NEUROSCIENCE

INTERNATIONAL SIXTH EDITION

NEUROSCIENCE

INTERNATIONAL SIXTH EDITION

EDITORS

Dale Purves • George J. Augustine
David Fitzpatrick • William C. Hall • Anthony-Samuel LaMantia
Richard D. Mooney • Michael L. Platt • Leonard E. White

This version of the text has been adapted and customized.
Not for sale in the USA or Canada.

NEW YORK OXFORD
OXFORD UNIVERSITY PRESS

Oxford University Press is a department of the University of Oxford. It furthers the University's objective of excellence in research, scholarship, and education by publishing worldwide.

Sinauer Associates is an imprint of Oxford University Press.

Published in the United States of America by Oxford University Press
198 Madison Avenue, New York, NY 10016, United States of America

Oxford is a registered trade mark of Oxford University Press.

ISBN 9781605358413
Printing number: 9 8 7 6 5 4 3 2

Printed in the United States of America

Contents in Brief

Contributors

George J. Augustine, Ph.D.
David Fitzpatrick, Ph.D.
William C. Hall, Ph.D.
Ben Hayden, Ph.D.
Anthony-Samuel LaMantia, Ph.D.
Richard D. Mooney, Ph.D.
Michael L. Platt, Ph.D.
Dale Purves, M.D.
Fan Wang, Ph.D.
Leonard E. White, Ph.D.

Unit Editors

UNIT I: George J. Augustine
UNIT II: David Fitzpatrick and Richard D. Mooney
UNIT III: Leonard E. White and William C. Hall
UNIT IV: Anthony-Samuel LaMantia
UNIT V: Dale Purves and Michael L. Platt

Contents

CHAPTER 8
Synaptic Plasticity 159

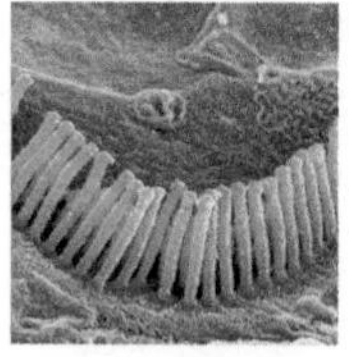

UNIT II Sensation and Sensory Processing 179

CHAPTER 9
The Somatosensory System: Touch and Proprioception 181

CHAPTER 10
Pain 201

CHAPTER 14

The Vestibular System 287

CHAPTER 15

The Chemical Senses 303

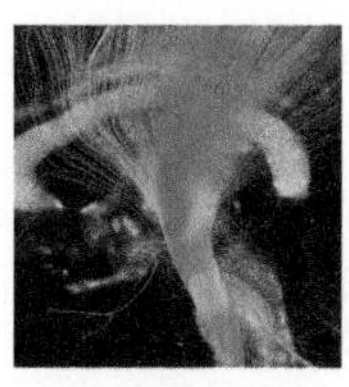

UNIT III Movement and Its Central Control 335

CHAPTER 16

Lower Motor Neuron Circuits and Motor Control 337

CHAPTER 17

Upper Motor Neuron Control of the Brainstem and Spinal Cord 359

CHAPTER 18

Modulation of Movement by the Basal Ganglia 385

CHAPTER 19

Modulation of Movement by the Cerebellum 403

CHAPTER 20

Eye Movements and Sensorimotor Integration 421

CHAPTER 21
The Visceral Motor System 437

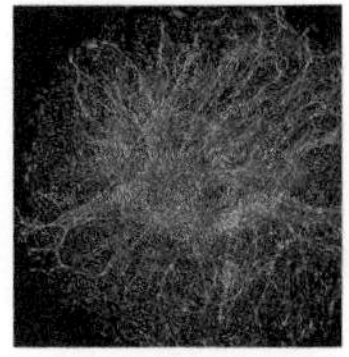

UNIT IV The Changing Brain 461

CHAPTER 22
Early Brain Development 463

CHAPTER 23
Construction of Neural Circuits 491

CHAPTER 24
Circuit Development: Intrinsic Factors and Sex Differences 521

CHAPTER 25
Experience-Dependent Plasticity in the Developing Brain 541

CHAPTER 26
Repair and Regeneration in the Nervous System 563

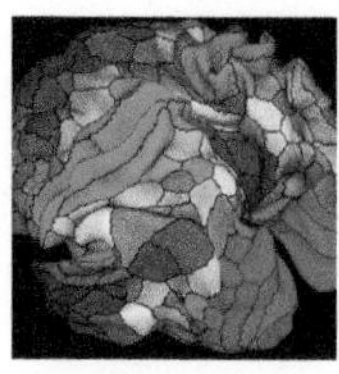

UNIT V Complex Brain Functions and Cognitive Neuroscience 591

CHAPTER 27
Cognitive Functions and the Organization of the Cerebral Cortex 593

CHAPTER 28
Cortical States 609

CHAPTER 29
Attention 633

CHAPTER 30
Memory 645

CHAPTER 31
Emotion 665

CHAPTER 32
Thinking, Planning, and Deciding 687

CHAPTER 33
Speech and Language 705

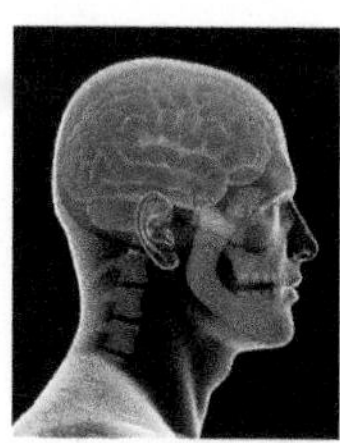

APPENDIX
Survey of Human Neuroanatomy A-1

ATLAS
The Human Central Nervous System AT-1

Preface

Whether judged in molecular, cellular, systemic, behavioral, or cognitive terms, the human nervous system is a stupendous piece of biological machinery. Given its accomplishments—all the artifacts of human culture, for instance—there is good reason for wanting to understand how the brain and the rest of the nervous system works. The debilitating and costly effects of neurological and psychiatric disease add a further sense of urgency to this quest. The aim of this book is to highlight the intellectual challenges and excitement—as well as the uncertainties—of what many see as the last great frontier of biological science. The information presented here is intended to serve as a starting point for undergraduates, medical students, students in other health professions, graduate students in the neurosciences, and many others who want insight into how the human nervous system operates.

Like any other great challenge, neuroscience should be, and is, full of debate, dissension, and considerable fun. All these ingredients have gone into the construction of this book's Sixth Edition; we hope they will be conveyed in equal measure to readers at all levels.

Acknowledgments

We are grateful to the many colleagues who provided helpful contributions, criticisms, and suggestions to this and previous editions. We particularly wish to thank Paul Adams, Ralph Adolphs, David Amaral, Dora Angelaki, Eva Anton, Gary Banker, the late Bob Barlow, Marlene Behrmann, Ursula Bellugi, Carlos Belmonte, Staci Bilbo, Dan Blazer, Alain Burette, Bob Burke, Roberto Cabeza, Jim Cavanaugh, Jean-Pierre Changeux, John Chapin, Milt Charlton, Michael Davis, Rob Deaner, Bob Desimone, Allison Doupe, Sasha du Lac, Jen Eilers, Chagla Eroglu, Anne Fausto-Sterling, Howard Fields, Elizabeth Finch, Nancy Forger, Jannon Fuchs, David Gadsby, Michela Gallagher, Dana Garcia, Steve George, the late Patricia Goldman-Rakic, Josh Gooley, Henry Greenside, Jennifer Groh, Mike Haglund, Zach Hall, Kristen Harris, Bill Henson, John Heuser, Bertil Hille, Miguel Holmgren, Jonathan Horton, Ron Hoy, Alan Humphrey, Jon Kaas, Kai Kaila, Jagmeet Kanwal, Herb Killackey, Len Kitzes, Marc Klein, Chieko Koike, Andrew Krystal, Arthur Lander, Story Landis, Simon LeVay, Darrell Lewis, Jeff Lichtman, Alan Light, Steve Lisberger, John Lisman, Arthur Loewy, Ron Mangun, Eve Marder, Robert McCarley, Greg McCarthy, Jim McIlwain, Daniel Merfeld, Steve Mitroff, Chris Muly, Vic Nadler, Sulochana Naidoo, Ron Oppenheim, Larysa Pevny, Franck Polleux, Scott Pomeroy, Rodney Radtke, Louis Reichardt, Sidarta Ribiero, Marnie Riddle, Jamie Roitman, Steve Roper, John Rubenstein, Ben Rubin, David Rubin, Josh Sanes, Cliff Saper, Lynn Selemon, Paul Selvin, Carla Shatz, Sid Simon, Bill Snider, Larry Squire, John Staddon, Peter Strick, Warren Strittmatter, Joe Takahashi, Stephen Traynelis, Christopher Walsh, Xiaoqin Wang, Richard Weinberg, Jonathan Weiner, Christina Williams, S. Mark Williams, Joel Winston, and Ryohei Yasuda. It is understood, of course, that any errors are in no way attributable to our critics and advisors.

Thanks are also due to our students at the several universities where the editors have worked, as well as to the many other students and colleagues who have suggested improvements and corrections. Finally, we owe special thanks to Andy Sinauer, Sydney Carroll, Martha Lorantos, Christopher Small, Jefferson Johnson, Joanne Delphia, Marie Scavotto, and the rest of the staff at Sinauer Associates for their outstanding work and the high standards they have maintained over six editions of this book.

Media and Supplements to Accompany *Neuroscience*, International Sixth Edition

For the Student

Companion Website (www.oup.com/uk/Purves6e)

The following resources are available to students free of charge:

- **Chapter Outlines:** The complete heading structure of each chapter.
- **Chapter Summaries:** Concise overviews of the important topics covered in each chapter.
- **Animations:** Topics such as synaptic transmission, resting membrane potential, information processing in the eye, the stretch reflex, and many others are presented in a dynamic manner that helps students visualize and better understand many of the complex processes of neuroscience.
- **Flashcards:** Flashcard activities help students master the extensive vocabulary of neuroscience.
- **Glossary:** The complete glossary, including all of the bold terms from the textbook.
- **Web Topics:** New for this edition, these provide novel or historical topics for special discussion.

For the Instructor

Ancillary Resource Center (www.oup.com/uk/Purves6e)

The Ancillary Resource Center (ARC) for *Neuroscience*, Sixth Edition includes a variety of resources to aid instructors in developing their courses and delivering their lectures:

- **Textbook Figures and Tables:** All of the figures and tables from the textbook are provided in JPEG format, reformatted for optimal readability, with complex figures provided in both whole and split formats.
- **PowerPoint Resources:** A PowerPoint presentation for each chapter includes all of the chapter's figures and tables, with titles and captions.
- **Atlas Images:** All of the images from the book's Atlas of the Human Central Nervous System are included in PowerPoint format, for use in lecture.
- **Animations:** All of the animations from the companion website are included for use in lecture and other course-related activities.
- **Test Bank:** A new complete Test Bank offers instructors a variety of questions for each chapter of the textbook. Available in Word, Diploma, and LMS formats.
- **Clinical Application Boxes:** New for this edition, these boxes examine common neural disorders and diseases.

CHAPTER

1 Studying the Nervous System

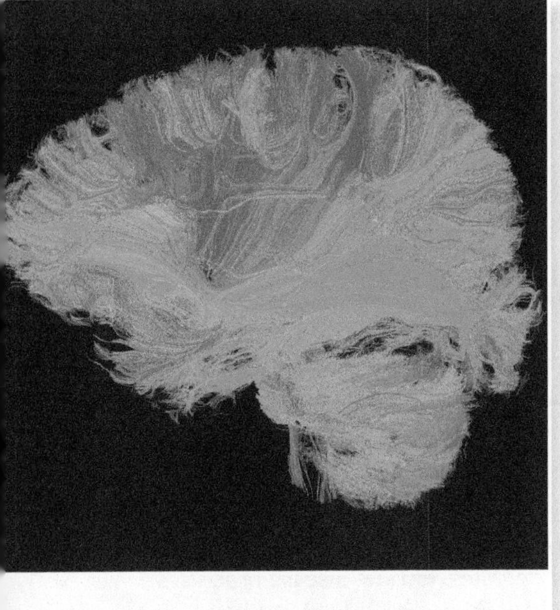

Overview

NEUROSCIENCE ENCOMPASSES A BROAD RANGE of questions about how the nervous systems of humans and other animals are organized, how they develop, and how they function to generate behavior. These questions can be explored using the tools of genetics and genomics, molecular and cell biology, anatomy, systems physiology, behavioral observation, psychophysics, and functional brain imaging. The major challenge facing students of neuroscience is to integrate the knowledge derived from these various levels and methods of analysis into a coherent understanding of brain structure and function. Many of the issues that have been explored successfully concern how the principal cells of all animal nervous systems—neurons and glia—perform their functions. Subsets of neurons and glia form ensembles called neural circuits, which are the primary components of neural systems that process different types of information. Neural systems in turn serve one of three general purposes: Sensory systems report information about the state of the organism and its environment; motor systems organize and generate actions; and associational systems provide "higher-order" brain functions such as perception, attention, memory, emotions, language, and thinking, all of which fall under the rubric of cognition. These latter abilities lie at the core of understanding human beings, their behavior, their history, and perhaps their future.

Genetics and Genomics

The nervous system, like all other organs, is the product of gene expression that begins at the outset of embryogenesis. A **gene** comprises both *coding* DNA sequences (exons) that are the templates for messenger RNA (mRNA) that will ultimately be translated into a protein, and *regulatory* DNA sequences (promoters and introns) that control whether and in what quantities a gene is expressed in a given cell type (i.e., transcribed into mRNA and then translated into a functional protein). **Genetic analysis** is thus fundamental to understanding the structure, function, and development of organs and organ systems. The advent of **genomics**, which focuses on the analysis of complete DNA sequences (both coding and regulatory) for a species or an individual, has provided insight into how nuclear DNA helps determine the assembly and operation of the brain and the rest of the nervous system.

Based on current estimates, the human genome comprises about 20,000 genes, of which some 14,000 (approximately 70%) are expressed in the developing or mature nervous system (Figure 1.1A). Of this subset, about 8000 are expressed in all cells and tissues, including the nervous system. The remaining 6000 genes of the 14,000 total are expressed only in the nervous system. Most "nervous system-specific" genetic information for the genes whose expression is shared among several tissues resides in the introns and regulatory sequences that control timing, quantity,

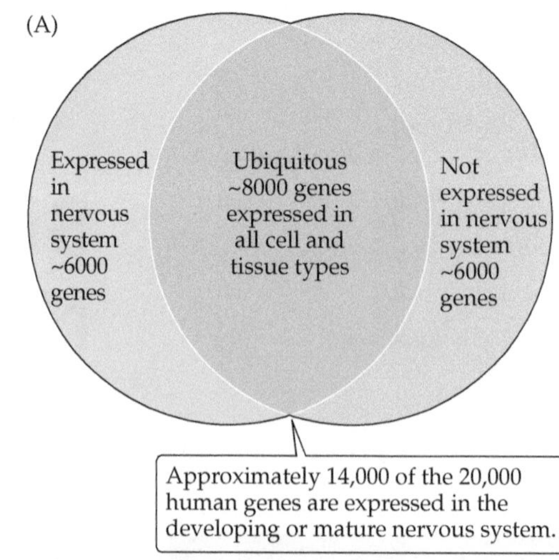

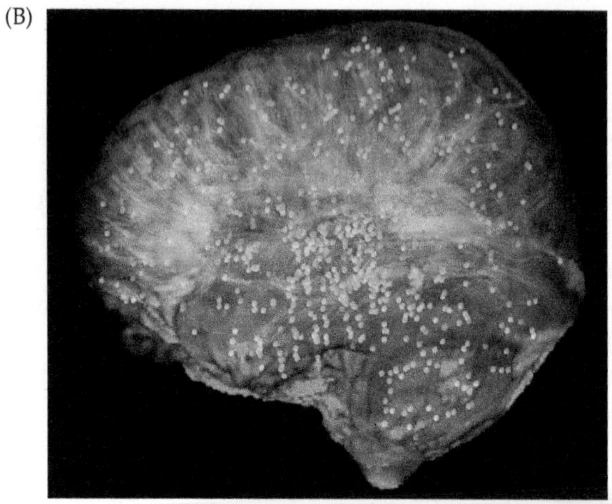

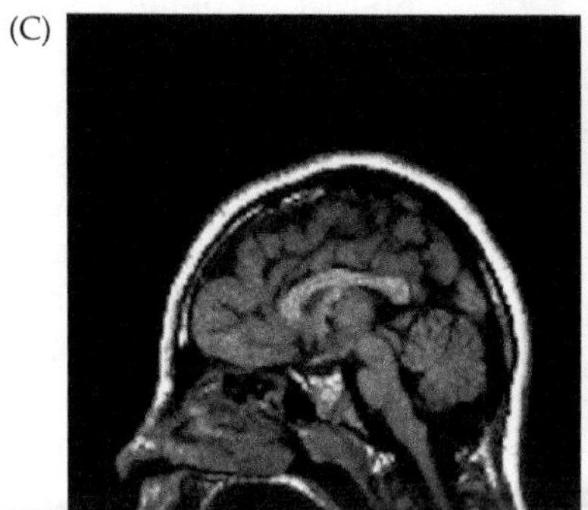

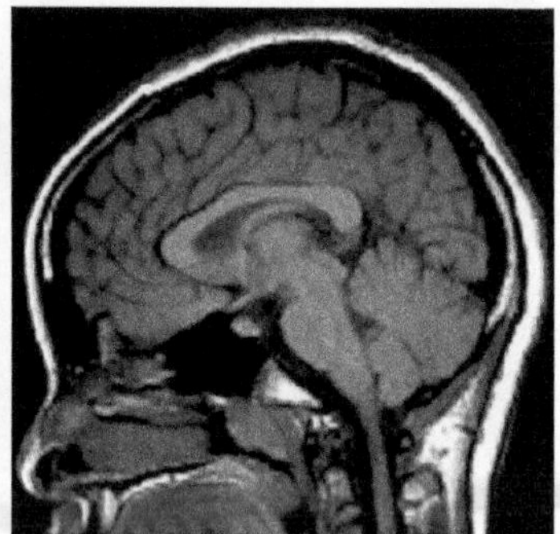

FIGURE 1.1 Genes and the nervous system. (A) In this Venn diagram of the human genome, the blue and purple regions represent genes that are expressed selectively in the nervous system along with those that are expressed in the nervous system as well as in all other tissues. (B) The locations and levels of expression of a single gene in the human brain. Dots indicate brain regions where mRNA for this particular gene are found, while their color (blue to orange) indicates the relative level (lower to higher) of mRNA detected at each location. (C) The consequence of a single-gene mutation for brain development. The gene, *ASPM* (*Abnormal Spindle-like Microcephaly-associated*), affects the function of a protein associated with mitotic spindles and results in microcephaly. In an individual carrying the *ASPM* mutation (left), the size of the brain is dramatically reduced and its anatomical organization is distorted compared with the brain of a typical control (right) of similar age and sex. (A data from Ramsköld et al., 2009; B courtesy of Allen Brain Atlas; C from Bond et al., 2002.)

variability, and cellular specificity of gene expression. Thus, despite the number of genes shared by the nervous system and other tissues, individual genes are regulated differentially throughout the nervous system, as measured by the amount of mRNA expressed from region to region and from one cell type to another (Figure 1.1B). Moreover, variable messages transcribed from the same gene, called **splice variants**, add diversity by allowing a single gene to encode information for a variety of related protein products. All these differences play a part in the diversity and complexity of brain structure and function.

A dividend of sequencing the human genome has been the realization that altered (mutated) genes, sometimes even one or a few, can underlie neurological and psychiatric disorders. For example, mutation of a single gene that regulates mitosis can result in microcephaly, a condition in which the brain and head fail to grow and brain function is dramatically diminished (Figure 1.1C; also see Chapter 22). In addition to genes that disrupt brain development, mutant genes can either cause (or are risk factors for) degenerative disorders of the adult brain, such as Huntington's and Parkinson's diseases. Using genetics and genomics to understand diseases of the developing and adult nervous system permits deeper insight into the pathology, and raises the hope for gene-based therapies.

The relationship between genotype and phenotype, however, is clearly not just the result of following genetic instructions, and genomic information on its own will not explain how the brain operates, or how disease processes disrupt normal brain functions. To understand how the brain and the rest of the nervous system work in health and disease, neuroscientists and clinicians must also understand the cell biology, anatomy, and physiology of the constituent cells, the neural circuits they form, and how the structure and function of such circuits change with use across the life span. Whereas understanding the operating principles of most other organ systems has long been clear, this challenge has yet to be met for the nervous system, and in particular the human brain.

Cellular Components of the Nervous System

Early in the nineteenth century, the cell was recognized as the fundamental unit of all living organisms. It was not until well into the twentieth century, however, that neuroscientists agreed that nervous tissue, like all other organs, is made up of these fundamental units. The major reason for this late realization was that the first generation of "modern" neuroscientists in the nineteenth century had

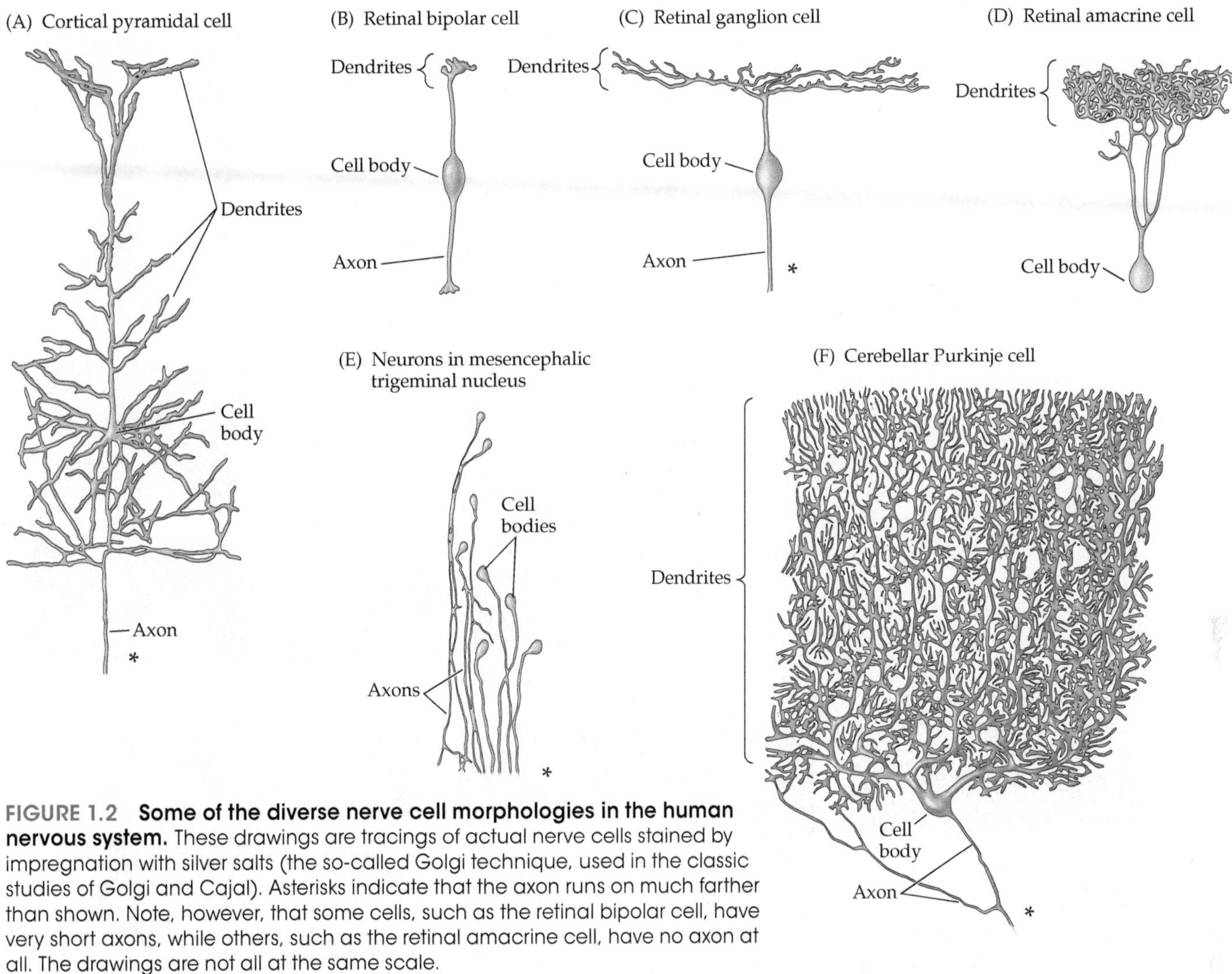

FIGURE 1.2 Some of the diverse nerve cell morphologies in the human nervous system. These drawings are tracings of actual nerve cells stained by impregnation with silver salts (the so-called Golgi technique, used in the classic studies of Golgi and Cajal). Asterisks indicate that the axon runs on much farther than shown. Note, however, that some cells, such as the retinal bipolar cell, have very short axons, while others, such as the retinal amacrine cell, have no axon at all. The drawings are not all at the same scale.

difficulty resolving the unitary nature of nerve cells with the microscopes and cell staining techniques then available. The extraordinarily complex shapes and extensive branches of individual nerve cells—all of which are packed together and thus difficult to distinguish from one another—further obscured their resemblance to the geometrically simpler cells of other tissues (Figure 1.2). Some biologists of that era even concluded that each nerve cell was connected to its neighbors by protoplasmic links, forming a continuous directly interconnected nerve cell network, or *reticulum* (Latin, "net"). The Italian pathologist Camillo Golgi articulated and championed this "reticular theory" of nerve cell communication. This mistake notwithstanding, Golgi made many important contributions to medical science, including identifying the cellular organelle eventually called the Golgi apparatus; developing the critically important cell staining technique that bears his name (see Figures 1.2 and 1.6); and contributing to the understanding of the pathophysiology of malaria. His reticular theory of the nervous system eventually fell from favor and was replaced by what came to be known as the "neuron doctrine." The major proponents of the neuron doctrine were the Spanish neuroanatomist Santiago Ramón y Cajal and the British physiologist Charles Sherrington.

The spirited debate occasioned by the contrasting views of Golgi and Cajal in the late nineteenth and early twentieth centuries set the course of modern neuroscience. Based on light microscopic examination of nervous tissue stained with silver salts according to Golgi's pioneering method, Cajal argued persuasively that nerve cells are discrete entities, and that they communicate with one another by means of specialized contacts that are not sites of continuity between cells. Sherrington, who had been working on the apparent transfer of electrical signals via reflex pathways, called these specialized contacts **synapses**. Despite the ultimate triumph of Cajal's view over Golgi's, both were awarded the 1906 Nobel Prize in Physiology or Medicine for their essential contributions to understanding the organization of the nervous system, and in 1932 Sherrington was likewise recognized for his contributions.

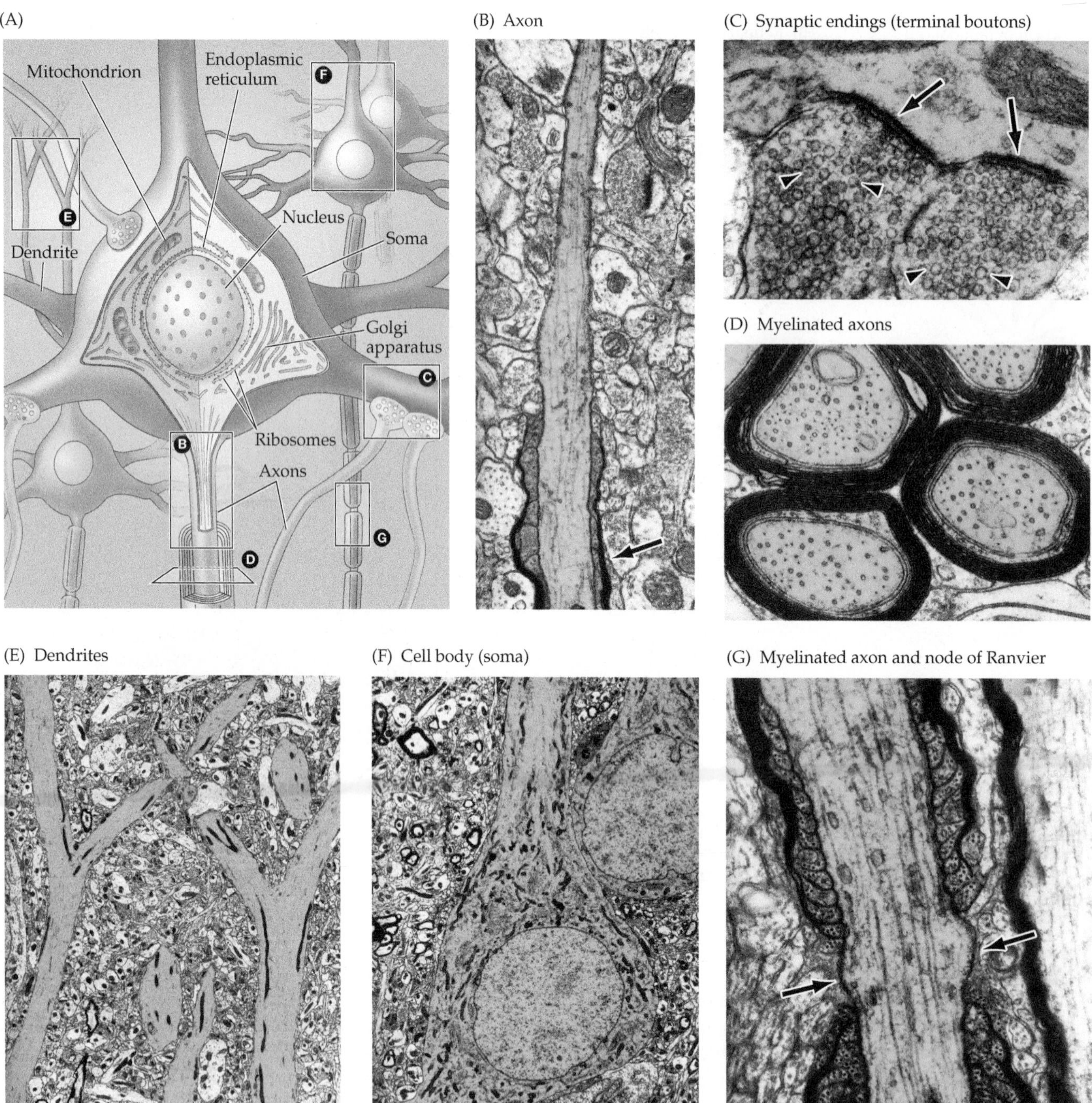

FIGURE 1.3 The major features of neurons visualized with electron microscopy. (A) Diagram of nerve cells and their component parts. The circled letters correspond to the micrographs in in the figure. (B) Axon initial segment (blue) entering a myelin sheath (gold). (C) Terminal boutons (blue) loaded with synaptic vesicles (arrowheads) forming synapses (arrows) with a dendrite (purple). (D) Transverse section of axons (blue) ensheathed by the processes of oligodendrocytes (gold); the surrounding myelin is black. (E) Apical dendrites (purple) of cortical pyramidal cells. (F) Nerve cell bodies (purple) occupied by large round nuclei. (G) Portion of a myelinated axon (blue) illustrating the intervals that occur between adjacent segments of myelin (gold and black) referred to as nodes of Ranvier (arrows). (Micrographs from Peters et al., 1991.)

The subsequent work of Sherrington and others demonstrating the transfer of electrical signals at synaptic junctions between nerve cells provided strong support for the neuron doctrine, although occasional challenges to the autonomy of individual neurons remained. It was not until the advent of electron microscopy in the 1950s that any lingering doubts about the discreteness of neurons were resolved. The high-magnification, high-resolution images obtained with the electron microscope (Figure 1.3) clearly established that nerve cells are functionally independent units; such micrographs also identified the junctions Sherrington had named synapses. As a belated consolation for

Golgi, however, electron microscopic studies also demonstrated specialized (albeit relatively rare) intercellular continuities between some neurons. These continuities, or **gap junctions**, are similar to those found between cells in epithelia such as the lung and intestine. Gap junctions do indeed allow for cytoplasmic continuity and the direct transfer of electrical and chemical signals between cells in the nervous system.

The histological studies of Cajal, Golgi, and a host of successors led to the consensus that the cells of the nervous system can be divided into two broad categories: **nerve cells**, or **neurons**, and supporting **glial cells** (also called **neuroglia**, or simply **glia**). Most, but not all, nerve cells are specialized for electrical signaling over long distances. Elucidating this process, which is the subject of Unit I, represents one of the more dramatic success stories in modern biology. In contrast to nerve cells, glial cells support the signaling functions of nerve cells rather than generating electrical signals themselves. They also serve additional functions in the developing and adult nervous system. Perhaps most important, glia are essential contributors to repairing nervous system damage, acting as stem cells in some brain areas where they promote regrowth of damaged neurons in regions where regeneration can usefully occur. In other regions, they prevent regeneration where uncontrolled regrowth might do more harm than good (see below and Unit IV).

Neurons and glia share the complement of organelles found in all cells, including endoplasmic reticulum, Golgi apparatus, mitochondria, and a variety of vesicular structures. In neurons and glia, however, these organelles are often more prominent in different regions of the cell. Mitochondria, for example, tend to be concentrated at synapses in neurons, while protein-synthetic organelles such as the endoplasmic reticulum are largely excluded from axons and dendrites. In addition to differing in the distribution of their organelles and subcellular components, neurons and glia differ in some measure from other cells in the specialized fibrillary or tubular proteins that constitute the cytoskeleton (see Figure 1.4). Although many of these proteins—isoforms of actin, tubulin, myosin, and several others—are found in other cells, their distinctive organization in neurons is critical for the stability and function of neuronal processes and synaptic junctions. Additional filament proteins characterize glial cells and contribute to their functions. The various filaments, tubules, subcellular motors, and scaffolding proteins of the neuronal and glial cytoskeleton orchestrate many functions, including the migration of nerve cells; the growth of axons and dendrites; the trafficking and appropriate positioning of membrane components, organelles, and vesicles; and the active processes of exocytosis and endocytosis underlying synaptic communication. Understanding the ways in which these molecular components are used to ensure the proper development and function of neurons and glia remains a primary focus of modern neurobiology.

Neurons

Most neurons are distinguished by their specialization for long-distance electrical signaling and intercellular communication by means of synapses. These attributes are apparent in the overall morphology of neurons, in the organization of their membrane components, and in the structural and functional intricacies of the synaptic contacts between neurons (see Figure 1.3C). The most obvious morphological sign of neuronal specialization for communication is the extensive branching of neurons. The two most salient aspects of this branching for typical nerve cells are the presence of an **axon** and the elaborate arborization of **dendrites** that arise from the neuronal cell body in the form of *dendritic branches* (or *dendritic processes*; see Figure 1.3E). Most neurons have only one axon that extends for a relatively long distance from the location of the cell body. Axons may have branches, but in general they are not as elaborate as those made by dendrites. Dendrites are the primary targets for synaptic input from the axon terminals of other neurons and are distinguished by their high content of ribosomes, as well as by specific cytoskeletal proteins.

The variation in the size and branching of dendrites is enormous, and of critical importance in establishing the information-processing capacity of individual neurons. Some neurons lack dendrites altogether, while others have dendritic branches that rival the complexity of a mature tree (see Figure 1.2). The number of inputs a particular neuron receives depends on the complexity of its dendritic arbor: Neurons that lack dendrites are innervated by the axons of just one or a few other neurons, which limits their capacity to integrate information from diverse sources, thus leading to more or less faithful relay of the electrical activity generated by the synapses impinging on the neurons. Neurons with increasingly elaborate dendritic branches are innervated by a commensurately larger number of other neurons, which allows for far greater integration of information. The number of inputs to a single neuron reflects the degree of **convergence**, while the *number of targets* innervated by any one neuron represents its **divergence**. The number of synaptic inputs received by each nerve cell in the human nervous system varies from 1 to about 100,000. This range reflects a fundamental purpose of nerve cells: to integrate and relay information from other neurons in a neural circuit.

The synaptic contacts made by axon endings on dendrites (and less frequently on neuronal cell bodies) represent a special elaboration of the secretory apparatus found in many polarized epithelial cells. Typically, the axon terminal of the **presynaptic** neuron is immediately adjacent to a specialized region of **postsynaptic** receptors on the target cell. For the majority of synapses, however, there is no physical continuity between these two elements. Instead, pre- and postsynaptic components communicate via the secretion of molecules from the presynaptic terminal that bind to receptors in the postsynaptic cell. These molecules, called **neurotransmitters**, must traverse an interval of extracellular space between

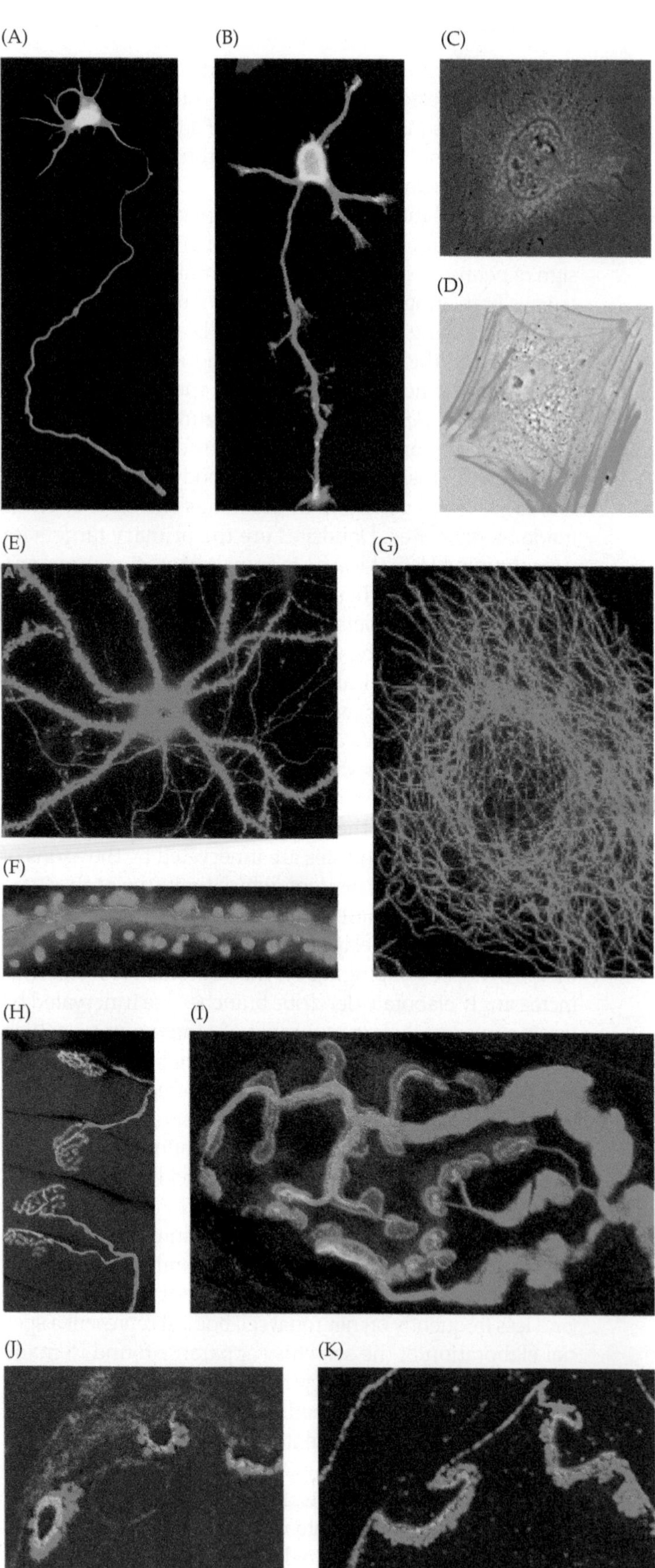

FIGURE 1.4 The diversity of cytoskeletal arrangements in neurons. (A) The cell body, the initial segment of the axon, and dendrites are distinguished by the distribution of tubulin (green). This distribution contrasts with the microtubule-binding protein tau (red), which is found in axons. (B) The localization of actin (red) to the growing tips of axonal and dendritic processes is shown here in a cultured neuron taken from the hippocampus. (C) In contrast, in a cultured epithelial cell, actin (red) is distributed in fibrils that occupy most of the cell body. (D) In astrocytes in culture, actin (red) is also seen in fibrillar bundles. (E) Tubulin (green) is found throughout the cell body and dendrites of neurons. (F) Although tubulin is a major component of dendrites, extending into small dendritic outgrowths called spines, the head of the spine is enriched in actin (red). (G) The tubulin component of the cytoskeleton in non-neuronal cells is arrayed in filamentous networks. (H–K) Synapses have a special arrangement of cytoskeletal elements, receptors, and scaffold proteins. (H) Two axons (green; tubulin) from motor neurons are seen issuing branches each to four muscle fibers. The red shows the clustering of postsynaptic receptors (in this case for the neurotransmitter acetylcholine). (I) A higher-power view of a single motor neuron synapse shows the relationship between the axon (green) and the postsynaptic receptors (red). (J) Proteins in the extracellular space between the axon and its target muscle are labeled green. (K) Scaffolding proteins (green) localize receptors (red) and link them to other cytoskeletal elements. The scaffolding protein shown here is dystrophin, whose structure and function are compromised in the many forms of muscular dystrophy. (A courtesy of Y. N. Jan; B from Kalil et al., 2000; C courtesy of D. Arneman and C. Otey; D courtesy of A. de Sousa and R. Cheney; E,F from Matus, 2000; G courtesy of T. Salmon; H–K courtesy of R. Sealock.)

pre- and postsynaptic elements called the **synaptic cleft**. The synaptic cleft is not simply an empty space, but is the site of extracellular proteins that influence the diffusion, binding, and degradation of the molecules, including neurotransmitters and other factors, secreted by the presynaptic terminal (see Chapter 5).

The information conveyed by synapses on the neuronal dendrites is integrated and generally "read out" at the origin of the axon (called the axon). The axon is the portion of the nerve cell specialized for relaying electrical signals over long distances (see Figure 1.3B). The axon is a unique extension from the neuronal cell body that may travel a few hundred micrometers or much farther, depending on the type of neuron and the size of the animal (some axons in large animals can be meters in length). The axon also has a distinct cytoskeleton whose elements are central for its functional integrity (Figure 1.4). Many nerve cells in the human brain have axons no more than a few millimeters long, and a few have no axon at all.

Relatively short axons are a feature of **local circuit neurons**, or **interneurons**, throughout the nervous system. In contrast, the axons of **projection neurons** extend to distant targets. For example, the axons that run from the human spinal cord to the foot are about a meter long. The axons of both interneurons and projection neurons often branch locally, resulting in the innervation of multiple post-synaptic sites on many post-synaptic neurons.

Axons convey electrical signals over such distances by a self-regenerating wave of electrical activity called an **action potential**. Action potentials (also referred to as "spikes" or "units") are all-or-nothing changes in the electrical potential (voltage) across the nerve cell membrane that conveys information from one place to another in the nervous system (see Chapter 2). The process by which the information encoded by action potentials is passed on at synaptic contacts to a target cell is called **synaptic transmission**, and its details are described in Chapter 5. Presynaptic terminals (also called *synaptic endings*, *axon terminals*, or *terminal boutons*; see Figure 1.3C) and their postsynaptic specializations are typically **chemical synapses**, the most abundant type of synapse in the mature nervous system. Another type, the **electrical synapse** (mediated by the gap junctions mentioned above), is relatively rare in the mature nervous system (but abundant in the developing CNS) and serves special functions, including the synchronization of local networks of neurons.

The secretory organelles in the presynaptic terminal of chemical synapses are called **synaptic vesicles** and are spherical structures filled with neurotransmitters and in some cases other neuroactive molecules (see Figure 1.3C). The positioning of synaptic vesicles at the presynaptic membrane and their fusion to initiate neurotransmitter release are regulated by a variety of proteins (including several cytoskeletal proteins) either in or associated with the vesicle. The neurotransmitters released from synaptic vesicles modify the electrical properties of the target cell by binding to receptors localized primarily at postsynaptic specializations. The intricate interplay of neurotransmitters, receptors, related cytoskeletal elements, and signal transduction molecules is the basis for communication among nerve cells and between nerve cells and effector cells in muscles and glands.

Glial Cells

Glial cells—usually referred to more simply as glia—are quite different from neurons, even though they are at least as abundant. Glia do not participate directly in synaptic transmission or in electrical signaling, although their supportive functions help define synaptic contacts and maintain the signaling abilities of neurons. Like nerve cells, many glial cells have complex processes extending from their cell bodies, but these are generally less prominent and do not serve the same purposes as neuronal axons and dendrites. Cells with glial characteristics appear to be the only stem cells retained in the mature brain, and are capable of giving rise both to new glia and, in a few instances, new neurons.

The word *glia* is Greek for "glue" and reflects the nineteenth-century presumption that these cells "held the nervous system together." The term has survived despite the lack of any evidence that glial cells actually bind nerve cells together. Glial functions that *are* well established include maintaining the ionic milieu of nerve cells; modulating the rate of nerve signal propagation; modulating synaptic action by controlling the uptake and metabolism of neurotransmitters at or near the synaptic cleft; providing a scaffold for some aspects of neural development; aiding (or in some instances impeding) recovery from neural injury; providing an interface between the brain and the immune system; and facilitating the convective flow of interstitial fluid through the brain during sleep, a process that washes out metabolic waste.

There are three types of differentiated glial cells in the mature nervous system: astrocytes, oligodendrocytes, and microglial cells. **Astrocytes**, which are restricted to the central nervous system (i.e., the brain and spinal cord), have elaborate local processes that give these cells a starlike ("astral") appearance (Figure 1.5A,F). A major function of astrocytes is to maintain, in a variety of ways, an appropriate chemical environment for neuronal signaling, including formation of the blood-brain barrier (see the Appendix). In addition, recent observations suggest that astrocytes secrete substances that influence the construction of new synaptic connections, and that a subset of astrocytes in the adult brain retains the characteristics of stem cells (Figure 1.5D; see below).

Oligodendrocytes, which are also restricted to the central nervous system, lay down a laminated, lipid-rich wrapping called **myelin** around some, but not all, axons (Figure 1.5B,G,H). Myelin has important effects on the speed of the transmission of electrical signals (see Chapter 3). In the peripheral nervous system, the cells that provide myelin are called **Schwann cells**. In the mature nervous system, subsets of oligodendrocytes and Schwann cells retain neural stem cell properties, and can generate new oligodendrocytes and Schwann cells in response to injury or disease (Figure 1.5E).

Microglial cells (Figure 1.5C,I) are derived primarily from hematopoietic precursor cells (although some may be derived directly from neural precursor cells). Microglia share many properties with macrophages found in other tissues: They are primarily scavenger cells that remove cellular debris from sites of injury or normal cell turnover. In addition, microglia, like their macrophage counterparts, secrete signaling molecules—particularly a wide range of cytokines that are also produced by cells of the immune system—that can modulate local inflammation and influence whether other cells survive or die. Indeed, some neurobiologists prefer to categorize microglia as a type of macrophage. Following brain damage, the number of microglia at the site of injury increases dramatically. Some of these cells proliferate from microglia resident in the brain,

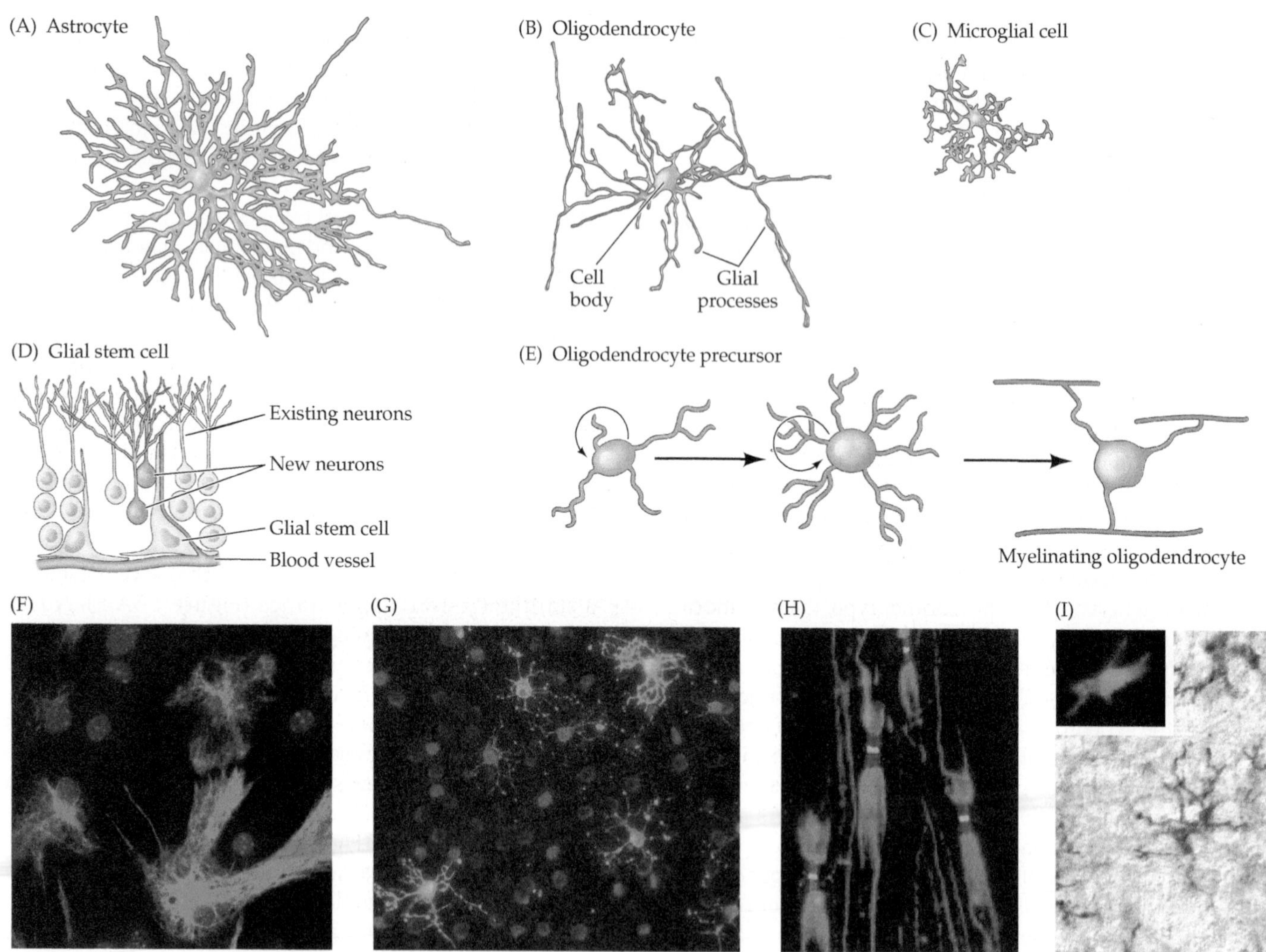

FIGURE 1.5 Glial cell types. (A–C) Tracings of differentiated glial cells in the mature nervous system visualized using the Golgi method include an astrocyte (A), an oligodendrocyte (B), and a microglial cell (C). The three tracings are at approximately the same scale. (D) Glial stem cells in the mature nervous system include stem cells with properties of astrocytes that can give rise to neurons, astrocytes, and oligodendrocytes. (E) Another class of glial stem cell, the oligodendrocyte precursor, has a more restricted potential, giving rise primarily to differentiated oligodendrocytes. (F) Astrocytes (red) in tissue culture are labeled with an antibody against an astrocyte-specific protein. (G) Oligodendrocytes (green) in tissue culture labeled with an antibody against an oligodendrocyte-specific protein. (H) Peripheral axons are ensheathed by myelin (labeled red) except at nodes of Ranvier (see Figure 1.3G). The green label indicates ion channels (see Chapter 4) concentrated in the node; the blue label indicates a molecularly distinct region called the paranode. (I) Microglial cells from the spinal cord labeled with a cell type–specific antibody. Inset: Higher-magnification image of a single microglial cell labeled with a macrophage-selective marker. (A–C after Jones and Cowan, 1983; D,E, after Nishiyama et al., 2009; F,G courtesy of A.-S. LaMantia; H from Bhat et al., 2001; I courtesy of A. Light, inset courtesy of G. Matsushima.)

while others come from macrophages that migrate to the injured area and enter the brain via local disruptions in the cerebral vasculature (the blood-brain barrier).

In addition to the three classes of differentiated glia, **glial stem cells** are also found throughout the adult brain. These cells retain the capacity to proliferate and generate additional precursors or differentiated glia, and in some cases neurons. Glial stem cells in the mature brain can be divided into two categories: a subset of astrocytes found primarily near the ventricles in a region called the subventricular zone (SVZ) or adjacent to ventricular zone blood vessels (see Figure 1.5D); and oligodendrocyte precursors scattered throughout the white matter and sometimes referred to as *polydendrocytes* (see Figure 1.5E). SVZ astrocytes, both in vivo and in vitro, can give rise to more stem cells, neurons, and mature astrocytes and oligodendrocytes. Thus, they have the key properties of all stem cells: proliferation, self-renewal, and the capacity to make all cell classes of a particular tissue. Oligodendrocyte precursors are more limited in their potential. They give rise primarily to mature oligodendrocytes as well as to some astrocytes, although under some conditions in vitro they can generate neurons.

The significance of stem cells that retain many molecular characteristics of glia in the mature brain remains unclear. They may reflect glial identity as the "default" for any proliferative cell derived from the embryonic precursors of the nervous system, or they may reflect distinctions in the differentiated state of neurons versus glia that allow proliferation only in cells with glial characteristics.

Cellular Diversity in the Nervous System

Although the cellular constituents of the human nervous system are in many ways similar to those of other organs, they are unusual in their extraordinary diversity. The human brain is estimated to contain about 86 billion neurons and at least that many glia. Among these two overall groups, the nervous system has a greater range of distinct cell types—whether categorized by morphology, molecular identity, or physiological role—than any other organ system (a fact that presumably explains why, as mentioned at the start of this chapter, so many different genes are expressed in the nervous system).

For much of the twentieth century, neuroscientists relied on the set of techniques developed by Cajal, Golgi, and other pioneers of histology (the microscopic analysis of cells and tissues) and pathology to describe and categorize the cell types in the nervous system. The staining method named for Golgi permitted visualization of individual nerve cells and their processes that had been impregnated, seemingly randomly, with silver salts (Figure 1.6A,B). More recently, fluorescent dyes and other soluble molecules injected into single neurons—often after physiological recording to identify the function of the cell—have provided more informative approaches to visualizing single nerve cells and their

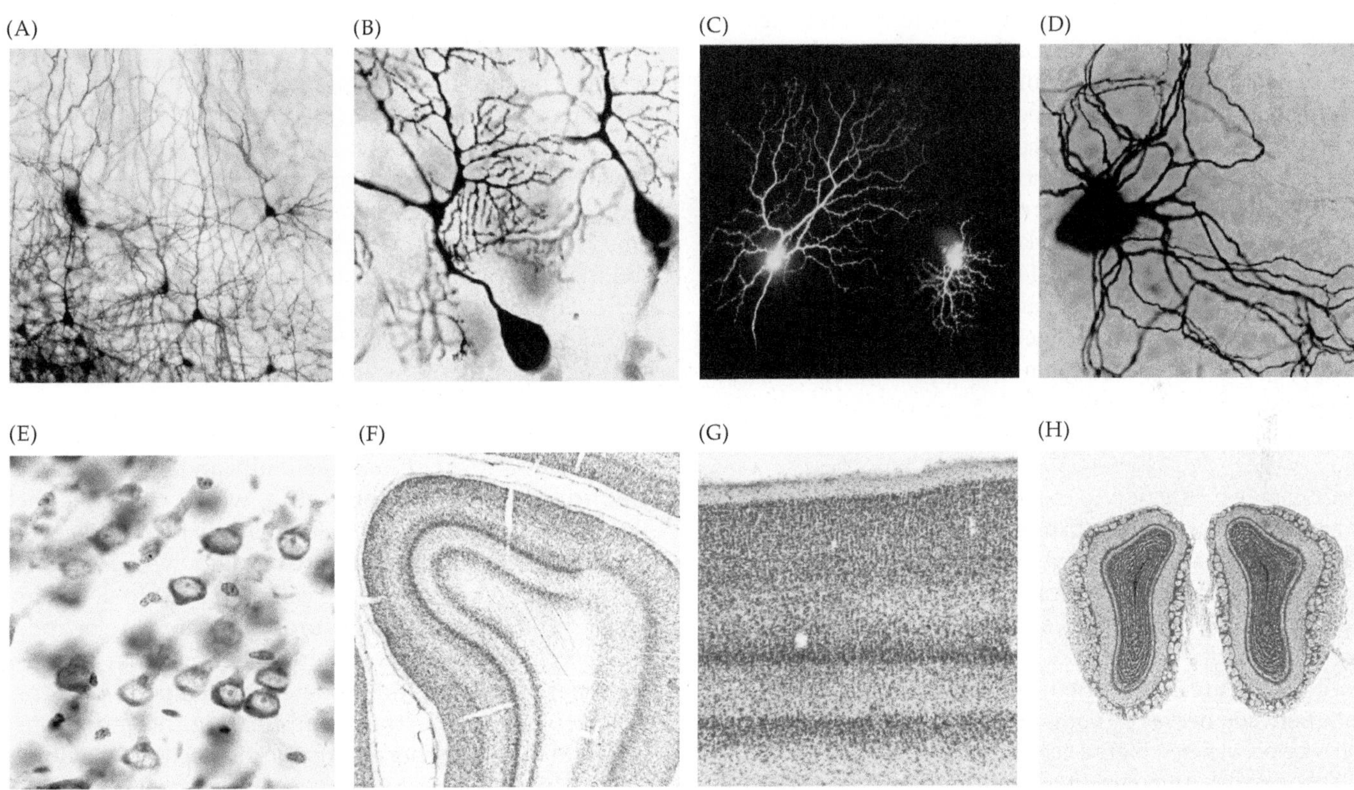

FIGURE 1.6 Visualizing nerve cells. (A) Cortical neurons stained using the Golgi method (impregnation with silver salts). (B) Golgi-stained Purkinje cells in the cerebellum. Purkinje cells have a single, highly branched apical dendrite (as diagrammed in Figure 1.2F). (C) Intracellular injection of fluorescent dye labels two retinal neurons that vary dramatically in the size and extent of their dendritic arborizations. (D) Intracellular injection of an enzyme labels a neuron in a ganglion of the autonomic (involuntary) nervous system. (E) The dye cresyl violet stains RNA in all cells in a tissue, labeling the nucleolus (but not the nucleus) as well as the ribosome-rich endoplasmic reticulum. Dendrites and axons are not labeled, which explains the "blank" spaces between these neurons. (F) Nissl-stained section of the cerebral cortex reveals lamination—cell bodies arranged in layers of differing densities. The different laminar densities define boundaries between cortical areas with distinct functions. (G) Higher magnification of the primary visual cortex, seen on the left side of panel (F). Differences in cell density define the laminae of the primary visual cortex and differentiate this region from other cerebral cortical areas. (H) Nissl stain of the olfactory bulbs reveals a distinctive distribution of cell bodies, particularly those cells arranged in rings on each bulb's outer surface. These structures, including the cell-sparse tissue contained within each ring, are called glomeruli. (C courtesy of C. J. Shatz; all others courtesy of A.-S. LaMantia and D. Purves.)

processes (Figure 1.6C,D). Today, many studies depend on molecular and genetic methods to introduce genes for fluorescent proteins that can fully label a neuron or glial cell and its processes. Additional methods use antibodies that label specific neuronal and glial components. Finally, nucleic acid probes with complimentary sequences can detect mRNAs that encode genes expressed in neurons or glia using a method called in situ hybridization (see Figure 1.6 and 1.14).

As a complement to these methods (which provide a sample of specific subsets of neurons and glia), other stains reveal the distribution of all cell bodies—but not their processes or connections—in neural tissue. The widely used Nissl method is one example; this technique stains the nucleolus and other structures (e.g., ribosomes) where DNA or RNA is found (Figure 1.6E). Such stains demonstrate that the size, density, and distribution of the total population of nerve cells are not uniform within the brain. In some regions, such as the cerebral cortex, cells are arranged in layers (Figure 1.6F,G), each of which is defined by differences in cell density. Structures such as the olfactory bulbs display even more complicated arrangements of cell bodies (Figure 1.6H).

Additional approaches, detailed later in the chapter, have further defined the differences among nerve cells from region to region. These include the identification of how subsets of neurons are connected to one another, and how molecular differences further distinguish classes of nerve cells in various brain regions (see Figure 1.14). The cellular diversity of the human nervous system apparent today presumably reflects the increasingly complex networks and behaviors that have arisen over the span of mammalian evolution.

Neural Circuits

Neurons never function in isolation; they are organized into ensembles called **neural circuits** that process specific kinds of information. The synaptic connections that underlie neural circuits are typically made in a dense tangle of dendrites, axon terminals, and glial cell processes that together constitute what is called **neuropil** (Greek *pilos*, "felt"; see Figure 1.3C). The neuropil constitutes the regions between nerve cell bodies where most synaptic connectivity occurs (see Figure 1.14D).

Although the arrangement of neural circuits varies greatly according to the function served, some features are characteristic of all such ensembles. Preeminent is the direction of information flow in any particular circuit, which is obviously essential to understanding its purpose. Nerve cells that carry information from the periphery *toward* the brain or spinal cord (or deeper centrally within the spinal cord and brain) are called **afferent neurons**; nerve cells that carry information *away* from the brain or spinal cord (or away from the circuit in question) are **efferent neurons**. Interneurons (local circuit neurons; see above) participate only in the local aspects of a circuit, based on the short distances over which their axons extend. These three functional classes—afferent neurons, efferent neurons, and interneurons—are the basic constituents of all neural circuits.

A simple example of a neural circuit is one that mediates the **myotatic reflex**, commonly known as the knee-jerk reflex (Figure 1.7). The afferent neurons that control the reflex are sensory neurons whose cell bodies lie in the **dorsal root ganglia** and send axons peripherally that terminate in sensory endings in skeletal muscles. (The ganglia that serve this same function for much of the head and neck are called **cranial nerve ganglia**; see the Appendix.) The central axons of these sensory neurons enter the spinal cord, where they terminate on a variety of central neurons concerned with the regulation of muscle tone—most obviously on the **motor neurons** that determine the activity of the related muscles. The motor neurons in the circuits are the efferent neurons, one group projecting to the flexor muscles in the limb, and the other to extensor muscles. Spinal cord interneurons are the third element of the circuit. The interneurons receive synaptic contacts from sensory afferent neurons and make synapses on the efferent motor neurons that project to the flexor muscles; thus, they are capable of modulating the input–output linkage. The excitatory synaptic connections between the sensory afferents and the extensor efferent motor neurons cause the extensor muscles to contract; at the same time, interneurons activated by the afferents are inhibitory, and their activation diminishes electrical activity in flexor efferent motor neurons and causes the flexor muscles to become less active. The result is a complementary activation and inactivation of the synergistic and antagonistic muscles that control the position of the leg.

A more detailed picture of the events underlying the myotatic or any other neural circuit can be obtained by **electrophysiological recording**, which measures the electrical activity of a nerve cell. There are two approaches to this method: **extracellular recording**, where an electrode is placed *near* the nerve cell of interest to detect its activity; and **intracellular recording**, where the electrode is placed *inside* the cell of interest. Extracellular recording is particularly useful for detecting temporal patterns of action potential activity and relating those patterns to stimulation by other inputs, or to specific behavioral events. Intracellular recording can detect the smaller, graded changes in electrical potential that trigger action potentials, and thus allows a more detailed analysis of communication among neurons within a circuit. These graded triggering potentials can arise at either sensory receptors or synapses and are called **receptor potentials** or **synaptic potentials**, respectively.

For the myotatic circuit, electrical activity can be measured both extracellularly and intracellularly, thus defining the functional relationships among the neurons in the circuit. With electrodes placed near—but still outside—individual cells, the pattern of action potential activity

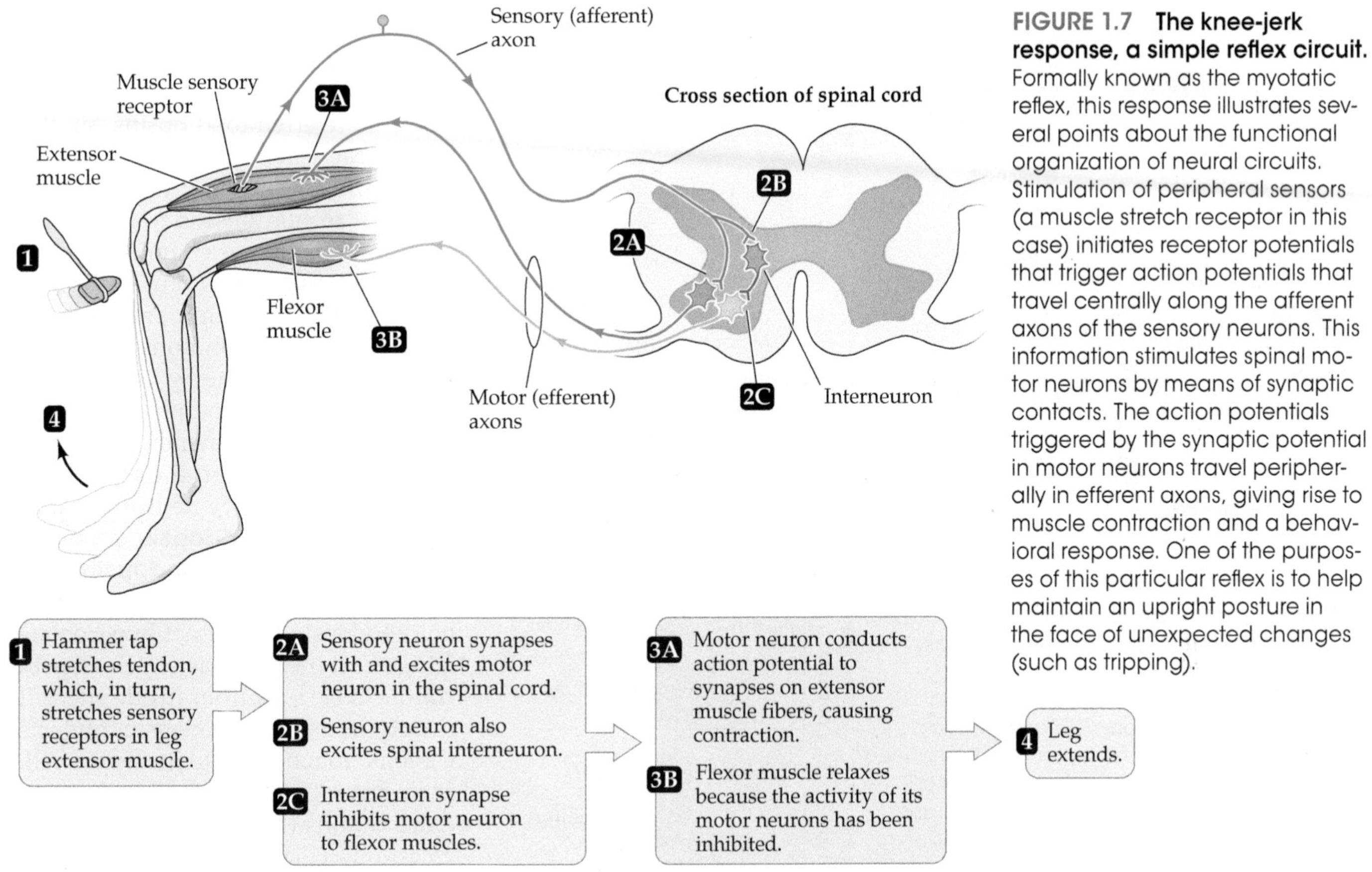

Sensory (afferent) axon
Muscle sensory receptor
Extensor muscle
Cross section of spinal cord
3A
2B
2A
1
Flexor muscle
3B
Motor (efferent) axons
2C
Interneuron
4
1 Hammer tap stretches tendon, which, in turn, stretches sensory receptors in leg extensor muscle.
2A Sensory neuron synapses with and excites motor neuron in the spinal cord.
2B Sensory neuron also excites spinal interneuron.
2C Interneuron synapse inhibits motor neuron to flexor muscles.
3A Motor neuron conducts action potential to synapses on extensor muscle fibers, causing contraction.
3B Flexor muscle relaxes because the activity of its motor neurons has been inhibited.
4 Leg extends.

FIGURE 1.7 The knee-jerk response, a simple reflex circuit. Formally known as the myotatic reflex, this response illustrates several points about the functional organization of neural circuits. Stimulation of peripheral sensors (a muscle stretch receptor in this case) initiates receptor potentials that trigger action potentials that travel centrally along the afferent axons of the sensory neurons. This information stimulates spinal motor neurons by means of synaptic contacts. The action potentials triggered by the synaptic potential in motor neurons travel peripherally in efferent axons, giving rise to muscle contraction and a behavioral response. One of the purposes of this particular reflex is to help maintain an upright posture in the face of unexpected changes (such as tripping).

can be recorded extracellularly for each element of the circuit (afferents, efferents, and interneurons) before, during, and after a stimulus (Figure 1.8). By comparing the onset, duration, and frequency of action potential activity in each cell, a functional picture of the circuit emerges. Using intracellular recording, it is possible to observe directly the changes in membrane potential underlying the synaptic connections of each element of the myotatic reflex (or any other) circuit (Figure 1.9).

Other Ways to Study Neural Circuits

Recent technological advances allow the activity of entire populations of neurons to be monitored. One approach, known as **calcium imaging**, records the transient changes in intracellular concentration of calcium ions (see Chapter 7) that are associated with action potential firing (Figure 1.10). Because calcium channels establish currents that lead to voltage changes in neurons, and because calcium

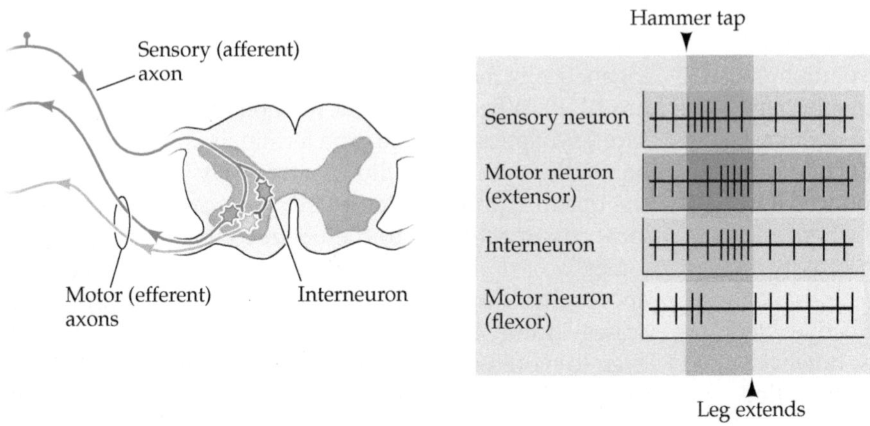

Sensory (afferent) axon
Motor (efferent) axons
Interneuron
Hammer tap
Sensory neuron
Motor neuron (extensor)
Interneuron
Motor neuron (flexor)
Leg extends

FIGURE 1.8 Extracellular recording shows the relative frequency and pattern of action potentials in neurons that form the neural circuits for the myotatic reflex. Action potentials are indicated by individual vertical lines. As a result of the stimulus, the sensory neuron is triggered to fire at higher frequency (i.e., more action potentials per unit of time). This increase triggers a higher frequency of action potentials in both the extensor motor neurons and the interneurons. Concurrently, the inhibitory synapses made by the interneurons onto the flexor motor neurons cause the frequency of action potentials in these cells to decline.

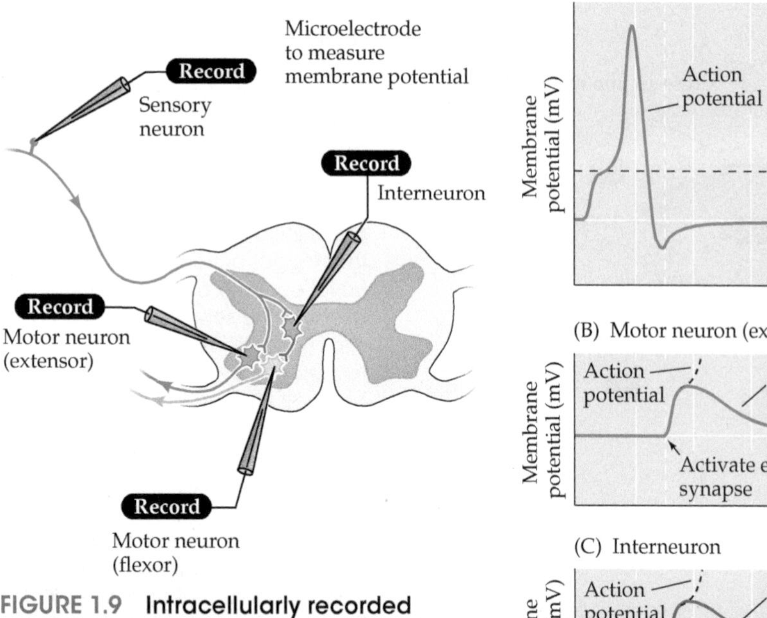

FIGURE 1.9 Intracellularly recorded responses underlying the myotatic reflex. (A) Action potential measured in a sensory neuron. (B) Postsynaptic potential recorded in an extensor motor neuron. (C) Postsynaptic potential recorded in an interneuron. (D) Postsynaptic potential recorded in a flexor motor neuron. Such intracellular recordings are the basis for understanding the cellular mechanisms of action potential generation, and the sensory receptor and synaptic potentials that trigger these conducted signals.

is an important second messenger, methods that rely on changes in fluorescence intensity caused by electrical activity can visualize neuronal activity in large numbers of individual cells based on calcium transients in the cells' cytoplasm. A related approach uses voltage-sensitive fluorescent dyes that insert into the neuronal plasma membrane and report on the transmembrane potential, thereby imaging the consequences of action potentials and other electrical signals in many neurons at once. Calcium indicators or voltage-sensitive dyes can be introduced directly into neurons in living slices or into primary cultured neurons based on their osmotic properties in solution. In addition, viral vectors can be used to transfect subpopulations of cells, either in living tissue slices or in the intact brain in a living animal. Finally, genes that encode calcium- or voltage-sensitive proteins can be introduced into transgenic animals for more precise control of where and when the proteins are available for measuring activity in the living animal.

The most specific and arguably the most effective way to manipulate the function of neural circuits, however, is to use molecular genetic tools, an approach called **optogenetics**. Optogenetic methods emerged as a consequence of the identification and cloning of bacterial channels referred to as opsins, similar to the opsins in animal retinas. Bacterial opsins use the same chromophore found in retinal opsins to transduce light energy into a chemical signal that activates channel proteins. Since opsins modulate membrane currents when they absorb photons, light can be used to control nerve cell activity when bacterial chromophores are incorporated into the membrane of any neuron. Three bacterial opsins have been used to modify neuronal excitability: **bacteriorhodopsin**, **halorhodopsin**, and **channelrhodopsin** (Figure 1.11A). Both bacteriorhodopsin and halorhodopsin have a net hyperpolarizing effect on cells: Bacteriorhodopsin conducts H^+ ions from inside the cell to outside, and halorhodopsin conducts Cl^- ions from outside to inside the cell. In contrast, channelrhodopsin conducts cations (Na^+, K^+Ca^{2+}, H^+) as well as anions (Cl^-), providing for either depolarizing or hyperpolarizing modulation, depending on the channelrhodopsin variant and the wavelengths of light used.

The genes for opsins can be introduced into neurons either in living brain slices or intact animals. In brain slices, a variety of viral transduction methods are used. In whole animals genetic methods are used (see "Genetic Analysis of Neural Systems"). Once the opsins are expressed in living neurons, these neurons can be illuminated by specific wavelengths of light, and neural activity can be manipulated with a high degree of spatial and temporal resolution, due to microscopic illumination of one or more opsin-labeled nerve cells. In awake and behaving animals, this approach can be used during the performance of specific tasks to evaluate the role of the optogenetically modified neurons in task performance (Figure 1.11B). When optogenetic methods are applied in brain slices, synaptic activity in axon terminals and dendrites can be modified locally by illuminating only those regions of the opsin-expressing nerve cell; the resulting change in local circuit activity can then be recorded electrophysiologically or optically (Figure 1.11C,D). Thus, optogenetic approaches can modify neuronal activity at a variety of scales—from single neurons to local neural circuits and even to more widely distributed neural networks that influence specific behaviors.

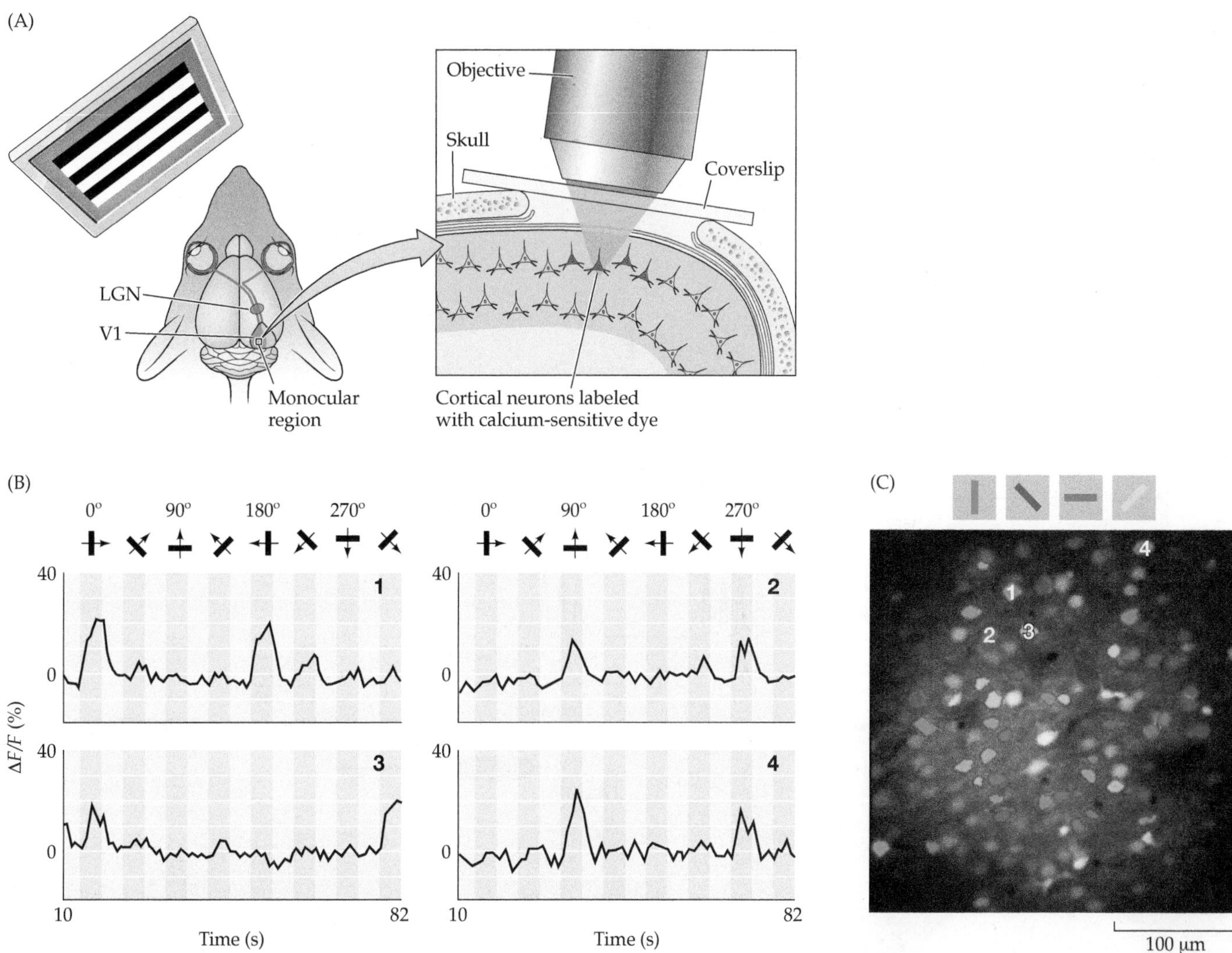

FIGURE 1.10 Imaging cortical neurons responding to visual stimuli using calcium-sensitive dyes. (A) The imaging was done in a live mouse presented with visual stimuli at different orientations (a horizontally oriented series of high-contrast stripes is shown here). A small "window" of bone was removed over the visual cortex for application of the dyes and subsequent imaging; changes in fluorescence intensity were detected using a microscope with the objective over the exposed cortical surface. (LGN = lateral geniculate nucleus; V1 = primary visual cortex.) (B) The change in fluorescence intensity ($\Delta F/F$ [%]) of four cells imaged this way while the mouse viewed stripes in the orientations shown at the top of the graphs, moving in directions indicated by the arrows (numbers 1–4 on the graphs identify the cells, which can be localized in C). Each separate graph shows the response over time of one cortical cell. In each graph, the peaks in fluorescence signal indicate robust responses when the cell's preferred stimulus orientation was presented; little response was elicited by nonpreferred stimuli. (C) The distribution of cells with preferred responses to stripes oriented at different angles (colors indicate preferred orientation). Activated cells preferring different orientations were interspersed, with each orientation represented by several cells in different positions within this small cortical area. (A from Mank, et al., 2008; B,C from Ohki et al., 2005.)

Organization of the Human Nervous System

Neural circuits that process similar types of information make up neural systems that serve broader purposes. The most general functional distinction divides such collections into **sensory systems** that acquire and process information from the internal and external environments (e.g., the visual system or the auditory system, both described in Unit II); and **motor systems** that respond to such information by generating movements (described in Unit III). There are, however, large numbers of cells and circuits that lie between these relatively well defined input and output systems. These are collectively referred to as **associational systems**, and they mediate the most complex and least well characterized brain functions (see Unit V).

In addition to recognizing these broad functional distinctions, neuroscientists and neurologists have

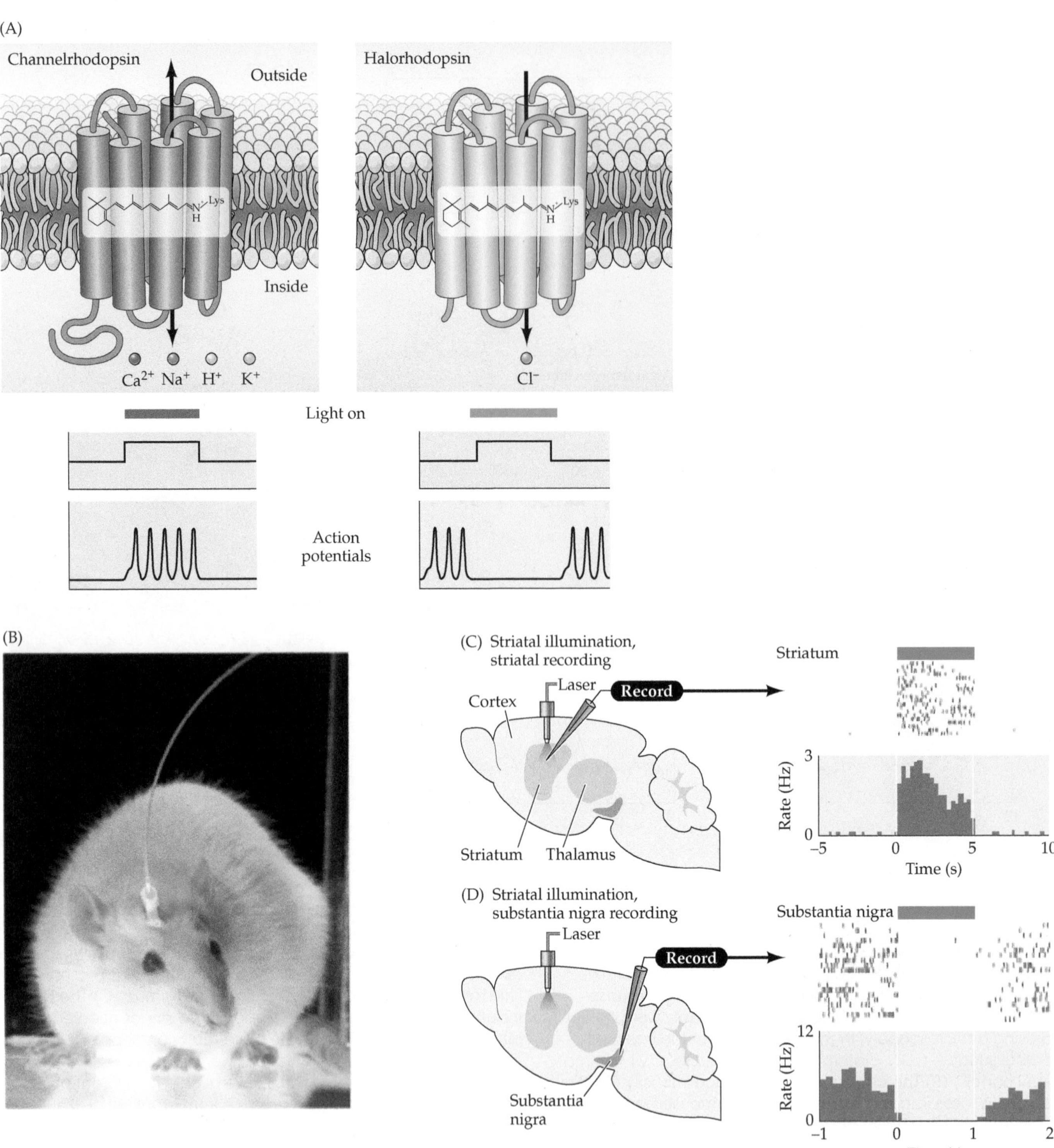

FIGURE 1.11 **Optogenetic methods used to control electrical activity in nerve cells.** (A) Two bacterial opsins, showing their 7-transmembrane domains. The light-sensing all-*trans* retinal transduces a change in illumination to transiently open the channels. (B) A fiber-optic probe, stabilized with a permanent head mount, can use a laser to deliver a narrow bandwidth of light to specific opsin-expressing neurons. (C) Illumination of bacterial opsins expressed in striatal neurons that regulate movement. Neurons expressing channelrhodopsin in the striatum, where neurons have little or no spontaneous action potential activity (regions on the graph with very few marks), fire robustly when illuminated (the histogram indicates action potential frequency when light is on and channelrhodopsin is activated). (D) Neurons in the substantia nigra pars reticulata, where neurons have a high frequency of spontaneous action potential activity (rasters and histograms in the left- and right-flanking regions), can be "silenced" transiently by illumination in the striatum. The striatal axons release the inhibitory neurotransmitter GABA. Thus, when the striatal neurons are stimulated optogenetically, the result of the "activation" is increased inhibition in the substantia nigra. Thus, optogenetic mechanisms can assess the physiology of neural circuits based on the activation of neuronal populations. (A after Zhang et al., 2011; C,D after Kravitz et al., 2010.)

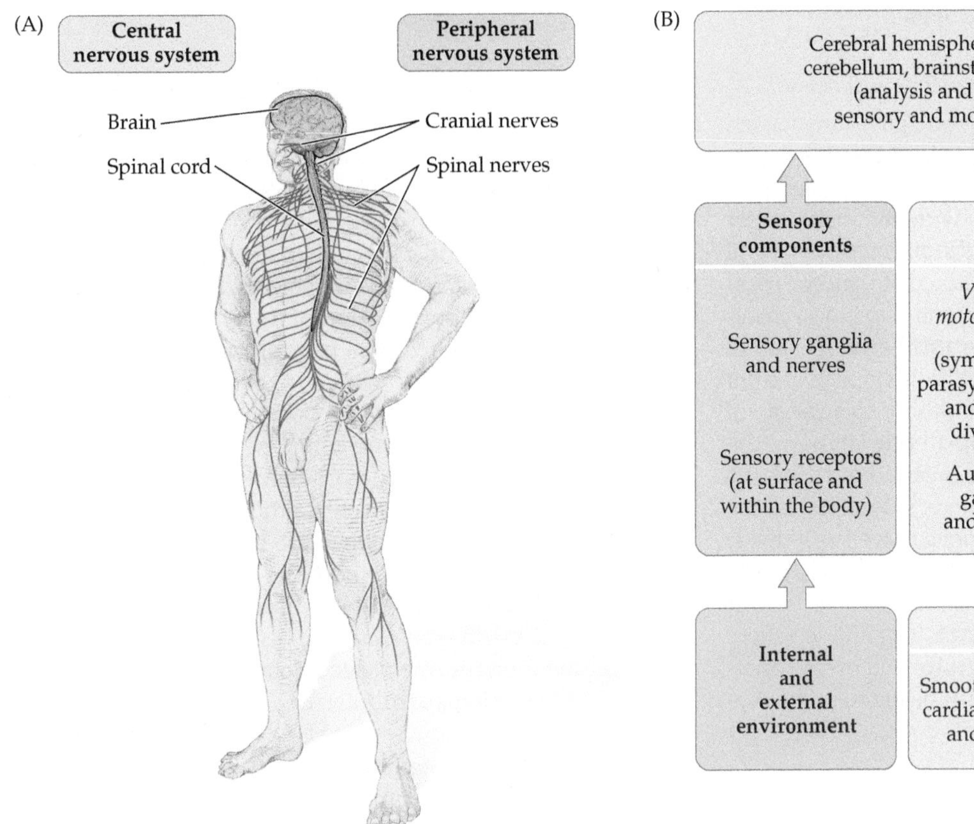

FIGURE 1.12 The major anatomical components of the nervous system and their functional relationships. (A) The CNS (brain and spinal cord) and PNS (spinal and cranial nerves). (B) Diagram of the major components of the CNS and PNS and their functional relationships. Stimuli from the environment convey information to processing circuits in the brain and spinal cord, which in turn interpret their significance and send signals to peripheral effectors that move the body and adjust the workings of its internal organs.

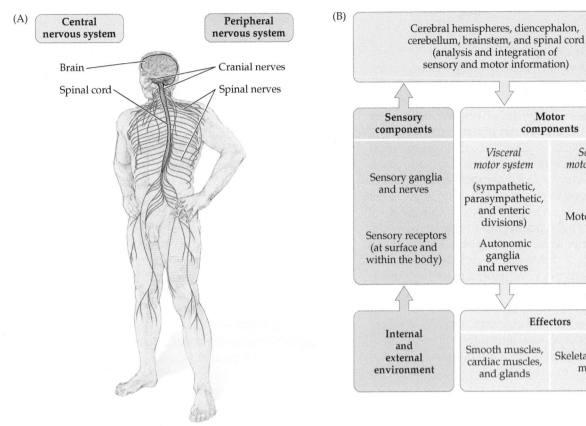

conventionally divided the vertebrate nervous system anatomically into central and peripheral components (Figure 1.12). The **central nervous system**, typically referred to as the **CNS**, comprises the **brain** (cerebral hemispheres, diencephalon, cerebellum, and brainstem) and the **spinal cord**. The **peripheral nervous system** (**PNS**) includes the sensory neurons that link sensory receptors on the body surface or deeper within it with relevant processing circuits in the CNS. The motor portion of the PNS in turn consists of two components. The motor axons that connect the brain and spinal cord to skeletal muscles make up the **somatic motor division** of the PNS, whereas the cells and axons that innervate smooth muscle, cardiac muscle, and glands make up the **visceral** or **autonomic motor division**.

Those nerve cell bodies that reside in the PNS are located in **ganglia**, which are simply local accumulations of nerve cell bodies and supporting cells. Peripheral axons are gathered into bundles called **nerves**, many of which are enveloped by the glial cells of the PNS; as mentioned earlier, these peripheral glia are called Schwann cells (see above), and they either myelinate these axons or provide a single glial covering that protects otherwise unmyelinated axons within peripheral nerves.

Two gross histological terms distinguish regions rich in neuronal cell bodies versus regions rich in axons. **Gray matter** refers to any accumulation of cell bodies and neuropil in the brain and spinal cord. **White matter** (named for its relatively light appearance, the result of the lipid content of myelin) refers to axon tracts and commissures. Within gray matter, nerve cells are arranged in two different ways. A local accumulation with neurons that have roughly similar connections and functions is called a **nucleus** (plural: *nuclei*, not to be confused with the nucleus of a cell); such collections are found throughout the cerebrum, diencephalon, brainstem, and spinal cord. In contrast, **cortex** (plural: *cortices*) describes sheetlike arrays of nerve cells. The cortices of the cerebral hemispheres and of the cerebellum provide the clearest examples of this organizational principle. Within the white matter of the CNS, axons are gathered into **tracts** that are more or less analogous to nerves in the periphery. Each tract contains axons that typically originate in the same gray matter structure, are organized in parallel, and often terminate in the same division of gray matter at some distance from their origin. Tracts that cross the midline of the brain, such as the corpus callosum that interconnects the cerebral hemispheres, are referred to as **commissures**. The sensory tracts of the dorsal spinal cord are referred to as **columns**.

The organization of the visceral motor division of the PNS (the nerve cells that control the functions of the visceral organs, including the heart, lungs, gastrointestinal tract, and genitalia) is a bit more complicated (see Chapter 21). Nearly all components of the PNS—neurons and glia—are derived from the neural crest (see Chapter 22). Visceral motor neurons in the brainstem and spinal cord—the so-called preganglionic neurons—form synapses with peripheral motor neurons that lie in the **autonomic ganglia**. The peripheral motor neurons in autonomic ganglia innervate smooth muscle, glands, and cardiac muscle, thus controlling most involuntary (visceral) behavior. In the **sympathetic** division of the autonomic motor system, the ganglia lie along or in front of the vertebral column and send their axons to a variety of peripheral targets. In the **parasympathetic** division, the ganglia are found in or adjacent to the organs they innervate. Another component of the visceral motor system, called the **enteric system**, comprises small ganglia as well as individual neurons scattered throughout the wall of the gut. These neurons and their intrinsic axonal connections comprise vast neural networks that influence enteric motility and secretion. More details about the physical structures and overall anatomy of the human nervous system can be found in the Appendix and the Atlas in the back of this book.

Neural Systems

Several characteristics distinguish neural systems within the complex array of anatomical components that make up any nervous system; primary among these are unity of function, representation of specific information, specific connectivity among select brain regions, and subdivision into subsystems for relaying and processing information in parallel. The most important of these is the unity of function evident in a selectively interconnected ensemble of neurons distributed over multiple ganglia in the PNS, or nuclei and cortices in the CNS. For example, the visual system is defined by all the neurons and connections primarily dedicated to vision, the auditory system by those dedicated to hearing, the somatosensory system by those dedicated to the sense of touch, the pyramidal motor system by the neurons and connections dedicated to voluntary movement, and so on for many other identifiable systems. In many instances, a system's components are distributed throughout the body and brain. Thus, sensory systems include peripheral sensory specializations in eye, ear, skin, nose, and tongue, while motor systems include the peripheral motor nerves and target muscles required to perform various actions. Both sensory and motor systems entail nerve pathways that connect the periphery with nuclei in the spinal cord, brainstem, and thalamus, as well as the relevant areas of the cerebral cortex.

Two other features of many neural systems are orderly *representation* of information at various levels, and a division of the function of the system into submodalities that are typically relayed and processed in *parallel pathways*. Parallel pathways arise because virtually all neural systems have identifiable subsystems. The human visual system, for instance, has subsystems that emphasize stimulus characteristics such as color, form, or motion, with each class of information relayed and processed separately to some degree. Similar segregation of information subtypes into parallel pathways is apparent in other sensory and motor systems.

For systems such as vision and somatic sensation (e.g., touch)—systems that function to distinguish differences between neighboring points in the visual field or on the body's surface—the representation of information is *topographic*. Such representations form **topographic maps** that reflect a point-to-point correspondence between the sensory periphery (the visual field or the body surface) and neighboring neurons within the central components of the system (in the spinal cord and brain). Motor systems also entail topographic representations of movements, although here the direction of information flow is from the CNS to the periphery.

For neural systems where the representation of information does not depend on discriminating neighboring points in a field—for example, hearing, smell, and taste—organizational principles compare, assess, and integrate multiple stimulus attributes in an orderly way that facilitates the extraction and processing of essential information. These representations, many of which remain only partially understood, are collectively referred to as **computational maps**. The organization of even more complex information such as perception, attention, emotions, and memories is also unclear, and presumably involves processing that engages additional networks beyond the relatively rudimentary level of topographic or computational maps in sensory and motor cortices (see Unit V).

Genetic Analysis of Neural Systems

Genetic analyses of nervous system function in health and neural disorders have been made by examining families in which a disease is inherited in a Mendelian fashion (i.e., due to both parents carrying one copy of a recessive gene, leading to homozygosity for this gene in offspring, or by one parent carrying a non-lethal dominant gene that is passed on to an offspring). An alternative approach is statistical correlation of likely disease genes drawn from analyses of large cohorts of individuals with the same clinical diagnoses (**genome-wide association studies**, or **GWAS**). The idea with GWAS is that if a genetic variant occurs with a greater than random frequency in patients with a clinically diagnosed condition such as Alzheimer's disease, schizophrenia, or autism, it probably contributes to that pathology. Once identified, human disease genes can be "modeled" in experimental animals that have **orthologous genes** (identical or similar genes based on sequence and chromosomal location). The identification of genes associated with diseases of the nervous system provides a direct connection between the brain's genomic foundations and brain structure and function.

Most of the analysis of the biological function of human disease genes has been done in mice, although some has been done in fruit flies (*Drosophila melangaster*), a nematode

BOX 1A ■ Model Organisms in Neuroscience

Much modern neuroscience focuses on understanding the organization and function of the human nervous system, as well as the pathological bases of neurological and psychiatric diseases. These issues, however, are difficult to address by studying the human brain; therefore, neuroscientists have relied on the nervous systems of other animals as a guide. A wealth of information about the anatomy, biochemistry, physiology, cell biology, and genetics of neural systems has been gleaned by studying the brains of a variety of species.

Often the choice of model species studied reflects the assumptions about enhanced functional capacity in that species; for example, from the 1950s through the 1970s, cats were the subjects of pioneering studies on visual function because they are highly "visual" animals, and therefore could be expected to have well-developed brain regions devoted to vision—regions similar to those found in primates, including humans. Much of what is currently known about human vision is based on studies carried out in cats. Studies on invertebrates such as the squid and the sea slug *Aplysia californica* yielded similarly critical insights into the basic cell biology of neurons, synaptic transmission, and synaptic plasticity (the basis of learning and memory). Both the squid and the sea slug were chosen because of certain exceptionally large nerve cells with stereotypic identity and connections that were well suited to physiological measurements. In each case, the advantages offered by these cells made it possible to perform experiments that helped answer key questions.

Biochemical, cellular, anatomical, physiological, and behavioral studies continue to be conducted on a wide range of animals. However, the complete sequencing of the genomes of invertebrate and vertebrate species, including mammals, has led to the informal adoption by many neuroscientists of four "model" organisms based on the ability to do genetic analysis and manipulation in each of these species. A majority of the genes in the human genome are expressed in the developing and adult nervous system. The same is true in the nematode worm *Caenorhabditis elegans*; the fruit fly *Drosophila melanogaster*; the zebrafish (*Danio rerio*); and the house mouse (*Mus musculus*)—the four species commonly used in modern genetics, and used increasingly in neuroscience. Despite certain limitations in each of these species, the availability of their genomes facilitates research on a range of questions at the molecular, cellular, anatomical, and physiological levels.

One advantage of these model species is that the wealth of genetic and genomic information for each one permits sophisticated manipulation of gene expression and function. Thus, once an important gene for brain development or later function is identified, it can be specifically manipulated in the worm, fly, fish, or mouse. Large-scale screens of mutant animals whose genomes have been modified randomly by chemical mutagens allow investigators to search for changes from typical development structure and function (the phenotype) and to identify genes related to specific aspects of brain architecture or behavior. Similar efforts, although more limited in scope, have identified spontaneous or induced mutations in the mouse that disrupt brain development or function. In addition, manipulations that result in so-called *transgenic animals* permit genes to be introduced into the genome ("knocked in"), or to be deleted or mutated ("knocked out") using the remarkable capacity of genomes to splice in new sequences that are similar to endogenous genes. This capacity, referred to as *homologous recombination*, allows DNA constructs that disrupt or alter the expression of specific genes to be inserted into the location of the normal gene in the host species. These approaches allow assessment of the consequences of eliminating or altering gene function.

Neuroscientists study the nervous systems and behaviors of other species as well, but with somewhat different aims. Crustaceans such as crayfish and lobsters and insects such as grasshoppers and cockroaches have been useful for discerning basic rules that govern neural circuit function. Avians and amphibians (chickens and frogs) continue to be useful for studying early neural development, and mammals such as rats are used extensively for neuropharmacological and behavioral studies of adult brain function. Finally, non-human primates (the rhesus monkey in particular) provide opportunities to study complex functions that closely approximate those carried out in the human brain. None of these species, however, is as amenable to genetic and genomic manipulations as are the four species mentioned above, each of which has made significant contributions to understanding the human brain.

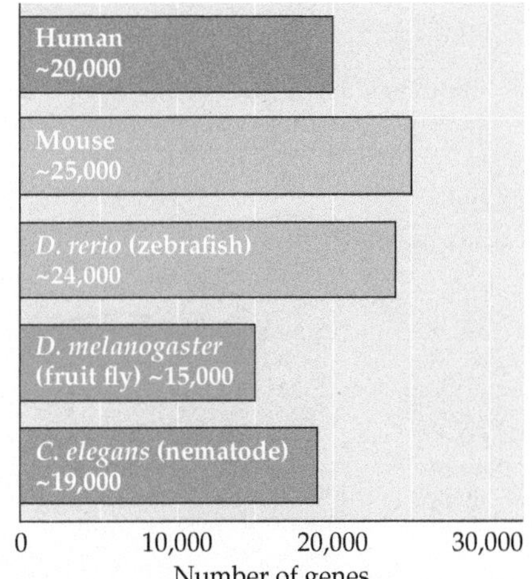

Estimated genome sizes of humans and several model species. Note that the number of genes in an organism's genome does not correlate with cellular or organismal complexity; the simple nematode *Caenorhabditis elegans*, for example, has almost the same number of genes as a human. Much genetic activity is dependent on transcription factors that regulate when and to what degree a given gene is expressed.

worm (*Caenorhabditis elegans*), and more recently, zebrafish (*Danio rerio*) (Box 1A). In the fly and worm, "forward" genetic analysis can be used in which flies or worms are randomly mutagenized using chemicals, ultraviolet light, or X-ray irradiation. Mutants with genes orthologous to human genes are then identified based on animal phenotypes that parallel those in humans.

An alternative to mutagenesis is **genetic engineering**, or "reverse" genetics. This approach is used in several model species; however, for neuroscience, genetically engineered mice have become particularly important, since mutations that parallel those found in humans can be made in orthologous mouse genes (Figure 1.13A). All of the techniques of genetic engineering rely on manipulations

(A)

Introduce a new gene, or modify or disable an existing gene, in mouse embryonic stem cells in vitro.

Gene

Isolation of stem cell line

Incorporation into inner cell mass

Surgical transfer of embryos into foster mother

Mosaic (chimera)

×

Wild type

If some manipulated cells produce an egg or sperm:

Heterozygous (F_1) carrying one copy of the manipulated gene

Homozygous carrier (F_2) formed when two F_1 mice are mated

(C)

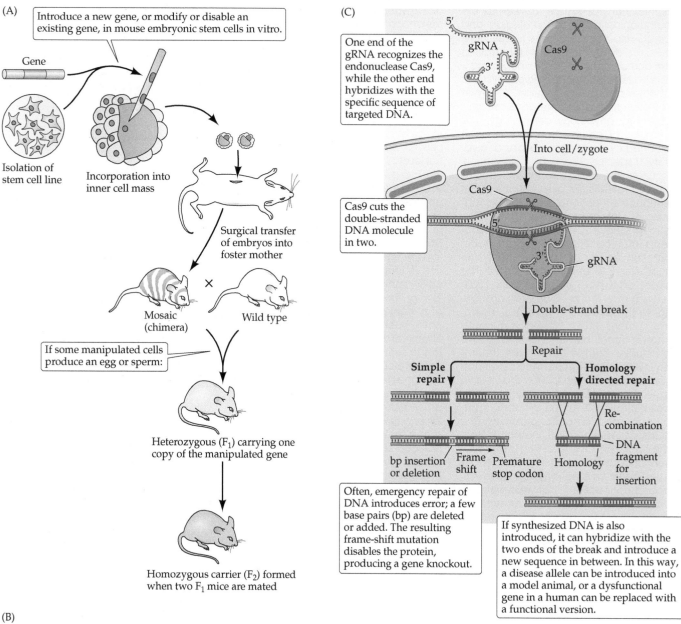

One end of the gRNA recognizes the endonuclease Cas9, while the other end hybridizes with the specific sequence of targeted DNA.

5′

gRNA

3′

Cas9

Into cell/zygote

Cas9

5′

Cas9 cuts the double-stranded DNA molecule in two.

3′

gRNA

Double-strand break

Repair

Simple repair

Homology directed repair

Re-combination

DNA fragment for insertion

bp insertion or deletion

Frame shift

Premature stop codon

Homology

Often, emergency repair of DNA introduces error; a few base pairs (bp) are deleted or added. The resulting frame-shift mutation disables the protein, producing a gene knockout.

If synthesized DNA is also introduced, it can hybridize with the two ends of the break and introduce a new sequence in between. In this way, a disease allele can be introduced into a model animal, or a dysfunctional gene in a human can be replaced with a functional version.

(B)

In most cells: No recombination

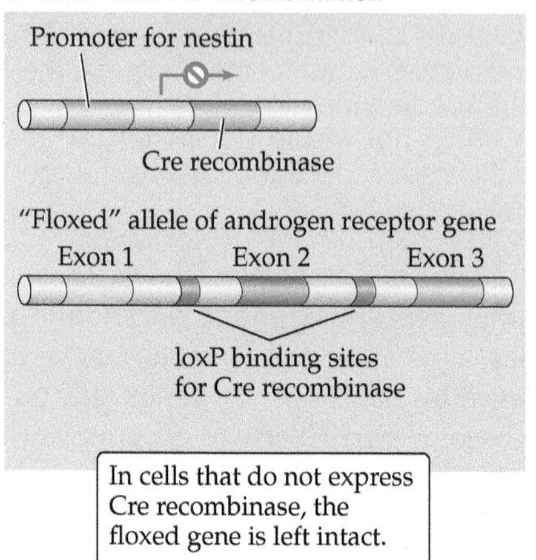

In cells that do not express Cre recombinase, the floxed gene is left intact.

In nervous system only (expressing nestin)

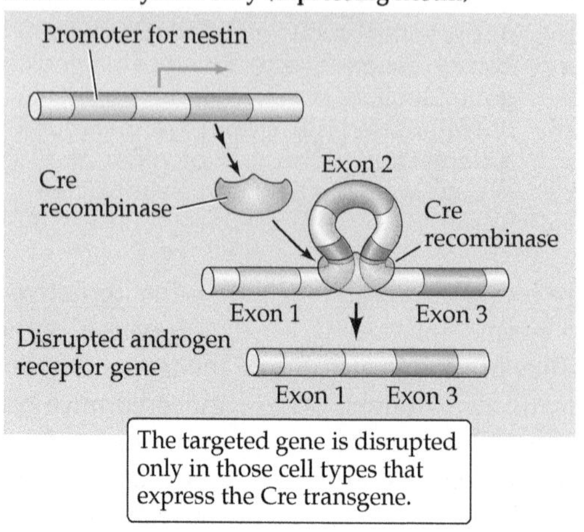

The targeted gene is disrupted only in those cell types that express the Cre transgene.

FIGURE 1.13 Genetic engineering. (A) Creation of a homozygous carrier mouse. (B) Conditional mutagenesis using the Cre/lox system, which excises a genetic sequence of interest. (C) DNA editing using CRISPR-Cas9. (A after Stewart and Mintz, 1981.)

either in newly fertilized mouse zygotes (transgenic mice are made this way) or in mouse embryonic stem (ES) cells ("knock-out" and "knock-in" mice are made this way). In each instance, the murine genome is altered so that all cells in the mouse will carry the mutation. Knock-in and knock-out mice are made by **homologous recombination**. This approach relies on endogenous cellular mechanisms for DNA replication and repair. DNA polymerases and ligases can mistake a synthetic DNA sequence targeted by homologous 3′ and 5′ sequences to a specific region of the genome and substitute ("recombine") the exogenous DNA sequence (at a low frequency) for the sequence normally found at that location. ES cells that have undergone homologous recombination for the targeted gene sequence are then injected into the blastocyst of a newly fertilized in vitro mouse embryo, where they integrate into the developing embryo, including germ cells (embryonic cells that give rise to egg or sperm precursors). This genetically engineered mouse can now pass the engineered gene to subsequent generations.

Knock-in and knock-out mice can be engineered for **conditional mutations** using the **Cre/lox** system, in which an exogenous recombinase recognizes unique DNA excision sequences (loxP sequences) that are introduced at the 5′ and 3′ ends of an endogenous gene and eliminates the intervening sequence (Figure 1.13B). The loxP sequences are not found in mammalian genomes but occur in genomes of bacteria targeted by certain viruses. The viruses use a unique DNA cutting enzyme, called **Cre recombinase** (Cre stands for Causes *re*combination), to cut pieces of DNA out of the bacterial genome, and then recombine the cut ends. Cre recombinase is also not found in any vertebrate genome, so in applying the Cre/lox system to murine models, the gene encoding Cre recombinase must first be introduced into the mouse genome. The Cre insert is engineered so that it has DNA sequences on the 5′ and 3′ ends that are homologous to an endogenous mouse gene. During mitotic DNA replication, the Cre DNA gets recombined into the genome at that locus and is then expressed under the control of the promoter and other regulatory sequences for that gene. With expression of the Cre DNA, the resulting Cre recombinase engages the loxP binding sites, and the intervening endogenous exon targeted for elimination (the so-called floxed sequence) is excised.

In a further refinement of this technique, Cre recombinase has been reengineered with a genetically modified estrogen receptor that cannot bind endogenous estrogen (the gonadal steroid) and can only be activated by an exogenous chemical (tamoxifen), a synthetic estrogen analog. This approach, referred to as the Cre:ERT method (ERT stands for *e*strogen *r*eceptor reengineered for *t*amoxifen activation), allows for temporal control of recombination. Tamoxifen is given at the time during development or in the adult that one wishes to assess gene function, and the target gene is excised or activated by the Cre recombinase only at that time.

An even newer approach to genetic engineering uses CRISPR-Cas9 DNA editing, which allows specific mutations to be inserted into targeted genes (Figure 1.13C). CRISPR-Cas9 relies on a specific RNA guide sequence (gRNA) that combines with **tracrRNA** recognized by the bacterial Cas9 DNA excision/repair enzyme. This RNA/enzyme complex cleaves the DNA at the genomic location recognized by the guide sequence. Following Cas9 excision, the DNA is repaired by non-homologous end joining, yielding a microdeletion mutation. Alternatively, a donor DNA sequence can be inserted following Cas9 cleavage via a mechanism similar to homologous recombination. The consequences of these modifications can then be studied.

Despite these remarkable techniques, identifying genes responsible for diseases has been much harder than anticipated, and whether mutations in the mouse are valid replicas of a human disease is always uncertain.

Structural Analysis of Neural Systems

Early observers of neural systems made inferences of functional localization (i.e., which region of the nervous system serves which function) by correlating behavioral deficits (e.g., paralysis of a specific extremity, difficulty speaking or understanding language) to damaged brain structures observed post mortem. Using structure to infer function was soon adapted to experimental animals, and much neuroscience rests on observations made by purposefully damaging a brain region, nerve, or tract and observing a subsequent loss of function. Indeed, such **lesion studies** have provided much current understanding of the anatomy of neural systems.

More detailed neuroanatomical studies that help define neural systems emerged with the advent of techniques that can trace neural connections from their source to their termination (**anterograde**), or from terminus to source (**retrograde**). Initially these techniques relied on injecting single neurons or a brain region with visible dyes or other molecules, which were then taken up by local cell bodies and transported to axon terminals, or taken up by local axon terminals and transported back to the parent cell body (Figure 1.14A,B). Such tracers can also demonstrate an entire network of axonal projections from nerve cells exposed to the tracer (Figure 1.14C). More recent approaches use viral vectors, made innocuous via genetic engineering, to insert tracer proteins into the nucleus of a target neuronal population. These viruses often can use the host cell to replicate and be released, thereby tracing circuitry beyond the direct target of any particular set of nerve cells. Together these approaches permit assessment of the extent of connections from a single population of nerve cells to their targets throughout the nervous system.

Analyses of connectivity have been augmented by molecular and histochemical techniques that demonstrate biochemical and genetic distinctions among nerve cells and their processes. Whereas conventional cell staining methods show differences in cell size and distribution, *antibody stains* recognize specific proteins in different

FIGURE 1.14 Cellular and molecular approaches for studying connectivity and molecular identity of nerve cells. (A–C) Tracing connections and pathways in the brain. (A) Radioactive amino acids can be taken up by one population of nerve cells (in this case, injection of a radioactively labeled amino acid into one eye) and transported to the axon terminals of those cells in the target region in the brain. (B) Fluorescent molecules injected into nerve tissue are taken up by the axon terminals at the site of the injection (the dark layers evident in the thalamus). The molecules are then transported, labeling the cell bodies and dendrites of the nerve cells that project to the injection site. (C) Tracers that label axons can reveal complex pathways in the nervous system. In this case, a dorsal root ganglion has been injected, showing the variety of axon pathways from the ganglion into the spinal cord. (D–F) Molecular differences among nerve cells. (D) A single glomerulus in the olfactory bulb (see Figure 1.6H) has been labeled with an antibody against the inhibitory neurotransmitter GABA. The label shows up as a red stain, revealing GABA to be localized in subsets of neurons around the glomerulus (arrowheads) as well as at synaptic endings in the neuropil of the glomerulus (asterisk). (E) The cerebellum has been labeled with an antibody that recognizes subsets of dendrites as well as cell bodies (green). (F) Here the cerebellum has been labeled with an RNA probe (blue) for a specific gene transcript that is expressed only by Purkinje cells. (A courtesy of P. Rakic; B courtesy of B. Schofield; C courtesy of W. D. Snider and J. Lichtman; D–F courtesy of A.-S. LaMantia, D. Meechan, and T. Maynard.)

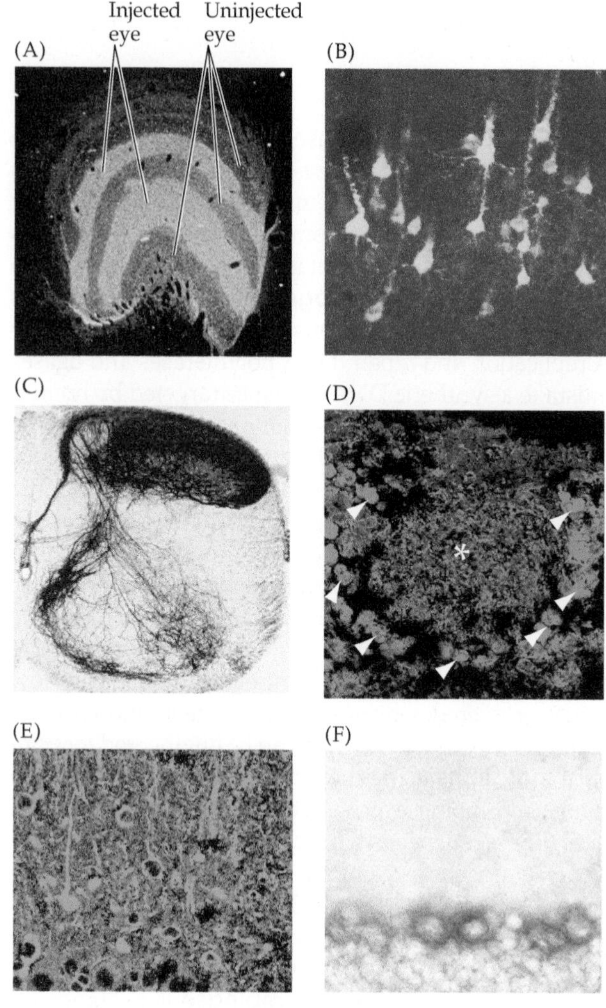

regions of a nerve cell, or molecular differences in classes of nerve cells. The use of antibody stains has clarified the functional distribution of synapses, dendrites, and other distinguishing features of nerve cell types in a variety of regions (Figure 1.14D,E). In addition, antibodies against various proteins, as well as probes for specific mRNA transcripts (which detect gene expression in the relevant cells), have been used for similar purposes (Figure 1.14F).

More recently still, molecular genetic and neuroanatomical methods have been combined to visualize the expression of fluorescent or other tracer molecules under the control of regulatory sequences of neural genes (Figure 1.15). This approach illuminates individual cells in fixed or living tissue in remarkable detail, allowing nerve cells and

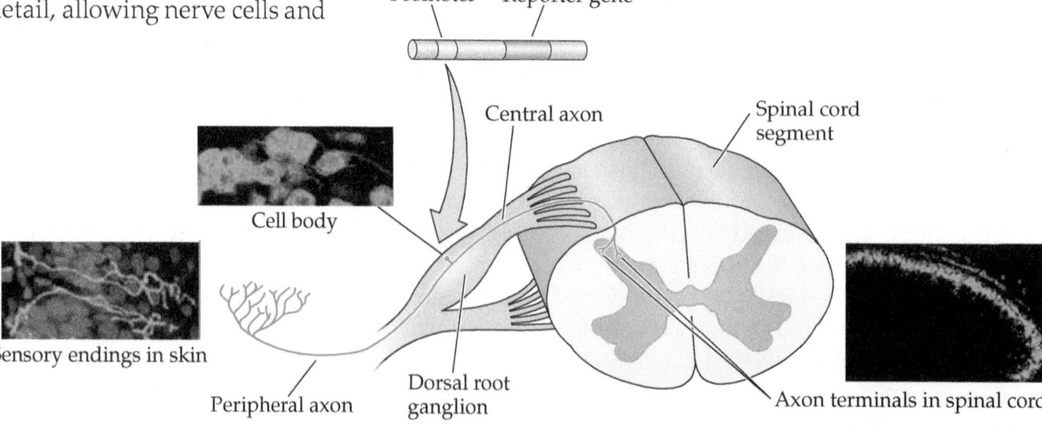

FIGURE 1.15 Genetic engineering used to show pathways in the nervous system. A reporter gene that codes for a visible protein (e.g., green fluorescent protein) is inserted into the genome under the control of a cell type–specific promoter (a mouse DNA sequence that turns the gene "on" in specific tissue and cell types). The reporter is expressed only in those cell types, revealing the cell bodies, axons, and dendrites of all cells in the nervous system that express the gene. Here the reporter is under the control of a promoter DNA sequence that is activated only in a subset of dorsal root ganglion neurons. The reporter labels neuronal cell bodies; the axons that project to the skin as free nerve endings; and the axon that projects to the dorsal root of the spinal cord to relay this sensory information from the skin to the brain. (Photographs from Zylka et al., 2005.)

their processes to be identified by their transcriptional state (i.e., which genes are being transcribed in that cell) as well as their structure and connections. These techniques enable molecular distinctions to be made between otherwise equivalent nerve cells.

In short, a variety of molecular and genetic engineering approaches now allow researchers to trace connections between defined populations of neurons and their targets. The use of pathway tracing and analyses of the molecular and genetic identity of nerve cells and connections in neural systems are now routine.

Functional Analysis of Neural Systems

Functional analyses provide another powerful way of understanding the organization of nervous systems. A variety of physiological methods are now available to evaluate the electrical and metabolic activity of the neuronal circuits that make up a neural system. Two approaches—electrophysiological recording and functional brain imaging—have been particularly useful in defining how neural systems convey and represent information.

The most widely used electrophysiological method, however, has been single-cell, or single-unit, electrophysiological recording with microelectrodes, as discussed earlier in the chapter (see Figures 1.8 and 1.9). This method often records from several nearby cells in addition to the one selected, providing further useful information. The use of microelectrodes to record action potential activity allows cell-by-cell analysis of the organization of topographic maps (see above), and can give specific insight into the type of stimulus to which the neuron is "tuned" (i.e., the stimulus that elicits a maximal change in neuronal activity from a baseline state). Such tuning defines a neuron's **receptive field**—the region in sensory space (e.g., the body surface or a specialized structure such as the retina) within which a specific stimulus elicits action potential response (Figure 1.16). This way of understanding neural systems was pioneered by Stephen Kuffler and Vernon Mountcastle in the early 1950s and has been used by several generations of neuroscientists to evaluate the relationship between peripheral stimuli (sensory) or actions (motor) and neuronal responses in both sensory and motor systems. Electrical recording techniques at the single-cell level have been extended and refined to include single and simultaneous multiple cell analysis in animals performing complex cognitive tasks, intracellular recordings in intact animals, and the use of patch electrodes to detect and monitor the activity of the individual membrane molecules that ultimately underlie neural signaling (see Chapters 2 and 3). For neurons that are not concerned with space as such, stimulus-selective responses further define **receptive field properties** that are used to categorize neurons. For example, the response to different odorant molecules can be used to define the preferences of individual neurons in the olfactory system. Such preferences can be identified for any neuron. Thus, a neuron in the visual system, for example, can be tested for preferences to color, motion, and so on, in addition to being tested for its response to stimuli presented in a distinct location in visual space.

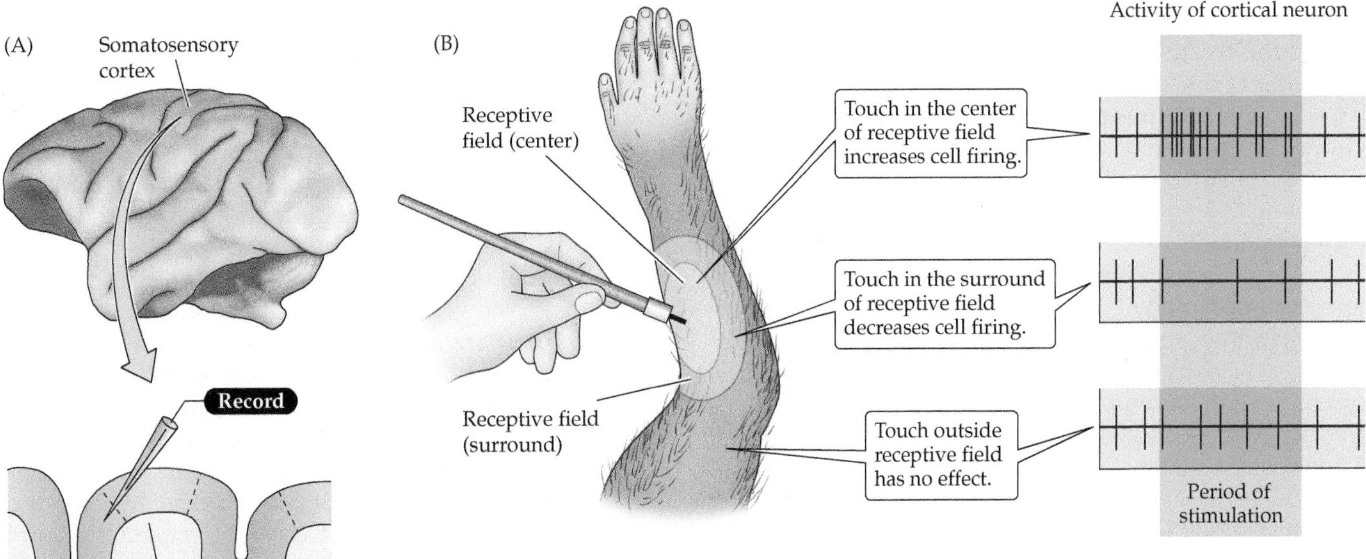

FIGURE 1.16 Single-unit electrophysiological recording in a monkey. This example is from a cortical pyramidal neuron, showing the firing pattern in response to a specific peripheral stimulus in the anesthetized animal. (A) Typical experiment setup, in which a recording electrode is inserted into the somatosensory cortex in the postcentral gyrus. (B) Defining neuronal receptive fields.

Functional brain imaging (see below) provides another, noninvasive approach for studying neural activity. Over the last three decades, techniques for brain imaging, especially in humans (and, to a lesser extent, experimental animals), have revolutionized our understanding of neural systems and the ability to describe and diagnose functional abnormalities in individuals with brain disorders. Unlike electrophysiological methods of recording neural activity, which require exposing the brain or some other part of the nervous system and inserting electrodes, brain imaging is thus applicable to both patients and healthy individuals who volunteer for neuroscientific investigations. Current functional imaging methods record local metabolic activity in relatively small volumes of brain tissue. These approaches allow the simultaneous evaluation of multiple brain structures, which is possible but difficult with electrical recording methods.

Analyzing Complex Behavior

Many advances in modern neuroscience have involved reducing the complexity of the brain to more readily analyzed components—genes, molecules, cells, and circuits. But, more complex brain functions such as perception, language, emotions, memory, and consciousness remain a challenge for contemporary neuroscientists. Over the last 30 years or so, a new field called **cognitive neuroscience** has emerged that is specifically devoted to understanding these issues (see Unit V). One approach to cognitive neuroscience is to design and validate specific behavioral tasks that can be used to assess aspects of human or animal information processing and behavior. These tasks can be used in humans and assessed based on correct versus incorrect responses, numbers of trials needed to learn the task, or the reaction time between the presentation of a stimulus and the individual's response. In addition, tasks can be adapted for presentation in a magnetic resonance scanner, and performed while the scanner records changes in patterns of activity (inferred from blood flow changes) in the human brain, as described in more detail in the following sections.

This evolution of cognitive neuroscience has also rejuvenated **neuroethology**, the field devoted to observing complex behaviors of animals in their native environments—for example, social communication in birds and non-human primates—and has encouraged the development of tasks to better evaluate the genesis of complex human behaviors. When used in combination with reductionist neuroscience methods, brain regions that are active when engaging in tasks that involve language, mathematics, music, aesthetics, and even abstract thinking and social appraisals can be evaluated. Carefully constructed behavioral tasks can also be used to study the pathology of complex neurological disorders that compromise cognition, such as Alzheimer's disease, schizophrenia, and depression. Although there is clearly a long way to go, these increasingly powerful approaches are beginning to unravel even the most complex aspects of human behavior in scientific terms.

Imaging the Living Human Brain

Until the late 1970s, most understanding of the structure and function of the human brain was derived from postmortem human specimens or inferred from animal studies. While these approaches were informative, the information they provided necessarily failed to convey the functional complexity of the human brain. Furthermore, most anatomical and functional correlations were based on the consequences of brain damage, raising additional uncertainties. The advent of techniques for imaging the anatomical and functional details of the human brain resulted in observations that supported many of the conclusions drawn from postmortem and animal studies. Their primary value was for clinical or diagnostic purposes. In the last several decades, however, newer approaches have been used in safe, non-invasive research in healthy human participants as well as individuals with brain damage or neurological diseases. In addition to confirming older conclusions in a more rigorous way, these methods opened new avenues for exploring how the brain carries out complex functions such as language, reading, math, music, and more.

Early Brain Imaging Using X-Rays

The first phase of human brain-imaging techniques included X-ray-based methods such as **pneumoencephalography** and **cerebral angiography** (Figure 1.17). In pneumoencephalography, air is injected into the subarachnoid space that contains cerebrospinal fluid, thus providing better X-ray contrast (see Figure 1.17A). This approach was especially useful in visualizing ventricular anomalies that cause hydrocephalus (dilation of one or several brain ventricles due to obstructions of the ventricular system). Nonetheless, the risks and discomfort produced by this method limited its use to patients whose provisional diagnoses could be confirmed in this way. Newer methods (described below) have replaced this approach.

In cerebral angiography, a contrast agent is introduced into the circulation via an arterial catheter. X-ray images are then taken of the head in different planes, revealing the blood vessels as a network within the brain (see Figure 1.17B). This approach allowed identification of arterio-venous malformations and other vascular anomalies such as arterial occlusions. Cerebral angiography could also inform clinical assessments of vascular accidents ("strokes") during a time when few other approaches were available. But again, the substantial risk of the procedure limits its use.

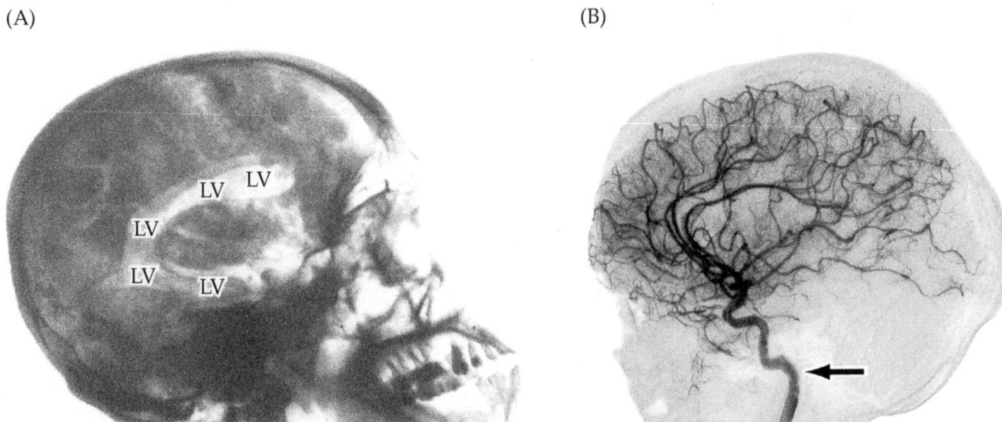

FIGURE 1.17 **Imaging basic features of the brain and cerebral vasculature using older X-ray-based methods.** (A) Sagittal view from a pneumoencephalogram of the head and brain of a typical individual. The ventricles in this individual have been injected with air, which makes them more translucent than the surrounding tissue. This allows the ventricular space, particularly the lateral ventricles (LV), which span the anterior as well as posterior regions of each cortical hemisphere, to be seen clearly. (B) Sagittal view from a cerebral angiogram. The carotid artery (arrow) is the large vessel leading to the reticulum of vessels that are seen on the lateral surface. The carotid artery gives rise to multiple branches on the lateral surface of the cerebral hemispheres that establish the blood supply for most of the cerebral cortex. Note that the brain tissue itself has little contrast when imaged by X-ray. (From Hoeffner et al., 2012.)

Early Functional Mapping Using Cortical Surface Stimulation and Electroencephalography

The advent of electrophysiological techniques that could be used in animals—particularly in non-human primates—brought about a new understanding of the functional organization of the brain, particularly the cerebral cortex. Over the late nineteenth century and on into the twentieth century, pioneering neurophysiologists demonstrated functional areas in the intact primate brain that process specific sensory modalities. More detailed studies of the brains of experimental animals in the mid-twentieth century showed that there are actually multiple cortical regions for various sensory modalities in each hemisphere. These studies also defined the basic properties of topographic maps, especially for the somatosensory and visual cortex. These areas implied a sophistication of processing beyond the low-resolution map based on postmortem analysis of lesions in humans, or even on lesion studies with more precisely localized cortical damage in non-human primates.

No doubt inspired by this work, research to develop techniques for recording the electrical activity of intact human brains proceeded apace. These methods initially were used as clinical tools, especially in the localization of epileptic foci while exploring sites for neurosurgical removal of epileptogenic cortical tissue. At the Montreal Neurological Institute, Herbert Jasper, one of the key developers of **electroencephalography** (**EEG**), worked with Wilder Penfield, who used non-penetrating surface electrodes to stimulate cortical regions exposed for neurosurgical purposes (EEG relies on scalp electrodes to detect changes in electrical activity; see below). Penfield's primary goal was to ensure that no critical cortical tissue would be removed from either hemisphere. For example, he carefully mapped the limits of the language areas in the left hemisphere to avoid any damage to these regions. Thus, Penfield (as well as subsequent neurosurgeons) exercised maximal caution when removing an "epileptic focus" identified by EEG to relieve otherwise intractable seizures in individuals with epilepsy. Penfield and Jasper's efforts also revealed both the somatosensory maps of the body surface (Figure 1.18A; also see Chapter 9) and the motor maps of movement intention expressed by the body's musculature (see Chapter 17) in humans (the *homunculus;* see Figure 9.11C). The patients being evaluated were alert and responsive (only local anesthesia was used). They could thus report where systematic stimulation elicited sensation on the body surface, and the surgeons could observe where stimulation caused muscle contraction. This work in neurosurgical patients was later extended to the mapping of other functions such as language (see Chapter 27). It was obvious, however, that there would be enormous benefit for both clinical practice and basic neuroscience if diagnosis and mapping could be done noninvasively in clinical patients as well as in nonclinical volunteers.

One useful, but low-resolution, approach to noninvasive activity mapping has been developed using a "cap" with scalp EEG electrodes placed in an ordered spatial array across the head. This approach, referred to as **event related potential** analysis or **ERP** (Figure 1.18B), uses neither radioactivity nor electrical stimulation. Instead, the

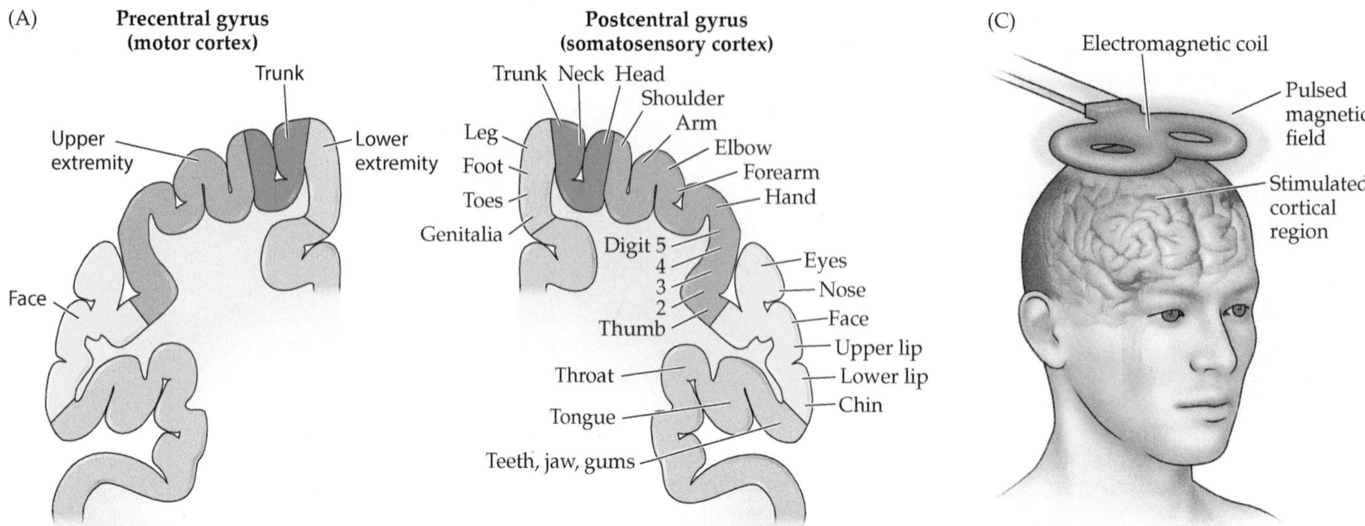

(B)
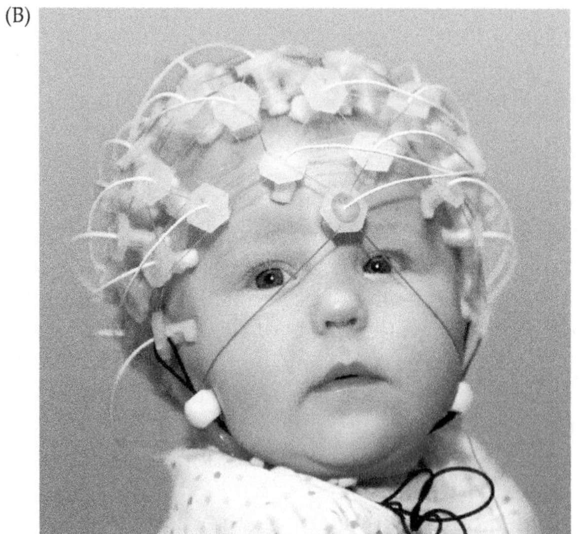

FIGURE 1.18 **Indirect "images" of functional localization and maps using invasive and noninvasive neurophysiological approaches.** (A) Penfield and Jasper defined the homunculus using intracranial surface electrode mapping. This approach relied on stimulation using an electrode placed carefully on the arachnoid surface to avoid damage to the underlying brain tissue. This early work resulted in the somatotopic map, as represented here. (B) Event related potential (ERP) recording in an awake, alert child. The array of scalp electrodes in the "cap" indicates the recording locations. Comparing the response intensity recorded at each site provides a low-resolution map of localized cortical activity. (C) Transcranial magnetic stimulation (TMS) relies on a handheld device that generates a magnetic field based on current flow through a magnetic coil. The device can deliver a pulse of current to the surface of the cerebral cortex, causing a brief disruption of electrical activity in that area. TMS has gained some acceptance as a clinical treatment for depression and other mood disorders, as well as being used to assess normal function. (B © Jon Wilson/Science Source.)

net electrical activity from each point in the scalp electrode array is detected, amplified, and mapped with reference to each electrode's position on the head. Individuals are presented with sensory stimuli or directed to execute a motor task, and time-locked electroencephalographic signals are averaged with respect to stimulus or task onset. ERPs can be recorded from adults as well as children, facilitating activity-based analysis of developmental behavioral and brain-based changes. These analyses allow, indirectly, for general localization during the performance of different tasks by examining anterior-posterior or medial-lateral differences in EEG activity. While ERP lacks the ability to define specific cortical areas, its relative ease of use makes it possible to perform experiments in standard laboratory settings. Thus, despite limited resolution, ERP analysis can be used in individuals performing a range of tasks.

Finally, a relatively novel, noninvasive approach for modifying local brain activity has been introduced into both clinical practice and behavioral and physiological research in humans. **Transcranial magnetic stimulation (TMS**; Figure 1.18C) uses magnetic pulses delivered by a paddlelike device held near the scalp. When the magnetic pulses are delivered locally in this way, activity in the underlying cortical tissue is briefly disrupted, leading to a transient change in behavioral performance. In effect, this transient "lesion" of activity causes no apparent harm to patients or to healthy volunteers. TMS has been adapted for use in standard nonclinical laboratory settings, often in combination with ERP analysis, to assess typical activity patterns during complex behavioral tasks. When stronger magnetic pulses are delivered, it is also possible to activate the output of the underlying cortical tissue. Therapeutic applications of TMS are under active investigation for a variety of neurological and neuropsychiatric conditions.

Computerized Tomography

In the 1970s, **computerized tomography**, or **CT**, ushered in a new era in noninvasive imaging using computer processing technology to probe the living brain. CT uses a narrow X-ray beam and a row of very sensitive detectors placed on opposite sides of the head to probe a small portion of tissue at a time with limited radiation exposure (Figure 1.19A). In order to form an image, the X-ray tube and detectors rotate around the head, collecting radiodensity information from every orientation around a narrow slice. Computer processing techniques then calculate the radiodensity of each point within the slice plane, producing a tomographic image (Greek *tomo,* "cut" or "slice"). If the individual is moved through the scanner slowly while the X-ray tube rotates in this manner, a three-dimensional radiodensity matrix can be created, allowing images corresponding to serial sections to be computed for any plane through the brain. CT scans can readily distinguish gray matter and white matter, differentiate the ventricles quite well, and show many other brain structures with a spatial resolution of several millimeters. Thus, major anatomical structures can be identified with relative confidence (Figure 1.19B), and lesions can be recognized if they are within the limits of the resolution of the CT (a few mm with newer techniques). Importantly, brain lesions that are not visible in standard X-ray methods can be resolved using CT. For example, metastatic lesions related to a distal malignancy can be localized fairly precisely, and can be correlated more definitively with functional loss as well as with subsequent response to treatments (Figure 1.19C). CT remains a valued diagnostic tool; however, its use in fundamental brain research on healthy individuals is limited due to the risks of unnecessary radiation exposure and its relatively low resolution of brain structure.

Magnetic Resonance Imaging

Brain imaging took a huge step forward in the 1980s with the development of **magnetic resonance imaging (MRI)**. Unlike CT, MRI is based on the physics of atomic motion. The nuclei of some atoms act as spinning magnets. If placed in a strong magnetic field, these atoms will line up with the field and spin at a frequency that is dependent on the field strength. If a brief radiofrequency pulse tuned to the atoms' spinning frequency is applied, the atoms are knocked out of alignment with the field and subsequently emit energy in an oscillatory fashion as they gradually realign themselves with the field. The strength of the emitted signal depends on how many atomic nuclei are affected by this process.

In MRI, the magnetic field is distorted slightly by imposing magnetic gradients along three different spatial axes so that only nuclei at certain locations are tuned to the detector's frequency at any given time. Almost all MRI scanners use detectors tuned to the radio frequencies of spinning hydrogen nuclei in water molecules, creating images based on the distribution of water in different tissues

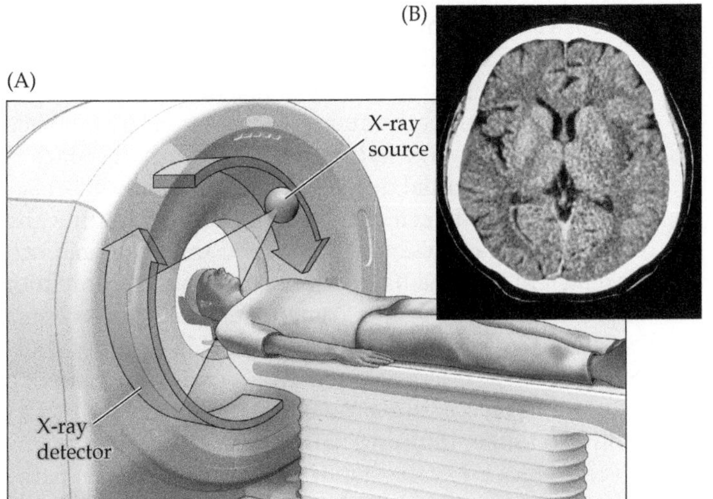

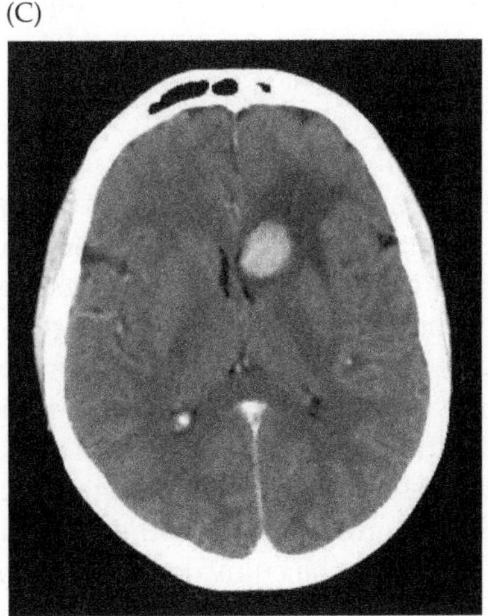

FIGURE 1.19 Computerized axial tomography. (A) In computerized tomography (CT), the X-ray source and detector are moved around the individual's head. (B) Horizontal CT section of a typical adult brain. (C) CT scan of an individual with multiple sites of a metastatic brain tumor (white spots throughout the cortical gray and white matter). CT scans are very useful in detecting brain lesions where the damage has a different tissue density than the normal brain tissue. (B © Puwadol Jaturawutthichai/ Alamy Stock Photo; C from Khairy et al., 2015.)

(A)

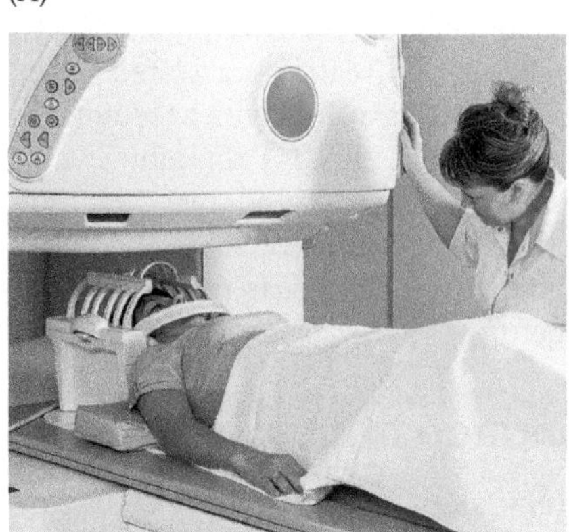

(B)

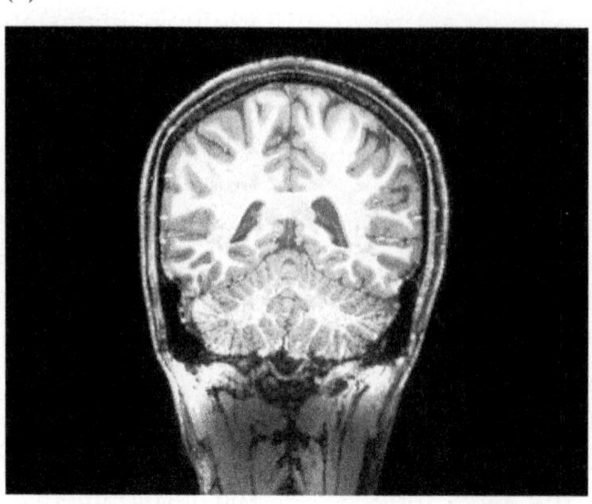

(C)

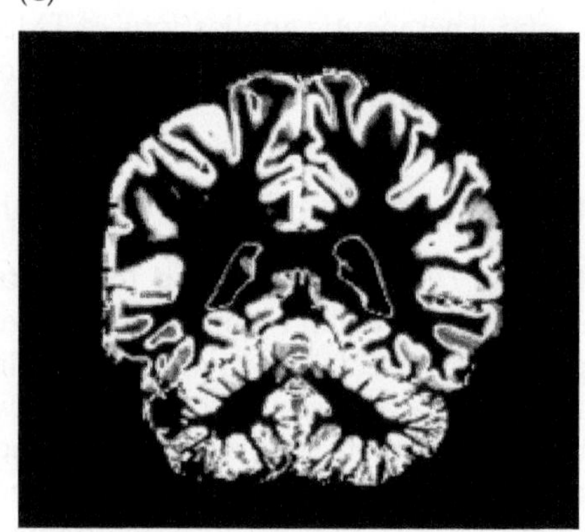

FIGURE 1.20 Magnetic resonance imaging. (A) The MRI scanner has a portal for the individual's head (or other region of the body to be imaged). A magnetic coil is placed around the head to activate and record magnetic resonance signal. Virtual reality goggles or earphones can be used to present visual or auditory stimuli. (B,C) MRI images obtained with two different pulse sequences. In (B), the pulse sequence yields data that record the white matter of the cerebral cortex as white and the gray matter as gray. In (C), the pulse sequence shows the cortical gray matter as a lighter gray with a white boundary at the external surface, and white matter is seen as a darker gray. (B,C from Seiger et al., 2015.)

(Figure 1.20A). Careful manipulation of magnetic field gradients and radiofrequency pulses makes it possible to construct extraordinarily detailed spatial images of the brain at any location and orientation, with millimeter resolution (Figure 1.20B,C; see also Figure 1.1C).

Safety (there is no high-energy radiation), noninvasiveness (no dyes are injected), and versatility (applicable to individuals in a variety of conditions) have made MRI the technique of choice for imaging brain structure in most applications. The strong magnetic field and radiofrequency pulses used in scanning are harmless (although ferromagnetic objects in or near the scanner are a safety concern). Moreover, MRI enables a suite of imaging modalities; by changing the scanning parameters, images based on a wide variety of different contrast mechanisms can be generated. For example, conventional MRI takes advantage of the fact that hydrogen in different types of tissue (e.g., gray matter, white matter, cerebrospinal fluid) has slightly different realignment rates, meaning that soft tissue contrast can be manipulated by adjusting the time at which the realigning hydrogen signal is measured. This results in remarkably detailed images that show structural features of the human brain and, for several important clinical conditions, the presence of pathophysiological processes. Another advantage is that, like CT scans but with far better resolution, the MRI data from each individual represent the equivalent of a "whole brain": Software has been developed so that from one detailed scan, it is possible to create detailed views in all the cardinal planes of section for the brain (see Appendix Figure A1), as well as three-dimensional renderings of volumes, such as those of the cortical surface or the intracerebral ventricles.

In addition, MRI can be used to detect changes in metabolic intermediates that may be related to ongoing neurotransmission. This application takes advantage of the well-established capacity of magnetic resonance methods (known as nuclear magnetic resonance, or NMR, in chemistry) to detect different organic molecules based on the spectra of their atomic properties. When this approach is used in human brain imaging, metabolites of excitatory amino acid neurotransmitters (glutamate) or inhibitory neurotransmitters (γ-aminobutyrate, or GABA; glycine) can be detected and their concentration and distribution mapped. The most commonly imaged metabolic intermediate is **N-acetyl aspartate** (**NAA**), which is thought to be primarily an indicator of glutamatergic transmission in the brain. Based on additional imaging routines (using differing field strengths and times), potential metabolites of GABA and glycine can be detected. Such approaches suggest that potential differences in neurotransmitter activity might be detected in brains of control or clinically compromised individuals. Moreover, this approach can give insight into the balance of excitatory and inhibitory activity needed for the performance of various behaviors.

The alignment of the magnetic fields of water molecules in axon tracts also makes it possible to visualize axon pathways using a variant of MRI referred to as **diffusion tensor imaging** (**DTI**) (see the image that opens this chapter). DTI can establish differences in axon pathway connectivity, making it possible to study individuals with genetic disorders that result in major alterations of axon projections. Additional settings can generate images in which gray matter and white matter are relatively invisible but the brain vasculature stands out in sharp detail. Thus, variations of MRI can clearly and safely visualize human neuroanatomy, including cerebral cortical regions, some subcortical structures, and axon tracts.

Functional Brain Imaging

Imaging specific functions in the brain has become possible with the development of techniques for detecting local changes in cerebral metabolism or blood flow. To conserve energy, the brain regulates its blood flow such that active neurons with relatively high metabolic demands receive more oxygen and nutrients than do relatively inactive neurons. Detecting and mapping these local changes in cerebral blood flow form the basis for four functional brain-imaging techniques: **positron emission tomography** (**PET**), **single-photon emission computerized tomography** (**SPECT**), **functional magnetic resonance imaging** (**fMRI**), and **magnetoencephalography** (**MEG**). PET and SPECT are less commonly chosen for either research or clinical applications due to the necessity of exposure to radiolabeled compounds (see below). fMRI has all of the advantages of safety enumerated for structural MRI imaging. MEG has excellent temporal resolution, allowing rapid acquisition of functional changes. Its limited spatial resolution, however, limits its utility in many applications. Thus, fMRI has emerged as the favored approach in most circumstances.

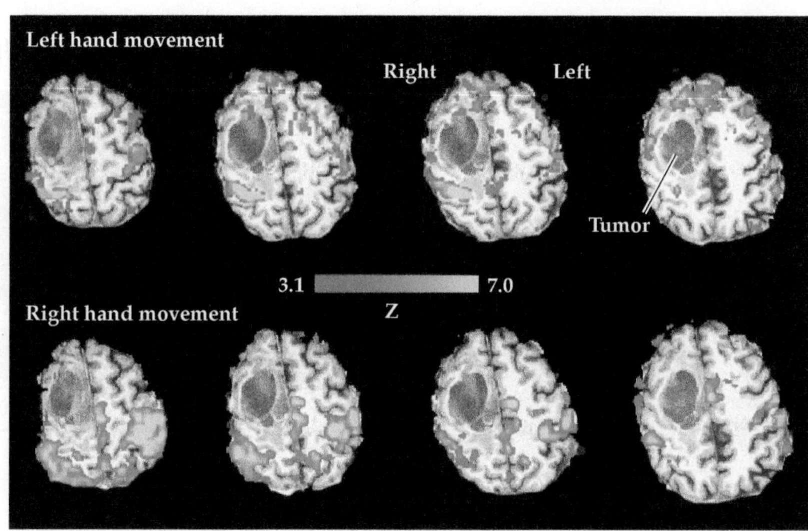

FIGURE 1.21 fMRI of a patient's brain that harbored a right frontal lobe glioma (gray area). When asked to move the left hand, greater activity was observed within the right motor and sensorimotor cortices compared to the resting state. Since the activity was not close to the tumor, the surgeon could operate with assurance that the motor control of the hand would not be affected. Red-to-yellow indicates increasing relative strength of brain activity. (From Goodyear et al., 2014.)

In PET scanning, unstable positron-emitting isotopes are incorporated into different reagents (including water, precursor molecules of specific neurotransmitters, or glucose) and injected into the bloodstream. Labeled oxygen and glucose quickly accumulate in more metabolically active areas. As the unstable isotope decays, it emits two positrons travelling in opposite directions. Gamma ray detectors placed around the head register a "hit" only when two detectors 180 degrees apart react simultaneously. Images of tissue isotope density can then be generated in much the way CT images are calculated, showing the location of active regions with a spatial resolution of about 4 mm (see Figures 33.6 and 33.7). Depending on the probe injected, PET imaging can be used to visualize activity-dependent changes in blood flow, tissue metabolism, or biochemical activity. SPECT is similar to PET in that it involves injection or inhalation of radiolabeled compound (for example, ^{133}Xe or ^{123}I-labeled iodoamphetamine), which produces photons that are detected by a gamma camera moving rapidly around the head.

fMRI offers the least invasive, most cost-effective approach for visualizing brain function based on local metabolism. fMRI relies on the fact that hemoglobin in blood slightly distorts the magnetic resonance properties of hydrogen nuclei in its vicinity, and the amount of magnetic distortion changes depending on whether the hemoglobin has oxygen bound to it. When a brain area is activated by a specific task, it begins to use more oxygen, and within seconds the brain microvasculature responds by increasing the flow of oxygen-rich blood to the active area. These changes in the concentration of oxygen and blood flow lead to localized **blood oxygenation level-dependent** (**BOLD**) changes in the magnetic resonance signal. Such fluctuations are detected using statistical image-processing techniques to produce maps of active brain regions (Figure 1.21). Because fMRI uses signals intrinsic in the brain, tracer injections are not necessary, and repeated observations can be made on the same individual—a major advantage over imaging methods such as PET. The spatial resolution (2 to 3 mm) and temporal resolution (a few seconds) of fMRI are superior to those of PET and SPECT, and can be combined with structural images generated by MRI using computational and statistical methods. fMRI has thus emerged as the technology of choice for functional brain imaging. It dominates the clinical as well as the cognitive science literature studying the functional basis of complex behaviors. The necessity of averaging fMRI signals over seconds, plus the statistical assumptions necessary to fit highly variable individual brains into standardized neuroanatomical templates, imposes some significant limitations, but these have not reduced the popularity of fMRI.

(A)

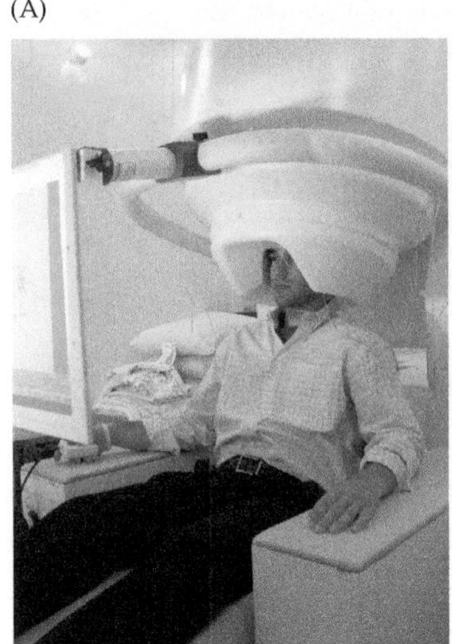

(B)

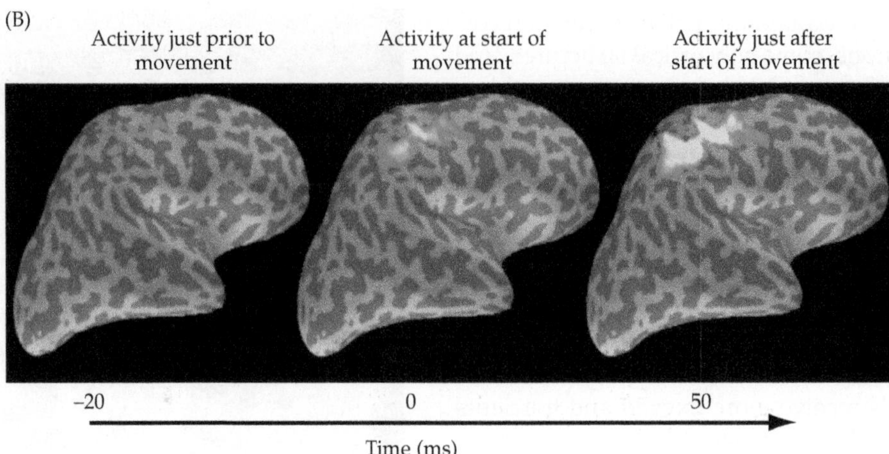

FIGURE 1.22 Magnetoencephalography (MEG) provides greater temporal resolution with useful spatial resolution. (A) The individual is fitted with a helmet that includes several magnetic detectors (a SQUID array) and then is placed in a biomagnetometer (the large cylindrical structure) that can amplify the small local changes in magnetic field orientation or strength that indicate fast temporal current flow changes in ensembles of neurons. MEG, like MRI and fMRI, is noninvasive. (B) The temporal resolution of MEG permits millisecond resolution of electrical activity in the human brain before, during, and after performance of a variety of tasks (displayed in color over a structural MRI of the same brain that has been "inflated" to reveal cortex folded into sulci and fissures). In this case, a digit was moved in response to a specific sensory cue. There was some "anticipatory" activity in the region of the somatosensory as well as motor cortices where the digit was represented. As the movement began, this baseline activity increased so that a wider region of both the somatosensory and motor cortices was activated. As the movement proceeded, the activity increased over the time of performance. (A courtesy of National Institute of Mental Health, National Institutes of Health, Department of Health and Human Services; B courtesy of Judith Schaechter, PhD, MGH Martinos Center for Biomedical Imaging.)

MEG provides a way of localizing brain function with better temporal resolution than fMRI. Thus, MEG is sometimes used to map changes in brain function with millisecond resolution rather than the lower temporal resolution of seconds afforded by fMRI. MEG, as its name suggests, records the magnetic consequences of brain electrical activity rather than the electrical signals themselves. Thus, unlike its close relative EEG, MEG detects independent sources of current flow, without reference to other currents. The magnetic signals that MEG recordings detect are quite local, and there is fairly good spatial resolution (a few mm) of the signal origin; thus, one can detect dynamic electrical activity in the brain with temporal resolution that approximates the key events in neuronal electrical signaling (i.e., action potentials and synaptic potentials) and can record the location of the source of that activity. For MEG recording, an array of individual detector devices called SQUIDs (superconducting *qu*antum *i*nterference *d*evices) is arranged as a helmet and fitted onto the individual. The individual is then placed in a biomagnetometer scanner that amplifies the signals detected by the SQUID (Figure 1.22).

These signals, like EEG signals, provide a map of current sources across the brain, using reference points (usually the ears and nose) to create a three-dimensional space. Given its millisecond (or even faster) temporal resolution combined with its spatial resolution, MEG can be used to evaluate local activity changes over time in individuals performing a variety of tasks. This allows for comparison of baseline activity before task initiation, changes during task performance, and activity after task completion (see Figure 1.22). In addition, MEG can be used to map the temporal characteristics as well as the brain localization of epileptic foci, reducing the need for intrasurgical brain surface electrode mapping. Even though MEG has reasonable spatial resolution, MEG maps alone often lack sufficient anatomical detail for some applications. Thus, MEG is often combined with structural MRI (see above), a combination referred to as **magnetic source imaging**, or **MSI**.

In sum, the use of modern structural and functional imaging methods has revolutionized human neuroscience. It is now possible to obtain images of the developing brain as it grows and changes, and of the living brain in action, assessing brain activity both in typical individuals and in individuals with neurological disorders.

Summary

The brain can be studied by methods ranging from genetics and molecular biology to behavioral testing and non-invasive imaging in healthy and neurologically compromised humans. Such studies have created ever-increasing knowledge about the organization of the nervous system. Many of the most notable successes in modern neuroscience have come from understanding nerve cells as the structural and functional units of the nervous system. The organization of neurons and glia into neural circuits provides the substrate for processing of sensory, motor, and cognitive information. The physiological characterization of the neurons and glia in such circuits, as well as more detailed analysis of the electrical properties of individual neurons, recorded with electrodes placed near or inserted inside the cell, provide insight into the dynamic nature of information processing, sensory maps, and the overall function of neural systems. Minimally invasive and non-invasive imaging methods can be used in humans with or without brain damage or disease. These methods can help integrate observations from animal models and human post-mortem analysis with structure and function in the living human brain. Among the goals that remain are understanding how basic molecular genetic phenomena are linked to cellular, circuit, and system functions; understanding how these processes go awry in neurological and psychiatric diseases; and understanding the especially complex functions of the brain that make us human.

ADDITIONAL READING

Reviews

Baker, M. (2014) Gene editing at CRISPR speed. *Nat. Biotechnol.* 32: 347–355.

Cell Migration Consortium (2014) Transgenic and knockout mouse—Approaches. www.cellmigration.org/resource/komouse/komouse_approaches.shtml.

Dennis, C., R. Gallagher and P. Campbell (eds.) (2001) Special issue on the human genome. *Nature* 409: 745–964.

Gao, X. and P. Zhang (2007) Transgenic RNA interference in mice. *Physiology* 22: 161–166.

Jasny, B. R. and D. Kennedy (eds.) (2001) Special issue on the human genome. *Science* 291: 1153.

Kim, H. and J.-S. Kim (2014) A guide to genome engineering with programmable nucleases. *Nat. Rev. Genet.* 5: 321–334.

Kravitz, A. V. and 6 others (2010) Regulation of parkinsonian motor behaviours by optogenetic control of basal ganglia circuitry. *Nature* 466: 622–626.

Ohki, K., S. Chung, Y. H. Ch'ng, P. Kara and R. C. Reid (2005) Functional imaging with cellular resolution reveals precise micro-architecture in visual cortex. *Nature* 433: 597–603.

Raichle, M. E. (1994) Images of the mind: Studies with modern imaging techniques. *Annu. Rev. Psychol.* 45: 333–356.

Stewart, T. A. and B. Mintz (1981) Successive generations of mice produced from an established culture line of euploid teratocarcinoma cells. *Proc. Natl. Acad. Sci. USA* 78: 6314–6318.

Wheless, J. W. and 6 others (2004) Magnetoencephalography (MEG) and magnetic source imaging (MSI). *Neurologist* 10: 138–153.

Zhang, F. and 12 others (2011) The microbial opsin family of optogenetic tools. *Cell* 147: 1446–1457.

Books

Brodal, P. (2010) *The Central Nervous System: Structure and Function*, 4th Edition. New York: Oxford University Press.

Gibson, G. and S. Muse (2009) *A Primer of Genome Science*, 3rd Edition. Sunderland, MA: Sinauer Associates.

Huettel, S. A., A. W. Song and G. McCarthy (2009) *Functional Magnetic Resonance Imaging*, 2nd Edition. Sunderland, MA: Sinauer Associates.

Oldendorf, W. and W. Oldendorf, Jr. (1988) *Basics of Magnetic Resonance Imaging.* Boston: Kluwer Academic Publishers.

Peters, A., S. L. Palay and H. de F. Webster (1991) *The Fine Structure of the Nervous System: Neurons and Their Supporting Cells*, 3rd Edition. New York: Oxford University Press.

Posner, M. I. and M. E. Raichle (1997) *Images of Mind*, 2nd Edition. New York: W. H. Freeman.

Purves, D. and 6 others (2013) *Principles of Cognitive Neuroscience*, 2nd Edition. Sunderland, MA: Sinauer Associates.

Ramón y Cajal, S. (1990) *New Ideas on the Structure of the Nervous System in Man and Vertebrates.* (Transl. by N. Swanson and L. W. Swanson.) Cambridge, MA: MIT Press.

Ropper, A. H. and N. Samuels (2009) *Adams and Victor's Principles of Neurology*, 9th Edition. New York: McGraw-Hill Medical.

Schild, H. (1990) *MRI Made Easy (...Well, Almost).* Berlin: H. Heineman.

Schoonover, C. (2010) *Portraits of the Mind: Visualizing the Brain from Antiquity to the 21st Century.* New York: Abrams.

Shepherd, G. M. (1991) *Foundations of the Neuron Doctrine.* History of Neuroscience Series, no. 6. Oxford, UK: Oxford University Press.

Go to the NEUROSCIENCE 6e Companion Website at **www.oup.com/uk/Purves6e** for Web Topics, Animations, Flashcards, and more.

UNIT I

Neural Signaling

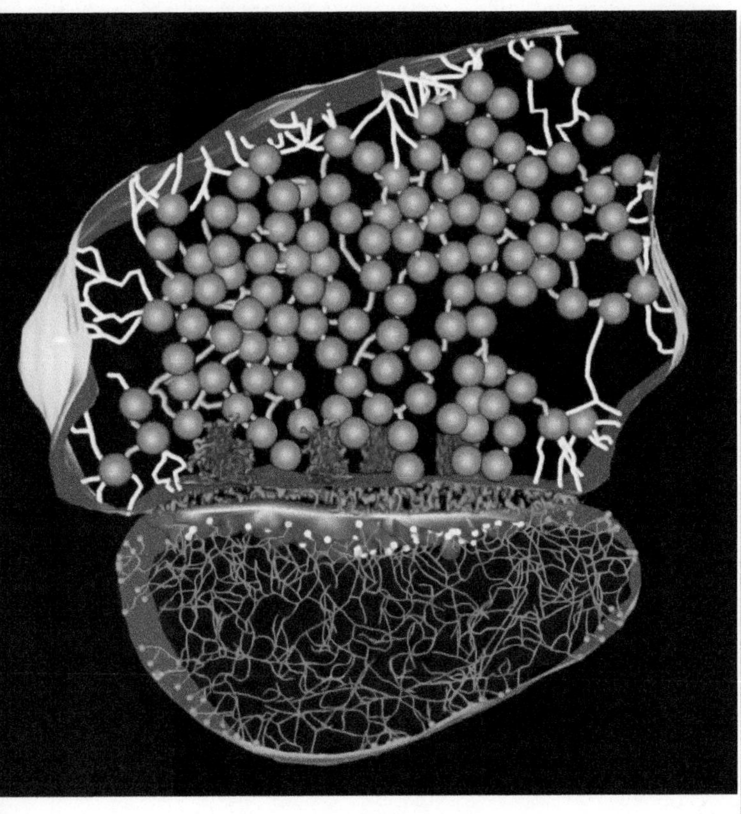

THE BRAIN IS REMARKABLY ADEPT at acquiring, coordinating, and disseminating information about the body and its environment. Such information must be processed within milliseconds, yet it also can be stored as memories that endure for years. Neurons perform these functions by generating sophisticated electrical and chemical signals. This unit describes these signals and how they are produced. It explains how one type of electrical signal, the action potential, allows information to travel along the length of a nerve cell. It also explains how other types of signals—both electrical and chemical—are generated at synaptic connections between nerve cells. Synapses permit information transfer by interconnecting neurons to form the circuitry on which brain information processing depends. Finally, this unit describes the intricate biochemical signaling events that take place within neurons and how such signaling can produce activity-dependent changes in synaptic communication. Understanding these fundamental forms of neuronal signaling provides a foundation for appreciating the higher-level functions considered in the rest of the book.

The cellular and molecular mechanisms that give neurons their unique signaling abilities are targets for disease processes that compromise the function of the nervous system, as well as targets for anesthetics and many other clinically important drugs. A working knowledge of the cellular and molecular biology of neurons is therefore fundamental to understanding a variety of brain pathologies, and for developing novel approaches to diagnosing and treating these all too prevalent problems.

On the previous page:

Structure of a chemical synapse within the cerebral cortex. A presynaptic terminal (top) forms a synapse with a dendritic spine of the postsynaptic neuron (bottom). Colors indicate different organelles found within these structures. (Courtesy of Alain Burette and Richard Weinberg; see Figure 5.4.)

CHAPTER

2 Electrical Signals of Nerve Cells

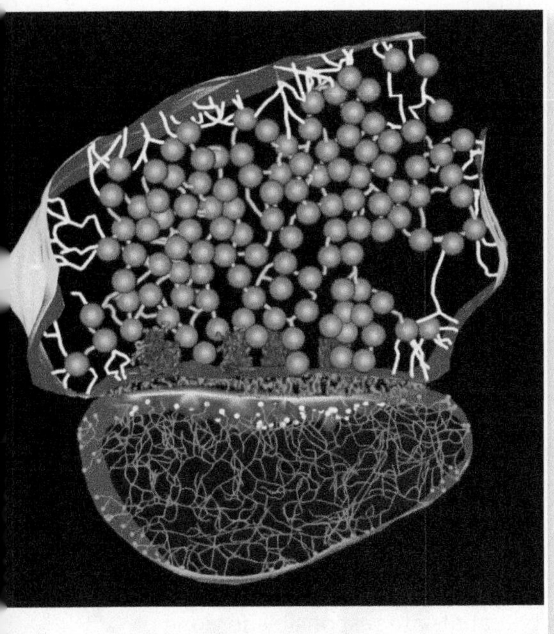

Overview

NERVE CELLS GENERATE A VARIETY of electrical signals that transmit and store information. Although neurons are not intrinsically good conductors of electricity, they have elaborate mechanisms that generate electrical signals based on the flow of ions across their plasma membranes. Ordinarily, neurons generate a negative potential, called the *resting membrane potential,* that can be measured by recording the voltage between the inside and outside of nerve cells. The action potential is a fundamental electrical signal that transiently abolishes the negative resting potential and makes the transmembrane potential positive. Action potentials are propagated along the length of axons and carry information from one place to another within the nervous system. Still other types of electrical signals are produced by the activation of synaptic contacts between neurons or by the actions of external forms of energy, such as light and sound, on sensory neurons. All of these electrical signals arise from ion fluxes brought about by the selective ionic permeability of nerve cell membranes, produced by ion channels, and the nonuniform distribution of these ions across the membrane, created by active transporters.

Electrical Signals of Nerve Cells

Neurons employ several different types of electrical signals to encode and transfer information. The best way to observe these signals is to use an intracellular microelectrode to measure the electrical potential across the neuronal plasma membrane. A typical microelectrode is a piece of glass tubing pulled to a very fine point (with an opening less than 1 μm in diameter) and filled with a good electrical conductor, such as a concentrated salt solution. This conductive core can then be connected to a voltmeter, typically a computer, that records the transmembrane voltage of the nerve cell.

The first type of electrical phenomenon can be observed as soon as a microelectrode is inserted through the membrane of the neuron. Upon entering the cell, the microelectrode reports a negative potential, indicating that neurons have a means of generating a constant voltage across their membranes when at rest. This voltage, called the **resting membrane potential**, depends on the type of neuron being examined, but it is always a fraction of a volt (typically –40 to –90 mV).

Neurons encode information via electrical signals that result from transient changes in the resting membrane potential. **Receptor potentials** are due to the activation of sensory neurons by external stimuli, such as light, sound, or heat. For example, touching the skin activated neve endings in Pacinian corpuscles, receptor neurons that sense mechanical disturbances of the skin. These neurons respond to touch with a receptor potential that changes the resting potential for a fraction of a second (Figure 2.1A). These transient changes in potential are the first step in generating the sensation of vibrations

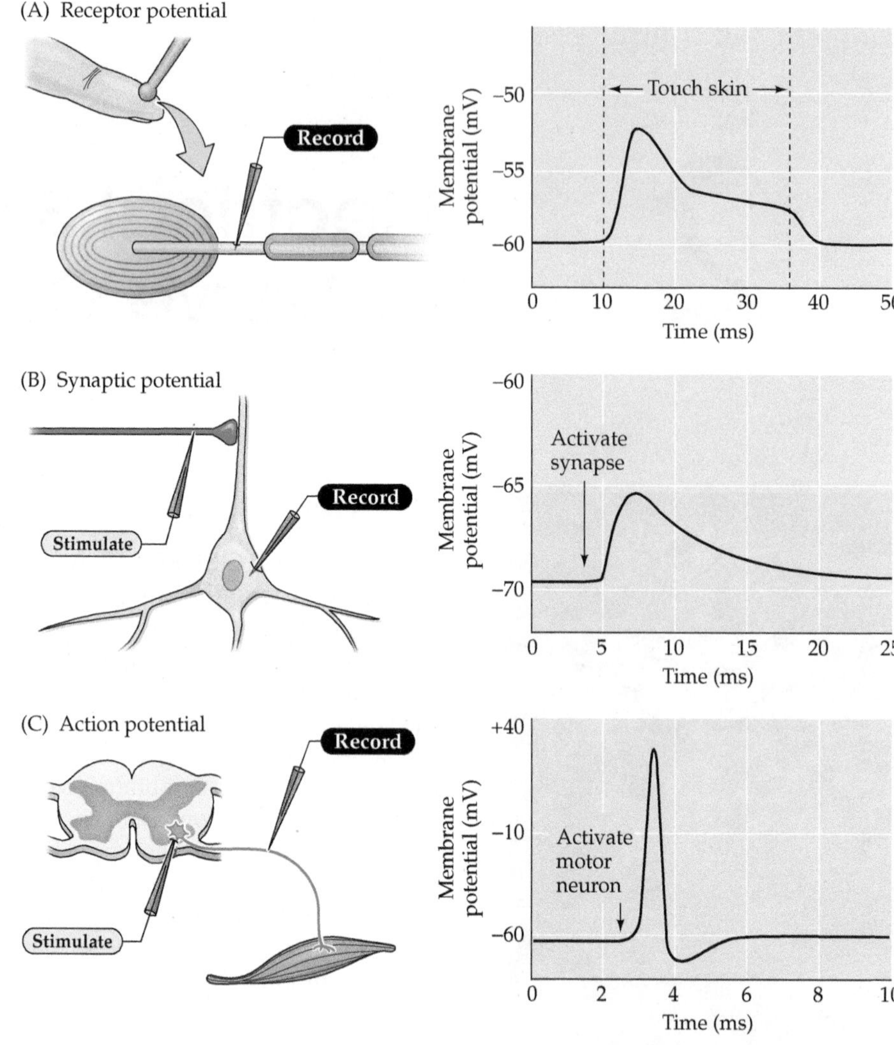

FIGURE 2.1 Types of neuronal electrical signals. In all cases, microelectrodes are used to measure changes in the resting membrane potential during the indicated signals. (A) A brief touch causes a receptor potential in a Pacinian corpuscle in the skin. (B) Activation of a synaptic contact onto a hippocampal pyramidal neuron elicits a synaptic potential. (C) Stimulation of a spinal reflex produces an action potential in a spinal motor neuron.

of the skin in the somatosensory system (see Chapter 9). Similar sorts of receptor potentials are observed in all other sensory neurons during transduction of sensory stimuli (see Unit II).

Another type of electrical signal is associated with communication between neurons at synaptic contacts. Activation of these synapses generates **synaptic potentials**, which allow transmission of information from one neuron to another. An example of such a signal is shown in Figure 2.1B. In this case, activation of a synaptic terminal innervating a hippocampal pyramidal neuron causes a very brief change in the resting membrane potential in the pyramidal neuron. Synaptic potentials serve as the means of exchanging information in the complex neural circuits found in both the central and peripheral nervous systems (see Chapter 5).

Finally, many neurons generate a special type of electrical signal that travels along their long axons. Such signals are called **action potentials** and are also referred to as *spikes* or *impulses*. An example of an action potential recorded from the axon of a spinal motor neuron is shown in Figure 2.1C. Action potentials are responsible for long-range transmission of information within the nervous system and allow the nervous system to transmit information to its target organs, such as muscle.

One way to elicit an action potential is to pass electrical current across the membrane of the neuron. In normal circumstances, this current would be generated by receptor potentials or by synaptic potentials. In the laboratory, however, electrical current suitable for initiating an action potential can be readily produced by inserting a microelectrode into a neuron and then connecting the electrode to a battery (Figure 2.2A). A second microelectrode can be inserted to measure the membrane potential changes produced by the applied current. If the current delivered in this way makes the membrane potential more negative (**hyperpolarization**), nothing very dramatic happens. The membrane potential simply changes in proportion to the magnitude of the injected current (Figure 2.2B, central part). Such hyperpolarizing responses do not require any unique property of neurons and are therefore called **passive electrical responses**. A much more interesting

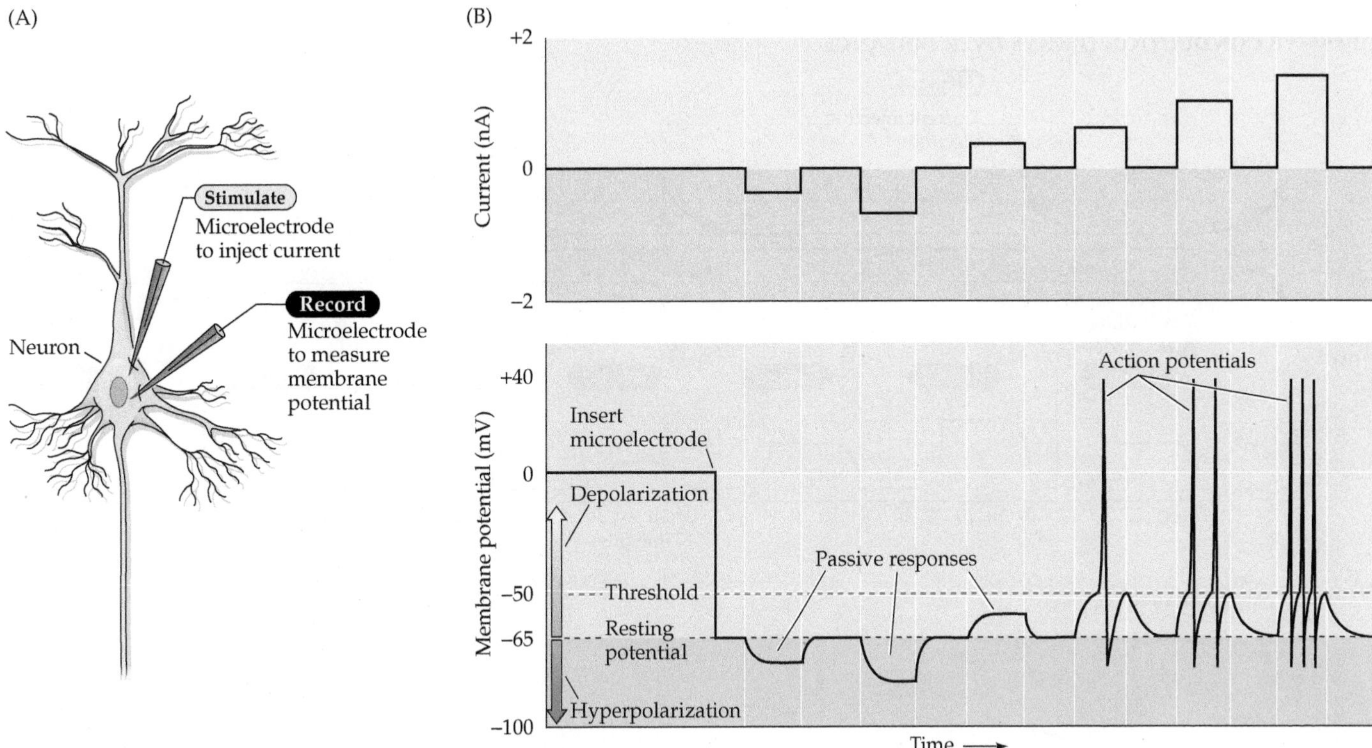

FIGURE 2.2 **Recording passive and active electrical signals in a nerve cell.** (A) Two microelectrodes are inserted into a neuron; one of these measures membrane potential while the other injects current into the neuron. (B) Inserting the voltage-measuring microelectrode into the neuron (bottom) reveals a negative potential, the resting membrane potential. Injecting current through the other microelectrode (top) alters the neuronal membrane potential. Hyperpolarizing current pulses produce only passive changes in the membrane potential. While small depolarizing currents also elicit only passive responses, depolarizations that cause the membrane potential to meet or exceed threshold additionally evoke action potentials.

phenomenon is seen if current of the opposite polarity is delivered, so that the membrane potential of the nerve cell becomes more positive than the resting potential (**depolarization**). In this case, at a certain level of membrane potential, called the **threshold potential**, action potentials occur (see Figure 2.2B, right side).

The action potential is an active response generated by the neuron and typically is a brief (about 1 ms) change from negative to positive in the transmembrane potential. Action potentials are considered active responses because they are generated by selective changes in the permeability of the neuronal membrane. Importantly, the amplitude of the action potential is independent of the magnitude of the current used to evoke it; that is, larger currents do not elicit larger action potentials. The action potentials of a given neuron are therefore said to be *all-or-none*—that is, they occur fully or not at all. If the amplitude or duration of the stimulus current is increased sufficiently, multiple action potentials occur, as can be seen in the responses to the three different current intensities shown in Figure 2.2B (right side). It follows, therefore, that the intensity of a stimulus is encoded in the *frequency* of action potentials rather than in their amplitude. This arrangement differs dramatically from that of receptor potentials, whose amplitudes are graded in proportion to the magnitude of the sensory stimulus; and from that of synaptic potentials, whose amplitudes vary according to the number of synapses activated, the strength of each synapse, and the previous amount of synaptic activity.

Long-Distance Transmission of Electrical Signals

The use of electrical signals—as in sending electricity over wires to provide power or information—presents a series of challenges in electrical engineering. A fundamental problem for neurons is that their axons, which can be quite long (remember that a spinal motor neuron can extend for a meter or more), are not good electrical conductors. Although neurons and wires are both capable of passively conducting electricity, the electrical properties of neurons compare poorly with that of an ordinary wire. This can be seen by measuring the passive electrical properties of a nerve cell axon by determining the voltage

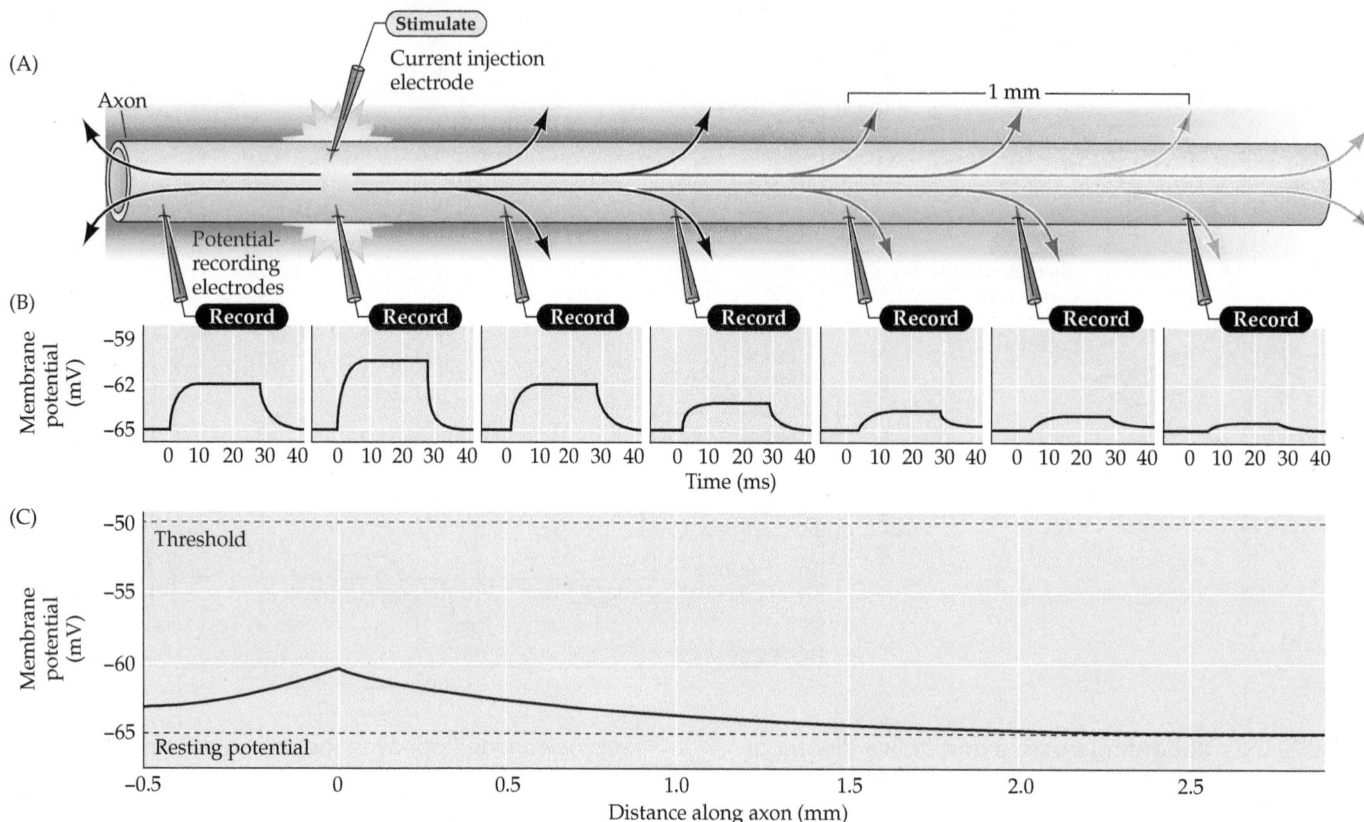
PASSIVE CONDUCTION DECAYS OVER DISTANCE
(A)
Stimulate
Current injection electrode
Axon
1 mm
Potential-recording electrodes
(B)
Record
Membrane potential (mV)
-59
-62
-65
0 10 20 30 40
Time (ms)
(C)
-50
Threshold
-55
-60
-65
Resting potential
-0.5
0
0.5
1.0
1.5
2.0
2.5
Distance along axon (mm)

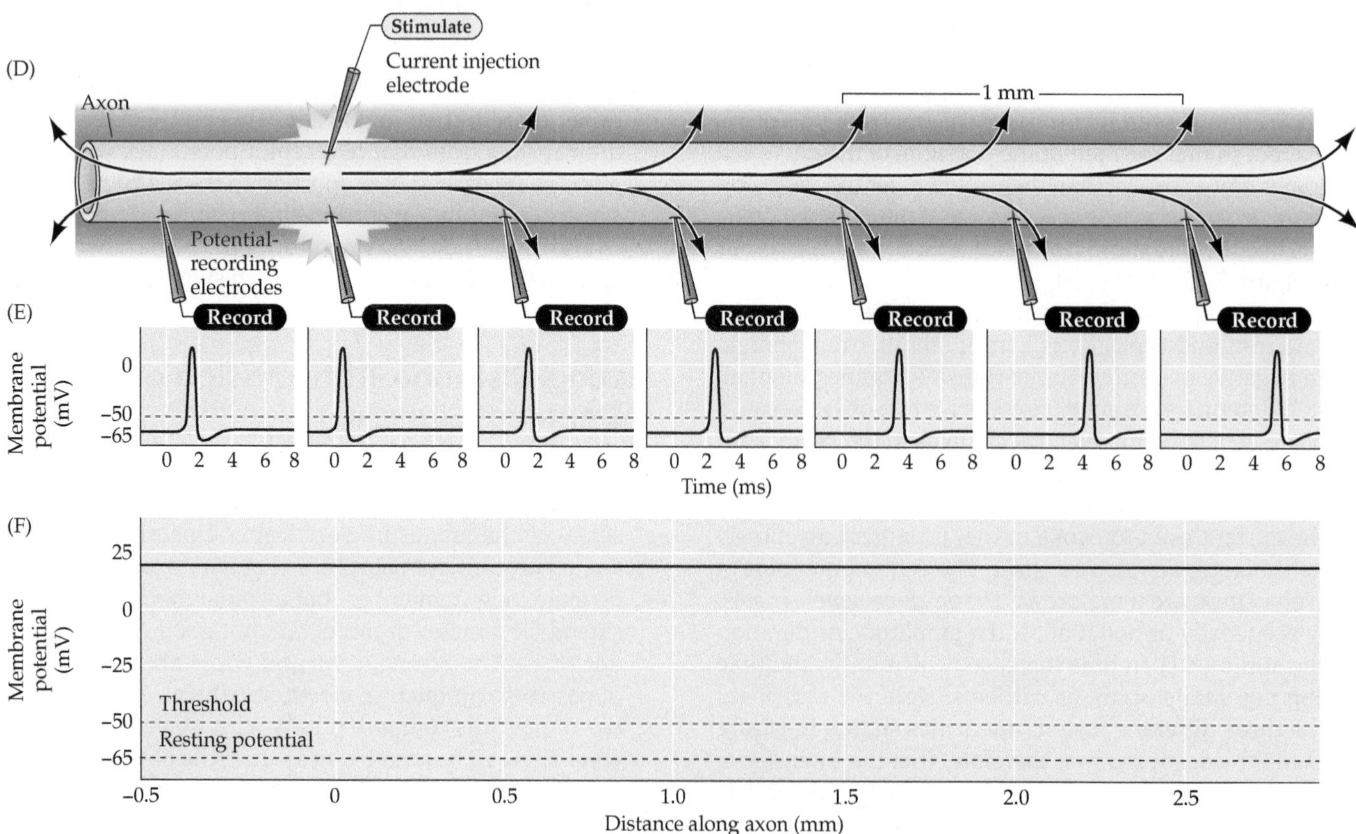
ACTIVE CONDUCTION IS CONSTANT OVER DISTANCE
(D)
Stimulate
Current injection electrode
Axon
1 mm
Potential-recording electrodes
(E)
Record
Membrane potential (mV)
0
-50
-65
0 2 4 6 8
Time (ms)
(F)
25
0
-25
Threshold
-50
Resting potential
-65
-0.5
0
0.5
1.0
1.5
2.0
2.5
Distance along axon (mm)

◀ FIGURE 2.3 **Passive and active current flow in an axon.** (A) Experimental arrangement for examining passive flow of electrical current in an axon. A current-passing electrode produces a current that yields a subthreshold change in membrane potential, which spreads passively along the axon. (B) Potential responses recorded by microelectrodes at the positions indicated. With increasing distance from the site of current injection, the amplitude of the potential change is attenuated as current leaks out of the axon. (C) Relationship between the amplitude of potential responses and distance. (D) If the experiment shown in (A) is repeated with a suprathreshold current, an active response, the action potential, is evoked. (E) Action potentials recorded at the positions indicated by microelectrodes. The amplitude of the action potential is constant along the length of the axon, although the time of appearance of the action potential is delayed with increasing distance. (F) The constant amplitude of an action potential (solid black line) measured at different distances. (After Hodgkin and Rushton, 1946.)

change resulting from a current pulse passed across the axonal membrane (Figure 2.3A). If this current pulse is below the threshold for generating an action potential, then the magnitude of the resulting potential change will decay with increasing distance from the site of current injection (Figure 2.3B). Typically, the potential falls to a small fraction of its initial value at a distance of no more than a few millimeters away from the site of injection (Figure 2.3C). For comparison, a wire would typically allow passive current flow over distances many thousands of times longer. The progressive decrease in the amplitude of the induced potential change occurs because the injected current leaks out across the axonal membrane; accordingly, farther along the axon less current is available to change the membrane potential. This leakiness of the axonal membrane prevents effective passive conduction of electrical signals along the length of all but the shortest axons (those 1 mm or less in length). To compensate for this deficiency, action potentials serve as a "booster system" that allows neurons to conduct electrical signals over great distances despite the poor passive electrical properties of axons.

The ability of action potentials to boost the spatial spread of electrical signals can be seen if the experiment shown in Figure 2.3A is repeated with a depolarizing current pulse that is large enough to produce an action potential (Figure 2.3D). In this case, the result is dramatically different. Now an action potential of constant amplitude is observed along the entire length of the axon (Figure 2.3E). The fact that electrical signaling now occurs without any decrement (Figure 2.3F) indicates that active conduction via action potentials is a very effective way to circumvent the inherent leakiness of neurons.

Action potentials are the basis of information transfer in the nervous system and are targets of many clinical treatments, including anesthesia. For these reasons, it is essential to understand how these and other neuronal electrical signals arise. Remarkably, all types of neuronal electrical signals are produced by similar mechanisms that rely on the movement of ions across the neuronal membrane. The remainder of this chapter addresses the question of how nerve cells use ions to generate electrical potentials. Chapter 3 explores more specifically the means by which action potentials are produced and how these signals solve the problem of long-distance electrical conduction within nerve cells. Chapter 4 examines the properties of membrane molecules responsible for electrical signaling. Finally, Chapters 5–8 consider how electrical signals are transmitted from one nerve cell to another at synaptic contacts.

How Ion Movements Produce Electrical Signals

Electrical potentials are generated across the membranes of neurons—and, indeed, across the membranes of all cells—because (1) there are *differences in the concentrations of specific ions* across nerve cell membranes, and (2) these *membranes are selectively permeable* to some of these ions. These two conditions depend in turn on two different kinds of proteins in the plasma membrane (Figure 2.4). The ion concentration gradients are established by proteins known as **active transporters**, which, as their name suggests, actively move ions into or out of cells against their concentration gradients. The selective permeability of membranes is due largely to **ion channels**, proteins that allow only certain kinds of ions to cross the membrane in the direction of their concentration gradients. Thus, channels and transporters basically work against each other, and in so doing they generate the resting membrane potential, action potentials, and the synaptic potentials and receptor potentials that trigger action potentials. Chapter 4 describes the structure and function of these channels and transporters.

To appreciate the role of ion gradients and selective permeability in generating a membrane potential, consider a simple system in which an artificial membrane separates two compartments containing solutions of ions. For comparison to the situation in neurons, we will refer to the left compartment as the inside and call the right compartment the outside. In such a system, it is possible to control the composition of the two solutions and thereby control the ion gradients across the membrane. It is also possible to control the ion permeability of the membrane.

As a first example, consider the case of a membrane that is permeable only to potassium ions (K^+). If the concentration of K^+ on each side of this membrane is equal,

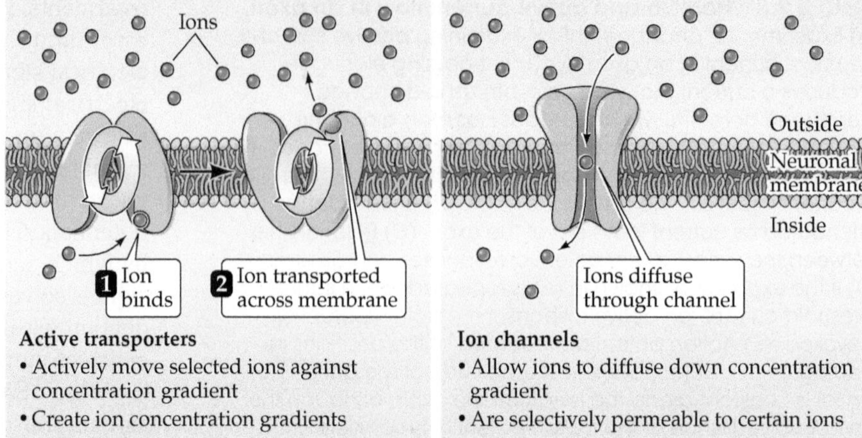

FIGURE 2.4 Active transporters and ion channels are responsible for ion movements across neuronal membranes. Transporters create ion concentration differences by actively transporting ions against their chemical gradients. Channels take advantage of these concentration gradients, allowing selected ions to move, via diffusion, down their chemical gradients.

then no electrical potential will be measured across it (Figure 2.5A). However, if the concentration of K^+ is not the same on the two sides, then an electrical potential will be generated. For instance, if the concentration of K^+ in the inside compartment is ten times higher than the K^+ concentration in the outside compartment, then the electrical potential of the inside will be negative relative to that of the outside (Figure 2.5B). This difference in electrical potential is generated because the potassium ions flow down their concentration gradient and take their electrical charge (one positive charge per ion) with them as they go. Because neuronal membranes contain pumps that accumulate K^+ in the cell cytoplasm, and because potassium-permeable channels in the plasma membrane allow a transmembrane flow of K^+, an analogous situation exists in living nerve cells. As will be proved later in the chapter, such an efflux of K^+ is responsible for the resting membrane potential.

In the hypothetical case just described, an equilibrium will quickly be reached. As K^+ moves from the inside

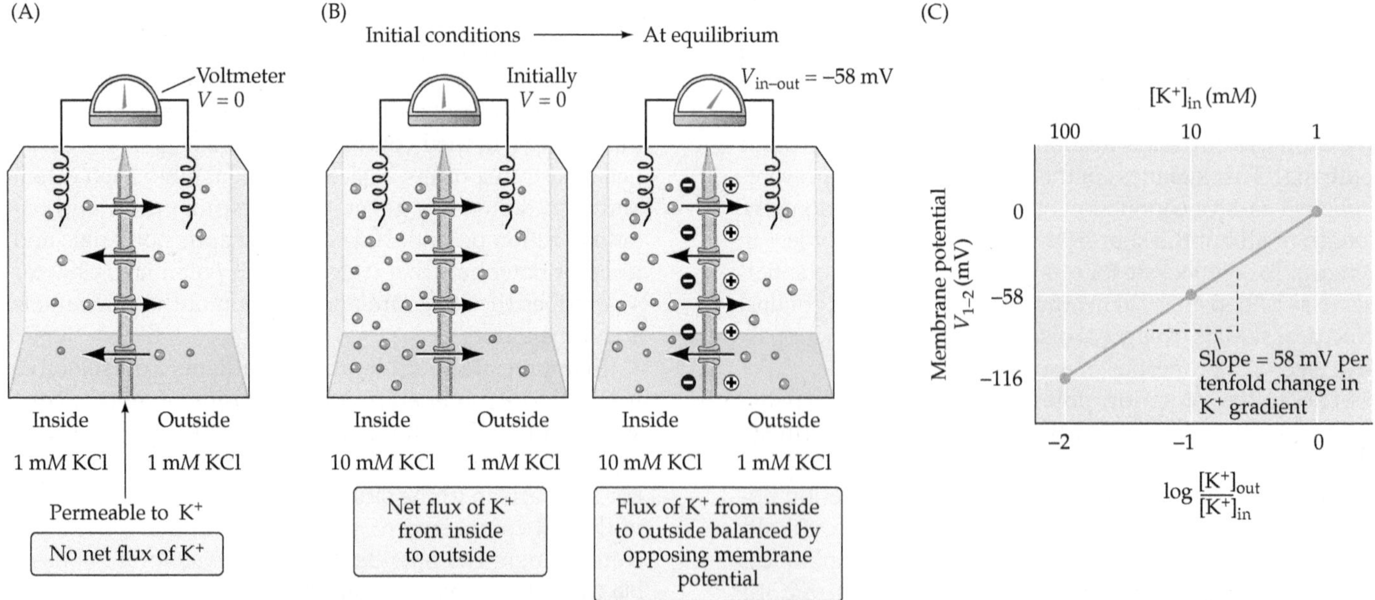

FIGURE 2.5 Electrochemical equilibrium. (A) A membrane permeable only to K^+ (gold spheres) separates the inside and outside compartments, which contain the indicated concentrations of KCl. (B) Increasing the KCl concentration of the inside compartment to 10 m*M* initially causes a small movement of K^+ into the outside compartment (initial conditions) until the electromotive force acting on K^+ balances the concentration gradient, and there is no further net movement of K^+ (at equilibrium). (C) The relationship between the transmembrane concentration gradient ($[K^+]_{out}/[K^+]_{in}$) and the membrane potential. As predicted by the Nernst equation, this relationship is linear when plotted on semilogarithmic coordinates, with a slope of 58 mV per tenfold difference in the concentration gradient.

compartment to the outside (the initial conditions; shown on the left in Figure 2.5B), a potential is generated that tends to impede further flow of K^+. This impediment results from the fact that the potential gradient across the membrane tends to repel the positive potassium ions that would otherwise move across the membrane. Thus, as the outside becomes positive relative to the inside, the increasing positivity makes the outside less attractive to the positively charged K^+. The net movement (or flux) of K^+ will stop at the point ("At equilibrium" in the right panel of Figure 2.5B) where the potential change across the membrane (the relative positivity of the outside compartment) exactly offsets the concentration gradient (the tenfold excess of K^+ in the inside compartment). At this **electrochemical equilibrium**, there is an exact balance between two opposing forces (see Figure 2.5B): (1) the concentration gradient that causes K^+ to move from inside to outside, taking along positive charge, and (2) an opposing electrical gradient that increasingly tends to stop K^+ from moving across the membrane. The number of ions that needs to flow to generate this electrical potential is very small (approximately 10^{-12} moles of K^+ per cm^2 of membrane, or less than one millionth of the K^+ ions present on each side). This last fact is significant in two ways. First, it means that the concentrations of permeant ions on each side of the membrane remain essentially constant, even after the flow of ions has generated the potential. Second, the tiny fluxes of ions required to establish the membrane potential do not disrupt chemical electroneutrality because each ion has an oppositely charged counter-ion (chloride ions in Figure 2.5) to maintain the neutrality of the solutions on each side of the membrane. The concentration of K^+ remains equal to the concentration of Cl^- in the solutions in both compartments, meaning that the separation of charge that creates the potential difference is restricted to the immediate vicinity of the membrane.

Forces That Create Membrane Potentials

The electrical potential generated across the membrane at electrochemical equilibrium—the **equilibrium potential**—can be predicted by a simple formula called the **Nernst equation**. This relationship is generally expressed as

$$E_X = \frac{RT}{zF} \ln \frac{[X]_{out}}{[X]_{in}}$$

where E_X is the equilibrium potential for any ion X, R is the gas constant, T is the absolute temperature (in degrees on the Kelvin scale), z is the valence (electrical charge) of the permeant ion, and F is the Faraday constant (the amount of electrical charge contained in one mole of a univalent ion). The brackets indicate the concentrations of ion X on each side of the membrane, with "in" referring to the inside compartment and "out" referring to the outside, and the symbol *ln* indicates the natural logarithm of the concentration gradient. Because it is easier to perform calculations using base 10 logarithms and to perform experiments at room temperature, this relationship is usually simplified to

$$E_X = \frac{58}{z} \log \frac{[X]_{out}}{[X]_{in}}$$

where log indicates the base 10 logarithm of the concentration ratio. (The constant of 58 becomes 61 mV at mammalian body temperatures.) Thus, for the example in Figure 2.5B, the potential across the membrane at electrochemical equilibrium is

$$E_K = \frac{58}{z} \log \frac{[K]_{out}}{[K]_{in}} = 58 \log \frac{1}{10} = -58\text{mV}$$

The equilibrium potential is conventionally defined in terms of the potential difference between the outside and inside compartments. Thus, when the concentration of K^+ is higher inside than out, an inside-negative potential is measured across the K^+-permeable neuronal membrane.

For a simple hypothetical system with only one permeant ion species, the Nernst equation allows the electrical potential across the membrane at equilibrium to be predicted exactly. For example, if the concentration of K^+ on the inside is increased to 100 m*M*, the membrane potential will be –116 mV. More generally, if the membrane potential is plotted against the logarithm of the K^+ concentration gradient ($[K^+]_{out}/[K^+]_{in}$), the Nernst equation predicts a linear relationship with a slope of 58 mV (actually 58/*z*) per tenfold change in the K^+ gradient (Figure 2.5C).

To reinforce and extend the concept of electrochemical equilibrium, consider some additional experiments on the influence of ion species and ion permeability that could be performed on the simple model system in Figure 2.5. What would happen to the electrical potential across the membrane (i.e., the potential of the inside relative to the outside) if the potassium on the outside were replaced with 10 m*M* sodium (Na^+) and the K^+ in the inside compartment were replaced by 1 m*M* Na^+? No potential would be generated because no Na^+ could flow across the membrane (which was defined as being permeable only to K^+). However, if under these ionic conditions (ten times more Na^+ outside) the K^+-permeable membrane were to be magically replaced by a membrane permeable only to Na^+, a potential of +58 mV would be measured at equilibrium. If 10 m*M* calcium (Ca^{2+}) were present outside and 1 m*M* Ca^{2+} inside, and a Ca^{2+}-selective membrane separated the two sides, what would happen to the membrane potential? A potential of +29 mV would develop—half that observed for Na^+, because the valence of calcium is +2. Finally, what would happen to the membrane potential if 10 m*M* Cl^- were

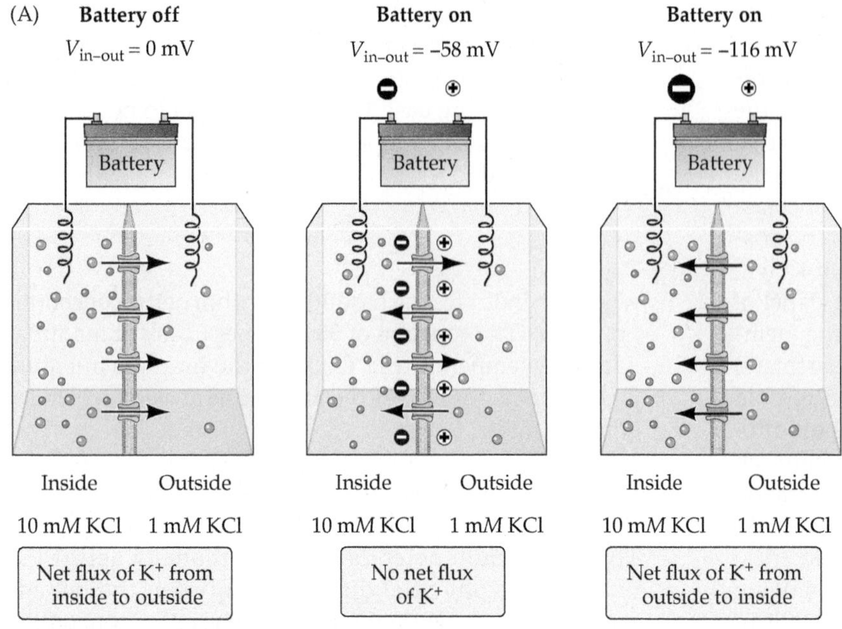

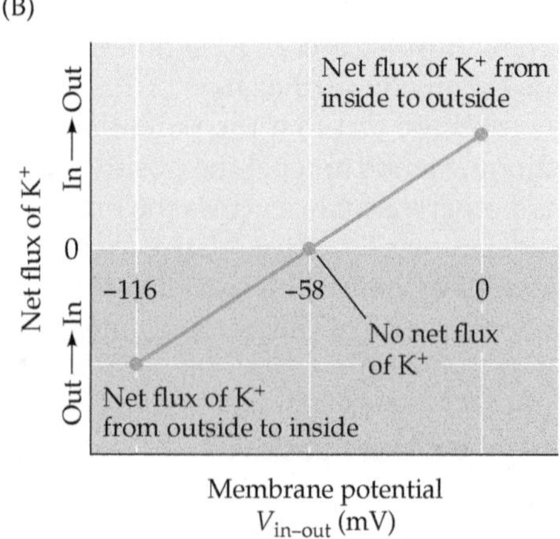

FIGURE 2.6 Membrane potential influences ion fluxes. (A) Connecting a battery across the K^+-permeable membrane allows direct control of membrane potential. When the battery is turned off (left), K^+ ions (gold) flow simply according to their concentration gradient. Setting the initial membrane potential (V_{in-out}) at the equilibrium potential for K^+ (center) yields no net flux of K^+, whereas making the membrane potential more negative than the K^+ equilibrium potential (right) causes K^+ to flow against its concentration gradient. (B) Relationship between membrane potential and direction of K^+ flux.

present inside and 1 m*M* Cl^- were present outside, with the two sides separated by a Cl^--permeable membrane? Because the valence of this anion is –1, the potential would again be +58 mV.

The balance of chemical and electrical forces at equilibrium means that the electrical potential can determine ion fluxes across the membrane, just as the ion gradient can determine the membrane potential. To examine the influence of membrane potential on ion flux, imagine connecting a battery across the two sides of the membrane to control the electrical potential across the membrane without changing the distribution of ions on the two sides (Figure 2.6). As long as the battery is off, things will be just as in Figure 2.5B, with the flow of K^+ from inside to outside causing a negative membrane potential (see Figure 2.6A, left). However, if the battery is used to make the inside compartment initially more negative relative to the outside, there will be less K^+ flux, because the negative potential will tend to keep K^+ in the inside compartment. How negative will the inside need to be before there is no net flux of K^+? The answer is –58 mV, the voltage needed to counter the tenfold difference in K^+ concentrations on the two sides of the membrane (see Figure 2.6A, center). If the inside is initially made more negative than –58 mV, then K^+ will actually flow from the outside (compartment 2, on the right) into the inside (compartment 1, on the left) because the positive ions will be attracted to the more negative potential of the inside (see Figure 2.6A, right). This example demonstrates that both the direction and magnitude of ion flux depend on the membrane potential. Thus, in some circumstances the electrical potential can overcome an ion concentration gradient.

The ability to alter ion flux experimentally by changing either the potential imposed on the membrane (see Figure 2.6B) or the transmembrane concentration gradient for an ion (see Figure 2.5C) provides convenient tools for studying ion fluxes across the plasma membranes of neurons, as will be evident in many of the experiments described in the chapters that follow.

Electrochemical Equilibrium in an Environment with More Than One Permeant Ion

Now consider a somewhat more complex situation in which both Na^+ and K^+ are unequally distributed across the membrane, as in Figure 2.7A. What would happen if 10 m*M* K^+ and 1 m*M* Na^+ were present inside, and 1 m*M* K^+ and 10 m*M* Na^+ were present outside? If the membrane were permeable only to K^+, the membrane potential would be –58 mV; if the membrane were permeable only to Na^+, the potential would be +58 mV. But what would the potential be if the membrane were permeable to both K^+ and Na^+? In this case, the potential would depend on the relative permeability of the

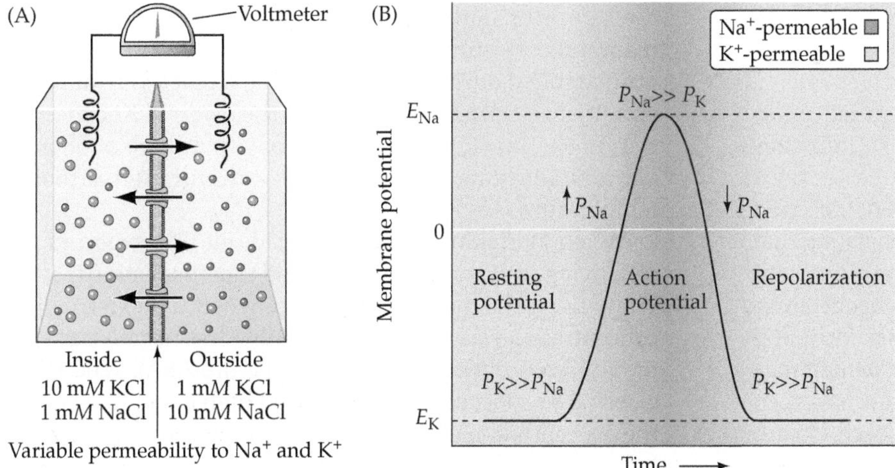

FIGURE 2.7 Resting and action potentials rely on permeabilities to different ions. (A) Hypothetical situation in which a membrane variably permeable to Na^+ (red) and K^+ (gold) separates two compartments that contain both ions. For simplicity, Cl^- ions are not shown in the diagram. (B) Schematic representation of the membrane ion permeabilities associated with resting and action potentials. At rest, neuronal membranes are more permeable to K^+ (gold) than to Na^+ (red); accordingly, the resting membrane potential is negative and approaches the equilibrium potential for K^+, E_K. During an action potential, the membrane becomes very permeable to Na^+ (red); thus, the membrane potential becomes positive and approaches the equilibrium potential for Na^+, E_{Na}. The rise in Na^+ permeability is transient, however, so the membrane again becomes primarily permeable to K^+, causing the potential to return to its negative resting value.

membrane to K^+ and Na^+. If it were more permeable to K^+, the potential would approach –58 mV, and if it were more permeable to Na^+, the potential would be closer to +58 mV. Because there is no permeability term in the Nernst equation, which considers only the simple case of a single permeant ion species, a more elaborate equation is needed. This equation must take into account both the concentration gradients of the permeant ions and the relative permeability of the membrane to each permeant species.

Such an equation was developed by David Goldman in 1943. For the case most relevant to neurons, in which K^+, Na^+, and Cl^- are the primary permeant ions at room temperature, the **Goldman equation** is written

$$V_m = 58 \log \frac{P_K[K]_{out} + P_{Na}[Na]_{out} + P_{Cl}[Cl]_{in}}{P_K[K]_{in} + P_{Na}[Na]_{in} + P_{Cl}[Cl]_{out}}$$

where V is the voltage across the membrane (again, the inside compartment, relative to the reference outside compartment) and P_x indicates the permeability of the membrane to each ion of interest. The Goldman equation is thus an extended version of the Nernst equation that takes into account the relative permeabilities of each of the ions involved. The relationship between the two equations becomes obvious in the situation where the membrane is permeable only to one ion, such as K^+; in this case, the Goldman expression collapses back to the simpler Nernst equation. In this context, it is important to note that the valence factor (z) in the Nernst equation has been eliminated; this is why the concentrations of negatively charged chloride ions, Cl^-, have been inverted relative to the concentrations of the positively charged ions [remember that $-\log(A/B) = \log(B/A)$].

If the membrane in Figure 2.7A is permeable only to K^+ and Na^+, the terms involving Cl^- drop out because P_{Cl} is 0. In this case, solution of the Goldman equation yields a potential of –58 mV when only K^+ is permeant, +58 mV when only Na^+ is permeant, and some intermediate value if both ions are permeant. For example, if K^+ and Na^+ were equally permeant, then the potential would be 0 mV.

With respect to neural signaling, it is particularly pertinent to ask what would happen if the membrane started out being permeable to K^+ and then temporarily switched to become most permeable to Na^+. In this circumstance, the membrane potential would start out at a negative level, become positive while the Na^+ permeability remained high, and then fall back to a negative level as the Na^+ permeability decreased again. As it turns out, this case essentially describes what goes on in a neuron during the generation of an action potential. In the resting state, P_K of the neuronal plasma membrane is much higher than P_{Na}; since, as a result of the action of ion transporters, there is always more K^+ inside the cell than outside, the resting potential is negative (Figure 2.7B). As the membrane potential is depolarized (by synaptic action, for example), P_{Na} increases. The transient increase in Na^+ permeability causes the membrane potential to become even more positive (red region in Figure 2.7B), because Na^+ rushes in (there is much more Na^+ outside a neuron than inside, again as a result of ion pumps). Because of this positive feedback relationship, an action potential occurs. The rise in Na^+ permeability during the action potential is transient, however; as the membrane permeability to K^+ is restored, the membrane potential quickly returns to its resting level.

Armed with an appreciation of these simple electrochemical principles, it will be much easier to understand the key experiments that proved how neurons generate resting and action potentials.

The Ionic Basis of the Resting Membrane Potential

The action of ion transporters creates substantial transmembrane gradients for most ions. Table 2.1 summarizes the ion concentrations measured directly in an exceptionally large neuron found in the nervous system of the squid (Box 2A). Such measurements are the basis for stating that there is much more K^+ inside the neuron than out, and much more Na^+ outside than in. Similar concentration gradients occur in the neurons of most animals, including humans. However, because the ionic strength of mammalian blood is lower than that of sea-dwelling animals such as squid, in mammals the concentrations of each ion are several times lower (Table 2.1). These transporter-dependent concentration gradients enable the resting membrane potential and the action potential of neurons.

Once the ion concentration gradients across various neuronal membranes are known, the Nernst equation can be used to calculate the equilibrium potential for K^+ and other major ions. Since the resting membrane potential of the squid neuron is approximately –65 mV, K^+ is the ion that is closest to being in electrochemical equilibrium when the cell is at rest. This fact implies that the resting membrane is more permeable to K^+ than to the other ions listed in Table 2.1, and that this permeability is the source of resting potentials.

It is possible to test this hypothesis, as Alan Hodgkin and Bernard Katz did in 1949, by asking what happens to the resting membrane potential if the concentration of K^+ outside the neuron is altered. If the resting membrane were permeable only to K^+, then the Goldman equation (or even the simpler Nernst equation) predicts that the membrane potential will vary in proportion to the logarithm of the K^+ concentration gradient across the membrane. Assuming that the internal K^+ concentration is unchanged during the experiment, a plot of membrane potential against the logarithm of the external K^+ concentration should yield a straight line with a slope of 58 mV per tenfold change in external K^+ concentration at room temperature (see Figure 2.5C).

When Hodgkin and Katz carried out this experiment on a living squid neuron, they found that the resting membrane potential did indeed change when the external K^+ concentration was modified, becoming less negative as external K^+ concentration was raised (Figure 2.8A). When the external K^+ concentration was raised high enough to equal the concentration of K^+ inside the neuron, thus making the K^+ equilibrium potential 0 mV, the resting membrane potential was also approximately 0 mV. In short, the resting membrane potential varied as predicted with the logarithm of the K^+ concentration, with a slope that approached 58 mV per tenfold change in K^+ concentration (Figure 2.8B). The value obtained was not exactly 58 mV because other ions, such as Cl^- and Na^+, are also slightly permeable, and thus influence the resting potential to a small degree. The

TABLE 2.1 ■ Extracellular and Intracellular Ion Concentrations

	Concentration (m*M*)	
Ion	**Intracellular**	**Extracellular**
Squid neuron		
Potassium (K^+)	400	20
Sodium (Na^+)	50	440
Chloride (Cl^-)	40-150	560
Calcium (Ca^{2+})	0.0001	10
Mammalian neuron		
Potassium (K^+)	140	5
Sodium (Na^+)	5-15	145
Chloride (Cl^-)	4-30	110
Calcium (Ca^{2+})	0.0001	1-2

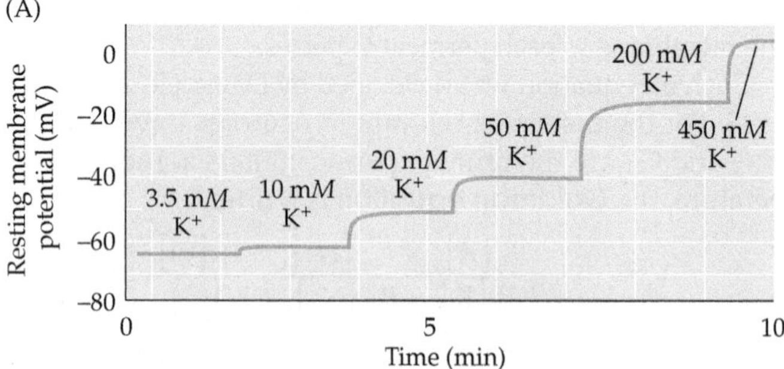

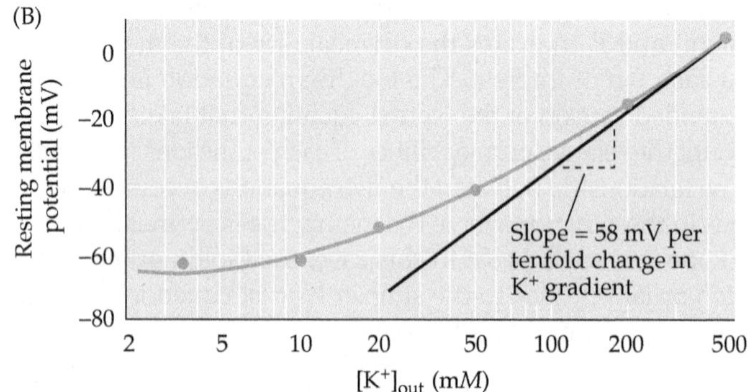

FIGURE 2.8 **The resting membrane potential of a squid giant axon is determined by the K^+ concentration gradient across the membrane.** (A) Increasing the external K^+ concentration depolarizes the resting membrane potential. (B) Relationship between resting membrane potential and external K^+ concentration, plotted on a semilogarithmic scale. The straight line represents a slope of 58 mV per tenfold change in concentration, as given by the Nernst equation. (After Hodgkin and Katz, 1949.)

BOX 2A ■ The Remarkable Giant Nerve Cells of Squid

Many of the initial insights into how ion concentration gradients and changes in membrane permeability produce electrical signals came from experiments performed on the extraordinarily large nerve cells of squid. The axons of these nerve cells can be up to 1 mm in diameter—100 to 1000 times larger than mammalian axons. Thus, squid axons are large enough to allow experiments that would be impossible on most other nerve cells. For example, it is not difficult to insert simple wire electrodes inside these giant axons and make reliable electrical measurements. The relative ease of this approach yielded the first direct recordings of action potentials from nerve cells and, as discussed in the next chapter, the first experimental measurements of the ion currents that produce action potentials. It also is practical to extrude the cytoplasm from giant axons and measure its ionic composition (see Table 2.1). In addition, some giant nerve cells form synaptic contacts with other giant nerve cells, producing very large synapses that have been extraordinarily valuable in understanding the fundamental mechanisms of synaptic transmission (see Chapter 5).

Giant neurons evidently evolved in squid because they enhanced the animal's survival. These neurons participate in a simple neural circuit that activates the contraction of the mantle muscle, producing a jet propulsion effect that allows the squid to move away from predators at a remarkably fast speed. As discussed in Chapter 3, larger axonal diameter allows faster conduction of action potentials. Thus, these huge nerve cells must help squid escape more successfully from their numerous enemies.

Today—more than 75 years after their discovery by John Z. Young at University College London—the giant nerve cells of squid remain useful experimental systems for probing basic neuronal functions.

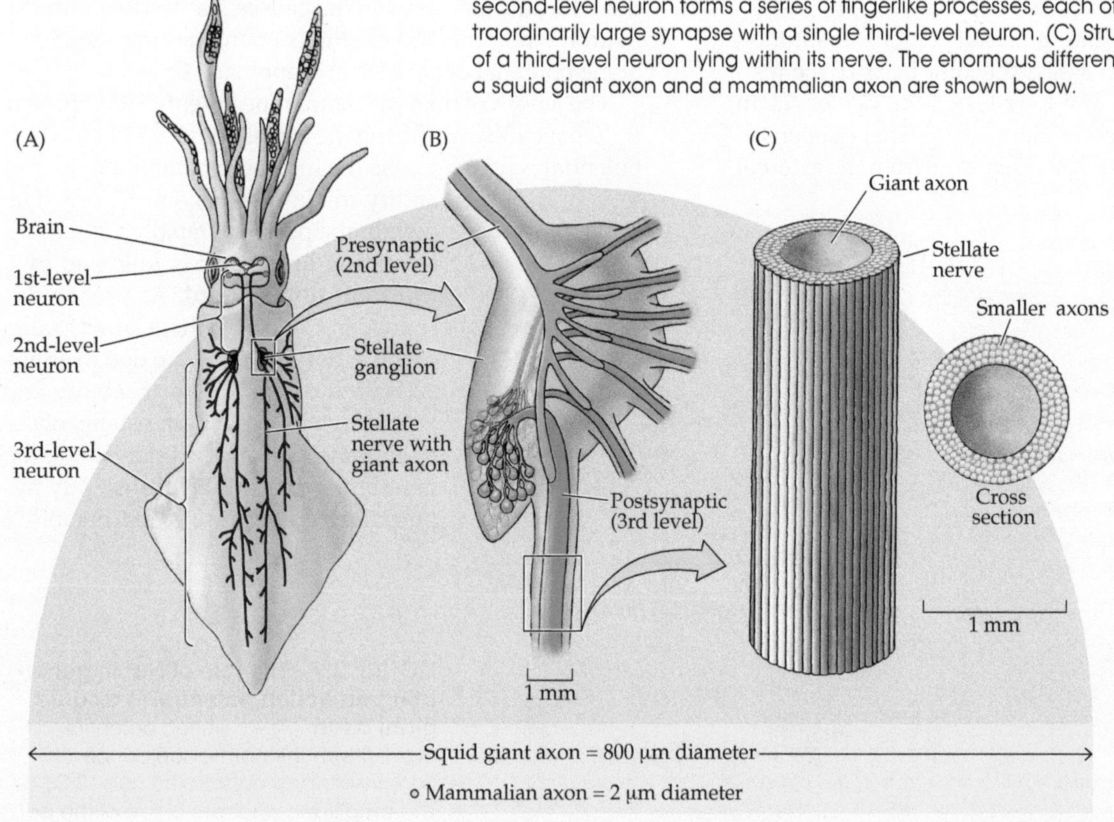

(A) Diagram of a squid, showing the location of its giant nerve cells. Different colors indicate the neuronal components of the escape circuitry. The first- and second-level neurons originate in the brain, while the third-level neurons are in the stellate ganglion and innervate muscle cells of the mantle. (B) Giant synapses within the stellate ganglion. The second-level neuron forms a series of fingerlike processes, each of which makes an extraordinarily large synapse with a single third-level neuron. (C) Structure of a giant axon of a third-level neuron lying within its nerve. The enormous difference in the diameters of a squid giant axon and a mammalian axon are shown below.

contribution of these other ions is particularly evident at low external K^+ levels, again as predicted by the Goldman equation. In general, however, manipulation of the external concentrations of these other ions has only a small effect (see Figure 2.9E), emphasizing that K^+ permeability is indeed the primary source of the resting membrane potential.

In summary, Hodgkin and Katz showed that the inside-negative resting potential arises because (1) the membrane of the resting neuron is more permeable to K^+ than to any of the other ions present, and (2) there is more K^+ inside the neuron than outside. The selective permeability to K^+ is caused by K^+-permeable membrane channels that are open in resting neurons, while the large K⁺ concentration gradient is produced by membrane transporters that selectively accumulate K⁺ within neurons. Many subsequent studies have confirmed the general validity of these principles.

The Ionic Basis of Action Potentials

What causes the membrane potential of a neuron to depolarize during an action potential? Although a general answer to this question has been given (i.e., increased permeability to Na⁺), it is well worth examining the most persuasive experimental support for this concept. Given the data in Table 2.1, one can use the Nernst equation to calculate that the equilibrium potential for Na⁺ (E_{Na}) in neurons, and indeed in most cells, is positive. Thus, if the membrane were to become highly permeable to Na⁺, the membrane potential would become positive. Based on these considerations, Hodgkin and Katz hypothesized that the action potential arises because the neuronal membrane becomes temporarily permeable to Na⁺.

Taking advantage of the same style of ion substitution experiment they used to assess the ionic basis of the resting potential, Hodgkin and Katz tested the role of Na⁺ in generating the action potential by asking what happens to the action potential when Na⁺ is removed from the external medium. They found that lowering the external Na⁺ concentration reduces both the rate of rise of the action potential and its peak amplitude (Figure 2.9A–C). Indeed, when they examined this Na⁺ dependence quantitatively, they found a more-or-less linear relationship between the amplitude of the action potential and the logarithm of the external Na⁺ concentration (Figure 2.9D). The slope of this relationship approached a value of 58 mV per tenfold change in Na⁺ concentration, as expected for a membrane selectively permeable to Na⁺. In contrast, lowering Na⁺ concentration had very little effect on the resting membrane potential (Figure 2.9E). Thus, while the resting neuronal membrane is only slightly permeable to Na⁺, the membrane becomes extraordinarily permeable to Na⁺ during the **rising phase** and the **overshoot phase** of an action potential. (Box 2B further explains action potential nomenclature.) This temporary increase in Na⁺ permeability results from the opening of Na⁺-selective channels that are closed in the resting state. Membrane pumps maintain a large electrochemical gradient for Na⁺, which is in much higher concentration outside the neuron than inside. This causes Na⁺ to flow into the neuron when the Na⁺ channels open, making the membrane potential depolarize and approach E_{Na}.

The length of time the membrane potential lingers near E_{Na} (about +50 mV) during the overshoot phase of an action potential is brief because the increased membrane permeability to Na⁺ itself is short-lived. The membrane potential rapidly repolarizes to resting levels and is followed by a transient **undershoot**. As will be described in Chapter 3, these latter phases of the action potential are due to an inactivation of the Na⁺ permeability and an increase in the K⁺ permeability of the membrane. During the undershoot, the membrane potential is transiently hyperpolarized because K⁺ permeability

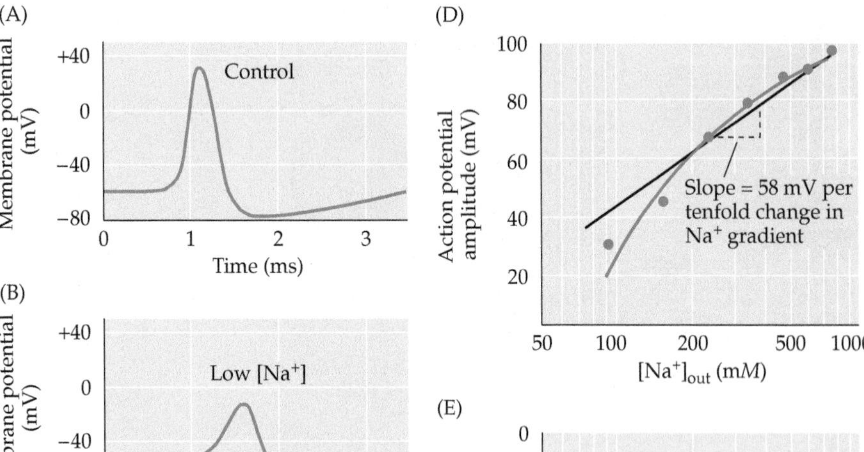

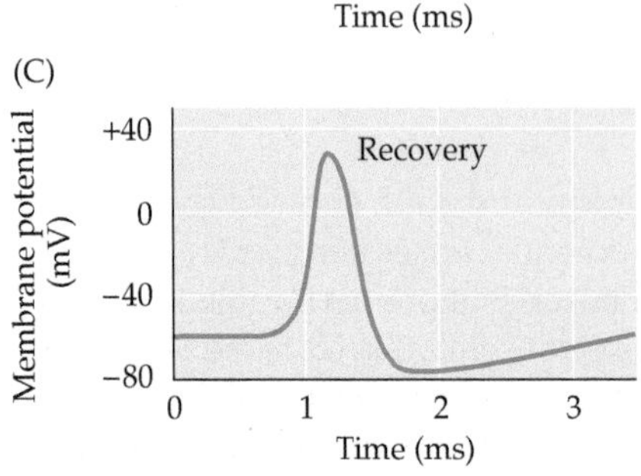

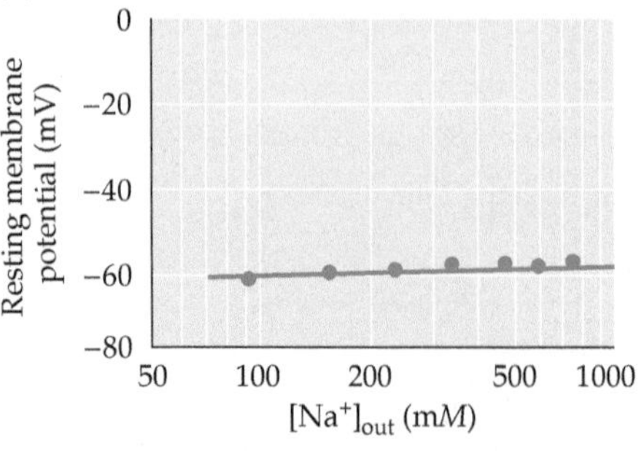

FIGURE 2.9 The role of Na⁺ in generating an action potential in a squid giant axon. (A) An action potential evoked with the normal ion concentrations inside and outside the cell. (B,C) The amplitude and rate of rise of the action potential (B) diminish when external Na⁺ concentration is reduced to one-third of normal, but recover (C) when the Na⁺ is replaced. (D,E) Although the amplitude of the action potential (D) is quite sensitive to the external concentration of Na⁺, the resting membrane potential (E) is little affected by changing the concentration of this ion. (After Hodgkin and Katz, 1949.)

BOX 2B ■ Action Potential Form and Nomenclature

The action potential of the squid giant axon has a characteristic shape, or waveform, with a number of different phases (Figure A). During the rising phase, the membrane potential rapidly depolarizes. In fact, action potentials cause the membrane potential to depolarize so much that the membrane potential transiently becomes positive with respect to the external medium, producing an overshoot. The overshoot of the action potential gives way to a falling phase, in which the membrane potential rapidly repolarizes. Repolarization takes the membrane potential to levels even more negative than the resting membrane potential for a short time; this brief period of hyperpolarization is called the undershoot.

Although the waveform of the squid action potential is typical, the detailed form of the action potential varies widely from neuron to neuron in different animals. In myelinated axons of vertebrate motor neurons (Figure B), the action potential is virtually indistinguishable from that of the squid axon. However, the action potential recorded in the cell body of this same motor neuron (Figure C) looks rather different. Thus, action potential waveform can vary even within the same neuron. More complex action potentials are seen in other central neurons. For example, action potentials recorded from the cell bodies of neurons in the mammalian inferior olive (a region of the brainstem involved in motor control) last tens of milliseconds (Figure D). These action potentials exhibit a pronounced plateau during their falling phase, and their undershoot lasts even longer than that of the motor neuron. One of the most dramatic types of action potentials occurs in cerebellar Purkinje neurons (Figure E). These potentials, termed *complex spikes*, are well named because they have several phases that result from the summation of multiple, discrete action potentials generated in different regions of the neuron.

The variety of action potential waveforms could mean that each type of neuron has a different mechanism of action potential production. Fortunately, however, these diverse waveforms all result from relatively minor variations in the scheme used by the squid giant axon. For example, plateaus in the repolarization phase result from the presence of ion channels that are permeable to Ca^{2+}, and long-lasting undershoots result from the presence of additional types of membrane K^+ channels. The complex action potential of the Purkinje cell results from these extra features plus the fact that different types of action potentials are generated in various parts of the Purkinje neuron—cell body, dendrites, and axons—and are summed together in recordings from the cell body. Thus, the lessons learned from the squid axon are applicable to, and indeed essential for, understanding action potential generation in all neurons.

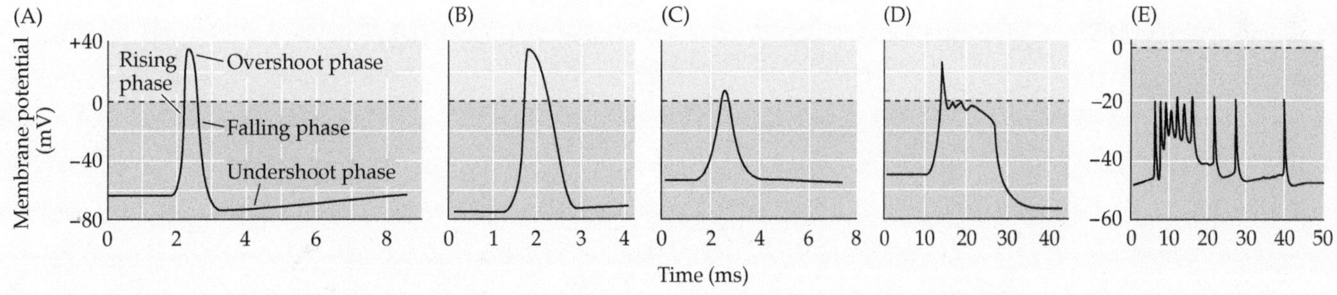

(A) The phases of an action potential of the squid giant axon. (B) Action potential recorded from a myelinated axon of a frog motor neuron. (C) Action potential recorded from the cell body of a frog motor neuron. The action potential is smaller and the undershoot prolonged in comparison with the action potential recorded from the axon of this same neuron (B). (D) Action potential recorded from the cell body of a neuron from the inferior olive of a guinea pig. This action potential has a pronounced plateau during its falling phase. (E) Action potential recorded from the dendrite of a Purkinje neuron in the cerebellum of a mouse. (A after Hodgkin and Huxley, 1939; B after Dodge and Frankenhaeuser, 1958; C after Barrett and Barrett, 1976; D after Llinás and Yarom, 1981; E after Chen et al., 2016.)

becomes even greater than it is at rest. The action potential ends when this phase of enhanced K^+ permeability subsides, and the membrane potential thus returns to its normal resting level.

The ion substitution experiments carried out by Hodgkin and Katz provided convincing evidence that (1) the resting membrane potential results from a high resting membrane permeability to K^+, and (2) depolarization during an action potential results from a transient rise in membrane Na^+ permeability. Although these experiments identified the ions that flow during an action potential, they did not establish *how* the neuronal membrane is able to change its ion permeability to generate the action potential, or what mechanisms trigger this critical change. The next chapter addresses these issues, documenting the surprising conclusion that the neuronal membrane potential itself affects membrane permeability.

Summary

Nerve cells generate electrical signals to convey information over substantial distances and to transmit it to other cells by means of synaptic connections. These signals ultimately depend on changes in the resting electrical potential across the neuronal membrane. A negative membrane potential at rest results from a net efflux of K^+ across neuronal membranes that are predominantly permeable to K^+. In contrast, an action potential occurs when a transient rise in Na^+ permeability allows a net influx of Na^+. The brief rise in membrane Na^+ permeability is followed by a secondary, transient rise in membrane K^+ permeability that repolarizes the neuronal membrane and produces a brief undershoot of the action potential. As a result of these processes, the membrane is depolarized in an all-or-none fashion during an action potential. When these active permeability changes subside, the membrane potential returns to its resting level because of the high resting membrane permeability to K^+.

ADDITIONAL READING

Reviews

Hodgkin, A. L. (1951) The ionic basis of electrical activity in nerve and muscle. *Biol. Rev.* 26: 339–409.

Hodgkin, A. L. (1958) The Croonian Lecture: Ionic movements and electrical activity in giant nerve fibres. *Proc. R. Soc. Lond.* (*B*) 148: 1–37.

Important Original Papers

Baker, P. F., A. L. Hodgkin and T. I. Shaw (1962) Replacement of the axoplasm of giant nerve fibres with artificial solutions. *J. Physiol.* (*London*) 164: 330–354.

Cole, K. S. and H. J. Curtis (1939) Electric impedence of the squid giant axon during activity. *J. Gen. Physiol.* 22: 649–670.

Goldman, D. E. (1943) Potential, impedence, and rectification in membranes. *J. Gen. Physiol.* 27: 37–60.

Hodgkin, A. L. and P. Horowicz (1959) The influence of potassium and chloride ions on the membrane potential of single muscle fibres. *J. Physiol.* (*London*) 148: 127–160.

Hodgkin, A. L. and B. Katz (1949) The effect of sodium ions on the electrical activity of the giant axon of the squid. *J. Physiol.* (*London*) 108: 37–77.

Hodgkin, A. L. and R. D. Keynes (1953) The mobility and diffusion coefficient of potassium in giant axons from *Sepia*. *J. Physiol.* (*London*) 119: 513–528.

Hodgkin, A. L. and W. A. H. Rushton (1946) The electrical constants of a crustacean nerve fibre. *Proc. R. Soc. Lond.* 133: 444–479.

Keynes, R. D. (1951) The ionic movements during nervous activity. *J. Physiol.* (*London*) 114: 119–150.

Nernst, W. (1888). Zur Kinetik der Lösung befindlichen Körper: Theorie der Diffusion. *Z. Phys. Chem.* 3: 613–637.

Books

Hodgkin, A. L. (1992) *Chance and Design*. Cambridge: Cambridge University Press.

Junge, D. (1992) *Nerve and Muscle Excitation*, 3rd Edition. Sunderland, MA: Sinauer Associates.

Katz, B. (1966) *Nerve, Muscle, and Synapse*. New York: McGraw-Hill.

Go to the NEUROSCIENCE 6e Companion Website at **www.oup.com/uk/Purves6e** for Web Topics, Animations, Flashcards, and more.

CHAPTER

3 Voltage-Dependent Membrane Permeability

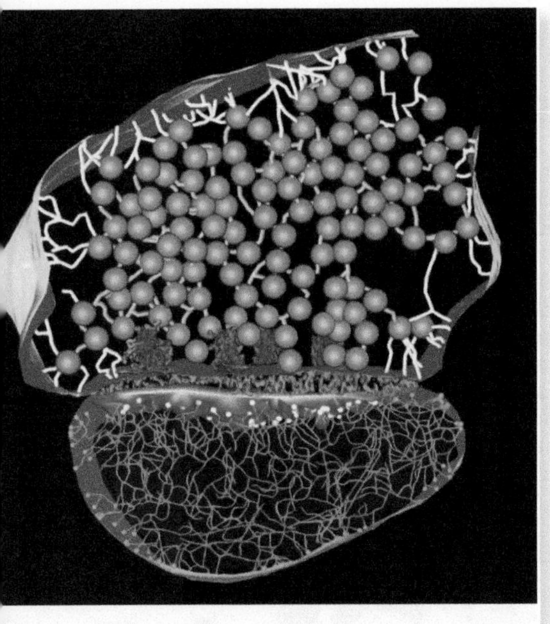

Overview

THE ACTION POTENTIAL IS A FUNDAMENTAL electrical signal generated by nerve cells and arises from changes in membrane permeability to specific ions. Present understanding of these changes in ion permeability is based on evidence obtained by the voltage clamp technique, which permits detailed characterization of permeability changes as a function of membrane potential and time. For most types of neurons, these changes consist of a rapid and transient rise in sodium (Na^+) permeability, followed by a slower but more prolonged rise in potassium (K^+) permeability. Both permeabilities are voltage-dependent, increasing as the membrane potential depolarizes. The measured kinetics and voltage dependence of Na^+ and K^+ permeabilities are sufficient to explain action potential generation. Depolarizing the membrane potential to the threshold level causes a rapid, self-sustaining increase in Na^+ permeability that produces the rising phase of the action potential; however, the Na^+ permeability increase is short-lived and is followed by a slower increase in K^+ permeability that restores the membrane potential to its usual negative resting level. A mathematical model that describes the behavior of these ion permeabilities accurately predicts the observed properties of action potentials. Importantly, voltage-dependent Na^+ and K^+ permeabilities also permit action potentials to be propagated along the length of axons, explaining how electrical signals are conveyed within neurons throughout the nervous system.

Ion Currents across Nerve Cell Membranes

The previous chapter introduced the idea that nerve cells generate electrical signals by virtue of a membrane that is differentially permeable to various ion species. In particular, a transient increase in the permeability of the neuronal membrane to Na^+ initiates the action potential. This chapter considers exactly how this increase in Na^+ permeability occurs. A key to understanding this phenomenon is the observation that action potentials are initiated *only* when the neuronal membrane potential becomes more positive than a threshold level. This observation suggests that the mechanism responsible for the increase in Na^+ permeability is sensitive to the membrane potential. Therefore, if one could understand how a change in membrane potential activates Na^+ permeability, it should be possible to explain how action potentials are generated.

The fact that the Na^+ permeability that generates the membrane potential change is itself sensitive to the membrane potential presents both conceptual and practical obstacles to studying the mechanisms underlying the action potential. A practical problem is the difficulty of systematically varying the membrane potential to study the permeability change, because such changes in membrane potential will produce an action potential, which causes further, uncontrolled changes in the membrane potential. Historically, then, it was not possible to understand action potentials until a technique was developed that allowed experimenters to control membrane potential *and* simultaneously measure the underlying permeability changes. This technique, the **voltage clamp method** (Box 3A), provides the information needed to define the ion permeability of the membrane at any level of membrane potential.

In the late 1940s, Alan Hodgkin and Andrew Huxley, from the University of Cambridge, used the voltage clamp technique to work out the permeability changes underlying the action potential. They again chose to use the giant axon of a squid because its large size (up to 1 mm in diameter; see Box 2A) allowed insertion of the electrodes necessary for voltage clamping. They were the first investigators to test directly the hypothesis that potential-sensitive Na^+ and K^+ permeability changes are both necessary and sufficient for the production of action potentials.

Hodgkin and Huxley's first goal was to determine whether neuronal membranes do, in fact, have voltage-dependent permeabilities. To address this issue, they asked whether ion currents flow across the membrane when its potential is changed. The result of one such experiment is shown in Figure 3.1. Figure 3.1A illustrates the currents produced by a squid axon when its membrane potential, V_m, is hyperpolarized from the resting level (–65 mV) to –130 mV. The initial response of the axon results from the

BOX 3A ■ The Voltage Clamp Method

Breakthroughs in scientific research often rely on the development of new technologies. In the case of the action potential, detailed understanding came only after of the invention of the voltage clamp technique by Kenneth Cole in the 1940s. This device is called a voltage clamp because it controls, or clamps, membrane potential (or voltage) at any level desired by the experimenter. The method measures the membrane potential with an electrode placed inside the cell (1) and electronically compares this voltage with the voltage to be maintained (called the *command voltage*) (2). The clamp circuitry then passes a current back into the cell though another intracellular electrode (3). This electronic feedback circuit holds the membrane potential at the desired level, even in the face of permeability changes that would normally alter the membrane potential (such as those generated during the action potential). Most importantly, the device permits the simultaneous measurement of the current needed to keep the cell at a given voltage (4). This current is exactly equal to the amount of current flowing across the neuronal membrane, allowing direct measurement of these membrane currents. Therefore, the voltage clamp technique can indicate how membrane potential influences ion current flow across the membrane. This information gave Hodgkin and Huxley the key insights that led to their model for action potential generation.

Today, the voltage clamp method remains widely used to study ion currents in neurons and other cells. The most popular contemporary version of this approach is the patch clamp technique, a method that can be applied to virtually any cell and has a resolution high enough to measure the minute electrical currents flowing through single ion channels (see Box 4A).

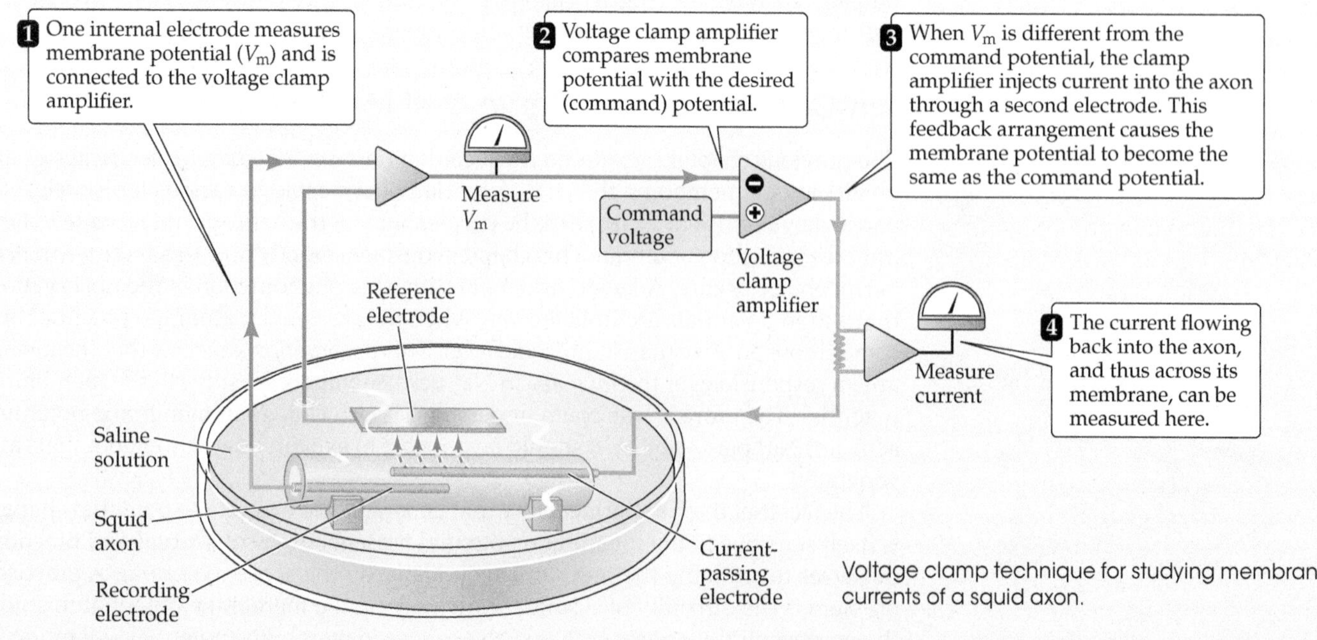

1 One internal electrode measures membrane potential (V_m) and is connected to the voltage clamp amplifier.
2 Voltage clamp amplifier compares membrane potential with the desired (command) potential.
3 When V_m is different from the command potential, the clamp amplifier injects current into the axon through a second electrode. This feedback arrangement causes the membrane potential to become the same as the command potential.
Measure V_m
Command voltage
Voltage clamp amplifier
Reference electrode
Measure current
4 The current flowing back into the axon, and thus across its membrane, can be measured here.
Saline solution
Squid axon
Recording electrode
Current-passing electrode

Voltage clamp technique for studying membrane currents of a squid axon.

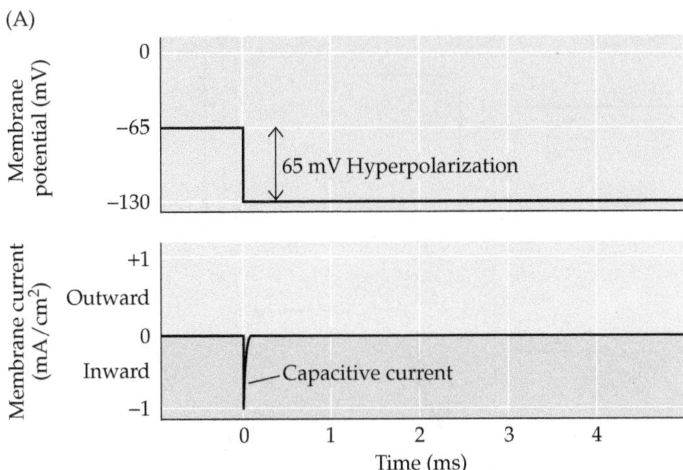

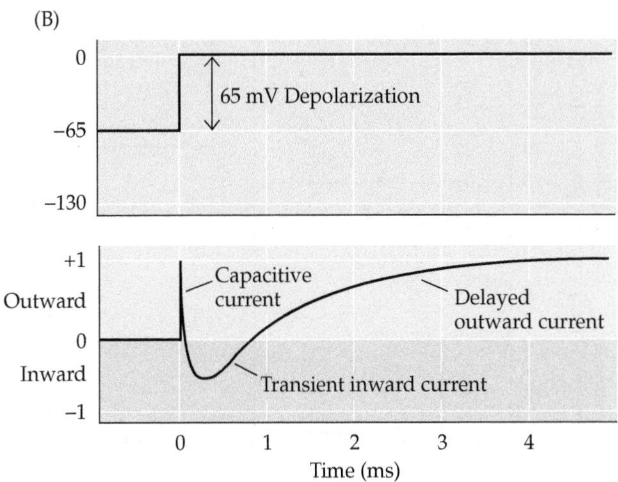

FIGURE 3.1 Current flow across a squid axon membrane during a voltage clamp experiment. (A) A 65 mV hyperpolarization of the membrane potential produces only a very brief capacitive current. (B) A 65 mV depolarization of the membrane potential also produces a brief capacitive current, which is followed by a longer lasting but transient phase of inward current and a delayed but sustained outward current. (After Hodgkin et al., 1952.)

redistribution of charge across the axonal membrane. This capacitive current is nearly instantaneous, ending within a fraction of a millisecond. Aside from this brief event, very little current flows when the membrane is hyperpolarized. However, when the membrane potential is depolarized from –65 mV to 0 mV, the response is quite different (see Figure 3.1B). Following the capacitive current, the axon produces a rapidly rising inward ion current (inward refers to a positive charge entering the cell—that is, cations in or anions out), which gives way to a more slowly rising, delayed outward current. The fact that membrane depolarization elicits these ion currents establishes that the membrane permeability of axons is indeed voltage-dependent.

Two Types of Voltage-Dependent Ion Currents

The results shown in Figure 3.1 demonstrate that the ion permeability of neuronal membranes is voltage-sensitive, but the experiments do not identify how many types of permeability exist, or which ions are involved. As discussed in Chapter 2 (see Figure 2.6), varying the potential across a membrane makes it possible to deduce the equilibrium potential for the ion fluxes through the membrane, and thus to identify the ions that are flowing. Because the voltage clamp method allows the membrane potential to be changed while measuring ion currents, it was a straightforward matter for Hodgkin and Huxley to determine ion permeability by examining how the properties of the initial inward and later outward currents changed as the membrane potential was varied (Figure 3.2). As already noted, no appreciable ion currents flow at membrane potentials more negative than the resting potential. At more positive potentials, however, the currents not only flow but change in magnitude. The early current has a U-shaped dependence on membrane potential, increasing over a range of depolarizations up to approximately 0 mV but decreasing as the potential is depolarized further. In contrast, the late current increases monotonically with increasingly positive membrane potentials. These different responses to membrane potential can be seen more clearly when the magnitudes of the two current components are plotted as a function of membrane potential, as in Figure 3.3.

The voltage sensitivity of the early current gives an important clue about the nature of the ions carrying the current—namely, that no current flows when the membrane potential is clamped at +52 mV. For the squid neurons studied by Hodgkin and Huxley, the external Na^+ concentration is 440 m*M*, and the internal Na^+ concentration is 50 m*M* (see Table 2.1). For this concentration gradient, the Nernst equation predicts that the equilibrium potential for Na^+ should be +55 mV. Recall further from Chapter 2 that at the Na^+ equilibrium potential there is no net flux of Na^+ across the membrane, even if the membrane is highly permeable to Na^+. Thus, the experimental observation that no early current flows at the membrane potential where Na^+ cannot flow is a strong indication that the current is carried by entry of Na^+ into the axon.

An even more demanding way to test whether Na^+ carries the early current is to examine the behavior of this current after *removing* external Na^+. Removing Na^+ outside the axon makes E_{Na} negative; if the permeability to Na^+ is increased under these conditions, current should flow outward as Na^+ leaves the neuron, due to the reversed electrochemical gradient. Hodgkin and Huxley performed this experiment, and found that removing external Na^+

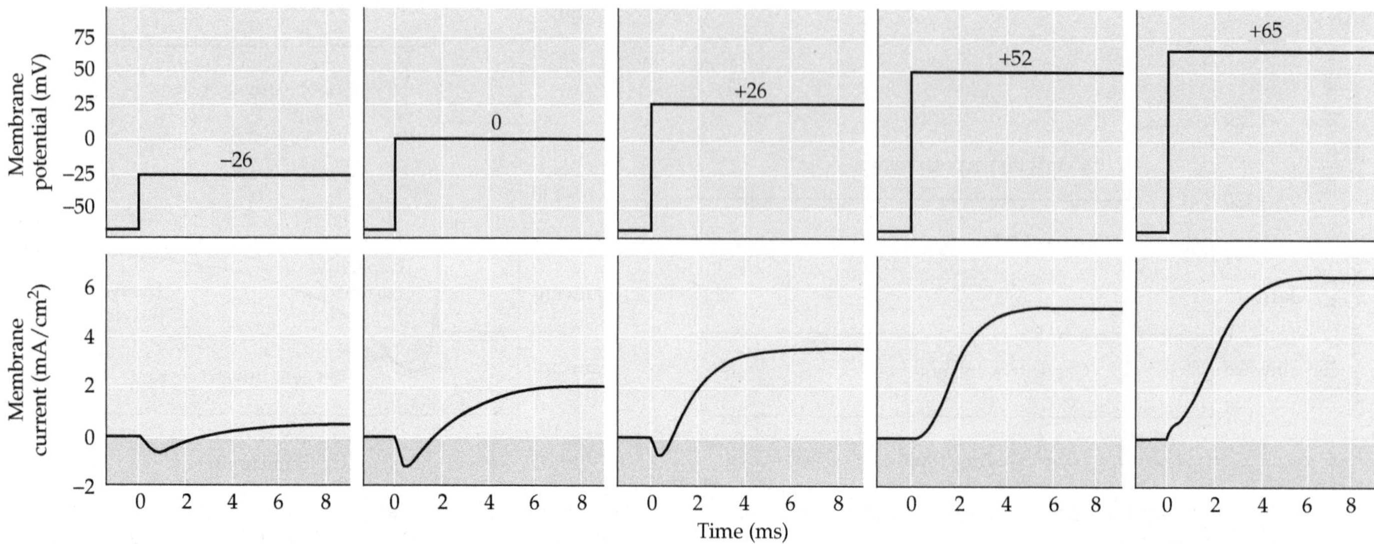

FIGURE 3.2 Currents produced by membrane depolarizations to several different potentials. The early current first increases, then decreases in magnitude as depolarization increases; note that this current reverses its polarity at potentials more positive than about +55 mV. The later outward current increases monotonically with increasing depolarization. (After Hodgkin et al., 1952.)

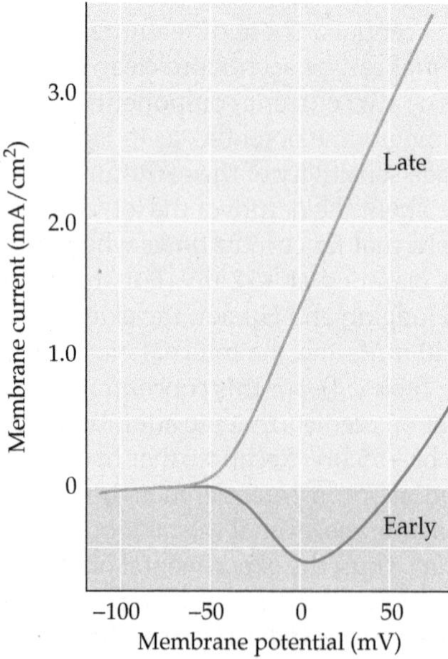

FIGURE 3.3 Relationship between current amplitude and membrane potential. Experiments such as the one shown in Figure 3.2 indicate that the late outward current increases steeply with increasing depolarization, whereas the early inward current first increases in magnitude but then decreases and reverses to outward current at about +55 mV (the sodium equilibrium potential). (After Hodgkin et al., 1952.)

caused the early current to reverse its polarity and become an outward current at a membrane potential that gave rise to an inward current when external Na^+ was present (Figure 3.4). This result demonstrates convincingly that the early inward current measured when Na^+ is present in the external medium must be due to Na^+ entering the neuron.

In the experiment shown in Figure 3.4, removal of external Na^+ has little effect on the outward current that flows after the neuron has been kept at a depolarized membrane voltage for several milliseconds. This further result shows that the late outward current must be due to the flow of an ion other than Na^+. Several lines of evidence presented by Hodgkin, Huxley, and others showed that this outward current is caused by K^+ exiting the neuron. Perhaps the most compelling demonstration of K^+ involvement is that the amount of K^+ efflux from the neuron (measured by loading the neuron with radioactive K^+) is closely correlated with the magnitude of the late outward current.

Taken together, these experiments show that changing the membrane potential to a level more positive than the resting potential produces two effects: an early influx of Na^+ into the neuron, followed by a delayed efflux of K^+. The early influx of Na^+ produces a transient inward current, whereas the delayed efflux of K^+ produces a sustained outward current. The differences in the time course and ion selectivity of the two fluxes suggest that two different ion permeability mechanisms are activated by changes in membrane potential. Confirmation that there are indeed two distinct mechanisms has come from pharmacological studies of drugs that specifically affect these two currents (Figure 3.5). *Tetrodotoxin*, an alkaloid neurotoxin found in certain puffer fish, tropical frogs, and salamanders, blocks the Na^+ current without affecting the K^+ current. Conversely, *tetraethylammonium* ions block K^+ currents without affecting Na^+

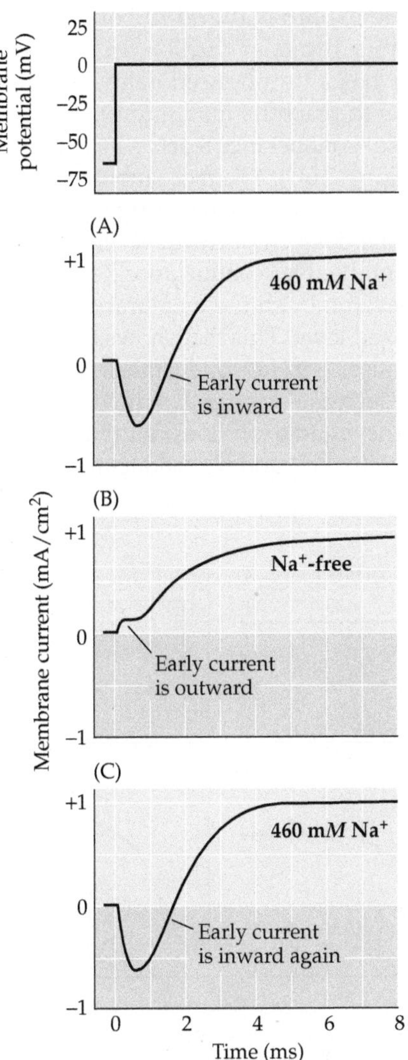

FIGURE 3.4 Dependence of the early current on sodium. (A) In the presence of normal external concentrations of Na^+, depolarization of a squid axon to 0 mV (top) produces an inward initial current. (B) Removal of external Na^+ causes this initial inward current to become outward, an effect that is reversed (C) by restoration of external Na^+. (After Hodgkin and Huxley, 1952a.)

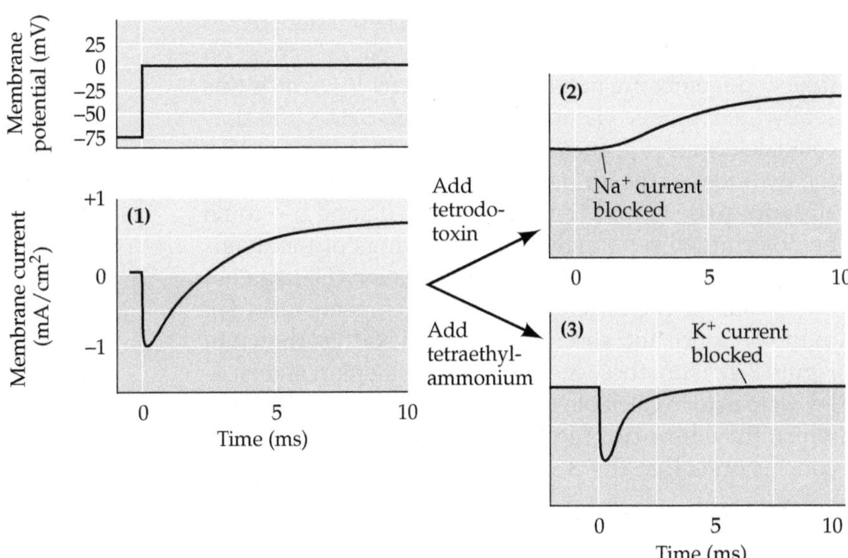

FIGURE 3.5 Pharmacological separation of Na^+ and K^+ currents into sodium and potassium components. Panel (1) shows the current that flows when the membrane potential of a squid axon is depolarized to 0 mV in control conditions. (2) Treatment with tetrodotoxin causes the early Na^+ current to disappear but spares the late K^+ current. (3) Addition of tetraethylammonium blocks the K^+ current without affecting the Na^+ current. (After Moore et al., 1967 and Armstrong and Binstock, 1965.)

currents. The differential sensitivity of Na^+ and K^+ currents to these drugs provides strong additional evidence that Na^+ and K^+ flow through independent permeability pathways. As discussed in Chapter 4, it is now known that these pathways are ion channels that are selectively permeable to either Na^+ or K^+. In fact, tetrodotoxin, tetraethylammonium, and other drugs that interact with specific types of ion channels have been extraordinarily useful tools in characterizing these channel proteins (see Box 4B).

Two Voltage-Dependent Membrane Conductances

The next goal Hodgkin and Huxley set for themselves was to describe Na^+ and K^+ permeability changes mathematically. To do this, they assumed that the ion currents are due to a change in **membrane conductance**, defined as the reciprocal of the membrane resistance. Membrane conductance is thus closely related, although not identical, to membrane permeability. When evaluating ion movements from an electrical standpoint, it is convenient to describe them in terms of ion conductances rather than ion permeabilities. For present purposes, permeability and conductance can be considered synonymous.

If membrane conductance (g) obeys Ohm's Law (which states that voltage is equal to the product of current and resistance), then the ion current that flows during an increase in membrane conductance is given by

$$I_{ion} = g_{ion}\,(V_m - E_{ion})$$

where I_{ion} is the ion current, V_m is the membrane potential, and E_{ion} is the equilibrium potential for the ion flowing through the conductance, g_{ion}. The difference between V_m and E_{ion} is the electrochemical driving force acting on the ion.

Hodgkin and Huxley used this simple relationship to calculate the dependence of Na^+ and K^+ conductances on time and membrane potential. They knew V_m, which was set by their voltage clamp device (Figure 3.6A), and they could determine E_{Na} and E_K from the ion concentrations on the two sides of the axonal membrane (see Table 2.1). The currents carried by Na^+ and K^+—that is, I_{Na} and I_K—could be determined separately from recordings of the membrane currents resulting from depolarization (Figure 3.6B) by measuring the difference between currents recorded in the presence and absence of external Na^+ (as shown in Figure 3.4). From these measurements, Hodgkin and Huxley were able to calculate g_{Na} and g_K (Figure 3.6C,D), from which they drew two fundamental conclusions. The first conclusion is that the Na^+ and K^+ conductances change over time. For example, both Na^+ and K^+ conductances require some time to activate, or turn on. In particular, the K^+ conductance has a pronounced delay, requiring several milliseconds to reach its maximum (see Figure 3.6D), whereas the Na^+ conductance reaches its maximum more rapidly (see Figure 3.6C). The more rapid activation of the Na^+ conductance allows the resulting inward Na^+ current to precede the delayed outward K^+ current (see Figure 3.6B). Although the Na^+ conductance rises rapidly, it quickly declines, even though the membrane potential is kept at a depolarized level. This fact shows that depolarization not only causes the Na^+ conductance to activate, but also causes it to decrease over time, or inactivate. The K^+ conductance of the squid axon does not inactivate in this way; thus, while the Na^+ and K^+ conductances share

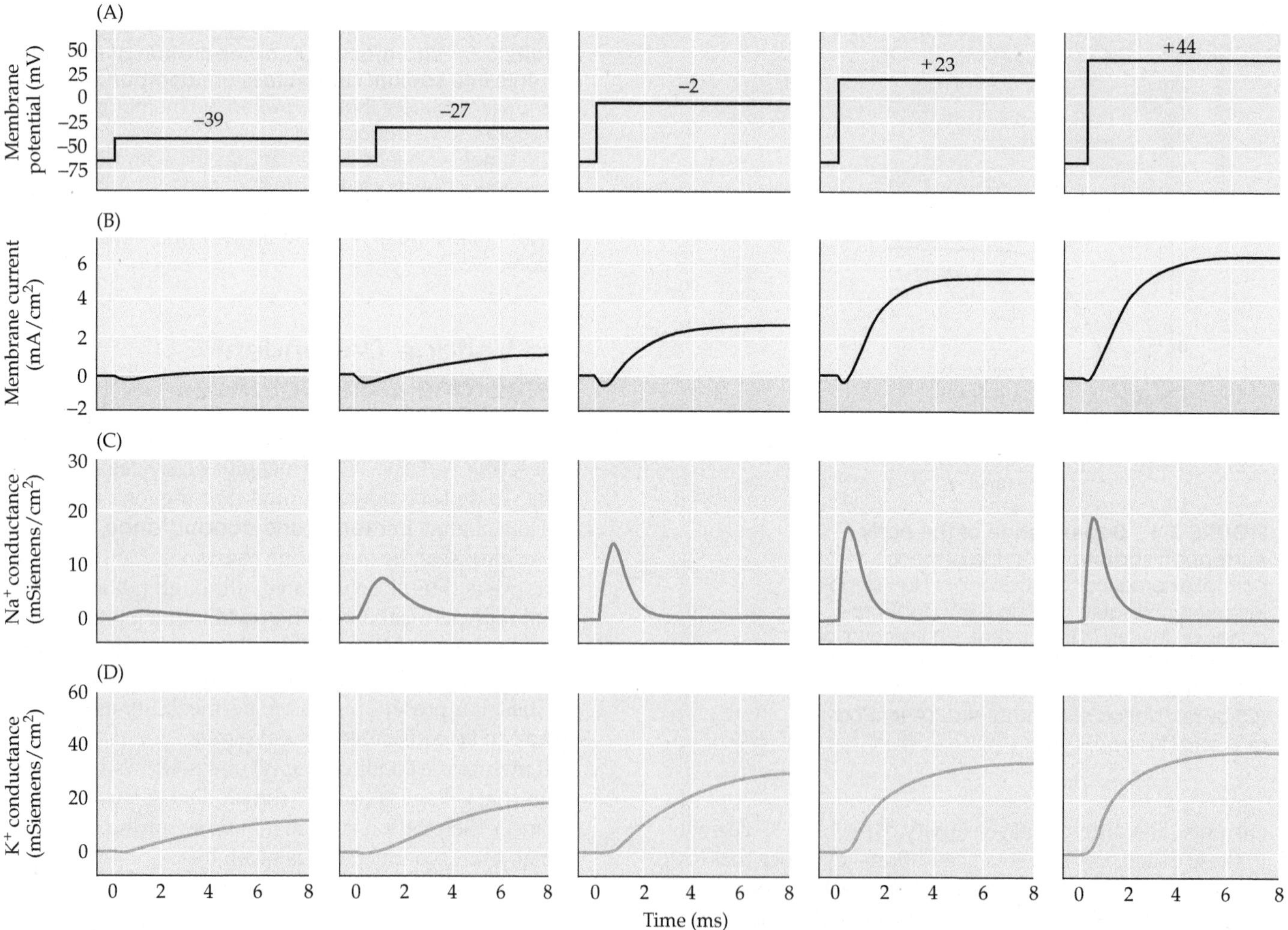

FIGURE 3.6 Membrane conductance changes underlying the action potential are time- and voltage-dependent. Depolarizations to various membrane potentials (A) elicit different membrane currents (B). Shown below are the conductances of Na^+ (C) and K^+ (D) calculated from these currents. Both peak Na^+ conductance and steady-state K^+ conductance increase as the membrane potential becomes more positive. In addition, the activation of both conductances, as well as the rate of inactivation of the Na^+ conductance, occurs more rapidly with larger depolarizations. (After Hodgkin and Huxley, 1952b.)

FIGURE 3.7 Depolarization increases Na^+ and K^+ conductances of the squid giant axon. The peak magnitude of Na^+ conductance and steady-state value of K^+ conductance both increase steeply as the membrane potential is depolarized. (After Hodgkin and Huxley, 1952b.)

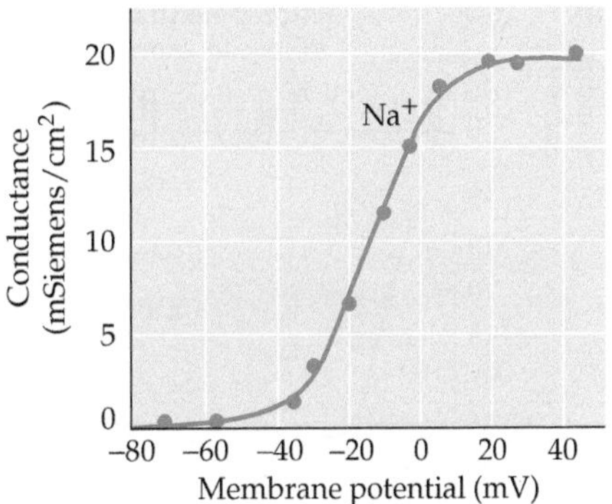

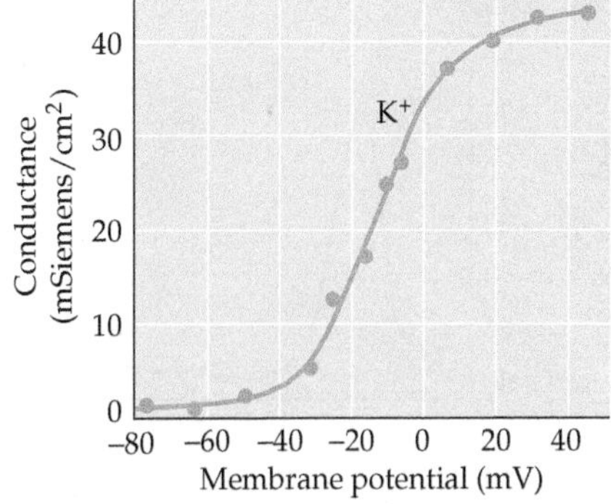

the property of time-dependent **activation**, only the Na^+ conductance exhibits **inactivation**. (Inactivating K^+ conductances have since been discovered in other types of nerve cells; see Chapter 4.) The time courses of the Na^+ and K^+ conductances are voltage-dependent, with the speed of both activation and inactivation increasing at more depolarized potentials. This accounts for the more rapid membrane currents measured at more depolarized potentials (see Figure 3.6B).

The second conclusion derived from Hodgkin and Huxley's calculations is that both the Na^+ and K^+ conductances are voltage-dependent—that is, both conductances increase progressively as the neuron is depolarized. Figure 3.7 illustrates this by plotting the relationship between peak value of the conductances (from Figure 3.6C,D) against the membrane potential. Note the similar voltage dependence for each conductance; both conductances are quite small at negative potentials, maximal at very positive potentials, and exquisitely dependent on membrane voltage at intermediate potentials. The observation that these conductances are sensitive to changes in membrane potential shows that the mechanism underlying the conductances somehow "senses" the voltage across the membrane.

All told, the voltage clamp experiments carried out by Hodgkin and Huxley showed that the ion currents that flow when the neuronal membrane is depolarized are due to three different time-dependent and voltage-sensitive processes: (1) activation of Na^+ conductance, (2) activation of K^+ conductance, and (3) inactivation of Na^+ conductance.

Reconstruction of the Action Potential

From their experimental measurements, Hodgkin and Huxley were able to construct a detailed mathematical model of the Na^+ and K^+ conductance changes. The goal of these modeling efforts was to determine whether the Na^+ and K^+ conductances alone are sufficient to produce an action potential. Using this information, they could in fact generate the form and time course of the action potential with remarkable accuracy (Figure 3.8A). The Hodgkin–Huxley model could simulate many other features of action potential behavior in the squid axon. For example, it was well known that, following an action potential, the axon becomes refractory to further excitation for a brief period of time, termed the **refractory period** (Figure 3.8B). The model was capable of closely mimicking such behavior (Figure 3.8C).

The Hodgkin–Huxley model also provided many insights into how action potentials are generated. Figure 3.8A compares the time courses of a reconstructed action potential and the underlying Na^+ and K^+ conductances. The coincidence of the initial increase in Na^+ conductance with the rapid rising phase of the action potential demonstrates that a selective increase in Na^+ conductance is responsible for action potential initiation. The increase in Na^+ conductance causes Na^+ to enter the neuron, thus depolarizing the membrane potential, which approaches E_{Na}. The rate of depolarization subsequently falls both because the electrochemical driving force on Na^+ decreases and because the Na^+ conductance inactivates. At the same time, depolarization slowly activates the voltage-dependent K^+ conductance, causing K^+ to leave the cell and repolarizing the membrane potential toward E_K. Because the K^+ conductance becomes temporarily higher than it is in the resting condition, the membrane potential briefly becomes more negative than the normal resting potential, to yield the undershoot. The hyperpolarization of the membrane potential causes the voltage-dependent K^+ conductance (and any Na^+ conductance not inactivated) to turn off, allowing the membrane potential to return to its resting level. The relatively slow time course of turning off the K^+ conductance, as well as the persistence of Na^+ conductance inactivation, is responsible for the refractory period (see also Figure 3.10).

This mechanism of action potential generation represents a positive feedback loop: Activating the voltage-dependent Na^+ conductance increases Na^+ entry into the neuron, which makes the membrane potential depolarize, which leads to the activation of still more Na^+ conductance, more Na^+ entry, and still further

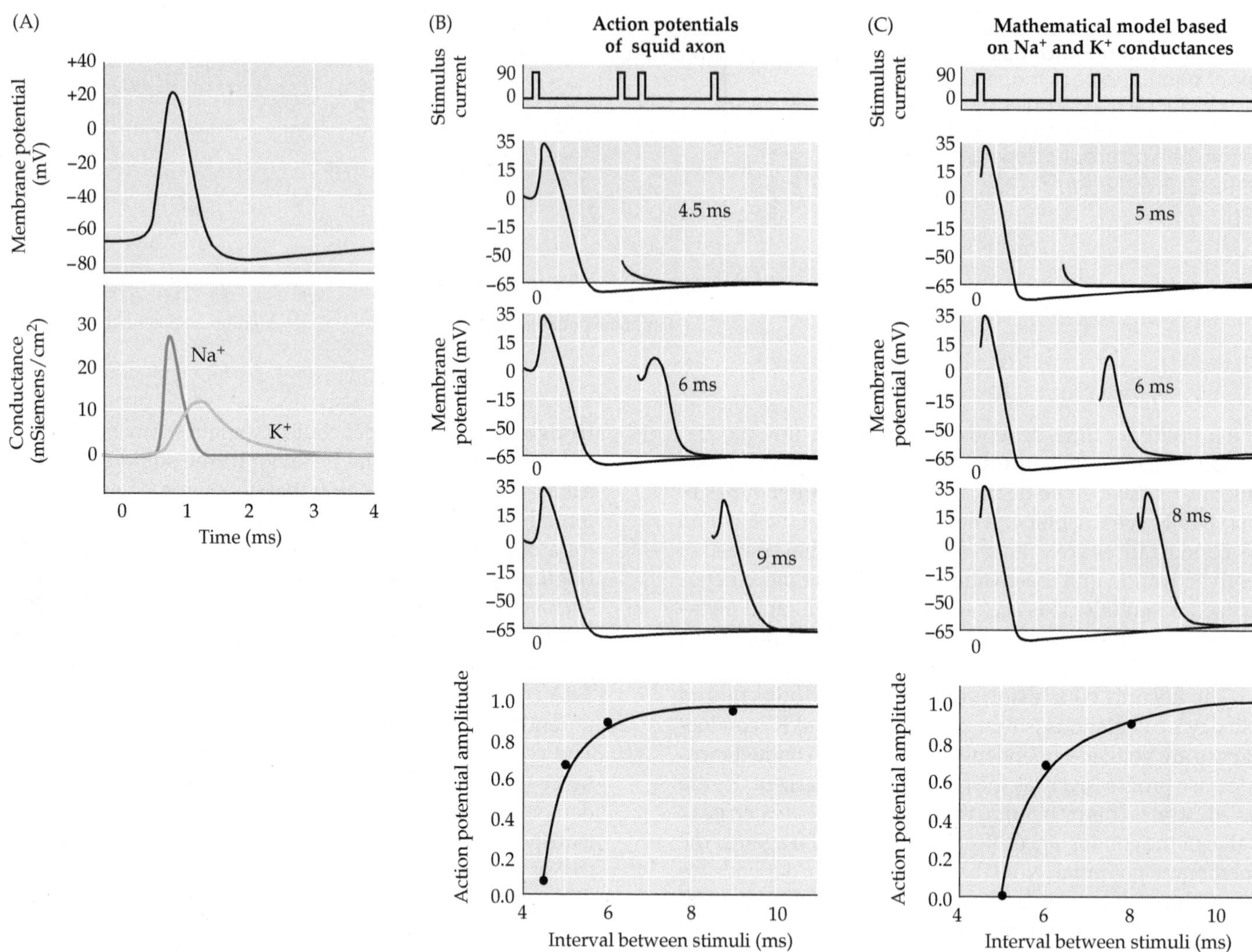

FIGURE 3.8 Mathematical simulation of the action potential. (A) Simulation of an action potential (black curve) together with the underlying changes in Na^+ (red curve) and K^+ (gold curve) conductance. The size and time course of the action potential were calculated using only the properties of g_{Na} and g_K measured in voltage clamp experiments. (B) The refractory period can be observed by stimulating an axon with two current pulses that are separated by variable intervals. Whereas the first stimulus reliably evokes an action potential, during the refractory period the second stimulus will generate only a small action potential or no response at all. (C) The mathematical model accurately predicts responses of the axon during the refractory period. (After Hodgkin and Huxley, 1952d.)

depolarization (Figure 3.9). Positive feedback continues unabated until Na^+ conductance inactivation and K^+ conductance activation restore the membrane potential to the resting level. Because this positive feedback loop, once initiated, is sustained by the intrinsic properties of the neuron—namely, the voltage dependence of the ion conductances—the action potential is self-supporting, or **regenerative**. This regenerative quality explains why action potentials exhibit all-or-none behavior (see Figure 2.2) and why they have a threshold. The delayed activation of the K^+ conductance represents a negative feedback loop that eventually restores the membrane to its resting state.

Hodgkin and Huxley's reconstruction of the action potential and all its features shows that the properties of the voltage-sensitive Na^+ and K^+ conductances, together with the electrochemical driving forces created by ion transporters, are sufficient to explain action potentials. Their use of both empirical and theoretical methods brought an unprecedented level of rigor to a long-standing problem, setting a standard of proof that is achieved only rarely in biological research.

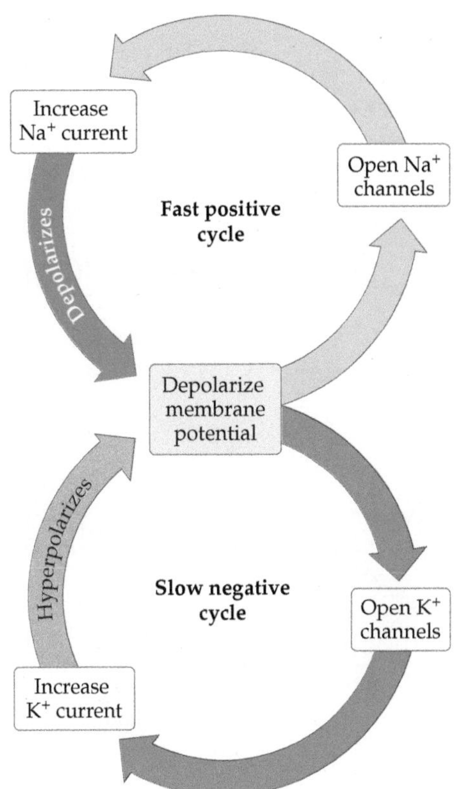

FIGURE 3.9 Feedback cycles are responsible for membrane potential changes during an action potential. Membrane depolarization rapidly activates a positive feedback cycle fueled by the voltage-dependent activation of Na^+ conductance. This phenomenon is followed by the slower activation of a negative feedback loop as depolarization activates a K^+ conductance, which helps to repolarize the membrane potential and terminate the action potential.

Long-Distance Signaling by Means of Action Potentials

The voltage-dependent mechanisms of action potential generation also explain the long-distance transmission of these electrical signals. Recall from Chapter 2 that neurons are relatively poor passive conductors of electricity, at least compared with a wire. Nonetheless, action potentials can traverse great distances along axons despite these poor passive properties. How does this occur? The mechanism of action potential propagation is easy to grasp once one understands how action potentials are generated and how current passively flows along an axon. A depolarizing stimulus—a synaptic potential or a receptor potential in an intact neuron, or an injected current pulse in an experiment such as the one depicted in Figure 3.10—locally depolarizes the axon, thus opening the voltage-sensitive Na^+ channels in that region. The opening of Na^+ channels causes inward movement of Na^+, and the resultant depolarization of the membrane potential generates an action potential at that site. Some of the local current generated by the action potential will then flow passively down the axon, in the same way that subthreshold currents spread along an axon (see Figure 2.3). Note that this passive current flow does not require the movement of Na^+ along the axon but instead occurs by a shuttling of charge, somewhat similar to what happens when wires passively conduct electricity by transmission of electron charge. This passive current flow depolarizes the membrane potential in the adjacent region of the axon, thus opening the Na^+ channels in the neighboring membrane. The local depolarization triggers an action potential in this region, which then spreads again in a continuing cycle until the action potential reaches the end of the axon. Thus, action potential propagation requires the coordinated action of two forms of current flow: the passive flow of current as well as active currents flowing through voltage-dependent ion channels. The regenerative properties of Na^+ channel opening allow action potentials to propagate in an all-or-none fashion by acting as a booster at each point along the axon, thus ensuring the long-distance transmission of electrical signals.

Recall that the axons are refractory following an action potential: generation of an action potential briefly makes it harder for the axon to produce subsequent action potentials (see Figure 3.8B). Refractoriness limits the number of action potentials that a neuron can produce per unit of time, with different types of neurons having different maximum rates of action potential firing due to different types and densities of ion channels. As described in a previous section, the refractory period arises because the depolarization that produces Na^+ channel opening also causes delayed activation of K^+ channels and Na^+ channel inactivation, which temporarily makes it more difficult for the axon to produce another action potential. This refractoriness also has important implications for action potential conduction along axons. As the action potential sweeps along the length of an axon, in its wake the action potential leaves the Na^+ channels inactivated and K^+ channels activated for a brief time. The resulting refractoriness of the membrane region where an action potential has been generated prevents subsequent re-excitation of this membrane as action potentials are generated in adjacent regions of the axon (see Figure 3.10). This important feature prevents action potentials from propagating backward, toward their point of initiation, as they travel along an axon. Thus, refractory behavior ensures polarized propagation of action potentials from their usual point of initiation near the neuronal cell body, toward the synaptic terminals at the distal end of the axon.

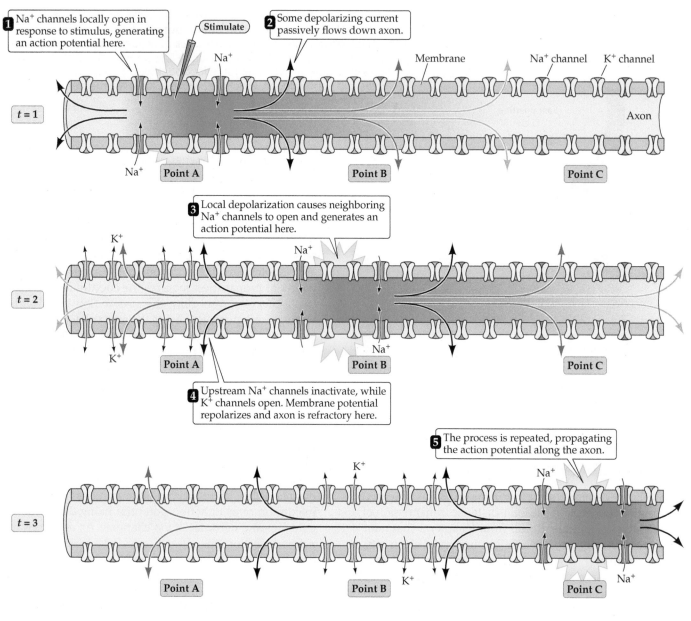

1 Na+ channels locally open in response to stimulus, generating an action potential here.

Stimulate

2 Some depolarizing current passively flows down axon.

Na+ Membrane Na+ channel K+ channel

t = 1 Axon

Na+ Point A Point B Point C

3 Local depolarization causes neighboring Na+ channels to open and generates an action potential here.

K+ Na+

t = 2

K+ Point A Na+ Point B Point C

4 Upstream Na+ channels inactivate, while K+ channels open. Membrane potential repolarizes and axon is refractory here.

5 The process is repeated, propagating the action potential along the axon.

K+ Na+

t = 3

Point A Point B K+ Point C Na+

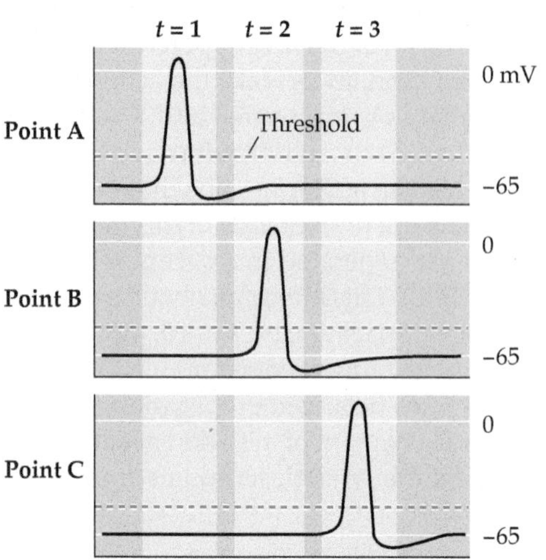

FIGURE 3.10 Action potential conduction requires both active and passive current flow. Depolarization opens Na^+ channels locally and produces an action potential at point A of the axon (time $t = 1$). The resulting inward current flows passively along the axon, depolarizing the adjacent region (point B) of the axon. At a later time ($t = 2$), the depolarization of the adjacent membrane has opened Na^+ channels at point B, resulting in the initiation of the action potential at this site and additional inward current that again spreads passively to an adjacent point (point C) farther along the axon. At a still later time ($t = 3$), the action potential has propagated even farther. This cycle continues along the full length of the axon. Note that as the action potential spreads, the membrane potential repolarizes due to K^+ channel opening and Na^+ channel inactivation, leaving a "wake" of refractoriness behind the action potential that prevents its backward propagation. The lower panel shows the time course of membrane potential changes at the points indicated.

Increased Conduction Velocity as a Result of Myelination

As a consequence of their mechanism of propagation, action potentials occur later and later at greater distances along the axon (see Figure 3.10, bottom left). Thus, the action potential has a measurable rate of propagation, called the **conduction velocity**. Conduction velocity is an important parameter because it defines the time required for electrical information to travel from one end of a neuron to another, and thus limits the flow of information within neural circuits. It is not surprising, then, that various mechanisms have evolved to optimize the propagation of action potentials along axons. Because action potential conduction requires passive and active flow of current, the rate of action potential propagation is determined by both of these phenomena. One way of improving passive current flow is to increase the diameter of an axon, which effectively decreases the internal resistance to passive current flow. For example, comparison of the conduction velocities of the axons of Aα and Aγ types of human motor neurons (Table 3.1) illustrates this point: increasing axon diameter only 2.5-fold yields a 20-fold increase in conduction velocity. The even larger giant axons of invertebrates such as squid presumably evolved because they increase action potential conduction velocity and thereby enhance the ability of these creatures to rapidly escape from predators.

A more efficient strategy to improve the passive flow of electrical current is to insulate the axonal membrane, reducing the ability of current to leak out of the axon and thus increasing the distance along the axon that a given local current can flow passively. This strategy is evident in the **myelination** of axons, a process by which oligodendrocytes in the central nervous system (and Schwann cells in the peripheral nervous system) wrap the axon in **myelin**, which consists of multiple layers of closely opposed glial membranes (Figure 3.11A; see also Chapter 1). By acting as an electrical insulator, myelin greatly speeds up action potential conduction (Figure 3.12). For example, Table 3.1 shows that whereas unmyelinated axon conduction velocities range from about 0.5 to 2 m/s, myelinated axons can conduct action potentials at velocities of up to 120 m/s (faster than a Formula 1 racing car). The major cause of this marked increase in speed is that the time-consuming process of action potential generation occurs only at specific points along the axon, called **nodes of Ranvier**, where there is a gap in the myelin wrapping (see Figure 1.3G). If the entire surface of an axon were insulated, there would be no place for current to flow out of the axon, and action potentials could not be generated. The voltage-gated Na^+ channels required for action potentials are found only at these nodes of Ranvier (Figure 3.11B). An action potential generated at one node of Ranvier elicits current that flows passively within the axon until the next few nodes are reached. This local current flow then generates an action potential in the neighboring nodes, and the cycle is repeated along the length of the axon. Because current flows across the neuronal membrane only at the nodes (Figure 3.11C), this type of propagation is called **saltatory**, meaning that the action potential jumps from node to node. Loss of myelin—as occurs in diseases such as multiple sclerosis, Guillain–Barré syndrome, and others—causes a variety of serious neurological problems because of the resulting defects in axonal conduction of action potentials.

TABLE 3.1 ■ Axon Conduction Velocities

Axon	Conduction velocity (m/s)	Diameter (μm)	Myelination
Squid giant axon	25	500	No
Human			
Motor axons			
Aα type	80–120	13–20	Yes
Aγ type	4–24	5–8	Yes
Sensory axons			
Aα type	80–120	13–20	Yes
Aβ type	35–75	6–12	Yes
Aδ type	3–35	1–5	Thin
C type	0.5–2.0	0.2–1.5	No
Autonomic			
preganglionic B type	3–15	1–5	Yes
postganglionic C type	0.5–2.0	0.2–1.5	No

Summary

The action potential and all its complex properties can be explained by time- and voltage-dependent changes in the Na^+ and K^+ permeabilities of neuronal membranes. This conclusion derives primarily from evidence obtained by a device called the voltage clamp. The voltage clamp technique is an electronic feedback method that allows control of neuronal membrane potential and, simultaneously, direct measurement of the voltage-dependent fluxes of Na^+ and K^+ that produce the action potential. Voltage clamp experiments show that a transient rise in Na^+ conductance activates rapidly and then inactivates during a sustained depolarization of the membrane potential. Such experiments also demonstrate a rise in K^+ conductance that activates in a delayed fashion and, in contrast to the Na^+ conductance, does not inactivate. Mathematical modeling of the properties of these conductances indicates that they, and they alone,

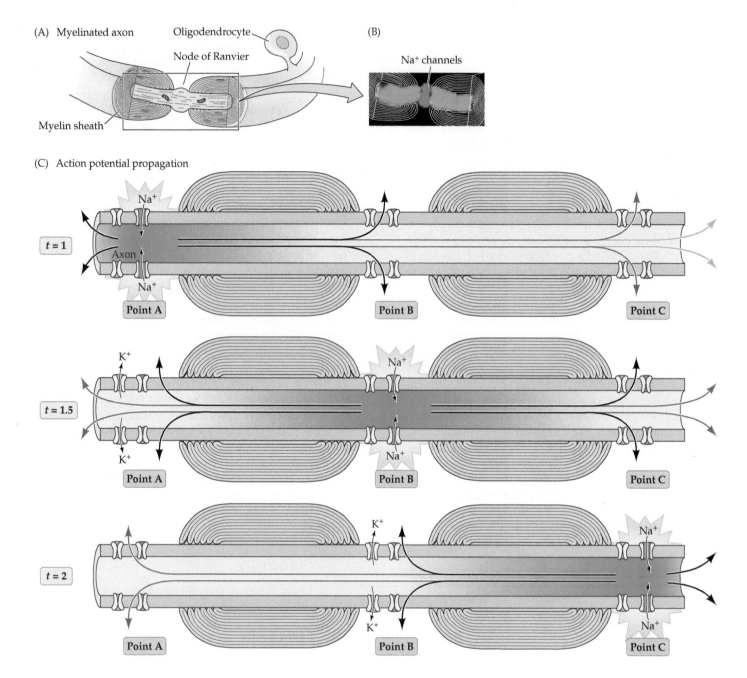

(A) Myelinated axon
Oligodendrocyte
Node of Ranvier
Myelin sheath
(B)
Na^+ channels

(C) Action potential propagation
t = 1
Na^+
Axon
Na^+
Point A
Point B
Point C
t = 1.5
K^+
K^+
Na^+
Na^+
Point A
Point B
Point C
t = 2
K^+
K^+
Na^+
Na^+
Point A
Point B
Point C

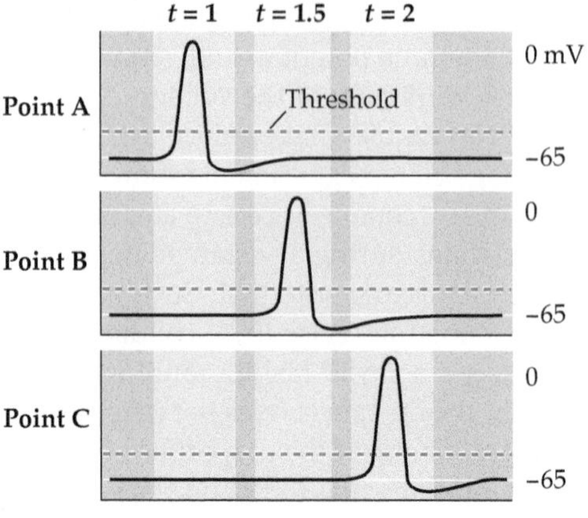

FIGURE 3.11 Saltatory action potential conduction along a myelinated axon. (A) Diagram of a myelinated axon. (B) Localization of voltage-gated Na^+ channels (red) at a node of Ranvier in a myelinated axon of the optic nerve. Green indicates the protein Caspr, which is located adjacent to the node of Ranvier. (C) Local current in response to action potential initiation at a particular site flows locally, as described in Figure 3.10. However, the presence of myelin prevents the local current from leaking across the internodal membrane; it therefore flows farther along the axon than it would in the absence of myelin, reaching several adjacent nodes (for clarity, only one node is shown). Moreover, voltage-gated Na^+ channels are present only at the nodes of Ranvier (voltage-gated K^+ channels are present at the nodes of some axons, but not others). This arrangement means that the generation of active, voltage-gated Na^+ currents need occur only at these unmyelinated regions. Bottom panel: The more rapid conduction of action potentials is illustrated by the more rapid timing of action potentials at the points indicated. (B from Chen et al., 2004.)

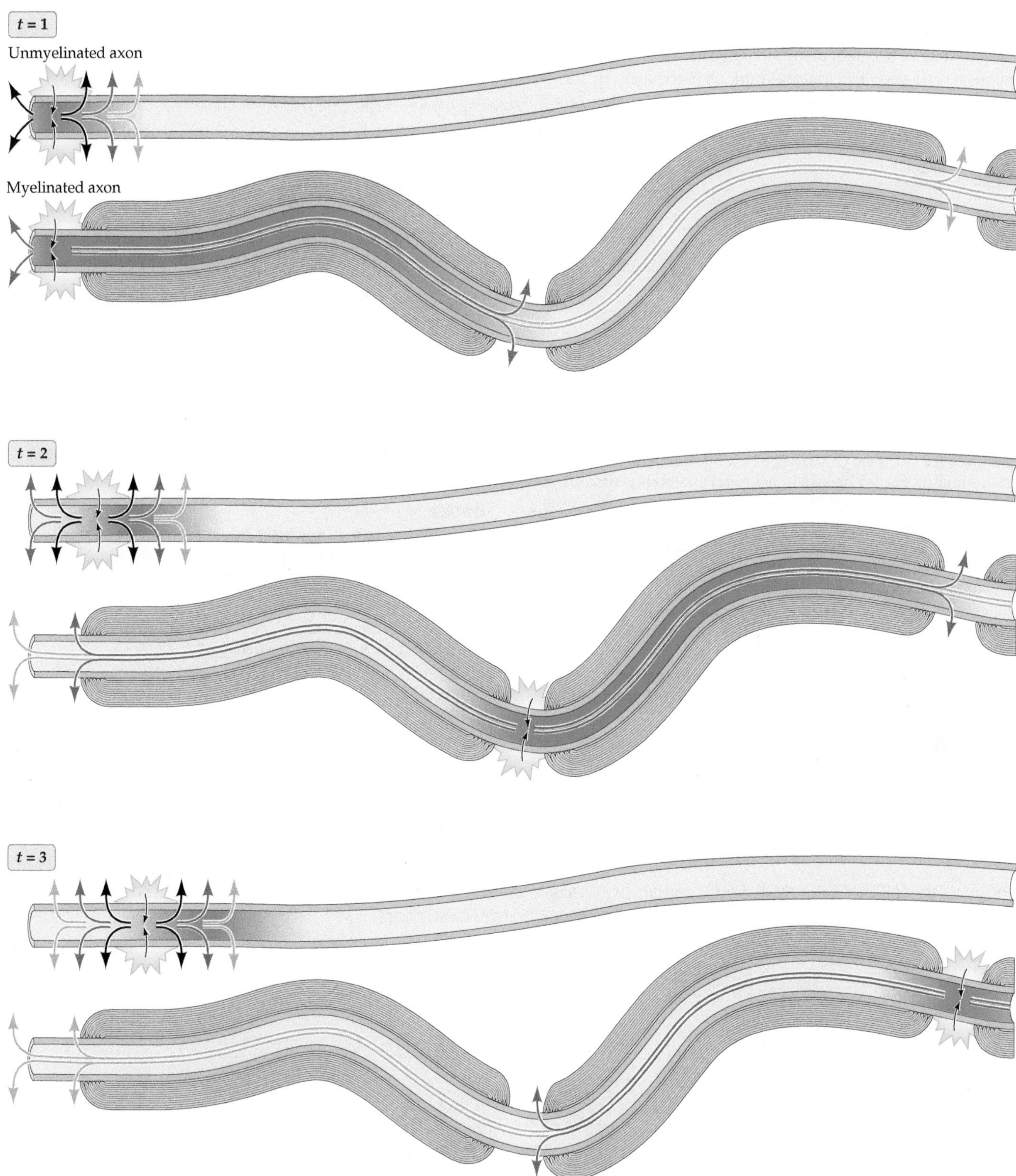

FIGURE 3.12 **Myelin increases action potential conduction speed.** The diagram compares the speed of action potential conduction in unmyelinated (upper panel in each pair) and myelinated (lower panels) axons. Passive conduction of current is shown by arrows.

are responsible for the production of all-or-none action potentials in the squid axon. Action potentials propagate along the nerve cell axons, initiated by the voltage gradient between the active and inactive regions of the axon by virtue of the local current flow. In this way, action potentials compensate for the relatively poor passive electrical properties of nerve cells and enable neural signaling over long distances. The molecular underpinnings of these signaling mechanisms will be revealed in the next chapter, which describes the properties of ion channels and transporters.

ADDITIONAL READING

Reviews

Armstrong, C. M. and B. Hille (1998) Voltage-gated ion channels and electrical excitability. *Neuron* 20: 371–380.

Salzer, J. L. (2003) Polarized domains of myelinated axons. *Neuron* 40: 297–318.

Important Original Papers

Armstrong, C. M. and L. Binstock (1965) Anomalous rectification in the squid giant axon injected with tetraethylammonium chloride. *J. Gen. Physiol.* 48: 859–872.

Chen, C. and 17 others (2004) Mice lacking sodium channel beta1 subunits display defects in neuronal excitability, sodium channel expression, and nodal architecture. *J. Neurosci.* 24: 4030–4042.

Hodgkin, A. L. and A. F. Huxley (1952a) Currents carried by sodium and potassium ions through the membrane of the giant axon of *Loligo*. *J. Physiol.* 116: 449–472.

Hodgkin, A. L. and A. F. Huxley (1952b) The components of membrane conductance in the giant axon of *Loligo*. *J. Physiol.* 116: 473–496.

Hodgkin, A. L. and A. F. Huxley (1952c) The dual effect of membrane potential on sodium conductance in the giant axon of *Loligo*. *J. Physiol.* 116: 497–506.

Hodgkin, A. L. and A. F. Huxley (1952d) A quantitative description of membrane current and its application to conduction and excitation in nerve. *J. Physiol.* 116: 507–544.

Hodgkin, A. L., A. F. Huxley and B. Katz (1952) Measurements of current–voltage relations in the membrane of the giant axon of *Loligo*. *J. Physiol.* 116: 424–448.

Hodgkin, A. L. and W. A. H. Rushton (1938) The electrical constants of a crustacean nerve fibre. *Proc. R. Soc. Lond.* 133: 444–479.

Huxley, A. F. and R. Stämpfli (1949) Evidence for saltatory conduction in peripheral myelinated nerve fibres. *J. Physiol.* 108: 315–339.

Moore, J. W., M. P. Blaustein, N. C. Anderson and T. Narahashi (1967) Basis of tetrodotoxin's selectivity in blockage of squid axons. *J. Gen. Physiol.* 50: 1401–1411.

Tasaki, I. and T. Takeuchi (1941) Der am Ranvierschen Knoten entstehende Aktionsstrom und seine Bedeutung für die Erregungsleitung. *Pflügers Arch.* 244: 696–711.

Books

Aidley, D. J. and P. R. Stanfield (1996) *Ion Channels: Molecules in Action*. Cambridge: Cambridge University Press.

Campenot, R. B. (2017) *Animal Electricity*. Cambridge, MA: Harvard University Press.

Hille, B. (2001) *Ion Channels of Excitable Membranes*, 3rd Edition. Sunderland, MA: Sinauer Associates.

Hodgkin, A. L. (1967) *The Conduction of the Nervous Impulse*. Springfield, IL: Charles C. Thomas.

Johnston, D. and S. M.-S. Wu (1995) *Foundations of Cellular Neurophysiology*. Cambridge, MA: MIT Press.

Junge, D. (1992) *Nerve and Muscle Excitation*, 3rd Edition. Sunderland, MA: Sinauer Associates.

Matthews, G. G. (2003) *Cellular Physiology of Nerve and Muscle*, 4th Edition. Malden, MA: Blackwell Publishing.

Go to the NEUROSCIENCE 6e Companion Website at **www.oup.com/uk/Purves6e** for Web Topics, Animations, Flashcards, and more.

CHAPTER

4 Ion Channels and Transporters

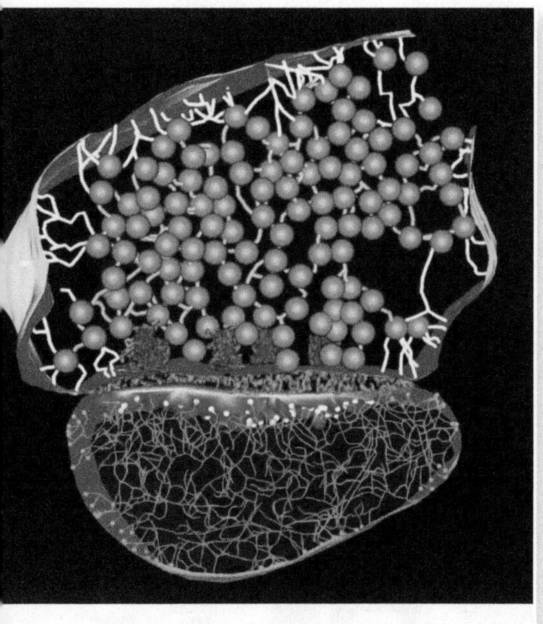

Overview

THE GENERATION OF ELECTRICAL SIGNALS in neurons requires that plasma membranes establish concentration gradients for specific ions and that these membranes undergo rapid and selective changes in their permeability to these ions. The membrane proteins that create and maintain ion gradients are called active transporters; other proteins, called ion channels, give rise to selective ion permeability changes. Ion channels are transmembrane proteins that contain a narrow pore that selectively permits particular ions to permeate the membrane. Some ion channels also contain voltage sensors that are able to detect the electrical potential across the membrane. Such voltage-gated channels open or close in response to the magnitude of the membrane potential, allowing the membrane's permeability to be regulated by changes in this potential. Other types of ion channels are gated by chemical signals, either by extracellular signals such as neurotransmitters or by intracellular signals such as second messengers. Still other channels respond to mechanical stimuli, temperature changes, or a combination of signals. Different combinations of ion channels are found in different cell types, yielding a wide spectrum of electrical characteristics. In contrast to ion channels, active transporters are membrane proteins that produce and maintain ion concentration gradients. The most important of these is the Na^+ pump, which hydrolyzes ATP to regulate the intracellular concentrations of both Na^+ and K^+. Other active transporters produce concentration gradients for the full range of physiologically important ions, including Cl^-, Ca^{2+}, and H^+. From the perspective of electrical signaling, active transporters and ion channels are complementary: Transporters create the concentration gradients that help drive ion fluxes through open ion channels, thereby generating electrical signals.

Ion Channels Underlying Action Potentials

Although Hodgkin and Huxley had no knowledge of the physical nature of the conductance mechanisms underlying action potentials, they nonetheless proposed that nerve cell membranes have channels that allow ions to pass selectively from one side of the membrane to the other (see Chapter 3). Based on ion conductances and currents measured in voltage clamp experiments, the postulated channels had to have several properties. First, because the ion currents are quite large, the channels had to be capable of allowing ions to move across the membrane at high rates. Second, because the currents depend on the electrochemical gradient across the membrane, the channels had to make use of these gradients. Third, because Na^+ and K^+ flow across the membrane independently of each other, different channel types had to be capable of discriminating between Na^+ and K^+, allowing only one of these ions to flow across the membrane under the relevant conditions. Finally, given that the ion conductances are voltage-dependent, the channels had to be able to sense the membrane potential, opening only when the voltage reached appropriate levels. While this concept of channels was highly speculative in the 1950s, later experimental work established beyond any doubt that transmembrane proteins called voltage-sensitive ion channels indeed exist and are responsible for action potentials and other types of electrical signals.

The first direct evidence for the presence of voltage-sensitive, ion-selective channels in nerve cell membranes came from measurements of the ion currents flowing through individual ion channels. The voltage clamp apparatus used by Hodgkin and Huxley could only resolve the *aggregate* current resulting from the flow of ions through many thousands of channels. A technique capable of measuring the currents flowing through single channels was devised in

BOX 4A ■ The Patch Clamp Method

A wealth of information about the function of ion channels has resulted from the invention of the patch clamp method. This technique is based on a very simple idea. A glass pipette with a very small opening is used to make tight contact with a tiny area, or patch, of neuronal membrane. After the application of a small amount of suction to the back of the pipette, the seal between the pipette and membrane becomes so tight that no ions can flow between the pipette and the membrane. Thus, all the current that flows when a single ion channel opens must flow into the pipette. This minute electrical current can be measured with an ultrasensitive electronic amplifier connected to the pipette. This arrangement is the *cell-attached patch clamp recording method*. As with the conventional voltage clamp method, the patch clamp method allows experimental control of the membrane potential to characterize the voltage dependence of membrane currents.

Minor manipulations allow other recording configurations. For example, if the membrane patch within the pipette is disrupted by briefly applying strong suction, the interior of the pipette becomes continuous with the cell cytoplasm. This arrangement allows measurements of electrical potentials and currents from the entire cell and is therefore called *whole-cell recording*. The whole-cell configuration also allows diffusional exchange between the solution in the pipette and the cytoplasm, producing a convenient way to inject substances into the interior of a "patched" cell.

Two other variants of the patch clamp method originate from the finding that once a tight seal has formed between the membrane and the glass pipette, small pieces of membrane can be pulled away from the cell without disrupting the seal; this yields a preparation that is free of the complications imposed by the rest of the cell. Simply retracting a pipette that is in the cell-attached configuration causes a small vesicle of membrane to remain attached to the pipette. By briefly exposing the tip of the pipette to air, the vesicle opens to yield a small patch of membrane with its (former) intracellular surface exposed. This arrangement, called the *inside-out patch recording configuration*, makes it possible to change the medium to which the intracellular surface of the membrane is exposed. Thus, the inside-out configuration is particularly valuable when studying the influence of intracellular molecules on ion channel function.

Alternatively, if the pipette is retracted while it is in the whole-cell configuration, the membrane patch produced has its extracellular surface exposed. This arrangement, called the *outside-out recording configuration*, is optimal for studying how channel activity is influenced by extracellular chemical signals such as neurotransmitters (see Chapter 5). This range of possible configurations makes the patch clamp method an unusually versatile technique for studies of ion channel function. Robotic versions of the patch clamp technique have made their way into industry, serving as a very sensitive and rapid means of screening therapeutic drugs that act on ion channels.

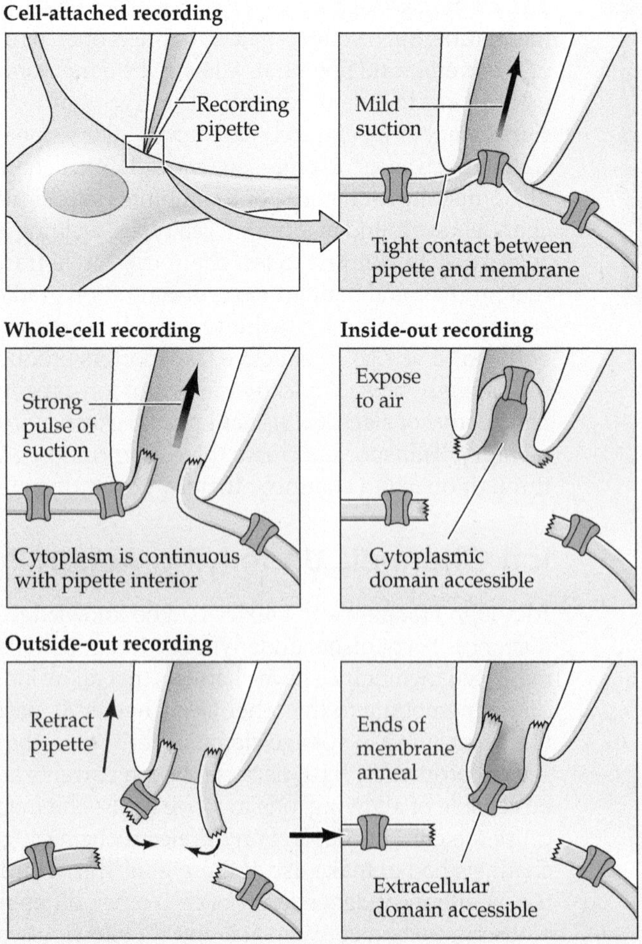

Four configurations in patch clamp measurements of ion currents.

1976 by Erwin Neher and Bert Sakmann at the Max Planck Institute in Germany. This remarkable approach, called a **patch clamp**, revolutionized the study of membrane currents (Box 4A). In particular, the patch clamp method provided the means to test directly Hodgkin and Huxley's deductions about the characteristics of ion channels.

Currents flowing through Na^+ channels are best examined in experimental circumstances that prevent the flow of current through other types of channels that are present in the membrane (e.g., K^+ channels). Under such conditions, depolarizing a patch of membrane from a squid giant axon causes tiny inward currents to flow, but only occasionally (Figure 4.1). The size of these currents is minuscule—approximately 1–2 pA (i.e., 10^{-12} ampere)—but is stereotyped, reaching discrete values that suggest ion flux though individual, open ion channels. These currents are orders of magnitude smaller than the Na^+ currents measured by voltage clamping the entire axon. The currents flowing through single channels are called **microscopic currents** to distinguish them from the **macroscopic currents** flowing through a large number of channels distributed over a much more extensive region of surface membrane. Although microscopic currents are certainly small, a current of 1 pA nonetheless reflects the flow of thousands of ions per millisecond. Thus, as predicted, a single channel can let many ions pass through the membrane in a very short time.

Several observations further proved that the microscopic currents in Figure 4.1B are due to the opening of single, voltage-activated Na^+ channels. First, the currents are carried by Na^+; thus, they are directed inward when the membrane potential is more negative than E_{Na}, reverse their polarity at E_{Na}, are outward at more positive potentials, and are reduced in size when the Na^+ concentration of the external medium is decreased. This behavior exactly parallels that of the macroscopic Na^+ currents described in Chapter 3 (see Figure 3.4). Second, the channels have a time course of opening, closing, and inactivating that matches the kinetics of macroscopic Na^+ currents. This correspondence is difficult to appreciate in the measurement of microscopic currents flowing through a single open channel, because individual channels open and close in a stochastic (random) manner, as can be seen by examining the individual traces in Figure 4.1B. However,

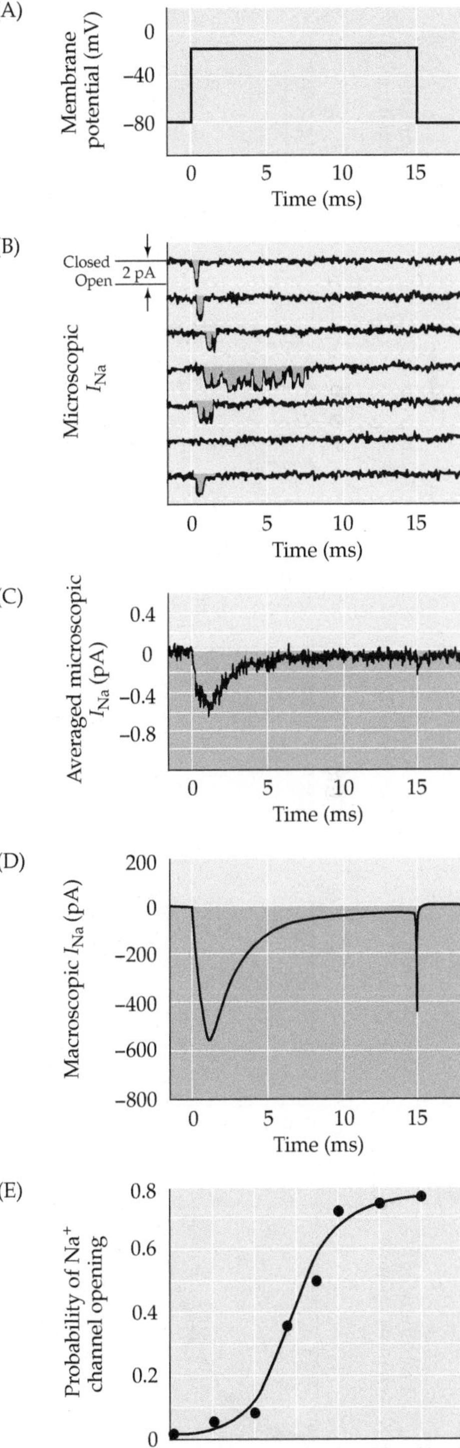

FIGURE 4.1 Patch clamp measurements of ion currents flowing through single Na^+ channels in a squid giant axon. In these experiments, Cs^+ was applied to the axon to block voltage-gated K^+ channels. Depolarizing voltage pulses (A) applied to a patch of membrane containing a single Na^+ channel result in brief currents (B, downward deflections) in the seven successive recordings of membrane current (I_{Na}). (C) The average of many such current records shows that most channels open in the initial 1–2 ms following depolarization of the membrane, after which the probability of channel openings diminishes because of channel inactivation. (D) A macroscopic current measured from another axon shows the close correlation between the time courses of microscopic and macroscopic Na^+ currents. (E) The probability of a Na^+ channel opening depends on the membrane potential, increasing as the membrane is depolarized. (B,C after Bezanilla and Correa, 1995; D after Vandenburg and Bezanilla, 1991; E after Correa and Bezanilla, 1994.)

repeated depolarization of the membrane potential causes each Na^+ channel to open and close many times. When the current responses to a large number of such stimuli are averaged together, the collective response has a time course that looks much like the macroscopic Na^+ current (see Figure 4.1C). In particular, the channels open mostly at the beginning of a prolonged depolarization, showing that they activate and subsequently inactivate, as predicted from the macroscopic Na^+ current (compare Figures 4.1C and D). Third, both the opening and closing of the channels are voltage-dependent; thus, the channels are closed at –80 mV but open when the membrane potential is depolarized. In fact, the probability that any given channel will be open varies with membrane potential (see Figure 4.1E), again as predicted from the macroscopic Na^+ conductance (see Figure 3.7). Finally, tetrodotoxin, which blocks the macroscopic Na^+ current (see Figure 3.5), also blocks microscopic Na^+ currents. Taken together, these results show that the macroscopic Na^+ current measured by Hodgkin and Huxley does indeed arise from the aggregate effect of many millions of microscopic Na^+ currents, each representing the opening of a single voltage-sensitive Na^+ channel.

Patch clamp experiments also revealed the properties of the channels responsible for the macroscopic K^+ currents associated with action potentials. When the membrane potential is depolarized (Figure 4.2A), microscopic outward currents (Figure 4.2B) can be observed under conditions that block Na^+ channels. The microscopic outward currents exhibit all the features expected for currents flowing through action potential-related K^+ channels. Thus, the microscopic currents (Figure 4.2C), like their macroscopic counterparts (Figure 4.2D), fail to inactivate during brief depolarizations. Moreover, these single-channel currents are sensitive to ionic manipulations and drugs that affect the macroscopic K^+ currents and, like the macroscopic K^+ currents, are voltage-dependent (Figure 4.2E). This and other evidence shows that macroscopic K^+ currents associated with action potential repolarization arise from the opening of many voltage-sensitive K^+ channels.

In summary, patch clamping has allowed direct observation of microscopic currents flowing through single ion channels, confirming that voltage-sensitive Na^+ and K^+ channels are responsible for the macroscopic conductances and currents that underlie the action potential. Measurements of the behavior of single ion channels have also provided insight into the molecular attributes of these

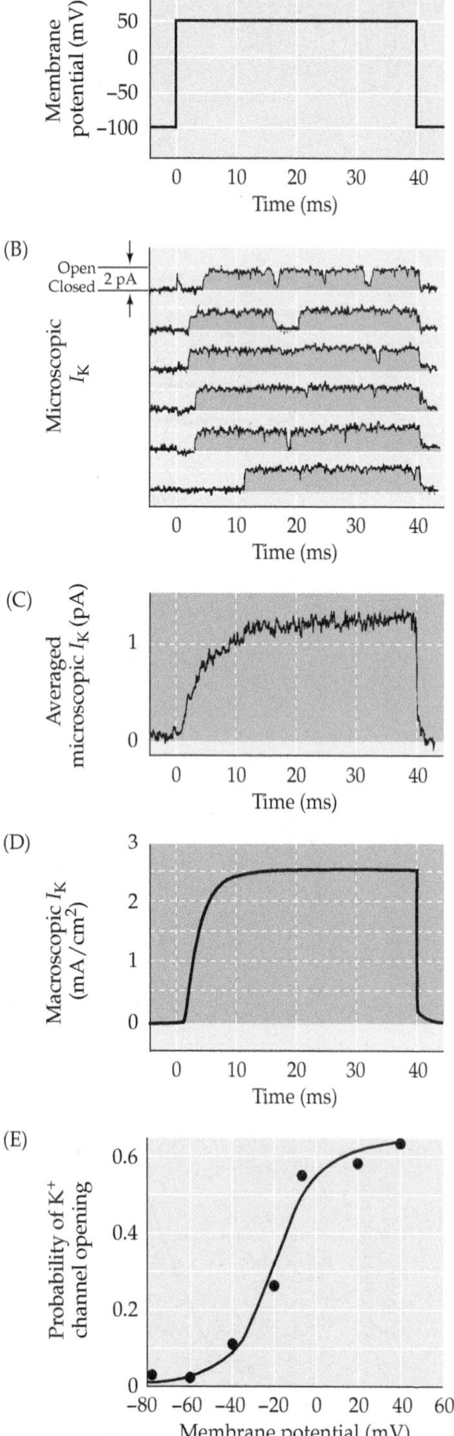

FIGURE 4.2 Patch clamp measurements of ion currents flowing through single K^+ channels in a squid giant axon. In these experiments, tetrodotoxin was applied to the axon to block voltage-gated Na^+ channels. Depolarizing voltage pulses (A) applied to a patch of membrane containing a single K^+ channel result in brief currents (B, upward deflections) whenever the channel opens. (C) The average of such current records shows that most channels open with a delay, but remain open for the duration of the depolarization. (D) A macroscopic current measured from another axon shows the correlation between the time courses of microscopic and macroscopic K^+ currents. (E) Membrane potential controls K^+ channel opening, with the probability of a channel opening increasing as the membrane is depolarized. (B,C after Augustine and Bezanilla, in Hille 2001; D after Augustine and Bezanilla, 1990; E after Perozo et al., 1991.)

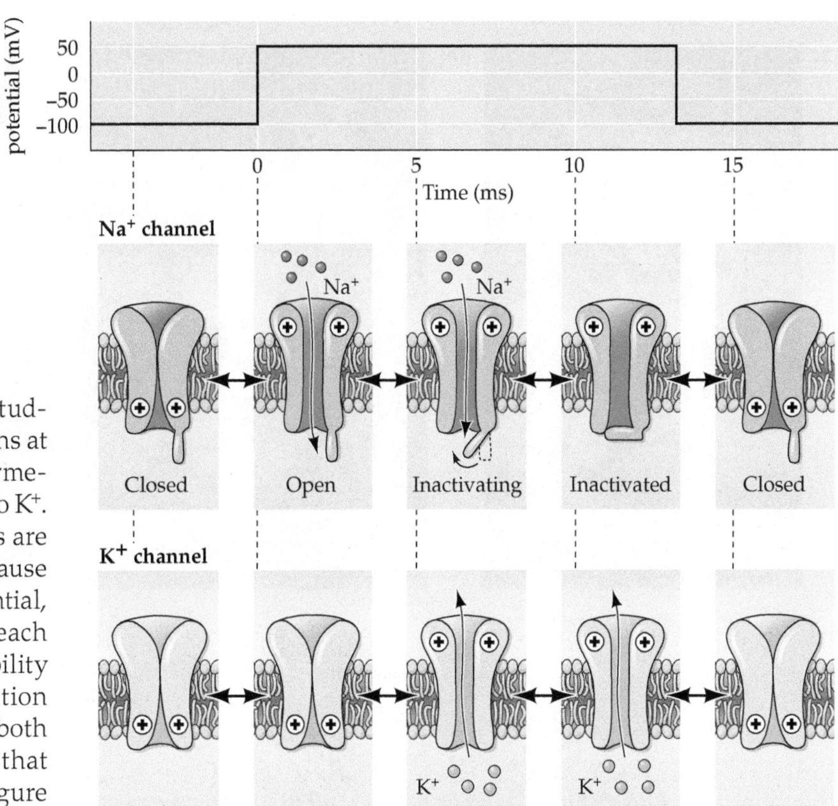

FIGURE 4.3 Functional states of voltage-gated Na^+ and K^+ channels. The gates of both channels are closed when the membrane potential is hyperpolarized. When the potential is depolarized, voltage sensors (indicated by +) allow the channel gates to open—first the Na^+ channels and then the K^+ channels. Na^+ channels also inactivate during prolonged depolarization, whereas many types of K^+ channels do not.

channels. For example, such single channel studies show that the squid axon membrane contains at least two types of channels—one selectively permeable to Na^+ and a second selectively permeable to K^+. This **ion selectivity** means that these channels are able to discriminate between Na^+ and K^+. Because their opening is influenced by membrane potential, both channel types are **voltage gated**. For each channel, depolarization increases the probability of channel opening, whereas hyperpolarization closes them (see Figures 4.1E and 4.2E). Thus, both channel types must have a **voltage sensor** that detects the potential across the membrane (Figure 4.3). However, these channels differ in important respects. In addition to differences in ion selectivity, the kinetic properties of the gating of the two channels differ as expected from the macroscopic behavior of the Na^+ and K^+ currents described in Chapter 3. Further, depolarization inactivates the Na^+ channel but not the K^+ channel, causing Na^+ channels to pass into a nonconducting state. The Na^+ channel must therefore have an additional molecular mechanism responsible for *inactivation*. Finally, these channel proteins provide unique binding sites for drugs and for various neurotoxins known to block specific subclasses of ion channels (Box 4B). This information about the physiology of ion channels set the stage for subsequent studies of their workings at the molecular level.

How Ion Channels Work

How can channels selectively conduct ions, such as Na^+ and K^+, across the membrane? How can these channels sense the transmembrane potential and use this information to regulate their ion fluxes? X-ray crystallography studies done by Rod MacKinnon at Rockefeller University and others have examined the molecular structure of ion channels and have answered these fundamental questions about channel function. Indeed, these studies have revealed that all channels are integral membrane proteins that span the plasma membrane repeatedly and share a common transmembrane architecture.

The ion permeability mechanism of channels was revealed by studies of a bacterial K^+ channel (Figure 4.4), which was chosen for analysis because the large quantity of channel protein needed for crystallography could be obtained by growing large numbers of bacteria. MacKinnon's results showed that the K^+ channel is formed by subunits that each cross the plasma membrane twice; between these two helical membrane-spanning structures is a **pore loop** that inserts into the plasma membrane (see Figure 4.4A). Four subunits are assembled together to form a single K^+ channel (see Figure 4.4B). In the center of the assembled channel, the four pore loops come together to form a **pore** that serves as a narrow tunnel that allows K^+ to flow through the protein and thus cross the membrane. The channel pore is formed by the pore loops of each subunit, as well as by adjacent membrane-spanning domains. The structure of the pore is well suited for conducting K^+; the narrowest part is near the outside mouth of the channel and is so constricted that only a nonhydrated K^+ ion can fit through the bottleneck (see Figure 4.4C). Larger cations, such as Cs^+, are too large to traverse this region of the pore, while smaller cations such as Na^+ cannot enter the pore because the pore "walls" are too far apart to stabilize a dehydrated Na^+ ion. This part of the channel complex is responsible for the selective permeability to K^+

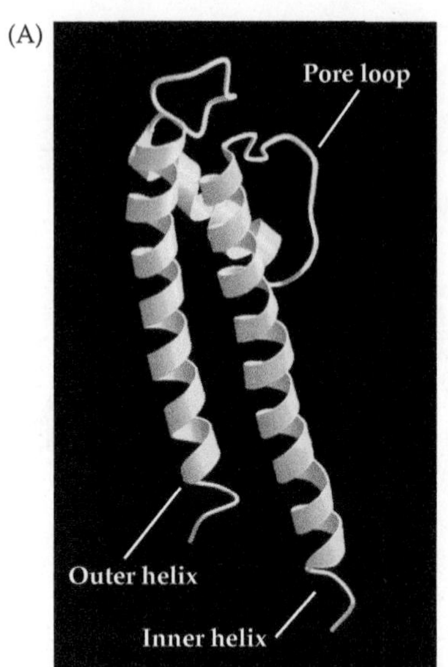

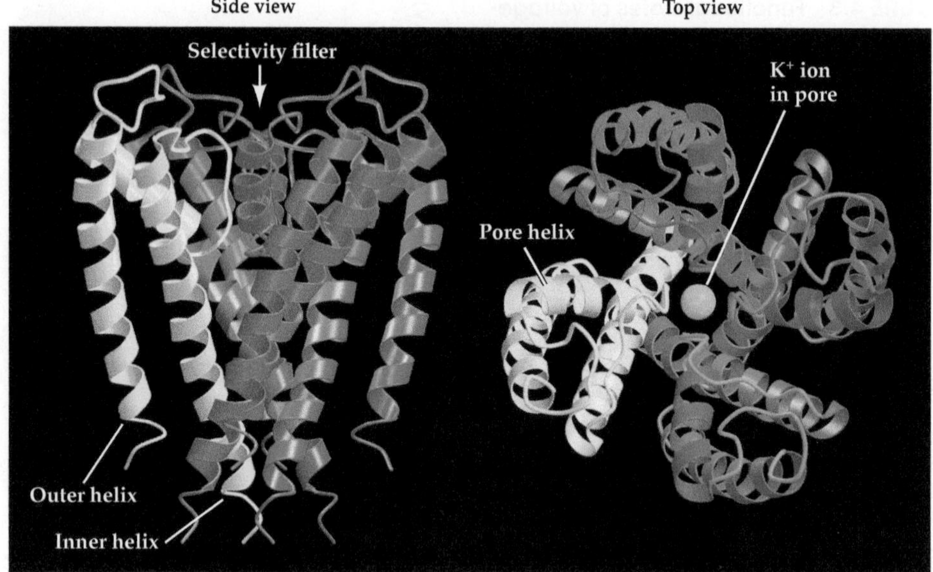

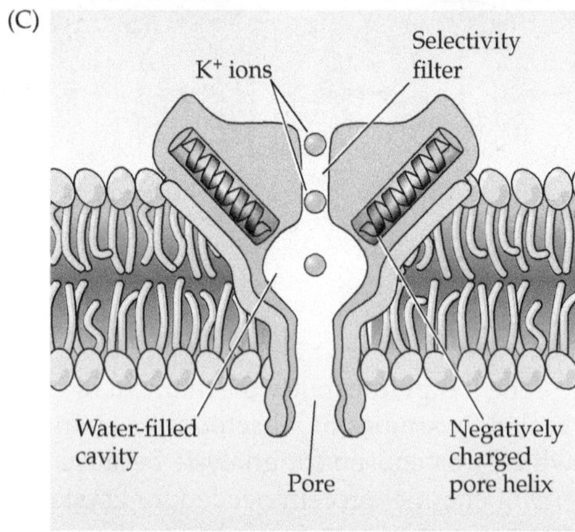

FIGURE 4.4 Structure of a simple bacterial K⁺ channel determined by crystallography. (A) Structure of one subunit of the channel, which consists of two membrane-spanning domains and a pore loop that inserts into the membrane. (B) Three-dimensional arrangement of four subunits (each in a different color) to form a K⁺ channel. The right-hand view is from the top of the channel, showing a K⁺ ion (yellow) within the channel pore. (C) The permeation pathway of the K⁺ channel consists of a large aqueous cavity connected to a narrow selectivity filter. Helical domains of the channel point negative charges (green) toward this cavity, allowing K⁺ ions (yellow) to become dehydrated and then move through the selectivity filter. (A,B From Doyle et al., 1998.)

and is therefore called the **selectivity filter**. Deeper within the channel is a water-filled cavity that connects to the interior of the cell. This cavity evidently collects K^+ from the cytoplasm and, using negative charges from the protein, allows K^+ ions to become dehydrated so they can enter the selectivity filter. These "naked" ions are then able to move through four K^+ binding sites within the selectivity filter to eventually reach the extracellular space (recall that the physiological electrochemical gradient normally drives K^+ out of cells). The presence of multiple (up to four) K^+ ions within the selectivity filter causes electrostatic repulsion between the ions that helps speed their transit through the selectivity filter, thereby permitting rapid ion flux through the channel. Thus, permeation of K^+ ions through these channels can be easily understood in structural terms. We now know that very similar principles also apply to ion permeation through other types of channels.

Structural insights into the voltage-dependent gating of ion channels emerged from further crystallographic studies of a mammalian voltage-gated K^+ channel. As is the case for the bacterial K^+ channel, four subunits assemble to form the voltage-gated K^+ channel (Figure 4.5A). The central pore region of this channel is very similar to that of the bacterial K^+ channel, confirming the generality of the ion permeation mechanism established from studies of the bacterial K^+ channel (compare Figure 4.5B and Figure 4.4B). This voltage-gated channel has additional structures on its cytoplasmic side, such as a regulatory β subunit and a T1 domain that links the β subunit to the channel (see Figure 4.5A). Most important, each channel subunit has four *additional* transmembrane structures that form the voltage sensors of this channel. These voltage sensors can be observed as separate domains that extend into the plasma membrane and are linked to the central pore of the channel (see Figure 4.5B). These voltage sensors contain positive charges that enable movement within the membrane in response to changes in membrane potential: Depolarization

BOX 4B ■ Toxins That Poison Ion Channels

Given the importance of Na^+ and K^+ channels for neuronal excitation, it is not surprising that channel-specific toxins have evolved in several organisms as mechanisms for self-defense or for capturing prey. A rich collection of natural toxins selectively target the ion channels of neurons and other cells. These toxins are valuable not only for survival, but also as tools for study of the function of cellular ion channels. The best-known channel toxin is *tetrodotoxin*, produced by certain puffer fish and other animals. Tetrodotoxin produces a potent and specific blockade of the pore of the Na^+ channels responsible for action potential generation, thereby paralyzing the animals unfortunate enough to ingest it. *Saxitoxin*, a chemical homologue of tetrodotoxin produced by dinoflagellates, has a similar action on Na^+ channels. The potentially lethal effects of eating shellfish that have ingested these "red tide" dinoflagellates are due to the potent neuronal actions of saxitoxin.

Fish-eating cone snails (see Box 6A) paralyze their prey by producing a potent venom consisting of tens or hundreds of peptide neurotoxins. One group of these toxins, called μ-conotoxins, produces paralysis by blocking the pore of voltage-gated Na^+ channels. Scorpions similarly paralyze their prey by injecting a potent mix of peptide toxins that also affect ion channels. Among these are the *α-toxins*, which slow the inactivation of Na^+ channels (Figure A1); exposure of neurons to α-toxins prolongs the action potential (Figure A2), thereby scrambling information flow within the nervous system of the soon-to-be-devoured victim. Other peptides in scorpion venom, called *β-toxins*, shift the voltage dependence of Na^+ channel activation (Figure B). These toxins cause Na^+ channels to open at potentials much more negative than normal, inducing uncontrolled action potential firing. *Batrachotoxin* is an alkaloid toxin, produced by a species of frog, that is used by some tribes of South American Indians to poison their arrow tips. This toxin works both by removing inactivation and shifting activation of Na^+ channels. Several plants produce similar toxins, including *aconitine*, from buttercups; *veratridine*, from lilies; and several insecticidal toxins (pyrethrins) produced by plants such as chrysanthemums and rhododendrons.

Potassium channels have also been targeted by toxin-producing organisms. Peptide toxins affecting K^+ channels include *dendrotoxin*, from wasps; *apamin*, from bees; and *charybdotoxin*, yet another toxin produced by scorpions. All of these toxins block K^+ channels as their primary action; no toxin is known to affect the activation or inactivation of these channels, although such agents may simply be awaiting discovery.

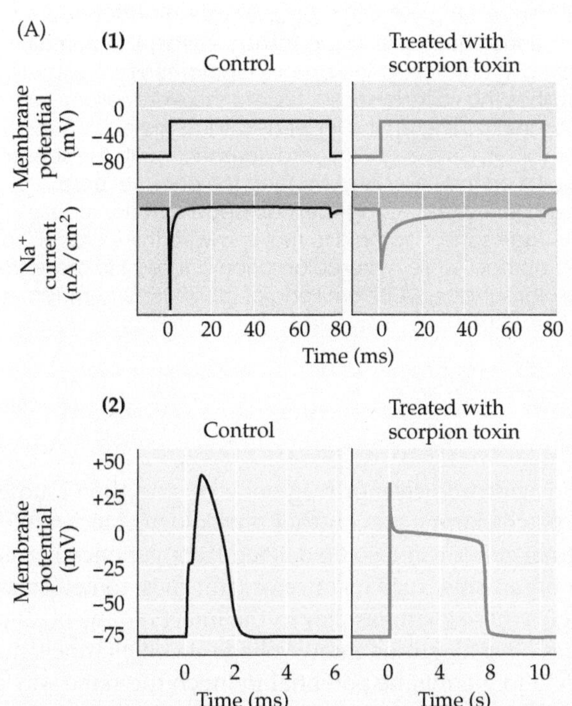

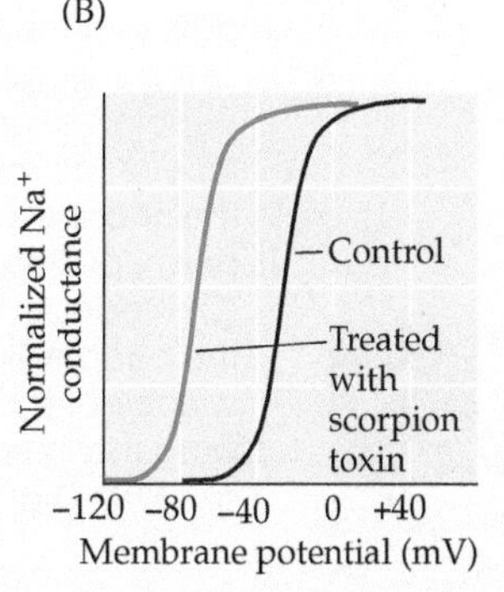

(A) Effects of toxin treatment on Na^+ channels of frog axons. (1) α-Toxin from the scorpion *Leiurus quinquestriatus* prolongs Na^+ currents recorded with the voltage clamp method. (2) As a result of the increased Na^+ current, α-toxin greatly prolongs the duration of the axonal action potential. Note the change in timescale after treating with toxin. (B) Treatment with β-toxin from another scorpion, *Centruroides sculpturatus*, shifts the activation of Na^+ channels, so that Na^+ conductance begins to increase at potentials much more negative than usual. (A after Schmidt and Schmidt, 1972; B after Cahalan, 1975.)

pushes the sensors outward, while hyperpolarization pulls them inward. Such movements of the sensors then exert force on the helical linkers connecting the sensors to the channel pore, pulling the channel pore open or pushing it closed (Figure 4.5C). Thus, voltage-dependent gating of ion channels can also be understood in structural terms. The precise movements of the voltage sensor that occur during membrane depolarization are not yet clear and are the subject of considerable debate. One proposal is that depolarization causes the paddle-like sensor to flip from one side of the membrane to the other (Figure 4.5D).

In summary, atomic-level structural characterization of K^+ channels has yielded considerable insight into how channels use pores that are permeable to a single type of ion to selectively conduct these ions from one side of the plasma membrane to the other, as well as illuminating how channels can sense changes in membrane voltage and use such changes to gate the opening and closing of their pores.

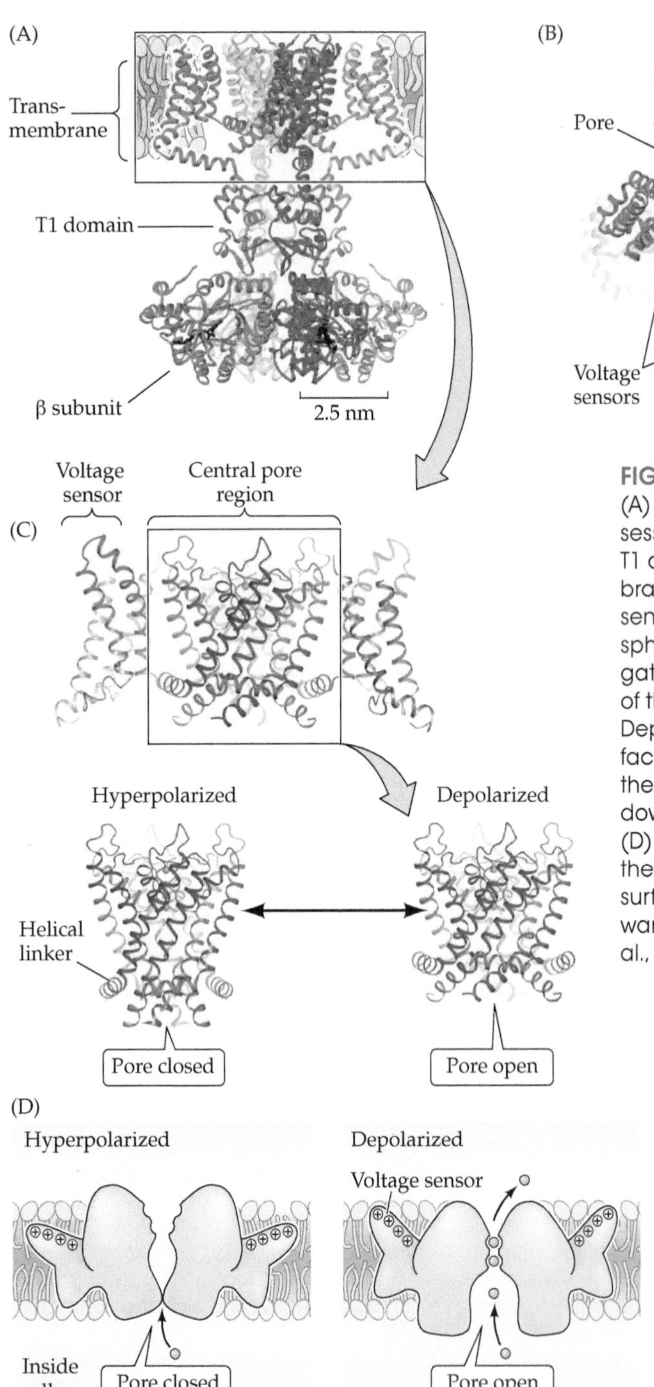

FIGURE 4.5 Structure of a mammalian voltage-gated K^+ channel. (A) The channel includes four subunits (in different colors), each possessing a transmembrane domain and a T1 domain. Attached to each T1 domain is a β subunit. (B) When viewed from above, the transmembrane domain can be seen to have separate domains for voltage sensing and for forming the K^+-conducting pore (K^+ indicated by black spheres in the middle of the pore). (C) Model for voltage-dependent gating of the K^+ channel. Upper: structure of the central pore domain of the voltage-gated K^+ channel in the open (depolarized) state. Lower: Depolarization pushes the voltage sensor toward the extracellular surface of the membrane, pulling on the linker (red) and thereby opening the channel pore (blue). Conversely, hyperpolarization pulls the sensor down, pushing down on this linker and shutting the channel pore. (D) Model for movement of the voltage sensor. Depolarization causes the paddle-like voltage sensor domain to move toward the extracellular surface of the membrane, while hyperpolarization causes it to move toward the intracellular surface. (A,B from Long et al., 2005; C from Tao et al., 2010; D after Lee, 2006.)

The Diversity of Ion Channels

Many additional insights have come from genetic studies of ion channels: More than 200 ion channel genes have been discovered, a remarkable number that could not have been anticipated from the work of Hodgkin and Huxley. To understand the functional significance of this multitude of ion channel genes, individual channel types can be selectively expressed in well-defined experimental systems, such as cultured cells or frog oocytes, and then studied with patch clamping and other physiological techniques. Channel genes can also be deleted from genetically tractable organisms, such as mice or fruit flies, to determine the roles these channels play in the intact organism. Such studies have identified many voltage-gated channels that respond to membrane potential in much the same way as do the Na^+ and K^+ channels that underlie the action potential. Voltage-gated ion channels also are involved in a broad range of neurological diseases. Other types of channels are insensitive to membrane voltage, instead being gated by chemical signals that bind to extracellular or intracellular domains on these proteins. Still others are sensitive to other types of physical stimuli, such as mechanical displacement or changes in temperature. Thus, although the basic electrical signals of the nervous system are relatively stereotyped, the proteins responsible for generating these signals are remarkably diverse, conferring specialized signaling properties to the many different types of neurons that populate the nervous system.

Voltage-Gated Ion Channels

Voltage-gated ion channels that are selectively permeable to each of the major physiological ions—Na^+, K^+, Ca^{2+}, and Cl^-—have been identified (Figure 4.6). Indeed, many different genes have been discovered for each type of voltage-gated channel; for example, there are ten human Na^+ channel genes. This finding was unexpected because Na^+ channels from many different cell types have similar functional properties, consistent with their origin from a single gene. It is now clear, however, that all of these Na^+ channel genes (called **SCN genes**) produce proteins that differ in their structure, function, and distribution in specific tissues. For instance, in addition to the rapidly inactivating Na^+ channels that underlie the action potential in many types of neurons, including the squid neurons studied by Hodgkin and Huxley, a voltage-sensitive Na^+ channel that does not completely inactivate has been identified in mammalian neurons. This channel gives rise to a "persistent" Na^+ current that helps regulate action potential threshold and repetitive firing. This channel serves as a target of local anesthetics such as procaine and lidocaine.

Na^+ channels consist of motifs of six membrane-spanning regions, similar to those on the voltage-gated K^+ channels shown in Figure 4.5, that are repeated four times, yielding a total of 24 transmembrane regions (see Figure

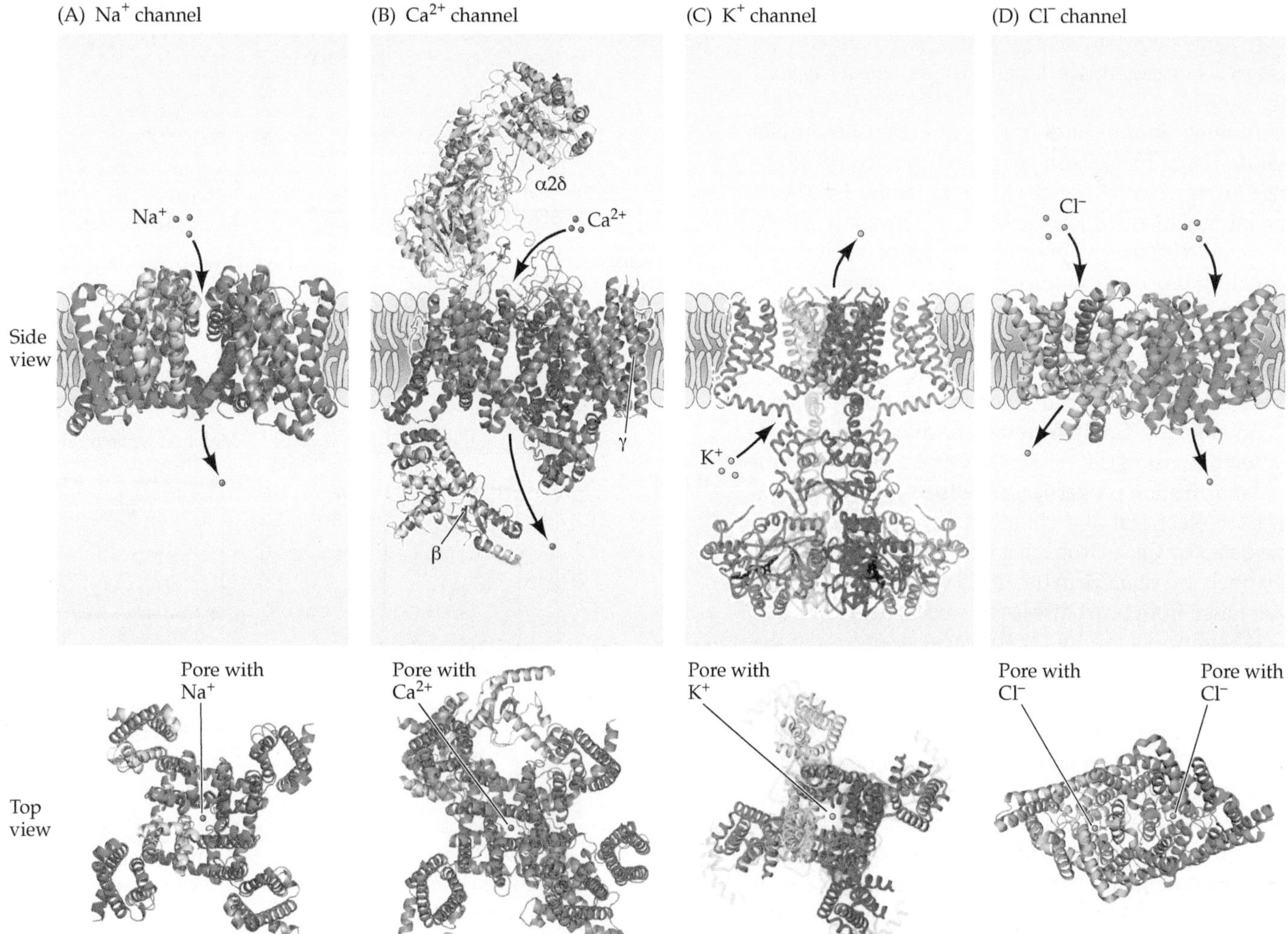

FIGURE 4.6 Types of voltage-gated ion channels. Examples of voltage-gated channels include those selectively permeable to Na^+ (A), Ca^{2+} (B), K^+ (C), and Cl^- (D). All channels include a number of transmembrane domains that form pores; the selectivity filters of these pores determine ion selectivity, while other structures are responsible for sensing the membrane potential and for gating pore opening and closing. Side view shows the channels within the plasma membrane; top view shows the transmembrane structures of the channels, emphasizing similarities between the Na^+, Ca^{2+}, and K^+ channels. γ, $\alpha 2\delta$, and β indicate accessory subunits of the Ca^{2+} channel. (A after Ahuja et al., 2015; B after Wu et al., 2015; C from Long et al., 2005; D after Dutzler et al., 2002.)

4.6A). Thus, one Na+ channel protein forms a structure very similar to that produced by four K+ channel subunits. Four of these transmembrane domains serve as voltage sensors that enable voltage-dependent gating of Na^+ channels. In the center of Na^+ channels is a pore that connects the extracellular and intracellular sides of the membrane. The selectivity filter of the Na^+ channel pore, like that of the K^+ channels shown in Figures 4.4 and 4.5, is formed by four pore loops. However, this selectivity filter is narrower to permit the smaller Na^+, but not the larger K^+, to diffuse through the Na^+ channel pore and permeate across the membrane. Accessory proteins, called β subunits, regulate the function of Na^+ channels.

Other electrical responses in neurons rely on voltage-gated Ca^{2+} channels (see Figure 4.6B). Ca^{2+} channels are similar in structure to Na^+ channels, consisting of six membrane-spanning regions that are repeated four times and include positively charged voltage sensors that enable voltage-dependent gating. Ca^{2+} channels also contain pore loops; these yield a pore that is selectively permeable to Ca^{2+} but otherwise look remarkably similar in structure to Na^+ and K^+ channels.

In some neurons, voltage-gated Ca^{2+} channels give rise to action potentials in much the same way as voltage-sensitive Na^+ channels. In other neurons, Ca^{2+} channels prolong the duration of action potentials whose rising phases are generated by currents flowing through Na^+ channels. More generally, by affecting intracellular Ca^{2+} concentrations, activation of Ca^{2+} channels regulates an enormous range of biochemical signaling processes within cells (see Chapter 7). Perhaps the most important brain process regulated by voltage-sensitive Ca^{2+} channels is the release of neurotransmitters at synapses (see Chapter 5). To mediate these diverse and crucial functions, ten different Ca^{2+} channel genes (**CACNA genes**) have been identified. Different types of Ca^{2+} channels vary in their activation and inactivation properties, allowing subtle variations in both electrical and chemical signaling processes mediated by Ca^{2+}. Drugs that block voltage-gated Ca^{2+} channels are valuable in treating a variety of conditions that range from heart disease to anxiety disorders.

K^+ channels are by far the largest and most diverse class of voltage-gated ion channels (see Figure 4.6C). Nearly 100 K^+ channel genes (**KCN genes**) are known, falling into several distinct groups that differ substantially in their activation, gating, and inactivation properties. Some take minutes to inactivate (Figure 4.7A), as found for the K^+ channels of squid axons. Others inactivate within milliseconds, reminiscent of most voltage-gated Na^+ channels (Figure 4.7B), and still others even faster (Figure 4.7C). These properties influence the duration and rate of action potential

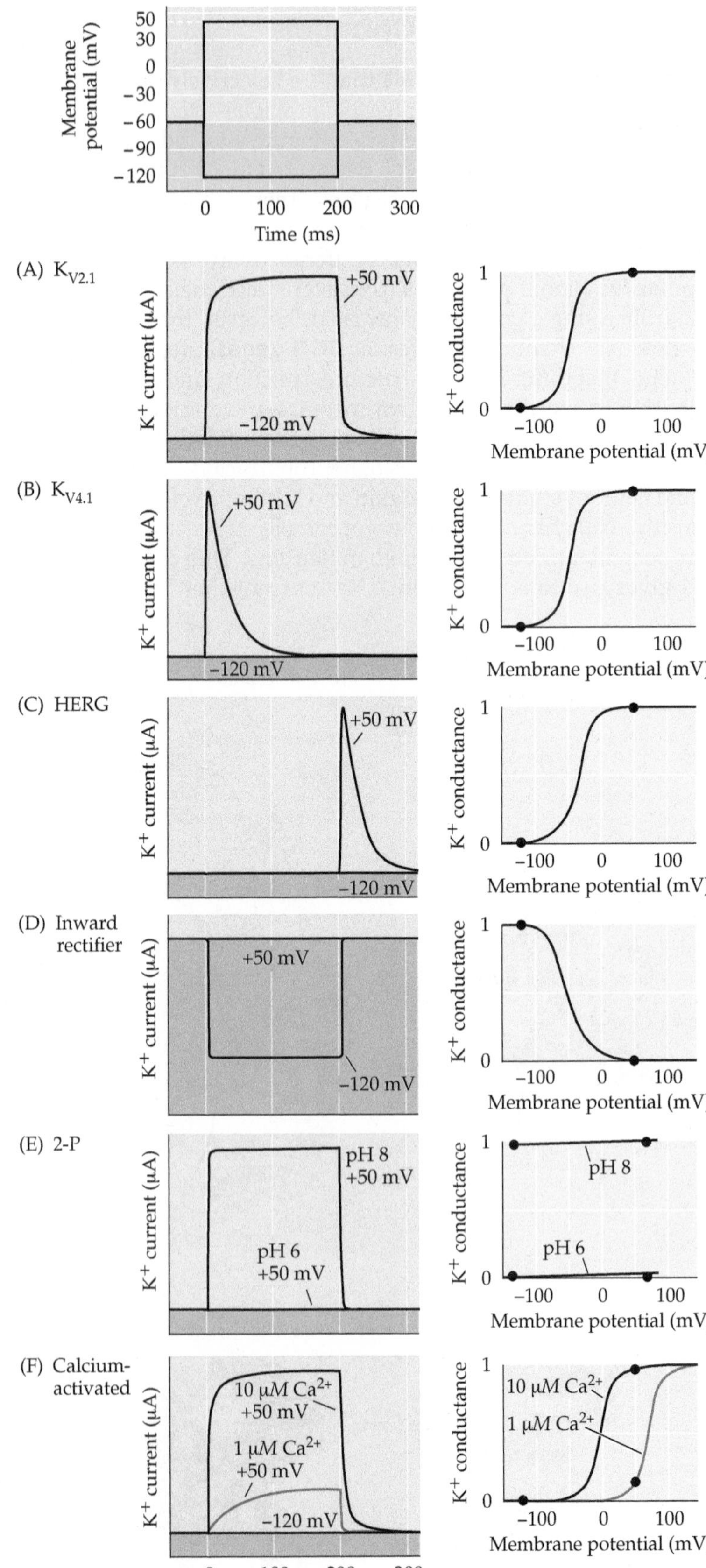

◀ FIGURE 4.7 **Diverse properties of K⁺ channels.** Different types of K^+ channels were expressed in *Xenopus* oocytes, and the voltage clamp method was used to change the membrane potential (top) and measure the resulting currents flowing through each type of channel. These K^+ channels vary markedly in their gating properties, as evident in their currents (left) and conductances (right). (A) $K_{V2.1}$ channels show little inactivation and are closely related to the delayed rectifier K^+ channels involved in action potential repolarization. (B) $K_{V4.1}$ channels inactivate during a depolarization and help regulate the interval between action potentials during repetitive firing. (C) HERG channels (*KCNH2* gene) inactivate so rapidly that current flows only when inactivation is rapidly removed at the end of a depolarization. (D) Inward rectifying K^+ channels allow more K^+ current to flow at hyperpolarized potentials than at depolarized potentials. (E) 2-P K^+ channels usually respond to chemical signals rather than to changes in membrane potential. In the case of the TASK channel (*KCNK9* gene) shown here, changes in extracellular pH regulate channel opening while membrane potential changes do not. (F) Ca^{2+}-activated K^+ channels open in response to intracellular Ca^{2+} ions and, in some cases, membrane depolarization.

firing, with important consequences for axonal conduction, synaptic transmission and information processing. These voltage-gated K^+ channels typically have the structural features described above: four subunits—each with six membrane-spanning domains, one pore loop, and a voltage sensor—that assemble to form a single functional ion channel (see Figure 4.6C).

Two other types of K^+ channels are responsible for the resting potential of neurons: the inward rectifier channels that respond to membrane hyperpolarization (Figure 4.7D) and the 2-P K^+ channels that are insensitive to membrane potential but are gated by chemical signals such as pH changes (Figure 4.7E). These resting K^+ channels deviate structurally from other K^+ channels in several ways, with neither having a voltage sensor domain. Furthermore, inward rectifier K^+ channels span the membrane only twice and have a single pore loop, while 2-P channels span the membrane four times and include two pore loops. Still other K^+ channels respond to intracellular Ca^{2+} levels (Figure 4.7F). These channels have a unique structure consisting of seven transmembrane domains, including a voltage sensor and pore loop, as well as a cytoplasmic domain responsible for sensing intracellular Ca^{2+} (see Figure 4.8C).

Finally, several types of Cl^- channel genes (**CLCN genes**), encoding different voltage-gated Cl^- channels, have been identified (see Figure 4.6D). These channels are present in every type of neuron, where they control excitability, contribute to the resting membrane potential, and help regulate cell volume. Chloride channels are structurally distinct from all other voltage-gated channels, consisting of dimers of two identical subunits that span the plasma membrane many times. Within each subunit is an ion-conducting pore, so that a complete Cl^- channel has two separate pores (see Figure 4.6D). The selectivity filter of these pores includes positive charges that coordinate the negative Cl^- as they diffuse from one side of the membrane to the other. Cl^- channels do not have the type of voltage sensor found in voltage-gated Na^+, Ca^{2+}, or K^+ channels. Instead, their voltage dependence seems to arise from voltage-dependent movement of a negatively charged amino acid near the selectivity filter.

In summary, voltage-gated ion channels are integral membrane proteins with characteristic structures that allow them to conduct ions and to open or close according to the transmembrane potential. Defects in these channel functions, associated with gene mutations, lead to a variety of neurological disorders.

Ligand-Gated Ion Channels

Many types of ion channels respond to chemical signals (ligands) rather than to changes in the membrane potential (Figure 4.8). The most important of these **ligand-gated ion channels** in the nervous system are neurotransmitter receptors, which are activated by binding of neurotransmitters to their extracellular domains (see Figure 4.8A). These channels are essential for synaptic transmission and other forms of cellular signaling phenomena discussed in Chapters 5–8. Whereas the voltage-gated ion channels underlying the action potential typically allow only one type of ion to permeate, channels activated by extracellular ligands are usually less selective, often allowing multiple types of ions to flow. For example, the neurotransmitter receptors involved in excitatory synaptic transmission typically are permeable to both Na^+ and K^+, as well as to other cations. The structures and gating mechanisms of neurotransmitter receptors are discussed in detail in Chapter 6. Another important class of channels activated by extracellular chemical signals is the acid-sensing ion channels (ASICs). These Na^+ channels are gated by external H^+, rather than by voltage, and are important for a wide range of functions including taste and pain sensation.

Other ligand-gated channels are sensitive to chemical signals arising within the cytoplasm of neurons (see Chapter 7), and can be selective for specific ions such as K^+ or Cl^-, or permeable to all physiological cations. Such channels are distinguished by ligand-binding domains on their *intracellular* surfaces that interact with second messengers such as Ca^{2+}, the cyclic nucleotides cAMP and cGMP, or protons. Examples of channels that respond to intracellular cues include Ca^{2+}-activated K^+ channels (see Figure 4.8C), and the cyclic nucleotide-gated cation channel (see Figure 4.8D). The main function of these channels is to convert intracellular chemical signals into electrical information. This process is particularly important in sensory transduction, where intracellular second messenger signals associated with sensory stimuli—such as odors and light—are transduced into electrical signals by cyclic nucleotide-gated channels.

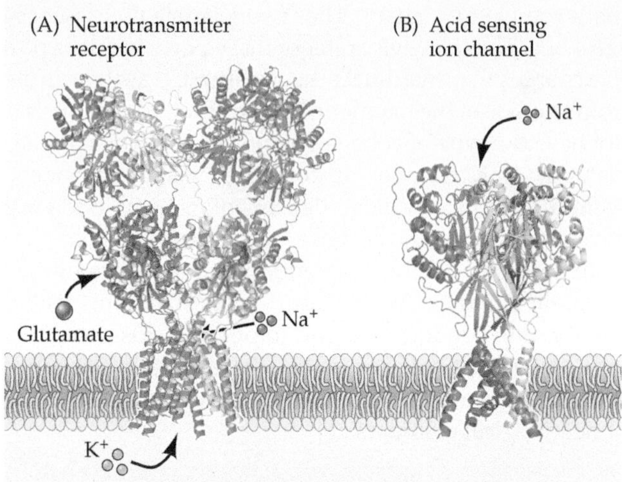

FIGURE 4.8 Ligand-gated ion channels. Some ligand-gated ion channels are activated by the extracellular presence of neurotransmitters, such as glutamate (A), while others are activated by extracellular H^+ (B). Still other channels are activated by intracellular second messengers, such as Ca^{2+} (C) or the cyclic nucleotides cAMP and cGMP (D). (A from Sobolevsky et al., 2009; B from Gonzalez et al., 2009; C from Hite et al., 2017; D from Li et al., 2017.)

Although many ligand-gated ion channels are located in the cell surface membrane, others are found in the membranes of intracellular organelles such as mitochondria or the endoplasmic reticulum. Some of these latter channels are selectively permeable to Ca^{2+} and regulate the release of Ca^{2+} from the lumen of the endoplasmic reticulum into the cytoplasm, where this second messenger can trigger a spectrum of cellular responses such as those described in Chapter 7.

Thermosensitive and Mechanosensitive Channels

Still other ion channels respond to other forms of stimuli, such as heat. **Thermosensitive** ion channels (Figure 4.9A), including some members of the transient receptor potential (TRP) gene family, contribute to the sensations of pain and body temperature. These cation channels open in response to specific temperature ranges, with some activated by cold temperatures rather than by heat. In some cases, channel gating is mediated by a unique mechanism based on a temperature-dependent displacement of membrane lipids (Figure 4.9B). Many thermosensitive TRP channels also are gated by ligands and are used to detect chemical signals. For example, the same TRP channel (called TRPV1; see Figure 4.9A) that responds to temperatures above 40°C is also sensitive to capsaicin (the ingredient that makes chili peppers spicy; see Box 10A); this channel thus transduces two different types of physical stimuli into the sensation of "hot." Thermosensitive TRP channels participate in numerous other functions, including pain sensation and inflammation (see Chapter 10).

Still other ion channels, including certain TRP channels and Piezo channels, respond to mechanical distortion of the plasma membrane. These **mechanosensitive** channels (Figure 4.9C) are the critical components of stretch receptors and neuromuscular stretch reflexes (see Chapters 9, 16, and 17). A specialized form of these channels apparently enables hearing by allowing auditory hair cells to respond to sound waves (see Chapter 13). The trimeric structure of Piezo channels is somewhat unusual compared with that of most other ion channels. Though still possessing a central pore, Piezo channels have extracellular blade structures that apparently serve as levers to open the channel pore in response to mechanical stimuli (see Figure 4.9D).

In summary, a tremendous variety of ion channels allows neurons to generate electrical signals in response to a broad range of stimuli, including changes in membrane potential, synaptic input, intracellular second messengers, light, odors, heat, sound, touch, pressure, pH, and many other stimuli.

Active Transporters Create and Maintain Ion Gradients

Up to this point, the discussion of the molecular basis of electrical signaling has taken for granted the remarkable fact that nerve cells maintain ion concentration gradients across their surface membranes: none of the ions of physiological importance (Na^+, K^+, Cl^-, H^+, and Ca^{2+}) are in electrochemical equilibrium. Further, because channels produce electrical effects by allowing one or more of these ions to diffuse down their electrochemical gradients, there would be a gradual dissipation of these concentration gradients unless nerve cells could restore ions displaced during the current flow that occurs as a result of both

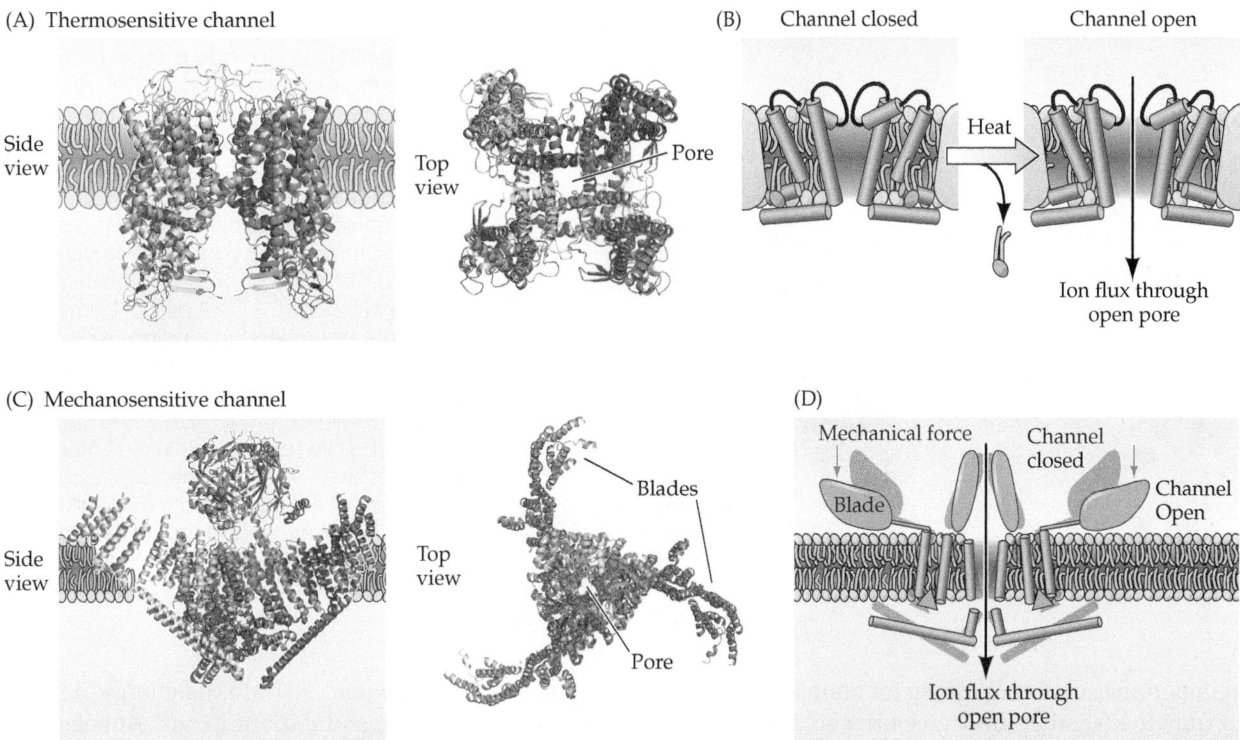

FIGURE 4.9 Thermosensitive and mechanosensitive channels. (A) The TRPV1 channel responds to heat as well as to chemical signals such as capsaicin. This channel consists of four subunits, which form a central cation-selective pore. (B)Heat is thought to open the TRPV1 channel pore by displacing membrane lipids closely associated with the channel, leading to a structural rearrangement that opens channel gates (red). (C) Piezo, an example of a mechanosensitive channel, has a unique structure that includes three subunits and large blades on the extracellular side of the channel. (D) The Piezo channel pore is thought to open when mechanical force displaces the blades, leading to rearrangement of the structure of the rest of the channel. (A after Gao et al., 2016; B after Ge et al., 2015.)

neural signaling and the continual leakage of ions that occurs even at rest. The work of generating and maintaining concentration gradients for particular ions is carried out by plasma membrane proteins known as active transporters. These proteins are called active transporters because they must consume energy as they transport ions uphill against their electrochemical gradients.

Active transporters carry out their task by forming complexes with the ions they are translocating. The process of ion binding and unbinding during transport typically requires several milliseconds. As a result, ion translocation by active transporters is orders of magnitude slower than the diffusion of ions through channel pores (recall that a single ion channel can conduct thousands of ions across a membrane each millisecond). In short, active transporters gradually store energy in the form of ion concentration gradients, whereas the opening of ion channels rapidly dissipates this stored energy during relatively brief electrical signaling events. Although the specific jobs of active transporters are highly diverse, active transporters can be sorted into two main types on the basis of the source of energy used for ion movement: **ATPase pumps** and **ion exchangers**.

ATPase Pumps

ATPase pumps acquire energy for ion translocation directly from the hydrolysis of ATP. The most prominent example of an ATPase pump is the Na^+ pump (or more properly, the Na^+/K^+ ATPase pump), which is responsible for maintaining transmembrane concentration gradients for both Na^+ and K^+ (Figure 4.10A). The Na^+ pump is a large integral membrane protein made up of at least two subunits, α and β. The α subunit is responsible for ion translocation and spans the membrane ten times, with most of the molecule found on the cytoplasmic side, whereas the β subunit spans the membrane only once and is predominantly extracellular.

Ca^{2+} pumps are another class of ATPase pump (Figure 4.10B). Ca^{2+} pumps are an important mechanism for removing Ca^{2+} from cells. Two different types of Ca^{2+} pumps have been identified. One, called PMCA, is found on the plasma membrane; the other, termed SERCA, is used to store Ca^{2+} in the endoplasmic reticulum (see Chapter 7). The structure of SERCA (see Figure 4.10B) is remarkably similar to that of the Na^+ pump, aside from having a binding site for Ca^{2+}.

(A) Na^+/K^+ pump

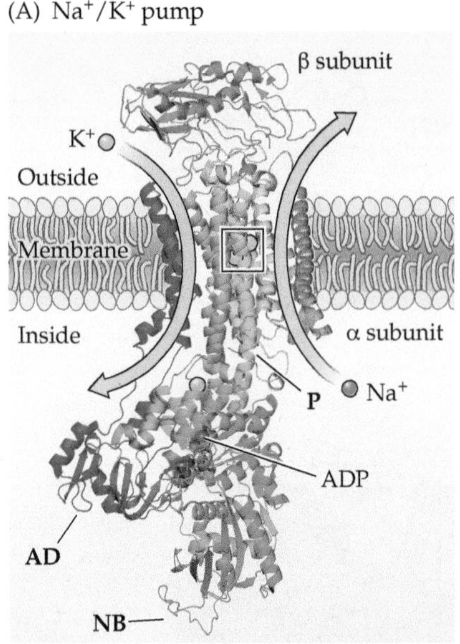

(B) Ca^{2+} pump

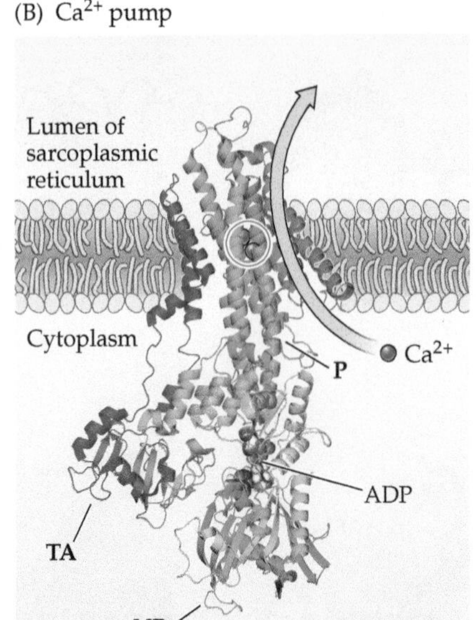

FIGURE 4.10 Examples of ATPase pumps. (A) Structure of the Na^+ pump. Domains responsible for nucleotide binding (NB), phosphorylation (P), and an actuator domain (AD) are evident. In this conformation, ADP occupies the NB domain of the pump, and two K^+ (inside square) can be observed in the transmembrane domain. Activity of the pump leads to transfer of Na^+ from inside to outside, and of K^+ in the opposite direction. (B) Structure of the SERCA Ca^{2+} pump. Domains responsible for nucleotide binding (NB), phosphorylation (P), and ion translocation activity (TA) are indicated. Shown is the structure of the pump when bound to ADP; in this state, two Ca^{2+} (purple spheres within red circle) are sequestered within the membrane-spanning regions of the pump. Note the similarity between this structure and that of the Na^+/K^+ pump shown in (A). (A after Shinoda et al., 2009; B after Toyoshima et al., 2004.)

The crucial importance of the Na^+ pump for brain function is evident from the fact that this transporter accounts for up to two-thirds of the brain's total energy consumption. The Na^+ pump of neurons was first discovered in the 1950s, when Richard Keynes at Cambridge University used radioactive Na^+ to demonstrate the energy-dependent efflux of Na^+ from squid giant axons. Keynes and his collaborators found that this efflux ceased when the axon's supply of ATP was interrupted by treatment with metabolic poisons (Figure 4.11, point 4). Other conditions that lowered intracellular ATP also prevented Na^+ efflux, proving that removal of intracellular Na^+ requires cellular metabolism. Further studies with radioactive K^+ demonstrated that Na^+ efflux is associated with the simultaneous ATP-dependent influx of K^+. These energy-dependent movements of Na^+ and K^+ generate transmembrane gradients for both ions. The opposing fluxes of Na^+ and K^+ are operationally inseparable—removal of external K^+ greatly reduces Na^+ efflux (see Figure 4.11, point 2), and vice versa. Subsequent work by Jens Christian Skou in Denmark established that these fluxes of Na^+ and K^+ are due to an ATP-hydrolyzing Na^+/K^+ pump. Quantitative studies indicate that Na^+ and K^+ are not pumped at identical rates: The rate of K^+ influx is only about two-thirds that of Na^+ efflux, indicating that the pump transports 2 K^+ into the cell for every 3 Na^+ that are removed.

FIGURE 4.11 Ion movements due to the Na^+ pump. Measurement of radioactive Na^+ efflux from a squid giant axon. This efflux depends on external K^+ and intracellular ATP. (After Hodgkin and Keynes, 1955.)

The Na^+ pump is thought to alternately shuttle Na^+ and K^+ across the membranes in a cycle fueled by binding of ATP and transfer of a phosphate group from ATP to the pump (Figure 4.12A). ATP binding promotes binding of intracellular Na^+ and release of K^+, while pump phosphorylation leads to extracellular release of Na^+ and binding of K^+.

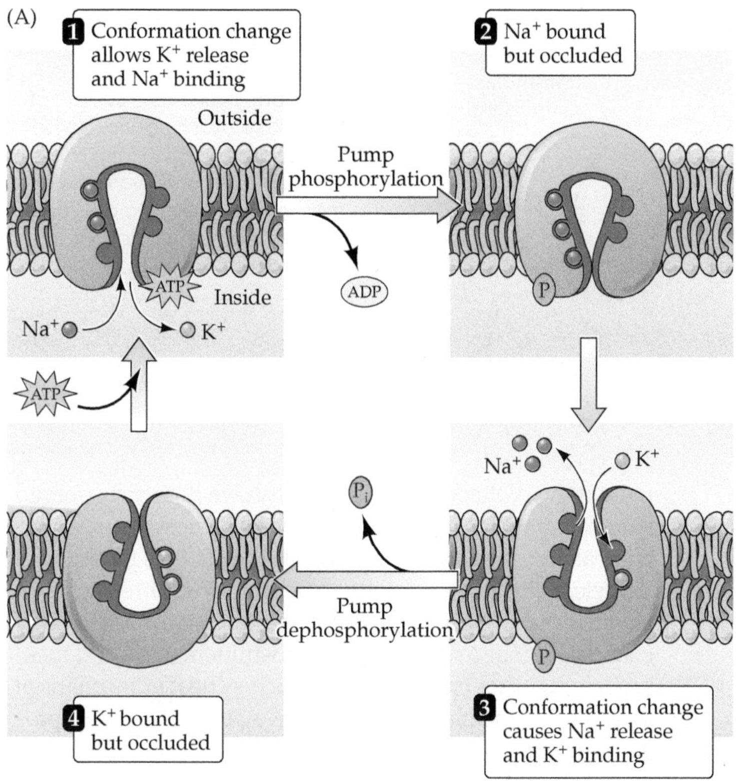

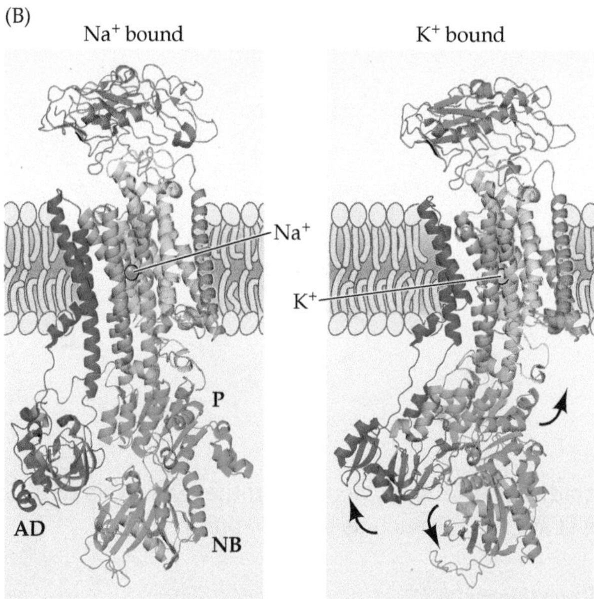

FIGURE 4.12 Translocation of Na^+ and K^+ by the Na^+ pump. (A) A model for the movement of ions by the Na^+ pump. Uphill movements of Na^+ and K^+ are driven by binding and hydrolysis of ATP, which phosphorylates the pump (indicated by P). These ion fluxes are asymmetrical, with 3 Na^+ carried out for every 2 K^+ brought in. (B) Comparison of structure of the Na^+ pump in Na^+-bound state (left), corresponding to step 2 in (A), and K^+-bound (right) state, corresponding to step 4. Changes in location of phosphorylation (P), nucleotide binding (NB), and activator (AD) domains (arrows) are associated with switch between Na^+-bound and K^+-bound conformations. (A after Lingrel et al., 1994; B after Nyblom et al., 2013.)

In between these two states of ion translocation are occluded states that prevent leakage of ions in the reverse direction, with subsequent hydrolysis of ATP leading to dissociation of ADP, which toggles the pump between accumulating intracellular K^+ and removing intracellular Na^+. While the precise mechanism of ion translocation is still being elucidated, many of the structures involved have been identified (Figure 4.12B). Na^+ and K^+ alternately bind to sites in the interior of the pump, within the transmembrane domain of the α subunit. A nucleotide-binding domain is responsible for binding ATP, with hydrolysis of ATP transferring a phosphate onto a phosphorylation domain that subsequently changes the location of an actuator domain thought to regulate ion binding to the pump (see Figure 4.12B). This ATP-dependent structural rearrangement explains how the pump is able to move Na^+ and K^+ uphill, against their steep electrochemical gradients. Very similar structural changes are also thought to be involved in removal of cytoplasmic Ca^{2+} by the SERCA Ca^{2+} pump.

Ion Exchangers

The second class of active transporter does not use ATP directly but instead uses the electrochemical gradients of other ions as an energy source. Such ion exchangers carry one or more ions *up* their electrochemical gradient while simultaneously taking another ion (most often Na^+) *down* its gradient. These transporters can be further subdivided into two types according to the direction of ion movement. As their name implies, **antiporters** exchange intracellular and extracellular ions. An example of an antiporter is the Na^+/Ca^{2+} exchanger, which shares with the Ca^{2+} pump the important job of keeping intracellular Ca^{2+} concentrations low (Figure 4.13A). Another antiporter, the Na^+/H^+ exchanger, regulates intracellular pH (Figure 4.13B). Ion exchangers of the second type, the **co-transporters**, work by carrying multiple ions in the same direction. Two such co-transporters regulate intracellular Cl^- concentration by translocating Cl^- along with extracellular Na^+ and/or K^+; these are the $Na^+/K^+/Cl^-$ co-transporter, which transports Cl^- along with Na^+ and K^+ into cells (Figure 4.13C), and the K^+/Cl^- co-transporter, which removes intracellular Cl^- (Figure 4.13D). As you will see in Chapter 6, neurotransmitters are transported into synaptic terminals and glial cells via other co-transporters (Figure 4.13E). Although the electrochemical gradient of Na^+ (or other counter-ions) is the proximate source of energy for both ion exchangers and co-transporters, these

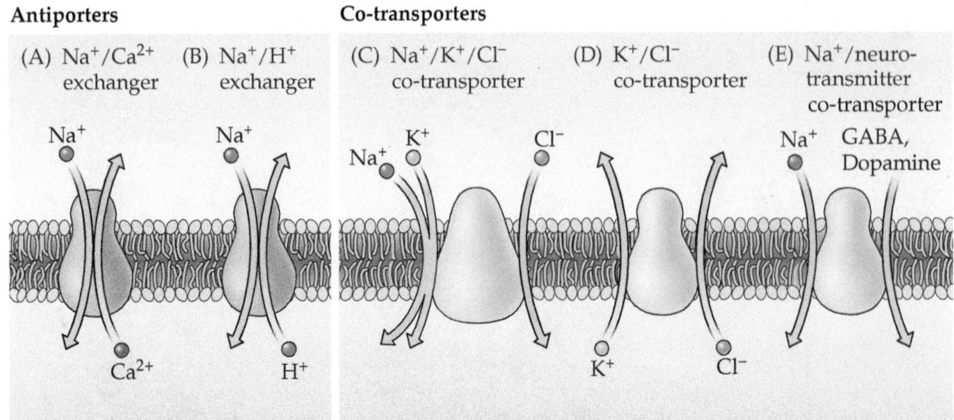

FIGURE 4.13 Examples of ion exchangers. Ion exchangers use the electrochemical gradients of co-transported ions as a source of energy. These exchangers can be further subdivided into antiporters that swap ions on the two sides of the membrane (A,B) and co-transporters that carry multiple ions in the same direction (C–E).

gradients ultimately depend on the hydrolysis of ATP by ATPase pumps such as the Na^+ pump.

Summary

Ion channels and active transporters have complementary functions. The primary purpose of transporters is to generate transmembrane concentration gradients, which are then exploited by ion channels to generate electrical signals. The flow of ions through single open channels can be detected as tiny electrical currents, and the synchronous opening of many channels generates the macroscopic currents that produce action potentials and other electrical signals. A large number of ion channel genes creates channels with a correspondingly wide range of functional characteristics, thus allowing different types of neurons to have a remarkable spectrum of electrical properties. Voltage-gated ion channels are responsible for the voltage-dependent conductances that underlie the action potential. These channels are integral membrane proteins that open or close ion-selective pores in response to the membrane potential, allowing specific ions to diffuse across the membrane. Voltage-gated channels have highly conserved structures that are responsible for features such as ion permeation and voltage sensing, as well as the features that specify ion selectivity and sensitivity to neurotoxins. Other types of channels have specialized structures that enable detection of chemical signals, such as neurotransmitters or second messengers, or other types of signals such as heat or mechanical distortion of the plasma membrane. Active transporter proteins are quite different from ion channels because they move ions against a concentration gradient. The energy required for ion translocation is provided either by the hydrolysis of ATP or by the electrochemical gradient of other ions, such as Na^+. The Na^+ pump produces and maintains the transmembrane gradients of Na^+ and K^+ by ATP-dependent phosphorylation of the pump, which causes structural changes that enable ion movement. Other transporters, both ATPase pumps and co-transporters, are responsible for the electrochemical gradients for other physiologically important ions, including Cl^-, Ca^{2+}, and H^+. Together, transporters and channels provide a comprehensive and satisfying molecular explanation for the ability of neurons to generate electrical signals.

ADDITIONAL READING

Reviews

Armstrong, C. M. and B. Hille (1998) Voltage-gated ion channels and electrical excitability. *Neuron* 20: 371–380.

Bezanilla, F. and A. M. Correa (1995) Single-channel properties and gating of Na^+ and K^+ channels in the squid giant axon. In *Cephalopod Neurobiology*, N. J. Abbott, R. Williamson and L. Maddock (eds.). New York: Oxford University Press, pp. 131–151.

Enyedi, P. and G. Czirják (2010) Molecular background of leak K^+ currents: two-pore domain potassium channels. *Physiol. Rev.* 90: 559–605.

Gouaux, E. and R. MacKinnon (2005) Principles of selective ion transport in channels and pumps. *Science* 310: 1461–1465.

Jentsch, T. J., M. Poet, J. C. Fuhrmann and A. A. Zdebik (2005) Physiological functions of CLC Cl^- channels gleaned from human genetic disease and mouse models. *Annu. Rev. Physiol.* 67: 779–807.

Lee, A. G. (2006) Ion channels: A paddle in oil. *Nature* 444: 697.

Vargas, E. and 10 others (2012) An emerging consensus on voltage-dependent gating from computational modeling and molecular dynamics simulations. *J. Gen. Physiol.* 140: 587–594.

Important Original Papers

Ahuja, S. and 34 others (2015) Structural basis of Nav1.7 inhibition by an isoform-selective small-molecule antagonist. *Science* 350: aac5464.

Caterina, M. J. and 5 others (1997) The capsaicin receptor: A heat-activated ion channel in the pain pathway. *Nature* 389: 816–824.

Doyle, D. A. and 7 others (1998) The structure of the potassium channel: Molecular basis of K^+ conduction and selectivity. *Science* 280: 69–77.

Dutzler, R., E. B. Campbell, M. Cadene, B. T. Chait and R. MacKinnon (2002) X-ray structure of a ClC chloride channel at 3.0 Å reveals the molecular basis of anion selectivity. *Nature* 415: 287–294.

Gao, Y., E. Cao, D. Julius and Y. Cheng (2016) TRPV1 structures in nanodiscs reveal mechanisms of ligand and lipid action. *Nature* 534: 347–351.

Ge, J. and 9 others (2015) Architecture of the mammalian mechanosensitive Piezo1 channel. *Nature* 527: 64–69.

Gonzalez, E. B., T. Kawate and Eric Gouaux (2009) Pore architecture and ion sites in acid sensing ion channels and P2X receptors. *Nature* 460: 599–604.

Hite, R.K., X. Tao and R. MacKinnon (2017) Structural basis for gating the high-conductance Ca2+-activated K+ channel. *Nature* 541: 52–57.

Hodgkin, A. L. and R. D. Keynes (1955) Active transport of cations in giant axons from *Sepia* and *Loligo*. *J. Physiol.* 128: 28–60.

Li, M. and 8 others (2017) Structure of a eukaryotic cyclicnucleotide-gated channel. *Nature* doi: 10.1038/nature20819

Llano, I., C. K. Webb and F. Bezanilla (1988) Potassium conductance of squid giant axon. Single-channel studies. *J. Gen. Physiol.* 92: 179–196.

Long, S. B., E. B. Campbell and R. MacKinnon (2005) Crystal structure of a mammalian voltage-dependent *Shaker* family K^+ channel. *Science* 309: 897–903.

Noda, M. and 6 others (1986) Expression of functional sodium channels from cloned cDNA. *Nature* 322: 826–828.

Nyblom, M. and 7 others (2013) Crystal structure of Na^+, K^+-ATPase in the Na^+-bound state. *Science* 342: 123–127.

Papazian, D. M., T. L. Schwarz, B. L. Tempel, Y. N. Jan and L. Y. Jan (1987) Cloning of genomic and complementary DNA from *Shaker*, a putative potassium channel gene from *Drosophila*. *Science* 237: 749–753.

Shinoda, T., H. Ogawa, F. Cornelius and C. Toyoshima (2009) Crystal structure of the sodium-potassium pump at 2.4 Å resolution. *Nature* 459: 446–450.

Sigworth, F. J. and E. Neher (1980) Single Na^+ channel currents observed in cultured rat muscle cells. *Nature* 287: 447–449.

Sobolevsky, A. I., M. P. Rosconi and E. Gouaux (2009) X-ray structure, symmetry and mechanism of an AMPA-subtype glutamate receptor. *Nature* 462: 745–756.

Tao, X., A. Lee, W. Limapichat, D. A. Dougherty and R. MacKinnon (2010) A gating charge transfer center in voltage sensors. *Science* 328: 67–73.

Toyoshima, C., H. Nomura and T. Tsuda (2004) Luminal gating mechanism revealed in calcium pump crystal structures with phosphate analogues. *Nature* 432: 361–368.

Vanderberg, C. A. and F. Bezanilla (1991) A sodium channel model based on single channel, macroscopic ionic, and gating currents in the squid giant axon. *Biophys. J.* 60: 1511–1533.

Waldmann, R., G. Champigny, F. Bassilana, C. Heurteaux and M. Lazdunski (1997) A proton-gated cation channel involved in acid-sensing. *Nature* 386: 173–177.

Wei, A. M. and 5 others (1990) K^+ current diversity is produced by an extended gene family conserved in *Drosophila* and mouse. *Science* 248: 599–603.

Wu, J. and 6 others (2015) Structure of the voltage-gated calcium channel CaV1.1 complex. *Science* 350: aad2395–1.

Books

Aidley, D. J. and P. R. Stanfield (1996) *Ion Channels: Molecules in Action*. Cambridge: Cambridge University Press.

Gribkoff, V. K. and L. K. Kaczmarek (2009) *Structure, Function and Modulation of Neuronal Voltage-Gated Ion Channels*. New York: Wiley.

Hille, B. (2001) *Ion Channels of Excitable Membranes*, 3rd Edition. Sunderland, MA: Sinauer Associates.

Zheng, J. and M. C. Trudeau (2015) *Handbook of Ion Channels*. Boca Raton, FL: CRC Press.

Go to the NEUROSCIENCE 6e Companion Website at **www.oup.com/uk/Purves6e** for Web Topics, Animations, Flashcards, and more.

CHAPTER

5 Synaptic Transmission

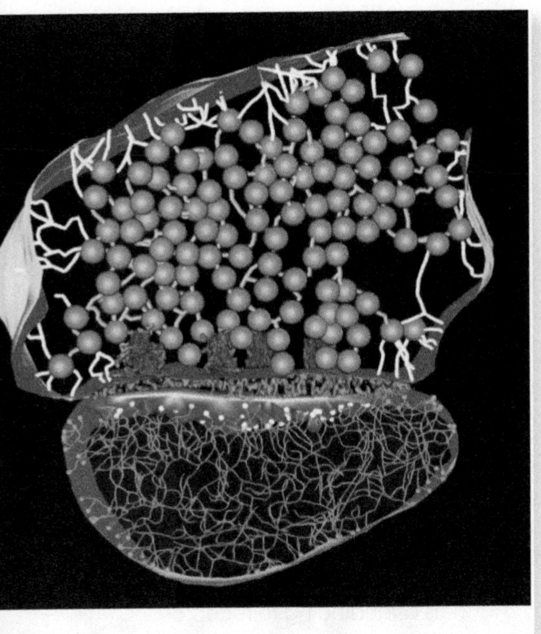

Overview

THE HUMAN BRAIN CONTAINS 86 billion neurons, each with the ability to influence many other cells. Clearly, sophisticated and highly efficient mechanisms are needed to enable communication among this astronomical number of elements. Such communication is made possible by synapses, the functional contacts between neurons. Two different types of synapses—electrical and chemical—can be distinguished on the basis of their mechanism of transmission. At electrical synapses, current flows through connexons, which are specialized membrane channels that connect two cells at gap junctions. In contrast, chemical synapses enable cell-to-cell communication via the secretion of neurotransmitters; these chemical agents released by the presynaptic neurons produce secondary current flow in postsynaptic neurons by activating specific neurotransmitter receptors. The total number of neurotransmitters is well over 100. Virtually all neurotransmitters undergo a similar cycle of use: synthesis and packaging into synaptic vesicles; release from the presynaptic cell; binding to postsynaptic receptors; and finally, rapid removal or degradation. The influx of Ca^{2+} through voltage-gated channels triggers the secretion of neurotransmitters; this, in turn, gives rise to a transient increase in Ca^{2+} concentration in the presynaptic terminal. The rise in Ca^{2+} concentration causes synaptic vesicles to fuse with the presynaptic plasma membrane and release their contents into the space between the pre- and postsynaptic cells. Proteins on the surface of the synaptic vesicle and the presynaptic plasma membrane mediate the triggering of exocytosis by Ca^{2+}. Neurotransmitters evoke postsynaptic electrical responses by binding to members of a diverse group of neurotransmitter receptors. There are two major classes of receptors: those in which the receptor molecule is also an ion channel, and those in which the receptor and ion channel are separate entities. These receptors give rise to electrical signals by transmitter-induced opening or closing of the ion channels. Whether the postsynaptic actions of a particular neurotransmitter are excitatory or inhibitory is determined by the ion permeability of the ion channel affected by the transmitter, and by the electrochemical gradient for the permeant ions.

Two Classes of Synapses

The many kinds of synapses in the human brain fall into two general classes: electrical synapses and chemical synapses. These two classes of synapses can be distinguished based on their structures and the mechanisms they use to transmit signals from the "upstream" neuron, called the **presynaptic** element, and the "downstream" neuron, termed **postsynaptic**.

The structure of an electrical synapse is shown schematically in Figure 5.1A. Electrical synapses permit direct, passive flow of electrical current from one neuron to

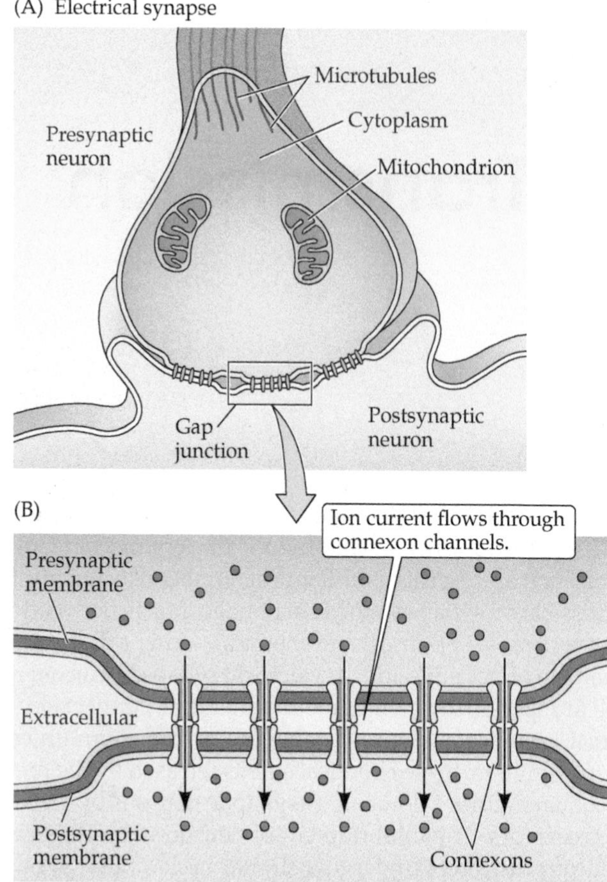

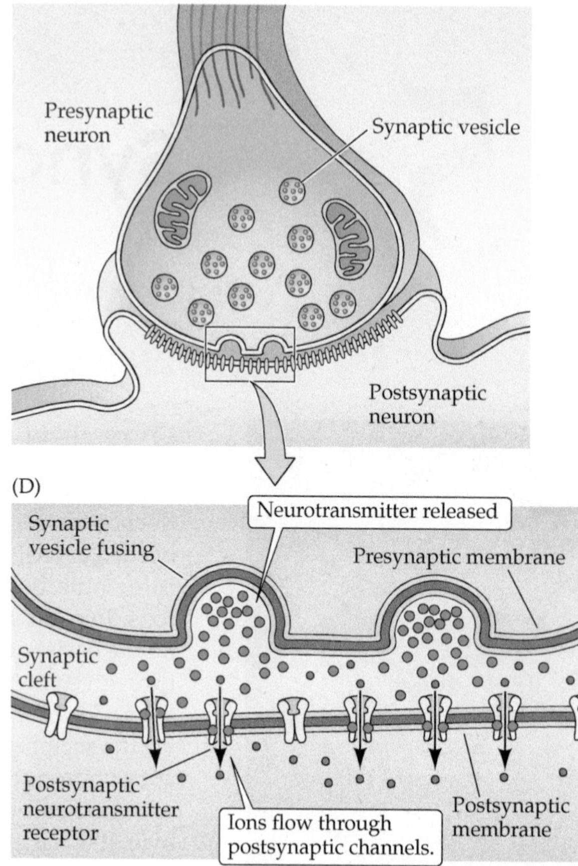

FIGURE 5.1 Electrical and chemical synapses differ fundamentally in their transmission mechanisms. (A) At electrical synapses, gap junctions occur between pre- and postsynaptic membranes. (B) Gap junctions contain connexon channels that permit current to flow passively from the presynaptic cell to the postsynaptic cell. (C) At chemical synapses, there is no intercellular continuity, and thus no direct flow of current from pre- to postsynaptic cell. (D) Synaptic current flows across the postsynaptic membrane only in response to the secretion of neurotransmitters, which open or close postsynaptic ion channels after binding to receptor molecules on the postsynaptic membrane.

another. The usual source of this current is the potential difference generated locally by the presynaptic action potential (see Chapter 3). Current flow at electrical synapses arises at an intercellular specialization called a **gap junction**, where membranes of the two communicating neurons come extremely close to one another and are linked together (Figure 5.1B). Gap junctions contain a unique type of channel, termed a **connexon**, which provides the path for electrical current to flow from one neuron to another (see Figure 5.2).

The general structure of a chemical synapse is shown schematically in Figure 5.1C. The space between the pre- and postsynaptic neurons is substantially greater at chemical synapses than at electrical synapses and is called the **synaptic cleft**. However, the most important structural feature of all chemical synapses is the presence of small, membrane-bounded organelles called **synaptic vesicles** within the presynaptic terminal. These spherical organelles are filled with one or more **neurotransmitters**, chemical signals that are secreted from the presynaptic neuron and detected by specialized receptors on the postsynaptic cell (Figure 5.1D). These chemical agents act as messengers between the communicating neurons and give this type of synapse its name.

Signaling Transmission at Electrical Synapses

Figure 5.2A shows an electron micrograph of an electrical synapse from a mammalian brain. As in the diagrams in Figure 5.1A and B, it can be seen that the processes of the presynaptic and postsynaptic neurons are connected via a gap junction (Figure 5.2B). The connexons contained within gap junctions are key to understanding how electrical

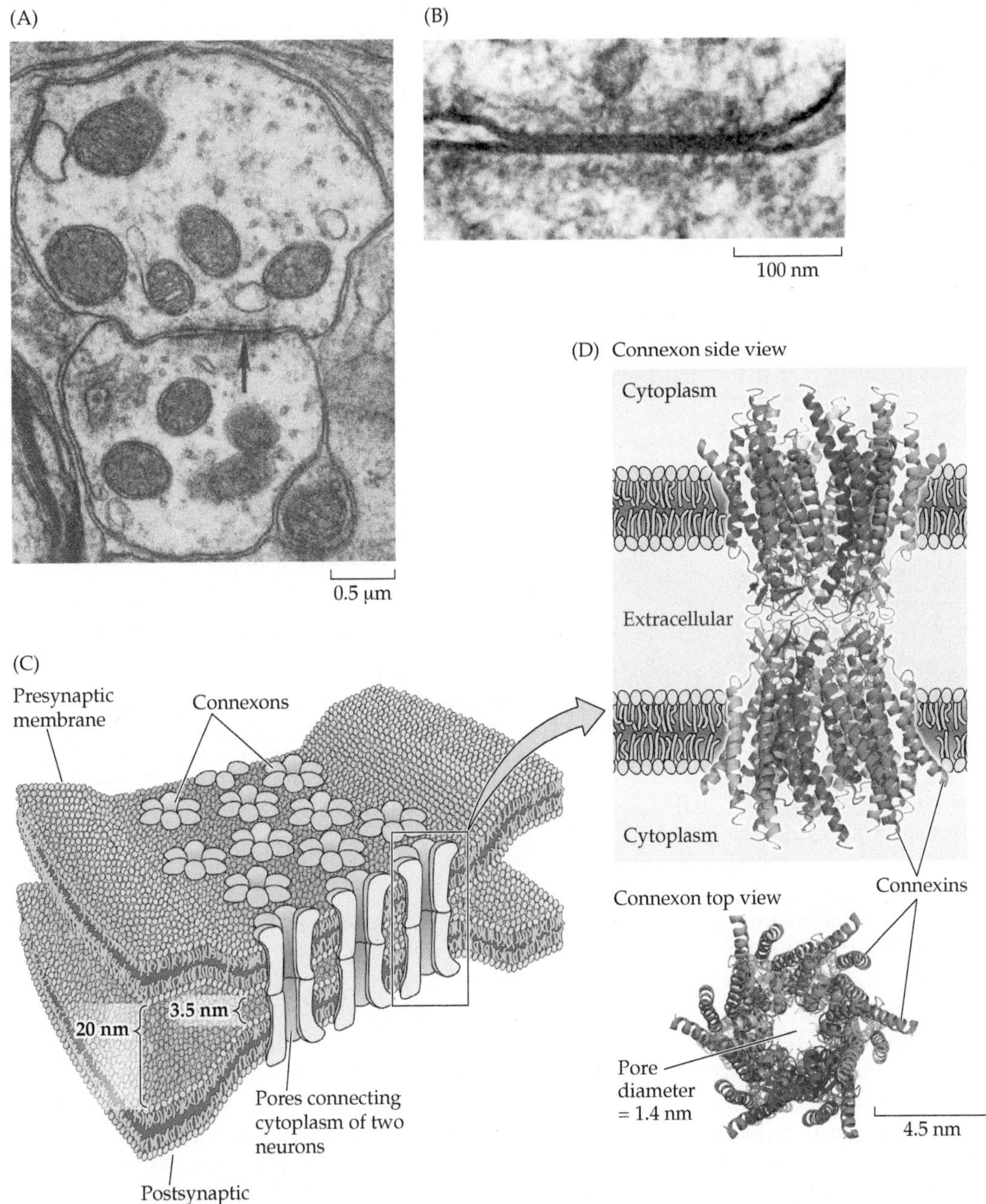

FIGURE 5.2 Structure of electrical synapses. (A) Electron micrograph of an electrical synapse (arrow) connecting two neurons within the inferior olive of a mammalian brain. (B) Higher-magnification electron micrograph of another electrical synapse, showing the gap junction structure characteristic of electrical synapses. (C) Gap junctions consist of connexons, hexameric complexes present in both the pre- and postsynaptic membranes. Channels assembled from connexons in these two membranes form pores that create electrical continuity between the two cells. (D) Crystallographic structure of connexons. Colors indicate individual connexins, integral membrane proteins that form the subunits of connexons. Side view shows the channels spanning the pre- and postsynaptic membranes; top view illustrates how six connexin subunits assemble in each membrane to form a channel with an exceptionally large pore. (A,B from Sotelo et al., 1974; D from Maeda et al., 2009.)

synapses work (Figure 5.2C). Connexons are composed of a unique family of ion channel proteins, the **connexins**, which serve as subunits to form connexon channels. There are twenty-one different types of human connexin genes (GJA–GJE) that are expressed in different cell types and yield connexons with diverse physiological properties. All connexins have four transmembrane domains, and all connexons consist of six connexins that come together to form a hemi-channel in both the pre- and postsynaptic neurons (Figure 5.2D). These hemi-channels are precisely aligned to form a pore that connects the two cells and permits electrical current to flow. The pore of a connexon channel is more than 1 nm in diameter, which is much larger than the pores of the voltage-gated ion channels described in Chapter 4. As a result, a variety of substances can simply diffuse between the cytoplasm of the pre- and postsynaptic neurons. In addition to ions, substances that diffuse through connexon pores include molecules with molecular weights as great as several hundred daltons. This permits important intracellular metabolites, such as ATP and second messengers (see Chapter 7), to be transferred between neurons.

Although they are a distinct minority, electrical synapses have several functional advantages. One is that transmission is extraordinarily fast: Because passive

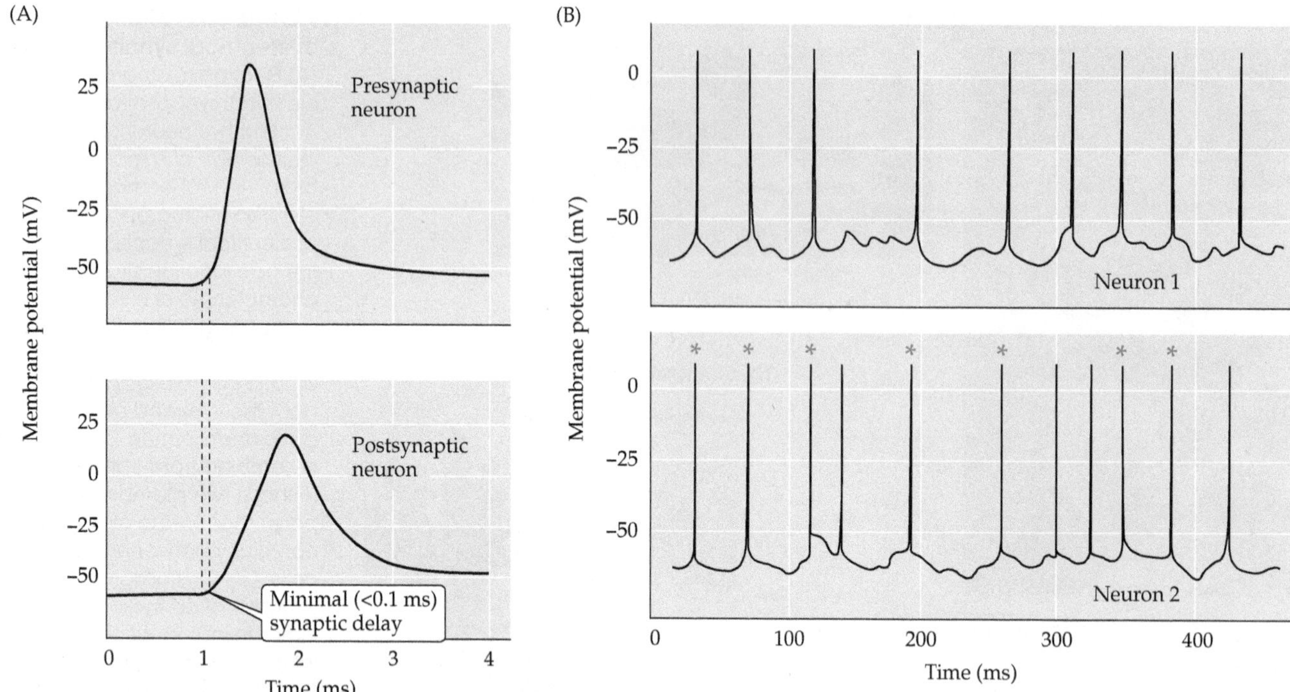

FIGURE 5.3 Function of gap junctions at electrical synapses. (A) Rapid transmission of signals at an electrical synapse in the crayfish. An action potential in the presynaptic neuron causes the postsynaptic neuron to be depolarized within a fraction of a millisecond. (B) Electrical synapses allow synchronization of electrical activity in hippocampal interneurons. In a pair of interneurons connected by electrical synapses, generation of an action potential in one neuron often results in the synchronized firing of an action potential in another neuron (asterisks). (A after Furshpan and Potter, 1959; B after Beierlein et al., 2000.)

current flow across connexons is virtually instantaneous, communication can occur without the delay that is characteristic of chemical synapses. The high speed of electrical synaptic transmission is apparent in the operation of the first electrical synapse to be discovered, which resides in the crayfish nervous system. A postsynaptic electrical signal is observed at this synapse within a fraction of a millisecond after the generation of a presynaptic action potential (Figure 5.3A). In fact, at least part of this brief synaptic delay is caused by propagation of the action potential into the presynaptic terminal, so there may be essentially no delay at all in the transmission of electrical signals across the synapse. Such synapses interconnect many of the neurons within the circuit that allows the crayfish to escape from its predators, thus minimizing the time between the presence of a threatening stimulus and a potentially lifesaving motor response.

Another unique advantage of electrical synapses is that transmission can be bidirectional; although some connexons have special features for unidirectional transmission, in most cases current can flow in either direction, depending on which member of the coupled pair is invaded by an action potential. This allows electrical synapses to synchronize electrical activity among populations of neurons. For example, the brainstem neurons that generate rhythmic electrical activity underlying breathing are synchronized by electrical synapses, as are populations of interneurons in the cerebral cortex, thalamus, cerebellum, and other brain regions (Figure 5.3B). Electrical transmission between vasopressin- and oxytocin-secreting neurons in the hypothalamus ensures that all cells fire action potentials at about the same time, thus facilitating a synchronized burst of secretion of these hormones into the circulation (see Box 21A). The fact that connexon pores are large enough to allow second messengers to diffuse between cells also permits electrical synapses to synchronize the intracellular signaling of coupled cells. This feature may be particularly important for glial cells, which form large intracellular signaling networks via their gap junctions.

Signaling Transmission at Chemical Synapses

Figure 5.4A shows an electron micrograph of a chemical synapse in the cerebral cortex. This image illustrates the presynaptic terminal, with its abundance of synaptic vesicles, as well as the postsynaptic cell separated by a synaptic cleft. A three-dimensional rendering of this chemical synapse, constructed from many images including the one in Figure 5.4A, reveals these features as well as many more structures, including filamentous elements in both pre- and postsynaptic processes, as well as structures in the synaptic cleft (Figure 5.4B). In the presynaptic terminal, dense projections (dark blue) are associated with the **active zone**, the place where synaptic vesicles discharge their neurotransmitters into the synaptic cleft, while the blue structure on the postsynaptic side represents the **postsynaptic density**, a structure important for postsynaptic signaling at excitatory synapses (see Box 7B).

Transmission at chemical synapses is based on the elaborate sequence of events depicted in Figure 5.4C. Prior to transmission, synaptic vesicles are formed and filled with neurotransmitter. Synaptic transmission is initiated when an action potential invades the terminal of the presynaptic neuron. The change in membrane potential caused by the arrival of the action potential leads to the opening of voltage-gated calcium channels in the presynaptic membrane. Because of the steep concentration gradient of Ca^{2+} across the presynaptic membrane (the external Ca^{2+} concentration is approximately 10^{-3} *M*, whereas the internal Ca^{2+} concentration is approximately 10^{-7} *M*), the opening of these channels causes a rapid influx of Ca^{2+} into the presynaptic terminal, with the result that the Ca^{2+} concentration of the cytoplasm in the terminal transiently rises to a much higher value. Elevation of the presynaptic Ca^{2+} concentration, in turn, allows synaptic vesicles to fuse with the plasma membrane of the presynaptic neuron. The Ca^{2+}-dependent fusion of synaptic vesicles with the terminal membrane causes their contents, most importantly neurotransmitters, to be released into the synaptic cleft, a process called **exocytosis**.

Following exocytosis, transmitters diffuse across the synaptic cleft and bind to specific receptors on the membrane of the postsynaptic neuron. The binding of neurotransmitter to the receptors causes channels in the postsynaptic membrane to open (or sometimes to close), thus changing the ability of ions to flow across the postsynaptic membrane. The resulting neurotransmitter-induced current flow alters the conductance and (usually) the membrane potential of the postsynaptic neuron, increasing or decreasing the probability that the neuron will fire an action potential. Subsequent removal of the neurotransmitter from the synaptic cleft, by uptake into glial cells or by enzymatic degradation, terminates the action of the neurotransmitter. In this way, information is transmitted transiently from one neuron to another.

Properties of Neurotransmitters

The notion that electrical information can be transferred from one neuron to the next by means of chemical signaling was the subject of intense debate through the first half of the twentieth century. In 1926, the German physiologist Otto Loewi performed a key experiment that supported this idea. Acting on an idea that allegedly came to him in the middle of the night, Loewi proved that electrical stimulation of the vagus nerve slows the heartbeat by releasing a chemical signal that was later shown to be **acetylcholine (ACh)**. ACh is now known to be a neurotransmitter that acts not only in the heart but also at a variety of postsynaptic targets in the central and peripheral nervous systems, preeminently at the neuromuscular junction of striated muscles and in the visceral motor system (see Chapters 6 and 21).

Formal criteria have been established to definitively identify a substance as a neurotransmitter. These criteria have led to the identification of more than 100 different neurotransmitters, which can be classified into two broad categories: small-molecule neurotransmitters, such as ACh, and neuropeptides (see Chapter 6). Having more than one transmitter diversifies the physiological repertoire of synapses. Multiple neurotransmitters can produce different types of responses on individual postsynaptic cells. For example, a neuron can be excited by one type of neurotransmitter and inhibited by another type of neurotransmitter. The speed of postsynaptic responses produced by different transmitters also differs, allowing control of electrical signaling over different timescales. In general, small-molecule neurotransmitters mediate rapid synaptic actions, whereas neuropeptides tend to modulate slower, ongoing neuronal functions. In some cases, neurons synthesize and release two or more different neurotransmitters; in this case, the molecules are called **co-transmitters**. Co-transmitters can be differentially released according to the pattern of synaptic activity, so that the signaling properties of such synapses change dynamically according to the rate of activity.

Effective synaptic transmission requires close control of the concentration of neurotransmitters within the synaptic cleft. Neurons have therefore developed a sophisticated ability to regulate the synthesis, packaging, release, and degradation (or removal) of neurotransmitters to achieve the desired levels of transmitter molecules. The synthesis of small-molecule neurotransmitters occurs locally within presynaptic terminals. The precursor molecules required to make new molecules of neurotransmitter are usually taken into the nerve terminal by transporters found in the plasma membrane of the terminal (see Figure 4.13E). The enzymes that synthesize these neurotransmitters are present in the cytoplasm of the presynaptic terminal, and the newly synthesized transmitters are then loaded into synaptic vesicles via another type of transporter located in the vesicular membrane. Most small-molecule neurotransmitters are packaged in vesicles 40 to 60 nm in diameter, the centers of which appear clear in electron micrographs (see Figure 5.4A); accordingly, these vesicles are referred to as **small clear-core vesicles**. Neuropeptides are synthesized in the cell body of a neuron, and peptide-filled

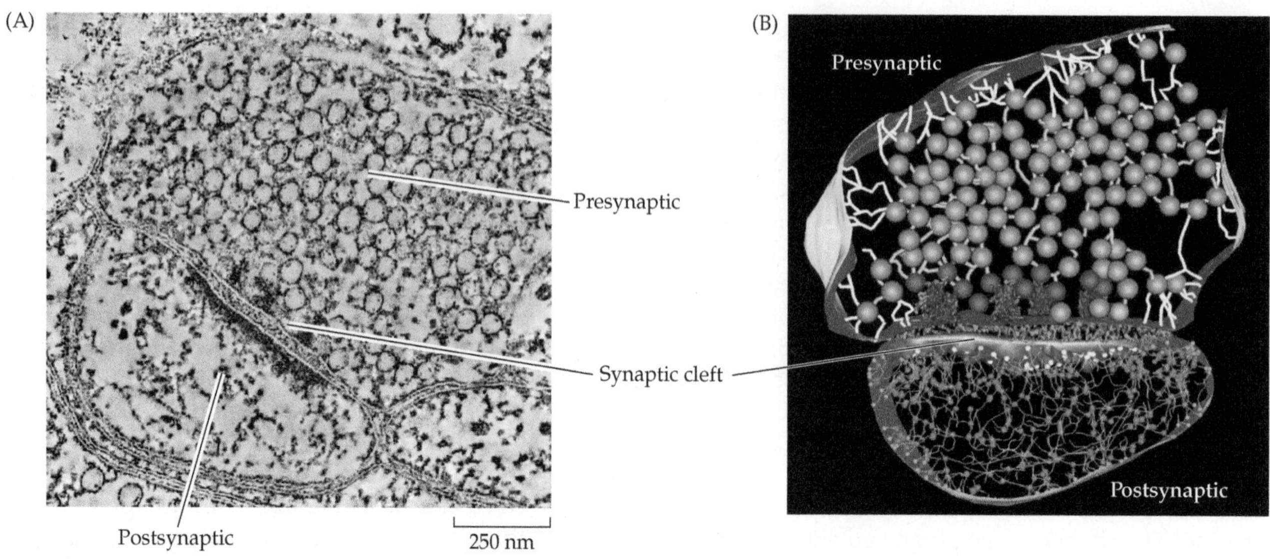

(C)

Myelin

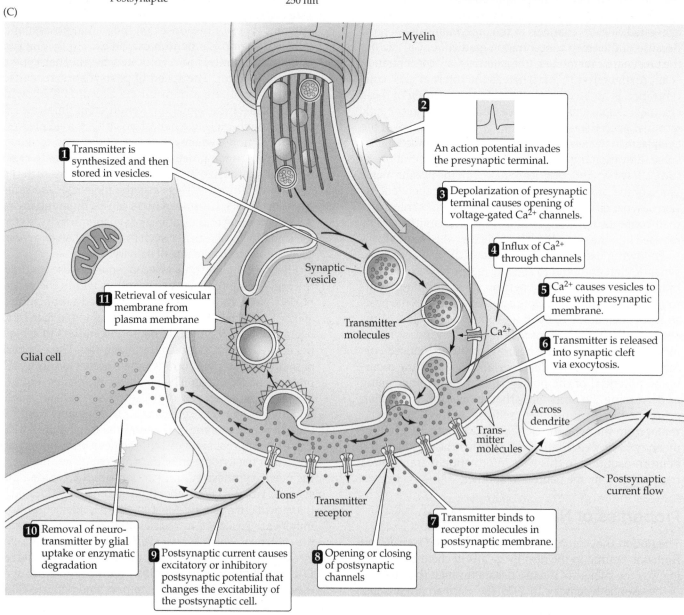

2 An action potential invades the presynaptic terminal.

1 Transmitter is synthesized and then stored in vesicles.

3 Depolarization of presynaptic terminal causes opening of voltage-gated Ca^{2+} channels.

4 Influx of Ca^{2+} through channels

5 Ca^{2+} causes vesicles to fuse with presynaptic membrane.

Synaptic vesicle

11 Retrieval of vesicular membrane from plasma membrane

Transmitter molecules

Ca^{2+}

6 Transmitter is released into synaptic cleft via exocytosis.

Glial cell

Across dendrite

Trans-mitter molecules

Postsynaptic current flow

Ions

Transmitter receptor

7 Transmitter binds to receptor molecules in postsynaptic membrane.

10 Removal of neuro-transmitter by glial uptake or enzymatic degradation

9 Postsynaptic current causes excitatory or inhibitory postsynaptic potential that changes the excitability of the postsynaptic cell.

8 Opening or closing of postsynaptic channels

◀ **FIGURE 5.4 Structure and function of chemical synapses.** (A) Structure of a chemical synapse in the cerebral cortex. A presynaptic terminal (pink) forms a synapse with a postsynaptic dendrite (green). (B) Three-dimensional reconstruction of the synapse shown in (A). Inside the presynaptic terminal, spheres indicate synaptic vesicles at various stages of their trafficking cycle, linear elements indicate intracellular filaments, and dark blue indicates dense projections associated with the active zone. Inside the postsynaptic neuron, the blue structure is the postsynaptic density, green structures represent filaments, red spheres indicate points where the filaments branch. Green material within the synaptic cleft indicates structures of unknown function. (C) Sequence of events involved in transmission at a typical chemical synapse. (A,B from Burette et al., 2012.)

vesicles are transported along an axon and down to the synaptic terminal via **axonal transport**. Neuropeptides are packaged into synaptic vesicles that range from 90 to 250 nm in diameter. Because the centers of these vesicles appear electron-dense in electron micrographs, they are referred to as **large dense-core vesicles**.

After a neurotransmitter has been secreted into the synaptic cleft, it must be removed to enable the postsynaptic cell to engage in another cycle of synaptic transmission. The removal of neurotransmitters involves diffusion away from the postsynaptic receptors, in combination with reuptake into nerve terminals or surrounding glial cells, degradation by specific enzymes, or a combination of these mechanisms. Specific transporter proteins remove most small-molecule neurotransmitters (or their metabolites) from the synaptic cleft, ultimately delivering them back to the presynaptic terminal for reuse (see Chapter 6).

Quantal Release of Neurotransmitters

Much of the evidence leading to the present understanding of chemical synaptic transmission was obtained from experiments examining the release of ACh at neuromuscular junctions. These synapses between spinal motor neurons and skeletal muscle cells are simple, large, and peripherally located, making them particularly amenable to experimental analysis. Such synapses occur at specializations called **end plates** because of the saucerlike appearance of the site on the muscle fiber where the presynaptic axon elaborates its terminals (Figure 5.5A). Most of the pioneering work on neuromuscular transmission was performed during the 1950s and 1960s by Bernard Katz and his collaborators at University College London. Although Katz worked primarily on the frog neuromuscular junction, numerous subsequent experiments have confirmed the applicability of his observations to transmission at all chemical synapses.

When an intracellular microelectrode is used to record the membrane potential of a muscle cell, an action potential in the presynaptic motor neuron can be seen to elicit a transient depolarization of the postsynaptic muscle fiber. This change in membrane potential, called an **end plate potential (EPP)**, is normally large enough to bring the membrane potential of the muscle cell well above the threshold

(A)

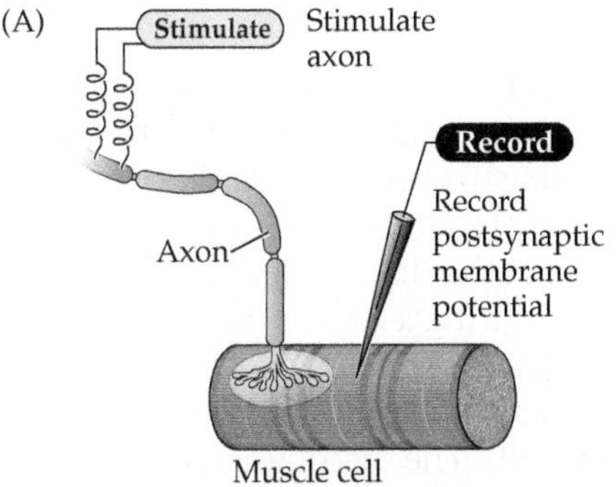

FIGURE 5.5 Synaptic transmission at the neuromuscular junction. (A) Experimental arrangement: The axon of the motor neuron innervating the muscle fiber is stimulated with an extracellular electrode, while an intracellular microelectrode is inserted into the postsynaptic muscle cell to record its electrical responses. (B) End plate potentials (shaded area) evoked by stimulation of a motor neuron are normally above threshold and therefore produce an action potential in the postsynaptic muscle cell. (C) Spontaneous miniature EPPs (MEPPs) occur in the absence of presynaptic stimulation. (D) When the neuromuscular junction is bathed in a solution that has a low concentration of Ca^{2+}, stimulating the motor neuron evokes EPPs whose amplitudes are reduced to about the size of MEPPs. (After Fatt and Katz, 1952.)

(B)

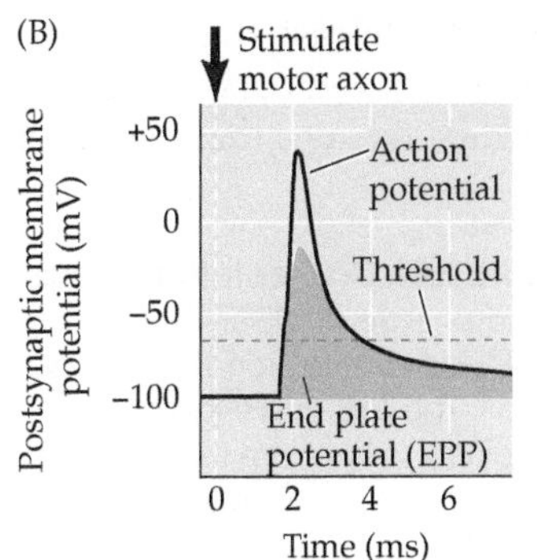

(C)

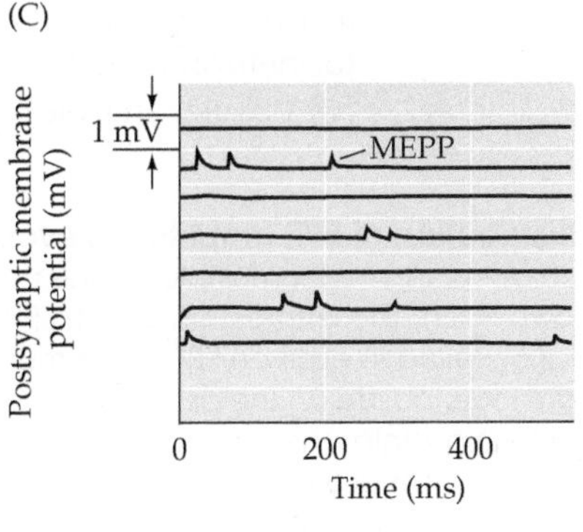

(D)

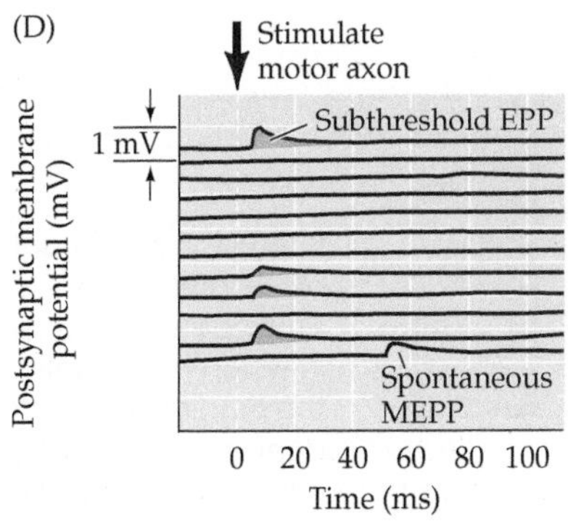

for producing a postsynaptic action potential (Figure 5.5B). The postsynaptic action potential triggered by the EPP causes the muscle fiber to contract. Unlike at electrical synapses, there is a pronounced delay between the time that the presynaptic motor neuron is stimulated and when the EPP occurs in the postsynaptic muscle cell. This synaptic delay is characteristic of all chemical synapses.

One of Katz's seminal findings, in studies carried out with Paul Fatt in 1951, was that spontaneous changes in muscle cell membrane potential occur even in the absence of stimulation of the presynaptic motor neuron (Figure 5.5C). These changes have the same shape as EPPs but are much smaller (typically less than 1 mV in amplitude, compared with an EPP of more than 50 mV). Both EPPs and these small, spontaneous events are sensitive to pharmacological agents that block postsynaptic acetylcholine receptors, such as curare (see Box 6A). These and other parallels between EPPs and the spontaneously occurring depolarizations led Katz and his colleagues to call these spontaneous events **miniature end plate potentials**, or **MEPPs**.

The relationship between the full-blown end plate potential and MEPPs was clarified by careful analysis of the EPPs. The magnitude of the EPP provides a convenient electrical assay of neurotransmitter secretion from a motor neuron terminal; however, measuring it is complicated by the need to prevent muscle contraction from dislodging the microelectrode. The usual means of eliminating muscle contractions is either to lower Ca^{2+} concentration in the extracellular medium or to partially block the postsynaptic ACh receptors with the drug curare. As expected from the scheme illustrated in Figure 5.4C, lowering the Ca^{2+} concentration reduces neurotransmitter secretion, thus reducing the magnitude of the EPP below the threshold for postsynaptic action potential production and allowing it to be measured more precisely. Under such conditions, stimulation of the motor neuron produces very small EPPs that fluctuate in amplitude from trial to trial (Figure 5.5D). These fluctuations give considerable insight into the mechanisms responsible for neurotransmitter release. In particular, the variable evoked response in low Ca^{2+} is now known to result from the release of unit amounts of ACh by the presynaptic nerve terminal. Indeed, the amplitude of the smallest evoked EPP response is strikingly similar to the size of single MEPPs (compare Figure 5.5C and D). Further supporting this similarity, increments in the EPP response (Figure 5.6A) occur in units about the size of single MEPPs (Figure 5.6B). These "quantal" fluctuations in the amplitude of EPPs indicated to Katz and his colleague Jose del Castillo that EPPs are made up of individual units, each equivalent to a MEPP.

The idea that EPPs represent the simultaneous release of many MEPP-like units can be tested statistically. A method of statistical analysis based on the independent occurrence of unitary events (called Poisson statistics) predicts what the distribution of EPP amplitudes would look like during a large number of trials of motor neuron stimulation, under the assumption that EPPs are built up from unitary events represented by MEPPs (see Figure 5.6B). The distribution of EPP amplitudes determined experimentally was found to be just that expected if transmitter release from the motor neuron is indeed quantal (the red curve in Figure 5.6A). Such analyses confirmed the idea that release of

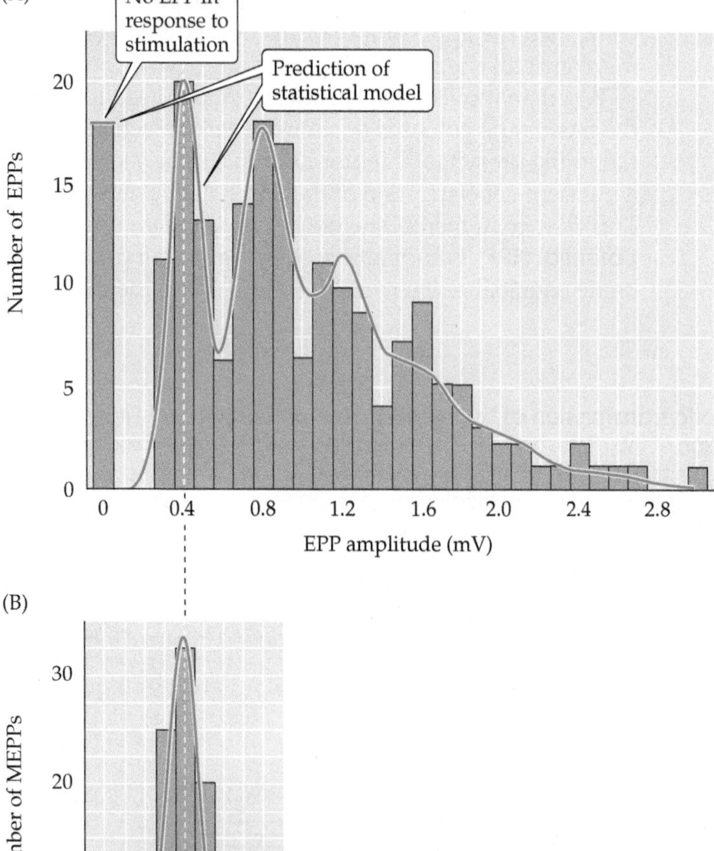

FIGURE 5.6 Quantized distribution of EPP amplitudes evoked in a low-Ca^{2+} solution. Peaks of EPP amplitudes (A) tend to occur in integer multiples of the mean amplitude of MEPPs, whose amplitude distribution is shown in (B). The leftmost bar in the EPP amplitude distribution shows trials in which presynaptic stimulation failed to elicit an EPP in the muscle cell. The red curve indicates the prediction of a statistical model based on the assumption that the EPPs result from the independent release of multiple MEPP-like quanta. The observed match, including the predicted number of failures, supports this interpretation. (After Boyd and Martin, 1955.)

acetylcholine does indeed occur in discrete packets, each equivalent to a MEPP. In short, a presynaptic action potential causes a postsynaptic EPP because it synchronizes the release of many transmitter quanta.

Release of Transmitters from Synaptic Vesicles

The discovery of the quantal release of packets of neurotransmitter immediately raised the question of how such quanta are formed and discharged into the synaptic cleft. At about the time Katz and his colleagues were using physiological methods to discover quantal release of neurotransmitter, electron microscopy revealed, for the first time, the presence of synaptic vesicles in presynaptic terminals. Putting these two discoveries together, Katz and others proposed that synaptic vesicles loaded with transmitter are the source of the quanta. Subsequent biochemical studies confirmed that synaptic vesicles are the repositories of transmitters. These studies have shown that ACh is highly concentrated in the synaptic vesicles of motor neurons, where it is present at a concentration of about 100 mM. Given the diameter of a small clear-core synaptic vesicle (~50 nm), approximately 10,000 molecules of neurotransmitter are contained in a single vesicle. This number corresponds quite nicely to the amount of ACh that must be applied to a neuromuscular junction to mimic a MEPP, providing further support for the idea that quanta arise from discharge of the contents of single synaptic vesicles.

To prove that quanta are caused by the fusion of individual synaptic vesicles with the plasma membrane, it is necessary to show that each fused vesicle produces a single quantal event in the postsynaptic cell. This challenge was met in the late 1970s, when John Heuser, Tom Reese, and colleagues correlated measurements of vesicle fusion with the quantal content of EPPs at the neuromuscular junction (Figure 5.7A). They used electron microscopy to determine the number of vesicles that fused with the presynaptic plasma membrane at the active zones of presynaptic terminals (Figure 5.7B). By treating terminals with different concentrations of a drug (4-aminopyridine, or 4-AP) that enhances the number of quanta released by single action potentials, it was possible to vary the amount of quantal release, determined from parallel electrical measurements of the quantal content of the EPPs. A comparison of the number of synaptic vesicle fusions observed with the electron microscope and

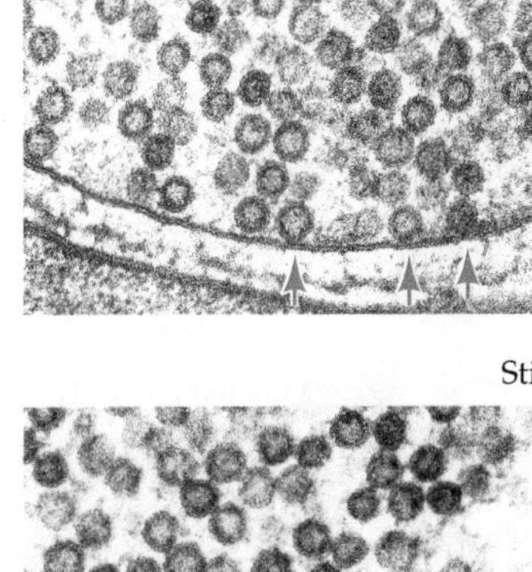

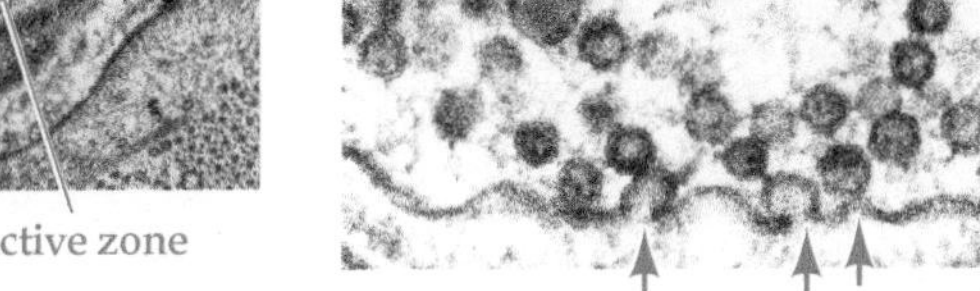

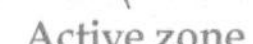

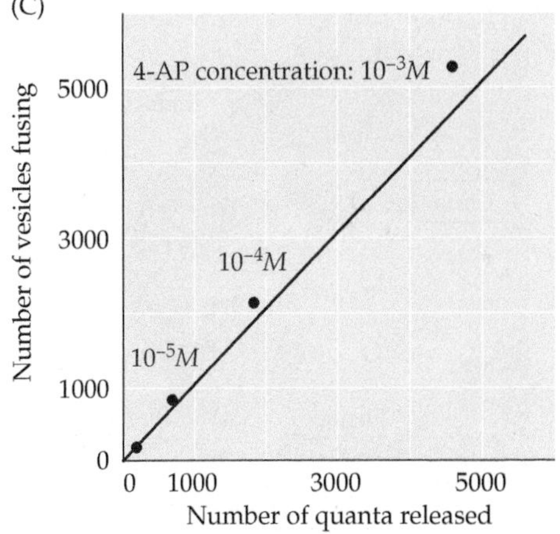

FIGURE 5.7 Relationship between synaptic vesicle exocytosis and quantal transmitter release. (A) Electron micrograph of a frog neuromuscular synapse. This synapse includes a presynaptic motor neuron that innervates a postsynaptic muscle cell and is covered by a type of glial cell called a Schwann cell. The active zone of the presynaptic terminal is the site of synaptic vesicle exocytosis. (B) Top: Active zone of an unstimulated presynaptic terminal. While many synaptic vesicles are present, including several that are docked at the active zone (arrows), none are fusing with the presynaptic plasma membrane. Bottom: Active zone of a terminal stimulated by an action potential; stimulation causes fusion (arrows) of synaptic vesicles with the presynaptic membrane. (C) Comparison of the number of observed vesicle fusion events with the number of quanta released by a presynaptic action potential. Transmitter release was varied by using different concentrations of a drug (4-AP) that affects the duration of the presynaptic action potential, thus changing the amount of calcium that enters during the action potential. The diagonal line is the 1:1 relationship expected if each vesicle that opened released a single quantum of transmitter. (C after Heuser et al., 1979; A and B courtesy of J. Heuser.).

the number of quanta released at the synapse showed a good correlation between these two measures (Figure 5.7C). These results remain one of the strongest lines of support for the idea that a quantum of transmitter release is due to fusion of a single synaptic vesicle with the presynaptic membrane. Subsequent evidence, based on other means of measuring vesicle fusion, has left no doubt about the validity of this interpretation, thereby establishing that chemical synaptic transmission results from the discharge of neurotransmitters from synaptic vesicles.

Local Recycling of Synaptic Vesicles

The fusion of synaptic vesicles causes new membrane to be added to the plasma membrane of the presynaptic terminal, but the addition is not permanent. Although a bout of exocytosis can dramatically increase the surface area of presynaptic terminals, this extra membrane is removed within a few minutes. Heuser and Reese performed another important set of experiments showing that the fused vesicle membrane is actually retrieved and taken back into the cytoplasm of the nerve terminal (a process called endocytosis). The experiments, again carried out at the frog neuromuscular junction, were based on filling the synaptic cleft with horseradish peroxidase (HRP), an enzyme that produces a dense reaction product that is visible in an electron microscope. Under appropriate experimental conditions, endocytosis could then be visualized by the uptake of HRP into the nerve terminal (Figure 5.8). To activate endocytosis, the presynaptic terminal was stimulated with a train of action potentials, and the subsequent fate of the HRP was followed by electron microscopy. Immediately following stimulation, the HRP was found in special endocytotic organelles called coated vesicles, which form from membrane budded off via coated pits (see Figure 5.8B). A few minutes later, however, the coated vesicles had disappeared, and the HRP was found in a different organelle, the endosome (see Figure 5.8C). Finally, within an hour after the terminal had been stimulated, the HRP reaction product appeared inside synaptic vesicles (see Figure 5.8D).

These observations indicate that synaptic vesicle membrane is recycled within the presynaptic terminal via the sequence summarized in Figure 5.8E. In this process, called the **synaptic vesicle cycle**, the retrieved vesicular

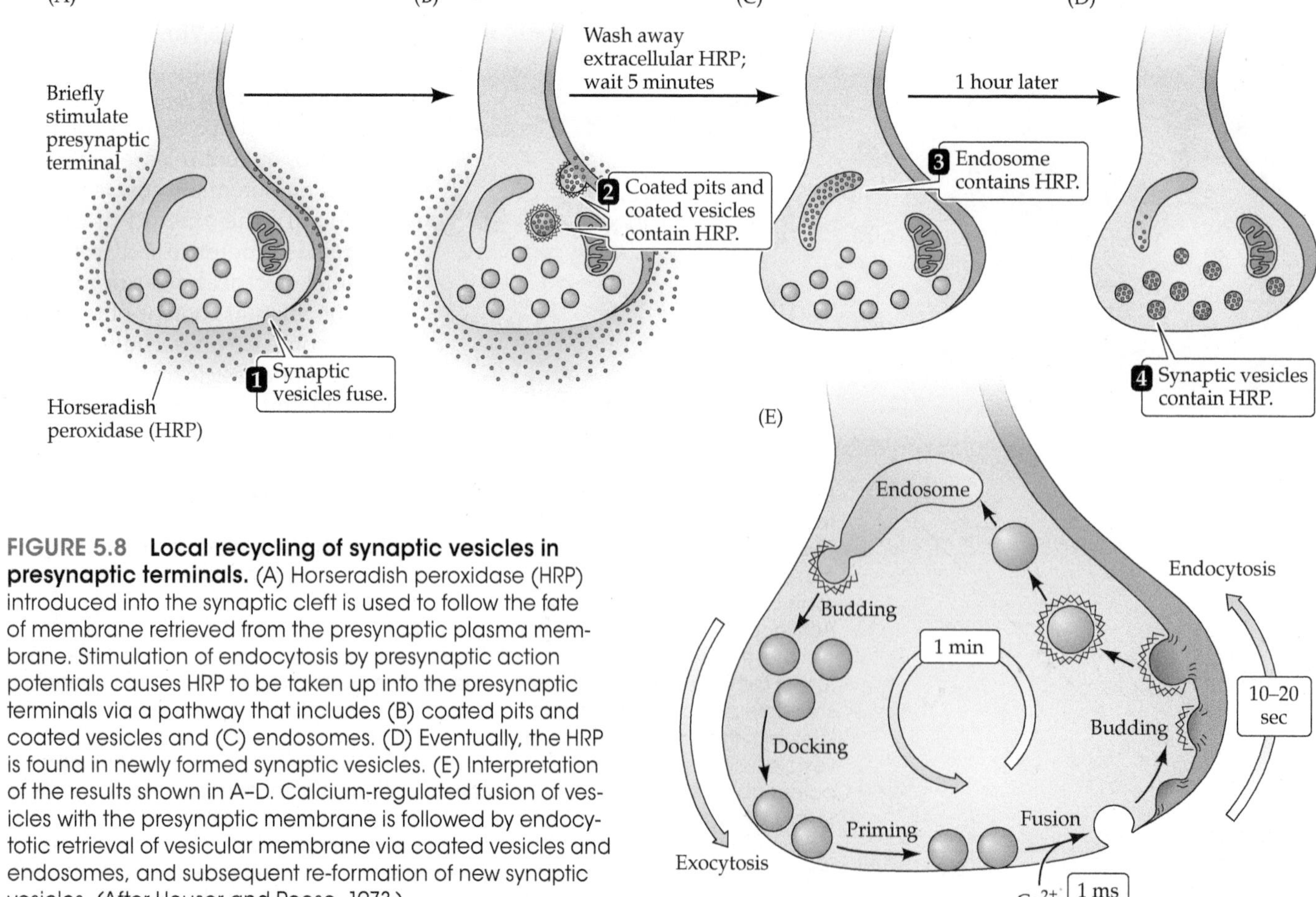

FIGURE 5.8 Local recycling of synaptic vesicles in presynaptic terminals. (A) Horseradish peroxidase (HRP) introduced into the synaptic cleft is used to follow the fate of membrane retrieved from the presynaptic plasma membrane. Stimulation of endocytosis by presynaptic action potentials causes HRP to be taken up into the presynaptic terminals via a pathway that includes (B) coated pits and coated vesicles and (C) endosomes. (D) Eventually, the HRP is found in newly formed synaptic vesicles. (E) Interpretation of the results shown in A–D. Calcium-regulated fusion of vesicles with the presynaptic membrane is followed by endocytotic retrieval of vesicular membrane via coated vesicles and endosomes, and subsequent re-formation of new synaptic vesicles. (After Heuser and Reese, 1973.)

membrane passes through several intracellular compartments—such as coated vesicles and endosomes—and is eventually used to make new synaptic vesicles. After synaptic vesicles are re-formed, they are stored in a reserve pool within the cytoplasm until they need to participate again in neurotransmitter release. These vesicles are mobilized from the reserve pool, docked at the presynaptic plasma membrane, and primed to participate in exocytosis once again. More recent experiments, employing a fluorescent label rather than HRP, have determined the time course of synaptic vesicle recycling. These studies indicate that the entire vesicle cycle requires approximately 1 minute, with membrane budding during endocytosis requiring 10 to 20 seconds of this time. As can be seen from the 1-millisecond delay in transmission following excitation of the presynaptic terminal (see Figure 5.5B), membrane fusion during exocytosis is much more rapid than budding during endocytosis. Thus, all of the recycling steps interspersed between membrane fusion and subsequent regeneration of a new vesicle are completed in less than a minute.

The precursors to synaptic vesicles *originally* are produced in the endoplasmic reticulum and Golgi apparatus in the neuronal cell body. Because of the long distance between the cell body and the presynaptic terminal in most neurons, transport of vesicles from the soma would not permit rapid replenishment of synaptic vesicles during continuous neural activity. Thus, local recycling is well suited to the peculiar anatomy of neurons, giving nerve terminals the means to provide a continual supply of synaptic vesicles.

The Role of Calcium in Transmitter Secretion

As was apparent in the experiments of Katz and others described in the preceding sections, lowering the concentration of Ca^{2+} outside a presynaptic motor nerve terminal reduces the size of the EPP (compare Figure 5.5B and D). Moreover, measurement of the number of transmitter quanta released under such conditions shows that the reason the EPP gets smaller is that lowering Ca^{2+} concentration decreases the number of vesicles that fuse with the plasma membrane of the terminal. An important insight into *how* Ca^{2+} regulates the fusion of synaptic vesicles was the discovery that presynaptic terminals have voltage-gated Ca^{2+} channels in their plasma membranes (see Chapter 4).

The first indication of presynaptic Ca^{2+} channels was provided by Katz and Ricardo Miledi. They observed that presynaptic terminals treated with tetrodotoxin (which blocks voltage-gated Na^+ channels; see Chapter 3) could still produce a peculiarly prolonged type of action potential. The explanation for this surprising finding was that current was still flowing through Ca^{2+} channels, substituting for the current ordinarily carried by the blocked Na^+ channels. Subsequent voltage clamp experiments, performed by Rodolfo Llinás and others at a giant presynaptic terminal of the squid (Figure 5.9A), confirmed the presence of voltage-gated Ca^{2+} channels in the presynaptic terminal (Figure 5.9B). Such experiments showed that the amount of neurotransmitter released is very sensitive to the exact amount of Ca^{2+} that enters. Furthermore, blockade of these Ca^{2+} channels with drugs also inhibits transmitter release (see Figure 5.9B, right). These observations establish that the voltage-gated Ca^{2+} channels are

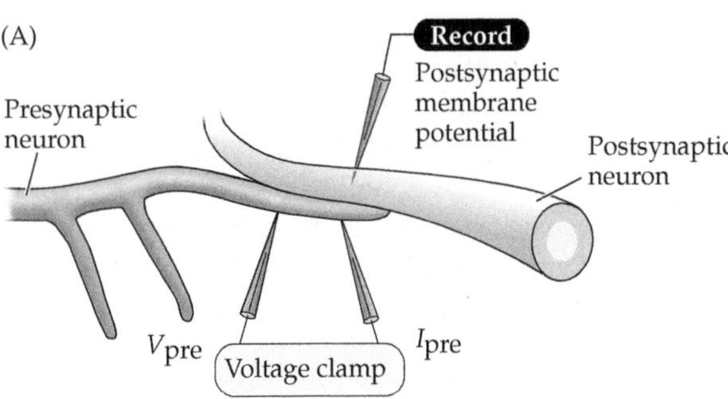

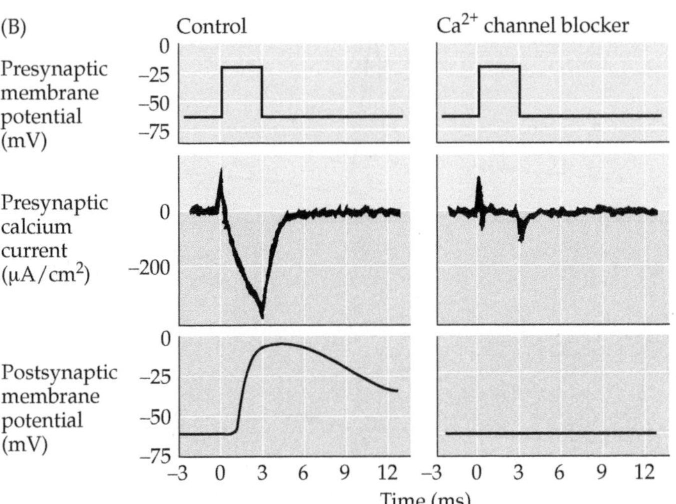

FIGURE 5.9 Entry of Ca^{2+} through presynaptic voltage-gated calcium channels causes transmitter release. (A) Experimental setup using an extraordinarily large synapse in the squid. The voltage clamp method detects currents flowing across the presynaptic membrane when the membrane potential is depolarized. (B) Pharmacological agents that block currents flowing through Na^+ and K^+ channels reveal a remaining inward current flowing through Ca^{2+} channels. This influx of calcium triggers transmitter secretion, as indicated by a change in the postsynaptic membrane potential. Treatment of the same presynaptic terminal with cadmium, a calcium channel blocker, eliminates both the presynaptic calcium current and the postsynaptic response. (After Augustine and Eckert, 1984.)

directly involved in synaptic transmission: Presynaptic action potentials open these Ca^{2+} channels, yielding an influx of Ca^{2+} into the presynaptic terminal.

As is the case for many other forms of neuronal signaling (see Chapter 7), Ca^{2+} serves as a second messenger during transmitter release. Ca^{2+} entering into presynaptic terminals accumulates within the terminal, as can be seen with microscopic imaging of terminals filled with Ca^{2+}-sensitive dyes (Figure 5.10A). The presynaptic second messenger function of Ca^{2+} has been directly shown in two complementary ways. First, microinjection of Ca^{2+} into presynaptic terminals triggers transmitter release even in the absence of presynaptic action potentials (Figure 5.10B). Second, presynaptic microinjection of calcium chelators (chemicals that bind Ca^{2+} and keep its concentration buffered at low levels) prevents presynaptic action potentials from causing transmitter secretion (Figure 5.10C). These results prove beyond any doubt that a rise in presynaptic Ca^{2+} concentration is both necessary and sufficient for neurotransmitter release. While Ca^{2+} is a universal trigger for transmitter release, not all transmitters are released with the same speed. For example, while secretion of ACh from motor neurons requires only a fraction of a millisecond (see Figure 5.5), release of neuropeptides requires high-frequency bursts of action potentials for many seconds. These differences in the rate of release probably arise from differences in the spatial arrangement of vesicles relative to presynaptic Ca^{2+} channels, yielding differences in the time course of local Ca^{2+} signaling.

FIGURE 5.10 Evidence that a rise in presynaptic Ca^{2+} concentration triggers transmitter release from presynaptic terminals. (A) Fluorescence microscopy measurements of presynaptic Ca^{2+} concentration at the squid giant synapse (see Figure 5.9A). A train of presynaptic action potentials causes a rise in Ca^{2+} concentration, as revealed by a dye (called fura-2) that fluoresces more strongly when the Ca^{2+} concentration increases (colors). (B) Microinjection of Ca^{2+} into a squid giant presynaptic terminal triggers transmitter release, measured as a depolarization of the postsynaptic membrane potential. (C) Microinjection of BAPTA, a Ca^{2+} chelator, into a squid giant presynaptic terminal prevents transmitter release. (A from Smith et al., 1993; B after Miledi, 1973; C after Adler et al., 1991.)

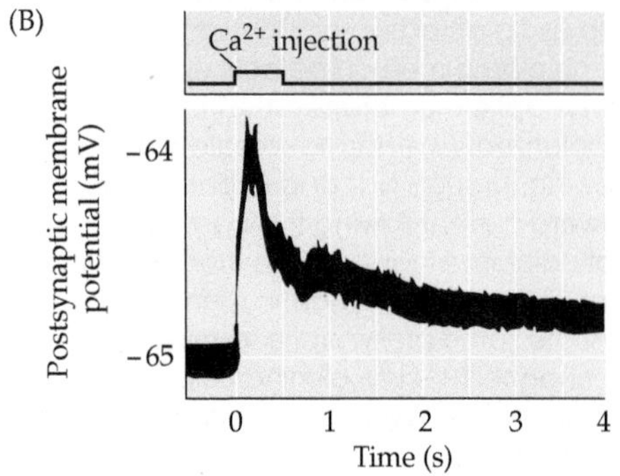

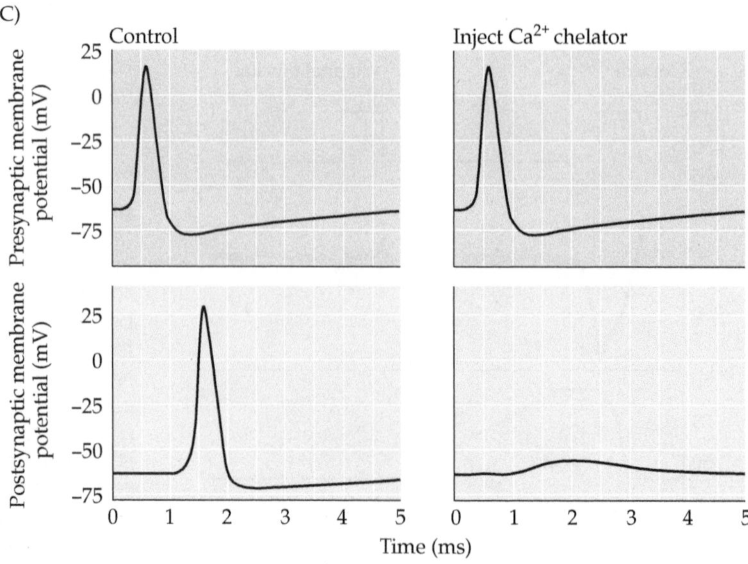

Molecular Mechanisms of Synaptic Vesicle Cycling

Precisely how an increase in presynaptic Ca^{2+} concentration goes on to trigger vesicle fusion and neurotransmitter release is not fully understood. Molecular studies have identified and characterized the proteins found on synaptic vesicles (Figure 5.11A) and their binding partners on the presynaptic plasma membrane and cytoplasm. Most, if not all, of these proteins act at one or more steps in the synaptic vesicle cycle (Figure 5.11B).

Several lines of evidence indicate that the protein **synapsin**, which reversibly binds to synaptic vesicles, may keep these vesicles tethered within the reserve pool by cross-linking vesicles to each other. Mobilization of these reserve pool vesicles is caused by phosphorylation of synapsin by proteins kinases, most notably the **Ca^{2+}/calmodulin-dependent protein kinase, type II** (CaMKII; see Chapter 7), which allows synapsin to dissociate from the vesicles. Once vesicles are free from their reserve pool tethers, they make their way to the plasma membrane and are then attached to this membrane by docking reactions that involve SNARE proteins (see below). A series of priming reactions then prepares the vesicular and plasma membranes for fusion. A large number of proteins are involved in priming, including some proteins that are also involved in other types of membrane fusion events common to all cells (see Figure 5.11B). For example, two proteins originally found to be important for the fusion of vesicles with membranes of the Golgi apparatus, the ATPase **NSF** (*N*EM-sensitive *f*usion protein) and **SNAPs** (soluble *N*SF-*a*ttachment *p*roteins), are also involved in priming synaptic vesicles for fusion. These two proteins work by regulating the assembly of other proteins that are called **SNAREs** (*SNAP re*ceptors). Many of the other proteins involved in priming—such as munc13, munc18, complexin, snapin, syntaphilin, and tomosyn—also interact with the SNAREs.

One of the main purposes of priming is to organize SNARE proteins into the correct conformation for membrane fusion. One of the SNARE proteins, **synaptobrevin**, is in the membrane of synaptic vesicles, while two other SNARE proteins called **syntaxin** and **SNAP-25** are found primarily on the plasma

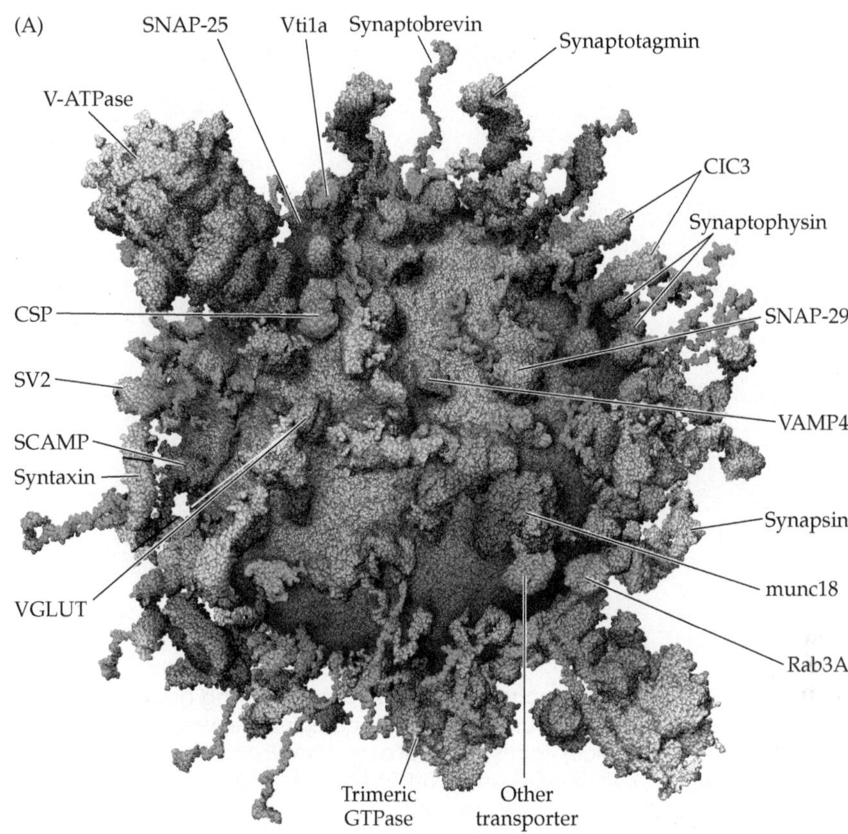

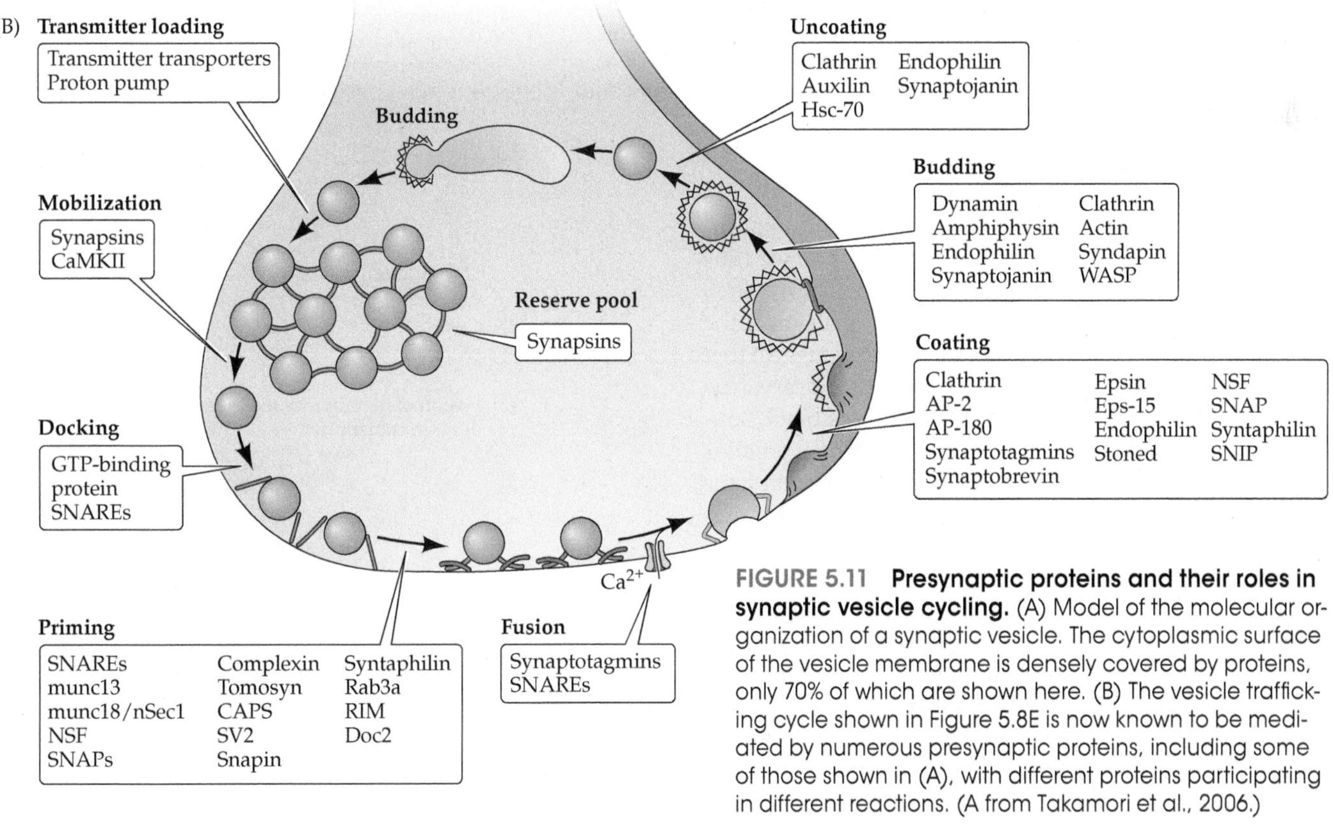

FIGURE 5.11 Presynaptic proteins and their roles in synaptic vesicle cycling. (A) Model of the molecular organization of a synaptic vesicle. The cytoplasmic surface of the vesicle membrane is densely covered by proteins, only 70% of which are shown here. (B) The vesicle trafficking cycle shown in Figure 5.8E is now known to be mediated by numerous presynaptic proteins, including some of those shown in (A), with different proteins participating in different reactions. (A from Takamori et al., 2006.)

(A)

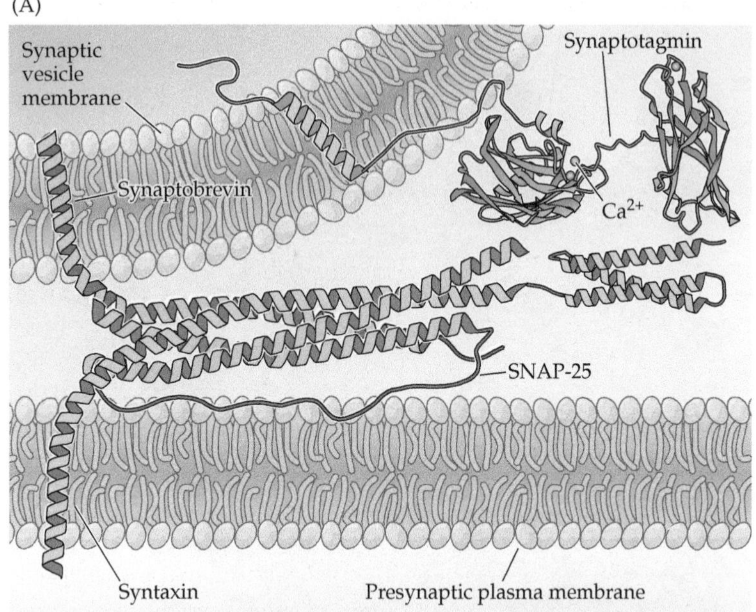

(B) (1) Free SNARES on vesicle and plasma membranes

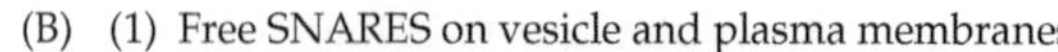

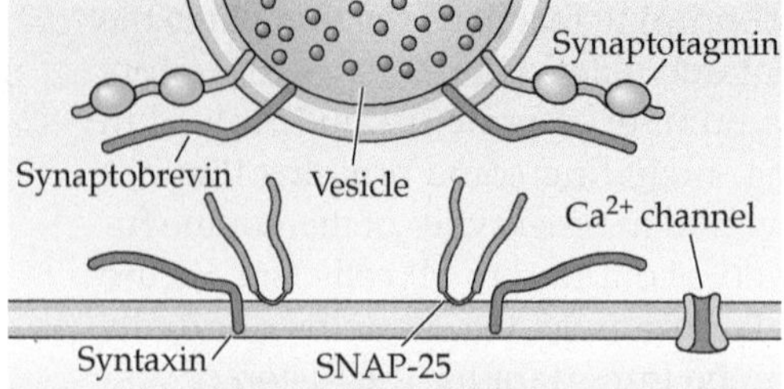

(2) SNARE complexes form as vesicle docks

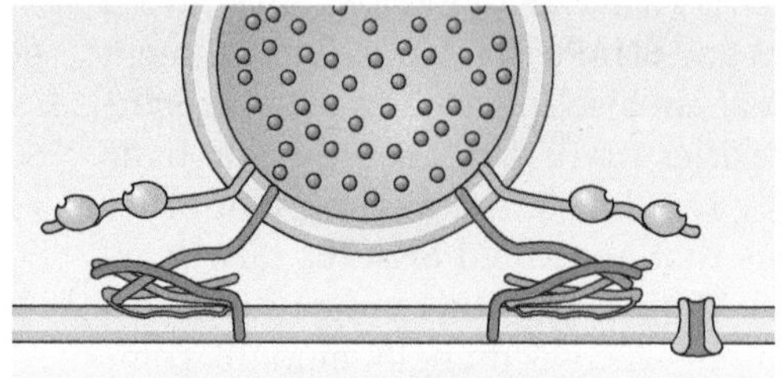

(3) Synaptotagmin binds to SNARE complex

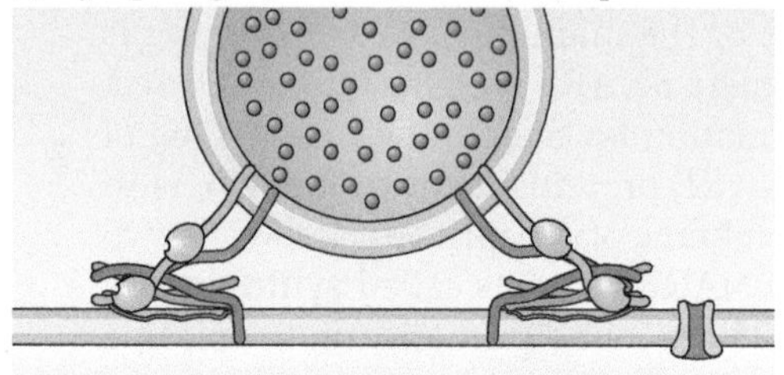

(4) Entering Ca^{2+} binds to synaptotagmin, leading to curvature of plasma membrane, which brings membranes together

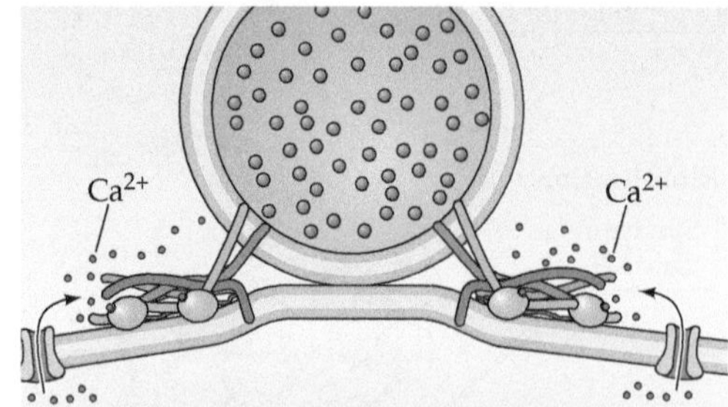

(5) Fusion of membranes leads to exocytotic release of neurotransmitter

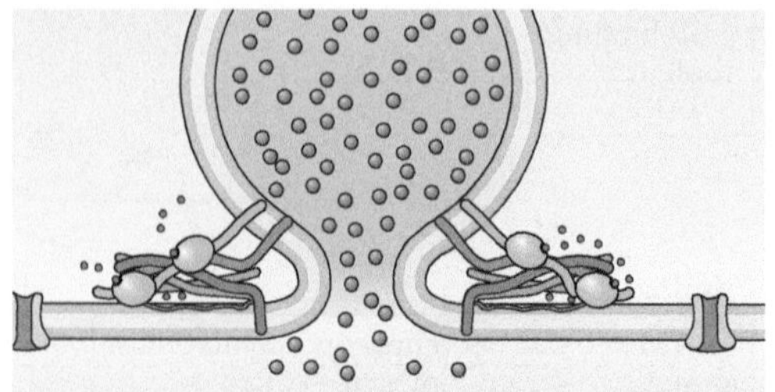

FIGURE 5.12 Molecular mechanisms of exocytosis during neurotransmitter release. (A) Structure of the SNARE complex. The vesicular SNARE, synaptobrevin (blue), forms a helical complex with the plasma membrane SNAREs syntaxin (red) and SNAP-25 (green). Also shown is the structure of synaptotagmin, a vesicular Ca^{2+}-binding protein, with bound Ca^{2+} indicated by spheres. (B) A model for Ca^{2+}-triggered vesicle fusion. During docking of the synaptic vesicle, SNARE proteins on the vesicle and plasma membranes form a complex (as in A) that brings together the two membranes. Synaptotagmin binds to this SNARE complex. Subsequent binding of Ca^{2+} to synaptotagmin causes the cytoplasmic region of this protein to insert into the plasma membrane to produce the membrane curvature that catalyzes membrane fusion. (A from Sutton et al., 1998 and Madej et al., 2014; B after Zhou et al., 2015.)

membrane. These SNARE proteins can form a macromolecular complex that spans the two membranes, thus bringing them into close apposition (Figure 5.12A). Such an arrangement is well suited to promote the fusion of the two membranes, and several lines of evidence suggest that this is what actually occurs. One important observation is that toxins that cleave the SNARE proteins block neurotransmitter release. In addition, putting SNARE proteins into artificial lipid membranes and allowing these proteins to form complexes with each other causes the membranes to fuse.

Because the SNARE proteins do not bind Ca^{2+}, still other molecules must be responsible for Ca^{2+} regulation of neurotransmitter release. Numerous presynaptic proteins, including calmodulin, CAPS, and munc-13, are capable of binding Ca^{2+}. However, it appears that Ca^{2+} regulation of neurotransmitter release usually is conferred by **synaptotagmins**, a family of proteins found in the membrane of synaptic vesicles (see Figure 5.12A). Synaptotagmin binds Ca^{2+} at concentrations similar to those required to trigger vesicle fusion within the presynaptic terminal, and this property allows synaptotagmin to act as a Ca^{2+} sensor that triggers vesicle fusion by signaling the elevation of Ca^{2+} within the terminal. In support of this idea, disruption of synaptotagmin in the presynaptic terminals of mice, fruit flies, squid, and other experimental animals impairs Ca^{2+}-dependent neurotransmitter release. In fact, deletion of only one of the 17 synaptotagmin genes (SYT) of mice is a lethal mutation, causing the mice to die soon after birth. It is thought that Ca^{2+} binding to synaptotagmin leads to exocytosis by changing the chemical properties of synaptotagmin, thereby allowing it to insert into the plasma membrane. This causes the plasma membrane to locally curve and leads to fusion of the two membranes. Thus, SNARE proteins bring the two membranes close together, while Ca^{2+}-induced changes in synaptotagmin then produce the final curvature that enables rapid fusion of these membranes (Figure 5.12B).

Still other proteins appear to be involved at the endocytosis steps of the synaptic vesicle cycle (Figure 5.13). The most important protein involved in endocytotic budding of vesicles from the plasma membrane is **clathrin**. Clathrin has a unique structure that is called a triskelion because of its three-legged appearance; these triskelia can assemble to form a cagelike coating around the vesicle membrane (see Figure 5.13A). Several adaptor proteins, such as AP-2 and AP-180, connect clathrin to the proteins and lipids of this membrane. These adaptor proteins, as well as other proteins such as amphiphysin, epsin, and Eps-15, help assemble individual triskelia into structures that resemble geodesic domes (see Figure 5.13A, bottom). Such domelike structures form coated pits that initiate membrane budding, increasing the curvature of the budding membrane until it forms a coated vesicle-like structure that remains connected to the plasma membrane via a narrow lipid stalk (Figure 5.13C). Another protein, called **dynamin**, forms a ringlike coil that surrounds the lipid stalk (see Figure 5.13B). This coil causes the final pinching-off of membrane that severs the stalk and completes the production of coated vesicles. Coated vesicles then

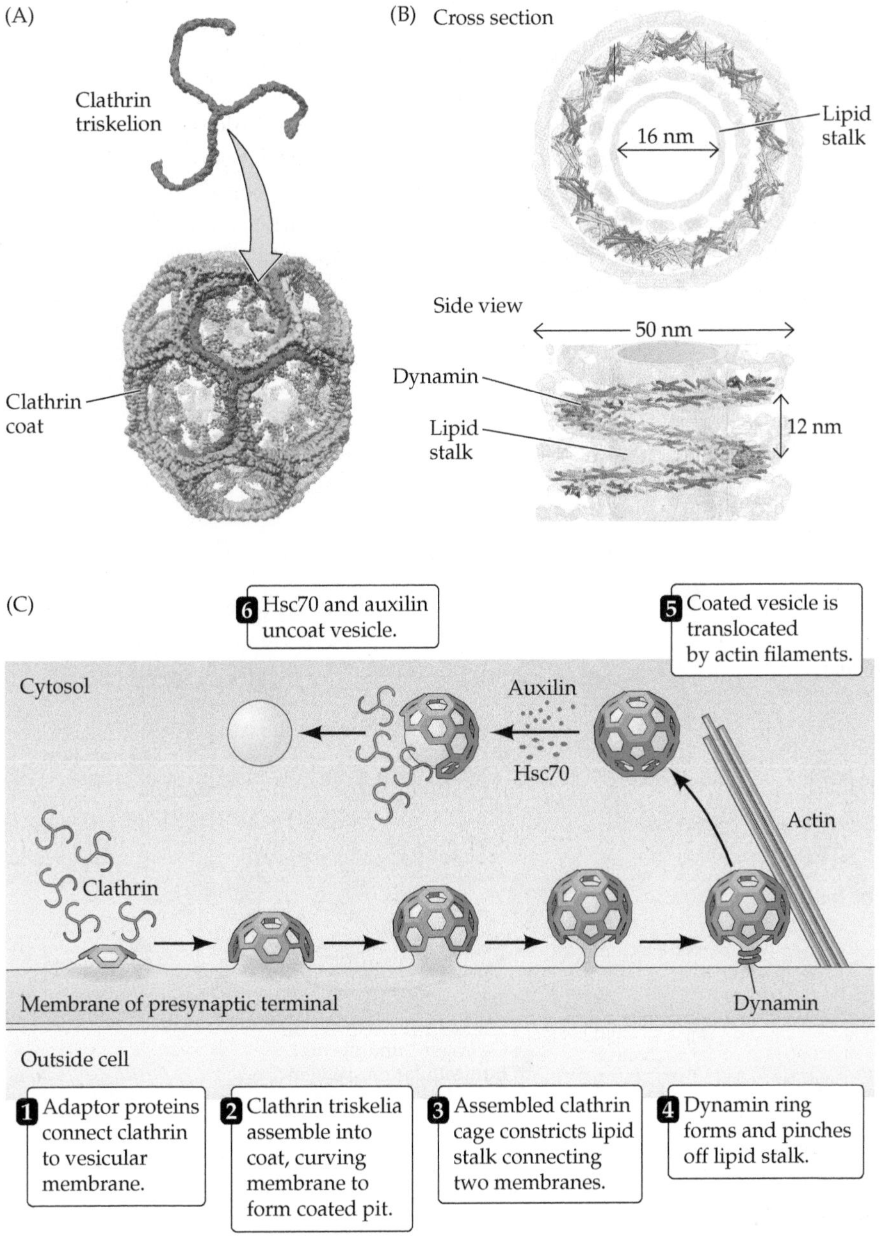

FIGURE 5.13 Molecular mechanisms of endocytosis following neurotransmitter release. (A) Individual clathrin triskelia assemble together to form membrane coats involved in membrane budding during endocytosis. (B) Dynamin forms ringlike coils around the lipid stalks of budding membranes; these rings disconnect vesicle membrane from plasma membrane during endocytosis. (C) A model for membrane budding during endocytosis. Following addition of synaptic vesicle membrane during exocytosis, clathrin triskelia attach to the vesicular membrane. Adaptor proteins, such as AP-2 and AP-180, aid their attachment. Polymerization of clathrin causes the membrane to curve and constrict, allowing dynamin to pinch off the coated vesicle. Subsequent uncoating of the vesicle, by Hsc70 and auxilin, yields a recycled synaptic vesicle. (A from Fotin et al., 2004; B from Ruebold et al., 2015; C after Shupliakov et al., 2010.)

are transported away from the plasma membrane by the cytoskeletal protein **actin**. This allows the clathrin coats to be removed by an ATPase, **Hsc70**, with another protein, **auxilin**, serving as a co-factor that recruits Hsc70 to the coated vesicle. Other proteins, such as **synaptojanin**, are also important for vesicle uncoating. Uncoated vesicles can then continue their journey through the recycling process, eventually becoming refilled with neurotransmitter due to the actions of neurotransmitter transporters in the vesicle membrane. These transporters exchange protons within the vesicle for neurotransmitter; the acidic interior of the vesicle is produced by a proton pump that also is located in the vesicle membrane.

In summary, a complex cascade of proteins, acting in a defined temporal and spatial order, allows neurons to secrete transmitters. This molecular cascade underlies the powerful ability of the brain to use its synapses to process and store information.

Neurotransmitter Receptors

The generation of postsynaptic electrical signals is also understood in considerable depth. Such studies began in 1907, when the British physiologist John N. Langley introduced the concept of **receptor molecules** to explain the specific and potent actions of certain chemicals on muscle and nerve cells. We now know that neurotransmitter receptors are proteins that are embedded in the plasma membrane of postsynaptic cells and have an extracellular neurotransmitter binding site that detects the presence of neurotransmitters in the synaptic cleft.

There are two broad families of receptor proteins that differ in their mechanism of transducing transmitter binding into postsynaptic responses. The receptors in one family contain a membrane-spanning domain that forms an ion channel (Figure 5.14A). These receptors combine transmitter-binding and channel functions into a single molecular entity and thus are called **ionotropic receptors** (Greek *tropos*, "to move in response to a stimulus") or ligand-gated ion channels.

The second family of neurotransmitter receptor is **metabotropic receptors**, so called because the eventual movement of ions through a channel depends on intervening metabolic steps. These receptors do not have ion channels as part of their structure; instead, they have an intracellular domain that indirectly affects channels though the activation of intermediate molecules called **G-proteins** (Figure 5.14B). Neurotransmitter binding to these receptors activates G-proteins, which then dissociate from the receptor and interact directly with ion channels or bind to other effector proteins, such as enzymes, that make intracellular messengers that open or close ion channels. Thus, G-proteins can be thought of as transducers that couple neurotransmitter binding to a receptor with regulation of

(A) Ligand-gated ion channels

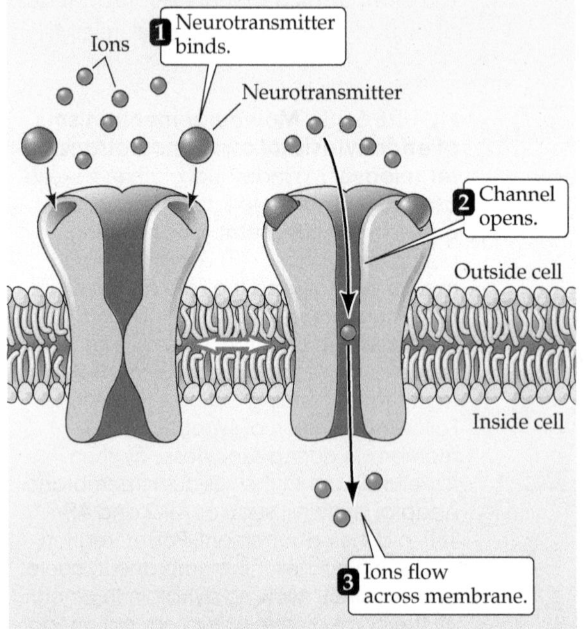

(B) G-protein-coupled receptors

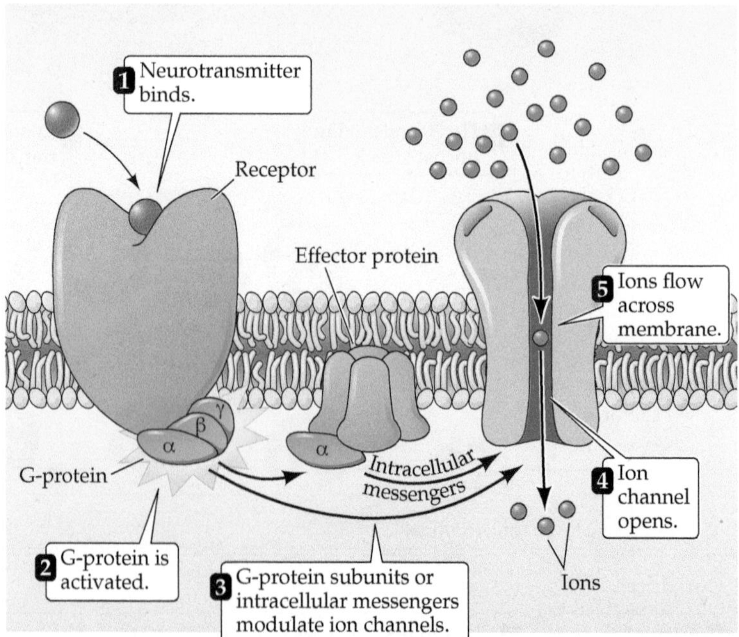

FIGURE 5.14 Two different types of neurotransmitter receptors. (A) Ligand-gated ion channels combine receptor and channel functions in a single protein complex. (B) Metabotropic receptors usually activate G-proteins, which modulate ion channels directly or indirectly through intracellular effector enzymes and second messengers.

postsynaptic ion channels. For this reason, metabotropic receptors are also called **G-protein-coupled receptors.** The postsynaptic signaling events initiated by metabotropic receptors are described in Chapter 7.

These two families of postsynaptic receptors give rise to postsynaptic actions that range from less than a millisecond to minutes, hours, or even days. Ionotropic receptors generally mediate rapid postsynaptic effects. Examples are the EPP produced at neuromuscular synapses by ACh (see Figure 5.5B), as well as the postsynaptic responses produced at certain glutamatergic synapses and GABAergic synapses (see Figure 5.19B). In these cases, the postsynaptic potentials arise within a millisecond or two of an action potential invading the presynaptic terminal and last for only a few tens of milliseconds or less. In contrast, the activation of metabotropic receptors typically produces much slower responses, ranging from hundreds of milliseconds to minutes or even longer. The comparative slowness of metabotropic receptor actions reflects the fact that multiple proteins need to bind to each other sequentially in order to produce the final physiological response. Importantly, many transmitters can activate both ionotropic and metabotropic receptors to produce both fast and slow postsynaptic potentials (sometimes even at the same synapse).

Postsynaptic Membrane Permeability Changes during Synaptic Transmission

Just as studies of the neuromuscular synapse paved the way for understanding neurotransmitter release mechanisms, this peripheral synapse has been equally valuable for understanding the mechanisms that allow neurotransmitter receptors to generate postsynaptic signals. The binding of ACh to postsynaptic receptors opens ion channels in the muscle fiber membrane. This effect can be demonstrated directly by using the patch clamp method (see Box 4A) to measure the minute postsynaptic currents that flow when two molecules of individual ACh bind to receptors, as Erwin Neher and Bert Sakmann first did in 1976. Exposure of the extracellular surface of a patch of postsynaptic membrane to ACh causes single-channel currents to flow for a few milliseconds (Figure 5.15A). This shows that ACh binding to its receptors opens ligand-gated ion channels,

FIGURE 5.15 Activation of ACh receptors at neuromuscular synapses. (A) Patch clamp measurement of single ACh receptor currents from a patch of membrane removed from the postsynaptic muscle cell. When ACh is applied to the extracellular surface of the membrane, the repeated brief opening of a single channel can be observed as downward deflections corresponding to inward current (i.e., positive ions flowing into the cell). (B) Synchronized opening of many ACh-activated channels at a synapse being voltage clamped. (1) If a single channel is examined during the release of ACh from the presynaptic terminal, the channel can be seen to open transiently. (2) If several channels are examined together, ACh release opens the channels almost synchronously. (3) The opening of a very large number of postsynaptic channels produces a macroscopic EPC. (C) In a normal muscle cell (i.e., not being voltage clamped), the inward EPC depolarizes the postsynaptic muscle cell, giving rise to an EPP. Typically, this depolarization generates an action potential (not shown; see Figure 5.5B).

(A) Patch clamp measurement of single ACh receptor current

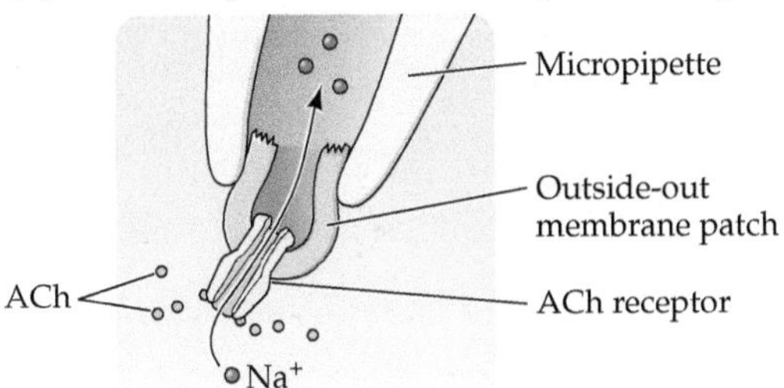

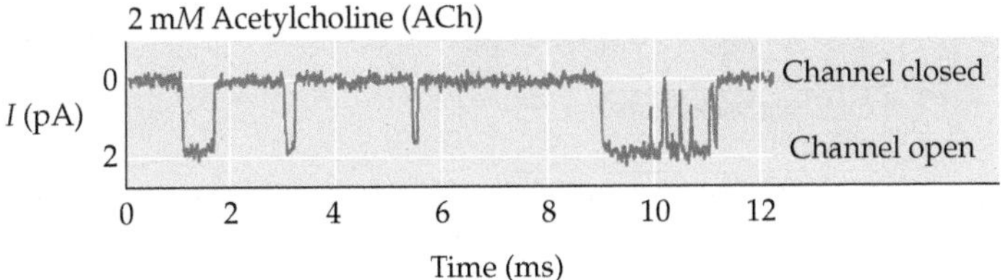

(B) Currents produced by:

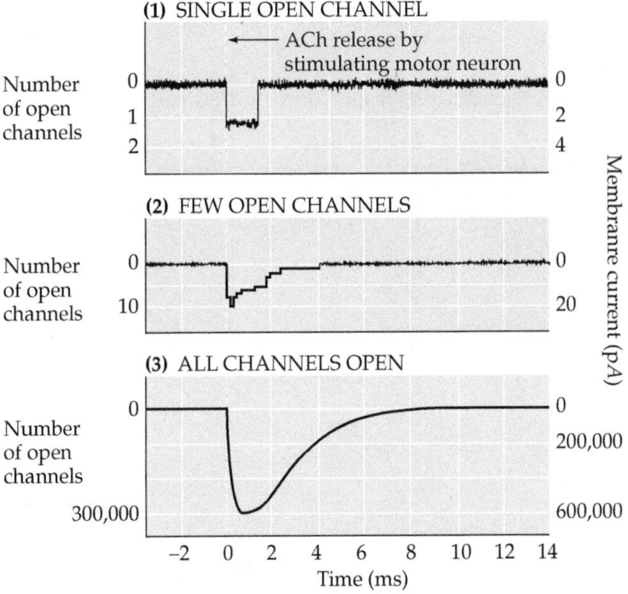

(C) Postsynaptic potential change (EPP) produced by EPC

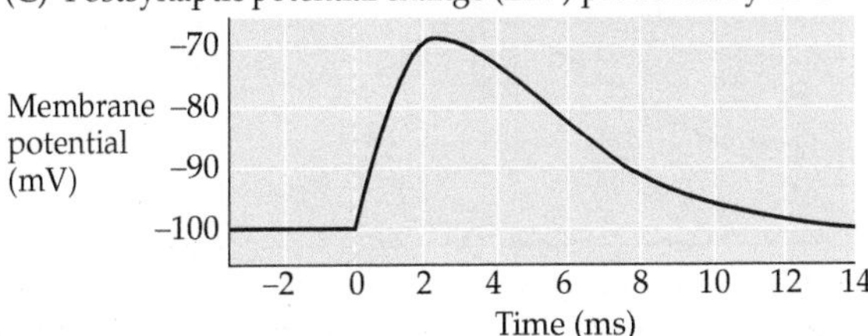

much in the way that changes in membrane potential open voltage-gated ion channels (see Chapter 4).

The electrical actions of ACh are greatly multiplied when an action potential in a presynaptic motor neuron causes the release of millions of molecules of ACh into the synaptic cleft. In this more physiological case, the transmitter molecules bind to many thousands of ACh receptors packed in a dense array on the postsynaptic membrane, transiently opening a very large number of postsynaptic ion channels. Although individual ACh receptors generate a microscopic current of only a few picoamperes (Figure 5.15B1), the opening of a large number of channels are opened synchronously when ACh is secreted from presynaptic terminals (Figure 5.15B2,3). The macroscopic current resulting from the summed opening of many ion channels is called the **end plate current**, or **EPC**. Because the EPC normally is inward, it causes the postsynaptic membrane potential to depolarize. This depolarizing change in

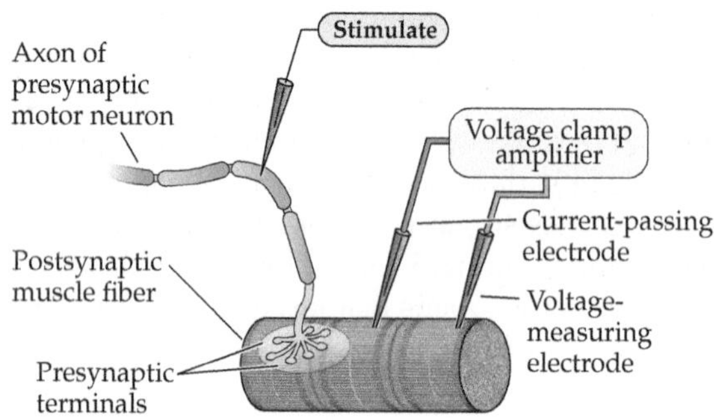

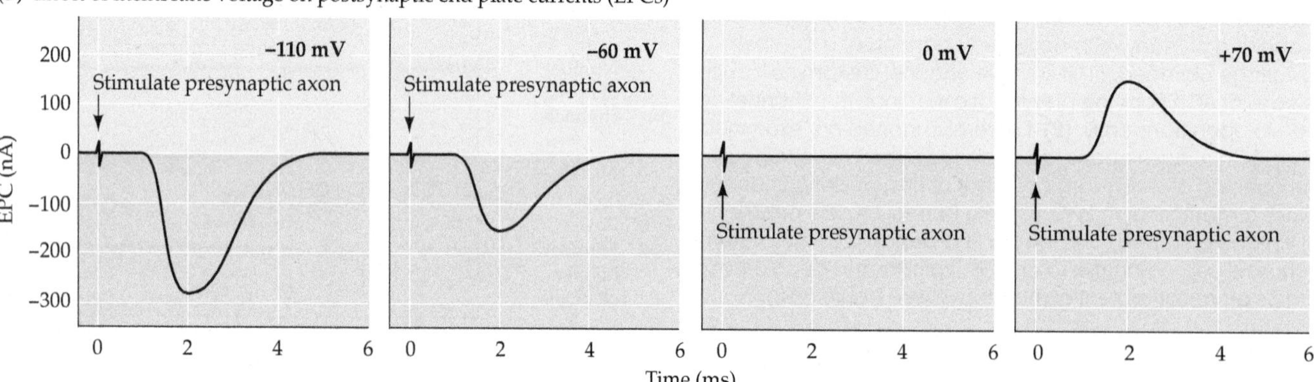

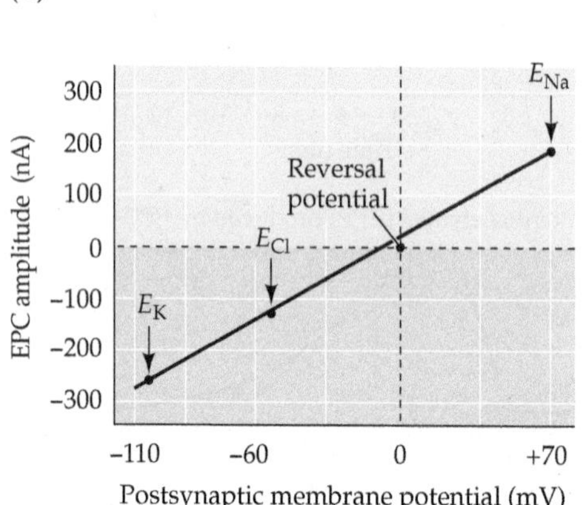

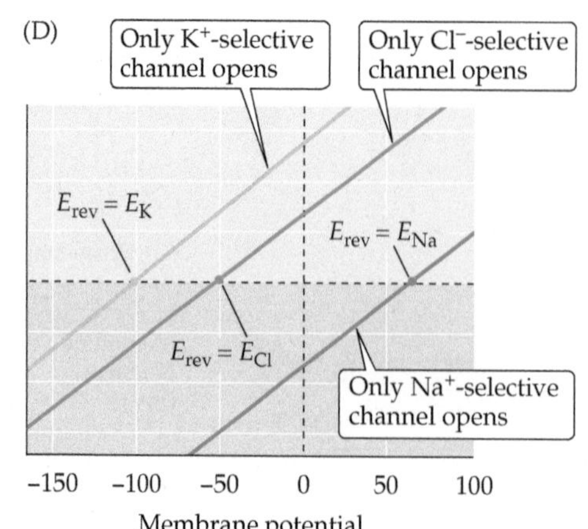

FIGURE 5.16 Influence of the postsynaptic membrane potential on end plate currents. (A) A postsynaptic muscle fiber is voltage clamped using two electrodes, while the presynaptic neuron is electrically stimulated to cause the release of ACh from presynaptic terminals. This experimental arrangement allows the recording of EPCs produced by ACh. (B) Amplitude and time course of EPCs generated by stimulating the presynaptic motor neuron while the postsynaptic cell is voltage clamped at four different membrane potentials. (C) The relationship between the peak amplitude of EPCs and postsynaptic membrane potential is nearly linear, with a reversal potential (the voltage at which the direction of the current changes from inward to outward) close to 0 mV. Also indicated on this graph are the equilibrium potentials of Na^+, K^+, and Cl^- ions. (D) The identity of the ions permeating postsynaptic receptors is revealed by the reversal potential (E_{rev}). Activation of postsynaptic channels permeable only to K^+ (yellow) results in currents reversing at E_K, near −100 mV, while activation of postsynaptic Na^+ channels results in currents reversing at E_{Na}, near +70 mV (red). Cl^--selective currents reverse at E_{Cl}, near −50 mV (green). (A–C after Takeuchi and Takeuchi, 1960.)

potential is the EPP (Figure 5.15C), which typically triggers a postsynaptic action potential by opening voltage-gated Na^+ and K^+ channels (see Figure 5.5B).

The identity of the ions that flow during the EPC can be determined via the same approaches used to identify the roles of Na^+ and K^+ fluxes in the currents underlying action potentials (see Chapter 3). Key to such an analysis is identifying the membrane potential at which no current flows during transmitter action. When the potential of the postsynaptic muscle cell is controlled by the voltage clamp method (Figure 5.16A), the magnitude of the membrane potential clearly affects the amplitude and polarity of EPCs (Figure 5.16B). Thus, when the postsynaptic membrane potential is made more negative than the resting potential, the amplitude of the EPC becomes larger, whereas this current is reduced when the membrane potential is made more positive. At approximately 0 mV, no EPC is detected, and at even more positive potentials, the current reverses its polarity, becoming outward rather than inward (Figure 5.16C). The potential where the EPC reverses, about 0 mV in the case of the neuromuscular junction, is called the **reversal potential**.

As was the case for currents flowing through voltage-gated ion channels (see Chapter 4), the magnitude of the EPC at any membrane potential is given by the product of the ion conductance activated by ACh (g_{ACh}) and the electrochemical driving force on the ions flowing through ligand-gated channels. Thus, the value of the EPC is given by the relationship

$$EPC = g_{ACh}(V_m - E_{rev})$$

where E_{rev} is the reversal potential for the EPC. This relationship predicts that the EPC will be an inward current at potentials more negative than E_{rev} because the electrochemical driving force, $V_m - E_{rev}$, is a negative number. Furthermore, the EPC will become smaller at potentials approaching E_{rev} because the driving force is reduced. At potentials more positive than E_{rev}, the EPC is outward because the driving force is reversed in direction (that is, positive). Because the channels opened by ACh are largely insensitive to membrane voltage, g_{ACh} will depend only on the number of channels opened by ACh, which depends in turn on the concentration of ACh in the synaptic cleft. Thus, the magnitude and polarity of the postsynaptic membrane potential determine the direction and amplitude of the EPC solely by altering the driving force on ions flowing through the receptor channels opened by ACh.

When V_m is at the reversal potential, $V_m - E_{rev}$ is equal to 0 and there is no net driving force on the ions that can permeate the receptor-activated channel. The identity of the ions that flow during the EPC can be deduced by observing how the reversal potential of the EPC compares with the equilibrium potential for various ion species (Figure 5.16D). For example, if ACh were to open an ion channel permeable only to K^+, then the reversal potential of the EPC would be at the equilibrium potential for K^+, which for a muscle cell is close to –100 mV. If the ACh-activated channels were permeable only to Na^+, then the reversal potential of the current would be approximately +70 mV, the Na^+ equilibrium potential of muscle cells; if these channels were permeable only to Cl^-, then the reversal potential would be approximately –50 mV. By this reasoning, ACh-activated channels cannot be permeable to only one of these ions, because the reversal potential of the EPC is not near the equilibrium potential for any of them (see Figure 5.16C). However, if these channels were permeable to both Na^+ and K^+, then the reversal potential of the EPC would be between +70 mV and –100 mV.

The fact that EPCs reverse at approximately 0 mV is therefore consistent with the idea that ACh-activated ion channels are almost equally permeable to both Na^+ and K^+. This hypothesis was tested in 1960, by the Japanese husband and wife team of Akira and Noriko Takeuchi, by altering the extracellular concentration of these two ions. As predicted, the magnitude and reversal potential of the EPC were changed by altering the concentration gradient of each ion. Lowering the external Na^+ concentration, which makes E_{Na} more negative, produces a negative shift in E_{rev} (Figure 5.17A), whereas elevating external K^+

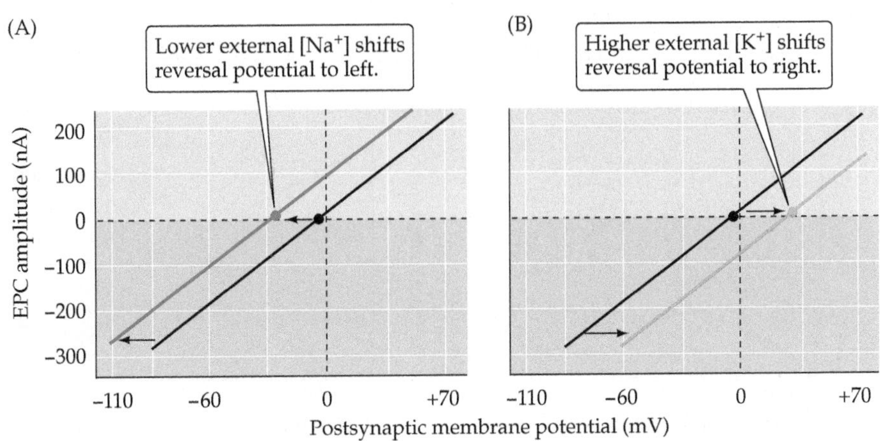

FIGURE 5.17 Reversal potential of the end plate current changes when ion gradients change. (A) Lowering the external Na^+ concentration causes EPCs to reverse at more negative potentials. (B) Raising the external K^+ concentration makes the reversal potential more positive. (After Takeuchi and Takeuchi, 1960.)

concentration, which makes E_K more positive, causes E_{rev} to shift to a more positive potential (Figure 5.17B). Such experiments establish that the ACh-activated ion channels are in fact permeable to both Na^+ and K^+.

Relationship between Ion Fluxes and Postsynaptic Potential Changes

Defining the ion fluxes occurring during the EPC permits understanding of how these ion fluxes generate the EPP. If the membrane potential of the muscle fiber is kept at E_K (approximately –100 mV), the EPC will arise entirely from an influx of Na^+ because at this potential there is no driving force on K^+. In the absence of a voltage clamp to prevent postsynaptic membrane potential changes, such an influx of Na^+ would cause a large depolarization and yield a large depolarizing EPP (Figure 5.18A). At the usual resting membrane potential of –90 mV, there is a small driving force on K^+, but a much greater one on Na^+. This means that much more Na^+ flows into the muscle cell than K^+ flows out (Figure 5.18B, left); the net influx of cations causes an EPC somewhat smaller than that measured at –100 mV and yields a depolarizing EPP that is also somewhat smaller than the EPP measured at –100 mV (Figure 5.18B, right). Thus, at the resting membrane potential, the EPP is generated primarily by Na^+ influx, along with a small efflux of K^+. At the reversal potential of 0 mV, Na^+ influx and K^+ efflux are exactly balanced, so no net current flows during the opening of channels by ACh binding (Figure 5.18C). This yields neither an EPC nor an EPP. At potentials more positive than E_{rev} the balance reverses; for example, at E_{Na} there

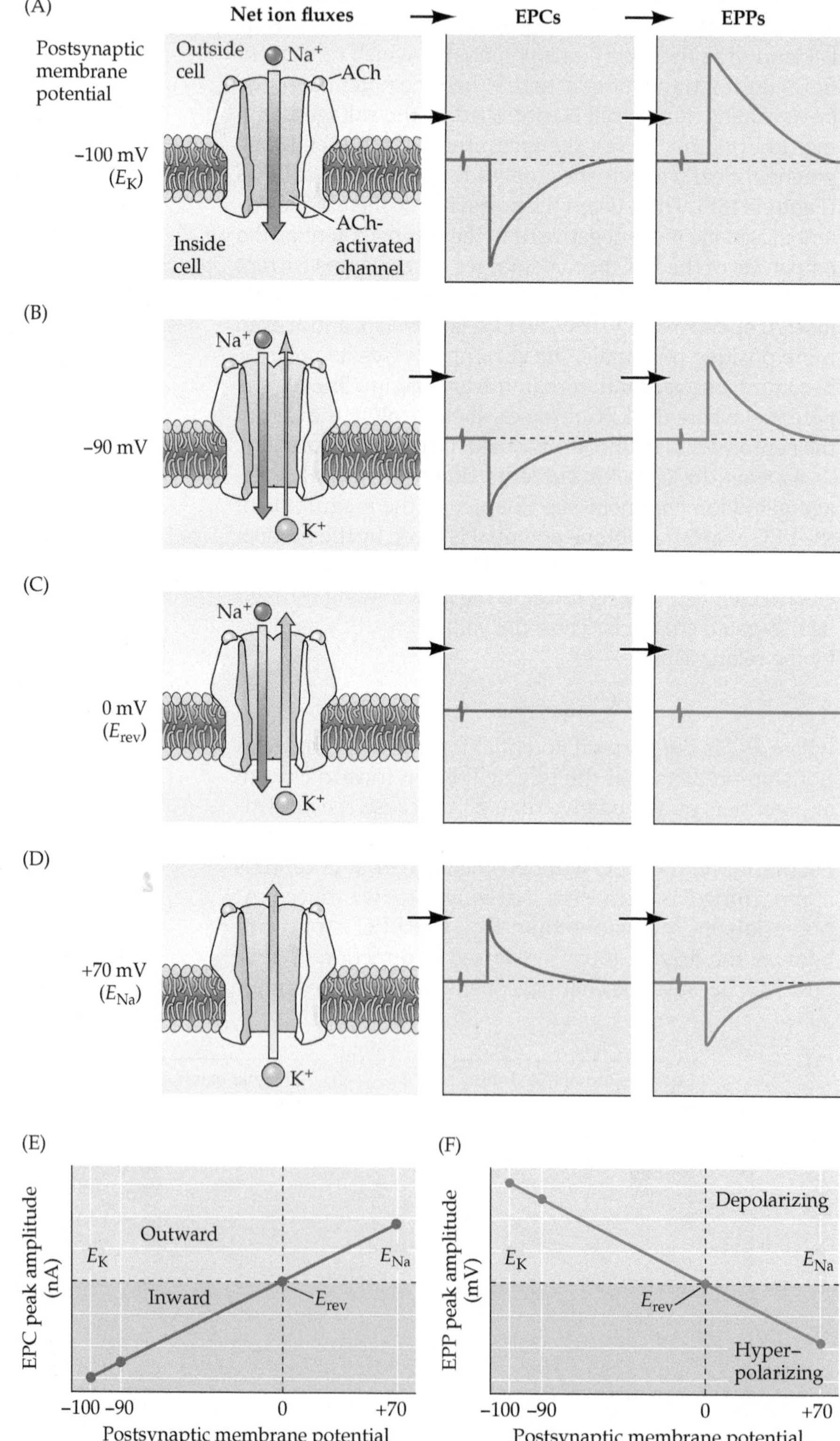

FIGURE 5.18 Na^+ and K^+ movements during EPCs and EPPs. (A–D) Each of the postsynaptic potentials indicated at the left results in different relative fluxes of Na^+ and K^+ (net ion fluxes). These ion fluxes determine the amplitude and polarity of the EPCs, which in turn determine the EPPs. Note that at about 0 mV the Na^+ flux is exactly balanced by an opposite K^+ flux, resulting in no net current flow, and hence no change in the membrane potential. (E) EPCs are inward at potentials more negative than E_{rev} and outward at potentials more positive than E_{rev}. (F) EPPs depolarize the postsynaptic cell at potentials more negative than E_{rev}. At potentials more positive than E_{rev}, EPPs hyperpolarize the cell.

is no influx of Na^+ and a large efflux of K^+ because of the large driving force on K^+ (Figure 5.18D). This produces an outward EPC and a hyperpolarizing EPP. In summary, the polarity and magnitude of the EPC (Figure 5.18E) depend on the electrochemical driving force on the permeant ions, which in turn determines the polarity and magnitude of the EPP (Figure 5.18F). EPPs will depolarize when the membrane potential is more negative than E_{rev}, and hyperpolarize when the membrane potential is more positive than E_{rev}. The general rule, then, is that *the action of a transmitter drives the postsynaptic membrane potential toward* E_{rev} *for the particular ion channels being activated.*

Although this discussion has focused on the neuromuscular junction, similar mechanisms generate postsynaptic responses at all chemical synapses: Transmitter binding to postsynaptic receptors produces a postsynaptic conductance change as ion channels are opened (or sometimes closed). The postsynaptic conductance is increased if—as at the neuromuscular junction—channels are opened, and is decreased if channels are closed. This conductance change typically generates an electrical current, the **postsynaptic current** (**PSC**), which in turn changes the postsynaptic membrane potential to produce a **postsynaptic potential** (**PSP**). As in the specific case of the EPP at the neuromuscular junction, PSPs are depolarizing if their reversal potential is more positive than the resting membrane potential and hyperpolarizing if their reversal potential is more negative.

The conductance changes and the PSPs that typically accompany them are the ultimate outcome of most chemical synaptic transmission, concluding a sequence of electrical and chemical events that begins with the invasion of an action potential into the terminals of a presynaptic neuron. In many ways, the events that produce PSPs at synapses are similar to those that generate action potentials in axons; in both cases, conductance changes produced by ion channels lead to ion current flow that changes the membrane potential.

Excitatory and Inhibitory Postsynaptic Potentials

PSPs ultimately alter the probability that an action potential will be produced in the postsynaptic cell. At the neuromuscular junction, synaptic action increases the probability that an action potential will occur in the postsynaptic muscle cell; indeed, the large amplitude of the EPP ensures that an action potential always is triggered. At many other synapses, PSPs similarly increase the probability of firing a postsynaptic action potential. However, still other synapses actually *decrease* the probability that the postsynaptic cell will generate an action potential. PSPs are called **excitatory** (or **EPSPs**) if they increase the likelihood of a postsynaptic action potential occurring, and **inhibitory** (or **IPSPs**) if they decrease this likelihood. Given that most neurons receive inputs from both excitatory and inhibitory synapses, it is important to understand more precisely the mechanisms that determine whether a particular synapse excites or inhibits its postsynaptic partner.

The principles of excitation just described for the neuromuscular junction are pertinent to all excitatory synapses. The principles of postsynaptic inhibition are much the same as for excitation, and they are also quite general. In both cases, neurotransmitters binding to receptors open or close ion channels in the postsynaptic cell. Whether a postsynaptic response is an EPSP or an IPSP depends on the type of channel that is coupled to the receptor, and on the concentration of permeant ions inside and outside the cell. In fact, the only distinction between postsynaptic excitation and inhibition is the reversal potential of the PSP in relation to the threshold voltage for generating action potentials in the postsynaptic cell.

Consider, for example, a neuronal synapse that uses glutamate as the transmitter. Many such synapses have receptors that, like the ACh receptors at neuromuscular synapses, open ion channels that are nonselectively permeable to cations (see Chapter 6). When these glutamate receptors are activated, both Na^+ and K^+ flow across the postsynaptic membrane, yielding an E_{rev} of approximately 0 mV for the resulting postsynaptic current. If the resting potential of the postsynaptic neuron is –60 mV, the resulting EPSP will depolarize by bringing the postsynaptic membrane potential toward 0 mV. For the hypothetical neuron shown in Figure 5.19A, the action potential threshold voltage is –40 mV. Thus, a glutamate-induced EPSP will increase the probability that this neuron produces an action potential, defining the synapse as excitatory.

As an example of inhibitory postsynaptic action, consider a neuronal synapse that uses GABA as its transmitter. At such synapses, the GABA receptors typically open channels that are selectively permeable to Cl^-, and the action of GABA causes Cl^- to flow across the postsynaptic membrane. Consider a case where E_{Cl} is –70 mV, as is the case for some neurons, so that the postsynaptic resting potential of –60 mV is less negative than E_{Cl}. The resulting positive electrochemical driving force ($V_m - E_{rev}$) will cause negatively charged Cl^- to flow into the cell and produce a hyperpolarizing IPSP (Figure 5.19B). This hyperpolarizing IPSP will take the postsynaptic membrane away from the action potential threshold of –40 mV, clearly inhibiting the postsynaptic cell.

Surprisingly, inhibitory synapses need not produce hyperpolarizing IPSPs. For instance, if E_{Cl} were –50 mV instead of –70 mV, then the negative electrochemical driving force would cause Cl^- to flow out of the cell and produce a depolarizing IPSP (Figure 5.19C). However, the synapse would still be inhibitory: Given that the reversal potential of the IPSP still is more negative than the action potential threshold (–40 mV), the depolarizing IPSP would inhibit

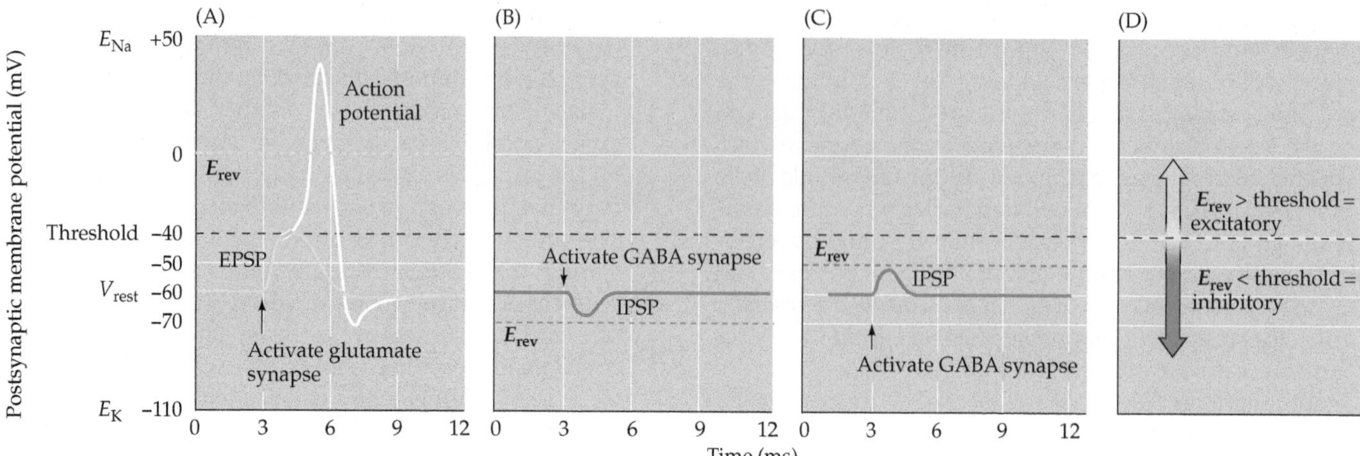

FIGURE 5.19 Reversal potentials and threshold potentials determine postsynaptic excitation and inhibition. (A) If the reversal potential for a PSP (0 mV) is more positive than the action potential threshold (–40 mV), the effect of a transmitter is excitatory, and it generates EPSPs. (B) If the reversal potential for a PSP is more negative than the action potential threshold, the transmitter is inhibitory and generates IPSPs. (C) IPSPs can nonetheless depolarize the postsynaptic cell if their reversal potential is between the resting potential and the action potential threshold. (D) The general rule of postsynaptic action is: If the reversal potential is more positive than threshold, excitation results; inhibition occurs if the reversal potential is more negative than threshold.

because the postsynaptic membrane potential would be kept more negative than the threshold for action potential initiation. Another way to think about this peculiar situation is that if another excitatory input onto this neuron brought the cell's membrane potential to –41 mV—just below the threshold for firing an action potential—the IPSP would then hyperpolarize the membrane potential toward –50 mV, bringing the potential away from the action potential threshold. Thus, while EPSPs depolarize the postsynaptic cell, IPSPs can either hyperpolarize or depolarize; indeed, an inhibitory conductance change may produce no potential change at all and still exert an inhibitory effect by making it more difficult for an EPSP to evoke an action potential in the postsynaptic cell.

Although the particulars of postsynaptic action can be complex, a simple rule distinguishes postsynaptic excitation from inhibition: An EPSP has a reversal potential more positive than the action potential threshold, whereas an IPSP has a reversal potential more negative than threshold (Figure 5.19D). Intuitively, this rule can be understood by realizing that an EPSP will tend to depolarize the membrane potential so that it exceeds threshold, whereas an IPSP will always act to keep the membrane potential more negative than the threshold potential.

Summation of Synaptic Potentials

The PSPs produced at most synapses in the brain are much smaller than those at the neuromuscular junction; indeed, EPSPs produced by individual excitatory synapses may be only a fraction of a millivolt and are usually well below the threshold for generating postsynaptic action potentials. How, then, can such synapses transmit information if their PSPs are subthreshold? The answer is that neurons in the central nervous system are typically innervated by thousands of synapses, and the PSPs produced by each active synapse can *sum together*—in space and in time—to determine the behavior of the postsynaptic neuron.

Consider the highly simplified case of a neuron that is innervated by two excitatory synapses, each generating a subthreshold EPSP, and an inhibitory synapse that produces an IPSP (Figure 5.20A). While activation of either one of the excitatory synapses alone (E1 or E2 in Figure 5.20B) produces a subthreshold EPSP, activation of both excitatory synapses at about the same time causes the two EPSPs to sum together. If the sum of the two EPSPs (E1 + E2) depolarizes the postsynaptic neuron sufficiently to reach the threshold potential, a postsynaptic action potential results. **Summation** thus allows subthreshold EPSPs to influence action potential production. Likewise, an IPSP generated by an inhibitory synapse (I) can sum (algebraically speaking) with a subthreshold EPSP to reduce its amplitude (E1 + I) or can sum with suprathreshold EPSPs to prevent the postsynaptic neuron from reaching threshold (E1 + I + E2).

In short, the summation of EPSPs and IPSPs by a postsynaptic neuron permits a neuron to integrate the electrical information provided by all the inhibitory and excitatory synapses acting on it at any moment. Whether the sum of active synaptic inputs results in the production of an action potential depends on the balance between excitation and inhibition. If the sum of all EPSPs and IPSPs results in a depolarization of sufficient amplitude to

(A)

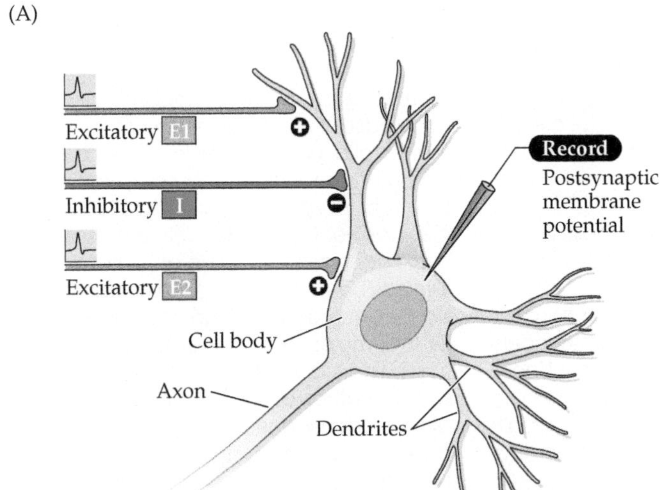

(B)

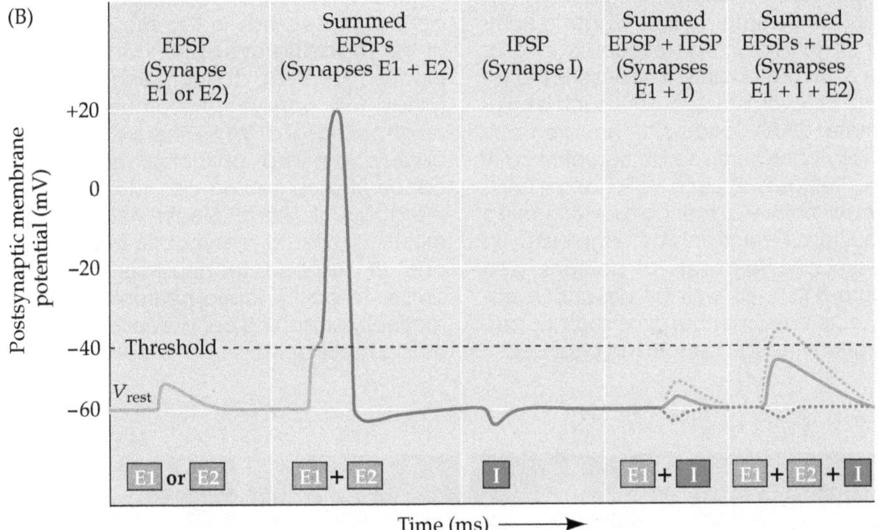

FIGURE 5.20 Summation of postsynaptic potentials. (A) A microelectrode records the postsynaptic potentials produced by the activity of two excitatory synapses (E1 and E2) and an inhibitory synapse (I). (B) Electrical responses to synaptic activation. Stimulating either excitatory synapse (E1 or E2) produces a subthreshold EPSP, whereas stimulating both synapses at the same time (E1 + E2) produces a suprathreshold EPSP that evokes a postsynaptic action potential (shown in blue). Activation of the inhibitory synapse alone (I) results in a hyperpolarizing IPSP. Summing this IPSP (dashed red line) with the EPSP (dashed yellow line) produced by one excitatory synapse (E1 + I) reduces the amplitude of the EPSP (solid orange line), while summing it with the suprathreshold EPSP produced by activating synapses E1 and E2 keeps the postsynaptic neuron below threshold, so that no action potential is evoked.

raise the membrane potential above threshold, then the postsynaptic cell will produce an action potential. Conversely, if inhibition prevails, then the postsynaptic cell will remain silent. Typically, the balance between EPSPs and IPSPs changes continually over time, depending on the number of excitatory and inhibitory synapses active at a given moment and the magnitude of the current at each active synapse. Summation is therefore a tug-of-war between all excitatory and inhibitory postsynaptic currents; the outcome of the contest determines whether or not a postsynaptic neuron fires an action potential and thereby becomes an active element in the neural circuits to which it belongs.

In conclusion, at chemical synapses neurotransmitter release from presynaptic terminals initiates a series of postsynaptic events that culminate in a transient change in the probability of a postsynaptic action potential occurring (Figure 5.21). Such synaptic signaling allows neurons to form the intricate synaptic circuits that play fundamental roles in brain information processing. Recent work further suggests that glial cells also contribute to synaptic signaling, adding another dimension to information processing in the brain (Box 5A).

Summary

Synapses communicate the information carried by action potentials from one neuron to the next in neural circuits. The mechanisms underlying postsynaptic potentials generated during synaptic transmission are closely related to the mechanisms that generate other types of neuronal electrical signals, namely, ionic flow through membrane channels. In the case of electrical synapses, these channels are connexons; direct but passive flow of current through connexons is the basis for transmission. In the case of chemical synapses, channels with smaller and more selective pores are activated by the binding of neurotransmitters to postsynaptic receptors following release of the neurotransmitters from the presynaptic terminal. The large number of neurotransmitters in the nervous system can be divided into two broad classes: small-molecule transmitters and neuropeptides. Neurotransmitters are synthesized from defined precursors by regulated enzymatic pathways, packaged into one of several types of synaptic vesicles, and released into the synaptic cleft in a Ca^{2+}-dependent manner. Transmitter agents are released presynaptically in units, or quanta, reflecting their storage within synaptic vesicles. Vesicles discharge their contents into the synaptic cleft when the presynaptic depolarization generated by the invasion of an action potential opens voltage-gated calcium channels, allowing Ca^{2+} to enter the presynaptic terminal. Calcium triggers neurotransmitter release

BOX 5A ■ The Tripartite Synapse

Up to this point, the discussion has considered signaling between presynaptic neurons and their postsynaptic targets to be a private dialogue between these two cells. However, recent work suggests that this synaptic conversation may involve glial cells as well.

As mentioned in Chapter 1, glial cells support neurons in several ways. For example, it is well established that glia regulate the extracellular environment by removing K^+ that accumulates during action potential generation and by removing neurotransmitters at the conclusion of synaptic transmission. Consistent with such roles, glia seem to occupy virtually all of the non-neuronal volume of the brain. This means that glia are found in very close association with synapses (see Figure A); a given synapse typically is no more than a few hundred nanometers away from a glial cell. Glia form exceedingly fine processes that completely envelop synapses (see Figure B), an intimate association that raises the possibility of a signaling role for glia at synapses.

The first support for such a role came from the discovery that glia respond to application of neurotransmitters. The list of neurotransmitters that elicit responses in glia now includes acetylcholine, glutamate, GABA, and many others. These responses are mediated by the same sorts of neurotransmitter receptors that are employed in synaptic signaling, most often the metabotropic receptors that are coupled to intracellular signaling cascades (see Figure 5.14B). In some cases, neurotransmitters produce changes in the membrane potential of glia. More often, these neurotransmitters cause transient changes in intracellular calcium concentration within the glial cell. These intracellular calcium signals often are observed to trigger calcium waves that spread both within a single glial cell and also propagate between glia (see Figure C).

These transient rises in intracellular calcium serve as second messenger signals (see Chapter 7) that trigger a number of physiological responses in glia. The most remarkable response is the release of several molecules—such as glutamate, GABA, and ATP—that are traditionally considered to be neurotransmitters. Release of such "gliotransmitters" occurs both via the calcium-triggered exocytotic mechanisms employed in presynaptic terminals of neurons (see Figure 5.4C), as well as via unconventional release mechanisms such as permeation through certain ion channels.

The ability both to respond to and to release neurotransmitters potentially makes glia participants in synaptic signaling. Indeed, release of neurotransmitters from a variety of presynaptic terminals has been found to elicit responses in glia. Furthermore, release of gliotransmitters has been found to regulate transmission at numerous synapses (see Figure D). In some cases, glia regulate release of transmitters from presynaptic terminals, while in other cases they alter postsynaptic responsiveness. Glia can also alter the ability of synapses to undergo activity-dependent, plastic changes in synaptic transmission (see Chapter 8).

This ability of glia to participate in synaptic signaling has led to the concept of the **tripartite synapse**, a three-way junction involving the presynaptic terminal, the postsynaptic process, and neighboring glia. While there is still debate about the physiological significance of such neuron–glia interactions, the ability of glia to release neurotransmitters, similar to presynaptic terminals, and to respond to neurotransmitters, similar to postsynaptic neurons, is dramatically changing our view of brain signaling mechanisms.

(A)

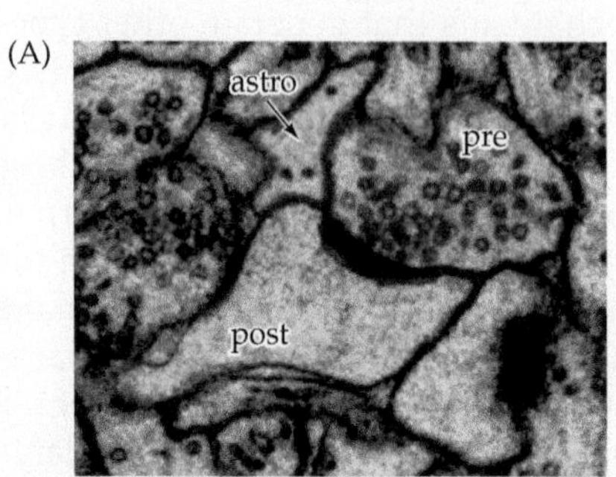

(B)

(C)

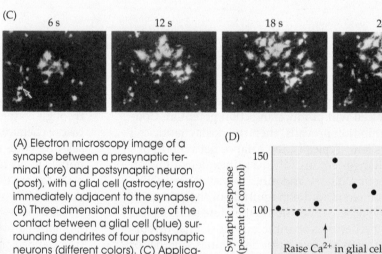

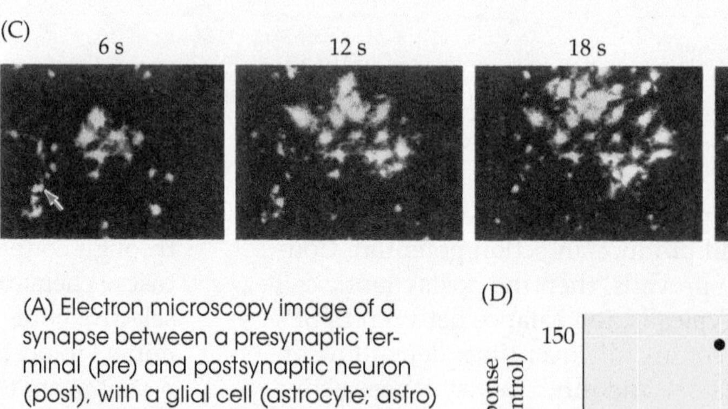

(D)

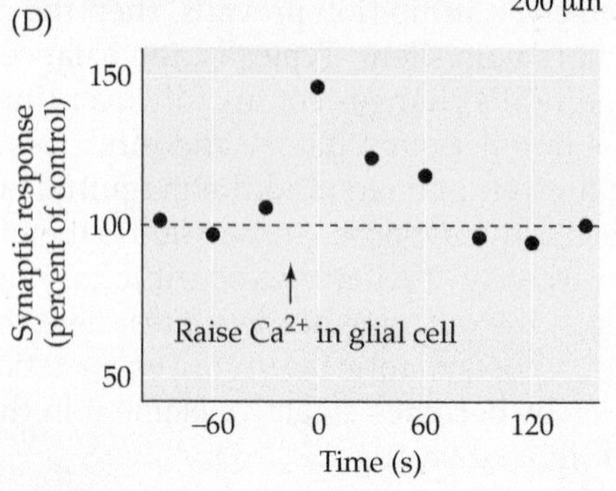

(A) Electron microscopy image of a synapse between a presynaptic terminal (pre) and postsynaptic neuron (post), with a glial cell (astrocyte; astro) immediately adjacent to the synapse. (B) Three-dimensional structure of the contact between a glial cell (blue) surrounding dendrites of four postsynaptic neurons (different colors). (C) Application of glutamate (at arrow) increases intracellular Ca^{2+} (white) in cultured glia. Examination of these cells at different times indicates that the calcium signals spread as a wave through neighboring glia. (D) Transient elevation of Ca^{2+} within a single glial cell (arrow) enhances transmission at an excitatory synapse in the hippocampus. (A,B from Witcher et al., 2007; C from Cornell-Bell et al., 1990; D after Perea and Araque, 2007.)

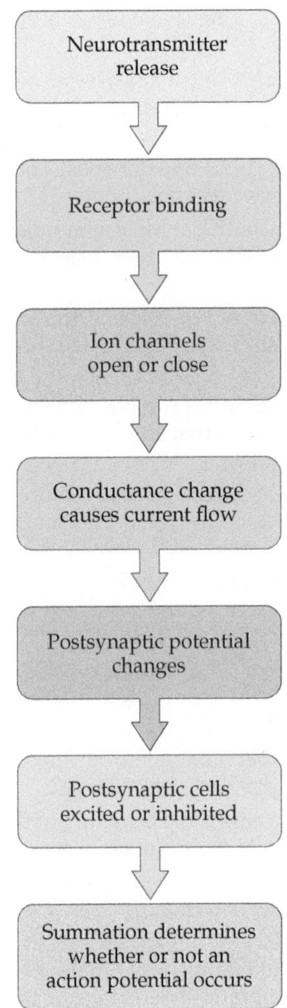

FIGURE 5.21 Overview of postsynaptic signaling. Neurotransmitter released from presynaptic terminals binds to its cognate postsynaptic receptor, which causes the opening or closing of specific postsynaptic ion channels. The resulting conductance change causes current to flow, which may change the membrane potential. The postsynaptic cell sums (or integrates) all of the EPSPs and IPSPs, resulting in moment-to-moment control of action potential generation.

by binding to the Ca^{2+} sensor protein, synaptotagmin, working in concert with SNARE proteins found on both vesicle and plasma membranes. Postsynaptic receptors are a diverse group of proteins that transduce binding of neurotransmitters into electrical signals by opening or closing postsynaptic ion channels. Two broadly different families of neurotransmitter receptors have evolved to carry out the postsynaptic signaling actions of neurotransmitters. The postsynaptic currents produced by the synchronous opening or closing of ion channels change the conductance of the postsynaptic cell, thus increasing or decreasing its excitability. Conductance changes that increase the probability of firing an action potential are excitatory, whereas those that decrease the probability of generating an action potential are inhibitory. Because postsynaptic neurons are usually innervated by many different inputs, the integrated effect of the conductance changes underlying all EPSPs and IPSPs produced in a postsynaptic cell at any moment determines whether or not the cell fires an action potential. The postsynaptic effects of neurotransmitters are terminated by the degradation of the transmitter in the synaptic cleft, by transport of the transmitter back into cells, or by diffusion out of the synaptic cleft. The response elicited at a given synapse depends on the type of neurotransmitter released and the postsynaptic complement of receptors and associated channels.

ADDITIONAL READING

Reviews

Ackermann, F., J. A. Gregory and L. Brodin (2012) Key events in synaptic vesicle endocytosis. Chapter 1 in *Molecular Regulation of Endocytosis*, B. Ceresa (ed.). Rijeka, Croatia: Intech.

Augustine, G. J. and H. Kasai (2006) Bernard Katz, quantal transmitter release, and the foundations of presynaptic physiology. *J. Physiol.* (Lond.) 578: 623–625.

Hestrin, S. and M. Galarreta (2005) Electrical synapses define networks of neocortical GABAergic neurons. *Trends Neurosci.* 28: 304–309.

Lefkowitz, R. J. (2007) Seven transmembrane receptors: something old, something new. *Acta Physiol.* (*Oxf.*) 190: 9–19.

Rizzoli, S. O. (2014) Synaptic vesicle recycling: steps and principles. *EMBO J.* 33: 788–822.

Important Original Papers

Adler, E., M. Adler, G. J. Augustine, M. P. Charlton and S. N. Duffy (1991) Alien intracellular calcium chelators attenuate neurotransmitter release at the squid giant synapse. *J. Neurosci.* 11: 1496–1507.

Augustine, G. J. and R. Eckert (1984) Divalent cations differentially support transmitter release at the squid giant synapse. *J. Physiol.* (Lond.) 346: 257–271.

Beierlein, M., J. R. Gibson and B. W. Connors (2000) A network of electrically coupled interneurons drives synchronized inhibition in neocortex. *Nature Neurosci.* 3: 904–910.

Boyd, I. A. and A. R. Martin (1955) The end-plate potential in mammalian muscle. *J. Physiol.* (Lond.) 132: 74–91.

Burette, A. C. and 6 others (2012) Electron tomographic analysis of synaptic ultrastructure. *J. Comp. Neurol.* 520: 2697–2711.

Curtis, D. R., J. W. Phillis and J. C. Watkins (1959) Chemical excitation of spinal neurons. *Nature* 183: 611–612.

Dale, H. H., W. Feldberg and M. Vogt (1936) Release of acetylcholine at voluntary motor nerve endings. *J. Physiol.* 86: 353–380.

del Castillo, J. and B. Katz (1954) Quantal components of the end plate potential. *J. Physiol.* (Lond.) 124: 560–573.

Fatt, P. and B. Katz (1951) An analysis of the end plate potential recorded with an intracellular electrode. *J. Physiol.* (Lond.) 115: 320–370.

Fatt, P. and B. Katz (1952) Spontaneous subthreshold activity at motor nerve endings. *J. Physiol.* (Lond.) 117: 109–128.

Fotin, A. and 6 others (2004) Molecular model for a complete clathrin lattice from electron cryomicroscopy. *Nature* 432: 573–579.

Furshpan, E. J. and D. D. Potter (1959) Transmission at the giant motor synapses of the crayfish. *J. Physiol.* (Lond.) 145: 289–325.

Harris, B. A., J. D. Robishaw, S. M. Mumby and A. G. Gilman (1985) Molecular cloning of complementary DNA for the alpha subunit of the G protein that stimulates adenylate cyclase. *Science* 229: 1274–1277.

Heuser, J. E. and T. S. Reese (1973) Evidence for recycling of synaptic vesicle membrane during transmitter release at the frog neuromuscular junction. *J. Cell Biol.* 57: 315–344.

Heuser, J. E. and 5 others (1979) Synaptic vesicle exocytosis captured by quick freezing and correlated with quantal transmitter release. *J. Cell Biol.* 81: 275–300.

Imig, C. and 8 others (2014) The morphological and molecular nature of synaptic vesicle priming at presynaptic active zones. *Neuron* 84: 416–431.

Loewi, O. (1921) Über humorale Übertragbarkeit der Herznerven-wirkung. *Pflügers Arch.* 189: 239–242.

Maeda, S. and 6 others (2009) Structure of the connexin 26 gap junction channel at 3.5 Å resolution. *Nature* 458: 597–602.

Miledi, R. (1973) Transmitter release induced by injection of calcium ions into nerve terminals. *Proc. R. Soc. Lond. B* 183: 421–425.

Neher, E. and B. Sakmann (1976) Single-channel currents recorded from membrane of denervated frog muscle fibres. *Nature* 260: 799–802.

Reubold, T. F. and 12 others (2015) Crystal structure of the dynamin tetramer. *Nature* 525: 404–408.

Smith, S. J., J. Buchanan, L. R. Osses, M. P. Charlton and G. J. Augustine (1993) The spatial distribution of calcium signals in squid presynaptic terminals. *J. Physiol.* (Lond.) 472: 573–593.

Sotelo, C., R. Llinas and R. Baker (1974) Structural study of inferior olivary nucleus of the cat: morphological correlates of electrotonic coupling. *J. Neurophysiol.* 37: 541–559.

Sutton, R. B., D. Fasshauer, R. Jahn and A. T. Brünger (1998) Crystal structure of a SNARE complex involved in synaptic exocytosis at 2.4 Å resolution. *Nature* 395: 347–353.

Takamori, S. and 21 others (2006) Molecular anatomy of a trafficking organelle. *Cell* 127: 831–846.

Takeuchi, A. and N. Takeuchi (1960) On the permeability of end-plate membrane during the action of transmitter. *J. Physiol.* (Lond.) 154: 52–67.

Wickman, K. and 7 others (1994) Recombinant $G_{\beta\gamma}$ activates the muscarinic-gated atrial potassium channel I_{KACh}. *Nature* 368: 255–257.

Zhou, Q. and 19 others (2015) Architecture of the synaptotagmin-SNARE machinery for neuronal exocytosis. *Nature* 525: 62–67.

Books

Katz, B. (1969) *The Release of Neural Transmitter Substances.* Liverpool: Liverpool University Press.

Nestler, E., S. Hyman, D. M. Holtzman and R. Malenka (2015) *Molecular Neuropharmacology: A Foundation for Clinical Neuroscience,* 3rd Edition. New York: McGraw Hill.

Peters, A., S. L. Palay and H. deF. Webster (1991) *The Fine Structure of the Nervous System: Neurons and Their Supporting Cells,* 3rd Edition. Oxford: Oxford University Press.

Go to the NEUROSCIENCE 6e Companion Website at **www.oup.com/uk/Purves6e** for Web Topics, Animations, Flashcards, and more.

CHAPTER

6 Neurotransmitters and Their Receptors

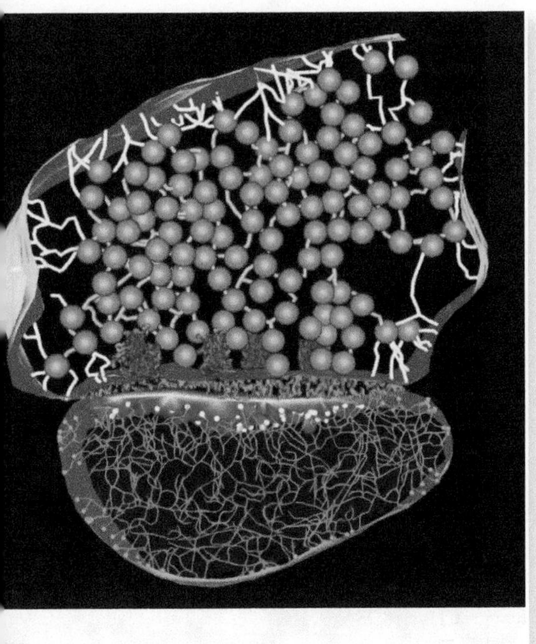

Overview

NEURONS IN THE HUMAN BRAIN COMMUNICATE with one another, for the most part, by releasing chemical messengers called neurotransmitters. A large number of neurotransmitters are known. The main excitatory neurotransmitter in the brain is the amino acid glutamate, while the main inhibitory neurotransmitter is γ-aminobutyric acid (GABA). These and all other neurotransmitters evoke postsynaptic responses by binding to and activating neurotransmitter receptors. Most neurotransmitters are capable of activating several different receptors, yielding many possible modes of synaptic signaling. After activating their postsynaptic receptors, neurotransmitters are removed from the synaptic cleft by neurotransmitter transporters or by degradative enzymes. Abnormalities in the function of neurotransmitter systems contribute to a wide range of neurological and psychiatric disorders; thus, many neuropharmacological therapies are based on drugs that affect neurotransmitters, their receptors, or the mechanisms responsible for removing neurotransmitters from the synaptic cleft.

Categories of Neurotransmitters

There are more than 100 different neurotransmitters. This large number of transmitters allows for tremendous diversity in chemical signaling between neurons. It is useful to separate this panoply of transmitters into two broad categories based simply on size (Figure 6.1). **Neuropeptides**, also called **peptide neurotransmitters**, are relatively large transmitter molecules composed of 3 to 36 amino acids. Individual amino acids, such as glutamate and GABA, as well as the transmitters acetylcholine, serotonin, and histamine, are much smaller than neuropeptides and have therefore come to be called **small-molecule neurotransmitters**. Within the category of small-molecule neurotransmitters, the **biogenic amines** (dopamine, norepinephrine, epinephrine, serotonin, and histamine) are often discussed separately because of their similar chemical properties and postsynaptic actions. The particulars of synthesis, packaging, release, and removal differ for each neurotransmitter (Table 6.1). This chapter describes some of the main features of these transmitters and their postsynaptic receptors.

Acetylcholine

As mentioned in Chapter 5, acetylcholine (ACh) was the first substance identified as a neurotransmitter. In addition to its function as the neurotransmitter at skeletal neuromuscular junctions (see Chapter 5), as well as at the neuromuscular synapse between the vagus nerve and cardiac muscle fibers, ACh serves as a transmitter at

SMALL-MOLECULE NEUROTRANSMITTERS

Acetylcholine $(CH_3)_3\overset{+}{N}-CH_2-CH_2-O-\overset{O}{\overset{\|}{C}}-CH_3$

AMINO ACIDS

Glutamate $H_3\overset{+}{N}-C(H)(CH_2-CH_2-COOH)-COO^-$

Aspartate $H_3\overset{+}{N}-C(H)(CH_2-COOH)-COO^-$

GABA $H_3\overset{+}{N}-CH_2-CH_2-CH_2-COO^-$

Glycine $H_3\overset{+}{N}-C(H)(H)-COO^-$

PURINES

ATP

$O^--P(=O)(O^-)-O-P(=O)(O^-)-O-P(=O)(O^-)-O-CH_2$ (ribose: H, H, OH, OH; adenine: NH_2, N, N, N, N)

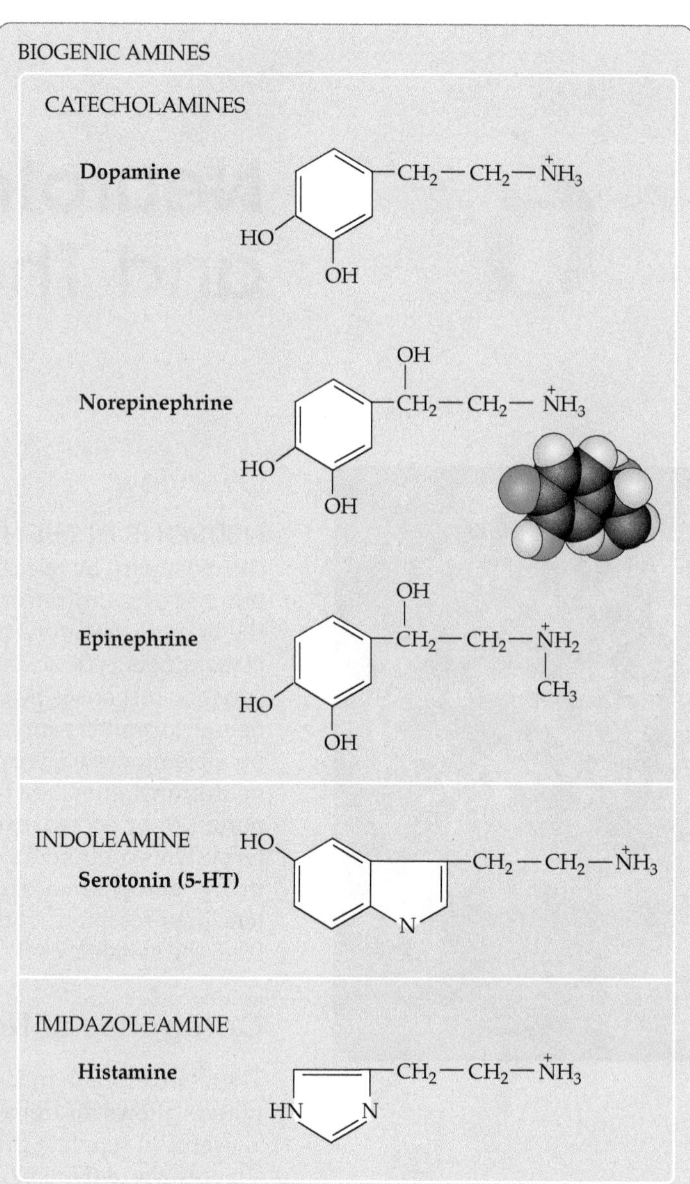

PEPTIDE NEUROTRANSMITTERS (more than 100 peptides, usually 3–36 amino acids long)

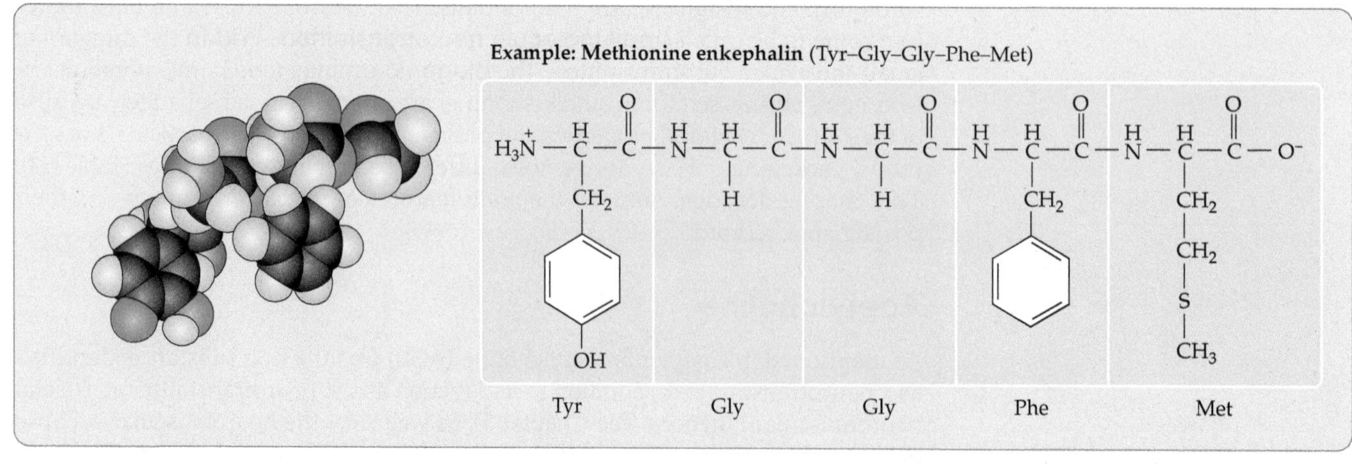

◀ FIGURE 6.1 **Examples of small-molecule and peptide neurotransmitters.** Small-molecule transmitters can be subdivided into acetylcholine, amino acids, purines, and biogenic amines. Size differences between the small-molecule neurotransmitters and the peptide neurotransmitters are indicated by the space-filling models for glycine, norepinephrine, and methionine enkephalin. (Carbon atoms are black, hydrogen gray, nitrogen blue, and oxygen red.)

synapses in the ganglia of the visceral motor system and at a variety of sites in the CNS. Whereas a great deal is known about the function of cholinergic transmission at neuromuscular junctions and ganglionic synapses, the actions of ACh in the CNS are not as well understood.

Acetylcholine is synthesized in nerve terminals from the precursors acetyl coenzyme A (acetyl CoA, which is synthesized from glucose) and choline, in a reaction catalyzed by choline acetyltransferase (ChAT; Figure 6.2). Choline is present in plasma at a high concentration (about 10 m*M*) and is taken up into cholinergic neurons by a high-affinity, Na^+-dependent choline co-transporter (ChT). After synthesis in the cytoplasm of the neuron, a vesicular ACh transporter (VAChT) loads approximately 10,000 molecules of ACh into each cholinergic vesicle. The energy required to concentrate ACh within the vesicle is provided by the acidic pH of the vesicle lumen, which allows the VAChT to exchange H^+ for ACh.

In contrast to most other small-molecule neurotransmitters, the postsynaptic actions of ACh at many cholinergic synapses (the neuromuscular junction in particular) are not terminated by reuptake but by a powerful hydrolytic enzyme, **acetylcholinesterase** (**AChE**). This enzyme is concentrated in the synaptic cleft, ensuring a rapid decrease in ACh concentration after its release from the presynaptic terminal. AChE has a very high catalytic activity (about 5000 molecules of ACh per AChE molecule per second) and rapidly hydrolyzes ACh into acetate and choline. The choline produced by ACh hydrolysis is recycled by being transported back into nerve terminals, where it is used to resynthesize ACh.

Among the many interesting drugs that interact with cholinergic enzymes are the organophosphates. This group includes some potent chemical warfare agents. One such compound is the nerve gas sarin, made notorious in 1995 when a group of terrorists released it in Tokyo's

TABLE 6.1 ■ Functional Features of the Major Neurotransmitters

Neurotransmitter	Postsynaptic effect[a]	Precursor(s)	Rate-limiting step in synthesis	Removal mechanism	Type of vesicle
ACh	Excitatory	Choline + acetyl CoA	ChAT	AChE	Small, clear
Glutamate	Excitatory	Glutamine	Glutaminase	Transporters	Small, clear
GABA	Inhibitory	Glutamate	GAD	Transporters	Small, clear
Glycine	Inhibitory	Serine	Phosphoserine	Transporters	Small, clear
Catecholamines (epinephrine, norepinephrine, dopamine)	Excitatory	Tyrosine	Tyrosine hydroxylase	Transporters, MAO, COMT	Small, dense-core or large, irregular dense-core
Serotonin (5-HT)	Excitatory	Tryptophan	Tryptophan hydroxylase	Transporters, MAO	Large, dense-core
Histamine	Excitatory	Histidine	Histidine decarboxylase	Transporters	Large, dense-core
ATP	Excitatory	ADP	Mitochondrial oxidative phosphorylation; glycolysis	Hydrolysis to AMP and adenosine	Small, clear
Neuropeptides	Excitatory and inhibitory	Amino acids (protein synthesis)	Synthesis and transport	Proteases	Large, dense-core
Endocannabinoids	Inhibits inhibition	Membrane lipids	Enzymatic modification of lipids	Hydrolysis by FAAH	None
Nitric oxide	Excitatory and inhibitory	Arginine	Nitric oxide synthase	Spontaneous oxidation	None

[a]The most common postsynaptic effect is indicated; the same transmitter can elicit postsynaptic excitation or inhibition, depending on the nature of the receptors and ion channels activated by transmitter.

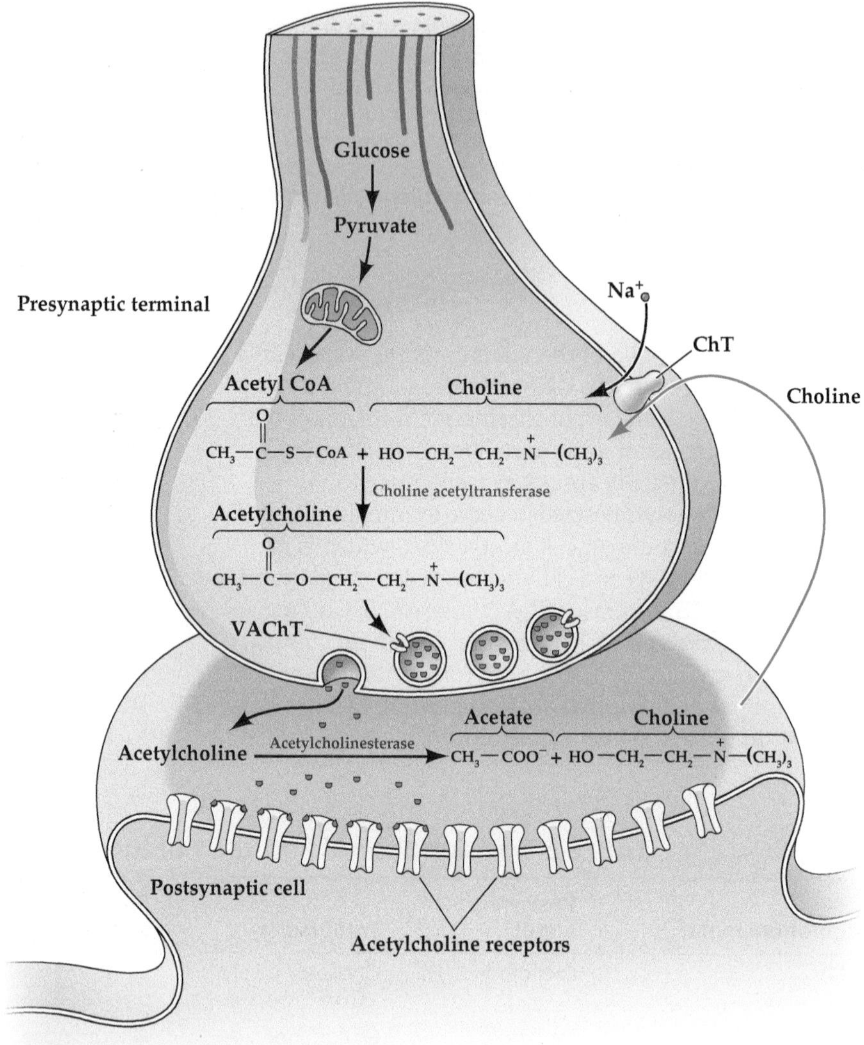

FIGURE 6.2 Acetylcholine metabolism in cholinergic nerve terminals. The synthesis of acetylcholine from choline and acetyl CoA requires choline acetyltransferase. Acetyl CoA is derived from pyruvate generated by glycolysis, while choline is transported into the terminals via an Na^+-dependent co-transporter (ChT). Acetylcholine is loaded into synaptic vesicles via a vesicular transporter (VAChT). After release, acetylcholine is rapidly metabolized by acetylcholinesterase, and choline is transported back into the terminal via the ChT.

underground rail system. Organophosphates can be lethal because they inhibit AChE, allowing ACh to accumulate at cholinergic synapses. This buildup of ACh depolarizes the postsynaptic muscle cell and renders it refractory to subsequent ACh release, causing neuromuscular paralysis and other effects. The high sensitivity of insects to AChE inhibitors has made organophosphates popular insecticides.

Many of the postsynaptic actions of ACh are mediated by the **nicotinic ACh receptor (nAChR)**, so named because the CNS stimulant nicotine also binds to these receptors. Nicotine consumption produces some degree of euphoria, relaxation, and eventually addiction, effects believed to be mediated by nAChRs. nAChRs are nonselective cation channels that generate excitatory postsynaptic responses, such as those illustrated in Figures 5.15 to 5.18. Several toxins, with remarkably diverse chemical structures, specifically bind to and block nicotinic receptors (Box 6A). The availability of these highly specific ligands—particularly a component of snake venom called α-bungarotoxin—has provided a valuable way to isolate and purify nAChRs. As a result, nAChRs are the best-studied type of ionotropic neurotransmitter receptor, and unraveling their molecular organization has provided deep insights into the workings of ionotropic receptors.

Nicotinic receptors are large protein complexes consisting of five subunits. At the neuromuscular junction, the nAChR contains two α subunits, each of which has a binding site that binds a single molecule of ACh. Both ACh binding sites must be occupied for the receptor to be activated, so only relatively high concentrations of ACh activate these receptors. These subunits also bind other ligands, such as nicotine and α-bungarotoxin. The two α subunits are combined with 3 other subunits from among the four other types of nAChR subunits—β, δ, and either γ or ε—in the ratio 2α:1β:1δ:1γ/ε. Neuronal nAChRs differ

BOX 6A ■ Neurotoxins That Act on Neurotransmitter Receptors

Poisonous plants and venomous animals are widespread in nature. The toxins they produce have been used for a variety of purposes, including hunting, healing, mind altering, and, more recently, research. Many of these toxins have potent actions on the nervous system, often interfering with synaptic transmission by targeting neurotransmitter receptors. The poisons found in some organisms contain a single type of toxin, whereas others contain a mixture of tens or even hundreds of toxins.

Consistent with the esential role of ACh receptors in mediating muscle contraction at neuromuscular junctions in numerous species, a large number of natural toxins interfere with signaling mediated by these receptors. In fact, the classification of nicotinic and muscarinic ACh receptors is based on the sensitivity of these receptors to the toxic plant alkaloids nicotine and muscarine, which activate nicotinic and muscarinic ACh receptors, respectively. Nicotine is derived from the dried leaves of the tobacco plant *Nicotinia tabacum*, and muscarine is from the poisonous red mushroom *Amanita muscaria*. Both toxins are stimulants that produce nausea, vomiting, mental confusion, and convulsions. Muscarine poisoning can also lead to circulatory collapse, coma, and death.

The poison α-bungarotoxin, one of many peptides that together make up the venom of the banded krait (*Bungarus multicinctus*) (Figure A), blocks transmission at neuromuscular junctions and is used by the snake to paralyze its prey. This 74-amino-acid toxin (Figure B) blocks neuromuscular transmission by irreversibly binding to nicotinic ACh receptors, thus preventing ACh from opening postsynaptic ion channels. Paralysis ensues because skeletal muscles can no longer be activated by motor neurons. As a result of its specificity and its high affinity for nACh receptors, α-bungarotoxin has contributed greatly to understanding the ACh receptor. Other snake toxins that block nicotinic ACh receptors are cobra α-neurotoxin and the sea snake peptide erabutoxin. The same strategy used by these snakes to paralyze prey was adopted by South American natives who used curare, a mixture of plant toxins from *Chondrodendron tomentosum*, as an arrowhead poison to immobilize their quarry. Curare also

Continued on the next page

(A)

(C)

(E)

(B) α-bungarotoxin

(D) α-conotoxin Vc1.1

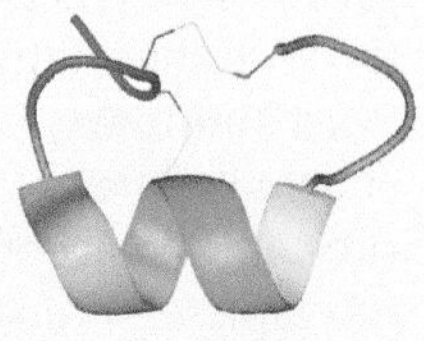

(F) Arecoline

(A) The banded krait (*Bungarus multicinctus*). (B) Structure of α-bungarotoxin produced by the banded krait. (C) A marine cone snail (*Conus* sp.) uses venomous darts to kill a small fish. (D) Structure of α-conotoxin Vc1.1, an nACh receptor blocker that is one of many neurotoxins produced by cone snails. (E) Betel nuts (*Areca catechu*) growing in Malaysia. (F) Structure of arecoline, contained in betel nuts. (A, Robert Zappalorti/Photo Researchers, Inc.; B from Tsetelin, 2015; C, © Alex Kerstitch/Getty Images; D from Lebbe et al., 2014; E, Fletcher & Baylis/Photo Researchers, Inc.)

BOX 6A ■ (*continued*)

blocks nACh receptors; the active agent is the alkaloid δ-tubocurarine.

Another interesting class of animal toxins that selectively block nACh and other receptors includes the peptides produced by fish-hunting marine cone snails (Figure C). These colorful snails kill small fish by "shooting" venomous darts into them. The venom contains hundreds of peptides, known as conotoxins, many of which target proteins that are important in synaptic transmission. Figure D illustrates α-conotoxin Vc1.1, a 14-amino-acid peptide that blocks nACh receptors. Other conotoxin peptides block numerous other types of channels and receptors, including Ca^{2+} channels, Na^+ channels, and glutamate receptors. The various physiological responses produced by these peptides all serve to immobilize any prey unfortunate enough to encounter a cone snail. Many other organisms, including other mollusks, corals, worms, and frogs, also utilize toxins containing specific blockers of ACh receptors.

Other natural toxins have mind- or behavior-altering effects and in some cases have been used for thousands of years by shamans and, more recently, physicians. Two examples are plant alkaloid toxins that block muscarinic ACh receptors: atropine from deadly nightshade (belladonna), and scopolamine from henbane. Because these plants grow wild in many parts of the world, exposure is not unusual, and poisoning by either toxin can be fatal.

Another postsynaptic neurotoxin that, like nicotine, is used as a social drug is found in the betel nut, the seed of the areca palm (*Areca catechu*) (Figure E). Betel nut chewing, although virtually unknown in the United States, is practiced by up to 25% of the population in India, Bangladesh, Ceylon, Malaysia, and the Philippines. Chewing these nuts produces a euphoria caused by arecoline, an alkaloid agonist of nACh receptors (Figure F). Like nicotine, arecoline is an addictive CNS stimulant.

Many other neurotoxins alter transmission at noncholinergic synapses. For example, amino acids found in certain mushrooms, algae, and seeds are potent glutamate receptor agonists. The excitotoxic amino acids kainate, from the red alga *Digenea simplex*, and quisqualate, from the seed of *Quisqualis indica*, are used to distinguish two families of glutamate receptors (see text). Other neurotoxic amino acid activators of glutamate receptors include ibotenic acid and acromelic acid, both found in mushrooms, and domoate, which occurs in algae, seaweed, and mussels. Another large group of peptide neurotoxins blocks glutamate receptors. These include the α-agatoxins from the funnel web spider, NSTX-3 from the orb weaver spider, jorotoxin from the Joro spider, and β-philanthotoxin from wasp venom, as well as many cone snail toxins.

All the toxins discussed so far target excitatory synapses. The inhibitory GABA and glycine receptors, however, have not been overlooked by the exigencies of survival. Strychnine, an alkaloid extracted from the seeds of *Strychnos nux-vomica*, is the only drug known to have specific actions on transmission at glycinergic synapses. Because the toxin blocks glycine receptors, strychnine poisoning causes overactivity in the spinal cord and brainstem, leading to seizures. Strychnine has long been used commercially as a poison for rodents, although alternatives such as the anticoagulant warfarin (Coumadin) are now more popular because they are safer for humans. Neurotoxins that block $GABA_A$ receptors include plant alkaloids such as bicuculline from Dutchman's breeches and picrotoxin from *Anamirta cocculus*. Dieldrin, a commercial insecticide, also blocks these receptors. Like strychnine, these agents are powerful CNS stimulants. Muscimol, a mushroom toxin that is a powerful depressant as well as a hallucinogen, activates $GABA_A$ receptors. A synthetic analogue of GABA, baclofen, is a $GABA_B$ receptor agonist that is used clinically to reduce the frequency and severity of muscle spasms.

Chemical warfare between species has thus given rise to a staggering array of molecules that target synapses throughout the nervous system. Although these toxins are designed to defeat normal synaptic transmission, they have also provided a set of powerful tools to understand postsynaptic mechanisms.

---

from those of muscle in that they (1) lack sensitivity to α-bungarotoxin and (2) comprise only two receptor subunit types (α and β), in a ratio of 3α:2β.

Each subunit of the receptor contains a large extracellular region (which in α subunits contains the ACh binding site) as well as four membrane-spanning domains (Figure 6.3A). The transmembrane domains of the five individual subunits together form a channel with a central membrane-spanning pore (Figure 6.3B,C). The width of this pore (Figure 6.3D) is substantially larger than that of the pores of voltage-gated ion channels (see Figure 4.6), consistent with the relatively poor ability of nACh receptors to discriminate between different cations. Within this pore is a constriction that may represent the gate of the receptor. Binding of ACh to the α subunits is thought to cause a twisting of the extracellular domains of the receptor, which causes some of the receptor transmembrane domains to tilt to open the channel gate and permit ions to diffuse through the channel pore (Figure 6.3E).

In summary, the nACh receptor is a ligand-gated ion channel. The intimate association of the ACh binding sites of this receptor with the pore of the channel permits the rapid response to ACh that is characteristic of nACh receptors. This general arrangement—several receptor subunits coming together to form a ligand-gated ion channel—is characteristic of *all* the ionotropic receptors at fast-acting synapses employing glutamate, GABA, serotonin, and other neurotransmitters. Thus, the nicotinic receptor has served as a paradigm for studies of other ionotropic receptors, at the same time leading to a much deeper appreciation of several neuromuscular diseases. The subunits used to make nAChRs and other types of ionotropic neurotransmitter receptors are summarized in Figure 6.3F.

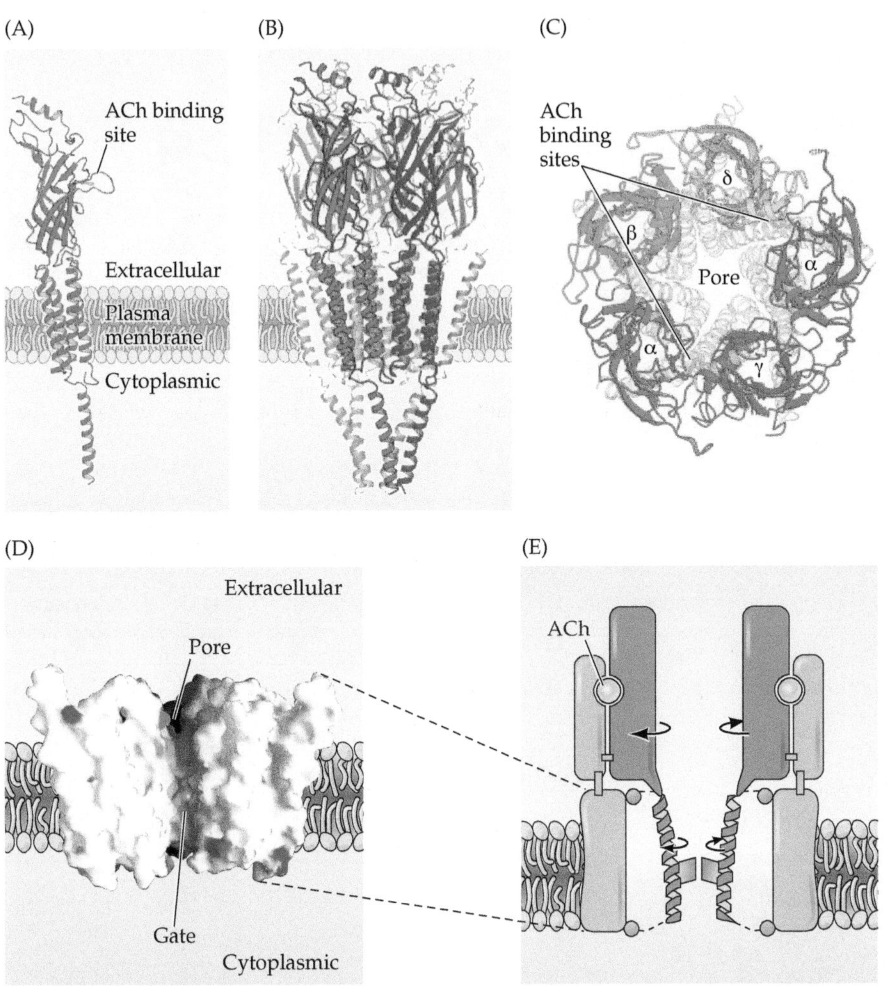

(F)

Receptor	nACh	AMPA	NMDA	Kainate	GABA	Glycine	Serotonin	Purines
Subunits (combination of 3–5 required for each receptor type)	α_{1-10}	GluA1	GluN1	GluK1	α_{1-6}	α_{1-6}	5-HT_{3A}	$P2X_1$
	β_{1-4}	GluA2	GluN2A	GluK2	β_{1-3}	β	5-HT_{3B}	$P2X_2$
	γ	GluA3	GluN2B	GluK3	γ_{1-3}		5-HT_{3C}	$P2X_3$
	δ	GluA4	GluN2C	GluK4	δ		5-HT_{3D}	$P2X_4$
	ε		GluN2D	GluK5	ε		5-HT_{3E}	$P2X_5$
			GluN3A		θ			$P2X_6$
			GluN3B		η			$P2X_7$
					ρ_{1-3}			

FIGURE 6.3 **Structure of the nicotinic ACh receptor.** (A) Structure of the α subunit of the receptor. Each subunit crosses the membrane four times; the α subunit additionally contains a binding site for ACh in its extracellular domain. (B) Five subunits come together to form a complete AChR. (C) View of the AChR from the perspective of the synaptic cleft. The arrangement of the five subunits is evident, with each subunit contributing one transmembrane helix that forms the channel pore. (D) Cross-section view of the transmembrane domain of the AChR. The openings at either end of the channel pore are very large, and the pore narrows at the channel gate. The turquoise sphere indicates the dimension of a sodium ion (0.3 nm diameter). (E) Model for gating of the AChR. Binding of ACh to its binding sites on the two α subunits causes a conformational change in part of the extracellular domain, which causes the pore-forming helices to move and open the pore gate. (F) A diversity of subunits come together to form ionotropic neurotransmitter receptors. (A–C from Unwin, 2005; D,E from Miyazawa et al., 2003.)

A second class of ACh receptors is activated by muscarine, a poisonous alkaloid found in some mushrooms (see Box 6A), and thus they are referred to as **muscarinic ACh receptors (mAChRs)**. mAChRs are metabotropic and mediate most of the effects of ACh in the brain. Like other metabotropic receptors, mAChRs have seven helical membrane-spanning domains (Figure 6.4A). ACh binds to a single binding site on the extracellular surface of the mAChR; this binding site is within a deep channel that is formed by several of the transmembrane helices (Figure 6.4B). Binding of ACh to this site causes a conformational change that permits G-proteins to bind to the cytoplasmic domain of the mAChR, which is only partially shown in Figure 6.4A.

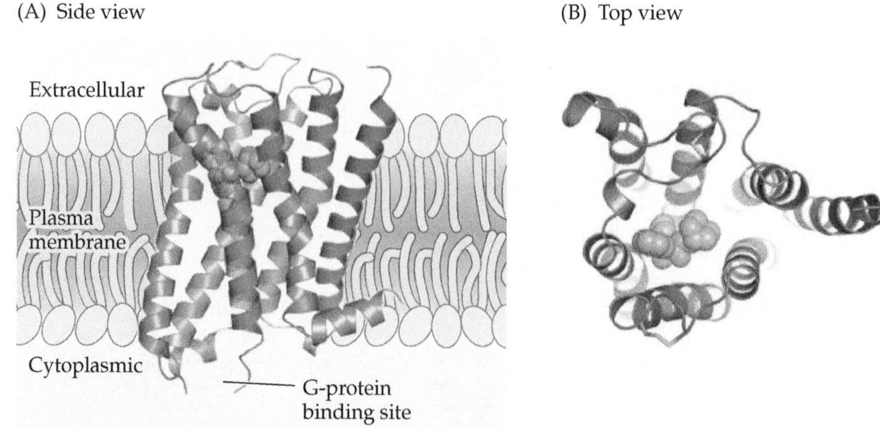

(C)

Receptor class	Muscarinic	Glutamate	$GABA_B$	Dopamine	Adrenergic	Histamine	Serotonin	Purines
Receptor subtype	M_1	Class I	$GABA_{B1}$	D1	Alpha	H_1	$5\text{-}HT_{1A}$	Adenosine
	M_2	$mGlu_1$	$GABA_{B2}$	D2	α_{1A}	H_2	$5\text{-}HT_{1B}$	A_1
	M_3	$mGlu_5$		D3	α_{1B}	H_3	$5\text{-}HT_{1D}$	A_{2A}
	M_4	Class II		D4	α_{1D}	H_4	$5\text{-}HT_{1E}$	A_{2B}
	M_5	$mGlu_2$		D5	α_{2A}		$5\text{-}HT_{1F}$	A_3
		$mGlu_3$			α_{2B}		$5\text{-}HT_{2A}$	P2Y
		Class III			α_{2C}		$5\text{-}HT_{2B}$	$P2Y_1$
		$mGlu_4$			Beta		$5\text{-}HT_{2C}$	$P2Y_2$
		$mGlu_6$			β_1		$5\text{-}HT_4$	$P2Y_4$
		$mGlu_7$			β_2		$5\text{-}HT_{5A}$	$P2Y_6$
		$mGlu_8$			β_3		$5\text{-}HT_6$	$P2Y_{11}$
							$5\text{-}HT_7$	$P2Y_{12}$
								$P2Y_{13}$
								$P2Y_{14}$

FIGURE 6.4 Muscarinic and other metabotropic receptors. (A,B) Structure of the human M_2 mAChR. (A) This receptor spans the plasma membrane seven times and has a cytoplasmic domain (only partially shown here) that binds to and activates G-proteins, as well as an extracellular domain that binds ACh. In this view, the ACh binding site is occupied by 3-quinuclidinyl-benzilate (QNB, colored spheres), a muscarinic receptor antagonist. (B) View of the extracellular surface of the mAChR showing QNB bound to the ACh binding site. (C) Muscarinic and other metabotropic neurotransmitter receptors. (A,B after Haga et al., 2012.)

Five subtypes of mAChRs are known (Figure 6.4C) and are coupled to different types of G-proteins, thereby causing a variety of slow postsynaptic responses. Muscarinic ACh receptors are highly expressed in the corpus striatum and various other forebrain regions, where they activate inward rectifier K^+ channels or Ca^{2+}-activated K^+ channels, thereby exerting an inhibitory influence on dopamine-mediated motor effects. In other parts of the brain, such as the hippocampus, mAChRs are excitatory and act by closing KCNQ-type K^+ channels. These receptors are also found in the ganglia of the peripheral nervous system. Finally, mAChRs mediate peripheral cholinergic responses of autonomic effector organs such as heart, smooth muscle, and exocrine glands and are responsible for the inhibition of heart rate by the vagus nerve. Numerous drugs act as mAChR agonists or antagonists; mAChR blockers that are therapeutically useful include atropine (used to dilate the pupil), scopolamine (effective in preventing motion sickness), and ipratropium (useful in the treatment of asthma).

Glutamate

Glutamate is the most important transmitter for normal brain function. Nearly all excitatory neurons in the CNS are glutamatergic, and it is estimated that more than half of all brain synapses release this neurotransmitter. During brain trauma, there is excessive release of glutamate that can produce *excitotoxic* brain damage.

Glutamate is a nonessential amino acid that does not cross the blood-brain barrier and therefore must be synthesized

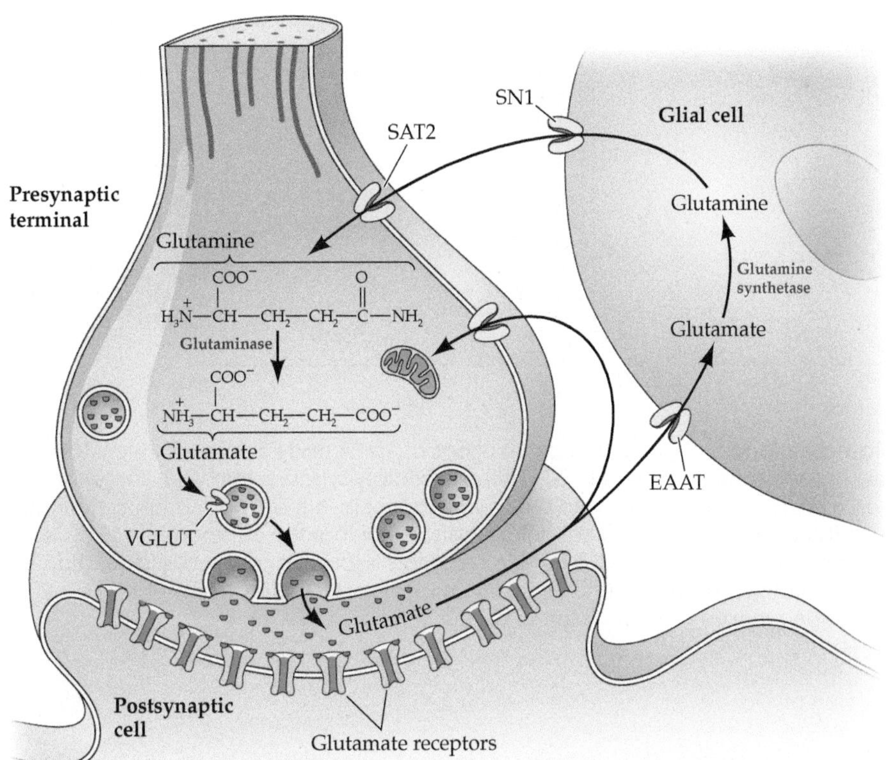

FIGURE 6.5 Glutamate synthesis and cycling between neurons and glia. The action of glutamate released into the synaptic cleft is terminated by uptake into surrounding glial cells (and neurons) via excitatory amino acid transporters (EAATs). Within glial cells, glutamate is converted to glutamine by glutamine synthetase and released by glial cells through the SN1 transporter. Glutamine is taken up into nerve terminals via SAT2 transporters and converted back to glutamate by glutaminase. Glutamate is then loaded into synaptic vesicles via vesicular glutamate transporters (VGLUTs) to complete the cycle.

in neurons from local precursors. The most prevalent precursor for glutamate synthesis is glutamine, which is taken up into presynaptic terminals by the system A transporter 2 (SAT2) and is then metabolized to glutamate by the mitochondrial enzyme glutaminase (Figure 6.5). Glucose metabolized by neurons also can be used to synthesize glutamate by transamination of 2-oxoglutarate, an intermediate of the tricarboxylic acid (Krebs) cycle. Glutamate synthesized in the presynaptic cytoplasm is packaged into synaptic vesicles by vesicular glutamate transporters (VGLUTs). At least three different VGLUT genes have been identified, with different VGLUTs involved in packaging glutamate into vesicles at different types of glutamatergic presynaptic terminals.

Once released, glutamate is removed from the synaptic cleft by the excitatory amino acid transporters (EAATs). EAATs are a family of five different Na^+-dependent glutamate co-transporters. Some EAATs are present in glial cells and others in presynaptic terminals. Glutamate transported into glial cells via EAATs is converted into glutamine by the enzyme glutamine synthetase. Glutamine is then transported out of the glial cells by a different transporter, the system N transporter 1 (SN1), and transported into nerve terminals via SAT2. This overall sequence of events is referred to as the **glutamate-glutamine cycle** (see Figure 6.5). This cycle allows glial cells and presynaptic terminals to cooperate both to maintain an adequate supply of glutamate for synaptic transmission and to rapidly terminate postsynaptic glutamate action.

There are several types of ionotropic glutamate receptors (see Figure 6.3F). **AMPA receptors**, **NMDA receptors**, and **kainate receptors** are named after the agonists that activate them: AMPA (α-amino-3-hydroxyl-5-methyl-4-isoxazole-propionate), NMDA (*N*-methyl-D-aspartate), and kainic acid. All of these receptors are glutamate-gated cation channels that allow the passage of Na^+ and K^+, similar to the nAChR. Hence AMPA, kainate, and NMDA receptor activation always produces excitatory postsynaptic responses.

Most central excitatory synapses possess both AMPA and NMDA receptors. Antagonist drugs that selectively block either AMPA or NMDA receptors are often used to identify synaptic responses mediated by each receptor type. Such experiments reveal that excitatory postsynaptic currents (EPSCs) produced by NMDA receptors are slower and last longer than those produced by AMPA receptors (Figure 6.6A). EPSCs generated by AMPA receptors usually are much larger than those produced by other types of ionotropic glutamate receptors, so that AMPA receptors are the primary mediators of excitatory transmission in the brain. The physiological roles of kainate receptors are less well defined; in some cases, these receptors are found on presynaptic terminals and serve as a feedback mechanism

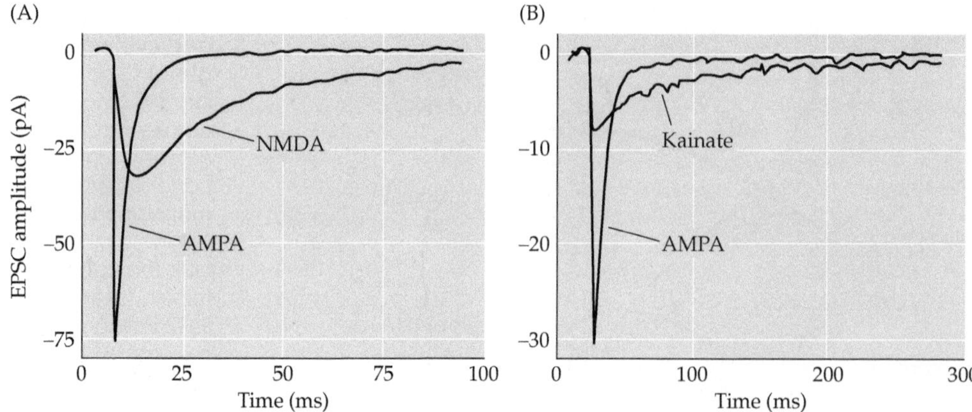

FIGURE 6.6 Postsynaptic responses mediated by ionotropic glutamate receptors. (A) Contributions of AMPA and NMDA receptors to EPSCs at a synapse between a presynaptic pyramidal cell and a postsynaptic interneuron in the visual cortex. Blocking NMDA receptors reveals a large and fast EPSC mediated by AMPA receptors, while blocking AMPA receptors reveals a slower EPSC component mediated by NMDA receptors. (B) Contributions of AMPA and kainate receptors to miniature EPSCs at the excitatory synapse formed between mossy fibers and CA3 pyramidal cells in the hippocampus. Pharmacological antagonists reveal that the component of EPSCs mediated by AMPA receptors is larger and decays faster than that mediated by kainate receptors. (A after Watanabe et al., 2005; B from Mott et al., 2008.)

to regulate glutamate release. When found on postsynaptic cells, kainate receptors generate EPSCs that rise quickly but decay more slowly than those mediated by AMPA receptors (Figure 6.6B).

Like all ionotropic receptors, AMPA receptors are composed of multiple subunits. There are four different AMPA receptor subunits, designated GluA1 to GluA4 (see Figure 6.3F), with each subunit conferring unique functional properties to AMPA receptors. AMPA receptor subunits have several different domains, including an extracellular ligand-binding domain that is responsible for binding glutamate, and a transmembrane domain that forms part of the ion channel (Figure 6.7A). These subunits are organized into the tetrameric structure shown in Figure 6.7B–D. The extracellular structure of AMPA receptors, unlike that of nAChRs, is asymmetrical and therefore looks different when viewed from its front and side surfaces (see Figure 6.7B,C). The AMPA receptor is Y-shaped (see Figure 6.7B), with the large extracellular domains of the subunits narrowing down as the receptor passes through the plasma membrane (see Figure 6.7D). The extracellular ligand-binding domains have a characteristic "clamshell" shape, with glutamate and other ligands binding within the opening of the clamshell (circle in Figure 6.7C). The transmembrane domain consists of helices that form both the channel pore and a gate that occludes the pore when glutamate is not bound to the receptor. Binding of glutamate causes the clamshell structure to "shut"; this movement then causes the gate helices within the transmembrane domain to move and thereby open the channel pore (Figure 6.7E).

NMDA receptors have physiological properties that set them apart from the other ionotropic glutamate receptors. Perhaps most significant is that the pore of the NMDA receptor channel allows the entry of Ca^{2+} in addition to Na^+ and K^+. As a result, excitatory postsynaptic potentials (EPSPs) produced by NMDA receptors increase the concentration of Ca^{2+} in the postsynaptic neuron, with Ca^{2+} then acting as a second messenger to activate intracellular signaling processes (see Chapter 7). Another key property is that Mg^{2+} blocks the pore of this channel at hyperpolarized membrane potentials, while depolarization pushes Mg^{2+} out of the pore (Figure 6.8A). This imparts a peculiar voltage dependence to current flow through the receptor (Figure 6.8B, red line); removing extracellular Mg^{2+} eliminates this behavior (blue line), which demonstrates that Mg^{2+} confers the voltage dependence. Because of this property, NMDA receptors pass cations (most notably Ca^{2+}) only when the postsynaptic membrane potential is depolarized, such as during activation of strong excitatory inputs and/or during action potential firing in the postsynaptic cell. This requirement for the coincident presence of both glutamate and postsynaptic depolarization to open NMDA receptors is widely thought to underlie some forms of synaptic information storage, such as long-term synaptic plasticity (see Chapter 8). Another unusual feature of NMDA receptors is that their gating requires a co-agonist—the amino acid glycine, which is present in the ambient extracellular environment of the brain.

NMDA receptors are tetrameric assemblies of subunits with many similarities to AMPA receptors. There are three groups of NMDA receptor subunits (GluN1, GluN2, and

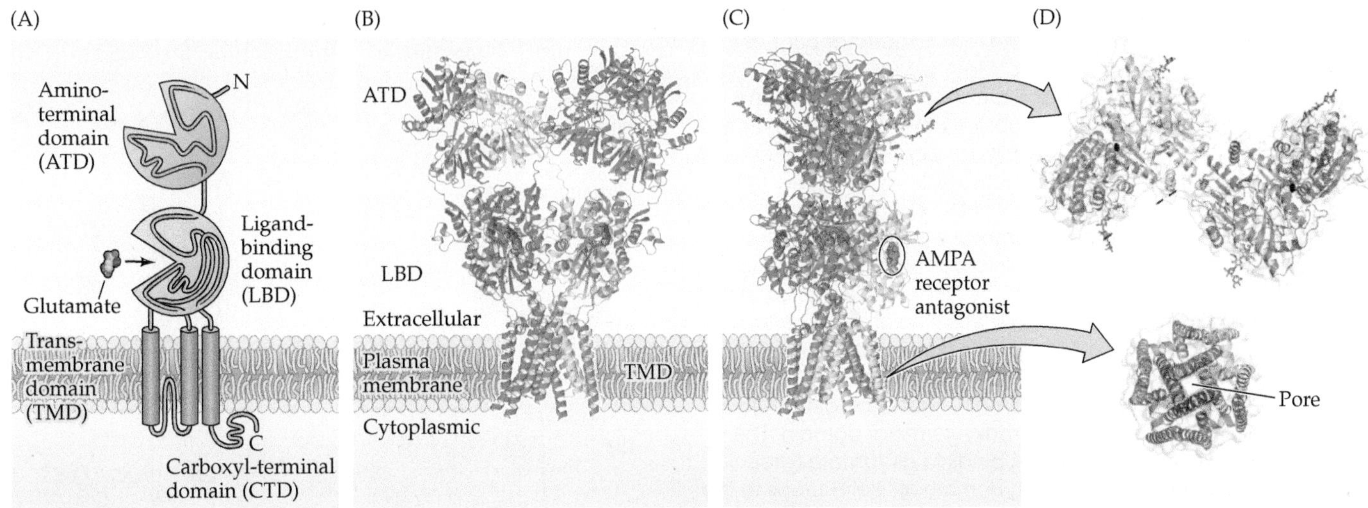

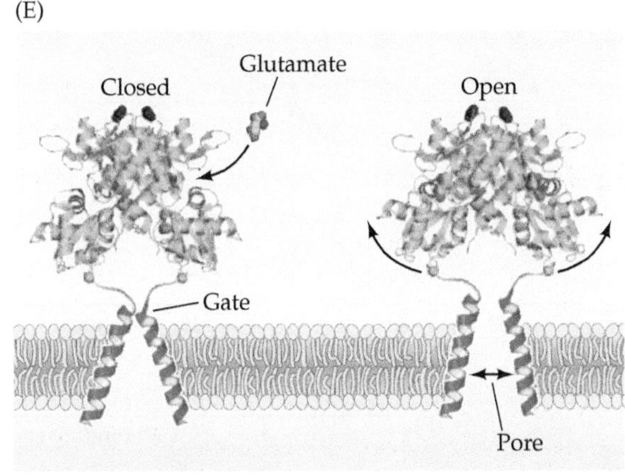

FIGURE 6.7 Structure of the AMPA receptor. (A) Domain structure of an AMPA receptor subunit. The largest part of each subunit is extracellular and consists of two domains, the amino-terminal domain (ATD) and the ligand-binding domain (LBD). In addition, a transmembrane domain (TMD) forms part of the ion channel pore, and an intracellular carboxyl-terminal domain (CTD) connects the receptor to intracellular proteins. (B–D) Crystallographic structure of the AMPA receptor. Each of the four subunits is indicated in a different color. (B) From this perspective, the Y shape of the AMPA receptor is evident. (C) After rotating the receptor 90 degrees, the asymmetrical dimensions of the receptor are evident. One ligand-binding domain is visible and is occupied by an antagonist drug (circled). (D) Cross-section views of the AMPA receptor at two different positions (gray arrows) reveal the spatial relationships between subunits and also illustrate the changes in shape that occur along the length of the receptor. (E) Model for gating of the AMPA receptor by glutamate. The transmembrane domain (blue helices) and part of the extracellular ligand-binding domain are shown. Binding of glutamate closes the clamshell structure of the ligand-binding domain (side arrows), leading to movement of the gate helices that opens the channel pore. (A,E from Traynelis et al., 2010; B–D from Sobolevsky et al., 2009.)

GluN3), with a total of seven different types of subunits (see Figure 6.3F). While GluN2 subunits bind glutamate, GluN1 and GluN3 subunits bind glycine. NMDA receptor tetramers typically comprise two glutamate-binding subunits (GluN2) and two glycine-binding subunits (GluN1). In some cases, GluN3 replaces one of the two GluN2 subunits. This mix of subunits ensures that the receptor binds both to glutamate released from presynaptic terminals and to the ambient glycine co-agonist.

The structure of an NMDA receptor resembles a hot-air balloon (Figure 6.8C,D). Similar to AMPA receptor subunits, NMDA receptor subunits possess clamshell-shaped ligand-binding domains that bind to glutamate and to glycine, as well as transmembrane domains that form the channel pore and gate. One unique feature of NMDA receptors is a structure in the extracellular vestibule, adjacent to the transmembrane domain, that is postulated to bind Ca^{2+} and may help confer Ca^{2+} permeability to NMDA receptors (Figure 6.8E). Gating of NMDA receptors is proposed to arise from closure of the ligand-binding domains upon binding of glutamate and glycine, leading to a conformational change that opens the channel pore; in contrast, binding of antagonists to the ligand-binding domains displaces them and prevents channel opening (Figure 6.8F). The site at which Mg^{2+} binds to block the pore of the NMRA receptor has not yet been identified.

In addition to these ionotropic glutamate receptors, there are three classes of metabotropic glutamate receptors (mGluRs; see Figure 6.4C). These receptors differ in their coupling to intracellular signal transduction pathways (see Chapter 7) and in their sensitivity to pharmacological agents. Activation of many of these receptors leads to inhibition of postsynaptic Ca^{2+} and Na^+ channels. Unlike the excitatory ionotropic glutamate receptors, mGluRs cause slower postsynaptic responses that can either excite or inhibit postsynaptic cells. As a result, the physiological roles of mGluRs are quite varied. Although they possess a transmembrane domain that spans the membrane seven times,

FIGURE 6.8 Function and structure of the NMDA receptor. (A) Voltage-dependent block of the NMDA receptor pore by Mg^{2+}. At hyperpolarized potentials, Mg^{2+} resides within the channel pore and blocks it (left). Depolarization of the membrane potential pushes Mg^{2+} out of the pore, so that current can flow through the NMDA receptor. (B) Voltage dependence of current flowing through NMDA receptors activated by glutamate. In the presence of Mg^{2+} (red), Mg^{2+} block of the channel pore prevents current flow at hyperpolarized membrane potentials. If extracellular Mg^{2+} is removed, there is no block of the channel pore. (C–E) Crystallographic structure of the NMDA receptor. Each of the four subunits is indicated in a different color: GluN1 subunits are orange and yellow, GluN2 subunits cyan and purple. (C) The structure of the NMDA receptor is similar to that of the AMPA receptor, with an amino-terminal domain, ligand-binding domain, transmembrane domain, and carboxyl-terminal domain. The ligand-binding domain of GluN2A binds to glutamate (green spheres), while the ligand-binding domain of GluN1 binds to the co-agonist, glycine (green spheres). (D) Rotating the receptor 90 degrees reveals the location of the putative binding site for Ca^{2+}. (E) Close-up view of the putative Ca^{2+} binding site (red and green mesh) in the extracellular vestibule of the receptor. (F) Model for gating of NMDA receptors. Proposed movements (arrows) of the amino-terminal domain and ligand-binding domain regions of the receptor when bound to antagonists, such as DCKA and D-APV (left), or agonists (glycine and glutamate, right). (C–E from Karakas and Furukawa, 2014; F from Zhu et al., 2016.)

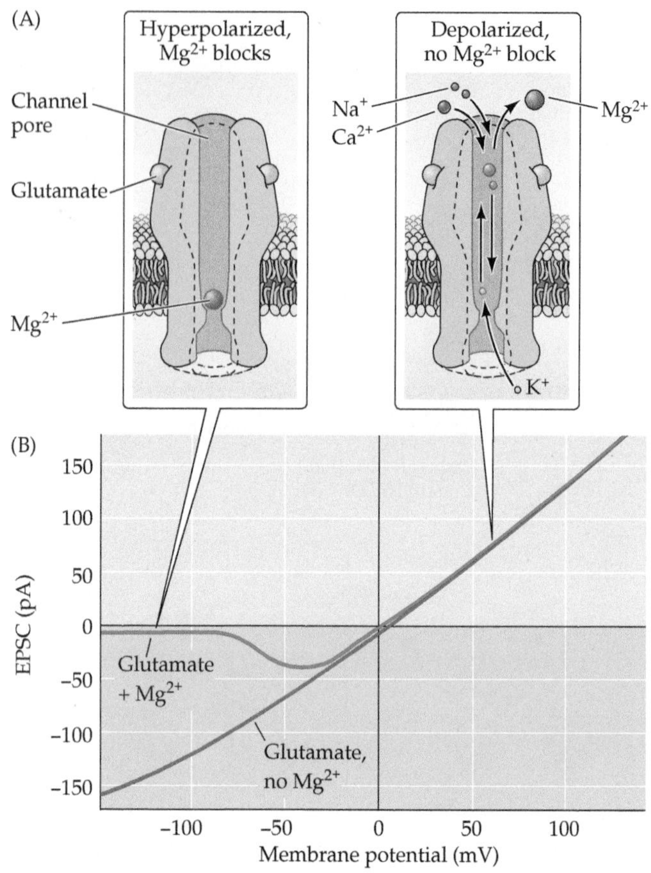

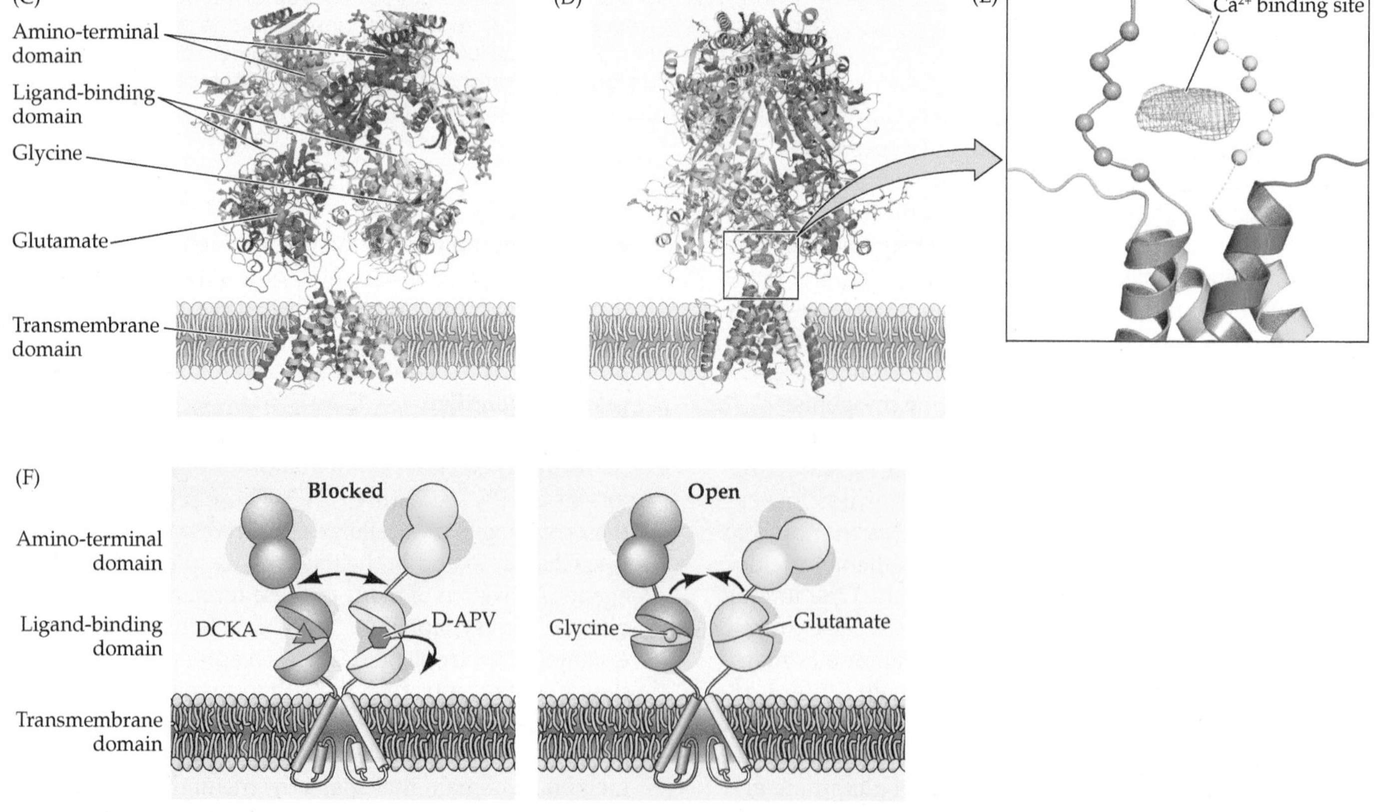

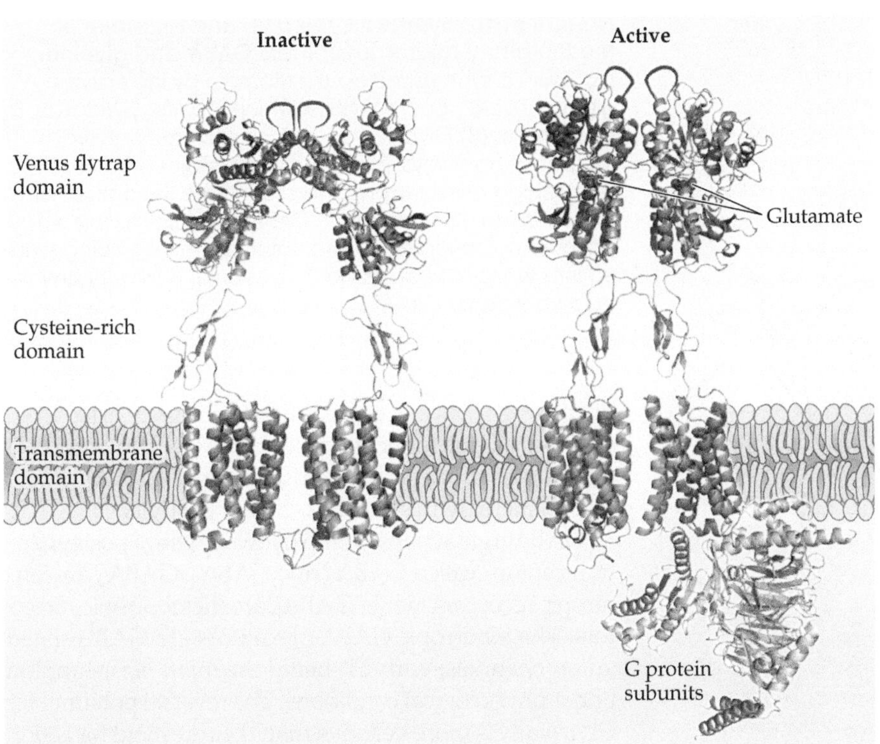

FIGURE 6.9 Structural model of metabotropic glutamate receptors. (Left) Metabotropic glutamate receptors consist of a pair of identical subunits, each containing a venus flytrap domain, a cysteine-rich linker domain, and a transmembrane domain consisting of the canonical seven membrane-spanning helices. (Right) Binding of glutamate (red spheres) to the venus flytrap domains causes the transmembrane domains to rotate and bind to G-proteins, thereby activating intracellular signaling processes. (From Pin and Bettler, 2016.)

characteristic of all G-protein-coupled receptors, mGluRs are structurally unique because they are dimers of two identical subunits (Figure 6.9). Each subunit possesses a *venus flytrap domain,* a glutamate-binding domain similar to the clamshell-shaped ligand-binding domains of ionotropic glutamate receptors (see Figures 6.7 and 6.8). This venus flytrap domain is connected to the transmembrane domain via a linker domain rich in the amino acid cysteine (see Figure 6.9, left). Binding of glutamate causes the venus flytrap domains to close, with the resulting movement causing the transmembrane domains to rotate and thereby activate the receptor (see Figure 6.9, right). Binding of G-proteins to the activated receptor then initiates intracellular signaling.

GABA and Glycine

Most inhibitory synapses in the brain and spinal cord use either γ-aminobutyric acid (GABA) or glycine as neurotransmitters (Figure 6.10). GABA was identified in brain tissue during the 1950s (as was glutamate). It is now known that as many as a third of the synapses in the brain use GABA as their inhibitory neurotransmitter. GABA is most commonly found in local circuit interneurons, although medium spiny neurons of the striatum (see Chapter 18) and cerebellar Purkinje cells (see Chapter 19) are examples of GABAergic projection neurons.

The predominant precursor for GABA synthesis is glucose, which is metabolized to glutamate by the tricarboxylic acid cycle enzymes. (Pyruvate and glutamine can also act as GABA precursors.) The enzyme glutamic acid decarboxylase (GAD), which is found almost exclusively in GABAergic neurons, catalyzes the conversion of glutamate to GABA (Figure 6.10A). GAD requires a co-factor, pyridoxal phosphate, for activity. Because pyridoxal phosphate is derived from vitamin B_6, a deficiency of this vitamin can lead to diminished GABA synthesis. The significance of this fact became clear after a disastrous series of infant deaths was linked to the omission of vitamin B_6 from infant formula. The absence of vitamin B_6 greatly reduced the GABA content of the brain, and the subsequent loss of synaptic inhibition caused seizures that in some cases were fatal. Once GABA is synthesized, it is transported into synaptic vesicles via a vesicular inhibitory amino acid transporter (VIAAT).

The mechanism of GABA removal is similar to that for glutamate: Both neurons and glia contain high-affinity Na^+-dependent co-transporters for GABA. These co-transporters are termed GATs, and several forms of GAT have been identified. Most GABA is eventually converted to succinate, which is metabolized further in the tricarboxylic acid cycle that mediates cellular ATP synthesis. Two mitochondrial enzymes are required for this degradation: GABA transaminase and succinic semialdehyde dehydrogenase. There are also other pathways for degradation of GABA, the most noteworthy of which results in the production of γ-hydroxybutyrate, a GABA derivative that has been abused as a "date rape" drug. Oral administration of γ-hydroxybutyrate can cause euphoria, memory deficits, and

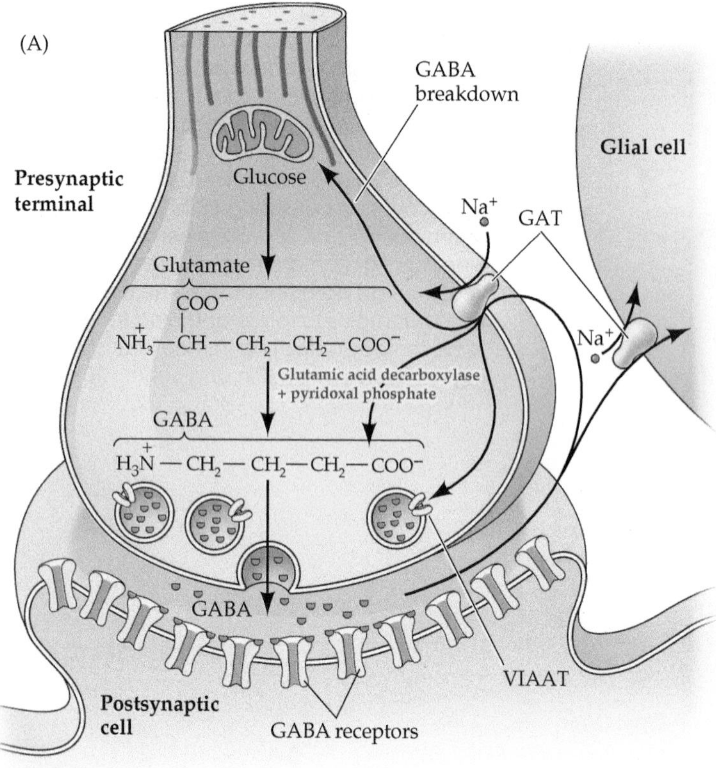

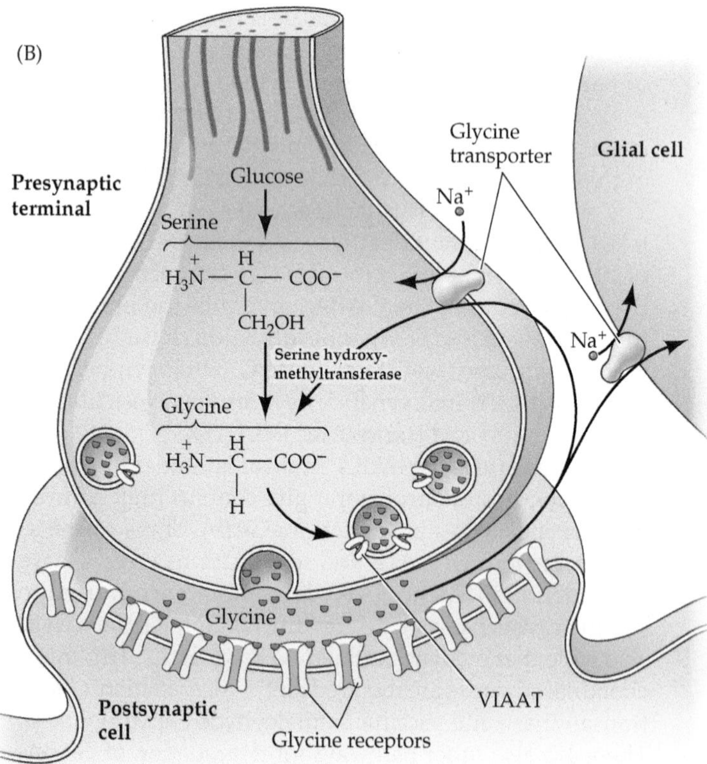

FIGURE 6.10 Synthesis, release, and reuptake of the inhibitory neurotransmitters GABA and glycine. (A) GABA is synthesized from glutamate by the enzyme glutamic acid decarboxylase, which requires pyridoxal phosphate. (B) Glycine can be synthesized by several metabolic pathways; in the brain, the major precursor is serine. High-affinity transporters terminate the actions of these transmitters and return GABA or glycine to the synaptic terminals for reuse, with both transmitters being loaded into synaptic vesicles via the vesicular inhibitory amino acid transporter (VIAAT).

unconsciousness. Presumably these effects arise from actions on GABAergic synapses in the CNS: Inhibition of GABA breakdown causes a rise in tissue GABA content and an increase in synaptic inhibitory activity.

GABAergic synapses employ two types of postsynaptic receptors, called $GABA_A$ and $GABA_B$. $GABA_A$ are ionotropic receptors, while $GABA_B$ are metabotropic receptors. The ionotropic $GABA_A$ receptors are GABA-gated anion channels, with Cl^- being the main permeant ion under physiological conditions. The reversal potential for Cl^- usually is more negative than the threshold for action potential firing (see Figure 5.19) due to the action of the K^+/Cl^- co-transporter (see Figure 4.13D), which keeps intracellular Cl^- concentration low. Thus, activation of these GABA receptors causes an influx of negatively charged Cl^- that inhibits postsynaptic cells (Figure 6.11A). In cases where postsynaptic Cl^- concentration is high—in developing neurons, for example—$GABA_A$ receptors can excite their postsynaptic targets (Box 6B).

Like nACh receptors, $GABA_A$ receptors are pentamers (Figure 6.11B). There are 19 types of $GABA_A$ subunits (see Figure 6.3F); this diversity of subunits causes the composition and function of $GABA_A$ receptors to differ widely among neuronal types. Typically, $GABA_A$ receptors consist of two α subunits, two β subunits, and one other subunit, most often a γ subunit. A specialized type of $GABA_A$ receptor, found exclusively in the retina, consists entirely of ρ subunits and is called the $GABA_{A\rho}$ receptor (formerly the $GABA_C$ receptor). The five $GABA_A$ receptor subunits are assembled into a structure quite similar to that of the nAChR (see Figure 6.11B). The transmembrane domains of the subunits form a central pore that includes a ring of positive charges that presumably serve as the binding site for Cl^- (Figure 6.11C). GABA binds in pockets found at the interface between the extracellular domains of the subunits; many other types of ligands also bind to these sites (Figure 6.11D). Benzodiazepines such as diazepam (Valium) and chlordiazepoxide (Librium) are anxiety-reducing drugs that enhance GABAergic transmission by binding to the extracellular domains of α and δ subunits of $GABA_A$ receptors. The same site

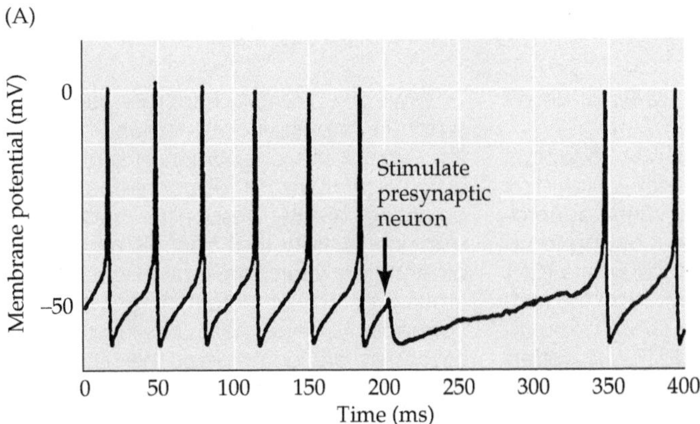

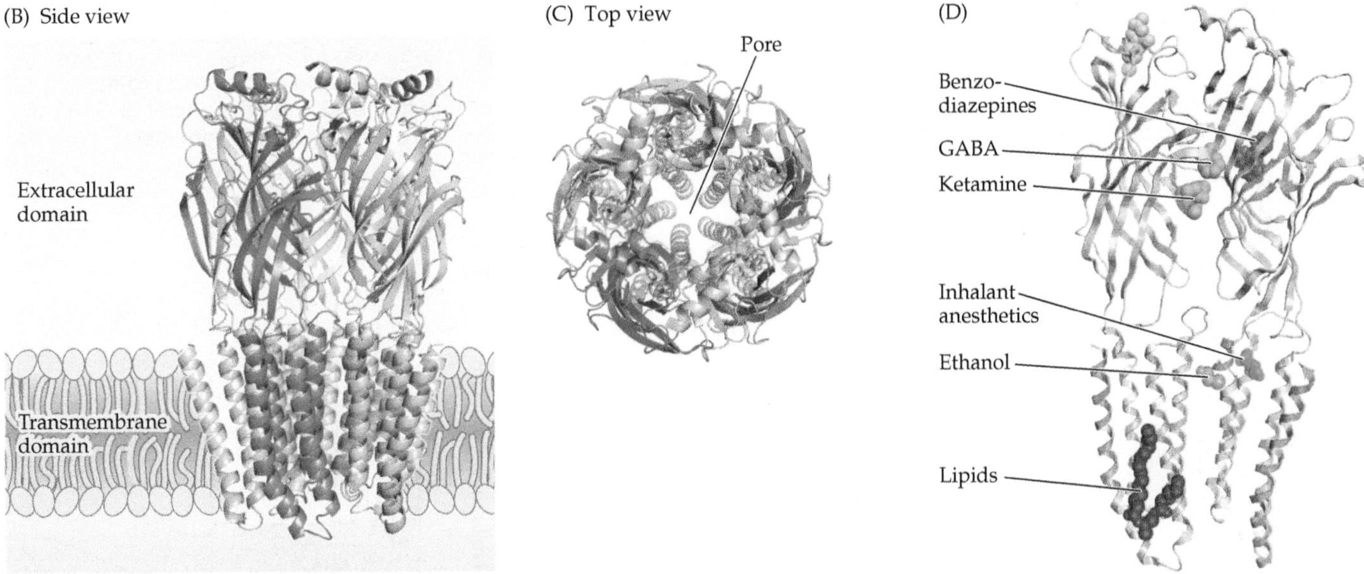

FIGURE 6.11 Ionotropic GABA receptors. (A) Stimulation of a presynaptic GABAergic interneuron, at the time indicated by the arrow, causes a transient inhibition of action potential firing in the postsynaptic target. This inhibitory response is caused by activation of postsynaptic $GABA_A$ receptors. (B–D) Crystallographic structure of a $GABA_A$ receptor. (B) The receptor is formed from five subunits, each containing an extracellular domain and a transmembrane domain. One subunit is highlighted in blue. (C) This extracellular perspective shows the pore formed by the transmembrane domains of the receptor subunits. (D) View of two receptor subunits, indicating the binding sites for numerous ligands. Here the GABA binding site is occupied by benzamidine, a $GABA_A$ receptor agonist. (A after Chavas and Marty, 2003; B,C from Miller and Aricescu, 2014; D from Puthenkalam et al., 2016.)

binds the hypnotic zolpidem (Ambien), which is widely used to induce sleep. Barbiturates such as phenobarbital and pentobarbital are other hypnotics that also bind to the extracellular domains of the α and β subunits of some GABA receptors and potentiate GABAergic transmission; these drugs are used therapeutically for anesthesia and to control epilepsy. The injection anesthetic ketamine also binds to the extracellular domain of GABA receptors (see Figure 6.11D). The transmembrane domains of $GABA_A$ receptors also serve as the targets for numerous ligands, such as inhalant anesthetics and steroids. Another drug that binds to the transmembrane domain of GABA receptors is ethanol; at least some aspects of drunken behavior are caused by ethanol-mediated alterations in ionotropic GABA receptors.

The metabotropic $GABA_B$ receptors are also widely distributed in the brain. Like the ionotropic GABA receptors, $GABA_B$ receptors are inhibitory. Rather than relying on Cl^--selective channels, however, $GABA_B$-mediated inhibition is often due to the activation of K^+ channels. A second action of $GABA_B$ receptors is to block Ca^{2+} channels, which also inhibits postsynaptic cells. The structure of $GABA_B$ receptors is similar to that of other metabotropic receptors, although $GABA_B$ receptors assemble as heterodimers of B1 and B2 subunits (Figure 6.12). Like the mGluRs, $GABA_B$ receptors possess venus flytrap domains (see Figure 6.12, left), but these bind GABA rather than glutamate. Binding of GABA to the venus flytrap domain of the B1 subunit causes this domain to close, leading to conformational changes in the transmembrane domains of both subunits that permit binding of G-proteins (see Figure 6.12, right).

The distribution of the neutral amino acid glycine in the CNS is more restricted than that of GABA. About half of the inhibitory synapses in the spinal cord use glycine; most other inhibitory synapses use GABA. Glycine is synthesized from serine by the mitochondrial isoform of serine hydroxymethyltransferase (see Figure 6.10B) and

BOX 6B ■ Excitatory Actions of GABA in the Developing Brain

Although GABA normally functions as an inhibitory neurotransmitter in the mature brain, in the developing brain GABA excites its target cells. This remarkable reversal of action arises from developmental changes in intracellular Cl^- homeostasis. The mechanisms involved in this switch have been studied most extensively in cortical neurons. In young neurons, intracellular Cl^- concentration is controlled mainly by the $Na^+/K^+/Cl^-$ co-transporter, which pumps Cl^- into the neurons and yields a high $[Cl^-]_i$ (Figure A, left). As the neurons continue to develop, they begin to express a K^+/Cl^- co-transporter that pumps Cl^- out of the neurons and lowers $[Cl^-]_i$ (see Figure A, right). Such shifts in Cl^- homeostasis can cause $[Cl^-]_i$ to drop several-fold over the first 1 to 2 postnatal weeks of development (Figure B).

Because ionotropic GABA receptors are Cl^--permeable channels, ion flux through these receptors varies according to the electrochemical driving force on Cl^-. In the young neurons, where $[Cl^-]_i$ is high, E_{Cl} is more positive than the resting potential. As a result, GABA depolarizes these neurons. In addition, E_{Cl} often is more positive than threshold, so GABA is able to excite these neurons to fire action potentials (Figure C). As described in the text, the lower $[Cl^-]_i$ of mature neurons causes E_{Cl} to be more negative than the action potential threshold (and often more negative than the resting potential), resulting in inhibitory responses to GABA.

Why does GABA undergo such a switch in its postsynaptic actions? While the logic of this phenomenon is not yet completely clear, it appears that depolarizing GABA responses produce electrical activity that controls neuronal proliferation, migration, growth, and maturation, as well as determining synaptic connectivity. Once these developmental processes are completed, the resulting neural circuitry requires inhibitory transmission that can then also be provided by GABA. Further work will be needed to fully appreciate the significance of the excitatory actions of GABA, as well as to understand the mechanisms underlying the expression of the K^+/Cl^- co-transporter that ends the brief career of GABA as an excitatory neurotransmitter.

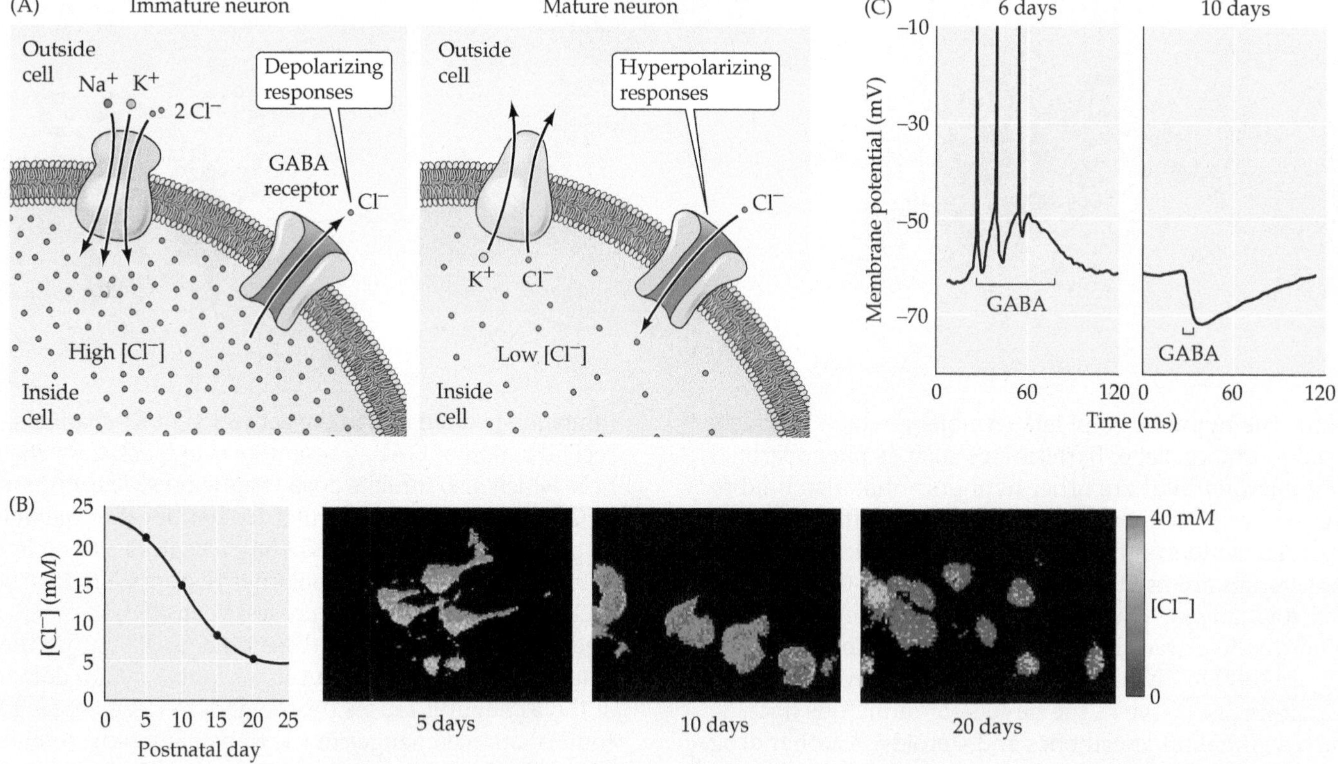

(A) The developmental switch in expression of Cl^- transporters lowers $[Cl^-]_i$, thereby reversing direction of Cl^- flux through GABA receptors. (B) Imaging $[Cl^-]_i$ between postnatal days 5 and 20 (right) demonstrates a progressive reduction in $[Cl^-]_i$ (left). (C) Developmental changes in $[Cl^-]_i$ cause GABA responses to shift from depolarizing in young (6-day-old) neurons (left) to hyperpolarizing in older (10-day-old) neurons (right) cultured from the chick spinal cord. (B from Berglund et al., 2006; C after Obata et al., 1978.)

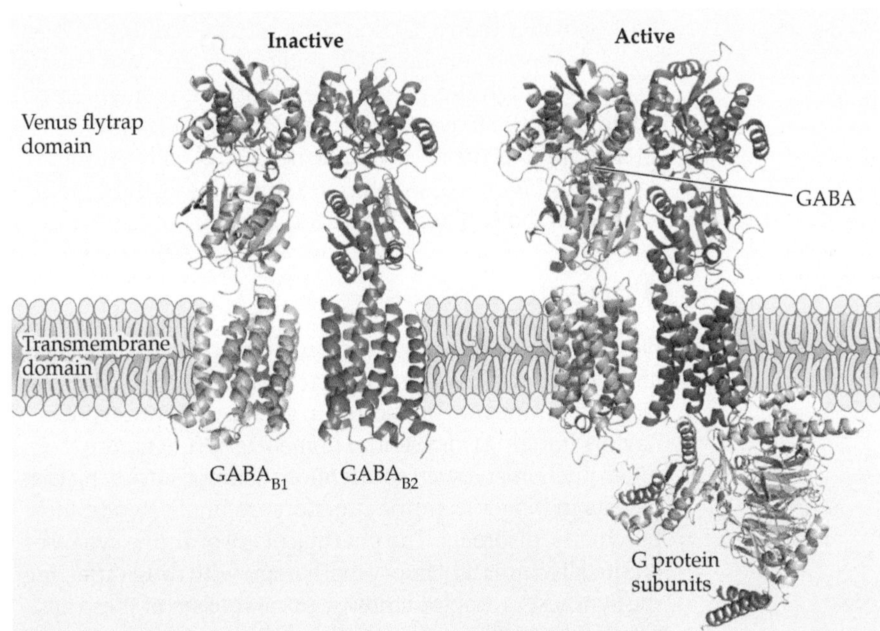

FIGURE 6.12 Structure of metabotropic $GABA_B$ receptors. (Left) Metabotropic $GABA_B$ receptors are heterodimers of B1 and B2 subunits, each containing a venus flytrap domain and a transmembrane domain. (Right) Binding of GABA (green spheres) to the venus flytrap domain of the B1 subunit causes the transmembrane domains to move and bind to G-proteins, thereby activating intracellular signaling processes. (From Pin and Bettler, 2016).

is transported into synaptic vesicles via the same vesicular inhibitory amino acid transporter that loads GABA into vesicles. Once released from the presynaptic cell, glycine is rapidly removed from the synaptic cleft by glycine transporters in the plasma membrane. Mutations in the genes coding for some of these transporters result in hyperglycinemia, a devastating neonatal disease characterized by lethargy, seizures, and mental retardation.

Glycine receptors are pentamers consisting of mixtures of four types of α subunits, along with an accessory β subunit (see Figure 6.3F). These receptors are potently blocked by strychnine, which may account for the toxic properties of this plant alkaloid (see Box 6A). Glycine receptors are ligand-gated Cl^- channels whose general structure closely mirrors that of the $GABA_A$ receptors (Figure 6.13). Gating of glycine receptors by ligands is well understood. Binding of glycine to a ligand-binding site on the extracellular domains causes a conformational change that opens the pore, increasing the pore radius from

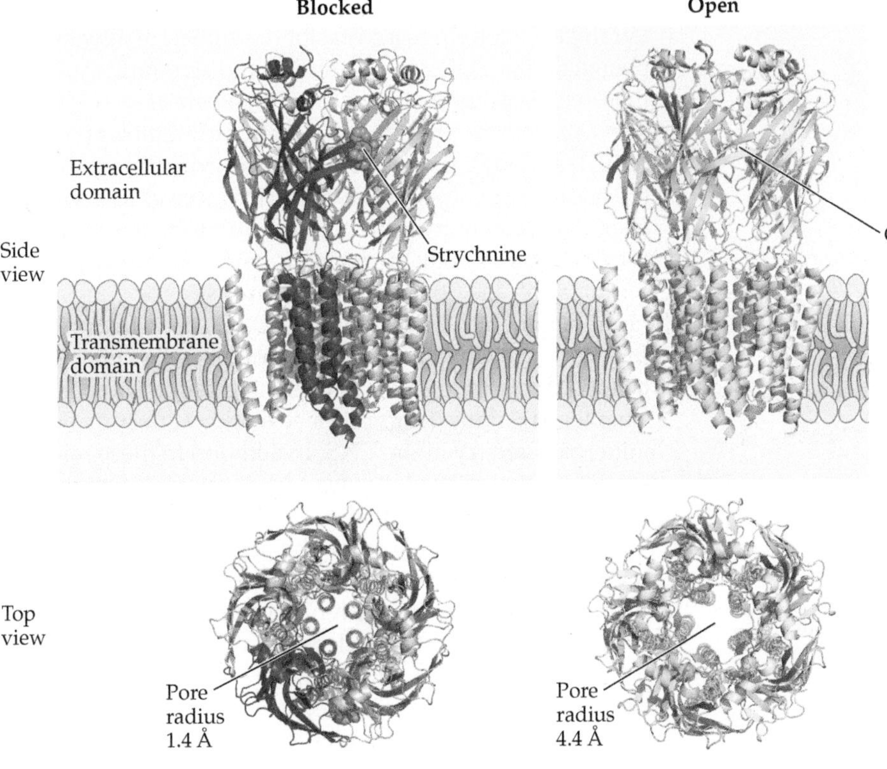

FIGURE 6.13 Gating of glycine receptors. Similar to $GABA_A$ receptors, glycine receptors are pentamers consisting of five subunits. Each subunit (one of which is highlighted) consists of an extracellular domain and a pore-forming transmembrane domain. (Left) Binding of strychnine (orange) to a ligand-binding site on the extracellular domain closes the channel pore. (Right) Binding of glycine to the same ligand-binding site causes a conformational change that opens the pore. (From Du et al., 2015.)

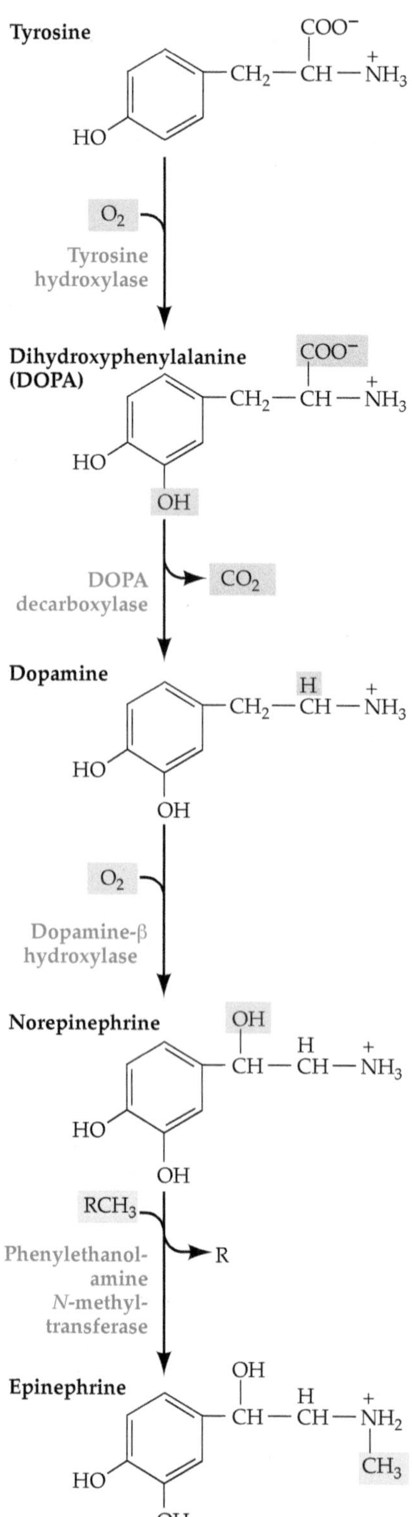

FIGURE 6.14 The biosynthetic pathway for the catecholamine neurotransmitters. The amino acid tyrosine is the precursor for all three catecholamines. The first step in this reaction pathway, catalyzed by tyrosine hydroxylase, is rate-limiting.

1.4 Å (smaller than a Cl^- ion, which has a radius of 1.8 Å) to 4.4 Å, thereby enabling Cl^- and other permeant anions to flow through the pore formed by the transmembrane domains of the five subunits (see Figure 6.13, right). This Cl^- flux inhibits the postsynaptic neuron. Blocking these receptors, by binding of strychnine to the same ligand-binding site, closes the pore (see Figure 6.13, left).

Biogenic Amines

Biogenic amine transmitters regulate many brain functions and are also active in the peripheral nervous system. Because biogenic amines are implicated in such a wide variety of behaviors (ranging from central homeostatic functions to cognitive phenomena such as attention), it is not surprising that defects in biogenic amine function are implicated in most psychiatric disorders. The pharmacology of amine synapses is critically important in psychotherapy, with drugs affecting the synthesis, receptor binding, or catabolism of these neurotransmitters being among the most important agents in the armamentarium of modern neuropharmacology. Many drugs of abuse also act on biogenic amine pathways.

There are five well-established biogenic amine neurotransmitters: the three **catecholamines—dopamine, norepinephrine** (**noradrenaline**), and **epinephrine** (**adrenaline**)—and **histamine** and **serotonin** (see Figure 6.1). All the catecholamines (so named because they share the catechol moiety) are derived from a common precursor, the amino acid tyrosine (Figure 6.14). The first step in catecholamine synthesis is catalyzed by tyrosine hydroxylase in a reaction requiring oxygen as a co-substrate and tetrahydrobiopterin as a co-factor to synthesize dihydroxyphenylalanine (DOPA). Histamine and serotonin are synthesized via other routes, as described below.

• *Dopamine* is present in several brain regions (Figure 6.15A), although the major dopamine-containing area of the brain is the corpus striatum, which receives major input from the substantia nigra and plays an essential role in the coordination of body movements. In Parkinson's disease, for instance, the dopaminergic neurons of the substantia nigra degenerate, leading to a characteristic motor dysfunction (see Chapter 18). Dopamine is also believed to be involved in motivation, reward, and reinforcement (see Chapter 31); many drugs of abuse work by affecting dopaminergic circuitry in the CNS. In addition to these roles in the CNS, dopamine also plays a poorly understood role in some sympathetic ganglia.

Dopamine is produced by the action of DOPA decarboxylase on DOPA (see Figure 6.14). Following its synthesis in the cytoplasm of presynaptic terminals, dopamine is loaded into synaptic vesicles via a vesicular monoamine transporter (VMAT). Dopamine action in the synaptic cleft is terminated by reuptake of dopamine into nerve terminals or surrounding glial cells by a Na^+-dependent dopamine co-transporter, termed DAT. Cocaine apparently produces its psychotropic effects by inhibiting DAT, thereby increasing dopamine concentrations in the synaptic cleft. Amphetamine, another addictive drug, also inhibits DAT as well as the transporter for norepinephrine (see below). The two major enzymes involved in the catabolism of dopamine are monoamine oxidase (MAO) and catechol *O*-methyltransferase (COMT). Both neurons and glia contain mitochondrial MAO and cytoplasmic COMT. Inhibitors of these enzymes, such as phenelzine and tranylcypromine, are used clinically as antidepressants.

Once released, dopamine acts exclusively by activating G-protein-coupled receptors. One of these, the D3 dopamine receptor, is shown in Figure 6.16A. The monomeric structure of this receptor closely parallels that of other metabotropic receptors, such as the mACh receptor (see Figure 6.4A), except that its ligand-binding site is optimized for binding to dopamine. Most dopamine receptor subtypes (see Figure 6.4C) act by either activating or inhibiting adenylyl cyclase (see Chapter 7). Activation of these receptors generally contributes to complex behaviors; for example, administration of dopamine receptor agonists causes hyperactivity and repetitive, stereotyped behavior in laboratory animals. Activation of another type of dopamine receptor in the medulla inhibits vomiting. Thus, antagonists of these receptors are used as emetics to induce vomiting after poisoning or a drug overdose. Dopamine receptor antagonists can also elicit catalepsy, a state in which it is difficult to initiate voluntary motor movement, suggesting a basis for this aspect of some psychoses.

• *Norepinephrine* (also called noradrenaline) is used as a neurotransmitter in the locus coeruleus, a brainstem nucleus that projects diffusely to a variety of forebrain targets (Figure 6.15B) and influences sleep and wakefulness, arousal, attention, and feeding behavior. Perhaps the most prominent noradrenergic neurons are sympathetic ganglion cells, which employ norepinephrine as the major peripheral transmitter in this division of the visceral motor system (see Chapter 21).

Norepinephrine synthesis requires dopamine β-hydroxylase, which catalyzes the production of norepinephrine from dopamine (see Figure 6.14). Norepinephrine is then loaded into synaptic vesicles via the same VMAT involved in vesicular dopamine transport. Norepinephrine is cleared from the synaptic cleft by the norepinephrine transporter (NET), an Na^+-dependent co-transporter that also is capable of taking up dopamine. As mentioned, NET is a molecular target of amphetamine, which acts as a stimulant by producing a net increase in the release of norepinephrine and dopamine. A mutation in the *NET* gene is a cause of orthostatic intolerance, a disorder that produces lightheadedness while standing up. Like dopamine, norepinephrine is degraded by MAO and COMT.

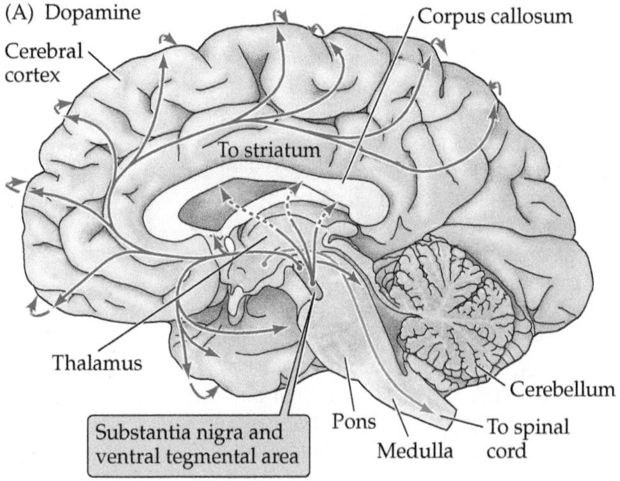

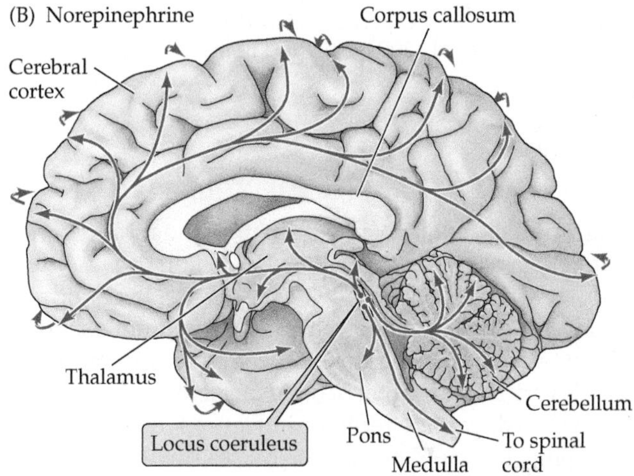

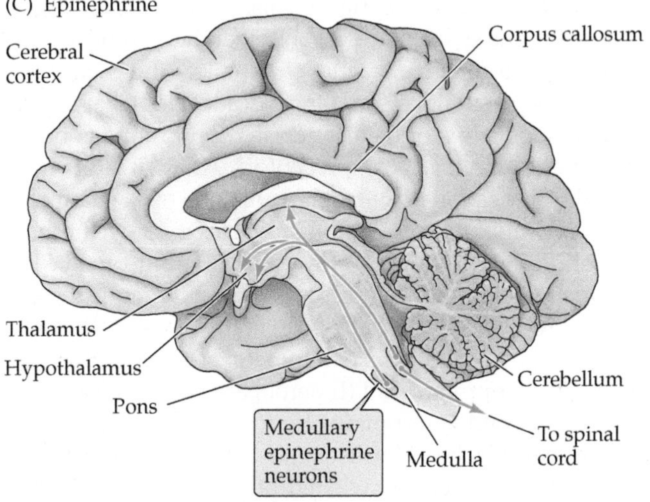

FIGURE 6.15 The distribution of catecholamine neurotransmitters in the human brain. Shown are neurons and their projections (arrows) that contain catecholamine neurotransmitters. Curved arrows along the perimeter of the cortex indicate the innervation of lateral cortical regions not shown in this midsagittal plane of section.

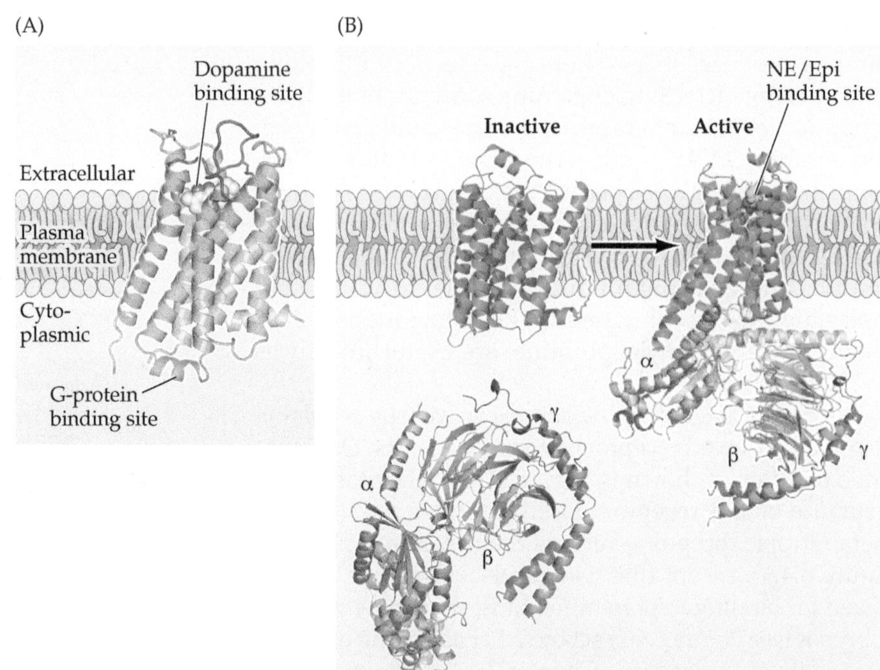

FIGURE 6.16 Metabotropic receptors for catecholamine neurotransmitters. (A) Structure of the D3 dopamine receptor. Like all metabotropic receptors, the D3 receptor spans the plasma membrane seven times and has a cytoplasmic domain that binds to and activates G-proteins, as well as an extracellular domain that binds dopamine. (B) Structure of the β_2-adrenergic receptor and its associated G-protein. (Left) In the absence of ligand, the cytoplasmic domain of the β_2 receptor is not bound to the G-protein (α, β, and γ subunits). (Right) Binding of ligand (β agonist, indicated by colored spheres) to the extracellular binding site for norepinephrine (NE) and epinephrine (Epi) causes the β_2 receptor to bind to the α subunit of the G-protein, which in turn induces a dramatic change in the structure of this subunit. (A from Chien et al., 2010; B from Rasmussen et al., 2007, 2011 and Betke et al., 2012.)

Both norepinephrine and epinephrine act on α- and β-adrenergic receptors (see Figure 6.4C). Both types of receptor are G-protein-coupled; in fact, the β-adrenergic receptor was the first identified metabotropic neurotransmitter receptor. As shown in Figure 6.16B, the structure of this receptor is very similar to that of other metabotropic receptors (such as the dopamine receptor in Figure 6.16A). Binding of norepinephrine or epinephrine causes small changes in the structure of this receptor, which permits the G-protein to bind (see Figure 6.16B, right). This, in turn, causes larger changes in the shape of the α subunit of the G-protein, the first step in a series of reactions that allow the G-protein to regulate intracellular signaling cascades (see Chapter 7).

Two subclasses of α-adrenergic receptors have been identified. Activation of α_1 receptors usually results in a slow depolarization linked to the inhibition of K^+ channels, while activation of α_2 receptors produces a slow hyperpolarization due to the activation of a different type of K^+ channel. There are three subtypes of β-adrenergic receptors, two of which are expressed in many types of neurons. Agonists and antagonists of adrenergic receptors, such as the β-blocker propranolol (Inderol), are used clinically for a variety of conditions ranging from cardiac arrhythmias to migraine headaches. However, most of the actions of these drugs are on smooth muscle receptors, particularly in the cardiovascular and respiratory systems (see Chapter 21).

• *Epinephrine* (also called adrenaline) is found in the brain at lower levels than the other catecholamines and also is present in fewer brain neurons than other catecholamines. Epinephrine-containing neurons in the CNS are primarily in the lateral tegmental system and in the medulla and project to the hypothalamus and thalamus (Figure 6.15C). These epinephrine-secreting neurons regulate respiration and cardiac function.

The enzyme that synthesizes epinephrine, phenylethanolamine-*N*-methyltransferase (see Figure 6.14), is present only in epinephrine-secreting neurons. Otherwise, the metabolism of epinephrine is very similar to that of norepinephrine. Epinephrine is loaded into vesicles via the VMAT. No plasma membrane transporter specific for epinephrine has been identified, although the NET is capable of transporting epinephrine. As already noted, epinephrine acts on both α- and β-adrenergic receptors.

• *Histamine* is found in neurons in the hypothalamus that send sparse but widespread projections to almost all regions of the brain and spinal cord (Figure 6.17A). The central histamine projections mediate arousal and attention, similar to central ACh and norepinephrine projections. Histamine also controls the reactivity of the vestibular system. Allergic reactions or tissue damage cause release of histamine from mast cells in the bloodstream. The close proximity of mast cells to blood vessels, together with the potent actions of histamine on blood vessels, raises the possibility that histamine may influence brain blood flow.

Histamine is produced from the amino acid histidine by a histidine decarboxylase (Figure 6.18A) and is transported into vesicles via the same VMAT as the catecholamines. No plasma membrane histamine transporter has been identified yet. Histamine is degraded by the combined actions of histamine methyltransferase and MAO.

The four known types of histamine receptors are all metabotropic receptors (see Figure 6.4C). Because of the role of

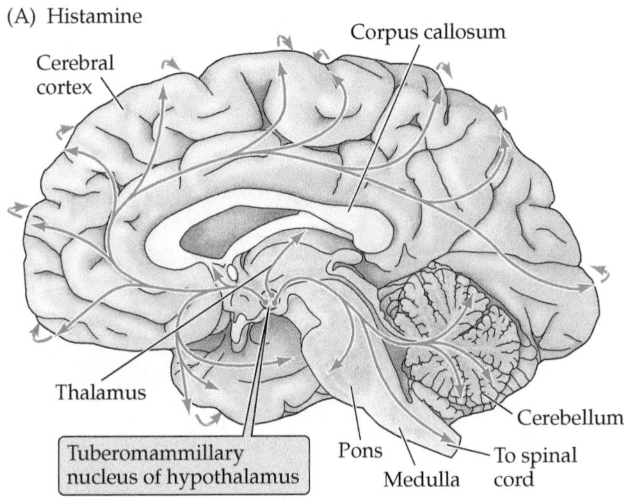

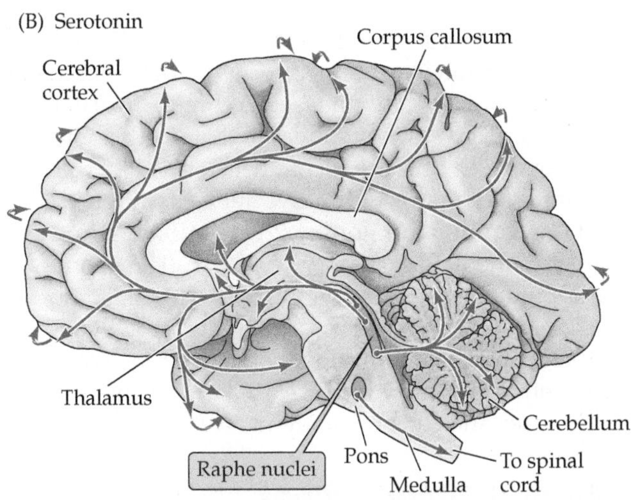

FIGURE 6.17 The distribution of histamine and serotonin neurotransmitters in the human brain. Diagrams show the distribution of neurons and their projections (arrows) containing histamine (A) or serotonin (B). Curved arrows along the perimeter of the cortex indicate the innervation of lateral cortical regions not shown in this midsagittal plane of section.

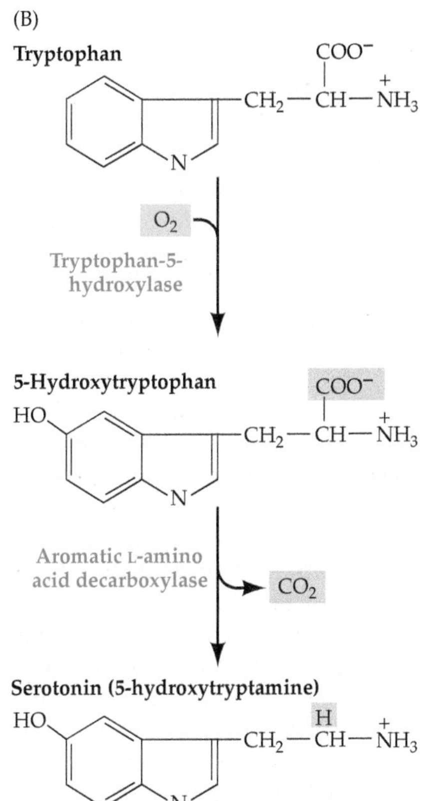

FIGURE 6.18 Synthesis of histamine and serotonin. (A) Histamine is synthesized from the amino acid histidine. (B) Serotonin is derived from the amino acid tryptophan by a two-step process that requires the enzymes tryptophan-5-hydroxylase and a decarboxylase.

histamine receptors in mediating allergic responses, many histamine receptor antagonists have been developed as antihistamine agents. Antihistamines that cross the blood-brain barrier, such as diphenhydramine (Benadryl), act as sedatives by interfering with the roles of histamine in CNS arousal. Antagonists of the H_1 receptor also are used to prevent motion sickness, perhaps because of the role of histamine in controlling vestibular function. H_2 receptors control the secretion of gastric acid in the digestive system, allowing H_2 receptor antagonists to be used in treating a variety of upper gastrointestinal disorders (e.g., peptic ulcers).

• *Serotonin*, or 5-hydroxytryptamine (5-HT), was initially thought to increase vascular tone by virtue of its presence in blood serum (hence the name serotonin). Serotonin is found primarily in groups of neurons in the raphe region of the pons and upper brainstem, which have widespread projections to the forebrain (Figure 6.17B) and regulate sleep and wakefulness (see Chapter 28). 5-HT occupies a place of prominence in neuropharmacology

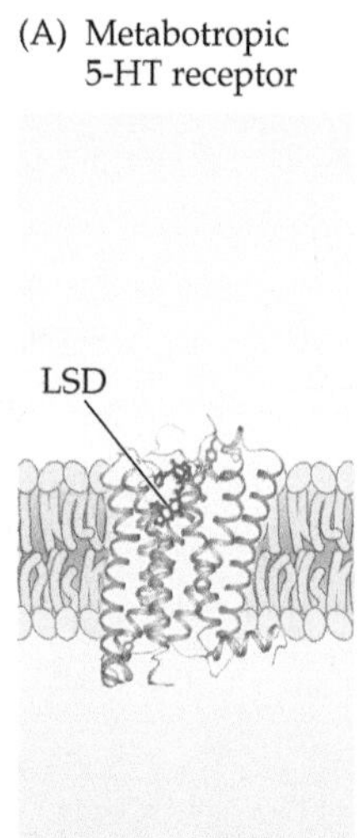

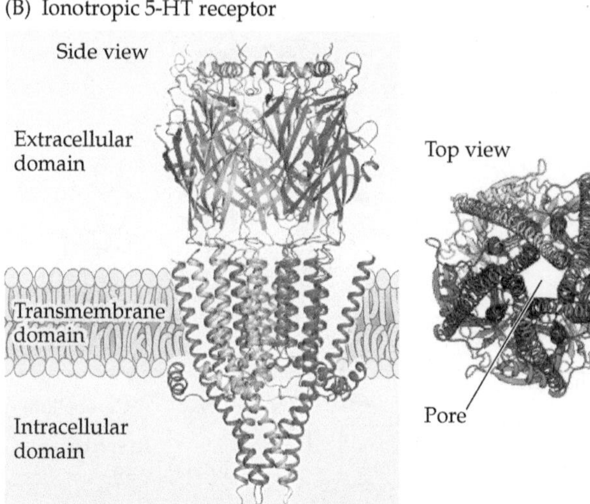

FIGURE 6.19 Serotonin receptors. (A) Structure of the human 5-HT$_{2B}$ receptor, a metabotropic 5-HT receptor. The pink structure indicates LSD bound to the 5-HT binding site of the receptor. (B) Structure of the human 5-HT$_3$ receptor, an ionotropic 5-HT receptor consisting of five subunits (each in a different color), each of which has an extracellular domain, transmembrane domain, and intracellular domain. An ion channel is formed by the transmembrane domains of the five subunits (right). (A from Wacker et al., 2017; B from Hassaine et al., 2014.)

because a large number of antipsychotic drugs that are valuable in the treatment of depression and anxiety act on serotonergic pathways.

5-HT is synthesized from the amino acid tryptophan, which is an essential dietary requirement. Tryptophan is taken up into neurons by a plasma membrane transporter and hydroxylated in a reaction catalyzed by the enzyme tryptophan-5-hydroxylase (Figure 6.18B), the rate-limiting step for 5-HT synthesis. Loading of 5-HT into synaptic vesicles is done by the VMAT that is also responsible for loading other monoamines into synaptic vesicles. The synaptic effects of serotonin are terminated by transport back into nerve terminals via a specific serotonin transporter (SERT) that is present in the presynaptic plasma membrane and is encoded by the 5HTT gene. Many antidepressant drugs are **selective serotonin reuptake inhibitors (SSRIs)** that inhibit transport of 5-HT by SERT. Perhaps the best-known example of an SSRI is the antidepressant drug Prozac. The primary catabolic pathway for 5-HT is mediated by MAO.

A large number of 5-HT receptors (encoded by HTR genes) have been identified. Most 5-HT receptors are metabotropic (see Figure 6.4C), with a monomeric structure typical of G-protein-coupled receptors (Figure 6.19A). Metabotropic 5-HT receptors have been implicated in a wide range of behaviors, including circadian rhythms, motor behaviors, emotional states, and state of mental arousal. Impairments in the function of these receptors have been implicated in numerous psychiatric disorders, such as depression, anxiety disorders, and schizophrenia (see Chapter 31), and drugs acting on serotonin receptors are effective treatments for several of these conditions. The psychedelic drug LSD (lysergic acid diethylamide) presumably causes hallucinations by activating multiple types of metabotropic 5-HT receptors (see Figure 6.19A). Activation of 5-HT receptors also mediates satiety and decreased food consumption, which is why serotonergic drugs are sometimes useful in treating eating disorders.

One group of serotonin receptors, the 5-HT$_3$ receptors, are ligand-gated ion channels formed from combinations of the five 5-HT$_3$ subunits (see Figure 6.3F). Their pentameric structure is very similar to that of other ionotropic receptors, with functional channels formed by the transmembrane domains of the five subunits (Figure 6.19B). 5-HT$_3$ receptors are nonselective cation channels and therefore mediate excitatory postsynaptic responses. Ligand-binding sites reside within the extracellular domains of these receptors and serve as targets for a wide variety of therapeutic drugs, including ondansetron (Zofran) and granisetron (Kytril), which are used to prevent postoperative nausea and chemotherapy-induced emesis.

ATP and Other Purines

All synaptic vesicles contain ATP, which is co-released with one or more "classic" neurotransmitters. This observation raises the possibility that ATP acts as a co-transmitter. It has been known since the 1920s that the extracellular application of ATP (or its breakdown products AMP and adenosine) can elicit electrical responses in neurons. The idea that some purines (so named because all these compounds contain a purine ring; see Figure 6.1) are also neurotransmitters is now well established. ATP acts as an excitatory neurotransmitter in motor neurons of the spinal cord, as well as in sensory and autonomic ganglia. Postsynaptic actions of ATP have also been demonstrated in the CNS, specifically for dorsal horn neurons and in a subset of hippocampal neurons. Extracellular enzymes degrade released ATP to adenosine, which has its own set of signaling actions. Thus, adenosine cannot be considered a classic neurotransmitter because it is not stored in synaptic vesicles or released in a Ca^{2+}-dependent manner. Several enzymes, including apyrase, ecto-5′ nucleotidase, and nucleoside transporters, are involved in the rapid catabolism and removal of purines from extracellular locations.

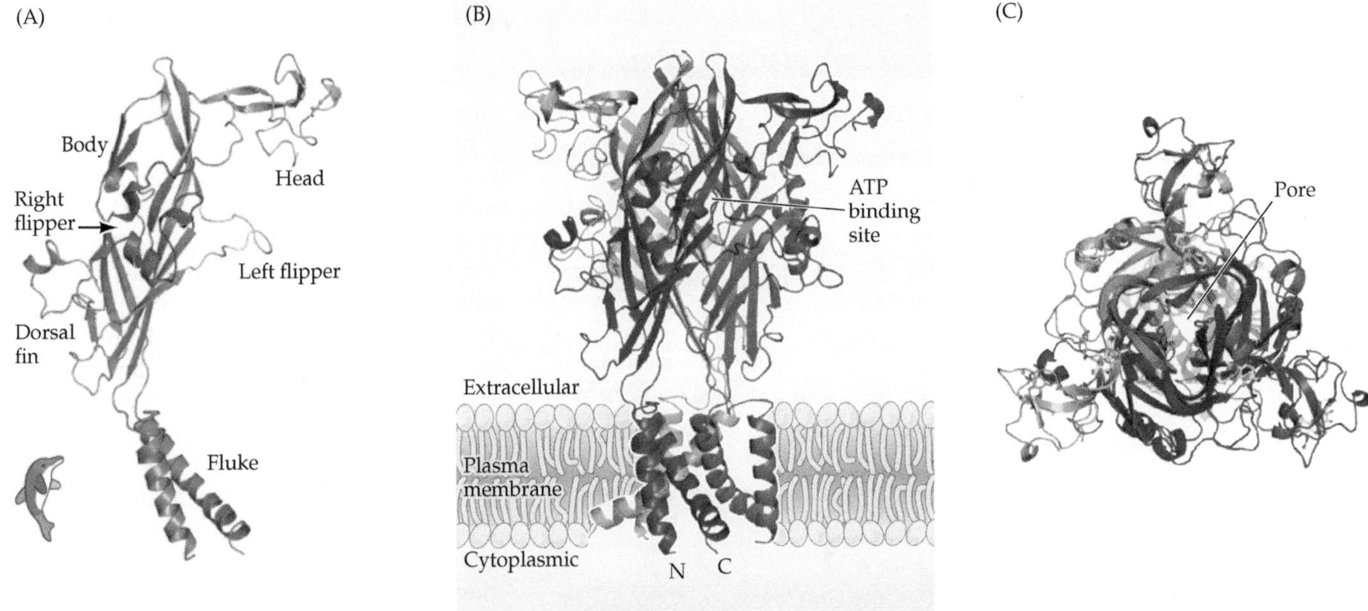

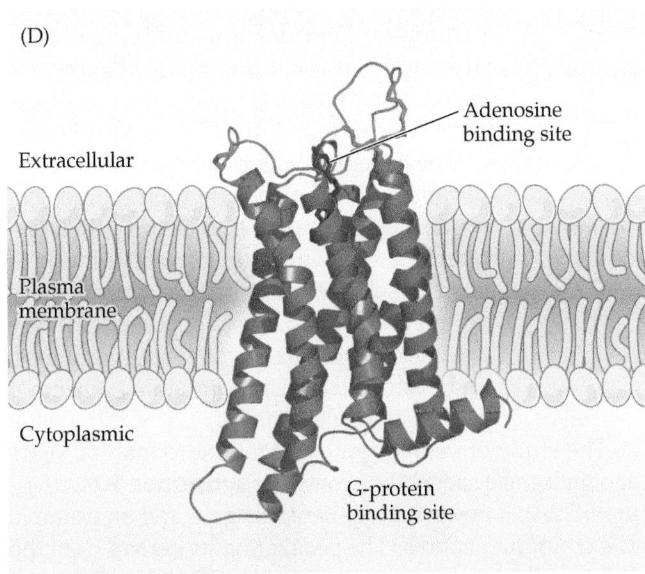

FIGURE 6.20 Purinergic receptors. (A) Subunit of an ionotropic $P2X_4$ receptor. Each subunit has a transmembrane domain consisting of two helical structures that form part of a channel, as well as a large extracellular domain that includes the ATP binding site. The shape of the subunit is reminiscent of a dolphin, with the structures color-coded as indicated in the inset. (B) Side view of a $P2X_4$ receptor; this receptor is a trimer of three subunits, with each subunit shown in a different color. The ATP binding site is proposed to be in the center of the extracellular domain. (C) Top view of the $P2X_4$ receptor, indicating the centrally located channel pore. (D) Structure of a metabotropic A_{2A} adenosine receptor. This receptor has the seven-membrane-spanning domain structure characteristic of metabotropic receptors and is shown with an antagonist drug (purple structure) occupying the adenosine binding site. (A–C from Kawate et al., 2009; D from Jaakola and Ijzerman, 2010.)

Receptors for both ATP and adenosine are widely distributed in the nervous system as well as in many other tissues. Three classes of these purinergic receptors are known. One class consists of ionotropic receptors called **P2X receptors** (see Figure 6.3F). The structure of these receptors is unique among ionotropic receptors because each subunit has a transmembrane domain that crosses the membrane only twice (Figure 6.20A). Furthermore, only three of these subunits are required to form a trimeric receptor (Figure 6.20B). As in all ionotropic receptors, a pore is located in the center of the P2X receptor (Figure 6.20C) and forms a nonselective cation channel. Thus, P2X receptors mediate excitatory postsynaptic responses. Ionotropic purinergic receptors are widely distributed in central and peripheral neurons. In sensory nerves, they evidently play a role in mechanosensation and pain; their function in most other cells, however, is not known.

The other two classes of purinergic receptors are G-protein-coupled metabotropic receptors (see Figure 6.4C). The two classes differ in their sensitivity to agonists—one type is preferentially stimulated by adenosine, whereas the other is preferentially activated by ATP. An example of the former, the A_{2A} adenosine receptor, is shown in Figure 6.20D. Both receptor types are found throughout the brain, as well as in peripheral tissues such as the heart, adipose tissue, and the kidney. Xanthines such as caffeine and theophylline block adenosine receptors, and this activity is thought to be responsible for the stimulant effects of these agents.

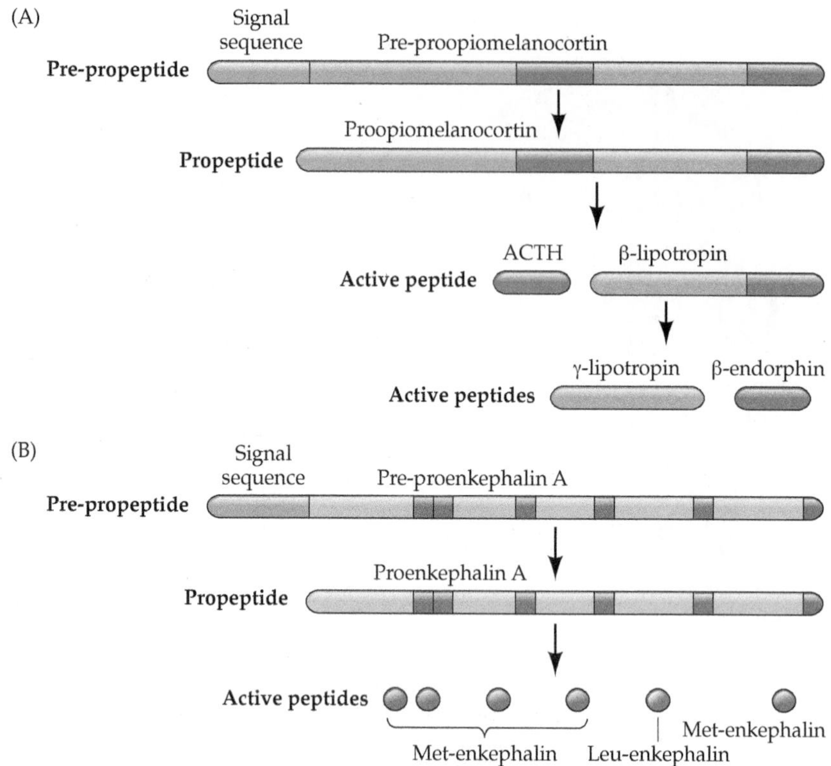

FIGURE 6.21 Proteolytic processing of pre-propeptides. Shown here are pre-proopiomelanocortin (A) and pre-proenkephalin A (B). For each pre-propeptide, the signal sequence is indicated at the left; the locations of active peptide products are indicated by darker colors. The maturation of the pre-propeptides involves cleaving the signal sequence and other proteolytic processing. Such processing can result in several different neuroactive peptides such as ACTH, γ-lipotropin, and β-endorphin (A), or multiple copies of the same peptide, such as methionine-enkephalin (B).

Peptide Neurotransmitters

Many peptides known to be hormones also act as neurotransmitters. Some peptide transmitters have been implicated in modulating emotions. Others, such as substance P and the opioid peptides, are involved in the perception of pain (see Chapter 10). Still other peptides, such as melanocyte-stimulating hormones, adrenocorticotropin, and β-endorphin, regulate complex responses to stress.

The mechanisms responsible for the synthesis and packaging of peptide transmitters are fundamentally different from those used for the small-molecule neurotransmitters and are much like those used for the synthesis of proteins that are secreted from non-neuronal cells (pancreatic enzymes, for instance). Peptide-secreting neurons generally synthesize polypeptides that are much larger than the final, "mature" peptide. Processing these polypeptides, which are called **pre-propeptides** (or pre-proproteins), takes place within the neuron's cell body by a sequence of reactions that occur in several intracellular organelles. Pre-propeptides are synthesized in the rough endoplasmic reticulum, where the signal sequence—that is, the sequence of amino acids indicating that the peptide is to be secreted—is removed. The remaining polypeptide, called a **propeptide** (or proprotein), then traverses the Golgi apparatus and is packaged into vesicles in the *trans*-Golgi network. The final stages of peptide neurotransmitter processing occur after packaging into vesicles and involve proteolytic cleavage, modification of the ends of the peptide, glycosylation, phosphorylation, and disulfide bond formation.

Propeptide precursors are typically larger than their active peptide products and can give rise to more than one species of neuropeptide (Figure 6.21), which means that multiple neuroactive peptides can be released from a single vesicle. In addition, neuropeptides often are co-released with small-molecule neurotransmitters. Thus, peptidergic synapses often elicit complex postsynaptic responses. Peptides are catabolized into inactive amino acid fragments by enzymes called peptidases, usually located on the extracellular surface of the plasma membrane.

The biological activity of the peptide neurotransmitters depends on their amino acid sequence (Figure 6.22). Based on their sequences, neuropeptide transmitters have been loosely grouped into five categories: the brain/gut peptides; opioid peptides; pituitary peptides; hypothalamic releasing hormones; and a catch-all category containing other, not easily classified, peptides.

The study of neuropeptides began more than 60 years ago with the accidental discovery of **substance P** (see Figure 6.22A), a powerful hypotensive agent and an example of a brain/gut peptide. (The peculiar name derives from the fact that this molecule was an unidentified component of *p*owder extracts from brain and intestine.) Substance P is an 11-amino-acid peptide present in high concentrations in the human hippocampus and neocortex and also in the gastrointestinal tract—hence its classification as a brain/gut peptide. It is also released from C fibers (see Tables 3.1 and 9.1), the small-diameter afferents in peripheral nerves that convey information about pain and temperature (as well as postganglionic autonomic signals). Substance P is a sensory neurotransmitter in the spinal cord, where its release can be inhibited by opioid peptides released from spinal cord interneurons, resulting in the suppression of pain (see Chapter 10). The diversity of neuropeptides is highlighted by the finding that the gene coding for substance P also encodes several other neuroactive peptides, including neurokinin A, neuropeptide K, and neuropeptide γ.

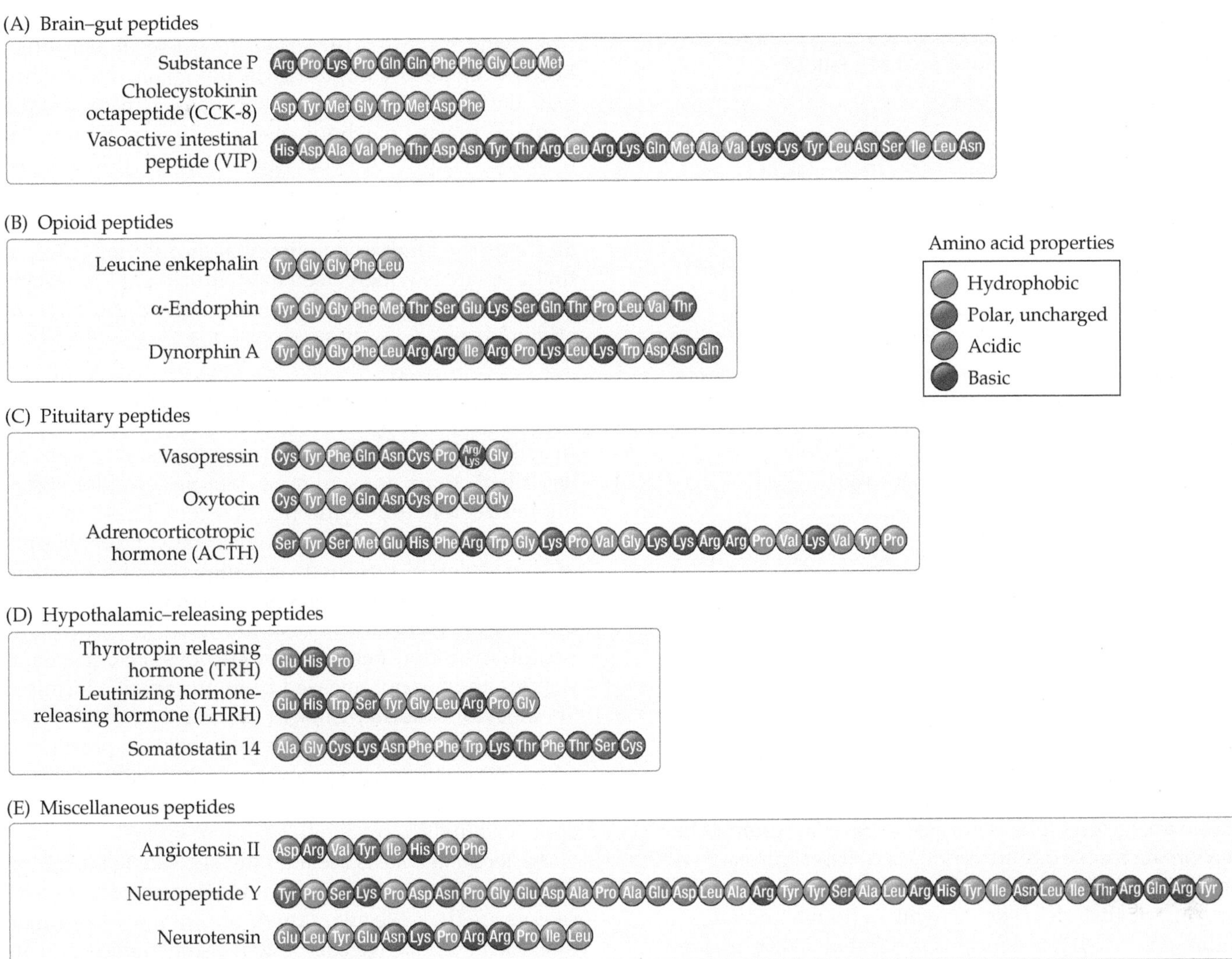

FIGURE 6.22 Amino acid sequences of neuropeptides. These neuropeptides vary in length, usually containing between 3 and 36 amino acids. The sequence of amino acids determines the biological activity of each peptide.

An especially important category of peptide neurotransmitters is the family of opioids (see Figure 6.22B), so named because they bind to the same postsynaptic receptors that are activated by opium. The opium poppy has been cultivated for some 5000 years, and its derivatives have been used as an analgesic since at least the Renaissance. The active ingredients in opium are a variety of plant alkaloids, predominantly morphine. Morphine, named for Morpheus, the Greek god of dreams, is still in use today and is one of the most effective analgesics, despite its addictive potential. Synthetic opiates such as meperidine, methadone, and fentanyl are also powerful analgesics.

The opioid peptides were discovered in the 1970s during a search for **endorphins**—*endo*genous compounds that mimicked the actions of m*orphine*. It was hoped that such compounds would be analgesics, and that understanding them would shed light on drug addiction. The endogenous ligands of the opioid receptors have now been identified as a family of more than 20 opioid peptides that fall into three classes: endorphins, enkephalins, and dynorphins (Table 6.2). Each class is liberated from an inactive pre-propeptide (pre-proopiomelanocortin, pre-proenkephalin A, and pre-prodynorphin) derived from distinct genes (see Figure 6.21). Opioid precursor processing is carried out by tissue-specific processing enzymes that are packaged into vesicles, along with the precursor peptide, in the Golgi apparatus.

Opioid peptides are widely distributed throughout the brain and are often co-localized with small-molecule neurotransmitters, such as GABA and 5-HT. In general, the opioids tend to be depressants. When injected intracerebrally in experimental animals, they act as analgesics; on the basis of this and other evidence, opioids are likely to be

TABLE 6.2 ■ Endogenous Opioid Peptides

Name	Amino acid sequence[a]
Endorphins	
α-Endorphin	*Tyr-Gly-Gly-Phe*-Met-Thr-Ser-Glu-Lys-Ser-Gln-Thr-Pro-Leu-Val-Thr
α-Neoendorphin	*Tyr-Gly-Gly-Phe*-Leu-Arg-Lys-Tyr-Pro-Lys
β-Endorphin	*Tyr-Gly-Gly-Phe*-Met-Thr-Ser-Glu-Lys-Ser-Gln-Thr-Pro-Leu-Val-Thr-Leu-Phe-Lys-Asn-Ala-Ile-Val-Lys-Asn-Ala-His-Lys-Gly-Gln
γ-Endorphin	*Tyr-Gly-Gly-Phe*-Met-Thr-Ser-Glu-Lys-Ser-Gln-Thr-Pro-Leu-Val-Thr-Leu
Enkephalins	
Leu-enkephalin	*Tyr-Gly-Gly-Phe*-Leu
Met-enkephalin	*Tyr-Gly-Gly-Phe*-Met
Dynorphins	
Dynorphin A	*Tyr-Gly-Gly-Phe*-Leu-Arg-Arg-Ile-Arg-Pro-Lys-Leu-Lys-Trp-Asp-Asn-Gln
Dynorphin B	*Tyr-Gly-Gly-Phe*-Leu-Arg-Arg-Gln-Phe-Lys-Val-Val-Thr

[a]Note the initial homology, indicated by italics.

involved in the mechanisms underlying acupuncture-induced analgesia. Opioids are also involved in complex behaviors such as sexual attraction, as well as aggressive and submissive behaviors. They have also been implicated in psychiatric disorders such as schizophrenia and autism, although the evidence for this is debated. Unfortunately, repeated administration of opioids leads to tolerance and addiction.

Virtually all neuropeptides initiate their effects by activating G-protein-coupled receptors. Studying these metabotropic peptide receptors in the brain has been difficult because few specific agonists and antagonists are known. Peptides activate their receptors at low (n*M* to μ*M*) concentrations compared with the concentrations required to activate receptors for small-molecule neurotransmitters. These properties allow the postsynaptic targets of peptides to be quite far removed from presynaptic terminals and to modulate the electrical properties of neurons that are simply in the vicinity of the site of peptide release. Neuropeptide receptor activation is especially important in regulating the postganglionic output from sympathetic ganglia and the activity of the gut (see Chapter 21). Peptide receptors, particularly the neuropeptide Y receptor, are also implicated in the initiation and maintenance of feeding behavior leading to satiety or obesity.

Other behaviors ascribed to peptide receptor activation include anxiety and panic attacks, and antagonists of cholecystokinin receptors are clinically useful in the treatment of these afflictions. Other useful drugs have been developed by targeting the opioid receptors. Three well-defined opioid receptor subtypes (μ, δ, and κ) play a role in reward mechanisms as well as addiction. The μ-opioid receptor has been specifically identified as the primary site for drug reward mediated by opiate drugs. Fentanyl, a selective agonist of μ-opioid receptors, has 80 times the analgesic potency of morphine. This synthetic opiate is widely used as a clinical analgesic agent to alleviate pain and is an increasingly popular recreational drug.

Unconventional Neurotransmitters

In addition to the conventional neurotransmitters already described, some unusual molecules are used for signaling between neurons and their targets. These chemical signals can be considered as neurotransmitters because of their roles in interneuronal signaling and because their release from neurons is regulated by Ca^{2+}. However, they are unconventional in comparison with other neurotransmitters because they are not stored in synaptic vesicles and are not released from presynaptic terminals via exocytotic mechanisms. In fact, these unconventional neurotransmitters need not be released from presynaptic terminals at all and are often associated with retrograde signaling (that is, from postsynaptic cells back to presynaptic terminals).

• *Endocannabinoids* are a family of related endogenous signals that interact with cannabinoid receptors. These receptors are the molecular targets of Δ^9-tetrahydrocannabinol, the psychoactive component of the marijuana plant, *Cannabis* (Box 6C). While some members of this emerging group of chemical signals remain to be determined, anandamide and 2-arachidonoylglycerol (2-AG) have been established as endocannabinoids. These signals are unsaturated fatty acids with polar head groups and are produced by enzymatic degradation of membrane lipids (Figure 6.23A,B). Production of endocannabinoids is stimulated by a second messenger within postsynaptic neurons, typically a rise in postsynaptic Ca^{2+} concentration, allowing these hydrophobic signals to diffuse through the postsynaptic membrane to reach cannabinoid receptors on other nearby cells. Endocannabinoid action is terminated by carrier-mediated transport of these signals back into the postsynaptic neuron, where they are hydrolyzed by the enzyme fatty acid hydrolase (FAAH).

At least two types of cannabinoid receptors have been identified, with most actions of endocannabinoids in the CNS mediated by the CB_1 type (see Box 6C). The CB_1 receptor is a G-protein-coupled receptor related to the metabotropic receptors for ACh, glutamate, and other conventional neurotransmitters. Several compounds that are

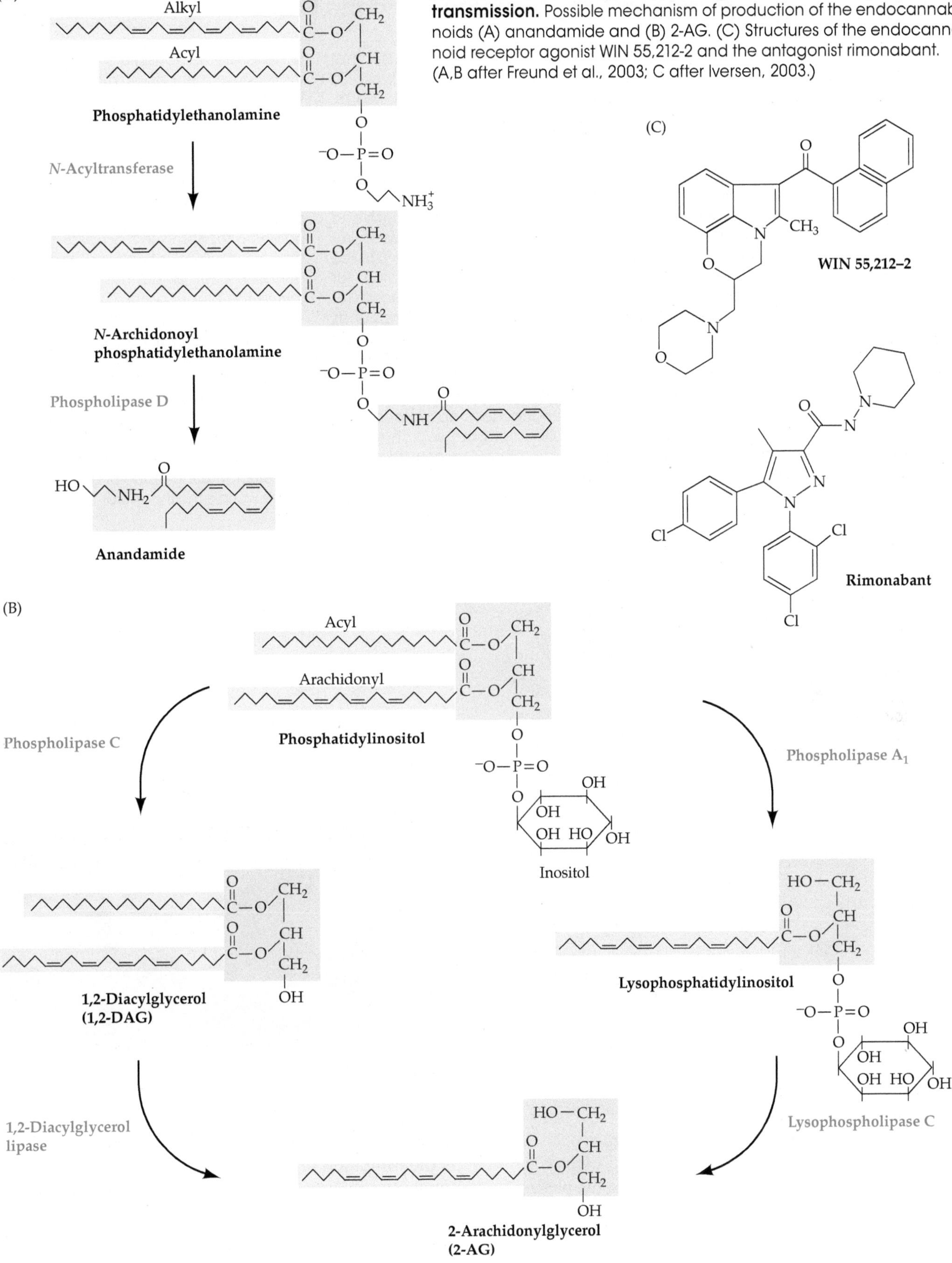

FIGURE 6.23 Endocannabinoid signals involved in synaptic transmission. Possible mechanism of production of the endocannabinoids (A) anandamide and (B) 2-AG. (C) Structures of the endocannabinoid receptor agonist WIN 55,212-2 and the antagonist rimonabant. (A,B after Freund et al., 2003; C after Iversen, 2003.)

BOX 6C ■ Marijuana and the Brain

Medicinal use of the marijuana plant (*Cannabis sativa*) (Figure A) dates back thousands of years. Ancient civilizations—including both Greek and Roman societies in Europe, as well as Indian and Chinese cultures in Asia—appreciated that this plant was capable of producing relaxation, euphoria, and several other psychopharmacological actions. In more recent times, medicinal use of marijuana has revived and the recreational use of marijuana has become so popular that some societies have decriminalized its use.

Understanding the brain mechanisms underlying the actions of marijuana was advanced by the discovery that a cannabinoid, Δ^9-tetrahydrocannabinol (THC; Figure B), is the active component of marijuana. This finding led to the development of synthetic derivatives, such as WIN 55,212-2 and rimonabant (see Figure 6.23), that have served as valuable tools for probing the brain actions of THC. Of particular interest is that receptors for these cannabinoids exist in the brain. The best studied of these receptors, called CB_1, is a metabotropic receptor that activates G-protein signaling pathways (Figure C). CB_1 exhibits marked regional variations in distribution, being especially enriched in brain areas—such as the substantia nigra and caudate putamen—that have been implicated in drug abuse (Figure D). The presence of these brain receptors for cannabinoids led in turn to a search for endogenous cannabinoid compounds in the brain, culminating in the discovery of endocannabinoids such as 2-AG and anandamide (see Figure 6.23). This path of discovery closely parallels the identification of endogenous opioid peptides, which resulted from the search for endogenous morphine-like compounds in the brain (see text and Table 6.2).

Given that THC interacts with brain endocannabinoid receptors, particularly the CB_1 receptor, it is likely that such actions are responsible for the behavioral consequences of marijuana use. Indeed, many of the well-documented effects of marijuana are consistent with the distribution and actions of brain CB_1 receptors. For example, marijuana's effects on perception could be due to CB_1 receptors in the neocortex, effects on psychomotor control due to endocannabinoid receptors in the basal ganglia and cerebellum, effects on short-term memory due to cannabinoid receptors in the hippocampus, and the well-known effects on stimulating appetite due to hypothalamic actions. While formal links between these behavioral consequences of marijuana and the underlying brain mechanisms are still being forged, studies of the actions of this drug have shed substantial light on basic synaptic mechanisms, which promise to further elucidate the mode of action of one of the world's most popular drugs.

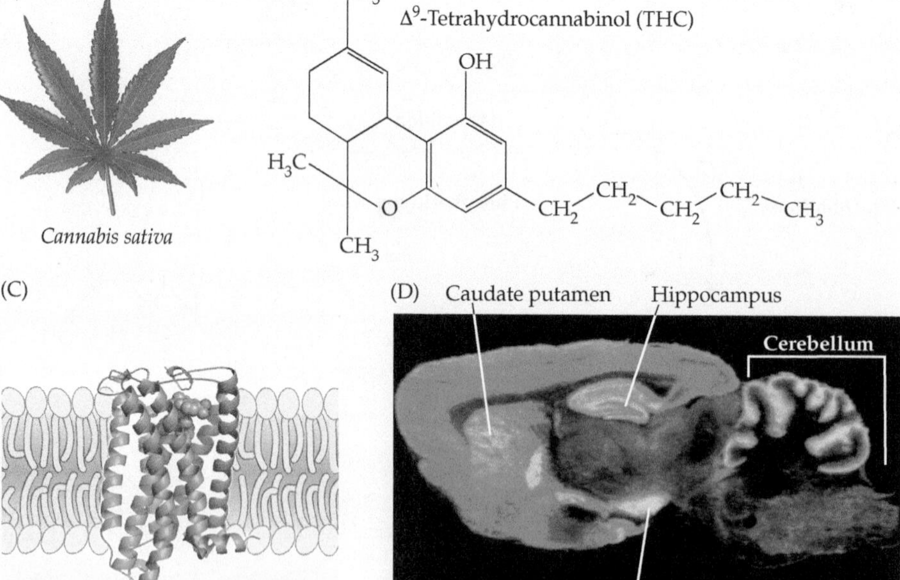

(A) Leaf of *Cannabis sativa*, the marijuana plant. (B) Structure of Δ^9-tetrahydrocannabinol (THC), the active ingredient of marijuana. (C) Structure of the human CB_1 receptor, bound to the ligand taranabant (colored spheres). (D) Distribution of brain CB_1 receptors, visualized by examining the binding of CP-55,940, a CB_1 receptor ligand. (B after Iversen, 2003; C from Shao et al., 2016; D courtesy of M. Herkenham, NIMH.)

structurally related to endocannabinoids and that bind to the CB_1 receptor have been synthesized (Figure 6.23C). These compounds act as agonists or antagonists of the CB_1 receptor and serve both as tools for elucidating the physiological functions of endocannabinoids and as targets for developing therapeutically useful drugs.

Endocannabinoids participate in several forms of synaptic regulation. The best-documented action of these agents is the inhibition of communication between presynaptic inputs and their postsynaptic target cells. In both the hippocampus and the cerebellum (among other brain regions), endocannabinoids serve as retrograde signals that regulate

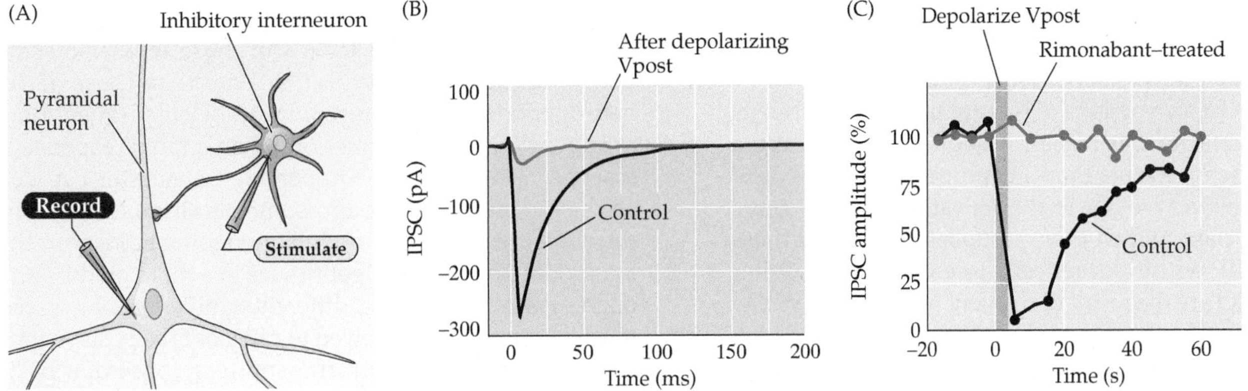

FIGURE 6.24 Endocannabinoid-mediated retrograde control of GABA release. (A) Experimental arrangement. Stimulation of a presynaptic interneuron causes release of GABA onto a postsynaptic pyramidal neuron. (B) Inhibitory postsynaptic currents (IPSCs) elicited by the inhibitory synapse (control) are reduced in amplitude following a brief depolarization of the postsynaptic neuron (Vpost). This reduction in the IPSC is due to less GABA being released from the presynaptic interneuron. (C) The reduction in IPSC amplitude produced by postsynaptic depolarization lasts a few seconds and is mediated by endocannabinoids, because it is prevented by the endocannabinoid receptor antagonist rimonabant. (B,C after Ohno-Shosaku et al., 2001.)

GABA release at certain inhibitory synapses. At such synapses, depolarization of the postsynaptic neuron causes a transient reduction in inhibitory postsynaptic responses (Figure 6.24). Depolarization reduces synaptic transmission by elevating the concentration of Ca^{2+} in the postsynaptic neuron; this rise in Ca^{2+} triggers synthesis and release of endocannabinoids from the postsynaptic cells. The endocannabinoids then bind to CB_1 receptors on presynaptic terminals, inhibiting the amount of GABA released in response to presynaptic action potentials, and thereby reducing inhibitory transmission. The mechanisms responsible for the reduction in GABA release are not entirely clear but probably involve effects on voltage-gated Ca^{2+} channels and/or K^+ channels in the presynaptic neurons.

• *Nitric oxide* (NO) is an unusual and especially interesting chemical signal. It is a gas produced by the action of nitric oxide synthase, an enzyme that converts the amino acid arginine into a metabolite (citrulline) and simultaneously generates NO (Figure 6.25). Within neurons, NO synthase is regulated by Ca^{2+} binding to the Ca^{2+} sensor protein calmodulin (see Chapter 7). Once produced, NO can permeate the plasma membrane, meaning that NO generated inside one cell can travel through the extracellular medium and act inside nearby cells. Thus, this gaseous signal has a range of influence that extends well beyond the cell of origin, diffusing a few tens of micrometers from its site of production before it is degraded. This property makes NO a potentially useful agent for coordinating the

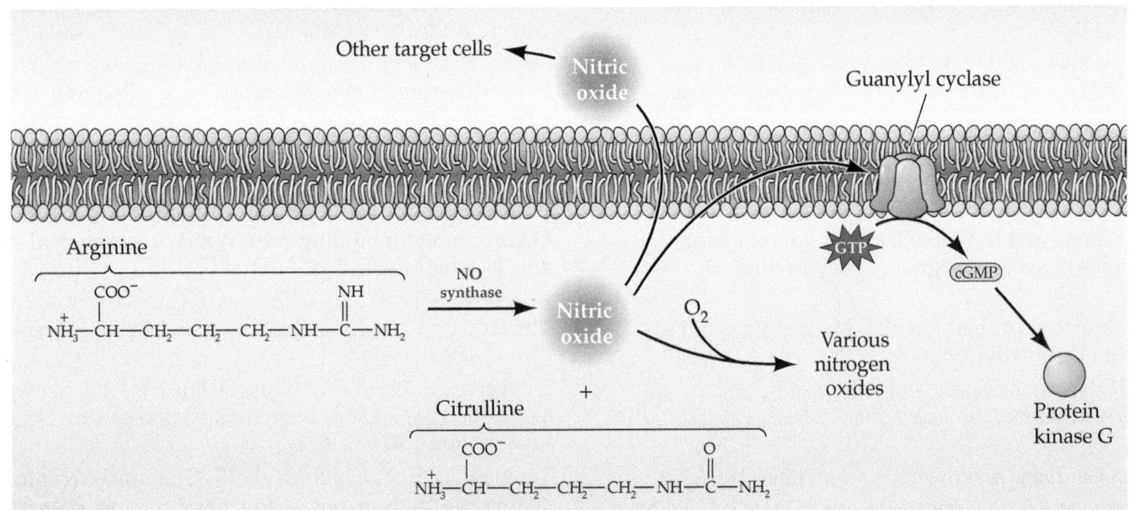

FIGURE 6.25 Synthesis, release, and termination of nitric oxide (NO).

activities of multiple cells in a localized region and may mediate certain forms of synaptic plasticity that spread within small networks of neurons.

All of the known actions of NO are mediated within its cellular targets; for this reason, NO often is considered a second messenger rather than a neurotransmitter. Some of the actions of NO are due to the activation of the enzyme guanylyl cyclase, which then produces the second messenger cGMP within target cells (see Chapter 7). Other actions of NO are the result of covalent modification of target proteins via nitrosylation, the addition of a nitryl group to selected amino acids within the proteins. NO decays spontaneously by reacting with oxygen to produce inactive nitrogen oxides; thus, its signals last for only a short time (seconds or less). NO signaling evidently regulates a variety of synapses that also employ conventional neurotransmitters; so far, presynaptic terminals that release glutamate are the best-studied NO targets in the CNS. NO may also be involved in some neurological diseases. For example, it has been proposed that an imbalance between nitric oxide and superoxide generation underlies some neurodegenerative diseases.

Summary

The complex synaptic computations occurring at neural circuits throughout the brain arise from the actions of a large number of neurotransmitters, which act on an even larger number of postsynaptic neurotransmitter receptors. Glutamate is the major excitatory neurotransmitter in the brain, whereas GABA and glycine are the major inhibitory neurotransmitters. The actions of these small-molecule neurotransmitters are typically faster than those of the neuropeptides. Thus, most small-molecule transmitters mediate synaptic transmission when a rapid response is essential, whereas the neuropeptide transmitters, as well as the biogenic amines and some small-molecule neurotransmitters, tend to modulate ongoing activity in the brain or in peripheral target tissues in a more gradual and ongoing way. Two broadly different families of neurotransmitter receptors have evolved to carry out the postsynaptic signaling actions of neurotransmitters. Ionotropic or ligand-gated ion channels combine the neurotransmitter receptor and ion channel in one molecular entity, and therefore give rise to rapid postsynaptic electrical responses. Metabotropic receptors regulate the activity of postsynaptic ion channels indirectly, usually via G-proteins, and induce slower and longer-lasting electrical responses. Metabotropic receptors are especially important in regulating behavior, and drugs targeting these receptors have been clinically valuable in treating a wide range of behavioral disorders. The postsynaptic response at a given synapse is determined by the combination of receptor subtypes, G-protein subtypes, and ion channels that are expressed in the postsynaptic cell. Because each of these features can vary both within and among neurons, a tremendous diversity of transmitter-mediated effects is possible. Drugs that influence transmitter actions have enormous importance in the treatment of neurological and psychiatric disorders, as well as in a broad spectrum of other medical problems.

ADDITIONAL READING

Reviews

Beaulieu, J. M. and R. R. Gainetdinov (2011) The physiology, signaling, and pharmacology of dopamine receptors. *Pharmacol. Rev.* 63: 182–217.

Betke, K. M., C. A. Wells and H. E. Hamm (2012) GPCR mediated regulation of synaptic transmission. *Prog. Neurobiol.* 96: 304–321.

Carlsson, A. (1987) Perspectives on the discovery of central monoaminergic neurotransmission. *Annu. Rev. Neurosci.* 10: 19–40.

Freund, T. F., I. Katona and D. Piomelli (2003) Role of endogenous cannabinoids in synaptic signaling. *Physiol. Rev.* 83: 1017–1066.

Hökfelt, T., O. Johansson, A. Ljungdahl, J. M. Lundberg and M. Schultzberg (1980) Peptidergic neurons. *Nature* 284: 515–521.

Hyland, K. (1999) Neurochemistry and defects of biogenic amine neurotransmitter metabolism. *J. Inher. Metab. Dis.* 22: 353–363.

Iversen, L. (2003) Cannabis and the brain. *Brain* 126: 1252–1270.

Jaakola, V. P. and A. P. Ijzerman (2010) The crystallographic structure of the human adenosine A_{2A} receptor in a high-affinity antagonist-bound state: Implications for GPCR drug screening and design. *Curr. Opin. Struct. Biol.* 20: 401–414.

Koob, G. F., P. P. Sanna and F. E. Bloom (1998) Neuroscience of addiction. *Neuron* 21: 467–476.

Pierce, K. L., R. T. Premont and R. J. Lefkowitz (2002) Seven-transmembrane receptors. *Nature Rev. Mol. Cell Biol.* 3: 639–650.

Pin, J.-P. and B. Bettler (2016) Organization and functions of mGlu and $GABA_B$ receptor complexes. *Nature* 540: 60–68.

Puthenkalam, R. and 6 others (2016) Structural studies of $GABA_A$ receptor binding sites: Which experimental structure tells us what? *Front. Mol. Neurosci.* 9: 44.

Rosenbaum, D. M., S. G. Rasmussen and B. K. Kobilka (2009) The structure and function of G-protein-coupled receptors. *Nature* 459: 356–363.

Schwartz, M. W., S. C. Woods, D. Porte Jr., R. J. Seeley and D. G. Baskin (2000) Central nervous system control of food intake. *Nature* 404: 661–671.

Traynelis, S. F. and 9 others (2010) Glutamate receptor ion channels: structure, regulation, and function. *Pharmacol. Rev.* 62: 405–496.

Important Original Papers

Charpak, S., B. H. Gähwiler, K. Q. Do and T. Knöpfel (1990) Potassium conductances in hippocampal neurons blocked by excitatory amino-acid transmitters. *Nature* 347: 765–767.

Chavas, J. and A. Marty (2003) Coexistence of excitatory and inhibitory GABA synapses in the cerebellar interneuron network. *J. Neurosci.* 23: 2019–2031.

Chien, E. Y. and 10 others (2010) Structure of the human dopamine D3 receptor in complex with a D2/D3 selective antagonist. *Science* 330: 1091–1095.

Curtis, D. R., J. W. Phillis and J. C. Watkins (1959) Chemical excitation of spinal neurons. *Nature* 183: 611–612.

Dale, H. H., W. Feldberg and M. Vogt (1936) Release of acetylcholine at voluntary motor nerve endings. *J. Physiol.* 86: 353–380.

Du, J., W. Lü, S. Wu, Y. Cheng and E. Gouaux (2015) Glycine receptor mechanism elucidated by electron cryo-microscopy. *Nature* 526: 224–229.

Gupta, S., S. Chakraborty, R. Vij and A. Auerbach (2017) A mechanism for acetylcholine receptor gating based on structure, coupling, phi, and flip. *J. Gen. Physiol.* 149: 85–103.

Haga, K. and 10 others (2012) Structure of the human M2 muscarinic acetylcholine receptor bound to an antagonist. *Nature* 482: 547–551.

Hassaine, G. and 14 others (2014) X-ray structure of the mouse serotonin 5-HT3 receptor. *Nature* 512: 276–281.

Hughes, J. and 5 others (1975) Identification of two related pentapeptides from the brain with potent opiate agonist activity. *Nature* 258: 577–580.

Karakas, E. and H. Furukawa (2014) Crystal structure of a heterotetrameric NMDA receptor ion channel. *Science* 344: 992–997.

Kawate, T., J. C. Michel, W. T. Birdsong and E. Gouaux (2009) Crystal structure of the ATP-gated $P2X_4$ ion channel in the closed state. *Nature* 460: 592–598.

Ledent, C. and 9 others (1997) Aggressiveness, hypoalgesia and high blood pressure in mice lacking the adenosine A_{2a} receptor. *Nature* 388: 674–678.

Miller, P. S. and A. R. Aricescu (2014) Crystal structure of a human $GABA_A$ receptor. *Nature* 512: 270–275.

Miyazawa, A., Y. Fujiyoshi and N. Unwin (2003) Structure and gating mechanism of the acetylcholine receptor pore. *Nature* 423: 949–955.

Mott, D. D., M. Benveniste and R. J. Dingledine (2008) pH-dependent inhibition of kainate receptors by zinc. *J. Neurosci.* 28: 1659–1671.

Ohno-Shosaku, T., T. Maejima and M. Kano (2001) Endogenous cannabinoids mediate retrograde signals from depolarized postsynaptic neurons to presynaptic terminals. *Neuron* 29: 729–738.

Rasmussen, S. G. and 12 others (2007) Crystal structure of the human beta2 adrenergic G-protein-coupled receptor. *Nature* 450: 383–387.

Rasmussen, S. G. and 19 others (2011) Crystal structure of the β_2 adrenergic receptor–Gs protein complex. *Nature* 477: 549–555.

Sobolevsky, A. I., M. P. Rosconi and E. Gouaux (2009) X-ray structure, symmetry and mechanism of an AMPA-subtype glutamate receptor. *Nature* 462: 745–756.

Thal, D. M. and 13 others (2016) Crystal structures of the M1 and M4 muscarinic acetylcholine receptors. *Nature* 531: 335–340.

Unwin, N. (2005) Refined structure of the nicotinic acetylcholine receptor at 4 Å resolution. *J. Mol. Biol.* 346: 967–989.

Wacker D. and 12 others (2017) Crystal structure of an LSD-bound human serotonin receptor. *Cell* 168: 377–389.

Wang, Y. M. and 8 others (1997) Knockout of the vesicular monoamine transporter 2 gene results in neonatal death and supersensitivity to cocaine and amphetamine. *Neuron* 19: 1285–1296.

Watanabe, J., A. Rozov and L. P. Wollmuth (2005) Target-specific regulation of synaptic amplitudes in the neocortex. *J. Neurosci.* 25: 1024–1033.

Zhu, S. and 6 others (2016) Mechanism of NMDA receptor inhibition and activation. *Cell* 165: 704–714.

Books

Cooper, J. R., F. E. Bloom and R. H. Roth (2003) *The Biochemical Basis of Neuropharmacology.* New York: Oxford University Press.

Nestler, E., S. Hyman, D. M. Holtzman and R. Malenka (2015) *Molecular Neuropharmacology: A Foundation for Clinical Neuroscience,* 3rd Edition. New York: McGraw Hill.

Siegel, G. J., R. W. Albers, S. Brady and D. Price (2012) *Basic Neurochemistry: Principles of Molecular, Cellular, and Medical Neurobiology.* Burlington, MA: Elsevier Academic Press.

Van Dongen, A. M. (2009) *Biology of the NMDA Receptor.* Boca Raton, FL: CRC Press.

Go to the NEUROSCIENCE 6e Companion Website at **www.oup.com/uk/Purves6e** for Web Topics, Animations, Flashcards, and more.

CHAPTER

7 Molecular Signaling within Neurons

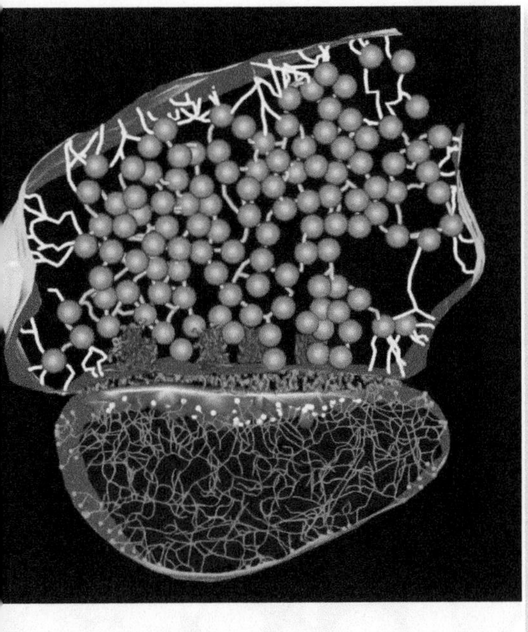

Overview

ELECTRICAL AND CHEMICAL SIGNALING MECHANISMS allow one nerve cell to receive and transmit information to another. This chapter focuses on the related events within neurons and other cells that are triggered by the interaction of a chemical signal with its receptor. This intracellular processing typically begins when extracellular chemical signals (such as neurotransmitters, hormones, and trophic factors) bind to specific receptors located either on the surface or in the cytoplasm or nucleus of the target cells. Such binding activates the receptors and in so doing stimulates cascades of intracellular reactions involving GTP-binding proteins, second-messenger molecules, protein kinases, ion channels, and many other effector proteins whose modulation temporarily changes the physiological state of the target cell. These same intracellular signal transduction pathways can also cause longer-lasting changes by altering the transcription of genes, thus affecting the protein composition of the target cells on a more permanent basis. The large number of components involved in intracellular signaling pathways allows precise temporal and spatial control over the function of individual neurons, thereby enabling control and coordination of the activity of neurons that comprise neural circuits and systems.

Strategies of Molecular Signaling

Chemical communication coordinates the behavior of individual nerve and glial cells in physiological processes that range from neural differentiation to learning and memory. Indeed, molecular signaling ultimately mediates and modulates all brain functions. To carry out such communication, a series of extraordinarily diverse and complex chemical signaling pathways has evolved. The preceding chapters have described in some detail the electrical signaling mechanisms that allow neurons to generate action potentials for conduction of information. Those chapters also described synaptic transmission, a special form of chemical signaling that transfers information from one neuron to another. But chemical signaling is not limited to synapses. Other well-characterized forms of chemical communication include **paracrine** signaling, which acts over a longer range than synaptic transmission and involves the secretion of chemical signals onto a group of nearby target cells, and **endocrine** signaling, which refers to the secretion of hormones into the bloodstream, where they can affect targets throughout the body (Figure 7.1).

Chemical signaling of any sort requires three components: a molecular *signal* that transmits information from one cell to another; a *receptor* molecule that transduces the information provided by the signal; and an *effector* molecule that mediates the cellular response (see Figure 7.1A). The part of this process that takes place within the confines

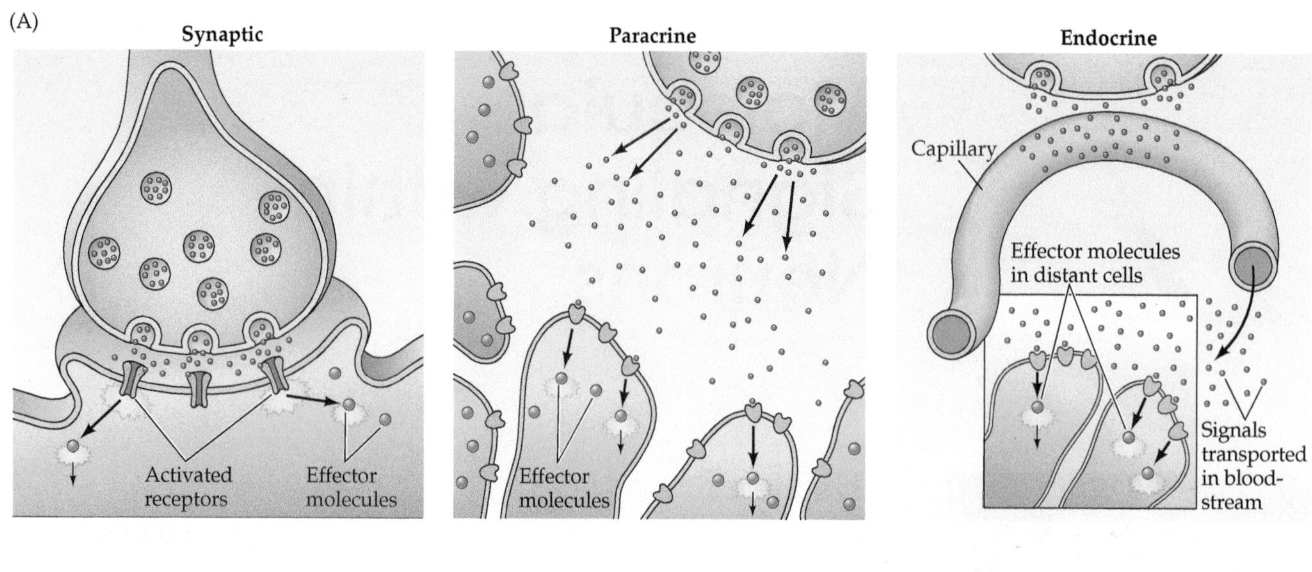

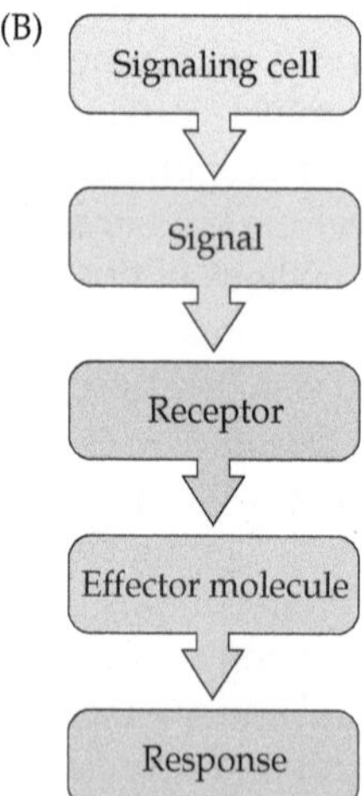

FIGURE 7.1 Chemical signaling. (A) Forms of chemical communication include synaptic transmission, paracrine signaling, and endocrine signaling. (B) The essential components of chemical signaling are: cells that initiate the process by releasing signaling molecules; specific receptors on target cells; intracellular effector molecules; and subsequent cellular responses.

of the target cell is called **intracellular signal transduction**. A good example of transduction in the context of *intercellular* communication is the sequence of events triggered by chemical synaptic transmission (see Figure 7.1A and Chapter 5): Neurotransmitters serve as the signal, neurotransmitter receptors serve as the transducing receptor, and the effector molecule is an ion channel that is opened or closed to produce the electrical response of the postsynaptic cell. In many cases, however, synaptic transmission activates additional *intracellular* pathways that have a variety of functional consequences. For example, the binding of the neurotransmitter norepinephrine to its receptor activates GTP-binding proteins (see Figure 6.16B), which produces second messengers within the postsynaptic target, activates enzyme cascades, and eventually changes the chemical properties of numerous effector molecules within the affected cell.

A general advantage of chemical signaling in both intercellular and intracellular contexts is **signal amplification**. Amplification occurs because individual signaling reactions can generate a much larger number of molecular products than the number of molecules that initiate the reaction. In the case of norepinephrine signaling, for example, a single norepinephrine molecule binding to its receptor can generate many thousands of second-messenger molecules (such as cyclic AMP), yielding an amplification of tens of thousands of phosphates transferred to effector proteins (Figure 7.2). Similar amplification occurs in all signal transduction pathways. Because the transduction processes often are mediated by a sequential set of enzymatic reactions, each with its own amplification factor, a small number of signal molecules ultimately can activate a very large number of effector molecules. Such amplification guarantees that a physiological response is evoked in the face of other, potentially countervailing, influences.

Another rationale for these complex signal transduction schemes is that they permit precise control of cell behavior over a wide range of times. Some molecular interactions allow information to be transferred rapidly, while others are slower and longer lasting. For example, the signaling cascades associated with synaptic transmission at neuromuscular junctions allow a person to respond to rapidly changing cues, such as the trajectory of a kicked ball, while

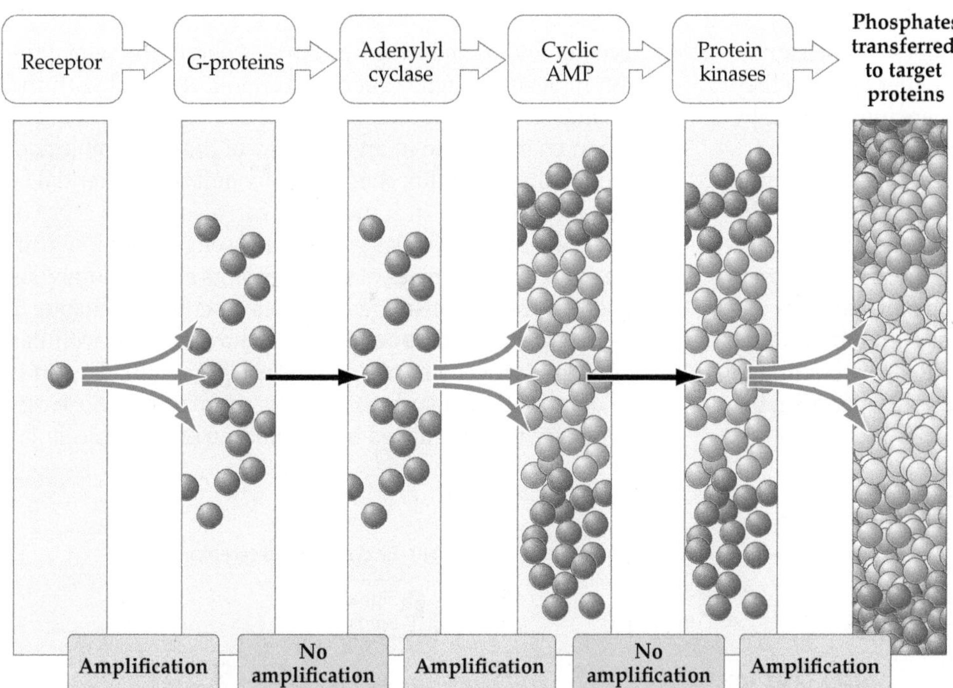

FIGURE 7.2 Amplification in signal transduction pathways. The activation of a single receptor by a signaling molecule, such as the neurotransmitter norepinephrine, can lead to the activation of numerous G-proteins inside cells. These activated proteins can bind to other signaling molecules, such as the enzyme adenylyl cyclase. Each activated enzyme molecule generates a large number of cAMP molecules. cAMP binds to and activates another family of enzymes—the protein kinases—that can phosphorylate many target proteins. Although not every step in this signaling pathway involves amplification, overall the cascade results in a tremendous increase in the potency of the initial signal.

the slower responses triggered by adrenal medullary hormones (epinephrine and norepinephrine) secreted during a challenging game produce slower and longer-lasting effects on muscle metabolism (see Chapter 21) and emotional state (see Chapter 31). To encode information that varies so widely over time, the concentration of the relevant signaling molecules must be carefully controlled. On one hand, the concentration of every signaling molecule within the signaling cascade must return to subthreshold values before the arrival of another stimulus. On the other hand, prolonged activation of the intermediates in a signaling pathway is critical for a sustained response. Having multiple levels of molecular interactions facilitates the intricate timing of these signaling events.

Activation of Signaling Pathways

The molecular components of intracellular signal transduction pathways are always activated by a chemical signaling molecule. Such signaling molecules can be grouped into three classes: **cell-impermeant**, **cell-permeant**, and **cell-associated signaling molecules** (Figure 7.3). The first two classes are secreted molecules and thus can act on target cells removed from the site of signal synthesis or release. Cell-impermeant signaling molecules typically bind to receptors associated with the plasma membrane. Hundreds of secreted molecules have now been identified, including the neurotransmitters discussed in Chapter 6; proteins such as neurotrophic factors (see Chapter 23); and peptide hormones such as glucagon,

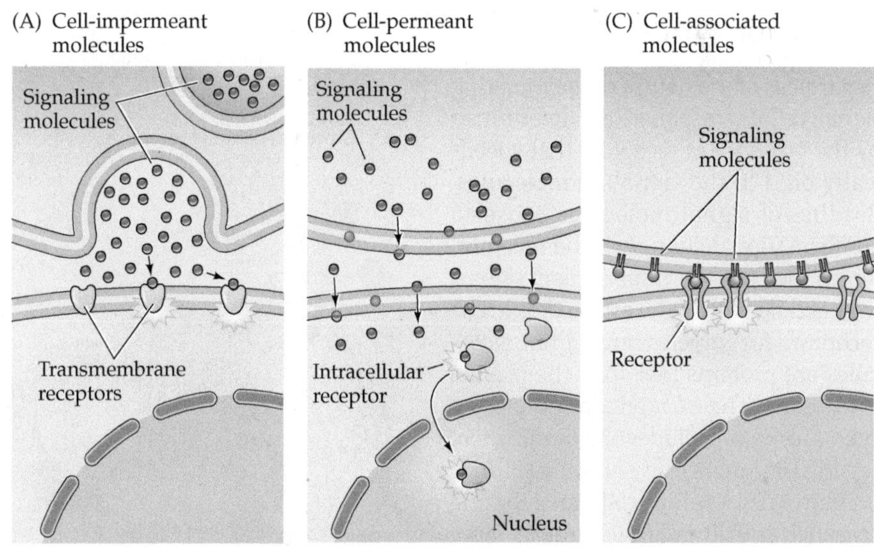

FIGURE 7.3 Three classes of cell signaling molecules. (A) Cell-impermeant molecules, such as neurotransmitters, cannot readily traverse the plasma membrane of the target cell and must bind to the extracellular portion of transmembrane receptor proteins. (B) Cell-permeant molecules are able to cross the plasma membrane and bind to receptors in the cytoplasm or nucleus of target cells. (C) Cell-associated molecules are presented on the extracellular surface of the plasma membrane. These signals activate receptors on target cells only if they are directly adjacent to the signaling cell.

insulin, and various reproductive hormones. These signaling molecules are typically short-lived, either because they are rapidly metabolized or because they are internalized by endocytosis once bound to their receptors.

Cell-permeant signaling molecules can cross the plasma membrane to act directly on receptors that are inside the cell. Examples include numerous steroid hormones (glucocorticoids, estradiol, and testosterone), thyroid hormones (thyroxin), and retinoids. These signaling molecules are relatively insoluble in aqueous solutions and are often transported in blood and other extracellular fluids by binding to specific carrier proteins. In this form, they may persist in the bloodstream for hours or even days.

The third group of chemical signaling molecules, cell-associated signaling molecules, is arrayed on the extracellular surface of the plasma membrane. As a result, these molecules act only on other cells that are physically in contact with the cell that carries such signals. Examples include proteins such as the integrins and neural cell adhesion molecules (NCAMs) that influence axonal growth (see Chapter 23). Membrane-bound signaling molecules are more difficult to study, but are clearly important in neuronal development and other circumstances where physical contact between cells provides information about cellular identities.

Receptor Types

Regardless of the nature of the initiating signal, cellular responses are determined by the presence of receptors that specifically bind to the signaling molecules. Binding of signal molecules causes a conformational change in the receptor, which then triggers a subsequent signaling cascade within the affected cell. The receptors for impermeant signal molecules are proteins that span the plasma membrane. The extracellular domain of such receptors includes the binding site for the signal, while the intracellular domain activates intracellular signaling cascades after the signal binds. A large number of these receptors have been identified and are grouped into families defined by the mechanism used to transduce signal binding into a cellular response (Figure 7.4).

Channel-linked receptors (see Figure 7.4A), also called ligand-gated ion channels (see Figure 4.8), have the receptor and transducing functions as part of the same protein molecule. Interaction of the chemical signal with the binding site of the receptor causes the opening or closing of an ion channel pore in another part of the same molecule. The resulting ion flux changes the membrane potential of the target cell and, in some cases, can also lead to entry of Ca^{2+} ions that serve as a second-messenger signal within the cell. Good examples of such receptors are the numerous ionotropic neurotransmitter receptors described in Chapter 6.

Enzyme-linked receptors also have an extracellular binding site for chemical signals (see Figure 7.4B). The intracellular domain of such receptors is an enzyme whose catalytic activity is regulated by the binding of an extracellular

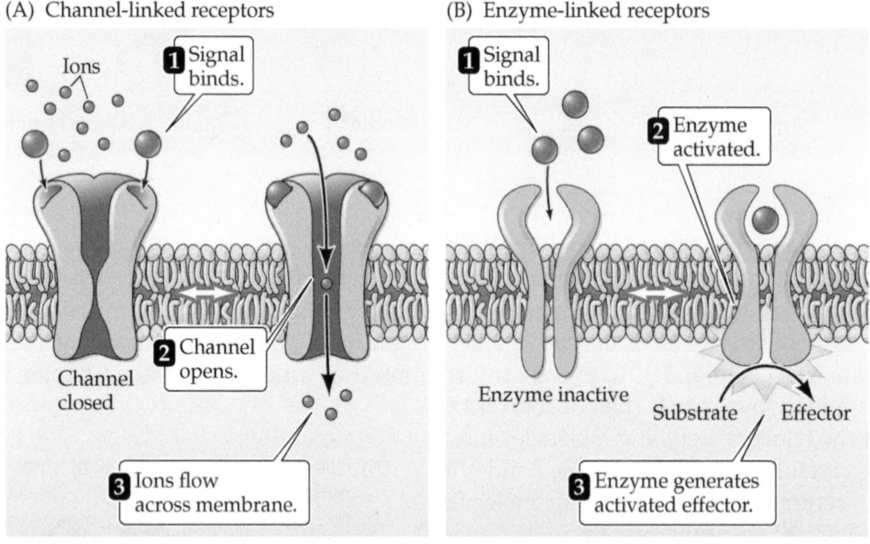

FIGURE 7.4 Categories of cellular receptors. Cell-impermeant signaling molecules can bind to and activate channel-linked receptors (A), enzyme-linked receptors (B), or G-protein-coupled receptors (C). Cell-permeant signaling molecules activate intracellular receptors (D).

signal. The great majority of these receptors are **protein kinases**, often tyrosine kinases, that phosphorylate intracellular target proteins, thereby changing the physiological function of the target cells. Noteworthy members of this group of receptors are the Trk family of neurotrophin receptors (see Chapter 23) and other receptors for growth factors.

G-protein-coupled receptors (see Figure 7.4C), also called metabotropic receptors (see Chapter 5), regulate intracellular reactions by an indirect mechanism involving an intermediate transducing molecule, called a **GTP-binding protein** (or **G-protein**). Hundreds of different G-protein-linked receptors have been identified. Well-known examples include the β-adrenergic receptor, the muscarinic type of acetylcholine receptor, and metabotropic glutamate receptors discussed in Chapter 6, as well as the receptors for odorants in the olfactory system, and many types of receptors for peptide hormones. Rhodopsins, the light-sensitive proteins of retinal photoreceptors, are another type of G-protein-linked receptor whose activating signal is photons of light, rather than a chemical signal (see Figure 11.9).

Intracellular receptors are activated by cell-permeant or lipophilic signaling molecules (see Figure 7.4D). Many of these receptors lead to the activation of signaling cascades that produce new mRNA and protein within the target cell. Often such receptors comprise a receptor protein bound to an inhibitory protein complex. When the signaling molecule binds to the receptor, the inhibitory complex dissociates to expose a DNA-binding domain on the receptor. This activated form of the receptor can then move into the nucleus and directly interact with nuclear DNA, resulting in altered transcription. Some intracellular receptors are located primarily in the cytoplasm, while others are in the nucleus. In either case, once these receptors are activated they affect gene expression by altering DNA transcription.

G-Proteins and Their Molecular Targets

Both G-protein-coupled receptors and enzyme-linked receptors can activate biochemical reaction cascades that ultimately modify the function of target proteins. For both of these receptor types, the coupling between receptor activation and their subsequent effects is provided by GTP-binding proteins. There are two general classes of GTP-binding proteins (Figure 7.5). **Heterotrimeric G-proteins** consist of three distinct subunits (α, β, and γ). There are many different α, β, and γ subunits, allowing a bewildering number of G-protein permutations. Regardless of the specific composition of the heterotrimeric G-protein, its α subunit binds to guanine nucleotides, either GTP or GDP. Binding of GDP then allows the α subunit to bind to the β and γ subunits to

(A) Heterotrimeric G-proteins

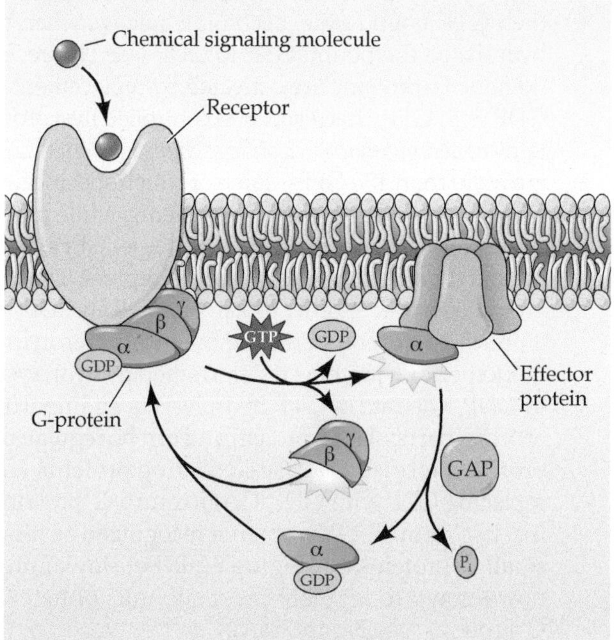

(B) Monomeric G-proteins

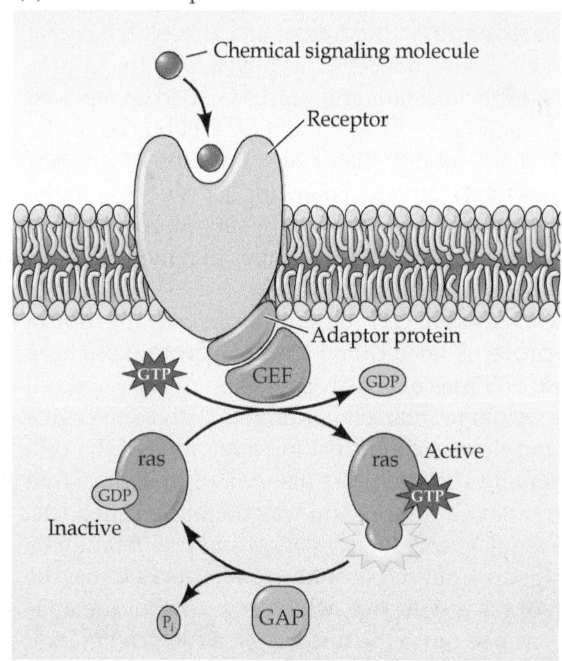

FIGURE 7.5 Types of GTP-binding proteins. (A) Heterotrimeric G-proteins consist of three distinct subunits (α, β, and γ). Receptor activation causes the binding of the G-protein and the α subunit to exchange GDP for GTP, leading to a dissociation of the α and βγ subunits. The biological actions of these G-proteins are terminated by hydrolysis of GTP, which is enhanced by GTPase-activating (GAP) proteins. (B) Monomeric G-proteins use similar mechanisms to relay signals from activated cell surface receptors to intracellular targets. The biological actions of these G-proteins depends on binding of GTP, which is activated by guanine nucleotide exchange factors (GEFs) that bind to the receptor in association with adaptor proteins, and their activity is terminated by hydrolysis of GTP, which is also regulated by GAP proteins.

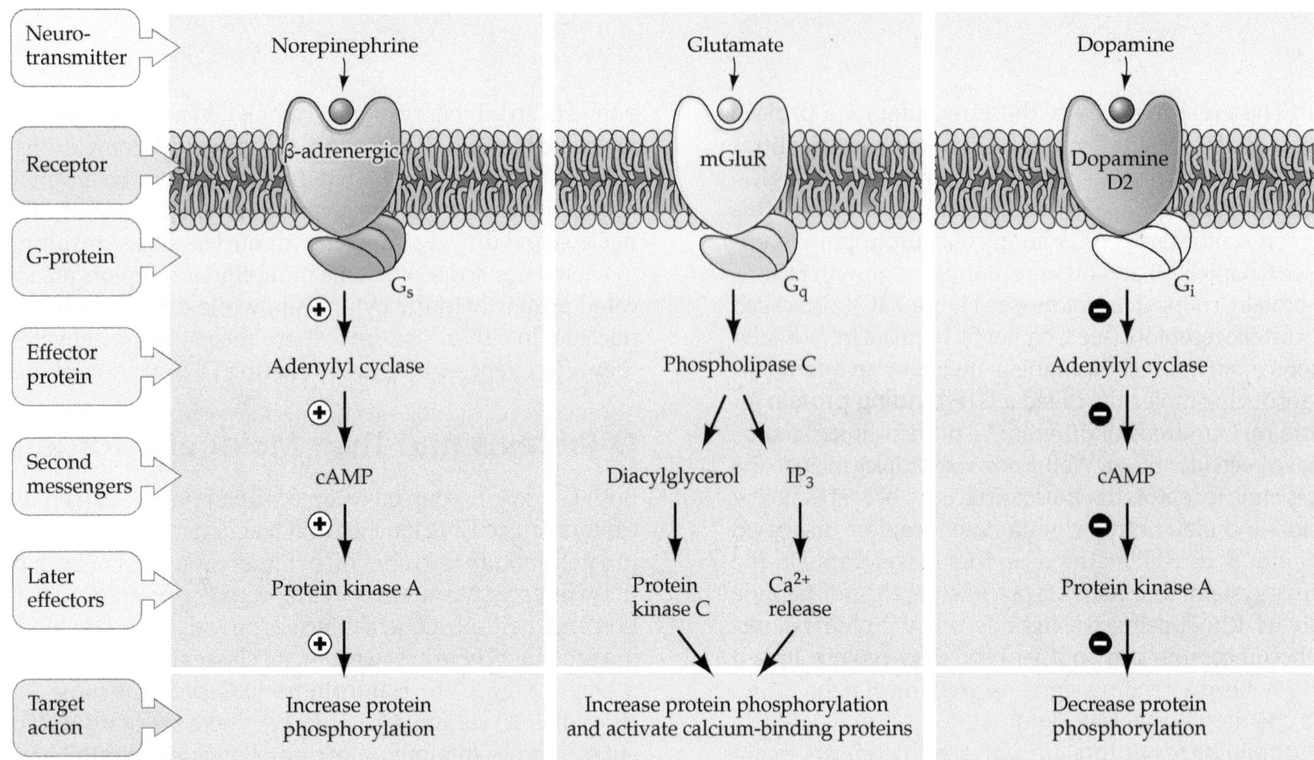

FIGURE 7.6 Effector pathways associated with G-protein-coupled receptors. In all three examples shown here, binding of a neurotransmitter to such a receptor leads to activation of a G-protein and subsequent recruitment of second-messenger pathways. G_s, G_q, and G_i refer to three different types of heterotrimeric G-proteins.

form an inactive trimer. Binding of an extracellular signal to a G-protein-coupled receptor in turn allows the G-protein to bind to the receptor and causes GDP to be replaced with GTP (see Figure 7.5A). When GTP is bound to the G-protein, the α subunit dissociates from the βγ complex and activates the G-protein. Following activation, both the GTP-bound α subunit and the free βγ subunit complex can bind to downstream effector molecules that mediate a variety of responses in the target cell.

The second class of GTP-binding proteins is the **monomeric G-proteins** (also called **small G-proteins**). These monomeric GTPases also relay signals from activated cell surface receptors to intracellular targets such as the cytoskeleton and the vesicle trafficking apparatus of the cell. The first small G-protein was discovered in a virus that causes *ra*t *s*arcoma tumors and was therefore called **ras**. Ras helps regulate cell differentiation and proliferation by relaying signals from receptor kinases to the nucleus; the viral form of ras is defective, which accounts for the ability of the virus to cause the uncontrolled cell proliferation that leads to tumors. Ras is known to be involved in many forms of neuronal signaling, including long-term synaptic potentiation (see Chapter 8). Since the discovery of ras, a large number of small GTPases have been identified and can be sorted into five different subfamilies with different functions. For instance, some are involved in vesicle trafficking in the presynaptic terminal or elsewhere in the neuron, while others play a central role in protein and RNA trafficking in and out of the nucleus.

Similar to heterotrimeric G-proteins, monomeric G-proteins function as molecular timers that are active in their GTP-bound state, becoming inactive when they have hydrolyzed the bound GTP to GDP (see Figure 7.5B). Monomeric G-proteins are activated by replacement of bound GDP with GTP; this reaction is controlled by a group of proteins called *guanine nucleotide exchange factors* (GEFs). There are more than 100 GEFs, some of which specifically activate single types of monomeric G-proteins, while other can activate multiple monomeric G-proteins. GEFs, in turn, are activated by binding to activated receptors. This binding is promoted by yet another protein, termed an adaptor protein.

Termination of signaling by both heterotrimeric and monomeric G-proteins is determined by hydrolysis of GTP to GDP. The rate of GTP hydrolysis is an important property of a particular G-protein and can be regulated by other proteins, termed GTPase-activating proteins (GAPs). By replacing GTP with GDP, GAPs return G-proteins to their inactive form. GAPs were first recognized as regulators of small G-proteins (see Figure 7.5B), but similar proteins are now known to regulate the α subunits of heterotrimeric G-proteins (see Figure 7.5A).

Activated G-proteins alter the function of many downstream effectors. Most of these effectors are enzymes that produce intracellular second messengers. Effector enzymes include adenylyl cyclase, guanylyl cyclase, phospholipase C, and others (Figure 7.6). The second messengers produced by these enzymes trigger the complex biochemical signaling cascades discussed in the next section. Because each of these cascades is activated by specific G-protein subunits, the pathways activated by a particular receptor are determined by the specific identity of the G-protein subunits associated with it.

G-proteins can also directly bind to and activate ion channels. For example, activation of muscarinic receptors by acetylcholine can open K^+ channels, thereby reducing the rate at which the neurons fire action potentials. Such inhibitory responses are believed to be the result of $\beta\gamma$ subunits of G-proteins binding to the K^+ channels. The activation of α subunits can also lead to the rapid closing of voltage-gated Ca^{2+} and Na^+ channels. Because these channels carry inward currents involved in generating action potentials, closing them also makes it more difficult for target cells to fire (see Chapters 3 and 4). Thus, by directly regulating the gating of ion channels, G-proteins can influence the electrical signaling of target cells.

In summary, the binding of chemical signals to their receptors activates cascades of signal transduction events in the cytosol of target cells. Within such cascades, G-proteins serve a pivotal function as the molecular transducing elements that couple membrane receptors to their molecular effectors within the cell. The diversity of G-proteins and their downstream targets leads to many types of physiological responses.

Second Messengers

Neurons use many different second messengers as intracellular signals. These messengers differ in the mechanism by which they are produced and removed, as well as in their downstream targets and effects (Figure 7.7A). This section summarizes the attributes of some of the principal second messengers.

• *Calcium*. The calcium ion (Ca^{2+}) is perhaps the most common intracellular messenger in neurons. Indeed, few neuronal functions are immune to the influence—direct or indirect—of Ca^{2+}. In all cases, information is transmitted by a transient rise in the cytoplasmic calcium concentration, which allows Ca^{2+} to bind to and activate a large number of Ca^{2+}-binding proteins that serve as molecular targets. One of the most thoroughly studied targets of Ca^{2+} is **calmodulin**, a Ca^{2+}-binding protein abundant in the cytosol of all cells. Binding of Ca^{2+} to calmodulin activates this protein, which then initiates its effects by binding to still other downstream targets, such as protein kinases. Another important family of intracellular Ca^{2+}-binding proteins are the synaptotagmins, which serve as Ca^{2+} sensors during neurotransmitter release and other forms of intracellular membrane fusion (see Chapter 5).

Ordinarily the concentration of Ca^{2+} ions in the cytosol is extremely low, typically 50 to 100 nanomolar (10^{-9} *M*). The concentration of Ca^{2+} ions outside neurons—in the bloodstream or cerebrospinal fluid, for instance—is several orders of magnitude higher, typically several millimolar ($10^{-3}M$). This steep Ca^{2+} gradient is maintained by several mechanisms (Figure 7.7B). Most important in this maintenance are two proteins that translocate Ca^{2+} from the cytosol to the extracellular medium: an ATPase called the **calcium pump**; and an **Na^+/Ca^{2+} exchanger**, which is a protein that replaces intracellular Ca^{2+} with extracellular sodium ions (see Chapter 4). In addition to these plasma membrane mechanisms, Ca^{2+} is also pumped into the endoplasmic reticulum and mitochondria. These organelles can thus serve as storage depots of Ca^{2+} ions that are later released to participate in signaling events. Finally, nerve cells contain other Ca^{2+}-binding proteins—such as **calbindin**—that serve as Ca^{2+} buffers. Such buffers reversibly bind Ca^{2+} and thus blunt the magnitude and slow the kinetics of Ca^{2+} signals within neurons.

The Ca^{2+} ions that act as intracellular signals enter the cytosol by means of one or more types of Ca^{2+}-permeable ion channels (see Chapter 4). These can be voltage-gated Ca^{2+} channels or ligand-gated channels in the plasma membrane, both of which allow Ca^{2+} to flow down the Ca^{2+} gradient and into the cell from the extracellular medium. In addition, other channels allow Ca^{2+} to be released from the interior of the endoplasmic reticulum into the cytosol. These intracellular Ca^{2+}-releasing channels are gated, so they can be opened or closed in response to various intracellular signals. One such channel is the **inositol trisphosphate (IP_3) receptor**. As the name implies, this channel is regulated by IP_3, a second messenger described in more detail below. A second type of intracellular Ca^{2+}-releasing channel is the **ryanodine receptor**, named after a drug that binds to and partially opens these receptors. Among the biological signals that activate ryanodine receptors are cytoplasmic Ca^{2+} and, at least in muscle cells, depolarization of the plasma membrane.

These various mechanisms for elevating and removing Ca^{2+} ions allow precise control of both the timing and location of Ca^{2+} signaling within neurons, which in turn permit Ca^{2+} to control many different signaling events. For example, voltage-gated Ca^{2+} channels allow Ca^{2+} concentrations to rise very rapidly and locally within presynaptic terminals to trigger neurotransmitter release, as described in Chapter 5. Slower and more widespread rises in Ca^{2+} concentration regulate a wide variety of other responses, including gene expression in the cell nucleus.

• *Cyclic nucleotides.* Another important group of second messengers is the cyclic nucleotides, specifically cyclic adenosine monophosphate (cAMP) and cyclic guanosine monophosphate (cGMP) (Figure 7.7C). Cyclic AMP is a derivative of the abundant cellular energy storage molecule ATP, and is produced when G-proteins activate adenylyl cyclase in the plasma membrane. Adenylyl cyclase converts

FIGURE 7.7 Neuronal second messengers. (A) Mechanisms responsible for producing and removing second messengers, and the downstream targets of these messengers. (B) Proteins involved in delivering calcium to the cytoplasm and in removing calcium from the cytoplasm. (C) Mechanisms for producing and degrading cyclic nucleotides. (D) Pathways involved in producing and removing diacylglycerol and IP_3.

(A)

Second messenger	Sources	Intracellular targets	Removal mechanisms
Ca^{2+}	Plasma membrane: Voltage-gated Ca^{2+} channels; Various ligand-gated channels; Endoplasmic reticulum: IP_3 receptors; Ryanodine receptors	Calmodulin; Protein kinases; Protein phosphatases; Ion channels; Synaptotagmins; Many other Ca^{2+}-binding proteins	Plasma membrane: Na^+/Ca^{2+} exchanger; Ca^{2+} pump; Endoplasmic reticulum: Ca^{2+} pump; Mitochondria
Cyclic AMP	Adenylyl cyclase acts on ATP	Protein kinase A; Cyclic nucleotide-gated channels	cAMP phosphodiesterase
Cyclic GMP	Guanylyl cyclase acts on GTP	Protein kinase G; Cyclic nucleotide-gated channels	cGMP phosphodiesterase
IP_3	Phospholipase C acts on PIP_2	IP_3 receptors on endoplasmic reticulum	Phosphatases
Diacylglycerol	Phospholipase C acts on PIP_2	Protein kinase C	Various enzymes

(B)

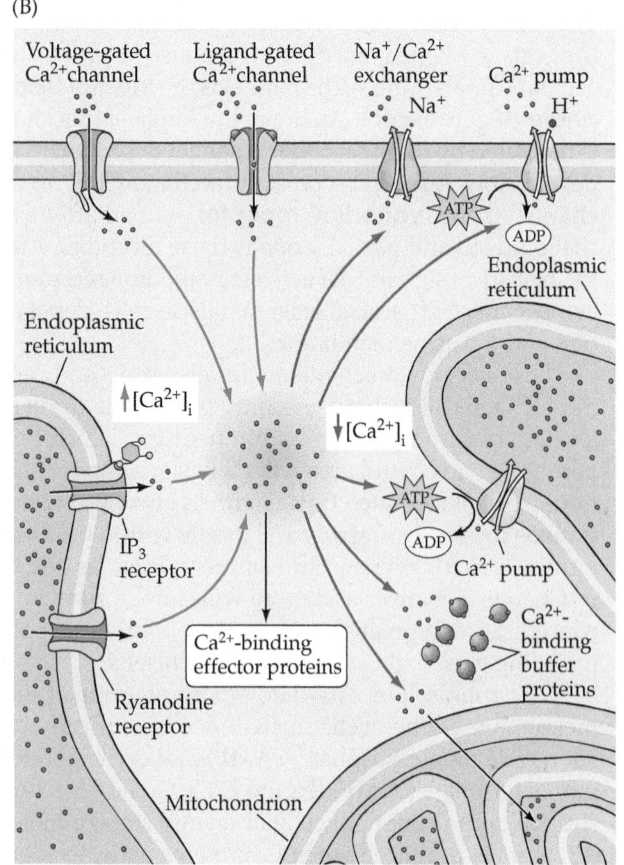

(C)

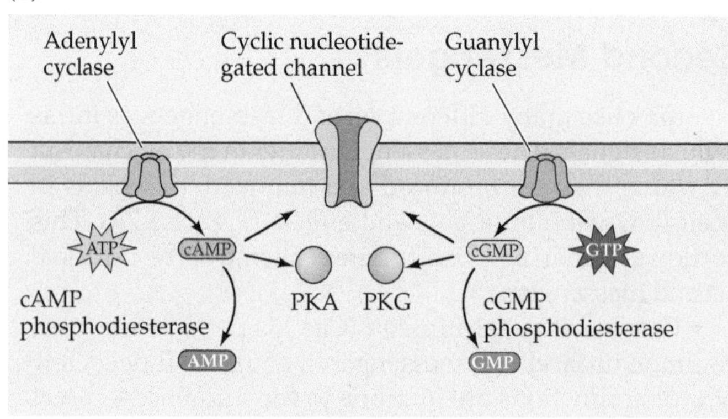

(D)

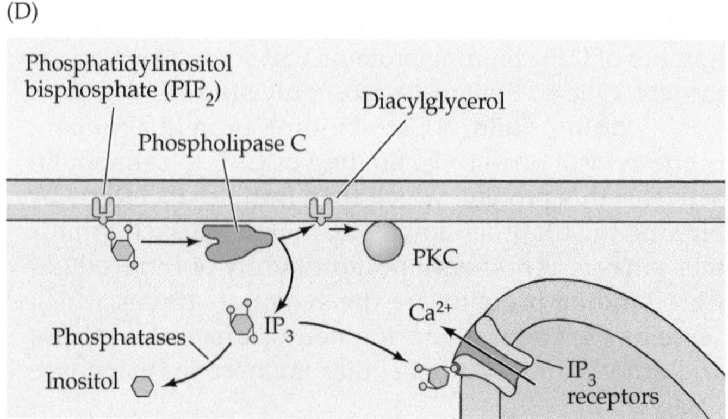

ATP into cAMP by removing two phosphate groups from the ATP. Cyclic GMP is similarly produced from GTP by the action of guanylyl cyclase. Once the intracellular concentration of cAMP or cGMP is elevated, these nucleotides can bind to two different classes of targets. The most common targets of cyclic nucleotide action are protein kinases, either the cAMP-dependent protein kinase (PKA) or the cGMP-dependent protein kinase (PKG). These enzymes mediate many physiological responses by phosphorylating target proteins, as described in the next section. In addition, cAMP and cGMP can influence neuronal signaling by binding to certain ligand-gated ion channels (see Figure 4.8D). These cyclic nucleotide-gated channels are particularly important in phototransduction and other sensory transduction processes, such as olfaction. Cyclic nucleotide signals are degraded by phosphodiesterases, enzymes that cleave phosphodiester bonds and convert cAMP into AMP or cGMP into GMP.

• *Diacylglycerol and* IP_3. Remarkably, membrane lipids can also be converted into intracellular second messengers (Figure 7.7D). The two most important messengers of this type are produced from phosphatidylinositol bisphosphate (PIP_2). This lipid component is cleaved by phospholipase C, an enzyme activated by certain G-proteins and by calcium ions. Phospholipase C splits the PIP_2 into two smaller molecules, each of which acts as a second messenger. One of these messengers is diacylglycerol (DAG), a molecule that remains within the membrane and activates protein kinase C, which phosphorylates substrate proteins in both the plasma membrane and elsewhere. The other messenger is IP_3, which leaves the cell membrane and diffuses within the cytosol. IP_3 binds to IP_3 receptors, channels that release calcium from the endoplasmic reticulum. Thus, the action of IP_3 is to produce yet another second messenger (perhaps a third messenger, in this case!) that triggers an entire spectrum of reactions in the cytosol. The actions of DAG and IP_3 are terminated by enzymes that convert these two molecules into inert forms that can be recycled to produce new molecules of PIP_2.

The intracellular concentration of these second messengers changes dynamically over time, allowing precise control over their downstream targets. These signals can also be localized to small compartments within single cells or can spread over great distances, some even spreading between cells via gap junctions (see Chapter 5). Understanding of the complex temporal and spatial dynamics of these second-messenger signals has been greatly aided by the development of imaging techniques that visualize second messengers and other molecular signals within cells (Box 7A).

BOX 7A Dynamic Imaging of Intracellular Signaling

Dramatic breakthroughs in our understanding of the brain often rely on development of new experimental techniques. This certainly has been true for our understanding of intracellular signaling in neurons, which has benefited enormously from the invention of imaging techniques that allow direct visualization of signaling processes within living cells. The first advance—and arguably the most significant—came from the development, by Roger Tsien and his colleagues, of the fluorescent dye fura-2 (Figure A). Calcium ions bind to fura-2 and cause the dye's fluorescence properties to change. When fura-2 is introduced inside cells and the cells are then imaged with a fluorescence microscope, this dye serves as a reporter of intracellular Ca^{2+} concentration. Fura-2 imaging has allowed investigators to detect the spatial and temporal dynamics of the Ca^{2+} signals that trigger innumerable process within neurons and glial cells; for example, fura-2 was used to obtain the image of Ca^{2+} signaling during neurotransmitter release shown in Figure 5.10A.

Subsequent refinement of the chemical structure of fura-2 has yielded many other fluorescent Ca^{2+} indicator dyes with different fluorescence properties and different sensitivities to Ca^{2+}. One of these dyes is Calcium Green, which was used to image the dynamic changes in Ca^{2+} concentration produced within the dendrites of cerebellar Purkinje cells by the intracellular messenger IP_3 (Figure B). Further developments have led to indicators for visualizing the spatial and temporal dynamics of other second-messenger signals, such as cAMP.

Another tremendous advance in the dynamic imaging of signaling processes came from the discovery of a green fluorescent protein that was first isolated from the jellyfish *Aequorea victoria* by Osamu Shimomura. Green fluorescent protein, or GFP, is (as its name indicates) a protein that is brightly fluorescent (Figure C). Molecular cloning of the *GFP* gene allows imaging techniques to visualize expression of gene products labeled with GFP fluorescence. The first such use of GFP was in experiments with the worm *Caenorhabditis elegans*, in which Martin Chalfie and his colleagues rendered neurons fluorescent by inducing GFP expression in these cells. Many subsequent experiments have used expression of GFP to image the structure of individual neurons in the mammalian brain (Figure D).

Molecular genetic strategies make it possible to attach GFP to almost any protein, thereby allowing fluorescence microscopy to image the spatial distribution of labeled proteins. In this way, it has been possible to visualize dynamic changes in the location of neuronal proteins during signaling events. Related techniques also allow visualization of the location of second-messenger signals or the biochemical activity of signaling proteins. For example, Figure 8.12 illustrates the use of this approach to monitor the activation of CaMKII during long-term synaptic potentiation.

As was the case with fura-2, subsequent refinement of GFP has led to numerous improvements. One significant improvement, also pioneered by Roger

Continued on the next page

BOX 7A (*continued*)

Tsien, was the production of proteins that fluoresce in colors other than green, thus permitting the simultaneous imaging of multiple types of proteins and/or neurons. A particularly vivid demonstration of the power of multicolor imaging of fluorescent proteins can be seen in Brainbow, a technique that uses differential expression of combinations of several fluorescent proteins to label neurons with one of nearly 100 different colors (Figure E). This permits the axons of individual neurons to be identified and followed, even through the complex tangle of neuronal processes typically found in the CNS, thereby defining the circuits formed by the labeled neurons.

Just as development of the Golgi staining technique opened our eyes to the cellular composition of the brain (see Chapter 1), study of intracellular signaling in the brain has been revolutionized by fura-2, GFP, and other fluorescent tools. There is no end in sight for the potential of such imaging methods to illuminate new and important aspects of brain signaling dynamics.

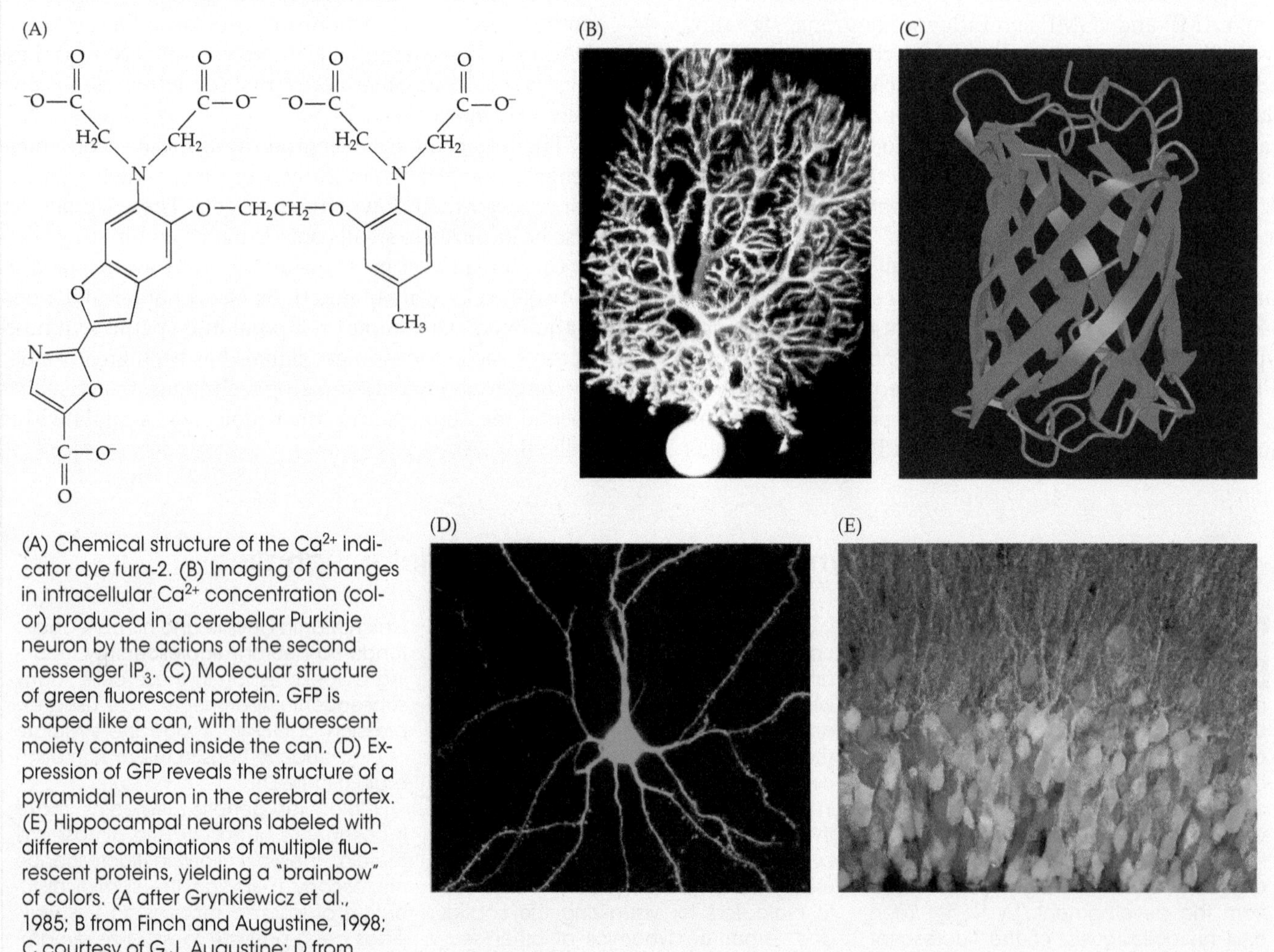

(A) Chemical structure of the Ca^{2+} indicator dye fura-2. (B) Imaging of changes in intracellular Ca^{2+} concentration (color) produced in a cerebellar Purkinje neuron by the actions of the second messenger IP_3. (C) Molecular structure of green fluorescent protein. GFP is shaped like a can, with the fluorescent moiety contained inside the can. (D) Expression of GFP reveals the structure of a pyramidal neuron in the cerebral cortex. (E) Hippocampal neurons labeled with different combinations of multiple fluorescent proteins, yielding a "brainbow" of colors. (A after Grynkiewicz et al., 1985; B from Finch and Augustine, 1998; C courtesy of G.J. Augustine; D from Vidal et al., 2016; E from Livet et al., 2007.)

Second-Messenger Signaling via Protein Phosphorylation

As already mentioned, second messengers typically regulate neuronal functions by modulating the phosphorylation state of intracellular proteins (Figure 7.8). Phosphorylation (the addition of phosphate groups) is a post-translational modification that rapidly and reversibly changes protein function. Proteins are phosphorylated by a wide variety of protein kinases; phosphate groups are removed by other enzymes called **protein phosphatases**. The importance of protein phosphorylation as a regulatory mechanism is emphasized by the fact that the human genome contains more than 500 protein kinase genes and approximately 200 protein phosphatase genes. This means that nearly 3% of the genome is directly dedicated to control of the phosphorylation state of proteins.

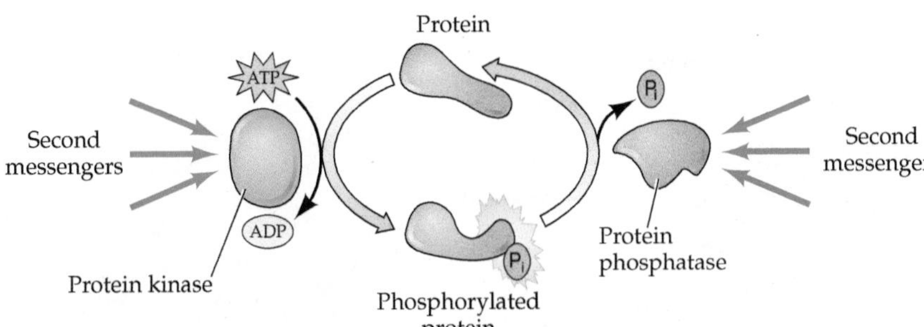

FIGURE 7.8 Regulation of cellular proteins by phosphorylation. Protein kinases transfer phosphate groups (P_i) from ATP to serine, threonine, or tyrosine residues on substrate proteins. This phosphorylation reversibly alters the structure and function of cellular proteins. Removal of the phosphate groups is catalyzed by protein phosphatases. Both kinases and phosphatases are regulated by a variety of intracellular second messengers.

The degree of phosphorylation of a target protein reflects a balance between the competing actions of protein kinases and phosphatases, thereby integrating a host of cellular signaling pathways. The substrates of protein kinases and phosphatases include enzymes, neurotransmitter receptors, ion channels, and structural proteins. Protein kinases and phosphatases typically act either on the serine and threonine residues (Ser/Thr kinases or phosphatases) or on the tyrosine residues (Tyr kinases or phosphatases) of their substrates. Some of these enzymes act specifically on only one or a handful of protein targets, while others are multifunctional and have a broad range of substrate proteins.

Protein Kinases

The activity of most protein kinases is regulated either by second messengers or by extracellular chemical signals such as growth factors (see Chapter 23). Typically, second messengers activate Ser or Thr kinases, whereas extracellular signals activate Tyr kinases. Each protein kinase has catalytic domains responsible for transferring phosphate groups to the relevant amino acids of their target proteins. Kinases that are regulated by second messengers typically have an additional regulatory domain that inhibits the catalytic site. Binding of second messengers (such as cAMP, DAG, and Ca^{2+}) to the regulatory domain removes the inhibition, allowing the catalytic domain to phosphorylate the substrate protein. Other kinases are activated by phosphorylation by another protein kinase; these kinases typically do not have inhibitory regulatory domains. Among the hundreds of protein kinases expressed in the brain, a relatively small number are the main regulators of neuronal signaling.

• *cAMP-dependent protein kinase (PKA).* PKA is the primary effector of cAMP action in neurons and consists of two catalytic subunits and two regulatory subunits (Figure 7.9A). cAMP activates PKA by binding to the regulatory subunits, releasing active catalytic subunits that can phosphorylate many different target proteins. Although the catalytic subunits are similar to the catalytic domains of other protein kinases, distinct amino acids allow PKA to bind to specific target proteins, thus allowing only those targets to be phosphorylated in response to intracellular cAMP signals. Signaling specificity is also achieved by using specific anchoring proteins, called A kinase anchoring proteins (AKAPs), to localize PKA activity to specific compartments within cells.

• *Ca^{2+}/calmodulin-dependent protein kinase, type II (CaMKII).* Ca^{2+} ions binding to calmodulin can regulate numerous protein kinases. In neurons, CaMKII is the predominant Ca^{2+}/calmodulin-dependent protein kinase; this kinase is the most abundant component of the postsynaptic density, a structure important for postsynaptic signaling (see Box 7B). CaMKII comprises 12 subunits that are connected by a central association domain to form a wheel-like structure (Figure 7.9B). Each subunit contains a catalytic domain and a regulatory domain. When intracellular Ca^{2+} concentration is low, binding of the regulatory domain to the catalytic domain inhibits kinase activity. Elevated Ca^{2+} allows Ca^{2+}/calmodulin to bind to the regulatory domain, relieving its inhibition of the catalytic domain. This allows the catalytic domain to extend to form a barrel-shaped structure that can phosphorylate substrate proteins. CaMKII substrates include ion channels and numerous proteins involved in intracellular signal transduction. The multimeric structure of CaMKII also allows neighboring subunits to phosphorylate each other, leading to sustained activation of CaMKII even after intracellular Ca^{2+} concentration returns to basal levels. This process of autophosphorylation is thought to serve as a cellular memory mechanism.

• *Protein kinase C (PKC).* Another important group of protein kinases is PKC, a diverse family of monomeric kinases activated by the second messengers DAG and Ca^{2+}. Ca^{2+} causes PKC to move from the cytosol to the plasma membrane, where the regulatory domains of PKC then bind to DAG and to membrane phospholipids (Figure 7.9C). These events separate the regulatory and catalytic domains, allowing the catalytic domain to phosphorylate various protein substrates. PKC also diffuses to sites other than the plasma membrane—such as the cytoskeleton, perinuclear sites, and the nucleus—where it phosphorylates still other substrate proteins. Tumor-promoting compounds called phorbol esters mimic DAG and cause a prolonged activation of PKC that is thought to trigger tumor formation.

• *Protein tyrosine kinases.* Two classes of protein kinases transfer phosphate groups to tyrosine residues on substrate proteins. Receptor tyrosine kinases are transmembrane proteins with an extracellular domain that binds to protein

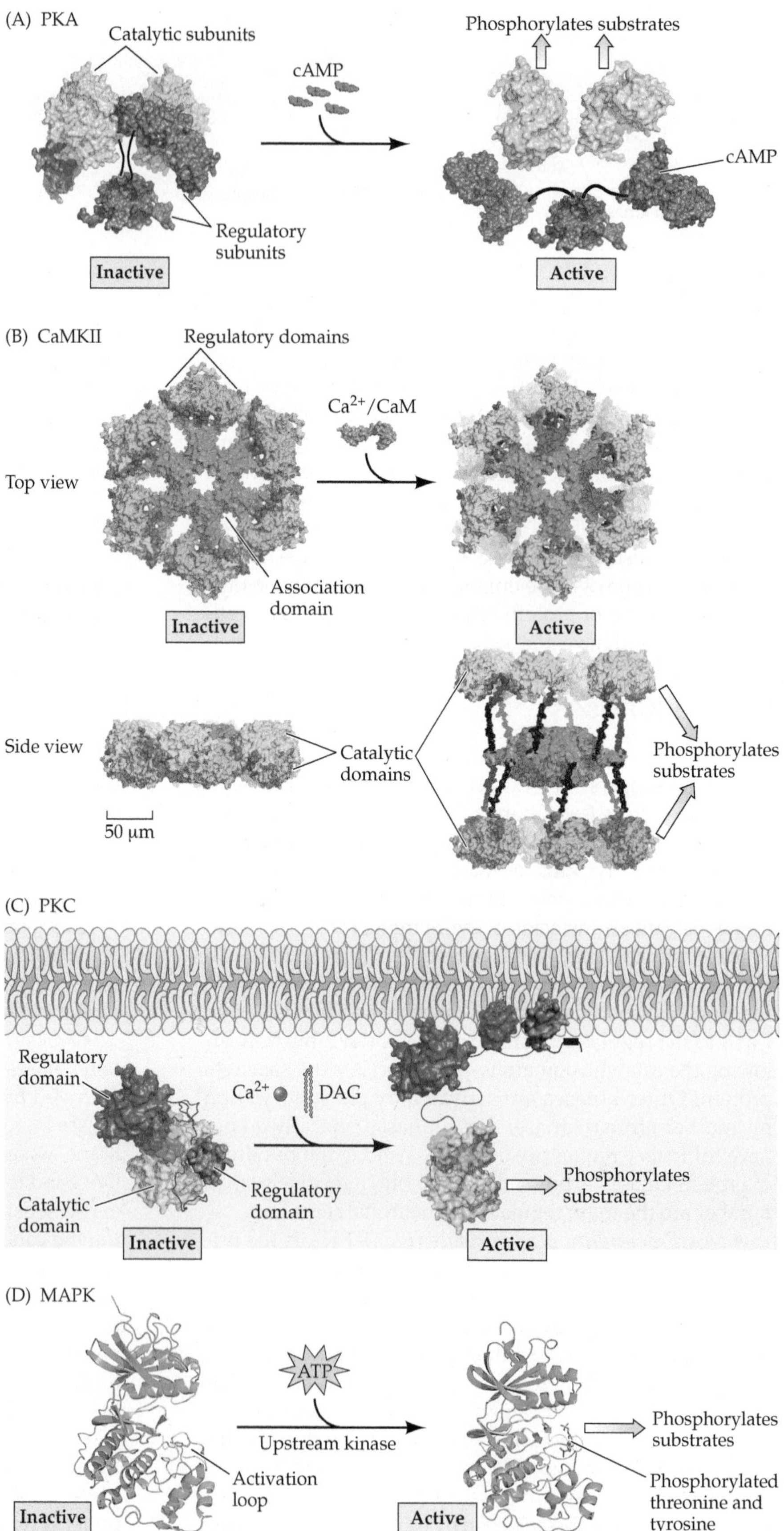

FIGURE 7.9 Activation of protein kinases. (A) In the inactive state (left), the catalytic subunits of PKA are inhibited by the regulatory subunits. Binding of cAMP to the regulatory subunits relieves the inhibition and frees the catalytic subunits to phosphorylate their targets. Black lines indicate flexible structures that connect parts of the regulatory subunit. (B) CaMKII is a large wheel-shaped structure comprising 12 subunits, each with a catalytic (tan/yellow) and a regulatory (blue) domain, held together by a central association domain (green). Binding of Ca^{2+}/calmodulin to the regulatory domain allows the catalytic domain to extend and phosphorylate its substrates. (C) Binding of Ca^{2+} allows lipid-binding domains of PKC (blue) to insert into the plasma membrane and bind to DAG and other membrane lipids. This change in structure and location displaces the regulatory domain (blue) allows the catalytic domain of PKC (yellow) to phosphorylate its substrates. Black lines indicate flexible structures that connect the various domains of PKC. (D) Activation of MAPK is caused by phosphorylation of an activation loop (yellow) by upstream kinases. Phosphorylation of the activation loop allows the catalytic domain of MAPK to assume its active conformation and phosphorylate downstream targets. (A from pdb101.rcsb.org; B after Craddock et al., 2012; C after Leonard et al., 2011; D from Turk, 2007.)

ligands (growth factors, neurotrophic factors, or cytokines) and an intracellular catalytic domain that phosphorylates the relevant substrate proteins. Non-receptor tyrosine kinases are cytoplasmic or membrane-associated enzymes that are indirectly activated by extracellular signals. Tyrosine phosphorylation is less common than Ser or Thr phosphorylation, and it often serves to recruit signaling molecules to the phosphorylated protein. Tyrosine kinases are particularly important for cell growth and differentiation (see Chapters 22 and 23).

• *Mitogen-activated protein kinase (MAPK).* MAPKs, also called extracellular signal-regulated kinases (ERKs), are important examples of protein kinases that are activated via phosphorylation by another protein kinase. MAPKs were first identified as participants in the control of cell growth but are now known to have many other signaling functions. MAPKs are inactive at rest but become activated when they are phosphorylated. In fact, MAPKs are part of a kinase cascade in which one protein kinase phosphorylates and activates the next protein kinase in the cascade. The signals that trigger these kinase cascades are often extracellular growth factors that bind to receptor tyrosine kinases that, in turn, activate monomeric G-proteins such as ras. However, MAPKs are also activated by other types of signals, such as osmotic stress and heat shock. Activation of MAPKs is caused by phosphorylation of one or more amino acids within a structure called the activation loop; phosphorylation changes the structure of the activation loop, enabling the catalytic domain of MAPK to phosphorylate downstream targets (Figure 7.9D). These targets include transcription factors—proteins that regulate gene expression (see Figure 7.12)—as well as various enzymes, including other protein kinases, and cytoskeletal proteins.

Protein Phosphatases

Among the several different families of protein phosphatases, the best characterized are the Ser/Thr phosphatases 1, 2A, and 2B. Like many protein kinases, these protein phosphatases consist of both catalytic and regulatory subunits (Figure 7.10). The catalytic subunits remove phosphates from proteins. Because of the remarkable similarity of their catalytic subunits, protein phosphatases display less substrate specificity than protein kinases. Furthermore, most regulatory subunits of phosphatases do not bind to second messengers, making most protein phosphatases constitutively active. Instead, expression of different regulatory subunits determines which substrates are dephosphorylated and where the phosphatase is located within cells.

• *Protein phosphatase 1 (PP1).* PP1 is a dimer consisting of one catalytic subunit and one regulatory subunit (see Figure 7.10A). PP1 activity is determined by which of the more than 200 different regulatory subunits binds to its catalytic subunit. PP1 also is regulated by phosphorylation of regulatory subunits by PKA. PP1 is one of the most prevalent Ser/Thr protein phosphatases in mammalian cells and dephosphorylates a wide array of substrate proteins. In addition to its well-studied regulation of metabolic enzymes, PP1 can also influence neuronal electrical signaling by dephosphorylating K^+ and Ca^{2+} channels, as well as neurotransmitter receptors such as AMPA-type and NMDA-type glutamate receptors.

• *Protein phosphatase 2A (PP2A).* PP2A is one of the most abundant enzymes in the brain and accounts for approximately 1% of the total amount of protein found within cells. PP2A is a multisubunit enzyme consisting of both catalytic and regulatory subunits, as well as an additional scaffold subunit that brings together the catalytic and regulatory subunits (see Figure 7.10B). There are two different versions of the catalytic and scaffold subunits and approximately 25 different regulatory subunits. Different combinations of these three subunits yield more than 80 different versions of PP2A. PP2A has a broad range of substrates that overlap with those of PP1. One of its best-studied substrates is tau, a protein associated with microtubules in the cytoskeleton. Alzheimer's disease is associated with excessive phosphorylation of tau, perhaps due to defects in PP2A. Alterations in PP2A activity also have been implicated in other neurodegenerative diseases, as well as in cancer and diabetes. Although PP2A is constitutively active, its activity can be regulated by phosphorylation and other post-translational modifications of both the catalytic and regulatory subunits.

• *Protein phosphatase 2B (PP2B).* PP2B, or calcineurin, is present at high levels in neurons and comprises a catalytic subunit and a regulatory subunit (see Figure 7.10C). Unlike the activity of PP1 and PP2A, PP2B activity is acutely controlled by intracellular Ca^{2+} signaling: Ca^{2+}/calmodulin activates PP2B by binding to the catalytic subunit and displacing the inhibitory regulatory domain, thereby activating PP2B. Even though both PP2B and CaMKII are activated by Ca^{2+}/calmodulin, they generally have different molecular targets. Substrates of PP2B include a transcriptional regulator, NFAT, and ion channels. Dephosphorylation of AMPA-type glutamate receptors by PP2B is thought to play a central role in signal transduction during long-term depression of hippocampal synapses (see Chapter 8).

In summary, activation of membrane receptors can elicit complex cascades of enzyme activation, resulting in second-messenger production and protein phosphorylation or dephosphorylation. These cytoplasmic signals produce a variety of physiological responses by transiently regulating enzyme activity, ion channels, cytoskeletal proteins, and many other cellular processes. Intracellular signaling mechanisms also serve as targets for numerous psychiatric disorders. At excitatory synapses, these signaling components are often contained within dendritic spines, which appear to serve as specialized signaling compartments within neurons (Box 7B). In addition, such signals can propagate to the nucleus and cause long-lasting changes in gene expression.

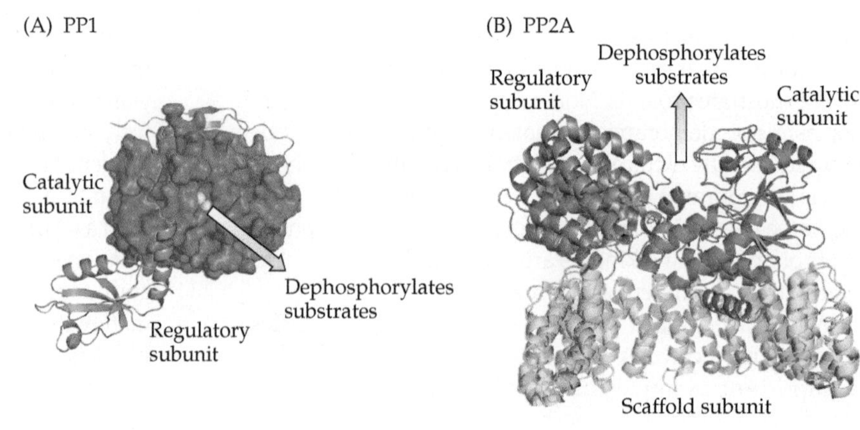

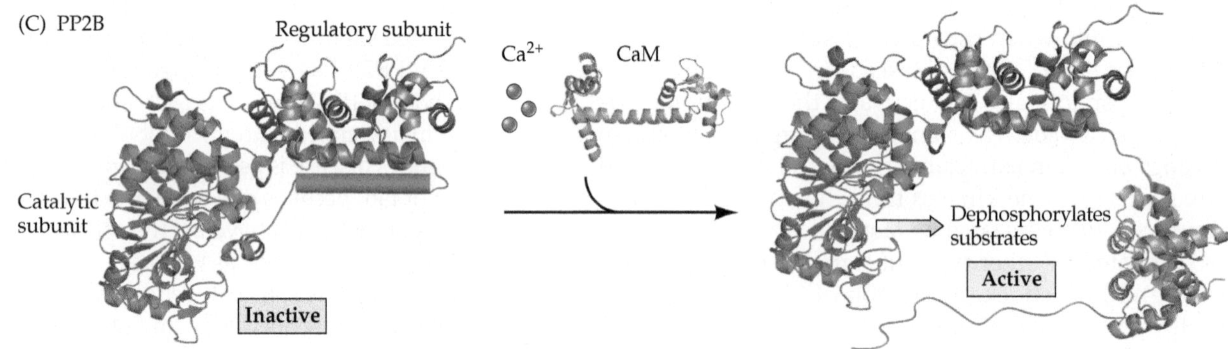

FIGURE 7.10 Types of protein phosphatases. (A) Structure of PP1, consisting of a catalytic subunit (red) bound to spinophilin (blue), a regulatory subunit found in dendritic spines. (B) PP2A is a trimeric enzyme that has a scaffold subunit (yellow), in addition to the catalytic (red) and regulatory (blue) subunits common to other phosphatases. (C) The activity of PP2B, or calcineurin, is regulated by Ca^{2+}/calmodulin. In the absence of Ca^{2+}/calmodulin (left), part of the regulatory subunit blocks the active site of the catalytic subunit, preventing phosphatase activity. Binding of Ca^{2+}/calmodulin to the regulatory subunit relieves this blockade, allowing PP2B to dephosphorylate its substrate proteins. (A from Bollen et al., 2010; B from Cho and Xu, 2007; C after Li et al., 2011.)

BOX 7B ■ Dendritic Spines

Many excitatory synapses in the brain involve microscopic dendritic protrusions known as spines (Figure A). Spines are distinguished by the presence of globular tips called spine heads, which serve as the postsynaptic site of innervation by presynaptic terminals. Spine heads are connected to the main shafts of dendrites by narrow links called spine necks (Figure B). Just beneath the site of contact between presynaptic terminals and spine heads are intracellular structures called postsynaptic densities. The number, size, and shape of spines are quite variable along some dendrites (Figure C). At least in some cases, spine shape can change dynamically over time (see Figure 8.15), and is altered in several neurodegenerative disorders (such as Alzheimer's disease) and psychiatric disorders (such as autism and schizophrenia).

Since the earliest description of these structures by Santiago Ramón y Cajal in the late 1800s, dendritic spines have fascinated generations of neuroscientists and have inspired much speculation about their function. One of the earliest conjectures was that the narrow spine

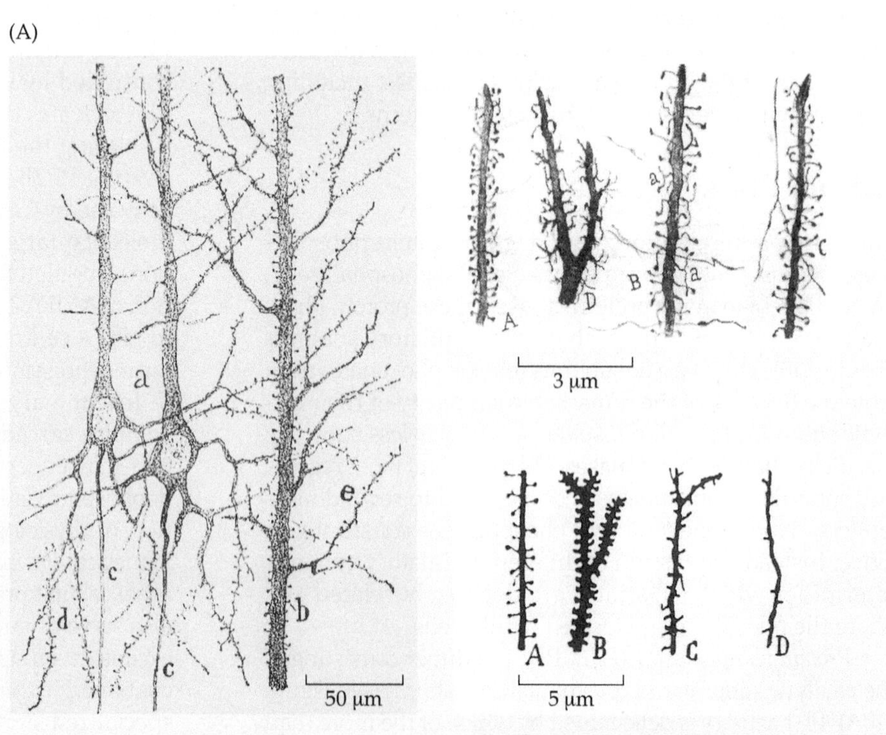

(A) Cajal's classic drawings of dendritic spines. Left, dendrites of cortical pyramidal neurons. Right, higher-magnification images of several different types of dendritic spines.

BOX 7B ■ (*continued*)

neck electrically isolates synapses from the rest of the neuron. However, the most recent measurements indicate that EPSPs can spread from some spine heads to dendrites with little attenuation.

Another theory—currently the most popular functional concept—postulates that spines create biochemical compartments. This idea is based on the idea that the spine neck could impede diffusion of biochemical signals from the spine head to the rest of the dendrite. Several observations are consistent with this notion. First, measurements show that the spine neck does indeed serve as a barrier to diffusion, in some cases slowing the rate of molecular movement by a factor of 100 or more. Second, spines are found at excitatory synapses, where it is known that synaptic transmission generates many diffusible signals, most notably the second messenger Ca^{2+}. Finally, fluorescence imaging shows that synaptic Ca^{2+} signals can indeed be restricted to dendritic spines in some circumstances (Figure D).

Nevertheless, there are counterarguments to the hypothesis that spines provide relatively isolated biochemical compartments. For example, it is known that other second messengers, such as IP_3, as well as other signaling molecules, can diffuse out of the spine head and into the dendritic shaft. This difference in diffusion presumably is due to the fact that such signals last longer than Ca^{2+} signals, allowing them sufficient time to overcome the diffusion barrier of the spine neck. Another relevant point is that postsynaptic Ca^{2+} signals are highly localized, even at excitatory synapses that do not have spines. Thus, in at least some instances, spines are neither

Continued on the next page

(B) Presynaptic terminal; Spine head; Spine neck; Dendrite

(C)

(B) Electron micrograph of an excitatory synapse in the hippocampus. Green arrow indicates postsynaptic density. (C) Reconstruction of a small region of the dendrite of a hippocampal pyramidal neuron, revealing a remarkable diversity in spine structure. Red structures indicate postsynaptic densities within each spine. (B from Harris and Weinberg, 2012; C from http://synapseweb.clm.utexas.edu/atlas; reprinted with permission from J. Spacek.)

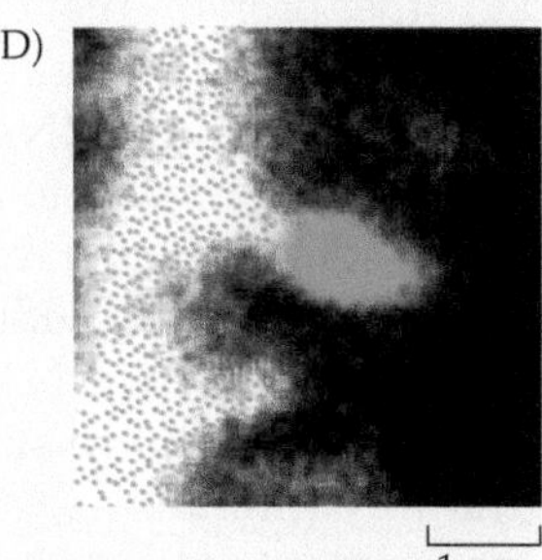

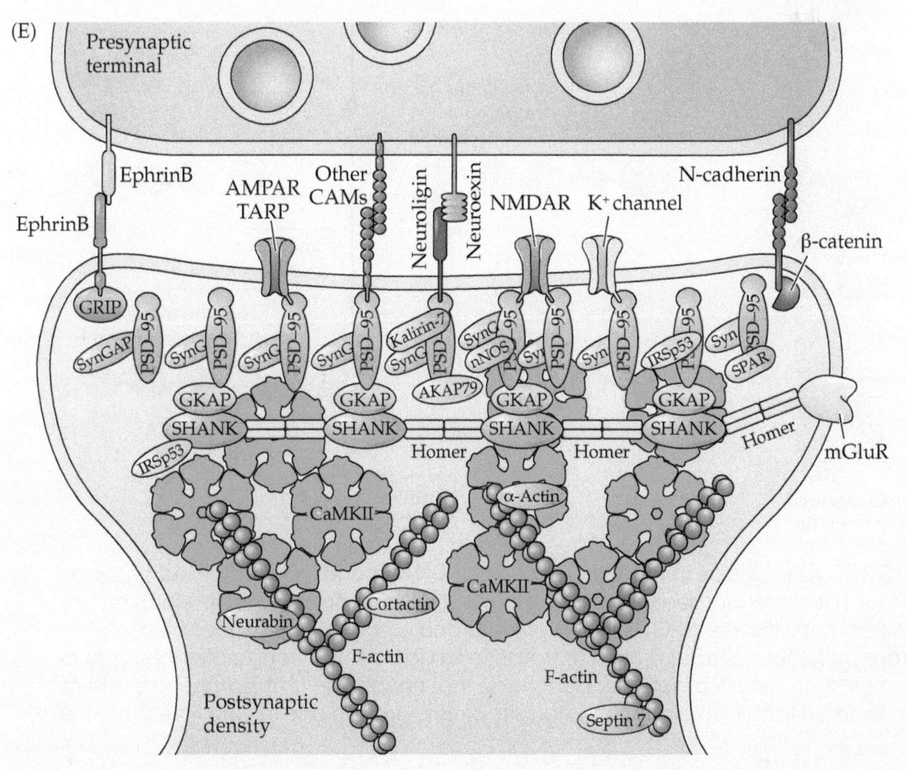

(D) Localized Ca^{2+} signal (green) produced in the spine of a hippocampal pyramidal neuron following activation of a glutamatergic synapse. (E) Postsynaptic densities include dozens of signal transduction molecules, including glutamate receptors (AMPA-type [AMPAR], NMDA-type [NMDAR], mGluR), other ion channels such as K^+ channels, and many intracellular signal transduction molecules, most notably the protein kinase CaMKII. Cytoskeletal elements, such as actin and its numerous binding partners, also are prominent and help create the structure of dendritic spines. (D from Sabatini et al., 2002; E after Sheng and Kim, 2011.)

BOX 7B ■ (continued)

necessary nor sufficient for localization of synaptic second-messenger signaling.

A final and less controversial idea is that the purpose of spines is to serve as reservoirs where signaling proteins—such as the downstream molecular targets of second-messenger signals—can be concentrated. Consistent with this possibility, glutamate receptors are highly concentrated on spine heads, and the postsynaptic density comprises dozens of proteins involved in intracellular signal transduction (Figure E). According to this view, the spine head is the destination for these signaling molecules during the assembly of synapses, as well as the target of the second messengers that are produced by the local activation of glutamate receptors. Spines also can trap molecules that are diffusing along the dendrite, which could be a means of concentrating these molecules within spines.

Although the function of dendritic spines remains enigmatic, Cajal undoubtedly would be pleased at the enormous amount of attention that these tiny synaptic structures continue to command, and the real progress that has been made in understanding the variety of tasks of which they are capable.

Nuclear Signaling

Second messengers elicit prolonged changes in neuronal function by promoting the synthesis of new RNA and protein. The resulting accumulation of new proteins requires at least 30 to 60 minutes, a time frame that is orders of magnitude slower than the responses mediated by ion fluxes or phosphorylation. Likewise, the reversal of such events requires hours to days. In some cases, genetic "switches" can be "thrown" to permanently alter a neuron, as occurs in neuronal differentiation (see Chapter 22).

The amount of protein present in cells is determined primarily by the rate of transcription of DNA into RNA (Figure 7.11). The first step in RNA synthesis is the decondensation of the structure of chromatin to provide binding sites for the RNA polymerase complex and for **transcriptional activator proteins**, also called **transcription factors**. Transcriptional activator proteins attach to binding sites that are present on the DNA molecule near the start of the target gene sequence; they also bind to other proteins that promote unwrapping of DNA. The net result of these actions is to allow RNA polymerase, an enzyme complex, to assemble on the **promoter** region of the DNA and begin transcription. In addition to clearing the promoter for RNA polymerase, activator proteins can stimulate transcription by interacting with the RNA polymerase complex or by interacting with other activator proteins that influence the polymerase.

Intracellular signal transduction cascades regulate gene expression by converting transcriptional activator proteins from an inactive state to an active state in which they are able to bind to DNA. This conversion comes about in several ways. The following sections briefly summarize three key activator proteins and the mechanisms that allow them to regulate gene expression in response to signaling events.

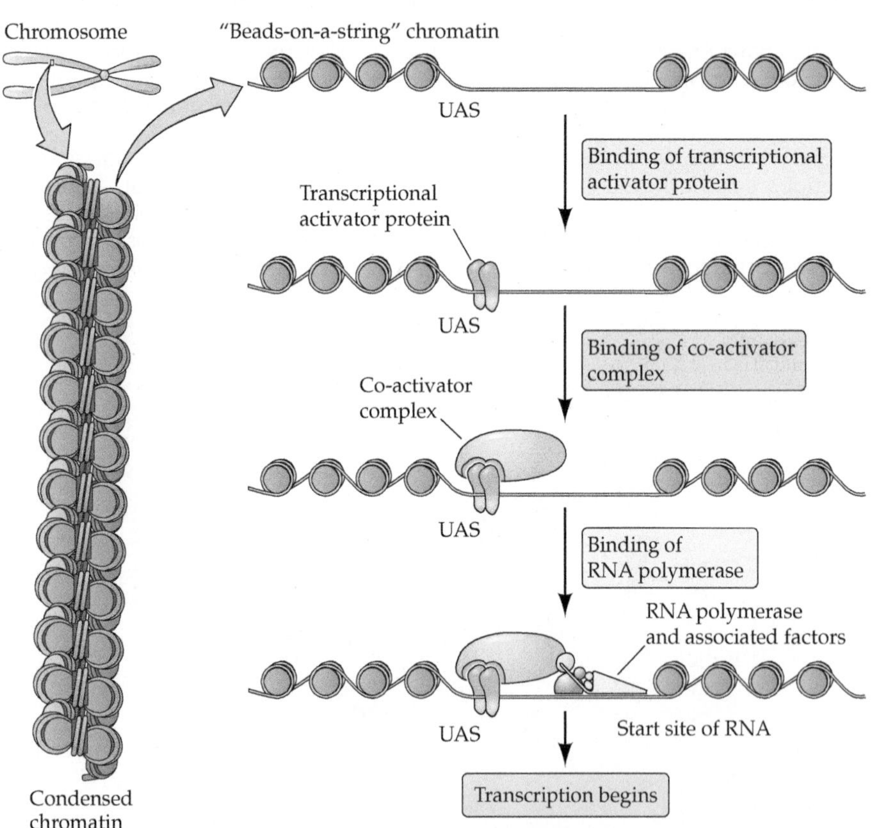

FIGURE 7.11 Steps in the transcription of DNA into RNA. Condensed chromatin (left) is decondensed into a beads-on-a-DNA-string array (right) in which an upstream activator site (UAS) is free of proteins and is bound by a sequence-specific transcriptional activator protein (a transcription factor). The transcriptional activator protein then binds co-activator complexes that enable the RNA polymerase with its associated factors to bind at the start site of transcription and initiate RNA synthesis.

• *CREB.* The **cAMP response element binding protein**, usually abbreviated **CREB**, is a ubiquitous transcriptional activator (Figure 7.12). CREB is typically bound to its binding site on DNA (called the cAMP response element, or CRE), either as a homodimer or bound to another, closely related transcription factor. In unstimulated cells, CREB is not phosphorylated and has little or no transcriptional activity. However, phosphorylation of CREB greatly potentiates transcription. Several signaling pathways are capable of causing CREB to be phosphorylated. Both PKA and the ras pathway, for example, can phosphorylate CREB. CREB can also be phosphorylated in response to increased intracellular calcium, in which case the CRE site is also called the CaRE (*ca*lcium *r*esponse *e*lement) site. The calcium-dependent phosphorylation of CREB is primarily caused by Ca^{2+}/calmodulin kinase IV (a relative of CaMKII) and by MAPK, which leads to prolonged CREB phosphorylation. CREB phosphorylation must be maintained long enough for transcription to ensue, even though neuronal electrical activity only transiently raises intracellular

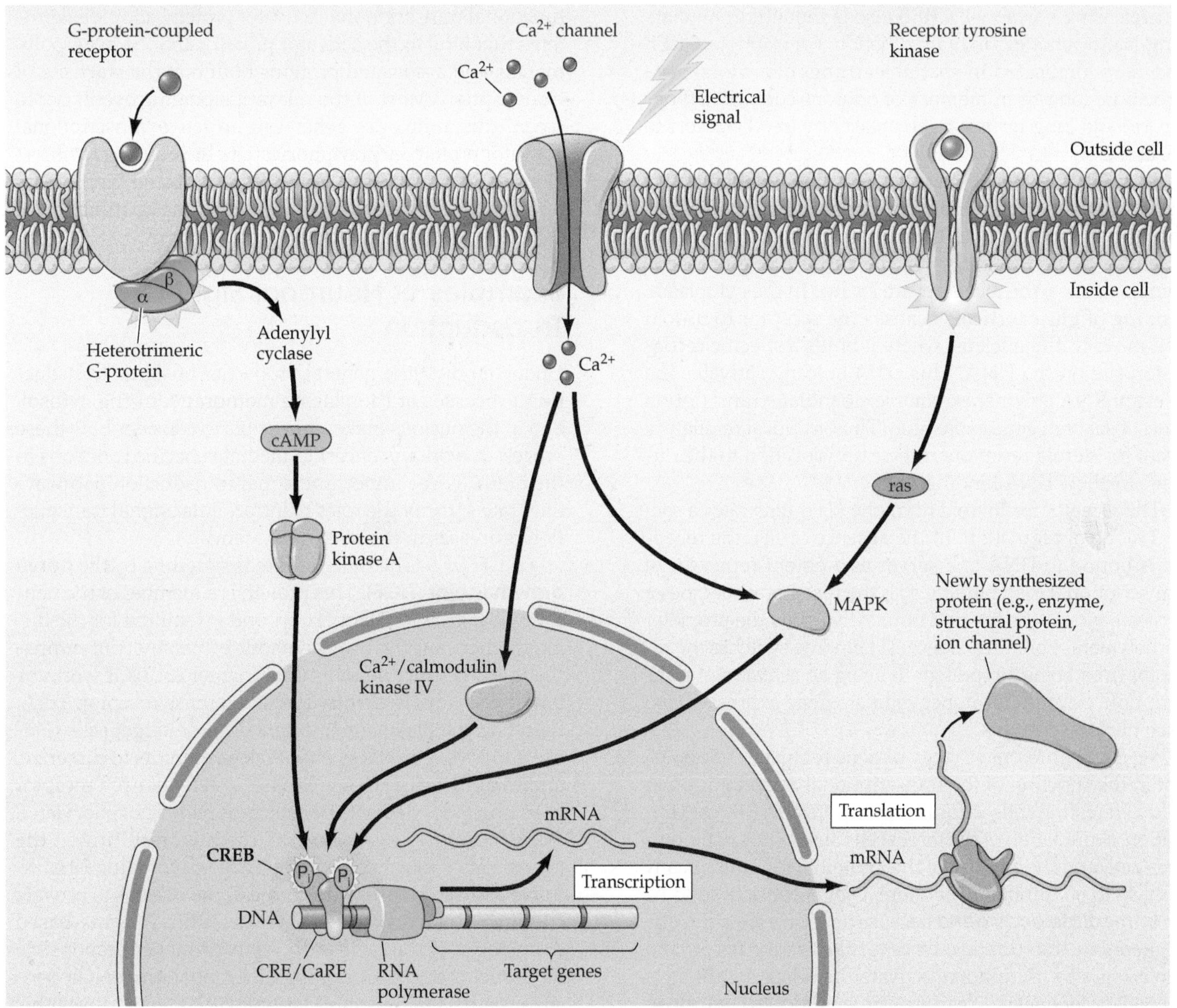

FIGURE 7.12 **Transcriptional regulation by CREB.** Multiple signaling pathways converge by activating kinases that phosphorylate CREB. These include PKA, Ca^{2+}/calmodulin kinase IV, and MAPK. Phosphorylation of CREB allows it to bind co-activators (not shown in the figure), which then stimulate RNA polymerase to begin synthesis of RNA. RNA is then processed and exported to the cytoplasm, where it serves as mRNA for translation into protein.

calcium concentration. Such signaling cascades can potentiate CREB-mediated transcription by inhibiting a protein phosphatase that dephosphorylates CREB. CREB is thus an example of the convergence of multiple signaling pathways onto a single transcriptional activator.

Many genes whose transcription is regulated by CREB have been identified. CREB-sensitive genes include the immediate early gene, *c-fos* (see below), the neurotrophin BDNF (see Chapter 23), the enzyme tyrosine hydroxylase (which is important for synthesis of catecholamine neurotransmitters; see Chapter 6), and many neuropeptides (including somatostatin, enkephalin, and corticotropin-releasing hormone). CREB also is thought to mediate long-lasting changes in brain function. For example, CREB has been implicated in spatial learning, behavioral sensitization, long-term memory of odorant-conditioned behavior, and long-term synaptic plasticity (see Chapters 8, 25, and 26).

• *Nuclear receptors.* Nuclear receptors for membrane-permeant ligands also are transcriptional activators. The receptor for glucocorticoid hormones illustrates one mode of action of such receptors. In the absence of glucocorticoid hormones, the receptors are located in the cytoplasm. Binding of glucocorticoids causes the receptor to unfold and move to the nucleus, where it binds a specific recognition site on the DNA. This DNA binding activates the relevant RNA polymerase complex to initiate transcription and subsequent gene expression. Thus, a critical regulatory event for steroid receptors is their translocation to the nucleus to allow DNA binding.

The receptor for thyroid hormone (TH) illustrates a second mode of regulation. In the absence of TH, the receptor is bound to DNA and serves as a potent repressor of transcription. Upon binding TH, the receptor undergoes a conformational change that ultimately opens the promoter for polymerase binding. Hence, TH binding switches the receptor from being a repressor to being an activator of transcription. Several hormones regulate gene expression via such nuclear receptors.

• *c-fos.* A different strategy of gene regulation is apparent in the function of the transcriptional activator protein **c-fos**. In resting cells, c-fos is present at a very low concentration. Stimulation of the target cell causes c-fos to be synthesized, and the amount of this protein rises dramatically over 30 to 60 minutes. Therefore, *c-fos* is considered to be an **immediate early gene** because its synthesis is directly triggered by the stimulus. Once synthesized, c-fos protein can act as a transcriptional activator to induce synthesis of second-order genes. These are termed **delayed response genes** because their activity is delayed by the fact that an immediate early gene—*c-fos,* in this case—must be activated first.

Multiple signals converge on *c-fos,* activating different transcription factors that bind to at least three distinct sites in the promoter region of the gene. The regulatory region of the *c-fos* gene contains a binding site that mediates transcriptional induction by cytokines and ciliary neurotropic factor. Another site is targeted by growth factors such as neurotrophins through ras and protein kinase C, and a CRE/CaRE that can bind to CREB and thereby respond to cAMP or calcium entry resulting from electrical activity. In addition to synergistic interactions among these *c-fos* sites, transcriptional signals can be integrated by converging on the same activator, such as CREB.

Nuclear signaling events typically result in the generation of a large and relatively stable complex composed of a functional transcriptional activator protein, additional proteins that bind to the activator protein, and the RNA polymerase and associated proteins bound at the start site of transcription. Most of the relevant signaling events act to "seed" this complex by generating an active transcriptional activator protein by phosphorylation, by inducing a conformational change in the activator upon ligand binding, by fostering nuclear localization, by removing an inhibitor, or simply by making more activator protein.

Examples of Neuronal Signal Transduction

Understanding the general properties of signal transduction processes at the plasma membrane, in the cytosol, and in the nucleus makes it possible to consider how these processes work in concert to mediate specific functions in the brain. Three important signal transduction pathways illustrate some of the roles of intracellular signal transduction processes in the nervous system.

• *NGF/TrkA.* The first of these is signaling by the **nerve growth factor (NGF)**. This protein is a member of the neurotrophin growth factor family and is required for the differentiation, survival, and synaptic connectivity of sympathetic and sensory neurons (see Chapter 23). NGF works by binding to a high-affinity tyrosine kinase receptor, TrkA, found on the plasma membrane of these target cells (Figure 7.13). NGF binding causes TrkA receptors to dimerize, and the intrinsic tyrosine kinase activity of each receptor then phosphorylates its partner receptor. Phosphorylated TrkA receptors trigger the ras cascade, resulting in the activation of multiple protein kinases, including MAPK. Some of these kinases translocate to the nucleus to activate transcriptional activators, such as CREB. This ras-based component of the NGF pathway is primarily responsible for inducing and maintaining differentiation of NGF-sensitive neurons. Phosphorylation of TrkA also causes this receptor to stimulate the activity of phospholipase C, which increases production of IP_3 and DAG. IP_3 induces release of Ca^{2+} from the endoplasmic reticulum, and DAG activates PKC. These two second messengers appear to target many of the same downstream effectors as ras. Finally, activation of TrkA receptors also causes activation of other protein kinases (such as Akt kinase) that inhibit cell death. This pathway, therefore, primarily mediates the NGF-dependent survival of sympathetic and sensory neurons described in Chapter 23.

• *Long-term synaptic depression (LTD).* The interplay between several intracellular signals can be observed at the excitatory synapses that innervate Purkinje cells in the cerebellum. These synapses are central to information flow through the cerebellar cortex, which coordinates motor movements (see Chapter 19). One of the synapses is between the parallel fibers (PFs) and their Purkinje cell targets: LTD is a form of synaptic plasticity that weakens these synapses (see Chapter 8). When PFs are active, they release the neurotransmitter glutamate onto the dendrites of Purkinje cells. This activates AMPA-type receptors, which are ligand-gated ion channels (see Chapter 6), and causes a small EPSP that briefly depolarizes the Purkinje cell. In addition to this electrical signal, the glutamate released by PFs also activates metabotropic glutamate receptors. These receptors stimulate phospholipase C to generate two second messengers in the Purkinje cell, IP_3 and DAG (Figure 7.14). When the PF synapses alone are active, these intracellular signals are insufficient to open IP_3 receptors or to stimulate PKC.

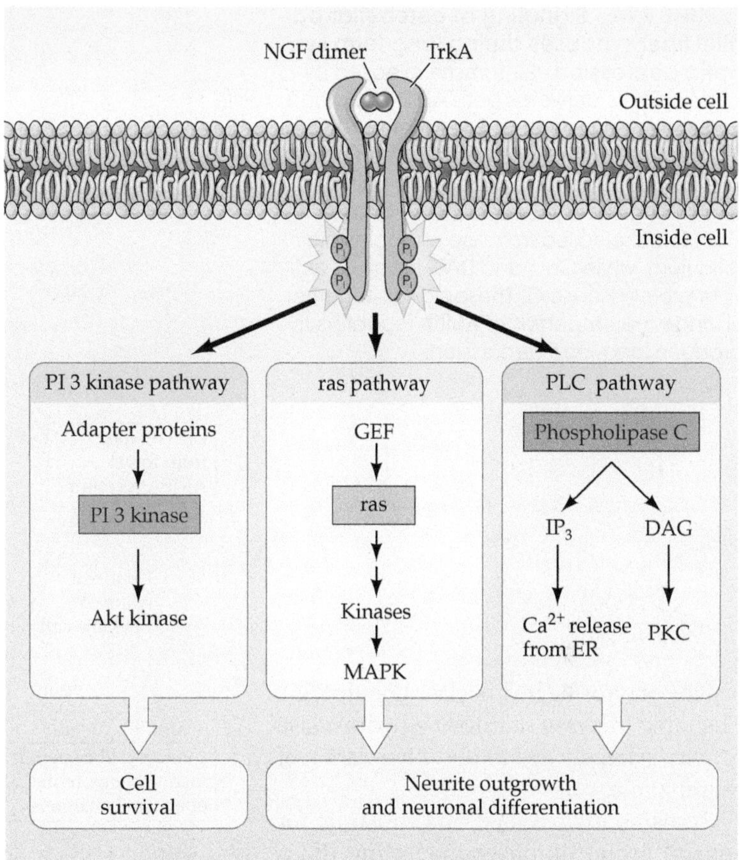

FIGURE 7.13 Mechanism of action of NGF. NGF binds to a high-affinity tyrosine kinase receptor, TrkA, on the plasma membrane to induce phosphorylation of TrkA at two different tyrosine residues. These phosphorylated tyrosines serve to tether various adapter proteins or phospholipase C (PLC), which in turn activate three major signaling pathways: the PI 3 kinase pathway leading to activation of Akt kinase, the ras pathway leading to MAPK, and the PLC pathway leading to release of intracellular Ca^{2+} from the endoplasmic reticulum and activation of PKC. The ras and PLC pathways primarily stimulate processes responsible for neuronal differentiation, whereas the PI 3 kinase pathway is primarily involved in cell survival.

LTD is induced when PF synapses are activated at the same time as the glutamatergic climbing fiber synapses that also innervate Purkinje cells. The climbing fiber synapses produce large EPSPs that strongly depolarize the membrane potential of the Purkinje cell. This depolarization allows Ca^{2+} to enter the Purkinje cell via voltage-gated Ca^{2+} channels. When both synapses are simultaneously activated, the rise in intracellular Ca^{2+} concentration caused by the climbing fiber synapse enhances the sensitivity of IP_3 receptors to the IP_3 produced by PF synapses and allows the IP_3 receptors in the Purkinje cell to open. This releases Ca^{2+} from the endoplasmic reticulum and further elevates Ca^{2+} concentration locally near the PF synapses. This larger rise in Ca^{2+}, in conjunction with the DAG produced by the PF synapses, activates PKC. PKC in turn phosphorylates several substrate proteins, including the AMPA-type receptors, ultimately altering trafficking of these receptors and leading to fewer AMPA-type receptors being located at the PF synapse (see Figure 8.17D). As a result, glutamate released from the PFs produces smaller EPSPs, causing LTD.

In short, transmission at Purkinje cell synapses produces brief electrical signals and chemical signals that last much longer. The temporal interplay between these signals allows LTD to occur only when both PF and climbing fiber synapses are active. The actions of IP_3, DAG, and Ca^{2+} also are restricted to small parts of the Purkinje cell dendrite, which is a more limited spatial range than the EPSPs, which spread throughout the entire dendrite and cell body of the Purkinje cell. Thus, in contrast to the electrical signals, the second-messenger signals can impart information about the location of active synapses and allow LTD to occur only in the vicinity of active PFs.

• *Phosphorylation of tyrosine hydroxylase.* A third example of intracellular signaling in the nervous system is the regulation of the enzyme tyrosine hydroxylase. Tyrosine hydroxylase governs the synthesis of the catecholamine neurotransmitters: dopamine, norepinephrine, and epinephrine (see Chapter 6). Several signals, including electrical activity, other neurotransmitters, and NGF, increase the rate of catecholamine synthesis by increasing the catalytic

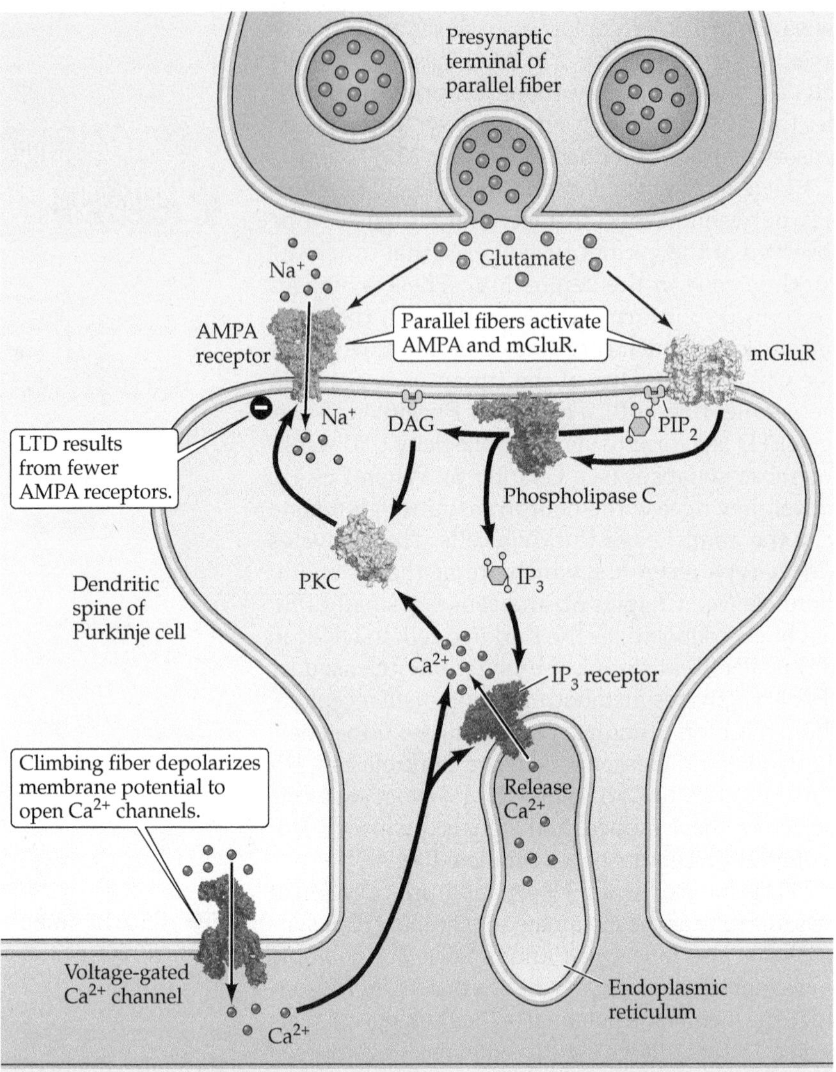

FIGURE 7.14 Signaling at cerebellar parallel fiber synapses during long-term synaptic depression. Glutamate released by parallel fibers activates both AMPA-type and metabotropic receptors. The latter produce IP_3 and DAG in the Purkinje cell. When paired with a rise in Ca^{2+} associated with activity of climbing fiber synapses, the IP_3 causes Ca^{2+} to be released from the endoplasmic reticulum, while Ca^{2+} and DAG together activate protein kinase C. These signals together change the properties of AMPA receptors to produce long-term depression.

activity of tyrosine hydroxylase (Figure 7.15). The rapid increase of tyrosine hydroxylase activity is largely due to phosphorylation of this enzyme.

Tyrosine hydroxylase is a substrate for several protein kinases, including PKA, CaMKII, MAPK, and PKC. Phosphorylation causes conformational changes that increase the catalytic activity of tyrosine hydroxylase. Stimuli that elevate cAMP, Ca^{2+}, or DAG can all increase tyrosine hydroxylase activity and thus increase the rate of catecholamine biosynthesis. This regulation by several different signals allows for close control of tyrosine hydroxylase activity and illustrates how several different pathways can converge to influence a key enzyme involved in synaptic transmission.

Summary

Diverse signal transduction pathways exist within all neurons. Activation of these pathways typically is initiated by chemical signals such as neurotransmitters and hormones, which bind to receptors that include ligand-gated ion channels, G-protein-coupled receptors, and tyrosine kinase receptors. Many of these receptors activate either heterotrimeric or monomeric G-proteins that regulate intracellular enzyme cascades and/or ion channels. A common outcome of the activation of these receptors is the production of second messengers, such as cAMP, Ca^{2+}, and IP_3, that bind to effector proteins. Particularly important effectors are protein kinases and phosphatases that regulate the phosphorylation state of their substrates, and thus their function. These substrates can be metabolic enzymes or other signal transduction molecules, such as ion channels, protein kinases, or transcription factors that regulate gene expression. Examples of such transcription factors include CREB, steroid hormone receptors, and c-fos. This plethora of molecular components allows intracellular signal transduction pathways to generate responses over a wide range of times and distances, greatly augmenting and refining the information-processing ability of neuronal circuits and, ultimately, brain systems.

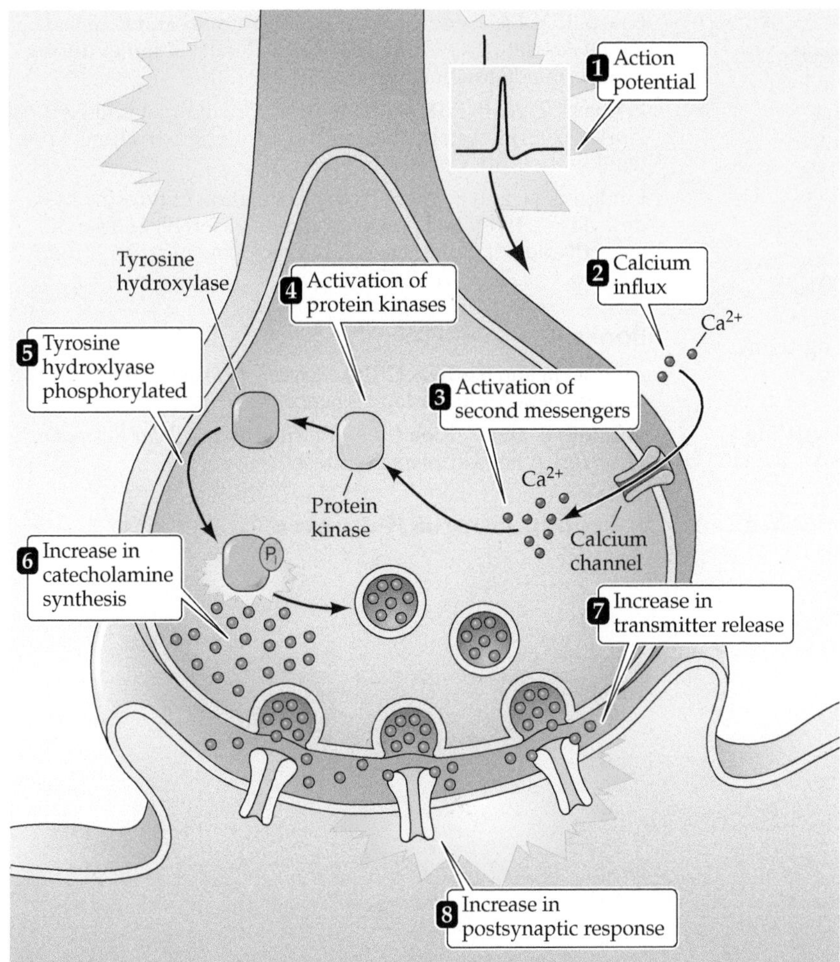

FIGURE 7.15 **Regulation of tyrosine hydroxylase by protein phosphorylation.** Tyrosine hydroxylase governs the synthesis of the catecholamine neurotransmitters and is stimulated by several intracellular signals. In the example shown here, neuronal electrical activity (1) causes influx of Ca^{2+} (2). The resultant rise in intracellular Ca^{2+} concentration (3) activates protein kinases (4), which phosphorylates tyrosine hydroxylase (5), stimulating catecholamine synthesis (6). This increased synthesis in turn increases the release of catecholamines (7) and enhances the postsynaptic response produced by the synapse (8).

ADDITIONAL READING

Reviews

Augustine, G. J., F. Santamaria and K. Tanaka (2003) Local calcium signaling in neurons. *Neuron* 40: 331–346.

Bollen, M., W. Peti, M. J. Ragusa and M. Beullens (2010) The extended PP1 toolkit: Designed to create specificity. *Trends Biochem. Sci.* 35: 450–458.

Greengard, P. (2001) The neurobiology of slow synaptic transmission. *Science* 294: 1024–1030.

Greer, P. L. and M. E. Greenberg (2008) From synapse to nucleus: Calcium-dependent gene transcription in the control of synapse development and function. *Neuron* 59: 846–860.

Ito, M., K. Yamaguchi, S. Nagao and T. Yamazaki (2014) Long-term depression as a model of cerebellar plasticity. *Prog. Brain Res.* 210: 1–30.

Kennedy, M. B., H. C. Beale, H. J. Carlisle and L. R. Washburn (2005) Integration of biochemical signalling in spines. *Nat. Rev. Neurosci.* 6: 423–434.

Kumer, S. and K. Vrana (1996) Intricate regulation of tyrosine hydroxylase activity and gene expression. *J. Neurochem.* 67: 443–462.

Li, H., A. Rao and P. G. Hogan (2011) Interaction of calcineurin with substrates and targeting proteins. *Trends Cell Biol.* 21: 91–103.

Reichardt, L. F. (2006) Neurotrophin-regulated signalling pathways. *Philos. Trans. R. Soc. London B* 361: 1545–1564.

Rosenbaum, D. M., S. G. Rasmussen and B. K. Kobilka (2009) The structure and function of G-protein-coupled receptors. *Nature* 459: 356–363.

Sangodkar, J. and 5 others (2016) All roads lead to PP2A: exploiting the therapeutic potential of this phosphatase. *FEBS J.* 283: 1004–1024.

Sheng, M. and E. Kim (2011) The postsynaptic organization of synapses. *Cold Spring Harb. Perspect. Biol.* 3: a005678.

Taylor, S. S., R. Ilouz, P. Zhang and A. P. Kornev (2012) Assembly of allosteric macromolecular switches: lessons from PKA. *Nature Rev. Mol. Cell. Biol.* 13: 646–658.

Turk, B. E. (2007) Manipulation of host signalling pathways by anthrax toxins. *Biochem. J.* 402: 405–417.

Important Original Papers

Burgess, G. M. and 5 others (1984) The second messenger linking receptor activation to internal calcium release in liver. *Nature* 309: 63–66.

Cho, U. S. and W. Xu (2007) Crystal structure of a protein phosphatase 2A heterotrimeric holoenzyme. *Nature* 445: 53–57.

Craddock, T. J. A., J. A. Tuszynski and S. Hameroff (2012) Cytoskeletal signaling: Is memory encoded in microtubule lattices by CaMKII phosphorylation? *PLoS Comput. Biol.* 8: e1002421.

De Zeeuw, C. I. and 6 others (1998) Expression of a protein kinase C inhibitor in Purkinje cells blocks cerebellar long-term depression and adaptation of the vestibulo-ocular reflex. *Neuron* 20: 495–508.

Finch, E. A. and G. J. Augustine (1998) Local calcium signaling by IP_3 in Purkinje cell dendrites. *Nature* 396: 753–756.

Lee, S. J., Y. Escobedo-Lozoya, E. M. Szatmari and R. Yasuda (2009) Activation of CaMKII in single dendritic spines during long-term potentiation. *Nature* 458: 299–304.

Leonard, T. A., B. Różycki, L. F. Saidi, G. Hummer and J. H. Hurley (2011) Crystal structure and allosteric activation of protein kinase C βII. *Cell* 144: 55–66.

Lindgren, N. and 8 others (2000) Regulation of tyrosine hydroxylase activity and phosphorylation at ser(19) and ser(40) via activation of glutamate NMDA receptors in rat striatum. *J. Neurochem.* 74: 2470–2477.

Books

Alberts, B. and 6 others (2014) *Molecular Biology of the Cell*, 6th Edition. New York: Garland Science.

Carafoli, E. and C. Klee (1999) *Calcium as a Cellular Regulator.* New York: Oxford University Press.

Go to the NEUROSCIENCE 6e Companion Website at **www.oup.com/uk/Purves6e** for Web Topics, Animations, Flashcards, and more.

CHAPTER

8 Synaptic Plasticity

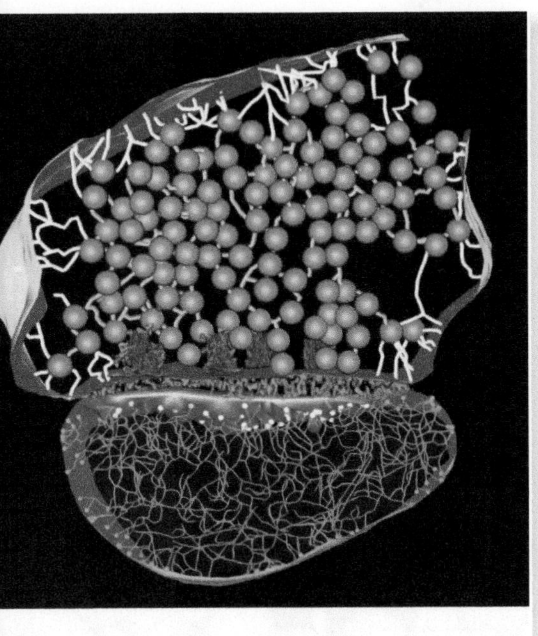

Overview

SYNAPTIC CONNECTIONS BETWEEN NEURONS provide the basic "wiring" of the brain's circuitry. However, unlike the wiring of an electronic device such as a computer, the strength of synaptic connections between neurons is a dynamic entity that is constantly changing in response to neural activity and other influences. Such changes in synaptic transmission arise from several forms of plasticity that vary in timescale from milliseconds to years. Most short-term forms of synaptic plasticity affect the amount of neurotransmitter released from presynaptic terminals in response to a presynaptic action potential. Several forms of short-term synaptic plasticity—including facilitation, augmentation, and potentiation—enhance neurotransmitter release and are caused by persistent actions of calcium ions within the presynaptic terminal. Another form of short-term plasticity, synaptic depression, decreases the amount of neurotransmitter released and appears to be due to an activity-dependent depletion of synaptic vesicles that are ready to undergo exocytosis. Long-term forms of synaptic plasticity alter synaptic transmission over timescales of 30 minutes or longer. Examples of such long-lasting plasticity include long-term potentiation and long-term depression. These long-lasting forms of synaptic plasticity arise from molecular mechanisms that vary over time: The initial changes in synaptic transmission arise from post-translational modifications of existing proteins, most notably changes in the trafficking of glutamate receptors, while later phases of synaptic modification result from changes in gene expression and synthesis of new proteins. These changes produce enduring changes in synaptic transmission, including synapse growth, that can yield essentially permanent modifications of brain function.

Short-Term Synaptic Plasticity

Chemical synapses are capable of undergoing plastic changes that either strengthen or weaken synaptic transmission. Synaptic plasticity mechanisms occur on timescales ranging from milliseconds to days, weeks, or longer. The short-term forms of plasticity—those lasting for a few minutes or less—are readily observed during repeated activation of any chemical synapse. There are several forms of short-term synaptic plasticity that differ in their time courses and their underlying mechanisms.

Synaptic facilitation is a rapid increase in synaptic strength that occurs when two or more action potentials invade the presynaptic terminal within a few milliseconds of each other (Figure 8.1A). By varying the time interval between presynaptic action potentials, it can be seen that facilitation produced by the first action potential lasts for tens of milliseconds (Figure 8.1B). Many lines of evidence indicate that facilitation is the result of prolonged elevation of presynaptic calcium levels following synaptic activity.

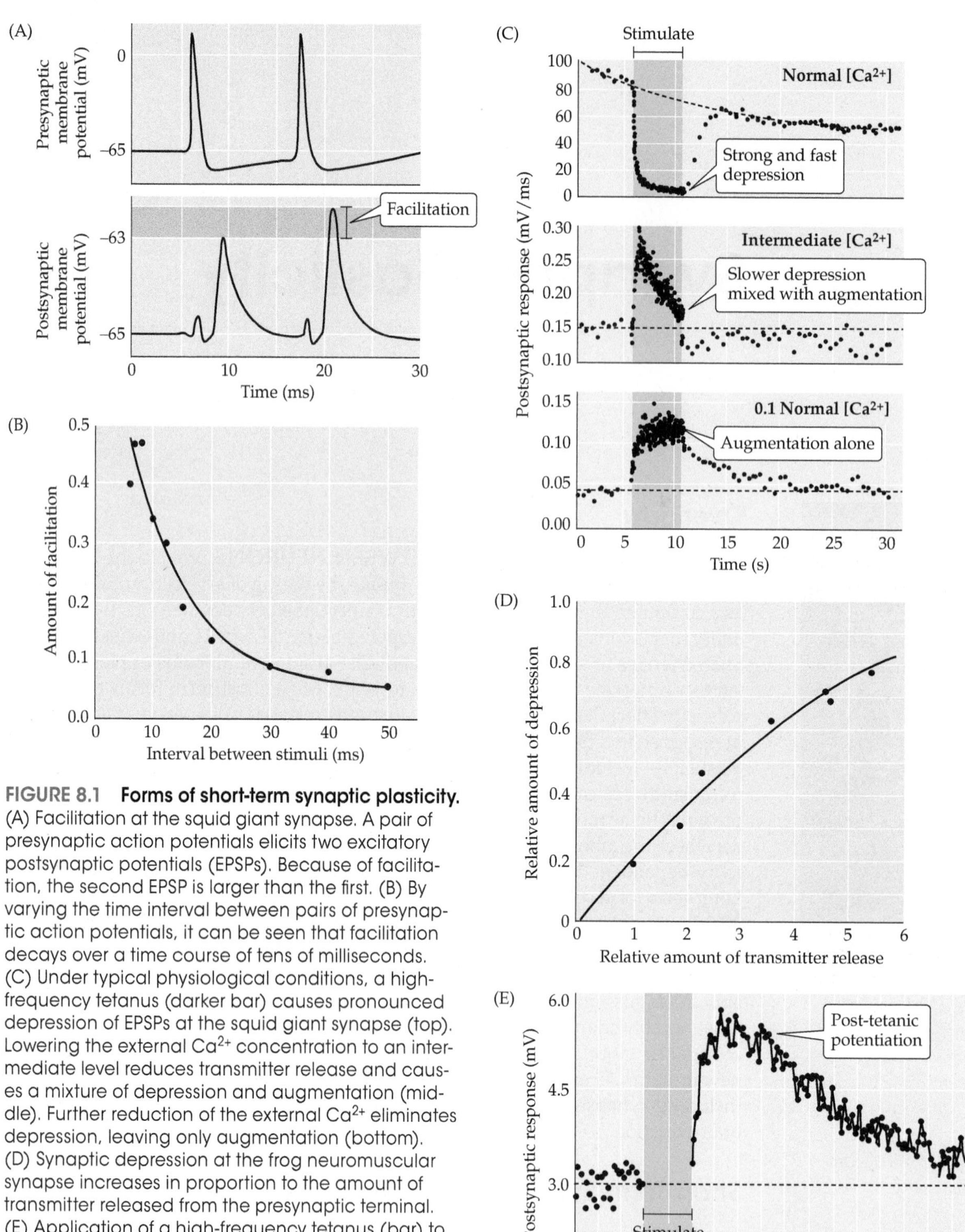

FIGURE 8.1 Forms of short-term synaptic plasticity. (A) Facilitation at the squid giant synapse. A pair of presynaptic action potentials elicits two excitatory postsynaptic potentials (EPSPs). Because of facilitation, the second EPSP is larger than the first. (B) By varying the time interval between pairs of presynaptic action potentials, it can be seen that facilitation decays over a time course of tens of milliseconds. (C) Under typical physiological conditions, a high-frequency tetanus (darker bar) causes pronounced depression of EPSPs at the squid giant synapse (top). Lowering the external Ca^{2+} concentration to an intermediate level reduces transmitter release and causes a mixture of depression and augmentation (middle). Further reduction of the external Ca^{2+} eliminates depression, leaving only augmentation (bottom). (D) Synaptic depression at the frog neuromuscular synapse increases in proportion to the amount of transmitter released from the presynaptic terminal. (E) Application of a high-frequency tetanus (bar) to presynaptic axons innervating a spinal motor neuron causes a post-tetanic potentiation that persists for a few minutes after the tetanus ends. (A,B after Charlton and Bittner, 1978; C after Swandulla et al., 1991; D from Betz, 1970; E after Lev-Tov et al., 1983.)

Although the entry of Ca^{2+} into the presynaptic terminal occurs within 1 to 2 milliseconds after an action potential invades (see Figure 5.9B), the mechanisms that return Ca^{2+} to resting levels are much slower. Thus, when action potentials arrive close together in time, calcium builds up in the terminal and allows more neurotransmitter to be released by a subsequent presynaptic action potential. Recent evidence indicates that the target of this residual Ca^{2+} signal is synaptotagmin 7, a Ca^{2+}-binding protein that is found on the plasma membrane and is related to the synaptotagmins that are present on synaptic vesicles and serve as Ca^{2+} sensors for triggering neurotransmitter release (see Chapter 5).

Opposing facilitation is **synaptic depression**, which causes neurotransmitter release to decline during sustained synaptic activity (Figure 8.1C, top). An important clue to the cause of synaptic depression comes from observations that depression depends on the amount of neurotransmitter that has been released. For example, lowering the external Ca^{2+} concentration, to reduce the number of quanta released by each presynaptic action potential, causes the rate of depression to be slowed (see Figure 8.1C). Likewise, the total amount of depression is proportional to the amount of transmitter released from the presynaptic terminal (Figure 8.1D). These results have led to the idea that depression is caused by progressive depletion of a pool of synaptic vesicles that are available for release: When rates of release are high, these vesicles deplete rapidly and cause a lot of depression; depletion slows as the rate of release is reduced, yielding less depression. According to this vesicle depletion hypothesis, depression causes the strength of transmission to decline until this pool is replenished by mobilization of vesicles from a reserve pool. Consistent with this explanation are observations that more depression is observed after the size of the reserve pool is reduced by impairing synapsin, a protein that maintains vesicles in the reserve pool (see Chapter 5).

Still other forms of synaptic plasticity, such as synaptic **potentiation** and **augmentation**, also are elicited by repeated synaptic activity and serve to increase the amount of transmitter released from presynaptic terminals. Both augmentation and potentiation enhance the ability of incoming Ca^{2+} to trigger fusion of synaptic vesicles with the plasma membrane, but the two processes work over different timescales. While augmentation rises and falls over a few seconds (see Figure 8.1C, bottom panel), potentiation acts over a timescale of tens of seconds to minutes (Figure 8.1E). As a result of its slow time course, potentiation can greatly outlast the tetanic stimulus that induces it, and is often called **post-tetanic potentiation** (**PTP**). Although both augmentation and potentiation are thought to arise from prolonged elevation of presynaptic calcium levels during synaptic activity, the mechanisms responsible for these forms of plasticity are poorly understood. It has been proposed that augmentation results from Ca^{2+} enhancing the actions of the presynaptic SNARE-regulatory protein munc-13 (see Figure 5.11), while potentiation may arise when Ca^{2+} activates presynaptic protein kinases that go on to phosphorylate substrates (such as synapsin) that regulate transmitter release.

During repetitive synaptic activity, the various forms of short-term plasticity can interact to cause synaptic transmission to change in complex ways. For example, at the peripheral neuromuscular synapse, repeated activity first causes an accumulation of Ca^{2+} in the presynaptic terminal that allows facilitation and then augmentation to enhance synaptic transmission (Figure 8.2). The ensuing depletion of synaptic vesicles then causes depression to dominate and weaken the synapse. Presynaptic action potentials that occur within 1 to 2 minutes after the end of the tetanus release more neurotransmitter because of the persistence of post-tetanic potentiation. Although their relative contributions vary from synapse to synapse, these forms of short-term synaptic plasticity collectively cause transmission at all chemical synapses to change dynamically as a consequence of the recent history of synaptic activity.

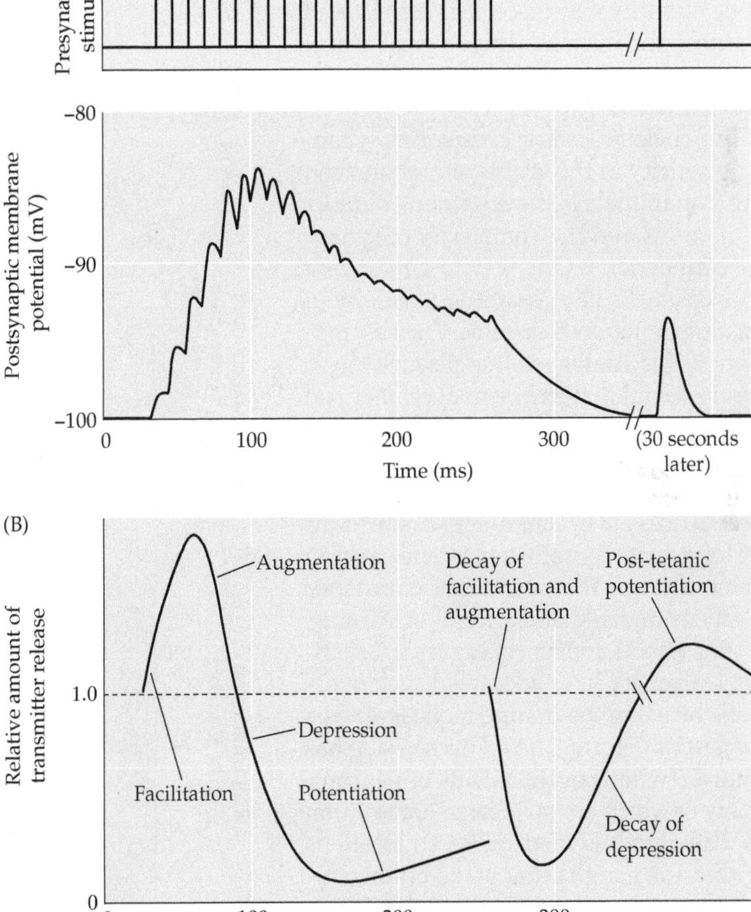

FIGURE 8.2 Short-term plasticity at the neuromuscular synapse. (A) A train of electrical stimuli (top) applied to the presynaptic motor nerve produces changes in the end plate potential (EPP) amplitude (bottom). (B) Dynamic changes in transmitter release caused by the interplay of several forms of short-term plasticity. Facilitation and augmentation of the EPP occurs at the beginning of the stimulus train and are followed by a pronounced depression of the EPP. Potentiation begins late in the stimulus train and persists for many seconds after the end of the stimulus, leading to post-tetanic potentiation. (A after Katz, 1966; B after Malenka and Siegelbaum, 2001.)

Long-Term Synaptic Plasticity and Behavioral Modification in *Aplysia*

Facilitation, depression, augmentation, and potentiation modify synaptic transmission over timescales of a few minutes or less. While these mechanisms probably are responsible for many short-lived changes in brain circuitry, they cannot provide the basis for changes in brain function that persist for weeks, months, or years. Many synapses exhibit long-lasting forms of synaptic plasticity that are plausible substrates for more permanent changes in brain function. Because of their duration, these forms of synaptic plasticity may be cellular correlates of learning and memory. Thus, a great deal of effort has gone into understanding how they are generated.

An obvious obstacle to exploring synaptic plasticity in the brains of humans and other mammals is the enormous number of neurons and the complexity of synaptic connections. One way to circumvent this dilemma is to examine plasticity in far simpler nervous systems. The assumption in this strategy is that plasticity is so fundamental that its essential cellular and molecular underpinnings are likely to be conserved in the nervous systems of very different organisms. This approach has been successful in identifying several forms of long-term synaptic plasticity and in demonstrating that such forms of synaptic plasticity underlie simple forms of learning.

Eric Kandel and his colleagues at Columbia University have addressed such questions by using the marine mollusk *Aplysia californica* (Figure 8.3A). This sea slug has only a few tens of thousands of neurons, many of which are quite large (up to 1 mm in diameter) and in stereotyped locations within the ganglia that make up the animal's nervous system (Figure 8.3B). These attributes make it practical to monitor the

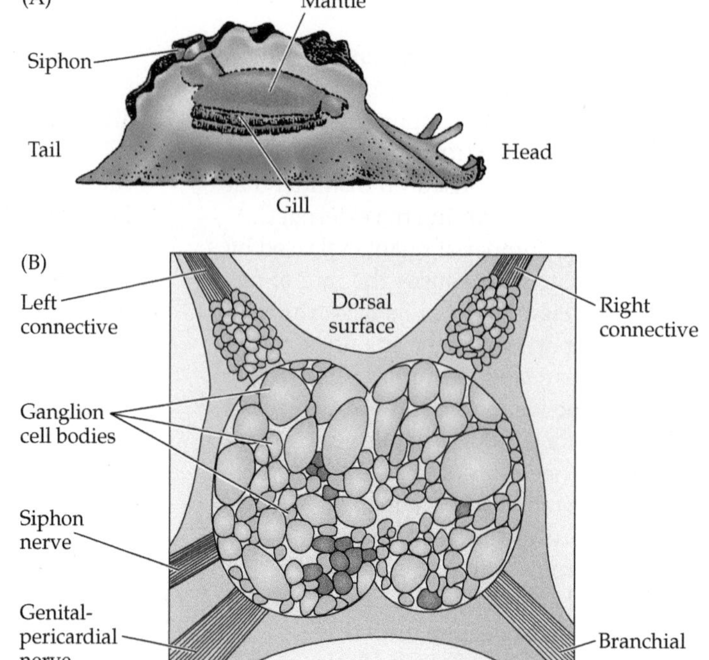

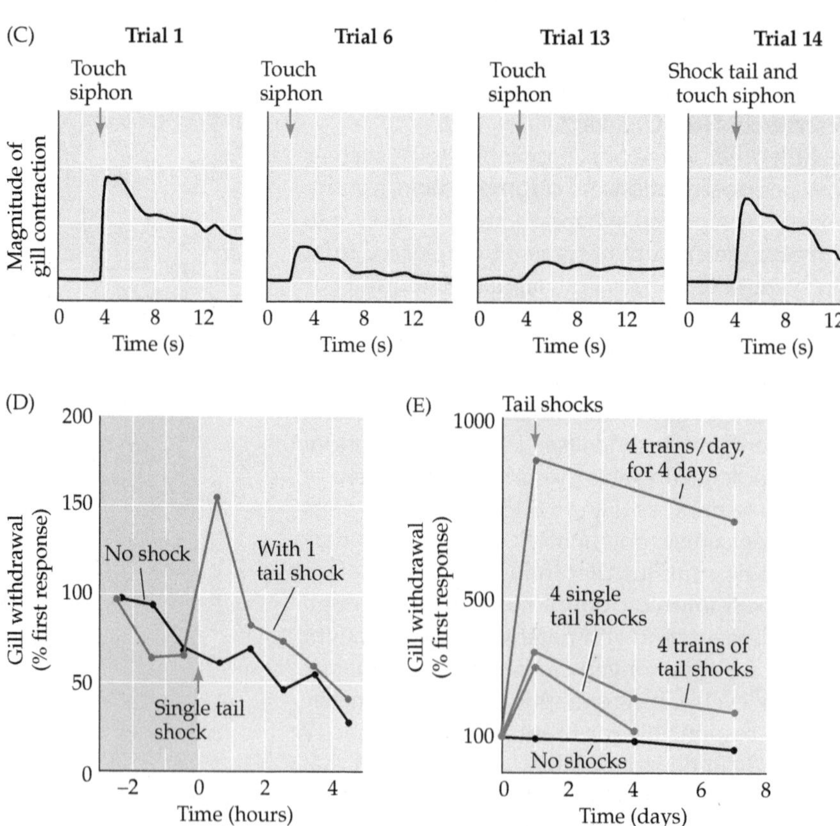

FIGURE 8.3 Short-term sensitization of the *Aplysia* gill withdrawal reflex. (A) Drawing of an *Aplysia* (commonly known as the sea slug). (B) The abdominal ganglion of *Aplysia*. The cell bodies of many of the neurons involved in gill withdrawal can be recognized by their size, shape, and position within this ganglion. (C) Changes in the gill withdrawal behavior due to habituation and sensitization. The first time the siphon is touched, the gill contracts vigorously. Repeated touches elicit smaller gill contractions due to habituation. Subsequently pairing a siphon touch with an electrical shock to the tail restores a large and rapid gill contraction, the result of short-term sensitization. (D) Time course of short-term sensitization of the gill withdrawal response following the pairing of a single tail shock with a siphon touch. (E) Repeated applications of tail shocks cause prolonged sensitization of the gill withdrawal response. (After Squire and Kandel, 1999.)

electrical activity of specific, identifiable nerve cells, and thus to define the synaptic circuits involved in mediating the limited behavioral repertoire of *Aplysia*.

Aplysia exhibit several elementary forms of behavioral plasticity. One form is **habituation**, a process that causes the animal to become less responsive to repeated occurrences of a stimulus. Habituation is found in many other species, including humans. For example, when dressing we initially experience tactile sensations due to clothes stimulating our skin, but habituation quickly causes these sensations to fade. Similarly, a light touch to the siphon of an *Aplysia* results in withdrawal of the animal's gill, but habituation causes the gill withdrawal to become weaker during repeated stimulation of the siphon (Figure 8.3C). The gill withdrawal response of *Aplysia* exhibits another form of plasticity called **sensitization**. Sensitization is a process that allows an animal to generalize an aversive response—elicited by a noxious stimulus—to a variety of other, non-noxious stimuli. In *Aplysia* that have habituated to siphon touching, sensitization of gill withdrawal is elicited by pairing a strong electrical stimulus to the animal's tail with another light touch to the siphon. This pairing causes the siphon stimulus to again elicit a strong withdrawal of the gill (see Figure 8.3C, right) because the noxious stimulus to the tail sensitizes the gill withdrawal reflex to light touch. Even after a single stimulus to the tail, the gill withdrawal reflex remains enhanced for at least an hour (Figure 8.3D). This can be viewed as a simple form of short-term memory. With repeated pairing of tail and siphon stimuli, this behavior can be altered for days or weeks (Figure 8.3E), thus demonstrating a simple form of long-term memory.

The small number of neurons in the *Aplysia* nervous system makes it possible to define the synaptic circuits involved in gill withdrawal and to monitor the activity of individual neurons in these circuits. Although hundreds of neurons are ultimately involved in producing this simple behavior, the activities of only a few different types of neurons can account for gill withdrawal and its plasticity during habituation and sensitization. These critical neurons include mechanosensory neurons that innervate the siphon, motor neurons that innervate muscles in the gill, and interneurons that receive inputs from a variety of sensory neurons (Figure 8.4A). Touching the siphon activates the mechanosensory neurons, which form

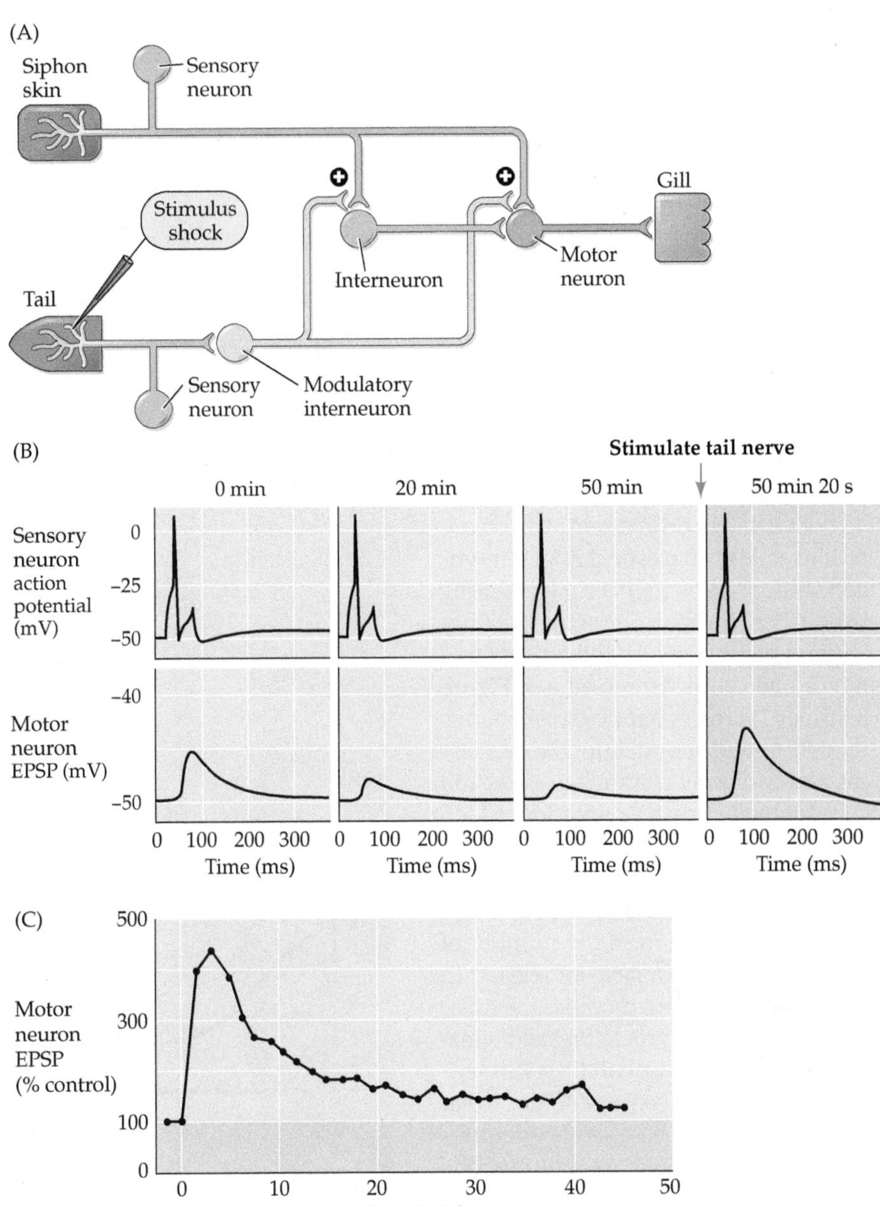

FIGURE 8.4 Synaptic mechanisms underlying short-term sensitization. (A) Neural circuitry involved in sensitization. Touching the siphon skin activates sensory neurons that excite interneurons and gill motor neurons, yielding a contraction of the gill muscle. A shock to the animal's tail stimulates modulatory interneurons that alter synaptic transmission between the siphon sensory neurons and gill motor neurons, resulting in sensitization. (B) Changes in synaptic efficacy at the sensory neuron–motor neuron synapse during short-term sensitization. Prior to sensitization, activating the siphon sensory neurons causes an EPSP to occur in the gill motor neurons. Repetitive activation of this synapse causes synaptic depression, indicated by a reduction in EPSPs in motor neurons. Activation of the serotonergic modulatory interneurons enhances release of transmitter from the sensory neurons onto the motor neurons, increasing the EPSP in the motor neurons and causing the motor neurons to more strongly excite the gill muscle. (C) Time course of the serotonin-induced facilitation of transmission at the sensory-motor synapse. (After Squire and Kandel, 1999.)

(A)

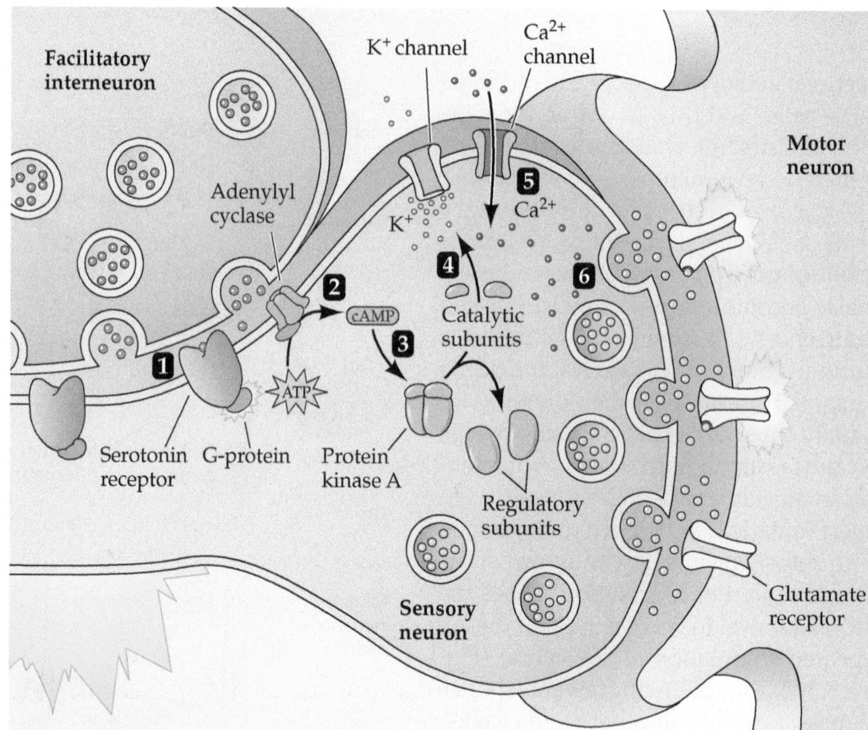

(B)

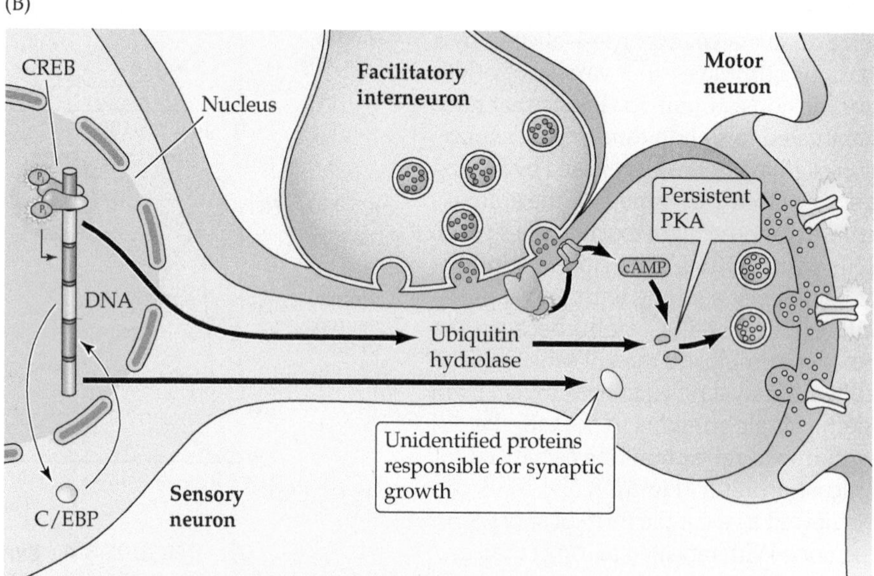

FIGURE 8.5 Mechanisms of presynaptic enhancement underlying behavioral sensitization. (A) Short-term sensitization is due to an acute, PKA-dependent enhancement of glutamate release from the presynaptic terminals of sensory neurons. See text for explanation. (B) Long-term sensitization is due to changes in gene expression, resulting in the synthesis of proteins that change PKA activity and lead to changes in synapse growth. (After Squire and Kandel, 1999.)

excitatory synapses that release glutamate onto both the interneurons and the motor neurons; thus, touching the siphon increases the probability that both of these postsynaptic targets will produce action potentials. The interneurons form excitatory synapses on motor neurons, further increasing the likelihood of the motor neurons firing action potentials in response to mechanical stimulation of the siphon. When the motor neurons are activated by the summed synaptic excitation of the sensory neurons and interneurons, they release acetylcholine that excites the muscle cells of the gill, producing gill withdrawal.

Both habituation and sensitization appear to arise from plastic changes in synaptic transmission in this circuit. During habituation, transmission at the glutamatergic synapse between the sensory and motor neurons is depressed (see Figure 8.4B, left). This synaptic depression is thought to be responsible for the decreasing ability of siphon stimuli to evoke gill contractions during habituation. Much like the short-term form of synaptic depression described in the preceding section, this depression is presynaptic and is due to a reduction in the number of synaptic vesicles available for release. In contrast, sensitization modifies the function of this circuit by recruiting additional neurons. The tail shock that evokes sensitization activates sensory neurons that innervate the tail. These sensory neurons in turn excite modulatory interneurons that release serotonin onto the presynaptic terminals of the sensory neurons of the siphon (see Figure 8.4A). Serotonin enhances transmitter release from the siphon sensory neuron terminals, leading to increased synaptic excitation of the motor neurons (Figure 8.4B). This modulation of the sensory neuron–motor neuron synapse lasts approximately 1 hour (Figure 8.4C), which is similar to the duration of the short-term sensitization of gill withdrawal produced by applying a single stimulus to the tail (see Figure 8.3D). Thus, the short-term sensitization apparently is due to recruitment of additional circuit elements, namely the modulatory interneurons that strengthen synaptic transmission in the gill withdrawal circuit.

The mechanism thought to be responsible for the enhancement of glutamatergic transmission during short-term sensitization is shown in Figure 8.5A. Serotonin released by the modulatory interneurons binds to G-protein-coupled receptors on the presynaptic terminals of the siphon sensory neurons (step 1), which stimulates production of the second messenger, cAMP (step 2). Cyclic AMP binds to the regulatory subunits of protein kinase A (PKA; step 3), liberating catalytic subunits of PKA that are then able to phosphorylate several proteins, probably including K^+ channels (step 4). The net effect of the action of PKA is to reduce the probability that the K^+ channels open during a presynaptic action potential. This effect prolongs the presynaptic action potential, thereby opening more presynaptic Ca^{2+} channels (step 5). There is evidence that the opening of presynaptic Ca^{2+} channels is also directly enhanced by serotonin. Finally, the enhanced influx of Ca^{2+} into the presynaptic terminals increases the amount of transmitter released onto motor neurons during a sensory neuron action potential (step 6). In summary, a signal transduction cascade that involves neurotransmitters, second messengers, protein kinases, and ion channels mediates short-term sensitization of gill withdrawal. This cascade ultimately causes a short-term enhancement of synaptic transmission between the sensory and motor neurons within the gill withdrawal circuit.

The same serotonin-induced enhancement of glutamate release that mediates short-term sensitization is also thought to underlie long-term sensitization. However, during long-term sensitization this circuitry is affected for up to several weeks. The prolonged duration of this form of plasticity is evidently due to changes in gene expression and thus protein synthesis (Figure 8.5B). With repeated training (i.e., additional tail shocks), the serotonin-activated PKA involved in short-term sensitization also phosphorylates—and thereby activates—the transcriptional activator CREB. As described in Chapter 7, CREB binding to the cAMP response elements (CREs) in regulatory regions of nuclear DNA increases the rate of transcription of downstream genes. Although the changes in genes and gene products that follow CRE activation have been difficult to sort out, several consequences of gene activation have been identified. First, CREB stimulates the synthesis of an enzyme, ubiquitin hydrolase, that stimulates degradation of the regulatory subunit of PKA. This causes a long-lasting increase in the amount of free catalytic subunit, meaning that some PKA is persistently active and no longer requires serotonin to be activated. CREB also stimulates another transcriptional activator protein called C/EBP. C/EBP stimulates transcription of other, unknown genes that cause addition of synaptic terminals, yielding a long-term increase in the number of synapses between the sensory and the motor neurons. Such structural increases are not seen following short-term sensitization. They may represent the ultimate cause of the long-lasting change in overall strength of the relevant circuit connections that produce a long-lasting enhancement in the gill withdrawal response. Another protein involved in the long-term synaptic facilitation is a cytoplasmic polyadenylation element binding protein, somewhat confusingly called CPEB. CPEB activates mRNAs and may be important for local control of protein synthesis. Most intriguing, CPEB has self-sustaining properties like those of prion proteins, which could allow CPEB to remain active in perpetuity and thereby mediate permanent changes in synaptic structure to generate long-term sensitization.

These studies of *Aplysia* and related work on other invertebrates, such as the fruit fly (Box 8A), have led to several generalizations about synaptic plasticity. First, synaptic plasticity clearly can lead to changes in circuit function and, ultimately, to behavioral plasticity. This conclusion has triggered intense interest in synaptic plasticity mechanisms.

BOX 8A ■ Genetics of Learning and Memory in the Fruit Fly

As part of a renaissance in the genetic analysis of simple organisms in the mid-1970s, several investigators recognized that the genetic basis of learning and memory might be effectively studied in the fruit fly *Drosophila melanogaster*. In the intervening 40 years, this approach has yielded some fundamental insights. Although the mechanisms of learning and memory have certainly been among the more difficult problems tackled by *Drosophila* geneticists, their efforts have been surprisingly successful. Several genetic mutations that alter learning and memory have been discovered, and the identification of these genes has provided a valuable framework for studying the cellular mechanisms of these processes.

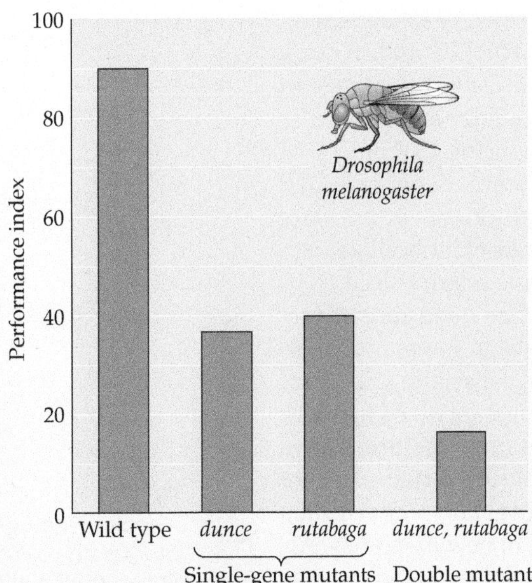

Performance of wild-type and mutant fruit flies (*Drosophila melanogaster*) on an olfactory learning task. The performance of *dunce* and of *rutabaga* mutants on this task is diminished by at least 50%. Flies that are mutant at both the *dunce* and *rutabaga* loci exhibit an even larger decrease in performance, suggesting that the two genes disrupt different but related aspects of learning. (After Tully, 1996.)

Continued on the next page

BOX 8A ■ *(continued)*

The initial problem in this work was to develop behavioral tests that could identify atypical learning or memory defects in large populations of flies. This challenge was met by Seymour Benzer and his colleagues Chip Quinn and Bill Harris at the California Institute of Technology, who developed the olfactory and visual learning tests that have become the basis for most subsequent analyses of learning and memory in the fruit fly (see figure). Behavioral paradigms pairing odors or light with an aversive stimulus allowed Benzer and his colleagues to assess associative learning in flies. The design of an ingenious testing apparatus controlled for non-learning-related sensory cues that had previously complicated such behavioral testing. Moreover, the apparatus allowed large numbers of flies to be screened relatively easily, expediting the analysis of mutagenized populations.

These studies led to the identification of an ever-increasing number of single gene mutations that disrupt learning or memory in flies. The behavioral and molecular studies of the mutants (given whimsical but descriptive names such as *dunce, rutabaga,* and *amnesiac*) suggested that a central pathway for learning and memory in the fly is signal transduction mediated by the cyclic nucleotide cAMP. Thus, the gene products of the *dunce, rutabaga,* and *amnesiac* loci are, respectively, a phosphodiesterase (which degrades cAMP), an adenylyl cyclase (which converts ATP to cAMP), and a peptide transmitter that stimulates adenylyl cyclase. This conclusion about the importance of cAMP has been confirmed by the finding that genetic manipulation of the CREB transcription factor also interferes with learning and memory in typical flies.

These observations in *Drosophila* accord with conclusions reached in studies of *Aplysia* and mammals (see text) and have emphasized the importance of cAMP-mediated learning and memory in a wide range of additional species. More generally, the genetic accessibility of *Drosophila* continues to make it a powerful experimental system for understanding the genetic underpinnings of learning and memory.

Second, these plastic changes in synaptic function can be either short-term effects that rely on post-translational modification of existing synaptic proteins, or they can be long-term changes that require changes in gene expression, new protein synthesis, and growth of new synapses (as well as enlarging or eliminating existing synapses). Thus, it appears that short- and long-term changes in synaptic function have different mechanistic underpinnings. As the following sections show, these generalizations apply to synaptic plasticity in the mammalian brain and have helped guide our understanding of these forms of synaptic plasticity.

Long-Term Potentiation at a Hippocampal Synapse

Long-term synaptic plasticity has also been identified in the mammalian brain. Here, some patterns of synaptic activity produce a long-lasting increase in synaptic strength known as **long-term potentiation (LTP)**, whereas other patterns of activity produce a long-lasting decrease in synaptic strength, known as **long-term depression (LTD)**. LTP and LTD are broad terms that describe only the direction of change in synaptic efficacy; in fact, different cellular and molecular mechanisms can be involved in producing LTP or LTD at different synapses throughout the brain. In general, LTP and LTD are produced by different histories of activity and are mediated by different complements of intracellular signal transduction pathways in the nerve cells involved.

Long-term synaptic plasticity has been most thoroughly studied at excitatory synapses in the mammalian hippocampus. The hippocampus is especially important in the formation and retrieval of some forms of memory (see Chapter 30). In humans, functional imaging shows that the hippocampus is activated during certain kinds of memory tasks and that damage to this brain region results in an inability to form certain types of new memories. Although many other brain areas are involved in the complex process of memory formation, storage, and retrieval, these observations from imaging and lesion (damage) studies have led many investigators to study long-term synaptic plasticity of hippocampal synapses.

Work on LTP began in the late 1960s, when Terje Lomo and Timothy Bliss, working in the laboratory of Per Andersen in Oslo, Norway, discovered that a few seconds of high-frequency electrical stimulation can enhance synaptic transmission in the rabbit hippocampus for hours or longer. More recently, however, progress in understanding the mechanism of LTP has relied heavily on in vitro studies of slices of living hippocampus. The arrangement of neurons

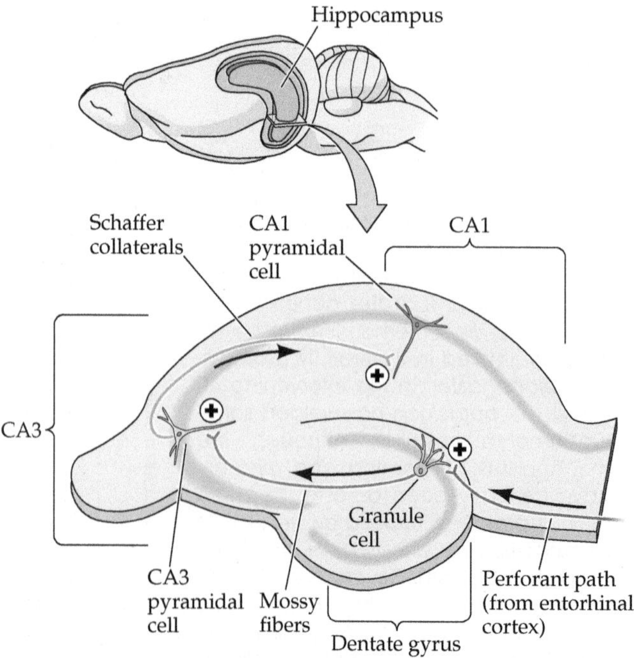

FIGURE 8.6 The trisynaptic circuit of the hippocampus. A section through the rodent hippocampus diagrams excitatory pathways and synaptic connections. Long-term potentiation (plus signs) has been observed at each of the three synaptic connections shown here.

allows the hippocampus to be sectioned such that most of the relevant circuitry is left intact. In such preparations, the cell bodies of the pyramidal neurons lie in a single densely packed layer that is readily apparent (Figure 8.6). This layer is divided into several distinct regions, the major ones being CA1 and CA3. "CA" refers to *cornu Ammonis,* Latin for Ammon's horn—the ram's horn that resembles the shape of the hippocampus. The dendrites of pyramidal cells in the CA1 region form a thick band (the stratum radiatum), where they receive synapses from Schaffer collaterals, the axons of pyramidal cells in the CA3 region.

Much of the work on LTP has focused on the synaptic connections between the Schaffer collaterals and CA1 pyramidal cells. Electrical stimulation of Schaffer collaterals generates EPSPs in the postsynaptic CA1 cells (Figure 8.7A,B). If the Schaffer collaterals are stimulated only two or three times per minute, the size of the EPSP elicited in the CA1 neurons remains constant. However, a brief, high-frequency train of stimuli to the same axons causes LTP, which is evident as a long-lasting increase in EPSP amplitude (Figure 8.7B,C). While the maximum duration of LTP is not known, in some cases LTP can last for more than a year (Figure 8.7D). The long duration of LTP shows that this form of synaptic plasticity is capable of serving as a mechanism for long-lasting storage of information. LTP occurs at each of the three excitatory synapses of the hippocampus

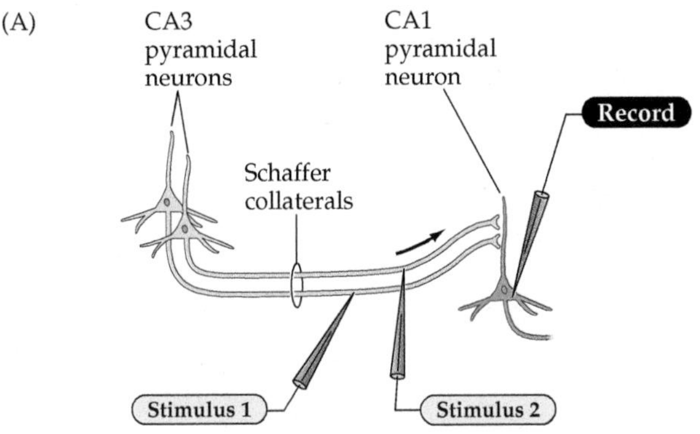

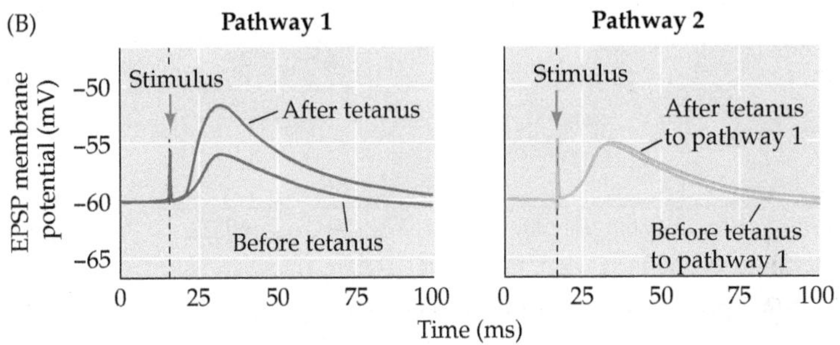

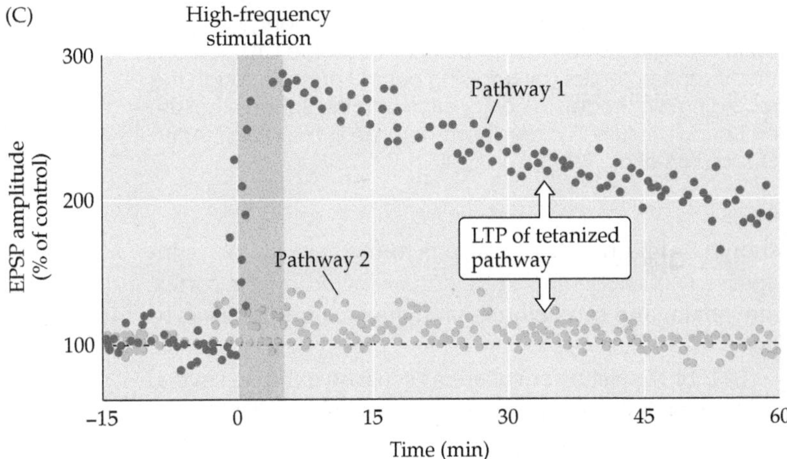

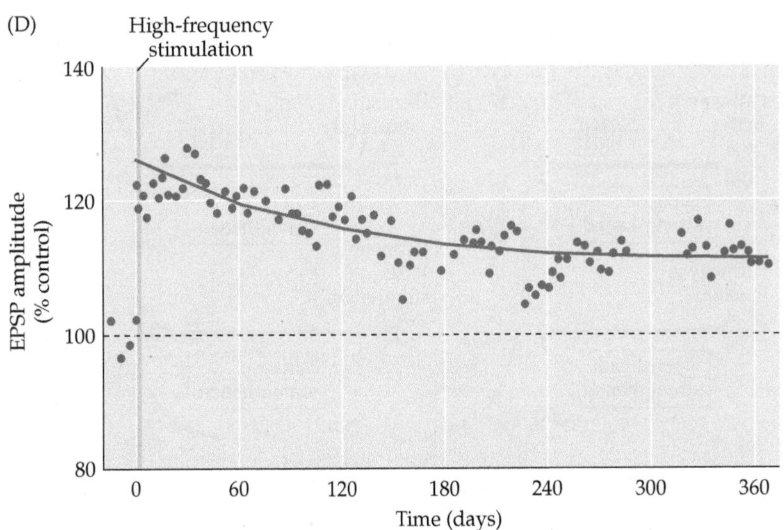

FIGURE 8.7 LTP of Schaffer collateral-CA1 synapses. (A) Arrangement for recording synaptic transmission. Two stimulating electrodes (1 and 2) activate separate populations of Schaffer collaterals, providing test and control synaptic pathways. (B) Left: Synaptic responses recorded in a CA1 neuron in response to single stimuli of synaptic pathway 1, minutes before and 1 hour after a high-frequency train of stimuli. The high-frequency stimulus train increases the size of the EPSP evoked by a single stimulus. Right: The size of the EPSP produced by stimulating synaptic pathway 2, which did not receive high-frequency stimulation, is unchanged. (C) Time course of changes in the amplitude of EPSPs evoked by stimulation of pathways 1 and 2. High-frequency stimulation of pathway 1 (darker bar) enhances EPSPs in this pathway (purple dots). This LTP of synaptic transmission in pathway 1 persists for more than an hour, while the amplitude of EPSPs produced by pathway 2 (orange dots) remains constant. (D) Recordings of EPSPs from the hippocampus in vivo reveal that high-frequency stimulation can produce LTP that lasts for more than 1 year. (A–C after Malinow et al., 1989; D from Abraham et al., 2002.)

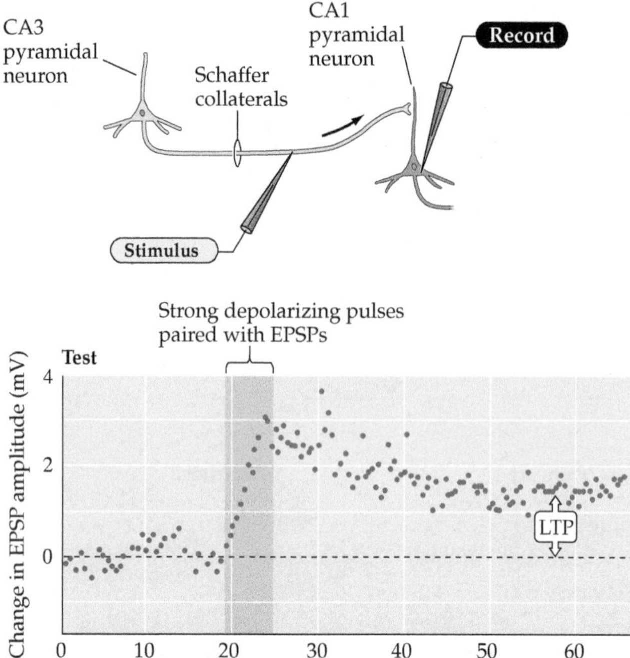

FIGURE 8.8 Pairing presynaptic and postsynaptic activity causes LTP. Single stimuli applied to a Schaffer collateral synaptic input evoke EPSPs in the postsynaptic CA1 neuron. These stimuli alone do not elicit any change in synaptic strength. However, brief polarization of the CA1 neuron's membrane potential (by applying current pulses through the recording electrode), in conjunction with the Schaffer collateral stimuli, results in a persistent increase in the EPSPs. (After Gustafsson et al., 1987.)

shown in Figure 8.6. LTP also is found at excitatory synapses in a variety of brain regions—including the cortex, amygdala, and cerebellum—and at some inhibitory synapses as well.

LTP of the Schaffer collateral synapse exhibits several properties that make it an attractive neural mechanism for information storage. First, LTP requires strong activity in both presynaptic and postsynaptic neurons. If action potentials in a small number of presynaptic Schaffer collaterals—which evoke transmitter release that produces subthreshold EPSPs that would not normally yield LTP—are paired with strong depolarization of the postsynaptic CA1 cell, the activated Schaffer collateral synapses undergo LTP (Figure 8.8). This increase in synaptic transmission occurs only if the paired activities of the presynaptic and postsynaptic cells are tightly linked in time, such that the strong postsynaptic depolarization occurs within about 100 milliseconds of transmitter release from the Schaffer collaterals. Such a requirement for coincident presynaptic and postsynaptic activity is the central postulate of a theory of learning devised by Donald Hebb in 1949. Hebb proposed that coordinated activity of a presynaptic terminal and a postsynaptic neuron would strengthen the synaptic connection between them, precisely as is observed for LTP. This indicates the involvement of a **coincidence detector** that allows LTP to occur only when both presynaptic and postsynaptic neurons are active. Hebb's postulate has also been useful in thinking about the role of neuronal activity in other brain functions, most notably development of neural circuits (see Chapter 25).

A second property that makes LTP a particularly attractive neural mechanism for information storage is that LTP is input specific: When LTP is induced by activation of one synapse, it does not occur in other, inactive synapses that contact the same neuron (see Figure 8.7C). Thus, LTP is restricted to activated synapses rather than to all of the synapses on a given cell (Figure 8.9A). Such specificity of LTP is consistent with its involvement in memory formation (or at least the selective storage of information at synapses). If activation of one set of synapses led to all other synapses—even inactive ones—being potentiated, it would be difficult to selectively enhance particular sets of inputs, as is presumably required to store specific information. The synapse specificity of LTP means that each of the tens of thousands of synapses on a hippocampal neuron can store information, thereby making it possible for the millions of neurons in the hippocampus to store a vast amount of information.

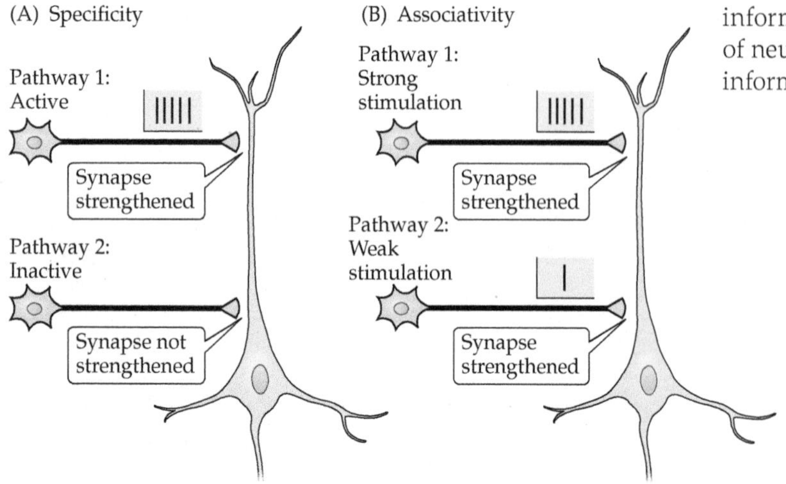

FIGURE 8.9 Properties of LTP at a CA1 pyramidal neuron receiving synaptic inputs from two independent sets of Schaffer collateral axons. (A) Strong activity initiates LTP at active synapses (pathway 1) without initiating LTP at nearby inactive synapses (pathway 2). (B) Weak stimulation of pathway 2 alone does not trigger LTP. However, when the same weak stimulus to pathway 2 is activated together with strong stimulation of pathway 1, both sets of synapses are strengthened.

Another important property of LTP is **associativity** (Figure 8.9B). As noted, weak stimulation of a pathway will not by itself trigger LTP. However, if one pathway is weakly activated at the same time that a neighboring pathway onto the same cell is strongly activated, both synaptic pathways undergo LTP. Associativity is another consequence of the coincidence detection feature of LTP, specifically pairing of the activity of the weak synapse with the coincident generation of action potentials by the strong synapse. The selective enhancement of conjointly activated sets of synaptic inputs is often considered a cellular analog of associative learning, where two stimuli are required for learning to take place. The best-known type of associative learning is classical (Pavlovian) conditioning. More generally, associativity is expected in any network of neurons that links one set of information with another.

Although there is clearly a gap between understanding LTP of hippocampal synapses and understanding learning, memory, or other aspects of behavioral plasticity in mammals, this form of long-term synaptic plasticity provides a plausible mechanism for producing enduring neural changes within a part of the brain that is known to be involved in certain kinds of memories.

Mechanisms Underlying Long-Term Potentiation

The molecular underpinnings of LTP now are well understood. A key advance was the discovery that antagonists of the NMDA-type glutamate receptor prevent LTP during high-frequency stimulation of the Schaffer collaterals, but have no effect on the synaptic response evoked by low-frequency stimulation. At about the same time, the unique biophysical properties of the NMDA receptor were first appreciated and provided a critical insight into how LTP is selectively induced by high-frequency activity. As described in Chapter 6, the NMDA receptor channel is permeable to Ca^{2+} but is blocked by Mg^{2+} at the normal resting membrane potential. The NMDA receptor is a molecular coincidence detector: The channel of this receptor opens (to induce LTP) only when glutamate is bound to the receptor, and the postsynaptic cell is depolarized to relieve the Mg^{2+} block of the channel pore. During low-frequency synaptic transmission, glutamate released by the Schaffer collaterals binds to both NMDA-type and AMPA-type glutamate receptors; however, Mg^{2+} blockade prevents current flow through the NMDA receptors, so that the EPSP is mediated entirely by the AMPA receptors (Figure 8.10, left). Because blockade of the NMDA receptor by Mg^{2+} is voltage-dependent, summation of EPSPs during high-frequency stimulation (as in Figure 8.7) leads to a prolonged depolarization that expels Mg^{2+} from the NMDA channel pore (see Figure 8.10, right). Removal of Mg^{2+} then allows Ca^{2+} to enter the dendritic spines of postsynaptic CA1 neurons.

These properties of the NMDA receptor can account for many of the characteristics of LTP. First, the requirement for strong coincident presynaptic and postsynaptic activity to induce LTP (see Figure 8.8) arises because presynaptic activity releases glutamate, while the coincident postsynaptic depolarization relieves the Mg^{2+} block of the NMDA receptor. The specificity of LTP (see Figure 8.9A) can be explained by the fact that NMDA channels will be opened only at synaptic inputs that are active and releasing glutamate, thereby confining LTP to these sites even though EPSPs generated at active synapses depolarize the postsynaptic neuron. With respect to associativity (see Figure 8.9B), a weakly stimulated input releases glutamate but cannot sufficiently depolarize the postsynaptic cell to relieve the Mg^{2+} block. If a large number of other inputs are strongly stimulated, however, they provide the "associative" depolarization

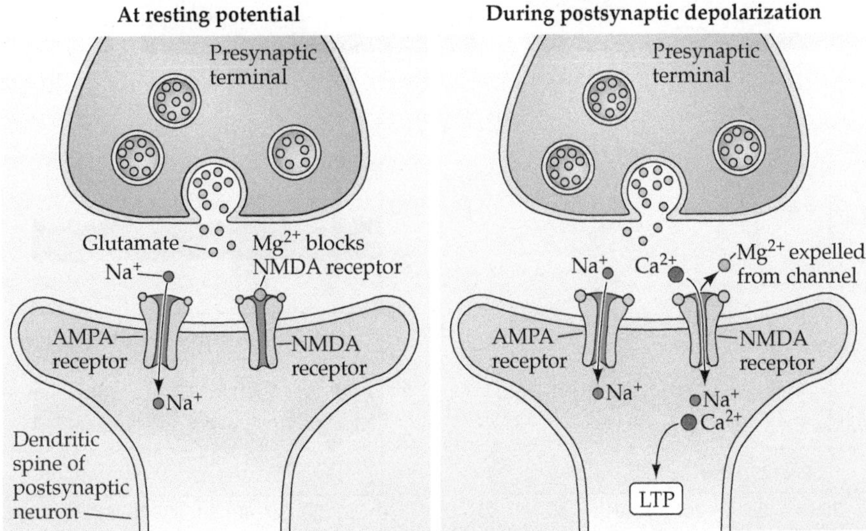

FIGURE 8.10 The NMDA receptor channel can open only during depolarization of the postsynaptic neuron from its normal resting potential. Depolarization expels Mg^{2+} from the NMDA channel, allowing current to flow into the postsynaptic cell. This leads to Ca^{2+} entry, which in turn triggers LTP. (After Nicoll et al., 1988.)

necessary to relieve the block. Thus, induction of LTP relies on activation of NMDA receptors.

Numerous observations have established that induction of LTP is due to accumulation of postsynaptic Ca^{2+} as a result of Ca^{2+} influx through NMDA receptors. Imaging studies have shown that activation of NMDA receptors increases postsynaptic Ca^{2+} levels and that these Ca^{2+} signals can be restricted to the dendritic spines of individual synapses. Furthermore, injection of Ca^{2+} chelators blocks LTP induction, whereas elevation of Ca^{2+} levels in postsynaptic neurons potentiates synaptic transmission. Thus, a rise in postsynaptic Ca^{2+} concentration serves as a second-messenger signal that induces LTP. The fact that these postsynaptic Ca^{2+} signals are highly localized (see Box 7B) can account for the input specificity of LTP (see Figure 8.9A).

The subsequent *expression* of LTP relies on dynamic changes in AMPA receptors. Excitatory synapses can dynamically regulate their postsynaptic AMPA receptors via the same sorts of membrane trafficking processes that occur in presynaptic neurons during neurotransmitter release (see Chapter 5). LTP apparently is due to synaptotagmin-mediated insertion of AMPA receptors into the postsynaptic membrane. These AMPA receptors come from an intracellular organelle known as the recycling endosome. The resulting increase in the number of AMPA receptors increases the response of the postsynaptic cell to released glutamate (Figure 8.11A), yielding a strengthening of synaptic transmission that can last for as long as LTP is maintained (Figure 8.11B). LTP does not affect the number of postsynaptic NMDA receptors; thus, while these receptors are crucial for induction of LTP, they do not play a major role in LTP expression. Stimulus-induced changes in AMPA receptor trafficking can even add new AMPA receptors to "silent" synapses that did not previously have postsynaptic AMPA receptors (Box 8B). At silent synapses, where synaptic activity generates no postsynaptic response at the normal resting potential, LTP adds AMPA receptors so that the synapse can produce postsynaptic responses (Figure 8.11C). Thus, the strengthening of synaptic transmission during LTP arises from an increase in the sensitivity of the postsynaptic cell to glutamate. Under some circumstances, LTP also can increase the ability of presynaptic terminals to release glutamate.

Ca^{2+} also activates complex postsynaptic signal transduction cascades that include at least two Ca^{2+}-activated protein kinases: Ca^{2+}/calmodulin-dependent protein kinase type II (CaMKII) and protein kinase C (PKC; see Chapter 7). CaMKII, which is the most abundant postsynaptic protein at Schaffer collateral synapses, seems to play an especially important role. This enzyme is activated by stimuli that induce

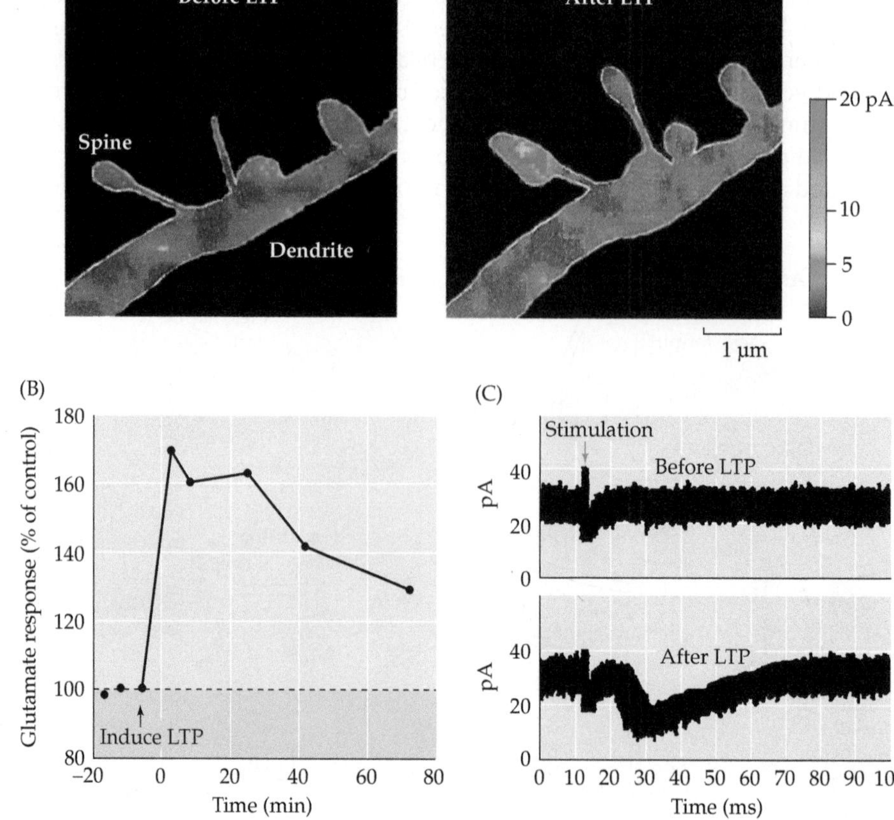

FIGURE 8.11 Addition of postsynaptic AMPA receptors during LTP. (A) Spatial maps of the glutamate sensitivity of a hippocampal neuron dendrite before and 120 minutes after induction of LTP. The color scale indicates the amplitude of responses to highly localized glutamate application. LTP causes an increase in the glutamate response of a dendritic spine due to an increase in the number of AMPA receptors on the spine membrane. (B) Time course of changes in glutamate sensitivity of dendritic spines during LTP. Induction of LTP at time = 0 causes glutamate sensitivity to increase for more than 60 minutes. (C) LTP induces AMPA receptor responses at silent synapses in the hippocampus. Prior to inducing LTP, no excitatory postsynaptic currents (EPSCs) are elicited at –65 mV at this silent synapse (upper trace). After LTP induction, the same stimulus produces EPSCs that are mediated by AMPA receptors (lower trace). (A,B from Matsuzaki et al., 2004; C after Liao et al., 1995.)

BOX 8B ■ Silent Synapses

Several observations indicate that postsynaptic glutamate receptors are dynamically regulated at excitatory synapses. Early insight into this process came from the finding that stimulation of some glutamatergic synapses generates no postsynaptic electrical signal when the postsynaptic cell is at its normal resting membrane potential (Figure A). However, once the postsynaptic cell is depolarized, these "silent synapses" can transmit robust postsynaptic electrical responses. The fact that transmission at such synapses can be turned on or off according to the postsynaptic membrane potential suggests an interesting and simple means of modifying neural circuitry.

Silent synapses are especially prevalent in development and have been found in many brain regions, including the hippocampus, cerebral cortex, and spinal cord. The silence of these synapses is evidently due to the voltage-dependent blockade of NMDA receptors by Mg^{2+} (see text and Chapter 6). At the normal resting membrane potential, presynaptic release of glutamate evokes no postsynaptic response at such synapses because their NMDA receptors are blocked by Mg^{2+}. However, depolarization of the postsynaptic neuron displaces the Mg^{2+}, allowing glutamate release to induce postsynaptic responses mediated by NMDA receptors.

Glutamate released at silent synapses evidently binds only to NMDA receptors. How, then, does glutamate release avoid activating AMPA receptors? The most likely explanation is that these synapses may have only NMDA receptors. Immunocytochemical experiments demonstrate that some excitatory synapses have only NMDA receptors (green spots in Figure B).

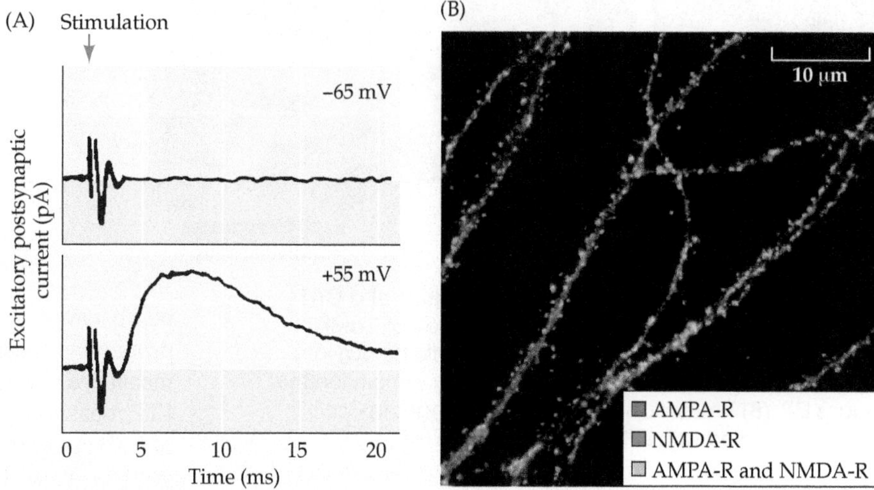

(A) Electrophysiological evidence for silent synapses. Stimulation of some axons fails to activate synapses when the postsynaptic cell is held at a negative potential (–65 mV, upper trace). However, when the postsynaptic cell is depolarized (+55 mV, lower trace), stimulation produces a robust response. (B) Immunofluorescent localization of NMDA receptors (green) and AMPA receptors (red) in a cultured hippocampal neuron. Many dendritic spines are positive for NMDA receptors but not AMPA receptors, indicating NMDA receptor-only synapses. (A after Liao et al., 1995; B courtesy of M. Ehlers.)

Such NMDA receptor-only synapses are particularly abundant early in postnatal development and decrease in adults (Figure C). Silencing of glutamatergic synapses has also been reported to result from the use of the recreational drug cocaine. Thus, at least some silent synapses are not a separate class of excitatory synapses that lack AMPA receptors, but rather an early stage in the ongoing maturation of the glutamatergic synapse (Figure D). Evidently, AMPA and NMDA receptors are not inextricably linked at excitatory synapses, but are targeted via independent cellular mechanisms. Such synapse-specific glutamate receptor composition implies sophisticated mechanisms for regulating the localization of each type of receptor. Dynamic changes in the trafficking of AMPA and NMDA receptors can strengthen or weaken synaptic transmission and are important in LTP and LTD, as well as in the maturation of glutamatergic synapses.

In summary, although silent synapses have begun to whisper their secrets, much remains to be learned about their physiological importance.

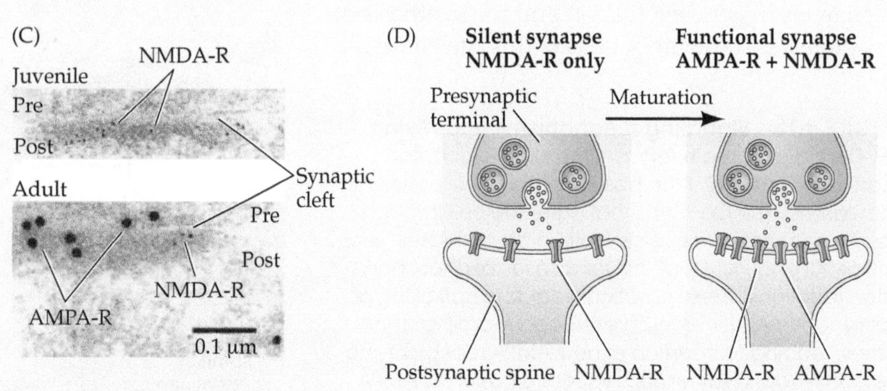

(C) Electron microscopy of excitatory synapses in CA1 stratum radiatum of the hippocampus from 10-day-old (juvenile) or 5-week-old (adult) rats double-labeled for AMPA receptors and NMDA receptors. The presynaptic terminal (pre), synaptic cleft, and postsynaptic spine (post) are indicated. AMPA receptors are abundant at the adult synapse but are absent from the younger synapse. (D) Diagram of glutamatergic synapse maturation. Early in postnatal development, many excitatory synapses contain only NMDA receptors. As synapses mature, AMPA receptors are recruited. (C from Petralia et al., 1999.)

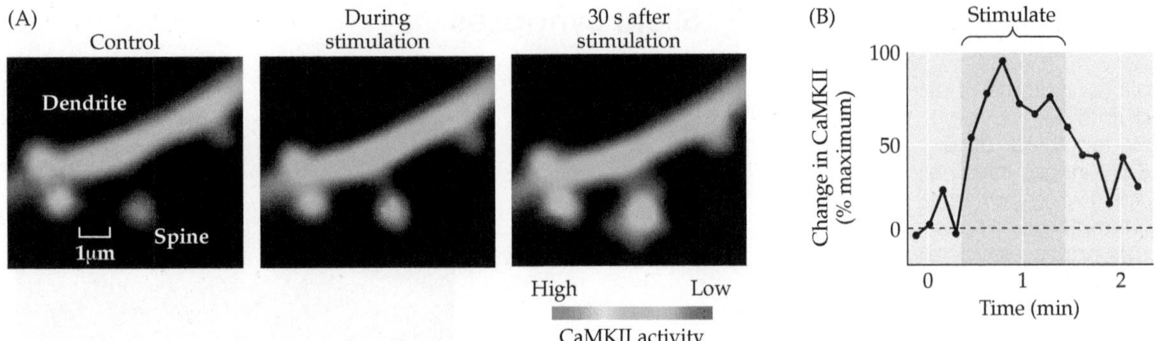

FIGURE 8.12 CaMKII activity in the dendrite of a CA1 pyramidal neuron during LTP. (A) The degree of CaMKII activation (indicated by the pseudocolor scale below) in a dendritic spine dramatically increases during stimulation that induces LTP. (B) Time course of transient changes in CaMKII activity associated with LTP. (From Lee et al., 2009.)

LTP (Figure 8.12), and pharmacological inhibition or genetic deletion of CaMKII prevents LTP. The autophosphorylation property of CaMKII (see Chapter 7) may also serve to prolong the duration of LTP. It is thought that CaMKII and PKC phosphorylate downstream targets, including both AMPA receptors and other targets, that collectively facilitate delivery of extrasynaptic AMPA receptors to the synapse.

In summary, the molecular signaling pathways involved in LTP at the Schaffer collateral–CA1 synapse are well understood. Ca^{2+} entering through postsynaptic NMDA receptors leads to activation of synaptotagmins and protein kinases that regulate trafficking of AMPA receptors, thereby enhancing the postsynaptic response to glutamate released from the presynaptic terminal (Figure 8.13). Still other forms of LTP are observed at other synapses and, in some cases, rely on signaling mechanisms different from those involved in LTP at the Schaffer collateral–CA1 synapse.

The scheme depicted in Figure 8.13 can account for the changes in synaptic transmission that occur over the first 1 to 2 hours after LTP is induced. However, there is also a later phase of LTP that depends on changes in gene expression and the synthesis of new proteins. The contributions of this late phase can be observed by treating synapses with drugs that inhibit protein synthesis: Blocking protein synthesis prevents LTP measured several hours after a stimulus but does not affect LTP measured at earlier times (Figure 8.14). This late phase of LTP appears to be initiated by protein kinase A, which goes on to activate transcription factors such as CREB, which stimulate the expression of other proteins. Although most of these newly synthesized proteins have not yet been identified, they include other transcriptional regulators, protein kinases, and AMPA receptors (Figure 8.15A). How these proteins contribute to the late phase of LTP is not yet known. There is evidence that the number and size of synaptic contacts increase during LTP (Figure 8.15B,C). Thus, it is likely that some of the proteins newly synthesized during the late phase of LTP are involved in construction of new synaptic

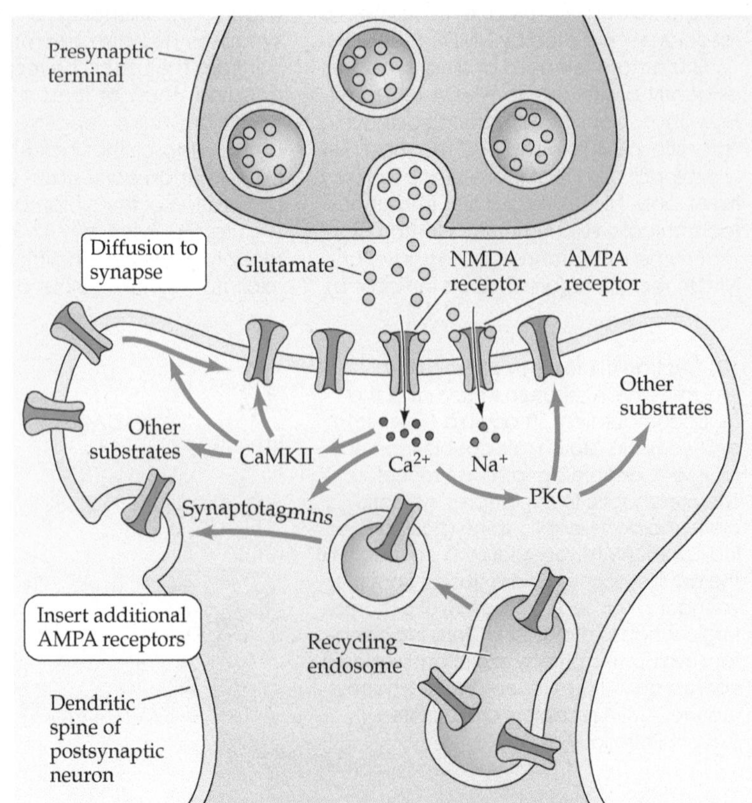

FIGURE 8.13 Signaling mechanisms underlying LTP. During glutamate release, the NMDA receptor channel opens only if the postsynaptic cell is sufficiently depolarized. The Ca^{2+} ions that enter the cell through the channel activate postsynaptic protein kinases, such as CaMKII and PKC, that trigger a series of phosphorylation reactions. These reactions regulate trafficking of postsynaptic AMPA receptors through recycling endosomes, leading to insertion of new AMPA receptors into the postsynaptic spine. Subsequent diffusion of AMPA receptors to the subsynaptic region yields an increase in the spine's sensitivity to glutamate, which causes LTP.

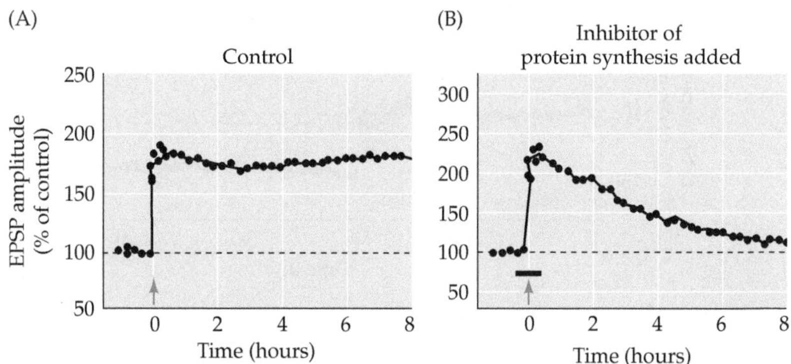

FIGURE 8.14 Role of protein synthesis in maintaining LTP. (A) Repetitive high-frequency stimulation (arrow) induces LTP that persists for many hours. (B) Treatment with anisomycin (at bar), an inhibitor of protein synthesis, causes LTP to decay within a few hours after the high-frequency stimulation (arrow). (After Frey and Morris, 1997.)

contacts that serve to make LTP essentially permanent (as in Figure 8.7D).

In conclusion, it appears that LTP in the mammalian hippocampus has many parallels to the long-term changes in synaptic transmission underlying behavioral sensitization in *Aplysia*. Both consist of an early, transient phase that relies on protein kinases to produce post-translational changes in membrane ion channels, and both have later, long-lasting phases that require changes in gene expression mediated by CREB. Both forms of long-term synaptic plasticity are likely to be involved in long-term storage of information, although the role of LTP in memory storage in the hippocampus is not firmly established.

Mechanisms Underlying Long-Term Depression

If synapses simply continued to increase in strength as a result of long-term potentiation, eventually they would reach some level of maximum efficacy, making it difficult to encode new information. Thus, to make synaptic strengthening useful, other processes must selectively weaken specific sets of synapses, and LTD is such a process. In the late 1970s, LTD was found to occur at the synapses between the Schaffer collaterals and the CA1 pyramidal cells in the hippocampus. Whereas LTP at these synapses requires brief, high-frequency stimulation, LTD occurs when the Schaffer collaterals are stimulated at a low rate—about 1 Hz—for long periods (10–15 minutes). This pattern of activity depresses the EPSP for several hours and, like LTP, is specific to the activated synapses (Figure 8.16A,B). Moreover, LTD can erase the increase in EPSP size due to LTP, and conversely, LTP can erase the decrease in EPSP size due to LTD. This complementarity suggests that

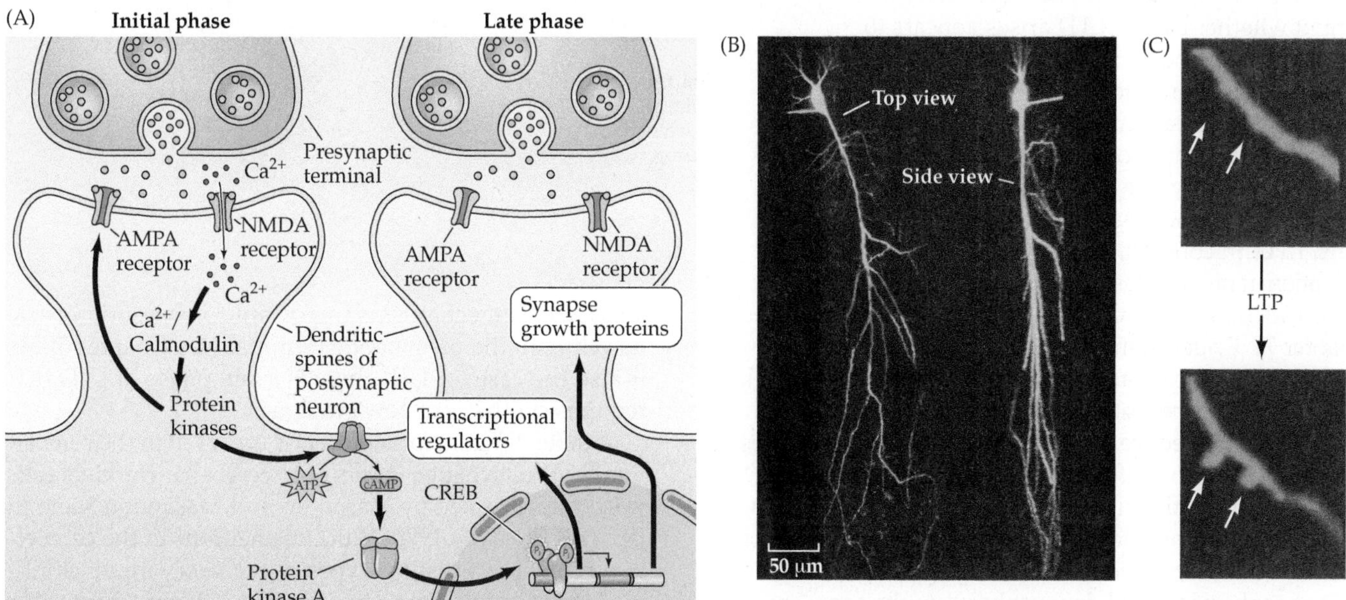

FIGURE 8.15 Mechanisms responsible for long-lasting changes in synaptic transmission during LTP. (A) The late component of LTP is the result of PKA activating the transcriptional regulator CREB, which turns on expression of several genes that produce long-lasting changes in PKA activity and synapse structure. (B,C) Structural changes associated with LTP in the hippocampus. (B) The dendrites of a CA1 pyramidal neuron were visualized by filling the cell with a fluorescent dye. (C) New dendritic spines (white arrows) can be observed approximately 1 hour after a stimulus that induces LTP. The presence of novel spines raises the possibility that LTP may arise, in part, from formation of new synapses. (A after Squire and Kandel, 1999; B and C after Engert and Bonhoeffer, 1999.)

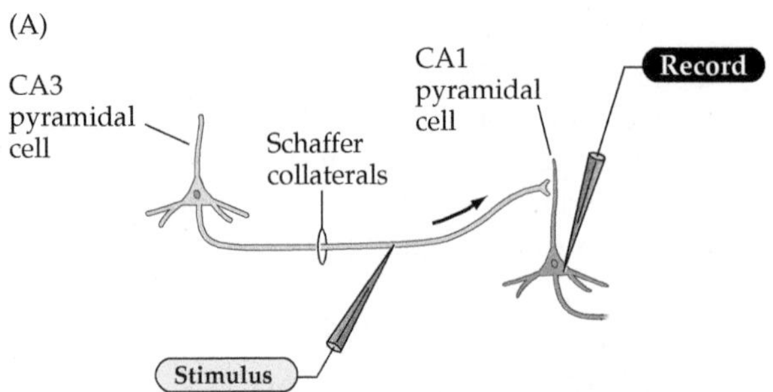

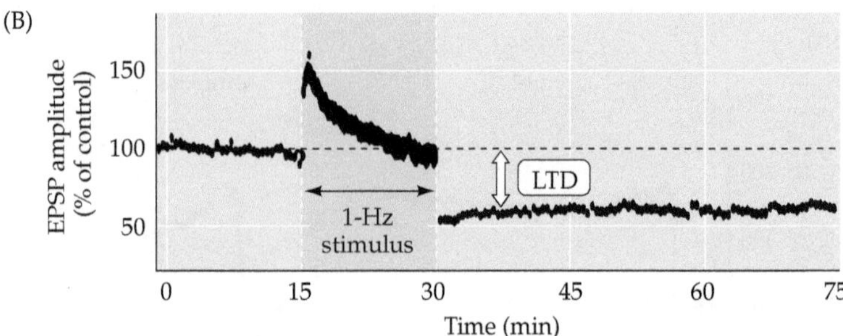

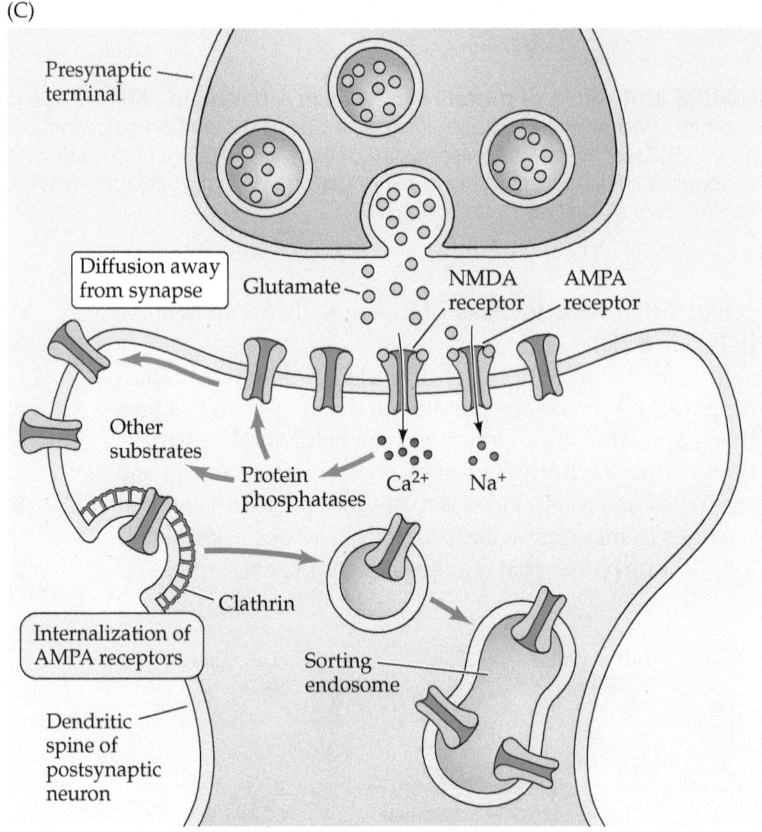

FIGURE 8.16 Long-term synaptic depression in the hippocampus. (A) Electrophysiological procedures used to monitor transmission at the Schaffer collateral synapses on CA1 pyramidal neurons. (B) Low-frequency stimulation (one per second) of the Schaffer collateral axons causes a long-lasting depression of synaptic transmission. (C) Mechanisms underlying LTD. A low-amplitude rise in Ca^{2+} concentration in the postsynaptic CA1 neuron activates postsynaptic protein phosphatases, which cause internalization of postsynaptic AMPA receptors, thereby decreasing the sensitivity to glutamate released from the Schaffer collateral terminals. (B after Mulkey et al., 1993.)

LTD and LTP reversibly affect synaptic efficiency by acting at a common site.

LTP and LTD at the Schaffer collateral–CA1 synapses actually share several key elements. Both require activation of NMDA-type glutamate receptors and the resulting entry of Ca^{2+} into the postsynaptic cell. The major determinant of whether LTP or LTD arises appears to be the nature of the Ca^{2+} signal in the postsynaptic cell: Small and slow rises in Ca^{2+} lead to depression, whereas large and fast increases in Ca^{2+} trigger potentiation. As noted above, LTP is at least partially due to activation of protein kinases, which phosphorylate their target proteins. LTD, in contrast, appears to result from activation of phosphatases, specifically PP1 and PP2B (calcineurin), a Ca^{2+}-dependent phosphatase (see Chapter 7). Evidence in support of this idea is that inhibitors of these phosphatases prevent LTD but do not block LTP. The different effects of Ca^{2+} during LTD and LTP may arise from the selective activation of protein phosphatases and kinases by the different types of postsynaptic Ca^{2+} signals occurring during these two forms of synaptic plasticity. Although the phosphatase substrates important for LTD have not yet been identified, it is possible that LTP and LTD phosphorylate and dephosphorylate the same set of regulatory proteins to control the efficacy of transmission at the Schaffer collateral–CA1 synapse. Just as LTP at this synapse is associated with insertion of AMPA receptors, LTD is often associated with a loss of synaptic AMPA receptors. This loss probably arises from internalization of AMPA receptors into sorting endosomes in the postsynaptic cell (Figure 8.16C), due to the same clathrin-dependent endocytosis mechanisms important for synaptic vesicle recycling in the presynaptic terminal (see Chapter 5). As is also the case for LTP, there is a late phase of LTD that requires synthesis of new proteins.

A quite different form of LTD is observed in the cerebellum. LTD of synaptic inputs onto cerebellar Purkinje cells was first described by Masao Ito and Masanobu Kano in Japan in the early 1980s. Purkinje neurons in the cerebellum receive two distinct types of excitatory input: climbing fibers and parallel fibers (Figure 8.17A; see Chapter 19). LTD reduces the strength of transmission at the parallel fiber synapse (Figure 8.17B) and subsequently was found to depress transmission at the climbing fiber synapse as well. This form of LTD has been implicated in the motor learning that mediates the coordination, acquisition, and storage of complex movements in the cerebellum. Although the role of LTD in cerebellar motor learning remains controversial, it

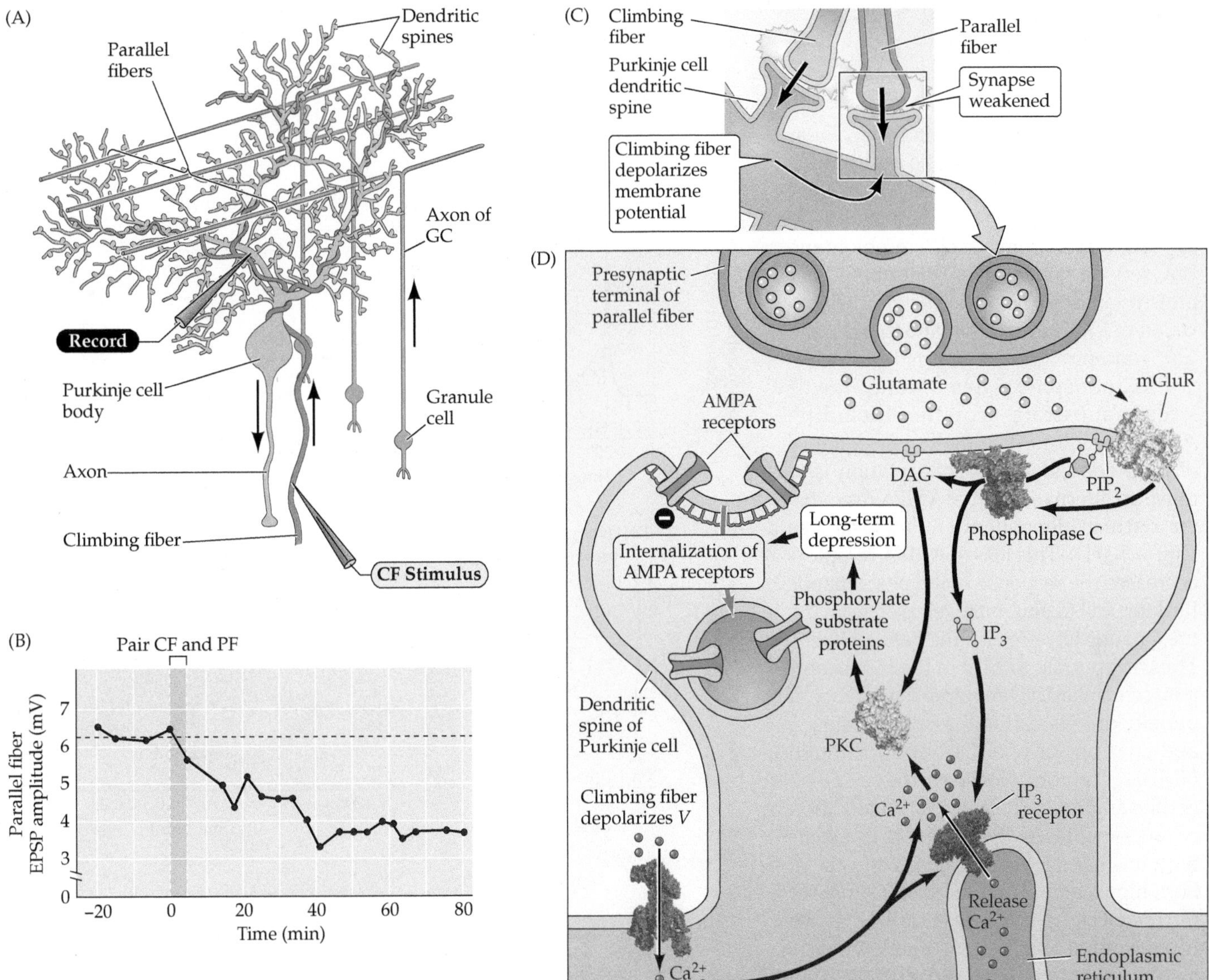

FIGURE 8.17 Long-term synaptic depression in the cerebellum. (A) Experimental arrangement. Synaptic responses were recorded from Purkinje cells following stimulation of parallel fibers and climbing fibers. (B) Pairing stimulation of climbing fibers (CF) and parallel fibers (PF) causes LTD that reduces the parallel fiber EPSP. (C) LTD requires depolarization of the Purkinje cell, produced by climbing fiber activation, as well as signals generated by active parallel fiber synapses. (D) Mechanism underlying cerebellar LTD. Glutamate released by parallel fibers activates both AMPA receptors and metabotropic glutamate receptors (mGluRs). The activated mGluRs produce two second messengers, DAG and IP_3, which interact with Ca^{2+} that enters when climbing fiber activity opens voltage-gated Ca^{2+} channels. The resultant release of Ca^{2+} from the endoplasmic reticulum leads to a further rise in intracellular Ca^{2+} concentration and the activation of PKC, which triggers clathrin-dependent internalization of postsynaptic AMPA receptors, weakening the parallel fiber synapse. (B after Sakurai, 1987.)

has nonetheless been a useful model system for understanding the cellular mechanisms of long-term synaptic plasticity.

Like many forms of motor learning, cerebellar LTD is associative because it occurs only when climbing fibers and parallel fibers are activated at the same time (Figure 8.17C). In this case, associativity arises from the combined actions of two distinct intracellular signal transduction pathways that are activated in the postsynaptic Purkinje cell in response to activity of the climbing fiber and parallel fiber synapses. In the first pathway, glutamate released from the parallel fiber terminals activates two types of receptors, the AMPA-type and metabotropic glutamate receptors (see Chapter 6). Glutamate binding to the AMPA receptor results in mild membrane depolarization, whereas binding to the metabotropic receptor produces the second messengers inositol triphosphate (IP_3) and diacylglycerol (DAG) (see Chapter 7). Glutamate released by climbing fibers also activates AMPA receptors, which strongly depolarizes the Purkinje cell membrane potential and initiates a second signal transduction pathway: an influx of Ca^{2+} through voltage-gated channels and a subsequent increase in intracellular Ca^{2+} concentration. These two second messengers—IP_3 and Ca^{2+}—cause an amplified rise in intracellular Ca^{2+} concentration by acting together on IP_3 receptors. This triggers release of Ca^{2+} from intracellular stores and leads to synergistic activation of PKC by Ca^{2+} and DAG (Figure 8.17D). Thus, the associative property of cerebellar LTD appears to arise from both IP_3 receptors and PKC serving as coincidence detectors. Other protein kinases also act to sustain the activation of PKC beyond the time that IP_3 and Ca^{2+} concentrations are elevated.

PKC phosphorylates several downstream substrate proteins, including AMPA receptors. The main consequence of PKC-dependent phosphorylation is to cause an internalization of AMPA receptors via clathrin-dependent endocytosis (see Figure 8.17D). This loss of AMPA receptors decreases the response of the postsynaptic Purkinje cell to glutamate released from the presynaptic terminals of the parallel fibers. Thus, in contrast to LTD in the hippocampus, cerebellar LTD requires the activity of protein kinases, rather than phosphatases, and does not involve Ca^{2+} entry through NMDA receptors. However, the net effect is the same in both cases: Internalization of postsynaptic AMPA receptors is a common mechanism for decreased efficacy of both hippocampal and cerebellar synapses during LTD. As is the case for LTP at the hippocampal Schaffer collateral synapse, as well as for long-term synaptic plasticity in *Aplysia*, CREB appears to be required for a late phase of cerebellar LTD. It is not yet known which proteins are synthesized as a consequence of CRE activation.

Spike Timing-Dependent Plasticity

The preceding sections show that LTP and LTD are preferentially initiated by different rates of repetitive synaptic activity, with LTP requiring high-frequency activity and LTD being induced by low-frequency activity. However, the precise temporal relationship between activity in the pre- and postsynaptic cells can also be an important determinant of the amount and direction of long-term synaptic plasticity. At a given (low) frequency of synaptic activity, LTD will occur if presynaptic activity is preceded by a postsynaptic action potential, while LTP will occur if the postsynaptic action potential follows presynaptic activity (Figure 8.18A,B). The relationship between the time interval and the magnitude of the synaptic change is a very sensitive function

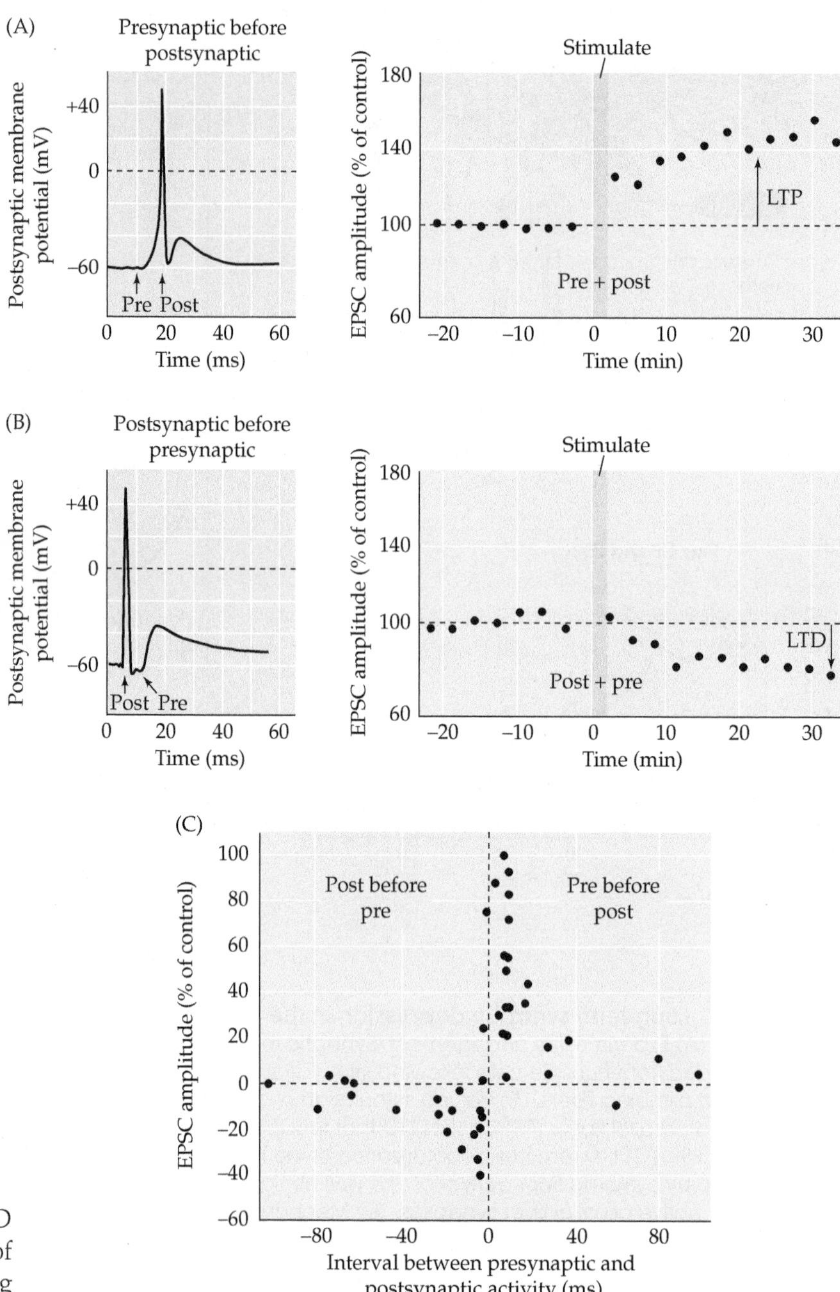

FIGURE 8.18 Spike timing-dependent synaptic plasticity in cultured hippocampal neurons. (A) Left: Stimulating a presynaptic neuron (Pre) causes an EPSP in the postsynaptic neuron; applying a subsequent stimulus to the postsynaptic neuron (Post) causes an action potential that is superimposed on the EPSP. Right: Repetitive application of this stimulus paradigm causes LTP of the EPSP. (B) Reversing the order of stimulation, so that the postsynaptic neuron is excited before the presynaptic neuron, causes LTD of the EPSP. (C) Complex dependence of STDP on the interval between presynaptic activity and postsynaptic activity. If the presynaptic neuron is activated 40 milliseconds or less before the postsynaptic neuron, then LTP occurs. Conversely, if the postsynaptic neuron is activated 40 milliseconds or less before the presynaptic neuron, LTD occurs. If the interval between the two events is longer than 40 milliseconds, no STDP is observed. (After Bi and Poo, 1998.)

of the time interval, with no changes observed if the presynaptic and postsynaptic activities are separated by 100 milliseconds or longer (Figure 8.18C).

Because precise timing of presynaptic and postsynaptic activity determines the polarity of these forms of long-lasting synaptic plasticity, they are called **spike timing-dependent plasticity (STDP)**. Although the mechanisms involved are not yet well understood, it appears that the properties of STDP arise from timing-dependent differences in postsynaptic Ca^{2+} signals. Specifically, if a postsynaptic action potential occurs *after* presynaptic activity, the resulting depolarization will relieve the block of NMDA receptors by Mg^{2+} and cause a relatively large amount of Ca^{2+} influx through postsynaptic NMDA receptors, yielding LTP. In contrast, if the postsynaptic action potential occurs *before* the presynaptic action potential, then the depolarization associated with the postsynaptic action potential will subside by the time an EPSP occurs. This sequence of events will reduce the amount of Ca^{2+} entry through the NMDA receptors, leading to LTD. It has been postulated that other signals, such as endocannabinoids (see Chapter 6), may also be required for LTD induction during STDP.

The requirement for a precise temporal relationship between presynaptic and postsynaptic activity means that STDP can perform several novel types of neuronal computation. STDP can provide a means of encoding information about causality. For example, if a synapse generates a suprathreshold EPSP, the resulting postsynaptic action potential would rapidly follow presynaptic activity, and the resulting LTP would encode the fact that the postsynaptic action potential resulted from the activity of that synapse. STDP could also serve as a mechanism for competition between synaptic inputs: Stronger inputs would be more likely to produce suprathreshold EPSPs and be reinforced by the resulting LTP, whereas weaker inputs would not generate postsynaptic action potentials that were correlated with presynaptic activity. There is evidence that STDP is important for neural circuit function, such as determining orientation preference in the visual system (see Chapter 12).

In summary, activity-dependent forms of synaptic plasticity cause changes in synaptic transmission that modify the functional connections within and among neural circuits. These changes in the efficacy and local geometry of synaptic connectivity can provide a basis for learning, memory, and other forms of brain plasticity. Activity-dependent changes in synaptic transmission may also be involved in some pathologies. Abnormal patterns of neuronal activity, such as those that occur in epilepsy, can stimulate abnormal changes in synaptic connections that may further increase the frequency and severity of seizures. Despite the substantial advances in understanding the cellular and molecular bases of some forms of plasticity, the means by which selective changes of synaptic strength encode memories or other complex behavioral modifications in the mammalian brain are simply not known.

Summary

Synapses exhibit many forms of plasticity that occur over a broad temporal range. At the shortest times (milliseconds to minutes), facilitation, augmentation, potentiation, and depression provide rapid but transient modifications in synaptic transmission. These forms of plasticity change the amount of neurotransmitter released from presynaptic terminals and are based on alterations in Ca^{2+} signaling and synaptic vesicle pools at recently active terminals. Longer-lasting forms of synaptic plasticity such as LTP and LTD are also based on Ca^{2+} and other intracellular second messengers. At least some of the synaptic changes produced by these long-lasting forms of plasticity are postsynaptic, caused by changes in neurotransmitter receptor trafficking, although alterations in neurotransmitter release from the presynaptic terminal can also occur. In these more enduring forms of plasticity, protein phosphorylation and changes in gene expression greatly outlast the period of synaptic activity and can yield changes in synaptic strength that persist for hours, days, or even longer. Long-lasting synaptic plasticity can serve as a neural mechanism for many forms of brain plasticity, such as learning new behaviors or acquiring new memories.

ADDITIONAL READING

Reviews

Chater, T. E. and Y. Goda (2014) The role of AMPA receptors in postsynaptic mechanisms of synaptic plasticity. *Front. Cell. Neurosci.* 8: 401.

Dan, Y. and M. M. Poo (2006) Spike timing-dependent plasticity: From synapse to perception. *Physiol. Rev.* 86: 1033–1048.

Huganir, R. L. and R. A. Nicoll (2013) AMPARs and synaptic plasticity: The last 25 years. *Neuron* 80: 704–717.

Ito, M., K. Yamaguchi, S. Nagao and T. Yamazaki (2014) Long-term depression as a model of cerebellar plasticity. *Prog. Brain Res.* 210: 1–30.

Jackman, S. L. and W. G. Regehr (2017) The mechanisms and functions of synaptic facilitation. *Neuron* 94: 447–464.

Kandel, E. R., Y. Dudai and M. R. Mayford (2014) The molecular and systems biology of memory. *Cell* 157: 163–186.

Kneussel, M. and T. J. Hausrat (2016) Postsynaptic neurotransmitter receptor reserve pools for synaptic potentiation. *Trends Neurosci.* 39: 170–182.

Malenka, R. C. and S. A. Siegelbaum (2001) Synaptic plasticity: Diverse targets and mechanisms for regulating synaptic efficacy. In *Synapses*, W. M. Cowan, T. C. Sudhof and C. F. Stevens (eds.). Baltimore, MD: Johns Hopkins University Press, pp. 393–413.

Song, S. H. and G. J. Augustine (2015) Synapsin isoforms and synaptic vesicle trafficking. *Mol. Cells.* 38: 936–940.

Important Original Papers

Abraham, W. C., B. Logan, J. M. Greenwood and M. Dragunow (2002) Induction and experience-dependent consolidation of stable long-term potentiation lasting months in the hippocampus. *J. Neurosci.* 22: 9626–9634.

Ahn, S., D. D. Ginty and D. J. Linden (1999) A late phase of cerebellar long-term depression requires activation of CaMKIV and CREB. *Neuron* 23: 559–568.

Betz, W. J. (1970) Depression of transmitter release at the neuromuscular junction of the frog. *J. Physiol. (Lond.)* 206: 629–644.

Bi, G. Q. and M. M. Poo (1998) Synaptic modifications in cultured hippocampal neurons: Dependence on spike timing, synaptic strength, and postsynaptic cell type. *J. Neurosci.* 18: 10464–10472.

Bliss, T. V. P. and T. Lomo (1973) Long-lasting potentiation of synaptic transmission in the dentate area of the anaesthetized rabbit following stimulation of the perforant path. *J. Physiol.* 232: 331–356.

Charlton, M. P. and G. D. Bittner (1978) Presynaptic potentials and facilitation of transmitter release in the squid giant synapse. *J. Gen. Physiol.* 72: 487–511.

Chung, H. J., J. P. Steinberg, R. L. Huganir and D. J. Linden (2003) Requirement of AMPA receptor GluR2 phosphorylation for cerebellar long-term depression. *Science* 300: 1751–1755.

Collingridge, G. L., S. J. Kehl and H. McLennan (1983) Excitatory amino acids in synaptic transmission in the Schaffer collateral-commissural pathway of the rat hippocampus. *J. Physiol.* 334: 33–46.

Engert, F. and T. Bonhoeffer (1999) Dendritic spine changes associated with hippocampal long-term synaptic plasticity. *Nature* 399: 66–70.

Enoki, R., Y. L. Hu, D. Hamilton and A. Fine (2009) Expression of long-term plasticity at individual synapses in hippocampus is graded, bidirectional, and mainly presynaptic: Optical quantal analysis. *Neuron* 62: 242–253.

Frey, U. and R. G. Morris (1997) Synaptic tagging and long-term potentiation. *Nature* 385: 533–536.

Gustafsson, B., H. Wigstrom, W. C. Abraham and Y. Y. Huang (1987) Long-term potentiation in the hippocampus using depolarizing current pulses as the conditioning stimulus to single volley synaptic potentials. *J. Neurosci.* 7: 774–780.

Junge, H. J. and 7 others (2004) Calmodulin and Munc13 form a Ca^{2+} sensor/effector complex that controls short-term synaptic plasticity. *Cell* 118: 389–401.

Katz, B. and R. Miledi (1968) The role of calcium in neuromuscular facilitation. *J. Physiol. (Lond.)* 195: 481–492.

Konnerth, A., J. Dreessen and G. J. Augustine (1992) Brief dendritic calcium signals initiate long-lasting synaptic depression in cerebellar Purkinje cells. *Proc. Natl. Acad. Sci. USA* 89: 7051–7055.

Lee, S. J., Y. Escobedo-Lozoya, E. M. Szatmari and R. Yasuda (2009) Activation of CaMKII in single dendritic spines during long-term potentiation. *Nature* 458: 299–304.

Lev-Tov, A., M. J. Pinter and R. E. Burke (1983) Posttetanic potentiation of group Ia EPSPs: Possible mechanisms for differential distribution among medial gastrocnemius motoneurons. *J. Neurophysiol.* 50: 379–398.

Liao, D., N. A. Hessler and R. Malinow (1995) Activation of postsynaptically silent synapses during pairing-induced LTP in CA1 region of hippocampal slice. *Nature* 375: 400–404.

Malinow, R., H. Schulman and R. W. Tsien (1989) Inhibition of postsynaptic PKC or CaMKII blocks induction but not expression of LTP. *Science* 245: 862–866.

Matsuzaki, M., N. Honkura, G. C. Ellis-Davies and H. Kasai (2004) Structural basis of long-term potentiation in single dendritic spines. *Nature* 429: 761–766.

Miyata, M. and 9 others (2000) Local calcium release in dendritic spines required for long-term synaptic depression. *Neuron* 28: 233–244.

Mulkey, R. M., C. E. Herron and R. C. Malenka (1993) An essential role for protein phosphatases in hippocampal long-term depression. *Science* 261: 1051–1055.

Murakoshi, H. and 5 others (2017) Kinetics of endogenous CaMKII required for synaptic plasticity revealed by optogenetic kinase inhibitor. *Neuron* 94: 37–47.

Sakurai, M. (1987) Synaptic modification of parallel fibre-Purkinje cell transmission in *in vitro* guinea-pig cerebellar slices. *J. Physiol. (Lond.)* 394: 463–480.

Tanaka, K. and G. J. Augustine (2008) A positive feedback signal transduction loop determines timing of cerebellar long-term depression. *Neuron* 59: 608–620.

Books

Bliss, T., G. Collingridge and R. Morris (eds.) (2004) *Long-term Potentiation: Enhancing Neuroscience for 30 Years.* New York: Oxford University Press.

Kandel, E. R. (2007) *In Search of Memory: The Emergence of a New Science of Mind.* New York: W. W. Norton.

Katz, B. (1966) *Nerve, Muscle, and Synapse.* New York: McGraw-Hill.

Squire, L. R. and E. R. Kandel (1999) *Memory: From Mind to Molecules.* New York: Scientific American Library.

Go to the NEUROSCIENCE 6e Companion Website at **www.oup.com/uk/Purves6e** for Web Topics, Animations, Flashcards, and more.

UNIT II

Sensation and Sensory Processing

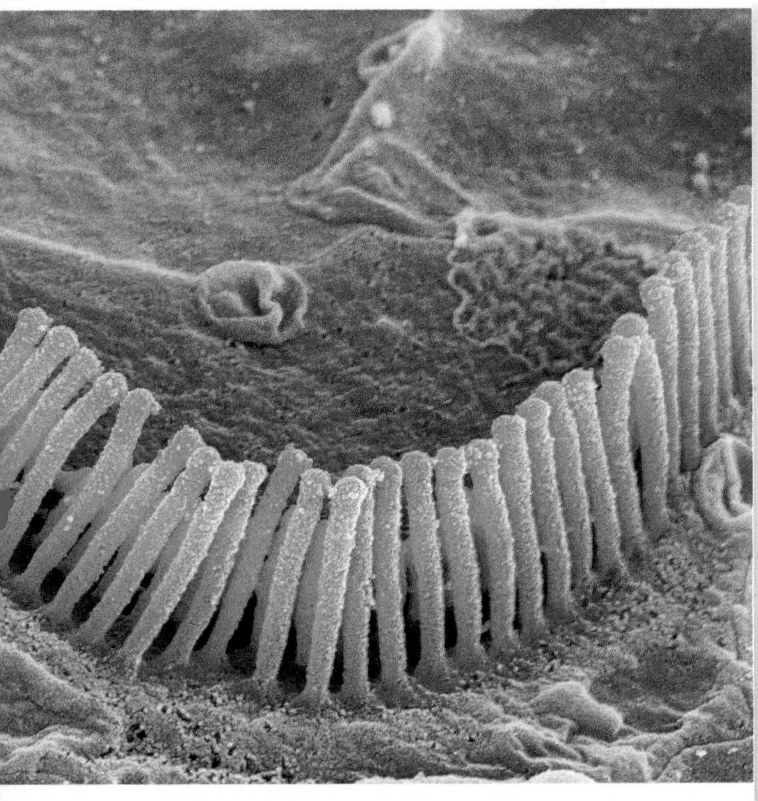

SENSATION ENTAILS THE ABILITY to transduce, encode, and ultimately perceive information generated by stimuli arising from both the external and internal environments. Much of the brain is devoted to these tasks. Although the basic senses—somatic sensation, vision, audition, vestibular sensation, and the chemical senses—are very different from one another, a few fundamental rules govern the way the nervous system deals with each of these diverse modalities. Highly specialized nerve cells called receptors convert the energy associated with mechanical forces, light, and sound waves—and the presence of odorant molecules or ingested chemicals—into neural signals that convey information about stimuli (afferent sensory signals) to the spinal cord and brain. Afferent sensory signals activate central neurons capable of representing both the qualitative and quantitative aspects of the stimulus (what it is and how strong it is) and, in some modalities (somatic sensation, vision, and audition) the location of the stimulus in space (where it is).

The clinical evaluation of patients routinely requires an assessment of the sensory systems to infer the nature and location of potential neurological problems. Knowledge of where and how the different sensory modalities are transduced, relayed, represented, and further processed to generate appropriate behavioral responses is therefore essential to understanding and treating a wide variety of diseases. Accordingly, these chapters on the neurobiology of sensation also introduce some of the major structure/function relationships in the sensory components of the nervous system.

On the previous page:

Scanning electron micrograph of outer hair cells (stereocilia) in the inner ear. © Steve Gschmeissner/Science Photo Library.

CHAPTER

9

The Somatosensory System: Touch and Proprioception

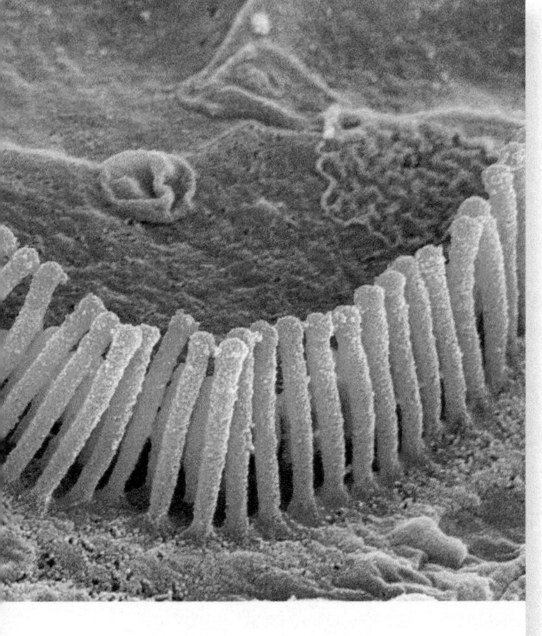

Overview

THE SOMATOSENSORY SYSTEM is arguably the most diverse of the sensory systems, mediating a range of sensations—touch, pressure, vibration, limb position, heat, cold, itch, and pain—that are transduced by receptors within the skin, muscles, or joints and conveyed to a variety of CNS targets. Not surprisingly, this complex neurobiological machinery can be divided into functionally distinct subsystems with distinct sets of peripheral receptors and central pathways. One subsystem transmits information from cutaneous mechanoreceptors and mediates the sensations of fine touch, vibration, and pressure. Another originates in specialized receptors that are associated with muscles, tendons, and joints and is responsible for proprioception—our ability to sense the position of our own limbs and other body parts in space. A third subsystem arises from receptors that supply information about painful stimuli and changes in temperature as well as non-discriminative (or sensual) touch. This chapter focuses on the tactile and proprioceptive subsystems; the mechanisms responsible for sensations of pain, temperature, and coarse sensual touch are considered in the following chapter.

Afferent Fibers Convey Somatosensory Information to the Central Nervous System

Somatic sensation originates from the activity of afferent nerve fibers whose peripheral processes ramify within the skin, muscles, or joints (Figure 9.1A). The cell bodies of afferent fibers reside in a series of ganglia that lie alongside the spinal cord and the brainstem and are considered part of the peripheral nervous system. Neurons in the dorsal root ganglia and in the cranial nerve ganglia (for the body and head, respectively) are the critical links supplying CNS circuits with information about sensory events that occur in the periphery.

Action potentials generated in afferent fibers by events that occur in the skin, muscles or joints propagate along the fibers and past the locations of the cell bodies in the ganglia until they reach a variety of targets in the CNS (Figure 9.1B). Peripheral and central components of afferent fibers are continuous, attached to the cell body in the ganglia by a single process. For this reason, neurons in the dorsal root ganglia are often called **pseudounipolar**. Because of this configuration, conduction of electrical activity through the membrane of the cell body is not an obligatory step in conveying sensory information to central targets. Nevertheless, cell bodies of sensory afferents play a critical role in maintaining the cellular machinery that mediates transduction, conduction, and transmission by sensory afferent fibers.

The fundamental mechanism of **sensory transduction**—the process of converting the energy of a stimulus into an electrical signal—is similar in all somatosensory afferents: A stimulus alters the permeability of cation channels in the afferent nerve

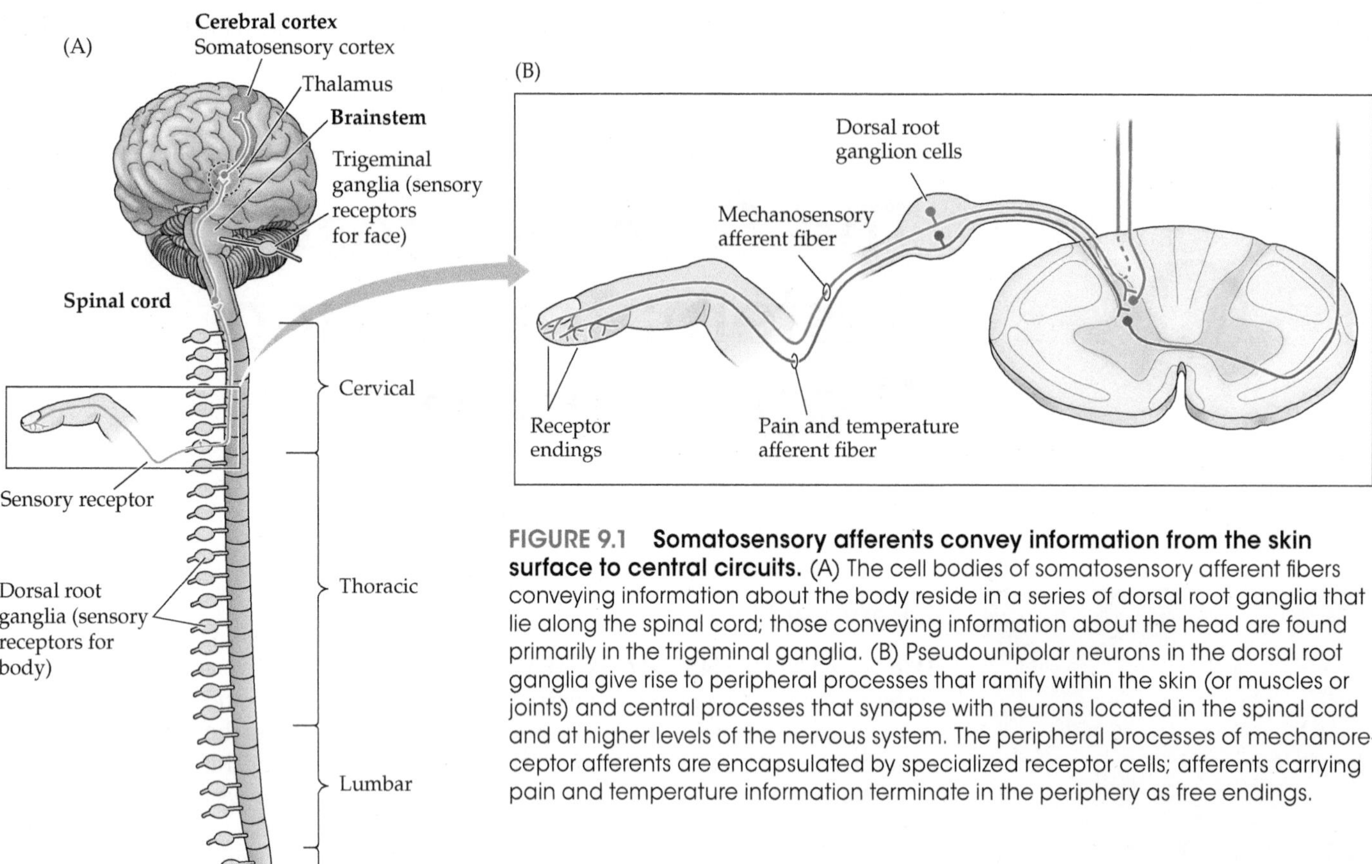

FIGURE 9.1 Somatosensory afferents convey information from the skin surface to central circuits. (A) The cell bodies of somatosensory afferent fibers conveying information about the body reside in a series of dorsal root ganglia that lie along the spinal cord; those conveying information about the head are found primarily in the trigeminal ganglia. (B) Pseudounipolar neurons in the dorsal root ganglia give rise to peripheral processes that ramify within the skin (or muscles or joints) and central processes that synapse with neurons located in the spinal cord and at higher levels of the nervous system. The peripheral processes of mechanoreceptor afferents are encapsulated by specialized receptor cells; afferents carrying pain and temperature information terminate in the periphery as free endings.

endings, generating a depolarizing current known as a **receptor** (or **generator**) **potential** (Figure 9.2). If sufficient in magnitude, the receptor potential reaches threshold for the generation of action potentials in the afferent fiber; the resulting rate of action potential firing is roughly proportional to the magnitude of the depolarization, as described in Chapters 2 and 3. Recently, the first family of mammalian mechanotransduction channels was identified. It consists of two members: Piezo1 and Piezo2 (Greek *piesi*, "pressure"; see Chapter 4). Piezo channels are predicted to have more than 30 transmembrane domains. Purified Piezo proteins reconstituted in artificial lipid bilayers form ion channels that transduce tension in the surrounding membrane. Importantly, Piezo2 is expressed in subsets of sensory afferent neurons as well as in other cells.

Afferent fiber terminals that detect and transmit touch sensory stimuli (**mechanoreceptors**) are often encapsulated by specialized receptor cells that help tune the afferent fiber to particular features of somatic stimulation. Afferent fibers that lack specialized receptor cells are referred to as **free nerve endings** and are especially important in the sensation of pain (see Chapter 10). Afferents that have encapsulated endings generally have lower thresholds for action potential generation and are thus are more sensitive to sensory stimulation than are free nerve endings.

Somatosensory Afferents Convey Different Functional Information

Somatosensory afferents differ significantly in their response properties. These differences, taken together, define distinct classes of afferents, each of which makes unique contributions to somatic sensation. Axon diameter is one factor that differentiates classes of somatosensory afferents (Table 9.1). The largest-diameter sensory afferents (designated Ia) are those that supply the sensory receptors in the muscles. Most of the information subserving touch is conveyed by slightly smaller diameter fibers (Aβ afferents), and information about pain and temperature is conveyed by even smaller diameter fibers (Aδ and C). The diameter of the axon determines the action potential conduction speed and is well matched to the properties of the central circuits and the various behavior demands for which each type of sensory afferent is employed (see Chapter 16).

Another distinguishing feature of sensory afferents is the size of the **receptive field**—for cutaneous afferents, the area of the skin surface over which stimulation results in a significant change in the rate of action potentials (Figure 9.3A). A given region of the body surface is served by sensory afferents that vary significantly in the size of their receptive fields. The size of the receptive field is largely a function of the branching characteristics of the afferent within the skin; smaller arborizations result in smaller receptive fields. Moreover, there are systematic regional variations in the average size of afferent receptive fields that reflect the density

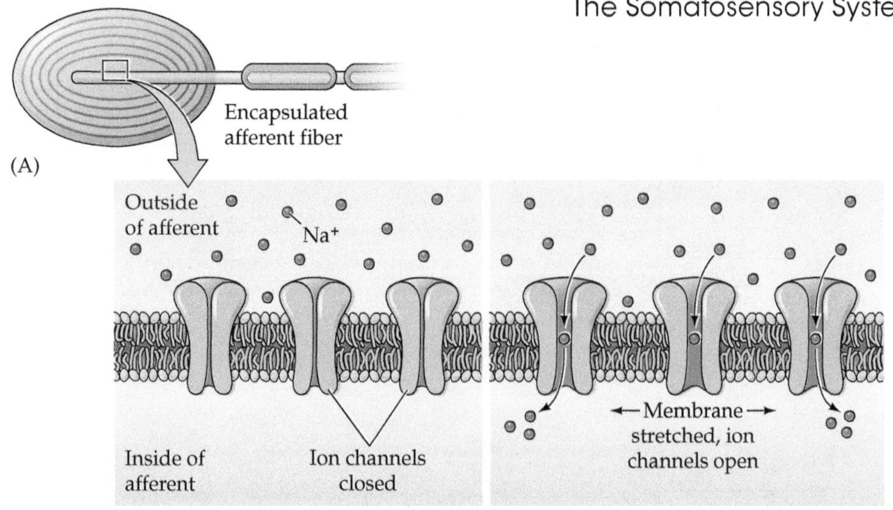

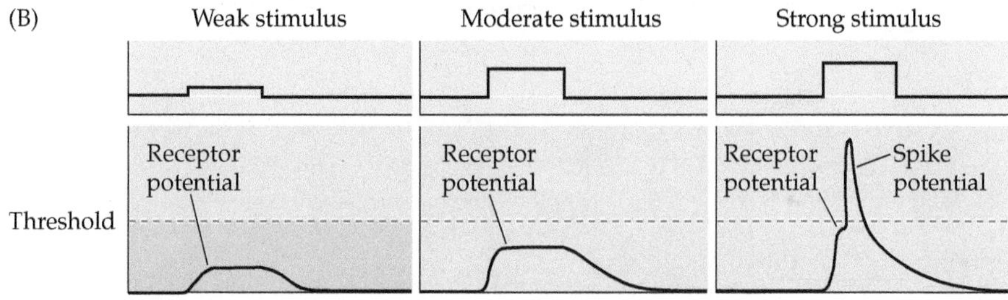

FIGURE 9.2 Transduction in a mechanosensory afferent. The process is illustrated here for a Pacinian corpuscle. (A) Deformation of the capsule leads to a stretching of the membrane of the afferent fiber, increasing the probability of opening mechanotransduction channels in the membrane. (B) Opening of these cation channels leads to depolarization of the afferent fiber (receptor potential). If the afferent is sufficiently depolarized, an action potential is generated and propagates to central targets.

of afferent fibers supplying the area. The receptive fields in regions with dense innervation (fingers, lips, toes) are relatively small compared with those in the forearm or back that are innervated by a smaller number of afferent fibers (Figure 9.3B).

Regional differences in receptive field size and innervation density are the major factors that limit the spatial accuracy with which tactile stimuli can be sensed. Thus, measures of **two-point discrimination**—the minimum interstimulus distance required to perceive two simultaneously applied stimuli as distinct—vary dramatically across the skin surface (Figure 9.3C). On the fingertips, stimuli (the indentation points produced by the tips of a caliper, for example) are perceived as distinct if they are separated by roughly 2 mm, but the same stimuli applied to the upper arm are not perceived as distinct until they are at least 40 mm apart.

Sensory afferents are further differentiated by the temporal dynamics of their response to sensory stimulation. Some afferents fire rapidly when a stimulus is first presented, then fall silent in the presence of continued stimulation; others generate a sustained discharge in the presence of an ongoing stimulus (Figure 9.4). **Rapidly adapting afferents** (those that become quiescent in the face of continued stimulation) are thought to be particularly effective in conveying information about changes in ongoing stimulation such as those produced by stimulus movement. In contrast, **slowly adapting afferents** are better suited to provide information about the

TABLE 9.1 Somatosensory Afferents That Link Receptors to the Central Nervous System

Sensory function	Receptor type	Afferent axon type[a]	Axon diameter	Conduction velocity
Proprioception	Muscle spindle	Ia, II	13–20 μm	80–120 m/s
Touch	Merkel, Meissner, Pacinian, and Ruffini cells	Aβ	6–12 μm	35–75 m/s
Pain, temperature	Free nerve endings	Aδ	1–5 μm	5–30 m/s
Pain, temperature, itch, non-discriminative touch	Free nerve endings (unmyelinated)	C	0.2–1.5 μm	0.5–2 m/s

[a]During the 1920s and 1930s, there was a virtual cottage industry classifying axons according to their conduction velocity. Three main categories were discerned, called A, B, and C. A comprises the largest and fastest axons, C the smallest and slowest. Mechanoreceptor axons generally fall into category A. The A group is further broken down into subgroups designated α (the fastest), β, and δ (the slowest). To make matters even more confusing, muscle afferent axons are usually classified into four additional groups—I (the fastest), II, III, and IV (the slowest)—with subgroups designated by lowercase roman letters! (After Rosenzweig et al., 2005.)

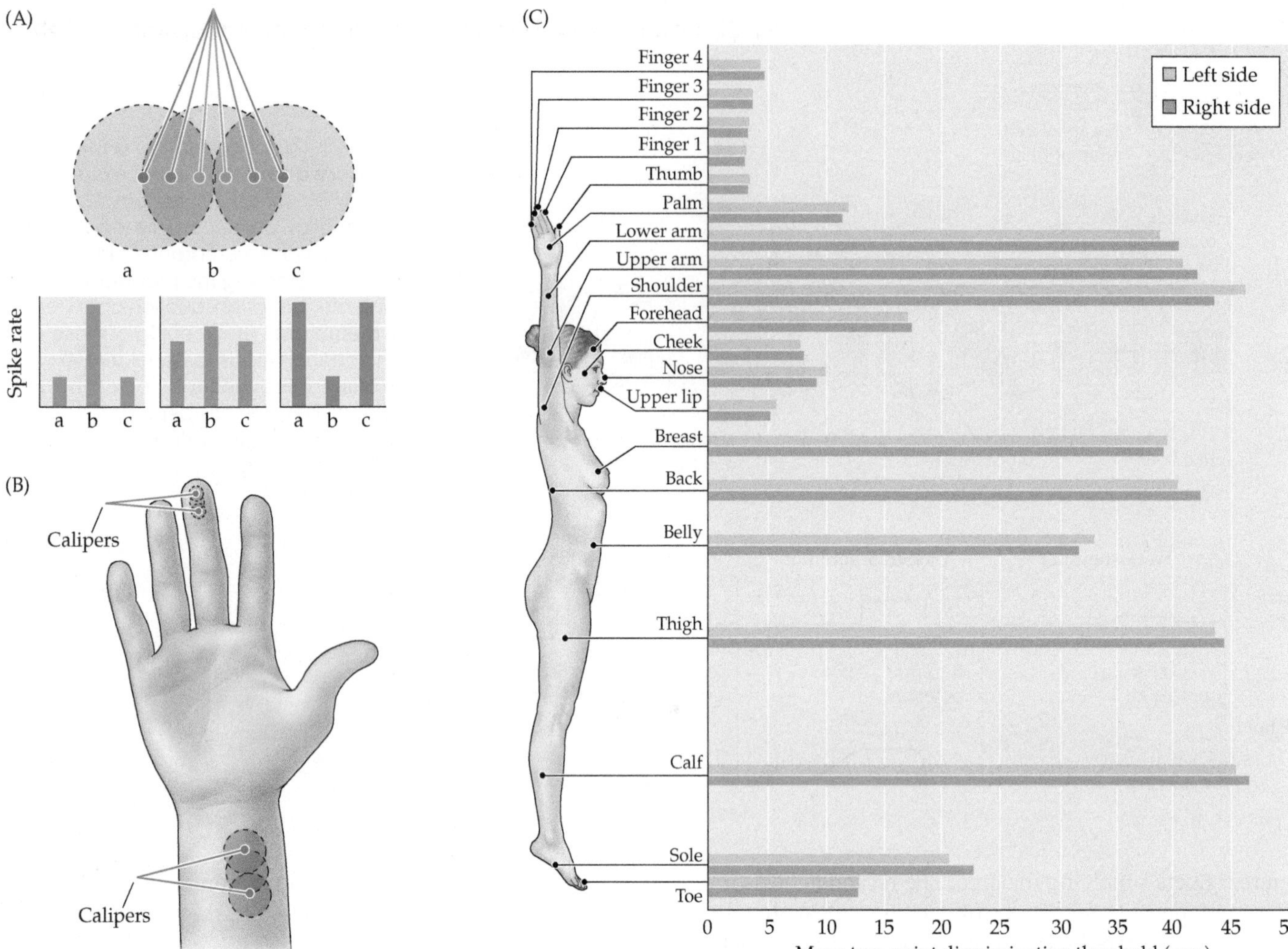

FIGURE 9.3 Receptive fields and the two-point discrimination threshold. (A) Patterns of activity in three mechanosensory afferent fibers with overlapping receptive fields a, b, and c on the skin surface. When two-point discrimination stimuli are closely spaced (green dots and histogram), there is a single focus of neural activity, with afferent b firing most actively. As the stimuli are moved farther apart (red dots and histogram), the activity in afferents a and c increases and the activity in b decreases. At some separation distance (blue dots and histogram), the activity in a and c exceeds that in b to such an extent that two discrete foci of stimulation can be identified. This differential pattern of activity forms the basis for the two-point discrimination threshold. Stimulation applied to the center of the receptive field tends to evoke stronger responses than stimuli applied at more eccentric locations within the receptive field (see Figure 1.14). (B) The two-point discrimination threshold in the fingers is much finer than that in the wrist because of differences in the sizes of afferent receptive fields—that is, the separation distance necessary to produce two distinct foci of neural activity in the population of afferents innervating the lower arm is much greater than that for the afferents innervating the fingertips. (C) Differences in the two-point discrimination threshold across the surface of the body. Somatic acuity is much higher in the fingers, toes, and face than in the arms, legs, or torso. (C after Weinstein, 1968.)

spatial attributes of the stimulus, such as size and shape. At least for some classes of afferent fibers, the adaptation characteristics are attributable to the properties of the receptor cells that encapsulate them. Rapidly adapting afferents that are associated with Pacinian corpuscles (see the following section) become slowly adapting when the corpuscle is removed.

Finally, sensory afferents respond differently to the qualities of somatosensory stimulation. Due to differences in the properties of the channels expressed in sensory afferents, or to the filter properties of the specialized receptor cells that encapsulate many sensory afferents, generator potentials are produced only by a restricted set of stimuli that impinge on a given afferent fiber. For example, the afferents encapsulated within specialized receptor cells in the skin respond vigorously to mechanical deformation of the skin surface, but not to changes in temperature or to the presence of mechanical forces or chemicals that are known to elicit painful sensations. The latter stimuli are especially effective in driving the responses of sensory afferents known as *nociceptors* (see Chapter 10) that terminate in the skin as free nerve endings. Further subtypes of mechanoreceptors and nociceptors are identified on the basis of their distinct responses to somatic stimulation.

While a given sensory afferent can give rise to multiple peripheral branches, the transduction properties of all the branches of a single fiber are identical. As a result,

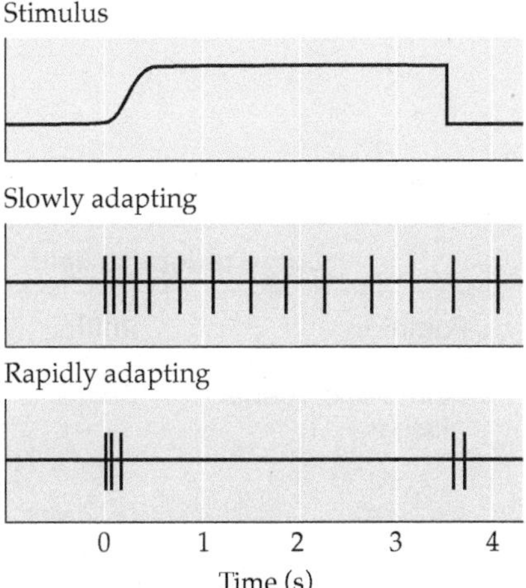

FIGURE 9.4 Slowly and rapidly adapting mechanoreceptors provide different information. Slowly adapting receptors continue responding to a stimulus, whereas rapidly adapting receptors respond only at the onset (and often the offset) of stimulation. These functional differences allow mechanoreceptors to provide information about both the static (via slowly adapting receptors) and dynamic (via rapidly adapting receptors) qualities of a stimulus.

somatosensory afferents constitute **parallel pathways** that differ in conduction velocity, receptive field size, dynamics, and effective stimulus features. As will become apparent, these different pathways remain segregated through several stages of central processing, and their activity contributes in unique ways to the extraction of somatosensory information that is necessary for the appropriate control of both goal-oriented and reflexive movements.

Mechanoreceptors Specialized to Receive Tactile Information

Our understanding of the contribution of distinct afferent pathways to cutaneous sensation is best developed for the glabrous (hairless) portions of the hand (i.e., the palm and fingertips). These regions of the skin surface are specialized to generate a high-definition neural image of manipulated objects. Active touching, or **haptics**, involves the interpretation of complex spatiotemporal patterns of stimuli that are likely to activate many classes of mechanoreceptors. Indeed, manipulating an object with the hand can often provide enough information to identify the object, a capacity called **stereognosis**. By recording the responses of individual sensory afferents in the nerves of humans and non-human primates, it has been possible to characterize the responses of these afferents under controlled conditions and gain insights into their contribution to somatic sensation. Here we consider four distinct classes of mechanoreceptive afferents that innervate the glabrous skin of the hand (Figure 9.5A; Table 9.2), as well as those that innervate the hair follicles in hairy skin (Figure 9.5B). An important aspect of the neurological assessment involves testing the functions of these different classes of mechanoreceptive afferents and noting

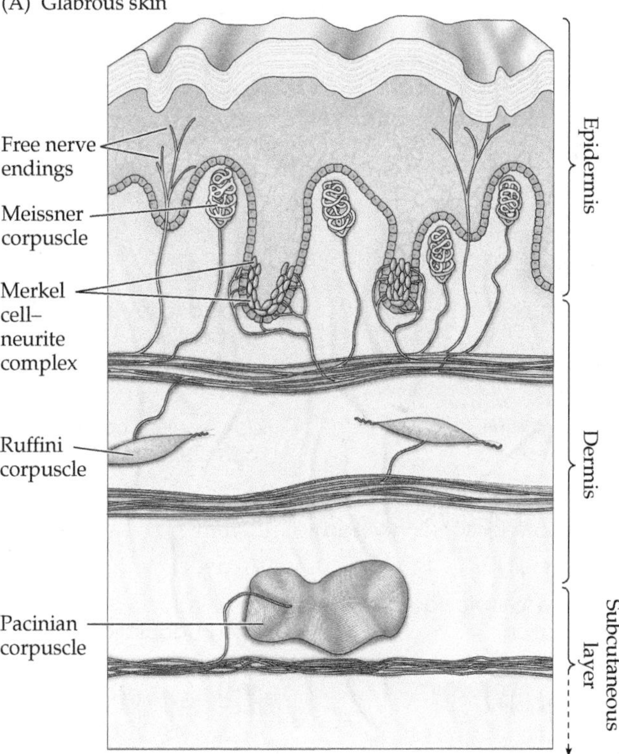

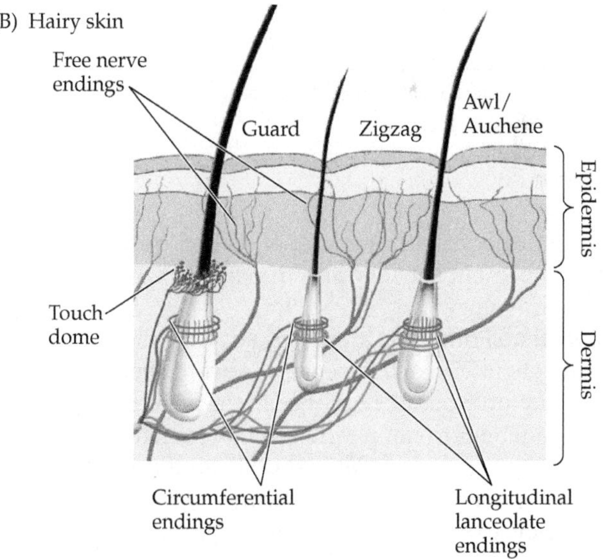

FIGURE 9.5 The skin harbors a variety of morphologically distinct mechanoreceptors. (A) This diagram represents the smooth, hairless (glabrous) skin of the fingertip. Table 9.2 summarizes the major characteristics of the various receptor types found in glabrous skin. (B) In hairy skin, tactile stimuli are transduced through a variety of mechanosensory afferents innervating different types of hair follicles. These arrangements are best known in mouse skin (illustrated here); see text for details. Similar mechanosensory afferents are believed to innervate hair follicles in human skin. (A after Johansson and Vallbo, 1983; B from Abraira and Ginty, 2013.)

TABLE 9.2 ■ Afferent Systems and Their Properties

	Small receptive field		Large receptive field	
	Merkel	**Meissner**	**Pacinian**	**Ruffini**
Location	Tip of epidermal sweat ridges	Dermal papillae (close to skin surface)	Dermis and deeper tissues	Dermis
Axon diameter	7–11 μm	6–12 μm	6–12 μm	6–12 μm
Conduction velocity	40–65 m/s	35–70 m/s	35–70 m/s	35–70 m/s
Sensory function	Shape and texture perception	Motion detection; grip control	Perception of distant events through transmitted vibrations; tool use	Tangential force; hand shape; motion direction
Effective stimuli	Edges, points, corners, curvature	Skin motion	Vibration	Skin stretch
Receptive field area[a]	9 mm²	22 mm²	Entire finger or hand	60 mm²
Innervation density (finger pad)	100/cm²	150/cm²	20/cm²	10/cm²
Spatial acuity	0.5 mm	3 mm	10+ mm	7+ mm
Response to sustained indentation	Sustained (slow adaptation)	None (rapid adaptation)	None (rapid adaptation)	Sustained (slow adaptation)
Frequency range	0–100 Hz	1–300 Hz	5–1000 Hz	0–? Hz
Peak sensitivity	5 Hz	50 Hz	200 Hz	0.5 Hz
Threshold for rapid indentation or vibration:				
Best	8 μm	2 μm	0.01μm	40 μm
Mean	30 μm	6 μm	0.08 μm	300 μm

[a]Receptive field areas as measured with rapid 0.5-mm indentation.

(After K. O. Johnson, 2002.)

geographically constrained zones, called **dermatomes**, that may present sensory loss in patients with nerve or spinal cord injury.

Merkel cell afferents are slowly adapting fibers that account for about 25% of the mechanosensory afferents in the hand. They are especially enriched in the fingertips, and are the only afferents to sample information from receptor cells located in the epidermis. **Merkel cell-neurite complexes** lie in the tips of the primary epidermal ridges—extensions of the epidermis into the underlying dermis that coincide with the prominent ridges ("fingerprints") on the finger surface. Both **Merkel cells** and their innervating sensory afferents express the mechanotransduction channel Piezo2. As a result, Merkel cells and their afferent axons can sense mechanical stimuli. Deleting Piezo2 selectively in Merkel cells significantly reduces the sustained and static firing of the innervating afferents. Thus, Merkel cells signal the static aspect of a touch stimulus, such as pressure, whereas the terminal portions of the Merkel afferents in these complexes transduce the dynamic aspects of stimuli. The slowly adapting character of the Merkel cell–neurite complexes depends on mechanotransduction. Merkel cells also play an active role in modulating the activity of their afferent axons by releasing neuropeptides on the neurites at junctions that resemble synapses, with the exocytosis of electron-dense secretory granules (see Chapter 6). Merkel cell afferents have the highest spatial resolution of all the sensory afferents—individual Merkel afferents can resolve spatial details of 0.5 mm. They are also highly sensitive to points, edges, and curvature, which makes them ideally suited for processing information about shape and texture.

Meissner afferents also express Piezo2. They are rapidly adapting fibers that innervate the skin even more densely than Merkel afferents, accounting for about 40% of the mechanosensory innervation of the human hand. Meissner corpuscles lie in the tips of the dermal papillae adjacent to the primary ridges and closest to the skin surface (see Figure 9.5A). These elongated receptors are formed by a connective tissue capsule that contains a set of flattened lamellar cells derived from Schwann cells and nerve terminals, with the capsule and the lamellar cells suspended from the basal epidermis by collagen fibers. The center of the capsule contains two to six afferent nerve fibers that terminate between and around the lamellar cells, a configuration thought to contribute to the transient response of these afferents to somatic stimulation. With indentation of the skin, the dynamic tension transduced by the collagen fibers provides the transient mechanical force that deforms the corpuscle and triggers generator potentials that may induce a volley of action potentials in the afferent fibers. When the stimulus is removed, the indented skin relaxes and the corpuscle returns to its resting configuration, generating another burst of action potentials. Thus, Meissner afferents display characteristic rapidly adapting, on–off responses (see Figure 9.4). Due at least in part to their close proximity to the skin surface, Meissner afferents are more than four times as sensitive to skin deformation as Merkel afferents; however, their receptive fields are larger than those of Merkel afferents, and thus they transmit signals with reduced spatial resolution (see Table 9.2).

Meissner corpuscles are particularly efficient in transducing information about the relatively low-frequency vibrations (3–40 Hz) that occur when textured objects are moved across the skin. Several lines of evidence suggest that the information conveyed by Meissner afferents is responsible for detecting slippage between the skin and an object held in the hand, essential feedback information for the efficient control of grip.

Pacinian afferents are rapidly adapting fibers that make up 10–15% of the mechanosensory innervation in the hand. Pacinian corpuscles are located deep in the dermis or in the subcutaneous tissue; their appearance resembles that of a small onion, with concentric layers of membranes surrounding a single afferent fiber (see Figure 9.5A). This laminar capsule acts as a filter, allowing only transient disturbances at high frequencies (250–350 Hz) to activate the nerve endings. Pacinian corpuscles adapt more rapidly than Meissner corpuscles and have a lower response threshold. The most sensitive Pacinian afferents generate action potentials for displacements of the skin as small as 10 nanometers. Because they are so sensitive, the receptive fields of Pacinian afferents are often large and their boundaries are difficult to define. The properties of Pacinian afferents make them well suited to detect vibrations transmitted through objects that contact the hand or are being grasped in the hand, especially when making or breaking contact. These properties are important for the skilled use of tools (e.g., using a wrench, cutting bread with a knife, writing).

Ruffini afferents are slowly adapting fibers and are the least understood of the cutaneous mechanoreceptors. Ruffini endings are elongated, spindle-shaped, capsular specializations located deep in the skin, as well as in ligaments and tendons (see Figure 9.5A). The long axis of the corpuscle is usually oriented parallel to the stretch lines in skin; thus, Ruffini corpuscles are particularly sensitive to the cutaneous stretching produced by digit or limb movements; they account for about 20% of the mechanoreceptors in the human hand. Although there is still some question as to their function, Ruffini corpuscles are thought to be especially responsive to skin stretches, such as those that occur during the movement of the fingers. Information supplied by Ruffini afferents contributes, along with muscle receptors, to providing an accurate representation of finger position and the conformation of the hand (see the following section on proprioception).

The different kinds of information that sensory afferents convey to central structures were first illustrated in experiments conducted by K. O. Johnson and colleagues, who compared the responses of different afferents as a fingertip was moved across a row of raised Braille letters (Figure 9.6). Clearly, all of the afferent types are activated by this stimulation, but the information supplied by each type varies enormously. The pattern of activity in the Merkel afferents is sufficient to recognize the details of the Braille pattern, and the Meissner afferents supply a slightly coarser version of this pattern. But these details are lost in the response of the Pacinian and Ruffini afferents; presumably these responses have more to do with tracking the movement and position of the finger than with the specific identity of the Braille characters. The dominance of Merkel afferents in transducing textural information is probably due to the fact that Braille letters are coarse. Human fingers are also exquisitely sensitive to fine textures. For example, we can easily distinguish silk from satin. The microgeometries of different fine textures produce different patterns of vibrations on the skin while the finger is scanning across the textured surface, which are best detected by the rapidly adapting afferents.

Finally, there are also several types of mechanoreceptive afferents that innervate the hair follicles in hairy skin (see Figure 9.5B). These include Merkel cell afferents innervating *touch domes* associated with the apical collars of hair follicles, and circumferential endings and longitudinal lanceolate endings surrounding the basal regions of the follicles. The longitudinal lanceolate endings form a palisade around the follicle that is exquisitely sensitive to the deflection of the hair by stroking the skin or simply the movement of air over the skin surface. These longitudinal lanceolate endings are derived from Aβ, Aδ, or C fibers, all of which form rapidly adapting low-threshold mechanoreceptors associated with the hairs. Interestingly, these lanceolate endings appear to be important for mediating forms of sensual touch, such as a gentle caress. These responses of longitudinal lanceolate endings should be distinguished from the responses of free nerve endings in the epidermis, which are also derived from Aδ and C axons in peripheral nerves. However, these free nerve endings and the distinct fibers from which they are derived have different physiological properties and respond (to painful stimuli) at much higher activation thresholds than touch-sensitive receptors associated with hair follicles (see Chapter 10).

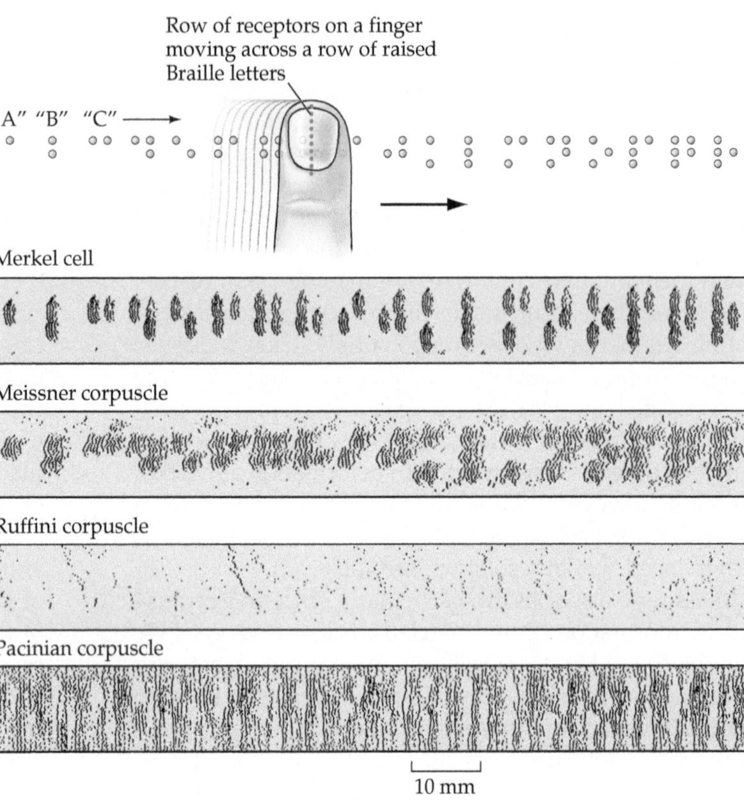

FIGURE 9.6 Simulation of activity patterns in different mechanosensory afferents in the fingertip. Each dot in the response records represents an action potential recorded from a single mechanosensory afferent fiber innervating the human finger as it moves across a row of Braille type. A horizontal line of dots in the raster plot represents the pattern of activity in the afferent as a result of moving the pattern from left to right across the finger. The position of the pattern (relative to the tip of the finger) was then displaced by a small distance, and the pattern was once again moved across the finger. Repeating this pattern multiple times produces a record that simulates the pattern of activity that would arise in a population of afferents whose receptive fields lie along a line in the fingertip (red dots). Only slowly adapting Merkel cell afferents (top panel) provide a high-fidelity representation of the Braille pattern—that is, the individual Braille dots can be distinguished only in the pattern of Merkel afferent neural activity. (After Phillips et al., 1990.)

Mechanoreceptors Specialized for Proprioception

While cutaneous mechanoreceptors provide information derived from external stimuli, another major class of receptors provides information about mechanical forces arising within the body itself, particularly from the musculoskeletal system. The purpose of these proprioceptors ("receptors for self") is primarily to give detailed and continuous information about the position of the limbs and other body parts in space. Low-threshold mechanoreceptors, including muscle spindles, Golgi tendon organs, and joint receptors, provide this kind of sensory information, which is essential to the accurate performance of complex movements. Information about the position and motion of the head is particularly important; in this case, proprioceptors are integrated with the highly specialized vestibular system, which we will consider in Chapter 14. (Specialized proprioceptors also exist in the heart and major blood vessels to provide information about blood pressure, but these neurons are considered to be part of the visceral motor system; see Chapter 21.)

The most detailed knowledge about proprioception derives from studies of **muscle spindles**, which are found in all but a few striated (skeletal) muscles. Muscle spindles consist of four to eight specialized **intrafusal muscle fibers** surrounded by a capsule of connective tissue. The intrafusal fibers are distributed among and in a parallel arrangement with the **extrafusal fibers** of skeletal muscle, which are the true force-producing fibers (Figure 9.7A). Sensory afferents are coiled around the central part of the intrafusal spindle, and when the muscle is stretched, the tension on the intrafusal fibers activates mechanically gated ion channels in the nerve endings, triggering action potentials. Innervation of the muscle spindle arises from two classes of fibers: primary and secondary endings. Primary endings arise from the largest myelinated sensory axons (group Ia afferents) and have rapidly adapting responses to changes in muscle length; in contrast, secondary endings (group II afferents) produce sustained responses to constant muscle lengths. Primary endings are thought to transmit information about limb dynamics—the velocity and direction of movement—whereas secondary endings provide information about the static position of limbs. Piezo2 is expressed by proprioceptors and is required for functional proprioception.

Changes in muscle length are not the only factors affecting the response of spindle afferents. The intrafusal fibers are themselves contractile muscle fibers and are controlled by a separate set of motor neurons (**γ motor neurons**) in the ventral horn of the spinal cord. Whereas intrafusal fibers do not add appreciably to the force of muscle contraction, changes in the tension of intrafusal fibers have significant impact on the sensitivity of the spindle afferents to changes in muscle length. Thus, in order for central circuits to provide an accurate account of limb position and movement, the level of activity in the γ system must be taken into account. (For a more detailed explanation of

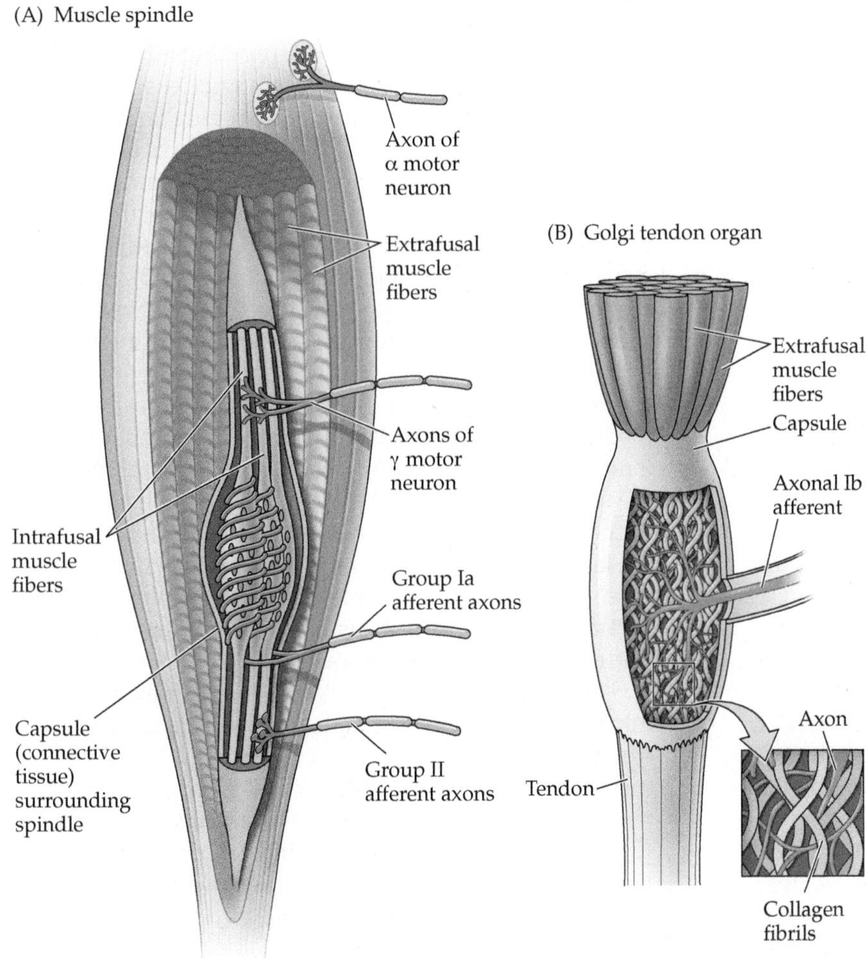

FIGURE 9.7 Proprioceptors in the musculoskeletal system. These "self-receptors" provide information about the position of the limbs and other body parts in space. (A) A muscle spindle and several extrafusal muscle fibers. The specialized intrafusal muscle fibers of the spindle are surrounded by a capsule of connective tissue. (B) Golgi tendon organs are low-threshold mechanoreceptors found in tendons; they provide information about changes in muscle tension. (A after Matthews, 1964.)

the interaction of the γ system and the activity of spindle afferents, see Chapter 16.)

The density of spindles in human muscles varies. Large muscles that generate coarse movements have relatively few spindles; in contrast, extraocular muscles and the intrinsic muscles of the hand and neck are richly supplied with spindles, reflecting the importance of accurate eye movements, the need to manipulate objects with great finesse, and the continuous demand for precise positioning of the head. This relationship between receptor density and muscle size is consistent with the generalization that the sensorimotor apparatus at all levels of the nervous system is much richer for the hands, head, speech organs, and other parts of the body that are used to perform especially important and demanding tasks. Spindles are lacking altogether in a few muscles, such as those of the middle ear, that do not require the kind of feedback that these receptors provide.

Whereas muscle spindles are specialized to signal changes in muscle length, low-threshold mechanoreceptors in tendons inform the CNS about changes in muscle tension. These mechanoreceptors, called **Golgi tendon organs**, are formed by branches of **group Ib afferents** distributed among the collagen fibers that form the tendons (Figure 9.7B). Each Golgi tendon organ is arranged in series with a small number (10–20) of extrafusal muscle fibers. Taken together, the population of Golgi tendon organs for a given muscle provides an accurate sample of the tension that exists in the muscle.

How each of these proprioceptive afferents contributes to the perception of limb position, movement, and force remains an area of active investigation. Experiments using vibrators to stimulate the spindles of specific muscles have provided compelling evidence that the activity of these afferents can give rise to vivid sensations of movement in immobilized limbs. For example, vibration of the biceps muscle leads to the illusion that the elbow is moving to an extended position, as if the biceps were being stretched. Similar illusions of motion have been evoked in postural and facial muscles. In some cases, the magnitude of the effect is so great that it produces a percept that is anatomically impossible; for example, when an extensor muscle of the wrist is vigorously vibrated, individuals report that the hand is hyperextended to the point where it is almost in contact with the back of the forearm. In all such cases, the illusion occurs only if the individual is blindfolded and cannot see the position of the limb, demonstrating that even though proprioceptive afferents alone can provide cues about limb position, under normal conditions both somatic and visual cues play important roles.

Prior to these studies, the primary source of proprioceptive information about limb position and movement was thought to arise from mechanoreceptors in and around joints. These **joint receptors** resemble many of the receptors found in the skin, including Ruffini endings and Pacinian corpuscles. However, individuals who have had artificial joint replacements were found to exhibit only minor deficits in judging the position or motion of limbs, and anesthetizing a joint such as the knee has no effect on judgments of the joint's position or movement. Although they make little contribution to limb proprioception, joint receptors appear to be important for judging position of the fingers. Along with cutaneous signals from Ruffini afferents and input from muscle spindles that contribute to fine representation of finger position, joint receptors appear to play a protective role in signaling positions that lie near the limits of normal finger joint range of motion.

Central Pathways Conveying Tactile Information from the Body: The Dorsal Column–Medial Lemniscal System

The axons of cutaneous mechanosensory afferents enter the spinal cord through the dorsal roots, where they bifurcate into ascending and descending branches. Both branches give off axonal collaterals that project into the gray matter of the spinal cord across several adjacent segments, terminating in the deeper layers (laminae III, IV, and V) in the dorsal horn. The main ascending branches extend ipsilaterally through the **dorsal columns** (also called the **posterior funiculi**) of the cord to the lower medulla, where they synapse on neurons in the dorsal column nuclei (Figure 9.8A). The term *column* refers to the gross columnar appearance of these fibers as they run the length of the spinal cord. These *first-order neurons* (primary sensory neurons) in the pathway can have quite long axonal processes: Neurons innervating the lower extremities, for example, have axons that extend from their peripheral targets through much of the length of the cord to the caudal brainstem. In addition to these so-called direct projections of the first-order neurons to the brainstem, projection neurons located in laminae III, IV, and V of the dorsal horn that receive inputs from mechanosensory collaterals project in parallel through the dorsal column to the same dorsal column nuclei. This indirect mechanosensory input to the brainstem is sometimes called the **postsynaptic dorsal column projection**.

The dorsal columns of the spinal cord are topographically organized such that the fibers conveying information from lower limbs lie most medial and travel in a circumscribed bundle known as the fasciculus gracilis (Latin *fasciculus*, "bundle"; *gracilis*, "slender"), or more simply, the **gracile tract**. Those fibers that convey information from the upper limbs, trunk, and neck lie in a more lateral bundle known as the fasciculus cuneatus ("wedge-shaped bundle") or **cuneate tract**. In turn, the fibers in these two tracts end in different subdivisions of the dorsal column nuclei: a medial subdivision, the nucleus gracilis or **gracile nucleus**; and a lateral subdivision, the nucleus cuneatus or **cuneate nucleus**.

The *second-order neurons* in the dorsal column nuclei send their axons to the somatosensory portion of the thalamus. The axons exiting from dorsal column nuclei are identified as the **internal arcuate fibers**. The internal arcuate fibers subsequently cross the midline and then form a dorsoventrally elongated tract known as the **medial lemniscus**. The word *lemniscus* means "ribbon"; the crossing of the internal arcuate fibers is called the *decussation* of the medial lemniscus, from the Roman numeral X, or *decem* (10). In a cross section through the medulla, such as the one shown in Figure 9.8A, the medial lemniscal axons carrying information from the lower limbs are located ventrally, whereas the axons related to the upper limbs are located dorsally. As the medial lemniscus ascends through the pons and midbrain, it rotates 90 degrees laterally, so that the fibers representing the upper body are eventually located in the medial portion of the tract and those representing the lower body are in the lateral portion. The axons of the medial lemniscus synapse with thalamic neurons located in the **ventral posterior lateral nucleus** (**VPL**). Thus, the VPL receives input from contralateral dorsal column nuclei.

Third-order neurons in the VPL send their axons via the **internal capsule** to terminate in the ipsilateral **postcentral gyrus** of the cerebral cortex, a region known as the **primary somatosensory cortex**, or **SI**. Neurons in the VPL also send axons to the **secondary somatosensory cortex** (**SII**), a smaller region that lies in the upper bank of the lateral sulcus. Thus, the somatosensory cortex represents mechanosensory signals first generated in the cutaneous surfaces of the contralateral body.

Central Pathways Conveying Tactile Information from the Face: The Trigeminothalamic System

Cutaneous mechanoreceptor information from the face is conveyed centrally by a separate set of first-order neurons that are located in the **trigeminal** (**cranial nerve V**) **ganglion** (Figure 9.8B). The peripheral processes of these neurons form the three main subdivisions of the trigeminal nerve (the *ophthalmic*, *maxillary*, and *mandibular* branches). Each branch innervates a well-defined territory on the face and head, including the teeth and the mucosa of the oral and nasal cavities. The central processes of trigeminal ganglion cells form the sensory roots of the trigeminal nerve; they enter the brainstem at the level of the pons to terminate on neurons in the **trigeminal brainstem complex**.

The trigeminal complex has two major components: the **principal nucleus** and the **spinal nucleus**. (A third component, the mesencephalic trigeminal nucleus, is considered below.) Most of the afferents conveying information from low-threshold cutaneous mechanoreceptors terminate in the principal nucleus. In effect, this nucleus corresponds to the

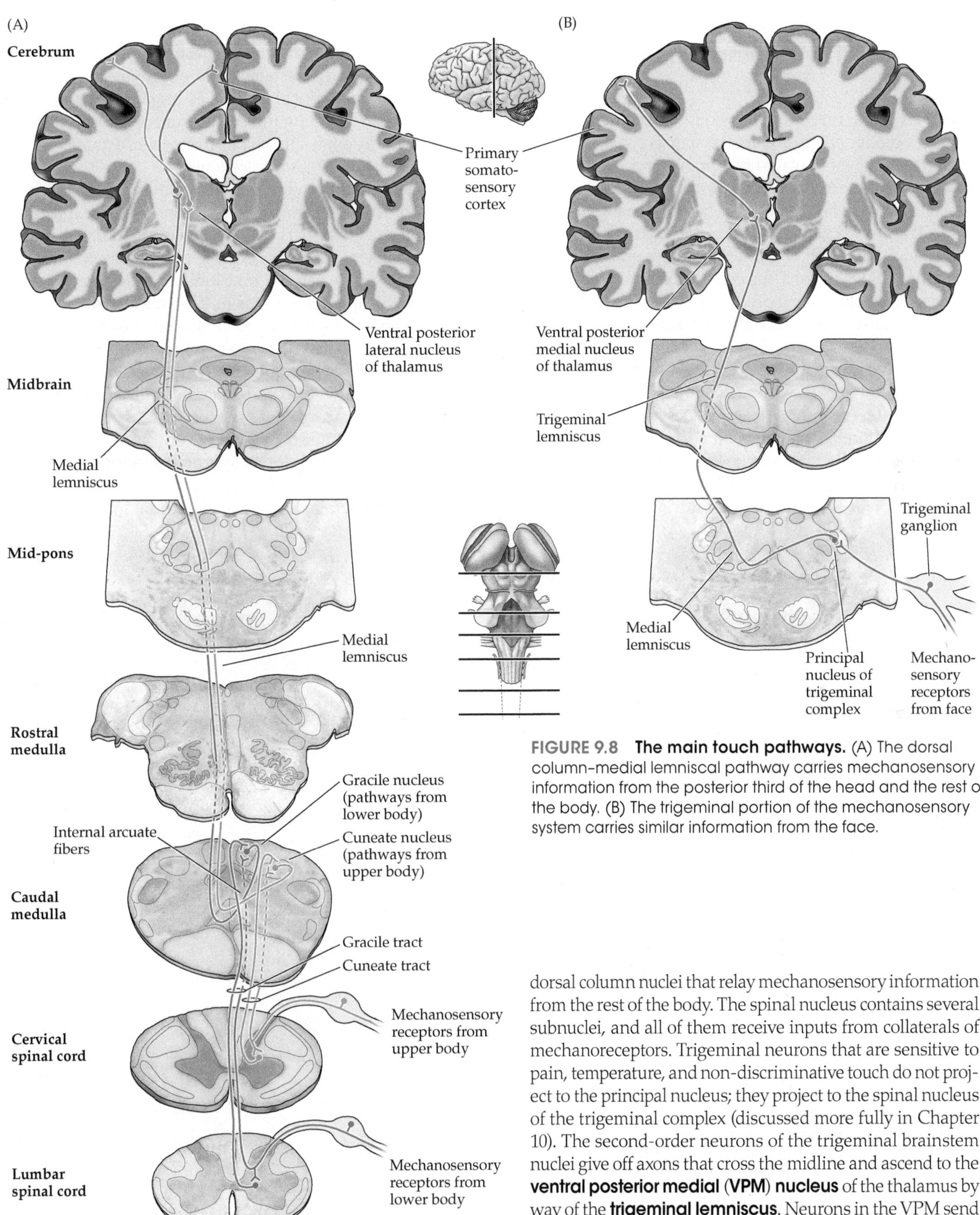

FIGURE 9.8 The main touch pathways. (A) The dorsal column-medial lemniscal pathway carries mechanosensory information from the posterior third of the head and the rest of the body. (B) The trigeminal portion of the mechanosensory system carries similar information from the face.

dorsal column nuclei that relay mechanosensory information from the rest of the body. The spinal nucleus contains several subnuclei, and all of them receive inputs from collaterals of mechanoreceptors. Trigeminal neurons that are sensitive to pain, temperature, and non-discriminative touch do not project to the principal nucleus; they project to the spinal nucleus of the trigeminal complex (discussed more fully in Chapter 10). The second-order neurons of the trigeminal brainstem nuclei give off axons that cross the midline and ascend to the **ventral posterior medial (VPM) nucleus** of the thalamus by way of the **trigeminal lemniscus**. Neurons in the VPM send their axons to ipsilateral cortical areas SI and SII.

Central Pathways Conveying Proprioceptive Information from the Body

Like their counterparts for cutaneous sensation, the axons of proprioceptive afferents enter the spinal cord through the dorsal roots, and many of the fibers from proprioceptive afferents also bifurcate into ascending and descending branches, which in turn send collateral branches to several spinal segments (Figure 9.9). Some collateral branches penetrate the dorsal horn of the spinal cord and synapse on neurons located there, as well as on neurons in the ventral horn. These synapses mediate, among other things, segmental reflexes such as the knee-jerk, or myotatic, reflex described in Chapters 1 and 16. The ascending branches of proprioceptive axons travel with the axons conveying cutaneous mechanosensory information through the dorsal column. However, there are also some differences in the spinal routes for delivering proprioceptive information to higher brain centers.

Specifically, the information supplied by proprioceptive afferents is important not only for our ability to sense limb position; it is also essential for the functions of the cerebellum, a structure that regulates the timing of muscle contractions necessary for the performance of voluntary movements. As a consequence, proprioceptive information reaches higher cortical circuits as branches of pathways that are also targeting the cerebellum, and some of these axons run through spinal cord tracts whose names reflect their association with this structure.

The association with cerebellar pathways is especially clear for the route that conveys proprioceptive information for the lower part of the body to the dorsal column nuclei. First-order proprioceptive afferents that enter the spinal cord between the mid-lumbar and thoracic levels (L2–T1) synapse on neurons in **Clarke's nucleus**, located in the medial aspect of the dorsal horn (see Figure 9.9, red pathway). Afferents that enter below this level ascend through the dorsal column and then synapse with neurons in Clarke's nucleus. Second-order neurons in Clarke's nucleus send their axons into the ipsilateral posterior lateral column of the spinal cord, where they travel up to the level of the medulla in the **dorsal spinocerebellar tract**. These axons continue into the cerebellum, but in their course, give off collaterals that synapse with neurons lying just

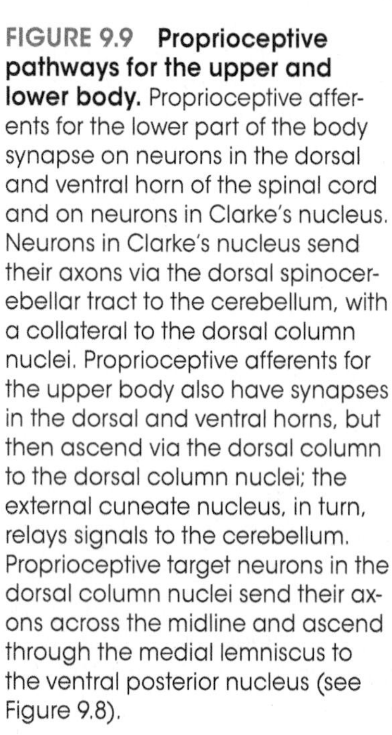

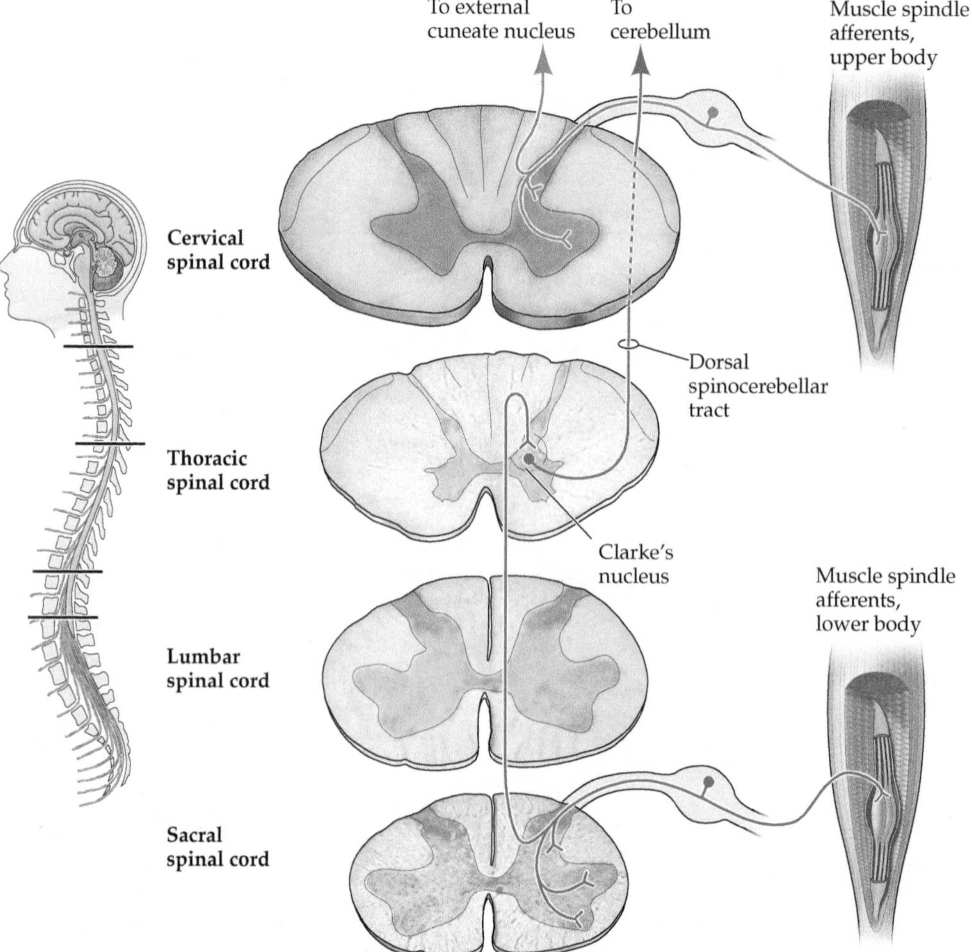

FIGURE 9.9 Proprioceptive pathways for the upper and lower body. Proprioceptive afferents for the lower part of the body synapse on neurons in the dorsal and ventral horn of the spinal cord and on neurons in Clarke's nucleus. Neurons in Clarke's nucleus send their axons via the dorsal spinocerebellar tract to the cerebellum, with a collateral to the dorsal column nuclei. Proprioceptive afferents for the upper body also have synapses in the dorsal and ventral horns, but then ascend via the dorsal column to the dorsal column nuclei; the external cuneate nucleus, in turn, relays signals to the cerebellum. Proprioceptive target neurons in the dorsal column nuclei send their axons across the midline and ascend through the medial lemniscus to the ventral posterior nucleus (see Figure 9.8).

outside the nucleus gracilis (for the present purpose, *proprioceptive neurons* of the dorsal column nuclei). Axons of these third-order neurons decussate and join the medial lemniscus, accompanying the fibers from cutaneous mechanoreceptors in their course to the VPL of the thalamus.

First-order proprioceptive afferents from the upper limbs have a course that is similar to that of cutaneous mechanoreceptors (see Figure 9.9, blue pathway). They enter the spinal cord and travel via the dorsal column (fasciculus cuneatus) up to the level of the medulla, where they synapse on proprioceptive neurons in the dorsal column nuclei, including a lateral nucleus among the tier of dorsal column nuclei in the caudal medulla called the **external cuneate nucleus**. Second-order neurons then send their axons into the ipsilateral cerebellum, while other branches cross the midline and join the medial lemniscus, ascending to the VPL of the thalamus.

Central Pathways Conveying Proprioceptive Information from the Face

Like the information from cutaneous mechanoreceptors, proprioceptive information from the face is conveyed through the trigeminal nerve. However, the cell bodies of the first-order proprioceptive neurons for the face have an unusual location: Instead of residing in the trigeminal ganglia, they are found within the CNS, in the **mesencephalic trigeminal nucleus**, a well-defined array of neurons lying at the lateral extent of the periaqueductal gray matter of the dorsal midbrain. Like their counterparts in the trigeminal and dorsal root ganglia, these pseudounipolar neurons have peripheral processes that innervate muscle spindles and Golgi tendon organs associated with facial musculature (especially the jaw muscles) and central processes that include projections to brainstem nuclei responsible for reflex control of facial muscles. Although the exact route is not clear, information from proprioceptive afferents in the mesencephalic trigeminal nucleus also reaches the thalamus and is represented in somatosensory cortex.

Somatosensory Components of the Thalamus

Each of the several ascending somatosensory pathways originating in the spinal cord and brainstem converges on the **ventral posterior complex** of the thalamus and terminates in an organized fashion (Figure 9.10). One of the organizational features of this complex instantiated by the pattern of afferent terminations is a complete and orderly somatotopic representation of the body and head. As already mentioned, the more laterally located ventral posterior lateral nucleus (VPL) receives projections from the medial lemniscus carrying somatosensory information from the body and posterior head, whereas the more medially located ventral posterior medial nucleus (VPM) receives axons from the trigeminal lemniscus conveying somatosensory information from the face. In addition, inputs carrying different types of somatosensory information—for example, those that respond to different types of mechanoreceptors, to muscle spindle afferents, or to Golgi tendon organs—terminate on separate populations of relay cells within the ventral posterior complex. Thus, the information supplied by different somatosensory receptors remains segregated in its passage to cortical circuits.

Primary Somatosensory Cortex

The majority of the axons arising from neurons in the ventral posterior complex of the thalamus project to cortical neurons located in layer 4 of the primary somatosensory cortex (see Box 27A for a description of cortical lamination). The primary somatosensory cortex in humans is located in the postcentral

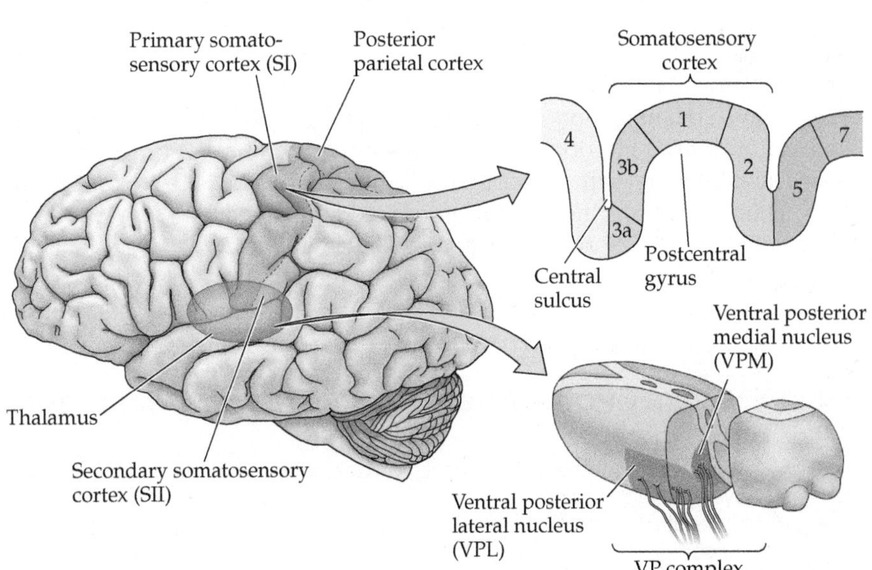

FIGURE 9.10 Somatosensory portions of the thalamus and their cortical targets in the postcentral gyrus. The ventral posterior nuclear complex comprises the VPM, which relays somatosensory information carried by the trigeminal system from the face, and the VPL, which relays somatosensory information from the rest of the body. The diagram at the upper right shows the organization of the primary somatosensory cortex in the postcentral gyrus, shown here in a section cutting across the gyrus from anterior to posterior. (After Brodal, 1992 and Jones et al., 1982.)

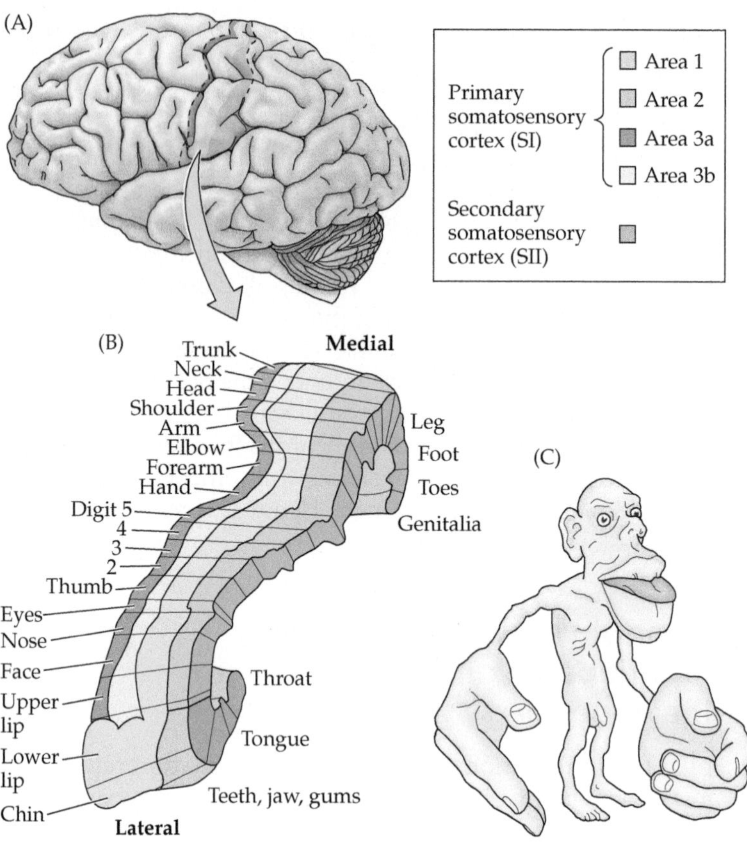

FIGURE 9.11 Somatotopic order in the human primary somatosensory cortex. (A) Diagram showing the region of the human cortex from which electrical activity is recorded following mechanosensory stimulation of different parts of the body. (The patients in the study were undergoing neurosurgical procedures for which such mapping was required.) Although modern imaging methods are now refining these classical data, the human somatotopic map defined in the 1930s has remained generally valid. (B) Diagram showing the somatotopic representation of body parts from medial to lateral. (C) Cartoon of the homunculus constructed on the basis of such mapping. Note that the amount of somatosensory cortex devoted to the hands and face is much larger than the relative amount of body surface in these regions. A similar disproportion is apparent in the primary motor cortex, for much the same reasons (see Chapter 17). (After Penfield and Rasmussen, 1950, and Corsi, 1991.)

gyrus of the parietal lobe and comprises four distinct regions, or fields, known as Brodmann's areas 3a, 3b, 1, and 2 (Figure 9.11A). Mapping studies in humans and other primates show further that each of these four cortical areas contains a separate and complete representation of the body. In these **somatotopic maps**, the foot, leg, trunk, forelimbs, and face are represented in a medial to lateral arrangement, as shown in Figure 9.11B.

A salient feature of somatotopic maps, recognized soon after their discovery, is their failure to represent the human body in its actual proportions. When neurosurgeons determined the representation of the human body in the primary sensory (and motor) cortex, the homunculus ("little man") defined by such mapping procedures had a grossly enlarged face and hands compared with the torso and proximal limbs (Figure 9.11C). These anomalies arise because manipulation, facial expression, and speech are extraordinarily important for humans and require a great deal of circuitry, both central and peripheral, to govern them. Thus, in humans the cervical spinal cord is enlarged to accommodate the extra circuitry related to the hand and upper limb, and as stated earlier, the density of receptors is greater in regions such as the hands and lips.

Such distortions are also apparent when topographical maps are compared across species. In the rat brain, for example, an inordinate amount of the somatosensory cortex is devoted to representing the large facial whiskers that are key components of the somatosensory input for rats and mice (Box 9A), while raccoons overrepresent their paws and the platypus its bill. In short, the sensory input (or motor output) that is particularly significant to a given species gets relatively more cortical representation.

BOX 9A ■ Patterns of Organization within the Sensory Cortices: Brain Modules

Observations over the last 45 years have made it clear that there is an iterated substructure within the somatosensory (and many other) cortical maps. This substructure takes the form of units called *modules*, each involving hundreds or thousands of nerve cells in repeating patterns. The advantages of these iterated patterns for brain function remain largely mysterious; for the neurobiologist, however, such iterated arrangements have provided important clues about cortical connectivity and the mechanisms by which neural activity influences brain development (see Chapter 25).

The observation that the somatosensory cortex comprises elementary units of vertically linked cells was first noted in the 1920s by the Spanish neuroanatomist Rafael Lorente de Nó, based on his studies in the rat. The potential importance of cortical modularity remained largely unexplored until the 1950s, however, when electrophysiological experiments indicated an arrangement of repeating units in the brains of cats and, later, monkeys. Vernon Mountcastle, a neurophysiologist at Johns Hopkins University School of Medicine, found that vertical microelectrode penetrations in the primary somatosensory cortex of these animals encountered cells that responded to the same sort of mechanical stimulus presented at the same location on the body surface. Soon after Mountcastle's pioneering

BOX 9A ■ *(continued)*

work, David Hubel and Torsten Wiesel discovered a similar arrangement in the cat primary visual cortex. These and other observations led Mountcastle to the general view that "the elementary pattern of organization of the cerebral cortex is a vertically oriented column or cylinder of cells capable of input-output functions of considerable complexity." Since these discoveries in the late 1950s and early 1960s, the view that modular circuits represent a fundamental feature of the mammalian cerebral cortex has gained wide acceptance, and many such entities have now been described in various cortical regions (see figure).

This wealth of evidence for such patterned circuits has led many neuroscientists to conclude, like Mountcastle, that modules are a fundamental feature of the cerebral cortex, essential for perception, cognition, and perhaps even consciousness. Despite the prevalence of iterated modules, there are some problems with the view that modular units are universally important in cortical function. First, although modular circuits of a given class are readily seen in the brains of some species, they have not been found in the same brain regions of other, sometimes closely related, animals. Second, not all regions of the mammalian cortex are organized in a modular fashion. And third, no clear function of such modules has been discerned, much effort and speculation notwithstanding. This salient feature of the organization of the somatosensory cortex and other cortical (and some subcortical) regions therefore remains a tantalizing puzzle.

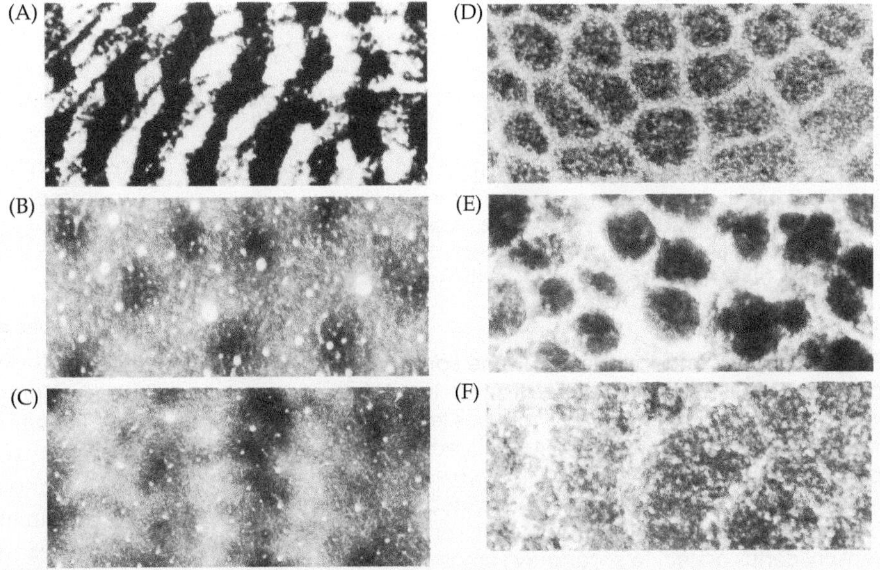

Examples of iterated modular substructures in the mammalian brain. (A) Ocular dominance columns in layer IV in the primary visual cortex (V1) of a rhesus monkey. (B) Repeating units called *blobs* in layers II and III in V1 of a squirrel monkey. (C) Stripes in layers II and III in V2 of a squirrel monkey. (D) Barrels in layer IV in primary somatosensory cortex of a rat. (E) Glomeruli in the olfactory bulb of a mouse. (F) Iterated units called *barreloids* in the thalamus of a rat. These and other examples indicate that modular organization is commonplace in the brain. These units are on the order of 100 to several hundred microns across. (From Purves et al., 1992.)

Although the topographic organization of the several somatosensory areas is similar, the functional properties of the neurons in each region are distinct. Experiments carried out in non-human primates indicate that neurons in areas 3b and 1 respond primarily to cutaneous stimuli, whereas neurons in 3a respond mainly to stimulation of proprioceptors; area 2 neurons process both tactile and proprioceptive stimuli. These differences in response properties reflect, at least in part, parallel sets of inputs from functionally distinct classes of neurons in the ventral posterior complex. In addition, a rich pattern of corticocortical connections between SI areas contributes significantly to the elaboration of SI response properties. Area 3b receives the bulk of the input from the ventral posterior complex and provides a particularly dense projection to areas 1 and 2. This arrangement of connections establishes a functional hierarchy in which area 3b serves as an obligatory first step in cortical processing of somatosensory information (Figure 9.12). Consistent with this view, lesions of area 3b in non-human primates result in profound deficits in all forms of tactile sensations mediated by cutaneous mechanoreceptors, while lesions limited to areas 1 or 2 result in partial deficits and an inability to use tactile information to discriminate either the texture of objects (area 1 deficit) or the size and shape of objects (area 2 deficit).

Even finer parcellations of functionally distinct neuronal populations exist within single cortical areas. Based on his analysis of electrode penetrations in primary somatosensory cortex, Vernon Mountcastle was the first to suggest that neurons with similar response properties might be clustered together into functionally distinct "columns" that traverse the depth of the cortex. Subsequent studies of finely spaced electrode penetrations in area 3b provided strong evidence in support of this idea, demonstrating that neurons with rapidly and slowly adapting properties were clustered into separate zones within the representation of a single digit

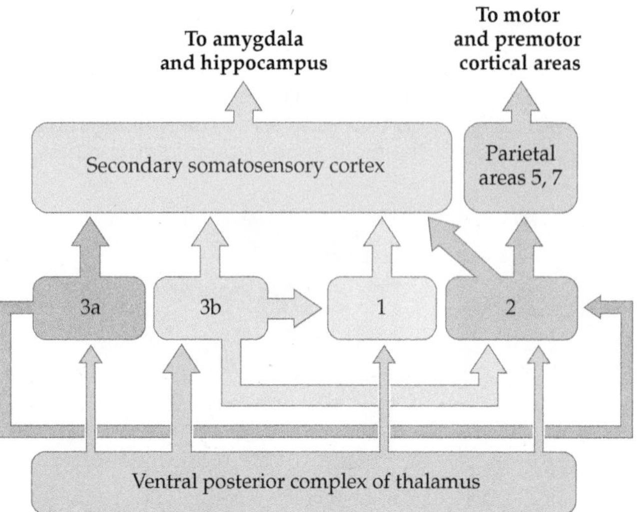

FIGURE 9.12 Connections within the somatosensory cortex establish functional hierarchies. Inputs from the ventral posterior complex of the thalamus terminate in Brodmann's areas 3a, 3b, 1, and 2, with the greatest density of projections in area 3b. Area 3b in turn projects heavily to areas 1 and 2, and the functions of these areas are dependent on the activity of area 3b. All subdivisions of primary somatosensory cortex project to secondary somatosensory cortex; the functions of SII are dependent on the activity of SI.

(Figure 9.13). In the past, it was assumed that the rapidly and slowly adapting cortical neurons receive segregated inputs from rapidly and slowly adapting mechanoreceptors, respectively. However, the cortical slowly adapting neurons all show a large touch-OFF response in addition to the sustained firing during contact. Such OFF responses are signaled only by rapidly adapting afferents in fingers. Furthermore, the rapidly adapting cortical neurons sometimes show sustained firing in response to stimuli of preferred directions. Thus, the cortical rapidly and slowly adapting columns reflect differential processing of convergent inputs from different peripheral receptors, rather than the strict segregation of afferent inputs that convey distinct physiological signals. This columnar organization of cortical areas, a fundamental feature of cortical organization throughout the neocortex (see Box 27A), is especially pronounced in visual cortical areas in primates (see Chapter 12). Slowly and rapidly adapting columns in somatosensory cortex are therefore more analogous to orientation columns in the visual cortex (reflecting cortical computations derived from converging input) than ocular dominance columns (which reflect strictly segregated thalamocortical inputs). Although these cortical patterns reflect specificity in the underlying patterns of thalamocortical and corticocortical connections, the functional significance of columns remains unclear (see Box 9A).

Beyond SI: Corticocortical and Descending Pathways

Somatosensory information is distributed from the primary somatosensory cortex to "higher-order" cortical fields. One of these higher-order cortical centers, the secondary somatosensory cortex, lies in the upper bank of the lateral sulcus (see Figures 9.10 and 9.11). SII receives convergent projections from all subdivisions of SI, and these inputs are necessary for the function of SII; lesions of SI eliminate the somatosensory responses of SII neurons. SII sends projections in turn to limbic structures such as the amygdala and hippocampus (see Chapters 30 and 31). This latter pathway is believed to play an important role in tactile learning and memory.

Neurons in SI also project to parietal areas posterior to area 2, especially areas 5a and 7b. These areas receive direct projections from area 2 and, in turn, supply inputs to neurons in motor and premotor areas of the frontal lobe. This is a major route by which information derived from proprioceptive afferents signaling the current state of muscle contraction gains access to circuits that initiate voluntary movements. More generally, the projections from parietal cortex to motor cortex are critical for the integration of sensory and motor information (see Chapters 17, 27, and 29 for discussion of sensorimotor integration in the parietal and frontal lobes).

Finally, a fundamental but often neglected feature of the somatosensory system is the presence of massive descending projections. These pathways originate in sensory cortical fields and run to the thalamus, brainstem, and spinal cord. Indeed, descending projections from the somatosensory cortex outnumber ascending somatosensory pathways. Although their physiological role is not well understood, it is generally thought that descending projections modulate the ascending flow of sensory information at the level of the thalamus and brainstem.

Plasticity in the Adult Cerebral Cortex

The analysis of maps of the body surface in primary somatosensory cortex and the responses to altered patterns of activity in peripheral afferents has been instrumental in understanding the potential for the reorganization of cortical circuits in adults. Jon Kaas and Michael Merzenich were the first to explore this issue, by examining the impact of peripheral lesions (e.g., cutting a nerve that innervates the hand, or amputation of a digit) on the topographic maps in somatosensory cortex. Immediately after the lesion, the corresponding region of the cortex was found to be unresponsive. After a few weeks, however, the unresponsive area became responsive to stimulation of neighboring regions of the skin (Figure 9.14). For example, if digit 3 was amputated, cortical neurons that formerly responded to stimulation of

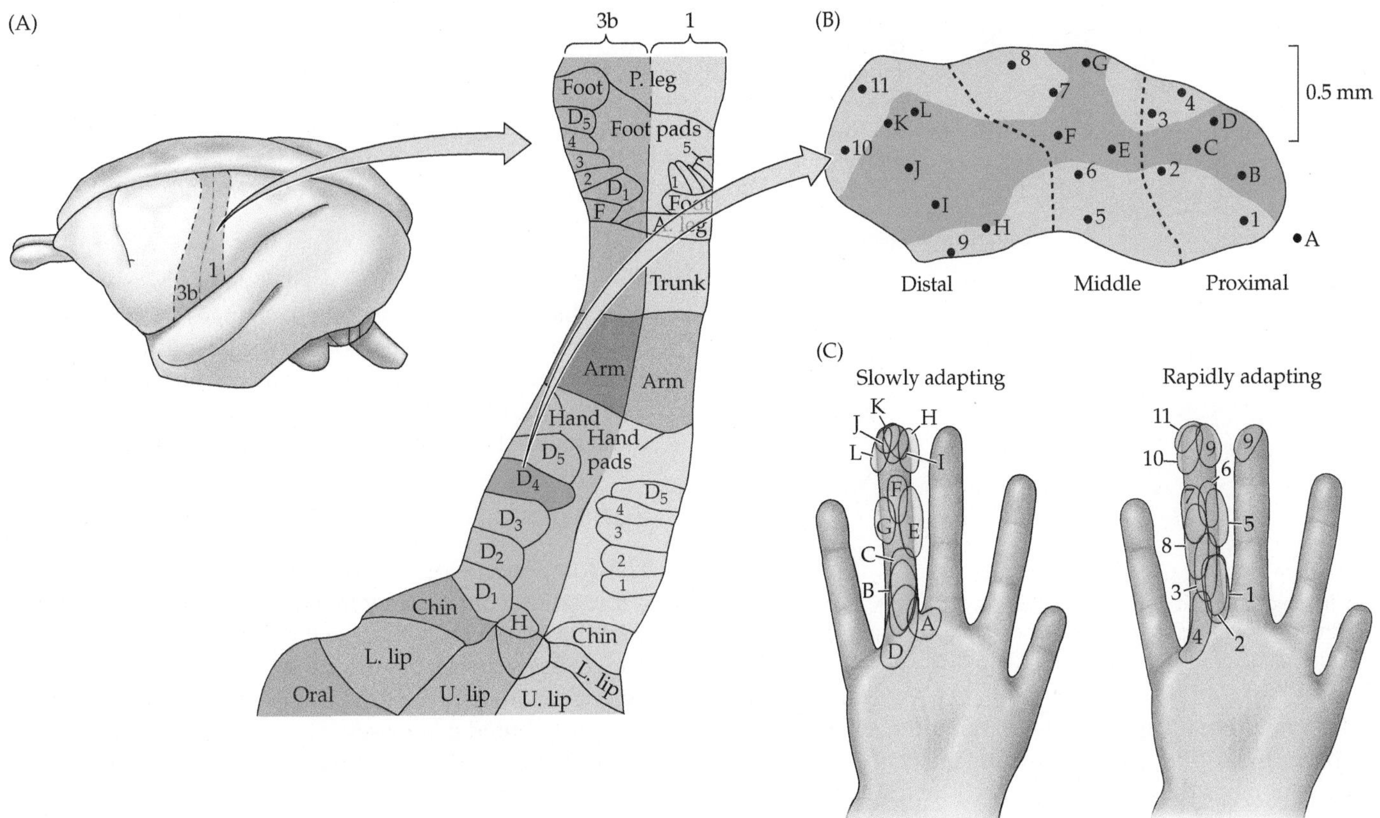

FIGURE 9.13 Neurons in the primary somatosensory cortex form functionally distinct columns. (A) Primary somatosensory map in the owl monkey based, as for the human in Figure 9.11, on the electrical responsiveness of the cortex to peripheral stimulation. The enlargement on the right shows Brodmann's areas 3b and 1, which process most cutaneous mechanosensory information. The arrangement is generally similar to that determined in humans. Note the presence of regions that are devoted to the representation of individual digits. (B) Modular organization of responses within the representation of a single digit, showing the location of electrode penetrations that encountered rapidly adapting (green) and slowly adapting (blue) responses within the representation of digit 4. (C) Distribution of slowly adapting and rapidly adapting receptive fields used to derive the plot in (B). Although the receptive fields of these different classes of afferents overlap on the skin surface, they are partially segregated within the cortical representation (A after Kaas, 1993; C after Sur et al., 1980.)

digit 3 now responded to stimulation of digits 2 or 4. Thus, the central representation of the remaining digits had expanded to take over the cortical territory that had lost its main input. Such "functional remapping" also occurs in the somatosensory nuclei in the thalamus and brainstem; indeed, some of the reorganization of cortical circuits may depend on this concurrent subcortical plasticity. This sort of adjustment in the somatosensory system may contribute to the altered sensation of phantom limbs after amputation. Similar plastic changes have been demonstrated in the visual, auditory, and motor cortices, suggesting that some ability to reorganize after peripheral deprivation or injury is a general property of the mature neocortex.

Appreciable changes in cortical representation also can occur in response to physiological changes in sensory or motor experience. For instance, if a monkey is trained to use a specific digit for a particular task that is repeated many times, the functional representation of that digit, determined by electrophysiological mapping, can expand at the expense of the other digits (Figure 9.15). In fact, significant changes in receptive fields of somatosensory neurons can be detected when a peripheral nerve is blocked temporarily by a local anesthetic. The transient loss of sensory input from a small area of skin induces a reversible reorganization of the receptive fields of both cortical and subcortical neurons. During this period, the neurons assume new receptive fields that respond to tactile stimulation of the skin surrounding the anesthetized region. Once the effects of the local anesthetic subside, the receptive fields of cortical and subcortical neurons return to their usual size. The common experience of an anesthetized area of skin feeling disproportionately large—as experienced, for example, following dental anesthesia—may be a consequence of this temporary change.

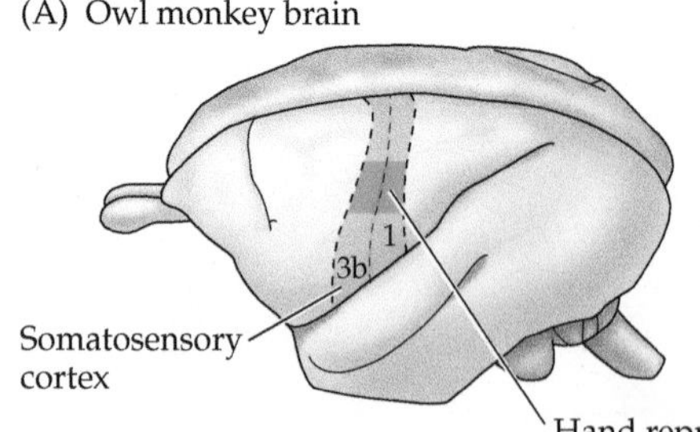

FIGURE 9.14 Functional changes in the somatosensory cortex following amputation of a digit. (A) Diagram of the somatosensory cortex in the owl monkey, showing the approximate location of the hand representation. (B) The hand representation in the animal before amputation; the numbers correspond to different digits. (C) The cortical map determined in the same animal 2 months after amputation of digit 3. The map has changed substantially; neurons in the area formerly responding to stimulation of digit 3 now respond to stimulation of digits 2 and 4. (After Merzenich et al., 1984.)

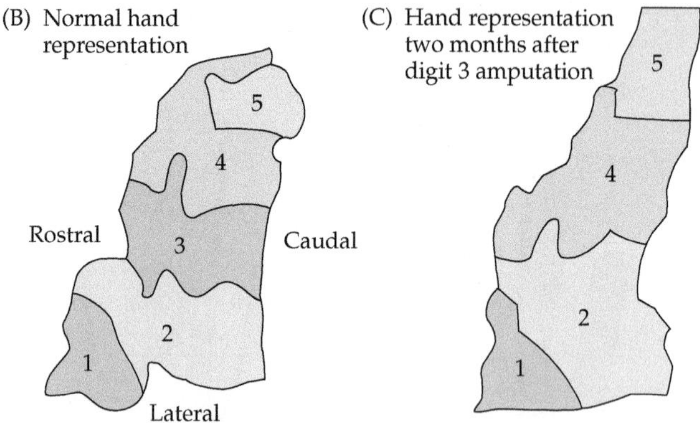

Despite these intriguing observations, the mechanism, purpose, and significance of the reorganization of sensory and motor maps that occurs in adult cortex are not known. Clearly, changes in cortical circuitry occur in the adult brain. Centuries of clinical observations, however, indicate that these changes may be of limited value for recovery of function following brain injury, and they may well lead to symptoms that detract from rather than enhance the quality of life following neural damage. Given their rapid and reversible character, most of these changes in cortical function probably reflect alterations in the strength of synapses already present. Indeed, finding ways to prevent or redirect the synaptic events that underlie injury-induced plasticity could reduce the long-term impact of acute brain damage.

Summary

The components of the somatosensory system process information conveyed by mechanical stimuli that either impinge on the body surface (cutaneous mechanoreception) or are generated within the body itself (proprioception). Somatosensory processing is performed by neurons distributed across several brain structures that are connected by both ascending and descending pathways. Transmission of afferent mechanosensory information from the periphery to the brain begins with a variety of receptor types that initiate action potentials. This activity is then conveyed centrally via a chain of nerve cells organized into distinct gray matter structures and white matter tracts. First-order neurons in this chain are the primary sensory neurons located in the dorsal root and cranial nerve ganglia. The next set of neurons conveying ascending mechanosensory signals is located in brainstem nuclei (although there are also projection neurons located in the spinal cord that project to the brainstem). The final link in the pathway from periphery to cerebral cortex consists of neurons found in the thalamus, which in turn project to the postcentral gyrus. These pathways are topographically arranged throughout the system, with the amount of cortical and subcortical space allocated to various body parts being proportional to the density of peripheral receptors. Studies of non-human primates show that specific cortical regions correspond to each functional submodality; area 3b, for example, processes information from low-threshold cutaneous receptors, while area 3a

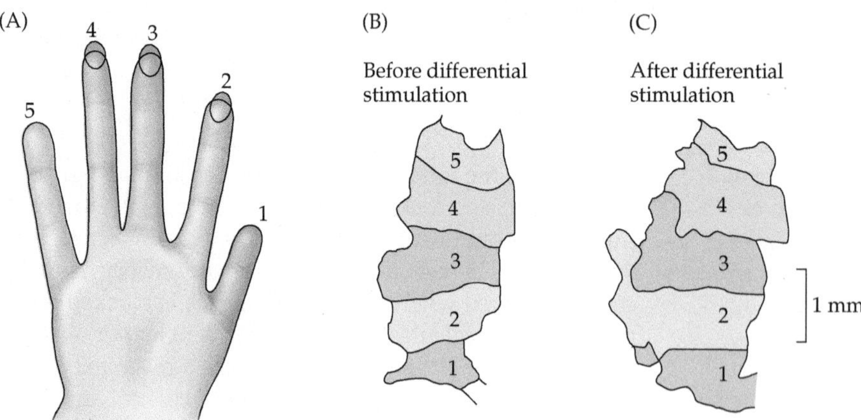

FIGURE 9.15 Functional expansion of a cortical representation by a repetitive behavioral task. (A) An owl monkey was trained in a task that required heavy usage of digits 2, 3, and occasionally 4. (B) The map of the digits in the primary somatosensory cortex prior to training. (C) After several months of "practice," a larger region of the cortex contained neurons activated by the digits used in the task. Note that the specific arrangements of the digit representations are somewhat different from those for the monkey shown in Figure 9.14, indicating the variability of the cortical representation in individual animals. (After Jenkins et al., 1990.)

processes inputs from proprioceptors. Thus, at least two broad criteria operate in the organization of the somatosensory system: modality and somatotopy. The end result of this complex interaction is the unified perceptual representation of the body and its ongoing interaction with the environment.

ADDITIONAL READING

Reviews

Abraira, V. E. and D. D. Ginty (2013) The sensory neurons of touch. *Neuron* 79: 618–639.

Barnes, S. J. and G. T. Finnerty (2010) Sensory experience and cortical rewiring. *Neuroscientist* 16: 186–198.

Chapin, J. K. (1987) Modulation of cutaneous sensory transmission during movement: Possible mechanisms and biological significance. In *Higher Brain Function: Recent Explorations of the Brain's Emergent Properties*, S. P. Wise (ed.). New York: John Wiley and Sons, pp. 181–209.

Darian-Smith, I. (1982) Touch in primates. *Annu. Rev. Psychol.* 33: 155–194.

Johansson, R. S. and J. R. Flanagan (2009) Coding and use of tactile signals from the fingertips in object manipulation tasks. *Nat. Rev. Neurosci.* 10: 345–359.

Johansson, R. S. and A. B. Vallbo (1983) Tactile sensory coding in the glabrous skin of the human. *Trends Neurosci.* 6: 27–32.

Johnson, K. O. (2002) Neural basis of haptic perception. In *Seven's Handbook of Experimental Psychology*, 3rd Edition, H. Pashler and S. Yantis (eds.). Vol 1: *Sensation and Perception.* New York: Wiley, pp. 537–583.

Kaas, J. H. (1990) Somatosensory system. In *The Human Nervous System*, G Paxinos (ed.). San Diego: Academic Press, pp. 813–844.

Kaas, J. H. (1993) The functional organization of somatosensory cortex in primates. *Ann. Anat.* 175: 509–518.

Kaas, J. H. and C. E. Collins (2003) The organization of somatosensory cortex in anthropoid primates. *Adv. Neurol.* 93: 57–67.

Mountcastle, V. B. (1975) The view from within: Pathways to the study of perception. *Johns Hopkins Med. J.* 136: 109–131.

Nicolelis, M. A. and E. E. Fanselow (2002) Thalamocortical optimization of tactile processing according to behavioral state. *Nature Neurosci.* 5 (6): 517–523.

Petersen, R. S., S. Panzeri and M. E. Diamond (2002) Population coding in somatosensory cortex. *Curr. Opin. Neurobiol.* 12: 441–447.

Ranade, S. S., R. Syeda and A. Patapoutian (2015) Mechanically activated ion channels. *Neuron.* 87: 1162–1179.

Saal, H. P. and S. J. Bensmaia (2014) Touch is a team effort: interplay of submodalities in cutaneous sensibility. *Trends Neurosci.* 37: 689–697.

Woolsey, C. (1958) Organization of somatosensory and motor areas of the cerebral cortex. In *Biological and Biochemical Bases of Behavior*, H. F. Harlow and C. N. Woolsey (eds.). Madison: University of Wisconsin Press, pp. 63–82.

Important Original Papers

Adrian, E. D. and Y. Zotterman (1926) The impulses produced by sensory nerve endings. II. The response of a single end organ. *J. Physiol.* 61: 151–171.

Friedman, R. M., L. M. Chen and A. W. Roe (2004) Modality maps within primate somatosensory cortex. *Proc. Natl. Acad. Sci. USA* 101: 12724–12729.

Johansson, R. S. (1978) Tactile sensibility of the human hand: Receptive field characteristics of mechanoreceptive units in the glabrous skin. *J. Physiol. (Lond.)* 281: 101–123.

Johnson, K. O. and G. D. Lamb (1981) Neural mechanisms of spatial tactile discrimination: Neural patterns evoked by Braille-like dot patterns in the monkey. *J. Physiol. (Lond.)* 310: 117–144.

Jones, E. G. and D. P. Friedman (1982) Projection pattern of functional components of thalamic ventrobasal complex on monkey somatosensory cortex. *J. Neurophysiol.* 48: 521–544.

Jones, E. G. and T. P. S. Powell (1969) Connexions of the somatosensory cortex of the rhesus monkey. I. Ipsilateral connexions. *Brain* 92: 477–502.

Lamotte, R. H. and M. A. Srinivasan (1987) Tactile discrimination of shape: Responses of rapidly adapting mechanoreceptive afferents to a step stroked across the monkey fingerpad. *J. Neurosci.* 7: 1672–1681.

Laubach, M., J. Wessber and M. A. L. Nicolelis (2000) Cortical ensemble activity increasingly predicts behavior outcomes during learning of a motor task. *Nature* 405: 567–571.

Moore, C. I. and S. B. Nelson (1998) Spatiotemporal subthreshold receptive fields in the vibrissa representation of rat primary somatosensory cortex. *J. Neurophysiol.* 80: 2882–2892.

Moore, C. I., S. B. Nelson and M. Sur (1999) Dynamics of neuronal processing in rat somatosensory cortex. *Trends Neurosci.* 22: 513–520.

Nicolelis, M. A. L., L. A. Baccala, R. C. S. Lin and J. K. Chapin (1995) Sensorimotor encoding by synchronous neural ensemble activity at multiple levels of the somatosensory system. *Science* 268: 1353–1359.

Ranade, S. S. and 16 others (2014) Piezo2 is the major transducer of mechanical forces for touch sensation in mice. *Nature* 516: 121–125.

Sur, M. (1980) Receptive fields of neurons in areas 3b and 1 of somatosensory cortex in monkeys. *Brain Res.* 198: 465–471.

Wall, P. D. and W. Noordenhos (1977) Sensory functions which remain in man after complete transection of dorsal columns. *Brain* 100: 641–653.

Weber, A. I. and 6 others (2013) Spatial and temporal codes mediate the tactile perception of natural textures. *Proc. Natl. Acad. Sci. USA* 110: 17107–17112.

Woo, S. H. and 11 others (2014) Piezo2 is required for Merkel-cell mechanotransduction. *Nature* 509: 622–626.

Zhu, J. J. and B. Connors (1999) Intrinsic firing patterns and whisker-evoked synaptic responses of neurons in the rat barrel cortex. *J. Neurophysiol.* 81: 1171–1183.

Books

Hertenstein, M. J. and S. J. Weiss (eds.) (2011) *The Handbook of Touch: Neuroscience, Behavioral, and Health Perspectives.* New York: Springer.

Linden, D. J. (2015) *Touch: The Science of Hand, Heart, and Mind.* New York: Viking Penguin.

Mountcastle, V. B. (1998) *Perceptual Neuroscience: The Cerebral Cortex.* Cambridge, MA: Harvard University Press.

Go to the NEUROSCIENCE 6e Companion Website at **www.oup.com/uk/Purves6e** for Web Topics, Animations, Flashcards, and more.

CHAPTER

10 Pain

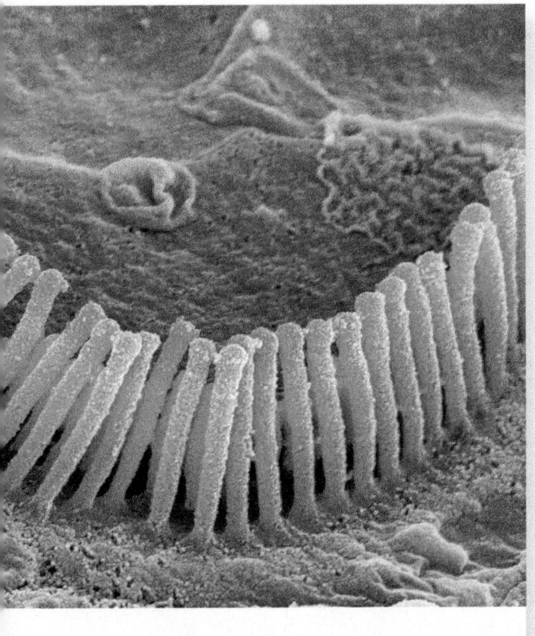

Overview

PAIN IS THE UNPLEASANT SENSORY AND EMOTIONAL EXPERIENCE associated with stimuli that cause tissue damage. Although it may be natural to assume that the sensations associated with injurious stimuli arise from excessive stimulation of the same receptors that generate other somatic sensations (i.e., those discussed in Chapter 9), this is not the case. The perception of injurious stimuli, called nociception, depends on specifically dedicated receptors and pathways. Moreover, the response of an organism to noxious stimuli is multidimensional, involving discriminative, affective, and motivational components. The central distribution of nociceptive information is correspondingly complex, involving multiple areas in the brainstem, thalamus, and forebrain. The overriding importance of pain in clinical practice (both as a diagnostic and as a focus of treatment) as well as the many aspects of pain physiology and pharmacology that remain imperfectly understood continue to make nociception an extremely active area of research.

Nociceptors

The relatively unspecialized nerve cell endings that initiate the sensation of pain are called **nociceptors** (Latin *nocere*, "to hurt"). Like other cutaneous and subcutaneous receptors, nociceptors transduce a variety of stimuli into receptor potentials, which in turn trigger afferent action potentials. Moreover, nociceptors, like other somatosensory receptors, arise from cell bodies in dorsal root ganglia (or in the trigeminal ganglion) that send one axonal process to the periphery and the other into the spinal cord or brainstem (see Figure 9.1).

Because peripheral nociceptive axons terminate in morphologically unspecialized "free nerve endings," it is conventional to categorize nociceptors according to the properties of the axons associated with them (see Table 9.1). As described in Chapter 9, the somatosensory receptors responsible for the perception of innocuous mechanical stimuli are associated with myelinated axons that have relatively rapid conduction velocities. The axons associated with nociceptors, in contrast, conduct relatively slowly, being only lightly myelinated or, more commonly, unmyelinated. Accordingly, axons conveying information about pain fall into either the **Aδ group** of myelinated axons, which conduct at 5 to 30 m/s, or into the **C fiber group** of unmyelinated axons, which conduct at velocities generally less than 2 m/s. Thus, even though the conduction of all nociceptive information is relatively slow, pain pathways can be either fast or slow.

Studies carried out in both humans and experimental animals demonstrated some time ago that the rapidly conducting axons that subserve somatic sensation are not involved in the transmission of pain. Figure 10.1 illustrates a typical experiment of this sort. The peripheral axons responsive to nonpainful mechanical or thermal

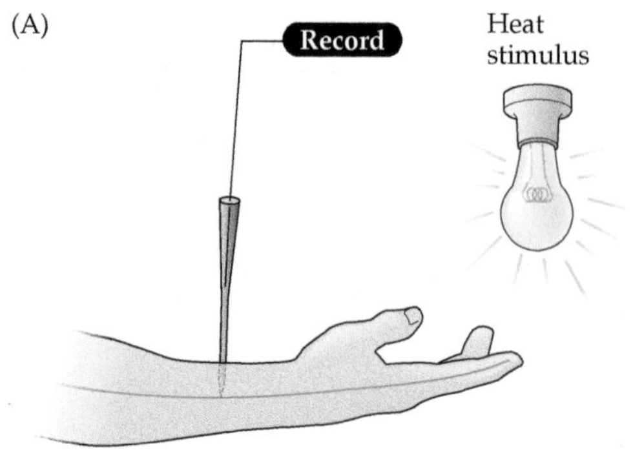

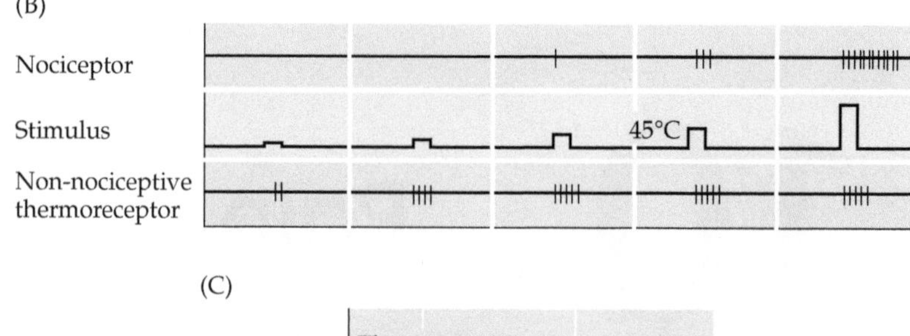

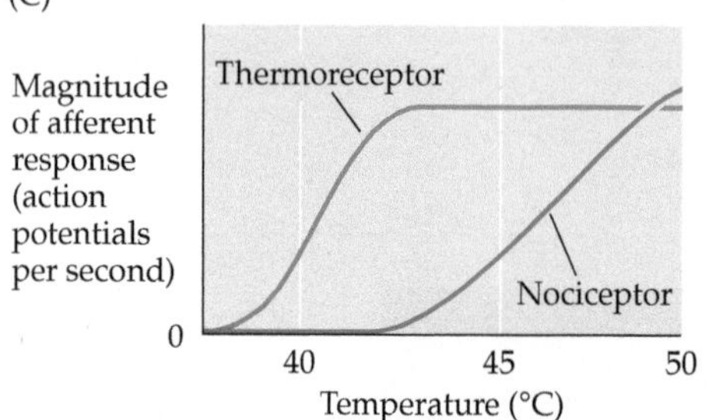

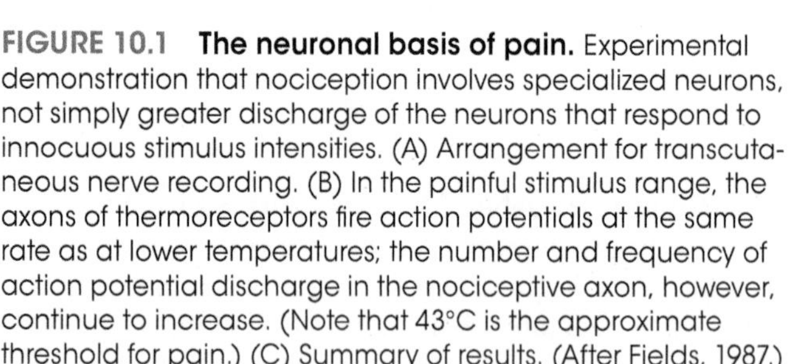

FIGURE 10.1 The neuronal basis of pain. Experimental demonstration that nociception involves specialized neurons, not simply greater discharge of the neurons that respond to innocuous stimulus intensities. (A) Arrangement for transcutaneous nerve recording. (B) In the painful stimulus range, the axons of thermoreceptors fire action potentials at the same rate as at lower temperatures; the number and frequency of action potential discharge in the nociceptive axon, however, continue to increase. (Note that 43°C is the approximate threshold for pain.) (C) Summary of results. (After Fields, 1987.)

stimuli do not discharge at a greater rate when painful stimuli are delivered to the same region of the skin surface. The nociceptive axons, by contrast, begin to discharge only when the strength of the stimulus (a thermal stimulus in the example in Figure 10.1) reaches high levels; at this same stimulus intensity, other thermoreceptors discharge at a rate no different from the maximum rate already achieved within the nonpainful temperature range, indicating the presence of both nociceptive and non-nociceptive thermoreceptors. Equally important, direct stimulation of the large-diameter somatosensory afferents at any frequency in humans does not produce sensations that subjects describe as painful. In contrast, the smaller diameter, more slowly conducting Aδ and C fibers are active when painful stimuli are delivered; when stimulated electrically in human subjects, these fibers produce sensations of pain. There are also C fibers that mediate nondiscriminative touch, as well as the sensations of warmth, coolness, and itch. These will be discussed later in this chapter.

How, then, do different classes of nociceptors lead to the perception of pain? As mentioned, one way of determining the answer has been to stimulate different nociceptors in human volunteers while noting the sensations reported. In general, two categories of pain perception have been described: a sharp **first pain** and a more delayed, diffuse, and longer-lasting sensation that is generally called **second pain** (Figure 10.2A). Stimulation of the large, rapidly conducting Aα and Aβ axons in peripheral nerves does not elicit the sensation of pain. When investigators raise the stimulus intensity to a level that activates a subset of Aδ fibers, however, a tingling sensation or, if the stimulation is intense enough, a feeling of sharp pain is reported. If the stimulus intensity is increased still further, so that the small-diameter, slowly conducting C-fiber axons are brought into play, then subjects report a duller, longer-lasting sensation of pain. It is also possible for researchers to selectively anesthetize C fibers and Aδ fibers; in general, these selective blocking experiments confirm that Aδ fibers are responsible for first pain and C fibers are responsible for the duller, longer-lasting second pain (Figure 10.2B,C).

The faster-conducting Aδ nociceptors are now known to fall into two main classes. Type I Aδ fibers respond to dangerously intense mechanical and chemical stimulation but have relatively high heat thresholds, while type II Aδ fibers have complementary sensitivities—that is, much lower thresholds for heat but very high thresholds for mechanical stimulation. Thus, the Aδ system has specialized pathways for the transmission of heat and mechanical nociceptive stimuli. Most of the slower-conducting, unmyelinated C-fiber nociceptors respond to all forms of nociceptive stimuli—thermal, mechanical, and chemical—and are therefore said to be polymodal. However, C-fiber nociceptors are also heterogeneous, with subsets that respond preferentially to heat or chemical stimulation rather than mechanical stimulation. Further subtypes of C-fiber nociceptors are especially responsive to chemical irritants, acidic substances, or cold. In short, each of the major classes of nociceptive afferents is composed of multiple subtypes with distinct sensitivity profiles.

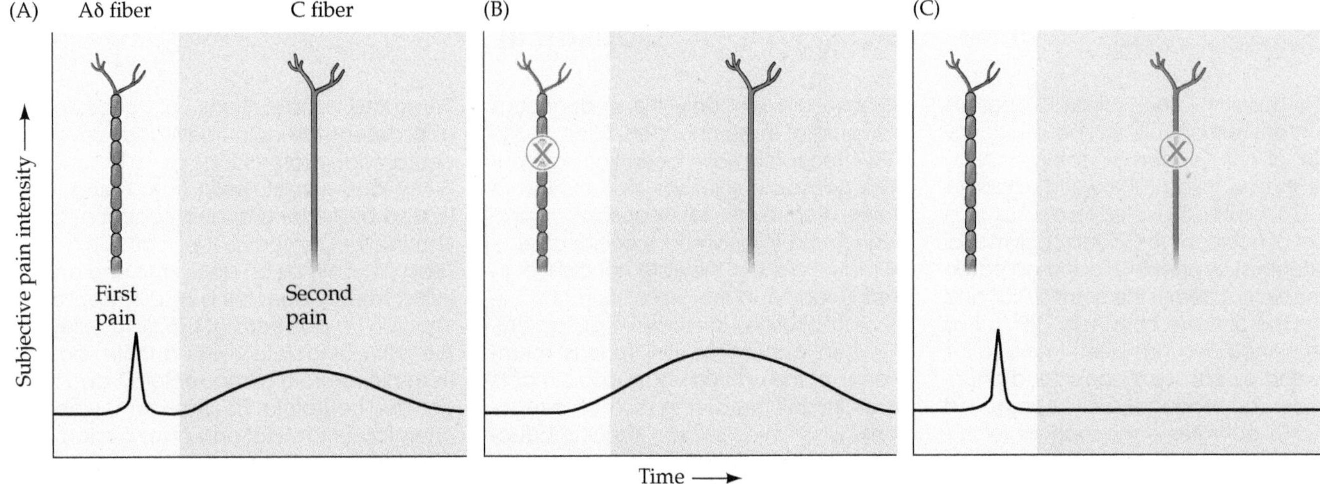

FIGURE 10.2 **First and second pain.** Pain can be separated into an early perception of sharp pain and a later sensation that is described as having a duller, burning quality. (A) First and second pain, as these sensations are called, are carried by different axons, as can be shown by (B) the selective blockade of the more rapidly conducting myelinated axons that carry the sensation of first pain, or (C) blockade of the more slowly conducting C fibers that carry the sensation of second pain. (After Fields, 1990.)

Transduction and Transmission of Nociceptive Signals

Given the variety of stimuli (mechanical, thermal, and chemical) that can give rise to painful sensations, the transduction of nociceptive signals is a complex task. While many puzzles remain, significant insights have come from the identification of a specific receptor associated with the sensation of noxious heat. The threshold for perceiving a thermal stimulus as noxious is around 43°C (110°F), and this pain threshold corresponds with the sensitivity of subtypes of Aδ- and C-fiber nociceptive endings. The receptor that confers this sensitivity to heat also confers sensitivity to capsaicin, the ingredient in chili peppers responsible for the tingling or burning sensation produced by spicy foods (Box 10A). The so-called vanilloid receptor (TRPV1), found in both C and Aδ fibers, is a member of the larger family of **transient receptor potential (TRP)** channels, first identified in studies of the phototransduction pathway in fruit flies and now known to comprise a large number of receptors sensitive to different ranges of heat and cold. Structurally, TRP channels resemble voltage-gated potassium or cyclic nucleotide-gated channels, having six transmembrane domains with a pore between domains 5 and 6. Under resting conditions, the pore of the channel is closed. In the open, activated state, these receptors allow an influx of sodium and calcium that initiates the generation of action potentials in the nociceptive fibers. Since the same receptor is responsive to heat as well as capsaicin, it is not surprising that many people experience the taste of chili peppers as "hot." A puzzle, however, is why the nervous system has evolved receptors that are sensitive to a chemical in chili peppers. As is the case with other plant compounds that selectively activate neural receptors (see the discussion of opioids in the section "The Physiological Basis of Pain Modulation" later in the chapter), it seems likely that TRPV1 receptors detect endogenous substances whose chemical structure resembles that of capsaicin. In fact, some recent evidence suggests that "endovanilloids" are produced by peripheral tissues in response to injury and that these substances, along with other factors, contribute to the nociceptive response to injury.

The receptors responsible for the transduction of mechanical and chemical forms of nociceptive stimulation are less well understood. Several different candidates for mechanotransducers have been identified, including other members of the TRP family (TRPV4), a rapidly adapting ion channel called Piezo2, and some members of the ASIC (*ac*id-sensing *i*on *c*hannels) family. TRP channels also appear to be responsible for the detection of chemical irritants in the environment. TRPA1 in particular has been shown to be sensitive to a diverse group of chemical irritants, including the pungent ingredients in mustard and garlic plants, as well as volatile irritants present in tear gas, vehicle exhaust, and cigarettes. The ASIC3 channel subtype is specifically expressed in nociceptors and is well represented in fibers that innervate skeletal and cardiac muscle. ASIC3 channels are thought to be responsible for the muscle or cardiac pain that results from changes in pH associated with ischemia. The complex molecular basis of mechano- and chemonociception is an area of active investigation and is critical for understanding the initial steps in the neural pathways that contribute to pain.

The graded potentials arising from receptors in the distal branches of nociceptive fibers must be transformed into

BOX 10A ■ Capsaicin

Capsaicin, the principle ingredient responsible for the pungency of hot peppers, is eaten daily by more than a third of the world's population. Capsaicin activates responses in a subset of nociceptive C fibers (polymodal nociceptors) by opening ligand-gated ion channels that permit the entry of Na^+ and Ca^{2+}. One of these channels, TRPV1, has been cloned and has been found to be activated by capsaicin, acid, and anandamide (an endogeneous compound that also activates cannabanoid receptors), or by heating the tissue to about 43°C. It follows that anandamide and temperature are probably the endogenous activators of these channels. Mice whose TRPV1 receptors have been knocked out drink capsaicin solutions as if they were water. Receptors for capsaicin have been found in polymodal nociceptors of all mammals, but they are not present in birds (leading to the production of squirrel-proof birdseed laced with capsaicin).

When applied to the mucus membranes of the oral cavity, capsaicin acts as an irritant, producing protective reactions. When injected into skin, it produces a burning pain and elicits hyperalgesia to thermal and mechanical stimuli. Repeated applications of capsaicin also desensitize pain fibers and prevent neuromodulators such as substance P, VIP, and somatostatin from being released by peripheral and central nerve terminals. Consequently, capsaicin is used clinically as an analgesic and anti-inflammatory agent; it is usually applied topically in a cream (0.075%) to relieve the pain associated with arthritis, postherpetic neuralgia, mastectomy, and trigeminal neuralgia. Thus, this remarkable chemical irritant not only gives gustatory pleasure on an enormous scale, but it is also a useful pain reliever.

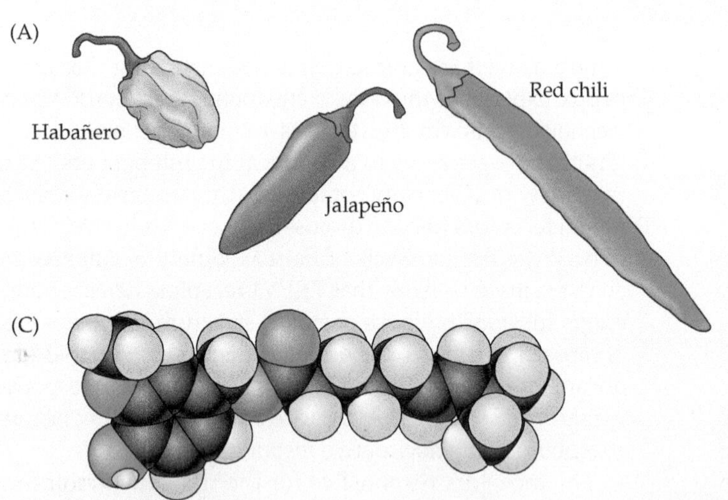

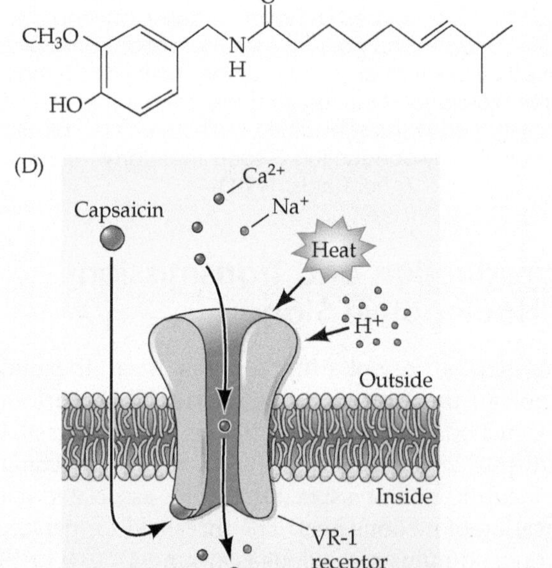

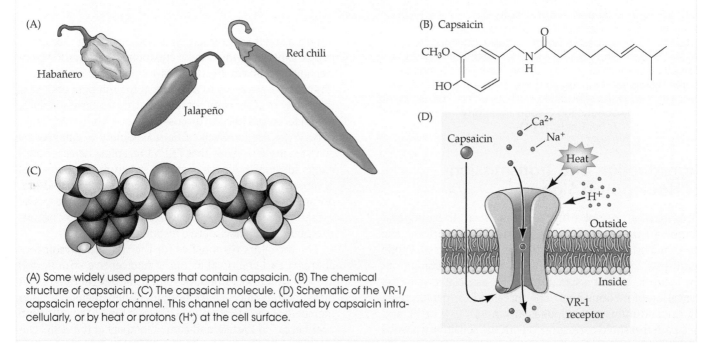

(A) Some widely used peppers that contain capsaicin. (B) The chemical structure of capsaicin. (C) The capsaicin molecule. (D) Schematic of the VR-1/capsaicin receptor channel. This channel can be activated by capsaicin intracellularly, or by heat or protons (H^+) at the cell surface.

action potentials in order to be conveyed to synapses in the dorsal horn of the spinal cord. Voltage-gated sodium and potassium channels are critical in this process (see Chapter 4), and one specific subtype of sodium channel—Nav1.7—appears to be especially important for the transmission of nociceptive information. Altered activity of Nav1.7 is responsible for a variety of human pain disorders. Mutations of the *NAV1.7* gene (known as *SCN9A* in humans) that lead to a loss of this channel's function result in an inability to detect noxious stimulation, while mutations leading to hyperexcitability of the channel are associated with pain disorders that cause intense burning sensations. The *NAV1.8* gene is highly expressed by most C-fiber nociceptors, and based on studies in mice, the protein has been associated with the transmission of noxious mechanical and thermal information. The development of local anesthetics specific to these subtypes of sodium channels may hold the key for treating a variety of intractable pain syndromes.

Central Pain Pathways Are Distinct from Mechanosensory Pathways

Pathways responsible for pain originate with other sensory neurons in dorsal root ganglia, and like other sensory nerve cells, the central axons of nociceptive nerve cells enter the spinal cord via the dorsal roots (Figure 10.3A). When these centrally projecting axons reach the dorsal horn of the spinal cord, they branch into ascending and descending collaterals, forming the **dorsolateral tract of Lissauer** (named after the German neurologist

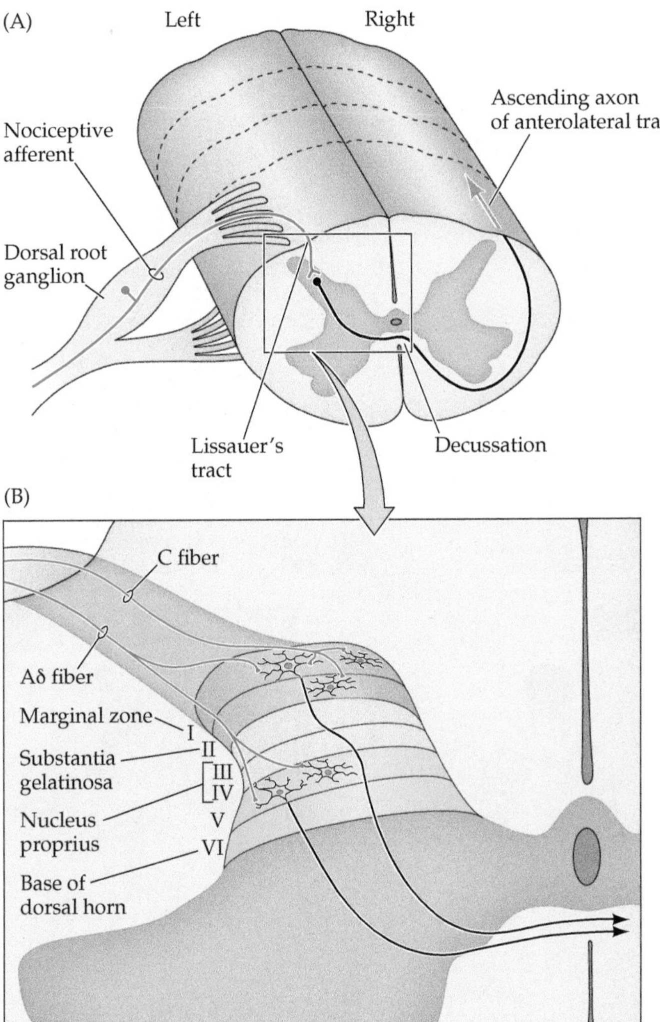

FIGURE 10.3 The anterolateral system. (A) Primary afferents in the dorsal root ganglia send their axons via the dorsal roots to terminate in the dorsal horn of the spinal cord. Afferents branch and course for several segments up and down the spinal cord in Lissauer's tract, giving rise to collateral branches that terminate in the dorsal horn. Second-order neurons in the dorsal horn send their axons (black) across the midline to ascend to higher levels in the anterolateral column of the spinal cord. (B) C-fiber afferents terminate in Rexed's laminae I and II of the dorsal horn, while Aδ fibers terminate in laminae I and V. The axons of second-order neurons in laminae I and V cross the midline and ascend to higher centers.

who first described this pathway in the late nineteenth century). Axons in Lissauer's tract typically run up and down for one or two spinal cord segments before they penetrate the gray matter of the dorsal horn. Once within the dorsal horn, the axons give off branches that contact second-order neurons located in Rexed's laminae I, II, and V. (Rexed's laminae are the descriptive divisions of the spinal gray matter in cross section, named after the neuroanatomist who described these details in the 1950s; see Table A1 and Figure A7 in the Appendix.) Laminae I and V contain projection neurons whose axons travel to brainstem and thalamic targets. While there are interneurons in all laminae of the spinal cord, they are especially abundant and diverse morphologically and histochemically in lamina II. These afferent terminations are organized in a lamina-specific fashion; for example, C fibers terminate exclusively in Rexed's laminae I and II, while Aδ fibers terminate in laminae I and V. Non-nociceptive (Aβ) afferents terminate primarily in laminae III, IV, and V, and a subset of the lamina V neurons receive converging inputs from nociceptive and non-nociceptive afferents. These multimodal lamina V neurons are called **wide-dynamic-range neurons**. Some of them receive visceral sensory input as well, making them a likely substrate for referred pain (i.e., pain that arises from damage to visceral organs but is misperceived as coming from a somatic location). The most common clinical example is angina, in which poor perfusion of the heart muscle is misperceived as pain in the chest wall, the shoulder, and the left arm and hand (Box 10B).

The axons of the second-order neurons in laminae I and V of the dorsal horn of the spinal cord cross the midline and ascend to the brainstem and thalamus in the anterolateral (also called ventrolateral) quadrant of the contralateral half of the spinal cord (Figure 10.3B). For this reason, the neural pathway that conveys pain and temperature information to higher centers is often referred to as the **anterolateral system**, to distinguish it from the dorsal column–medial lemniscal system that conveys mechanosensory information (see Chapter 9).

Axons conveying information for the anterolateral system and the dorsal column–medial lemniscal system travel in different parts of the spinal cord white matter. This difference provides a clinically relevant sign that is useful for defining the locus of a spinal cord lesion. Axons of the first-order neurons for the dorsal column–medial lemniscal system enter the spinal cord, turn, and ascend in the ipsilateral dorsal columns all the way to the medulla, where they synapse on neurons in the dorsal column nuclei (Figure 10.4, left panel). The axons of neurons in the dorsal column nuclei then cross the midline and ascend to the contralateral thalamus. In contrast, the crossing point for information conveyed by the anterolateral system lies within the spinal cord. First-order neurons contributing to the anterolateral system terminate in the dorsal horn, and second-order neurons in the dorsal horn send their axons across the midline and ascend on the contralateral side of the cord (in the anterolateral column) to their targets in the thalamus and brainstem.

Because of this anatomical difference in the site of decussation, a unilateral spinal cord lesion results in dorsal column–medial lemniscal symptoms (loss of sensation of touch, pressure, vibration, and proprioception) on the side

BOX 10B ■ Referred Pain

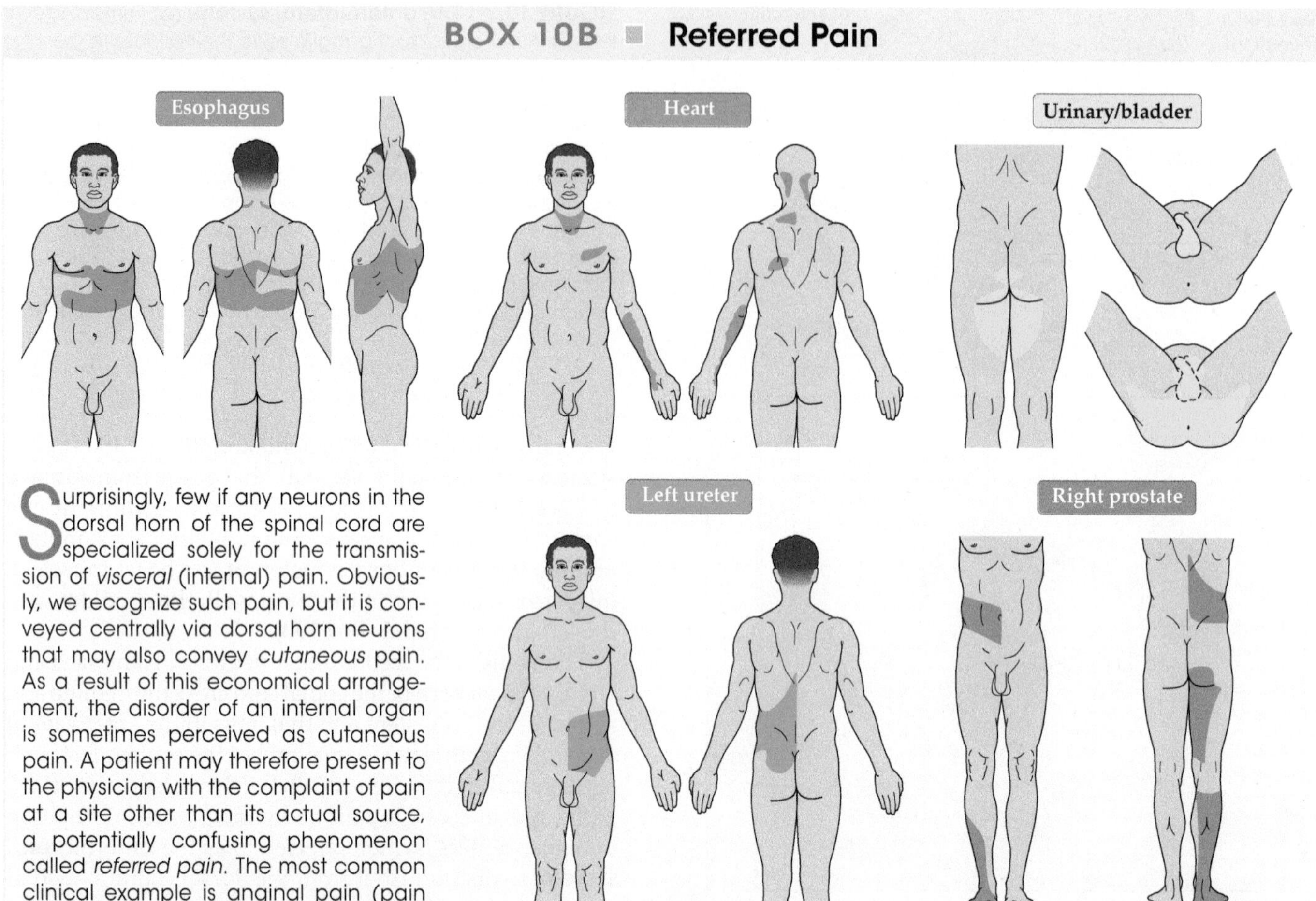

Examples of pain arising from a visceral disorder referred to a cutaneous region (color).

Surprisingly, few if any neurons in the dorsal horn of the spinal cord are specialized solely for the transmission of *visceral* (internal) pain. Obviously, we recognize such pain, but it is conveyed centrally via dorsal horn neurons that may also convey *cutaneous* pain. As a result of this economical arrangement, the disorder of an internal organ is sometimes perceived as cutaneous pain. A patient may therefore present to the physician with the complaint of pain at a site other than its actual source, a potentially confusing phenomenon called *referred pain*. The most common clinical example is anginal pain (pain arising from heart muscle that is not being adequately perfused with blood) referred to the upper chest wall, with radiation into the left arm and hand. Other important examples are gallbladder pain referred to the scapular region, esophageal pain referred to the chest wall, ureteral pain (e.g., from passing a kidney stone) referred to the lower abdominal wall, bladder pain referred to the perineum, and the pain from an inflamed appendix referred to the anterior abdominal wall around the umbilicus. Understanding referred pain can lead to an astute diagnosis that might otherwise be missed.

of the body *ipsilateral* to the lesion, and anterolateral symptoms (deficits of pain and temperature perception) on the *contralateral* side of the body (Figure 10.4, right panel). The deficits are due to the interruption of fibers ascending from lower levels of the cord; for this reason they include all regions of the body (on either the contralateral or ipsilateral side) that are innervated by spinal cord segments that lie below the level of the lesion. This pattern of **dissociated sensory loss** (contralateral pain and temperature, ipsilateral touch and pressure) is a signature of spinal cord lesions and, together with local dermatomal signs, can be used to define the level of the lesion. (Box 10C discusses an important exception to the functional dissociation of the dorsal column–medial lemniscal and anterolateral systems for visceral pain.)

Parallel Pain Pathways

Second-order fibers in the anterolateral system project to several different structures in the brainstem and forebrain, making it clear that pain is processed by a diverse and distributed network of neurons. While the full significance of this complex pattern of connections remains unclear, these central destinations are likely to mediate different aspects of the sensory and behavioral response to a painful stimulus.

One component of this system, the spinothalamic tract, mediates the **sensory-discriminative** aspects of pain: the location, intensity, and quality of the noxious stimulation. These aspects of pain are thought to depend on information relayed through the ventral posterior lateral nucleus (VPL) to neurons in the primary and secondary somatosensory

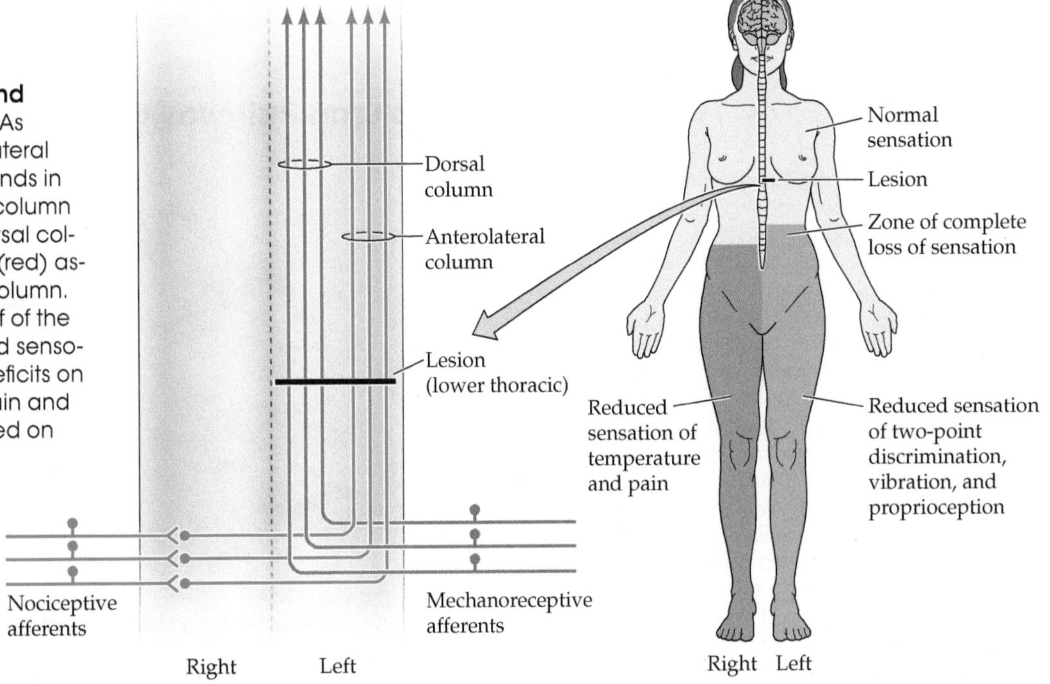

FIGURE 10.4 Nociceptive and mechanosensory pathways. As diagrammed here, the anterolateral system (blue) crosses and ascends in the contralateral anterolateral column of the spinal cord, while the dorsal column–medial leminiscal system (red) ascends in the ipsilateral dorsal column. A lesion restricted to the left half of the spinal cord results in dissociated sensory loss and mechanosensory deficits on the left half of the body, with pain and temperature deficits experienced on the right.

Dorsal column
Anterolateral column
Lesion (lower thoracic)
Nociceptive afferents
Mechanoreceptive afferents
Right
Left
Normal sensation
Lesion
Zone of complete loss of sensation
Reduced sensation of temperature and pain
Reduced sensation of two-point discrimination, vibration, and proprioception
Right Left

cortex (Figures 10.5 and 10.6A). (The pathway for relay of information from the face to the ventral posterior medial nucleus, or VPM, is considered in the next section.) Although axons from the anterolateral system overlap those from the dorsal column system in the ventral posterior nuclei, these axons contact different classes of relay neurons, so that nociceptive information remains segregated up to the level of cortical circuits. Consistent with mediating the discriminative aspects of pain, electrophysiological recordings from nociceptive neurons in the primary somatosensory cortex (SI) show that these neurons have small, localized receptive fields—properties commensurate with behavioral measures of pain localization.

Other parts of the system convey information about the **affective-motivational** aspects of pain: the unpleasant feeling, the fear and anxiety, and the autonomic activation that accompany exposure to a noxious stimulus (the classic fight-or-flight response; see Chapter 21). Targets of these projections include several subdivisions of the reticular formation, the periaqueductal gray, the deep layers of the superior colliculus, and the parabrachial nucleus in the rostral pons (see Figure 10.5). The parabrachial nucleus processes and relays second pain signals to the amygdala, hypothalamus, and a distinct set of thalamic nuclei that lie medial to the ventral posterior nucleus, which we group together here as the medial thalamic

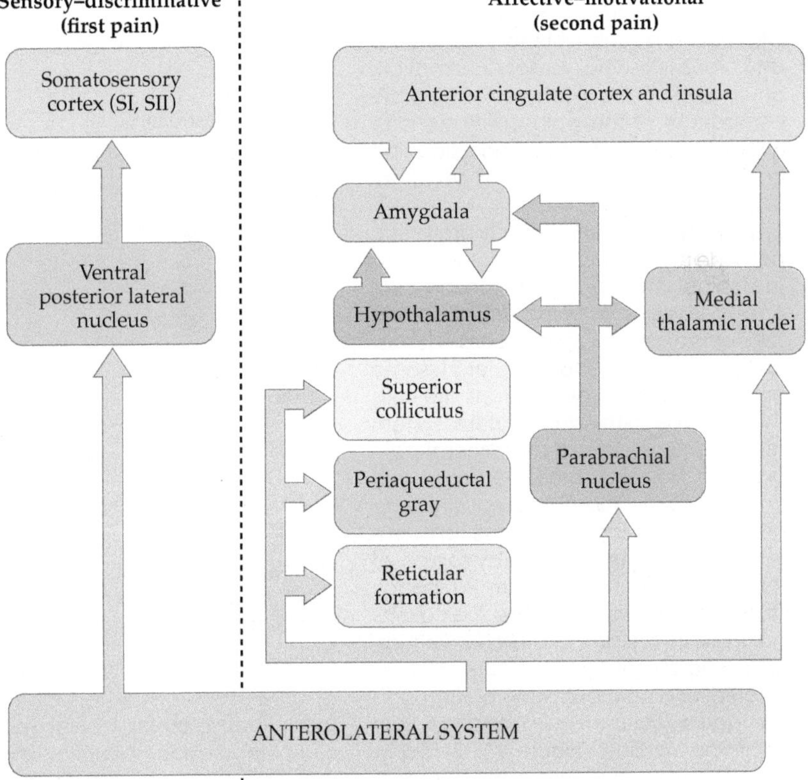

FIGURE 10.5 Two distinct aspects of the experience of pain. The anterolateral system supplies information to different structures in the brainstem and forebrain that contribute to different aspects of the experience of pain. The spinothalamic tract (left of dashed line) conveys signals that mediate the sensory discrimination of first pain. The affective and motivational aspects of second pain are mediated by complex pathways that reach integrative centers in the limbic forebrain.

BOX 10C ■ A Dorsal Column Pathway for Visceral Pain

Chapters 9 and 10 present a framework for considering the central neural pathways that convey innocuous mechanosensory signals and painful signals from cutaneous and deep somatic sources. Considering just the signals derived from the body below the head, discriminative mechanosensory and proprioceptive information travels to the ventral posterior thalamus via the dorsal column-medial lemniscal system (see Figure 10.4A), while nociceptive information travels to the same (and additional) thalamic relays via the anterolateral system (see Figure 10.6A).

But how do painful signals that arise in the visceral organs of the pelvis, abdomen, and thorax enter the central nervous system and ultimately reach an individual's consciousness? The answer is via a newly discovered component of the dorsal column-medial lemniscal pathway that conveys visceral nociception. Although Chapter 21 will present more information on the systems that receive and process visceral sensory information, at this juncture it is worth considering this component of the pain pathways and the way in which a better understanding of this particular pathway has begun to affect clinical medicine.

Primary visceral afferents from the pelvic and abdominal viscera enter the spinal cord and synapse on second-order neurons in the dorsal horn of the lumbar-sacral spinal cord. As discussed in Box 10B and Chapter 21, some of these second-order neurons are cells that give rise to the anterolateral system and contribute to referred visceral pain patterns. However, other neurons—perhaps primarily those that give rise to nociceptive signals—synapse on neurons in the intermediate gray region of the spinal cord near the central canal. These neurons, in turn, send their axons not through the anterolateral white matter of the spinal cord (as might be expected for a pain pathway) but through the dorsal columns in a position very near the midline (Figure A). Similarly, second-order neurons in the thoracic spinal cord that convey nociceptive signals from thoracic viscera send their axons rostrally through the dorsal columns along the dorsal intermediate septum, near the division of the gracile and cuneate tracts. These second-order axons then synapse on the dorsal column nuclei of the caudal medulla, where neurons give rise to arcuate fibers that form the contralateral medial lemniscus and eventually synapse on thalamocortical projection neurons in the ventral posterior thalamus.

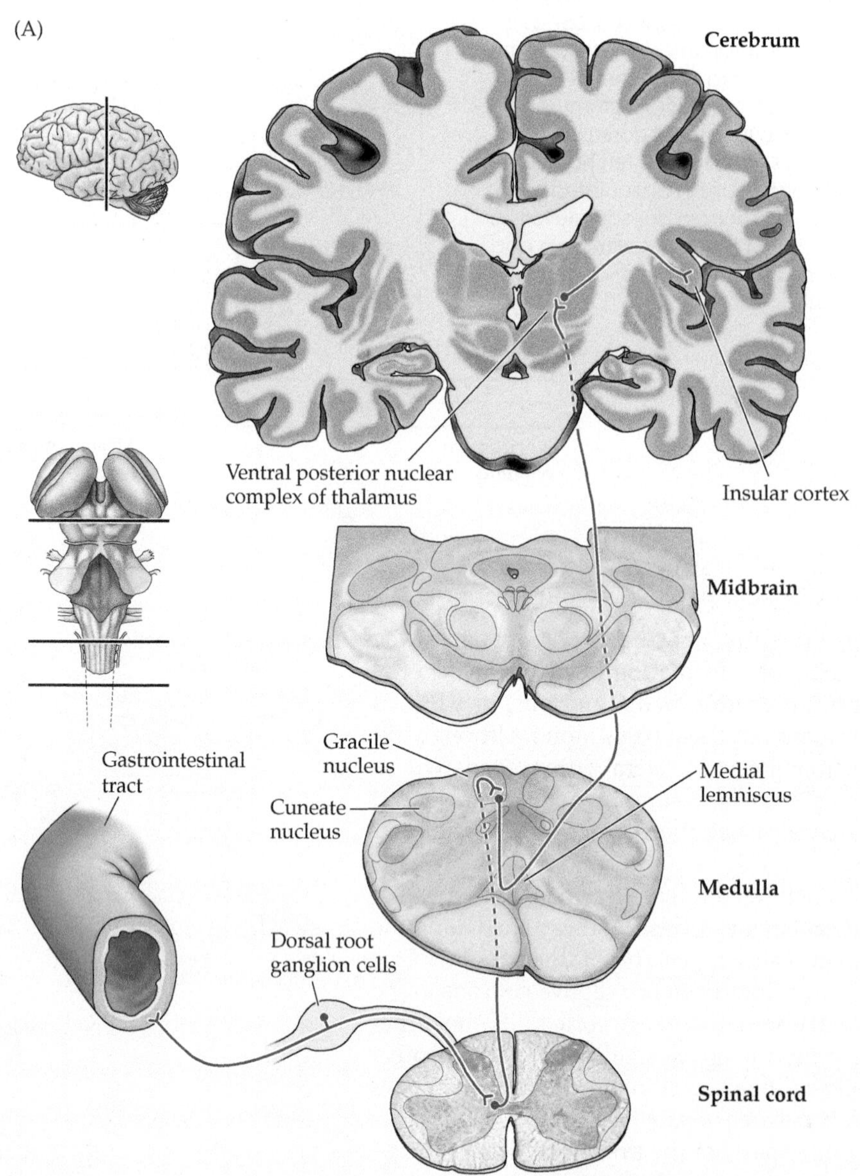

(A) A visceral pain pathway in the dorsal column–medial lemniscal system. For simplicity, only the pathways that mediate visceral pain from the pelvis and lower abdomen are illustrated.

This dorsal column visceral sensory projection now appears to be the principal pathway by which painful sensations arising in the viscera are detected and discriminated. Several observations support this conclusion: (1) neurons in the ventral posterior lateral nucleus, gracile nucleus, and near the central canal of the spinal cord all respond to noxious visceral stimulation; (2) responses of neurons in the ventral posterior lateral

BOX 10C ■ (continued)

nucleus and gracile nucleus to such stimulation are greatly reduced by spinal lesions of the dorsal columns (Figure B), but not by lesions of the anterolateral white matter; and (3) infusion of drugs that block nociceptive synaptic transmission into the intermediate gray region of the sacral spinal cord blocks the responses of neurons in the gracile nucleus to noxious visceral stimulation, but not to innocuous cutaneous stimulation.

The discovery of this visceral sensory component in the dorsal column–medial lemniscal system has helped explain why surgical transection of the axons that run in the medial part of the dorsal columns (a procedure termed *midline myelotomy*) generates significant relief from the debilitating pain that can result from visceral cancers in the abdomen and pelvis. Although the initial development of this surgical procedure preceded the elucidation of this visceral pain pathway, these new discoveries have renewed interest in midline myelotomy as a palliative neurosurgical intervention for cancer patients whose pain is otherwise unmanageable. Indeed, precise knowledge of the visceral sensory pathway in the dorsal columns has led to further refinements that permit a minimally invasive (*punctate*) surgical procedure that attempts to interrupt the second-order axons of this pathway within just a single spinal segment (typically, at mid- or lower-thoracic level; Figure C). In so doing, this procedure offers some hope to patients who struggle to maintain a reasonable quality of life in extraordinarily difficult circumstances.

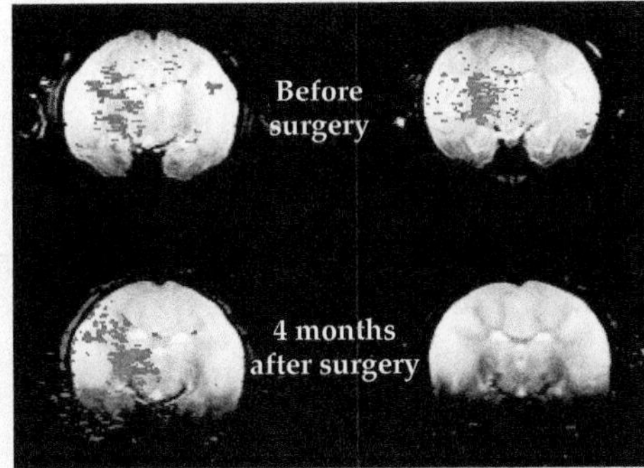

(B) Empirical evidence supporting the existence of the visceral pain pathway shown in (A). Increased neural activity was observed with fMRI techniques in the thalamus of monkeys that were subjected to noxious distention of the colon and rectum, indicating the processing of visceral pain. This activity was abolished by lesion of the dorsal columns at T10, but not by "sham" surgery.

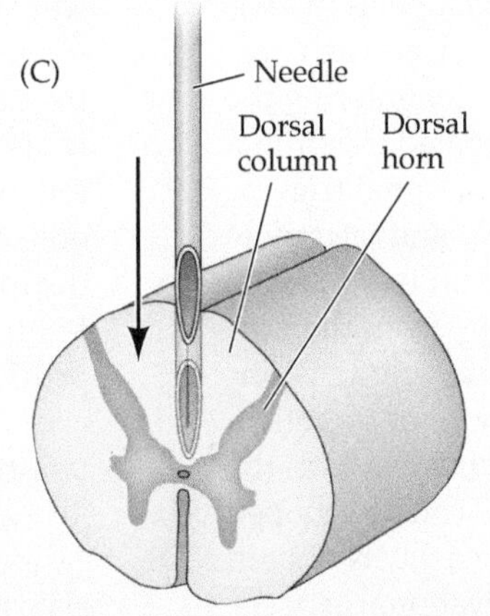

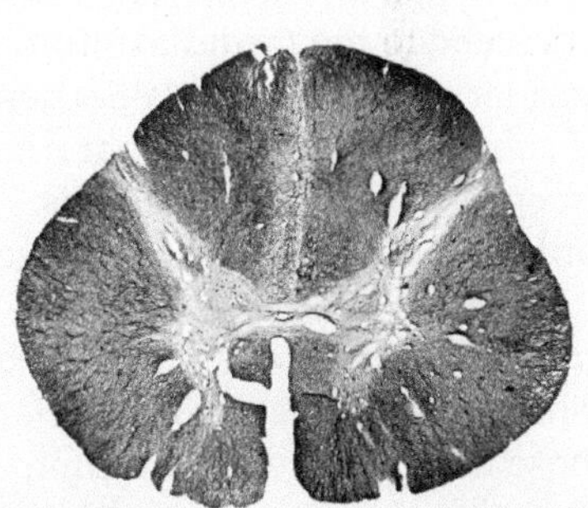

(C) Left: One method of punctate midline myelotomy for the relief of severe visceral pain. Right: Myelin-stained section of the thoracic spinal cord (T10) from a patient who underwent midline myelotomy for the treatment of colon cancer pain that was not controlled by analgesics. After surgery, the patient experienced relief from pain during the remaining 3 months of his life. (B from Willis et al., 1999; C from Hirshberg et al., 1996; drawing after Nauta et al., 1997.)

nuclei. These medial thalamic nuclei, which also receive input from anterolateral system axons, play an important role in transmitting nociceptive signals to both the anterior cingulate cortex and to the insula. Together with the amygdala and hypothalamus, which are also interconnected with the cingulate cortex and insula, these limbic forebrain structures elaborate affective-motivational aspects of pain (see Chapter 31). Electrophysiological recordings in human patients, which show that cingulate neurons respond to noxious stimuli, support the role of the anterior cingulate cortex in the perception of pain. Moreover, patients who have undergone cingulotomies report an attenuation of the unpleasantness that accompanies pain.

Evidence from functional imaging studies in humans supports the view that different brain regions mediate the sensory–discriminative and affective–motivational aspects of pain. The presentation of a painful stimulus results in the activation of both primary somatosensory cortex and anterior cingulate cortex; however, by using hypnotic suggestion to selectively increase or decrease the unpleasantness of a painful stimulus, it has been possible to tease apart the neural response to changes in the intensity of a painful stimulus versus changes in its unpleasantness. Changes in intensity are accompanied by changes in the activity of neurons in somatosensory cortex, with little change in the activity of cingulate cortex, whereas changes in unpleasantness are highly correlated with changes in the activity of neurons in cingulate cortex.

From this description, it should be evident that the full experience of pain involves the cooperative action of an extensive network of forebrain regions whose properties we are only beginning to understand. Indeed, brain-imaging studies frequently refer to the broad array of areas whose activity is associated with the experience of pain—including the somatosensory cortex, insular cortex, amygdala, and anterior cingulate cortex—as the **pain matrix**. In retrospect, the distributed nature of pain representation should not be surprising given that pain is a multidimensional experience with sensory, motor, affective, and cognitive effects. A distributed representation also explains why ablations of the somatosensory cortex do not usually alleviate chronic pain, even though they severely impair contralateral mechanosensory perception.

Pain and Temperature Pathways for the Face

Information about noxious and thermal stimulation of the face originates from first-order neurons located in the trigeminal ganglion and from ganglia associated with cranial nerves VII, IX, and X (Figure 10.6B). After entering the pons, these small myelinated and unmyelinated trigeminal fibers descend to the medulla, forming the spinal trigeminal tract (or spinal tract of cranial nerve V) and terminate in two subdivisions of the spinal trigeminal nucleus: the pars interpolaris and pars caudalis. Axons from the second-order neurons in these two trigeminal subdivisions cross the midline and terminate in a variety of targets in the brainstem and thalamus. Like their counterparts in the dorsal horn of the spinal cord, these targets can be grouped into those that mediate the discriminative aspects of pain and those that mediate the affective–motivational aspects. The discriminative aspects of facial pain are thought to be mediated by projections to the contralateral ventral posterior medial nucleus (via the trigeminothalamic tract) and projections from the VPM to primary and secondary somatosensory cortex. Affective–motivational aspects are mediated by connections to various targets in the reticular formation and parabrachial nucleus, as well as by the medial nuclei of the thalamus, which supply the cingulate and insular regions of cortex.

Other Modalities Mediated by the Anterolateral System

While the anterolateral system plays a critical role in mediating nociception, it is also responsible for transmitting a variety of other innocuous information to higher centers. For example, in the absence of the dorsal column system, the anterolateral system appears to be capable of mediating what is commonly called *nondiscriminative touch*, a form of tactile sensitivity that lacks the fine spatial resolution that can be supplied only by the dorsal column system. The C fiber low-threshold mechanoreceptors mediate this nondiscriminative (sensual) touch. Thus, following damage to the dorsal column–medial lemniscal system, a crude form of tactile sensation remains, one in which two-point discrimination thresholds are increased and the ability to identify objects by touch alone (*stereognosis*) is markedly impaired.

As already alluded to, the anterolateral system is responsible for mediating innocuous temperature sensation. The sensation of warmth and cold is thought to be subserved by separate sets of primary afferents: warm fibers that respond with increasing spike discharge rates to increases in temperature, and cold fibers that respond with increasing spike discharge to decreases in temperature. Neither of these afferents responds to mechanical stimulation, and they are distinct from afferents that respond to temperatures that are considered painful (noxious heat, above 43°C; or noxious cold, below –17°C). The recent identification of TRP channels with sensitivity to temperatures in the innocuous range—TRPV3 and TRPV4, which respond to warm temperatures, and TRPM8, which responds to cold temperatures—raises the possibility of labeled lines for the transmission of warmth and cold, beginning at the level of transduction and continuing within central pathways. Consistent with this idea, the information supplied by innocuous warm and cold afferents is relayed to higher centers by distinct classes of secondary neurons that reside in lamina I of the spinal cord. In addition, subsets of C fibers, called pruriceptors, are activated by prurigenic (itch-inducing) chemicals. Interestingly, many pruriceptors also respond to painful stimuli, and how circuitry in the spinal cord decode itch versus pain is an active area of research.

Indeed, the emerging view of lamina I is that it consists of several distinct classes of modality-selective neurons that convey noxious and innocuous types of sensory information into the anterolateral system. These include individual classes of neurons that are sensitive to a variety of stimuli: sharp (first) pain, burning (second) pain, innocuous warmth, innocuous cold, the sense of itch, slow mechanical stimulation (sensual touch), and a class of inputs that innervates muscles and senses lactic acid and other metabolites that are released during muscle contraction. The latter could contribute to the "burn" or ache that can accompany strenuous exercise.

Is lamina I merely an eclectic mixture of cells with different properties, or might a unifying theme account for this diversity? It has been proposed that the lamina I system functions as the sensory input to a network that is responsible for representing the physiological condition of the body—a modality that has been called **interoception**, to distinguish it from **exteroception** (touch and pressure) and proprioception. These inputs drive the homeostatic mechanisms that maintain an optimal internal state. Some of these mechanisms are automatic, and the changes necessary to maintain homeostasis can be mediated by reflexive adjustment of the autonomic nervous system (see Chapter 21). For example, changes in temperature evoke autonomic reflexes (e.g., sweating or shivering) that

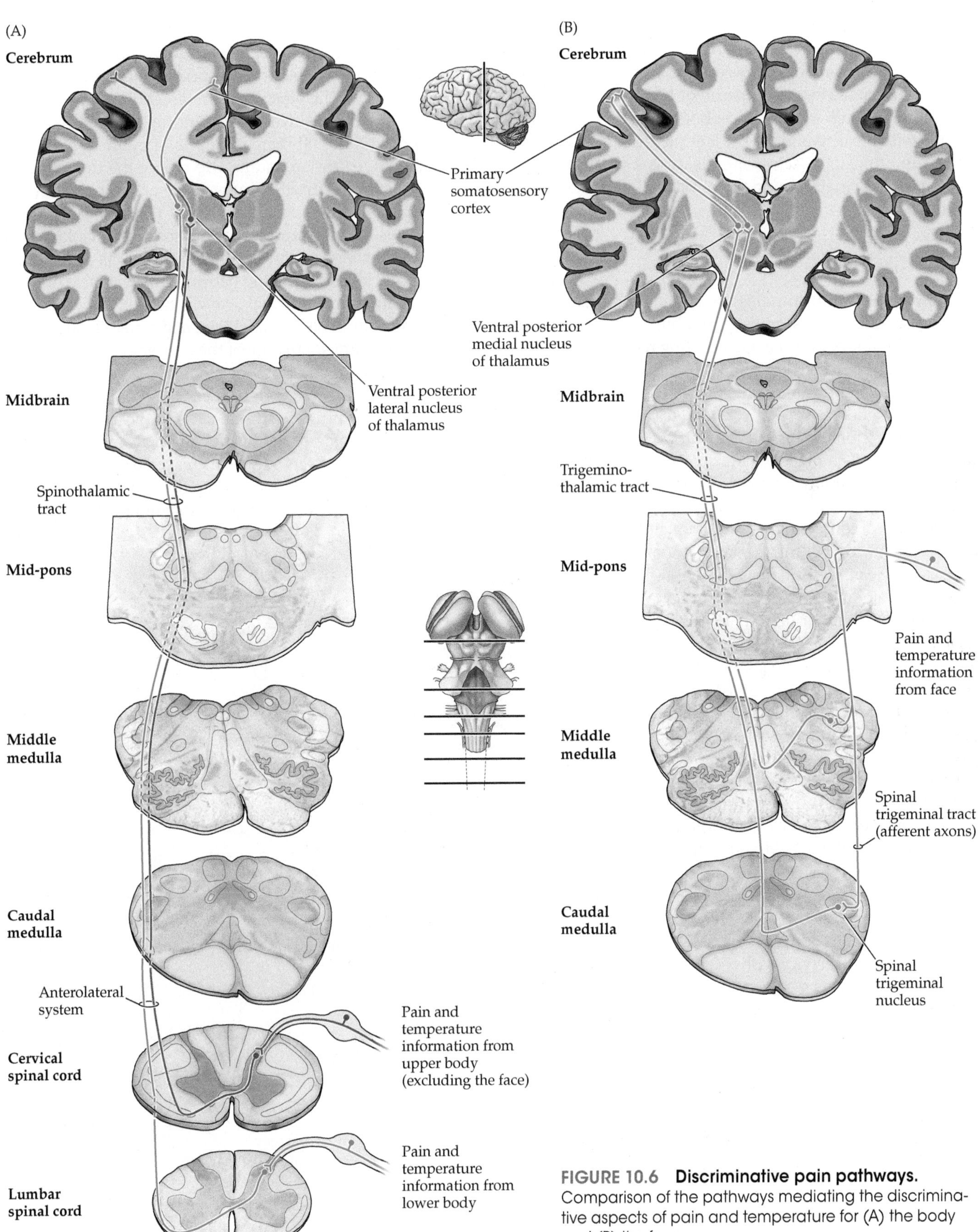

FIGURE 10.6 Discriminative pain pathways. Comparison of the pathways mediating the discriminative aspects of pain and temperature for (A) the body and (B) the face.

counter a disturbance to the body's optimal temperature. Homeostatic disturbances are sometimes too great to be mediated by autonomic reflexes alone and require behavioral adjustments (e.g., putting on or taking off a sweater) to restore balance. In this conception, the sensations associated with the activation of the lamina I system—whether pleasant or noxious—motivate the initiation of behaviors appropriate to maintaining the physiological homeostasis of the body.

Sensitization

Following a painful stimulus associated with tissue damage (e.g., cuts, scrapes, bruises, and burns), stimuli in the area of the injury and the surrounding region that would ordinarily be perceived as slightly painful are perceived as significantly more so, a phenomenon referred to as **hyperalgesia**. A good example of hyperalgesia is the increased sensitivity to temperature that occurs after sunburn. This effect is due to changes in neuronal sensitivity that occur at the level of peripheral receptors as well as their central targets.

Peripheral sensitization results from the interaction of nociceptors with the "inflammatory soup" of substances released when tissue is damaged. These substances arise from activated nociceptors or from non-neuronal cells that reside within, or migrate to, the injured area. Nociceptors release peptides and neurotransmitters such as substance P, calcitonin gene-related peptide (CGRP), and ATP, all of which further contribute to the inflammatory response (vasodilation, swelling, and the release of histamine from mast cells). The list of non-neuronal cells that contribute to this inflammatory soup includes mast cells, platelets, basophils, macrophages, neutrophils, endothelial cells, keratinocytes, and fibroblasts. These cells are responsible for releasing extracellular protons, arachidonic acid and other lipid metabolites, bradykinin, histamine, serotonin, prostaglandins, nucleotides, nerve growth factor (NGF), and numerous cytokines, chief among them interleukin-1β (IL-1β) and tumor necrosis factor α (TNF-α). Most of these substances interact directly with receptors or ion channels of nociceptive fibers, augmenting their response (Figure 10.7). For example, the responses of the TRPV1 receptor to heat can be potentiated by direct interaction of the channel with extracellular protons or lipid metabolites. NGF and bradykinin also potentiate the activity of the TRPV1 receptors, but do so indirectly through the actions of separate cell surface receptors (TrkA and bradykinin receptors, respectively) and their associated intracellular signaling pathways. The prostaglandins are thought to contribute to peripheral sensitization by binding to G-protein-coupled receptors that increase levels of cyclic AMP within nociceptors. Prostaglandins also reduce the threshold depolarization required for generating action potentials via phosphorylation of a specific class of TTX-resistant sodium channels that are expressed in nociceptors. Cytokines can directly increase sodium channel activity, via activation of the MAP kinase signaling pathway, and can also potentiate the inflammatory response via increased production of prostaglandins, NGF, bradykinin, and extracellular protons.

The presumed purpose of the complex chemical signaling cascade arising from local damage is not only to protect the injured area (as a result of the painful perceptions produced by ordinary stimuli close to the site of damage), but also to promote healing and guard against infection by means of local effects such as increased blood flow and the migration of white blood cells to the site, and by the production of factors (e.g., resolvins) that reduce inflammation

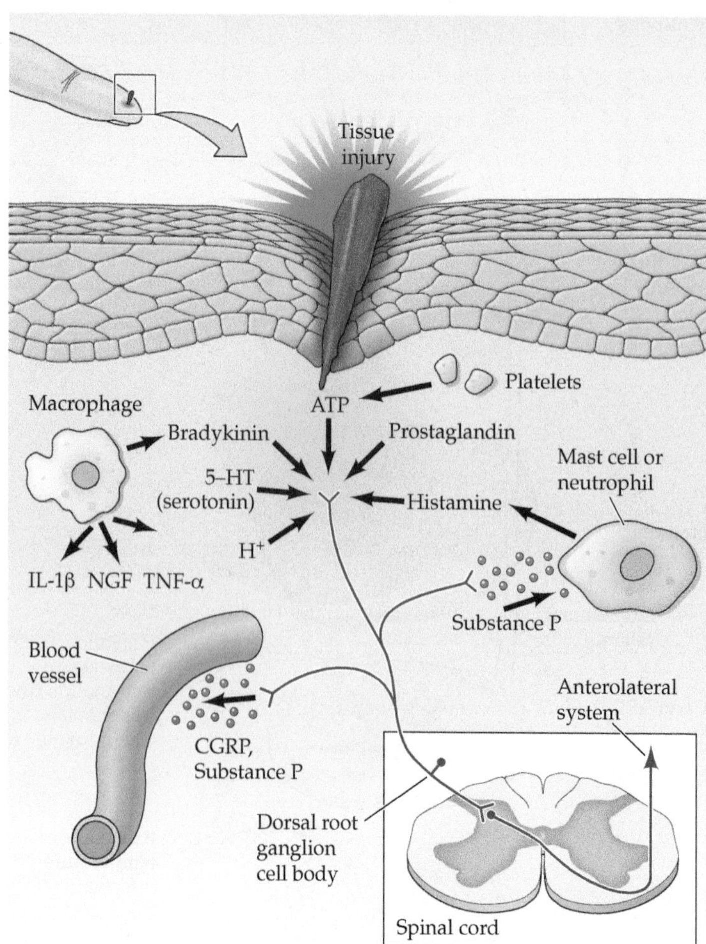

FIGURE 10.7 Inflammatory response to tissue damage. Substances released by damaged tissues augment the response of nociceptive fibers. In addition, electrical activation of nociceptors causes the release of peptides and neurotransmitters that further contribute to the inflammatory response.

and resolve pain. Indeed, identifying the components of the inflammatory soup and their mechanisms of action is a fertile area of exploration in the search for potential analgesics (compounds that reduce pain's intensity). For example, NSAIDs (*n*onsteroidal *a*nti-*i*nflammatory *d*rugs), which include aspirin and ibuprofen, act by inhibiting cyclooxygenase (COX), an enzyme important in the biosynthesis of prostaglandins. Interfering with neurotrophin or cytokine signaling has become a major strategy for controlling inflammatory disease and the resulting pain. Blocking the action of TNF-α with a neutralizing antibody has been significantly effective in the treatment of autoimmune diseases, including rheumatoid arthritis and Crohn's disease, leading to dramatic reduction in both tissue destruction and the accompanying hyperalgesia. Likewise, anti-NGF antibodies have been shown to prevent and to reverse the behavioral signs of hyperalgesia in animal models.

Central sensitization refers to a rapid onset, activity-dependent increase in the excitability of neurons in the dorsal horn of the spinal cord following high levels of activity in the nociceptive afferents. As a result, activity levels in nociceptive afferents that were subthreshold prior to the sensitizing event become sufficient to generate action potentials in dorsal horn neurons, contributing to an increase in pain sensitivity. Although central sensitization is triggered in dorsal horn neurons by activity in nociceptors, the effects generalize to other inputs that arise from low-threshold mechanoreceptors. Thus, stimuli that under normal conditions would be innocuous (such as brushing the surface of the skin) activate second-order neurons in the dorsal horn that receive nociceptive inputs, giving rise to a sensation of pain. The induction of pain by a normally innocuous stimulus is referred to as **allodynia**. This phenomenon typically occurs immediately after the painful event and can outlast the pain of the original stimulus by several hours.

As in peripheral sensitization, several different mechanisms contribute to central sensitization. One form of central sensitization, called *windup*, involves a progressive increase in the discharge rate of dorsal horn neurons in response to repeated low-frequency activation of nociceptive afferents. A behavioral correlate of the windup phenomenon has been studied by examining the perceived intensity of pain in response to multiple presentations of a noxious stimulus. Although the intensity of the stimulation is constant, the perceived intensity increases with each stimulus presentation. Windup lasts only during the period of stimulation and arises from the summation of the slow synaptic potentials evoked in dorsal horn neurons by nociceptive inputs. The sustained depolarization of the dorsal horn neurons results in part from the activation of voltage-dependent L-type calcium channels, and in part from the removal of the Mg^{2+} block of NMDA receptors. Removing the Mg^{2+} block increases the sensitivity of the dorsal horn neuron to glutamate, the neurotransmitter in nociceptive afferents.

Other forms of central sensitization that last longer than the period of sensory stimulation (e.g., allodynia) are thought to involve an LTP-like enhancement of postsynaptic potentials much like that described for the hippocampus (see Chapter 8). These effects are dependent on NMDA receptor-mediated elevations of Ca^{2+} in spinal cord neurons postsynaptic to nociceptors. Reduction in the level of GABAergic or glycinergic inhibition in spinal cord circuits is also thought to contribute to persistent pain syndromes by increasing the excitability of dorsal horn projection neurons. One mechanism affecting GABA-mediated inhibition is the dysregulation of intracellular chloride. In conditions promoting central sensitization, the function and/or expression of a potassium–chloride co-transporter (KCC2) in dorsal horn neurons may become impaired. Consequently, the concentration of intracellular chloride may become significantly elevated and the reversal potential of the GABA-A receptor channel may drift in the depolarizing direction past the threshold for generating action potentials. Thus, dorsal horn neurons postsynaptic to GABAergic interneurons may be depolarized by GABA, rather than hyperpolarized, similar to what is common in immature neurons (see Box 6B). Microglia and astrocytes also contribute to the central sensitization process, especially when there is injury to the nerve, or in other chronic pain conditions associated with arthritis, chemotherapy, and cancer. For example, pro-inflammatory cytokines such as IL-1β released from microglia promote the widespread transcription of the COX-2 enzyme and ensuing production of prostaglandins in dorsal horn neurons. As described for nociceptive afferents, increased levels of prostaglandins in CNS neurons augment neuronal excitability. Thus, the analgesic effects of drugs that inhibit COX-2 transcription are due to actions in both the periphery and within the dorsal horn. Microglia also produce TNF-α and BDNF (*b*rain-*d*erived *n*eurotrophic *f*actor), which enhance excitatory synaptic transmission and suppress inhibitory synaptic transmission in nociceptive circuitry. Furthermore, astrocytes also produce chemokines such as CCL2 and CXCL1 to enhance pain transmission in the spinal cord. Finally, while microglia are activated after injury to the nerve in males and females, drugs that inhibit microglial activation are effective mainly in males, suggesting sex-specific effects of certain drugs after nerve damage.

As injured tissue heals, the sensitization induced by peripheral and central mechanisms typically declines and the threshold for pain returns to pre-injury levels. However, when the afferent nerve fibers or central pathways themselves are damaged—a frequent complication in pathological conditions, including diabetes, shingles, AIDS, multiple sclerosis, trauma, and stroke—these processes can persist. The resulting condition is referred to as **neuropathic pain**: a chronic, intensely painful experience that is difficult to treat with conventional analgesic medications. Neuropathic pain can arise spontaneously (i.e., without any stimulus), or it can be produced by mild stimuli that are common to everyday experience, such as the gentle touch and pressure of clothing, or warm and cool temperatures. Patients often describe their experience as a constant burning sensation interrupted by episodes of shooting, stabbing, or electric shock–like jolts. Because the disability and psychological stress associated with chronic neuropathic pain can be severe, much present research is being devoted to better understanding the mechanisms of peripheral and central sensitization, as well as glial activation and neuroinflammation, with the hope of developing more effective therapies for this debilitating syndrome.

Descending Control of Pain Perception

With respect to the *interpretation* of pain, observers have long commented on the difference between the objective reality of a painful stimulus and the subjective response to it. Modern studies of this discrepancy have provided considerable insight into how circumstances affect pain perception and, ultimately, into the anatomy and pharmacology of the pain system.

During World War II, Henry Beecher and his colleagues at Harvard Medical School made a fundamental observation. In the first systematic study of its kind, they found that soldiers suffering from severe battle wounds often experienced little or no pain. Indeed, many of the wounded expressed surprise at this odd dissociation. Beecher, an anesthesiologist, concluded that the perception of pain depends on its context. For instance, the pain of an injured soldier on the battlefield would presumably be mitigated by the imagined benefits of being removed from danger, whereas a similar injury in a domestic setting would present quite a different set of circumstances that could exacerbate the pain (loss of work, financial liability, and so on). Such observations, together with the well-known placebo effect (discussed in the next section), make clear that the perception of pain is subject to central modulation (also discussed below); indeed, all sensations are subject to at least some degree of this kind of modification.

The Placebo Effect

The word *placebo* means "I will please," and the **placebo effect** is defined as a physiological response following the administration of a pharmacologically inert "remedy." The placebo effect has a long history of use (and abuse) in medicine, but its reality is undisputed. In one classic study, medical students were given one of two different pills, one said to be a sedative and the other a stimulant. In fact, both pills contained only inert ingredients. Of the students who received the "sedative," more than two-thirds reported feeling drowsy, and students who took two such pills felt sleepier than those who took only one. Conversely, a large fraction of the students who took the "stimulant" reported that they felt less tired. Moreover, about one-third of the entire group reported side effects ranging from headaches and dizziness to tingling extremities and a staggering gait. Only 3 of the 56 students in the group reported that the pills they took had no appreciable effect.

In another study of this general sort, 75% of patients suffering from postoperative wound pain reported satisfactory relief after an injection of sterile saline. The researchers who carried out this work noted that the responders were indistinguishable from the non-responders, both in the apparent severity of their pain and in their psychological makeup. Most tellingly, this placebo effect in postoperative patients could be blocked by naloxone, a competitive antagonist of opioid receptors, indicating that there is a substantial physiological basis for the pain relief experienced (see the next section). In addition, imaging studies show that the administration of a placebo with the expectation that it represents an analgesic agent is associated with activation of endogeneous opioid receptors in cortical and subcortical brain regions that are part of the pain matrix, including the anterior cingulate and insular regions of cortex and the amygdala.

A common misunderstanding about the placebo effect is the view that patients who respond to a therapeutically meaningless reagent are not suffering real pain, but only "imagining" it. This certainly is *not* the case. Among other things, the placebo effect may explain some of the efficacy of acupuncture anesthesia and the analgesia that can sometimes be achieved by hypnosis. In China, surgery is often carried out under the effect of a needle (often carrying a small electrical current) inserted at locations dictated by ancient acupuncture charts. Before the advent of modern anesthetic techniques, operations such as thyroidectomies for goiter were commonly done without extraordinary discomfort, particularly among populations where stoicism was the cultural norm. At least in rodents, acupuncture also reduces nociceptive responses by releasing adenosine at the site of needle stimulation, suggesting that this ancient treatment engages an endogenous anti-nociceptive (adenosine) mechanism in addition to a placebo effect.

The mechanisms of pain amelioration on the battlefield, in acupuncture anesthesia, and with hypnosis are presumably related. Although the mechanisms by which the brain affects the perception of pain are only beginning to be understood, the effect is neither magical nor a sign of a suggestible intellect. In short, the placebo effect is quite real.

The Physiological Basis of Pain Modulation

Understanding the central modulation of pain perception (on which the placebo effect is presumably based) was greatly advanced by the finding that electrical or pharmacological stimulation of certain regions of the midbrain produces relief of pain. This analgesic effect arises from activation of descending pain-modulating pathways that project to the dorsal horn of the spinal cord (as well as to the spinal trigeminal nucleus) and regulate the transmission of information to higher centers. One of the major brainstem regions that produce this effect is located in the periaqueductal gray matter of the midbrain. Electrical stimulation at this site in experimental animals not only produces analgesia by behavioral criteria, but also demonstrably inhibits the activity of nociceptive projection neurons in the dorsal horn of the spinal cord.

Further studies of descending pathways to the spinal cord that regulate the transmission of nociceptive information have shown that they arise from several brainstem sites, including the parabrachial nucleus, dorsal raphe, locus coeruleus, and medullary reticular formation (Figure 10.8A). The analgesic effects of stimulating the periaqueductal gray are mediated through these brainstem sites. These centers employ a wealth of different neurotransmitters (e.g., noradrenaline, serotonin, dopamine, histamine, acetylcholine) and can exert both facilitatory and inhibitory effects on the activity of neurons in the dorsal horn. The complexity of these interactions is made even greater by the fact that descending projections can exert their effects on a variety of sites within the dorsal horn, including the synaptic terminals of nociceptive afferents, excitatory and inhibitory interneurons, and the synaptic terminals of the other descending pathways, as well as by contacting the projection neurons themselves. Although these descending projections were originally viewed as a mechanism that served primarily to inhibit the transmission of nociceptive signals, it is now evident that these projections provide a balance of facilitatory and inhibitory influences that ultimately determines the efficacy of nociceptive transmission.

In addition to descending projections, local interactions between mechanoreceptive afferents and neural circuits within the dorsal horn can modulate the transmission of nociceptive information to higher centers (Figure 10.8B). These interactions are thought to explain the ability to reduce the sensation of sharp pain by activating low-threshold mechanoreceptors—for example, if you crack your shin or stub a toe, a natural (and effective) reaction is to vigorously rub the site of injury for a minute or two. Such observations, buttressed by experiments in animals, led Ronald Melzack and Patrick Wall to propose that the flow of nociceptive information through the spinal cord is modulated by concomitant activation of the large myelinated fibers associated with low-threshold mechanoreceptors. Even though further investigation led to modification of some of the original propositions in Melzack and Wall's **gate theory of pain**, the idea stimulated a great deal of work on pain modulation and has emphasized the importance of synaptic interactions within the dorsal horn for modulating the perception of pain intensity.

The most exciting advance in this longstanding effort to understand central mechanisms of pain regulation has been the discovery of **endogenous opioids**. For centuries, opium derivatives such as morphine have been known to be powerful analgesics—indeed, they remain a mainstay of analgesic therapy today. In the modern era, animal studies have shown that a variety of brain regions are susceptible to the action of opioid drugs, particularly—and significantly—the periaqueductal gray matter and other sources of descending projections. There are, in addition, opioid-sensitive neurons within the dorsal horn of the spinal cord. In other words, the areas that produce analgesia when stimulated are also responsive to exogenously administered opioids. It seems likely, then, that opioid drugs act at most or all of the sites shown in Figure 10.8 in producing their dramatic pain-relieving effects.

The analgesic action of opioids led researchers to suspect the existence of specific brain and spinal cord receptors for these drugs long before the 1960s and '70s, when such receptors were actually identified. Since these receptors are unlikely to have evolved in response to the exogenous administration of opium and its derivatives, the conviction grew that *endogenous* opioid-like compounds must exist in order to explain the evolution of opioid receptors (see Chapter 6). Several categories of endogenous opioids have been isolated from the brain and intensively studied. These agents are found in the same regions involved in the modulation of nociceptive afferents, although each of the families of endogenous opioid peptides has a somewhat different distribution. All three of the major groups—**enkephalins**, **endorphins**, and **dynorphins** (see Table 6.2)—are present in the periaqueductal gray matter. Enkephalins and dynorphins have also been found in the rostral ventral medulla and in those spinal cord regions involved in pain modulation.

One of the most compelling examples of the mechanism by which endogenous opioids modulate transmission of nociceptive information occurs at the first synapse in the pain pathway between nociceptive afferents and projection neurons in the dorsal horn of the spinal cord (see Figure 10.8C). A class of enkephalin-containing local circuit neurons within the dorsal horn synapses with the axon terminals of nociceptive afferents, which in turn synapse with dorsal horn projection neurons. The release of enkephalin onto the nociceptive terminals inhibits their release of neurotransmitter onto the projection neuron, thus reducing the level of activity that is passed on to higher centers. Enkephalin-containing local circuit neurons are themselves the targets of descending projections, providing a powerful mechanism by which higher centers can decrease the activity relayed by nociceptive afferents.

In a similar fashion, the analgesic effects of marijuana (*Cannabis*) led to the discovery of **endocannabinoids** (see Chapter 6). Exogenously administered cannabinoids

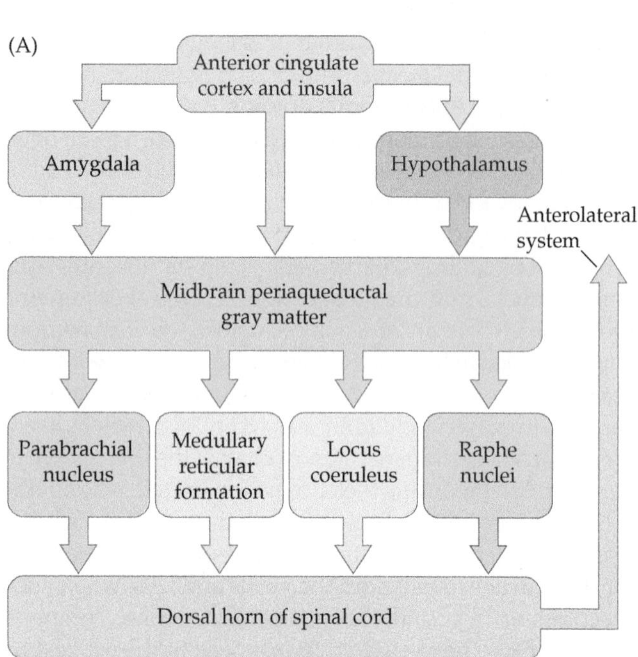

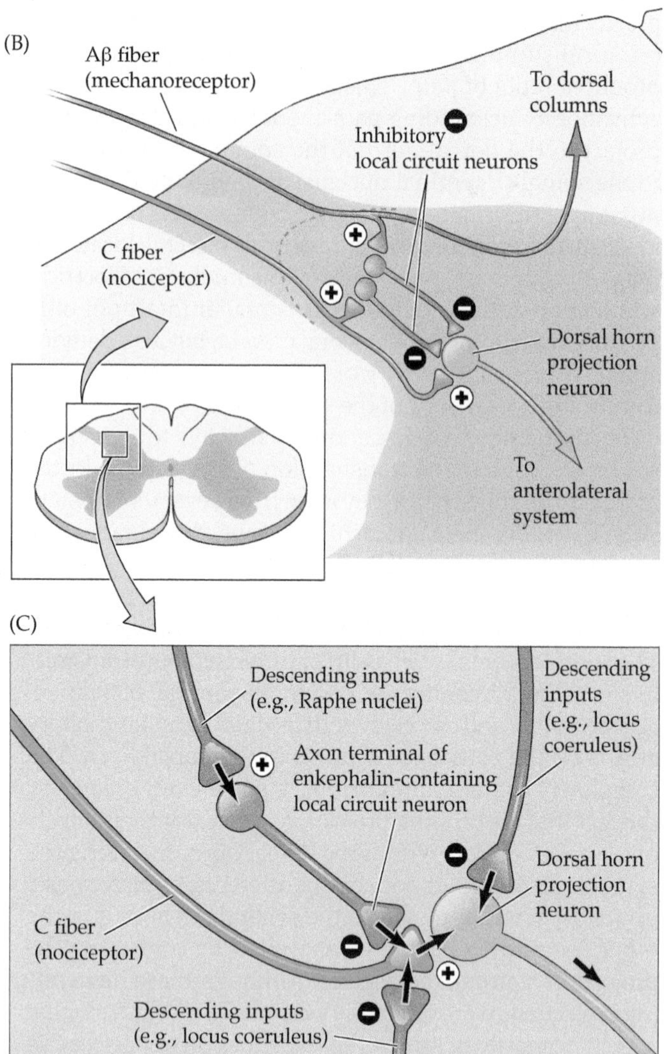

FIGURE 10.8 Descending systems modulate the transmission of ascending pain signals. (A) These modulatory systems originate in the anterior cingulate cortex and insula, the amygdala, the hypothalamus, the midbrain periaqueductal gray, the raphe nuclei, and other nuclei of the pons and rostral medulla. Complex modulatory effects occur at each of these sites, as well as in the dorsal horn. (B) Gate theory of pain. Activation of mechanoreceptors modulates the transmission of nociceptive information to higher centers. (C) Descending inputs from the brainstem modulate the transmission of pain signals in the dorsal horn. Some inputs interact directly with dorsal horn projection neurons or the presynaptic terminals of C fibers. Others interact indirectly via enkephalin-containing local circuit neurons.

are known to suppress nociceptive neurons in the dorsal horn of the spinal cord without altering the activity of non-nociceptive neurons. We now know that endogenous cannabinoids in the CNS act as neurotransmitters; they are released from depolarized neurons and travel to presynaptic terminals, where they activate cannabinoid receptors (CB_1) through a retrograde signaling mechanism. The actions of endocannabinoids are thought to *decrease* the release of neurotransmitters such as GABA and glutamate, thus modulating neuronal excitability. Evidence for a direct effect of endocannabinoids on the transmission of nociceptive signals comes from studies showing that analgesic effects induced by electrical stimulation of the periaqueductal gray can be blocked if CB_1 antagonists are administered. In addition, it appears that exposure to noxious stimuli increases the level of endocannabinoids in the periaqueductal gray matter, a finding that supports a major role for these molecules in the descending control of pain transmission. Cannabinoids also activate CB_2 receptors in microglia, resulting in reduced activation of glial cells, which may further diminish pain transmission.

The story of endogenous anti-nociceptive compounds is impressive in its wedding of physiology, pharmacology, and clinical research to yield a richer understanding of the intrinsic modulation of pain. This information has finally begun to explain the subjective variability of painful stimuli and the striking dependence of pain perception on the context of experience. Many laboratories are exploring the precise mechanisms by which pain is modulated, motivated by the tremendous clinical (and economic) benefits that would accrue from a deeper understanding of the pain system and its molecular underpinnings in the spinal cord and throughout the forebrain—wherever pain processing modifies cognition and behavior.

Summary

Whether studied from a structural or from a functional perspective, pain is an extraordinarily complex sensory modality. Because pain is an important means of warning an animal of dangerous circumstances, the mechanisms and pathways that subserve nociception are widespread and redundant. A distinct set of pain afferents with membrane receptors known as nociceptors transduces noxious stimulation and conveys this information to neurons in the dorsal horn of the spinal cord. The major central pathway responsible for transmitting the discriminative aspects of pain (location, intensity, and quality) differs from the mechanosensory pathway primarily in that the central axons of dorsal root ganglion cells synapse on second-order neurons in the dorsal horn; the axons of the second-order neurons then cross the midline in the spinal cord and ascend to thalamic nuclei that relay information to the somatosensory cortex of the postcentral gyrus. Additional pathways involving a number of centers in the brainstem, thalamus, and limbic forebrain mediate the affective and motivational responses to painful stimuli. Descending pathways interact with local circuits in the spinal cord to regulate the transmission of nociceptive signals to higher centers. Researchers have made tremendous progress in understanding pain in the last several decades, including transduction and sensitization of pain, pain-modulating neural circuits, network connectivity in chronic pain, and chronic pain modulation by glial cells and neuroinflammation. Much more progress seems likely, given the importance of the problem. Few patients are more distressed—or more difficult to treat—than those with chronic pain, a devastating by-product of the protective function of this vital sensory modality.

ADDITIONAL READING

Reviews

Basbaum, A. I., D. M. Bautista, G. Scherrer and D. Julius (2009) Cellular and molecular mechanisms of pain. *Cell* 139: 267–284.

Braz J., C. Solorzano, X. Wang and A. I. Basbaum (2014) Transmitting pain and itch messages: a contemporary view of the spinal cord circuits that generate gate control. *Neuron* 82: 522–536.

Cregg, R., A. Momin, F. Rugiero, J. N. Wood and J. Zhao (2010) Pain channelopathies. *J. Physiol.* 588: 1897–1904.

Di Marzo, V., P. M. Blumberg and A. Szallasi (2002) Endovanilloid signaling in pain. *Curr. Opin. Neurobiol.* 12: 372–379.

Fields, H. L. and A. I. Basbaum (1978) Brainstem control of spinal pain transmission neurons. *Annu. Rev. Physiol.* 40: 217–248.

Gold, M. S. and G. F. Gebhart (2010) Nociceptor sensitization in pain pathogenesis. *Nature Med.* 16: 1248–1257.

Guindon, J. and A. G. Hohmann (2009) The endocannabinoid system and pain. *CNS Neurol. Disord. Drug Targets* 8: 403–421.

Hunt, S. P. and P. W. Mantyh (2001) The molecular dynamics of pain control. *Nat. Rev. Neurosci.* 2: 83–91.

Ji, R. R., T. Kohno, K. A. Moore and C. J. Woolf (2003) Central sensitization and LTP: Do pain and memory share similar mechanisms? *Trends Neurosci.* 26: 696–705.

Millan, M. J. (2002) Descending control of pain. *Prog. Neurobiol.* 66: 355–474.

Neugebauer, V., V. Galhardo, S. Maione and S. C. Mackey (2009) Forebrain pain mechanisms. *Brain Res. Rev.* 60: 226–242.

Patapoutian, A., A. M. Peier, G. M. Story and V. Viswanath (2003) ThermoTRP channels and beyond: Mechanisms of temperature sensation. *Nat. Rev. Neurosci.* 4: 529–539.

Rainville, P. (2002) Brain mechanisms of pain affect and pain modulation. *Curr. Opin. Neurobiol.* 12: 195–204.

Scholz, J. and C. J. Woolf (2002) Can we conquer pain? *Nat. Rev. Neurosci.* 5 (Suppl): 1062–1067.

Taves, S., T. Berta, G. Chen and R. R. Ji (2013) Microglia and spinal cord synaptic plasticity in persistent pain. *Neural Plast.* 2013: 753656.

Trang, T. and 5 others (2015) Pain and poppies: The good, the bad, and the ugly of opioid analgesics. *J. Neurosci.* 35: 13879–13888.

Treede, R. D., D. R. Kenshalo, R. H. Gracely and A. K. Jones (1999) The cortical representation of pain. *Pain* 79: 105–111.

Zubieta, J.-K. and S. Christian (2009) Neurobiological mechanisms of placebo responses. *Ann. N.Y. Acad. Sci.* 1156: 198–210.

Important Original Papers

Basbaum, A. I. and H. L. Fields (1979) The origin of descending pathways in the dorsolateral funiculus of the spinal cord of the cat and rat: Further studies on the anatomy of pain modulation. *J. Comp. Neurol.* 187: 513–522.

Beecher, H. K. (1946) Pain in men wounded in battle. *Ann. Surg.* 123: 96.

Blackwell, B., S. S. Bloomfield and C. R. Buncher (1972) Demonstration to medical students of placebo response and non-drug factors. *Lancet* 1: 1279–1282.

Caterina, M. J. and 8 others (2000) Impaired nociception and pain sensation in mice lacking the capsaicin receptor. *Science* 288: 306–313.

Craig, A. D., E. M. Reiman, A. Evans and M. C. Bushnell (1996) Functional imaging of an illusion of pain. *Nature* 384: 258–260.

Hunt, S. P. and P. W. Mantyh (2001) The molecular dynamics of pain control. *Nat. Rev. Neurosci.* 2: 83–91.

LaMotte, R. H., X. Dong and M. Ringkamp (2014) Sensory neurons and circuits mediating itch. *Nat. Rev. Neurosci.* 15: 19–31.

Lavertu, G., S. L. Côté and Y. De Koninck (2014) Enhancing K–Cl co-transport restores normal spinothalamic sensory coding in a neuropathic pain model. *Brain* 137: 724–738.

Levine, J. D., H. L. Fields and A. I. Basbaum (1993) Peptides and the primary afferent nociceptor. *J. Neurosci.* 13: 2273–2286.

Sorge, R. E. and 19 others (2015) Different immune cells mediate mechanical pain hypersensitivity in male and female mice. *Nat. Neurosci.* 18: 1081–1083.

Books

Fields, H. L. (1987) *Pain*. New York: McGraw-Hill.

Fields, H. L. (ed.) (1990) *Pain Syndromes in Neurology*. London: Butterworths.

Kolb, L. C. (1954) *The Painful Phantom*. Springfield, IL: Charles C. Thomas.

Skrabanek, P. and J. McCormick (1990) *Follies and Fallacies in Medicine*. New York: Prometheus Books.

Wall, P. D. and R. Melzack (1989) *Textbook of Pain*. New York: Churchill Livingstone.

Go to the NEUROSCIENCE 6e Companion Website at **www.oup.com/uk/Purves6e** for Web Topics, Animations, Flashcards, and more.

CHAPTER

11 Vision: The Eye

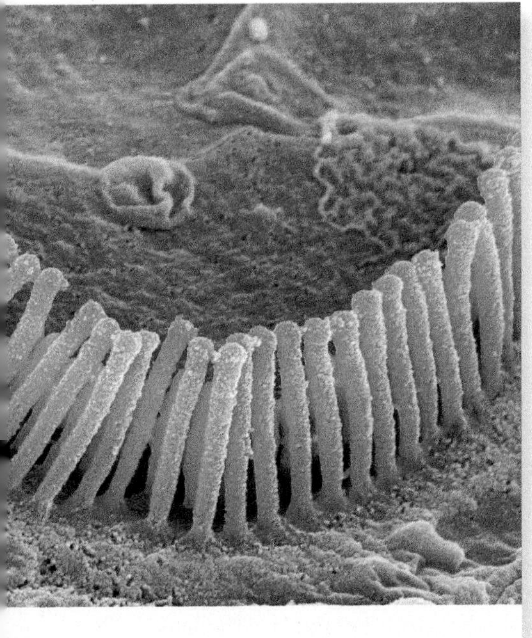

Overview

THE HUMAN VISUAL SYSTEM IS EXTRAORDINARY in the quantity and quality of information it supplies about the world. A glance is sufficient to describe the location, size, shape, color, and texture of objects and, if the objects are moving, their direction and speed. Equally remarkable is the fact that viewers can discern visual information over a wide range of stimulus intensities, from the faint light of stars at night to bright sunlight. This chapter and the next one describe the molecular, cellular, and higher order mechanisms that allow us to see. The first steps in the process of seeing involve transmission and refraction of light by the optics of the eye, the transduction of light energy into electrical signals by photoreceptors, and the refinement of these signals by synaptic interactions within the neural circuits of the retina.

Anatomy of the Eye

The eye is a fluid-filled sphere enclosed by three layers of tissue (Figure 11.1). The innermost layer of the eye, the **retina**, contains neurons that are sensitive to light and transmit visual signals to central targets. The immediately adjacent layer of tissue includes three distinct but continuous structures collectively referred to as the **uveal tract**. The largest component of the uveal tract is the choroid, which is composed of a rich capillary bed that nourishes the retinal photoreceptors; a major feature is a high concentration of the light-absorbing pigment melanin in what is called the *pigment epithelium* (discussed later in the chapter). Extending from the choroid near the front of the eye is the **ciliary body**, a ring of tissue that encircles the lens and consists of two parts: a muscular component that adjusts the refractive power of the lens, and a vascular component (the so-called ciliary processes) that produces the fluid that fills the front of the eye. The most anterior component of the uveal tract is the iris, the colored portion of the eye that can be seen through the cornea. It contains two sets of muscles with opposing actions, which allow the size of the **pupil** (the opening in its center) to be adjusted under neural control. The **sclera** forms the outermost tissue layer of the eye and is composed of a tough, white, fibrous tissue. At the front of the eye, however, this opaque outer layer is transformed into the **cornea**, a highly specialized transparent tissue that permits light rays to enter the eye.

Once beyond the cornea, light rays pass through two distinct fluid environments before striking the retina. In the **anterior chamber**, just behind the cornea and in front of the lens, is the **aqueous humor**, a clear, watery liquid that supplies nutrients to both of these structures. Aqueous humor is produced by the ciliary processes in the **posterior chamber** (the region between the lens and the iris) and flows into the anterior chamber through the pupil. The amount of fluid produced by the ciliary

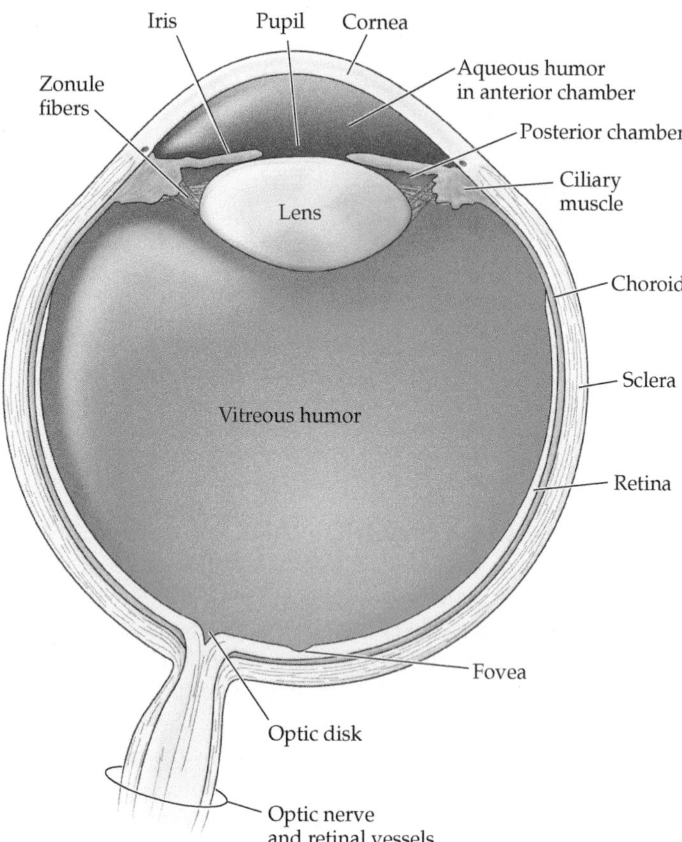

FIGURE 11.1 Anatomy of the human eye.

processes is substantial: the entire volume of fluid in the anterior chamber is replaced about 12 times a day. Thus, the rate of aqueous humor production must be balanced by a comparable rate of drainage from the anterior chamber in order to ensure a constant intraocular pressure. A meshwork of cells lying at the junction of the iris and the cornea is responsible for aqueous drainage. Failure of this drainage causes **glaucoma**, a disorder in which high levels of intraocular pressure reduce the blood supply to the eye and eventually damage retinal neurons.

The space between the back of the lens and the surface of the retina is filled with a thick, gelatinous substance called the **vitreous humor**, which accounts for about 80% of the volume of the eye. In addition to maintaining the shape of the eye, the vitreous humor contains phagocytic cells that remove blood and other debris that might otherwise interfere with light transmission. The housekeeping abilities of the vitreous humor are limited, however, as many middle-aged and elderly individuals with vitreal "floaters" will attest. Floaters are collections of debris too large for phagocytic consumption that therefore remain, casting annoying shadows on the retina. Floaters typically arise when the aging vitreous membrane pulls away from the overly long eyeball of myopic individuals (Box 11A).

Image Formation on the Retina

Normal vision requires that the optical media of the eye be transparent, and both the cornea and the lens are remarkable examples, achieving a level of transparency that rivals that found in inorganic materials such as glass. Not surprisingly, alterations in the composition of the cornea or the lens can significantly reduce their transparency and have serious consequences for visual perception. Indeed, opacities in the lens known as **cataracts** account for roughly half the cases of blindness in the world, and almost everyone over the age of 70 experiences some loss of transparency in the lens that ultimately degrades the quality of vision. Fortunately, successful surgical treatments for cataracts can restore vision in most cases.

In addition to efficiently transmitting light energy, the primary function of the optical components of the eye is to generate a focused image on the surface of the retina. The cornea and lens are primarily responsible for the refraction (bending) of light necessary for focused images on retina (Figure 11.2). The cornea contributes most of the necessary refraction, as can be appreciated by considering the hazy, out-of-focus images experienced when swimming underwater. Water, unlike air, has a refractive index close to that of the cornea; as a result, immersion in water virtually eliminates the refraction that normally occurs at the air–cornea interface; thus, the image is no longer focused on the retina. The lens has considerably less refractive power than the cornea; however, the refraction supplied by the lens is adjustable, allowing the observer to bring objects at various distances into sharp focus.

Dynamic changes in the shape of the lens are referred to as **accommodation**. When viewing distant objects, the lens is made relatively thin and flat and has the least

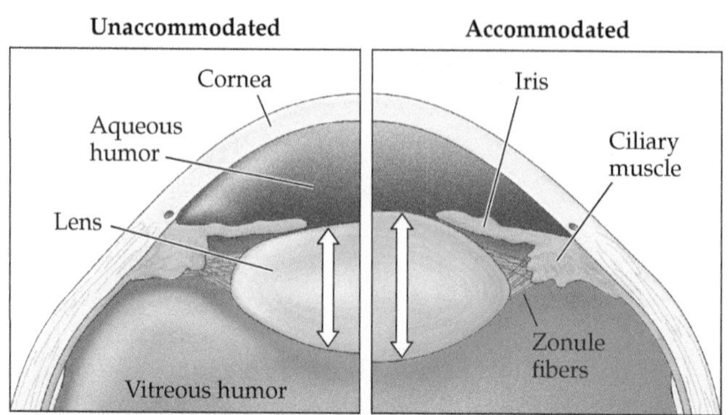

FIGURE 11.2 Accommodation in the human eye. The diagram shows the anterior part of the eye in the unaccommodated (left) and accommodated (right) states. Accommodation for focusing on near objects involves contraction of the ciliary muscle, which reduces tension in the zonule fibers and allows the elasticity of the lens to increase its curvature (white arrows).

BOX 11A ■ Myopia and Other Refractive Errors

Discrepancies of the various physical components of the eye cause a majority of the human population to have some form of refractive error, called ametropia. People who are unable to bring distant objects into clear focus are said to be nearsighted, or **myopic** (Figure B). Myopia can be caused by a corneal surface that is too curved or by an eyeball that is too long. In either case, with the lens as flat as it can be, the image of distant objects focuses in front of, rather than on, the retina.

People who are unable to focus on near objects are said to be farsighted, or **hyperopic**. An eyeball that is too short or a refracting system that is too weak can cause hyperopia (Figure C). Even with the lens in its most rounded-up state, the image is out of focus on the retinal surface (focusing at some point behind it). Both myopia and hyperopia are correctable by appropriate lenses—concave (minus) and convex (plus), respectively—or by the increasingly popular technique of corneal surgery (laser-assisted in situ keratomileusis or *Lasik* for short).

Myopia is by far the most common ametropia; it affects an estimated 50% of the United States population. Given the large number of people who need glasses, contact lenses, or surgery to correct this refractive error, one naturally wonders how nearsighted people coped before spectacles were invented only a few centuries ago. From what is now known about myopia, most people's vision may have been considerably better in ancient times. The basis for this assertion is the surprising finding that the growth of the eyeball is strongly influenced by focused light falling on the retina. This phenomenon was first described in monkeys reared with their lids sutured (the same approach used to demonstrate the effects of visual deprivation on cortical connections in the visual system; see Chapter 25), a procedure that deprives the eye of focused retinal images. Animals growing to maturity under these conditions show an elongation of the eyeball. The effect of focused light deprivation appears to be a local one, since the abnormal growth of the eye occurs in experimental animals even if the optic nerve is cut. Indeed, if only a portion of the retinal surface is deprived of focused light, then only that region of the eyeball grows abnormally. Although the mechanism of light-mediated control of eye growth is not understood, many experts believe that the modern prevalence of myopia may be due to some aspect of modern civilization—perhaps learning to read and write at an early age—that interferes with the normal feedback control of vision on eye development, leading to abnormal elongation of the eyeball.

Even people with normal (**emmetropic**) vision as young adults eventually experience difficulty focusing on near objects. One of the many consequences of aging is that the lens loses its elasticity; as a result, the maximum curvature the lens can achieve when the ciliary muscle contracts is gradually reduced. The near point (the closest point that can be brought into clear focus) thus recedes, and objects (such as this book) must be held farther and farther away from the eye in order to focus them on the retina. At some point, usually during early middle age, the accommodative ability of the eye is so reduced that near-vision tasks such as reading become difficult or impossible (Figure D). This condition is referred to as **presbyopia**. Presybyopia can be corrected by convex lenses for near-vision tasks, or by bifocal lenses if myopia (a circumstance that requires a negative correction) is also present. Replacement with an artificial lens after cataract removal also solves this problem.

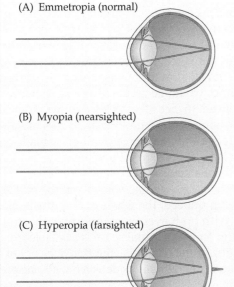

(A) Emmetropia (normal)

(B) Myopia (nearsighted)

(C) Hyperopia (farsighted)

Refractive errors. (A) In the normal eye, with ciliary muscles relaxed, an image of a distant object is focused on the retina. (B) In myopia, light rays are focused in front of the retina. (C) In hyperopia, images are focused at a point beyond the retina.

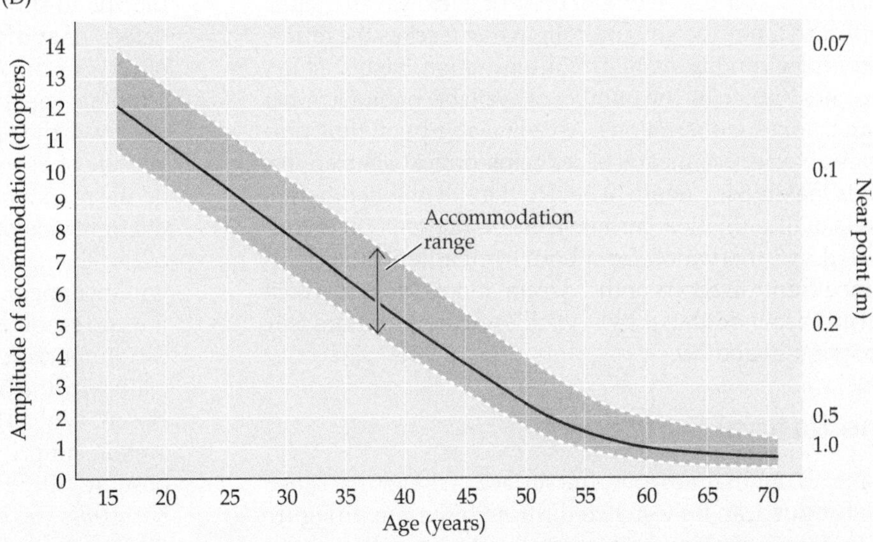

(D) Changes in the ability of the lens to round up (accommodate) with age. The graph also shows how the near point (the closest point to the eye that can be brought into focus) changes. Accommodation, which is an optical measurement of the refractive power of the lens, is given in diopters. (After Westheimer, 1974.)

refractive power. For near vision, the lens becomes thicker and rounder and has the most refractive power (see Figure 11.2). These changes arise from the tension of the **ciliary muscle** that surrounds the lens. The lens is held in place by radially arranged connective tissue bands called **zonule fibers** that are attached to the ciliary muscle. The shape of the lens is determined by two opposing forces: the elasticity of the lens, which tends to keep it rounded up (removed from the eye, the lens becomes spheroidal); and the tension exerted by the zonule fibers, which tends to flatten it. When viewing distant objects, the force exerted by the zonule fibers is greater than the elasticity of the lens, and the lens assumes the flatter shape appropriate for distance viewing. Focusing on closer objects requires relaxing the tension in the zonule fibers, allowing the inherent elasticity of the lens to increase its curvature. This relaxation is accomplished by the sphincter-like contraction of the ciliary muscle. Because the ciliary muscle forms a ring around the lens, when the muscle contracts, the attachment points of the zonule fibers move toward the central axis of the eye, thus reducing the tension on the lens.

Adjustments in the size of the pupil also contribute to the clarity of images formed on the retina. Like the images formed by other optical instruments, those generated by the eye are affected by spherical and chromatic aberrations, which tend to blur the retinal image. Since these aberrations are greatest for light rays that pass farthest from the center of the lens, narrowing the pupil reduces both spherical and chromatic aberration, just as closing the iris diaphragm on a camera lens improves the sharpness of a photographic image. Reducing the size of the pupil also increases the depth of field—that is, the distance within which objects are seen without blurring. However, a small pupil also limits the amount of light that reaches the retina, and under conditions of dim illumination, visual acuity becomes limited by the number of available photons rather than by optical aberrations. An adjustable pupil thus provides an effective means of reducing optical aberrations, while maximizing depth of field to the extent that different levels of illumination permit. The size of the pupil is controlled by innervation from both sympathetic and parasympathetic divisions of the visceral motor system, which in turn are modulated by several brainstem centers (see Chapters 20 and 21).

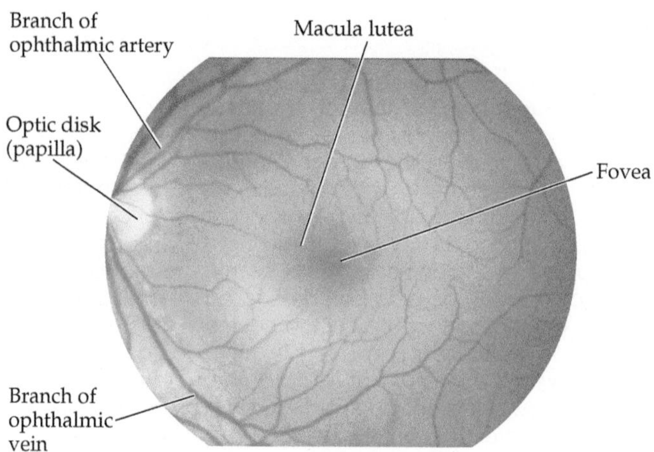

FIGURE 11.3 The inner surface of the retina, viewed with an ophthalmoscope. The optic disk is the region where the ganglion cell axons leave the retina to form the optic nerve; it is also characterized by the entrance and exit, respectively, of the ophthalmic arteries and veins that supply the retina. The macula lutea can be seen at the center of the optical axis (the optic disk lies nasally); the macula is the region of the retina that has the highest visual acuity. The fovea is a depression, or pit, about 1.5 mm in diameter that lies at the center of the macula. (Courtesy of National Eye Institute, National Insitutes of Health.)

The Retinal Surface

Using an ophthalmoscope, the surface of the retina, called the **fundus**, can be visualized through the pupil (Figure 11.3). Numerous blood vessels, both arteries and veins, fan out over the inner surface of the retina. These blood vessels arise from the ophthalmic artery and vein, which enter or leave the eye through a whitish circular area known as the **optic disk**, or **optic papilla**. The optic disk is also where retinal axons leave the eye to reach targets in the thalamus and midbrain via the optic nerve. This region of the retina contains no photoreceptors and, because it is insensitive to light, produces the perceptual phenomenon known as the "blind spot." It is logical to suppose that a visual field defect, or **scotoma**, would be seen. A scotoma, however, often goes unnoticed. In fact, all of us have a physiological scotoma, i.e., the blind spot. To find the blind spot of right eye cover the left eye and fixate on a small mark made on left side of a sheet of paper. Then, without breaking fixation on the mark, move the tip of a pencil slowly toward it from the right side of the page. At some point about the middle of the sheet, the tip (indeed, the whole end) of the pencil will disappear. With both eyes open, information about the corresponding region of visual space is available from the temporal retina of the other eye. But this does not explain why the blind spot remains undetected with one eye closed. When the world is viewed monocularly, the visual system simply "fills in" the missing part of the scene. To observe filling in, notice what happens when the pencil lies *across* the optic disk representation, at about 5–8° in diameter—remarkably, it looks complete.

In addition to being a conspicuous retinal landmark, the appearance of the optic disk provides a useful gauge of intracranial pressure. The subarachnoid space surrounding the optic nerve is continuous with that of the brain; as a result, increases in intracranial pressure—a sign of serious neurological problems such as space-occupying lesions or brain swelling due to trauma—can be detected as a protuberance of the optic disk.

Another prominent feature of the fundus is the **macula lutea**, a circular region containing yellow pigment (xanthophyll) roughly 3 mm in diameter that is located near the center of the retina. The macula is the region of the retina that supports high visual acuity (the ability to resolve fine details). Acuity is greatest at the center of the macula, a small depression or pit in the retina called the **fovea**. The pigment xanthophyll has a protective role, filtering ultraviolet wavelengths that could be harmful to the photoreceptors. Damage to this region of the retina, as occurs in age-related macular degeneration, has a devastating impact on visual perception.

Retinal Circuitry

Despite its peripheral location, the retina, which is the neural portion of the eye, is actually part of the central nervous system. During development, the retina forms as an outpocketing of the diencephalon called the optic vesicle (Figure 11.4). Consistent with its status as a full-fledged part of the central nervous system, the retina exhibits complex neural circuitry that converts the graded electrical activity of specialized photosensitive neurons—the photoreceptors—into action potentials that travel to central targets via axons in the optic nerve. Although it has the same types of functional elements and neurotransmitters found in other parts of the central nervous system, the retina comprises fewer broad classes of neurons, and these are arranged in a manner that has been less difficult to unravel than the circuits in other areas of the brain.

There are five basic classes of neurons in the retina: **photoreceptors**, **bipolar cells**, **ganglion cells**, **horizontal cells**, and **amacrine cells**. The cell bodies and processes of these neurons are stacked in alternating layers, with the cell bodies located in the inner nuclear, outer nuclear, and ganglion cell layers, and the processes and synaptic contacts located in the inner plexiform and outer plexiform layers (Figure 11.5A,B). The photoreceptors comprise two types, **rods** and **cones** (Figure 11.5B,C). Both types have an outer segment adjacent to the pigment epithelium that contains membranous disks with light-sensitive photopigment, and an inner segment that contains the cell nucleus and gives rise to synaptic terminals that contact bipolar or horizontal cells.

A three-neuron chain—photoreceptor cell to bipolar cell to ganglion cell—is the most direct path of information flow from photoreceptors to the optic nerve. Absorption of light by the photopigment in the outer segment of the photoreceptors initiates a cascade of events that changes the membrane potential of the receptor, and therefore the amount of neurotransmitter released by the photoreceptor terminals. (This process, called **phototransduction**, is discussed later in the chapter.) The synapses between photoreceptor terminals and bipolar cells (and horizontal cells) occur in the outer plexiform layer, the cell bodies of photoreceptors making up the outer nuclear layer, and the cell bodies of bipolar cells making up the inner nuclear layer. The short axonal processes of bipolar cells make synaptic contacts in turn on the dendritic processes of ganglion cells in the inner plexiform layer. The much larger axons of the ganglion cells form the **optic nerve** and carry information about retinal stimulation to the rest of the central nervous system.

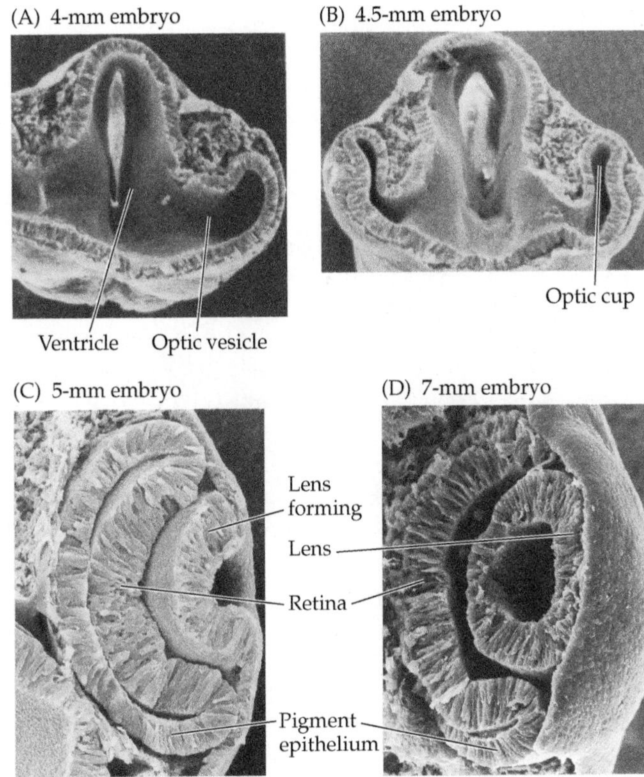

FIGURE 11.4 Development of the human eye. (A) The retina develops as an outpocketing from the neural tube, called the optic vesicle. (B) The optic vesicle invaginates to form the optic cup. (C,D) The inner wall of the optic cup becomes the neural retina, while the outer wall becomes the pigment epithelium. (A–C from Hilfer and Yang, 1980; D courtesy of K. Tosney.)

The two other types of neurons in the retina, horizontal cells and amacrine cells, have their cell bodies in the inner nuclear layer and have processes that are limited to the outer and inner plexiform layers, respectively (see Figure 11.5B). The processes of horizontal cells enable lateral interactions between photoreceptors and bipolar cells that maintain the visual system's sensitivity to contrast, over a wide range of light intensities, or **luminance**. The processes

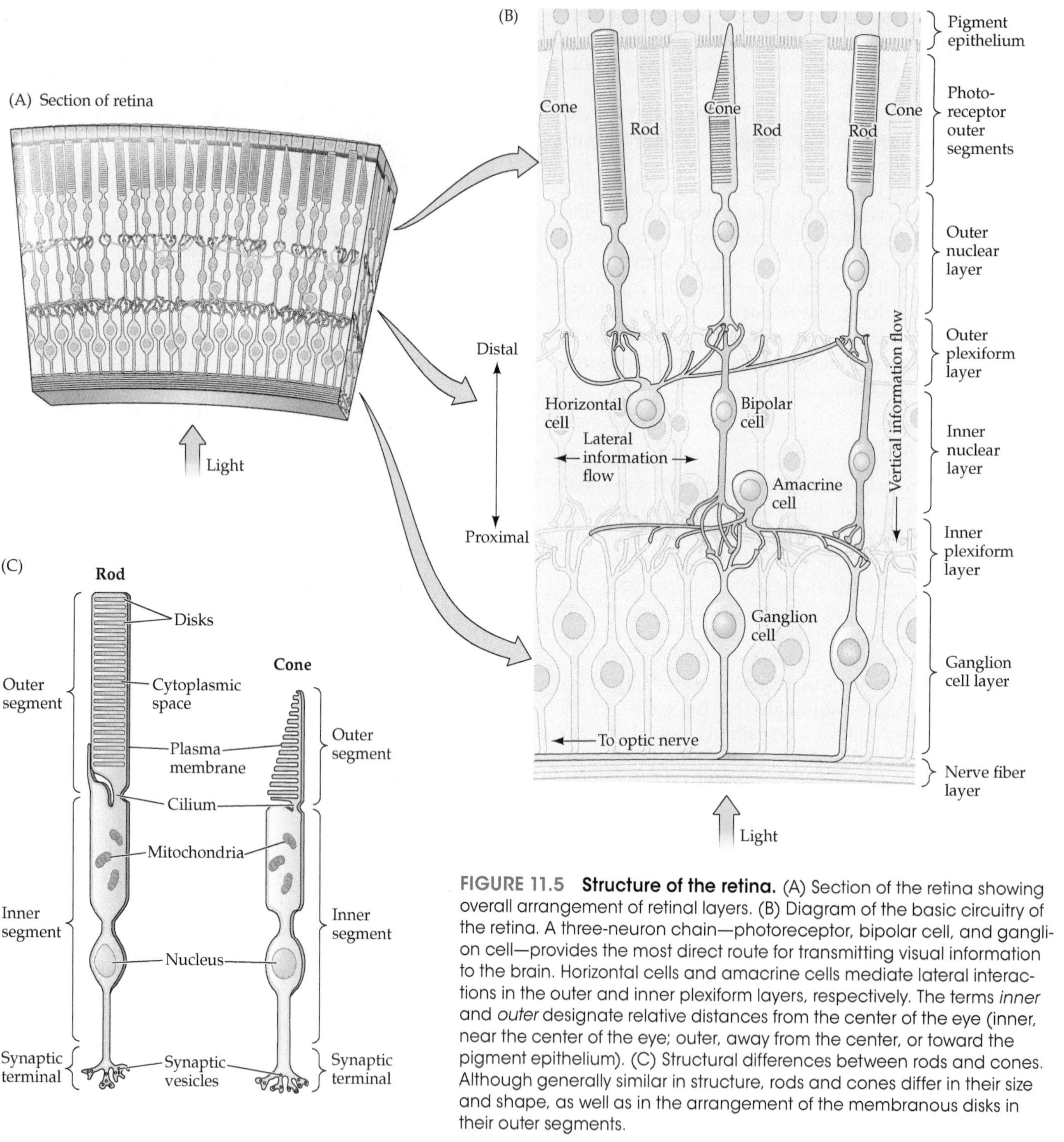

FIGURE 11.5 Structure of the retina. (A) Section of the retina showing overall arrangement of retinal layers. (B) Diagram of the basic circuitry of the retina. A three-neuron chain—photoreceptor, bipolar cell, and ganglion cell—provides the most direct route for transmitting visual information to the brain. Horizontal cells and amacrine cells mediate lateral interactions in the outer and inner plexiform layers, respectively. The terms *inner* and *outer* designate relative distances from the center of the eye (inner, near the center of the eye; outer, away from the center, or toward the pigment epithelium). (C) Structural differences between rods and cones. Although generally similar in structure, rods and cones differ in their size and shape, as well as in the arrangement of the membranous disks in their outer segments.

of amacrine cells are postsynaptic to bipolar cell terminals and presynaptic to the dendrites of ganglion cells. Different subclasses of amacrine cells are thought to make distinct contributions to visual function. For example, one amacrine cell type has an obligatory role in the pathway that transmits information from rod photoreceptors to retinal ganglion cells. Another type is critical for generating the direction-selective responses exhibited by a specialized subset of ganglion cells.

The variety of amacrine cell subtypes illustrates the more general rule that, even with only five basic classes of retinal neurons, there can be considerable diversity within a given cell class. This diversity is also a hallmark of retinal ganglion cells and the basis for pathways that convey different sorts of information to central targets in a parallel manner, a topic that we consider in more detail in Chapter 12.

The Pigment Epithelium

The spatial arrangement of retinal layers at first seems counterintuitive: light rays must pass through various non-light-sensitive elements of the retina as well as the retinal vasculature before reaching the outer segments of the photoreceptors where photons are absorbed (see Figure 11.5A,B). The reason for this curious feature of retinal organization is the special relationship that exists among the outer segments of the photoreceptors and the pigment epithelium. The cells that make up the retinal pigment epithelium have long processes that extend into the photoreceptor layer, surrounding the tips of the outer segments of each photoreceptor (Figure 11.6A).

The pigment epithelium plays two roles that are critical to the function of retinal photoreceptors. First, the membranous disks in the outer segment, which house the light-sensitive photopigment and other proteins involved in phototransduction, have a life span of only about 12 days. New outer segment disks are continuously being formed near the base of the outer segment, while the oldest disks are removed (Figure 11.6B). During their life span, disks move progressively from the base of the outer segment to the tip, where the pigment epithelium plays an essential role in removing the expended receptor disks. This shedding involves the "pinching off" of a clump of receptor disks by the outer segment membrane of the photoreceptor. This enclosed clump of disks is then phagocytosed by the pigment epithelium (Figure 11.6C). The epithelium's second role is to regenerate photopigment molecules after they have been exposed to light. Photopigment is cycled continuously between the outer segment of the photoreceptor and the pigment epithelium.

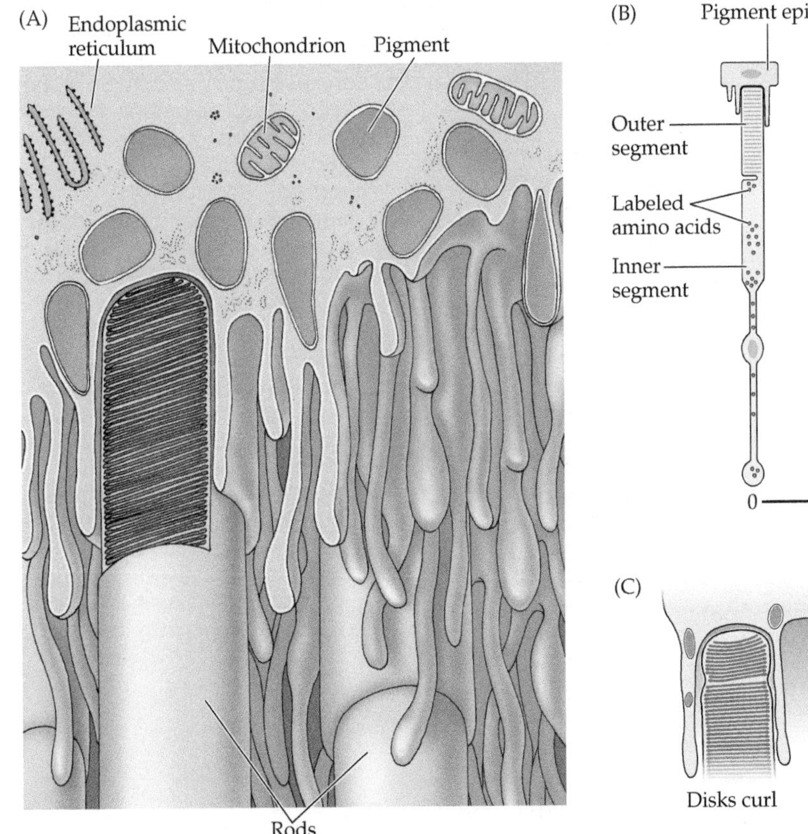

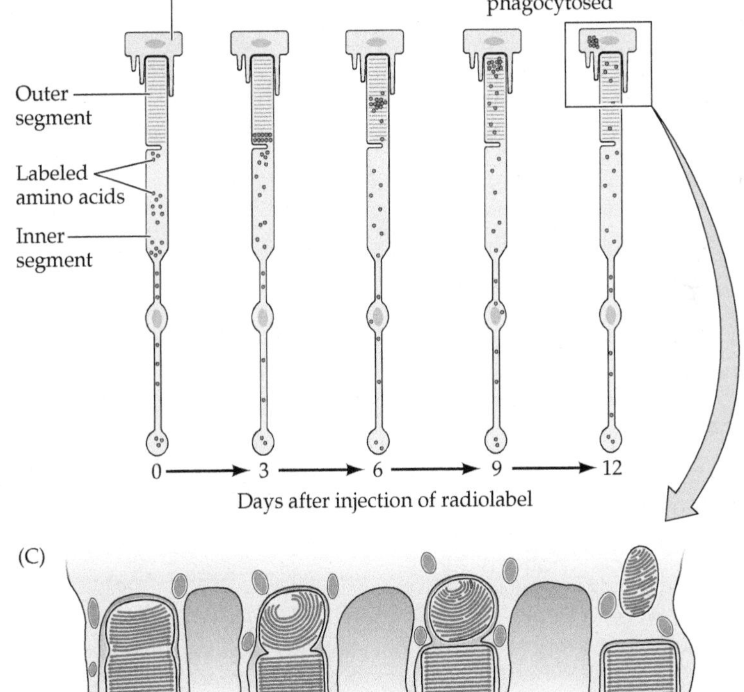

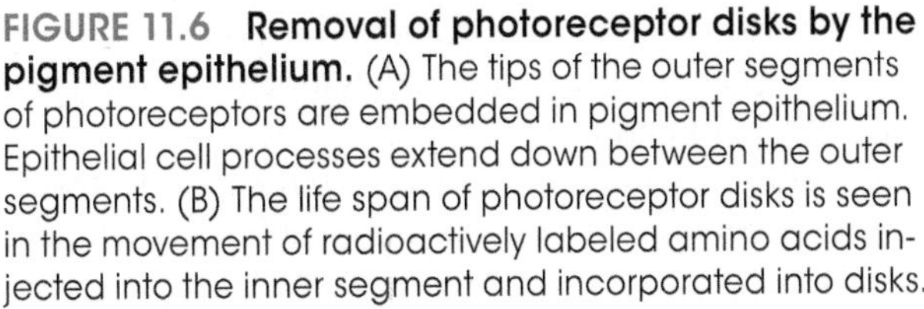

FIGURE 11.6 Removal of photoreceptor disks by the pigment epithelium. (A) The tips of the outer segments of photoreceptors are embedded in pigment epithelium. Epithelial cell processes extend down between the outer segments. (B) The life span of photoreceptor disks is seen in the movement of radioactively labeled amino acids injected into the inner segment and incorporated into disks. The labeled disks migrate from the inner to the outer portion of the outer segment over a 12-day period. (C) Expended disks are shed from the outer segment and phagocytosed. Photopigment from the disks enters the pigment epithelium, where it will be biochemically cycled back to "newborn" photoreceptor disks. (A after Oyster, 1999; B,C after Young, 1971.)

BOX 11B ■ Retinitis Pigmentosa

Retinitis pigmentosa (RP) refers to a heterogeneous group of hereditary eye disorders characterized by progressive vision loss due to a gradual degeneration of photoreceptors. An estimated 100,000 people in the United States have RP. In spite of the name, inflammation is not a prominent part of the disease process; instead, the photoreceptor cells appear to die by apoptosis (determined by the presence of DNA fragmentation).

Classification of this group of disorders under one rubric is based on the clinical features commonly observed. The hallmarks of RP are night blindness, a reduction of peripheral vision, narrowing of the retinal vessels, and the migration of pigment from disrupted retinal pigment epithelium into the retina, forming clumps of various sizes, often next to retinal blood vessels.

Typically, patients first notice difficulty seeing at night due to the loss of rod photoreceptors; the remaining cone photoreceptors then become the mainstay of visual function. Over many years, the cones also degenerate, leading to a progressive loss of vision. In most RP patients, visual field defects begin in the midperiphery, between 30° and 50° from the point of foveal fixation. The defective regions gradually enlarge, leaving islands of vision in the periphery and a constricted central field—a condition known as tunnel vision. When the visual field contracts to 20° or less and/or central vision is 20/200 or worse, the person is considered to be legally blind.

Inheritance patterns indicate that RP can be transmitted in an X-linked (XLRP), autosomal dominant (ADRP), or recessive (ARRP) manner. In the United States, the percentage of these genetic types is estimated to be 9%, 16%, and 41%, respectively. When only one member of a pedigree has RP, the case is classified as "simplex," which accounts for about a third of all cases.

Among the three genetic types of RP, ADRP is the mildest. These patients often retain good central vision until they are 60 years of age or older. In contrast, patients with the XLRP form of the disease are usually legally blind by 30 to 40 years of age. However, the severity and age of onset of the symptoms varies greatly among patients with the same type of RP, and even within the same family (when, presumably, all the affected members have the same genetic mutation).

To date, RP-inducing mutations of 30 genes have been identified. Many of these genes encode photoreceptor-specific proteins, several being associated with phototransduction in the rods. Among the latter are genes for rhodopsin, subunits of the cGMP phosphodiesterase, and the cGMP-gated channel. Multiple mutations have been found in each of these cloned genes. For example, in the case of the rhodopsin gene, 90 different mutations have been identified among ADRP patients.

The heterogeneity of RP at all levels, from genetic mutations to clinical symptoms, has important implications for understanding its pathogenesis and for designing therapies. Given the complex molecular etiology of RP, it is unlikely that a single cellular mechanism will explain all cases. Regardless of the specific mutation or causal sequence, the vision loss that is most critical is the result of gradual degeneration of cones. But in many cases, the RP-causing mutation affects proteins that are not even expressed in the cones—the prime example is the rod-specific visual pigment rhodopsin—and the loss of cones may be an indirect result of a rod-specific mutation. Identifying the cellular mechanisms that directly cause cone degeneration should lead to a better understanding of this pathology.

Characteristic appearance of the retina in patients with retinitis pigmentosa. Note the dark clumps of pigment that are the hallmark of this disorder. (Photo © 2006 C. Hamel.)

Note: The Foundation Fighting Blindness of Hunt Valley, MD, maintains a website that provides updated information about many forms of retinal degeneration (www.blindness.org). RetNet provides updated information, including references to original articles, on genes and mutations associated with retinal diseases (www.sph.uth.tmc.edu/RetNet).

These considerations, along with the fact that the capillaries in the choroid underlying the pigment epithelium are the primary source of nourishment for retinal photoreceptors, presumably explain why rods and cones are found in the outermost rather than the innermost layer of the retina. Indeed, disruptions in this normal relationship between the pigment epithelium and retinal photoreceptors have severe consequences for vision (Box 11B).

Phototransduction

In most sensory systems, activation of a receptor by the appropriate stimulus causes the cell membrane to depolarize, ultimately stimulating an action potential and transmitter release onto the neurons it contacts. In the retina, however, photoreceptors do not exhibit action potentials; rather, light activation causes a graded change in membrane potential and a corresponding change in the rate of transmitter release onto postsynaptic neurons. Indeed, much of the processing within the retina is mediated by graded potentials, largely because action potentials are not required to transmit information over the relatively short distances involved.

Perhaps even more surprising is that shining light on a photoreceptor, either a rod or a cone, leads to membrane *hyperpolarization* rather than depolarization (Figure 11.7). In the dark, the receptor is in a depolarized state, with a membrane potential of roughly –40 mV (including those portions of the cell that release transmitters). Progressive increases in the intensity of illumination cause the potential across the receptor membrane to become more negative, a response that saturates when the membrane potential reaches about –65 mV. Although the direction of the potential change may seem odd, the only logical requirement for subsequent visual processing is a consistent relationship between luminance changes and the rate of transmitter release from the photoreceptor terminals. As in other nerve cells, transmitter release from the synaptic terminals of the photoreceptor is dependent on voltage-sensitive Ca^{2+} channels in the terminal membrane. Thus, in the dark, when photoreceptors are relatively depolarized, the number of open Ca^{2+} channels in the synaptic terminal is high, and the rate of transmitter release is correspondingly great; in the light, when receptors are hyperpolarized, the number of open Ca^{2+} channels is reduced, and the rate of transmitter release is also reduced. The reason for this unusual arrangement compared with that in other sensory receptor cells is not known, but it may have to do with the challenge of responding to both increases and decreases of luminance (see below).

In the dark, cations (both Na^+ and Ca^{2+}) flow into the outer segment through membrane channels that are gated by the nucleotide cyclic guanosine monophosphate (cGMP) similar to other second-messenger systems (see Chapter 7). This inward current is opposed by an outward current that is mediated by potassium-selective channels in the inner segment. Thus, the depolarized state of the photoreceptor in the dark reflects the net contribution of Na^+ and Ca^{2+} influx, which acts to depolarize the cell, and K^+ efflux, which acts to hyperpolarize the cell (Figure 11.8A). Absorption of light by the photoreceptor reduces the concentration of cGMP in the outer segment, leading in turn to a closure of the cGMP-gated channels in the outer segment membrane and, consequently, a reduction in the inward flow of Na^+ and Ca^{2+}. As a result, positive charge (carried by K^+) flows out of the cell more rapidly than positive charge (carried by Na^+ and Ca^{2+}) flows in, and the cell becomes hyperpolarized (Figure 11.8B).

The series of biochemical changes that ultimately leads to a reduction in cGMP levels begins when a photon is absorbed by the photopigment in the receptor disks. The photopigment contains the light-absorbing chromophore **retinal** (an aldehyde of vitamin A) coupled to one of several possible proteins called **opsins**. The different opsins tune the molecule's absorption of light to a particular region of the light spectrum; indeed, it is the differing protein components of the photopigments in rods and cones that allow the functional specialization of these two receptor types.

Most of what is known about the molecular events of phototransduction has been gleaned from experiments in rods, in which the photopigment is **rhodopsin**. The seven transmembrane domains of the opsin molecule traverse the membrane of the disks in the outer segment, forming a pocket in which the retinal molecule resides (Figure 11.9A). When retinal absorbs a photon of light, one of the double bonds between the carbon atoms in the retinal molecule breaks, and its configuration changes from the 11-*cis* isomer to all-*trans* retinal (Figure 11.9B); this change triggers a series of alterations in the opsin component of the molecule. The changes in opsin lead, in turn, to the activation of an intracellular messenger called **transducin**, which activates a phosphodiesterase (PDE) that hydrolyzes cGMP. All of these events take place within the disk membrane. The hydrolysis by PDE at the disk membrane lowers cGMP concentration throughout the outer segment, thus reducing the number of cGMP molecules available to bind to the channels in the surface of the outer segment membrane and leading in turn to channel closure (Figure 11.9C).

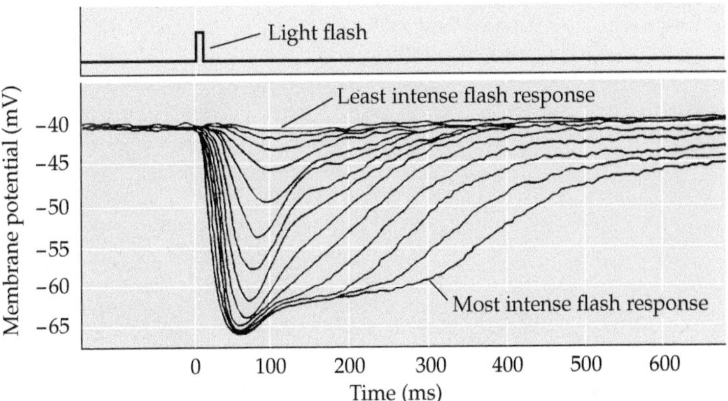

FIGURE 11.7 Hyperpolarization of a photoreceptor. This intra-cellular recording is from a single cone stimulated with different amounts of light (the cone has been taken from a turtle retina, which accounts for the relatively long time course of the response). Each trace represents the response to a brief flash that was varied in intensity. At the highest light levels, the response amplitude saturates (at about –65 mV). The hyperpolarizing response is characteristic of vertebrate photoreceptors. (After Baylor, 1987.)

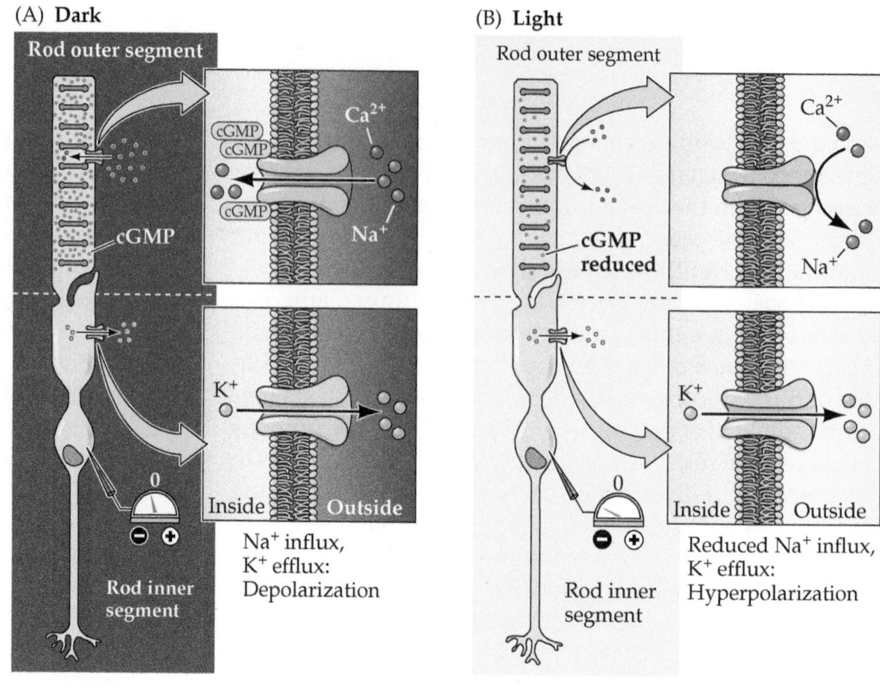

FIGURE 11.8 Cyclic GMP-gated channels and light-induced changes in the electrical activity of photoreceptors. This simplified diagram shows a rod, but the same scheme applies to cones. (A) In the dark, cGMP levels in the outer segment membrane are high; cGMP binds to the Na^+-permeable channels in the membrane, keeping them open and allowing sodium and other cations to enter, thus depolarizing the cell. (B) Absorption of photons leads to a decrease in cGMP levels, closing the cation channels and resulting in receptor hyperpolarization.

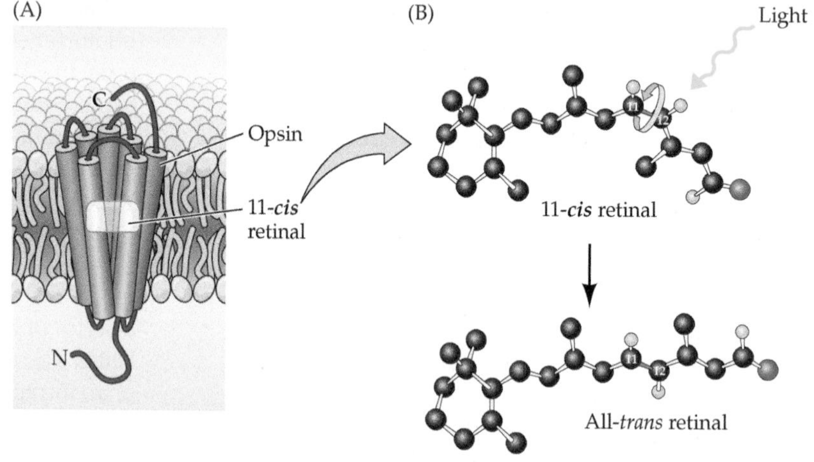

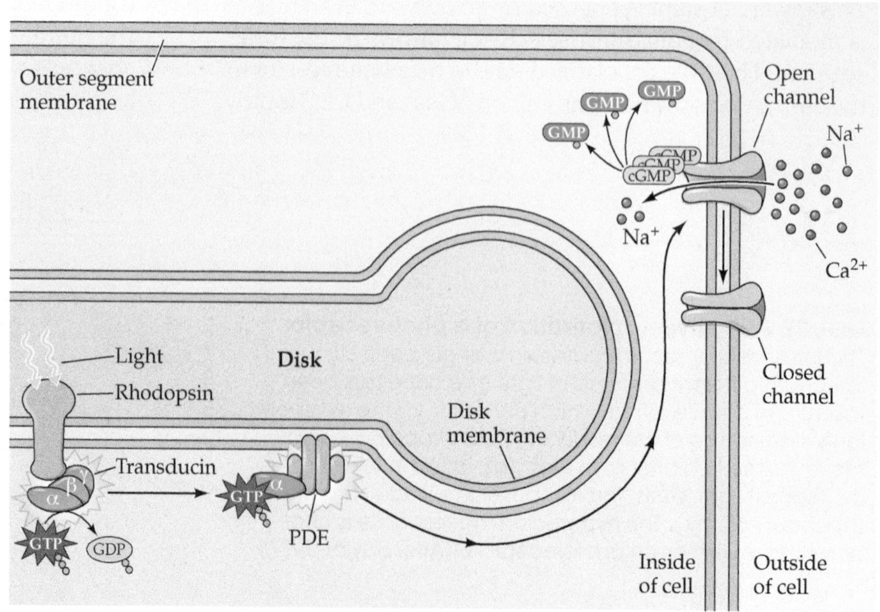

FIGURE 11.9 Phototransduction in rod photoreceptors. (A) Rhodopsin resides in the disk membrane of the photoreceptor outer segment. The seven transmembrane domains of the opsin molecule enclose the light-sensitive retinal molecule. (B) Absorption of a photon of light by retinal leads to a change in configuration from the 11-*cis* to the all-*trans* isomer. (C) The second messenger cascade of phototransduction. The change in the retinal isomer activates transducin, which in turn activates a phosphodiesterase (PDE). The PDE then hydrolyzes cGMP, reducing its concentration in the outer segment and leading to the closure of channels in the outer segment membrane. (A,B after Oyster, 1999; A after Stryer, 1986; B after Stryer, 1987.)

One of the important features of this complex biochemical cascade initiated by photon capture is that it provides enormous signal amplification. A single light-activated rhodopsin molecule can activate as many as 800 transducin molecules—roughly 8% of the transducin molecules on the disk surface. Although each transducin molecule activates only one PDE molecule, each PDE is capable of catalyzing the breakdown of as many as 6 cGMP molecules. As a result, the absorption of a single photon by a rhodopsin molecule results in the closure of approximately 200 ion channels, or about 2% of the number of channels in each rod that are open in the dark. This number of channel closures causes a net change in the membrane potential of about 1 mV.

Once initiated, additional mechanisms limit the duration of this amplifying cascade and restore the various molecules to their inactivated states. Activated rhodopsin is rapidly phosphorylated by rhodopsin kinase, which permits the protein **arrestin** to bind to rhodopsin. Bound arrestin blocks the ability of activated rhodopsin to activate transducin, thus effectively truncating the phototransduction cascade.

The restoration of retinal to a form capable of signaling photon capture is a complex process known as the **retinoid cycle** (Figure 11.10A). The all-*trans* retinal dissociates from opsin and diffuses into the cytosol of the outer segment; there it is converted to all-*trans* retinol and transported into the pigment epithelium via a chaperone protein, **interphotoreceptor retinoid binding protein** (**IRBP**; see Figure 11.10A), where appropriate enzymes ultimately convert it to 11-*cis* retinal. After being transported back into the outer segment via IRBP, 11-*cis* retinal recombines with opsin in the receptor disks. The retinoid cycle is critically important for maintaining the light sensitivity of photoreceptors. Even under intense illumination, the rate of retinal regeneration is sufficient to maintain a significant number of active photopigment molecules.

The magnitude of the phototransduction amplification varies with the prevailing level of illumination, a phenomenon known as **light adaptation**. Photoreceptors are most sensitive to light at lower levels of illumination. As levels of illumination increase, sensitivity decreases, preventing the receptors from saturating and thereby greatly extending the range of light intensities over which they operate. The concentration of Ca^{2+} in the outer segment appears to play a key role in the light-induced modulation of photoreceptor sensitivity. The cGMP-gated channels in the outer segment are permeable to both Na^+ and Ca^{2+} (Figure 11.10B); thus, light-induced closure of these channels leads to a net decrease in the internal Ca^{2+} concentration. This decrease triggers a number of changes in the phototransduction cascade, all of which tend to reduce the sensitivity of the receptor to light. For example, the decrease in Ca^{2+} increases the activity of guanylate cyclase, the cGMP-synthesizing enzyme, leading to increased cGMP levels. The decrease in Ca^{2+} also increases the activity of rhodopsin kinase, permitting more arrestin to bind to rhodopsin. Finally, the decrease in Ca^{2+} increases the affinity of the cGMP-gated channels for cGMP, reducing the impact of the light-induced reduction of cGMP levels. The regulatory effects of Ca^{2+} on the phototransduction cascade are only one part of the mechanism that adapts retinal sensitivity to background levels of illumination; another important contribution comes from neural interactions between horizontal cells and photoreceptor terminals, as discussed later in the chapter.

Functional Specialization of the Rod and Cone Systems

The two types of photoreceptors, rods and cones, are distinguished by their shape (from which they derive their names), the type of photopigment they contain, their distribution across the retina, and their pattern of synaptic connections. These properties reflect the fact that the rod and cone systems (i.e., the receptor cells and their connections within the retina) are specialized for different aspects of vision. The rod system has very low spatial resolution but is extremely sensitive to light; it is therefore specialized for sensitivity at the expense of seeing detail. Conversely, the cone system has very high spatial resolution but is relatively insensitive to light; it is specialized for acuity at the expense of sensitivity. The properties of the cone system also allow humans and many other animals to see color.

Figure 11.11 shows the range of illumination over which the rods and cones operate. At the lowest levels of illumination, only the rods are activated. Such rod-mediated perception is called **scotopic vision**. The difficulty of making fine visual discriminations under very low light conditions where only the rod system is active is a common experience. The problem is primarily the poor resolution of the rod system (and to a lesser extent, the fact that there is no perception of color because in dim light there is no significant involvement of the cones). Although cones begin to contribute to visual perception at about the level of starlight, spatial discrimination at this light level is still very poor.

As illumination increases, cones become more and more dominant in determining what is seen, and they are the major determinant of perception under conditions such as normal indoor lighting or sunlight. The contributions of rods to vision drops out nearly entirely in **photopic vision** because their response to light saturates—that is, the membrane potential of individual rods no longer varies as a function of illumination because all of the membrane channels are closed (see Figure 11.9). **Mesopic vision** occurs in levels of light at which both rods and cones contribute—at twilight, for example. From these considerations it

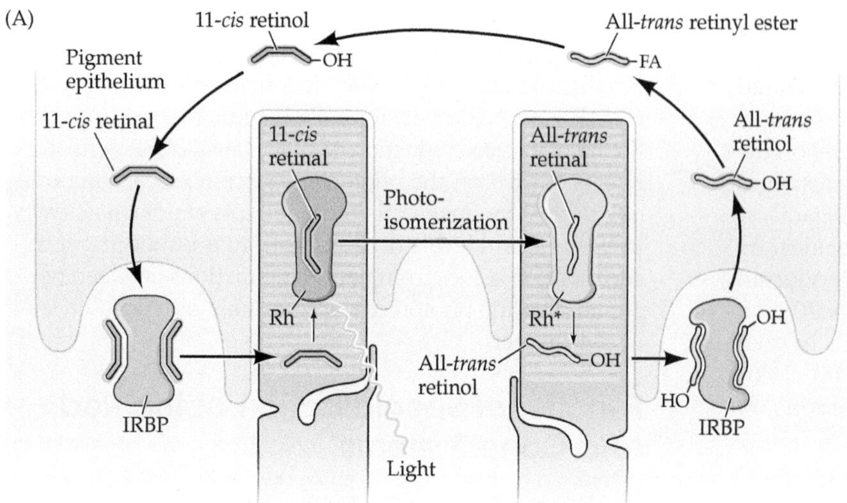

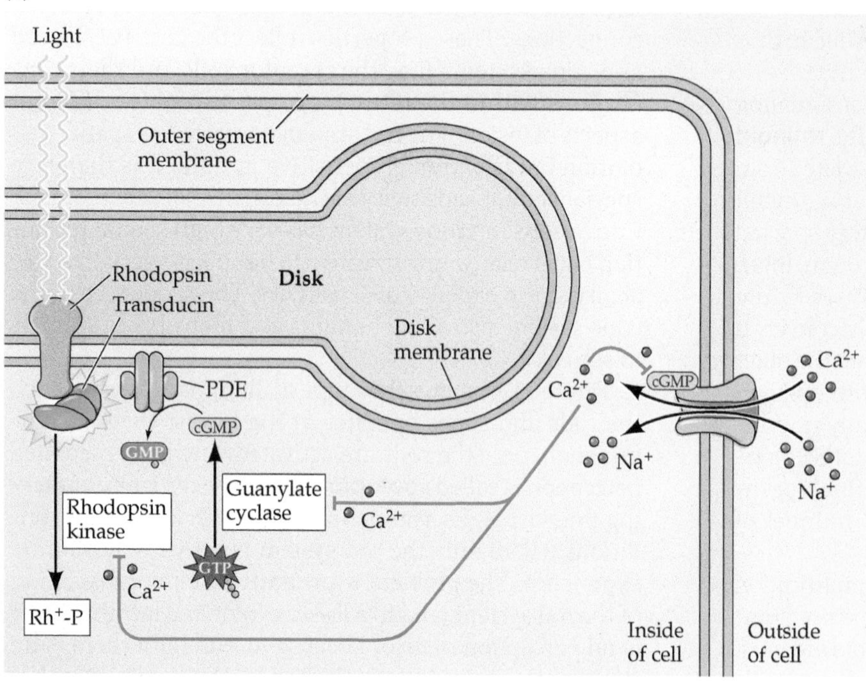

FIGURE 11.10 The retinoid cycle and photoadaptation. (A) Following photoisomerization, all-*trans* retinal is converted into all-*trans* retinol and is transported by the chaperone protein IRBP into the pigment epithelium. There, in a series of steps, it is converted to 11-*cis* retinal and transported back to the outer segment (again via IRBP), where it recombines with opsin. (B) Photoreceptor adaptation. Calcium in the outer segment inhibits the activity of guanylate cyclase and rhodopsin kinase, and reduces the affinity of cGMP-gated channels for cGMP. Light-induced closure of channels in the outer segment membrane leads to a reduction in Ca^{2+} concentration and a reduction in Ca^{2+}-mediated inhibition of these elements of the cascade. As a result, the photoreceptor's sensitivity to photon capture is reduced.

should be clear that most of what we think of as normal "seeing" is mediated by the cone system, and that loss of cone function is devastating, as occurs in individuals suffering from macular degeneration (see Box 11B). People who have lost cone function are legally blind, whereas those who have lost rod function only experience difficulty seeing at low levels of illumination (night blindness).

Differences in the transduction mechanisms utilized by the two receptor types are a major factor in the ability of rods and cones to respond to different ranges of light intensity. For example, rods produce a reliable response to a single photon of light, whereas more than 100 photons are required to produce a comparable response in a cone. It is not true, however, that cones fail to effectively capture photons. Rather, the change in current produced by single photon capture in cones is comparatively small and difficult to distinguish from background noise.

Another difference is that the response of an individual cone does not saturate at high levels of steady illumination, as the rod response does. Although both rods and cones adapt to operate over a range of luminance values, the adaptation mechanisms of the cones are more effective. This difference in adaptation is apparent in the time course of the response of rods and cones to light flashes. The response of a cone, even to a bright light flash that produces the maximum change in photoreceptor current, recovers in about 200 ms, more than four times faster than rod recovery (Figure 11.12A).

The arrangement of the circuits that transmit rod and cone information to retinal ganglion cells also contributes to the different characteristics of scotopic and photopic vision. In most parts of the retina, rod and cone signals converge on the same ganglion cells; that is, individual ganglion cells respond to both rod and cone inputs, depending on the level of illumination. The early stages of the pathways that link rods and cones to ganglion cells, however, are largely independent. For example, the pathway from rods to ganglion cells involves a distinct class of rod bipolar cells that, unlike cone bipolar cells, do not contact retinal ganglion cells. Instead, rod bipolar cells synapse with the dendritic processes

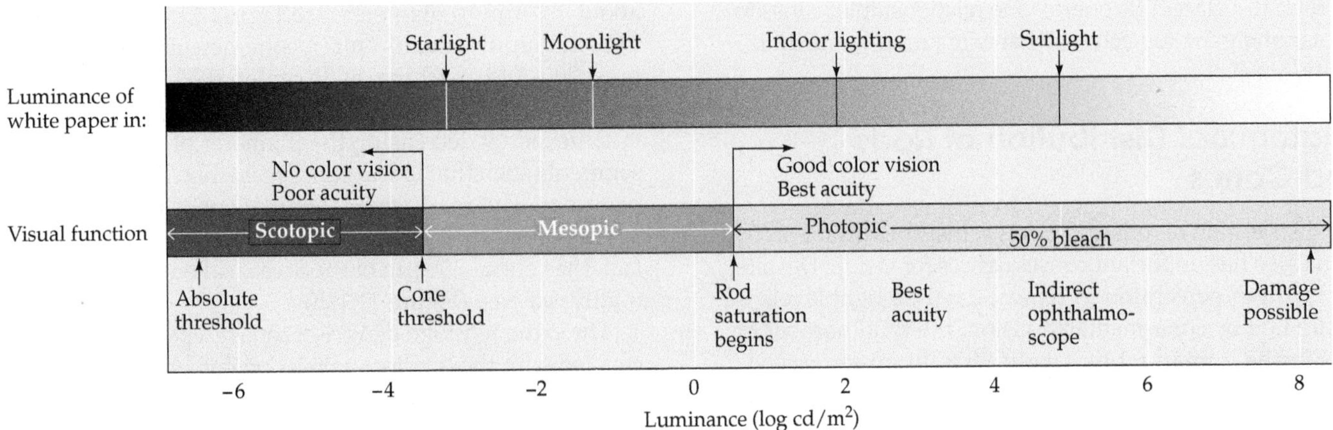

FIGURE 11.11 The range of luminance values over which the visual system operates. At the lowest levels of illumination, only rods are activated. Cones begin to contribute to perception at about the level of starlight and are the only receptors that function under relatively bright conditions.

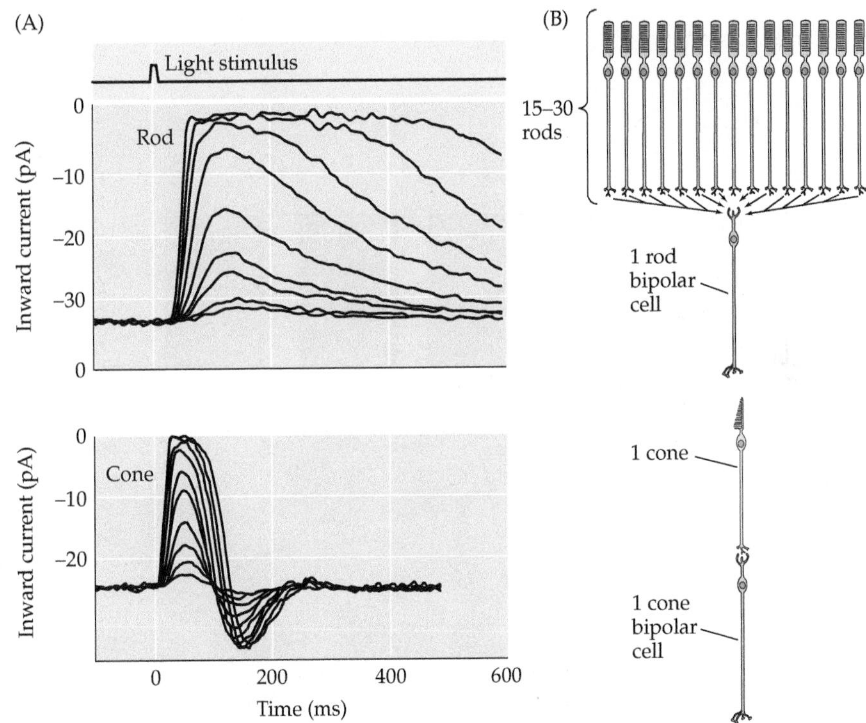

FIGURE 11.12 Differential responses of primate rods and cones. (A) Suction electrode recordings of the reduction in inward current produced by flashes of successively higher light intensity. For moderate to long flashes, the rod response continues for more than 600 ms, whereas even for the brightest flashes tested, the cone response returns to baseline (with an overshoot) in roughly 200 ms. (B) Difference in the amount of convergence in the rod and cone pathways. Each rod bipolar cell receives synapses from 15 to 30 rods. Additional convergence occurs at downstream sites in the rod pathway (see text). In contrast, in the center of the fovea, each bipolar cell receives its input from a single cone and synapses with a single ganglion cell. (A after Baylor, 1987.)

of a specific class of amacrine cells that makes gap junctions and chemical synapses with the terminals of cone bipolars; these processes, in turn, make synaptic contacts on the dendrites of ganglion cells in the inner plexiform layer. Another dramatic difference between rod and cone circuitry is their degree of convergence (Figure 11.12B). Each rod bipolar cell is contacted by a number of rods, and many rod bipolar cells contact a given amacrine cell. In contrast, the cone system is much less convergent. Thus, each of the retinal ganglion cells that dominate central vision (called midget ganglion cells) receives input from only one cone bipolar cell, which, in turn, is contacted by a single cone. Convergence makes the rod system a better detector of light, because small signals from many rods are pooled to generate a large response in the bipolar cell. At the same time, convergence reduces the spatial resolution of the rod system, since the source of a signal in a rod bipolar cell or retinal ganglion cell could have come from anywhere within a relatively large area of the retinal surface. The one-to-one relationship of cones to bipolar and ganglion cells is what is required to maximize acuity.

Anatomical Distribution of Rods and Cones

The distribution of rods and cones across the surface of the retina also has important consequences for vision. Despite the fact that perception in typical daytime light levels is dominated by cone-mediated vision, the total number of rods in the human retina (about 90 million) far exceeds the number of cones (roughly 4.5 million). As a result, the density of rods is much greater than that of cones throughout most of the retina (Figure 11.13A). However, this relationship changes dramatically in the fovea, the highly specialized region in the center of the macula that measures about 1.2 mm in diameter (see Figure 11.1). In the fovea (which literally means "pit"), cone density increases almost 200-fold, reaching, at its center, the highest receptor packing density anywhere in the retina. This high density is achieved by decreasing the diameter of the cone outer segments such that foveal cones resemble rods in their appearance. The increased density of cones in the fovea is accompanied by a sharp decline in the density of rods. In fact, the central 300 μm of the fovea, called the **foveola**, is totally rod-free (Figure 11.13B).

The extremely high density of cone receptors in the fovea, coupled with the one-to-one relationship with bipolar cells and retinal ganglion cells (see Figure 11.12), endows this component of the cone system with the capacity to mediate the highest levels of visual acuity. As cone density declines with eccentricity and the degree of convergence onto retinal ganglion cells increases, acuity is markedly reduced. Just 6

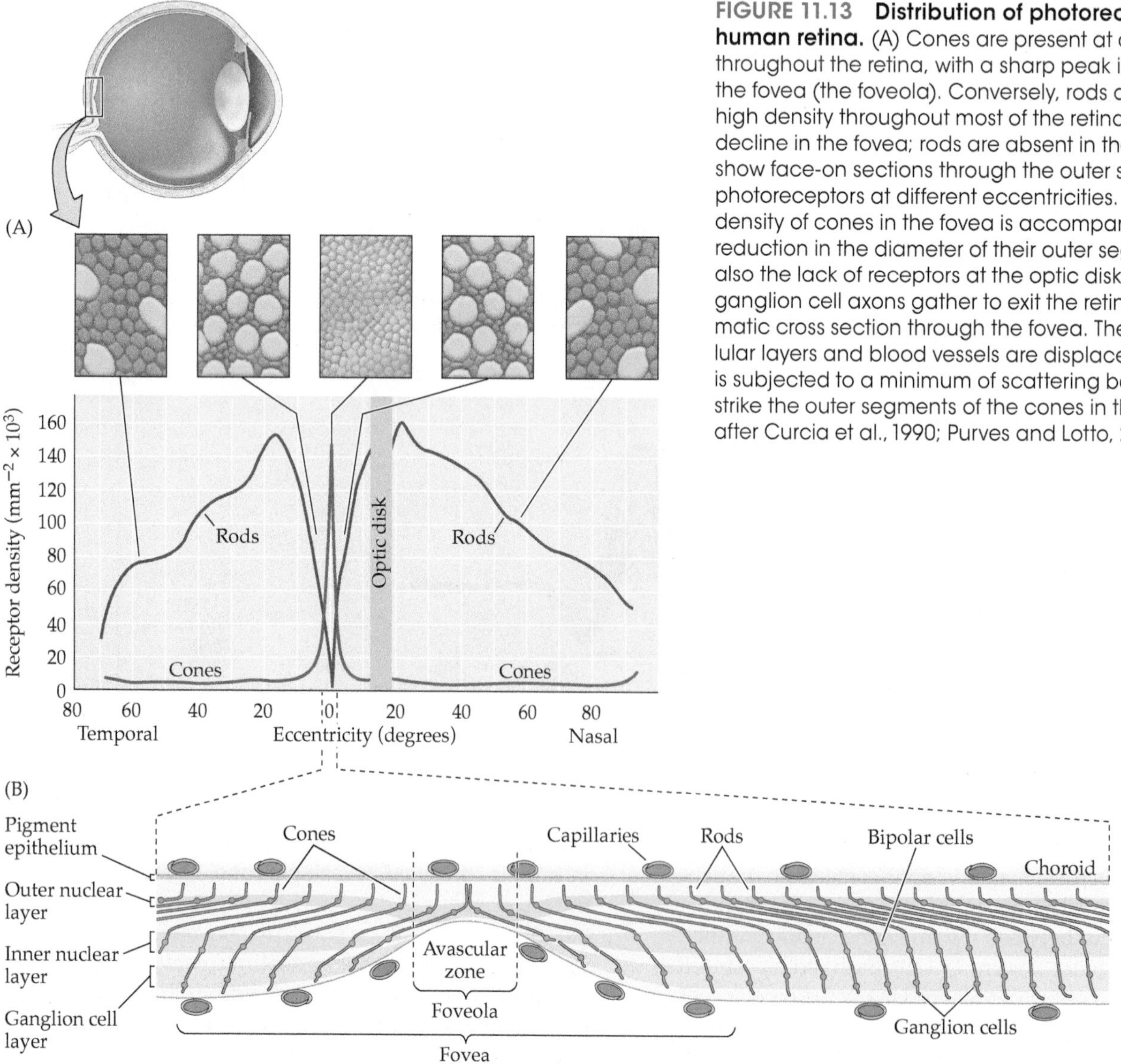

FIGURE 11.13 Distribution of photoreceptors in the human retina. (A) Cones are present at a low density throughout the retina, with a sharp peak in the center of the fovea (the foveola). Conversely, rods are present at high density throughout most of the retina, with a sharp decline in the fovea; rods are absent in the foveola. Boxes show face-on sections through the outer segments of the photoreceptors at different eccentricities. The increased density of cones in the fovea is accompanied by a striking reduction in the diameter of their outer segments. Note also the lack of receptors at the optic disk, where retinal ganglion cell axons gather to exit the retina. (B) Diagrammatic cross section through the fovea. The overlying cellular layers and blood vessels are displaced so that light is subjected to a minimum of scattering before photons strike the outer segments of the cones in the foveola. (A after Curcia et al., 1990; Purves and Lotto, 2011.)

degrees eccentric to the line of sight, acuity is reduced by 75%, a fact that can be readily appreciated by trying to read the words on any line of this page beyond the word being fixated on. The restriction of highest acuity vision to such a small region of the retina is the main reason humans spend so much time moving their eyes (and heads) around—in effect directing the foveae of the two eyes to objects of interest (see Chapter 20). It is also the reason why disorders that affect the functioning of the fovea have such devastating effects on sight (see Box 11B). Conversely, the exclusion of rods from the fovea, and their presence in high density away from the fovea, explains why the threshold for detecting a light stimulus is lower outside the region of central vision. It is easier to see a dim object (such as a faint star) by looking slightly away from it, so that the stimulus falls on the region of the retina that is richest in rods (see Figure 11.13A).

Another anatomical feature of the fovea that contributes to the superior acuity of the cone system is the displacement of the inner layers of the retina such that photons are subjected to a minimum of scattering before they strike the photoreceptors. The fovea is also devoid of another source of optical distortion that lies in the light path to the receptors elsewhere in the retina—the retinal blood vessels (see Figure 11.13B). This avascular central region of the fovea is thus dependent on the underlying choroid and pigment epithelium for oxygenation and metabolic sustenance.

Cones and Color Vision

Perceiving color allows humans and many other animals to discriminate objects on the basis of the distribution of the wavelengths of light that they reflect to the eye. While differences in luminance are often sufficient to distinguish objects, color adds another perceptual dimension that is especially useful when differences in light intensity are subtle or nonexistent. Color obviously gives us a quite different way of perceiving and describing the world we live in, and our color vision is the result of special properties of the cone system.

Unlike rods, which contain a single photopigment, the three types of cones are defined by the photopigment they contain. Each photopigment is differentially sensitive to light of different wavelengths, and for this reason cones are referred to as *blue*, *green*, and *red* or, more appropriately, as short- (S-), medium- (M-), and long- (L-) wavelength cones—terms that more or less describe their spectral sensitivities (Figure 11.14A). This nomenclature can be a bit misleading in that it seems to imply that individual cones provide color information for the wavelength of light that excites them best. In fact, individual cones, like rods, are entirely color-blind in that their response is simply a reflection of the number of photons they capture, regardless of the wavelength of the photon (or more properly, its vibrational energy). It is impossible, therefore, to determine whether the change in the membrane potential of a particular cone has arisen from exposure to many photons at wavelengths to which the receptor is relatively insensitive, or fewer photons at wavelengths to which it is most sensitive. This ambiguity can be resolved only by *comparing* the activity in different classes of cones. Based on the responses of individual ganglion cells and cells at higher levels in the visual pathway (see Chapter 12), comparisons of this type are clearly involved in how the visual system

(A)

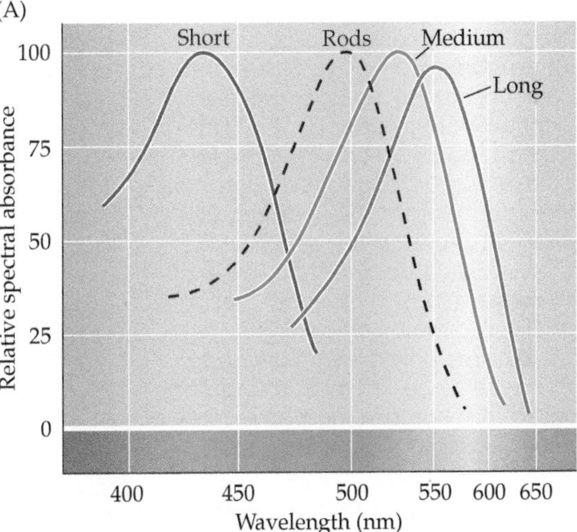

(B)

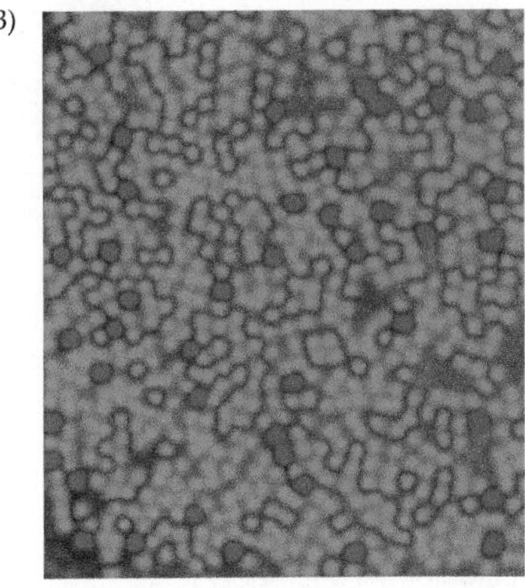

FIGURE 11.14 Absorption spectra and distribution of cone opsins. (A) Light absorption spectra of the four photopigments in normal human retina. (Recall that *light* is defined as electromagnetic radiation having wavelengths between ~400 and 700 nm. *Absorbance* is defined as the log value of the intensity of incident light divided by intensity of transmitted light.) Solid curves represent the three cone opsins; the dashed curve shows rod rhodopsin for comparison. (B) Using a technology known as adaptive optics and clever "tricks" of light adaptation, it is possible to map with great precision the distribution of different cone types within the living retina. Pseudocolor has been used to identify the short- (blue), medium- (green), and long- (red) wavelength cones. (A after Schnapf et al., 1987; B from Hofer et al., 2005.)

extracts color information from spectral stimuli, although a full understanding of the neural mechanisms that underlie color perception has been elusive (Box 11C).

While it might seem natural to assume that the three cone types are present in roughly the same number, this is clearly not the case. S cones make up only about 5% to 10% of the cones in the retina, and they are virtually absent from the center of the fovea. Although the M and L types are the predominant retinal cones, the ratio of M to L varies considerably from individual to individual, as has been

BOX 11C ■ The Importance of Context in Color Perception

Seeing color logically demands that retinal responses to different wavelengths in some way be compared. The three human cone types and their different absorption spectra are therefore correctly regarded as the basis for human color vision. Nevertheless, how these human cone types and the higher-order neurons they contact (see Chapter 12) produce the sensations of color is still unclear. Indeed, some of the greatest minds in science (Hering, Helmholtz, Maxwell, Schroedinger, and Mach, to name only a few) have debated this issue since Thomas Young first proposed that humans must have three different receptive "particles"—that is, the three cone types.

A fundamental problem has been that, although the relative activities of three cone types can more or less explain the colors perceived in color-matching experiments performed in the laboratory, the perception of color is strongly influenced by context. For example, a patch returning the exact same spectrum of wavelengths to the eye can appear quite different depending on its surround, a phenomenon called *color contrast* (Figure A). Moreover, test patches returning different spectra to the eye can appear to be the same color, an effect called *color constancy* (Figure B). Although these phenomena were well known in the nineteenth century, they were not accorded a central place in color vision theory until Edwin Land's work in the 1950s. In his most famous demonstration, Land (who, among other achievements, founded the Polaroid Corporation and became a billionaire) used a collage of colored papers that have been referred to as *the Land Mondrians* because of their similarity to the work of the Dutch artist Piet Mondrian.

Using a telemetric photometer and three adjustable illuminators generating short-, middle-, and long-wavelength light, Land showed that two patches that in white light appeared quite different in color (e.g., green and brown) continued to look their respective colors even when the three illuminators were adjusted so that the light being returned from the "green" surfaces produced exactly the same readings on the three telephotometers as had previously come from the "brown" surface—a striking demonstration of color constancy.

For Land, the answer lay in a series of ratiometric equations that could integrate the spectral returns of different regions over the entire scene. It was recognized even before Land's death in 1991, however, that his so-called retinex theory did not work in all circumstances and was in any event a description rather than an explanation. An alternative explanation of these

(A)

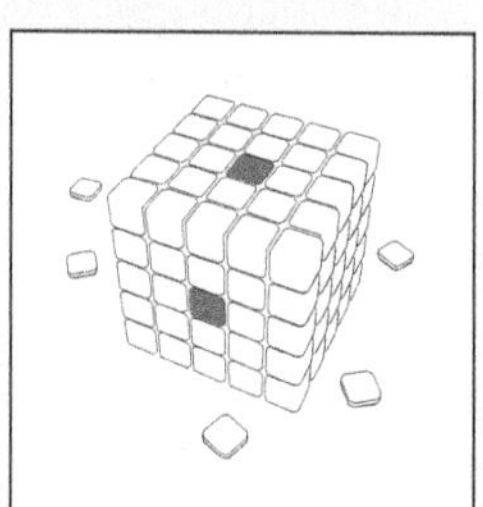

(B)

The genesis of contrast and constancy effects by exactly the same context. The two panels demonstrate the effects on apparent color when two *similarly* reflective target surfaces (A) or two *differently* reflective target surfaces (B) are presented in the *same* context in which all the information provided is consistent with illumination that differs only in intensity. The appearances of the relevant target surfaces in a neutral context are shown in the insets below. (From Purves and Lotto, 2011.)

The phenomena of color contrast and color constancy have led to a heated modern debate about how color percepts are generated that now spans several decades. For Land, the answer contextual aspects of color vision is that color, like brightness, is generated empirically according to what spectral stimuli have typically signified in past experience (see Box 11D).

shown using optical techniques that permit visualization of identified cone types in the intact human retina (Figure 11.14B). Interestingly, large differences in the ratio of M and L cone types (from nearly 4:1 to 1:1) do not appear to have much impact on color perception.

Thus, normal human color vision is fundamentally **trichromatic**, based on the relative levels of activity in three sets of cones that have different absorption spectra. The trichromatic nature of color vision is supported by perceptual studies showing that any color stimulus can be matched to a second stimulus composed of three superimposed light sources (long, medium, and short wavelengths), provided the intensity of the light sources can be independently adjusted. But about 8% of the male population in the United States (and a much smaller percentage of the female population) have a deficiency in color vision (commonly referred to as *color blindness*) that manifests as difficulty in distinguishing colors that are easily perceived by individuals with normal trichromatic vision. For some of these individuals, color vision is **dichromatic**: only two bandwidths of light are needed to match all the colors that can be perceived. The two most prevalent forms of dichromacy are *protanopia*, characterized by impairment in perception of long wavelengths, and *deuteranopia*, impairment in the perception of medium wavelengths. Although there are differences in the color discrimination capabilities of protanopes and deuteranopes, both have difficulties with the discrimination of red and green, and for this reason dichromacy is commonly called red–green color blindness (Figure 11.15). The other form of dichromacy, *tritanopia*, is extremely rare. Tritanopes have impaired perception of short wavelengths, a condition commonly called blue–yellow color blindness.

The majority of individuals with color vision deficiencies, however, are **anomalous trichromats**. In this condition, three light sources (i.e., short, medium, and long wavelengths) are needed to make all possible color matches; but the matches are made using intensity values significantly different from those used by most individuals. Some anomalous trichromats require higher intensity long-wavelength stimulation to make color matches (protanomalous trichromats), while other require higher intensity medium-wavelength (deuteranomalous trichromats) or short-wavelength (tritanomalous trichromats) stimulation. Anomalous trichromats may not be aware that they have a color vision deficiency and often pass as normal observers in everyday activities. Dichromats, however, can be severely color deficient, creating safety concerns in some contexts that can limit career opportunities. In professions where color discriminations

(A) Normal (trichromat)

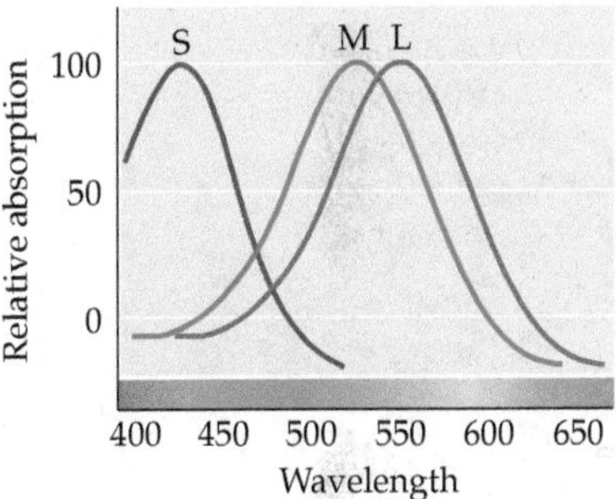

(B) Protanopia

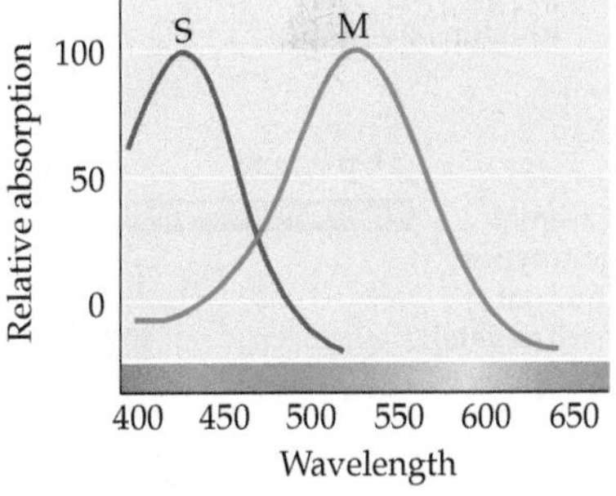

(C) Deuteranopia

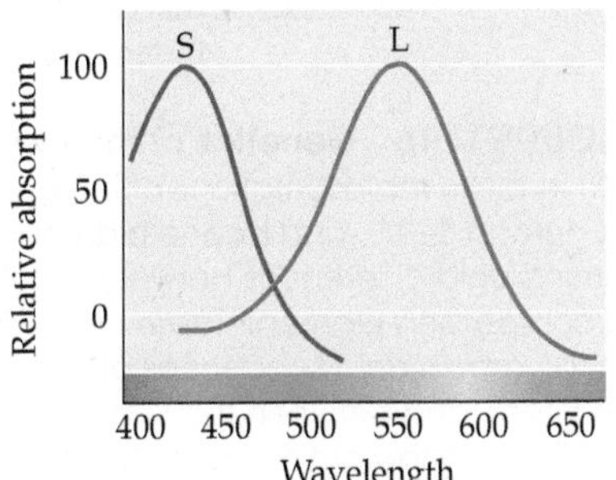

FIGURE 11.15 Abnormalities of color vision. Simulation of the image of a flower as it would appear to (A) an observer with normal color vision; (B) an observer with protanopia (loss of long-wavelength-sensitive cones); and (C) an observer with deuteranopia (loss of medium-wavelength-sensitive cones). The graphs show the corresponding absorption spectra of retinal cones in normal males and in males with defective color vision. (Photos by M. H. Siddall.)

play a critical role in performance (airline pilots, firefighters, police officers, operators of public transportation, etc.), normal color vision capabilities are often a requirement.

Color vision deficiencies result either from the inherited failure to make one or more of the cone pigments, or from an alteration in the absorption spectra of cone pigments (or rarely, from lesions in the central stations that process color information; see Chapter 12). Jeremy Nathans and his colleagues at Johns Hopkins University have provided a deeper understanding of color vision deficiencies by identifying and sequencing the genes that encode the three human cone pigments (Figure 11.16A). The genes that encode the red and green pigments show a high degree of sequence homology and lie adjacent to each other on the X chromosome, thus explaining the prevalence of red–green color deficiency in males. Normal trichromats have one gene for the red pigments and can have anywhere from one to five genes for green pigments. In contrast, the blue-sensitive pigment gene, found on chromosome 7, is quite different in its amino acid sequence. These facts suggest that the red and green pigment genes evolved relatively recently, perhaps as a result of the duplication of a single ancestral gene; they also explain why most color vision abnormalities involve the red and green cone pigments.

Human dichromats lack one of the three cone pigments, either because the corresponding gene is missing or because it exists as a hybrid of the red and green pigment genes (Figure 11.16B,C). For example, some dichromats lack the green pigment gene altogether, while others have a hybrid gene that is thought to produce a redlike pigment in the "green" cones. Anomalous trichromats also possess hybrid genes, but these genes elaborate pigments whose spectral properties lie between those of the normal red and green pigments. Thus, although most anomalous trichromats have distinct sets of medium- and long-wavelength cones, there is more overlap in their absorption spectra

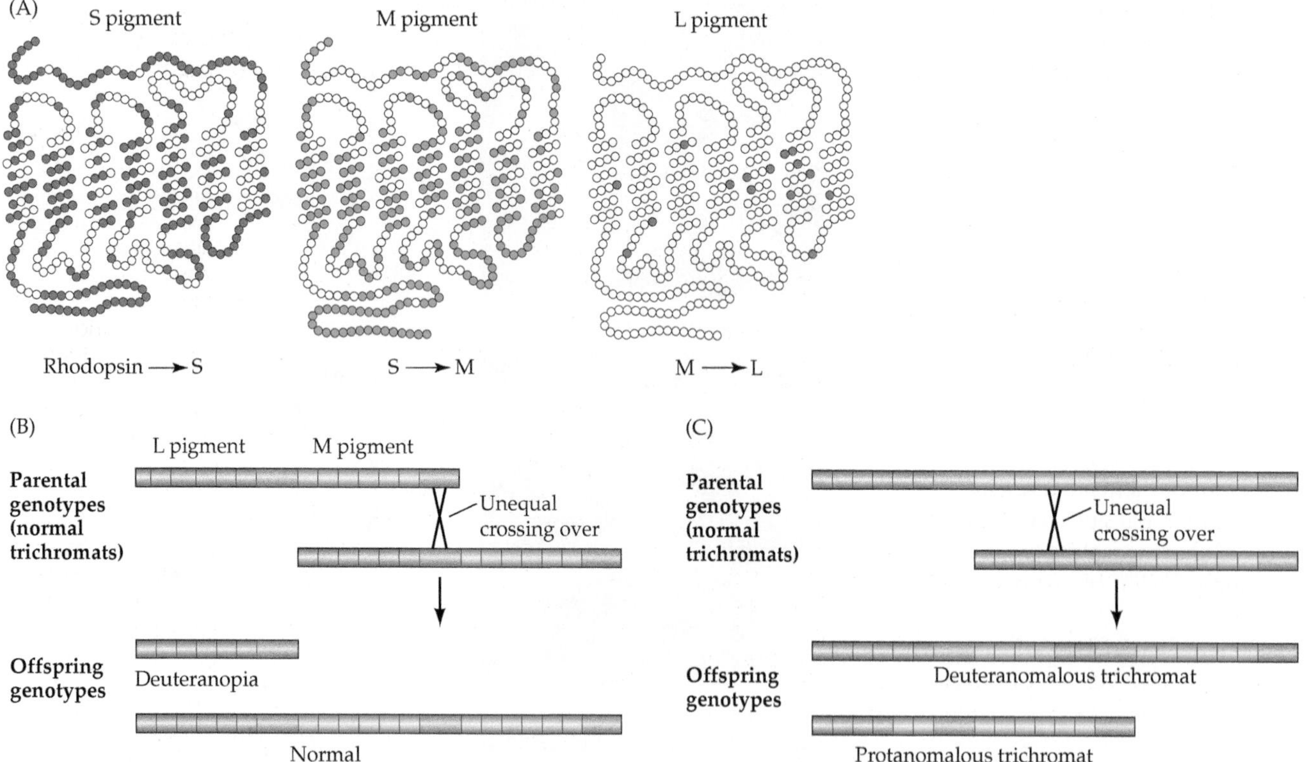

FIGURE 11.16 Genetics of the cone pigments. (A) In these representations of the amino acid sequences of human S-, M-, and L-cone pigments, colored dots identify amino acid differences between each photopigment and a comparison pigment. There are substantial differences in the amino acid sequences of rhodopsin and the S-cone pigment, and between the S- and M-cone pigments; however, only a few amino acid differences separate the M- and L-cone pigment sequences. (B,C) Many deficiencies of color vision arise from alterations in the M- or L-cone pigment genes as a result of chromosomal crossing over during meiosis. Colored squares represent the six exons of the L and M genes. (B) Unequal recombination in the intergenic region results in loss of a gene (or duplication of a gene). Loss of a gene results in dichromatic color capabilities (protanopia or deuteranopia). (C) Intragenic recombination results in hybrid genes that code for photopigments with abnormal absorption spectra, consistent with the color vision capabilities of anomalous trichromats. (A after Nathans, 1987; B,C after Deeb, 2005.)

than in normal trichromats, and thus less difference in how the two sets of cones respond to a given wavelength, with resulting anomalies in color perception.

At present there is no cure for color deficiencies. However, experiments in animals that are natural dichromats (in mammals, trichromatic vision is limited to Old World primates) have shown that it is possible to use molecular genetic techniques to express an additional pigment in a subset of cone photoreceptors, and that this manipulation is sufficient to support trichromatic color discriminations. These observations offer the hope that the most serious color deficiencies may eventually be ameliorated by gene therapy.

Retinal Circuits for Light and Dark

Despite the aesthetic pleasure inherent in having color vision, most of the information in visual scenes consists of spatial variations in light intensity; a black-and-white movie, for example, has most of the information a color version has, although it is deficient in some respects and may be less interesting to watch. The mechanisms by which central targets decipher the spatial patterns of light and dark that fall on the photoreceptors have been a vexing problem (Box 11D). To understand what the complex neural circuits within the retina accomplish during this process, it is useful to begin by considering the responses of individual retinal ganglion cells to small spots of light.

Stephen Kuffler, working at Johns Hopkins University, pioneered this approach in mammals early in the 1950s when he characterized the responses of single ganglion cells in the cat retina. He found that each ganglion cell responds to stimulation of a small circular patch of the retina, which defines the cell's receptive field (see Chapter 9 for a discussion of receptive fields). Based on these responses, Kuffler distinguished two classes of ganglion cells, ON-center and OFF-center. Turning on a spot of light in the receptive field center of an **ON-center neuron** produces a burst of action potentials. The same stimulus applied to the receptive field center of an **OFF-center neuron** reduces the rate of discharge, and when the spot of light is turned off, the cell responds with a burst of action potentials (Figure 11.17A). Complementary patterns of activity are found for each cell type when a dark spot is placed in the receptive field center (Figure 11.17B). Thus, ON-center cells increase their discharge rate to luminance *increments* in the receptive field center, whereas OFF-center

BOX 11D ■ The Perception of Light Intensity

Understanding the link between retinal stimulation and what we see (perception) is arguably the central problem in vision, and the relation of luminance (a physical measurement of light intensity) and brightness (the sensation elicited by light intensity) is probably the simplest place to consider this challenge.

As indicated in the text, how we see the brightness differences (i.e., contrast) between adjacent territories having distinct luminances depends in part on the relative firing rate of retinal ganglion cells, modified by lateral interactions. However, there is a problem with the assumption that the central nervous system simply "reads out" these relative rates of ganglion cell activity to sense brightness. The difficulty, as in perceiving color, is that the brightness of a given target is markedly affected by its context in ways that are difficult or impossible to explain in terms of the retinal output as such. The accompanying figures, illustrate two simultaneous brightness contrast illusions that make this point. In Figure A, two photometrically identical (equiluminant) gray squares appear differently bright as a function of the background in which they are presented. A conventional interpretation of this phenomenon is that the receptive field properties illustrated in Figures 11.14 through 11.17 cause ganglion cells to fire differently depending on whether the surround of the equiluminant target is dark or light. The demonstration in Figure B, however, undermines this explanation, since in this case the target surrounded by more dark area actually looks darker than the same target surrounded by more light area.

An alternative interpretation of luminance perception that can account for these puzzling phenomena is that brightness percepts are generated on the basis of experience as a means of contending with the fact that biological vision does not have the ability to measure the physical parameters of objects and conditions in the world (in this case, surface reflectance and illumination). Since to be successful an observer has to respond to the real-world sources of luminance and not to light intensity as such, this ambiguity of the retinal stimulus presents a quandary. A plausible solution to the inherent uncertainty of the relationship between luminance values and their actual sources would be to generate the sensation of brightness elicited by a given luminance (e.g., in the brightness of the identical test patches in the figure) on the basis of what the luminance of the test patches had typically turned out to be in the past experience of human observers. To get the understand of need for this sort of strategy consider Figure C, which shows that the equiluminant target patches in Figure A could have been generated by two differently painted surfaces in different illuminants, as in a comparison of the target patches on the left and middle cubes, or two similarly reflecting surfaces in similar amounts of light, as in a comparison of the target patches on the middle and right cubes. An expedient—and perhaps the only—way the visual system can cope with this profound uncertainty is to generate the perception of the stimul1 in Figure A and B empirically—that is, based on what the target patches typically turned out to signify in the past. Since the equiluminant targets will have arisen from a variety of possible

continued on next page

BOX 11D (continued)

sources, it makes sense to have the lightness elicited by the patches determined by the relative frequency of occurrence of that luminance in the particular context in which it is presented. The advantage of seeing luminance according to the relative probabilities of the possible sources of the stimulus is that percepts generated in this way give the observer the best chance of making appropriate behavioral responses to real world sources that can't be measured.

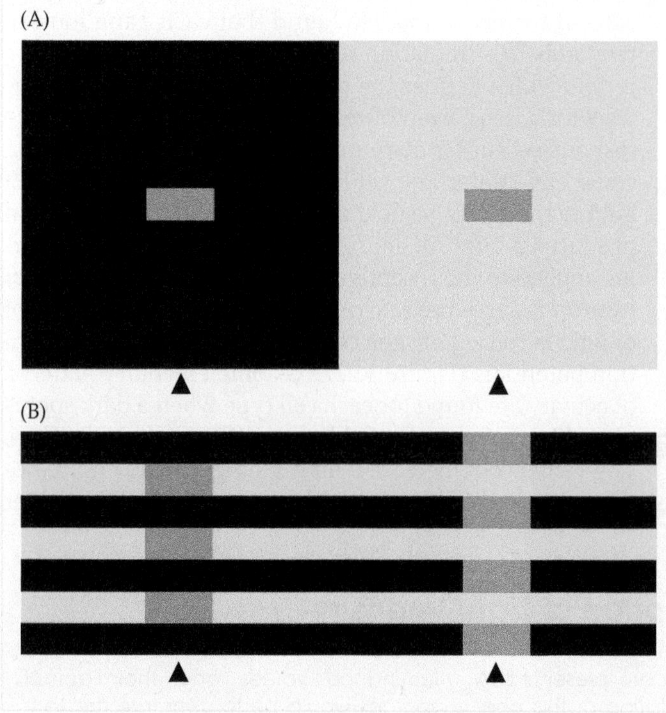

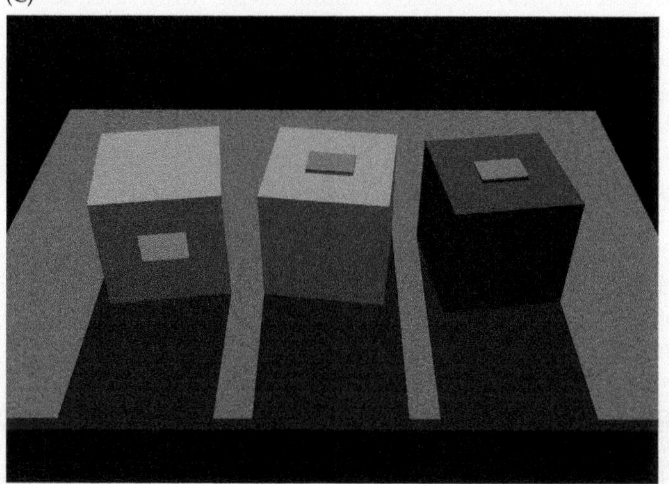

(A) Standard illusion of simultaneous brightness contrast. (B) Another illusion of simultaneous brightness contrast that is difficult to explain in conventional terms. (C) Cartoons of some possible sources of the standard simultaneous brightness contrast illusion in (A). (Courtesy of R. B. Lotto and D. Purves.)

cells increase their discharge rate to luminance *decrements* in the receptive field center.

The receptive fields of ON- and OFF-center ganglion cells have overlapping distributions in visual space, so several ON-center and several OFF-center ganglion cells analyze every point on the retinal surface (i.e., every part of visual space). Peter Schiller and his colleagues at the Massachusetts Institute of Technology, who examined the effects of pharmacologically inactivating ON-center ganglion cells on a monkey's ability to detect visual stimuli, suggested a rationale for having two distinct types of retinal ganglion cells. After silencing ON-center ganglion cells, the animals showed a deficit in their ability to detect stimuli that were brighter than the

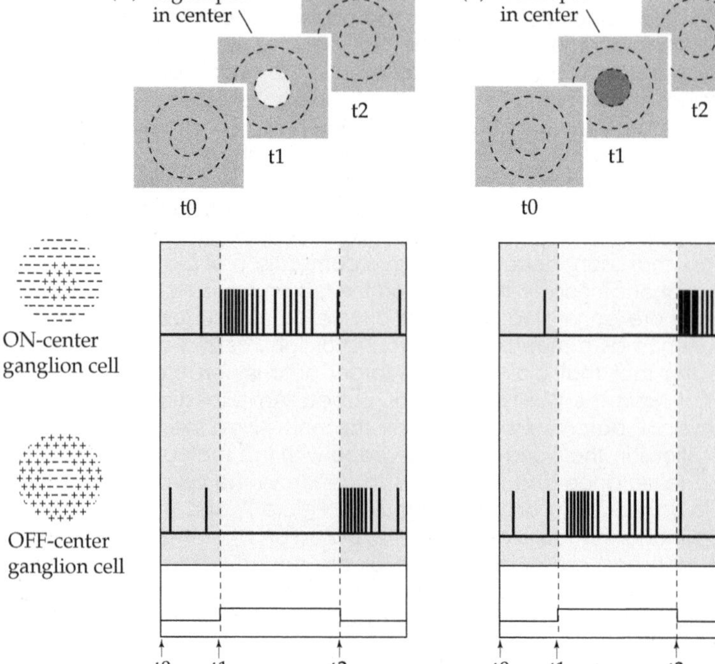

FIGURE 11.17 ON- and OFF-center retinal ganglion cell responses to stimulation of different regions of their receptive fields. Upper panels indicate the time sequence of stimulus changes; note the overlapping receptive fields. (A) Effects of light spot in the receptive field center. (B) Effects of dark spot in the receptive field center.

background; however, they could still see objects that were darker than the background.

These observations imply that information about increases or decreases in luminance is carried separately to the brain by reciprocal changes in the activity of these two types of retinal ganglion cells. Why there are separate "channels" for light and dark remains unclear, although there is increasing evidence that there are additional differences in the morphology and the response properties of ON- and OFF-center ganglion cells that are important to consider. Compared with ON-center ganglion cells, OFF-center ganglion cells are more numerous and have smaller dendritic fields, endowing them with a capacity for greater spatial resolution. Combined with evidence from statistical analysis of natural visual scenes showing that dark edges are more numerous, the presence of separate ON-center and OFF-center pathways may be another example of the remarkable power of natural selection to optimize the coding of information in sensory pathways to match the properties of the physical environment.

The functional differences between these two cell types can be understood in terms of both their anatomy and their physiological properties and relationships. ON- and OFF-center ganglion cells have dendrites that arborize in separate strata of the inner plexiform layer, forming synapses selectively with the terminals of ON- and OFF-center bipolar cells that respond to luminance increases and decreases, respectively (Figure 11.18A). As mentioned previously, the principal difference between ganglion cells and bipolar cells lies in the nature of their electrophysiological responses. Like most other cells in the retina, bipolar cells have graded potentials rather than action potentials. Graded depolarization of bipolar cells leads to an increase in transmitter release (glutamate) at their synapses and consequent depolarization of the ON-center ganglion cells that they contact via AMPA, kainate, and NMDA receptors.

The selective response of ON- and OFF-center bipolar cells to light increments and decrements is explained by the fact that they express different types of glutamate receptors (see Figure 11.18A). OFF-center bipolar cells have ionotropic receptors (AMPA and kainate) that cause the cells to depolarize in response to glutamate released from photoreceptor terminals. In contrast, ON-center bipolar cells express a G-protein-coupled metabotropic glutamate receptor (mGluR6). When bound to glutamate, these receptors activate an intracellular cascade that closes cGMP-gated Na^+ channels, reducing inward current and hyperpolarizing the cell. Thus, glutamate has opposite effects on these two classes of cells, depolarizing OFF-center bipolar cells and hyperpolarizing ON-center cells. Photoreceptor synapses with OFF-center bipolar cells are described as *sign-conserving,* since the sign of the change in membrane potential of the bipolar cell (depolarization or hyperpolarization) is the same as that in the photoreceptor. Photoreceptor synapses with ON-center bipolar cells are called *sign-inverting* because the change in the membrane potential of the bipolar cell is the opposite of that in the photoreceptor. In order to understand the response of ON- and OFF-center bipolar cells to changes in light intensity, recall that photoreceptors hyperpolarize in response to light increments, decreasing their release of neurotransmitter (Figure 11.18B). Under these conditions, ON-center bipolar cells contacted by the photoreceptors are freed from the hyperpolarizing influence of the photoreceptor's transmitter, and they depolarize. In contrast, for OFF-center cells, the reduction in glutamate represents the withdrawal of a depolarizing influence, and these cells hyperpolarize. Decrements in light intensity naturally have the opposite effect on these two classes of bipolar cells, hyperpolarizing ON-center cells and depolarizing OFF-center ones (Figure 11.18C).

The Adjustable Operating Range of Retinal Ganglion Cells

As noted, the polarity of luminance change (light versus dark) is encoded by different populations of ganglion cells, but these ganglion cells must be sensitive to these changes over the extremely broad range of luminance values that occurs during the typical day and night cycle (almost 10^9). We have already considered the specializations in the rod and cone systems and the adaptation mechanisms in the phototransduction cascade that adjust the sensitivity of the retina to these dramatically different light intensities. But even within a single visual scene there are significant differences in light intensity that must be accommodated quickly by the visual system as we make rapid and frequent eye movements to objects of interest. The problem is well illustrated by considering the experience one has with digital photography, where even when adjusting exposure for the average level of light in a scene, it is often the case that details in some regions of the photographic image are lost because the light levels exceed or fall below the camera sensor's operating range. Circuits within the retina do a remarkable job of adjusting the operating range of retinal ganglion cell responses so that they continue to supply information despite rapid changes in luminance.

Figure 11.19 shows how the response rate of an ON-center ganglion cell to a small spot of light turned on in its receptive field center varies as a function of the spot's intensity. The response rate is roughly proportional to the spot's intensity over a range of about 1 log unit. However, the intensity of spot illumination required to evoke a given discharge rate is dependent on the level of illumination in the receptive field center prior to the onset of the spot. Increases in background level of illumination are accompanied by adaptive shifts in the ganglion cell's operating range such that greater stimulus intensities are required to achieve the same discharge rate. By scaling the ganglion

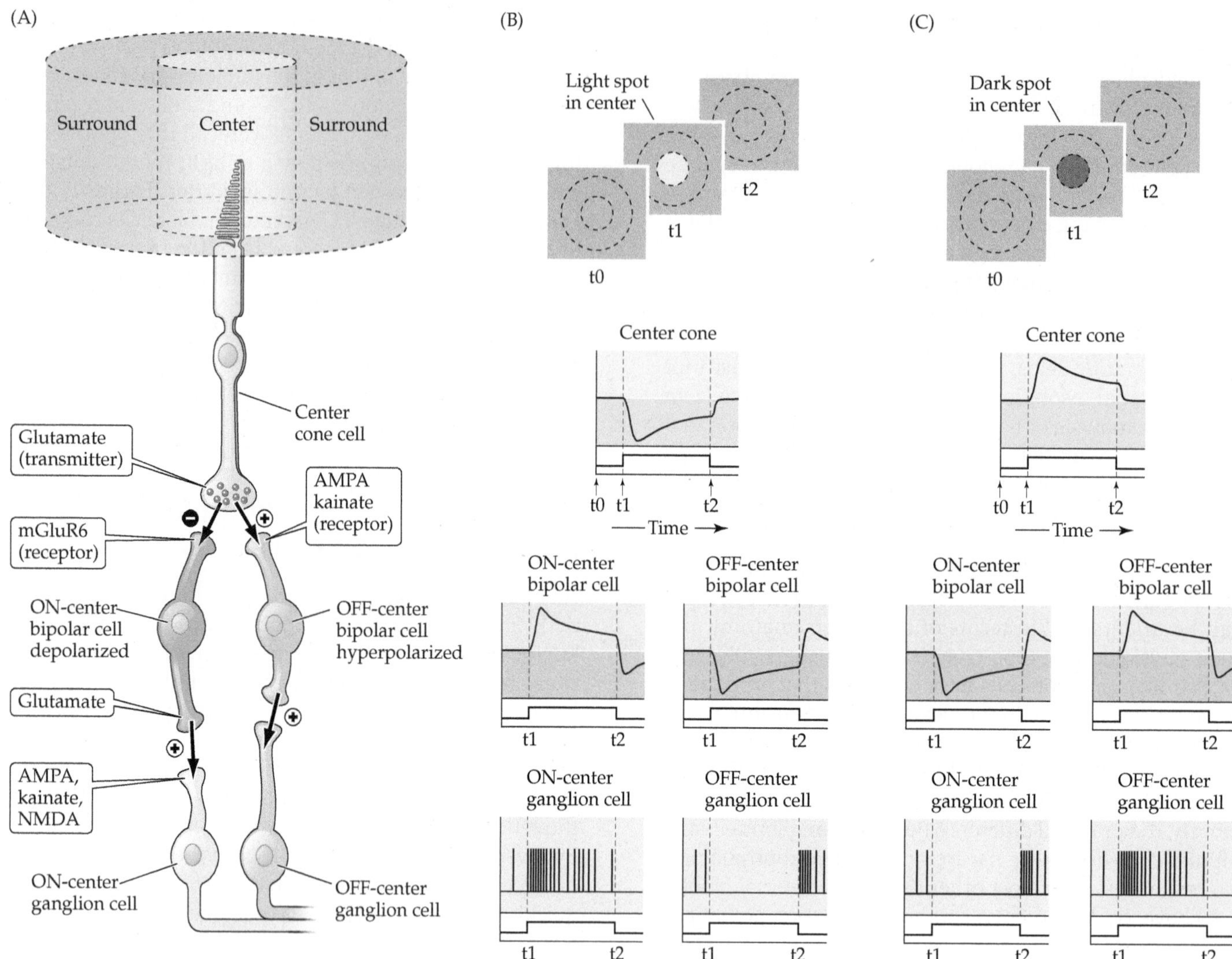

FIGURE 11.18 Circuitry responsible for generating receptive field center responses of retinal ganglion cells. (A) Functional anatomy of cone inputs to the center of a ganglion cell receptive field. A plus indicates a sign-conserving synapse; a minus represents a sign-inverting synapse. (B) Responses of the various cell types to the presentation of a light spot in the center of the ganglion cell receptive field. (C) Responses of the various cell types to the presentation of a dark spot in the center of the ganglion cell receptive field.

cell's response to prevailing levels of illumination (adjusting the gain), the entire dynamic range of a ganglion cell's firing rate can be used to encode information about intensity differences over the range of luminance values that are relevant for a given part of the visual scene. Thus, ganglion cell firing rate is not an absolute measure of light intensity, but a value that reflects the prevailing luminance conditions.

Several lines of evidence indicate that adjustments in the gain of ganglion cell response are due to changes that occur beyond the level of the photoreceptor. In particular, dynamic regulation of neurotransmitter release within the bipolar cell terminal is thought to play a major role in ganglion cell adaptation. Other factors such as synaptic inputs from amacrine cells and mechanisms intrinsic to ganglion cells' spike generation mechanism are also implicated.

Luminance Contrast and Receptive Field Surrounds

Kuffler's work also called attention to the fact that retinal ganglion cells do not act as simple photodetectors. Indeed, most ganglion cells are relatively poor at signaling differences in the level of diffuse illumination. Instead, they are sensitive to differences between the level of illumination that falls on the receptive field center and the level of illumination that falls on the surround—that is, to **luminance contrast**. The center of a ganglion cell receptive field is

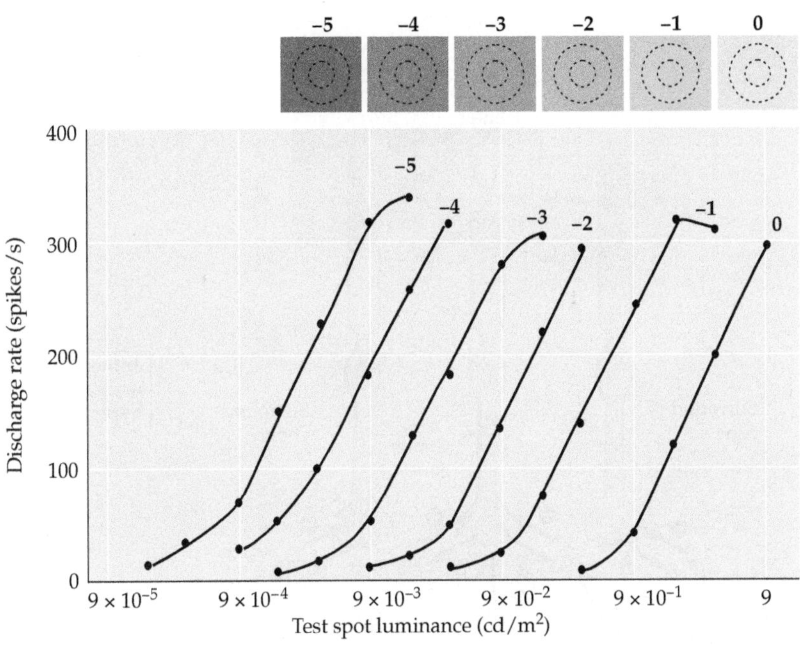

FIGURE 11.19 Adaptive changes in ganglion cell operating range. A series of curves illustrating the discharge rate of a single ON-center ganglion cell to the onset of a small test spot of light in the center of its receptive field. Each curve represents the discharge rate evoked by spots of varying intensity at a constant background level of receptive field center illumination, which is given by the numbers at the top of each curve. The highest background level is 0, the lowest –5. Examples at top of figure depict the receptive field center at different background levels of illumination. The response rate is proportional to stimulus intensity over a range of 1 log unit, but the operating range shifts to the right as the background level of illumination increases. (After Sakmann and Creutzfeldt, 1969.)

(A)

(B)

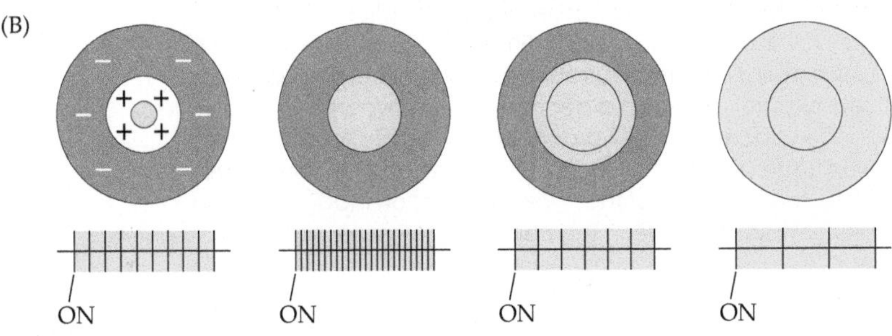

FIGURE 11.20 Responses of ON-center ganglion cells to different light conditions. (A) The rate of discharge of an ON-center cell to a spot of light as a function of the distance of the spot from the receptive field center. Zero on the x axis corresponds to the center; at a distance of 5°, the spot falls outside the receptive field. (B) Response of an ON-center ganglion cell to the increase in the size of a spot of light placed in the receptive field center. As the spot increases in size to fill the center, the response of the ganglion cell increases, but as it extends into the receptive field surround, the response of the ganglion cell decreases. (B after Hubel and Wiesel, 1961.)

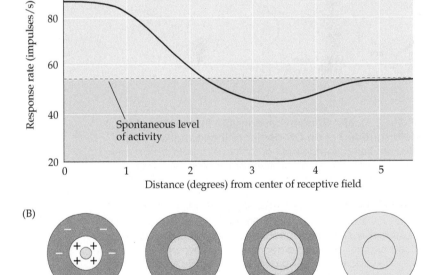

surrounded by a concentric region that, when stimulated, antagonizes the response to stimulation of the receptive field center (see Figure 11.20). For example, presentation of a small spot of light in the center of the receptive field of an ON-center ganglion cell generates a response that is enhanced as the spot size increases. But once the size of the spot exceeds the receptive field center and enters the surround, further increases in the diameter of the stimulus lead to a progressive decrease in the cell's response. OFF-center ganglion cells exhibit a similar surround antagonism with reversed polarity: the increase in response that occurs to the presentation of a dark spot that fills the center of an OFF-center ganglion cell receptive field is reduced when the dark spot extents into the receptive field surround. Because of their antagonistic surrounds, most ganglion cells respond much more vigorously to small spots

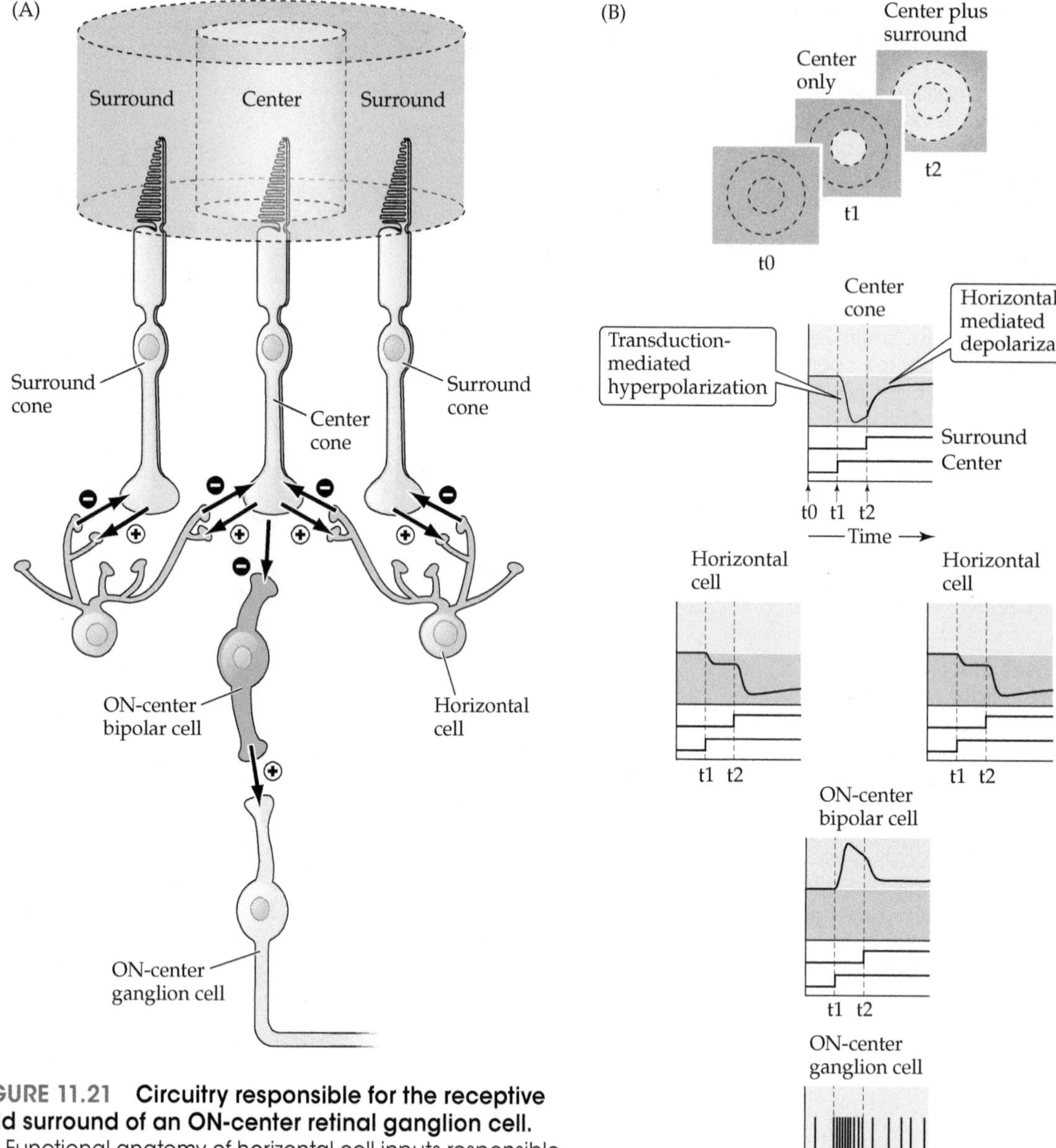

FIGURE 11.21 Circuitry responsible for the receptive field surround of an ON-center retinal ganglion cell. (A) Functional anatomy of horizontal cell inputs responsible for generating surround antagonism. A plus indicates a sign-conserving synapse; a minus represents a sign-inverting synapse. (B) Responses of various cell types to the presentation of a light spot in the center of the receptive field (t1) followed by the addition of light stimulation in the surround (t2). Light stimulation of the surround leads to hyperpolarization of the horizontal cells and a decrease in the hyperpolarizing influence of horizontal cell processes on the photoreceptor terminals. The net effect is to depolarize the center cone terminal, offsetting much of the hyperpolarization induced by the transduction cascade in the center cone's outer segment.

of light confined to their receptive field centers than to either large spots or to uniform illumination of the visual field. Thus, the information supplied by the retina to central visual stations for further processing does not give equal weight to all regions of the visual scene; rather, it emphasizes the regions where there are spatial differences in luminance: that is, object boundaries.

Like the mechanism responsible for generating the ON- and OFF-center response, the antagonistic surround of ganglion cells is thought to be a product of interactions that occur at the early stages of retinal processing. Much of the antagonism is believed to arise via lateral connections established by horizontal cells and receptor terminals (Figure 11.21). Horizontal cells receive synaptic inputs from photoreceptor terminals and are linked via gap junctions with a vast network of other horizontal cells distributed over a wide area of the retinal surface. As a result, the activity in horizontal cells reflects levels of illumination over a broad area of the retina. Although the details of their actions are not entirely clear, horizontal cells are thought to exert their influence via action on photoreceptor terminals, regulating the amount of transmitter that the photoreceptors release onto bipolar cell dendrites.

Glutamate release from photoreceptor terminals has a depolarizing effect on horizontal cells (sign-conserving synapse), while horizontal cells have a hyperpolarizing influence on photoreceptor terminals (sign-inverting synapse) (see Figure 11.21A). As a result, the net effect of inputs from the horizontal cell network is to oppose changes in the membrane potential of the photoreceptor that are induced by phototransduction events in the outer segment. Figure 11.21B illustrates how these events lead to surround suppression in an ON-center ganglion cell. A small spot of light centered on a photoreceptor supplying input to the center of the ganglion cell's receptive field produces a strong hyperpolarizing response in the photoreceptor. Under these conditions, changes in the membrane potential of the horizontal cells that synapse with the photoreceptor terminal are relatively small, and the response of the photoreceptor to light is largely determined by its phototransduction cascade. With the addition of light to the surround, however, the impact of the horizontal network becomes significantly greater; the light-induced reduction in the release of glutamate from the photoreceptors in the surround leads to a strong hyperpolarization of the horizontal cells whose processes converge on the terminal of the photoreceptor in the receptive field center. The reduction in activity of the horizontal cells has a depolarizing effect on the membrane potential of the central photoreceptor, reducing the light-evoked response, and ultimately reducing the firing rate of the ON-center ganglion cell.

Thus, even at the earliest stages in visual processing, neural signals do not represent the absolute numbers of photons that are captured by a receptor, but rather the relative intensity of stimulation—how much the current level of stimulation differs from previous stimulation levels, and how much it differs from the activity of neurons in adjacent areas of the retina (see Box 11D). These network mechanisms allow retinal circuits to reliably convey the most salient aspects of luminance changes to the central stages of the visual system described in Chapter 12.

Summary

Light falling on photoreceptors is transformed by retinal circuitry into a pattern of action potentials that ganglion cell axons convey to the visual centers in the rest of the brain. This process begins with phototransduction, a biochemical cascade that ultimately regulates the opening and closing of ion channels in the membrane of the photoreceptor's outer segment, and thereby the amount of neurotransmitter the photoreceptor releases. Two systems of photoreceptors—rods and cones—allow the visual system to meet the conflicting demands of sensitivity and acuity, respectively. Retinal ganglion cells operate quite differently from the photoreceptor cells. Two distinct classes of ganglion cells convey information about luminance increments and decrements (light and dark). Retinal circuits dynamically regulate the sensitivity of retinal ganglion cells and adjust their operating range to permit changes in activity to convey information over a broad range of stimulus conditions. The center–surround arrangement of ganglion cell receptive fields enhances the activity of those ganglion cells that carry the most information about object boundaries in the visual scene.

ADDITIONAL READING

Reviews

Arshavsky, V. Y., T. D. Lamb and E. N. Pugh Jr. (2002) G proteins and phototransduction. *Annu. Rev. Physiol.* 64: 153–187.

Burns, M. E. and D. A. Baylor (2001) Activation, deactivation, and adaptation in vertebrate photoreceptor cells. *Annu. Rev. Neurosci.* 24: 779–805.

Deeb, S. S. (2005) The molecular basis of variation in human color vision. *Clin. Genet.* 67: 369–377.

Euler, T., S. Haverkamp, T. Schubert and T. Baden (2014) Retinal bipolar cells: elementary building blocks of vision. *Nat. Rev. Neurosci.* 15: 507–519.

Lamb, T. D. and E. N. Pugh Jr. (2004) Dark adaptation and the retinoid cycle of vision. *Prog. Retin. Eye Res.* 23: 307–380.

Masland, R. H. (2012) The neuronal organization of the retina. *Neuron.* 76: 266–280.

Nathans, J. (1987) Molecular biology of visual pigments. *Annu. Rev. Neurosci.* 10: 163–194.

Rieke, F. and M. E. Rudd (2009) The challenges natural images pose for visual adaptation. *Neuron* 64: 605–616.

Schnapf, J. L. and D. A. Baylor (1987) How photoreceptor cells respond to light. *Sci. Amer.* 256 (April): 40–47.

Sterling, P. (1990) Retina. In *The Synaptic Organization of the Brain*, G. M. Shepherd (ed.). New York: Oxford University Press, pp. 170–213.

Stryer, L. (1986) Cyclic GMP cascade of vision. *Annu. Rev. Neurosci.* 9: 87–119.

Thoreson ,W. B. and S. C. Mangel (2012) Lateral interactions in the outer retina. *Prog. Retin. Eye Res.* 31: 407–441.

Wassle, H. (2004) Parallel processing in the mammalian retina. *Nat. Rev. Neurosci.* 5: 747–757.

Important Original Papers

Baylor, D. A., M. G. F. Fuortes and P. M. O'Bryan (1971) Receptive fields of cones in the retina of the turtle. *J. Physiol. (Lond.)* 214: 265–294.

Dowling, J. E. and F. S. Werblin (1969) Organization of the retina of the mud puppy, *Necturus maculosus*. I. Synaptic structure. *J. Neurophysiol.* 32: 315–338.

Enroth-Cugell, C. and R. M. Shapley (1973) Adaptation and dynamics of cat retinal ganglion cells. *J. Physiol.* 233: 271–309.

Fasenko, E. E., S. S. Kolesnikov and A. L. Lyubarsky (1985) Induction by cyclic GMP of cationic conductance in plasma membrane of retinal rod outer segment. *Nature* 313: 310–313.

Gilbert, C. D. (1992) Horizontal integration and cortical dynamics. *Neuron* 9: 1–13.

Hofer, H., J. Carroll, J. Neitz, M. Neitz and D. R. Williams (2005) Organization of the human trichromatic cone mosaic. *J. Neurosci.* 25: 9669–9679.

Kuffler, S. W. (1953) Discharge patterns and functional organization of mammalian retina. *J. Neurophysiol.* 16: 37–68.

Mancuso, K. and 7 others (2009) Gene therapy for red-green colorblindness in adult primates. *Nature* 461: 784–787.

Nathans, J., T. P. Piantanida, R. Eddy, T. B. Shows and D. S. Hogness (1986) Molecular genetics of inherited variation in human color vision. *Science* 232: 203–211.

Nathans, J., D. Thomas and D. S. Hogness (1986) Molecular genetics of human color vision: The genes encoding blue, green, and red pigments. *Science* 232: 193–202.

Ramachandran, V. S. and T. L. Gregory (1991) Perceptual filling in of artificially induced scotomas in human vision. *Nature* 350: 699–702.

Roorda, A. and D. R. Williams (1999) The arrangement of the three cone classes in the living human eye. *Nature* 397: 520–522.

Sakmann, B. and O. D. Creutzfeldt (1969) Scotopic and mesopic light adaptation in the cat's retina. *Pflügers Arch.* 313: 168–185.

Schiller, P. H., J. H. Sandell and J. H. R. Maunsell (1986) Functions of the "on" and "off" channels of the visual system. *Nature* 322: 824–825.

Werblin, F. S. and J. E. Dowling (1969) Organization of the retina of the mud puppy, *Necturus maculosus*. II. Intracellular recording. *J. Neurophysiol.* 32: 339–354.

Young, R. W. (1978) The daily rhythm of shedding and degradation of rod and cone outer segment membranes in the chick retina. *Invest. Ophthalmol. Vis. Sci.* 17: 105–116.

Books

Barlow, H. B. and J. D. Mollon (1982) *The Senses*. London: Cambridge University Press.

Dowling, J. E. (1987) *The Retina: An Approachable Part of the Brain*. Cambridge, MA: Belknap Press.

Fain, G. L. (2003) *Sensory Transduction*. Sunderland, MA: Sinauer Associates.

Hart, W. M. J. (ed.) (1992) *Adler's Physiology of the Eye: Clinical Application*, 9th Edition St. Louis, MO: Mosby Year Book.

Hogan, M. J., J. A. Alvarado and J. E. Weddell (1971) *Histology of the Human Eye: An Atlas and Textbook*. Philadelphia: Saunders.

Hubel, D. H. (1988) *Eye, Brain, and Vision*. Scientific American Library Series. New York: W. H. Freeman.

Hurvich, L. (1981) *Color Vision*. Sunderland, MA: Sinauer Associates, pp. 180–194.

Ogle, K. N. (1964) *Researches in Binocular Vision*. New York: Hafner.

Oyster, C. (1999) *The Human Eye: Structure and Function*. Sunderland, MA: Sinauer Associates.

Polyak, S. (1957) *The Vertebrate Visual System*. Chicago: University of Chicago Press.

Rodieck, R. W. (1973) *The Vertebrate Retina*. San Francisco: W. H. Freeman.

Rodieck, R. W. (1998) *First Steps in Seeing*. Sunderland, MA: Sinauer Associates.

von Helmholtz, H. L. F. (1924) *Helmholtz's Treatise on Physiological Optics*, Vol. I–III. Menasha, WI: George Banta Publishing Co. (Translated from the 3rd German edition by J. P. C. Southall.)

Wandell, B. A. (1995) *Foundations of Vision*. Sunderland, MA: Sinauer Associates.

Go to the NEUROSCIENCE 6e Companion Website at **www.oup.com/uk/Purves6e** for Web Topics, Animations, Flashcards, and more.

CHAPTER

12 Central Visual Pathways

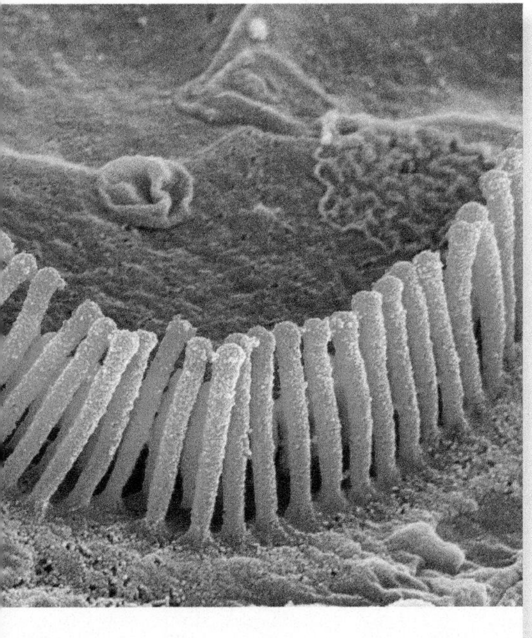

Overview

INFORMATION SUPPLIED BY THE RETINA initiates interactions among multiple subdivisions of the brain; these interactions eventually lead to perception of the visual scene, whether consciously or not. At the same time, this information activates more conventional reflexes such as adjustment of the pupil, direction of the eyes toward targets of interest, and regulation of homeostatic behaviors that are tied to the day–night cycle and circadian rhythmicity. The pathways and structures mediating this range of functions are necessarily diverse. Of these, the primary visual pathway from the retina to the dorsolateral geniculate nucleus in the thalamus and on to the primary visual cortex is the most important and certainly the most thoroughly studied component of the visual system or any other. Different classes of neurons within this pathway encode the variety of visual information—luminance, spectral differences, orientation, and motion—that we ultimately "see." The parallel processing of different categories of visual information continues in cortical pathways that extend beyond the primary visual cortex, supplying visual and other areas in the occipital, parietal, and temporal lobes. Visual areas in the temporal lobe are primarily involved in object recognition, whereas those in the parietal lobe are concerned with motion and location. Normal vision depends on the integration of information in all these cortical areas and many more. The processes underlying visual perception are not well understood and remain one of the central challenges of modern neuroscience.

Central Projections of Retinal Ganglion Cells

As was indicated in Chapter 11, ganglion cell axons exit the retina through a circular region in its nasal part called the optic disk (or optic papilla), where they bundle together to form the optic nerve. Axons in the optic nerve run a straight course to the **optic chiasm** at the base of the diencephalon (Figure 12.1). In humans, about 60% of these fibers cross in the chiasm; the other 40% continue toward thalamus and midbrain targets on the same side.

Once past the optic chiasm, the ganglion cell axons on each side form the **optic tract**. Thus, the optic tract, unlike the optic nerve, contains fibers from *both* eyes. The partial crossing (*decussation*) of ganglion cell axons at the optic chiasm allows information from corresponding points on the two retinas to be processed by approximately the same cortical site in each hemisphere, an important feature considered in the next section.

The ganglion cell axons in the optic tract reach a number of structures in the diencephalon and midbrain (see Figure 12.1). The major target in the diencephalon is the **dorsolateral geniculate nucleus** of the thalamus. Neurons in the lateral geniculate nucleus, like their counterparts in the thalamic relays of other sensory

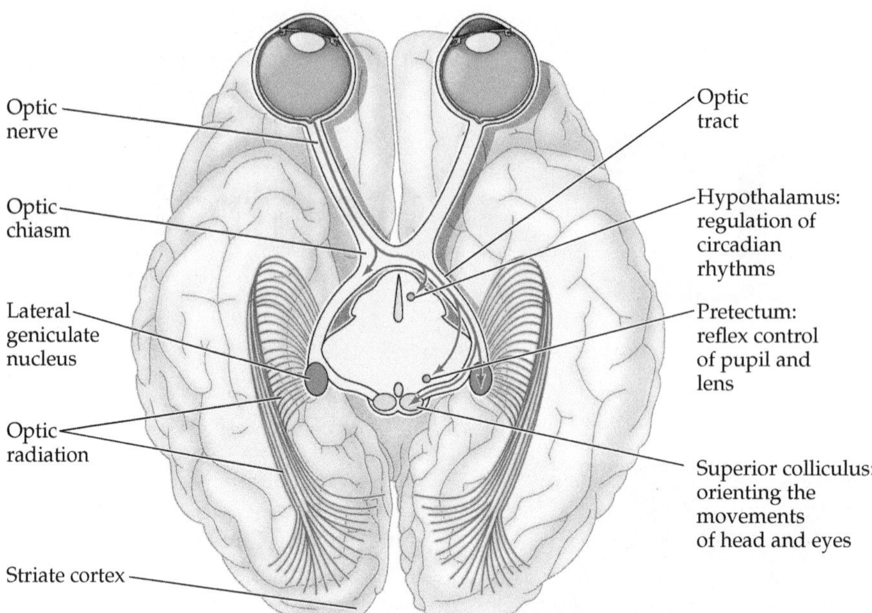

FIGURE 12.1 Central projections of retinal ganglion cells. Ganglion cell axons terminate in the lateral geniculate nucleus of the thalamus, the superior colliculus, the pretectum, and the hypothalamus. For clarity, only the crossing axons of the right eye are shown (view is looking up at the inferior surface of the brain).

systems, send their axons to the cerebral cortex via the internal capsule. These axons pass through a portion of the internal capsule called the **optic radiation** and terminate in the **primary visual cortex (V1)**, or **striate cortex** (also referred to as **Brodmann's area 17**), which lies largely along and within the calcarine fissure in the occipital lobe. The **retinogeniculostriate pathway**, or **primary visual pathway**, conveys information that is essential for most of what is thought of as seeing; damage anywhere along this route results in serious visual impairment.

A second major target of ganglion cell axons is a collection of neurons that lies between the thalamus and the midbrain in a region known as the **pretectum**. Although small in size compared with the lateral geniculate nucleus, the pretectum is particularly important as the coordinating center for the **pupillary light reflex** (i.e., the reduction in the diameter of the pupil that occurs when sufficient light falls on the retina; Figure 12.2). The initial component of the pupillary light reflex pathway is a bilateral projection from the retina to the pretectum.

Pretectal neurons, in turn, project to the **Edinger-Westphal nucleus**, a small group of nerve cells that lies close to the nucleus of the oculomotor nerve (cranial nerve III) in the midbrain. The Edinger–Westphal nucleus contains the preganglionic parasympathetic neurons that send their axons via the oculomotor nerve to terminate on

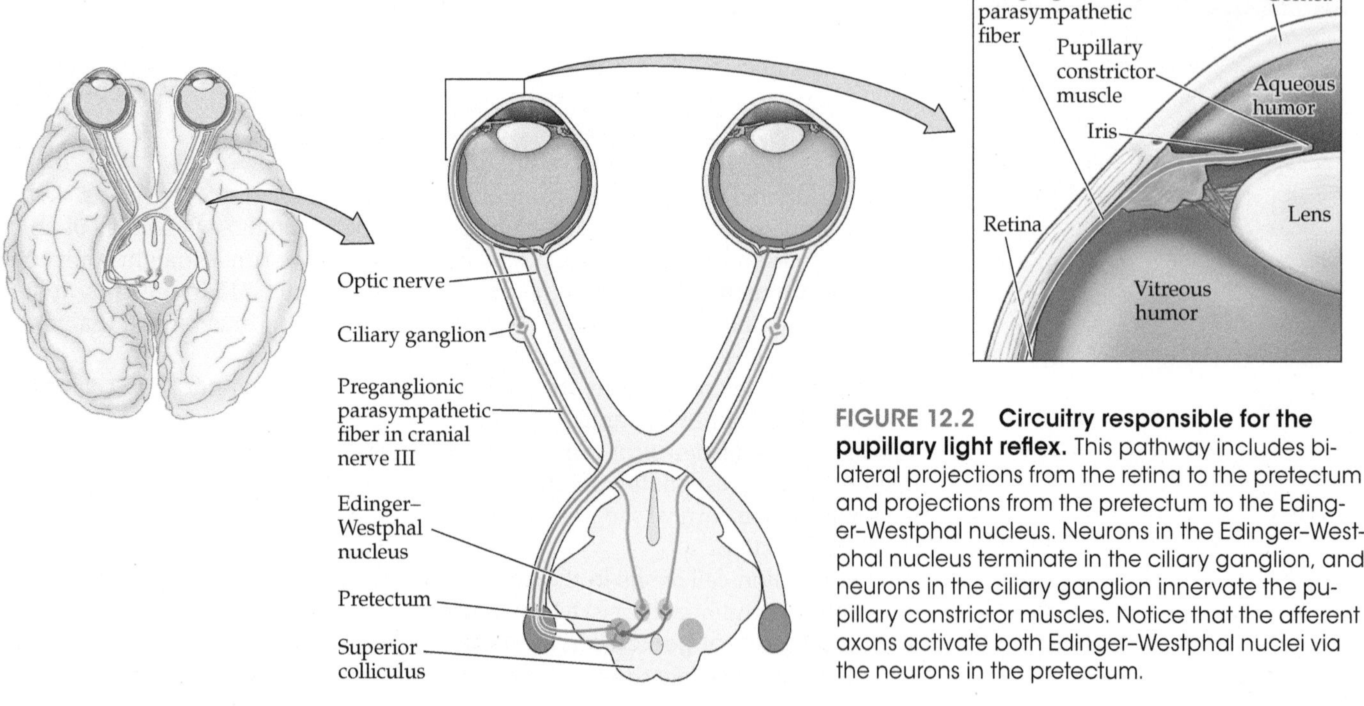

FIGURE 12.2 Circuitry responsible for the pupillary light reflex. This pathway includes bilateral projections from the retina to the pretectum and projections from the pretectum to the Edinger-Westphal nucleus. Neurons in the Edinger-Westphal nucleus terminate in the ciliary ganglion, and neurons in the ciliary ganglion innervate the pupillary constrictor muscles. Notice that the afferent axons activate both Edinger-Westphal nuclei via the neurons in the pretectum.

neurons in the ciliary ganglion (see Chapter 20). Neurons in the ciliary ganglion innervate the constrictor muscle in the iris, which decreases the diameter of the pupil when activated. Shining light in the eye leads to an increase in the activity of pretectal neurons, which stimulates the Edinger–Westphal neurons and the ciliary ganglion neurons they innervate, thus constricting the pupil.

In addition to its normal role in regulating the amount of light that enters the eye, the pupillary reflex provides an important diagnostic tool that allows the physician to test the integrity of the visual sensory apparatus, the motor outflow to the pupillary muscles, and the central pathways that mediate the reflex. Under normal conditions, the pupils of both eyes respond identically, regardless of which eye is stimulated; that is, light in one eye produces constriction of both the stimulated eye (the direct response) and the unstimulated eye (the consensual response). Comparing the responses in the two eyes is often helpful in localizing a lesion. For example, a direct response in the left eye without a consensual response in the right eye suggests a problem with the visceral motor outflow to the right eye, possibly as a result of damage to the oculomotor nerve or Edinger–Westphal nucleus in the brainstem. Failure to elicit a response (either direct or indirect) to stimulation of the left eye if both eyes respond normally to stimulation of the right eye suggests damage to the sensory input from the left eye, possibly to the left retina or optic nerve.

Retinal ganglion cell axons have several other important targets. One is the **suprachiasmatic nucleus** of the hypothalamus, a small group of neurons at the base of the diencephalon (see Box 21A). The **retinohypothalamic pathway** is the route by which variation in light levels influences a spectrum of visceral functions that are entrained to the day–night cycle (see Chapter 28). Another target is the **superior colliculus**, a prominent structure visible on the dorsal surface of the midbrain (see Figures 12.1 and 12.2). The superior colliculus coordinates head and eye movements to visual (as well as other) targets; its functions are considered in Chapter 20.

Functionally Distinct Types of Retinal Ganglion Cells

The types of visual information required to perform the functions of these different retinal targets are quite varied. Reading the text on this page, for example, requires a high-resolution sampling of the retinal image, whereas regulating circadian rhythms and adjusting the pupil accordingly require only a measure of the overall changes in light levels, and little or no information about the features of the image. It should come as no surprise, then, that a diversity of ganglion cell types provide information appropriate to the functions of different targets. Indeed, it is estimated that there are at least 30 morphologically and physiologically distinct retinal ganglion cell types, each of which has a distinct pattern of projection to central visual targets. The availability of transgenic mice with fluorescent reporters under the control of specific promoters has made it possible to gain a wealth of information on ganglion cell types, their response properties, central projections, and contribution to behavior. Many of these cell types in mice are analogous to cell types in other species, including primates, but the properties of cell types in these other species and their relation to those in the mouse are as yet unclear.

The retinal organization underlying for these distinct classes of retinal ganglion cells are only beginning to be identified; they include not only differences in ganglion cell synaptic connections, but in the locus of the phototransduction event itself. Unlike the majority of ganglion cells, which depend on rods and cones for their sensitivity to light, the ganglion cells that project to the hypothalamus and pretectum express their own light-sensitive photopigment (*melanopsin*) and can modulate their response to changes in light levels in the absence of signals from rods and cones. The presence of light sensitivity within this class of ganglion cells explains why normal circadian rhythms are maintained in animals that have lost form vision as the result of complete degeneration of rod and cone photoreceptors.

Retinotopic Representation of the Visual Field

The spatial relationships among the ganglion cells in the retina are maintained in most of their central targets as orderly representations, or "maps," of visual space. Most of these structures receive information from both eyes, requiring that these inputs be integrated to form a coherent map of individual points in space. As a general rule, information from the left half of the visual world, whether it originates from the left or right eye, is represented in the right half of the brain, and vice versa.

Understanding the neural basis for the appropriate arrangement of inputs from the two eyes requires considering how images are projected onto the two retinas, and the central destination of the ganglion cells located in different parts of the retina. Each eye sees a part of visual space that defines its **visual field** (Figure 12.3A). For descriptive purposes, each retina and its corresponding visual field are divided into quadrants. In this scheme, vertical and horizontal lines that intersect at the center of the fovea subdivide the surface of the retina (Figure 12.3B). The vertical line divides the retina into **nasal** and **temporal divisions**, and the horizontal line divides the retina into **superior** and **inferior divisions**. Corresponding vertical and horizontal lines in visual space (also called *meridians*) intersect at the **point of fixation** (the point in visual space that falls on the fovea) and define the quadrants of the visual field. The

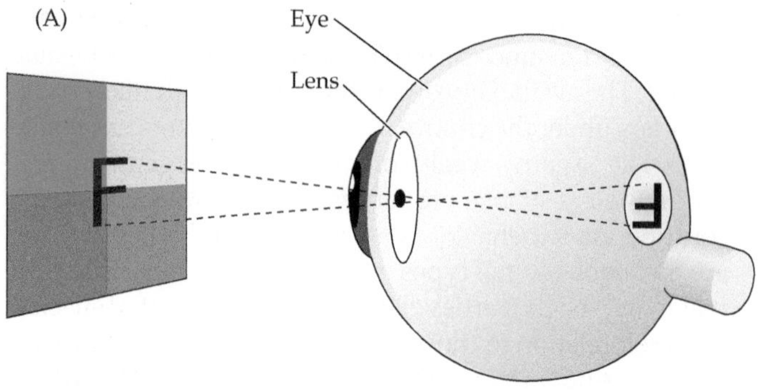

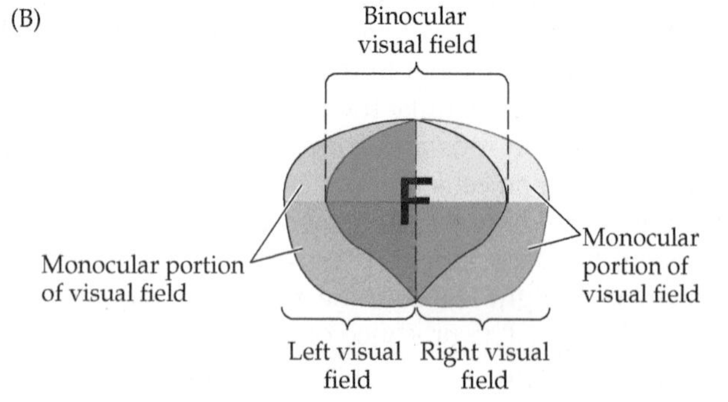

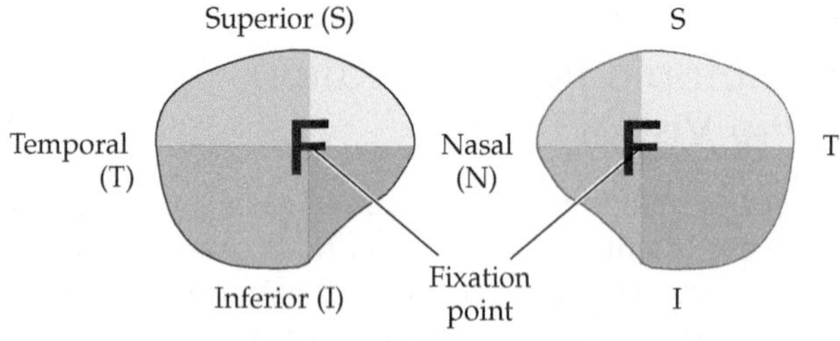

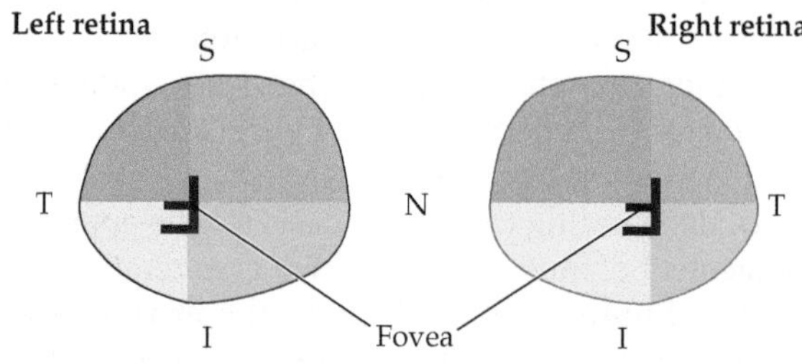

FIGURE 12.3 Projection of the visual fields onto the left and right retinas. (A) Projection of an image onto the surface of the retina. The passage of light rays through the pupil of the eye results in images that are inverted and left-right reversed on the retinal surface. (B) Retinal quadrants and their relation to the organization of monocular and binocular visual fields, as viewed from the back surface of the eyes. Vertical and horizontal lines drawn through the center of the fovea define retinal quadrants (bottom). Comparable lines drawn through the point of fixation define visual field quadrants (center). Color coding illustrates corresponding retinal and visual field quadrants. The overlap of the two monocular visual fields is shown at the top.

crossing of light rays diverging from different points on an object at the pupil causes the images of objects in the visual field to be inverted and left–right reversed on the retinal surface. As a result, objects in the temporal part of the visual field are seen by the nasal part of the retina, and objects in the superior part of the visual field are seen by the inferior part of the retina. (It may help in understanding Figure 12.3B to imagine that you are looking at the back surfaces of the retinas, with the corresponding visual fields projected onto them.)

With both eyes open, the two foveas normally align on a single target in visual space, causing the visual fields of both eyes to overlap extensively (Figure 12.4; also see Figure 12.3B). This **binocular field** consists of two symmetrical visual hemifields (left and right). The left binocular hemifield includes the nasal visual field of the right eye and the temporal visual field of the left eye; the right hemifield includes the temporal visual field of the right eye and the nasal visual field of the left eye. The temporal visual fields are more extensive than the nasal visual fields, reflecting the sizes of the nasal and temporal retinas, respectively. As a result, vision in the periphery of the field of view is strictly monocular, mediated by the most medial portion of the nasal retina. Most of the rest of the field of view can be seen by both eyes; that is, individual points in visual space lie in the nasal visual field of one eye and the temporal visual field of the other. It is worth noting, however, that the shape of the face and nose affect the extent of this region of binocular vision. In particular, the inferior nasal visual fields are less extensive than the superior nasal fields, and consequently the binocular field of view is smaller in the lower visual field than in the upper (see Figure 12.3B).

Ganglion cells that lie in the nasal division of each retina give rise to axons that cross in the optic chiasm, while those that lie in the temporal retina give rise to axons that remain on the same side (see Figure 12.4). The boundary, or line of decussation, between contralaterally and ipsilaterally projecting ganglion cells runs through the center of the fovea and defines the border between the nasal and temporal hemiretinas. Images of objects in the left visual hemifield (such as point B in Figure 12.4) fall on the nasal retina of the left eye and the temporal retina of the right eye, and the axons from ganglion cells in these regions of the two retinas project through the right optic tract. Objects in the right visual hemifield (such as point C in Figure 12.4) fall on the nasal retina of the right eye and the temporal retina of the left eye; the axons from ganglion cells in these regions project through the left optic tract. As mentioned previously, objects in the monocular portions of the visual hemifields (points A and D in Figure 12.4) are seen only by the most peripheral nasal retina of each eye; the axons of ganglion cells in these regions (like the rest of the nasal retina) run in the contralateral optic tract. Thus,

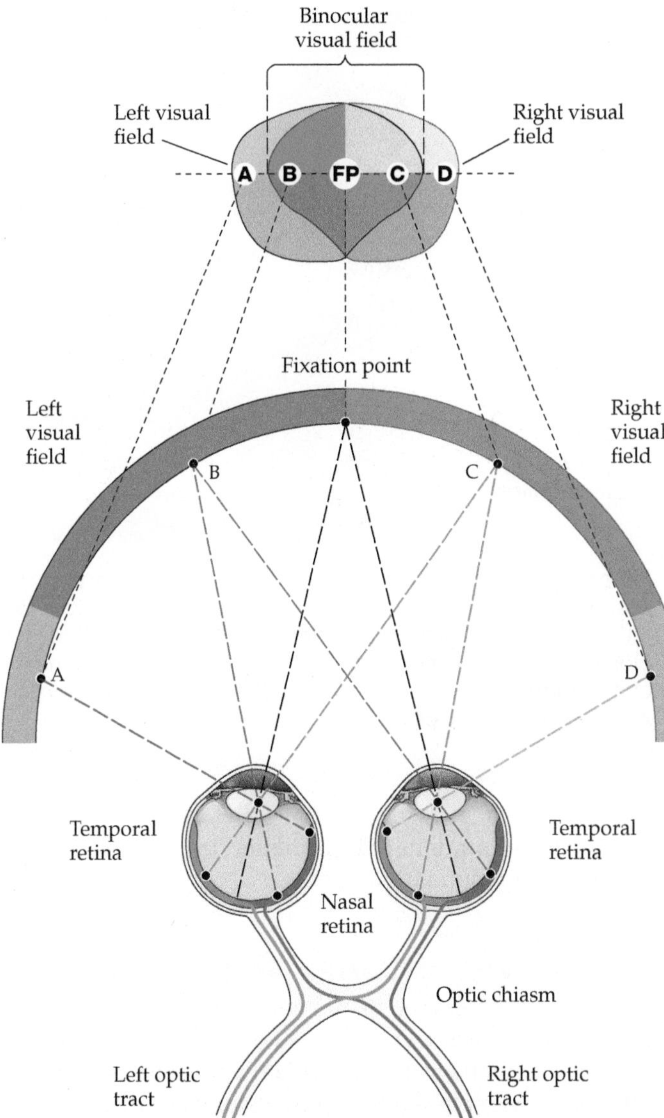

FIGURE 12.4 Binocular vision. The diagram illustrates the projection of the binocular field of view onto the two retinas and its relation to the crossing of fibers in the optic chiasm. Points in the binocular portion of the left visual field (B) fall on the nasal retina of the left eye and the temporal retina of the right eye. Points in the binocular portion of the right visual field (C) fall on the nasal retina of the right eye and the temporal retina of the left eye. Points that lie in the monocular portions of the left and right visual fields (A and D) fall on the left and right nasal retinas, respectively. The axons of ganglion cells in the nasal retina cross in the optic chiasm, whereas those from the temporal retina do not. As a result, the right optic tract carries information from the left visual field, and the left optic tract carries information from the right visual field.

unlike the optic nerve, the optic tract contains the axons of ganglion cells that originate in both eyes and represent the contralateral field of view.

Optic tract axons terminate in an orderly fashion within their target structures, thus generating well-ordered maps of the contralateral hemifield. For the primary visual pathway, the map of the contralateral hemifield established in the lateral geniculate nucleus is maintained in the projections of the lateral geniculate nucleus to the striate cortex (Figure 12.5). Thus, the fovea is represented in the posterior part of the striate cortex, whereas the more peripheral regions of the retina are represented in progressively more anterior parts of the striate cortex. The upper visual field is mapped below the calcarine sulcus, and the lower visual field is mapped above it. As in the somatosensory system, the amount of cortical area devoted to each unit of area of the sensory surface is not uniform, but reflects the density of receptors and sensory axons that supply the peripheral region. Like the representation of the hand region in the somatosensory cortex, the representation of the macula is therefore disproportionately large, occupying most of the caudal pole of the occipital lobe (this discrepancy is called "cortical magnification").

Spatiotemporal Tuning Properties of Neurons in Primary Visual Cortex

Much of the current understanding of the functional organization of visual cortex had its origin in the pioneering studies of David Hubel and Torsten Wiesel at Harvard Medical School. The method they used was primarily microelectrode recordings in anesthetized animals that reported the responses of individual neurons in the lateral geniculate nucleus and the cortex to various patterns of retinal stimulation (Figure 12.6A). The responses of neurons in the lateral geniculate nucleus were found to be remarkably similar to those in the retina, with a center–surround receptive field organization and selectivity for luminance increases or decreases. However, the small spots of light that were so effective at stimulating neurons in the retina and lateral geniculate nucleus were largely ineffective in visual cortex. Instead, most cortical neurons in cats and monkeys responded vigorously to light–dark bars or edges, and only if the bars were presented at a particular range of orientations within the cell's receptive field (Figure 12.6B). The responses of cortical neurons are thus tuned to the orientation of edges, much as cone photoreceptors are tuned to the wavelength of light; the peak in the tuning curve (the orientation to which a cell is most responsive) is referred to as the neuron's preferred orientation (Figure 12.6C). By sampling the responses of a large number of single cells, Hubel and Wiesel demonstrated that all edge orientations were roughly equally represented in visual cortex. As a result, a given orientation in a visual scene appears to be "encoded" in the activity of a distinct population of orientation-selective neurons.

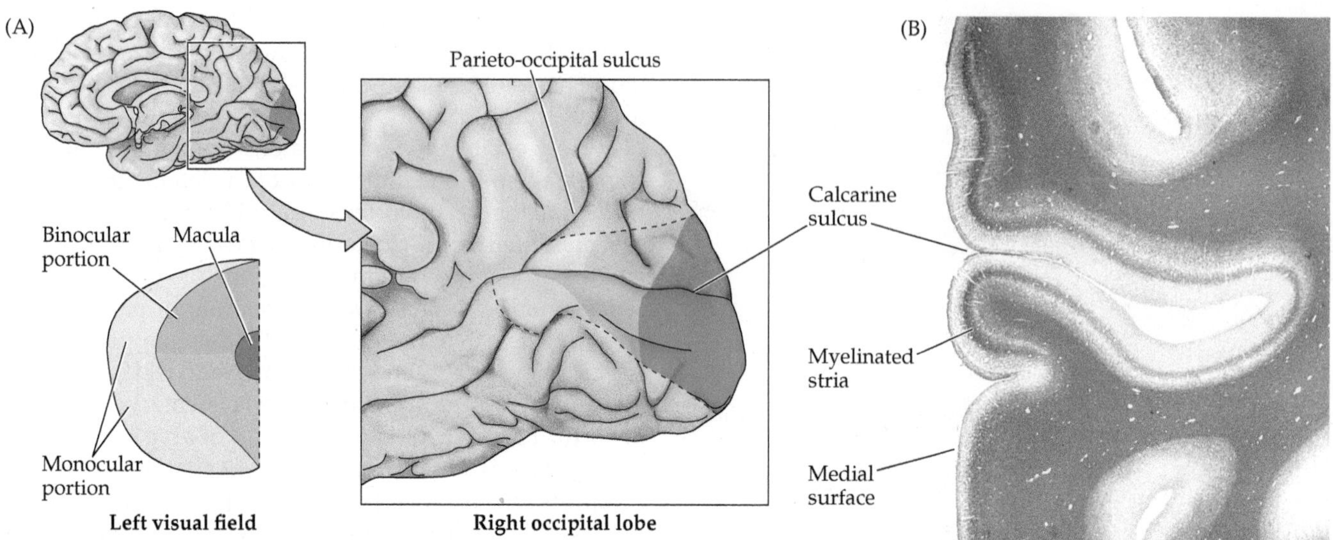

FIGURE 12.5 Visuotopic organization of the striate cortex in the right occipital lobe. (A) Seen in midsagittal view, the primary visual cortex occupies a large part of the occipital lobe. The area of central vision (the fovea) is represented over a disproportionately large part of the caudal portion of the lobe, whereas peripheral vision is represented more anteriorly. The upper visual field is represented below the calcarine sulcus, the lower field above the calcarine sulcus. (B) Coronal section of the human striate cortex, showing the characteristic myelinated band, or *stria*, that gives this region of the cortex its name. The calcarine sulcus on the medial surface of the occipital lobe is indicated. (B courtesy of T. Andrews and D. Purves.)

To appreciate how the properties of an image might be represented by populations of neurons that are tuned to different orientations, an image can be decomposed into its frequency components (an analytical approach discovered by the French mathematician Joseph Fourier) and then filtered to create a set of images whose spectral composition simulates the information that would be conveyed by neurons tuned to different orientations (Figure 12.7). Each class of orientation-selective neuron transmits only a fraction of the information in the scene—the part that matches its filter properties—but the information from these different filters contains all the spatial information necessary to generate a faithful representation of the original image.

Orientation preference is only one of the qualities that define the filter properties of neurons in primary visual cortex. A substantial fraction of cortical neurons are also tuned to the direction of stimulus motion, for example, responding much more vigorously when a stimulus moves to the right than when it moves to the left. Neurons can also be characterized by their preference for spatial frequency (the coarseness or fineness of the variations in contrast that fall within their receptive fields) as well as temporal frequency (rate of change in contrast). Why should cortical neurons show selectivity for these particular stimulus dimensions? Computational analyses suggest that receptive fields with properties such as these are well matched to the statistical structure of natural scenes and would therefore maximize the amount of information transferred with a minimum of redundancy.

Primary Visual Cortex Architecture

Like all neocortex, the visual cortex is a sheet approximately 2 mm thick and composed of two broad classes of neurons: spiny neurons (pyramidal and stellate) that exhibit dendritic spines, and aspinous or smooth dendritic neurons. Pyramidal neurons employ the excitatory neurotransmitter glutamate and are the principal source of axonal projections that leave the cortex to target subcortical and other cortical areas. Most smooth dendritic neurons have local axonal arbors and are the principal source of cortical inhibition, employing the neurotransmitter GABA (see Chapter 26 for general overview of cortical structure and cell types). Neocortex exhibits a conspicuous laminar structure in preparations stained to reveal the density and size of neuronal cell bodies. By convention, neocortex is divided into six cellular layers (layers 1–6; Figure 12.8). To accommodate the laminar complexity exhibited by primate visual cortex, the layers can be further subdivided using Latin and Greek lettering (e.g., layer 4Cβ).

Although the organization of intracortical circuits is complex and not fully understood, it is useful for didactic purposes to outline the basic input–output organization of the visual cortex (see Figure 12.8B). Lateral geniculate axons terminate primarily in cortical layer 4C, which is

(A) Experimental setup

Light bar stimulus projected on screen

Record

Recording from visual cortex

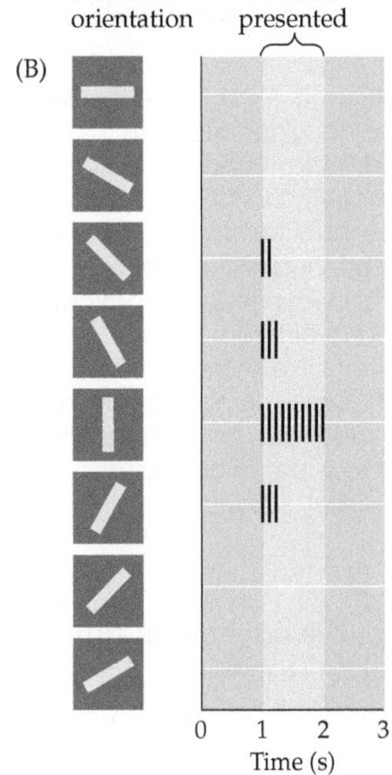

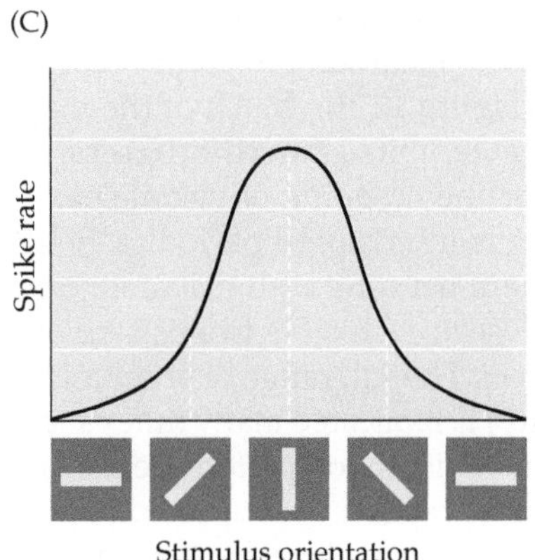

FIGURE 12.6 Neurons in the primary visual cortex respond selectively to oriented edges. (A) An anesthetized animal is fitted with contact lenses to focus the eyes on a screen, where images can be projected; an extracellular electrode records the neuronal responses. (B) Neurons in the primary visual cortex typically respond vigorously to a bar of light oriented at a particular angle and less strongly—or not at all—to other orientations. (C) Orientation tuning curve for a neuron in primary visual cortex. In this example, the highest rate of action potential discharge occurs for vertical edges—the neuron's "preferred" orientation.

composed of spiny stellate neurons whose axons convey the activity supplied by the lateral geniculate nucleus to other cortical layers. Pyramidal neurons in the superficial layers of visual cortex project to extrastriate cortical areas, while those in the deeper cortical layers send their axons to subcortical targets, including the lateral geniculate nucleus and the superior colliculus. Thus, the laminar organization of visual cortex, like that of other cortical areas, segregates populations of neurons having distinct patterns of connections.

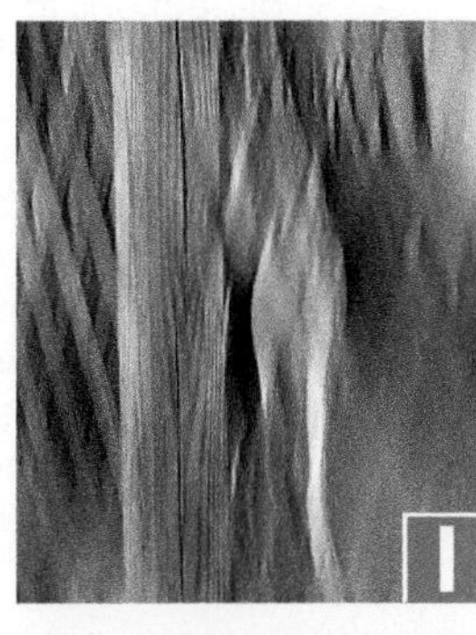

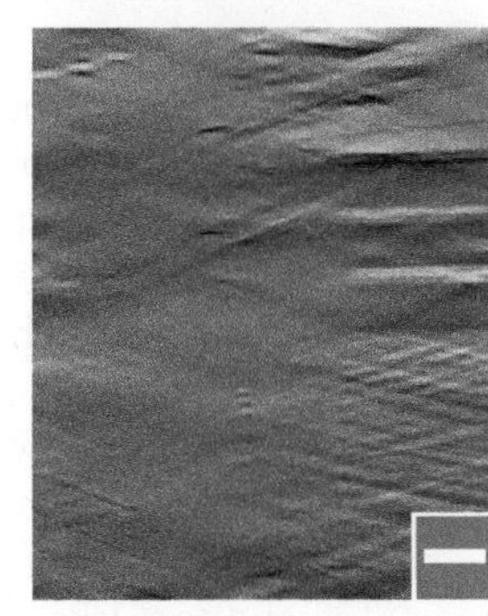

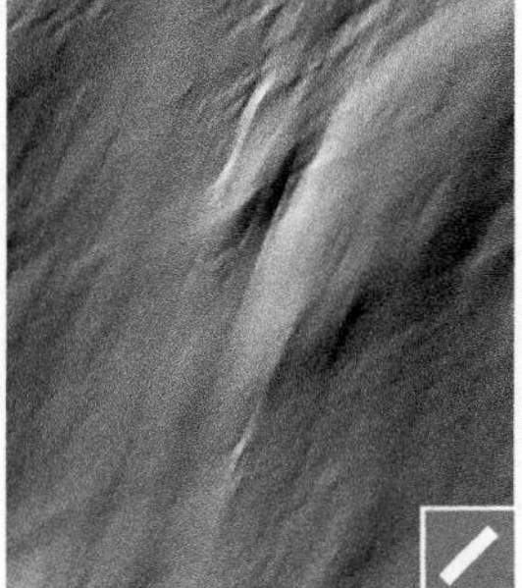

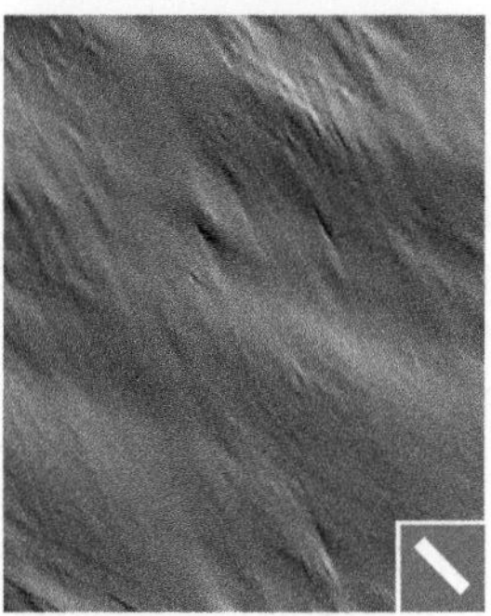

FIGURE 12.7 Representation of a visual image by neurons selective for different stimulus orientations. This simulation uses image mathematics (selective filtering of the two-dimensional Fourier transform of the image) to illustrate the attributes of a visual image (greyhound and fence) that would be represented in the responses of populations of cortical neurons tuned to different preferred orientations. The panels surrounding the image illustrate the components of the image that would be detected by neurons tuned to vertical, horizontal, and oblique orientations (blue boxes). In ways that are still not understood, the activity in these different populations of neurons is integrated to yield a coherent representation of the image features. (Photos courtesy of Steve Van Hooser and Elizabeth Johnson.)

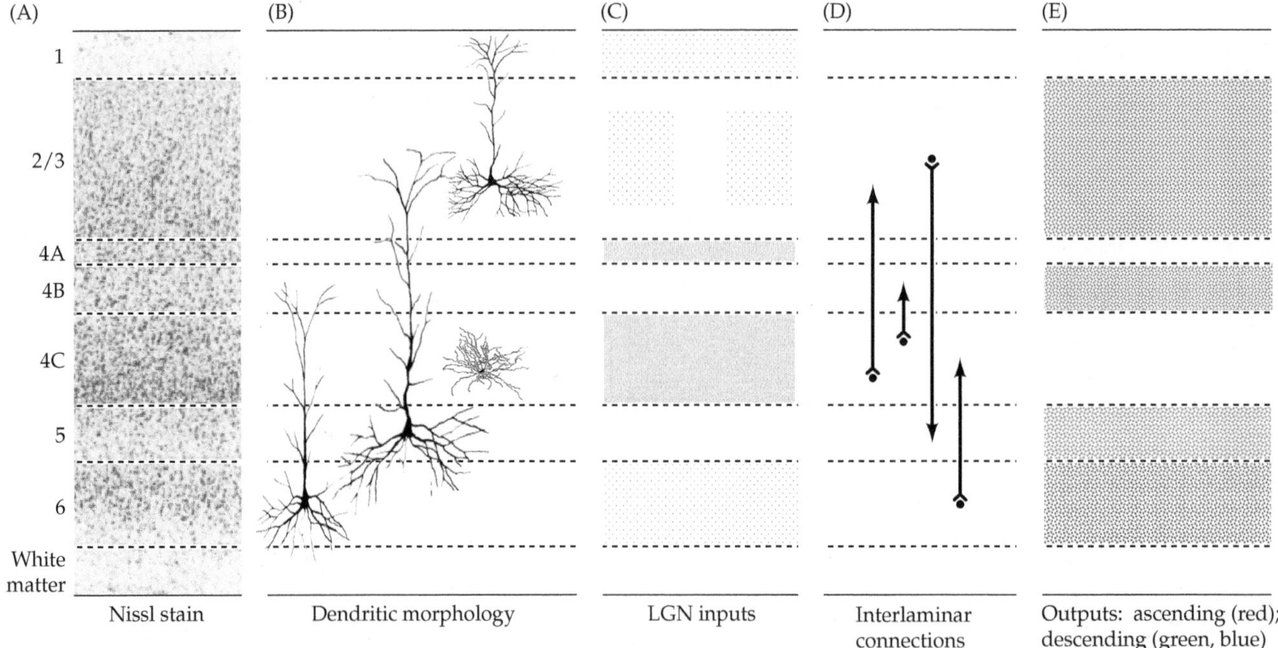

FIGURE 12.8 Organization of primary visual (striate) cortex. Striate cortex is divided into six principal cellular layers that differ in cell packing density, cellular morphology, and connections. (A) Primary visual cortex visualized using a histological stain that reveals neuronal cell bodies. In primates, layer 4 has several subdivisions (4A, 4B, and 4C; see also Figure 12.5). (B) Pyramidal cells with prominent apical and basilar dendrites are the most numerous cell type in the neocortex; they are located in all layers except 4C. Layer 4C is dominated by spiny stellate neurons, whose dendrites are confined to this layer. (C) Laminar organization of inputs from the lateral geniculate nucleus (LGN). Lateral geniculate axons terminate most heavily in layers 4C and 4A, with less dense projections to layers 1, 2/3, and 6; the terminations in layer 2/3 are "patchy." (D) Laminar organization of major intracortical connections. Neurons in layer 4C give rise to axons that terminate in more superficial layers (4B and 2/3). Axons of layer 2/3 neurons terminate heavily in layer 5. Axons of layer 6 neurons terminate in layer 4C. (E) Laminar organization of neurons projecting to different targets. Connections with extrastriate cortex arise primarily from neurons in layers 2/3 and 4B (red). Descending projections to the lateral geniculate nucleus arise from layer 6 neurons (blue), while those projecting to the superior colliculus reside in layer 5 (green). (A from Hubel, 1988.)

What cannot be discerned from a cursory examination of anatomical sections is that the cortex also exhibits a much more detailed organization. Thus microelectrode penetrations perpendicular to the cortical surface encounter columns of neurons that have similar receptive field properties, responding, for example, to stimulation arising from the same region of visual space and exhibiting preferences for similar stimulus properties, such as edge orientation and direction of motion (Figure 12.9). The uniformity in response along the radial axis raises the obvious question of how response properties change across adjacent columns. From the previous description of the mapping of visual space in primary visual cortex, it should come as no surprise that adjacent columns have similar but slightly shifted receptive field locations, such that tangential electrode penetrations encounter columns of neurons whose receptive field locations overlap significantly, but shift progressively in a fashion that is consistent with the global mapping of visual space.

Like receptive field location, the orientation tuning curves of neurons in adjacent columns also overlap significantly, but tangential penetrations frequently reveal an orderly progression in orientation preference (see Figure 12.9B). The availability of functional imaging techniques has made it possible to visualize the two-dimensional layout of the map of orientation preference on the surface of visual cortex (Figure 12.10). Much of the map of orientation preference exhibits smooth progressive change, like that seen for the mapping of visual space. This smooth progression is interrupted periodically by point discontinuities, where neurons with disparate orientation preferences lie close to each other in a pattern resembling a child's pinwheel. The full range of orientation preferences (0–180 degrees) is replicated many times such that neurons with the same orientation preference are arrayed in an iterated fashion, repeating at approximately 1-mm intervals across the primary visual cortex. This iteration ensures that the full range of orientation values

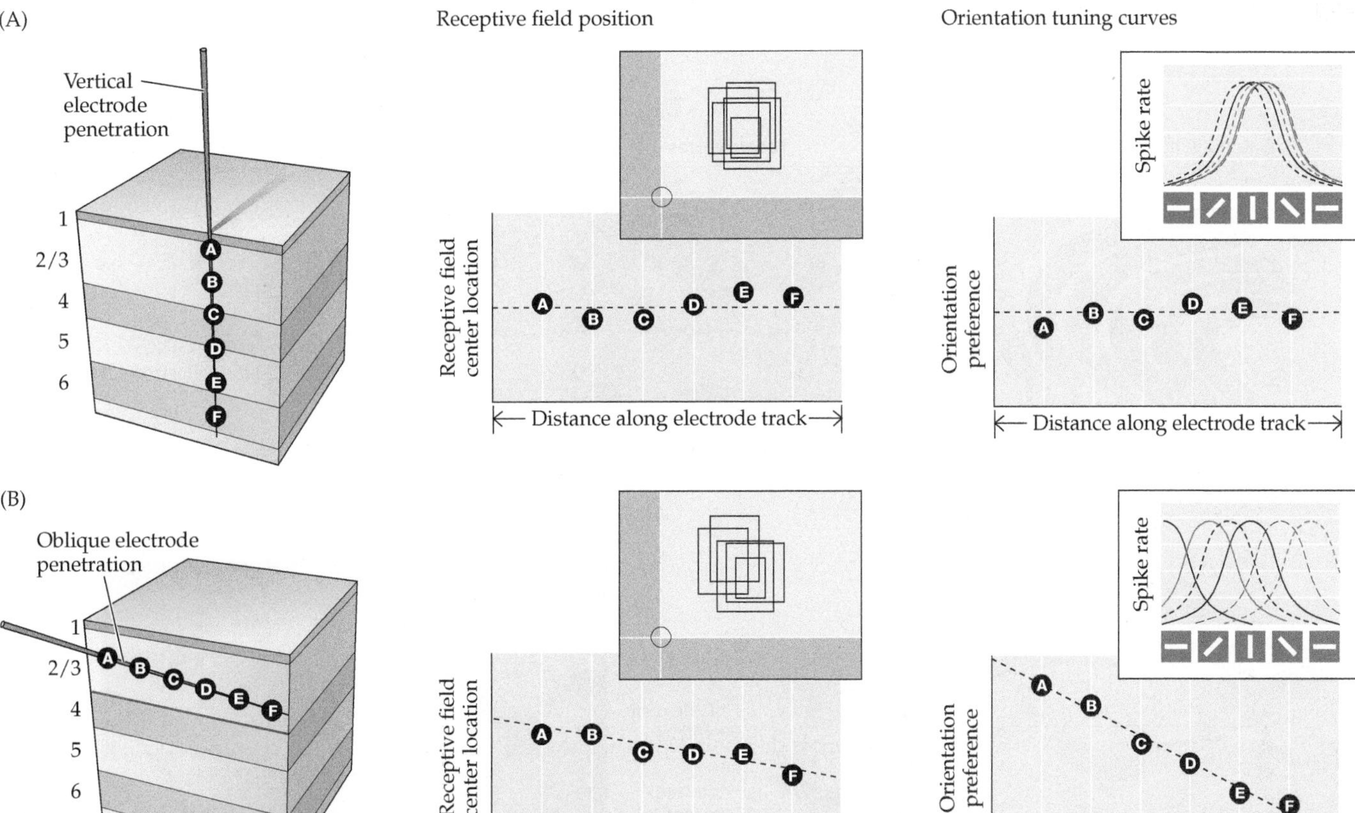

FIGURE 12.9 Orderly progression of columnar response properties forms the basis of functional maps in primary visual cortex. (A) Neurons displaced along the radial axis of the cortex have receptive fields that are centered on the same region of visual space and exhibit similar orientation preferences. At left is a depiction of a vertical microelectrode penetration into primary visual cortex. Neuronal receptive fields encountered along the electrode track are located in the upper part of the right visual field (center panel, top; intersection of axes represents center of gaze). Note that there is little variation in the location of the receptive field centers (center panel, bottom). The orientation tuning curves (right panel, top) and preferred orientation (right panel, bottom) for neurons encountered along the electrode track show that there is little variation in the orientation preference of the neurons. (B) Neurons displaced along the tangential axis of the cortex exhibit an orderly progression of receptive field properties. Neurons encountered along the electrode penetration have receptive field centers (center panel) and orientation preferences (right panel) that shift in a progressive fashion.

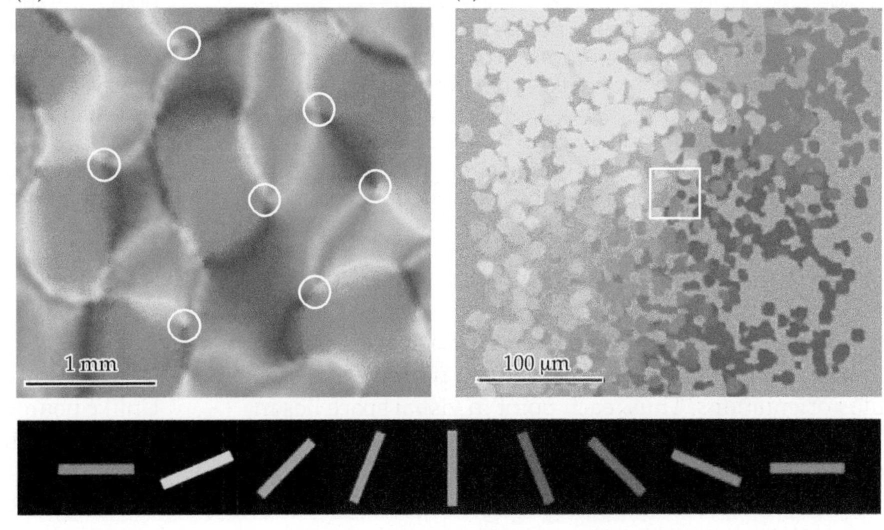

FIGURE 12.10 Functional imaging reveals orderly mapping of orientation preference in the primary visual cortex. (A) Surface view of the striate cortex, using intrinsic signal imaging tec hniques to visualize the map of preferred orientation. Colors indicate the average preferred orientation of columns at a given location; red indicates the location of columns that respond preferentially to horizontal orientations, blue those that respond preferentially to vertical orientations; other colors represent intermediate orientations. The smooth progression of preferred orientations indicated by the gradations in color is interrupted by point discontinuities (pinwheel centers; white circles). (B) Single-cell view of a "pinwheel" visualized using two-photon imaging of calcium signals. Note that adjacent cells have similar preferred orientations except at the very center, where nearby cells exhibit nearly orthogonal orientation preferences. (A courtesy of D. Fitzpatrick; B modified from Ohki et al., 2006.)

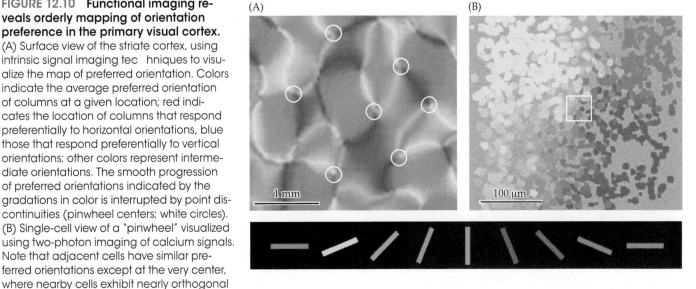

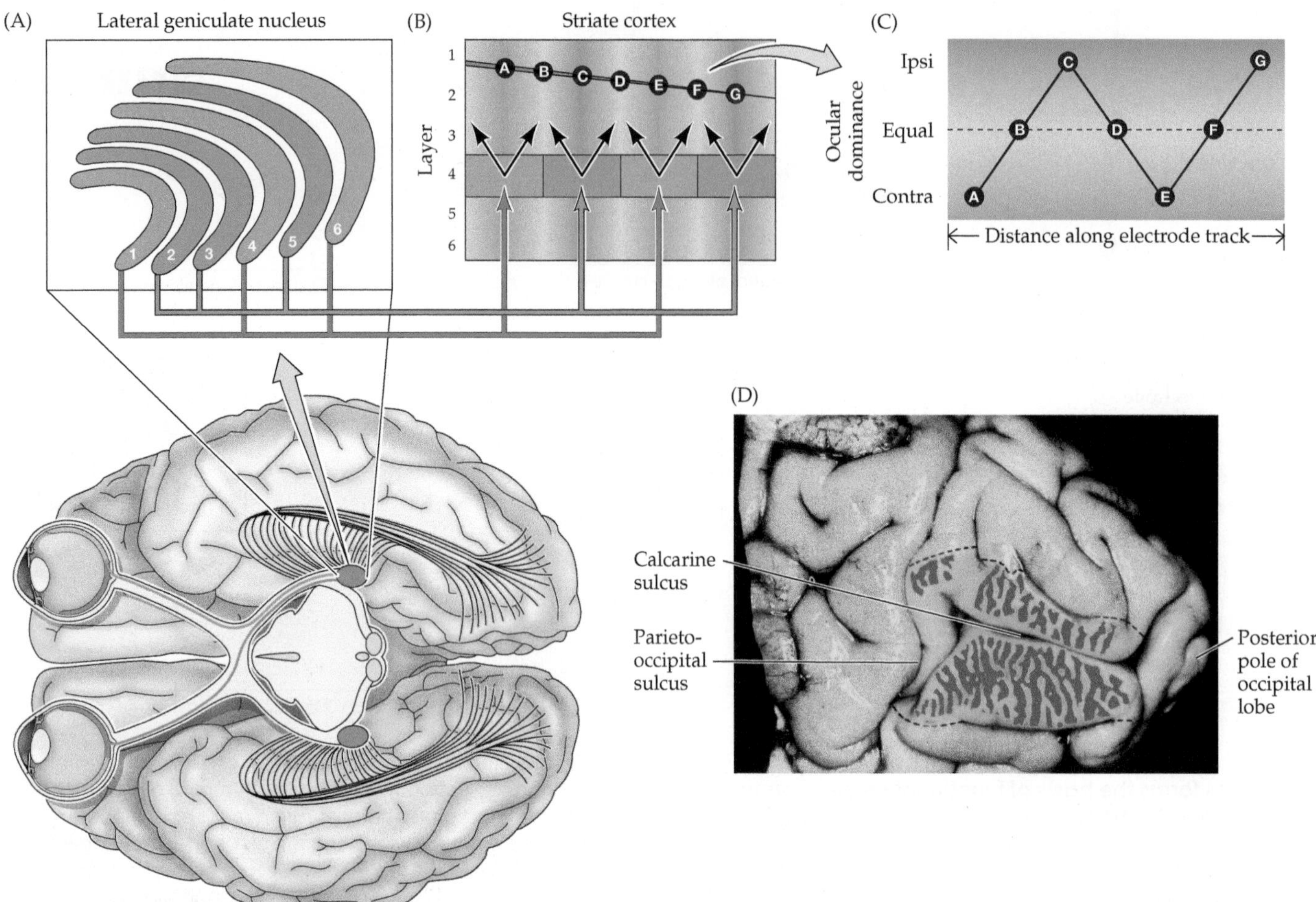

FIGURE 12.11 Mixing of the pathways from the two eyes first occurs in the striate cortex. (A) Although the lateral geniculate nucleus receives inputs from both eyes, the inputs are segregated in separate layers. (B) In many species, including most primates, inputs from the two eyes remain segregated in the ocular dominance columns of layer 4. Layer 4 neurons send their axons to other cortical layers; it is at this stage that the information from the two eyes converges onto individual neurons. (B,C) Physiological demonstration of columnar organization of ocular dominance in primary visual cortex. Cortical neurons vary in the strength of their response to the inputs from the two eyes, from complete domination by one eye to equal influence of the two eyes. Neurons encountered in a vertical electrode penetration (other than those neurons that lie in layer 4) tend to have similar ocular dominance. Tangential electrode penetration across the superficial cortical layers reveals a gradual shift in the strength of response to the inputs from the two eyes, from complete domination by one eye to equal influence of the two eyes. (D) Pattern of ocular dominance columns in human striate cortex. The alternating left and right eye columns in layer 4 have been reconstructed from tissue sections and projected onto a photograph of the medial wall of the occipital lobe. (D from Horton and Hedley-Whyte, 1984.)

is represented for each region of visual space—that is, there are no "holes" in the capacity to perceive all stimulus orientations. Thus, each point in visual space lies in the receptive fields of a large population of neurons that collectively occupy several millimeters of cortical surface area, an area that contains neurons having the full range of orientation preferences. As described in Box 9A, many other cortical regions show a similar columnar arrangement of their processing circuitry.

Combining Inputs from Two Eyes

Unlike neurons at earlier stages in the primary visual pathway, most neurons in striate cortex are binocular, responding to stimulation of both the left and right eyes. Inputs from both eyes are present at the level of the lateral geniculate nucleus, but contralateral and ipsilateral retinal axons terminate in separate layers, so that individual geniculate neurons are strictly monocular, driven by either the left or right eye, but not by both (Figure 12.11A–C). Activity arising from the left and right eyes that is conveyed by geniculate axons continues to be segregated at the earliest stages of cortical processing as the axons of geniculate neurons terminate in alternating eye-specific **ocular dominance columns** in cortical layer 4 (Figure 12.11D). Beyond this point, however, signals from the two eyes converge as the axons from layer 4 neurons in adjacent monocular stripes synapse on individual neurons in other cortical layers. While most neurons outside of layer 4 are binocular, the relative strength of the inputs from the two eyes varies from neuron to neuron in a columnar fashion that reflects the pattern of ocular dominance stripes in layer 4. Thus, neurons that are located over the centers of layer 4 ocular dominance columns respond almost exclusively to the left or right eye, while those that lie over the borders between ocular dominance columns in layer 4 respond equally well to stimulation of either eye. Similar to the mapping of orientation preference, tangential electrode penetrations across the superficial layers reveal a gradual, continuous shift in the ocular dominance of the recorded neurons (see Figure 12.11B,C). With the exception of layer 4, which is strictly monocular, vertical penetrations encounter neurons with similar ocular preferences.

Bringing together the inputs from the two eyes at the level of the striate cortex provides a basis for **stereopsis**, the sensation of depth that arises from viewing nearby objects with two eyes instead of one. Because the two eyes look at the world from slightly different angles, objects that lie in front of or behind the plane of fixation project to non-corresponding points on the two retinas. To convince yourself of this fact, hold your hand at arm's length and fixate on the tip of one finger. Maintain fixation on the finger as you hold a pencil in your other hand about half an arm's length away from you. At this distance, the image of the pencil falls on non-corresponding points on the two retinas and will therefore be perceived as two separate pencils (a phenomenon called double vision, or **diplopia**). If you move the pencil toward the finger (the point of fixation), the two images of the pencil fuse and you see a single pencil in front of the finger. Thus, for a small distance on either side of the plane of fixation, where the disparity between the two views of the world remains modest, a single image is perceived; the disparity between the two eye views of objects nearer or farther than the point of fixation is perceived as *depth* (Figure 12.12). While disparity cues normally arise from specific objects in the visual scene, disparity cues alone can give rise to a sense of depth that a scene is nothing more than random dots, meaning that no object is apparent in monocular viewing (Box 12A).

Some binocular neurons in the striate cortex and in other visual cortical areas have receptive field properties that make them good candidates for extracting information about binocular disparity. In these neurons, the receptive fields driven by the left and right eyes are slightly offset either in their position in visual space or in their internal organization so that the cell is maximally activated by stimuli that fall on non-corresponding parts of the retinas. Some of these neurons (so-called **far cells**) discharge to retinal disparities that arise from points beyond the plane of fixation, while others (**near cells**) respond to retinal disparities that arise from points in front of the plane of fixation. A third class of neurons (**tuned zero**) responds selectively to points that lie on the plane of fixation. The relative activity in these different classes of neurons is thought to mediate sensations of stereoscopic depth.

Interestingly, the presence of binocular responses in cortical neurons is contingent on normal patterns of activity from the two eyes during early postnatal life. Any factor that creates an imbalance in the activity of the two

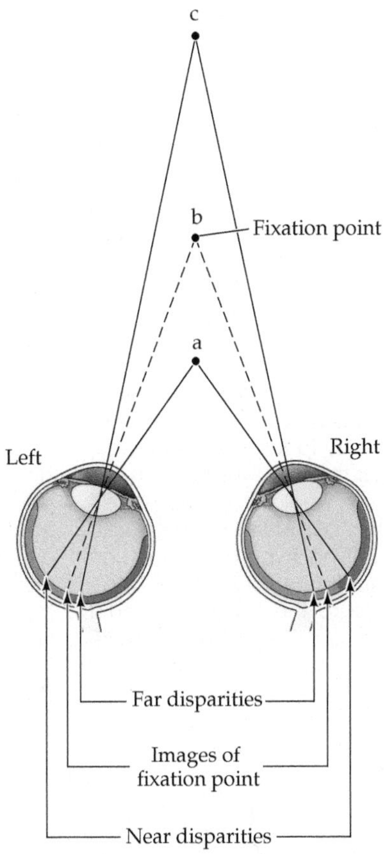

FIGURE 12.12 Binocular disparities are thought to be the basis of stereopsis. When the eyes are fixated on point b, points that lie beyond the plane of fixation (point c) or in front of the point of fixation (point a) project to non-corresponding points on the two retinas. When these disparities are small, the images are fused, and the brain interprets the disparity as small differences in depth. When the disparities are greater, double vision occurs (although this normal phenomenon is generally unnoticed).

BOX 12A ■ Random Dot Stereograms and Related Amusements

An important advance in studies of stereopsis occurred in 1959 when Bela Julesz, then working at the Bell Laboratories in Murray Hill, New Jersey, discovered an ingenious way of showing that stereoscopy depends on matching information seen by the two eyes without any prior recognition of what object(s) such matching might generate. Julesz, a Hungarian whose background was in engineering and physics, was working on the problem of how to "break" camouflage. He surmised that the brain's ability to fuse the slightly different views of the two eyes to bring out new information would be an aid in overcoming military camouflage. Julesz also realized that, if his hypothesis was correct, a hidden figure in a random pattern presented to the two eyes should emerge when a portion of the otherwise identical pattern was shifted horizontally in the view of one eye or the other. A horizontal shift in one direction would cause the hidden object to appear in front of the plane of the background, whereas a shift in the other direction would cause the hidden object to appear in back of and the hidden figure appears (in this case, a square that occupies the middle portion of the figure). The random dot stereogram has been widely used in stereoscopic research for about 45 years, although how such stimuli elicit depth remains very much a matter of dispute.

An impressive—and extraordinarily popular—derivative of the random dot stereogram is the autostereogram (Figure C). The nineteenth-century British physicist David Brewster was the first to discern the possibility of autostereograms. While staring at a Victorian wallpaper with an iterated but offset pattern, he noticed that when the patterns were fused he perceived two different planes. The plethora of autostereograms that can be seen today in posters, books, and newspapers are close cousins of the random dot stereogram in that computers are used to shift patterns of iterated information with respect to each other. The result is that different planes emerge from what appears to be a meaningless array of visual information (or depending on the taste of the creator, an apparently "normal" scene in which the iterated and displaced information is hidden). Some autostereograms are designed to reveal the hidden figure when the eyes diverge, others when the eyes converge. (Looking at a plane more distant than the plane of the surface causes divergence; looking at a plane in front of the picture causes the eyes to converge; see Figure 12.12.)

The elevation of the autostereogram to a popular art form should probably be attributed to Chris W. Tyler, a student of Julesz's and a visual psychophysicist, who was among the first to create commercial

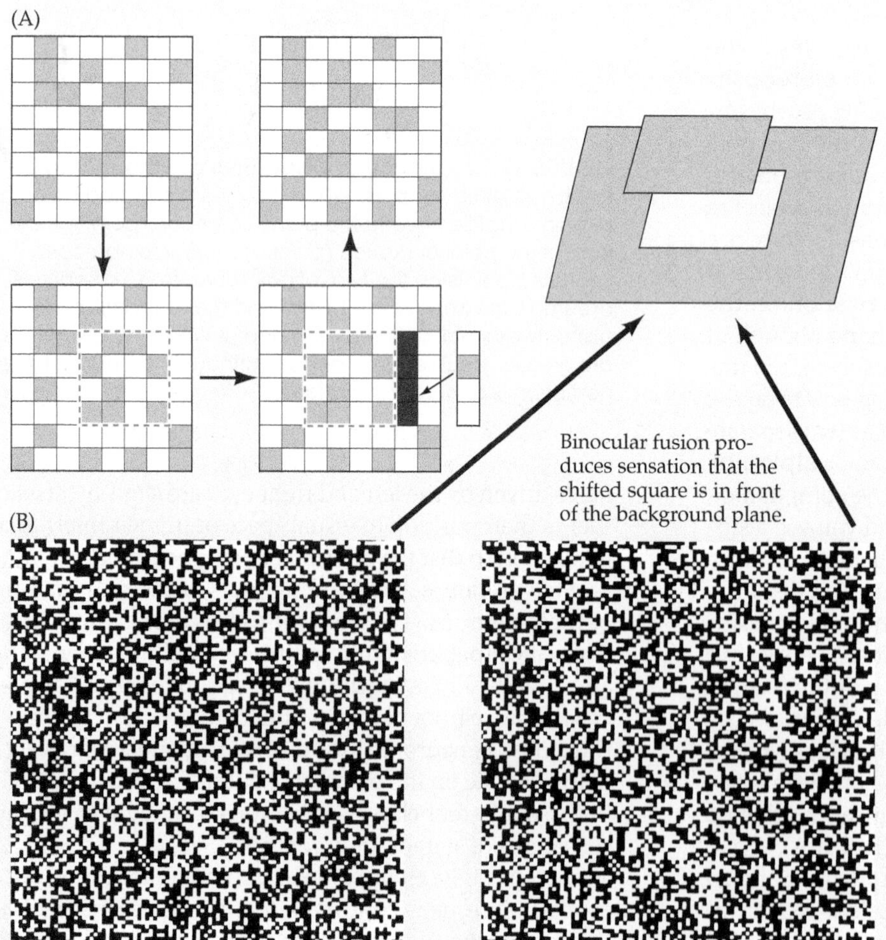
(A)
(B)
Binocular fusion produces sensation that the shifted square is in front of the background plane.

Random dot stereograms. (A) First, a random dot pattern, to be observed by one eye, is created. The stimulus for the other eye is then created by copying the first image, displacing a particular region horizontally, and then filling in the gap with a random sample of dots. (B) When the right and left images are viewed simultaneously but independently by the two eyes (either by using a stereoscope or by fusing the images by converging or diverging the eyes), the plane of the shifted region (a square) appears to be different from the plane of the other dots. (A from Wandell, 1995; B from Juksz, 1964.)

BOX 12A (continued)

autostereograms. Numerous graphic artists—preeminently in Japan, where the popularity of the autostereogram has been enormous—have generated many such images. As with the random dot stereogram, the task in viewing the autostereogram is not clear to the observer. Nonetheless, the hidden figure emerges, often after minutes of effort in which the brain automatically tries to make sense of the occult information.

(C)

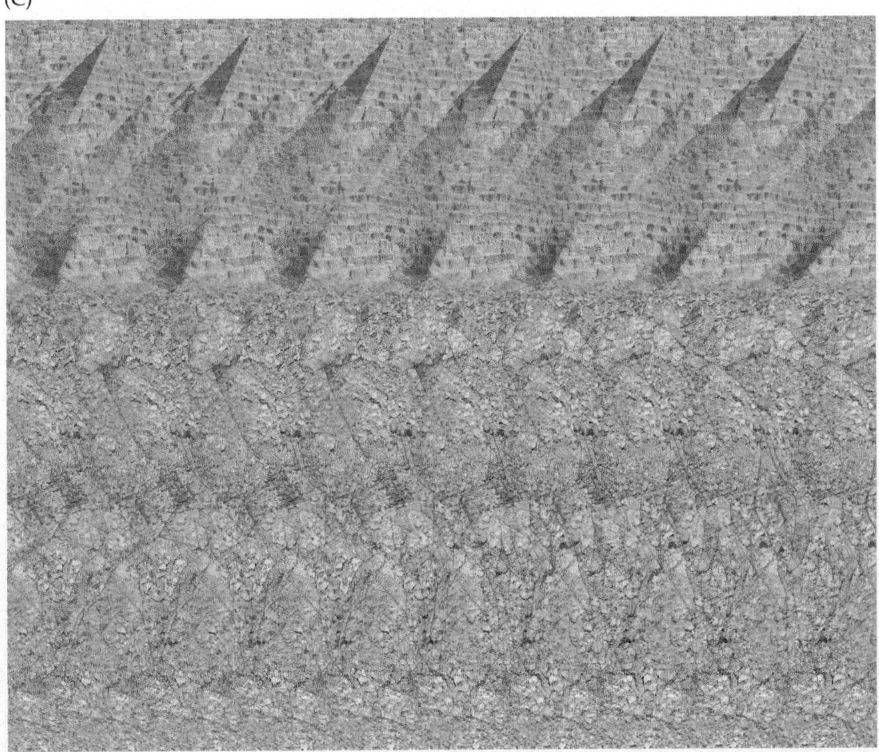

(C) An autostereogram. In this case, the hidden figure (two camels and pyramids) emerges by diverging the eyes. (© 3 Dimka/Shutterstock.)

eyes—for example, the clouding of one lens or the abnormal alignment of the eyes during infancy (*strabismus*)—can permanently reduce the effectiveness of one eye in driving cortical neurons, thus impairing the ability to use binocular information as a cue for depth. Early detection and correction of visual problems are therefore essential for normal visual function in maturity.

Division of Labor within the Primary Visual Pathway

In addition to being specific for input from one eye or the other, the layers in the lateral geniculate are also distinguished on the basis of cell size. Two ventral layers, which are composed of large neurons, are referred to as the **magnocellular layers**; the four more dorsal layers, composed of small neurons, are referred to as the **parvocellular layers**. The magno- and parvocellular layers receive inputs from distinct populations of ganglion cells that exhibit corresponding differences in cell size. M ganglion cells that terminate in the magnocellular layers have larger cell bodies, more extensive dendritic fields, and larger-diameter axons than the P ganglion cells that terminate in the parvocellular layers (Figure 12.13A,B). Moreover, the axons of relay cells in the magno- and parvocellular layers of the lateral geniculate nucleus terminate on distinct populations of neurons located in separate strata within layer 4C of primary visual cortex: magnocellular axons terminate in the upper part of layer 4C (4Cα), while parvocellular axons terminate in the lower part of layer 4C (4Cβ) (Figure 12.13C). Thus, the retinogeniculate pathway is composed of parallel magnocellular and parvocellular pathways that convey distinct types of information to the initial stages of cortical processing.

The response properties of the M and P ganglion cells provide clues about the contributions of the magno- and parvocellular pathways to visual perception. Both M and P ganglion cells exhibit the ON-center and OFF-center organization described in Chapter 11, but M ganglion cells have larger receptive fields than P cells, and their axons have faster conduction velocities. M and P ganglion cells also differ in ways that are not so obviously

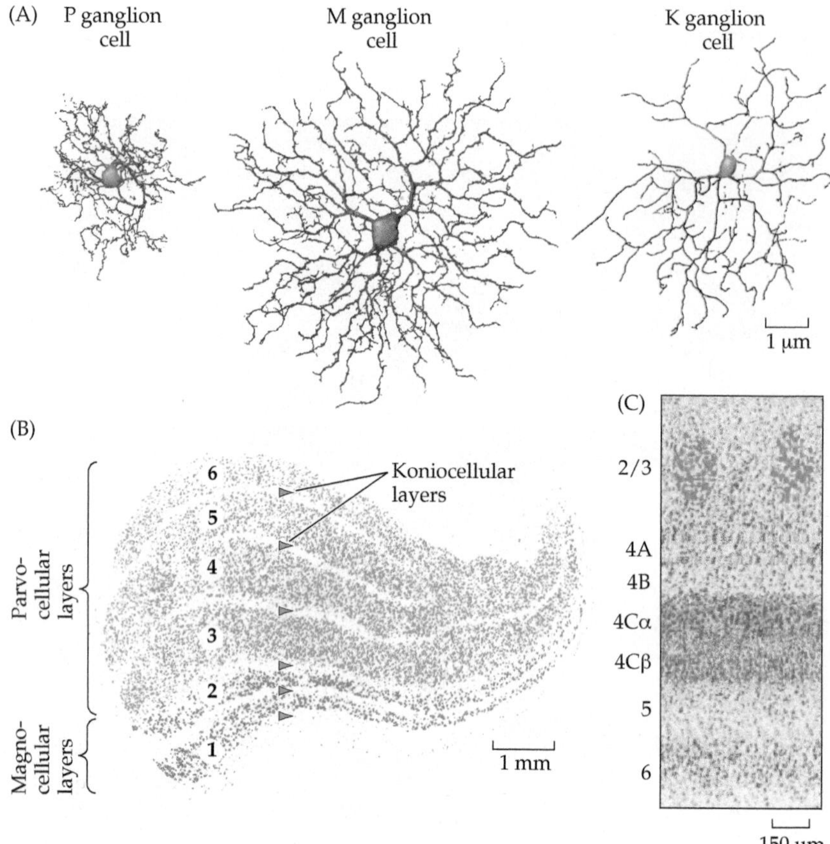

FIGURE 12.13 Magno-, parvo-, and koniocellular pathways. (A) Tracings of M, P, and K ganglion cells as seen in flat mounts of the retina. M cells have large-diameter cell bodies and large dendritic fields. They supply the magnocellular layers of the lateral geniculate nucleus. P cells have smaller cell bodies and dendritic fields. They supply the parvocellular layers of the lateral geniculate nucleus. K cells have small cell bodies and intermediate-sized dendritic fields. They supply the koniocellular layers of the lateral geniculate nucleus. (B) The human lateral geniculate nucleus showing the magnocellular, parvocellular, and koniocellular layers. (C) Termination of lateral geniculate axons in striate cortex. Magnocellular layers terminate in layer 4Cα, parvocellular layers terminate in layer 4Cβ, and koniocellular layers terminate in a patchy pattern in layers 2 and 3. Inputs to other layers have been omitted for simplicity (see Figure 12.8). (A after Watanabe and Rodieck, 1989; B courtesy of T. Andrews and D. Purves.)

related to their morphology. M cells respond transiently to the presentation of visual stimuli, while P cells respond in a sustained fashion. Moreover, P ganglion cells can transmit information about color, whereas M cells cannot. P cells convey color information because their receptive field centers and surrounds are driven by different classes of cones (i.e., cones responding with greatest sensitivity to either short-, medium-, or long-wavelength light). For example, some P ganglion cells have centers that receive inputs from long-wavelength-sensitive cones and surrounds that receive inputs from medium-wavelength-sensitive cones. Others have centers that receive inputs from medium wavelength-sensitive cones and surrounds from long-wavelength-sensitive cones (see Chapter 11). As a result, P cells are sensitive to differences in the wavelengths of light striking their receptive field center and surround. Although M ganglion cells also receive inputs from cones, there is no difference in the type of cone input to the receptive field center and surround; the center and surround of each M cell receptive field is driven by all cone types. The absence of cone specificity to center–surround antagonism makes M cells largely insensitive to differences in the wavelengths of light that strike their receptive field centers and surrounds, and they are thus unable to transmit color information to their central targets.

The contribution of the magno- and parvocellular pathways to visual perception has been tested experimentally by examining the visual capabilities of monkeys after selectively damaging either the magno- or parvocellular layers of the lateral geniculate nucleus. Damage to the magnocellular layers has little effect on visual acuity or color vision but sharply reduces the ability to perceive rapidly changing stimuli. In contrast, damage to the parvocellular layers has no effect on motion perception but severely impairs visual acuity and color perception. These observations suggest that the visual information conveyed by the parvocellular pathway is particularly important for high spatial resolution—the detailed analysis of the shape, size, and color of an object. The magnocellular pathway, by contrast, appears critical for tasks that require high temporal resolution, such as evaluating the location, speed, and direction of a rapidly moving object.

In addition to the magno- and parvocellular pathways, a third distinct anatomical pathway—the **koniocellular**, or **K-cell, pathway**—has been identified in the lateral geniculate nucleus (see Figure 12.13). Neurons contributing to the K-cell pathway reside in the interlaminar zones that separate lateral geniculate layers; these neurons receive inputs from fine-caliber retinal axons and project in a patchy fashion to the superficial layers (layers 2 and 3) of primary visual cortex. Although the contribution of the K-cell pathway to perception is not understood, it appears that some aspects of color vision, especially information derived from short-wavelength-sensitive cones, may be transmitted via the K-cell rather than the P-cell pathway. Why short-wavelength-sensitive cone signals should be processed differently from middle- and long-wavelength information is not

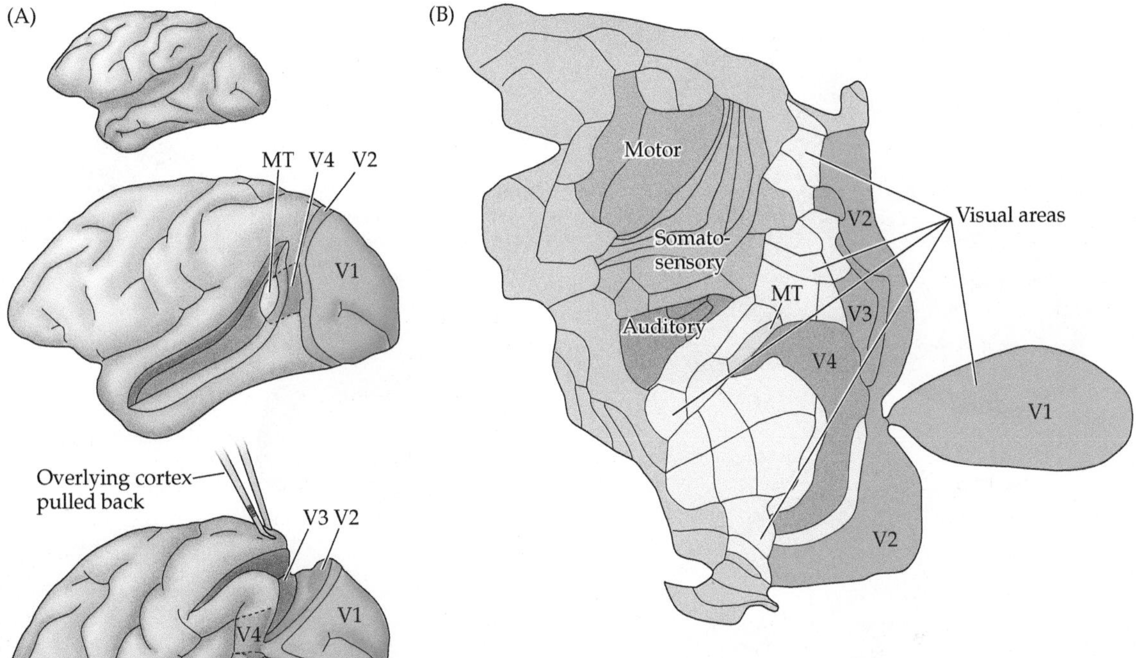

FIGURE 12.14 Subdivisions of the extrastriate cortex in the macaque monkey. (A) Each of the subdivisions indicated in color contains neurons that respond to visual stimulation. Many are buried in sulci, and the overlying cortex must be removed in order to expose them. Some of the more extensively studied extrastriate areas are specifically identified (V2, V3, V4, and MT). V1 is the primary visual cortex; MT is the middle temporal area. (B) The arrangement of extrastriate and other areas of neocortex in a flattened view of the monkey neocortex. At least 25 areas are predominantly or exclusively visual in function, and 7 additional areas are suspected to play a role in visual processing. (A after Maunsell and Newsome, 1987; B after Felleman and Van Essen, 1991.)

clear, but the distinction may reflect an earlier evolutionary origin of the K-cell pathway (see Chapter 11).

Functional Organization of Extrastriate Visual Areas

Anatomical and electrophysiological studies in monkeys have led to the discovery of a multitude of areas in the occipital, parietal, and temporal lobes that are involved in processing visual information (Figure 12.14). Each of these areas contains a map of visual space, and each is largely (but not exclusively) dependent on the primary visual cortex for its activation. The response properties of the neurons in some of these regions suggest that they are specialized for different aspects of the visual scene. For example, the **middle temporal area** (**MT**) contains neurons that respond selectively to the direction of a moving edge without regard to its color. In contrast, area **V4** contains a high percentage of neurons that respond selectively to the color of a visual stimulus without regard to its direction of movement. These physiological findings are supported by behavioral evidence; thus, damage to the MT leads to a specific impairment in a monkey's ability to perceive the direction of motion in a stimulus pattern, while other aspects of visual perception remain intact.

Functional imaging studies have indicated a similar arrangement of visual areas in human extrastriate cortex. Using retinotopically restricted stimuli, it has been possible to localize at least ten separate representations of the visual field (Figure 12.15). One of these areas exhibits a large motion-selective signal, suggesting that it is the homologue of the motion-selective middle temporal area described in monkeys. Another area exhibits color-selective responses, suggesting that it may be similar to V4 in non-human primates. A role for these areas in the perception of motion and color, respectively, is further supported by evidence for increases in activity not only during the presentation of the relevant stimulus, but also during periods when subjects experience motion or color afterimages.

The clinical description of selective visual deficits after localized damage to various regions of extrastriate cortex also supports functional specialization of these visual areas in humans. For example, a well-studied patient who suffered a stroke that damaged the extrastriate region thought to be comparable to the MT in the monkey was unable to appreciate the motion of objects, a rare disorder called **cerebral akinetopsia**. The neurologist who treated her noted that she had difficulty in pouring tea into a cup because the

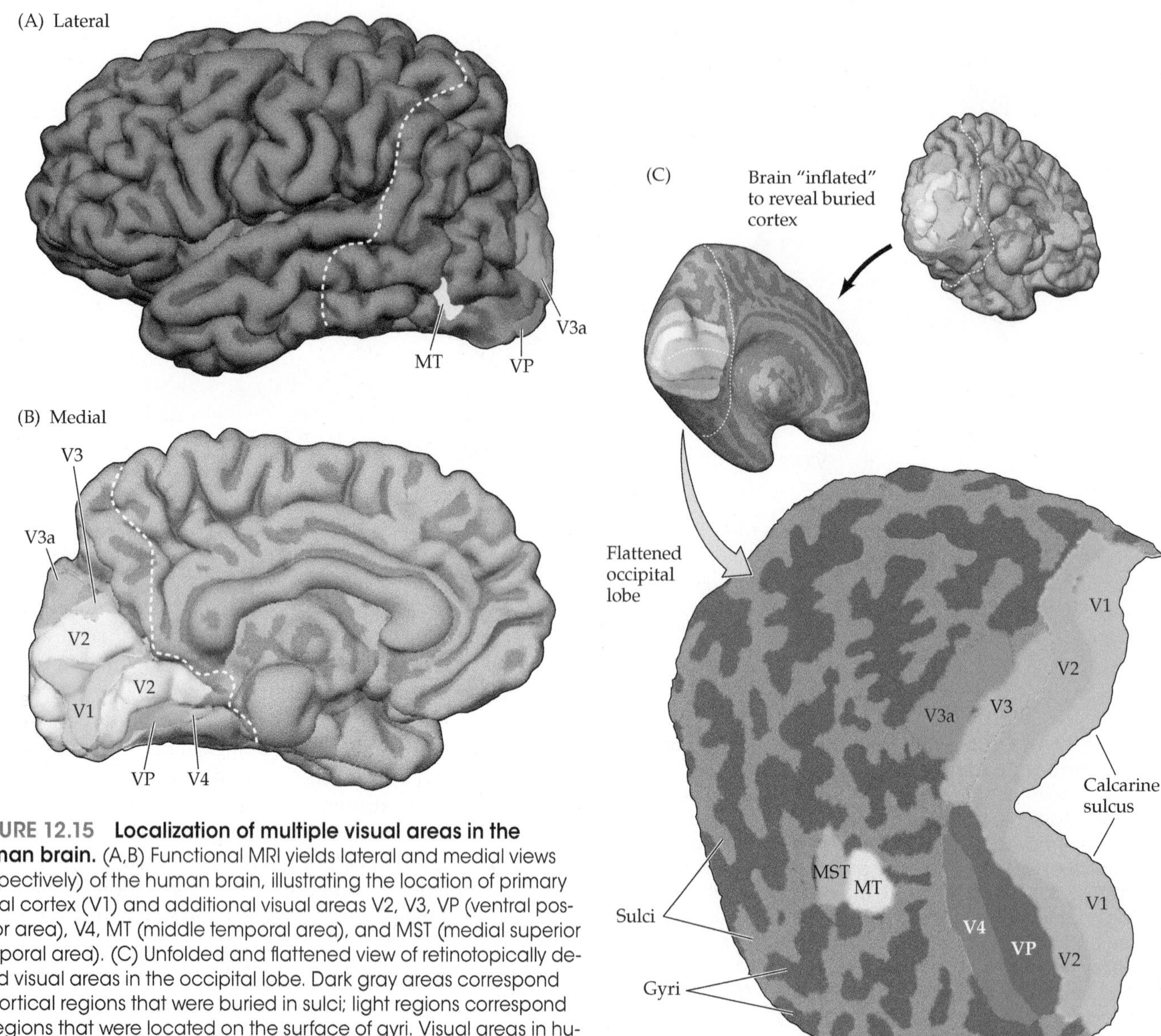

FIGURE 12.15 Localization of multiple visual areas in the human brain. (A,B) Functional MRI yields lateral and medial views (respectively) of the human brain, illustrating the location of primary visual cortex (V1) and additional visual areas V2, V3, VP (ventral posterior area), V4, MT (middle temporal area), and MST (medial superior temporal area). (C) Unfolded and flattened view of retinotopically defined visual areas in the occipital lobe. Dark gray areas correspond to cortical regions that were buried in sulci; light regions correspond to regions that were located on the surface of gyri. Visual areas in humans show a close resemblance to visual areas originally defined in monkeys (compare with Figure 12.14). (After Sereno et al., 1995.)

fluid seemed to be "frozen." In addition, she could not stop pouring at the right time because she was unable to perceive when the fluid level had risen to the brim. The patient also had trouble following a dialog because she could not follow the movements of the speaker's mouth. Crossing the street was potentially terrifying because she couldn't judge the movement of approaching cars. As the patient related, "When I'm looking at the car first, it seems far away. But then, when I want to cross the road, suddenly the car is very near." Her ability to perceive other features of the visual scene, such as color and form, was intact.

Another example of a specific visual deficit as a result of damage to extrastriate cortex is **cerebral achromatopsia**. These patients lose the ability to see the world in color, although other aspects of vision remain in good working order. The normal colors of a visual scene are described as being replaced by "dirty" shades of gray, much like looking at a poor-quality black-and-white movie. Achromatopsic individuals know the normal colors of objects—that a school bus is yellow, an apple red—but can no longer see them. When asked to draw objects from memory, they have no difficulty with shapes but are unable to appropriately

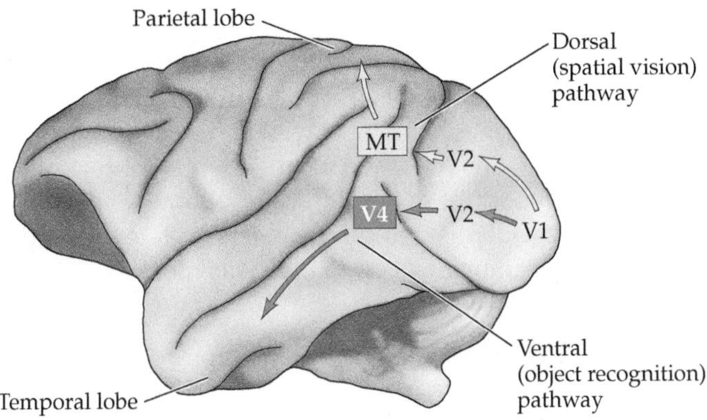

FIGURE 12.16 Visual areas beyond the striate cortex. Outside the occipital lobe, visual areas are broadly organized into two pathways: a ventral pathway that leads to the temporal lobe, and a dorsal pathway that leads to the parietal lobe. The ventral pathway plays an important role in object recognition, the dorsal pathway in spatial vision.

color the objects they have represented. It is important to distinguish this condition from the color blindness that arises from the congenital absence of one or more cone pigments in the retina (see Chapter 11). In achromatopsia, the three types of cones are functioning normally; it is damage to specific extrastriate cortical areas that renders the patient unable to use the information supplied by the retina.

Based on the anatomical connections between visual areas, differences in electrophysiological response properties, and the effects of cortical lesions, a consensus has emerged that extrastriate cortical areas are organized into two largely separate systems that eventually feed information into cortical association areas in the temporal and parietal lobes (see Chapter 27). One system, called the ventral stream, includes area V4 and leads from the striate cortex into the inferior part of the temporal lobe. This system is thought to be responsible for high-resolution form vision and object recognition. The dorsal stream, which includes the middle temporal area, leads from striate cortex into the parietal lobe. This system is thought to be responsible for spatial aspects of vision, such as the analysis of motion, and positional relationships between objects in the visual scene (Figure 12.16).

The functional dichotomy between these two streams is supported by observations on the response properties of neurons and the effects of selective cortical lesions. Neurons in the ventral stream exhibit properties that are important for object recognition, such as selectivity for shape, color, and texture. At the highest levels in this pathway, neurons exhibit even greater selectivity, responding preferentially to faces and objects (see Chapter 27). In contrast, those in the dorsal stream are not tuned to these properties, but show selectivity for direction and speed of movement. Consistent with this interpretation, lesions of the parietal cortex severely impair a monkey's ability to distinguish objects on the basis of their position, while having little effect on its ability to perform object-recognition tasks. In contrast, lesions of the inferior temporal cortex produce profound impairments in the ability to perform recognition tasks but no impairment in spatial tasks. These effects are remarkably similar to symptoms found after damage to the parietal and temporal lobe in humans (see Chapters 28 and 29).

What, then, is the relationship between these "higher order" extrastriate visual streams and the magno-, parvo-, and koniocellular pathways that supply the input to primary visual cortex? Despite the initial segregation of magno-, parvo-, and koniocellular inputs in primary visual cortex, at subsequent stages of processing these inputs at least partially converge. The extrastriate areas in the ventral stream clearly have access to the information conveyed by all three pathways, and the dorsal stream, while dominated by inputs from the magnocellular pathway, also receives inputs from the parvo- and koniocellular pathways. Even within primary visual cortex there is ample evidence for the convergence of information from these different lateral geniculate pathways. Thus, it would appear that the functions of higher visual areas involve the integration of information derived from distinct geniculocortical pathways, as well as input form other cortical regions.

Summary

Distinct populations of retinal ganglion cells send their axons to a number of central visual structures that serve different functions. The most important projections are to the pretectum for mediating the pupillary light reflex; to the hypothalamus for the regulation of circadian rhythms; to the superior colliculus for the regulation of eye and head movements; and—most important of all—to the lateral geniculate nucleus for mediating vision and visual perception. The retinogeniculostriate projection (the primary visual pathway) is arranged topographically such that central visual structures contain an organized map of the contralateral visual field. Damage anywhere along the primary visual pathway, which includes the optic nerve, optic tract, lateral geniculate nucleus, optic radiation, and striate cortex, results in a loss of vision confined to a predictable region of visual space. Compared with retinal ganglion cells, neurons at higher levels of the visual pathway become increasingly selective in their stimulus requirements. Thus, most neurons in the striate cortex respond to light–dark edges only if they are presented at a certain orientation at a particular locus in visual space, or to movement of the edge in a specific direction. The neural circuitry in the striate cortex also brings together information from the two eyes; most cortical neurons (other than those in layer 4, which are segregated into eye-specific columns) have binocular responses. Binocular convergence is presumably essential for the detection of binocular disparity, a key factor depth perception. The primary visual pathway is composed of separate functional pathways that convey information from different types of retinal ganglion cells to the initial stages of cortical processing. The magnocellular pathway conveys information that is critical for the detection of rapidly changing stimuli; the parvocellular pathway mediates high-acuity vision and appears to share responsibility for color vision with the koniocellular pathway. Finally, beyond striate cortex, parcellation of function continues in the ventral and dorsal streams that lead to the extrastriate and association areas in the temporal and parietal lobes, respectively. Areas in the inferotemporal cortex are especially important in object recognition, whereas areas in the parietal lobe are critical for understanding the spatial relationships among objects in the visual field.

ADDITIONAL READING

Reviews

Berson, D. M. (2003) Strange vision: Ganglion cells as circadian photoreceptors. *Trends Neurosci.* 26: 314–320.

Callaway, E. M. (2005) Neural substrates within primary visual cortex for interactions between parallel visual pathways. *Prog. Brain Res.* 149: 59–64.

Courtney, S. M. and L. G. Ungerleider (1997) What fMRI has taught us about human vision. *Curr. Opin. Neurobiol.* 7: 554–561.

Dhande, O. S., B. K. Stafford, J.-H. A. Lim and A. D. Huberman (2015) Contributions of retinal ganglion cells to subcortical visual processing and behaviors. *Annu. Rev. Vis. Sci.* 1: 291–328.

Do, M. T. and K. W. Yau (2010) Intrinsically photosensitive retinal ganglion cells. *Physiol. Rev.* 90: 1547–1581.

Felleman, D. J. and D. C. Van Essen (1991) Distributed hierarchical processing in primate cerebral cortex. *Cerebral Cortex* 1: 1–47.

Felsen, G. and Y. Dan (2005) A natural approach to studying vision. *Nat. Neurosci.* 8: 1643–1646.

Grill-Spector, K. and R. Malach (2004) The human visual cortex. *Annu. Rev Neurosci.* 27: 649–677.

Hendry, S. H. and R. C. Reid (2000) The koniocellular pathway in primate vision. *Annu. Rev. Neurosci.* 23: 127–153.

Hubel, D. H. and T. N. Wiesel (1977) Functional architecture of macaque monkey visual cortex. *Proc. R. Soc. (Lond.) B* 198: 1–59.

Maunsell, J. H. R. (1992) Functional visual streams. *Curr. Opin. Neurobiol.* 2: 506–510.

Nassi, J. J. and E. M. Callaway (2009) Parallel processing strategies of the primate visual system. *Nat. Rev. Neurosci.* 10: 360–372.

Olshausen, B. A. and D. J. Field (2004) Sparse coding of sensory inputs. *Curr. Opin. Neurobiol.* 14: 481–487.

Sanes, J. R. and R. H. Masland (2015) The types of retinal ganglion cells: current status and implications for neuronal classification. *Annu. Rev. Neurosci.* 38: 221–246.

Schiller, P. H. and N. K. Logothetis (1990) The color-opponent and broad-band channels of the primate visual system. *Trends Neurosci.* 13: 392–398.

Sincich, L. C. and J. C. Horton (2005) The circuitry of V1 and V2: Integration of color, form, and motion. *Annu. Rev. Neurosci.* 28: 303–326.

Tootell, R. B., A. M. Dale, M. I. Sereno and R. Malach (1996) New images from human visual cortex. *Trends Neurosci.* 19: 481–489.

Ungerleider, J. G. and M. Mishkin (1982) Two cortical visual systems. In *Analysis of Visual Behavior,* D. J. Ingle, M. A. Goodale and R. J. W. Mansfield (eds.). Cambridge, MA: MIT Press, pp. 549–586.

Important Original Papers

Basole, A., L. E. White and D. Fitzpatrick (2003) Mapping multiple features in the population response of visual cortex. *Nature* 423: 986–990.

Brincat. S. L. and C. E. Connor (2004) Underlying principles of visual shape selectivity in posterior inferotemporal cortex. *Nat. Neurosci.* 7: 880–886.

Chong, E., A. M. Familiar and W. M. Shim (2016) Reconstructing representations of dynamic visual objects in early visual cortex. *Proc. Natl. Acad. Sci. USA* 113: 1453–1258.

Feinberg, E. H. and M. Meister (2015) Orientation columns in the mouse superior colliculus. *Nature* 519: 229–232.

Glasser, M. F. and 11 others (2016). A multi-modal parcellation of human cerebral cortex. *Nature* 536: 171–178.

Hattar, S., H. W. Liao, M. Takao, D. M. Berson and K. W. Yau (2002) Melanopsin-containing retinal ganglion cells: architecture, projections, and intrinsic photosensitivity. *Science* 295: 1065–1070.

Horton, J. C. and D. L. Adams (2005). The cortical column: a structure without a function. *Phil. Trans. R. Soc. B* 360: 837–862.

Hubel, D. H. and T. N. Wiesel (1962) Receptive fields, binocular interaction and functional architecture in the cat's visual cortex. *J. Physiol.* (Lond.) 160: 106–154.

Hubel, D. H. and T. N. Wiesel (1968) Receptive fields and functional architecture of monkey striate cortex. *J. Physiol.* (Lond.) 195: 215–243.

Hung, C. P., G. Kreiman, T. Poggio and J. J. DiCarlo (2005) Fast readout of object identity from macaque inferior temporal cortex. *Science* 310: 863–866.

Huth, A. G., W. A. de Heer, T. L. Griffiths, F. E. Theunissen and J. L. Gallant (2016) Natural speech reveals the semantic maps that tile human cerebral cortex. Nature 532: 453–458.

Michel, M. M., Y. Chen, W. S. Geisler and E. Seidemann (2013) An illusion predicted by V1 population activity implicates cortical topography in shape perception. *Nat. Neurosci.* 16: 1477–1483.

Nadler, J. W. and 5 others (2013) Joint representation of depth from motion parallax and binocular disparity cues in macaque area MT. *J. Neurosci.* 33: 14061–14074.

Ohki, K. and 5 others (2006) Highly ordered arrangement of single neurons in orientation pinwheels. *Nature* 442: 925–928.

Sereno, M. I. and 7 others (1995) Borders of multiple visual areas in humans revealed by functional magnetic resonance imaging. *Science* 268: 889–893.

Zihl, J., D. von Cramon and N. Mai (1983) Selective disturbance of movement vision after bilateral brain damage. *Brain* 106: 313–340.

Books

Brodmann K (1909). *Vergleichende Lokalisationslehre der Grosshirnrinde* (in German). Leipzig: Johann Ambrosius Barth.

Chalupa, L. M. and J. S. Werner (eds.) (2004) *The Visual Neurosciences.* Cambridge, MA: MIT Press.

Horton, J. C. (1992) The central visual pathways. In *Alder's Physiology of the Eye.* W. M. Hart (ed.). St. Louis: Mosby Yearbook.

Hubel, D. H. (1988) *Eye, Brain, and Vision.* New York: Scientific American Library.

Rodieck, R. W. (1998) *The First Steps in Seeing.* Sunderland, MA: Sinauer Associates.

Schwartz, S. H. (2009) *Visual Perception: A Clinical Orientation,* 4th Edition. New York: McGraw-Hill.

Sherrington, S. C. (1947) *The Integrative Action of the Nervous System.* 2nd edition. New Haven, CT: Yale University Press.

Zeki, S. (1993) *A Vision of the Brain.* Oxford: Blackwell Scientific Publications.

Go to the NEUROSCIENCE 6e Companion Website at **www.oup.com/uk/Purves6e** for Web Topics, Animations, Flashcards, and more.

CHAPTER

13 The Auditory System

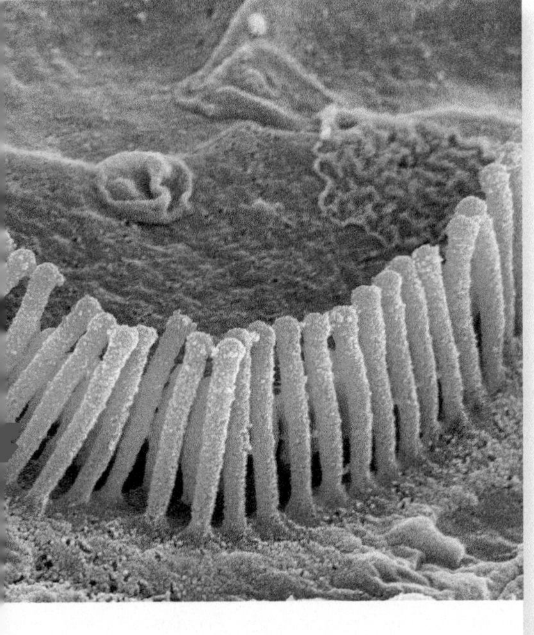

Overview

THE AUDITORY SYSTEM IS ONE OF THE engineering masterpieces of the human body. At the periphery of the system is an array of miniature acoustical detectors packed into a space no larger than a pea. These detectors can transduce vibrations as small as the diameter of an atom, and they respond 1000 times faster than visual photoreceptors. Such rapid responses to acoustical cues, which are paralleled by rapid signaling in the auditory brainstem, facilitate the initial orientation of the head and body to novel stimuli, especially those stimuli that are not initially within the field of view. Human social communication is largely mediated by the auditory system, making the auditory system at least as critical for well being as the visual system; indeed, loss of hearing can be more socially debilitating than blindness. From a cultural perspective, the auditory system is essential not only to understanding speech, but also to perceiving music, one of the most aesthetically sophisticated forms of human expression. For these and other reasons, audition represents a fascinating and especially important mode of sensation.

Sound

In physical terms, *sound* refers to pressure waves generated by vibrating air molecules; in more casual usage (and somewhat confusingly), *sound* refers to an auditory percept. Physical sound waves radiate in three dimensions, creating concentric spheres of alternating compression and rarefaction. Like all wave phenomena, sound waves have four major features: waveform, phase, amplitude (usually expressed in logarithmic units known as decibels, abbreviated dB), and frequency (expressed in cycles per second or Hertz, abbreviated Hz). For a human listener, the amplitude and frequency of a sound pressure change at the ear roughly correspond to that listener's experience of **loudness** and **pitch**, respectively.

The waveform of a sound stimulus is its amplitude plotted against time. To understand this concept, it helps to begin by visualizing an acoustical waveform as a sine wave. Figure 13.1 diagrams the behavior of air molecules near a tuning fork that vibrates sinusoidally when struck. The vibrating tines of the tuning fork produce local displacements of the surrounding molecules, such that when the tine moves in one direction, there is molecular condensation; when it moves in the other direction, there is rarefaction. These changes in density of the air molecules are equivalent to local changes in air pressure. Such regular, sinusoidal cycles of compression and rarefaction can be thought of as a form of circular motion, with one complete cycle equivalent to one full revolution (360°). This point can be illustrated with two sinusoids of the same frequency projected onto a circle, a strategy that also makes it easier to understand the concept of phase (Figure 13.2). Imagine that two tuning forks, both of which resonate at the same frequency, are struck at slightly different times. At a given time $t = 0$, one wave is at position P and the other

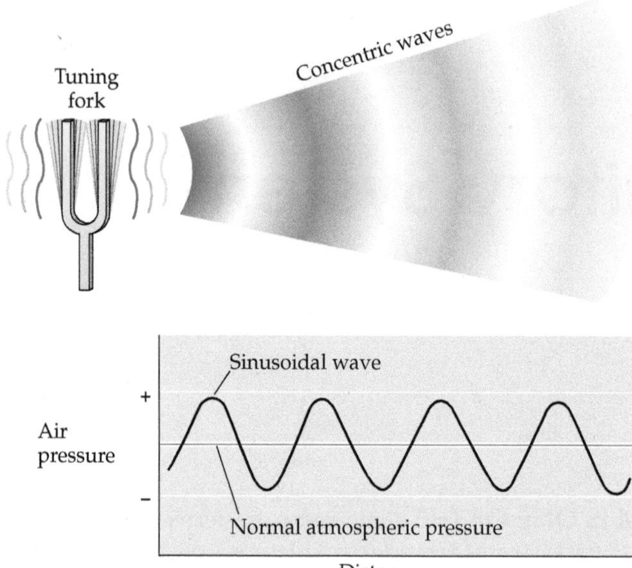

FIGURE 13.1 A sine wave. Diagram of the periodic condensation and rarefaction of air molecules produced by the vibrating tines of a tuning fork. The molecular disturbance of the air is pictured as if frozen at the instant the constituent molecules responded to the resultant pressure wave. Shown below is a plot of the air pressure versus distance from the fork. Note its sinusoidal quality.

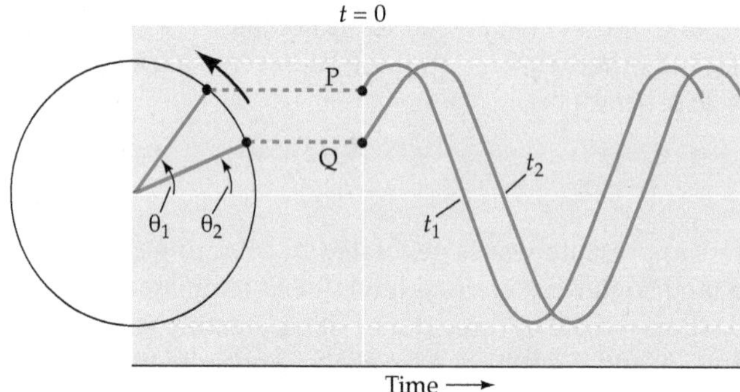

FIGURE 13.2 A sine wave and its projection as circular motion. The two sinusoids shown are at different phases, such that point P corresponds to phase angle θ_1 and point Q corresponds to phase angle θ_2.

at position Q. By projecting P and Q onto the circle, their respective phase angles, θ_1 and θ_2, are apparent. The sine wave that begins at P reaches a particular point on the circle, say 180°, at time t_1, whereas the wave that begins at Q reaches 180° at time t_2. Thus, phase differences have corresponding time differences, a concept that is important in appreciating how the auditory system locates sounds in space.

Although useful in a didactic sense, it must be kept in mind that sounds composed of single sine waves (i.e., pure tones) are rarely found in nature. At any instant, most sounds in speech, for example, contain energy distributed across a broad frequency spectrum, while some other environmental stimuli (e.g., a babbling brook, wind in trees) have little or no periodicity. Spectrally complex waveforms (such as those produced in speech) that have a periodic quality can be modeled as the sum of sinusoidal waves of varying amplitudes, frequencies, and phases. In engineering applications, an algorithm called the *Fourier transform* decomposes a complex signal into its sinusoidal components. In the auditory system, as will be apparent later in the chapter, the inner ear acts as a sort of acoustical prism, decomposing complex sounds into a myriad of constituent frequencies. In addition to complex spectral features, many biologically important sounds, including animal vocalizations, speech, and music, are characterized by temporal modulations in the sound's amplitude envelope (Figure 13.3). Along with its ability to encode spectral features, the auditory system's ability to encode these temporal features is important to speech and melody perception and more generally to the perceptual grouping of auditory objects in complex acoustical scenes (i.e., such as the soprano melody in a four-voice fugue).

The human ear is extraordinarily sensitive to changes in sound pressure. At the threshold of hearing, air molecules are displaced an average of only 10 picometers (10^{-11} m), and the intensity of such a sound is about one-trillionth of a watt per square meter! Even dangerously high sound pressure levels (>100 dB) have power at the eardrum that is only in the milliwatt range.

The Audible Spectrum

Humans with normal hearing are able to detect sounds that fall within a frequency range from about 20 Hz to 20 kHz, with the upper limit dropping off somewhat in adulthood. Not all mammalian species are sensitive to the same range of frequencies, and most small mammals are sensitive to very high frequencies but not to low frequencies. For instance, some species of bats are sensitive to tones as high as 200 kHz, but their lower limit is around 20 kHz—the upper limit for young people with normal hearing. Different animal species also tend to emphasize certain frequency bandwidths in both their vocalizations and their range of hearing. In general, vocalizations, by virtue of their periodicity, can be distinguished from the noise "barrier" created by environmental sounds, such as babbling brooks and rustling leaves. Animals that echolocate, such as bats and dolphins, rely on very high frequency vocal sounds to maximally resolve spatial features of the target, while animals intent on avoiding predation have auditory systems "tuned" to the low-frequency vibrations that approaching predators transmit through the substrate. These behavioral differences are mirrored by a wealth of anatomical and functional specializations throughout the auditory system.

A Synopsis of Auditory Function

The auditory system transforms sound stimuli into distinct patterns of neural activity. These patterns are then integrated with information from other sensory systems and

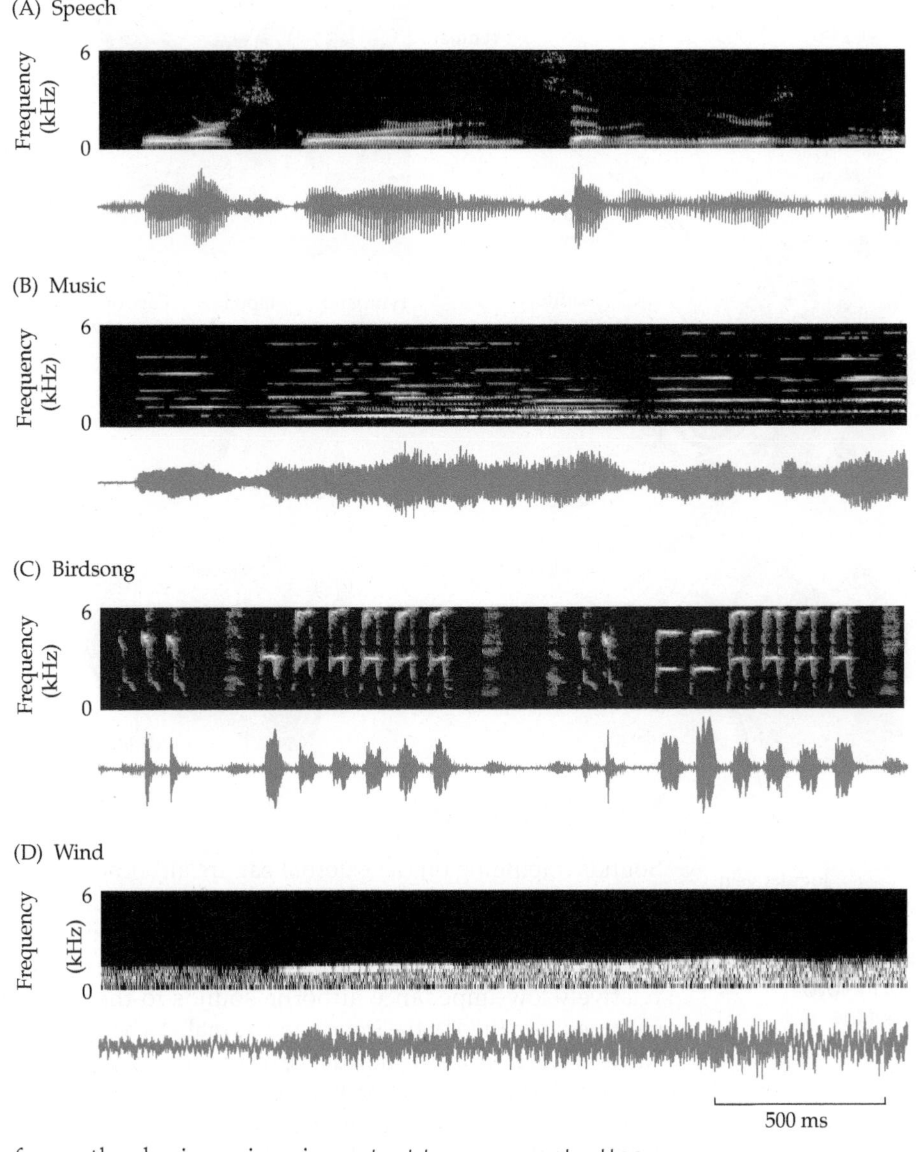

FIGURE 13.3 Examples of different sounds. In each case, the top panel is a sonogram (frequency versus time with increasing intensity represented by hotter colors) and the bottom panel is an oscillogram (amplitude versus time plot). Note that animal vocalizations, speech, and music can contain highly periodic (tonal and harmonic) elements, whereas environmental sounds such as wind lack such periodic structure. (Courtesy of Timothy Warren.)

from other brain regions important to movement, attention, and arousal to guide behaviors that include orienting to acoustical stimuli, engaging in intraspecies communication, and distinguishing self-generated sounds from other sounds in the environment. The first stage of this transformation occurs at the external and middle ears, which collect sound waves and amplify their pressure so that the sound energy in the air can be successfully transmitted to the fluid-filled cochlea of the inner ear. In the inner ear, a series of biomechanical processes enable the frequency, amplitude, and phase of the original signal to be transduced by the sensory **hair cells** and encoded by the electrical activity of the auditory nerve fibers. One product of this process of acoustical decomposition is the systematic representation of sound frequency along the length of the cochlea, referred to as **tonotopy**, which is an important organizational feature preserved throughout the central auditory pathways.

The earliest stage of central processing occurs at the cochlear nucleus, where the peripheral auditory information diverges into several parallel central pathways. Accordingly, the output of the cochlear nucleus has several targets. One of these is the superior olivary complex, the first point at which information from the two ears interacts and the site of the initial processing of the cues that allow listeners to localize sound in space. The cochlear nucleus also projects to the inferior colliculus of the midbrain, a major integrative center and the first point at which auditory information can interact with the motor system. The inferior colliculus relays auditory information to the thalamus and cortex, the latter of which plays a prominent role in the perception of speech and music. The large number of stations between the auditory periphery and the cortex far exceeds those in other sensory systems, providing a hint that the perception of communication and environmental sounds is an especially critical and intensive neural process. Furthermore, both the peripheral and central auditory systems are "tuned" to conspecific communication vocalizations, pointing to the interdependent evolution of the entire system for perceiving these signals.

The External Ear

The external ear, which consists of the **pinna**, **concha**, and **auditory meatus**, gathers sound energy and focuses it on the eardrum, or **tympanic membrane** (Figure 13.4). One consequence of the configuration of the human auditory meatus is that it selectively boosts the sound pressure 30- to 100-fold for frequencies around 3 kHz via passive resonance effects. This amplification makes humans especially sensitive to frequencies in the range of 2–5 kHz—and also explains why we are particularly prone to hearing loss near this frequency following exposure to high-intensity broadband noise, such as that generated by heavy machinery or explosives. In humans, the sensitivity to this frequency range appears to be directly related to speech perception. Although human speech is a broadband signal,

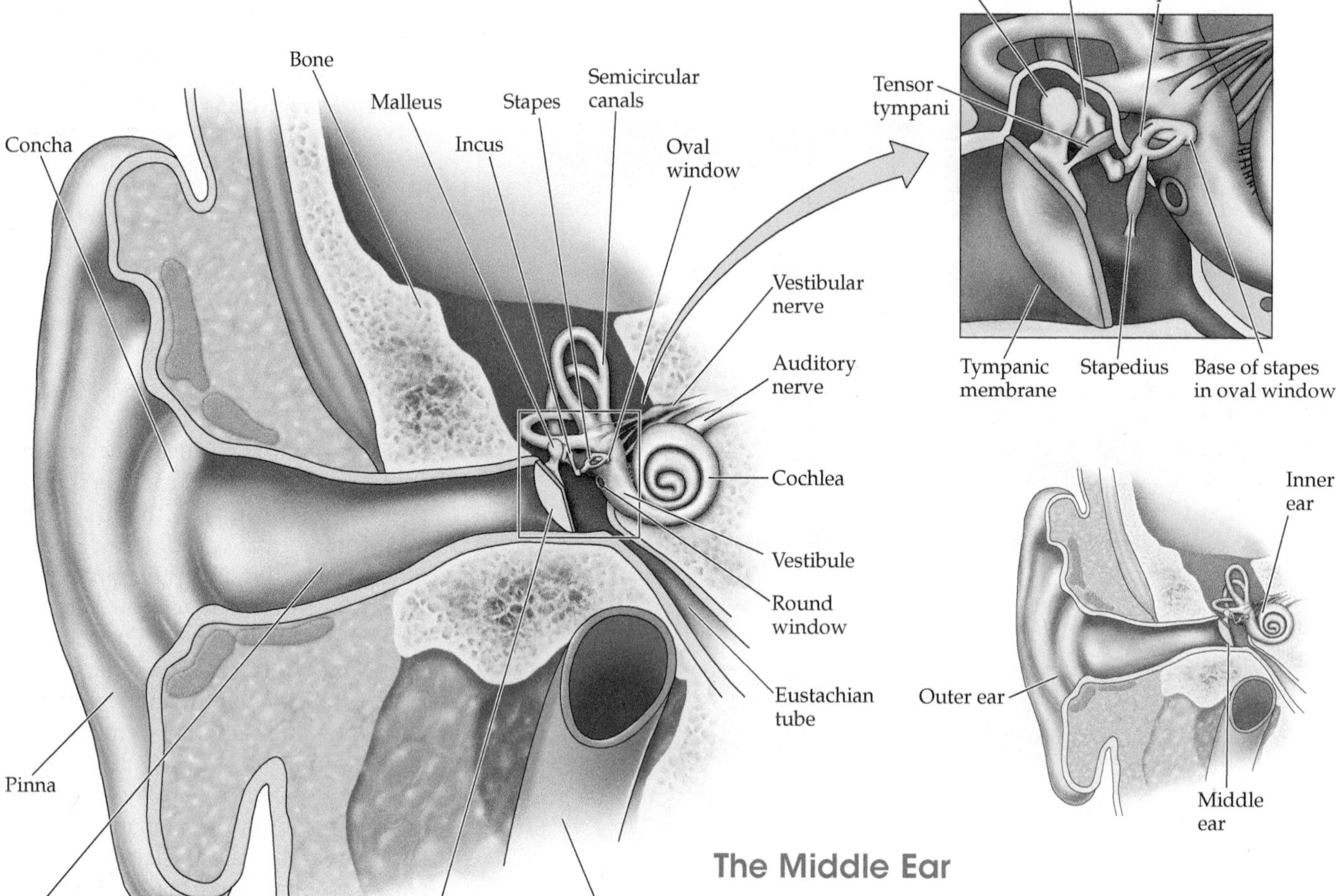

FIGURE 13.4 The human ear. Note the large surface area of the tympanic membrane (eardrum) relative to the oval window. This feature, along with the lever action of the malleus, incus, and stapes, facilitates transmission of airborne sounds to the fluid-filled cochlea.

important spectral cues used for discriminating different speech sounds, including plosive consonants (e.g., *ba* and *pa*), are concentrated around 3 kHz (see Box 33A). Therefore, selective hearing loss in the range of 2–5 kHz disproportionately degrades speech recognition.

A second important function of the pinna and concha is to selectively filter different sound frequencies in order to provide cues about the elevation of the sound source. The vertically asymmetrical convolutions of the pinna are shaped so that the external ear transmits more high-frequency components from an elevated source than from the same source at ear level. This effect can be demonstrated by recording identical sounds from different elevations after they have passed through an "artificial" external ear; when the recorded sounds are played back via earphones, so that the whole series is presented from a source at the same elevation relative to the listener, the recordings from higher elevations are perceived as coming from positions higher in space than the recordings from lower elevations.

The Middle Ear

Sounds impinging on the external ear are airborne; however, the environment within the inner ear, where the sound-induced vibrations are converted to neural impulses, is aqueous. The major function of the middle ear is to match relatively low-impedance airborne sounds to the higher-impedance fluid of the inner ear. Normally, when sound waves travel from a low-impedance medium such as air to a much higher-impedance medium such as water, almost all (>99.9%) of the acoustical energy is reflected. The middle ear (see Figure 13.4) overcomes this problem and ensures transmission of the sound energy across the air–fluid boundary by boosting the pressure measured at the tympanic membrane almost 200-fold by the time it reaches the inner ear.

Two mechanical processes occur within the middle ear to achieve this large pressure gain. The first and major boost is achieved by focusing the force impinging on the relatively large-diameter tympanic membrane onto the much smaller-diameter **oval window**, the site where the bones of the middle ear contact the inner ear. A second and related process relies on the mechanical advantage gained by the lever action of the three small, interconnected middle ear bones, or **ossicles** (i.e., the malleus, incus, and stapes; see Figure 13.4), which connect the tympanic membrane to the oval window. **Conductive hearing loss**, which involves damage to the external or middle ear, lowers the efficiency at which sound energy is transferred to the inner ear and can be partially overcome by artificially boosting sound pressure levels with an external hearing aid. In normal hearing, the efficiency of sound transmission to the inner ear also is regulated by two small muscles in the middle ear, the tensor tympani, innervated by cranial nerve V, and the stapedius, innervated by cranial nerve VII (see the Appendix). Contraction of these muscles, which is triggered automatically by loud noises or during self-generated vocalization, counteracts the movement of the ossicles and reduces the amount of sound energy transmitted to the cochlea, serving to protect the inner ear. Conversely, conditions that lead to flaccid paralysis of either of these muscles, such as Bell's palsy (nerve VII), can trigger a painful sensitivity to moderate or even low-intensity sounds known as **hyperacusis**.

Bony tissues and soft tissues, including those surrounding the inner ear, have impedance values close to that of water. Therefore, even without an intact tympanic membrane or middle ear ossicles, acoustical vibrations of sufficient energy, such as those arising from a tuning fork directly touching the head, can still be transferred directly through the bones and tissues of the head to the inner ear. In the clinic, the Weber test uses a tuning fork placed against the scalp to determine whether hearing loss is due to conductive problems or to damage either to the hair cells of the inner ear or to the auditory nerve itself (**sensorineural hearing loss**).

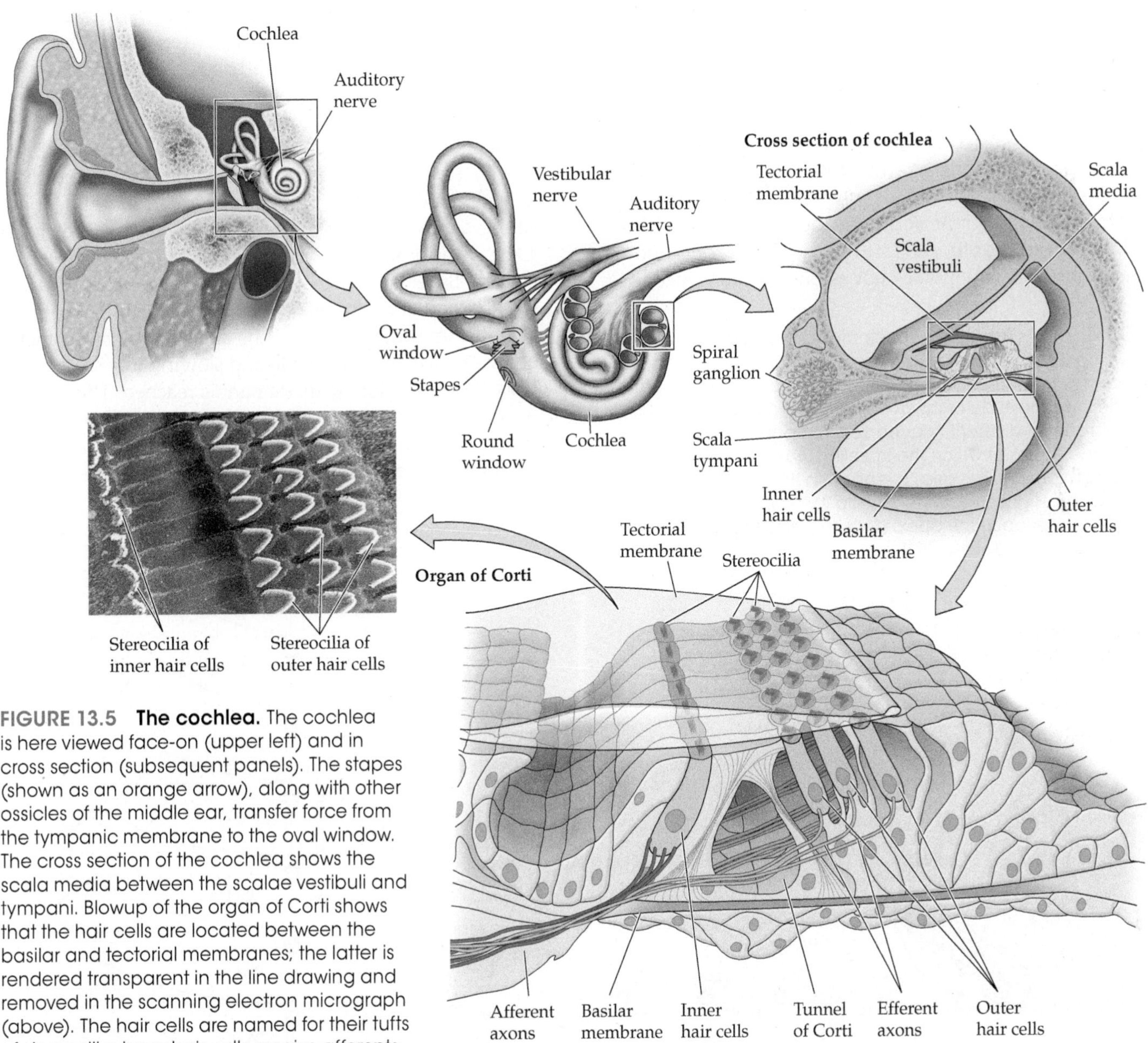

FIGURE 13.5 **The cochlea.** The cochlea is here viewed face-on (upper left) and in cross section (subsequent panels). The stapes (shown as an orange arrow), along with other ossicles of the middle ear, transfer force from the tympanic membrane to the oval window. The cross section of the cochlea shows the scala media between the scalae vestibuli and tympani. Blowup of the organ of Corti shows that the hair cells are located between the basilar and tectorial membranes; the latter is rendered transparent in the line drawing and removed in the scanning electron micrograph (above). The hair cells are named for their tufts of stereocilia; inner hair cells receive afferents from cranial nerve VIII, whereas outer hair cells receive mostly efferent innervation. (Micrograph from Counter et al., 1991.)

The Inner Ear

The **cochlea** of the inner ear is the site where the energy from sonically generated pressure waves is transformed into neural impulses. The cochlea not only amplifies sound waves and converts them into neural signals, but it also acts as a mechanical frequency analyzer, decomposing complex acoustical waveforms into simpler elements. Many features of auditory perception accord with aspects of the physical properties of the cochlea; hence, it is important to consider this structure in some detail.

The cochlea (from the Latin for "snail") is a small (~10 mm wide) coiled structure, which, were it uncoiled, would form a tube about 35 mm long (Figures 13.5 and 13.6). Both the oval window and the **round window**, another region where the bone is absent surrounding the cochlea, are at the basal end of this tube. The cochlea is bisected from its basal end almost to its apical end by the cochlear partition, a flexible structure that supports the **basilar membrane** and the **tectorial membrane**. There are fluid-filled chambers on each side of the cochlear partition, called the **scala vestibuli** and the **scala tympani**. A distinct channel, the **scala media**, runs within the cochlear partition. The cochlear partition does not extend all the way to the apical end of the cochlea; instead, an opening known as the **helicotrema** joins the scala vestibuli to the scala tympani, allowing their fluid, known as **perilymph**, to mix. One consequence of this structural arrangement is that inward movement of the oval window displaces the fluid of the inner ear, causing the round window to bulge out slightly and deforming the cochlear partition.

The manner in which the basilar membrane vibrates in response to sound is the key to understanding how hearing is initiated. Measurements of the vibration of different parts of the basilar membrane, as well as the discharge rates of individual auditory nerve fibers that terminate along its length, show that both of these features are tuned; that is, although they respond to a broad range of frequencies, they respond most intensely to a specific frequency. Frequency tuning within the inner ear is attributable in part to the geometry of the basilar membrane, which is wider and more flexible at the apical end and narrower and stiffer at the basal end. Georg von Békésy, working at Harvard University, showed that a membrane that varies systematically in its width and flexibility vibrates maximally at different positions as a function of the stimulus frequency (see Figure 13.6). Using models and human cochleas taken from cadavers, von Békésy found that an acoustical stimulus such as a sine tone initiates a **traveling wave** in the cochlea that propagates from the base toward the apex of the basilar membrane, growing in amplitude and slowing in velocity until a point of maximum displacement is reached. The point of maximum displacement is determined by the frequency of the stimulus and persists vibrating in that pattern as long as the tone endures. The points responding to high frequencies

FIGURE 13.6 Traveling waves along the cochlea. A traveling wave is shown at a given instant along the cochlea, which has been uncoiled for clarity. The graphs on the right profile the amplitude of the traveling wave along the basilar membrane for different frequencies. The position (labeled 1–7 in the figure) at which the traveling wave reaches its maximum amplitude varies directly with the frequency of stimulation: Higher frequencies map to the base, and lower frequencies map to the apex. (Drawing after Dallos, 1992; graphs after von Békésy, 1960.)

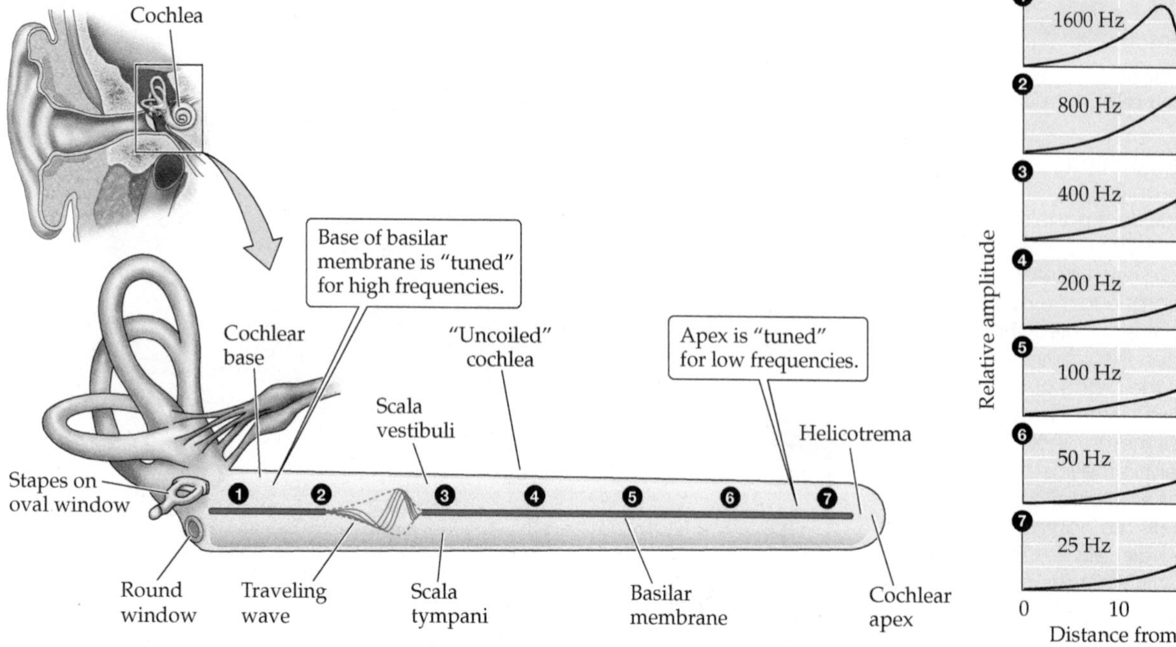

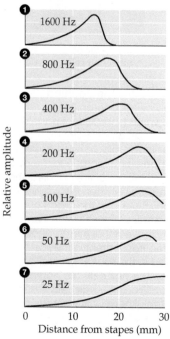

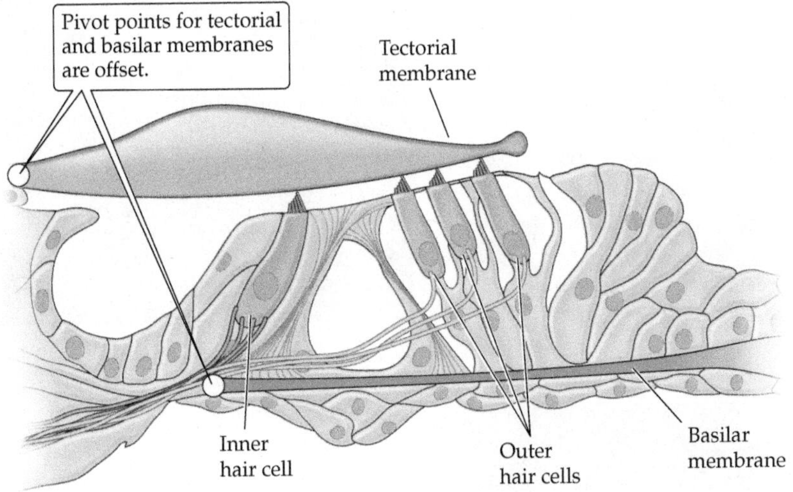

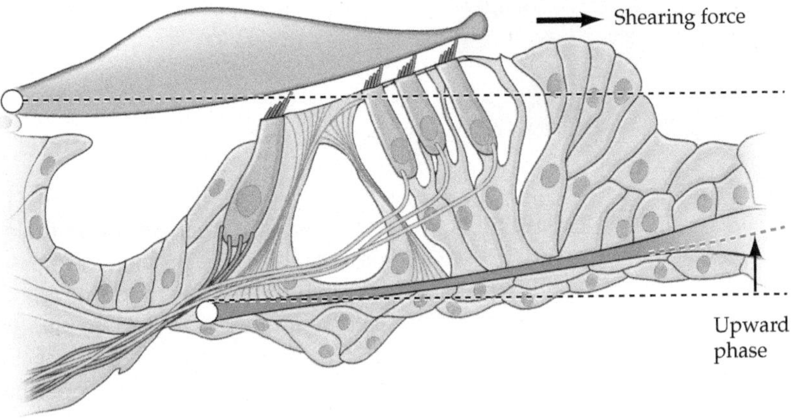

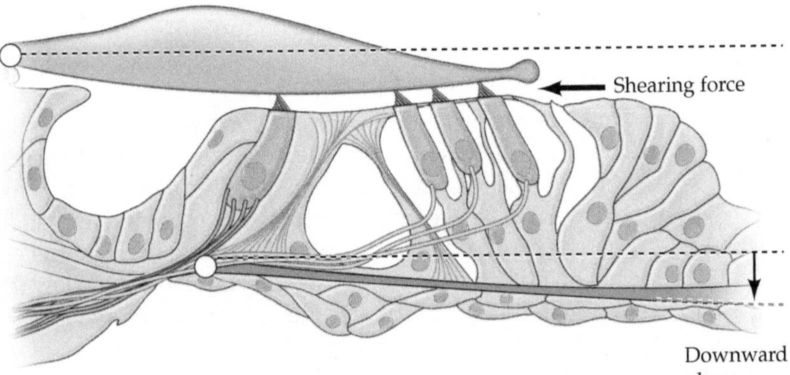

FIGURE 13.7 Traveling waves initiate auditory transduction. Vertical movement of the basilar membrane is translated into a shearing force that bends the stereocilia of the hair cells. The pivot point of the basilar membrane is offset from the pivot point of the tectorial membrane so that when the basilar membrane is displaced, the tectorial membrane moves across the tops of the hair cells, bending the stereocilia.

are at the base of the basilar membrane, and the points responding to low frequencies are at the apex, giving rise to a topographical mapping of frequency (i.e., tonotopy). Spectrally complex stimuli cause a pattern of vibration equivalent to the superposition of the vibrations generated by the individual tones making up that complex sound, thus accounting for the decompositional aspects of cochlear function mentioned earlier. This process of spectral decomposition appears to be an important strategy for detecting the various harmonic combinations that distinguish natural sounds that have a periodic character, such as animal vocalizations, including vowels and some consonants in speech.

The traveling wave initiates sensory transduction by displacing the sensory hair cells that sit atop the basilar membrane. Because the basilar membrane and the overlying tectorial membrane are anchored at different positions, the vertical component of the traveling wave is translated into a shearing motion between these two membranes (Figure 13.7). This motion bends the tiny processes, called **stereocilia**, that protrude from the apical ends of the hair cells, leading to voltage changes across the hair cell membrane. How the bending of stereocilia leads to receptor potentials in hair cells is considered in the following section.

Hair Cells and the Mechanoelectrical Transduction of Sound Waves

The cochlear hair cells in humans consist of one row of inner hair cells and three rows of outer hair cells (see Figures 13.5 and 13.7). The inner hair cells are the sensory receptors, and 95% of the fibers of the auditory nerve that project to the brain arise from this subpopulation. The terminations on the outer hair cells are almost all from efferent axons that arise from cells in the superior olivary complex. Current evidence suggests that the outer hair cells play an important role in modulating basilar membrane motions and that they function as an important component of the cochlear amplifier (see below).

The hair cell is a flask-shaped epithelial cell named for the bundle of hairlike processes that protrude from its apical end into the scala media. Each hair bundle contains anywhere from 30 to a few hundred stereocilia, with one taller **kinocilium** (Figure 13.8A). Despite their names,

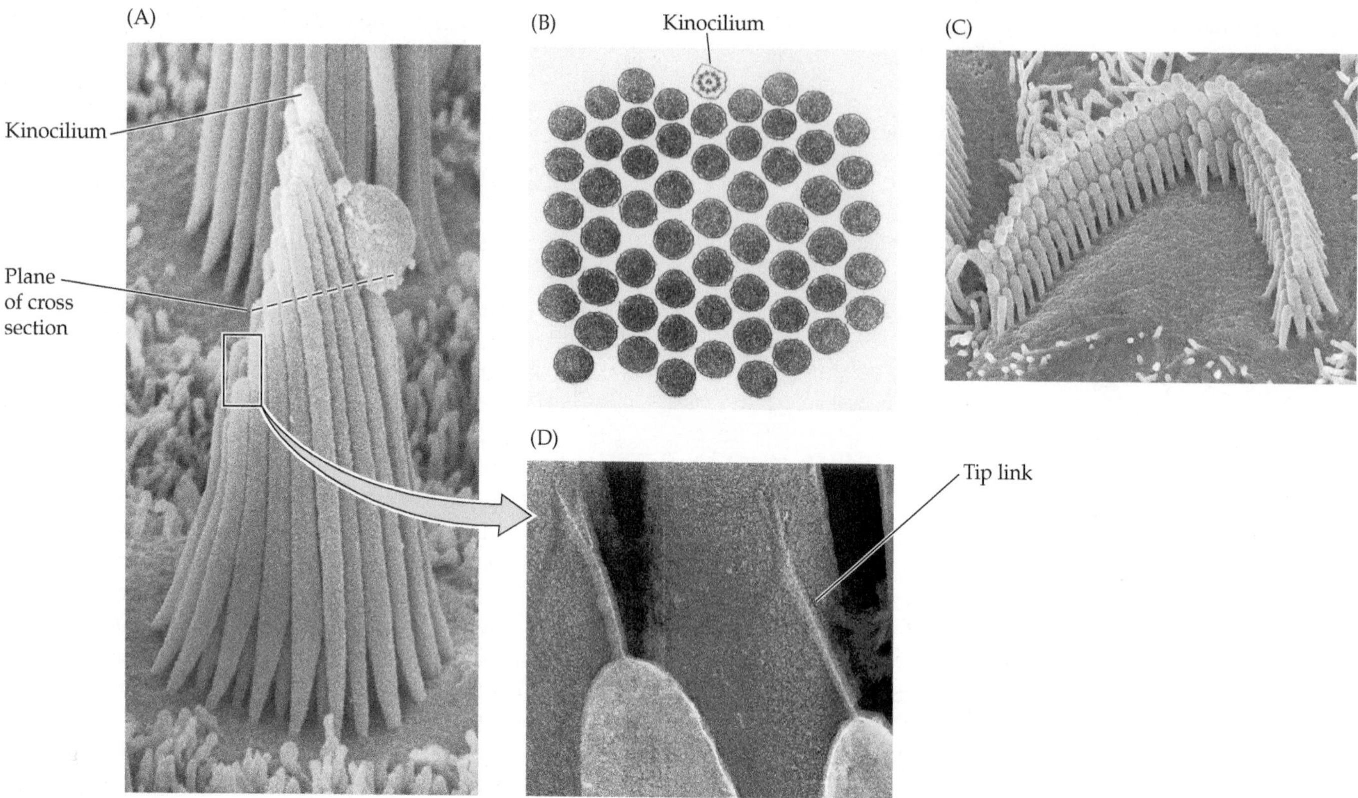

FIGURE 13.8 The structure of the hair bundle in cochlear and vestibular hair cells. (A) The hair bundle of a frog vestibular hair cell. This view shows the increasing height leading to the kinocilium. (B) Cross section through the vestibular hair bundle shows the kinocilium at the top, with its 9 + 2 array of microtubules, which contrasts with the simpler actin filament structure of the stereocilia. (C) Scanning electron micrograph of a cochlear outer hair cell bundle viewed along the plane of mirror symmetry. Note the graded lengths of the stereocilia and the absence of a kinocilium. (D) Tip links that connect adjacent stereocilia are believed to be mechanical linkages that open and close transduction channels. (A from Hudspeth, 2014; B from Hudspeth, 1983. © 1983 Scientific American, a division of Nature America, Inc. All rights reserved. C courtesy of David Furness and Carole Hackney, Keele University, UK; D from Kachar et al., 2000. © 2000 National Academy of Sciences, USA.)

only the kinocilium is a true ciliary structure (Figure 13.8B), and in mammalian cochlear hair cells it disappears shortly after birth (Figure 13.8C). The stereocilia are simpler, containing only an actin cytoskeleton. Each stereocilium tapers where it inserts into the apical membrane, forming a hinge about which each stereocilium pivots. The stereocilia are graded in height and are arranged in a bilaterally symmetrical fashion (see Figure 13.8C); in vestibular hair cells, this plane runs through the kinocilium (see Figure 13.8A). Fine filamentous structures, known as tip links, run in parallel to the plane of bilateral symmetry, connecting the tips of adjacent stereocilia (Figure 13.8D).

The tip links, which consist of the cell adhesion molecules cadherin 23 and protocadherin 15, provide the means for rapidly translating hair bundle movement into a receptor potential. Displacement of the hair bundle parallel to the plane of bilateral symmetry in the direction of the tallest stereocilia stretches the tip links, directly opening cation-selective mechanoelectrical transduction (hair cell MET or hcMET) channels located at the end of the link and depolarizing the hair cell (Figure 13.9). Movement in the opposite direction compresses the tip links, closing the hcMET channels and hyperpolarizing the hair cell. As the linked stereocilia pivot back and forth, the tension on the tip link varies, modulating the ionic flow and resulting in a graded receptor potential that follows the movements of the stereocilia. This receptor potential in turn leads to transmitter release from the basal end of the hair cell, which triggers action potentials in cranial nerve VIII fibers that follow the up-and-down vibration of the basilar membrane at relatively low frequencies (see the following section).

The transduction of mechanical forces by hair cells is both fast and remarkably sensitive. The hair bundle movements at the threshold of hearing are approximately 0.3 nm—about the diameter of an atom of gold. Hair cells can convert the displacement of the stereociliary bundle into an electrical potential in as little as 10 μs; as described below, such speed is required for the accurate localization of the source of the sound. The need for microsecond resolution

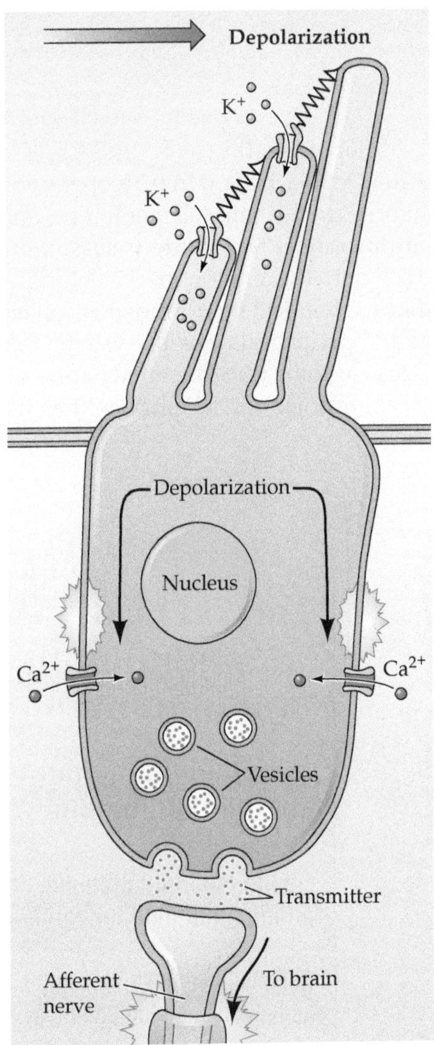

FIGURE 13.9 Mechanoelectrical transduction mediated by hair cells. When the hair bundle is deflected toward the tallest stereocilium, cation-selective hcMET channels open near the tips of the stereocilia, allowing K^+ to flow into the hair cell down their electrochemical gradient (see text for the explanation of this peculiar situation). The resulting depolarization of the hair cell opens voltage-gated Ca^{2+} channels in the cell soma, allowing calcium entry and release of neurotransmitter onto the nerve endings of the auditory nerve. (After Lewis and Hudspeth, 1983.)

requires direct mechanical gating of the transduction channel, rather than the relatively slow second-messenger pathways used in visual and olfactory transduction (see Chapters 11 and 15). Although mechanotransduction in hair cells is extremely fast, springiness of the tip link introduces distortion effects that, in some cases, are audible (Box 13A). Moreover, the exquisite mechanical sensitivity of the stereocilia also presents substantial risks. High-intensity sounds can break the tip links and destroy the hair bundle, resulting in profound hearing deficits. Because human stereocilia, unlike those in fish and birds, do not regenerate, such damage is irreversible. The small number of hair cells (a total of ~30,000 in a human, or 15,000 per ear) further compounds the sensitivity of the inner ear to environmental and genetic insults. An important goal of current research is to identify stem cells and factors that could contribute to the regeneration of human hair cells, thus affording a possible therapy for some forms of sensorineural hearing loss.

BOX 13A ■ The Sweet Sound of Distortion

As early as the first half of the eighteenth century, musical composers such as G. Tartini and W. A. Sorge discovered that, upon playing pairs of tones, other tones not present in the original stimulus are also heard. These combination tones, *fc*, are mathematically related to the played tones, f_1 and f_2 ($f_2 > f_1$), by the formula

$$fc = mf_1 \pm nf_2$$

where *m* and *n* are positive integers. Combination tones have been used for a variety of compositional effects, as they can strengthen the harmonic texture of a chord. Furthermore, organ builders sometimes use the difference tone ($f_2 - f_1$) created by two smaller organ pipes to produce the extremely low tones that would otherwise require building one especially large pipe.

Modern experiments suggest that this distortion product is due at least in part to the nonlinear properties of the inner ear. M. Ruggero and his colleagues placed small glass beads (10–30 nm in diameter) on the basilar membrane of an anesthetized animal and then determined the movement of the basilar membrane in response to different combinations of tones by measuring the Doppler shift of laser light reflected from the beads. When two tones were played into the ear, the basilar membrane vibrated not only at those two frequencies but also at other frequencies predicted by the above formula. Related experiments on hair cells studied in vitro suggest that these nonlinearities result from the properties of the mechanical linkage of the transduction apparatus. By moving the hair bundle sinusoidally with a metal-coated glass fiber, A. J. Hudspeth and his coworkers found that the hair bundle exerts a force at the same frequency. However, when two sinusoids were applied simultaneously, the forces exerted by the hair bundle occurred not only at the primary frequencies but at several combination frequencies as well. These distortion products are due to the transduction apparatus, because blocking the transduction channels causes the forces exerted at the combination frequencies to disappear, even though the forces at the primary frequencies remain unaffected. It seems that the tip links add a certain extra springiness to the hair bundle in the small range of motions over which the transduction channels are changing between closed and open states. If nonlinear distortions of basilar membrane vibrations arise from the properties of the hair bundle, then it is likely that hair cells can indeed influence basilar membrane motion, thereby accounting for the cochlea's extreme sensitivity. When we hear difference tones, we may be paying the price in distortion for an exquisitely fast and sensitive transduction mechanism.

The Ionic Basis of Mechanotransduction in Hair Cells

Intracellular recordings reveal that the hair cell has a resting potential between −45 and −60 mV relative to the fluid that bathes the basal end of the cell. At the resting potential, only a small fraction of the transduction channels are open. When the hair bundle is displaced in the direction of the tallest stereocilium, more transduction channels open, which causes depolarization as K^+ and Ca^{2+} enter the cell (Figure 13.10). Depolarization in turn opens voltage-gated calcium channels in the hair-cell membrane, and the resultant Ca^{2+} influx causes transmitter release from the basal end of the cell onto the auditory nerve endings (see Figure 13.9), similar to chemical neurotransmission elsewhere in the central and peripheral nervous systems (see Chapters 5 and 6). Because some transduction channels are open at rest, the receptor potential is biphasic: Movement toward

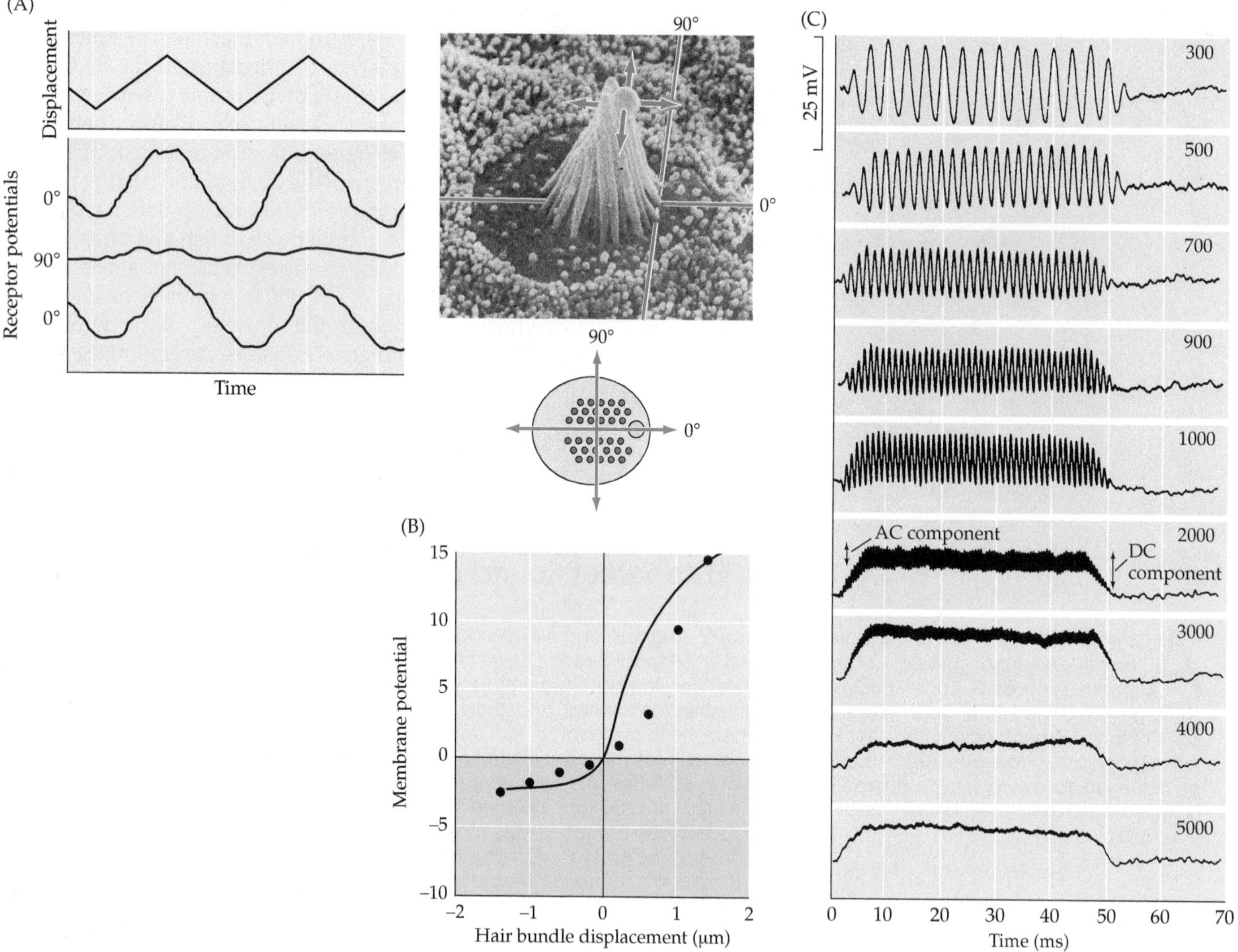

FIGURE 13.10 Frequency-dependent characteristics of receptor potentials in hair cells. (A) Vestibular hair cell receptor potentials (bottom three traces; blue) measured in response to symmetrical displacement (top trace; yellow) of the hair bundle about the resting position, either parallel (0°) or orthogonal (90°) to the plane of bilateral symmetry. (B) The asymmetrical stimulus-response (*x*-axis/*y*-axis) function of the hair cell. Equivalent displacements of the hair bundle generate larger depolarizing responses than hyperpolarizing responses, because most transduction channels are closed while "at rest" (i.e., 0 μm). (C) Receptor potentials generated by an individual hair cell in the cochlea in response to pure tones (indicated in Hz, right). Note that the hair cell potential faithfully follows the waveform of the stimulating sinusoids for lower frequencies (< 3 kHz) but still responds with a direct current offset to higher frequencies due to the asymmetrical stimulus-response function and the electrical filtering properties of the hair cells. (A from Shotwell et al., 1981; B after Hudspeth and Corey, 1977; C after Palmer and Russell, 1986.)

the tallest stereocilia depolarizes the cell, whereas movement in the opposite direction leads to hyperpolarization. This situation allows the hair cell to generate a sinusoidal receptor potential in response to a sinusoidal stimulus, thus preserving the temporal information present in the original signal, up to frequencies of around 3 kHz (see Figure 13.10). Hair cells still can signal at frequencies above 3 kHz, although without preserving the exact temporal structure of the stimulus; the cell's membrane time-constant filters the asymmetrical displacement-receptor current function of the hair cell bundle to produce a tonic depolarization of the soma, augmenting transmitter release and thus exciting auditory nerve terminals.

The high-speed demands of mechanoelectrical transduction have resulted in some impressive specializations of the ion fluxes within the inner ear. An unusual adaptation of the hair cell in this regard is that K^+ serves both to depolarize *and* repolarize the cell, enabling the hair cell's K^+ gradient to be largely maintained by passive ion movement alone. As with other epithelial cells, the basal and apical surfaces of the hair cell are separated by tight junctions, allowing separate extracellular ionic environments at these two surfaces. The apical end (including the stereocilia) protrudes into the scala media and is exposed to the K^+-rich, Na^+-poor **endolymph** produced by dedicated ion-pumping cells in the **stria vascularis** (Figure 13.11). The basal end of the hair cell body is bathed in perilymph—the same fluid that fills the scala tympani. Perilymph resembles other extracellular fluids in that it is K^+-poor and Na^+-rich. However, the compartment containing endolymph is about 80 mV more positive than the perilymph compartment (this difference is known as the *endocochlear potential*), while the inside of the hair cell is about 45 mV more negative than the perilymph and about 125 mV more negative than the endolymph. The resulting electrical gradient across the membrane of the stereocilia (~125 mV) drives K^+ through open transduction channels into the hair cell, which depolarizes the hair cell, opening voltage-gated K^+ and Ca^{2+} channels located in the membrane of the hair cell soma (see Web Topic 14.2). The opening of *somatic* K^+ channels favors K^+ efflux, and thus repolarization, whereas Ca^{2+} entry triggers transmitter release and opens Ca^{2+}-dependent K^+ channels, which provide another avenue for K^+ to enter the perilymph. Indeed, the interaction of Ca^{2+} influx and Ca^{2+}-dependent K^+ efflux can lead to electrical resonances that enhance the tuning of response properties of hair cells.

In essence, the hair cell operates as two distinct compartments, each dominated by its own Nernst equilibrium potential for K^+; this arrangement ensures that the hair cell's ionic gradient does not run down, even during prolonged stimulation. The rupture of Reissner's membrane (which normally separates the scalae media and vestibuli) or the presence of compounds such as ethacrynic acid that

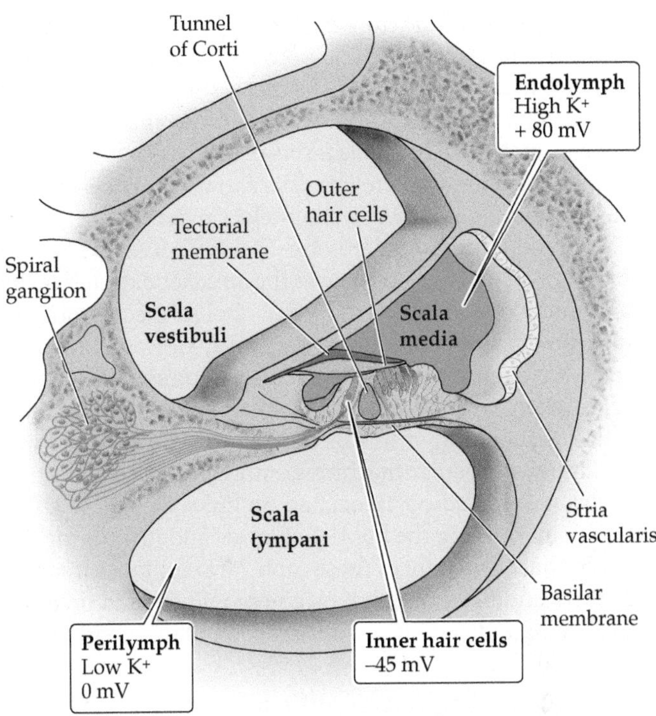

FIGURE 13.11 **Depolarization and repolarization of hair cells are mediated by K^+.** The stereocilia of the hair cells protrude into the endolymph, which is high in K^+ and has an electrical potential of +80 mV relative to the perilymph. This endocochlear potential drives K^+ into open transduction channels located at the apical ends of the stereocilia; the resulting depolarization of the hair cell body opens somatic K^+ channels. The negative resting potential of the hair cell and the low K^+ concentration in the surrounding perilymph result in an outward K^+ current through these somatic channels.

selectively poison the ion-pumping cells of the stria vascularis can cause the endocochlear potential to dissipate, resulting in a sensorineural hearing deficit. In short, the hair cell exploits the different ionic milieus of its apical and basal surfaces to provide extremely fast and energy-efficient repolarization.

The Hair Cell Mechanoelectrical Transduction Channel

The hcMET channel has yet to be isolated and the genes that encode for it have yet to be identified, despite extensive knowledge of the hcMET's physiological properties and intensive genetic analyses that have led to the promotion of several potential candidates. Sheer paucity of material is one of the major challenges to isolating the hcMET protein: A single hair bundle may possess as few as 200 functional channels, which represent a tiny fraction (<0.001%) of all hair bundle proteins, factors that have rendered biochemical purification impractical. A further challenge is the complexity of the transduction apparatus, with current evidence indicating that the pore-forming molecule must interact with a variety of other accessory proteins to enable mechanotransduction. Despite these challenges, the genetic analysis of heritable forms of deafness has identified numerous genes that are important to normal hearing, including candidate hcMET channels. Currently, four especially promising candidates are *TMC1, TMC2, TMIE,* and *LHFPL5,* all of which localize to the apical end of stereocilia and mutations in which reduce or abolish mechanotransduction currents in auditory hair cells. However, none of these has been shown to sustain mechanotransduction currents when expressed in heterologous systems, which may reflect that mechanotransduction in hair cells is the product of a multi-molecular machine comprising these and other molecules, including associated tip links. Despite the difficulty in isolating the hcMET channel and identifying the genes that encode for it, these topics are of intense interest, as fully understanding how we hear will depend on a full molecular and genetic characterization of this channel.

The Cochlear Amplifier

Von Békésy's model of cochlear mechanics was a passive one, resting on the premise that the basilar membrane acts like a series of linked resonators, much as a concatenated set of tuning forks. More recent studies made from the intact, living cochlea, however, indicate that normal hearing depends on the activity of a biological amplifier located within the cochlea. The rationale for this is based on three observations. First, the tuning of the auditory periphery, whether measured at the basilar membrane or recorded as the electrical activity of auditory nerve fibers, is too sharp to be explained by passive mechanics alone. Second, at very low sound intensities, the basilar membrane vibrates 100-fold more than would be predicted by linear extrapolation from the motion measured at high intensities. Third, the ear can generate sounds under certain conditions. These otoacoustic emissions, which can be detected by placing a sensitive microphone at the eardrum and monitoring the response after briefly presenting a tone or click, provide a useful means to assess cochlear function in the newborn, and this test is now done routinely to rule out congenital deafness. Such emissions, called **cochlear microphonics**, can also occur spontaneously, and are thus one possible source of **tinnitus** (ringing in the ears; most cases of tinnitus, however, arise centrally). These various observations indicate that the ear's sensitivity arises from an active biomechanical process as well as from its passive resonant properties (see Box 13A).

Although the basis for this active process remains a matter of debate, evidence suggests that the outer hair cells are an essential component of the cochlear amplifier. First, the high sensitivity "notch" of auditory nerve tuning curves is lost when the outer hair cells are selectively inactivated. Second, mutant mice lacking inner hair cells, although deaf, nonetheless produce otoacoustic emissions. Further, isolated outer hair cells contract and expand in response to small electrical currents, thus providing a potential source of energy to drive an active process within the cochlea. Finally, the opening and closing of hcMET channels may provide another source of energy driving basilar membrane motions, raising the possibility that inner hair cells also contribute to amplification. Ultimately, active processes are needed to explain the remarkable sensitivity of the auditory system to faint sounds.

Tuning and Timing in the Auditory Nerve

The rapid response time of the transduction apparatus allows the membrane potential of the hair cell to follow deflections of the hair bundle up to moderately high frequencies of oscillation (~3 kHz). As a result, spectral components as well as temporal modulations in the sound envelope that fall below 3 kHz can be encoded by the temporal patterns of activity of hair cells and their associated auditory nerve fibers. Even these extraordinarily rapid processes, however, fail to follow frequencies above 3 kHz (see above and Figure 13.10). Accordingly, some other mechanism must be used to transmit auditory information at higher frequencies. The tonotopically organized basilar membrane provides an alternative to temporal coding—namely a *labeled-line* coding mechanism.

In labeled-line coding, frequency information is specified by preserving tonotopy at higher levels in the auditory pathway. Because a single auditory nerve fiber innervates only a single inner hair cell (although several or more auditory nerve fibers synapse on a single hair cell), each auditory nerve fiber transmits information about only a small part of the audible frequency spectrum. As a result, auditory nerve fibers related to the apical end of the cochlea respond to low frequencies, and fibers that are related to the basal end respond to high frequencies (see Figure 13.6). The properties of specific fibers can be seen in electrophysiological recordings of responses to sound (Figure 13.12). These threshold functions are called **tuning curves**, and the lowest threshold of the tuning curve is called the **characteristic frequency**. Since the topographical order of the characteristic frequency of neurons is retained throughout the system, information about frequency is also preserved. Cochlear implants exploit the tonotopic organization of the cochlea, and particularly its auditory nerve afferents, to roughly recreate the patterns of auditory nerve activity elicited by sounds. In patients with damaged hair cells, such implants can effectively bypass the impaired transduction apparatus, and thus restore some degree of auditory function.

The other prominent feature of hair cells—their ability to follow the waveform of low-frequency sounds—is also important in other, more subtle aspects of auditory coding. As mentioned earlier, hair cells have biphasic response properties. Because hair cells release transmitter only when depolarized, auditory nerve fibers fire only during the positive phases of low-frequency sounds (see Figure 13.12). The resultant "phase-locking" provides temporal information from the two ears to neural centers that compare interaural time differences for frequencies up to 3 kHz. The evaluation of interaural time differences provides a critical cue for sound localization and the perception of auditory "space." That auditory space can be perceived is remarkable, given that the cochlea, unlike the retina, cannot represent space directly.

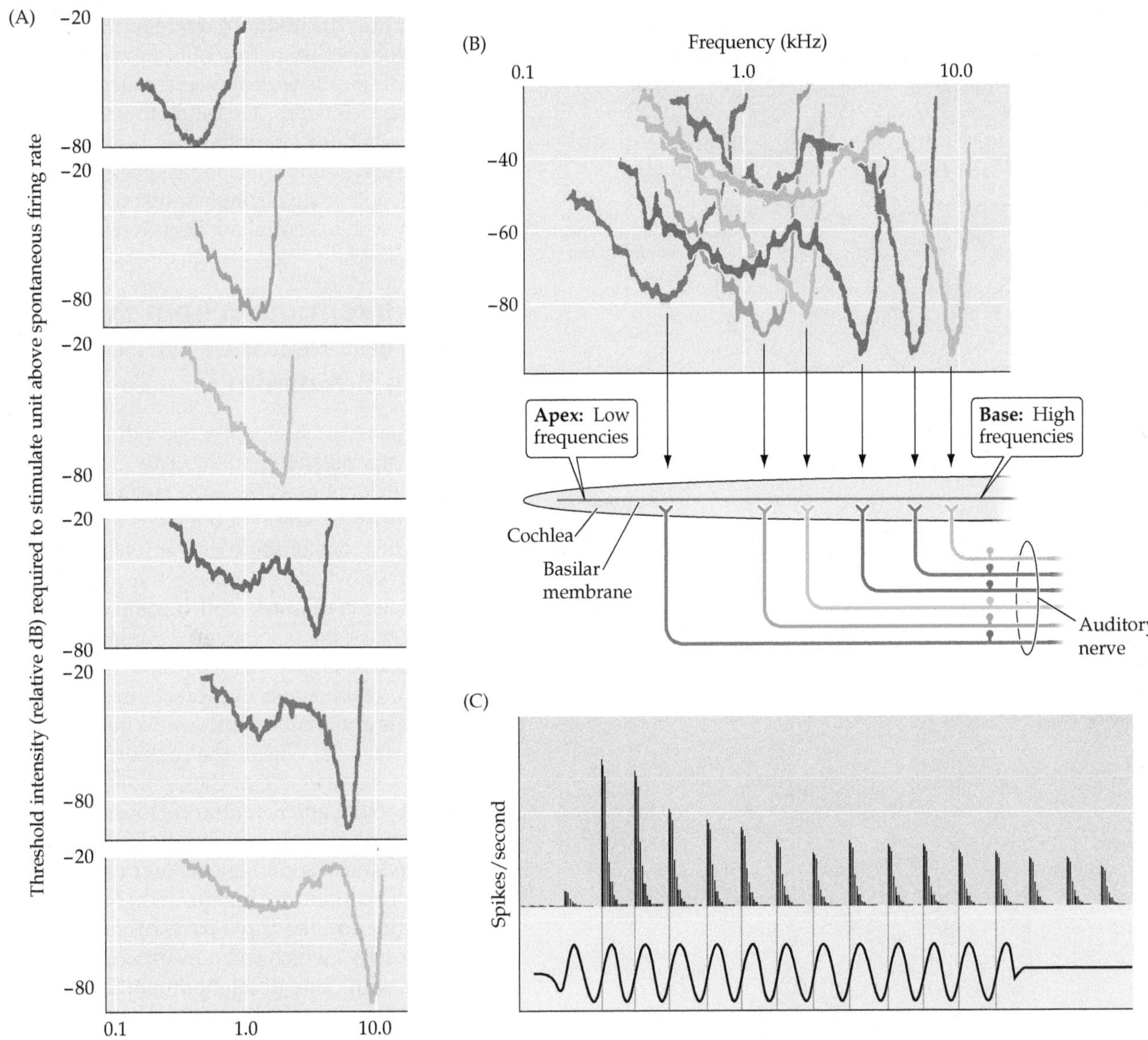

FIGURE 13.12 Response properties of auditory nerve fibers. (A) Frequency tuning curves of six different fibers in the auditory nerve. Each graph plots the minimum sound level required to increase the fiber's firing rate above its spontaneous firing level, across all frequencies to which the fiber responds. The lowest point in the plot is the weakest sound intensity to which the neuron will respond. The frequency at this point is called the neuron's characteristic frequency. (B) The frequency tuning curves of auditory nerve fibers superimposed and aligned with their approximate relative points of innervation along the basilar membrane. (In the side-view schematic below, the basilar membrane is represented as a straight line within the unrolled cochlea.) (C) Temporal response patterns of a low-frequency axon in the auditory nerve. The stimulus waveform is indicated beneath the histograms, which show the phase-locked responses to a 50-ms tone pulse of 260 Hz. Note that the spikes are all locked to the same phase of the sinusoidal stimulus. (A,B after Kiang and Moxon, 1972; C after Kiang, 1984.)

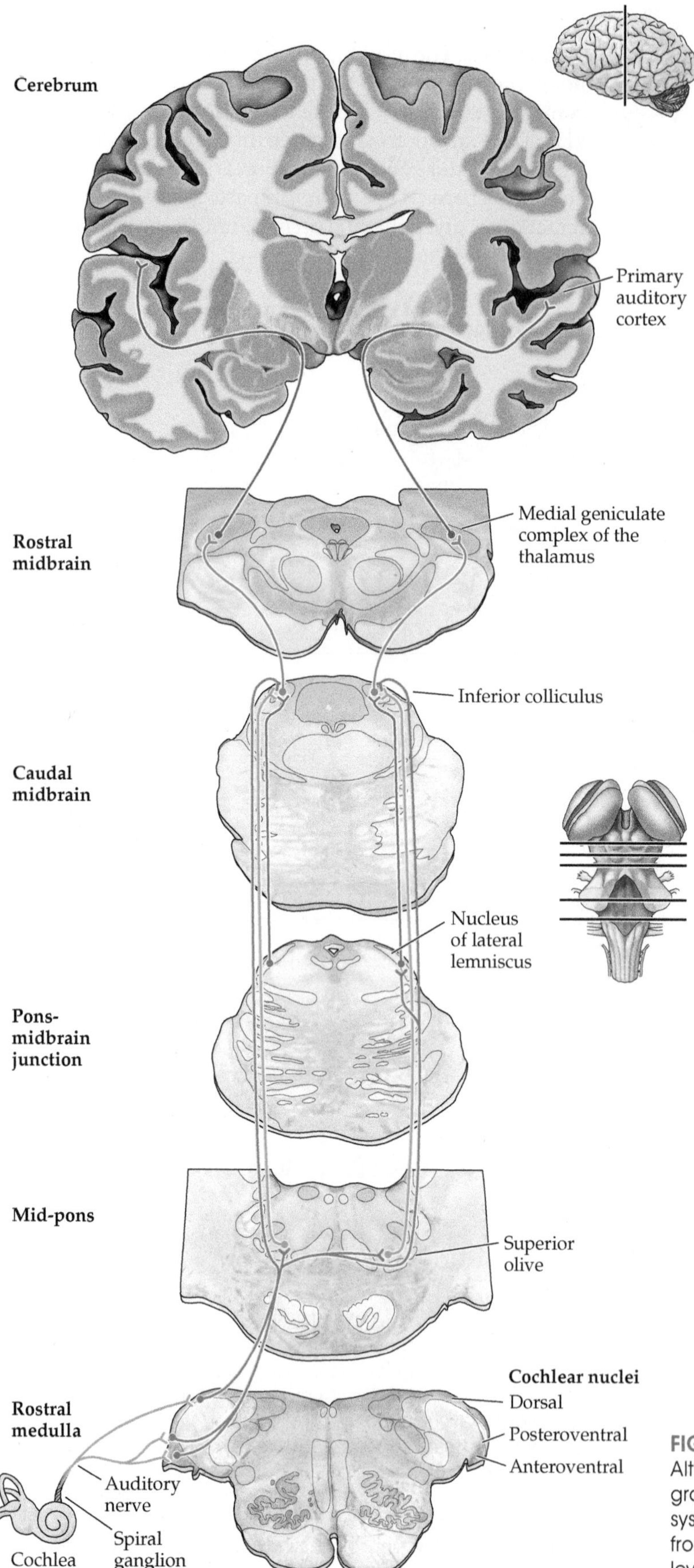

Much of the information described in these sections indicates that auditory nerve activity patterns are not simply neural replicas of the auditory stimulus itself. This conclusion accords with what is known about other sensory systems, all of which are specialized to process information that is particularly important for the species in question. An example in hearing is work by William Bialek and his colleagues at Princeton University, who have shown that the auditory nerve in the bullfrog encodes conspecific mating calls more efficiently than artificial sounds with similar frequency and amplitude characteristics. This and other studies suggest that the auditory periphery is optimized to transmit natural sounds, including species-typical vocal sounds, rather than simply transmitting all sounds equally well to central auditory areas.

How Information from the Cochlea Reaches Targets in the Brainstem

As in the visual system, the ascending auditory system is organized in parallel. This arrangement becomes evident as soon as the auditory nerve enters the brainstem, where it branches to innervate the three divisions of the cochlear nucleus. The auditory nerve (which, along with the vestibular nerve, constitutes cranial nerve VIII) comprises the central processes of the bipolar spiral ganglion cells in the cochlea (see Figure 13.5); each of these cells sends a peripheral process to contact one or a few inner hair cells and a central process to innervate the cochlear nucleus. Within the cochlear nucleus, each auditory nerve fiber branches, sending an ascending branch to the anteroventral cochlear nucleus and a descending branch to the posteroventral cochlear nucleus and the dorsal cochlear nucleus (Figure 13.13). The tonotopic organization of the cochlea is maintained in the three parts of the cochlear nucleus, each of which contains different populations of cells with quite different properties. In addition, the patterns of termination of the auditory nerve axons differ in density and type; thus, there are several opportunities at this level for transformation of the information from the hair cells.

FIGURE 13.13 Diagram of the major auditory pathways. Although many details are missing from this simplified diagram, two important points are evident: (1) The auditory system entails several parallel pathways, and (2) information from each ear reaches both sides of the system, even at the level of the brainstem.

Integrating Information from the Two Ears

Just as the auditory nerve branches to innervate several different targets in the cochlear nuclei, the neurons in these nuclei give rise to several different pathways (see Figure 13.13). One clinically relevant feature of the ascending projections of the auditory brainstem is a high degree of bilateral connectivity, which means that damage to central auditory structures is almost never manifested as a monaural hearing loss. (Sound arriving at one ear only is referred to as monaural.) Indeed, a monaural hearing loss strongly implicates unilateral peripheral damage, either to the middle or inner ear or to the auditory nerve itself. Given the relatively byzantine organization already present at the level of the auditory brainstem, it is useful to consider these pathways in the context of their functions.

The best-understood function mediated by the auditory brainstem nuclei, and certainly the one most intensively studied, is sound localization. Humans use at least two different strategies to localize the horizontal position of sound sources, depending on the frequencies in the stimulus. For frequencies below 3 kHz (which auditory nerve fibers can follow in a phase-locked manner), interaural *time* differences are used to localize the source; above these frequencies, interaural *intensity* differences are used as cues. Parallel pathways originating from the cochlear nucleus serve each of these strategies for sound localization.

The human ability to detect interaural time differences is remarkable. The longest interaural time differences, which are produced by sounds arising directly lateral to one ear, are on the order of only 700 μs (a value given by the width of the head divided by the speed of sound in air, or about 340 m per s). Psychophysical experiments show that humans can detect interaural time differences as small as 10 μs; two sounds presented through earphones separated by such small interaural time differences are perceived as arising from the side of the leading ear. This sensitivity translates into accuracy for sound localization to within about 1 degree.

The neural circuitry that computes these tiny interaural time differences consists of binaural inputs to the **medial superior olive** (**MSO**) that arise from the right and left anteroventral cochlear nuclei (Figure 13.14; see also Figure 13.13). The MSO contains cells with bipolar dendrites that extend both medially and laterally. The lateral dendrites receive input from the ipsilateral anteroventral cochlear nucleus, and the medial dendrites receive input

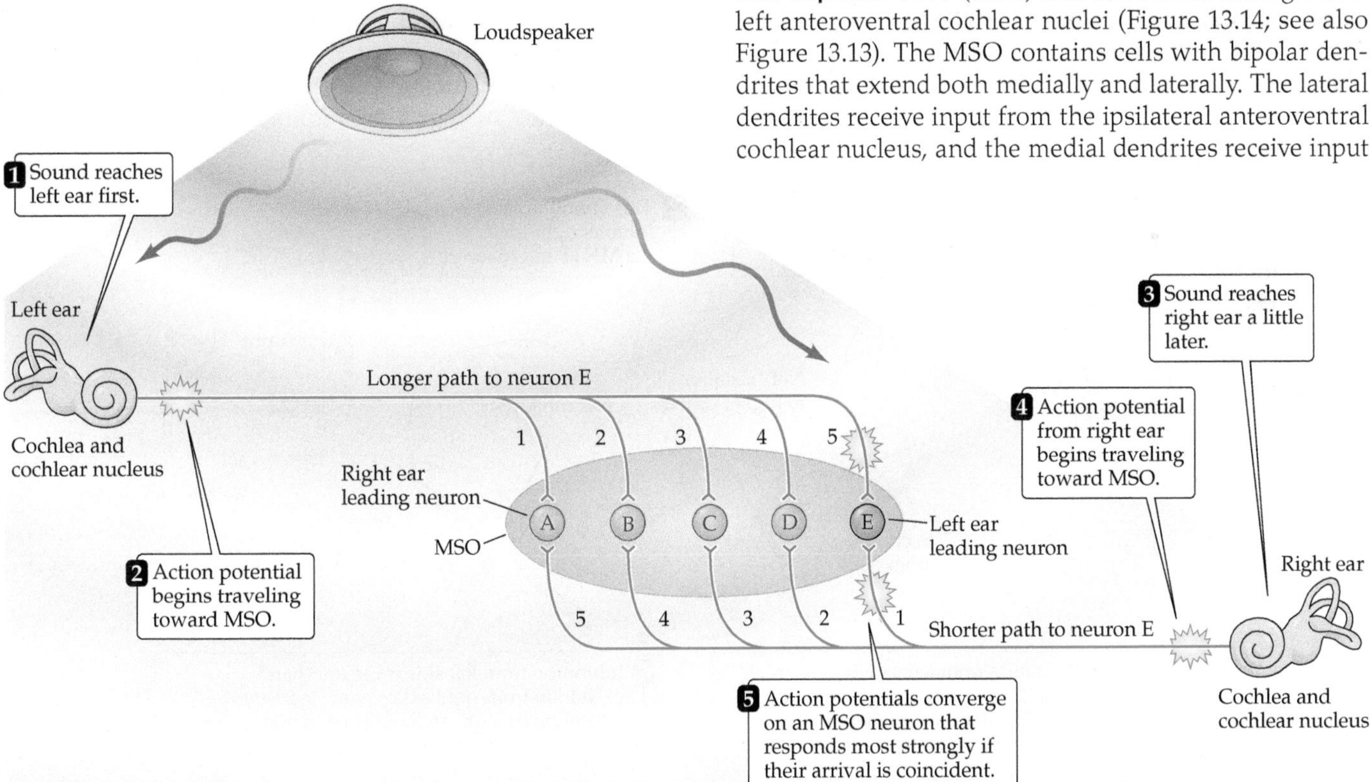

FIGURE 13.14 **A model of how the MSO computes the location of a sound by interaural time differences.** A given MSO neuron responds most strongly when the two inputs arrive simultaneously, as occurs when the contralateral and ipsilateral inputs precisely compensate (via their different lengths) for differences in the time of arrival of a sound at the two ears. The systematic (and inverse) variation in the delay lengths of the two inputs creates a map of sound location. In this model, neuron E in the MSO would be most sensitive to sounds located to the left, and neuron A to sounds from the right; neuron C would respond best to sounds coming from directly in front of the listener. (After Jeffress, 1948.)

from the contralateral anteroventral cochlear nucleus (both inputs are excitatory). An influential model of sound localization, first formally described by J. Jeffress, posits that the neurons of the MSO work as **coincidence detectors**, responding when both excitatory signals arrive at the same time. For a coincidence mechanism to be useful in localizing sound, different neurons must be maximally sensitive to different interaural time delays. The axons that project from the anteroventral cochlear nucleus evidently vary systematically in length to create delay lines. These anatomical differences compensate for sounds arriving at slightly different times at the two ears, so that the resultant neural impulses arrive at a particular MSO neuron simultaneously, making each cell especially sensitive to sound sources in a particular place. Experimental evidence of delay lines and coincidence detectors has been found in the brainstem of the barn owl, an animal that is highly adept at locating and capturing prey (e.g., mice) by the sounds they make as they scamper along the ground. However, experiments in gerbils point to a role for inhibition in shaping MSO neuronal responses, suggesting that some mammals may employ additional or alternate mechanisms to compute interaural time differences.

Sound localization perceived on the basis of interaural time differences requires phase-locked information from the periphery, which, as already emphasized, is available to humans only for frequencies below 3 kHz. Therefore, a second mechanism must come into play at higher frequencies. One clue to the solution is that at frequencies higher than about 2 kHz, the human head begins to act as an acoustical obstacle, because the wavelengths of the sounds are too short to bend around it. As a result, when high-frequency sounds are directed toward one side of the head, an acoustical "shadow" of lower intensity is created at the far ear. These intensity differences provide a second cue about the location of a sound. The circuits that compute the position of a sound source on this basis are found in the **lateral superior olive** (**LSO**) and the **medial nucleus of the trapezoid body** (**MNTB**) (Figure 13.15). Excitatory axons project directly from the ipsilateral anteroventral cochlear nucleus to the LSO (as well as to the MSO; see Figure 13.14). Note that the LSO also receives inhibitory input from the contralateral ear via an inhibitory neuron in the MNTB. This excitatory–inhibitory interaction results in a net excitation of the LSO on the same side of the head as the sound source. For sounds arising directly lateral to the listener's head, firing rates will be highest in the LSO on that side; in this circumstance, the excitation via the ipsilateral anteroventral cochlear nucleus will be maximal, and inhibition from the contralateral MNTB minimal. In contrast, sounds arising closer to the listener's midline will elicit lower firing rates in the ipsilateral LSO because of increased inhibition arising from the contralateral MNTB. For sounds arising at the midline, or from the other side,

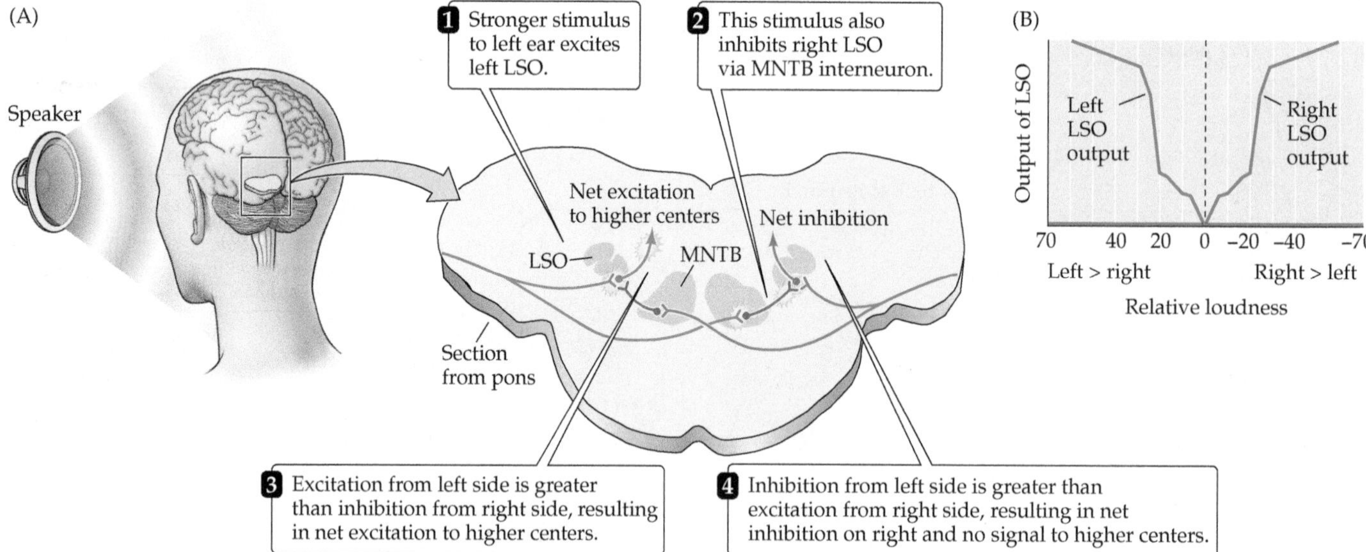
(A)
Speaker
1 Stronger stimulus to left ear excites left LSO.
2 This stimulus also inhibits right LSO via MNTB interneuron.
Net excitation to higher centers
Net inhibition
LSO
MNTB
Section from pons
3 Excitation from left side is greater than inhibition from right side, resulting in net excitation to higher centers.
4 Inhibition from left side is greater than excitation from right side, resulting in net inhibition on right and no signal to higher centers.
(B)
Output of LSO
Left LSO output
Right LSO output
70 40 20 0 −20 −40 −70
Left > right
Right > left
Relative loudness

FIGURE 13.15 LSO neurons encode sound location through interaural intensity differences. (A) LSO neurons receive direct excitation from the ipsilateral cochlear nucleus; input from the contralateral cochlear nucleus is relayed via inhibitory interneurons in the MNTB. (B) This arrangement of excitation–inhibition makes LSO neurons fire most strongly in response to sounds arising directly lateral to the listener on the same side as the LSO, because excitation from the ipsilateral input will be great and inhibition from the contralateral input will be small. In contrast, sounds arising from in front of the listener, or from the opposite side, will silence the LSO output, because excitation from the ipsilateral input will be minimal, but inhibition driven by the contralateral input will be great. Note that LSOs are paired and bilaterally symmetrical; each LSO encodes only the location of sounds arising from the ipsilateral hemifield.

the increased inhibition arising from the MNTB is powerful enough to completely silence LSO activity. Note that each LSO encodes only sounds arising in the ipsilateral hemifield; therefore, it takes both LSOs to represent the full range of horizontal positions.

In summary, two separate pathways—and two separate mechanisms—enable us to localize sound along the azimuth. Humans process interaural time differences in the MSO and interaural intensity differences in the LSO. These two pathways eventually merge in the midbrain auditory centers. The elevation of sound sources is determined by spectral filtering mediated by the external pinnae. Experimental evidence suggests that the spectral "notches" created by the shape of the pinnae are detected by neurons in the dorsal cochlear nucleus. Thus, binaural cues play an important role in localizing the azimuthal position of sound sources, whereas spectral cues are used to localize the elevation of sound sources.

Monaural Pathways from the Cochlear Nucleus to the Nuclei of the Lateral Lemniscus

The binaural pathways for sound localization are only part of the output of the cochlear nucleus. This fact is hardly surprising, given that auditory perception involves much more than locating the position of the sound source. A second major set of pathways from the cochlear nucleus bypasses the superior olive and terminates in the **nuclei of the lateral lemniscus** on the contralateral side of the brainstem (see Figure 13.13). These particular pathways respond to sound arriving at one ear only and are thus referred to as monaural. Some cells in the lateral lemniscus nuclei signal the onset of sound, regardless of its intensity or frequency. Other cells in the lateral lemniscus nuclei process other temporal aspects of sound, such as duration. The role of these pathways in processing temporal features of sound is not yet known. As with the outputs of the superior olivary nuclei, the pathways from the nuclei of the lateral lemniscus converge at the midbrain.

Integration in the Inferior Colliculus

Auditory pathways ascending via the olivary and lemniscal complexes, as well as other projections that arise directly from the cochlear nucleus, project to the midbrain auditory center, also known as the **inferior colliculus**. In examining how integration occurs in the inferior colliculus, it is again instructive to turn to the most completely analyzed auditory mechanism, the binaural system for localizing sound. As already noted, space is not mapped on the auditory receptor surface; thus, the perception of auditory space must somehow be synthesized by circuitry in the lower brainstem and midbrain. Experiments in the barn owl show that the convergence of binaural inputs in the midbrain produces something entirely new, relative to the periphery—namely a computed topographical representation of auditory space. Neurons within this auditory space map in the colliculus respond best to sounds originating in a specific region of space; thus, they have both a preferred elevation and a preferred azimuthal location. Although the circuit mechanisms underlying sound localization in humans are unknown, humans have a clear perception of both the elevational and azimuthal components of a sound's location, indicating that our brains contain a neural representation of auditory space.

Another important property of the inferior colliculus is its ability to process sounds with complex temporal patterns. Many neurons in the inferior colliculus respond only to frequency-modulated sounds, while others respond only to sounds of specific durations or in specific temporal sequences. Such sounds are typical components of biologically relevant sounds, such as those made by predators, or intraspecific communication sounds, which in humans include speech. In summary, the high degree of convergence in the inferior colliculus of inputs conveying information about simpler cues, such as timing, intensity, and frequency, results in more integrative and complex response properties that are likely to be important to the representation of auditory objects.

The Auditory Thalamus

Despite the parallel pathways in the auditory stations of the brainstem and midbrain, the **medial geniculate complex (MGC)** in the thalamus is an obligatory relay for all ascending auditory information destined for the cortex (see Figure 13.13). Most input to the MGC arises from the inferior colliculus, although a few auditory axons from the lower brainstem bypass the inferior colliculus to reach the auditory thalamus directly. The MGC has several divisions, including the ventral division, which projects to the core region of the auditory cortex, and the medial and dorsal divisions, which are organized like a belt around the ventral division and project to the belt regions that surround the core region of the auditory cortex.

In some mammals, the strictly maintained tonotopy of the lower brainstem areas is exploited by convergence onto MGC neurons, generating specific responses to certain spectral combinations. The original evidence for this statement came from research on the response properties of cells in the MGC of echolocating bats. Some cells in the so-called belt regions of the bat MGC respond only to combinations of widely spaced frequencies that are specific components of the bat's echolocation signal and of the echoes that are reflected from objects in the bat's environment. In the mustached bat, where this phenomenon has been most thoroughly studied, the echolocation pulse has a changing frequency (frequency-modulated, or FM) component that includes a fundamental frequency and one or more harmonics. The fundamental frequency (FM_1) has low intensity and sweeps from 30 to 20 kHz. The second harmonic (FM_2) is the most intense component and sweeps from 60 to 40 kHz. Note that these frequencies do not overlap. Most of the echoes are from the intense FM_2 sound, and virtually none arise from the weak FM_1, even though the emitted FM_1 is loud enough for the bat to hear. Apparently, the bat assesses the distance to an object by measuring the delay between the FM_1 emission and the FM_2 echo. Certain MGC neurons respond when FM_2 follows FM_1 by a specific delay, providing a mechanism for sensing such frequency combinations. Because each neuron responds best to a particular delay, the population of MGC neurons encodes a range of distances.

Bat sonar illustrates two important points about the function of the auditory thalamus. First, the MGC displays pronounced selectivity for frequency combinations. The mechanism responsible for this selectivity is presumably the ultimate convergence of many inputs from cochlear regions with different spectral sensitivities. Second, cells in the MGC are selective not only for frequency combinations, but also for specific time intervals between the two frequencies. The principle is the same as that described for binaural neurons in the medial superior olive, but in this instance, two monaural signals with different frequency sensitivities coincide, and the time difference is in the millisecond rather than the microsecond range. For echolocating bats, time differences in the millisecond range correspond to target distances of catchable prey. Interestingly, speech sounds change continuously over the range of a few milliseconds, suggesting that MGC neurons in the human capable of integrating acoustical information over the millisecond timescale could facilitate speech perception.

In summary, neurons in the MGC receive convergent inputs from spectrally and temporally separate pathways. This complex, by virtue of its convergent inputs, mediates the detection of specific spectral and temporal combinations of sounds. In many species, including humans, varying spectral and temporal cues are especially important features of communication sounds. It is not known whether cells in the human MGC are selective to combinations of sounds, but the processing of speech certainly requires both spectral and temporal combination sensitivity.

The Auditory Cortex

The auditory cortex is the major target of the ascending fibers from the MGC, and it plays an essential role in our conscious perception of sound, including recognition of speech and music. Indeed, auditory cortical lesions in humans are often accompanied by deficits in speech and music perception. Despite the obvious importance of auditory cortical processing for sound perception, the auditory cortex has been less completely studied than the auditory periphery and is less well understood than visual cortex (although one should hasten to add that the mechanisms of both peripheral auditory processing and visual cortical processing are far from clear).

Although the auditory cortex has several subdivisions, as in the visual and somatosensory systems a broad distinction can be made between primary (i.e., core) and secondary (i.e., belt and parabelt) regions. The core region in macaque monkeys comprises three divisions, including **auditory area 1** (**A1**), **rostral** (**R**), and **rostrotemporal** (**RT**), all of which are located on the lower bank of the lateral sulcus in the medial and posterior part of the superior temporal gyrus (STG) in the temporal lobe. Imaging studies in humans indicate that the core region is located in the transverse temporal gyri (Heschl's gyri, or Brodmann's areas 41 and 42), buried in the lateral sulcus. The core region receives point-to-point input from the ventral division of the MGC; thus, the three divisions of the core region contain precise tonotopic maps. The **belt** and **parabelt regions** of the auditory cortex receive more diffuse input from the belt division of the MGC, as well as input from the primary auditory cortex, and are less precise in their tonotopic organization. Additionally, these various auditory cortical areas are strongly interconnected, with reciprocal connections between the core and belt regions, between the belt and parabelt regions, and between these latter two regions and auditory-related cortical areas in the ST and the STS (superior temporal sulcus), suggestive of a processing hierarchy.

The core divisions of the auditory cortex each contain a topographical map of the cochlea (Figure 13.16), just as the primary visual cortex (V1) and the primary somatosensory cortex (S1) have topographical maps of their respective sensory epithelia. Just as the visual and somatosensory systems represent their peripheral receptor surfaces in central maps, so the organization of the cochlea is laid out in a central map. Since frequencies are arrayed tonotopically along the length of the basilar membrane, this organization as reflected in A1 is said to comprise a tonotopic map, as do most of the ascending auditory structures between the cochlea and the cortex. Orthogonal to the frequency axis of the tonotopic map are irregular patches of neurons that are excited by both ears (and are therefore called EE cells) interspersed with patches of cells that are excited by one ear and inhibited by the other ear (EI cells). The EE and EI stripes alternate, an arrangement that is reminiscent of the ocular dominance columns in V1 (see Chapter 12).

The auditory cortex obviously does much more than provide a tonotopic map and respond differentially to ipsilateral and contralateral stimulation. The sorts of sensory processing that occur in the auditory cortex are not fully

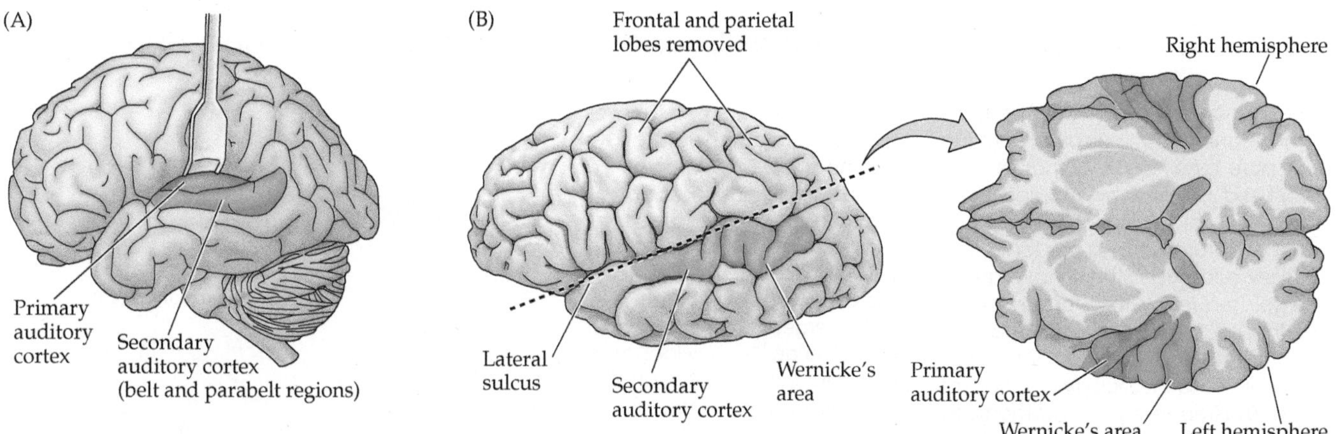

FIGURE 13.16 The human auditory cortex. (A) Diagram showing the brain in left lateral view, including the depths of the lateral sulcus, where part of the auditory cortex occupying the superior temporal gyrus normally lies hidden. The core region is shown in blue; the surrounding belt regions of the auditory cortex are in red. (B) Diagram of the brain in left lateral view, showing locations of human auditory cortical regions related to processing speech sounds in the intact hemisphere. Right: An oblique section (plane of dashed line) shows the cortical areas on the superior surface of the temporal lobe. Note that Wernicke's area, a region important in comprehending speech, is just posterior to the primary auditory cortex.

understood, but they are likely to be important to higher-order processing of natural sounds, especially those used for communication (see Box 13B and Chapter 33). One clue about such processing comes from work in marmosets, small New World monkeys with a complex vocal repertoire. The A1 and belt areas of these animals are indeed organized tonotopically, but they also contain neurons that are strongly responsive to spectral combinations that characterize certain vocalizations. The responses of these neurons to the tonal stimuli do not accurately predict their responses to the spectral combinations, suggesting that, in accordance with peripheral optimization, cortical processing is, in part, dedicated to detecting particular intraspecific vocalizations. Recent studies in marmosets and humans also implicate secondary regions of the auditory cortex in the perception of pitch. This percept is especially important to our musical sense and to vocal communication, because it enables us to hear two speech sounds as distinct even when they have overlapping spectral content and arise from the same location. A curious feature of pitch perception is that, for the harmonically complex sounds that typify speech and music, pitch corresponds to the fundamental frequency, even when it is absent from the actual stimulus. This ability of pitch processing to "fill in" a missing frequency further underscores the idea that the auditory cortex is doing far more than faithfully representing what the auditory periphery provides as input.

Another clue about the role of the auditory cortex in speech processing comes from electrocorticographic recordings made in human epilepsy patients from parts of the STG that most likely correspond to parabelt regions in non-human primates (i.e., Brodmann's area 22). These recordings reveal that neural responses to speech sounds spread from posterior to anterior regions of the STG, consistent with a processing hierarchy. Moreover, STG population activity strongly correlates with syllable onsets and offsets, which are especially important to speech intelligibility, and is sensitive to acoustical cues, such as voice onset times, that are important to the perceptual categorization of different speech sounds (e.g., distinguishing between *ba* and *pa*). Neural activity in the STG also depends strongly on context and attention: When subjects are instructed to attend to only one of two people speaking simultaneously, their STG neurons robustly encode fine spectrotemporal features of the attended voice but display little or no responsiveness to the other voice. Thus, auditory cortical activity is strongly influenced both by linguistic features and cognitive context, consistent with an influence of experience and task demands on auditory cortical processing of speech.

Sounds that are especially important for intraspecific communication often have a highly ordered temporal structure. In humans, the best example of such time-varying signals is speech, where different phonetic sequences are perceived as distinct syllables and words. Behavioral studies in cats and monkeys show that the auditory cortex is especially important for processing temporal sequences of sound. If the auditory cortex is ablated in these animals, they lose the ability to discriminate between two complex sounds that have the same frequency components but differ in temporal sequence. Thus, without the auditory cortex, monkeys cannot discriminate one conspecific communication sound from another. The physiological basis of such temporal sensitivity likely requires neurons that are

BOX 13B ■ Representing Complex Sounds in the Brains of Bats and Humans

Most natural sounds are complex, meaning that they differ from the pure tones or clicks that are frequently used in neurophysiological studies of the auditory system. Only a small minority of natural sounds are tonal, but when they occur the stimuli have a fundamental frequency that largely determines the *pitch* of the sound, and one or more harmonics of different intensities that contribute to the quality or *timbre* of a sound. The frequency of a harmonic is, by definition, an integer multiple of the fundamental frequency, and both may be modulated over time. Such *frequency-modulated* (*FM*) sweeps can rise or fall in frequency, or they can change in a sinusoidal (or some other) fashion. Occasionally, multiple nonharmonic frequencies may be simultaneously present in some communication or musical sounds. In some sounds, a level of spectral splatter, or *broadband noise*, is embedded within tonal or frequency-modulated sounds. The variations in the sound spectrum are typically accompanied by a modulation of the amplitude envelope of the complex sound as well. All of these features can be visualized by spectrographic analysis.

How does the brain represent such complex natural sounds? Cognitive studies of complex sound perception provide some understanding of how a large but limited number of neurons in the brain can dynamically represent an infinite variety of natural stimuli in the sensory environment of humans and other animals. In bats, specializations for processing complex sounds are apparent. Studies in echolocating bats show that both communication and echolocation sounds (Figure A) are processed not only within some of the same areas, but also within the same neurons in the auditory cortex. In humans, multiple modes of processing are also likely, given the large overlap within the superior and middle temporal gyri in the temporal lobe for the representation of different types of complex sounds.

Asymmetrical representation is another common principle of complex sound processing that results in lateralized (though largely overlapping) representations of natural stimuli. Thus, speech sounds that are important for communication are lateralized to the left in the belt regions of the auditory cortex, whereas environmental sounds that are important for recognizing aspects of the

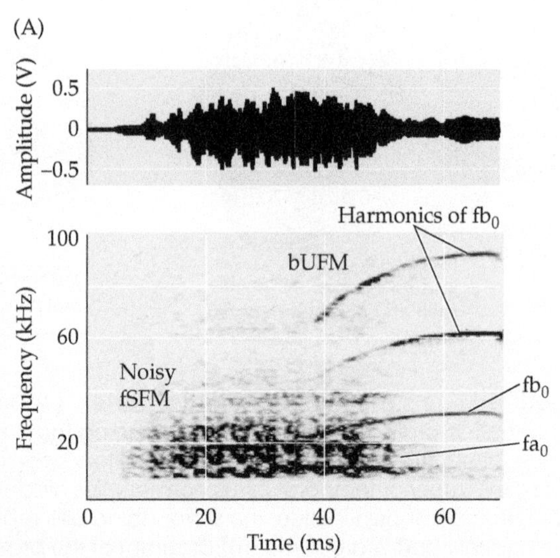

(A) Amplitude envelope (above) and spectrogram (below) of a composite syllable emitted by mustached bats for social communication. This composite consists of two simple syllables: a fixed Sinusoidal FM (fSFM), and a bent Upward FM (bUFM) that emerges from the fSFM after some overlap. Each syllable has its own fundamental (fa_0 and fb_0) and multiple harmonics. (Courtesy of Jagmeet Kanwal.)

sensitive to time-varying cues in communication sounds. Indeed, electrophysiological recordings from the primary auditory cortices of both marmosets and bats show that some neurons that respond to intraspecific communication sounds do not respond as strongly when the sounds are played in reverse, indicating sensitivity to the sounds' temporal features. Studies of human patients with bilateral damage to the auditory cortex also reveal severe problems in processing the temporal order of sounds. It seems likely, therefore, that specific regions of the human auditory cortex are specialized for processing elementary speech sounds, as well as other temporally complex acoustical signals, such as music. Thus, Wernicke's area, which is critical to the comprehension of human language, is contiguous with the secondary auditory area (see Figure 13.16 and Chapter 33).

Rather than serving as the end point of a vertical sensory hierarchy, the auditory cortex integrates a wide range of nonauditory information from other cortical and subcortical regions. One especially important type of nonauditory information is motor related. In fact, auditory cortical activity is modulated by a variety of movements, especially vocal and manual gestures, and motor and auditory cortical regions are reciprocally connected, providing a pathway for motor-related signals to influence auditory cortical activity. Moreover, electrical or transmagnetic stimulation of the motor cortex can interfere with speech perception, and EEG recordings from humans and single-unit recordings in freely vocalizing marmosets reveal that auditory cortical activity decreases immediately before and during vocalization, consistent with a motor-related suppressive signal. Such movement-related suppression may serve as part of an active filter to dampen responsiveness to the anticipated acoustical consequences of movements, such as those that produce speech, while also heightening sensitivity to unexpected sounds, including errors in vocal performance. Consistent with this idea, experiments in which marmosets listen to the auditory feedback from their vocalizations through headphones show that auditory cortical neurons are strongly excited

BOX 13B (continued)

auditory environment are represented in each hemisphere (Figure B). The very different types of musical sounds that can motivate us to march in war or to relax when coping with stress are to a considerable degree lateralized to the right in the belt regions of the auditory cortex. The extent of lateralization for speech, and possibly music, may vary with sex, age, and training. In some species of bats, mice, and primates, processing of natural communication sounds appears to be lateralized to the left hemisphere. In summary, natural sounds are complex, and their representation in the sensory cortex tends to be asymmetrical across the two hemispheres.

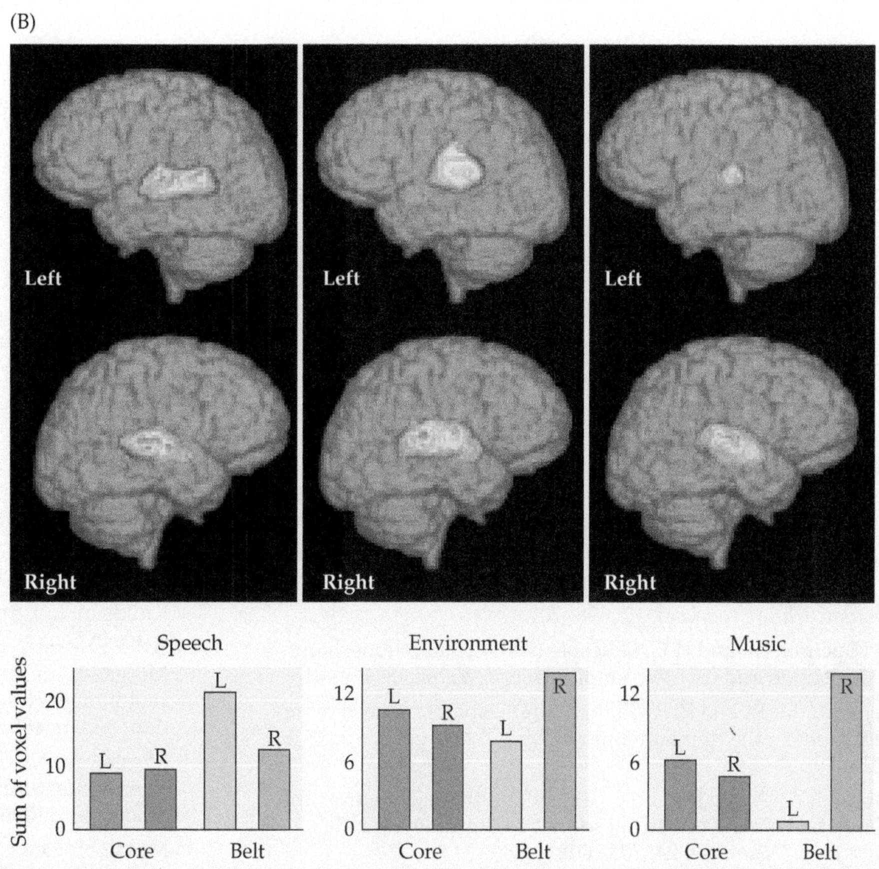

(B) Reconstructed fMRI of BOLD (blood oxygen level-dependent) contrast signal change (average for eight subjects) showing significant ($p < 0.001$) activation elicited by speech, environmental, and musical sounds on surface views of the left versus the right side of the human brain. Bar graphs show the total significant activation to each category of complex sounds in the core and belt areas of the auditory cortex for the left versus the right side. (Courtesy of Jagmeet Kanwal.)

rather than suppressed when the frequency of feedback is artificially shifted outside the normal range. More broadly, motor-based predictions of sensory feedback figure prominently in forward models of speech learning, which posit that differences in predicted and actual feedback lead to the generation of error signals important to speech learning and maintenance.

Summary

Sound waves are transmitted via the external and middle ear to the cochlea of the inner ear, which exhibits a traveling wave when stimulated. For high-frequency sounds, the amplitude of the traveling wave reaches a maximum at the base of the cochlea; for low-frequency sounds, the traveling wave reaches a maximum at the apical end. The associated motions of the basilar membrane are transduced primarily by the inner hair cells, while the basilar membrane motion is itself actively modulated by the outer hair cells. Damage to the outer or middle ear results in conductive hearing loss, while hair cell damage results in a sensorineural hearing deficit. The tonotopic organization of the cochlea is retained at all levels of the central auditory system. Projections from the cochlea travel via the auditory nerve to the three main divisions of the cochlear nucleus. The targets of the cochlear nucleus neurons include the superior olivary complex and nuclei of the lateral lemniscus, where the binaural cues for sound localization are processed. The inferior colliculus, the target of nearly all of the auditory pathways in the lower brainstem, carries out important integrative functions, such as processing of sound frequencies and integration of the cues for localizing sound in space. The primary auditory cortex, which is also organized tonotopically, supports basic auditory functions, such as frequency discrimination and sound localization, and also plays an important role in processing of intraspecific communication sounds. Populations of neurons in belt areas of the auditory cortex, which have a less strict tonotopic organization, display activity patterns that correlate with speech intelligibility and that are strongly modulated by linguistic features and cognitive context. In the human brain, the major speech comprehension areas reside in the zone immediately adjacent to the auditory cortex, and motor-related activity can strongly modulate auditory cortical responses to vocalization-related auditory feedback, suggestive of a predictive sensorimotor mechanism.

ADDITIONAL READING

Reviews

Fettiplace, R. and K. X. Kim (2014) The physiology of mechanoelectrical transduction in hearing. *Phys. Rev.* 94: 951–986.

Grothe, B., M. Pecka and D. McApline (2010) Mechanisms of sound localization in mammals. *Phys. Rev.* 90: 983–1012.

Hackett, T. A. (2015) Anatomic organization of the auditory cortex. In *Handbook of Clinical Neurology*, vol. 129, G. G. Celesia and G. Hickok (eds.). New York: Elsevier, pp. 27–53.

Hudspeth, A. J. (2001–2002) How the ear's works work: Mechanoelectrical transduction and amplification by hair cells of the internal ear. *Harvey Lect.* 97: 41–54.

Hudspeth, A. J. (2008) Making an effort to listen: Mechanical amplification in the ear. *Neuron* 59: 530–545.

King, A. J. and I. Nelken (2009) Unraveling the principles of auditory cortical processing: Can we learn from the visual system? *Nature Neurosci.* 12: 698–701.

LeMasurier, M. and P. G. Gillespie (2005) Hair-cell mechanotransduction and cochlear amplification. *Neuron* 48: 403–415.

Leonard, M. K. and E. F. Chang (2014) Dynamic speech representations in the human temporal lobe. *Trends Cogn. Sci.* 18 (9): 472–479.

Mizrahi, A., A. Shalev and I. Nelken (2014) Single neuron and population coding of natural sounds in the auditory cortex. *Curr. Opin. Neurobiol.* 24: 103–110.

Nelken, I. (2002) Feature detection by the auditory cortex. In *Integrative Functions in the Mammalian Auditory Pathway*, D. Oertel, R. Fay and A. N. Popper (eds.). *Springer Handbook of Auditory Research*, vol. 15. New York: Springer-Verlag, pp. 358–416.

Nelken, I. (2008) Processing of complex sounds in the auditory system. *Curr. Opin. Neurobiol.* 18: 413–417.

Pickles, J. O. (2015) Auditory pathways: anatomy and physiology. In *Handbook of Clinical Neurology, vol. 129,* G. G. Celesia and G. Hickok (eds.). New York: Elsevier, pp. 3–25.

Vollrath, M. A., K. Y. Kwan and D. P. Corey (2007) The micromachinery of mechanotransduction in hair cells. *Annu. Rev. Neurosci.* 30: 339–365.

Zhao, B. and U. Müller (2015) The elusive mechanotransduction machinery of hair cells. *Curr. Opin. Neurobiol.* 34: 172–179.

Important Original Papers

Barbour, D. L. and X. Wang (2005) The neuronal representation of pitch in primate auditory cortex. *Nature* 436: 1161–1165.

Brand, A., O. Behrand, T. Marquardt, D. McAlpine and B. Grothe (2002) Precise inhibition is essential for microsecond interaural time difference coding. *Nature* 417: 543–547.

Chen, X., U. Leischner, N. L. Rochefort, I. Nelken and A. Konnerth (2011) Functional mapping of single spines in cortical neurons in vivo. *Nature* 475: 501–505.

Corey, D. P. and A. J. Hudspeth (1979) Ionic basis of the receptor potential in a vertebrate hair cell. *Nature* 281: 675–677.

Crawford, A. C. and R. Fettiplace (1981) An electrical tuning mechanism in turtle cochlear hair cells. *J. Physiol.* 312: 377–413.

Eliades, S. J. and X. Wang (2008) Neural substrates of vocalization feedback monitoring in primate auditory cortex. *Nature* 453: 1102–1106.

Fitzpatrick, D. C., J. S. Kanwal, J. A. Butman and N. Suga (1993) Combination-sensitive neurons in the primary auditory cortex of the mustached bat. *J. Neurosci.* 13: 931–940.

Jeffress, L. A. (1948) A place theory of sound localization. *J. Comp. Physiol. Psychol.* 41: 35–39.

Knudsen, E. I. and M. Konishi (1978) A neural map of auditory space in the owl. *Science* 200: 795–797.

Mesgarani, N. and E. F. Chang (2012) Selective cortical representation of attended speaker in multi-talker speech perception. *Nature* 485: 233–236.

Mesgarani, N., C. Cheung, K. Johnson and E. F. Chang (2014) Phonetic feature encoding in human superior temporal gyrus. *Science* 343: 1006–1010.

Middlebrooks, J. C., A. E. Clock, L. Xu and D. M. Green (1994) A panoramic code for sound location by cortical neurons. *Science* 264: 842–844.

Rieke, F., D. A. Bodnar and W. Bialek (1995) Naturalistic stimuli increase the rate and efficiency of information transfer by primary auditory afferents. *Proc. Biol. Sci.* 262 (1365): 259–265.

Rothschild, G., I. Nelken and A. Mizrahi (2010) Functional organization and population dynamics in the mouse primary auditory cortex. *Nature Neurosci.* 13: 353–360.

Schneider, D. M., A. Nelson and R. Mooney (2014) A synaptic and circuit basis for corollary discharge in the auditory cortex. *Nature* 513: 189–194.

Suga, N., W. E. O'Neill and T. Manabe (1978) Cortical neurons sensitive to combinations of information-bearing elements of biosonar signals in the mustache bat. *Science* 200: 778–781.

von Békésy, G. (1960) *Experiments in Hearing.* New York: McGraw-Hill. (A collection of von Békésy's original papers.)

Books

Moore, B. C. J. (2003) *An Introduction to the Psychology of Hearing.* London: Academic Press.

Pickles, J. O. (2013) *An Introduction to the Physiology of Hearing*, 4th Edition. Leiden: Brill.

Schnupp, J., I. Nelken and A. King (2011) *Auditory Neuroscience.* Cambridge, MA: MIT Press.

Go to the NEUROSCIENCE 6e Companion Website at **www.oup.com/uk/Purves6e** for Web Topics, Animations, Flashcards, and more.

CHAPTER

14 The Vestibular System

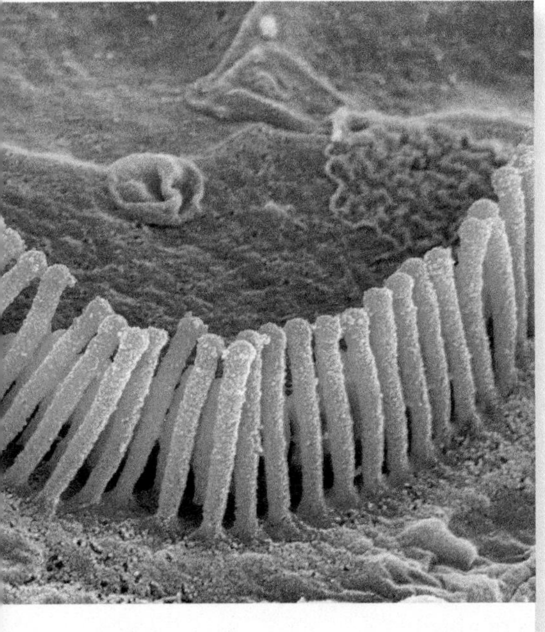

Overview

THE VESTIBULAR SYSTEM PROCESSES sensory information underlying motor responses to (and perceptions of) self-motion, head position, and spatial orientation relative to gravity, helping to stabilize gaze, head, and posture. The peripheral portion of the vestibular system includes inner ear structures that function as an inertial guidance device made up of small linear accelerometers and angular velocity sensors, continuously reporting information about the motions and position of the head to integrative centers in the brainstem, cerebellum, and cerebral cortices. The central portion of the system includes the vestibular nuclei, which make extensive connections with brainstem and cerebellar structures. The vestibular nuclei also innervate motor neurons controlling extraocular, cervical, and postural muscles, thus mediating stabilization of gaze, head orientation, and posture during movement. Interestingly, the vestibular nuclei also receive visual input, resulting in multisensory integration at the earliest point in central vestibular processing. Although we are normally unaware of its functioning, the vestibular system is a key component in postural reflexes and eye movements and, in concert with the visual and proprioceptive systems, plays a central role in distinguishing self-generated "active" movements of the head and body from passive movements resulting from externally applied forces. This multimodal integration is critical to our perception of self-motion, spatial orientation, and body representation. In summary, the vestibular system gives rise to a "sixth sense" that is critical both to automatic behaviors and to perception, with the consequence that balance, gaze stabilization during head movement, and sense of orientation in space are all adversely affected if the system is damaged.

The Vestibular Labyrinth

The main peripheral component of the vestibular system is an elaborate set of interconnected chambers—the **labyrinth**—that has much in common, and is in fact continuous, with the cochlea (see Chapter 13). Like the cochlea, the labyrinth is derived from the otic placode of the embryo and uses the same specialized set of sensory cells—hair cells—to transduce physical motion into neural impulses. In the cochlea, the motion arises from airborne sound stimuli; in the labyrinth, the pertinent motions arise from the effects of gravity and from translational and rotational movements of the head.

The labyrinth is buried deep in the temporal bone and consists of the two **otolith organs**—the **utricle** and **saccule**—and three **semicircular canals** (Figure 14.1). The elaborate architecture of these components explains why this part of the vestibular system is called the labyrinth. The utricle and saccule are specialized primarily to respond to **translational movements** of the head and static head position relative

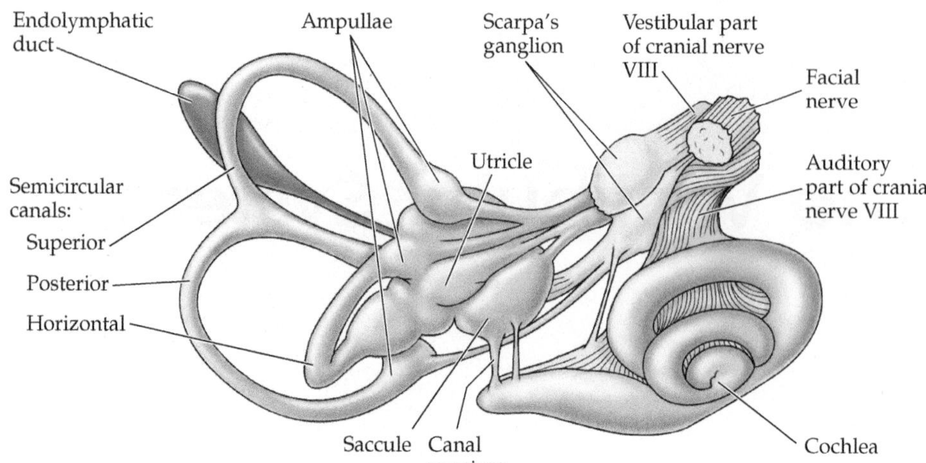

FIGURE 14.1 The labyrinth and its innervation. The vestibular and auditory portions of cranial nerve VIII are shown; the small connection from the vestibular nerve to the cochlea contains auditory efferent fibers. General orientation within the head is shown in Figure 13.4; see also Figure 14.8.

to the gravitational axis (i.e., head tilts), whereas the semicircular canals, as their shapes suggest, are specialized for responding to rotations of the head.

The intimate relationship between the cochlea and the labyrinth goes beyond their common embryonic origin. Indeed, the cochlear and vestibular spaces are actually joined (see Figure 14.1), and the specialized ionic environments of the vestibular end-organ are much like those of the cochlea. The membranous sacs within the bone are filled with fluid (endolymph) and are collectively called the membranous labyrinth. Between the bony walls (the osseous labyrinth) and the membranous labyrinth is another fluid, the perilymph, which is similar in composition to cerebrospinal fluid (see Chapter 13). The vestibular hair cells are located in the utricle and saccule and in three jug-like swellings called **ampullae**, located at the base of the semicircular canals next to the utricle. As in the cochlea, tight junctions seal the apical surfaces of the vestibular hair cells, ensuring that endolymph selectively bathes the hair cell bundle while remaining separate from the perilymph surrounding the basal portion of the hair cell.

Vestibular Hair Cells

The vestibular hair cells, which, like cochlear hair cells, transduce minute displacements into behaviorally relevant receptor potentials, provide the basis for vestibular function. Vestibular and auditory hair cells are quite similar; Chapter 13 gave a detailed description of hair cell structure and function. As in the case of auditory hair cells, movement of the stereocilia toward the kinocilium in the vestibular end-organs opens mechanically gated transduction channels located at the tips of the stereocilia, depolarizing the hair cell and causing neurotransmitter release onto (and excitation of) the vestibular nerve fibers (Figure 14.2A,B). Movement of the stereocilia in the direction away from the kinocilium closes the channels, hyperpolarizing the hair cell and thus reducing vestibular nerve activity. The biphasic nature of the receptor potential means that some transduction channels are open in the absence of stimulation, with the result that hair cells tonically release transmitter, thereby generating considerable spontaneous activity in vestibular nerve fibers (see Figure 14.6). One consequence of these spontaneous action potentials is that the firing rates of vestibular fibers can increase or decrease in a manner that faithfully mimics the receptor potentials produced by the hair cells. Adaptation in vestibular hair cells, mediated by calcium entering through mechanoelectrical transduction (MET) and voltage-gated calcium channels, is especially important to vestibular function, as it allows hair cells to continue to signal small changes in head position despite much larger tonic forces of gravity.

Importantly, the hair cell bundles in each vestibular organ have specific orientations (Figure 14.2C). As a result, the organ as a whole is responsive to displacements in all directions. In a given semicircular canal, the hair cells in the ampulla are all polarized in the same direction, and the three semicircular canals are oriented to detect rotational motion around the three cardinal (i.e., x, y, and z) axes. In the utricle and saccule, a specialized area called the **striola** divides the hair cells into two populations having opposing polarities (see Figures 14.2C and 14.4C). The directional polarization of the receptor surfaces is a principle of organization in the vestibular system, as will become apparent in the following descriptions of the individual vestibular organs.

Otolith Organs: The Utricle and Saccule

The two otolith organs, the utricle and the saccule, detect tilting and translational (i.e, linear, as opposed to rotational) movements of the head. Both of these organs

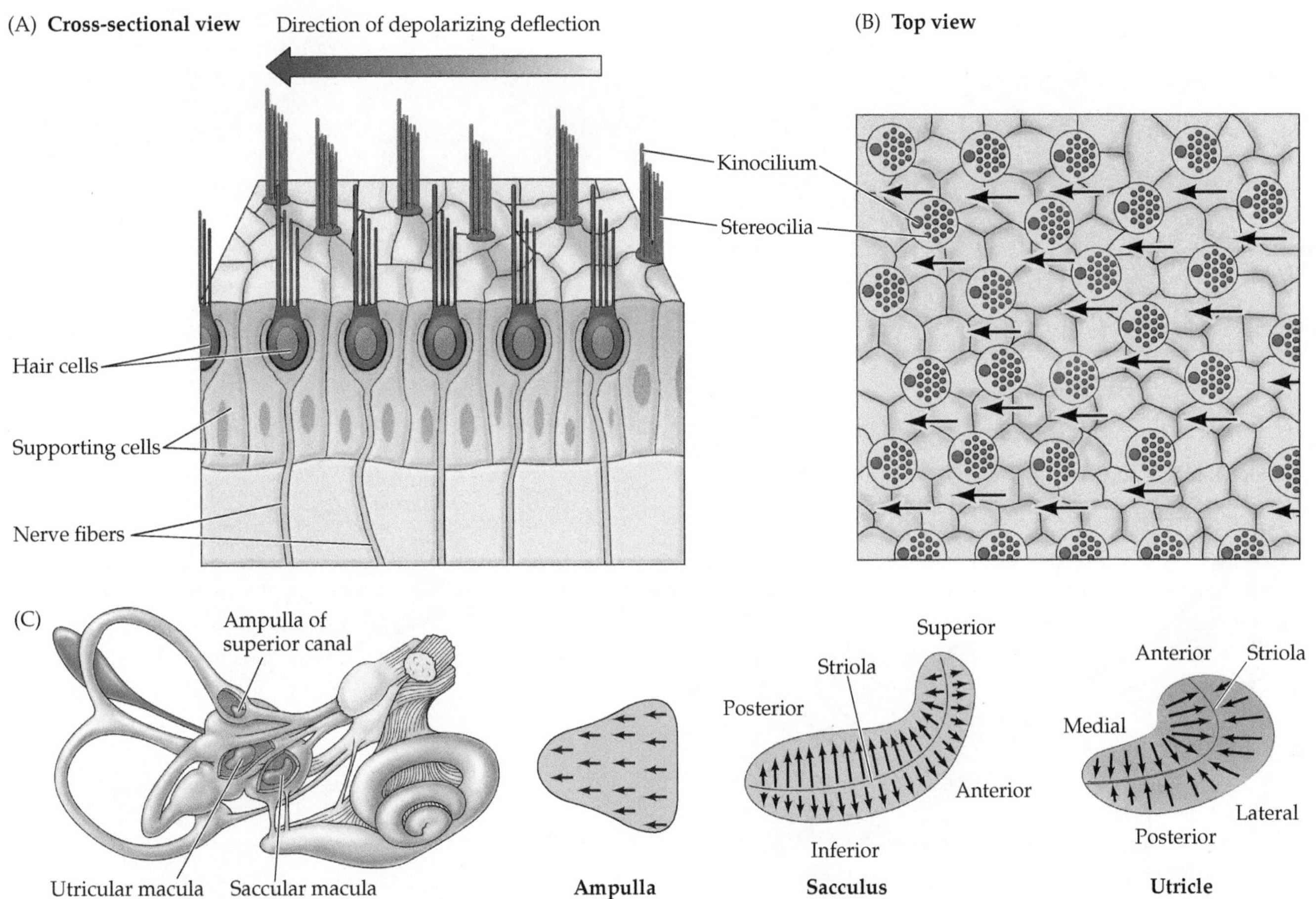

FIGURE 14.2 The morphological polarization of vestibular hair cells and polarization maps of the vestibular organs. (A) A cross section of hair cells shows that the kinocilia of a group of hair cells are all located on the same side of the hair cell. The arrow indicates the direction of deflection that depolarizes the hair cell. (B) View looking down on the hair bundles. (C) In the ampulla located at the base of each semicircular canal, the hair bundles are oriented in the same direction. In the sacculus and utricle, the striola divides the hair cells into populations with opposing hair bundle polarities.

contain a sensory epithelium, the **macula**, which consists of hair cells and associated supporting cells. Overlying the hair cells and their hair bundles is a gelatinous layer; above this layer is a fibrous structure, the **otolithic membrane**, in which are embedded crystals of calcium carbonate called **otoconia** (Figure 14.3). The crystals give the otolith organs their name (*otolith* is Greek for "ear stones"). The otoconia make the otolithic membrane heavier than the structures and fluids surrounding it; thus, when the head tilts, gravity causes the membrane to shift relative to the macula (Figure 14.4). The resulting shearing motion between the otolithic membrane and the macula displaces the hair bundles, which are embedded in the lower, gelatinous surface of the membrane. This displacement of the hair bundles generates a receptor potential in the hair cells. A shearing motion between the macula and the otolithic

FIGURE 14.3 Calcium carbonate crystals (otoconia) in the utricular macula of a quail. Each otoconium in this scanning electron micrograph is about 50 μm long. (From Dickman et al., 2004.)

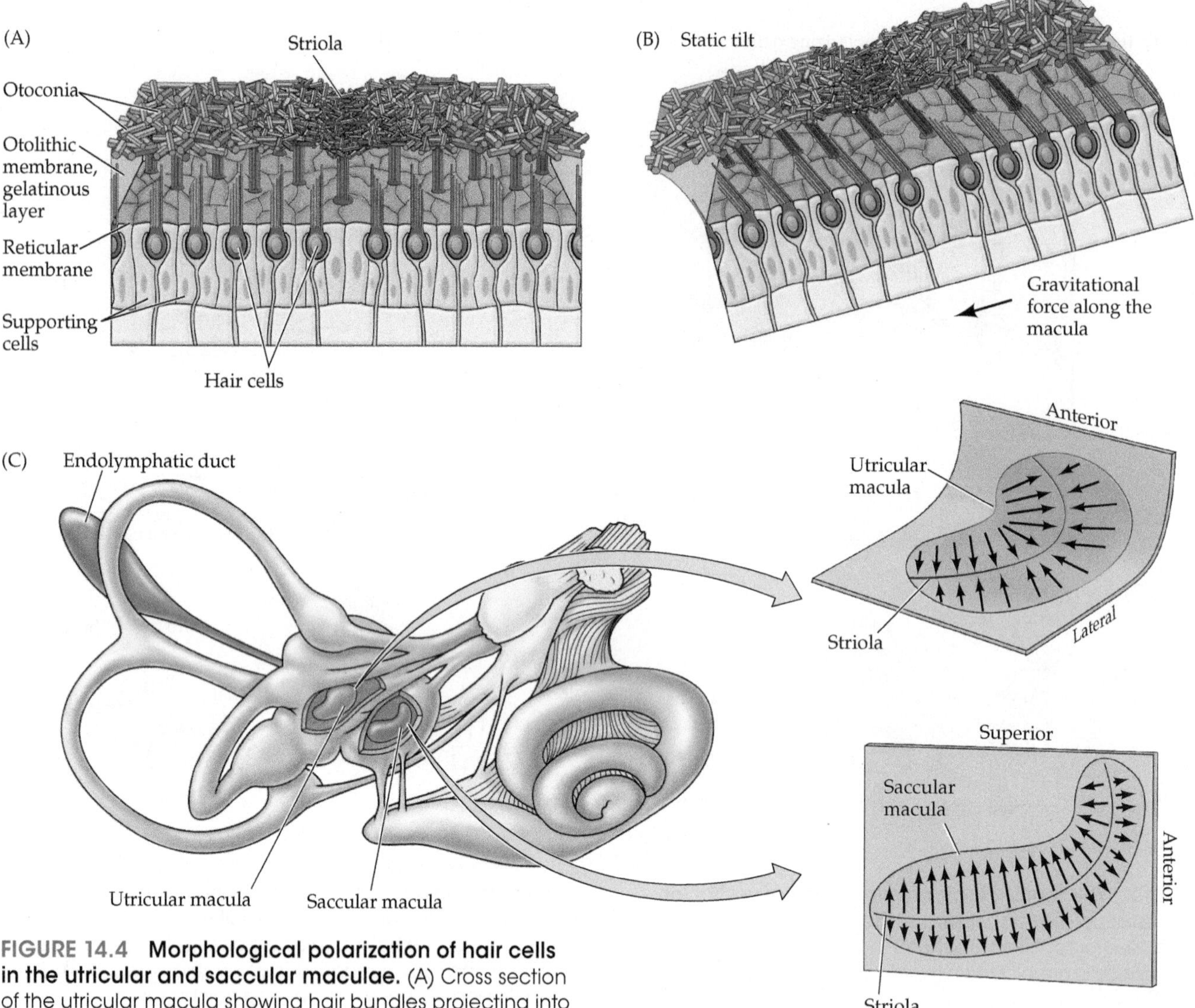

FIGURE 14.4 Morphological polarization of hair cells in the utricular and saccular maculae. (A) Cross section of the utricular macula showing hair bundles projecting into the gelatinous layer when the head is level. (B) Cross section of the utricular macula when the head is tilted. The hair bundles are deflected by the otoconia in the direction of the gravitational force along the macular plane. An equivalent linear acceleration opposite to this force would induce the same deflection of the otoconia and is referred to as the *equivalent acceleration.* (C) Orientation of the utricular and saccular maculae in the head; arrows show orientation of the kinocilia, as in Figure 14.2. The saccules on either side are oriented more or less vertically, and the utricles more or less horizontally. The striola is a structural landmark consisting of small otoconia arranged in a narrow trench that divides each otolith organ. In the utricular macula, the kinocilia are directed toward the striola. In the saccular macula, the kinocilia point away from the striola. Note that, given the utricle and sacculus on both sides of the body, there is a continuous representation of all directions of head movement.

membrane also occurs when the head undergoes translational movements (Figure 14.5); the greater relative mass of the otolithic membrane causes it to lag behind the macula temporarily, leading to transient displacement of the hair bundle.

One consequence of the similar effects exerted on otolithic hair cells by certain head tilts and translational movements is that otolith afferents cannot convey information that distinguishes between these two types of stimuli. Consequently, one might expect that these different stimuli would be rendered perceptually equivalent when visual feedback is absent, as occurs in the dark or when the eyes are closed. Nevertheless, blindfolded subjects can discriminate between these two stimulus categories, a feat that depends on the integration of information from the otolith organs and the semicircular canals in the central vestibular system, as described in more detail in the section "Central Pathways for Stabilizing Gaze, Head, and Posture."

As already mentioned, the orientation of the hair cell bundles is organized relative to the striola, which demarcates the overlying layer of otoconia (see Figure 14.4A). The

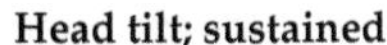

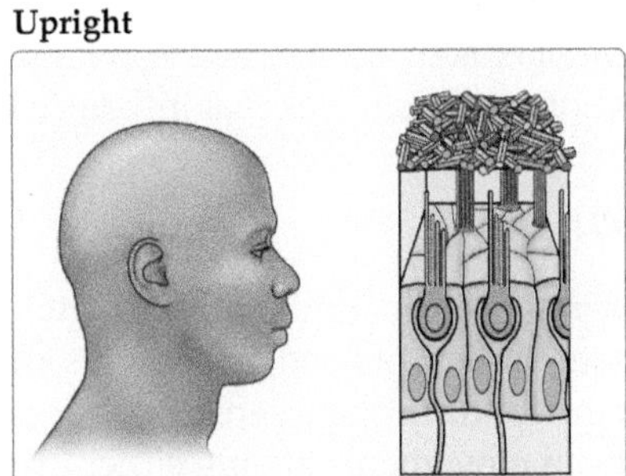

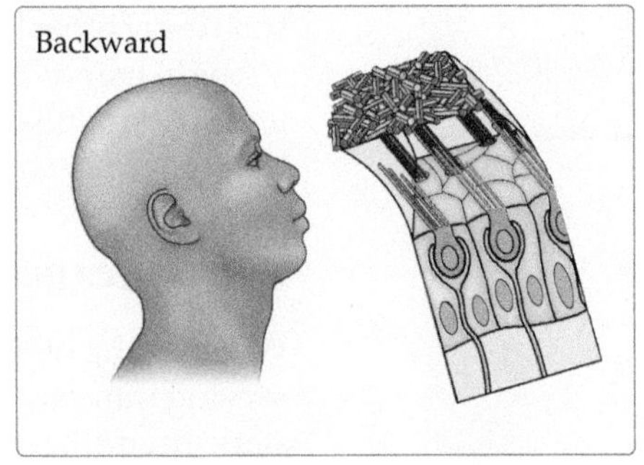

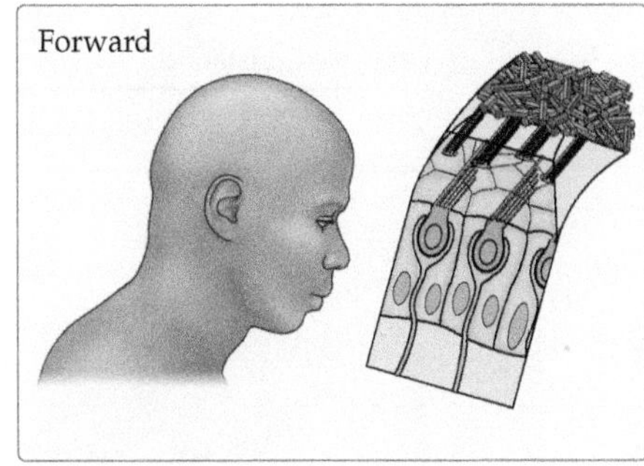

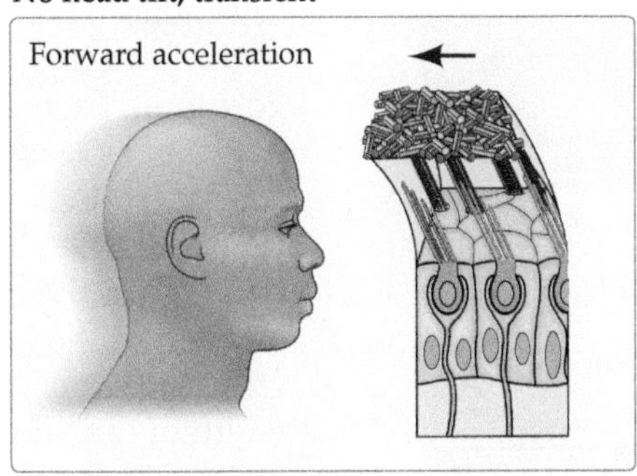

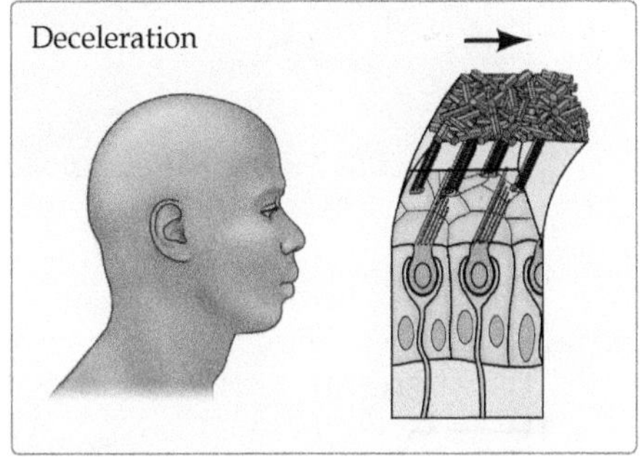

FIGURE 14.5 **Forces acting on the head displace the otolithic membrane of the utricular macula.** For each of the positions and accelerations due to translational movements, some set of hair cells will be maximally excited, while another set will be maximally inhibited. Note that head tilts produce displacements similar to certain accelerations.

striola forms an axis of mirror symmetry such that hair cells on opposite sides of the striola have opposing morphological polarizations. Thus, a head tilt along the axis of the striola will excite the hair cells on one side while inhibiting the hair cells on the other side. The saccular macula is oriented vertically and the utricular macula horizontally, with continuous variation in the morphological polarization of the hair cells located in each macula (as shown in Figure 14.4C, where the black arrows indicate the direction of movement that produces excitation). Inspection of the excitatory orientations in the maculae indicates that the utricle responds to translational movements of the head in the horizontal plane and to sideways head tilts, whereas the saccule responds to vertical translational movements of the head and to upward or downward head tilts.

Note that the saccular and utricular maculae on one side of the head are mirror images of those on the other side. Thus, a tilt of the head to one side has opposite effects on corresponding hair cells of the two utricular maculae. This concept is important in understanding how the central connections of the vestibular periphery mediate the interaction of inputs from the two sides of the head.

How Otolith Neurons Sense Head Tilts and Translational Head Movements

The structure of the otolith organs enables them to sense both static displacements, as would be caused by head tilts, and translational components of head movements. Figure 14.5 illustrates some of the forces produced by head tilts and translational movements on the utricular macula.

The mass of the otolithic membrane relative to the surrounding endolymph, as well as the otolithic membrane's physical uncoupling from the underlying macula, means that hair bundle displacement will occur transiently in response to translational head movements, and tonically in response to tilting of the head. The resulting hair bundle displacements are reflected in the responses of the vestibular nerve fibers that innervate the otolith organs. As mentioned, these nerve fibers have a steady and relatively high firing rate when the head is upright. Figure 14.6 shows these responses recorded from an otolith afferent fiber, or axon, in a monkey seated in a chair that could be tilted for several seconds to produce a steady force on the head. Prior to the tilt, the axon has a high firing rate, which increases or decreases depending on the direction of the tilt. Notice also that the response remains at a high level as long as the tilting force remains constant; thus, such neurons faithfully encode the static force being applied to the head (see Figure 14.6A). When the head is returned to the original position, the firing level of the neurons returns to baseline value. Conversely, when the tilt is in the opposite direction, the neurons respond by decreasing their firing rate below the

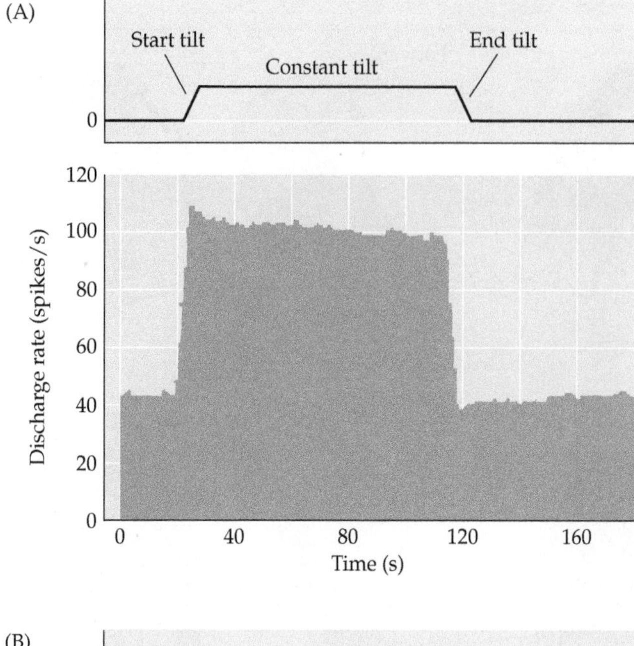

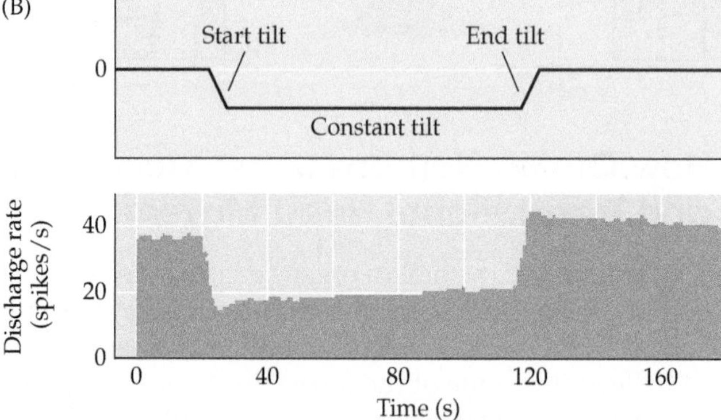

FIGURE 14.6 Response of a vestibular nerve axon from an otolith organ. The utricle is the example shown here. (A) The stimulus (top) is a change in head tilt. The spike histogram shows the neuron's response to tilting in a particular direction. (B) A response of the same fiber to tilting in the opposite direction. (After Fernández and Goldberg, 1976.)

resting level (see Figure 14.6B) and remain depressed as long as the static force continues. In a similar fashion, transient increases or decreases in firing rate from spontaneous levels signal the direction of translational movements of the head.

The range of orientations of hair bundles within the otolith organs enables them to transmit information about linear forces in every direction the head might move (see Figure 14.4C). The utricle, which is primarily concerned with motion in the horizontal plane, and the saccule, which is concerned with vertical motion, combine to effectively gauge the linear forces acting on the head at any instant, in three dimensions. Tilts of the head off the horizontal plane and translational movements of the head in any direction stimulate a distinct subset of hair cells in the saccular and utricular maculae, while simultaneously suppressing the responses of other hair cells in these organs. Ultimately, variations in hair cell polarity in the otolith organs produce patterns of vestibular nerve fiber activity that, at a population level, encode head position and the forces that influence it.

Semicircular Canals

Whereas the otolith organs are concerned primarily with sensing translational components of head movements and static head tilts, the semicircular canals sense head rotational components of head movements arising either from self-induced head rotations or from angular accelerations of the head imparted by external forces, such as a merry-go-round. Each of the three semicircular canals has at its base a bulbous expansion—the ampulla—that houses the sensory epithelium, or **crista**, that contains the hair cells (Figure 14.7). The structure of the canals suggests how they detect the angular accelerations that arise through rotation of the head. The hair bundles extend out of the crista into a gelatinous mass, the **cupula**, that bridges the width of the ampulla, forming a viscous barrier through which endolymph cannot circulate. As a result, movements

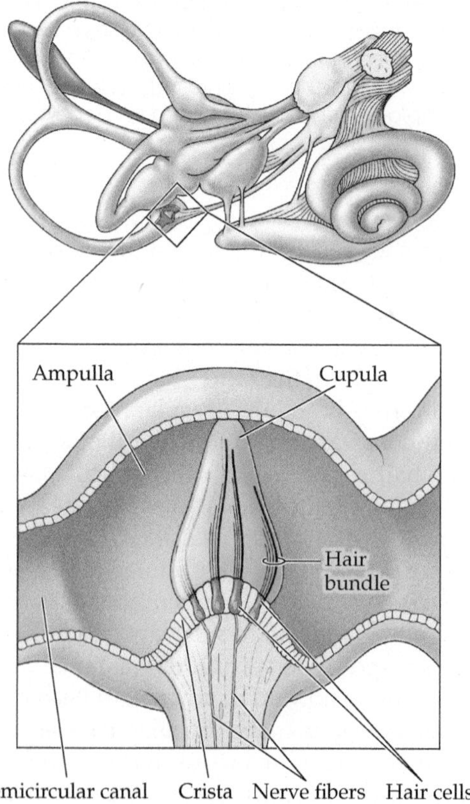

FIGURE 14.7 The ampulla of the posterior semicircular canal. The crista, hair bundles, and cupula are diagrammed. When the head rotates, fluid in the membranous canal distorts the cupula.

of the endolymphatic fluid distort the relatively compliant cupula. When the head turns in the plane of one of the semicircular canals, the inertia of the endolymph produces a force across the cupula, distending it away from the direction of head movement and causing a displacement of the hair bundles within the crista (Figure 14.8A,B). Note that semicircular canals can be excited by rotations that occur during the initiation of a head tilt as well as rotations resulting from other active or passive movements of the head. In contrast, translational movements of the head produce equal forces on the two sides of the cupula, so the hair bundles within the ampulla are not displaced.

Unlike the saccular and utricular maculae, all of the hair cells in the crista within each semicircular canal are organized with their kinocilia pointing in the same direction (see Figure 14.2B). Thus, when the cupula moves in the appropriate direction, the entire population of hair cells is depolarized and activity in all of the innervating axons increases. When the cupula moves in the opposite direction, the population is hyperpolarized and neuronal activity decreases. Deflections orthogonal to the excitatory–inhibitory direction produce little or no response.

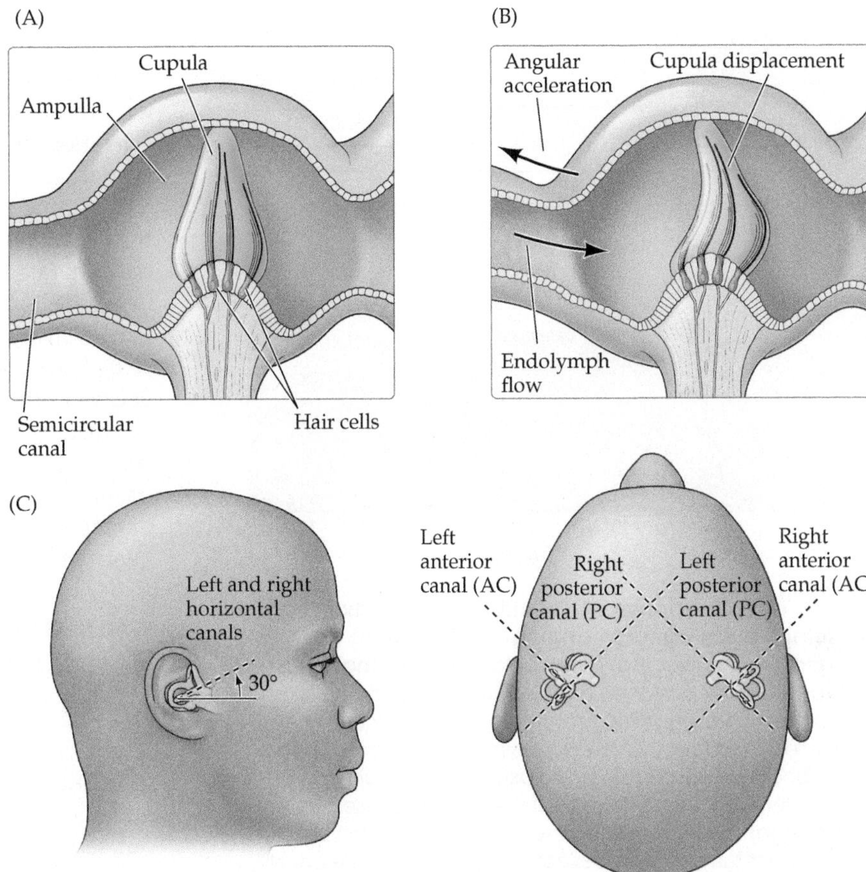

FIGURE 14.8 Functional organization of the semicircular canals. (A) The position of the cupula without angular acceleration. (B) Distortion of the cupula during angular acceleration. When the head is rotated in the plane of the canal (arrow outside canal), the inertia of the endolymph creates a force (arrow inside canal) that displaces the cupula. (C) Arrangement of the canals in pairs. The two horizontal canals form a pair; the right anterior canal (AC) and the left posterior canal (PC) form a pair; and the left AC and the right PC form a pair.

Each semicircular canal works in concert with the partner located on the other side of the head that has its hair cells aligned oppositely. There are three such pairs: the two (right and left) horizontal canals, and the anterior canal on each side working with the posterior canal on the other side (Figure 14.8C). Head rotation deforms the cupula in opposing directions for the two partners, resulting in opposite changes in their firing rates. Thus, the orientation of the horizontal canals makes them selectively sensitive to rotation in the horizontal plane. More specifically, the hair cells in the canal toward which the head is turning are depolarized, while those on the other side are hyperpolarized.

For example, when the head rotates to the left, the cupula is pushed toward the kinocilium in the left horizontal canal, and the firing rate of the relevant axons in the left vestibular nerve increases. In contrast, the cupula in the right horizontal canal is pushed away from the kinocilium, with a concomitant decrease in the firing rate of the related neurons. If the head rotation is to the right, the result is just the opposite. This push–pull arrangement operates for all three pairs of canals; the pair whose activity is modulated is in the plane of the rotation, and the member of the pair whose activity is increased is on the side toward which the head is turning. The net result is a system that provides information about the rotation of the head in any direction.

How Semicircular Canal Neurons Sense Head Rotations

Like axons that innervate the otolith organs, the vestibular fibers that innervate the semicircular canals exhibit a high level of spontaneous activity. As a result, they can transmit information by either increasing or decreasing their firing rate, thus more effectively encoding rotational head movements (see above). The bidirectional responses of fibers innervating the hair cells of the semicircular canal have been studied by recording the axonal firing rates in a monkey's vestibular nerve. Seated in a chair, the monkey was rotated

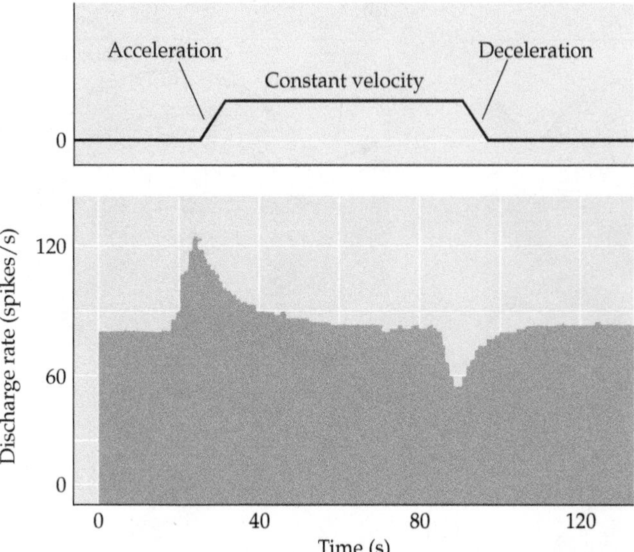

FIGURE 14.9 Response of a vestibular nerve axon from the semicircular canal to angular acceleration. The stimulus (top) is a rotation that first accelerates, then maintains constant velocity, and then decelerates the head. The stimulus-evoked change in firing rate of this vestibular unit (bottom) reflects the fact that the endolymph has viscosity and inertia and that the cupola has elasticity. Thus, during the initial acceleration, the deflection of the cupula causes the unit activity to rapidly increase. During constant angular velocity, the cupula returns to its undeflected state over a time course related to its elasticity and the viscosity of the fluid, and the unit activity returns to the baseline rate. During deceleration, the cupula is deflected in the opposite direction, causing a transient decrease in the unit firing rate. This behavior can be thought of as the cupula–endolymph system dynamic; the inertia of the fluid plays a minor role in this dynamic, coming into play only at very high frequencies of head movement. (After Fernández and Goldberg, 1976.)

continuously in one direction during three phases: an initial period of acceleration, then a period of several seconds at constant velocity, and finally a period of sudden deceleration to a stop (Figure 14.9). The maximum firing rates observed correspond to the period of acceleration, when the cupula is deflected; the minimum firing rate corresponds to the period of deceleration, when the cupula is deflected in the opposite direction. During the constant-velocity phase, firing rates return to a baseline level as the cupula returns to its undeflected state over a time course that is related to the cupular elasticity and the viscosity of the endolymph (about 15 seconds). Note that the time it takes the cupula to return to its undistorted state (and for the hair bundles to return to their undeflected position) can occur while the head is still turning, as long as a constant angular velocity is maintained. Such constant forces are rarely found in nature, although they are encountered aboard ships, airplanes, space vehicles, and amusement park rides, where prolonged acceleratory arcs can occur.

An interesting aspect of the cupula–endolymph system dynamic is that it "smooths" the transduction of head accelerations into neural signals. For example, when the head is angularly accelerated to a constant velocity rather rapidly (corresponding to high-frequency rotational movements of the head), vestibular units associated with the affected canal generate a velocity signal; note that the axon firing rate in Figure 14.9 rises linearly during the acceleration phase. However, when the head is moving at a constant angular velocity (i.e., low-frequency rotational movements), the rate decays to the spontaneous level (corresponding to an acceleration of zero). This transduction process results in a velocity signal at high frequencies and an acceleration signal at low frequencies, a behavior that can be seen clearly in response to sinusoidal stimuli applied over a wide frequency range.

Central Pathways for Stabilizing Gaze, Head, and Posture

The vestibular system contributes to rapid automatic behaviors, such as reflexive eye movements that stabilize gaze and rapid postural adjustments to maintain balance, and also to higher-order processes that are important to our sense of spatial orientation and self-motion. The organization of the central vestibular pathways reflects this multifunctional role; these pathways also display two features that distinguish them from the pathways that convey information important to sight, hearing, and touch. First, *central vestibular processing is inherently multisensory*, because many neurons in the vestibular nuclei—the earliest point in central vestibular processing—receive visual input. Second, *many neurons in the vestibular nuclei function as premotor neurons in addition to giving rise to ascending sensory projections*, providing a very short-latency sensorimotor arc that can drive extremely rapid (~10 ms) compensatory eye and head movements in response to vestibular stimulation.

The vestibular end-organs communicate, via the vestibular branch of cranial nerve VIII, with targets in the brainstem and the cerebellum that process much of the information necessary to compute head position and motion. As with the cochlear nerve, the vestibular nerves arise from a population of bipolar neurons, the cell bodies of which in this instance reside in the **vestibular nerve ganglion** (also called **Scarpa's ganglion**; see Figure 14.1). The distal processes of these cells innervate the semicircular canals and the otolith organs, while the central processes project via the vestibular portion of the **vestibulocochlear nerve** (cranial nerve VIII) to the **vestibular nuclei** (and also directly to the cerebellum; Figure 14.10). Although the canal and otolith afferents are largely segregated in the periphery, a large amount of canal–otolith convergence is found in the vestibular nuclei, a feature that ultimately enables the unambiguous encoding of head orientation and motion through the environment. Indeed, although head tilts and translational movements of the head can similarly excite otolith organs, the semicircular canals are excited only by rotations that accompany head tilts and not by purely translational movements. Therefore, integration of information from the otolith organs and semicircular canals in the vestibular nuclei and cerebellum can be used to distinguish head tilts from translational head movements. The vestibular nuclei also integrate a broad range of vestibular and non-vestibular information, receiving input from the vestibular nuclei of the opposite side as well as from the cerebellum and the visual and somatosensory systems.

The central projections of the vestibular system participate in two major classes of reflexes: (1) those responsible for maintaining equilibrium and gaze during movement, and (2) those responsible for maintaining posture. The first helps coordinate head and eye movements to keep the gaze fixed on objects of interest during movement. (Other functions include protective or escape reactions; see Box 14A.) The **vestibulo-ocular reflex (VOR)** in particular is a mechanism for producing eye movements that counter head movements, thus permitting the gaze to remain fixed on a particular point (see Chapter 20). For example, activity in the left horizontal canal induced by leftward rotary acceleration of the head excites neurons in the left vestibular nucleus and results in compensatory eye movements to the right.

Figure 14.10 illustrates the circuits mediating this reflex. Vestibular nerve fibers originating in the left horizontal semicircular canal project to the medial and superior vestibular nuclei. Excitatory fibers from the medial vestibular nucleus cross to the contralateral abducens nucleus, which has two outputs. One of these is a motor pathway that causes the lateral rectus of the right eye to contract; the other is an excitatory projection that crosses the midline and ascends via the **medial longitudinal fasciculus** to the left oculomotor nucleus, where it activates neurons that cause the medial rectus of the left eye to contract. Finally, inhibitory neurons project from the

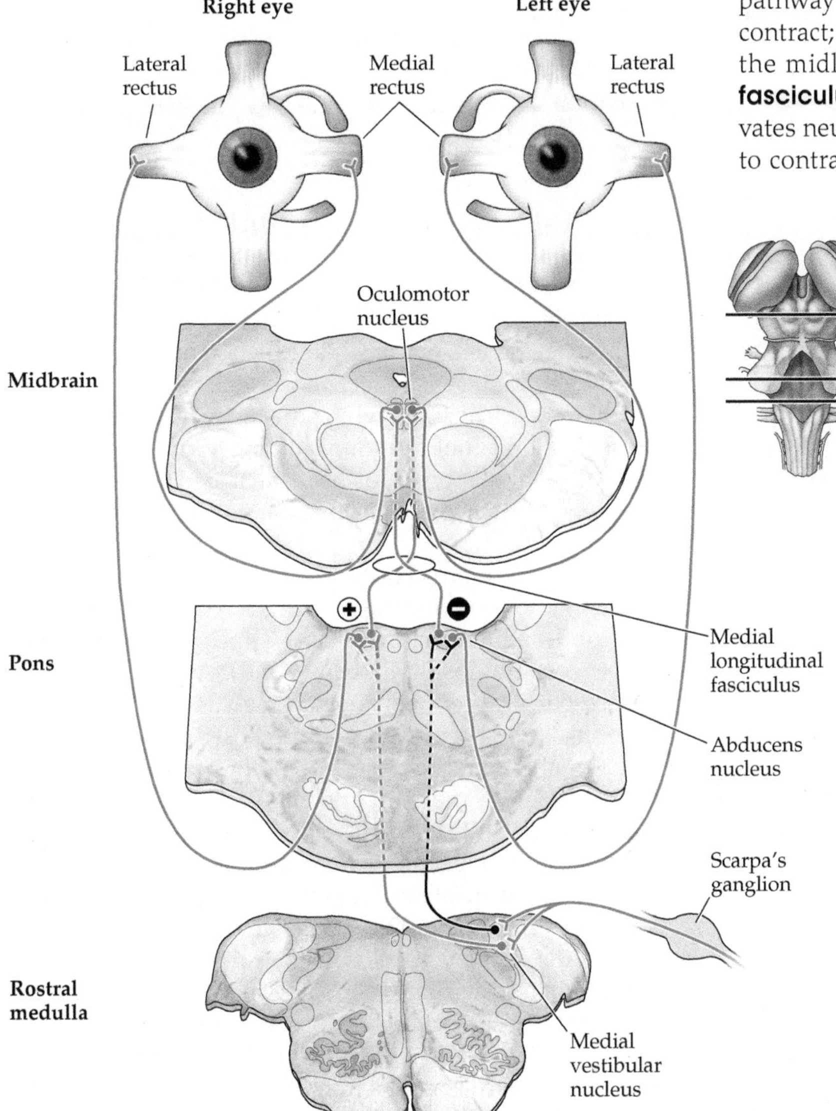

FIGURE 14.10 Connections underlying the vestibulo-ocular reflex. Projections of the vestibular nucleus to the nuclei of cranial nerves III (oculomotor) and VI (abducens). The connections to the oculomotor nucleus and to the contralateral abducens nucleus are excitatory (red), whereas the connections to ipsilateral abducens nucleus are inhibitory (black). There are connections from the oculomotor nucleus to the medial rectus of the left eye and from the abducens nucleus to the lateral rectus of the right eye. This circuit moves the eyes to the right—that is, in the direction away from the left horizontal canal—when the head rotates to the left. Turning to the right, which causes increased activity in the right horizontal canal, has the opposite effect on eye movements. The projections from the right vestibular nucleus are omitted for clarity.

BOX 14A ■ Mauthner Cells in Fish

A primary function of the vestibular system is to provide information about the direction and speed of ongoing movements, ultimately enabling rapid, coordinated reflexes to compensate for both self-induced and externally generated forces. One of the most impressive and speediest vestibular-mediated reflexes is the tail-flip escape behavior of fish (and larval amphibians), a stereotyped response that allows a potential prey to elude its predators (Figure A; tap on the side of a fish tank if you want to observe the reflex). In response to a perceived risk, fish flick their tail and are thus propelled laterally away from the approaching threat.

The circuitry underlying the tail-flip escape reflex includes a pair of giant medullary neurons called Mauthner cells, their vestibular inputs, and the spinal cord motor neurons to which the Mauthner cells project. (Most fish possess one pair of Mauthner cells in a stereotypic location. Thus, these cells can be consistently visualized and studied from animal to animal.) Movements in the water, such as might be caused by an approaching predator, excite saccular hair cells in the vestibular labyrinth. These receptor potentials are transmitted via the central processes of vestibular ganglion cells in cranial nerve VIII to the two Mauthner cells in the brainstem. As in the vestibulospinal pathway in humans, the Mauthner cells project directly to spinal motor neurons. The small number of synapses intervening between the receptor cells and the motor neurons is one of the ways that this circuit has been optimized for speed by natural selection, an arrangement evident in humans as well. The large size of the Mauthner axons is another; the axons from these cells in a goldfish are about 50 μm in diameter.

(A) Bird's-eye view of the sequential body orientations of a fish engaging in a tail-flip escape behavior, with time progressing from left to right. This behavior is mediated largely by vestibular inputs to Mauthner cells. (After Eaton et al., 1977.)

medial vestibular nucleus to the left abducens nucleus, directly causing the motor drive on the lateral rectus of the left eye to decrease and also indirectly causing the right medial rectus to relax. The consequence of these several connections is that excitatory input from the horizontal canal on one side produces eye movements toward the opposite side. Therefore, turning the head to the left causes eye movements to the right. In a similar fashion, head turns in other planes activate other semicircular canals, causing other appropriate compensatory eye movements. Thus, the VOR also plays an important role in vertical gaze stabilization in response to the linear vertical head oscillations that accompany locomotion and in response to vertical angular accelerations of the head, as can occur when riding on a swing. Notably, voluntary movements to redirect gaze transiently diminish the VOR, preventing vestibular reflexes from interfering with goal-directed movements. In the clinic, caloric testing provides a useful way to activate the VOR without moving the head, and it is a valuable tool for diagnosing peripheral and central lesions to the vestibular system.

Loss of the VOR can have severe consequences. A patient with vestibular damage finds it difficult or impossible to fixate on visual targets while the head is moving, a condition called **oscillopsia** ("bouncing vision"). If the damage is unilateral, the patient usually recovers the ability to fixate objects during head movements. However, a patient with bilateral loss of vestibular function has the persistent and disturbing sense that the world is moving when the head moves. The underlying problem in such cases is that information about head movements normally generated by the vestibular organs is not available to the oculomotor centers, so that compensatory eye movements cannot be made.

Descending projections from the vestibular nuclei are essential for postural adjustments of the head, mediated by the vestibulocervical reflex (VCR), and body, mediated by the vestibulospinal reflex (VSR). As with the VOR, these postural reflexes are extremely fast, in part due to the small number of synapses interposed between the vestibular organ and the relevant motor neurons (Box 14A). Like the VOR, the VCR and the VSR are both compromised in patients with bilateral damage to the vestibular periphery. Such patients exhibit diminished head and postural stability, resulting in gait deviations; they also have difficulty balancing. These balance defects become more pronounced in low light or while walking on uneven surfaces, indicating that balance normally is the product of vestibular, visual, and proprioceptive inputs.

The anatomical substrate for the VCR involves the medial vestibular nucleus; axons from this nucleus descend in the medial longitudinal fasciculus to reach the upper cervical

BOX 14A (*continued*)

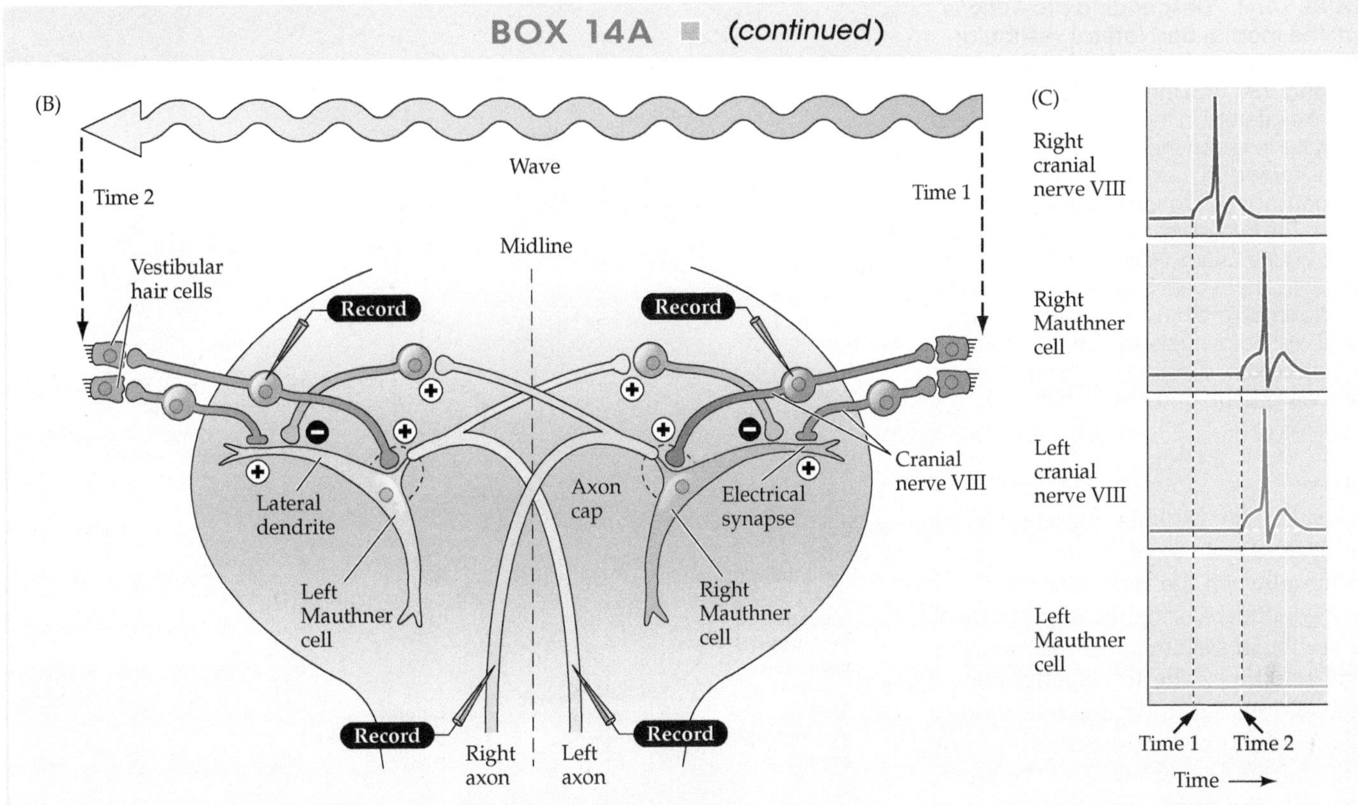

(B) Diagram of synaptic events in the Mauthner cells of a fish in response to a disturbance in the water coming from the right. (C) Complementary responses of the right and left Mauthner cells mediating the escape response. Times 1 and 2 correspond to those indicated in Figure B. (After Furshpan and Furukawa, 1962.)

The optimization for speed and direction in the escape reflex also is reflected in the synapses that vestibular nerve afferents make on each Mauthner cell (Figure B). These connections are electrical synapses that allow rapid and faithful transmission of the vestibular signal.

An appropriate direction for escape is promoted by two features: (1) each Mauthner cell projects only to contralateral motor neurons; and (2) a local network of bilaterally projecting interneurons inhibits activity in the Mauthner cell away from the side on which the vestibular activity originates. In this way, the Mauthner cell on one side faithfully generates action potentials that command contractions of contralateral tail musculature, thus moving the fish out of the path of the oncoming predator. Conversely, the local inhibitory network silences the Mauthner cell on the opposite side during the response (Figure C).

The Mauthner cells in fish are analogous to the reticulospinal and vestibulospinal pathways that control balance, posture, and orienting movements in mammals. The equivalent behavioral responses in humans are evident in a friendly game of tag, or more serious escape endeavors.

levels of the spinal cord (Figure 14.11). This pathway regulates head position by reflex activity of neck muscles in response to stimulation of the semicircular canals caused by rotations of the head. For example, during a downward pitch of the body (e.g., tripping), the superior canals are activated and the head muscles reflexively pull the head up. The dorsal flexion of the head initiates other reflexes, such as forelimb extension and hindlimb flexion, to stabilize the body and protect against a fall (see Chapter 17).

The VSR is mediated by a combination of pathways, including the **lateral** and **medial vestibulospinal tracts** and the reticulospinal tract. The inputs from the otolith organs project mainly to the lateral vestibular nucleus, which in turn sends axons in the lateral vestibulospinal tract to the ipsilateral ventral horn of the spinal cord (see Figure 14.11). These axons terminate monosynaptically on extensor motor neurons, and they disynaptically inhibit flexor motor neurons; the net result is a powerful excitatory influence on the extensor (antigravity) muscles. When hair cells in the otolith organs are activated, signals reach the medial part of the ventral horn. By activating the ipsilateral pool of motor neurons innervating extensor muscles in the trunk and limbs, this pathway mediates balance and the maintenance of upright posture.

Decerebrate rigidity, characterized by rigid extension of the limbs, arises when the brainstem is transected above

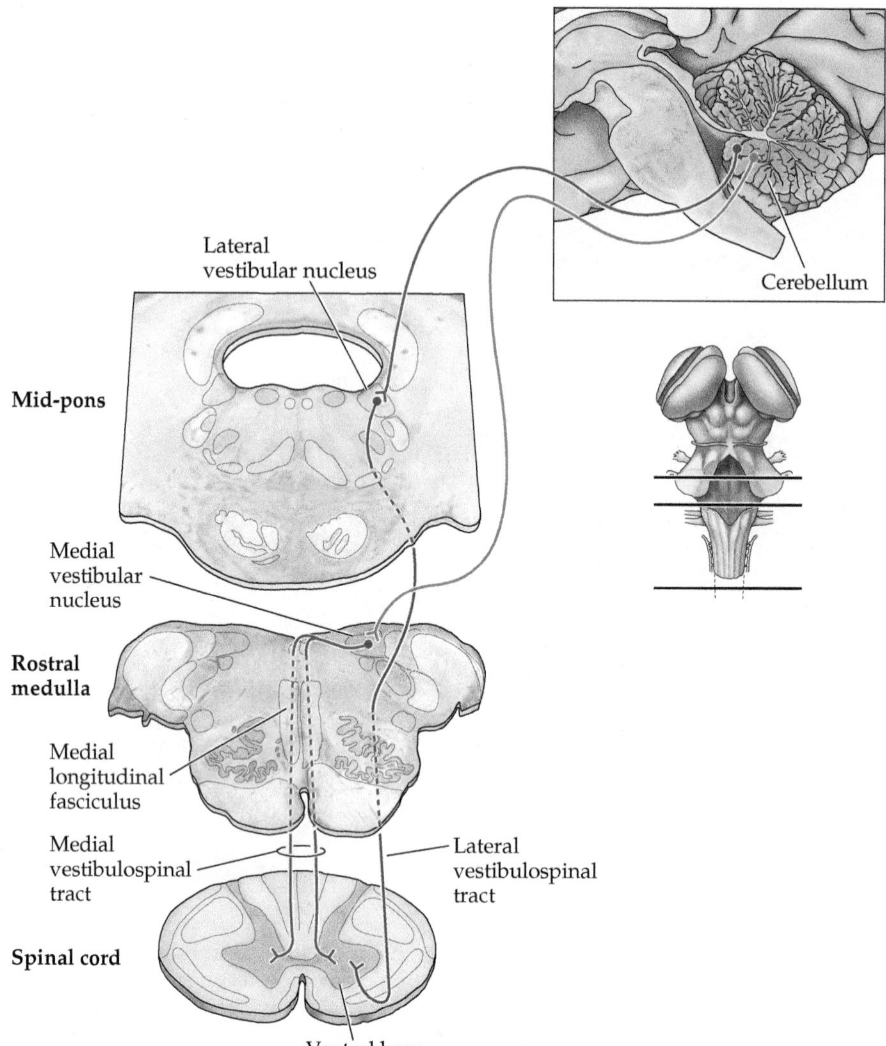

FIGURE 14.11 Descending projections from the medial and lateral vestibular nuclei to the spinal cord underlie the VCR and VSR. The medial vestibular nuclei project bilaterally in the medial longitudinal fasciculus to reach the medial part of the ventral horns and mediate head reflexes in response to activation of semicircular canals. The lateral vestibular nucleus sends axons via the lateral vestibular tract to contact ventral horn cells innervating the axial and proximal limb muscles. Neurons in the lateral vestibular nucleus receive input from the cerebellum, allowing the cerebellum to influence posture and equilibrium.

the level of the vestibular nucleus. Decerebrate rigidity in experimental animals is relieved when the vestibular nuclei are lesioned, underscoring the importance of the vestibular system to the maintenance of muscle tone. The tonic activation of extensor muscles in decerebrate rigidity suggests further that descending projections from higher levels of the brain, especially the cerebral cortex, normally suppress the vestibulospinal pathway (see also Chapter 17).

Vestibular–Cerebellar Pathways

The cerebellum is a major target of ascending vestibular pathways and also provides descending input to the vestibular nuclei, resulting in a recurrent circuit architecture that plays an important role in modulating vestibular activity. These vestibular–cerebellar circuits play a critical role in integrating and modulating vestibular signals to enable adaptive changes to the VOR, distinguish head tilts from translational movements, and distinguish passive movements of the head and body from those that are self-generated. Major vestibular targets in the cerebellum include the flocculus, paraflocculus, nodulus, uvula, and rostral fastigial nucleus, all of which play distinct roles in vestibular plasticity and multimodal integration.

The VOR displays remarkable plasticity in response to altered visual or vestibular input, a feature that has advanced the VOR as a major subject of sensorimotor learning studies. This plasticity manifests after unilateral damage to the vestibular periphery and can also be triggered experimentally by placing optics over the eyes that magnify, miniaturize, or even invert the optical image on the retina. Subjects wearing such optics adapt by changing the gain of the VOR to maintain a stable image on the retina and enable coordinated movements under visual control. In a famous series of experiments Ivo Kohler conducted in the 1950s, he found that after wearing inverting prisms for a week he could function well enough to ride a bicycle down a busy street. The flocculus and paraflocculus contain Purkinje cells that transmit inhibitory signals to the vestibular nucleus, and these neurons change their firing properties in a manner that accounts for adaptation of the VOR. Although most people do not elect to wear prism glasses except for experiments, VOR plasticity comes into play during development, when the rapid growth of the head changes the transform between head and eye movements, and also when people are fitted with corrective lenses.

A previously mentioned, integration of signals from the otolith organs and semicircular canals is needed to disambiguate head tilts from purely translational movements. Recordings made in the nodulus and uvula reveal that individual Purkinje cells integrate signals from these two vestibular sources to unambiguously encode either head tilts or translational movements, suggesting that the nodulus and uvula are critical sites of computation for making this distinction. Another major function

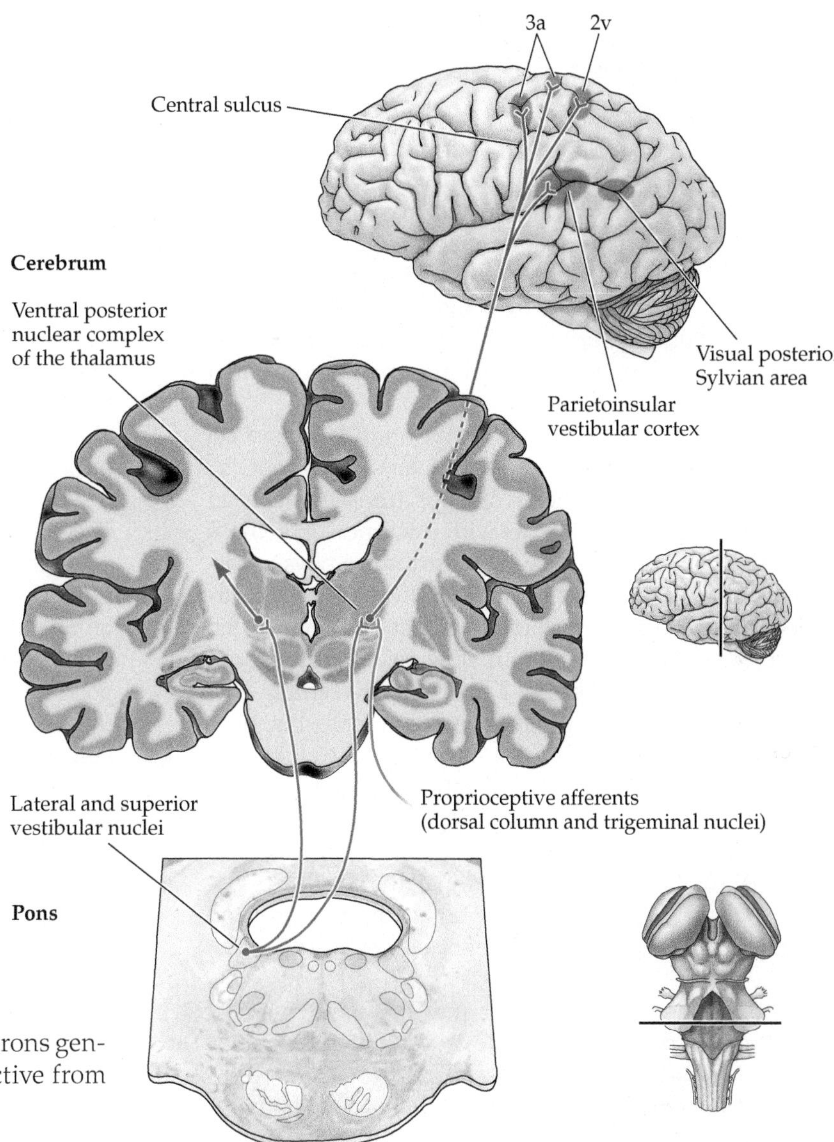

FIGURE 14.12 Thalamocortical pathways carrying vestibular information. Unlike somatosensory, visual, and auditory systems, there is no single, canonical "primary vestibular cortex." Rather, there is a "vestibular cortical system" involving a distributed set of cortical areas in the parietal and posterior insular regions (purple ovals). Each of these areas contains neurons that are modulated by vestibular signals, interconnected across areas, and give rise to subcortical connections with the vestibular nuclear complex of the brainstem. Of particular importance is the parietoinsular vestibular cortex, which integrates multimodal proprioceptive signals and generates a "head-in-space" frame for body orientation and motor control. Other areas contributing to this vestibular cortical system include the ventral premotor cortex and the cingulate motor area (not shown; see Chapter 17).

of vestibular–cerebellar circuitry is helping distinguish vestibular signals that arise from self-generated motion from those that are triggered by external forces. An interesting feature of neurons in the rostral fastigial nucleus is that they do not respond to self-generated head or body movements, despite receiving both vestibular and proprioceptive signals from the head and body. An influential idea to account for this specialized responsiveness is that predictive signals generated in the cerebellum cancel out vestibular and proprioceptive signals in rostral fastigial neurons generated from self-motion, helping distinguish active from passive movements of the head.

Vestibular Pathways to the Thalamus and Cortex

In addition to the several projections of the vestibular nuclei already mentioned, the superior and lateral vestibular nuclei send axons to the ventral posterior nuclear complex of the thalamus. This, in turn, projects to a variety of cortical areas relevant to the perceptions arising from the processing of vestibular information, including Brodmann's area 2v just posterior of the face representation in the somatosensory cortex, two regions in Brodmann's area 3a in the fundus of the central sulcus, and the parietoinsular vestibular cortex (PIVC), which may be especially important to our sense of self-motion and orientation in space (Figure 14.12). Indeed, in the 1950s Wilder Penfield found that electrically stimulating the PIVC could elicit strong vestibular sensations, and more modern imaging studies indicate that this region is activated by vestibular stimulation. Electrophysiological studies of individual neurons in these various cortical areas show that the relevant cells respond to proprioceptive and visual stimuli as well as to vestibular stimuli, reflecting the multisensory nature of central vestibular processing. Many of these cortical neurons are activated by moving visual stimuli as well as by rotation of the body (even with the eyes closed), suggesting that these cortical regions are involved in the perception of body orientation in extrapersonal space. Consistent with this interpretation, patients with lesions of the right parietal cortex, including the PIVC, suffer altered perception of personal and extrapersonal space, as discussed in greater detail in Chapter 26.

Spatial Orientation Perception and Multisensory Integration

Although the vestibular system contributes to many automatic reflexes and most of us remain unaware of its functioning unless it is damaged, it also plays an important role in our perception of spatial orientation and self-motion. Indeed, the vestibular system underlies our ability to detect the direction and magnitude of self-generated motion and distinguish this motion from the movement of objects around us. Working in concert with the visual system, the vestibular system provides critical information for spatial orientation and navigation.

As might be expected based on the stimuli that act on the vestibular end-organs, the vestibular system contributes to our perception of head rotations, translational head movements, and head tilts. As with other percepts, these vestibular percepts do not simply mirror the physical attributes of the associated stimulus. For example, a blindfolded subject riding on a chair rotating at a constant speed will perceive that the rotation is slowing and, after about 30 seconds, that it has stopped entirely. Interestingly, the time course of this perceptual decay is similar to, but more prolonged than, the decrement of the signal transmitted from the semicircular canal to the brain, suggesting that the brain is somehow compensating for the diminishing signal to generate a percept that more closely approximates the actual rotation. In a similar manner, the vestibular system also contributes to our perception of the direction and velocity of translational movements, as well as the direction and magnitude of static tilts.

As mentioned, the vestibular and visual systems normally work in concert to provide an estimate of self-motion. As we walk along a forest trail, for example, translational movements activate the vestibular system and also generate visual flow across the retina. The contributions of visual flow to self-motion can be very compelling, as witnessed by riders waiting for their train to leave the station. If a train on an adjacent track starts to move, the resultant visual flow generates a strong sense of self-motion, a perceptual process known as **vection**. More systematic tests of vestibular–visual interactions using controlled visual stimulation suggest that the normal function of the vestibular system is to suppress the sense of self-generated motion that can arise from visual cues. Consistent with this idea, people with vestibular damage—as well as astronauts working in space, who lack gravitational cues from the otolith organs—report heightened vection in response to visual flow information.

Summary

The vestibular system provides information about the orientation of the head with respect to gravity and head motion. The sensory receptor cells are located in the otolith organs and the semicircular canals of the inner ear. The otolith organs provide information for ocular reflexes and postural adjustments when the head tilts in various directions or undergoes translational movements, and for our perception of these tilts and translations. The semicircular canals, in contrast, provide information about head rotations; these stimuli initiate reflex movements that adjust the eyes, head, and body during motor activities. Among the best studied of these reflexes are eye movements that compensate for head movements, thereby stabilizing the visual scene when the head moves. Information from the vestibular system also plays a central role in our perception of spatial orientation and our ability to navigate through the environment. Vestibular processing is inherently multisensory: Input from all the vestibular organs is integrated with input from the visual and somatosensory systems to provide perceptions of body position and orientation in space.

ADDITIONAL READING

Reviews

Angelaki, D. E. and K. E. Cullen (2008) Vestibular system: The many facets of a multimodal sense. *Annu. Rev. Neurosci.* 31: 125–150.

Benson, A. (1982) The vestibular sensory system. In *The Senses*, H. B. Barlow and J. D. Mollon (eds.). New York: Cambridge University Press, pp. 333–368.

Brandt, T. (1991) Man in motion: Historical and clinical aspects of vestibular function. A review. *Brain* 114: 2159–2174.

Cullen, K. E. (2011) The neural encoding of self-motion. *Curr. Opin. Neurobiol.* 21: 587–595.

Cullen, K. E. (2012) The vestibular system: multimodal integration and encoding of self-motion for motor control. *Trends Neurosci.* 35: 185–196.

Cullen, K. E. and J. X. Brooks (2015) Neural correlates of sensory prediction errors in monkeys: evidence for internal models of voluntary self-motion in the cerebellum. *Cerebellum* 14: 31–34.

Eatock, R. A. and J. E. Songer (2011) Vestibular hair cells and afferents: two channels for head motion signals. *Annu. Rev. Neurosci.* 34: 501–534.

Furman, J. M. and R. W. Baloh (1992) Otolith-ocular testing in human subjects. *Ann. N. Y. Acad. Sci.* 656: 431–451.

Goldberg, J. M. (2000) Afferent diversity and the organization of the central vestibular pathways. *Exp. Brain Res.* 130: 277–297.

Goldberg, J. M. and C. Fernandez (1984) The vestibular system. In *Handbook of Physiology. Section 1: The Nervous System, Volume III: Sensory Processes, Part II*, J. M. Brookhart, V. B. Mountcastle,

I. Darian-Smith and S. R. Geiger (eds.). Bethesda, MD: American Physiological Society, pp. 977–1022.

Green, A. M. and D. E. Angelaki (2010) Multisensory integration: resolving sensory ambiguities to build novel representations. *Curr. Opin. Neurobiol.* 20: 353–360.

Hess, B. J. (2001) Vestibular signals in self-orientation and eye movement control. *News Physiolog. Sci.* 16: 234–238.

Raphan, T. and B. Cohen (2002) The vestibulo-ocular reflex in three dimensions. *Exp. Brain Res.* 145: 1–27.

Important Original Papers

Angelaki, D. E., A. G. Shaikh, A. M. Green and J. D. Dickman (2004) Neurons compute internal models of the physical laws of motion. *Nature* 430: 560–564.

Brooks, J. X., J. Carriot and K. E. Cullen (2015) Learning to expect the unexpected: rapid updating in primate cerebellum during voluntary self-motion. *Nat. Neurosci.* 18: 1310–1317.

Goldberg, J. M. and C. Fernandez (1971) Physiology of peripheral neurons innervating semicircular canals of the squirrel monkey, Parts 1, 2, 3. *J. Neurophysiol.* 34: 635–684.

Goldberg, J. M. and C. Fernandez (1976) Physiology of peripheral neurons innervating otolith organs of the squirrel monkey, Parts 1, 2, 3. *J. Neurophysiol.* 39: 970–1008.

Laurens, J., H. Meng and D. E. Angelaki (2013) Neural representation of orientation relative to gravity in the macaque cerebellum. *Neuron* 80: 1508–1518.

Lindeman, H. H. (1973) Anatomy of the otolith organs. *Adv. Otorhinolaryngol.* 20: 405–433.

Merfeld, D. M. (1995) Modeling the vestibular-ocular reflex of the squirrel monkey during eccentric rotation and roll tilt. *Exp. Brain. Res.* 106: 123–134.

Books

Baloh, R. W. (1998) *Dizziness, Hearing Loss, and Tinnitus.* Philadelphia: F. A. Davis Company.

Baloh, R. W. and V. Honrubia (2001) *Clinical Neurophysiology of the Vestibular System,* 3rd Edition. New York: Oxford University Press.

Go to the NEUROSCIENCE 6e Companion Website at **www.oup.com/uk/Purves6e** for Web Topics, Animations, Flashcards, and more.

CHAPTER

15 The Chemical Senses

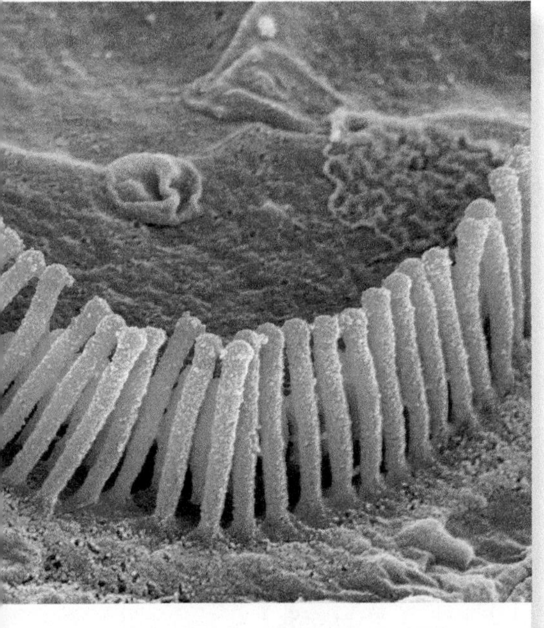

Overview

THREE SENSORY SYSTEMS ARE ASSOCIATED with the nose and mouth: the olfactory (smell), vomeronasal (pheromone sensation), and gustatory (taste) systems. Each detects chemicals in the environment. The olfactory system detects airborne molecules called odorants that provide information about animals and plants, help identify food and noxious substances, as well as signals regarding self versus others. Olfactory information thus influences social interactions, reproduction, defensive responses, and feeding. In most mammals—except, notably, humans—a second chemosensory system, the vomeronasal system, detects airborne odors from predators, prey, and potential mates. The gustatory system detects ingested tastants (primarily water- or fat-soluble molecules) that provide information about the quality, quantity, and safety of ingested food. For smell, vomeronasal chemosensation, and most aspects of taste, the initiation of sensory transduction relies on G-protein-coupled receptors (GPCRs) and second-messenger-mediated signaling. In each system, there are a large number of receptor genes that encode GPCRs with capacity to bind specific odorants or tastants. The stimulus selectivity for olfactory receptor molecules remains mostly unknown; however, specific vomeronasal receptor molecules are selective either for sex-specific odorants in males versus females or for odorants of predators versus prey. Specific dimeric combinations of taste receptor molecules are selective for the five basic classes of taste stimuli. Information from primary sensory receptors in the nose or tongue is relayed to CNS regions that guide a broad range of social and defensive behaviors. Despite this anatomical knowledge, the central representation of conscious olfactory perception remains uncertain. The central representation of vomeronasal information activates hypothalamic and amygdala circuitry to influence sexual, homeostatic, or predator-avoidance behaviors. The central representation of taste is by far the best understood of the chemical senses: Information about each of the five major taste categories is preserved as five representations from the tongue in the insular cortex, where taste is processed. From an evolutionary perspective, the chemical senses—particularly olfaction and vomeronasal sensation—are deemed to be the "oldest" or "most primitive," yet they are in many ways the least understood of the sensory modalities.

Organization of the Olfactory System

The olfactory system—the most thoroughly studied component of the chemosensory triad—processes information about the identity, concentration, and quality of a wide range of airborne, volatile chemical stimuli called **odorants**. Odorants interact with olfactory receptor neurons found in an epithelial sheet, the **olfactory**

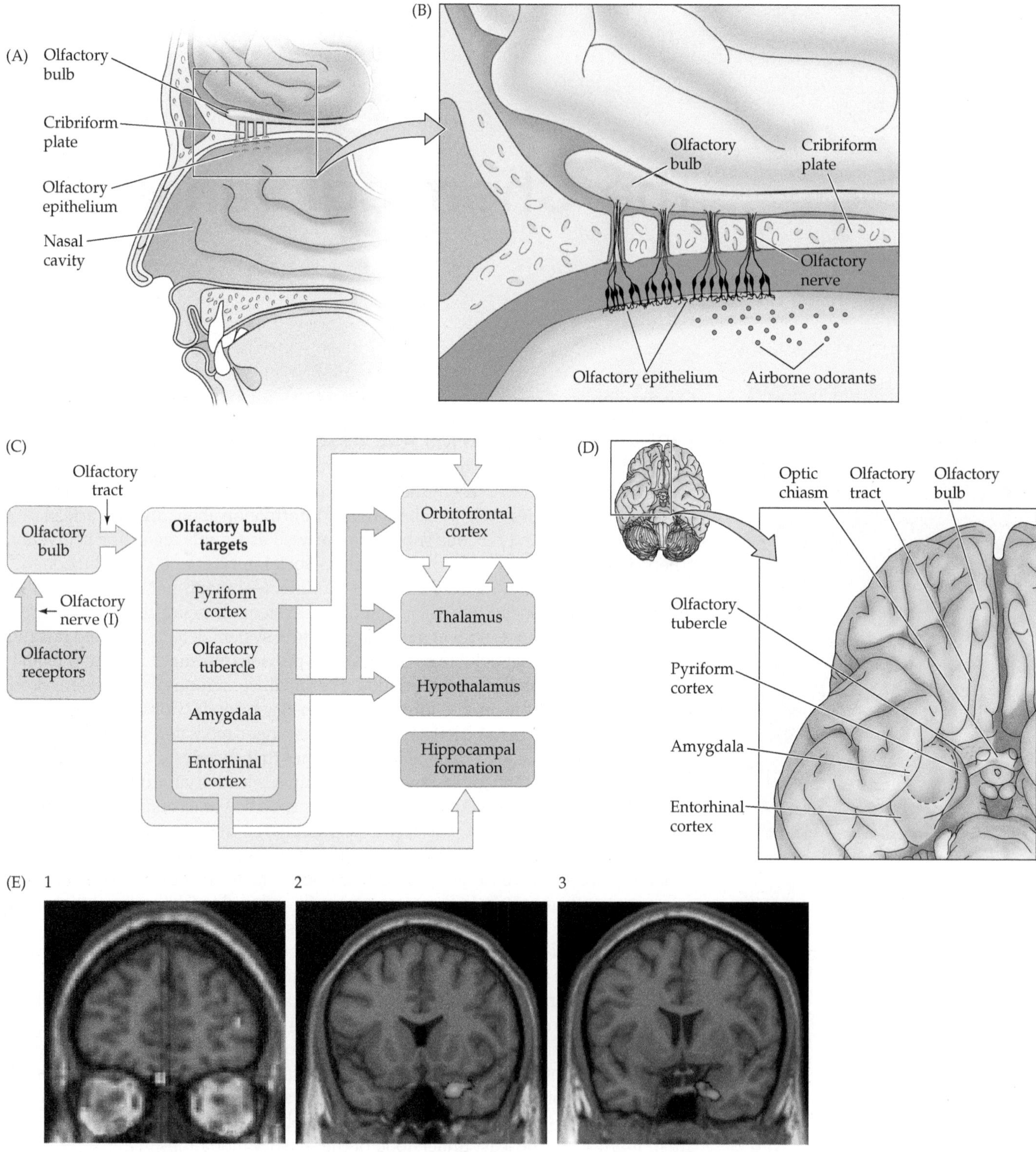

FIGURE 15.1 Organization of the human olfactory system. (A) Peripheral and central components of the primary olfactory pathway. (B) Enlargement of region boxed in (A), showing the relationship between the olfactory epithelium (which contains the ORNs) and the olfactory bulb (the central target of ORNs). (C) The basic pathways for processing olfactory information. (D) Central components and basic connections of the olfactory system. (E) Functional MRI images of coronal sections through the human brain at the level of (1) the orbitofrontal cortex, (2) the pyriform cortex and olfactory bulbs, and (3) the amygdala. Maximum focal activation in response to odor presentation (in this case correlated either with pleasantness [1] or intensity [2,3]), is seen in the orbitofrontal and pyriform cortices as well as in the amygdala. (E from Rolls et al., 2003.)

epithelium, that lines the interior of the nose (Figure 15.1A,B). Axons arising from receptor cells project through the **cribiform plate** (a thin perforated region of the skull that separates the olfactory epithelium from the brain) directly to neurons in the **olfactory bulb**, which in turn sends projections to the pyriform cortex in the temporal lobe, as well as to other structures in the forebrain, via an axon pathway known as the **olfactory tract** (Figure 15.1C,D).

The olfactory system is unique among the sensory systems in that it does not include a thalamic relay from primary receptors en route to a cortical region that processes the sensory information. Instead, after synapses are made by axon terminals of olfactory receptor neurons onto projection neurons in the olfactory bulb, olfactory sensory information is relayed to and then processed in the **pyriform cortex**, a three-layered archicortex dedicated to olfaction and considered to be phylogenetically older than the six-layered neocortex. Although this initial pathway bypasses the thalamus, the thalamus does play an important role in subsequent stages of olfaction. Olfactory information from pyriform cortex is relayed to the thalamus en route to association areas in the neocortex, where further processing occurs (see Figure 15.1C). Together, the pyriform cortex and the multi-modal sensory association areas of the neocortex are thought to be essential for the conscious appreciation of odorants as well as the association of odors with other sensory characteristics of environmental stimuli. The olfactory bulb relays information directly to several other targets in addition to the pyriform cortex, including the hypothalamus and amygdala (see Figure 15.1C,D). The neural computations that occur in these regions influence motor, visceral, and emotional reactions to chemosensory stimuli, particularly those relevant to feeding, reproduction, and aggression (Figure 15.1E).

Despite its phylogenetic age and the unusual trajectory of olfactory information to the neocortex, the olfactory system abides by the same principle that governs other sensory modalities: Sensory stimuli—in this case, airborne chemicals—interact with receptors at the periphery and are transduced and encoded into electrical signals, which are relayed via synaptic transmission to higher-order centers. Unlike other sensory systems, very little is known about the neural representation of olfactory information in the central nervous system. For example, the somatosensory and visual cortices described in the preceding chapters feature topographic maps of the relevant receptor surface, and the auditory cortex features a computational map of frequencies. Whether analogous maps exist in the pyriform cortex (or the olfactory bulb) is not yet known. Indeed, until recently it was difficult to imagine how sensory qualities might be represented in an orderly olfactory map (e.g., odor identity, intensity, or behavioral significance), or what features of chemosensory stimuli might be processed in parallel (as occurs in other sensory systems).

Olfactory Perception in Humans

In humans, olfaction is often considered the least acute of the senses, and many animals obviously possess far superior olfactory abilities. The greater chemosensory sophistication of such animals may be explained by increased numbers of olfactory receptor neurons and odorant receptor proteins in an expanded olfactory epithelium, as well as by a relatively larger portion of the forebrain devoted to olfaction (Figure 15.2). In a 70-kg human, the surface area of the olfactory epithelium is approximately 10 cm^2; in contrast, in a rat it is 15 cm^2, in a 3-kg cat it is about 20 cm^2, in dogs it is 150 to 170 cm^2, and in bloodhounds bred for their increased olfactory sensitivity it increases to 380 cm^2.

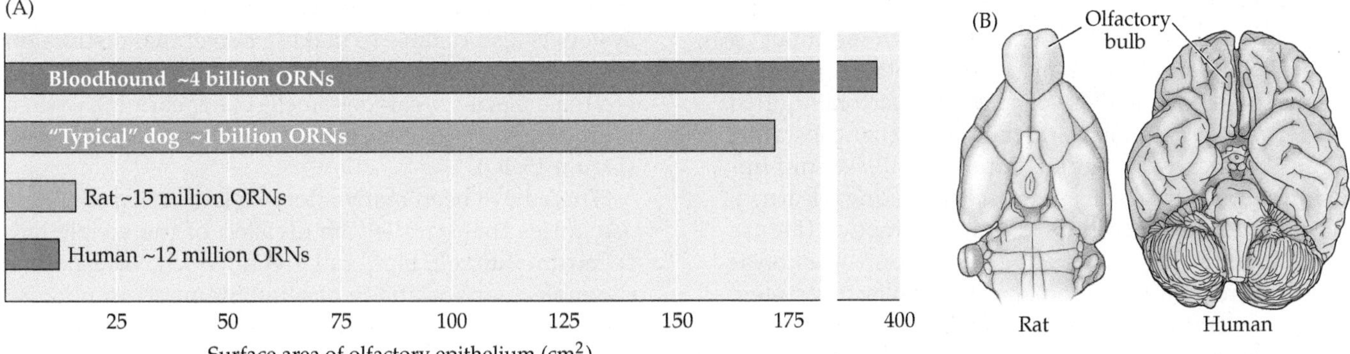

FIGURE 15.2 Odorant perception in mammals. (A) Comparison of the surface area (bars) of the olfactory epithelium and the number of ORNs in a human, a rat, a "typical" dog, and a bloodhound (bred for maximum olfactory discrimination). (B) Proportional sizes of the olfactory bulb in rat and human brains; the bulbs comprise relatively more of the forebrain in rats than they do in humans. (A, data from Shier et al., 2004.)

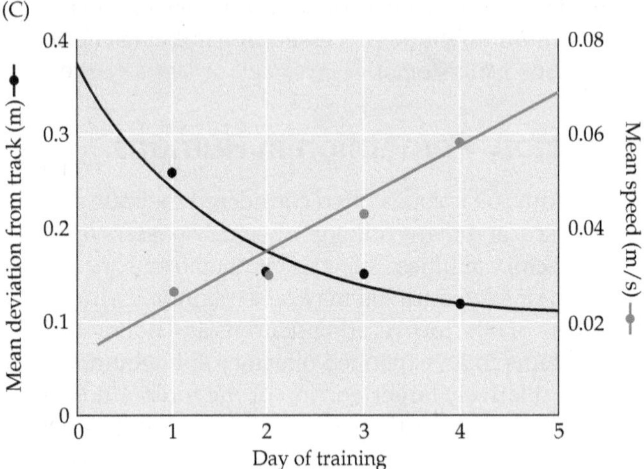

FIGURE 15.3 Humans can track scents at low concentrations over long distances. (A) The yellow line indicates the scent trail established by dragging a pheasant through a field (the pheasant, immobilized, is seen at the bottom of the picture). The red line indicates the path of a pointer tracking the scent. The dog's tracking includes several orthogonal digressions, which are common when a new scent is present at relatively low concentrations in a complex odor environment. (B) The yellow line shows a scent trail established with chocolate essential oil; the red line indicates the trail of a human tracking the scent. Like the dog, the human makes orthogonal digressions. (C) Learning curves for human scent tracking. Over a short training period, humans can acquire greater skill and accuracy tracking a scent at low concentrations (black trace). This improvement indicates that the olfactory system has the capacity for enhanced performance, perhaps by increased sensitivity to a learned signal over background "odor noise." Humans also acquire greater speed in tracking scents at low concentration over a small number of trials (red trace). Apparently, olfactory sensation, like other sensory modalities, can be used in complex tasks in which performance speed as well as accuracy can be enhanced by repetition. (After Porter et al., 2007.)

With such a quantitative disadvantage, the human nose seems ill-suited for certain tasks, such as following a scent to a specific target, that are second nature to a cat or dog.

Nevertheless, humans can, when challenged, use their somewhat modest olfactory endowment to "sniff out" a scent trail. Moreover, we seem to use scent-tracking strategies that are similar to those of our more olfactorally gifted counterparts: We pursue a tracking path that constantly bisects the linear scent trail (Figure 15.3A,B); we sniff frequently; our sniffing increases as scent tracking is learned; and our performance improves with practice (Figure 15.3C). Thus, although humans do not rely on olfaction as a major source of information, the human olfactory system has the capacity to use chemosensory information to track targets and locate items of interest in space.

Humans are also quite good at detecting and identifying individual airborne odorants with a wide range of aesthetic (unpleasant/pleasant) and behavioral (irritant/attractant) significance. Ozone (the smell that accompanies lightning and electrical arcing) becomes an irritant above relatively low concentrations. The human olfactory system can detect ozone reliably at approximately 10 molecules per *billion* in room air. Similarly, humans can identify D-limonene, the major element of citrus smells, fairly reliably at 15 molecules per billion in room air (Figure 15.4A). Other molecules are detected only at much higher concentrations. For example, some estimates place human sensitivity to the odor of ethanol at 2000 molecules per billion.

A further complication in rationalizing the perception of odors is that their quality may change with odorant concentration. For example, at low concentrations the molecule indole has a "floral" odor, whereas at higher concentrations it smells "putrid" (Figure 15.4B). The human olfactory system is also capable of making perceptual distinctions based on small changes in molecular structure; for example, the molecule D-carvone smells like spearmint, whereas L-carvone smells like the caraway seeds found in rye bread (Figure 15.4C).

There have been many attempts to classify odors into categories that parallel the division of the visible light spectrum into red, blue, and green (which correspond to the molecular specificity of photopigments in photoreceptors). Despite these efforts, there is no indication that any currently available arbitrary scheme reflects biologically significant categories of odorants. However, one of the most consistent aspects of olfactory perception is the classification of odors as either pleasant and attractive or

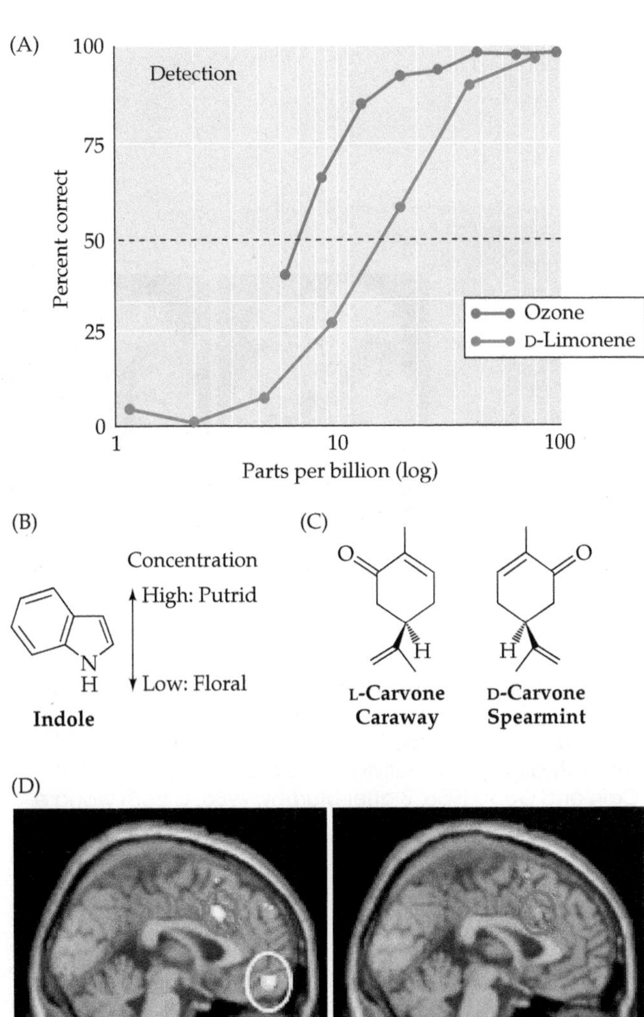

FIGURE 15.4 Human sensitivity to odors. (A) In a controlled setting where room air is presented in precise mixtures with single odors, the threshold for detection reflects the concentration at which a human correctly identifies the presence of the odor above chance (50%). Humans can detect ozone, a somewhat unpleasant odor, at approximately 10 parts per billion. The pleasant and nutritionally significant odor of D-limonene can be identified at approximately 15 parts per billion. (B) Perception of the molecule indole is concentration-dependent. At low concentrations, it is perceived as a pleasant floral smell. At high concentrations, the same molecule (which is produced by bacteria in decomposing organic material) is experienced as putrid. (C) The D and L enantiomers of carvone produce very different olfactory perceptions (spearmint versus caraway) when present at similar concentrations. (D) Functional MRI analysis in typical humans indicates that odorants perceived as "pleasant" versus "unpleasant" elicit maximum activity in distinct regions of the orbitofrontal (white oval) and cingulate (red ovals) cortex. (A, data from Cain et al., 2007; D from Rolls et al., 2003.)

unpleasant and repulsive. These basic properties of olfactory stimuli—their "aesthetic" qualities (or lack thereof)—are apparently represented in distinct cortical regions that mediate olfactory perception (Figure 15.4D). This suggests that perceived aesthetic properties of odorants have distinct representations in the forebrain, including in distinct regions of the cerebral cortex. Most naturally occurring odors are blends of several odorant molecules, even though they are typically perceived as a single smell (such as the scent of a particular perfume or the bouquet of a wine). Thus, it remains to be determined whether animals "map" odors or their attractive or repellent qualities based on single perceptual attributes.

Assessing Olfactory Function in the Laboratory or Clinic

Most people are able to consistently identify a broad range of odorants, and they can distinguish distinct odors from one another. Indeed, many clinicians use uniquely scented "probes"—such as coffee grounds or soap—to test the function of the **olfactory nerve** (cranial nerve I) as part of the standard cranial nerve examination. But some individuals consistently fail to identify one or more common odors (Figure 15.5A). Such chemosensory deficits, known as **anosmias**, are often restricted to a single odorant, suggesting that a specific element in the olfactory system—either an olfactory receptor gene (see below) or genes that control expression or function of specific odorant receptor genes—is inactivated. Genetic analysis of anosmic individuals has yet to confirm this possibility. Thus, unlike blindness and deafness, olfactory loss is difficult to classify as either peripheral or central in its origins.

Anosmias can be congenital, or they may be acquired following chronic sinus infection or inflammation, traumatic head injury, or exposure to toxins. Olfactory loss is also a common consequence of aging (see below). In some cases, such disruption is not a source of great concern (e.g., the transient anosmia that occurs with a severe cold). Nevertheless, it can diminish the enjoyment of food and, if sustained, can lead to decreased appetite, weight loss, and eventual malnutrition (especially in aged individuals). If an anosmia is particularly specific and severe, it can affect a person's ability to identify and respond appropriately to potentially dangerous odors such as spoiled food, toxic chemicals, or smoke. Anosmias often target perception of distinct noxious odorants. For example, approximately 1 person in 1000 is insensitive to butyl mercaptan, the foul-smelling odorant released by skunks. More serious is the inability to detect hydrogen cyanide (1 person in 10), which can be lethal, or ethyl mercaptan, the chemical added to natural gas to enable people to detect gas leaks.

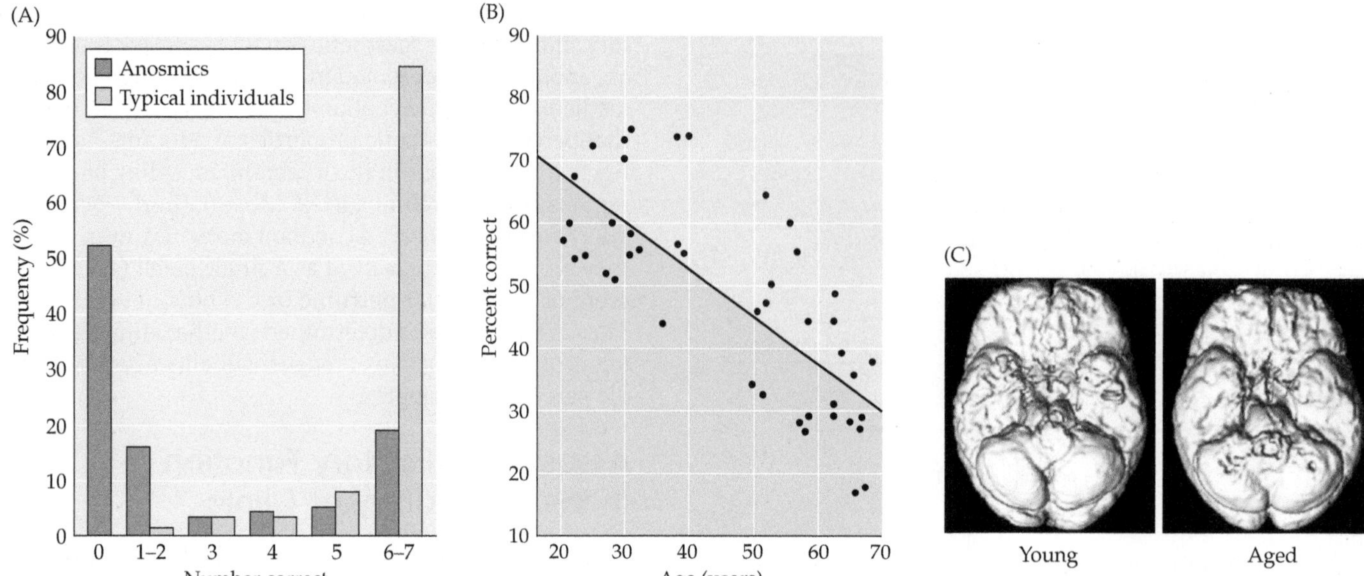

FIGURE 15.5 Loss of olfactory sensitivity. (A) Anosmia is the inability to identify common odors. The majority of typical individuals presented with seven common odors (a test frequently used by neurologists) can identify all seven correctly (in this case, baby powder, chocolate, cinnamon, coffee, mothballs, peanut butter, and soap). Persons who are anosmic have difficulty identifying even these common scents. (B) The ability to identify 80 common odorants declines markedly between ages 20 and 70. Such loss of sensory acuity is normal. (C) Maximum activation (red) of orbitofrontal and medial (pyriform cortex/amygdala) cerebral cortex by familiar odors in young and typical (i.e., without dementia) aged individuals. Areas of focal activation remain similar, but there is clearly diminished activity in the older individuals. (A after Cain and Gent, 1986; B after Murphy, 1986; C from Wang et al., 2005.)

Like other sensory modalities, human olfactory capacity typically decreases with age. If otherwise healthy individuals are challenged to identify a large battery of common odorants, people 20 to 40 years of age can ordinarily identify 50% to 75% of the odors, whereas those between ages 50 and 70 correctly identify only 30% to 45% (Figure 15.5B,C). These changes may reflect either diminished peripheral sensitivity or altered activity of central olfactory structures in otherwise typical aging individuals. A more radically diminished or distorted sense of smell often accompanies neurodegenerative conditions associated with aging, especially Alzheimer's disease. In fact, odor discrimination (the ability to tell two odors apart, usually measured by a standardized "scratch and sniff" test known as the University of Pennsylvania Smell Identification Test) is often part of a battery of diagnostic tests administered at the early stages of age-related dementia and other neurodegenerative diseases.

In addition to normal and pathological age-related changes in olfaction, olfactory sensation and perception can be disrupted by chemotherapy, eating disorders, diabetes, neurological disorders (olfaction is often compromised early in the course of Parkinson's disease), and psychotic disorders (especially schizophrenia). Thus, in people with schizophrenia, olfactory hallucinations (i.e., perception of a stimulus that is not actually present in the environment) are among the earliest symptoms of psychosis. One of the earliest signs of dysfunction in individuals ultimately diagnosed with Parkinson's disease is diminished olfactory function. In individuals on the autistic spectrum, odorant detection thresholds can be lowered, and the experience of neutral or even attractive odors can be reported as unpleasant. The causes of olfactory deficits in this broad range of disorders are not known. Some dysfunction may reflect lost capacity of the olfactory epithelium to maintain neural stem cells or the newly generated neurons that normally replace damaged olfactory receptor neurons over the course of a lifetime (see Figure 15.6)—perhaps an early sign of more general pathogenic deficits in maintenance of optimally fuctioning neurons and circuits.

Physiological and Behavioral Responses to Olfactory Stimuli

In addition to conscious olfactory perceptions, odorants can elicit a variety of physiological responses. Examples are the visceral motor responses to the aroma of appetizing food (salivation and increased gastric motility) or to a noxious smell (gagging and, in extreme cases, vomiting). Olfaction can also influence reproductive and endocrine functions. For instance, there is some evidence that women living in single-sex dormitories tend to have synchronized menstrual cycles, apparently mediated by olfaction. This evidence was reinforced by studies with volunteer women exposed to gauze pads from the underarms of other women at different stages of their menstrual cycles also experience synchronized menses. In addition this synchronization can be disrupted by exposure to analogous gauze pads from men. These responses are thought to reflect in part detection of gender-specific odorants (see below). In more recent work, these studies have not been fully replicated, and there is some contention of whether this particular phenomenon is relevant to understanding human olfactory function. Surprisingly, studies of parallel responses in female animals including rodents and non-human primates are similarly inconclusive and controversial. Thus, possibility of human sex-specific olfactory influence on reproduction remains uncertain. Olfaction also influences mother–child interactions. Infants recognize their mother within hours after birth by smell, preferentially orienting toward their mother's breasts and showing increased rates of suckling when fed by their mother compared with being fed by other lactating females, or when presented experimentally with their mother's odor versus that of an unrelated female. A mother's recognition ability matches that of her infant, and mothers can reliably discriminate their own infant's odor from that of other infants of similar age. These observations are supported by much more detailed analysis of maternal–offspring bonding and subsequent maternal–pup behavior in rodents and other species.

Olfactory Epithelium and Olfactory Receptor Neurons

The transduction of olfactory information—a series of neural events that ultimately results in the conscious sense of smell—begins in the olfactory epithelium, the sheet of neurons and supporting cells that lines approximately half of the surface of the nasal cavity (see Figure 15.1A). The remaining intranasal surface is lined by respiratory epithelium similar to that in the trachea and lungs. Respiratory epithelium primarily maintains appropriate temperature and moisture for inhaled air (which may be important for the presentation of odorants) and provides an immune barrier that protects the nasal cavity from irritation and infection. A layer of thick mucus lines the nasal cavity and protects the exposed neurons, respiratory epithelial cells, and supporting cells of the olfactory epithelium. The mucus also controls the ionic milieu of the olfactory cilia, the primary site of odorant transduction (see below). Mucus is produced by secretory specializations called **Bowman's glands** that are distributed throughout the olfactory epithelium. When the mucus layer thickens, as during a cold, olfactory acuity decreases significantly. Two other cell classes, **basal cells** and **sustentacular** (supporting) **cells**, are also present in the olfactory epithelium. The mucus secreted by Bowman's glands traps and neutralizes some potentially harmful agents. In both the respiratory and olfactory epithelium (Figure 15.6B), immunoglobulins in mucus provide an initial line of defense against harmful antigens. The sustentacular cells also contain enzymes (cytochrome P-450s and others) that catabolize organic chemicals and other potentially damaging molecules. In addition, macrophages found throughout the nasal mucosa isolate and remove harmful material—as well as the remains of degenerating cells of the olfactory epithelium. This entire apparatus—mucus layer and epithelium with neural and supporting cells—is called the **nasal mucosa**. Given the unusual direct exposure of the olfactory epithelium to the external environment (outdoor or room air), immune protection is especially important.

The neural portion of the olfactory epithelium is primarily defined by the **olfactory receptor neurons** (**ORNs**). These bipolar cells give rise to small-diameter, unmyelinated axons at their basal surface that transmit olfactory information centrally. At the apical surface, an ORN has a single dendritic process that expands into a knoblike protrusion from which several microvilli, called **olfactory cilia**, extend into a thick layer of mucus (Figure 15.6A). The microvilli are not actual "cilia" based upon the distinctive cytoskeletal arrangement that defines a cilium (see Chapter 22). Instead they are actin based cellular protrusions (more like filopodia seen in growth cones; see Chapter 23) that have scaffolding proteins that localize odorant receptors and signal transduction molecules within the ORN apical dendrite. ORNs have direct access to odorant molecules as air is inspired through the nose into the lungs; however, this access exposes these neurons to airborne pollutants, allergens, microorganisms, and other potentially harmful substances, subjecting them to more or less continual damage. The ultimate solution to the vulnerability of ORNs is to maintain a healthy population by a normal cycle of degeneration and regeneration, analogous to that in other exposed epithelia (such as the intestine and lung). This constant process of degeneration and regeneration is found in all vertebrates, including mammals. ORN regeneration relies on maintaining among the basal cells in the mature olfactory epithelium a population of neural stem cells that divide to give rise to new receptor neurons (Figure 15.6C; see also Chapter 26).

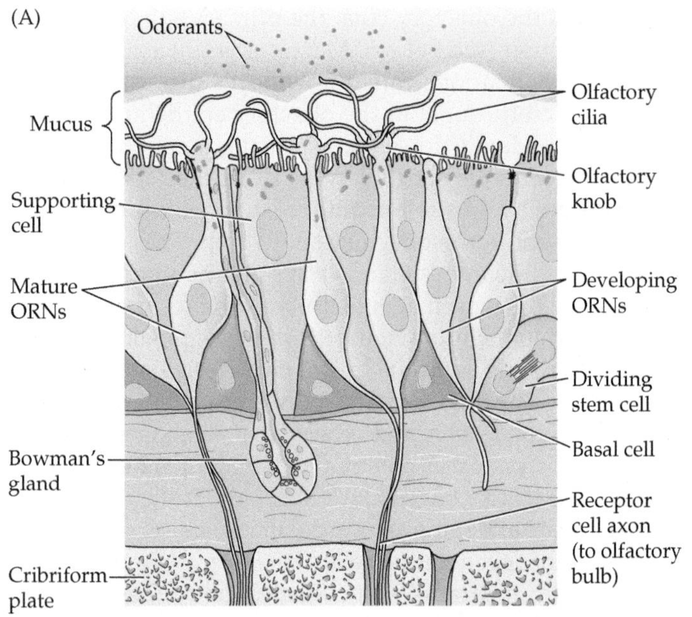

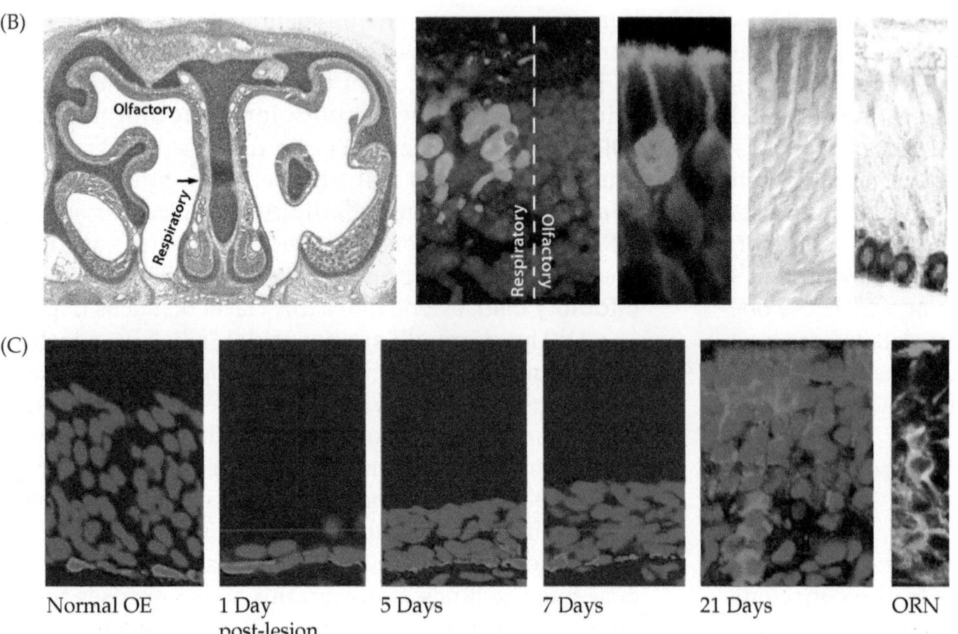

FIGURE 15.6 Structure and function of the olfactory epithelium. (A) Diagram of the olfactory epithelium showing the major cell types: ORNs and their cilia, sustentacular (supporting) cells that detoxify potentially dangerous chemicals, and basal cells. Bowman's glands produce mucus. Bundles of unmyelinated axons and blood vessels run in the basal part of the mucosa (the lamina propria). ORNs are generated continuously from dividing stem cells maintained among the basal cells of the olfactory epithelium. (B) Distinctions between respiratory and olfactory (neural) epithelium in the nasal cavity. At far left, the nasal cavity of a juvenile mouse, composed of a fairly thin respiratory epithelium and much thicker olfactory epithelium. The arrow indicates the approximate location of the image in the next panel, which shows a sharp boundary between respiratory epithelium (labeled green here, based on expression of the transcription factor forkhead1) and olfactory epithelium. The remaining three panels show distinct cell classes in the olfactory epithelium. ORNs are labeled with olfactory marker protein (OMP; green), a molecule expressed uniquely in these neurons. Supporting (sustentacular) cells express another cell-class-specific molecule (light brown). At far right, basal cells, the stem cells of the adult olfactory epithelium, are recognized by their expression of the filament protein cytokeratin 5 (dark brown). (C) Olfactory epithelium (OE) regeneration depends on basal cells. A reporter protein genetically labels these cells and their descendants so that they appear red. When the olfactory epithelium is undisturbed, basal cells are seen in their appropriate position. Immediately after a lesion, basal cells begin to proliferate; their progeny are additional basal cells. Within 21 days, all cells in the regenerated epithelium have arisen from basal cells. Regenerated cells include ORNs, which are double-labeled (green for OMP expression as well as the red indicating the neuron's basal-cell derivation). (A after Anholt, 1987; B adapted from Rawson and LaMantia, 2006; C from Leung et al., 2007.)

In rodents, most if not all olfactory neurons are renewed every 6 to 8 weeks. This extended time period (6 to 8 weeks represents a significant portion of a mouse or rat's typical 1.5- to 2-year life span) suggests that neural regeneration is a gradual process. The period for complete turnover has not been defined in humans. Nevertheless, it is clear that ORNs will be regenerated if large populations of extant neurons, but not the neural stem cells that are maintained throughout life in the OE (see Figure 5.6) are eliminated at one time. This can happen due to environmental exposure, viral or bacterial infection (colds and sinus infections), or traumatic head injuries such as whiplash that occurs in automobile accidents (the axons are sheared by the force of the impact due to differential movement of the neural tissue versus the cribiform plate). Unfortunately, this sort of large-scale regeneration does not fully restore normal function. In such individuals, after a period of complete anosmia, odor discrimination and identification as well as olfactory-guided behavior are often permanently altered.

ORNs are the only neurons with long axons (referred to generally as projection neurons) that project to a distal target and are constantly newly generated from neural stem cells found in the OE in mature individuals. The newly generated ORNs can then grow an axon that establishes new synaptic connections with appropriate targets (see Figure 15.13). In the mature olfactory system, many of the molecules that influence initially neuronal differentiation, axon outgrowth, and synapse formation during development (see Chapters 22 and 23) are apparently retained or reactivated to perform similar functions for regenerating ORNs. Understanding how new ORNs differentiate, extend axons to the brain, and reestablish appropriate functional synaptic connections is obviously relevant to stimulating regeneration of functional connections elsewhere in the brain after injury or disease (see Chapter 26). Indeed, other specialized cell classes in the mature olfactory system are adapted to facilitate constant regeneration. In adults, glial cells called **olfactory ensheathing cells** surround axons in the olfactory nerve and bulb. These glial cells, which are derived initially from the olfactory epithelium, are believed to support the growth of new axons through a mature nervous system. In experimental therapies following damage to other regions of the CNS (e.g., the spinal cord), olfactory glial cells have been used to construct cellular "bridges" across sites of axonal damage to promote regeneration. Thus, the regenerative capacity of ORNs, with the assistance of other cell types in the olfactory epithelium provides a potentially instructive model for understanding how regeneration of neurons or axons can be stimulated throughout the nervous system. This is a fundamental issue since in all mammals, including humans, cellular regeneration and related functional recovery after central nervous system damage does not occur to a useful degree (see Chapter 26).

Odor Transduction and Odorant Receptor Proteins

Odor transduction in the olfactory epithelium begins with odorant binding to specific odorant receptor proteins concentrated on the external surface of olfactory cilia. Prior to the identification of odorant receptor proteins, the compartmental sensitivity of the cilia to odors was demonstrated in physiological experiments (Figure 15.7). Odorants presented to the *cilia* of an isolated ORN elicit a robust electrical response; those presented to the *cell body* do not. Despite their external appearance, olfactory cilia do not have the cytoskeletal features of motile cilia (i.e., the 9+2 arrangement of microtubules). Instead, the actin-rich olfactory cilia more closely resemble microvilli of other epithelia (such as the lung and gut) and thus have a greatly expanded cellular surface to which odorants can bind. Many molecules that are crucial for olfactory transduction are either enriched or found exclusively in the cilia, including the odorant receptor proteins that bind and transduce chemosensory information.

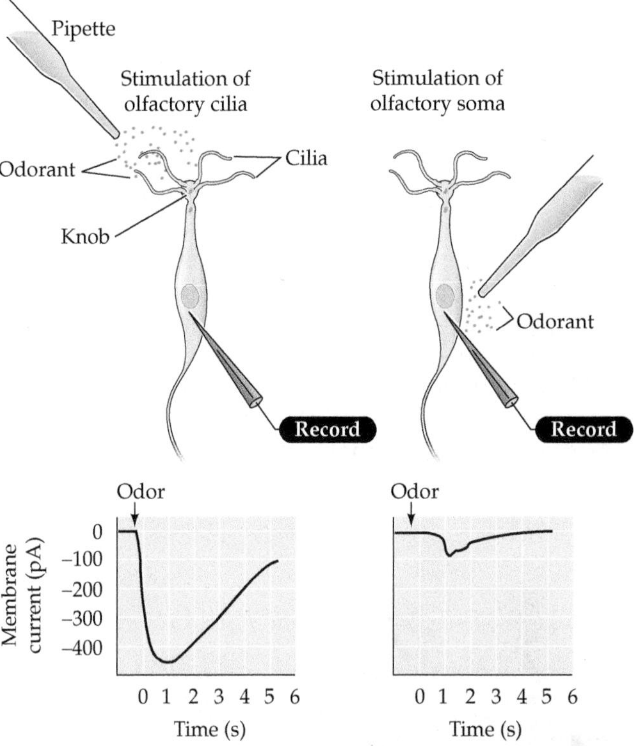

FIGURE 15.7 **Receptor potentials are generated in the cilia of receptor neurons.** Odorants evoke a large inward (depolarizing) current when applied to the cilia (left), but only a small current when applied to the cell body (right). (After Firestein et al., 1991.)

The central role of odorant receptor proteins in the encoding of olfactory information was acknowledged in 2004 when a Nobel Prize was awarded to Richard Axel and Linda Buck for their discovery of the **odorant receptor gene family**. Olfactory receptor molecules are homologous to G-protein-coupled receptors, a category that includes β-adrenergic and muscarinic acetylcholine receptors, the light-responsive rhodopsin, and the cone opsins (see Chapters 6 and 11). In all invertebrates and vertebrates examined thus far, odorant receptor proteins have seven membrane-spanning hydrophobic domains, potential odorant binding sites in the extracellular domain of the protein, and the ability to interact with G-proteins at the carboxyl terminal region of their cytoplasmic domain (Figure 15.8A). The amino acid sequences for these molecules show substantial variability in several of the membrane-spanning regions, as well as in the extracellular and cytoplasmic domains. The specificity of odorant recognition and signal transduction is presumably the result of this molecular variety of odorant receptor proteins in the nasal epithelium; however, the molecular mechanism by which individual receptors bind specific odorants remains poorly understood.

The number of odorant receptor genes, though substantial in all species, varies widely. Nevertheless, in all mammals, odorant receptors are the largest known single gene family, representing 3% to 5% of the genome. Analysis of the human genome has identified approximately 950 odorant receptor genes (Figure 15.8B); the number is similar in other primates, including chimpanzees, which have approximately 1100. Analysis of the mouse genome indicates about 1500 different odorant receptor genes, and in certain dogs, including those noted for their olfactory abilities (Box 15A), the number is around 1200. Additional sequence analysis of apparent mammalian odorant receptor genes, however, suggests that many of these genes—around 60% in humans and chimps versus 15% to 20% in mice and dogs—are not transcribed due to changes that have rendered them pseudogenes.* Thus, the number of functional odorant receptor proteins encoded by stably transcribed and translated genes is estimated to be around 400 in humans and chimps versus about 1200 in mice and 1000 in dogs. In mammals, the number of expressed odorant receptors apparently is correlated with the olfactory capacity of different species. Similar analyses

*A pseudogene is a sequence of DNA that contains a promoter and a transcription initiation site, but because of sequence changes, the DNA either cannot be transcribed into a stable mRNA, or the transcript cannot be translated into a protein.

(A)

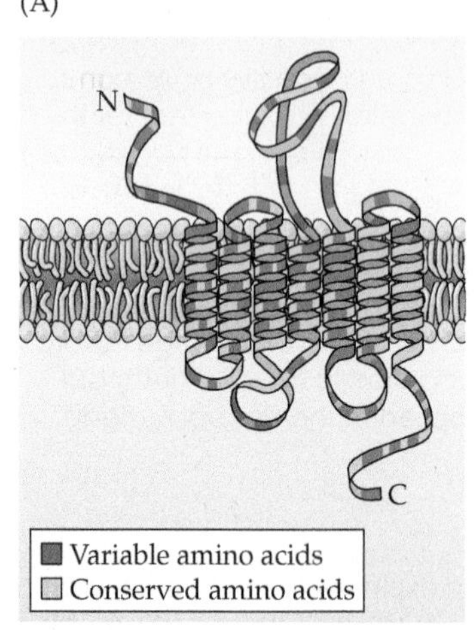

(B)

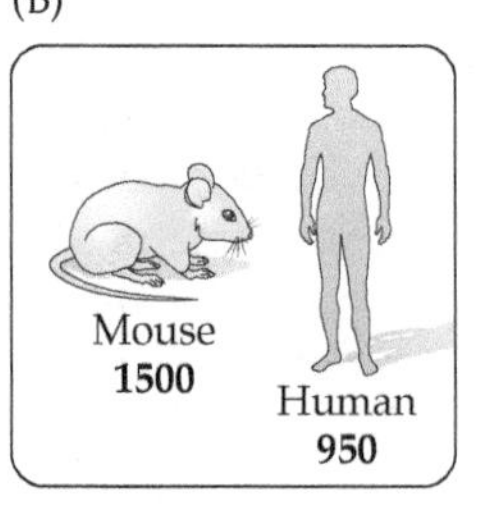

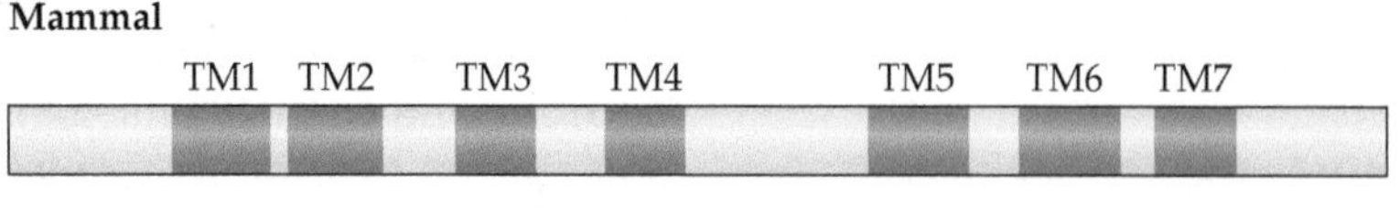

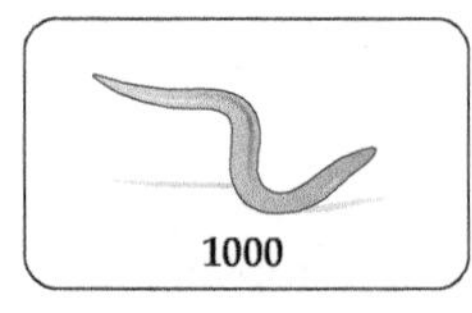

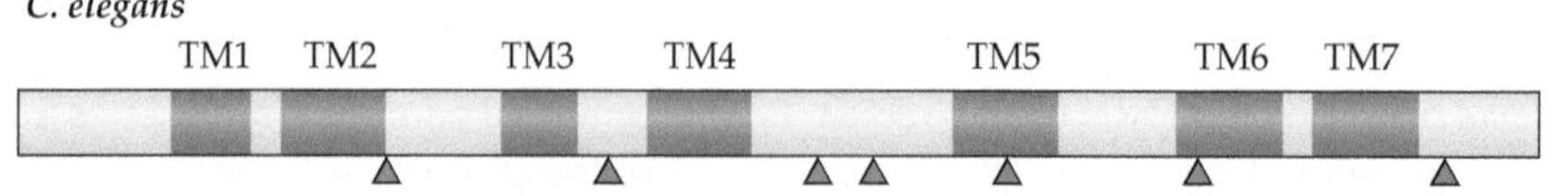

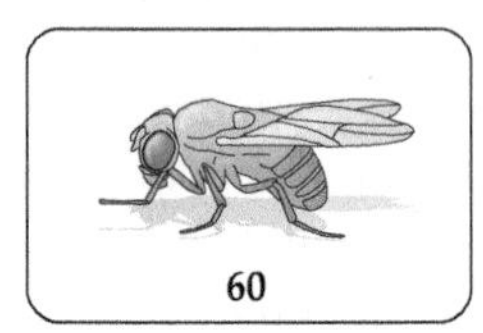

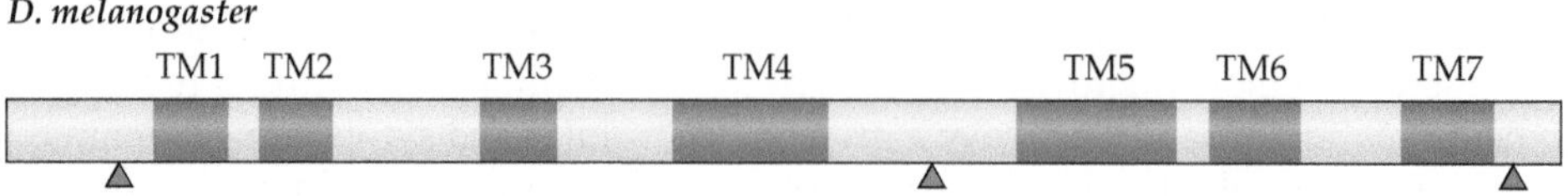

FIGURE 15.8 Odorant receptor proteins. (A) The generic structure of putative olfactory odorant receptors. These proteins have seven transmembrane domains, plus a variable cell surface region and a cytoplasmic tail that interacts with G-proteins. As many as 1000 genes encode proteins of similar inferred structure in several mammalian species, including humans. Each gene presumably encodes a receptor protein that detects a particular set of odorant molecules. (B) Regions encoding the seven transmembrane domains characteristic of G-protein-coupled receptors are shown in green on maps of receptor genes in mammals, the nematode *C. elegans*, and the fruit fly *D. melanogaster*. The comparative size of each domain, as well as the size of the intervening cytoplasmic or cell surface domains (beige), varies from species to species. In addition, splice sites (red arrowheads) reflect introns in the genomic sequences of the two invertebrates; genes for mammalian odorant receptors lack introns. The number of genes that encode odorant receptors in each of the four species is indicated in the corresponding boxes. (A after Menini, 1999; B after Dryer, 2000.)

BOX 15A ■ The "Dogtor" Is In

Conventional wisdom holds that having a pet, particularly a dog, is good for your health. Most of us assume that the primary benefits come from the companionship, as well as the daily exercise, that a dog provides. However, there may be more critical benefits of pet ownership that reflect the remarkable acuity of the canine olfactory system. The family dog may in fact be a reliable source of early diagnosis for several cancers—albeit a diagnostician that likes to chew shoes and has a wet nose.

In the late 1980s, anecdotal reports emerged that claimed family dogs could use smell to identify moles and other skin blemishes on their owners that turned out to be malignant. In recounting this seemingly strange capacity of several dogs, H. Williams, one of the original discoverers, reported "a patient whose dog constantly sniffed at a mole on her leg. On one occasion, the dog even tried to bite the lesion off. ... [The] constant attention [of the dog] prompted her to seek medical advice. The lesion was excised and histology showed the lesion to be a malignant melanoma."

Subsequently, similar diagnoses by individual pets for their owners were reported, including a Labrador retriever that detected a basal cell carcinoma that had developed from an eczema lesion on its master's skin. A slightly less anecdotal study relied on techniques used to train explosive-sniffing dogs for airport security. In this instance, George, a schnauzer, was trained to distinguish malignant melanomas in cell culture from their nonmalignant melanocyte counterparts. George was then introduced to a person who had several moles. One mole caused George to "go crazy"; a biopsy proved that the mole was indeed an early malignant melanoma.

Over the ensuing years, further anecdotal evidence suggested that dogs could recognize lung, breast, and bladder cancer using olfaction. These reports remained isolated anecdotes until 2006, when a truly systematic analysis of this apparent diagnostic capacity was published. In this study, five ordinary adult dogs were trained to distinguish exhaled breath samples from patients with lung or breast cancer versus controls who did not have cancer. The dogs were then tested for their ability to distinguish patients from controls in an entirely novel sample population. In this instance, the specificity and sensitivity of the dogs' ability to detect lung cancer from early to late stages was 99% as accurate as that of conventional biopsy diagnosis. The accuracy of breast cancer detection was slightly lower—approximately 90% that of conventional methods.

A similar study that challenged dogs to discriminate urine from patients with and without bladder cancer had parallel but somewhat less robust results. During the course of this study, however, the dogs consistently identified a presumed "control" sample as that from a patient with cancer. Clinicians were sufficiently alerted to perform further diagnostic tests, and in fact discovered a kidney carcinoma in this individual.

Aside from writing a new chapter in the saga of the salutatory relationship between humans and dogs, these observations have several implications for understanding the mechanisms and biological significance of olfactory acuity and selectivity. First, there is evidence that the concentration of alkanes and other volatile organic compounds is increased in air exhaled from patients with lung cancer. Thus, as indicated by preliminary studies of odorant receptor molecule sensitivity, 7-transmembrane G-protein-coupled odorant receptors may be specialized to detect and discriminate a wide—and biologically significant—spectrum of volatile organic compounds at low concentrations. Second, the discrimination made between patients and controls, either by untrained individual dogs or the trained group of dogs, suggests that subtle distinctions in olfactory perception are clearly represented and can guide behavior. The apparent heightened olfactory ability in dogs may reflect a somewhat larger number of odorant receptors and/or relatively larger olfactory periphery that allow increased sensitivity, or specialized circuitry in the olfactory bulb, pyriform cortex, or other brain regions that assign cognitive significance to distinct olfactory stimuli. Whether this ability has adaptive significance for dogs or is just the ultimate smart pet trick is unclear.

Does this mean the term *pet scan* will soon take on a new meaning in clinical medicine? Clearly, the complexity of making critical diagnoses and the potential lack of reliability of dogs—however well trained—render routine use of diagnostic dogs difficult to imagine. Nevertheless, the remarkable olfactory capacity of these animals provides a starting point for an understanding of the molecular specificity of odorant receptors, as well as processing capacity and representations of olfactory information in the CNS. Such understanding may not only illuminate the functional characteristics of the olfactory system; it may provide a natural guide to specific molecules associated with disease states and the design of better diagnostic tools—or at least diagnoses that don't rely on cold, wet noses.

of complete genome sequences from the worm *C. elegans* and the fruit fly *D. melanogaster* indicate that the worm has approximately 1000 odorant receptor genes, whereas the fruit fly has only about 60. The functional significance of these disparate numbers is not known.

Expression in ORNs has been confirmed for only a limited subset of the huge number of odorant receptor genes. Messenger RNAs for different odorant receptor genes are expressed in subsets of ORNs that occur in bilaterally symmetrical zones of olfactory epithelium. Additional evidence for restricted patterns of odorant receptor gene expression in spatially restricted subsets of ORNs comes from molecular genetic experiments (primarily in mice and fruit flies) in which reporter proteins such as β-galactosidase or green fluorescent protein (GFP) are inserted into odorant receptor gene locus (Figure 15.9; see

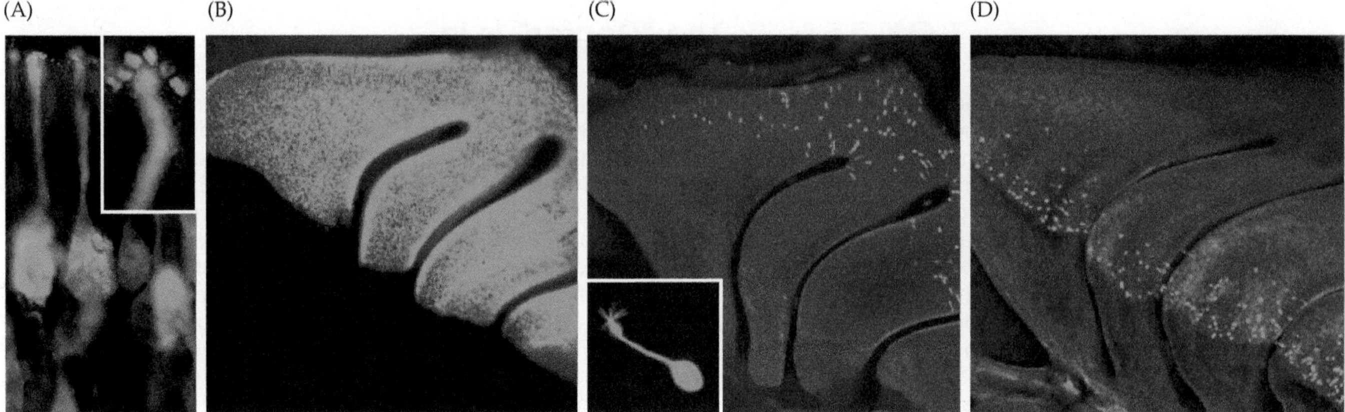

FIGURE 15.9 Odorant receptor gene expression. (A) Individual ORNs labeled immunohistochemically with olfactory marker protein (OMP, green; OMP is selective for *all* ORNs) and the ORN-specific adenylyl cyclase III (red) that is limited to olfactory cilia (inset). The labels are in register with the segregation of signal transduction components to this domain. (B) The distribution of OMP-expressing ORNs throughout the entire nasal epithelium of an adult mouse, demonstrated with an OMP-GFP reporter transgene. The protuberances oriented diagonally from left to right represent individual turbinates in the olfactory epithelium. The remaining bony and soft-tissue structures of the nose have been dissected away. (C) The distribution of ORNs expressing the I7 odorant receptor. These cells are restricted to a distinct domain or zone in the epithelium. The inset shows that odorant receptor-expressing cells are indeed cilia-bearing ORNs. (D) ORNs expressing the M71 odorant receptor are limited to a zone that is completely distinct from that of the I7 receptor. (A courtesy of A.-S. LaMantia; B–D from Bozza et al., 2002.)

also Chapter 1 for a general summary of this approach). Genetic as well as cell biological analyses show that most mammalian ORNs express only one odorant receptor gene. Moreover, the mRNA that is translated to generate the single odorant receptor protein expressed in each ORN is apparently transcribed from only one of the two allelic copies of each odorant receptor gene in the genome (one allele from maternal chromosomes, one allele from paternal chromosomes). The mechanism of this **allelic silencing** is thought to reflect local chromatin confirmation changes in regions of the genome where odorant receptor genes are clustered. Local transcriptional feedback within each ORN reinforces the selection of one allele and the exclusion of the other so that this allelic choice is maintained. Remarkably, this allelic silencing must be maintained in all of the ORNs that are generated throughout the lifetime of all vertebrates. How the mechanism is established and maintained in the ORN progeny of stem cells resident in the OE remains unknown.

Different odors activate molecularly and spatially distinct subsets of ORNs. Furthermore, because only one of the two copies of each odorant receptor gene is expressed in any particular receptor neuron, one of the two alleles for each gene must be silenced in each ORN. Thus, molecular diversity, along with the complex genomic regulation of odorant receptors and the resulting cellular diversity for ORNs, certainly mediates the capacity of olfactory systems to detect and encode a wide range of complex and novel odors in the environment.

Molecular and Physiological Mechanisms of Olfactory Odor Transduction

Once an odorant is bound to an odor receptor protein, several additional steps are required to generate a receptor potential that converts chemical information into electrical signals that can be interpreted by the brain. In mammals, the principal pathway for generating electrical activity in olfactory receptors involves cyclic nucleotide-gated ion channels similar to those found in rod photoreceptors (see Chapter 11). The ORNs express an olfactory-specific heterotrimeric G-protein, $\mathbf{G_{olf}}$, whose α subunit dissociates upon odorant binding to receptor proteins and then activates **adenylyl cyclase III** (**ACIII**), an olfactory-specific adenylate cyclase (Figure 15.10A). Both of these proteins are restricted to the olfactory knob and cilia, consistent with the idea that odor transduction occurs in these domains of the ORN (see Figure 15.6A). Stimulation of odorant receptor molecules leads to an increase in cyclic AMP (cAMP), which opens **cyclic nucleotide-gated channels** that permit the entry of Na^+ and Ca^{2+} (mostly Ca^{2+}), thus depolarizing the neuron. This depolarization, amplified by a Ca^{2+}-activated Cl^- current, is conducted passively from the cilia to the axon hillock region of the ORN, where action potentials are generated via voltage-regulated Na^+ channels and transmitted to the olfactory bulb.

There are also distinct signaling mechanisms for repolarization, recovery, and adaptation in response to odorants.

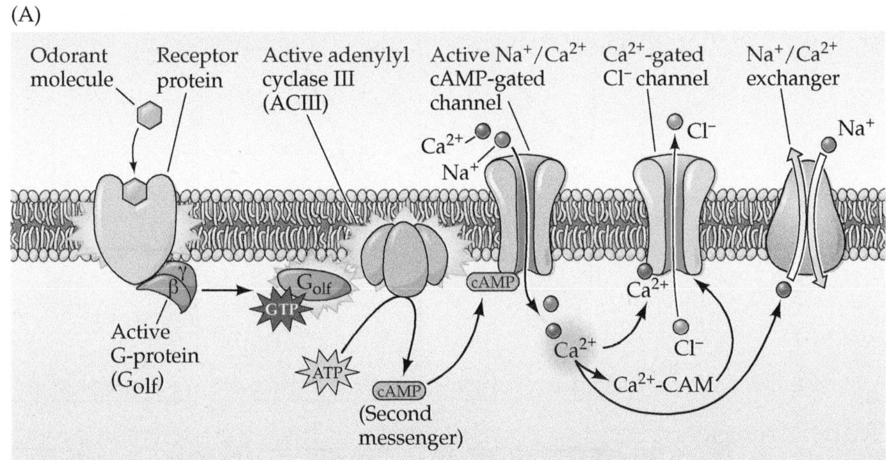

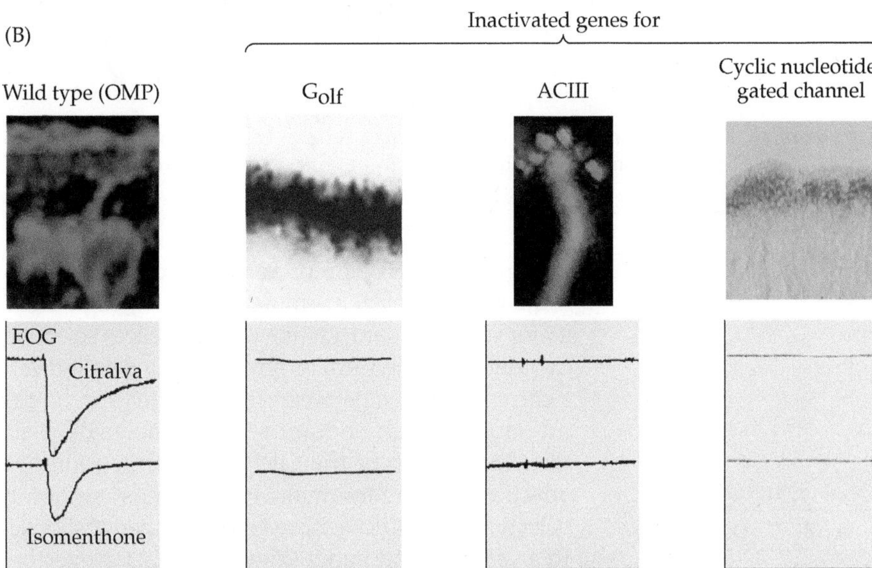

FIGURE 15.10 Molecular mechanisms of odorant transduction. (A) Odorants in the mucus bind directly (or are shuttled via odorant binding proteins) to one of many receptor molecules located in the membranes of the cilia. This association activates an odorant-specific G-protein (G_{olf}) that, in turn, activates an adenylate cyclase (ACIII), resulting in the generation of cyclic AMP (cAMP). One target of cAMP is a cation-selective channel that, when open, permits the influx of Na^+ and Ca^{2+} into the cilia, resulting in depolarization. The ensuing increase in intracellular Ca^{2+} opens Ca^{2+}-gated Cl^- channels that provide most of the depolarization of the olfactory receptor potential. The receptor potential is reduced in magnitude when cAMP is broken down by specific phosphodiesterases to reduce its concentration. At the same time, Ca^{2+} complexes with calmodulin (Ca^{2+}-CAM) and binds to the channel, reducing its affinity for cAMP. Finally, Ca^{2+} is extruded through the Ca^{2+}/Na^+ exchange pathway. (B) Consequences of inactivation of critical molecules in the odorant signal transduction cascade. The images of ORNs show expression of G_{olf}, ACIII, and the cyclic nucleotide-gated channel. The traces below show odorant-elicited electrical activity in the olfactory epithelium, measured extracellularly using the electro-olfactogram (EOG). In the wild type, a robust response results when either pleasant (citralva) or pungent (isomenthone) odors are presented. Inactivating any of the major signal transduction molecules linked to the 7-transmembrane odorant receptors abolishes these responses.(A after Menini, 1999; B from Wong et al., 2000 [wild type OMP]; Belluscio et al., 1998 [G_{olf}]; courtesy of A.-S. LaMantia [ASCII]; Brunet et al., 1996 [cyclic nucleotide-gated channel].)

Most of these mechanisms reflect concurrent increases in Ca^{2+} and activation of calcium/calmodulin-dependent kinase II. In response to elevated Ca^{2+}, an Na^+/Ca^{2+} exchanger extrudes Ca^{2+} and transports Na^+ to repolarize the membrane. Recovery relies on calcium/calmodulin kinase II-mediated mechanisms that restore the heterotrimeric G_{olf} and diminish cAMP levels via activation of phosphodiesterases. Finally, adaptation relies on cAMP-regulated phosphorylation of intracellular domains of the odorant receptor proteins, as well as engagement of β-arrestin (which serves a similar role in photoreceptor adaptation) to modify receptor sensitivity. These mechanisms for adaptation likely play a role in perceived changes in sensitivity to smells, such as initially noticing, but later not sensing, the smell of cigarette smoke in a "smoking" hotel room.

In genetically engineered mice, inactivation of any one of the major signal transduction elements associated with G-protein-coupled oderant receptors (G_{olf}, ACIII, or the cyclic nucleotide-gated channel) results in a loss of receptor response to odorants in ORNs—neurons that otherwise appear normal in these animals (Figure 15.10B). There is also complete loss of behavioral response to most odorants; in other words, the mice are anosmic. This common end point following loss of function of each molecule demonstrates that each signaling step—receptor-mediated G-protein activation, adenylyl cyclase-mediated elevation of cAMP levels, and Ca^{2+}-mediated activation of the cyclic nucleotide-gated channel—contributes to the transduction of odorants. Nevertheless, in mice a few ORNs may use a different G-protein-mediated transduction pathway, perhaps explaining some residual chemosensory function following inactivation of G_{olf}, ACIII, or the cyclic nucleotide-gated channel. These ORNs, however, represent a small portion of the full complement of chemosensory

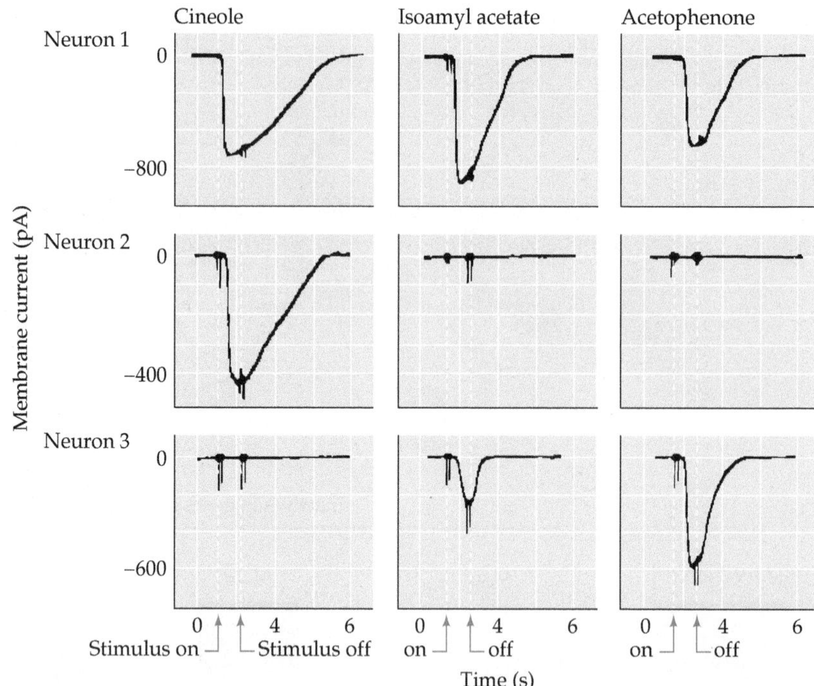

FIGURE 15.11 Responses of receptor neurons to selected odorants. Neuron 1 responds similarly to three different odorants. In contrast, neuron 2 responds to only one of these odorants. Neuron 3 responds to two of the three odorants. The responses of these receptor neurons were recorded by whole-cell patch clamp recording (see Box 4A); downward deflections represent inward currents measured at a holding potential of –55 mV. (From Firestein et al., 1992.)

neurons. Moreover, in invertebrates (including *Drosophila*), some odorant receptors may act as ion channels, directly influencing depolarization (and thus odor transduction) without activating G-proteins. Thus, although the overall molecular structure of odorant receptor proteins has been conserved, these proteins function in diverse ways in different species.

Like other sensory receptor cells, individual ORNs are sensitive to subsets of stimuli—that is, there is receptor specificity. Some ORNs exhibit marked selectivity to a single chemically defined odorant. Several different odorant molecules, however, may activate other ORNs (Figure 15.11). Presumably, these differences in odorant sensitivity parallel the expression of a single odorant receptor gene in each ORN. Thus, some receptor proteins must have fairly high affinity for some odorants, while others are less selective. Currently, no chemical or physiological data indicate a correspondence between high-affinity binding of an odorant to an odorant receptor molecule, electrical activation of the ORN, and perception of a specific odor in mammals. Indeed, most randomly selected single ORNs have broadly tuned responses to a variety of odorants (Figure 15.12A). Nevertheless, genetic labeling studies have demonstrated that consistent relationships exist between classes of odorants and the responses they elicit in individual cells isolated from olfactory epithelium. Response specificity reflects chemical differences in subsets of

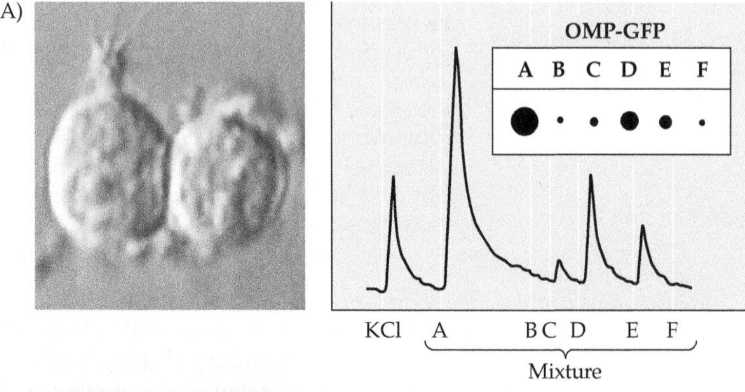

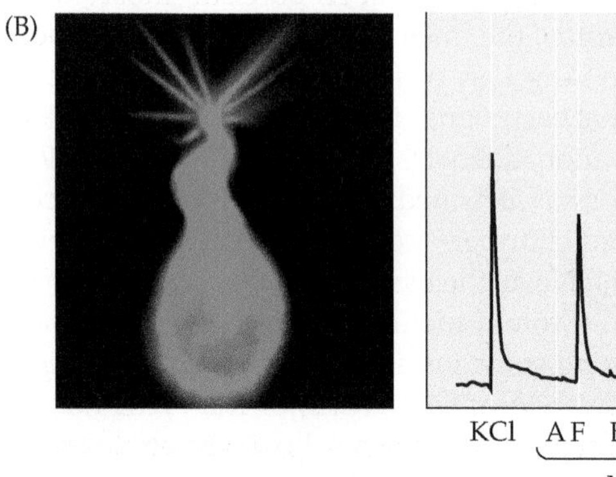

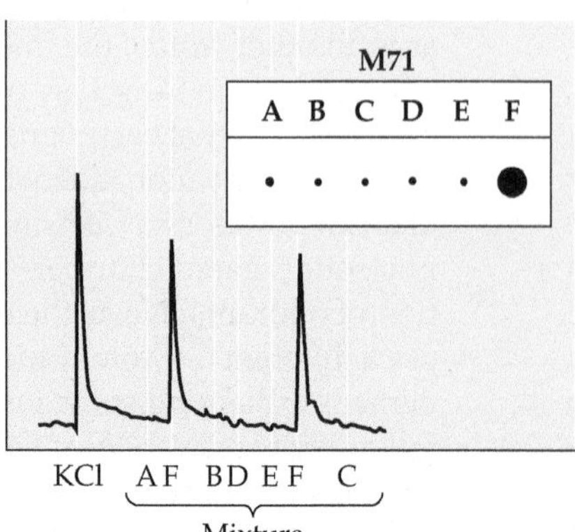

FIGURE 15.12 Odorant receptor neuron selectivity. Odorant receptor neurons (ORNs) were isolated and tested for their responses to six different odorant molecule mixtures (indicated here as A–F). KCl was used as a control to demonstrate the capacity of the cell to generate an action potential. The size of the dots and magnitude of the spikes in the graphs indicate the strength of the electrical response to each odorant mixture. (A) Randomly chosen ORNs responded to several of the six mixtures. (B) The M71R-expressing cell was isolated by linking its gene to green fluorescent protein. M71-selected ORNs responded preferentially only to mixture F. (Graphs after Bozza et al., 2002; A micrograph from Bozza and Kauer, 1998; B micrograph courtesy of T. Bozza.)

odorant mixtures (defined by differences in carbon chain length of the molecular "backbone" of an odorant). These responses can be fairly broad (for example, a single ORN responds relatively well to several different mixtures; see Figure 15.12A). If, however, an ORN that expresses a single odorant receptor protein that binds an identified ligand is isolated (for example, the M71 odorant receptor, which is known to bind acetophenone and benzaldehyde; Figure 15.12B), it responds specifically to the mixture that includes that odorants. It is not known, however, whether any of these odorant molecules represent the "best" (i.e., highest affinity) or most environmentally relevant odorants for any given receptor protein.

The Olfactory Bulb

The transduction of odorants in the olfactory cilia and the subsequent changes in electrical activity in the ORN are only the first steps in olfactory information processing. Unlike other primary sensory receptor cells (e.g., photoreceptors in the retina or hair cells in the cochlea), ORNs have axons, and these axons relay odorant information directly to the rest of the brain via action potentials. As the axons leave the olfactory epithelium, they coalesce to form a large number of bundles that together make up the olfactory nerve (cranial nerve I). Each olfactory nerve projects ipsilaterally to the olfactory bulb, which in humans lies on the ventral anterior aspect of the ipsilateral cerebral hemisphere. The most distinctive feature of the olfactory bulb is the array of **glomeruli**—more or less spherical accumulations of neuropil 100 to 200 μm in diameter. Glomeruli lie just beneath the surface of the bulb and are the synaptic target of the primary olfactory axons (Figure 15.13). In vertebrates, ORN axons make excitatory glutamatergic synapses within the glomeruli. Remarkably, this relationship between the olfactory periphery (the nose, or insect antennae—that is, the structures where ORNs are found) and glomeruli in the CNS is maintained across the animal kingdom (see Figure 15.13A, inset).

In mammals, including humans, within each glomerulus the axons of the receptor neurons contact apical dendrites of **mitral cells**, which are the principal projection neurons of the olfactory bulb. The cell bodies of the mitral cells are located in a distinct layer of the olfactory bulb deep within the glomeruli. A mitral cell extends its primary dendrite into a single glomerulus, where the dendrite gives rise to an elaborate tuft of branches onto which the axons of ORNs synapse (see Figure 15.13B,D). In the mouse, in which glomerular connectivity has been studied quantitatively, each glomerulus includes the apical dendrites of approximately 25 mitral cells, which in turn receive input from approximately 25,000 olfactory receptor axons. Remarkably, most if not all 25,000 of these axons come from ORNs that express the same, single odorant receptor gene (see Figure 15.13E). This degree of convergence presumably increases the sensitivity of mitral cells to ensure maximum fidelity of odor detection. It may also maximize the signal strength from the convergent olfactory receptor neuron input by averaging out uncorrelated "background" noise. Each glomerulus also includes dendritic processes from two other classes of local circuit neurons: Approximately 50 *tufted cells* and 25 *periglomerular cells* contribute to each glomerulus (see Figure 15.13D). Although it is generally assumed that these neurons sharpen the sensitivity of individual glomeruli to specific odorants, their function is unclear.

Finally, *granule cells*, which constitute the innermost layer of the vertebrate olfactory bulb, synapse primarily on the basal dendrites of mitral cells in the external plexiform layer (see Figure 15.13C,D). Granule cells lack an identifiable axon, and instead make reciprocal dendrodendritic synapses with mitral cells. Granule cells are thought to establish local lateral inhibitory circuits with mitral cells as well as participating in synaptic plasticity in the olfactory bulb. Olfactory granule cells and periglomerular cells are among the few classes of neurons in the forebrain that can be replaced throughout life in some mammals (see Chapter 26). In humans, however, the available evidence suggests that these cells are not lost and regenerated in adulthood.

The relationship between ORNs expressing one odorant receptor and small subsets of glomeruli (see Figure 15.13E) suggests that individual glomeruli respond specifically (or at least selectively) to distinct odorants. The selective (but not singular) responses of subsets of glomeruli to particular odorants have been confirmed physiologically in invertebrates such as *Drosophila*, as well as in mice, using single and multiunit recordings, metabolic mapping, voltage-sensitive dyes, genetically encoded sensors of electrical activity, or intrinsic signals that depend on blood oxygenation. Such studies show that increasing the odorant concentration increases the activity of individual glomeruli, as well as the number of glomeruli activated. In addition, different single odorants, or odorants with distinct chemical structures (e.g., length of the carbon chain in the backbone of the odorant molecule) maximally activate one or a few glomeruli (Figure 15.14A).

It is still not clear how (or whether) odor identity and concentration are mapped across the entire array, or reflect the activation of smaller subsets of glomeruli. Given the response of small numbers of glomeruli to single odorants, one might expect that complex natural odors such as those of coffee, fruits, cheeses, or spices—each of which is composed of more than 100 compounds—would activate a very large number of olfactory glomeruli. Surprisingly, this is not the case. In mice, natural odorants presented at their normal concentrations activate a relatively small number of glomeruli (up to 20), each of which responds selectively to one or two molecules that characterize the complex odor (Figure 15.14B). Thus, to solve the problem of representing

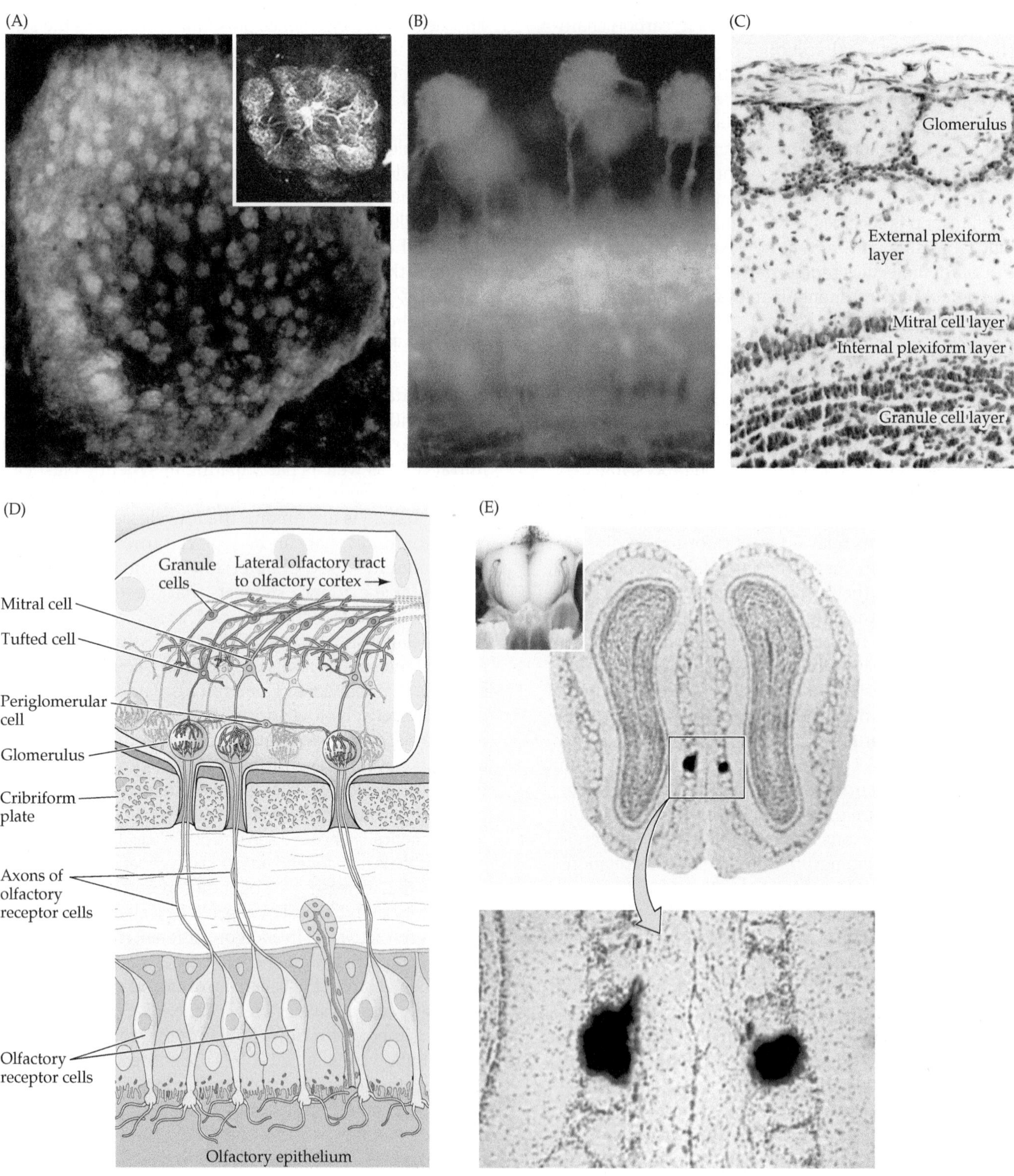

complex odorants, the olfactory system appears to employ a sparse coding mechanism that cues in on a small number of dominant chemicals within a mixture (see Figure 15.12) and represents that mixture over a relatively small subset of glomeruli. One useful metaphor is to envision the sheet of glomeruli in the olfactory bulb as an array of lights on a movie marquee; the spatial distribution of the active and inactive glomeruli ("lit and unlit lights") produces a message that is unique for a given odorant at a particular concentration.

◀ FIGURE 15.13 **Organization of the mammalian olfactory bulb.** (A) When the bulb is viewed from its dorsal surface (visualized here in a living mouse in which the overlying bone has been removed), olfactory glomeruli can be seen. The dense accumulation of dendrites and synapses that constitutes glomeruli are stained here with a vital fluorescent dye that recognizes neuronal processes. The inset shows a similar arrangement of glomeruli in the mushroom body (the equivalent of the olfactory bulb) in *Drosophila*. (B) Among the major neuronal components of each glomerulus are the apical tufts of mitral cells, which project to the pyriform cortex and other bulb targets (see Figure 15.1C). In this image of a coronal section through the bulb, the mitral cells have been labeled retrogradely by placing the lipophilic tracer Di-I in the lateral olfactory tract. (C) The cellular structure of the olfactory bulb, shown in a Nissl-stained coronal section. The five layers of the bulb are indicated. The glomerular layer includes the tufts of mitral cells, the axon terminals of ORNs, and periglomerular cells that define the margins of each glomerulus. The external plexiform layer is made up of lateral dendrites of mitral cells, cell bodies and lateral dendrites of tufted cells, and dendrites of granule cells that make dendrodendritic synapses with the other dendritic elements. The mitral cell layer is defined by the cell bodies of mitral cells, and mitral cell axons are found in the internal plexiform layer. Finally, granule cell bodies are densely packed into the granule cell layer. (D) Laminar and circuit organization of the olfactory bulb, shown diagrammatically in a cutaway view from its medial surface. Olfactory receptor cell axons synapse with mitral cell apical dendritic tufts and periglomerular cell processes within glomeruli. Granule cells and mitral cell lateral dendrites constitute the major synaptic elements of the external plexiform layer. (E) Axons from ORNs that express a particular odorant receptor gene converge on a small subset of bilaterally symmetrical glomeruli. These glomeruli, indicated in the boxed area in the upper panel, are shown at higher magnification in the lower panel. The projections from the olfactory epithelium have been labeled by a reporter transgene inserted by homologous recombination ("knocked in") into the genetic locus that encodes the particular receptor. (A courtesy of D. Purves and A.-S. La Mantia; inset from Wang et al., 2003; B from Blanchart et al., 2006; C from Pomeroy et al., 1990; E from Mombaerts et al., 1996; inset from Tadenev et al., 2011.)

Pyriform Cortical Processing of Information Relayed from the Olfactory Bulb

Mitral cell axons, as well as those from tufted cells (a less frequent class of olfactory bulb projection neurons), provide the only relay for olfactory information to the rest of the brain. The mitral cell axons from each olfactory bulb form a bundle—the **lateral olfactory tract**—that projects to the accessory olfactory nuclei, the olfactory tubercle, the pyriform and entorhinal cortices, as well as to portions of the amygdala (see Figure 15.1; Figure 15.15A). Most projections of the lateral olfactory tract are ipsilateral; however, a subset of mitral cell axons cross the midline, presumably initiating bilateral processing of some aspects of olfactory information. In humans, the major target of the lateral olfactory tract is the three-layered pyriform cortex in the ventromedial aspect of the temporal lobe, near the optic chiasm. Mitral cell inputs from glomeruli that receive

(A)

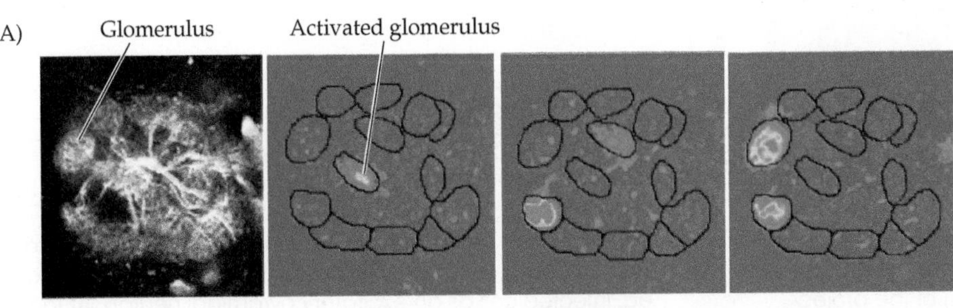

(B)

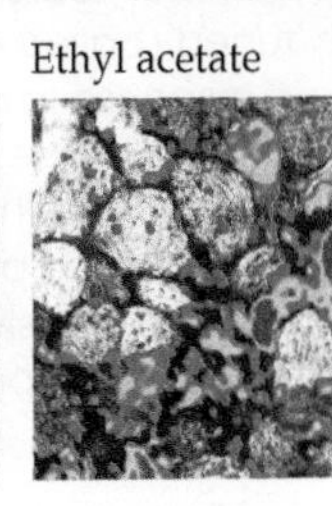

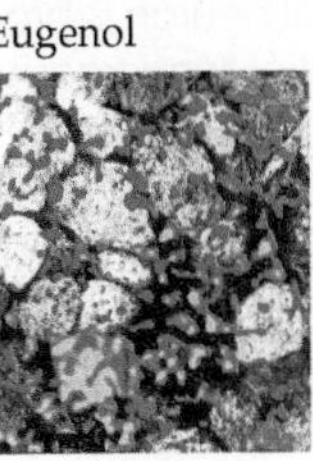

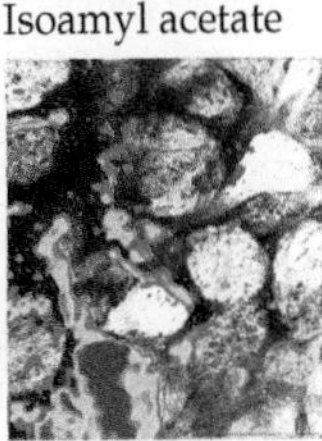

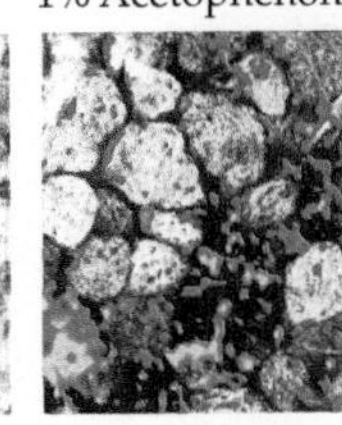

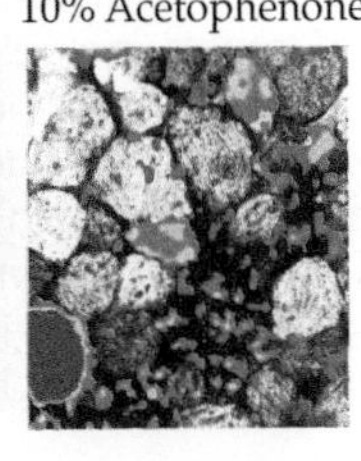

FIGURE 15.14 **Responses of chemically distinct odorants in individual glomeruli.** (A) At left, the array of glomeruli in the *Drosophila melanogaster* olfactory lobe (the equivalent of the mammalian olfactory bulb) is visualized with a fluorescent protein expressed under the genetic control of an olfactory lobe-specific gene. Subsequent panels (left to right) show that three distinct odorants—1-octan-3-ol, an insect attractant; hexane, which has a chemical smell to humans; and isoamyl acetate, the major molecular constituent of the aroma given off by bananas—activate different glomeruli. In each case, activation (yellow and the maximum red) is limited to one or two distinct glomeruli. (B) Surface images of glomeruli from a mouse olfactory bulb. An overlying response-intensity map shows the response to odorants with different carbon backbones; red represents high-intensity response. The glomerulus that responds specifically to acetophenone shows different levels of response to 1% and 10% concentrations in room air. (A from Wang et al., 2003; B from Fleischmann et al., 2008.)

odorant receptor-specific projections are distributed across the pyriform cortex. Accordingly, neurons in pyriform cortex respond to odors based upon the relay of odorant information from ORNs through the olfactory bulb via mitral cell projections.

Recent work suggests that the segregation of projections based upon the relationship between ORNs expressing a single odorant receptor protein and specific subsets of glomeruli in the olfactory bulb is far less constrained in the pyriform cortex (Figure 15.15A). Furthermore, pyriform cortical neurons have a variety of responses to multiple versus single odors (Figure 15.15B). In fact, some individual pyriform cortical cells seem to be more broadly tuned to different odors than are cells in the olfactory bulb, and the neurons that respond to single odors are distributed throughout extended regions of the pyriform cortex (Figure 15.16). Apparently, the segregation of information seen in the olfactory bulb is not maintained in the pyriform cortex; however, there may be some segregation of glomerular inputs to the amygdala. Thus, transformation

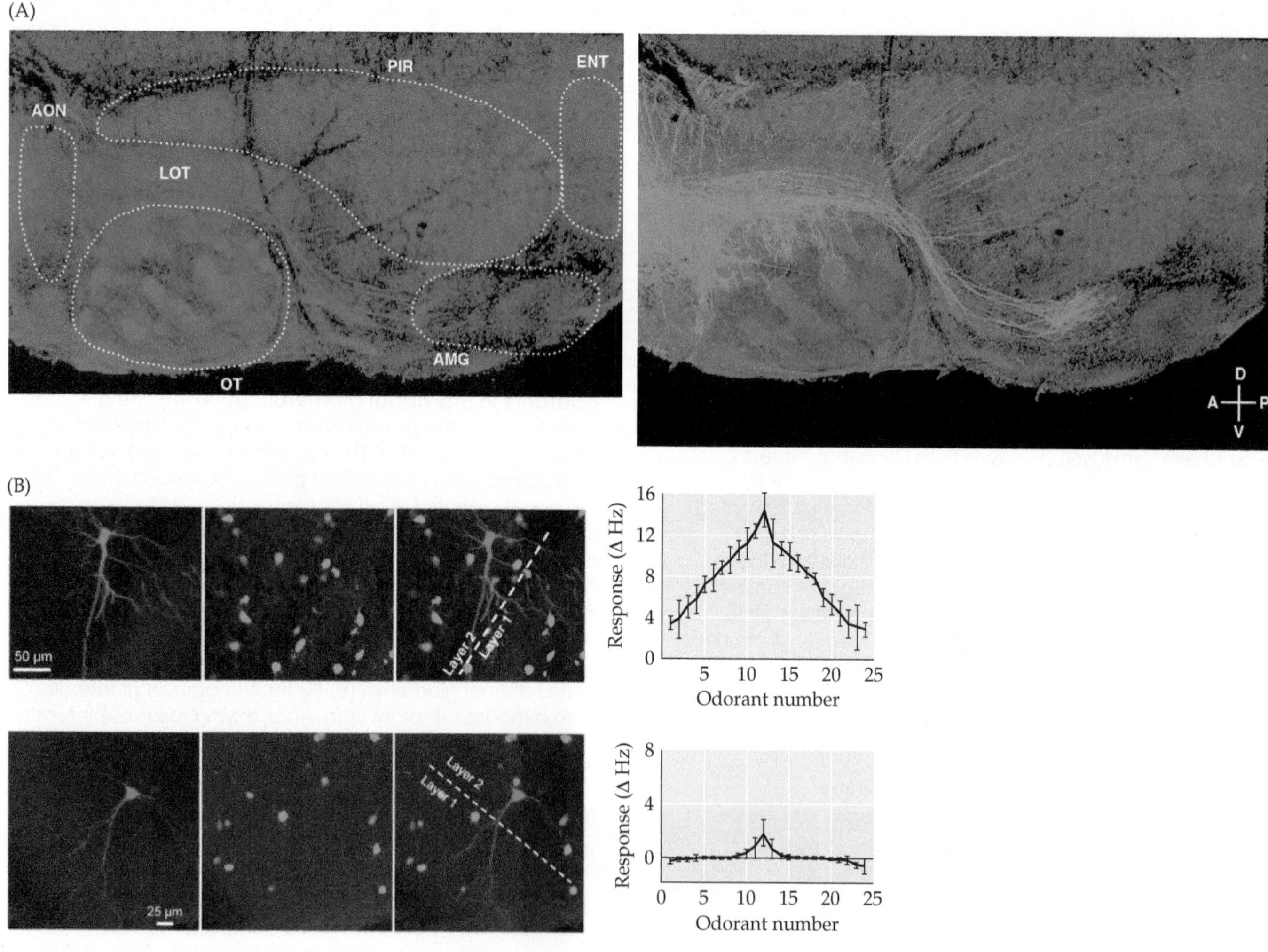

FIGURE 15.15 Single glomeruli exhibit divergent projections to olfactory recipient areas. (A) Anatomical tracing (using focal electroporation of TMR-dextran) of the projections of a single olfactory glomerulus via the lateral olfactory tract (LOT) to multiple olfactory bulb targets, including the pyriform cortex (PIR), entorhinal cortex (ENT), amygdala (AMG), olfactory tubercle (OT), and accessory olfactory nucleus (AON). (B) Distinct "tuning" of pyramidal neurons in the pyriform cortex to 24 distinct odorants in awake, behaving mice. The red label (far right) reflects single-cell injection into the neuron whose tuning curve is shown at the far left. The green co-label is for GABAergic interneurons in the pyriform cortex. The lack of co-localization confirms the identity of the recorded, injected cell as an excitatory pyramidal neuron. At top, a single pyramidal neuron is broadly tuned (based on changes in action potential firing frequency in response to odorant presentation: ΔHZ) so that it has detectable responses to all 24 odors, with some modest selectivity for a subset that elicits the highest change in firing frequency. At bottom, a pyramidal neuron from the pyriform cortex that is unresponsive to all but 3 to 5 of the 24 odorants, with a distinct peak response. (A from Sosulski et al., 2011; B from Zhan and Luo, 2010.)

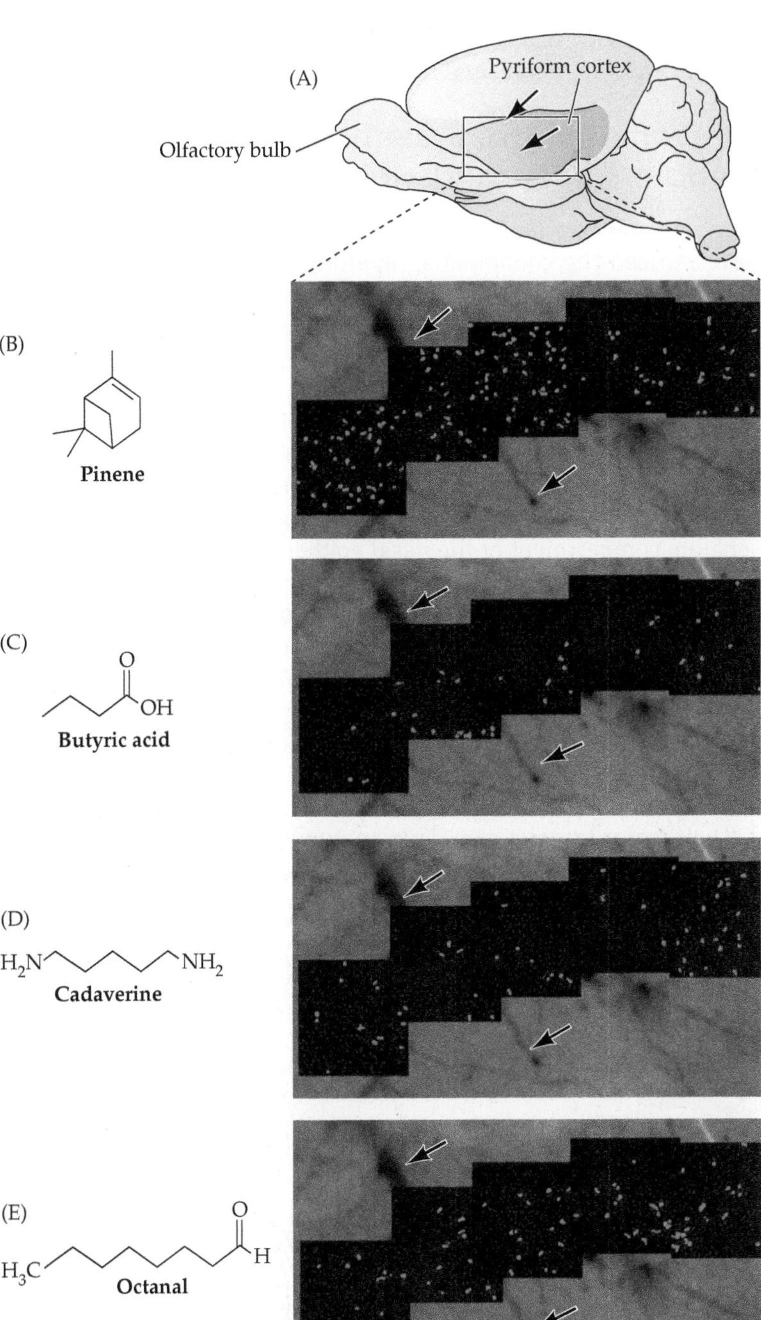

FIGURE 15.16 Differential activation of widely distributed ensembles of neurons in pyriform cortex. (A) Lateral surface of the mouse brain showing the pyriform cortex and outlining the region from which optical recordings of the electrical activity of single neurons were made. (B–E) Four distinct odors recruited different subsets of cells across the pyriform cortex. Activated cells (bright red) were recorded based on local change in fluorescence signal emitted by each cell. The fluorescence signal was due to a Ca^{2+}-sensitive dye introduced into all cells in the pyriform cortex prior to the recording session. The arrows in each panel indicate a blood vessel, which provides a landmark to compare patterns of activation of the multiple odorants in this single animal. (From Stettler and Axel, 2009.)

of odorant information from the olfactory epithelium through the bulb to the pyriform cortex that may build a representation of olfactory sensation remains somewhat difficult to discern based on patterns of connectivity alone.

The pyriform cortex has pyramidal neurons that project to a variety of forebrain targets. Thus, olfactory information is distributed broadly to forebrain regions, where it can influence a wide range of behaviors. Significant numbers of neurons in the pyriform cortex innervate directly a variety of areas in the neocortex, including the orbitofrontal cortex in humans and other primates, where multimodal responses to complex stimuli—particularly food—include an olfactory component. Pyriform cortical neurons also project to the thalamus, hippocampus, hypothalamic nuclei, and amygdala. The connections between the pyriform cortex and the mediodorsal nucleus of the thalamus, which is implicated in human memory (see Chapter 30), are thought to influence olfactory-guided "declarative" memory (see Chapter 30) via mediodorsal connections with the frontal cortex. Projections to the hippocampus are similarly thought to play a role in olfactory-guided memory, but there is very little indication of how olfaction and declarative, or conscious mnemonic information are integrated. Finally, connections between the pyriform cortex, hypothalamus, and amygdala are thought to influence visceral, appetitive, and sexual behaviors.

The details of central olfactory processing are unclear, in large measure because of the difficulty of studying the processing of specific olfactory stimuli experimentally. Identifying key odor stimuli, presenting them to the ORNs with consistent concentrations and without environmental "noise" from other molecules in air that can act as odorants, and quantifying receptor responses are daunting tasks that even now have not been completely solved for the olfactory system (however, see below for how these issues have been approached for vomeronasal sensation). Nevertheless, these pathways ensure that information about odors reaches a variety of forebrain regions, allowing olfactory perception to influence cognitive, visceral, emotional, and homeostatic behaviors.

The Vomeronasal System: Predators, Prey, and Mates

Many dog owners (and the occasional brave cat owner) have noticed the conspicuous openings in the mucus membranes above the upper gum line of their pet—usually while trying to pry an especially chewable household item from a recalcitrant pet's jaws. These modest supra-lingual openings represent a second division of the olfactory system that is prominent in carnivores (including dogs and cats) and rodents, but less robust or absent in primates (especially humans). This **vomeronasal system** encompasses a distinct receptor cell population in a separate compartment of the nasal epithelium called the **vomeronasal organ (VNO)**, as well as a separate region of the olfactory bulb—called the **accessory olfactory bulb (AOB)**—where axons from chemosensory receptor cells in the vomeronasal organ synapse (Figure 15.17).

The projections of the accessory olfactory bulb are distinct from those of the remainder of the olfactory bulb (referred to as the "main" olfactory bulb in rodents and carnivores) and include the hypothalamus and amygdala as their major target zones. This anatomical distinction provides an important clue to the primary function of the vomeronasal system: It is believed to encode and process information about odorants from conspecifics or predators and to mediate sexual, reproductive, homeostatic, and aggressive responses. The specific stimuli detected and represented by the vomeronasal system that mediate behaviors with conspecifics (i.e., mating, parental, and other social behaviors) are referred to as **pheromones**. The specific stimuli that mediate behaviors with other animals are referred to as **kairomones** (airborne molecular stimuli from a predator [e.g., an owl] or prey [e.g., a mouse]). The existence of pheromones and kairomones, distinct from consciously perceived odors, remains a focus of research for a variety of purposes, including animal population control and assisted reproduction.

The fate of the vomeronasal system in primates, especially humans, is mysterious. The vomeronasal organ is absent in most primates, as is a region of the olfactory bulb that corresponds to the accessory olfactory bulb. There are few recognizable vomeronasal receptor genes in the human genome, and those that have some homology are pseudogenes—they are not expressed and do not appear to encode functional proteins. Nevertheless, primates, including humans, have behavioral responses that can be attributed to stimuli similar to the pheromones that activate the vomeronasal system in other animals. These include control of the menstrual cycle in women exposed to either same-sex or opposite-sex individuals (e.g., same sex or coed dormitories—these studies, however, remain controversial, see above). There are equally controversial studies of male- and female-specific responses to odorants in distinct regions of the hypothalamus (presumably relayed via the olfactory bulb) as well as the reversal of these responses in the hypothalamus of individuals of different sexual orientations (Figure 15.18; see also Chapter 24). Thus, in some mammals the vomeronasal system provides

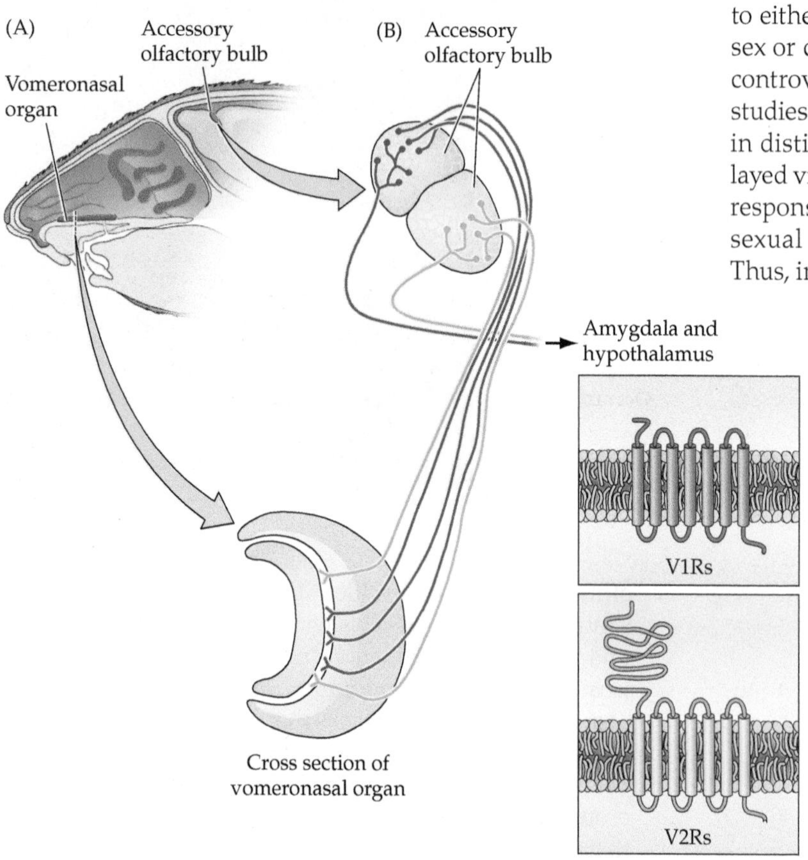

(A) Accessory olfactory bulb
Vomeronasal organ
(B) Accessory olfactory bulb
Amygdala and hypothalamus
V1Rs
V2Rs
Cross section of vomeronasal organ

FIGURE 15.17 The vomeronasal system. (A) A midsagittal section through the head of a mouse shows the location of the vomeronasal organ in the nasal cavity, and the accessory olfactory bulb located in the dorsal posterior region of the main olfactory bulb. (B) As diagrammed here, the two divisions of the accessory olfactory bulb each have glomeruli (spherical units of neuropil where synapses take place) that receive input from only one of two classes of vomeronasal receptor neurons, VR1 or VR2 (vomeronasal receptor class 1, shown here as dark blue, or class 2, light blue). (B after Pantages and Dulac, 2000.)

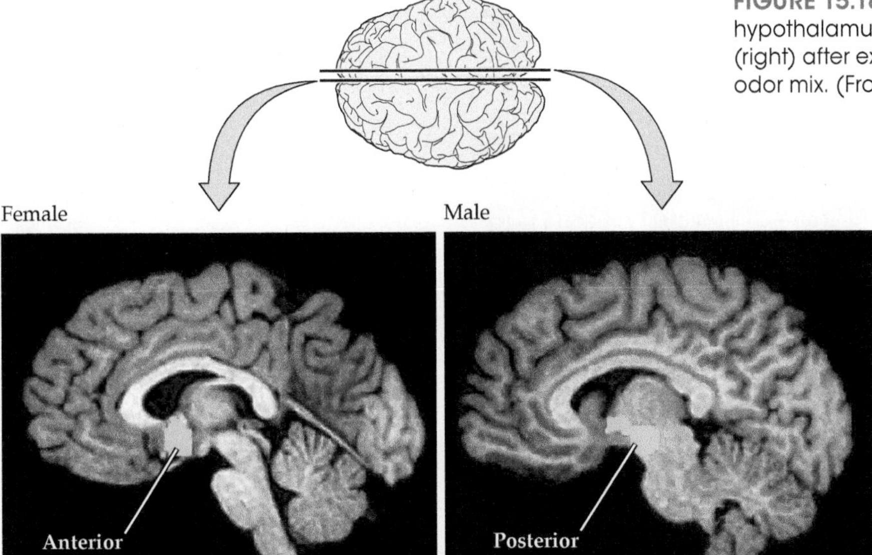

FIGURE 15.18 Differential patterns of activation in the hypothalamus of a typical human female (left) and male (right) after exposure to an estrogen- or androgen-containing odor mix. (From Savic et al., 2001.)

a distinct chemosensory parallel pathway for detecting and processing chemosensory signals about reproduction, social interactions, predator threats and prey opportunities, For other mammals, including humans, the representation of such information—if indeed it is specifically represented—remains obscure.

The Vomeronasal System: Molecular Mechanisms of Sensory Transduction

In the late 1990s, the separate identity of the vomeronasal system was confirmed at the molecular level with the cloning of a family of **vomeronasal receptors** (**VRs**). The genomic identity and expression of VRs is specific for the chemosensory neurons of the VNO. The VRs are therefore distinct from their counterparts found in ORNs. The VRs are a large class (as many as 250 individual receptor genes in the mouse) of 7-transmembrane G-protein-coupled receptors (Figure 15.19A) expressed uniquely in **vomeronasal receptor neurons** (**VRNs**). They fall into two major classes, **V1Rs** and **V2Rs**. V1Rs have a limited extracellular domain, whereas V2Rs have an extracellular domain that is quite extensive. V1Rs and V2Rs use different G-protein-coupled cascades to activate signaling (see Figure 15.19A). Thus, although VRNs in the VNO look much like their ORN counterparts in the olfactory epithelium (and share expression of some molecules), their G-protein-coupled receptors are genetically, molecularly, structurally, and functionally different. Moreover, signal transduction is accomplished via a different set of second messengers and cyclic nucleotide-gated ion channels. The V1Rs use the G-protein Gαi2, and the V2Rs use the G-protein Gαo (versus the olfactory-selective G_{olf} that is activated by odorant binding to ORs in ORNs). The activity of **transient receptor potential** (**TRP**) channels, specifically TRP2, in vomeronasal receptor neurons is regulated by phospholipase C (PLC) and diacylglycerol (DAG) in response to G-protein stimulation via V1Rs and V2Rs, whereas a cyclic nucleotide-gated ion channel and a Ca^{2+}-gated Cl^- channel are the primary molecular mediators of excitability in ORNs via activation of adenylate cyclase and the generation cAMP (see Figure 15.10). In addition, some V2Rs are coexpressed with and may interact with **major histocompatibility complex** (**MHC**) gene families for M10 and M1. The functional significance of the V2R–MHC interaction remains uncertain. The MHC protein may contribute to pheromone detection by subsets of V2Rs and also is thought to regulate the trafficking of the V2R to the vomeronasal receptor cell membrane.

The molecular map of pheromones (airborne molecular stimuli from conspecifics: e.g., mouse to mouse) and kairomones (airborne molecular stimuli from predators or prey) is now fairly well understood. Based on cellular and physiological analyses, there is a fairly broad tuning of the vomeronasal receptors (Figure 15.19B), with some specificity for distinct V1Rs and V2Rs. Thus, several V1Rs and V2Rs respond robustly and differentially to pheromones as sex-specific or predator–prey-specific cues (urine, etc.) in male versus female mice. Elimination of the TRP channels that mediate vomeronasal signal transduction (or elimination, replacement, or mutation of the receptors themselves) leads to changes in sexual or reproductive behavior, often in a **dimorphic** (see Chapter 24) manner (that is, males and females are differentially compromised). In contrast, several V2Rs and fewer V1Rs respond robustly and differentially to predator cues, including those of snakes, owls, rats, and ferrets. This selectivity is assumed to mediate avoidance responses to these threatening species. V1Rs and V2Rs are functionally partially segregated—V1Rs apparently participate more in pheromone sensing, and V2Rs in kairomone sensing.

The physiology of central targets of the vomeronasal system, relayed via the AOB—which has a cellular

(A) VNO transduction model

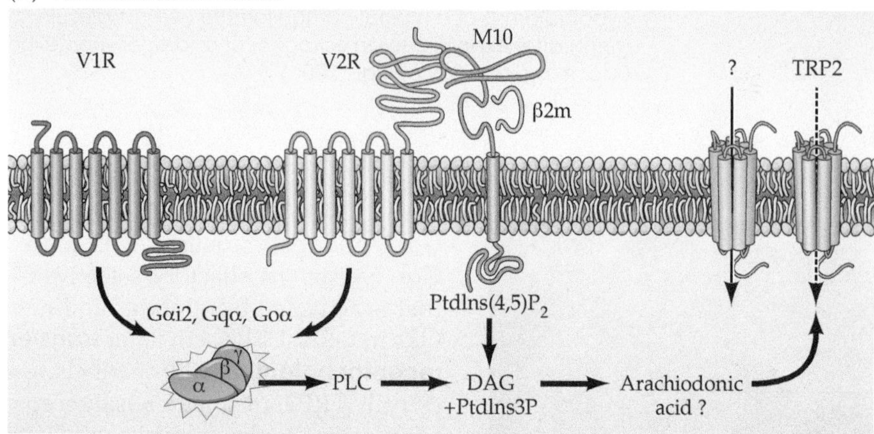

(B)

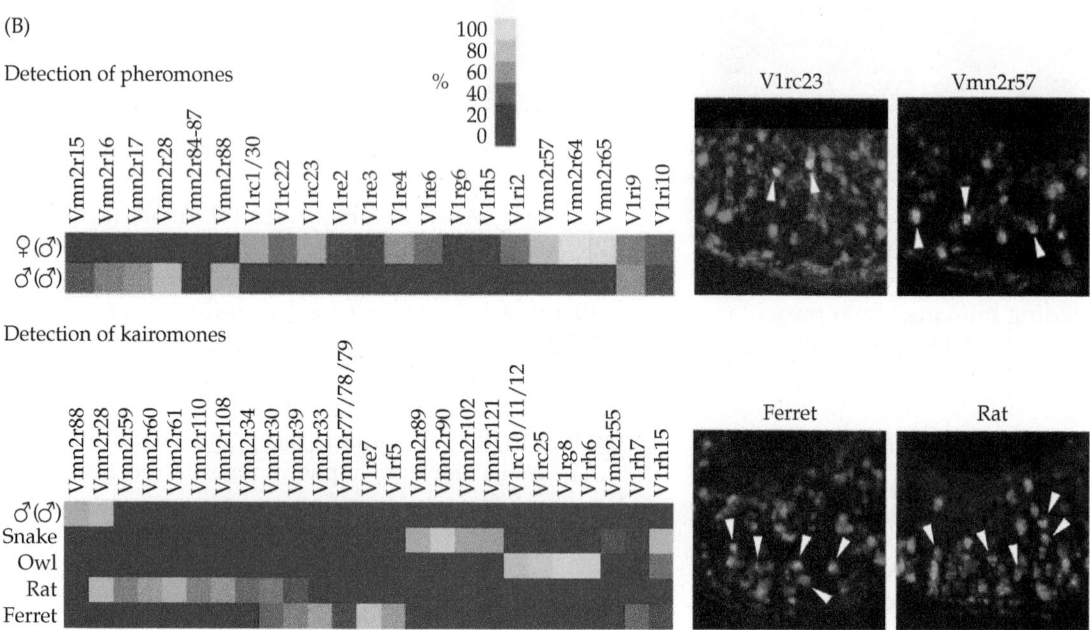

FIGURE 15.19 The two basic classes of vomeronasal receptor (VR) proteins. (A) Both V1Rs and V2Rs are 7-transmembrane G-protein-coupled receptors; however, the signal transduction mechanisms that activate the transient receptor potential (TRP) channel in the two classes of cells are distinct. (B) Selectivity of single VR proteins for sex-specific pheromonal cues (top) and species-specific kairomonal cues. The selectivity is not exclusive; however, based on the heat maps, it is clear that some receptor proteins (e.g., Vmn2r64 for sex-specific pheromones and V1rg8 for species-specific kairomones) respond dimorphically (Vmn2r64) or selectively (V1rg8), to distinct cues. These estimates of response selectivity are based on frequency of co-localization of single VRs (green) with expression of Egr1 protein (red) that is rapidly expressed in response to neuronal activity. (A after Dulac and Torello, 2003; B from Isogai et al., 2011.)

structure similar to that of the OB—is consistent with the selectivity of the V1Rs and V2Rs. Electrical recordings from single neurons in the hypothalamus and amygdala (presumed to be the targets of accessory olfactory bulb inputs) in rodents show that these cells respond specifically to chemical constituents of urine or other excreta thought to contain pheromones that elicit stereotypical reproductive or aggressive behaviors.

Organization of the Taste System

The third chemosensory system, the taste system, represents the chemical as well as physical qualities of ingested substances, primarily food. In concert with the olfactory and trigeminal systems, taste reflects the aesthetic and nutritive qualities of food as well as indicating whether or not a food item is safe to be ingested. Once in the mouth, the chemical constituents of food interact

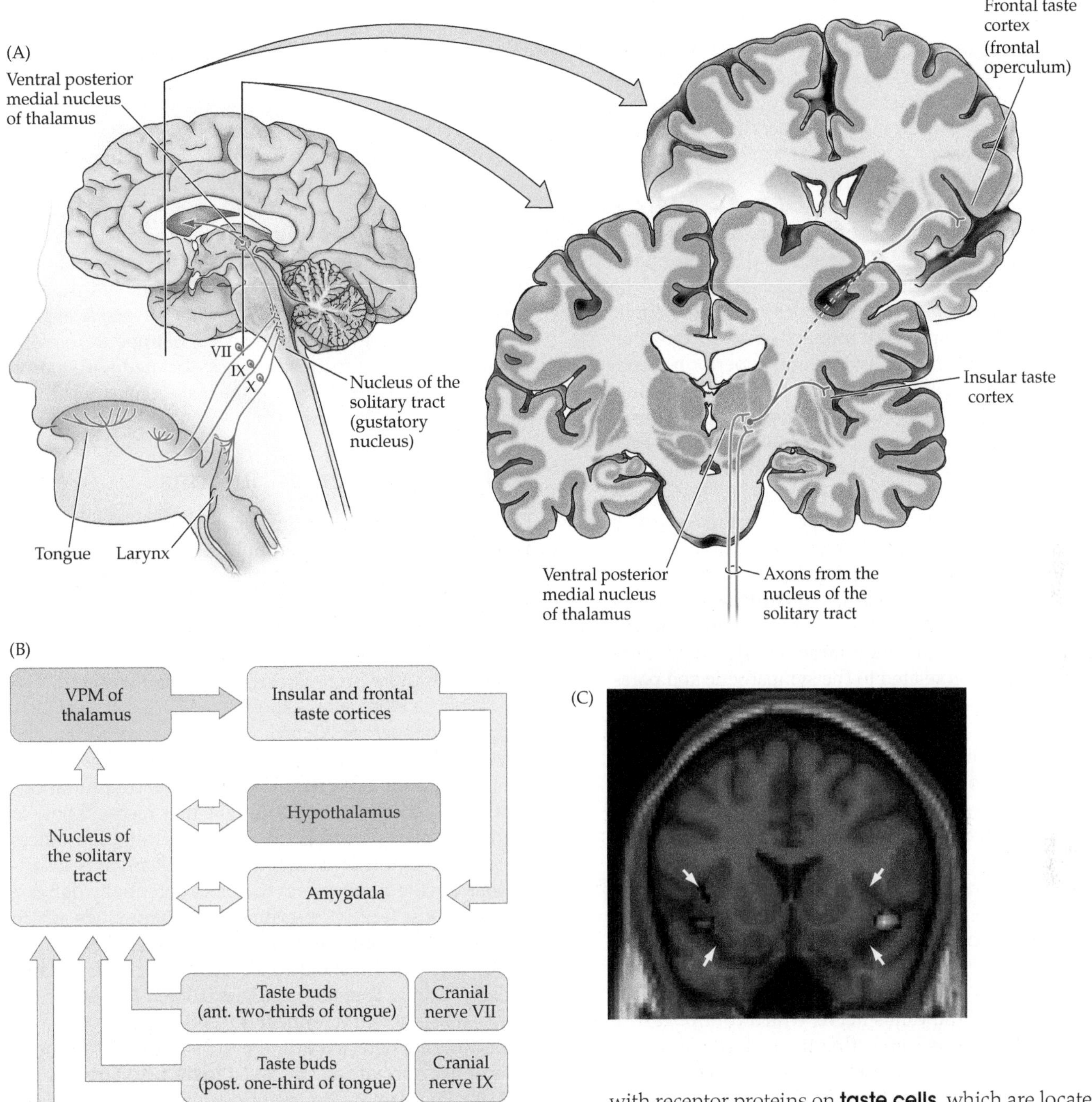

FIGURE 15.20 The human taste system. (A) The drawing shows the relationship between receptors in the mouth and upper alimentary canal, and the nucleus of the solitary tract in the medulla. The coronal section shows the ventral posterior medial (VPM) nucleus of the thalamus and its connection with gustatory regions of the cerebral cortex. (B) Basic pathways for processing taste information. (C) Functional MRI of a typical person consuming food. Note bilateral focal activation (red) in the insular cortex (arrows), with a bias for greater activation in the dominant hemisphere (left in most humans). (C from Schoenfeld et al., 2004.)

with receptor proteins on **taste cells**, which are located in epithelial specializations called **taste buds** in the tongue. Taste cells transduce chemical stimuli to encode information about the identity, concentration, and qualities (pleasant, unpleasant, or potentially harmful) of the substance. This information also prepares the gastrointestinal system to receive and digest food by causing salivation and swallowing—or if the substance is noxious, gagging and regurgitation. Information about the temperature and texture of food (including viscosity and fat content) is transduced and relayed from the tongue and mouth via somatosensory receptors from the trigeminal and other sensory cranial nerves to the thalamus and somatosensory cortices (see Chapters 9 and 10). Of course, food is not eaten simply for nutritional value or avoided because of unpleasant or potentially harmful qualities; "taste" also depends on cultural and psychological factors. How else can one explain why so many people enjoy consuming hot peppers or bitter-tasting liquids such as beer?

Like the olfactory system, the taste system is defined by its specialized peripheral receptors as well as by several central pathways that relay and process taste information (Figure 15.20). Taste cells (the peripheral receptors) are found in taste buds distributed on the dorsal surface of the tongue, soft palate, pharynx, and upper part of the esophagus. Taste cells synapse with primary sensory axons that run in the chorda tympani and greater superior petrosal branches of the facial nerve (cranial nerve VII), the lingual branch of the glossopharyngeal nerve (cranial nerve IX), and the superior laryngeal branch of the vagus nerve (cranial nerve X) to innervate the taste buds in the tongue, palate, epiglottis, and esophagus, respectively. The central axons of these primary sensory neurons in the respective cranial nerve ganglia project to rostral and lateral regions of the **nucleus of the solitary tract** in the medulla (see Figure 15.20A), also known as the **gustatory nucleus** of the solitary tract complex. (The posterior region of the solitary tract nucleus is the main target of afferent visceral sensory information related to the sympathetic and parasympathetic divisions of the visceral motor system; see Chapter 21.)

The distribution of the cranial nerves that innervate taste buds in the oral cavity is topographically represented along the rostral–caudal axis of the rostral portion of the gustatory nucleus; the terminations from the facial nerve are rostral, those from the glossopharyngeal are in the midregion, and those from the vagus nerve are more caudal in the nucleus (see Figure 15.20A). Integration of taste and visceral sensory information is presumably facilitated by this arrangement. The caudal part of the nucleus of the solitary tract also receives innervation from subdiaphragmatic branches of the vagus nerve, which control gastric motility. Interneurons connecting the rostral and caudal regions of the nucleus represent the first interaction between visceral and gustatory stimuli, and these connections can be thought of as the sensory limb of a gustatory–visceral reflex arc. This close relationship between gustatory and visceral information makes sense, since an animal must quickly recognize if it is eating something that is likely to make it sick and respond accordingly.

Axons from the rostral (gustatory) part of the solitary nucleus project to the ventral posterior complex of the thalamus, where they terminate in the medial half of the **ventral posterior medial nucleus**. This nucleus projects in turn to several regions of the neocortex, including the anterior insula in the temporal lobe (the **insular taste cortex**) and the operculum of the frontal lobe (see Figure 15.20B). There is also a secondary neocortical taste area in the caudolateral orbitofrontal cortex; here neurons respond to combinations of visual, somatosensory, olfactory, and gustatory stimuli. Interestingly, in the monkey, when a given food is consumed to the point of satiety, specific orbitofrontal neurons diminish their activity to that tastant, suggesting that these neurons are involved in the conscious motivation to eat (or not to eat) particular foods. Finally, reciprocal projections connect the nucleus of the solitary tract via nuclei in the pons to the hypothalamus and amygdala. These projections presumably influence affective aspects (e.g., pleasurable versus aversive experience of food; food-seeking behavior) of appetite, satiety, and other homeostatic responses associated with eating (recall that the hypothalamus is the major center governing homeostasis; see Chapter 21).

Taste Perception in Humans

The taste system encodes information about the quantity as well as the identity of stimuli. Most taste stimuli are nonvolatile, hydrophilic molecules that are soluble in saliva. In general, the perceived intensity of taste is directly proportional to the concentration of the taste stimulus. In humans, threshold concentrations for most ingested tastants are quite high. For example, the threshold concentration for citric acid is about 2 m*M*; for salt (NaCl) 10 m*M*; and for sucrose 20 m*M*. (In contrast, recall that the perceptual threshold for some odorants is as low as 0.01 n*M*.) Because the body requires substantial concentrations of salt and carbohydrates, taste cells may respond only to relatively high concentrations of these essential substances in order to promote an adequate intake. Clearly, it is advantageous for the taste system to detect potentially dangerous substances (e.g., bitter-tasting plant compounds, which may be noxious or poisonous) at much lower concentrations. Thus, the threshold concentration for such tastants is relatively low: That for quinine is 0.008 m*M*, and for the deadly substance strychnine it is 0.0001 m*M*.

Tastants are detected over the full surface of the tongue in receptive specializations called **taste papillae** (Figure 15.21A). Papillae are defined by multicellular protuberances surrounded by local invaginations in the tongue epithelium. These invaginations form a trench to concentrate solubilized tastants. Taste buds are distributed along the lateral surfaces of the papillar protuberance as well as in the trench walls. They consist of specialized neuroepithelial receptor cells called taste cells, some supporting cells, and occasional basal cells (Figure 15.21B). In humans, approximately 4000 taste buds are distributed throughout the surface of the tongue as well as the palate, epiglottis, and esophagus. Taste cells are clustered around a 1-mm opening called a taste pore in the taste bud near the surface of the tongue (see Figure 15.21B). Solubilized tastants are further concentrated and are presented directly to the exposed taste receptor cells in the relatively small region of the taste pore. Like ORNs (and presumably for the same reason—because they are exposed to infectious agents and environmental toxins), taste cells have a lifetime of about 2 weeks. Taste cells are apparently regenerated from basal cells, which constitute a local stem cell population that is retained in the mature tongue.

There are three types of papillae: **fungiform** (which contain about 25% of the total number of taste buds), **circumvallate** (50%), and **foliate** (the remaining 25%). The three classes are distributed discontinuously on the surface of the tongue. Fungiform papillae are found only on the anterior two-thirds of the tongue; the highest density (about 30 per square centimeter) is at the tip. Fungiform papillae have a mushroom-like structure (hence their name) and typically have about three taste buds at their apical surface. Nine circumvallate papillae form a chevron at the rear of the tongue. Each consists of a circular trench containing about 250 taste buds along the trench walls (see Figure 15.21A). Two foliate papillae are present on the posterolateral tongue, each having about 20 parallel ridges with about 600 taste buds in their walls. Thus, chemical stimuli on the tongue first stimulate receptors in the fungiform papillae and then in the foliate and circumvallate papillae. Tastants subsequently stimulate scattered taste buds in the pharynx, larynx, and upper esophagus.

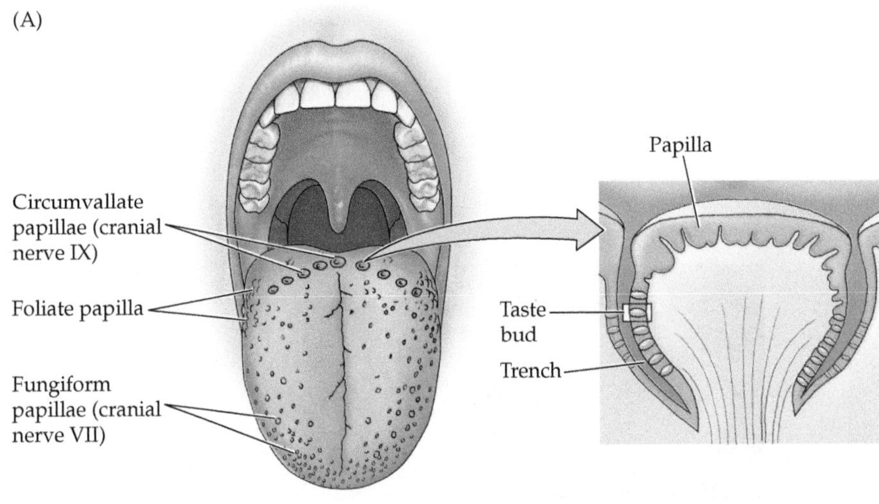

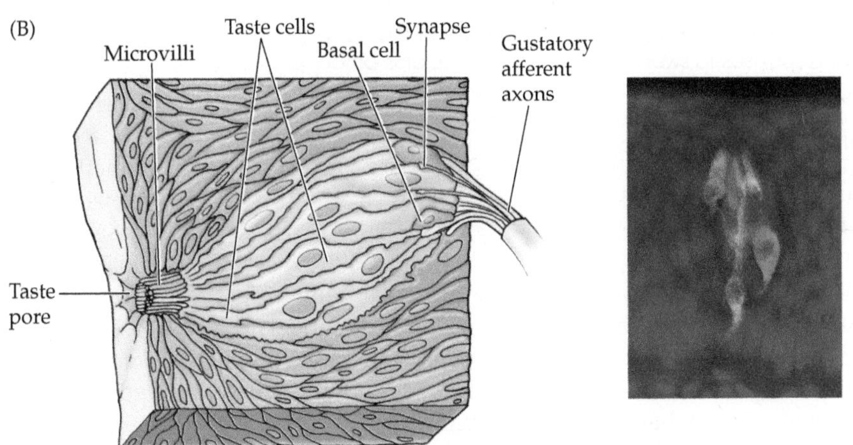

FIGURE 15.21 Taste buds and taste papillae. (A) Distribution of taste papillae on the dorsal surface of the tongue. The blowup shows the location of individual taste buds on a circumvallate papilla. (B) Diagram and light micrograph of a taste bud, showing various types of taste cells and the associated gustatory nerves. The taste cells make synapses on the gustatory afferent axons. The apical surface of the receptor cells has microvilli that are oriented toward the taste pore. (Micrograph courtesy of M. Tizzano, T. Finger and colleagues.)

Based on general agreement across cultures, the taste system detects five perceptually distinct categories of tastants: **salt**, **sour**, **sweet**, **bitter**, and **umami**. (From the Japanese word for "delicious," *umami* refers to savory tastes, including monosodium glutamate and other amino acids that provide the flavor in cooked meat and other protein-rich foods.) These five perceptual categories have dietary and metabolic significance: Salt tastes include NaCl, which is needed for electrolyte balance; sour tastes, associated with acidity and thus protons (H^+), indicate the palatability of various foods (e.g., the citric acid in oranges); sugars such as glucose and other carbohydrates are needed for energy; bitter-tasting molecules, including plant alkaloids such as atropine, quinine, and strychnine, indicate foods that may be poisonous; and essential amino acids such as glutamate are needed for protein synthesis.

There are obvious limitations to this classification. People experience a variety of gustatory or ingestive sensations in addition to these five, including astringent (cranberries, tea), pungent (hot peppers, ginger), fat, starch, and various metallic tastes, to name only a few. In addition, mixtures of chemicals may elicit entirely new taste sensations. Finally, at low concentrations the protective response to aversive tastes can be overridden, leading to acquired tastes for foods having sour or bitter flavor, such as lemons (sour) and quinine (bitter).

Although all tastes can be detected over the entire surface of the tongue, different regions of the tongue have different thresholds for various tastes (Figure 15.22A). These discontinuities in taste sensitivity may be

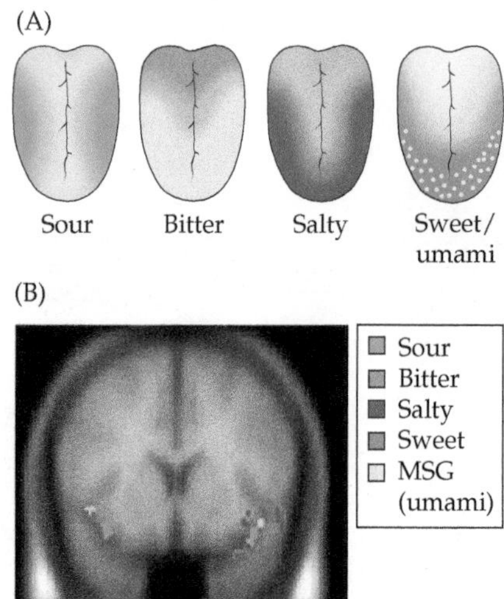

FIGURE 15.22 Peripheral innervation of the tongue. (A) Responses to sweet/umami, salty, sour, and bitter tastants recorded in the three cranial nerves that innervate the tongue and epiglottis. (B) Composite fMRI showing the different locations of focal activation in the insular cortex in response to each of the tastes encoded by taste receptors. (B from Schoenfeld et al., 2004.)

related to the aesthetic, metabolic, and potentially toxic qualities detected by the taste receptors in the tongue. The tip of the tongue is most responsive to sweet, umami, and salty compounds, all of which produce pleasurable sensations at somewhat higher concentrations. Thus, tastes encountered by this region—the initial point of contact for most ingested foods—activate feeding behaviors such as mouth movements, salivary secretion, insulin release, and swallowing. The acquisition of foods high in carbohydrates and amino acids is beneficial (in moderation), and thus it is not surprising that the most exposed region of the tongue is especially sensitive to these tastes.

Sour and bitter taste sensitivity is lowest toward the tip and greatest on the sides and back of the tongue. It seems reasonable that, once it has analyzed for **nutrient** content, the receptor surface might next evaluate aesthetic characteristics like acidity and bitterness that indicate lack of palatability (excessive sourness) or even toxicity (bitterness). Sour-tasting compounds elicit grimaces, puckering, and massive salivary secretion to dilute the tastant. Activation of the rear of the tongue by bitter-tasting substances elicits protrusion of the tongue and other protective reactions (expectoration and gagging) that prevent ingestion.

Each of the primary tastes represented over the surface of the tongue corresponds to a distinct class of receptor molecules expressed in subsets of taste cells (see the next section). Thus, representation in taste buds of the five primary categories of taste perception is closely linked to the molecular biology of taste transduction. These taste categories are also maintained in the representation of taste information in the CNS, including in the insular taste cortex (Figure 15.22B). Mapping of responses to sweet, bitter, salty, sour, and umami in typical humans shows that each of these tastes elicits focal activity in the taste cortex, suggesting that information about each taste category remains somewhat segregated throughout the taste system.

As with olfaction, gustatory sensitivity declines with age. An obvious index of this decline is the tendency of adults to add more salt and spices to food than children do. The decreased sensitivity to salt can be problematic for older people with hypertension as well as electrolyte and/or fluid balance problems. Unfortunately, a safe and effective substitute for table salt (NaCl) has not yet been developed.

Taste Receptor Proteins and Transduction

Within the taste buds, only the taste cells are specialized for sensory transduction, and their basic structure and function are uniform across all classes of papillae and their constituent taste buds. Taste cells have distinct apical and basal domains, reflecting their epithelial character (Figure 15.23). Chemosensory transduction is initiated in the apical domain of the taste cells, and electrical signals are generated at the basal domain via graded receptor potentials (and corresponding secretion of neurotransmitters). The specific neurotransmitters released by taste cells remain uncertain but are thought to include serotonin, ATP, and GABA. Taste receptor proteins and related signaling molecules, like those in ORNs, are concentrated on microvilli that emerge from the taste cell apical surface. The basal domain is specialized for synaptic activation in response to tastant binding on apical receptor proteins. There are voltage-regulated ion channels as well as channels controlled by second messengers—especially members of the transient receptor potential, or TRP, family. In addition, local endoplasmic reticulum acts as a store that provides Ca^{2+} to facilitate synaptic vesicle fusion and neurotransmitter release at synapses made onto gustatory afferents at the basal surface. These synapses are made onto primary afferent axons from branches of three cranial nerves: the facial (VII), glossopharyngeal (IX), and vagus (X) nerves (see Figure 15.20A,B).

Five distinct classes of taste receptor molecules represent tastants in the major perceptual categories—salty, sour, sweet, bitter, and umami. These receptor molecules are thought to be concentrated primarily in the apical microvilli of taste cells. Salty and sour tastes are elicited by ionic stimuli such as the positively charged ions in salts (e.g., Na^+ from NaCl), or the H^+ in acids (e.g., acetic acid, which gives vinegar its sour taste). Thus, the ions in salty and sour tastants initiate sensory transduction via specific ion channels, most likely an amiloride-sensitive Na^+ channel for salty tastes (Figure 15.24A) and, for sour, an H^+-permeant, nonselective cation channel that is a member of the TRP family (Figure 15.24B). The sour receptor channel is related to a similar channel protein that is mutated in polycystic kidney disease; thus, the channel is referred to as PKD.

The sour receptor is expressed in a subset of taste cells, similar to the segregated expression of receptor proteins for sweet, umami, and bitter. The receptor potentials generated by the positive inward current carried either by Na^+ for salty or H^+ for sour directly depolarize the relevant taste cell. The initial depolarization leads to the activation of voltage-gated Na^+ channels in the basolateral aspect of the taste cell. This additional depolarization activates voltage-gated Ca^{2+} channels, leading to the release of neurotransmitter from the basal aspect of the taste cell and the activation of action potentials in ganglion cell axons (see Figure 15.23).

In humans and other mammals, sweet and umami receptors are heterodimeric G-protein-coupled receptors that share a common 7-transmembrane receptor subunit called T1R3, paired with the T1R2 7-transmembrane receptor for perception of sweet, or with the T1R1 receptor for amino acids (Figure 15.24C,D). The T1R2 and T1R1 receptors are expressed in different subsets of taste cells, indicating the presence of, respectively, sweet- and amino acid-selective cells in the taste buds. Upon binding sugars or other sweet stimuli, the T1R2/T1R3 receptor heterodimer initiates a G-protein-mediated signal transduction cascade that leads to activation of the phospholipase C isoform **PLCβ2**, leading in turn to increased concentrations of inositol triphosphate (IP_3) and to the opening of TRP channels (specifically the $TRPM_5$ channel), which depolarizes the taste cell via increased intracellular Ca^{2+}. Similarly, the T1R1/T1R3 receptor is broadly tuned to the 20 standard L-amino acids found in proteins (but not to their D-amino acid enantiomers). Transduction of amino acid stimuli via the T1R1/T1R3 receptor also reflects G-protein-coupled intracellular signaling leading to PLCβ2-mediated activation of the $TRPM_5$ channel and depolarization of the taste cell (see Figure 15.24D).

Another family of G-protein-coupled receptors known as T2R receptors transduces bitter tastes. Approximately 30 genes in humans and other mammals encode 30 T2R subtypes, and single taste cells express multiple T2R subtypes. Indeed, in humans a well-known mutation for the perception of a specific bitter tastant—phenothylcarbamide (PTC)—was originally discovered in the early 1930s and identified as a simple Mendelian trait shortly thereafter;

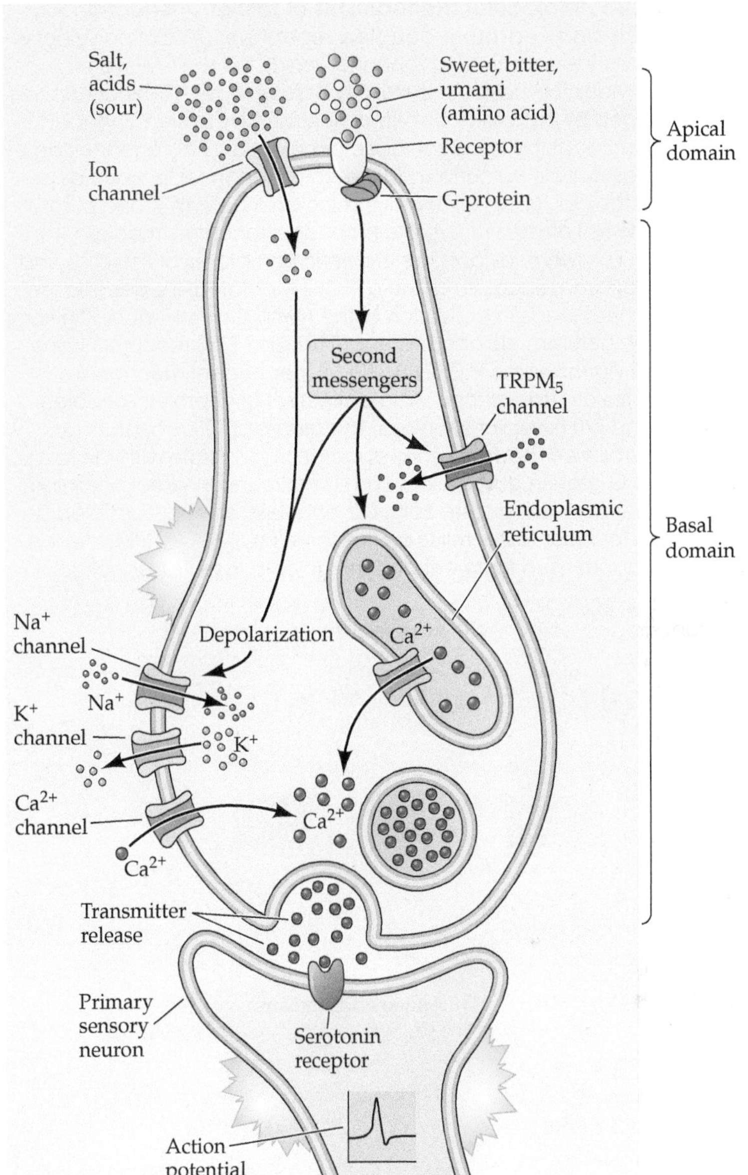

FIGURE 15.23 **Sensory transduction in taste cells.** Taste cells are polarized epithelial cells with an apical and a basal domain separated by tight junctions. Tastant-transducing channels (salt and sour) and G-protein-coupled receptors (sweet, amino acid, and bitter) are limited to the apical domain. Intracellular signaling components that are coupled to taste receptor molecules (G-proteins and various second-messenger-related molecules) are also enriched in the apical domain. Voltage-regulated Na^+, K^+, and Ca^{2+} channels mediate release of neurotransmitter from presynaptic specializations at the base of the cell onto terminals of peripheral sensory afferents. These channels are limited to the basolateral domain, as is endoplasmic reticulum that also modulates intracellular Ca^{2+} concentration and contributes to the release of neurotransmitter. The neurotransmitter serotonin, among others, is found in taste cells, and serotonin receptors are found on the sensory afferents. Finally, the $TRPM_5$ channel, which facilitates G-protein-coupled receptor-mediated depolarization, is expressed in taste cells. Its localization to apical versus basal domains is not yet known.

FIGURE 15.24 Molecular mechanisms of taste transduction via ion channels and G-protein-coupled receptors. (A) Cation selectivity of the amiloride-sensitive Na^+ channel versus the H^+-sensitive proton channel provides the basis for specificity of salty tastes. (B) Sour tastants are transduced by a proton-permeant, nonselective cation channel that is a member of the transient receptor potential (TRP) channel family. In both cases, positive current via the cation channel leads to depolarization of the cell. (C–E) For sweet, amino acid (umami), and bitter tastants, different classes of G-protein-coupled receptors mediate transduction. (C) For sweet tastants, heteromeric complexes of the T1R2 and T1R3 receptors transduce stimuli via a PLCβ2-mediated, IP_3-dependent mechanism that leads to activation of the $TRPM_5$ Ca^{2+} channel. (D) For amino acids, heteromeric complexes of T1R1 and T1R3 receptors transduce stimuli via the same PLCβ2/IP_3/$TRPM_5$-dependent mechanism. (E) Bitter tastes are transduced via a distinct set of G-protein-coupled receptors, the T2R receptor subtypes. The details of T2R receptors are less well established; however, they apparently associate with the taste cell-specific G-protein gustducin, which is not found in sweet or amino acid receptor-expressing taste cells. Nevertheless, stimulus-coupled depolarization for bitter tastes relies on the same PLCβ2/IP_3/$TRPM_5$-dependent mechanism used for sweet and amino acid taste transduction.

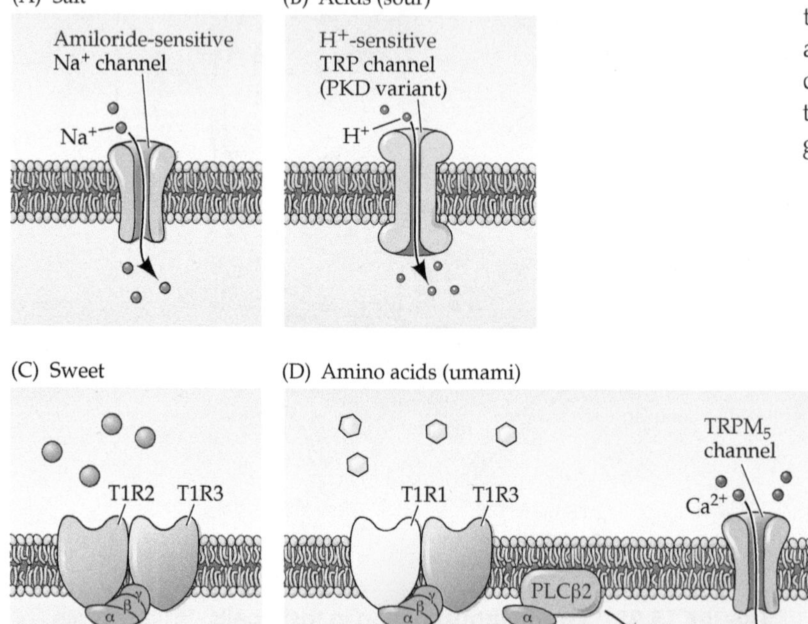

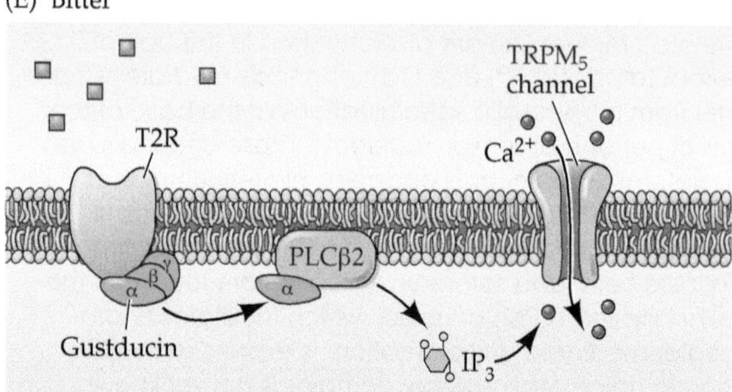

this has proven to be a mutation of a human T2R gene. The observation of a selective single gene mutation for bitter taste indicates that this taste category is distinct and encoded specifically in taste receptor cells. The distribution of T2R receptors among taste cells supports this view. T2Rs are not expressed in the same taste cells as T1R1, T1R2, and T1R3 receptors. Thus, the receptor cells for bitter tastants are presumably completely distinct from those for sweet and umami, which share at least one heterodimeric G-protein-coupled receptor subunit.

Although the transduction of bitter stimuli relies on a mechanism similar to that for sweet and amino acid tastes, the taste cell-specific G-protein **gustducin**, found primarily in T2R-expressing taste cells, apparently contributes to the transduction of bitter tastes (Figure 15.24E). The role of gustducin versus that of the other G-proteins for transduction of sweet and umami/amino acid tastes remains unclear. The remaining steps in bitter transduction are similar to those for sweet and amino acids: PLCβ2-mediated activation of $TRPM_5$ channels depolarizes the taste cell, resulting in the release of neurotransmitter at the synapse between the taste cell and sensory ganglion cell axon.

Neural Coding in the Taste System

In the taste system, *neural coding* refers to the way that the identity, concentration, and "hedonic" (pleasurable or aversive) value of tastants is represented in the pattern of action potentials relayed to the brain from the taste buds. Neurons in the taste system might be specifically "tuned" to respond with a maximum change in electrical activity to a single taste stimulus. Such tuning might rely on specificity at the level of the receptor cells, as well as on the maintenance of separate channels for the relay of this information from the periphery to the brain. This sort of coding scheme is often referred to as a *labeled line code,* since responses in specific cells at multiple points in the pathway presumably correspond to distinct stimuli. The segregated expression of sour, sweet, amino acid, and bitter receptors in different taste cells (Figure 15.25A–C) and the maintenance of focal activation for each class of taste in the insular taste cortex (see Figure 15.20C) are consistent with labeled line coding.

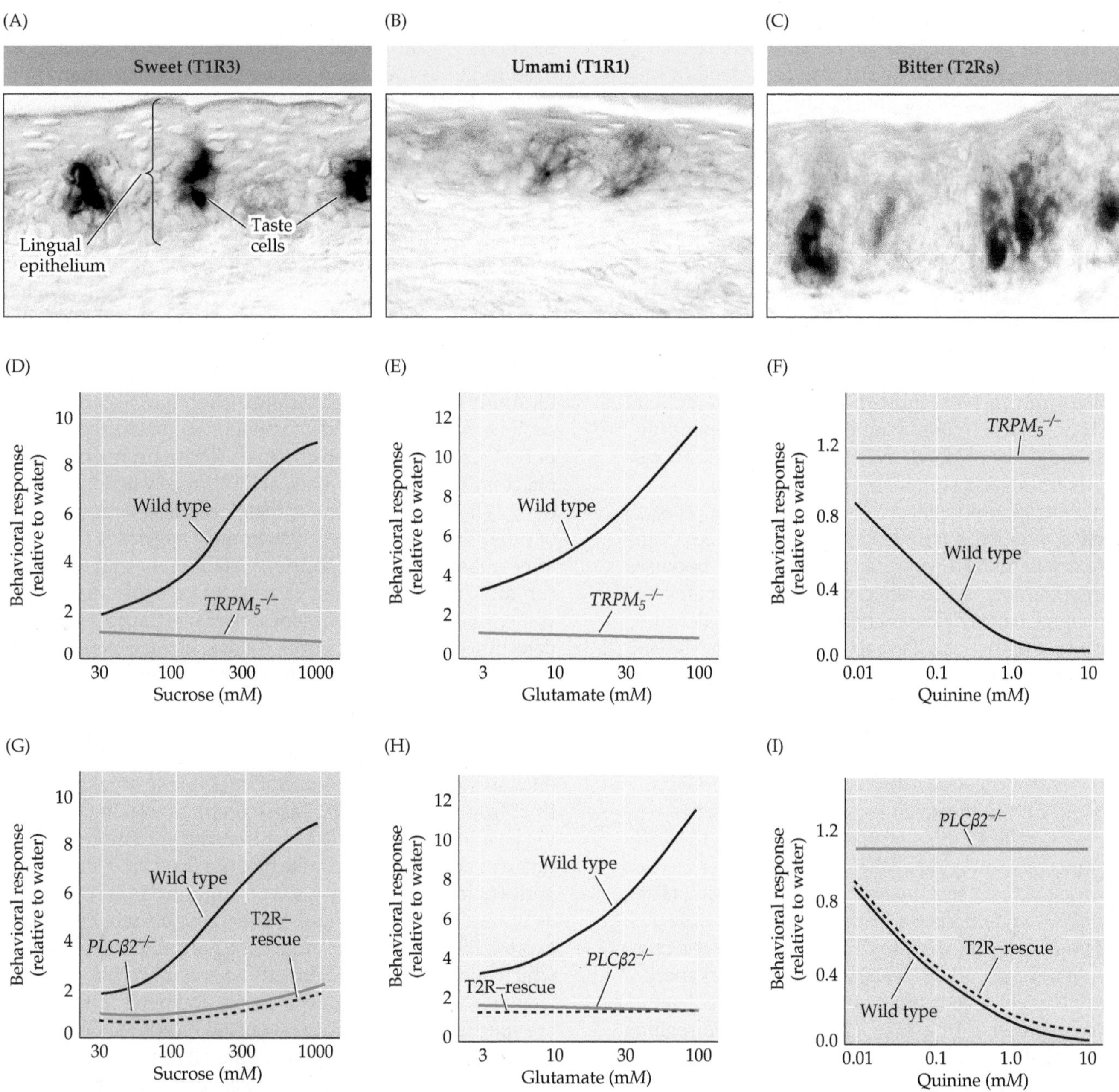

FIGURE 15.25 Specificity in peripheral taste coding supports the labeled line hypothesis. (A–C) Sweet (A), amino acid (B), and bitter (C) receptors are expressed in different subsets of taste cells. (D–E) The gene for the $TRPM_5$ channel can be inactivated, or "knocked out," in mice (*$TRPM_5^{-/-}$*) and behavioral responses measured with a taste preference test. The mouse is presented with two drinking spouts, one with water and the other with a tastant; behavioral responses are measured as the frequency of licking of the two spouts. For pleasant tastes such as sweet (sucrose; D) or umami (glutamate; E), control (i.e., wild-type) mice lick the spout with the tastant more frequently, and higher concentrations of tastant lead to increased response (black lines). In *$TRPM_5^{-/-}$* mice, this behavioral response (i.e., a preference for the tastant versus water) is eliminated at all concentrations (red lines). (F) For an aversive tastant like bitter quinine, control mice prefer water. This behavioral response—which is initially low—is further diminished with higher quinine concentrations (black line). Inactivation of *$TRPM_5$* also eliminates this behavioral response, regardless of tastant concentration (red line). (G–I) When the *PLCβ2* gene is knocked out, the behavioral response to sucrose (G), glutamate (H), and quinine (I) is eliminated (red lines). When *PLCβ2* is reexpressed only in T2R-expressing taste cells, behavioral responses to sucrose and glutamate are not rescued (dashed black lines in G and H); however, the behavioral response to quinine is restored to normal levels (compare the solid and dashed black lines in I). (A,B from Nelson et al., 2001; C from Adler et al., 2000; D-I after Zhang et al., 2003.)

Molecular genetic experiments in mice indicate that some perceptual taste categories are established based on the identity of T1Rs and T2Rs that are expressed in individual taste cells. Initial support came from studies in which the genes that specify the sweet and amino acid heteromeric receptors (T1R2 and T1R1) were inactivated in mice. Such mice lack behavioral responses to a broad range of sweet or amino acid stimuli, depending on the gene that has been inactivated. Moreover, recordings of electrical activity in the relevant branches of cranial nerves VII, IX, or X showed that action potentials in response to sweet or amino acid stimuli were lost in parallel with the genetic mutation and behavioral change. Finally, these deficits in transduction and perception were unchanged at a broad range of concentrations, indicating that the molecular specificity of each receptor is quite rigid—the remaining receptors could not respond, even at high concentrations of sweet or amino acid stimuli (Figure 15.25D–H).

These observations suggest that sweet and amino acid transduction and perception depend on labeled lines from the periphery. Bitter taste proved harder to analyze because of the larger number of T2R bitter receptors. To circumvent this challenge, Charles Zuker, Nicholas Ryba, and colleagues took advantage of the shared aspects of intracellular signaling for sweet, amino acid, and bitter tastes. If the genes for either the $TRPM_5$ channel or PLCβ2 are inactivated, behavioral and physiological responses to sweet, amino acid, and bitter stimuli are all abolished (see Figure 15.25), while salty and sour perceptions—which do not rely on the G-protein-coupled, PLCβ2-mediated transduction mechanism—remain. To evaluate whether taste cells expressing the T2R family of receptors provide a labeled line for bitter tastes, PLCβ2 was selectively reexpressed in T2R-expressing taste cells in a *PLCβ2* mutant mouse. Thus, in these mice only the taste cells that normally express the T2R subset of receptor genes could tranduce taste signals. If these receptors specifically and uniquely encode bitter tastes, the "rescued" mice (i.e., those expressing PLCβ2 in T2R cells) should regain their perceptual and physiological responses to bitter, but not sweet or amino acid, tastes. This was indeed the result of the experiment: Behavioral and physiological responses to bitter tastes, but not sweet or amino acid tastes, were restored to normal levels (Figure 15.25I). Evidently, receptor proteins uniquely expressed in subsets of taste cells encode sweet, amino acid, and bitter, as judged by taste perception, action potential activity in peripheral nerves, and behavioral responses. This specificity and segregation of receptor cells at the periphery can be considered to establish labeled lines that relay the information to the CNS, where information about the identity of the five primary taste categories remains segregated. Given the clear distinctions one makes between sweet, salty, sour, bitter, and umami tastes, it seems likely that this perceptual clarity established by peripheral receptors is maintained by central representations and is used to guide specific ingestive (sweet, salty, umami) or aversive (sour, bitter) behaviors.

Summary

The chemical senses—olfaction, vomeronasal sensation, and taste—all contribute to sensing airborne or soluble molecules from a variety of sources. Humans and other mammals rely on this information for behaviors as diverse as attraction, reproduction, feeding, and avoiding potentially dangerous circumstances. Receptor neurons in the olfactory epithelium transduce chemical stimuli into neuronal activity by stimulation of a large family of G-protein-coupled receptors that elicit second-messenger-mediated regulation of Na^+, Ca^{2+}, and Cl^- ion channels. These events generate olfactory receptor potentials, and ultimately action potentials, in the afferent axons of these cells. The large number of odorant receptor molecules in most species is believed to establish sensitivity to the myriad odors that animals can discriminate. In most vertebrates except humans, the vomeronasal pathway provides a parallel pathway for the detection of pheromones from conspecifics (attractive cues) and kairomones from predators or prey (aversive or attractive cues). The peripheral vomeronasal receptor neurons appear similar to ORNs, but they express a different family of G-protein-coupled receptors and the intercellular transduction mechanisms for generating action potentials is distinct from that in ORNs. Taste receptor cells, in contrast, use a variety of mechanisms for transducing a more limited range of chemical stimuli. Each of the five perceptual categories of taste—salty, sour, sweet, amino acids (also known as umami), and bitter—are encoded by receptor cells that express distinct receptor proteins. Salts and protons (acids, which elicit a sour taste) directly activate two different ion channels, and for sweet, amino acid, and bitter tastes there are specific sets of G-protein-coupled receptors. Olfaction, vomeronasal sensation, and taste are all relayed via specific pathways in the CNS. ORNs project directly from the periphery to the olfactory bulb, and the olfactory bulb projects primarily to the pyriform cortex. Vomeronasal receptor neurons project directly to the accessory olfactory bulb, which in turn projects to targets in the hypothalamus and amygdala. In the taste system, cranial sensory ganglion neurons relay information from taste cells to the solitary nucleus in the brainstem. The solitary nucleus projects, via the thalamus, to the taste area of the cerebral cortex, where each of the five taste categories is represented in a distinct, non-overlapping domain. Each of these systems ultimately processes chemosensory information in ways that give rise to some of the most sublime pleasures humans can experience.

ADDITIONAL READING

Reviews

Axel, R. (2005) Scents and sensibility: A molecular logic of olfactory perception (Nobel lecture). *Angew Chem.*, Int. Ed. (English) 44 (38): 6110–6127.

Buck, L. B. (2000) The molecular architecture of odor and pheromone sensing in mammals. *Cell* 100: 611–618.

Chandreshekar, J., M. A. Hoon, N. J. Ryba and C. S. Zuker (2006) The receptors and cells for mammalian taste. *Nature* 444: 288–294.

Hildebrand, J. G. and G. M. Shepherd (1997) Mechanisms of olfactory discrimination: Converging evidence for common principles across phyla. *Annu. Rev. Neurosci.* 20: 595–631.

Lindemann, B. (1996) Taste reception. *Physiol. Rev.* 76: 719–766.

Mombaerts, P. (2004) Genes and ligands for odorant, vomeronasal and taste receptors *Nat. Rev. Neurosci.* 5: 263–278.

Scott, K. (2004) The sweet and the bitter of mammalian taste. *Curr. Opin. Neurobiol.* 14: 423–427.

Zufall, F. and T. Leinders-Zufall (2000) The cellular and molecular basis of odor adaptation. *Chem. Senses* 25: 473–481.

Important Original Papers

Adler, E. and 5 others (2000) A novel family of mammalian taste receptors. *Cell* 100: 693–702.

Astic, L. and D. Saucier (1986) Analysis of the topographical organization of olfactory epithelium projections in the rat. *Brain Res. Bull.* 16: 455–462.

Avanet, P. and B. Lindemann (1988) Amiloride-blockable sodium currents in isolated taste receptor cells. *J. Memb. Biol.* 105: 245–255.

Bozza, T., P. Feinstein, C. Zheng and P. Mombaerts (2002) Odorant receptor expression defines functional units in the mouse olfactory system. *J. Neurosci.* 22: 3033–3043.

Buck, L. and R. Axel (1991) A novel multigene family may encode odorant receptors: A molecular basis for odor recognition. *Cell* 65: 175–187.

Caterina, M. J. and 8 others (2000) Impaired nociception and pain sensation in mice lacking the capsaicin receptor. *Science* 288: 306–313.

Chaudhari, N., A. M. Landin and S. D. Roper (2000) A metabotropic glutamate receptor variant functions as a taste receptor. *Nature Neurosci.* 3: 113–119.

DuLac, C. and A. T. Torello (2003) Molecular detection of pheromone signals in mammals: from genes to behaviour. *Nat. Rev. Neurosci.* 4: 551–562.

Fleischmann, A. and 10 others (2008) Mice with a "monoclonal nose": Perturbations in an olfactory map impair odor discrimination. *Neuron* 60: 1068–1081.

Graziadei, P. P. C. and G. A. Monti-Graziadei (1980) Neurogenesis and neuron regeneration in the olfactory system of mammals. III. Deafferentation and reinnervation of the olfactory bulb following section of the fila olfactoria in rat. *J. Neurocytol.* 9: 145–162.

Isogai, Y. and 5 others (2011) Molecular organization of vomeronasal chemoreception *Nature* 478: 241–245.

Kay, L. M. and G. Laurent (2000) Odor- and context-dependent modulation of mitral cell activity in behaving rats. *Nat. Neurosci.* 2: 1003–1009.

Lin, D. Y., S. D. Shea and L. D. Katz (2006) Representation of natural stimuli in the rodent main olfactory bulb. *Neuron* 50: 937–949.

Malnic, B., J. Hirono, T. Sato and L. B. Buck (1999) Combinatorial receptor codes for odors. *Cell* 96: 713–723.

Mombaerts, P. and 7 others (1996) Visualizing an olfactory sensory map. *Cell* 87: 675–686.

Nelson, G. and 5 others (2001) Mammalian sweet taste receptors. *Cell* 106: 381–390.

Nelson, G. and 6 others (2002) An amino-acid taste receptor. *Nature* 416: 199–202.

Sosulski, D. L., M. L. Bloom, T. Cutforth, R. Axel and S. R. Datta (2011) Distinct representations of olfactory information in different cortical centers. *Nature* 472: 213–216.

Stettler, D. D. and R. Axel (2009) Representations of odor in the piriform cortex. *Neuron* 63: 854–864.

Vassar, R. and 5 others (1994) Topographic organization of sensory projections to the olfactory bulb. *Cell* 79: 981–991.

Wong, G. T., K. S. Gannon and R. F. Margolskee (1996) Transduction of bitter and sweet taste by gustducin. *Nature* 381: 796–800.

Zhan, C. and M. Luo (2010) Diverse patterns of odor representation by neurons in the anterior piriform cortex of awake mice. *J. Neurosci.* 30: 16662–16672.

Zhang, Y. and 7 others (2003) Coding of sweet, bitter, and umami tastes: Different receptor cells sharing similar signaling pathways. *Cell* 112: 293–301.

Zhao, G. Q. and 6 others (2003) The receptors for mammalian sweet and umami taste. *Cell* 115: 255–266.

Books

Barlow, H. B. and J. D. Mollon (1989) *The Senses.* Cambridge, UK: Cambridge University Press, chapters 17–19.

Doty, R. L. (ed.) (1995) *Handbook of Olfaction and Gustation.* New York: Marcel Dekker.

Farbman, A. I. (1992) *Cell Biology of Olfaction.* New York: Cambridge University Press.

Getchell, T. V., L. M. Bartoshuk, R. L. Doty and J. B. Snow, Jr. (1991) *Smell and Taste in Health and Disease.* New York: Raven Press.

Shier, D., J. Butler, and R. Lewis (2004) *Hole's Human Anatomy and Physiology.* Boston: McGraw-Hill.

Go to the NEUROSCIENCE 6e Companion Website at **www.oup.com/uk/Purves6e** for Web Topics, Animations, Flashcards, and more.

UNIT III

Movement and Its Central Control

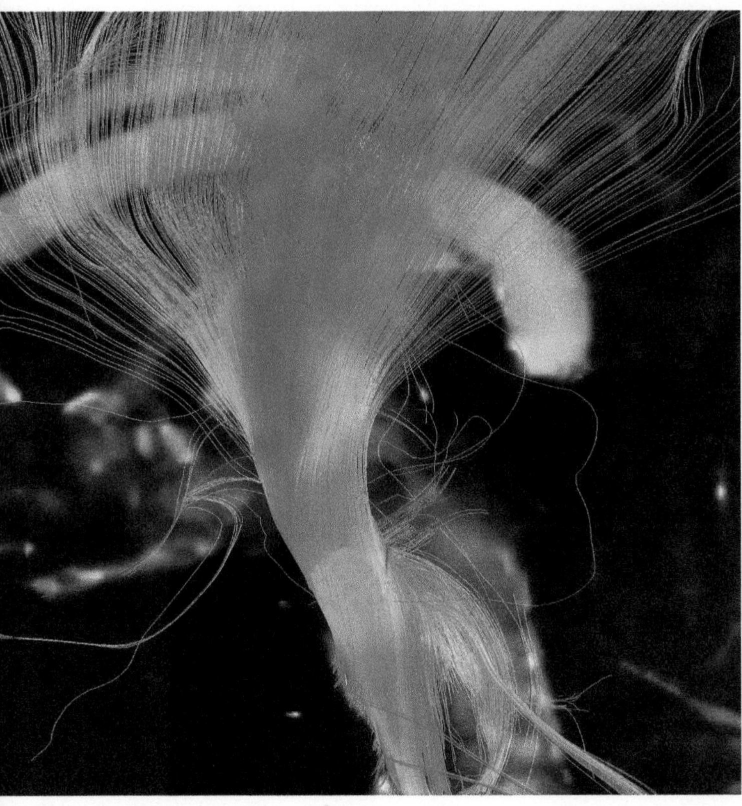

MOVEMENTS, WHETHER VOLUNTARY or involuntary, are produced by spatial and temporal patterns of muscular contractions orchestrated by neural circuits in the brain and spinal cord. Analysis of these circuits is fundamental to an understanding of both typical behavior and the etiology of a variety of neurological disorders. This unit considers the brainstem, and spinal cord circuitry that makes elementary reflex movements possible, as well as the circuits in the forebrain and cerebellum that organize the intricate patterns of neural activity responsible for more complex motor acts.

The "lower" motor neurons in the spinal cord and brainstem directly innervate skeletal muscles. These lower motor neurons are controlled directly by local circuits within the spinal cord and brainstem and indirectly by "upper" motor neurons in the cerebral cortex and brainstem. Circuits in the basal ganglia and cerebellum regulate upper motor neurons, facilitating the initiation and performance of movement with spatial and temporal precision. The autonomic divisions of the visceral motor system organize the innervation of visceral smooth muscles, cardiac muscle, and glandular secretions by a similar network of neurons in both lower and higher CNS centers. All of this circuitry works together to enable and coordinate complex sequences of bodily movements and ensure appropriate autonomic activity in support of these movements.

The various symptoms of movement disorders often signify damage to particular brain regions. For example, amyotrophic lateral sclerosis, Parkinson's disease, and Huntington's disease are the result of pathological changes in different parts of the motor system. Thus, knowledge of the various levels of motor control is essential for understanding, diagnosing, and treating these diseases.

On the previous page:

Demonstration of projections from the cerebral cortex to the brainstem and spinal cord (lavender fibers) in a living person constructed from data obtained via diffusion tensor imaging (DTI). DTI measures the direction of water diffusion, which is highly constrained in white matter pathways by the organization of axons into parallel fascicles. Computational methods, called DTI tractography, are employed to generate images of fiber bundles that represent in 3D views the course of axonal projections in cerebral white matter. © Sherbrooke Connectivity Imaging Lab/Science Photo Library.

CHAPTER

16 Lower Motor Neuron Circuits and Motor Control

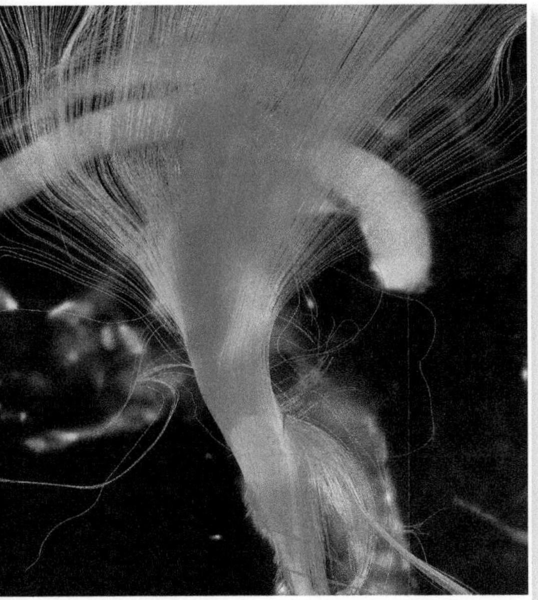

Overview

SKELETAL MUSCLE CONTRACTION is initiated by "lower" motor neurons in the spinal cord and brainstem. The cell bodies of the lower neurons are located in the ventral horn of the spinal cord gray matter and in the motor nuclei of the cranial nerves in the brainstem. These neurons (also called α motor neurons) send axons directly to skeletal muscles via the ventral roots and spinal peripheral nerves or, in the case of brainstem motor nuclei, via cranial nerves. The spatial and temporal patterns of activation of lower motor neurons are determined primarily by local circuits located within the spinal cord and brainstem. The local circuit neurons receive direct input from sensory neurons and mediate sensorimotor reflexes; they also maintain precise interconnections that enable the coordination of a rich repertoire of rhythmical and stereotyped behaviors. The local circuit neurons also receive input from descending pathways from higher centers. These descending pathways comprise the axons of "upper" motor neurons that modulate the activity of lower motor neurons by influencing the local circuitry. The cell bodies of the upper motor neurons are located in brainstem centers, such as the vestibular nuclei, superior colliculus, and reticular formation, as well as in the cerebral cortex. These diverse sources of upper motor neurons initiate and guide a wide variety of both involuntary and voluntary movements. The axons of the upper motor neurons typically synapse on the local circuit neurons in the brainstem and spinal cord, which, via relatively short axons, in turn make synaptic connections with the appropriate combinations of lower motor neurons. Lower motor neurons, therefore, are the final common pathway for transmitting information from a variety of sources to the skeletal muscles. Comparable circuits of interneurons and lower motor neurons may be recognized within the divisions of the visceral motor system, but consideration of these motor circuits will be reserved for Chapter 21. Until then, the principal context for our exploration of the central control of movement will be those movements that are executed by musculoskeletal systems.

Neural Centers Responsible for Movement

The neural centers responsible for the control of movement can be divided into four distinct but highly interactive subsystems, each of which makes a unique contribution to motor control (Figure 16.1). The first of these subsystems is located within the gray matter of the spinal cord and the tegmentum of the brainstem. The relevant cells include the lower motor neurons, which send their axons out of the brainstem and spinal cord to innervate the skeletal muscles of the head and body, respectively, and the local circuit neurons, which are the major source of synaptic input to all lower motor neurons. Commands for movement, whether reflexive or voluntary, are

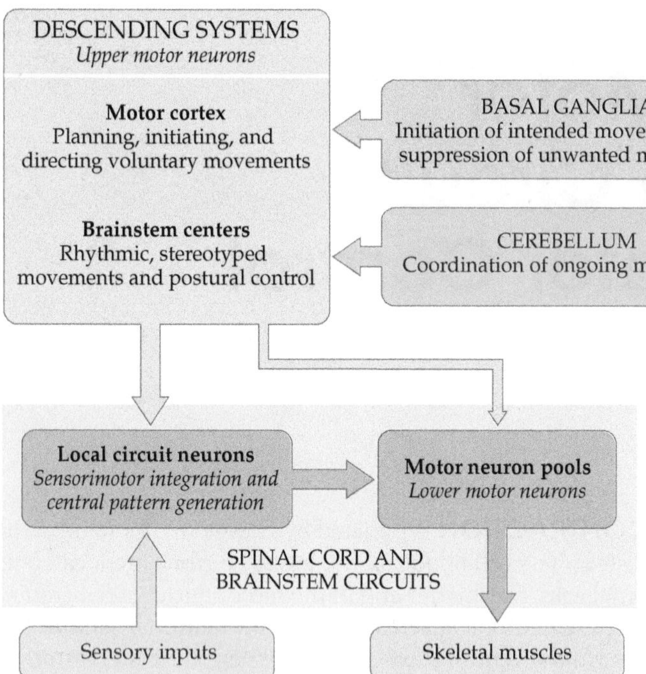

FIGURE 16.1 Organization of neural structures involved in the control of movement. Four systems—local spinal cord and brainstem circuits, descending control centers in the cerebral cortex and brainstem, the cerebellum, and the basal ganglia—make essential and distinct contributions to motor control.

ultimately conveyed to the muscles by the activity of the lower motor neurons; thus, these neurons comprise, in the words of the great British neurophysiologist Charles Sherrington, the "final common path" for initiating movement. The local circuit neurons that innervate the lower motor neurons receive sensory inputs as well as descending projections from higher centers. The circuits they form provide much of the coordination between different muscle groups that is essential for organized movement. Even after the spinal cord is disconnected from the brain in an experimental animal, appropriate stimulation of local circuits in the isolated spinal cord can elicit involuntary but highly coordinated limb movements that resemble walking.

The second motor subsystem consists of the **upper motor neurons**, whose cell bodies lie in the brainstem or cerebral cortex, and whose axons descend to synapse with the local circuit neurons or (more rarely) with the lower motor neurons directly. The upper motor neuron pathways that arise in the cortex are essential for the initiation of voluntary movements and for complex spatiotemporal sequences of skilled movements. In particular, descending upper motor neuron pathways from cortical areas in the frontal lobe, including the **primary motor cortex** (Brodmann's area 4) and several divisions of the **premotor cortex** (mainly Brodmann's area 6), are essential for planning, initiating, and directing sequences of voluntary movements involving the head, trunk, and limbs. The frontal lobe also contains cortical areas that play a similar role in the control of eye movements (Brodmann's area 8). In addition, cortical areas in the anterior cingulate gyrus (Brodmann's area 24) govern the expression of emotions, especially with respect to the facial musculature. The posterior portion of inferior frontal gyrus—typically in the left hemisphere (referred to as Broca's area or Brodmann's areas 44 and 45; see Chapter 33)—is a division of premotor cortex that plays a critical role in the production of speech. Upper motor neurons originating in the brainstem are responsible for regulating muscle tone and for orienting the eyes, head, and body with respect to vestibular, somatic, auditory, and visual sensory information. Their contributions also are critical for basic navigational movements and for the control of posture.

The third and fourth subsystems are massive, complex neural circuits with output pathways that have no direct access to either the local circuit neurons or the lower motor neurons. Instead, they control movement indirectly by regulating the activity of the upper motor neurons in the cerebral cortex and brainstem.

The larger of these latter two subsystems, the **cerebellum**, overlies the pons and fourth ventricle in the posterior cranium (see Chapter 19). The cerebellum functions via its efferent pathways to the upper motor neurons as a servomechanism, detecting and attenuating the difference, or "motor error," between an intended movement and the movement actually performed. The cerebellum mediates both real-time and long-term reductions in these inevitable motor errors (the latter being a form of motor learning). Patients with cerebellar damage exhibit incoordination with persistent errors in controlling the direction and amplitude of ongoing movements.

Last, embedded in the depths of the forebrain, is a group of structures collectively referred to as the **basal ganglia**. The basal ganglia prevent upper motor neurons from initiating unwanted movements and prepare the motor circuits for the initiation of movements. The problems with movements associated with disorders of basal ganglia, such as Parkinson's disease and Huntington's disease, attest to the importance of this subsystem in the regulation of transitions from one pattern of voluntary movements to another (see Chapter 18).

Despite much effort, the sequence of events that lead from thought and emotion to movement is still poorly understood. The picture is clearest, however, at the level of control of the skeletal muscles themselves. It therefore makes sense to begin an account of motor behavior by considering the anatomical and physiological relationships between lower motor neurons and the striated muscle fibers they innervate.

Motor Neuron–Muscle Relationships

An orderly relationship between the locations of motor neuron pools and the muscles they innervate is evident both along the length of the spinal cord and across the medial-to-lateral dimension of the cord, an arrangement that, in effect, provides a spatial map of the body's musculature. This map can be demonstrated in animal experiments by injecting individual muscle groups with visible tracers that are transported by the axons of the lower motor neurons in a retrograde direction from their terminals back to their cell bodies. The lower motor neurons that innervate each of the body's skeletal muscles can then be seen in histological sections of the ventral horns of the spinal cord. Each lower motor neuron innervates muscle fibers within a single muscle, and all the motor neurons innervating a single muscle (called the motor neuron pool for that muscle) are grouped together into a rod-shaped cluster that runs parallel to the long axis of the spinal cord for one or more spinal cord segments (Figure 16.2). For example, the motor neuron pools that innervate the arm are located in the cervical enlargement of the cord, and those that innervate the leg are located in the lumbar enlargement (see Appendix Figure A3). There is also a map, or *topography*, of motor neuron pools in the medial-to-lateral dimension of the spinal cord. Motor neurons that innervate the axial musculature (i.e., the postural muscles of the trunk) are located most medially in the ventral horn of the spinal cord, whereas neurons that innervate the muscles of the shoulders (or pelvis in the lumbar spinal cord; see Figure 16.2) are lateral to the axial neurons. Lower motor neurons that innervate the proximal muscles of the arm are the next most lateral, while those that innervate the distal parts of the extremities, including the hands and fingers, lie farthest from the midline (Figure 16.3).

This spatial organization of motor neuron pools in the ventral horn provides a framework for understanding

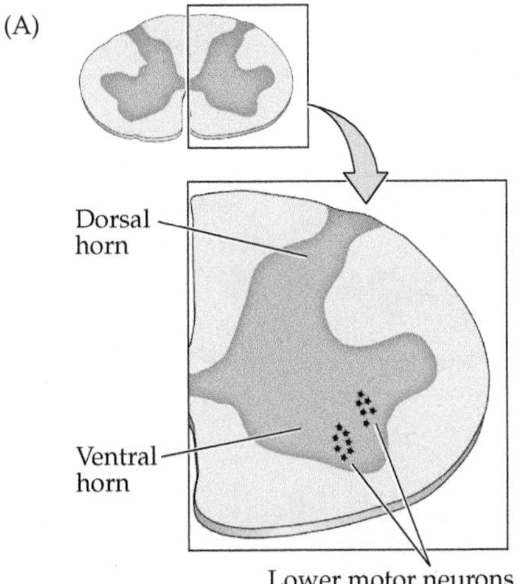

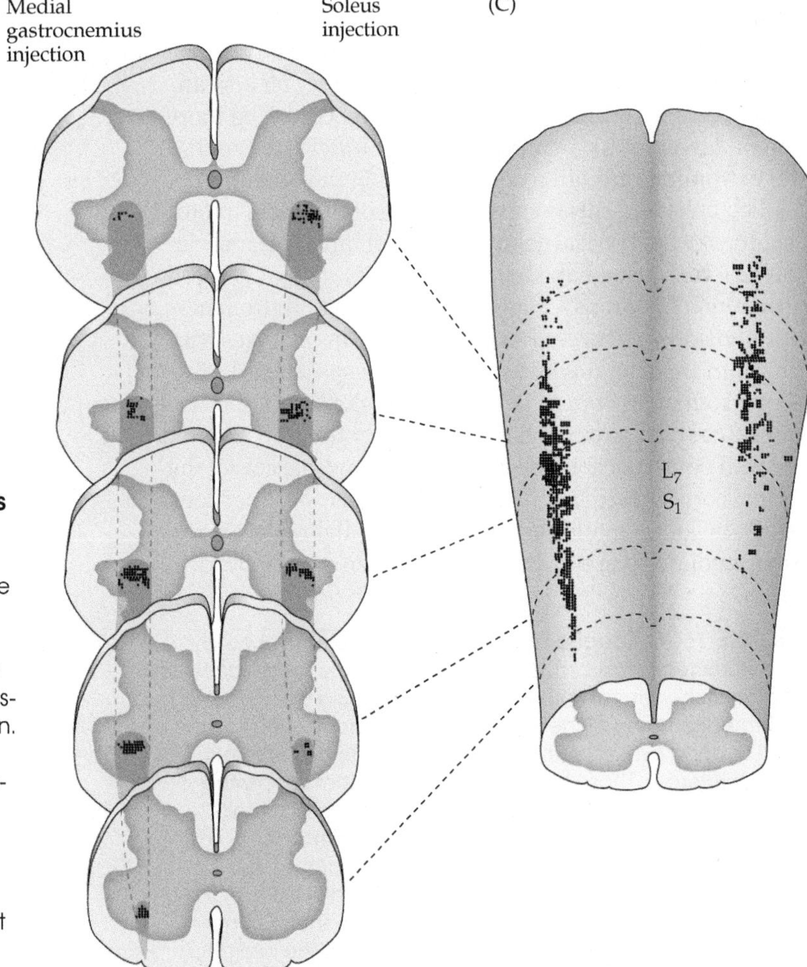

FIGURE 16.2 Distribution of lower motor neurons in the ventral horn of the spinal cord. Motor neurons were identified by injecting a retrograde tracer into either the medial gastrocnemius or soleus muscle of the cat, thus labeling neuronal cell bodies and revealing their spatial distribution. A transverse section through the lumbar level of the spinal cord (A) shows lower motor neurons forming distinct, rod-shaped clusters (motor neuron pools) in the ipsilateral ventral horn. Spinal cord cross sections (B) and a reconstruction seen from the dorsal surface (C) illustrate the distribution of motor neurons innervating individual skeletal muscles in both axes of the cord. The rodlike shape and distinct distribution of different motor neuron pools are especially evident in the dorsal view of the reconstructed cord. The dashed lines in (C) represent the locations of individual lumbar and sacral spinal cord sections shown in (B). (After Burke et al., 1977.)

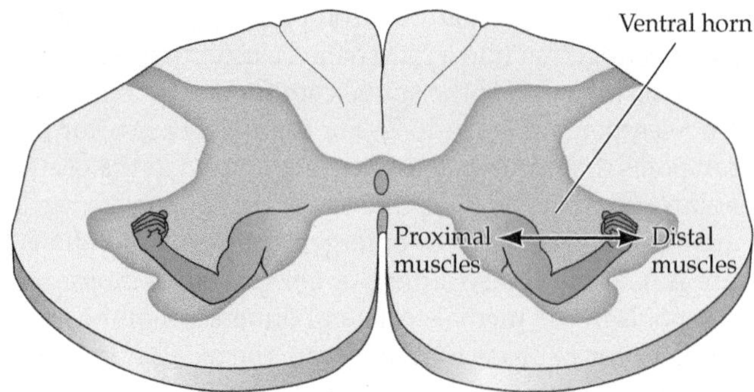

FIGURE 16.3 Somatotopic organization of lower motor neuron pools. A cross section of the ventral horn at the cervical level of the spinal cord, illustrating that the motor neurons innervating the axial musculature are located medially, whereas those innervating the distal musculature are located more laterally.

how descending projections of upper motor neurons and intersegmental spinal cord circuits control posture and modulate movement. Thus, medial lower motor neuron pools that govern postural control and the maintenance of balance receive input from upper motor neurons in the brainstem vestibular nuclei and reticular formation. They comprise long pathways that run in the medial and anterior (ventral) white matter of the spinal cord. The more lateral lower motor neuron pools that innervate the distal extremities are often concerned with the execution of skilled behavior; this is especially true of the lateral motor neurons of the cervical enlargement that innervate muscles of the forearm and hand in primates. These laterally placed lower motor neurons are governed by projections from motor divisions of the cerebral cortex that, in primates, run through the lateral white matter of the spinal cord. This same somatotopic plan is reflected in the location of local spinal cord circuits that interconnect the lower motor neuron pools distributed along the longitudinal axis of the spinal cord (Figure 16.4). Thus, the patterns of connections made by local circuit neurons in the medial region of the intermediate zone are different from those made by local circuit neurons in the lateral region, and these differences are related to their respective functions. The medial local circuit neurons, which supply the lower motor neurons in the medial ventral horn, have axons that project to many spinal cord segments. Indeed, some projections run between the cervical and lumbar enlargements and participate in the coordination of rhythmic movements of the upper and lower limbs (see the section "Spinal Cord Circuitry and Locomotion" later in this chapter), while other axons terminate along the entire length of the cord and help mediate posture. Moreover, many of these neurons have axonal branches that cross the midline in the ventral commissure of the spinal cord to innervate lower motor neurons in the medial part of the contralateral hemicord. This arrangement ensures that groups of axial muscles on both sides of the body act in concert to maintain and adjust motor activity that requires synchronous bilateral coordination of muscles, such as maintenance of posture or breathing. In contrast, local circuit neurons in the lateral region of the intermediate zone have shorter axons that typically extend fewer than five segments and are predominantly ipsilateral. This more restricted pattern of connectivity provides the finer and more differentiated control that is exerted over the muscles of the distal extremities on

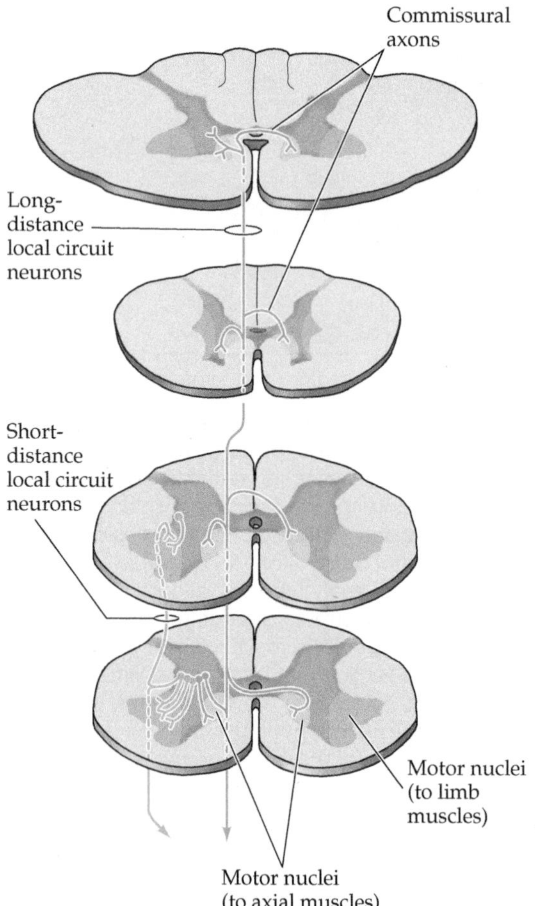

FIGURE 16.4 Local circuit neurons in the spinal cord gray matter. Local circuit neurons that supply the medial region of the ventral horn are situated medially within the intermediate zone of the spinal cord gray matter. Their axons (red) extend over several spinal cord segments and terminate bilaterally. Those local circuit neurons that supply the lateral parts of the ventral horn are located more laterally; their axons (orange) extend over just a few spinal cord segments, always terminating on the same side of the cord as the cell body. Pathways that contact the medial parts of the spinal cord gray matter are involved primarily in the control of posture and locomotion; those that contact the lateral parts are involved in the fine control of the distal extremities.

one side, such as that required for the independent movement of individual fingers during typing, picking up small objects, or playing a musical instrument.

Two types of lower motor neurons are found in the motor neuron pools of the ventral horn. Large motor neurons are called **α motor neurons**; they innervate the striated muscle fibers that actually generate the forces needed for posture and movement. Interspersed among the α-motor neurons are smaller **γ motor neurons**, which innervate specialized muscle fibers that, in combination with the nerve fibers that innervate them, are actually sensory receptors arranged in parallel with the force-generating striated muscle fibers. These specialized muscle fibers, called **muscle spindles** (see Chapter 9), are embedded within connective tissue capsules in the muscle and are thus referred to as intrafusal muscle fibers (*fusal* means capsular or spindle-shaped, in contrast to the surrounding uncapsulated striated muscle fibers, which are termed extrafusal). The intrafusal muscle fibers are innervated by sensory axons that send information to the spinal cord and brainstem about the length of the muscle. The function of the γ motor neurons is to regulate this sensory input by setting the intrafusal muscle fibers to an appropriate length (see the section "The Spinal Cord Circuitry Underlying Muscle Stretch Reflexes" later in this chapter). The output of both types of lower motor neurons is coordinated to optimize movement, particularly when the lengths of active muscles change and the forces acting on the body are dynamic.

Although the following discussion focuses on the lower motor neurons in the spinal cord, comparable sets of motor neurons responsible for the control of muscles in the head, eyes, and neck are located in the brainstem. The lower motor neurons in the brainstem are distributed in the eight somatic and branchial motor nuclei of the cranial nerves that are located in the medulla, pons, and midbrain (see Appendix Figures A9 and A11). Their activity is controlled by analogous patterns of connections with local circuit and upper motor neurons (see Chapter 17).

The Motor Unit

Most extrafusal skeletal muscle fibers in mature mammals are innervated by only a single α motor neuron (immature muscle fibers are innervated by several α motor neurons; see Chapter 23). Since there are, by far, more muscle fibers than motor neurons, individual motor axons branch within muscles to synapse on multiple extrafusal fibers. These fibers are typically distributed over a relatively wide area within the muscle, presumably to ensure that the contractile force is spread evenly (Figure 16.5). In addition, this arrangement reduces the chance that damage to one or a few α motor neurons will significantly alter a muscle's action. Because an action potential generated by a motor neuron typically brings to contraction threshold all of the muscle fibers the neuron contacts, the single α motor neuron and its associated muscle fibers constitute the smallest unit of force that can be activated by the muscle. Sherrington was again the first to recognize this fundamental relationship between an α motor neuron and the muscle fibers it innervates, for which he coined the term **motor unit**.

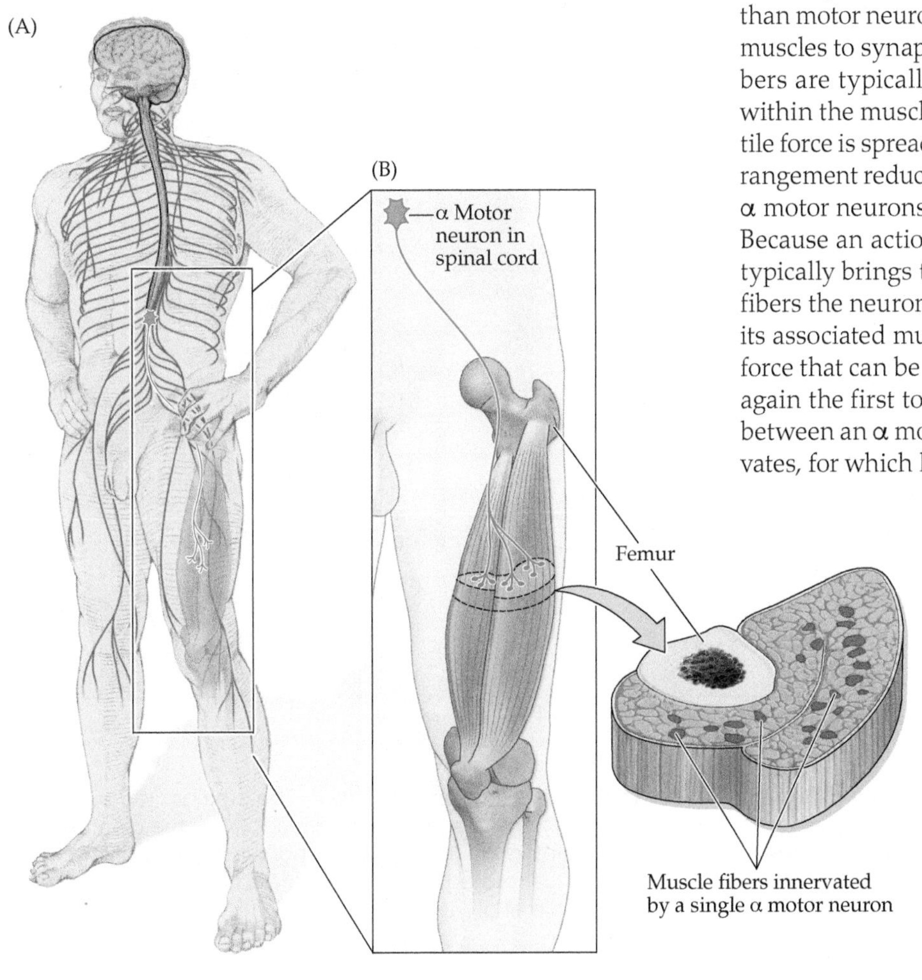

FIGURE 16.5 The motor unit. (A) Diagram showing a lower motor neuron in the spinal cord and the course of its axon to its target muscle. (B) Each α motor neuron synapses with multiple fibers in the muscle. The α motor neuron and the muscle fibers it contacts define the motor unit. The cross section through the muscle shows the relatively diffuse distribution of muscle fibers (dark red) contacted by a single α motor neuron.

Both motor units and the α motor neurons themselves vary in size. Small α motor neurons innervate relatively few muscle fibers to form motor units that generate small forces, whereas large motor neurons innervate larger, more powerful motor units. Motor units also differ in the types of muscle fibers they innervate. In most skeletal muscles, the smaller motor units comprise small "red" muscle fibers that contract slowly and generate relatively small forces; but because of their rich myoglobin content, plentiful mitochondria, and rich capillary beds, these small red fibers are resistant to fatigue. These small units are called **slow (S) motor units** and are especially important for activities that require sustained muscular contraction, such as maintaining an upright posture. Larger α motor neurons innervate larger, pale muscle fibers that generate more force; however, these fibers have sparse mitochondria and are therefore easily fatigued. These units are called **fast fatigable (FF) motor units** and are especially important for brief exertions that require large forces, such as running or jumping. A third class of motor unit has properties in between those of the other two. These **fast fatigue-resistant (FR) motor units** are of intermediate size and are not quite as fast as FF motor units. They generate about twice the force of a slow motor unit and, as the name implies, are resistant to fatigue (Figure 16.6).

These distinctions among different types of motor units explain how the nervous system produces movements appropriate for different circumstances. In most muscles, small, slow motor units have lower thresholds for activation than do the larger units and are tonically active during motor acts that require sustained effort (standing, for instance). The thresholds for the large, fast motor units are reached only during rapid movements requiring great force, such as jumping.

The functional distinctions between the various classes of motor units also explain some structural differences among muscles. For example, a motor unit in the soleus (a muscle important for posture that comprises mostly small motor units) has an average innervation ratio of 180 muscle fibers for each motor neuron. In contrast, the gastrocnemius, a muscle that comprises both small and larger motor units, has an innervation ratio of 1000 to 2000 muscle fibers per motor neuron and can generate forces needed for sudden changes in body position. Other differences are related to the highly specialized functions of particular muscles. For instance, the rotation of the eyes in the orbits requires rapid, precise movements that are generated by small forces; in consequence, extraocular muscle motor units are extremely small (with an average innervation ratio of only three fibers per unit) and have a very high proportion of muscle fibers capable of contracting with maximum velocity. More subtle motor unit variations are present in athletes on different training regimens; indeed, both the myofibril and neuronal properties of motor units

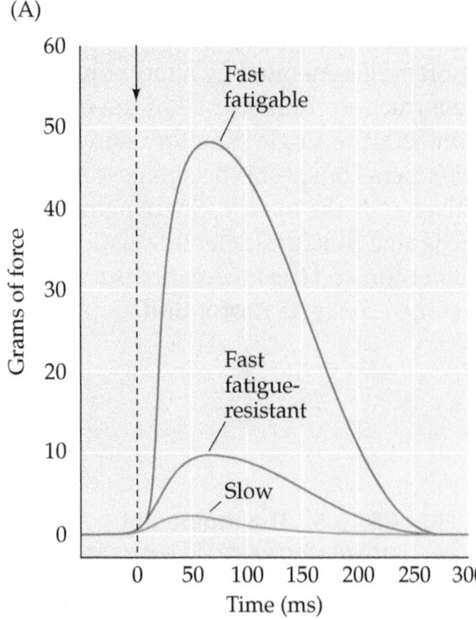

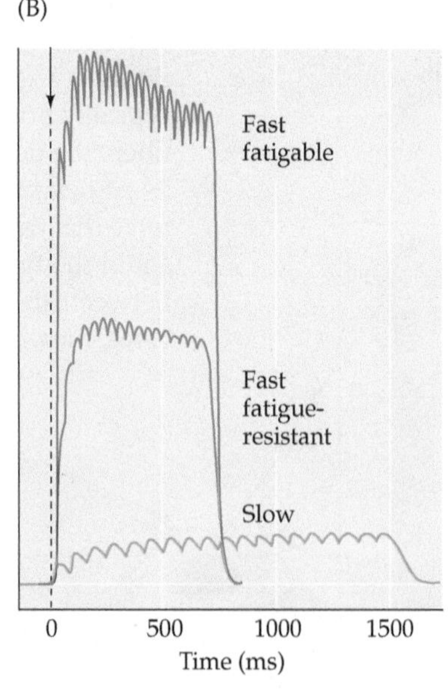

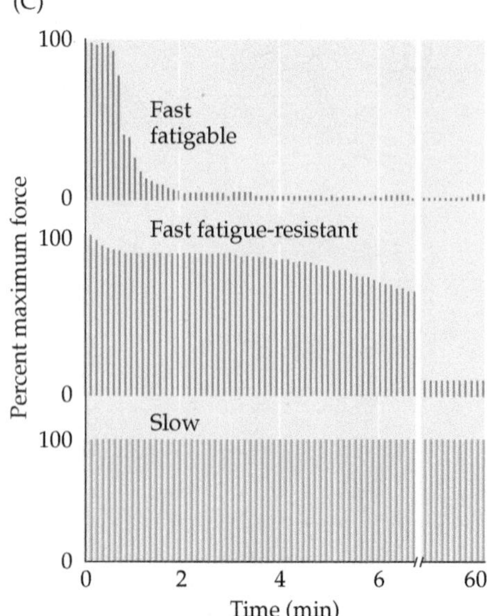

FIGURE 16.6 Force and fatigability of the three different types of motor units. In each case, the response reflects stimulation of a single α motor neuron. (A) Change in muscle tension in response to a single action potential. (B) Tension in response to repetitive stimulation of each type of motor unit. (C) Response to repeated stimulation at a level that initially evokes maximum tension. The ordinate represents the force generated by each stimulus. Note the different time scales in the three panels and the strikingly different tensions generated and fatigue rates among motor units. (After Burke et al., 1973.)

are subject to use-dependent plasticity. This potential for change in part underlies neuromuscular adaptations to physical exercise and training (Box 16A). Thus, muscle biopsies show that sprinters have a larger proportion of powerful, but rapidly fatiguing, pale fibers in their leg muscles than do marathon runners.

Regulation of Muscle Force

Increasing or decreasing the number of motor units active at any one time changes the amount of force produced by a muscle. In the 1960s, Elwood Henneman and his colleagues at Harvard Medical School found that progressive increases in muscle tension could be produced by progressively increasing the activity of axons that provide input to the relevant pool of lower motor neurons. This gradual increase in tension results from the recruitment of motor units in a fixed order, according to their size. By stimulating either sensory nerves or upper motor pathways that project to a lower motor neuron pool while measuring the tension changes in the muscle, Henneman found that, in experimental animals, only the smallest motor units in the pool are activated by weak synaptic input. When synaptic input to the motor pool increases, progressively larger motor units that generate larger forces are recruited. Thus, as the synaptic activity driving a motor neuron pool increases, low-threshold S motor units are recruited first, then FR motor units, and finally, at the highest levels of activity, the FF motor units. Since these original experiments were performed, evidence for the orderly recruitment of motor units has been found in a variety of voluntary and reflexive movements, including exercise activities. This systematic relationship has come to be known as the **size principle**.

BOX 16A ■ Motor Unit Plasticity

Organisms with complex nervous systems demonstrate an astounding ability to acquire new motor skills and modify the strength and endurance of motor function. The neural basis of these abilities depends heavily on the operations of supraspinal motor centers (i.e., neural centers above the spinal cord) whose functions in volitional motor behavior and motor learning are described in Chapters 17–19. But what role—if any—do the motor units themselves play in the functional changes that underlie such abilities? Are motor units subject to use-dependent plasticity, and if so, how are the anatomical and physiological properties of motor units changeable? To address these questions, it is necessary to consider in more detail the range of phenotypes expressed by motor units.

When considering the structure and function of skeletal muscle, it is convenient to classify the constituent motor units into one of three categories: slow (S), fast fatigable (FF), or fast fatigue-resistant (FR) (see Figure 16.6). However, with increasingly sophisticated means of characterizing the intrinsic architecture, biochemistry, and physiology of muscle fibers, it has become clear that most skeletal muscles possess a broader spectrum of fiber phenotypes that vary in speed of contraction, tension generation, oxidative capacity, and endurance. These variations among muscle fibers combine with corresponding variations in the morphological and biophysical properties of α motor neurons to determine the physiological function of motor units (Figure A). Thus, the features of α motor neurons that serve small motor units explain why such neurons are easily depolarized to firing threshold but typically maintain only slow, steady rates of firing—properties that are well suited for the control of slow muscle fibers that mediate postural stability, for example. By contrast, α motor neurons that serve large motor units are more difficult to depolarize to threshold but are capable of achieving high frequencies of firing—properties consistent with the force-generating potential of FF muscle fibers that are recruited for production of maximum tension. Not surprisingly, muscle fibers that are intermediate in their functional properties are supplied by α motor neurons whose phenotypes are midway between these extremes.

An early clue to the nature of the mechanisms underlying motor unit plasticity came from a classic series of "cross-innervation" experiments performed by the Australian Nobel laureate J. C. Eccles and his colleagues, most notably A. J. Buller. The results demonstrated that the physiological properties of slow and fast muscle fibers could be reversed when the innervation to these fibers was surgically altered so that slow muscle fibers were innervated by a nerve that typically supplies fast fibers, and vice versa. Subsequent studies by other investigators demonstrated that the actual pattern of neural activity in a motor nerve, in addition to—or perhaps

(A)

With increased motor unit size, α motor neurons exhibit:	
Increased	**Decreased**
Cell body size	Input resistance
Dendritic complexity	Excitability
Short-term EPSP potentiation with repeated activation	Ia EPSP amplitude
Axonal diameter (i.e., faster conduction)	PSP decay constant
Number of axonal branches (i.e., more muscle fibers innervated)	Duration of after-hyperpolarization

(A) Morphological and biophysical properties of α motor neurons that scale proportionally with the size of motor units

Continued on the next page

BOX 16A ■ *(continued)*

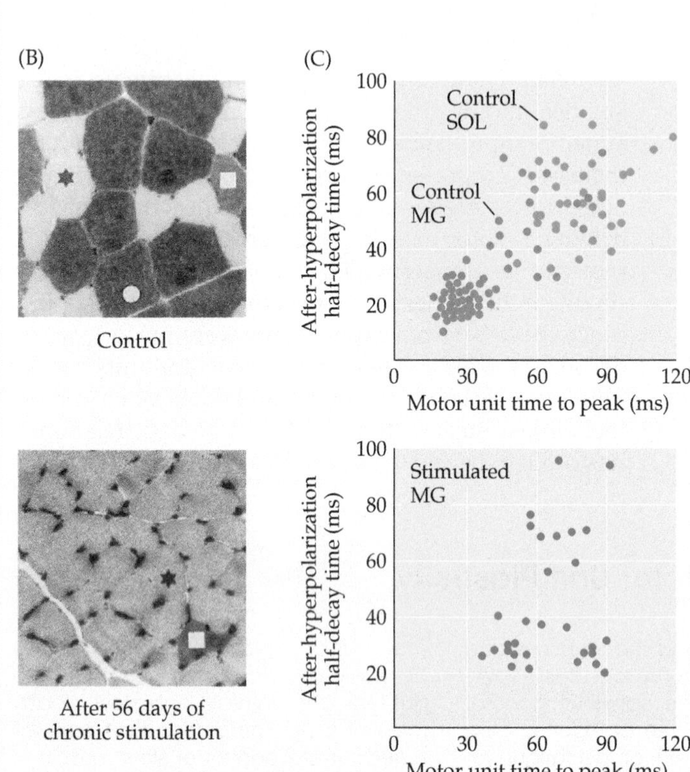

(B) Photomicrographs of muscle fibers in cat medial gastrocnemius stained to demonstrate the presence of myosin ATPase activity in alkaline conditions. In control muscle, fast fatigable fibers (circle) and fast fatigue-resistant fibers (square) stain darkly, but slow oxidative fibers (star) stain very lightly. Following 56 days of chronic electrical nerve stimulation, nearly all fibers acquired the histochemical phenotype of slow oxidative fibers. (C) The electrophysiological properties of the α motor neurons supplying the stimulated nerve also shifted toward those more characteristic of the slower motor units of the soleus muscle (SOL). The upper graph shows control data in which the faster motor units of the medial gastrocnemius (MG) are differentiated from the slower motor units of the soleus muscle by shorter neuronal after-hyperpolarizations and time-to-peak tension in the supplied muscle fibers. The lower graph shows the impact of chronic stimulation, which shifts the properties of MG motor neurons toward those seen in SOL motor neurons. (B from Gordon et al., 1997; C after Munson et al., 1997.)

instead of—the molecular identity of the innervating motor neurons themselves, provides an instructive signal that can influence the expression of muscle fiber phenotype. For example, chronic electrical nerve stimulation transforms the metabolic and contractile properties of FF fibers to those consistent with S fibers (Figures B and C). Corresponding changes were also observed in the biophysical properties of the α motor neurons whose axons were stimulated. Although the effects were more subtle, stimulated α motor neurons were modified toward slow, fatigue-resistant motor units, with increased excitability, lengthened after-hyperpolarizations, and short-term depression of excitatory postsynaptic potential (EPSP) amplitudes following high-frequency activation.

It is much more difficult to control and interpret studies of exercising organisms; nevertheless, the same general principles of motor unit plasticity that were derived from nerve stimulation studies apply to neuromuscular adaptation in more naturalistic contexts, including resistance and endurance training. Thus, the nature and degree of muscle adaptation following exercise are functions of the tensions exerted by the muscle fibers and the duration of their increased activity. Most commonly, exercise regimes can "slow" the contractile properties of motor units while increasing the endurance and strength of muscle fibers. Moreover, the impact of exercise is distributed proportionally to motor units in order of their recruitment during training activities, with S motor units being affected most at low exertion levels, and FR and FF motor units being affected only if recruited by higher intensities of exercise.

Interestingly, neural contributions to exercise-induced changes in performance are not limited to alterations of motor unit phenotype. Indeed, the increase in strength achieved in the early phases of resistance training often exceeds what can be attributable to changes in the structure and function of

An illustration of how the size principle operates for the motor units of the medial gastrocnemius muscle in the cat is shown in Figure 16.7. When the animal is standing quietly, the force measured directly from the muscle tendon is only a small fraction (about 5%) of the total force that the muscle can generate. The force is provided by the S motor units, which make up about 25% of the motor units in this muscle. When the cat begins to walk, larger forces are necessary. Locomotor activities that range from slow walking to fast running require up to 25% of the muscle's total force capacity. This additional need is met by the recruitment of FR motor units. Only movements such as galloping and jumping, which are performed infrequently and for short periods, require the full power of the muscle; such demands are met by the additional recruitment of the FF motor units. Thus, the size principle provides a simple solution to the problem of grading muscle force. The combination of motor units activated by such orderly recruitment optimally matches the physiological properties of different motor unit types with the range of forces required to perform different motor tasks.

The frequency of the action potentials generated by motor neurons also contributes to the regulation of muscle tension. The increase in force that occurs with increased

BOX 16A ■ (*continued*)

muscle fibers, implying the operation of spinal and/or supraspinal neural mechanisms that mediate increased motor function. At the motor unit level, these neural adaptations include an increase in the instantaneous discharge rate, a reduction in discharge rate variability, and a marked decrease in interspike interval at the onset of contraction, all of which facilitate the rapid generation of tension (Figures D and E). Furthermore, studies of unilateral exercise (e.g., training one arm and not the other) have shown appreciable gains in the non-exercised limb, indicating the recruitment and adaptation of central neural circuits that have access to contralateral motor units. There are even documented gains in muscle strength with *imagined* exercise—a provocative result that may eventually have profound implications for athletic training and rehabilitation science.

There is still much to be learned about how motor units respond to alterations in strength and endurance training, and scientists are only beginning to probe the neurobiological and neuromuscular mechanisms that underlie skill acquisition. Pursuit of these aims will surely lead to a better understanding of how to maximize motor performance in exercising human (and non-human) individuals, as well as in the rehabilitation of patients coping with neuromuscular or other physical impairment.

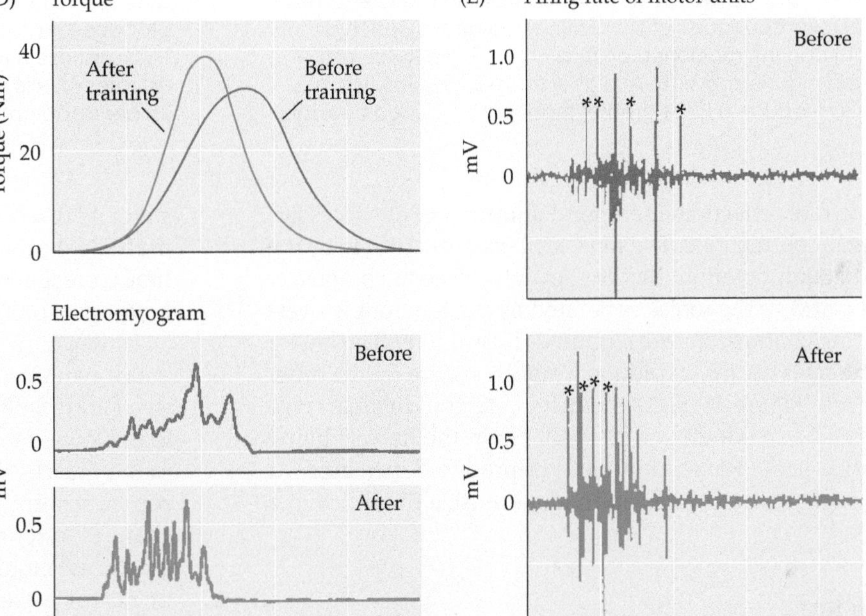

(D) Comparison of torque and electromyogram (EMG) activity during ballistic contractions of ankle dorsiflexor muscles in humans before and after dynamic training. Note the increased rate of tension development after training and the accompanying increase in rectified, surface EMG activity in the early phase of contraction. (E) These changes are associated with an increase in the instantaneous firing rate of motor units recorded from intramuscular electrodes; asterisks mark repetitive discharges of the same motor unit. (D,E after Van Cutsem et al., 1998.)

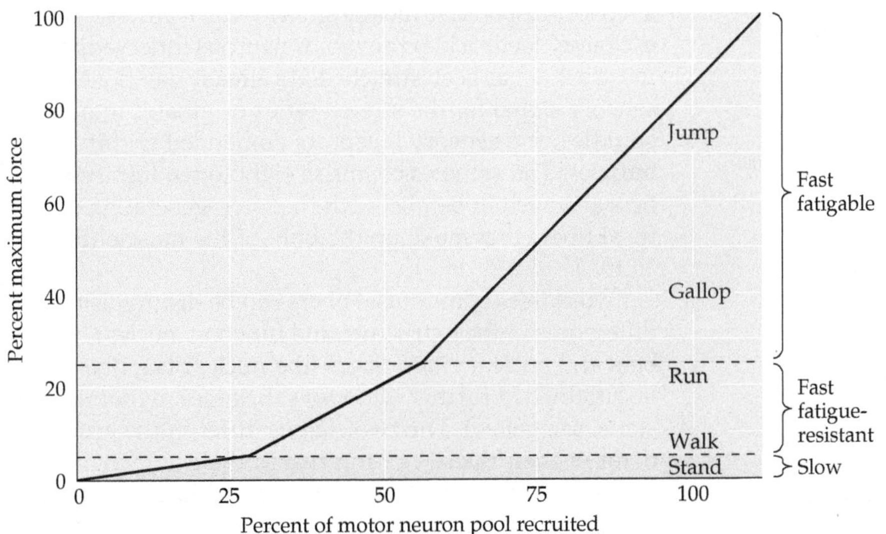

FIGURE 16.7 Motor neuron recruitment in the cat medial gastrocnemius muscle under different behavioral conditions. Slow (S) motor units provide the tension required for standing. Fast fatigue-resistant (FR) motor units provide the additional force needed for walking and running. Fast fatigable (FF) motor units are recruited for the most strenuous activities, such as jumping. (After Walmsley et al., 1978.)

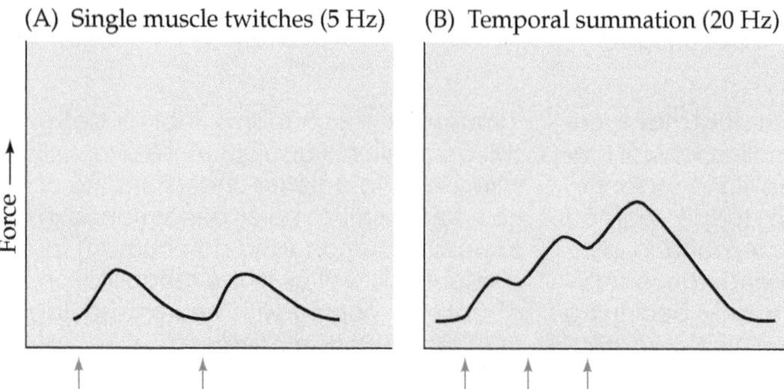

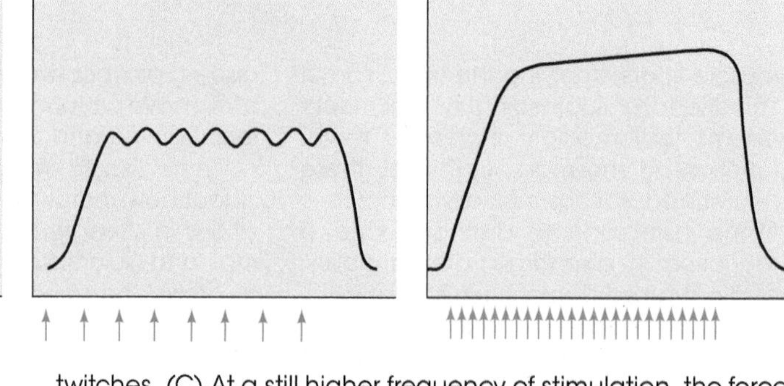

FIGURE 16.8 Effect of stimulation rate on muscle tension. (A) At low frequencies of stimulation (red arrows), each action potential in the motor neuron results in a single twitch of the related muscle fibers. (B) At higher frequencies, the twitches sum to produce a force greater than that produced by single twitches. (C) At a still higher frequency of stimulation, the force produced is greater, but individual twitches are still apparent. This response is referred to as unfused tetanus. (D) At the highest rates of motor neuron activation, individual twitches are no longer apparent—a condition called fused tetanus.

firing rate reflects the temporal summation of successive muscle contractions. The muscle fibers are activated by the next action potential before they have time to completely relax, and so the forces generated by the temporally overlapping contractions are summed (Figure 16.8). The lowest firing rates during a voluntary movement are on the order of 8 Hz (Figure 16.9). As the firing rate of individual units rises (to a maximum of 20 to 25 Hz in the muscle being studied here), the amount of force produced increases. At the highest firing rates, individual muscle fibers are in a state of "fused tetanus"—that is, the tension produced in individual motor units no longer has peaks and troughs that correspond to the individual twitches evoked by the motor neuron's action potentials.

Under normal conditions, the maximum firing rate of motor neurons is less than that required for fused tetanus (see Figure 16.8). However, the asynchronous firing of different lower motor neurons provides a steady level of input to the muscle, which causes the contraction of a relatively constant number of motor units, and averages out the changes in tension due to contractions and relaxations of individual motor units. All this allows the resulting movements to be executed smoothly.

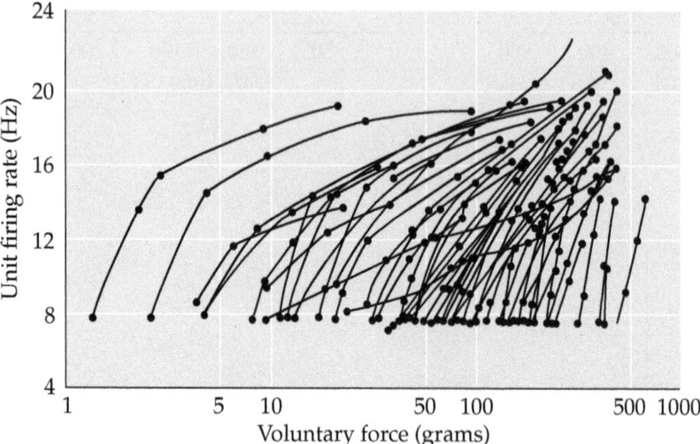

FIGURE 16.9 The number of active motor units and their rate of firing both increase with voluntary force. Motor unit firing in a muscle of the human hand (each unit is represented here by a single trace) was recorded transcutaneously as the amount of voluntary force produced by the individual progressively increased. The lowest-threshold motor units generate the least amount of voluntary force and are recruited first. As the individual generates more and more force, both the number and the rate of firing of active motor units increase. (Note that all motor units fire initially at about 8 Hz.) (After Monster and Chan, 1977.)

The Spinal Cord Circuitry Underlying Muscle Stretch Reflexes

Local circuitry within the spinal cord mediates several sensorimotor reflexes. The simplest of these reflex arcs entails a sensory response to muscle stretch, which provides direct excitatory feedback to the motor neurons innervating the muscle that has been stretched. As already mentioned, the sensory signal for the stretch reflex originates in muscle spindles, the sensory receptors embedded within most muscles. The spindles comprise eight to ten intrafusal fibers arranged in parallel with the force-generating extrafusal fibers that make up the bulk of the muscle (Figure 16.10A).

Two classes of intrafusal fibers can be distinguished by differences in their structure and function: nuclear bag fibers and nuclear chain fibers (the nuclear bag fibers can be subdivided further into two subclasses, dynamic and static; see below). The two classes differ in the arrangement of their nuclei (giving rise to their nomenclature, bag and chain fibers), the intrinsic architecture of their myofibrils, and their dynamic sensitivity to stretch. Most

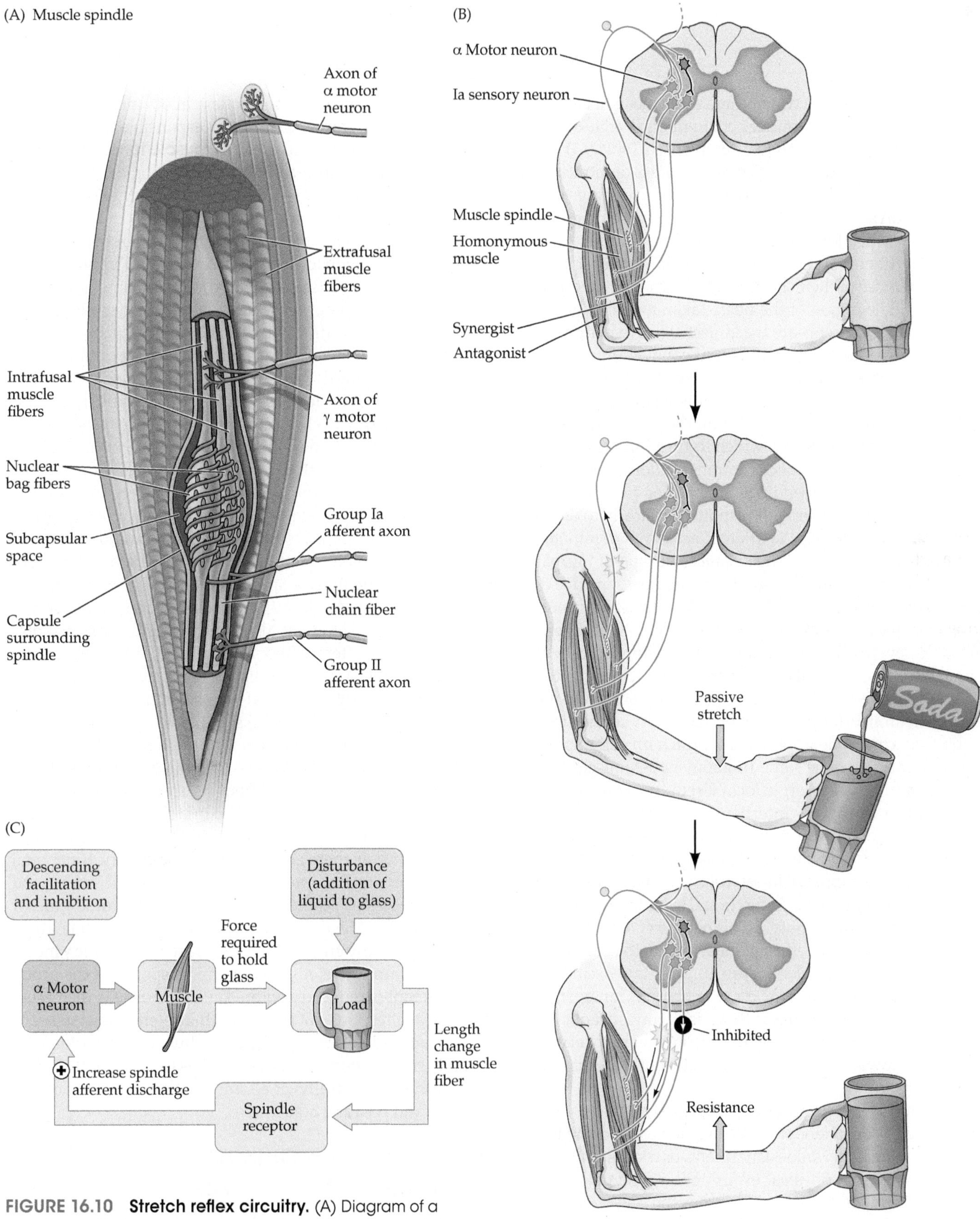

FIGURE 16.10 Stretch reflex circuitry. (A) Diagram of a muscle spindle, the sensory receptor that initiates the stretch reflex. (B) Stretching a muscle spindle leads to increased activity in group Ia afferents and an increase in the activity of α motor neurons that innervate the same muscle. Group Ia afferents also excite the motor neurons that innervate synergistic muscles, and they indirectly inhibit the motor neurons that innervate antagonists via intervening reciprocal-Ia-inhibitory interneurons (gray neurons; see also Figures 1.7–1.9). (C) The stretch reflex operates as a negative feedback loop to regulate muscle length.

muscle spindles contain two or three nuclear bag fibers and at least twice that many nuclear chain fibers. Large-diameter sensory axons (group Ia afferents; see Table 9.1) are coiled around the middle region of each class of intrafusal fiber, forming so-called annulospiral primary endings (see Figure 16.10A). Nearly as large in diameter are the group II afferents, which form secondary endings, mainly on nuclear chain fibers; these are referred to as "flower-spray" endings because of their short, petal-like contacts just outside the middle region of the fiber. Taken together, group Ia and group II afferents are the largest axons in peripheral nerves, and because action potential conduction velocity is a direct function of axon diameter (see Chapter 3), they mediate very rapid reflex adjustments when the muscle is stretched. The stretch imposed on the muscle deforms the intrafusal muscle fibers, which in turn initiates action potentials by activating mechanotransduction channels in the group I and II axon endings innervating the spindle.

Group Ia afferents tend to respond phasically to small stretches. This is because Ia afferent activity is dominated by signals transduced by the *dynamic* subtype of nuclear bag fiber whose biomechanical properties are sensitive to the *velocity* of fiber stretch. Group II afferents, which innervate *static* nuclear bag fibers and the nuclear chain fibers, signal the level of *sustained* fiber stretch by firing tonically at a frequency proportional to the degree of stretch, with little dynamic sensitivity. The centrally projecting branch of the sensory neuron forms monosynaptic excitatory connections with those α motor neurons in the ventral horn of the spinal cord that innervate the same (homonymous) muscle and, via intervening GABAergic local circuit neurons (called reciprocal-Ia-inhibitory interneurons), forms inhibitory connections with those α motor neurons that innervate antagonistic (heteronymous) muscles. This arrangement is an example of **reciprocal innervation** and results in rapid contraction of the stretched muscle and simultaneous relaxation of the antagonist muscle. This pattern of activity leads to especially rapid and efficient adjustments to changes in the length of the muscle (Figure 16.10B). The excitatory pathway from a spindle to the α motor neurons innervating the same muscle is unusual in that it is a monosynaptic reflex; in most cases, sensory neurons from the periphery do not contact lower motor neurons directly but instead exert their effects through local circuit neurons.

This monosynaptic reflex arc is variously referred to as the "stretch," "deep tendon," or "myotatic" reflex, and it is the basis of the knee, ankle, jaw, biceps, or triceps response tested in a routine physical examination. The tap of the reflex hammer on the tendon stretches the muscle, which evokes an afferent volley of activity in the Ia sensory axons that innervate the muscle spindles. The afferent volley is relayed to the α motor neurons in the brainstem or spinal cord, which then deliver an efferent volley to the same muscle (see Figure 1.7). Since muscles are always under some degree of stretch, this reflex circuit, mediated largely by group II afferents, is typically responsible for the steady level of tension in muscles called **muscle tone**. Changes in muscle tone occur in a variety of pathological conditions, and these changes are assessed by examination of deep tendon reflexes (see Box 17D).

In terms of engineering principles, the stretch reflex arc is a negative feedback loop used to maintain muscle length at a desired value (Figure 16.10C). In the context of motor control, the appropriate muscle length is specified by the activity of descending upper motor neuron pathways that influence the lower motor neuron pool. Deviations from the desired length are detected by the muscle spindles, since increases or decreases in the stretch of the intrafusal fibers alter the level of activity in the sensory axons that innervate the spindles. These changes lead, in turn, to adjustments in the activity of the α motor neurons, returning the muscle to the desired length by contracting the stretched muscle and relaxing the opposing muscle group, and by restoring the level of spindle activity and sensitivity.

Smaller γ motor neurons control the functional characteristics of the muscle spindles by modulating their level of excitability. As described earlier, when the muscle is stretched, the spindle is also stretched and the rate of discharge in the afferent fibers is increased. When the muscle shortens, the spindle is relieved of tension ("unloaded"), and the sensory axons that innervate the spindle might therefore be expected to fall silent, but in fact they remain active. The γ motor neurons terminate on the contractile poles of the intrafusal fibers, and the activation of these neurons causes intrafusal fiber contraction—in this way, maintaining the tension on the middle, or equatorial region, of the intrafusal fibers where the sensory axons terminate. Just as there are the dynamic and static functional classes of intrafusal muscle fibers, there are dynamic and static classes of γ motor neurons. When dynamic γ motor neurons fire, the dynamic response of the group Ia afferent is markedly enhanced. In contrast, when static γ motor neurons are activated, the dynamic response of the group Ia afferent is reduced and the static response is increased; the static response of the group II afferent likewise is enhanced under these conditions. Thus, co-activation of α and γ motor neurons allows spindles to function (i.e., send information centrally) at all muscle lengths during movement and postural adjustment.

Modifying the Gain of Muscle Stretch Reflexes

The level of γ motor neuron activity is often referred to as γ *bias*, or *gain*, and can be adjusted by upper motor neuron pathways as well as by local reflex circuitry. The gain of the myotatic reflex refers to the amount of muscle force generated in response to a given stretch of the intrafusal fibers. If the gain of the reflex is high, then a small amount of stretch applied to the intrafusal fibers will produce a large increase in the number of α motor neurons recruited and a large increase in their firing rates; this, in turn, will lead to a large increase in the amount of tension produced by the extrafusal fibers. If the gain is low, a greater stretch will be required to generate the same amount of tension in the extrafusal muscle fibers. In fact, the gain of the stretch reflex is continuously adjusted to meet different functional requirements. For example, while standing in a moving bus, the gain of the stretch reflex can be modulated by upper motor neuron pathways to compensate for the variable changes that occur as the bus stops and starts or progresses relatively smoothly. During voluntary stretching, such as warming up for athletic performance, the gain of myotatic reflexes must be reduced to facilitate the lengthening of muscle fibers and other elastic elements of the musculotendinous system that is desirable under these circumstances. Thus, under the various demands of voluntary (and involuntary) movement, α and γ motor neurons are often co-activated by higher centers to prevent muscle spindles from being unloaded or overactivated (Figure 16.11).

In addition, the level of γ motor neuron activity can be modulated independently of α motor neuron activity to allow fine adjustments in movements. In general, the baseline activity level of γ motor neurons is high if a movement is relatively difficult and demands rapid and precise execution. For example, recordings from cat hindlimb muscles show that γ motor neuron activity is high when the animal has to perform a difficult movement, such as walking across a narrow beam. Unpredictable conditions, as when the animal is picked up or handled, also lead to marked increases in γ motor neuron activity and greatly increased spindle responsiveness.

However, γ motor neuron activity is not the only factor that sets the gain of the stretch reflex. The gain also depends on the level of excitability of the α motor neurons that serve as the efferent side of this reflex loop. Thus, in addition to the influence of descending upper motor neuron projections, local circuits in the spinal cord can change the gain of the stretch reflex by excitation or inhibition of either α or γ motor neurons. In addition, there are inhibitory interneurons that form axo-axonal synapses on the terminals of Ia afferents and are thus positioned to selectively suppress the transfer of excitatory drive to specific subpopulations of lower motor neurons. The activities of local circuits in the spinal cord are themselves influenced by the projections of upper motor neurons in the brainstem and cerebral cortex, as well as by neuromodulatory

(A) α Motor neuron activation without γ

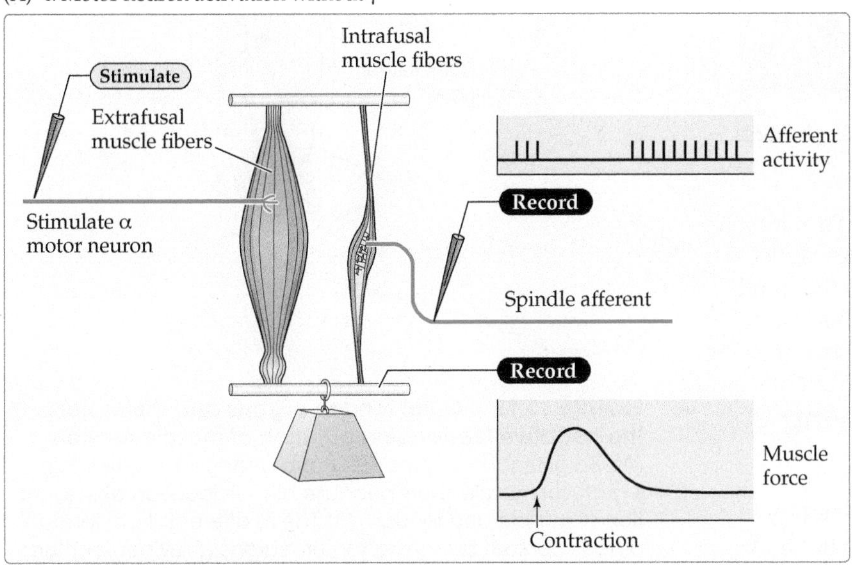

(B) α Motor neuron activation with γ

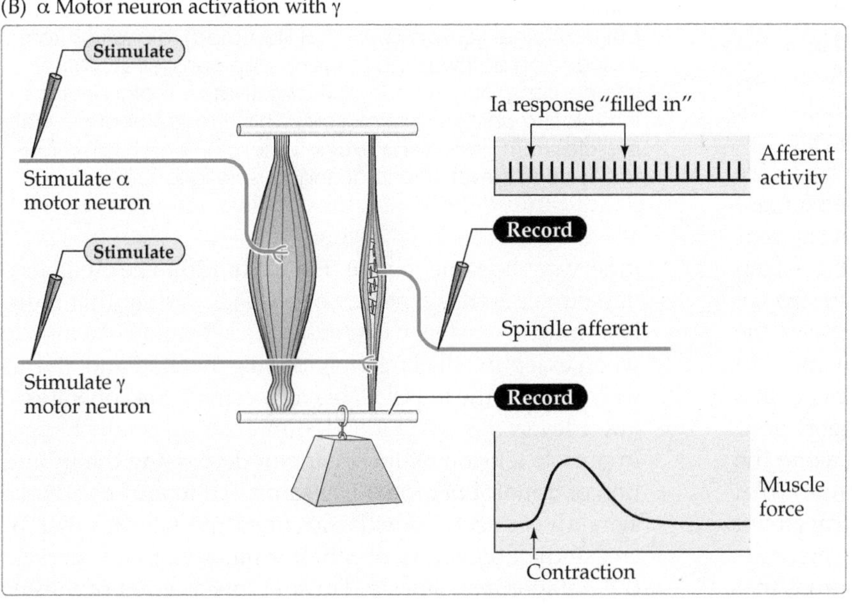

FIGURE 16.11 The role of γ motor neurons in regulating muscle spindle responses. (A) When α motor neurons are stimulated without activation of γ motor neurons, the response of the Ia fiber decreases as the muscle contracts. (B) When both α and γ motor neurons are activated, there is no decrease in Ia firing during muscle shortening. Thus, the γ motor neurons can regulate the gain of muscle spindles so they can operate efficiently at any length of the parent muscle. (After Hunt and Kuffler, 1951.)

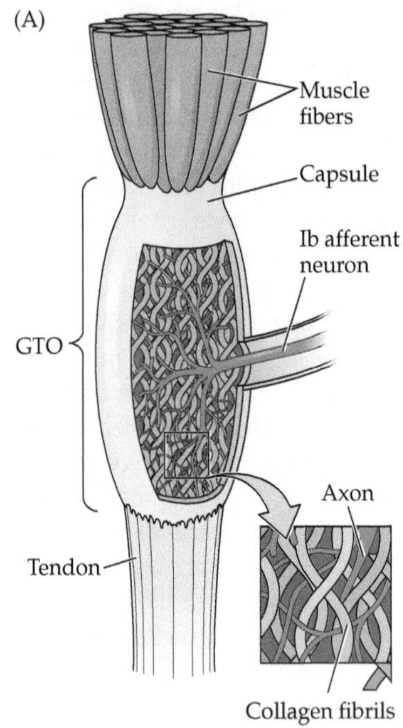

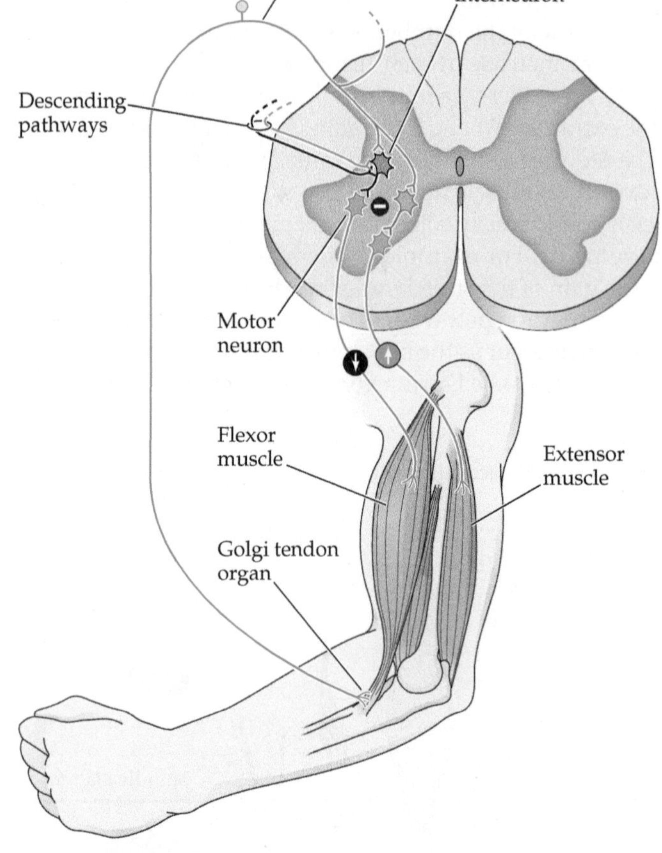

FIGURE 16.12 **Golgi tendon organs and their role in the negative feedback regulation of muscle tension.** (A) Golgi tendon organs (GTO) are arranged in series with extrafusal muscle fibers because of their location at the junction of muscle and tendon. (B) The Ib afferents from tendon organs contact Ib inhibitory interneurons (gray neuron) that decrease the activity of α motor neurons innervating the same muscle. The Ib inhibitory interneurons also receive input from other sensory fibers (not illustrated), as well as from descending pathways. Ib afferents also contact excitatory interneurons (purple neuron) that activate α motor neurons innervating antagonistic muscles. This arrangement prevents muscles from generating excessive tension and helps maintain a steady level of tone during muscle fatigue.

systems that originate in the brainstem reticular formation (see Chapter 17). Many of these neuromodulatory projections release biogenic amine neurotransmitters that bind to G-protein-coupled receptors and mediate long-lasting effects on the gain of segmental circuits in the spinal cord.

The Spinal Cord Circuitry Underlying the Regulation of Muscle Force

Another sensory receptor that is important in the reflexive regulation of motor unit activity is the Golgi tendon organ. Golgi tendon organs are encapsulated afferent nerve endings located at the junction of a muscle and a tendon (Figure 16.12A). Each tendon organ is innervated by a single group Ib sensory axon (Ib axons are slightly smaller than the Ia axons that innervate the muscle spindles; see Table 9.1). In contrast to the parallel arrangement of extrafusal muscle fibers and spindles, Golgi tendon organs are in series with the extrafusal muscle fibers. When a muscle actively contracts, the force acts directly on the tendon, leading to an increase in the tension of the collagen fibrils in the tendon organ and consequent compression of the intertwined sensory nerve endings. Activation of the non-selective, cationic mechanosensitive ion channels in the nerve endings of the Golgi tendon organ results in a generator potential that, if suprathreshold, triggers generation of action potentials that are propagated along the group Ib axon to the spinal cord. The Ib axons from Golgi tendon organs contact GABAergic inhibitory local circuit neurons in the spinal cord (called Ib inhibitory interneurons) that synapse, in turn, with the α motor neurons that innervate the same muscle (Figure 16.12B). The Golgi tendon circuit is thus a negative feedback system that regulates muscle tension; it decreases the activation of a muscle when exceptionally large forces are generated and, in this way, protects the muscle. This reflex circuit also operates at lower levels of muscle force, counteracting small changes in muscle tension by increasing or decreasing the inhibition of α motor neurons. The same Ib afferents also make synaptic connections with excitatory interneurons that increase the excitability of α motor neurons that innervate the antagonistic muscle. Thus, at lower levels of muscle force, the Golgi tendon system tends to maintain a steady level of tension and a stable joint angle, counteracting effects that diminish muscle force (such as fatigue).

Like the muscle spindle system, the Golgi tendon organ system is subject to a variety of influences. The Ib inhibitory interneurons receive synaptic inputs from several other sources, including upper motor neurons, cutaneous receptors, muscle spindles, and joint receptors. The joint receptors comprise several types of receptors resembling Ruffini's and Pacinian corpuscles that are located in joint capsules (see Chapter 9). Joint receptors signal hyperextension or hyperflexion of the joint, thereby contributing to the protective functions mediated by Ib inhibitory interneurons when the risk of injury is markedly increased. Acting in concert, these diverse inputs regulate the responsiveness of Ib interneurons to activity arising in Golgi tendon organs.

Complementary Functions of Muscle Spindles and Golgi Tendon Organs

From the preceding discussion, it should be evident that muscle spindles and Golgi tendon organs serve in a complementary fashion to help regulate motor performance through the operations of distinct spinal cord reflexes. Consider the circumstance of passively stretching a muscle. With passive stretch, most of the change in length occurs in the muscle fibers, since they are more elastic than the fibrils of the tendon. Thus, activity increases in spindle afferents with stretch while there is little change in the firing rate of Golgi tendon organ afferents (Figure 16.13A). Now, consider active muscle contraction. The generated force is transmitted to the tendon and transduced by the Golgi tendon organ, leading to an increase in the firing

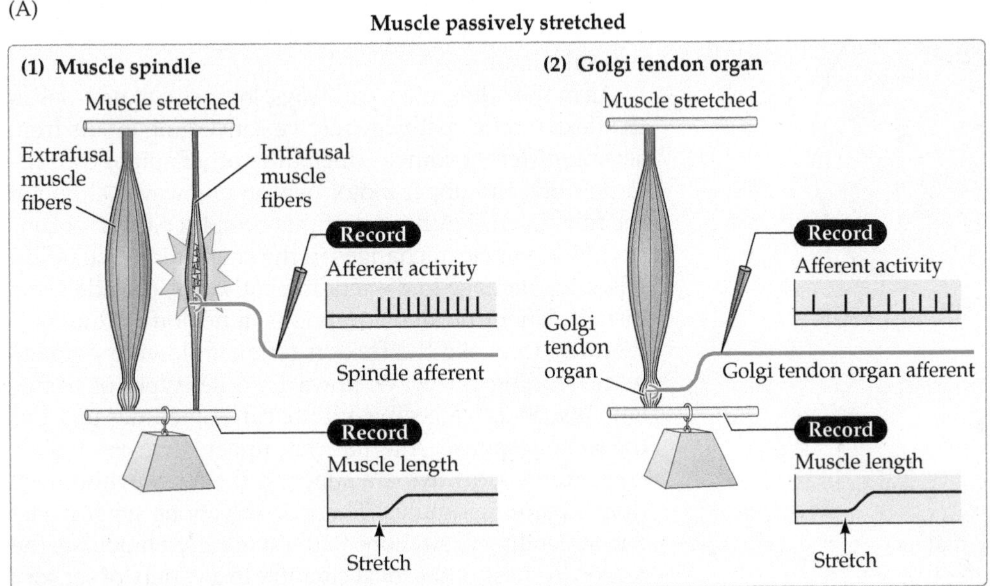

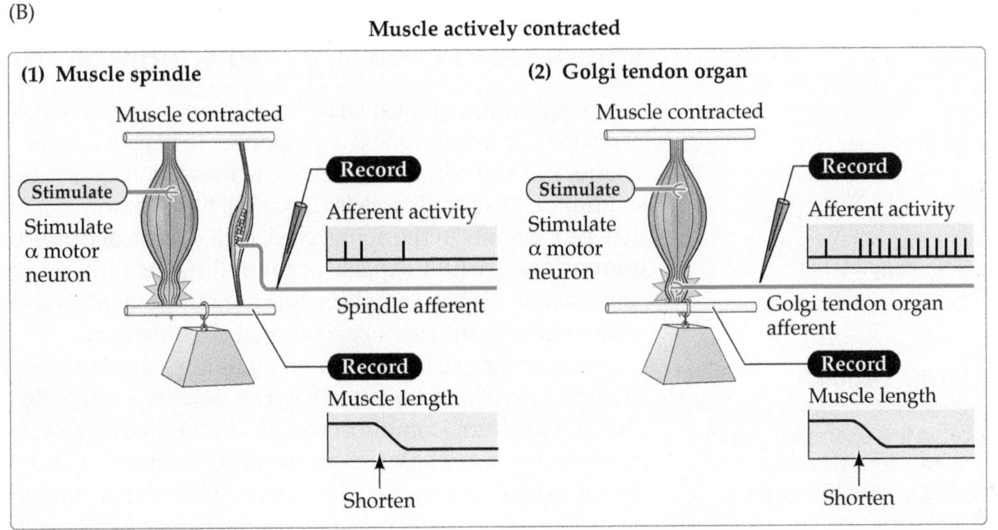

FIGURE 16.13 **Muscle spindles and Golgi tendon organs.** The two types of muscle receptors, the muscle spindles (1) and the Golgi tendon organs (2), have different responses to passive muscle stretch (A) and active muscle contraction (B). Both afferents discharge in response to passively stretching the muscle, although the Golgi tendon organ discharge is much less than that of the spindle. When the extrafusal muscle fibers are made to contract by stimulation of their α motor neurons, however, the spindle is unloaded and its activity decreases, whereas the rate of Golgi tendon organ firing increases. (After Patton, 1965.)

rate of Ib afferents (Figure 16.13B). Thus, Golgi tendon organs are exquisitely sensitive to increases in muscle *tension* that arise from muscle contraction but, unlike spindles, are relatively insensitive to *passive stretch*. During active contraction, if it were not for a compensatory increase in the output of the relevant γ motor neurons, the muscle spindle would be unloaded and the activity of the associated Ia afferents would decrease (see Figure 16.13B). In short, the muscle spindle system is a feedback system that monitors and maintains muscle *length*, and the Golgi tendon system is a feedback system that monitors and maintains muscle *force*.

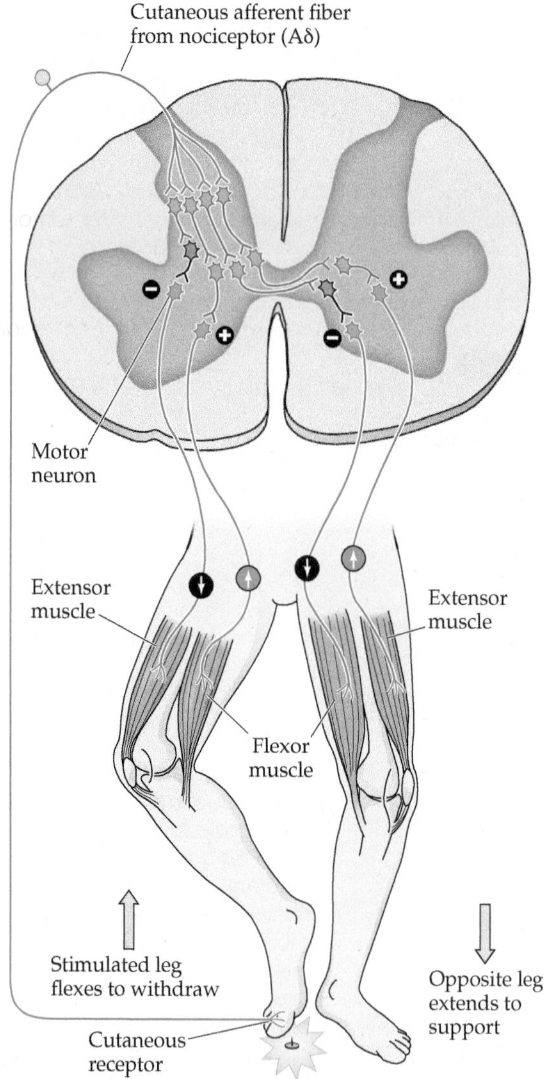

FIGURE 16.14 Spinal cord circuitry for the flexion-crossed extension reflex. Stimulation of cutaneous receptors in the foot (by stepping on a tack, in this example) leads to activation of spinal cord local circuits that serve to withdraw (flex) the stimulated extremity and extend the other extremity to provide compensatory support.

Flexion Reflex Pathways

So far, this discussion has focused on reflexes driven by sensory receptors located within muscles or tendons. Other reflex circuitry mediates the withdrawal of a limb from a painful stimulus, such as a pinprick or the heat of a flame. Contrary to what might be imagined given the speed with which we are able to withdraw from a painful stimulus, this flexion reflex involves slowly conducting afferent axons and several synaptic links (Figure 16.14). As a result of activity in this circuitry, stimulation of nociceptive sensory fibers leads to withdrawal of the limb from the source of pain by excitation of ipsilateral flexor muscles and reciprocal inhibition of ipsilateral extensor muscles. Flexion of the stimulated limb is also accompanied by an opposite reaction in the contralateral limb (i.e., the contralateral extensor muscles are excited while flexor muscles are inhibited). This crossed extension reflex provides postural support during withdrawal of the affected limb from the painful stimulus.

As in the other reflex pathways, local circuit neurons in the flexion reflex pathway receive converging inputs from several different sources, including other spinal cord interneurons and upper motor neuron pathways. Although the functional significance of this complex pattern of connectivity is unclear, changes in the character of the reflex, following damage to descending pathways, provide some insight. Under normal conditions, a noxious stimulus is required to evoke the flexion reflex; following damage to descending pathways, however, other types of stimulation, such as squeezing a limb, can sometimes produce the same response. Alternatively, under some conditions, descending pathways can suppress the reflex withdrawal from a painful stimulus. These observations suggest that the descending projections to the spinal cord modulate the responsiveness of the local circuitry to a variety of sensory inputs.

Spinal Cord Circuitry and Locomotion

The contribution of local circuitry to motor control is not, of course, limited to reflexive responses to sensory inputs. Studies of rhythmic movements, such as locomotion and swimming in animal models (Box 16B), have demonstrated that local circuits in the spinal cord, called **central pattern generators**, are fully capable of controlling the timing and coordination of such complex patterns of movement, and of adjusting them in response to altered circumstances.

A good example is locomotion (walking, running, etc.). In quadrupeds and bipeds alike, the movement of a single limb during locomotion can be thought of as a cycle consisting of two phases: a *stance phase*, during which the limb is extended and placed in contact with the ground to propel the animal forward; and a *swing phase*, during which

BOX 16B ■ Locomotion in the Leech and the Lamprey

All animals must coordinate body movements so they can navigate successfully in their environment. All vertebrates, including mammals, use local circuits in the spinal cord (central pattern generators) to control the coordinated movements associated with locomotion. The cellular basis of organized locomotor activity, however, is best understood in an invertebrate, the leech, and in a simple vertebrate, the lamprey.

Both the leech and the lamprey lack the peripheral appendages for locomotion possessed by many vertebrates (limbs, wings, flippers, fins, or their equivalent). Furthermore, their bodies comprise repeating muscle segments (as well as repeating skeletal elements in the lamprey). Thus, in order to move through the water, both animals must coordinate the movements generated by each segment. They do this by orchestrating a sinusoidal displacement of each body segment in sequence, so that the animal is propelled forward through the water.

The leech is particularly well suited for studying the circuitry responsible for coordinated movement. The nervous system in the leech consists of a series of interconnected segmental ganglia, each with motor neurons that innervate the corresponding segmental muscles (Figure A). These segmental ganglia facilitate electrophysiological studies of the circuitry because there is a limited number of neurons in each ganglion and each neuron has a distinct identity. Specific neurons can thus be recognized and studied from animal to animal and their electrical activity correlated with the sinusoidal swimming movements.

A central pattern generator circuit coordinates this undulating motion. In the leech, the circuit is an ensemble of sensory neurons, interneurons, and motor neurons repeated in each segmental ganglion that controls the local sequence of contraction and relaxation in each segment of the body wall musculature (Figure B). The sensory neurons detect the stretching and contraction of the body wall associated with the sequential swimming movements. Dorsal and ventral motor neurons in the circuit provide innervation to dorsal and ventral longitudinal muscles, whose phasic contractions propel the leech forward. The sensory and motor neuron signals are coordinated by interneurons that fire rhythmically, setting up phasic patterns of activity in the dorsal and ventral cells that lead to sinusoidal movement. The intrinsic swimming rhythm is established by a variety of membrane conductances that mediate periodic bursts of suprathreshold action potentials generated by depolarization, which are followed by well-defined periods of hyperpolarization.

The lamprey, one of the simplest vertebrates, is distinguished by its clearly segmented musculature and by its lack of bilateral fins or other appendages. In order to move through the water, the lamprey contracts and relaxes each muscle segment in sequence (Figure C), which produces a sinusoidal motion, much like that of the leech. Again, a central pattern generator coordinates this sinusoidal movement.

Unlike the leech with its segmental ganglia, the lamprey has a continuous spinal cord that gives rise to nerves that connect each spinal level with the adjacent muscle segments. The lamprey spinal cord is simpler than that of other vertebrates, and several classes of identified neurons occupy stereotyped positions. This orderly arrangement again facilitates the identification and analysis of the neurons that constitute the central pattern generator circuit.

Continued on the next page

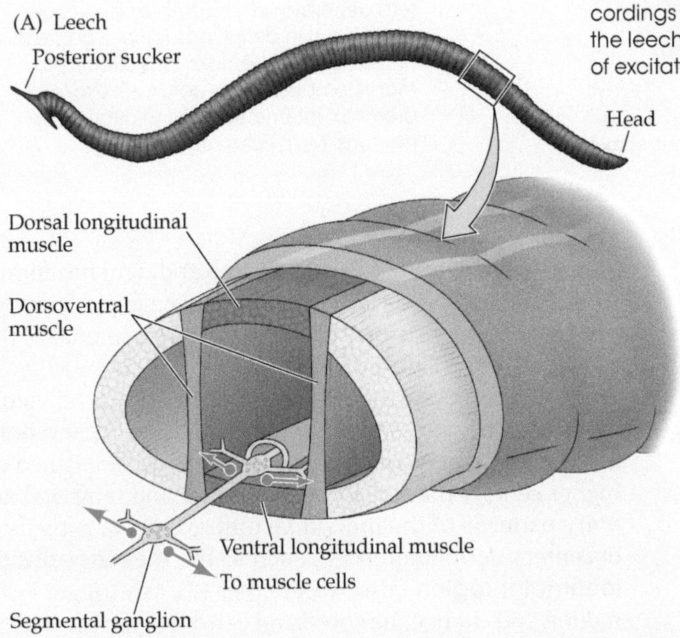

(A) Leech
Posterior sucker
Head
Dorsal longitudinal muscle
Dorsoventral muscle
Ventral longitudinal muscle
To muscle cells
Segmental ganglion

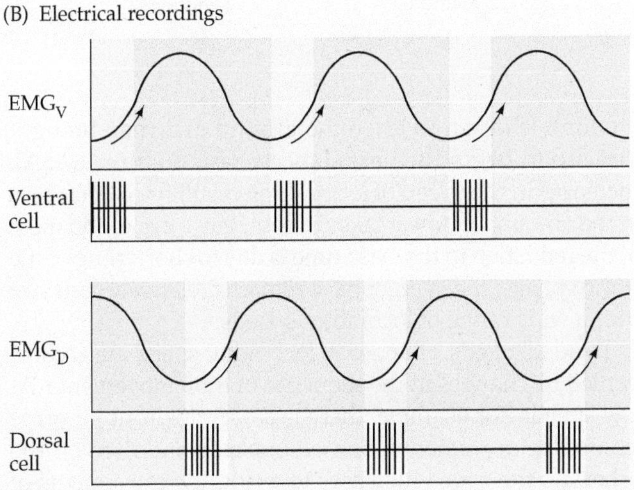

(B) Electrical recordings
EMG_V
Ventral cell
EMG_D
Dorsal cell

(A) The leech propels itself through the water by sequential contraction and relaxation of the body wall musculature of each segment. The segmental ganglia in the ventral midline coordinate swimming, with each ganglion containing a population of identified neurons. (B) Electromyographic recordings from the ventral (EMG_V) and dorsal (EMG_D) longitudinal muscles in the leech and the corresponding motor neurons show a reciprocal pattern of excitation for the dorsal and ventral muscles of a given segment.

BOX 16B ■ (continued)

In the lamprey spinal cord, the intrinsic firing pattern of a set of interconnected sensory neurons, interneurons, and motor neurons establishes the pattern of undulating muscle contractions that underlie swimming (Figure D). The patterns of connectivity between neurons, the neurotransmitters used by each class of cell, and the physiological properties of the elements in the lamprey pattern generator are now known. One set of interneurons—known as excitatory premotor interneurons—release glutamate as their neurotransmitter and thereby excite one another, as well as nearby inhibitory interneurons. A local pool of these excitatory premotor interneurons generates a burst of activity for segmental motor output. One class of inhibitory interneurons makes reciprocal connections across the midline that coordinate the pattern-generating circuitry on each side of the spinal cord.

This circuitry in the lamprey provides a basis for understanding similar circuits that control locomotion in more complex vertebrates. Thus, investigations of pattern-generating circuits for locomotion in relatively simple animals have guided studies that have identified similar central pattern generators in the spinal cords of terrestrial mammals. Although different in detail, terrestrial locomotion ultimately relies on sequential movements similar to those that propel the leech and the lamprey through aquatic environments. Simple aquatic organisms and more complex terrestrial vertebrates likely share many key features that facilitate central pattern generation, including the intrinsic physiological properties of spinal cord neurons that establish rhythmicity and the modulation by descending monoaminergic input from the brainstem and hypothalamus that can alter rhythmic patterns and cycle frequency.

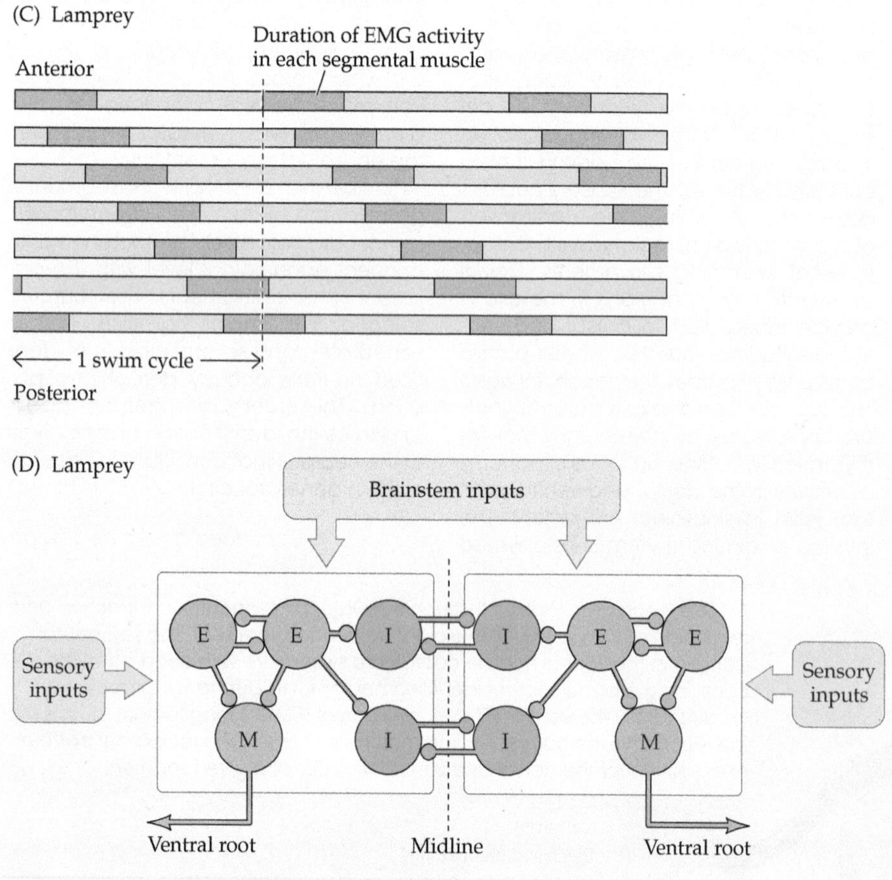

(C) In the lamprey, the pattern of activity across segments is also highly coordinated. (D) The elements of the central pattern generator in the lamprey have been worked out in detail, providing a guide to understanding homologous circuitry in more complex spinal cords. (E: excitatory premotor interneurons; I: inhibitory interneurons; M: lower motor neurons).

the limb is flexed to leave the ground and then brought forward to begin the next stance phase (Figure 16.15A). Increases in the speed of locomotion result from decreases in the amount of time it takes to complete a cycle, and most of the reduction in the cycle time is due to shortening of the stance phase; the swing phase remains relatively constant over a wide range of locomotor speeds.

In quadrupeds, changes in locomotor speed are accompanied by changes in the sequence of limb movements. At low speeds, for example, there is a back-to-front progression of leg movements, first on one side and then on the other. As the speed increases to a trot, the movements of the right forelimb and left hindlimb are synchronized (as are the movements of the left forelimb and right hindlimb). At the highest speeds (a gallop), the movements of the two front legs are synchronized, as are the movements of the two hindlimbs (Figure 16.15B).

Given the precise timing of the movements of individual limbs and the necessity of coordinating these movements, it is natural to assume that locomotion is accomplished by higher centers that organize the spatial and temporal activity patterns of the individual limbs. Indeed, activation of centers in the brainstem, such as the **mesencephalic locomotor region** (also see Chapter 17), can trigger locomotion and change the speed and pattern of the movement by changing the level of activity delivered to the spinal cord.

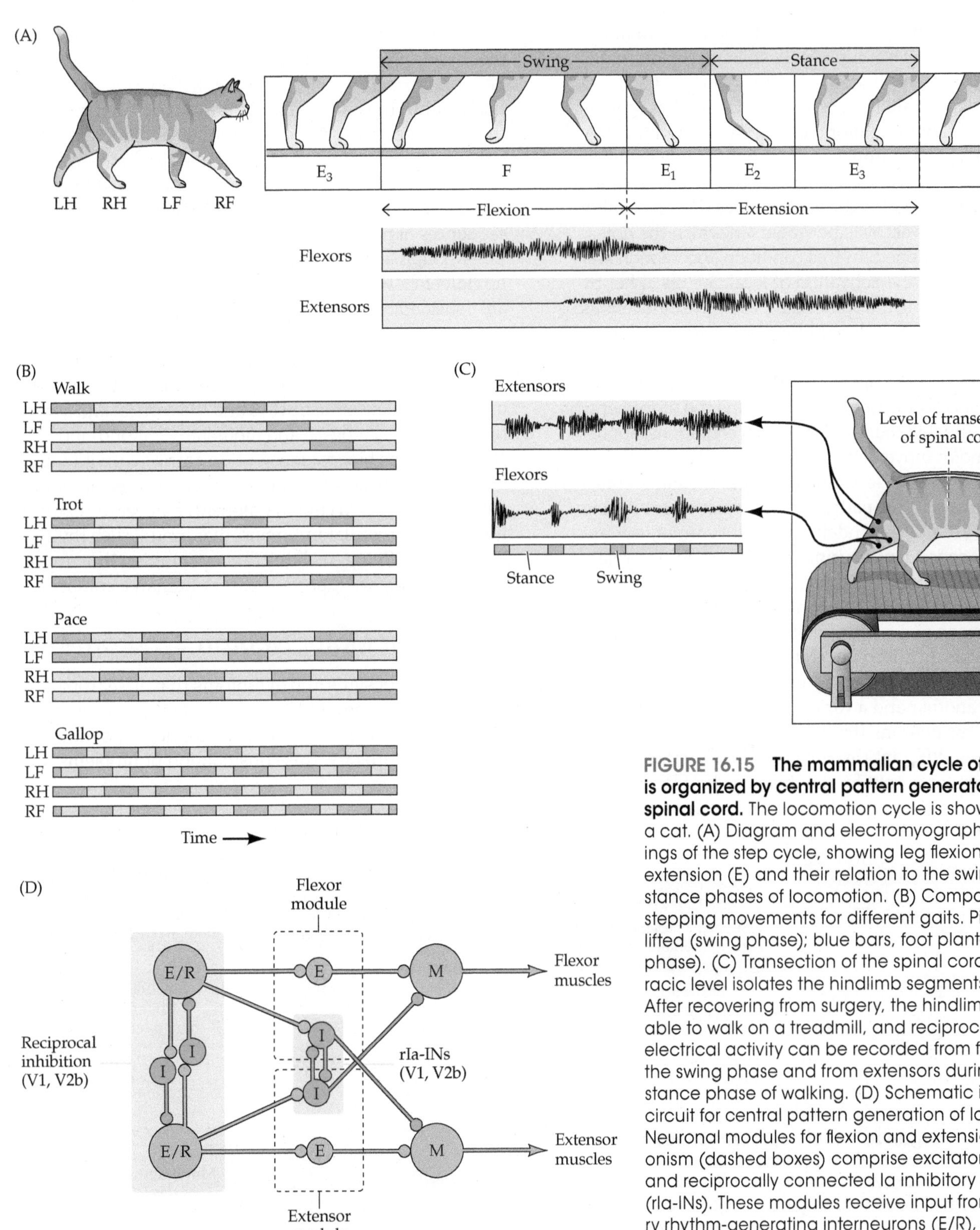

FIGURE 16.15 The mammalian cycle of locomotion is organized by central pattern generators in the spinal cord. The locomotion cycle is shown here for a cat. (A) Diagram and electromyographic recordings of the step cycle, showing leg flexion (F) and extension (E) and their relation to the swing and stance phases of locomotion. (B) Comparison of the stepping movements for different gaits. Pink bars, foot lifted (swing phase); blue bars, foot planted (stance phase). (C) Transection of the spinal cord at the thoracic level isolates the hindlimb segments of the cord. After recovering from surgery, the hindlimbs are still able to walk on a treadmill, and reciprocal bursts of electrical activity can be recorded from flexors during the swing phase and from extensors during the stance phase of walking. (D) Schematic illustrating a circuit for central pattern generation of locomotion. Neuronal modules for flexion and extension antagonism (dashed boxes) comprise excitatory neurons and reciprocally connected Ia inhibitory interneurons (rIa-INs). These modules receive input from excitatory rhythm-generating interneurons (E/R), which are reciprocally inhibited by interneurons belonging to the V1 and V2b classes of spinal cord interneurons (expressing distinct transcription factors and derived from distinct embryonic lineages); rIa-INs also belong to the V1 and V2b neuronal classes. (A–C after Pearson, 1976; D after Kiehn, 2016.)

However, following transection of the spinal cord at the thoracic level, a cat's hindlimbs will still make coordinated locomotor movements if the animal is supported and placed on a moving treadmill (Figure 16.15C). Under these conditions, the speed of locomotor movements is determined by the speed of the treadmill, suggesting that the movement is nothing more than a reflexive response to the sensory input initiated by stretching the limb muscles. This possibility is ruled out, however, by experiments in which the dorsal roots are also sectioned. In this condition, locomotion still can be induced by the activation of local circuits either by the act of transecting the spinal cord, or by the intravenous injection of L-DOPA (a dopamine precursor), which may serve to release neurotransmitter from the axon terminals of the now-transected upper motor neuron pathways. Although the speed of walking is slowed and the movements are less coordinated than under normal conditions, appropriate locomotor movements are still observed.

These and other observations in experimental animals show that the rhythmic patterns of limb movement during locomotion are not dependent on sensory input, nor are they wholly dependent on input from descending projections from higher centers. Rather, local circuitry provides for each limb a central pattern generator responsible for the alternating flexion and extension of the limb during locomotion. This central pattern generator comprises local circuit neurons that include excitatory glutamateric neurons coupled to one another and a variety of inhibitory GABAergic and glycinergic neurons (Figure 16.15D). The mechanisms for generating different rhythms that characterize central pattern generator output for varying speeds of locomotion are not yet well understood. However, current evidence indicates that rhythmogenesis depends on both intrinsic membrane properties of excitatory local circuit neurons and network properties reflecting the distribution of connections within the circuit. The central pattern generators for the limbs are coupled to each other by additional modular circuits that coordinate left–right and forelimb–hindlimb activities (such as the long-distance local circuit neurons illustrated in Figure 16.4) in order to achieve the different sequences of movements that occur at different speeds.

Although locomotor movements can also be elicited in humans following damage to descending pathways, these are considerably less effective than the movements seen in the cat. The reduced ability of the transected spinal cord to mediate rhythmic stepping movements in humans presumably reflects an increased dependence of local circuitry on upper motor neuron pathways and the cortical and subcortical circuits that govern and modulate their output (Figure 16.16). Perhaps bipedal locomotion carries with it requirements for postural control greater than can be accommodated by spinal cord circuitry alone. Whatever the explanation, the basic oscillatory circuits that control such rhythmic behaviors as flying, walking, and swimming in many animals also play an important part in human locomotion.

The Lower Motor Neuron Syndrome

The complex of signs and symptoms that arise from damage to the lower motor neurons of the brainstem and spinal cord is referred to as the "lower motor neuron syndrome." In clinical neurology, this constellation of problems must

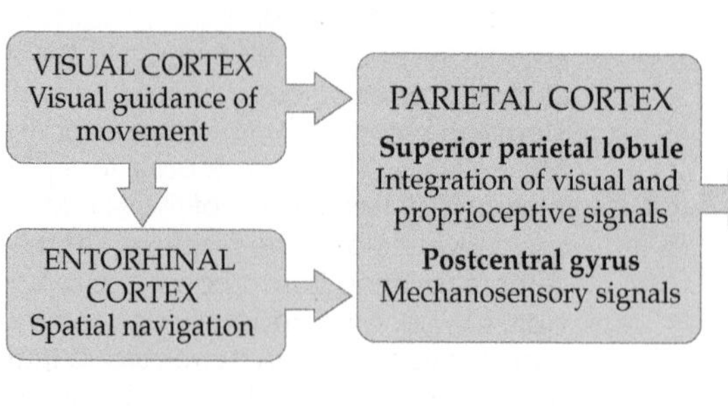

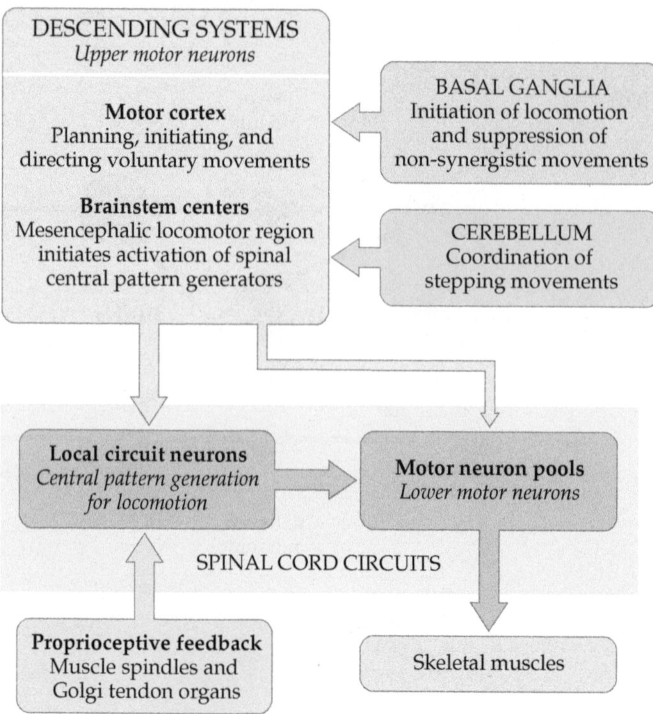

FIGURE 16.16 Organization of neural structures involved in the control of locomotion. Schematic representation of how the four motor subsystems illustrated in Figure 16.1 activate and govern the activities of spinal circuits for central pattern generation. Once activity is initiated, cortical systems play a relatively minor role in sustaining central pattern generation. Cortical control is most relevant for conveying motor intention (e.g., spatial navigation; walking or running) and visual guidance of locomotion through complex environments (e.g., stepping over or avoiding obstacles). (After Drew and Marigold, 2015; Kiehn, 2016.)

be distinguished from the "upper motor neuron syndrome" that results from damage to the descending upper motor neuron pathways (see Table 17.1).

Damage to lower motor neuron cell bodies, or their peripheral axons, results in paralysis (loss of movement) or paresis (weakness) of the affected muscles, depending on the extent of the damage. In addition to paralysis and/or paresis, the lower motor neuron syndrome includes a loss of reflexes (areflexia) due to interruption of the efferent (motor) limb of the sensorimotor reflex arcs. Damage to lower motor neurons also entails a loss of muscle tone, since tone is dependent in part on the monosynaptic reflex arc that links the muscle spindles to the lower motor neurons (see Box 17D). The muscles involved may also exhibit fibrillations and fasciculations, which are spontaneous twitches characteristic of single denervated muscle fibers or motor units, respectively. These phenomena arise from changes in the excitability of single denervated muscle fibers in the case of fibrillation, and from pathological activity of injured α motor neuron units in the case of fasciculations. These spontaneous contractions can be readily recognized in an electromyogram, providing an especially helpful clinical tool in diagnosing lower motor neuron disorders. A somewhat later effect of lower motor neuron damage is atrophy of the affected muscles due to long-term denervation and disuse.

Summary

Four distinct but highly interactive motor subsystems make essential contributions to motor control: local circuits in the spinal cord and brainstem, descending upper motor neuron pathways that control these circuits, and the basal ganglia and cerebellum, which modulate the activity of upper motor neuronal circuits. Alpha motor neurons in the spinal cord and in the cranial nerve nuclei of the brainstem directly link the nervous system and muscles, with each motor neuron and its associated muscle fibers constituting a functional entity called a motor unit. Motor units vary in size, amount of tension produced, speed of contraction, and degree of fatigability. Graded increases in muscle tension are mediated by both the orderly recruitment of different types of motor units and an increase in lower motor neuron firing frequency.

Local circuitry involving sensory inputs, local circuit neurons, and α and γ motor neurons is especially important in the reflexive control of muscle activity. The stretch reflex is a monosynaptic circuit with connections between sensory fibers arising from muscle spindles and the α motor neurons that innervate the same, or synergistic, muscles. The γ motor neurons regulate the gain of the stretch reflex by adjusting the level of tension in the intrafusal muscle fibers of the muscle spindle. This mechanism sets the baseline level of activity in α motor neurons and helps regulate muscle length and tone. Other reflex circuits provide feedback control of muscle tension and mediate essential functions, such as the rapid withdrawal of limbs from painful stimuli.

Much of the spatial organization and timing of muscle activation required for complex rhythmic movements such as locomotion is provided by specialized local circuits called central pattern generators. Because of the essential role of lower motor neurons in all of these circuits, damage to them results in paralysis or paresis of the associated muscle and other changes, including loss of reflex activity, loss of muscle tone, and eventual muscle atrophy.

ADDITIONAL READING

Reviews

Burke, R. E. (1981) Motor units: Anatomy, physiology and functional organization. In *Handbook of Physiology*, V. B. Brooks (ed.). Section 1: *The Nervous System*, vol. 1, part 1. Bethesda, MD: American Physiological Society, pp. 345–422.

Grillner, S. and P. Wallen (1985) Central pattern generators for locomotion, with special reference to vertebrates. *Annu. Rev. Neurosci.* 8: 233–261.

Henneman, E. (1990) Comments on the logical basis of muscle control. In *The Segmental Motor System*, M. C. Binder and L. M. Mendell (eds.). New York: Oxford University Press, pp. 7–10.

Henneman, E. and L. M. Mendell (1981) Functional organization of the motoneuron pool and its inputs. In *Handbook of Physiology*, V. B. Brooks (ed.). Section 1: *The Nervous System*, vol. 1, part 1. Bethesda, MD: American Physiological Society, pp. 423–507.

Kiehn, O. (2016) Decoding the organization of spinal circuits that control locomotion. *Nat. Rev. Neurosci.* 17: 224–238.

Lundberg, A. (1975) Control of spinal mechanisms from the brain. In *The Nervous System*, vol. 1: *The Basic Neurosciences*, D. B. Tower (ed.). New York: Raven Press, pp. 253–265.

Nistri, A., K. Ostoumov, E. Sharifullina and G. Taccola (2006) Tuning and playing a motor rhythm: How metabotropic glutamate receptors orchestrate generation of motor patterns in the mammalian central nervous system. *J. Physiol.* 572: 323–334.

Patton, H. D. (1965) Reflex regulation of movement and posture. In *Physiology and Biophysics*, 19th Ed., T. C. Rugh and H. D. Patton (eds.). Philadelphia: Saunders, pp. 181–206.

Prochazka, A., M. Hulliger, P. Trend and N. Durmuller (1988) Dynamic and static fusimotor set in various behavioral contexts. In *Mechanoreceptors: Development, Structure, and Function*, P. Hnik, T. Soulup, R. Vejsada and J. Zelena (eds.). New York: Plenum, pp. 417–430.

Important Original Papers

Barker, D. (1948) The innervation of the muscle-spindle. *Q. J. Microsc. Sci.* 89: 143–186.

Burke, R. E., D. N. Levine, P. Tsairis and F. E., III, Zajac (1973) Physiological types and histochemical profiles in motor units of the cat gastrocnemius. *J. Physiol.* 234: 723–748.

Burke, R. E., D. N. Levine, M. Salcman and P. Tsaires (1974) Motor units in cat soleus muscle: Physiological, histochemical, and morphological characteristics. *J. Physiol.* 238: 503–514.

Burke, R. E., P. L. Strick, K. Kanda, C. C. Kim and B. Walmsley (1977) Anatomy of medial gastrocnemius and soleus motor nuclei in cat spinal cord. *J. Neurophysiol.* 40: 667–680.

Drew, T. and D. S. Marigold (2015) Taking the next step: cortical contributions to the control of locomotion. *Curr. Opin. Neurobiol.* 33: 25–33.

Goetz, L. and 5 others (2016) On the role of the pedunculopontine nucleus and mesencephalic reticular formation in locomotion in nonhuman primates. *J. Neurosci.* 36: 4917–4929.

Henneman, E., E. Somjen, and D. O. Carpenter (1965) Excitability and inhibitability of motoneurons of different sizes. *J. Neurophysiol.* 28: 599–620.

Hunt, C. C. and S. W. Kuffler (1951) Stretch receptor discharges during muscle contraction. *J. Physiol.* 113: 298–315.

Liddell, E. G. T. and C. S. Sherrington (1925) Recruitment and some other factors of reflex inhibition. *Proc. R. Soc. London* 97: 488–518.

Lloyd, D. P. C. (1946) Integrative pattern of excitation and inhibition in two-neuron reflex arcs. *J. Neurophysiol.* 9: 439–444.

Monster, A. W. and H. Chan (1977) Isometric force production by motor units of extensor digitorum communis muscle in man. *J. Neurophysiol.* 40: 1432–1443.

Walmsley, B., J. A. Hodgson and R. E. Burke (1978) Forces produced by medial gastrocnemius and soleus muscles during locomotion in freely moving cats. *J. Neurophysiol.* 41: 1203–1215.

Books

Lieber, R. L. (2002) *Skeletal Muscle Structure, Function, and Plasticity,* 2nd Edition. Baltimore, MD: Lippincott Williams & Wilkins.

Sherrington, C. (1947) *The Integrative Action of the Nervous System,* 2nd Edition. New Haven, CT: Yale University Press.

CHAPTER

17 Upper Motor Neuron Control of the Brainstem and Spinal Cord

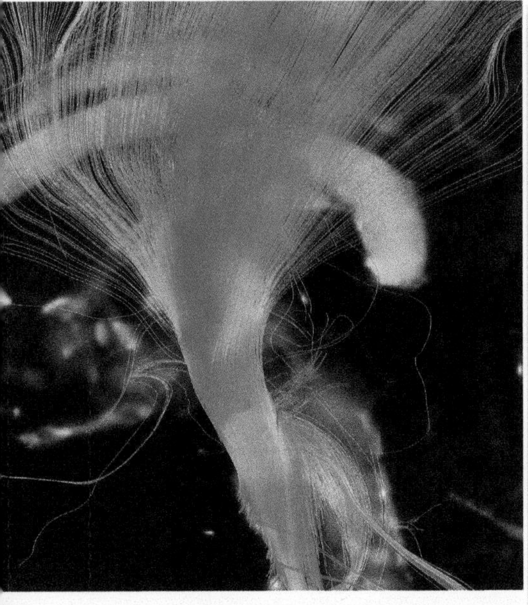

Overview

THE AXONS OF UPPER MOTOR NEURONS arise from cell bodies in higher centers and descend to influence the local circuits in the brainstem and spinal cord. These local circuits organize movements by coordinating the activity of the lower motor neurons that innervate different muscles. The sources of these upper motor neuron pathways include several brainstem centers and multiple cortical areas in the frontal lobe. The motor control centers in the brainstem are especially important in postural control, orientation toward sensory stimuli, locomotion, and orofacial behavior, with each center having a distinct influence. The mesencephalic locomotor region initiates locomotion. Two other centers, the vestibular nuclear complex and the reticular formation, make widespread contributions to the maintenance of body posture and position. The reticular formation also contributes to a variety of somatic and visceral motor circuits that govern the expression of autonomic and stereotyped somatic motor behavior. Also in the brainstem, the superior colliculus contains upper motor neurons that initiate orienting movements of the head and eyes. The primary motor cortex and a mosaic of "premotor" areas in the posterior frontal lobe, in contrast, are responsible for the planning, initiation, and control of complex sequences of voluntary movements, as well as mediating the somatic expression of emotional states. Most upper motor neurons, regardless of their source, influence the generation of movements by modulating the activity of the local circuits in the brainstem and spinal cord. Upper motor neurons in the cortex also control movement indirectly, via pathways that project to motor control centers in the brainstem, which in turn project to the local organizing circuits in the brainstem and spinal cord. These indirect pathways mediate the automatic adjustments in the body's posture that occur during cortically initiated voluntary movements.

Organization of Descending Motor Control

The spatial arrangement of the lower motor neurons and local circuit neurons within the spinal cord—the ultimate targets of the upper motor neurons—provides insight into the functions of different sets of upper motor neurons. As described in Chapter 16, lower motor neurons in the ventral horn of the spinal cord are distributed in a somatotopic fashion: The most medial part of the ventral horn contains lower motor neuron pools that innervate axial muscles or proximal muscles of the limbs, whereas the more lateral parts contain lower motor neurons that innervate the distal muscles of the limbs (Figure 17.1). The local circuit neurons, which lie primarily in the intermediate zone of the spinal cord and supply much of the direct input to the lower motor neurons, are also topographically arranged. Thus, the medial region of the intermediate zone of the spinal cord gray matter contains the local circuit

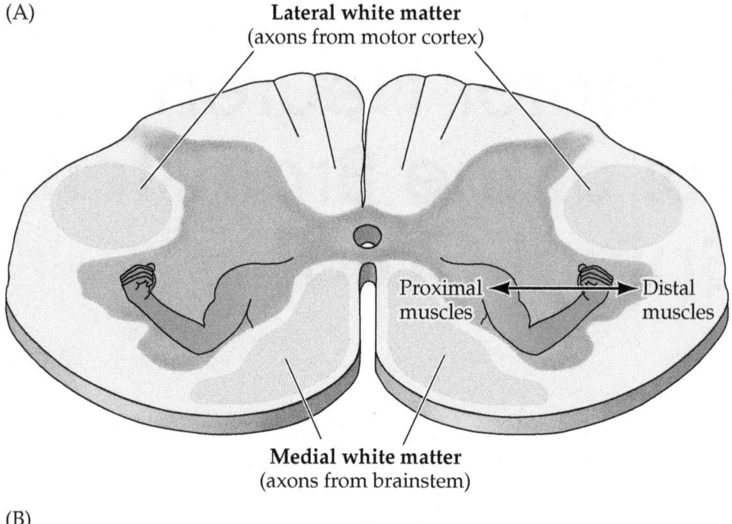

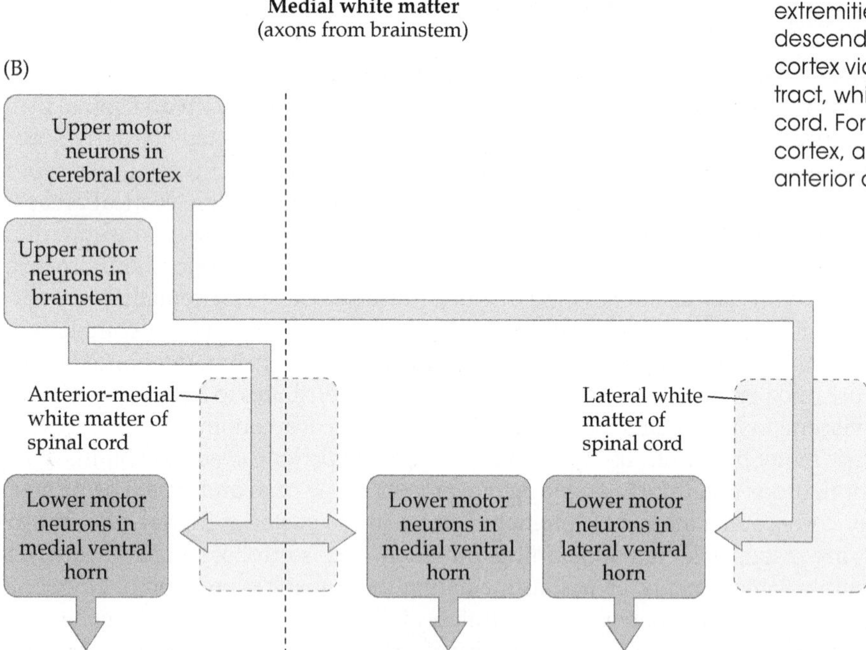

FIGURE 17.1 **Overview of descending motor control.** (A) Somatotopic organization of the ventral horn in the cervical enlargement. The locations of descending projections from the motor cortex in the lateral white matter and from the brainstem in the anterior-medial white matter are shown. (B) Schematic illustration of the major pathways for descending motor control. The medial ventral horn contains lower motor neurons that govern posture, balance, locomotion, and orienting movements of the head and neck during shifts of visual gaze. These medial motor neurons receive descending input from pathways that originate mainly in the brainstem, course through the anterior-medial white matter of the spinal cord, and then terminate bilaterally. The lateral ventral horn contains lower motor neurons that mediate the expression of skilled voluntary movements of the distal extremities. These lateral motor neurons receive a major descending projection from the contralateral motor cortex via the main (lateral) division of the corticospinal tract, which runs in the lateral white matter of the spinal cord. For simplicity, only one side of the brainstem, motor cortex, and lateral ventral horn is shown, and the minor anterior corticospinal tract is not illustrated.

neurons that synapse primarily with lower motor neurons in the medial part of the ventral horn, whereas the lateral regions of the intermediate zone contain local neurons that synapse primarily with lower motor neurons in the lateral ventral horn. This somatotopic organization of the ventral and intermediate spinal cord gray matter provides an important framework for understanding the control of the body's musculature in posture and movement, as well as how the descending projections of different groups of upper motor neurons are organized to influence movements.

Differences in where upper motor neuron pathways from the cortex and brainstem terminate in the spinal cord conform to the functional distinctions between the local circuits that organize the activity of axial and distal muscle groups. Thus, most upper motor neurons that project to the medial part of the ventral horn also project to the medial region of the intermediate zone. The axons of these upper motor neurons course through the anterior-medial white matter of the spinal cord and give rise to collateral branches that terminate over many spinal cord segments among medial cell groups on both sides of the spinal cord. The sources of these projections are located primarily in the brainstem, and as their terminal zones in the medial spinal cord gray matter suggest, they are concerned primarily with proximal muscles that control posture, balance, orienting mechanisms, and the initiation and regulation of stereotyped, rhythmic behavior (see Figure 17.1B). In contrast, the large majority of axons that project from the motor cortex to the spinal cord course through the lateral white matter of the spinal cord and terminate in lateral parts of the ventral horn, with terminal fields that are restricted to only a few spinal cord segments. The major component of this corticospinal pathway is concerned with the voluntary expression of precise, skilled movements involving more distal parts of the limbs.

The Corticospinal and Corticobulbar Tracts

The upper motor neurons in the cerebral cortex reside in several adjacent and highly interconnected areas in the posterior frontal lobe, which together mediate the planning and initiation of complex temporal sequences of voluntary movements. These cortical areas all receive regulatory input from the basal ganglia and cerebellum via relays in the ventrolateral thalamus (see Chapters 18 and 19), as well as inputs from sensory regions of the parietal lobe (see Chapter 9). Although the label "motor cortex" is sometimes used to refer to these frontal areas collectively, the term is more commonly restricted to the primary motor cortex located in the precentral gyrus and the paracentral lobule (Figure 17.2; see the Appendix and Plates 1–4 of the Atlas for annotated photographs of the gyral formations). The primary motor cortex can be distinguished from a complex mosaic of adjacent "premotor" areas both cytoarchitectonically (it is area 4 in Brodmann's nomenclature; see Figure 27.1) and by the low intensity of current necessary to elicit movements by electrical stimulation in this region. The low threshold for eliciting movements is an indicator of a relatively large and direct pathway from the primary area to the lower motor neurons of the brainstem and spinal cord.

The pyramidal cells of cortical layer 5 are the upper motor neurons of the primary motor cortex. Among these neurons are the conspicuous Betz cells, which are the largest neurons (by soma size) in the human CNS (Figure 17.3). Although it is often assumed that Betz cells are the principal upper motor neurons of the motor cortex, there are far too few of them to account for the number of axons that project from the motor cortex to the brainstem and spinal cord; indeed, in the human CNS they account for no more than 5% of the axons that project from the motor cortex to the spinal cord. Despite their small numbers, Betz cells play an important role in the activation of lower motor

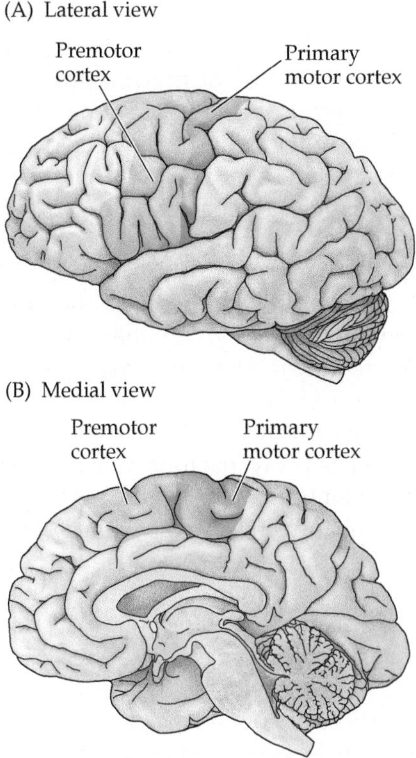

FIGURE 17.2 Primary motor cortex and premotor areas in the human cerebral cortex. Seen here in lateral (A) and medial (B) views, the primary motor cortex is located in the precentral gyrus. The mosaic of premotor areas is more rostral.

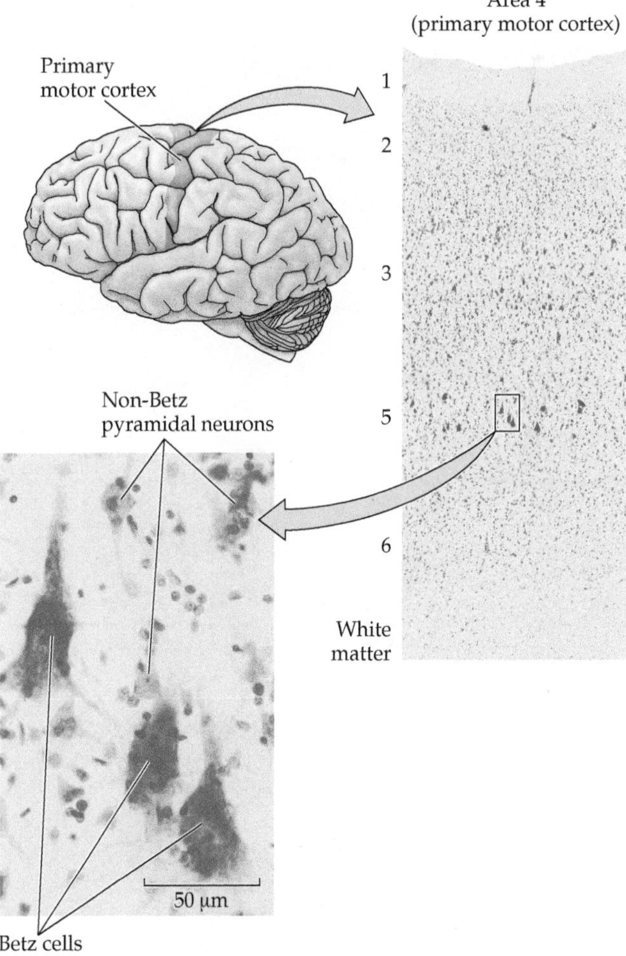

FIGURE 17.3 Cytoarchitectonic appearance of the primary motor cortex in the human brain. Histological photomicrographs show Nissl-stained sections that demonstrate cell bodies; note the presence of Betz cells among the pyramidal neurons of cortical layer 5. (Micrographs courtesy of L.E. White.)

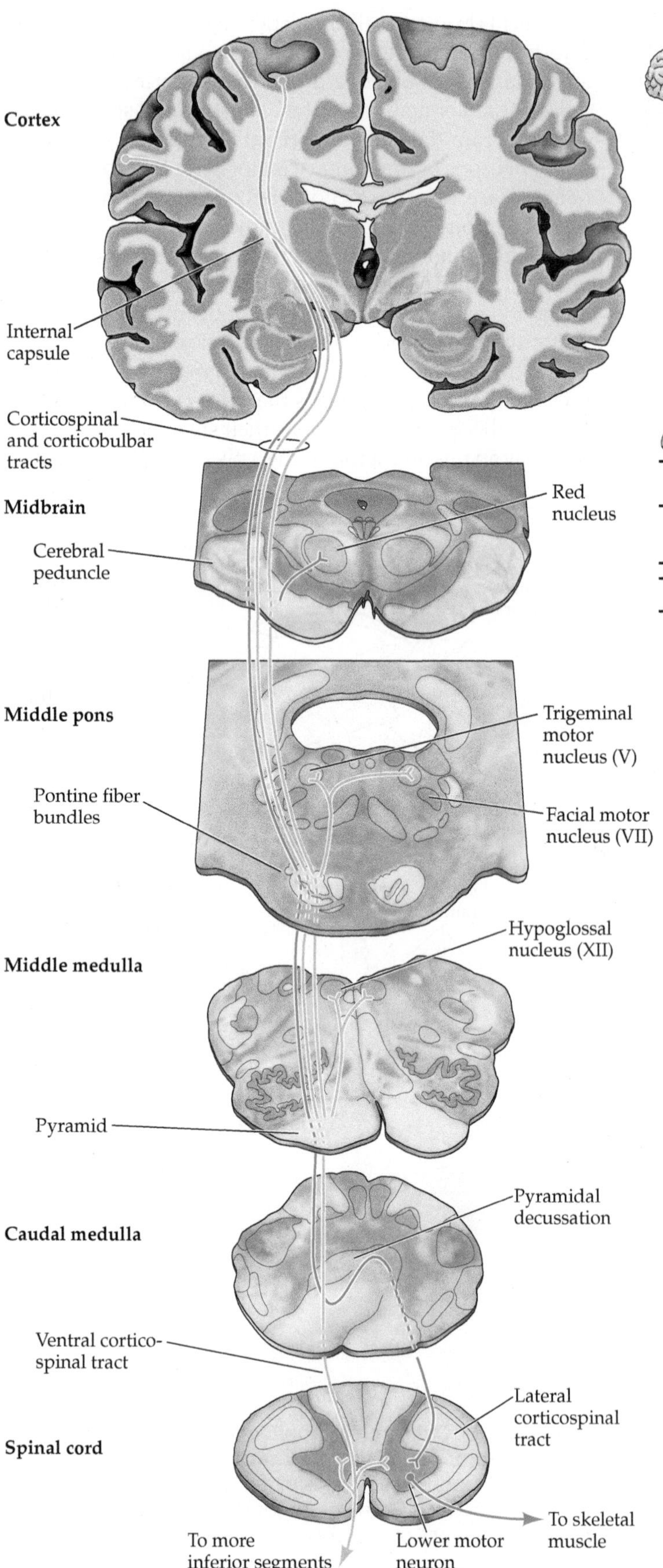

FIGURE 17.4 The corticospinal and corticobulbar tracts. Neurons in the motor cortex give rise to axons that travel through the internal capsule and coalesce on the ventral surface of the midbrain, within the cerebral peduncle. These axons continue through the pons and come to lie on the ventral surface of the medulla, giving rise to the medullary pyramids. As they course through the brainstem, corticobulbar axons (gold) give rise to bilateral collaterals that innervate brainstem nuclei (only collaterals to the trigeminal motor nuclei and the hypoglossal nuclei are shown). Most of the corticospinal fibers (dark red) cross in the caudal part of the medulla to form the lateral corticospinal tract in the spinal cord. Those axons that do not cross (light red) form the ventral corticospinal tract, which terminates bilaterally.

neurons that control muscle activities in the distal extremities. The remaining upper motor neurons are the smaller, non-Betz pyramidal neurons of layer 5 that are found in the primary motor cortex and in each division of the premotor cortex. The axons of these upper motor neurons descend in the **corticobulbar** and **corticospinal tracts**, terms that are used to distinguish axons that terminate in the brainstem ("bulbar" refers to the brainstem) or spinal cord. Along their course, these axons pass through the posterior limb of the internal capsule in the forebrain to enter the cerebral peduncle at the base of the midbrain (Figure 17.4). They then pass through the base of the pons, where they are scattered among the transverse pontine fibers and nuclei of the basal pontine gray matter. They coalesce again on the ventral surface of the medulla, where they form the **medullary pyramids**. The components of this upper motor neuron pathway that innervate cranial nerve nuclei, the reticular formation, and the red nucleus (that is, the corticobulbar tract) leave the pathway at the appropriate levels of the brainstem (see Figure 17.4). There is also a massive corticobulbar projection that terminates among nuclei in the base of the pons that project in turn to the cerebellum; this projection is often called the corticopontine tract and will be discussed in Chapter 19.

Most corticobulbar axons that govern the cranial nerve motor nuclei (see the Appendix) terminate *bilaterally* on local circuit neurons embedded in the brainstem reticular formation (see the section "Motor Control Centers in the Brainstem" later in this chapter), rather than directly on the lower motor neurons in the motor nuclei. These local circuit neurons, in turn, coordinate the output of different groups of lower motor neurons in the cranial nerve motor nuclei. The consequence of the bilateral corticobulbar innervation is that damage to the corticobulbar fibers on only one side does not result in dramatic deficits in function.

There are three notable exceptions to the pattern of symmetrical, bilateral cortical innervation of the local circuits controlling cranial nerve motor nuclei. For each of these exceptions, corticobulbar inputs to the relevant local circuits arise from both cerebral hemispheres; but there is significant bias in favor of inputs from the *contralateral* motor cortex. Specifically, the local circuits that organize the output of lower motor neurons in the hypoglossal nucleus (which governs tongue protrusion), the trigeminal motor nucleus (which governs chewing), and the part of the facial motor nucleus that innervates the lower face each receive corticobulbar input primarily from the contralateral motor cortex. The part of the facial motor nucleus that innervates the upper face is supplied more equally by the corticobulbar inputs from the two sides; this is an important clinical point that is also relevant for understanding facial expressions of emotion. Essentially, lower facial movements that may be performed unilaterally—such as pushing the tongue against one cheek, biting on one side of the mouth, or raising or lowering one corner of the mouth—are governed primarily by the contralateral motor cortex. Most other motor functions governed by cranial nerve nuclei in which the movements of the two sides are largely in synchrony (e.g., vocalization, salivation, tearing, swallowing) are subject to symmetrical, bilateral upper motor neuronal control.

Near the caudal end of the medulla, nearly all of the fibers in the medullary pyramids are corticospinal axons. Just before entering the spinal cord, about 90% of these axons cross the midline—*decussate*—to enter the lateral columns of the spinal cord on the opposite side, where they form the **lateral corticospinal tract**. The remaining 10% of the pyramidal tract fibers enter the spinal cord without crossing; these axons, which constitute the **ventral (anterior) corticospinal tract**, terminate bilaterally. Collateral branches of these axons cross the midline via the ventral white commissure of the spinal cord to reach the opposite ventral horn. The ventral corticospinal pathway arises primarily from dorsal and medial regions of the motor cortex that serve trunk and proximal limb muscles—the same divisions of the motor cortex that give rise to projections to the reticular formation (see the section "Motor Control Centers in the Brainstem").

The lateral corticospinal tract forms a direct pathway from the cortex to the spinal cord and terminates primarily in the lateral portions of the ventral horn and intermediate gray matter. Some of these axons (including those derived from Betz cells) synapse directly on α motor neurons that govern the distal extremities (see Figures 17.1 and 17.4). However, this privileged synaptic contact on lower motor neurons is restricted to a subset of α motor neurons that supply the muscles of the forearm and hand; most axons of the lateral corticospinal tract, in contrast, terminate among pools of local circuit neurons that coordinate the activities of the lower motor neurons in the lateral cell columns of the ventral horn that innervate different muscles. This difference in terminal distribution implies a special role for the lateral corticospinal tract in the control of the hands. Although selective damage to this pathway in humans is rarely seen, evidence from experimental studies in non-human primates indicates that direct projections from the motor cortex to the spinal cord are essential for the performance of discrete finger movements. This evidence helps explain the limited recovery in humans after damage to the motor cortex or some component of this pathway. Immediately after such an injury, such patients are typically paralyzed on the affected side. With time, however, some ability to perform voluntary movements reappears. These movements, which are presumably mediated by residual corticospinal inputs and by motor centers in the brainstem, are crude for the most part. The ability to perform fractionated finger movements, such as those required for writing, typing, playing a musical instrument, or buttoning clothes, typically remains impaired.

Finally, some components of the corticobulbar and corticospinal projections do not participate directly in upper motor control of lower motor neurons. These components are derived from layer 5 neurons in somatosensory regions of the anterior parietal lobe and terminate among local circuit neurons near the sensory trigeminal nuclei and dorsal column nuclei of the brainstem, and in the dorsal horn of the spinal cord. They are likely involved in modulating the transmission of proprioceptive signals and other mechanosensory inputs relevant to sensory perception and the monitoring of body movements. Interestingly, the corticospinal projection to the ventral horn is largest in vertebrates that have the most complex repertoire of fractionated movements with their hands or forepaws. In animals with little ability to execute skilled movements with their forepaws, the corticospinal projection is predominantly directed toward the dorsal horn, where it modulates sensory input to the brain and spinal cord.

Functional Organization of the Primary Motor Cortex

Clinical observations and experimental work dating back 100 years or more provide a foundation for understanding the functional organization of the motor cortex. By the end of the nineteenth century, experimental work in animals by the German physiologists G. Theodor Fritsch and Eduard Hitzig had shown that electrical stimulation of the motor cortex elicits contractions of muscles on the contralateral side of the body. Around the same time, the British neurologist John Hughlings Jackson surmised that the motor cortex contains a complete spatial representation, or map, of the body's musculature. Jackson reached this conclusion from his observation that the movements accompanying certain types of epileptic seizures often began locally and proceeded to "march" systematically from one part of the body to another. For instance, partial motor seizures may begin with twitches and other non-purposeful movements of a finger, then involve the entire hand, and progressively affect the forearm, the arm, the shoulder, and finally the face.

This early evidence for motor maps in the cortex was confirmed shortly after the turn of the nineteenth century when the great British neurophysiologist Sir Charles Sherrington published his classic maps of the organization of the motor cortex in great apes. These maps were created using focal electrical stimulation applied to the surface of the cortex. During the 1930s, one of Sherrington's students, the renowned neurosurgeon Wilder Penfield, extended this work by demonstrating that the human motor cortex also contains a spatial map of the contralateral body. By correlating the location of muscle contractions with the site of electrical stimulation on the surface of the motor cortex (the same method used by Sherrington), Penfield mapped the motor representation in the precentral gyrus in more than 400 neurosurgical patients. He found that this motor map shows the same general disproportions observed in the somatosensory maps in the postcentral gyrus (see Chapter 9). Thus, the musculature used in tasks requiring fine motor control (such as movements of the face and hands) is represented by a greater area of motor cortex than is the musculature requiring less precise motor control (such as that of the trunk) (Figure 17.5).

The introduction in the 1960s of intracortical microstimulation (a more refined method of cortical activation than cortical surface stimulation) allowed a more detailed understanding of motor maps. Microstimulation entails the delivery of brief electrical currents an order of magnitude smaller than those used by Sherrington and Penfield. By passing the current through the sharpened tip of a metal microelectrode inserted into the cortex, the upper motor neurons in layer 5 that project to lower motor neuron circuitry could be stimulated more focally. Although intracortical stimulation generally confirmed Penfield's spatial map in the motor cortex, it also showed that the finer organization of the map is rather different from what most neuroscientists had imagined. For example, when microstimulation was combined with recordings of muscle electrical activity, even the smallest currents capable of eliciting a response initiated the excitation of several muscles (and the simultaneous suppression of others), suggesting that organized movements rather than individual muscles are represented in the map (Box 17A). Furthermore, within major subdivisions of the map (e.g., forearm or face regions), a particular movement could be elicited by stimulation of widely separated sites, supporting the argument that neurons in nearby regions are linked by local circuits in the cortex and spinal cord to organize specific movements. This interpretation has been supported by the observation that the regions responsible for initiating different movements overlap substantially. The conclusion that movements—or action goals—rather than the contractions of individual muscles are encoded in the cortex also applies to the motor areas of the frontal cortex that control eye movements, where focal stimulation elicits binocular shifts in the direction of gaze (see Chapter 20).

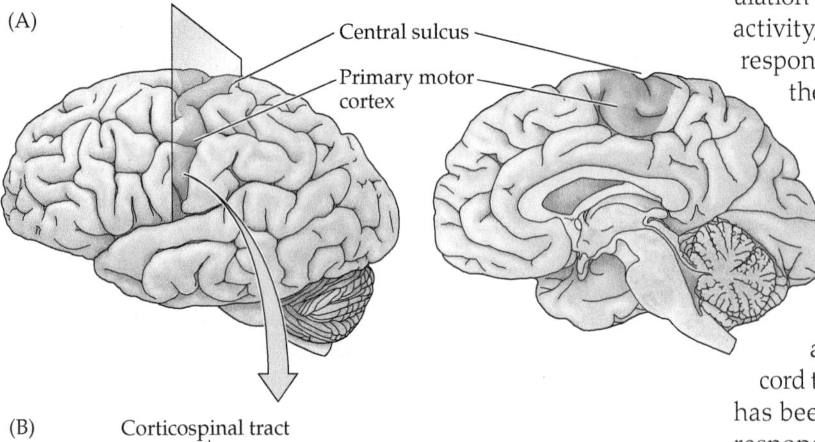

FIGURE 17.5 Topographic map of movement in the primary motor cortex. (A) Location of primary motor cortex in the precentral gyrus. (B) Section along the precentral gyrus, illustrating the somatotopic organization of the motor cortex. In contrast to the precise and detailed representation of the contralateral body in the primary somatosensory cortex (see Figure 9.11), the somatotopy of the primary motor cortex is much more coarse.

BOX 17A ■ What Do Motor Maps Represent?

Electrical stimulation studies carried out in human patients by the neurosurgeon Wilder Penfield and his colleagues (and in experimental animals by Sherrington and, later, by Clinton Woolsey and his colleagues) clearly demonstrated a systematic motor map in the precentral gyrus (see text). The fine structure of this map and the nature of its representation, however, have been a continuing source of controversy. Is the map in the motor cortex a map of *musculature* that operates like a "piano keyboard" for the control of individual muscles? Is it a map of *movements*, in which specific sites control multiple muscle groups that contribute to the generation of particular actions? Is it a map of *intentions*, with the goal of the movement having preeminence over the means by which the goal is achieved?

Initial experiments implied that the map in the motor cortex is a fine-scale representation of individual muscles. Thus, stimulation of small regions of the map activated single muscles, suggesting that vertical columns of cells in the motor cortex were responsible for controlling the actions of particular muscles, much as columns in the somatosensory map are thought to analyze stimulus information from particular locations in the body (see Chapter 9).

More recent studies using anatomical and physiological techniques, however, have shown that the map in the motor cortex is far more complex than a columnar representation of particular muscles. Individual pyramidal tract axons are now known to terminate on sets of spinal motor neurons that innervate different muscles. This relationship is evident even for neurons in the hand representation of the motor cortex, the region that controls the most discrete, fractionated movements.

Furthermore, cortical microstimulation experiments have shown that contraction of a single muscle can be evoked by stimulation over a wide region of the motor cortex (about 2 to 3 mm in macaque monkeys) in a complex, mosaic fashion. It seems likely that horizontal connections within the motor cortex and local circuits in the spinal cord create ensembles of neurons that coordinate the pattern of firing in the population of ventral horn cells that ultimately generate a given movement.

Thus, while the somatotopic maps in the motor cortex generated by early studies are correct in their overall topography, the fine structure of the map is far more abstract. As discussed in the text, it is now widely accepted that the functional maps in the primary motor and premotor cortex are maps of movement. Although coarse somatotopy provides one means of understanding the organization of these motor maps (see Figure 17.5), Michael Graziano and his colleagues at Princeton University have proposed another scheme. Their microstimulation studies of awake, behaving monkeys suggest that the topographic representations of movement in the motor cortex are organized around ethologically relevant categories of motor behavior.

For example, microstimulation of sites in the arm region of the primary motor cortex often invoked movements of the arm that brought the monkey's hand into central space, where the animal might visually inspect and manipulate a held object (see figure; see also Figure 17.7B). Stimulation of more lateral regions (toward the face representation) often led to hand-to-mouth motions and mouth opening, whereas more medial stimulation sites (toward the trunk and leg representations) evoked climbing- or leaping-like postures. These observations suggests that the posterior regions of the motor cortex, including the primary motor cortex, are most concerned with manual and oral behaviors that occur in central, personal space. Just anterior to this cortical region (in the premotor cortex) are sites that, when stimulated, elicit reaching motions and other outward arm movements directed away from the body. Other anterior sites may evoke coordinated, defensive postures as well, perhaps reflecting the integration of threatening sensory signals derived from extrapersonal space. New studies of the mirror motor system raise the intriguing possibility that what is actually represented in the motor cortex is the intention of movement or action goal, rather than movement per se.

Obviously, these are exciting times for investigators studying the cortical governance of movement. Unraveling the details of what is represented in motor maps still holds the key to understanding how patterns of activity in the primate motor cortex generate the rich repertoire of volitional movement.

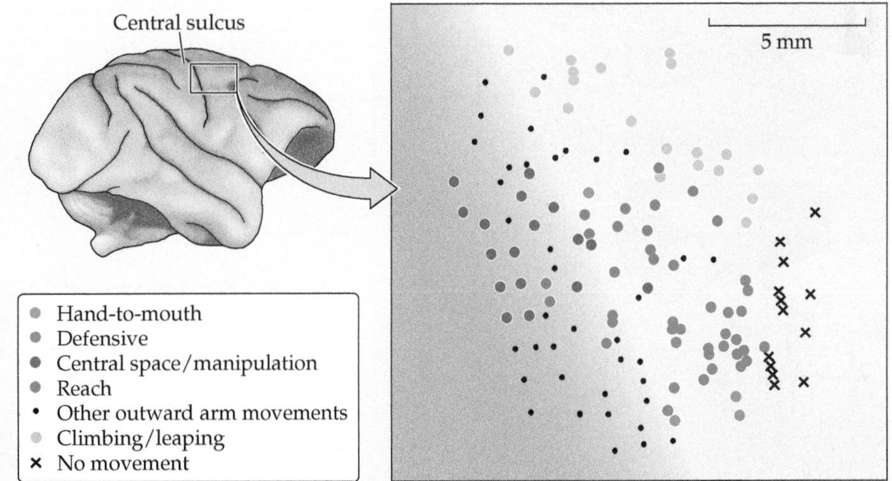

Topographic distribution of microstimulation sites that evoke behaviorally relevant movements in a macaque monkey. The rectangular region on the brain (left) shows the portion of the motor cortex under investigation. The shaded region in the map of stimulation sites indicates cortex folded into the anterior bank of the central sulcus. (After Graziano et al., 2005.)

At about the same time that these microstimulation studies were undertaken, Edward Evarts and his colleagues at the National Institutes of Health were pioneering a technique in which implanted microelectrodes were used to record the electrical activity of individual motor neurons in awake, behaving monkeys. The monkeys were trained to perform a variety of motor tasks during the cortical recording, thus providing a means of correlating neuronal activity with voluntary movements. Evarts and his group found that the force generated by contracting muscles changed as a function of the firing rate of upper motor neurons. Moreover, the firing rates of the active neurons often changed *prior* to movements involving very small forces. Evarts therefore proposed that the primary motor cortex contributes to the initial phase of recruitment of lower motor neurons involved in the generation of finely controlled movements. Additional experiments showed that the activity of primary motor neurons is correlated not only with the magnitude, but also with the direction of the force produced by muscles. Thus, some neurons show progressively less activity as the vector of the movement deviates from the neuron's "preferred direction."

A further advance was made in the 1970s with the introduction of *spike-triggered averaging* (Figure 17.6). By correlating the timing of a single cortical neuron's discharges with the onset times of the contractions generated by the various muscles used in a movement, this method provides a way of measuring the influence of the single neuron on a population of lower motor neurons in the spinal cord. Recording such activity from different muscles as monkeys performed wrist flexion or extension demonstrated that the activity of multiple different muscles is directly facilitated by the discharges of a given upper motor neuron. This peripheral muscle group is referred to as the *muscle field* of the upper motor neuron.

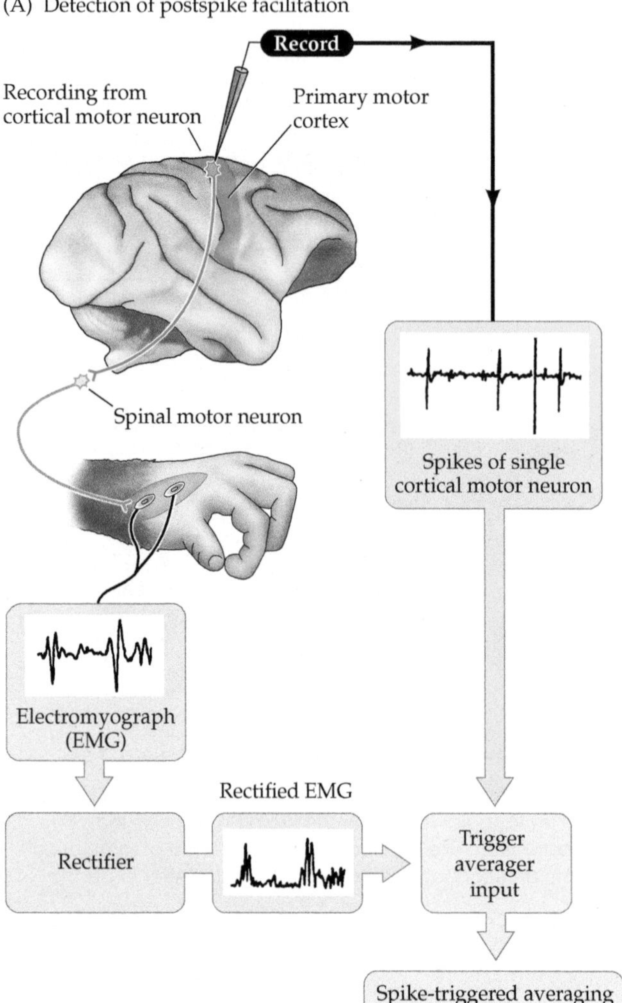

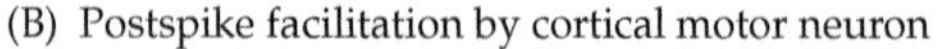

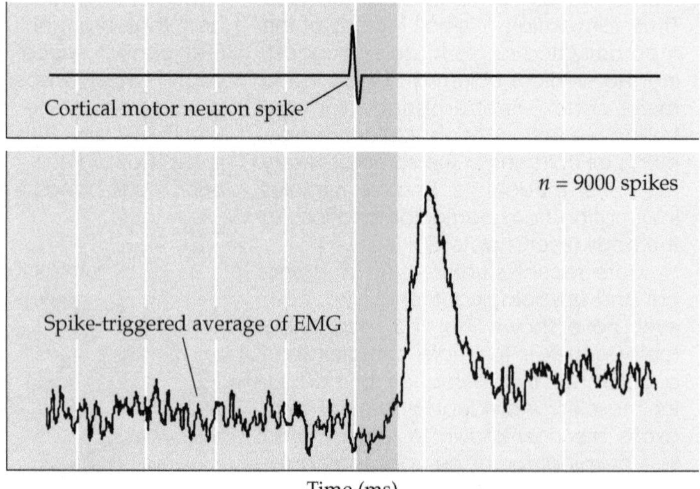

FIGURE 17.6 The influence of single cortical upper motor neurons on muscle activity. (A) Diagram illustrating the spike-triggered averaging method for correlating muscle activity with the discharges of single upper motor neurons. (B) The response of a thumb muscle (bottom trace) follows by a fixed latency the single spike discharge of a pyramidal tract neuron (top trace). This technique can be used to determine all the muscles that are influenced by a given motor neuron. (After Porter and Lemon, 1993.)

On average, the size of the muscle field in the wrist region is two or three muscles per upper motor neuron. These observations, which confirmed that single upper motor neurons contact several lower motor neuron pools, are consistent with the general conclusion that the activity of the upper motor neurons in the cortex controls *movements*, rather than individual muscles.

For the several reasons discussed above, the motor map in the precentral gyrus is much less precise than the somatotopic map in the postcentral gyrus, where the receptive fields of adjacent cortical neurons overlap in a smooth and continuous progression across the surface of the primary somatosensory cortex. Indeed, it is problematic to represent the motor map in the form of a homunculus cartoon that would be analogous to the somatosensory homunculus in the postcentral gyrus (see Figure 9.11), since the representation of muscle movement is not organized at the level of individual muscles or body parts, and the distribution of muscle fields among neighboring cortical neurons is neither spatially continuous nor temporally fixed. However, this apparent imprecision in the motor map does not indicate a degenerate representation of the body's musculature in the motor cortex. Rather, it suggests a dynamic and flexible means for encoding higher order movement parameters that entail the coordinated activation of multiple muscle groups across several joints to perform behaviorally useful actions.

This principle of upper motor neural control has been demonstrated by Michael Graziano and his colleagues at Princeton University, who extended the duration of cortical microstimulation in behaving monkeys to a timescale that more closely corresponds to the duration of volitional movements (from hundreds of milliseconds to several seconds). When such stimuli are applied to the precentral gyrus, the resulting movements are sequentially distributed across multiple joints and are strikingly purposeful (Figure 17.7). Examples of motor patterns frequently elicited with prolonged microstimulation of the precentral gyrus are movements of the hand to the mouth as if to feed, movements that bring the hand to central space as if to inspect an object of interest, and defensive postures as if to protect the body from an impending collision. These findings reinforce the current view that purposeful movements are organized by the circuitry of the primary motor cortex and that their somatotopic organization is best understood in the context of ethologically relevant behaviors (see Box 17A and the next section).

Finally, the commands to perform precise movement patterns are encoded in the activity of a large population of upper motor neurons integrated by intracortical circuitry. One well-studied paradigm for exploring the

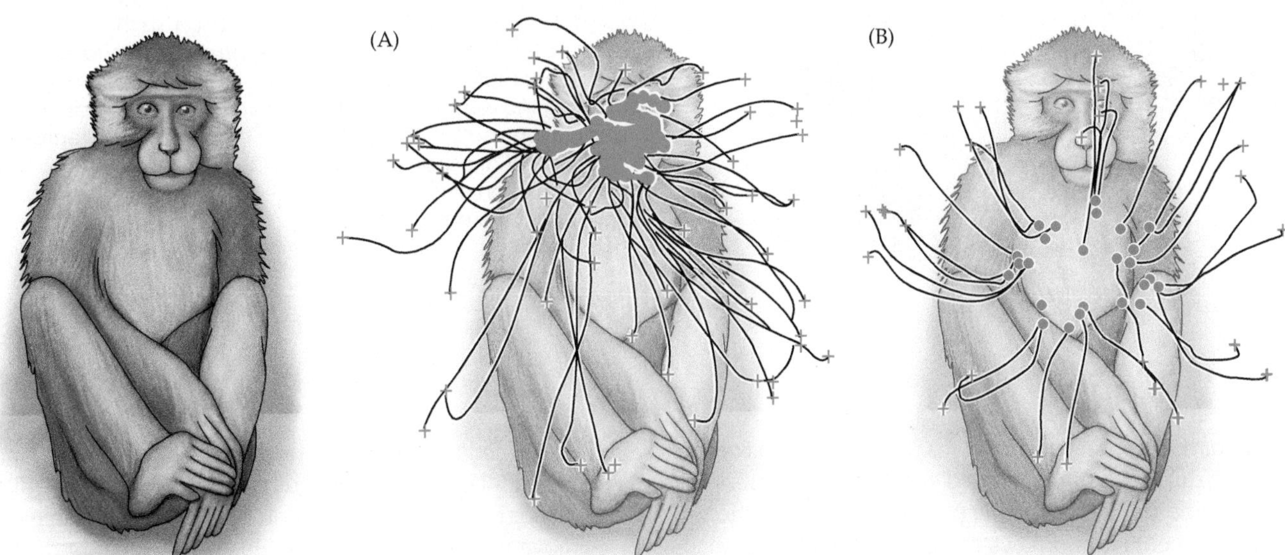

FIGURE 17.7 Purposeful movements of the contralateral arm and hand in a macaque monkey. Prolonged microstimulation of primary motor cortex sites near the middle of the precentral gyrus elicits coordinated movements of the hand and mouth (A) or movements of the arm that bring the hand toward central space, as if to visually inspect and manipulate a held object (B). The starting positions of the contralateral hand are indicated by the blue crosses, the elicited movements are illustrated with the curved black lines, and the final positions of the hand at the end of microstimulation are indicated by the red dots. (After Graziano et al., 2005.)

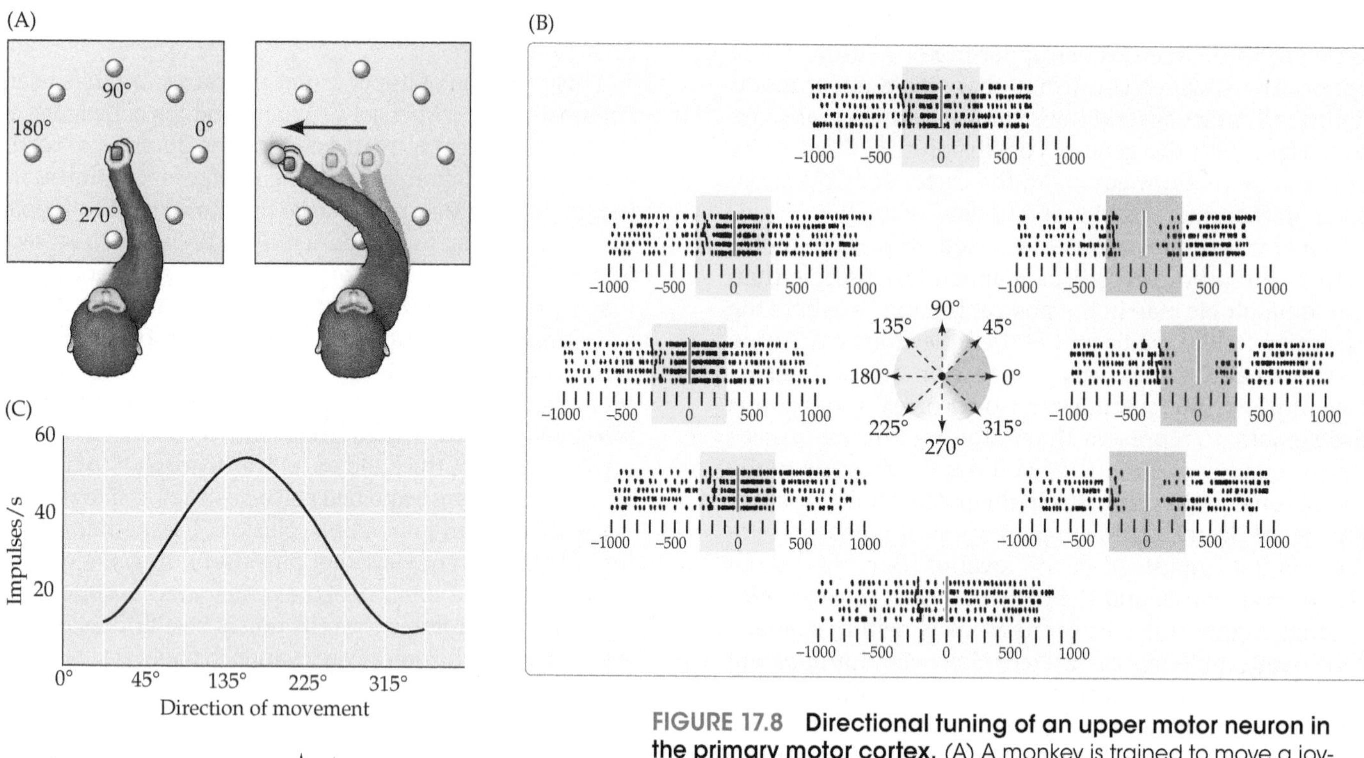

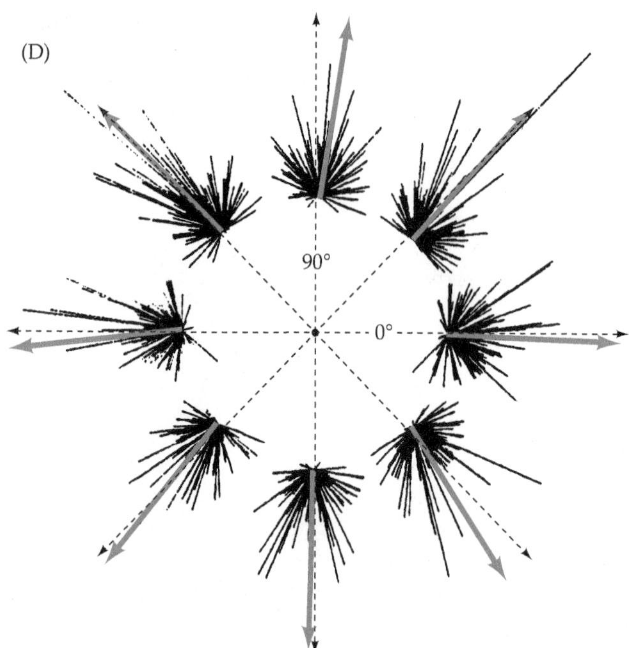

FIGURE 17.8 Directional tuning of an upper motor neuron in the primary motor cortex. (A) A monkey is trained to move a joystick in the direction indicated by a light. (B) The activity of a single neuron was recorded during arm movements in each of eight different directions (0 indicates the time of movement onset; each short vertical line in this raster plot represents an action potential). The activity of the neuron increased before movements between 90° and 225° (yellow zone) but decreased in anticipation of movements between 45° and 315° (blue zone). (C) Plot showing that the neuron's discharge rate was greatest before movements in a particular direction, which defines the neuron's "preferred direction" in this experimental paradigm. (D) The black lines indicate the discharge rates of individual upper motor neurons prior to each direction of movement. By combining the responses of all the neurons in the recording session, a "population vector" (red arrows) can be derived that represents the movement direction encoded by the simultaneous activity of the entire population of recorded units. (A–C after Georgeopoulos et al., 1982; D after Georgeopoulos, 1983.)

nature of this neural code involves recording from cortical neurons during visually guided reaching movements of the arm and hand. Using this paradigm, the direction of arm movements in monkeys could be predicted by calculating a "neuronal population vector" derived simultaneously from the discharges of a population of upper motor neurons that are "broadly tuned" in the sense that each neuron discharges prior to movements in many directions (Figure 17.8). These observations showed that the discharges of individual upper motor neurons cannot specify the direction of an arm movement, simply because they are tuned too broadly (likely reflecting the summed tuning of inputs from other upper motor neurons). Rather, each arm movement must be encoded by the concurrent discharges of a large population of such functionally linked neurons. The fact that the same site in the primary motor cortex can encode different trajectories of motion depending on the starting position of the limb (see Figure 17.7) suggests that multiple parameters of movement may be selected by the relevant ensemble of upper motor neurons to achieve a behaviorally useful action. Thus, microstimulation experiments use exogenous electrical currents to engage populations of upper motor neurons whose output encodes not simply the trajectory of arm motion, but also the final position of the hand in the context of an action goal.

The Premotor Cortex

A complex mosaic of interconnected frontal lobe areas that lie rostral to the primary motor cortex also contributes to motor functions (see Figure 17.2). This functional division of the motor cortex includes Brodmann's areas 6, 8, and 44/45 on the lateral surface of the frontal lobe and parts of areas 23 and 24 on the medial surface of the hemisphere. Although the organization of this premotor mosaic is best understood in the macaque monkey (Figure 17.9), recent functional brain imaging studies, as well as structural brain imaging studies in patients with frontal lobe injuries, suggest that a comparable distribution of premotor areas is present in humans. Each of the divisions of the premotor cortex receives extensive multisensory input from regions of the inferior and superior parietal lobules, as well as more complex signals related to motivation and intention from the rostral ("prefrontal") divisions of the frontal lobe. The upper motor neurons in this premotor cortex influence motor behavior both indirectly, through extensive reciprocal connections with the primary motor cortex, and directly, via axons that project through the corticobulbar and corticospinal pathways to influence the local circuits that organize the output of lower motor neurons in the brainstem and spinal cord.

Indeed, over 30% of the axons in the corticospinal tract arise from neurons in the premotor cortex. Thus, past arguments that the premotor cortex occupies a higher position in a cortical hierarchy of motor control by operating through feedforward signals to the primary motor cortex are no longer tenable. Rather, a variety of experiments indicate that the premotor cortex uses information from other cortical regions to select movements appropriate to the context and goal of the action (see Chapter 32). The principal difference between the premotor cortex and primary motor cortex lies in the strength of their connections to lower motor neurons, with more upper motor neurons in the primary motor cortex making monosynaptic connections to α motor neurons, especially those in the ventral horn of the cervical spinal cord that control precise movements of the distal upper extremities. Recent evidence suggests that other differences may reflect the mapping of purposeful movements relative to personal and extrapersonal space and the nature of the signals that lead to the initiation of motor commands in the context of action goals. The action goals encoded by the primary motor cortex tend to be localized to personal space (within arm's length), while the action goals encoded by premotor cortex

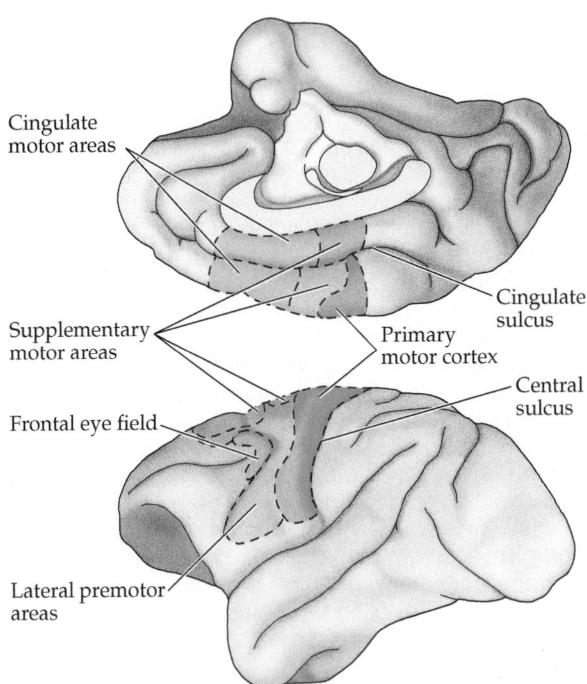

FIGURE 17.9 Divisions of the motor cortex in the macaque monkey brain. As in humans, the primary motor cortex resides in the anterior bank of the central sulcus. Anterior to this region is a complex mosaic of premotor areas that extends from the frontal operculum on the lateral surface of the frontal lobe to the cingulate gyrus on the medial hemispheric surface. The lateral premotor and supplementary motor areas are involved in selecting and organizing purposeful movements of the limbs and face; the frontal eye fields organize voluntary gaze shifts (see Chapter 20), and the cingulate motor areas are involved in the expression of emotional somatic behavior (see Chapter 31). Current evidence supports the existence of comparable premotor areas in the human motor cortex. (After Geyer et al., 2000.)

are more typically oriented toward extrapersonal space (beyond arm's length; see Box 17A). In the exciting new field of neuroengineering, the neural engrams of such action goals are now serving to drive machines and a variety of computerized systems through so-called brain–machine interfaces (Box 17B).

The functions of the premotor cortex may be understood in terms of differences between the lateral and medial components of this region. As many as 65% of the neurons in the lateral premotor cortex have responses that are linked in time to the occurrence of movements; as in the primary motor area, many of these cells fire most strongly before and during movements made in a specific direction. However, these neurons are especially important in conditional ("closed loop") motor tasks. That is, in contrast to the neurons in the primary motor area, when a monkey is trained to reach in different directions depending on the nature of

BOX 17B ■ Minds and Machines

Science fiction has long imagined the melding of the human mind and mechanical or digital machines that would enact our thoughts without the obligatory actions of our evolved musculoskeletal effectors—that is, our physical bodies. In recent years, a consortium of neuroscientists, computer scientists, material scientists, and electrical, mechanical, and biomedical engineers have boldly envisioned the means for realizing what was once mere fantasy. In research laboratories and in some neurorehabilitation clinics around the world, such scientists are teaming up with neurologists, neurosurgeons, and physical therapists to translate *brain-machine interface* technology to clinical practice in the hopes of restoring function lost to neurological injury and disease.

Brain-machine interface (BMI; also known as brain-computer interface, BCI) refers to the systems and technologies that enable thought-controlled operation of virtual or real actuators for communication, movement, and the remote operation of a variety of computer-based systems to enable activities of daily living. The basic design of BMI systems involves (1) the acquisition of brain-generated signals that reflect information processing and the encoding of action goals; (2) the processing and decoding of brain signals using artificial neural networks to extract salient features and translate them into pragmatic control signals; (3) the implementation of control signals for the operation of digital and mechanical systems; and (4) the generation of sensory-based feedback signals to promote adaptive plasticity and improved brain control of BMI technology (Figure A).

Continued on the next page

(A) General design of a brain-machine interface (BMI) system based on the invasive or noninvasive acquisition of brain-derived signals. Artificial neural networks decode brain activity and generate control signals that drive relatively simple, intermediate, or complex BMI systems. Visual, proprioceptive, or haptic feedback is provided to enhance brain control of BMI performance. (BOLD, blood oxygenation level-dependent; ECoG, electrocorticography; NIRS, near-infrared spectroscopy.) (After Leuthardt, Washington University School of Medicine.)

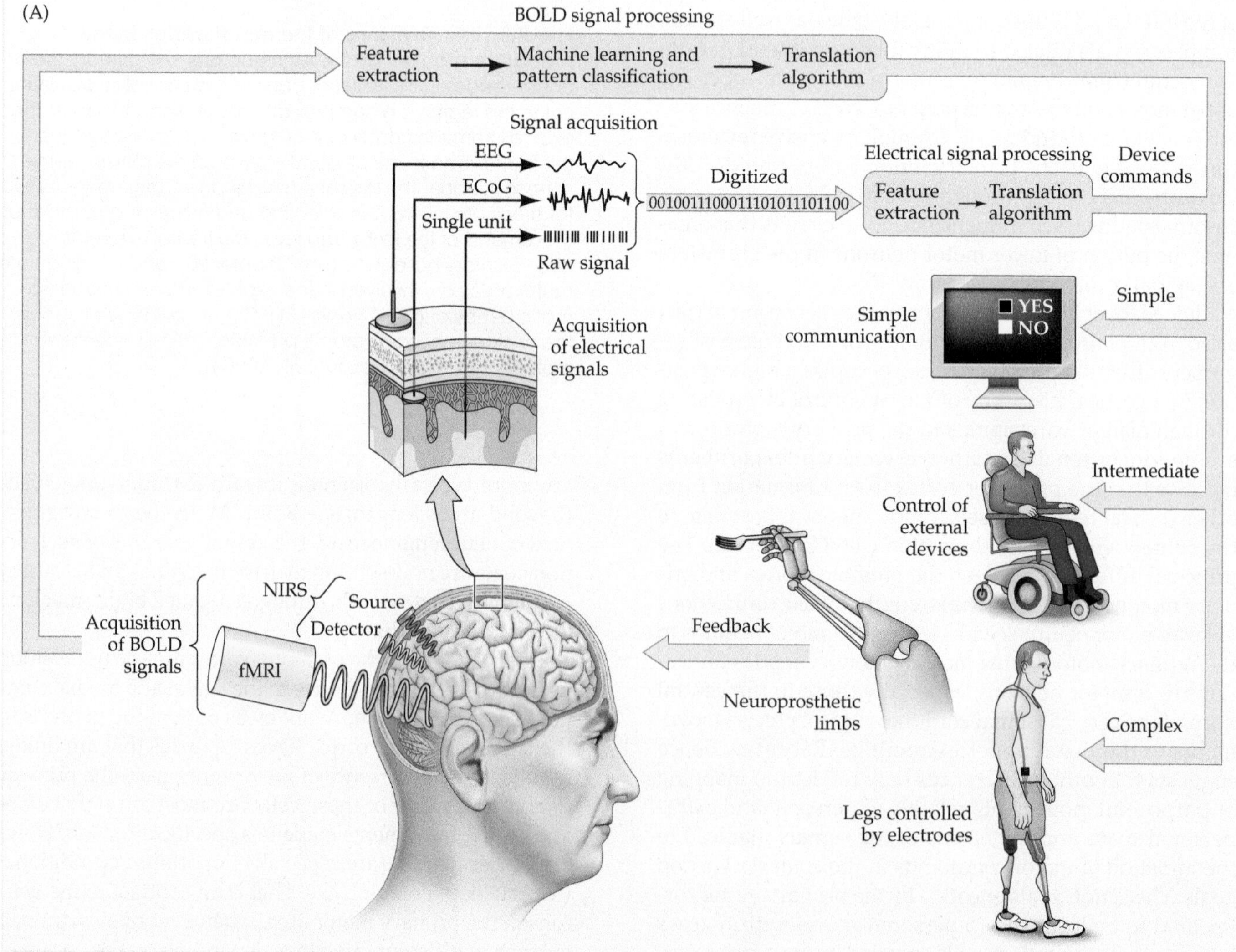

BOX 17B (continued)

Brain-derived signals for driving BMI technologies may be sampled invasively by methods such as single- or multi-unit recording of neuronal action potentials, local field potentials, or electrocorticography (ECoG); or noninvasively, using approaches such as electroencephalography (EEG) and blood oxygenation level–dependent (BOLD) fMRI or near-infrared spectroscopy (NIRS). Invasive means for acquiring brain-generated signals have the advantage of dense information content that would provide high-fidelity signals for processing and decoding in artificial neural networks, but with the obvious disadvantage of neurosurgical intervention and the risk of attending medical and postsurgical complications. Noninvasive means for acquiring brain-generated signals obviate the risks of neurosurgery; however, until recently the information content of signals recorded from outside the cranium had been considered too impoverished to be useful for driving BMI technologies. Improvements in signal processing and the performance of artificial neural networks have proved sufficient for the implementation of BMI systems based on brain-generated signals acquired noninvasively.

One striking demonstration of the promise of such BMI systems was showcased in the opening ceremony for the 2014 FIFA (Fédération Internationale de Football Association) World Cup in São Paulo, Brazil. In a fleeting moment—amidst the pageantry and spectacle of the global celebration of the "beautiful game"—a 29-year-old Brazilian man who had suffered a complete spinal cord injury in the upper thoracic region 6 years earlier executed a simple kicking motion, sending the ball toward the referee in a ceremonial first kick. That was indeed "one small step for [a] man, one giant leap for mankind," (reminiscent of the interdisciplinary science supporting the Apollo 11 mission and the moment of Neil Armstrong's famous declaration) as this simple motor action was performed using a noninvasive BMI system driving a wearable exoskeleton to enable brain-controlled body-weight support, posture, and locomotion.

Quite unexpectedly, several members of the cohort of individuals who were training for this groundbreaking public demonstration of BMI technology experienced neurological improvement and some measure of clinically significant functional recovery. Eight individuals who were 3 to 13 years post spinal cord injury underwent 12 months of training with a multistage, BMI-based gait neurorehabilitation program that entailed immersive virtual reality training, enriched visual-tactile feedback, and extensive training with an EEG-controlled robotic exoskeleton. By the conclusion of the training period, all of the individuals experienced improvements in somatosensation across multiple dermatomes, and most regained some measure of voluntary muscle contraction below the level of the injury. Moreover, half of the individuals were upgraded to an incomplete paraplegia classification (Figure B). It remains to be determined which components of this complex, intensive training paradigm were most effective in promoting improved neurological function. Likewise, it is not known if these individuals have achieved a recovery "ceiling," or if ongoing BMI-assisted neurorehabilitation might promote even further functional gains years after injury.

The neurobiological mechanisms that underlie these functional improvements remain a matter of speculation. Perhaps some individuals classified as having complete spinal cord injury retain some latent corticospinal connections that may be "awakened" through long-term potentiation and synaptic sprouting (see Chapter 8) during intensive neurorehabilitation with virtual and real, brain-controlled BMI technologies. Ongoing synaptic and circuit plasticity at the cortical and spinal levels may further consolidate functional gains and promote ongoing adaptation to BMI-assisted neurorehabilitation.

Such dramatic demonstrations notwithstanding, BMI technology remains in its infancy, and the barriers to greater efficacy and widespread implementation are daunting. However, the pace of advancement in the multidisciplinary domains of science and technology that support BMI systems holds great promise for a future where minds and machines seamlessly integrate thought, feeling, and action.

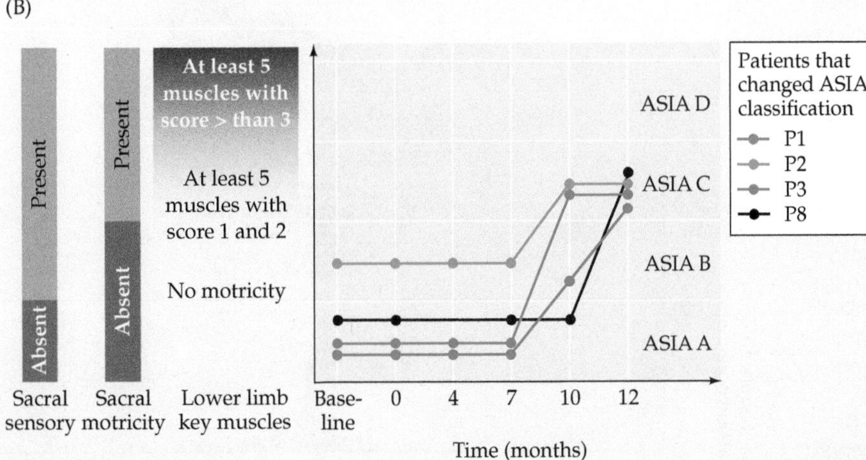

(B) Functional improvements in four of eight patients, indicated by upgrades in American Spinal Injury Association (ASIA) classification, during BMI-assisted neurorehabilitation. Three patients were upgraded from ASIA A (complete spinal cord injury with no sensory or motor function in sacral segments S4–S5) to ASIA C (incomplete spinal cord injury with motor function observable below neurologic level, with majority of affected muscles graded < 3 of 5). One patient was upgraded from ASIA B (incomplete spinal cord injury with sensory, but not motor, function preserved below neurologic level) to ASIA C. Four other patients graded ASIA A showed functional gains in somatosensation and motor control but did not change ASIA classification. (After Donati et al., 2016.)

a visual cue, the appropriately tuned lateral premotor neurons begin to fire at the appearance of the cue, well before the monkey receives a signal to actually make the movement. As the animal learns to associate a new visual cue with the movement, appropriately tuned neurons begin to increase their rate of discharge in the interval between the cue and the onset of the signal to perform the movement. Rather than directly commanding the initiation of a movement, these neurons appear to encode the monkey's *intention* to perform a particular movement; thus, they seem to be particularly involved in the selection of movements based on external events.

A ventrolateral subdivision of the premotor cortex has received considerable attention in recent years, after the discovery that a subset of its neurons responds not just in preparation for the execution of particular movements, such as a precision grip to retrieve a morsel of food, but also when the same action is *observed*, being performed by another individual (monkey or human). For example, these premotor neurons fire action potentials when a monkey observes the hand of a human trainer engaging in the same or similar action that would activate these same neurons during self-initiated movements (Figure 17.10). However, these so-called **mirror motor neurons** respond much less well when the same

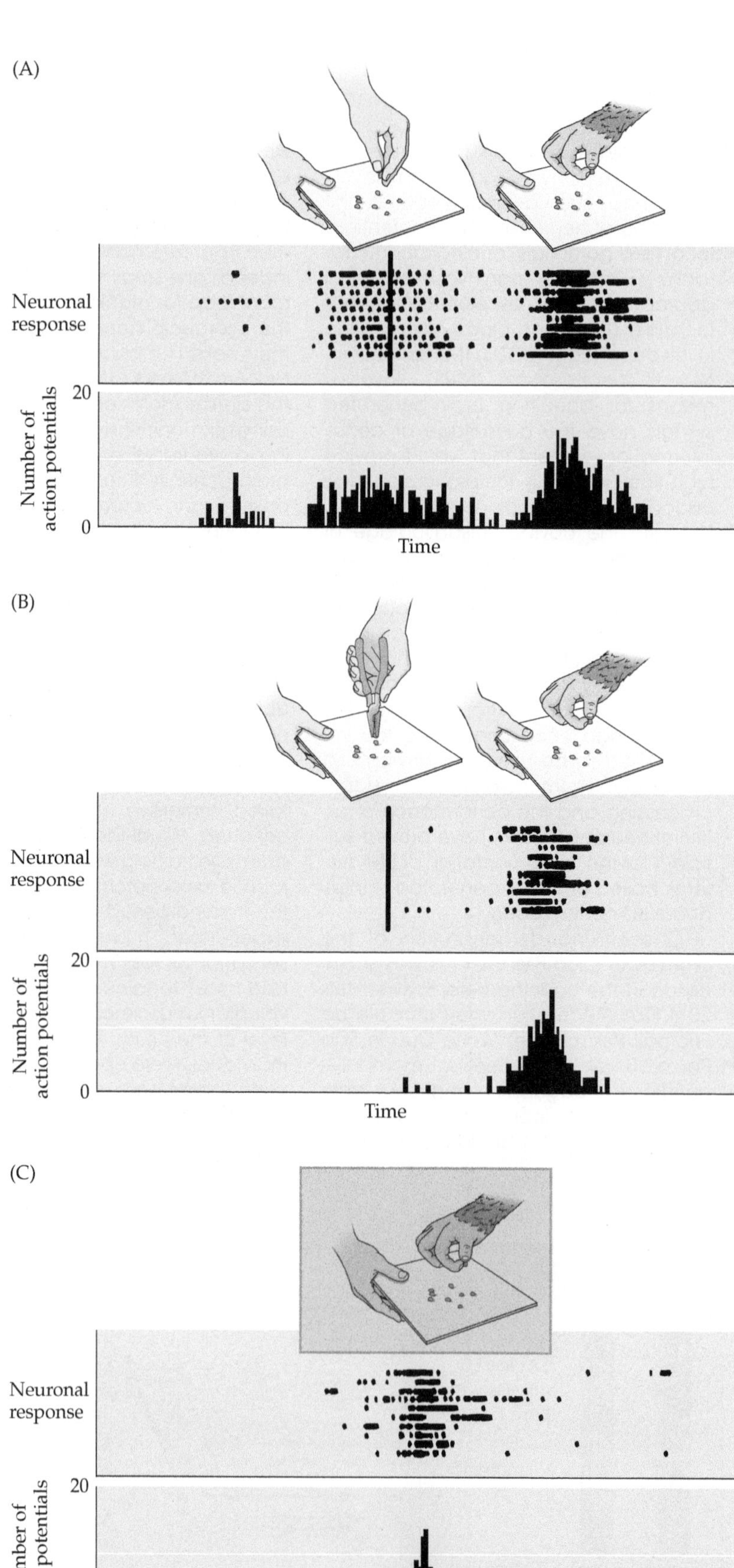

FIGURE 17.10 Mirror motor neuron activity in a ventral-anterior sector of the lateral premotor cortex. In the panels, the upper graphics illustrate the monkey's view of the hand of the trainer placing a food morsel on a tray and the monkey's own hand extending to retrieve the morsel. The middle graphics illustrate raster plots that show the firing of the neuron relative to the observed and executed movements (each tick mark indicates an action potential, and each row represents one trial). The lower graphs are peristimulus response histograms aligned to the overlying raster plots. The mirror motor neuron fires during the passive observation of a human hand placing the morsel of food on the tray (A), as well as during the execution of a similar action to retrieve the food. (The vertical line in the raster plots indicates the time at which the food was placed on the tray; 1 to 2 s later, the monkey reaches to retrieve the morsel.) The same neuron does not respond when the food is placed with the aid of pliers (B), but it does fire during the monkey's reaching and retrieval movements when the monkey is allowed to observe its reach (B) and when the behavior is executed behind a barrier (C). These findings suggest that this division of the premotor cortex plays a role in encoding the observed actions of others. (After Rizzolatti et al., 1996.)

actions are pantomimed without the explicit presence of an action goal, such as an object to be grasped. Furthermore, they respond during the observation of goal-directed behavior even when the final stage of the action is hidden from view—for example, the grasping of an object known to the monkey to have been placed behind a small barrier. Recent studies have demonstrated that some mirror motor neurons show suppression of firing during action observation, even if the same neurons fire during action execution. Such neuronal activities may contribute to the suppression of imitation. Taken together, these findings suggest that the mirror motor system is involved in encoding the intention to make a specific movement based on the observation of the behaviorally relevant actions of others. Evidently, this system participates in an extended network of parietal and frontal regions that subserve action understanding and imitation learning, whether or not observed behavior is "mirrored" in one's own actions (Figure 17.11). The functions of the mirror motor system are among the most actively studied and debated domains of motor and cognitive neuroscience, but the full scope of this system's contributions to motor control, motor learning, and more complex brain functions such as social communication, language, theory of mind, and empathy remain to be elucidated.

Further evidence that the lateral premotor area is concerned with movement selection comes from the effects of cortical damage on motor behavior. Lesions in this region severely impair the ability of monkeys to perform visually cued conditional tasks, even though they can still respond to the visual stimulus and can perform the same movement in a different setting. Similarly, patients with frontal lobe damage have difficulty learning to select a particular movement to be performed in response to a visual cue, even though they understand the instructions and can perform the movements. Individuals with lesions in the premotor cortex may also have difficulty performing movements in response to verbal commands.

Finally, a rostral division of the lateral premotor cortex in the human brain, especially in the left hemisphere, has evolved to play a special role in the production of speech sounds. This region, called **Broca's area** (which typically corresponds to Brodmann's areas 44 and 45, but may be localized to adjacent area 6 in some individuals), is critical for the production of speech and will be considered in detail

(A)
Dorsal premotor cortex
Central sulcus
Superior parietal lobule
Intraparietal sulcus
Ventral premotor cortex
Middle temporal visual areas
Lateral (Sylvian) fissure
Inferior parietal lobule
Superior temporal sulcus

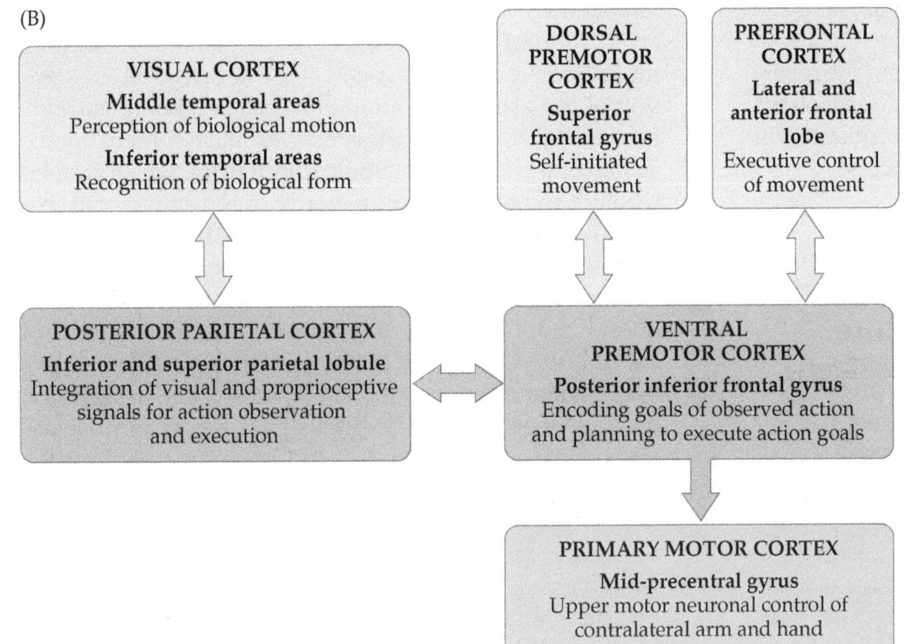

FIGURE 17.11 Cortical network for visually guided reach-to-grasp. (A) Illustration of significant activation in humans during observation of reach-to-grasp. These same cortical areas are recruited during execution of reach-to-grasp with the right arm and hand, with the addition of activation in the mid-precentral gyrus (arm and hand region of the primary motor cortex). (B) Schematic summary of the cortical network for visually guided reach-to-grasp. Areas enriched in neurons with mirroring properties are colored red (mirror motor neurons are also found in other areas not illustrated). This network is proposed to encode the action goal of observed or executed reach-to-grasp. (A from Rizzolatti and Sinigaglia, 2016.)

in Chapter 33. The evolution of this premotor division in primates and its functional relation to semantic processing regions in the parietal and temporal lobe are areas of active investigation.

The medial division of the premotor cortex extends onto the medial aspect of the frontal lobe (including a division that has been referred to as the "supplementary motor area"). Like the lateral area, the medial premotor cortex mediates the selection of movements. However, this region appears to be specialized for initiating movements specified by *internal* rather than *external* cues ("open-loop" conditions). In contrast to lesions in the lateral premotor area, removal of the medial premotor area in a monkey reduces the number of self-initiated or "spontaneous" movements the animal makes, whereas the ability to execute movements in response to external cues remains largely intact. Imaging studies suggest that this cortical region in humans functions in much the same way. For example, functional brain imaging studies show that the medial region of the premotor cortex is activated when individuals perform motor sequences from memory (i.e., without relying on an external instruction). In accord with this evidence, single-unit recordings in monkeys indicate that many neurons in the medial premotor cortex begin to discharge 1 to 2 s before the onset of a self-initiated movement. Among the areas of the medial premotor cortex are two divisions that will be considered in more detail elsewhere: a frontal eye field (see Figure 17.9) involved in directing visual gaze toward a location of interest (see also Chapter 20); and a set of areas in the depths of the cingulate sulcus (see Figure 17.9) that plays a role in the expression of emotional behavior (see also Chapter 31).

In summary, both the lateral and medial areas of the premotor cortex are intimately involved in selecting a specific movement or sequence of movements from the repertoire of possible behaviorally relevant actions. The functions of the areas differ, however, in the relative contributions of external and internal cues to the selection process.

Motor Control Centers in the Brainstem: Upper Motor Neurons That Maintain Balance, Govern Posture, Initiate Locomotion, and Orient Gaze

Several structures in the brainstem contain circuits of upper motor neurons whose activities serve to organize a variety of somatic movements involving the axial musculature of the trunk and the proximal musculature of the limbs. These movements include the maintenance of balance, the regulation of posture, the initiation and regulation of locomotion, and the orientation of visual gaze. They are governed by upper motor neurons in the nuclei of the vestibular complex, the reticular formation, and the superior colliculus (Figure 17.12). Such movements are usually necessary to support the expression of skilled motor behaviors involving the more distal parts of the extremities or, in the case of visual gaze, when attention is directed toward a particular sensory stimulus. Indeed, the relevant brainstem circuits are competent to direct many motor activities without supervision by higher motor centers in the cerebral cortex. However, these brainstem motor centers usually work in concert with divisions of the motor cortex that organize volitional movements, which always entail both skilled (voluntary) and supporting (reflexive) motor activities.

As described in Chapter 14, the vestibular nuclei are the major destination of the axons that form the vestibular division of the eighth cranial nerve; as such, they receive sensory information from the semicircular canals and the otolith organs that specifies the position of the head and its rotational and translational movements. Many of the cells in the vestibular nuclei that receive this information are upper motor neurons with descending axons that terminate in the medial region of the spinal cord gray matter, although some extend more laterally to contact the neurons that control the proximal muscles of the limbs. The projections from the vestibular nuclei that control axial muscles and those that influence proximal limb muscles originate from different cells and take somewhat different routes to the spinal cord (see Figure 17.12A).

Neurons in the medial vestibular nucleus give rise to a **medial vestibulospinal tract** that terminates bilaterally in the medial ventral horn of the cervical cord. There, the medial vestibulospinal tract regulates head position by reflex activation of neck muscles in response to the stimulation of the anterior semicircular canals resulting from unexpected rapid, downward rotation of the head. For example, when an individual falls forward, the medial vestibulospinal tract mediates reflexive dorsiflexion of the neck as well as extension of the arms in an attempt to protect the upper body from injury. Neurons in the lateral vestibular nucleus are the source of the **lateral vestibulospinal tract**, which courses through the anterior white matter of the spinal cord in a slightly more lateral position relative to the medial vestibulospinal tract. Despite the modifier in its name, the lateral vestibulospinal tract terminates ipsilaterally among medial lower motor neuron pools that govern proximal muscles of the limbs. As discussed in more detail in Chapter 14, this tract facilitates the activation of limb extensor (antigravity) muscles when the otolith organs signal deviations from stable balance and upright posture. Other upper motor neurons in the vestibular nuclei project to local circuit neurons and lower motor neurons in the cranial nerve nuclei that control eye movements (the third, fourth, and sixth cranial nerve nuclei). This pathway produces the eye movements that maintain fixation while the head is moving (the vestibulo-ocular reflex; see Chapters 14 and 20).

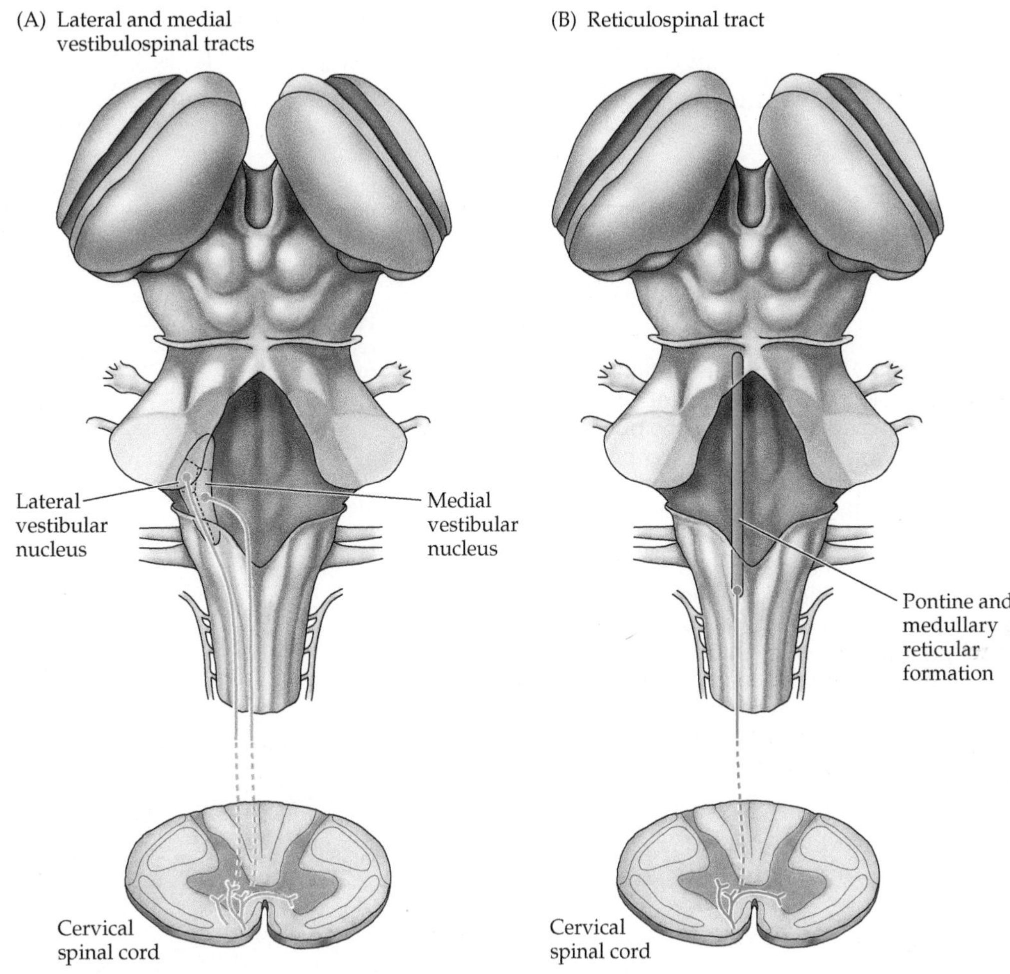

FIGURE 17.12 Descending projections from the brainstem to the spinal cord. Pathways that influence motor neurons in the medial part of the ventral horn originate in the vestibular nuclei (A) and the reticular formation (B).

The **reticular formation** is a complicated network of circuits in the core of the brainstem that extends from the rostral midbrain to the caudal medulla; it is similar in structure and function to the local circuitry in the intermediate gray matter of the spinal cord (Figure 17.13 and Box 17C). Unlike the well-defined sensory and motor nuclei of the cranial nerves, the reticular formation comprises numerous clusters of neurons scattered among a welter of interdigitating axon bundles; it is therefore difficult to subdivide anatomically. The neurons within the reticular formation serve a disparate variety of functions, including cardiovascular and respiratory control (see Chapter 21), governance of myriad sensorimotor reflexes (see Chapters 16 and 21), coordination of eye movements (see Chapter 20), regulation of sleep and wakefulness (see Chapter 28), and most important for the purpose of this discussion, the temporal and spatial coordination of limb and trunk movements, particularly those that control rhythmic, stereotypical behaviors such as locomotion. The descending motor control pathways from the reticular formation to the spinal cord are similar to those of the vestibular nuclei; they terminate primarily in the medial parts of the gray matter, where they influence the local circuit neurons that coordinate axial and proximal limb muscles (see Figure 17.12B). With few exceptions, reticulospinal projections are distributed bilaterally to the medial ventral horns.

Both the vestibular nuclei and the reticular formation provide information to the spinal cord that maintains posture in response to environmental (or self-induced) disturbances of body position and stability. Direct projections from the vestibular nuclei to the spinal cord ensure a rapid compensatory *feedback* response to any postural instability detected by the vestibular labyrinth (see Chapter 14). In contrast, the motor centers in the reticular formation are controlled largely by motor centers in the cerebral cortex, hypothalamus, or brainstem. The relevant neurons in the reticular formation initiate *feedforward* adjustments that stabilize posture during ongoing movements.

The way neurons of the reticular formation maintain posture can be appreciated by analyzing their activity during voluntary movements. Even the simplest movements are accompanied by the activation of muscles that at first glance seem to have little to do with the primary

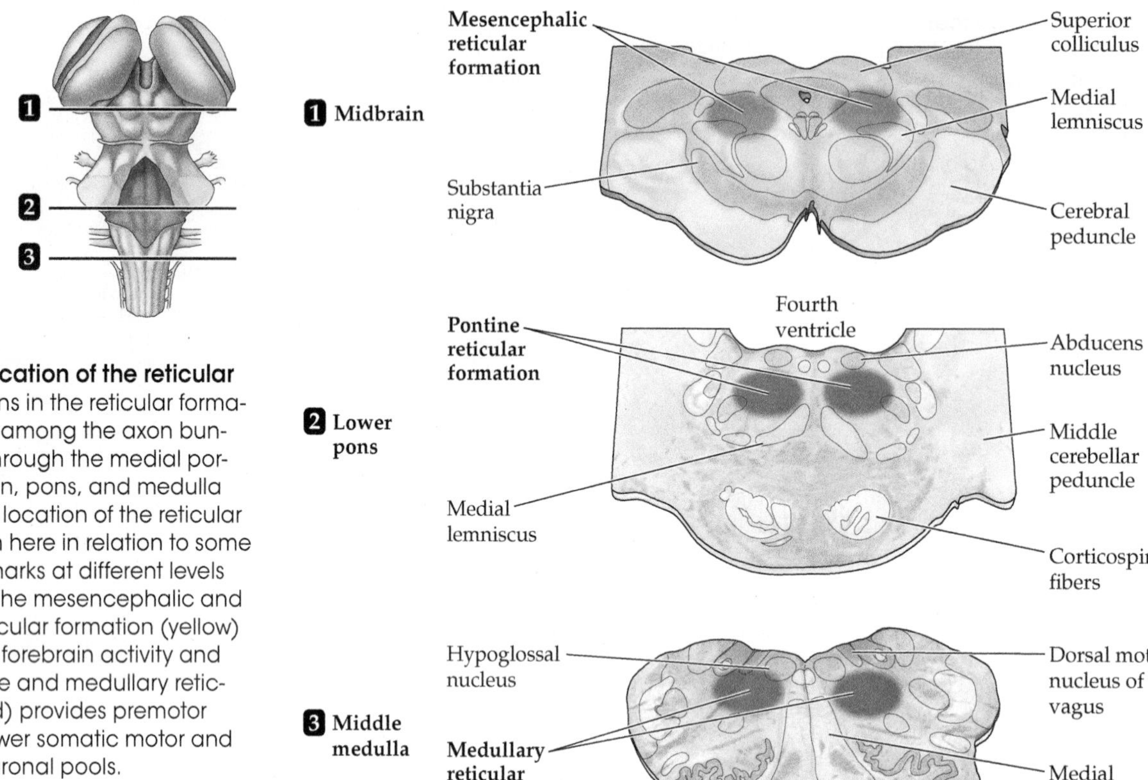

FIGURE 17.13 **Location of the reticular formation.** Neurons in the reticular formation are scattered among the axon bundles that course through the medial portion of the midbrain, pons, and medulla (see Box 17C). The location of the reticular formation is shown here in relation to some other major landmarks at different levels of the brainstem. The mesencephalic and rostral pontine reticular formation (yellow) mainly modulates forebrain activity and the caudal pontine and medullary reticular formation (red) provides premotor coordination of lower somatic motor and visceral motor neuronal pools.

BOX 17C ■ The Reticular Formation

If one were to exclude from the structure of the brainstem the cranial nerve nuclei, the nuclei that provide input to the cerebellum, the long ascending and descending tracts that convey explicit sensory and motor signals, and the structures that lie dorsal and lateral to the ventricular system, what would be left is a central core region known as the *tegmentum* (Latin, "covering structure"), so named because it "covers" the ventral part of the brainstem. Scattered among the diffuse fibers that course through the tegmentum are small clusters of neurons that are collectively known as the reticular formation. With few exceptions, these clusters of neurons are difficult to recognize as distinct nuclei in standard histological preparations. Indeed, the modifying term *reticular* ("netlike") was applied to this loose collection of neuronal clusters because early histologists envisioned these neurons as part of a sparse network of diffusely connected cells that extends from the intermediate gray regions of the cervical spinal cord to the lateral regions of the hypothalamus and certain nuclei along the midline of the thalamus.

These early anatomical concepts were influenced by lesion experiments in animals and clinical observations in human patients made in the 1930s and 1940s. These studies showed that damage to the upper brainstem tegmentum produced coma, suggesting the existence of a neural system in the midbrain and rostral pons that supported typical conscious brain states and transitions between sleep and wakefulness. These ideas were articulated most influentially by G. Moruzzi and H. Magoun when they proposed a "reticular activating system" to account for these functions and the critical role of the brainstem reticular formation.

Current evidence generally supports the notion of an activating function of the rostral reticular formation; however, neuroscientists now recognize the complex interplay of a variety of neurochemical systems (with diverse postsynaptic effects) comprising distinct cell clusters in the rostral tegmentum, and myriad other functions performed by neuronal clusters in more caudal parts of the reticular formation. Thus, with the advent of more precise means of demonstrating anatomical connections, as well as more sophisticated means of identifying neurotransmitters and the activity patterns of individual neurons, the concept of a "sparse network" engaged in a common function is now obsolete. Nevertheless, the term *reticular formation* remains, as does the daunting challenge of understanding the anatomical complexity and functional heterogeneity of this intricate brain region. Fortunately, two simplifying generalizations can be made. First, the functions of the different clusters of neurons in the reticular formation can be grouped into two broad

BOX 17C ■ (continued)

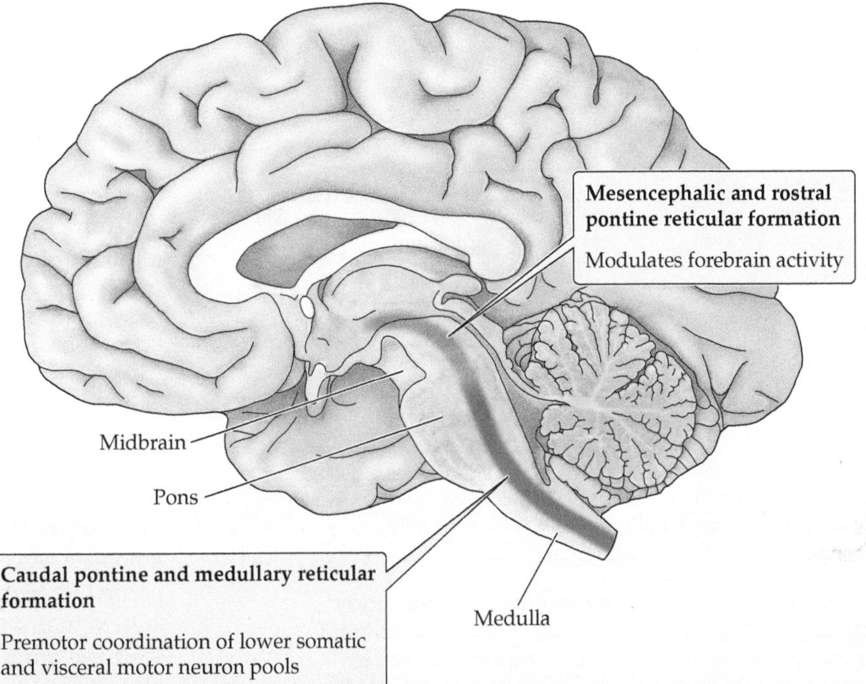

Midsagittal view of the brain showing the longitudinal extent of the reticular formation and highlighting the broad functional roles performed by neuronal clusters in its rostral (gold) and caudal (red) sectors.

categories: *modulatory functions* and *premotor functions*. Second, the modulatory functions are found primarily in the rostral sector of the reticular formation, whereas most of the premotor functions are localized in more caudal regions.

Several clusters of large (magnocellular) neurons in the midbrain and rostral pontine reticular formation participate—together with certain diencephalic nuclei—in the modulation of conscious states (see Chapter 28). These effects are accomplished by long-range, diencephalic projections of cholinergic neurons near the superior cerebellar peduncle, as well as the more widespread forebrain projections of noradrenergic neurons in the locus coeruleus and serotonergic neurons in the raphe nuclei. Generally speaking, these biogenic amine neurotransmitters function as neuromodulators (see Chapter 6) that alter the membrane potential and thus the firing patterns of thalamocortical and cortical neurons (the details of these effects are explained in Chapter 28). Also included in this category are the dopaminergic systems of the ventral midbrain that modulate corticostriatal interactions in the basal ganglia (see Chapter 18) and the responsiveness of neurons in the prefrontal cortex and limbic forebrain (see Chapter 32). However, not all modulatory projections from the rostral reticular formation are directed toward the forebrain. Although they are not always considered part of the reticular formation, it is helpful to include in this functional group certain neuronal columns in the periaqueductal gray (surrounding the cerebral aqueduct) that project to the dorsal horn of the spinal cord and modulate the transmission of nociceptive signals (see Chapter 10). Reticular formation neurons in the caudal pons and medulla oblongata generally serve a premotor function in the sense that they integrate feedback sensory signals with executive commands from upper motor neurons and deep cerebellar nuclei and, in turn, organize the efferent activities of lower visceral motor and certain somatic motor neurons in the brainstem and spinal cord. Examples of this functional category include the smaller (parvocellular) neurons that coordinate a broad range of motor activities, including the gaze centers discussed in Chapter 20 and local circuit neurons near the somatic motor and branchiomotor nuclei that organize mastication, facial expressions, and a variety of reflexive orofacial behaviors such as sneezing, hiccupping, yawning, and swallowing. In addition, autonomic centers organize the efferent activities of specific pools of primary visceral motor neurons. Included in this subgroup are distinct clusters of neurons in the ventrolateral medulla that generate respiratory rhythms, and others that regulate the cardioinhibitory output of neurons in the nucleus ambiguus. Still other clusters organize more complex activities that require the coordination of both somatic motor and visceral motor outflow, such as gagging and vomiting, and even laughing and crying.

One set of neuronal clusters that does not fit easily into this rostrocaudal framework is the set of neurons that give rise to the reticulospinal projections. As described in the text, these neurons are distributed in both rostral and caudal sectors of the reticular formation, and they give rise to long-range projections that innervate lower motor neuron pools in the medial ventral horn of the spinal cord. The reticulospinal inputs serve to modulate the gain of segmental reflexes involving the muscles of the trunk and proximal limbs and to relay initiation signals for certain stereotypical patterns of limb movement, such as locomotion.

In summary, the reticular formation is best viewed as a heterogeneous collection of distinct neuronal clusters in the brainstem tegmentum. These neuronal clusters either modulate the excitability of distant neurons in the forebrain and spinal cord or coordinate the firing patterns of more local lower motor neuron pools engaged in reflexive or stereotypical somatic motor and visceral motor behaviors.

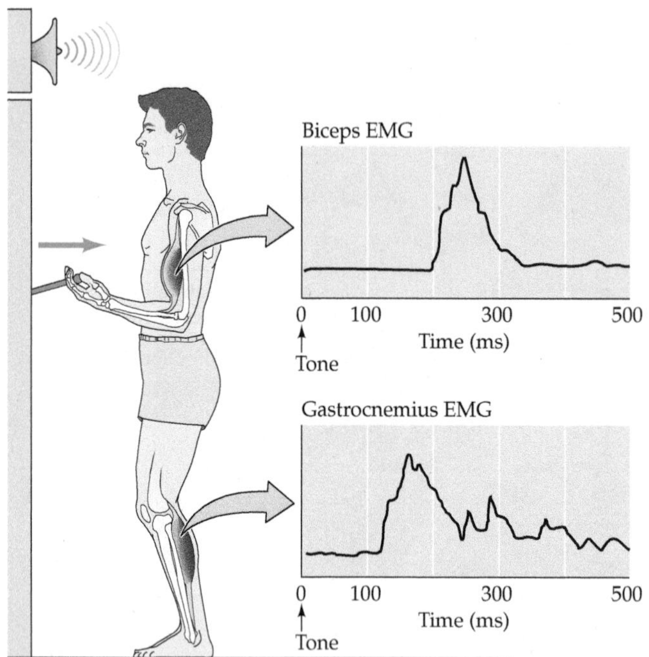

FIGURE 17.14 Anticipatory maintenance of body posture. At the onset of an audible tone, the individual pulls on a handle, contracting the biceps muscle. To ensure postural stability, contraction of the gastrocnemius muscle precedes that of the biceps. EMG refers to the electromyographic recording of muscle activity. (After Nashner, 1979.)

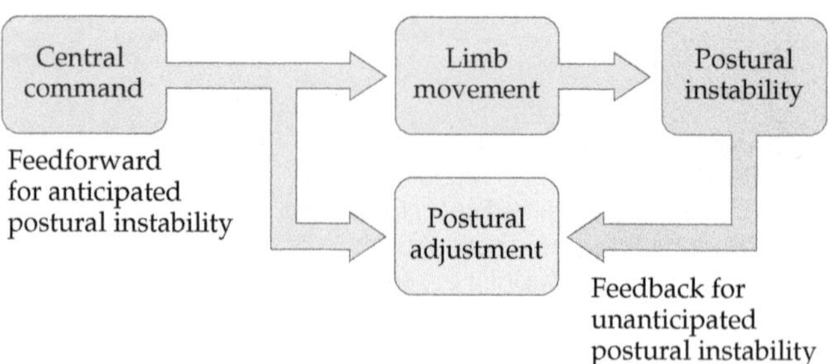

FIGURE 17.15 Feedforward and feedback mechanisms of postural control. Feedforward postural responses are "preprogrammed" and typically precede the onset of limb movement (see Figure 17.14). Feedback responses are initiated by sensory inputs that detect postural instability.

purpose of the movement. For example, Figure 17.14 shows the pattern of muscle activity that occurs as an individual uses his arm to pull on a handle in response to an auditory tone. Activity in the biceps muscle begins about 200 ms after the tone. However, as the records show, the contraction of the biceps is accompanied by a significant increase in the activity of a proximal leg muscle, the gastrocnemius (as well as many other muscles not monitored in the experiment). In fact, contraction of the gastrocnemius muscle begins well before contraction of the biceps. These observations show that postural control during movement entails an anticipatory, or feedforward, mechanism (Figure 17.15). As part of the motor plan for moving the arm, the effect of the impending movement on body stability is predicted and used to generate a change in the activity of the gastrocnemius muscle. This change actually precedes and provides preparatory postural support for the movement of the arm. In the example given in Figure 17.14, contraction of the biceps would tend to pull the entire body forward, an action that is opposed by the contraction of the gastrocnemius muscle. In short, this feedforward mechanism predicts the resulting disturbance in body stability and generates an appropriate stabilizing response.

The importance of the reticular formation for feedforward mechanisms of postural control has been explored in more detail in cats trained to use a forepaw to strike an object. As expected, the forepaw movement is accompanied by feedforward postural adjustments in the other legs to maintain the animal upright. These adjustments shift the animal's weight from an even distribution over all four feet to a diagonal distribution pattern, in which the weight is carried mostly by the contralateral, nonreaching forelimb and the ipsilateral hindlimb. Lifting of the forepaw and postural adjustments in the other limbs can also be induced in an alert cat by electrical stimulation of the motor cortex. After pharmacological inactivation of the reticular formation, however, electrical stimulation of the motor cortex evokes only the forepaw movement, without the feedforward postural adjustments that normally accompany them.

The results of this experiment can be understood in terms of the fact that the upper motor neurons in the motor cortex influence the spinal cord circuits by two routes: direct projections to the spinal cord (as discussed above) and indirect projections to brainstem centers that in turn project to the spinal cord. The reticular formation is one of the major destinations of these latter projections from the motor cortex; thus, cortical upper motor neurons initiate both the reaching movement of the forepaw and also the postural adjustments in the other limbs necessary to maintain body stability. The forepaw movement is initiated by the direct pathway from the cortex to the spinal cord, whereas the postural adjustments are mediated via pathways from the motor cortex that reach the spinal cord indirectly, after an intervening relay in the reticular formation (the so-called cortico-reticulospinal pathway; Figure 17.16).

Further evidence for the contrasting functions of the direct and indirect pathways from the motor cortex to the spinal cord comes from experiments carried out by the Dutch neurobiologist Hans Kuypers, who examined the behavior

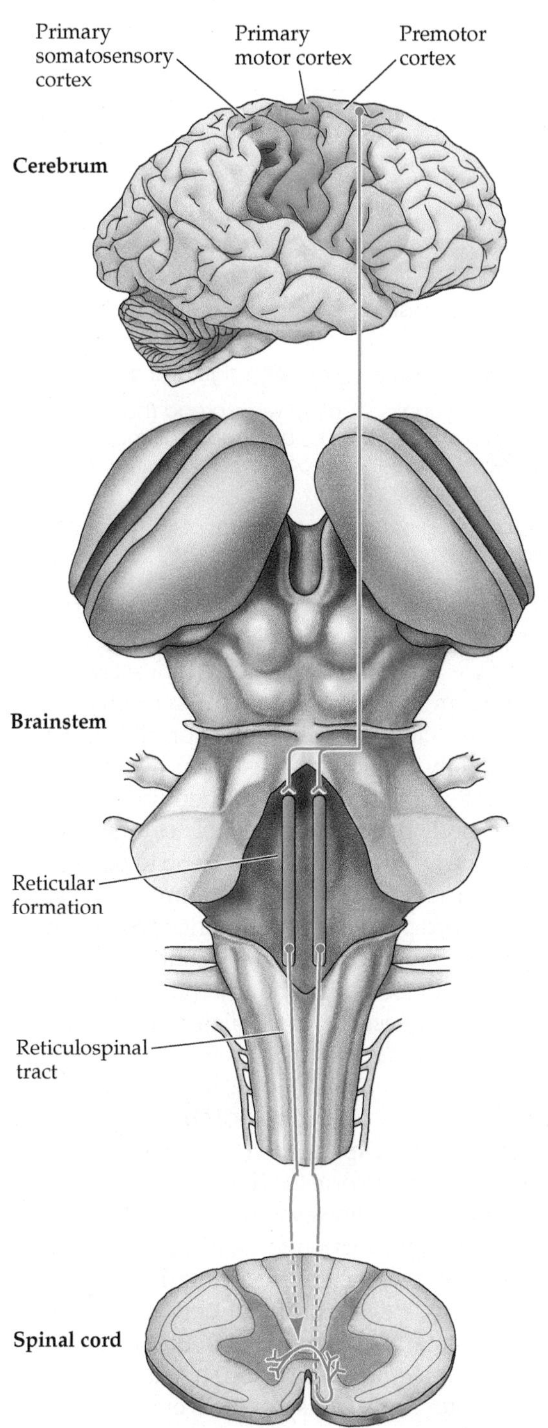

FIGURE 17.16 **Indirect pathways from the motor cortex to the spinal cord.** Neurons in the motor cortex that supply the lateral part of the ventral horn to initiate movements of the distal limbs (see Figure 17.4) also terminate on neurons in the reticular formation to mediate postural adjustments that support the movement. The reticulospinal pathway terminates in the more medial parts of the ventral horn, where lower motor neurons that innervate axial and proximal muscles are located. Thus, the motor cortex can influence the activity of spinal cord neurons via both direct and indirect routes.

of rhesus monkeys that had the direct pathway to the spinal cord transected at the level of the medulla, leaving only the indirect pathways to the spinal cord via the brainstem centers intact. Immediately after the surgery, the animals were able to use axial and proximal muscles to stand, walk, run, and climb, but they had great difficulty using the distal parts of their limbs (especially their hands) independently of other body movements. For example, the monkeys could cling to the cage but were unable to reach toward and pick up food with their fingers; rather, they used the entire arm to sweep the food toward them. After several weeks, the animals recovered some independent use of their hands and were again able to pick up objects of interest, but this action involved the concerted closure of all of the fingers. The ability to make independent, fractionated movements of the fingers, as in opposing the movements of the fingers and thumb to pick up an object, never returned.

These observations show that following damage to the direct projections from the motor cortex to the spinal cord at the level of the medulla, the indirect projections to the spinal cord from the motor cortex via the brainstem centers (or from brainstem centers alone) are capable of sustaining motor behavior that involves primarily the use of proximal muscles. In contrast, the direct projections from the motor cortex to the spinal cord provide the speed and agility of movements, and they enable a higher degree of precision in fractionated finger movements than is possible using the indirect pathways alone.

An additional brainstem structure, the **superior colliculus**, which is located in the dorsal midbrain, also contributes upper motor neuron pathways that govern lower motor neurons in the spinal cord. Although most mammals are likely to have direct projections from neurons in deep layers of the superior colliculus to the spinal cord (comprising a so-called colliculospinal or tectospinal tract), the major output of the superior colliculus to the spinal cord is mediated by the reticular formation. Thus, upper motor neurons in the superior colliculus innervate neural circuits in the reticular formation, which in turn give rise to reticulospinal projections that supply medial cell groups in the cervical cord. Functionally, this pathway plays a role in controlling axial musculature in the neck. These projections are particularly important in generating orienting movements of the head (Chapter 20 provides a detailed description of the role of the superior colliculus in the generation of head and eye movements).

Also in the midbrain is the mesencephalic locomotor region, which is involved in the initiation of locomotion (see Chapter 16). This region comprises a set of nuclei embedded in the reticular formation of the midbrain, just ventral and lateral to the periaqueductal gray matter. The mesenencephalic locomotor region projects to reticulospinal neurons in the medulla that, in turn, mediate the initiation and regulation of locomotion via connections with central pattern generators in the spinal cord.

In non-human primates and other mammals, a large nucleus in the tegmentum of the midbrain, termed the red nucleus, projects via the **rubrospinal tract** to the cervical level of the spinal cord (*rubro*—Latin, "red"—refers to the reddish color of this nucleus in fresh tissue, presumably due to the enrichment of its neurons with iron–protein complexes). Unlike the other projections from the brainstem to the spinal cord discussed thus far, the rubrospinal tract is located in the lateral white matter of the spinal cord; its axons terminate in lateral regions of the ventral horn and intermediate zone, where circuits of lower motor neurons governing the distal musculature of the upper extremities reside. Presumably, this projection participates together with the direct pathway from the motor cortex in the control of the arms (or forepaws). The limited distribution of rubrospinal projections may seem surprising, given the large size of the red nucleus in most mammals. However, the rubrospinal tract arises from especially large (magnocellular) neurons in the caudal pole of the red nucleus, which account for a relatively small fraction of the total number of neurons in the nucleus. In the human midbrain, there are few—if any—large neurons in the red nucleus; thus, if the rubrospinal tract exists in humans (which may not be the case in some individuals), its significance for motor control is dubious. Indeed, nearly all of the neurons in the red nucleus in humans are small (parvocellular) and do not project to the spinal cord at all; instead, many of these neurons relay information to the inferior olive, an important source of learning signals for the cerebellum (this role of the red nucleus will be discussed in Chapter 19).

TABLE 17.1 ■ **Signs and Symptoms of Lower and Upper Motor Neuron Lesions**

	Lower motor neuron syndrome	Upper motor neuron syndrome
Strength	Weakness or paralysis	Weakness
Muscle bulk	Severe atrophy develops	Mild or no atrophy develops
Reflexes	Hypoactive superficial and deep reflexes	Hyperactive deep reflexes after initial period of spinal shock
Special signs and symptoms	Initial signs and symptoms persist	Initial period of spinal shock, then spasticity ensues
	Fasciculations and fibrillations	Babinski's sign and clonus
	Geographic distribution of impairment (reflecting distribution of affected spinal segments, cranial nuclei, or spinal/cranial nerves)	More widespread (nongeographic) distribution of impairment in body regions
	Impairments of reflexive and gross and/or fine voluntary movements	Impairment of fine voluntary movements; gross movements relatively unimpaired

Damage to Descending Motor Pathways: The Upper Motor Neuron Syndrome

Injury to upper motor neurons is common because of the large amount of cortex occupied by the motor areas and because their pathways extend all the way from the cerebral cortex to the lower end of the spinal cord. Damage to the descending motor pathways anywhere along this trajectory gives rise to a set of symptoms called the **upper motor neuron syndrome**. This clinical picture is important for diagnosis of neurological problems because it differs markedly from the lower motor neuron syndrome described in Chapter 16 and entails a characteristic set of motor deficits (Table 17.1).

Damage to the motor cortex or the descending upper motor axons in the internal capsule typically causes an immediate flaccidity of the muscles on the contralateral side of the body and lower face. Given the topographical arrangement of the motor system, identifying the specific parts of the body that are affected helps localize the site of the injury. The acute manifestations tend to be most severe in the arms and legs. If the affected limb is elevated and released, it drops passively, and all reflex activity on the affected side is abolished. In contrast, control of trunk muscles is usually preserved, either by the remaining brainstem pathways or because of the bilateral projections of the corticospinal pathway to local circuits that control midline musculature. This initial period of "hypotonia" after upper motor neuron injury is called **spinal shock** and reflects the decreased activity of spinal circuits suddenly deprived of input from the motor cortex and brainstem.

After several days, however, the spinal cord circuits regain much of their function for reasons that are not fully understood, but may include the strengthening of remaining connections, the sprouting of new connections, and other homeostatic reactions that promote sustained neural activity in local segmental circuits. Thereafter, a consistent pattern of motor signs and symptoms emerges, including:

- *The Babinski sign.* The typical response in an adult to sharply stroking the sole of the foot is flexion of the big toe, and often the other toes. Following damage to descending upper motor neuron pathways, however, this stimulus may elicit extension of the big toe and a fanning of the other toes (Figure 17.17). A similar response occurs in human infants before the maturation of the corticospinal pathway and presumably indicates incomplete upper motor neuron control of local motor neuronal circuitry.

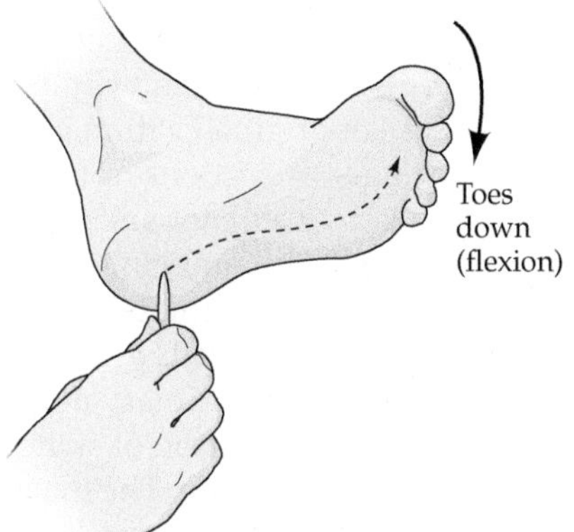

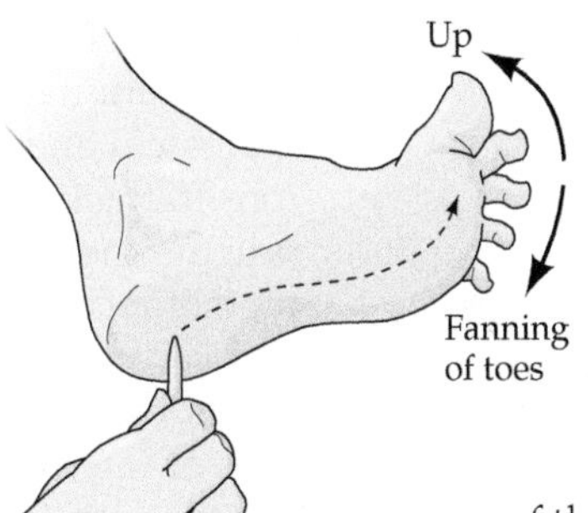

FIGURE 17.17 The Babinski sign. Following damage to descending corticospinal pathways, stroking the sole of the foot may cause an atypical fanning of the toes and the extension of the big toe.

- *Spasticity*. Spasticity is increased muscle tone (Box 17D), hyperactive stretch reflexes, and clonus (oscillatory contractions and relaxations of muscles in response to muscle stretching). Extensive upper motor neuron lesions may be accompanied by rigidity of the extensor muscles of the leg and the flexor muscles of the arm (called decerebrate rigidity; see below). Spasticity is probably caused by disruption of the regulatory influences exerted by the cortex on the postural centers of the vestibular nuclei and reticular formation, which in turn serve to govern the excitability of segmental circuits in the spinal cord. In experimental animals, for instance, lesions of the vestibular nuclei ameliorate the spasticity that follows damage to the corticospinal tract. Spasticity is also eliminated by sectioning the dorsal roots, suggesting that it represents an atypical increase in the *gain* of the spinal cord stretch reflexes due to loss of descending suppression (see Chapter 16). This increased gain of segmental circuits is also thought to explain clonus.

BOX 17D ■ Muscle Tone

Muscle tone refers to the resting level of tension in a muscle. In general, maintaining an appropriate level of tone allows a muscle to respond optimally to voluntary or reflexive commands in a given context. Tone in the extensor muscles of the legs, for example, helps maintain posture while standing. By keeping the muscles in a state of readiness to resist stretch, tone in the leg muscles prevents the amount of sway that typically occurs while standing from becoming too large. During activities such as walking or running, the "background" level of tension in leg muscles also helps store mechanical energy, in effect enhancing the muscle tissue's springlike qualities.

Muscle tone depends on the resting level of discharge of α motor neurons. Activity in muscle spindle afferents—the neurons responsible for the stretch reflex—is the major contributor to this tonic level of firing. As described in Chapter 16, the γ efferent system (by its action on intrafusal muscle fibers) regulates the resting level of activity in spindle afferents and thus establishes the baseline level of α motor neuron activity in the absence of muscle stretch.

Clinically, muscle tone is assessed by judging the resistance of a patient's limb to passive stretch. Damage to either the α motor neurons or the spindle afferents carrying sensory information to the α motor neurons results in a decrease in muscle tone, called *hypotonia*. In general, damage to descending pathways that terminate in the spinal cord has the opposite effect, leading to an increase in muscle tone, or *hypertonia* (except during the initial phase of spinal shock; see text). The neural changes responsible for hypertonia following damage to higher centers are not well understood; however, at least part of this change is due to an increase in the responsiveness of α motor neurons to spindle afferent inputs. Thus, in experimental animals in which descending inputs have been severed, the resulting hypertonia can be eliminated by sectioning the dorsal roots.

Increased resistance to passive movement following damage to higher centers is called *spasticity* and is associated with two other characteristic signs: the clasp-knife phenomenon and clonus. When first stretched, a spastic muscle provides a high level of resistance to the stretch and then suddenly yields, much like the blade of a pocket knife (or clasp knife, in old-fashioned terminology). Hyperactivity of the stretch reflex loop is the reason for the increased resistance to stretch in the clasp-knife phenomenon. The physiological basis for the inhibition that causes the sudden collapse of the stretch reflex (and loss of muscle tone) may involve the activation of Golgi tendon organs and/or inhibitory Ib interneurons in the spinal cord (see Chapter 16).

Clonus refers to a rhythmic pattern of contractions (3 to 7 per second) due to the alternate stretching and unloading of the muscle spindles in a spastic muscle. Clonus can be demonstrated in the flexor muscles of the leg by pushing up on the sole of an individual's foot to dorsiflex the ankle. If there is damage to descending upper motor neuron pathways, holding the ankle loosely in this position generates rhythmic contractions of both the gastrocnemius and soleus muscles. Both the increase in muscle tone and the pathological oscillations seen after damage to descending pathways are very different from the tremor at rest and cogwheel rigidity present in basal ganglia disorders such as Parkinson's disease, phenomena discussed in Chapter 18.

- *Loss of the ability to perform fine movements.* If the lesion involves the descending pathways that control the lower motor neurons to the distal upper limbs, the ability to execute fine movements (such as independent movements of the fingers) may be severely impaired.

Although these upper motor neuron signs and symptoms may arise from damage anywhere along the descending pathways, the spasticity that follows damage to descending pathways in the spinal cord is less marked than the spasticity that follows damage to the cortex or internal capsule. For example, the spastic extensor muscles in the legs of a patient with spinal cord damage cannot support the individual's body weight, whereas those of a patient with damage at the cortical level often can. However, lesions that interrupt the descending pathways in the brainstem above the level of the vestibular nuclei but below the level of the red nucleus cause even greater extensor tone than that which occurs after damage to higher regions. Sherrington, who first described this phenomenon, called the increased tone **decerebrate rigidity**. In the cat, the extensor tone in all four limbs is so great after lesions that spare the vestibulospinal tracts that the animal can stand without support. Patients with severe brainstem injury at the level of the pons may exhibit similar signs of decerebration: arms and legs stiffly extended, jaw clenched, and neck retracted. The relatively greater hypertonia following damage to the nervous system above the level of the medulla oblongata is presumably explained by the remaining activity of the intact descending pathways from the vestibular nuclei and reticular formation, which evidently have a net excitatory influence on the gain of segmental reflexes that contribute to posture and equilibrium in the context of impaired cortico-reticular regulation.

Summary

Two sets of upper motor neuron pathways make distinct contributions to the control of the local circuitry in the brainstem and spinal cord. One set originates from neurons in the frontal lobe and includes projections from the primary motor cortex and the nearby premotor areas. The premotor cortices are responsible for planning, initiating, and controlling complex sequences of voluntary movements, especially movements that are triggered by sensory cues or internal motivations, whereas the primary motor cortex is especially involved with the execution of skilled movements of the limb and facial musculature. The motor cortex influences movements *directly* by contacting lower motor neurons and local circuit neurons in the spinal cord and brainstem; and *indirectly* by innervating neurons in brainstem centers (mainly the reticular formation) that in turn project to lower motor neurons and circuits. The other major upper motor neuron pathways originate from brainstem centers—primarily the reticular formation and the vestibular nuclei—and are responsible for postural regulation. The reticular formation is especially important in *feedforward* control of posture (i.e., movements that occur in anticipation of changes in body stability). In contrast, the neurons in the vestibular nuclei that project to the spinal cord are especially important in *feedback* postural mechanisms (i.e., in producing movements that are generated in response to sensory signals that indicate an existing postural disturbance). Although the brainstem pathways can independently organize gross motor control, direct projections from the motor cortex to local circuit neurons in the brainstem and spinal cord are essential for the fine, fractionated movements of the face and the distal parts of the limbs that are especially important in activities of daily living and the expression of motor skill.

ADDITIONAL READING

Reviews

Dum, R. P. and P. L. Strick (2002) Motor areas in the frontal lobe of the primate. *Physiol. Behav.* 77: 677–682.

Gahery, Y. and J. Massion (1981) Coordination between posture and movement. *Trends Neurosci.* 4: 199–202.

Georgeopoulos, A. P., M. Taira and A. Lukashin (1993) Cognitive neurophysiology of the motor cortex. *Science* 260: 47–52.

Geyer, S., M. Matelli and G. Luppino (2000) Functional neuroanatomy of the primate isocortical motor system. *Anat. Embryol.* 202: 443–474.

Graziano, M. S. A. (2016) Ethological action maps: a paradigm shift for the motor cortex. *Trends. Cog. Sci.* 20: 121–132.

Kuypers, H. G. J. M. (1981) Anatomy of the descending pathways. In *Handbook of Physiology*, V. B. Brooks (ed.). Section 1: *The Nervous System*, vol. II, part 1. Bethesda, MD: American Physiological Society, pp. 597–666.

Nashner, L. M. (1979) Organization and programming of motor activity during posture control. In *Reflex Control of Posture and Movement*, R. Granit and O. Pompeiano (eds.). *Prog. Brain Res.* 50: 177–184.

Nashner, L. M. (1982) Adaptation of human movement to altered environments. *Trends Neurosci.* 5: 358–361.

Rizzolatti, G. and C. Sinigaglia (2016) The mirror mechanism: a basic principle of brain function. *Nat. Rev. Neurosci.* 17: 757–765.

Sherrington, C. and S. F. Grunbaum (1901) Observations on the physiology of the cerebral cortex of some of the higher apes. *Proc. R. Soc.* 69: 206–209.

Important Original Papers

Caspers, S., K. Zilles, A. R. Laird and S. B. Eickhoff (2010) ALE meta-analysis of action observation and imitation in the human brain. *NeuroImage* 50: 1148–1167.

Evarts, E. V. (1981) Functional studies of the motor cortex. In *The Organization of the Cerebral Cortex,* F. O. Schmitt, F. G. Worden, G. Adelman and S. G. Dennis (eds.). Cambridge, MA: MIT Press, pp. 199–236.

Fetz, E. E. and P. D. Cheney (1978) Muscle fields of primate corticomotoneuronal cells. *J. Physiol.* (*Paris*) 74: 239–245.

Fetz, E. E. and P. D. Cheney (1980) Postspike facilitation of forelimb muscle activity by primate corticomotoneuronal cells. *J. Neurophysiol.* 44: 751–772.

Georgeopoulos, A. P., A. B. Swartz and R. E. Ketter (1986) Neuronal population coding of movement direction. *Science* 233: 1416–1419.

Graziano, M. S. A., T. N. S. Aflalo and D. F. Cooke (2005) Arm movements evoked by electrical stimulation in the motor cortex of monkeys. *J. Neurophysiol.* 94: 4209–4223.

Kuypers, H. G. J. M. (1958) Corticobulbar connexions to the pons and lower brain-stem in man. *Brain* 81: 364–388.

Lawrence, D. G. and H. G. J. M. Kuypers (1968) The functional organization of the motor system in the monkey. I. The effects of bilateral pyramidal lesions. *Brain* 91: 1–14.

Mitz, A. R., M. Godschalk and S. P. Wise (1991) Learning-dependent neuronal activity in the premotor cortex: Activity during the acquisition of conditional motor associations. *J. Neurosci.* 11: 1855–1872.

Rizzolatti, G., L. Fadiga, V. Gallese and L. Fogassi (1996) Premotor cortex and the recognition of motor actions. *Cogn. Brain Res.* 3: 131–141.

Roland, P. E., B. Larsen, N. A. Lassen and E. Skinhof (1980) Supplementary motor area and other cortical areas in organization of voluntary movements in man. *J. Neurophysiol.* 43: 118–136.

Sanes, J. N. and W. Truccolo (2003) Motor "binding": Do functional assemblies in primary motor cortex have a role? *Neuron* 38: 115–125.

Schieber, M. H. and L. S. Hibbard (1993) How somatotopic is the motor cortex hand area? *Science* 261: 489–492.

Books

Asanuma, H. (1989) *The Motor Cortex.* New York: Raven Press.

Nicolelis, M. A. L. (2011) *Beyond Boundaries: The New Neuroscience of Connecting Brains with Machines—and How It Will Change Our Lives.* New York: Times Books.

Passingham, R. (1993) *The Frontal Lobes and Voluntary Action.* Oxford, UK: Oxford University Press.

Penfield, W. and T. Rasmussen (1950) *The Cerebral Cortex of Man: A Clinical Study of Localization of Function.* New York: Macmillan.

Porter, R. and R. Lemon (1993) *Corticospinal Function and Voluntary Movement.* Oxford, UK: Oxford University Press.

Sherrington, C. (1947) *The Integrative Action of the Nervous System,* 2nd Edition. New Haven, CT: Yale University Press.

Sjölund, B. and A. Björklund (1982) *Brainstem Control of Spinal Mechanisms.* Amsterdam: Elsevier.

Go to the NEUROSCIENCE 6e Companion Website at **www.oup.com/uk/Purves6e** for Web Topics, Animations, Flashcards, and more.

CHAPTER

18

Modulation of Movement by the Basal Ganglia

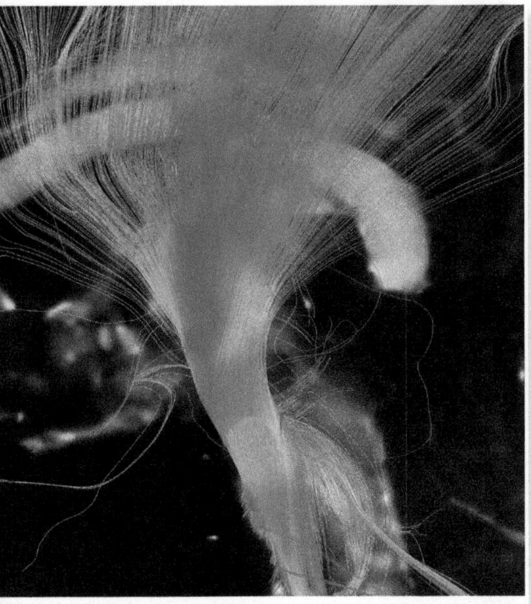

Overview

IN CONTRAST TO THE UPPER MOTOR NEURONS IN THE MOTOR REGIONS of the cerebral cortex and brainstem (discussed in Chapter 17), the basal ganglia and cerebellum do not directly influence lower motor neuronal circuitry; instead, these brain regions influence movement by regulating the activity of upper motor neuronal circuits. The term *basal ganglia* refers to a large and functionally diverse set of nuclei that lies deep within the cerebral hemispheres. The subset of these nuclei relevant to this account of motor function includes the caudate, the putamen, and the globus pallidus. Two additional structures, the substantia nigra in the base of the midbrain and the subthalamic nucleus in the ventral thalamus, are closely associated with the motor functions of these basal ganglia nuclei. The motor components of the basal ganglia, together with the substantia nigra and the subthalamic nucleus, comprise a subcortical loop that links most areas of the cerebral cortex with upper motor neurons in the primary motor and premotor cortices and in the brainstem. The neurons in this loop modulate their activity mainly at the beginning and ending of movement sequences, and their influences on upper motor neurons are required for functional regulation of voluntary movements. When one of these components of the basal ganglia or associated structures is compromised, the motor systems cannot switch smoothly between commands that initiate and maintain a movement and those that terminate the movement. The disordered movements that result can be understood as a consequence of maladaptive upper motor neuron activity that results from dysregulation of the control provided by the basal ganglia.

Projections to the Basal Ganglia

The motor nuclei of the basal ganglia that modulate the movements of the body are divided into several functionally distinct groups. The first and larger of these groups is called the **striatum**, which includes two principle nuclei, the **caudate** and the **putamen** (Figure 18.1). An older term for these nuclei (plus additional components of the basal ganglia described below) is *corpus striatum*, which means "striped body," reflecting the fact that the caudate and the dorsal part of the putamen are joined by slender bridges of gray matter that extend through the internal capsule and confer a striped appearance in parasagittal sections through this area. These two subdivisions of the corpus striatum comprise the *input zone* of the basal ganglia, since their neurons are the destinations of most of the pathways that reach this complex from other parts of the brain (Figure 18.2). The destinations of the incoming axons from the cerebral cortex are the dendrites of a class of cells in the corpus striatum called **medium spiny neurons** (Figure 18.3). The large dendritic trees of these neurons

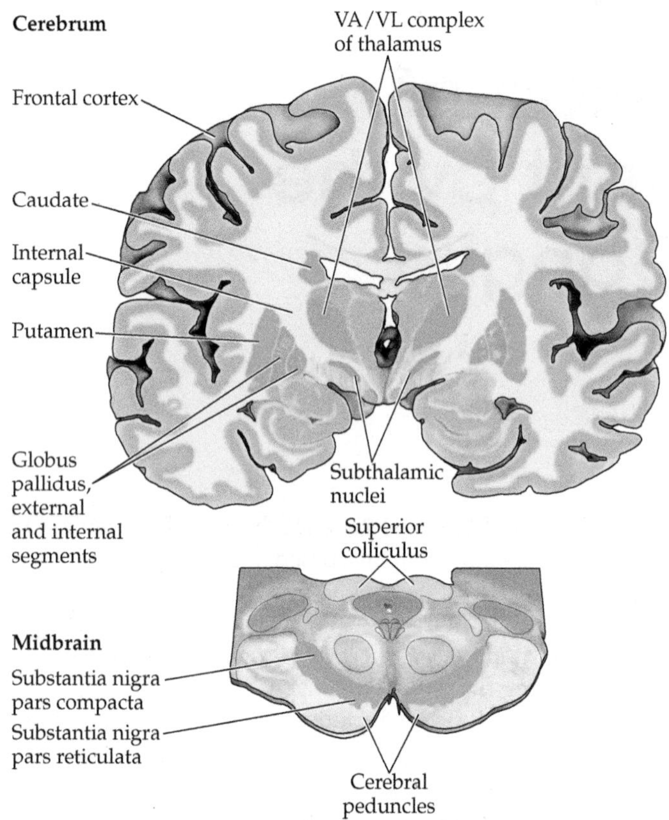

FIGURE 18.1 Motor components of the basal ganglia. The human basal ganglia comprise a set of gray matter structures, most of which are buried deep in the telencephalon, although some are found in the diencephalon and midbrain. The major components that receive and process movement-related signals are the striatum (caudate and putamen) and the pallidum (globus pallidus and substantia nigra pars reticulata). These structures border the internal capsule in the forebrain and midbrain (the cerebral peduncle is a caudal extension of the internal capsule). Smaller but functionally significant components of the basal ganglia system are the substantia nigra pars compacta and the subthalamic nucleus, which provide input to the striatum and pallidum, respectively. For the control of limb movements, output from the basal ganglia arises in the internal segment of the globus pallidus and is sent to the ventral anterior and ventral lateral nuclei (VA/VL complex) of the thalamus, which interact directly with circuits of upper motor neurons in frontal cortex. The substantia nigra pars reticulata projects to upper motor neurons in the superior colliculus and controls orienting movements of the eyes and head.

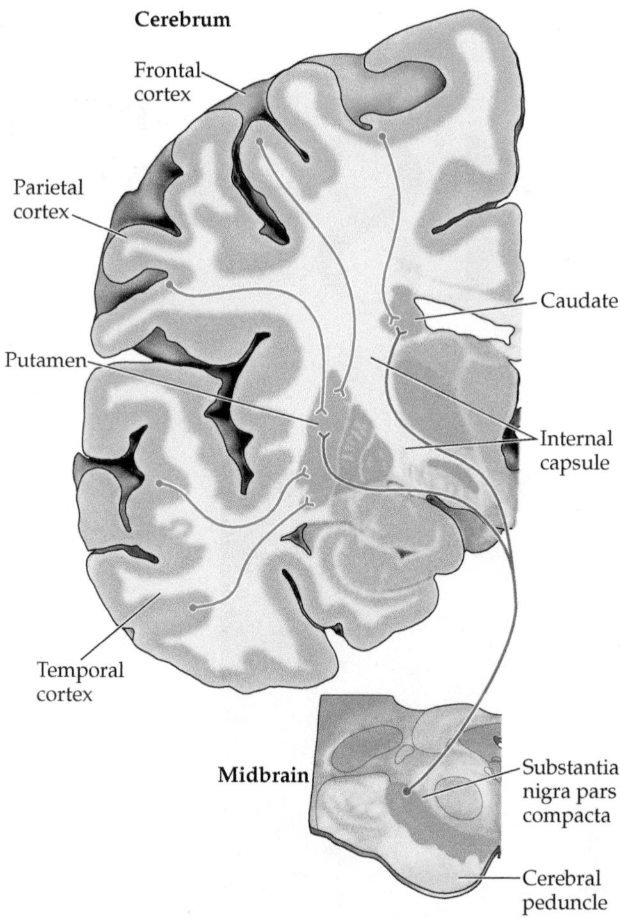

FIGURE 18.2 Anatomical organization of the inputs to the basal ganglia. Idealized coronal sections through the human forebrain and midbrain, showing the projections from the cerebral cortex and substantia nigra pars compacta to the caudate and putamen.

allow them to collect and integrate input from a variety of cortical, thalamic, and brainstem structures. The axons arising from the medium spiny neurons converge on neurons in the **pallidum**, which includes the **globus pallidus** and the **substantia nigra pars reticulata**. The globus pallidus and substantia nigra pars reticulata are the main sources of *output* from the basal ganglia complex to other parts of the brain (see Figure 18.4).

Historically, the globus pallidus had been recognized as a component of the corpus striatum; however, given the important neurochemical, anatomical, and physiological distinctions between striatum and pallidum (discussed in detail below), it is important to distinguish the globus pallidus from striatal divisions of the corpus striatum. Therefore, to avoid confusion that often attends this terminology, we will hereafter avoid the term *corpus striatum* in favor of more specific reference to components of the striatum or the pallidum.

Nearly all regions of the cerebral cortex project directly to the striatum, making the cortex the source of the largest input to the basal ganglia. The majority of these projections are from association areas in the frontal and parietal lobes, but substantial contributions also arise from the temporal, insular, and cingulate cortices. All of these projections,

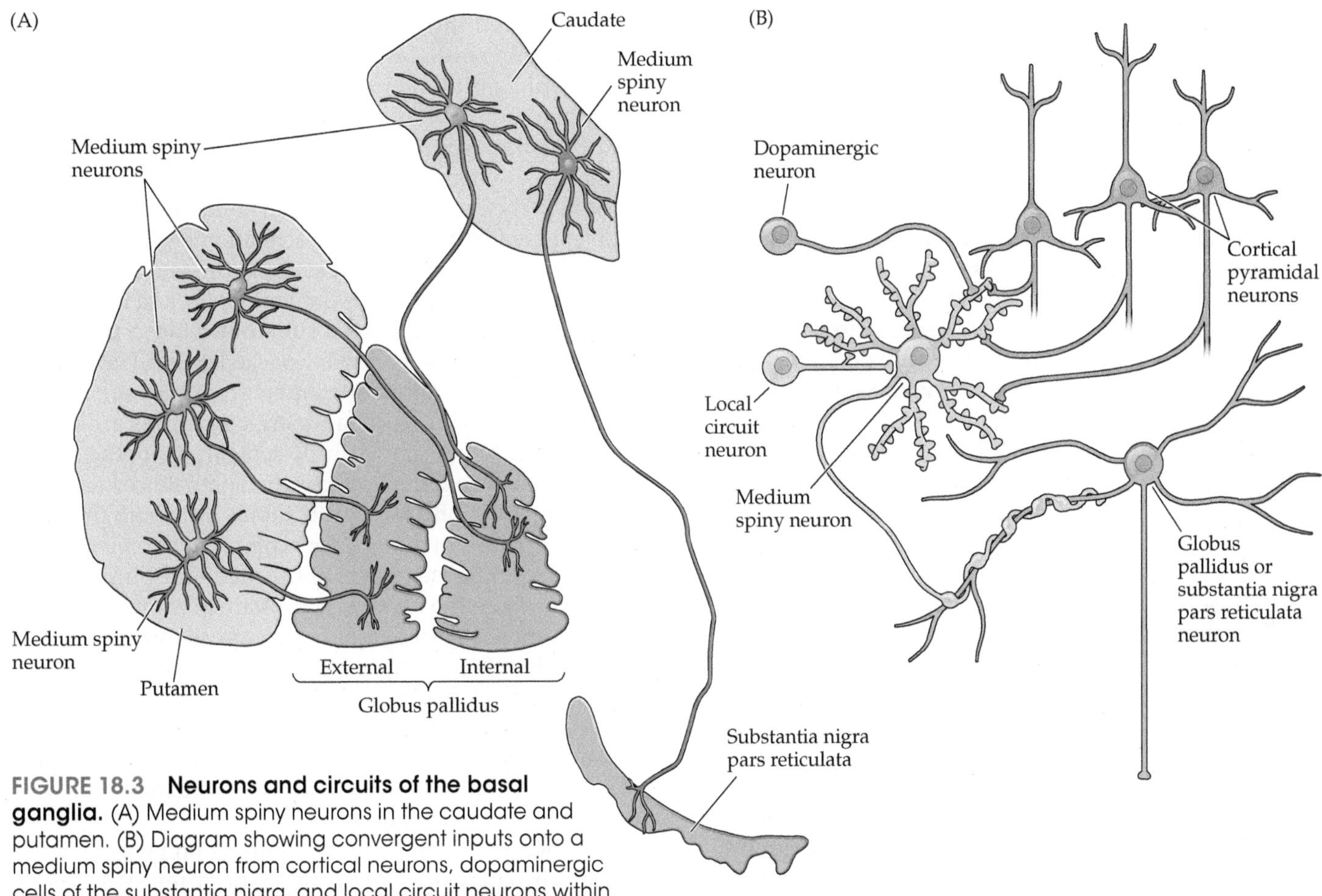

FIGURE 18.3 Neurons and circuits of the basal ganglia. (A) Medium spiny neurons in the caudate and putamen. (B) Diagram showing convergent inputs onto a medium spiny neuron from cortical neurons, dopaminergic cells of the substantia nigra, and local circuit neurons within the striatum. The arrangement of these synapses indicates that the response of the medium spiny neurons to their principal input, derived from the cerebral cortex, can be modulated by dopamine and the inputs of local circuit neurons. The primary output of the medium spiny cells is to neurons in the globus pallidus and substantia nigra pars reticulata.

referred to collectively as the **corticostriatal pathway**, travel through the subcortical white matter on their way to the caudate and putamen (see Figure 18.2).

The cortical inputs to the caudate and putamen are not equivalent, however, and the differences in these inputs reflect functional differences between the two nuclei. The caudate receives cortical projections primarily from multimodal association cortices and from motor areas in the frontal lobe that control eye movements. As their name implies, the association cortices do not process any one type of sensory information; rather, they receive input from several primary and secondary sensory cortices and their associated thalamic nuclei (see Chapter 27). The putamen, by contrast, receives input from the primary and secondary somatosensory cortices in the parietal lobe, the higher order (extrastriate) visual cortices in the occipital and temporal lobes, the premotor and primary motor cortices in the frontal lobe, and the auditory association areas in the temporal lobe. The fact that different cortical areas project to different regions of the striatum implies that the corticostriatal pathway consists of multiple parallel pathways serving different functions. This interpretation is supported by the observation that the segregation is maintained in the output structures that receive projections from the striatum and in the output pathways that project from the basal ganglia to other brain regions.

The distribution of parallel corticostriatal pathways within the striatum reflects the functional organization of the cerebral cortex. For example, visual and somatosensory cortical projections are topographically mapped within different regions of the putamen. Moreover, the cortical areas that are functionally interconnected at the level of the cortex give rise to projections that overlap extensively in the striatum. Anatomical studies by Ann Graybiel and her colleagues at the Massachusetts Institute of Technology have shown that different cortical areas concerned with the hand (see Chapter 9) send projections that converge in specific rostrocaudal bands within the striatum; conversely, cortical areas concerned with the leg give rise to projections that converge in other striatal bands. These rostrocaudal bands therefore appear to be functional units concerned with the movement of particular body parts. Another study by the same group shows that the more extensive the interconnections of cortical areas by corticocortical pathways, the greater the overlap in their projections to the striatum. Thus, the specialization of functional units within the striatum reflects the specialization of the cortical areas that provide their input.

A further indication of functional subdivision within the striatum is evident when tissue sections obtained post mortem are stained for the presence of different neurotransmitters and their related enzymes. For example, when the striatum is stained for the enzyme acetylcholinesterase, which inactivates acetylcholine (see Chapter 6), a compartmental organization is revealed within the striatum. The compartments are defined by lightly stained regions, called *patches* or *striosomes*, surrounded by densely stained tissue, called *matrix* or *matrisomes*. Subsequent studies of the distributions of other neurochemicals, including peptide neurotransmitters, have cataloged a variety of neuroactive substances that localize to the patch or matrix compartments. Tract-tracing experiments in animals have likewise shown differences between these striatal compartments in the sources of their inputs from the cortex and in the destinations of their projections to other parts of the basal ganglia. For example, the matrix makes up the bulk of the striatum; it receives input from most areas of the cerebral cortex and sends projections to the globus pallidus and the substantia nigra pars reticulata. The patches in the caudate receive most of their input from the prefrontal cortex (see Chapter 27) and project preferentially to a different subdivision of the substantia nigra (the dopaminergic neurons of the pars compacta; see below). Distinct patterns of projection from medium spiny neurons in the patches and matrix further support the conclusion that functionally distinct pathways project in parallel from the cerebral cortex to the striatum.

The nature of the information transmitted to the caudate and putamen from the cerebral cortex is not understood. It is known, however, that collateral axons of corticocortical, corticothalamic, and corticospinal pathways all form excitatory glutamatergic synapses on the dendritic spines of medium spiny neurons (see Figure 18.3B). The number of contacts established between an individual cortical axon and a single medium spiny cell is very small, whereas the number of spiny neurons contacted by a single axon is extremely large. This divergence of the inputs from corticostriatal axons allows a single medium spiny neuron to integrate the influences of thousands of cortical cells.

The medium spiny cells also receive inputs from several sources besides the cerebral cortex, including other medium spiny neurons via their local axon collaterals, local circuit interneurons of the striatum, neurons in the midline and intralaminar nuclei of the thalamus, and neurons in several nuclei of the brainstem that produce biogenic amine neurotransmitters. In contrast to the cortical inputs that synapse on the dendritic spines of the medium spiny neurons, the local circuit neuron and thalamic synapses are made on the dendritic shafts and close to the cell soma, where they can modulate the effectiveness of cortical synaptic activation of the more distal dendrites. One important set of brainstem inputs to the medium spiny neurons is dopaminergic, and it originates in a subdivision called the **substantia nigra pars compacta** because of its densely packed cells. (The striatum also receives serotonergic inputs from the raphe nuclei; see Chapter 6.) The dopaminergic synapses are located on the base of the spine, in close proximity to the cortical synapses, where they selectively modulate cortical input (see Figure 18.3B). As a result, inputs from both the cortex and the substantia nigra pars compacta are relatively far from the initial segments of the medium spiny neurons' axons, where the nerve impulses are generated. Furthermore, medium spiny neurons express inward-rectifier potassium conductances that tend to remain open near resting membrane potentials, but close with depolarization. Accordingly, these neurons exhibit very little spontaneous activity and must simultaneously receive many excitatory inputs to overcome the stabilizing influence of this potassium conductance.

When the medium spiny neurons do become active, their firing is associated with the occurrence of a movement. Extracellular recordings show that these neurons typically increase their rate of discharge before an impending movement. Neurons in the putamen tend to discharge in anticipation of limb and trunk movements, whereas caudate neurons fire prior to eye movement. These anticipatory discharges are evidently part of a movement selection as well as a movement initiation process; in fact, they can precede the initiation of movement by as much as several seconds. Similar recordings have also shown that the discharges of some striatal neurons vary according to the location in space of the *destination* of a movement, rather than with the starting position of the limb relative to the destination. Thus, the activity of these cells may encode the *decision to move* toward a goal rather than the direction and amplitude of the actual movement necessary to reach the goal. Furthermore, medium spiny neurons increase their firing rate at the termination of a movement sequence, which routinely coincides with the initiation of a subsequent motor program (e.g., the reinitiation of stationary, stable posture following a sequence of steps). This temporal relationship between the firing of medium spinal neurons and the initiation and termination of movement sequences has implicated the basal ganglia in the selection of action plans and the instantiation of habitual patterns of movement (Box 18A).

BOX 18A ■ Making and Breaking Habits

To one degree or another, we are all "creatures of habit," which is to say that patterns of thought and movement often display repetitive stereotypes that may serve to increase the efficiency of goal-oriented behavior. Indeed, habitual patterns of movement become "second nature" as motivated behavior increasingly loses dependence on explicit outcomes (attainment of reward) and component movements are consolidated into stereotyped, automated sequences. It has been long suspected that this process of associative sensorimotor learning and action automation involves basal ganglia circuitry. Presumably, one function of motor circuits in the basal ganglia is to acquire information related to stimulus-response associations and to initiate efficient patterns of movement driven by stimulus-response contingencies.

Research from the laboratory of Ann Graybiel at the Massachusetts Institute of Technology has shed considerable light on the contributions of striatal neurons and their through pathways in habit formation and the execution of habitual behavior. These studies show that the firing patterns of medium spiny neurons in the dorsolateral aspect of the striatum (primate putamen and head of the caudate nucleus) serve to "chunk" action sequences by accentuating the initiation and termination of overlearned patterns of movement. For example, as macaque monkeys freely viewed a visual display of possible targets

Continued on the next page

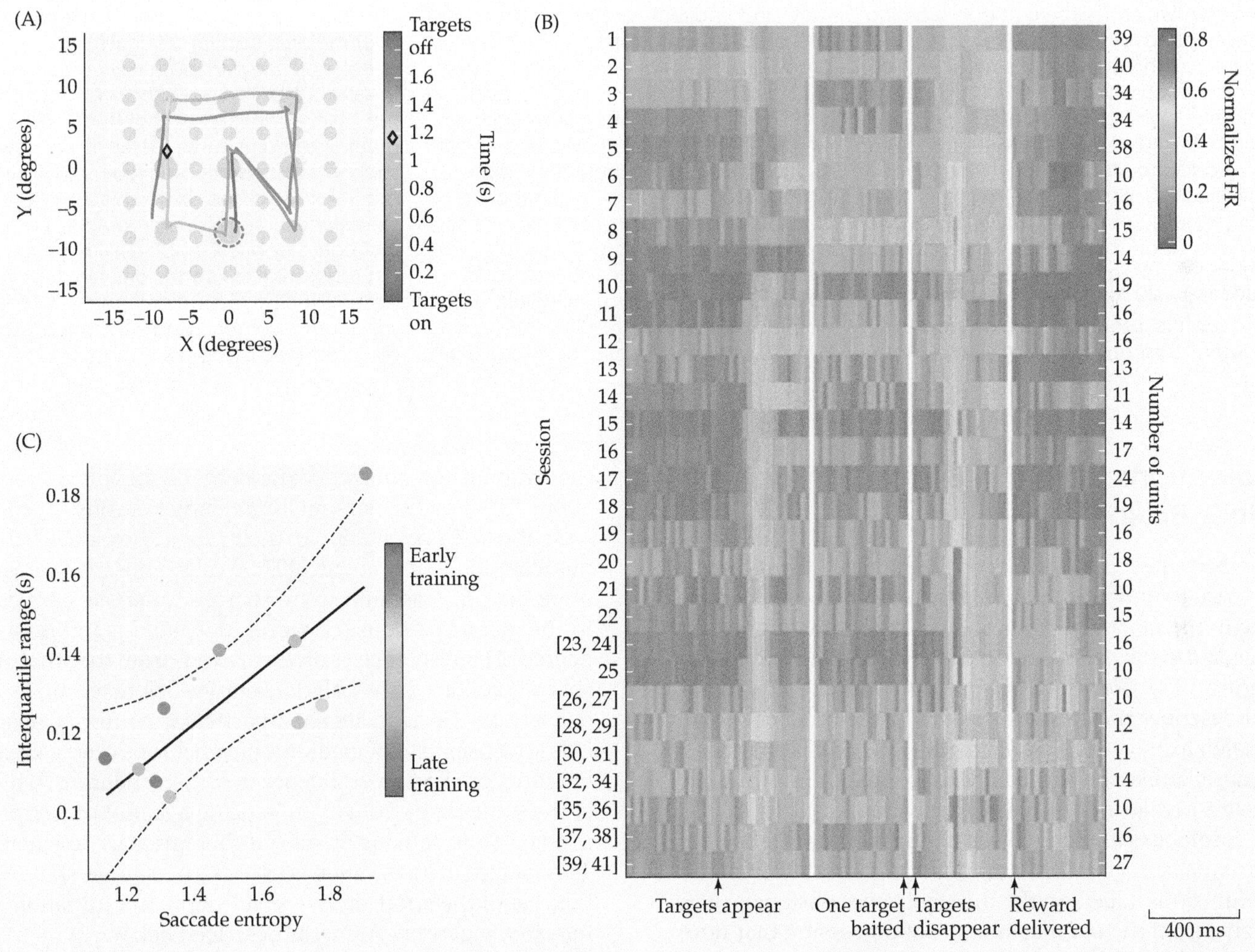

(A) Sample sequence of free visual scanning. Green targets appeared among an array of gray spots and a monkey began to scan the green targets until the scan path passed through a randomly chosen target (indicated by red dashed circle); the grid was then turned off and a reward provided. The monkey was not cued about the location of the bated target; it simply continued to scan across the green targets until they turned off at the conclusion of a rewarded trial. Black diamond indicates time (on color bar) and position (on grid) of monkey's gaze when the target became bated with reward. (B) All units recorded from the caudate nucleus in one monkey displayed across sessions. The firing rate (FR) of each unit was normalized and units were binned (20-ms bins), with each row representing the average activity of all units in a session. White vertical lines divide the phases of each session. Note the progressive increase in neuronal firing and the development of "sharpness" (increased tuning) of neuronal discharges across sessions at the start and especially at the termination of the visual scans. (C) Direct correlation between the tuning of neuronal activity in the striatum and the repetitiveness of visual scanning (saccade entropy). Smaller values of interquartile range (*y*-axis) indicate sharper tuning of neuronal activity, and lower values of saccade entropy (*x*-axis) indicate the formation of habitual patterns of scanning across the visual targets. Across sessions (color code), striatal neurons became more sharply tuned at the start and end of visual scans, and scan paths became more stereotyped. (A–C after Desrochers et al., 2015.)

BOX 18A ■ (continued)

for successive fixations (saccades; see Chapter 20), the discharges of medium spiny neurons in the head of the caudate nucleus became increasingly well tuned to the initiation and termination of stereotyped scan paths through the visual array as eye movements became more refined and habitual (see figure). This result suggests that striatal neurons encode an integrated cost-benefit signal by which reinforcement learning drives behaviors that minimize costs (in this case, the number of saccades necessary for completing a scan through the targets) and signal outcomes (completing a rewarded movement sequence).

Additional studies recently performed by Nicole Calakos, Henry Yin, and their colleagues at Duke University have teased apart the contributions of direct and indirect pathway striatal neurons in making and breaking habits. These investigators used two-photon scanning laser microscopy to perform calcium imaging in mouse brain slices. Their goal was to record simultaneously the evoked activities of both types of medium spiny neurons (and associated interneuons and glia) as a function of recent habitual behavior. The results indicated that plasticity mechanisms operating at the level of cortical inputs to striatal neurons are sufficient to drive habit formation, but with differential contributions of direct and indirect pathway projection neurons. As habits formed, there was a broadly distributed increase in the gain of striatal neuronal responses to cortical input, with a tendency for direct pathway striatal neurons to fire in advance of indirect pathway striatal neurons. Evidently, a timing competition between the direct and indirect pathways mediates the formation and expression of habitual patterns of movement. Plasticity advancing direct pathway activation would favor habit formation and reduce the probability of action cancellation associated with activation of the indirect pathway. Interestingly, habit breaking was mediated by weakening the response of direct pathway neurons to cortical activation, rather than by strengthening indirect pathway connections. This implies that habit suppression is a manifestation of a reduced drive for volitional movement.

Taken together, these studies indicate that broadly distributed plastic changes in corticostriatal connections alter the propagation of activity through direct and indirect pathways and bias the output of the basal ganglia toward the consolidation of habitual movement. It remains to be determined how such mechanisms of circuit plasticity are related to the excessive or overly stereotyped movement routines that are commonly associated with certain neuropsychiatric conditions, including obsessive-compulsive disorder, autism spectrum disorders, and substance use disorders (see Chapters 31 and 34).

Projections from the Basal Ganglia to Other Brain Regions

The medium spiny neurons of the caudate and putamen give rise to inhibitory GABAergic projections that terminate in the globus pallidus and the substantia nigra pars reticulata in the pallidal nuclei of the basal ganglia complex (Figure 18.4). *Globus pallidus* means "pale body," a name that describes the appearance of the large number of myelinated axons in this nucleus; *pars reticulata* is so named because, unlike the pars compacta, axons passing through give it a netlike, or reticulated, appearance.

The globus pallidus and substantia nigra pars reticulata share the same types of neurons and perform comparable functions, albeit on the different types of signals they receive from the parallel streams of processing that flow through the basal ganglia. In fact, the pars reticulata may be understood as being a part of the globus pallidus that, during early brain development, became separated from the rest of the pallidum by the formation of the posterior limb of the internal capsule and cerebral peduncle. The striatal projections to these two nuclei resemble the corticostriatal pathways in that they terminate in rostrocaudal bands, the locations of which vary with the locations of sources in the striatum. A striking feature of these projections is the degree of convergence from the medium spiny neurons to the neurons of the globus pallidus and substantia nigra pars reticulata. In humans, for example, the striatum contains approximately 100 million neurons, about 75% of which are medium spiny neurons. In contrast, the main destination of their axons, the globus pallidus, comprises only about 700,000 cells. Thus, on average, more than 100 medium spiny neurons innervate each cell in the globus pallidus. However, despite this impressive degree of convergence, individual axons from the striatum sparsely contact many pallidal neurons before terminating densely on the dendrites of a particular neuron. Consequently, ensembles of medium spiny neurons exert a broad but functionally weak influence over many neurons, while at the same time strongly influencing a subset of neurons in the globus pallidus or substantia nigra pars reticulata. This pattern of innervation is important for understanding the role of the striatum in the selection and initiation of intended motor programs, as described below.

The efferent neurons of the globus pallidus and substantia nigra pars reticulata together give rise to the major output pathways that allow the basal ganglia to influence the activity of upper motor neurons located in the motor cortex and in the brainstem (see Figure 18.4). The pathway to the cortex arises primarily in the medial division of the globus pallidus, called the **internal segment**, and reaches the motor cortex via a relay in the **ventral anterior** and **ventral lateral nuclei** of the dorsal thalamus. These thalamic nuclei project directly to motor areas of the cerebral cortex, thus completing a vast loop of circuitry that originates in multiple areas of the cortex and terminates in the

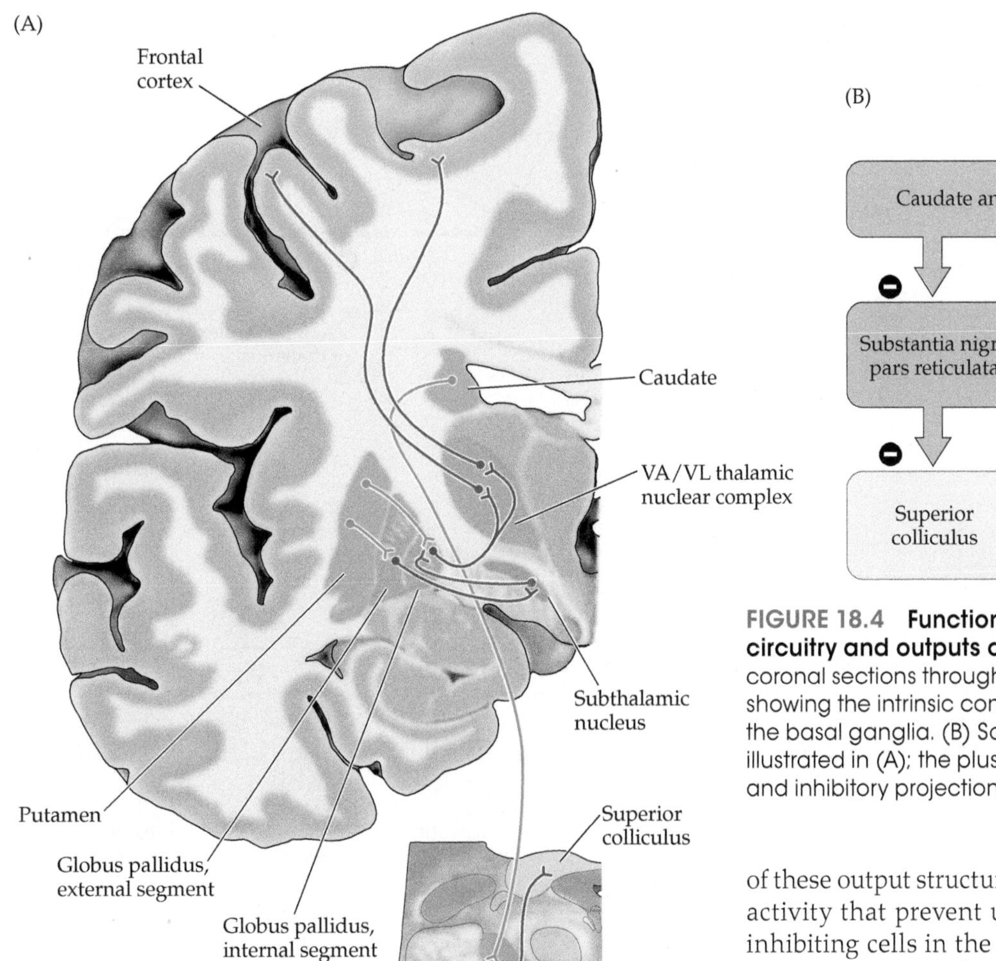

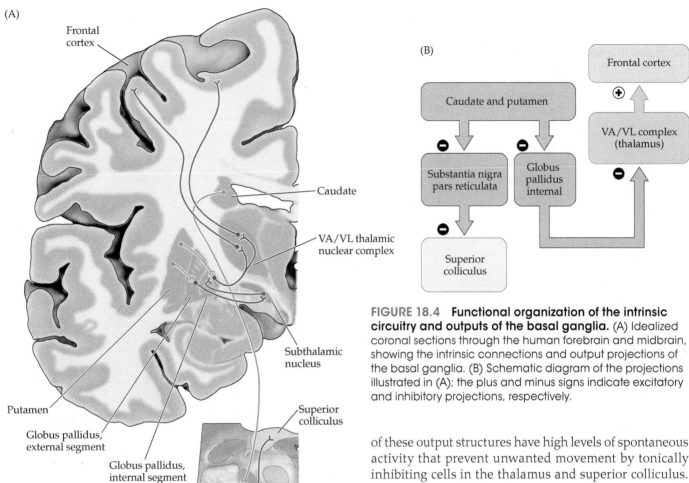

FIGURE 18.4 Functional organization of the intrinsic circuitry and outputs of the basal ganglia. (A) Idealized coronal sections through the human forebrain and midbrain, showing the intrinsic connections and output projections of the basal ganglia. (B) Schematic diagram of the projections illustrated in (A); the plus and minus signs indicate excitatory and inhibitory projections, respectively.

motor areas of the frontal lobe, after successive stages of processing in the basal ganglia and thalamus. In contrast, many efferent axons from substantia nigra pars reticulata have more direct access to upper motor neurons by synapsing on neurons in the superior colliculus that command head and eye movements, without an intervening relay in the thalamus. This difference between the globus pallidus and substantia nigra pars reticulata is not absolute, however, since many reticulata axons also project to the thalamus (mediodorsal and ventral anterior nuclei), where they contact relay neurons that project to the frontal eye fields of the premotor cortex (see Chapter 20). The thalamic relay is a mechanism for facilitating or suppressing inputs to circuits of upper motor neurons in the cortex—a level of organization that is not shared by the superior colliculus.

Because the efferent cells of both the globus pallidus and substantia nigra pars reticulata are GABAergic, the main output of the basal ganglia is *inhibitory*. In contrast to the quiescent medium spiny neurons, the neurons in both of these output structures have high levels of spontaneous activity that prevent unwanted movement by tonically inhibiting cells in the thalamus and superior colliculus. Because the medium spiny neurons of the striatum also are GABAergic and inhibitory, the net effect of the phasic excitatory inputs that reach the striatum from the cortex is to open a physiological gate by inhibiting the tonically active inhibitory cells of the globus pallidus and substantia nigra pars reticulata (Figure 18.5). For example, in the absence of volitional body movements (and the intention to make movements), the globus pallidus neurons provide tonic inhibition to the relay cells in the ventral lateral and ventral anterior nuclei of the thalamus. When the pallidal cells are inhibited by activation of the medium spiny neurons (as signals for volitional movement converge on the striatum), the thalamic neurons are *disinhibited* and can trigger the activation of upper motor neurons in the cortex. This disinhibition allows the upper motor neurons to send commands to local circuit neurons and lower motor neurons that in turn initiate movement.

Evidence from Studies of Eye Movements

The permissive, or gating, role of the basal ganglia in the initiation of movement is perhaps most clearly demonstrated by studies of eye movements carried out by Okihide Hikosaka and Robert Wurtz at the National Institutes of Health (Figure 18.6). As described in the previous section,

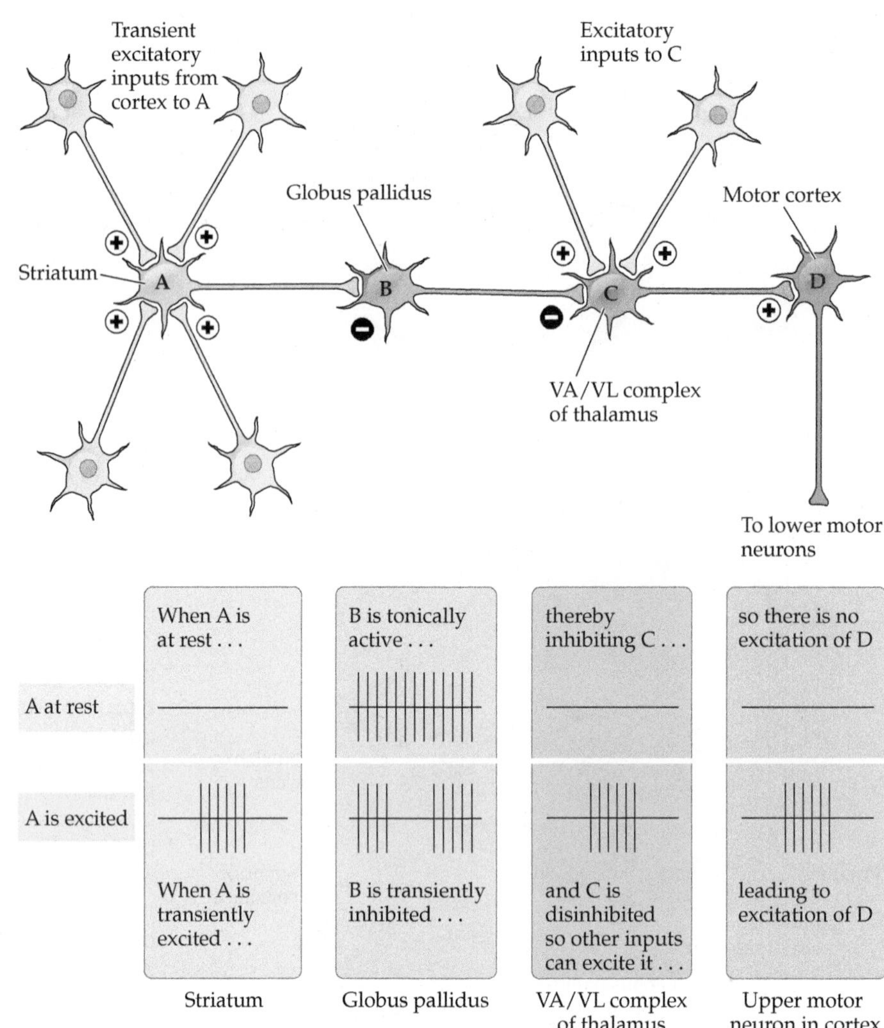

FIGURE 18.5 A chain of nerve cells arranged in a disinhibitory circuit. At the top is a diagram of the connections between neurons A and B and an excitatory neuron, C, which activates D, an upper motor neuron in the cortex. The colored boxes below diagram the pattern of action potential activity in A, B, C, and D both when neuron A is at rest and when neuron A fires transiently as a result of excitatory input. Such circuits are central to the gating operations of the basal ganglia.

the substantia nigra pars reticulata is part of the output circuitry of the basal ganglia. Instead of projecting to the ventral anterior and ventral lateral nuclei of the thalamus, however, it sends axons mainly to the deep layers of the superior colliculus. The upper motor neurons in these layers command the rapid, orienting movements of the eyes called *saccades* (see Chapter 20). When the eyes are fixating a visual target, these upper motor neurons are tonically inhibited by the spontaneously active reticulata cells, thus preventing unwanted saccades. Shortly before the onset of a saccade, the tonic discharge rate of the reticulata neurons is sharply reduced by input from the GABAergic medium spiny neurons of the caudate, which have been activated by signals from the cortex. The subsequent reduction in the tonic discharge from reticulata neurons disinhibits the upper motor neurons of the superior colliculus, allowing them to generate the bursts of action potentials that command the saccade. Thus, the projections from the substantia nigra pars reticulata to the upper motor neurons act as a physiological "gate" that must be "opened" to allow either sensory or other higher order signals from cognitive centers to activate the upper motor neurons and initiate a saccade.

This brief account of the genesis of saccadic eye movements provides an important illustration of the principal functions of the basal ganglia in motor control: The basal ganglia facilitate the *initiation* of motor programs that express movement and the *suppression* of competing or non-synergistic motor programs that would otherwise interfere with the expression of sensory-driven or goal-directed behavior (see Box 18A). Chapter 20 provides a more complete account of sensorimotor integration and the origins of eye movements; the remaining sections of this chapter will explain how the intrinsic and accessory circuits of the basal ganglia accomplish these principal functions in motor control and why disease that afflicts elements of these circuits can lead to devastating movement disorders.

Circuits within the Basal Ganglia System

The projections from the medium spiny neurons of the caudate and putamen to the internal segment of the globus pallidus constitute the so-called *direct pathway* through the basal ganglia and, as illustrated in Figure 18.4, serve to release from tonic inhibition the thalamic neurons that drive cortical circuits of upper motor neurons. Thus, this direct pathway provides a means for the basal ganglia to facilitate the initiation of volitional movement. Figure 18.7A summarizes the functional organization of the direct pathway.

Additional circuits of the basal ganglia constitute a so-called *indirect pathway* linking the caudate and putamen to the internal segment of the globus pallidus (Figure 18.7B). This second pathway increases the level of tonic inhibition mediated by the projection neurons of the internal segment (and the substantia nigra pars reticulata). In the indirect pathway, a distinct population of medium spiny neurons

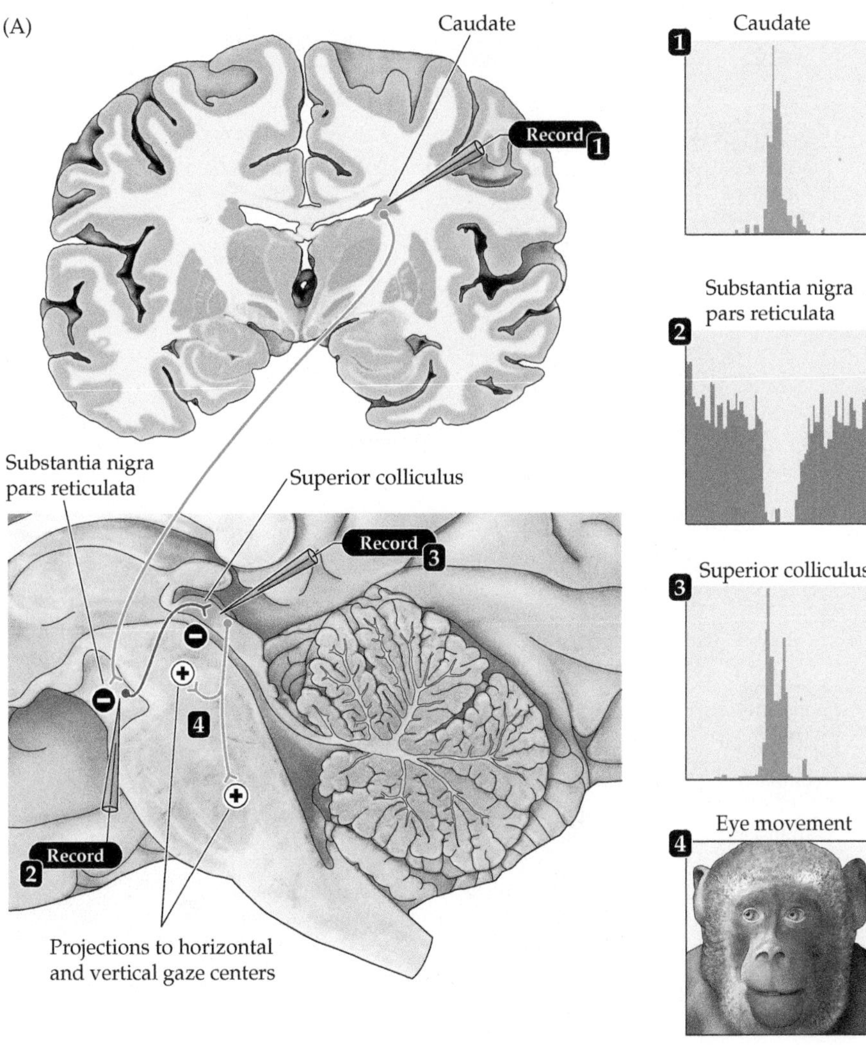

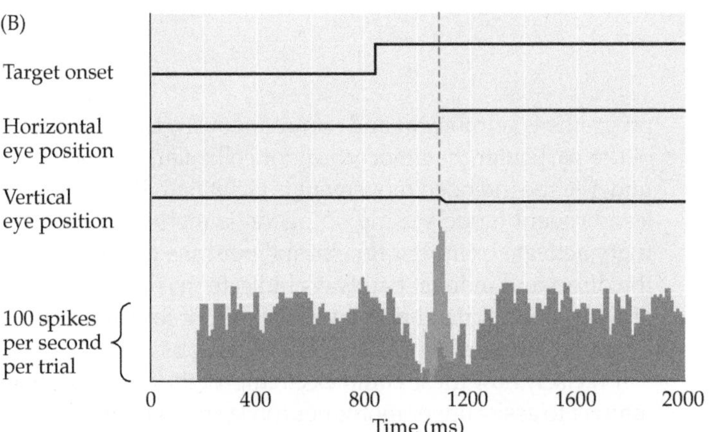

FIGURE 18.6 The role of basal ganglia disinhibition in the generation of saccadic eye movements. (A) Medium spiny cells in the caudate respond with a transient burst of action potentials to an excitatory input from the cerebral cortex (1). Spiny cells inhibit tonically active GABAergic cells in the substantia nigra pars reticulata (2). As a result, the upper motor neurons in the deep layers of the superior colliculus are no longer tonically inhibited and can generate the bursts of action potentials that command a saccade (3, 4). (B) The graph shows the temporal relationship between inhibition in the substantia nigra pars reticulata (purple) and disinhibition in the superior colliculus (light blue) preceding a saccade to a visual target. (A1 after Hikosaka and Wurtz, 1986; A2–3,B after Hikosaka and Wurtz, 1983.)

projects to the lateral division of the globus pallidus, called the **external segment.** The external segment of the globus pallidus sends projections to both the adjacent internal segment and the **subthalamic nucleus** of the ventral thalamus (see Figure 18.1). The subthalamic nucleus also receives excitatory projections from the cerebral cortex (sometimes referred to as the *hyperdirect pathway*) that work synergistically with the disinhibitory effect mediated by the projections from the external segment of the globus pallidus. In turn, the subthalamic nucleus projects diffusely back to the internal segment of the globus pallidus and to the substantia nigra pars reticulata. Thus, the indirect pathway feeds back onto the output nuclei that provide the means by which the basal ganglia gain access to upper motor neurons. But as will become clear in the following discussion, *the indirect pathway antagonizes the activity of the direct pathway;* together, they function to open or shut the physiological gates that initiate and terminate movements.

The indirect pathway through the basal ganglia modulates the disinhibitory actions of the direct pathway. The subthalamic nucleus neurons that project to the internal segment of the globus pallidus and substantia nigra pars reticulata use glutamate as their neurotransmitter and are excitatory. When signals from the cortex activate the indirect pathway, the striatal medium spiny neurons discharge and inhibit the tonically active GABAergic neurons of the external globus pallidus. As a result of the removal of this tonic inhibition and the simultaneous arrival of excitatory inputs from the cerebral cortex, the subthalamic cells become more active, and by virtue of their excitatory synapses with the GABAergic cells of the internal segment of the globus pallidus and substantia nigra pars reticulata, they increase the inhibitory outflow of the basal ganglia. In contrast to the direct pathway, which, when activated, releases thalamocortical and collicular circuits from tonic inhibition, the indirect pathway has the net effect of increasing the inhibitory influences of the basal ganglia. The balance of activity mediated by the

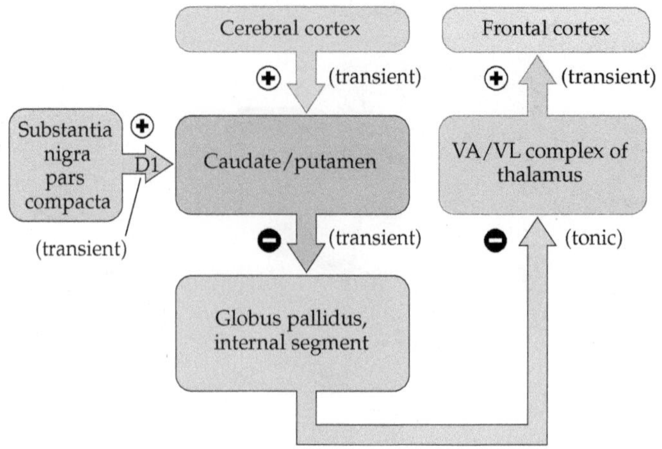

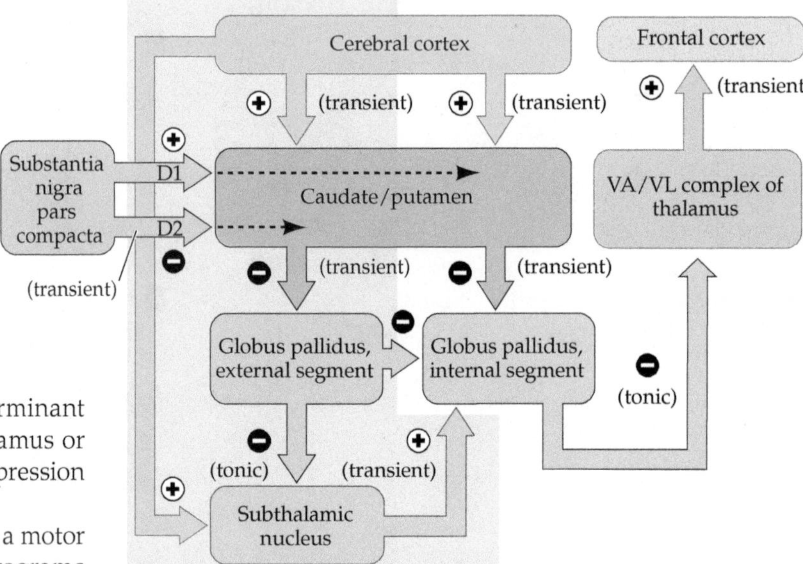

FIGURE 18.7 Disinhibition in the direct and indirect pathways through the basal ganglia. (A) In the direct pathway, transiently inhibitory neurons in the caudate and putamen project to tonically active inhibitory neurons in the *internal* segment of the globus pallidus, which project in turn to the VA/VL complex of the thalamus. Transiently excitatory inputs to the caudate and putamen from the cortex and substantia nigra are also shown, as is the transiently excitatory input from the thalamus back to the cortex. (B) In the indirect pathway (shaded), transiently active inhibitory neurons from the caudate and putamen project to tonically active inhibitory neurons of the *external* segment of the globus pallidus. Note that the influence of nigral dopaminergic input to neurons in the indirect pathway is inhibitory. The globus pallidus (external segment) neurons project to the subthalamic nucleus, which also receives a strong excitatory input from the cortex. The subthalamic nucleus in turn projects to the globus pallidus (internal segment), where its transiently excitatory drive acts to oppose the disinhibitory action of the direct pathway. In this way, the indirect pathway modulates the effects of the direct pathway.

direct and indirect pathways is the principal determinant of whether output from the pallidum to the thalamus or superior colliculus will select and facilitate the expression of the intended motor program.

These circuits not only facilitate the selection of a motor program; they also suppress competing motor programs that could interfere with the expression of sensory-driven or goal-oriented behavior. A concept called *focused selection* has increased understanding of this antagonistic interaction. According to this concept, the direct and indirect pathways are functionally organized in a center–surround fashion within the output nuclei of the basal ganglia (Figure 18.8). The influence of the direct pathway is tightly focused on particular functional units in the internal segment of the globus pallidus (and the substantia nigra pars reticulata), whereas the influence of the indirect pathway is much more diffuse, covering a broader range of functional units. Recall that individual axons from the striatum to the internal segment of the globus pallidus tend to synapse densely on single pallidal neurons (despite making sparse contacts on numerous pallidal cells); this provides a means for the direct pathway to focus its input on a "central" functional unit at the output stage of the basal ganglia. In contrast, afferents from the subthalamic nucleus are distributed much more evenly throughout the internal segment, providing a means for the indirect pathway to suppress the activity of a broader "surrounding" set of functional units. Accordingly, when basal ganglia systems receive and process cortical signals, the suppression of competing or recently activated motor programs is reinforced, and simultaneously, the activation of the particular thalamocortical (or collicular) circuits that underlie the intended movement is facilitated. Recent studies in rodent models using optogenetic methods to selectivity activate or inhibit the striatal neurons giving rise to the direct and indirect pathways indicate that co-activation of both sets of striatal neurons is important for the smooth initiation and execution of new motor actions.

Precisely how these complex circuits of the basal ganglia interact to assist upper motor neuron systems in the execution of volitional behavior remains poorly understood, and this simplified description will undoubtedly be subject to revision as further anatomical and physiological details become available. Nevertheless, this account serves as a useful model for understanding the architecture and function of neural systems that achieve fine control of their output by an interplay between neural excitation and inhibition (recall, for example, the center–surround antagonism of ganglion cell receptive fields in the retina; see Chapter 11). Furthermore, this model provides an instructive framework for understanding disorders of movement that result

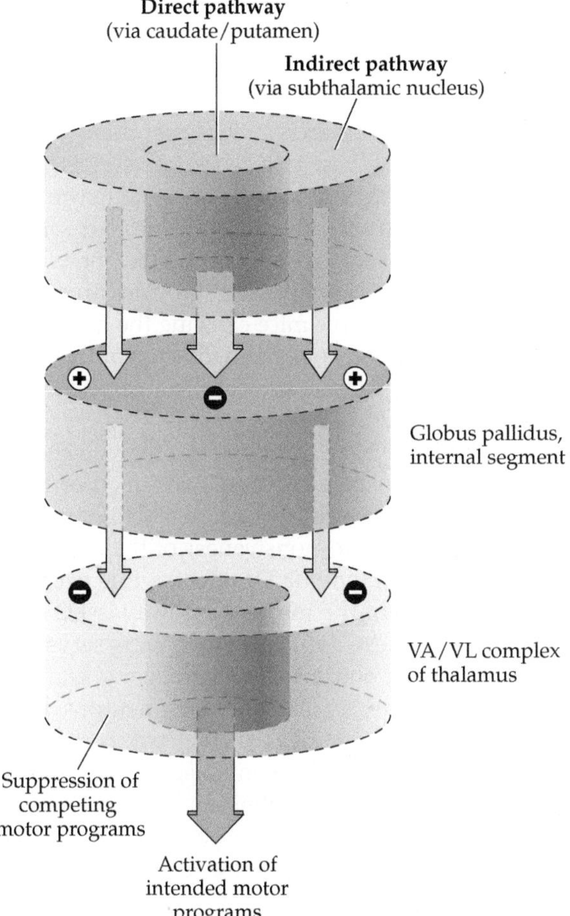

FIGURE 18.8 Center-surround functional organization of the direct and indirect pathways. Integration of cortical input by the striatum leads to the activation of the direct and indirect pathways. With activation of the indirect pathway, neurons in a "surround" region of the internal segment of the globus pallidus are driven by excitatory inputs from the subthalamic nucleus; this reinforces the suppression of a broad set of competing motor programs. Simultaneously, activation of the direct pathway leads to the focal inhibition of a more restricted "center" cluster of neurons in the internal segment; this in turn results in the disinhibition (bottom arrow) of the VA/VL complex and the expression of the intended motor program.

from injury or disease that afflicts one or more components of the basal ganglia system (see below).

Dopamine Modulates Basal Ganglia Circuits

As described earlier in this chapter, an important circuit within the basal ganglia system involves the dopaminergic cells in the pars compacta subdivision of the substantia nigra. Although this circuit derives from a relatively small pool of dopaminergic neurons, it exerts a profound influence over the integration of cortical input in the striatum. The medium spiny neurons of the striatum (especially from striosome compartments) project directly to the substantia nigra pars compacta, which in turn sends widespread dopaminergic projections back to the medium spiny neurons. The effects of dopamine on the spiny neurons are complex; they illustrate the principle that the action of a neurotransmitter is determined by the types of receptors expressed in postsynaptic neurons and by the downstream signaling pathways to which the receptors are linked (see Chapter 6). In this case, the same nigral neurons can provide excitatory inputs to the spiny cells that project to the internal globus pallidus (the direct pathway) and inhibitory inputs to the spiny cells that project to the external globus pallidus (the indirect pathway). This duality is achieved by the differential expression of two types of dopamine receptors—types D1 and D2—by the medium spiny neurons.

Both D1 and D2 dopamine receptors are members of the 7-transmembrane G-protein-coupled family of cell surface receptors. The major functional difference between them is that the D1 receptors mediate the activation of G-proteins that *increase* cAMP, while D2 receptors act through different G-proteins that *decrease* cAMP. For both types of receptors, the dopaminergic synapses on medium spiny neurons tend to be located on the shafts of the spines that receive synaptic input from the cerebral cortex. This arrangement suggests that dopamine exerts its effects on the spiny neurons by modulating their responses to cortical input, with D1 receptors positioned to enhance the excitatory input from cortex and D2 receptors positioned to suppress this excitation. Since the actions of the direct and indirect pathways on the output of the basal ganglia are antagonistic, these different influences of dopamine on medium spiny neurons have the same effects—a decrease in the inhibitory outflow of the basal ganglia, and the consequent release of projections from the thalamus to the frontal cortex or projections from the superior colliculus to circuits of lower motor neurons in the brainstem.

This dopaminergic input to the striatum may contribute to reward-related modulation of behavior. For example, in monkeys the latencies of saccades toward a target are shorter when the goal of the movement is associated with a larger reward. This effect is eliminated by caudate injections of the dopamine D1 receptor antagonist and enhanced by injections in the same site of the D2 receptor antagonist. These results suggest that the influence of motivation on motor performance may be modulated by circuits in the basal ganglia that recruit dopaminergic input from the midbrain. The role of dopamine in motivated behavior and the deleterious impact of addictive substances on dopaminergic modulation of basal ganglia function will be discussed in more detail in Chapter 31.

Hypokinetic Movement Disorders

The modulatory influences of this dopaminergic circuit may also help explain many of the manifestations of basal ganglia disorders, especially those characterized by decreased voluntary movement (*hypokinesia*). For example, **Parkinson's disease** is the second most common degenerative disease of the nervous system (Alzheimer's disease being the leader). Described by James Parkinson in 1817, this disorder is characterized by tremor at rest, slowness of movement (*bradykinesia*), rigidity of the extremities and neck, and minimal facial expressions. Walking entails short steps, stooped posture, and a paucity of associated movements such as arm swinging. In some individuals, these abnormalities of motor function are associated with dementia. Following a gradual onset, typically between the ages of 50 and 70, the disease progresses slowly and culminates in death some 10 to 20 years later.

Unlike in other neurodegenerative diseases (such as Alzheimer's disease and amyotrophic lateral sclerosis), in Parkinson's disease the spatial distribution of the degenerating neurons is largely restricted to the substantia nigra pars compacta. Thus, idiopathic Parkinson's disease is caused by the loss of the nigrostriatal dopaminergic neurons (Figure 18.9A). Although the cause of the progressive deterioration of these dopaminergic neurons is not known, genetic investigations provide clues to the etiology and pathogenesis. Whereas the majority of cases of Parkinson's disease are sporadic, there may be specific forms of susceptibility genes that confer increased risk of acquiring the disease, just as the e4 allele of the *ApoE* gene increases the risk of Alzheimer's disease (see Chapter 30). Familial forms of Parkinson's caused by single gene mutations account for fewer than 10% of all cases; identification of these rare genes, however, is likely to provide insight into molecular pathways that may underlie the disease. Mutations of three distinct genes—*α-synuclein*, *Parkin*, and *DJ-1*—have been implicated in rare forms of Parkinson's disease. Their identification provides an opportunity to generate transgenic mice carrying the mutant form of the human gene, potentially providing an animal model in which the pathogenesis can usefully be elucidated and therapies can be tested.

As described above, activation of the nigrostriatal projection leads to opposite but synergistic effects on the direct and indirect pathways: The release of dopamine in the striatum increases the responsiveness of the direct pathway to corticostriatal input (a D1 effect) while decreasing the responsiveness of the indirect pathway (a D2 effect). Typically, both of these dopaminergic effects serve to decrease the inhibitory outflow of the basal ganglia and thus to increase the excitability of upper motor neurons. In contrast, when the dopaminergic cells of the pars compacta are destroyed, as occurs in Parkinson's disease, the inhibitory outflow of the basal ganglia is abnormally high, and timely thalamic activation of upper motor neurons in the motor cortex is therefore less likely (Figure 18.9B).

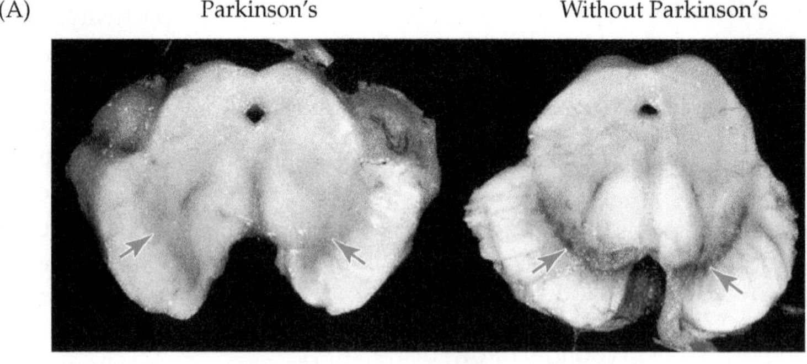

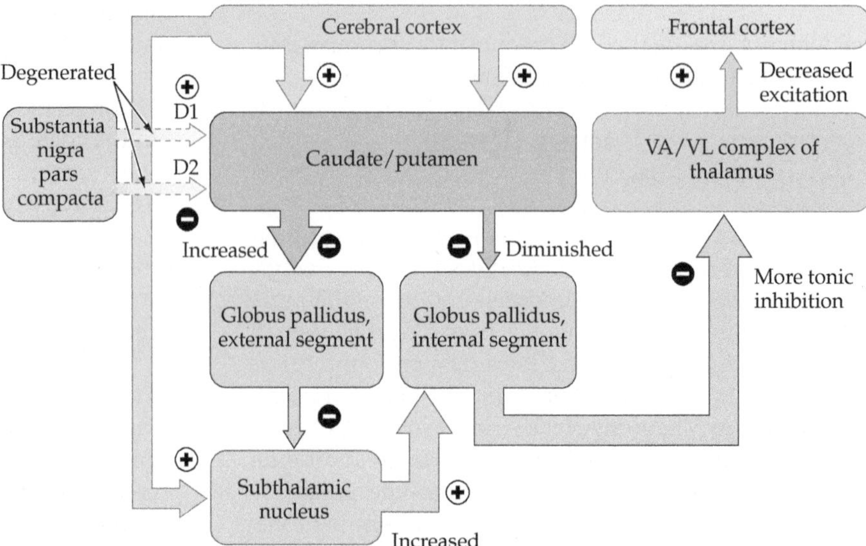

FIGURE 18.9 Degeneration of dopaminergic neurons reduces voluntary movement in Parkinson's disease. (A) In the midbrain of an individual with Parkinson's disease, the substantia nigra (pigmented area) is largely absent in the region above the cerebral peduncles. The midbrain from an individual without Parkinson's disease shows intact substantia nigra (cf. regions indicated with red arrows). (B) In Parkinson's disease, the dopaminergic inputs provided by the substantia nigra pars compacta are diminished (dashed arrows), making it more difficult to generate the transient inhibition from the caudate and putamen. The result of this change in the direct pathway is to sustain or increase the tonic inhibition from the internal segment of the globus pallidus to the thalamus (thicker arrow than corresponding arrow in Figure 18.7B), making thalamic excitation of the motor cortex less likely (thinner arrow from thalamus to frontal cortex). (© 2010, European Association for Predictive, Preventive and Personalised Medicine; B after DeLong, 1990.)

In fact, many of the symptoms seen in Parkinson's disease and other hypokinetic movement disorders reflect a failure of the disinhibition normally mediated by the basal ganglia. Thus, individuals with Parkinson's tend to have diminished facial expressions and reduced amplitude of movement, such as diminished arm swinging during walking. Indeed, any movement is difficult to initiate and, once initiated, is often difficult to terminate. Disruption of the same circuits also increases the discharge rate of the inhibitory cells in the substantia nigra pars reticulata. The resulting increase in tonic inhibition reduces the excitability of upper motor neurons in the superior colliculus, thus reducing the frequency and amplitude of saccades.

Support for this explanation of hypokinetic movement disorders such as Parkinson's disease comes from studies of monkeys in which degeneration of the dopaminergic cells of the substantia nigra pars compacta has been induced by the neurotoxin 1-methyl-4-phenyl-1,2,3,6-tetrahydropyridine (MPTP). Monkeys (or humans) exposed to MPTP develop symptoms that are very similar to those of individuals with Parkinson's disease. Furthermore, a second lesion placed in the subthalamic nucleus results in significant improvement in the ability of these animals to initiate movement, as would be expected based on the circuitry of the indirect pathway (see Figure 18.9B). In humans, rather than creating lesions, neurologists and neurosurgeons are now increasingly using deep brain stimulation to normalize permissive patterns of neural activity in basal ganglia circuits, with the subthalamic nucleus emerging as a strategic target for neuromodulation.

Other novel and promising therapeutic approaches include gene therapy and stem cell grafts. Gene therapy involves correcting a disease phenotype by introducing new genetic information into the affected organism. Although still in its infancy, this approach has the potential to revolutionize treatment of human disease. One such therapy proposed for Parkinson's disease would enhance release of dopamine in the caudate and putamen. In principle, this could be accomplished by implanting cells that have been genetically modified to express tyrosine hydroxylase, the enzyme that converts tyrosine to L-DOPA, which in turn is converted by a nearly ubiquitous decarboxylase into the neurotransmitter dopamine (see Figure 6.14). An alternative strategy involves "neural grafts" using stem cells. Stem cells are self-renewing, multipotent progenitors with broad developmental potential (see Chapters 22 and 26). This approach entails identifying and isolating stem cells, and identifying the growth factors needed to promote differentiation into the desired phenotype (i.e., dopaminergic neurons). The identification and isolation of multipotent mammalian stem cells has already been accomplished, and several factors likely to be important in differentiation of midbrain precursors into dopamine neurons have now been identified. Establishing the efficacy of this approach for individuals with Parkinson's disease would increase the possibility of its application to other neurodegenerative diseases.

Hyperkinetic Movement Disorders

Given the preceding discussion of how an overactive pallidum can suppress voluntary movement, it should not be surprising to learn that insufficient tonic output from the pallidum permits the expression of unwanted movement. Thus, knowledge of the architecture and neurophysiology of basal ganglia circuits also helps explain the motor abnormalities seen in *hyperkinetic* movement disorders, such as **Huntington's disease**.

In 1872, a physician named George Huntington described a group of patients seen by his father and grandfather in their practice in East Hampton, Long Island. The disease he defined, which became known as Huntington's disease (HD), is characterized by the gradual onset of defects in behavior, cognition, and movement beginning in the fourth and fifth decades of life. The disorder is inexorably progressive, resulting in death within 10 to 20 years. HD is inherited in an autosomal dominant pattern, a feature that has led to a much better understanding of its cause in molecular terms.

One of the more common inherited neurodegenerative diseases, HD usually presents as an alteration in mood (especially depression) or a change in personality that often takes the form of increased irritability, suspiciousness, and impulsive or eccentric behavior. Defects of memory and attention may also occur. The hallmark of the disease, however, is a movement disorder consisting of rapid, jerky motions, with no clear purpose. These *choreiform* ("dancelike") movements may be confined to a finger or may involve a whole extremity, the facial musculature, or even the vocal apparatus. The movements themselves are involuntary, but the patient often incorporates them into apparently deliberate actions, presumably in an effort to obscure the problem. There is no weakness, ataxia, or deficit of sensory function.

A distinctive neuropathology is associated with these clinical manifestations: a profound but selective atrophy of the caudate and putamen, with some associated degeneration of the frontal and temporal cortices (Figure 18.10A). However, not all striatal neurons are equally susceptible, especially early in the course of disease. In patients with HD, medium spiny neurons that project to the external segment of the globus pallidus degenerate. In the absence of their normal inhibitory input from the spiny neurons, the external globus pallidus cells become abnormally active; this activity reduces in turn the excitatory output of the subthalamic nucleus to the internal segment of the globus pallidus, and the inhibitory outflow of the basal ganglia is reduced (Figure 18.10B). Without the restraining influence of the basal ganglia, upper motor neurons can be activated

(A)

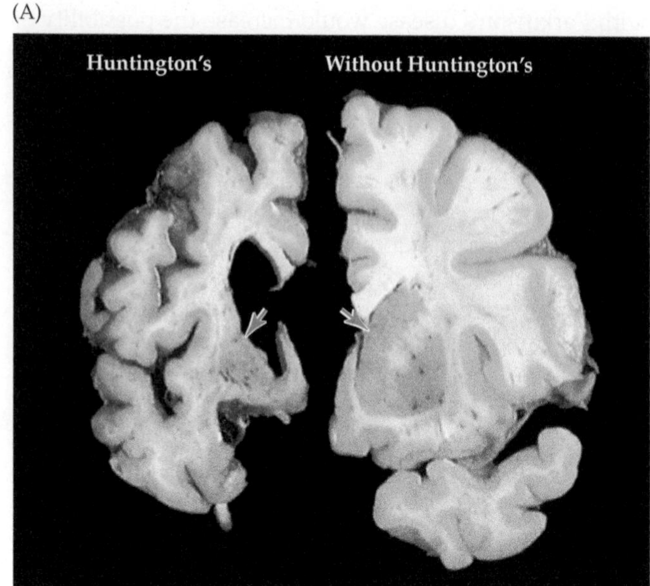

(B) Huntington's disease (hyperkinetic)

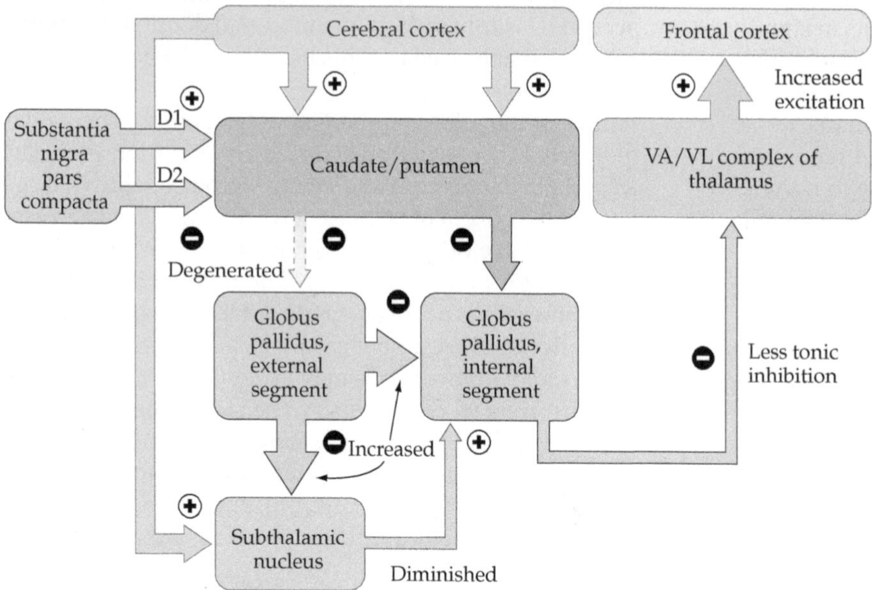

FIGURE 18.10 Degeneration of medium spiny neurons increases involuntary movement in Huntington's disease. (A) The size of the caudate and putamen (the striatum) is dramatically reduced in patients with advanced Huntington's disease. (B) In Huntington's disease, the projection from the caudate and putamen to the external segment of the globus pallidus is diminished (dashed arrow). This effect increases the tonic inhibition from the globus pallidus to the subthalamic nucleus (thicker arrow), making the excitatory subthalamic nucleus less effective in opposing the action of the direct pathway (thinner arrow). Thus, thalamic excitation of the cortex is increased (thicker arrow), leading to the expression of unwanted motor activity. (A courtesy of Harvard Brain Tissue Resource Center; B after DeLong, 1990.)

by inappropriate signals, resulting in the undesired ballistic (flailing or jerking) and choreiform movements that characterize the disease.

The availability of extensive HD pedigrees has allowed geneticists to decipher the molecular cause of this disease. HD was one of the first human diseases in which DNA polymorphisms were used to localize the mutant gene, which in 1983 was mapped to the short arm of chromosome 4. This discovery led to an intensive effort to identify the HD gene within this region by positional cloning. Ten years later, these efforts culminated in identification of the gene (named *Huntingtin*) responsible for the disease. The *Huntingtin* mutation is an unstable triplet repeat present within the coding region of the gene consisting of a DNA segment (CAG) that codes for the amino acid glutamine. In typical individuals, *Huntingtin* contains between 15 and 34 repeats, whereas in HD patients the gene contains from 42 to more than 66 repeats. The mechanism by which the increased number of polyglutamine repeats injures neurons is not clear. The leading hypothesis is that the increased numbers of glutamines alter protein folding, which somehow triggers a cascade of molecular events culminating in dysfunction and neuronal death.

Just as in Huntington's disease, imbalances in the fine control mechanism represented by the convergence of the

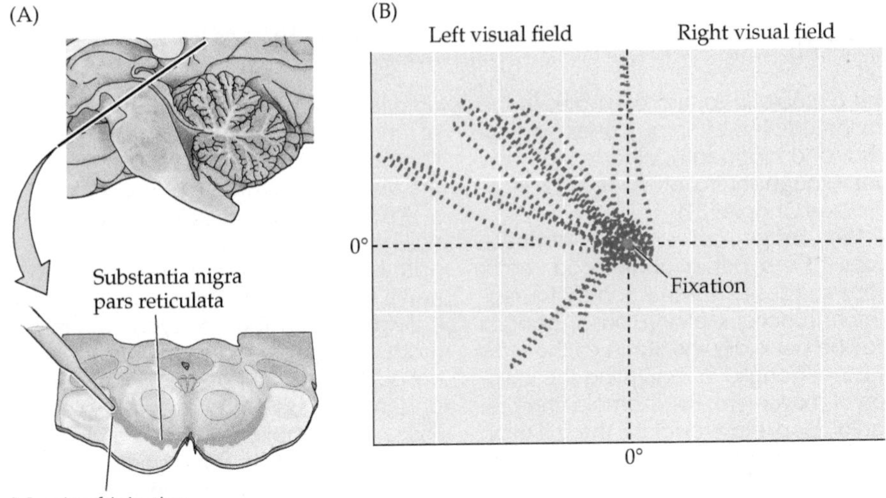

FIGURE 18.11 A GABA agonist produces involuntary movements resembling hyperkinesia. When the tonically active cells of the substantia nigra pars reticulata are inactivated by an intranigral injection of the GABA agonist muscimol (A), the upper motor neurons in the deep layers of the superior colliculus are disinhibited and the monkey generates spontaneous irrepressible saccades (B). Because the cells in both the substantia nigra pars reticulata and the deep layers of the superior colliculus are arranged in spatially organized motor maps of saccade vectors (see Chapter 20), the direction of the involuntary saccades—in this case toward the upper left quadrant of the visual field—depends on the precise location of the injection site within the substantia nigra.

direct and indirect pathways in the pallidum are apparent in other hyperkinetic movement disorders caused by diseases that affect primarily the subthalamic nucleus. The pathophysiology renders dysfunctional a source of excitatory input to the internal segment of the globus pallidus and substantia nigra pars reticulata, thus abnormally reducing the inhibitory outflow of the basal ganglia. A basal ganglia syndrome called **hemiballismus**, which is characterized by ballistic involuntary movements of the limbs, is the result of damage to the subthalamic nucleus. As in Huntington's disease, the involuntary movements of hemiballismus are initiated by the abnormal discharges of upper motor neurons that are receiving less than adequate governance via the tonic inhibition that the basal ganglia normally exerts over the motor nuclei of the thalamus.

As predicted by these accounts of hypokinetic and hyperkinetic movement disorders, GABA agonists and antagonists applied to substantia nigra pars reticulata of monkeys produce symptoms similar to those seen in human basal ganglia disease. For example, intranigral injection of bicuculline, which blocks the GABAergic inputs from the striatal medium spiny neurons to the reticulata cells, increases the amount of tonic inhibition on the upper motor neurons in the deep collicular layers. These animals exhibit fewer and slower saccades, reminiscent of patients with Parkinson's disease. In contrast, injection of the GABA agonist muscimol into the substantia nigra pars reticulata decreases the tonic GABAergic inhibition of the upper motor neurons in the superior colliculus, with the result that the injected monkeys generate spontaneous, irrepressible saccades that resemble the involuntary movements characteristic of basal ganglia diseases such as hemiballismus and Huntington's disease (Figure 18.11).

The central theme of this unit is "Movement and Its Central Control," and the focus of this account of the basal ganglia has been its role in the modulation of movement. However, several parallel streams of processing flow through different sectors of the basal ganglia, including functional loops that modulate the expression of cognitive and affective behavior (Box 18B). Studies of the anatomical and physiological organization of the better understood motor and oculomotor loops have provided the foundation for investigations of the anterior and ventral circuitry of the basal ganglia that subserves a variety of non-motor functions (see Chapter 31). Thus, each functional loop through the basal ganglia is likely to exert a similar influence on the selection, initiation, and suppression of motor or non-motor programs, with equally significant clinical implications should injury, disease, or neurochemical imbalance impair the function of one or more components of the diverse basal ganglia loops.

Summary

The contributions of the basal ganglia to motor control are apparent from the deficits that result from damage to the component nuclei. Such lesions compromise the initiation and performance of voluntary movements, as exemplified by the paucity of movement typical of Parkinson's disease and in the inappropriate "release" of movements characteristic of Huntington's disease. The organization of the basic circuitry of the basal ganglia indicates how this constellation of nuclei modulates movement. With respect to motor function, the system forms a loop that originates in almost every area of the cerebral cortex and eventually terminates, after enormous convergence within the basal ganglia, on the upper motor neurons in the motor and premotor areas of the frontal lobe and in the superior colliculus. The efferent neurons of the basal ganglia influence the upper motor neurons in the cortex by gating the flow

BOX 18B ■ Basal Ganglia Loops and Non-Motor Brain Functions

The basal ganglia traditionally have been regarded as motor structures that regulate the initiation of voluntary movements, such as those involving the limbs and eyes. However, the basal ganglia are also central structures in anatomical circuits or loops that are involved in modulating non-motor aspects of behavior. These parallel loops originate in different regions of the cerebral cortex, engage specific subdivisions of the basal ganglia and thalamus, and ultimately affect areas of the frontal lobe outside the primary motor and premotor cortices. The most prominent of these non-motor loops are a dorsolateral prefrontal loop, involving the dorsolateral sector of the prefrontal cortex and the head of the caudate (see Chapter 32); and a *limbic loop* that originates in the orbitomedial prefrontal cortex, amygdala, and hippocampal formation and runs through ventral divisions of the striatum (see Chapter 31).

The anatomical similarity of these loops to the better understood motor loops suggests that the non-motor regulatory functions of the basal ganglia may be generally the same as the roles of basal ganglia in regulating the initiation of movement. For example, the prefrontal loop may regulate the initiation and termination of cognitive processes such as planning, short-term memory, and attention. Likewise, the limbic loop may regulate emotional and motivated behavior, as well as the transitions from one mood state to another. Indeed, the deterioration of cognitive and emotional function in both Parkinson's and Huntington's diseases could be the result of the disruption of these non-motor loops.

In fact, a variety of other disorders are now thought to be caused, at least in part, by damage to non-motor components of the basal ganglia. For example, individuals with Tourette syndrome may produce inappropriate utterances and obscenities as well as unwanted vocal and motor tics and repetitive grunts. These manifestations may be a result of excessive activity in basal ganglia loops that regulate the cognitive circuitry of the prefrontal speech areas. Another example is schizophrenia, which some investigators have argued is associated with aberrant activity within the limbic

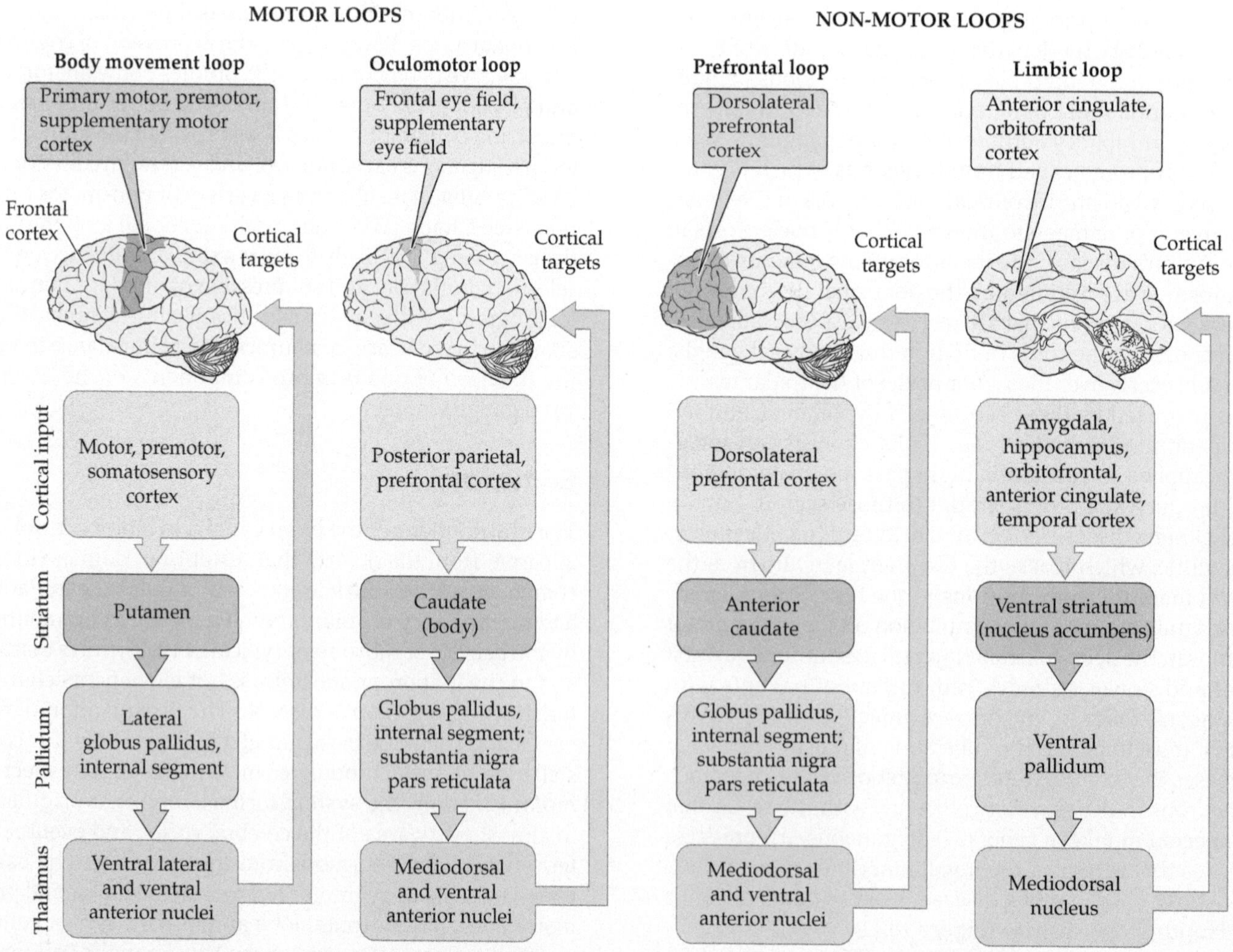

Comparison of motor and non-motor basal ganglia loops.

BOX 18B (continued)

and prefrontal loops, resulting in hallucinations, delusions, disordered thoughts, and loss of emotional expression. In support of the argument for a basal ganglia contribution to schizophrenia, antipsychotic drugs are known to act on dopaminergic receptors, which are found in high concentrations in the striatum.

Still other psychiatric disorders, including obsessive-compulsive disorder, depression, and chronic anxiety, may also involve dysfunctions of the limbic loop. Indeed, one particular component of the limbic loop in a ventral division of the striatum is the nucleus accumbens. This structure is implicated in both the neuropharmacology of addiction to drugs of abuse and of the expression of addictive reward-seeking behavior (see Chapter 31). A challenge for future research is to understand more fully the relationships between these clinical conditions and the functions of the basal ganglia.

of information through relays in the ventral nuclei of the thalamus. The upper motor neurons in the superior colliculus that initiate saccadic eye movements are controlled by monosynaptic projections from the substantia nigra pars reticulata. In each case, the basal ganglia loops regulate movement by a process of disinhibition that results from the serial interaction within the basal ganglia circuitry of two sets of GABAergic neurons. Internal circuits within the basal ganglia system modulate the amplification of the signals that are transmitted through the loops.

ADDITIONAL READING

Reviews

Alexander, G. E. and M. D. Crutcher (1990) Functional architecture of basal ganglia circuits: Neural substrates of parallel processing. *Trends Neurosci.* 13: 266–271.

Cattaneo, E., C. Zuccato and M. Tartari (2005) Normal huntingtin function: An alternative approach to Huntington's disease. *Nat. Rev. Neurosci.* 6: 919–930.

DeLong, M. R. (1990) Primate models of movement disorders of basal ganglia origin. *Trends Neurosci.* 13: 281–285.

Gerfen, C. R. and C. J. Wilson (1996) The basal ganglia. In *Handbook of Chemical Neuroanatomy,* L. W. Swanson, A. Björklund and T. Hokfelt (eds.). Vol. 12: *Integrated Systems of the CNS,* part III. New York: Elsevier Science Publishers, pp. 371–468.

Goldman-Rakic, P. S. and L. D. Selemon (1990) New frontiers in basal ganglia research. *Trends Neurosci.* 13: 241–244.

Graybiel, A. M. and C. W. Ragsdale (1983) Biochemical anatomy of the striatum. In *Chemical Neuroanatomy,* P. C. Emson (ed.). New York: Raven Press, pp. 427–504.

Grillner, S., J. Hellgren, A. Ménard, K. Saitoh and M. A. Wikström (2005) Mechanisms for selection of basic motor programs: Roles for the striatum and pallidum. *Trends Neurosci.* 28: 364–370.

Hardy, J. (2010) Genetic analysis of pathways to Parkinson disease. *Neuron* 68: 201–206.

Hikosaka, O. and R. H. Wurtz (1989) The basal ganglia. In *The Neurobiology of Eye Movements,* R. H. Wurtz and M. E. Goldberg (eds.). New York: Elsevier Science Publishers, pp. 257–281.

Kaji, R. (2001) Basal ganglia as a sensory gating devise for motor control. *J. Med. Invest.* 48: 142–146.

Ledonne, A. and N. B. Mercuri (2017) Current concepts on the physiopathological relevance of dopaminergic receptors. *Front. Cell. Neurosci.* 11: 27. doi: 10.3389/fncel.2017.00027

Mink, J. W. and W. T. Thach (1993) Basal ganglia intrinsic circuits and their role in behavior. *Curr. Opin. Neurobiol.* 3: 950–957.

Pollack, A. E. (2001) Anatomy, physiology, and pharmacology of the basal ganglia. *Neurol. Clin.* 19: 523–534.

Schapira, A. H. V., K. R. Chaudhuri and P. Jenner (2017) Non-motor features of Parkinson disease. *Nat. Rev. Neurosci.* 18: 435–450.

Shepherd, G. M. G. (2013) Corticostriatal connectivity and its role in disease. *Nat. Rev. Neurosci.* 14: 278–291.

Slaght, S. J. and 5 others (2002) Functional organization of the circuits connecting the cerebral cortex and the basal ganglia. Implications for the role of the basal ganglia in epilepsy. *Epileptic Disord.* Suppl 3: S9–S22.

Wilson, C. J. (1990) Basal ganglia. In *Synaptic Organization of the Brain,* G. M. Shepherd (ed.). Oxford, UK: Oxford University Press, chapter 9.

Important Original Papers

Anden, N.-E. and 5 others (1966) Ascending monoamine neurons to the telencephalon and diencephalon. *Acta Physiol. Scand.* 67: 313–326.

Brodal, P. (1978) The corticopontine projection in the rhesus monkey: Origin and principles of organization. *Brain* 101: 251–283.

Crutcher, M. D. and M. R. DeLong (1984) Single cell studies of the primate putamen. *Exp. Brain Res.* 53: 233–243.

DeLong, M. R. and P. L. Strick (1974) Relation of basal ganglia, cerebellum, and motor cortex units to ramp and ballistic movements. *Brain Res.* 71: 327–335.

DiFiglia, M., P. Pasik and T. Pasik (1976) A Golgi study of neuronal types in the neostriatum of monkeys. *Brain Res.* 114: 245–256.

Huntington, G. (1872) On chorea. *Med. Surg. Reporter* 26: 317.

Huntington's Disease Collaborative Research Group (1993) A novel gene containing a trinucleotide repeat that is expanded and unstable on Huntington's disease chromosomes. *Cell* 72: 971–983.

Kemp, J. M. and T. P. S. Powell (1970) The cortico-striate projection in the monkey. *Brain* 93: 525–546.

Kim, R., K. Nakano, A. Jayaraman and M. B. Carpenter (1976) Projections of the globus pallidus and adjacent structures: An autoradiographic study in the monkey. *J. Comp. Neurol.* 169: 217–228.

Kocsis, J. D., M. Sugimori and S. T. Kitai (1977) Convergence of excitatory synaptic inputs to caudate spiny neurons. *Brain Res.* 124: 403–413.

Mink, J. W. (1996) The basal ganglia: Focused selection and inhibition of competing motor programs. *Prog. Neurobiol.* 50: 381–425.

Nakamura, K. and O. Hikosaka (2006) Role of dopamine in the primate caudate nucleus in reward modulation of saccades. *J. Neurosci.* 26: 5360–5369.

Smith, Y., M. D. Bevan, E. Shink and J. P. Bolam (1998) Microcircuitry of the direct and indirect pathways of the basal ganglia. *Neuroscience* 86: 353–387.

Tecuapetla F., X. Jin, S. Q. Lima and R. M. Costa (2016) Complementary contributions of striatal projection pathways to action initiation and execution. *Cell* 166: 703–715.

Books

Bradley, W. G., R. B. Daroff, G. M. Fenichel and C. D. Marsden (eds.) (1991) *Neurology in Clinical Practice.* Boston: Butterworth-Heinemann, chapters 29 and 77.

Donaldson, I., C. D. Marsden, K. P. Bhatia and S. A. Schneider (2012) *Marsden's Book of Movement Disorders.* Oxford, UK: Oxford University Press.

Klawans, H. L. (1989) *Toscanini's Fumble and Other Tales of Clinical Neurology.* New York: Bantam, chapters 7 and 10.

Go to the NEUROSCIENCE 6e Companion Website at **www.oup.com/uk/Purves6e** for Web Topics, Animations, Flashcards, and more.

CHAPTER

19

Modulation of Movement by the Cerebellum

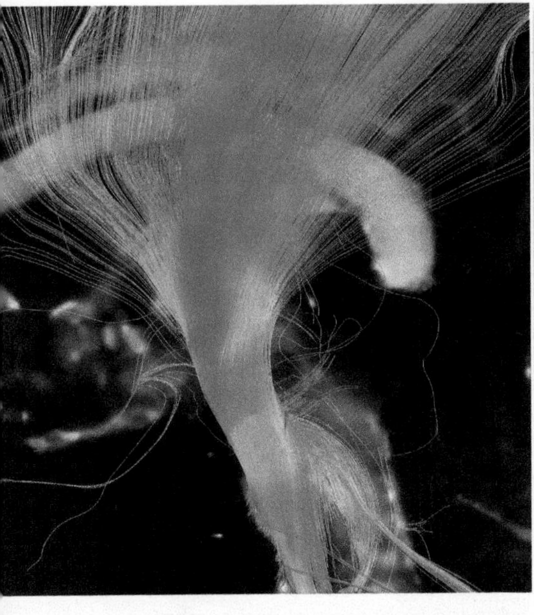

Overview

IN CONTRAST TO THE UPPER MOTOR NEURONS described in Chapter 17, the efferent cells of the cerebellum do not project directly to the local circuits of the brainstem and spinal cord that organize movement, nor do they directly contact the lower motor neurons that innervate muscles. Instead—like the basal ganglia—the cerebellum influences movements primarily by modifying the activity patterns of upper motor neurons. In fact, the cerebellum sends prominent projections to virtually all circuits that govern upper motor neurons. Anatomically, the cerebellum has two main gray matter structures: a laminated cortex on its surface and clusters of cells in nuclei buried deep in the white matter of the cerebellum. Pathways that reach the cerebellum from other brain regions (in humans, the largest contribution arises in the cerebral cortex) project to both components by means of afferent axons that send branches to both the deep nuclei and the cortex. Neurons in the deep nuclei are the main source of the output from the cerebellum. Their spatiotemporal patterns of activity are sculpted by descending input from the overlying cortex. In this way, the output of the cerebellum is integrated before being sent to circuits of upper motor neurons in the cerebral cortex, by way of thalamic relays, and in the brainstem. A primary function of the cerebellum is to detect the difference, or "motor error," between an intended movement and the actual movement, and through its influence over upper motor neurons, to reduce the error. These corrections can be made both during the course of the movement and as a form of motor learning when the correction is stored. When this feedback loop is damaged, as occurs in many cerebellar disorders and injuries, the afflicted individual makes persistent errors when executing movement. The specific pattern of incoordination depends on the location of the damage.

Organization of the Cerebellum

The cerebellar hemispheres can be subdivided into three main parts based on differences in their sources of input (Figure 19.1A). By far, the largest subdivision in humans is the **cerebrocerebellum**. It occupies most of the lateral part of the cerebellar hemisphere and receives input, indirectly, from many areas of the cerebral cortex (Figure 19.2). This region of the cerebellum is especially well developed in primates and is particularly prominent in humans. The cerebrocerebellum is concerned with the regulation of highly skilled movements, especially the planning and execution of complex spatial and temporal sequences of movement (including speech). Just medial to the cerebrocerebellum is the **spinocerebellum**. The spinocerebellum occupies the median and paramedian zones of the cerebellar hemispheres and is the only part that receives input directly from the spinal cord. The more lateral

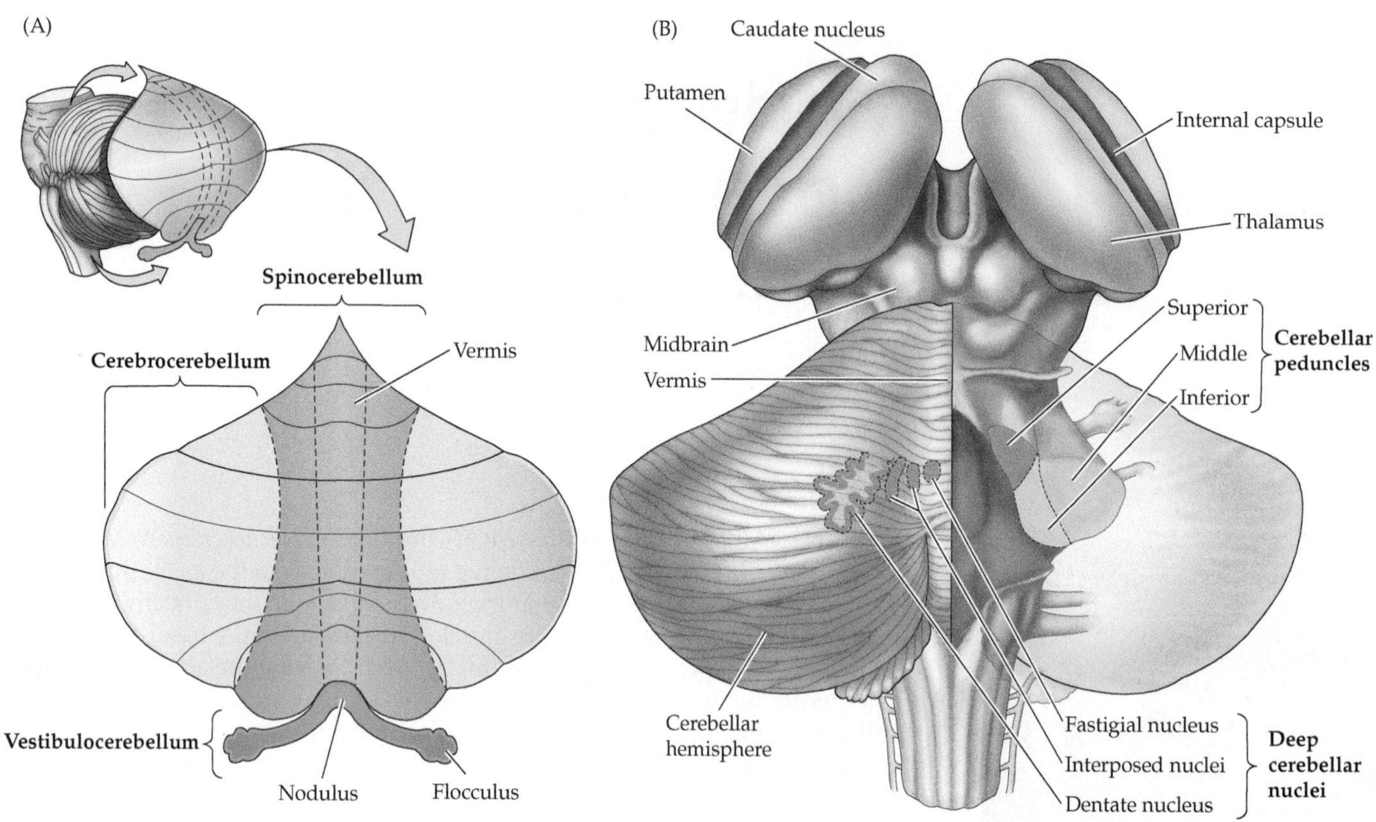

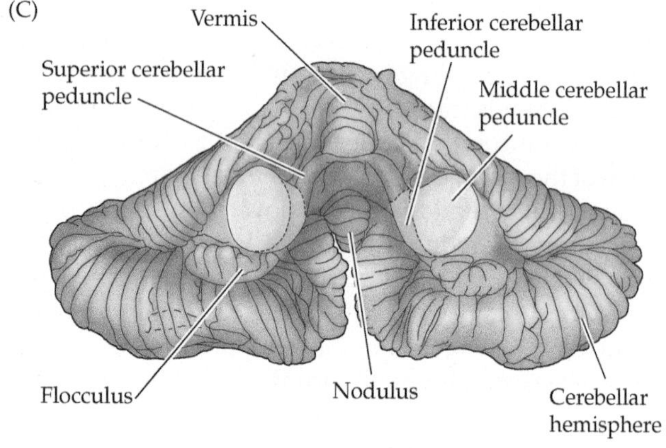

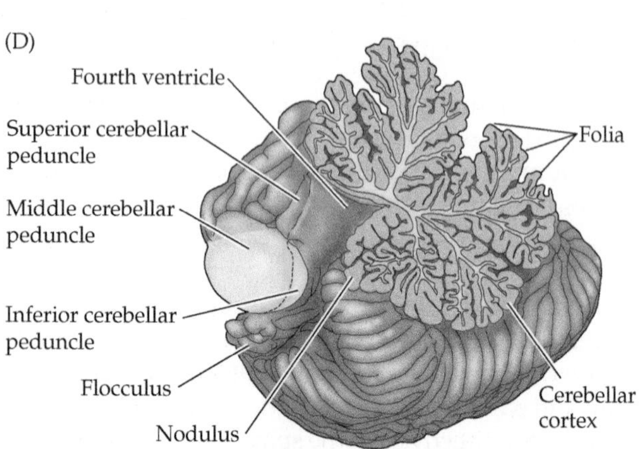

FIGURE 19.1 Organization and subdivisions of the cerebellum. (A) Flattened view of the cerebellar surface illustrating the three major subdivisions. (B) Dorsal views of the cerebellum. This view shows the left cerebellar hemisphere and illustrates the location of the deep cerebellar nuclei. The right hemisphere has been removed to show the cerebellar peduncles. (C) Removal from the brainstem reveals the cerebellar peduncles on the anterior aspect of the inferior surface. (D) Paramedian sagittal section through the left cerebellar hemisphere showing the highly convoluted cerebellar cortex. The small gyri in the cerebellum are called *folia*.

(paramedian) part of the spinocerebellum is concerned primarily with movements of distal muscles. The most median strip of cerebellar hemisphere lies along the midline and is called the **vermis**. The vermis is concerned primarily with movements of proximal muscles; it also regulates certain types of eye movements (see Chapter 20). The third major subdivision is the **vestibulocerebellum**, the phylogenetically oldest part of the cerebellum. This portion comprises the caudal-inferior lobes of the cerebellum and includes the **flocculus** and **nodulus** (see Figure 19.1A). As its name suggests, the vestibulocerebellum receives input from the vestibular nuclei in the brainstem and is concerned primarily with the vestibulo-ocular reflex (see Chapter 14) and with the regulation of movements that maintain posture and equilibrium.

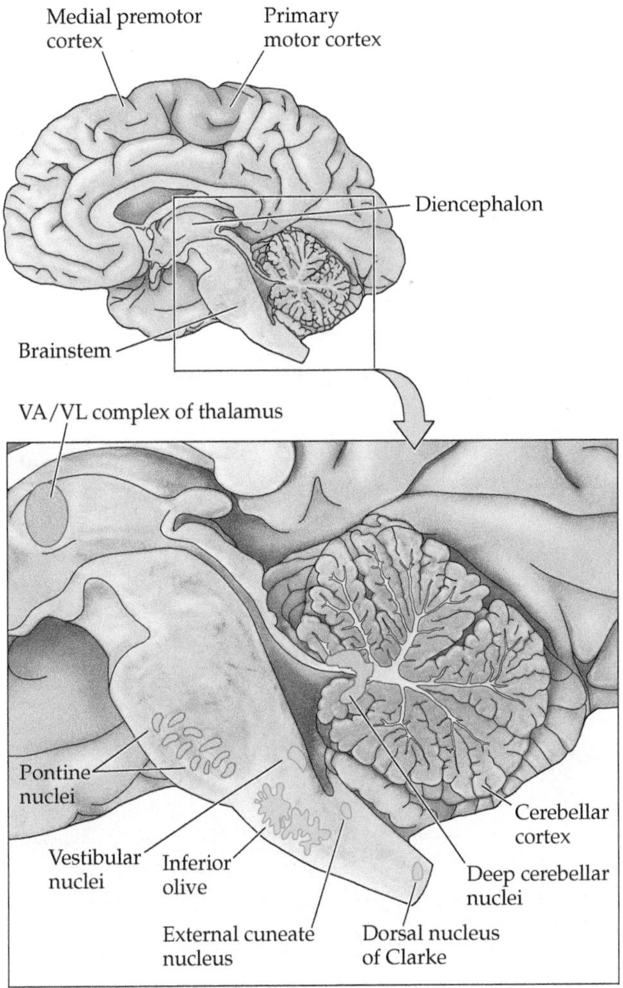

FIGURE 19.2 Components of the brainstem and diencephalon related to the cerebellum. This sagittal section shows the major structures of the cerebellar system, including the cerebellar cortex, the deep cerebellar nuclei, and the ventral anterior and ventral lateral (VA/VL) complex (which is an important target of some deep cerebellar nuclei). Also shown are brainstem and spinal cord nuclei that provide input to the cerebellum.

The connections between the cerebellum and other parts of the nervous system are made by three large pathways called **cerebellar peduncles** (Figure 19.1B–D; also see Figure 19.3). The **superior cerebellar peduncle** (or **brachium conjunctivum**) is almost entirely an efferent pathway. The neurons that give rise to this pathway are located in the **deep cerebellar nuclei** (Table 19.1). Their axons project to the motor nuclei of the thalamus, which in turn relay signals to circuits of upper motor neurons in the primary motor and premotor divisions of the cerebral cortex. Efferent axons in the superior peduncle also project directly to upper motor neurons in the deep layers of the superior colliculus that control orienting movements of the head and eyes. In non-human species, neurons in the deep cerebellar nuclei also provide input to upper motor neurons in the caudal part of the red nucleus. The **middle cerebellar peduncle** (or **brachium pontis**) is an afferent pathway to the cerebellum; most of the cell bodies that give rise to this pathway are in the base of the contralateral pons, where they form the **pontine nuclei** (see Figure 19.2). Finally, the **inferior cerebellar peduncle** (or **restiform body**) is the smallest but most complex of the cerebellar peduncles, containing multiple afferent and efferent pathways. Afferent pathways in the inferior peduncle include axons from the vestibular nuclei, the spinal cord, and several regions of the brainstem tegmentum, while efferent pathways project to the vestibular nuclei and the reticular formation.

TABLE 19.1 ■ Major Components of the Cerebellum

Cerebellar cortex	Deep cerebellar nuclei	Cerebellar peduncles
Cerebrocerebellum	Dentate nucleus	Superior and middle peduncle
Spinocerebellum	Interposed nuclei	Inferior peduncle
Vestibulocerebellum	Fastigial nucleus	Inferior peduncle

Projections to the Cerebellum

The cerebral cortex is by far the source of the largest input to the cerebellum, and the major destination of this input is the cerebrocerebellum. These cortical axons do not project directly into the cerebellum. Rather, they synapse on neurons in the ipsilateral pontine nuclei (i.e., on the same side of the brainstem as their hemisphere of origin). These pontine nuclei receive input from a wide variety of sources, including almost all areas of the cerebral cortex and the superior colliculus. The axons of the cells in the pontine nuclei, called **transverse pontine fibers** (or pontocerebellar fibers), cross the midline and enter the contralateral cerebellum via the middle cerebellar peduncle. Each of the two middle cerebellar peduncles contains more than 20 million axons, making them among the largest pathways in the brain. (In comparison, the optic nerves each contain about 1 million axons, and the pyramidal tracts about 0.5 million each.) In fact, the size of the cerebral peduncles in the ventral portion of the human midbrain (each of which also contains about 20 million axons) is primarily due to the magnitude of the projection from the cerebral cortex that, via the pontine nuclei, enters the cerebellum. (In comparison, the corticospinal projection, which is often assumed incorrectly to account for the bulk of the cerebral peduncles, comprises less than

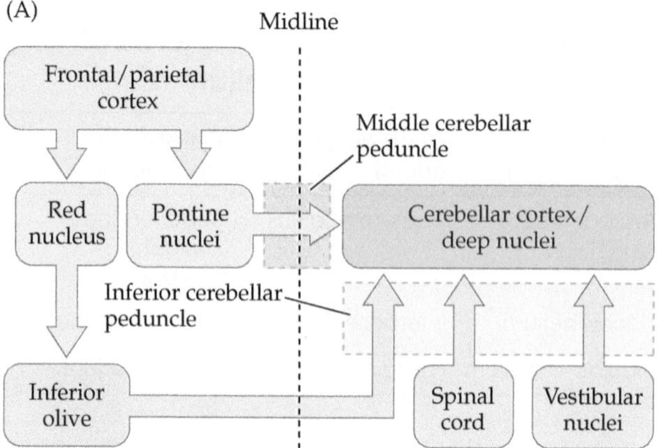

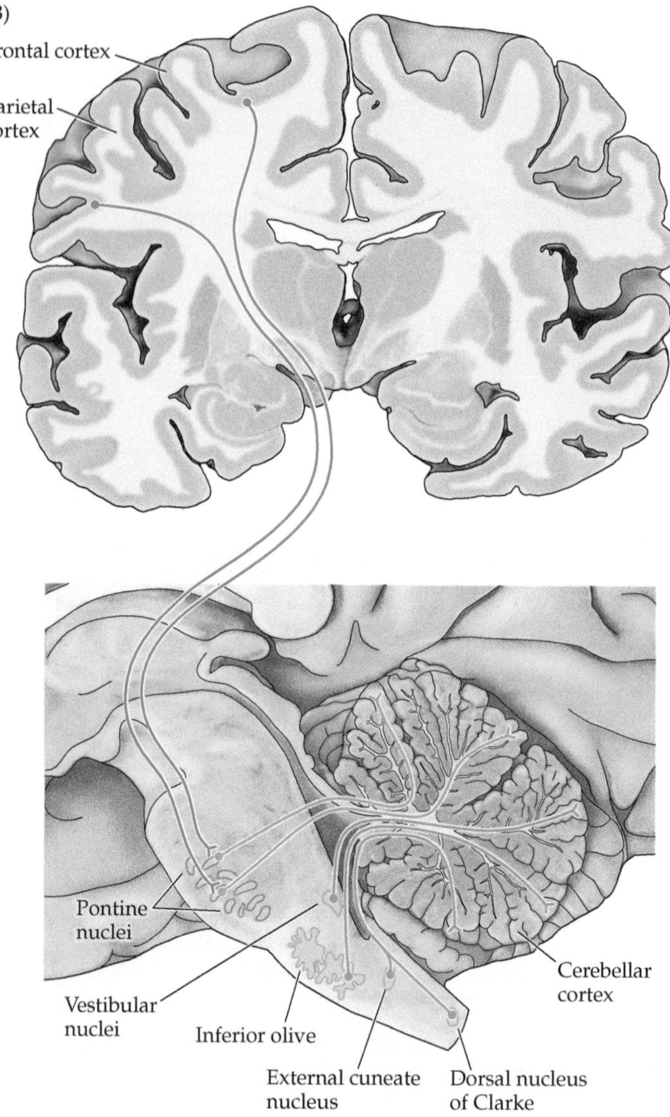

FIGURE 19.3 Functional organization of the inputs to the cerebellum. (A) Diagram of the major inputs. (B) Idealized coronal and sagittal sections through the human brainstem and cerebrum, showing inputs to the cerebellum from the cerebral cortex, vestibular system, spinal cord, and brainstem. The cortical projections to the cerebellum are made via relay neurons in the pons. These pontine axons then cross the midline within the pons and project to the cerebellum via the middle cerebellar peduncle. Axons from the inferior olive, spinal cord, and vestibular nuclei enter via the inferior cerebellar peduncle.

5% of the total number of axons in each cerebral peduncle.) This massive, crossed projection of transverse pontine fibers into the cerebellum via the middle cerebellar peduncle is the means by which signals originating in one *cerebral* hemisphere are sent to neural circuits in the opposite *cerebellar* hemisphere (Figure 19.3A).

Sensory pathways also project to the cerebellum (Figure 19.3B). Vestibular axons in the eighth cranial nerve as well as axons from the vestibular nuclei in the pons and medulla project to the vestibulocerebellum. In addition, somatosensory relay neurons in the **dorsal nucleus of Clarke** in the spinal cord and the **external** (or accessory) **cuneate nucleus** of the caudal medulla send their axons to the spinocerebellum (recall that these nuclei comprise groups of relay neurons innervated by proprioceptive axons from the lower and upper parts of the body, respectively; see Chapter 9). Proprioceptive signals from the face are likewise relayed via the **mesencephalic trigeminal nucleus** to the spinocerebellum. The vestibular, spinal, and trigeminal inputs provide the cerebellum with information from the labyrinth in the ear, muscle spindles, and other mechanoreceptors that monitor the position and motion of the body. Visual and auditory signals are relayed via brainstem nuclei to the cerebellum; they provide the cerebellum with additional sensory signals that supplement the proprioceptive information regarding body position and motion.

The somatosensory input is topographically mapped in the spinocerebellum, providing the basis for orderly representations of the body within the cerebellum (Figure 19.4). However, these maps are "fractured"; that is, fine-grain electrophysiological analysis indicates that each small area of the body is represented multiple times by spatially separated clusters of cells, rather than by a specific site within a single continuous somatotopic map. The vestibular and spinal inputs remain ipsilateral as they pass through the inferior cerebellar peduncle and enter the cerebellum, having their origin on the same side of the brainstem and spinal cord (see Figure 19.3A). This arrangement ensures that the right cerebellum is concerned with the right half of the body and the left cerebellum with the left half. Thus, while many areas of the brain are concerned with *contralateral* representations (of the body and external space), the cerebellum is concerned with *ipsilateral* representations.

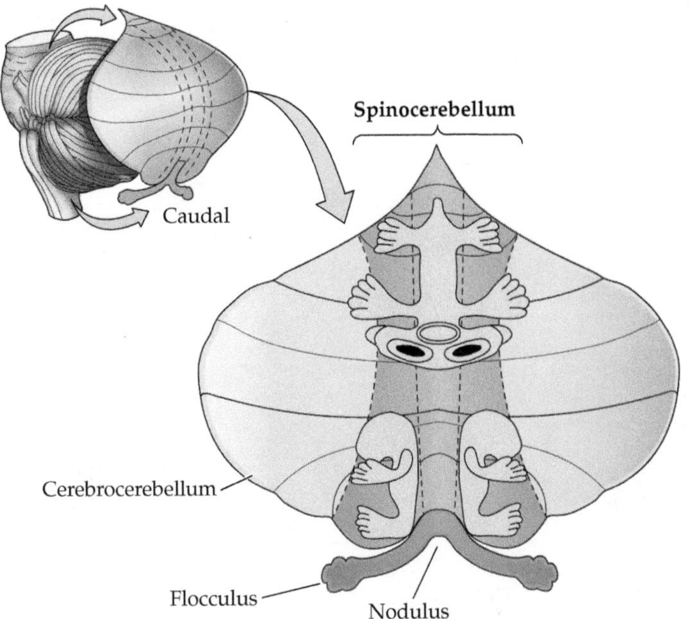

FIGURE 19.4 **Somatotopic maps of the body surface in the cerebellum.** The spinocerebellum contains at least two maps of the body.

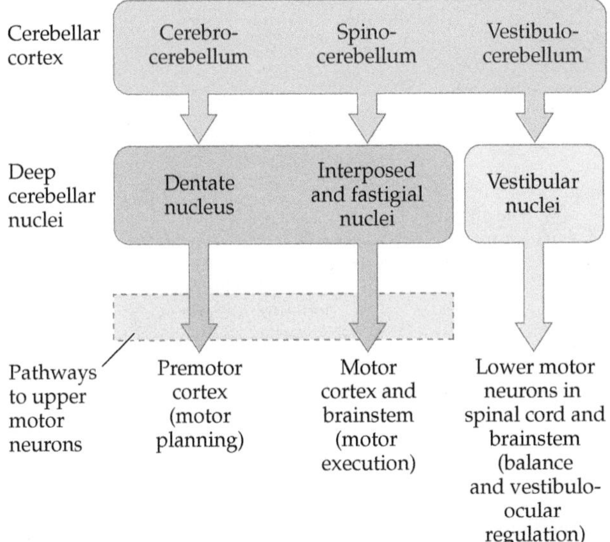

FIGURE 19.5 **Functional organization of cerebellar outputs.** The three major functional divisions of the cerebellar hemispheres project to corresponding deep cerebellar nuclei and the vestibular nuclei, which in turn provide input to neural circuits that govern different aspects of motor control.

Finally, the entire cerebellum receives modulatory inputs from the **inferior olivary nucleus** (or **inferior olive**) in the medulla oblongata. These inputs participate in the learning and memory functions served by cerebellar circuitry. The inferior olive receives input from a wide variety of structures, including the cerebral cortex (via a relay in the parvocellular, or small-celled, division of the red nucleus), the reticular formation, and the spinal cord. The so-called olivo-cerebellar axons exit medially from the inferior olive, cross the midline, and enter the cerebellum on the opposite side via the inferior cerebellar peduncle (see Figure 19.3A). Electrotonic gap junctions are abundant among neurons in the inferior olive, and these evidently play an important role in the timing and spatial distribution of cerebellar responses to olivary inputs.

Projections from the Cerebellum

The efferent neurons of the **cerebellar cortex** project to the deep cerebellar nuclei and to the vestibular complex; these structures project, in turn, to upper motor neurons in the brainstem and to thalamic nuclei that innervate upper motor neurons in the motor cortex (Figure 19.5). In each cerebellar hemisphere, there are four major deep nuclei: the **dentate nucleus** (by far the largest in humans), two **interposed nuclei**, and the **fastigial nucleus**. Each receives input from a different region of the cerebellar cortex. Although the borders are not distinct, the cerebrocerebellum projects primarily to the dentate nucleus, and the spinocerebellum to the interposed and fastigial nuclei. The vestibulocerebellum projects directly to the vestibular complex in the brainstem. As discussed in Chapter 17, parts of the vestibular complex are sources of upper motor neurons that influence posture, equilibrium, and vestibulo-ocular eye movements.

Pathways originating in the dentate nucleus project primarily to the premotor and associational cortices of the frontal lobe, which function in planning and initiating volitional movements. The pathways reach these cortical areas after a relay in the ventral lateral nuclear complex of the thalamus (Figure 19.6A). Since each cerebellar hemisphere is concerned with the ipsilateral side of the body, this pathway crosses the midline so the motor cortex in each hemisphere, which governs contralateral musculature, receives information from the appropriate cerebellar hemisphere. For this reason, the dentate axons that exit the cerebellum via the superior cerebellar peduncle cross the midline at the decussation of the superior cerebellar peduncle in the caudal midbrain, and then ascend to the contralateral thalamus. Along its course to the thalamus, this pathway also sends axons to eye-movement-related upper motor neurons in the superior colliculus and, in addition, sends collaterals to the parvocellular division of the **red nucleus** in the midbrain (which accounts for virtually the entire red nucleus in the human midbrain) (Figure 19.6B). This division of the red nucleus projects, in turn, to the inferior olive, thus providing a means for cerebellar output to feed back on a major source of cerebellar input. This feedback is crucial for the adaptive functions of cerebellar circuits in motor learning.

Anatomical studies using viruses to trace chains of connections between nerve cells have shown that, in addition

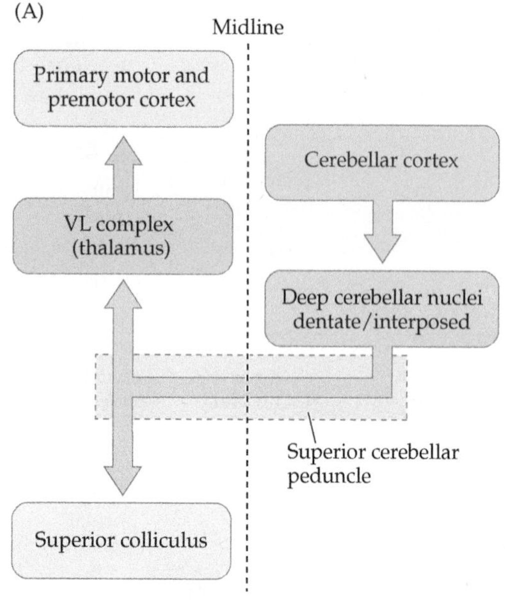

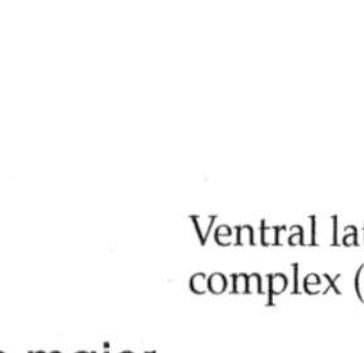

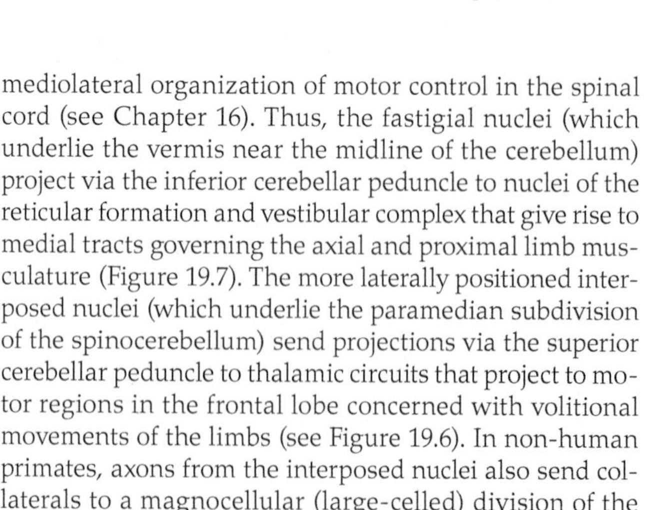

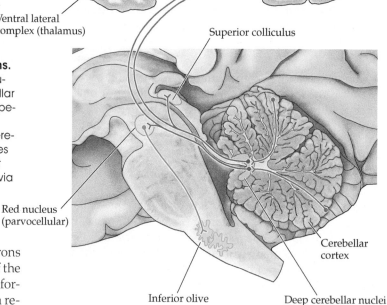

FIGURE 19.6 Functional organization of the major outputs from the cerebellum to cortical motor systems. (A) Diagram of major outputs that affect upper motor neurons in the cerebral cortex. The axons of the deep cerebellar nuclei cross in the midbrain—in the decussation of the superior cerebellar peduncle—before reaching the thalamus. (B) Idealized coronal and sagittal sections through the cerebrum and brainstem, showing the location of the structures and pathways diagrammed in (A) and a feedback circuit by which cerebellar output is directed to the inferior olive via the red nucleus.

to sending ascending projections to upper motor neurons concerned with the control of movement, large parts of the cerebrocerebellum form "closed loops" by sending information back to non-motor areas of the cortex. That is, a region of the cerebellum sends projections back to the same cortical areas (via thalamic projections) from which (via the pontine nuclei) its input signals originated. Such closed loop cerebellar circuits provide a mechanism for the cerebellum to modulate its own input. In the case of cerebellar circuits that modulate the prefrontal cortex, these closed loops may also influence the coordination of non-motor programs—such as problem solving—in a manner that is analogous to their modulation of movement-related signals. The closed loops run parallel to the more commonly recognized "open loops" that receive input from multiple cortical areas and funnel output back to upper motor neurons in specific regions of the motor and premotor cortices.

Spinocerebellar pathways are directed toward circuits of upper motor neurons that govern the execution of movement. The somatotopic organization of the spinal subdivision of the cerebellum is reflected in the organization of its efferent projections, both of which conform to the mediolateral organization of motor control in the spinal cord (see Chapter 16). Thus, the fastigial nuclei (which underlie the vermis near the midline of the cerebellum) project via the inferior cerebellar peduncle to nuclei of the reticular formation and vestibular complex that give rise to medial tracts governing the axial and proximal limb musculature (Figure 19.7). The more laterally positioned interposed nuclei (which underlie the paramedian subdivision of the spinocerebellum) send projections via the superior cerebellar peduncle to thalamic circuits that project to motor regions in the frontal lobe concerned with volitional movements of the limbs (see Figure 19.6). In non-human primates, axons from the interposed nuclei also send collaterals to a magnocellular (large-celled) division of the red nucleus that gives rise to the rubrospinal tract, a lateral tract of the spinal cord that functions synergistically with

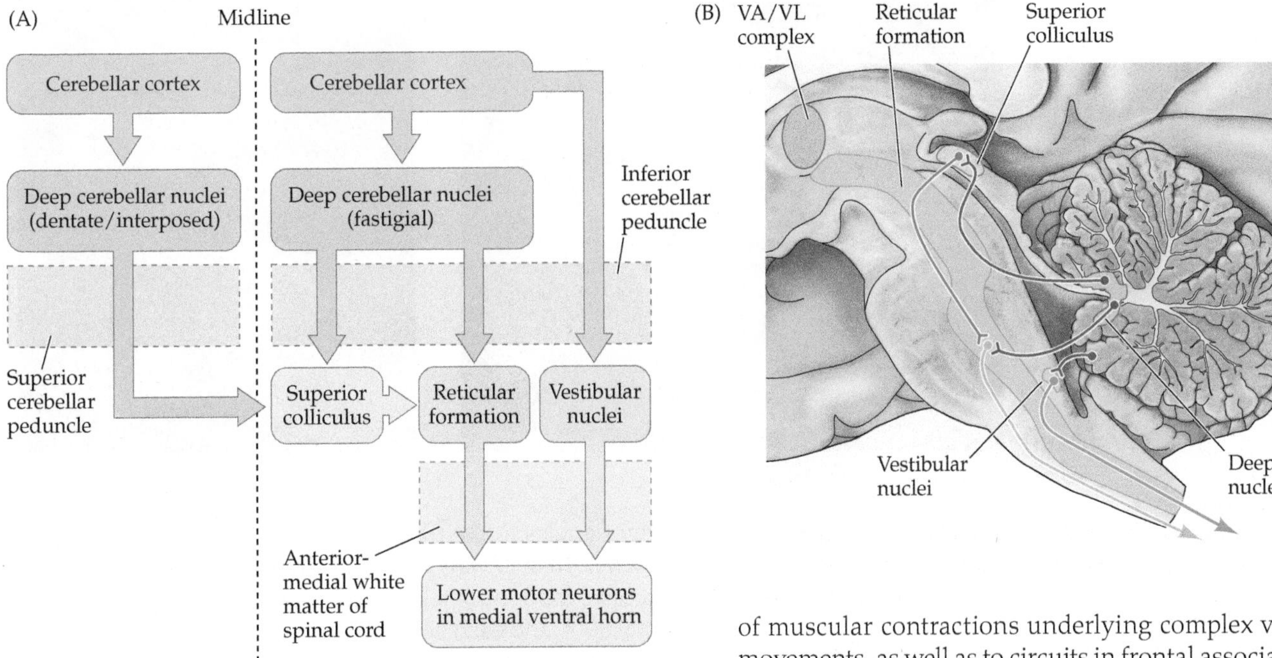

FIGURE 19.7 **Functional organization of the major outputs from the cerebellum to brainstem motor systems.** (A) Diagram of major outputs that affect upper motor neurons in the brainstem. The axons of the deep cerebellar nuclei and the vestibulocerebellar cortex project to upper motor neurons that contribute to the control of axial and proximal limb musculature in the medial ventral horn of the spinal cord. (B) Idealized sagittal section through the brainstem, showing the location of the structures diagrammed in (A).

the lateral corticospinal tract. (As discussed in Chapter 17, this division of the red nucleus and its spinal projection are vestigial in humans relative to other primates.)

Most of the projection from the cerebellum to the eye-movement-related upper motor neurons in the superior colliculus arises in the dentate and interposed nuclei, which receive their input from the lateral portions of the cerebellar cortex. The output pathway travels in the superior cerebellar peduncle and crosses the midline to terminate on upper motor neurons in the deep layers of the superior colliculus on the contralateral side. So, for example, the right cerebellar hemisphere projects to the left superior colliculus, which in turn controls saccades toward the right half of the visual field (see Chapter 20).

The thalamic nuclei that receive projections from the cerebrocerebellum (dentate nuclei) and spinocerebellum (interposed nuclei) are segregated in two distinct subdivisions of the ventral lateral nuclear complex: the oral, or anterior, part of the posterolateral segment, and a region simply called "area X." Both of these thalamic relays project directly to primary motor and premotor association cortices. Through these pathways, the cerebellum has access to the upper motor neurons that organize the sequence of muscular contractions underlying complex voluntary movements, as well as to circuits in frontal association cortex that exert executive control over planning movements (see Chapter 32).

Projections from the vestibulocerebellum course through the inferior cerebellar peduncle and terminate in nuclei of the vestibular complex in the brainstem. These nuclei govern the movements of the eyes, head, and neck that compensate for linear and rotational accelerations of the head (see Figure 19.5).

Circuits within the Cerebellum

The ultimate destination of the afferent pathways to the cerebellar cortex is a distinctive cell type called the Purkinje cell (see Figures 19.8 and 19.9). The largest of these afferent pathways arises in widespread areas of the cerebral cortex and terminates in the pontine nuclei of the basal pons, as described above. The pontine nuclei, in turn, project to the contralateral cerebellum. The axons from the pontine nuclei—and most other sources of cerebellar input from the brainstem and spinal cord—are called **mossy fibers** because of the appearance of their synaptic terminals. Mossy fibers send collateral branches that synapse both on neurons in the deep cerebellar nuclei and on granule cells in the granule cell layer of the cerebellar cortex. Cerebellar granule cells, which are widely held to be the most abundant class of neurons in the human brain, give rise to axons called **parallel fibers** that ascend to the outermost molecular layer of the cerebellar cortex. The parallel fibers bifurcate in the molecular layer to form T-shaped branches that extend for several millimeters parallel to the orientation of the small cerebellar gyri (called folia). There, they form excitatory synapses with the dendritic spines of the underlying Purkinje cells.

(A)

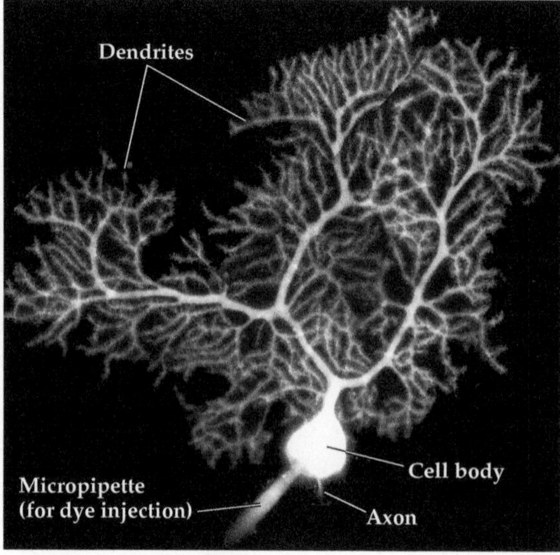

(B)

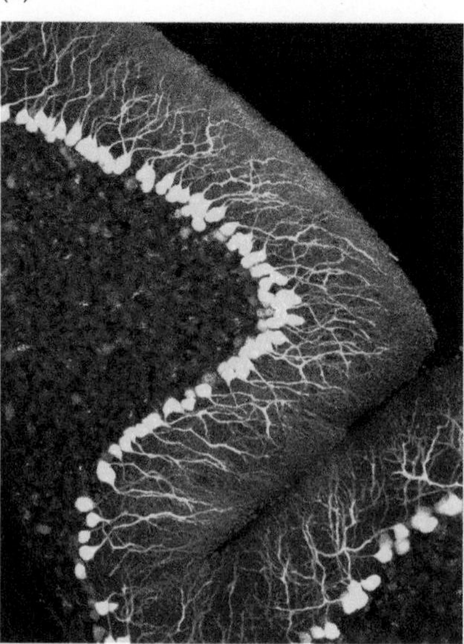

FIGURE 19.8 Cerebellar cortical neurons. (A) A cerebellar Purkinje neuron in a living slice from mouse cerebellum. The neuron has been visualized by infusing a fluorescent dye that indicates Ca^{2+} concentrations, via a micropipette inserted into the cell body. (B) Histological preparation of mouse cerebellar cortex. Purkinje neurons were engineered to fluoresce yellow, while granule cells fluoresce blue. (A courtesy of K. Tanaka and G. Augustine; B courtesy of A. Agmon.)

The Purkinje cells (Figure 19.8A) are the most distinctive histological feature of the cerebellum. Their elaborate dendrites extend into the molecular layer from a single subjacent layer of giant Purkinje cell bodies (called the Purkinje cell layer; Figure 19.8B). In the molecular layer, the Purkinje cell dendrites branch extensively but in a plane restricted at right angles to the trajectory of the parallel fibers (Figure 19.9A). In this way, each Purkinje cell is in a position to receive input from a large number of parallel fibers (about 200,000), and each parallel fiber can contact a vast number of Purkinje cells (on the order of tens of thousands). The Purkinje cells also receive a direct input on their dendritic

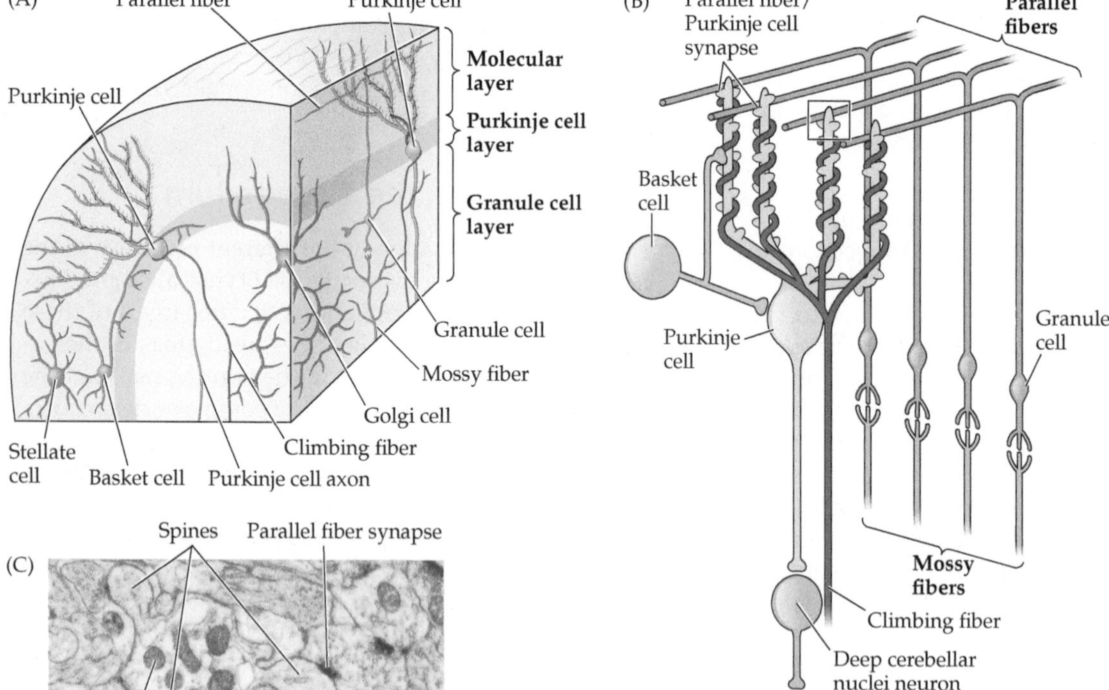

FIGURE 19.9 Neurons and circuits of the cerebellum. (A) Neuronal types in the cerebellar cortex. Note that the various neuron classes are found in distinct layers. (B) Diagram showing convergent inputs onto the Purkinje cell from climbing and parallel fibers and from local circuit neurons (climbing and mossy fiber inputs to deep cerebellar neurons are omitted for clarity). The boxed region is shown at higher magnification in (C). The output of the Purkinje cells is to the deep cerebellar nuclei. (C) Electron micrograph showing a Purkinje cell dendritic shaft with spines contacted by parallel fibers in rhesus macaque cerebellum. (C courtesy of A.-S. La Mantia and P. Rakic.)

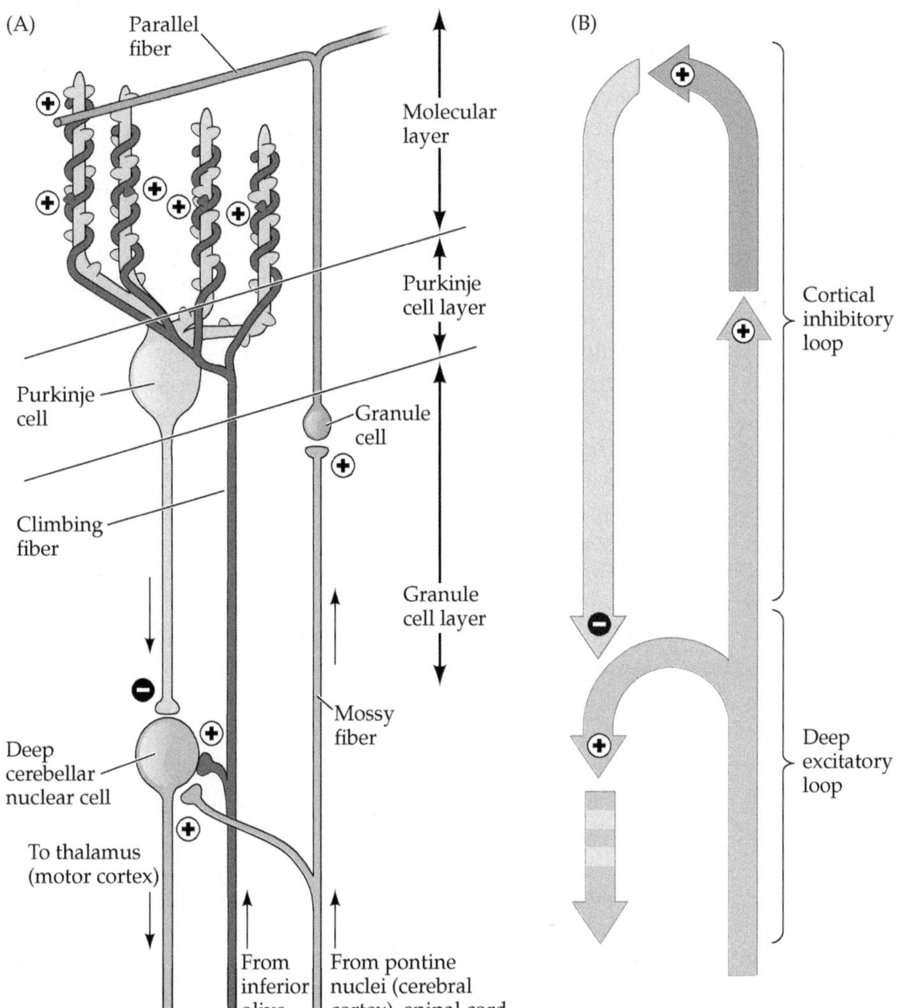

FIGURE 19.10 Excitatory and inhibitory connections in the cerebellar cortex and deep cerebellar nuclei. (A) The excitatory input from mossy fibers and climbing fibers to Purkinje cells and deep nuclear cells is basically the same. Additional convergent input onto the Purkinje cell from local circuit neurons (basket and stellate cells) and other Purkinje cells (not illustrated) establishes a basis for the comparison of ongoing movement and sensory feedback derived from it. The output of the Purkinje cell onto the deep cerebellar nuclear cell is inhibitory. (B) Conceptual diagram of the circuitry illustrated in (A). The deep cerebellar nuclei and their excitatory afferents constitute a *deep excitatory loop* whose output is shaped by a *cortical inhibitory loop* that inverts the "sign" of the input signals. The Purkinje neuron output to the deep cerebellar nuclear cell thus generates an error correction signal that can modify movements. The climbing fibers modify the efficacy of the parallel fiber–Purkinje cell connection, producing long-term changes in cerebellar output. (A after Stein, 1986.)

shafts from **climbing fibers**, all of which arise in the contralateral inferior olive (Figure 19.9B). Each Purkinje cell receives numerous synaptic contacts from a single climbing fiber. The climbing fibers provide a "training" signal that modulates the synaptic strength of the parallel fiber connection with the Purkinje cells (see below).

The Purkinje cells project in turn to the deep cerebellar nuclei and comprise the only output cells of the cerebellar cortex. Since Purkinje cells are GABAergic, the output of the cerebellar cortex is wholly inhibitory. However, the neurons in the deep cerebellar nuclei also receive excitatory input from the collaterals of the mossy and climbing fibers. The inhibitory projections of Purkinje cells serve to sculpt the discharge patterns that deep nuclei neurons generate in response to their direct mossy and climbing fiber inputs (Figure 19.10).

Inputs from GABAergic interneurons modulate the inhibitory activity of Purkinje cells. The most powerful of these local inputs are inhibitory nests of synapses made with the Purkinje cell bodies by **basket cells** (see Figure 19.9). Another type of local circuit neuron, the **stellate cell**, receives input from the parallel fibers and provides an inhibitory input to the Purkinje cell dendrites. Finally, the molecular layer contains the apical dendrites of inhibitory interneurons called **Golgi cells**; these neurons have their cell bodies in the granular cell layer. Golgi cells receive input from the parallel fibers and provide an inhibitory feedback to the cells of origin of the parallel fibers (the granule cells). There are other classes of interneurons in the cerebellar cortex (including excitatory unipolar brush cells and inhibitory Lugaro cells; not illustrated in Figure 19.9), whose functions are less well understood.

This circuit module of excitatory and inhibitory cells is repeated over and over throughout every subdivision of the cerebellum in all vertebrates, and suggests that—in spite of the differences in the sources of their inputs and in the destinations of their outputs—all of these subdivisions share a similar function. That is, in each subdivision, transformation of signal flow through these modules provides the basis for both the real-time regulation of movement and the long-term changes in regulation that underlie motor learning.

Description of the flow of signals through these complex modules may be simplified by distinguishing the two basic stages of cerebellar processing, beginning with the deep cerebellar nuclei. Mossy fiber and climbing fiber collaterals drive the activation of neurons in the deep cerebellar nuclei; this constitutes a *deep excitatory loop* in which input signals converge on the final output stage of cerebellar processing. However, as suggested above, the spatiotemporal patterns of the output activity are not simply faithful replications of the input patterns. The response patterns of the deep cerebellar nuclei to their direct inputs are modified by the descending inhibitory inputs of Purkinje cells, which are driven by these same two afferent pathways (i.e., the mossy and climbing fiber projections to the cerebellar cortex). For their part, Purkinje cells integrate these principal inputs and invert their "sign" by responding to excitatory inputs with an inhibitory output (see Figure 19.10B). Thus, Purkinje cells convey the product of computations performed by a *cortical inhibitory loop* that comprises the circuitry of the cerebellar cortex, including the interneurons of the granule and molecular layers, as well as the Purkinje cells themselves. The interneurons control the flow of information through the cerebellar cortex. For example, the Golgi cells form an inhibitory feedback circuit that controls the temporal properties of the granule cell input to the Purkinje cells, whereas the basket cells provide lateral inhibition that may focus the spatial distribution of Purkinje cell activity.

The modulation of cerebellar output by the cerebellar cortex may be responsible for motor learning. According to a model proposed by Masao Ito and his colleagues at Tokyo University, the climbing fibers from the inferior olive relay the message of a motor error to the Purkinje cells. This message is derived from inputs that the inferior olive receives from multiple structures (including the cerebral cortex and spinal cord) as well as from feedback signals from the cerebellum via the red nucleus, as described above. The 1000 or so synapses made by a single climbing fiber with the proximal dendrites of a single Purkinje cell constitute one of the most powerful excitatory connections in the CNS. Activation of the climbing fibers induces a strong excitatory postsynaptic potential in Purkinje cells that generates an initial action potential followed by series of smaller "spikelets." This postsynaptic response is termed a *complex spike*; it typically occurs infrequently (1 to 2 Hz), depending on the context and demands of the concurrent movements. In contrast, the parallel fiber input to the Purkinje cells gives rise to individual action potentials, called *simple spikes*, that typically fire at a much higher rate (30 to 100 Hz). The impact of the climbing fiber input on Purkinje cell output is further enhanced by the gap junctions that electronically join and synchronize the activity of neurons in the inferior olive. These ensembles of olivary neurons both drive the activation of cerebellar circuits and also promote adaptive plasticity in the inhibitory output of the cerebellar cortex. The plasticity results from long-term reductions in the Purkinje cell responses to their parallel fiber inputs. The mechanism for this long-term depression is a complex chain of cellular events leading from the climbing fiber input to the Purkinje cells to the endocytosis of AMPA receptors at parallel fiber–Purkinje cell synapses. (Recall that AMPA receptors mediate fast, excitatory responses at glutamatergic synapses; for an account of the cellular mechanisms responsible for this long-term reduction in the efficacy of the parallel fiber synapse on Purkinje cells, see Chapter 8.)

The reduction in the efficacy of parallel fiber input to Purkinje cells has the effect of increasing the response of neurons in the deep cerebellar nuclei to afferent activity (by weakening the influence of the inhibitory loop). Thus, the feedback signals from the cerebellum to circuits of upper motor neurons in the motor cortex and brainstem are altered as a consequence of climbing fiber activation. It is not yet understood at the circuit level how this alteration mediates a "correction" of movement error. Nevertheless, it is clear from studies of animal models and of humans with damage to the inferior olive that both short-term sensorimotor adaptation (error correction) and long-term motor learning require the modulation of cerebellar processing by climbing fiber activation.

Despite the consistency with which these basic structural and functional features of cerebellar circuitry are replicated throughout the cortex of the cerebellum, recent molecular, genetic, anatomical, and physiological studies have revealed longitudinal compartments in the cerebellar cortex (Figure 19.11). For example, subsets of Purkinje cells show variable expression of zebrin II, which is an antigen localized to aldolase C (an enzyme in the glycolysis pathway). Purkinje cells that express zebrin II are clustered together into rostrocaudal bands that are interleaved with bands lacking zebrin II expression (reminiscent of modules found elsewhere in the CNS; see Box 9A). A variety of other molecular markers are co-localized to zebrin II–positive or –negative bands, including molecules related to glutamatergic neurotransmission and second-messenger systems within postsynaptic processes. One physiological consequence of this distinction is that long-term depression (as discussed above) is more prominent at synapses between parallel fibers and Purkinje neurons lacking zebrin II. These zebrin II-negative Purkinje neurons also tend to exhibit higher basal rates of simple spike activity. In contrast, long-term potentiation is more likely at parallel fiber–Purkinje cell synapses in zebrin II–positive bands where the rate of simple cell spikes is relatively low. These findings indicate that the canonical circuits of the cerebellar cortex may yet be

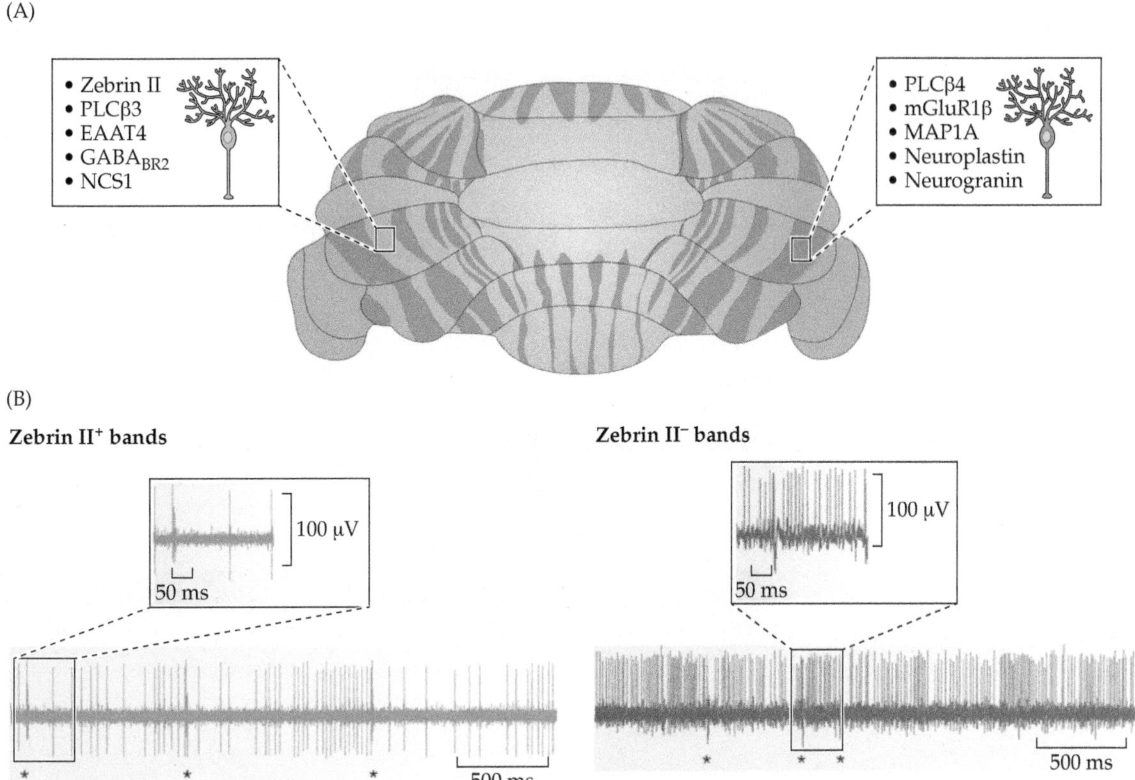

FIGURE 19.11 Compartments in the cerebellar cortex. (A) Illustration of the mouse cerebellum showing alternating bands of zebrin II expression in Purkinje cells and co-localization of a variety of molecular markers. EAAT4, excitatory amino acid transporter 4; $GABA_{BR2}$, $GABA_B$ receptor subtype 2; MAP1A, microtubule-associated protein 1A; mGluR1β, metabotropic glutamate receptor 1β; NCS1, neuronal calcium sensor 1; PLCβ3, phospholipase Cβ3; PLCβ4, phospholipase Cβ4; neuroplastin is a member of the immunoglobulin superfamily that functions as a cell adhesion molecule, and neurogranin is a calmodulin-binding protein. (B) Extracellular recordings of Purkinje cells in zebrin II-positive and -negative bands of rat cerebellar cortex. Zebrin II-negative Purkinje cells show higher rates of sustained simple spike activity, while zebrin II-positive Purkinje cells show stronger suppression of simple spikes following complex spikes (asterisks). (From Cerminara et al., 2015.)

further differentiated as the tools of contemporary neuroscience continue to reveal previously unrecognized order. It remains a challenge to determine how such regional variations in gene expression and physiological phenotype convey different capabilities in information processing across the cerebellar cortex.

Cerebellar Circuitry and the Coordination of Ongoing Movement

As one would expect for a structure that monitors and adjusts motor behavior, neuronal activity in the cerebellum changes continually during the course of a movement. For instance, the execution of a relatively simple task such as flexing and extending the wrist back and forth elicits a dynamic pattern of activity in both the Purkinje cells and the deep cerebellar nuclear cells that closely follows the ongoing movement (Figure 19.12). Both types of cells are tonically active at rest and change their frequency of firing as movements occur. The neuronal responses are influenced by various aspects of movement, including relaxation or contraction of specific muscles, the position of the joints, and the direction of the next movement that will occur. All this information is encoded by changes in the discharge pattern of Purkinje cells, and these changes modulate the ongoing output of the deep cerebellar nuclear cells.

As these neuronal response properties predict, cerebellar injuries and disease tend to disrupt the modulation and coordination of ongoing movements, and the specific movements that are disrupted vary with the location of the damage. The hallmark of individuals with cerebellar damage is difficulty producing smooth, well-coordinated, multi-jointed movements. Instead,

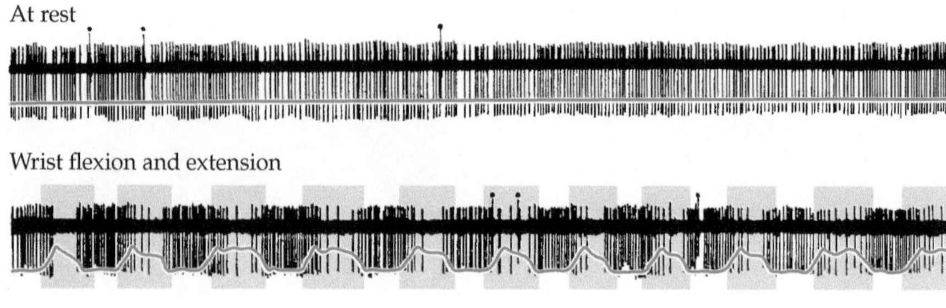

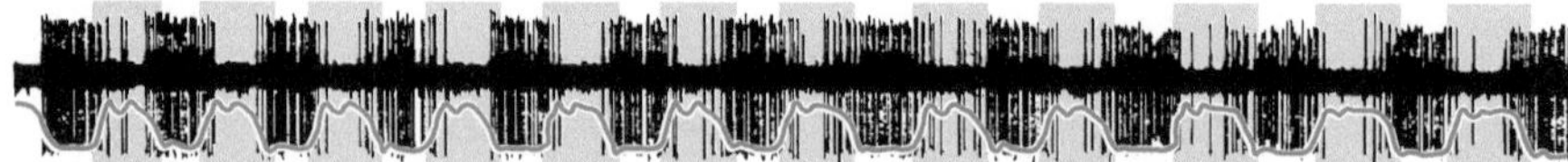

FIGURE 19.12 Activity of Purkinje cells and cells of the deep cerebellar nuclei. Neuronal activity is shown for a Purkinje cell (A) and a deep cerebellar nuclear cell (B) both at rest (upper records) and during movement of the ipsilateral wrist (lower records). The red traces represent wrist movement; up is flexion, down is extension. The durations of the wrist movements are indicated by the colored blocks. Both classes of cells are tonically active at rest. Rapid alternating movements result in the transient inhibition of the tonic activity of both cell types. (After Thach, 1968.)

movements tend to be decomposed into jerky and imprecise elements, a condition referred to as **cerebellar ataxia**. Many of these difficulties in performing movements can be explained as disruption of the cerebellum's role in correcting errors in ongoing movements, since the cerebellar error correction mechanism normally ensures that movements are modified to cope with changing circumstances. As described earlier, the Purkinje cells and deep cerebellar nuclear cells recognize potential errors by comparing patterns of convergent activity that are concurrently available to both cell types. The deep nuclear cells then send corrective signals to the upper motor neurons in order to maintain or improve the accuracy of the movement.

As in the case of the basal ganglia, studies of the oculomotor system (saccades in particular) have contributed greatly to understanding the contribution that the cerebellum makes to motor error reduction. For example, cutting part of the tendon to the lateral rectus muscles in one eye of a monkey weakens horizontal eye movements by that eye (Figure 19.13). When a patch is then placed over the normal eye to force the animal to use its weak eye, the saccades performed by the weak eye are initially *hypometric*—they fall short of visual targets. Over the next few days, however, the amplitude of the saccades gradually increases until they again become accurate. If the patch is then switched to cover the weakened eye, the saccades performed by the normal eye are now *hypermetric*. In other words, over a period of a few days the nervous system corrects the errors in the saccades made by the weak eye by increasing the gain in a region of the saccade motor system that controls both eyes (see Chapter 20). Lesions in the vermis of the spinocerebellum (see Figure 19.1) eliminate this ability to reduce the motor error.

Similar evidence of the cerebellar contribution to movement has come from studies of the vestibulo-ocular reflex (VOR) in monkeys and humans. The VOR keeps the eyes trained on a visual target during head movements (see Chapter 14). The relative simplicity of this reflex has made it possible to analyze some of the mechanisms that enable motor learning as a process of error reduction. When the head moves, the eyes must move at the same velocity in the opposite direction to maintain a stable representation of the visual image on the retina. In these studies, the adaptability of the VOR to changes in the nature of incoming sensory information is challenged by fitting individuals (either monkeys or humans) with magnifying or minifying spectacles (Figure 19.14). Because the glasses alter the size of the visual image on the retina, the compensatory eye movements, which would normally maintain a stable image of an object on the retina, are either too large or too small. Over time, individuals (whether monkeys or humans) learn to adjust the distance the eyes must move in response to head movements to accord with the artificially altered size of the visual field. Moreover, this change is retained for significant periods after the spectacles are removed and can be detected electrophysiologically in the responses that

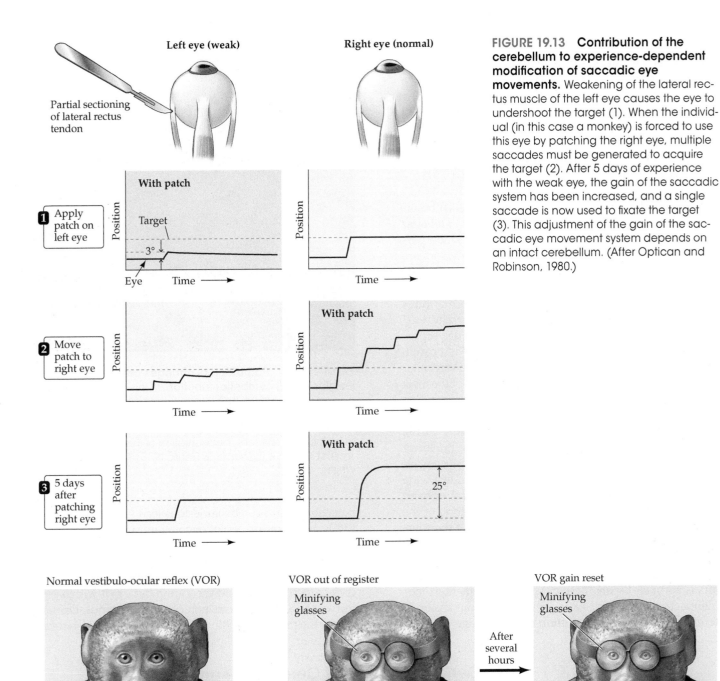

FIGURE 19.13 Contribution of the cerebellum to experience-dependent modification of saccadic eye movements. Weakening of the lateral rectus muscle of the left eye causes the eye to undershoot the target (1). When the individual (in this case a monkey) is forced to use this eye by patching the right eye, multiple saccades must be generated to acquire the target (2). After 5 days of experience with the weak eye, the gain of the saccadic system has been increased, and a single saccade is now used to fixate the target (3). This adjustment of the gain of the saccadic eye movement system depends on an intact cerebellum. (After Optican and Robinson, 1980.)

Head and eyes move in a coordinated manner to keep image on retina.

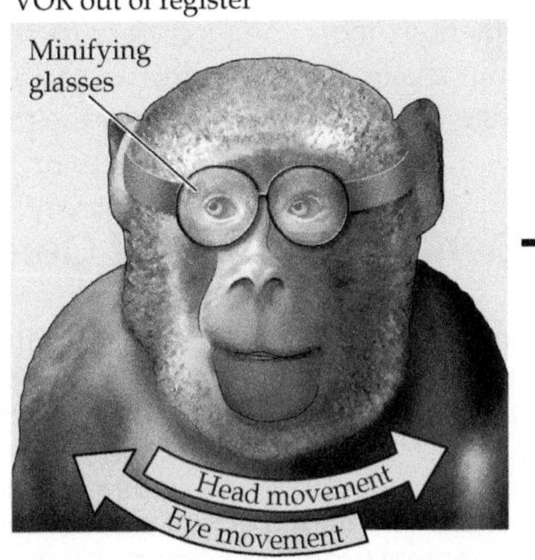

Eyes move too far in relation to image movement on the retina when the head moves.

Eyes move smaller distances in relation to head movement to compensate.

FIGURE 19.14 Learned changes in the vestibulo-ocular reflex in monkeys. Normally, the VOR operates to move the eyes as the head moves, so that the retinal image remains stable. When the animal observes the world through minifying spectacles, the eyes initially move too far with respect to the "slippage" of the visual image on the retina. After some experience, however, the gain of the VOR is reset and the eyes move an appropriate distance in relation to head movement, thus compensating for the altered size of the visual image.

can be recorded from Purkinje cells and neurons in the deep cerebellar nuclei. Once again, if the cerebellum is damaged or removed, the ability of the VOR to adapt to the new conditions is lost. These observations support the conclusion that the cerebellum is critically important in error reduction during motor learning.

Further Consequences of Cerebellar Lesions

A basic circuit module is present throughout all subdivisions of the cerebellum and individuals with cerebellar damage, regardless of the cause or location, exhibit persistent errors in movement. These movement errors are always on the same side of the body as the damage to the cerebellum, reflecting the cerebellum's unusual status as a brain structure in which sensorimotor information is represented ipsilaterally rather than contralaterally. Furthermore, somatic, visual, and other inputs are represented topographically within the cerebellum; as a result, the movement deficits following circumscribed cerebellar damage may be quite specific. For example, one of the most common cerebellar syndromes is caused by degeneration in the anterior portion of the cerebellar cortex in individuals with a long history of alcohol abuse (Figure 19.15). Such damage specifically affects movement in the lower limbs, which are represented in the anterior spinocerebellum (see Figure 19.4). The consequences include a wide and staggering gait but little impairment of arm or hand movements. Thus, the topographical organization of the cerebellum allows cerebellar damage to disrupt the coordination of movements performed by some muscle groups but not others.

The implication of these pathologies is that the cerebellum is normally capable of integrating the moment-to-moment actions of muscles and joints throughout the body to ensure the smooth execution of a full range of motor behaviors. Thus, cerebellar lesions lead first and foremost to a lack of coordination of ongoing movements (Box 19A). For example, damage to the vestibulocerebellum impairs the ability to stand upright and maintain the direction of gaze. The eyes have difficulty maintaining fixation; they drift from the target and then jump back to it with a corrective saccade, a phenomenon called **nystagmus**. Disruption of the pathways to the vestibular nuclei may also result

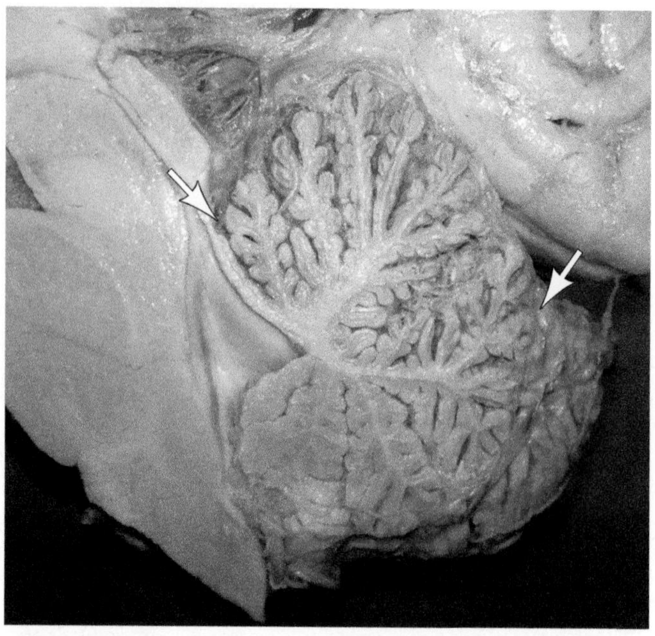

FIGURE 19.15 **Pathological changes can provide insights about cerebellar function.** In this example, chronic alcohol abuse has caused degeneration of the anterior vermis (arrows), while leaving other cerebellar regions intact. The individual had difficulty walking but little impairment of arm movements or speech. The orientation of this paramedian sagittal section is the same as Figure 19.1D. (Courtesy of L. E. White.)

BOX 19A ■ Genetic Analysis of Cerebellar Function

Since the early 1950s, investigators interested in motor behavior have identified and studied strains of mutant mice in which movement is compromised. These mutant mice are easy to spot: Following induced or spontaneous mutagenesis, the "screen" is simply to look for animals that have difficulty moving.

Genetic analysis suggested that some of these abnormal behaviors could be explained by single autosomal recessive or semidominant mutations, in which homozygotes are most severely affected. The strains were given names such as *reeler, weaver, lurcher, staggerer*, and *leaner* that reflected the nature of the motor dysfunction they exhibited (see table). The relatively large number of mutations that compromise movement suggested it might be possible to understand some aspects of motor circuits and function at the genetic level.

A common feature of the mutants is ataxia resembling that associated with cerebellar dysfunction in humans. Indeed, all the mutations are associated with some form of cerebellar pathology. The pathologies associated with the *reeler* and *weaver* mutations are particularly striking (see figure). In the *reeler* cerebellum, Purkinje cells, granule cells, and interneurons are all displaced from their usual laminar positions, and there are fewer granule cells than normal. In *weaver*, most of the granule cells are lost prior to their migration from the external granule layer (a proliferative region where cerebellar granule cells are generated during development), leaving only Purkinje cells and local

BOX 19A ■ *(continued)*

interneurons to carry on the work of the cerebellum. Thus, these mutations causing deficits in motor behavior impair the development and final disposition of the neurons that make up the major processing circuits of the cerebellum (see Figures 19.8 and 19.9).

Efforts to characterize the cellular mechanisms underlying these motor deficits were unsuccessful, and the molecular identity of the affected genes remained obscure until recently. In the past few decades, however, both the *reeler* and *weaver* genes have been identified and cloned.

The *reeler* gene was cloned through a combination of good luck and careful observation. In the course of making transgenic mice by inserting DNA fragments in the mouse genome, investigators in Tom Curran's laboratory created a new strain of mice that behaved much like *reeler* mice and had similar cerebellar pathology. This "synthetic" *reeler* mutation was identified by finding the position of the novel DNA fragment—which turned out to be on the same chromosome as the original *reeler* mutation. Further analysis showed that the same gene had indeed been mutated, and the *reeler* gene was subsequently identified. Remarkably, the protein encoded by this gene is homologous to known extracellular matrix proteins such as tenascin, laminin, and fibronectin (see Chapter 23). This finding makes sense, since the pathophysiology of the *reeler* mutation entails altered cell migration, resulting in misplaced neurons in the cerebellar cortex as well as the cerebral cortex and hippocampus.

Molecular genetic techniques have also led to cloning the *weaver* gene. Using linkage analysis and the ability to clone and sequence large pieces of mammalian chromosomes, Andy Peterson and his colleagues "walked" (i.e., sequentially cloned) several kilobases of DNA in the chromosomal region to find where the *weaver* gene mapped. By comparing normal and mutant sequences within this region, they determined *weaver* to be a mutation in an inward rectifier K^+ channel (see Chapter 4). How this particular molecule influences the development of granule cells or causes their death in the mutants is not yet clear. Nevertheless, the story of the proteins encoded by the *reeler* and *weaver* genes indicates both the promise and the challenge of a genetic approach to understanding cerebellar function.

In recent years, such investigative opportunity has arisen with respect to the putative role of the cerebellum in the expression of disorder in the neurocognitive domain. Thus, genetic, behavioral, and clinical investigations have suggested that dysfunctional cerebellar circuits may contribute to the development of neurocognitive disorders, including autism spectrum disorders. Early disruption of the cerebellar circuitry has been shown to be positively correlated with autism; indeed, cerebellar injury conveys the largest nonheritable risk for the emergence of autism spectrum

Continued on the next page

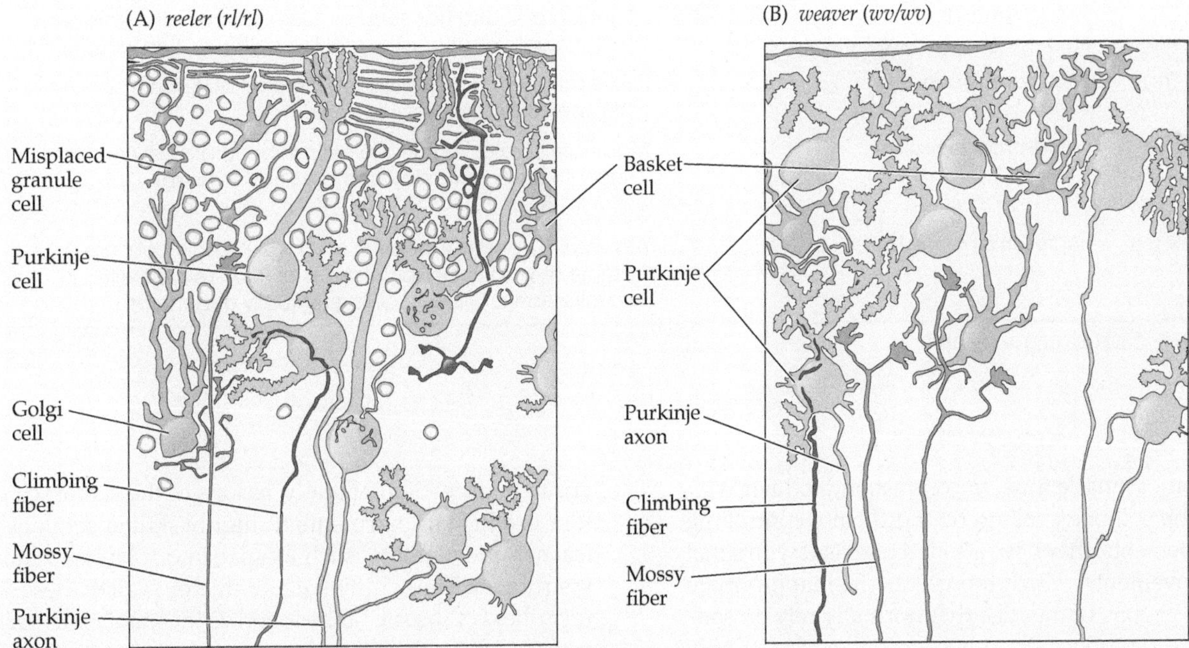

The cerebellar cortex is disrupted in both the *reeler* and *weaver* mutations. (A) The cerebellar cortex in homozygous *reeler* mice. The *reeler* mutation causes the major cell types of the cerebellar cortex to be displaced from their normal laminar positions. Despite the disorganization of the cerebellar cortex in *reeler* mutants, the major inputs—mossy fibers and climbing fibers—find appropriate targets. (B) The cerebellar cortex in homozygous *weaver* mice. The granule cells are missing, and the major cerebellar inputs synapse inappropriately on the remaining neurons. (After Caviness and Rakic, 1978.)

BOX 19A ■ (*continued*)

disorders. Among heritable factors, recent work using mouse models has shown that mutations in the genes that encode the SHANK (SH3 and multiple ankyrin repeat domains) family of postsynaptic scaffolding proteins produce impairments in cerebellar-dependent sensorimotor learning and alterations in Purkinje cell dendritic morphology. Perhaps such mutations induce synaptic modifications that alter the role of cerebellar circuitry in mediating neural activities that depend on precise temporal control. Whatever the explanation, it has become clear that early disturbances in cerebellar structure and function have broad implications for the ongoing (postnatal) construction and refinement of circuits in the cerebral hemispheres, including those that are involved in governing motor behavior, and also those involved in the expression of cognition and affect.

Motor Mutations in Mice

Mutation	Inheritance	Chromosome affected	Behavioral and morphological characteristics
reeler (*rl*)	Autosomal recessive	5	Reeling ataxia of gait, dystonic postures, and tremors. Systematic malposition of neuron classes in the forebrain and cerebellum. Small cerebellum, reduced number of granule cells.
weaver (*wv*)	Autosomal recessive	?	Ataxia, hypotonia, and tremor. Cerebellar cortex reduced in volume. Most cells of external granular layer degenerate prior to migration.
leaner (*tgla*)	Autosomal recessive	8	Ataxia and hypotonia. Degeneration of granule cells, particularly in the anterior and nodular lobes of the cerebellum. Degeneration of a few Purkinje cells.
lurcher (*lr*)	Autosomal semi-dominant	6	Homozygote dies. Heterozygote is ataxic with hesitant, lurching gait and has seizures. Cerebellum half normal size; Purkinje cells degenerate; granule cells reduced in number.
nervous (*nr*)	Autosomal recessive	8	Hyperactivity and ataxia. Ninety percent of Purkinje cells die between 3 and 6 weeks of age.
Purkinje cell degeneration (*pcd*)	Autosomal recessive	13	Moderate ataxia. All Purkinje cells degenerate between the fifteenth embryonic day and third month of age.
staggerer (*sg*)	Autosomal recessive	9	Ataxia with tremors. Dendritic arbors of Purkinje cells are simple (few spines). No synapses of Purkinje cells with parallel fibers. Granule cells eventually degenerate.

Adapted from Caviness and Rakic, 1978.

in a reduction of muscle tone. In contrast, individuals with damage to the spinocerebellum have difficulty controlling walking movements; they have a wide-based gait with small shuffling movements, which represents the inappropriate operation of groups of muscles that normally rely on sensory feedback to produce smooth, concerted actions. These individuals also have difficulty performing rapid alternating movements, a sign referred to as **dysdiadochokinesia**. Over- and underreaching, or **dysmetria**, may also occur. Tremors, known as **action** or **intention tremors**, accompany over- and undershooting of a movement, due to disruption of the mechanism for detecting and correcting movement errors (Figure 19.16). Finally, lesions of the cerebrocerebellum produce impairments in highly skilled sequences of learned movements, such as speech or playing a musical instrument. The common denominator of all of these signs, regardless of the site of the lesion, is the inability to perform smooth, precisely directed movements.

Summary

The cerebellum receives input from regions of the cerebral cortex that plan and initiate complex and highly skilled movements; it also receives input from sensory systems

(A)

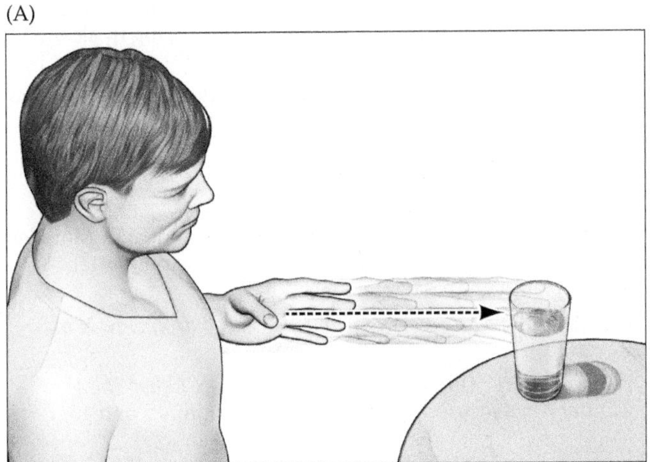

(B)

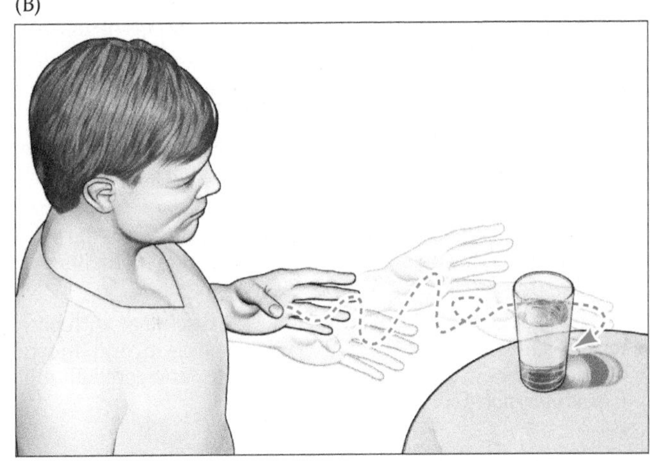

FIGURE 19.16 Illustration of appendicular ataxia with cerebellar damage. (A) Smooth execution of a visually guided reach in an individual with typical cerebellum function. (B) Poorly coordinated visually guided reach (appendicular ataxia) in an individual with cerebellar damage. The hand takes a much less straight trajectory to the target, with irregular movements that typically overshoot or undershoot the visual target and so require frequent corrective movements to execute the intended motor task.

that monitor the course of movements. This arrangement enables a comparison of an intended movement with the actual movement and a reduction in the difference, or "motor error." The corrections of motor error produced by the cerebellum occur in real time and are stored over longer periods as a form of motor learning. Error correction is mediated by climbing fibers that ascend from the inferior olive to contact the dendrites of the Purkinje cells in the cerebellar cortex. Information provided by the climbing fibers modulates the effectiveness of the second major input to the Purkinje cells, which arrives via the parallel fibers from the granule cells. The granule cells receive information about the intended movement—and the actual performance of the movement—from the vast number of mossy fibers that enter the cerebellum from multiple sources. As might be expected, the output of the cerebellum from the deep cerebellar nuclei projects to circuits that govern all the major sources of upper motor neurons described in Chapter 17. The effects of cerebellar disease provide strong support for the idea that the cerebellum regulates the performance of movements. Thus, individuals with cerebellar disorders show severe ataxias in which the site of the lesion determines the particular movements affected, with the incoordination of movement on the same side of the body as the site of the lesion.

ADDITIONAL READING

Reviews

Allen, G. and N. Tsukahara (1974) Cerebrocerebellar communication systems. *Physiol. Rev.* 54: 957–1006.

Apps, R. and R. Hawkes (2009) Cerebellar cortical organization: a one-map hypothesis. *Nature Rev. Neurosci.* 10: 670–681.

Cerminara, N. L., E. J. Lang, R. V. Sillitoe and R. Apps (2015) Redefining the cerebellar cortex as an assembly of non-uniform Purkinje cell microcircuits. *Nat. Rev. Neurosci.* 16: 79–93.

Glickstein, M. and C. Yeo (1990) The cerebellum and motor learning. *J. Cog. Neurosci.* 2: 69–80.

Lisberger, S. G. (1988) The neural basis for learning of simple motor skills. *Science* 242: 728–735.

Ohyama, T., W. L. Nores, M. Murphy and M. D. Mauk (2003) What the cerebellum computes. *Trends Neurosci.* 26: 222–227.

Robinson, F. R. and A. F. Fuchs (2001) The role of the cerebellum in voluntary eye movements. *Annu. Rev. Neurosci.* 24: 981–1004.

Thach, W. T. (2007) On the mechanism of cerebellum contributions to cognition. *Cerebellum* 6: 163–167.

Thach, W. T., H. P. Goodkin and J. G. Keating (1992) The cerebellum and adaptive coordination of movement. *Annu. Rev. Neurosci.* 15: 403–442.

Therrien, A. S. and A. J. Bastian (2015) Cerebellar damage impairs internal predictions for sensory and motor function. *Curr. Opin. Neurobiol.* 33: 127–133.

Important Original Papers

Asanuma, C., W. T. Thach and E. G. Jones (1983) Distribution of cerebellar terminals and their relation to other afferent terminations in the ventral lateral thalamic region of the monkey. *Brain Res. Rev.* 5: 237–265.

Brodal, P. (1978) The corticopontine projection in the rhesus monkey: Origin and principles of organization. *Brain* 101: 251–283.

DeLong, M. R. and P. L. Strick (1974) Relation of basal ganglia, cerebellum, and motor cortex units to ramp and ballistic movements. *Brain Res.* 71: 327–335.

Eccles, J. C. (1967) Circuits in the cerebellar control of movement. *Proc. Natl. Acad. Sci. USA* 58: 336–343.

McCormick, D. A., G. A. Clark, D. G. Lavond and R. F. Thompson (1982) Initial localization of the memory trace for a basic form of learning. *Proc. Natl. Acad. Sci. USA* 79: 2731–2735.

Thach, W. T. (1968) Discharge of Purkinje and cerebellar nuclear neurons during rapidly alternating arm movements in the monkey. *J. Neurophysiol.* 31: 785–797.

Thach, W. T. (1978) Correlation of neural discharge with pattern and force of muscular activity, joint position, and direction of intended next movement in motor cortex and cerebellum. *J. Neurophysiol.* 41: 654–676.

Victor, M., R. D. Adams and E. L. Mancall (1959) A restricted form of cerebellar cortical degeneration occurring in alcoholic patients. *Arch. Neurol.* 1: 579–688.

Yang, Y. and S. G. Lisberger (2014) Purkinje-cell plasticity and cerebellar motor learning are graded by complex-spike duration. *Nature* 510: 529–532.

Books

Bradley, W. G., R. B. Daroff, G. M. Fenichel and C. D. Marsden (eds.) (1991) *Neurology in Clinical Practice.* Boston: Butterworth-Heinemann, chapters 29 and 77.

Ito, M. (1984) *The Cerebellum and Neural Control.* New York: Raven Press.

Klawans, H. L. (1989) *Toscanini's Fumble and Other Tales of Clinical Neurology.* New York: Bantam, chapters 7 and 10.

Go to the NEUROSCIENCE 6e Companion Website at **www.oup.com/uk/Purves6e** for Web Topics, Animations, Flashcards, and more.

CHAPTER

20 Eye Movements and Sensorimotor Integration

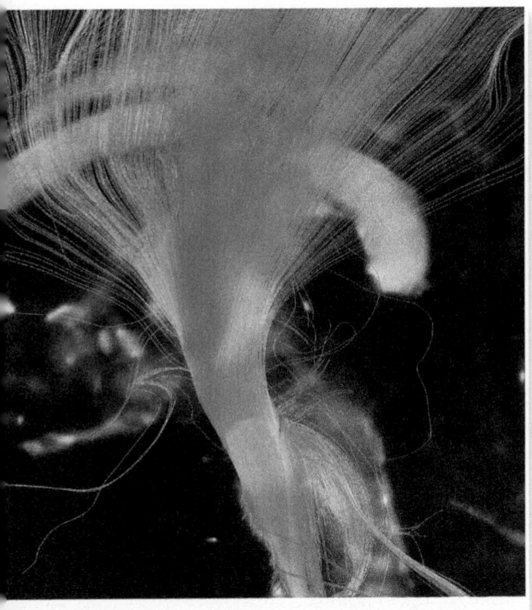

Overview

EYE MOVEMENTS ARE EASIER TO STUDY than movements of other parts of the body. This fact arises in part from the relative simplicity of muscle actions on the eyeball. There are only six extraocular muscles, each of which has a specific role in adjusting eye position. Moreover, there is a limited set of stereotyped eye movements, and the central circuits governing each one are partially distinct. Eye movements have therefore been a useful model for understanding the mechanisms of motor control. Indeed, much of what is known about the regulation of movements by the vestibular system, basal ganglia, and cerebellum has come from the study of eye movements (see Chapters 14, 18, and 19). In this chapter, the major features of eye movement control are used to illustrate principles of sensorimotor integration that also apply to more complex motor behaviors.

What Eye Movements Accomplish

Eye movements are important in humans because high visual acuity is restricted to the fovea, the small circular region (about 1.2 mm in diameter) in the central retina that is densely packed with cone photoreceptors (see Chapter 11). Eye movements can direct the fovea to new objects of interest in the visual field—a process called **foveation**—or compensate for disturbances that cause the fovea to be displaced from an object already foveated.

Several decades ago, the Russian physiologist Alfred Yarbus demonstrated that eye movements reveal a good deal about the strategies used to inspect a scene. Yarbus used contact lenses with small mirrors on them to document (by the position of a reflected beam on photosensitive paper) the pattern of eye movements made while individuals examined a variety of objects and scenes. Figure 20.1 shows the changes in the direction of an individual's gaze while viewing a photograph. The thin, straight lines represent the quick, ballistic eye movements (**saccades**) used to align the foveae with particular parts of the scene. Little or no visual perception occurs during a saccade, which occupies only a few tens of milliseconds. The denser spots along these lines represent points of fixation where the observer paused for a variable period to take in visual information from the area of interest. These results obtained by Yarbus, and subsequently by many others, showed that vision is an active process in which eye movements typically shift the view several times each second to direct the foveae toward selected parts of the scene to examine especially interesting or informative features. The selection of areas of interest as targets of the saccades indicates that non-foveal areas of the retina have sufficient resolution to guide the foveae toward these areas for closer examination. In the figure, the spatial distribution of the fixation points is not random, and indicates that the individual spent much

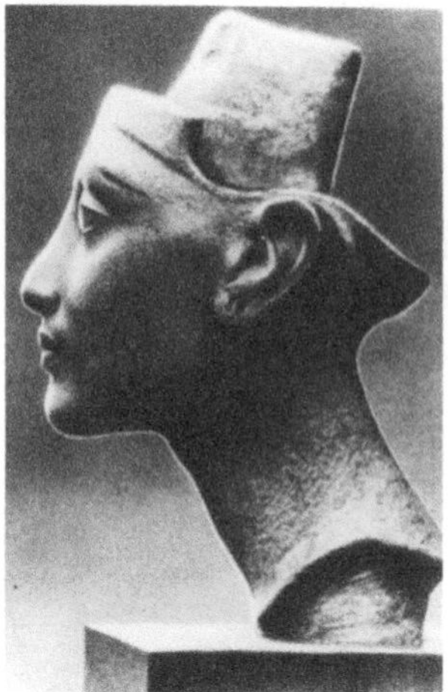

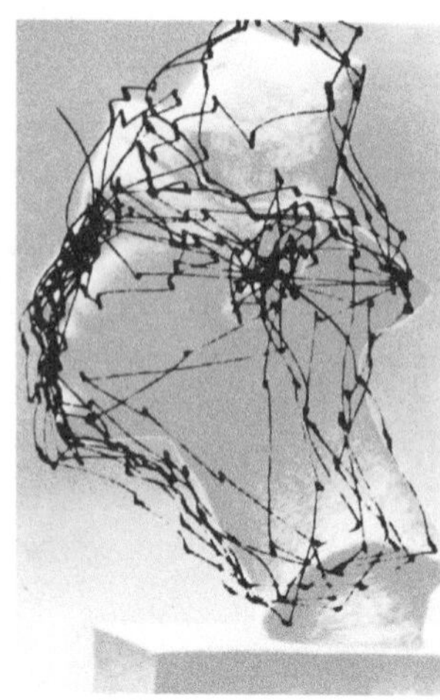

FIGURE 20.1 **Eye movements of an individual viewing a photograph.** The individual was shown this photograph (left) of the famous bust of Queen Nefertiti. The diagram on the right shows the individual's eye movements over a 2-min viewing period. (From Yarbus, 1967.)

more time scrutinizing Nefertiti's eye, nose, mouth, and ear than examining the middle of her cheek or neck. Thus, eye movements allow us to scan the visual field, pausing to focus attention on the portions of the scene that convey the most significant information. It follows from Figure 20.1 that tracking eye movements can be used to determine which aspects of a scene are particularly arresting; in fact, today's corporate advertisers can use modern versions of Yarbus's method to determine which pictures and scene arrangements will best sell their products.

The importance of eye movements for visual perception has also been demonstrated by experiments in which a visual image is stabilized on the retina, either by paralyzing the extraocular eye muscles or by moving a scene in exact register with eye movements so that the different features of the image always fall on exactly the same parts of the retina (Box 20A). Such stabilized visual images rapidly disappear, for reasons that remain poorly understood. Nonetheless, observations on motionless images make it plain that eye movements are essential for visual perception.

Actions and Innervation of Extraocular Muscles

Three antagonistic pairs of muscles control eye movements: the **lateral** and **medial rectus muscles**; the **superior** and **inferior rectus muscles**; and the **superior** and **inferior oblique muscles**. These muscles are responsible for movements of the eye along three different axes: *horizontal*, either toward the nose (adduction) or away from the nose (abduction); *vertical*, either elevation or depression; and *torsional*, movements that bring the top of the eye toward the nose (intorsion) or away from the nose (extorsion). Horizontal movements are controlled entirely by the medial

BOX 20A ■ The Perception of Stabilized Retinal Images

Visual perception depends critically on frequent changes of scene. Normally, our view of the world is changed by saccades, and tiny saccades that continue to move the eyes abruptly over a fraction of a degree of visual arc occur even when the observer stares intently at an object of interest. Moreover, continual drift of the eyes during fixation progressively shifts the image onto a nearby but different set of photoreceptors. As a consequence of these several sorts of eye movements (Figure A), our point of view changes more or less continually.

The importance of a continually changing scene for normal vision is dramatically revealed when the retinal image is stabilized. If a small mirror is attached to the eye by means of a contact lens and an image is reflected off the mirror onto a screen, then the individual necessarily sees the same thing, whatever the position of the eye—every time the eye moves, the projected image moves by exactly the same amount (Figure B). Under these circumstances, the stabilized image actually disappears from perception within a few seconds!

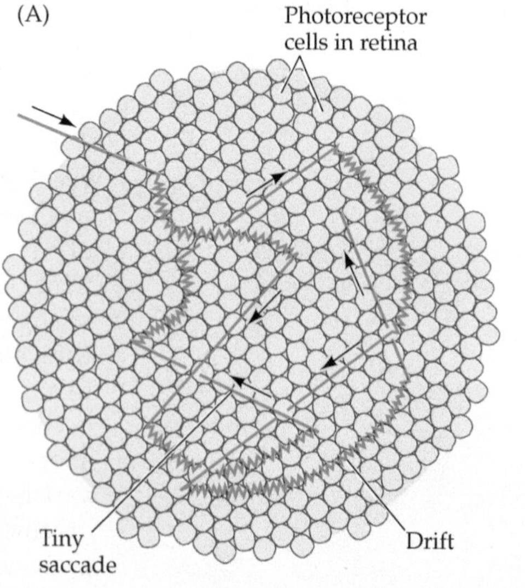

(A) Diagram of the types of eye movements that continually change the retinal stimulus during fixation. The straight lines indicate microsaccades, and the zigzag lines drift; the structures in the background are photoreceptor cells drawn approximately to scale. The normal scanning movements of the eyes (saccades) are much too large to be shown here but obviously contribute to the changes of view that we continually experience, as do slow tracking eye movements (although the fovea tracks a particular object, the scene nonetheless changes). (After Pritchard, 1961.)

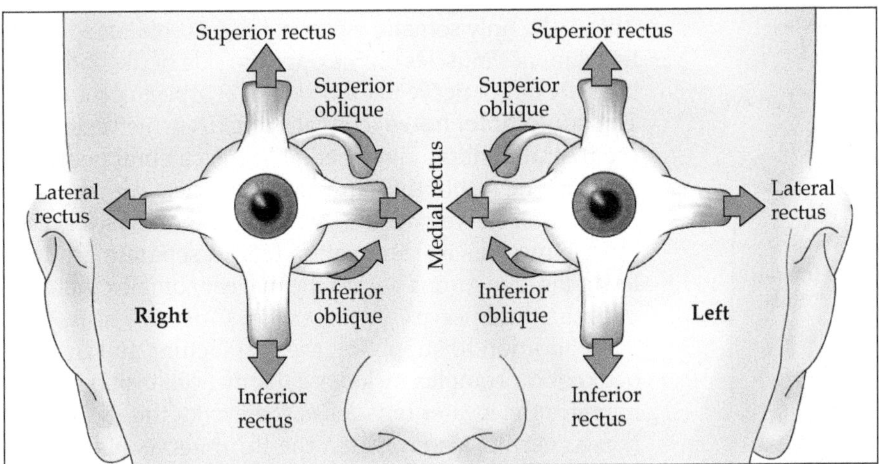

FIGURE 20.2 Extraocular muscles contribute to vertical and horizontal eye movements. Horizontal movements are mediated by the medial and lateral rectus muscles, while vertical movements are mediated by the superior and inferior rectus and the superior and inferior oblique muscles.

and lateral rectus muscles; the medial rectus muscle is responsible for adduction, the lateral rectus muscle for abduction (Figure 20.2). Vertical movements require the coordinated action of the superior and inferior rectus muscles, as well as the oblique muscles. The relative contributions of the rectus and oblique muscles depend on the horizontal position of the eye. In the primary position (eyes straight ahead), both of these muscle groups contribute to vertical movements. Elevation is due to the action of the superior rectus and inferior oblique muscles, while depression is due to the action of the inferior rectus and superior oblique muscles. When the eye is abducted, the rectus muscles are the prime vertical movers; elevation is due to the action of the superior rectus, and depression is due to the action of the inferior rectus. When the eye is adducted, the oblique muscles are the prime vertical movers. In this position, elevation is due to the action of the inferior oblique muscle, while depression is due to the action of the superior oblique muscle. The oblique muscles are also primarily responsible for torsional movements.

The extraocular muscles are innervated by lower motor neurons whose axons form three cranial nerves: the abducens, the trochlear, and the oculomotor (Figure 20.3). The **abducens nerve** (cranial nerve VI) exits the brainstem

BOX 20A (*continued*)

A simple way to demonstrate the rapid disappearance of a stabilized retinal image is to visualize one's own retinal blood vessels. The blood vessels, which lie in front of the photoreceptor layer, cast a shadow on the underlying receptors. Although normally invisible, the vascular shadows can be seen by moving a source of light across the eye, a phenomenon first noted by J. E. Purkinje more than 150 years ago. This perception can be elicited with an ordinary penlight pressed gently against the lateral side of the closed eyelid. When the light is wiggled vigorously, a rich network of black blood vessel shadows (called a "Purkinje tree") appears against an orange background. (The vessels appear black because they are shadows.) By starting and stopping the movement, it is readily apparent that the image of the blood vessel shadows disappears within a fraction of a second after the light source is stilled.

The conventional interpretation of the rapid disappearance of stabilized images is retinal adaptation. In fact, the phenomenon is at least partly of central origin. Stabilizing the retinal image in one eye, for example, diminishes perception through the other eye, an effect known as interocular transfer. Although the explanation of these remarkable effects is not entirely clear, they emphasize the point that the visual system is designed to deal with novelty.

(B) One means of producing stabilized retinal images. By attaching a small mirror to the eye, the scene projected onto the screen will always fall on the same set of retinal points, no matter how the eye is moved. (After Riggs et al., 1953.)

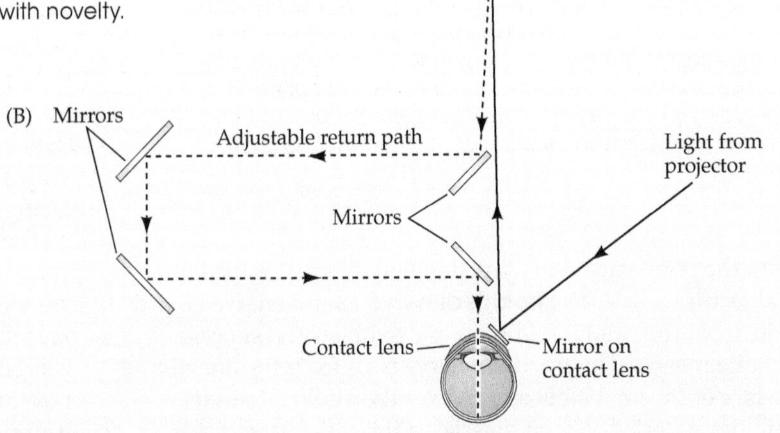

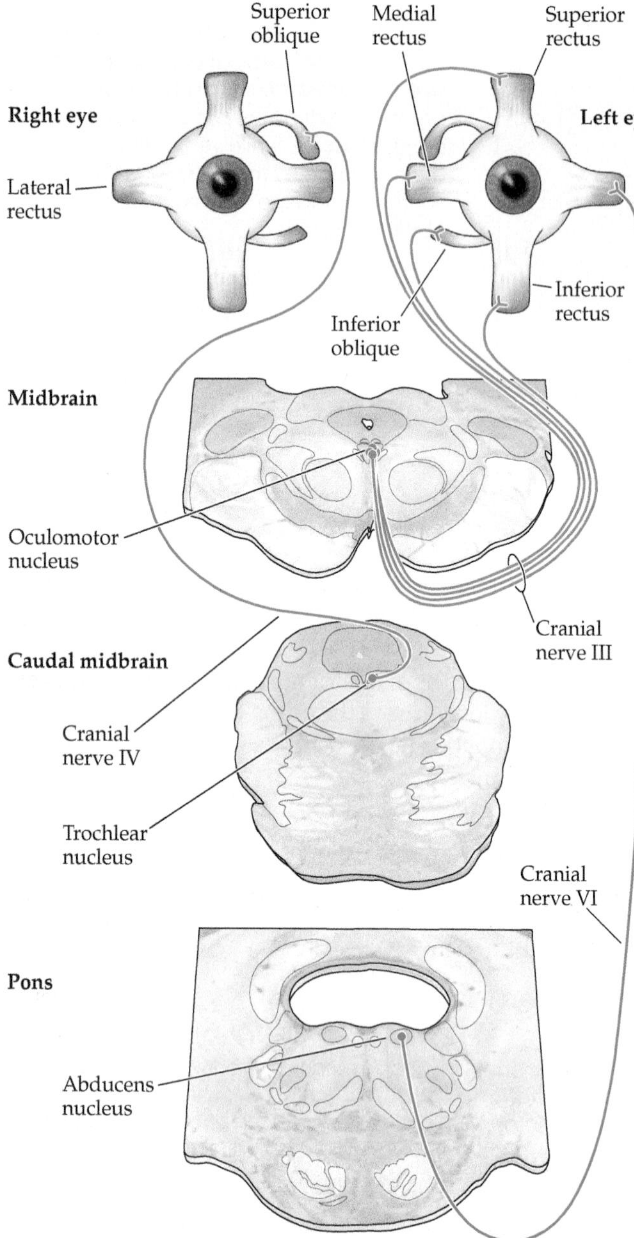

FIGURE 20.3 Innervation of the extraocular muscles by the cranial nerve nuclei governing eye movements. The abducens nucleus innervates the ipsilateral lateral rectus muscle; the trochlear nucleus innervates the contralateral superior oblique muscle; and the oculomotor nucleus innervates all the rest of the ipsilateral extraocular muscles (medial rectus, inferior rectus, superior rectus, and inferior oblique).

from the pons–medullary junction and innervates the lateral rectus muscle. The **trochlear nerve** (cranial nerve IV) exits from the caudal midbrain and supplies the superior oblique muscle. The trochlear nerve exits from the dorsal surface of the brainstem and crosses the midline to innervate the superior oblique muscle on the contralateral side (this is the only somatic motor nerve—cranial or spinal—that supplies muscles on the opposite side of the body, and the only motor nerve to exit the dorsal aspect of the CNS). The **oculomotor nerve** (cranial nerve III), which exits from the rostral midbrain just medial to the cerebral peduncle, supplies the rest of the extraocular muscles. Although the oculomotor nerve governs several different muscles, each muscle receives its innervation from a separate group of lower motor neurons within the nuclear complex that supplies the third nerve.

In addition to supplying the extraocular muscles, the oculomotor complex includes a distinct cell group that innervates the levator muscles of the eyelid; the axons from these neurons also travel in the third nerve. Finally, the third nerve carries preganglionic parasympathetic axons that are responsible for pupillary constriction (see Chapters 12 and 21) from the nearby Edinger–Westphal nucleus. Thus, damage to the third nerve results in three characteristic deficits: impairment of eye movements, drooping of the eyelid (a clinical sign called **ptosis**), and pupillary dilation, due to the unopposed action of sympathetic inputs to the dilator muscles of the iris.

Types of Eye Movements and Their Functions

The five basic types of eye movements can be grouped into two functional categories: those that serve to *shift* the direction of gaze, and those that serve to *stabilize* gaze. Shifts in eye position are necessary to foveate new targets and to follow foveated targets as they move in visual space. Stabilizing movements of the eyes are used to maintain foveation when the head moves and when there are large-scale movements of the visual field. Thus, saccades, smooth pursuit movements, and vergence movements shift the direction of gaze, and vestibulo-ocular and optokinetic movements stabilize gaze. The functions of each type of eye movement are introduced here; in subsequent sections, the neural circuitry responsible for movements that shift the direction of gaze is presented in more detail (see Chapters 14 and 19 for further discussion of the neural circuitry underlying gaze-stabilizing movements).

As noted earlier, saccades are rapid, ballistic movements of the eyes that abruptly change the direction of fixation. They range in amplitude from the small movements made while reading to the much larger movements made while gazing around a room. Saccades can be elicited voluntarily, but they occur reflexively whenever the eyes are open, even when they are fixated on a target (see Box 20A). The rapid eye movements (REM) that occur during an important phase of sleep (see Chapter 28) also are saccades.

Figure 20.4 shows the time course of a saccadic eye movement. After the onset of a target for a saccade (in this example, the stimulus was the movement of an already

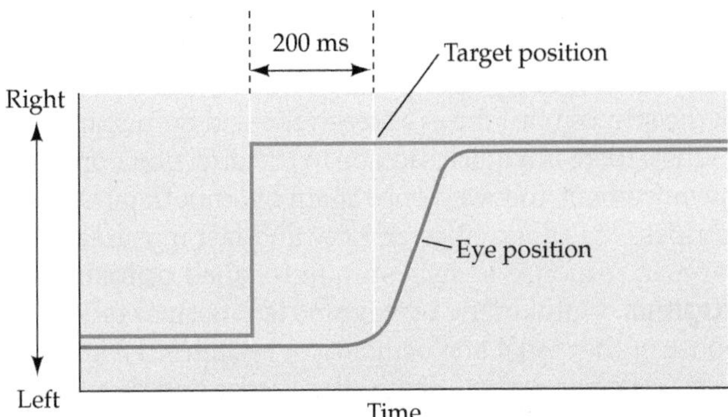

FIGURE 20.4 **The metrics of a saccadic eye movement.** The red line indicates the position of a fixation target and the blue line the position of the fovea. When the target moves suddenly to the right, there is a delay of about 200 ms before the eye begins to move to the new target position. (After Fuchs, 1967.)

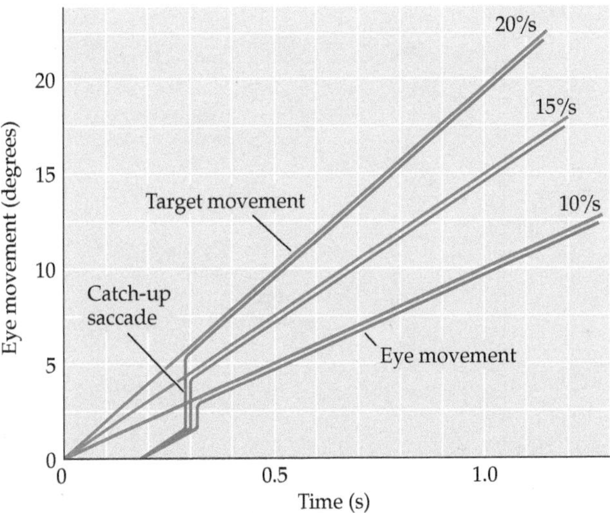

FIGURE 20.5 **The metrics of smooth pursuit eye movements.** These traces show eye movements (blue lines) tracking a stimulus moving at one of three different velocities (red lines). After a quick saccade to foveate the target, the eye movement attains a velocity that matches the velocity of the target. (After Fuchs, 1967.)

fixated target), it takes about 200 ms for eye movement to begin. During this delay, the position of the target with respect to the fovea (that is, how far the eye has to move) is computed, and the difference between the initial and intended position is converted into a motor command that activates the extraocular muscles to move the eyes the correct distance in the appropriate direction. Saccadic eye movements are said to be ballistic because the saccade-generating system usually does not respond to subsequent changes in the position of the target during the course of the eye movement. If the target moves again during this time (which is on the order of 15 to 100 ms), the saccade will miss the target, and a second saccade must be made to correct the error.

Smooth pursuit movements are much slower tracking movements of the eyes designed to keep a moving stimulus on the fovea once foveation is achieved. Such movements are under voluntary control in the sense that the observer can choose whether or not to track a moving stimulus (Figure 20.5). Surprisingly, only highly trained observers can make a smooth pursuit movement in the absence of a moving target. Most people who try to move their eyes in a smooth fashion without a moving target to track simply make saccades.

Vergence movements align the fovea of each eye with targets located at different distances from the observer. Although vergence movements are required to track a visual target that may be moving closer or farther away, they are more commonly employed when abruptly shifting the direction of gaze, for example, from a near object to one that is more distant. Unlike other types of eye movements, in which the two eyes move in the same direction (**conjugate eye movements**), vergence movements are **disconjugate** (or **disjunctive**); they involve either a convergence or divergence of the lines of sight of each eye to foveate an object that is nearer or farther away. Convergence is one of the three reflexive visual responses elicited together to shift gaze from a distant to a near object. The other components of the so-called **near reflex triad** are accommodation of the lens, which by increasing the curvature of the lens brings the close object into focus, and pupillary constriction, which by reducing spherical aberration increases the depth of field and sharpens the image on the retina (see Chapter 11).

Vestibulo-ocular movements and **optokinetic eye movements** operate together to move the eyes and stabilize gaze relative to the external world, thus compensating for head movements. These reflexive responses prevent visual images from "slipping" on the surface of the retina as head position varies and, more rarely, when confronted with large-scale movements of the visual scene (such as a flowing river or a passing train).

The action of vestibulo-ocular movements can be appreciated by fixating an object and moving the head from side to side; the eyes automatically compensate for the head movement by moving the same distance and at the same velocity but in the opposite direction, thus keeping the image of the object at more or less the same place on the retina. The vestibular system detects brief, transient changes in head position and produces rapid, corrective eye movements using the pathways described in Chapter 14. Sensory information from the semicircular canals directs the eyes to move in a direction opposite to the head movement. Although the vestibular system operates

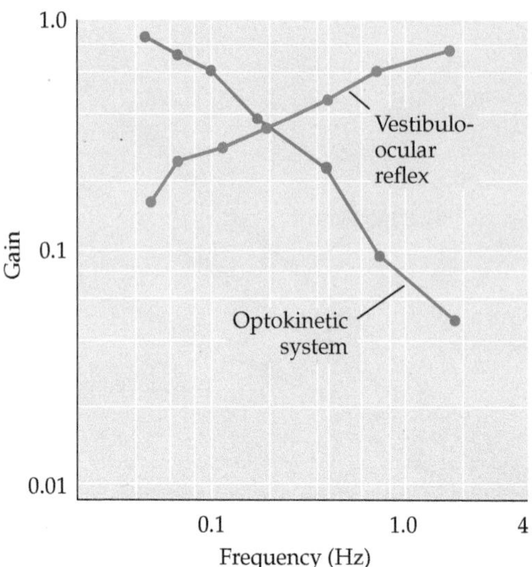

FIGURE 20.6 Operational ranges of the vestibulo-ocular and optokinetic systems. Functions of the vestibulo-ocular and optokinetic systems were assessed independently in rabbits by rotating the animals with their eyes closed (to isolate the vestibulo-ocular reflex) or following recovery from bilateral labyrinthectomy (to isolate the optokinetic system). At low frequencies of movement (below 1 Hz or one back-and-forth cycle of stimulation per second), the gain of the vestibulo-ocular reflex (the ratio of eye movement to head movement) diminishes below unity. However, the gain of the optokinetic system (the ratio of eye movement to retinal slip) approaches unity at such low frequencies of stimulation. Thus, the vestibulo-ocular and optokinetic systems act in a complementary, frequency-dependent fashion to stabilize gaze over a broad range of stimulation frequencies. (After Baarsma and Collewijn, 1974.)

effectively to counteract rapid movements of the head, it is relatively insensitive to slow movements (below 1 Hz) or to persistent rotation of the head. For example, if the vestibulo-ocular reflex is tested with continuous rotation of an individual and without visual cues about the movement of the image (i.e., with eyes closed or in the dark), the compensatory eye movements cease after only about 30 seconds of rotation. However, if the same test is performed with visual cues, eye movements persist. The compensatory eye movements in this case are due to the activation of another system that relies not on vestibular information, but on visual cues indicating motion of the visual field. This optokinetic system is especially sensitive to slow movements (below 1 Hz) of large areas of the visual field, and its response builds up slowly. These features complement the properties of the vestibulo-ocular reflex, especially as head movements slow down and vestibular signals decay (Figure 20.6). Thus, should a visual image slowly "slip" across the retina, the optokinetic system will respond by inducing compensatory movements of the eyes at the same speed and in the opposite direction.

The optokinetic system can be tested by seating an individual in front of a screen on which a series of horizontally moving vertical bars is presented. The eyes automatically track the stripes until the eyes reach the end of their excursion. Then there is a quick saccade in the direction opposite to the movement, followed once again by smooth pursuit of the stripes. This alternation of slow and fast movement of the eyes in response to such stimuli is called **optokinetic nystagmus**. Optokinetic nystagmus is a normal reflexive response of the visual and oculomotor systems in response to large-scale movements of the visual scene and should not be confused with the pathological nystagmus that can result from certain kinds of brain injury (for example, damage to the vestibular system or the cerebellum; see Chapters 14 and 19, respectively). Indeed, clinicians have long regarded eye movements to be key indicators of neurological function and dysfunction.

Neural Control of Saccadic Eye Movements

The problem of moving the eyes to fixate a new target in space (or indeed any other movement) involves two separate tasks: controlling the *amplitude* of movement (how far), and controlling the *direction* of the movement (which way). The amplitude of a saccadic eye movement is encoded by the duration of neuronal activity in the lower motor neurons of the oculomotor nuclei. For instance, as shown in Figure 20.7, neurons in the abducens nucleus fire a burst of action potentials just prior to abducting the eye (by causing the lateral rectus muscle to contract) and are silent when the eye is adducted. The amplitude of the movement is correlated with the duration of the burst of action potentials in abducens neurons. Following each saccade, abducens neurons reach a new baseline level of discharge that is correlated with the position of the eye in the orbit. The steady baseline level of firing generates the muscle force needed to hold the eye in its new position.

The direction of the movement is determined by which eye muscles are activated. Although in principle any given direction of movement could be specified by independently adjusting the activity of individual eye muscles, the complexity of the task would be overwhelming. Instead, the direction of eye movement is controlled by the local circuit neurons in two **gaze centers** in the reticular formation (see Box 17C), each of which is responsible for generating movements along a particular axis. The **paramedian pontine reticular formation** (**PPRF**), also called the horizontal gaze center, is a collection of local circuit neurons near the midline in the pons. These neurons are responsible for generating horizontal eye movements. The **rostral interstitial nucleus**, or vertical gaze center, is located in the rostral part of the midbrain reticular formation and is responsible for vertical movements. Activation of each gaze center separately results

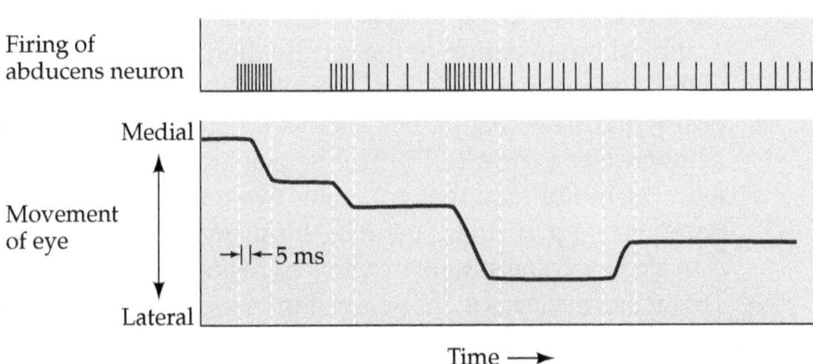

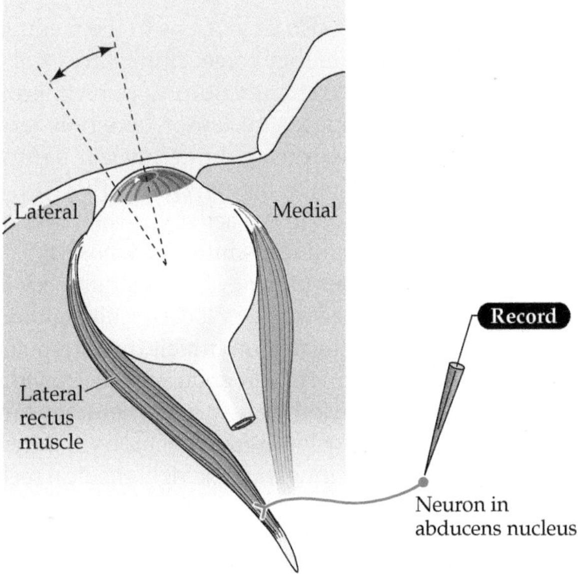

FIGURE 20.7 Motor neuron activity in relation to saccadic eye movements. The experimental setup is shown on the right. In this example, an abducens lower motor neuron fires a burst of activity (upper trace) that precedes and extends throughout the movement (solid line). An increase in the tonic level of firing is associated with more lateral displacement of the eye. Note also the decline in firing rate during a saccade in the opposite direction. (After Fuchs and Luschei, 1970.)

in movements of the eyes along a single axis, either horizontal or vertical. Activation of the gaze centers in concert results in oblique movements whose trajectories are specified by the relative contribution of each center.

An example of how the PPRF works with the abducens and oculomotor nuclei to generate a horizontal saccade to the right is shown in Figure 20.8. Neurons in the PPRF innervate cells in the abducens nucleus on the same side of the brain. The abducens nucleus contains two types of neurons. One type comprises the lower motor neurons that innervate the lateral rectus muscle on the same side. The other type, called internuclear neurons, send their axons across the midline. These axons ascend in a fiber tract called the **medial longitudinal fasciculus** and terminate in the portion of the oculomotor nucleus that contains lower motor neurons that innervate the medial rectus muscle. As a result of this arrangement, activation of PPRF neurons in the right side of the brainstem causes horizontal movements of both eyes to the right; likewise, activation of PPRF neurons in the left half of the brainstem induces horizontal movements of both eyes to the left.

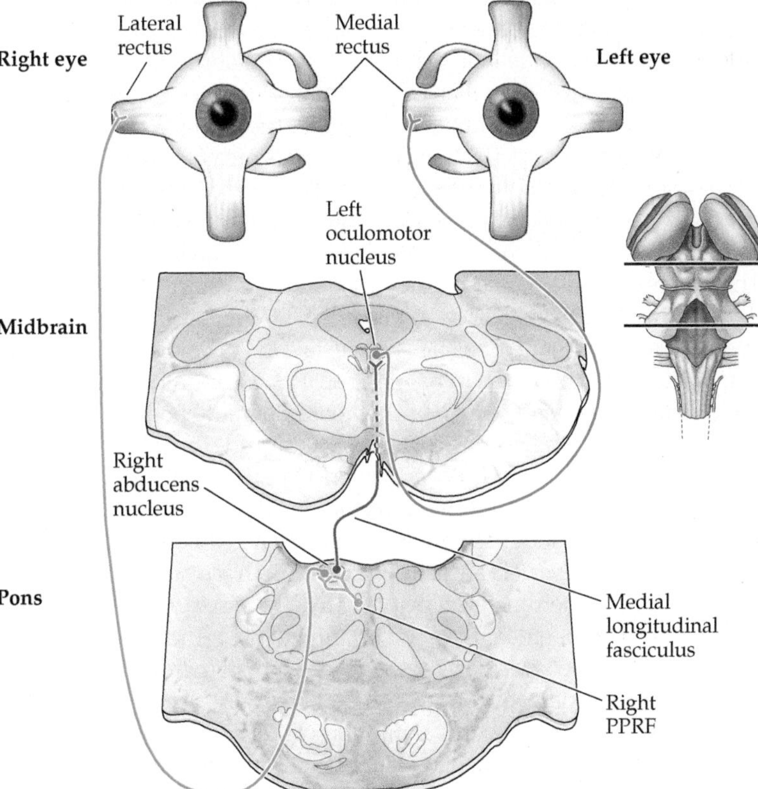

FIGURE 20.8 Synaptic circuitry responsible for horizontal movements of the eyes to the right. This simplified diagram depicts how activation of local circuit neurons in the right horizontal gaze center (the right PPRF; orange) leads to increased activity of lower motor neurons and internuclear neurons in the right abducens nucleus. The lower motor neurons (green) innervate the lateral rectus muscle of the right eye. The internuclear neurons (blue) project via the medial longitudinal fasciculus to the contralateral oculomotor nucleus, where they activate lower motor neurons (red) that in turn innervate the medial rectus muscle of the left eye. The coordinated action of the right lateral rectus and left medial rectus muscles rotates the eyes to the right. Inhibitory local circuit neurons in the medullary reticular formation (not illustrated) inhibit activity in the left abducens nucleus, which has the effect of decreasing tone in the antagonistic muscles.

Neurons in the PPRF also send axons to the medullary reticular formation, where they contact inhibitory local circuit neurons. These local circuit neurons, in turn, project to the contralateral abducens nucleus, where they terminate on lower motor neurons and internuclear neurons. In consequence, activation of neurons in the PPRF on the right results in a reduction in the activity of the lower motor neurons in the left abducens nucleus, whose muscles would oppose movements of the eyes to the right (see Figure 20.7). Likewise, these inhibitory local circuit neurons in the medullary reticular formation inhibit the internuclear neurons that project from the left abducens nucleus to the right oculomotor nucleus, thus assuring a commensurate reduction in the activity of lower motor neurons in the right oculomotor nucleus that innervate the right medical rectus. This inhibition of antagonists resembles the strategy used by local circuit neurons in the spinal cord to control limb muscle antagonists (see Chapter 16).

Although saccades can occur in complete darkness, they are often elicited when something in the visual field attracts attention and the observer directs the foveae toward the object of interest for more detailed examination. How, then, is sensory information about the location of a salient target in space transformed into an appropriate pattern of activity in the horizontal and vertical gaze centers? Two regions of the brain that project to the gaze centers are demonstrably important for the initiation and accurate targeting of saccadic eye movements: the superior colliculus of the midbrain (called the **optic tectum** in non-mammalian vertebrates) and several areas in the frontal and parietal cortex. Especially well studied is a region of the frontal lobe that lies in a rostral portion of the premotor cortex, known as the **frontal eye field** (classically, Brodmann's area 8, although in humans the frontal eye field may encroach posteriorly into Brodmann's area 6). Upper motor neurons in both the superior colliculus and frontal eye fields, each of which contains a topographical map of eye movement vectors, discharge immediately prior to saccades. Thus, activation of a particular site in the superior colliculus or in the frontal eye field elicits saccadic eye movements in a specified direction and for a specified distance. This movement is independent of the initial position of the eyes in the orbit. However, when the eyes are in the same initial position, the direction and distance of the elicited saccades are always the same for a given site of activation. Consistent with a topographic map of eye movement vectors, the direction and distance of the saccade change systematically when different sites in the frontal eye field are activated. Since each saccade is produced by the coordinated activity of all of the extraocular muscles, this arrangement is a good example of the principle that the activation of specific movements rather than individual muscles is encoded by the upper motor neurons.

Both the superior colliculus and the frontal eye field also contain cells that are activated by visual stimuli; however, the relationship between the sensory and motor responses of individual cells is better understood for the superior colliculus. An orderly map of visual space is established by the topographical organization of the termination of retinal axons within the superior colliculus, as well as by inputs from cortical visual areas that participate in the dorsal spatial vision pathway (see Figure 12.16). This sensory map is in register with the motor map that generates eye movements. Thus, neurons in a particular region of the superior colliculus are activated by visual stimuli in a limited region of visual space. This activation leads to the generation of a saccade by activating neighboring upper motor neurons that move the eye by an amount just sufficient to align the foveae with the region of visual space that provided the stimulation (Figure 20.9).

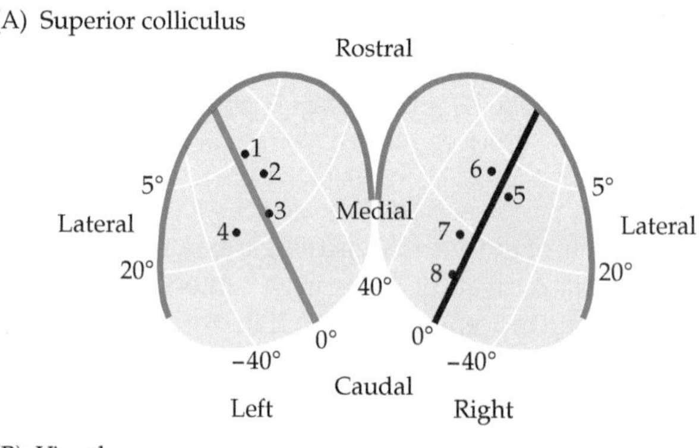

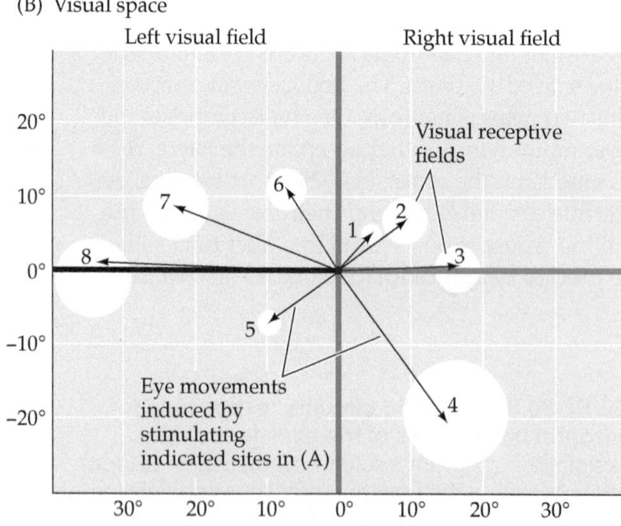

FIGURE 20.9 The sensory map of visual space in the superior colliculus is in register with the motor map that generates eye movements. Evidence for this registration has been obtained from electrical recording and stimulation. (A) Surface views of the superior colliculus illustrating the location of eight separate electrode recording and stimulation sites. (B) Map of visual space showing the visual receptive field location of the sites in (A) (white circles), and the amplitude and direction of the eye movements elicited by stimulating these sites electrically (arrows). In each case, electrical stimulation results in eye movements that align the foveae with a region of visual space that corresponds to the visual receptive fields of neurons at that site. (After Schiller and Stryker, 1972.)

Neurons in the superior colliculus also respond to auditory and somatosensory stimuli. Indeed, location in space for these other modalities is mapped in register with the visual and motor maps in the colliculus. Topographically organized maps of auditory space and of the body surface in the superior colliculus can orient the eyes (and the head, via output projections from the superior colliculus to neurons that give rise to the reticulospinal tract; see Chapter 17) in response to a variety of different sensory stimuli. This registration of the sensory and motor maps in the colliculus illustrates an important principle of topographical maps: They provide an efficient mechanism for the transformation of sensory signals into the movements that are guided by these signals (in this case, the extraocular muscles and muscles of the posterior head and neck; Box 20B). However, the motor map in the deep layers of the superior colliculus is not simply organized in the framework established by the spatial distribution of the sensory inputs. Rather, the input signals must also be

BOX 20B ■ Sensorimotor Integration in the Superior Colliculus

The superior colliculus is a laminated structure in which differences between the layers provide clues about how sensory and motor maps interact to produce appropriate movements. As discussed in the text, the superficial or "visual" layer of the colliculus receives input from retinal axons that form a topographic map. Thus, each site in the superficial layer is activated maximally by the presence of a stimulus at a particular point of visual space. In contrast, neurons in the deeper or "motor" layers generate bursts of action potentials that command saccades, effectively generating a motor map; thus, activation of different sites generates saccades having different vectors (see Figure 20.9). The visual and motor maps are *in register*, so that visual cells responding to a stimulus in a specific region of visual space are located directly above the motor cells that command eye movements toward that same region (Figure A).

The registration of the visual and motor maps suggests a simple strategy for how the eyes might be guided toward an object of interest in the visual field. When an object appears at a particular location in the visual field, it activates neurons in the corresponding part of the visual map. As a result, bursts of action potentials are generated by the subjacent motor cells to command a saccade that rotates the two eyes just the right amount to direct the foveae toward that same location in the visual field. This behavior is called "visual grasp" because successful sensorimotor integration results in the accurate foveation of a visual target.

This seemingly simple model, formulated in the early 1970s when the collicular maps were first found, assumes point-to-point connections between the visual and motor maps. In practice, however, these connections have been difficult to demonstrate. Neither the anatomical nor the physiological methods available at the time were sufficiently precise to establish these postulated synaptic connections. At about the same time, motor neurons were found to command saccades to nonvisual stimuli; moreover, spontaneous saccades occur in the dark. Thus, it was clear that visual-layer activity is not always necessary for saccades. To confuse matters further,

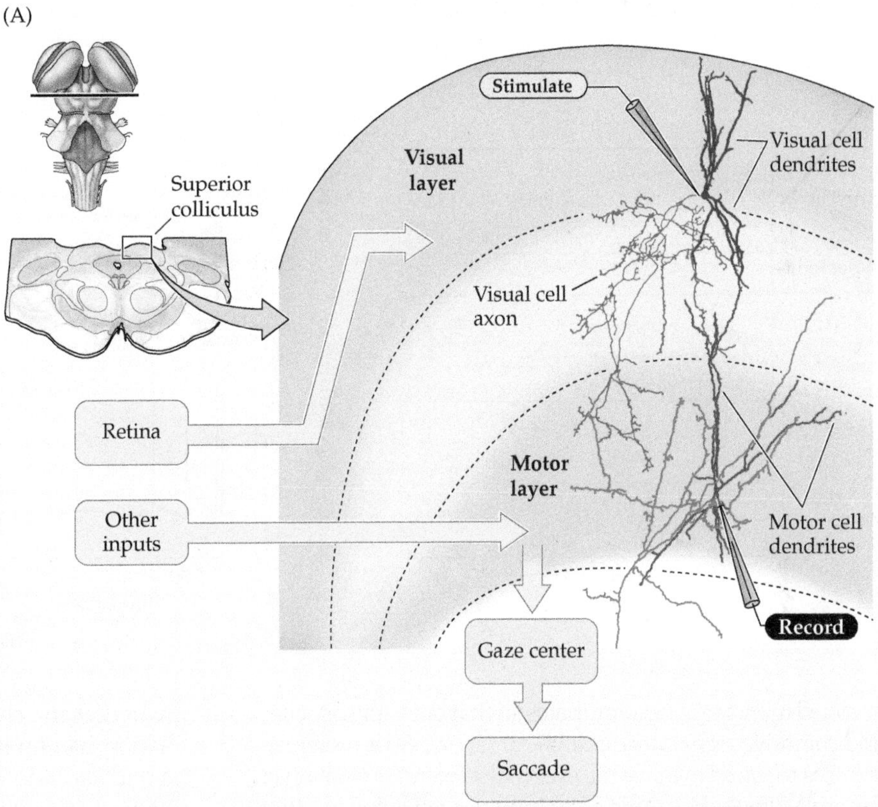

(A) The superior colliculus receives visual input from the retina and sends a command signal to the gaze centers to initiate a saccade. In the experiment illustrated here, a stimulating electrode activates cells in the visual layer, and a patch clamp pipette records the response evoked in a neuron in the subjacent motor layer. The cells in the visual and motor layers were subsequently labeled with a tracer called biocytin. This experiment demonstrates that the terminals of the visual neuron are located in the same region as the dendrites of the motor neuron.

Continued on the next page

BOX 20B (continued)

animals could be trained *not* to make a saccade when an object appeared in the visual field, showing that the activation of visual neurons is sometimes insufficient to command saccades. The fact that activity of neurons in the visual map is *neither necessary nor sufficient* for eliciting saccades led investigators away from the simple model of direct connections between corresponding regions of the two maps, toward models that linked the layers indirectly through pathways that detoured through the cortex.

Eventually, however, new and better methods resolved this uncertainty. Techniques for filling single cells with axonal tracers showed an overlap between descending visual-layer axons and ascending motor-layer dendrites, in accord with direct anatomical connections between corresponding regions of the maps. At the same time, in vitro whole-cell patch clamp recording (see Box 4A) permitted more discriminating functional studies that distinguished excitatory and inhibitory inputs to the motor cells. These experiments showed that the visual and motor layers do indeed have the functional connections required to initiate the command for a visually guided saccadic eye movement. A single brief electrical stimulus delivered to the superficial layer generates a prolonged burst of action potentials that resembles the command bursts that normally occur just before a saccade (Figures B and C).

These direct connections presumably provide the substrate for the very short-latency, reflex-like express saccades that are unaffected by destruction of the frontal eye fields. Other visual and nonvisual inputs to the deep layers probably explain why activation of the retina is neither necessary nor sufficient for the production of saccades.

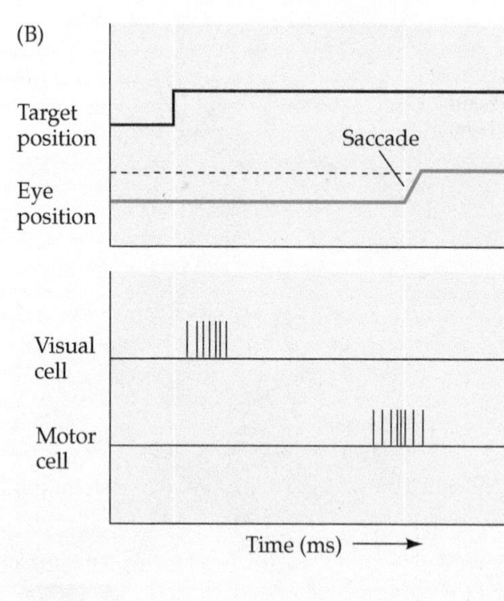

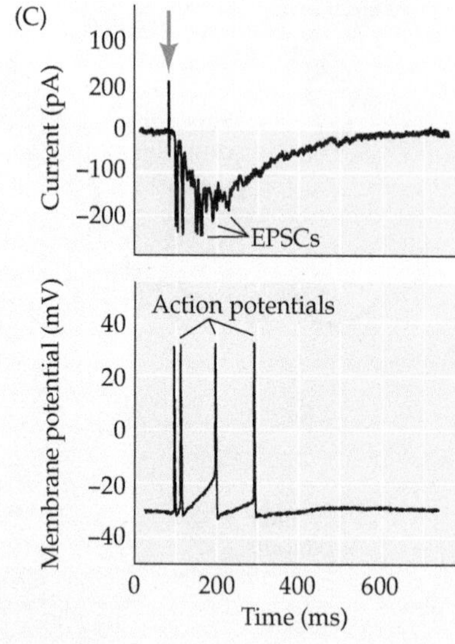

(B) The onset of a target in the visual field (top trace) is followed, after a short interval, by a saccade to foveate the target (second trace). In the superior colliculus, the visual cell responds shortly after the onset of the target, while the motor cell responds later, just before the onset of the saccade. (C) Bursts of excitatory postsynaptic currents (EPSCs) recorded from a motor-layer neuron in response to a brief (0.5 ms) current stimulus applied via a steel wire electrode in the visual layer (top; see arrow). These synaptic currents generate bursts of action potentials in the same cell (bottom). (B after Wurtz and Albano, 1980; C after Ozen et al., 2000.)

encoded in movement coordinates so that both sensory cues and cognitive signals (see below) can activate the motor responses required to move the eyes to the intended position in the orbits. Thus, the output of the superior colliculus specifies movement intention rather than movements to fixed positions in external space or on the body surface.

The organizing framework of this motor map was demonstrated in an ingenious series of studies performed by David Sparks and his colleagues at the University of Alabama. They showed that retinal error signals (i.e., the distance and direction of the retinal projection of the target from the fovea) in retinotopic coordinates are often not sufficient to localize saccade targets. Using trained monkeys, the investigators cued a voluntary saccade with a brief flash of light, but before the saccade could be initiated, they stimulated a site in the deep layers of the superior colliculus that induced a saccade away from the point of fixation. They recorded eye movements to determine whether the change in eye position induced by the stimulation had an impact on the direction and distance of the cued saccade (Figure 20.10). If the saccade vectors were determined simply by the retinotopic coordinates of the target, then the monkey would be expected to make a saccade of the cued direction and distance (about 10° in the upward direction in this example). However, because of the deviated starting position, the saccade should systematically miss the target position by the amount of the stimulation-induced deviation (indicated by the dashed arrow pointing upward to the dash-encircled T on the left side of Figure 20.10A). The results consistently showed, however, that this was not the case. The animals compensated for the stimulation-induced shift by performing a compensatory saccade (a saccade indicated by the oblique, black dashed arrow to the T that appears within a black circle—the actual target location—in Figure 20.10A). This compensatory action was based on stored information about

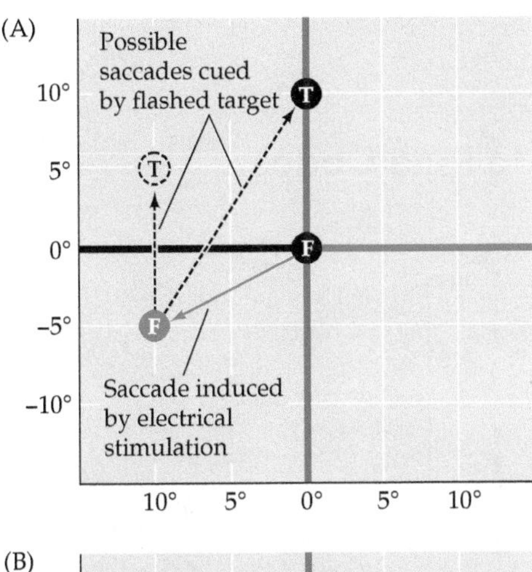

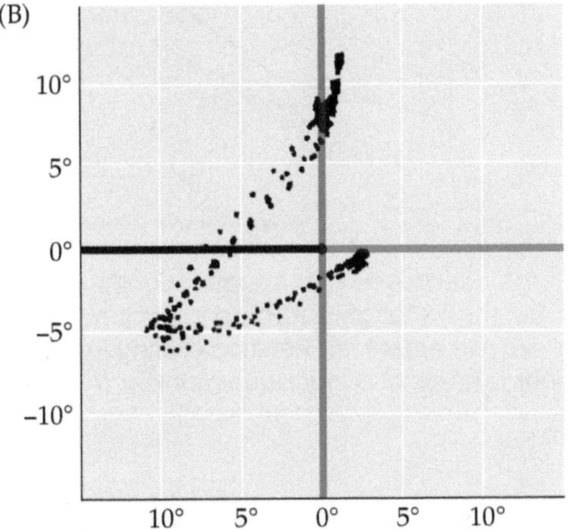

FIGURE 20.10 Saccades are encoded in movement coordinates, not retinotopic coordinates. (A) Map of visual space illustrating experimental design. Monkeys were trained to fixate a central location (F, in black) and then perform a saccade to a remembered target location cued by a brief flash at a location 10° above the starting position (T, in black). After cueing, but before expression of the cued saccade, an electrical stimulus was applied to a site in the superior colliculus that induced a saccade down and to the left (to the location marked by the F in red). If the cued saccade was encoded in retinotopic coordinates, the monkey should move its eyes 10° above the stimulus-induced position of foveation (F, in red) to a location marked by the dash-encircled T. If the saccade was encoded in movement coordinates, then a compensatory saccade to the cued target location (T, in black) would be expected. (B) Consistent with the encoding of saccades in movement coordinates, the monkeys performed compensatory saccades upward and to the right, toward the location of the cued target. Dots represent eye movements sampled at 500 Hz. (After Sparks and Mays, 1983.)

the location of the retinal image and current information about the position of the eyes in the orbit. The upper motor neurons that initiate the compensatory saccade are located at the expected site in the motor map of saccade vectors, but their activation depends on information in addition to the retinotopic location of the target. This information may be provided by circuits in the cerebral cortex that integrate this information and, in turn, activate the site in the superior colliculus that initiates the compensatory saccade (see Figure 20.10B).

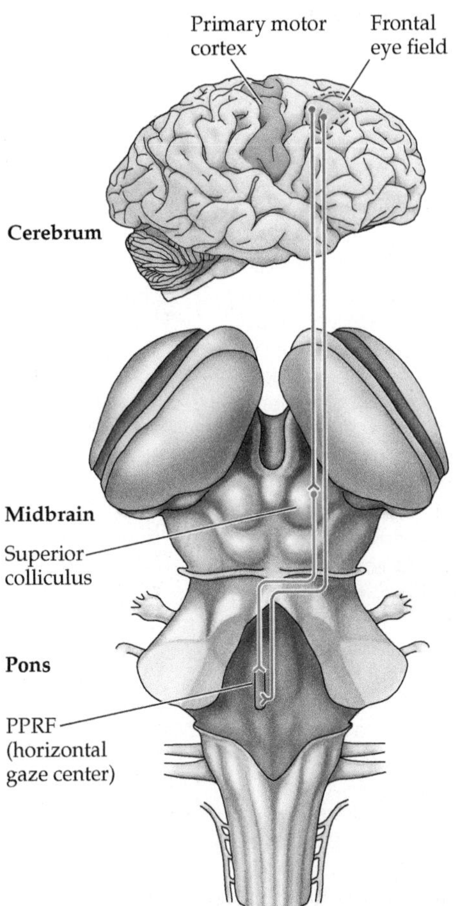

FIGURE 20.11 Neurons in the frontal eye field collaborate with cells in the superior colliculus to control eye movements. The projections shown here are from the frontal eye field in the right cerebral hemisphere (Brodmann's area 8) to the superior colliculus and the horizontal gaze center (PPRF). In humans, the frontal eye field can influence eye movements by either of two routes: indirectly, by projections to the ipsilateral superior colliculus, which in turn projects to the contralateral PPRF; and directly, by projections to the contralateral PPRF.

This study and several that followed showed that signals from different sensory modalities are integrated and transformed into a common motor frame of reference that encodes the direction and distance of the eye movements necessary to foveate an intended target. This "place code" for intended eye position generated in the upper motor neurons of the superior colliculus is then translated into a "rate code" by downstream gaze centers in the reticular formation that can then direct the activity of lower motor neurons in the ocular motor nuclei (Box 20C).

The eye movement regions of the cerebral cortex collaborate with the superior colliculus in controlling saccades. Thus, the frontal eye field projects to the superior colliculus, and the superior colliculus projects to the PPRF on the contralateral side (Figure 20.11). (The superior colliculus also projects to the vertical gaze center, but for simplicity the discussion here is limited to the PPRF.) The frontal eye field can thus control eye movements by

BOX 20C ■ From Place Codes to Rate Codes

How does the pattern of activity in the superior colliculus get translated into a motor command that can be delivered to muscle fibers? Recall that neurons in the superior colliculus have "movement fields," discharging in conjunction with saccadic eye movements of a particular direction and amplitude. Movement fields are conceptually similar to the receptive fields that occur in various sensory areas of the brain. Across the entire population of collicular neurons, all possible saccade vectors are represented (Figure A). Because the movement fields are topographically organized, the superior colliculus forms a *motor map* of saccade vectors (or movement intentions; see text).

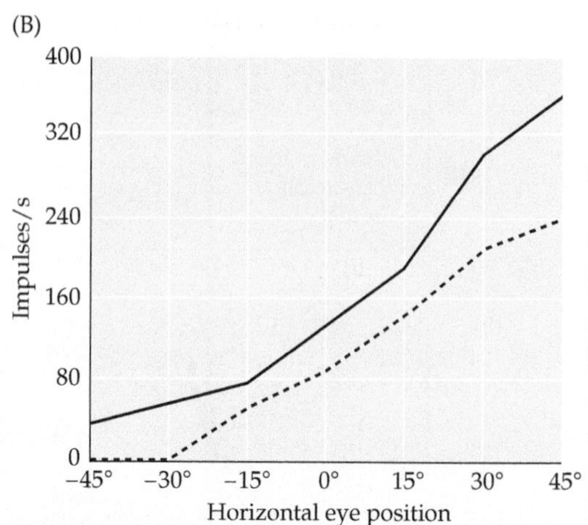

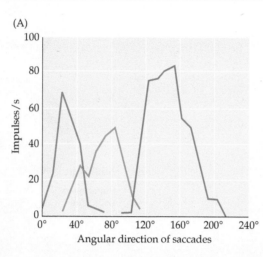

(A) Direction tuning of three neurons recorded from the deep layers of the superior colliculus in macaque monkeys. Each neuron is broadly tuned, but most strongly activated in conjunction with a particular direction (and amplitude) of saccadic eye movement. (B) Relation of firing frequency to steady eye position in two neurons in the abducens nucleus of a macaque monkey. (A after Sparks, 1975; B after Fuchs and Luschei, 1970.)

The direction and amplitude of eye movements are encoded quite differently by the extraocular muscles (Figure B). *Direction* is controlled by the ratio of activation of the different muscles, and *amplitude* is controlled by the magnitude of the activation of those muscles. In other words, to make a saccade go farther, the muscle pulling the eye must pull harder and longer than it would for a shorter saccade. Amplitude is therefore a *monotonic* function of muscle activation.

The pattern of activity must be transformed from a code in which collicular neurons are tuned for particular saccade amplitudes to a code in which most or all α motor neurons respond, regardless of saccade amplitude, but the level or duration of their activity varies monotonically with saccade amplitude. This transformation occurs before signals from the superior colliculus reach the α motor neurons that activate the extraocular muscles.

Various models have been proposed to explain this transformation. The basic idea, shared by all models, is that the saccade vector, as signaled by the locus of activity in the superior colliculus, is decomposed into two monotonic amplitude signals corresponding roughly to the horizontal and vertical components of the saccade vector. The weights of the projections from the

activating selected populations of upper motor neurons in the superior colliculus. This cortical area also projects directly to the contralateral PPRF; as a result, the frontal eye field can also control eye movements independently of the superior colliculus. The parallel inputs to the PPRF from the frontal eye field and superior colliculus are reflected in the different deficits that result from damage to these structures. Injury to the frontal eye field results in an inability to make saccades to the contralateral side and a deviation of the eyes to the side of the lesion. These effects are transient, however; in monkeys with experimentally induced lesions of this cortical region, recovery is virtually complete in 2 to 4 weeks. Lesions of the superior colliculus increase the latency and decrease the accuracy, frequency, and velocity of saccades; yet saccades still occur, and the deficits also improve with time. These results suggest that the frontal eye fields and the superior colliculus provide complementary pathways for the control of saccades. Moreover, one of these structures appears to be able to compensate (at least partially) for the loss of the other. In support of this interpretation, combined lesions of the frontal eye field and the superior colliculus produce a dramatic and permanent loss in the ability to make saccadic eye movements.

These observations do not, however, imply that the frontal eye fields and the superior colliculus have the same functions. Superior colliculus lesions produce a permanent deficit in the ability to perform very short-latency, reflex-like eye movements called **express saccades**. Express saccades are evidently mediated by direct pathways to the superior colliculus from the retina or visual cortex that can access the upper motor neurons in the colliculus

BOX 20C (continued)

superior colliculus to the horizontal and vertical gaze control centers are thought to be tuned to accomplish this. For example, a site in the superior colliculus where the movement fields encode 5° rightward movements would project to the rightward horizontal gaze control center with a modest strength. A site encoding 10° rightward saccades would send a stronger projection to that center. A site encoding an oblique saccade with a 10° horizontal and a 5° vertical component would project to both the horizontal and vertical centers, with weights proportional to the required contribution along each direction (Figure C).

This model is too simple to account for all the relevant experimental findings. However, it gives a general idea of how the brain might convert information encoded in one kind of format into another. This kind of transformation is a likely requirement of sensorimotor integration in many behavioral contexts where sensory cues guide movement.

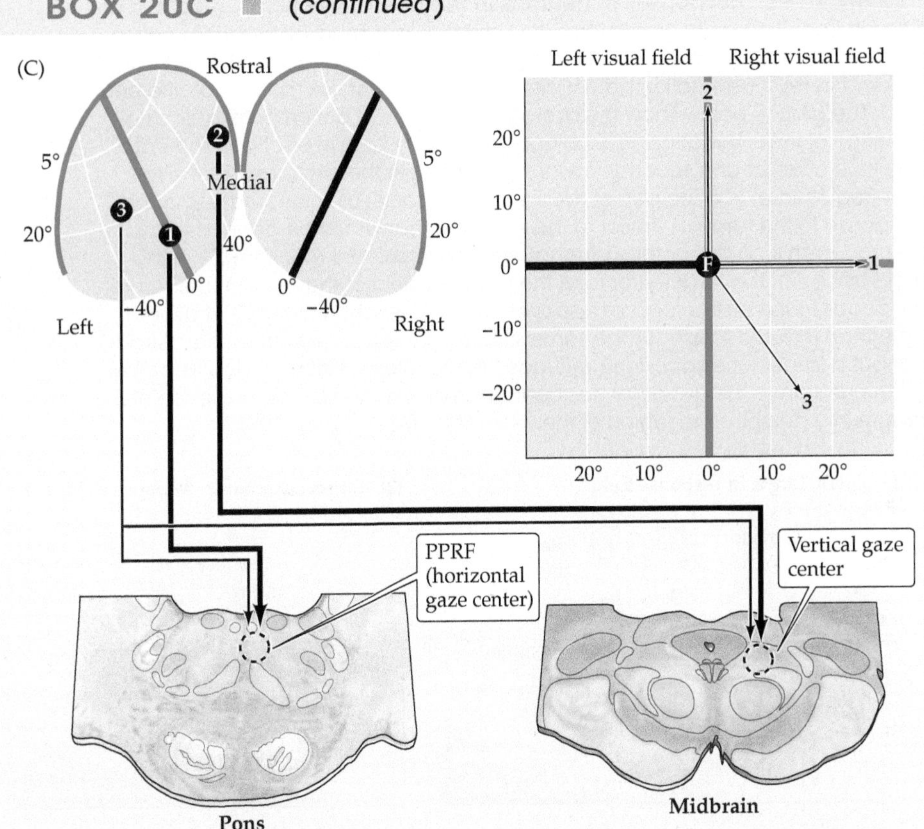

(C) Projections from the deep layers of the superior colliculus to the vertical and horizontal gaze centers in the mesencephalic and pontine reticular formation, respectively. Sites in the colliculus that encode horizontal movements (site 1) project mainly to the paramedian pontine reticular formation (PPRF, the horizontal gaze center), while sites that encode vertical movements (site 2) project mainly to the vertical gaze center in the mesencephalic reticular formation. Other sites that encode oblique saccades project to both gaze centers with weights proportional to the required horizontal and vertical displacements (thinner arrows projecting from site 3 to both gaze centers).

without extensive, and more time-consuming, processing in the frontal cortex (see Box 20B). In contrast, frontal eye field lesions produce permanent deficits in the ability to make saccades that are not guided by an external target. For example, people (or monkeys) with a lesion in the frontal eye field cannot voluntarily direct their eyes *away* from a stimulus in the visual field; this type of eye movement is called an "anti-saccade." Such lesions also eliminate the ability to make a saccade to the remembered location of a target that is no longer visible.

Finally, the frontal eye fields are essential for systematically scanning the visual field to locate an object of interest within an array of distracting objects (see Figure 20.1). Figure 20.12 shows the responses of a frontal eye field neuron during a visual task in which a monkey was required to foveate a target located within an array of distracting objects. This frontal eye field neuron discharges at different levels to the same stimulus, depending on whether the stimulus is the target of the saccade or a "distractor," and on the location of the distractor relative to the actual target. For example, the differences between the middle and the left and right traces in Figure 20.12 demonstrate that the response to the distractor is much reduced if it is located close to the target in the visual field. Results such as these suggest that lateral interactions within the frontal eye fields enhance the neuronal responses to stimuli that will be selected as saccade targets. They also suggest that such interactions suppress the responses to uninteresting and potentially distracting stimuli. These sorts of interactions presumably reduce the occurrence of unwanted saccades to distracting stimuli in the visual field.

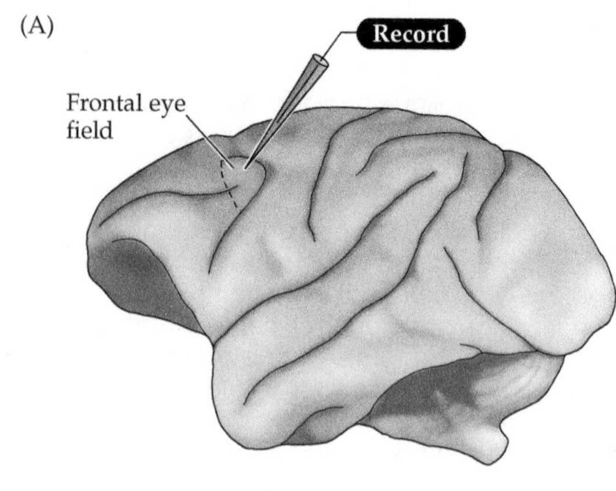

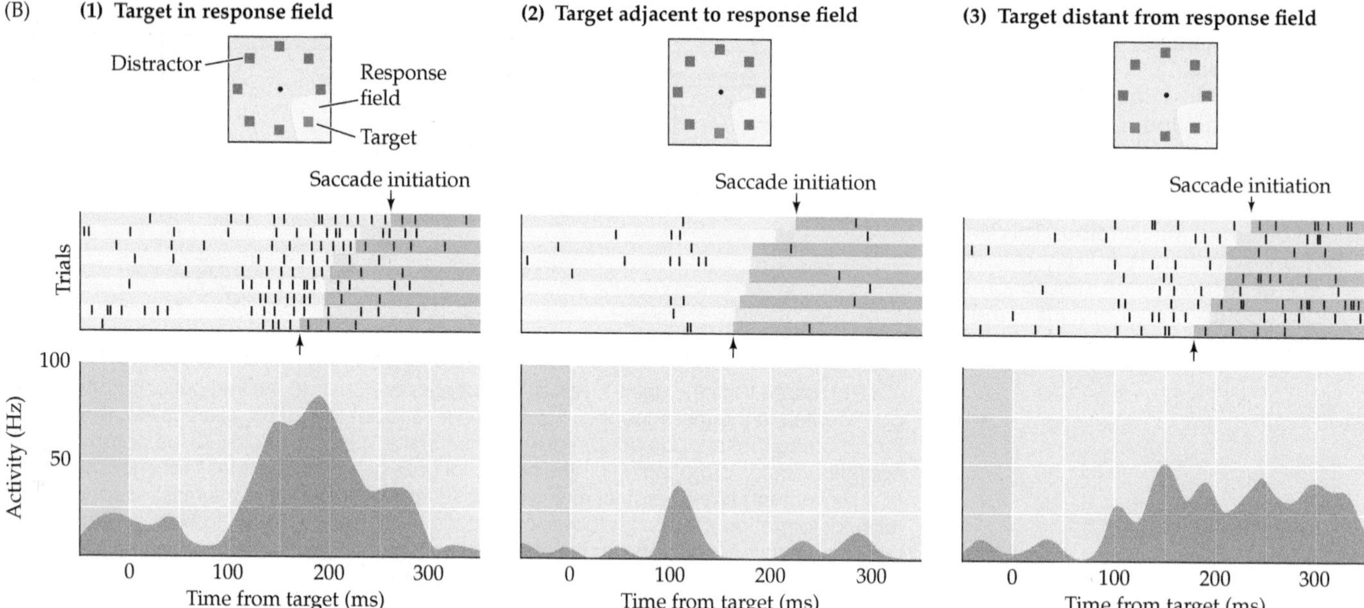

FIGURE 20.12 Responses of neurons in the frontal eye fields. (A) Locus of the left frontal eye field on a lateral view of the rhesus monkey brain. (B) Activation of a frontal eye field neuron during visual search for a target. The vertical tick marks represent action potentials, and each row of tick marks is a different trial. The graphs below show the average frequency of action potentials as a function of time. The change in color from beige to blue in each row indicates the time of onset of a saccade toward the target. In the left trace (1), the target (red square) is in the part of the visual field "seen" by the neuron, and the response to the target is similar to the response that would be generated by the neuron even if no distractors (blue squares) were present (not shown). In the right trace (3), the target is far from the response field of the neuron. The neuron responds to the distractor in its response field. However, it responds at a lower rate than it would to exactly the same stimulus if the square were not a distractor but a target for a saccade (left trace). In the middle trace (2), the response of the neuron to the distractor has been sharply reduced by the presence of the target in a neighboring region of the visual field. (After Schall, 1995.)

Neural Control of Smooth Pursuit Movements

Until recently, smooth pursuit and saccades were considered to be mediated by different structures, but studies such as those carried out by Richard Krauzlis at the Salk Institute for Biological Studies indicate that these two types of eye movements involve many of the same structures. Not only are smooth pursuit movements mediated by neurons in the PPRF, they also are under the influence of motor control centers in the rostral superior colliculus and subareas within the frontal eye fields, both of which receive sensory input from the dorsal spatial vision pathway in the parietal and temporal lobes. The exact routes by which visual information reaches the PPRF to generate smooth pursuit movements are not known, but pathways from the cortex to the superior colliculus and PPRF similar to those that mediate saccades may play a role; an indirect pathway through the cerebellum also has been suggested (Figure 20.13). It is clear, however, that neurons in the striate and extrastriate visual areas provide sensory information that is essential for the initiation and accurate guidance of smooth pursuit movements. In monkeys, neurons in the middle temporal area (which is largely concerned with the perception of moving stimuli; see Chapter 12) respond selectively to targets moving in a specific direction, and damage to this area disrupts smooth pursuit movements. In humans, damage of comparable areas in the parietal and occipital lobes also results in abnormalities of smooth pursuit movements. Finally, a pathway from the retina that detects movements of the visual stimulus on the retina (retinal drift) terminates in the cerebellum after relays in the pretectum and inferior olive (see Chapter 19), and adjusts the gain of this system to ensure that the velocity of the eye movements matches that of the movement of the visual target.

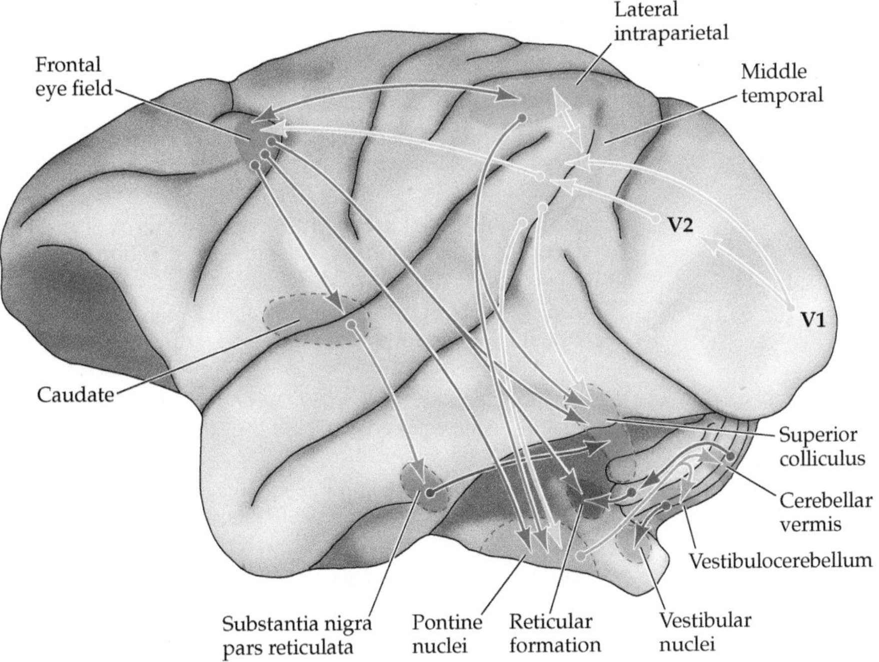

FIGURE 20.13 Sensory and motor structures and the connections that govern saccadic and smooth pursuit eye movements. This illustration summarizes data from studies on the rhesus macaque brain. Although these two types of eye movements were once thought to be controlled by separate circuits in the forebrain and brainstem, it is now recognized that they are governed by similar networks of cortical and subcortical structures. Visual signals are processed by the dorsal spatial vision pathway, including the middle temporal and lateral intraparietal areas. Sensory and attentional signals then guide motor planning areas in the frontal eye field. These cortical areas interact with subcortical structures, including basal ganglia (caudate and substantia nigra pars reticulata) and pontine-cerebellar structures (pontine nuclei, cerebellar vermis, and vestibulocerebellum), that modulate the initiation and coordination of eye movements by the superior colliculus and downstream oculomotor centers in the reticular formation and vestibular nuclei. The eye movements regulated by this complex circuitry are guided by a variety of sensory and cognitive signals, including perception, attention, memory, and reward expectation. (After Krauzlis, 2005.)

Neural Control of Vergence Movements

When a person wishes to look from one object to another object that is located at a different distance from the eyes, a saccade is made that shifts the direction of gaze toward the new object, and the eyes either diverge or converge until the object falls on the fovea of each eye. The structures and pathways responsible for mediating such vergence movements are not well understood, but they appear to include several extrastriate areas in the occipital lobe. Information about the location of retinal activity is relayed through the two lateral geniculate nuclei to the cortex, where the information from the two eyes is integrated. The appropriate command to diverge or converge the eyes, which is based largely on information from the two eyes about the amount of binocular disparity (see Chapter 12), is then sent from the occipital cortex to "vergence centers" in the brainstem. One such center is a population of local circuit neurons located in the midbrain near the oculomotor nucleus. These neurons generate a burst of action potentials that initiate a vergence movement, and the frequency of the burst determines its velocity. There is a division of labor within the vergence center, so that some neurons command convergence movements while others command divergence movements. These neurons also coordinate vergence movements of the eyes with accommodation of the lens and pupillary constriction to maximize the clarity of images formed on the retina, as discussed in Chapter 11.

Summary

Despite their specialized function, the systems that control eye movements have much in common with the motor systems that govern movements of other parts of the body. Just as the spinal cord provides the basic circuitry for coordinating the actions of muscles around a joint, the reticular formation of the pons and midbrain provides the basic circuitry that mediates movements of the eyes. Descending projections from upper motor neurons in the superior colliculus and the frontal eye field innervate gaze centers in the brainstem, providing a basis for integrating eye movements with sensory information that indicates the location of objects in space. The superior colliculus and the frontal eye field are organized in a parallel as well as a hierarchical fashion, enabling one of these structures to compensate for the loss of the other. Eye movements, like other movements, are also under the control of the basal ganglia and cerebellum; this control ensures the proper initiation and successful execution of these relatively simple motor behaviors, thus allowing observers to interact efficiently with the visual environment.

ADDITIONAL READING

Reviews

Foulsham, T. (2015) Eye movements and their functions in everyday tasks. *Eye* 29: 196–199.

Fuchs, A. F., C. R. S. Kaneko and C. A. Scudder (1985) Brainstem control of eye movements. *Annu. Rev. Neurosci.* 8: 307–337.

Hikosaka, O. and R. H. Wurtz (1989) The basal ganglia. In *The Neurobiology of Saccadic Eye Movements: Reviews of Oculomotor Research*, vol. 3, R. H. Wurtz and M. E. Goldberg (eds.). Amsterdam: Elsevier, pp. 257–281.

Krauzlis, R. J. (2005) The control of voluntary eye movements: New perspectives. *Neuroscientist* 11: 124–137.

May, P. J. (2006) The mammalian superior colliculus: Laminar structure and connections. *Prog. Brain Res.* 151: 321–378.

Robinson, D. A. (1981) Control of eye movements. In *Handbook of Physiology*, V. B. Brooks (ed.). Section 1: *The Nervous System*, vol. II: *Motor Control*, part 2. Bethesda, MD: American Physiological Society, pp. 1275–1320.

Schall, J. D. (1995) Neural basis of target selection. *Rev. Neurosci.* 6: 63–85.

Sparks, D. L. and L. E. Mays (1990) Signal transformations required for the generation of saccadic eye movements. *Annu. Rev. Neurosci.* 13: 309–336.

Spering, M. and M. Carrasco (2015) Acting without seeing: eye movements reveal visual processing without awareness. *Trends Neurosci.* 38: 247–258.

Zee, D. S. and L. M. Optican (1985) Studies of adaption in human oculomotor disorders. In *Adaptive Mechanisms in Gaze Control: Facts and Theories.* A Berthoz and G. Melvill Jones (eds.). Amsterdam: Elsevier, pp. 165–176.

Important Original Papers

Baarsma, E. and H. Collewijn (1974) Vestibulo-ocular and optokinetic reactions to rotation and their interaction in the rabbit. *J. Physiol.* 238: 603–625.

Fuchs, A. F. and E. S. Luschei (1970) Firing patterns of abducens neurons of alert monkeys in relationship to horizontal eye movements. *J. Neurophysiol.* 33: 382–392.

Optican, L. M. and D. A. Robinson (1980) Cerebellar-dependent adaptive control of primate saccadic system. *J. Neurophysiol.* 44: 1058–1076.

Schiller, P. H. and M. Stryker (1972) Single unit recording and stimulation in superior colliculus of the alert rhesus monkey. *J. Neurophysiol.* 35: 915–924.

Schiller, P. H., S. D. True and J. L. Conway (1980) Deficits in eye movements following frontal eye-field and superior colliculus ablations. *J. Neurophysiol.* 44: 1175–1189.

Sparks, D. L. and L. E. Mays (1983) Spatial localization of saccade targets. I. Compensation for stimulation-induced perturbations in eye position. *J. Neurophysiol.* 49: 45–63.

Sun, Z., A. Smilgin, M. Junker, P. W. Dicke and P. Their (2017) The same oculomotor vermal Purkinje cells encode the different kinematics of saccades and of smooth pursuit eye movements. *Sci. Rep.* 7: 40613; doi: 10.1038/srep40613

Books

Hall, W. C. and A. Moschovakis (eds.) (2004) *The Superior Colliculus: New Approaches for Studying Sensorimotor Integration.* Methods and New Frontiers in Neuroscience Series. New York: CRC Press.

Leigh, R. J. and D. S. Zee (1983) *The Neurology of Eye Movements.* Contemporary Neurology Series. Philadelphia, PA: F. A. Davis.

Schor, C. M. and K. J. Ciuffreda (eds.) (1983) *Vergence Eye Movements: Basic and Clinical Aspects.* Boston: Butterworth.

Yarbus, A. L. (1967) *Eye Movements and Vision* (trans. B. Haigh). New York: Plenum Press.

Go to the NEUROSCIENCE 6e Companion Website at **www.oup.com/uk/Purves6e** for Web Topics, Animations, Flashcards, and more.

CHAPTER

21 The Visceral Motor System

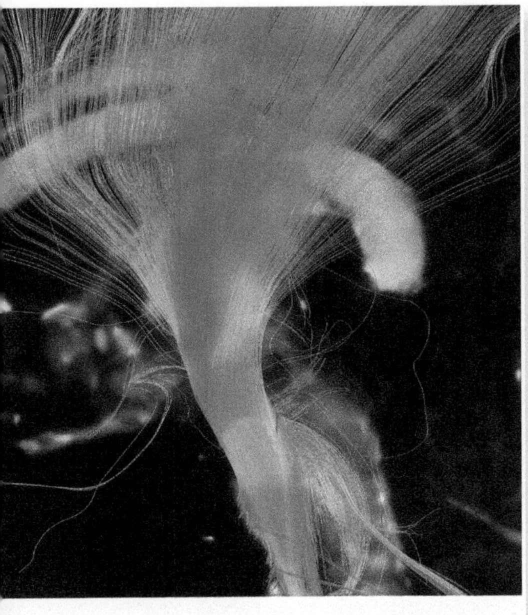

Overview

THE VISCERAL, OR AUTONOMIC, MOTOR SYSTEM controls involuntary functions mediated by the activity of smooth muscle fibers, cardiac muscle fibers, and glands. The system comprises two major divisions, the sympathetic and parasympathetic subsystems. The specialized innervation of the gut is a further, semi-independent division that is usually referred to as the enteric nervous system. Although these divisions are always active at some level, the sympathetic division mobilizes the body's resources for dealing with challenges of one sort or another. Conversely, parasympathetic activity predominates during states of relative quiescence, so that energy sources previously expended can be restored. This continuous neural regulation of the expenditure and replenishment of the body's resources is crucial for the overall physiological balance of bodily functions called homeostasis. Whereas the major controlling centers for somatic motor activity are the motor cortex in the frontal lobes and related subcortical nuclei, the major locus of central control in the visceral motor system is the hypothalamus, which in turn is modulated by activity in the amygdala, hippocampus, insula, and other cortical regions in the ventral and medial aspects of the frontal lobes. The function of both principal divisions of the visceral motor system is governed by descending pathways from the hypothalamus and the reticular formation of the brainstem to preganglionic neurons in the brainstem and spinal cord, which in turn determine the activity of the primary, or lower, visceral motor neurons in autonomic ganglia located outside the CNS. The autonomic regulation of several organ systems of particular importance in clinical practice (including cardiovascular function, control of the bladder, and governance of the reproductive organs) is considered in more detail as specific examples of visceral motor control and the importance of central integration for the coordination of somatic motor and visceral motor function.

Early Studies of the Visceral Motor System

Although humans must always have been aware of involuntary motor reactions to stimuli in the environment (e.g., narrowing of the pupil in response to bright light, constriction of superficial blood vessels in response to cold or fear, increased heart rate in response to exertion or anxiety), it was not until the late nineteenth century that the neural control of these and other visceral functions came to be understood in modern terms. The researchers who first rationalized the workings of the **visceral motor system** were Walter Gaskell and John Langley, British physiologists at Cambridge University. Gaskell's work preceded that of Langley and established the overall anatomy of the system as he carried out early physiological experiments that demonstrated some of its salient functional characteristics (e.g., that the heartbeat of an experimental animal is accelerated by stimulating the outflow of the upper thoracic spinal cord segments). Based on these and other observations, Gaskell concluded in 1866 that "every tissue is innervated by two sets of nerve fibers of opposite characters," and he further surmised that these actions showed "the characteristic signs of opposite chemical processes."

Using similar electrical stimulation techniques in experimental animals, Langley went on to establish the function of **autonomic ganglia** (which harbor the lower visceral motor neurons), defined the terms *preganglionic* and *postganglionic* (see below), and coined the phrase **autonomic nervous system**, which is commonly used as a synonym for visceral motor system (although some somatic motor actions related to emotion are closely tied to autonomic motor actions such as facial expressions; see Chapter 31). Langley's work on the pharmacology of the autonomic system initiated the classic studies indicating the roles of acetylcholine and the catecholamines in visceral motor function, and in neurotransmitter function more generally (see Chapter 6). In short, Langley's ingenious physiological and anatomical experiments established in detail the general proposition put forward by Gaskell on more circumstantial grounds.

The third major figure in the pioneering studies of the visceral motor system was Walter Cannon at Harvard Medical School, who during the early to middle 1900s devoted his career to understanding visceral motor functions in relation to homeostatic mechanisms, emotions, and other complex brain functions (see Chapter 31). Like Gaskell and Langley before him, Cannon based his work primarily on electrical stimulation in experimental animals, including activation of the hypothalamus, brainstem, and peripheral components of the system. He also established the effects of denervation in the visceral motor system, laying some of the foundation for current understanding of neuronal plasticity (see Chapters 8 and 25).

Distinctive Features of the Visceral Motor System

Chapters 16 and 17 discussed in detail the organization of lower motor neurons in the CNS, their relationships to striated muscle fibers, and the means by which their activities are governed by higher motor centers. With respect to the efferent systems that govern the actions of smooth muscle fibers, cardiac muscle fibers, and glands, it is instructive to recognize the anatomical and functional features of the visceral motor system that distinguish it from the somatic motor system.

First, the lower motor neurons of the visceral motor system are located outside the CNS (Figure 21.1). The cell bodies of these primary visceral motor neurons are found in autonomic ganglia that are either close to the spinal cord (sympathetic division) or embedded in a neural **plexus**—a network of intersecting nerves—very near or in the target organ (parasympathetic and enteric divisions).

Second, the contacts between visceral motor neurons and the viscera are much less differentiated than the neuromuscular junctions of the somatic motor system. Visceral motor axons tend to be highly branched and give rise to many synaptic terminals at varicosities (swellings) along the length of the terminal axonal branch. Moreover, the surfaces of the visceral muscle usually lack the highly ordered structure of the motor end plates that characterizes postsynaptic target sites on striated muscle fibers. As a consequence, the neurotransmitters released by visceral motor terminals often diffuse for hundreds of microns before binding to postsynaptic receptors—a far greater distance than at the synaptic cleft of the somatic neuromuscular junction.

Third, whereas the principal actions of the somatic motor system are governed by motor cortical areas in the posterior frontal lobe (see Chapter 17), the activities of the visceral motor system are coordinated by a distributed set of cortical and subcortical structures in the ventral and medial parts of the forebrain and in the brainstem; collectively, these structures comprise a central autonomic network.

Finally, visceral motor terminals release a variety of neurotransmitters, including primary small-molecule neurotransmitters (which differ depending on whether the motor neuron in question is sympathetic or parasympathetic) and one or more of a variety of co-neurotransmitters that may be a different small-molecule neurotransmitter or a neuropeptide (see Chapter 6). These neurotransmitters interact with a diverse set of postsynaptic receptors that mediate myriad postsynaptic effects in smooth and cardiac muscle and glands. It should be clear, then, that whereas the major effect of somatic motor activation on striated muscle is nearly the same throughout the body, the effects of visceral motor activation are remarkably varied (Table 21.1). This fact should come as no surprise, given the challenge of maintaining homeostasis across the many organ systems of the body in the face of variable environmental conditions and ever-changing behavioral contingencies.

The Sympathetic Division of the Visceral Motor System

Activity of the neurons that make up the sympathetic division of the visceral motor system ultimately prepares individuals for "fight or flight," as Cannon famously put it. Cannon meant that, in extreme circumstances, heightened levels of sympathetic neural activity allow the body to make maximum use of its resources (particularly its metabolic resources), thereby increasing the chances of survival or success in threatening or otherwise challenging situations. Thus, during high levels of sympathetic activity, the

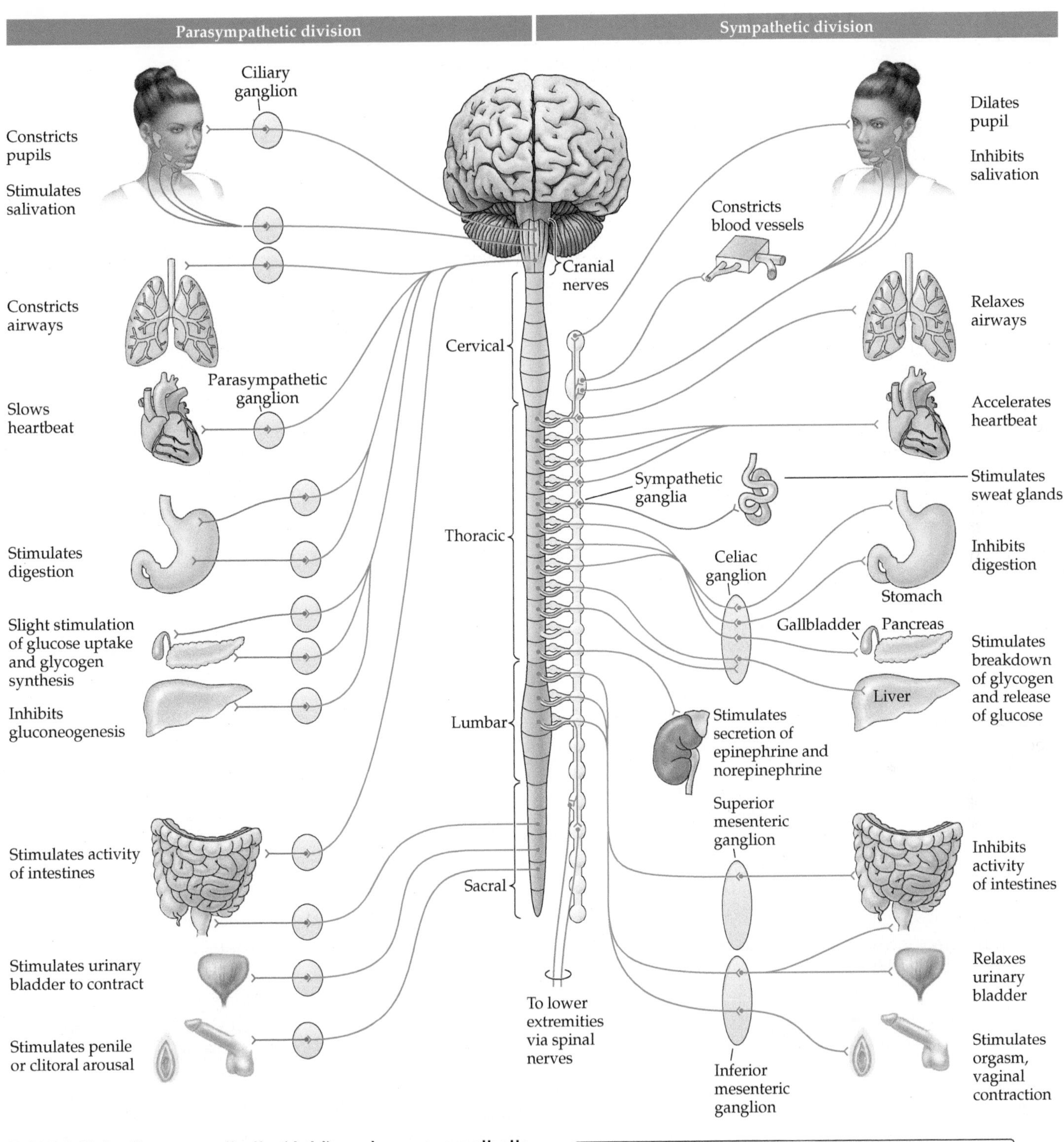

FIGURE 21.1 The sympathetic (right) and parasympathetic (left) divisions of the visceral motor system. The figure identifies neurons that are understood to be sympathetic or parasympathetic by conventional anatomical, physiological, and pharmacological criteria. By such criteria, the sacral visceral motor outflow has been considered parasympathetic (lower left side of illustration), but new evidence in mice suggests sympathetic ontogeny, with genetic and molecular phenotypes consistent with thoracic sympathetic outflow.

Noradrenergic neurons	Cholinergic neurons	
Postganglionic	Preganglionic	Postganglionic

TABLE 21.1 ■ **Major Functions of the Visceral Motor System**

Parasympathetic Division			
Target organ	**Location of preganglionic neurons**	**Location of ganglionic neurons**	**Actions**
Eye	Edinger-Westphal nucleus	Ciliary ganglion	Pupillary constriction, accommodation
Lacrimal gland	Superior salivatory nucleus	Pterygopalatine ganglion	Secretion of tears
Submandibular and sublingual glands	Superior salivatory nucleus	Submandibular ganglion	Secretion of saliva, vasodilation
Parotid gland	Inferior salivatory nucleus	Otic ganglion	Secretion of saliva, vasodilation
Head, neck (blood vessels, sweat glands, piloerector muscles)	None	None	None
Upper extremity	None	None	None
Heart	Nucleus ambiguus and dorsal motor nucleus of the vagus nerve	Cardiac plexus	Reduced heart rate
Bronchi, lungs	Dorsal motor nucleus of the vagus nerve	Pulmonary plexus	Bronchial constriction and secretion
Stomach	Dorsal motor nucleus of the vagus nerve	Myenteric and submucous plexus	Peristaltic movement and secretion
Pancreas	Dorsal motor nucleus of the vagus nerve	Pancreatic plexus	Secretion of insulin and digestive enzymes
Ascending small intestine, transverse large intestine	Dorsal motor nucleus of the vagus nerve	Ganglia in the myenteric and submucous plexus	Peristaltic movement and secretion
Descending large intestine, sigmoid, rectum	S3–S4	Ganglia in the myenteric and submucous plexus	Peristaltic movement and secretion
Adrenal gland	None	None	None
Ureter, bladder	S2–S4	Pelvic plexus	Contraction of bladder wall and inhibition of internal sphincter
Lower extremity	None	None	None

TABLE 21.1 ■ (*continued*)

Sympathetic Division			
Target organ	**Location of preganglionic neurons**	**Location of ganglionic neurons**	**Actions**
Eye	Upper thoracic spinal cord (C8–T7)	Superior cervical ganglion	Pupillary dilation
Lacrimal gland			Protein secretion in tears
Submandibular and sublingual glands			Vasoconstriction
Parotid gland			Vasoconstriction
Head, neck (blood vessels, sweat glands, piloerector muscles)			Sweat secretion, vasoconstriction, piloerection
Upper extremity	T3–T6	Stellate and upper thoracic ganglia	Sweat secretion, vasoconstriction, piloerection
Heart	Middle thoracic spinal cord (T1–T5)	Superior cervical and upper thoracic ganglia	Increased heart rate and stroke volume, dilation of coronary arteries
Bronchi, lungs		Upper thoracic ganglia	Vasodilation, bronchial dilation
Stomach	Lower thoracic spinal cord (T6–T10)	Celiac ganglion	Inhibition of peristaltic movement and gastric secretion, vasoconstriction
Pancreas		Celiac ganglion	Vasoconstriction, inhibition of insulin secretion
Ascending small intestine, transverse large intestine		Celiac, superior, and inferior mesenteric ganglia	Inhibition of peristaltic movement and secretion
Descending large intestine, sigmoid, rectum		Inferior mesenteric hypogastric and pelvic plexus	Inhibition of peristaltic movement and secretion
Adrenal gland	T9–L2	Cells of gland are modified neurons	Catecholamine secretion
Ureter, bladder	T11–L2	Hypogastric and pelvic plexus	Relaxation of bladder wall muscle and contraction of internal sphincter
Lower extremity	T10–L2	Lower lumbar and upper sacral ganglia	Sweat secretion, vasoconstriction, piloerection

pupils dilate and the eyelids retract (allowing more light to reach the retina and the eyes to move more efficiently); the blood vessels of the skin and gut constrict (rerouting blood to muscles, thus allowing them to extract a maximum of available energy); the hairs stand on end (which made our hairier ancestors look more fearsome); the bronchi dilate (increasing oxygenation); the heart rate accelerates and the force of cardiac contraction is enhanced (maximally perfusing skeletal muscles and the brain); and digestive and other vegetative functions become quiescent (thus diminishing activities that are temporarily unnecessary) (see Figure 21.1). At the same time, sympathetic activity stimulates the adrenal medulla to release epinephrine and norepinephrine into the bloodstream and causes the release of glucagon from the pancreas, further enhancing energy-mobilizing (or catabolic) functions. These coordinated responses illustrate an important principle of visceral motor function: There are circumstances that necessitate a departure from homeostatic set points in the regulation of the body's physiological systems (Box 21A), and such

BOX 21A ■ The Hypothalamus

The hypothalamus is located at the base of the forebrain, bounded by the optic chiasm rostrally and the midbrain tegmentum caudally. It forms the floor and ventral walls of the third ventricle and is continuous through the infundibular stalk with the posterior pituitary gland, as illustrated in Figure A. Given its central position in the brain and its proximity to the pituitary, it is not surprising that the hypothalamus integrates information from the forebrain, brainstem, spinal cord, and various intrinsic chemosensitive neurons.

What is surprising about this structure is the remarkable diversity of homeostatic functions that are governed by this relatively small region of the forebrain. The diverse functions in which hypothalamic involvement is at least partially understood include: *the control of blood flow* (by promoting adjustments in cardiac output, vasomotor tone, blood osmolarity, and renal clearance, and by motivating drinking and salt consumption); the *regulation of energy metabolism* (by monitoring blood glucose levels and regulating feeding behavior, digestive functions, metabolic rate, and temperature); the *regulation of reproductive activity* (by influencing gender identity, sexual orientation, and mating behavior, and in females, by governing menstrual cycles, pregnancy, and lactation); and the *coordination of responses to threatening conditions* (by governing the release of stress hormones, modulating the balance between sympathetic and parasympathetic tone, and influencing the regional distribution of blood flow).

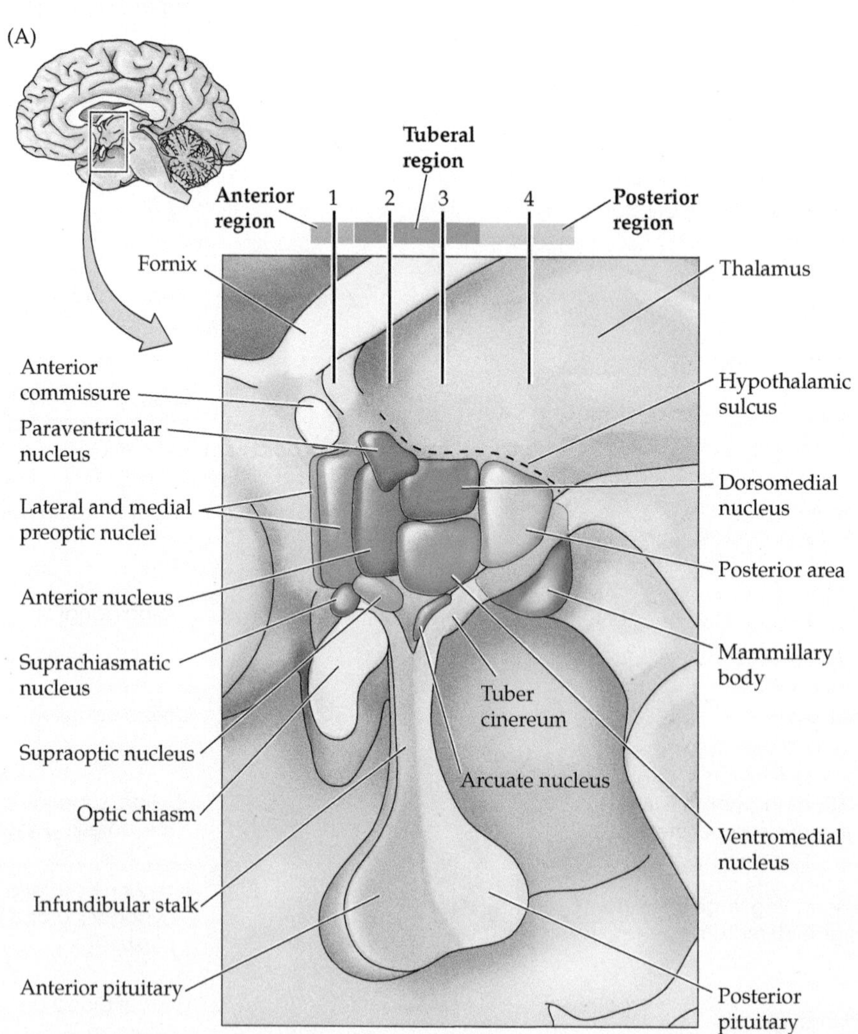

(A) Diagram of the human hypothalamus, illustrating its major nuclei.

Despite the impressive scope of hypothalamic control, the individual components of the hypothalamus use similar physiological mechanisms to exert their influence over these many functions (Figure B). Thus, hypothalamic circuits receive sensory and contextual information, compare that information with biological set points, and activate relevant visceral motor, neuroendocrine, and somatic motor effector systems that restore homeostasis or elicit appropriate behavioral responses.

Like the overlying thalamus—and consistent with the scope of hypothalamic functions—the hypothalamus comprises a large number of distinct nuclei, each with its own specific pattern of connections and functions. The nuclei, most

BOX 21A ■ (continued)

of which are intricately interconnected, can be grouped in three longitudinal regions referred to as *periventricular, medial,* and *lateral* (Figure C). The nuclei can also be grouped along the anterior-posterior dimension into the *anterior* (or preoptic), *tuberal*, or *posterior* region (see Figure A). The anterior-periventricular group contains the suprachiasmatic nucleus, which receives direct retinal input and drives circadian rhythms (see Chapter 28). More scattered neurons in the periventricular region (located along the wall of the third ventricle) manufacture peptides known as releasing or inhibiting factors, which control the secretion of a variety of hormones by the anterior pituitary. The axons of these neurons project to the median eminence, a region at the junction of the hypothalamus and pituitary stalk, where the peptides are secreted into the portal circulation that supplies the anterior pituitary.

Nuclei in the medial-tuberal region (*tuberal* refers to the tuber cinereum, the anatomical name given to the middle portion of the inferior surface of the hypothalamus) include the paraventricular and supraoptic nuclei, which contain the neurosecretory neurons whose axons extend into the

Continued on the next page

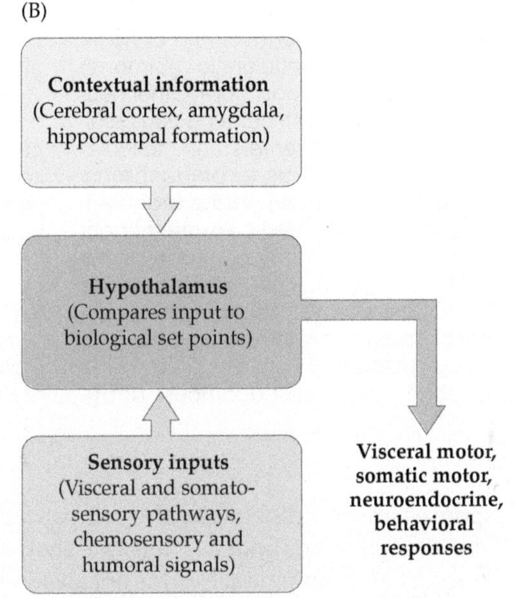

(B) Physiological mechanisms underlying hypothalamic function.

(C)

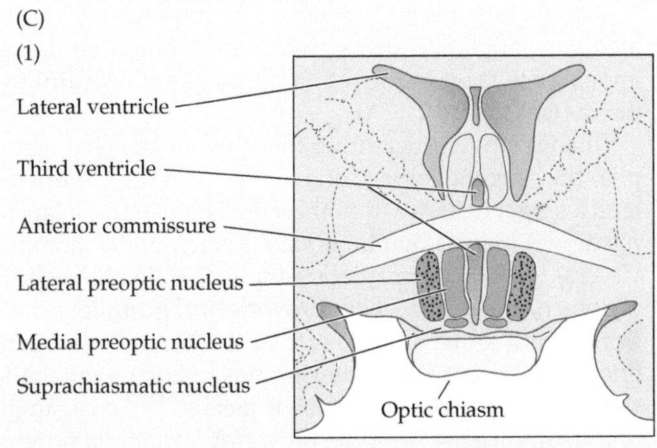

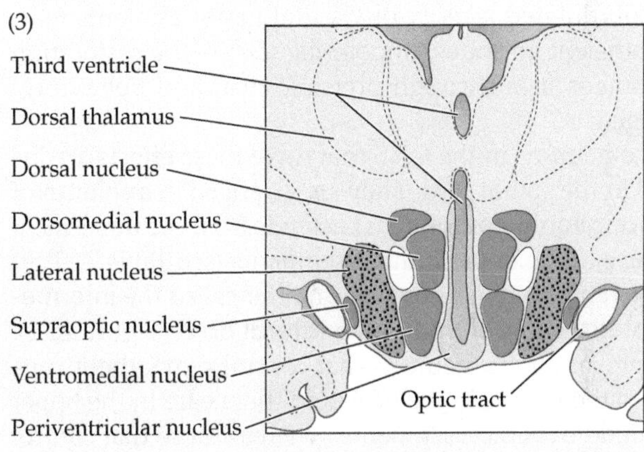

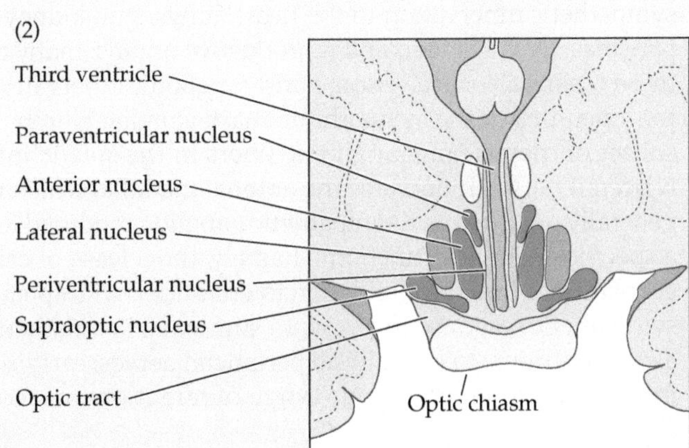

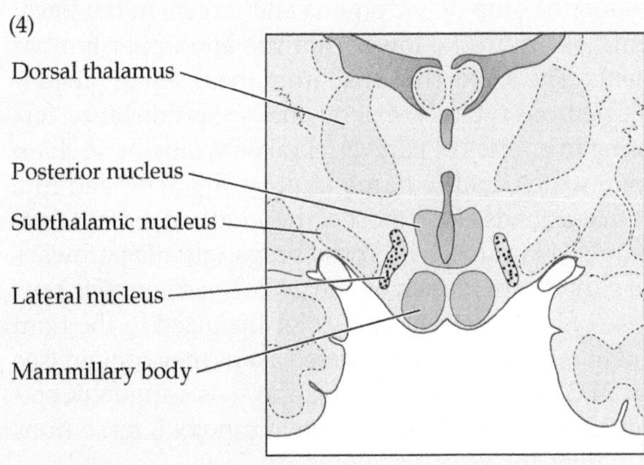

(C) Coronal sections through the human hypothalamus (see Figure A for location of sections 1–4). Color coding of the nuclei illustrates the two dimensions by which hypothalamic nuclei are subdivided (see text). Blue, red, and green illustrate nuclei in the anterior, tuberal, and posterior regions, respectively. The relative shading of these hues illustrates the three mediolateral zones: Lighter shading represents nuclei in the periventricular zone, whereas darker shades represent medial zone nuclei. Nuclei in the lateral zone are stippled. (1) Section through the anterior region illustrating the preoptic and suprachiasmatic nuclei. (2) Rostral tuberal region. (3) Caudal tuberal region. (4) Section through the posterior region illustrating the mammillary bodies.

BOX 21A ■ (continued)

posterior pituitary. With appropriate stimulation, these neurons secrete oxytocin or vasopressin (antidiuretic hormone) directly into the bloodstream. Other neurons in the paraventricular nucleus project to autonomic centers in the reticular formation, as well as to preganglionic visceral motor neurons in the brainstem and spinal cord; these cells are thought to exert hypothalamic control over the visceral motor outflow throughout the body. The paraventricular nucleus receives inputs from other hypothalamic zones, which are in turn related to the cerebral cortex, hippocampus, amygdala, and other central structures, all of which are capable of influencing visceral motor function.

Also in this region of the hypothalamus are the dorsomedial and ventromedial nuclei, which are involved in feeding, reproductive and parenting behavior, thermoregulation, and water balance. These nuclei receive inputs from structures in the limbic forebrain, as well as from visceral sensory nuclei in the brainstem (e.g., the nucleus of the solitary tract).

Finally, the lateral region of the hypothalamus is really a rostral continuation of the midbrain reticular formation (see Box 17C). Thus, the neurons of the lateral region are not grouped into nuclei, but are scattered among the fibers of the medial forebrain bundle, a prominent collection of axonal projections that run through the lateral hypothalamus. Cells in this lateral region control behavioral arousal and shifts of attention.

In summary, the hypothalamus regulates an enormous range of physiological and behavioral activities and serves as the key controlling center for visceral motor activity and homeostatic functions that are essential for survival.

responses are coordinated by the sympathetic division of the visceral motor system. Thus, the short-term functional goal of autonomic activity is not always homeostasis (the maintenance of a constant internal state). Rather, the coordinated activity of visceral motor efferents may, for transient episodes, impose *allostasis*—the restoration of homeostasis through physiological and behavioral change.

The neurons in the CNS that drive these effects are located in the spinal cord. They are arranged in a column of **preganglionic neurons** that extends from the uppermost thoracic to the upper lumbar segments (see Table 21.1) in a region of the spinal cord gray matter called the **intermediolateral cell column** in the **lateral horn** (Figure 21.2). The preganglionic neurons that control sympathetic outflow to the organs in the head and thorax are in the upper and middle thoracic segments, whereas those that control the abdominal and pelvic organs and targets in the lower extremities are in the lower thoracic and upper lumbar segments. The axons that arise from these spinal preganglionic neurons typically extend only a short distance, terminating in a series of paravertebral or sympathetic chain ganglia, which, as the name implies, are arranged in a chain that extends along most of the length of the vertebral column (see Figure 21.1). These preganglionic pathways to the ganglia are known as the *white communicating rami* because of the relatively light color imparted to the rami (singular, *ramus*) by the myelinated axons they contain (see Figure 21.2A). Roughly speaking, these preganglionic spinal neurons are comparable to somatic motor interneurons (see Chapter 16).

The neurons in **sympathetic ganglia** are the primary or lower motor neurons of the sympathetic division in that they directly innervate smooth muscles, cardiac muscle, and glands. The **postganglionic axons** arising from these **paravertebral sympathetic chain** neurons travel to various targets in the body wall, joining the segmental spinal nerves of the corresponding spinal segments by way of the *gray communicating rami*. The gray rami are another set of short linking nerves, so named because the unmyelinated postganglionic axons give them a somewhat darker appearance than the myelinated preganglionic linking nerves (see Figure 21.2A).

In addition to innervating the sympathetic chain ganglia, the preganglionic axons that govern the viscera extend a longer distance from the spinal cord in the splanchnic nerves (nerves that innervate thoracic and abdominal viscera) to reach sympathetic ganglia that lie in the chest, abdomen, and pelvis. These **prevertebral ganglia** include sympathetic ganglia in the cardiac plexus; the celiac ganglion; the superior and inferior mesenteric ganglia; and sympathetic ganglia in the pelvic plexus. The postganglionic axons arising from the prevertebral ganglia provide sympathetic innervation to the heart, lungs, gut, kidneys, pancreas, liver, bladder, and reproductive organs; many of these organs also receive some postganglionic innervation from neurons in the sympathetic chain ganglia. Finally, a subset of thoracic preganglionic fibers in the splanchnic (visceral) nerves innervates the adrenal medulla, which is generally regarded as a sympathetic ganglion modified for a specific endocrine function—namely, the release of catecholamines into the circulation to enhance a widespread sympathetic response to stress. In summary, sympathetic axons contribute to virtually all peripheral nerves, carrying innervation to an enormous range of target organs (see Table 21.1).

Cannon's memorable truism that the sympathetic activity prepares the animal for "fight or flight" notwithstanding, the sympathetic division of the visceral motor system is tonically active to maintain sympathetic function at appropriate levels, whatever the circumstances. Nor should the sympathetic system be thought of as responding in an

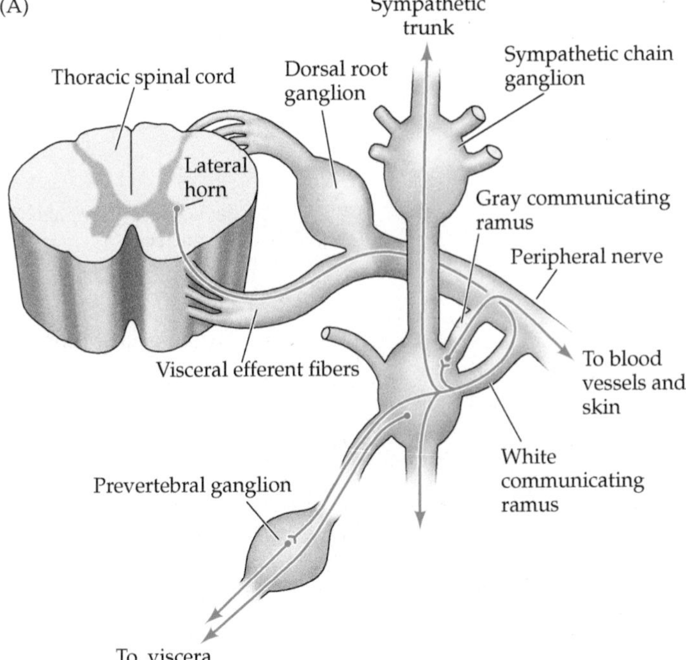

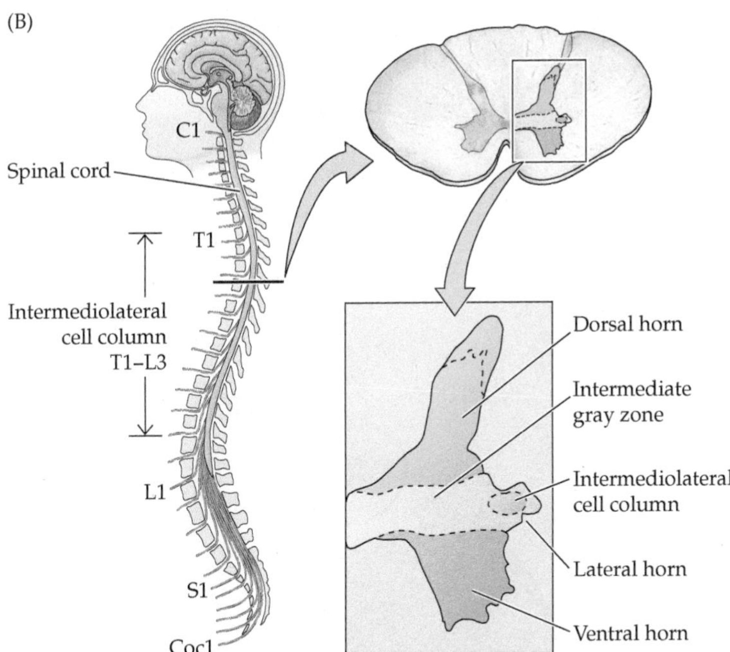

FIGURE 21.2 Organization of the preganglionic spinal outflow to sympathetic ganglia. (A) General organization of the sympathetic division of the visceral motor system in the spinal cord and the preganglionic outflow to the sympathetic ganglia that contain the primary visceral motor neurons. (B) Cross section of the thoracic spinal cord at the level indicated, showing location of the sympathetic preganglionic neurons of the intermediolateral cell column in the lateral horn.

all-or-none fashion; many sympathetic reflexes operate more or less independently, as might be expected from the obvious need to specifically control various organ functions (e.g., the heart during exercise, the bladder during urination, and the reproductive organs during sexual intercourse).

The Parasympathetic Division of the Visceral Motor System

The preganglionic outflow from the CNS to the ganglia of the parasympathetic division stems from neurons whose distribution is limited to the brainstem and (according to long-standing convention, but see below) the sacral part of the spinal cord (Figure 21.3; see also Figure 21.1). The cranial preganglionic innervation arising from the brainstem, which is analogous to the preganglionic sympathetic outflow from the spinal cord, includes the **Edinger-Westphal nucleus** in the midbrain (which innervates the ciliary ganglion via the oculomotor nerve and mediates construction of the pupil in response to increased light; see Chapter 12); the **superior** and **inferior salivatory nuclei** in the pons and medulla (which innervate the salivary glands and tear glands, mediating salivary secretion and the production of tears); a visceral motor division of the **nucleus ambiguus** in the medulla; and the **dorsal motor nucleus of the vagus nerve**, also in the medulla. Neurons in the ventrolateral part of the nucleus ambiguus provide an important source of cardio-inhibitory innervation to the cardiac ganglia via the vagus nerve. The more dorsal part of the dorsal motor nucleus of the vagus nerve primarily governs glandular secretion via the parasympathetic ganglia located in the viscera of the thorax and abdomen, whereas the more ventral part of the nucleus controls the motor responses of the heart, lungs, and gut elicited by the vagus nerve (e.g., constricting the bronchioles). In addition, other preganglionic neurons in the nucleus ambiguus innervate parasympathetic ganglia in the submandibular salivary glands and the mediastinum (a different division of the nucleus ambiguus provides branchiomotor innervation of striated muscle in the pharynx and larynx; see the Appendix). The location of the parasympathetic brainstem nuclei is shown in Figure 21.3.

The sacral preganglionic innervation arises from neurons in the lateral gray matter of the sacral segments of the spinal cord, which are located in much the same position as the sympathetic preganglionic neurons in the intermediolateral cell column of the thoracic cord (see Figure 21.3C,D). The axons from these neurons travel in the splanchnic nerves to innervate ganglia in the lower third of the colon, rectum, bladder, and reproductive organs.

The **parasympathetic ganglia** innervated by preganglionic outflow from both cranial and sacral levels are in or near the end organs they serve. In this way they differ from the ganglionic targets of the sympathetic system (recall that both the paravertebral chain and prevertebral

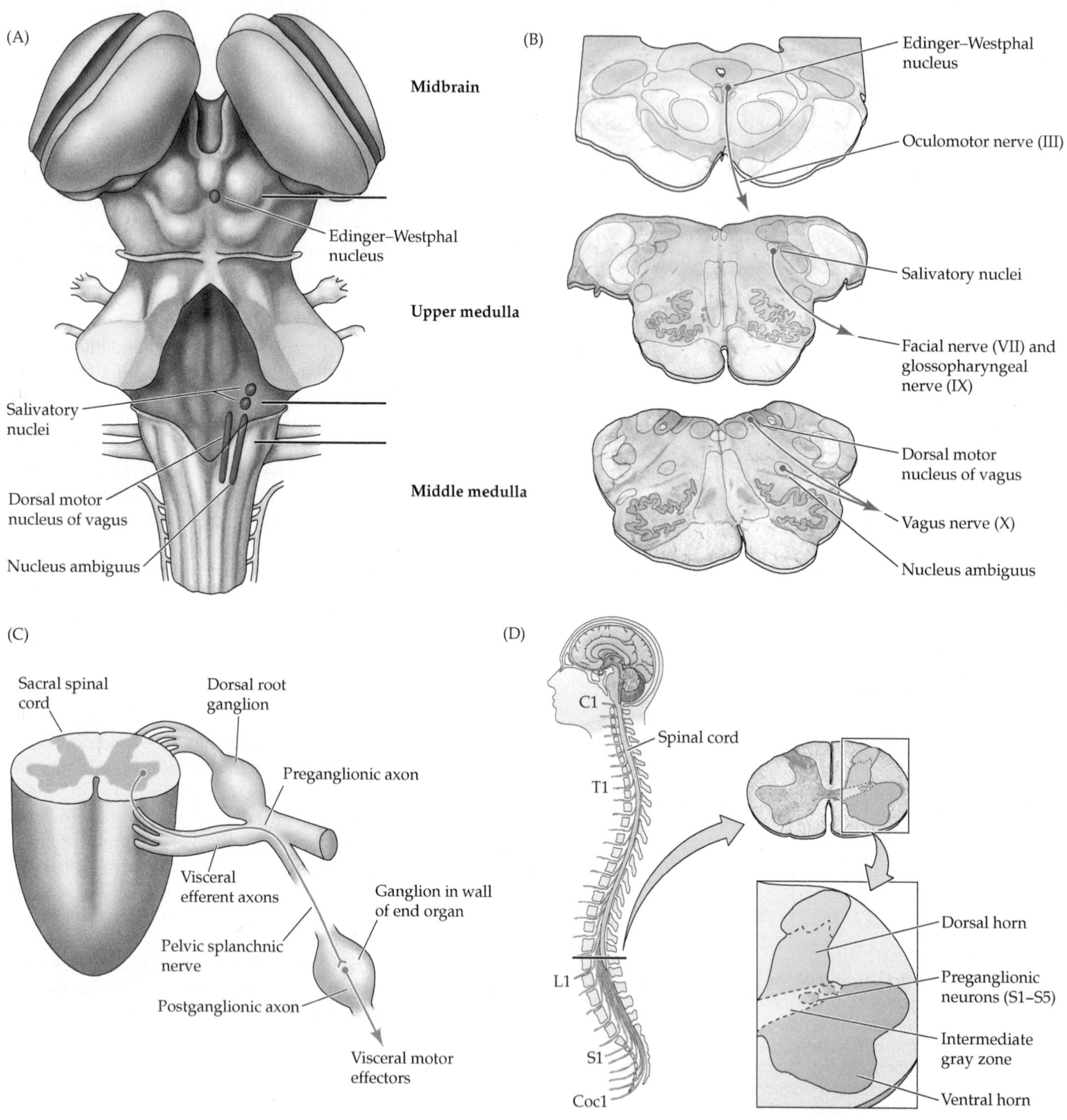

FIGURE 21.3 Organization of the cranial preganglionic outflow to parasympathetic ganglia and sacral visceral motor outflow. (A) Dorsal view of brainstem showing the locations of the nuclei of the cranial part of the parasympathetic division of the visceral motor system. (B) Cross sections of the brainstem at the relevant levels [horizontal lines in (A)] showing the locations of these parasympathetic nuclei. (C) Main features of the preganglionic visceral motor outflow in the sacral segments of the spinal cord. Until recently, this outflow has been considered parasympathetic (as in Figure 21.1); new evidence suggests sympathetic ontogeny (see text for details). (D) Cross section of the sacral spinal cord showing the location of sacral preganglionic neurons.

ganglia are located relatively far from their target organs; see Figure 21.1). An important anatomical difference between sympathetic and parasympathetic ganglia at the cellular level is that sympathetic ganglion cells tend to have extensive dendritic arbors and are, as might be expected from this arrangement, innervated by a large number of preganglionic fibers. Parasympathetic ganglion cells have few if any dendrites and consequently are each innervated by only one or a few preganglionic axons. This arrangement implies a greater diversity of converging influences on sympathetic ganglion neurons compared with parasympathetic ganglion neurons.

The overall function of the parasympathetic system, as Gaskell, Langley, and Cannon demonstrated, is generally opposite to that of the sympathetic system, serving to increase metabolic and other resources during periods when the animal's circumstances allow it to "rest and digest." In contrast to the sympathetic functions enumerated earlier, the activity of the parasympathetic system constricts the pupils, slows the heart rate, increases the peristaltic activity of the gut, and promotes emptying of urine from the bladder (or *voiding*, as clinicians often call this process). At the same time, diminished activity in the sympathetic system allows the blood vessels of the skin and gut to dilate, the piloerector muscles to relax, and the outflow of catecholamines from the adrenal medulla to decrease.

Although most organs receive innervation from *both* the sympathetic and parasympathetic divisions of the visceral motor system (as Gaskell surmised), some receive only sympathetic innervation. These exceptional targets include the sweat glands, adrenal medulla, piloerector muscles of the skin, and most arterial blood vessels (see Table 21.1).

Until very recently, there was little reason to challenge the conventional schema presented above for cranial-sacral parasympathetic outflow (see Figures 21.1 and 21.3). The standard classification of sacral visceral motor outflow as *parasympathetic* has been based on (1) anatomy—similarities in organization to vagal innervation (long preganglionic axons, short ganglionic axons); (2) physiology—presumed actions that oppose the effects mediated by the thoraco-lumbar (sympathetic) visceral motor outflow; and (3) pharmacology—general antagonism of end-organ action by blockade of muscarinic cholinergic receptors (see below). However, recent molecular and genetic analysis of the mouse nervous system has questioned this understanding of sacral visceral motor outflow. Studies by J.-F. Brunet and colleagues at the École Normale Supérieure in Paris suggest that the sacral division of the visceral motor system, which provides innervation to the pelvic organs, should now be considered *sympathetic*. The basis for this proposed reclassification is an analysis of 15 phenotypic and ontogenetic features of preganglionic and ganglionic elements showing that the sacral outflow and the thoraco-lumbar outflow share all 15 features. Furthermore, these shared features are distinct from features expressed by the parasympathetic outflow derived from the brainstem. Thus, this new work suggests that visceral motor outflow from the CNS may be understood in simple, bipartite terms comprising a cranial parasympathetic division and a spinal sympathetic division. While these molecular and genetic studies are compelling, it remains to be determined if these findings in mice generalize to all mammals. Furthermore, questions remain as to whether and how this "sacral sympathetic" concept might be reconciled with the anatomical, physiological, and pharmacological criteria outlined above that support the more conventional schema for cranial-sacral parasympathetic outflow.

The Enteric Nervous System

An enormous number of neurons are specifically associated with the gastrointestinal tract to control its many functions; indeed, it is likely that more neurons reside in the human gut than in the entire spinal cord. As already noted, the activity of the gut is modulated by both the sympathetic and parasympathetic divisions of the visceral motor system. However, the gut also has an extensive system of nerve cells in its wall (as do its accessory organs such as the pancreas and gallbladder) that do not fit neatly into the sympathetic or parasympathetic divisions of the visceral motor system (Figure 21.4A). To a surprising degree, these neurons and the complex enteric networks in which they are found operate more or less independently according to their own reflex rules; as a result, many gut functions continue perfectly well without sympathetic or parasympathetic supervision (peristalsis, for example, occurs in isolated gut segments in vitro). Thus, most investigators prefer to classify the enteric nervous system as a unique, autonomic component of the visceral motor system.

The neurons in the gut wall include local and centrally projecting sensory neurons that monitor mechanical and chemical conditions in the gut, local circuit neurons that integrate this information, and motor neurons that influence the activity of the smooth muscles in the wall of the gut and glandular secretions (e.g., of digestive enzymes, mucus, stomach acid, and bile). This complex arrangement of nerve cells intrinsic to the gut is organized into (1) the **myenteric** (or **Auerbach's**) **plexus**, which is specifically concerned with regulating the musculature of the gut; and (2) the **submucous** (or **Meissner's**) **plexus**, which is located, as the name implies, just beneath the mucus membranes of the gut and is concerned with chemical monitoring and glandular secretion (Figure 21.4B).

As already mentioned, the preganglionic parasympathetic neurons that influence the gut are primarily in the dorsal motor nucleus of the vagus nerve in the brainstem and the intermediate gray zone in the sacral spinal cord segments. The preganglionic sympathetic innervation that

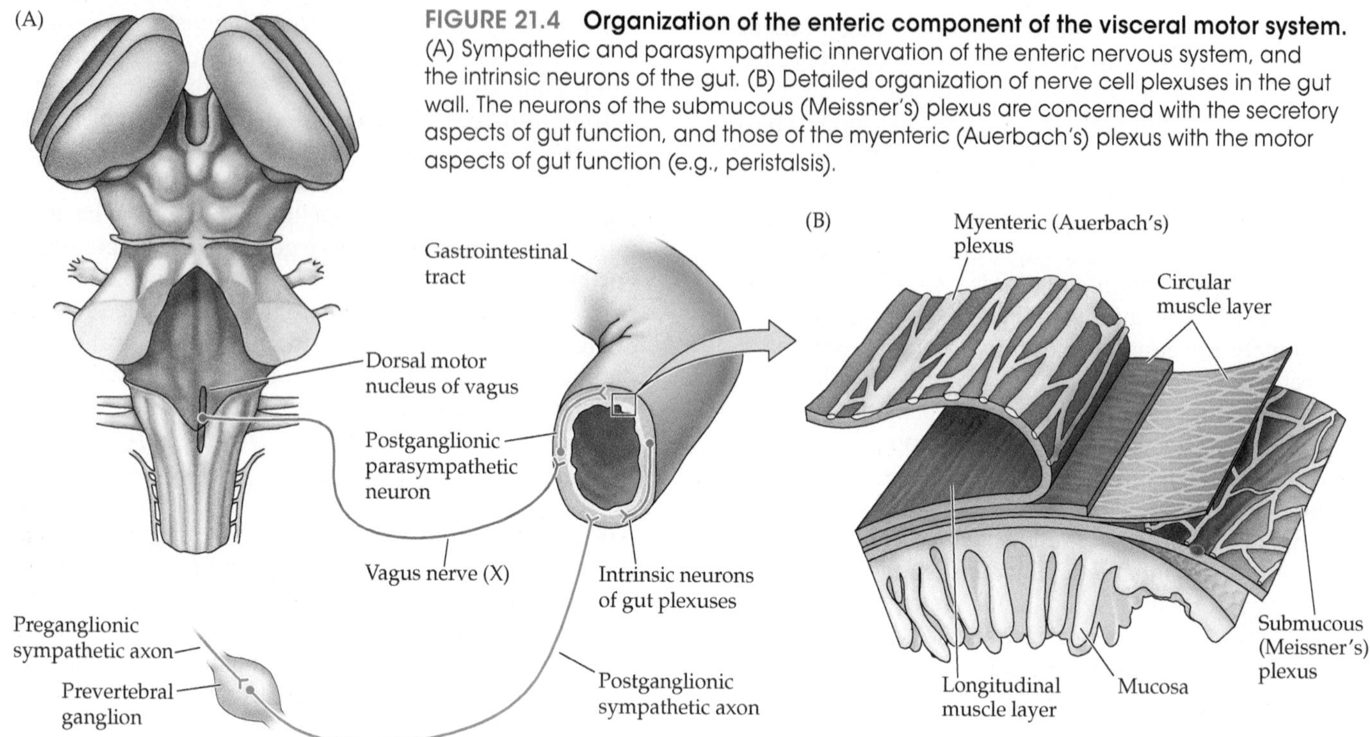

FIGURE 21.4 Organization of the enteric component of the visceral motor system. (A) Sympathetic and parasympathetic innervation of the enteric nervous system, and the intrinsic neurons of the gut. (B) Detailed organization of nerve cell plexuses in the gut wall. The neurons of the submucous (Meissner's) plexus are concerned with the secretory aspects of gut function, and those of the myenteric (Auerbach's) plexus with the motor aspects of gut function (e.g., peristalsis).

modulates the action of the gut plexuses derives from the thoraco-lumbar cord, primarily by way of the celiac, superior, and inferior mesenteric ganglia.

Sensory Components of the Visceral Motor System

Although the focus of this unit is movement and its central control, it is important to understand the sources of visceral sensory information and the means by which this input becomes integrated with visceral motor networks in the CNS. Generally speaking, afferent activity arising from the viscera serves two important functions. First, it provides feedback to local reflexes that modulate moment-to-moment visceral motor activity within individual organs. Second, it informs higher integrative centers of more complex patterns of stimulation that may signal potentially threatening conditions or require the coordination of more widespread visceral motor, somatic motor, neuroendocrine, and behavioral activities (Figure 21.5). The **nucleus of the solitary tract** in the medulla is the central structure in the brain that receives visceral sensory information and distributes it accordingly to serve both purposes.

The afferent fibers that provide these visceral sensory inputs arise from cell bodies in the dorsal root ganglia (as is the case of somatosensory modalities; see Chapters 9 and 10) and the sensory ganglia associated with the glossopharyngeal and vagus cranial nerves. However, far fewer visceral sensory neurons (by a factor of about 10) innervate

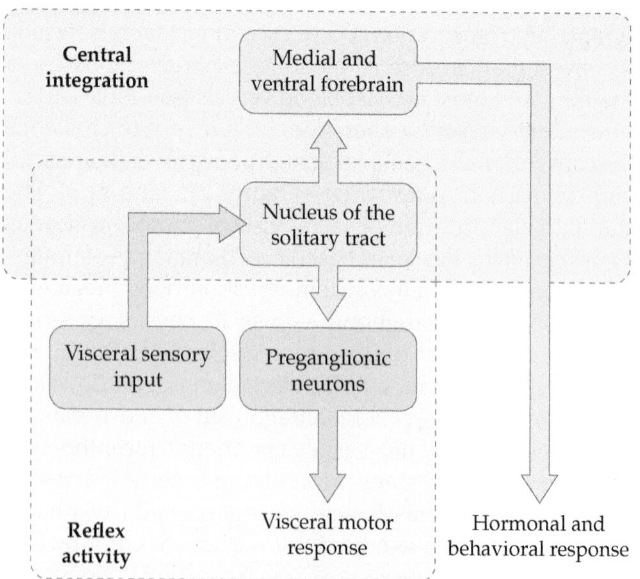

FIGURE 21.5 Distribution of visceral sensory information by the nucleus of the solitary tract. Sensory information transduced via this pathway serves either local reflex responses or more complex hormonal and behavioral responses via integration within a central autonomic network. As expanded upon in Figure 21.7, forebrain centers also provide input to visceral motor effector systems in the brainstem and spinal cord.

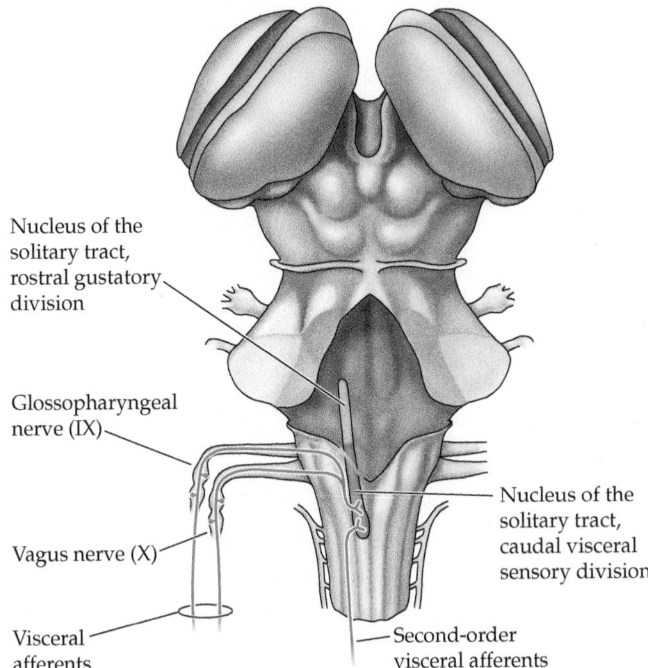

FIGURE 21.6 Organization of sensory input to the visceral motor system. Afferent input from the cranial nerves relevant to visceral sensation (as well as afferent input ascending from second-order visceral afferents in the spinal cord) converges on the caudal division of the nucleus of the solitary tract (the rostral division is a gustatory relay; see Chapter 15).

the skin and deeper somatic structures in comparison with the number of mechanosensory neurons that do so. This relative sparseness of peripheral visceral sensory innervation accounts in part for why most visceral sensations are diffuse and difficult to localize precisely.

The spinal visceral sensory neurons in the dorsal root ganglia send axons peripherally, through sympathetic nerves, ending in sensory receptor specializations such as nerve endings that are sensitive to pressure or stretch (in the walls of the heart, bladder, and gastrointestinal tract); endings that innervate specialized chemosensitive cells (oxygen-sensitive cells in the carotid bodies); or nociceptive endings that respond to damaging stretch, ischemia, or the presence of irritating chemicals. The central axonal processes of these dorsal root ganglion neurons terminate on second-order neurons and local interneurons in the dorsal horn and in intermediate gray regions of the spinal cord. Some primary visceral sensory axons terminate near the lateral horn, where the preganglionic neurons of sympathetic and parasympathetic divisions are located; these terminals mediate visceral reflex activity in a manner not unlike that of the segmental sensorimotor reflexes described in Chapter 16.

In the dorsal horn, many of the second-order neurons that receive visceral sensory inputs are actually neurons of the anterolateral system, which also receives nociceptive or crude mechanosensory inputs from more superficial sources (see Chapter 10). As described in Box 10B, this is one means by which painful visceral sensations may be referred to more superficial somatic territories. Axons of these second-order visceral sensory neurons travel rostrally in the ventrolateral white matter of the spinal cord and the lateral sector of the brainstem and eventually reach the ventral posterior complex of the thalamus. However, the axons of other second-order visceral sensory neurons terminate before reaching the thalamus; the principal target of these axons is the nucleus of the solitary tract (Figure 21.6). Other brainstem targets of second-order visceral sensory axons are visceral motor centers in the medullary reticular formation (see Box 17C).

In the last decade, it has become clear that visceral sensory information, especially that related to painful visceral sensations, also ascends the CNS by another spinal pathway. Second-order neurons whose cell bodies are located near the central canal of the spinal cord send their axons through the dorsal columns to terminate in the dorsal column nuclei, where third-order neurons relay visceral nociceptive signals to the ventral posterior thalamus. Although the existence of this visceral pain pathway in the dorsal columns complicates the simplistic view of the dorsal column–medial lemniscal pathway as a discriminative mechanosensory projection and the anterolateral system as a pain pathway, mounting empirical and clinical evidence highlights the importance of this newly discovered dorsal column pain pathway in the central transmission of visceral nociception (see Box 10C).

In addition to these spinal visceral afferents, general visceral sensory inputs from thoracic and upper abdominal organs, as well as from viscera in the head and neck, enter the brainstem directly via the glossopharyngeal and vagus cranial nerves (see Figure 21.6). These glossopharyngeal and vagal visceral afferents terminate in the nucleus of the solitary tract. This nucleus, as described in the next section, integrates a wide range of visceral sensory information and transmits this information directly (and indirectly) to relevant visceral motor nuclei, to the brainstem reticular formation, and to several regions in the medial and ventral forebrain that coordinate visceral motor activity (see Figure 21.5).

Finally, unlike in the somatosensory system (where virtually all sensory signals gain access—albeit gated access—to conscious neural processing), sensory fibers related to the viscera convey only limited information to consciousness. For example, most of us are completely unaware of the subtle changes in peripheral vascular resistance that raise or lower our mean arterial blood pressure, yet such covert visceral afferent information is essential for the functioning of autonomic reflexes and the maintenance of homeostasis. Typically, only painful visceral sensations enter conscious awareness (see Chapter 31).

Central Control of Visceral Motor Functions

The caudal part of the nucleus of the solitary tract is a key integrative center for reflexive control of visceral motor function and an important relay for visceral sensory information that reaches other brainstem nuclei and forebrain structures (Figure 21.7; see also Figure 21.5). The rostral part of this nucleus is a gustatory relay, receiving input from primary taste afferents (cranial nerves VII, IX, and X) and sending projections to the gustatory nucleus in the ventral posterior thalamus (see Chapter 15). The caudal visceral sensory part of the nucleus of the solitary tract provides input to primary visceral motor nuclei, such as the dorsal motor nucleus of the vagus nerve and the nucleus ambiguus. It also projects to premotor autonomic centers in the medullary reticular formation and to higher integrative centers in the amygdala (specifically, the central group of amygdaloid nuclei; see Box 31B) and hypothalamus (see below). In addition, the nucleus of the solitary tract projects to the **parabrachial nucleus** (so named because it envelopes the superior cerebellar peduncle, also known by the Latin name *brachium conjunctivum*). The parabrachial nucleus, in turn, relays visceral sensory information to the hypothalamus, amygdala, thalamus, and medial prefrontal and insular cortex (see Figure 21.7; for clarity, the cortical projections of the parabrachial nucleus are omitted).

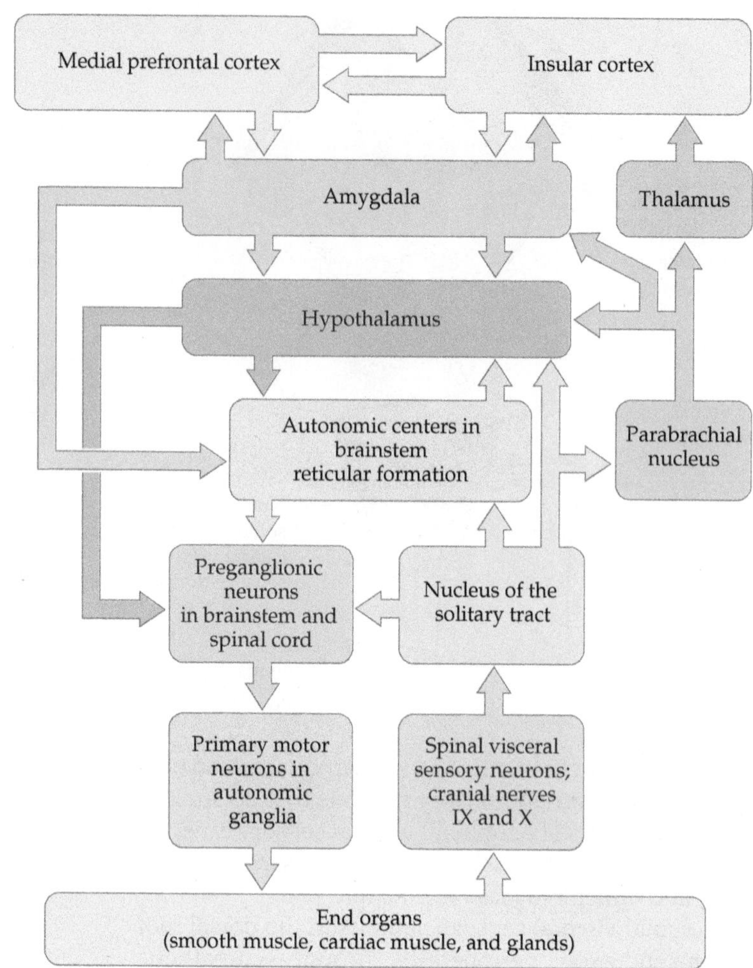

FIGURE 21.7 **A central autonomic network for the control of visceral motor function.** The distribution of visceral sensory information within this network is illustrated on the right side of the figure; the generation of visceral motor commands is shown on the left. However, extensive interconnections among autonomic centers in the forebrain (e.g., between the amygdala and associated cortical regions or hypothalamus) militate against a strict parsing of this network into afferent and efferent limbs. The hypothalamus, a key structure in this network, integrates visceral sensory input and higher-order visceral motor signals (see Box 21A).

Although one might argue that the posterior insular cortex is the primary visceral sensory area and the medial prefrontal cortex is the primary visceral motor area, it is more accurate to emphasize the interactions among these cortical areas and related subcortical structures. Taken together, they constitute a **central autonomic network**. This network accounts for the integration of visceral sensory information with input from other sensory modalities and from higher cognitive centers that process emotional experiences. Involuntary visceral reactions such as blushing in response to consciously embarrassing stimuli, vasoconstriction and pallor in response to fear, and autonomic responses to sexual situations are examples of the integrated activity of this network. Indeed, autonomic function is intimately related to emotional processing, as emphasized in Chapter 31.

The hypothalamus is a key component of this central autonomic network that deserves special consideration. The **hypothalamus** is a heterogeneous collection of nuclei in the base of the diencephalon that plays an important role in the coordination and expression of visceral motor activity (see Box 21A). The major outflow from the relevant hypothalamic nuclei is directed toward autonomic centers in the reticular formation; these centers can be thought of as dedicated premotor circuits that coordinate the efferent activity of preganglionic visceral motor neurons. They organize specific visceral functions such as cardiac reflexes, bladder control reflexes, sexual function reflexes, and critical reflexes underlying respiration and vomiting (see Box 17C).

In addition to these important projections to the reticular formation, hypothalamic control of visceral motor function is also exerted more directly by projections to the cranial nerve nuclei that contain parasympathetic preganglionic neurons, and to the sympathetic and parasympathetic preganglionic neurons in the spinal cord. Nevertheless, the autonomic centers of the reticular formation and the preganglionic visceral motor neurons that they control are competent to function autonomously should disease or injury impede the ability of the hypothalamus to govern the many bodily systems that maintain homeostasis. Figure 21.7 summarizes the general organization of this central autonomic control. Box 21B discusses the relevance of this central control to the feeding system and the problem of obesity.

BOX 21B ■ Obesity and the Brain

Obesity—and its relationship to a broad range of diseases, including diabetes, cardiovascular disease, and cancer—has become a major public health concern in most developed countries, particularly the United States. Whereas the signature of obesity is obviously an excess of body fat, the underlying cause or causes are generally thought to lie in abnormal regulation by the brain circuits that control appetite and satiety (the feeling of fullness following a meal). This fact makes weight loss particularly difficult for many obese individuals. Thus, understanding the CNS mechanisms that regulate food intake and metabolism is essential for developing effective strategies to combat this serious health problem.

The brain regulates appetite and satiety via the neural activity that is modulated by chemical signals. These chemical signals are secreted into the circulation by fat-storing adipose tissues throughout the body. This feedback loop entails some of the central components of the visceral motor system, in addition to endocrine mechanisms via insulin, growth hormone, and a growing list of factors that signal metabolic state, adiposity (amount of body fat), and nutrient balance.

The peptide **ghrelin** is secreted by the stomach prior to feeding, presumably as a signal of hunger; adipocytes (the cells that concentrate lipids in fatty tissues) increase their secretion of **leptin** into the circulation following feeding, presumably one of several signals for satiety (Figure A). The receptors for these peptides are concentrated in small groups of neurons in the ventrolateral and anterior hypothalamus (see Box 21A), which interact with additional hypothalamic neurons in the arcuate region. These ghrelin- and leptin-sensitive cells modulate the activity of neurons expressing the opiomelanocortin propeptide (POMC) and the subsequent secretion of α-melanocyte secreting hormone (α-MSH), one of the peptides encoded by the POMC transcript. This hormone evidently regulates appetite and satiety by acting on specific receptors (particularly the melanocortin receptor subtype called MCR-4) located on additional populations of hypothalamic and brainstem neurons (particularly those in the nucleus of the solitary tract), as well as by endocrine mechanisms that remain poorly understood.

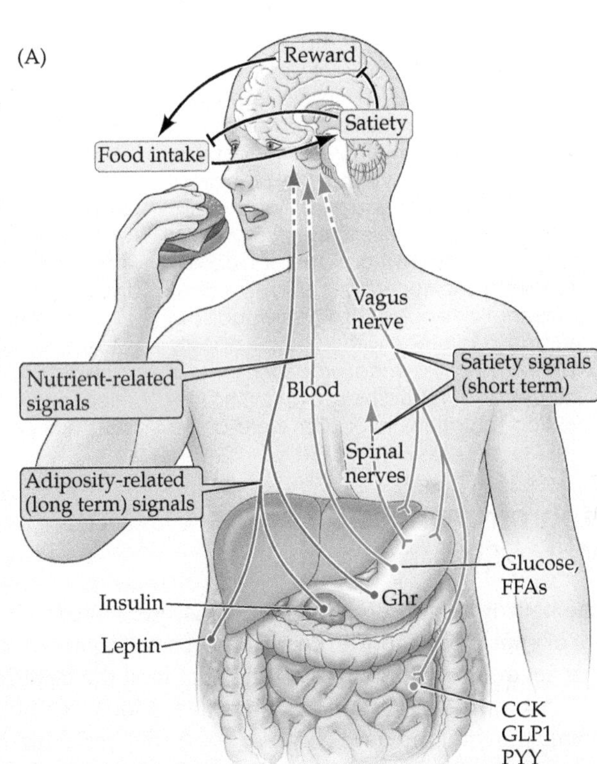

(A) Body-brain dynamics in energy homeostasis. The CNS integrates longer-term, state-dependent signals (leptin, ghrelin [Ghr], insulin) and shorter-term, feeding-dependent signals related to nutrient content (glucose, free fatty acids [FFAs]), satiety (peptide YY [PYY], glucagon-like peptide 1 [GLP1], cholecystokinin [CCK]), and visceral motor activity of the gut. Central integration of such signals regulates food intake and energy expenditure. (After Marx, 2003; Morton et al., 2014.)

The interactions of leptin, ghrelin, α-MSH, and MCR-4 were determined in animal models. Two recessive mutations in mice—obese (*ob/ob*) and diabetic (*db/db*)—were identified based on excessive body weight and failure to regulate food intake, respectively. When each mutation was cloned, the mutant gene in *ob* mice turned out to be the gene for leptin, and the *db* gene that for the leptin receptor. Mutations in the *POMC* (Figure B) and *MCR4* genes also lead to obesity in mice. The results of inactivation of the *ghrelin*

Continued on the next page

(B)

(C)

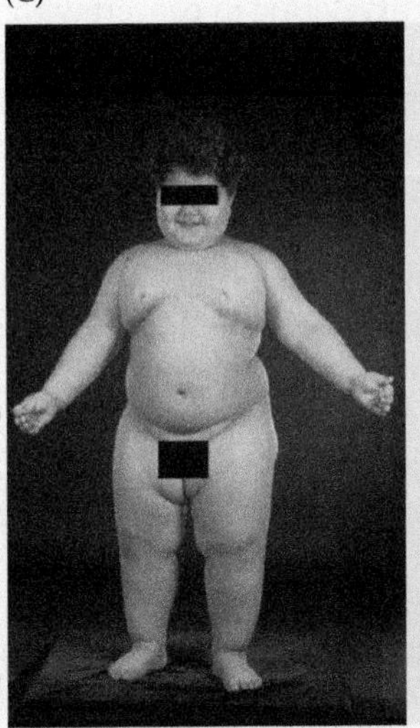

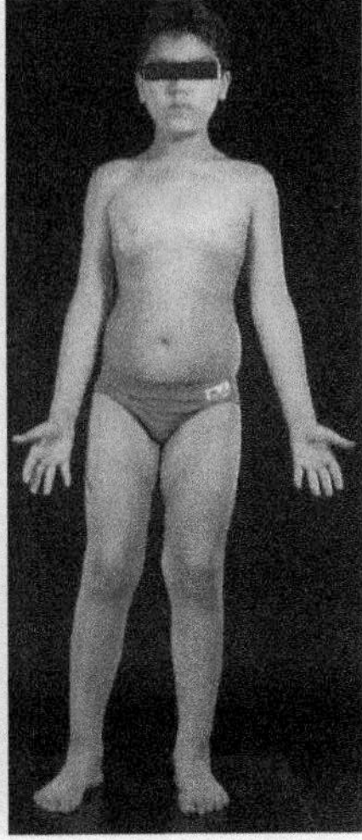

(B) A *POMC* knockout mouse (left) and a wild-type littermate (right). (C) The effect of leptin treatment in a human. At age 3 years, the child weighed 42 kg (left); at age 7 years, following treatment, the same child weighed 32 kg (right). (B from Yaswen et al., 1999; C from O'Rahilly et al., 2003.)

BOX 21B ■ (continued)

gene are less clear; however, pharmacological and physiological studies associate changes in ghrelin levels with altered feeding patterns and weight loss. Studies in mice have thus provided a solid framework for examining the physiological mechanisms regulating food intake in humans. Nonetheless, the relevance of the mice studies to morbid human obesity remained unclear until recently.

Genetic analysis of individuals in human pedigrees with extreme obesity (determined via measured body mass indices and weight/height ratios) has revealed mutations in one or more of the leptin, leptin receptor, or *MCR4* genes. As a result, these individuals have little sense of satiety after eating and thus fail to regulate food intake based on signals other than gastric distension, pain, and plasma osmolality. How this pathophysiology is related to less extreme degrees of obesity is not yet known, but it is being intensely studied because of its implications for normal weight control.

The emerging understanding of body weight regulation by hypothalamic circuits that are modulated by feedback elicited by hormonal signals from fat tissues has provided new ways of thinking about pharmacological therapies for weight control. Although leptin mimetics have proven generally ineffective, leptin administration in people with leptin deficiencies does reduce food intake and obesity (Figure C). Currently, there is great interest in drugs that modulate α-MSH signaling via MCR-4. Although no effective pharmacological therapies presently exist, clinical investigators hope that such drugs, when combined with behavioral changes in dietary practices and physical activity, will effectively combat this often intractable and increasingly common health problem.

Neurotransmission in the Visceral Motor System

The neurotransmitters used by the visceral motor system are of enormous importance in clinical practice, and drugs that act on the autonomic system are among the most important in the clinical armamentarium. Moreover, autonomic transmitters have played a major role in the history of efforts to understand synaptic function.

Acetylcholine is the primary neurotransmitter of both sympathetic and parasympathetic preganglionic neurons. Nicotinic receptors on autonomic ganglion cells are ligand-gated ion channels that mediate a fast EPSP (much like nicotinic receptors at the neuromuscular junction). In contrast, muscarinic acetylcholine receptors on ganglion cells are members of the 7-transmembrane G-protein-coupled receptor family (see Chapters 6 and 7), and they mediate slower synaptic responses. The primary action of muscarinic receptors in autonomic ganglion cells is to close K^+ channels, making the neurons more excitable and generating a prolonged EPSP. Acting in concert with the muscarinic activities are neuropeptides that serve as co-neurotransmitters at the ganglionic synapses. As described in Chapter 6, peptide neurotransmitters also tend to exert slowly developing but long-lasting effects on postsynaptic neurons. As a result of these two acetylcholine receptor types and a rich repertoire of neuropeptide transmitters, ganglionic synapses mediate both rapid excitation and a slower modulation of autonomic ganglion cell activity.

The postganglionic effects of autonomic ganglion cells on their smooth muscle, cardiac muscle, or glandular targets are mediated by two primary neurotransmitters: norepinephrine (NE) and acetylcholine (ACh). For the most part, sympathetic ganglion cells release norepinephrine onto their targets (a notable exception is the cholinergic sympathetic innervation of sweat glands), whereas parasympathetic ganglion cells typically release acetylcholine. As might be expected from the foregoing account, these two neurotransmitters usually have opposing effects on their target tissue—contraction versus relaxation of smooth muscle, for example.

As described in Chapters 6 and 7, the specific effects of ACh and NE are determined by the type of receptor expressed in the target tissue and the downstream signaling pathways to which these receptors are linked. Peripheral sympathetic targets generally have two subclasses of noradrenergic receptors in their cell membranes, referred to as α and β receptors. Like muscarinic ACh receptors, both α and β receptors and their subtypes belong to the 7-transmembrane G-protein-coupled class of cell surface receptors. The different distribution of these receptors in sympathetic targets allows for a variety of postsynaptic effects mediated by norepinephrine released from postganglionic sympathetic nerve endings (Table 21.2).

The effects of acetylcholine released by parasympathetic ganglion cells onto smooth muscles, cardiac muscle, and glandular cells also vary according to the subtypes of muscarinic cholinergic receptors found in the peripheral target (Table 21.3). The two major subtypes are known as M_1 and M_2 receptors, M_1 receptors being found primarily in the gut and M_2 receptors in the cardiovascular system. Another subclass of muscarinic receptor, M_3, occurs in both smooth muscle and glandular tissues. Muscarinic receptors are coupled to a variety of intracellular signal transduction mechanisms that modify K^+ and Ca^{2+} channel conductances. They can also activate nitric oxide synthase, which promotes the local release of nitric oxide in some parasympathetic target tissues (see, for example, the section later in this chapter that discusses autonomic control of sexual function).

In contrast to the relatively restricted responses generated by norepinephrine and acetylcholine released by sympathetic and parasympathetic ganglion cells, respectively, neurons of the enteric nervous system achieve an enormous diversity of effects by virtue of many different

TABLE 21.2 ■ Adrenergic Receptor Types and Some of Their Effects in Sympathetic Targets

Receptor	G-protein	Tissue	Response
α_1	G_q	Smooth muscle of blood vessels, iris, ureter, urethra, hairs, uterus	Contraction of smooth muscle
		Heart muscle	Positive inotropic effect ($\beta_1 >> \alpha_1$)
		Salivary gland	Secretion
		Adipose tissue	Glycogenolysis, gluconeogenesis
		Sweat glands	Secretion
		Kidney	Na^+ reabsorbed
α_2	G_i	Adipose tissue	Inhibition of lipolysis
		Pancreas	Inhibition of insulin release
		Smooth muscle of blood vessels	Contraction
β_1	G_s	Heart muscle	Positive inotropic effect; positive chronotropic effect
		Adipose tissue	Lipolysis
		Kidney	Renin release
β_2	G_s	Liver	Glycogenolysis, gluconeogenesis
		Skeletal muscle	Glycogenolysis, lactate release
		Smooth muscle of bronchi, uterus, gut, blood vessels	Relaxation
		Pancreas	Stimulates insulin secretion
		Salivary glands	Thickened secretions
β_3	G_s	Adipose tissue	Lipolysis
		Smooth muscle of gut	Modulation of intestinal mobility
		Smooth muscle of bladder	Bladder filling

neurotransmitters, most of which are neuropeptides associated with specific cell groups in either the myenteric or submucous plexuses mentioned earlier. The details of these agents and their actions are beyond the scope of this introductory account.

Many examples of specific autonomic functions could be used to illustrate in more detail how the visceral motor system operates. The three outlined here—control of cardiovascular function, control of the bladder, and control of sexual function—have been chosen primarily because of their importance in human physiology and clinical practice.

Autonomic Regulation of Cardiovascular Function

The cardiovascular system is subject to precise reflex regulation so that an appropriate supply of oxygenated blood can reliably be provided to different body tissues under a wide range of circumstances. The sensory monitoring for this critical homeostatic process entails primarily mechanical (*barosensory*) information about pressure in the arterial system and, secondarily, chemical (*chemosensory*) information about the levels of oxygen and carbon dioxide

TABLE 21.3 ■ Cholinergic Receptor Types and Some of Their Effects in Parasympathetic Targets

Receptor	G-protein	Tissue	Response
Nicotinic	None (ionotropic receptor)	Most parasympathetic targets (and all autonomic ganglion cells)	Relatively fast postsynaptic response
Muscarinic (M_1)	G_q	Smooth muscles and glands of the gut	Smooth muscle contraction and glandular secretion (relatively slow response)
Muscarinic (M_2)	G_i	Smooth and cardiac muscle of cardiovascular system	Reduction in heart rate, smooth muscle contraction
Muscarinic (M_3)	G_q	Smooth muscles and glands of all targets	Smooth muscle contraction, glandular secretion

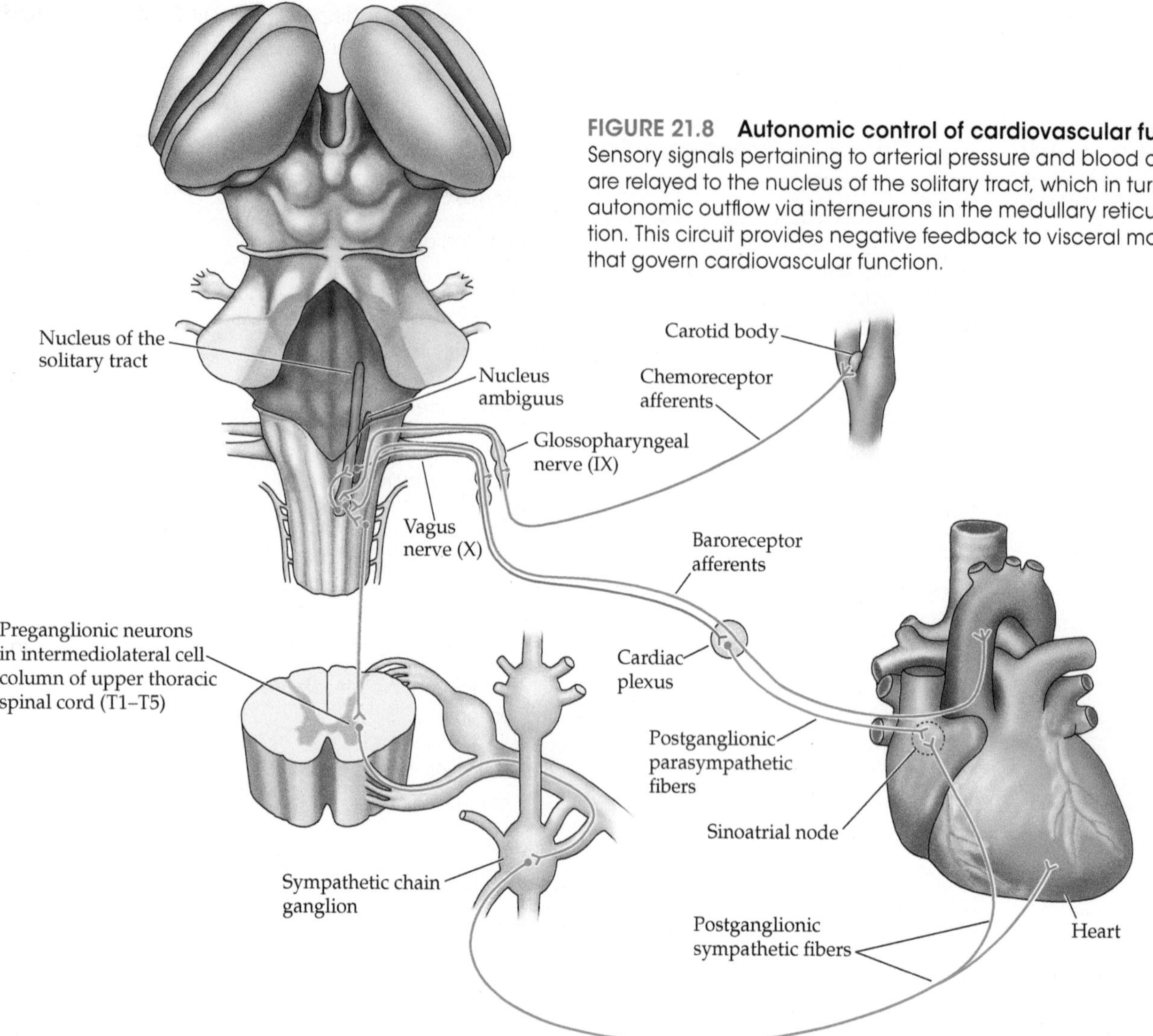

FIGURE 21.8 Autonomic control of cardiovascular function. Sensory signals pertaining to arterial pressure and blood oxygenation are relayed to the nucleus of the solitary tract, which in turn organizes autonomic outflow via interneurons in the medullary reticular formation. This circuit provides negative feedback to visceral motor neurons that govern cardiovascular function.

in the blood. The parasympathetic and sympathetic activities relevant to cardiovascular control are determined by the information supplied by these sensors.

The mechanoreceptors, called baroreceptors, are located in the heart and major blood vessels; the chemoreceptors are located primarily in the carotid bodies, which are small, highly specialized organs located at the bifurcation of the common carotid arteries (some chemosensory tissue is also found in the aorta). The nerve endings in baroreceptors are activated by deformation as the elastic elements of the vessel walls expand and contract. The chemoreceptors in the carotid bodies and aorta respond directly to the partial pressure of oxygen and carbon dioxide in the blood. Visceral afferents from the aortic arch and carotid bifurcation reach the brainstem via the vagus nerve and glossopharyngeal nerve, respectively. Both afferent systems convey their signals to the nucleus of the solitary tract, which relays this information to the hypothalamus and the relevant autonomic centers in the reticular formation (Figure 21.8).

The afferent information derived from changes in arterial pressure and blood gas levels reflexively modulates the activity of the relevant visceral motor pathways and, ultimately, smooth and cardiac muscles and other more specialized structures. For example, a rise in blood pressure activates baroreceptors that, via the pathway illustrated in Figure 21.8, inhibit the tonic activity of sympathetic preganglionic neurons in the spinal cord. In parallel, the pressure increase stimulates the activity of the parasympathetic preganglionic neurons in the nucleus ambiguus and the dorsal motor nucleus of the vagus that influence heart rate and contractility. The carotid chemoreceptors also have some influence, but less than that stemming from the baroreceptors.

This shift in the balance of sympathetic and parasympathetic activity results in reduction of the stimulatory noradrenergic effects of postganglionic sympathetic innervation on the cardiac pacemaker and cardiac musculature. These effects are abetted by the decreased output of catecholamines from the adrenal medulla and the decreased vasoconstrictive effects of sympathetic innervation on the peripheral blood vessels. At the same time, activation of the cholinergic parasympathetic innervation of the heart decreases the discharge rate of the cardiac pacemaker in the sinoatrial node and slows the ventricular conduction system. These parasympathetic influences are mediated by an extensive series of parasympathetic ganglia in and near the heart, which release acetylcholine onto cardiac pacemaker cells and cardiac muscle fibers. As a result of this combination of sympathetic and parasympathetic effects, heart rate and the effectiveness of atrial and ventricular myocardial contraction are reduced and the peripheral arterioles dilate, thus lowering the blood pressure.

In contrast to this sequence of events in response to raised blood pressure, a fall in blood pressure (as might occur from blood loss) has the opposite effect—it inhibits parasympathetic activity while increasing sympathetic activity. As a result, norepinephrine is released from sympathetic postganglionic terminals, increasing the rate of cardiac pacemaker activity and enhancing cardiac contractility, at the same time increasing release of catecholamines from the adrenal medulla (which further augments these and many other sympathetic effects that enhance the response to this threatening situation). Norepinephrine released from the terminals of sympathetic ganglion cells also acts on the smooth muscles of the arterioles to increase the tone of the peripheral vessels, particularly those in the skin, subcutaneous tissues, and muscles, thus shunting blood away from these tissues to those organs where oxygen and metabolites are urgently needed to maintain function (e.g., brain, heart, and kidneys in the case of blood loss). If these reflex sympathetic responses fail to raise the blood pressure sufficiently (in which case the patient is said to be in shock), the vital functions of these organs begin to fail, often catastrophically.

A more mundane circumstance that requires a reflex autonomic response to a fall in blood pressure is standing up. Rising quickly from a prone position produces a shift of some 300 to 800 mL of blood from the thorax and abdomen to the legs, resulting in a sharp (approximately 40%) decrease in the output of the heart. The adjustment to this normally occurring drop in blood pressure (called **orthostatic hypotension**) must be rapid and effective, as evidenced by the dizziness sometimes experienced in this situation. Indeed, anyone can briefly lose consciousness as a result of blood pooling in the lower extremities, which is the usual cause of fainting when standing still for exceptionally long periods.

The sympathetic innervation of the heart arises from the preganglionic neurons in the intermediolateral cell column of the spinal cord, extending from roughly the first through fifth thoracic segments (T1–T5; see Table 21.1). The primary visceral motor neurons are in the adjacent thoracic paravertebral and prevertebral ganglia of the cardiac plexus. The parasympathetic preganglionics, as already mentioned, are in the nucleus ambiguus and the dorsal motor nucleus of the vagus nerve and project to parasympathetic ganglia in and around the heart and great vessels.

Autonomic Regulation of the Bladder

The autonomic regulation of the bladder provides an especially instructive example of the interplay between components of the somatic motor system that are subject to volitional control (we usually have voluntary control over urination), and the sympathetic and parasympathetic divisions of the visceral motor system, which operate involuntarily. This should not be surprising given that, for many mammals, the act of urination (and of defecation) places the individual at increased risk for attack, since the capacity for immediate fight or flight is reduced. In addition, for many mammals, urine contains chemical signals (pheromones) that mediate complex social behaviors. The neural control of bladder function therefore involves the coordination of relevant autonomic, somatic motor, and cognitive faculties that inhibit or promote urination.

The arrangement of afferent and efferent innervation of the bladder is shown in Figure 21.9A. The sympathetic innervation of the bladder originates in the lower thoracic and upper lumbar spinal cord segments (T10–L2), with preganglionic axons running to primary sympathetic neurons in the inferior mesenteric ganglion and the ganglia of the pelvic plexus. The postganglionic fibers from these ganglia travel in the hypogastric and pelvic nerves to the bladder, where sympathetic activity is believed to cause the smooth muscle of the bladder wall to relax and the internal urethral sphincter to close (postganglionic sympathetic fibers also innervate the blood vessels of the bladder). Stimulation of this sympathetic pathway in response to a modest increase in bladder pressure from the accumulation of urine thus allows the bladder to fill and prevents leakage of urine. At the same time, moderate distension of the bladder inhibits its sacral outflow, which causes contraction of the bladder musculature and bladder emptying. This is promoted by preganglionic neurons in the sacral spinal cord segments (S2–S4) that innervate visceral motor neurons in ganglia in or near the bladder wall. (Conventionally, this is considered parasympathetic innervation of the bladder. However, as noted above, recent molecular and genetic studies are suggesting that this sacral outflow should be reclassified as sympathetic.)

The afferent limb of this reflexive circuit is supplied by mechanoreceptors in the bladder wall that convey visceral afferent information to second-order neurons in the dorsal horn of the spinal cord. In addition to local connections within spinal cord circuitry, these neurons project to higher integrative centers in the periaqueductal gray of the midbrain. This midbrain region (which is also involved in the descending control of nociception; see Chapter 10) receives input from the hypothalamus, amygdala, and orbital-medial prefrontal cortex. These forebrain structures participate in limbic networks that evaluate risk and the emotional significance of contextual cues (see Chapter 31);

in the context of bladder filling, they signal when it is safe and socially appropriate to urinate.

When the bladder is full, sacral visceral motor outflow increases and thoraco-lumbar motor outflow decreases, causing the bladder to contract and the internal sphincter muscle to relax. However, urine is held in check by the voluntary somatic motor innervation of the external urethral sphincter muscle. The voluntary control of the external sphincter is mediated by motor neurons of the ventral horn in sacral spinal cord segments (S2–S4), which cause the striated muscle fibers of the sphincter to contract. During bladder filling (and subsequently, until circumstances permit urination) these neurons are active, keeping the external sphincter closed and preventing voiding. During urination, this tonic activity is temporarily inhibited, leading to relaxation in the external sphincter muscle. Normally, this is only possible when integrative signals derived from the periaqueductal gray activate a collection of premotor neurons in the dorsal pontine reticular formation, known as the "pontine micturition center" (or Barrington's nucleus). The pontine micturition center (*micturition* is clinical term for urination) projects to preganglionic neurons and inhibitory local circuit neurons in the sacral spinal cord; the net result is increased visceral motor outflow from the sacral cord (leading to stronger contraction of the bladder wall) and inhibition of the somatic lower motor neurons that innervate the external sphincter muscle (allowing for voiding; Figure 21.9B). Thus, urination results from the coordinated activation of sacral visceral motor neurons and temporary inactivation of motor neurons of the somatic motor system; this coordination is ultimately governed by the integration of visceral sensory, emotional, social, and contextual cues.

Importantly, individuals who are paraplegic or otherwise have lost descending control of the sacral spinal cord continue to exhibit reflexive, autonomic regulation of bladder function. Unfortunately, this reflex is not fully efficient in the absence of descending motor control, resulting in a variety of problems in paraplegics and others with diminished or defective central control of bladder function. The

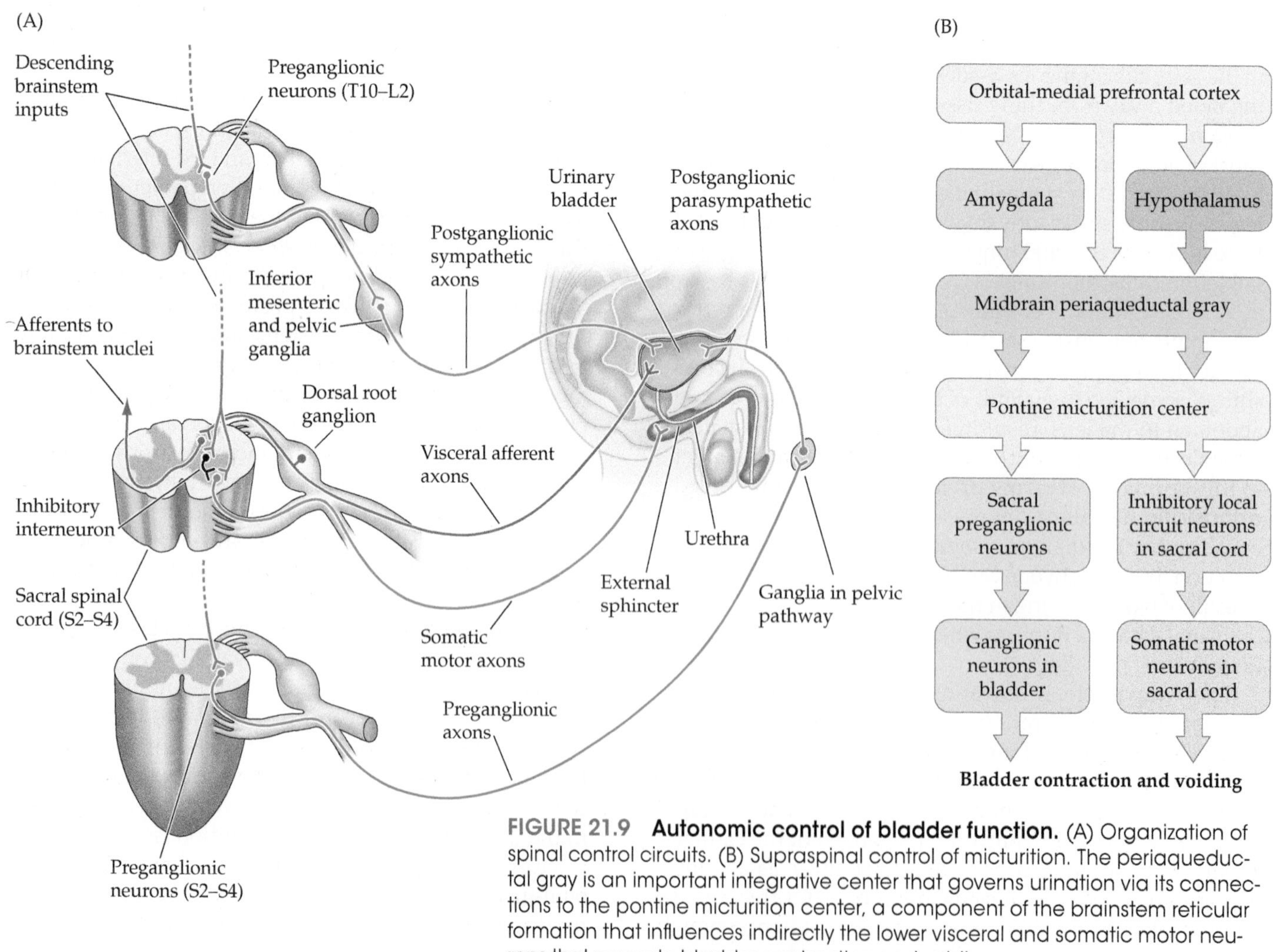

FIGURE 21.9 Autonomic control of bladder function. (A) Organization of spinal control circuits. (B) Supraspinal control of micturition. The periaqueductal gray is an important integrative center that governs urination via its connections to the pontine micturition center, a component of the brainstem reticular formation that influences indirectly the lower visceral and somatic motor neurons that promote bladder contraction and voiding.

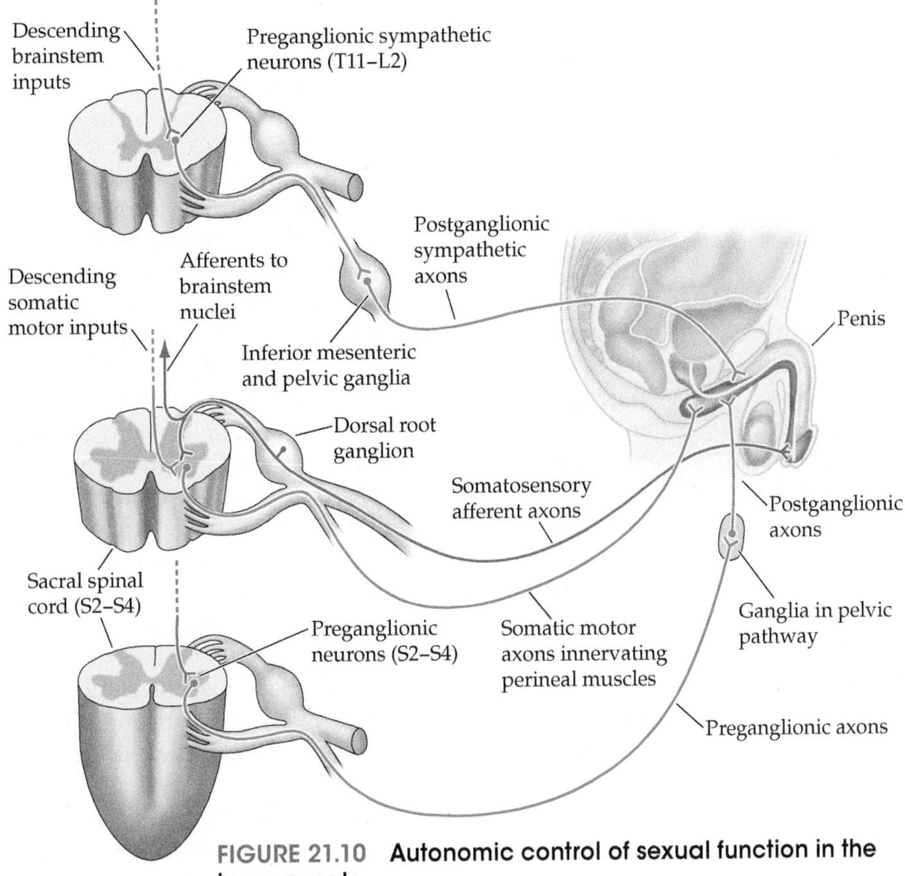

FIGURE 21.10 Autonomic control of sexual function in the human male.

major difficulty in these cases is incomplete bladder emptying, which often leads to chronic urinary tract infections from the culture medium provided by retained urine, and thus the need for an indwelling catheter to ensure adequate drainage. Indeed, urinary tract morbidity is recognized as the second leading cause of death in patients with spinal cord injury. In other individuals with urge incontinence and overactive bladder disorders, leakage of urine and an "absence of warning" are the problems. Mounting evidence obtained from functional and structural studies of the brains of such individuals implicates structural lesions or dysfunction that impair the integrative activity of the periaqueductal gray and its control over the pontine micturition center.

Autonomic Regulation of Sexual Function

Much like control of the bladder, sexual responses are mediated by the coordinated activity of visceral motor and somatic motor innervation, both of which are governed by complex cognitive, emotional, and contextual cues processed in the limbic forebrain. Although these reflexes differ in detail in males and females, basic similarities, not only in humans but in mammals generally, allow the autonomic sexual responses of the two sexes to be considered together. These similarities include: (1) the mediation of vascular dilation, which causes penile or clitoral erection; (2) stimulation of prostatic or vaginal secretions; (3) smooth muscle contraction of the vas deferens during ejaculation in males or rhythmic vaginal contractions during orgasm in females; and (4) contractions of the somatic pelvic muscles that accompany orgasm in both sexes.

Like the urinary tract, the reproductive organs receive preganglionic visceral motor innervation from the sacral spinal cord and from the lower thoracic and upper lumbar spinal cord segments, and somatic motor innervation from α motor neurons in the ventral horn of the lower spinal cord segments (Figure 21.10). The sacral visceral motor pathway controlling the sexual organs in both males and females originates in the sacral segments S2–S4 and reaches the target organs via the pelvic nerves. Activity of the postganglionic neurons in the relevant ganglia causes dilation of penile or clitoral arteries, and a corresponding relaxation of the smooth muscles of the venous (cavernous) sinusoids, which leads to expansion of the sinusoidal spaces. As a result, the amount of blood in the tissue is increased, leading to a sharp rise in the pressure and an expansion of the cavernous spaces (i.e., erection). The mediator of the smooth muscle relaxation leading to erection is not acetylcholine (as in most postganglionic parasympathetic actions), but nitric oxide (NO; see Chapter 6). The drug sildenafil (Viagra), for instance, acts by inhibiting PDE-5, the predominant phosphodiesterase (PDE) expressed in erectile tissue, which leads to an increase in the intracellular concentration of cyclic GMP. This second messenger mediates the activity of endogenous NO; thus, PDE-5 inhibitors enhance the relaxation of the venous sinusoids and promote erection in males during sexual stimulation (when NO is released in erectile tissue). Sacral visceral motor outflow also provides excitatory input to the vas deferens, seminal vesicles, and prostate in males, or vaginal glands in females.

In contrast, visceral motor outflow from the thoraco-lumbar spinal cord causes vasoconstriction and loss of erection. This pathway to the sexual organs originates in spinal segments T1–L2 and reaches the target organs via the corresponding sympathetic chain ganglia and the inferior mesenteric and pelvic ganglia.

The afferent effects of genital stimulation are conveyed centrally from somatosensory endings via the dorsal roots of S2–S4, eventually reaching the somatosensory cortex (reflex sexual excitation may also occur by local stimulation, as is evident in individuals with spinal cord injuries sparing the sacral cord). The reflex effects of such stimulation are increased visceral motor outflow from the sacral cord, which, as noted, causes relaxation of the smooth muscles in the wall of the sinusoids and subsequent erection.

Finally, the somatic motor component of reflex sexual function arises from α motor neurons in the lumbar and sacral spinal cord segments. These neurons provide excitatory innervation to the bulbocavernosus and ischiocavernosus muscles, which are active during ejaculation in males and mediate the contractions of the perineal (pelvic floor) muscles that accompany orgasm in both males and females.

Sexual functions are governed centrally by the anteromedial and mediotuberal zones of the hypothalamus, which contain a variety of nuclei pertinent to visceral motor control and reproductive behavior (see Box 21A). Although they remain poorly understood, these nuclei appear to act as integrative centers for sexual responses and are also thought to be involved in more complex aspects of sexuality, such as sexual preference and gender identity (see Chapter 24). The relevant hypothalamic nuclei receive inputs from several areas of the brain, including—as one might imagine—the cortical and subcortical structures concerned with emotion, hedonic reward, and memory (see Chapters 30 and 31).

Summary

Sympathetic and parasympathetic ganglia, which contain the visceral lower motor neurons that innervate smooth muscles, cardiac muscle, and glands, are controlled by preganglionic neurons in the spinal cord and brainstem. The sympathetic preganglionic neurons that govern ganglion cells in the sympathetic division of the visceral motor system arise from neurons in the thoracic and upper lumbar segments of the spinal cord; parasympathetic preganglionic neurons, in contrast, are located in the brainstem. Preganglionic neurons in the sacral spinal cord have long been considered parasympathetic, but recent evidence shows that they share many features with preganglionic neurons in the thoraco-lumbar cord. Sympathetic ganglion cells are distributed in the sympathetic chain (paravertebral) and prevertebral ganglia, whereas the parasympathetic motor neurons are more widely distributed in ganglia that lie in or near the organs they control. Most visceral structures receive inputs from both the sympathetic and parasympathetic systems, which act in a generally antagonistic fashion. The diversity of autonomic functions is achieved primarily by different types of receptors for the two primary classes of postganglionic autonomic neurotransmitters, norepinephrine in the case of the sympathetic division and acetylcholine in the parasympathetic division. The visceral motor system is regulated by sensory feedback provided by dorsal root and cranial nerve sensory ganglion cells that make local reflex connections in the spinal cord or brainstem and project to the nucleus of the solitary tract in the brainstem. The visceral motor system is also regulated by descending pathways from the hypothalamus and brainstem reticular formation, the major control centers of homeostasis more generally. The importance of the visceral motor control of organs such as the heart, bladder, and reproductive organs—and the many pharmacological means of modulating autonomic function—have made visceral motor control a central theme in clinical medicine.

ADDITIONAL READING

Reviews

Andersson, K.-E. and G. Wagner (1995) Physiology of penile erections. *Physiol. Rev.* 75: 191–236.

Brown, D. A. and 8 others (1997) Muscarinic mechanisms in nerve cells. *Life Sciences* 60 (13–14): 1137–1144.

Costa, M. and S. J. H. Brookes (1994) The enteric nervous system. *Am. J. Gastroenterol.* 89: S129–S137.

Craig, A. D. (2009) How do you feel—now? The anterior insula and human awareness. *Nat. Rev. Neurosci.* 10: 59–70.

Dampney, R. A. L. (1994) Functional organization of central pathways regulating the cardiovascular system. *Physiol. Rev.* 74: 323–364.

Fowler, C. J., D. Griffiths and W. C. de Groat (2008) The neural control of micturition. *Nat. Rev. Neurosci.* 9: 453–466.

Gershon, M. D. (1981) The enteric nervous system. *Annu. Rev. Neurosci.* 4: 227–272.

Holstege, G. (2005) Micturition and the soul. *J. Comp. Neurol.* 493: 15–21.

Mundy, A. R. (1999) Structure and function of the lower urinary tract. In *Scientific Basis of Urology*, A. R. Mundy, J. M. Fitzpatrick, D. E. Neal and N. J. R. George (eds.). Oxford, UK: Isis Medical Media Ltd., pp. 217–242.

Patton, H. D. (1989) The autonomic nervous system. In *Textbook of Physiology: Excitable Cells and Neurophysiology*, vol. 1, section VII: *Emotive Responses and Internal Milieu*, H. D. Patton, A. F. Fuchs, B. Hille, A. M. Scher and R. Steiner (eds.). Philadelphia, PA: Saunders, pp. 737–758.

Pryor, J. P. (1999) Male sexual function. In *Scientific Basis of Urology*, A. R. Mundy, J. M. Fitzpatrick, D. E. Neal and N. J. R. George (eds). Oxford, UK: Isis Medical Media, pp. 243–255.

Important Original Papers

Espinosa-Medina, I. and 6 others (2016) The sacral autonomic outflow is sympathetic. *Science* 354: 893–897.

Jansen, A. S. P., X. V. Nguyen, V. Karpitskiy, T. C. Mettenleiter and A. D. Loewy (1995) Central command neurons of the sympathetic nervous system: Basis of the fight or flight response. *Science* 270: 644–646.

Langley, J. N. (1894) The arrangement of the sympathetic nervous system chiefly on observations upon pilo-erector nerves. *J. Physiol.* (*Lond.*) 15: 176–244.

Langley, J. N. (1905) On the reaction of nerve cells and nerve endings to certain poisons chiefly as regards the reaction of striated muscle to nicotine and to curare. *J. Physiol.* (*Lond.*) 33: 374–473.

Lichtman, J. W., D. Purves and J. W. Yip (1980) Innervation of sympathetic neurones in the guinea-pig thoracic chain. *J. Physiol.* 298: 285–299.

Rubin, E. and D. Purves (1980) Segmental organization of sympathetic preganglionic neurons in the mammalian spinal cord. *J. Comp. Neurol.* 192: 163–174.

Books

Appenzeller, O. (1997) *The Autonomic Nervous System: An Introduction to Basic and Clinical Concepts*, 5th Edition. Amsterdam: Elsevier Biomedical Press.

Blessing, W. W. (1997) *The Lower Brainstem and Bodily Homeostasis.* New York: Oxford University Press.

Brading, A. (1999) *The Autonomic Nervous System and Its Effectors.* Oxford: Blackwell Science.

Burnstock, G. and C. H. V. Hoyle (1995) *The Autonomic Nervous System,* vol. 1: *Autonomic Neuroeffector Mechanism.* London: Harwood Academic.

Cannon, W. B. (1932) *The Wisdom of the Body.* New York: Norton.

Furness, J. B. and M. Costa (1987) *The Enteric Nervous System.* Edinburgh: Churchill Livingstone.

Gabella, G. (1976) *Structure of the Autonomic Nervous System.* London: Chapman and Hall.

Jänig, W. (2006) *The Integrative Action of the Autonomic Nervous System: Neurobiology of Homeostatis.* Cambridge, UK: Cambridge University Press.

Langley, J. N. (1921) *The Autonomic Nervous System.* Cambridge, UK: Heffer & Sons.

Loewy, A. D. and K. M. Spyer (eds.) (1990) *Central Regulation of Autonomic Functions.* New York: Oxford University Press.

Pick, J. (1970) *The Autonomic Nervous System: Morphological, Comparative, Clinical and Surgical Aspects.* Philadelphia, PA: J. B. Lippincott Co.

Randall, W. C. (ed.) (1984) *Nervous Control of Cardiovascular Function.* New York: Oxford University Press.

Go to the NEUROSCIENCE 6e Companion Website at **www.oup.com/uk/Purves6e** for Web Topics, Animations, Flashcards, and more.

UNIT IV
The Changing Brain

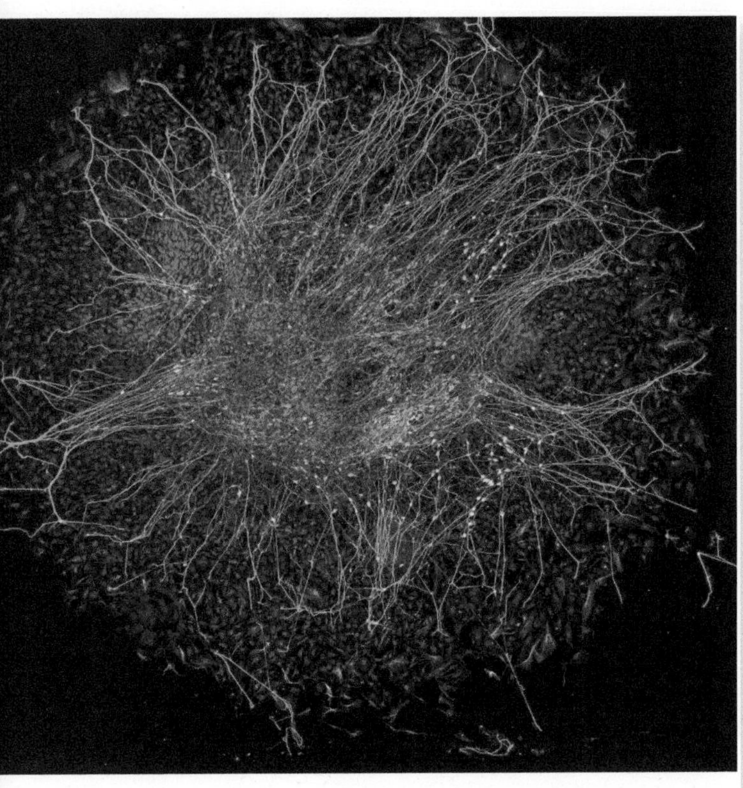

THE STRUCTURE AND FUNCTIONAL CAPACITY of the brain change dramatically over the human life span. As soon as the nervous system is established, the coordinated gene expression, neuronal genesis, axonal and dendritic growth, and formation of connections yields a brain whose shape, size, and cellular architecture are continually transformed. These events rely on the transcriptional regulators, secreted signals and their receptors, and adhesion and recognition molecules that determine the appropriate neuronal identities, positions, and connections. The neural circuits that emerge eventually mediate a remarkably complex array of behaviors. Nevertheless, they must be refined by subsequent experience during postnatal life to be maximally efficient for each individual. This refinement occurs via activity-dependent mechanisms that translate experience into altered synaptic efficacy, gene expression, neuronal growth or, in some cases, elimination of neuronal processes or synapses. Such changes are most pronounced during "windows" in early life, called critical periods. Finally, like any other organ, the brain is subject to traumatic insults and disease. Some injuries activate repair mechanisms that resemble those used during development, but the capacity of the mature brain for repair and regeneration is limited. Some new neurons can be generated, new axons and dendrites extended, and new connections made. These new elements, however, rarely completely restore brain structure or function. Diseases such as amyotrophic lateral sclerosis, Parkinson's, and Alzheimer's disease all reflect pathologies of processes that typically contribute to neuronal development and subsequent maintenance of neural circuitry. The degeneration of brain circuits in these diseases ultimately accounts for the behavioral deficits seen in individuals with these disorders.

On the previous page:

A trigeminal ganglion from an E10.5 mouse embryo grown in cell culture for 24 hours. The green is neuron-selective label (β-tubulin III), the red is a transcription factor marker (Six1), and the blue is a nuclear DNA label (DAPI). (Courtesy of A.-S. LaMantia.)

CHAPTER

22 Early Brain Development

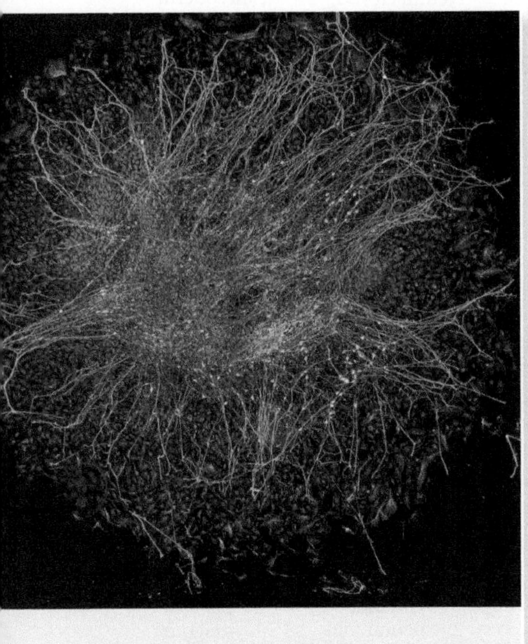

Overview

THE ELABORATE ARCHITECTURE OF THE ADULT BRAIN is the product of cell-to-cell signals, genetic instructions, and their consequences for the stem cells in the embryo that generate the entire nervous system. These early events include the establishment of the primordial nervous system and its constituent cells in the embryo, the initial formation of the major brain regions, the generation of neurons and glial cells from undifferentiated neural stem or precursor cells, and the migration of neurons from sites of generation to their final positions. These processes set the stage for the subsequent differentiation of local dendrites, axons, and synapses, as well as long distance axon pathways and synaptic connections. When any of these processes goes awry—because of genetic mutation, disease, or exposure to drugs or other chemicals—the consequences can be disastrous. Indeed, many well-studied congenital brain defects result from interference with the typical mechanisms of early nervous system development, prior to the formation of synaptic connections. With the aid of cell biological, molecular, and genetic tools, investigators are beginning to understand the machinery underlying these extraordinarily complex events.

Formation of the Nervous System: Gastrulation and Neurulation

The cells that will generate the nervous system become distinct early in the generation of a vertebrate embryo, concurrent with the establishment of the midline and the basic body axes: anterior–posterior (mouth–anus), dorsal–ventral (back–belly), and medial–lateral (midline–periphery). These axes are foundational for proper generation of every organ in the body, including the brain. In addition, the unique curvature of the human CNS generates a distinctive rostral–caudal (Latin, "nose–tail") axis in the developing brain (see Figure A1 in the Appendix). The axes, and thus the initiation of neural development, are critically dependent on the process of **gastrulation**. Gastrulation begins as the local invagination of a subset of cells in the very early embryo (which starts out as a single sheet of cells). By the time invagination is complete, the embryo consists of three layers of cells called the **germ layers**: an outer **ectoderm**; a middle **mesoderm** (these cells initiate the invagination that defines gastrulation); and an inner **endoderm** (Figure 22.1A). Based on the position of the invaginating mesoderm and endoderm, gastrulation defines the midline as well as the anterior–posterior and dorsal–ventral axes of all vertebrate embryos. These axes then determine the position of all organ systems including the peripheral and CNS, as well as facial structures and appendages.

The formation of the **notochord** at the midline of the gastrulating embryo is a central event for the development of the nervous system. The notochord is a distinct cylinder of mesodermal cells that condenses at the midline as the mesoderm invaginates and extends from the mid-anterior to the posterior aspect of the embryo. It is generated at the site of a singular surface indentation called the **primitive pit**, which subsequently elongates to form the **primitive streak**. As a result of these cell movements, the notochord defines the embryonic midline, and thus the axis of symmetry for the entire body. The ectoderm that lies immediately above the notochord, called **neuroectoderm**, gives rise to the entire nervous system (see Figure 22.1A). The notochord itself, however, is a transient structure that disappears once early development is complete. The notochord specifies the basic topography of the embryo by defining the midline and axis of symmetry, determines the position of the nervous system, and is required for subsequent early neural differentiation. Along with cells that define the primitive pit, the notochord sends inductive signals (see below) to the overlying ectoderm that cause a subset of cells to differentiate into **neuroectodermal precursor cells**. During this process, called **neurulation**, the midline ectoderm that contains these cells thickens into a distinct columnar epithelium called the **neural plate** (Figure 22.1B). The lateral margins of the neural plate, referred to as the alar plate, then fold inward, transforming the neural plate into a tube called the **neural tube** (Figure 22.1C,D). The portion of the neural plate at the midline, referred to as the basal plate, becomes the ventral region of the neural tube. The neural tube is not a uniform structure. The cells at the ventral midline of the neural tube differentiate into a specialized strip of epithelial-like cells called the **floorplate** (reflecting their position at the ventral and medial part of the neural tube, above the notochord; see Figure 22.1). Molecular signals from the floorplate as well as from the notochord specify position and fate for the neuroectodermal precursors of the spinal cord and hindbrain. In the forebrain, non-floorplate ventral midline structures as well as neural crest-derived mesenchyme immediately adjacent to the prosencephalic vesicle provide similar signals. The multipotent **neural stem cells** within the neural tube subsequently give rise to the entire brain

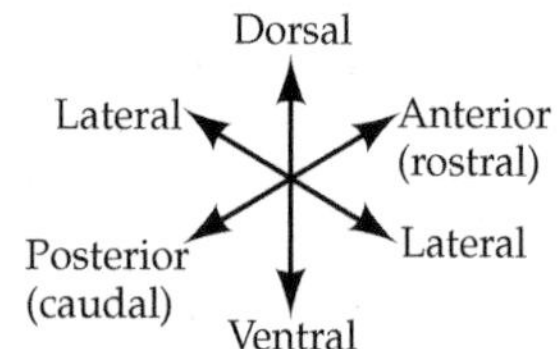

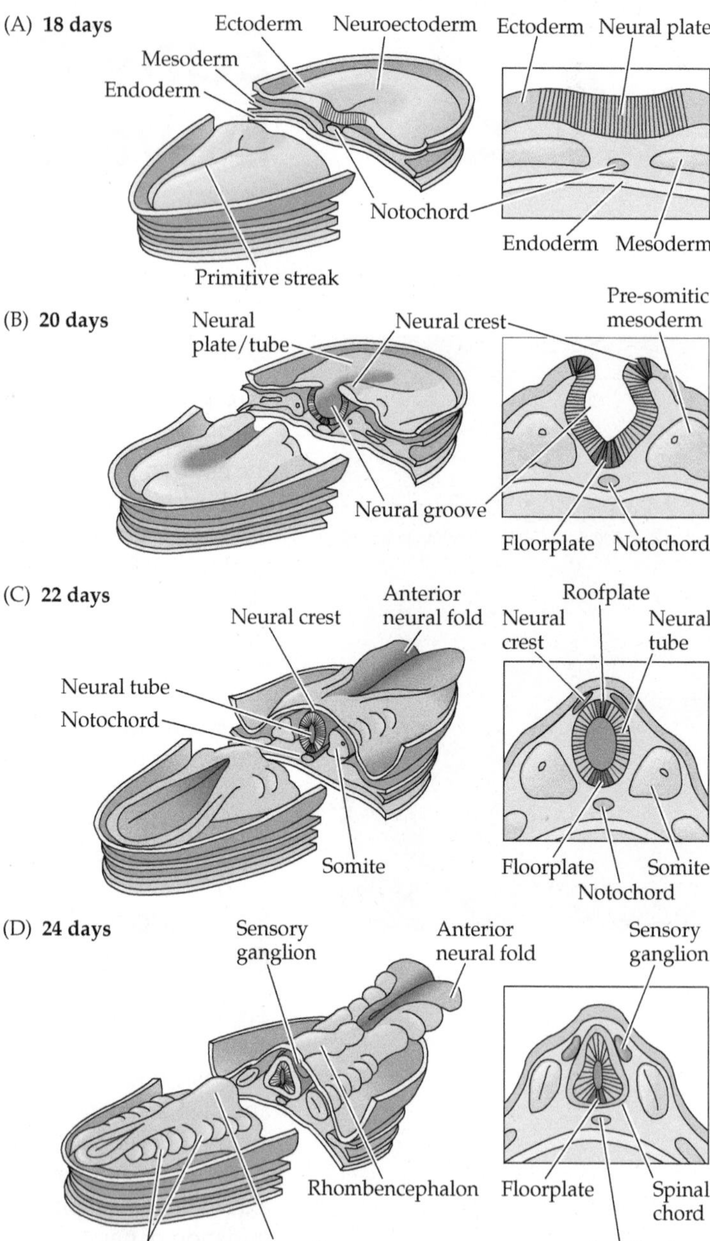

FIGURE 22.1 Neurulation in the mammalian embryo. On the left are dorsal views of a human embryo at several different stages of early development; each boxed view on the right is a midline cross section through the embryo at the same stage. (A) During late gastrulation and early neurulation, the notochord forms by invagination of the mesoderm in the region of the primitive streak. The ectoderm overlying the notochord becomes defined as the neural plate. (B) As neurulation proceeds, the neural plate begins to fold at the midline (adjacent to the notochord), forming the neural groove and, ultimately, the neural tube. The neural plate immediately above the notochord differentiates into the floorplate, whereas the neural crest emerges at the lateral margins of the neural plate (farthest from the notochord). (C) Once the edges of the neural plate meet in the midline, the neural tube is complete. The mesoderm adjacent to the tube then thickens and subdivides into structures called somites—the precursors of the axial musculature and skeleton. (D) As development continues, the neural tube adjacent to the somites becomes the rudimentary spinal cord, and the neural crest gives rise to sensory and autonomic ganglia (the major elements of the peripheral nervous system). Finally, the anterior ends of the neural plate (anterior neural folds) grow together at the midline and continue to expand, eventually giving rise to the brain.

and spinal cord, as well as most of the peripheral nervous system. Thus, these cells are neural stem cells, which in turn divide to produce more precursor cells (self-renewal being a hallmark of all stem cells) with the capacity to give rise to the full range of cell classes found in mature tissues (another defining feature of stem cells). Neural stem cells produce neurons, astrocytes, and oligodendroglial cells (Box 22A). Eventually, subsets of the neuroectodermal precursor cells generate region- and fate-specified neural progenitors that differentiate into specific classes of neurons in distinct brain structures. Signals from the floorplate, the somites, and the cranial neural crest for the forebrain lead to differentiation of cells in the ventral neural tube that eventually give rise to spinal and hindbrain motor neurons and related interneurons, closest to the ventral midline, or basal forebrain structures as well as related interneurons, once again closest to the ventral midline. Precursor cells farther away from the ventral midline give rise to sensory relay neurons and related interneurons in more dorsal regions of the spinal cord and hindbrain. The differentiation of these dorsal cell

BOX 22A ■ Stem Cells: Promise and Peril

One of the most highly publicized issues in biology has been the potential use of stem cells to treat a variety of neurodegenerative conditions, including Parkinson's, Huntington's, and Alzheimer's diseases. Amidst the social, political, and ethical debate set off by the promise of stem cell therapies, the question of what exactly is a stem cell tends to get lost.

Neural stem cells are an example of a broader class of **somatic stem cells** found in various tissues, both during development and in the adult. All somatic stem cells share two fundamental characteristics: they are self-renewing; and, upon terminal division and differentiation, they can give rise to the full range of cell classes within the relevant tissue. Thus, a neural stem cell can give rise to another neural stem cell, or to any of the differentiated cell types found in the central and peripheral nervous systems (i.e., inhibitory and excitatory neurons, astrocytes, and oligodendrocytes). A neural stem cell is distinct from a *neural progenitor cell*, which is incapable of continuous self-renewal and usually has the capacity to give rise to only one class of differentiated progeny. An oligodendroglial progenitor cell, for example, gives rise to oligodendrocytes until its mitotic capacity is exhausted; a neural stem cell, in contrast, can generate more neural stem cells as well as a full range of differentiated neural and glial cell classes, presumably indefinitely.

Neural stem cells, and indeed all classes of somatic stem cells, are different from *embryonic* stem cells. **Embryonic stem cells (ES cells)** are derived from pre-gastrula embryos. Like somatic stem cells, ES cells have the potential for infinite self-renewal. However, ES cells can give rise to *all* tissue and cell types of the organism—including the germ cells that undergo meiosis and generate haploid gametes as well as neural and other somatic stem cells (Figure A). Somatic stem cells, on the other hand, generate only diploid, tissue-specific cell types. In some cases, however, somatic cells such as skin cells or fibroblasts have been induced to acquire stem cell properties—including the ability to generate all tissues, as ES cells do. These *induced pleuripotential stem cells*, or *IPSCs*, are produced in vitro by introducing the genes for several transcription factors associated with stem cells into somatic cells (e.g., fibroblasts). IPSCs hold out the possibility of generating stem cells from mature individuals for therapeutic uses in regeneration and tissue repair.

The ultimate therapeutic promise of stem cells—neural or other types—is their ability to generate newly differentiated cells and tissues to replace those that may have been lost due to disease or injury. Such therapies have been imagined for some forms of diabetes (replacement of islet cells that secrete insulin) and some hematopoietic diseases. In the nervous system, stem cell therapies have been suggested for replacement of dopaminergic cells lost to

Continued on the next page

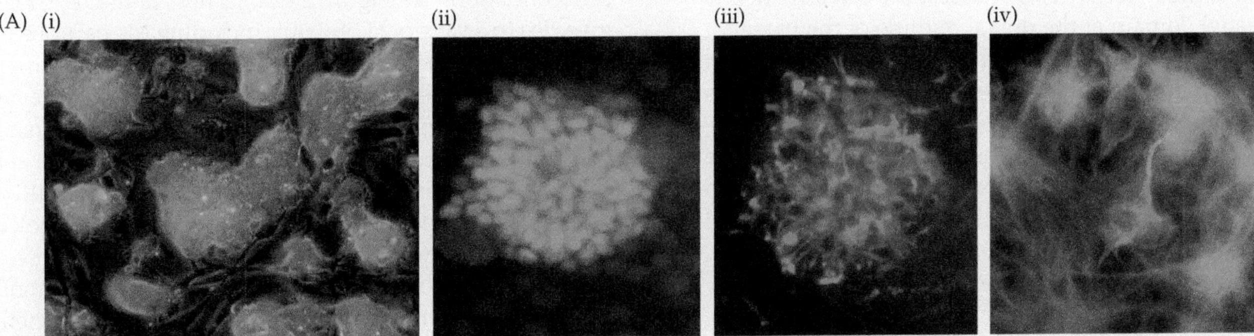

(A) Embryonic stem cells differentiate into various neuronal cell types. (i) Colonies of ES cells prior to differentiation. (ii, iii) After exposure to neuralizing signals, individual stem cell colonies express markers associated with different neural precursor cells. Cells in this colony express both Sox2 (green), a marker of early neural precursors, and nestin (red), a marker for later neural progenitor cells. (iv) After several days in culture, both neurons (red, labeled for neuron-specific tubulin) and astrocytes (green, labeled for glial fibrillary protein) have been generated from ES cells. (Photographs courtesy of L. Pevny.)

BOX 22A ■ (continued)

Parkinson's disease and replacing lost neurons in other degenerative disorders.

While intriguing, this projected use of stem cell technology raises some significant perils. These include ensuring the controlled division of stem cells when they are introduced into mature tissue, and identifying the appropriate molecular instructions to achieve differentiation of the desired cell class. Clearly, the latter challenge will need to be met with a fuller understanding of the signaling and transcriptional regulatory steps used during development to guide differentiation of relevant neuron classes in the embryo.

At present, there is no clinically validated use of stem cells for human therapeutic applications in the nervous system. Nevertheless, some promising work in mice and other experimental animals indicates that both somatic and ES cells can acquire distinct identities if given appropriate instructions in vitro (i.e., prior to introduction into the host), and if delivered into a supportive host environment. For example, ES cells grown in the presence of platelet-derived growth factor, which biases progenitors toward glial fates, have generated oligodendroglial cells that can myelinate axons in myelin-deficient rats. Similarly, ES cells pretreated with retinoic acid matured into motor neurons when introduced into the developing spinal cord (Figure B). While such experiments suggest that a combination of proper instruction and correct placement can lead to appropriate differentiation of embryonic or somatic stem cells, there are still many issues to be resolved before the promise becomes reality.

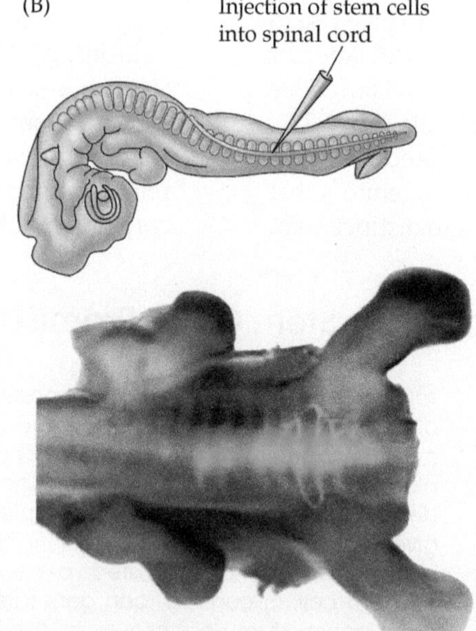

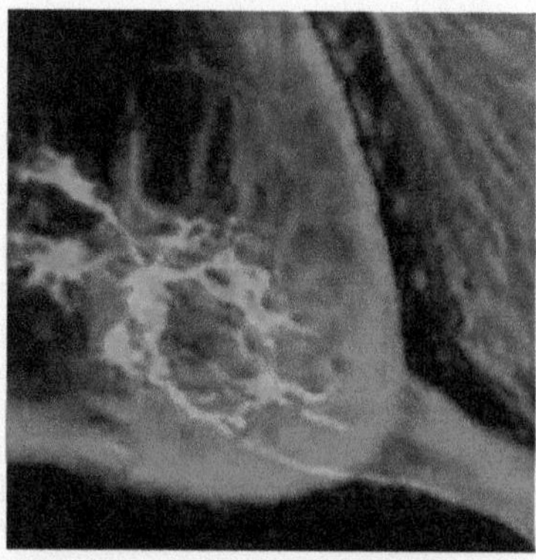

(B) Left top: Injection of fluorescently labeled embryonic stem cells into the spinal cord of a host chicken embryo shows that ES cells integrate into the host spinal cord and apparently extend axons. Below: the progeny of the grafted ES cells are seen in the ventral horn of the spinal cord. They have motor neuron-like morphologies, and their axons extend into the ventral root. (From Wichterle et al., 2002.)

groups is also facilitated by a narrow strip of neuroepithelial cells at the dorsal midline of the neural tube referred to as the **roofplate** in the spinal cord. In the forebrain, this signaling is mediated by two sources: the cortical "hem," an epithelial domain at the dorsal midline of the prosencphalic–telencephalic vesicles, and the choroid plexus, which secretes signals directly into the developing ventricles to modulate forebrain stem cell proliferation and differentiation (see Figure 22.12D,E). Like the notochord, the floorplate and roofplate are transient structures that provide signals to the developing neural tube and all but disappear once initial nervous system development is complete.

At the lateral edges of the neural plate (the alar plate, which subsequently become the dorsal and medial aspect of the neural tube), a third population of precursor cells emerges that constitutes the **neural crest**, the region where the edges of the folded neural plate come together (Figure 22.2). These **neural crest cells** migrate away from the neural tube through a matrix of loosely packed mesenchymal cells that fill the spaces between the neural tube, embryonic epidermis, and somites. Subsets of neural crest cells follow different pathways, along which they are exposed to additional signals that influence their specific differentiation (see Figure 22.2B). Thus, neural crest cells give rise to a variety of progeny, including the neurons and glia of the sensory and visceral motor (autonomic) ganglia, the neurosecretory cells of the adrenal gland, and the neurons of the enteric nervous system. Neural crest cells also contribute to non-neural structures such as pigment cells, cartilage, and bone, particularly in the face and skull.

The past decades have witnessed an explosion of molecular biological knowledge about the inductive signaling events and their consequences for gene expression and differentiation that transform neuroectodermal precursors and neural stem cells into the diverse cell and tissue types of the nervous system, including the neural crest. The establishment of cellular diversity, outlined in detail in subsequent sections of this chapter, occurs in parallel with the formation of the anatomical structures that will define the gross subdivisions of the brain: the spinal cord, brainstem, midbrain, and forebrain.

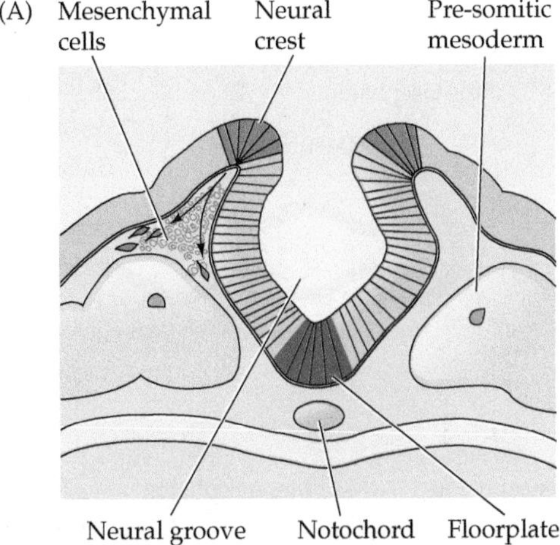

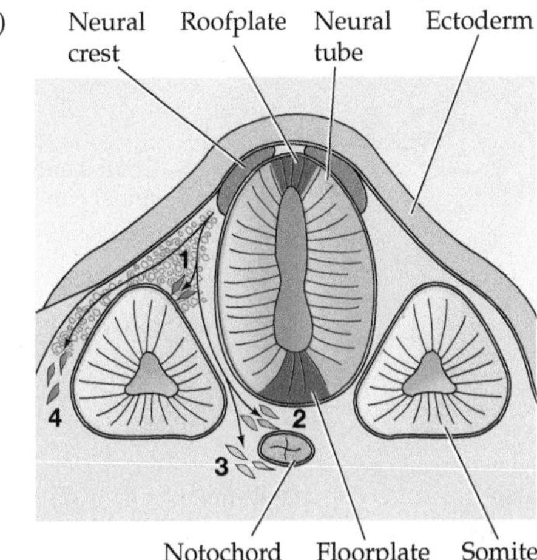

FIGURE 22.2 The neural crest. (A) Cross section through a developing mammalian embryo at a stage similar to that seen in Figure 22.1B. The neural crest cells are established based on their position at the boundary of the embryonic epidermis and neuroectoderm. Arrows indicate the initial migratory route of undifferentiated neural crest cells. (B) Four distinct migratory paths lead to differentiation of neural crest cells into specific cell types and structures. Cells that follow pathways 1 and 2 give rise to sensory and autonomic ganglia, respectively. The precursors of adrenal neurosecretory cells migrate along pathway 3 and eventually aggregate around the dorsal portion of the kidney. Cells destined to become non-neural tissues (for example, melanocytes) migrate along pathway 4. Each pathway permits the migrating cells to interact with different kinds of cellular environments, from which they receive inductive signals (see Figure 22.14). (After Sanes, 1989.)

Formation of the Major Brain Subdivisions

Soon after the neural tube forms, the forerunners of the major brain regions become apparent as a result of morphogenetic movements that bend, fold, and constrict the tube. Initially, the anterior end of the tube forms a crook, or "cane handle" (Figure 22.3A). The end of the "handle" nearest the sharpest bend is the **cephalic flexure**, which balloons out to form the **prosencephalon**, which in turn gives rise to the forebrain. The midbrain (**mesencephalon**) forms as a bulge above the cephalic flexure. The hindbrain (**rhombencephalon**) forms in the long, relatively straight stretch between the cephalic flexure and the more caudal cervical flexure. Caudal to the cervical flexure, the neural tube forms the precursor of the spinal cord. This bending and folding diminish or enlarge the different regions of the lumen enclosed by the developing neural tube. These lumenal spaces eventually become the ventricles of the mature brain (Figure 22.3B; also see the Appendix).

Once the primitive brain regions are established, they undergo at least two more rounds of partitioning, each of which elaborates the developing brain regions into forerunners of adult structures (Figure 22.3C). Thus, the lateral aspects of the rostral prosencephalon form the **telencephalon**. The two bilaterally symmetrical telencephalic vesicles include dorsal and ventral territories. The dorsal territory gives rise to the rudiments of the cerebral cortex and hippocampus, while the ventral territory gives rise to the basal ganglia (derived from embryonic structures called the **ganglionic eminences**), basal forebrain nuclei, and olfactory bulb. The more caudal portion of the prosencephalon forms the **diencephalon**, which contains the rudiments of the thalamus and hypothalamus, as well as a pair of lateral outpocketings—the **optic vesicles**—from which the neural portion of the retina will form. The dorsal portion of the mesencephalon gives rise to the superior and inferior colliculi, while the ventral portion gives rise to a collection of nuclei known as the midbrain tegmentum. The rostral part of the rhombencephalon becomes the **metencephalon** and gives rise to the adult cerebellum and pons. Finally, the caudal part of the rhombencephalon becomes the **myelencephalon** and gives rise to the adult medulla.

How can a simple tube of neuronal precursor cells produce such a variety of brain structures? At least part of the answer comes from the observation, made early in the twentieth century, that much of the neural tube is organized into repeating units called **neuromeres**. This discovery led to the idea that the process of **segmentation**—seen in most animal embryos (as well as in angiosperm plant embryos) to establish regional identity in the body by dividing the embryo into repeated units, or segments—might also establish regional identity in the developing brain. Enthusiasm for this hypothesis was stimulated by observations of the development of the body plan of the fruit fly *Drosophila*. In the fly, early expression of a class of genes called **homeotic** or **homeobox genes** guides the differentiation of the embryo into distinct segments that give

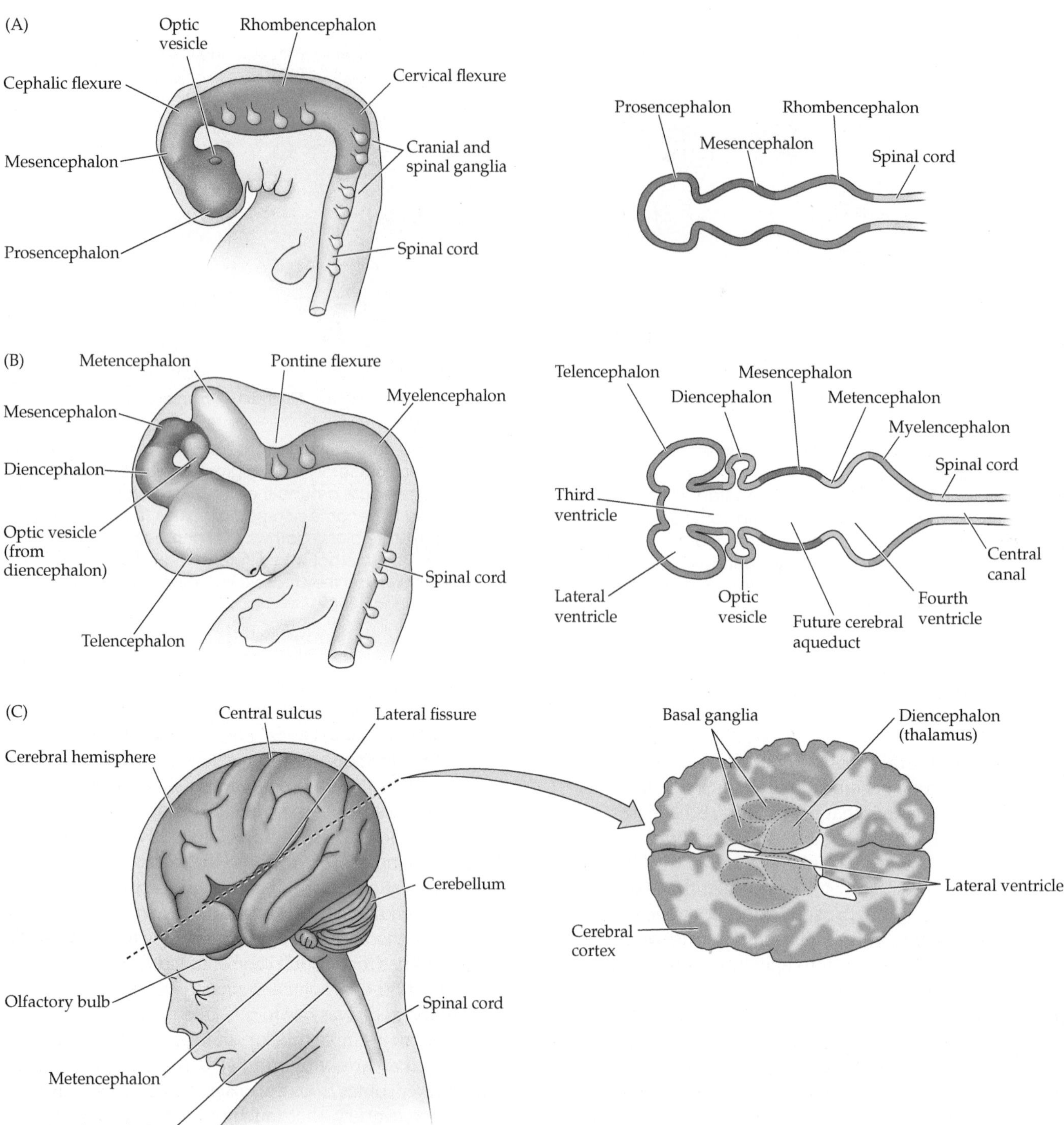

FIGURE 22.3 Regional specification of the developing brain. (A) Early in gestation the neural tube becomes subdivided into the prosencephalon (at the anterior end of the embryo), mesencephalon, and rhombencephalon. The spinal cord differentiates from the more posterior region of the neural tube. The initial bending of the neural tube at its anterior end leads to a cane shape. At right is a longitudinal section of the neural tube at this stage, showing the position of the major brain regions. (B) Further development distinguishes the telencephalon and diencephalon from the prosencephalon; two other subdivisions—the metencephalon and myelencephalon—derive from the rhombencephalon. These subregions give rise to the rudiments of the major functional subdivisions of the brain, while the spaces they enclose eventually form the ventricles of the mature brain. At right is a longitudinal section of the embryo at the developmental stage shown in (B). (C) The fetal brain and spinal cord are clearly differentiated by the end of the second trimester. Several major subdivisions, including the cerebral cortex and cerebellum, are clearly seen from the lateral surfaces. At right is a cross section through the forebrain at the level indicated showing the nascent sulci and gyri of the cerebral cortex, as well as the differentiation of the basal ganglia and thalamic nuclei.

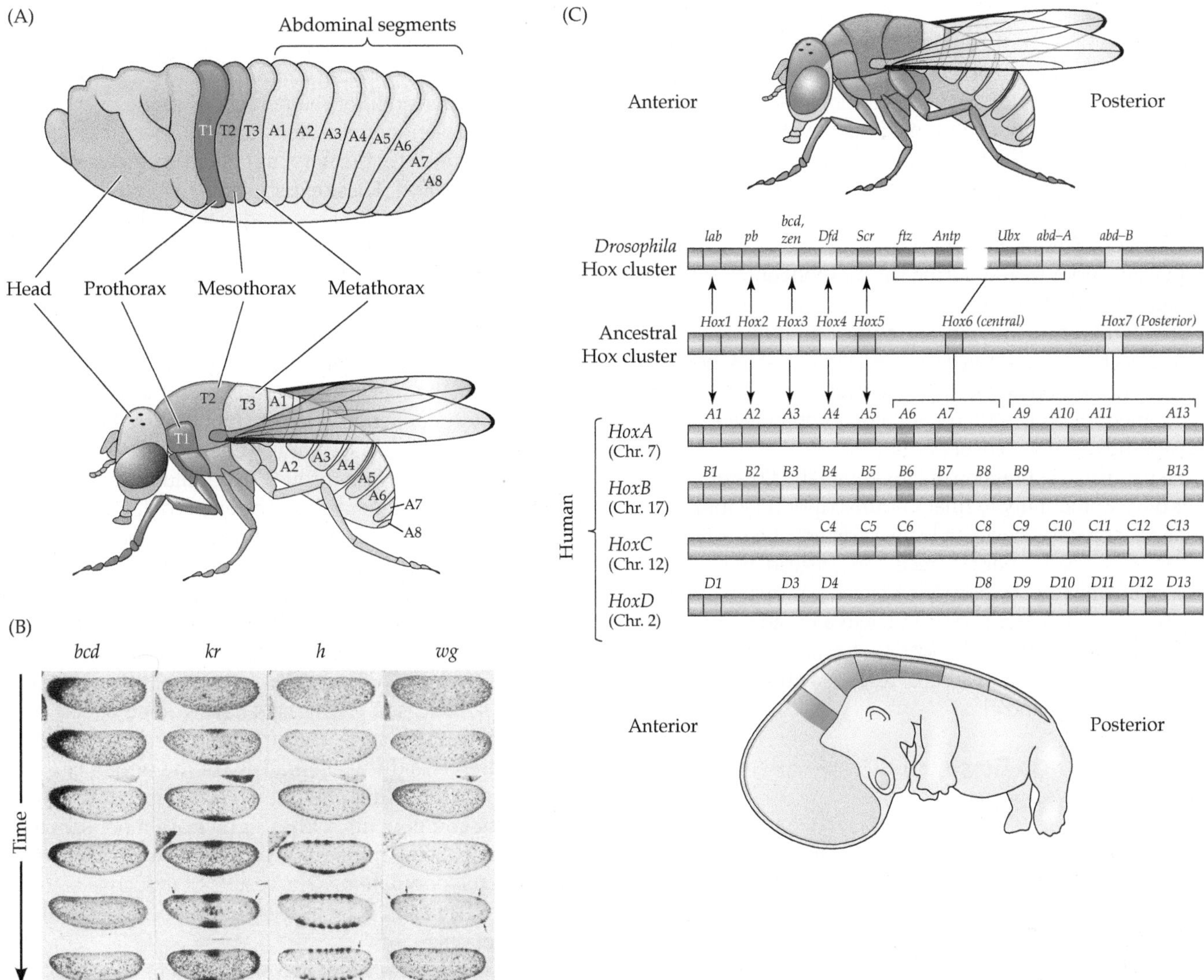

FIGURE 22.4 Sequential gene expression divides the embryo into regions and segments. (A) The relationship of the embryonic segments in the *Drosophila* larva, defined by sequential gene expression, to the body plan of the mature fruit fly. (B) Temporal pattern of expression of four genes that influence the establishment of the body plan in *Drosophila*. A series of sections through the anterior-posterior midline of the embryo is shown from early to later stages of development (top to bottom in each row). Initially, expression of the gene *bicoid* (*bcd*) helps define the anterior pole of the embryo. Next, *krüppel* (*kr*) is expressed in the middle and then at the posterior end of the embryo, defining the anterior-posterior axis. Then *hairy* (*h*) expression helps delineate the domains that will eventually form the mature segmented body of the fly. Finally, the *wingless* (*wg*) gene is expressed, further refining the organization of individual segments. (C) Parallels between *Drosophila* segmental genes (the inferred "ancestral" homeobox genes from which invertebrate and vertebrate segmental genes evolved) and human Hox genes. Hox genes of humans (and of most mammals) have apparently been duplicated twice, leading to four independent groups, each on a distinct chromosome. The anterior-to-posterior pattern of Hox gene expression in both flies and mammals (including humans) follows the 3′-to-5′ orientation of these genes on their respective chromosomes. (A after Gilbert, 1994, and Lawrence, 1992; B from Ingham, 1988; C after Veraksa and McGinnis, 2000.)

rise to the head, thorax, and abdomen (Figure 22.4A,B). *Drosophila* homeobox genes code for DNA-binding proteins that modulate the expression of other genes that mediate morphogenesis. Similar genes have been identified in vertebrates, including mammals, and are referred to as **Hox genes.** Rather than having one copy of each segmentally essential homeobox gene, as seen in the fly, vertebrate Hox genes have undergone multiple duplications so that today's vertebrates have four "clusters" of homologous genes with similar functions. In most mammals, including humans, each Hox gene cluster is located on a distinct chromosome, and their anterior-to-posterior expression and function are reflected in their position 5′→3′ on that chromosome. In some cases in the developing nervous system, the pattern of Hox gene expression coincides with, or even precedes, the formation of morphological features—that is, the various bends, folds, and constrictions—that underlie the progressive regionalization of the developing neural tube, particularly in the hindbrain and spinal cord (Figure 22.4C). In vertebrates, Hox gene expression does not extend into the midbrain or forebrain; however, regional differences in expression of other transcription factors are seen in these subdivisions prior to and during the morphogenetic events that define them (see Figure 22.7). The genes seen in the vertebrate midbrain and forebrain (including members of the *distaless* [*DLX*] and *paired box* [*PAX*] families) are homologs of transcription factors in *Drosophila* that influence the development of body structures such as appendages, head and mouthparts, and sensory organs.

The patterned expression of Hox genes and the genes for other developmentally regulated transcription factors and signaling molecules does not by itself determine the fate of a group of embryonic neural precursors. Instead, this aspect of regionally distinct transcription factor expression during early brain development contributes to a broader series of cellular and molecular processes that eventually produce distinct fully differentiated brain regions with appropriate classes of neurons and glia.

The Molecular Basis of Neural Induction

Neural stem cells in the early neural plate and tube, and subsequently in each nascent brain region, must acquire instructions that establish their capacity to make nerve cells specific to each region. It is clear that these instructions come from neighboring cells or tissues. During the first half of the twentieth century, the instructions needed for gastrulation and neurulation were defined using a variety of classic experiments based on the removal or transfer of embryonic tissues to assess the capacity of ectodermal, mesodermal, and endodermal cells to form organs composed of differentiated cell types. Cells that are moved either acquire the identity of the new region in which they are placed (thus receiving instructions based on their new location), or they retain an identity that reflects their original position (and thus possess immutable instructions from their original location). Cells that are removed are either compensated by local cell proliferation, causing little noticeable disruption of subsequent development (a process referred to as embryonic regulation), or their absence disrupts subsequent development. Finally, in some cases, relocation of cells causes a complete change of the local developmental program. (For example, transplantation of the equivalent of the cells that define the primitive pit to another location on an early embryo can cause a second notochord to form and a second nervous system to develop.)

Taken together, these experiments showed that *interactions* between cells in adjacent germ layers (e.g., mesoderm adjacent to ectoderm) are essential for regional and cellular identity in the developing embryo. By the early 1920s, it was clear that apposition of the notochord to the overlying ectoderm is essential for establishing neuroectodermal stem cells and thus the entire nervous system, a process known as **neural induction**. These experiments suggested that neural induction relies on signals provided locally by adjacent cells or tissues; however, the ultimate proof of this conjecture did not emerge until the early 1990s.

In the past 30 years, molecular and genetic approaches have demonstrated that the generation of cell identity and diversity—of which neural induction is only one example—results from the spatial and temporal control of different sets of genes by endogenous signaling molecules. Most of these molecular signals are secreted by one embryonic cell class or tissue and then diffuse through extracellular space to act on an adjacent cell class or tissue. The embryonic structures that are critical for the induction and patterning of the CNS (including the notochord, floorplate, roofplate, and neuroectoderm itself, as well as adjacent mesodermally derived tissues such as somites) all secrete these inductive signals (Figure 22.5A). Different classes of receptors transduce the signals in the neuroectoderm to drive further cellular differentiation. In some cases, the signals have graded effects based on the distance of target cells from the source. These effects may represent a diffusion gradient of the signal, or graded activity due to the distribution pattern of receptors or other signaling components. Other signals are more specific in their action, being most effective at the boundaries between distinct cell populations. The results of inductive signaling include changes in gene expression, shape, and motility in the target cells.

One of the first inductive signals to be identified was **retinoic acid** (**RA**), a derivative of vitamin A and a member of the steroid/thyroid superfamily of hormones (Figure 22.5B). Retinoic acid is a small, lipophilic molecule synthesized via metabolic enzymes, similar to gonadal steroids and small-molecule neurotransmitters. RA activates a unique class of transcription factors that are also receptors for RA and related ligands—the **retinoid receptors**. When activated by RA or related retinoids, the ligand-receptor complex modulates the expression of several target genes. The capacity of RA receptors, when bound by RA, to stimulate or repress gene expression depends on co-activators or co-repressors that form complexes with the receptors when bound to nuclear DNA. RA signaling drives cellular differentiation, regulating the transitions between various classes of neural stem cells leading up to their terminal division for neurogenesis. Excess retinoid signaling (due to excessive or insufficient dietary intake of vitamin A or exposure to retinoid-based medications) can cause severe birth defects, including incomplete neural tube closure and other disruptions of early brain morphogenesis.

Most inductive signaling molecules, however, are peptide hormones. The **fibroblast growth factor** (**FGF**) family

(A)

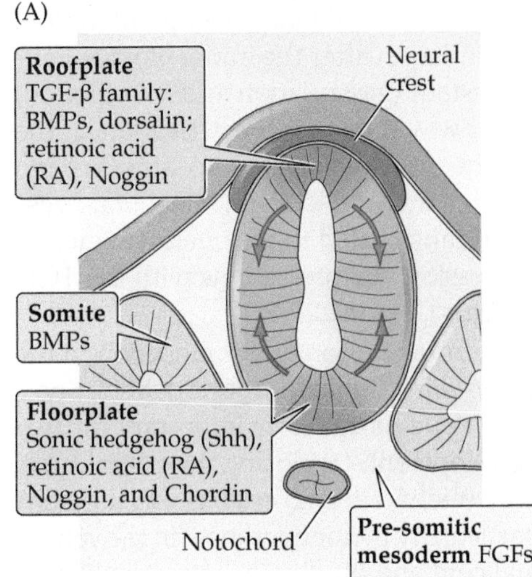

(B) Retinoic acid (RA)

(C) Fibroblast growth factor (FGF)

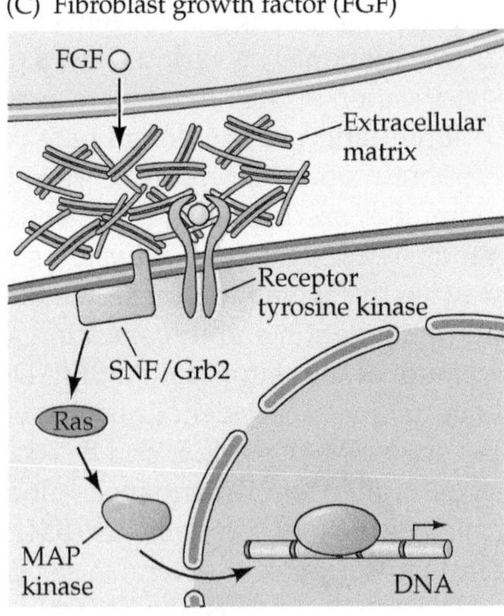

(D) Bone morphogenetic protein (BMP)

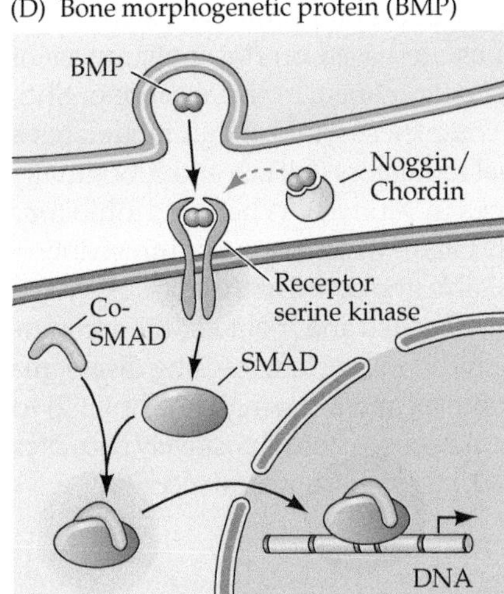

(E) Noncanonical Wnt

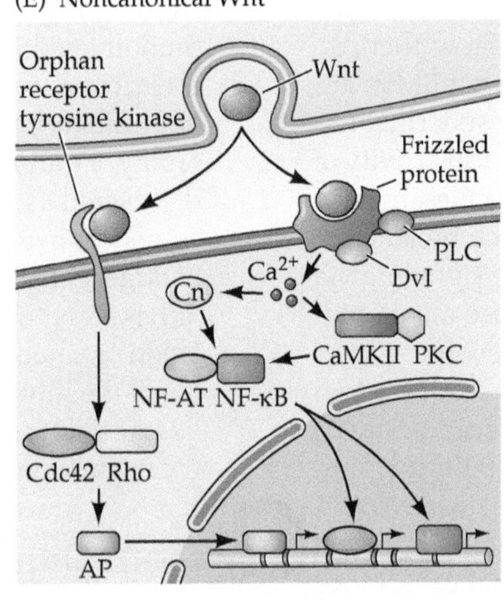

(F) Canonical Wnt

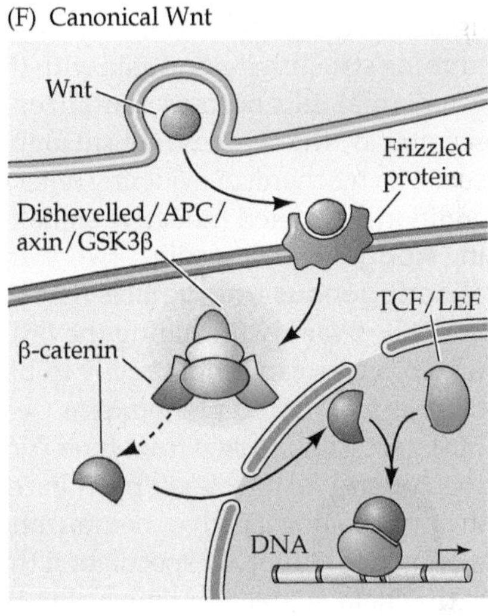

(G) Sonic hedgehog (Shh)

FIGURE 22.5 Major inductive signaling pathways in vertebrate embryos. (A) The embryonic notochord, floorplate, and neural ectoderm, as well as adjacent tissues such as somites, produce the molecular signals that induce cell and tissue differentiation in the vertebrate embryo. (B–G) Schematics of ligands, receptors, and primary intracellular signaling molecules for retinoic acid (RA); members of the FGF and TGF-β (BMP) superfamilies of peptide hormones; the Wnt family of signals; and Sonic hedgehog (Shh). Each of these pathways contributes to the initial establishment of the neural ectoderm, as well as to the subsequent differentiation of distinct classes of neurons and glia throughout the brain.

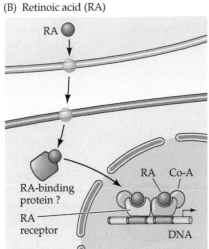

of peptide hormones is among the largest sets of inductive signals. There are 22 different FGF ligands encoded by 22 genes in the human genome. All of these ligands bind to the same receptor tyrosine kinases that initiate a phosphorylation-based signaling cascade via the ras-MAP kinase pathway (Figure 22.5C). MAP kinase activation can lead to altered expression of several target genes, especially those that modulate cell proliferation and differentiation. Among mammals, including humans, FGF8 has emerged as a particularly important regulator of forebrain and midbrain development. In addition, FGFs from the pre-somitic mesoderm (from which somites form) regulate spinal cord neurogenesis.

The **bone morphogenetic proteins** (**BMPs**), members of the TGF-β family of peptide hormones, are particularly important for a variety of events in neural induction and differentiation. The various BMPs play roles in the initial specification of the neural plate as well as the subsequent differentiation of the dorsal part of the spinal cord and hindbrain and the cerebral cortex. In humans and other mammals, six distinct genes encode six different BMP ligands. These ligands all activate a singular signaling pathway via the same receptor serine kinases, resulting in the phosphorylation and translocation to the nucleus of transcriptional regulators called SMADs (Figure 22.5D). Following BMP-dependent phosphorylation, three different phospho-SMADs—1, 5, and 8—translocate into the nucleus, bind to specific enhancer/supressor DNA sequences called BMP response elements (BMPre) and thereby influence transcription of several target genes.

In *mesodermal* cells, the BMPs (as their name suggests) elicit osteogenesis (bone cell formation). When *ectodermal* cells are exposed to BMPs, they assume an epidermal fate, forming structures associated with the skin. How, then, do ectodermal cells become neuralized, given that BMPs are secreted by the somites and surrounding mesodermal tissue? The mechanism evidently relies on the local activity of additional secreted inductive signaling molecules, including **Noggin** and **Chordin**—two members of a broad class of **endogenous antagonists** that modulate signaling via the TGF-β family (including the BMPs). These antagonistic molecules can bind directly to BMPs, preventing their binding to BMP receptor proteins (see Figure 22.5D). When BMPs are thus blocked from their "normal" receptors, neuroectoderm is "rescued" from becoming epidermis and continues along a path of neuralization. This negative regulation has reinforced speculation that becoming a neuron is the "default" fate for embryonic ectodermal cells. Once neuronal precursor identity is firmly established, however, BMPs can act on neural precursors or differentiating neurons to further influence their identity and fate.

Members of the **Wnt** family of secreted signals also modulate several aspects of nervous system morphogenesis and neuronal differentiation, including some aspects of neural crest differentiation. In contrast to several other inductive signaling pathways, the 19 human Wnt ligands (encoded by 19 separate genes) can activate two distinct signal transduction cascades, the "canonical" and "noncanonical" pathways. Although researchers elucidated the canonical pathway first, the noncanonical pathway is featured during early nervous system morphogenesis.

- The *noncanonical Wnt pathway*, also known as the planar cell polarity (PCP) pathway, regulates cell movements necessary for lengthening the neural plate and neural tube. Here Wnt ligands activate receptor proteins (Frizzled), leading to changes in intracellular Ca^{2+} levels; alternatively, the Wnt ligands can bind an orphan receptor tyrosine kinase, leading to activation of a Jun kinase (Jnk) signaling pathway that can phosphorylate several intracellular targets, leading to changes in cell shape and polarity (Figure 22.5E).
- The *canonical Wnt pathway* influences cell proliferation, adhesion, and differentiation after the initial morphogenesis of the nervous system (gastrulation and neurulation) is complete. This pathway relies on the activation of the Frizzled receptor in the presence of a co-receptor (Lrp5/6) which leads to stabilization of β-catenin, a cellular messenger which is then translocated to the nucleus, where it influences gene expression via interactions with the TCF/LEF transcription factor (Figure 22.5F).

Another peptide hormone essential for induction in the developing nervous system is **Sonic hedgehog (Shh)**. Sonic hedgehog is thought to be particularly important for two phases of neural development: (1) closing the neural tube, especially the anterior midline, and (2) establishing the identity of neurons—particularly motor neurons—in the ventral portion of the spinal cord and hindbrain. The transduction of signals via Shh (Figure 22.5G) requires the cooperative binding of two surface receptor proteins, Patched and Smoothened (the names are based on the appearances of their respective *Drosophila* mutants). In the absence of Shh, an inhibitory protein complex assembles that modulates a family of transcriptional regulators (Gli1, 2, and 3, originally discovered as oncogenes in *gli*omas). When this inhibitory complex is in place, only Gli3—which represses transcription of target genes—is available and active in the nucleus. When Shh is present, it binds to Patched and promotes the accumulation of Smoothened on the cell surface, causing disassembly of the inhibitory complex and allowing Gli1 (or Gli2) to be translocated into the nucleus, where it positively regulates the expression of genes that establish neural identity.

Stem Cells

Stem cell biology has emerged as a promising field for establishing new therapies to repair degenerating or injured organs and to understand the pathogenesis of serious diseases such as several cancers. Stem cell biology has also had an impact on understanding the development of the nervous system as well as prospects (still quite speculative) for repair of degenerating or damaged neural tissue (see Box 22A and Chapter 26). Thus, understanding stem cells in embryos, in tissues found in maturing or mature animals, or in their in vitro counterparts is vital for understanding vertebrate development and pathology.

The foundational experiment that defined stem cell biology was carried out in the early 1960s. In this pioneering work (for which Sir John Gurdon shared the 2012 Nobel Prize in physiology and medicine with Shinya Yamanaka), a nucleus from a somatic tissue in the frog—in this case the intestinal epithelium—was transplanted to an unfertilized frog oocyte whose nucleus had been removed (Figure 22.6A). The result, although obtained at low frequency (~10/750 attempts), showed that a single germ-line cell (in this case, an egg cell whose nucleus with one rather than two copies of each chromosome) with a somatic nucleus

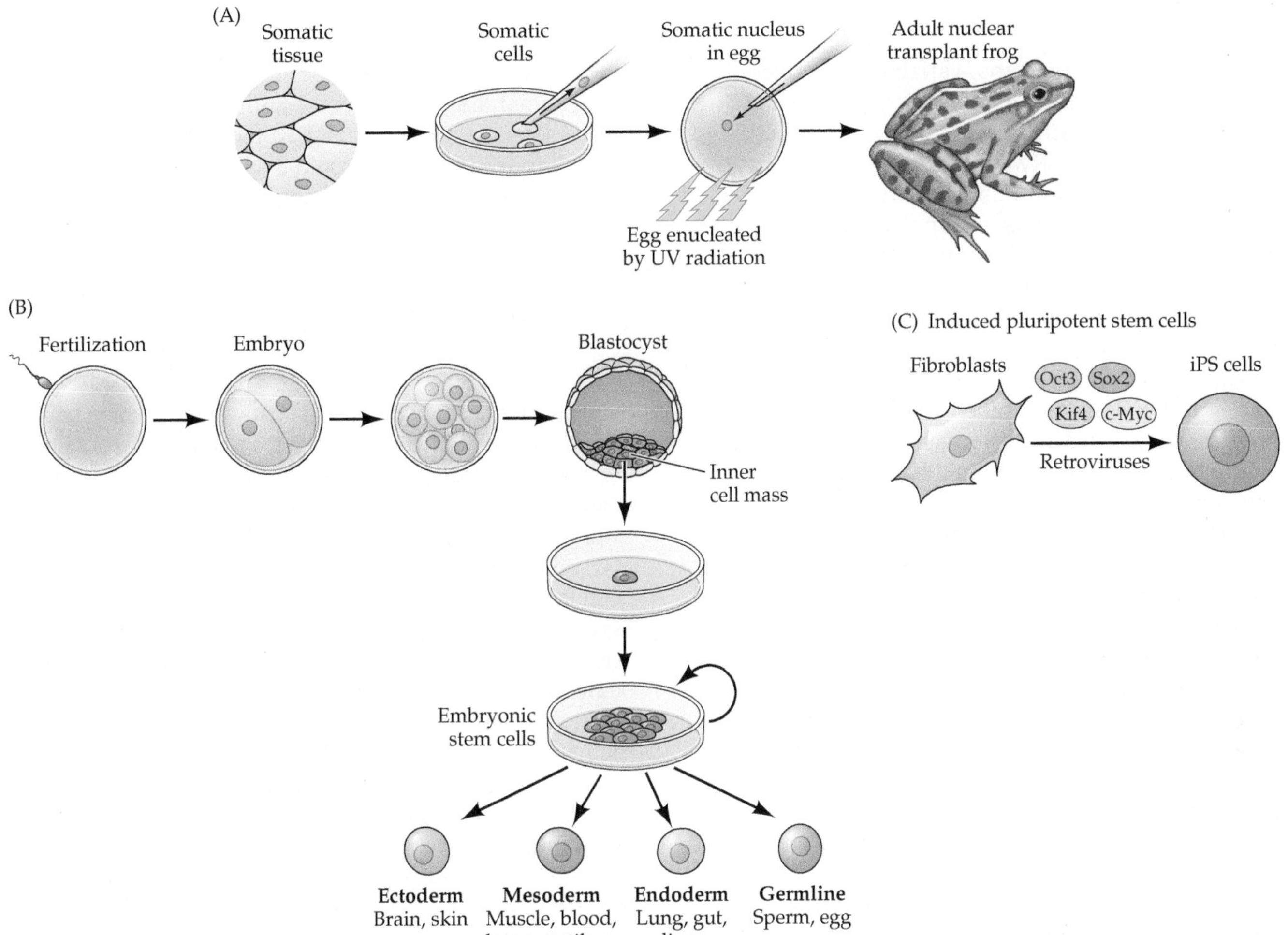

FIGURE 22.6 **Stem cell specification accounts for development of all somatic tissues, including the nervous system.** (A) The first demonstration that the somatic nucleus of an adult tissue had the complete genomic information necessary to generate an entire organism. In this experiment, the nucleus of a somatic cell (from the intestine, in this example) was removed from its parent cell. This nucleus was then injected into a UV irradiated oocyte (the UV irradiation completely inactivates the oocyte nucleus). This combination of somatic nucleus and oocyte was capable of generating an entire adult frog. Thus, the oocyte (in this case with a somatic nucleus, normally post-fertilization when the DNA content restored to diploid from monoploid), with a full complement of genomic DNA from the somatic nucleus was the ultimate stem cell capable of giving rise to all somatic tissues in the organism. (B) In vitro experiments, in which cells from the inner cell mass of the mammalian blastocyst (the totipotent stem cells generated by the first series of cell divisions of the fertilized oocyte) are isolated and maintained in cell culture conditions that preserve their capacity to generate all classes of somatic tissues. Such cultured embryonic stem cells (ES cells) when injected into a host blastocyst, can integrate into an embryo and contribute to every tissue in the mature organism, including the germ line. (C) Somatic cells can be "reprogrammed" to become the equivalent of ES cells. Such cells are referred to as induced pluripotent stem cells. The reprogramming relies upon the introduction of four transcription factors that are thought to be essential for establishing and maintaining the ES cell state. In a small number of somatic cells in which these transcription factors are expressed, those cells will re-acquire their ES cell properties, including the capacity to be introduced into a blastocyst and contribute to all tissues in the adult organism that originates from the chimeric embryo.

had the potential to generate a whole healthy frog, with all somatic tissues including the central and peripheral nervous systems. Thus, these oocytes with a transplanted nucleus were said to be **totipotent**: capable of generating all somatic cells plus gametes that define the germline (oocytes and sperm) that constitute the tissues of the living adult organism. Not only did this experiment suggest that somatic nuclei had not been irrevocably altered through DNA excision or other mechanisms, and retained the full embryonic genome, it also defined a cell type—albeit experimentally created—that could generate all other cells and tissues, including the nervous system, in the organism. This was the first demonstration of a true embryonic stem (ES) cell (Figure 22.6B). Subsequent work showed that the oocyte with transplanted nuclei could also give rise to the germline (gamete precursors), confirming the true totipotency of these hybrid cells. Indeed, the hybrid cell had the ploidy (chromosome number) of a somatic cell (two copies of each autosome and, variably, two copies of the X chromosome in females, and one copy of the X and Y chromosomes for males) without the driving force of fertilization and nuclear fusion between oocyte and sperm; nevertheless, it could generate an entire organism capable of reproduction. Subsequently, it was shown that these animals had germline cells (gametes). Thus, these oocyte–somatic nuclei hybrids were truly totipotent—capable of giving rise to all cells in the organism, including gametes and the entire nervous system.

The next step for defining stem cell biology, including neural stem cell biology, was taken when Martin Evans and his colleagues isolated and propagated embryonic stem cells in vitro by harvesting cells from the inner-cell mass of the blastocyst in mice and other mammals. (Other groups, using discarded human blastulae from in vitro fertilization procedures, isolated—with great controversy—human stem cells.) These cells, when maintained under rigorous conditions in vitro, can differentiate into all somatic cell types in the organism. Moreover, when injected into the blastocyst of an early-stage embryo (mouse embryo in most cases), they can create a chimeric mouse, including integrating into the germline. It is through this mechanism of germline integration that "knock-in" and "knock-out" mice are made in the laboratory from genetically modified ES cells.

Shinya Yamanaka and his colleagues provided the ultimate molecular validation of these embryonic and in vitro phenomena that apparently define embryonic stem cells. Based on extensive studies of transcription factors expressed early in developing embryos, Yamanaka and colleagues arrived at a combination of four—Sox2, Oct4, Kif4, and c-Myc—that could reprogram somatic cells to become pluripotent stem cells capable of generating an entire organism (in this case a mouse). These cells, called induced pluripotent stem (iPS) cells, like the hybrid somatic cells for the frog or ES cells from mice, are capable of generating a chimeric mouse that is reproduction-competent (i.e., they can integrate into the germline) when injected into a host blastocyst. Clearly reprogramming via four transcription factors expressed at high levels at the zygote and blastula stage is effective for establishing pluripotency in these ES cell-like iPS cells that can establish the germline as well as somatic tissue (Figure 22.6C).

The current status of stem cell biology is fraught with uncertainty and contention. As biological tools, stem cells have provided assay systems to evaluate the activity of a variety of soluble factors and transcriptional regulators in the differentiation of stem cells to a variety of somatic cell classes. In addition, the stem cell concept has been applied to cancer, where it is thought that tumors arise from a "cancer stem cell" that then gives rise clonally to tumor cells. Nevertheless, there are no effective therapies using stem cells to engineer replacement organs or tissues after traumatic injury or degeneration. Moreover, the prospect of whole-organism cloning (realized in mice and a few other mammalian species) has raised ethical issues regarding the prospect of using human ES cells to clone a human. The technical details of potential human cloning would make such attempts difficult indeed (and one would likely generate a chimeric human clone). Moreover, the moral issues militate strongly against the use of such technology to circumvent other conventional means of human reproduction.

Integrated Inductive Signals Establish Neuron Identity

Stem cells in the embryo are guided toward becoming neural stem cells, and ultimately neurons via a wide range of signaling molecules that are either secreted by other cells nearby, or available on the cell surfaces of neighboring cells or in the extracellular matrix. The combined activities of retinoic acid, FGFs, BMPs (antagonized by Noggin and Chordin), Sonic hedgehog, and Wnts specify a mosaic of transcription factor expression in subsets of precursor cells throughout the developing neuroectoderm, then the neural plate, and then the neural tube. The earliest signaling molecules are thought to be essential for distinguishing stem cells destined to become the neural crest (see below) at the margins of the neural plate from stem cells in the presumptive neural tube ectoderm that will give rise to neural stem cells that generate the brain and spinal cord. The subsequent activity of inductive signals and their transcription factor targets for neural stem cell specification and differentiation in the CNS is best understood in the spinal cord, where they interact to establish differences in gene expression in the ventral, intermediate, and dorsal spinal cord (Figure 22.7A–C). This mechanism of signaling and transcriptional control sequentially limits downstream gene expression in neural precursors, leading to specific patterns of gene expression in postmitotic neuronal progeny. These specific patterns of transcription factor expression will provide a foundation for the differentiation of motor neurons, interneurons, and sensory relay neurons in the ventral, intermediate, and dorsal domains of the nascent spinal cord. Additional transcription factors, also influenced by local inductive signals, are thought to specify neuronal identity in immature postmitotic neuroblasts; they also support the position-specific final differentiation of motor neurons, interneurons, and sensory relay neurons in the mature ventral, intermediate, and dorsal spinal cord. This general mechanism—local signaling leading to local variation in transcription factor expression in distinct precursor cells or early postmitotic neuroblasts—operates throughout the developing central and peripheral nervous systems. The combination of transcription factors necessary to establish identity of specific neuron classes is often referred to as a *transcriptional code*. Indeed, although some of the signaling molecules and transcription factors are different, a similar mechanism establishes identity for neuronal progenitors in the forebrain (Figure 22.7D). In the forebrain, inductive signaling via molecules identical or analogous to those in

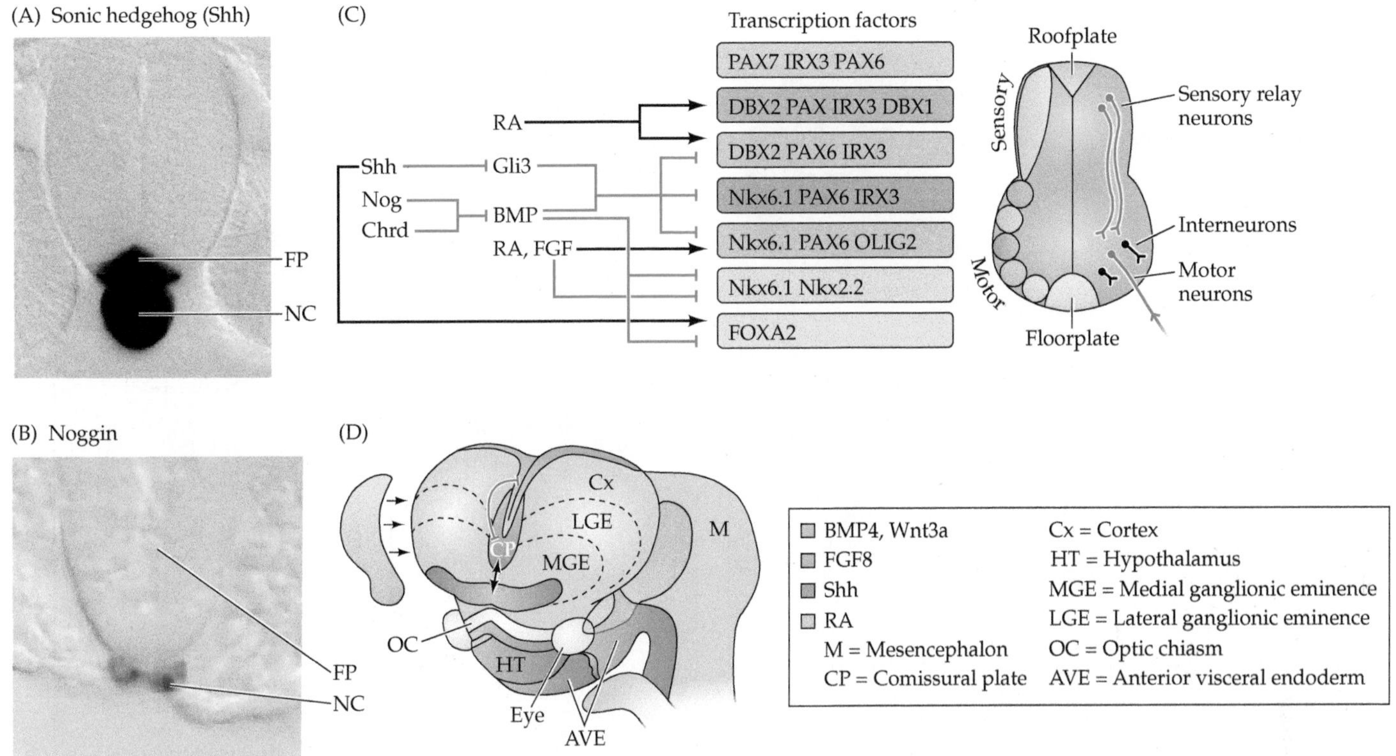

FIGURE 22.7 An integrated network of local signals specifies neural identity. (A) A section through the embryonic chick spinal cord shows the distribution of the Sonic hedgehog signal (purple label) in the notochord (NC) and floorplate (FP). (B) The BMP antagonist Noggin (light blue label), which helps preserve the neural identity of ectoderm by preventing BMP signaling, is available from the roofplate (see part C and Figure 22.5A) as well as the floorplate and notochord. This image is a section through the embryonic mouse spinal cord; the mouse notochord is a somewhat smaller structure than that of the chick. (C) Interactions between Shh (via Gli3 repression), Noggin/Chordin, BMP, RA, and FGF lead to either expression (black) or repression (red) of a set of transcription factors that distinguish different precursors. These distinct precursors, based on their dorsal-to-ventral position in the spinal cord, will go on to become sensory relay neurons (dorsal), interneurons (intermediate), or motor neurons (ventral). (D) A similar mechanism establishes identity for neuronal progenitors in the forebrain. (A from Dodd et al., 1998; B from Fausett et al., 2014.)

the spinal cord as well as transcriptional differences among cells prefigure morphogenesis and cellular differentiation in ventral, dorsal, and intermediate forebrain subdivisions. These domains then give rise to the olfactory bulb, basal ganglia, basal forebrain (amygdala and other ventrolateral structures), hippocampus, and neocortex. Thus, inductive signals, their receptors, and the resulting regulation of gene expression (particularly of locally expressed transcription factors) can specify cell identity as well as influence other aspects of neural development.

Initial Differentiation of Neurons and Glia

The numbers of neurons versus glia in the human brain are not precisely known; estimates place both numbers at 86 *billion* or more. Despite the uncertainty of numbers—one way or another there are a lot of cells—all of these cells must be generated over the course of a few months from a small population of neural stem cells in the early embryo. Neurogenesis begins after the initial patterning of the neural tube is complete. At this time, precursor cells in various brain regions have distinct signatures of gene expression that assign basic identities. These precursor cells are located in the ventricular zone: the innermost cell layer surrounding the lumen of the neural tube, and a region of extraordinary proliferative activity during gestation. It has been estimated that a developing human generates about 250,000 new neurons *each minute* during the peak of cell proliferation. Except for a few specialized cases (see Chapter 26), the entire neuronal complement of the adult brain is produced during a time window that closes before birth; thereafter, precursor cells mostly disappear, and in most brain regions, few if any new neurons can be added to replace those lost by age or injury.

Dividing precursor cells in the ventricular zone undergo a stereotyped pattern of cell movements as they progress through the cell cycle, leading to the formation of either new stem or precursor cells or postmitotic **neuroblasts** (immature nerve cells) that differentiate into neurons (Figure 22.8). The distinction between different modes of cell division that give rise to either new stem cells or to neuroblasts is an essential aspect of the process of neurogenesis. Thus,

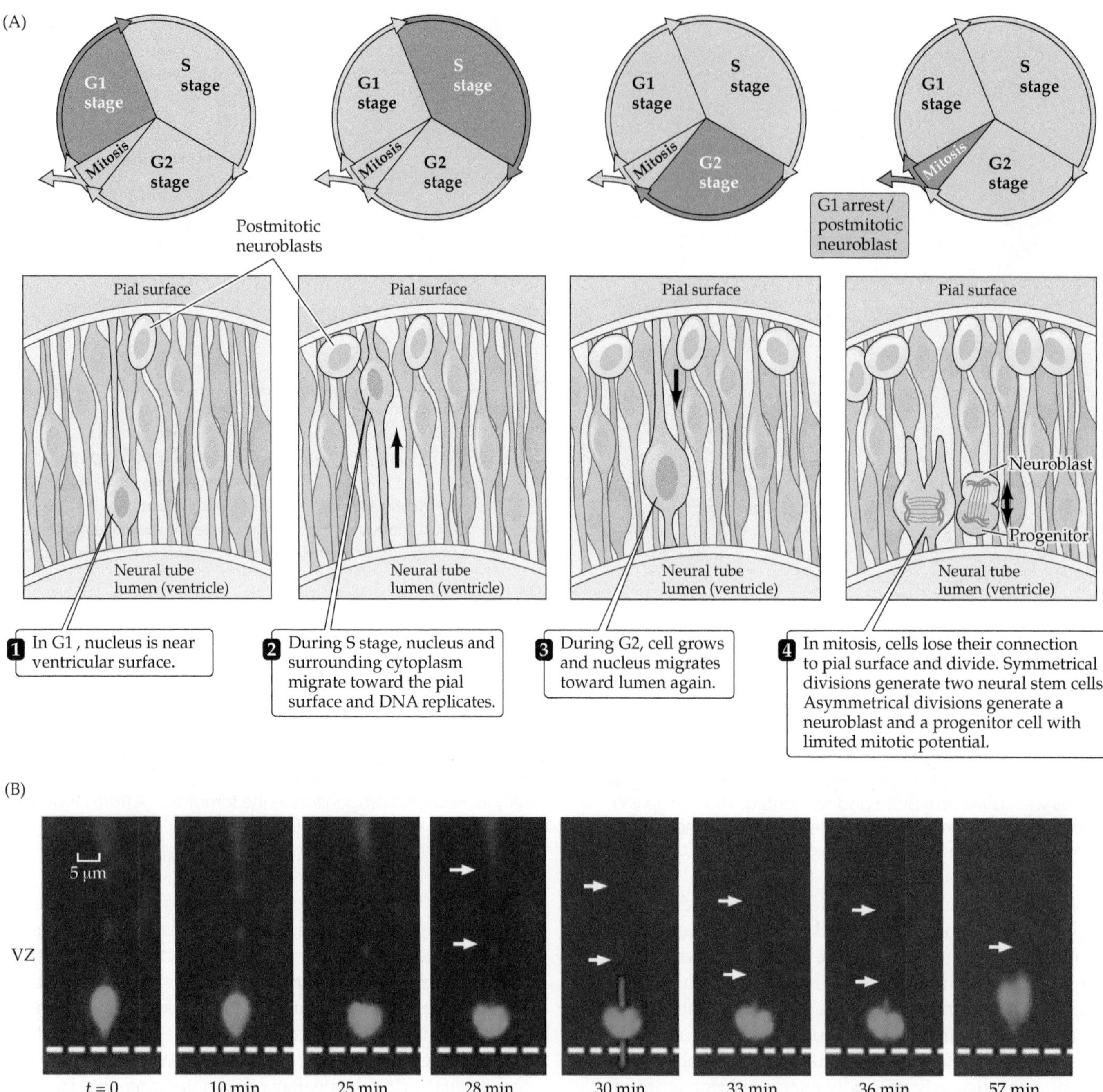
(A)
G1 stage
S stage
G2 stage
Mitosis
G1 arrest/ postmitotic neuroblast
Postmitotic neuroblasts
Pial surface
Neural tube lumen (ventricle)
Neuroblast
Progenitor
1 In G1, nucleus is near ventricular surface.
2 During S stage, nucleus and surrounding cytoplasm migrate toward the pial surface and DNA replicates.
3 During G2, cell grows and nucleus migrates toward lumen again.
4 In mitosis, cells lose their connection to pial surface and divide. Symmetrical divisions generate two neural stem cells. Asymmetrical divisions generate a neuroblast and a progenitor cell with limited mitotic potential.
(B)
5 μm
VZ
t = 0
10 min
25 min
28 min
30 min
33 min
36 min
57 min

FIGURE 22.8 Neural precursor cells undergo mitosis in the ventricular zone. (A) Precursor cells in the vertebrate neuroepithelium are attached both to the pial (outside) surface of the neural tube and to its ventricular (lumenal) surface. The nucleus of the cell translocates between these limits within a narrow cylinder of cytoplasm (the ventricular zone, VZ). When cells are closest to the outer surface of the tube, they enter the DNA synthesis phase (S stage) of the cell cycle. Once the nucleus moves back to the ventricular surface (G2 stage), the precursor cells lose their connection to the outer surface and enter mitosis. When mitosis is complete, the two daughter cells extend processes back to the outer surface of the neural tube, and the new precursor cells enter a resting (G1) phase of the cell cycle. At some point a precursor cell generates either another progenitor cell that will go on dividing and a daughter cell—a neuroblast—that will not divide further, or two postmitotic daughter cells. (B) Time-lapse microscopy permits visualization of symmetrical, vertically oriented division (red line) of a single radial glial stem cell in the cortex. The cell body is seen at the ventricular surface (dashed line); the arrows indicate the radially oriented process of the cell, which is mostly out of the focal plane necessary for visualizing the cell body. The radial processes are retained once the cell has divided. (B from Noctor et al., 2008.)

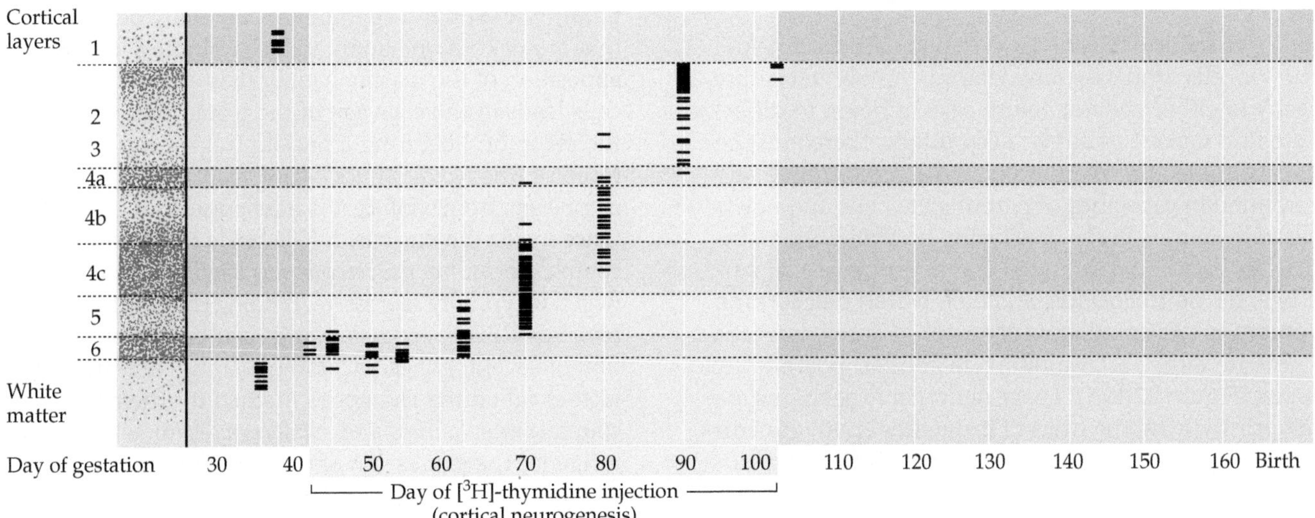

FIGURE 22.9 Generation of cortical neurons. The graph covers a span of about 165 days during the gestation of a rhesus monkey. The final cell divisions of the neuronal precursors, determined by maximal incorporation of radioactive thymidine administered to the pregnant mother, occur primarily during the first half of pregnancy and are complete on or about embryonic day 105. Each short horizontal line represents the position of a neuron heavily labeled by maternal injection of radiolabeled thymidine at the time indicated by the corresponding vertical line. The numerals on the left designate the cortical layers. The earliest-generated cells are found in a transient layer called the subplate (a few of these cells survive in the white matter) and in cortical layer 1 (the Cajal–Retzius cells). (After Rakic, 1974.)

the regulation of cell division, as well as the regulation of expression of distinct transcription factors, is a key determinant of the fate of any cell in the developing nervous system.

New stem cells arise from *symmetrical* divisions of neuroectodermal cells. These cells divide relatively slowly and can renew themselves indefinitely. Surprisingly, neural stem cells seem to acquire and retain many of the molecular characteristics of glial cells. Thus, in the developing brain some multipotent neural precursors are indistinguishable from the **radial glial cells** that also act as a substrate for migration of postmitotic neurons in the cerebral cortex (see below). Postmitotic neurons, in contrast, are generated from cells that divide *asymmetrically*: One of the two daughter cells becomes a postmitotic neuroblast while the other reenters the cell cycle to give rise to yet another postmitotic progeny via asymmetrical division (see Figure 22.8A). Such asymmetrically dividing progenitors are molecularly distinct from the slowly dividing radial glial stem cells (see Figure 22.11), and although they tend to divide more rapidly, they have a limited capacity for division over time. They are sometimes known as *transit amplifying cells* because they are a *transitional* form between stem cells and differentiated neurons, and they account for the *amplification* in numbers of differentiated cells due to their rapid mitotic kinetics and serial asymmetrical divisions.

Different populations of spinal cord neurons, as well as nuclei of the brainstem, thalamus, and cerebral cortex, are distinguished by the times when their component neurons are generated. Some of these distinctions are influenced by local differences in the signaling molecules and transcription factors that characterize the precursors (see Figure 22.7D). In the cerebral cortex, most neurons of the six cortical layers are generated in an inside-out manner; each layer consists of a cohort of cells "born"—and which therefore undergo their terminal cell divisions—at a distinct time. The firstborn cells are eventually located in the deepest layers; later generations of neurons migrate radially from the site of their final division in the ventricular zone, traveling through the older cells and coming to lie superficial to them (Figure 22.9). These differences in time of cell origin are matched by differences in gene expression in distinct cohorts of cells. In the cortex, early-born neurons in the lower cortical layers express transcription factors distinct from those of the later-born neurons in the upper cortical layers. Indeed, in most regions of the CNS where neurons are arranged into layered or laminar structures (the hippocampus, cerebellum, superior colliculus, and retina), there is a systematic relationship between the layers, time of cell origin, and additional molecular properties, including transcription factor expression. In these brain regions, neuroblasts from the ventricular zone either migrate or are passively displaced radially and outward, thus establishing a systematic relationship between the time of a cell's final division and its laminar position. The implication of this phenomenon is that organized periods of neurogenesis are important for the development of the cell types and connections that characterize the distinct brain regions.

Molecular Regulation of Neurogenesis

Clearly, the mode of cell division in populations of neural precursor cells is an essential determinant of their future capacity to either generate more cells or begin to differentiate into mature neurons. Accordingly, there must be extensive molecular regulation of the proliferative and differentiation capacities of neural stem cells and newly generated neurons in the developing nervous system. Interactions between a family of transmembrane cell surface ligands, **delta ligands**, and their **Notch cell surface receptors**, are key regulators of neural stem cell decisions to generate either additional stem cells or postmitotic neurons (Figure 22.10A). These neurogenic decisions are made primarily on the basis of influences from immediately neighboring cells. Thus, signaling via Delta ligands binding to Notch receptors happens only between cells that are next to one another. Delta binding to Notch via apposition of the plasma membranes of two neighboring cells leads to the cleavage of the intracellular domain of the receptor, liberating a protein fragment of the Notch receptor (the *Notch intracellular domain*, or *NICD*) into the cytoplasm, from whence it is transported into the nucleus. Once inside the nucleus, NICD binds to a transcriptional complex including the *recombining binding protein J* (*RBP-J*), which is typically repressive. Binding the NICD, however, reverses RBP-J repression and results in the transcription of several genes, including a family of transcription factors called the *Hes* genes (named for their *Drosophila* counterparts, *hairless* and *enhancer-of-split*), which in turn influence the expression of transcription factors involved

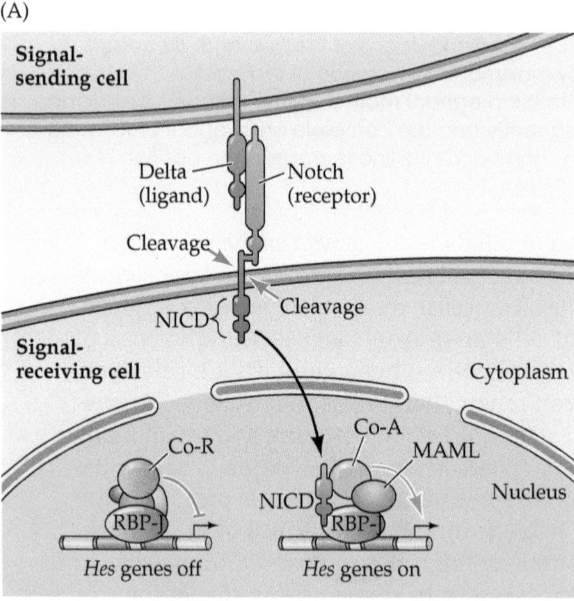

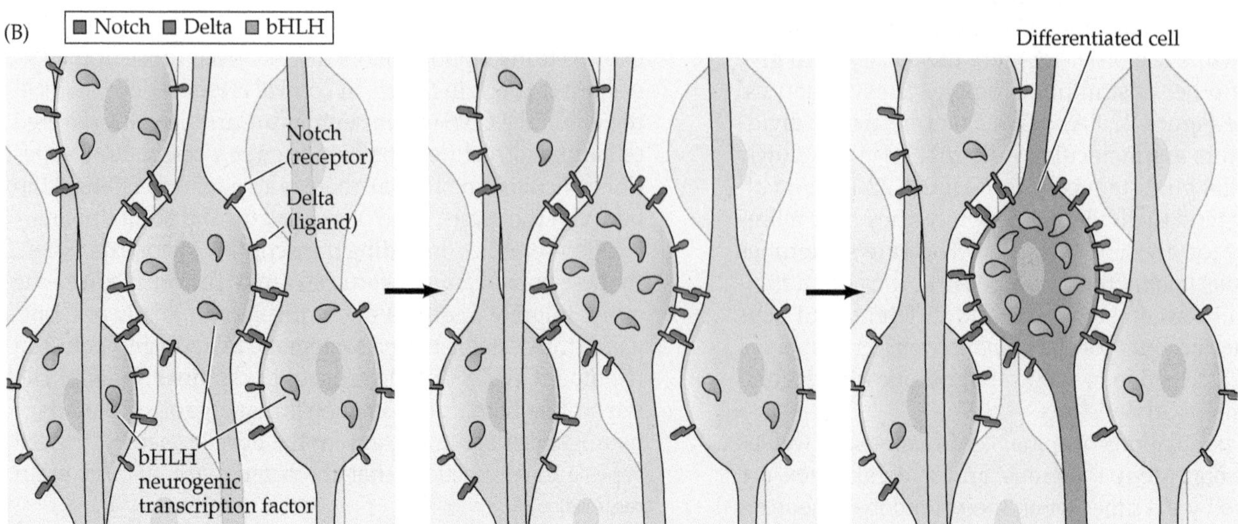

FIGURE 22.10 Delta-Notch signaling leads to neuronal differentiation. Interaction between the Delta cell surface ligands and their Notch receptors on neural progenitor cells in close proximity to one another regulates transcription factors necessary for the generation of differentiated neurons. (A) Delta binding cleaves a protein fragment of the receptor (the Notch intracellular domain, or NICD). When the NICD is transported into the nucleus, it binds to RBP-J, inhibiting RBP-J repression and resulting in the transcription of (among others) the *Hes* family of transcription factors, including the bHLH neurogenic factors responsible for neuronal differentiation. (B) Delta, Notch, and bHLH proteins are expressed at similar levels in a cluster of progenitor cells and neuroblasts. A stochastic increase in Delta ligands on a particular cell leads to downregulation of Delta in neighboring cells, while in the Delta-upregulated cell, bHLH gene expression is also upregulated, and the cell becomes primed for neuronal differentiation.

in the terminal differentiation of neural cells. Aside from the *Hes* genes, the most important homologues of these factors in the vertebrate genome are referred to as the **bHLH (basic helix-loop-helix)** neurogenic factors. Local Delta-Notch signaling among neighboring cells leads to downregulation of Delta in several cells (thus diminishing signaling capacity), but to its upregulation in one or a few of the neighboring cells. In Delta-upregulated cells, bHLH gene expression is also upregulated, and the cell becomes primed for neuronal differentiation (Figure 22.10B). Cells whose Delta levels have been diminished remain as neural stem cells. Similar mechanisms, regulated by Notch signaling and bHLH neurogenic transcription factors, are thought to influence the generation of oligodendrocytes and astrocytes. Thus, although influenced by secreted patterning factors, the ultimate decision to exit the cell cycle and embark on terminal neurogenesis or gliogenesis appears to be regulated by local, cell contact-mediated interactions that depend initially on Delta-Notch signaling.

Generation of Neuronal Diversity

Neural precursor cells that look and act more or less the same give rise to postmitotic cells that are enormously diverse. On the most basic level, these precursors produce neurons and glial cells, two cell types with markedly different properties and functions. Neuronal stem cells then differentiate to produce the diverse neurons and glia of each of the brain regions specified during early morphogenetic events that guide initial steps of brain development. Thus, morphology, neurotransmitter synthesis, cell surface molecules, and the types of synapses made or received distinguish neuron classes of the spinal cord, brainstem, cerebellum, cerebral cortex, hippocampus, and subcortical nuclei, including the basal ganglia and thalamus.

The bulk of evidence favors the view that neuronal differentiation is based primarily on local cell–cell interactions followed by distinct histories of transcriptional regulation via a "code" of transcription factors expressed in each cell, specified by diffusible as well as local cell–cell signals (Figure 22.11; see also Figures 22.5, 22.7, and 22.10). A balance between cell lineage (i.e., who a cell's "parent" is) and cell–cell interactions (i.e., who a cell's neighbors are) has been invoked to explain differentiation of neuronal and glial classes in the developing vertebrate brain. Many of the signaling molecules that are essential for the initial steps of neural induction and regionalization—BMPs, Sonic hedgehog, and Wnts, as well as Delta and Notch—also influence the genesis of specific classes of neurons and glia via local cell–cell interactions.

Among the targets of all of these pathways, different bHLH genes have emerged as central to subsequent differentiation of distinct neural or glial fates (see Figures 22.10 and 22.11). There are multiple bHLH genes, and their restricted expression in distinct rudimentary brain regions exerts a powerful influence on cell identity in those domains (see Figure 22.11). Some of these bHLH genes are homologs of genes originally discovered in developing fruit flies, while others have been identified based on their predicted amino acid sequences inferred from genomic sequences.

These molecular details of cell signaling and subsequent lineage relationships provide an outline of how general cell classes are established. Nevertheless, there is presently no clear and complete explanation for the means by which any specific neuronal class achieves final mature identity. Neuronal identity must ultimately be defined not only by a history of signaling and transcriptional regulation but as current gene expression. In addition, each neuron acquires identity based upon its connections and projections—thus the circuitry of which it is part. This gap in knowledge presents a problem in using neural stem cells to generate replacements for specific cell classes lost to disease or injury, as well as to efforts to understand how neural precursors can transform into the tumorogenic cells that cause medulloblastoma gliomas and other cancers of the developing or mature nervous system.

Molecular and Genetic Disruptions of Early Neural Development

New awareness of the molecules involved in neural induction, neurogenesis, and the generation of neuronal diversity has led to a more informed way of thinking about the etiology of several congenital disorders of the nervous system. Anomalies such as *spina bifida* (failure of the posterior neural tube to close completely), *anencephaly* (failure of the anterior neural tube to close at all), *holoprosencephaly* (disrupted regional differentiation of the forebrain), and other brain malformations can result from environmental insults that disrupt inductive signaling, and from the mutation of genes that participate in this process.

Maternal dietary insufficiency of substances such as folic acid can disrupt embryonic neural tube formation by compromising cellular mechanisms essential for DNA replication and for typical cell division and motility. Conversely, excessive maternal intake of vitamin A—the metabolic precursor of retinoic acid—can impede neural tube closure and differentiation, or it can disrupt later aspects of neuronal differentiation due to an excess of ectopic RA signaling. Embryonic exposure to a variety of other drugs—alcohol and thalidomide are notable examples—can also elicit pathological differentiation of the embryonic nervous system by providing or blocking inductive signals at inappropriate times or places. Altered cholesterol metabolism can compromise Sonic hedgehog signaling because cholesterol molecules play a role in modulating the interaction of Shh with its receptor, Patched. Such metabolic disruptions, as well as rare mutations in the human *SHH*

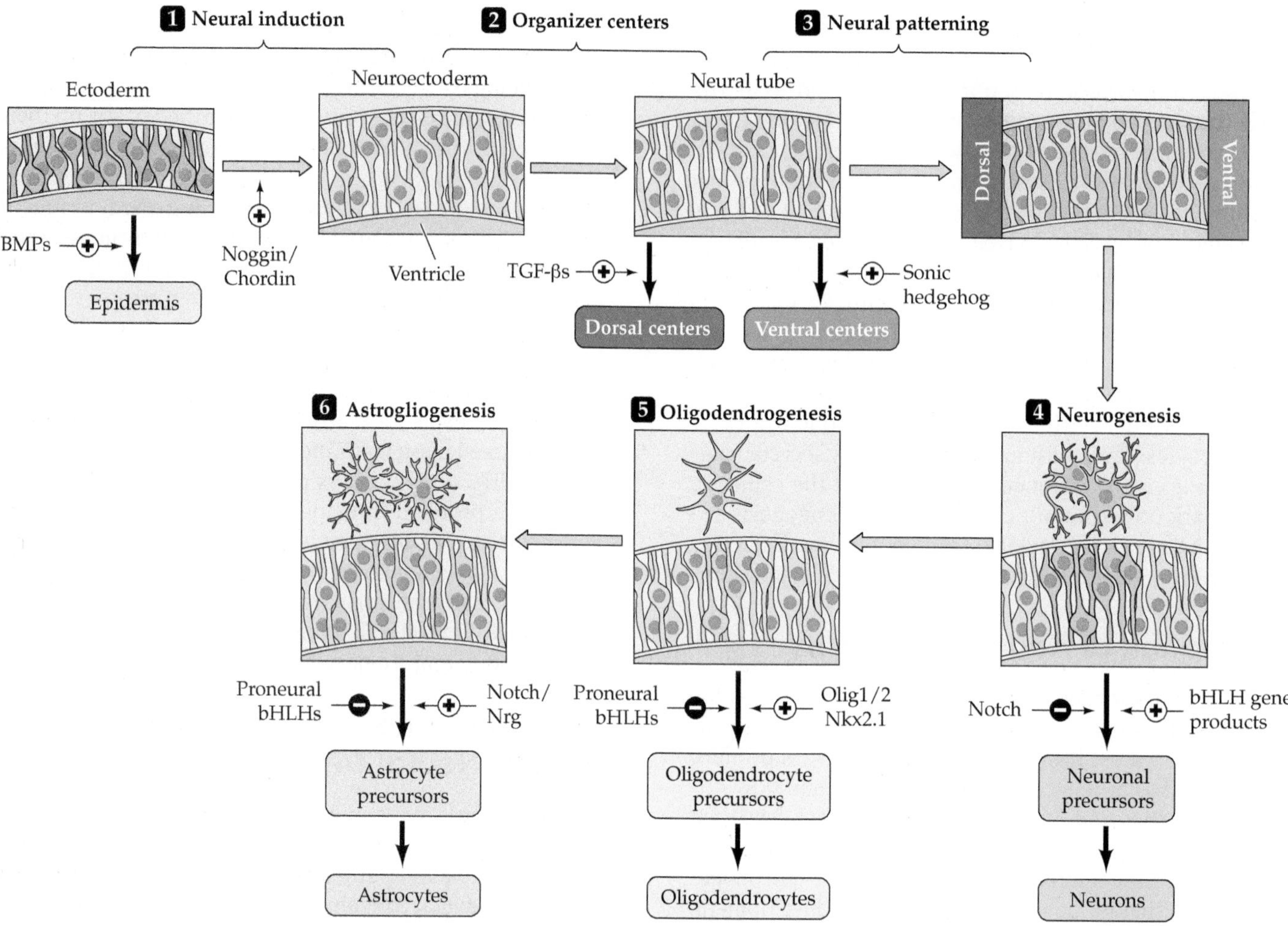

FIGURE 22.11 **Molecular and cellular mechanisms that guide neuronal and glial differentiation.** (1-3) The steps by which ectoderm acquires its identity as neural ectoderm. Generation of neural precursors, or stem cells, relies first on the balance of BMP and its endogenous antagonists (e.g., Noggin and Chordin) in the developing embryo. Next, local sources of inductive signals, including TFG-β family members and Sonic hedgehog, establish gradients that influence subsequent neural precursor identities, as well as identifying local "organizers" (such as the floorplate and roofplate) that define the cellular identity of the inductive signaling centers. (4-6) Steps thought to define neurons, oligodendroglia, and astrocytes from multipotent neuronal precursors. Balanced signaling activity of Notch and transcriptional control of the bHLH proneural genes (named based on their ability to bias neural progenitor cells toward a differentiated neural fate) influence neurogenesis. Similarly, antagonistic transcriptional regulation via either the products of the bHLH genes or three additional transcription factors—Olig1, Olig2, and Nkx2.1—influences the generation of oligodendroglia. Continued antagonism between bHLH proteins, Notch signaling proteins, and the signaling molecule neuregulin (Nrg) is thought to influence the generation of mature astrocytes. Finally, in the adult brain, cells adjacent to the ventricles (which apparently have avoided becoming differentiated) remain as ependymal cells. These may include a subpopulation of neural stem cells (see Box 22A). (After Kintner, 2002.)

gene, are associated with a small proportion of the recorded cases of holoprosencephaly. Mutations in the genes for Shh and other proteins in the Shh pathway—particularly Patched—are also associated with the most prevalent childhood brain tumor, *medulloblastoma*.

Some forms of *hydrocephalus* (a condition in which impeded flow of cerebrospinal fluid increases pressure and results in enlarged ventricles and eventual cortical atrophy from compression) can be traced to mutations of genes on the X chromosome, especially those for the L1 cell adhesion molecule (see Chapter 23). Similarly, fragile-X syndrome, the most common form of congenital mental retardation in males, is associated with triplet repeats in a subset of genes on the X chromosome, particularly the gene for fragile-X protein, which is involved in stabilizing dendritic processes and synapses.

Mutations in additional single genes with distinct functions including regulation of DNA methylation (Methly

CpG-binding Protein 2: MECP2) are associated with rare X-linked genetic disorders. MECP2 loss-of-function mutations are the primary cause of an "autistic-like" disorder, Rett Syndrome. This syndrome is seen nearly exclusively in girls, who are mosaic for a mutant (on one X chromosome) as well as functional MECP2 allele (on the other) due to X-inactivation. Males with a mutant MECP2 gene do not survive through birth. Girls with Rett Syndrome begin life similar to other typically developing children, and then around age 2 to 3 begin to regress, losing language and cognitive function, and developing motor disabilities that are devastating for further intellectual and social behaviors.

Some disorders that compromise the nervous system reflect single-gene mutations in homeobox-like transcription factors. *Aniridia* (characterized by loss of the iris in the eye and mild mental retardation) and *Waardenburg's syndrome* (characterized by craniofacial abnormalities, spina bifida, and hearing loss) are the result of mutations in the *PAX6* and *PAX3* genes, respectively, both of which encode transcription factors. Many additional rare single-gene mutations have been identified and correlated with behavioral deficits that also are seen in broader categories of neurodevelopmental disorders such as intellectual disability or autism. In addition, aneuploid disorders, which are defined by the deletion or duplication of a small region or large segment of an entire chromosome, are associated with neurodevelopmental disorders. Perhaps the best-known example of this class of disorders is *Down syndrome,* or *trisomy 21,* which is caused by the duplication of part or all of chromosome 21, most often due to failure of meiosis during the final stages of oogenesis. This duplication results in three copies of all the genes on chromosome 21. Although the connections between aberrant gene dosage and the resulting anomalies of neural induction, patterning, and neurogenesis are not yet understood, such correlations provide a starting point for exploring the molecular pathogenesis of many congenital disorders of the nervous system.

Neuronal Migration in the Peripheral Nervous System

Cell migration, a ubiquitous feature in the embryo, brings distinct classes of cells into appropriate spatial relationships within differentiating tissues. In the developing nervous system, migration brings different classes of neurons together so they can interact, and it ensures the final position of many postmitotic neurons. The capacity of a neural precursor or an immature nerve cell to move and the path of its transit through a changing cellular environment are essential for its subsequent differentiation. The final location of a postmitotic neuron is presumably especially critical, since neural function depends on precise connections made by these cells and their targets. The developing neuron must be in the right place at the right time so that it can be properly integrated into a functional circuit that can mediate behavior.

As is the case for inductive events during initial formation of the nervous system, stereotyped movements during neuronal migration bring different classes of cells into contact with one another (often transiently), thereby providing a means of constraining cell–cell signaling to specific times and places. Such effects are most thoroughly documented for the migration and differentiation of neural crest cells from the neural tube to the periphery of the embryo; these particular migratory paths are influenced by the initial positions of the neural crest cells at distinct anterior–posterior locations in the neural tube. Their initial positional identity is reflected in the final locations of neural crest cells from various anterior–posterior levels in distinct parts of the body (Figure 22.12A). Thus, the final fates of neural crest cells—including becoming sensory, sympathetic, parasympathetic, and enteric neurons of the peripheral nervous system—are critically dependent on their proper exit from the epithelium of the neural tube, and on their subsequent migration through terrain that provides instructive as well as trophic signals.

The neural crest arises from the dorsal neural tube along the entire length of the spinal cord and hindbrain (see Figure 22.2). Thus, as neural crest cells begin their journeys, they carry with them information about their point of origin, including expression of distinct Hox genes (see Figure 22.4C) that are limited to various spinal cord and hindbrain domains. Regardless of where they originate, all neural crest cells must undergo an essential transition in order to begin migration. They all start out as neuroepithelial cells, with all of the intercellular junctions and adhesive interactions that keep epithelial cells in place. To move, then, neural crest cells must downregulate expression of these adhesive genes and undergo an **epithelial-to-mesenchymal transition** (epithelial cells being bound in a sheet, mesenchymal cells being loosely packed and tending to migrate freely). Presumptive neural crest cells express several transcription factors, including the bHLH family members Snail1 and Snail2 (Figure 22.12B), which repress expression of intercellular junctional proteins and epithelial adhesion molecules.

This process of delamination and migration in the neural crest is similar to the process of metastasis when transformed oncogenic cells escape their epithelial confines, assume a mesenchymal character, and are able to migrate freely (with grave pathological consequences). Indeed, some of the genes that modulate delamination in the neural crest—particularly the Snail genes—can become oncogenes if mutations arise that allow them to be expressed in mature epithelial tissues. When motile neural crest cells reach their final destinations, they cease to express Snail and other transcription factors that induce

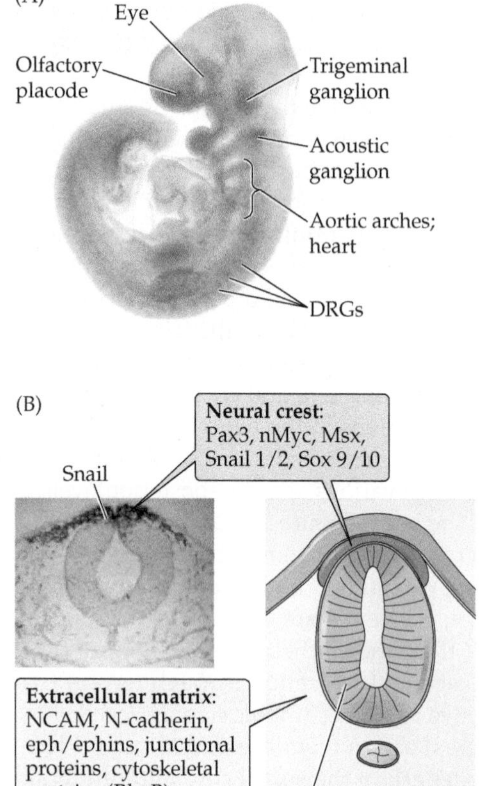

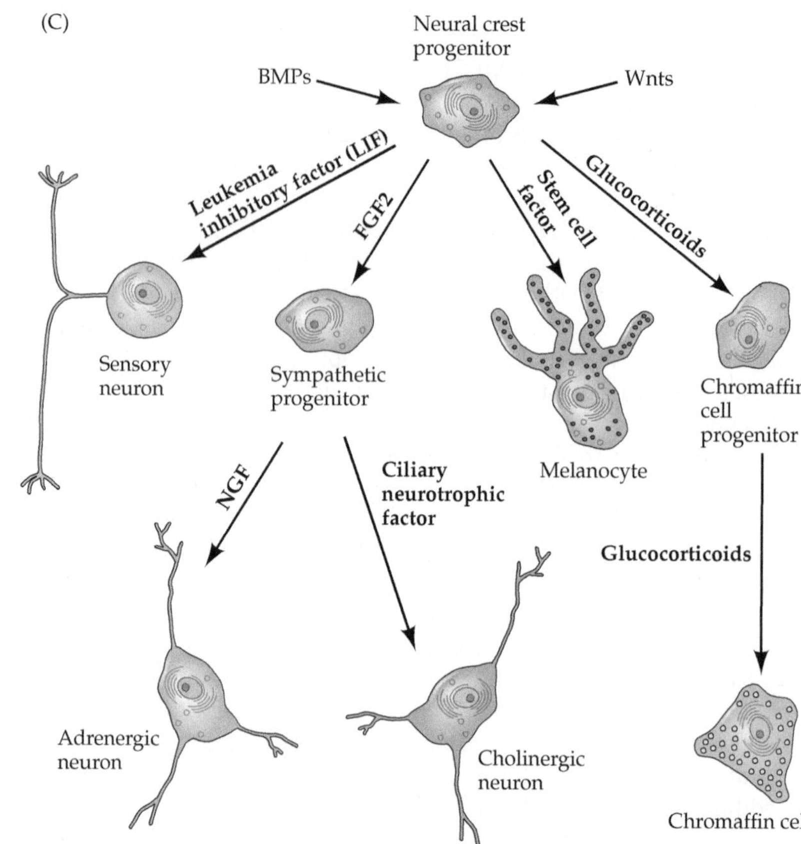

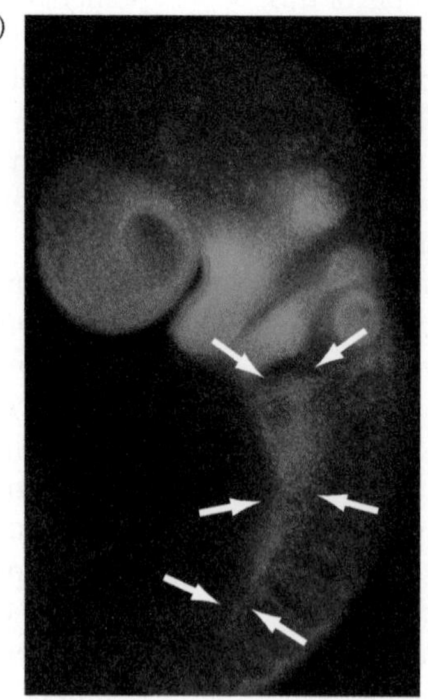

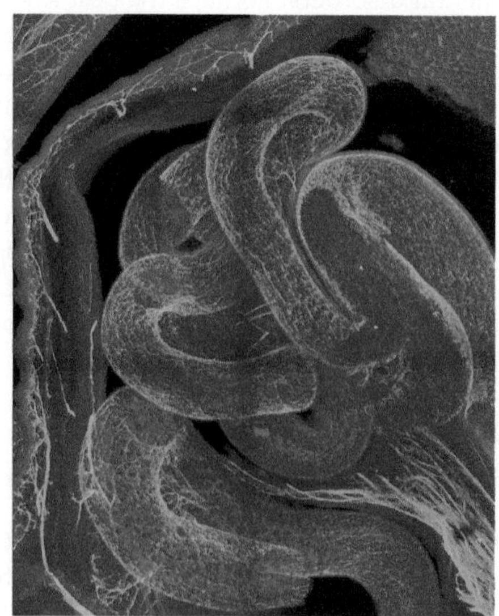

the mesenchymal, migratory state. This change is thought to reflect the integration of several signals the cells encounter along their migratory route. Understanding this normal event has implications for cancer biology, since inducing transformed, migratory cancerous cells to revert to a stationary state would have clear therapeutic value.

Neural crest cells are largely guided along different migratory pathways by signals from non-neural peripheral structures such as somites (which eventually form the axial muscles and skeleton) and other rudimentary musculoskeletal or visceral tissues. The signals along these pathways can be secreted molecules (including some of the same peptide hormones used at earlier times for neural induction), cell surface ligands and receptors (adhesion molecules and other signals), or extracellular matrix molecules; most of these molecules also are used at later stages of development to guide axonal growth and targeting, as described in Chapter 23. Thus, the surfaces of cells in the embryonic periphery display specialized adhesion molecules in the extracellular matrix, such as laminin or fibronectin, or adhesion molecules, such as neural cadherin (N-cadherin), neural cell adhesion molecule (NCAM), or the Eph receptors and their ligands, the ephrins (see Figure 22.12B). In addition, secreted signals including neurotrophic molecules may influence neural crest cell migration. Of particular significance is the fact that specific peptide hormone growth factors available in

◀ FIGURE 22.12 **Migration and differentiation of the neural crest rely on regulation of gene expression and cell-cell signaling.** (A) The migrating neural crest is visualized in a mouse embryo in which a reporter has been inserted into a gene that is typically expressed in the neural crest. Individual neural crest cells (blue) can be seen accumulating around the olfactory placode, the eye (where they will contribute to the pigment epithelium of the retina), the trigeminal (V) and acoustic (VIII) ganglia, the aortic arches and heart, and the immature dorsal root ganglia (DRGs). (B) In a section through the spinal cord (left), neural crest cells labeled for the bHLH transcription factor Snail exit the dorsal spinal cord and stream toward the periphery. The schematic (right) indicates sites of action of molecules that favor the mesenchymal, migratory (for the neural crest), or neuroepithelial state. (C) Cell signaling during migration of neural crest cells influences progenitor identity and terminal differentiation. Each signal is available along a specific migratory route taken by subsets of neural crest cells. (D) Migrating neural crest cells during early mid-gestation (embryonic day 10 of 21 in the mouse embryo, shown here) include a substantial contingent that targets the developing viscera (white arrows). These neural cells crest are called vagal neural crest cells. The mature neurons and other cell types generated by these crest cells will be either innervated by the vagal motor neurons in the brainstem, or give rise to the vagal sensory nerve from the periphery that relays visceral sensation. In and around the heart, the vagal crest contributes directly to the great vessels of the heart, and in the more ventral visceral regions it forms the enteric nervous system. (E) The enteric nervous system of any particular vertebrate species includes more neurons than are found in the spinal cord of that same species. This mature enteric nervous system, primarily the portion that innervates the small intestine, forms a plexus, or complex network of cells and axons (green labeled processes) that envelop the outer wall of the gut. The mature enteric neurons secrete neurotransmitters and neuropeptides that regulate gut motility, and modulate gut immune responses and other physiological functions of the intestine. (A courtesy of A.-S. LaMantia; B courtesy of M. A. Nieto; D courtesy of B. Karpinski; E courtesy of M. Howard.)

particular peripheral targets cause neural crest cells to differentiate into distinct phenotypes (Figure 22.12C). These cues modulate the expression of bHLH genes in neural crest cells during the transition from migratory precursor to postmitotic neuroblast. Thus, the balance of migratory capacity, instructive cues, and modification of gene expression seen during the transit of the neural crest from the neural tube to the periphery illustrates the influence of migration on the establishment of neuronal identity.

The neural crest can become non-neuronal cells including melanocytes (pigment cells in skin), chromaffin cells of the adrenal gland, or cardiovascular cell classes in the great arteries of the heart. There are several serious cardiovascular developmental defects that reflect anomalous migration or differentiation of the component of the crest referred to as the **vagal crest** (Figure 22.12D). In addition, the neural crest gives rise to one of the largest divisions of the entire nervous system, the enteric nervous system (Figure 22.12E). The **enteric nervous system** (**ENS**) consists of multiple neuron classes that secrete neurotransmitters and neuropeptides to regulate gut motility, immune function and local secretion of digestive enzymes. The ENS is in turn regulated by autonomic pathways and ultimately by the CNS via the hypothalamus and its influence on vagal motor neurons and pre-ganglionic thoracic motor neurons. Disrupted migration of the neural crest (also part of the vagal crest; see Figure 22.12D) that forms the enteric nervous system can result in serious digestive disorders including Hirschprung's disease. Individuals with Hirschprung's disease—particularly newborn infants—develop distensions of the small intestine and colon, have difficulty excreting waste, fail to gain weight, and thus can be under-nourished. This disorder is treated surgically, by removing the intestinal tissue that lacks proper enteric nervous system innervation and function.

Neuronal Migration in the Central Nervous System

Neuronal migration is not limited to the periphery. Neurons generated in several locations in the CNS must also move from the site of their initial genesis to a distant site, where they differentiate and are integrated into mature neural circuits. The mechanisms of central neuronal migration are diverse, and its successful completion is essential for many aspects of typical brain function.

A minority of nerve and glial cells in the CNS (and a few in the periphery) use existing axon pathways as migratory guides. These include subsets of cranial nerve nuclei in the hindbrain; nuclei in the pons that project to the cerebellum; Schwann cells that myelinate peripheral axons; and a small population of neurons that migrate during embryonic development from the olfactory epithelium in the nose to the hypothalamus, where they secrete gonadotropin-releasing hormone (GnRH), which is essential for regulating reproductive functions in the mature animal.

The most prominent form of neuroblast migration in the CNS, however, is that guided by glial cells. Many neuroblasts that migrate long distances in the CNS—including those in the cerebral cortex, cerebellum, and hippocampus—are guided to their final destinations by following the long processes of radial glia (or similar glia in the cerebellum, called Bergmann glia). Radial glial cells have multiple functions in the developing brain. In addition to acting as migratory guides, they are currently thought to be neuronal progenitor cells in the developing CNS (see Figure 22.8 and Chapter 25).

Histological observations of embryonic brains made by Wilhelm His and Ramón y Cajal during the nineteenth and early twentieth centuries suggested that neuroblasts in the developing cerebral cortical hemispheres followed glial guides to their final locations (Figure 22.13A). These light microscopic observations were supported by analyses of electron microscopic images of fixed tissue in the 1960s and 1970s as well as molecular labeling that identified the radial glial cells and migrating neurons as distinct cell classes (Figure 22.13B). More recent in vivo microscopic observation has confirmed these inferences, and also has shown that radial glia do indeed serve dual functions, being both stem cells and migration guides. The apparent scaffold for radial movement of postmitotic neurons established by the radial glia fits well with the orderly relationship between birth dates and final position of distinct cell types in the cerebral cortex as well as the cerebellum (see Figure 22.9). By adhering to the glial process, a cell can move past dividing cells in the ventricular and subventricular zone as well as already differentiating neurons in lower cortical layers (layers 5 and 6, for example) toward the cortical surface, where it will contact the end feet of radial glial cells and disengage from the glial surface. In the cerebellum, the direction of this process is reversed. The precursors for cerebellar granule cells are actually on the cerebellar surface, and the processes of Bergmann glia guide postmitotic granule cells past the already generated Purkinje cells into the nascent internal granule cell layer. Thus, glia-guided migration appears to ensure translocation of newly generated neurons past other neurons already in place in the developing nervous system.

Molecular Mechanisms of Neuronal Migration and Cortical Migration Disorders

Increased understanding of molecular mechanisms and direct observation of migrating neurons and their glial guides in live developing brains indicate that the process of neuronal migration, particularly in the cortex, may be vulnerable to the effects of genetic mutations that disrupt either the ability of the nerve cell to move, the ability of radial glial cells to support migration, or both. This inference received initial support from the characterization of several single-gene mutants in the mouse that disrupted the orderly placement of neurons based on their time of origin. Subsequent work identified several proteins whose function is vital for normal migration along radial glia. Some of these proteins are found

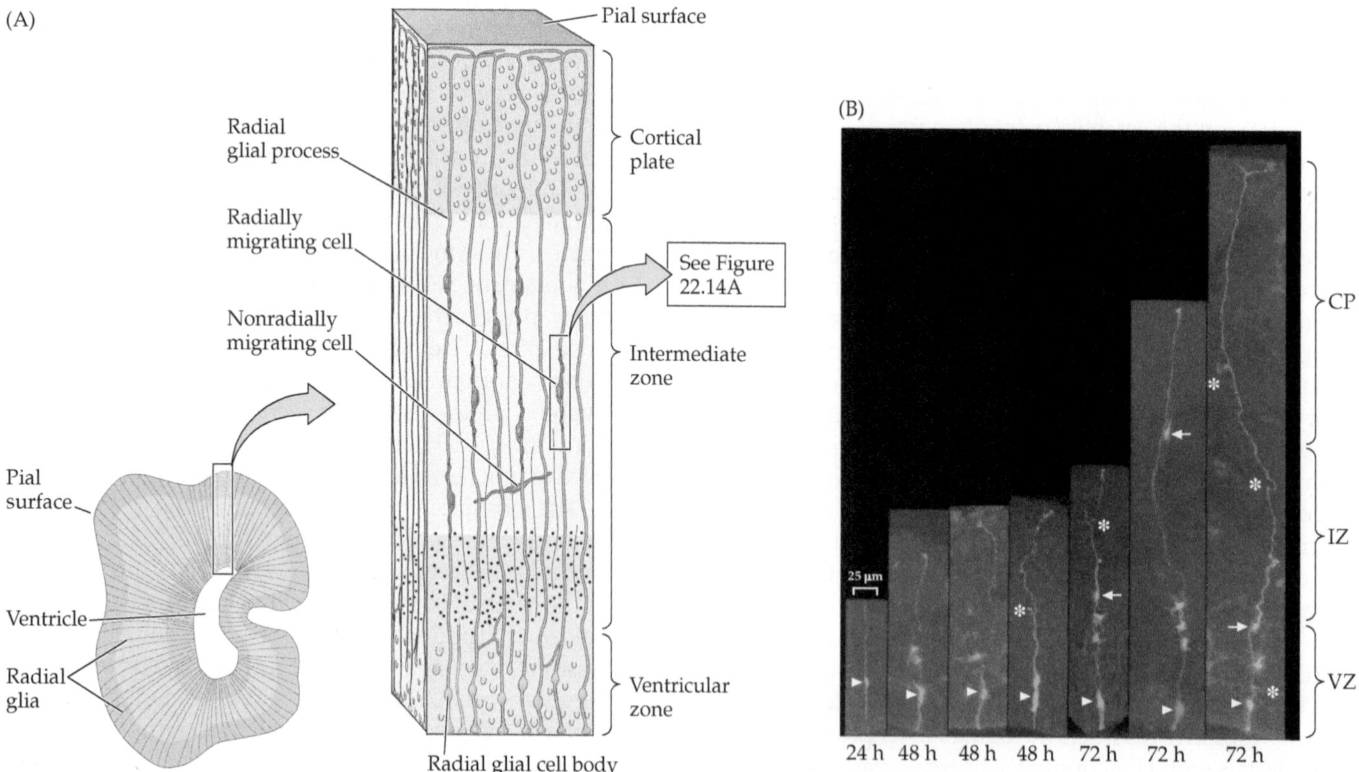

FIGURE 22.13 Radial glia guide migrating neurons. (A) Section through the developing forebrain showing radial glial processes from the ventricular to the pial surfaces. Migrating neurons are intimately apposed to radial glial cells, which guide them to their final position in the cortex. Some cells take a nonradial migratory route, which can lead to wide dispersion of neurons derived from the same precursor. (B) Time-lapse micrography showing radially migrating (arrows) and nonradially migrating (asterisks) neurons. (A after Cowan, 1979, based on Rakic, 1971; B from Noctor et al., 2001.)

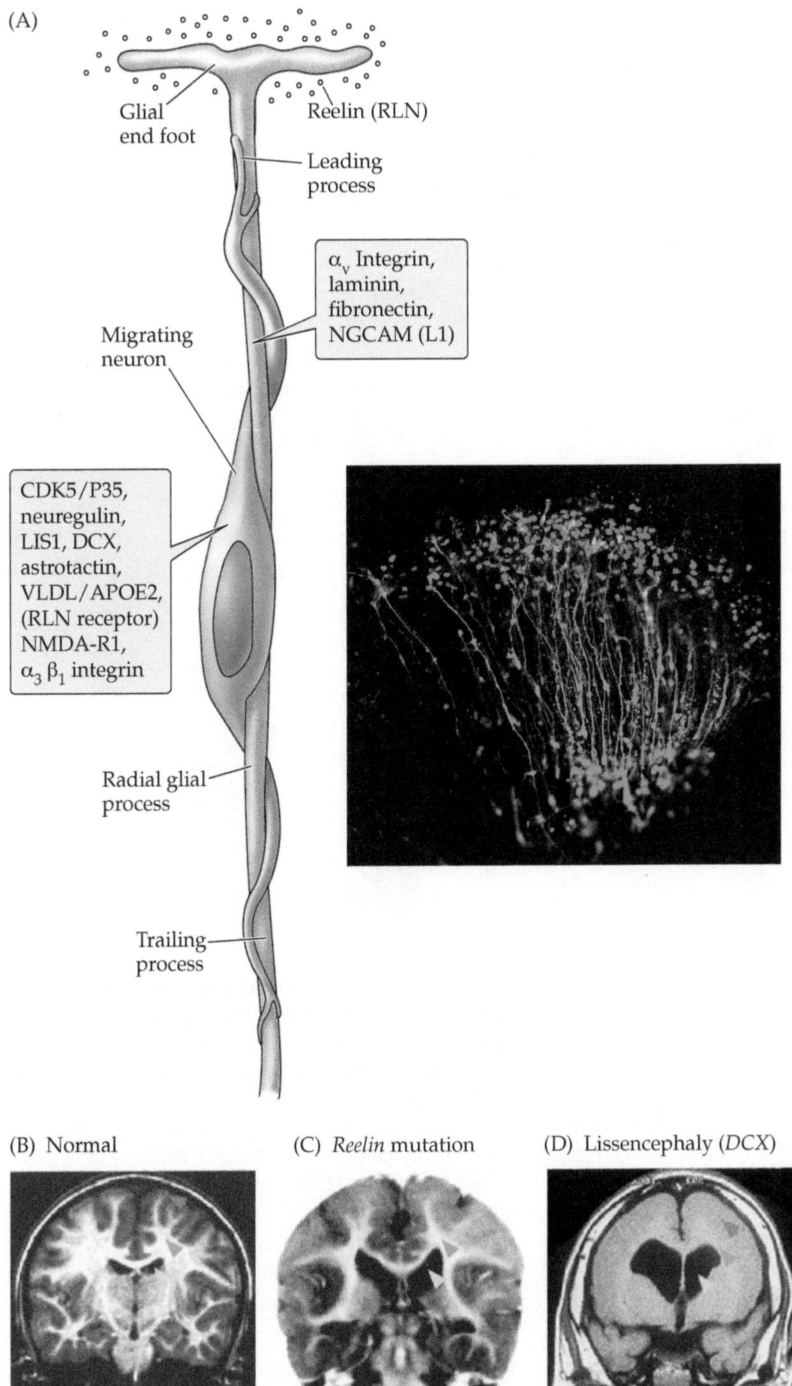

FIGURE 22.14 Radial migration in the developing cortex. (A) A single neuroblast migrates on a radial glial process (based on serial reconstruction of EM sections as well as in vitro assays of migration). Cell adhesion and other signaling molecules or receptors found on the surface of either the neuron (green) or the radial glial process (tan) are indicated in the respective boxes. The micrograph shows migrating neurons labeled with an antibody to neuregulin, a protein specific to migrating cortical neurons. (B–D) Mutations in genes that influence neuronal migration cause malformations of the human cerebral cortex. In these MRI images, yellow arrows point to the lateral ventricle, green arrows indicate the subcortical white matter in the internal capsule (a "thoroughfare" for axons entering and exiting the cerebral cortex), and red arrows highlight the typical appearance of sulci and gyri. (B) Typical cerebral cortex. (C) Individual with a mutation of the gene encoding Reelin, a protein that influences radial neuronal migration in the cortex. The lateral ventricles are enlarged, subcortical white matter is diminished, and the pattern of sulci and gyri is disrupted. (D) In an individual with a mutation in the *DCX* gene, the ventricles are dramatically enlarged, subcortical white matter is nearly absent, and sulci and gyri are completely absent. This dramatic cortical malformation is known clinically as lissencephaly ("smooth brain"). (A Cowan, 1979, based on Rakic, 1971; micrograph courtesy of E. S. Anton and P. Rakic; B,C from Hong et al., 2000; D courtesy of C. A. Walsh.)

on the surface or in the cytoplasm of the migrating neuron itself; others are found on the glial cell surface (Figure 22.14A). When the function of any of these proteins is disrupted, neurogenesis, migration, and cortical lamination can be compromised.

A particularly compelling demonstration of the reliance of cortical neurogenesis and migration on specific genes and their protein products has come from analysis of human patients with a variety of brain malformations that can be visualized using magnetic resonance imaging (Figure 22.14B–D). Genetic analysis of these patients has identified mutations in several genes that can disrupt cortical migration. In some cases (such as that of the molecule Reelin, which is available near the glial end feet and is thought to influence the detachment of neurons from the radial glia; see Figure 22.14A), mutations in humans result in a cortical phenotype very similar to that seen in mice in which the analogous gene is experimentally deleted (see Figure 22.14C). In patients with lissencephaly—"smooth brain," a condition in which the cortex has no sulci or gyri (see Figure 22.14D)—several novel genes have been identified, including *Lissencephaly 1*, or *LIS1*, and *Doublecortin*, or *DCX*. The LIS1 protein interacts with the cellular motor protein dynein, and mutations may disrupt dynein-mediated aspects of cell division in cortical progenitors or organelle transport in migrating cortical neurons. The DCX protein interacts with microtubules in migrating neurons and is thought to influence both the integrity of the cytoskeleton in these cells and the appropriate transport of organelles during initial differentiation. In other individuals, including those with "small brains," or microcephaly, other genes are mutated (including those for the mitotic spindle–associated protein ASPM and the protein kinase CDK5RAP2), leading to reduced neurogenesis as well as altered migration.

Neuropathological observations as well as molecular and genetic studies have confirmed that several neurological problems, including epilepsy and some forms of mental retardation, arise from the abnormal migration of cerebral cortical neurons. Several additional molecules that influence cell-cell signaling or cell adhesion are associated with disorders that are thought to reflect, in part, disrupted neuronal migration in the developing cerebral cortex. In particular, neuregulin (a secreted signal), NCAM, and *DISC1* (a gene mutated or deleted in a small number of cases of schizophrenia) are highly associated with risk for psychiatric diseases. Thus, disrupted cell migration (and its consequences for subsequent development of brain synapses and circuits) may underlie the pathology of several serious brain disorders.

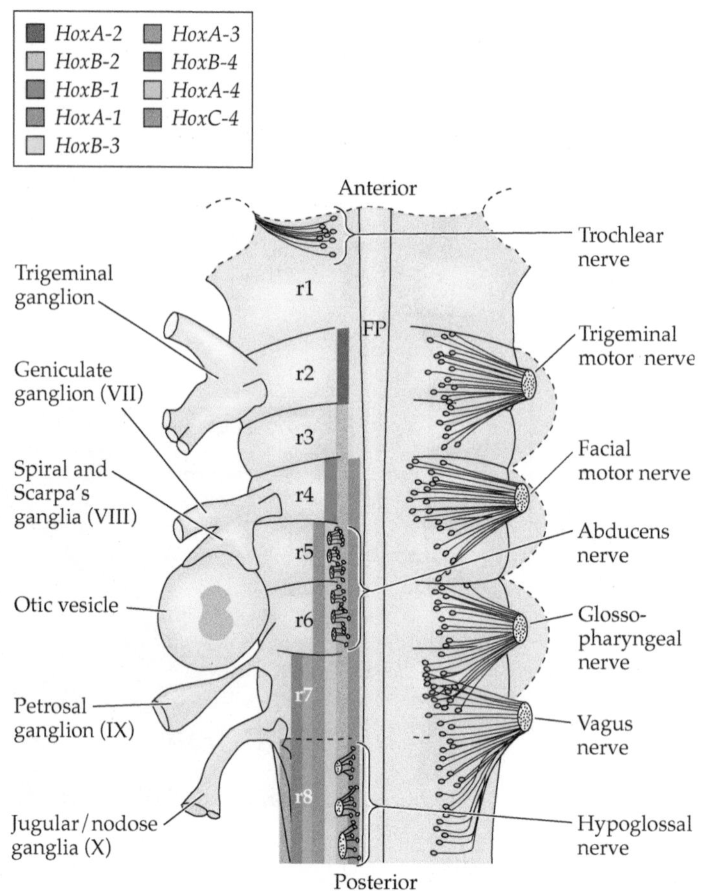

FIGURE 22.15 The relationship between early segmental hindbrain patterning and the differentiation of mature cranial sensory and motor nerves. Rhombomeres in the developing hindbrain are defined by restricted patterns of Hox gene expression (colored bands). The subsequent specification of neural crest as well as neural progenitors that remain in the hindbrain guide the differentiation of the cranial sensory ganglia and nerves (left) as well as cranial motor or mixed nerves (right). (Courtesy of A. Lumsden.)

Cranial Placodes: The Critical Connection between the Brain and the Outside World

An additional critical event occurs during early mammalian neural development that is essential for establishing a functional nervous system: the specification and differentiation of the **cranial** and **epibranchial placodes**. These patches of primarily neurogenic ectoderm are established on the surface of the embryo, and subsequently undergo neuronal differentiation, distinguishing them from the surrounding ectoderm that will presumably differentiate to epidermis. Thus, there are cranial or epibranchial placodes that give rise to the olfactory epithelium, the otic vesicle and organ of Corti, the lens and cornea (but not the neural part of the retina), and the mechanosensory neurons of the cranial sensory ganglia (Figure 22.15): the trigeminal, the geniculate, the petrosal/jugular, the superior vagal, and the nodose. Once the placodes are differentiated from the surface ectoderm, all of them undergo inductive interactions with neural crest cells, which drive further differentiation. In most cases, with the exception of the olfactory epithelium, the resulting cranial sensory structure will contain both placode- and neural crest-derived neurons. Each of these placode- or crest-derived sensory neuron populations has the unique capacity to transduce specific classes of sensory stimuli directly from the environment, leading to the initial encoding of information to be further processed in the brain. Accordingly, the result of placode formation is the establishment of the "lines into" the CNS regions that mediate sensory processing.

The cranial placodes themselves are initially apparent as ectodermal thickenings during early development of the vertebrate head (Figure 22.16A,B). They can be recognized molecularly by their expression of signaling molecules such as FGF8 (see Figure 22.16A) as well as transcription factors. Adjacent to these thickenings are primarily populations of neural crest cells that remain as mesenchyme (loosely arrayed cells that are not organized into a "sheet" or epithelium). The interaction of the placodes with the adjacent neural crest is key. In some cases (olfactory placode, lens placode, otic placode), the neural crest makes no contribution or only a negligible contribution to the sensory neuron populations (see Figure 22.16B, top). In the case of the cranial ganglia a significant contributor to the peripheral portions of the cranial nerves (trigeminal, vagal, etc.; see Appendix), the placodal cells are transformed to neural progenitors and generate sensory—primarily mechanosensory—neurons (touch, pressure, proprioception; see Chapter 9). The neural crest also contributes to the cranial ganglia by generating the majority of nociceptive or pain-sensing sensory neurons (see Figure 22.16B, bottom).

All of the cranial placodes that generate either special sensory (olfactory, auditory) or somatosensory

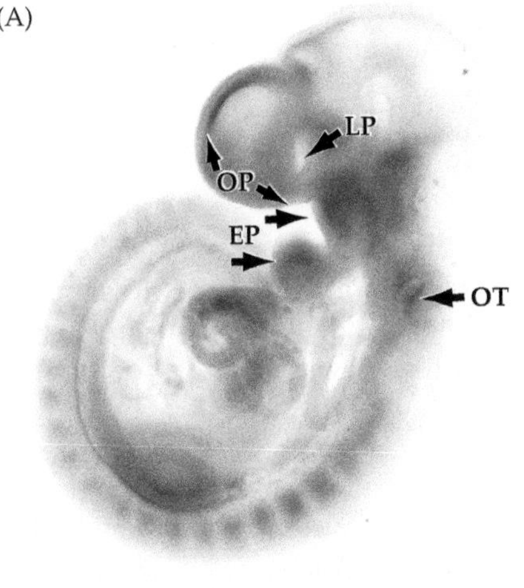

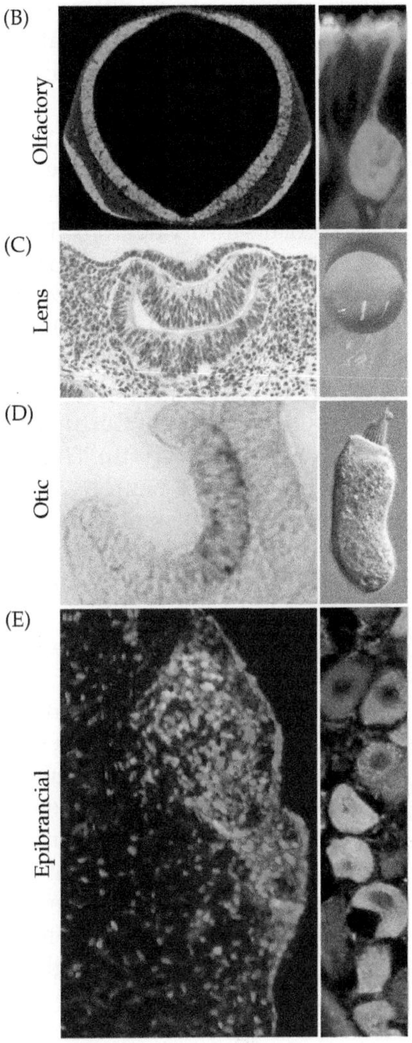

FIGURE 22.16 The cranial ectodermal placodes are essential for establishing peripheral sensory receptors and related structures. (A) The location of the cranial placodes in an E9.5 mouse embryo. This embryo has been hybridized for the *Dgcr8* gene which encodes a protein, Pasha, that is part of the Drosha/Pasha micro RNAase microprocessor complex. This gene is expressed at all of the sites of the cranial placodes: the olfactory placode (OP) which will generate the olfactory sensory epithelium, the lens placode (LP) which will generate the lens of the eye, the epibranchial placodes (EP) which are associated with the branchial arches and contribute neural precursors that generate cranial somatosensory mechanoreceptive and proprioceptive neurons, and the otic placode which will generate the hair cells of the inner ear. (B) The olfactory placode labeled with an antibody to β-tubulin, a marker of early neuronal identity. The two placodes, now still thickenings on the ventro-lateral aspect of the head, will eventually invaginate and then continue to grow until the mature olfactory epithelium (see Chapter 15) is in place. At right is an image of an olfactory receptor neuron. These neurons are generated from stem cells resident in the olfactory placode. (C) The lens placode gives rise to the non-neural lens in the mature eye. The placode, seen here as a thickening of the surface ectoderm that has begun to invaginate, will eventually invaginate completely. The placode-dervied epithelium then forms an epithelial sphere that encloses a lumen. The mature lens cells, which are optically clear, will differentiate from these progenitors. (D) The otic placode gives rise to structures of the ear, including the sensory organs: the cochlea and the vestibular labyrinth. Aside from the physical structures of the outer and inner ear, these progenitors will also generate hair cells (both auditory and vestibular; an auditory hair cell is shown at right) that transduce air molecule displacement ("sound waves") into what is perceived as sound. (E) The epibranchial placodes are thickenings of the ectoderm in the region of the branchial arches that also form the main structures of the lower face (jaws, teeth, etc.). These thickenings of the branchial arch ectoderm, shown here labeled with an antibody to the Six1 protein, a specific marker for placode cells and their progeny, will eventually undergo an epithelial to mesenchymal transition. The delaminated placodal cells are then displaced into the underlying mesenchyme where they coalesce with neural crest derived cells to form the cranial ganglia (the nascent trigeminal ganglion-V; and geniculate or facial ganglion-VII). Eventually, the placodal cells will generate somatic sensory receptor cells (at right), particularly those that transduce information about mechanosensory stimuli. (A courtesy of A.-S. LaMantia and T. M. Maynard; B,E courtesy of A.-S. LaMantia and B. A. Karpinski; B right courtesy of C. Balmer and A.-S. LaMantia; C from Carmona et al., 2008; C right courtesy of M. Fickett; D from Birol et al., 2016; D right from Hudspeth, 1985.)

(trigeminal ganglion and others) neurons are mediated by molecular signals between the placodal ectoderm and the underlying neural crest–derived mesenchyme (see Figure 22.16A). These interactions were first characterized for non-neural cranial structures, including the teeth, and then the observations were extended to the establishment of sensory neurons. The olfactory placode is perhaps the clearest example of the placodal ectoderm–neural crest interactions (see Figure 22.16A,B). In *Pax6* mutant mice, the neural crest fails to migrate properly into the olfactory placode, and it does not produce retinoic acid, and presumably, other neural crest–mesenchyme-derived factors. Consequently, the olfactory epithelium (and olfactory bulb, since the neural crest mesenchyme also influences ventral forebrain induction) fails to differentiate. If one recapitulates the basic mesenchymal-epithelial interactions in the olfactory primordium (placodal ectoderm plus mesenchyme) in vitro, by separating the layers and then either culturing them independently or reapposing them to one another (Figure 22.16C), an olfactory epithelium only forms when mesenchyme and epithelium are apposed to one another. These mesenchymal–epithelial interactions establish the appropriate medial–lateral and anterior–posterior axes. A coherent, fasiculated olfactory nerve forms, and grows anteriorly toward its absent forebrain target (see Figure 22.16C). Similarly, for the otic placode and vesicle, local neural crest interacts with placodal ectoderm to drive axial patterning and final differentiation of cochlear sensory cells (Figure 22.16D).

Subsequent placodal differentiation is distinct for each site. Somatosensory cranial ganglion neurons generate mechanosensory (mostly placode) as well as pain- and temperature-sensitive neurons, each of which has distinct endings for somatosensory transduction in the periphery (see Chapters 9 and 10). In the olfactory placode, ectoderm cells generate local neural stem cells that then generate bipolar sensory neurons. These neurons express the gene for single-allele single-odorant receptor molecules (7-transmembrane G-protein-coupled receptors, of which there are between 600 and 2000 in vertebrate and invertebrate genomes; see Chapter 15). In the otic placode, the placodal ectoderm produces hair cells for the auditory system (in the organ of Corti) and the vestibular system (in the vestibular labyrinth). The lens placode is distinct—it does not produce neurons. Instead, it invaginates to produce nearly transparent cells that secrete an equally transparent matrix to create the lens, which transmits and focuses light on the neural retina, and is flexible so that it can be changed to accommodate different focal planes. The epibranchial placodes are ectodermal thickenings that arise at the boundaries of each branchial arch. The thickened ectodermal cells become neural (influenced by local FGF signaling, as well as BMPs) and then delaminate from the placodal epithelium and translocate a small distance into an adjacent nascent cranial ganglion (Figure 22.16E). These neurons then coalesce with the neural crest cells to form the ganglion.

The establishment of placodes is often the target of genetic diseases that cause extreme disruption of sensory neuron development along with craniofacial anomalies. In addition, it is likely that more subtle changes in placode induction and differentiation could modulate the efficiency of sensory transduction and thus account for individual differences—including some that impair olfaction, audition, somatosensation, and taste—in sensory acuity.

Summary

The initial development of the nervous system depends on an intricate interplay of inductive signals, stem cell lineage progression from ectodermal to neural stem cells, cell proliferation, and cellular movements. In addition to the early establishment of regional identity, cellular identity, and cellular position within the brain, substantial migration of neuronal precursors is necessary for the subsequent differentiation of classes of neurons and the eventual formation of specialized patterns of synaptic connections (see Chapters 8 and 23). The fate of individual precursor cells is not determined simply by their mitotic history; rather, the information required for differentiation arises from interactions between the developing cells, local signaling molecules, and the subsequent activity of distinct transcriptional regulators. A final step in the initial differentiation of the nervous system is the specification and differentiation of the sensory placodes. These "neural outposts" in the periphery of the early embryo are key for establishing communication between the outside sensory world and the nervous system. They engage both local ectoderm and neural crest cells to generate peripheral sensory neurons such as the somatosensory neurons of the dorsal root and cranial sensory ganglia, the olfactory receptor neurons, and the hair cells and supporting structures of the inner ear. Although the neural portion of the retina is derived from the CNS, supporting structures such as the lens and cornea arise from placodal tissues. All of these events depend on the same categories of molecular and cellular phenomena: cell-cell signaling, changes in motility and adhesion, transcriptional regulation, and ultimately, cell-specific changes in gene expression. The molecules that participate in signaling during early brain development are the same as the signals used by mature cells: hormones, transcription factors, and second messengers (see Chapter 7), as well as cell adhesion molecules. The importance of signals in the progression from multipotent ectodermal stem cell is essential for the appropriate development of the central and peripheral nervous system. The identification and characterization of these molecules in the developing brain, the regions they specify and the stem cells they influence have begun to explain a variety of congenital neurological defects, as well as providing initial insight into the genetic and cellular basis of a number of developmental disorders. These associations with brain pathologies reflect the vulnerability of signaling and transcriptional regulation during early neural development to the effects of genetic mutations, as well as to the actions of the many drugs and other chemicals that can compromise the elaboration of a typical nervous system.

ADDITIONAL READING

Reviews

Anderson, D. J. (1993) Molecular control of cell fate in the neural crest: The sympathoadrenal lineage. *Annu. Rev. Neurosci.* 16: 129–158.

Caviness, V. S., Jr. and P. Rakic (1978) Mechanisms of cortical development: A view from mutations in mice. *Annu. Rev. Neurosci.* 1: 297–326.

Francis, N. J. and S. C. Landis (1999) Cellular and molecular determinants of sympathetic neuron development. *Annu. Rev. Neurosci.* 22: 541–566.

Hatten, M. E. (1993) The role of migration in central nervous system neuronal development. *Curr. Opin. Neurobiol.* 3: 38–44.

Ingham, P. (1988) The molecular genetics of embryonic pattern formation in *Drosophila*. *Nature* 335: 25–34.

Jessell, T. M. and D. A. Melton (1992) Diffusible factors in vertebrate embryonic induction. *Cell* 68: 257–270.

Kessler, D. S. and D. A. Melton (1994) Vertebrate embryonic induction: Mesodermal and neural patterning. *Science* 266: 596–604.

Keynes, R. and R. Krumlauf (1994) Hox genes and regionalization of the nervous system. *Annu. Rev. Neurosci.* 17: 109–132.

Kintner, C. (2002) Neurogenesis in embryos and in adult neural stem cells. *J. Neurosci.* 22: 639–643.

Lewis, E. M. (1992) The 1991 Albert Lasker Medical Awards. Clusters of master control genes regulate the development of higher organisms. *JAMA* 267: 1524–1531.

Linney, E. and A. S. LaMantia (1994) Retinoid signaling in mouse embryos. *Adv. Dev. Biol.* 3: 73–114.

Rice, D. S. and T. Curran (1999) Mutant mice with scrambled brains: Understanding the signaling pathways that control cell positioning in the CNS. *Genes Dev.* 13: 2758–2773.

Rubenstein, J. L. R. and P. Rakic (1999) Genetic control of cortical development. *Cerebral Cortex* 9: 521–523.

Sanes, J. R. (1989) Extracellular matrix molecules that influence neural development. *Annu. Rev. Neurosci.* 12: 491–516.

Selleck, M. A., T. Y. Scherson and M. Bronner-Fraser (1993) Origins of neural crest cell diversity. *Dev. Biol.* 159: 1–11.

Zipursky, S. L. and G. M. Rubin (1994) Determination of neuronal cell fate: Lessons from the R7 neuron of *Drosophila*. *Annu. Rev. Neurosci.* 17: 373–397.

Important Original Papers

Anchan, R. M., D. P. Drake, C. F. Haines, E. A. Gerwe and A. S. LaMantia (1997) Disruption of local retinoid-mediated gene expression accompanies abnormal development in the mammalian olfactory pathway. *J. Comp. Neurol.* 379: 171–184.

Angevine, J. B. and R. L. Sidman (1961) Autoradiographic study of cell migration during histogenesis of cerebral cortex in the mouse. *Nature* 192: 766–768.

Bulfone, A. and 5 others (1993) Spatially restricted expression of *Dlx-1*, *Dlx-2* (*Tes-1*), *Gbx-2*, and *Wnt-3* in the embryonic day 12.5 mouse forebrain defines potential transverse and longitudinal segmental boundaries. *J. Neurosci.* 13: 3155–3172.

Eksioglu, Y. Z. and 12 others (1996) Periventricular heterotopia: An X-linked dominant epilepsy locus causing aberrant cerebral cortical development. *Neuron* 16: 77–87.

Ericson, J., S. Morton, A. Kawakami, H. Roelink and T. M. Jessell (1996) Two critical periods of sonic hedgehog signaling required for the specification of motor neuron identity. *Cell* 87: 661–673.

Galileo, D. S., G. E. Gray, G. C. Owens, J. Majors and J. R. Sanes (1990) Neurons and glia arise from a common progenitor in chicken optic tectum: Demonstration with two retroviruses and cell type-specific antibodies. *Proc. Natl. Acad. Sci. USA* 87: 458–462.

Gray, G. E. and J. R. Sanes (1991) Migratory paths and phenotypic choices of clonally related cells in the avian optic tectum. *Neuron* 6: 211–225.

Hafen, E., K. Basler, J. E. Edstroem and G. M. Rubin (1987) *Sevenless*, a cell-specific homeotic gene of *Drosophila*, encodes a putative transmembrane receptor with a tyrosine kinase domain. *Science* 236: 55–63.

Hemmati-Brivanlou, A. and D. A. Melton (1994) Inhibition of activin receptor signaling promotes neuralization in *Xenopus*. *Cell* 77: 273–281.

Liem, K. F., Jr., G. Tremml and T. M. Jessell (1997) A role for the roof plate and its resident TGFβ-related proteins in neuronal patterning in the dorsal spinal cord. *Cell* 91: 127–138.

McMahon, A. P. and A. Bradley (1990) The *wnt-1* (*int-1*) protooncogene is required for the development of a large region of the mouse brain. *Cell* 62: 1073–1085.

Noden, D. M. (1975) Analysis of migratory behavior of avian cephalic neural crest cells. *Dev. Biol.* 42: 106–130.

Patterson, P. H. and L. L. Y. Chun (1977) The induction of acetylcholine synthesis in primary cultures of dissociated rat sympathetic neurons. *Dev. Biol.* 56: 263–280.

Rakic, P. (1971) Neuron–glia relationship during granule cell migration in developing cerebral cortex: A Golgi and electron microscopic study in *Macacus rhesus*. *J. Comp. Neurol.* 141: 283–312.

Rakic, P. (1974) Neurons in rhesus monkey visual cortex: Systematic relation between time of origin and eventual disposition. *Science* 183: 425–427.

Sauer, F. C. (1935) Mitosis in the neural tube. *J. Comp. Neurol.* 62: 377–405.

Spemann, H. and H. Mangold (1924) Induction of embryonic primordia by implantation of organizers from a different species. Translated by V. Hamburger and reprinted in *Foundations of Experimental Embryology*, B. H. Willier and J. M. Oppenheimer (eds.) (1974). New York: Hafner Press.

Stemple, D. L. and D. J. Anderson (1992) Isolation of a stem cell for neurons and glia from the mammalian neural crest. *Cell* 71: 973–985.

Walsh, C. and C. L. Cepko (1992) Widespread dispersion of neuronal clones across functional regions of the cerebral cortex. *Science* 255: 434–440.

Yamada, T., M. Placzek, H. Tanaka, J. Dodd and T. M. Jessell (1991) Control of cell pattern in the developing nervous system. Polarizing activity of the floor plate and notochord. *Cell* 64: 635–647.

Zimmerman, L. B, J. M. De Jesus-Escobar and R. M. Harland (1996) The Spemann organizer signal Noggin binds and inactivates bone morphogenetic protein 4. *Cell* 86: 599–606.

Books

Gilbert, S. F. and M. Barresi (2016) *Developmental Biology*, 11th Edition, Chs. 9–15. Sunderland, MA: Sinauer Associates.

Lawrence, P. A. (1992) *The Making of a Fly: The Genetics of Animal Design*. Oxford: Blackwell Scientific Publications.

Moore, K. L. (1988) *The Developing Human: Clinically Oriented Embryology*, 4th Edition. Philadelphia: W. B. Saunders Company.

CHAPTER

23 Construction of Neural Circuits

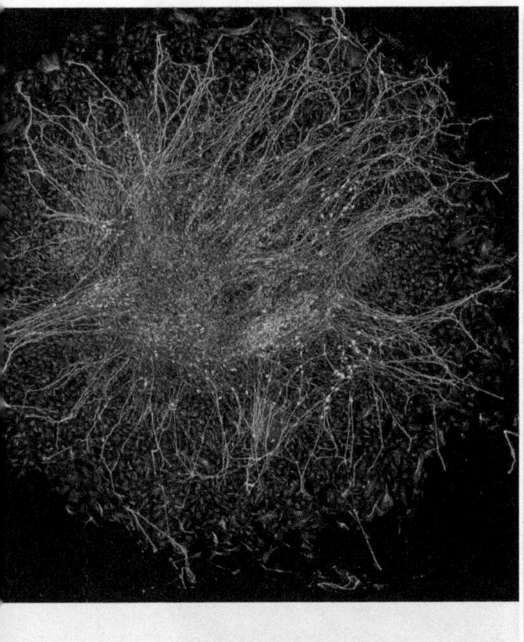

Overview

ONCE NERVE CELLS HAVE BEEN GENERATED, groups of neurons must become interconnected to form the neural circuits that mediate brain function. The first step in this process is to establish axons and dendrites in the newly generated neurons. The initial distinction of the characteristic single axon from the multiple dendrites of a neuron depends on cellular polarization, a process that reflects interactions among proteins found in different regions of a neuron's cytoskeleton. Once established, axons grow to reach appropriate target cells, which can be local or distant, and begin to make the synaptic connections that will define neural circuits. The directed growth of axons and recognition of appropriate synaptic targets depend on growth cones—the specialized endings of growing axons. The dynamic behavior of growth cones depends in turn on adhesive, attractive, and repulsive molecular signals in the embryonic environment. Once axons find their way to appropriate targets and form synapses, molecular neurotrophic factors influence neuron survival. In many regions of the developing nervous system, the death of some neurons helps match the numbers of innervating neurons to the needs of the target group. Cell adhesion, neurotrophic and other signals also regulate the subsequent growth of axons and dendrites and the addition of synapses to match the strength of connections to target and circuit needs. As in other instances of intercellular communication, a variety of receptors and second-messenger molecules transduce adhesion and neurotrophic signals as synapses and circuits mature. Their targets include proteins that stabilize the molecular architecture of both pre- and post-synaptic specializations. These cellular mechanisms establish topographic maps and other orderly representations of information underlying the complex neural circuits that allow animals to behave in increasingly sophisticated ways as they mature.

Neuronal Polarization: The First Step in Neural Circuit Formation

Neurons are especially elaborate examples of **polarized epithelial cells**, a fundamental cell class found in most tissues. Epithelial cells assemble into sheets that absorb molecules from the environment in one domain and secrete proteins and other cell products in another (Figure 23.1A). The tissues of the gut, lung, kidney, and pancreas are all epithelial cell sheets enclosing lumens from which molecules are absorbed adjacent to structures, especially blood vessels, and into which molecules are released. The apical surface of an epithelial cell faces the lumen and may have specializations such as cilia or microvilli that increase the surface area for taking in and releasing specific molecules. The basolateral surface is typically specialized for intercellular communication. Junctions that hold epithelial cells together are found on this surface, as are ion channels, small signaling molecules, and the machinery for intercellular protein exchange. These apical and basolateral distinctions arise from the differential distribution of proteins

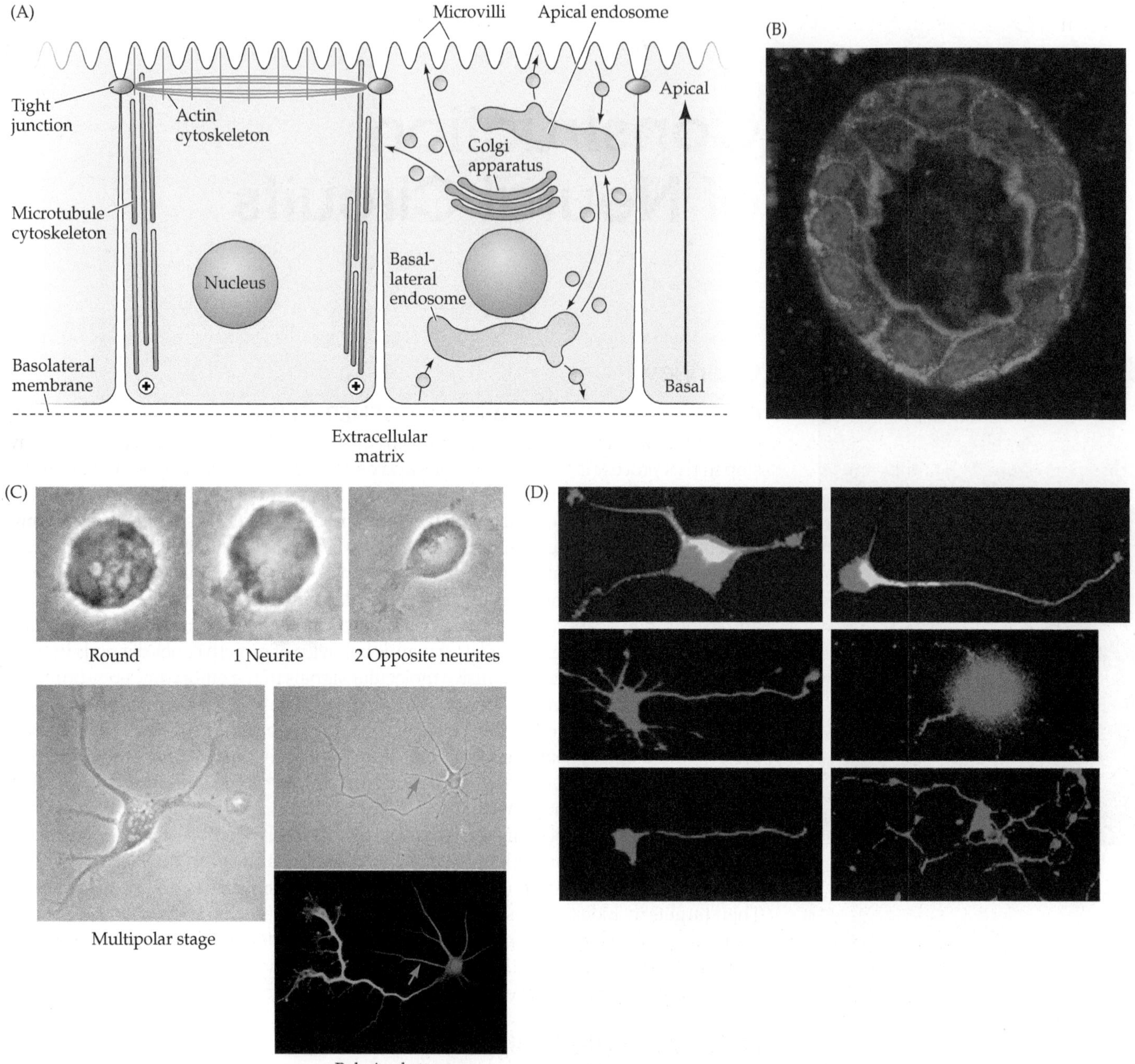

FIGURE 23.1 Cell polarity and the differentiation of axons and dendrites. (A) Apical/basal distinctions in a simple epithelium. The apical domain has a distinctive actin cytoskeleton and membrane extensions (villi); there are tight junctions; the Golgi apparatus is oriented toward the apical membrane; secretory vesicles fuse; and vesicular endocytosis targets endosomal compartments within the cell. The basolateral domain makes contact with the extracellular matrix; it has specialized adhesion contacts; and the plus ends of microtubules are oriented toward the basal membrane, which is the site of endosomal traffic. (B) Image of a simple polarized epithelial tube. The red labeling is actin filaments, which segregate to the apical domain, and the green is an adhesion molecule e-cadherin, which is found at specialized adhesion contacts in the basolateral domain. (C) Images of a postmitotic neuroblast initiating the process of neurite growth and establishing one of the neurites as an axon (red arrow). This process reflects establishing polarity that is striking similar to that of epithelial cells. (D) The polarity scaffolding protein Par-3 (red) is enriched in the growing tips of all neurites in an isolated neuron (top). When Par-3 is functioning normally, one of the processes of the neuron elongates and becomes an axon (middle left). When Par-3 function is disrupted, none of the processes elongate (middle right). The axon-specific cytoskeleton protein tau (red; bottom left) is usually seen only in the single developing axon. When Par-3 function is disrupted, tau is seen in all of the neurites (bottom right). (B courtesy of Natalie Elia and Jennifer Lippincott-Schwartz; C courtesy of Annette Gärtner; D from Shi et al., 2003.)

that constitute the cellular cytoskeleton (Figure 23.1B). Because neurons are highly specialized epithelial cells, a first step in their development is distinguishing their polarity. For the neuron, the fundamental polarity reflects the distinction between the dendrites (specialized for signal transduction) and the axon (specialized for secretion) (Figure 23.1C).

Once neurogenesis is complete and the neuroblast has entered a fully committed postmitotic state, the outgrowth of neuronal processes begins. Initially, several apparently equivalent small extensions (referred to as *neurites*, since at first they have neither axonal nor dendritic identities) protrude from the immature neuron. Soon after, microtubule and actin components of the cytoskeleton as well as other proteins are redistributed among the neurites so that a single process is identified as the axon. The remaining processes become dendrites (see Figures 23.1 and 23.7). Several studies, initially done in cell culture and more recently confirmed in developing embryos, indicate that many proteins, particularly members of the PAR family, are distributed preferentially in the nascent axon (Figure 23.1D). (PAR stands for "*par*titioning-defective"; PAR proteins were originally identified in the worm *Caenorhabditis elegans* based on their control of the axis of cell division and distribution of daughter cell proteins.) These proteins interact with cytoskeletal elements and signaling molecules, including Rho and other protein kinases; signal transduction molecules activated by secreted Wnts (see Chapter 22); neurotrophins; and cell surface-bound cell adhesion molecules. When the function of PAR proteins or related signaling molecules is disrupted, the specification of a single axon does not occur (see Figure 23.1D). PAR proteins and other polarity regulators also play a role in defining the regions of dendrites that receive synapses. In sum, in developing neurons, the molecular mechanisms that establish epithelial cell polarity are adapted to generate axons and dendrites, which in turn grow and travel, eventually making the connections that define a neural circuit.

The Axon Growth Cone

Once the axon has been specified, it navigates over millimeters or even centimeters, through complex embryonic terrain, to find appropriate synaptic partners. In 1910, Ross G. Harrison first observed this phenomenon in a living tadpole and wrote, "The growing fibers are clearly endowed with considerable energy and have the power to make their way through the solid or semi-solid protoplasm of the cells of the neural tube. But we are at present in the dark with regard to the conditions which guide them to specific points." More recent efforts, using increasingly sophisticated vital dyes, molecular labels, and optical techniques with improved resolution in living specimens, have confirmed Harrison's initial description of the remarkable navigation of growing axons through the embryo (Figure 23.2A), and defined many of the

(A)

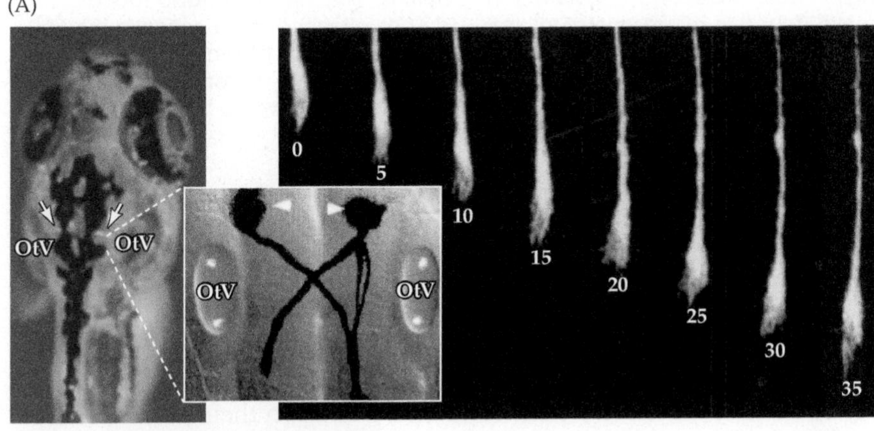

(B)

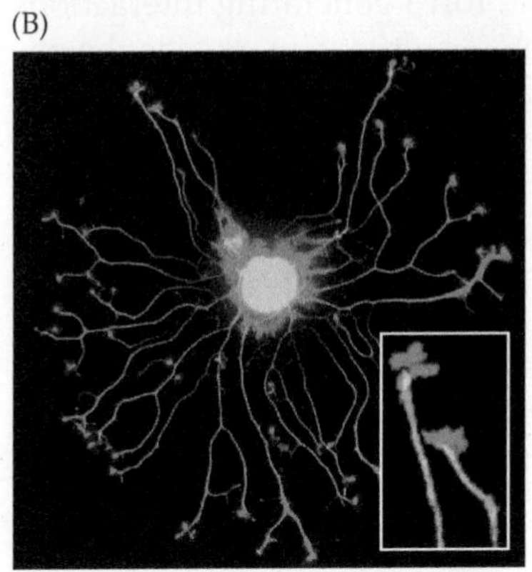

(C)

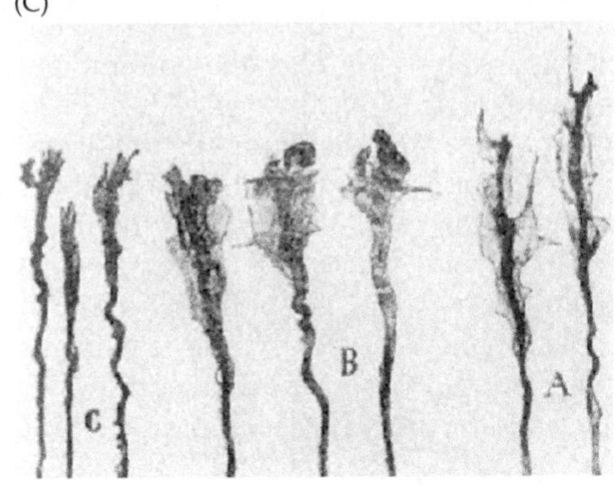

FIGURE 23.2 Growth cones guide axons in the developing nervous system. (A) Mauthner neurons (arrows) in the hindbrain of a zebrafish embryo, adjacent to the otic vesicles (OtV), give rise to the sensory neurons of the inner ear. The inset shows a higher magnification of the hindbrain. The Mauthner neuron axons can be seen crossing the midline and extending down to the spinal cord. Right: A single Mauthner axon labeled with a fluorescent dye in a living zebrafish embryo, led through the spinal cord by a relatively simple growth cone. Over 35 minutes, the axon advances approximately 50 μm. (B) A single dorsal root ganglion cell isolated in culture extends many processes called neurites. Each neurite has a long shaft in which microtubules (green) predominate, tipped by a growth cone in which actin (red) is the major molecular constituent. (C) Growth cone shape varies at "decision" regions. Here, growth cones from sensory relay neurons in the dorsal spinal cord are relatively simple near their origin (C in this drawing). They change in shape as they approach and cross the ventral midline to form the spinothalamic tract (A and B). Once across the midline, they reassume the simpler morphology. (A from Takahashi et al., 2002 and Jontes et al., 2000; B courtesy of F. Zhou and W. D. Snider; C after S. Ramón y Cajal, courtesy of C. A. Mason.)

BOX 23A ■ Choosing Sides: Axon Guidance at the Optic Chiasm

The functional requirement that a subset of axons from retinal ganglion cells in each eye must cross while the remaining axons project to the ipsilateral side of the brain was predicted based on optical principles—most notably by Sir Isaac Newton in the seventeenth century—and confirmed (much later) by neuroanatomists and neurophysiologists (see Chapter 12). The partial crossing, or *decussation*, of retinal axons is most striking in primates, including humans, where approximately half of the axons cross and the other half do not. Although all other mammals also have crossed and uncrossed retinal projections, the percentage of uncrossed axons diminishes from 20%–30% in carnivores to less than 5% in most rodents. The frequency of uncrossed axons decreases even more in other vertebrates; thus, in amphibians, fish, and birds most or all of the retinal projection is crossed. For both functional and evolutionary reasons, the partial decussation of the retinal pathways and its variable extent in different species has engaged the imagination of biologists and others interested in vision.

For developmental neurobiologists, this phenomenon raises an obvious question: How do retinal ganglion cells "choose sides" such that some project contralaterally and others ipsilaterally? This question is central to understanding how the peripheral visual projection is organized to construct two accurate visual hemifield maps that superimpose points of space seen jointly by the two eyes (see Chapter 12). It also speaks to the more general issue in neural development of how axons distinguish between ipsilateral and contralateral targets.

It is clear that the laterality of retinal axons is determined by initial cell identity and axon guidance mechanisms rather than by regressive processes that subsequently select or sculpt these projections. Thus, the distinction between the nasal and temporal retinal regions that project ipsilaterally and contralaterally is already apparent in the retina—as well as in axon trajectories at the midline and in the developing optic tract—long before the axons reach their targets. In the retina, this specificity is seen as a "line of decussation," or border, between ipsilaterally and contralaterally projecting retinal ganglion cells.

The line of decussation can be detected experimentally by injecting a retrograde tracer into the nascent optic tract of very young embryos. In the retinas of such embryos, there is a distinct boundary between the population of retinal ganglion cells projecting ipsilaterally in one eye (found in the temporal retina), as well as a complementary boundary for contralaterally projecting cells in the other eye (see figure). A molecular basis for this specificity was initially suggested by studies of albino mammals, including mice and humans. In albinos, where single gene mutation disrupts melanin synthesis throughout the animal, including in the pigment epithelium of the retina, the ipsilateral component of the retinal projection from each eye is dramatically reduced, the line of decussation in the retina is disrupted, and the distribution of glia and other cells in the vicinity of the optic chiasm is altered. These and other observations suggested that identity of retinal axons with respect to decussation is established in the retina, and further reinforced by axonal "choices" influenced by cues provided by cells within the optic chiasm.

Analysis of growth cone cell morphologies has shown that the chiasm is indeed a region where growth cones explore the molecular environment in a particularly detailed way, presumably to make choices pertinent to directed growth. Furthermore, molecular analysis reveals specialized neuroepithelial cells in and around the chiasm that express several cell adhesion molecules associated with axon guidance. Interestingly, some of these molecules—particularly netrins, slits, and their robo receptors—do not influence decussation in the chiasm as they do at other regions of the nervous system. Instead, they are expressed in cells where the chiasm forms,

"conditions" that guide axons to specific points. In addition, it is now clear that dendrites, especially primary dendrites such as those of cortical pyramidal cells or cerebellar Purkinje cells, must also extend over relatively long distances. The manner in which they do so is similar to that of axons.

Harrison recognized two fundamental features of axonal growth that continue to motivate efforts to understand the assembly of neural circuits. First, the "considerable energy… and power" of growing axons reflect the cellular properties of the **growth cone**, a specialized structure at the tip of the extending axon. Growth cones are highly motile. They explore the extracellular environment, determine the direction of growth, and then guide the extension of the axon in that direction. The primary morphological characteristic of a growth cone is a sheetlike expansion of the growing axon at its tip called a **lamellipodium**. When growth cones are examined in vitro, lamellipodia can be seen clearly, as can numerous fine processes called **filopodia** that extend from each lamellipodium (Figure 23.2B). Filopodia rapidly form and disappear from the terminal expansion, like fingers reaching out to sense the environment. The lamellipodium and filopodia are distinguished from the axon shaft by different cytoskeletal molecules (see Figure 23.1B and the following section). They have the capacity to localize or concentrate molecular receptors on their cell surfaces, based upon their distinctive cytoskeleton. ATP-dependent, force generating interactions between cytoskeletal proteins ultimately provide the energy and power to propel the growth cone and its axon to its target. Thus, the growth cone is a distinct, if transient, neuronal specialization whose activity is critical for the formation of tracts and circuits in the developing brain. Once a growth cone reaches and recognizes an appropriate target—relying on cues or "conditions which guide them to specific points" that Harrison rightly stated were "in the dark" and remained so until late in the last century—it is gradually transformed into either a presynaptic ending for an axon or the terminal domain of a dendrite.

BOX 23A (continued)

apparently constraining its location on the ventral surface of the diencephalon. The establishment of ipsilateral versus contralateral identity is evidently more dependent on the zinc finger transcription factor Zic2, as well as on cell adhesion molecules of the ephrin family. Zic2, which is expressed specifically in the temporal retina, is associated with the expression of a distinct receptor, EphB1, in the axons arising from temporal retinal ganglion cells. The ephrin B2 ligand, which is recognized as a repellent of EphB1 axons, is found in midline glial cells in the optic chiasm. In support of the functional importance of these molecules, disrupting *Zic2*, *EphB1*, or *ephrin B2* gene function diminishes the degree of ipsilateral projection in developing mice; in accord with this finding, neither the *Zic2* nor the *ephrin B2* gene is expressed in vertebrate species that lack ipsilateral projections.

These observations provide a molecular framework for the identification of retinal ganglion cells and the sorting of their projections at the optic chiasm. How this sorting is related to the topography of tectal, thalamic, and cortical representations is not yet known. Most observations suggest that retinal topography is not faithfully preserved among axons in the optic tracts. The identity and position of axons from nasal and temporal retinas whose retinal ganglion cells "see" a common point in the binocular hemifield must therefore be restored in the thalamus and subsequently retained or reestablished in the thalamic projections to cortex. Choosing sides at the chiasm is only a first step in establishing maps of visual space.

(A) +/+

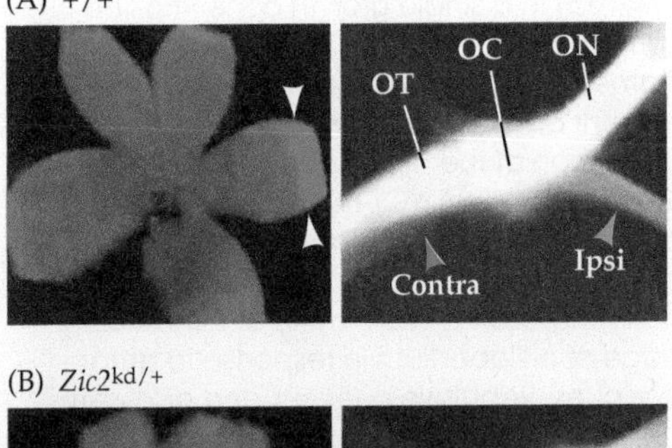

(B) *Zic2*kd/+

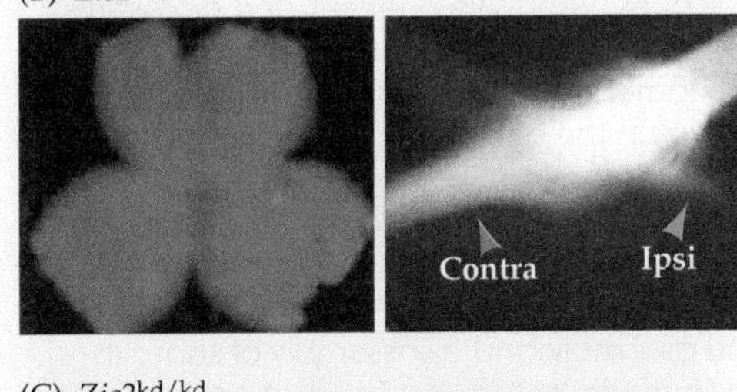

(C) *Zic2*kd/kd

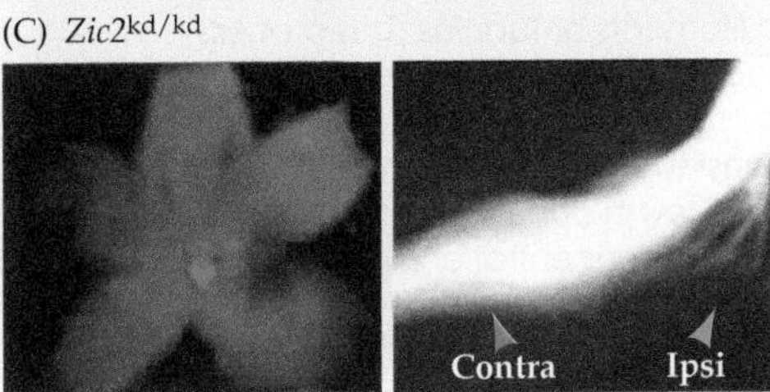

(A) A small population of Zic2-expressing retinal ganglion cells (arrowheads) is seen in the ventrotemporal region of the normal retina (at left, mounted flat by making several radial cuts). At right, the normal projection of one eye via the optic nerve (ON), through the optic chiasm (OC), and into the optic tract (OT) has been traced using a lipophilic dye placed in one eye. After the chiasm, labeled axons can be seen both in the contralateral (Contra) as well as the ipsilateral (Ipsi) optic tract. (B) When Zic2 function is diminished in a mouse heterozygous for a *Zic2* "knock-out" mutation (in which expression of Zic2 protein is diminished but not eliminated), the number of ipsilateral axons in the optic tract is similarly diminished. (C) When Zic2 function is further diminished in homozygous *Zic2* knock-out mice, the ipsilateral projection can no longer be detected in the optic tract; thus, each optic tract consists of contralateral axons. (From Herrera et al., 2003.)

Santiago Ramón y Cajal, a contemporary of Harrison's, observed in sections of fixed specimens that growth cones in several established axon pathways—pathways presumably "pioneered" by one or a few early growing axons—tend to be simple in shape when in "familiar" territory close to the neuronal cell body of its origin. In contrast, when a pioneer axon (i.e., the first axon to extend through a given region, thus establishing that terrain as hospitable to axon growth) extends in a new direction through a territory not previously innervated, or reaches a region where a choice must be made about which direction to take, the growth cone changes dramatically (Figure 23.2C). The lamellipodium of the growth cone expands as it encounters a potential target and extends numerous filopodia, actions that suggest an active search for appropriate cues to direct subsequent growth. These changes of growth cone shape at "decision points" have been observed in both the peripheral and central nervous systems (PNS and CNS). In the periphery, the growth cones of both motor and sensory neurons change shape as they enter the developing muscles of immature limbs, presumably facilitating the selection of appropriate targets in the musculature. Distinct molecules on or around the muscle surface—including cell surface adhesion molecules, receptors for secreted signals, and specialized extracellular matrix molecules—mediate these local cellular changes (see below). Growth cones in all peripheral as well as central axon pathways examined thus far (including the olfactory nerve, optic chiasm, and cerebral commissures) change shape when they reach critical points in their trajectories (see Figure 23.2C and Box 23A).

The Molecular Basis of Growth Cone Motility

Growth cone motility reflects rapid, controlled rearrangement of the cytoskeleton. The force to move the axon is generated by ATP-dependent modification of the actin and microtubule cytoskeletons. The **actin cytoskeleton**

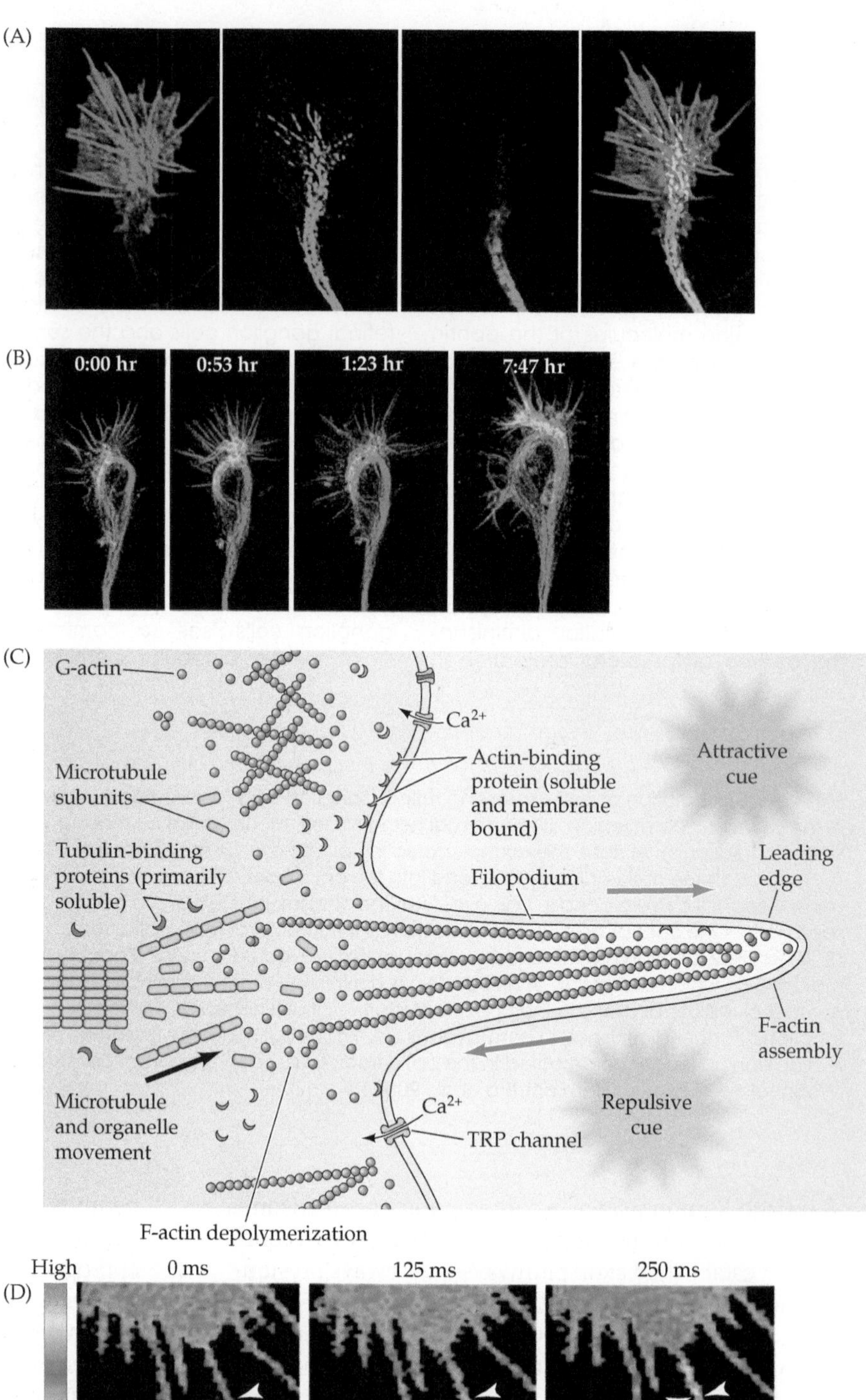

FIGURE 23.3 The structure and action of growth cones. (A) Distinct types of actin and tubulin are seen in discrete regions of the growth cone. Filamentous actin (F-actin, red) is seen in the lamellipodium and filopodia. Tyrosinated microtubules are the primary tubular constituents of the lamellar region (green), and acetylated microtubules are restricted to the axon itself (blue). (B) Dynamics of actin cytoskeleton (yellow, red) in a single growth cone imaged over an 8-hour interval. The distribution of filamentous actin, labeled red with a fluorescent actin-binding protein, changes in the region of the lamellipodium as well as in the filopodium. (C) The distribution and dynamics of cytoskeletal elements in the growth cone. Globular actin (G-actin) can be incorporated into F-actin at the leading edge of a filopodium in response to attractive cues. Repulsive cues support disassembly and retrograde flow of G-actin toward the lamellipodium. Organized microtubules make up the cytoskeletal core of the axon, while more broadly dispersed microtubule subunits are found at the transition between the axon shaft and the lamellipodium. Actin- and tubulin-binding proteins regulate the assembly and disassembly of subunits to filaments or tubules. This process is influenced by changes in intracellular Ca^{2+} via voltage-regulated Ca^{2+} channels as well as transient receptor potential (TRP) channels. (D) Growth cones show rapid changes in Ca^{2+} concentration. In this example, a single filopodium (white arrowheads) undergoes a rapid increase in Ca^{2+}. (A from Dent and Gertler, 2003; B from Dent and Kalil, 2001; D from Gomez and Zheng, 2006.)

regulates changes in lamellipodial and filopodial shape for directed growth, while the **microtubule cytoskeleton** is responsible for elongation of the axon itself (Figure 23.3A,B). The molecular composition of both the actin and microtubule cytoskeleton changes in distinct regions of the growing axon, suggesting a great deal of dynamism within growing neural processes (Figure 23.3B,C). Thus, the ways in which the actin and microtubule cytoskeletons are modified in growing axons are key to understanding how growth cones and axons extend.

Actin is the primary molecular constituent of a network of cellular filaments found in the lamellipodia and filopodia of growth cones. Tubulin is the primary molecular constituent of the microtubules that run parallel to the axis of the axon and give it both structural integrity and a means for transporting proteins from the nerve cell body to the axon terminal (axons do not contain significant protein synthesis machinery). Actin and tubulin are found in two forms in the growth cone and axon: freely soluble monomers in the cytoplasm, and polymers that form filaments (actin) or microtubules (tubulin) in more organized bundles within filopodia or the axon itself. The dynamic polymerization and depolymerization of actin at the membrane of the lamellipodium, as well as within the filopodium, sets the direction of growth cone movement, in part by generating local forces that orient the growth cone toward or away from substrates. Similarly, the polymerization and depolymerization of tubulin into microtubules consolidate the direction of movement of the growth cone by stabilizing the axon shaft. The interface between the actin and microtubule cytoskeletons regulates the balance of active growth versus stability.

Several proteins regulate the polymerization and depolymerization of actin and tubulin by binding to these molecules and catalyzing post-translational modifications, or by recruiting enzymes that modify the primary molecular elements of the cytoskeleton. Actin-binding proteins are found throughout the growth cone cytoplasm. Most either bind actin directly, or modify actin monomers by phosphorylation and other post-translational modifications. These molecules are particularly enriched at the inner surface of the growth cone plasma membrane, presumably to mediate assembly and membrane anchoring of actin filaments. The local anchoring of actin filaments is essential to generate the forces that extend the membrane and direct the movement of the lamellipodium or filopodia (see Figure 23.3C). In addition, the actin cytoskeleton is the anchor for multiple "protein scaffolding" molecules that in turn localize or concentrate receptors and channels to the lamellopodial or filopodial membrane. Microtubule-binding proteins are more concentrated in the axon shaft, and they modulate post-translational modifications of monomeric and polymerized tubulin. Other microtubule-binding proteins ("motors") are essential for moving molecules and organelles ("cargo") up and down neuronal processes. These proteins include dynein and several kinesins. They rely upon ATP hydrolysis to generate force to "walk" along the microtubule with their cargo proteins. The interaction of these motor proteins with tubulin allows the axon to remain relatively stable in the face of the expanding exploration of the growth cone and extension of the axon.

The constant flux between monomeric actin and tubulin versus the polymerized actin filaments and microtubules is regulated via binding proteins that rely upon enzymatic cleavage of second messengers like cAMP and cGMP to generate energy. This signaling is also thought to explain the higher concentration of mitochondria in the lamellopodium of a growth cone as well as the growing axon. Some of the signals ultimately enhance mitochondrial ATP genesis. Thus, directed growth is elicited in response to signals from the environment. These signals are transduced by receptors and channels on the growth cone membrane surface, structures that influence the levels of intracellular messengers, particularly that of Ca^{2+} (Figure 23.3D). Indeed, the regulation of intracellular Ca^{2+} levels, either through voltage-regulated Ca^{2+} channels, transient receptor potential (TRP) channels activated by second messengers, or second-messenger pathways that mobilize intracellular Ca^{2+} stores, is thought to be a major mediator of actin and microtubule dynamics in the growing axon. These fluctuations of Ca^{2+} are quite localized in the growth cone—sometimes in a single filopodium (see Figure 23.3D)—and are thought to influence decisions on direction of growth. Thus, the conditions that Ross Harrison suggested "guide [growth cones] to specific points" are now understood to be changes in the cytoskeleton and local composition of membrane proteins of the growth cone and axon, mediated by intracellular signal transduction in response to adhesion molecules and diffusible signals in the embryonic terrain through which the growth cone extends.

Non-Diffusible Signals for Axon Guidance

The complex behavior of growth cones during axon extension suggests the presence of specific cues that cause the growth cone to move in a particular direction, and of an array of receptors and transduction mechanisms that respond to these cues. These presumed molecular cues remained elusive for nearly 75 years following Harrison and Cajal's initial observations, but over the past 35 years many of the relevant molecules have been identified. The cues comprise a large group of proteins associated with cell adhesion and cell–cell recognition throughout the organism. These molecules initiate intracellular signaling cascades that can alter the actin or microtubule cytoskeleton, the complement of receptors and channels on the cell surface, or gene expression. The association of specific cell adhesion molecules with axon growth is based on experiments either in vitro, where addition or removal of a particular molecule results in modifying the relevant behavior of growing axons; or in vivo, where genetic mutation, deletion, or manipulation disrupts the growth, guidance, or targeting of a particular axon pathway (see Box 23A).

Even though there are a daunting number of them, the molecules known to influence axon growth and guidance can be grouped into families of ligands and their receptors (Figure 23.4). The major classes of non-diffusible axon guidance molecules are: the extracellular matrix molecules and their integrin receptors; the Ca^{2+}-independent cell adhesion molecules (CAMs); the Ca^{2+}-dependent cell adhesion molecules, or cadherins; and the ephrins and Eph receptors.

The **extracellular matrix cell adhesion molecules** were the first to be associated with axon growth. The most prominent members of this group are the **laminins**, the **collagens**, and **fibronectin**. As their family name indicates, all three of these adhesion molecules are found in the extracellular matrix (ECM), an adhesive macromolecular complex outside the cell. The ECM's components can be secreted by the cell itself or by its neighbors; however,

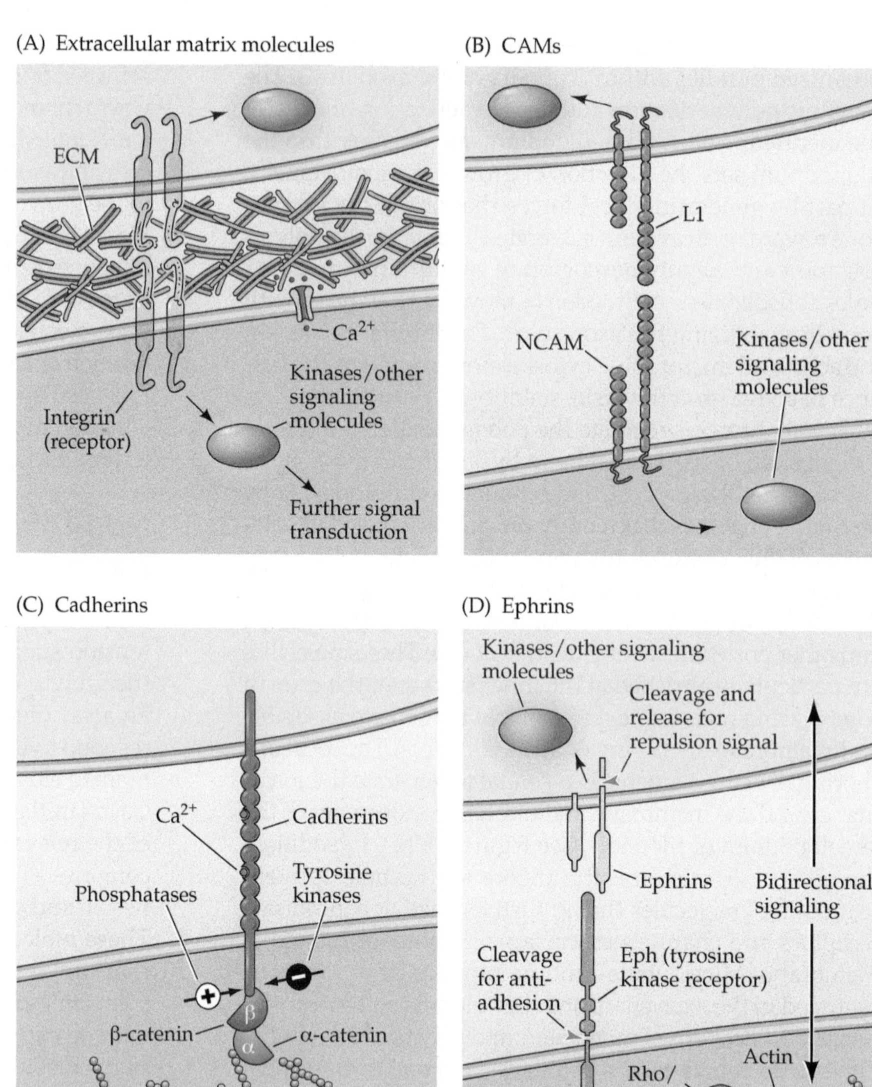

FIGURE 23.4 Several families of ligands and receptors constitute the major classes of axon guidance molecules. These ligand-receptor pairs can be either attractive or repulsive, depending on the identity of the molecules and the context in which they signal the growth cone. (A) Extracellular matrix (ECM) molecules including fibronectin and several isoforms of both laminin and collagen serve as the ligands for multiple integrin receptors. Integrins transduce ECM signals by interacting with cytoplasmic protein kinases and activating Ca^{2+} channels. (B) Homophilic, Ca^{2+}-independent cell adhesion molecules (CAMs) are at once ligands and receptors. Homophilic binding activates intracellular kinases, leading to cytoskeletal changes. (C) Ca^{2+}-dependent adhesion molecules, or cadherins, are also capable of homophilic binding. They signal via activation of β-catenin, which influences gene expression. (D) Ephrins, which can be either transmembrane or membrane-associated, signal via the Eph receptors, which are receptor tyrosine kinases.

rather than diffusing away from the cell after secretion, these molecules form polymers and create a durable local extracellular substance. A broad class of cell surface receptors known as **integrins** binds specifically to these ECM molecules (see Figure 23.4A). Integrins do not have kinase activity or any other direct signaling capacity. Instead, the binding of laminin, collagen, or fibronectin to integrins triggers a cascade of events—perhaps via interactions between the cytoplasmic domains of integrins with non-receptor cytoplasmic kinases (like the SRC family kinases—SRC indicates the initial identification of this kinase family in sarcoma tumor tissue) and other signaling molecules, as well as Ca^{2+} channels—that can stimulate axon growth and elongation.

The role of ECM molecules in axon guidance is particularly clear in the embryonic PNS. Axons extending through peripheral tissues grow through loosely arrayed mesenchymal cells that fill the interstices of the embryo (between sheets of epithelia), and the spaces between these mesenchymal cells are rich in ECM molecules. Peripheral axons also grow along the interface of mesenchyme and epithelial tissues, including the boundaries between the neural tube and mesenchyme and between the mesenchyme and the epidermis, where organized sheets of ECM components called **basal lamina** provide a supportive substrate. In tissue culture as well as in the periphery of the embryo, the different ECM molecules have different capacities to stimulate axon growth. The role of these molecules in the CNS is less clear. Some of the same molecules are present in the extracellular spaces of the CNS, but they are not organized into a basal lamina as in the periphery, and therefore have been more difficult to study.

The **CAMs** and **cadherins** are found on growth cones and growing axons as well as on surrounding cells and targets (see Figure 23.4B,C). Moreover, both CAMs and cadherins have dual functions as ligands and receptors, usually via homophilic ("like with like") binding. Some of the CAMs, especially the L1 CAM, have been associated with the bundling, or *fasciculation*, of groups of axons. Cadherins have been suggested as important determinants of final target selection in the transition from growing axon to synapse (see below). For both CAMs and cadherins, the unique ability of each class to function as both ligand and receptor (L1 CAM, for example, is its own receptor) may be important for recognition between specific sets of axons and targets. Both CAMs and cadherins rely on a somewhat indirect route of signal transduction. The Ca^{2+}-independent CAMs interact with cytoplasmic kinases to initiate cellular responses, whereas the Ca^{2+}-dependent cadherins engage the β-catenin pathway (also activated by Wnts; see Chapter 22).

A final class of non-diffusible axon guidance molecules includes a large family of **ephrin** ligands and their tyrosine kinase receptors (**Eph receptors**, or **Ephs**) that constitute a cell–cell recognition code in a variety of tissues (see Figure 23.4D). In the developing nervous system, immature axons use ephrins and Eph receptors to recognize appropriate pathways for growth as well as appropriate sites for synaptogenesis (see below). Although identified as ligands and receptors, the binding of ephrins with Ephs can initiate "reverse" signaling via the ephrins, which can interact with cytoplasmic protein kinases. Ephrins and Ephs activate a variety of signaling pathways and, depending on the nature of signal transduction, can be either growth-promoting or growth-limiting. To limit axon growth, the extracellular domain of an ephrin ligand can be proteolytically cleaved, or Eph receptors can be removed via selective endocytosis, thus terminating signaling. The dependence of axon growth and guidance on adhesive interactions and the signal transduction pathways activated by these interactions is underscored by the pathogenesis of several inherited human developmental or neurological disorders. These syndromes include X-linked hydrocephalus; MASA (an acronym for *m*ental retardation, *a*phasia, *s*huffling gait, and *a*dducted thumbs); Kallmann syndrome (which compromises reproductive and chemosensory function); X-linked spastic paraplegia; and several even rarer disorders. Some are consequences of mutations in genes encoding the Ca^{2+}-independent CAMs, while others compromise secreted guidance signals or additional mechanisms that influence axon growth. These mutations can also lead to the partial absence of the corpus callosum that connects the two cerebral hemispheres (referred to as *callosal agenesis*), or of the corticospinal tract, which carries cortical information to the spinal cord. These rare congenital anomalies are now understood to arise from mutations in genes that result in loss of function of the encoded protein. The absence or diminished activity of these proteins lead to errors in signaling mechanisms normally responsible for axon navigation via cell surface adhesion molecules.

Chemoattraction and Chemorepulsion

A growing axon must eventually find an appropriate target while avoiding inappropriate ones. With remarkable foresight, Cajal proposed that target-derived signals, most likely released by the target cells themselves, selectively attract growth cones to useful destinations. In addition to this *chemoattraction* predicted by Cajal, it was long supposed that there might also be *chemorepellent* signals that discourage axon growth toward inappropriate regions (Figure 23.5). Despite the importance of chemoattraction and chemorepulsion in constructing pathways and circuits, the identity of the signals themselves remained uncertain until the early 1990s. One problem was the vanishingly small amounts of such factors expressed in the developing embryo. Another was that of distinguishing *tropic* molecules (which *guide* growing axons toward a source) from *trophic* molecules (which *support* the survival and growth of neurons and their processes once an appropriate target has been contacted). These problems were solved by laborious biochemical purification and analysis of attractive or repulsive activities from vertebrate (chick) embryos and genetic analysis of axon growth in both *Drosophila* and *C. elegans*. Remarkably, the identity and function of chemoattractants and chemorepellents across phyla are highly conserved.

One of the first families of chemoattractant molecules to be identified was the **netrins** (Sanskrit, "to guide"; see Figure 23.5B). In chick embryos, netrins were identified as proteins with chemoattractant activity following biochemical purification. In *C. elegans*, an activity analogous to that of the netrins was first recognized as the product of a gene that influenced axon growth and guidance—mutation of this gene caused altered axon growth and disrupted behavior. The first such gene to be isolated was named *Unc* for "uncoordinated," which describes the behavioral phenotype of the mutant worms; the cause is misrouted axons, misplaced or absent synapses, and disrupted circuits due to the absence of netrin activity. The netrins themselves have high homology to ECM molecules such as laminin, and in some cases may actually interact with the ECM to influence directed axon growth. Netrin chemoattractant signals are transduced by specific receptors, including the molecule DCC (*d*eleted in *c*olorectal *c*ancer). A different receptor, Unc5 (yet another *C. elegans* protein whose name is also used for its vertebrate orthologs), mediates netrin-dependent chemorepulsion. Like many cell surface adhesion molecules, netrin receptors have repeated amino acid motifs in their extracellular domains—in this case, a transmembrane domain; and an intracellular domain with no known enzymatic activity. Thus, in order for netrin binding to stimulate changes in the target cell, other proteins with catalytic activity must interact with the

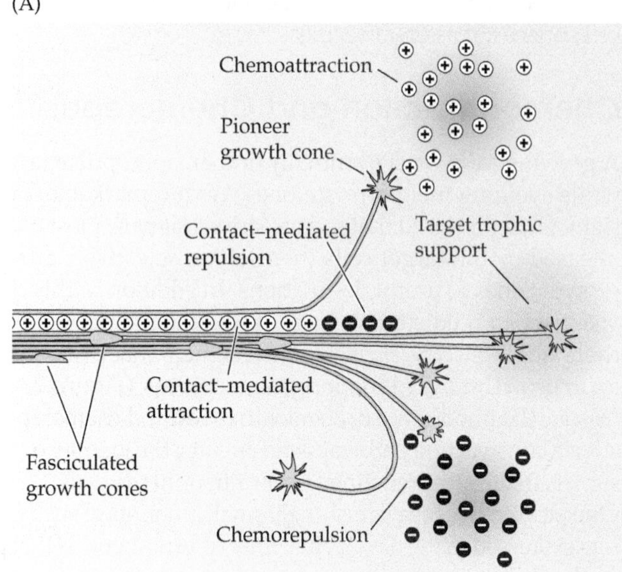

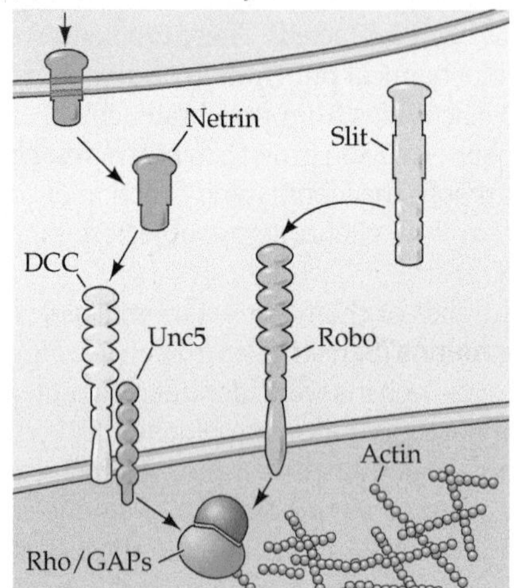

cytoplasmic domain. In this way, they are particularly similar to the integrin receptors for ECM molecules, especially laminin, homologous to the transmembrane netrin. The Rho/GAP family of signaling proteins, all of which modulate second-messenger-mediated cytoskeletal modification, are thought to provide a final step in netrin signaling.

The initial characterization of netrin in vertebrates was guided by the observation that there seemed to be a chemoattractive signal in the spinal cord ventral midline that influenced the directed growth of spinothalamic axons (Figure 23.6A,B). Netrins are localized to the floorplate of the neural tube (see Figure 22.2), which defines the ventral midline in the developing spinal cord as well as sites where axons must cross from one side of the cord to the other. Indeed, mutation of the *netrin-1* gene (whose expression is dramatically enhanced in the floorplate and other midline brain regions where axons cross) disrupts the development of axon pathways that cross the midline, including those of the spinothalamic tract (Figure 23.6C) and the cerebral commissures (the corpus callosum and anterior commissure).

Even though netrin signaling can be downregulated by proteolytic cleavage of the DCC receptor, additional mechanisms ensure that once netrin directs axons across the midline, they do not cross back. The secreted factor **slit** and its receptor **robo** (named for the phenotypes of *Drosophilia* mutants in which these genes were first identified) are important for preventing an axon from straying back over the midline once it has crossed initially in response to netrin (see Figures 23.5B and 23.6B). Slit and robo are available immediately off

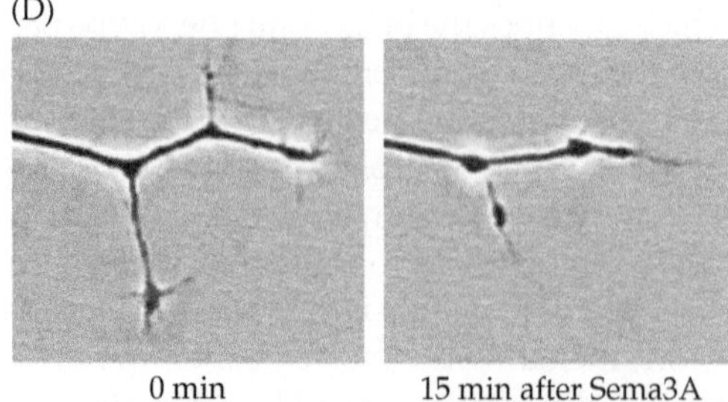

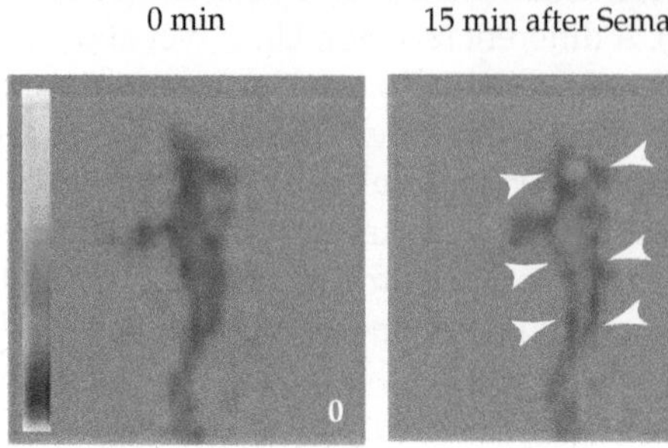

FIGURE 23.5 Signals provided by the embryonic environment. (A) Basic signal classes. Chemoattractant, or tropic, signals (pluses) can operate from a distance and reorient growth toward the source of the cue, often by acting on a pioneer growth cone that sets out a course distinct from that of the fasciculated followers. Similar cues acting on or near the surface of the axon shaft help maintain groups of axons as fascicles, which is essential for formation of coherent nerves and tracts. Chemorepellant signals (minuses) can also act from a distance, or they act at regions where axons must defasciculate from a nascent nerve in order to change their trajectory or avoid an inappropriate target. Trophic signals (gold) support growth and differentiation of the axon and its parent nerve cell. (B) The netrin/slit family of attractive and repulsive secreted signals acts through two distinct receptors, DCC (*deleted in colorectal cancer*), which binds netrin, and robo, the receptor for slit. (C) Semaphorins are primarily repulsive cues that can either be bound to the cell surface or secreted. Their receptors (the plexins and neuropilin) are found on growth cones. (D) Growth cone–collapsing activity of semaphorin (upper panels) depends on Ca^{2+} signaling, which is initiated by semaphorin signaling (lower panels). (A after Huber et al., 2003; D from Gomez and Zheng, 2006; Dontechev and Letourneau, 2002.)

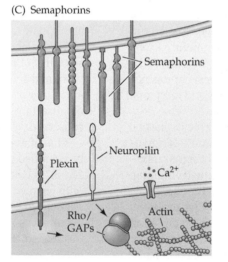

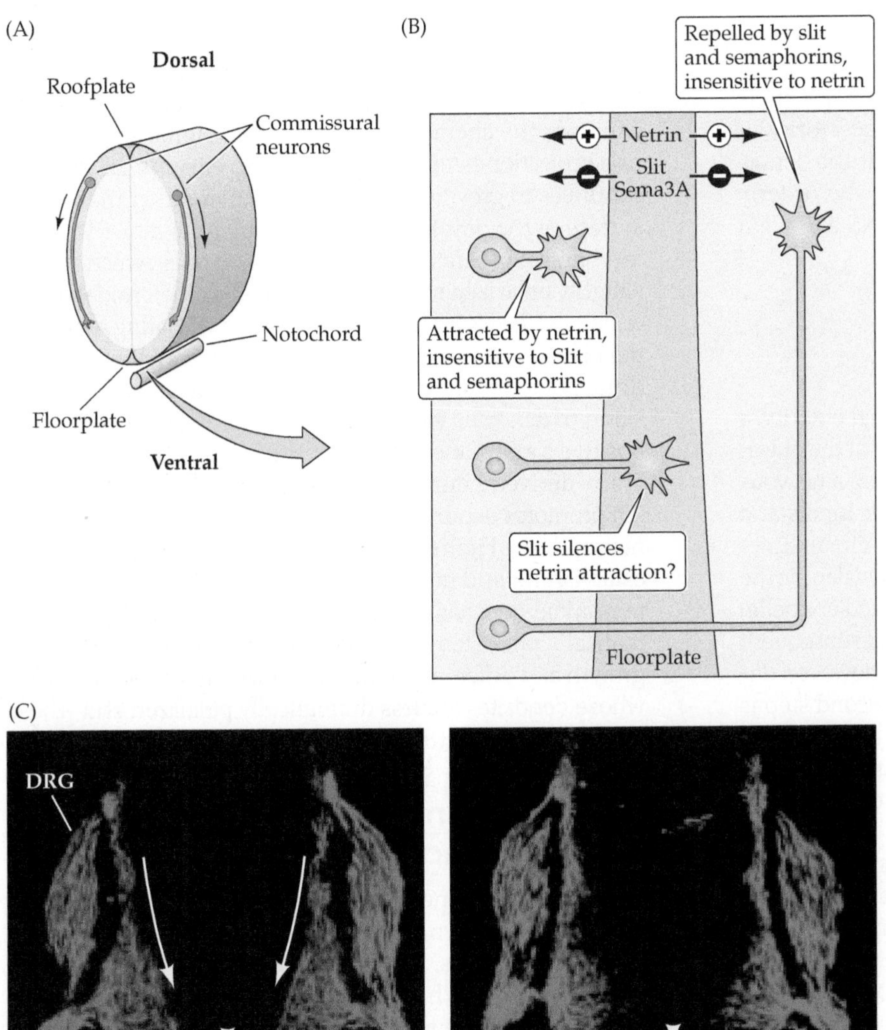

FIGURE 23.6 Netrins chemotropically regulate pathway formation in the developing spinal cord. (A) Commissural neurons send axons to the ventral region of the spinal cord, including the floorplate. (B) Opposing activities of netrin and slit at the ventral midline of the spinal cord. This molecular guidance system ensures that the axons relaying pain and temperature via the anterolateral pathway cross the midline at appropriate levels of the spinal cord and remain on the contralateral side until they reach their targets in the thalamus. (C) At left, labeled commissural axons (red) descend through the spinal cord, pass the motor column (MC), and cross the midline (arrowhead) into the anterior (ventral) commissure of the spinal cord. (DRG = dorsal root ganglion.) At right, the *netrin* gene of a mouse has been homozygously inactivated, and the commissural axons do not fasciculate, nor do they cross at the ventral midline (arrowhead). (A after Serafini et al., 1994; B after Dickinson, 2001; C from Serafini et al., 1996.)

the midline and are thought to terminate the growth cone's sensitivity to netrin once it has crossed from one side of the CNS to the other. Thus, slit and robo (and probably several other molecules) and the signaling pathways they activate orchestrate the unidirectional crossing of axons at the midline. The successful completion of such crossing by axons is essential for the construction of all major sensory, motor, and associational pathways in the mammalian brain (see Box 23A).

Beyond the specific challenge of crossing the midline—and then not crossing back—constructing the nervous system generally requires telling axons where *not* to grow. Two broad classes of chemorepellent molecules (in addition to slit and robo, which function primarily at the midline) have been described. The first class is associated with CNS myelin; molecules of this class are particularly important for regulating axon regrowth following injury in the mature nervous system (see Chapter 26). Molecules belonging to the second class of chemorepellents, the **semaphorins** (Greek, "signal bearer") are active primarily during neural development (see Figure 23.5C). Semaphorins are bound to cell surfaces or to the ECM, where they prevent the extension of nearby axons. Their receptors, like those for cell surface adhesion molecules, are transmembrane proteins (including the plexins and a protein called neuropilin) whose cytoplasmic domains have no known catalytic activity. Nevertheless, semaphorin signaling leads to changes in Ca^{2+} concentration that presumably activate intercellular kinases and other signaling molecules to modify the growth cone cytoskeleton. Studies using cultured vertebrate neurons indicate that semaphorins cause growth cones to collapse and axon extension to cease (see Figure 23.5D). Much of the initial characterization of semaphorin chemorepellent activity emerged from studies of invertebrates (particularly *Drosophila*) in which mutation or manipulation of these genes caused axons to grow abnormally. In developing vertebrates, however, the activity of semaphorins has been harder to demonstrate. Mutations in semaphorins, plexins, or neuropilin in mammals do not yield dramatic phenotypes. Although none of these signals alone explains initial choices and resulting trajectories of developing axons, it is nonetheless clear that the semaphorins make an important contribution to the orderly construction of axon pathways in both the PNS and CNS of vertebrates.

Directed Dendritic Growth: Ensuring Polarity

The establishment and maintenance of neuronal polarity, a phenomenon first recognized by Ramon y Cajal at the outset of the twentieth century, is essential for building a nervous system that can manage integration of synaptic inputs and generation of synaptic outputs to define neural circuit function. The first step identifies the axon as the equivalent of the basolateral secretory domain of simpler polarized epithelial cells (see above) via activity of PAR proteins and subsequent stabilizers of the axon cytoskeleton. This step, however, defines only the synaptic output domain. The second step is the maintenance, guided growth, and local branching of dendrites. In the CNS, several classes of neurons, particularly projection neurons that extend axons relatively long distances from the location of neuronal cell bodies, have striking **dendritic polarization** that underlies their unique information-processing capacity. Retinal ganglion cells, olfactory bulb mitral cells, cerebellar Purkinje cells, and cortical pyramidal neurons represent major neuron classes whose dendritic polarity is essential for appropriate circuit function (see Chapters 1, 15, 19, and 33). Divergent cytological differentiation of dendritic versus axonal processes in these neurons requires molecular mechanisms that, quite literally, can tell one end of a neuron from the other.

In most cases, a relatively common mechanism is used, and it is one that relies on chemoattractive and chemorepellant signaling molecules that also influence axon growth and guidance: semaphorins, neuropilins, and related signaling intermediates (Figure 23.7). This mechanism has been best characterized for cortical neurons, but it also operates for cerebellar Purkinje cells and retinal ganglion cells. The generally chemorepellant cell surface molecule Semaphorin 3A (Sema3A) is central in this mechanism. Sema3A simultaneously repels the axons of developing cortical pyramidal, retinal ganglion, and cerebellar Purkinje neurons, while acting as a chemoattractant for the dendrites of the same cells. Thus, in the developing cerebral cortex, secreted Sema3A is found at high concentrations in the marginal zone, the outermost layer of the nascent cortex. This Sema3A attracts developing dendrites via neuropilin receptors; however, polarized differences in the distribution of downstream signaling intermediates, including soluble guanylyl cyclase (sGC), convert the normally repulsive Sema3A/neuropilin signal into a chemoattractant signal. This downstream conversion does not happen in the axon, where sGC is absent, and thus the axon grows in the opposite direction of the dendrite. In addition, the chemorepellant secreted signal slit1 also repels projection neuron axon growth to ensure that the axon continues to grow toward the ventricular rather than apical surface of the developing cortex (see Figure 23.7A). Similar mechanisms operate for retinal ganglion cells, which acquire a highly polarized morphology soon after their initial differentiation (see Figure 23.7B). Local Notch signaling reinforces the primary consequences of semaphorin signaling for dendritic polarization in the cortex. At this stage in the acquisition of dendritic versus axonal polarity, slit1 also acts as a positive signal for cortical dendritic branching. In addition, brain-derived neurotrophic factor, or BDNF (see Chapter 25) promotes dendritic growth in the differentiating cortical neuropil (see Figure 23.7C). While these signals are key for establishing and maintaining polarized growth of cortical pyramidal neurons as well as Purkinje cells and other dramatically polarized neuronal cell types, they also influence growth and differentiation of local GABAergic interneurons, whose dendrites are less dramatically polarized and whose axons ramify near the dendrites of the same interneuron.

Dendritic Tiling: Defining Synaptic Space

One last step of neuronal differentiation precedes the major phase of synapse formation and establishment of functional circuits. This step ensures proper modulation of dendritic growth so that each dendritic arbor occupies appropriate space to accommodate incoming axons that will synapse on it. This mechanism, referred to as *dendritic tiling*, was first observed in the PNS of *Drosophila* and has subsequently been analyzed in the developing CNS of mammals, especially mice. There are two key outcomes of this aspect of control of initial dendritic differentiation: First, developing dendrites are regulated so they do not grow toward nearby dendrites from the same neuron; and second, developing dendrites from different neurons are repelled from one another to a greater or lesser degree to ensure that each neuron's dendritic arbor provides adequate "coverage" for a particular region of neuronal space in a developing structure such as the retina.

The mechanism for dendritic tiling therefore must accomplish an intriguing feat: dual repulsion. First, each neuron must respond to molecular signals, presumably from its own dendrites. These signals prevent dendrites of the same neuron from growing on top of each other. Next, each neuron must respond to cues, in this case from the dendrites of other neurons, that restrict the territories of dendritic arbors from neighboring neurons. A novel molecular mechanism has evolved to mediate this essential aspect of growth and avoidance of dendrites in space. The molecular basis for dendritic tiling was first defined in the fly, based on the remarkable

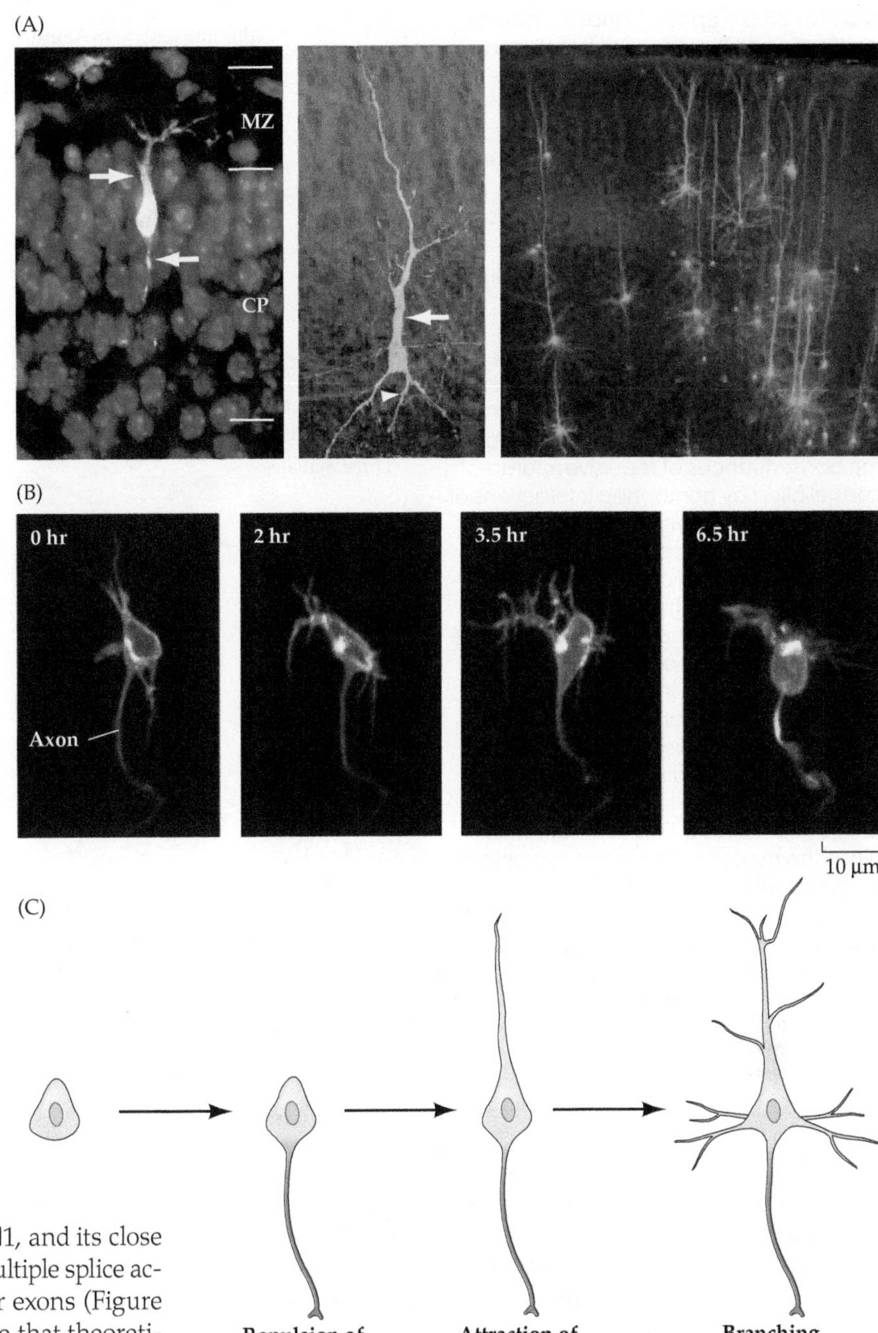

FIGURE 23.7 Polarized dendritic growth relies on secreted signals. (A) At left, the initial polarized differentiation of the axon (arrowhead) and dendrite (arrow) of a cortical pyramidal or projection neuron. This neuron, which is migrating through the cortical plate (CP) on its way to its final destination at the outermost region of the developing cortex, has already begun to respond to directional cues in the marginal zone (MZ). Subsequently (middle image), the apical dendrite (oriented toward the MZ) begins to grow and branch. In addition, at this time the basal dendrites (arrowhead) of the neuron begin to differentiate, although they grow only in the local territory close to the cell body and apical dendrite. (B) A time-lapse series of images of a single retinal ganglion cell (imaged in an intact but ex vivo retina) shows initial branching of dendrites in several directions that are then modified so that the dendritic arbor extends laterally from the cell body, and at the opposite pole from the axon. (C) A schematic of the signals necessary for polarized, opposite growth of dendrites and axons. In this instance, signaling for a cortical pyramidal cell is outlined. This general mechanism also operates for retinal ganglion cells, cerebellar Purkinje cells, and several other classes of neurons with distinctly polarized axons and dendrites. (A,C from Whitford et al., 2002; B from Choi et al., 2010.)

genomic structure of a single gene, **DSCAM**1, and its close relative DSCAM2. In the fly, *DSCAM* has multiple splice acceptor and donor sites distributed over four exons (Figure 23.8A). This genomic structure yields a gene that theoretically can encode at least 37,000 variants, and this flexibility is key for *Drosophila* dendritic tiling. When the *DSCAM* gene is unperturbed, dendrites grow in a normal distribution, avoiding growing on their sibling dendrites and establishing reasonable spacing between dendrites from neighboring neurons (Figure 23.8B). When the *DSCAM* gene is mutated so that fewer splice variants are possible, or so it is not expressed at all, dendritic tiling is impaired; sibling dendrites grow closer or on top of one another, and neighboring dendrites no longer keep their distance (Figure 23.8C). This has been seen for *Drosophila* sensory neurons whose dendrites cover the body wall, usually in an evenly distributed pattern, as well as for *Drosophila* retinal neurons whose connections must be precisely matched to the omatidial structure (columnar units defined by photoreceptors and adjacent processing neurons) of the fly's eye. The fundamental mechanism for dendritic tiling relies on distinctions between homophilic binding between DSCAM isoforms of the same identity, which leads, perhaps counterintuitively, to repulsion, so that dendrites from the same neuron recognize one another and

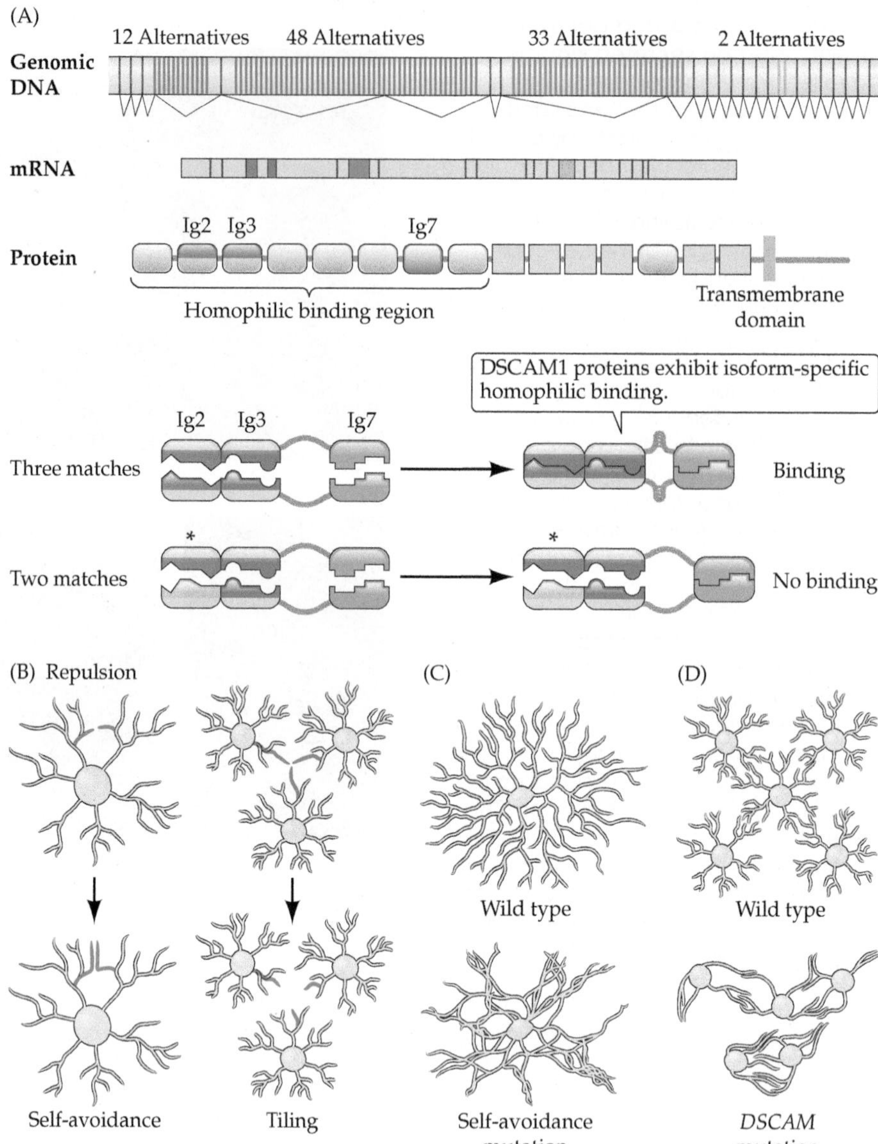

FIGURE 23.8 Splice variants, homophilic recognition or repulsion, and dendritic growth and tiling. (A) Top panel: In *Drosophila*, the *DSCAM1* gene has 12 alternative splice variants in exon 4, 48 variants in exon 6, 33 in exon 9, and 2 in exon 17. Middle panel: The variable mRNAs transcribed from this gene give rise to DSCAM proteins with variable homophilic immunoglobulin (Ig) adhesive domains. Bottom panel: If the three homophilic variable Ig domains are matched, there is recognition and binding. If the protein variability of one or more Ig domains is not matched (asterisk), recognition and binding do not occur. (B) A schematic of the consequences of repulsive interactions elicited by homophilic interactions of DSCAM1 variants. Within a single neuron (left panel), these repulsive interactions, based on homophilic binding, prevent branches of the same neuron from growing on top of one another, while the same mechanism for sets of different neurons (right panel) results in dendritic avoidance by each neuron of dendrites from other neurons, leading to dendritic tiling. (C) An example of disrupted dendritic branching, leading to bundled, dysmorphic dendrites, in a *DSCAM1* mutant fly. (D) In the mouse retina, amacrine cells in a normal wild-type mouse are tiled, based on murine DSCAM function. In the case of a *DSCAM* loss-of-function mutation, tiling fails; amacrine cell processes grow on top of one another, and cell bodies are no longer evenly spaced. (A,B,D from Hattori et al., 2008; C from Lefebvre et al., 2015.)

don't grow on top of each other (Figure 23.8C). Heterophilic binding, or the absence of a DSCAM variant on a nearby process, results in permissive growth that can lead to apposition, fasciculation, and other forms of dendritic contact.

Mammals also have a *DSCAM* gene. In fact, the acronym for *DSCAM* is derived from the location of the human orthologue, found on human chromosome 21, the chromosome duplicated in Down syndrome: the human *DSCAM* stands for Down syndrome *c*ell *a*dhesion *m*olecule. The mammalian *DSCAM* gene, including that in humans, does not encode the large number of splice isoforms encoded by its *Drosophila* counterpart. Nevertheless, *DSCAM* does mediate some forms of dendritic tiling. This is particularly recognizable in the retinas of mice that carry a homozygous loss-of-function mutation in the *DSCAM* gene. In the retinas of wild-type mice, a key class of intraretinal relay neuron, amacrine cells (*amacrine* means "star-shaped"), is tiled across the retinal surface. In contrast, in the retinas of *DSCAM* mutant mice, dendrites of individual amacrine cells grow on top of one another, and the regular tiled distribution is completely disrupted (Figure 23.8D). This mechanism of homophilic repulsion and heterophilic permission may be further modulated by alternately spliced variants of mammalian protocadherin (PCDH) genes, whose splice architecture is reminiscent of that of *DSCAM* in *Drosophila* (see Figure 23.11). This speculation, however, remains to be confirmed by experimental observation. The sum of these established and speculative repulsion/permission growth mechanisms is to constrain neuronal differentiation so that the distribution of dendrites is initially appropriate for the volume of local and afferent synapses that will be made. These mechanisms may also act as a substrate for further activity-mediated specification and stabilization of synapses (see Chapter 25).

Formation of Topographic Maps

In the somatosensory, visual, and motor systems, neuronal connections are arranged such that neighboring points in the periphery are represented by adjacent locations in the appropriate regions of the CNS (see Chapters 9, 11, and 16). In other systems (e.g., the auditory and olfactory systems), there are also orderly representations of various stimulus attributes such as frequency or sensory receptor identity in at least some central regions (see Chapters 13 and 15). In all of these examples, there must be some sort of fairly precise recognition mechanisms during initial development to guide peripheral axons from distinct locations to appropriately "mapped" target locations in the brain. In the early 1960s, Roger Sperry (who also did pioneering work on the functional specialization of the cerebral hemispheres; see Chapter 33) articulated the **chemoaffinity hypothesis**. This hypothesis, based primarily on Sperry's work on the visual systems of frogs and goldfish, explains how topographic maps arise during development. In these animals, the terminals of retinal ganglion cells form a precise topographic map in the optic tectum (the tectum is homologous to the mammalian superior colliculus; Figure 23.9A). When Sperry crushed the optic nerve and allowed it to regenerate (fish and amphibians, unlike mammals, can regenerate CNS axon tracts; see Chapter 26), he found that retinal axons reestablished the original topographic pattern of connections in the tectum. Even if the eye was rotated 180 degrees, the regenerating axons grew back to their original tectal destinations (causing some behavioral confusion for the frog; Figure 23.9B). Accordingly, Sperry proposed that each tectal cell carries a chemical "identification tag" and that the growing terminals of retinal ganglion cells have complementary tags such that the retinal cells seek out a specific location in the tectum. In a non-perturbed frog, such tags would therefore match the topography of the sensory surface, in this case the retina, with connections within the CNS target—the tectum. The "identification tags" were assumed to be cell adhesion or cell recognition molecules, and the "affinity" they engendered was presumed to be due to the selective binding of receptor molecules on retinal ganglion cell growth cones to corresponding molecules on the tectal cells at appropriate relative positions. It is easy to imagine that such a mechanism might account for the full range of topographic maps seen in the nervous systems of many animals, including humans.

Further experiments in the amphibian and avian visual systems made the strictest form of the chemoaffinity hypothesis—that is, the labeling of each tectal location by a distinctive recognition molecule—untenable. Rather than displaying a precise "lock-and-key" affinity, the behavior of growing axons suggested that there are *gradients* of cell surface molecules to which growing axons respond. Normally, axons from the temporal region of the retina innervate the anterior pole of the tectum and avoid the posterior pole. Temporal retinal axons, when presented with a choice of cell membranes derived from anterior or posterior tectal regions as a substrate, grow exclusively on anterior membranes, avoiding membranes derived from the "wrong" region of the tectum (Figure 23.9C). A likely candidate for the negative-guidance signal for temporal axons in the posterior tectum was purified and its gene cloned. The protein—initially called RAGS (*r*epulsive *a*xon *g*uidance *s*ignal)—turned out to belong to the Eph family of cell surface-bound adhesion and signaling molecules (see Figure 23.4D). The Ephs had been previously identified and characterized as cell–cell recognition molecules in tumor cells. In the eye and the tectum, ephrins and Eph receptors are distributed in complementary gradients across the temporonasal and dorsoventral retina and the anterior–posterior and medio-lateral tectum. These gradients result in matching levels of specific ephrin ligands and receptors; thus facilitating topographic mapping of the nasal and temporal retina along the anterior–posterior axis of the tectum. Subsequent work has associated several ephrins and Ephs with topographic mapping in numerous systems, including the mammalian visual system (see Box 23A). These observations accord with the idea that chemoaffinity operates not by one-to-one or lock-and-key recognition, but by gradients of affinities that provide axons and their targets with markers of general position within a system of coordinates (like north, south, east, and west on a map). Additional sharpening of topography may rely on activity-dependent mechanisms that continue to modify patterns of synaptic connections (see Chapter 25).

Selective Synapse Formation

Once an axon reaches its target region, additional cell–cell interactions dictate which target cells to innervate from among a variety of potential synaptic partners. There are some absolute restrictions to synaptic associations. For example, neurons do not make synapses on nearby glial cells in the CNS or connective tissue cells in the PNS. Furthermore, instances have been described in which nerve and target cell types that are not normally interconnected show little or no inclination to establish synaptic partnerships with one another when confronted with the possibility by experimental manipulation, either in vivo or in vitro. When synaptogenesis does proceed, however, neurons and their targets in both the CNS and PNS appear to associate according to a continuously variable graded system of preferences. Such biases guide the pattern of innervation that arises in development (or reinnervation during regeneration; see Chapter 26) without limiting it in an absolute way. The target cells residing in muscles, autonomic ganglia, or the brain are certainly not equivalent, but neither are they unique with respect to the innervation they can receive. This relative promiscuity most likely reflects the fact that potential pre- and postsynaptic sites may share many molecules

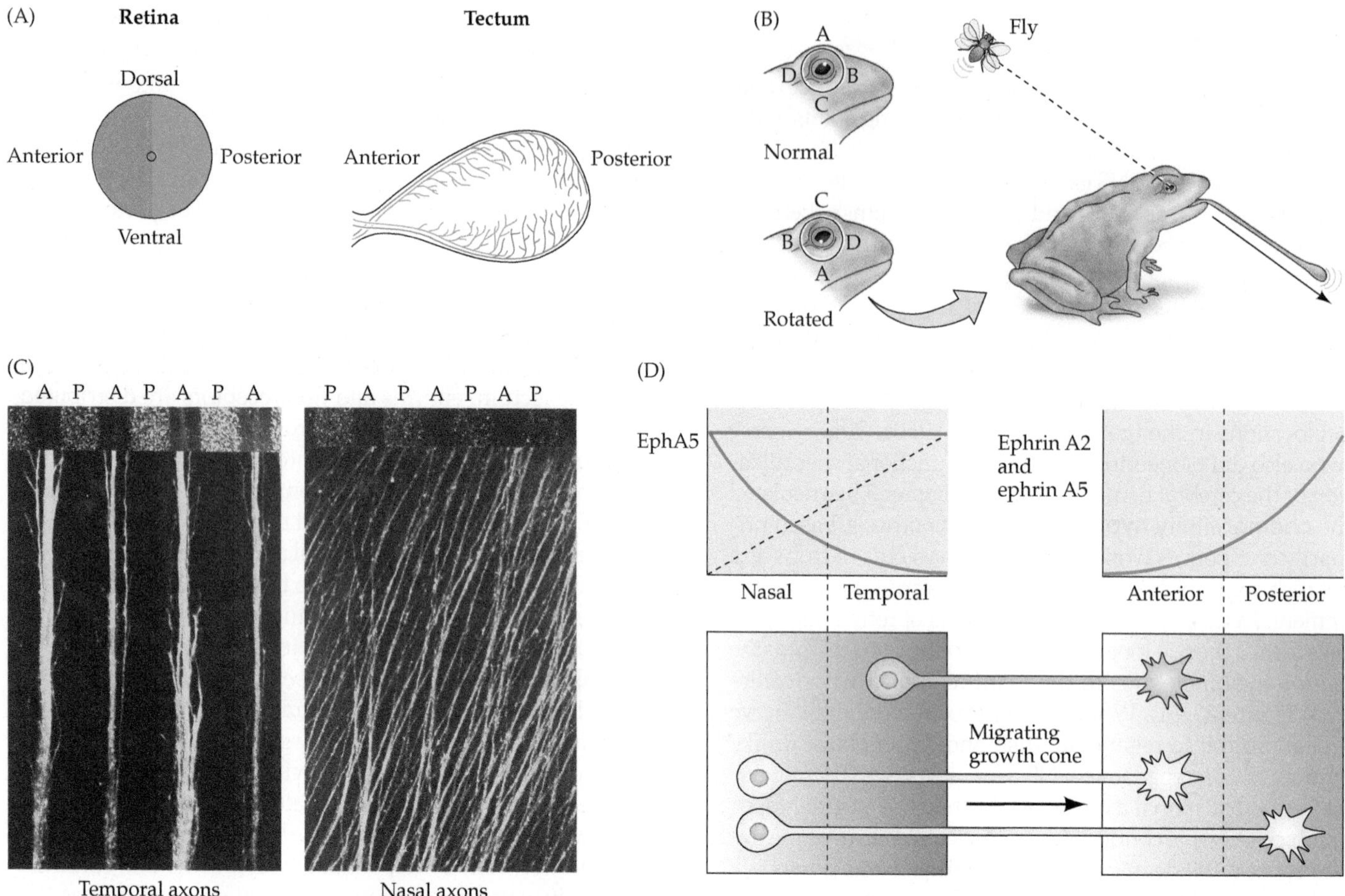

FIGURE 23.9 Mechanisms of topographic mapping in the vertebrate visual system. (A) Posterior retinal axons project to the anterior tectum, and anterior retinal axons project to the posterior tectum. When the optic nerve of a frog is surgically interrupted, the axons regenerate with the appropriate specificity. (B) Even if the eye is rotated after severing the optic nerve, the axons regenerate to their original position in the tectum. This topographic constancy is evident from the frog's behavior: When a fly is presented above, the frog consistently strikes downward, and vice versa. (C) In vitro assay for cell surface molecules that contribute to topographic specificity in the optic tectum. A set of alternating, 90-μm-wide stripes of membranes from anterior (A) and posterior (P) optic tectum of chicks were laid down on a glass coverslip. The posterior membranes have fluorescent particles added to make the boundaries of the stripes apparent (top of panels). Explants of retina from either nasal or temporal retina were placed on the stripes. Temporal axons prefer to grow on anterior membranes and are repulsed by posterior membranes. In contrast, nasal retinal axons grow equally well on both stripes. (D) Complementary gradients of Eph receptors (in afferent cells and their growth cones) and ephrins (in the target cells) lead to differential affinities and topographic mapping. In this model, retinal growth cones with a high concentration of Eph receptors (for a mouse retina, EphA5) would be more likely to recognize a lower concentration of ligand (for the mouse superior colliculus, ephrins A2 and A5), whereas a growth cone with low Eph receptor concentration would recognize a higher concentration of ligand. (A,B after Sperry, 1963; C from Walter et al., 1987; D after Wilkinson, 2001.)

that identify them as potential locations for making a connection. Thus, if there is a not a strong specific affinity, more generic recognition events occur, and anomalous connections will be made. This can result in problems following neural injury, since regenerated patterns of innervation are not always appropriate (see Figure 23.9B and Chapter 26).

Much of this imprecision may reflect the overlapping subset of molecules that regulate general aspects of individual synapse formation as well as axonal and dendritic growth. Several observations show that many of the same adhesion molecules that participate in axon guidance contribute to the identification and stabilization of generic synaptic sites on target cells, as well as to the ability of a growing axon to recognize specific sites as optimal. Thus, in the first stages of synapse formation (Figure 23.10A), the ephrins, the Ca^{2+}-independent CAMs, and the cadherin families of Ca^{2+}-dependent adhesion molecules are all thought to influence recognition of any suitable postsynaptic positions on dendrites,

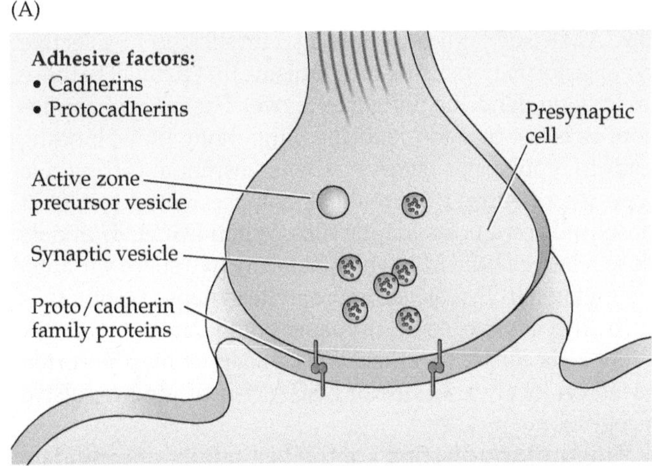

FIGURE 23.10 Molecular mechanisms involved in synapse formation. (A) Initiation of a synapse depends centrally on local recognition between the presumptive pre- and postsynaptic membranes mediated by members of the cadherin and protocadherin family of Ca^{2+} cell adhesion molecules (cadherins and protocadherins). This local recognition is accompanied by the initial accumulation of synaptic vesicles as well as transport vesicles that contain molecular components that contribute to the presynaptic active zone. (B) Once the initial specialization is established, additional adhesion molecules are recruited, including synaptic cell adhesion molecule (SynCAM), a member of the Ca^{2+}-independent, homophilic binding adhesion molecule family (like NCAM; see Figure 23.4B), neurexin and neuroligin (see panel C), and the ephrin B ligands and their EphB receptors. Adhesive signaling between these molecules initiates differentiation of the presynaptic active zone and the postsynaptic density. The presynaptic terminal also releases molecules (e.g., neuregulin) that influence the expression and clustering of postsynaptic receptors and associated proteins. (C) The interaction of neurexin (a presynaptic transmembrane adhesion protein) with neuroligin (a postsynaptic adhesion protein) is central for recruiting and retaining cytoskeletal elements that localize synaptic vesicles to the presynaptic terminal and mediate their fusion. In addition, neurexin is important for localizing voltage-gated Ca^{2+} channels to ensure local vesicle release. Neuroligin, upon binding neurexin, is essential for localizing neurotransmitter receptors and postsynaptic proteins to the postsynaptic specialization. Neuregulin is released via local proteolytic cleavage and binds to dimeric ErbB receptor kinases or to dimeric ErbB/epidermal growth factor (EGF) receptor kinases. (A,B after Waites et al, 2005; C after Dean and Dreshbach, 2006.)

cell bodies, or other appropriate targets (i.e., muscle fibers) by a nascent presynaptic process (derived by the conversion of a growth cone to an immature presynaptic terminal). In the next step, pre- and postsynaptic specializations must be elaborated for appropriate cellular specializations for synaptic communication (Figure 23.10B,C). Several soluble or secreted signals have been implicated in this process, including growth factors and neurotransmitters themselves. Subsequently, specialized adhesion molecules link pre- and postsynaptic domains so that the synapse emerges as a discrete, relatively stable intracellular specialization for local electrical or chemical signaling.

Among this list of molecules that initiate synaptogenesis, **neuregulin1** (**Nrg1**) has emerged as an essential regulator of expression and localization of postsynaptic receptors and other proteins. Nrg1 is a transmembrane protein usually made in presynaptic cells and can be released following proteolytic cleavage of the ectodomain (outside portion) of the protein. This cleaved, "mature" form of Nrg1 then diffuses and binds to specific receptors: the ErbB family of epidermal growth factor-like receptors, found on the surfaces of many developing central neurons as well as on muscle cells and other targets of peripheral neurons. Nrg1 signaling is thought to elicit increased synthesis and insertion of neurotransmitter receptors at a nascent postsynaptic site. Intriguingly, the human *NRG1* gene is a site of multiple polymorphic changes (altered DNA sequences that differ from those of most individuals sequenced) associated with schizophrenia and other behavioral disorders thought to alter development or maintenance of synaptic connections. Individuals with these altered DNA sequences have a slightly (but significantly) increased probability of developing schizophrenia compared with individuals without the polymorphisms. The DNA sequence changes, however, do not result in altered amino acid sequences for Nrg1 protein (i.e., they occur in noncoding regions of the *NRG1* gene), and it remains uncertain how they might alter the gene's expression or the protein's activity.

Two families of adhesion molecules are particularly central to the construction of all synapses: **neurexins**, found in the presynaptic membrane; and their binding partners the **neuroligins**, found in the postsynaptic membrane. In addition to the ability of neurexins and neuroligins to bind one another and promote adhesion between the pre- and postsynaptic membranes, neurexins have a specialized transmembrane domain that helps localize synaptic vesicles, docking proteins, and fusion molecules contributed by active zone vesicles in the presynaptic terminal. Neuroligins have similar functions for the postsynaptic site, where they interact with specialized postsynaptic proteins to promote the clustering of receptors and channels of the postsynaptic density as the synapse matures (see Figure 23.10C). Neurexins and neuroligins are shared by all developing synapses, perhaps explaining why some cells, when confronted by targets different from those they normally innervate, can make connections with the available, if unusual, target. The association of polymorphisms in neurexin and neuroligin genes with increased risk for behavioral disorders such as autism and schizophrenia has reinforced the hypothesis that these molecules are key for establishing appropriate connectivity. There is some indication that different neuroligins are deployed at postsynaptic specializations in glutamatergic, excitatory synapses versus GABAergic, inhibitory synapses. Nevertheless, there must be additional mechanisms that sort out specific individual synapses from one another to achieve the ultimate precision required to establish functional neural circuits.

The mechanisms by which neighboring synapses are sorted out remain unclear; however, some common themes and compelling candidate molecules have emerged. First, the diversity of ephrin ligands and Eph receptors, along with their established roles in topographic map formation, indicates that these molecules likely contribute to synapse specificity. Second, genes encoding additional candidates—all of which are cell adhesion molecules—have multiple sites for alternative splicing of transcripts, and thus can encode a large number of variants of the same basic protein. Third, some of these variants tend to be distributed in different pre- and postsynaptic sites, sometimes in a single neuron. Finally, when these genes are mutated, patterns of connectivity are disrupted in subtle but informative ways.

In the fly, the gene for the cell adhesion molecule DSCAM1 (see Figure 23.8) is also implicated in selective synapse formation, using mechanisms that parallel to those underlying dendritic tiling (see above). For synapse formation, as is the case for dendritic tiling, homophilic binding leads to repulsion, suggesting that an essential sorting rule prevents a neuron from making synapses with itself, or at closely adjacent postsynaptic sites. When *DSCAM1* and its close relative *DSCAM2* (which has only two splice variants) are mutated in *Drosophila*, appropriate synapse sorting fails in the insect's eye, so that synapses of like rather than different cells cluster together. Thus, synapses may be sorted based on local variations of DSCAM-mediated repulsive recognition.

The **protocadherins** are the last family of candidate genes thought to underlie synapse specificity. At the molecular level, protocadherins resemble the general cadherin family of cell adhesion molecules (see Figure 23.4C). The single gene encoding a large number of mammalian protocadherins, however, is remarkably similar to *DSCAM* in the fly. Thus, there are three regions (α, β, and γ) consisting of multiple alternatively spliced exons that encode the extracellular and transmembrane domains of individual protocadherin variants (there are 58) and a conserved domain that encodes the intracellular portion of all protocadherin isoforms (Figure 23.11A). Protocadherin isoforms on opposing cells bind to each other with varying affinity based on their degree of similarity (i.e., more hemophilic; high binding) or divergence (i.e., more heterophilic; lower binding). Moreover, protocadherins are not uniformly expressed at neighboring synaptic sites in cultured neurons (Figure 23.11B) or in several regions of the CNS. Thus, isoforms of protocadherins may invest synaptic sites in the mammalian nervous system with distinct identities. Analysis of mutant mice in which protocadherin gene function is eliminated indicates that these genes are crucial for synapse formation generally; however, their specific role in the development of synaptic specificity remains uncertain. Thus, despite potential candidate molecular families, it is still not clear whether molecular differences alone can explain the final synaptic pattern that emerges in a given neural circuit (see Chapter 25).

Regulation of Neuronal Connections by Trophic Interactions

Once synaptic contacts are established and the initial distribution of synapses is set, neurons become dependent on the presence of their targets for continued survival as well as the further growth and differentiation of axons and dendrites. In the absence of synaptic partners, the axons and dendrites of developing neurons typically atrophy and often die. This long-term dependency between neurons and their targets is referred to as **trophic interaction** (from the Greek *trophé*, meaning, roughly, "nourishment"). The nourishment

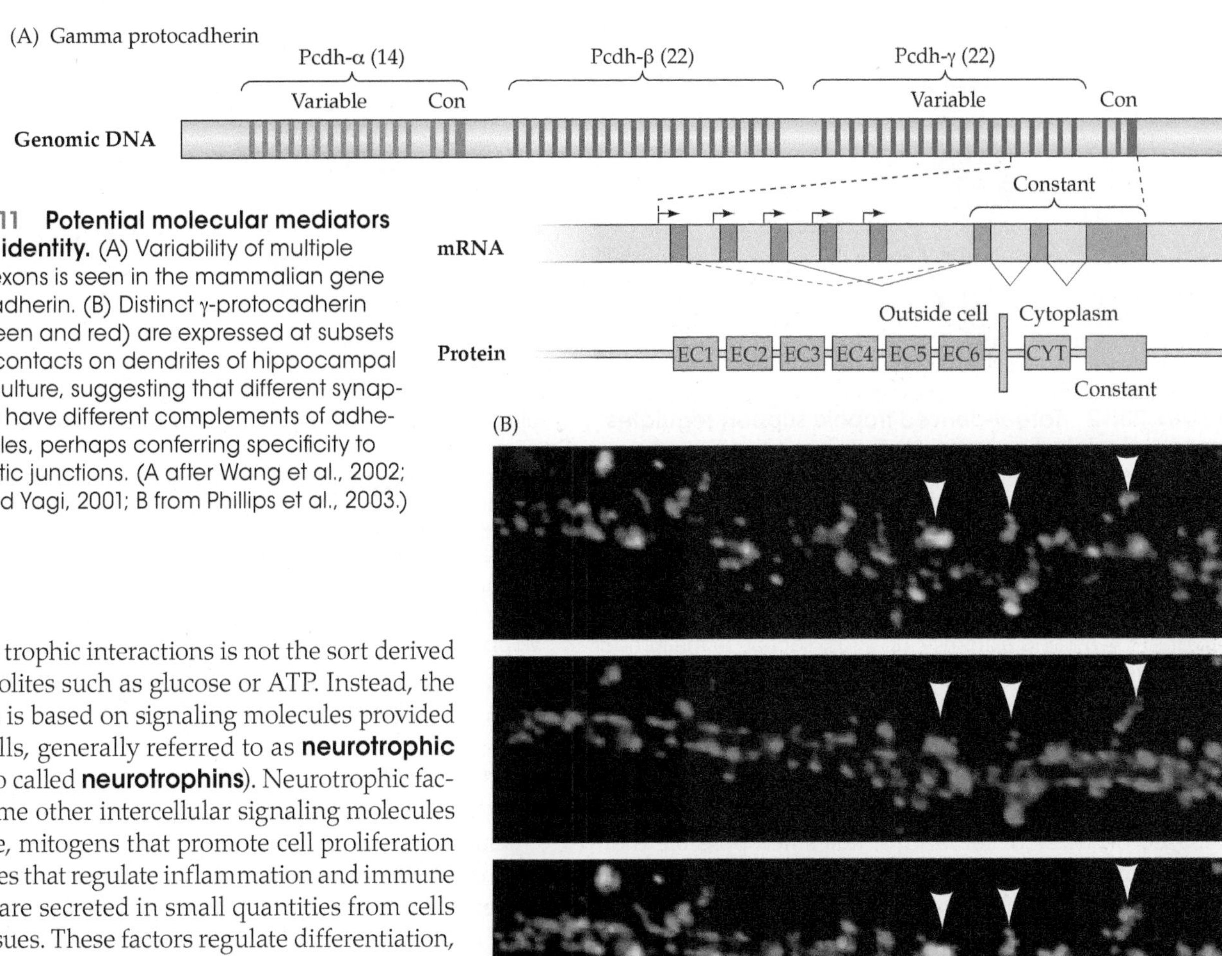

FIGURE 23.11 Potential molecular mediators of synapse identity. (A) Variability of multiple alternative exons is seen in the mammalian gene for γ-protocadherin. (B) Distinct γ-protocadherin isoforms (green and red) are expressed at subsets of synaptic contacts on dendrites of hippocampal neurons in culture, suggesting that different synaptic sites may have different complements of adhesion molecules, perhaps conferring specificity to those synaptic junctions. (A after Wang et al., 2002; Hamada and Yagi, 2001; B from Phillips et al., 2003.)

provided by trophic interactions is not the sort derived from metabolites such as glucose or ATP. Instead, the dependence is based on signaling molecules provided by target cells, generally referred to as **neurotrophic factors** (also called **neurotrophins**). Neurotrophic factors, like some other intercellular signaling molecules (for example, mitogens that promote cell proliferation and cytokines that regulate inflammation and immune responses), are secreted in small quantities from cells in target tissues. These factors regulate differentiation, growth, and ultimately survival in nearby cells. Neurotrophic factors are unique in that, unlike inductive signaling molecules and cell adhesion molecules, their expression is limited to neurons and a few non-neural neuronal targets such as muscles. These factors are first detected after the initial populations of postmitotic neurons have been generated in the nascent CNS and PNS, and they help regulate the phase of neural development that begins once neurogenesis has concluded.

Why do developing neurons depend so strongly on their targets, and what specific cellular and molecular interactions mediate this dependence? The answer to the first question lies in the changing scale of the developing nervous system and the body it serves, and the related need to precisely match the number of neurons in particular populations with the size and functional demands of their targets. A general—and surprising—strategy in vertebrate development is the production of an initial surplus of nerve cells (on the order of two- or threefold); the final population is subsequently established by the death of those neurons that fail to interact successfully with their intended targets. The elimination of supernumerary neurons, particularly the initiation of **apoptosis**—the highly regulated processes that result in cell death—is mediated by a specific group of neurotrophic factors, the neurotrophins (see below and Chapter 26).

A series of studies dating from the early twentieth century showed that targets play a major role in determining the size of the neuronal populations that innervate them, presumably based on the targets' provision of neurotrophic factors. The pioneering neuroembryologists Viktor Hamburger and Rita Levi-Montalcini made these seminal observations, first independently and then collaboratively, in the 1930s and 1940s. A critical finding was that the removal of a limb bud from a chick embryo results, at later embryonic stages, in a striking reduction in the number of nerve cells (in this case α motor neurons) in corresponding portions of the spinal cord (Figure 23.12A,B). Apparently, signals in the limb bud (the target cells) ensure the survival of α motor neurons. Furthermore, in typical embryos, an apparent surplus of motor neurons is generated prior to the growth and innervation of the limb, and disappears later in development. Initially, these surplus motor neurons innervate the immature limb, but then die—presumably due to a lack of trophic support—as the motor neuron population becomes matched to the needs of the developing limb musculature.

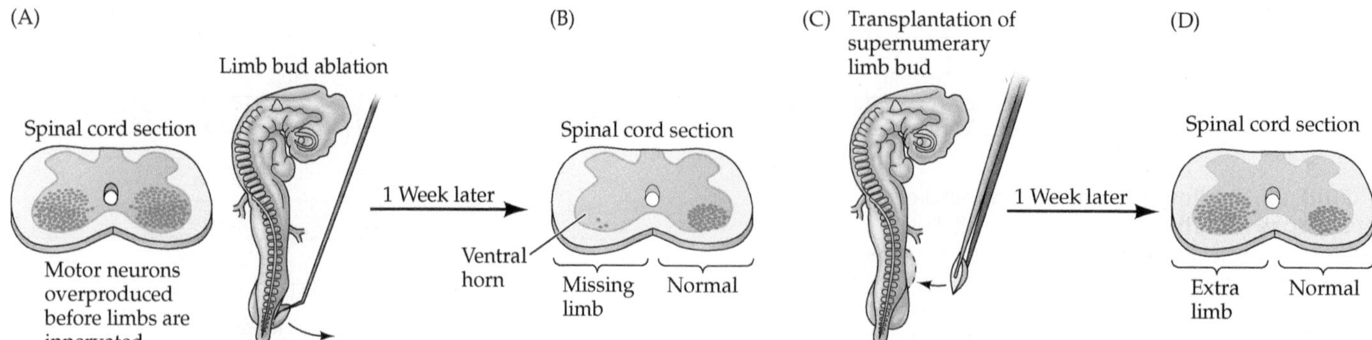

FIGURE 23.12 Target-derived trophic support regulates survival of related neurons. (A) The chick spinal cord generates an excess of neurons (green) prior to the differentiation and innervation of the limb. Normally, some of these neurons are lost once the appropriate level of innervation is established in the developing limb bud. Limb bud amputation in a chick embryo at the appropriate stage of incubation (about 2.5 days) further depletes the pool of motor neurons that would have innervated the missing extremity. (B) Cross section of the lumbar spinal cord in an embryo that underwent this surgery about a week earlier. The motor neurons (red dots) in the ventral horn that would have innervated the hindlimb have degenerated almost completely. A normal complement of motor neurons is present on the other side; most of the normally supernumerary neurons have been lost. (C) Adding an extra limb bud before the normal period of cell death rescues early-generated neurons that normally would have died. (D) Such augmentation leads to an abnormally large number of limb motor neurons on the side related to the extra limb, and these neurons are recruited from the pool of cells overproduced at an earlier stage of development (green dots) rather than generated de novo through cell proliferation elicited by the added target. (After Hamburger, 1958, 1977 and Hollyday and Hamburger, 1976.)

Based on these observations, it seemed possible that when a limb bud is present, innervating neurons in the spinal cord compete with one another for a resource available in limited supply in the developing limb. In support of this idea, many neurons that would normally die can be rescued by augmenting the amount of target tissue available, thereby providing extra trophic support—in this example, by experimentally adding to the embryo a limb bud that can be innervated by the same spinal segments that innervate the normal limb (Figure 23.12C,D). Careful monitoring of cell proliferation versus cell death shows that the extra cells are not generated de novo (i.e., they do not arise from precursors in response to a mitogenic signal from the extra target). Instead, they are "rescued" from a neuron population that is overproduced in early development and normally winnowed based on the limited trophic support from the targets. Thus, the size of nerve cell populations in the adult is not fully determined by a rigid genetic program of cell proliferation followed by highly specified innervation of the target. The connections between nerve cells and their targets can be modified by neuron–target interactions in each developing individual.

Competitive Interactions and the Formation of Neuronal Connections

Once the size of a neuronal population is established by trophic regulation, trophic interactions continue to modulate the formation of synaptic connections, beginning in embryonic life and extending far beyond birth. Certain problems must be solved during the establishment of innervation. These include ensuring that the "right" number of remaining axons innervates each target cell ("right" suggests sufficient innervation to allow the target cell to generate electrical signals and integrate or process information vital to the operation of its neural circuit) and ensuring that each axon innervates the "right" number of target cells. Getting these numbers correct is another major achievement of trophic interactions between developing neurons and target cells, and it is necessary for establishing appropriate circuits to support specific functional demands of each individual organism.

Studying synaptic refinement in the complex circuitry of the cerebral cortex or other regions of the CNS is a formidable challenge. Many fundamental ideas about the ongoing modification of developing brain circuitry have come from simpler, more accessible parts of the nervous system, most notably the vertebrate neuromuscular junction and autonomic ganglion cells (Figure 23.13). Adult skeletal muscle fibers and neurons in some classes of autonomic ganglia (parasympathetic neurons) are each innervated by a single axon. Initially, however, each of these target cells is innervated by axons from several central motor neurons (the pre-ganglionic autonomic neurons in the brainstem and spinal cord are considered analogous to α motor neurons), a condition called **polyneuronal innervation**. In such cases, inputs are gradually lost during early postnatal development until only one remains. This process of loss

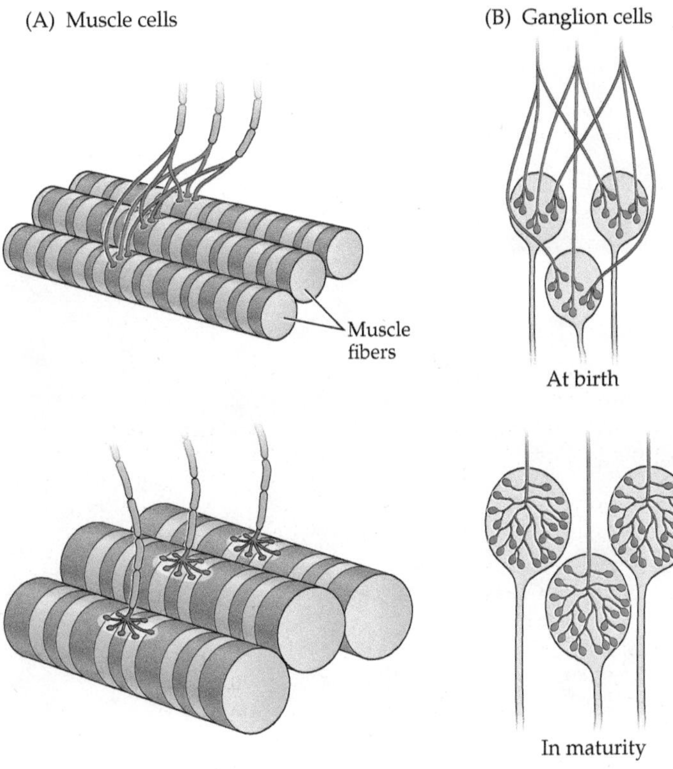

FIGURE 23.13 Synapse number and pattern in the mammalian PNS are adjusted during early postnatal life. In muscles (A) and in peripheral ganglia whose neurons have no dendrites (B), each axon innervates a larger number of target cells at birth than in maturity. Most of this rudimentary multiple innervation is eliminated shortly after birth. In both muscles and ganglia, however, the size and complexity of the terminal arbor that remains on each mature target cell increase. Thus, each axon elaborates more and more terminal branches and synaptic endings on the target cells it will innervate in maturity. The common denominator of this process is not a net loss of synapses, but the removal of immature contacts from all but one or a few axons on each target, and the focus on fewer target cells by a progressively increasing amount of synaptic machinery for each axon that remains. (After Purves and Lichtman, 1980.)

is generally referred to as **synapse elimination**, although here "elimination" refers to a reduction in the number of different axonal inputs to the target cells, not to a reduction in the overall number of synaptic contacts (the individual specialized junctions between pre- and postsynaptic cells) made on postsynaptic cells.

In fact, the overall number of synaptic contacts and the complexity of pre- and post-synaptic specializations in the PNS increases steadily during the course of development, as is also the case throughout the brain (see Chapter 25). The elimination of some initial minimally functional inputs (i.e., an individual axon making one or a few synaptic contacts) to muscle and ganglion cells is a process in which synapses originating from different neurons compete with one another for "ownership" of an individual target cell. Patterns of electrical activity in the pre- and postsynaptic partners are thought to influence this competition for target space and neurotrophic support. For example, if acetylcholine receptors at the neuromuscular junction are blocked by the potent AChR antagonist curare, polyneuronal innervation persists. Blocking presynaptic action potentials in the motor neuron axons (by silencing the nerve with TTX, a sodium channel blocker; see Box 6A) also prevents the reduction of polyneuronal innervation. Blocking neural activity therefore reduces or delays competitive interactions, presumably mediated by neurotrophic signaling, and the associated synaptic rearrangements.

Many useful insights into the nature of synaptic rearrangement during development have come from direct observations of competition between the presynaptic endings of two motor neuron axons for a single synaptic site on a developing muscle fiber. Using different colored fluorescent dyes or genetically encoded reporters that label either each presynaptic terminal or the postsynaptic receptors, the same neuromuscular junction can be followed over days, weeks, and even longer (Figure 23.14A). Competition between synapses arising from different motor neurons does not involve the active displacement of the "losing" input by the eventual "winner." Instead, it appears that the inputs of the two competitors initially occupy the same subregion of a nascent postsynaptic specialization, but then gradually segregate. The post-synaptic territory of the "losing" axon diminishes over time, and eventually the presynaptic ending atrophies and the axon retracts from the synaptic site. Neurotransmitter receptors beneath the terminal branches that eventually will be eliminated are also lost. This receptor loss occurs before the nerve terminal has withdrawn and presumably reduces the synaptic strength of the input, which results in further loss of postsynaptic receptors, leading to further reduction in the strength of the input. This downward spiral of synaptic efficacy and its molecular and cellular foundations presumably results in withdrawal of the presynaptic terminal. The remaining terminals continue to enlarge and strengthen as the end plate region expands during postnatal muscle growth.

Input elimination occurs in a variety of other PNS and CNS regions. In autonomic ganglia in the PNS, the number

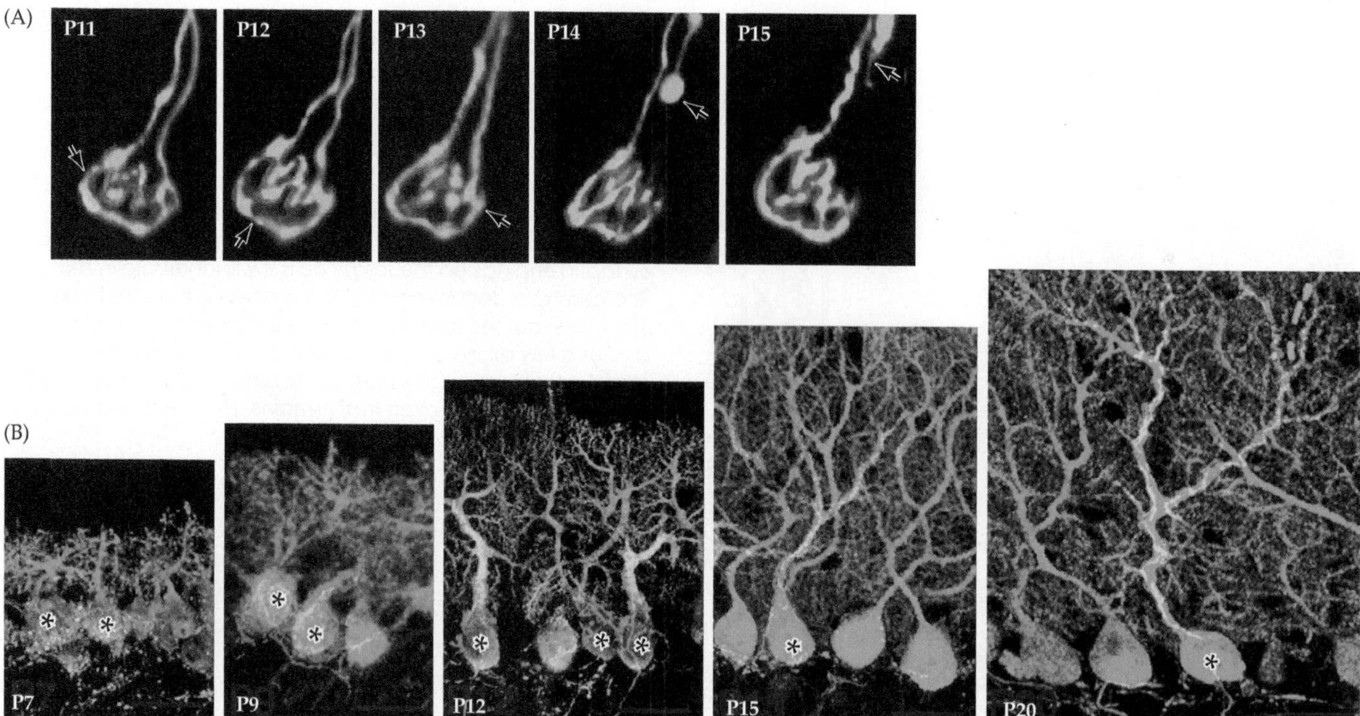

FIGURE 23.14 The elimination of multiple innervation in the PNS (neuromuscular junction) and CNS (cerebellum). (A) In this series of images, the same neuromuscular junction in a neonatal mouse has been imaged repeatedly, beginning on postnatal day 11 (P11). Initially, two axons (green and blue) innervate the muscle fiber (the local clustering of postsynaptic ACh receptors is shown in red). The arrow indicates the boundary between the postsynaptic territory of the green and blue axons. By P12, the proportion of territory occupied by the green and blue axons has begun to shift, with the blue axon terminal apparently losing postsynaptic space while the green axon terminal expands. This process continues at P13, and by P14 the blue axon has fully retreated, its synaptic terminal transformed into a large retraction bulb (arrow at P14). Within an additional day, the retracting axon is almost fully withdrawn from the synaptic site. (B) Single climbing fiber axons (red) in the developing cerebellum initially innervate multiple Purkinje cells (asterisks) during early postnatal development (postnatal day 7, P7). Over the course of approximately 2 weeks, individual climbing fibers retract immature axon branches from all but one Purkinje cell. The climbing fiber then elaborates an extensive axon that innervates most of the dendritic arbor of one (and only one) Purkinje cell. (A from Walsh and Lichtman, 2003; B from Hashimoto et al., 2009.)

of presynaptic axons innervating target neurons also decreases as the number of cellular synaptic specializations (pre-synaptic terminals arising from axon branches) from the remaining axons *increases*. A similar process has been described in the CNS. In the cerebellum, each adult Purkinje cell is innervated by a single axon called a climbing fiber (see Chapter 19). During early development, however, each Purkinje cell receives multiple climbing fiber inputs. During postnatal life, each climbing fiber axon retracts branches from neighboring Purkinje cells and elaborates multiple branches and synaptic contacts on the apical dendrite of a single Purkinje cell target (Figure 23.14B). Finally, in the visual cortex, initial binocular innervation of layer 4 cortical neurons is eliminated to establish segregated, molecularly driven inputs (see Chapter 25).

Thus, the pattern of synaptic connections that emerges in the adult is not simply a consequence of the biochemical identities of synaptic partners or other rigid developmental rules. Rather, the mature wiring plan is the result of a much more flexible process in which local neuronal connections are formed, removed, and remodeled according to local circumstances that reflect molecular constraints, the detailed structure and size of the target, and ongoing electrical activity. These interactions guarantee that every target cell is innervated—and continues to be innervated—by the right number of inputs and synapses, and that every innervating axon contacts the right number of target cells with an appropriate number of synaptic endings. This regulation of **convergence** (the number of inputs to a target cell) and **divergence** (the number of connections made by a neuron) in the developing nervous system is another key consequence of trophic interactions among neurons and their targets. The regulation of convergence and divergence by neurotrophic interactions is also influenced by the shape and size of neurons, particularly the dendrites (Box 23B).

BOX 23B ■ Why Do Neurons Have Dendrites?

Perhaps the most striking feature of neurons is their diverse morphology. Some classes of neurons have no dendrites at all; others have a modest dendritic arborization; still others have an arborization that rivals the complex branching of a fully mature tree (see Figures 1.2 and 1.6). Why should this be? Although there are many reasons for this diversity, neuronal geometry influences the number of different inputs that a target neuron receives by modulating competitive interactions among the innervating axons.

Evidence that the number of inputs a neuron receives depends on its geometry has come from studies of the peripheral autonomic system, where it is possible to stimulate the full complement of axons innervating an autonomic ganglion and its constituent neurons. This approach is not usually feasible in the CNS because of the anatomical complexity of most central circuits. Since individual postsynaptic neurons can also be labeled via an intracellular recording electrode, electrophysiological measurements of the number of different axons innervating a neuron can routinely be correlated with target cell shape. In both parasympathetic and sympathetic ganglia, the degree of preganglionic convergence onto a neuron is proportional to its dendritic complexity. Thus, neurons that lack dendrites altogether are generally innervated by a single input, whereas neurons with increasingly complex dendritic arborizations are innervated by a proportionally greater number of different axons (see figure). This correlation of neuronal geometry and input number holds within a single ganglion, among different ganglia in a single species, and among homologous ganglia across a range of species. Since ganglion cells that have few or no dendrites are initially innervated by several different inputs (see text), confining inputs to the limited arena of the developing cell soma evidently enhances competition among them, whereas the addition of dendrites to a neuron allows multiple inputs to persist in peaceful coexistence. Importantly, the dendritic complexity of at least some classes of autonomic ganglion cells is influenced by neurotrophins.

A neuron innervated by a single axon will clearly be more limited in the scope of its responses than a neuron innervated by 100,000 inputs (1 to 100,000 is the approximate range of convergence in the mammalian brain). By regulating the number of inputs that neurons receive, dendritic form greatly influences function.

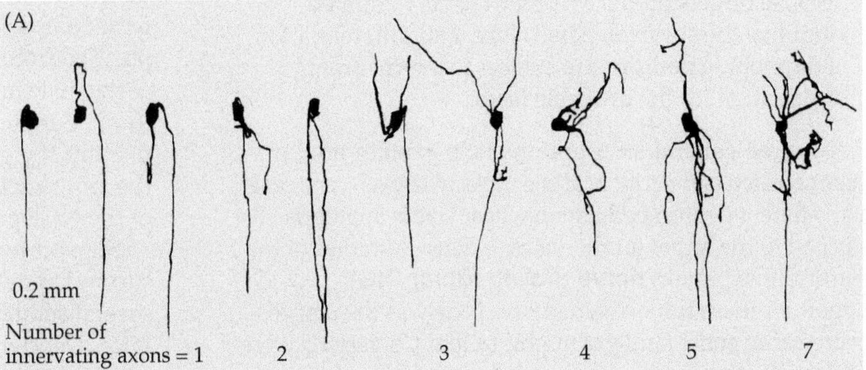

(A) Axons innervating ciliary ganglion cells in adult rabbits. Neurons studied electrophysiologically and then labeled by intracellular injection of a marker enzyme have been arranged in order of increasing dendritic complexity. The number of axons innervating each neuron is indicated. (From Purves and Hume, 1981.)

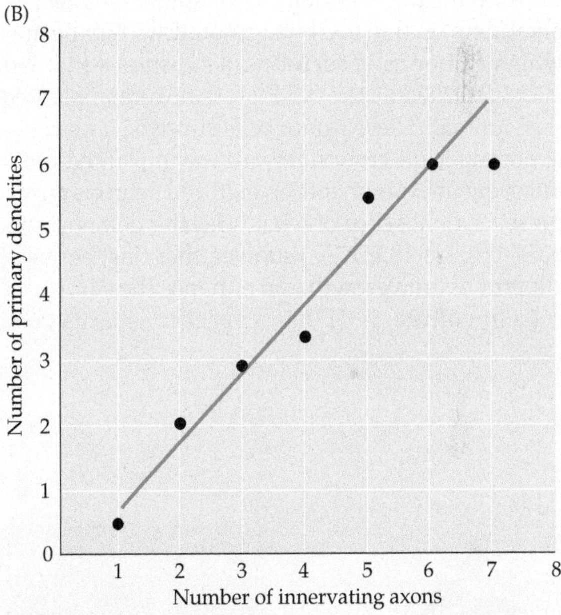

(B) This graph summarizes observations of the number of innervating axons on a large number of rabbit ciliary ganglion cells. There is a strong correlation between dendritic geometry and input number. (After Purves and Hume, 1981.)

The Molecular Basis of Trophic Interactions

Trophic interactions regulate three essential steps in the formation of mature neural circuits. These three functions—survival of a subset of neurons from a considerably larger population, formation and maintenance of appropriate numbers of connections, and the elaboration of axonal and dendritic branches to support these connections—can be rationalized at least in part by the supply and availability of molecular trophic factors. This perspective yields at least three general assumptions about neurons and their targets (which may be other neurons, or muscles and other peripheral structures):

1. Neurons depend on the availability of some minimum amount of trophic factor for the survival and subsequent persistence of appropriate numbers of target connections.

2. Target tissues synthesize appropriate trophic factors and make them available to developing neurons.
3. Because targets produce trophic factors in limited amounts, the survival, persistence, and differentiation of developing neurons are subject to interneuronal competition for the available factor.

These three general assumptions help explain how trophic-mediated competition, at the molecular level, can result in an afferent neuron population whose size is appropriately matched to the target it innervates. Extensive studies of the neurotrophic protein **nerve growth factor** (**NGF**) provide support for these three assumptions. NGF was the first-discovered of an entire family of trophic factors. Characterization of its activity and regulation provide a model for understanding how neurotrophins provided by targets influence the survival and connections of the nerve cells that innervate them.

NGF was discovered as an "activity" that elicited robust growth of neuronal processes both in the animal and in cell culture. The idea that a specific molecule might elicit neuronal growth originated with experiments in which tumor cells secreting a substance that would be later identified as the NGF protein were implanted into a host animal. These tumor cells survived, and they caused abnormal axon growth toward the implanted tumor cells. Subsequent experiments on cultured neurons grown in the presence or absence of what was later demonstrated to be NGF (Figure 23.15A–C) showed that this "activity" could support neurite growth and enhance the survival of nerve cells in culture. NGF has a specific capacity to support sympathetic neurons in vitro; accordingly, much of the subsequent in vivo analysis of NGF activity was focused on the sympathetic portion of the autonomic nervous system. NGF was eventually identified as a protein and was purified from a rich biological source, the salivary glands of the male mouse (the reasons for the enrichment of NGF in male mouse salivary glands remain obscure). Subsequently, the gene for NGF was identified and cloned, and the amino acid sequence and three-dimensional structure of the NGF protein were determined.

Support for the idea that NGF is important for neuronal survival in more physiological circumstances emerged from several additional observations made primarily in rats and mice. Depriving developing mice of NGF (by the chronic administration of an NGF antiserum or by eliminating the gene for NGF selectively) resulted in adult mice that lacked most NGF-dependent neurons (Figure 23.15D). Conversely, injection of exogenous NGF into newborn rodents caused enlargement of sympathetic ganglia (which are particularly NGF-dependent) due to additional cells as well as more extensive axonal and dendritic growth, an effect opposite to that of NGF deprivation. Finally, once the protein and gene for NGF were identified, it was possible to show that NGF was indeed made available by the targets of the nerve cells that it acted on (a variety of smooth muscle, vascular, and glandular tissues; see Chapter 21). The dramatic influence of NGF on afferent neuron survival and its secretion from specific neuronal targets, together with what was known about the significance of neuronal death and regulation of neurite growth in development, suggested that NGF is indeed a target-derived

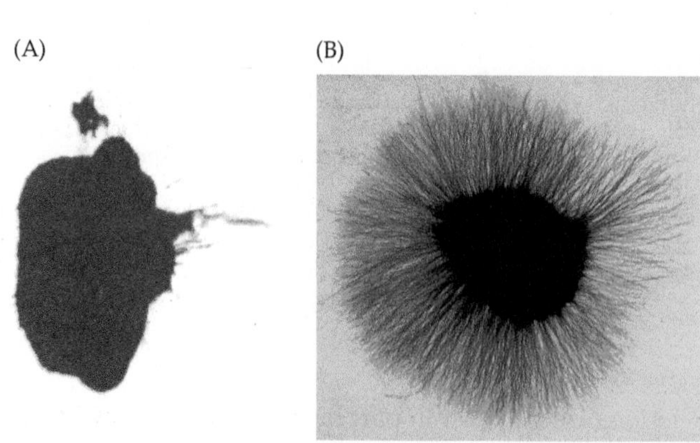

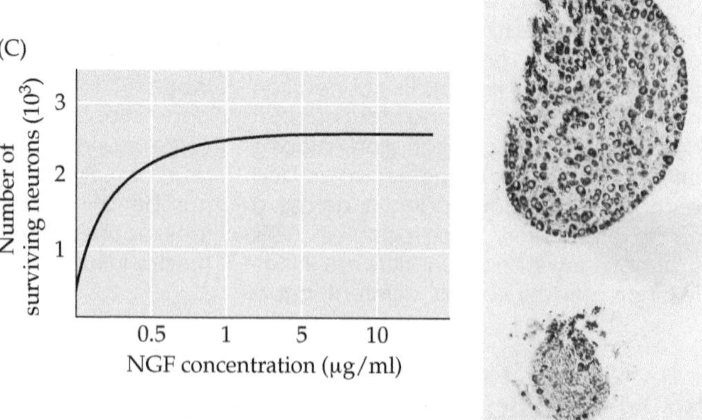

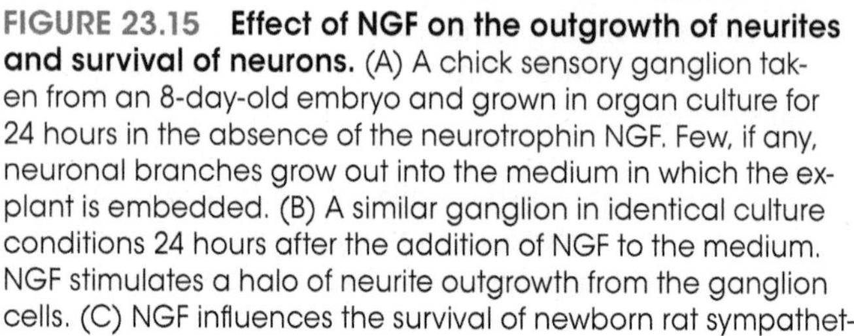

FIGURE 23.15 Effect of NGF on the outgrowth of neurites and survival of neurons. (A) A chick sensory ganglion taken from an 8-day-old embryo and grown in organ culture for 24 hours in the absence of the neurotrophin NGF. Few, if any, neuronal branches grow out into the medium in which the explant is embedded. (B) A similar ganglion in identical culture conditions 24 hours after the addition of NGF to the medium. NGF stimulates a halo of neurite outgrowth from the ganglion cells. (C) NGF influences the survival of newborn rat sympathetic ganglion cells grown in culture for 30 days. Dose-response curves confirm the dependence of these neurons on the availability of NGF. (D) Cross section of a superior cervical ganglion from a normal 9-day-old mouse (top) compared with a similar section from a littermate that was injected daily since birth with NGF antiserum (bottom). The ganglion of the treated mouse shows marked atrophy, with obvious loss of nerve cells. (A,B from Purves and Lichtman, 1985, courtesy of R. Levi-Montalcini; C after Chun and Patterson, 1977; D from Levi-Montalcini, 1972.)

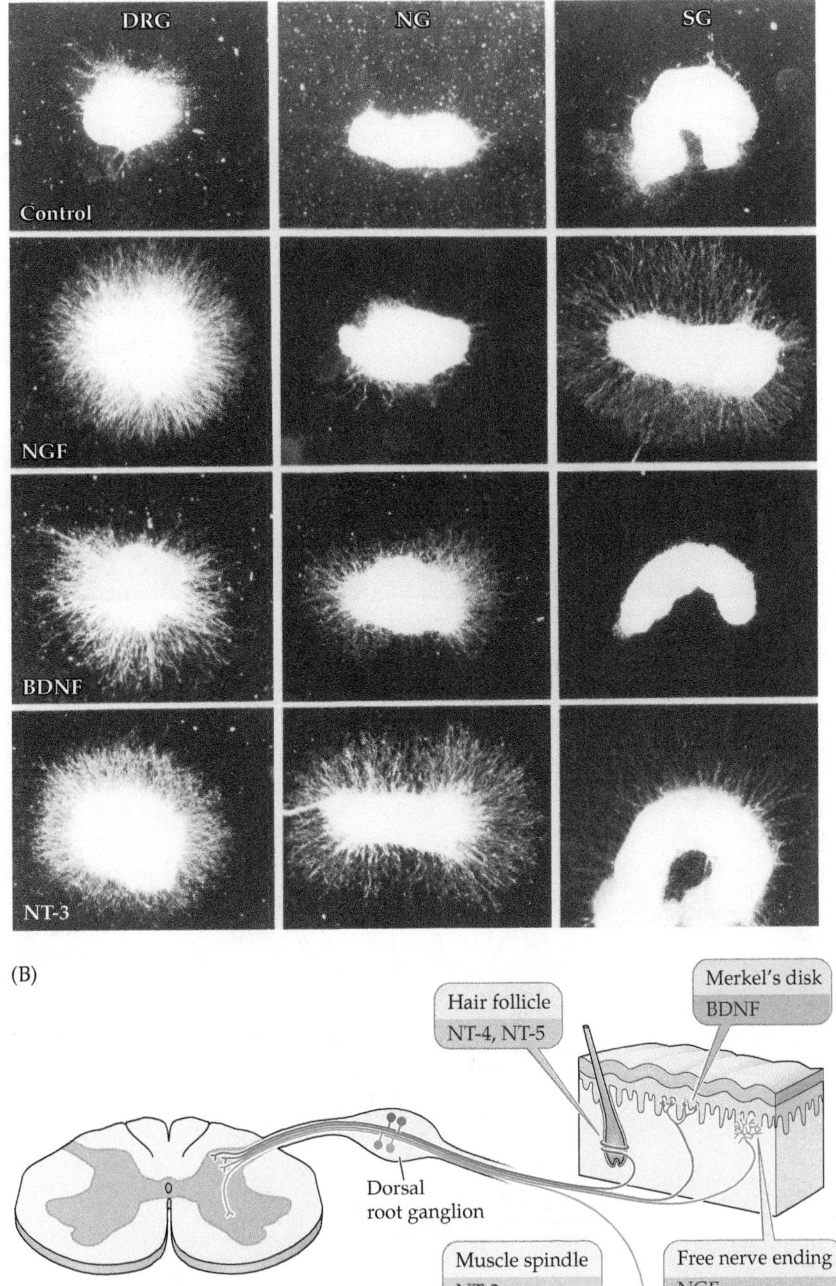

FIGURE 23.16 Neurotrophins have distinct effects on different target neurons. (A) Effect of NGF, BDNF, and NT-3 on the outgrowth of neurites from explanted dorsal root ganglia (left column), nodose ganglia (middle column), and sympathetic ganglia (right column). The specificities of these several neurotrophins are evident in the ability of NGF to induce neurite outgrowth from sympathetic and dorsal root ganglia, but not from nodose ganglia (which are cranial nerve sensory ganglia that have a different embryological origin from dorsal root ganglia); of BDNF to induce neurite outgrowth from dorsal root and nodose ganglia, but not from sympathetic ganglia; and of NT-3 to induce neurite outgrowth from all three types of ganglia. (B) Specific influence of neurotrophins in vivo. Distinct classes of peripheral somatosensory receptors and the dorsal root ganglion cells that give rise to these sensory endings depend on different trophic factors in specific target tissues. (A from Maisonpierre et al., 1990; B after Bibel and Barde, 2000.)

signal that serves to match the number of nerve cells to the number of target cells.

From the outset, it was apparent that only certain classes of nerve cells respond to NGF. Furthermore, diversity of trophic factors most certainly would facilitate greater specificity between pre- and postsynaptic partners during development (or in regeneration; see Chapter 26). In the 1980s and 1990s, work from several laboratories showed that NGF belongs to the family of related trophic molecules referred to as neurotrophins. Neurotrophins are functionally similar to a broader class of signaling molecules found throughout the organism and referred to generically as *growth factors*. The expression and activity of the neurotrophins, however, are mostly limited to neurons and their targets. At present, there are four well-characterized members of the neurotrophin family in addition to NGF: **brain-derived neurotrophic factor** (**BDNF**), **neurotrophin-3** (**NT-3**), and **neurotrophins 4** and **5** (**NT-4/5**) (Figure 23.16). In addition, two other ligand families, those that include the ciliary neurotrophic factor (CNTF) and the glial derived neurotrphic factor (GDNF), also have neurotrophic signaling capacity (see below).

Although the members of the neurotrophin family are homologous in amino acid sequence and structure, they are encoded by distinct genes and are very different in their specificity (see Figure 23.16A). For example, NGF supports the survival of (and neurite outgrowth from) sympathetic neurons, while another family member—BDNF—cannot. Conversely, BDNF, but not NGF, can support the survival of certain sensory ganglion neurons, which have a different embryonic origin. NT-3 supports both of these populations. Given the diverse systems whose growth and connectivity must be coordinated during neural development, this specificity makes good sense. Indeed, different neurotrophins are selectively available in different targets. For example, the different receptor specializations in the skin that transduce somatosensory information (touch and proprioception versus pain and temperature) express different neurotrophins, and this specificity is matched by expression of neurotrophin receptors (see the next section) that distinguish the sensory neurons that innervate each distinct structure (see Figure 23.16B). All of these neurotrophins also act in the CNS to regulate the number and growth of neurons; however, the local specificities of their actions in the CNS are not as well established as in the PNS.

In addition to the neurotrophins, other secreted molecules also have neurotrophic influences. These include the ciliary neurotrophic factor (CNTF), which is considered a cytokine because of its role in inflammation and immune responses beyond neurotrophic interactions; leukemia inhibitory factor (LIF), also a cytokine; and glial-derived neurotrophic factor (GDNF) and related proteins (referred to as the GDNF family of ligands). GDNF ligands activate the RET tyrosine kinase receptor; they influence kidney development and spermatogenesis as well as neuronal differentiation. Thus, a variety of factors—some specific to the nervous system, others used for purposes beyond neural development—influence the survival and growth of developing nerve cells and therefore the elaboration of neural circuits.

Neurotrophin Signaling

Neurotrophic factors are clearly key regulators for three distinct cellular mechanisms: neural process growth or retraction; synapse stabilization or elimination; and cell survival or death. The ways in which these ligands influence different aspects of neuronal differentiation reflect the details of signal transduction in response to neurotrophic molecules. The specific role of NGF in axon growth was most dramatically demonstrated by experiments that isolated NGF availability to subsets of neural processes without exposing the cell body to the factor (Figure 23.17A). The results of these experiments indicated that NGF acts locally to stimulate neurite growth—even while other processes of the same cell, deprived of NGF, are retracting. Apparently, general availability of NGF to an individual neuron does not support growth of all of that neuron's processes. Similar observations have been made for other neurotrophins. For example, BDNF can influence neurite growth by altering local Ca^{2+} signaling in growth cones (Figure 23.17B). In addition, physiological experiments indicate that NGF, BDNF, and other neurotrophins influence synaptic activity and plasticity (see Chapter 25), again independent of their effects on cell survival. Finally, neurotrophins engage signaling pathways that protect cells from programmed cell death (apoptosis), and their absence activates signaling that leads to expression and activity of apoptotic genes. These and related effects depend on relaying the neurotrophic signal from the axon terminal to the cell body. To accomplish this relay, neurotrophins bound to their transmembrane receptors (primarily the Trk subset of receptors; see below) are selectively internalized by assembling a **signaling endosome** that also

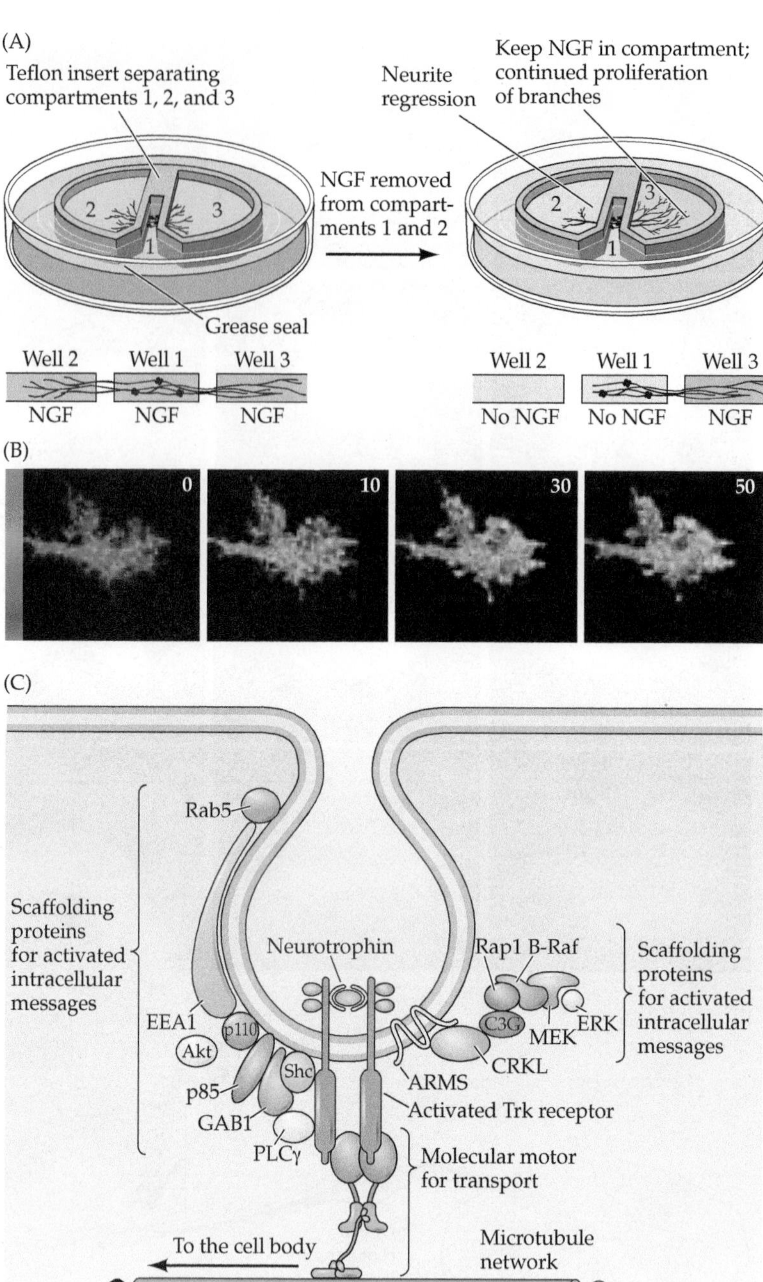

FIGURE 23.17 Neurotrophins influence neurite growth by local action. (A) Three compartments ("wells") of a culture dish are separated by a Teflon divider sealed to the bottom of the dish with grease. A magnified view looking down on the wells is shown below. Isolated rat sympathetic ganglion cells plated in well 1 can grow through the grease seal into wells 2 and 3. Growth into a lateral well occurs as long as the well contains an adequate concentration of NGF. This local application does not influence the neurites in the other lateral well. Subsequent removal of NGF from a well causes local regression of neurites without affecting neurite survival in the other wells. (B) Ca^{2+} signaling within a growth cone from a developing spinal cord neuron in culture. This time-lapse sequence illustrates that neuritic growth can be locally controlled by neurotrophins that stimulate Ca^{2+} release. (C) Trk-mediated neurotrophin signaling at the axon is maintained and propagated by endocytic internalization of the ligand-receptor complex with several scaffolding proteins that bind one of three intracellular effectors: the Akt kinase, phospholipase Cγ (PLCγ), or extracellular signal-regulated kinase (ERK). This "signaling endosome" can also bind molecular motors that engage the microtubule cytoskeleton. The signaling endosome, activated via neurotrophin binding, is then transported back to the cell body to activate downstream targets, including modifying gene expression. (A after Campenot, 1981; B from Li. et al., 2005; C after Zweifel et al., 2005.)

includes neurotrophins bound to the now-activated neurotrophin receptor/kinase, as well as an assembly of related signaling proteins (Figure 23.17C). This signaling endosome is transported back to the cell body. The binding of the neurotrophin at a distal target, and the subsequent endocytic internalization and retrograde transport of the ligand/receptor kinase complex (with the catalytic domain of the Trk receptor still facing the cytoplasm) facilitates a key aspect of neurotrophic signaling. In this way, a neurotrophic ligand can activate and maintain signaling from the site of availability to the neuronal cell body, which is often a substantial distance. The ongoing maintenance of trophic support provided by an afferent by a target neuron may be a point of vulnerability for initiating pathology in a number of neurodegenerative diseases.

The selective actions of neurotrophins reflect their interactions with two classes of neurotrophin receptors: the **tyrosine kinase (Trk)** receptors and the **p75** receptor. There are three Trk receptors, each of which is a single transmembrane protein with a cytoplasmic tyrosine kinase domain. TrkA is primarily a receptor for NGF, TrkB is a receptor for BDNF and NT-4/5, and TrkC is a receptor for NT-3 (Figure 23.18). In addition, all neurotrophins activate the p75 receptor protein. The interactions between neurotrophins, Trks, and p75 demonstrate another level of selectivity and specificity of neurotrophin signaling. All neurotrophins are secreted in an unprocessed form that undergoes subsequent proteolytic cleavage. The Trk receptors have high affinity for processed ligands, while the p75 receptor has high affinity for unprocessed neurotrophins but low affinity for the processed ligands. The expression of a particular Trk receptor subtype or p75 therefore confers the capacity for distinct neurotrophin responses. Since Trk and p75 receptors are expressed only in subsets of neurons, and neurotrophins are available from different classes of targets (see Figure 23.16), selective binding between ligand and receptor likely accounts for some of the specificity of neurotrophic interactions.

Trk receptors, via stimulation of their intercellular tyrosine kinase domain and subsequent phosphorylation of target proteins, engage three distinct second-messenger pathways that alter functions of proteins (via phosphorylation and other post-translational modifications) or that change gene expression in the target cell (Figure 23.19). These receptors activate the small GTPase ras, which then activates a family of cytoplasmic protein kinases—the *m*itogen *a*ctivated *p*rotein (MAP) kinases—to elicit a variety of cellular responses, including changes in gene expression. Signaling from the Trk receptors via ras and MAP kinases is thought to regulate neurite outgrowth in response to neurotrophin signaling at the cell surface. Trk receptors also activate two enzymes that modify or release phospholipid second messengers—phospholipase C (PLC) and phosphoinositol 3 (PI 3) kinase—that influence function of existing proteins in the cell or cause changes in gene expression. Phospholipase C-dependent signaling preferentially influences cellular responses that lead to activity-dependent synapse plasticity (see Chapter 25). PI3 kinase interacts with pathways that regulate the activity of Akt kinase, a cytoplasmic kinase that modulates proteins that either prevent or promote cell death. The last of the neurotrophin receptors, p75, also engages three different intercellular signaling pathways. One of these is mediated by Rho GTPases (which function in much the same way as ras) that influence neurite growth. The second p75-dependent pathway leads to activation of the c-Jun

(A)

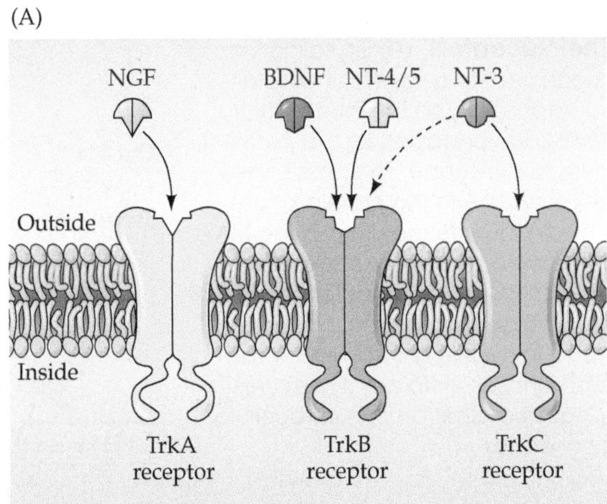

(B)

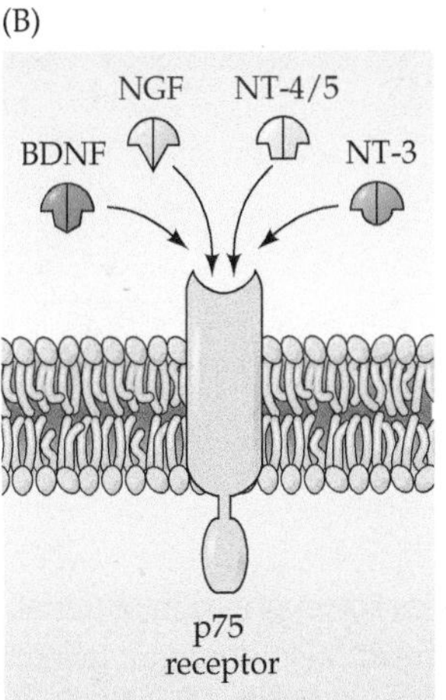

FIGURE 23.18 Neurotrophin receptors and their specificity. (A) The Trk family of receptor tyrosine kinases for the neurotrophins. TrkA is primarily a receptor for NGF, TrkB a receptor for BDNF and NT-4/5, and TrkC a receptor for NT-3. Because of the high degree of structural homology among both the neurotrophins and the Trk receptors, there is some degree of cross-activation between factors and receptors. For example, NT-3 can bind to and activate TrkB under some conditions, as indicated by the dashed arrow. These distinct receptors allow various neurons to respond selectively to the different neurotrophins. (B) The p75 low-affinity neurotrophin receptor binds all neurotrophins at low affinities (as its name implies). This receptor confers the ability to respond to a broad range of neurotrophins on fairly broadly distributed classes of neurons in the PNS and CNS.

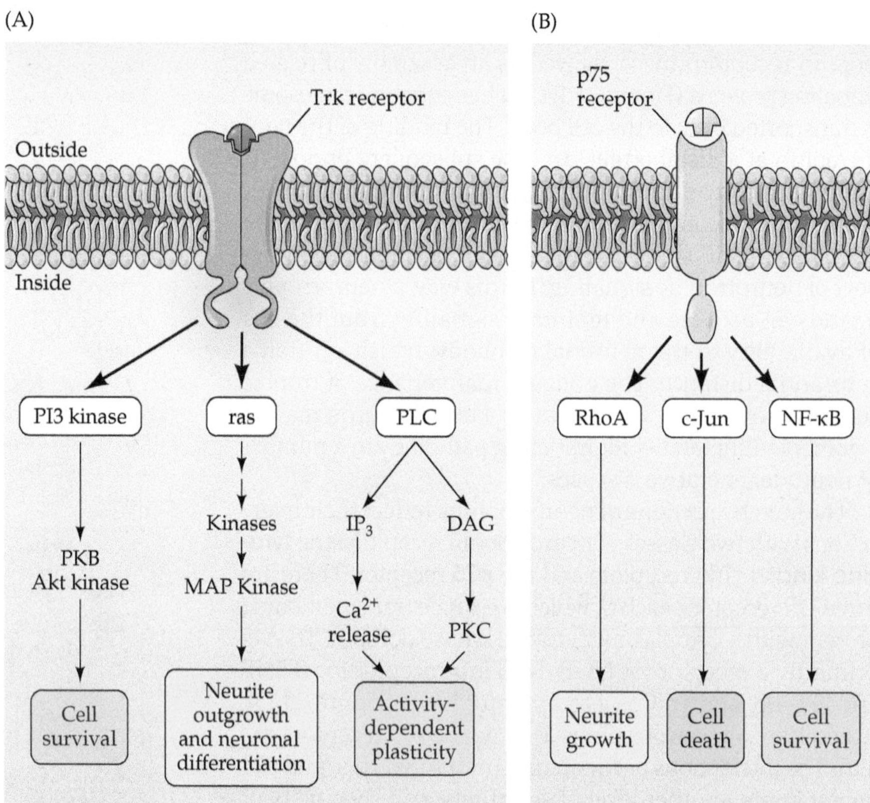

FIGURE 23.19 Signaling via neurotrophins and their receptors. (A) Signaling via Trk dimers can lead to a variety of cellular responses, depending on the intracellular signaling cascade engaged by the receptor after binding to the ligand. The possibilities include cell survival (via the protein kinase C/Akt pathway); neurite growth (via the MAP kinase pathway); and activity-dependent plasticity (via the Ca^{2+}/calmodulin and PKC pathways). (B) Signaling via the p75 pathway can lead to neurite growth via interaction with Rho kinases, or to cell cycle arrest and cell death via other distinct intracellular signaling cascades.

transcription factor, which regulates the expression of genes involved in apoptosis. Finally, p75 can also engage nuclear factor κB (NF-κB), which enhances expression of genes that promote cell survival. Thus, the final functional influence of any neurotrophin depends on (1) the receptor it encounters, and (2) that receptor's likelihood of engaging diverse intercellular signaling pathways that influence growth, change, or cell death.

The subtlety and diversity of neuronal circuits are set during later stages of neural development by different combinations of neurotrophins, their receptors, and signal transduction mechanisms. These mechanisms in concert determine the numbers of neurons, their shape, and their patterns of connections. Presumably, disruption of these neurotrophin-dependent processes, either during development or in the adult brain, can result in neurodegenerative conditions in which neurons die due to lack of appropriate trophic support, or fail to make and maintain appropriate connections due to failure to grow. Such disruption has devastating consequences for the circuits that the cells define and for the behaviors controlled by those circuits. Indeed, the pathogenic mechanisms of developmental disorders such as autism and schizophrenia, as well as neurodegenerative diseases as diverse as amyotrophic lateral sclerosis (ALS) and Parkinson's, Huntington's, and Alzheimer's diseases, may all reflect abnormalities of neurotrophic regulation.

Summary

Neurons in the developing brain must integrate a variety of signals to determine where to send their axons, what cells to form synapses on, how many synapses to make and retain, and whether to live or die. A remarkable transient cellular specialization, the growth cone, is responsible for axon growth and guidance. Growth cones explore the embryonic environment and determine the direction of axon growth as well as recognize appropriate targets. Their motile properties allow growth cones to approach, select, or avoid a target according to modulation of the actin and microtubule cytoskeleton by numerous signaling mechanisms, many of which involve changes in intracellular Ca^{2+}. The instructions that elicit growth cone responses come from adhesive, chemotropic, chemorepellant, and trophic molecules. These molecules are embedded in the extracellular matrix, found on cell surfaces, or secreted into extracellular spaces. Their cues ensure that coherent axon pathways are formed and prevent inappropriate connections. Initial growth of dendrites is influenced by similar adhesion and recognition signaling mechanisms, resulting in appropriate dendritic orientation, branching, and distribution. Adhesive, attractive, and repulsive molecules also influence the differentiation of growth cones and dendritic domains into pre-and post-synaptic specializations that define a synapse. Further signals that specify synaptic partners, and stabilize or destabilize nascent synapses are transmitted by neurotrophins, molecules made by neuronal targets in small quantities that bind to a variety of receptors to elicit distinct cellular responses. Neurotrophic influences—cell survival or death, process growth, and modulation of synaptic activity—help determine which neurons remain in a neural circuit, how they are connected, and how they continue to change. Defects in the early guidance of axons or subsequent trophic regulation of synaptogenesis have been implicated in a variety of congenital neurological syndromes and developmental disorders, and neurotrophic dysfunction in the adult CNS may underlie degenerative pathologies such as Alzheimer's and Parkinson's diseases.

ADDITIONAL READING

Reviews

Culotti, J. G. and D. C. Merz (1998) DCC and netrins. *Curr. Opin. Cell Biol.* 10: 609–613.

Huber, A. B., A. L. Kolodkin, D. D. Ginty and J. F. Cloutier (2003) Signaling at the growth cone: Ligand-receptor complexes and the control of axon growth and guidance. *Annu. Rev. Neurosci.* 26: 509–563.

Kolodkin, A. L. and M. Tessier-Lavigne (2010) Mechanisms and molecules of neuronal wiring: A primer. *Cold Spring Harb. Perspect. Biol.* (epub).

Mei, L. and W.-C. Xiong (2008) Neuregulin 1 in neural development, synaptic plasticity and schizophrenia. *Nature Rev. Neurosci.* 9: 437–452.

Raper, J. A. (2000) Semaphorins and their receptors in vertebrates and invertebrates. *Curr. Opin. Neurobiol.* 10: 88–94.

Reichardt, L. F. (2006) Neurotrophin-regulated signalling pathways. *Philos. Trans. R. Soc. Lond. B Biol. Sci.* 361: 1545–1564.

Reichardt, L. F. and K. J. Tomaselli (1991) Extracellular matrix molecules and their receptors: Functions in neural development. *Annu. Rev. Neurosci.* 14: 531–570.

Rutishauser, U. (1993) Adhesion molecules of the nervous system. *Curr. Opin. Neurobiol.* 3: 709–715.

Sanes, J. R. and J. W. Lichtman (1999) Development of the vertebrate neuromuscular junction. *Annu. Rev. Neurosci.* 22: 389–442.

Segal, R. A. (2003) Selectivity in neurotrophin signaling: Theme and variations. *Annu. Rev. Neurosci.* 26: 299–330.

Wiggin, G. R., J. P. Fawcett and T. Pawson (2005). Polarity proteins in axon specification and synaptogenesis. *Dev. Cell* 8 (6): 803–816.

Zipursky, S. L. and J. R. Sanes (2010) Chemoaffinity revisited: dscams, protocadherins, and neural circuit assembly. *Cell* 143: 343–353.

Important Original Papers

Baier, H. and F. Bonhoeffer (1992) Axon guidance by gradients of a target-derived component. *Science* 255: 472–475.

Brown, M. C., J. K. S. Jansen and D. Van Essen (1976) Polyneuronal innervation of skeletal muscle in new-born rats and its elimination during maturation. *J. Physiol.* (*Lond.*) 261: 387–422.

Campenot, R. B. (1977) Local control of neurite development by nerve growth factor. *Proc. Natl. Acad. Sci. USA* 74: 4516–4519.

Drescher, U. and 5 others (1995) In vitro guidance of retinal ganglion cell axons by RAGS, a 25 kDa tectal protein related to ligands for Eph receptor tyrosine kinases. *Cell* 82: 359–370.

Farinas, I., K. R. Jones, C. Backus, X. Y. Wang and L. F. Reichardt (1994) Severe sensory and sympathetic deficits in mice lacking neurotrophin-3. *Nature* 369: 658–661.

Kaplan, D. R., D. Martin-Zanca and L. F. Parada (1991) Tyrosine phosphorylation and tyrosine kinase activity of the *trk* proto-oncogene product induced by NGF. *Nature* 350: 158–160.

Kennedy, T. E., T. Serafini, J. R. de la Torre and M. Tessier-Lavigne (1994) Netrins are diffusible chemotropic factors for commissural axons in the embryonic spinal cord. *Cell* 78: 425–435.

Kolodkin, A. L., D. J. Matthes and C. S. Goodman (1993) The *semaphorin* genes encode a family of transmembrane and secreted growth cone guidance molecules. *Cell* 75: 1389–1399.

Levi-Montalcini, R. and S. Cohen (1956) In vitro and in vivo effects of a nerve growth-stimulating agent isolated from snake venom. *Proc. Natl. Acad. Sci. USA* 42: 695–699.

Luo, Y., D. Raible and J. A. Raper (1993) Collapsin: A protein in brain that induces the collapse and paralysis of neuronal growth cones. *Cell* 75: 217–227.

Oppenheim, R. W., D. Prevette and S. Homma (1990) Naturally occurring and induced neuronal death in the chick embryo in vivo requires protein and RNA synthesis: Evidence for the role of cell death genes. *Dev. Biol.* 138: 104–113.

Serafini, T. and 6 others (1996) Netrin-1 is required for commissural axon guidance in the developing vertebrate nervous system. *Cell* 87: 1001–1014.

Sperry, R. W. (1963) Chemoaffinity in the orderly growth of nerve fiber patterns and connections. *Proc. Natl. Acad. Sci. USA* 50: 703–710.

Walter, J., S. Henke-Fahle and F. Bonhoeffer (1987) Avoidance of posterior tectal membranes by temporal retinal axons. *Development* 101: 909–913.

Books

Letourneau, P. C., S. B. Kater and E. R. Macagno (eds.) (1991) *The Nerve Growth Cone.* New York: Raven Press.

Loughlin, S. E. and J. H. Fallon (eds.) (1993) *Neurotrophic Factors.* San Diego, CA: Academic Press.

Purves, D. (1988) *Body and Brain: A Trophic Theory of Neural Connections.* Cambridge, MA: Harvard University Press.

Ramón y Cajal, S. (1928) *Degeneration and Regeneration of the Nervous System.* R. M. May (ed.). New York: Hafner Publishing.

CHAPTER 24

Circuit Development: Intrinsic Factors and Sex Differences

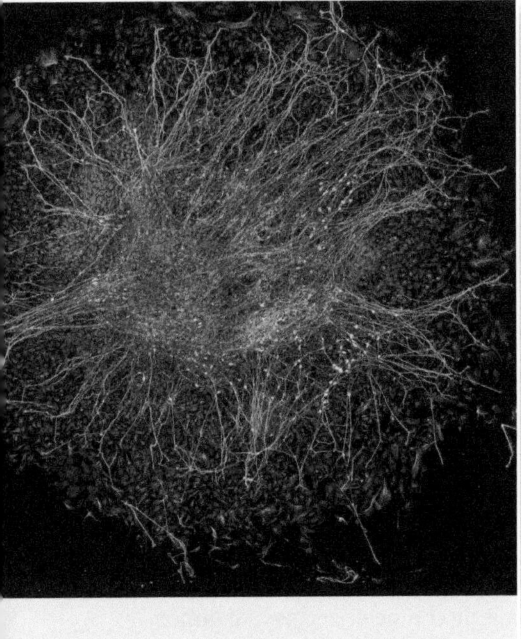

Overview

SEVERAL ESSENTIAL ASPECTS of neuronal differentiation and circuit development rely upon the activity of intrinsic factors. These factors are molecular signals produced by the embryo itself and elicit cellular changes that drive differentiation of specific brain regions and connections. These developmental events are key for establishing circuits that mediate some of the most fundamental, and adaptive behaviors of any organism. The best example of the influence of intrinsic factors is the establishment of different (dimorphic) structures and circuits in the brains of females and males of both invertebrate and vertebrate species, including mammals. The level and temporal pattern of secretion of gonadal steroids—estrogen and testosterone—once the developing ovaries or testes have differentiated, influences multiple programs of brain differentiation. These sex-specific patterns of intrinsic signals are transduced via multiple steroid receptors. The receptors are selectively expressed in neural precursors or brain regions that then go onto develop differently in females versus males. These regions and connections become specialized in register with different peripheral structures, particularly genitalia, mammary tissue. In addition, intrinsic signaling via gonadal steroids can mediate differentiation of sensory and motor pathways that serve male versus female behaviors for communication, mate identification or selection, and interactions with offspring. The mechanisms that establish sexually dimorphic circuits in the brain rely on the capacity of gonadal steroids and their receptors to modify gene expression and thus establish divergent developmental trajectories based on availability, concentration, and enzymatic processing of testosterone and estrogen—the two steroid hormones produced by the developing gonads. These mechanisms of circuit differentiation result in the specific behaviors of males and females that facilitate reproduction and parenting.

Sexual Dimorphisms and Sexually Dimorphic Behaviors

Sexual dimorphisms—clear and consistent physical differences between females and males of the same species—are seen throughout the animal kingdom and are usually associated with functional differences that promote reproduction or rearing of offspring. These physical differences are often recognized in peripheral structures found in one sex but not the other. Usually, different neural circuits in female and male brains parallel differences in female and male bodies. For example, in the tobacco hawk moth (*Manduca sexta*), the female and male antennae are strikingly different in size and structure: The female antennae are small and smooth, while the male antennae are larger and lined with rows of ciliated structures (Figure 24.1A). These anatomical specializations are essential for female versus male reproductive behaviors; female antennae sense specific odorants from the tobacco plants that are

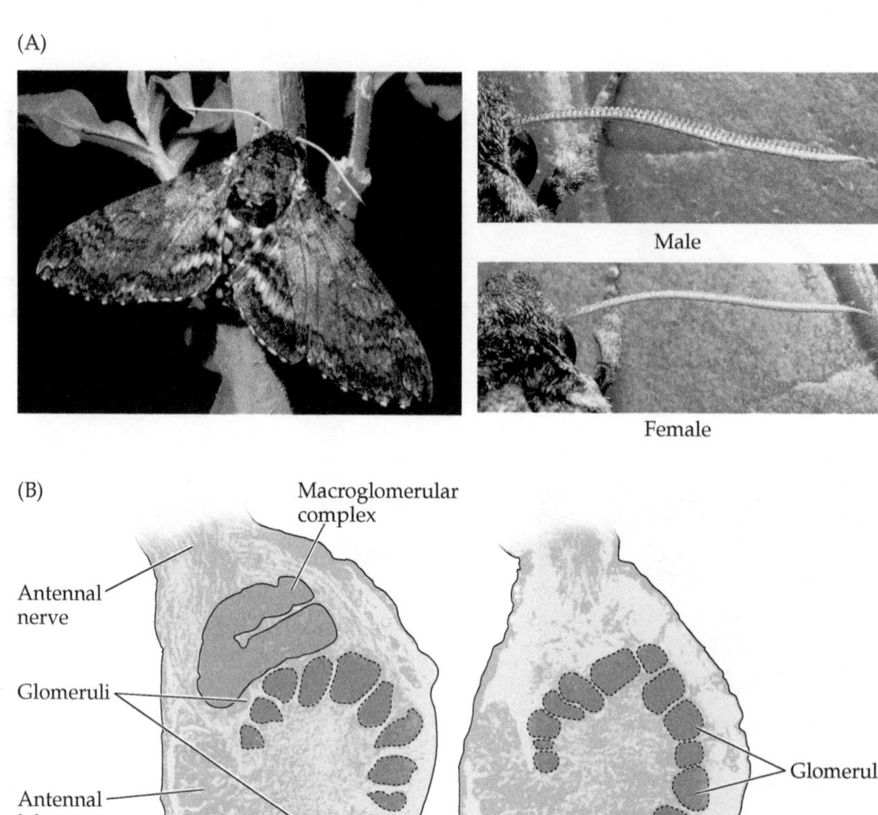

FIGURE 24.1 Sexually dimorphic anatomy in the tobacco hawk moth. (A) The antennae of male and female *Manduca sexta* are specialized for their different roles in courtship and mating behavior. (B) The physical dimorphism of the moth antennae is matched by dimorphism in the olfactory glomeruli of the brain's antennal lobe, which are specialized for odorant-mediated, sex-specific behaviors. The male-specific macroglomerular complex is essential for processing the female pheromone. (A, photos by C. Hedgcock, © Arizona Board of Regents.)

optimal egg-laying sites, while male antennae can detect extremely low concentrations of an airborne pheromone (see Chapter 15) that identifies a nearby female as a potential mate. This physical dimorphism is matched by dimorphic circuitry in the brain. Male and female antennae have olfactory receptor neurons whose axons project to glomerular structures in the antennal lobe (the equivalent of the vertebrate olfactory bulb; see Chapter 15), where the organization of these glomeruli corresponds to the peripheral dimorphism (Figure 24.1B). The large glomeruli of male moths (referred to as the macroglomerular complex) match the larger, more complex antennae of the male, and have neural circuitry specialized for responses to female pheromone. Similar glomeruli in females are smaller in register with the diminished size of female antennae, and their constituent cells respond more robustly to molecules released by tobacco plants. The development of these glomeruli depends on the antennae that project to it. If a male antenna is transplanted onto the head of a female at a late larval stage, a male macroglomerular complex develops in the female brain.

Among the most thoroughly characterized sexual dimorphisms of body, behavior, and brain in non-mammalian vertebrates are those of several songbird species, including the zebra finch (*Taeniopygia guttata;* Figure 24.2A,B), in which males produce complex songs and females do not. A male zebra finch's song is critical for attracting a mate and establishing territorial dominion. It is also an important aspect of parenting: Male hatchlings learn to sing from adult male "tutors" whose vocalizations they hear and mimic, thus determining their reproductive success. The peripheral pharyngeal structure that produces songs (the *syrinx,* a series of membranes around the trachea that vibrate based on air flow in the trachea and contraction of throat muscles) is larger and more differentiated in male birds than in females. Accordingly, in the songbird brain, the nuclei that control motor as well as sensory aspects of song production via the syrinx are larger in males. In both the body (in this instance, the syrinx) and the brain, many of these structural dimorphisms are under the control of circulating steroid hormones produced by the male and female gonads, particularly during development. Thus,

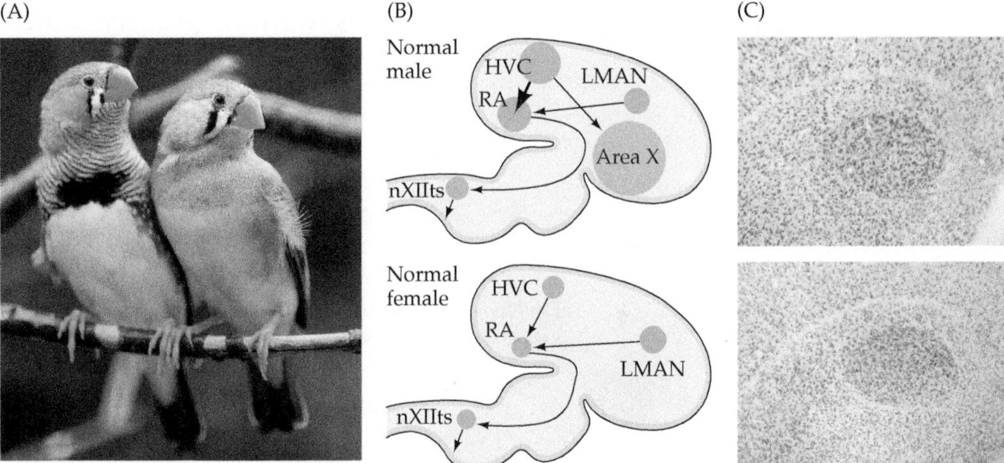

FIGURE 24.2 Sexual dimorphism in the zebra finch. (A) The male zebra finch (left) is larger than the female (right), has different feather patterns (note the chestnut cheek patch and white-spotted chestnut flanks, both of which are gray in the female), and has a red beak (versus yellower in the female). (B) Forebrain regions that control song in the zebra finch vary between sexes. The key areas shown here are both pallial structures: area X, which is similar to the mammalian medial striatum (but not exactly parallel), and HVC which is also a striatal nucleus. In a male, area X can be easily recognized and HVC is quite prominent. In the female, area X is absent and the HVC is substantially smaller than that in the male. (C) The HVC in a female (lower panel) can be "masculinized" by early posthatching exposure to estradiol (in males, normally derived from testosterone and then aromatized to estradiol in the brain) and results in an HVC that is similar in size to that of a male control (top panel). (B after Arnold, 1980; C from Grisham et al. 2011.)

female zebra finches treated with male-specific levels of gonadal steroids during development acquire a highly differentiated, hypertrophic syrinx. The song-control nuclei in their brains also become masculinized, and these female birds—masculinized in the peripheral sexually dimorphic structures as well as in the brain—sing like their male counterparts (Figure 24.2C).

The relationships between brain and peripheral sexual dimorphism in *Manduca sexta* (tobacco moths) and *Taeniopygia guttata* (zebra finches) illustrate a fundamental concept in the development of sex differences: Dimorphisms in body structure that are essential for reproductive and parenting behaviors are paralleled by the sex-specific prenatal or early postnatal differentiation of distinct brain structures and circuits that ultimately control those peripheral specializations and behaviors. For the nervous systems of moths, birds, and several other species, including humans, these dimorphisms reflect differential growth due to distinct patterns of gonadal steroids that leads to divergent organization of brain structures that are nevertheless present in both sexes.

Sex, Gonads, Bodies, and Brains

Chromosomal sex is a biological term that refers specifically to an individual's sex chromosomes. Most species have two types of chromosomes: **autosomes**, which are identical in the two sexes; and **sex chromosomes**, the number and/or identity of which determine **chromosomal sex**. In some species, males have three copies of the sex chromosomes, whereas females have only two. In others, including humans, there are different chromosomal identities; most commonly, a male-specific chromosome is present or absent.

Not surprisingly, the genes critical for the development of **primary sex characteristics**—the gonadal tissues that support either male or female gametes in humans and most other mammals—are found on sex chromosomes X (female and male) and Y (male-specific; see below). The physical state of the gonads and external genitalia is the primary determinant of **phenotypic sex**—the fundamental physical attributes that define sex. A range of **secondary sex characteristics** further defines an individual's phenotypic sex; these include mammary glands in females, sex-specific hair patterns in males and females, and musculoskeletal as well as organismal size differences. These phenotypic characteristics are more or less related to distinct reproductive and parenting functions in females or males and are regulated primarily by hormonal secretions from the gonads.

In humans, the letters X and Y identify the sex chromosomes, in contrast to the 22 pairs of autosomes, which are identified by numbers. With few exceptions (see the discussion of intersex conditions in the section "Human Genetic Disorders of Genotypic and Phenotypic Sex" later in this

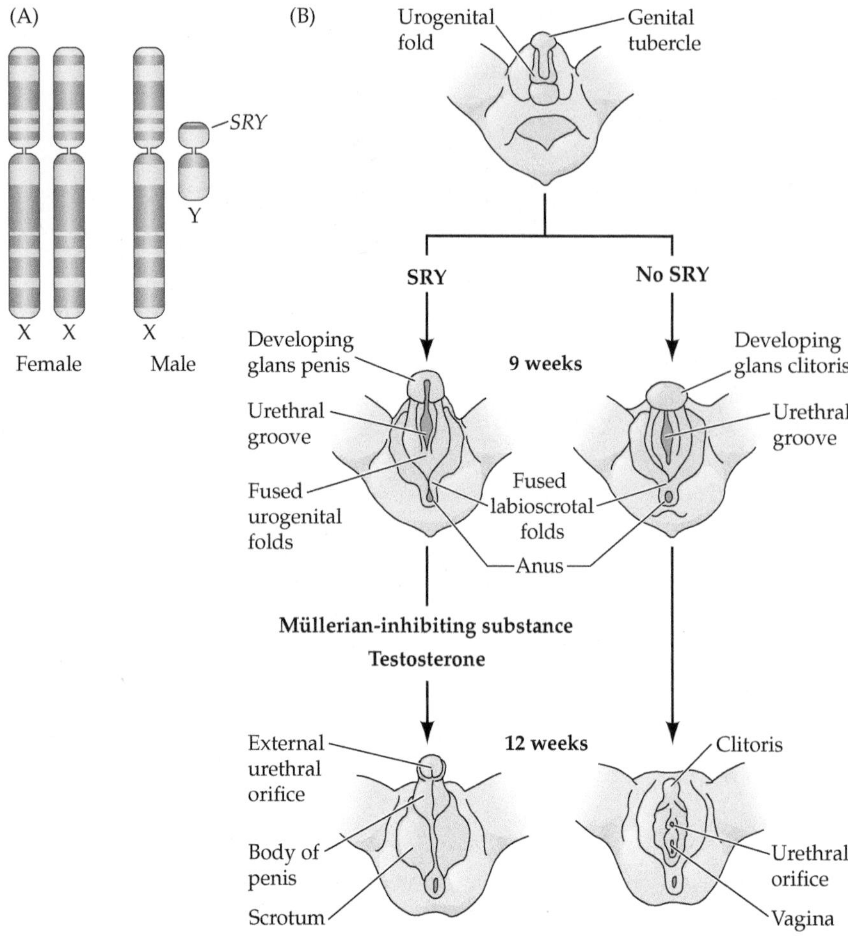

FIGURE 24.3 Chromosomal sex and primary sex determination in humans. (A) The *SRY* gene, located on the short arm of the Y chromosome, initiates a cascade of gene expression and hormonal signaling that results in masculinization of human genitalia. (B) The genitalia of the early human embryo (weeks 4–7 of gestation) are sexually indifferent. Under the influence of *SRY* gene product, testes develop and produce hormones that result in male genitalia (left diagrams) between weeks 9 and 12 of gestation. Without the influence of the Y chromosome and its *SRY* gene, the human gonads become ovaries, whose hormonal cascade results in female external genitalia (right). (After Moore, 1977.)

chapter), individuals with two X chromosomes are female, while those with one X and one Y chromosome are male (Figure 24.3A). The correlations between human genotype (i.e., chromosomal sex) and phenotype (i.e., phenotypic sex) for primary sex characteristics are best understood for the male genotype, XY. The defining gene for the development of male-specific genotypic and phenotypic sex is found on the Y chromosome and is a single gene for a transcription factor known as **testis-determining factor** (**TDF**) or **SRY** (**sex-reversal gene on the Y chromosome**). Remarkably, a translocation of the *SRY* gene from the Y chromosome to the X chromosome in a male parent can result in the complete masculinization of that male's XX offspring.

In most instances, the XY genotype leads to a male phenotype (hence the term phenotypic sex) with testicles, epididymis, vas deferens, seminal vesicles, penis, and scrotum; the XX genotype results in the development of ovaries, oviducts, uterus, cervix, clitoris, labia, vagina, and mammary glands that define female phenotypic sex (Figure 24.3B). Rare apparently male individuals (and mutant mice, in which the *Sry* gene was first cloned and sequenced) have two X chromosomes, but one of these carries the translocated *SRY* gene (transferred to the paternal X chromosome from the paternal Y chromosome during meiotic recombination). Despite the overwhelming majority of dual copies of X-chromosome genes in these individuals (and the corresponding lack of any additional Y genes), *SRY* alone is sufficient to completely masculinize them, and they become phenotypic males. In fact, whether *SRY* is translocated as an individual gene or is transferred to an XX individual along with the entire Y chromosome (resulting in an aneuploid XXY phenotype known as Klinefelter's syndrome; see below), the result is the same—a phenotypic male. Other rare individuals carry a deletion of *SRY* on an otherwise intact Y chromosome and are phenotypic females. *SRY* is thus believed to be at the apex of a genetic network that mediates the differentiation of male primary and secondary sex characteristics (see Figure 24.3B). These genetic details of gonadal and secondary sex characteristic specification are crucial for brain development because, as described in detail below, the organization of female versus male brains depends in large measure on the initial divergent differentiation of gonadal tissue. Surprisingly, however, *SRY* is not expressed in the brain. In mammalian brains, including humans, as is the case for the sexually dimorphic brains of songbirds and hawk moths, sexual dimorphisms arise secondarily, in response to primary peripheral distinctions.

This biological account of genotypic and phenotypic sex, while compelling, does not explain the full range of cognitive, emotional, and cultural experiences that emerge around sexual attraction and behavior during human development (Box 24A). Accordingly, additional vocabulary is needed to describe various facets of human sexuality. The phrase **sexual identity** is used to describe an individual's

BOX 24A ■ The Science of Love (or, Love As a Drug)

For much of recorded history, romantic love has been the province of poets, painters, and musicians. Scientists, it seemed, had little to add—and perhaps it was feared that scientists might negate the charm of this intoxicating human experience. Nevertheless, over the past decade neuroscientists have added their interpretation to the countless couplets, cupids, and cantatas that celebrate coupling. The results seem to confirm the notion of songwriters and rock stars that love is a drug.

Beyond inspiring poetry and lyrics, the biological purpose of romantic love seems to be to reinforce mate selection and bonding for maximum parental effectiveness. Thus, the brain systems that are the most engaged by "being in love" are also associated with reward and reinforcement—the very same systems that are activated, often with disastrous consequences, by alcohol and drugs. In addition, in humans the state of romantic love (and also maternal love) relies on the activity of two peptide hormone neurotransmitters, oxytocin and vasopressin. These peptides are associated with social recognition and maternal-offspring bonding in diverse species, regardless of whether monogamy and romantic love (or its biological counterpart, pair bonding) are part of the behavioral repertoire. Finally, when a person is in love, the brain regions that normally regulate social interactions—especially those in the cerebral cortex and basal forebrain that enhance social vigilance and caution—are diminished. Thus, love may actually be "blind" to the behavioral cues that in other social encounters inspire caution.

Monogamy and pair bonding were the first behaviors related to romantic love to be understood biologically. These two behaviors emerge in only a few mammalian species, including the prairie vole (*Microtus ochrogaster*). Prairie voles bond with a single mate and remain monogamous for life. In contrast, individuals of the closely related species montane vole (*Microtus montanus*) are promiscuous and do not form lifelong mating pairs. It turns out that these differences in mate selection and preference are matched by differences in the distribution of oxytocin and vasopressin receptors in the nucleus accumbens, caudate-putamen, and ventral pallidum (Figures A,B)—all regions that are associated with reward, reinforcement, and addictive behaviors (see Chapter 31). These

(A) Oxytocin receptors

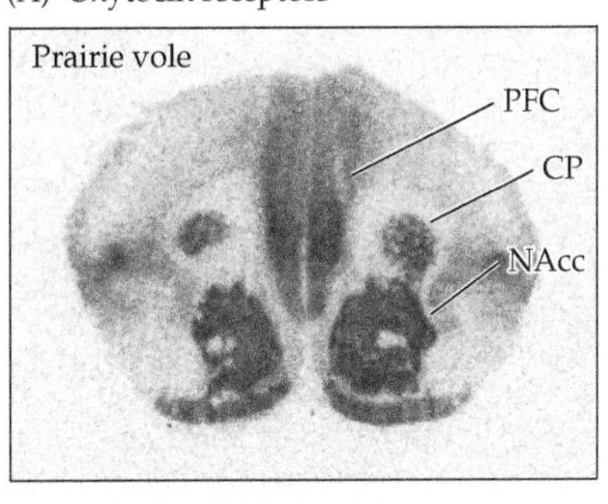

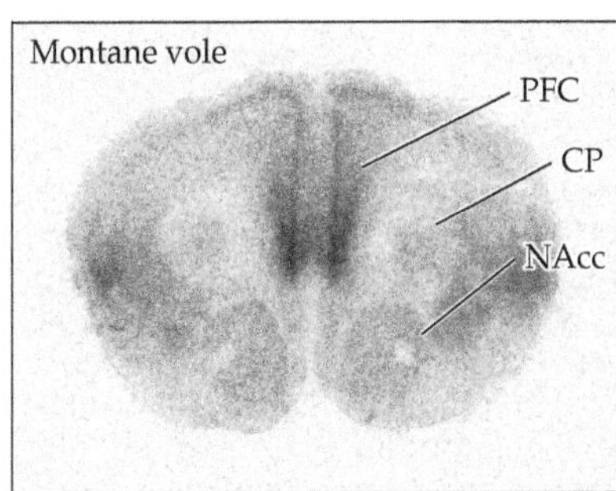

(B) Vasopressin receptors

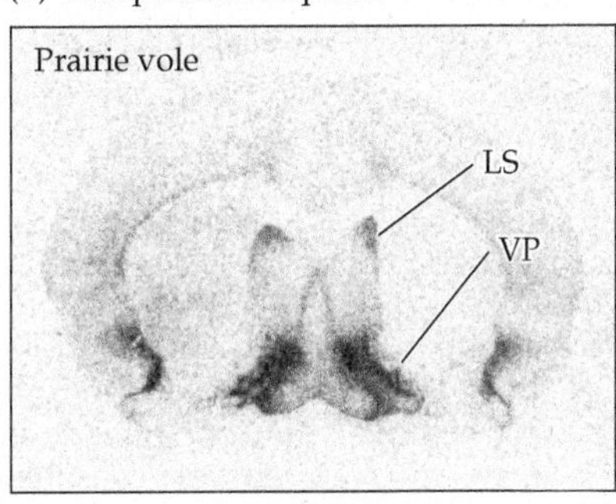

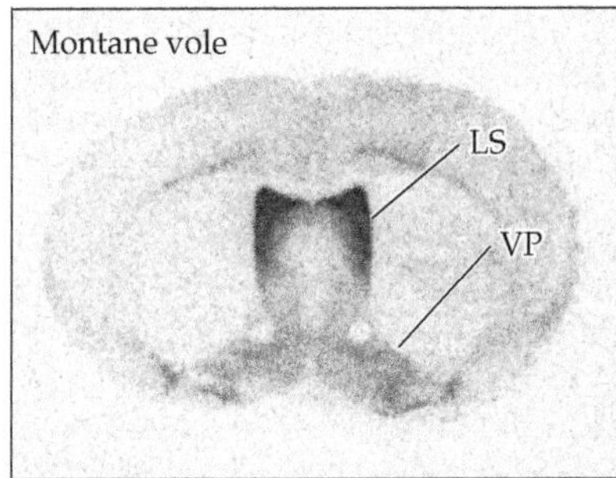

(C)

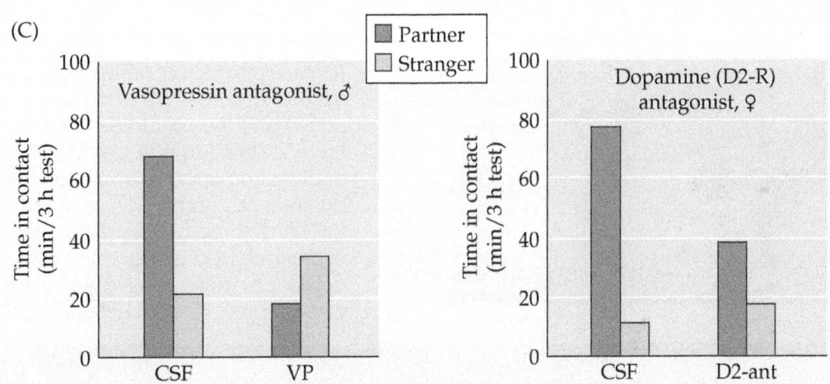

(A) Prairie voles (left), which form lifelong mating pair bonds, have a high density of oxytocin receptors in the nucleus accumbens (NAcc), caudate putamen (CP), and prefrontal cortex (PFC), all associated with reward and reinforcement. Promiscuous montane voles (right) lack a high density of receptors in the NAcc and CP. (B) Male prairie voles (left) have enhanced density of vasopressin receptors in their ventral pallidum (VP, part of the striatum) and lateral septal nuclei (LS). In the montane vole (right), there is a high density of receptors in the LS, but the density is diminished in the VP. (C) Typical male prairie voles (left) presented with their pair-bonded partner versus a stranger prefer to mate with the partner, based on the amount of time spent in contact. If a vasopressin antagonist is infused into the ventral pallidum versus a cerebrospinal fluid (CSF) control, this preference diminishes substantially. Typical female prairie voles (right) also prefer to mate with their pair-bonded partner; however, if a dopamine antagonist is infused into the NAcc, this preference is decreased. (From Young and Wang, 2004.)

Continued on the next page

regions are also sites of enriched dopaminergic neurotransmission, which is known to contribute to reinforcement, including the maladaptive reinforcement seen in drug-taking behavior. Perhaps not surprisingly, vasopressin and dopamine both influence monogamous mate selection in prairie voles. Thus, when vasopressin antagonists are injected into the ventral pallidum of male prairie voles, monogamy is disrupted, and when dopamine antagonists are injected into the nucleus accumbens of female prairie voles, single-partner preference induced by mating and pair bonding is no longer seen (Figure C).

The contributions of the nucleus accumbens and of the ventral pallidum and its dopaminergic innervation from the ventral tegmental area (VTA) have led to a hypothesis for studying human romantic love: Individuals newly in love should have elevated brain activity in all of these regions in response to considering or seeing the object of their affections versus neutral stimuli. Indeed, when males and females monitored using fMRI were presented with images of their beloved versus images of friends or acquaintances, maximum brain activity was observed in the caudate putamen and VTA (Figure D). Moreover, since "love is blind," it was reasoned that the regions of the brain essential for social vigilance and caution, including several cortical regions and the amygdala, should be diminished in activation, which was also the case (Figure E). These observations placed love somewhat outside the brain systems that mediate emotion—particularly limbic regions (see Chapter 31). Instead, brain regions essential for reward, risk taking, and social cognition seem to be featured.

Subsequent studies suggest that the ups and downs of love recruit additional brain circuits. Individuals recently rejected by a love interest retain maximum activation in the VTA and caudate nucleus in response to viewing an image of the lost love, but add several regions, including cortical regions also associated with motivation, calculation of gain and loss (such as in gambling or risk taking), emotional regulation, and even drug craving that are not activated in the "in love" state. Conversely, when men and women who have been married an average of more than 20 years are presented with images of their spouse versus other familiar individuals, regions associated with attachment—particularly those activated by mother-child bonding—are differentially activated.

While one is unlikely to see Valentine cards reading "You've been carried away my VTA" any time soon, these results drain neither the passion nor mystery from love. Indeed, while we can identify the neural consequences of attraction and romance, the reasons why the beloved elicits these responses in the first place remain unknown. Nevertheless, these observations point to our subjective experience of romantic love as a key aspect of the overall biology of sex, reproduction, and parenting. The human experience of romantic love activates regions of the brain that favor reinforcement of a connection with a sexual partner, promote exclusivity in mating, and enhance cooperation in the rearing of offspring. These desires—passion, constancy, and domesticity—offer depth of feeling, whether expressed by poets or glimpsed by fMRI. Thus, love may be like a drug in the best sense, in that it can be highly therapeutic to those who find it.

(D)

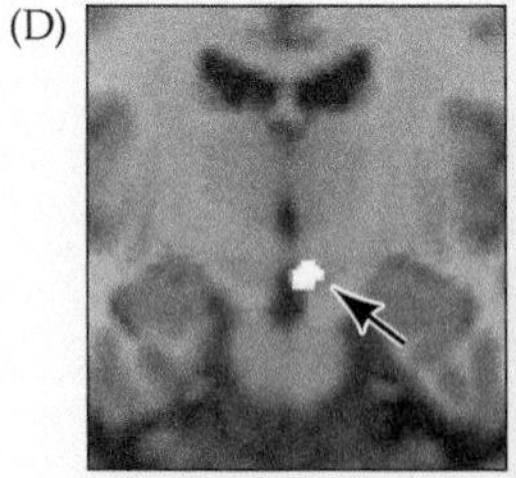

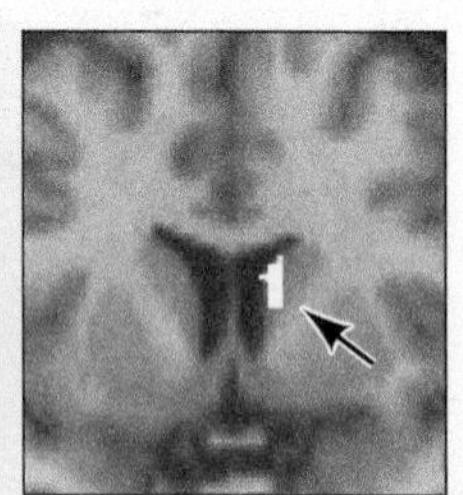

(E)

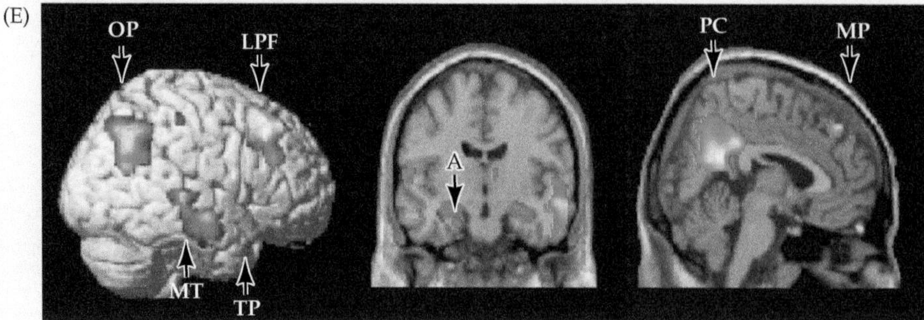

(D) Maximum focal activation in response to seeing a picture of one's beloved versus that of a friend or acquaintance is detected in the ventral tegmental area (left, arrow) and the caudate nucleus (right, arrow) in both young men and women in the early to middle stages of self-identified love affairs. (E) Brain regions that are deactivated upon seeing a picture of one's beloved versus that of a friend or acquaintance. Cortical areas in the occipital/parietal junction (OP), medial temporal (MT), temporal pole (TP), lateral prefrontal (LPF), posterior cingulate (PC), medial prefrontal/paracingulate (MP), and the amygdala (A) are all sites of diminished activation. These same areas are deactivated when mothers view pictures of their own children. (D from Fisher et al. 2005; E from Bartels and Zeki, 2004.)

conscious perception of his or her phenotypic sex. **Sexual orientation** refers to the cognitive experience of emotions and attractions that are associated with sexual relationships. Sexual orientation is not simply coordinated with obvious genotypic, phenotypic, or gender-associated characteristics—homosexual women are not phenotypically masculinized (i.e., genotypic and phenotypic sex remain female), nor are homosexual men phenotypically feminized. In clinical studies of transgendered individuals, most transgendered men and women report having had deeply held identities that oppose their genotypic and phenotypic sex from very early in life.

Hormonal Influences on Sexually Dimorphic Development

The differentiation of male versus female gonadal tissue specified by chromosomal sex sets in motion a series of events that define major phenotypic dimorphisms both for peripheral sexually dimorphic structures, particularly gonads and genitalia, and in the brain. The steroid hormones **testosterone** and **estrogen** (Figure 24.4A), secreted by the developing testes and ovaries, respectively, influence most aspects of sexual dimorphism. Steroids are a distinct class of circulating signaling molecules. In contrast to protein signals (see Chapter 22), steroids are not directly produced as polypeptides from transcribed genes. Instead, a metabolic precursor, cholesterol, is enzymatically modified to produce steroid variants with different activities and receptor specificities. The initial differentiation of male and female gonads is the central event for the translation of chromosomal sex into phenotypic sex. Although both the ovaries (XX) and testes (XY, dependent on a functional *SRY* gene) secrete both estrogens and androgens, the two gonadal tissue types secrete very different levels of these hormones at distinct developmental times. At critical junctures, differing hormone levels influence the undifferentiated primordial gonadal structures, resulting in divergent developmental paths of gonads themselves, the genitalia and, later, secondary sex characteristics of males and females. This early hormonal influence on the development of sexually dimorphic structures is sometimes referred to as the *organizational effects* of the gonadal steroids, reflecting their actions in guiding distinct male versus female differentiation in a variety of tissues, including the brain.

Genotypic males experience an early surge of testosterone, which, along with the peptide hormone Müllerian-inhibiting substance (MIS), masculinizes the genitalia (Figure 24.4B). Paradoxically, many of the effects of testosterone in the male brain are in fact due to estrogens during midgestation (the second trimester of pregnancy in humans) and result from the conversion of testosterone to estrogen in the brain. Vertebrate neurons, especially those in brain regions that will develop divergently in males and females, contain an enzyme called **aromatase** that converts testosterone to **estradiol**, an "active" form of estrogen that binds with high affinity to the estrogen receptor (see Figure 24.4A). Thus, a surge of testosterone in developing males also generates a surge of estradiol that affects developing neurons with appropriate receptors. In these instances, estrogen and estradiol in the developing brain via local estrogen receptors are assumed to mediate subsequent sexual dimorphisms that arise due to increased testosterone. There are, however, instances where testosterone acts directly (i.e., through its own receptors) on developing as well as mature neurons.

Estrogen, testosterone, and other steroids are highly lipophilic molecules and are usually transported via carrier proteins circulating in the blood. Thus, the fetuses of placental mammals are exposed to estrogens generated by the maternal ovary and placenta delivered via maternal circulation. So why doesn't this

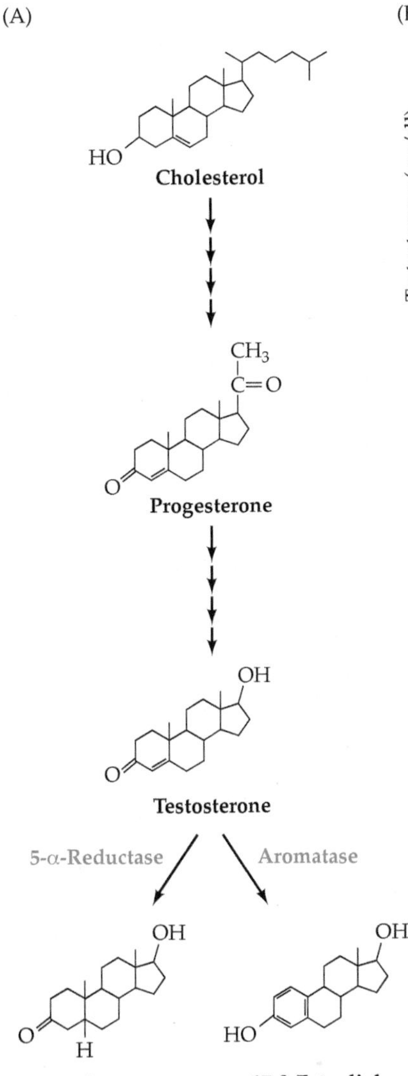

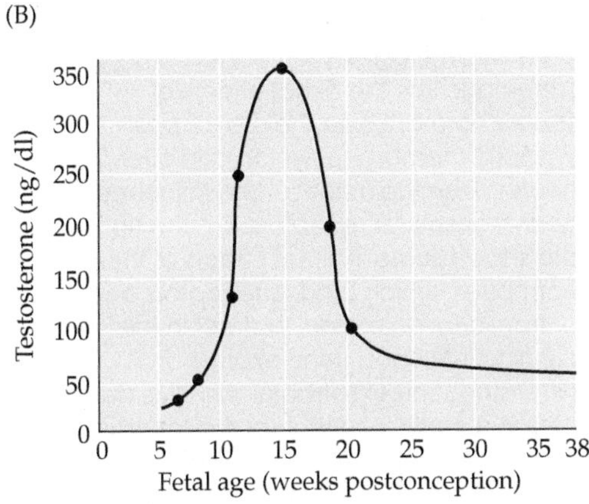

FIGURE 24.4 Gonadal sex steroids and their influence. (A) All sex steroids are synthesized from cholesterol, which is converted to progesterone, the common precursor, by four enzymatic reactions (represented by four arrows). Progesterone can then be converted into testosterone via another series of enzymatic reactions; testosterone in turn is converted to 5-α-dihydrotestosterone via 5-α-reductase, or to 17-β-estradiol via an aromatase. 17-β-estradiol mediates most of the known hormonal effects in the brains of both female and male rodents. (B) In human male fetuses, masculinization of the genitalia reflects increased secretion of testosterone by the immature testes between weeks 7 and 20 of gestation. (B after Gustafson and Donahoe, 1994.)

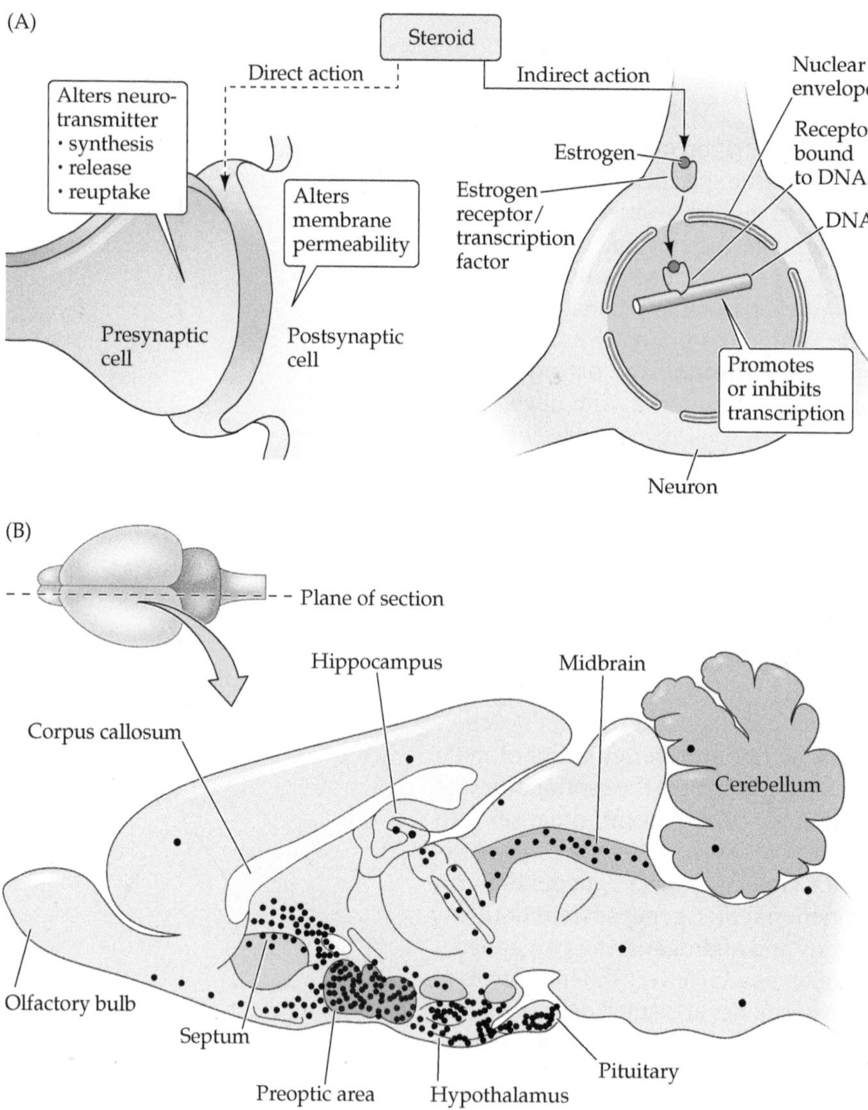

FIGURE 24.5 Effects of sex steroids on neurons. (A) The left panel of this schematic lists the direct effects of steroid hormones on the pre- or postsynaptic membrane, which can alter neurotransmitter release and influence neurotransmitter receptors. On the right are shown some indirect effects of these hormones, which bind to receptors and/or transcription factors that act in the nucleus to influence gene expression. (B) Distribution of estradiol-sensitive neurons in a sagittal section of the rat brain. Animals were given radioactively labeled estradiol; dots represent regions where the label accumulated. In the rat, most estradiol-sensitive neurons are located in the basal diencephalon and telencephalon, with a high concentration in the preoptic area, hypothalamus, and more laterally in the amygdala, which is not shown in this mid-sagittal section. (A after McEwen et al., 1979; B after McEwen, 1976.)

maternal estrogen interfere with sex differentiation in the offspring? Developing fetuses have high levels of **α-fetoprotein**, a protein that binds circulating estrogens including those from the maternal blood supply. In both sexes, the brain is apparently protected from early exposure because estrogens secreted by the gonads into the bloodstream are bound by α-fetoprotein. Testosterone does not bind to α-fetoprotein, however, and is aromatized to estradiol only once inside neurons; therefore the male brain is exposed to an early dose of masculinizing steroids—including 17-β-estradiol generated from testosterone by aromatase—from the testosterone surge generated by the embryonic testes (see Figure 24.4B). The consequences of this difference in steroid metabolism, subsequent steroid signaling and dimorphic brain development can be seen in mice in which the α-fetoprotein gene has been knocked out. Female mice that lack α-fetoprotein are infertile due to a failure of the hypothalamic control of ovulation, presumably the result of masculinizing overexposure to estrogen during embryonic development and resulting changes in the hypothalamus (see below). Surprisingly, there is no information available on ovulation or fertility for the rare cases of human females that lack the gene for α-fetoprotein. Male mice and men lacking this gene are fully fertile.

Gonadal steroids act by binding to specific receptors for either testosterone or estrogen (Figure 24.5A). These receptors are distributed in fairly limited populations of cells in the mammalian brain, being concentrated at sites where reproductive and parenting functions are centrally represented and sexual dimorphisms are seen (Figure 24.5B). Gonadal steroid receptors belong to a larger family of proteins called the **steroid-thyroid nuclear receptors**, which includes the receptors for vitamin A (retinoic acid; see Chapter

22), vitamin D, and glucorticoids. Unlike most receptors for peptide hormones (or neurotransmitters), which are on the cell surface, most steroid–thyroid receptors are found in the cytoplasm. Thus, their ligands (testosterone and estrogen) must diffuse through the membrane to reach their receptors. When the active forms of testosterone (as 5-α-dihydrotestosterone) or estrogen (as 17-β-estradiol) bind to their respective receptors, the receptors move from the cytoplasm to the nucleus, where they are able to bind to DNA recognition sites (response elements) and regulate gene expression.

Development of Dimorphisms in the Central Nervous System

Testosterone and estrogen each have specific tissue or cellular targets, both in the body and in the brain. Many of the brain targets are neural structures (particularly nuclei in the hypothalamus; see below) that control the external genitalia, gonads, and other dimorphic structures (such as mammary glands) and mediate sex-specific behaviors. A well-studied example of central nervous system dimorphism beyond the hypothalamus related to motor control of sex-specific reproductive behavior is the difference in size of the **spinal nucleus of the bulbocavernosus (SNB)** primarily located in the fifth lumbar segment of the rodent spinal cord, which innervates muscles of the genitalia. The motor neurons of the SNB innervate two striated muscles of the male perineum (the bulbocavernosus and levator ani) that attach at the base of the penis and are involved in both penile erection and urination (Figure 24.6A). In female rats, the bulbocavernosus muscle is absent and the levator ani is dramatically reduced in size. The SNB, though present in both sexes, is significantly larger in males and contains more and larger neurons. Thus, the spinal cords of male and female rodents are sexually dimorphic, in parallel with the genitalia.

The SNB dimorphism in rodents is established via the influence of testosterone around the time of birth; however, in this case the primary target of gonadal steroid hormone action is not in the brain, but in the perineal muscles that control the genitalia. At birth, undeveloped bulbocavernosus muscles of similar size are present in male and female rodents (Figure 24.6B), and there are equivalent numbers of neurons in the SNB. The muscles in both males and females have testosterone receptors (as do all muscles, much to the chagrin, in the case of humans, of the governing bodies of major sports organizations), but only males have sufficient endogenous levels of testosterone to activate these receptors. The activation of testosterone receptors in the male perineal muscles spares these muscles from the apoptotic cell death for which the female bulbocavernosus muscle is destined shortly after birth. Subsequently, the male peripheral target supports, via trophic interactions (see Chapter 25), the survival and differentiation of significantly more SNB neurons. In contrast, the loss of a major target muscle in the female leads to trophic factor deprivation, cell atrophy, and apoptosis. If a female rodent is artificially exposed to testosterone, the muscles are rescued and the number of SNB neurons approaches that of the male. This mechanism for the development of a sexual dimorphism represents a special case of a more general rule: During development, structures in the CNS are matched to the periphery based on the level of trophic support provided by their targets. In addition to this trophic mechanism, SNB neurons in the male spinal cord express testosterone receptors, and activation of these receptors is thought to secondarily regulate the enhanced growth of SNB neurons in males. The functional significance of this hormonally induced dimorphism in neuronal size is unknown.

As with most sexual dimorphisms, the situation in humans is considerably less clear than its analog in experimental animals. The human spinal cord structure that corresponds to the rodent SNB is **Onuf's nucleus**, which consists of two cell groups in the sacral spinal cord (the dorsomedial and ventrolateral groups). The dorsomedial group is not sexually dimorphic; however, human females have fewer neurons in the ventrolateral group than do males (Figure 24.6C). In contrast to rodents, adult human females retain a bulbocavernosus muscle (which serves to constrict the vagina), but the muscle is smaller than in males. The difference in nuclear size and cell number in humans, as in rats, presumably reflects the difference in the number of bulbous cavernosus muscle fibers that Onuf's nucleus motor neurons innervate.

Brain Dimorphisms and the Establishment of Reproductive Behaviors

In addition to the primary motor neurons that directly innervate dimorphic genital muscles, there are dimorphic structures in the CNS that do not project directly to the genitalia. Nevertheless, their divergent cellular composition and size reflect differences in reproductive function. A major site of these dimorphisms is the hypothalamus, presumably because of its central role in the control of visceral motor function (see Chapter 21)—which includes the secretory, vascular, and smooth muscle control necessary for sexual function in both males and females. The concentration of gonadal steroid receptors in the hypothalamus (see Figure 24.5B) reinforces this conclusion. Neurons in the anteroventral and medial preoptic areas of the anterior hypothalamus (where estrogen and androgen receptors are concentrated) mediate key sexual and reproductive behaviors, and in most mammals there are sexually dimorphic differences in size and cell number of subsets of nuclei in this region.

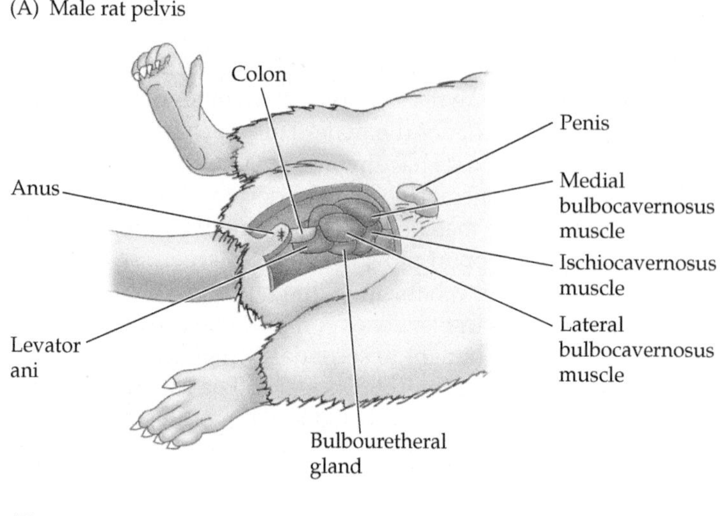

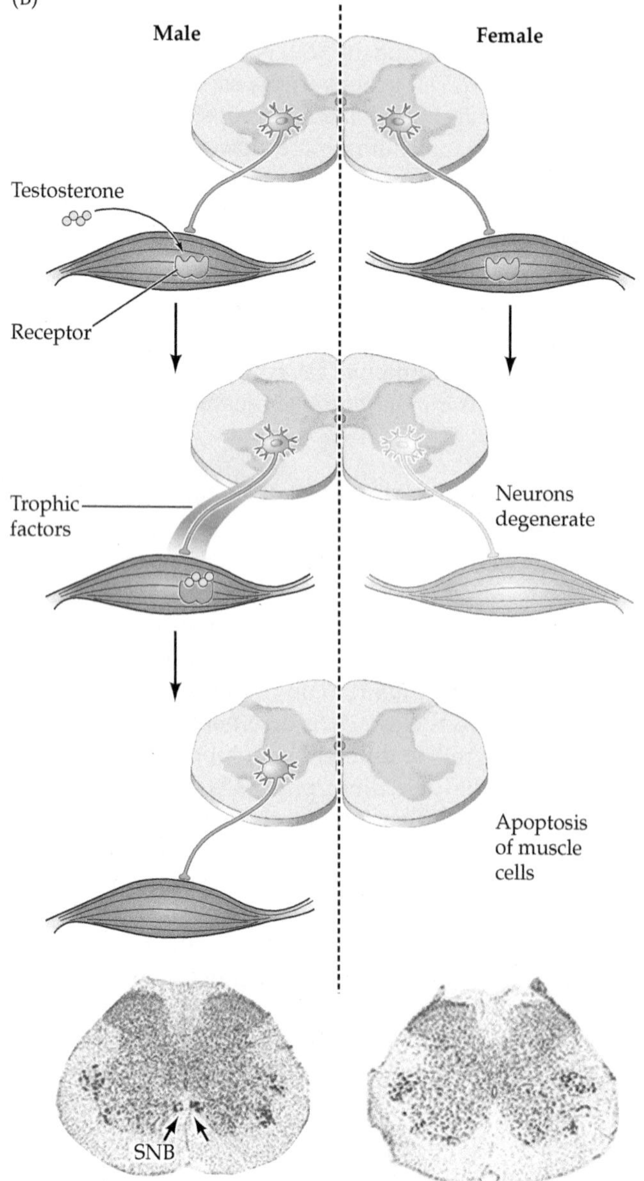

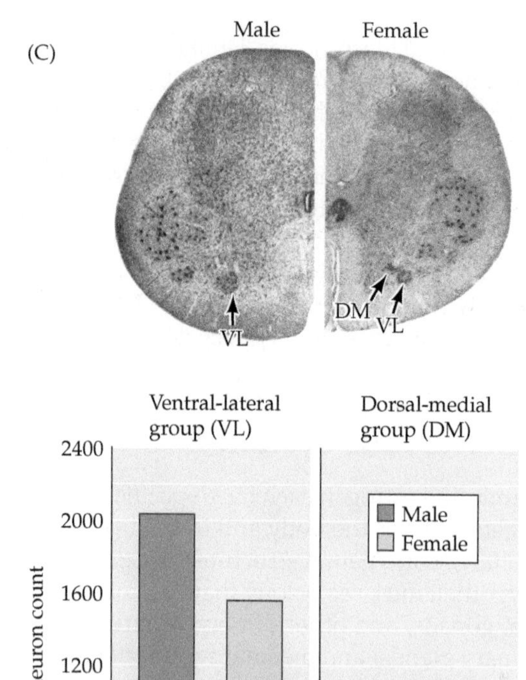

FIGURE 24.6 The number of spinal motor neurons related to the perineal muscles is sexually dimorphic. (A) The perineal region of a male rat. (B) Comparison of the developmental sequence for the bulbocavernosus muscle and the spinal motor neurons that innervate it, in male and female rats. Histological cross sections show the dimorphic spinal nucleus of the bulbocavernosus (SNB), found in the fifth lumbar spinal cord segment. Arrows point to the SNB in the male; there is no equivalent grouping of densely stained neurons in the female. (C) The micrograph shows Onuf's nucleus in the lumbar spinal cord of a human male, and at right, in a female. Histograms show motor neuron counts in the dorsomedial (DM) and ventrolateral (VL) groups of Onuf's nucleus in human males and females. The dorsomedial group is not visible in the section through the male spinal cord. (A after Breedlove and Arnold, 1981; B, diagram after Morris et al., 2004; photographs from Breedlove and Arnold, 1983; C from Forger and Breedlove, 1986.)

Among the fundamental differences between adult male and female mammals, the cycle for ovulation is perhaps the most distinctive: Female gametes (eggs) mature at distinct times, whereas male gametes (sperm) are constantly generated. A specific group of cells in the hypothalamus, the **anteroventral paraventricular nucleus (AVPV)**, regulates the ovulatory cycle. AVPV cells are far

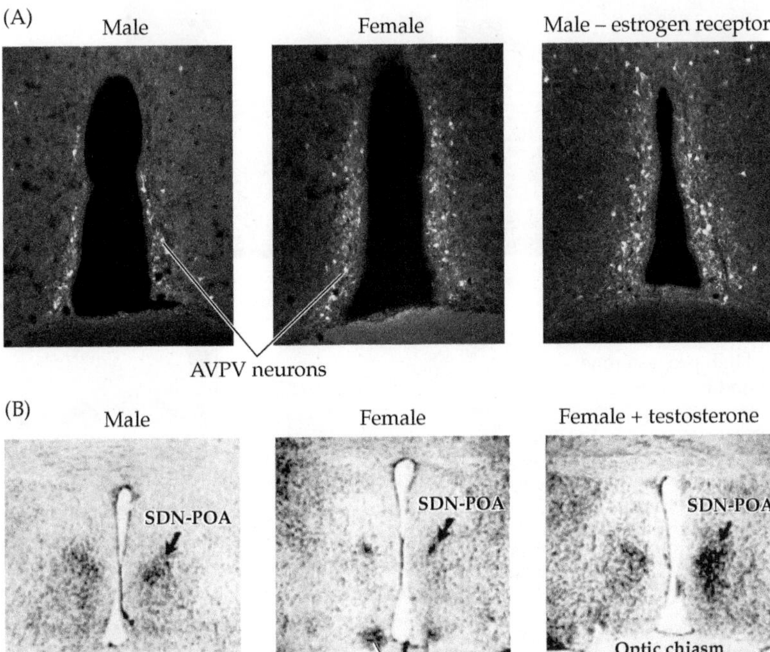

FIGURE 24.7 Sexually dimorphic hypothalamic nuclei associated with sexual behaviors. (A) The anteroventral paraventricular nucleus (AVPV) is a collection of dopaminergic neurons (labeled here with an antibody against the dopamine-synthetic enzyme tyrosine hydroxylase) that is larger in females than in males (the AVPV from male and female mice is shown here) and is responsible for the control of cyclic ovulation. When the estrogen receptor gene is inactivated in male mice (thus preventing the masculinizing effects of testosterone, converted in the brain to estradiol by aromatase), the AVPV is similar in size to that of the female (far right). (B) The sexually dimorphic nucleus of the preoptic area (SDN-POA) is larger in male rats than in female rats. This size difference can be approximated in genotypically female rats that are given testosterone perinatally (far right). (A courtesy of R. Simerly; B from Arnold and Gorski, 1984, and Gorski, 1983.)

more numerous in females than in males, and they project to other cell groups in the hypothalamus (Figure 24.7A). These projections modulate the systemic release of gonadotropin-releasing hormone as well as prolactin secretion, both of which are key for the cyclic control of ovulation. Sexual dimorphism in the AVPV arises during development under the negative influence of elevated levels of testosterone in males. Transient elevated levels of testosterone (converted to estrogen by aromatase in the hypothalamus) induce cell death in the developing male AVPV; the absence of testosterone in females ensures AVPV cell survival. As is the case for many other sexually dimorphic cell groups, altering hormone levels during development can alter the AVPV dimorphism. Elevated testosterone at the critical stage of development in the female will cause AVPV cell death and lack of ovulation, and lack of testosterone signaling in males (via aromatized 17-β-estradiol binding to estrogen receptors) will rescue the AVPV cells that would normally die (see Figure 24.7A). In male mice in which the estrogen receptor has been knocked out, the AVPV neurons that would normally die are rescued as well.

In rodents, another nuclear group in the hypothalamus, the **sexually dimorphic nucleus of the preoptic area** (**SDN-POA**), is consistently larger and has more neurons in males than in females (Figure 24.7B). The size and cell number of this nucleus are regulated by testosterone during early postnatal development. In the male, estradiol (once again derived from testosterone via aromatase) influences anti-apoptotic genes and stabilizes SDN-POA neurons. In the female, the lack of significant levels of circulating testosterone leads to cell death. Females exposed to elevated levels of androgens during early postnatal life develop an enlarged SDN-POA with more neurons, presumably due to diminished cell death. In contrast to what happens in the spinal cord nucleus of the bulbous cavernosus, however, the effects of testosterone on cell survival in the SDN-POA are direct, via its conversion to estrogen in the brain and (presumably) the subsequent actions of estrogen via its receptors on SND-POA neurons. This dimorphism, like most others, can ultimately be linked to the differing ability of male and female gonadal tissues to provide distinct levels of testosterone and estrogen. Similar dimorphisms have been reported for several other nuclei in the preoptic area of the human hypothalamus; however, their consistency and relationship to sexual behavior remain controversial.

In a range of laboratory animals, the preoptic area in general, including the SDN-POA, has been implicated in dimorphic sexual behaviors. In male rats, lesions of the entire preoptic area abolish all copulatory behavior, while more discrete lesions of the SDN-POA diminish the frequency of mounting and copulation. In female rats, such lesions yield individuals that avoid male partners and do not display female-specific copulatory behaviors. Thus, the preoptic area is thought to mediate mate selection and the preparatory behaviors for copulation, as well as some of the motor and visceral aspects of male intromission and ejaculation, and female copulatory responses.

Cellular and Molecular Basis of Sexually Dimorphic Development

The establishment, maintenance, and plasticity of sexually dimorphic structures and behaviors clearly depend on differences in organizational as well as activational effects of circulating levels of gonadal steroids. Thus, an important goal is to understand how gonadal steroids, their receptors and synthetic enzymes, and related genes influence neuronal structure and function.

One essential target for estrogen and testosterone must be the molecular pathways that regulate cell survival and cell death, since many sexually dimorphic nuclei achieve their distinct cell numbers and cell sizes through apoptotic cell death and trophically mediated stabilization and growth of the surviving cells. It is not clear how estrogen and testosterone initially stimulate mechanisms that favor either apoptosis or cell survival; however, recent evidence implies that the ultimate targets of the sex steroids are genes that regulate apoptotic cell death. Male or female mice overexpressing the anti-apoptotic gene *Bcl2* have more neurons in the SNB, while mice in which the pro-apoptotic gene *Bax* has been inactivated do not display the sexual dimorphisms normally seen in the basal forebrain. Thus, the regulation of structural sexual dimorphisms that match phenotypic sex is most likely dependent on gonadal steroid regulation of apoptosis, perhaps further modulated by trophic factors during brain development (see Figure 24.6).

Besides influencing cell death, in certain instances gonadal steroids can act as trophic factors, directly regulating neuronal size as well as process growth. During development, and to some extent throughout life, estradiol (the aromatized form of estrogen; see Figure 24.4A) stimulates brain dimorphisms by increasing cell size, nuclear volume, dendritic length, dendritic branching, dendritic spine density, and synaptic connectivity of the sensitive neurons, independent of cell survival or apoptosis (Figure 24.8A,B). Testosterone can also influence neuronal size and differentiation, at least in vitro, in neurons that express testosterone receptors (Figure 24.8C); however, the extent to which these effects rely on the direct action of testosterone on its receptor versus aromatized 17-β-estradiol via estrogen receptors is unclear.

(A)

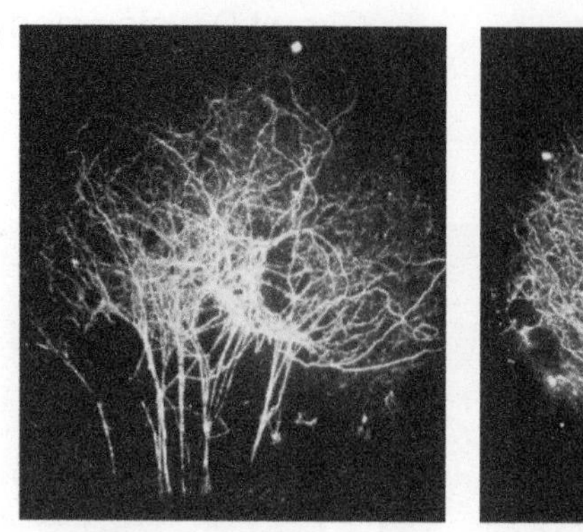

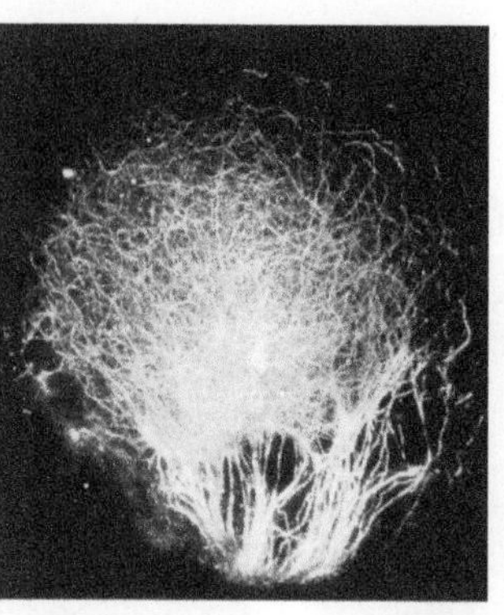

(B)

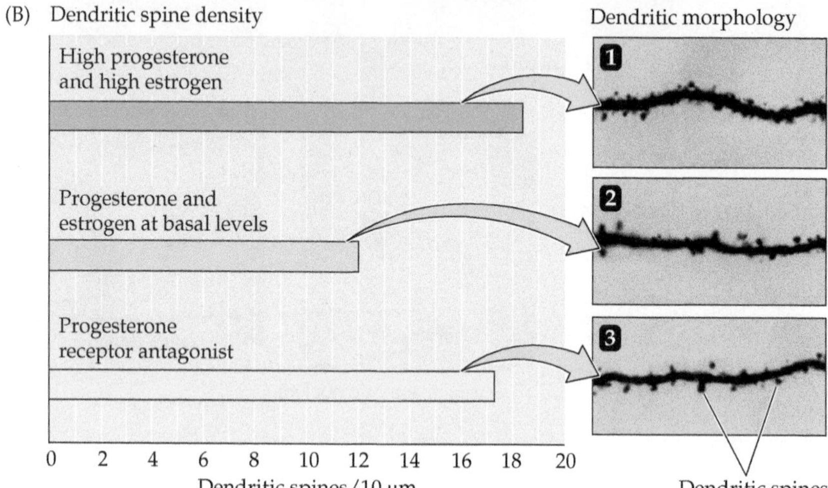

(C)

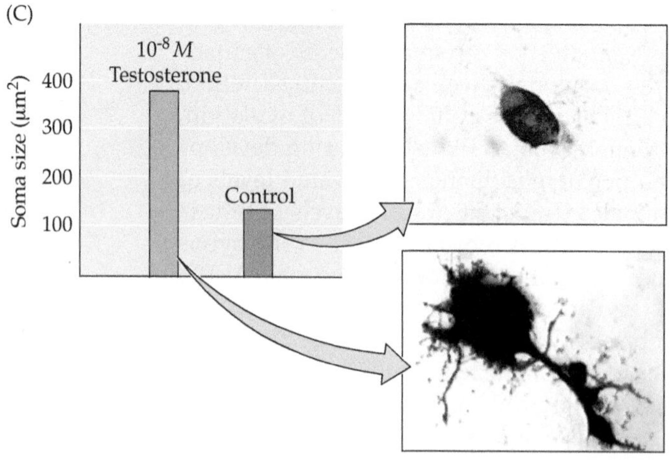

FIGURE 24.8 Estrogen and testosterone influence neuronal growth and differentiation. (A) A control explant (left) from the mouse hypothalamus shows only a few silver-impregnated processes; an estradiol-treated explant (right) has many more neurites growing from its center. (B) Dendritic spine density in female rat hippocampal neurons in response to progesterone (a precursor of both estrogen and testosterone; see Figure 24.4A) and to estrogen. Recall that dendritic spines, which are small extensions from the dendritic shaft, are sites of synapses. Tracings at right are of representative apical dendrites from hippocampal pyramidal neurons: (1) after administration of progesterone and estrogen in high dosage; (2) after administration of progesterone and estrogen at basal levels; and (3) after administration of a progesterone receptor antagonist. (C) Effects of testosterone on embryonic rat pelvic ganglion in cell culture. In response to testosterone, processes become thicker and more highly branched, and the cell body (soma) grows in size. (A from Toran-Allerand, 1976; B after Woolley and McEwen, 1993; C from Meusberger and Keast, 2001.)

Estradiol can also stimulate an increase in the number of synaptic contacts in adult animals. For example, during periods of high circulating estrogen in the estrous cycle of female rodents (or after administration of exogenous estrogen), there is an increase in the density of spines (and presumably synapses) on the apical dendrites of pyramidal neurons in the hippocampus (see Figure 24.8B). These apparent changes in neuronal circuitry might contribute to differences in learning and memory during the course of the estrous cycle. Such differences have been observed in rodents using tests of spatial navigation and memory; however, the relevance of these laboratory behaviors to significant functional differences in reproductive behavior engendered by estrous-dependent hippocampal changes is not understood.

Gonadal steroids can also modify electrical signaling between neurons in a variety of brain regions. Perhaps the most compelling example of this phenomenon is in the periventricular nucleus (PVN) of the hypothalamus, where fluctuating steroid levels facilitate the formation of gap junctions by regulating the transcription of relevant proteins. The resulting increase in gap junctions allows for neuronal synchrony correlated with lactation and maternal behavior. In addition, the influences of gonadal steroids, particularly estrogen, on neuronal activity have been evaluated in the hippocampus. The hippocampus is an established site of neuronal plasticity (see Chapters 25 and 30) and is sensitive to hormonal fluctuations, including those seen during estrous. Estrogen receptors are expressed by mature neurons and often are localized to the cytoplasm at synapses, as well as in the cytoplasm of the cell body (Figure 24.9A). Estrogen can modify excitable properties of hippocampal neurons, including K^+ and Ca^{2+} conductances and the rate of action potential firing. Estrogen can also influence hippocampal synaptic signaling and plasticity. Estrogen at relatively high concentrations (arguably higher than those seen physiologically) can increase the amplitude of excitatory postsynaptic currents over minutes to hours and, when coupled with high-frequency stimulation that elicits long-term potentiation (LTP; see Chapter 25) results in a sustained change in excitatory postsynaptic potentials (Figure 24.9B,C). It is tempting to speculate that such changes underlie some of the learned behaviors and memories associated with fluctuating gonadal steroid levels, and therefore are associated with reproductive behaviors. There is no solid evidence as yet that supports this speculation.

Steroid Receptors and Responses in the Adult Brain

Estrogens and androgens can influence neuronal and glial structure and function throughout life in brain regions beyond the hypothalamus, and in circumstances beyond

(A)

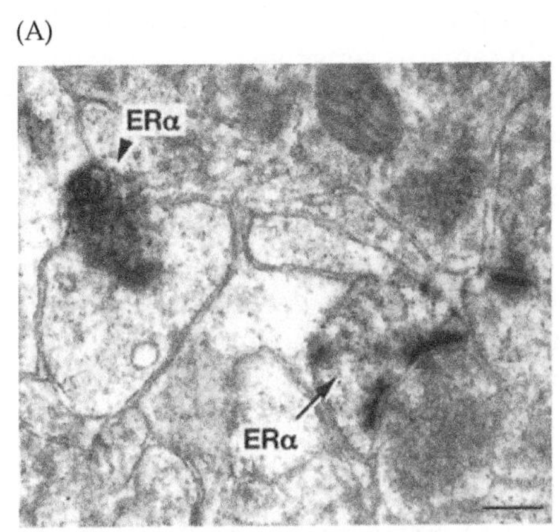

(B)

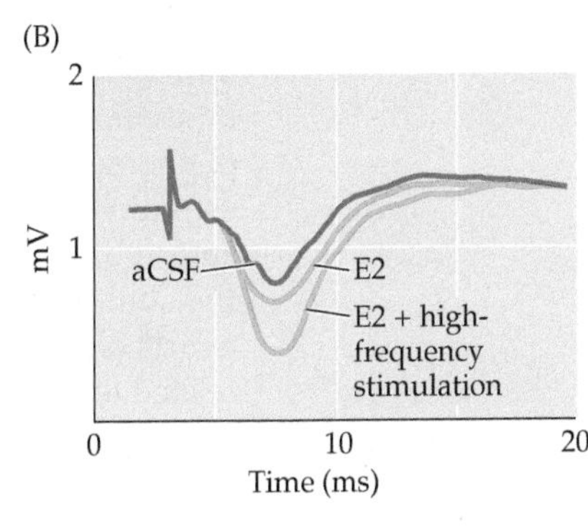

(C)

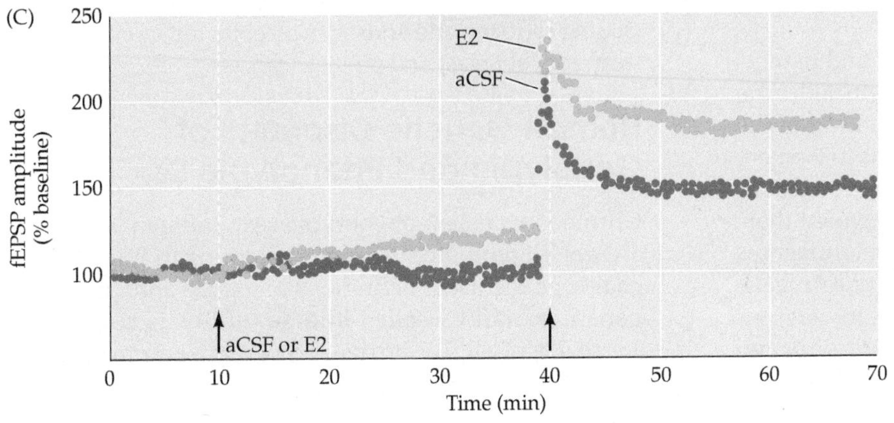

FIGURE 24.9 **Estrogen influences synaptic transmission.** (A) Electron micrograph showing localization of the estrogen receptor α (ERα; the dark, "electron-dense" label) in postsynaptic processes (presumably spines) in the rat hippocampus. (B) Estrogen (E2) increases the amplitude of excitatory postsynaptic potentials in individual hippocampal neurons (the same physicological measurement done in artificial cerebrospinal fluid—aCSF—is shown as a control). High-frequency stimulation further enhances the effects of E2, suggesting that estrogen may modulate use-dependent plasticity in hippocampal synapses. (C) High-frequency stimulation in the presence of E2 in rat hippocampal slices (see Chapter 8) results in enhanced long-term potentiation consistent with a role for estrogen in synaptic and circuit plasticity in the hippocampus. The data shown here plots the frequency of EPSP values (fEPSP) over time. E2 alone results in a clear increase in EPSP values, and this effect is magnified and maintained after high frequency stimulation, denoted by the second arrow. (From Woolley, 2007.)

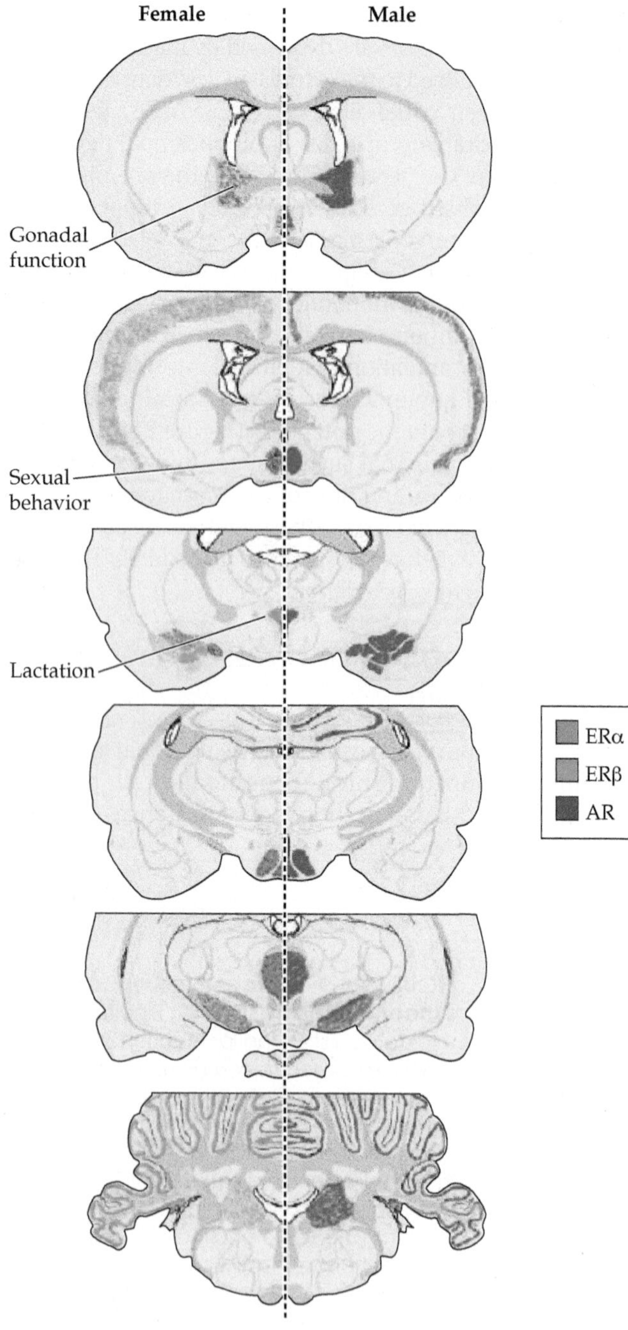

FIGURE 24.10 Estrogen and androgen receptor distribution in the brain is widespread. Distribution in the rat brain of the three major receptor/transcription factors that bind estrogen (ERα and ERβ) and androgen/testosterone (AR), eliciting corresponding changes in gene expression. ERα, ERβ, and AR tend to be expressed in the same subsets of brain structures. However, these structures are not restricted to the hypothalamic nuclei that control gonadal function, sexual behavior, and parenting behavior; they also include large regions of the cerebral cortex, amygdala, hippocampus, thalamus, substantia nigra, and cerebellum. The significance of gonadal steroid receptor expression and activity at sites beyond the hypothalamus is less well understood than their reproduction-specific functions. They may provide a substrate for the influence of these hormones on behaviors beyond reproduction and parenting, including cognition (cortex), learning and memory (cortex, hippocampus, amygdala), aggression and stress (hippocampus, amygdala), pain sensation (thalamus, brainstem), and motor control (substantia nigra, cerebellum). (From Patchev et al., 2004.)

those directly associated with reproduction and parenting. The evidence of this influence in humans includes the consequences for behavior of therapeutic removal of the gonads (e.g., hysterectomy for medical reasons in women, or orchiotomy—removal of the testis—to treat testicular cancer in men), followed by hormone replacement therapy. Perhaps best known among these apparent influences are the suggested "neuroprotective" effects of estrogens for ischemic and other degenerative changes in neurons, including those associated with age- or stress-related cognitive decline and Alzheimer's disease. The significance of these effects remains highly controversial due to difficulties in evaluating the large, heterogeneous population of women who have received estrogen replacement therapy.

The consequences of testosterone treatment following castration, or the effects of illicit testosterone use on mood and behavior (especially aggressive behavior) in athletes, indicate that the brain remains sensitive to this particular androgen throughout life. It is not certain whether these phenomena represent direct actions of testosterone (or of estradiol following aromatization in the brain). The widespread distribution of estrogen and androgen receptors in the adult brain most likely mediates ongoing activational effects on a wide range of behaviors beyond those directly related to reproduction and parenting (Figure 24.10). Aside from their high concentrations in the hypothalamus (see Figure 24.5B), there are significant numbers of both estrogen and androgen receptors in the cerebral cortex, amygdala, and substantia nigra. These observations inform several important clinical issues, in particular the potential side effects of therapies that manipulate gonadal steroid signaling, including estrogen and androgen receptor antagonists for the treatment of breast and prostate cancer.

Human Genetic Disorders of Genotypic and Phenotypic Sex

Chromosomal sex, phenotypic sex, and gender are not always aligned, and genetic variations in humans can challenge the usual definitions of female and male. Such genetic variation, called **intersexuality**, is apparent in 1% to 2% of all live births and can arise from a variety of mutations associated with sex chromosomes as well as autosomes (Table 24.1). The most common genomic variations result in misaligned chromosomal and phenotypic sex. XXY individuals (Klinefelter's syndrome) have male genitalia (due to the presence of *SRY* on the Y chromosome; see Figure 24.3) but some female secondary sex characteristics (e.g., mammary tissue), presumably due to a "double dose" of the genes on the X chromosome. This increased dosage may reflect disruption of the typical process of X inactivation due to the presence of a Y chromosome. (In XX females, one of each copy of most X-chromosome genes is inactivated via DNA modifications that ensure appropriate expression levels). XO individuals (Turner syndrome) are small in stature, have rudimentary gonadal development and underdeveloped external genitalia (which usually appear female), and are sterile. XYY individuals are the least compromised. Their gonadal tissues and external genitalia are male (although they are sterile); their primary identifying physical characteristic is slightly increased height.

Other genetic disorders that result in intersexuality are the result of mutations in genes that encode metabolic enzymes for steroid hormone production. One of the most prevalent examples is **congenital adrenal hyperplasia (CAH)**, usually the result of mutations in the gene encoding 21-hydroxylase, an enzyme that, using testosterone as the precursor, synthesizes cortisol and aldosterone in the adrenal gland. In affected individuals, failure of cortisol and aldosterone synthesis leads to increased testosterone, and XY genotypes with CAH are dramatically masculinized, often very tall at an early age, and undergo precocious puberty. In XX genotypes, CAH leads to overactive adrenal secretion of testosterone during development, resulting in an ambiguous, masculinized sexual phenotype.

Androgen insensitivity syndrome (AIS) (sometimes called *testicular feminization*) illustrates the results of genetic disruption of receptor-mediated responses to gonadal steroids. The best-studied cases of AIS are males who carry mutations in the gene encoding the receptors for testosterone and/or Müllerian-inhibiting hormone (see Figure 24.3B). In these XY individuals, testes form and secrete androgens; however, the testes are hypotrophic, and the deficiency of androgen receptors leads to the development of female external genitalia. Thus, XY individuals with AIS look like females and self-identify as female, even though they have a Y chromosome and testicular tissues. Individuals with AIS present one of the strongest arguments that brain circuits in primates are masculinized mainly by the action of androgens rather than by estrogens, which are the masculinizing agent in rodents.

XY infants carry two copies of a recessive mutation that makes them deficient in one of two forms of 5-α-reductase. 5-α-reductase catalyzes the conversion of testosterone to the biologically active dihydrotestosterone during fetal and early postnatal life, the genitalia resemble those of females. These individuals are generally perceived and raised as females. At puberty, however, testicular secretion of testosterone increases and dihydrotestosterone synthesis is regulated by the second 5-α-reductase variant (encoded by a separate gene). At this point the apparent clitoris enlarges into a penis and the testes descend. In the Dominican Republic, where this recessive syndrome has been thoroughly studied in particularly consanguineous families, the condition is referred to colloquially as *guevedoces*, roughly translated "testes-at-12." Genetic testing can identify affected children, most of whom are then raised from birth in a manner consistent with their genotypic sex. Anecdotal reports indicate that most such young men retain their masculine gender identity as well as a heterosexual orientation.

Due to the long-range psychological and social consequences of conflict between genotypic sex, phenotypic sex, and sexual and gender identity, the rapid diagnosis of intersexuality and a variety of adjustments to assure appropriate matching of the sex of an individual's body, brain, and gender

TABLE 24.1 ■ Genetic Disorders Resulting in Intersexuality in Humans

Syndrome	Mutation	Frequency	Phenotype
Klinefelter's	XXY	1/2500 live births	Male secondary sex characteristics
Turner	XO	1/10,000 live births	Incomplete female development
47-XYY	XYY	1/1000 live births	Male secondary sex characteristics
Congenital adrenal hyperplasia (CAH)	21-hydroxylase (chrom. 6)	1/5000 live births	Hypermasculinization of XY individuals; masculinization of XX individuals
Androgen insensitivity (AIS)	Testosterone receptor (X chrom.); Müllerian-inhibiting substance (MIS) (chrom. 19)	Rare (1/100,000 live births)	Hypotrophic testicular tissue, female syndrome secondary sex characteristics in XY individuals
Testes-at-12	5-α-reductase (chrom. 2)	Exceedingly rare	Incomplete male genital development prior to puberty; brain and behavior remain masculinized

identity as soon as possible after birth are now standard clinical practice.

Sexual Orientation and Human Brain Structure

In the early 1990s, several high-profile studies of postmortem brain samples reported anatomical dimorphisms between the brains of homosexual and heterosexual men. This issue was approached primarily in males, most likely because of the increased availability of postmortem brain samples from self-identified homosexual men who suffered AIDS-related deaths—a complicating factor for interpretation of these studies. Such analyses of anatomical dimorphisms are based on the notion that mechanisms resulting in a "homosexual brain" (if such a singular entity exists) would tend to make dimorphic structures more feminized in homosexual men and masculinized in homosexual women.

Initial studies by Simon LeVay in a much-cited analysis of a sample of postmortem tissue from heterosexual and homosexual men indicated that such differences might exist. These observations, however, were only modestly significant and not absolutely predictive. They suggested that INAH3, a sexually dimorphic subnucleus of the **interstitial nuclei of the hypothalamus (INAH)**, the homolog of the sexually dimorphic SDN-POA in rodents (see Figure 24.7B), was smaller in homosexual than in heterosexual men, being similar in size to that of heterosexual women. Nevertheless, the size of INAH3 alone was not a reliable indicator of sexual orientation in the sample reported by LeVay. Subsequent studies that have taken into account a significant tendency for degenerative changes in brain tissue from individuals who died of AIDS-related illnesses (now uncommon due to improved anti-retroviral treatment), regardless of sexual orientation, failed to replicate this suggested dimorphism. Indeed, analyses of several other anatomical dimorphisms of the hypothalamus in heterosexual and homosexual men (women remain largely unstudied) have failed to generate consistent results. Non-invasive anatomical imaging approaches such as MRI (see Chapter 1) lack the resolution to identify these cellular differences in the hypothalamus. Given the current evidence, it would seem that the volume and number of neurons in sexually dimorphic hypothalamic nuclei do not, taken alone, reliably predict sexual orientation.

The application of functional imaging techniques to this question has added some clarity by mapping differences in activation of potentially functionally dimorphic regions in the brains of heterosexual and homosexual men and women in response to behaviorally relevant stimuli. In these studies, individuals have been carefully selected for comparable age, consistent sexual orientation (both heterosexual and homosexual), relationship status (a balanced percentage of both the heterosexual and homosexual individuals were in committed relationships), age, and HIV status (none of the individuals were HIV-positive). Heterosexual men and women show differential patterns of hypothalamic activation when presented with estrogens and androgens as odorants (see Chapter 15). In homosexual men, androgens maximally activate the anterior hypothalamus as they do in heterosexual women; and in homosexual women, estrogens maximally activate the anterior hypothalamus, as in heterosexual men (Figure 24.11). The behavioral significance of this functional dimorphism is not clear. Nevertheless, these observations provide the best evidence to date for the existence of differences in the brain correlated with sexual orientation. Moreover, these differences suggest that there may be some acquisition of female characteristics in the brains of homosexual men and male characteristics in the brains of homosexual women.

(A) Androgen administered

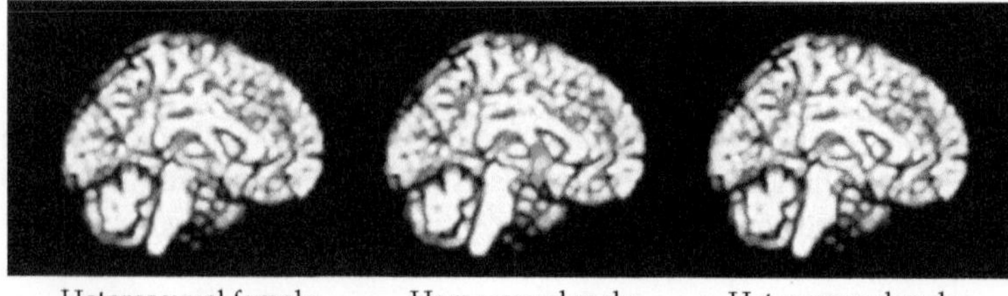

Heterosexual female Homosexual male Heterosexual male

(B) Estrogen administered

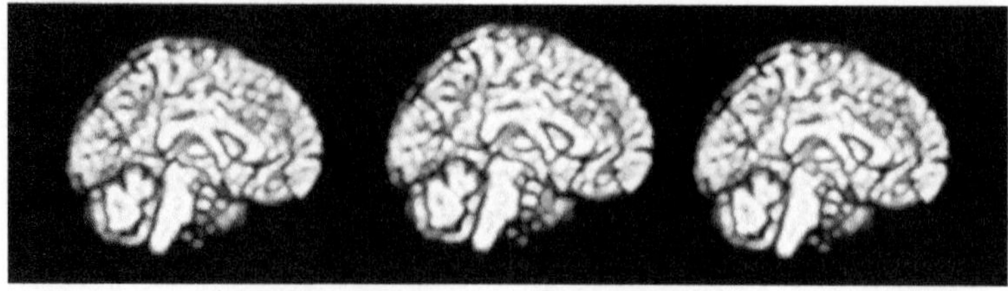

Heterosexual female Homosexual female Heterosexual male

FIGURE 24.11 **Hypothalamic activation by estrogen and androgens in heterosexual and homosexual women and men.** (A) Inhaling androgens elicits focal activation of the hypothalamus (red) in heterosexual women and homosexual men; there is no activation in the hypothalamus of heterosexual men. (B) Inhaled estrogen activates the cingulate cortex, but not the hypothalamus, in heterosexual women. In homosexual women, estrogen elicits some activation in the hypothalamus, similar to that seen in heterosexual men. (From Savic et al., 2005.)

Sex-Based Differences in Cognitive Functions

Aside from conventional wisdom (or prejudice) that boys and girls and/or men and women have greater aptitudes for, or interest in, certain tasks, there is little scientific evidence that the cognitive abilities of men and women differ in ways that correlate strictly

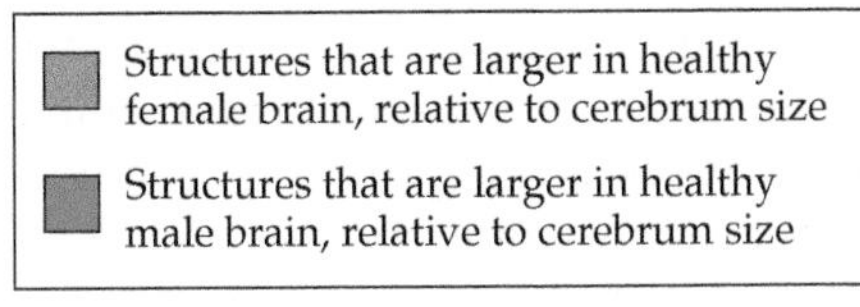

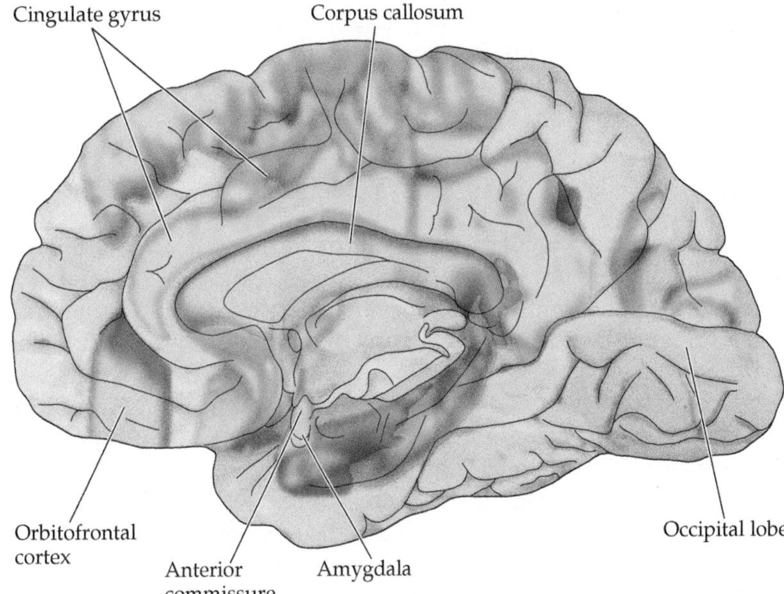

FIGURE 24.12 Potentially dimorphic brain regions beyond the hypothalamus. In the human brain, the average size of these regions differs in the females versus males. It is important to note, however, that these representations are based on mean estimates; individual variability makes it impossible to predict an individual's sex based solely on the sizes of the brain regions depicted here. (After Cahill, 2006.)

with either genotypic or phenotypic sex. Many apparent sex differences in cognitive tasks such as language, learning, memory, or visuospatial ability reflect influences not directly related to genetically established sexual dimorphisms. Thus, statistically significant differences in the performance of men and women on a variety of tasks are at least as likely to represent social or cultural influences that result in different patterns of learned behavior. The relationship of these behaviors to genotypic or phenotypic sex is not known.

The issue of brain structural differences beyond quantitative cell biological dimorphisms detected via histological analysis in the spinal cord and hypothalamus has proven even more difficult to evaluate. Postmortem as well as structural MRI analyses suggest that several forebrain structures (including cerebral commissures such as the corpus callosum and anterior commissure, nuclei such as the amygdala, and many cortical regions) may differ in size or shape in men and women (Figure 24.12). Many of the relevant anatomical studies, however, are complicated by small samples and highly derived analyses that reveal only small differences in size and shape, and often are not replicated in additional independent samples.

A great deal of interest has focused on the amygdala as the most likely central site of sexual dimorphism beyond the spinal cord and hypothalamus, perhaps because of its established role in regulating the output of hypothalamic nuclei, as well as the hypersexuality observed in animals or patients with bilateral damage to the amygdala. Some MRI studies suggest that the male amygdala has a larger volume than the female; however, additional observations indicate that these differences are not significant when measurements are corrected for brain size or cranial volume that differ in men and women. Studies of intersex individuals (see Table 24.1) suggest that the amygdala may be sexually dimorphic, perhaps due to the influence of altered gonadal steroid levels in these conditions. These studies, however, are complicated by the generally smaller total brain size in genetically intersex individuals.

In contrast to these equivocal anatomical data, there is a fairly robust functional distinction in the amygdalas of men and women performing emotional memory tasks. In such tasks, individuals view aversive or frightening films or images that elicit an emotional response. Several weeks later, the individuals are evaluated for their memory of these images. In recalling emotionally charged content, the right amygdala is maximally activated in men, while the left is maximally activated in women (Figure 24.13). These

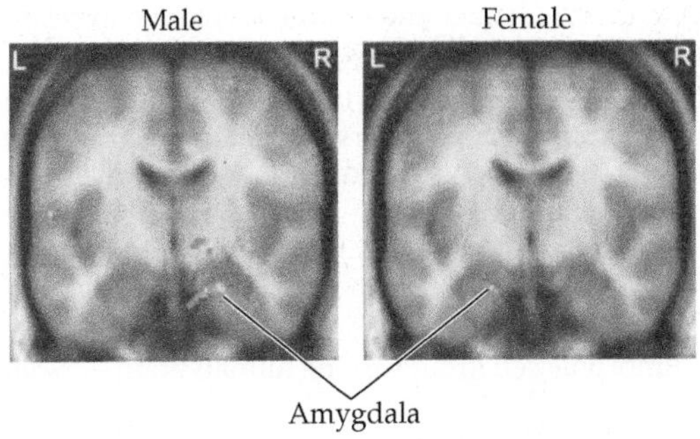

FIGURE 24.13 Laterality of activation in the amygdala. When men are presented with a previously viewed image depicting a strong negative emotion, the right amygdala is focally activated. Women presented with the same image, again seen previously, show focal activation in the left amygdala. (From Cahill et al., 2004.)

functional differences suggest that *laterality of activation*, rather than size of the nucleus, is the most robust sexual dimorphism in the amygdala. The functional significance of this observation, however, again is not known.

Summary

Throughout the animal kingdom, the brains of male and female individuals become specialized during pre- and postnatal development for the division of behavioral tasks that deal with reproduction and the rearing of offspring. These differences reflect the consequences of intrinsically produced signals provided primarily by the embryo itself. In mammals, the strongest determinant of these differences is the initial differentiation of gonadal tissues, under the control of the masculinizing transcription factor SRY. SRY determines an individual's genetic sex, and usually the phenotypic sex as well, but it is not expressed in the brain. The influence of SRY on the nervous system is thus indirect: SRY-mediated masculinization leads to differential production of gonadal tissues during fetal development and thus to sex-specific levels of circulating gonadal hormones (estrogen and testosterone in particular over the course of development). These hormones profoundly influence the development of brain structures that subserve the peripheral structures (genitalia, mammary glands) directly related to reproduction and parenting. Some of these dimorphisms reflect trophic regulation of cell survival and death in relevant structures based on parallel development of peripheral organs that these cells innervate or regulate (male and female genitalia, mammary glands in females, and related muscles). Differences in size, presence or absence of muscles or glandular tissue results in distinct levels of target derived trophic support. The origins, existence, and functional significance of dimorphisms related to distinctions in gender identity and sexual orientation remain controversial. These differences in the brains of women and men must ultimately be due to brain organization and function. Whether they arise via learned behaviors that are used to define gender roles in society or intrinsic developmental mechanisms is not known.

ADDITIONAL READING

Reviews

Blackless, M. and 5 others (2000) How sexually dimorphic are we? Review and synthesis. *Amer. J. Human Biol.* 12: 151–166.

MacLusky, N. J. and F. Naftolin (1981) Sexual differentiation of the central nervous system. *Science* 211: 1294–1302.

McEwen, B. S. (1999) Permanence of brain sex differences and structural plasticity of the adult brain. *Proc. Natl. Acad. Sci. USA* 96: 7128–7129.

Morris, J. A., C. L. Jordan and S. M. Breedlove (2004) Sexual differentiation of the vertebrate nervous system. *Nat, Neurosci.* 7: 1034–1039.

Smith, C. L and B. W. O'Malley (1999) Evolving concepts of selective estrogen receptor action: From basic science to clinical applications. *Trends Endocrinol. Metab.* 10: 299–300.

Swaab, D. F. (1992) Gender and sexual orientation in relation to hypothalamic structures. *Horm. Res.* 38 (Suppl. 2): 51–61.

Swaab, D. F. and M. A. Hofman (1984) Sexual differentiation of the human brain: A historical perspective. In *Progress in Brain Research,* vol. 61. G. J. De Vries (ed.). Amsterdam: Elsevier, pp. 361–374.

Important Original Papers

Allen, L. S., M. Hines, J. E. Shryne and R. A. Gorski (1989) Two sexually dimorphic cell groups in the human brain. *J. Neurosci.* 9: 497–506.

Allen, L. S., M. F. Richey, Y. M. Chai and R. A. Gorski (1991) Sex differences in the corpus callosum of the living human being. *J. Neurosci.* 11: 933–942.

Beyer, C., B. Eusterschulte, C. Pilgrim and I. Reisert (1992) Sex steroids do not alter sex differences in tyrosine hydroxylase activity of dopaminergic neurons in vitro. *Cell Tissue Res.* 270: 547–552.

Breedlove, S. M. and A. P. Arnold (1981) Sexually dimorphic motor nucleus in the rat lumbar spinal cord: Response to adult hormone manipulation, absence in androgen-insensitive rats. *Brain Res.* 225: 297–307.

Byne, W. and 5 others (2000) The interstitial nuclei of the human anterior hypothalamus: Assessment for sexual variation in volume and neuronal size, density, and number. *Brain Res.* 856: 254–258.

Byne, W. and 8 others (2002) The interstitial nuclei of the human anterior hypothalamus: An investigation of variation with sex, sexual orientation, and HIV status. *Horm. Behav.* 40: 86–92.

Cooke, B. M., G. Tabibnia and S. M. Breedlove (1999) A brain sexual dimorphism controlled by adult circulating androgens. *Proc. Natl. Acad. Sci. USA* 96: 7538–7540.

De Vries, G. J. and 9 others (2002) A model system for study of sex chromosome effects on sexually dimorphic neural and behavioral traits. *J. Neurosci.* 22: 9005–9014.

Forger, N. G. and S. M. Breedlove (1987) Motoneuronal death during human fetal development. *J. Comp. Neurol.* 264: 118–122.

Frederikse, M. E., A. Lu, E. Aylward, P. Barta and G. Pearlson (1999) Sex differences in the inferior parietal lobule. *Cerebral Cortex* 9: 896–901.

Gorski, R. A., J. H. Gordon, J. E. Shryne and A. M. Southam (1978) Evidence for a morphological sex difference within the medial preoptic area of the rat brain. *Brain Res.* 143: 333–346.

Gron, G., A. P. Wunderlich, M. Spitzer, R. Tomczak and M. W. Riepe (2000) Brain activation during human navigation: Gender different neural networks as substrate of performance. *Nat. Neurosci.* 3: 404–408.

Lasco, M. S., T. J. Jordan, M. A. Edgar, C. K. Petito and W. Byne (2002) A lack of dimorphism of sex or sexual orientation in the human anterior commissure. *Brain Res.* 936: 95–98.

LeVay, S. (1991) A difference in hypothalamic structure between heterosexual and homosexual men. *Science* 253: 1034–1037.

Meyer-Bahlburg, H. F. L., A. A. Ehrhardt, L. R. Rosen and R. S. Gruen (1995) Prenatal estrogens and the development of homosexual orientation. *Dev. Psych.* 31: 12–21.

Modney, B. K. and G. I. Hatton (1990) Motherhood modifies magnocellular neuronal interrelationships in functionally meaningful ways. In *Mammalian Parenting*, N. A. Krasnegor and R. S. Bridges (eds.). New York: Oxford University Press, pp. 306–323.

Raisman, G. and P. M. Field (1973) Sexual dimorphism in the neuropil of the preoptic area of the rat and its dependence on neonatal androgen. *Brain Res.* 54: 1–29.

Rossell, S. L., E. T. Bullmore, S. C. R. Williams and A. S. David (2002) Sex differences in functional brain activation during a lexical visual field task. *Brain Lang.* 80: 97–105.

Swaab, D. F. and E. Fliers (1985) A sexually dimorphic nucleus in the human brain. *Science* 228: 1112–1115.

Wallen, K. (1996) Nature needs nurture: The interaction of hormonal and social influences on the development of behavioral sex differences in Rhesus monkeys. *Horm. Behav.* 30: 364–378.

Woolley, C. S. and B. S. McEwen (1992) Estradiol mediates fluctuation in hippocampal synapse density during the estrous cycle in the adult rat. *J. Neurosci.* 12: 2549–2554.

Xerri, C., J. M. Stern and M. M. Merzenich (1994) Alterations of the cortical representation of the rat ventrum induced by nursing behavior. *J. Neurosci.* 14: 1710–1721.

Zhou, J.-N., M. A. Hofman, L. J. G. Gooren and D. F. Swaab (1995) A sex difference in the human brain and its relation to transsexuality. *Nature* 378: 68–70.

Books

Fausto-Sterling, A. (2000) *Sexing the Body*. New York: Basic Books.

Goy, R. W. and B. S. McEwen (1980) *Sexual Differentiation of the Brain*. Cambridge, MA: MIT Press.

LeVay, S. (1993) *The Sexual Brain*. Cambridge, MA: MIT Press.

LeVay, S. and J. Baldwin (2012) *Human Sexuality*, 4th Edition. Sunderland, MA: Sinauer Associates.

Go to the NEUROSCIENCE 6e Companion Website at **www.oup.com/uk/Purves6e** for Web Topics, Animations, Flashcards, and more.

CHAPTER

25 Experience-Dependent Plasticity in the Developing Brain

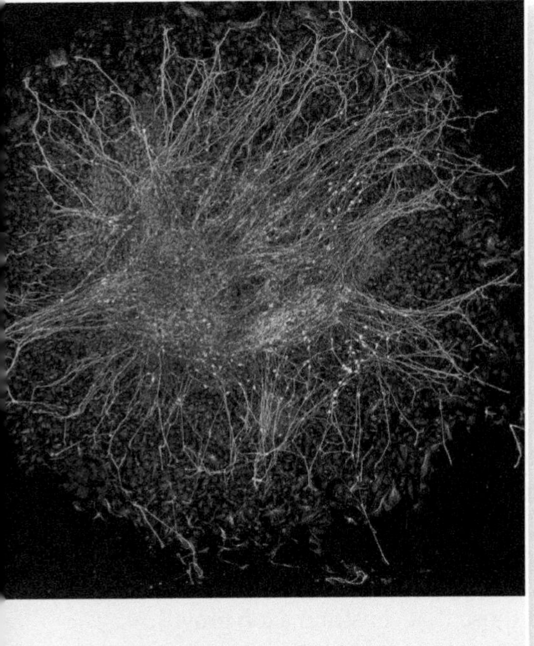

Overview

ONCE THE BASIC FRAMEWORK of neuronal diversity, axon projections, dendritic arborization, and synapse formation is established in the nervous system, a final phase of developmental change begins. This phase of development—during late pre-natal and early post-natal life—relies on cellular changes driven by neural activity, most often elicited by environmental stimuli that reflect the newborn's experience of her environment. These limited times of postnatal developmental change are referred to as critical periods. As humans (and other mammals) mature, the cellular mechanisms that modify neural connectivity become less effective, and the dramatic changes in neural circuits and related behaviors seen during critical periods are no longer possible. Thus, critical periods are thought to optimize the brain of each individual so that circuitry is adapted to specific demands that confront that individual throughout the balance of life. There are, however, some parallels between cellular mechanisms that mediate activity-dependent developmental change and those that mediate synaptic modification that underlies learning and memory. Many of the effects of activity depend on secreted signals, including neurotrophins and neurotransmitters, transduced via second messengers and their effectors. These activity-elicited changes influence local gene expression, neurotrophic interactions leading to final adjustments of axon or dendrite growth, as well as synapse growth and stability. Much of our knowledge of the influence of activity on developing neural circuits comes from studies of the mammalian visual system, where differences in input from each eye influence patterns of connections in the visual cortex. Beyond the primary visual areas, changes in the structure and size of the human cerebral cortex from birth through late adolescence indicate that experience-dependent critical periods likely influence connectivity and complex behaviors. In humans, postnatal changes in cortical size may reflect the consequences of experience and activity for the differential growth of some cortical areas but not others. These mechanisms may be compromised in disorders that result in intellectual disability, developmental delay, autism, or psychiatric diseases such as schizophrenia.

Neural Activity and Brain Development

In 1949, the psychologist D. O. Hebb hypothesized that the coordinated electrical activity of a presynaptic terminal and a postsynaptic neuron strengthens the synaptic connection between them. **Hebb's postulate**, as it has come to be known, was originally formulated to explain the cellular basis of learning and memory, but the concept has been widely applied to situations involving long-term modifications in synaptic strength or distribution, including those that occur during development of neural circuits. In this context, Hebb's postulate implies that synaptic terminals strengthened by correlated activity during development will be retained or sprout new branches,

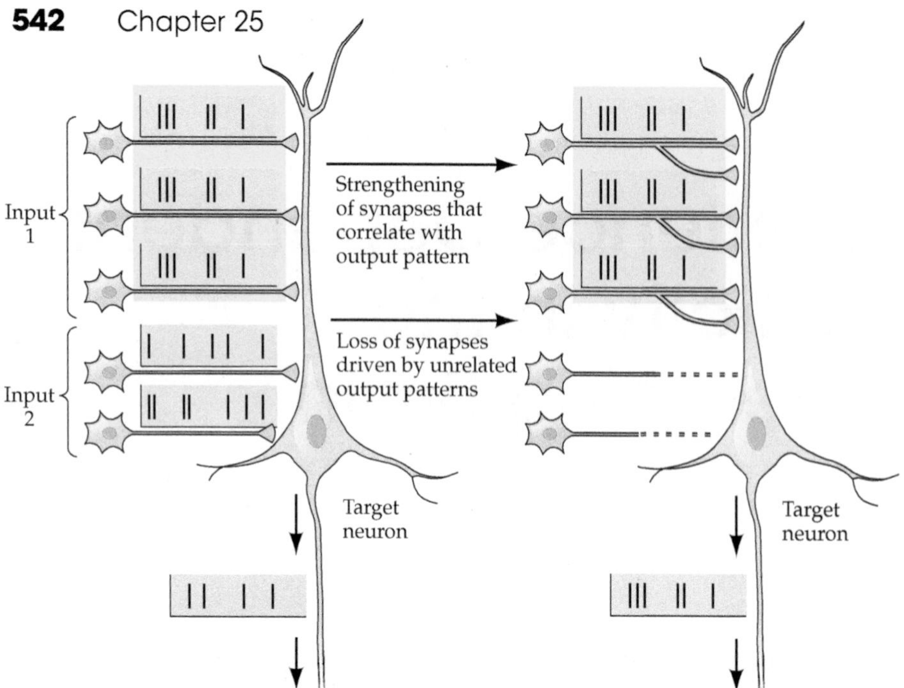

FIGURE 25.1 Hebb's postulate and the development of synaptic inputs. In this drawing, a postsynaptic neuron is shown with two sets of presynaptic inputs, each with a different pattern of electrical activity. Activity patterns, corresponding to action potential frequency, are represented by the short vertical bars. In the example here, the three correlated inputs at the top are better able to activate the postsynaptic cell. These inputs cause the postsynaptic cell to fire a pattern of action potentials that follows the pattern seen in the input. As a result, the activity of the presynaptic terminals and the postsynaptic neuron is highly correlated. According to Hebb's postulate, these synapses are therefore strengthened. The two additional inputs relay a different pattern of activity that is less well correlated with the majority of the activity elicited in the postsynaptic cell. These synapses gradually weaken and are eventually eliminated (right-hand side of figure), while the correlated inputs form additional synapses.

whereas those terminals that are persistently weakened by uncorrelated activity will eventually lose their hold on the postsynaptic cell, either leading to the death of the cell that gives rise to those synapses, or to the stabilization and growth of synapses from that cell on another target (Figure 25.1; see also Chapter 22).

This Hebbian formulation of the role of activity in influencing the ongoing organization of neural circuits provided an initial clue to understanding three obvious phenomena in brain development. First, behaviors not initially present in newborns emerge and are shaped by experience throughout early life; second, there is superior capacity for acquiring complex skills and cognitive abilities during early life; and third, the brain continues to grow after birth, roughly in parallel with the emergence and acquisition of increasingly complex behaviors and the addition of pre-and postsynaptic processes (dendritic and axonal branches). The parallel growth of dendritic and axonal branches and the addition of synaptic connections early in life must account for a significant portion of postnatal brain growth (Figure 25.2A,B) and perhaps provide a substrate for enhanced behavioral capacities. Not surprisingly—especially if Hebb's postulate is correct—there is a subsequent decline in synapse number during adolescence (Figure 25.2C). Accordingly, synaptic connections and related neuronal growth are the ultimate targets of activity-dependent change during early postnatal life—initially for progressive construction, and then selective elimination of connections. During the initial phases of progressive construction the brain gets larger because of post-natal growth of dendrites, axons and synapse—but not (with few exceptions) the addition of neurons. During the elimination phase, counterintuitively, the brain continues to grow. This reflects the continued elaboration of the synapses that remain, and the neurons that are their targets. From birth through early adulthood, these events occur in synchrony with the acquisition of sensory and motor abilities, the capacity for social interaction, and increasingly sophisticated cognitive behaviors including spoken, signed, or written language in humans (see below and Chapter 33). These coincidences suggest that the combination of activity-dependent modification of connections initially suggested by Hebb and corresponding brain growth and behavioral changes during early life must underlie how each individual's brain ultimately develops to meet the challenges of adapting to a dynamic environment.

The electrical activity initiated by exposure to and interaction with the environment following birth is a key element in the construction of any individual's nervous system. Intrinsic mechanisms (including those that constrain axon growth, initiate the formation of topographic maps, and establish the first synapses; see Chapter 23) establish the general circuitry required for most behaviors. These cellular and molecular mechanisms, however, do not yield a final pattern of connectivity. "Typical" experiences evidently validate initial wiring and preserve, augment, or adjust the initial arrangement that is established by intrinsic developmental mechanisms. In the case of "diminished" experiences due to lack of sensory exposure or disrupted sensory transduction and relay, these adjustments do not occur appropriately and brain connectivity and behavioral capacity can be altered. These alterations can have some adaptive advantage—for example, in the context of compromised capacity of the individual to acquire information from the environment (e.g., sensory impairments such as blindness or deafness)—and sometimes to the detriment of optimal brain function as the individual matures (e.g., sensory deprivation or trauma due to life circumstances for an otherwise intact infant or child). The eventual decline of the capacity to remodel cortical (and subcortical) connections most likely explains changes in the capacity of the brain to acquire new information and direct

(A)

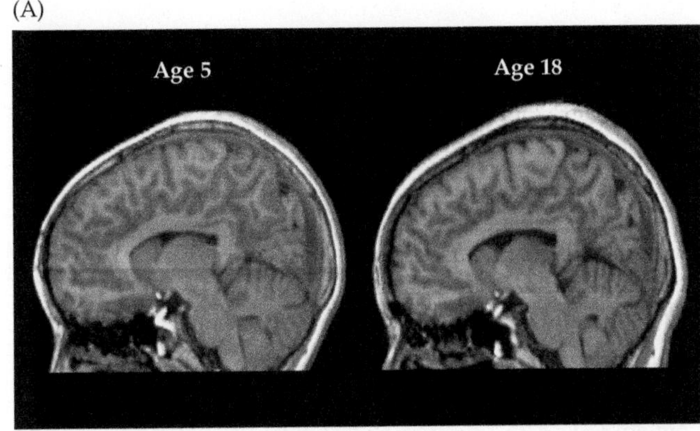

(B)

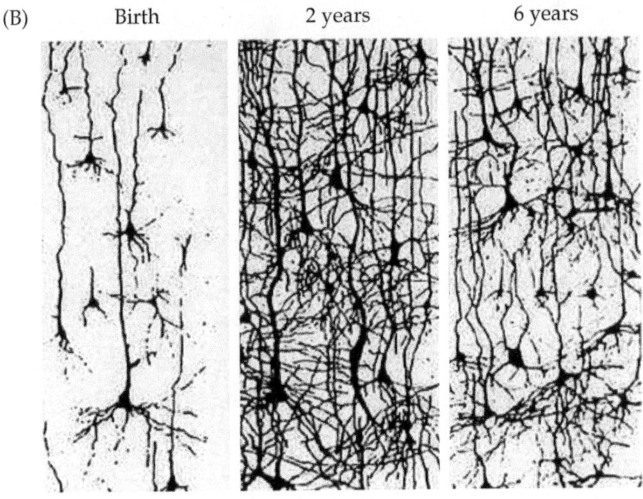

(C)

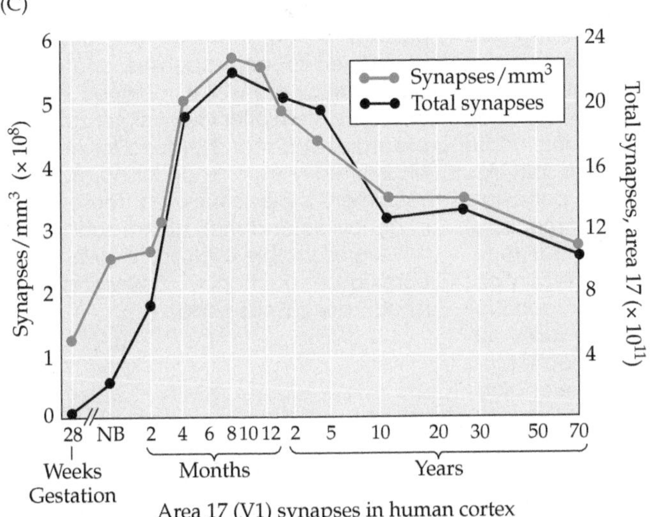

FIGURE 25.2 The human brain changes significantly during postnatal life. (A) MRI images of a brain from the same individual at 5 years and 18 years of age. The brain at 18 years is slightly larger, and the pattern of sulci and gyri in the cerebral cortex is more prominent. (B) Axons and dendrites in the human cerebral cortex at birth, 2 years, and 6 years of age. These drawings are based upon Golgi-stained neurons in the cerebral cortex from individual post-mortem samples of different ages. (C) Addition and then elimination of synapses in the human brain. This data is based upon electron microscopic analysis of synapse density in the human primary visual cortex. (A courtesy of Jay Giedd and P. Shaw; B from Conel, 1939–1967; C after Huttenlocher et al., 1982).

new behaviors over a lifetime—a concept with obvious educational, psychiatric, and social implications.

Critical Periods

For most animals, the behavioral repertoire for basic survival, including foraging, fighting, and mating strategies, largely relies on patterns of connectivity established by intrinsic developmental mechanisms. Indeed, embryonic mechanisms and their developmental consequences are sufficient to create some remarkably sophisticated innate or "instinctual" behaviors, including the complex repertoire for parental identification, feeding, and responding to predators seen in newborn birds and mammals (see Chapter 24). Nevertheless, the nervous systems of animals with increasingly complex repertoires of behaviors, including humans, clearly adapt to and are influenced by the particular circumstances of an individual's environment beyond innate behavioral capacities. These environmental factors are especially influential in early life, during temporal windows called **critical periods**—the time when experience and the neural activity that reflects that experience have maximal effect on the acquisition or skilled execution of a particular behavior. Some behaviors, such as parental imprinting in hatchling birds (the event by which the hatchling recognizes its "parent" based on initial recognition after emerging from the egg), are expressed only if animals have certain specific experiences during a sharply restricted time (hours or days) in early postnatal (or post-hatching) development (Box 25A). By contrast, critical periods for sensorimotor skills and complex behaviors end far less abruptly and provide far more time for environmentally acquired experience. In some cases, such as the acquisition communication skills in songbirds (see Chapter 24), or language in humans, detailed instructive influences from the environment (i.e., exposure to complex calls in songbirds or words and sentences in humans) and opportunities for mimicry and repetition are required for an extended period to ensure normal development of the behavior. The availability of instructive experiences from the environment, as well as the neural capacity to respond to them, is key for successful completion of the critical period. These instructive influences are important for territorial and reproductive behaviors in a subset of non-human species. In some songbirds, both male and female birds acquire the capacity to produce species-specific song by mimicking tutor birds during a limited period of postnatal life (see Chapter 24). If this essential instruction is withheld or disrupted, the birds are not effective in using communication to define their territories and compete for mates.

Although they vary widely in both their duration and the behaviors affected, all critical periods share some basic properties. Each critical period encompasses the time during which a given behavior is especially susceptible to—indeed, requires—specific environmental influences

BOX 25A ■ Built-In Behaviors

The idea that animals possess an innate set of behaviors appropriate for a world not yet experienced has always been difficult to accept. However, the preeminence of instinctual responses is obvious to any biologist who looks at what animals actually do. Perhaps the most thoroughly studied examples occur in young birds. Hatchlings emerge from the egg with an elaborate set of innate behaviors. First, of course, is the complex behavior that allows the chick to escape from the egg. Once the chick has emerged, a variety of additional abilities indicate how much early behavior is "preprogrammed."

In a series of seminal observations based on his work with geese, Konrad Lorenz showed that goslings follow the first large, moving object they see and hear during their first day of life. Although this object is typically the mother goose, Lorenz found that goslings will imprint on a wide range of animate and inanimate objects presented during this period, including Lorenz himself. The window for imprinting in goslings is less than a day; if animals are not exposed to an appropriate stimulus during this time, they will never form the appropriate parental relationship. Once imprinting occurs, however, it is irreversible, and geese will continue to follow inappropriate objects (male conspecifics, people, or even inanimate objects).

In many mammals, auditory and visual systems are poorly developed at birth, and maternal imprinting relies on olfactory or gustatory cues. For example, during the first week of life (but not later), infant rats develop a lifelong preference for odors associated with their mother's nipples. As in birds, this filial imprinting plays a role in the rats' social development and later sexual preferences.

Imprinting is a two-way street, with parents (especially mothers) rapidly forming exclusive bonds with their offspring. This phenomenon is especially important in animals such as sheep that live in large groups or herds in which all the females produce offspring at about the same time of year. Ewes have a critical period 2 to 4 hours after giving birth during which they imprint on the scent of their own lamb. After about 4 hours, they rebuff approaches by other lambs.

Work by Harry Harlow and his colleagues at the University of Wisconsin in the 1950s underscored the relevance of these studies of avian imprinting to primates. Harlow isolated monkeys within a few hours of birth and raised them in the absence of either a natural mother or a human substitute. In the best known of these experiments, the baby monkeys had one of two maternal surrogates: a "mother" constructed of a wooden frame covered with wire mesh that supported a nursing bottle, or a similarly shaped object covered with soft terrycloth but without any source of nourishment for the young monkey. When presented with this choice, the baby monkeys preferred the terrycloth mother and spent much of their time clinging to it, even though the feeding bottle was with the wire mother. Harlow took this to mean that newborn monkeys have a built-in need for maternal care and have at least some innate idea of what a mother should feel like. Several other endogenous behaviors have been studied in infant monkeys, including a naïve monkey's fear reaction to the presentation of certain objects (e.g., a snake) and the "looming" response (fear elicited by the rapid approach of any formidable object). Most of these built-in behaviors have analogs in human infants.

Konrad Lorenz followed by imprinted geese. (Photograph courtesy of H. Kacher.)

Taken together, these observations make plain that many complicated behaviors, emotional responses, and other predilections are well established in the nervous system prior to any significant experience, and that the need for certain kinds of early experience for normal development is predetermined. These built-in behaviors and their neural substrates have presumably evolved to give newborns a better chance of surviving in a reliably dangerous world.

in order to develop normally. Environmental influence elicits neural activity in the relevant sensory or motor pathway, and the nature of this activity—its frequency, amplitude, duration, and correlation—ultimately drives changes in synaptic connections. These experience or activity influences can be as subtle as the ongoing stimuli of light or sound encountered by an infant, or the precisely articulated instruction in one's native (or a foreign) language required to achieve fluent speech and accurate comprehension. Once a critical period ends, the core features of the behavior are largely unaffected by subsequent experience. This suggests that the cellular and molecular mechanisms that are influenced by experience via neural activity must also change. Failure to be exposed to appropriate stimuli during the critical period is difficult or impossible to remedy subsequently—likely because the biological mechanisms needed to change connections during the critical period are no longer available in an older individual (however, see Box 26B). In most mammals, including humans, critical periods seem to rely particularly on changes in organization and function of circuits in the cerebral cortex. Thus, much of our subsequent discussion of critical periods will focus on the consequences of experience and activity for influencing cortical growth, connections, and function from birth through young adulthood.

The Role of Oscillations in Establishing Critical Periods

The evidence for critical periods initially relied on behavioral observations, then on electrophysiological data that measured action potential activity elicited by relevant stimuli—especially visual stimuli—coupled with anatomical assessment of changes in patterns of connections. These approaches provided definitive proof of the critical period for the visual system. It is beyond doubt that "elicited" sensory experience—that driven by external stimuli resulting in action potential activity—is key. Nevertheless, there are additional forms of subthreshold as well as action potential-mediated physiological activity that occur prior to experience-driven activity. These forms of electrical activity establish a framework within which sensory experience can further influence patterns and function of connectivity. Local **oscillations**, or "waves," of electrical activity that are initially beneath the threshold for action potential generation are now known to be essential for shaping circuit networks so they are prepared for optimal experience-driven activity.

Most of the insights into the developmental importance of neuronal oscillations have come from analyses of the mammalian visual system (see Chapter 11). The initial evidence for oscillatory activity came from analysis of activity in the retina that begins long before birth in most mammalian species and is referred to as **retinal waves** (Figure 25.3A,B). It had long been known, based on anatomical tracing (see below), that a scaffold of segregated inputs from the lateral geniculate nucleus (LGN), driven in turn by segregated afferents from retinal ganglion cells in the left versus right eye, is present in many animals, including rhesus monkeys, cats, and ferrets either slightly before or after birth. It was unclear, however, how this segregation, prior to experience-driven competition

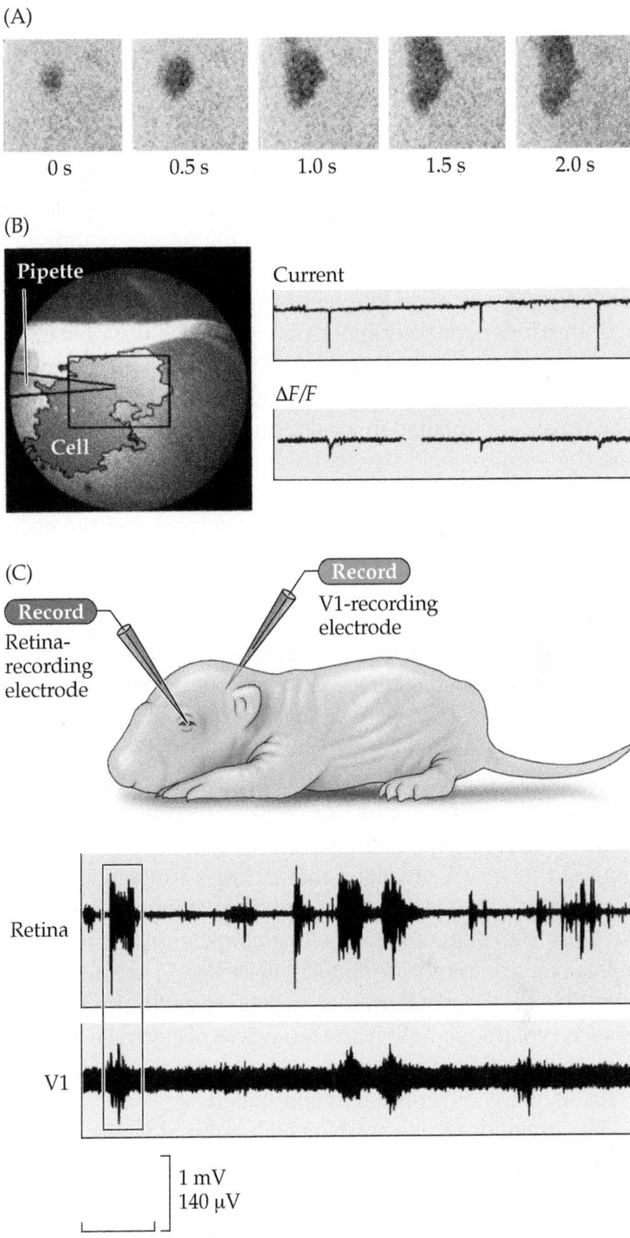

FIGURE 25.3 **Spontaneous activity establishes rudimentary patterns of connectivity in the retinogeniculocortical pathway.** (A) Retinal waves, measured here by imaging calcium transients (proportional to action potential activity) in a flattened retina in vitro. The activity that defines a single wave, depicted using a grey scale that maps the local change of Ca^{2+} influx over time, measured by the change in fluorescence intensity ($\Delta F/F$) spreads across the retinal surface. The small gray spot in the upper left quadrant of the far left panel indicates the initiation of the wave at time 0 (0s). The subsequent images, taken every 0.5 second, show the spreading excitation across a sub-region of the retinal surface that defines the wave (which is ultimately relayed to the thalamus via action potential activity in retinal ganglion cells) until the wave abates (not shown). (B) The relationship between single cell excitation measured intracellularly (the downward deflections in the top trace indicate excitatory postsynaptic potentials) and calcium transients that define each retinal wave. Within the wave region outlined at left, the firing of a cell recorded at the position (shown by the pipette cartoon and box is synchronized to the maximal depolarization events that define the wave, shown as the downward traces in the fluorescence intensity record (bottom trace). (C) Simultaneous extracellular recordings are made in the retina and primary visual cortex of an early postnatal rat pup, prior to eye opening and the capacity of the retina to relay light information from rods and cones to the ganglion cells. The spontaneous activity that defines retinal waves is shown as bursts of action potential activity. These bursts in the retina are highly correlated with bursts of action potential activity recorded in primary visual cortex in the same pup. (A,B from Feller et al., 1996, images courtesy of C. Shatz; C from Hanganu et al., 2006.)

between the two eyes, was put into place. There are some molecular affinities reflecting the origin of ipsilateral projecting versus contralateral projecting retinal ganglion cells in the nasal versus temporal retina (see Box 23A) that biases them toward distinct target regions in the LGN. Nevertheless, these affinities alone are not sufficient to explain the initial pre- or perinatal segregation of the inputs from the two eyes. A substantial amount of this initial segregation is due to organized electrical activity in the retinas of individual animals before they are born or their eyes have opened. Each retina in an individual fetus or newborn (prior to eye opening) independently generates a pattern of waves of electrical activity (usually measured as Ca^{2+} influx) that moves across large populations of retinal cells in an orderly fashion. The waves are initiated in local retinal cells (amacrine cells), and this subthreshold activity leads to action potential firing by ganglion cells that is then relayed to the lateral geniculate nucleus (LGN). These waves, though coherent in each eye, are asynchronous between the two eyes. The lack of correlated activity establishes a modest competitive interaction—leading to Hebbian synaptic reinforcement—between the two eyes for target space in the LGN and via the LGN to the primary visual cortex. Thus, afferents driven by one eye are likely to segregate, at least partially, from those of the other eye. Retinal waves can be abolished via pharmacological blockers of various neurotransmitters, or in mice in which a variety of receptors or neurotransmitter transporters have been inactivated by mutation. These experiments suggest that a nascent circuit that includes acetylcholinergic, GABAergic, and glycinergic synaptic transmission from retinal amacrine cells combined with glutamatergic release from bipolar cells onto retinal ganglian cells (see Chapter 6) is responsible for the generation of retinal waves. In the absence of such synaptic signaling and the waves of electrical activity that result, segregation in the LGN by left eye versus right eye is substantially diminished or eliminated.

The excitatory drive established by retinal waves via the synaptic signals delivered to ganglion cells and transmitted through LGN relay neurons that project to the primary visual cortex, establishes related oscillatory activity in the visual cortex. These oscillations can be recognized with multi-unit recordings from the neonatal primary visual cortex in animals. Such activity in developing cortical networks reflects voltage changes—synaptic as well as action potential activity—in large numbers of neurons can be recorded using extracellular electrodes placed in the cortex or thalamus in early postnatal animals such as a rat or mouse, or can be detected using electroencephalography (EEG) in pre-term as well as full-term human infants (Figure 25.3C; also see Chapter 1). In early postnatal rat visual cortex, bursts of action potential activity in the cortex are highly correlated with the wave-induced bursts of action potential activity from the retina. The absence of this cortical activity delays or abolishes appropriate maturation of visually evoked responses in the cortex. In humans, similar activity, detected as bursts of population electrical activity (generally called spindles) via EEG recordings, indicates that even before extensive visual experience, there is subthreshold oscillatory activity as well as bursts of action potential signaling that influence synapse and circuit maturation in the cortex. This activity is thought to be essential to prepare the cortex for visual-experience-dependent critical period plasticity.

Critical Periods in Visual System Development

Fundamental understanding of how changes in activity and connectivity might contribute to critical periods and behavioral capacity has come from studies of the developing visual system in animals with highly developed visual abilities, particularly cats and monkeys. The visual system is extremely amenable to the sorts of experimental manipulations necessary to test the relationship between experience, activity, and circuitry. It is relatively easy to either deprive or augment visual experience in an experimental animal; eyes can be sutured shut, or animals can be reared in illumination conditions ranging from total darkness to maximal light and pattern. Such control of sensory experience is almost impossible in any other modality—it is much harder to deprive an animal of auditory, somatosensory, olfactory, or taste stimuli. Moreover, the organization of the visual pathways provides ideal opportunities to evaluate how experience influences ongoing function and connections.

Information from the two eyes is first integrated in the primary visual (striate) cortex (see Chapters 11 and 12), where most afferents from the LGN of the thalamus terminate. In some mammals—carnivores, anthropoid primates, and humans—the afferent terminals form an alternating series of eye-specific domains in cortical layer 4 called **ocular dominance columns** (Figure 25.4). Ocular dominance columns can be visualized by injecting tracers such as radioactive amino acids into one eye; the tracer is then transported along the visual pathway to specifically label the geniculocortical terminals (i.e., synaptic terminals in the visual cortex) corresponding to that eye. In the adult macaque, the domains representing the LGN input from the two eyes are stripes of about equal width (0.5 mm) that occupy roughly equal areas of layer 4 of the primary visual cortex. Electrical recordings confirm that the cells in layer 4 of macaques respond strongly or exclusively to stimulation of either the left or the right eye, while neurons in layers above and below layer 4 integrate inputs from both the left and right eyes and respond to visual stimuli seen by both eyes. Complete ocular dominance is thus seen in the domains (stripes) in cortical layer 4, where all neurons are driven exclusively by one eye or the other reflecting convergent information from LGN axons driven by one or the other eye. Ocular dominance can be seen beyond layer 4. It is measured based upon the extent to which one or both eyes activate individual cortical neurons (i.e., in layers 2 to

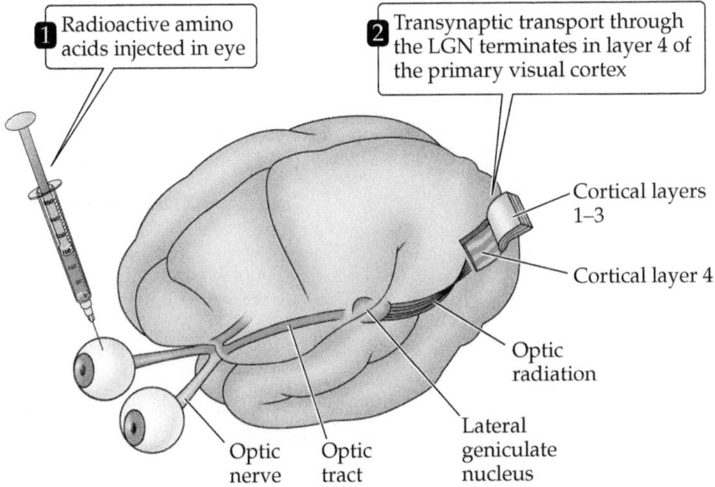
1 Radioactive amino acids injected in eye
2 Transynaptic transport through the LGN terminates in layer 4 of the primary visual cortex
Cortical layers 1–3
Cortical layer 4
Optic radiation
Lateral geniculate nucleus
Optic nerve
Optic tract

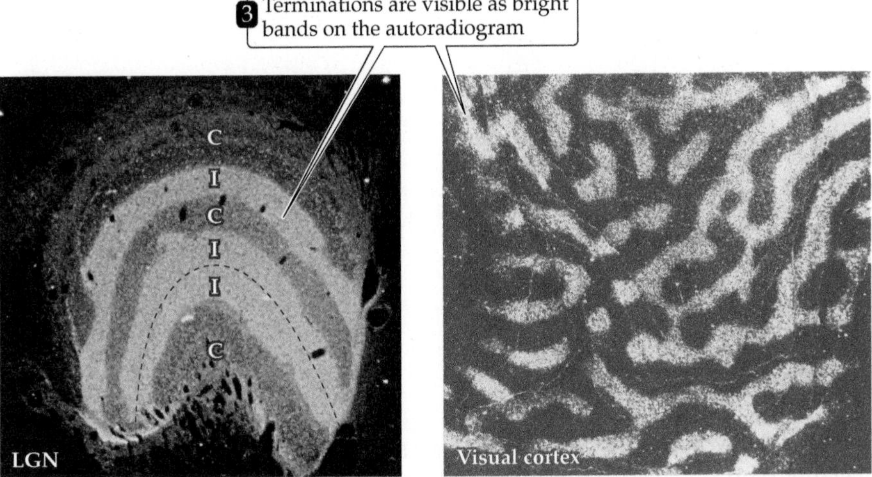
3 Terminations are visible as bright bands on the autoradiogram
C
I
C
I
I
C
LGN
Visual cortex

FIGURE 25.4 **Ocular dominance columns in layer 4 of the primary visual cortex of an adult macaque.** The diagram illustrates the labeling procedure. Following transynaptic transport of the radioactive label, the distribution of ipsilateral (I) versus contralateral (C) retinal ganglion cell axon terminals is seen in the lateral geniculate nucleus (LGN; lower left). Geniculocortical terminals (from the labeled LGN layers) related to the injected eye are visible as a pattern of light stripes in an autoradiogram (lower right) of a section through layer 4 in the plane of the cortex (that is, as if looking down on the cortical surface). The dark areas are the zones occupied by geniculocortical terminals related to the unlabeled eye. (LGN micrograph courtesy of P. Rakic; ocular dominance columns from LeVay, et al., 1980.)

3 and 5 to 6). The clarity of these patterns of anatomical and functional connectivity, and the precision by which experience via the two eyes can be manipulated led to a series of experiments that defined the relationship between activity, experience and circuitry. These experiments greatly clarified the neurobiological processes underlying critical periods. It was clear that the critical period for vision relied upon correlated activity due to visually evoked stimuli delivered to each eye, changes in functional innervation, that underlie changes in patterns of connectivity measured both physiologically and anatomically.

Effects of Visual Deprivation on Ocular Dominance

The clear distinction between activation of visual cortical neurons by one or the other eye via the retinogeniculocortical pathways was the primary measure of how peripheral experience influences visual development. As described in Chapter 12, if an electrode is passed at a shallow angle through the cortex while the responses of individual neurons to light stimulation of one or the other eye are being recorded, detailed assessment of ocular dominance can be made at the level of individual cells that share, more or less, the same laminar location (see Figure 12.11). The original studies of visual cortical plasticity and critical periods were performed using a similar physiological approach. Accordingly, visual cortical neurons within individual cortical layers were divided into seven "ocular dominance" groups based on their degree of response to either the contralateral or ipsilateral eye (Figure 25.5). Group 1 cells are driven only by stimulation of the contralateral eye; group 7 cells are driven entirely by the ipsilateral eye; and neurons driven equally well by either eye are assigned to group 4. Based on this measurement scheme, ocular dominance distribution in all layers but layer 4 in primary visual cortex is roughly Gaussian in a normal adult (cats were used in these experiments). Most cells were activated to some degree by both eyes (distributed around a mean defined by "group 4" cells), and a substantial minority was more activated by either the contralateral or ipsilateral eye (see Figure 25.5A).

This normal distribution of ocular dominance at the level of single cortical neurons can be altered by visual

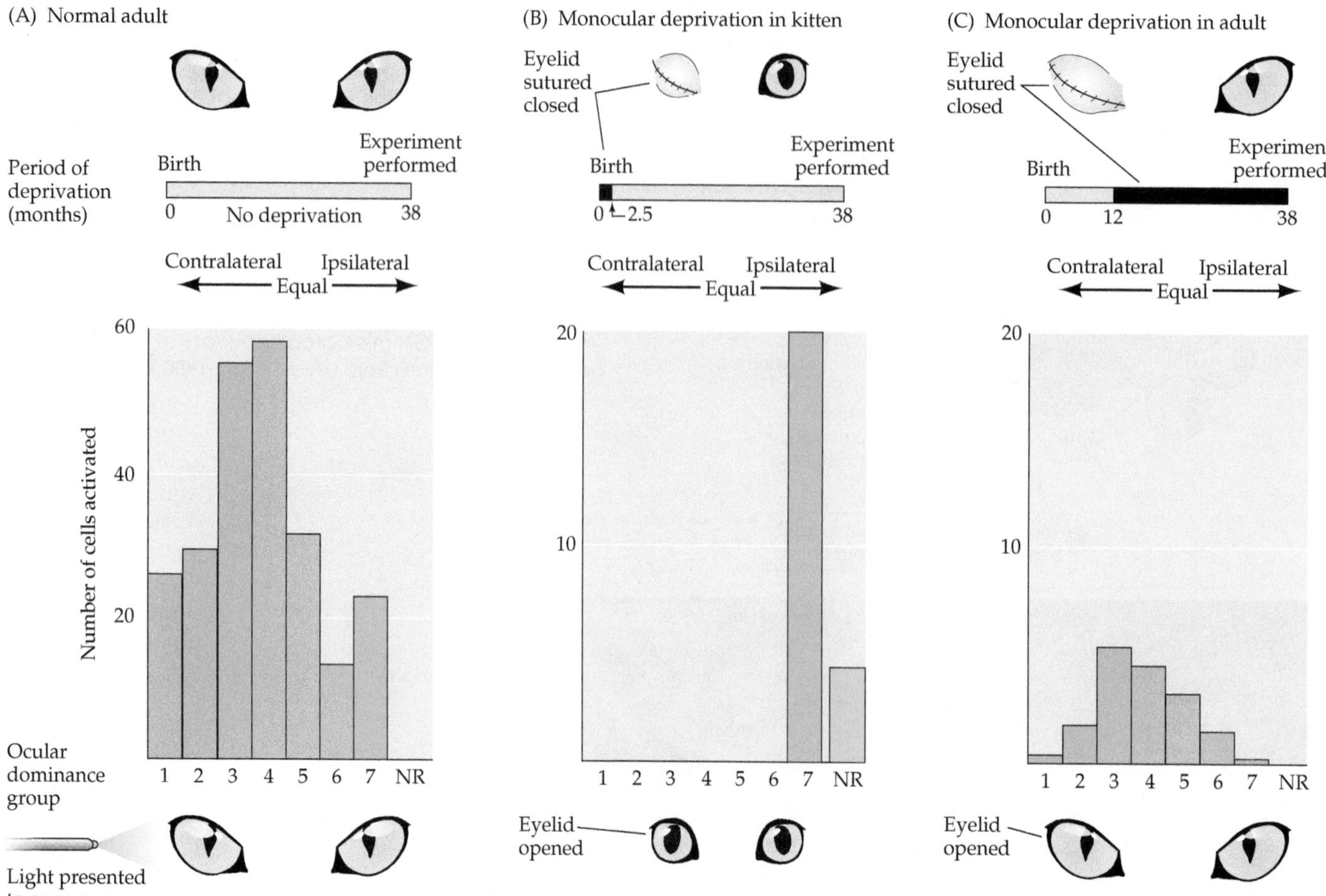

FIGURE 25.5 Effect of early eyelid closure of one eye on the distribution of cortical neurons driven by stimulation of both eyes. Histograms plot the number of cells that fall into one of the seven ocular dominance categories, defined based on the frequency of action potential activity elicited from visual cortical neurons following illumination in the relevant eye. In all three cases, the experimental readings were performed on cats 38 months after birth, at which time a penlight was shined into both wide-open eyes to elicit responses in the visual cortex. (A) Ocular dominance distribution of single-unit recordings from a large number of neurons in the primary visual cortex of normal adult cats. Cells in group 1 were activated exclusively by the contralateral eye, cells in group 7 by the ipsilateral eye. There were no cells that were not responsive (NR) to light stimulation in the retina. (B) One eye of a newborn kitten was closed from 1 week after birth until 2.5 months of age. After 2.5 months, the eye was opened and the kitten matured normally to 38 months. Note that the deprivation was relatively brief—the sutured eye had been open for 35.5 months of the cat's life. Even so, light presented to the open but transiently deprived eye elicited no electrical responses in visual cortical neurons. The only visually responsive cells responded to the ipsilateral (non-deprived) eye. (C) A much longer period of monocular deprivation in an adult cat showed little effect on ocular dominance, although overall cortical activity was diminished. Most of the responsive cells were driven by both eyes. In addition, some cells (groups 1 and 2) were uniquely or mostly responsive to the deprived eye. (A after Hubel and Wiesel, 1962; B after Wiesel and Hubel, 1963; C after Hubel and Wiesel, 1970.)

experience. When one eye of a kitten was sutured closed early in life and the animal then matured to adulthood (which takes about 6 months), a remarkable change was observed. Once the eyelid was opened, electrophysiological recordings showed that very few cortical cells could be driven from the deprived (previously sutured) eye. Recordings from the retina and LGN layers in response to direct electrical stimulation in the deprived eye (as opposed to light-elicited electrical responses) indicated that these more peripheral stations in the visual pathway worked quite normally. Nevertheless, the ocular dominance distribution in the visual cortex had shifted; the eye that remained open was uniquely able to drive most cortical cells (see Figure 25.5B). Thus, the absence of cortical cells that responded to stimulation of the closed eye was not a result of retinal degeneration or a loss of retinal connections to

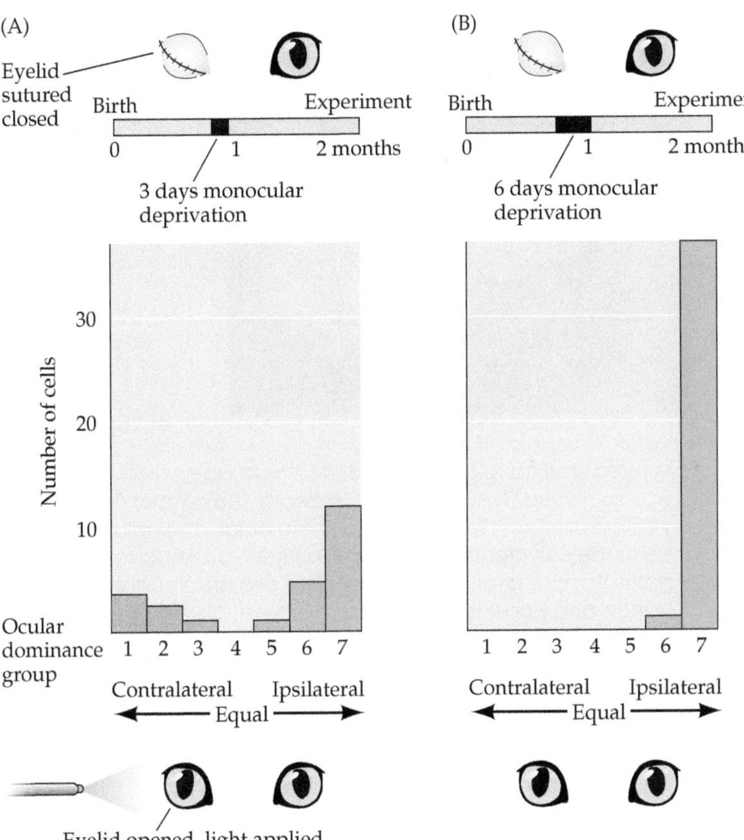

FIGURE 25.6 Consequences of a short period of monocular deprivation at the height of the critical period in the cat. (A) Just 3 days of deprivation produced a significant shift of cortical activation in favor of the non-deprived eye. (B) Six days of deprivation produced a shift of cortical activation in favor of the non-deprived eye that was almost as complete as that elicited by 2.5 months of deprivation (see Figure 25.5B). (After Hubel and Wiesel, 1970.)

the thalamus. Rather, the deprived eye had been functionally disconnected from the visual cortex. Consequently, such animals are behaviorally blind in the deprived eye. This "cortical blindness," or *amblyopia*, is permanent (see below). Even if the formerly deprived eye remains open, little or no recovery occurs.

Remarkably, the same manipulation—closing one eye—performed in adulthood has no effect on the responses of cells in the mature visual cortex (see Figure 25.5C). If one eye of an adult cat was closed for a year or more, both the ocular dominance distribution measured across all cortical layers and the animal's visual behavior were indistinguishable from normal when tested through the reopened eye. Thus, sometime between when a kitten's eyes open (about a week after birth) and 1 year of age, visual experience determines how the visual cortex is wired with respect to eye dominance. At the height of this critical period (about 4 weeks of age in a cat), as little as 3 or 4 days of eye closure profoundly alters the ocular dominance profile of the striate cortex (Figure 25.6). After this time, deprivation or manipulation has little or no permanent, detectable effect. In fact, eye closure is effective only if the deprivation occurs during the first 3 months of a kitten's life. In keeping with the ethological observations described earlier in the chapter, David Hubel and Torsten Wiesel called this period of susceptibility to visual deprivation the critical period for ocular dominance development. Similar experiments in monkeys have shown that the same phenomenon occurs in primates, although the critical period is longer (up to about 6 months of age).

Visual deprivation during the critical period must result in some sort of cortical connectivity changes that influence the functional response properties of individual neurons—especially since neither retinal nor retinogeniculate activity is altered. Indeed, the segregation of afferents by eye of origin in the LGN, and the separation of the single eye recipient LGN layers is present at birth, and does not change in deprivation experiments. Subsequent anatomical studies established that physiological changes were due to changes in patterns of connections in the visual cortex, especially those made by geniculo-cortical axons. In monkeys, the stripelike pattern of geniculocortical axon terminals in layer 4 that defines ocular dominance columns is already present at birth, and this pattern reflects the functional segregation of inputs from the two eyes (Figure 25.7A; also see Figure 25.4). Indeed, the early formation of this pattern in layer 4 reflects a significant amount of segregation of the LGN axons that relay information from one eye or the other, and this early segregation occurs even in the absence of meaningful visual experience (see the earlier section on retinal waves and early oscillatory activity). Subsequent observations have confirmed this initial experience-independent segregation, and there is some indication that specific molecular signals as well as pre-experience physiological activity may distinguish LGN cells innervated by one eye or the other. Thus, the visual cortex is clearly not a blank slate on which the effects of experience are inscribed. Nevertheless, animals deprived from birth of vision in one eye develop abnormal patterns of ocular dominance stripes in the visual cortex, presumably due to the altered patterns of activity caused by deprivation (Figure 25.7B). The stripes related to the open eye are substantially wider, and the stripes representing the deprived eye are correspondingly diminished. The absence of cortical neurons that respond physiologically to the deprived eye is not simply a result of the relatively inactive inputs withering away. If this were the case, one would expect to see areas of layer 4 devoid of

(A)

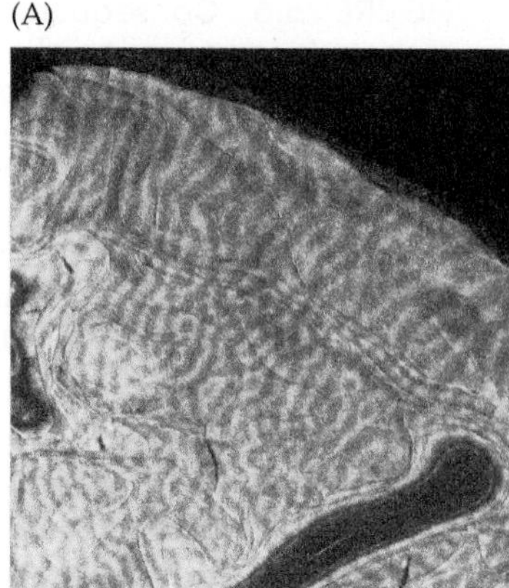

(B)

FIGURE 25.7 Effect of monocular deprivation on the pattern of ocular dominance columns in the macaque. (A) In normal monkeys, ocular dominance columns are seen as alternating stripes of roughly equal width. (B) The picture is quite different after monocular deprivation. This dark-field autoradiograph shows a reconstruction of several sections through layer 4 of the primary visual cortex of a monkey whose right eye was sutured shut from 2 weeks to 18 months of age, when the animal was sacrificed. Two weeks before death, the normal (left) eye was injected with radiolabeled amino acids. The columns related to the non-deprived eye (white stripes) are much wider than normal; those related to the deprived eye are shrunken. (A from Horton and Hocking, 1999; B from Hubel et al., 1977.)

any thalamic innervation. Instead, inputs from the active (open) eye take over some—but not all—of the territory that formerly belonged to the inactive (closed) eye. These inputs then dominate the physiological responses of the target cortical neurons.

These results suggest **competitive interaction** for post-synaptic space between afferent axons driven by each of the two eyes during the critical period, reminiscent of Hebb's description of synaptic plasticity but in the context of development. In normal animals, an equivalent amount of synaptic territory driven by each eye is retained (and sharpened in terms of segregation of ocular dominance stripes in layer 4 of the cortex) if both eyes experience roughly comparable levels of visual stimulation. However, when an imbalance in visual experience is induced by monocular deprivation, the active eye gains a competitive advantage and replaces many of the synaptic inputs from the closed eye. In this case, even though LGN axons arising from neurons innervated by the closed eye are retained in the cortex (albeit with much less extensive terminals and fewer functional synapses), few if any neurons fire action potentials when light is presented to the deprived eye. These observations in experimental animals have important implications for children with birth defects or ocular injuries that result in an imbalance of inputs from the two eyes. Unless the imbalance is corrected during the critical period, the child may ultimately have poor binocular fusion, diminished depth perception, and degraded acuity. In addition, recent work (see below) indicates that this mechanism is essential for organizing the orientation selectivity of cortical neurons that are driven by some degree by both eyes. Thus, a significant disruption of ocular competition and its consequences may permanently impair many aspects of a child's vision.

The idea that a competitive imbalance underlies the altered distribution of inputs after deprivation has been confirmed by suturing shut *both* eyes shortly after birth. This manipulation equally deprives all visual cortical neurons of normal experience during the critical period. The arrangement of ocular dominance recorded some months later is, by either electrophysiological or anatomical criteria, much closer to normal than if just one eye is closed. Thus, the balance of inputs, not the absolute level of activity, is a key feature for shaping the normal pattern of connections. This also reinforces the conclusion that modest correlation of activity between the two eyes prior to visual experience (i.e., retinal waves) can influence initial ocular segregation. Although several peculiarities in the response properties of cortical cells deprived of normal light-dependent vision are apparent, roughly normal proportions of neurons responsive to the two eyes are found in binocularly deprived animals after the period of deprivation is finished. Because there is no added imbalance in visual activity favoring one eye or the other (both sets of cortical inputs being deprived), both eyes retain their territory in the cortex. If disuse atrophy of the closed-eye inputs were the main effect of monocular deprivation on visual cortical function, then binocular deprivation during the critical period would cause the visual cortex to be largely unresponsive. Furthermore, in dark reared animals, even when the animals are reintroduced to light after the close of the typical critical period, visual activity and acuity returns to

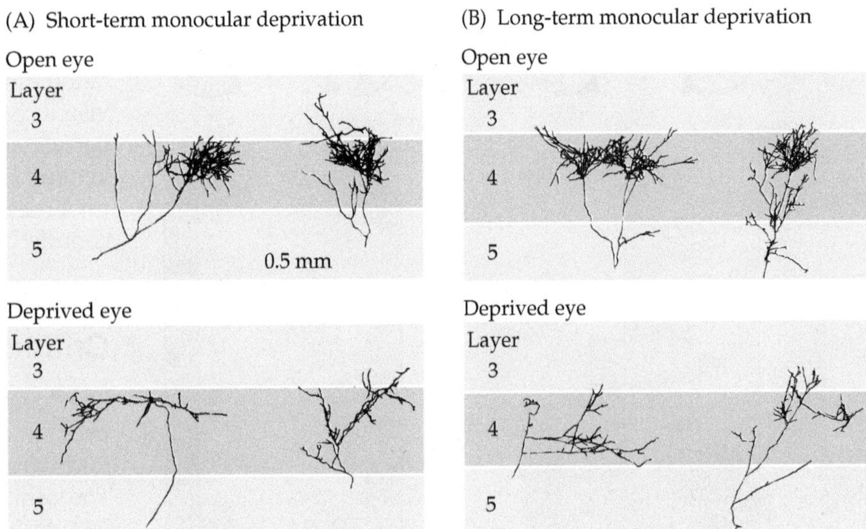

FIGURE 25.8 Effects of monocular deprivation on arborizations of LGN axons in the visual cortex. (A) After only a week of monocular deprivation during the critical period, axons terminating in layer 4 of the primary visual cortex from LGN neurons driven by the deprived eye have greatly reduced numbers of branches compared with those from the open eye. (B) Deprivation for longer periods does not result in appreciably larger changes in the arborization of geniculate axons. (After Antonini and Stryker, 1993.)

levels that approximate animals reared in a typical environment (neither dark reared nor monocularly deprived).

Experiments using techniques that label individual axons from distinct layers in the LGN have shown in greater detail what happens to the arborizations of individual LGN neurons in layer 4 of visual cortex after visual deprivation (Figure 25.8). At the level of single axons, loss of cortical territory related to the deprived eye and concomitant expansion of the open eye's territory are reflected in decreased size and complexity of the arborizations of LGN neurons related to the deprived eye, and increased growth and complexity of the arborizations related to the open eye (i.e., those arising from LGN cells innervated by the open eye). Individual axon arbors can be substantially altered after as little as 1 week of unequal deprivation. This finding highlights the ability of developing thalamic and cortical neurons to rapidly remodel their connections (presumably making and breaking synapses) in response to environmental circumstances during the critical period.

Manipulating Competition

To specifically test the role of correlated activity in driving the competitive postnatal rearrangement of cortical connections, it is necessary to create a situation in which activity levels in each eye remain the same but the correlations between the two eyes are altered. This circumstance can be created in experimental animals by cutting one of the extraocular muscles in one eye. This condition, in which the two eyes can no longer be aligned, is called **strabismus** (a condition that is also recognized clinically in children; see below). The major consequence of strabismus is that objects in the same location in visual space no longer stimulate corresponding points on the two retinas at the same time. As a result, differences in the visually evoked patterns of activity between the two eyes are far greater than normal. Unlike monocular deprivation, however, the overall amount of activity in each eye remains roughly the same; only the correlation of activity arising from corresponding retinal points is changed. Accordingly, the anatomical pattern of ocular dominance columns in layer 4 of cats in which input from both eyes remains active but highly asynchronous is *sharper* than normal. Apparently, the total independence of the two eyes further enhances the correlations between ipsilateral versus contralateral activity. In addition, ocular asynchrony prevents the binocular convergence that normally occurs in cells above and below layer 4; ocular dominance histograms from such animals show that cells in *all* layers are driven exclusively either by one eye or the other (Figure 25.9). Evidently, strabismus not only accentuates the competition between the two sets of thalamic inputs in layer 4, but also prevents binocular interactions in the other layers, which are mediated by local connections originating from cells in layer 4.

Binocular Competition and Orientation Tuning for Binocular Vision

The relationship between action potential activity correlated by eye of origin and the acquisition of other physiological properties that define neuronal identity and capacity for information processing in the primary visual cortex has recently been better defined in experiments that demonstrate the influence of binocular competition on establishing orientation preference in individual cortical neurons (see Chapter 12). Prior to the onset of the critical period, there is little or no correlation between relatively broad orientation sensitivities in cortical neurons driven by both eyes: The maximal response to apparent preferred

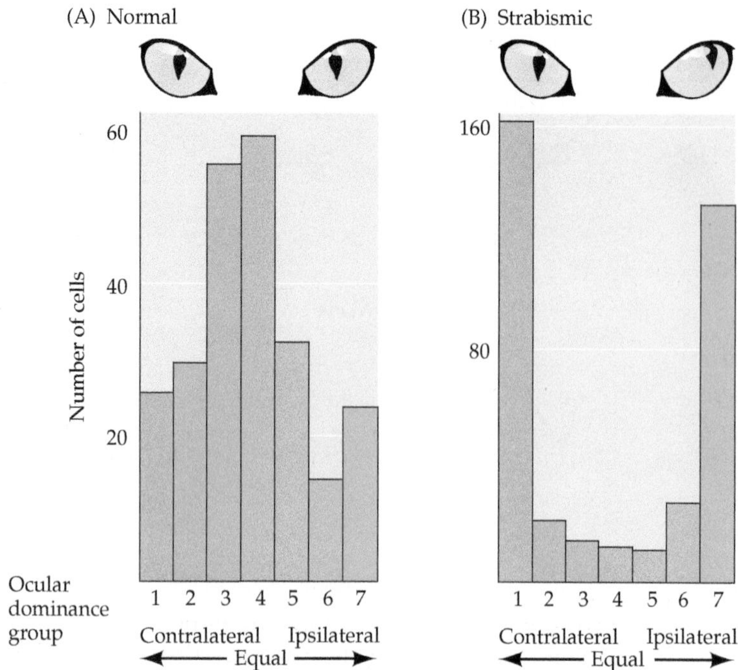

FIGURE 25.9 Ocular asynchrony prevents binocular convergence. Ocular dominance histograms obtained by electrophysiological recordings in normal adult cats (A) and adult cats in which strabismus was induced during the critical period (B). The number of binocularly driven cells (groups 3, 4, and 5) is sharply decreased as a consequence of strabismus; most of the cells are driven exclusively by stimulation of one eye or the other. This enhanced segregation of the inputs presumably results from the greater discrepancy in the patterns of activity between the two eyes as a result of surgically interfering with normal conjugate vision. This pathological state is thought to enhance the relative degree of correlation within inputs from each eye and to decrease the possibility of correlation between eye inputs. (After Hubel and Wiesel, 1965.)

orientations is fairly low, and the orientations themselves are dissimilar. As visual system development progresses toward the start of the critical period for visual-experience-evoked plasticity, the magnitude and frequency of the maximal responses to oriented stimuli in a single cell driven by both the right and left eyes increase dramatically. Their orientation preferences, however, remain dissimilar. The increased correlation of visually evoked stimuli, presumably correlated based upon coincident activation of retinal ganglion cells by binocularly coherent stimuli with identical orientations, leads to the matching of orientation tuning of the right and left eye inputs to single cortical binocularly driven neurons (Figure 25.10). This process of shaping the circuitry for binocular neurons so that orientation information is singularly represented can be modified by the same peripheral manipulations that change ocular dominance. Thus, if one eye or the other is closed during the critical period (monocular deprivation; see above), the matching of orientation tuning of binocular inputs does not occur, and cannot be restored once the closed eye is opened after the end of the critical period. In contrast, closing one eye for an extended time after the critical period is over has no effect on the matching of orientation tuning of the right and left eye inputs to single visual cortical binocular neurons. This suggests that correlated activity is a key determinant of not only general binocular circuitry, but of the circuitry that underlies feature detection (such as oriented bars or contrast and edges) in the visual system.

Amblyopia, Strabismus, and Critical Periods for Human Vision

Developmental phenomena in the visual systems of experimental animals accord with clinical problems seen in children who have experienced similar deprivation. The loss of acuity, diminished stereopsis, and problems with fusion that arise from early deficiencies of visual experience are called **amblyopia** (Greek, "dim sight"). These functional difficulties are all believed to reflect the essential contribution made by normal binocular input and competition for defining, in an experience-dependent manner, the cortical circuitry necessary for binocular vision and depth perception, during the critical period for visual cortical development.

In humans, amblyopia is most often the result of strabismus. Depending on the extraocular muscles affected, the misalignment can produce convergent strabismus, called **esotropia** ("crossed eyes"); or divergent strabismus, called **exotropia** ("wall eyes"). These alignment errors, both of which produce double vision (which describes a lack of proper binocular fusion as well as depth perception), are surprisingly common, affecting about 5% of children. In some of these individuals, the response of the visual system is to suppress input from one eye by mechanisms that are not completely understood but which are thought to reflect competitive interactions during the critical period. Presumably, the inputs to the LGN from the eye that is more optimally aligned are competitively advantaged. Thus, the corresponding LGN inputs are accorded more territory in the visual cortex. Functionally, the suppressed eye eventually comes to have very low acuity that may render an individual effectively blind in that eye. Early surgical correction of ocular misalignment in strabismic children (by adjusting the lengths of extraocular muscles during the critical period) has become an essential treatment to correct strabisimus and preserve normal vision.

Another cause of visual deprivation in humans is cataracts, which can be caused by several congenital conditions and which render the lens or cornea opaque. Diseases such as onchocerciasis ("river blindness," the result of infection by the parasitic nematode *Onchocerca volvulus*) and trachoma (caused by the parasitic bacterium *Chlamydia trachomatis*) affect millions of people in tropical regions. These diseases can lead to cataracts. A cataract or corneal occlusion in one

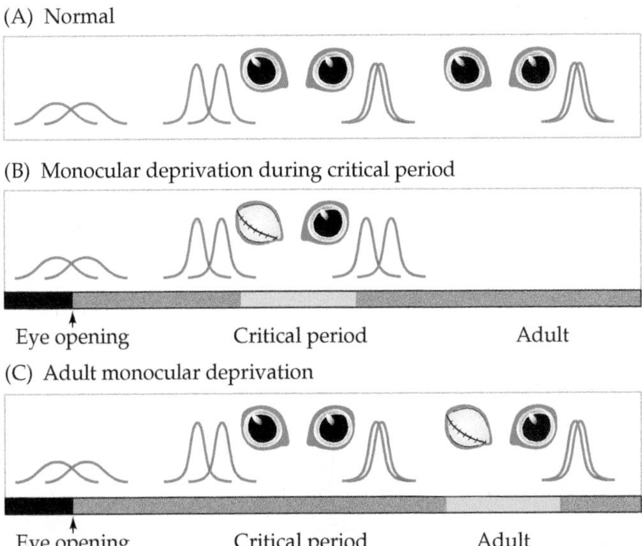

FIGURE 25.10 Binocular competition during the critical period aligns orientation tuning in binocularly innervated cortical neurons. (A) At eye opening and the onset of visual experience in a typically developing animal are the orientation tuning curves for responses in a single cell to excitation in a normal left eye (red) and right eye (blue). The responses are completely uncorrelated. Thus, this binocularly innervated primary visual cortical neuron is sensitive to two very different orientations depending on which eye is stimulated. Such incongruent sensitivity would likely make information processing much less efficient. As postnatal development proceeds after eye opening, the responses of the cortical neuron increase in amplitude; however, the incongruence in their orientation tuning remains as the critical period begins. After the critical period, if binocular competition proceeds uninterrupted, the orientation tuning of binocularly driven single cortical neurons now matches. (B) When normal binocular competition is disrupted during the critical period (but not before or after) using transient monocular deprivation (MD, shown here as reversible lid closure of the left eye), the lack of binocular competition results in a failure to match orientation tuning of binocular inputs to single visual cortical neurons. (C) As a control, after the same transient monocular deprivation in an adult after the close of the critical period, the orientation congruence of binocular inputs to single cortical neurons does not change. (After Wang et al., 2010.)

eye is functionally equivalent to monocular deprivation in experimental animals. Left untreated in children, this ocular defect also results in irreversible damage to visual acuity in the deprived eye. If either the cataract or corneal occlusion is remedied by removing and replacing the biological lens with an artificial lens or removing the damaged cornea and providing a transplant before about 4 months of age, however, the consequences of this monocular deprivation are largely avoided. As expected from Hubel and Wiesel's work, bilateral cataracts or corneal occlusions—which are similar to binocular deprivation in experimental animals—produce less dramatic deficits even if treatment is delayed. As predicted by detailed animal experiments, unequal competition during the critical period for normal vision is more deleterious than the complete disruption of visual input that occurs with binocular deprivation. Correction of unequal competitive interactions during the critical period based upon the pioneering work of Hubel and Wiesel and their colleagues has restored or preserved visual function in countless children.

In keeping with the findings in experimental animals, the visual abilities of individuals monocularly deprived of vision after the close of the critical period (for example, by cataracts or corneal scarring common in adults) are much less compromised, even after decades of deprivation. When vision is restored in such individuals, there may be difficult psychological consequences described by the neurologist Oliver Sacks and then adapted as the play *Molly Sweeney* by the Irish playwright Brian Friel. Nor is there any evidence of anatomical change in this circumstance. For instance, a patient whose eye was surgically removed in adulthood showed normal ocular dominance columns when his brain was examined post mortem many years later. Thus, all of the predictions of decades of basic experiments defining critical period mechanisms and how they can be disrupted has led to a clear understanding of several visual disorders and their prevention or treatment.

Cellular and Molecular Regulation of Critical Periods

Critical periods must ultimately be defined by molecular mechanisms that can transduce activity-mediated synaptic signaling and modify connections. These molecular events must make activity and experience-dependant modifications permanent within a distinct developmental window. Furthermore, these mechanisms must be regulated so that, the system is refractory to additional substantial changes once the critical period has ended. Clearly, the steps that initiate these processes must rely on signals generated by synaptic activity associated with sensory experience, perceptual and cognitive processing, or motor performance. Neurotransmitters and several other signaling molecules, including neurotrophic factors, are obvious candidates for initiating the changes that occur with correlated or repeated activity, including that driven by experience. Indeed, mice that lack genes for several neurotransmitter-synthesizing or degrading enzymes exhibit changes in experience-dependent plasticity of the visual cortex and other cortical regions. Without doubt, the effects of correlated excitatory neurotransmission, particularly that via glutamate transduced by NMDA receptors. NMDA-Rs are capable of detecting local correlated synaptic activity, initiating Ca^{2+} influx at the postsynaptic specialization, and thus modifying signaling in the postsynaptic cell (see Chapter 8). Indeed, NMDA-R signaling in the context of the developing nervous system

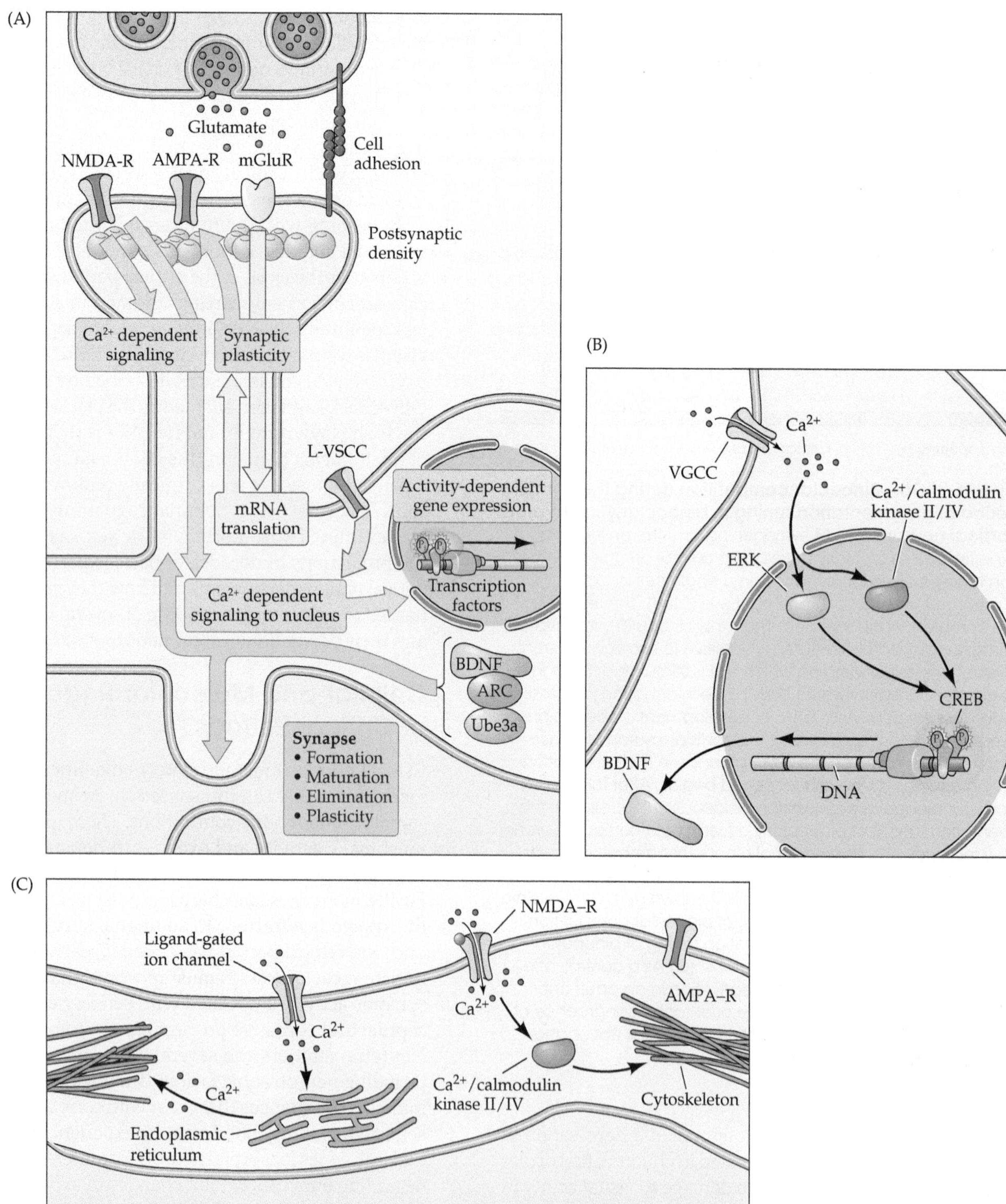

has been shown to be a key mechanism for the molecular transduction of activity and experience driven changes during critical periods. Additional ionotropic glutamate receptors—particularly the AMPA receptor (AMPA-R) which is permeable to Na^+ and K^+ but not Ca^{2+} as well as members of the mGluR metabotropic glutamate receptors (see Chapter 8)—also establish or reinforce depolarization due to correlated activity. This additional depolarization further modifies other voltage-gated channels (including the NMDA-R and L-type voltage-sensitive Ca^{2+} channels

◀ **FIGURE 25.11 Transduction of electrical activity into cellular change via Ca^{2+} signaling.** (A) Summary of the signal transduction essential for critical period synaptic plasticity, primarily at the postsynaptic specialization. Excitatory activity, which is experience-dependent after birth and relies on the appropriate maturation of peripheral sensory relay pathways, is proportional to the level of glutamate release from the presynaptic terminal. In turn, glutamate binds to the ionotropic receptors NMDA-R and AMPA-R, as well as to the metabotropic glutamate receptor, mGluR. The consequence of NMDA-R and AMPA-R activation via glutamate binding is depolarization that favors the influx of Ca^{2+} via the NMDA-R and the initiation of Ca^{2+}-dependent signaling that can influence local cytoskeletal integrity and receptor distribution and stability. This aspect of structural modulation to translate electrical activity to cellular change also includes the modulation of Ca^{2+}-dependent cell adhesion to either maintain or disrupt the relationship between pre-and postsynaptic sites. Signaling through mGluR activates second-messenger cascades that rely on mTOR activation to modulate mRNA translation into protein, or to activate ERK signaling, leading to altered nuclear gene transcription. (B) Correlated or sustained activity leads to increased Ca^{2+} conductances and increased intracellular Ca^{2+} concentration, which results in activation of CaMKII or CaMKIV as well as ERK, and their subsequent translocation to the nucleus. ERK and CaMKII/IV then activate Ca^{2+}-regulated transcription factors such as CREB, as well as other chromatin-binding proteins (not shown). The target genes for activated CREB may include neurotrophic signals such as BDNF, which when secreted by a cell may help stabilize or promote the growth of active synapses on that cell. (C) Local increases in Ca^{2+} signaling in distal dendrites due to correlated or sustained activity may lead to local increases in Ca^{2+} concentration that modify cytoskeletal elements (actin- or tubulin-based structures), perhaps through the activity of kinases such as CaMKII/IV operating in the cytoplasm rather than the nucleus. Changes in these cytoskeletal elements lead to local changes in dendritic structure. In addition, increased local Ca^{2+} concentration may influence local translation of transcripts in the endoplasmic reticulum, including transcripts for neurotransmitter receptors and other modulators of postsynaptic responses. Increased Ca^{2+} may also influence the trafficking of these proteins, their insertion into the postsynaptic membrane, and their interaction with local scaffolds for cytoplasmic proteins (see Figure 23.10). (A after Ebert and Greenberg, 2013; B,C after Wong and Ghosh, 2002.)

[L-VSCCs]), regulates Ca^{2+} influx, initiates Ca^{2+}-dependent second-messenger cascades that lead to altered cell adhesion signaling, influences cytoskeletal integrity, and ultimately changes gene expression that can reinforce or weaken a synaptic contact (Figure 25.11A). Thus, Hebb's postulate anticipated the discovery of a molecular signal transduction mechanism that is the cellular determinant of activity-dependent changes in connectivity.

The signaling intermediates that respond to the altered Ca^{2+} concentration associated with synaptic activity during critical periods are key mediators of the cellular and molecular basis of critical period plasticity. Ca^{2+}/calmodulin-dependent protein kinase type II (CaMKII) phosphorylates the amino acids serine and threonine (thus CaMKII is referred to as a serine-threonine kinase) on target proteins at the postsynaptic site. These target proteins include the AMPA-R, and this modification further reinforces depolarization sensitivity that mediates additional synaptic strengthening. In addition, the ubiquitin ligase, Ube3a, plays an important role in degrading ARC, the *a*ctivity *r*egulated *c*ytoskeletal protein, which regulates, among other targets, membrane turnover and trafficking of the AMPA-R. Cell adhesion signaling through neurexins (presynaptic) and neuroligins (postsynaptic) regulates the integrity of cytoskeletal scaffolding proteins, including postsynaptic density protein 95 (PSD95), the guanylate kinase protein GKAP, the scaffolding protein SHANK (*SH*3- and *ank*yrin repeat domains protein), and the activity-related cystokeletal protein HOMER (Figure 25.11B). Cytoskeletal integrity at the pre- and post-synaptic membrane is essential for retaining the local accumulation of proteins for synaptic vesicle fusion (pre-synaptic) and receptor concentration (post-synaptic and pre-synaptic) at the synapse. If the synaptic cytoskeleton is disrupted the efficiency and fidelity of synaptic transmission declines. Such changes can substantially diminish the effectiveness of activity and experience in shaping maximally adaptive neural circuits and behaviors. Finally, signaling mediated by metabotropic glutamate receptor (mGluR) engages the tuberosclerosis complex genes 1 and 2 (*TSC1* and *TSC2*), originally identified as tumor suppressor genes. *TSC1* and *TSC2* ultimately regulate M-TOR (mammalian target of rapamycin), a protein that controls bioenergetic homeostasis, and thus influence protein synthesis, stability, and turnover. The consequence of the activity of this combination of molecular mediators of excitatory synaptic transmission is to modify postsynaptic integrity and thus either strengthen the physical synaptic specialization or weaken it for eventual elimination.

A second key pathway for the molecular regulation of critical periods is via neurotrophins (see Chapter 23), particularly the brain-derived neurotrophin BDNF. BDNF in this instance is primarily available via secretion from the afferent synaptic input rather than the synaptic target (this can also be the case for other forms of neurotrophic signaling; see Chapter 23). BDNF, via the TrkB receptor tyrosine kinase, initiates a signaling cascade that depends on ras and RAF intermediates. The RAF serine-threonine kinase phosphorylates targets that lead to activation of the extracellular signal-regulated kinase (ERK) pathway (Figure 25.11C), while ras (a small GTPase) binds and catalyzes the hydrolysis of GTP in its active form. Ultimately, these two second-messenger-regulated signaling cascades can activite the CREB transcription factor as well as related activity-dependent transcriptional regulators. This activation leads to changes in gene expression that record activity-dependent plasticity as an altered transcriptional state in the target cell.

The genes for many critical period–related signal transduction molecules as well as cytoskeletal and synaptic scaffold proteins (SHANK, neurexins and neuroligins, TSC1 and TSC2, as well as the Fragile X-mental retardation protein [FMRP]) are also, perhaps not surprisingly, targets for mutation associated with a range of neurodevelopmental disorders, including intellectual disability and autistic spectrum disorders. In addition, local and global effects of modified neurotrophin signaling in genetically engineered mice (see Chapter 1) include altered synaptic plasticity during critical periods; however, a genetic association between neurotrophic signaling and neurodevelopmental disorders in humans has not been established. Evidence from studies in a number of animal models shows that when these signaling molecules are manipulated genetically or pharmacologically, some critical period phenomena are altered. One major target of all these molecular signaling processes is apparently the network of local inhibitory connections made by GABAergic neurons. Regulation of the number and placement of local inhibitory synapses, as well as the expression of GABA receptors at postsynaptic sites, seems to be exquisitely sensitive to changes in levels of electrical activity during early postnatal life (see Figure 25.3). Thus, the activity- and experience-dependent regulation of inhibitory connectivity, and the excitatory/inhibitory balance (E/I balance) it establishes in neuronal circuits, has become a major focus for understanding the molecular and cellular mechanisms of activity- and experience-dependent postnatal construction of neural circuits.

Evidence for Critical Periods in Other Sensory Systems

Although the neural basis of critical periods has been most thoroughly studied in the mammalian visual system, similar phenomena exist in the auditory, somatosensory, and olfactory systems as well as in motor pathways. Experiments on the role of auditory experience and neural activity in owls (which use auditory information to localize prey) indicate that neural circuits for auditory localization, like mammalian visual circuits, are shaped by experience (Table 25.1). Thus, deafening an owl or altering its neural

TABLE 25.1 Critical Periods and Molecular Regulators for Some Neural Systems

System	Species[a]	Critical period (postnatal)[b]	Confirmed molecular regulators[c]
Neuromuscular junction	Mouse	Prior to day 12	ACh
Cerebellum	Mouse	Days 15–16	NMDA, mGluR1, G_q, PLCβ, PKCγ
Lateral geniculate	Mouse, ferret, cat	Prior to day 10	ACh, cAMP, MAOA, NO, MHC1, CREB nucleus layers
Ocular dominance	Cat, rat, mouse, ferret	3 weeks–months	GABA, NMDA, PKA, ERK, CaMKII, CREB, BDNF, tPA, protein synthesis, NE, ACh
Orientation bias	Cat, mouse	Prior to day 28	NR1, NR2A, PSD95
Somatosensory map	Mouse, rat	Prior to day 7–16	NR1, MAOA, $5HT_{1B}$, cAMP, mGluR5, PLCβ, FGF8
Tonotopic map (cortex)	Rat	Days 16–50	ACh
Absolute pitch	Human	Before 7 years	Unknown
Taste, olfaction	Mouse	None	GABA, mGLuR2, NO, neurogenesis
Imprinting	Chick	14–42 hours	Catecholamines
Stress, anxiety	Rat, mouse	Prior to day 21	Hormones, $5HT_{1A}$
Slow-wave sleep	Cat, mouse	Days 40–60	NMDA
Sound localization	Barn owl	Prior to day 200	GABA, NMDA
Birdsong	Zebra finch	Prior to day 100	GABA, hormones, neurogenesis
Language	Human	0–12 years	Unknown

[a]Primary research species for elucidation of molecular mechanisms.

[b]Although the details vary from system to system and from species to species, all critical periods are limited to a definite window of time during early postnatal (or posthatching) life and are complete before the onset of sexual maturation.

[c]Molecules known to regulate critical periods include neurotransmitters, their receptors, and related signaling proteins.

After Hensch, 2004.

activity during early postnatal development compromises the bird's ability to localize sounds and capture prey and can alter the neural circuits that mediate these capacities. In addition, this auditory critical period in the owl is coordinated with critical periods for vision so that the two sensory modalities can operate efficiently together to enhance detection and capture of prey. The development of song in many bird species provides another example of a critical period for auditory function, as well as critical periods for motor control of a complex behavior (see Chapter 24). In the somatosensory system, there is a critical period during which cortical maps can be changed by experience. In mice or rats, for instance, the anatomical patterns of "whisker barrels" in the somatosensory cortex (see Chapter 9) can be altered by abnormal sensory experience (or by removing subsets of sensory receptors such as whiskers) during a narrow window in early postnatal life. In the olfactory system, behavioral studies (outlined in Chapter 15) indicate that exposure to maternal odors for a limited period can alter the ability to respond to such odorants—a change that can persist throughout life.

Clearly, critical periods are common in the development of sensory perception and related motor skills. The primary evidence for these critical periods comes from deprivation experiments analogous to those done in the visual system, complemented by analysis using pharmacological approaches or genetically modified animals in which major neurotransmitter synthetic pathways are disabled, essential neurotransmitter receptors such as NMDA-Rs are lost, or major signaling molecules (e.g., the Ca^{2+}/calmodulin kinases, BDNF, the neurotrophin receptors, or in the case of sexually dimorphic song behaviors in birds, gonadal steroids such as estrogen and testosterone) have been disrupted. In each instance, these molecular modifications of synaptic signaling result in changes in the duration or efficiency of critical period-dependent plasticity.

Language Development: A Critical Period for a Distinctly Human Behavior

The series of experimental and clinical observations in the visual system of cats, monkeys, and humans described above provides the most complete description of the phenomenon of the critical period and its underlying physiological mechanisms. These studies raise the question of whether critical periods can be documented for even more complex behaviors, including cognitive function and, in humans, language. While cellular and physiological evidence is not yet available, several behavioral observations have defined a critical period for language acquisition and production.

Exposure to language from birth onward is essential for the development of appropriate capacity to comprehend and produce meaningful communication. The various forms of early language exposure, including the "baby talk" that parents and other adults often use to communicate with infants and small children, may actually serve to emphasize important perceptual distinctions that facilitate proper language production and comprehension. To be effective, this linguistic experience must occur in early life. The requirement for perceiving and practicing language (as opposed to specific auditory, visual, or motor skills) during a critical period is apparent in studies of language acquisition in congenitally deaf children, whose language acquisition (i.e., sign language) relies on seeing and moving the hands and fingers (the equivalent of spoken and heard language) rather than on listening and moving the lips, tongue, and larynx. Whereas most hearing and speaking babies begin producing speechlike sounds ("babbling") at about 7 months, congenitally deaf infants show obvious deficits in their early vocalizations and fail to develop language if not provided with an alternative form of symbolic expression such as sign language (see Chapter 33). If, however, deaf children are exposed to sign language at an early age (from approximately 6 months onward, which is particularly likely for the children of deaf, signing parents), they begin to "babble" with their hands, just as a hearing infant babbles audibly (Figure 25.12). This manual babbling suggests that, regardless of the modality, early experience shapes language behavior. Children who have acquired speech but lose their hearing before puberty also

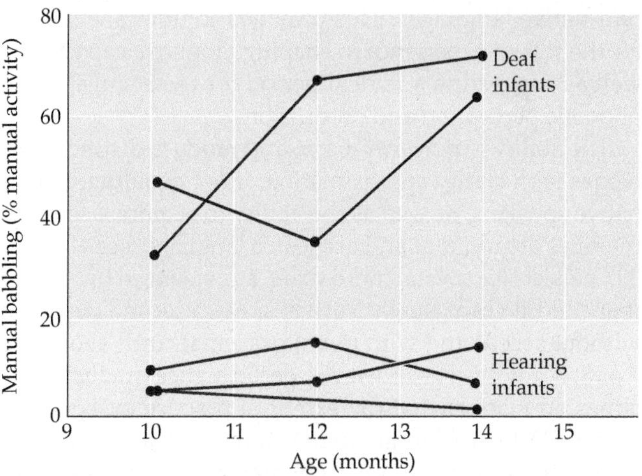

FIGURE 25.12 Manual "babbling" in deaf infants. The two deaf infants studied here were raised by deaf, signing parents. Babbling was judged by scoring hand positions and shapes that showed some resemblance to the components of American Sign Language and comparing these scores with those of manual babble in three hearing infants. In these deaf infants with signing parents, meaningful hand shapes increased as a percentage of manual activity between ages 10 and 14 months. Hearing children raised by hearing, speaking parents do not produce similar hand shapes. (After Petitto and Marentette, 1991.)

suffer a substantial decline in spoken language, presumably because they are unable to hear themselves or others talk and thus lose the opportunity to refine their speech by auditory feedback during the final stages of the critical period for language.

The auditory details of the language an individual hears during early life shape both the perception and production of speech. Many of the thousands of human languages and dialects use appreciably different speech sounds called *phonemes* to produce spoken words (examples are the phonemes *ba* and *pa* in English; see Chapter 33). Very young human infants can perceive and discriminate between differences in *all* human speech sounds, and are not innately biased toward phonemes characteristic of any particular language. However, this universal perceptual capacity does not persist. For example, adult Japanese speakers cannot reliably distinguish between the *r* and *l* sounds in English, presumably because this phonemic distinction is not made in Japanese and thus is not reinforced by experience during the critical period. Nonetheless, 4-month-old Japanese infants can make this discrimination as reliably as 4-month-olds raised in English-speaking households (as indicated by increased suckling frequency or head turning in the presence of a novel stimulus). By 6 months of age, however, infants begin to show preferences for phonemes in their native language over those in foreign languages, much as deaf infants do for moving digits that suggest signs. By the end of their first year, infants no longer respond robustly to phonetic elements that are peculiar to non-native languages. This provides additional evidence for the role of experience in shaping language capacity, as well as suggesting a critical period for the acquisition of phonetic perception and production.

The ability to perceive, learn, and produce distinct phonemes with clarity approximating, if not equaling, that of native speakers, as well as the ability to acquire a sense of the rules of grammar and usage in a language (see Chapter 33), persists for several more years, as evidenced by the fact that children can usually learn to speak a second language without accent and with fluent grammar until about age 7 or 8. After this age, however, performance gradually declines no matter what the extent of practice or exposure (Figure 25.13A). Changes in the patterns of activity in language regions of the brain in children versus adults suggest that the relevant neural circuits may undergo functional or structural modifications during the critical period for language (Figure 25.13B). Comparisons of patterns of activity in children ages 7 to 10 with the patterns of adults performing the same specific word-processing tasks suggest that different brain regions are activated for the same task in children versus adults. While the significance of such differences is not clear—they may reflect anatomical plasticity associated with critical periods, or distinct modes of performing language tasks in children versus adults—there is nevertheless an indication that brain circuits change to accommodate language function during early life, with very different patterns of activity seen in adulthood. Thus, although the cellular basis is difficult to study, it is likely that activity elicited by language experience leads to rearrangement of connectivity, analogous to the much better documented changes shaped by experience during the critical period in the visual cortex.

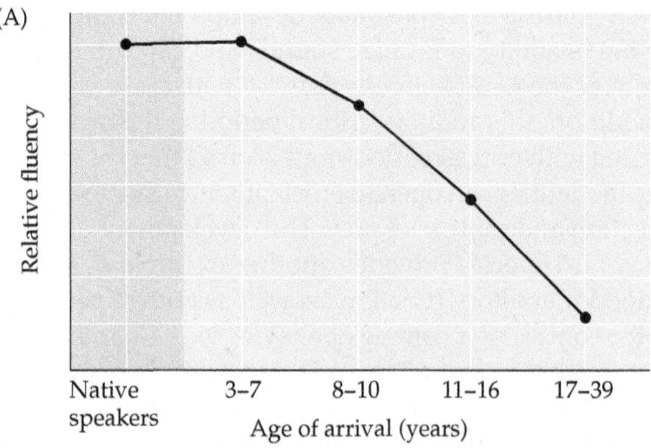

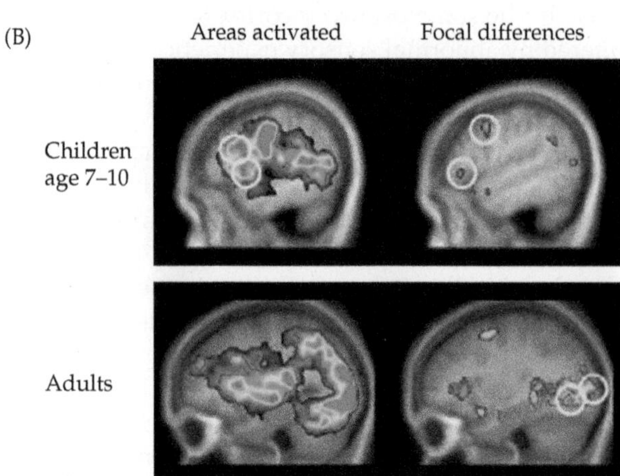

FIGURE 25.13 Learning language. (A) A critical period for learning language is demonstrated by the decline in language ability (fluency) of non-native speakers of English as a function of their age upon arrival in the United States. The ability to score well on tests of English grammar and vocabulary declines from approximately age 7 onward. (B) Maps derived from fMRI of adults and children performing visual word-processing tasks. Images are sagittal sections, with the front of the brain toward the left. The top row shows the range of active areas (left) and foci of activity based on group averages (right) for children ages 7 to 10. The bottom row shows analogous results for adults performing the same task. The differences in regions of maximal activation (shown in red in the images at left; highlighted by white circles in the right-hand images) indicate changes in either the circuitry or the mode of processing and performing the same task in children versus adults. (A after Johnson and Newport, 1989; B after Schlaggar et al., 2002.)

Human Brain Development, Activity-Dependent Plasticity, and Critical Periods

The advent of high-resolution, noninvasive imaging techniques has made it possible to reevaluate some basic aspects of human brain development in the context of physiological and behavioral understanding of critical periods and accompanying changes in neuronal growth that have been discerned in animal experiments. In the late 1980s, studies in multiple cortical areas in the rhesus monkey demonstrated what had been suggested by less detailed analysis of synapses in the human visual cortex: The number of synapses throughout the cortex (not just in the visual cortex) increased during prenatal and a limited period of postnatal life, declined during a protracted period that included much of adolescence, and reached a steady state in early adulthood (Figure 25.14). This pattern of initial increase followed by a decline in synapse numbers indicated that critical periods may be mediated first by local growth of neural elements in an activity-dependent manner, followed by a subsequent elimination of some synapses—perhaps the cellular equivalent of the consequences of Hebbian competition—and the selective growth and stabilization of other synapses (see Figure 25.1). These quantitative observations suggest a cellular basis for activity-dependent plasticity and critical period phenomena throughout the cerebral cortex.

This suggestion received compelling support from a remarkable series of studies begun in the late 1990s and not completed until the end of the first decade of the new millennium. A group of investigators at the National Institute of Mental Health did longitudinal MRI scans of the developing brain in 13 children starting at age 4 and ending at age 20 to measure the growth of the gray matter (the location of cell bodies, axon terminal branches, dendrites, and synapses) throughout the cortical mantle in individual children over time (see Figure 25.2). The study's results paralleled those predicted from the analyses of synapses in rhesus monkeys: Gray matter (the location of synapses in the cerebral cortex) grows throughout the cortex during early life, then declines slightly over a protracted period of late childhood and early adolescence (Figure 25.15A). There are some important regional distinctions in this overall trajectory of early growth and subsequent loss of gray matter volume. Primary sensory cortices appear to

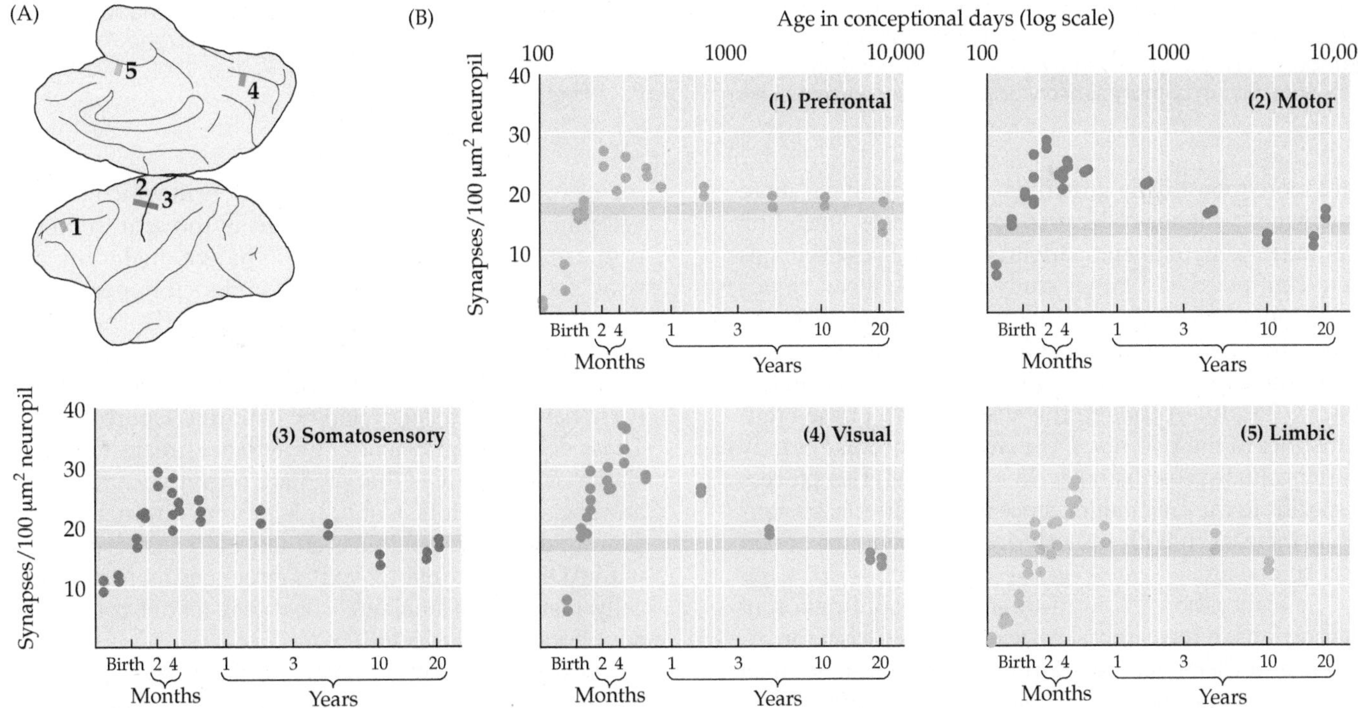

FIGURE 25.14 Synapse addition and elimination in rhesus monkey cortex. (A) Location of brain regions where synapse density was measured between mid-gestation and 20 years of age. (B) Rapid addition followed by gradual decline of synapse density in the cerebral cortex. Age has been converted to a logarithmic scale of "conceptional days" in order to fit the entire life span onto one graph. Synapse addition apparently continues through early life, gradually declines throughout most of adolescence, and reaches a steady state (shaded horizontal bar) after puberty (between 2 and 3 years of age in the rhesus monkey). (After Rakic et al., 1986.)

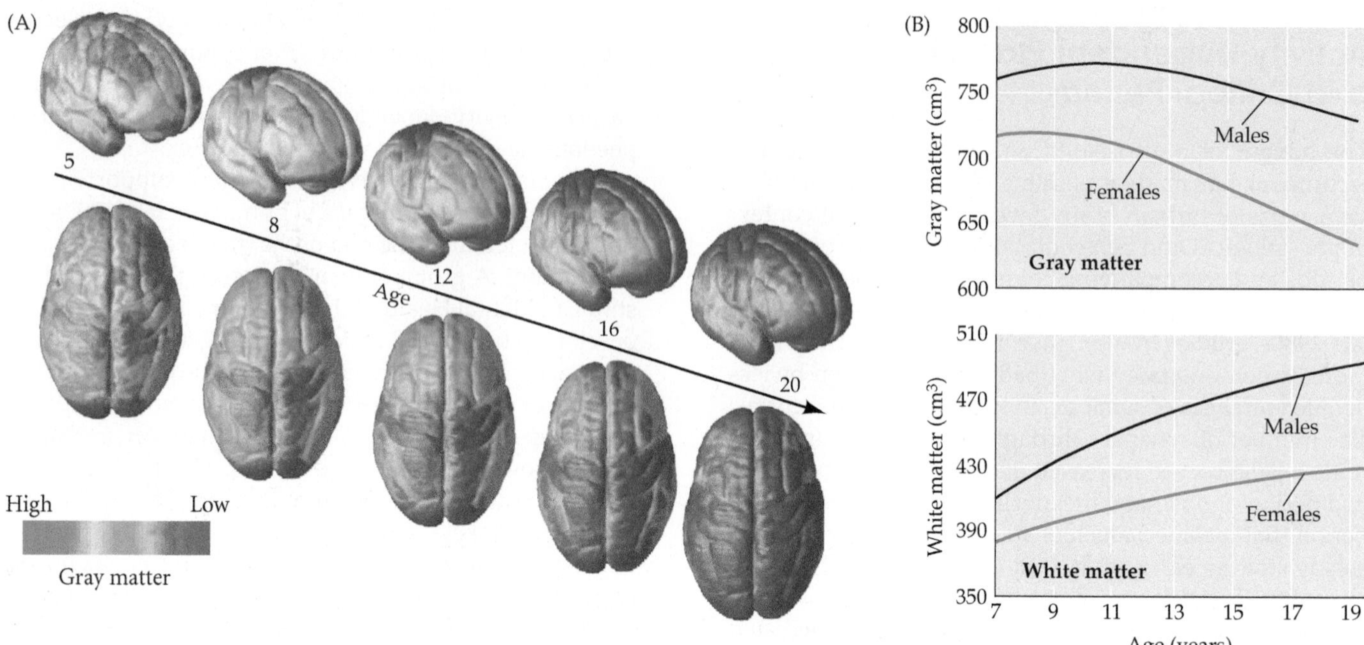

FIGURE 25.15 Increased and decreased gray matter volumes parallel critical periods in humans. (A) A composite map of cortical gray matter volume growth (red/yellow/green) and decline (blue/purple), based on longitudinal MRI scans of 13 typical individuals from 5 years of age until 20 years of age. There is initial growth of gray matter throughout the cortex, especially in primary sensory and motor regions, followed by gradual decline. There is some heterogeneity in the timing and rate of decline in primary sensory and motor versus association areas. (B) Mean cortical gray matter (above) and white matter (below) volume in males and females from a cross-sectional study (i.e., means computed from multiple individuals rather than from the same individual at each age). Although the absolute growth of the male and female brain differs, gray matter volume increases and then decreases in roughly the same way. White matter volume, in contrast, increases throughout early childhood and adolescence. (A from Gogtay et al., 2004; B after Lenroot et al., 2007.)

have more robust early growth,; however, decline is more prolonged in higher-order association cortices, including prefrontal, temporal, and parietal regions. These changes were confirmed in a larger study in which mean values were generated from multiple individuals at each age rather than for one individual followed longitudinally. In these studies, males and females were analyzed separately (Figure 25.15B). Although the absolute size of the brain is distinct in the two sexes, the overall trajectory of gray matter volume growth and decline is parallel. This process is specific to the gray matter, since white matter has a continual increase in both sexes during the same time period of postnatal life.

The elaboration followed by selective elimination of connections in the cerebral cortex—inferred from the increase and then decrease in gray matter volumes—may indeed underlie the remarkable capacity of the human brain to acquire and refine behavioral capacity from birth through early adulthood. The cessation of this process is intriguingly coincident with the time in life when the process of learning new information and skills becomes increasingly difficult. Moreover, this prolonged process of experience-driven construction of cortical circuits seems to be altered in several disorders in which the development of cortical connections is thought to be the primary target for pathological change, including autism, schizophrenia, and attention deficit hyperactivity disorder (ADHD). In children with ADHD, the rate of cortical growth during early postnatal life is delayed and the overall magnitude of growth is diminished compared with that of typically developing children (Figure 25.16A). These deficits in gray matter growth are greatest in association cortical areas that mediate cognitive, emotional, and social behaviors. Not only is growth diminished and slowed; decline in volume is enhanced. Thus, in children with ADHD, gray matter volume declines more dramatically, resulting in smaller cortical gray matter volumes in adulthood (Figure 25.16B).

These observations of postnatal brain development in living humans allow one to infer (but do not prove) that experience- and activity-dependent mechanisms during critical periods are primarily responsible for much of what is recognized as typical behavioral development, social development, and learning. This implies that the final target (although not necessarily the initial cause) of several behavioral and psychiatric disorders may include the

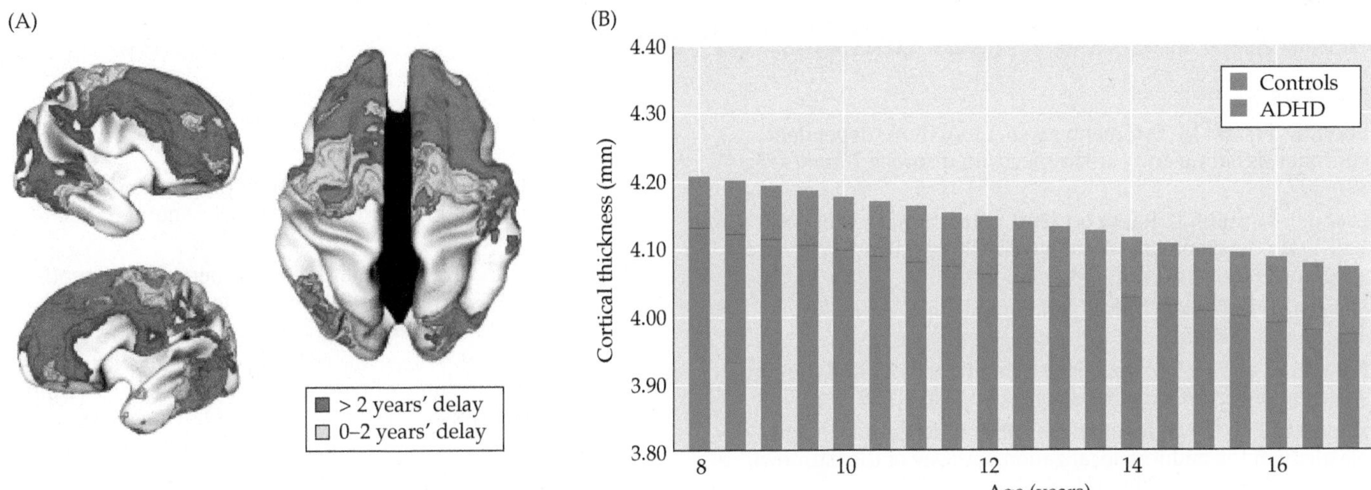

FIGURE 25.16 A behavioral disorder accompanied by altered addition of gray matter volume. (A) Map of cortical regions in which gray matter volume increases more slowly in children with attention deficit hyperactivity disorder (ADHD) than in typically developing children. (B) The rate of decline of gray matter volume is equivalent for children with ADHD and typically developing children. The net result is lower gray matter volumes in adulthood for individuals with ADHD. (After Shaw et al., 2006.)

activity- and experience-dependent processes that shape and—under typical circumstances—optimize those connections that mediate complex behaviors.

Summary

An individual animal's history of interaction with its environment—its "experience"—helps shape its neural circuitry and thus determines subsequent behavior. Experience during specific times in early life, referred to as critical periods, helps shape behaviors as diverse as maternal bonding and the acquisition of language. Correlated patterns of activity are thought to mediate critical periods by stabilizing concurrently active synaptic connections and weakening or eliminating connections whose activity is divergent. Some of this correlated activity depends on excitable properties of neurons that emerge prior to the influence of sensory-evoked, experience-dependent electrical signals, while other correlated activity is established by patterns of excitable changes elicited by sensory inputs or motor behaviors. The cellular and molecular mechanisms implicated in critical periods rely on the activity of several neurotransmitters, receptors, and intracellular signaling cascades that modify cytoskeletal integrity, receptor function and stability, and ultimately gene expression in response to changes in synaptic activity in a target cell. Receptors for the excitatory neurotransmitter glutamate, including the NMDA-R, AMPA-R, and metabotropic glutamate receptor (mGluR), are essential for the transduction of excitatory activity into postsynaptic changes that underlie critical period changes in circuits and behaviors. This transduction is often mediated by Ca^{2+} dependent second messenger systems in the post-synaptic cell activated by ligand or voltage gated channels. The intracellular signaling consequences rely upon the concerted activity of scaffolding proteins, protein kinases, and their targets that regulate the integrity of individual synapses as well as gene transcription, translation and protein stability. Genes for neurotrophins such as BDNF, extracellular matrix components, and neurotransmitter receptors are all targets for altered expression in response to synaptic activity during critical periods. Signaling via BDNF from the pre- to post-synaptic sites is especially important for the modification of gene expression that records activity-dependent change more permanently as transcriptional change. The most accessible and thoroughly studied example of a critical period is that responsible for the establishment of normal vision in mammals, including humans. When typical patterns of activity are disturbed during the critical period in early life (experimentally in animals or by pathology in humans), connectivity in the visual cortex is altered, as is visual function. If not reversed before the end of the critical period, these structural and functional alterations of brain circuitry are difficult or impossible to change. Observations of the addition and elimination of synapses throughout the cerebral cortex in animals, and parallel analysis of the increase and decrease of cortical gray matter volumes where such synapses are made in the brains of children and adolescents, indicate that a full range of human behaviors—including those compromised in conditions such as autism, schizophrenia, and ADHD—may be shaped by activity- and experience-dependent addition and subsequent elimination of synaptic connections during critical periods that begin at birth and end in early adulthood.

ADDITIONAL READING

Reviews

Ebert, D. H. and M. E. Greenberg (2013) Activity-dependent neuronal signaling and autism spectrum disorder. *Nature* 493: 327–337.

Giedd, J. N. and J. L. Rapoport (2010) Structural MRI of pediatric brain development: What have we learned and where are we going? *Neuron* 67: 728–734.

Hensch, T. K. (2004) Critical period regulation. *Annu. Rev. Neurosci.* 27: 549–579.

Katz, L. C. and C. J. Shatz (1996) Synaptic activity and the construction of cortical circuits. *Science* 274: 1133–1138.

Knudsen, E. I. (1995) Mechanisms of experience-dependent plasticity in the auditory localization pathway of the barn owl. *J. Comp. Physiol.* 184: 305–321.

Sherman, S. M. and P. D. Spear (1982) Organization of visual pathways in normal and visually deprived cats. *Physiol. Rev.* 62: 738–855.

Wiesel, T. N. (1982) Postnatal development of the visual cortex and the influence of environment. *Nature* 299: 583–591.

Wong, W. O. and A. Ghosh (2002) Activity-dependent regulation of dendritic growth and patterning. *Nature Rev. Neurosci.* 10: 803–812.

Important Original Papers

Antonini, A. and M. P. Stryker (1993) Rapid remodeling of axonal arbors in the visual cortex. *Science* 260: 1819–1821.

Cabelli, R. J., A. Hohn and C. J. Shatz (1995) Inhibition of ocular dominance column formation by infusion of NT-4/5 or BDNF. *Science* 267: 1662–1666.

Feller, M. B., D. P. Wellis, D. Stellwagen, F. S. Werblin and C. J. Shatz (1996) Requirement for cholinergic synaptic transmission in the propagation of spontaneous retinal waves. *Science* 272: 1182–1187.

Gogtay, N. and 11 others (2004) Dynamic mapping of human cortical development during childhood through early adulthood. *Proc. Natl. Acad. Sci. USA* 101: 8174–8179.

Hanganu, I. L., Y. Ben-Ari and R. Khazipov (2006) Retinal waves trigger spindle bursts in the neonatal rat visual cortex. *J. Neurosci.* 26: 6728–6736.

Horton, J. C. and D. R. Hocking (1999) An adult-like pattern of ocular dominance columns in striate cortex of newborn monkeys prior to visual experience. *J. Neurosci.* 16: 1791–1807.

Huang, Z. J. and 7 others (1999) BDNF regulates the maturation of inhibition and the critical period of plasticity in mouse visual cortex. *Cell* 98: 739–755.

Hubel, D. H. and T. N. Wiesel (1965) Binocular interaction in striate cortex of kittens reared with artificial squint. *J. Neurophysiol.* 28: 1041–1059.

Hubel, D. H. and T. N. Wiesel (1970) The period of susceptibility to the physiological effects of unilateral eye closure in kittens. *J. Physiol.* 206: 419–436.

Hubel, D. H., T. N. Wiesel and S. LeVay (1977) Plasticity of ocular dominance columns in monkey striate cortex. *Phil. Trans. R. Soc. Lond. B* 278: 377–409.

Kuhl, P. K., K. A. Williams, F. Lacerda, K. N. Stevens and B. Lindblom (1992) Linguistic experience alters phonetic perception in infants by 6 months of age. *Science* 255: 606–608.

LeVay, S., T. N. Wiesel and D. H. Hubel (1980) The development of ocular dominance columns in normal and visually deprived monkeys. *J. Comp. Neurol.* 191: 1–51.

Rakic, P. (1977) Prenatal development of the visual system in the rhesus monkey. *Phil. Trans. R. Soc. Lond. B* 278: 245–260.

Rakic, P., J. P. Bourgeois, M. F. Eckenhoff, N. Zecevic, and P. S. Goldman-Rakic (1986) Concurrent overproduction of synapses in diverse regions of the primate cerebral cortex. *Science* 232: 232–235.

Stryker, M. P. and W. Harris (1986) Binocular impulse blockade prevents the formation of ocular dominance columns in cat visual cortex. *J. Neurosci.* 6: 2117–2133.

Wang, B.-S., R. Sarnaik and J. Cang (2010) Critical period plasticity matches binocular orientation preference in the visual cortex. *Neuron* 65: 246–256.

Wiesel, T. N. and D. H. Hubel (1965) Comparison of the effects of unilateral and bilateral eye closure on cortical unit responses in kittens. *J. Neurophysiol.* 28: 1029–1040.

Books

Curtiss, S. (1977) *Genie: A Psycholinguistic Study of a Modern-Day "Wild Child."* New York: Academic Press.

Hubel, D. H. (1988) *Eye, Brain, and Vision*. Scientific American Library Series. New York: W. H. Freeman.

Purves, D. (1994) *Neural Activity and the Growth of the Brain.* Cambridge, UK: Cambridge University Press.

Go to the NEUROSCIENCE 6e Companion Website at **www.oup.com/uk/Purves6e** for Web Topics, Animations, Flashcards, and more.

CHAPTER

26 Repair and Regeneration in the Nervous System

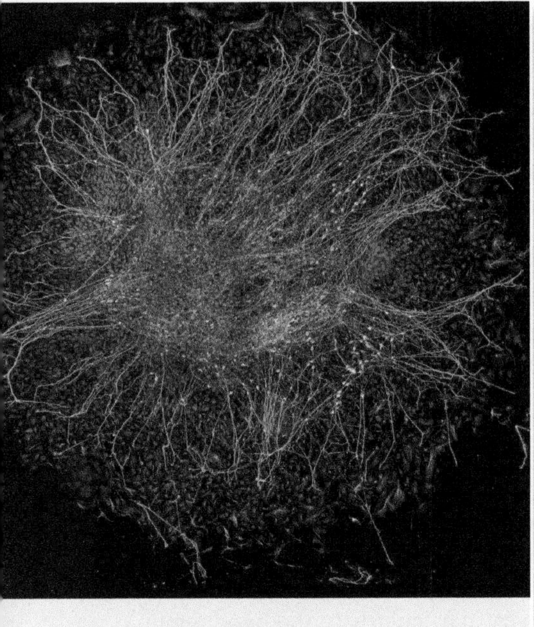

Overview

THE ABILITY OF BRAIN TISSUE TO ALTER, RENEW, OR REPAIR itself (beyond the ongoing molecular and cellular changes associated with synaptic plasticity) is limited. Unlike many other organs—notably the skin, lungs, intestine, and liver—that continuously generate new cells, human brains do not produce many new neurons once the initial complement is established during mid-gestation through early postnatal life. Nevertheless, some nervous system repair does occur after injury. In the periphery, axons can regrow through vacated peripheral nerve sheaths. Using these sheathes as guides, peripheral sensory neurons can eventually reinnervate sensory specializations in the skin, and central motor neuron axons can regrow through the periphery to reinervate synaptic sites on muscles. In contrast, modest recovery seen after brain injury—if there is recovery at all—is usually attributed to reorganization of function using remaining, intact circuits rather than repair of damaged brain tissue. Few brain neurons can grow a new axon if the original axon is severed or injured, nor can neurons replace dendrites lost due to local tissue damage or degenerative disease. At least four barriers impede central nervous system regeneration. First, local injury of brain tissue often leads to neuronal death. Second, several other cell classes, particularly glial cells, actively inhibit axon growth. Third, although neural stem cells are retained in the adult brain, most are constrained in their ability to divide, migrate, and differentiate. And fourth, immune responses in the nervous system, mediated by microglia, astrocytes, and oligodendrocytes, release cytokines that further inhibit extensive regrowth. Nevertheless, in some vertebrate species, these impediments are circumvented in specific regions of the central nervous system. Efforts to understand these exceptions provide a foundation for ongoing research into potential therapies for brain repair following traumatic injury, hypoxia/ischemia, stroke, or degeneration due to pathologies such as Parkinson's, Huntington's, and Alzheimer's diseases.

The Damaged Brain

Many organs are capable of repair and regeneration. The epithelial cells of the epidermis and intestinal lining are constantly lost and replaced, as are blood cells. Broken bones knit, and wounds heal. Transplant surgeries have demonstrated that the adult liver has regenerative abilities. However, the brain, especially in mammals, is generally refractory to repair. This deficiency was noted clinically at the dawn of medical history in ancient Egypt (Figure 26.1), and the understanding of the structure and function of nervous tissue gained in the intervening two millennia has not provided much encouragement. The realization that nervous tissue is made up of many classes of highly branched, interconnected nerve cells that communicate via specialized, specified cellular junctions to transmit electrical impulses made it clear that repairing nervous tissue

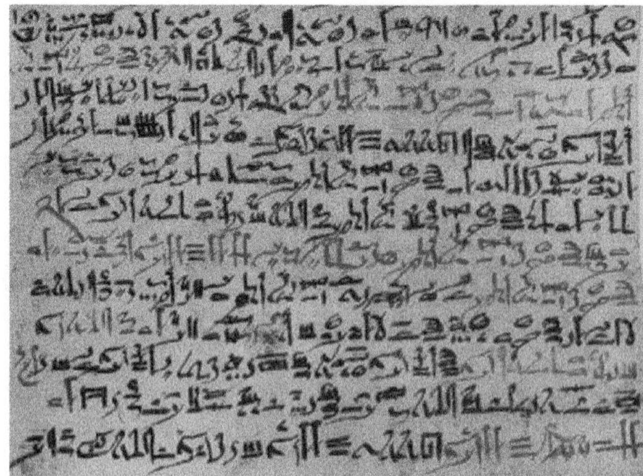

FIGURE 26.1 **An ancient Egyptian papyrus acknowledges the difficulty of repairing the brain and spinal cord.** The symbols in brown translate to: *When you examine a man with a dislocation of a vertebra of his neck, and you find him unable to move his arms, and his legs... Then you have to say: a disease one cannot treat.* (From Case and Tessier-Lavigne, 2005; courtesy of N. Y. Academy of Medicine Library.)

presents a far greater challenge than regenerating the liver. Moreover, as postmortem analyses and histological studies progressed, it became clear that specifically localized lesions accompany the behavioral deficits seen after brain damage, and that these lesions remain visible even many years after the injury. In such patients, there appears to be little repair of damaged tissue; the injured region is characterized either by a fluid-filled cyst or by "scar tissue" that lacks the histological integrity of neighboring brain regions. Thus, any adaptive changes in brain function following injury are unlikely to reflect wholesale replacement or remodeling of the cellular constituents of the brain.

Although still widely held, the view that the brain, once damaged, can never be restored to its previous state has been challenged for more than a century and continues to be debated. Nevertheless, nearly all modern studies of functional recovery after brain injury indicate that improvement seen in some patients following neural trauma primarily reflects the reorganization of intact circuits rather than the regrowth or replacement of damaged neurons.

Functional Reorganization without Repair

Neurologists have long recognized that, over time, patients who suffer strokes or sustain limited injuries to distinct brain regions often recover some of the deficits seen immediately after the trauma. Movement in paralyzed limbs can improve (especially if physical therapy is included in the treatment plan), and problems with verbal communication may diminish with intensive speech therapy. Such recovery is not thought to reflect significant regrowth or replacement of damaged neurons. Instead, the available evidence indicates that undamaged brain regions eventually become activated and reorganized to support, at least in part, functions whose primary representation was disrupted. The best understanding of functional recovery comes from studies of the primary motor cortex, where force generation and accuracy of movement can be measured reliably over time after focal damage.

Observations made in animals have shown that circuits in the adult primary motor cortex retain some capacity for use-dependent plasticity, suggesting a biological mechanism for the reorganization and recovery seen in patients. The plasticity of the primary motor cortex reflects that region's rich array of horizontally spreading axonal connections. Thus, connections that might not be active when the system is intact can be "unmasked" when there is damage nearby. In addition, plastic changes that favor functional recovery—similar to long-term potentiation (LTP) or long-term depression (LTD; see Chapter 8)—may occur at synapses between intact excitatory or inhibitory neurons, perhaps resetting the excitatory/inhibitory balance to maximize circuit function. There may also be some modest local growth of axon branches or dendrites as well as new synaptogenesis from intact neurons that further strengthens remaining connections. Finally, altered activity in the ipsilateral motor cortex may provide activation that can pattern the appropriate movement via spared contralateral pathways.

The consequences of motor cortex plasticity can be seen in stroke patients using fMRI in parallel with observing the patient's progress during rehabilitation. The most thorough studies have been of patients with focal strokes in the subcortical white matter that results in specific deficits in hand movement and grip strength. In these individuals, the amount of cortical activity, particularly in the hand region of the motor map, is *increased* and broadened shortly after injury, and *declines* with improved function (Figure 26.2A). Indeed, those with poorer outcomes did not show as extensive a decline in ectopic cortical activation. This surprising finding can be best appreciated by following differences in cortical activity in a single patient over time (Figure 26.2B,C). As hand movement improved in a patient with focal pontine stroke (compromising axons from upper motor neurons in the hand representation), focal activity in the hand representation, which increased substantially shortly after the stroke, declined as functional recovery progressed. Ectopic or increased activity in primary and supplementary motor cortex, as well as in the primary and secondary visual cortices both ipsilateral and contralateral to the lesion, also diminished (see Figure 26.2B). Whether these changes represent diminished excitatory neuronal activity or increased inhibitory neuronal activity is not known.

Similar observations have been made in patients with strokes that compromise complex functions such as language. Activation of remaining brain circuits changes over

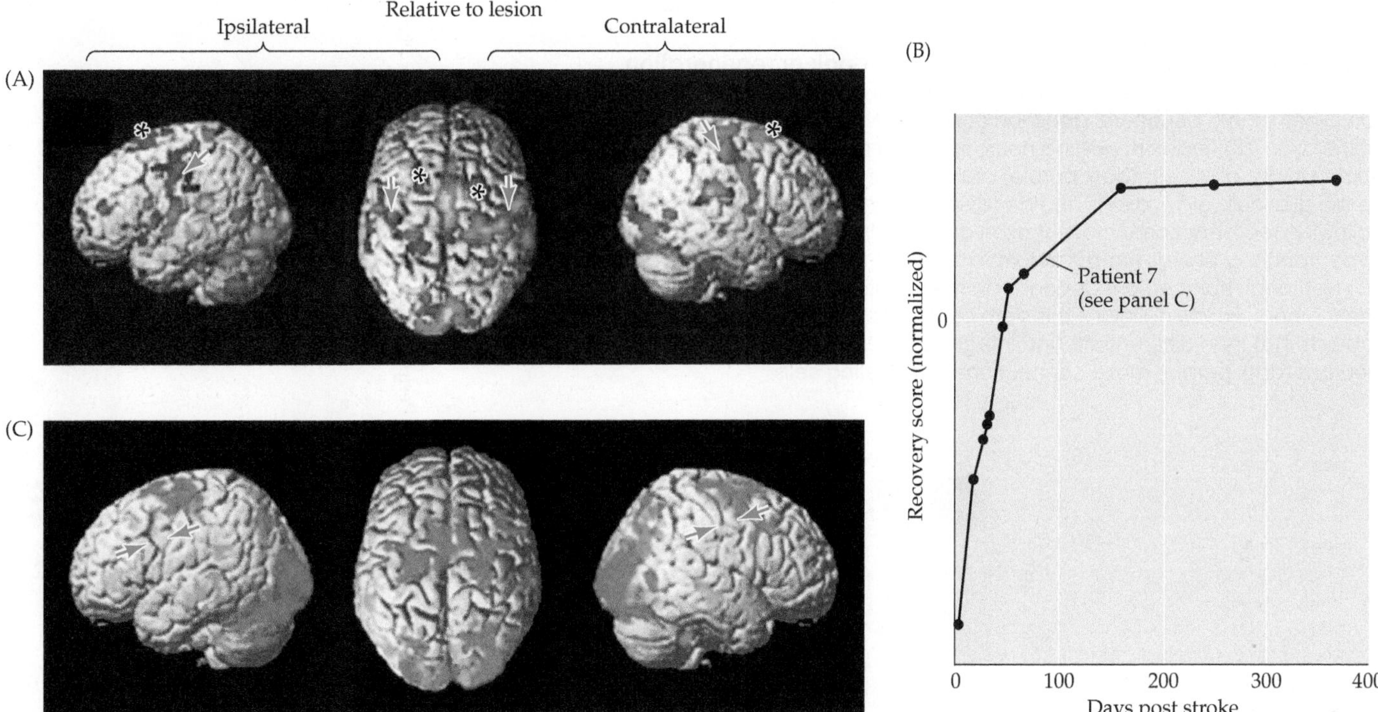

FIGURE 26.2 Altered cortical activity patterns are correlated with functional recovery after a focal stroke. (A) An fMRI map of *diminished* activity compared with age-matched control individuals following a middle cerebral artery (MCA) stroke. Regions of maximally diminished activity are red; orange and yellow reflect moderate changes. The data reflect the average of changes from six patients (from a total of 20 studied—the others had unilateral infarcts either in the internal capsule or pons) with comparable MCA strokes projected onto an idealized brain surface. There is a statistically significant bilateral decline in activity in much of the precentral gyrus (primary motor cortex; blue arrows); the decline is slightly enhanced (red) contralateral to the lesion. There is also decline in activity in supplementary motor areas (asterisks). Finally, visual cortex activity is diminished. (B) Improvement of motor performance over time for a single patient. Improvement is seen as an increased "recovery score," which measures performance on a battery of tasks that require hand-grasping and controlled movement. (C) Areas of diminished (red) and increased (green) cortical activity over time in a single patient scanned multiple times. This patient (the individual whose recovery is documented in panel B) had a pontine stroke that compromised axons carrying motor commands from the "hand" region of the primary motor cortex. This composite image shows bilateral diminished activity in the primary and supplementary motor areas, as well as in the visual cortex. The changes include diminished activity in the region of the primary motor cortex associated with the hand ipsilaterally and contralaterally (arrowheads). (From Ward et al., 2003a,b.)

time following rehabilitation and functional recovery; however, the magnitude, direction and localization of these changes are more variable than changes that occur in the motor cortex. In sum, neural circuits that remain following focal brain damage can reorganize, based on changing patterns of activation, to accommodate functional recovery, even if rules for reorganization remain elusive. The limited nature of this reorganization and its absence in patients with more profound impairments indicate the challenge of regaining normal function after the brain has been damaged.

Three Types of Neuronal Repair

Neuronal regeneration in a damaged brain (rather than functional reorganization of intact neural circuits) reflects capacities and limitations of three types of repair. The first type is the regrowth of axons, either from nerve cells in peripheral ganglia or from otherwise intact central nerve cells of which peripherally projecting axons have been severed, mostly aplha motor neurons of the spinal cord and brainstem (Figure 26.3A). This scenario requires a reactivation of the developmental processes for axon growth and guidance, as well as those for synapse formation. In addition, such repair entails activity-dependent competitive mechanisms similar to those used during development to ensure proper quantitative matching of newly regrown afferents to temporarily denervated targets. This first type of repair is seen primarily when sensory or motor nerves are damaged in the periphery, leaving the nerve cell bodies in the relevant sensory and autonomic ganglia or the spinal cord intact. This *peripheral nerve regeneration* is the most readily accomplished type of repair in the nervous system, and the most clinically successful.

The second type of repair is *restoration of damaged central nerve cells.* Although the axons or dendrites of these neurons may be injured, the cells themselves nevertheless

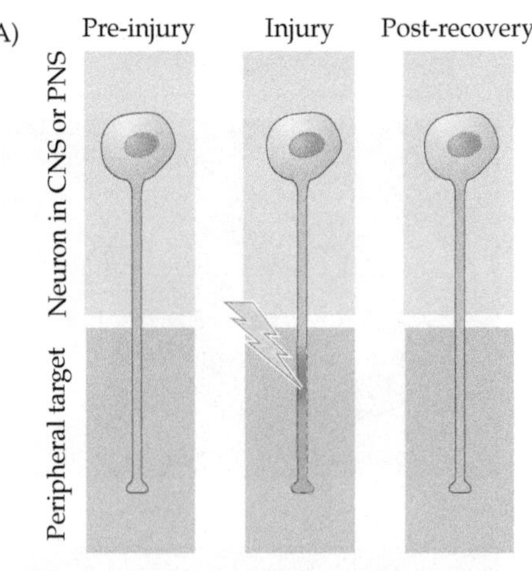

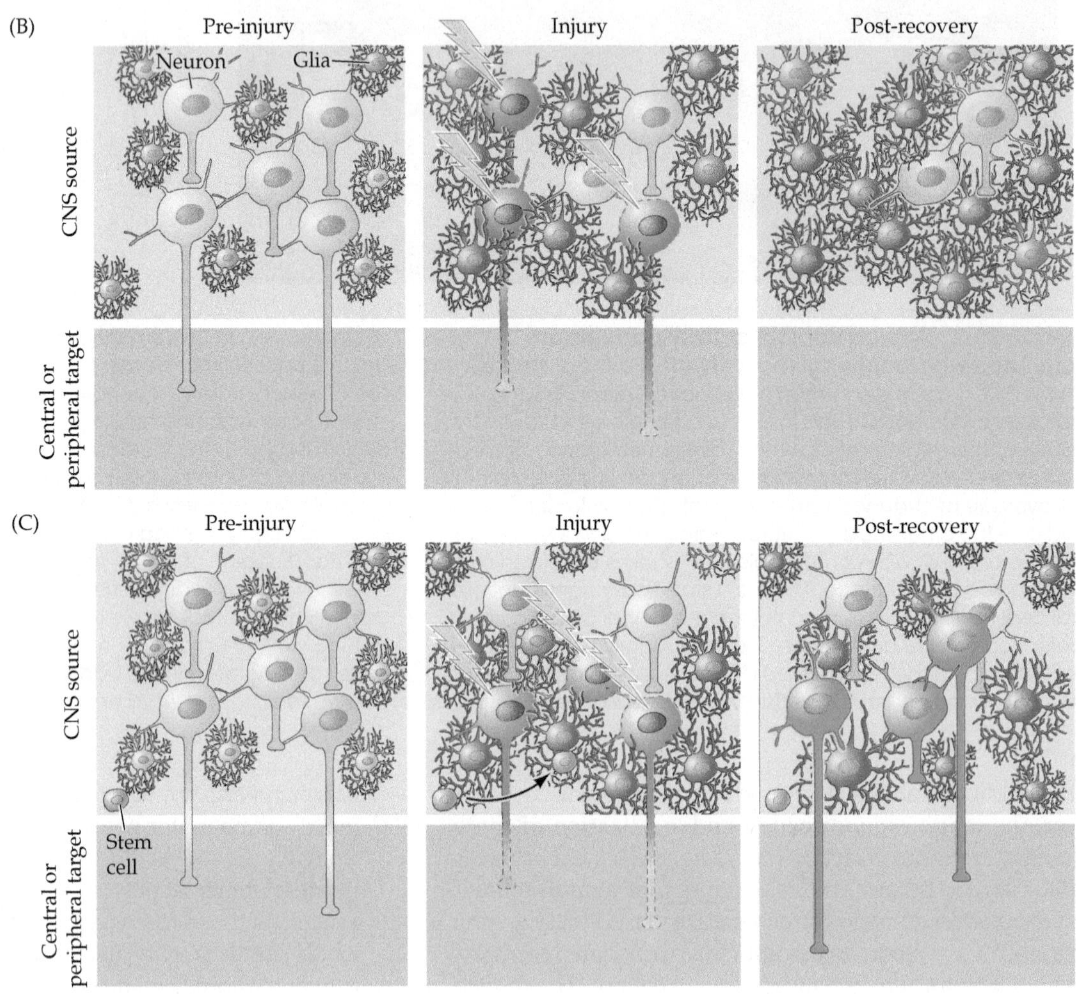

FIGURE 26.3 Three types of nervous system repair or regeneration. (A) In peripheral nerve regeneration, when peripheral axons are severed, the neuron, whether in a peripheral ganglion or in the CNS, regenerates the distal portion of the axon. (B) Repair of existing neurons at and around a site of injury in the CNS. Prior to injury, glial cells (dark purple) are quiescent. Immediately following the injury, the glial cells grow, axons and dendrites degenerate, and connections are lost. Following recovery, some modest axon and dendrite growth may be seen, but the hypertrophic glial cells remain to form a "scar" at the site of the tissue damage. (C) Neuronal replacement depends on the maintenance of a neural stem cell (dark green). Following injury, this stem cell proliferates and gives rise to new neuroblasts that then differentiate and integrate into the damaged tissue. These new neurons (dark purple) make connections with existing cells.

survive (Figure 26.3B). This response requires first that nerve cells are capable of restoring their damaged processes and connections to some level of functional integrity. Thus, new dendrites, axons, and synapses must grow from an existing cell body—a phenomenon sometimes referred to as *sprouting*. To achieve this, several developmental mechanisms must be reengaged, including appropriate regulation of cell polarity to distinguish dendrites and axons; adhesion signals to direct process extension; and trophic signaling to support growth (see Chapter 23). This type of repair requires the cooperative regrowth of existing neuronal and glial elements in a more complex environment—the local neuropil—than that of the peripheral nervous system (PNS). It generally fails in the injured CNS except over limited distances, most likely because of local overgrowth of glial cells and their production of signals that inhibit neuron growth (see below). A possible reason for glial overgrowth is the delicate balance of immune system-mediated clearance of damaged tissue versus maintenance of a small number of surviving neurons. Damage to brain tissue elicits a local inflammatory response that supports glial rather than neuronal growth. This glial growth insulates surrounding intact tissue from further damage due to inflammation. The subsequent loss of trophic support to damaged axons and dendrites due to inaccessibility of targets obscured by glial overgrowth, along with the action of inflammatory cytokines (released by macrophages and other immune cells in response to tissue damage) may suppress reactivation of cellular mechanisms for dendritic and axonal regrowth as well as those for synapse formation. Thus, this second type of repair is quite limited.

The third type of repair is the *genesis of new neurons* to replace those that have been lost, whether through normal wear and tear or as a result of damage (Figure 26.3C). Such adult neuronal genesis occurs rarely, and its mechanisms are controversial. The genesis of peripheral olfactory receptor neurons throughout adult life (see Chapter 15) is an example of this type of repair that is ongoing and gradual; however, the growth of newly generated axons from new olfactory receptor neurons resembles other instances of peripheral nerve regeneration. Indeed, the olfactory ensheathing cells that support ongoing olfactory receptor neuron axon growth to the olfactory bulb resemble peripheral Schwann cells more than central glia, even though their derivation from the neural crest (which generates Schwann cells) versus cranial placodes (which generate the neurons and supporting cells of the olfactory epithelium) remains controversial. If repair is accomplished by genesis of new neurons in an adult brain, several criteria must be met. First, nervous tissue must retain a population of *multipotent neural stem cells* (see Box 22A) that can give rise to all of the cell types of the relevant brain region. Second, these neural stem cells must be present in a region that retains an appropriate environment for the genesis and differentiation of new nerve cells and glia. Third, the regenerating tissue must preserve the capacity to recapitulate (or closely parallel) the migration, process outgrowth, and synapse formation necessary to reconstitute local functional networks of connections as well as long-distance connections.

The remainder of the chapter considers each of these three types of brain repair and regeneration. For the most part, peripheral neurons seem to be quite capable of extensive axonal regeneration leading to functional recovery; however, the extent of similar repair possible in the mammalian brain is limited. Some other vertebrate species (and many more invertebrate species) exhibit the capacity for axon regrowth, neuron replacement, and wholesale regeneration of lost or damaged neural tissue. These examples in insects, fish, frogs, and birds suggest to some that a better understanding of axon growth and guidance, neuronal mechanisms of use-dependent plasticity, and neural stem cell biology (see Chapters 22, 23, and 25) could eventually indicate strategies to promote the repair of damaged neural tissue in humans.

Peripheral Nerve Regeneration

In the early 1900s, the British neurologist Henry Head provided a particularly dramatic account of repair in the PNS. By this time, it had become clear that damage to a peripheral nerve resulted in a gradual but usually incomplete restoration of sensory and motor function. The speed and precision of this recovery could be facilitated by the surgical reapposition of the two ends of the severed nerve. Apparently, if regeneration occurred in an environment that restored continuity of the existing nerve sheath, functional recovery could be far greater. Head's interest in this possibility culminated in a somewhat idiosyncratic approach to documenting the extent and functional precision of regeneration in damaged peripheral sensory and motor nerves. Rather than evaluating functional recovery in patients whose traumatic injuries were variable in location and extent of tissue damage, he chose to undergo a precise nerve transection and reapposition experiment *himself*, documenting the results as a personal narrative. In his 1905 paper, Head wrote:

> *On April 25, 1903, the radial (ramus cutaneus radialis) and external cutaneous nerves were divided (cut) in the neighborhood of my elbow, and after small portions had been excised, the ends were united with silk sutures. Before the operation the sensory condition of the arm and back of the hand had been minutely examined and the distances at which two points of the compass could be discriminated had been everywhere measured.*
>
> Head, et al., 1905, *Brain* 28: 99–115

Head went on to carefully monitor the return of sensation to the parts of his hand that had been rendered insensitive by the lesion (Figure 26.4). His observations emphasized several important aspects of peripheral nerve regeneration in this one-of-a-kind experiment. The first indication of recovery was a difference in the return of general sensitivity to pressure and touch that was not well localized (a sensitivity he called "protopathic"), beginning at approximately 6 weeks and lasting about 13 weeks. Based on this observation, Head suggested that the "protopathic system regenerates more rapidly and with greater ease. It can triumph over want of apposition and the many disadvantages that are liable to follow traumatic division of a nerve."

Head also experienced a set of sensations that recovered more slowly and with less fidelity to his recollection of his original sensory state. These phenomena included sensitivity to light touch, temperature discrimination, pinprick, and two-point discrimination, as well as fine motor control (which he referred to as "epicritic" abilities). In fact, these faculties were not fully recovered over the 2 years between Head's surgical injury and publication of the paper he wrote. He noted "the fibers of this system are more easily injured, and regenerate more slowly, than those of the protopathic

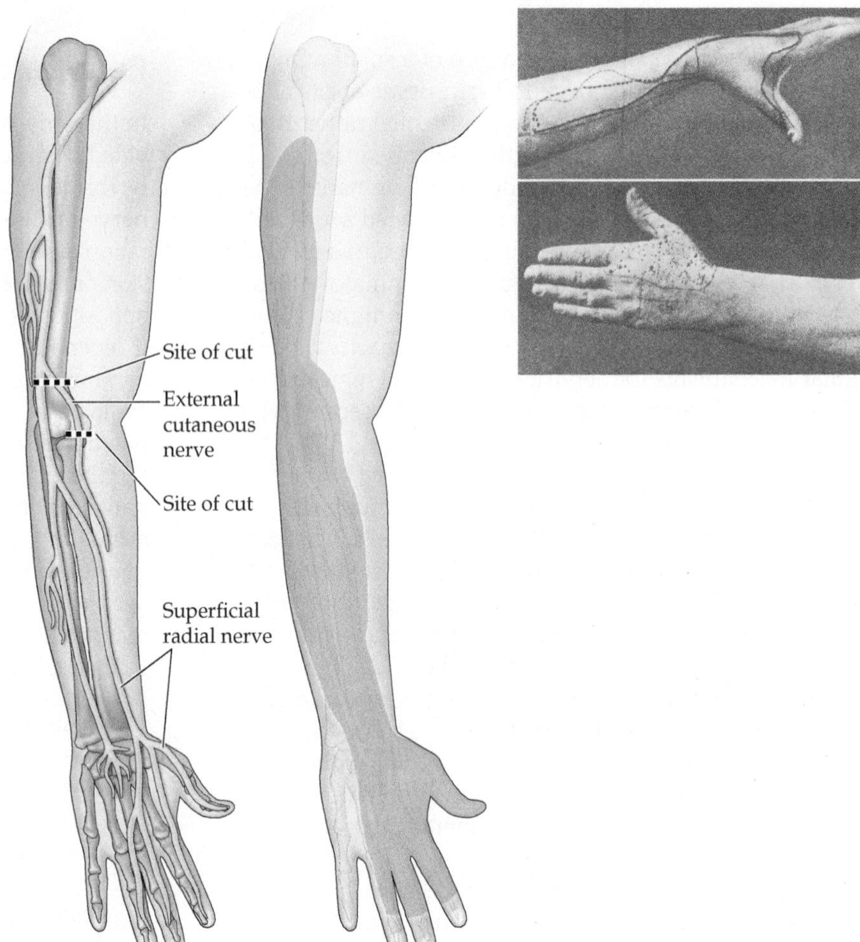

FIGURE 26.4 Henry Head's peripheral nerve regeneration experiment. (A) Diagram of the human arm showing the location of the radial nerve (left), which was severed and the proximal and distal ends surgically reapposed in Head's experiment on his own arm; and the territory normally innervated by the radial nerve (purple, right). (B) Two photographs taken from Head's 1905 paper on his recovery from the peripheral nerve cut. The top panel shows the outlines of regions of Head's lower arm and hand that were insensitive to painful stimuli (e.g., a pinprick), and the dotted line shows regions that were insensitive to light touch (e.g., a wisp of cotton). The bottom panel shows the region of Head's hand and thumb that regained sensation after an initial period of recovery (2 to 6 months). The various marks within the resensitized region indicate "hot" and "cold" spots that were more or less sensitive to stimulation. (B from Head, et al., 1905.)

system. They are evidently more highly developed and approach more nearly to the motor fibres that supply voluntary muscle in the time required for their regeneration."

These observations —despite their unusual experimental source, and limited number of subjects (only 1!)—were the first to distinguish between the regenerative abilities of various classes of dorsal root ganglion cells and spinal motor neurons during the process of peripheral reinnervation. They also suggested, for the first time, that there may be a biological difference between sensory neurons that detect general, less well-localized somatosensory information such as pressure and touch and more specific, localized information such as two-point discrimination and pain (see Chapters 8 and 9). Of course, the actual process of regeneration was inferred based on the recovery of peripheral sensory and motor functions. The distinct schedules for "protopathic" versus "epicritic" abilities are presumably related to the initial specificity between different classes of dorsal root ganglion cell and motor axons, the environment of the re-apposed, but denervated nerve sheath, and their targets during development. This specificity is now known to rely upon a variety of molecular signals, including several different neurotrophins (see Figure 23.19). The importance of these cues, and their retention in the adult PNS to facilitate regeneration, is reviewed in the following section.

The Cellular and Molecular Basis of Peripheral Nerve Repair

The cellular basis of peripheral nerve regeneration provides perhaps the clearest example of the relationship between the mechanisms that repair neural damage and those used to promote initial axon growth and synapse formation during development. Even though the molecular mechanisms are similar, the environment for adult peripheral nerve repair (Figure 26.5A) is far different than that in the embryo. Obviously, the distances that must be accommodated by the growing axon are larger, and the synaptic targets have already been specified. The major cellular elements that contribute to peripheral axon regrowth and the reinnervation of targets in the mature PNS are *Schwann cells*, the glial cells that myelinate peripheral axons; and *macrophages*, immune system cells that clear the degenerating remains of severed axons (Figure 26.5B). In addition to their respective roles in supporting intact axons and removing debris, both of these adult cell types secrete molecules that are essential for successful regeneration. In this process, the Schwann cell plays a dominant role in ensuring the appropriate cellular and molecular milieu for regeneration following injury.

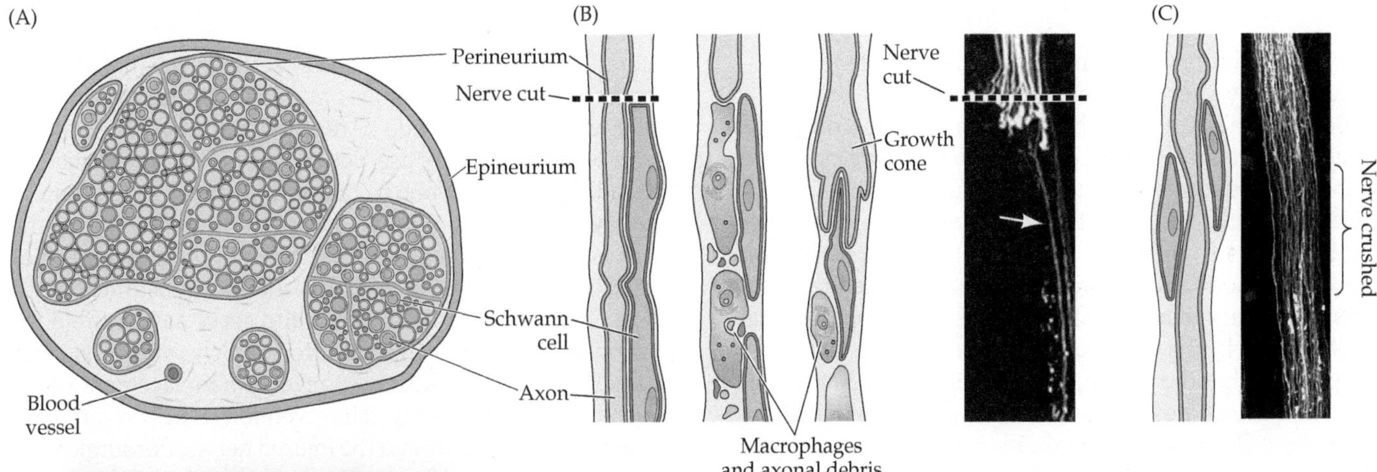

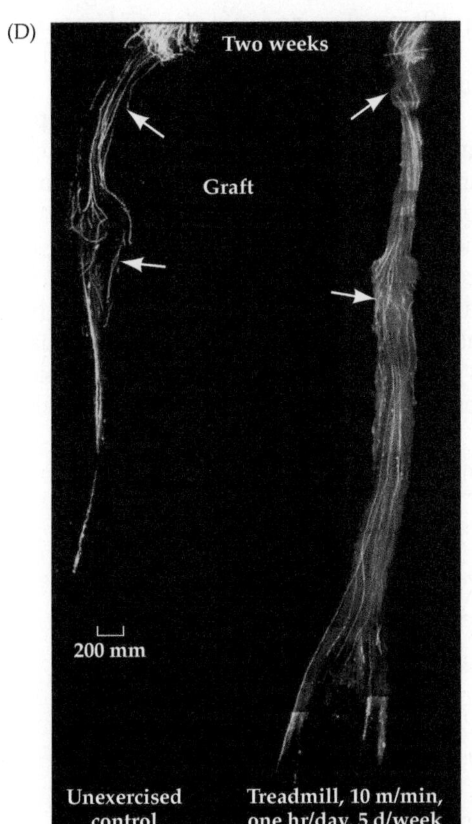

FIGURE 26.5 Regeneration in peripheral nerves. (A) Cross section through a peripheral nerve showing the connective tissue sheath of the epineurium and the extracellular matrix-rich perineurium that immediately surrounds the axons and Schwann cells. (B) Degeneration and regeneration in an idealized single peripheral nerve "tube" of perineurium/basal lamina. Once the axon is cut, the distal portion degenerates and is phagocytosed by macrophages. After the debris is mostly cleared, the proximal axon stump transforms into a growth cone, and this growth cone interacts with the adjacent Schwann cells. The image next to the drawing shows this step in peripheral nerve regeneration imaged in a live mouse after peripheral nerve damage. The nascent growth cones can be seen at the site of the cut, and in a few cases (arrow) growing beyond it into the distal stump. (C) Regeneration is more efficient after crushing versus cutting a nerve. This panel shows a nerve adjacent to that which was cut in the same animal shown in B. The remaining axons have recovered far more rapidly, and there is extensive regeneration across the site of the crush. (D) Activity and use can influence peripheral axon regrowth. The fibular nerve in the leg of a mouse was transected and a significant length of nerve removed. Subsets of axons are labeled using a fluorescent reporter transgene as in panel C. A heterologous graft of peripheral nerve from a donor mouse was positioned as a "bridge" (between the arrows in each nerve shown). After the graft was complete, over the course of 2 additional weeks the mice were either left to recover while pursuing their normal cage activities (left) or given a regime of exercise on a treadmill (right). Exercised mice showed significantly more nerve growth through the graft and recovered function more rapidly. (B,C photos from Pan et al., 2003; D from Sabatier et al., 2009.)

When a peripheral axon is severed, the axon segment distal to the site of the cut degenerates, and the debris left by the dead axon is cleared by macrophages. When the axon is crushed rather than severed, more rapid recovery occurs because the damaged segments, some of which have not fully degenerated, provide a guide to the regenerating proximal axons (Figure 26.5C). Since severed distal axons are cleared rapidly, only the basal lamina components secreted by Schwann cells, and the Schwann cells themselves, are available to stimulate and guide regeneration of intact proximal axons into the distal stump of a severed nerve (see Figures 26.5C and Figure 26.6). The extracellular matrix, within the spaces defined by Schwann cell processes, provides a conduit of sorts for the regenerating axons. These remaining "channels" in the distal peripheral nerve, referred to as *bands of Bungner*, are composed of a relatively orderly array of extracellular matrix components, Schwann cells, immune cells, and connective tissue. The extracellular matrix associated with axon fascicles in the nerve before damage (also referred to as the basal lamina, although a peripheral nerve is not really an epithelium) is more or less continuous to the axon's target. This is presumably why precise reapposition of distal and proximal nerve segments facilitates better recovery of function, especially of fine touch and movement (Henry Head's "epicritic" abilities; see above). Precise surgical reapposition is now done routinely using microscopic guidance to maximize recovery after peripheral nerve injury.

Both regenerating peripheral sensory and motor axons express *integrins* (see Figure 23.4A) that mediate recognition of the matrix and subsequent intracellular signaling that facilitates growth. The capacity of the motor neurons, whose cell bodies are in the CNS, to reactivate the expression of growth-promoting factors suggests that mature central neurons can modify gene expression and respond to appropriate cues, at least in the periphery, to accommodate regrowth during adulthood. If an injury is extensive and much of the nerve is lost, new Schwann cells can be generated by Schwann cell precursors remaining in the damaged proximal nerves. These new cells can then provide an appropriate environment to support the extension of a newly generated growth cone from the axon stump. Nevertheless, the absence of a well-aligned distal stump makes surgical repair more difficult, limits the precision of regeneration, and dramatically diminishes the recovery of function.

In some cases, damage is so extensive that the proximal and distal nerves cannot be directly rejoined, and recovery cannot be achieved without additional surgical intervention. In such cases, surgeons attach both proximal and distal ends to a nearby intact nerve in the hope that growth will occur via the undamaged nerve. The other option is to take a length of nerve from another, less important nerve (e.g., a cutaneous nerve in the leg) and attach the proximal and distal ends to this heterologous graft. In experimental animals, the rate and magnitude of regrowth through such grafts can be accelerated by use-dependent mechanisms, such as treadmill running done by rats with peripheral nerve grafts following peripheral nerve damage (Figure 26.5D). Additional approaches include development of biomaterials that approximate extracellular matrix that can substitute for the lost length of peripheral nerve and serve as a scaffold for the migration and differentiation of Schwann cells from the proximal nerve stump to provide a replacement nerve sheath. The success of such interventions relies on control of mechanical forces around the injured nerve after surgical repair, minimizing inflammation and infection and immune response to the engrafted biomaterial, perhaps enhanced by rehabilitative training (see Figure 26.5D).

The Schwann cell is the essential cellular mediator of peripheral axon regrowth through any conduit. Schwann cells provide molecular support that facilitates regeneration by recreating an environment similar to the milieu that supports axon guidance and growth during early development (Figure 26.6). Schwann cells secrete additional extracellular matrix molecules such as laminin, fibronectin, and collagens that provide a substrate for axon growth via activation of signaling that supports growth cone pathfinding and re-extension of the axon. In response to axon injury, Schwann cells also

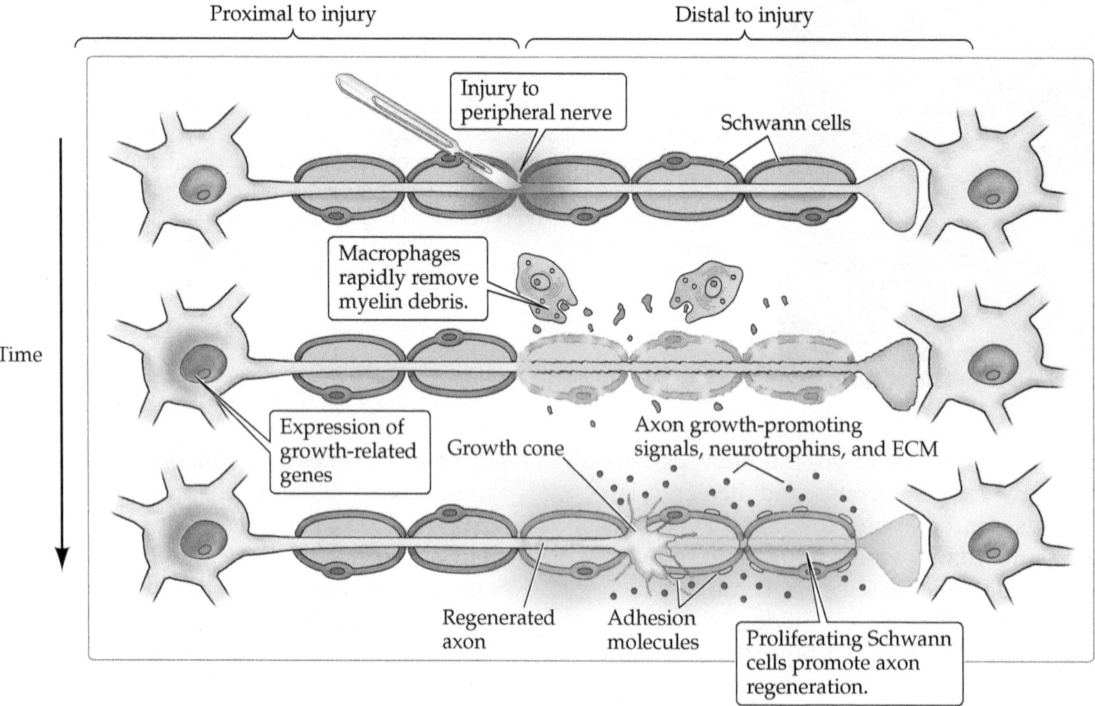

FIGURE 26.6 Molecular and cellular responses that promote peripheral nerve regeneration. The Schwann cell is essential for this process. Once the macrophages have cleared the debris from the degenerating peripheral stump, the Schwann cells proliferate, express adhesion molecules on their surface, and secrete neurotrophins and other growth-promoting signaling molecules. In parallel, the parent neuron of the regenerating axon expresses genes that restore it to a growth state. The gene products are often receptors, or signal transduction molecules, that allow the cell to respond to the factors provided by the Schwann cell. (ECM = extracellular matrix.)

increase the amount of transmembrane- or membrane-associated cell adhesion molecules such as NCAM, L1, and N-cadherin on their surfaces. At the same time, the regenerating axons express complementary cell surface adhesion molecules and co-receptors. These adhesion molecules likely mediate signaling that facilitates growth cone motility, force generation, and microtubule assembly in the newly generated axon. Schwann cells near the site of injury in the distal end of the nerve increase expression and secretion of several neurotrophins, including NGF and BDNF (thought to be especially crucial for motor axon growth). Trk and p75 neurotrophin receptors are elevated following injury on the newly generated growth cones of regenerating peripheral axons. There is some specificity in timing of the availability of NGF, P75 and TrkA at an earlier stage (both are primarily associated with nociceptive axon growth) followed by expression of other Trks that support mechanoreceptive and proprioceptive axon growth (see Chapter 23). The local availability of neurotrophins may promote a "growth" state (i.e., reactivate the capacity for *trophic* signaling) for the damaged axons, as well as attract the growing axons to appropriate local targets distal to the site of damage (i.e., *tropic* effects). Its specificity and timing provides a molecular explanation of the early observations of Head et al. that showed the initial recovery of nociception (Head's "protopathic" sense) followed by proprio- and mechanoreception (Head's "epicritic" sense).

In anticipation of reactivation of nerve growth immediately after injury or in response to these changes orchestrated by the Schwann cell after debris is cleared and Schwann cells have begun to migrate into the distal stump, the regenerating peripheral neurons also change. Injured and actively regenerating sensory or motor neurons modify gene expression for cytoskeletal reassembly, signal transduction, and protein trafficking in the cell body to accommodate regrowth. Some of this modification relies upon early re-expression of developmentally regulated transcription factors including those in the bHLH family (see Chapter 22). The dynamics of the actin and microtubule cytoskeleton must be restored to a growth state so that growth cone navigation and axon extension can occur. Changes in gene expression enable this, and the genes that are switched on include several associated with proteins that modulate the cytoskeleton during developmental axon growth—actin- and microtubule-binding proteins, Ca^{2+}-regulated modulators, and molecular "motors" such as kinesins and dynein. In addition, genes selectively associated with the growth state, including *GAP43*, are reexpressed. The protein GAP43 (*g*rowth-*a*ssociated *p*rotein-*43*) is normally found in growing axons in the embryo (see Chapter 23), but is also present at high levels following axotomy, as a result of increased gene expression. As regrowth reaches its conclusion, and synapses are formed, these growth-promoting genes are downregulated.

It is possible to provide a milieu and elicit a growth state that promotes CNS axon outgrowth from adult neurons.; however, it depends upon an impractical experimental approach, and its effects are sharply limited. In experimental animals, severed axons in the optic nerve (recall that the optic nerve and retina are components of the CNS) or spinal cord can be provided with a peripheral nerve graft that offers the Schwann cell, basal lamina, and connective tissue components that normally support peripheral nerve regeneration (Figure 26.7A). Central axons grow readily through

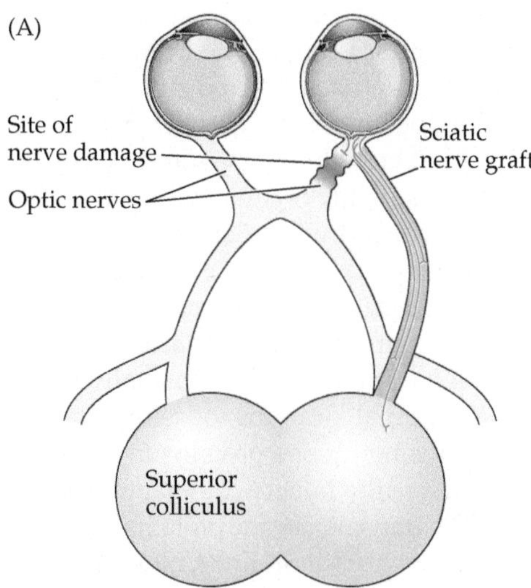

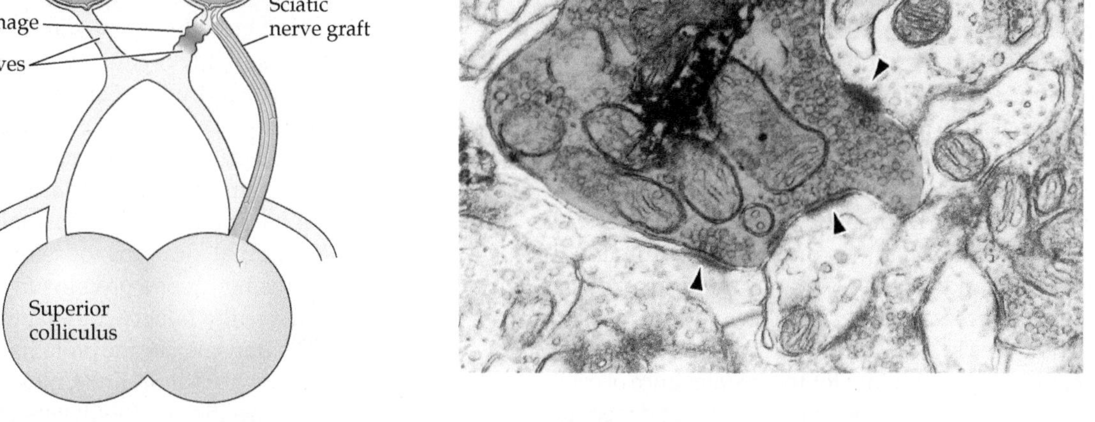

FIGURE 26.7 Growth-promoting properties of peripheral nerve sheaths and Schwann cells facilitate growth of damaged axons in the CNS. (A) Severed axons from the optic nerve are apposed to a peripheral nerve graft. The axons, which would normally not regenerate through the optic nerve (recall that the optic nerve, while physically in the periphery, is wholly within the CNS), now grow through the peripheral nerve graft to reach the superior colliculus, a normal target for retinal ganglion cells. (B) Regenerated axons make synapses with targets in the superior colliculus. The dark material is an electron-dense, intracellularly transported tracer that identifies specific synaptic terminals (arrowheads) as emanating from a regenerated retinal axon. (A after So and Aguyao, 1985; B from Aguayo et al., 1991.)

the peripheral nerve graft. Some of these axons can even make synapses in the target territory to which the distal end of the graft is connected (Figure 26.7B). The numbers of synapses, and their functional capacity to restore vision; however, are very limited. Nevertheless, such experiments show that Schwann cells define an environment in the peripheral nerve sheath that is particularly well adapted to initiate and support the regrowth of damaged adult axons, whether they project to the periphery as do motor neurons during peripheral nerve repair or remain within the CNS. Perhaps more importantly, they indicate that CNS neurons, beyond motor neurons when damaged, can return to a transcriptional state that will support growth. Nevertheless, without this very artificial solution to CNS regenerative axon growth, the net effect of CNS damage apparently overcomes any change in surviving neurons and prevents their regrowth.

Regeneration of Peripheral Synapses

The extension of axons from mature sensory, autonomic, or motor neurons is only the first step in peripheral nerve regeneration. The next essential event in successful recovery of function is reinnervation of appropriate target tissues and reestablishment of synaptic connections. This process must occur for all three classes of peripheral axons (sensory, motor, and autonomic); however, it has been most thoroughly characterized at the neuromuscular junction (NMJ) and in the peripheral autonomic system (Box 26A). Because of the relative ease of identifying and visualizing synaptic sites on muscle fibers, the regenerating NMJ has been studied in great detail (Figure 26.8A,B). The ability to define major molecular constituents—synaptic extracellular matrix, postsynaptic receptors, and related proteins—gives insight into the stability of a denervated synaptic site after damage, as well as the changes that accompany reinneveration (Figure 26.8C,D).

When skeletal muscle fibers are denervated, the original neuromuscular synaptic sites remain intact for weeks. Indeed, even specialized components of the basal lamina that mark synaptic sites remain in the absence of the synapse or the muscle itself. At these or nearby sites, many secreted signaling molecules are either increased or decreased in both the muscle cells and the Schwann cells near the denervated end plate sites. Presumably, the neurotrophins whose expression is increased at the denervated neuromuscular junction—NGF and BDNF—enhance the tropic and trophic signaling necessary to recapitulate target recognition and synaptogenesis. Those that are diminished—NT3 and NT4—may be more important for the maintenance of established synapses. Their decline may ensure that denervated synaptic sites accept innervation from regenerating axons.

The clustering of acetylcholine receptors (AChRs) that defines the postsynaptic membrane specialization also remains, as does the local scaffold of proteins that retain the AChRs at the synaptic specialization on the muscle fiber (see Figure 26.8D). The secreted factor neuregulin and its receptors, involved in the initiation of receptor clustering (see Chapter 23), are also expressed at the denervated synapse. In addition, the extracellular matrix components that distinguish the synaptic portion of the muscle basal lamina are maintained when mature muscle fibers are denervated. This matrix includes specialized forms of laminin (synaptic, or S laminin) normally found at the neuromuscular synapse. In addition, the proteoglycan agrin is key for initiating or maintaining the synaptic site on muscle fibers by binding to the low-density lipoprotein receptor 4 (Lrp4), a transmembrane receptor that then activates the muscle-specific receptor kinase MuSK. The activation of MuSK, via phosphorylation, leads to the recruitment of a transmembrane complex that includes the AChR itself (a multimeric ligand-gated ion channel/receptor), the transmembrane scaffolding protein dystrobrevin, the cytoplasmic scaffold dystrophin, and an additional cytoplasmic membrane–associated kinase, rapsyn. The immobilization of these AChrR complexes in the postsynaptic NMJ cluster, particularly at the top of the postsynaptic membrane foldings, closet to the sites of motorneuron synaptic vesicle release, maximize the initiation of neuromuscular synaptic transmission. Perhaps it is not surprising that the genes for many of these scaffolds are causal mutations in muscular dystrophy; indeed, first such disease-causing mutation was found in the dystrophin gene (named for its association with muscular dystrophy). Finally, molecules that are normally bound to the specialized extracellular matrix (ECM), produced by perisynaptic Schwann cells at the synapse, are maintained at denervated sites. In particular, acetylcholinesterase (AchE), which mediates degradation of acetylcholine at the NMJ, is concentrated in the synaptic cleft by the collagen-related protein ColQ. ColQ is localized to the synaptic cleft via binding to perlecan, an ECM protein. The ECM at the synapse also localizes and concentrates secreted growth factors—not only neurotrophins, but additional growth-promoting molecules such as fibroblast growth factors—produced on denervation. Accordingly, this highly organized region of the basement membrane, in concert with secreted factors, postsynaptic receptor proteins, and Schwann cells, defines the site of reinnervation. In fact, even if the muscle fiber is eliminated, motor axons recognize these specialized sites (maintained in "tubes" of basal lamina that normally surround muscle fibers) as optimal locations for reinnervation. Finally, a complex of transmembrane proteins in the muscle plasma membrane ensure the fidelity of synaptogenesis during regeneration. The transmembrane proteins Agrin as well as the muscle specific kinase MuSK link ColQ, AChE, and perlican in the synapse-specific ECM with clustered nicotinic acetyl choline receptors (AchRs) in the muscle post-synaptic membrane (see Figure 26.8D). This ECM-postsynaptic complex is reinforced by two scaffolding proteins, dystroglycan and rapsyn, which the anchor it to the actin cytoskeleton of the muscle cell.

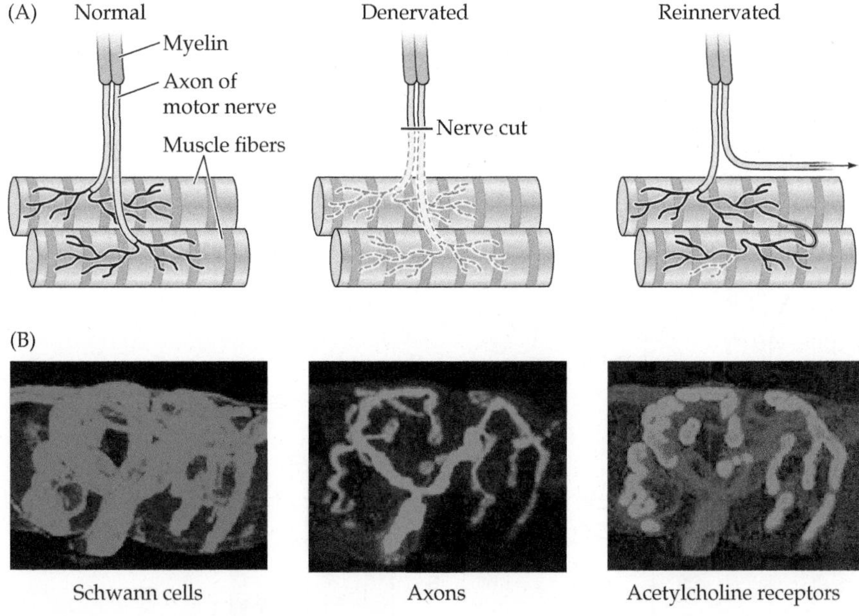

(B)

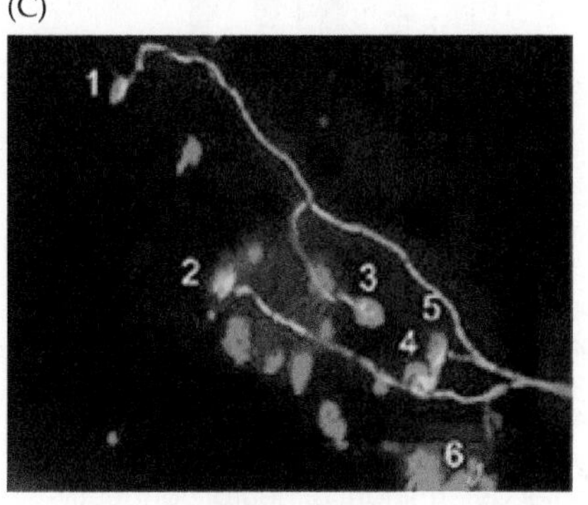

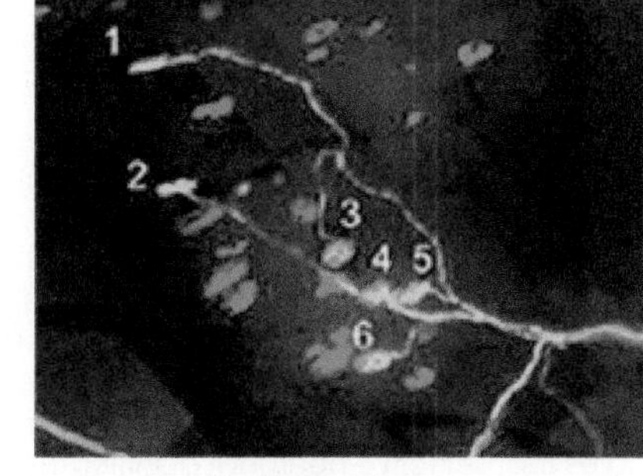

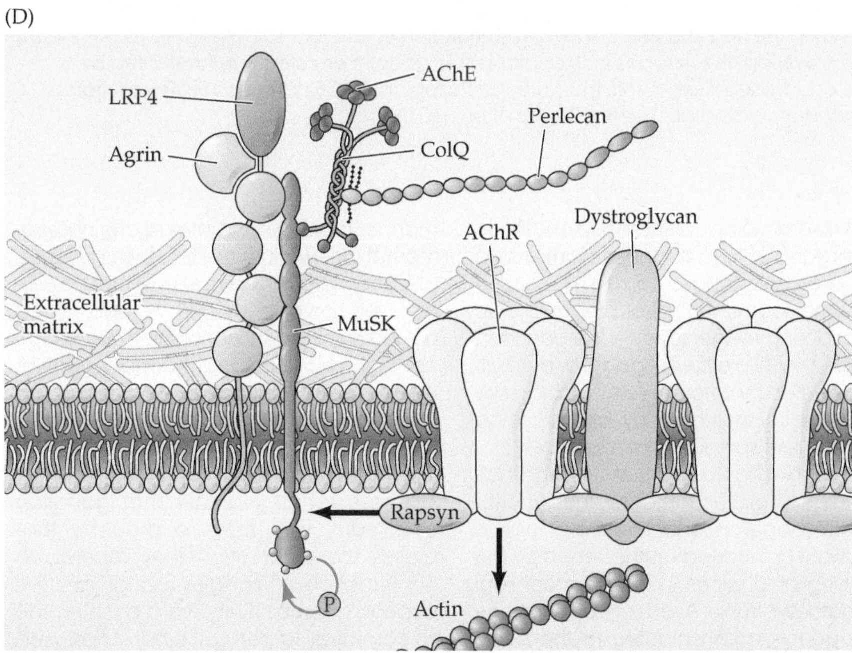

FIGURE 26.8 Reinnervation of muscles following peripheral motor nerve damage. (A) Schematic showing the degeneration of a distal motor axon, its regrowth, and the maintenance of the postsynaptic specialization on the muscle surface during the period of denervation. (B) The cellular components usually found at the neuromuscular junction (NMJ). When the axon degenerates, the Schwann cells and acetylcholine receptors (AChRs) remain in place. (C) The pattern of motor innervation in an isolated muscle before an injury and 25 days after, imaged in a single live mouse. The postsynaptic specializations of the NMJs on each individual muscle fiber are labeled in red (for AChR). The axon fluoresces green. This axon makes six synapses on six muscle fibers. Twenty-five days later, the axon has reinnervated all six sites, and the basic pattern is similar to that prior to the injury. (D) A specific molecular architecture in the synaptic cleft and postsynaptic membrane of the muscle preserves neuromuscular synaptic sites and facilitates reinnervation and functional recovery. The synaptic cleft is the site of several specific extracellular matrix (ECM) molecules, including specific classes of laminin (e.g., the NMJ-specific variant S laminin) and special variants of collagen, including ColQ, which specifically binds perlecan, another ECM molecule found in the NMJ matrix. ColQ in turn binds and localizes acetylcholinesterase (AChE) to the synaptic cleft. The postsynaptic muscle membrane itself is key for maintaining ongoing NMJ repair and regeneration following denervation. Two key transmembrane receptor proteins, low-density lipoprotein receptor 4 (Lrp4) and the muscle-specific tyrosine kinase MuSK, bind the protein agrin (secreted by the presynaptic terminal of the ingrowing motor neuron axon). The secretion of agrin into the ECM triggers the muscle cell to establish or maintain AChR clusters that rely on additional scaffold proteins, including dystrobrevin and rapsyn. (B from Pitts et al., 2006; C from Nguyen et al., 2002; D after Campanari et al., 2016.)

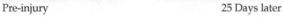

BOX 26A ■ Specific Regeneration of Synaptic Connections in Autonomic Ganglia

The peripheral autonomic system has been studied in detail for more than a century with respect to neural regeneration. The accessibility and regenerative properties of peripheral axons allow a variety of investigations to be done with relative ease. Most of these studies have been done in the mammalian sympathetic system. Preganglionic sympathetic fibers, like other peripheral axons, regenerate when they are severed. Toward the end of the nineteenth century, the English physiologist John Langley, working at Cambridge University, found that sympathetic end-organ responses (e.g., blood vessel constriction, piloerection, and pupillary dilation) recovered a few weeks after cutting the preganglionic nerve to the superior cervical ganglion. As indicated in Figure A, the normal innervation of this and other sympathetic ganglia is selectively organized in that preganglionic axons arising from different spinal segments innervate particular functional classes of cells in the ganglion. Langley found that after reinnervation the end-organ responses were organized much as before; thus, stimulation of T1 elicited its particular constellation of largely nonoverlapping end-organ effects compared with stimulation of the preganglionic axons arising from T4.

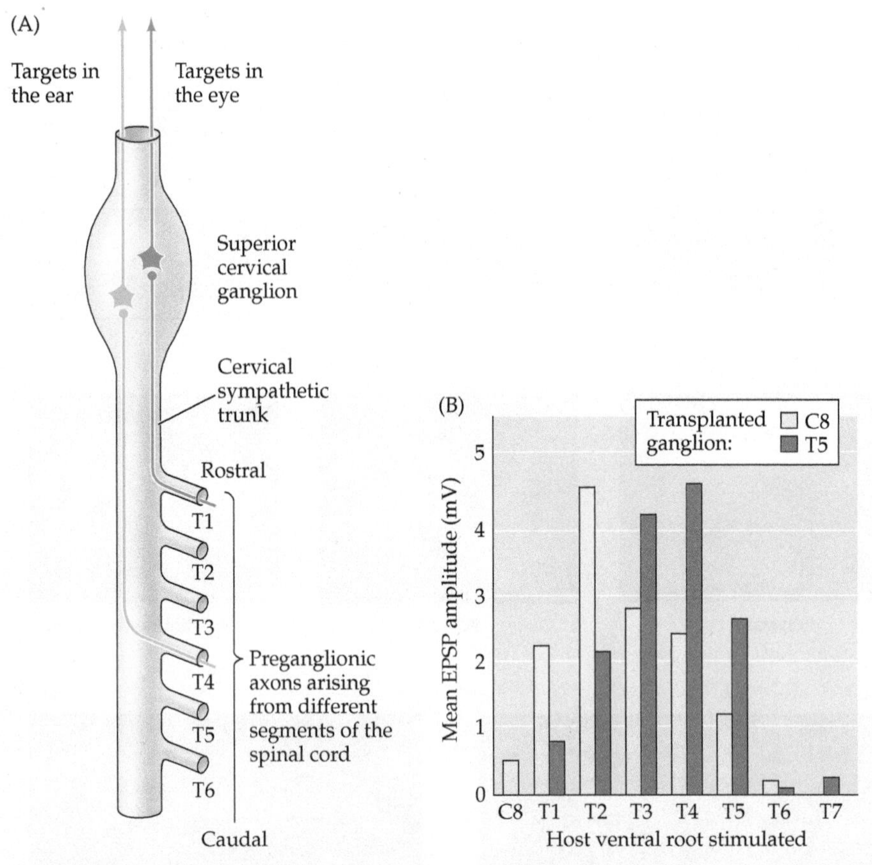

Evidence that synaptic connections between mammalian neurons form according to specific affinities between different classes of pre- and postsynaptic cells. (A) In the superior cervical ganglion, preganglionic neurons located in particular spinal cord segments (T1, for example) innervate ganglion cells that project to particular peripheral targets (the eye, for example). Establishment of these preferential synaptic relationships indicates that selective neuronal affinities are a major determinant of neural connectivity. (B) In a transplantation experiment, guinea pig donor ganglia C8 (superior cervical ganglion; control) and T5 (fifth thoracic ganglion) were transplanted into the superior cervical bed of a host animal. The graph shows the average postsynaptic response of neurons in the transplanted ganglia to stimulation of different spinal segments. Although there is overlap, the neurons in transplanted T5 ganglia are clearly reinnervated by a more caudal set of segments than are the transplanted C8 neurons. (EPSP = excitatory postsynaptic potential.) (B after Purves et al., 1981.)

Modern experiments confirmed Langley's observations and showed further that the normal pattern of innervation observed with intracellular recording is indeed reestablished following regeneration of the preganglionic axons. Selective reinnervation also occurs in parasympathetic ganglia. The chick ciliary ganglion has two functionally and anatomically distinct populations of ganglion cells: the ciliary cells and the choroidal cells. Because these ganglion cell types can be separately identified and are in turn innervated by preganglionic axons with different conduction velocities, one can ask whether these two populations are reinnervated by the same classes of axons that contacted them in the first place. As in the mammalian sympathetic system, appropriate contacts are reestablished during reinnervation.

The accurate reinnervation of different classes of sympathetic neurons is especially remarkable because the ganglion cells innervated by a particular spinal segment (and that innervate a particular target) are distributed more or less randomly through the ganglion. This arrangement implies that recognition of the pre- and postsynaptic elements must occur at the level of the target cells. One way to explore the implication that ganglion cells bear some more or less permanent identity is to transplant different sympathetic chain ganglia from a donor animal to a host where the ganglia can be exposed to the same segmental set of preganglionic axons during reinnervation. One can then ask whether two different ganglia, normally innervated by different sets of axons, are selectively reinnervated by axons arising from different spinal segments.

As seen in Figure B, different sympathetic chain ganglia (in this case, the superior cervical ganglion and the fifth thoracic ganglion) are indeed distinguished by the preganglionic axons in the host cervical sympathetic trunk. A donor superior cervical ganglion transplanted to the cervical sympathetic trunk is reinnervated in a manner that approximates its original segmental innervation; the fifth thoracic ganglion transplanted to that position, however, is reinnervated by an overlapping but caudally shifted subset of the thoracic spinal cord segments that normally contribute to the cervical sympathetic trunk. This more caudal innervation approximates the original segmental innervation of the fifth thoracic ganglion.

These results indicate that ganglion cells carry with them a property that biases the innervation they receive, in confirmation of Langley's original concept of "chemoaffinity" as a basis for the selectivity of target cell innervation (see Chapter 23).

It is not surprising that many of these molecules, if mutated or attacked by auto-antibodies or general inflammatory responses, can compromise ongoing maintenance and repair of the NMJ, and thus neurodegenerative diseases, including muscular dystrophies, myasthenia gravis, and other conditions where muscle weakness and wasting reflect failure of normal NMJ maintenance.

This molecular specificity, however, provides only one part of the instructions needed to reestablish synapses after peripheral nerve regeneration. Activity-dependent processes similar to those that eliminate polyneuronal innervation at neuromuscular synaptic sites during development are also essential for restoring function after peripheral nerve damage. There is a fair degree of imprecision in the reinnervation of specific targets, as is evident in Henry Head's description of the slowness and imprecision of his recovery of fine motor and sensory function. This result has been confirmed by more recent studies in which reinnervation is observed over time in adult animals with experimentally damaged peripheral motor nerves (see Figure 26.8C). The subsequent regeneration can be fairly faithful to the original pattern, or not. Imprecision is due not only to a lack of fidelity of the regenerating axon and its original target site on the muscle. Instead, the return of polyneuronal innervation of neuromuscular synapses returns during regeneration and reinnervation. Much of this innervation is eventually eliminated, presumably via the same activity-dependent mechanisms that operate during the early postnatal period, when supernumerary axons are eliminated from the developing synapse (see Figure 23.13). If electrical activity is blocked during regeneration, either in the muscle fiber or in the afferent nerve, multiple innervation of the end plate sites remains. This return of activity dependent mechanisms that facilitate elimination of multiple innervation may be used to revalidate the pattern of singly innervated motor units and thus optimize restoration of function.

Regeneration in the Central Nervous System

With the exception of spinal cord and brainstem motor neurons, whose axons project to peripheral muscles and therefore have access to instructions for relatively successful peripheral regeneration, there is very little long-distance axon growth or reestablishment of functional connections within the CNS following injury. The limited regrowth of damaged CNS axons, even those whose cell bodies remain intact, largely accounts for the relatively poor prognosis following brain or spinal cord damage.

Damage to the CNS can occur in several ways. The brain or spinal cord can be injured acutely by *physical trauma*. Another type of damage is caused by a lack of oxygen created by locally diminished blood flow (ischemia) due to a vascular occlusion or local bleeding (stroke). Damage can also arise from global deprivation of oxygen, as in drowning or cardiac arrest. A different type of damage arises from *neurodegenerative diseases* such as Alzheimer's and Parkinson's diseases and amyotrophic lateral sclerosis (ALS). In all of these cases, axonal and dendritic neuronal loss occurs for reasons that are still poorly understood. Because central axons have little ability to regenerate, the key to whatever recovery is possible after brain injury lies with the complex cellular events that support the survival of those neurons that have not been killed outright and whose processes remain relatively intact.

Cellular and Molecular Responses to Injury in the Central Nervous System

There are two major reasons for the differences between successful peripheral regeneration and the limited regeneration in the CNS. First, damage to brain tissue tends to engage the mechanisms that lead to necrotic and apoptotic cell death for nearby neurons whose process have been severed. Second, the neuronal changes at the site of injury do not recapitulate developmental signaling that supports the initial establishment of brain and spinal cord circuits. Instead, a combination of glial growth and proliferation along with microglial activity (microglia have immune functions that lead to local inflammation) actively inhibits growth. Finally, there is an upregulation of growth-inhibiting molecules related to the chemorepellent factors that influence axon trajectories during development (see Figure 23.5C,D).

One of the striking differences in the consequences of central versus peripheral nerve cell damage is the extent of neuronal cell death that occurs after direct damage to the brain. Neuronal cell death in the CNS has been studied extensively in brains where hypoxia has occurred due to local vascular occlusion (a common form of stroke). In such cases, a loss begins rapidly in the hypoxic region and continues for an extended period, often eliminating most or all of the local neurons (Figure 26.9A). Where such cell loss has occurred, there is enhanced activation of caspase-3, an enzyme that, when activated, causes a cell to die by apoptosis (Figure 26.9B). This genetically regulated mechanism can be elicited by growth factor deprivation, by the hypoxia itself, by DNA damage, or cellular stress (diminished oxidative phosphorylation in mitochondria). The transection of axons presumably can lead to growth factor deprivation by removing the target source for the parent neurons. DNA damage and cellular stress (including changes in oxidative metabolism and damage from oxygen free radicals, which also trigger apoptosis) have been suggested to underlie some neurodegenerative diseases.

In addition to apoptosis, **autophagy** has emerged as an intermediate response to cellular stress that can either prevent, or ultimatey lead to neuronal degeneration. Autophagic responses operate normally to target cellular contents for lysosomal degeneration. When neurons (or other

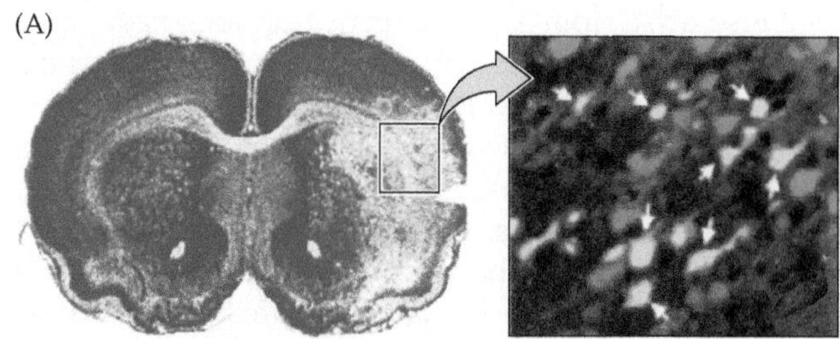

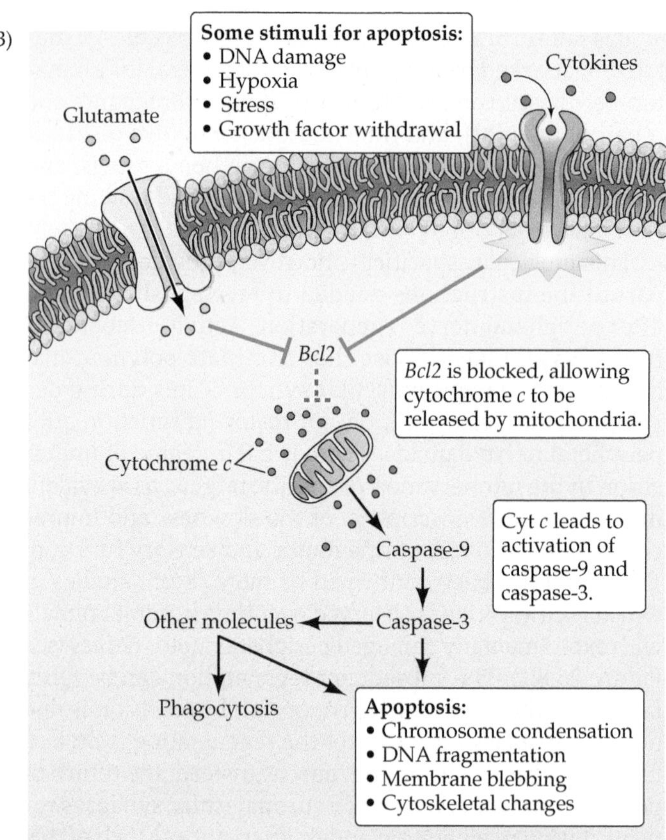

FIGURE 26.9 Consequences of hypoxia/ischemia in the mammalian brain. (A) Section through the brain of a 7-day-old mouse in which the carotid artery was transiently constricted. Nissl stain (see Chapter 1) was used to visualize cell bodies. The lighter region (i.e., little or no staining) shows the extent of cell damage and loss caused by this brief deprivation of oxygen. Cells in the higher-magnification image were stained for the neuronal marker Neu-N (red) and for activated caspase-3 (yellow), indicative of neurons undergoing apoptosis. (B) Model of the primary mechanism for neuronal apoptosis after injury. Apoptosis can be elicited by excitotoxicity via excess glutamate, or by the binding of inflammatory cytokines to receptors in the neuronal membrane. In addition, loss of neuronal connections to a target and resultant deprivation of trophic support can initiate apoptosis. Any or all of these stimuli result in the removal of the anti-apoptotic gene *Bcl2*. Cytochrome *c* is then released from mitochondria, activating caspase-3 and obligating the cell to apoptotic death as caspase-3 stimulates destructive changes in downstream molecules. (A from Manabat et al., 2003.)

cells) are under transient metabolic stress due to hypoxia or other bioenergetic fluctuations, the autophagic pathway will alter activity to accommodate the metabolic demands and move the as close as possible to a state that does not activate apoptosis. Disruptions of autophagy are implicated in a number of neurodegenerative disorders including Parkinson's and Alzheimer's diseases. In addition, autophagic responses may enhance neuronal survival in the face of acute damage to the nervous system and subsequent neuroinflammatory activity.

A major source of cellular stress is glutamatergic overstimulation caused by bursts of abnormal activity arising after local brain damage. Such overstimulation can also arise from epileptogenic foci and seizure activity generated at sites of damage. Elevated neuronal activity and its consequences are referred to as *excitotoxicity* and, if unchecked, can lead to neuron death. Following injury or seizure, large amounts of neurotransmitters are released. This enhanced signaling modifies the effectiveness of members of the Bcl2 family of anti-apoptotic molecules, which normally oppose changes in mitochondrial function that reflect oxidative stress. Diminished Bcl2 activity allows cytochrome *c* to be liberated from mitochondria into the cytoplasm, where it facilitates activation of caspase-3. Activated caspase-3 can then cause the fragmentation of nuclear DNA, membrane and cytoskeletal changes, and ultimately cell death (see Figure 26.9B). Thus, one of the key determinants of the long-term effects of damage to adult neural tissue is the extent to which the damage activates apoptosis, either directly due to immediate cell damage, or due to excitoxicity following excess neurotransmitter release in response to brain injury.

Glial Scar Formation in the Injured Brain

As might be expected from cellular mechanisms of peripheral nerve injury, glial cells found at the site of injury contribute to the degenerative and regenerative processes that occur after brain damage; however, central glia differ from Schwann cells in their responses. Brain injury elicits responses from all three glial classes—astrocytes, oligodendrocytes, and microglia—that actively oppose neuronal regrowth (Figure 26.10). Such cells, referred to as *activated glia,* are less susceptible to the stimuli that result in neuronal apoptosis after injury. Thus, local growth of glial cells is preserved while neighboring neurons die. Most brain lesions cause local proliferation of otherwise quiescent glial precursors, as well as extensive growth of processes from existing glial cells in or around the site of injury. These reactions lead to **glial scarring** (see Figures 26.10 and 26.11A), which reflects a local overgrowth and

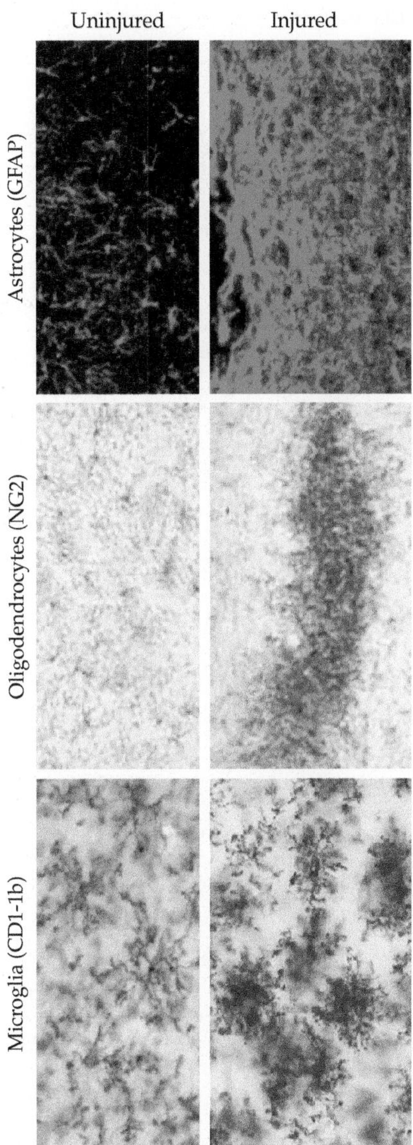

FIGURE 26.10 The reaction of the three major classes of glia in the CNS to local tissue damage. In each case, there is growth and change in expression of molecules normally associated with each cell class. (Top) Astrocytes labeled to visualize glial fibrillary acidic protein (GFAP) both before and after injury. (Center) The molecule NG2, notably present in glial scar tissue, is visualized here in oligodendroglial precursors and immature oligodendrocytes. (Bottom) CD1-1b, a marker for microglia. (Top from McGraw et al., 2001; center from Tan et al., 2005; bottom from Ladeby et al., 2005.)

sustained concentration of astrocytes as well as oligodendrocytes and their processes. In addition, macrophages undergo significant, efflorescent growth, adding to the physical barrier established by the glial scar. Glial scars are thought to be a major barrier for axon and dendrite regrowth in the CNS.

Several inhibitory molecules, most of which are related to brain myelin, prevent axon growth in the CNS. Most of these molecules are produced by central oligodendrocytes that contribute to glial scars. The protein components of brain myelin produced by oligodendrocytes inhibit axon growth, including the diminished ability of axons to grow on substrates enriched in myelin-associated proteins such as myelin-associated glycoprotein (MAG; Figure 26.11B). This observation presents a puzzle since MAG is also produced by Schwann cells but does not impede peripheral regeneration. Several transmembrane receptors expressed on injured CNS axons interact with myelin proteins including MAG, as well as ECM molecules, secreted signals for cytokines including tumor necrosis factor (TNF) and other cytokines (see below).

Immune Activation and Inflammation Following Brain Injury

Perhaps one of the most important challenges to CNS neuronal regeneration is the specific immune response initiated by glial cells before, during and after formation of a glial scar following injury. In some cases, the immune-inflammatory response is reinforced by additional cells of the monocyte lineage after damage to the *blood-brain barrier* (see Appendix), a cellular interface between blood vessels and neural tissue that keeps most molecules and immune cells from entering the brain (Figure 26.12A). Damage to brain tissue disrupts the blood-brain barrier by disrupting the tight junctions that link endothelial cells and thus prevent large molecules and circulating blood cells from directly entering CNS tissue. This allows invasion of neutrophils and monocytes, the subsequent activation of microglia as well as astrocytes, and finally the invasion of T and B cells (Figure 26.12B). This inflammatory response is initiated by the release of damage-associated proteins that include the contents of lysed neurons, mitochondrial enzymes, and non-protein signaling intermediates such as reactive oxygen species and pyrinergic ligands. These proteins initiate a cascade of cellular responses that elicit secretion of pro-inflammatory cytokines (Table 26.1), including cytokines that attract neutrophils and other monocytes, cytokines that mediate T and B cell maturation, and perhaps most important, the pro-inflammatory cytokine interleukin 1 (IL-1). IL-1 modulates expression of several immune mediators that reinforce the inflammatory state. The end result of this neuroinflammatory response is glial scarring that encapsulates the site of inflammation and forms a protective barrier for adjacent healthy brain tissue. Since inflammatory mediators are targets of many drugs, the inflammatory state could potentially be manipulated to minimize additional damage and improve the efficiency of repair after brain injury.

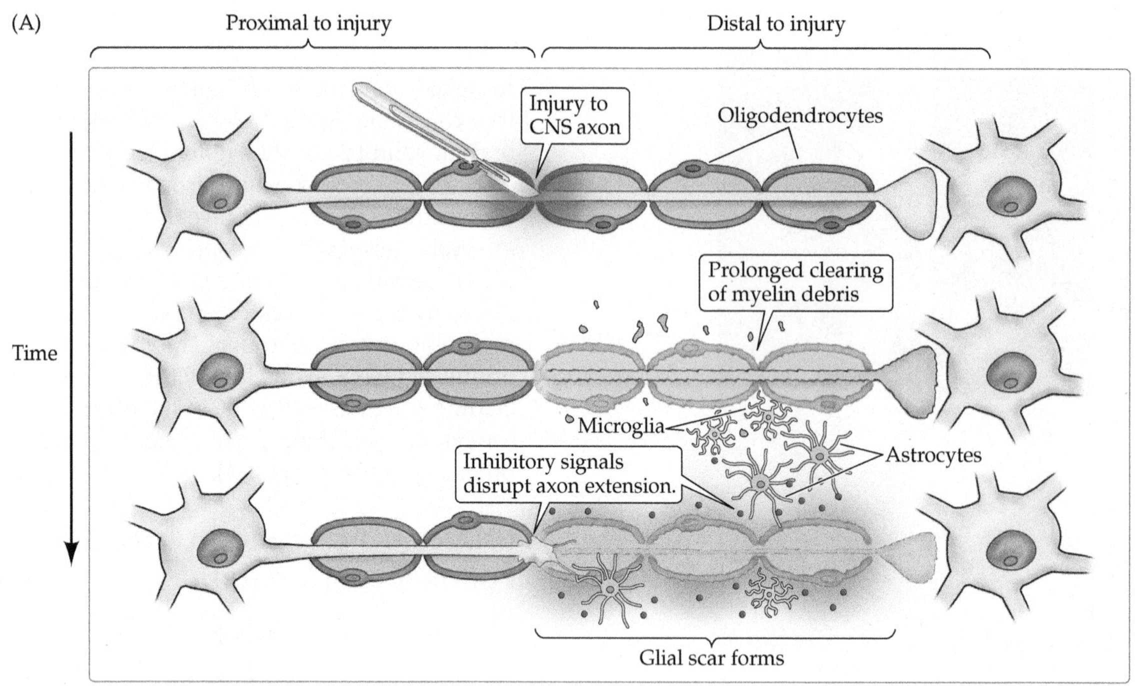

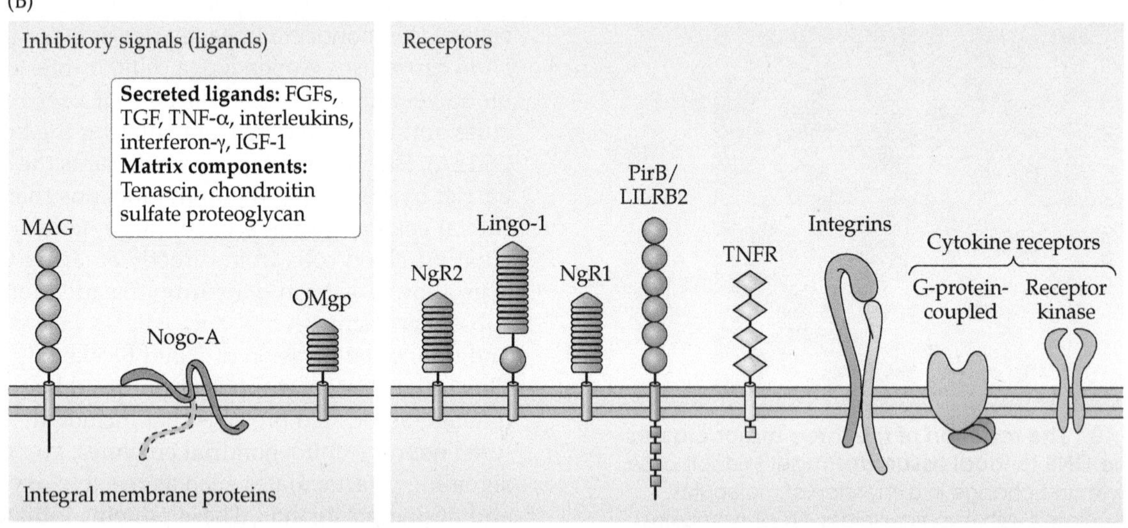

FIGURE 26.11 Cellular response to injury in the CNS. (A) Local cellular changes at or near an injured site include the degeneration of myelin and other cellular elements; the clearing of this debris by microglia, which act as phagocytic cells in the CNS; local production of inhibitory factors by astrocytes, oligodendroglia, and microglia; and glial scar formation. (B) Oligodendroglial cells and other glia either provide (on their cell surfaces) or secrete inhibitory signals that limit axon growth. Receptors for these signals are expressed on the newly generated axons of neurons whose original axons have been severed. NgR1 and 2 are receptors for the integral membrane protein NogoA. Lingo1 and PirB/LILRB2 are receptors for the myelin-associated proteins MAG (*my*elin-*a*ssociated *g*lycoprotein) and OMgp (*o*ligodendrocyte *m*yelin *g*lyco*p*rotein). Tissue necrosis factor receptors (TNFRs) bind the inflammatory cytokine TNF-α. Integrins bind several extracellular matrix proteins, and two classes of cytokine receptors bind pro- and anti-inflammatory cytokines released by astrocytes, oligodendroglia, or macrophages at sites of injury. (B after Kolodkin and Tessier-Lavigne, 2011.)

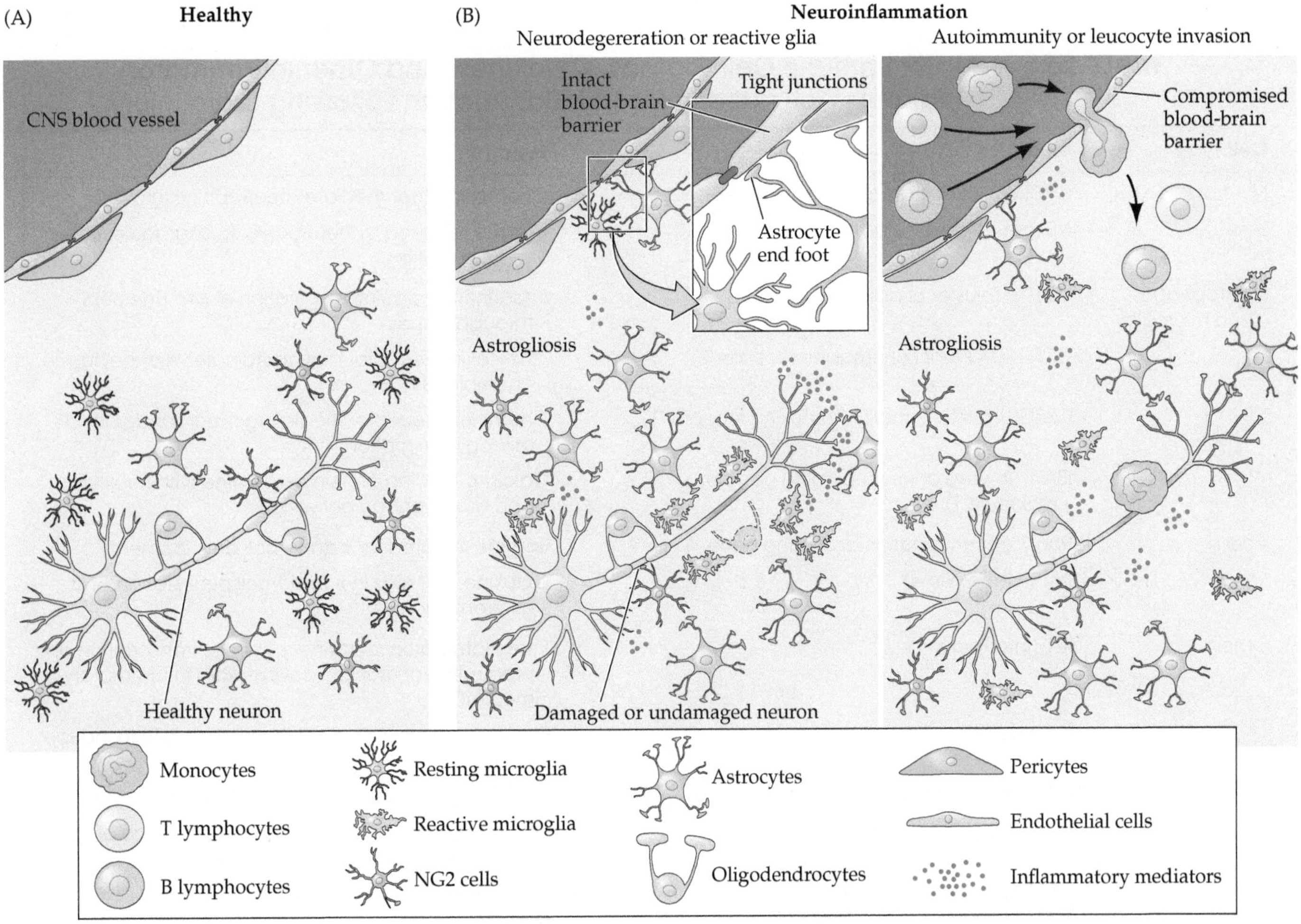

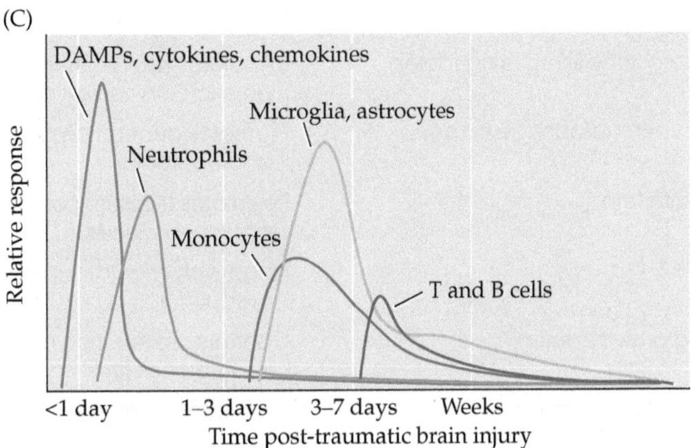

FIGURE 26.12 Immune-mediated inflammatory responses that drive glial responses and resistance to neuronal regrowth and repair. (A) The distribution and state of glial cells in an undamaged brain in which the blood-brain barrier is intact. Quiescent microglia, a few activated microglia, astrocytes, oligodendrocytes, and presumed oligodendrocyte precursors (Ng2 cells) are distributed throughout the tissue. (B) When local damage occurs in brain tissue (middle) or when the blood-brain barrier is disrupted (right), cytokines and other signaling molecules activate microglia and astrocytes. In turn, astrocytic processes as well as the astrocytes themselves increase in frequency, providing the scaffold for a local glial scar. Once the blood-brain barrier is compromised, neutrophils and other monocytes rapidly infiltrate the brain tissue and release additional pro-inflammatory cytokines that elicit a more robust astrocyte response and reinforce local inflammation. This leads to an additional decrease in the potential for preserving neuronal survival, tissue integrity, as well as the possibility of even modest regrowth and repair. (C) The timing of immune responses and activation, proliferation or invasion of cells that mediate inflammation in CNS tissue following acute damage to the CNS and disruption of the blood/brain barrier. This process begins with the release of damage-associated proteins, local cytokines and chemokines (produced by activated glia). (A,B from Waisman et al., 2015; C from McKee and Lukens, 2016.)

TABLE 26.1 ■ **Major Immune Cell Classes, Cytokines, and Other Inflammatory Mediators Activated in Neuroinflammation Following Brain Injury**

Cell type	Mediator	Function
Neutrophils	CXCR2 (C-X-G motif chemokine receptor 2)	Chemokine that mediates neutrophil migration
	NE (neutrophil elastase)	Enzyme released by neutrophils to degrade extra-cellular matrix
Macrophages and microglia	CD1-1b (cluster of differentiation 1-1b)	Integrin that regulates migration of immune cells through tissues
	CCR2 (C-G motif chemokine receptor 2)	Chemokine receptor that coordinates monocyte chemotaxis
	CX3CR1 (C-X3-G motif chemokine receptor 1)	Chemokine receptor mediating macrophage and microglia migration
	IBA 1 (ionized calcium-binding adapter molecule 1)	Calcium-binding protein associated with microglia and macrophage activation
T cells	Rag 1 (recombination activating gene 1)	Enzyme required for B and T cell development
	IL-4 (interleukin-4)	Cytokine that aids in B and T cell proliferation and differentiation
Others	IL-1 (interleukin-1)	Pro-inflammatory cytokine regulates transcription and production of multiple downstream inflammatory mediators
	Caspase-1	Enzyme that cleaves pro-IL-1β and pro-IL-18 to induce inflammation
	IL-18 (interleukin-18)	Pro-inflammatory cytokine that activates NK and T cells
	IL-6 (interleukin-6)	Pleiotropic cytokine that induces a multitude of inflammatory responses
	GFAP (glial fibrillary acidic protein)	Intermediate filament protein expressed by astrocytes
	TNF-α (tumor necrosis factor α)	Pleotropic cytokine that can promote cell death, inflammatory cytokine production, and cell proliferation
	G-CSF (granulocyte colony-stimulating factor)	Stimulates proliferation and differentiation of hematopoietic cells as well as neural progenitors
	GM-CSF (granulocyte-macrophage colony-stimulating factor)	Promotes generation and activation of myeloid cells and neurons
	Type 1 IFN (type 1 interferon)	Regulates transcription of pro-inflammatory cytokines and chemokines
	IL-10 (interleukin-10)	Negatively regulates pro-inflammatory cytokine production
	TGF-β (transforming growth factor β)	Controls proliferation and differentiation of multiple immune cell types
	TREM2 (triggering receptor expressed on myeloid cells 2)	Stimulates proliferation and differentiation of hematopoietic cells as well as neural progenitors

From McKee and Lukens, 2016.

Axon Growth after Brain Injury

One unfortunate consequence of glial scarring is its impediment to axon and dendrite growth. The reasons for this blockage (beyond physical obstruction) are not well understood. It is clear, however, that astroctyes within the glial scar produce molecules that inhibit axon growth. These include semaphorin 3A (causes growth cone collapse and axon withdrawl), several ephrins, and slit (see Figures 23.4 and 23.5). The receptors for each of these molecules are upregulated on the growth cones of axons that approach the glial scar, resulting in local distortions on the direction of growth: the axons turn away from the glial scar. In addition, matrix components that inhibit axon growth (particularly tenascin and chondroitin sulfate proteoglycan) are enriched in the extracellular spaces within the glial scar.

Thus, the proliferation and hypertrophy of glial cells, and their expression of chemorepellent or growth-inhibiting molecules, dominate the tissue reaction to focal brain injury (see Figures 26.11 and 26.10).

Neurogenesis in the Mature Central Nervous System

Few issues in modern neuroscience have engendered as much controversy as the ability of the adult central nervous system to generate new neurons in response to acute or degenerative damage to neural tissue. There was no reliable way to assess the extent of cell proliferation and mitotic activity in the mature brain until the advent of neuronal birth-dating techniques. These approaches use analogs to the DNA specific nucleotide thymidine. When a cell divides in the presence of such analogs, including radioactive-labeled thymidine (H^3-T) or BromodeoxyUridine (BrdU) and other Uracil variants, the analog is incorporated into the nuclear DNA. If the division is a terminal division or that of a slowly dividing stem cell, the nuclear DNA will be heavily labeled, and detectable by autoradiography (H^3-T) or immunohistochemistry (BrdU). Indeed, the neurogenesis that occurs in the peripheral olfactory epithelium of all mammals (see Chapter 15) remained the only unchallenged example for decades. Clinical experience, buttressed by animal studies primarily in mammals, indicated that the mature brain was unlikely to undergo significant genesis of new functional nerve cells; however, some vertebrates, such as fish and birds, do have a capacity for neurogenesis in response to brain injury.

The ability to label cells undergoing DNA replication and presumably mitosis and to track their progeny led to a clear but discouraging assessment of the potential for generating new neurons in the adult mammalian brain. Quite simply, there did not seem to be any extensive addition of neurons after the completion of early postnatal development. While this conclusion had extensive support, some reservations were raised as early as the mid-1960s. Joseph Altman and his colleagues, then at the Massachusetts Institute of Technology, found that small numbers of apparent granule neurons in the hippocampus and olfactory bulb in guinea pigs and rats could be heavily labeled with H^3-T injected at adult ages. Their work suggested that these local inhibitory interneurons with short axons that remained within the hippocampus or olfactory bulb, respectively, might be added to the brain during adulthood, either replacing or augmenting the cohort generated during development. At the time, however, the lack of additional markers that could identify these cells securely as neurons led other investigators to conclude that many, if not all, of these cells were newly generated glia rather than neurons. This conclusion was reached based upon more secure evidence available at the time that glial cells continued to divide in the CNS throughout life. Moreover, it seemed relatively unlikely that proliferative activity of already differentiated neurons might generate such cells.

Modern approaches have led to a much clearer understanding of the identity of the sources and progeny of mitotically active cells in the adult brain of many vertebrates, including mammals. A low level of glial cell proliferation—for both oligodendrocytes and astrocytes—does indeed continue throughout life; however, these glia are not the primary source of new neurons in mature brains. Moreover, it is clear that existing, differentiated neurons do not de-differentiate and divide. Instead, at a few locations in the brains of several species—including humans—there are regions where neural stem cells are maintained in the adult brain. As discussed in Chapter 22, the neural stem cells of the embryonic nervous system give rise to the full complement of cell classes found in neural tissue—that is, neurons, astrocytes, and oligodendroglia (see Box 22A)—as well as to more neural stem cells. Apparently, mature central nervous systems in several vertebrates provide an environment, or niche, that supports the maintenance of such multipotent neural precursors. These adult neural stem cells, like their embryonic counterparts, express primarily glial markers, but are distinguished by their location and their proliferative characteristics. The extent to which these stem cells give rise to neurons that replace or augment existing populations varies depending on species, brain region, and the conditions (e.g., growth, seasonal change, injury) that influence neurogenesis in the adult brain. Whether these limited examples of neurogenesis play a normal role in brain function is not yet clear.

Adult Neurogenesis in Non-Mammalian Vertebrates

Observations in several non-mammalian vertebrate species, particularly in teleost fish and songbirds, make a strong case for the capacity of adult vertebrate brains to add new neurons and incorporate them into functional circuits that guide behavior. One of the first thoroughly characterized examples of ongoing vertebrate adult neurogenesis was the goldfish. Goldfish, like many other fish, continue to grow throughout their lifetimes. Their body growth is matched by the growth of sensory structures in the periphery, particularly the eye. By the early 1970s, several investigators had recognized that growth of the eye was accompanied by the generation of new retinal neurons. Subsequent work showed that these neurons are generated from a subset of stem cells that form a ring around the entire margin of the goldfish retina (Figure 26.13A). These cells are capable of generating all of the cell classes in the goldfish retina (with the exception of rod photoreceptors, which are regenerated by a distinct adult precursor cell; see Chapter 11 for a review of retinal cell classes). They retain many of the

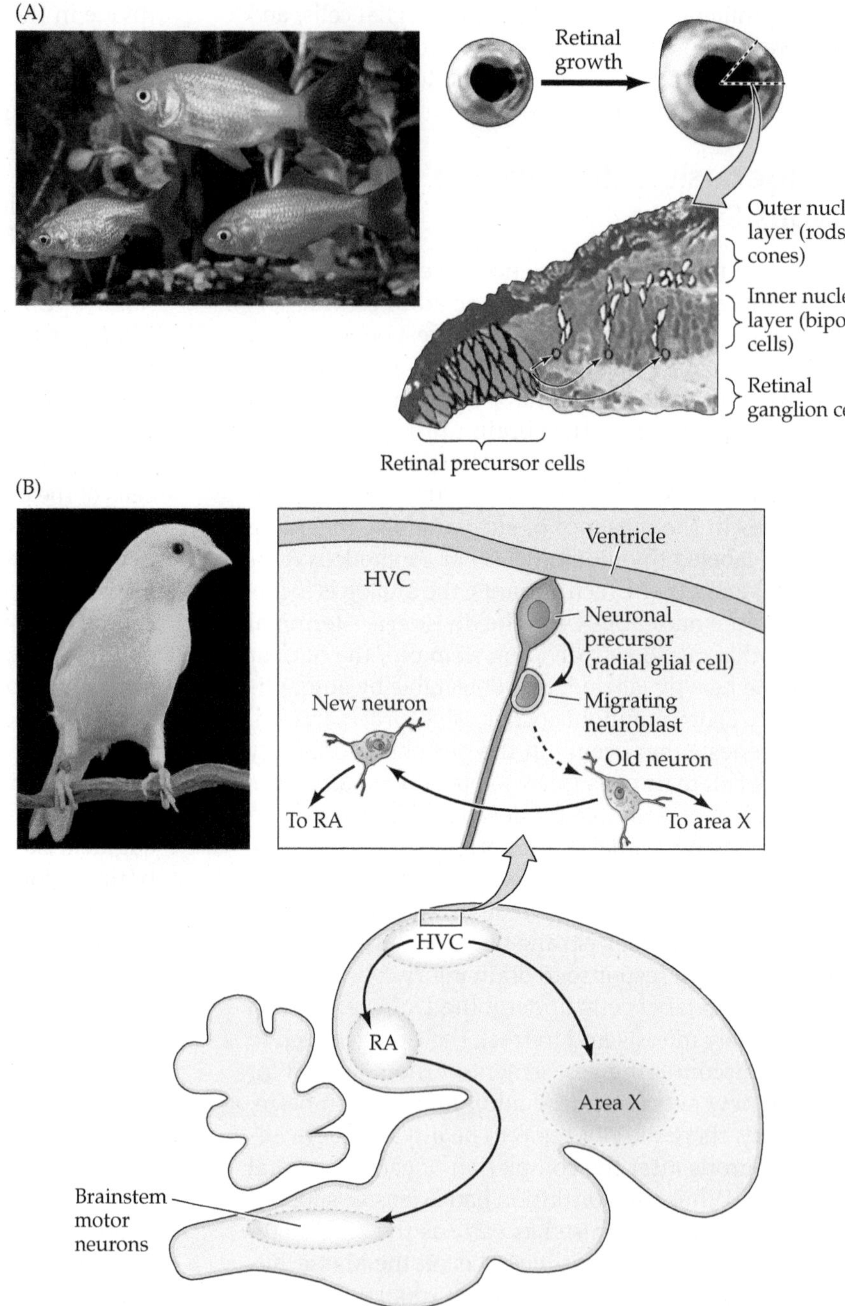

FIGURE 26.13 Adult neurogenesis in non-mammalian vertebrates. (A) Given favorable environmental conditions, teleost fish such as goldfish grow throughout their entire adult lives; the growth of the fish's body is matched by the growth of its eyes and brain. The retina grows by adding new neurons generated from a population of stem cells distributed in a ring at the very margin of the retina (red). These stem cells give rise to all retinal cell types except the rods (which are regenerated from precursors found in the existing differentiated region of the retina). (B) In a process of ongoing adult neurogenesis, male songbirds such as the canary lose and replace significant numbers of neurons in forebrain nuclei that control the production and perception of song. These song-control centers include the HVC (higher vocal center), RA (robustus archistriatus), and "area X," which is the equivalent of the caudate nucleus in the mammalian brain. In the HVC, a population of radial stem cells is maintained. The cell bodies are adjacent to the ventricular space, and their processes extend into the neuropil of the nucleus. A subset of neurons is retained in the nucleus as new ones are added. Neuroblasts migrate from the ventricular zone along the radial processes of the precursor cells and then integrate into circuits with existing neurons. (A after Otteson and Hitchcock, 2003, photo © Juniors Bildarchiv/Alamy; B after Goldman, 1998, photo © Eric Isselée/istockphoto.com.)

molecular characteristics of the stem cells in the embryo that initially generate the retina. The new neurons generated by these adult retinal stem cells integrate into the existing retina in rings between the precursor cells at the periphery and the existing differentiated retina—much like yearly growth rings being added to tree trunks. The axons of the new retinal ganglion cells enter and grow through the optic nerve and tract to reinnervate the optic tectum. Most of this axon growth happens along the extracellular matrix deposited at the glial limiting membrane of the optic nerve—the equivalent of the basal lamina conduits made by Schwann cells in the periphery. Regeneration of all retinal cell classes can also happen in response to a local injury of existing retinal tissue; however, the details of how this limited, localized repair is completed are somewhat different than that for ongoing addition of cells. As surprising as these remarkable capacities are, it is perhaps equally surprising that the optic tectum in the brain adds new neurons to accommodate the quantitative expansion of the retinal projection. These cells are not added in complementary rings; instead, populations of new neurons are added as crescents to the back of the tectum. Accordingly, the geometry of adult neurogenesis in the periphery and brain is mismatched. This divergence requires that new retinal inputs be constantly remapped along with existing retinal projections. Thus, there must be a great deal of dynamism in synaptic connections in the optic tectum of the adult goldfish. Connections must be made, broken, and remade to maintain the integrity of the retinotopic map as the fish grows throughout its lifetime.

An equally striking example of adult neurogenesis is found in several species of songbirds, including the canary and the zebra finch. This ongoing neurogenesis occurs in several regions of the avian brain; however, it has been most thoroughly studied in the structures that control vocalization and perception of song (Figure 26.13B; see also Figure 24.2). In most male songbirds, there is continual loss and addition of neurons in these regions. In some, the cycle of loss and regeneration follows mating seasons and is under the control of gonadal steroids (see Chapter 24), while in others it occurs continuously. Although it is tempting to speculate that the new neurons are a substrate for flexible acquisition or production of songs, no definitive evidence confirms this speculation. Moreover, many birds that have significant amounts of adult neurogenesis in song-control regions show very little flexibility in their song after the critical period for song learning is complete in early life.

Regardless of the behavioral consequences, it is estimated that birds replace most of the neurons in several song-control centers of their brain several times over a lifetime. The new neurons are generated from neural stem cells found in a limited region of the neural tissue immediately adjacent to the forebrain lateral ventricles. Stem cell bodies are found in this zone, and their radial processes (much like the radial glia that are the stem cells of the developing mammalian neocortex; see Chapter 22) extend into song-control centers (see Figure 26.13B). These cells function as precursors, generating new neurons via asymmetrical divisions, and as migration guides that constrain the translocation of new neurons from the ventricular zone to the song-control nuclei. Many of the new neurons integrate themselves into behaviorally relevant circuits and have functional properties that are consistent with a contribution to song production or perception. Nevertheless, a significant number die before they can fully differentiate, suggesting that there may be limits to the capacity of newly generated neurons to establish sufficient trophic support and activity-dependent validation to survive. Such effects would be similar to the processes that limit neuron numbers during initial development in many species (see Chapter 23). A key feature of adult neurogenesis in the avian brain is that there is always a balance of existing, long-lived neurons and newly generated neurons. Thus, even this compelling example of adult neurogenesis occurs in the context of significant stability in the mature brain.

Neurogenesis in the Adult Mammalian Brain

Over the past decade, the mechanisms and extent of neurogenesis in the adult mammalian brain have been reexamined in mice, rats, monkeys, and humans. Clearly, the capacity for ongoing neuronal replacement in specific CNS regions would provide a model for how regeneration might be elicited following brain injury or to combat neurodegenerative disease. The results of these efforts to understand neurogenesis in the adult brain are quite clear. New nerve cells in the CNS are generated reliably in just two regions, the olfactory bulb and the hippocampus (Figure 26.14A; see also Figure 15.6).

In the CNS, new nerve cells are primarily interneurons: granule cells and periglomerular cells in the olfactory bulb (see Chapter 15), or granule cells in the hippocampus (see Chapters 8 and 30). These newly generated olfactory or hippocampal interneurons are apparently the progeny of precursor or stem cells located close to the surface of the lateral ventricles, relatively near either the bulb or hippocampus. For the mature olfactory bulb, neurogenic stem cells are found in the anterior subventricular zone (SVZ) of the forebrain and in the hippocampus they are found in the subgranular zone (SGZ) within the hippocampal formation. The neural stem cells in the SVZ and SGZ do not give rise to projection neurons with long axons: only interneurons are generated at each site. At least some of these new neurons become integrated into functional synaptic circuits (Figure 26.14B,C); however, most new neurons generated in the adult brain die before being integrated into existing circuitry. No one has yet explained the functional significance of restricting neurogenesis to just these few regions in the adult brain, the exclusive GABAergic interneuron identity of the newly generated neurons, and the ultimate behavioral consequences for the addition of such cells. Moreover, the death of most newly generated neurons suggests that there may be a premium placed on stability in the mammalian brain—even in regions like the olfactory bulb and hippocampus—thus limiting opportunities for new neurons to join existing circuits. Nevertheless, the fact that new neurons can be generated and incorporated in at least a few adult brain regions shows that this phenomenon can occur in the mammalian CNS.

Cellular and Molecular Mechanisms of Adult Neurogenesis

In regenerating tissues such as intestine or lung, stem cells are found in a distinct location. Presumably there is a local environment within the stem cell **niche** that is conducive to the maintenance of stem cells as well as to the division and initial differentiation of cells that will reconstitute the adult tissue. The SVZ is a cell-dense region adjacent to the ventricular space found in the anterior aspect of the

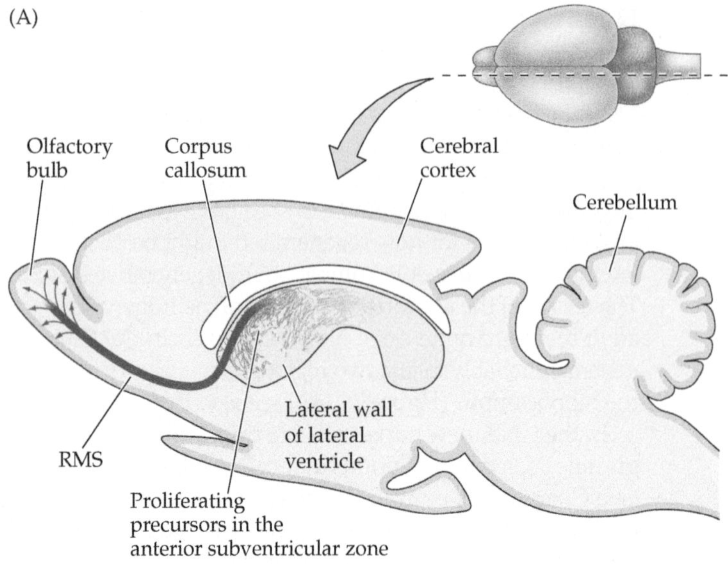

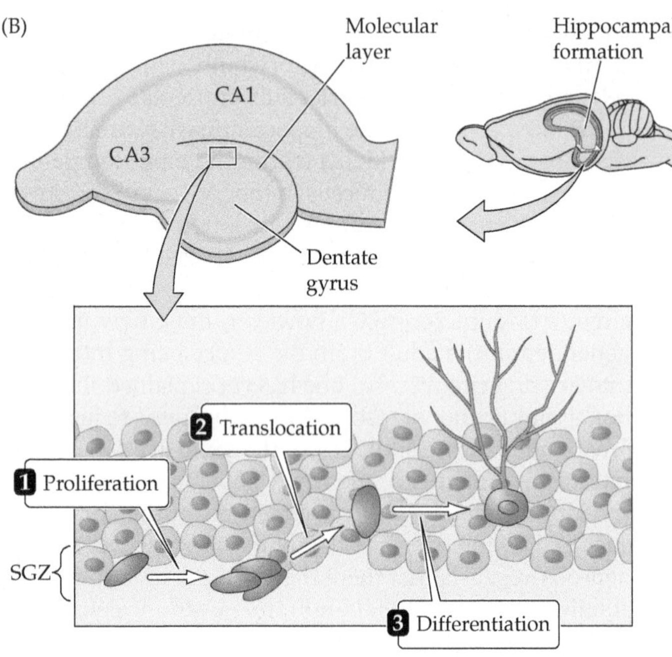

FIGURE 26.14 Neurogenesis in the adult mammalian brain. (A) Neural precursors in the epithelial lining of the anterior lateral ventricles in the forebrain (a region called the anterior subventricular zone, or SVZ) give rise to postmitotic neuroblasts that migrate to the olfactory bulb via a distinctive pathway known as the rostral migratory stream (RMS). Neuroblasts that migrate to the bulb via the RMS become either olfactory bulb granule cells or periglomerular cells; both cell types function as interneurons in the bulb. (B) In the mature hippocampus, a population of neural precursors is resident in the basal aspect of the granule cell layer of the dentate gyrus (the subgranular zone, or SGZ). These precursors give rise to postmitotic neuroblasts that translocate from the basal aspect of the granule cell layer to more apical levels. In addition, some of these neuroblasts elaborate dendrites and a local axonal process and apparently become GABAergic interneurons within the dentate gyrus. (C) A newly generated granule neuron (labeled green using a genetic marker) is contacted by GABAergic synapses (labeled red) in the granule cell layer of the adult hippocampus. (A,B after Gage, 2000; C from Kelsch et al., 2010.)

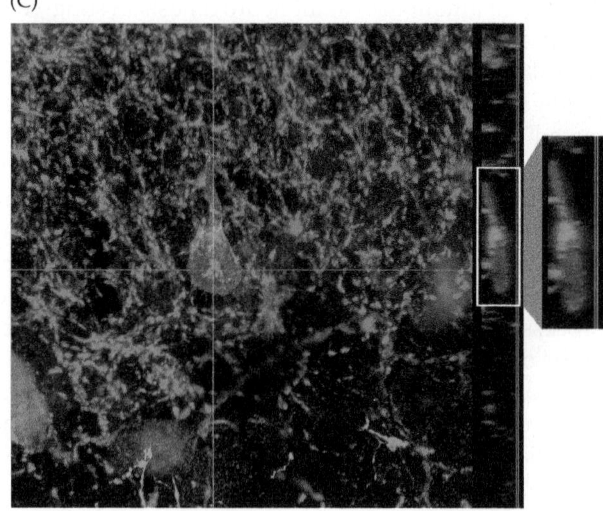

cortical hemispheres, adjacent to the ventricular space that is filled with cerebrospinal fluid (Figure 26.15A; also see Figure 26.14A). The SGZ is a region of cells immediately adjacent to the granule cell layer of the hippocampus (see Figure 26.14B). The SVZ and SGZ are the only regions that give rise to post-mitotic neurons in the adult brain that can integrate into existing circuits. Nevertheless, adult stem cells can be isolated not only from the SVZ (Figure 26.15B) and dentate gyrus, but also from subventricular regions of the cerebellum, midbrain, and spinal cord. When isolated and maintained in culture, these stem cells can give rise to additional stem cells or to neurons and glia, depending on the experimental conditions (see Figure 26.15B). In vivo, however, at sites beyond the olfactory bulb and hippocampus, adult neural stem cells do not appear to produce new neurons that are integrated into adjacent CNS regions.

In both the SVZ and SGZ, neural stem cells have characteristics similar to those of astrocytes, including the expression of multiple molecules that are also seen in astrocytes. Thus, similar to the developing brain (as well as in avian brains; see Figure 26.14B), the multipotent neural stem cell has an apparent glial rather than neuronal identity (see Chapter 22). Furthermore, neural stem cells are often found in close proximity to blood vessels, suggesting that they may be regulated by circulating as well as local signaling molecules. Although the route of delivery

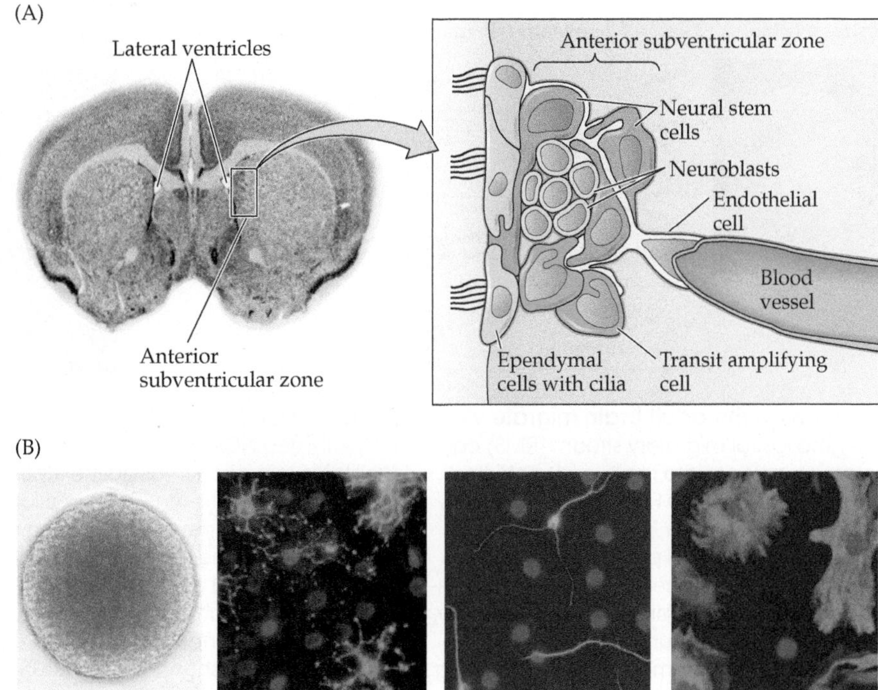

FIGURE 26.15 The forebrain's anterior subventricular zone provides a stem cell niche. (A) Coronal section through the mouse brain indicating the location of the anterior subventricular zone in the anterior part of the lateral ventricles. The schematic shows the arrangement of cells at the ventricular surface. The ciliated ependymal cells form a tight epithelial boundary separating cerebrospinal fluid from brain tissue. Immediately adjacent are neural stem cells, whose processes are intercalated between the ependyma and all other cell types. These stem cells are also often seen in proximity to blood vessels. Transit amplifying cells are also found in this domain, and their progeny—the neuroblasts—are often clustered close by before adhering to glia that guide them into the rostral migratory stream (RMS; see Figure 26.16). (B) Isolation and propagation in vitro of neural stem cells from the anterior subventricular zone. At left is a "neurosphere," a ball of neural stem and transit amplifying cells that has been generated clonally from a single founder dissociated from the adult subventricular zone epithelium. The adjacent three panels show differentiated cell types generated from the neurosphere. From left to right: oligodendroglia, neurons, and astrocytes. (A after Alvarez-Buylla and Lim, 2004; B from Councill et al., 2006.)

is different, this proximity to signals suggests that these precursors, like their developing counterparts, rely on cell-cell signaling to guide proliferation and differentiation. It is also possible that circulating signals relay information about the overall physiological state of the animal, thus linking broader homeostatic mechanisms with neurogenesis in the adult brain.

In order to generate differentiated neurons and glia, the neural stem cell must give rise to an intermediate precursor cell class, generally referred to as a **transit amplifying cell**. These cells retain the ability to divide; however, their cell cycles are much faster than those of stem cells, and they divide asymmetrically. After each cell division, a transit amplifying cell gives rise to a postmitotic daughter cell, plus another transit amplifying cell that reenters the cell cycle for an additional round of asymmetrical division. Transit amplifying cells are limited in their number of divisions, and eventually their potential for generating postmitotic blast cells is exhausted—thus yielding a terminal symmetric division. For neurogenesis to proceed constitutively, the transit amplifying cells must be replenished from the stem cell population. These newly generated, still undifferentiated neurons and glial cells—neuroblasts and glioblasts—are no longer competent to divide, and they move away from the SGZ or SVZ into regions of the olfactory bulb or hippocampus where mature neurons or glia are found. In the hippocampus, this distance is relatively small, and the cells undergo a modest local displacement to reach a final position relatively close to their site of generation. For newly generated neurons destined for the olfactory bulb, however, the distance from the SVZ, which is in the anterior ventricular region adjacent to the cerebral cortical hemisphere, to the olfactory bulb (which has no recognizable ventricular space) is considerable. A specific migratory route, defined by a distinct subset of glial cells, facilitates migration of newly generated neurons from the anterior SVZ to the bulb. This route is referred to as the **rostral migratory stream**

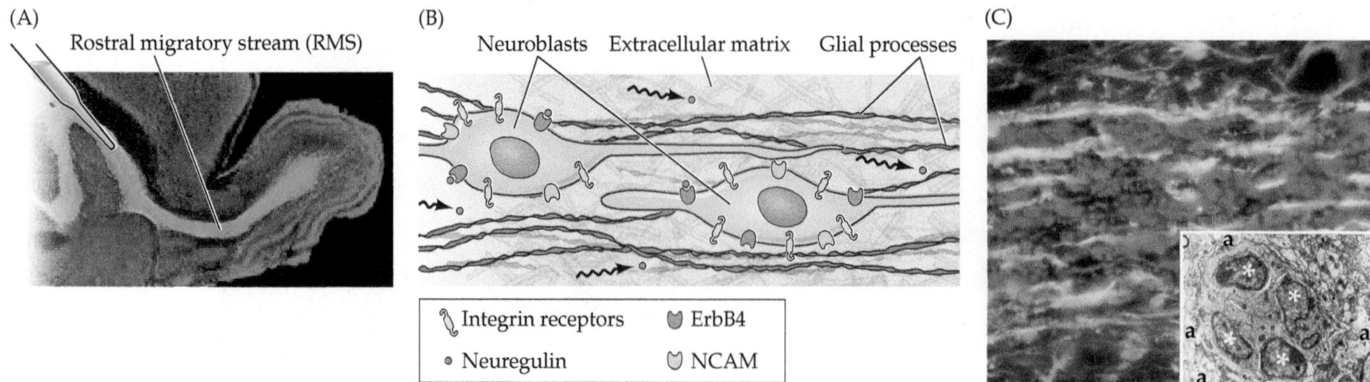

FIGURE 26.16 New neurons in the adult brain migrate via a specific pathway. (A) The rostral migratory stream (RMS) can be demonstrated by injecting a tracer into the lateral ventricle. The cells in the SVZ take up the tracer (see Figure 26.14), and the labeled cells enter the forebrain tissue in a "stream" of migrating neurons. (B) A schematic of the RMS. Glial processes (red) form conduits for migrating neurons. The extracellular matrix (ECM) associated with these processes influences migration, mediated by integrin receptors for ECM components found on migrating neurons. Secreted neuregulin also influences motility of the migrating neurons in the RMS, via the ErbB4 neuregulin receptor. Finally, polysialyated NCAM on the surfaces of newly generated neurons facilitates migration through the RMS. (C) The glial cells of the rostral migratory stream (RMS) in a longitudinal view are labeled in green, and the neurons migrating from the SVZ via the RMS are labeled red. The neurons are constrained within apparent tubes composed of glial cells and processes. The inset shows a cross section of one such "tube" with glial processes (a) encapsulating the migrating neurons (asterisks). (A from Ghashghaei et al., 2006; B after Ghashghaei et al., 2007; C from Peretto et al., 1999.)

(Figure 26.16; also see Figure 26.14A); within it, neuroblasts move along channels defined by the surfaces of elongated glial cells (these glial cells do not have stem cell properties). In this migratory pathway, as at other sites of cell migration or axon growth, an extracellular matrix, presumably secreted by the glial cells, facilitates migration. In addition, the migrating cells express the polysialyated form of the neural cell adhesion molecule NCAM, which promotes cell-cell interactions that facilitate migration. Secreted signaling molecules also influence migration in the rostral migratory stream. In this case, neuregulin and its ErbB receptors, which also influence axon guidance and synapse formation in the periphery, facilitates motility, particularly via interaction with the ErbB4 neuregulin receptor. Thus, in the rostral migratory stream, several developmentally regulated adhesion molecules and secreted signals mediate migration of new neurons through otherwise mature brain tissue.

Identification of molecular mediators of adult neurogenesis remains a major focus of current research. The most attractive hypothesis, and one that has support from the available data, is that the signaling molecules and transcriptional regulators used to define neural stem cells early in development are either retained or reactivated to facilitate neurogenesis in the adult. Accordingly, many of the inductive signaling molecules described in Chapter 22 as mediators of the specification of neural precursors and their progeny in the neural plate and neural tube are also active in adult SVZs. These include Sonic hedgehog (Shh), members of the fibroblast growth factor (FGF) family, TGF-β family members (including the bone morphogenetic proteins, or BMPs), and retinoic acid. The essential transcription factors associated with neural stem cells during initial CNS development, including Sox2, and those that identify transit amplifying cells and newly generated neurons in the embryonic brain, like the bHLH neurogenic genes (see Chapter 22) are expressed in adult neural stem and transit amplifying cells. Finally, many of the adhesion molecules that influence cell migration and dendritic and axonal outgrowth during neural development (see Figures 23.3–23.5) also influence the migration and differentiation of newly generated neurons.

Adult Neurogenesis, Stem Cells, and Brain Repair in Humans

The addition or replacement of neurons in the adult brains of fish, birds, rats, and mice provide clear examples of how new neurons may be integrated into existing circuits, presumably preserving, replacing, or augmenting function. In most cases, neuron replacement is gradual and likely related to ongoing low-level neurogenesis rather than to wholesale reconstitution of brain tissue in response to injury. Nevertheless, the limited capacity to replace neurons in an adult brain has offered some promise that, under the right conditions, neuron replacement might be used to repair the injured brain. In humans, the entire subventricular zone of the cerebral hemispheres provides a supportive environment for neural stem cells. At present, however, there is no evidence that adult neurogenesis occurs outside the hippocampus. The rostral migratory stream is absent in humans, suggesting that ongoing neurogenesis in the ventricular zone does not produce new neurons that migrate to olfactory bulb. Moreover, the available evidence suggests that there is no neurogenesis in the adult cerebral cortex (Box 26B)—an observation that establishes a much higher threshold for the possibility of stem cell-mediated repair of cortical circuitry damaged by trauma, hypoxia, or neurodegenerative disease. Cell replacement therapies have been attempted in a relatively small number of patients with Parkinson's disease (see Chapter 18), but the overall effectiveness of such treatments has been poor. Furthermore, the extent to which truly undifferentiated human neural stem cells can be made to acquire and maintain characteristics of dopaminergic neurons of the substantia nigra (or any other distinctive neuronal identity) as well as an appropriate pattern of synaptic connections is unclear. Thus, although there is some promise that understanding the maintenance of neural stem cells and their neurogenic potential in limited regions of the adult brain for repairing the damaged brain, the fulfillment of this promise may be extremely challenging.

Perhaps the key question in all this is why peripheral neural repair is broadly successful in mammals whereas central repair fails. A sensible speculation is that the CNS puts a premium on the stability of connections to ensure that learned behaviors are maintained. Since learning by experience plays a limited role peripherally, repair in the periphery is preeminent.

BOX 26B ■ Nuclear Weapons and Neurogenesis

The presence of stockpiles of nuclear weapons in an ever-increasing number of nations has always been a heavy burden for world affairs. It therefore may come as a surprise that the nuclear weapons testing carried out during the height of the Cold War (from the early 1950s through 1963) might play a positive, if unanticipated, role in resolving a major conflict in neuroscience.

In the late 1990s, a consensus emerged that the hippocampus and olfactory bulb are sites of gradual, limited addition of new neurons in the adult brain of all mammalian species, including humans. This consensus, however, did not extend to the question of whether new neurons are added to the cerebral cortex in adulthood. If adults do indeed add new neurons in significant numbers, such a mechanism would demand revision of current notions of plasticity, learning, and memory; it would also offer new avenues for treating traumatic, hypoxic, and neurodegenerative cortical damage. Several reports in the mid-1990s, including some from experiments on non-human primates, suggested that there might be substantial addition of neurons to the adult cortex. Despite the provocative nature and exciting implications of these findings, several other laboratories had difficulty replicating the surprising and controversial results. The disparities engendered a polarized debate with no easy resolution. Clearly, what was needed was an independent means of assessing neurogenesis in the adult cortex, preferably in humans.

In a unique approach—and in one of the more successful searches for evidence of weapons of mass destruction in the last several years—Jonas Friesén and colleagues at the Karolinska Institute in Stockholm took advantage of fluctuations in environmental exposure to radioisotopes from nuclear weapons testing to evaluate when cortical neurons are indeed generated over an individual's lifetime. Their method relied on the recognition that the normally steady state of the isotope carbon-14 (^{14}C) in Earth's atmosphere had been dramatically altered for a brief period between the mid-1950s and early 1960s. During this time, many countries conducted multiple tests of nuclear weapons, introducing large amounts of ionizing radiation into the atmosphere and nearly doubling the atmospheric concentration of ^{14}C. The Nuclear Test Ban Treaty of 1963 (which was adhered to by most countries until fairly recently) put a fairly abrupt end to this frightening period in human history, and atmospheric ^{14}C levels declined exponentially (Figure A).

This change in atmospheric ^{14}C concentration provided a natural version of experimental birth-dating techniques. Rather than injecting a bolus of tritiated thymidine or BrdU into the individual, people of varying ages had been naturally exposed to a "bolus" of ^{14}C that was incorporated into DNA being synthesized at the time. Thus, regardless of the age of the individual, cortical neurons generated between 1955 and 1963 should have a higher concentration of ^{14}C in their nucleus than those generated before or after that time frame.

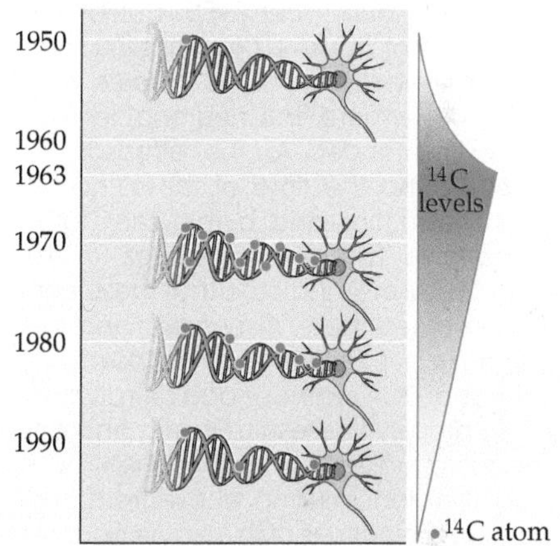

(A) Changes in atmospheric ^{14}C levels (right) and the availability of the isotope for incorporation into mitotic neurons at different times between 1955 (the start of frequent nuclear testing) and 1963 (when the Nuclear Test Ban Treaty was put in place). Neurons generated between 1963 and 1970, whether in adults or in individuals who underwent gestation and birth during that time frame, would incorporate significant amounts of ^{14}C into their nuclear DNA. (After Au and Fishell, 2006.)

To assess neurogenesis in this ingenious way, the researchers obtained autopsy samples from the cerebral cortices of seven individuals born between 1933 and 1973. The logic was that those born before 1955 would have significant numbers of ^{14}C-labeled neurons due to exposure as adults *only if there was indeed*

Continued on the next page

BOX 26B ■ (continued)

adult cortical neurogenesis. If there was no adult neurogenesis, only those individuals born during (or shortly after) 1955–1963 should have ^{14}C-containing cortical neurons. To ensure that only cortical neurons were assessed, dissociated cortical cells were labeled fluorescently with a neuron-specific marker and then sorted so that a comparison could be made between fluorescent-tagged neurons and the non-fluorescent glia and supporting cells (non-neurons). ^{14}C levels were then measured using accelerator mass spectroscopy.

The results were clear: Individuals born before 1955 had no cortical neurons with elevated ^{14}C levels; thus, no neurons had been generated in their adult cortices (Figure B, top). In contrast, individuals born after 1955, but before the return of ^{14}C levels to baseline, had significant numbers of ^{14}C-labeled neurons; furthermore, the neuronal level of ^{14}C corresponded to the atmospheric level around the time of these individuals' gestations and births (see Figure B, bottom). The different values for the non-neural cells indicate that these cells turn over, so their ^{14}C content declines because of dilution by subsequent rounds of DNA synthesis and cell division.

To amplify this result, Friesén and colleagues studied a group of patients who had skin cancers and whose treatment included injections of thymidine analogs such as BrdU. They then examined cells heavily labeled with BrdU in the cortex post mortem. BrdU does not label neurons (recognized by staining for Neu-N, a neural marker) or neuron-specific neurofilaments; however, BrdU does label glial cells, recognized with antibodies against glial fibrillary protein (Figure C). This observation, especially when taken together with the ^{14}C study, argues strongly against significant neurogenesis in the adult cerebral cortex.

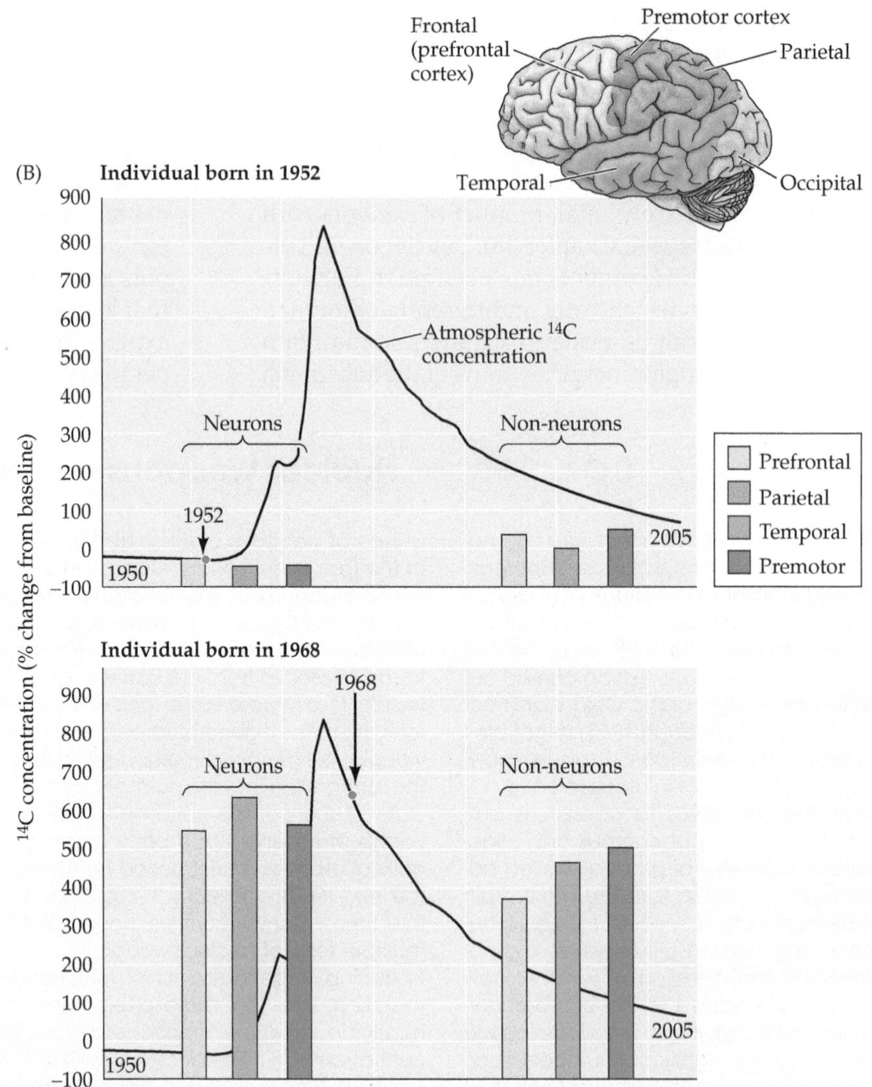

(B) Autopsy results for the individual born in 1952 showed no cortical neurons with elevated ^{14}C levels; thus, no neurons had been generated in the adult cortices. In contrast, the individual born in 1968 had significant numbers of ^{14}C-labeled neurons. Furthermore, in both cases, the neuronal level of ^{14}C corresponded to the atmospheric level at around the time of the individuals' gestations and births. The slightly elevated levels for non-neural cells in the 1952 individual, and the lower levels in the 1968 individual, indicate that these cells turn over, and their ^{14}C content is altered by subsequent rounds of DNA synthesis and cell division. (After Bhardwaj et al., 2006.)

(C)

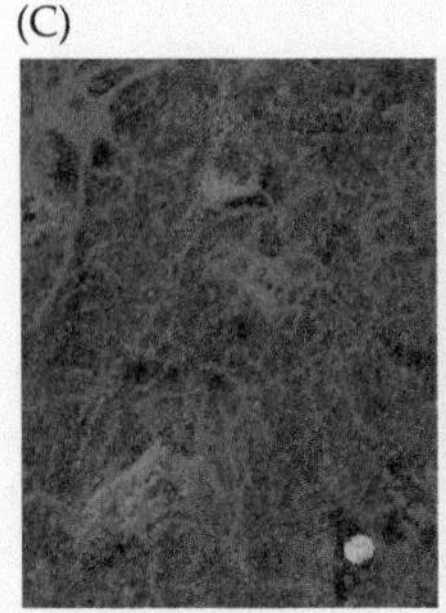

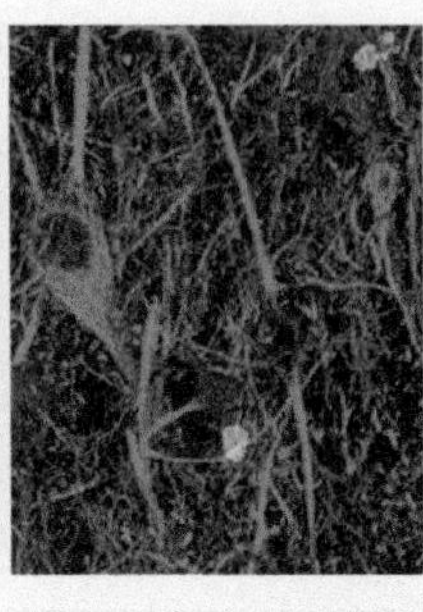

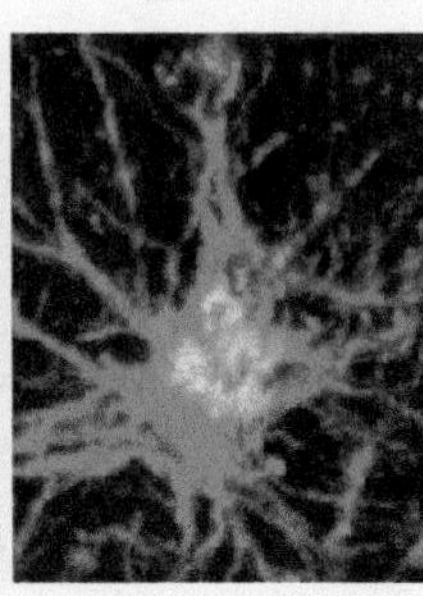

(C) The cortices from patients receiving BrdU during their lifetime were processed for BrdU histochemistry combined with neuronal and glial markers post mortem. Left: A BrdU nucleus (green) is distinct from cells labeled with Neu-N (red), a neuronal marker. Center: A similar distinction between neurons labeled for neurofilaments (red) and BrdU-labeled cells (green). Right: Cells labeled for GFAP are coincident with the BrdU-labeled nuclei, suggesting that only glia are generated in the adult brain. (From Bhardwaj et al., 2006.)

Summary

There are three types of cellular repair in the adult nervous system, in addition to the functional reorganization of surviving neurons and circuits that typically follows brain damage. The first and most effective is the regrowth of severed peripheral axons either from peripheral sensory neurons or central motor neurons, usually via the peripheral nerve sheaths once occupied by their forerunners. After regrowth, these axons reestablish sensory and motor synapses on muscles or other targets. During this regeneration, mature Schwann cells provide many of the molecules that regulate axon regrowth and targeting; these molecules are mostly those used for the same purpose during initial development. A second, and far more limited, type of repair is local sprouting or longer extension of axons and dendrites at sites of traumatic damage or degenerative pathology in the brain or spinal cord. Major impediments to such local repair include formation of glial scars; the death of damaged neurons due to trophic deprivation or other stress; inhibition of axon growth by protein components of myelin inhibition of neuronal growth by cytokines released during the immune response to brain tissue damage; and the formation of a glial scar by extensive hypertrophy of existing glial cells plus proliferation of glia at the site of the injury. The role of immune-mediated inflammation in establishing an anti-regenerative state in brain tissue is central. Molecular mediators of inflammation, including cytokines, their receptors, and related signaling intermediates, drive this process and establish barriers to neuronal regrowth. A third type of repair is generation of new neurons in the adult brain. Although there is no evidence for wholesale replacement of neurons and circuits in most vertebrate brains, the capacity for limited ongoing neuronal replacement exists in some species—sometimes in register with ongoing growth of the animal or due to seasonal variations. In most mammals, the olfactory bulb and the hippocampus are the only sites of adult neurogenesis. In both of these brain regions, new neurons are generated by neural stem cells retained in specific restricted locations in the adult brain. Many of the molecules that regulate the maintenance, proliferation, and differentiation of adult neural stem cells and their progeny are used for similar purposes for neural stem cells in the embryonic brain. The challenge of developing this capacity to generate new neurons and circuits as a strategy for repair following brain injury or degenerative disease continue to capture the imagination of patients, physicians, and many neuroscientists.

ADDITIONAL READING

Reviews

Alvarez-Buylla, A. and D. A. Lim (2004) For the long run: Maintaining germinal niches in the adult brain. *Neuron* 41: 683–686.

Boyd, J. G. and T. Gordon (2003) Neurotrophic factors and their receptors in axonal regeneration and functional recovery after peripheral nerve injury. *Mol. Neurobiol.* 27: 277–324.

Case, L. C. and M. Tessier-Lavigne (2005) Regeneration of the adult central nervous system. *Curr. Biol.*15: 749–753.

Deshmukh, M. and E. M. Johnson Jr. (1997) Programmed cell death in neurons: Focus on the pathway of nerve growth factor deprivation-induced death of sympathetic neurons. *Mol. Pharmacol.* 51: 897–906.

Gage, F. H. (2000) Mammalian neural stem cells. *Science* 287: 1433–1488.

Goldman, S. A. (1998) Adult neurogenesis: From canaries to the clinic. *J. Neurobiol.* 36: 267–286.

Johnson, E. O., A. B. Zoubos and P. N. Soucacos (2005) Regeneration and repair of peripheral nerves. *Injury* 36S: S24–S29.

Otteson, D. C. and P. F. Hitchcock (2003) Stem cells in the teleost retina: Persistent neurogenesis and injury-induced regeneration. *Vision Res.* 43: 927–936.

Rossini, P. M., C. Calautti, F. Pauri and J.-C. Baron (2003) Post-stroke plastic reorganisation in the adult brain. *Lancet Neurol.* 2: 493–502.

Sanes, J. N. and J. P. Donoghue (2000) Plasticity and primary motor cortex. *Annu. Rev. Neurosci.* 23: 393–415.

Silver, J. and J. H. Miller (2004) Regeneration beyond the glial scar. *Nature Rev. Neurosci.* 5: 146–156.

Song, Y., J. A. Panzer, R. M. Wyatt and R. J. Balice-Gordon (2006) Formation and plasticity of neuromuscular synaptic connections. *Internatl. Anesthesiol. Clinics* 44: 145–178.

Terenghi, G. (1999) Peripheral nerve regeneration and neurotrophic factors. *J. Anat.* 194: 1–14.

Wieloch, T. and K. Nikolich (2006) Mechanisms of neural plasticity following brain injury. *Curr. Opin. Neurobiol.* 16: 258–264.

Important Original Papers

Altman, J. (1969) Autoradiographic and histological studies of postnatal neurogenesis. IV. Cell proliferation and migration in the anterior forebrain, with special reference to persisting neurogenesis in the olfactory bulb. *J. Comp. Neurol.* 136: 269–293.

Altman, J. and G. D. Das (1967) Postnatal neurogenesis in the guinea-pig. *Nature* 214: 1098–1101.

Bregman, B. S. and 5 others (1995) Recovery from spinal cord injury mediated by antibodies to neurite growth inhibitors. *Nature* 378: 498–501.

David, S. and A. J. Aguayo (1981) Axonal elongation into peripheral nervous system "bridges" after central nervous system injury in adult rats. *Science* 214: 931–933.

Easter, S. S. Jr. and C. A. Stuermer (1984) An evaluation of the hypothesis of shifting terminals in goldfish optic tectum. *J. Neurosci.* 4: 1052–1063.

Eriksson, P. S. and 6 others (1998) Neurogenesis in the adult human hippocampus. *Nat. Med.* 4: 1313–1317.

Goldman, S. A. and F. Nottebohm (1983) Neuronal production, migration, and differentiation in a vocal control nucleus of the adult female canary brain. *Proc. Natl. Acad. Sci. USA* 80: 2390–2394.

Graziadei, G. A. and P. P. Graziadei (1979) Neurogenesis and neuron regeneration in the olfactory system of mammals. II. Degeneration and reconstitution of the olfactory sensory neurons after axotomy. *J. Neurocytol.* 8: 197–213.

Head, H., W. H. R. Rivers and J. Sherren. (1905) The afferent nervous system from a new aspect. *Brain* 28: 99–111.

Kim, J. E., S. Li, T. GrandPre, D. Qiu and S. M. Strittmatter (2003) Axon regeneration in young adult mice lacking Nogo-A/B. *Neuron* 38: 187–199.

Lois, C., J. M. Garcia-Verdugo and A. Alvarez-Buylla (1996) Chain migration of neuronal precursors. *Science* 271: 978–981.

Luskin, M. B. (1993) Restricted proliferation and migration of postnatally generated neurons derived from the forebrain subventricular zone. *Neuron* 11: 173–189.

Marshall, L. M., J. R. Sanes and U. J. McMahan (1977) Reinnervation of original synaptic sites on muscle fiber basement membrane after disruption of the muscle cells. *Proc. Natl. Acad. Sci. USA* 74: 3073–3077.

Nguyen, Q. T., J. R. Sanes and J. W. Lichtman (2002) Pre-existing pathways promote precise projection patterns. *Nat. Neurosci.* 5: 861–867.

Sanai, N. and 11 others (2004) Unique astrocyte ribbon in adult human brain contains neural stem cells but lacks chain migration. *Nature* 427: 740–744.

Skene, J. H. and M. Willard (1981) Axonally transported proteins associated with axon growth in rabbit central and peripheral nervous systems. *J. Cell Biol.* 89: 96–103.

Suhonen, J. O., D. A. Peterson, J. Ray and F. H. Gage (1996) Differentiation of adult hippocampus-derived progenitors into olfactory neurons in vivo. *Nature* 383: 624–627.

Zheng, B. and 5 others (2003) Lack of enhanced spinal regeneration in Nogo-deficient mice. *Neuron* 38: 213–224.

Go to the NEUROSCIENCE 6e Companion Website at **www.oup.com/uk/Purves6e** for Web Topics, Animations, Flashcards, and more.

UNIT V

Complex Brain Functions and Cognitive Neuroscience

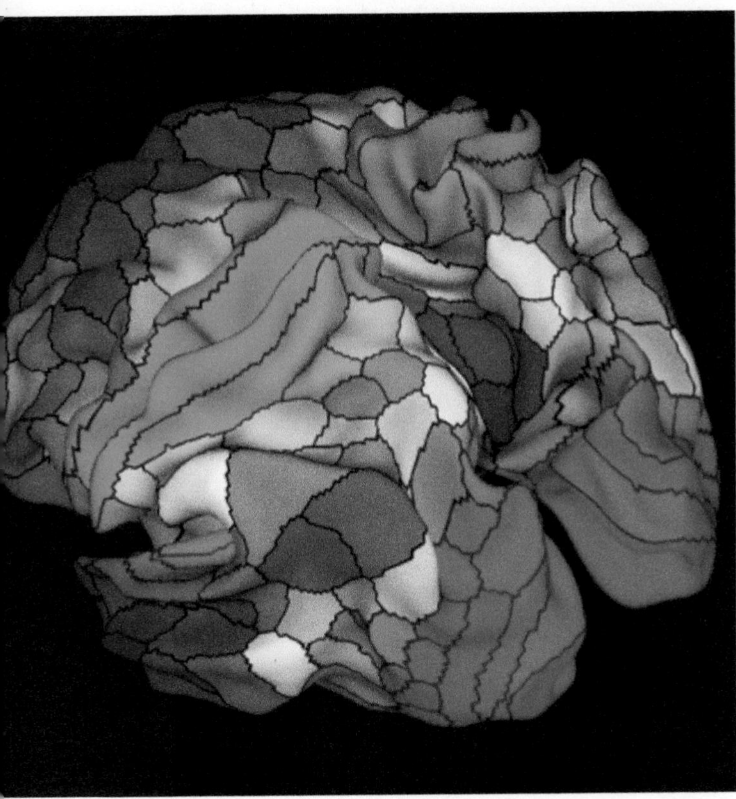

THE PERCEPTION OF PHYSICAL and social circumstances, attending to those that are especially important, remembering the past, planning for the future, experiencing emotions, and using language to express these aspects of our subjective world all rank among the most scientifically challenging functions of the human brain. The intrinsic interest of these aspects of brain function is unfortunately equaled by the difficulty—both technical and conceptual—involved in unraveling their neurobiological underpinnings. Nonetheless, over the last century or more much progress has been made in deciphering some of the principles involved, as well as the structural and functional organization of many of the relevant brain regions. The older approach of clinical evaluation of brain damaged patients and postmortem correlation is now complemented by noninvasive brain imaging and other techniques. And these methods can of course be used to study healthy individuals as well as patients. At the same time, electrophysiological experiments in non-human primates and other experimental animals have begun to elucidate the cellular correlates of many of these functions. This general domain is called *cognitive neuroscience*. The goal of this Unit is to review the progress made in this relatively new field, pointing out some of the challenges that remain.

On the previous page:

A map of 360 areas in human cerebral cortex that differ from one another in structure, function, and/or connectivity, as determined using in vivo neuroimaging data from the Human Connectome Project. (From Glasser et al., 2016.)

CHAPTER 27

Cognitive Functions and the Organization of the Cerebral Cortex

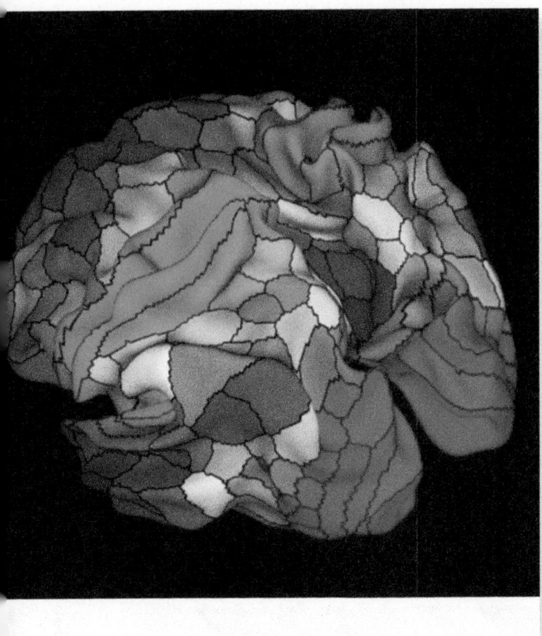

Overview

GIVEN THE IMPORTANCE OF COGNITIVE FUNCTIONS for human behavior and culture, it is hardly surprising that much of the human brain is devoted to the operations listed in the Unit Outline. The admittedly vague descriptor for the brain regions that carry out these functions is the *association cortex* (*or cortices*), which brings together sensory and much other information to produce useful behavior, whatever the circumstances might be. The term *association* refers to the fact that these regions of the cortex integrate (associate) information derived from other brain regions. This strategy makes sense, since any animal would be advantaged by tapping into the full range of available information before responding. The inputs to the association cortices are projections from the primary and secondary sensory and motor cortices, the hippocampus, the thalamus, and the brainstem. Outputs from the association cortices reach the hippocampus, the basal ganglia and cerebellum, and the thalamus, as well as other cortical areas. Although insight into the function of different cortical regions initially came from observations of patients with damage to one or more of these areas, noninvasive brain imaging of normal subjects, functional mapping at neurosurgery, and electrophysiological analysis of comparable brain regions in non-human animals have confirmed and greatly extended clinical deductions. Together, these studies show that the parietal association cortex is especially important for attending to stimuli in the external and internal environment; the temporal association cortex is especially important for recognizing objects and conditions; and the frontal association cortex is especially important for selecting and planning appropriate behavioral responses. The importance of the occipital/parietal association cortex for vision and of temporal association cortex for audition has already been discussed in earlier chapters.

A Primer on Cortical Structure

Before delving into a more detailed account of the functions of the different regions of the association cortices (Figure 27.1A), some general understanding of cortical structure and the organization of its canonical circuitry is useful. Most of the cortex that covers the cerebral hemispheres is *neocortex*, defined as cortex that has six cellular layers, or **laminae**. Each layer comprises more or less distinctive populations of cells based on their different densities, sizes, shapes, inputs, and outputs. The laminar organization and basic connectivity of the human cerebral cortex are summarized in Figure 27.1B and Table 27.1. Despite an overall uniformity, regional differences based on these laminar features have long been apparent (Box 27A), allowing investigators to identify numerous subdivisions of the cerebral cortex (Figure 27.1C,D). These histologically defined subdivisions are referred to

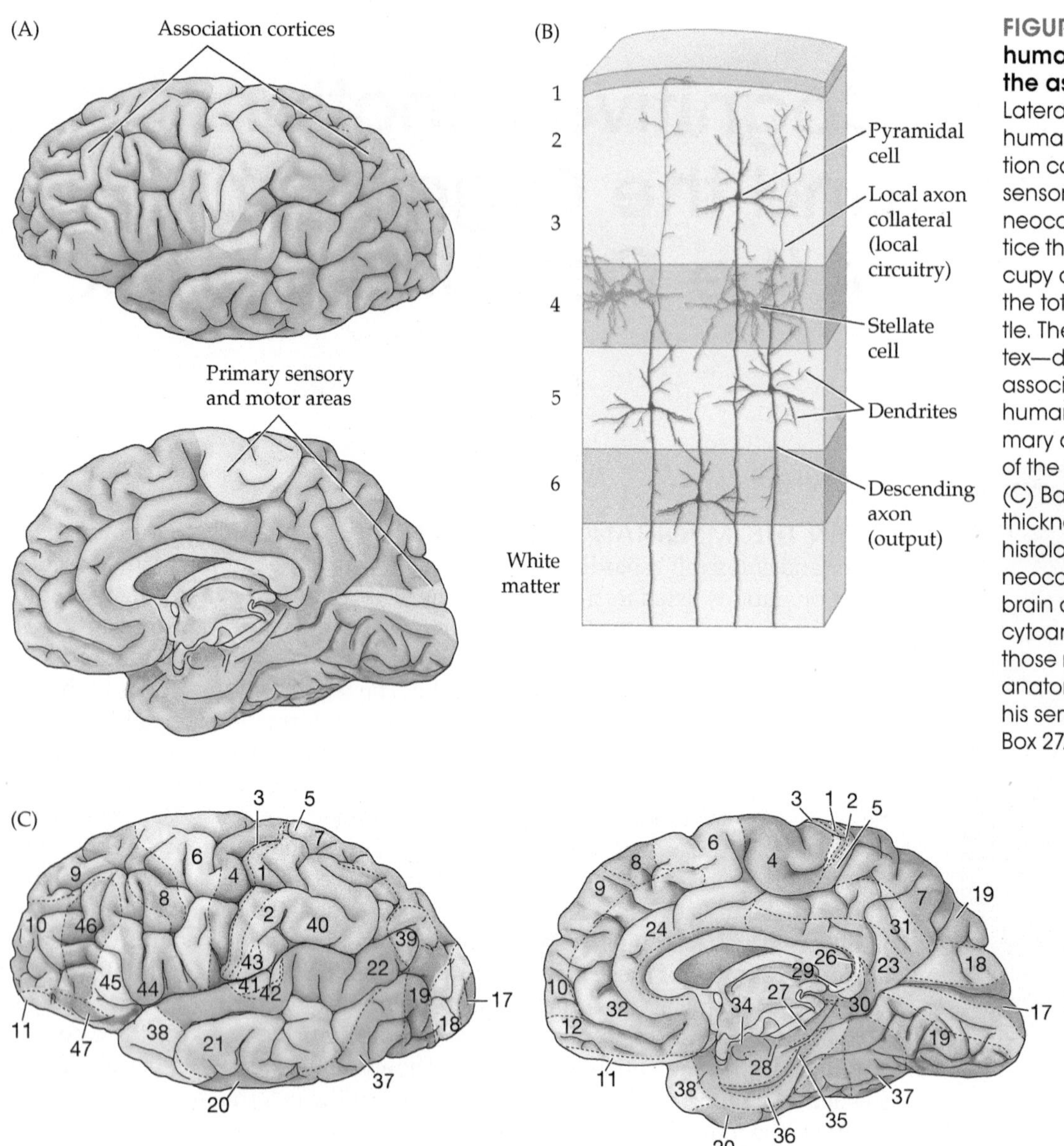

FIGURE 27.1 Structure of the human neocortex, including the association cortices. (A) Lateral and medial views of the human brain show the association cortices in blue. The primary sensory and motor regions of the neocortex are shown in yellow. Notice that the primary cortices occupy a relatively small fraction of the total area of the cortical mantle. The remainder of the neocortex—defined by exclusion as the association cortices—is the seat of human cognitive ability. (B) Summary of the cellular composition of the six layers of the neocortex. (C) Based on variations in the thickness, cell density, and other histological features of the six neocortical laminae, the human brain can be divided into some 50 cytoarchitectonic areas, typically those recognized by the neuroanatomist Korbinian Brodmann in his seminal 1909 monograph (see Box 27A).

as **cytoarchitectonic areas**, and over the years a zealous band of neuroanatomists has painstakingly mapped these areas in humans and in many widely used laboratory animals.

Early in the twentieth century, cytoarchitectonically distinct regions were identified with little or no knowledge of their functional significance. Eventually, studies of patients in whom one or more of these cortical areas had been damaged, supplemented by electrophysiological mapping in both laboratory animals and neurosurgical patients, supplied this information. This work showed that many of the regions neuroanatomists had distinguished on histological grounds are also functionally distinct. Thus, cytoarchitectonic areas can sometimes be identified by the physiological

TABLE 27.1 ■ Major Connections of the Neocortex

Sources of cortical input	Targets of cortical output
Other cortical regions	Other cortical regions
Hippocampal formation	Hippocampal formation
Amygdala	Amygdala
Thalamus	Thalamus
Brainstem modulatory systems	Caudate and putamen (striatum)
	Brainstem
	Spinal cord

BOX 27A ■ Cortical Lamination

Much knowledge about the cerebral cortex is based on descriptions of differences in cell number and density throughout the cortical mantle. Nerve cell bodies, because of their high metabolic rate, are rich in basophilic substances (RNA, for instance) and therefore tend to stain darkly with reagents such as cresyl violet acetate. These *Nissl stains* (named after Franz Nissl, who first described this technique when he was a medical student in nineteenth-century Germany) provide a dramatic picture of brain structure at the histological level. The most striking feature revealed in this way is the distinctive lamination of the cortex in humans and other mammals, as seen in the figure. Humans have three to six cortical layers, depending on the area of cortex. These layers, or *laminae*, are designated with the numerals 1–6 (or with Roman numerals I–VI). Laminar subdivisions are indicated with letters (layers 4a, 4b, and 4c of the visual cortex, for example).

Each of the cortical laminae in the so-called *neocortex* (which covers the bulk of the cerebral hemispheres and is defined by six layers) has characteristic functional and anatomical features (see Figures 27.1 and 27.2). For example, cortical layer 4 is typically rich in stellate neurons with locally ramifying axons; in the primary sensory cortices, these neurons receive input from the thalamus, the major sensory relay from the periphery. Layer 5, and to a lesser degree layer 6, contain pyramidal neurons whose axons typically leave the cortex. The generally smaller pyramidal neurons in layers 2 and 3 (which are not as distinct as their numerical assignments suggest) have primarily corticocortical connections, and layer 1 contains mainly neuropil.

Early in the twentieth century, Korbinian Brodmann devoted his career to an analysis of brain regions distinguished in this way, describing some 50 distinct cortical regions, or *cytoarchitectonic areas* (see Figure 27.1C). These structural features of the cerebral cortex continue to figure importantly in discussions of the brain, particularly in structural-functional correlation of intensely studied regions such as the primary sensory and motor cortices.

Not all cortex is six-layered neocortex. The hippocampal cortex, which lies deep in the temporal lobe and has been implicated in acquisition of declarative memories (see Chapter 30), has only three or four laminae. The hippocampal cortex is regarded as evolutionarily more primitive and is therefore called *archicortex* (*archi*, "first") to distinguish it from the six-layered neocortex. Another, presumably even more primitive, type of cortex is the *paleocortex* (*paleo*, "ancient"); paleocortex generally has three layers and is found on the ventral surface of the cerebral hemispheres and along the parahippocampal gyrus in the medial temporal lobe.

The functional significance of different numbers of laminae in neocortex, archicortex, and paleocortex is not known, although it seems likely that the greater number of layers in neocortex reflects more complex information processing. The general similarity of neocortical structure across the entire cerebrum suggests a common denominator of cortical operation, although no one has yet deciphered what it is.

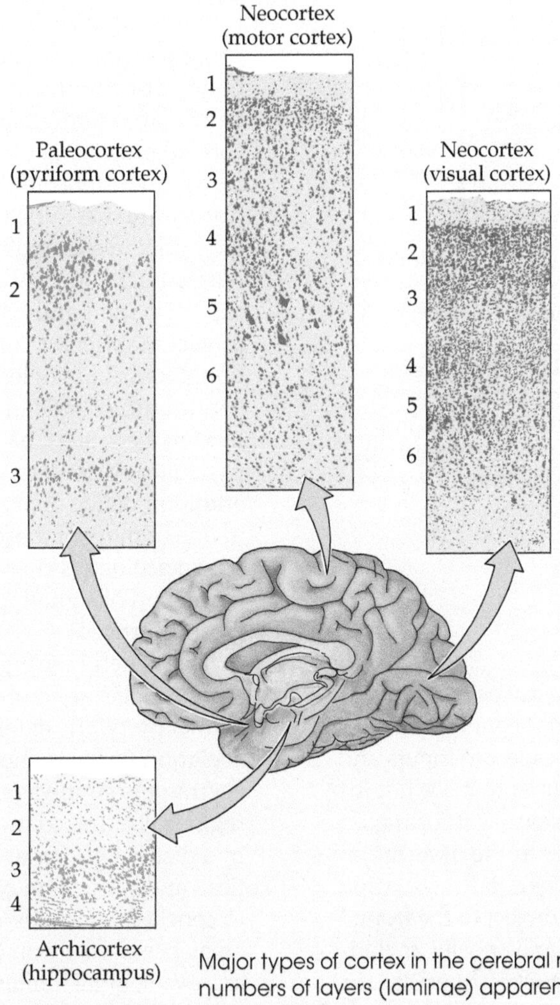

Major types of cortex in the cerebral mantle, based primarily on the different numbers of layers (laminae) apparent in histological sections.

response properties of their constituent cells, and often by their patterns of local and long-distance connections.

Despite significant variations among different cytoarchitectonic areas, the circuitry of all cortical regions has some common features (Figure 27.2). First, each cortical layer has a primary source of inputs and a primary output target. Second, each area has connections in the vertical axis (called *columnar* or *radial* connections) and connections in the horizontal axis (called *lateral* or *horizontal* connections). Third, cells with similar functions tend to be arrayed in radially aligned groups that span all the cortical layers and receive inputs that are often segregated into radial bands or columns. Finally, interneurons within specific cortical layers give rise to extensive local axons that extend horizontally in the cortex, often linking functionally similar groups of cells. The particular circuitry of any cortical region is a variation on this canonical pattern of inputs, outputs, and vertical and horizontal patterns of connectivity.

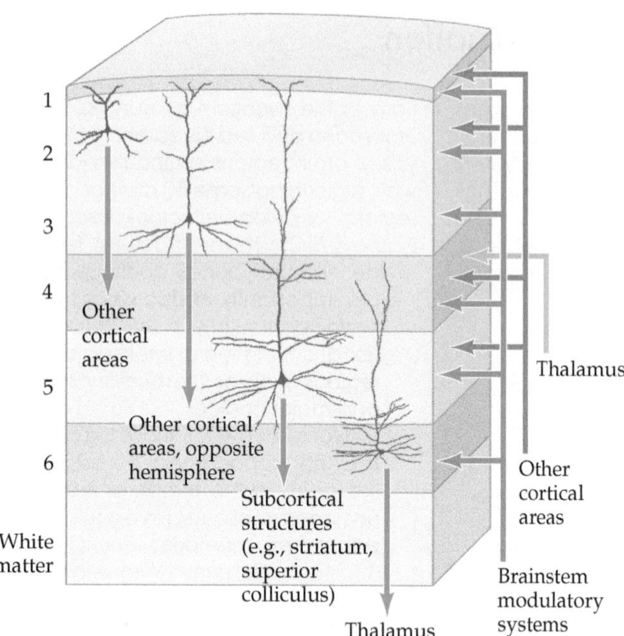

FIGURE 27.2 Canonical neocortical circuitry. Green arrows indicate outputs to the major targets of each of the neocortical layers in humans; orange arrow indicates thalamic input (primarily to layer 4); purple arrows indicate input from other cortical areas; blue arrows indicate input from the brainstem modulatory systems to each layer.

Unique Features of the Association Cortices

The connectivity of the association cortices is appreciably different from that of primary and secondary sensory and motor cortices, particularly with respect to inputs and outputs. For instance, two thalamic nuclei that are not involved in relaying primary motor or sensory information provide much of the subcortical input to the association cortices: the pulvinar projects to the parietal association cortex, while the medial dorsal nuclei project to the frontal association cortex. Several other thalamic nuclei, including the anterior and ventral anterior nuclei, innervate the association cortices as well. In consequence, the signals coming into the association cortices via the thalamus reflect sensory and motor information that has *already* been processed in the primary sensory and motor areas of the cerebral cortex and is being fed back to the association regions. The primary sensory cortices, in contrast, receive thalamic information that is more directly related to peripheral sense organs (see Unit II). Similarly, much of the thalamic input to primary motor cortex is derived from the thalamic nuclei related to the basal ganglia and cerebellum rather than to other cortical regions (see Unit III).

A second major difference in the sources of innervation to the association cortices is their enrichment in projections from other cortical areas, called **corticocortical connections** (see Figure 27.2). Indeed, these connections form the majority of the input to the association cortices. Ipsilateral corticocortical connections arise from primary and secondary sensory and motor cortices, and from other association cortices within the same hemisphere. Corticocortical connections also arise from both corresponding and noncorresponding cortical regions in the opposite hemisphere via the corpus callosum and anterior commissure, which together are referred to as **interhemispheric connections**. In the association cortices of humans and other primates, corticocortical connections often form segregated radial bands or columns in which interhemispheric projection bands are interdigitated with bands of ipsilateral corticocortical projections.

Another important source of innervation to the association areas is subcortical, arising from the dopaminergic nuclei in the midbrain, the noradrenergic and serotonergic nuclei in the brainstem, and cholinergic nuclei in the brainstem and basal forebrain. These diffuse inputs project to different cortical layers and, among other functions, contribute to learning, motivation, and arousal (the continuum of mental states that ranges from deep sleep to high alert; see Chapter 28). A variety of behavioral and psychiatric disorders, including addiction, depression, and attention deficit disorder, are associated with dysfunction in one or more of these neuromodulatory circuits. Current pharmacological therapies for these diseases rely on manipulation of signaling via modulatory inputs to the association cortices.

The general wiring plan for the association cortices is summarized in Figure 27.3. Despite this degree of interconnectivity, the extensive inputs and outputs of the association cortices should not be taken to imply that everything is simply connected to everything else in these regions. On the contrary, each association area is defined by a distinct, if overlapping, subset of thalamic, corticocortical, and subcortical connections. It is nonetheless difficult to conclude much about the role of these different cortical areas based solely on connectivity. This information is, in any event, quite limited for the human association cortices; most of the evidence comes from anatomical tracing studies in non-human primates, supplemented by the limited pathway tracing that can be done in human brain tissue postmortem as well as the newer noninvasive technique called **diffusion tensor imaging**, or **DTI**, which can identify large bundles of axons connecting brain areas in living humans. DTI forms the core technique driving the Human Connectome Project, an ambitious effort to map all the connections within the human brain.

Inferences about the function of human association areas continue to depend critically on observations of patients with cortical lesions. Damage to the association cortices in the parietal, temporal, and frontal lobes,

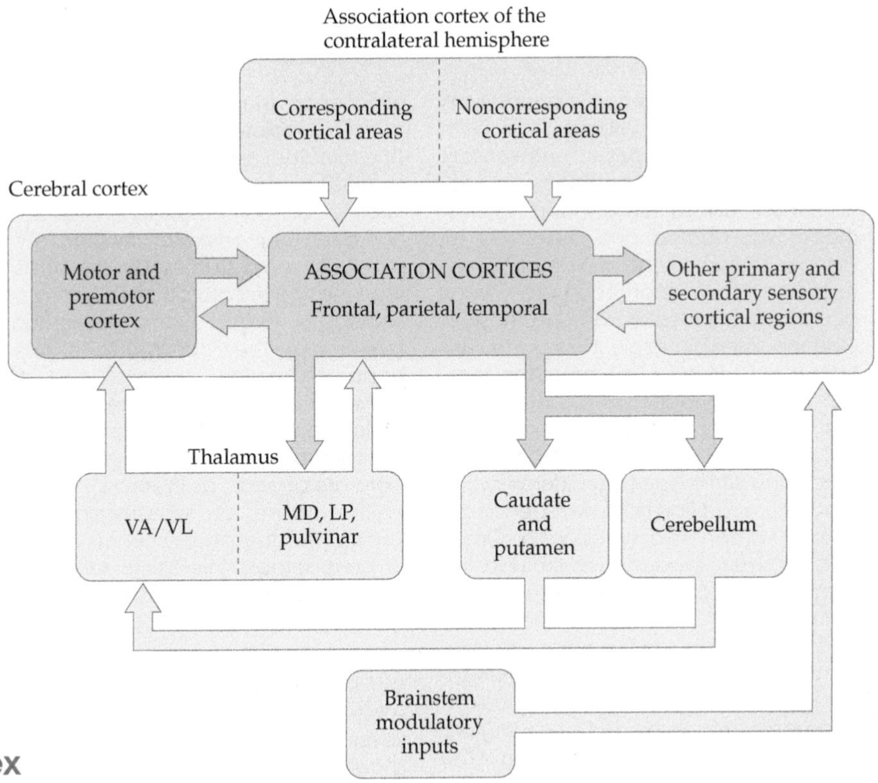

Association cortex of the contralateral hemisphere
Corresponding cortical areas
Noncorresponding cortical areas
Cerebral cortex
Motor and premotor cortex
ASSOCIATION CORTICES
Frontal, parietal, temporal
Other primary and secondary sensory cortical regions
Thalamus
VA/VL
MD, LP, pulvinar
Caudate and putamen
Cerebellum
Brainstem modulatory inputs

FIGURE 27.3 Summary of the overall connectivity of the association cortices. VA = ventral anterior nucleus, VL = ventral lateral nucleus, MD = medial dorsal nucleus, LP = lateral posterior nucleus.

respectively, results in specific cognitive deficits that indicate much about the operations and purposes of each of these regions, as can apparent "miswiring" in conditions such as **synesthesia** (Box 27B). These deductions have largely been corroborated by patterns of neural activity observed in the homologous regions of the brains of experimental animals, as well as in humans using noninvasive imaging techniques. The following sections provide an overview of the major association cortices, which are critical for many of the cognitive functions considered in detail in the rest of Unit V.

The Parietal Association Cortex

In 1941, the British neurologist W. R. Brain reported three patients with unilateral parietal lobe lesions in whom the primary problem was varying degrees of difficulty paying attention to objects and events contralateral to the lesion. Brain described their peculiar deficiency in the following way:

> *Though not suffering from a loss of topographical memory or an inability to describe familiar routes, they nevertheless got lost in going from one room to another in their own homes, always making the same error of choosing a right turning instead of a left, or a door on the right instead of one on the left. In each case there was a massive lesion in the right parieto-occipital region, and it is suggested that this ... resulted in an inattention to or neglect of the left half of external space.*
>
> *The patient who is thus cut off from the sensations which are necessary for the construction of a body scheme may react to the situation in several different ways. He may remember that the limbs on his left side are still there, or he may periodically forget them until reminded of their presence. He may have an illusion of their absence, i.e. they may "feel absent" although he knows that they are there; he may believe that they are absent but allow himself to be convinced by evidence to the contrary; or, finally, his belief in their absence may be unamenable to reason and evidence to the contrary and so constitute a delusion.*
>
> W. R. Brain, 1941 (*Brain* 64, pp. 257 and 264)

This description is generally considered to be the first account of the link between parietal lobe lesions and deficits in attention or perceptual awareness. Based on a large number of patients studied since Brain's pioneering work, these deficits are now referred to as **contralateral neglect syndrome**. The hallmark of contralateral neglect syndrome is an inability to attend to objects, or even one's own body, in a portion of space, despite the fact that visual acuity, somatic sensation, and motor ability remain intact. Affected individuals fail to report, respond to, or even orient to stimuli presented to the side of the body (or visual space) opposite the lesion.

Importantly, contralateral neglect syndrome is typically associated with damage to the *right* parietal cortex (see Chapter 29). The unequal distribution of this particular cognitive function between the hemispheres is thought to arise because the right parietal cortex mediates attention to both left and right halves of the body and extrapersonal space, whereas the left hemisphere mediates attention primarily to the right; this hemispheric bias is thought to arise from specialization of the *left* hemisphere for language, thereby driving attentive functions into the right hemisphere. Thus, left parietal lesions tend to be compensated by the intact right hemisphere. In contrast, when the right parietal cortex is damaged, there is little or no compensatory capacity in the left hemisphere to mediate attention to the left side of the body or extrapersonal space. This

BOX 27B ■ Synesthesia

A remarkable sensory anomaly is evident in individuals who conflate experiences in one sensory domain with those in another—a phenomenon called synesthesia. Synesthesia was named and described by Francis Galton in the nineteenth century, and the phenomenon received a good deal of attention among those in Galton's scientific circle in England. The term means literally "mixing of the senses," and its best-understood expression is in individuals who see specific numerals, letters, or similar shapes printed in black and white as being differently colored; this condition is known specifically as color-graphemic synesthesia. Other, less common synesthesias include the experience of colors in response to musical notes, and specific tastes elicited by certain words and/or numbers. The list of famous synesthetes includes painter David Hockney, novelist Vladimir Nabokov, composer and musician Duke Ellington, and physicist Richard Feynman. The experience of synesthetes is not in any sense metaphorical. Nor do they consider it "abnormal"; it is simply the way they experience the world. People who experience color-graphemic synesthesia (the form that has been most thoroughly studied) perceive numbers as being differently colored; the reality of their ability has been demonstrated in a variety of psychophysical studies. On the basis of the synesthetic colors they see, they can segregate targets from backgrounds (see figure), they can group targets in apparent motion displays, and they show the Stroop effect (the slowed reaction time that everyone exhibits when the printed ink and the spelling of a color word are at odds, as in yellow).

The cause of synesthesia is not known, but the phenomenon is clearly of considerable interest to researchers trying to sort out how information inputs from different sensory modalities are integrated. A number of cognitive neuroscientists have used functional MRI and other modern methods to study synesthesia, but so far, without leading to any definite conclusions. The influence of synesthetic color perception on the various psychophysical tasks shows that the phenomenon occurs at the level of the cerebral cortex. Numerous neurobiological theories have been put forward, the most plausible of which entail some form of aberrant wiring during early development. A good deal of novel synaptic connectivity is required as a person becomes literate, numerate, or musically trained, and it may be during this period of plasticity that "miswiring" occurs.

Improved performance on a visual search task by a color-grapheme synesthete, "subject W. O." (A) The physical stimulus presented to W. O. and to a nonsynesthete control subject. The task was to find the numeral 2 among the multiple numeral 5's. (B) The same stimulus with synesthetic colors assigned to the two numbers tested, which presumably shows how W. O. perceives the physical stimulus. (C) The graph reveals that W. O.'s reaction time in the task was faster than that of the control subject. W. O.'s performance is presumably better because the differently colored 2 "pops out" for him, whereas it doesn't for the control subject. (From Palmeri et al., 2002.)

(A) Physical stimulus as presented

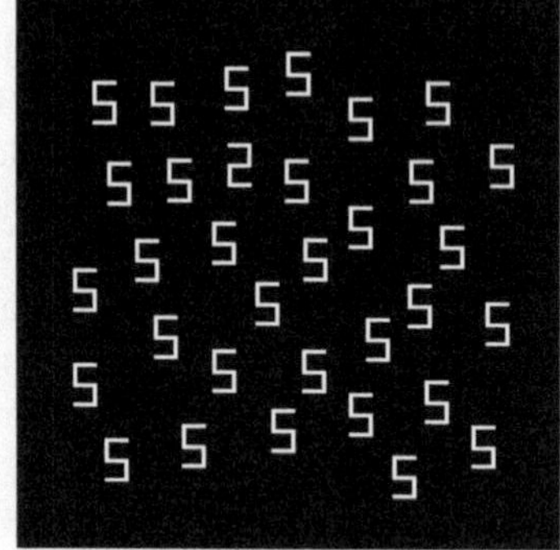

(B) Presumed synesthete perception

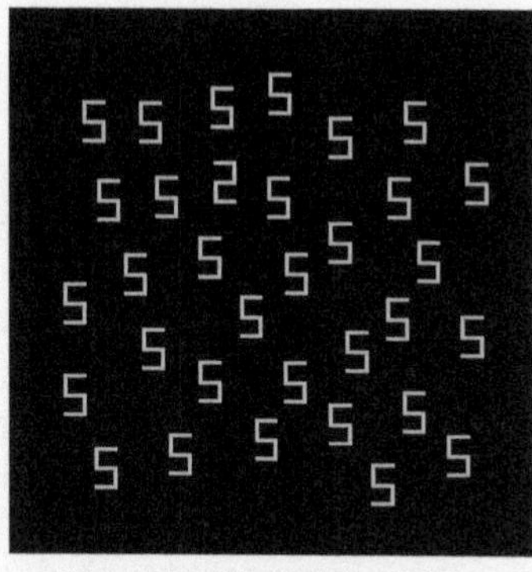

(C)

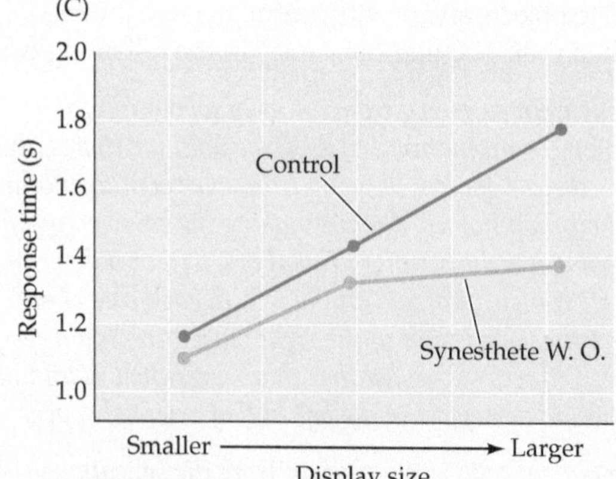

interpretation has been confirmed by noninvasive imaging of parietal lobe activity during specific attention tasks carried out by normal subjects. Such studies show that neural activity is increased in *both* the right and left parietal cortices when subjects are asked to perform tasks in the *right* visual field requiring selective attention to distinct aspects of a visual stimulus such as its shape, velocity, or color. However, when a similar challenge is presented in the *left* visual field, typically the right parietal cortex is activated, although left parietal cortical activity is often observed as well. There is also evidence of increased activity in the right frontal cortex during such tasks. This observation suggests that multiple regions contribute to attentive behavior, and perhaps to some aspects of the pathology of neglect syndromes. Overall, however, brain-imaging data are consistent with the clinical fact that contralateral neglect typically arises from a right parietal lesion, and these data endorse the broader idea of hemispheric specialization

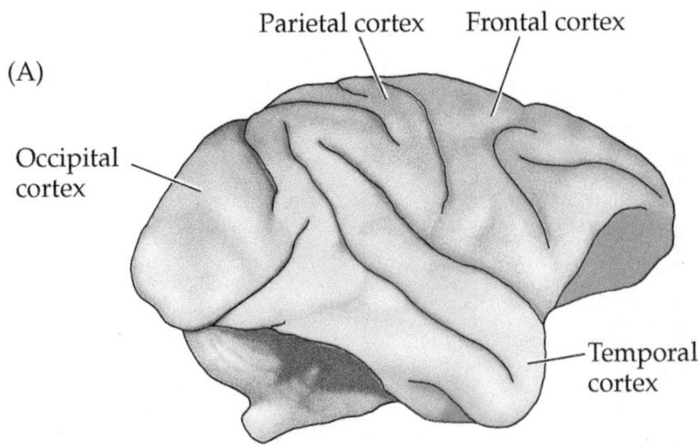

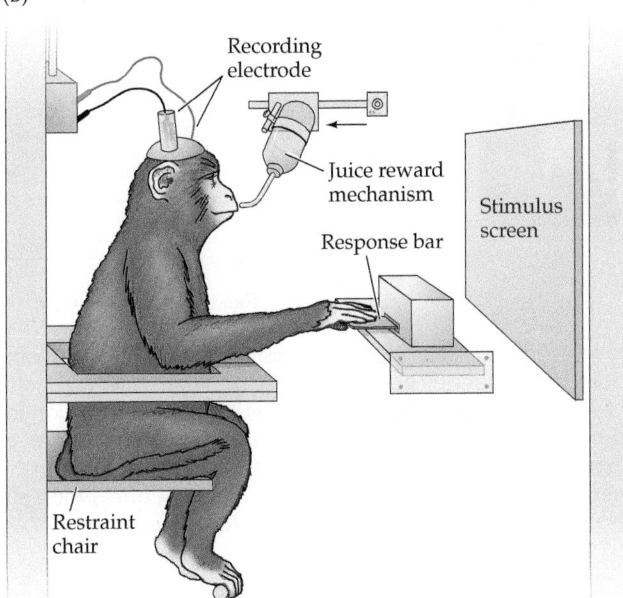

FIGURE 27.4 Recording from single neurons in the brain of an awake, behaving rhesus monkey. (A) Lateral view of the rhesus monkey brain showing the parietal, temporal, and frontal cortices. (B) The animal is seated in a chair and gently restrained. Several weeks before data collection begins, a recording well is placed through the skull by means of a sterile surgical technique. For electrophysiological recording experiments, a tungsten microelectrode is inserted through the dura and arachnoid and into the cortex. The screen and response bar in front of the monkey are for behavioral testing. In this way, individual neurons can be monitored while the monkey performs specific cognitive tasks to gain a fruit juice reward.

for attention, in keeping with the concept of hemispheric specialization for a number of other cognitive functions.

These clinical and brain-imaging observations do not, however, provide much insight into how the nervous system represents cognitive information in nerve cells and their interconnections. The apparent functions of the association cortices implied by clinical observations stimulated a wealth of informative electrophysiological studies in non-human primates, particularly macaque (usually rhesus) monkeys.

As in humans, a wide range of cognitive abilities in monkeys is mediated by the association cortices of the parietal, temporal, and frontal lobes (Figure 27.4A). Moreover, these functions can be tested using behavioral paradigms that assess attention, identification, and planning capabilities—the broad functions assigned to the parietal, temporal, and frontal association cortices, respectively, in humans. By means of implanted electrodes, recordings from single neurons in the brains of awake, behaving monkeys are used to assess the activity of individual cells in the brain as various tasks are performed (Figure 27.4B).

Neurons in the parietal cortex of monkeys have been studied using this approach. The studies take advantage of the fact that monkeys can be trained to selectively attend to particular objects or events and report their experience in a variety of nonverbal ways, typically by looking at a response target (thus allowing their eye movements to be monitored) or manipulating a joystick. Attention-sensitive neurons can be identified by recording electrophysiological changes in neuronal activity associated with simultaneous changes in the attentive behavior of the animal. As might be expected from the clinical evidence in humans, neurons in specific regions of the parietal cortex of the rhesus monkey are activated when the animal attends to a target, but not when the same stimulus is ignored (Figure 27.5A). In another study, monkeys were rewarded with different amounts of fruit juice (a highly desirable treat) for attending to each of a pair of simultaneously illuminated targets (Figure 27.5B). Not surprisingly, the frequency with which monkeys attended to each target varied with the amount of juice they learned to expect for doing so. Moreover, the activity of some neurons in parietal cortex varied systematically as a function of the amount of juice associated with each target, and therefore the amount of attention paid by the monkey to the target. Thus, the primate parietal cortex contains neurons that respond specifically when the animal attends to a behaviorally meaningful stimulus, and the vigor of the response reflects the amount of attention paid to the stimulus.

The Temporal Association Cortex

Clinical evidence from patients with lesions of the association cortex in the temporal lobe indicates that one of the major functions of this part of the brain is the recognition and identification of stimuli that are attended to, particularly complex stimuli. Thus, damage to either temporal lobe can result in difficulty recognizing, identifying, and

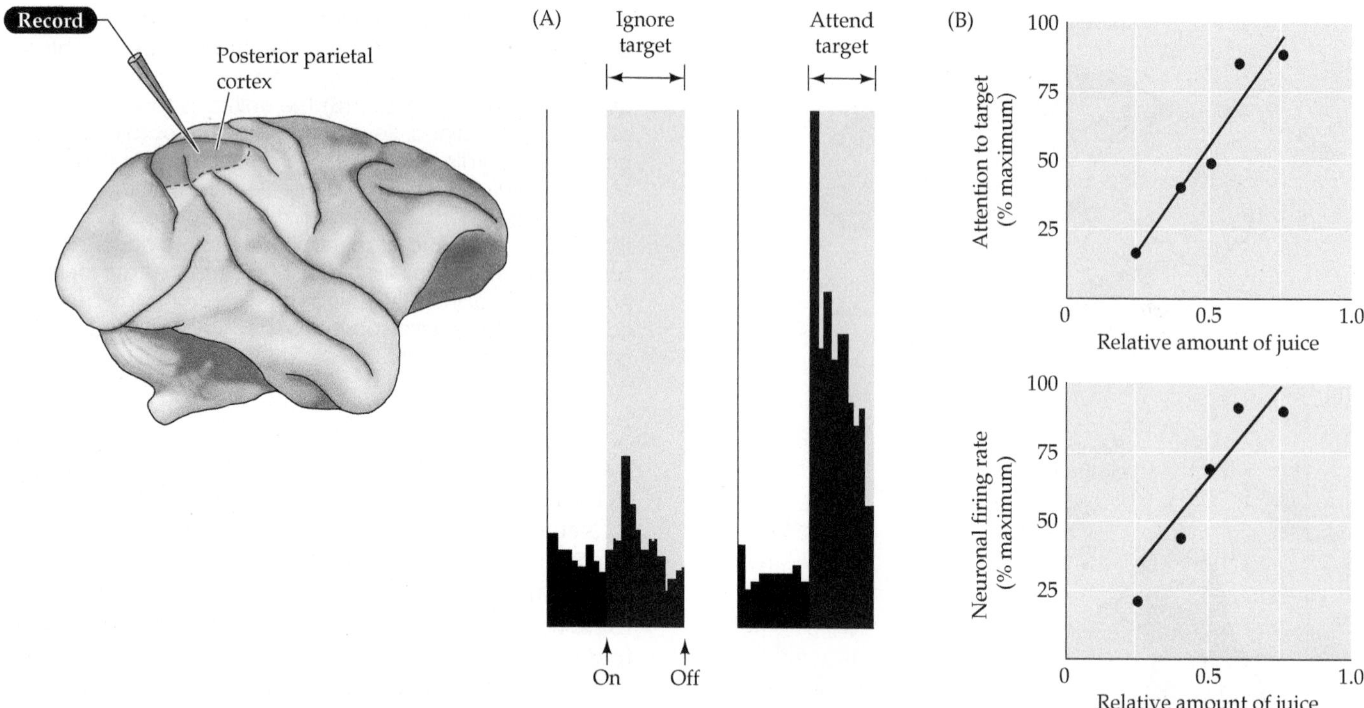

FIGURE 27.5 Selective activation of neurons in the monkey parietal cortex as a function of attention. In this case, a rhesus monkey's attention is directed to a light associated with a fruit juice reward. (A) Although the baseline level of activity of the neuron being studied here remains unchanged when the monkey ignores a visual target (left), firing rate increases dramatically when the monkey attends to the same stimulus (right). The histograms indicate action potential frequency per unit of time. (B) When given a choice of where to attend, the monkey pays increasing attention to a particular visual target when more fruit juice reward can be expected for doing so (left), and the firing rate of the parietal neuron under study increases accordingly (A after Lynch et al., 1977; B after Platt and Glimcher, 1999.)

naming different categories of objects. These disorders, collectively called **agnosias** (Greek, "unknown"), are quite different from the neglect syndromes. As noted, patients with right parietal lobe damage often deny awareness of sensory information in the left visual field (and are less attentive to the left sides of objects generally), despite the fact that the sensory systems are intact (an individual with contralateral neglect syndrome typically withdraws his left arm in response to a pinprick, even though he may not admit to the arm's existence). Patients with agnosia, by contrast, acknowledge the presence of a stimulus but are unable to report what it is. Agnosias have both a lexical aspect (a mismatching of verbal or other cognitive symbols with sensory stimuli) and a mnemonic aspect (a failure to recall stimuli when confronted with them again).

One of the most thoroughly studied agnosias following damage to the temporal association cortex in humans is the inability to recognize and identify faces. This disorder, called **prosopagnosia** (Greek *prosopon*, "face" or "person"), was recognized by neurologists in the late nineteenth century and remains an area of intense investigation. After damage to the inferior temporal cortex, typically on the right, patients are often unable to identify familiar individuals by their facial characteristics, and in some cases cannot recognize a face at all. Nonetheless, such individuals are perfectly aware that some sort of visual stimulus is present and can describe particular aspects or elements of it without difficulty.

An example is the case of L. H., a patient described by the neuropsychologist N. L. Etcoff and colleagues. L. H. (the use of initials to identify neurological patients in published reports is standard practice) was a 40-year-old minister and social worker who sustained a severe head injury in an automobile accident when he was 18. After recovery, he could not recognize familiar faces, report that they were familiar, or answer questions about faces from memory. He could identify common objects, could discriminate subtle shape differences, and could recognize the sex, age, and even the "likeability" of faces. Moreover, he could identify particular people by nonfacial cues such as voice, body shape, and gait. The only other category of visual stimuli he had trouble recognizing was animals and their expressions, though these impairments were not as severe as for human faces, and he was able to lead a fairly normal and productive life.

Noninvasive brain imaging showed that L. H.'s prosopagnosia was the result of damage to the right temporal lobe.

Prosopagnosia and related agnosias involving objects are specific instances of a broad range of functional deficits that have as their hallmark the inability to recognize a complex sensory stimulus as familiar, and to identify and name that stimulus as a meaningful entity in the environment. Depending on the laterality, location, and size of the lesion in temporal cortex, agnosias can be as specific as an inability to recognize human faces or as general as an inability to name most familiar objects. In general, lesions of the right temporal cortex lead to agnosia for faces and objects, whereas lesions of the corresponding regions of the left temporal cortex tend to result in difficulties with language-related material. (Recall that the primary auditory cortex is on the superior aspect of the temporal lobe; as described in Chapter 33 the cortex adjacent to the auditory cortex in the left temporal lobe is specifically concerned with language.) The lesions that typically cause recognition deficits, particularly for faces, are in the inferior temporal cortex, in or near the fusiform gyrus; those that cause language-related problems in the left temporal lobe tend to be on the lateral surface of the cortex. Consistent with these conclusions, direct cortical stimulation in subjects whose temporal lobes are being mapped for neurosurgery (typically removal of a seizure focus) may induce transient prosopagnosia as a consequence of this abnormal activation of the relevant regions of the right temporal cortex.

More recently, brain imaging and direct electrophysiological recording studies in normal subjects have confirmed that the inferior temporal cortex, particularly the fusiform gyrus, mediates face recognition and that nearby regions are responsible for categorically different recognition functions. The special role of faces in social behavior in our own species is highlighted by the fact that there appear to be several discrete "patches" of neurons in the temporal lobe that are activated when participants identify faces, dynamic facial gestures, and the direction in which another individual is looking. Recognizing these stimuli provides important clues to another person's emotional state and intentions, and the existence of brain regions specialized for processing this information has been taken as evidence for the important role of social behavior and cognition during human evolution.

In keeping with human deficits of recognition following temporal lobe lesions, neurons with responses that correlate with the recognition of specific stimuli are present in the temporal cortex of rhesus monkeys. The behavior of these neurons is generally consistent with one of the major functions ascribed to the human temporal cortex, namely, the recognition and identification of complex stimuli. For example, some neurons in the inferior temporal gyrus of the rhesus monkey cortex respond specifically to the presentation of a monkey face. These cells are often quite selective; some respond only to the frontal view of a face, others only to profiles. Furthermore, the cells are not easily deceived. When parts of faces or generally similar objects are presented, such cells typically fail to respond; in fact, the only things that confuse face-selective neurons are round or fuzzy objects such as apples, clock faces, or toilet brushes—all of which are vaguely facelike in appearance.

In principle, it is unlikely that such "face cells" are tuned to specific faces or objects. However, it is not hard to imagine that populations of neurons differently responsive to various features of faces or other objects could act in concert to enable the recognition of such complex sensory stimuli. The notion of such "population coding" of objects is supported by the recent observation that face-selective neuronal responses in the temporal cortex of monkeys vary in intensity with respect to an average face. Both monkeys and humans are better at recognizing faces having extreme features—caricatures—than they are at recognizing less distinctive faces, suggesting that faces are identified by comparison with a mental standard or norm. Similarly, neurons in the inferior temporal cortex of monkeys respond much more strongly to caricatures of human faces than to an "average" human face, which is represented by the average activity of the neuronal population. Such norm-based tuning has also been reported for neuronal responses to shapes in the inferior temporal cortex.

Recent studies suggest that such complex response properties may be based on a columnar anatomical arrangement similar to that in the primary visual cortex (see Chapter 12). Each column has been taken to represent different arrangements of complex features making up an object, the overall spatial pattern of neuronal activity representing the object in view. In keeping with this general idea, optical imaging of the surface of the temporal cortex shows that large populations of neurons are activated when monkeys view an object comprising several different geometric features. The locus of this activity in the upper layers of the cortex shifts systematically when object features, such as the orientation of a face, are systematically altered. Taken together, these further observations suggest that object identification relies on graded signals carried by a population of neurons rather than on the specific output of one or a few cells that are selective for a particular object.

The Frontal Association Cortex

The functional deficits that result from damage to the human frontal lobe are diverse and devastating, particularly if both hemispheres are involved. This broad range of clinical effects stems from the fact that the frontal cortex has a wider repertoire of functions than any other neocortical region (consistent with the fact that the frontal lobes in humans and other primates are the largest of the brain's lobes and comprise a greater number of cytoarchitectonic areas).

The particularly devastating nature of the behavioral deficits after frontal lobe damage reflects the role of this part of the brain in maintaining what is normally thought of as an individual's "personality." The frontal cortex integrates complex information from sensory and motor cortices, as well as from the parietal and temporal association cortices. The result is an appreciation of self in relation to the world that allows behaviors to be planned and executed normally. When this ability is compromised, the afflicted individual often has difficulty carrying out complex behaviors that are appropriate to the circumstances. These deficiencies in the normal ability to match ongoing behavior to present or future demands are, not surprisingly, interpreted as a change in the patient's "character."

The case that first called attention to the consequences of frontal lobe damage was that of Phineas Gage, a laborer on the Rutland and Burlington Railroad in mid-nineteenth-century Vermont. In that era, the conventional method of blasting rock involved tamping powder into a hole with a heavy metal rod. Gage, a popular and respected crew foreman, was undertaking this procedure one day in 1848 when his tamping rod sparked the powder, setting off an explosion that drove the rod—which was about 1 meter long and 4 or 5 centimeters in diameter—through his left eye socket, destroying much of the frontal part of his brain in the process. Gage, who never lost consciousness, was promptly taken to a local doctor who treated his wound. An infection set in, presumably destroying additional frontal lobe tissue, and Gage was an invalid for several months. Eventually, he recovered and was to outward appearances well again. Those who knew Gage, however, were profoundly aware that he was not the "same" individual he had been before. A temperate, hardworking, and altogether decent person had, by virtue of this accident, been turned into an inconsiderate, intemperate lout who could no longer cope with normal social intercourse or the kind of practical planning that had allowed Gage the social and economic success he enjoyed before.

The physician who treated Gage followed his story until his Gage's death in 1863 and summarized his initial impressions of his patient as follows:

> *[Gage is] fitful, irreverent, indulging at times in the grossest profanity (which was not previously his custom), manifesting but little deference for his fellows, impatient of restraint or advice when it conflicts with his desires, at times pertinaciously obstinate, yet capricious and vacillating, devising many plans of future operations, which are no sooner arranged than they are abandoned in turn for others appearing more feasible. A child in his intellectual capacity and manifestations, he has the animal passions of a strong man. Previous to his injury, although untrained in the schools, he possessed a well-balanced mind, and was looked upon by those who knew him as a shrewd, smart businessman, very energetic and persistent in executing all his plans of operation. In this regard his mind was radically changed, so decidedly that his friends and acquaintances said he was "no longer Gage."*
>
> J. M. Harlow, 1868 (*Publications of the Massachusetts Medical Society* 2: 339–340)

Another classic case of frontal lobe deficits was that of a patient followed throughout the 1920s and 1930s by the neurologist R. M. Brickner. Joe A., as Brickner referred to his patient, was a stockbroker who at age 39 underwent bilateral frontal lobe resection to remove a large tumor. After the operation, Joe A. had no obvious sensory or motor deficits—he could speak and understand verbal communication and was aware of people, objects, and temporal order in his environment. He acknowledged his illness and retained a high degree of intellectual power, as judged from an ongoing ability to play an expert game of checkers. Nonetheless, Joe A.'s personality underwent a dramatic change. This formerly restrained, modest man became boastful of professional, physical, and sexual prowess, showed little restraint in conversation, and was unable to match the appropriateness of what he said to his audience. Like Gage, his ability to plan for the future was largely lost, as was much of his earlier initiative and creativity. Even though he retained the ability to learn complex procedures, he was unable to return to work and had to rely on his family for support and care.

The effects of widespread frontal lobe damage documented by these case studies encompass a wide range of cognitive disabilities, including impaired restraint, disordered thought, perseveration (repetition of the same behavior), and the inability to plan appropriate action. Recent studies of patients with focal damage to particular regions of the frontal lobe suggest that some of the processes underlying these deficits may be localized. Short-term memory functions (see Chapter 30) are situated more dorsolaterally, and planning and social restraint functions are located more ventromedially. Some of these functions can be clinically assessed using standardized tests such as the Wisconsin Card Sorting Task for planning (Box 27C), the delayed response task for short-term memory, and the "go–nogo" task for inhibition of inappropriate responses. All these observations are consistent with the idea that the common denominator of the cognitive functions subserved by the frontal cortex is the selection, planning, and execution of appropriate behavior, particularly in social contexts.

More recent brain-imaging studies have confirmed the important role of the frontal association cortex in planning, decision making, and behavioral inhibition; these studies corroborate the allocation of these functions to specific areas of the frontal lobes hinted at by studies of

BOX 27C ■ Neuropsychological Testing

Long before PET scanning and functional MRI were available to evaluate normal and abnormal cognitive brain function, several "low-tech" methods provided reliable means of assessing these operations in human subjects. From the late 1940s onward, psychologists and neurologists developed a battery of behavioral tests—generally called neuropsychological tests—to evaluate the integrity of cognitive function and to help localize lesions.

One of the most frequently used measures is the Wisconsin Card Sorting Task illustrated here. In this test, the examiner places four cards with symbols that differ in number, shape, or color before the subject, who is given a set of response cards with similar symbols on them. The subject is then asked to place an appropriate response card in front of the stimulus card based on a sorting rule established, but not stated, by the examiner (i.e., sort by either color, number, or shape). The examiner then indicates whether the response is "right" or "wrong." After ten consecutive correct responses, the examiner changes the sorting rule simply by saying "wrong." The subject must then ascertain the new sorting rule and perform ten correct trials. The sorting rule is then changed again, until six cycles have been completed.

In 1963, the neuropsychologist Brenda Milner at the Montreal Neurological Institute showed that patients with frontal lobe lesions have consistently poor performance in the Wisconsin Card Sorting Task. By comparing patients with known brain lesions as a result of surgery for epilepsy or tumor, Milner was able to demonstrate that this impairment is fairly specific for frontal lobe damage. Particularly striking is the inability of frontal lobe patients to use previous information to guide subsequent behavior. A widely accepted explanation for the sensitivity of the Wisconsin Card Sorting Task to frontal lobe deficits is the "planning" aspect of this test. To respond correctly, the subject must retain information about the previous trial, which is then used to guide behavior on future trials. Processing this sort of information is characteristic of normal frontal lobe function. In keeping with the conservation of neural function between humans and non-human primates, rhesus monkeys can also be taught to perform a computerized variant of the Wisconsin Card Sort Task. Lesion studies in monkeys confirm that the dorsolateral prefrontal cortex is crucial for performing this task, specifically for representing the rule (i.e., sort by shape, color, or number). Single-unit recordings in monkeys showing neurons whose activity specifically encodes particular rules endorse this conclusion.

A variety of other neuropsychological tests have been devised to evaluate the functional integrity of other cognitive functions. These include tasks in which a patient is asked to identify familiar faces in a series of pictures, and others in which "distractors" interfere with the patient's ability to attend to salient stimulus features. An example of the latter is the Stroop Test, in which patients are asked to read the names of colors presented in color-conflicting print (for example, the word *green* printed in red ink). This sort of challenge evaluates abilities related to both attention and identification.

The simplicity, economy, and accumulated experience with such tests as the Wisconsin Card Sorting Task continue to make them a valuable means of evaluating cognitive functions.

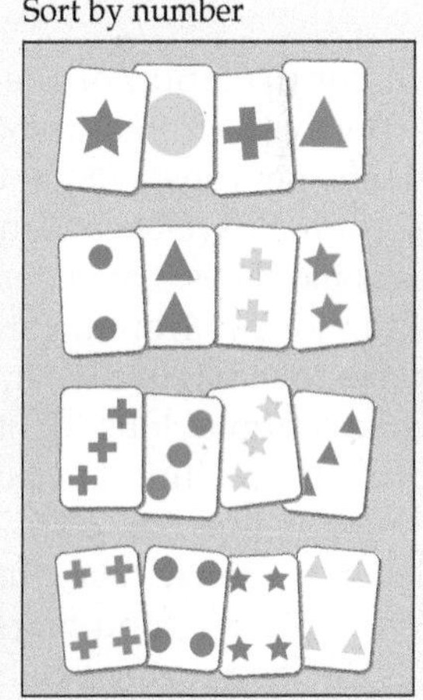

brain-damaged patients. For example, the dorsal and lateral aspects of the frontal cortex are activated when normal subjects actively suppress a behavioral response when an expected pattern of events is violated in order to generate the appropriate behavior. These findings endorse a role for this part of the brain in maintaining information in short-term memory (see Chapter 30) and using that information to override automatic reactions to anticipated events. By contrast, personal preferences for different types of rewards are correlated with individual differences in brain activation within the ventromedial prefrontal cortex, suggesting this part of the brain signals the value

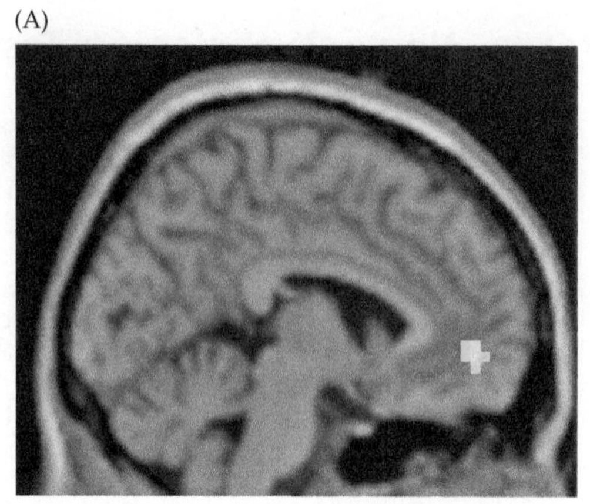

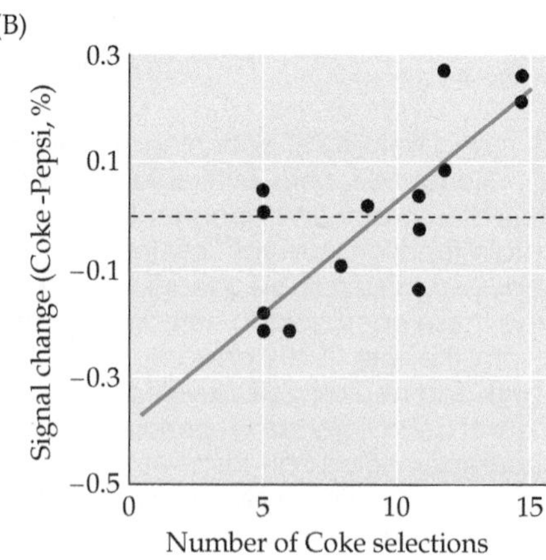

FIGURE 27.6 Activation in the ventromedial prefrontal cortex correlates with subjective preferences for soft drinks. Subjects were given a taste-test to determine whether they preferred Coke or Pepsi. The participants then underwent functional MRI scans while they received squirts of each soft drink. Activation in the ventromedial prefrontal cortex in response to squirts of Coke compared with activation in response to squirts of Pepsi (A) was correlated with the frequency with which subjects selected Coke over Pepsi in the taste-test (B). (From McClure et al., 2004.)

individuals place on a reward independent of its physical attributes (Figure 27.6). Such signals are crucial for making adaptive decisions, and the effects of disrupted reward processing in ventromedial prefrontal cortex on decision making are all too apparent in the disordered behavior of patients (such as Phineas Gage) with damage to this area.

Sadly, the effects of damage to the frontal lobes have also been documented by the many thousands of frontal lobotomies performed in the 1930s and 1940s as a means of treating mental illness. The rise and fall of the popularity of this "psychosurgery" provides a compelling example of the frailty of human judgment in medical practice and of the conflicting approaches of neurologists, neurosurgeons, and psychiatrists in that era to the treatment of mental disease.

In confirmation of the clinical evidence about the function of the frontal association cortices in neurological patients, neurons that appear to be specifically involved in planning have been identified in the frontal cortices of rhesus monkeys. One behavioral test used to study cells in the monkey frontal cortex is called the **delayed response task** (Figure 27.7A). Variants of this task are used to assess frontal lobe function in a variety of situations, including the clinical evaluation of frontal lobe function in humans (see Box 27C). In the simplest version of the delayed response task, the monkey watches an experimenter place a food morsel in one of two wells; both wells are then covered. Subsequently, a screen is lowered for an interval of a few seconds to several minutes (the delay). When the screen is raised, the monkey gets only one chance to uncover the well containing food and retrieve the reward. Thus, the animal must decide that it wants the food, remember where it is placed, recall that the cover must be removed to obtain the food, and keep all this information available during the delay so that it can be used to get the reward. The monkey's ability to carry out this short-term memory task is diminished or abolished if the area anterior to the motor region of the frontal cortex—called the prefrontal cortex—is destroyed bilaterally; this result is in accord with clinical findings in human patients.

Some neurons in the prefrontal cortex, particularly those in and around the principal sulcus (Figure 27.7B), are activated when monkeys perform computerized variants of the delayed response task, and they are maximally active during the period of the delay, as if their firing represented information about the location of the food morsel maintained from the presentation part of the trial (i.e., the cognitive information needed to guide behavior when the screen is raised; Figure 27.7C,D). Such neurons return to a low level of activity during the actual motor phase of the task, suggesting that they represent short-term memory and planning (see Chapter 30) rather than the actual movement itself. Delay-specific neurons in the prefrontal cortex are also active in monkeys that have been trained to perform a variant of the delayed response task in which well-learned movements are produced in the absence of any cue. Evidently, these neurons are equally capable of using remembered information to guide behavior. Thus, if a monkey is trained to associate looking at a particular target with a delayed reward, the delay-associated neurons in the prefrontal cortex will fire during the delay, even if the monkey shifts its gaze to the appropriate region of the visual field in the absence of the target.

In addition to maintaining cognitive information during short delays, some neurons in prefrontal cortex also appear to participate directly in longer range planning of sequences of movements. When monkeys are trained to perform a motor sequence, such as turning a joystick to the left, then right, then left again, some neurons in prefrontal cortex fire at a particular point in the sequence (such as the third response), regardless of which movement (e.g., left or right) is made. Prefrontal neurons have also been found that are selective for each position in a learned motor sequence, thus ruling out the possibility that these neurons

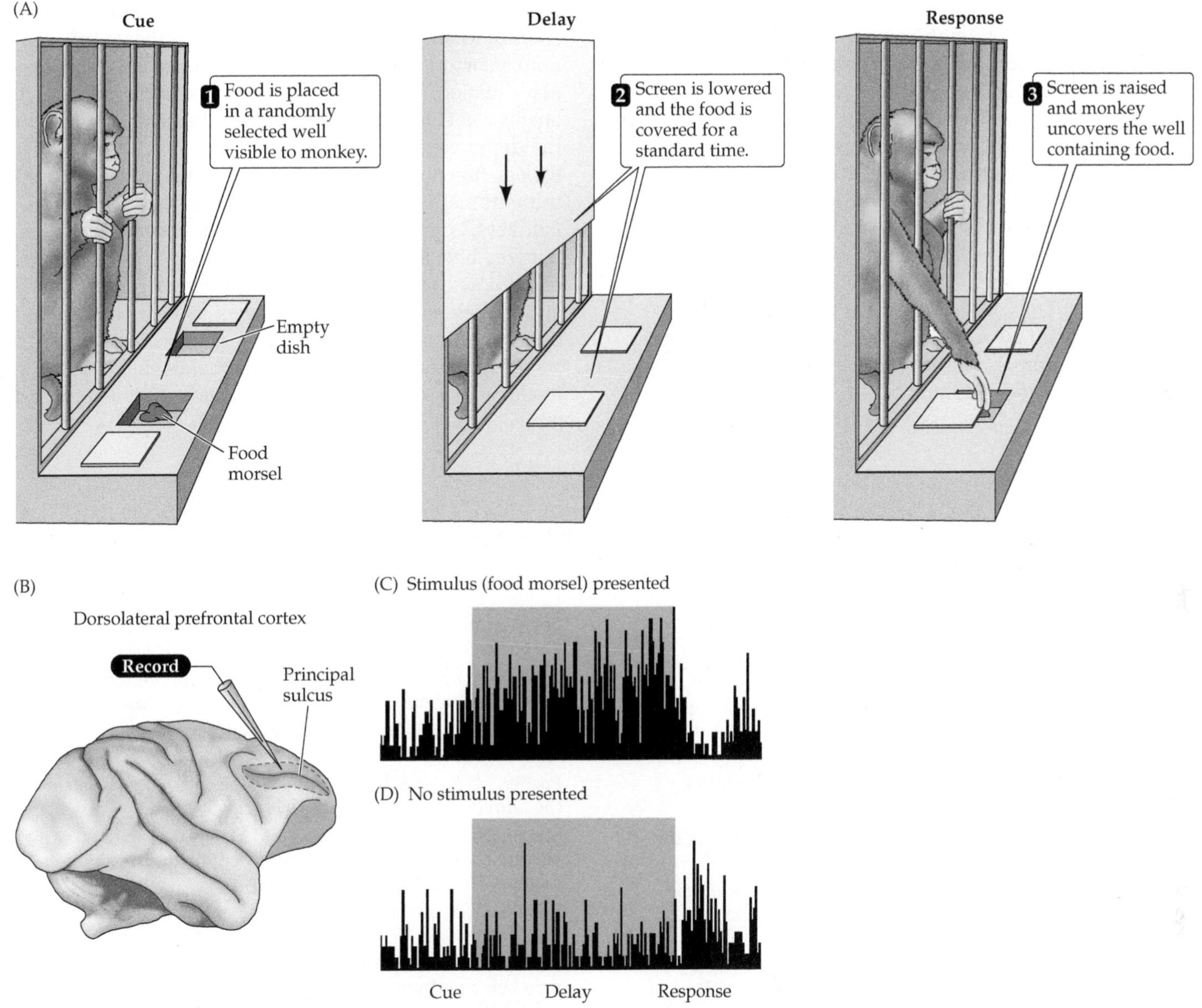

FIGURE 27.7 Activation of neurons near the principal sulcus of the frontal lobe during the delayed response task. (A) Illustration of task. The experimenter randomly varies the well in which the food is placed. The monkey watches the morsel being covered, and then the screen is lowered for a standard time. When the screen is raised, the monkey is allowed to uncover only one well to retrieve the food. Normal monkeys learn this task quickly, usually performing at a level of 90% correct after fewer than 500 training trials, whereas monkeys with frontal lesions perform poorly. (B) Region of recording. (C) Activity of a delay-specific neuron in the prefrontal cortex of a rhesus monkey recorded during the delayed response task shown in (A). The histograms show the number of action potentials during the cue, delay, and response periods. The neuron begins firing more rapidly when the screen is lowered and remains active throughout the delay period. (D) When the screen is lowered and raised but no food is presented, the same neuron is less active. (After Goldman-Rakic, 1987.)

merely encode task difficulty or proximity to reward as the monkey nears the end of the series of responses. When these regions of prefrontal cortex are inactivated pharmacologically, monkeys lose the ability to execute sequences of movements from memory. These observations endorse the notion, first inferred from studies of brain-damaged individuals such as Phineas Gage, that the frontal lobe contributes specifically to the cognitive functions that use remembered information to plan and guide appropriate sequences of behavior.

Summary

The majority of the human cerebral cortex is devoted to tasks that transcend encoding primary sensations or commanding motor actions. Collectively, the association cortices mediate these cognitive functions of the brain—broadly defined as the ability to attend to, identify, and act meaningfully in response to complex external or internal stimuli. Descriptions of patients with cortical lesions, functional brain imaging of normal subjects, and behavioral and electrophysiological studies of non-human primates have established the general purpose of the major association areas. Thus, parietal association cortex is involved in attention and awareness of the body and the stimuli that act on it; temporal association cortex is involved in the recognition and identification of highly processed sensory information; and frontal association cortex is importantly involved in guiding complex behavior by planning responses to ongoing stimulation (or remembered information), matching such behaviors to the demands of a particular situation. The especially extensive association areas in our species compared with those of other primates support the cognitive processes that define human culture. Nonetheless, functional localization, whether inferred by examining human patients or by recording single neurons in monkeys, is an imprecise business. The observations summarized here are merely a rudimentary guide to thinking about how complex cognitive information is represented and processed in the brain, as well as how the relevant brain areas and their constituent neurons contribute to such important but still ill-defined qualities as personality, intelligence, and other cognitive functions that define what it means to be human.

ADDITIONAL READING

Reviews

Behrmann, M. (1999) Spatial frames of reference and hemispatial neglect. In *The Cognitive Neurosciences*, 2nd Edition. M. Gazzaniga (ed.). Cambridge, MA: MIT Press, pp. 651–666.

Bisley, J. W. and M. E. Goldberg (2010) Attention, intention, and priority in the parietal lobe. *Annu. Rev. Neurosci.* 33: 1–21.

Buschman, T. J. and S. Kastner (2015) From behavior to neural dynamics: An integrated theory of attention. *Neuron* 88: 127–144.

Carter, R. M. K. and S. A. Huettel (2013) A nexus model of the temporal–parietal junction. *Trends Cogn. Sci.* 17: 328–336

Damasio, A. R. (1985) The frontal lobes. In *Clinical Neuropsychology*, 2nd Edition. K. H. Heilman and E. Valenstein (eds.). New York: Oxford University Press, pp. 409–460.

Damasio, A. R., H. Damasio and G. W. Van Hoesen (1982) Prosopagnosia: Anatomic basis and behavioral mechanisms. *Neurology* 32: 331–341.

Desimone, R. (1991) Face-selective cells in the temporal cortex of monkeys. *J. Cog. Neurosci.* 3: 1–8.

Filley, C. M. (1995) Right hemisphere syndromes. Chapter 8 in *Neurobehavioral Anatomy*. Boulder: University of Colorado Press, pp. 113–130.

Goldman-Rakic, P. S. (1987) Circuitry of the prefrontal cortex and the regulation of behavior by representational memory. In *Handbook of Physiology*. Section 1: The Nervous System. Vol. 5, Higher Functions of the Brain, Part I. F. Plum (ed.). Bethesda, MD: American Physiological Society, pp. 373–417.

Grill-Spector, K. and K. S. Weiner (2014) The functional architecture of the ventral temporal cortex and its role in categorization. *Nature Rev. Neurosci.* 15: 536–548.

Halligan, P. W. and J. C. Marshall (1994) Toward a principled explanation of unilateral neglect. *Cog. Neuropsych.* 11: 167–206.

Koechlin, E. (2016) Prefrontal executive function and adaptive behavior in complex environments. *Curr. Opin. Neurobiol.* 37: 1–6.

Làdavas, E., A. Petronio and C. Umilta (1990) The deployment of visual attention in the intact field of hemineglect patients. *Cortex* 26: 307–317.

Lara, A. H. and J. D. Wallis (2015) The role of prefrontal cortex in working memory: A mini review. *Front. Sys. Neurosci.* 9: 173.

Macrae, D. and E. Trolle (1956) The defect of function in visual agnosia. *Brain* 77: 94–110.

Posner, M. I. and S. E. Petersen (1990) The attention system of the human brain. *Annu. Rev. Neurosci.* 13: 25–42.

Sreenivasan, K. K., C. E. Curtis and M. D'Esposito (2014) Revisiting the role of persistent neural activity during working memory. *Trends Cogn. Sci.* 18: 82–89

Suzuki, W. A. and Y. Naya (2014) The perirhinal cortex. *Annu. Rev. Neurosci.* 37: 39–53.

Szczepanski, S. M. and R. T. Knight (2014) Insights into human behavior from lesions to the prefrontal cortex. *Neuron* 83: 1002–1018.

Vallar, G. (1998) Spatial hemineglect in humans. *Trends Cog. Sci.* 2: 87–96.

Important Original Papers

Baldauf, S. and R. Desimone (2014) Neural mechanisms of object-based attention. *Science* 344: 424–427.

Brain, W. R. (1941) Visual disorientation with special reference to lesions of the right cerebral hemisphere. *Brain* 64: 224–272.

Colby C. L., J. R. Duhamel and M. E. Goldberg (1996) Visual, presaccadic, and cognitive activation of single neurons in monkey lateral intraparietal area. *J. Neurophysiol.* 76: 2841–2852.

Crowe, D. A. and 6 others (2013) Prefrontal neurons transmit signals to parietal neurons that reflect executive control of cognition. *Nature Neurosci.* 16: 1484–1491.

Desimone, R., T. D. Albright, C. G. Gross and C. Bruce (1984) Stimulus-selective properties of inferior temporal neurons in the macaque. *J. Neurosci.* 4: 2051–2062.

Desrochers, T. M., C. H. Chatham and D. Badre (2015) The necessity of rostrolateral prefrontal cortex for higher-level sequential behavior. *Neuron* 87: 1357–1368.

Etcoff, N. L., R. Freeman and K. R. Cave (1991) Can we lose memories of faces? Content specificity and awareness in a prosopagnosic. *J. Cog. Neurosci.* 3: 25–41.

Funahashi, S., M. V. Chafee and P. S. Goldman-Rakic (1993) Prefrontal neuronal activity in rhesus monkeys performing a delayed anti-saccade task. *Nature* 365: 753–756.

Fuster, J. M. (1973) Unit activity in prefrontal cortex during delayed-response performance: Neuronal correlates of transient memory. *J. Neurophysiol.* 36: 61–78.

Geschwind, N. (1965) Disconnexion syndromes in animals and man. Parts I and II. *Brain* 88: 237–294.

Harlow, J. M. (1868) Recovery from the passage of an iron bar through the head. *Publications of the Massachusetts Medical Society* 2: 327–347.

Jacob, S. M. and A. Nieder (2014) Complementary roles for primate frontal and parietal cortex in guarding working memory from distractor stimuli. *Neuron* 83: 226–237.

Kim, H., S. Ährlund-Richter, X. Wang, K. Deisseroth and M. Carlén (2016) Prefrontal parvalbumin neurons in control of attention. *Cell* 164: 208–218.

Leopold, D. A., I. V. Bondar and M. A. Giese (2006) Norm-based encoding by single neurons in the monkey inferotemporal cortex. *Nature* 442: 572–575.

Mormann, F. and 9 others. (2017) Scene-selective coding by single neurons in the human parahippocampal cortex. *P. Natl. Acad. Sci. U.S.A.* (in press).

Mountcastle, V. B., J. C. Lynch, A. Georgopoulous, H. Sakata and C. Acuna (1975) Posterior parietal association cortex of the monkey: Command function from operations within extrapersonal space. *J. Neurophysiol.* 38: 871–908.

Platt, M. L. and P. W. Glimcher (1999) Neural correlates of decision variables in parietal cortex. *Nature* 400: 233–238.

Quiroga, R. Q., A. Kraskov, F. Mormann, I. Fried and C. Koch (2014) Single-cell responses to face adaptation in the human medial temporal lobe. *Neuron* 84: 363–369.

Sprague, T. C. and J. T. Serences (2013) Attention modulates spatial priority maps in the human occipital, parietal and frontal cortices. *Nature Neurosci.* 16: 1879–1887.

Tanji, J. and K. Shima (1994) Role for supplementary motor area cells in planning several movements ahead. *Nature* 371: 413–416.

Wang, G., K. Tanaka and M. Tanifuji (1996) Optical imaging of functional organization in the monkey inferotemporal cortex. *Science* 272: 1665–1668.

Books

Brickner, R. M. (1936) *The Intellectual Functions of the Frontal Lobes.* New York: Macmillan.

Damasio, A. R. (1994) *Descartes' Error: Emotion, Reason and the Human Brain.* New York: Grosset/Putnam.

DeFelipe, J. and E. G. Jones (1988) *Cajal on the Cerebral Cortex: An Annotated Translation of the Complete Writings.* New York: Oxford University Press.

Garey, L. J. (1994) *Brodmann's "Localisation in the Cerebral Cortex."* London: Smith-Gordon. (Translation of K. Brodmann's 1909 book. Leipzig: Verlag von Johann Ambrosius Barth.)

Glimcher, P. W. (2003) *Decisions, Uncertainty, and the Brain: The Science of Neuroeconomics.* Cambridge, MA: MIT Press.

Heilman, H. and E. Valenstein (1985) *Clinical Neuropsychology,* 2nd Edition. New York: Oxford University Press, Chapters 8, 10, and 12.

Klawans, H. L. (1988) *Toscanini's Fumble, and Other Tales of Clinical Neurology.* Chicago: Contemporary Books.

Klawans, H. L. (1991) *Newton's Madness.* New York: Harper Perennial Library.

Posner, M. I. and M. E. Raichle (1994) *Images of Mind.* New York: Scientific American Library.

Purves, D. and 5 others (2013) *Principles of Cognitive Neuroscience.* Sunderland, MA: Sinauer.

Sacks, O. (1987) *The Man Who Mistook His Wife for a Hat.* New York: Harper Perennial Library.

Sacks, O. (1995) *An Anthropologist on Mars.* New York: Alfred A. Knopf.

Go to the NEUROSCIENCE 6e Companion Website at **www.oup.com/uk/Purves6e** for Web Topics, Animations, Flashcards, and more.

CHAPTER

28 Cortical States

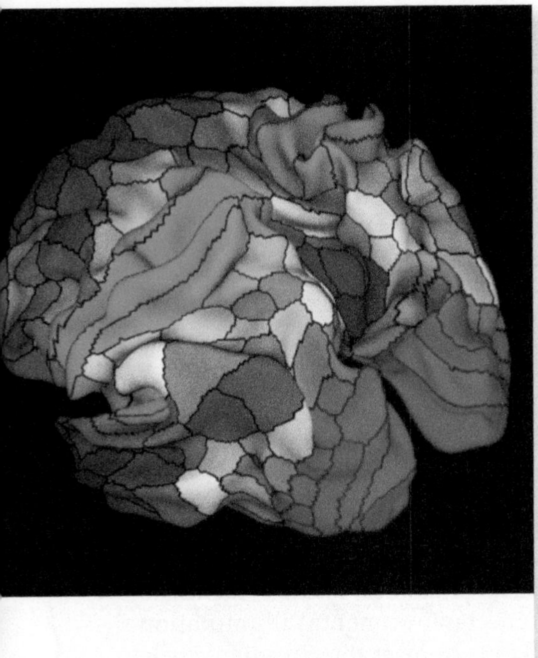

Overview

THE BRAIN, EVEN IN SLEEP, VARIES IN ITS ACTIVITY in ways and for reasons that remain unclear. Most progress has been made in understanding the daily changes in brain activity under the rubric of circadian rhythmicity. The most obvious aspect of these daily rhythms is sleep and wakefulness. About a third of our lives is spent in various stages of sleep, which, except for periods of dreaming, is unconscious and fundamentally different from wakefulness. Any consideration of cortical states thus leads to the knotty problem of consciousness and its neural basis, which is in turn an important feature of many cognitive functions. But even when we are fully awake, our awareness of the world around us and of our internal states (e.g., feelings, thoughts) varies greatly. What follows is a review of how the brain and the rest of the nervous system regulate changes from alert wakefulness to deep sleep, and the bearing this physiological regulation has on the neural underpinnings of consciousness and perceptual awareness.

The Circadian Cycle

Human physiology varies with **circadian** (Latin, "about a day") periodicity, and biologists have explored several questions about this 24-hour cycle. What happens, for example, when individuals are prevented from sensing the cues they normally use to distinguish night and day? This question was addressed by placing volunteers in an environment that lacks external time cues (Figure 28.1A). In a typical experiment of this sort, participants undergo a 5- to 8-day period of acclimatization that includes normal social interactions, meals at the usual times, and temporal cues (e.g., radio and television). During this period, the individuals typically arise and go to sleep at the usual times and maintain a 24-hour sleep–wake cycle. When the cues are removed, however, the volunteers awaken later each day, and under these conditions the clock is said to be "free running." Nonetheless, the cycle remains about 24 hours in length. Thus, humans (and many other animals) have an internal "clock" that operates even in the absence of external information about the time of day.

Presumably, circadian clocks evolved to maintain appropriate daily rhythms of homeostatic functions in spite of the variable amount of daylight and darkness in different seasons and at different locations on the planet (Figure 28.1B). To photoentrain physiological processes with this day–night cycle, the biological clock must be able to detect variations in light levels. The receptors that sense these light changes are, not surprisingly, in the retina, as demonstrated by the fact that removing or covering the eyes abolishes photoentrainment. In mammals, however, the most important retinal detectors are not rod or cone cells, but neurons that lie within the ganglion cell layer of the retina. Unlike rods and cones, which are hyperpolarized

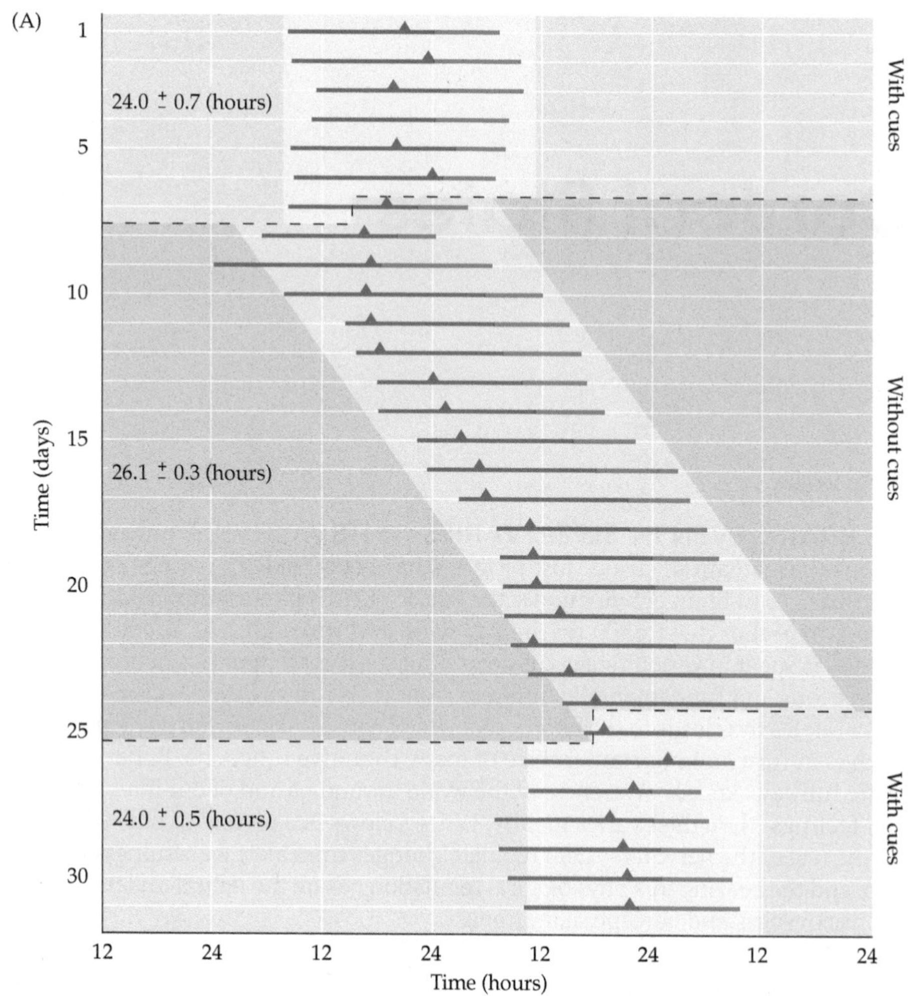

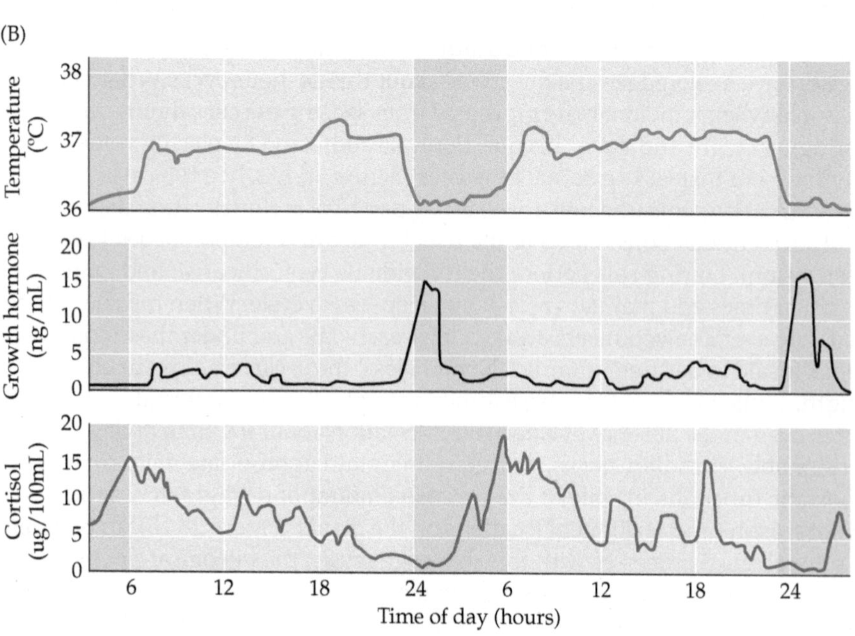

FIGURE 28.1 Circadian rhythmicity. (A) Circadian rhythm in the absence of cues. The illustration graphs the waking (blue) and sleeping (red) status of a volunteer in an isolation chamber with and without cues about the day-night cycle. Numbers represent the mean ± standard deviation of a complete wake-sleep cycle in each condition. Triangles represent times when the rectal temperature was maximum. (B) Circadian rhythmicity of homeostatic regulation. Core body temperature and blood levels of growth hormone and cortisol all show a rhythmic pattern of roughly 24 hours. In the early evening, core temperature begins to decrease whereas growth hormone begins to increase. The level of cortisol, which reflects stress, begins to increase toward morning and stays elevated for several hours. (A after Aschoff, 1965, reproduced in Schmidt et al., 1983; B after Hobson, 1989.)

when activated by light (see Chapter 11), these photosensitive ganglion cells contain a novel photopigment called **melanopsin** and are *depolarized* by light (Figure 28.2A,B). The function of these unusual photoreceptors is evidently to encode environmental illumination and thus to reset the circadian clock, although rods and cones are still able to mediate some circadian entrainment in melanopsin-knockout mice.

The axons of these melanopsin-containing neurons run in the retinohypothalamic tract, which projects to the **suprachiasmatic nucleus** (**SCN**) of the anterior hypothalamus, the central site of the circadian control of homeostatic functions (Figure 28.2C). Activation of the SCN via this pathway evokes responses in the paraventricular nucleus of the hypothalamus and, ultimately, the preganglionic sympathetic neurons in the intermediolateral zone of the lateral horns of the thoracic spinal cord. As described in Chapter 21, these preganglionic neurons modulate neurons in the superior cervical ganglia, some

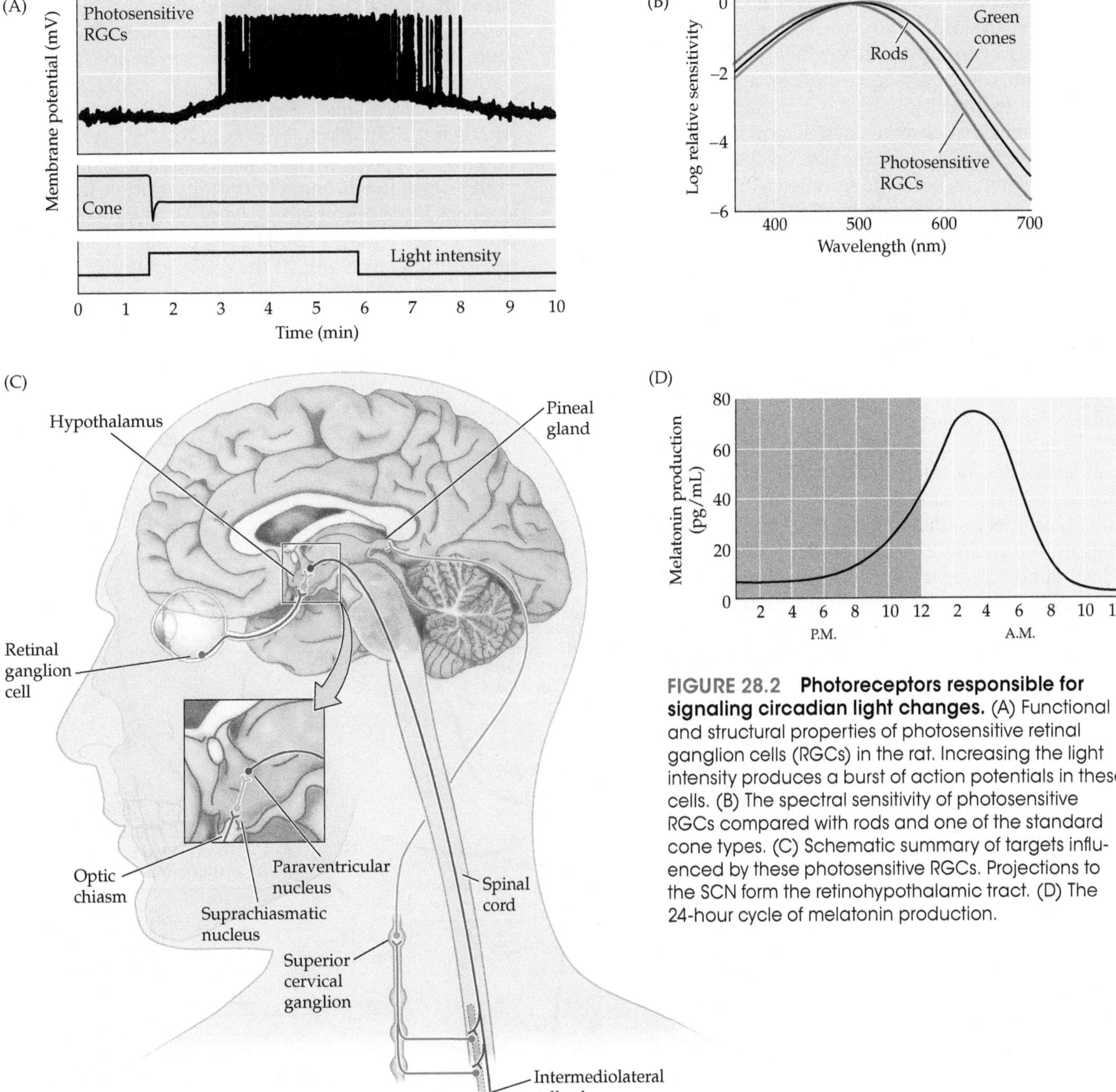

FIGURE 28.2 Photoreceptors responsible for signaling circadian light changes. (A) Functional and structural properties of photosensitive retinal ganglion cells (RGCs) in the rat. Increasing the light intensity produces a burst of action potentials in these cells. (B) The spectral sensitivity of photosensitive RGCs compared with rods and one of the standard cone types. (C) Schematic summary of targets influenced by these photosensitive RGCs. Projections to the SCN form the retinohypothalamic tract. (D) The 24-hour cycle of melatonin production.

of whose postganglionic axons project to the **pineal gland** (*pineal* means "pinecone-like") in the midline near the dorsal thalamus. The pineal gland synthesizes the sleep-promoting neurohormone **melatonin** (*N*-acetyl-5-methoxytryptamine) from tryptophan. When secreted into the bloodstream, melatonin modulates neural activity by interacting with melatonin receptors on neurons in the SCN that in turn influence the sleep–wake cycle. Melatonin synthesis increases as light from the environment decreases, reaching a maximum between 2 and 4 A.M. (Figure 28.2D). In the elderly, the pineal gland produces less melatonin, perhaps explaining why older people sleep less at night. Melatonin supplements have been used to promote sleep in elderly insomniacs and to reduce the disruption of biological clocks that occurs with jet lag, but it remains unclear whether these therapies are effective.

Most sleep researchers consider the suprachiasmatic nucleus to be a "master clock." Evidence for this conclusion is that removal of the SCN in experimental animals abolishes their circadian sleep–wake cycle. Furthermore, when SCN cells are placed in culture, they exhibit characteristic circadian rhythms of activity. Some other isolated cell types also show this rhythmicity, implying that the SCN is the apex of a hierarchy that governs physiological timing functions that are synchronized with the sleep–wake cycle, including body temperature, hormone secretion (e.g., cortisol), blood pressure, and urine output. In adults, urine production is reduced at night because of a circadian upregulation of antidiuretic hormone (ADH, also called vasopressin).

Molecular Mechanisms of Biological Clocks

Virtually all animals, plants, and other organisms adjust their physiology and behavior to the 24-hour day–night cycle under the influence of circadian variations. Recent studies have revealed much about the genes and proteins that are evidently the molecular machinery underlying these effects. This work began in the early 1970s, when Ron Konopka and Seymour Benzer at the California Institute of Technology discovered three mutant strains of fruit flies whose circadian rhythms were abnormal. Analysis showed the mutations to be alleles that differed at a single locus, which Konopka and Benzer called the *period* or *per* gene. In the absence of normal environmental cues (that is, in constant light or dark), wild-type flies have periods of activity geared to a 24-hour cycle; per^S mutants have 19-hour rhythms, per^1 mutants have 29-hour rhythms, and per^0 mutants have no apparent circadian rhythm.

Michael Young at Rockefeller University and Jeffrey Hall and Michael Rosbash at Brandeis University independently cloned the first of the three *per* genes in the early 1990s. Cloning a gene does not necessarily reveal its function, however, and so it was in this case. Nonetheless, the gene product Per, a nuclear protein, is found in many *Drosophila* cells pertinent to the production of the fly's circadian rhythms. Moreover, normal flies show a circadian variation in the amount of *per* mRNA and Per protein, whereas per^0 flies do not show circadian rhythmicity of gene expression.

Many of the genes and proteins responsible for circadian rhythms in fruit flies have now been discovered in mammals (Figure 28.3). In mice, the circadian clock arises from the temporally regulated activity of proteins (given here in capital letters) and genes (both abbreviations and full names in italics), including CRY (*Cry, cryptochrome*), CLOCK (*Clk, circadian locomotor output cycles kaput*), BMAL1 (*Bmal1, brain and muscle, ARNT-like*), PER1 (*Per1, Period1*), PER2 (*Per2, Period2*), and PER3 (*Per3, Period3*). These genes and the proteins they express give rise to regulatory transcription/translation feedback loops with both excitatory and inhibitory components (see Figure 28.3). The key points in this complex regulatory scheme are: (1) the concentrations of BMAL1 and the three PER proteins cycle in counterphase; (2) PER2

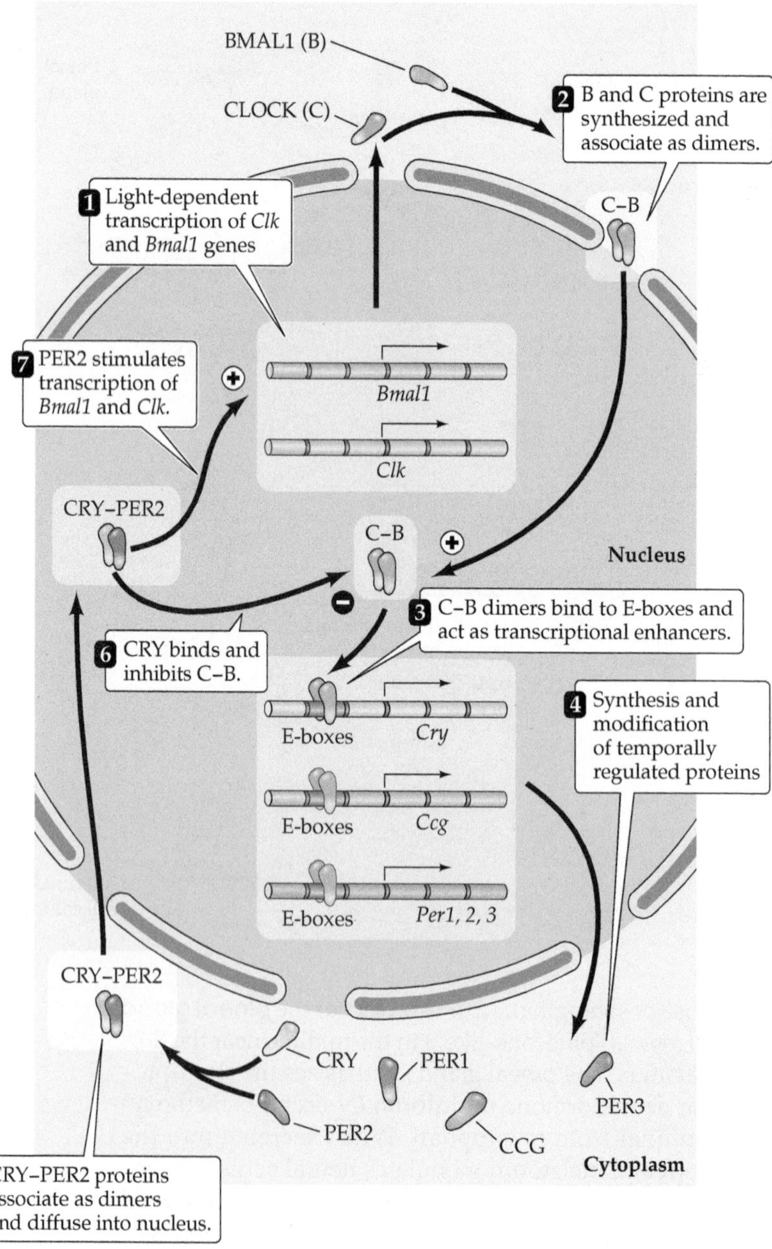

FIGURE 28.3 The molecular feedback loop believed to govern circadian clocks in mammals. (After Okamura et al., 1999.)

is a positive regulator of the BMAL1 loop; and (3) CRY is a negative regulator of the period and cryptochrome loops. The two positive components of this scheme are influenced, albeit indirectly, by light and/or temperature. The cycle begins when CLOCK and BMAL1 drive transcription of *Cry* and the *Per* genes and *Rev-erb* Per proteins and CRY heterodimerize and translocate to the nucleus, where they inhibit the transcriptional activity of CLOCK and BMAL1. This cycle takes about 24 hours. In a second feedback loop, REV-ERB competes with ROR to cyclically repress transcription of *Bmal1*. In the most recent update of this classic model, transcription of another gene, *Bmal1*, is regulated by the transcription factors ROR (which activates *Bmal1* transcription) and REV-ERB (which represses transcription). CLOCK and BMAL1 proteins rhythmically drive direct transcriptional activation of *Rev-erb*, resulting in rhythmic expression of *Bmal1* that is antiphase to the *Per* and *Cry* genes. A complicated story to be sure, but an important one for understanding the molecular machinery that mediates the internal circadian "clock."

Sleep

Sleep is defined behaviorally by the normal suspension of consciousness and electrophysiologically by specific brain wave criteria (Box 28A). It occurs in all mammals, and probably in all vertebrates. We crave sleep when deprived of it, and continued sleep deprivation is highly deleterious and can even be fatal. Surprisingly, however, sleep is not the result of a simple diminution of brain activity. Indeed, in rapid eye movement (REM) sleep, the brain is about as active as it is when people are awake. Rather, sleep is a series of precisely controlled physiological states, the sequence of which is governed by a group of brainstem nuclei that project widely throughout the brain and spinal cord. The reason for the high levels of brain activity during REM sleep, the significance of dreaming, and the basis of the restorative effect of sleep are all topics that remain poorly understood.

BOX 28A ■ Electroencephalography

Although electrical activity recorded from the exposed cerebral cortex of a monkey was reported in 1875, it was not until 1929 that Hans Berger, a psychiatrist at the University of Jena, Germany, first made scalp recordings of this activity in humans. Since then, the electroencephalogram, or EEG, has received a mixed press. Touted by some as a unique opportunity to understand human thinking, the EEG is denigrated by others as too complex and poorly resolved to allow anything more than a superficial glimpse of what the brain is actually doing. The truth probably lies somewhere in between, but no one disputes that electroencephalography has provided a valuable tool to both researchers and clinicians, particularly in the fields of sleep physiology and epilepsy.

The major advantage of electroencephalography, which involves the application of a set of electrodes to standard positions on the scalp (Figure A), is its great simplicity. Its most serious limitation is poor spatial resolution, allowing localization of an active site only to within several centimeters. Four basic EEG phenomena have been defined in humans (albeit somewhat arbitrarily). The alpha rhythm is typically recorded in awake individuals with their eyes closed. By definition, the frequency of the alpha rhythm is 8 to 13 Hz, with amplitudes that are typically 10 to 50 mV. Lower-amplitude beta activity is defined by frequencies of 14 to 60 Hz and is indicative of mental activity and attention. The theta and delta waves, which are characterized by frequencies of 4 to 7 Hz and less than 4 Hz, respectively, imply drowsiness, sleep, or one of a variety of pathological conditions; these slow waves in individuals with normal brain activity are the signature of stage IV non-REM sleep. The way these phenomena are generated is shown in Figures B and C.

Continued on the next page

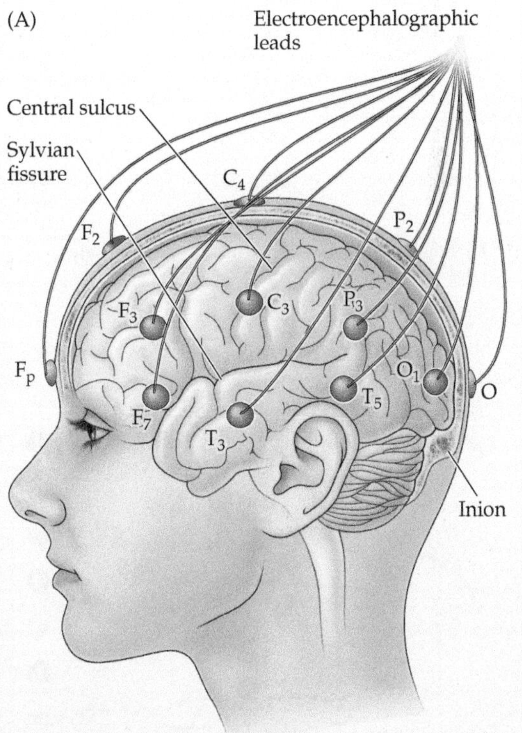

(A) The electroencephalogram represents the voltage recorded between two electrodes applied to the scalp. The placement of the electrodes depends on the application. In the example here, 19 pairs of electrodes are placed in standard positions distributed over the head. Letters indicate position (F = frontal, P = parietal, T = temporal, O = occipital, C = central). The recording obtained from each pair of electrodes is somewhat different because each samples the activity of a population of neurons in a different brain region.

BOX 28A ■ (continued)

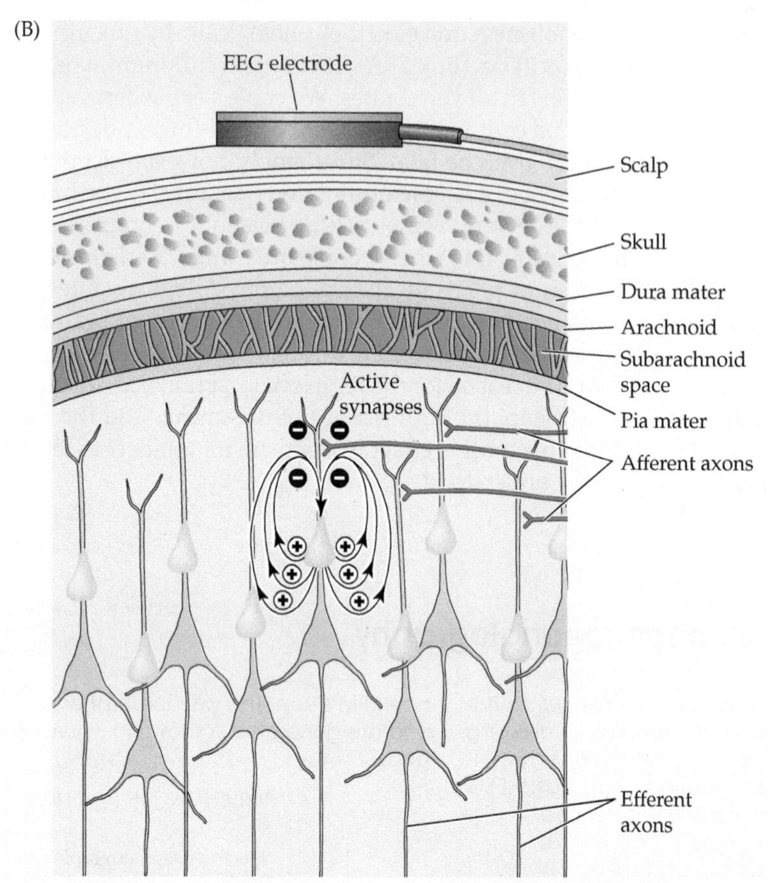

(B) An electrode on the scalp measures the activity of a very large number of neurons in the underlying regions of the brain, each of which generates a small electrical field that changes over time. This activity (which is thought to be mostly synaptic) makes the more superficial extracellular space negative with respect to deeper cortical regions. The EEG electrode measures a synchronous signal because many thousands of cells are responding in the same manner at more or less the same time. (After Bear et al., 2001.)

Far and away the most obvious component of these various oscillations is the alpha rhythm. Its prominence in the occipital region—and its modulation by eye opening and closing—implies that it is somehow linked to visual processing, as was first proposed in 1935 by the British physiologist Edgar Adrian. In fact, evidence from large numbers of subjects suggests that at least several different regions of the brain have their own characteristic rhythms; for example, within the alpha band (8 to 13 Hz), one rhythm, the classic alpha rhythm, associated with visual cortex, one (the is mu rhythm) with the sensory motor cortex around the central sulcus, and yet another (the kappa rhythm) with the auditory cortex.

In the 1940s, Edward Dempsey and Robert Morrison showed that these EEG rhythms depend in part on activity in the thalamus, since thalamic lesions can reduce or abolish the oscillatory

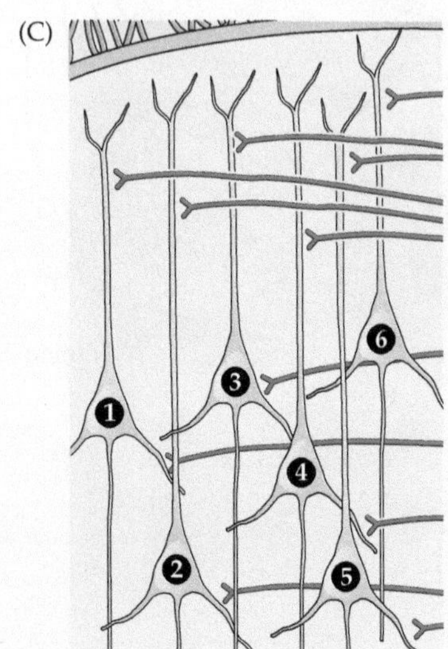

(C) Generation of the synchronous activity that characterizes deep sleep. In the pyramidal cell layer below the EEG electrode, each neuron receives thousands of synaptic inputs. If the inputs are irregular or out of phase, their algebraic sum will have a small amplitude, as occurs in the waking state. If, however, the neurons are activated at approximately the same time, then the EEG waves will tend to be in phase and the amplitude will be much greater, as occurs in the delta waves that characterize stage IV sleep. (After Bear et al., 2001.)

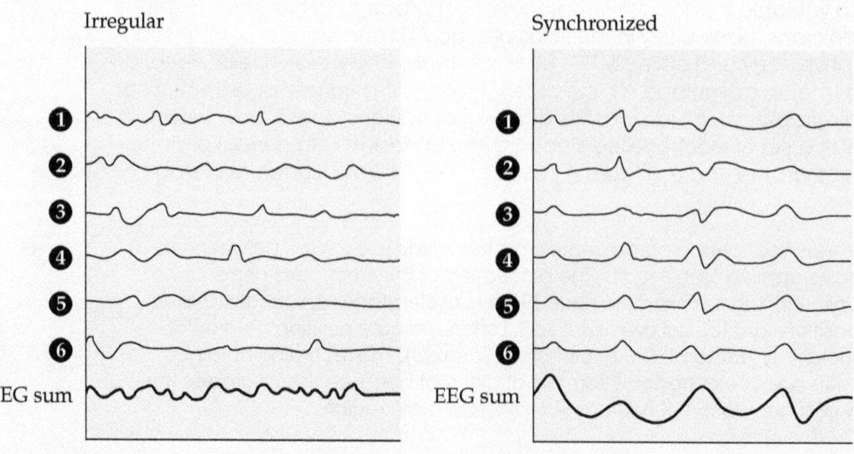

BOX 28A (continued)

cortical discharge (although some oscillatory activity remains even after the thalamus has been inactivated). At about the same time, Horace Magoun and Giuseppe Moruzzi showed that the reticular activating system in the brainstem is also important in modulating EEG activity. For example, activation of the reticular formation changes the cortical alpha rhythm to beta activity, in association with greater behavioral alertness. In the 1960s, Per Andersen and his colleagues in Sweden further advanced these studies by showing that virtually all areas of the cortex participate in these oscillatory rhythms, which reflect a feedback loop between neurons in the thalamus and cortex (see Figure 28.9).

The cortical origin of EEG activity has been clarified by animal studies, which have shown that the source of the current that causes the fluctuating scalp potential is primarily the pyramidal neurons and their synaptic connections in the deeper layers of the cortex (see Figures B and C). (This conclusion was reached by noting the location of electrical field reversal upon passing an electrode vertically through the cortex from surface to white matter.) In general, oscillations come about either because membrane voltage of thalamocortical cells fluctuates spontaneously or as a result of the reciprocal interaction of excitatory and inhibitory neurons in circuit loops. The oscillations of the EEG are thought to arise from the latter mechanism.

Despite these intriguing observations, the functional significance of these cortical rhythms is not known. The purpose of the brain's remarkable oscillatory activity is a puzzle that has defied electroencephalographers and neurobiologists for more than 75 years.

To feel rested, most adults require 7 to 8 hours of sleep, although this number varies among individuals (Figure 28.4A). For infants, the requirement is much higher (17 hours a day or more), and teenagers need on average about 9 hours of sleep. Thus, we spend a substantial fraction of our lives in this mysterious state. As people age, they tend to sleep more lightly and for shorter times (Figure 28.4B). Older adults often "make up" for shorter and lighter nightly sleep periods by napping during the day. Getting too little sleep creates a "sleep debt" that must be repaid in the following days. In the meantime, judgment, reaction time, and other functions are impaired in varying degrees. Poor sleep therefore has a price, and sometimes tragic consequences. In the United States, fatigue is estimated to contribute to more than 100,000 highway accidents each year, resulting in some 70,000 injuries and 1500 deaths.

The clinical importance of sleep is obvious from the prevalence of sleep disorders. In any given year an estimated 40 million Americans suffer from chronic sleep disorders, and an additional 30 million experience occasional sleep problems that are severe enough to interfere with their daily activities. In consequence, the phenomenology of sleep presents major challenges in both basic neurobiology and clinical medicine.

Sleep (or at least a physiological period of quiescence) is a highly conserved behavior found in animals ranging from fruit flies to humans and goes on even in the absence of cues.

The Purpose of Sleep

Despite the prevalence of sleep, *why* we sleep is not well understood. Because an animal is particularly vulnerable while sleeping, there must be evolutionary advantages that

FIGURE 28.4 The duration of sleep. (A) In adults, the duration of sleep each night is normally distributed with a mean of 7.5 hours and a standard deviation of about 1.25 hours. Thus, each night about two-thirds of the population sleeps between 6.25 and 8.75 hours. (B) The duration of daily sleep as a function of age. (After Hobson, 1989.)

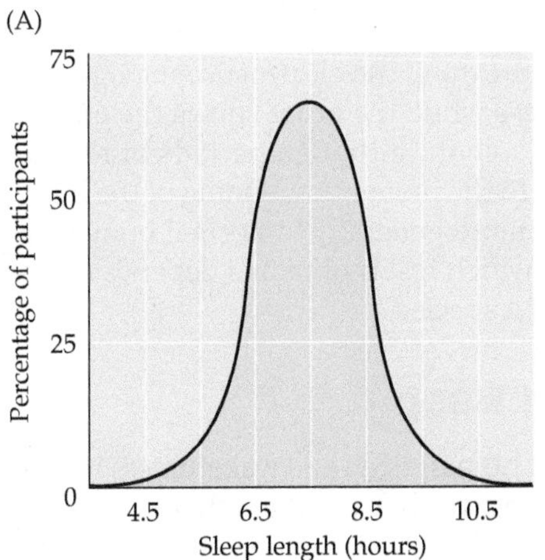

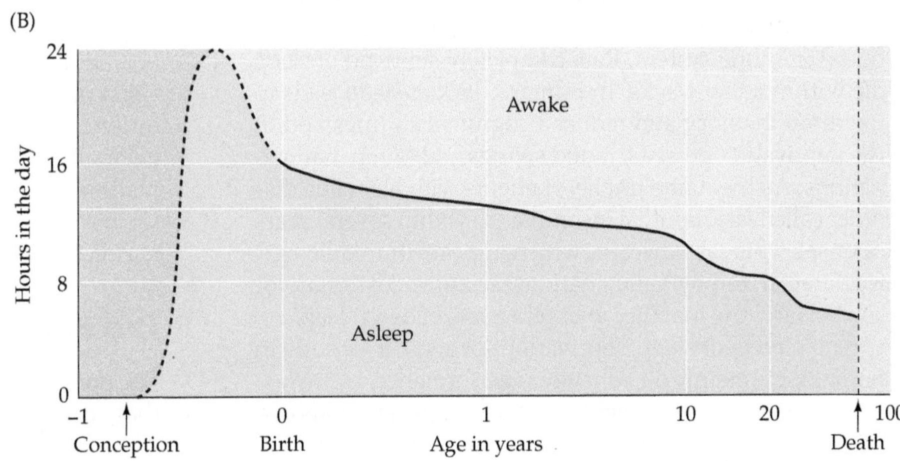

BOX 28B ■ Sleep and Memory

Studies in recent years have indicated that memory consolidation is enhanced during sleep. The idea that memories are replayed during sleep was suggested by behavioral evidence that sleeping soon after learning improves subsequent recall. Electroencephalographic studies further showed that REM sleep is more important for nondeclarative memory consolidation, whereas slow-wave sleep is more important for declarative memory consolidation. For example, one study implanted electrodes in the hippocampus and visual cortex of rats and measured neuronal firing in these brain regions while the rats ran in a figure-eight maze, as well as during periods of slow-wave sleep before and after maze running. The simultaneous hippocampal-cortical reactivation observed was proposed to strengthen cortical-cortical synapses, leading to memory consolidation.

In another study, investigators showed that stimuli presented during sleep and associated with a learning experience can promote replay. Participants learned object-location associations while a specific odor ("rose") was present in the room. Re-presenting the same smell during slow-wave sleep increased hippocampal activity and improved memory for object locations the next day. In another study, participants learned associations between objects and spatial locations while listening to sound associated with each object. Spatial memory after a nap improved. This range of evidence suggests yet another benefit of sleep.

outweigh this considerable disadvantage. From a perspective of energy conservation, one function of sleep is to replenish brain glycogen levels, which fall during the waking hours. In addition, since it is generally colder at night, more energy would have to be expended to keep warm were we nocturnally active. Human body temperature has a 24-hour cycle (as do many other indices of activity and stress; see Figure 28.1B), reaching a minimum at night and thus reducing heat loss. Metabolism measured by oxygen consumption decreases during sleep. Another plausible reason is that humans and many other animals that sleep at night are highly dependent on visual information to find food and avoid predators. A recent idea about an advantage of sleep proposes that memories, in the form of changes in the strength of synaptic connections induced by experiences during waking hours, are consolidated while we sleep (Box 28B). An even more provocative idea, and not necessarily mutually exclusive, is that metabolic wastes produced by neurons active during wakefulness are cleared from the brain during sleep, thus accounting for the restorative effects of sleep.

In mammals, sleep is evidently necessary for successful behavior and even survival. Sleep-deprived rats lose weight despite increased food intake and progressively fail to regulate body temperature as it increases several degrees. They also develop infections, suggesting a compromised immune system. Rats completely deprived of sleep die within a few weeks. In humans, lack of sleep leads to impaired memory and reduced cognitive abilities and, if the deprivation persists, mood swings and, often, hallucinations. As the name implies, patients with a genetic disorder called fatal familial insomnia die within several years of onset. This rare disease, which appears in middle age, is characterized by hallucinations, seizures, loss of motor control, and the inability to enter a state of deep sleep.

The effects of sleep deprivation for shorter periods in humans are mainly on behavioral performance, as anyone who has experienced a night or two of sleeplessness knows. The longest documented period of voluntary sleeplessness in humans is 264 hours (approximately 11 days)—a record achieved without any pharmacological stimulation. The young man involved recovered within a few days, during which he slept more than normal but otherwise seemed none the worse for wear.

Sleep in Different Species

A wide variety of animals have a rest–activity cycle that often (but not always) occurs in a daily, or circadian, rhythm. Even among mammals, however, the organization of sleep depends on the species in question. As a general rule, predatory animals can indulge, as humans do, in long, uninterrupted periods of sleep that can be nocturnal or diurnal, depending on the time of day when the animal acquires food, mates, cares for its young, and deals with life's other necessities. The survival of animals that are preyed on depends much more critically on continued vigilance. Such species—as diverse as rabbits and giraffes—sleep during short intervals that usually last no more than a few minutes. Shrews, the smallest mammals, hardly sleep at all.

Dolphins and seals show an especially remarkable solution to the problem of maintaining vigilance during sleep. Their sleep alternates between the two cerebral hemispheres. EEG tracings taken simultaneously from left and right cerebral hemispheres of dolphins show that one hemisphere can exhibit the electroencephalographic signs of wakefulness while the other shows the characteristics of sleep. In short, although periods of rest are evidently essential to the proper functioning of the mammalian brain, and more generally to normal homeostasis, the manner in which rest is obtained depends on the particular needs of a species.

The Stages of Sleep

The normal cycle of human sleep and wakefulness implies that, at specific times, various neural systems are active while others are turned off. For centuries—indeed, up until the 1950s—most researchers considered sleep a unitary phenomenon whose physiology was essentially passive and whose purpose was simply restorative. In 1953, however, Nathaniel Kleitman and Eugene Aserinksy showed by means of electroencephalographic (EEG) recordings from healthy individuals that sleep actually comprises different stages that occur in a characteristic sequence.

Over the first hour after retiring, humans descend into successive stages of sleep (Figure 28.5). These characteristic stages are defined primarily by electroencephalographic criteria (see Box 28A). Initially, during "drowsiness," the frequency spectrum of the electroencephalogram shifts toward lower values and the amplitude of the cortical waves increases slightly. This drowsy period, called *stage I sleep,* eventually gives way to light *stage II sleep,* which is characterized by a further decrease in the frequency of the EEG waves and an increase in their amplitude, together with intermittent high-frequency spike clusters called **sleep spindles**. Sleep spindles are periodic bursts of activity at about 10 to 12 Hz that generally last 1 to 2 seconds and arise as a result of interactions between thalamic and cortical neurons (see below). In *stage III sleep,* which represents moderate to deep sleep, the number of spindles decreases, whereas the amplitude of EEG activity increases further and the frequency continues to fall. In the deepest level of sleep, *stage IV sleep,* the predominant EEG activity consists of very low frequency (0.5 to 4 Hz), high-amplitude fluctuations called **delta waves**, the characteristic slow waves for which this phase of sleep is named; stages III and IV together are known as **slow-wave sleep**. (Note that delta waves can also be thought of as reflecting synchronized electrical activity of cortical neurons.) The entire sequence from drowsiness to stage IV sleep usually takes about an hour.

Taken together, sleep stages I–IV are called **non-rapid eye movement**, or **non-REM sleep**. The most prominent feature of non-REM sleep is the slow-wave stage IV. It is more difficult to awaken people from slow-wave sleep, which is therefore considered to be the deepest stage of sleep. Following a period of slow-wave sleep, however, EEG recordings show that participants enter a quite different state called **rapid eye movement (REM) sleep**. EEG recordings of REM sleep are remarkably similar to those of individuals in the awake state (see Figure 28.5). After about 10 minutes in REM sleep, the brain typically cycles back through the four non-REM sleep stages. Slow-wave sleep is usually most pronounced early in an 8-hour sleep episode, although it occurs periodically during any given night (see Figure 28.6). On average, four additional periods of REM sleep occur, each having a longer duration.

In summary, the typical 8 hours of sleep experienced each night actually comprise several cycles that alternate between non-REM and REM sleep, and the brain is quite active during much of this supposedly dormant, restful time. The amount of daily REM sleep decreases from about 8 hours at birth to 2 hours at 20 years to only about 45

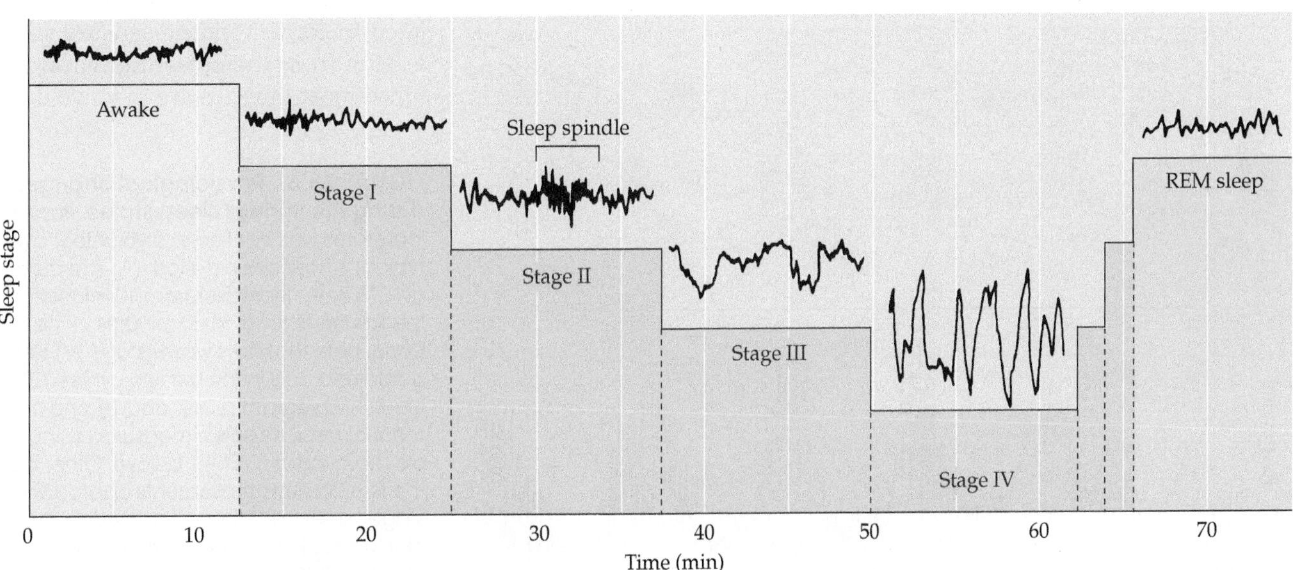

FIGURE 28.5 EEG recordings during the first hour of sleep in humans. The waking state with the eyes open is characterized by high-frequency (15 to 60 Hz), low-amplitude activity (~30 μV) activity. This pattern is called beta activity. Descent into stage I non-REM sleep is characterized by decreasing EEG frequency (4 to 8 Hz) and increasing amplitude (50 to 100 μV), called theta waves. Descent into stage II non-REM sleep is characterized by 10 to 12 Hz oscillations (50 to 150 μV) called sleep spindles, which occur periodically and last for a few seconds. Stages III and IV of non-REM sleep are characterized by slower waves (also called delta waves) at 0.5 to 4 Hz (100 to 150 μV). After reaching this level of deep sleep, the sequence changes and a period of rapid eye movement sleep, or REM sleep, ensues. REM sleep is characterized by low-voltage, high-frequency activity similar to the EEG activity of individuals who are awake. Although the diagram makes it appear that the amounts of sleep in each stage are about the same, they are not (see text). (After Hobson, 1989.)

minutes at 70 years of age. The reasons for this change over the human lifespan are not known.

The Physiological Changes in Sleep States

A variety of additional physiological changes take place during the different stages of sleep (Figure 28.6). Stage I sleep is characterized by slow, rolling eye movements followed by decreases in muscle tone, body movements, heart rate, breathing, blood pressure, metabolic rate, and temperature. All these parameters reach their lowest values during stage IV sleep. Periods of REM sleep, in contrast, are accompanied by increases in blood pressure, heart rate, and metabolism to levels almost as high as those found in the awake state. REM sleep, as the name implies, is also characterized by rapid, ballistic eye movements, as well as by pupillary constriction, paralysis of many large muscle groups (although obviously not the diaphragm and other muscles used for breathing), and the twitching of the smaller muscles in the fingers, toes, and middle ear. Spontaneous penile erection also occurs during REM sleep, a fact that is clinically important in determining whether a complaint of impotence has a physiological or psychological basis. REM sleep has been observed in all mammals and in at least some birds; certain reptiles also have periods of increased brain activity during sleep that are arguably homologous to REM sleep in mammals.

Despite the similarity of EEG recordings obtained in REM sleep and in wakefulness, the two brain states are clearly not equivalent. For one thing, REM sleep is characterized by a greater prevalence of **dreaming** (Box 28C), a unique state of awareness that entails some features of memory and hallucinations in the sense that the experience of dreams is not related to corresponding sensory stimuli arising from the present environment. Since most muscles are inactive during

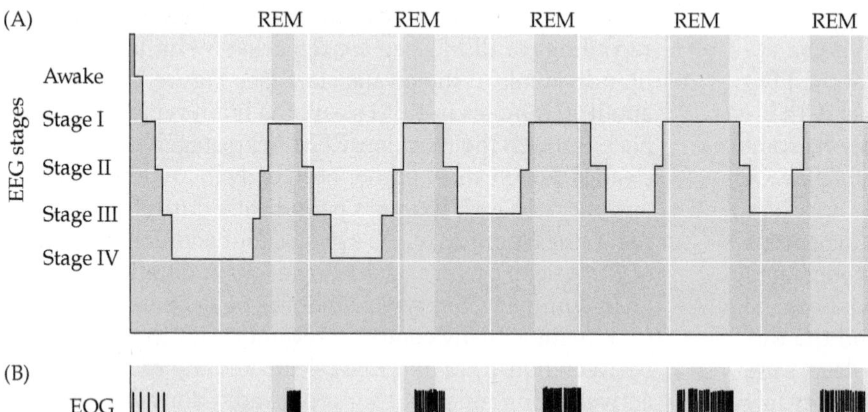

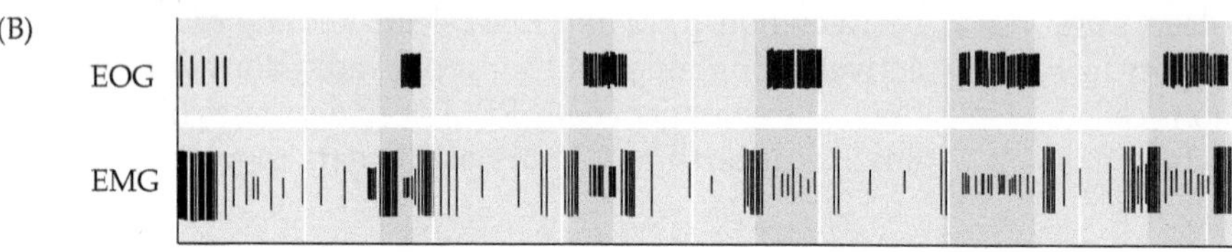

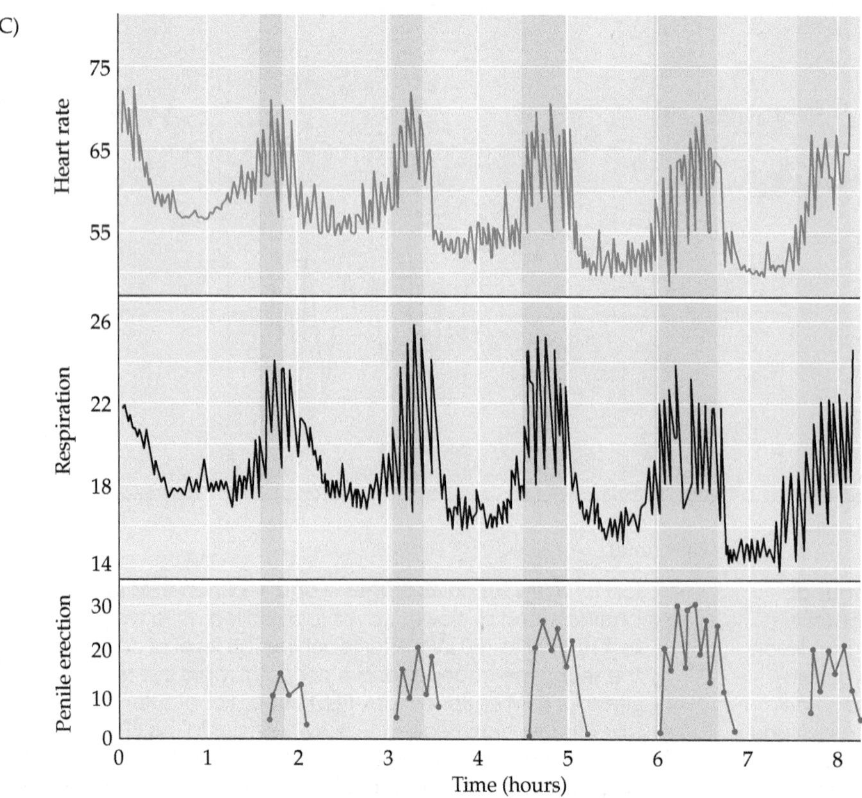

FIGURE 28.6 Physiological changes during the various sleep states. Physiological status was tracked in a volunteer over a typical 8-hour sleep period. (A) The duration of REM sleep increases from 10 minutes in the first cycle to up to 50 minutes in the final cycle; note that slow-wave (stage IV) sleep is attained only in the first two cycles. (B) The electrooculogram (EOG, above) and movement of neck muscles, measured using an electromyogram (EMG, below). Other than the few slow eye movements approaching stage I sleep, all other eye movements evident in the EOG occur in REM sleep. The greatest EMG activity occurs during the onset of sleep and just prior to awakening. (C) The heart rate (beats per minute) and respiration (breaths per minute) slow in non-REM sleep, but increase almost to the waking levels in REM sleep. Penile erection (strain gauge units) occurs only during REM sleep. (A after Woods and Greenhouse, 1974; B,C after Funkhouser, 2014; Jovanovic, 1971.)

BOX 28C ■ Dreaming

Despite the wealth of descriptive information about the stages of sleep and an intense research effort over the last 50 years, the functional purposes of the various sleep states remain poorly understood. Whereas most sleep researchers accept the idea that the purpose of non-REM sleep is at least in part restorative, the function of REM sleep and sleep in general remains a matter of considerable controversy.

A clue about additional purposes of sleep concerns dreams. The time of occurrence of dreams during sleep has been determined by waking volunteers during either non-REM or REM sleep and asking them if they were dreaming. Individuals awakened from REM sleep usually recall elaborate, vivid, and often emotional dreams; those awakened during non-REM sleep report fewer dreams, and when dreams do occur, they are more conceptual, less vivid, and tend to be less emotion-laden. Dreaming during light non-REM sleep tends to be more prevalent near the onset of sleep and before awakening. In any event, dreaming is not limited to REM sleep.

Dreams have been studied in a variety of ways, perhaps most notably within the psychoanalytic framework aimed at revealing unconscious thought processes considered to be at the root of neuroses. Sigmund Freud's *The Interpretation of Dreams*, published in 1900, speaks eloquently to the complex relationship between conscious and unconscious mentation. Freud thought that during dreaming the conscious "ego" relaxes its hold on the "id," or subconscious. These ideas have been out of fashion in recent decades, but to give Freud his due, at the time he made these speculations little was known about neurobiology of the brain in general and sleep in particular. In fact, some recent evidence supports Freud's idea that dreams often reflect events and conflicts of the day (the "day residue," in his terminology) and may play a role in memory. Several investigators have suggested that dreams help consolidate learned tasks, perhaps by further strengthening synaptic changes associated with recent experiences. The more general hypothesis that sleep is important in consolidating memories has been supported by studies of remembered spatial location in rodents in which ensembles of hippocampal neurons activated during a spatial memory task are reactivated during subsequent sleep, and by experiments in humans that show a sleep-dependent improvement in learning.

A quite different idea is that dreaming has evolved to dispose of unwanted memories that accumulate during the day. Francis Crick, for example, suggested that dreams might reflect a mechanism for expunging "parasitic" modes of thought that would otherwise become overly intrusive, as occurs in compulsive thought disorders. Finally, some experts such as Allan Hobson have taken the more skeptical view that dream content may be "as much dross as gold, as much cognitive trash as treasure, as much informational noise as a signal of something."

Adding to the uncertainty about the purposes of sleep and dreaming is the fact that depriving human subjects of REM sleep for as much as 2 weeks has little or no obvious effect on their behavior. Furthermore, patients taking serotonin reuptake inhibitors for depression have markedly less REM sleep. The apparent innocuousness of REM sleep deprivation contrasts markedly with the devastating effects of total sleep deprivation mentioned earlier. The implication of these findings is that we can get along without REM sleep, but we need non-REM sleep in order to survive.

REM sleep, the motor responses to dreams are relatively minor. (Sleepwalking, which is most common in children ages 4 to 12, and "sleeptalking" actually occur during non-REM sleep and are not usually accompanied or motivated by dreams.)

Taken together, these observations have led to the aphorism that non-REM sleep is characterized by an inactive brain in an active body, whereas REM sleep is characterized by an active brain in an inactive body. In any event, it is apparent that a variety of sensory and motor systems are sequentially activated and inactivated during the different stages of sleep.

The Neural Circuits Governing Sleep

From the descriptions of the various physiological states that occur during sleep, it is clear that periodic changes in the balance of excitation and inhibition must occur in many neural circuits. What follows is a brief overview of these still incompletely understood circuits and the interactions among them that govern sleep and wakefulness.

In 1949, Horace Magoun and Giuseppe Moruzzi provided one of the first pieces of evidence about the circuits involved in the sleep–wake cycle. They found that electrically stimulating a group of neurons near the junction of the pons and midbrain causes a state of wakefulness and arousal. This region of the brainstem was given the name **reticular activating system** (Figure 28.7A; see also Box 17D). Their work implied that wakefulness requires specialized activating circuitry—that is, wakefulness is not just the result of adequate sensory experience. At about the same time, the Swiss physiologist Walter Hess found that stimulating the thalamus in an awake cat with low-frequency pulses produced a slow-wave sleep (Figure 28.7B). These key experiments showed that sleep entails a patterned interaction between the brainstem, thalamus, and cortex.

Further evidence was provided by work on the circuitry underlying REM sleep. The saccade-like eye movements that define REM sleep are now known to arise because, in the absence of external visual stimuli, endogenously generated signals from the **pontine reticular formation** are transmitted to the motor region of the superior colliculus. As described in Chapter 20, collicular neurons project to the **paramedian pontine reticular formation** (**PPRF**) and the **rostral interstitial nucleus**, which coordinates

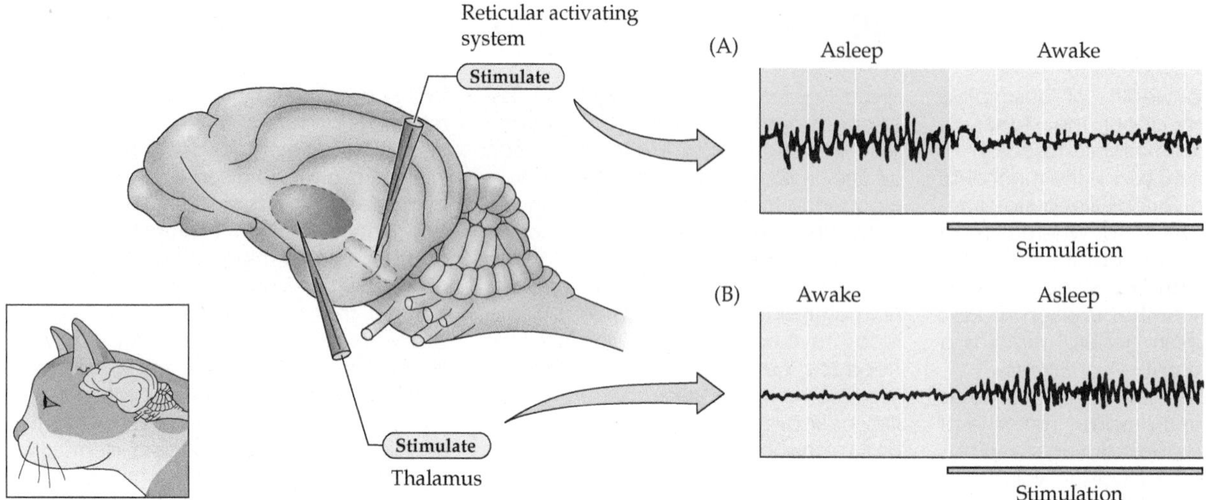

FIGURE 28.7 Activation of specific neural circuits triggers sleep and wakefulness. (A) Electrical stimulation of the cholinergic neurons near the junction of pons and midbrain (the reticular activating system) causes a sleeping cat to awaken. (B) Slow electrical stimulation of the thalamus causes an awake cat to fall asleep. Graphs show EEG recordings before and during stimulation. (After Magoun, 1952.)

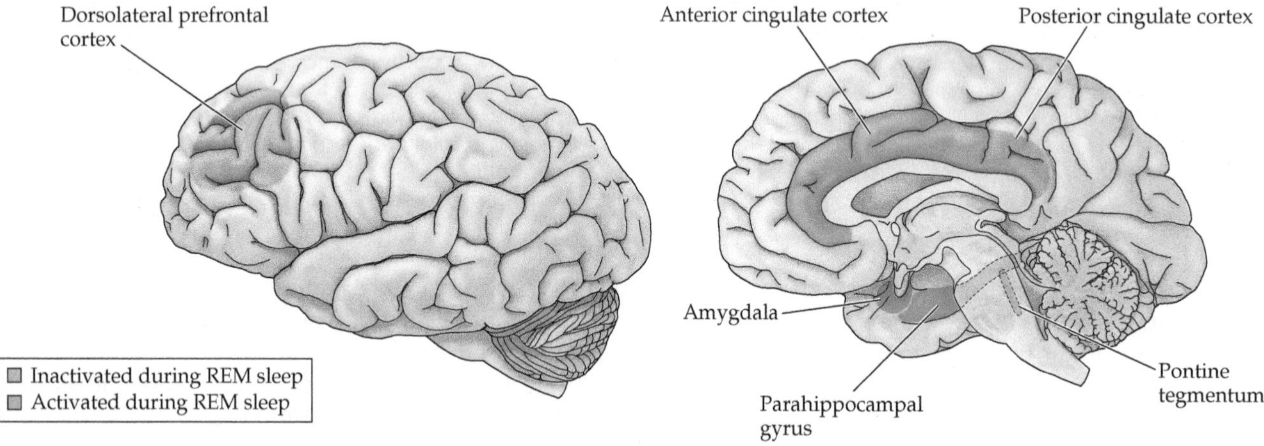

FIGURE 28.8 Cortical activity during REM sleep. The diagram shows cortical regions whose activity is increased or decreased during REM sleep. (After Hobson et al., 1989.)

timing and direction of eye movements. REM sleep is also characterized by EEG waves that originate in the pontine reticular formation and propagate through the lateral geniculate nucleus of the thalamus to the occipital cortex. These **pontine-geniculo-occipital (PGO) waves** provide a useful marker for the beginning of REM sleep; they also indicate yet another neural network by which brainstem nuclei can activate the cortex.

A further advance has come from fMRI and PET studies that have compared human brain activity in the awake state and in REM sleep, as well as studies of the phenomenon of consciousness more generally (see the section "Consciousness" later in the chapter). Activity in the amygdala, parahippocampus, pontine tegmentum, and anterior cingulate cortex increases during REM sleep, whereas activity in the dorsolateral prefrontal and posterior cingulate cortices decreases (Figure 28.8). The increase in limbic system activity, coupled with a marked decrease in the influence of the frontal cortex during REM sleep, presumably explains some characteristics of dreams (e.g., their heightened emotionality in REM sleep and their often inappropriate social content).

On the basis of these discoveries and other studies using neuronal recording in experimental animals, it is now generally agreed that a key component of the reticular activating system is a group of **cholinergic nuclei** near the pons–midbrain junction that project to thalamocortical neurons (Figure 28.9). The relevant neurons in the nuclei

(A)

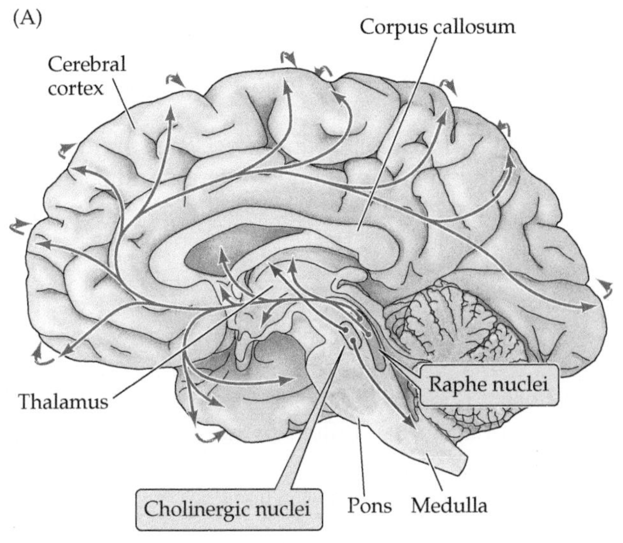

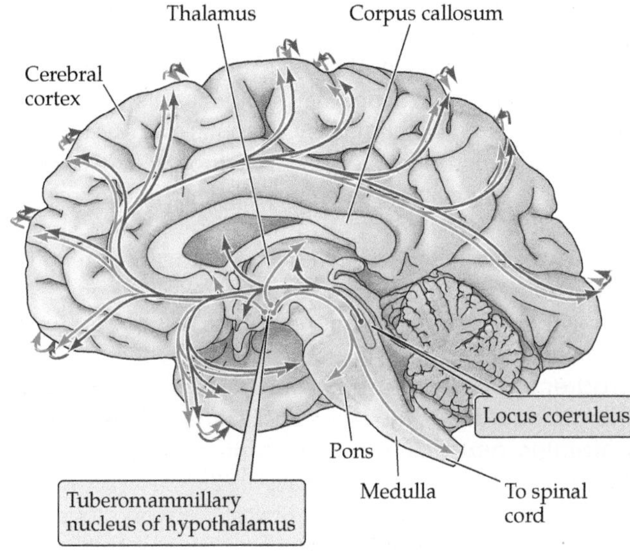

(B)

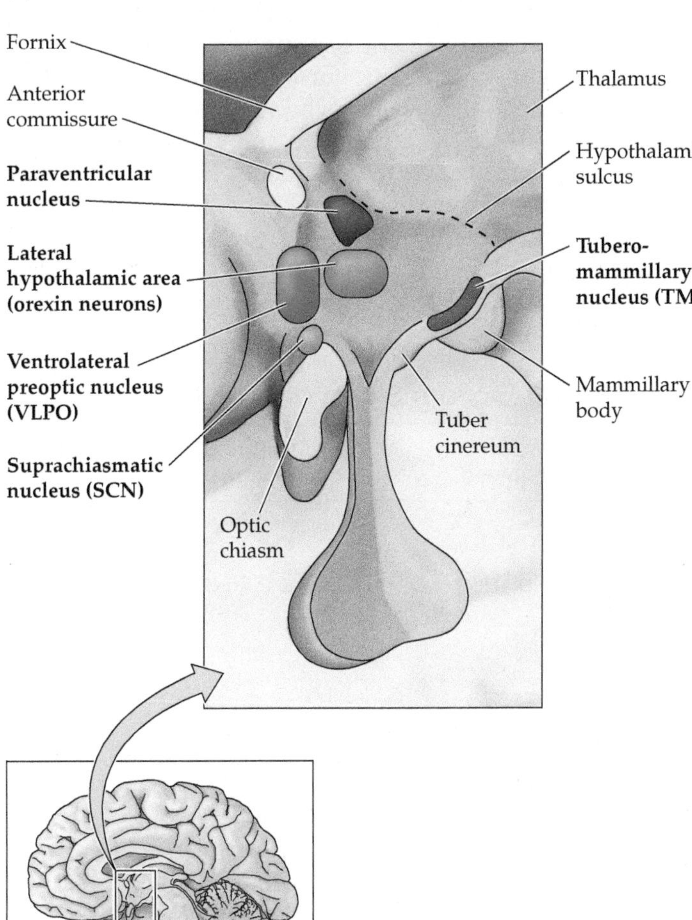

(C)

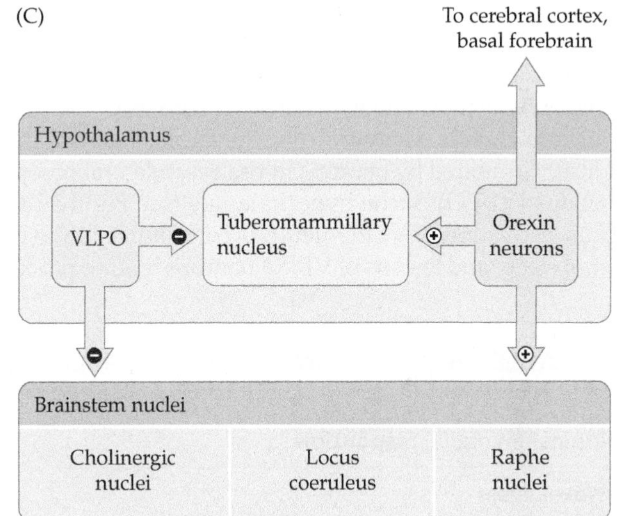

FIGURE 28.9 Important nuclei in regulation of the sleep-wake cycle. (A) A variety of brainstem nuclei using several different neurotransmitters determines mental status on a continuum that ranges from deep sleep to a high level of alertness. These nuclei include: (left) the cholinergic nuclei of the pons–midbrain junction and the raphe nuclei; and (right) the locus coeruleus and the tuberomammillary nucleus. All have widespread ascending and descending connections to other regions (arrows), which explains their numerous effects. Curved arrows along the perimeter of the cortex indicate the innervation of lateral cortical regions not shown in this plane of section. (B) Location of hypothalamic nuclei involved in sleep. (C) Activation of VLPO induces sleep. Orexin-containing neurons project to different nuclei and produce arousal.

are characterized by high discharge rates during both waking and REM sleep, and by quiescence during non-REM sleep. When stimulated, these nuclei cause "desynchronization" of the electroencephalogram (i.e., a shift of EEG activity from high-amplitude, synchronized waves to lower-amplitude, higher-frequency, desynchronized waves; see Box 28A). These features imply that activity of cholinergic neurons in the reticular activating system is a primary cause of wakefulness and REM sleep and that their relative inactivity is important for producing non-REM sleep.

Activity of these cholinergic neurons is not, however, the only neuronal basis of wakefulness; also involved are the **noradrenergic neurons** of the locus coeruleus; the **serotonergic neurons** of the raphe nuclei; and the **histamine-containing neurons** in the tuberomammillary nucleus (TMN) of the hypothalamus (see Figure 28.9). The activation of these cholinergic and monoaminergic networks together produces the awake state. The locus coeruleus and raphe nuclei are modulated by the TMN neurons located near the tuberal region. The TMN is activated by neurons in the lateral hypothalamus that secrete the peptide **orexin** (also called **hypocretin**), which promotes waking. Conversely, antihistamines inhibit the histamine-containing TMN network, which is why they tend to make people drowsy.

These circuits responsible for the awake state are periodically inhibited by neurons in the ventrolateral preoptic nucleus (VLPO) of the hypothalamus (see Figure 28.9). Thus, activation of VLPO neurons contributes to the onset of sleep, and lesions of VLPO neurons tend to produce insomnia. As if all this were not complicated enough, recent work suggests that adenosine neurotransmission in the basal forebrain is also involved in the regulation of sleep.

These complex interactions and effects are summarized in Table 28.1. In brief, both monoaminergic and cholinergic systems are active during the waking state, and their decreased activity leads to the onset of non-REM sleep. In REM sleep, monoaminergic and serotonergic neurons are markedly less active, while cholinergic levels of activity increase to approximately the levels found in the awake state. Given that so many systems and transmitters are involved in the different phases of sleep, it is not surprising that a wide variety of drugs can influence the sleep cycle.

Thalamocortical Interactions in Sleep

The effects of neuronal activity (or its absence) in these brainstem nuclei are achieved by modulating the interactions between the thalamus and the cortex. Thus, the activity of several ascending systems from the brainstem decreases both the rhythmic bursting of the thalamocortical neurons and the related synchronized activity of cortical neurons (hence the diminution and ultimate disappearance of high-voltage, low-frequency slow waves during waking and during REM sleep).

To appreciate how different sleep states reflect modulation of thalamocortical activity, consider the electrophysiological responses of the relevant neurons.

TABLE 28.1 ■ **Summary of the Cellular Mechanisms That Govern Sleep and Wakefulness**

Brainstem nuclei responsible	Neurotransmitter involved	Activity state of the relevant brainstem neurons
Wakefulness		
Cholinergic nuclei of pons-midbrain junction	Acetylcholine	Active
Locus coeruleus	Norepinephrine	Active
Raphe nuclei	Serotonin	Active
Tuberomammillary nuclei	Histamine	Active
Lateral hypothalamus	Orexin	Active
Non-REM sleep		
Cholinergic nuclei of pons-midbrain junction	Acetylcholine	Decreased
Locus coeruleus	Norepinephrine	Decreased
Raphe nuclei	Serotonin	Decreased
REM sleep		
Cholinergic nuclei of pons-midbrain junction	Acetylcholine	Active (PGO waves)
Raphe nuclei	Serotonin	Inactive
Locus coeruleus	Norepinephrine	Inactive

Thalamocortical neurons receive ascending projections from the locus coeruleus (noradregeneric), raphe nuclei (serotonergic), reticular activating system (cholinergic), and TMN (histaminergic), and as their name implies, project to cortical pyramidal cells. The primary characteristic of thalamocortical neurons is that they can be in one of two stable electrophysiological states: an intrinsic bursting (or oscillatory) state; or a tonically active state. The latter is generated when the thalamocortical neurons are depolarized, such as occurs when the reticular activating system generates wakefulness (Figure 28.10). In the tonically active state, thalamocortical neurons transmit information to the cortex that is correlated with the spike trains encoding peripheral stimuli. In contrast, when thalamocortical neurons are in the bursting state, the neurons in the thalamus become synchronized with those in the cortex, essentially "disconnecting" the cortex from the outside world. The disconnection is maximal during slow-wave sleep, when EEG recordings show the lowest frequency and the highest amplitude.

The oscillatory, or bursting, state of thalamocortical neurons can be transformed into the tonically active state by activity in the cholinergic or monoaminergic projections from the brainstem nuclei (Figure 28.11). Moreover, the bursting state is stabilized by hyperpolarizing the relevant thalamic cells. Such hyperpolarization occurs as a consequence of stimulation by GABAergic neurons in the thalamic reticular nucleus. These neurons receive ascending information from the brainstem and descending projections from cortical neurons, and they contact the thalamocortical neurons. When neurons in the reticular nucleus undergo a burst of activity, they cause thalamocortical neurons to generate short bursts of action potentials, which in turn generate spindle activity in cortical EEG recordings (indicating a lighter sleep state).

In sum, the control of sleep and wakefulness depends on brainstem modulation of the thalamus and cortex. It is this modulation of the thalamocortical loop that generates the EEG signature of brain function along the continuum of deep sleep to high alert. The major components of the modulatory system are the cholinergic nuclei of the pons–midbrain junction; the noradrenergic cells of the locus coeruleus in the pons; the serotonergic raphe nuclei; and GABAergic neurons in the VLPO of the hypothalamus. All of these nuclei exert both direct and indirect effects on the overall cortical activity that determines sleep and wakefulness. Figure 28.12 summarizes the relationships among the various sleep–wake states.

Consciousness

Although being awake is clearly a prerequisite to being conscious in the sense of being normally aware of the world and the self, these functions are not equivalent. Most definitions of consciousness thus refer to two different aspects or meanings of the term: a physiological meaning that describes consciousness in terms of the brain state we think of as wakefulness, and a more abstract meaning that refers to a subjective awareness of the world—a brain state that must have a more subtle signature than wakefulness, since one can be awake and yet be unaware of some or even most aspects of the external and internal environments.

This latter meaning of consciousness refers not to the neuroanatomical and physiological bases of wakefulness, but to the deeply puzzling ability we all have to be subjectively aware of the world and of ourselves as actors in it. Whereas consciousness as wakefulness

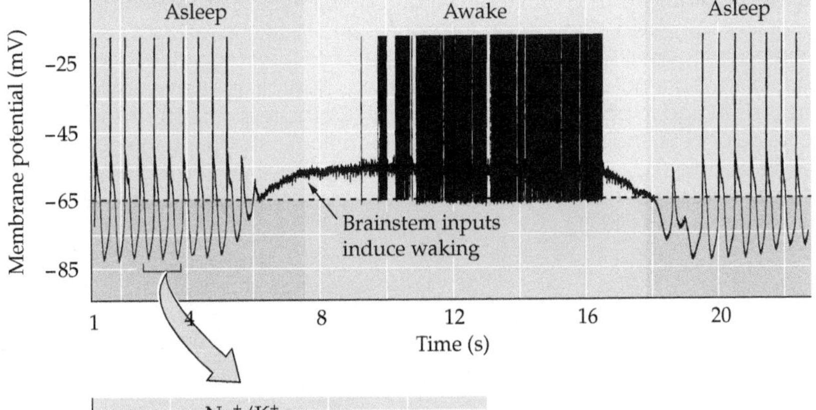

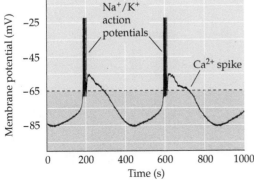

FIGURE 28.10 **Thalamocortical neurons and the sleep cycle.** Recordings from a thalamocortical neuron, showing the oscillatory mode corresponding to a sleep state and the tonically active mode corresponding to an awake state. An expanded view of oscillatory phase is shown at left. Bursts of action potentials are evoked only when the thalamocortical neuron is hyperpolarized sufficiently to activate low-threshold calcium channels. These bursts account for the spindle activity seen in EEG recordings in stage II sleep (see Figures 28.5 and 28.11). Depolarizing the cell either by injecting current or by stimulating the reticular activating system transforms this oscillatory activity into a tonically active mode. (After McCormick and Pape, 1990.)

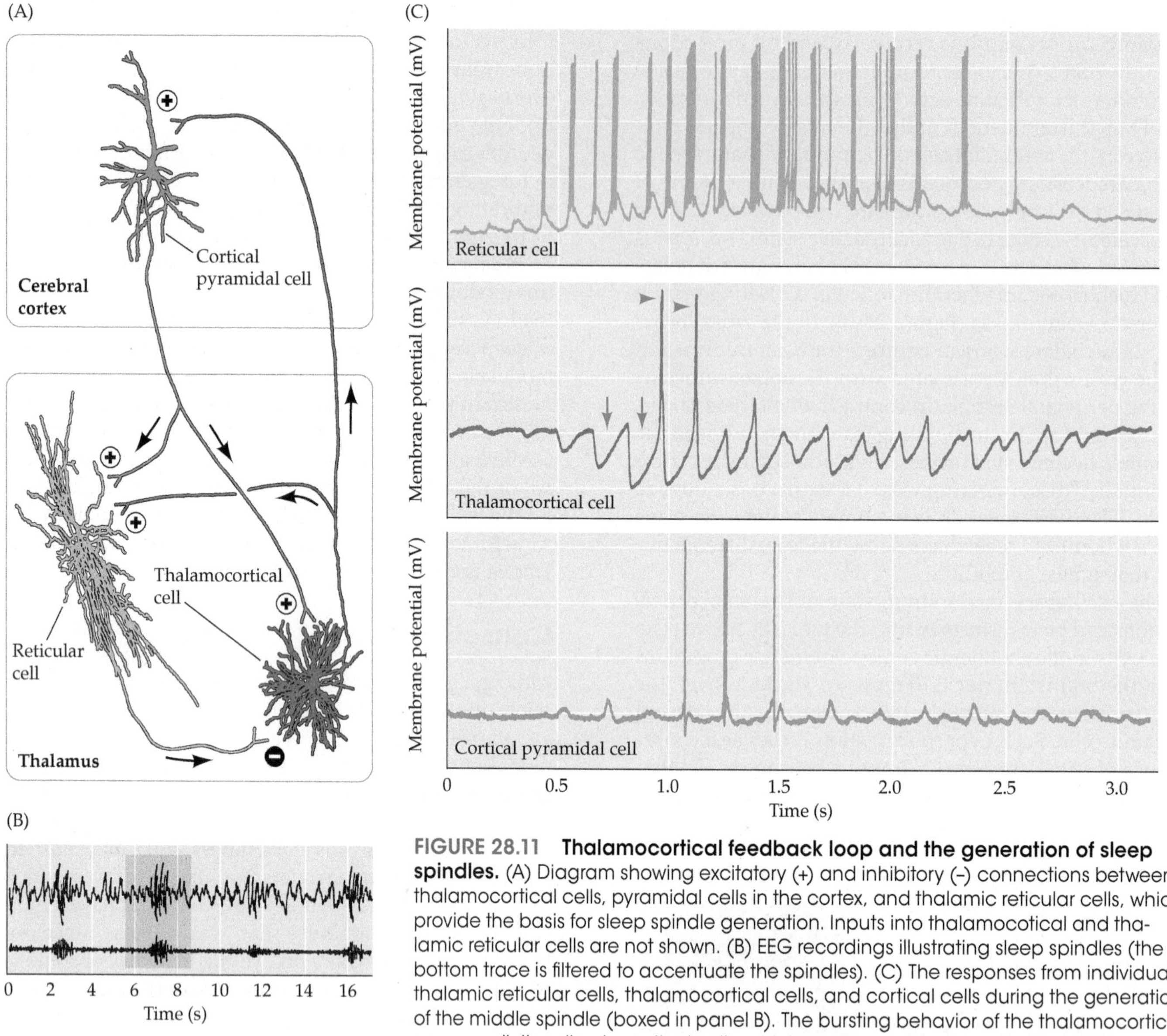

FIGURE 28.11 Thalamocortical feedback loop and the generation of sleep spindles. (A) Diagram showing excitatory (+) and inhibitory (–) connections between thalamocortical cells, pyramidal cells in the cortex, and thalamic reticular cells, which provide the basis for sleep spindle generation. Inputs into thalamocotical and thalamic reticular cells are not shown. (B) EEG recordings illustrating sleep spindles (the bottom trace is filtered to accentuate the spindles). (C) The responses from individual thalamic reticular cells, thalamocortical cells, and cortical cells during the generation of the middle spindle (boxed in panel B). The bursting behavior of the thalamocortical neurons elicits spikes in cortical cells, which are then evident as spindles in EEG recordings. (After Steriade et al., 1993.)

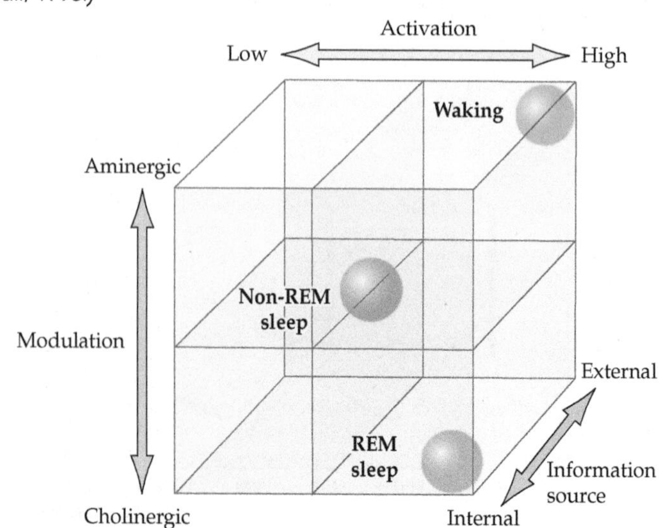

FIGURE 28.12 A summary scheme of sleep-wake states. In the waking state, activation is high, modulation is aminergic, and the information source is external. In REM sleep, activation is also high, the modulation is cholinergic, and the information source is internal. The other states can likewise be remembered in terms of this general diagram. (After Hobson, 1989.)

and its physiological basis fit easily in the conventional framework of neurobiology, consciousness as awareness, or self-awareness, raises more difficult and contentious philosophical issues.It has thus been far more difficult to investigate consciousness with the techniques used in cognitive neuroscience or, for that matter, with any other methods.

Investigating consciousness is further complicated by the fact that being *aware* is not the same as being *self-aware*: one can imagine an animal that is aware of sensory input without having the integrated sense of a separate self that humans possess. Whereas laboratory animals such as rats and pigeons seem aware of the world, whether they are self-aware is debatable. However, some non-human primates seem to be both aware of the world and (at least to some degree) aware of themselves as actors in it, although the level, or even existence, of such awareness in controversial. Going further down the scale of brain (or nervous system) complexity, very simple animals presumably lack awareness altogether, at least in any conventional sense. Indeed, the most neurally simple animals would appear to operate as automatons, although it is not clear at what taxonomic level this occurs. Moreover, it is important to note that even in the human nervous system most neural processing is largely automatic, operating below the threshold of awareness. Think, for example, of all the homeostatic neural mechanisms that insure your well-being in innumerable ways even while you are pondering the meaning of this sentence.

These daunting issues notwithstanding, several cognitive neuroscientists have sought to address the basis of consciousness by ferreting out some signature of the neural processing that occurs when we are aware of something (e.g., a visual stimulus) compared with when we are not (e.g., presentation of the same stimulus under circumstances in which it does not elicit a reportable percept). Studies of both normal subjects and patients with various pathological conditions have shed some light on this challenge. In normal subjects, the approach most commonly taken has been to assess the nature and location of neural activity while a particular sensory percept moves in and out of awareness. By asking the person (or experimental animal) to report these perceptual transitions (typically verbally or by a button press), the investigator can compare neural activity during awareness in response to a stimulus with that when the subject is unaware of the stimulus. In such paradigms, the physical stimulus remains unchanged and thus serves as its own control. One paradigm for this purpose has been binocular rivalry. Binocular rivalry refers to the fact that when a particular stimulus is presented to one eye while a discordant stimulus is presented to the other, the visual percept is of either one stimulus or the other, and alternates back and forth every few seconds, rather than being a combination or blending of the views coming from the two eyes (Figure 28.13).

Using humans or monkeys trained to report what they are seeing at any given moment, electrophysiological methods and fMRI can assess changes in brain activity that occur when there is a switch of conscious content. For example, when the monocular inputs are faces and houses, recordings of fMRI activity show increases in the fusiform face area of the temporal lobe when a face is perceived, and in the parahippocampal place area when houses are seen. Using similar paradigms, single neurons in at least some regions of the visual cortex in monkeys tend to show increased activity when the animal is perceiving the view with one eye but not the other.

A different fMRI paradigm with the same aim of probing activity changes in the low-level visual cortical regions during perception versus its absence takes advantage of visual aftereffects. For example, orientation aftereffects can be induced by exposing participants to a series of lines (referred to as *gratings*) in a particular orientation (say, 45°) for a minute or two; following this exposure, "neutral" line stimuli (vertical lines) are perceived for several seconds as being slightly tilted in the direction opposite the angle of the inducing exposure. The same inducing stimulus can be presented without awareness by masking it during the presentation. One can then ask whether lack of awareness of the inducing stimulus abolishes the aftereffect. It does not, implying that the low-level visual cortical neurons sensitive to orientation are about as active when subjects are aware of the inducing stimulus as when they are not. This lack of effect has been further confirmed in non-human primates by

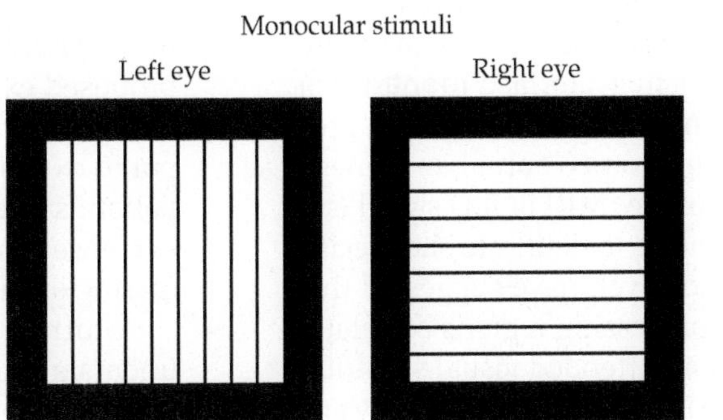

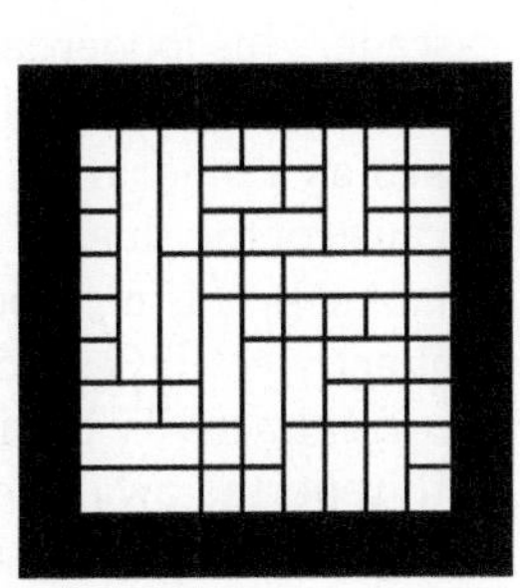

FIGURE 28.13 Binocular rivalry. The perceptual phenomenon of binocular rivalry is illustrated here by the presentation of vertical lines to the left eye and horizontal lines to the right eye. A grid pattern is not seen, indicating that the views of the two eyes are not simply brought together in V1 by the activity of binocular neurons in the visual cortex.

examining the activity of depth-sensitive (disparity-tuned) neurons in the primary visual cortex (V1) that are active in response to stereoscopic stimuli. Again, little or no difference in activity is apparent when the monkeys indicate behaviorally that they experienced a perception of depth compared with when they do not. In short, studies of activity in at least the earliest stages of visual cortical processing have not led to any clear hallmark of visual awareness.

It may help in rationalizing these somewhat confusing results to distinguish lower-order visual processing from higher-order visual processing in the visual association cortices. Recall from Chapter 12 that the higher-order visual areas are broadly divisible into a ventral stream running to the temporal lobe that is concerned with the recognition of visual stimuli (the "what" pathway), and a dorsal stream that is more concerned with stimulus location and spatial relations (the "where" pathway). Correlations of neural activity with stimulus awareness in monkeys tend to increase as recording sites move away from V1 along the ventral pathway, toward the temporal lobe. This interpretation accords with the majority of the studies mentioned above, showing that awareness of objects such as houses and faces activate the relevant cortical regions.

A further point is that the activity of some cortical regions not generally thought of as visual are also correlated with visual awareness. For example, neuroimaging studies in humans have shown that perceptual changes in binocular rivalry and other bistable-image paradigms are associated with activity in frontal and parietal cortical regions (i.e., there is activity in these cortical regions that is time-locked to the subjects' reports of the perceptual changes). Consistent with this evidence, other perceptual effects such as "pop-out" of a visual stimulus, or the awareness of a previously missed stimulus feature, are also correlated with activity changes in frontal and parietal areas. Transient functional disruption of processing in these areas by transcranial magnetic stimulation (TMS) also disturbs perception. Disturbances in perception following damage to these higher-level brain regions further confirm that they are involved in perceptual awareness and consciousness, although their roles in these phenomena remain unclear.

Yet another idea concerning possible neural mechanisms underlying awareness involves a longer-latency return of activity back to the relevant sensory processing regions, but much later in time than the initial feedforward cascade. This longer-latency activity, termed **reentrant** or **recurrent neural activation**, has specifically been proposed as a mechanism leading to perceptual awareness. Because of the sluggishness of the fMRI BOLD signal (see Chapter 1), this method is unable to delineate such recurrent activity. EEG studies, however, have suggested that recurrent activity in visual processing regions correlates with reported awareness of an attended visual stimulus, although the tightness of the link between the activity and awareness is still unsettled.

Considered together, this evidence is consistent with the generally accepted idea that awareness of responses to stimuli (i.e., perception and thinking) is based on altered activity of populations of cortical neurons in the regions of the association cortices that process the relevant stimuli, integrating the result with information arising from other modalities and contextual influences. With respect to vision, the observed correlations described here suggest that activity in the visual association cortices is *necessary* for visual awareness, and that continued activity in these areas may be important. However, such activity does not appear to be a *sufficient* cause of awareness, and no defining neural signature of awareness has been discerned.

Clinical evidence has also contributed to understanding the neural basis of awareness and consciousness. A pathological phenomenon of particular interest is **blindsight**. As described in Chapter 12, patients with damage to the primary visual cortex are blind in the affected area of the contralateral visual field (recall that as a result of the visual system anatomy, V1 in the right occipital lobe processes information arising from the left visual field, and V1 in the left occipital lobe processes input from the right visual field). Objects presented within the area of blindness in the visual field (i.e., the scotoma) are not seen, at least according to the patient's verbal reports. Nevertheless, when some blindsight patients are forced by the experimenter to make a response to simple stimuli presented within their scotoma, the responses are often significantly above chance performance. For instance, if patients are asked to guess whether a stimulus line within their scotoma is vertical or horizontal, they answer correctly much of the time even though they claim to have seen nothing. Blindsight can also be simulated in normal individuals by transient inactivation of V1 by TMS applied over the occipital lobes. TMS creates a temporary scotoma for a specific region of the visual field, and again the features of simple stimuli presented in the unseen region tend to be guessed correctly at levels well above chance.

Functional neuroimaging and electrophysiological studies of patients with blindsight show that the unseen stimuli elicit some activity in extrastriate regions beyond V1, implying that these cortical areas are needed for successful behavior in the absence of awareness. This evidence underscores the conclusion above that extrastriate activity may be necessary for awareness, but is not sufficient. One proposed explanation of blindsight is that subcortical visual processing of the stimulus, or coarse subcortical input to extrastriate cortex that bypasses V1, influences the patient's guesses. This interpretation accords with other evidence that subliminal (unconscious) information processing influences behavior of all sorts.

Another example of neural pathology pertinent to understanding awareness is the experience of *split-brain patients* (see Chapter 33). When the corpus callosum is cut as a treatment for otherwise intractable epileptic seizures, direct communication between the right and left hemispheres is no longer possible. The perceptual consequences of this surgery, first studied by Roger Sperry and Michael Gazzaniga in the 1960s, showed that the divided hemispheres in these patients function relatively independently, and that awareness generated by neural processing in one hemisphere is largely unavailable to the other. For example, when simple written instructions such as "laugh" or "walk" are presented visually to the left visual field—and thus to the right brain—of a split-brain patient, many subjects have enough rudimentary verbal understanding in the right hemisphere to execute the commanded action. However, when asked to report *why* they laughed or walked, they typically confabulate a response using the superior language skills in the left hemisphere, saying, for instance, that something the experimenter said struck them as funny, or that they were tired of sitting and needed to walk a bit. Thus, the same individual would appear, under these circumstances, to harbor two relatively independent domains of awareness. This evidence raises the provocative question of whether awareness is really the unified function we generally take it to be, as well as what the role of the corpus callosum is in engendering such unity.

Whereas blindsight and hemispheric division lead to circumstances in which patients are unaware of stimulus processing that nonetheless influences their behavior, it is also possible to be aware (or to believe one is aware) of something that doesn't actually exist. Perhaps the most striking illustration of this sort of phenomenon is the *phantom limb* experiences of amputees. Recall that a common experience following amputation is the patient's subjective awareness of the missing arm or leg, despite the fact that the physical limb and its peripheral sensory input are absent. Although the interpretation of this bizarre condition is debated, the awareness of the missing limb and sensations arising from it seem quite real to the patient (pain is especially problematic for some amputees), emphasizing that processing of peripheral information is an active process in which the cortex constructs the percepts we experience. Hallucinations make the same point.

The Neural Correlates of Consciousness

Thus, efforts to identify a neural basis for awareness and consciousness have not provided the sort of answers sought by many investigators in this field. The neurological aspects of consciousness described in the previous section are insufficient to satisfy philosophers, theologians, and many neuroscientists interested in the broader issues that the phenomenon of consciousness raises. Surely a phenomenon as important to us as consciousness should have a prominent signature. A prominent correlate would answer a number of questions such as whether other animals are conscious in the same way we are, and whether machines could ever be self-aware.

With respect to the first of these issues, despite a long-standing debate about consciousness in other animals, it would be unwise to assert that humans are alone in possessing this obviously useful biological attribute. However, from a purely logical vantage it is impossible, strictly speaking, to know whether any being other than one's own self is conscious; as philosophers have long pointed out, we must inevitably take the consciousness of others on faith (or at least on the basis of common sense). Nonetheless, it is reasonable to assume that animals with brains structured much like ours (other primates and, to a considerable degree, mammals in general) have, in some measure, the same ability to be self-aware as we do. The ability to reflect on the past and plan for the future, made possible by self-awareness, is surely an advantage that evolution would have inculcated. A reasonable supposition would be that consciousness is present in animals in proportion to the complexity of their brains and behaviors—particularly those behaviors that are sophisticated enough to benefit from reflecting on past outcomes and future eventualities.

The question of whether machines can ever be conscious is a much more contentious issue. If one rejects dualism—the Cartesian proposition that consciousness, or "mind," is an entity beyond the ken of physics, chemistry, and biology and therefore is not subject to the rules of these disciplines—it follows that a structure could be built that either mimicked human consciousness (by being effectively isomorphic with brains) or achieved consciousness using physically different elements (e.g., computer elements). There is no logical reason to assume that such constructions could not be conscious,

Although some contemporary scientists continue to believe that neurobiology may soon reveal the *basis* of consciousness, such revelations are not likely. A more plausible scenario is that as information grows about the nature of other animals, about computers, and indeed about the brain, the question "What is consciousness?" will fade from center stage in much the same way that the question "What is life?" (which stirred up a similar debate for much of the twentieth century) was asked less and less frequently as biologists and others recognized it to be an ill-posed question that admits no definite answer. We can already create many aspects of living organisms in the laboratory, and there seems to be no reason in principle why artificial neural networks capable of self-awareness should not someday be possible.

Comatose States

An important pathological condition pertinent to consciousness is *coma*. **Coma** (Greek for "deep sleep") is the term applied to the brain state of individuals who have suffered brain injury that leaves them in a deeply unconscious state defined by apparent unresponsiveness to sensory stimuli. The condition typically involves compromised function of the brainstem and other deep brain structures, such that the normal interaction of these structures with the cerebral cortex is interrupted.

Coma can arise from varying degrees of brain damage, and the prognosis is often uncertain. Most comatose individuals recover consciousness within a few days or weeks as the compromised neurons and the circuits they contribute to gradually regain their functions. Impairment can persist for much longer, however, if neural damage is more profound. Some patients recover consciousness after months, although typically with residual effects, and extremely rare cases have been reported in which consciousness is regained after some years. This variability has led to social, religious, and political controversy over the point at which an unresponsive patient should be considered to be in a **persistent vegetative state**, a diagnosis that raises ethical issues surrounding decisions about withholding life-support. This sensitive issue means there has been, and will continue to be, interest in techniques that could contribute to a better understanding a given patient's prognosis. Electroencephalography has been fundamental in diagnosing *irreversible brain death*, which occurs when brain trauma is so severe that no EEG activity can be recorded (i.e., a flat electrical trace). More recently, functional neuroimaging has also been used to evaluate persistent vegetative state, sometimes with surprising results. For example, when the brain of a 23-year-old woman who had been uncommunicative for 5 months following a traffic accident was functionally imaged, fMRI showed that her brain responded to simple verbal commands by generating appropriate patterns of activity.

The "Default State" of the Brain

A final question in this chapter concerns the purpose of high levels of activity in many brain regions when people are awake and aware, but without pursuing any particular goal, thought, or task. On the basis of neuroimaging studies, neuroscientists have come to call this the *default* state of brain activity. Some brain regions, including the posterior cingulate cortex, the ventral anterior cingulate cortex, the medial prefrontal cortex, and the cortex at the junction of the temporal and parietal lobes, show consistently *greater* activity during resting states than during the processing of specific cognitive tasks. When the activity during a baseline or rest condition is subtracted from the activity during cognitive tasks, relative decreases of activity (generally called *deactivations*) are observed in these areas. Because cognitive tasks induce activation increases in other brain areas carrying out the relevant processing, this finding led to the proposal that these regions constitute a network supporting a default mode of brain function that is engaged in the *absence* of any particular cognitive task (Figure 28.14). Further supporting the hypothesis that these areas constitute a network, analysis of the functional connectivity of these areas (i.e., how much their activity covaries across time) has shown strong coupling during the resting state.

Single-unit recording in monkeys has supported the inverse activation pattern of the default-mode network. For example, neurons in the posterior cingulate cortex, a posterior brain region considered to be part of the default-mode network, show changes in firing rates that closely match the fMRI activation patterns observed in humans. In particular, their fluctuations are anticorrelated with activity in dorsal parietal regions associated with the attentional control network and predict lapses in attention and the ability to switch from one task to another. Moreover, neuroimaging studies also have shown abnormal activity patterns in the default-mode network in several major neurological and psychiatric disorders, including being less active in autism and more so in schizophrenia.

The obvious question is what purpose neural activity in a default-mode network serves—that is, why should these regions be active if and when the brain is doing nothing in particular? Although the default-mode network activity might be related to mental "idling," another possibility is that this network is activated when attention is inwardly focused, the "standard" attentional control system being activated primarily when a person is focused on events and stimuli in the external environment (see Chapter 29). Whatever this network does, its activation pattern also occurs in monkeys, so it presumably evolved to carry out some relatively basic yet important function. The inverse relationship of activity in the default network and in the dorsal frontal-parietal attentional control network during focused attention suggests that these complementary systems may play an interactive role in system-wide brain function related to engaging and disengaging attention and other cognitive functions.

Summary

The most obvious cortical states are sleep and wakefulness. All animals exhibit a restorative cycle of rest following daily activity, but only mammals divide the period of rest into distinct phases of non-REM and REM sleep. Why humans (and many other animals) need a restorative phase of suspended consciousness accompanied by decreased metabolism and lowered body temperature is not known. Even more mysterious is why the human brain is periodically active during sleep at levels not appreciably different from those of the waking state (that is, the neural activity during REM sleep). The highly organized sequence of human sleep states is actively generated by nuclei in the brainstem, most importantly the cholinergic nuclei of the pons–midbrain junction, the noradrenergic cells of the locus coeruleus, and

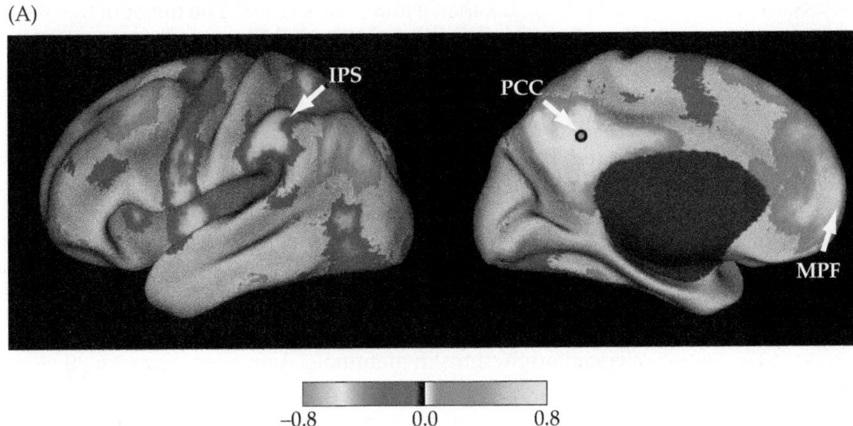

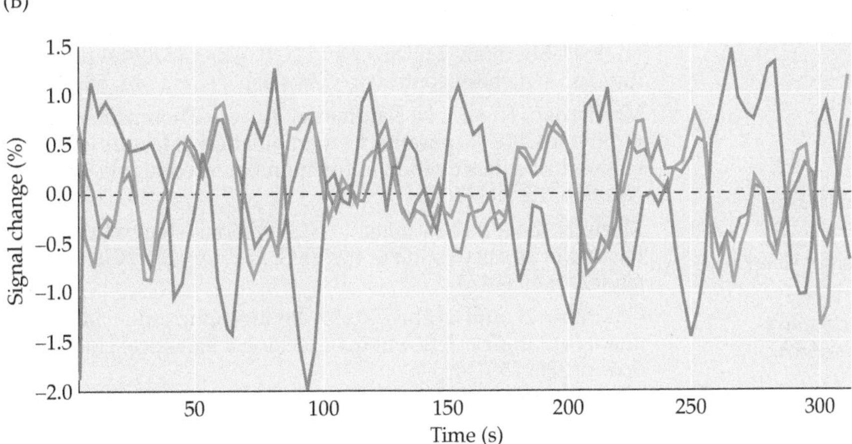

FIGURE 28.14 Default-mode network activity is anticorrelated with frontal-parietal attentional control regions. (A) fMRI data were collected from a subject at rest, and one part of the default-mode network, the posterior cingulate cortex (PCC), was chosen as a seed region to help identify other parts of the network. The time course of activity in the seed region was correlated with every other part of the brain, and those correlation coefficients are shown plotted on an inflated representation of the brain surface. Using this simple method, the rest of the default-mode network shows up as warm-colored, indicating a similar activity profile as that of the PCC, whereas areas of the frontal-parietal attentional control network show up in cooler colors, indicating an anticorrelation with the PCC. (B) fMRI activity time courses over one run are plotted from two regions of the default network (PCC and medial prefrontal cortex, or MPF) and from one region of the attentional network (intraparietal sulcus, or IPS). The time courses clearly show that when activity in one network goes up, activity in the other goes down. (From Fox et al., 2005.)

the serotonergic neurons of the raphe nuclei. This control by the brainstem of the cortical states is mediated by modulation of activity in a thalamocortical loop. This complex physiological interplay involving brainstem, thalamus, and cortex controls the degree of mental alertness on a continuum from deep sleep to waking attentiveness. A circadian clock located in the suprachiasmatic nucleus and the hypothalamus in turn influences these systems, adjusting cortical and other physiological states to appropriate durations during the 24-hour cycle of light and darkness that is fundamental to life on Earth. Exploring consciousness as something that cannot be reduced to wakefulness raises the more difficult questions including its neural correlates, if they exist as distinct from wakefulness and attention; whether animals are conscious in the same way we are; and whether machines can be conscious. More tractable is the puzzling question of why large regions of the cortex are more active at rest than when a subject is carrying a task.

ADDITIONAL READING

Reviews

Brown, R. E., R. Basheer, J. T. McKenna, R. E. Strecker and R. W. McCarley (2012). Control of sleep and wakefulness. *Physiol. Rev.* 92 (3): 1087–1187. doi:10.1152/physrev.00032.2011

Davidson, A. J. and M. Menaker (2003) Birds of a feather clock together—sometimes: Social synchronization of circadian rhythms. *Curr. Opin. Neurobiol.* 13: 765–769.

Green, C., J. Takahashi and J. Bass (2008) The meter of metabolism. *Cell* 134: 727–742.

Hobson, J. A. (1990) Sleep and dreaming. *J. Neurosci.* 10: 371–382.

McCarley, R. W. (1995) Sleep, dreams and states of consciousness. In *Neuroscience in Medicine*, P. M. Conn (ed.). Philadelphia: J. B. Lippincott, pp. 535–554.

McCormick, D. A. (1989) Cholinergic and noradrenergic modulation of thalamocortical processing. *Trends Neurosci.* 12: 215–220.

McCormick, D. A. (1992) Neurotransmitter actions in the thalamus and cerebral cortex. *J. Clin. Neurophysiol.* 9: 212–223.

Posner, M. I. and S. Dehaene (1994) Attentional networks. *Trends Neurosci.* 17: 75–79.

Rees, G., G. Kreiman and C. Koch (2002) Neural correlates of consciousness in humans. *Nat. Rev. Neurosci.* 3: 261–270.

Saper, C. B. and F. Plum (1985) Disorders of consciousness. In *Handbook of Clinical Neurology*, Vol. 1 (45): *Clinical Neuropsychology*, J. A. M. Frederiks (ed.). Amsterdam: Elsevier Science Publishers, pp. 107–127.

Siegel, J. M. (2000) Brainstem mechanisms generating REM sleep. In *Principles and Practice of Sleep Medicine*, 3rd Edition. M. H. Kryger, T. Roth and W. C. Dement (eds.). New York: W. B. Saunders.

Steriade, M. (1992) Basic mechanisms of sleep generation. *Neurology* 42: 9–18.

Steriade, M. (1999) Coherent oscillations and short-term plasticity in corticothalamic networks. *Trends Neurosci.* 22: 337–345.

Steriade, M., D. A. McCormick and T. J. Sejnowski (1993) Thalamocortical oscillations in the sleeping and aroused brain. *Science* 262: 679–685.

Stoerig, P. and A. Cowey (1997) Blindsight in man and monkey. *Brain* 120: 535–559.

Tong, F. (2004) Primary visual cortex and visual awareness. *Nat. Rev. Neurosci.* 4: 219–228.

Important Original Papers

Allison, T. and D. V. Cicchetti (1976) Sleep in mammals: Ecological and constitutional correlates. *Science* 194: 732–734.

Allison, T. H. and H. Van Twyver (1970) The evolution of sleep. *Natural History* 79: 56–65.

Allison, T., H. Van Twyver and W. R. Goff (1972) Electrophysiological studies of the echidna, *Tachyglossus aculeatus. Arch. Ital. Biol.* 110: 145–184.

Aschoff, J. (1965) Circadian rhythms in man. *Science* 148: 1427–1432.

Aserinsky, E. and N. Kleitman (1953) Regularly occurring periods of eye motility, and concomitant phenomena, during sleep. *Science* 118: 273–274.

Cashmore, A. R. (2003) Cryptochromes: Enabling plants and animals to determine circadian time. *Cell* 114: 537–543.

Churchland, P. M. and P. S. Churchland (1990) Could a machine think? *Sci. Amer.* 262: 32–37.

Colwell, C. S. and S. Michel (2003) Sleep and circadian rhythms: Do sleep centers talk back to the clock? *Nature Neurosci.* 10: 1005–1006.

Crick, F. (1995) *The Astonishing Hypothesis: The Scientific Search for the Soul.* New York: Touchstone.

Crick, F. and C. Koch (1998) Consciousness and neuroscience. *Cerebral Cortex* 8: 97–107.

Czeisler, C. A. and 11 others (1999) Stability, precision, and near-24-hour period of the human circadian pacemaker. *Science* 274: 2177–2181.

Dement, W. C. and N. Kleitman (1957) Cyclic variations in EEG during sleep and their relation to eye movements, body motility, and dreaming. *Electroenceph. Clin. Neurophysiol.* 9: 673–690.

Dunlap, J. C. (1993) Genetic analysis of circadian clocks. *Annu. Rev. Physiol.* 55: 683–727.

Fox, M. D. and 5 others. (2005) The human brain is intrinsically organized into dynamic, anticorrelated function networks. *Proc. Natl. Acad. Sci. USA* 102: 9673–9678.

Green, C., J. Takahashi and J. Bass (2008) The meter of metabolism. *Cell* 134: 727–742.

Hardin, P. E., J. C. Hall and M. Rosbash (1990) Feedback of the *Drosophila period* gene product on circadian cycling of its messenger RNA levels. *Nature* 348: 536–540.

Hayden, B. Y., D. V. Smith and M. E. Platt (2009). Electrophysiological correlates of default-mode in macacque posterior cingulate cortex. *Proc. Natl. Acad. Sci. USA* 106: 5948–5953.

Hobson, J. A., R. Strickgold and E. F. Pace-Schott (1998) The neuropsychology of REM sleep and dreaming. *NeuroReport* 9: R1–R14.

King, D. P. and J. S. Takahashi (2000) Molecular mechanism of circadian rhythms in mammals. *Annu. Rev. Neurosci.* 23: 713–742.

Lu J., M. A. Greco, P. Shiromani and C. B. Saper (2000) Effect of lesions of the ventrolateral preoptic nucleus on NREM and REM sleep. *J. Neurosci.* 20: 3830–3842.

Lu, J., D. Sherman, M. Devor and C. B. Saper (2006) A putative flip-flop switch for control of REM sleep. *Nature* 441: 589–594.

McKiernan, K. A., J. N. Kaufman, J. Kucera-Thompson and J. R. Binder (2003) A parametric manipulation of factors affecting task-induced deactivation in functional neuroimaging. *J. Cogn. Neurosci.* 15: 394–408.

Moruzzi, G. and H. W. Magoun (1949). Brain stem reticular formation and activation of the EEG. *Electroenceph. Clin. Neurophysiol.* 1: 455–473.

Okamura, H. and 8 others (1999) Photic induction of *mPer1* and *mPer2* in *Cry*-deficient mice lacking a biological clock. *Science* 286: 2531–2534.

Owen, A. M. and 5 others (2006) Detecting awareness in the vegetative state. *Science* 313: 1402.

Provencio, I. and 5 others (2000) A novel human opsin in the inner retina. *J. Neurosci.* 20: 600–605.

Raichle, M. E. and 5 others (2001) A default mode of brain function. *Proc. Natl. Acad. Sci. USA* 98: 676–682.

Ren, D. and J. D. Miller (2003) Primary cell culture of suprachiasmiatic nucleus. *Brain Res. Bull.* 61: 547–553.

Ribeiro, S. and 7 Others (2004) Long-lasting, novelty-induced neuronal reverberation during slow-wave sleep in multiple forebrain areas. *PLoS Biology* January 20: E24.

Roffwarg, H. P., J. N. Muzio and W. C. Dement (1966) Ontogenetic development of the human sleep–dream cycle. *Science* 152: 604–619.

Searle, J. R. (2000) Consciousness. *Annu. Rev. Neurosci.* 23: 557–578.

Shearman, L. P. and 10 others (2000) Interacting molecular loops in the mammalian circadian clock. *Science* 278: 1013–1019.

Takahashi, J. S. (1992) Circadian clock genes are ticking. *Science* 258: 238–240.

Tononi, G. and G. Edelman (1998) Consciousness and complexity. *Science* 272: 1846–1851.

Vitaterna, M. H. and 9 others (1994) Mutagenesis and mapping of a mouse gene, *clock*, essential for circadian behavior. *Science* 264: 719–725.

von Schantz, M. and S. N. Archer (2003) Clocks, genes, and sleep. *J. Roy. Soc. Med.* 96: 486–489.

Willie, J. T. and 13 others (2003) Distinct narcolepsy syndromes in *orexin receptor-2* and *orexin*-null mice: Molecular genetic dissection of non-REM and REM sleep regulatory processes. *Neuron* 38: 715–730.

Wilson, M. A. (2002) Hippocampal memory formation, plasticity, and the role of sleep. *Neurobiol. Learn. Mem.* 3: 565–569.

Wilson, M. A. and B. L. McNaughton (1994) Reactivation of hippocampal ensemble memories during sleep. *Science* 265: 603–604.

Xie, L. and 12 others (2013) Sleep drives metabolite clearance from the adult brain. *Science.* 342: 373–377. doi: 10.1126/science.1241224

Books

Foulkes, D. (1999) *Children's Dreaming and the Development of Consciousness.* Cambridge, MA: Harvard University Press.

Gardner, H. (1974) *The Shattered Mind: The Person After Brain Damage.* New York: Vintage.

Hobson, J. A. (1989) *Sleep.* New York: Scientific American Library.

Hobson, J. A. (2002) *Dreaming.* New York: Oxford University Press.

Lavie, P. (1996) *The Enchanted World of Sleep* (translated by A. Barris). New Haven, CT: Yale University Press.

Lenneberg, E. (1967) *The Biological Foundations of Language.* New York: Wiley.

McNamara, P., R. A. Barton and C. L. Nunn (2010) *Evolution of sleep: Phylogenetic and functional perspectives.* Cambridge: Cambridge University Press.

Penrose, R. (1996) *Shadows of the Mind: A Search for the Missing Science of Consciousness.* Oxford: Oxford University Press.

Pinker, S. (1994) *The Language Instinct: How the Mind Creates Language.* New York: William Morrow and Company.

Posner, M. I. and M. E. Raichle (1994) *Images of Mind.* New York: Scientific American Library.

Weiskrantz, L. (1986) *Blindsight: A Case Study and Its Implications.* Oxford: Oxford University Press.

Go to the NEUROSCIENCE 6e Companion Website at **www.oup.com/uk/Purves6e** for Web Topics, Animations, Flashcards, and more.

CHAPTER

29 Attention

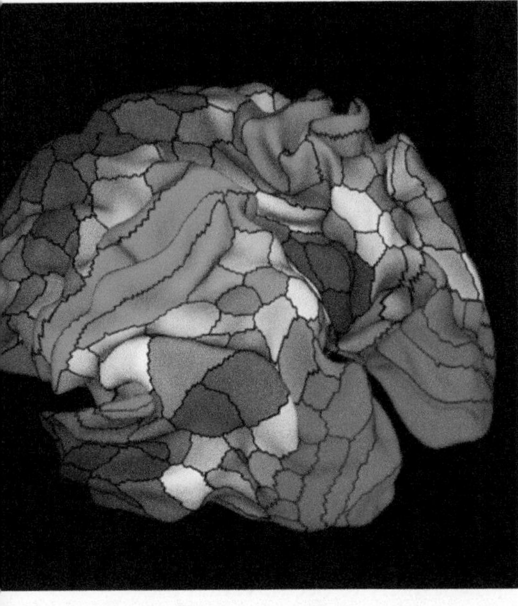

Overview

ATTENTION IS THE CONSCIOUS OR UNCONSCIOUS FUNCTION that focuses on some external or internal stimulus, presumably at the expense of fully processing other information. This idea can be traced back to William James, who stated:

> *Everyone knows what attention is. It is the taking possession by the mind, in clear and vivid form, of one out of what seem several simultaneously possible objects or trains of thought. ... It implies withdrawal from some things in order to deal effectively with others...*
>
> James, 1890 (*The Principles of Psychology*, pp. 403–404)

A problem, however, is the implication of a subjective "I" that decides to attend to one thing or another, as in "I paid attention to what he said." Who or what "I" might be raises broad philosophical as well as key neurobiological questions. To complicate matters further, attention must be distinguished from other terms that are related to it but not synonymous, for example, the difference between attention and other forms of arousal. As described in Chapter 28, wakefulness is a continuum of brain states that range from inattentiveness to a fully alert and aroused state when one is specifically paying attention to something. Studies of attention consider how and why we focus on particular aspects of the flood of information available in the internal and external environments. Although such studies were initially limited to the behavioral measures such as reaction time and processing accuracy, or the behavioral consequences of brain damage, over the last few decades neuroscientists have directly measured the influence of attention on brain activity during pertinent tasks. The goal of this chapter is to review the phenomenonology of attention, its implications, and the puzzles that it entails.

Attention as "Selective" Processing

Attention as "selective" refers to the allocation of neural resources to the analysis of particular information at the expense of resources that might have been allocated to other concurrent information that is in principle available. The quotation marks around "selective" are needed because it is unclear who or what the implied agent is, a quandary considered later in the chapter.

The usual example demonstrating selective attention is the *cocktail party effect*, wherein a listener can attend to one voice in a noisy conversation and "tune out" other simultaneous sound signals (Figure 29.1). An analysis of this effect was carried out in the 1950s by psychologist Colin Cherry who presented different dialogues to

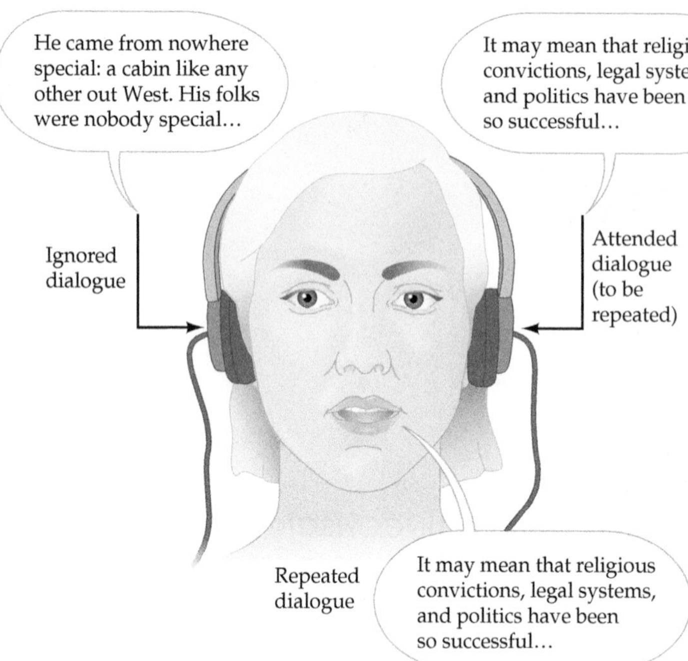
He came from nowhere special: a cabin like any other out West. His folks were nobody special...
It may mean that religious convictions, legal systems, and politics have been so successful...
Ignored dialogue
Attended dialogue (to be repeated)
Repeated dialogue
It may mean that religious convictions, legal systems, and politics have been so successful...

FIGURE 29.1 Attention as selective filtering. Colin Cherry's experiment using two voices speaking different dialogues presented separately to the left and right ears. See text for explanation.

each ear at the same time. Subjects were instructed to attend to only one of these inputs, and to immediately repeat the content to ensure that they were indeed attending to that stream of speech. Cherry then tested subjects' ability to report the content of the other, unattended input stream. Whereas the subjects could accurately report the content of the attended channel, they were unable to provide more than rudimentary information about the unattended stream. Cherry concluded that an attentional mechanism was filtering out unattended information at a relatively low level in sensory processing.

Although the idea that early attentional processing actively determines the information passed on to other parts of the nervous system for further analysis seems plausible, it soon became clear that at least some information in an unattended channel was being processed. A familiar example is attending to one's name when it is mentioned in otherwise unattended conversations. Consistent with this finding, other investigators proposed a "late-selection model" in which information filtering occurs relatively late in sensory processing pathways. According to this theory, only after higher-level processing is complete does an attentional mechanism determine what input information enters consciousness or influences behavior. At about the same time, psychologist Anne Treisman suggested a filtering system that could attenuate the inputs from concurrent sensory channels in a flexible manner. In an unattended channel, only especially salient or relevant information (e.g., one's name) would reach threshold for further processing and ultimately entry into consciousness. Psychological theories of attention were revised still further by incorporating the influence of the "perceptual load" imposed by a task—that is, the difficulty of a task arising from stimulus complexity and/or presentation brevity. The idea was that when the load is low, more processing would be available for other inputs. Thus, depending on the allocation of neural resources, information would (or would not) be allowed to reach higher processing levels where late-selection mechanisms presumably operate. The merits of these various concepts of attentional filtering are still being studied today.

Endogenous versus Exogenous Attention

Many behavioral studies, like those of Cherry, entail *voluntary* attentional tasks. Following an experimenter's instruction (or subjectively fulfilling a self-generated desire), subjects consciously direct attention to a particular aspect of the environment, such as an individual voice or a location in visual space. This type of attention is called **endogenous attention**. A particularly useful paradigm for studying endogenous attention was developed by Michael Posner and colleagues in the late 1970s (Figure 29.2A). While subjects maintain visual fixation on a central point, a trial begins with a centrally presented cue, such as an arrow pointing to the left or right, indicating where an upcoming target stimulus is most likely to occur. When the target appears, regardless of whether it occurs in the cued location, the subject must perform a discrimination task, such as indicating whether the target is a circle or an oval. In most of the trials the target is presented at the cued location, but sometimes it is presented at another location. Figure 29.2B shows the typical behavioral results: subjects respond faster to targets appearing at the cued location ("valid") than those appearing away from the cued location ("invalid"). An advantage of this and related paradigms is quantification of attentional effects.

In contrast, stimuli arising from events or conditions in the environment that attract attention automatically trigger **exogenous attention**. Exogenous (or involuntary) attention thus refers to the situation in which an unexpected noise, flash of light, movement, or other salient stimulus causes a shift of focus, trumping whatever else one happens to be attending at that moment. In conjunction with this shift in attention, the unexpected stimulus also facilitates the processing of information in that region, at the same time diminishing the efficacy of processing elsewhere. Like endogenous attention, exogenous attention has been

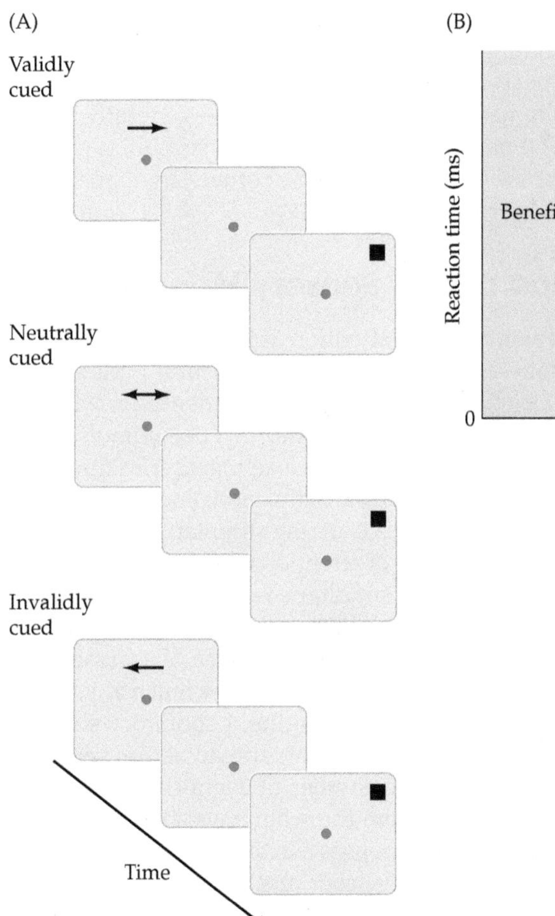

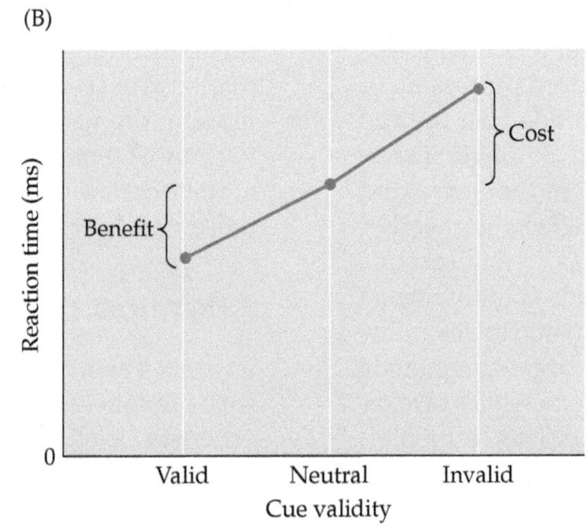

FIGURE 29.2 A cueing paradigm for studying endogenous visual spatial attention. (A) In this paradigm, a centrally presented instructional cue indicates where a target will most likely be presented (validly cued), where it will be less likely to be presented (invalidly cued), and where the cue provides no information as to the likely target location (neutrally cued). (B) Typical results show the benefits and costs in the reaction time for target detection after valid and invalid cueing, relative to the neutral-cue condition. (After Posner et al., 1980.)

studied in a variety of behavioral experiments. One approach has used trial-by-trial cueing in which a sensory cue such as a flash of light is presented at a particular location shortly before a target stimulus is presented either in that location or elsewhere (Figure 29.3). In such circumstances, subjects are again quicker to respond to a target presented in the cued location compared with an uncued location.

On the face of it, this sort of paradigm and the behavioral effects observed are similar to studies of endogenous cueing. In both instances, a cue induces a shift in the focus of attention that facilitates the sensory processing of stimuli in the attended region, and diminishes the efficacy of processing elsewhere. Despite these similarities, endogenous cueing and exogenous cueing have important functional differences. In endogenous cueing, information about the likelihood that the target stimulus will occur in the cued location is provided by prior knowledge (e.g., being informed where the target is likely to occur). In contrast, exogenous attention is not driven by any explicit information about a likely target location. That is, even if an exogenous cue (e.g., a flash) is presented randomly in the two possible locations from trial to trial, and thus has no predictive value about where a target will occur, processing in the cued location is nonetheless facilitated, presumably because the flash preceding the appearance of a target automatically draws attention to that location.

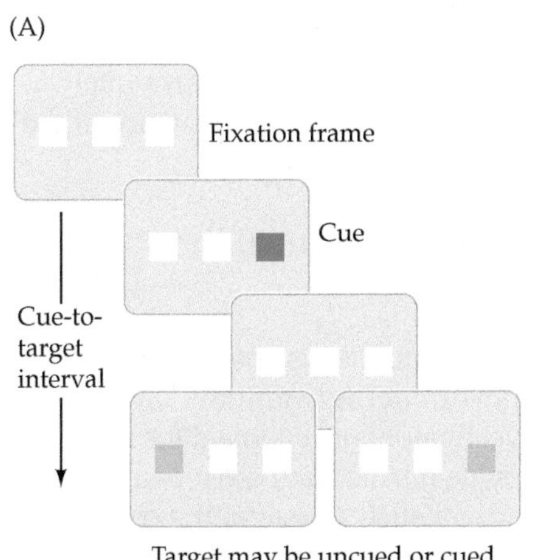

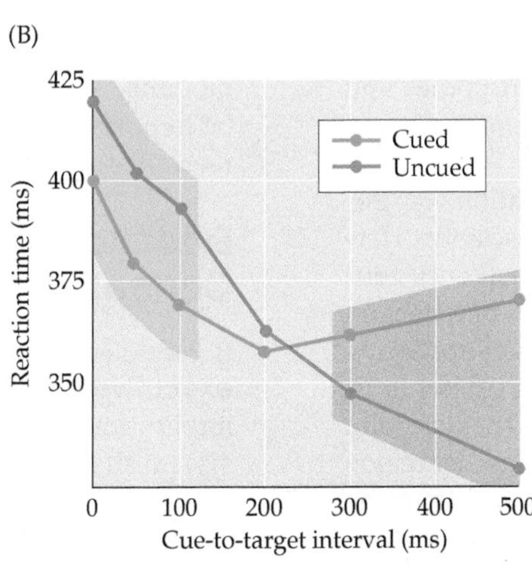

FIGURE 29.3 Exogenously triggered attention. (A) In this paradigm a brief flash is presented in one of two possible target locations, serving as an exogenous cue for a target that might follow at that location or at the other location. The occurrence of a target at the cued versus uncued location is random, with the probability at each location being 50%. (B) Shortly after the exogenous cue (green-shaded time period), stimulus processing at that location is facilitated, as indicated by faster response times to cued relative to uncued targets. At longer intervals (orange-shaded time period), however, there is a decrement in performance for the cued targets, known as *inhibition of return*. (From Klein, 2000; data from Posner and Cohen, 1984.)

Endogenous and exogenous attention also differ in the time courses of their influence on target processing. For endogenously cued attention, the improved processing of a cued target begins about 300 milliseconds after the cue and can last for some seconds afterward, or longer if subjects maintain their focus of attention on the instructed location. In contrast, exogenous cueing effects start earlier and are short-lived, beginning as early as 75 milliseconds after the cue and lasting only a few hundred milliseconds or so. Moreover, at still longer intervals (~400 to 800 milliseconds after the cue), the effect of the target cuing tends to reverse, with subjects actually being somewhat slower at responding to targets in the cued location. This "inhibition of return" probably reflects the reasonable redeployment of attention to other locations when a target fails to appear within a short time at the cued location. In any event, it is clear that the pattern of effects on behavioral task performance differs between attentional shifts that are triggered endogenously and those that are driven by exogenous factors. One important question emerging from these findings is whether or not different neural systems mediate exogenous and endogenous attention—a topic taken up in more detail below.

Covert Attention

Another way of categorizing attention is whether it is overt or covert. **Overt attention** involves orienting the head and eyes to a stimulus, thereby aligning visual (and auditory) processing with it and improving perception. **Covert attention** involves somehow directing attention to a stimulus without moving the head or eyes. The Russian psychologist and cyberneticist Alfred Yarbus first quantified overt attention by measuring subjects' patterns of gaze in response to viewing paintings and sculptures, using an ingenious system of small mirrors glued to the eyes of participants which redirected light to photo-tracing paper. He found that, in the absence of any instructions, participants tended to look at the faces and eyes of individuals in the artwork. But when instructed to ascertain the ages or wealth of individuals depicted in a painting, their gaze patterns shifted to focus on their bodies and clothing, reflecting the importance of this information for solving the task.

An experimental example of covert attention was described by the German physicist and vision scientist Hermann von Helmholtz at the end of the nineteenth century. When Helmholtz briefly flashed arrays of letters on a screen and asked subjects to report the letter appearing at a particular location, he observed that if a subject (typically himself) steadily fixated gaze on a particular point in the visual field but directed attention to another region of the field (that is, without moving the eyes), then the stimuli presented in the attended location were reported better than stimuli in the rest of the field. These and many related findings have established that attention to particular aspects of the environment—whether overtly by moving the head and eyes or covertly—generally leads to improved processing of the attended stimuli, typically at the expense of the processing of other, simultaneously presented information.

Attention across Sensory Modalities

It would be surprising if attention did not work across sensory modalities and other sources of neural inputs, and other studies have indeed provided evidence for what is often called **supramodal attention**. When stimuli from two different modalities occur close together in time, attention to the stimulus in one of the modalities will tend to encompass concurrently occurring stimulation in another modality. This *spreading* of attention makes sense in terms of the tendency to link simultaneously occurring stimulation from different modalities into a *multisensory object*, such as a barking dog, the look and voice of a friend, the odor and appearance of a pizza, or pretty much any other ordinary situation one can imagine. Laboratory studies have explored how attention can spread across sensory modalities. For instance, studies of event-related potentials (ERPs) recorded from the scalp have shown that the electrophysiological responses elicited by auditory stimuli are enhanced when they occur in a visually attended location, even when they are task-irrelevant. Correspondingly, ERP responses to task-irrelevant visual stimuli are enhanced when they occur in a location being attended for auditory stimuli. Similar results are observed between the visual and tactile modalities, as well as between tactile and auditory modalities. Complementary studies using fMRI have indicated that these enhanced responses to stimuli in the task-irrelevant modality include increased activity at relatively low-level processing areas in the sensory cortices. The biological value of this supramodal linkage during spatial attention is easy to understand: if important stimuli pertinent to one modality arise from a particular location, stimulus information from another modality arising from the same location is also likely to be important and key to understanding the nature of the relevant object or event.

Evidence for a Brain System That Controls Attention

It would seem to follow from all this that some form of **executive control** must underlie attention. Given this implication, many investigators have looked for a neural system that might provide it. Although most neuroscientists accept the evidence for such a system, the concept of executive control remains problematic, as explained below.

Notably, there is a strong overlap in the circuits that control attention and those that govern movements of the head and eyes. This overlay suggests attentional control builds upon systems that originally evolved to orient organisms to objects and events in the world that were crucial for successful behavior, belying the notion of a privileged central executive in control of attention.

Evidence from Neurological Patients: Hemispatial Neglect Syndrome

The earliest evidence for an anatomically definable system of attentional control came from neurological patients. A relatively common lesion is injury to the right inferior parietal lobe and adjacent regions (Figure 29.4). As introduced in Chapter 27, such lesions, first described in the 1940s, cause deficits in attention to the left side of personal and extrapersonal space (i.e., the side contralateral to the lesion). Right parietal, right superior temporal, or right frontal brain damage can all lead to difficulty attending to the left side of visual space and/or the left side of objects.

Since objects in the left visual field stimulate the visual system normally in such patients, the primary problem is attentional, not sensory. Depending on the extent and severity of the lesion, when an object in the left visual field is specifically pointed out or the object is made particularly salient (for example, by presenting food to a hungry individual), patients then tend to report being able to see it. Thus, these impairments are quite different from the visual deficit that follows a lesion in the visual cortex (see Chapter 12). Patients with visual cortical lesions are effectively blind in specific corresponding parts of the contralateral visual field. In contrast, the underlying problem with lesions to right parietal cortex appears to be an attentional deficit, not a sensory one; the patients can apparently *see* stimuli in the left visual field, but they tend not to notice them or be able to orient their attention to them effectively. Although these deficits are often most obvious in vision, they are evident in other sensory modalities. For instance, many patients with right hemisphere damage are less able to attend to the left side of their own body, as shown by the tendency to shave or apply makeup on only the right side of the face, or to dress only the right side of the body. This constellation of signs and symptoms is called **hemispatial** (or **contralateral**) **neglect syndrome**.

A possible reason why attentional deficits are most often associated with right parietal lesions is that this region influences mechanisms of attention in *both* hemispheres, whereas the corresponding left parietal area influences mainly those on the right. There are, however, alternative explanations, and the anatomy is a good deal more complex than implied here. For instance, some investigators have suggested that the relatively greater importance of

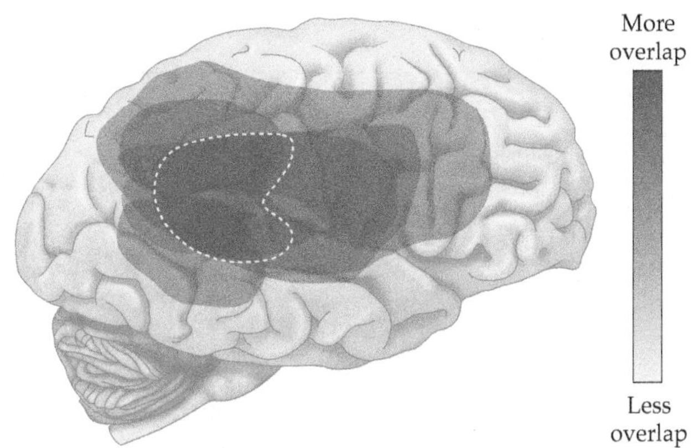

FIGURE 29.4 **Cortical lesions leading to left hemispatial neglect syndrome.** This composite diagram shows the distribution of right-hemisphere damage in eight patients with left hemispatial neglect. The degree of overlap of damaged brain areas across patients is indicated by shading level. Although some of the lesions include parietal and frontal lobes, as well as parts of the temporal lobe, the region most commonly affected is in the right inferior parietal lobe (dashed line). (After Heilman and Valenstein, 1985.)

right parietal cortex in neglect syndromes reflects right hemispheric lateralization for vigilance or alertness. Interestingly, in non-human primates damage to the inferior parietal lobe on either the left or right side induces neglect of contralateral space.

As shown in Figure 29.5, the left-sided neglect evident in these patients can be demonstrated clinically by one of several simple tests. In the single-line bisection test (see Figure 29.5A), patients are asked to mark the center of a horizontal line. Patients with neglect tend to ignore the left side of the line and thus their estimate of the center is displaced to the right. In the line cancellation test (see Figure 29.5B), patients are asked to draw a line through each of several lines scattered across a page; in this test, patients cancel lines mainly on the right side of the page. In addition, the left-sided neglect is not limited to ignoring objects in the left hemispace; patients with this syndrome also tend to ignore the left sides of objects *wherever* they are in visual space. For example, if asked to draw a copy of an object, these patients tend to draw only its right side (see Figure 29.5C). Such patients even tend to ignore the left side of their visual imagery and memory. So if asked to draw a clock from memory, they are likely to draw half a clock, sometimes remembering to include all 12 numbers in the drawing, but placing all of the numbers on the right (see Figure 29.5D).

Assessments of hemispatial neglect patients also elicit *extinction*. This phenomenon is revealed when the neurologist stands in front of the patient with arms outstretched and moves a finger on either the right or the left hand. If

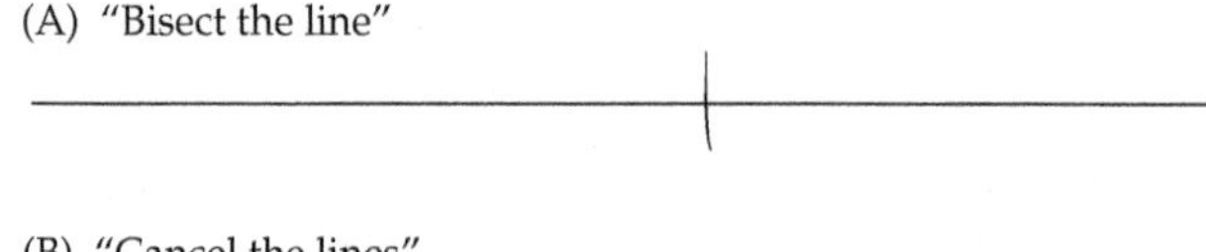

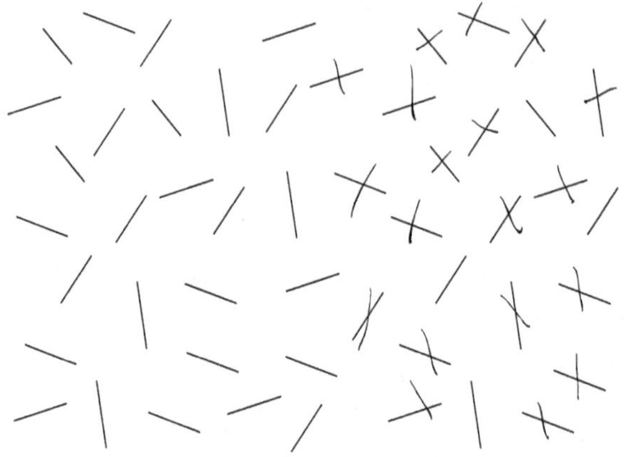

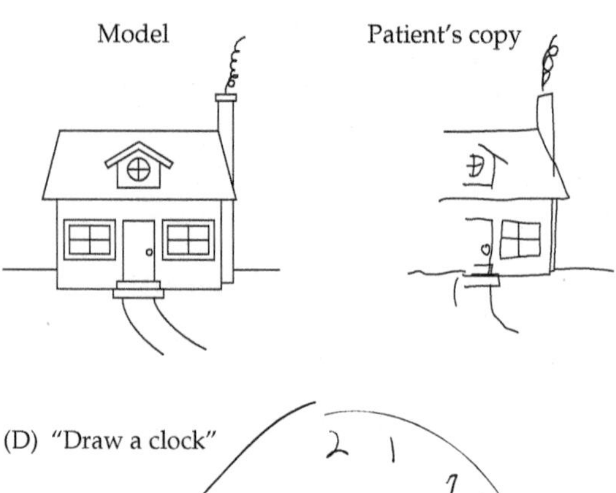

FIGURE 29.5 **Clinical tests of left hemispatial neglect caused by damage to the right inferior parietal lobe.** The performances in the single-line bisection test (A) and the line cancellation test (B) shown here are characteristic of hemispatial neglect patients. (C) An example of a visual copying task as performed by a hemispatial neglect patient. (D) A drawing of a clock face from memory by a hemispatial neglect patient. (A,C after Posner and Raichle, 1994; B after Blumenfeld, 2010; D after Grabowecky et al., 1993.)

a finger on either side is moved by itself, the patient generally reports the presence of the moving finger correctly, presumably because a moving stimulus is a particularly strong attractor of attention, even for these patients. If both fingers are moved at the same time, however, the patient typically reports seeing only the one on the right. This test suggests that the normal competition between the stimulus inputs from the two sides is now dominated by the right visual field, which "extinguishes" the input from the left. Extinction emphasizes again that the underlying problem is an attentional deficit, not a sensory one.

Other Brain Regions That Affect Attention

Lesions in parts of the frontal cortex that are connected to the parietal cortex can also cause attentional deficits. In particular, lesions to the **frontal eye fields** disrupt the ability both to initiate eye movements to targets in the contralateral visual field and to direct attention toward that side (Box 29A). Moreover, lesions in these frontal regions can interfere with attentional functions such as task switching and ignoring irrelevant information. As a rule, unilateral frontal lesions tend to have a greater effect on motor-related aspects of attention, compromising the ability to initiate or direct eye or limb movements toward contralateral space.

Brainstem lesions can also affect attentional control. The interactions between the superior colliculi and the parietal cortex are apparent in the so-called **Sprague effect**, in which the hemispatial neglect induced by a parietal lesion can be mostly compensated by a lesion of the superior colliculus on the other side. The proposed explanation is that the parietal lesion-induced neglect results not from the cortical damage itself, but from an imbalance of activity between the two parietal lobes in attentional control. According to this theory, a lesion of the contralateral colliculus helps to restore the appropriate balance, because of its connections to the parietal lobe on the same side. Regardless, this effect underscores an important role for the superior colliculus in attentional control, possibly via functional interactions with parietal cortex. Experimental studies in monkeys have demonstrated that these interactions are mediated by the pulvinar, the thalamic relay that connects the parietal cortex and the superior colliculus.

In sum, clinical evidence indicates that damage to a variety of brain areas can lead to deficits of attention with different characteristics and possible mechanisms. These findings are generally inconsistent with the idea of

BOX 29A ■ Attention and the Frontal Eye Fields

One of the main issues in understanding attention is demonstrating the effect of one brain region on another as a way of supporting the concept of a specific attention network. In the example here, the effect of frontal eye field (FEF) stimulation on the response of a single neuron in the extrastriate visual cortex of a monkey is evaluated. The rationale of the experiment is that the normal function of the FEF is to generate saccadic eye movements to locations in visual space that warrant attention. Thus, to test the presence of a link and the effect of attention, a stimulating electrode is placed in the FEF in a locus that would evoke saccades to a given location with respect to the center of gaze; a second electrode records the activity of a visual cortical neuron responsive to the same location in visual space. As shown, FEF stimulation while the monkey attended the fixation point caused a saccade to the expected location and increased neuronal activity at the recoding site. The implication is that the FEF and this region of visual cortex (V4) comprise part of an attention network.

Microstimulation of sites within the FEF, below the threshold for eliciting a saccade, was carried out while the visual stimulus responses of single V4 neurons were recorded in monkeys performing a fixation task. (A) The stimulating electrode was positioned so that suprathreshold stimulation would evoke a saccade into the receptive field (RF) of the V4 cell under study. (B) This example shows the effect of subthreshold FEF microstimulation on the response of a single V4 neuron to an oriented bar presented in the cell's receptive field. The mean response during control trials is shown in black; the enhanced response arising from the FEF microstimulation is shown in red. (C) On trials in which the visual stimulus was presented outside the receptive field of the V4 neuron, no enhancement is seen. (After Moore et al., 2003.)

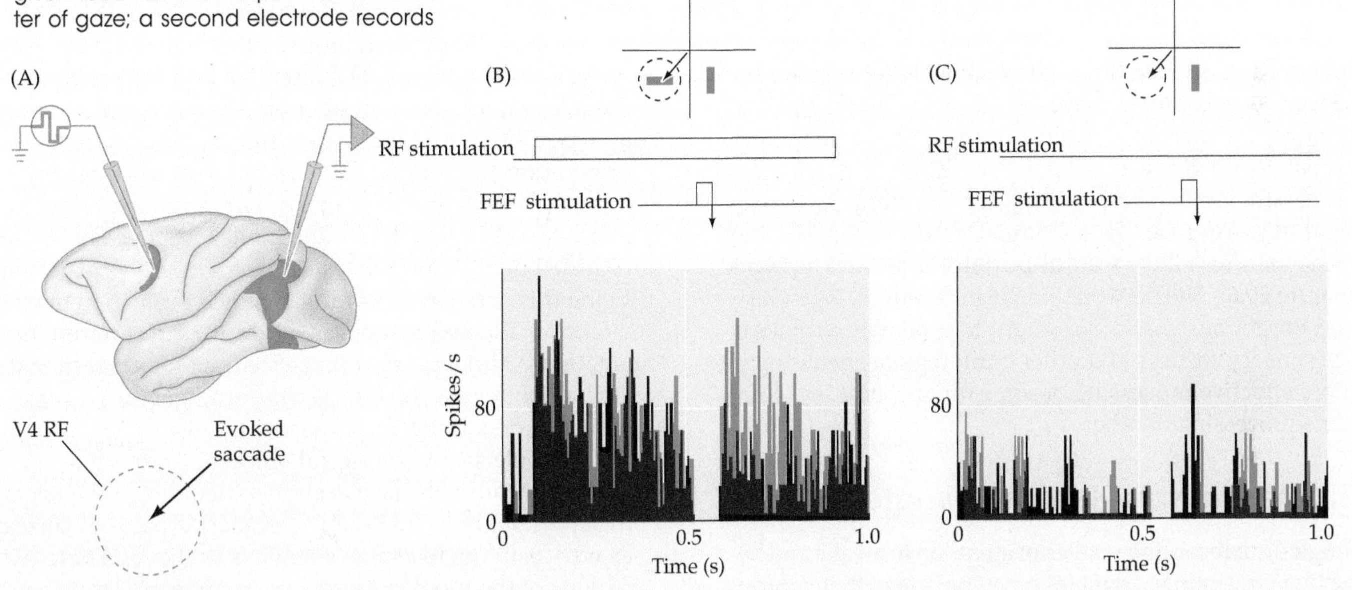

a specific brain region devoted to attention and its control, and instead suggest that attention depends on a large number of brain regions working together to determine what we attend to.

Evidence from Normal Subjects

The advent of noninvasive brain-imaging methods has complemented clinical observations with studies of attention in normal participants. In accord with studies of neglect patients, tasks involving attention reliably activate a set of brain regions in the dorsal parietal and dorsolateral frontal cortices. Much the same regions are activated by attending voluntarily, be it to spatial locations or nonspatial features like color or orientation. Involuntary shifts in attention elicited by unexpected events are associated with additional activity in ventral areas of the parietal cortex in the vicinity of the right temporal–parietal junction, which in turn activates more dorsal regions of the frontal cortex.

Based on this and other evidence, Maurizio Corbetta and Gordon Shulman have proposed that attention is mediated by two interacting systems that carry out different functions. In this interpretation, one system consisting of regions of the intraparietal cortex and superior frontal cortex serves endogenous attention; the other system, consisting mainly of the cortex near the temporal–parietal junction and ventral frontal cortex (mainly in the right hemisphere), is specialized for the detection of unexpected or salient exogenous stimuli. Both systems are

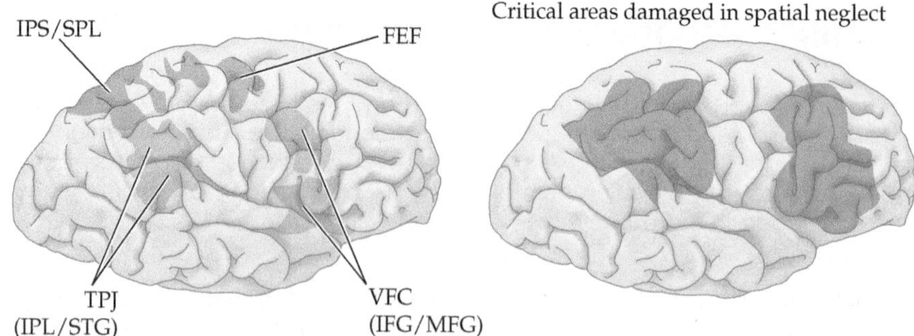

FIGURE 29.6 A postulated attentional control network, illustrated in the right hemisphere. The areas in blue indicate the dorsal frontal-parietal regions that tend to be activated by endogenous stimuli; the areas in yellow indicate the more ventral regions that tend to be activated during reorienting, and by exogenous stimuli. IPS/SPL = interparietal sulcus/superior parietal lobule; FEF = frontal eye fields; TPJ = temporal-parietal junction; IPL/STG = inferior parietal lobule/superior temporal gyrus; VFC = ventral frontal cortex; IFG/MFG = inferior frontal gyrus/middle frontal gyrus. (After Corbetta and Shulman, 2002.)

proposed to operate by sending signals that prepare the sensory cortical regions that need to deal with expected or ongoing objects and events in the environment. Such preparatory activity would presumably result in enhanced processing and corresponding improvements in behavioral performance. This complex of brain regions has come to be called the **frontal-parietal attention network** (Figure 29.6). The network is activated both endogenously and exogenously, and is thought to modulate activity in the sensory cortices and other brain regions, resulting in more effective processing of some inputs and a less complete processing of others.

Studies in Non-Human Primates

Single-unit recording studies of attention in awake and behaving non-human primates have confirmed that neurons in parietal and frontal regions homologous to the putative attention network in humans tend to be active when animals are attending to a task. These findings, mostly in macaque monkeys, have allowed researchers to study attention and the possible mechanisms of attentional selection at the neuronal level, most often using visual attention as the experimental paradigm.

In general, visual cortical neurons respond strongly only if a stimulus is presented within the cell's receptive field. The firing rate signals the optimal stimulus for that neuron (i.e., the characteristics to which the cell is tuned, such as a particular orientation, direction of movement, color, etc.). Once a cortical neuron is located and its receptive field characterized, the animal's attention can be manipulated to investigate its effects on neuronal responsiveness. When effective and ineffective stimuli (i.e., stimuli that matched or did not match the neuron's tuning curve) were presented together within a neuron's receptive field in visual area V4, the cell fired strongly only when the effective stimulus was being attended. When the monkey attended the ineffective stimulus, the neuron responded weakly, even though the stimulus had not changed (Figure 29.7). These observations indicate that neuronal responses depend on the locus of attention within a neuron's receptive field, at least for cells in these areas of cortex. In the later stages of visual processing—that is, in the ventral pathway leading to the inferior temporal cortex—the observed pattern was different. At this level, attention modulated the neuronal responses even if the ignored stimulus was relatively far away from the attended one, presumably because the receptive fields in this region are much larger.

As shown in Figure 29.7, attention enhances activity of the relevant neurons. Another non-human primate study assessed how the locus of spatial attention affects the orientation tuning curves of visual neurons after training monkeys to attend to one of two gratings. When the monkey attended to the stimuli in the receptive field of the recorded neuron, responses were enhanced, as expected. Using this paradigm, however, allowed evaluation of how attention affected responses to gratings of different orientations. Although attention enhanced neural responses at all orientations, the effects were stronger for a neuron's preferred orientation (see Chapter 12).

Other electrophysiological studies of attention in experimental animals have examined neurons in the lateral intraparietal area (LIP) of the posterior parietal cortex, as well as in the frontal eye fields (see above). These two regions of the monkey cortex are assumed to correspond with the parietal and frontal areas in humans where damage causes neglect syndromes, and where neuroimaging studies have shown activity related to attentional control. One interest in these particular areas is the possibility of "integrating centers" within a broader set of brain areas involved in attention. For instance, the firing rates of LIP neurons in response to a stimulus in their receptive field are greater when the task is to make a saccade to a target in the receptive field rather than simple fixation. Neuronal responses are also enhanced when a monkey attends to the stimulus in the receptive field but does not make a saccade, or when the saccade is delayed. These results suggest that enhanced neuronal responsiveness is not due to saccade preparation per se, but to the allocation of attention to the spatial location of the target in the neuron's receptive field, which occurs when the monkey plans to shift his gaze there.

A related question in experimental animals is how activity in these regions could lead to enhanced stimulus processing in the sensory cortices. Some relevant information had already been provided by fMRI studies in humans. When participants direct sustained attention to a particular visual-field location expecting the onset of a visual stimulus there, increased activity is elicited not only in the frontal and parietal cortices but in extrastriate cortex as well. The implication is that the increased activity in visual cortex in the absence of visual stimulation reflects preparatory signals from the frontal-parietal network that favors the attended location. In accord with this idea, microstimulation of the frontal eye fields in monkeys improves performance in attentional tasks and simultaneously increases the activity of neurons in V4 with receptive aligned with the retinotopic locus of stimulation. Saccade-related activity in the frontal eye fields has also supported this "premotor theory" of attention, although other interpretations have also been suggested.

This wealth of studies in patients, normal subjects, and non-human primates shows that many interrelated brain areas are involved in attention. These facts, plus the reliable activation of this network irrespective of the sensory modality (e.g., visual, auditory) or stimulus category (e.g., location, feature, or object type) examined, *seem* to validate calling these interconnected areas an attentional system or network (see Figure 29.6).

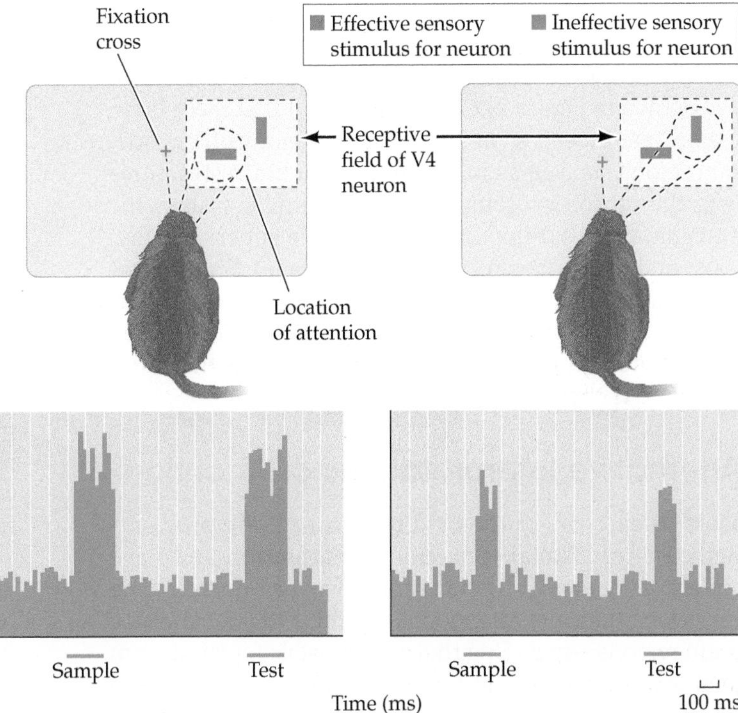

FIGURE 29.7 Effects of attention on the firing rates of single neurons in the visual cortex. At the attended location (circled), two stimuli—sample and test—were presented sequentially; the monkey had to discriminate whether they were the same or different. Irrelevant stimuli were presented simultaneously with the sample and test but at a separate location in the receptive field. Stimuli could either be effective stimuli for the neuron (red bars in this example) or ineffective stimuli (green bars). When both an effective stimulus and an ineffective stimulus were presented within the receptive field and the monkey attended to the effective stimulus, the neural responses were robust. When the monkey attended to the ineffective stimulus, however, the responses were much reduced, despite the presence of an effective stimulus in the receptive field. In short, the locus of attention has a clear effect on the activity the relevant neurons. (After Moran and Desimone 1985.)

Problems with the Concept of Attention as Executive Control

Many aspects of the brain can be understood in terms of systems and subsystems; the visual system, the auditory system, the skeletal motor system, and the visceral motor system, to name a few, are anatomically and functionally specialized regions that carry out relatively specific tasks. Attention, however, does not fit easily into this category. The anatomical regions of association cortex illustrated in Figure 29.6 participate in many different neural functions and have no defining characteristic other than their enhanced activity when human subjects or other animals are focused on something. Although the putative attention network could be interpreted in functional rather than anatomical terms, this concept would still entail a network charged with assessing what is required for any given task, selecting the most relevant components of concurrent stimulus information and directing their flow. Thus, even a functional characterization of attention leaves the concept of executive control on center stage.

At a subjective level, the notion of executive control as the foundation of attention obviously has its uses. For example, we routinely make statements such as "I decided to pay attention" and assume a personal "I" that makes decisions and directs the focus of endogenous attention. But in terms of neuroscience, the concept of an executive controller that monitors the rest of the brain and makes decisions about its operation is hard to credit. An executive system would have to receive input from all the systems and subsystems the nervous system uses to sense the external and internal environment, take into account relevant information from memory, and integrate the affective and motivational state of the individual. To complicate matters further, the system would have to direct processing resources to any location in space, to any specifiable feature of an object or a scene, and to any point in time. In short, the neural processes that an attentional system would need to have to monitor and direct include the full gamut of information-processing circuitry entailed in normal brain function.

In response, advocates of the dedicated attentional system shown in Figure 29.6 would point out that these brain regions are indeed widely connected to many other cortical regions, as well as many subcortical brain structures (e.g., the subcortical components of the limbic system, the amygdala, the subthalamic nucleus, the superior colliculus, and the striatum, to name just some). But the fact remains that the only "system" fully capable of integrating the range of external and internal information needed to generate optimal behavior at any moment is the entire brain and the rest of the nervous system.

Alternative Interpretations of Attention

Most researchers in the field recognize these problems with the traditional conception of attention and have sought to deal with them in one way or another. One variation on idea of attentional control is that a variety of brain regions—including those responsible for short-term memory, emotional state, motivation, motor planning, and other areas—issue *biasing signals* to specialized nodal areas (e.g., the LIP and frontal eye fields) where signal integration and control occur. Although these biasing signals are not themselves taken to be attentional, they would nonetheless give rise to the privileged processing of some information at the expense of other input determined by the output of these nodes. So construed, the integration of biasing signals in specialized nodes behaves *as if* there were attentional control, but without an overarching executive.

Despite the appeal of this sort of explanation, the decentralization of attentional control among a set of integrating nodes assumes much the same theoretical burden as the traditional concept of attention. Unless one understands how and in what sense input acting on nodes can privilege some information, the problem of attentional selection has only been pushed back a step. Instead of attributing selection to a single executive, selection is now carried out by a plurality of minor executives, each charged with monitoring, winnowing, and redirecting incoming information. Although abolishing the idea of a dedicated executive system seems a step in the right direction, replacing it with a set of distributed executives does not resolve the basic problem inherent in the traditional concept of attention.

Another proposal for understanding attentional control is the idea of **saliency maps**. The interpretation in this case is that cortical areas such as the LIP and frontal eye fields—as well as subcortical structures such as the superior colliculus—contain maps defining corresponding regions of space or some other key quality. When incoming signals activate different locations within these maps, that region of space (or its equivalent) is given processing priority. Thus, at any given moment the highest level of activity in the combination of postulated saliency maps creates a de facto priority map that determines what we attend to. The activity in saliency maps is imagined to be affected by biasing signals from sensory input, the relevance of objects and locations for current goals, motor planning, differing motivations to respond to objects and locations, the emotional valence of stimuli, and so on. As a result, attention in one domain (e.g., visual) can be influenced by a range of additional (e.g., nonvisual) factors. While the idea of replacing a dedicated attentional system with saliency maps is also attractive, this approach suffers from a reliance on preparatory signals that arise in a manner that is not well defined in terms of locus or mechanism. Rather than resolving the flaws inherent in the traditional concept of attention, saliency maps may also be reformulating the basic problem in the concept of executive control, albeit in a manner consistent with newer evidence from brain-imaging and other studies.

Other conceptions of how attention could operate without a dedicated executive are based on computational principles. For example, redirection of attention could arise from attractor states found in dynamic systems (the brain being one such system). Thus, when an experimenter or naturally occurring stimuli "cue attention," the rules governing "attractor states" would redirect processing. Attention would thus be construed as an emergent phenomenon determined by the computational demands of efficient neural processing.

Another way that some investigators in engineering and computer science have thought about brain operations without executive controllers is based on robotics. Instead of programming a robotic system with fixed behavioral capabilities directed by a central executive, "decisions" are made in a non-representational manner based on information accumulated and instantiated in the robot's circuitry by trial and error. The behavior that ensues from this approach provides the appearance of central control when in fact there is none. Although the performance of such mobile robots is still in its infancy, the framework they provide applies to attention since the behaviors elicited from these design principles could be construed in attentional terms.

Thus a more radical possibility is that attention is generated automatically (reflexively) according to the full range of internal and external stimuli that affect the activity in the nervous system's circuitry at any given time. The responses would, like the robot, have been generated by trial and learning over evolutionary and individual time. As a result there would no executive system, no overarching control, and no selectivity in the usual sense of the word. In these terms, whatever is attended to is simply the result of preexisting neural circuitry shaped empirically by the mechanisms of natural selection and individual learning that favor responses that promoted successful behavior in the past. From this perspective, the term *attention* would simply be a useful colloquialism that expresses a subjective sense of what are, in fact, predetermined responses based on existing neural circuitry. The idea that the internal mental processes that we think of as voluntary outcomes determined be a neural "I" would be no different in kind from the involuntary shifts of attention elicited by a flash of light or a novel sound.

Summary

Attention research seeks to understand how processing resources are directed to deal effectively with ever-changing internal and external environments. Endogenous attention refers to the ability to voluntarily direct attention based on one's goals, expectations, or knowledge. Exogenous attention refers to involuntary shifts of attention triggered by salient stimuli in the environment. Both lead to enhanced processing of the information to which attention has been directed. Insight into both the psychological and the neural mechanisms of attention has been greatly advanced in recent years by combining older behavioral approaches with EEG and fMRI that can evaluate brain activity while humans or other animals are engaged in attentional tasks. The widely accepted idea based on these studies of a dedicated attentional system that monitors brain activity and makes decisions about the allocation of neural resources is nonetheless problematic. Many attention researchers recognize this problem and have sought to provide other ways of conceptualizing attention and interpreting the relevant experimental results, but so far without any general agreement among them.

ADDITIONAL READING

Reviews

Buxbaum, L. J. (2006) On the right (and left) track: Twenty years of progress in studying hemispatial neglect. *Cog. Neuropsych.* 23: 184–201. doi: 10.1080/02643290500202698

Corbetta, M. and G. L. Shulman (2002) Control of goal-directed and stimulus-driven attention in the brain. *Nat. Rev. Neurosci.* 3: 201–215.

Desimone, R. and J. Duncan (1995) Neural mechanisms of selective visual attention. *Annu. Rev. Neurosci.* 18: 193–222.

Driver, J. (2001) A selective review of selective attention research from the past century. *Br. J. Psychol.* 92: 53–78.

Hannula, D. E., D. J. Simons and N. J. Cohen (2005) Imaging implicit perception: Promise and pitfalls. *Nat. Rev. Neurosci.* 6: 247–255.

Husain, M. and P. Nachev (2007). Space and the parietal cortex. *Trends Cogn. Sci.* 11: 30–36.

Kastner, S. and L. G. Ungerleider (2000) Mechanisms of visual attention in the human cortex. *Annu. Rev. Neurosci.* 23: 315–341.

Knudsen, E. I. (2007) Fundamental components of attention. *Annu. Rev. Neurosci.* 30: 57–78.

Maunsell, J. H. R. and S. Treue (2006) Feature-based attention in visual cortex. *Trends Neurosci.* 29: 317–322.

Miller, E. K. and J. D. Cohen (2001) An integrative theory of prefrontal cortex function. *Annu. Rev. Neurosci.* 24: 167–202.

Posner, M. I. and S. E. Petersen (1990) The attention system of the human brain. *Annu. Rev Neurosci.* 13: 25–42.

Shamma, S. A., M. Elhilali and C. Micheyl (2011) Temporal coherence and attention in auditory scene analysis. *Trends Neurosci.* 34: 114–123.

Talsma, D., D. Senkowski, S. Soto-Faraco and M. G. Woldorff (2010) The multifaceted interplay between attention and multisensory integration. *Trends Cogn. Sci.* 14: 400–410.

Important Original Papers

Baldauf, D. and R. Desimone (2014) Neural mechanisms of object-based attention. *Science.* 344: 424–427.

Buschman, T. J. and E. K. Miller (2007) Top-down versus bottom-up control of attention in the prefrontal and posterior parietal cortices. *Science* 315: 1860–1862.

Cerf, M. and 6 others (2010) On-line voluntary control of human temporal lobe neurons. *Nature* 467: 1104–1108.

Colby, C. L., J. R. Duhamel and M. E. Goldberg (1996) Visual, presaccadic, and cognitive activation of single neurons in monkey lateral intraparietal area. *J. Neurophysiol.* 76: 641–652.

Collet, C., M. Morel, A. Chapon and C. Petit (2009) Physiological and behavioral changes associated to the management of secondary tasks while driving. *Appl. Ergon.* 40: 1041–1046. doi: 10.1016/j.apergo.2009.01.00

Cooper, A. A. and G. W. Humphreys (2000) Coding space within but not between objects: Evidence from Balint's syndrome. *Neuropsychologia* 38: 723–733.

Corbetta, M. and 10 others (1998) A common network of functional areas for attention and eye movements. *Neuron* 21: 761–773.

Corbetta, M., F. M. Miezin, G. L. Shulman and S. E. Petersen (1993) A PET study of visuospatial attention. *J. Neurosci.* 13: 602–626.

De Weerd, P., M. R. Peralta III, R. Desimone and L. G. Ungerleider (1999) Loss of attentional stimulus selection after extrastriate cortical lesions in macaques. *Nat. Neurosci.* 2: 753–758.

Friedman-Hill, S. R., L. C. Robertson and A. Treisman (1995) Parietal contributions to visual feature binding: Evidence from a patient with bilateral lesions. *Science* 269: 853–855.

Fries, P., J. H. Reynolds, A. E. Rorie and R. Desimone (2001) Modulation of oscillatory neuronal synchronization by selective visual attention. *Science* 291: 1560–1563.

Gazzaley, A., J. W. Cooney, K. McKevoy, R. T. Knight and M. D'Esposito (2005) Top-down enhancement and suppression of the magnitude and speed of neural activity. *J. Cogn. Neurosci.* 17: 507–517.

Grabowecky, M., L. C. Robertson and A. Treisman (1993) Preattentive processes guide visual search: evidence from patients with unilateral visual neglect. *J. Cogn. Neurosci.* 5: 288–302.

Gregoriou, G. G., S. J. Gotts, H. Zhou and R. Desimone (2009) High-frequency, long-range coupling between prefrontal and visual cortex during attention. *Science* 324: 1207–1210.

Heinze, H. J. and 11 others (1994) Combined spatial and temporal imaging of brain activity during visual selective attention in humans. *Nature* 372: 543–546.

Hillyard, S. A., R. F. Hink, V. L. Schwent and T. W. Picton (1973) Electrical signs of selective attention in the human brain. *Science* 182: 177–180.

Kastner, S., M. A. Pinsk, P. De Weerd, R. Desimone and L. G. Ungerleider (1999) Increased activity in human visual cortex during directed attention in the absence of visual stimulation. *Neuron* 22: 751–761.

Knight, R. T., S. A. Hillyard, D. L. Woods and H. J. Neville (1981) The effects of frontal cortex lesions on event-related potentials during auditory selective attention. *Electroencephalogr. Clin. Neurophysiol.* 52: 571–582.

Lakatos, P., G. Karmos, A. D. Mehta, I. Ulbert and C. E. Schroeder (2008) Entrainment of neuronal oscillations as a mechanism for attentional selection. *Science* 320: 110–113.

Lavie, N. (1995) Perceptual load as a necessary condition for selective attention. *J. Exp. Psychol. Hum. Percept. Perform.* 21: 451–468.

Liu, T., S. D. Slotnick, J. T. Serences and S. Yantis (2003) Cortical mechanisms of feature-based attentional control. *Cereb. Cortex* 13: 1334–1343.

Luck, S. J., L. Chelazzi, S. A. Hillyard and R. Desimone (1997) Neural mechanisms of spatial selective attention in areas V1, V2, and V4 of macaque visual cortex. *J. Neurophysiol.* 77: 24–42.

Macaluso, E., C. D. Frith and J. Driver (2000) Modulation of human visual cortex by crossmodal spatial attention. *Science* 289: 1206–1208.

McAdams, C. J. and J. H. R. Maunsell (1999) Effects of attention on orientation-tuning functions of single neurons in macaque cortical area V4. *J. Neurosci.* 19: 431–441.

Mesulam, M. M. (1981) A cortical network for directed attention and unilateral neglect. *Ann. Neurol.* 10: 309–325.

Moore, T. and K. M. Armstrong (2003) Selective gating of visual signals by microstimulation of frontal cortex. *Nature* 421: 370–373.

Moran, J. and R. Desimone (1985) Selective attention gates visual processing in the extrastriate cortex. *Science* 229: 782–784.

Motter, B. C. (1993) Focal attention produces spatially selective processing in visual cortical areas V1, V2, and V4 in the presence of competing stimuli. *J. Neurophysiol.* 70: 909–919.

O'Connor, D. H, M. M. Fukui, M. A. Pinsk and S. Kastner (2002) Attention modulates responses in the human lateral geniculate nucleus. *Nat. Neurosci.* 5: 1203–1209.

O'Craven, K. M., P. E. Downing and N. Kanwisher (1999) fMRI evidence for objects as the units of attentional selection. *Nature* 401: 584–587.

Petkov, C. I. and 5 others (2004) Attentional modulation of human auditory cortex. *Nat. Neurosci.* 7: 658–663.

Posner, M. I., C. R. R. Snyder and B. J. Davidson (1980) Attention and the detection of signals. *J. Exp. Psychol. Gen.* 59: 160–174.

Ptak, R. and A. Schnider (2010) The dorsal attention network mediates orienting toward behaviorally relevant stimuli in spatial neglect. *J. Neurosci.* 30: 12557–12565.

Rees, G., C. D. Frith and N. Lavie (1997) Modulating irrelevant motion perception by varying attentional load in an unrelated task. *Science* 278: 1616–1619.

Reynolds, J. H. and D. Heeger (2009) The normalization model of attention. *Neuron* 61: 168–185.

Reynolds, J. H., T. Pasternak and R. Desimone (2000) Attention increases sensitivity of V4 neurons. *Neuron* 26: 703–714.

Schoenfeld, M. A. and 6 others (2003) Dynamics of feature binding during object selective attention. *Proc. Natl. Acad. Sci. USA* 100: 11806–11811.

Serences, J. T. and 5 others (2005) Coordination of voluntary and stimulus-driven attentional control in human cortex. *Psychol. Sci.* 16: 114–122.

Thompson, K. G., K. L. Biscoe and T. R. Sato (2005) Neuronal basis of covert spatial attention in the frontal eye field. *J. Neurosci.* 25: 9479–9487.

Treisman, A. (1960) Contextual cues in selective listening. *Q. J. Exp. Psychol.* 12: 242–248.

Treisman, A. and G. Gelade (1980) A feature integration theory of attention. *Cogn. Psychol.* 12: 97–136.

Treue, S. and J. C. Martinez Trujillo (1999) Feature-based attention influences motion processing gain in macaque visual cortex. *Nature* 399: 575–579.

Wojciulik, E. and N. Kanwisher (1999) The generality of parietal involvement in visual attention. *Neuron* 23: 747–764.

Woodman, G. F. and S. J. Luck (1999) Electrophysiological measurement of rapid shifts of attention during visual search. *Nature* 400: 867–869.

Worden, M. S., J. J. Foxe, N. Wang and G. V. Simpson (2000) Anticipatory biasing of visuospatial attention indexed by retinotopically specific alpha-band electroencephalography increases over occipital cortex. *J. Neurosci.* 20: 1–28.

Yantis, S. and 6 others (2002) Transient neural activity in human parietal cortex during spatial attention shifts. *Nat. Neurosci.* 5: 995–1002.

Books

Humphreys, G., J. Duncan and A. Treisman (eds.) (1999) *Attention, Space and Action: Studies in Cognitive Neuroscience.* Oxford: Oxford University Press.

Itti, L., G. Rees and J. K. Tsotsos (eds.) (2005) *Neurobiology of Attention.* Amsterdam: Elsevier.

James, W. (1890) *The Principles of Psychology.* New York: Henry Holt and Company.

Näätänen, R. (1992) *Attention and Brain Function.* Hillsdale, NJ: Lawrence Erlbaum.

Parasuraman, R. (ed.) (1998) *The Attentive Brain.* Cambridge, MA: MIT Press.

Posner, M. I. (1978) *Chronometric Explorations of Mind.* Hillsdale; NJ: Lawrence Erlbaum.

CHAPTER

30 Memory

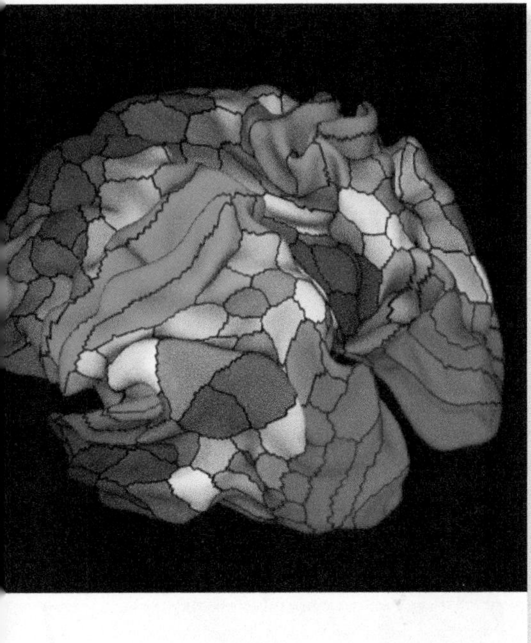

Overview

IT SHOULD BE OBVIOUS THAT ONE OF THE MOST IMPORTANT of the brain's complex functions is the ability to store information from past experience and retrieve it, either consciously or unconsciously. Without this ability, access to the past and imagination of the future would be lost. *Learning* is the name given to the processes through which new information is acquired by the nervous system, and has already been addressed at the cellular level in the chapters on the mechanisms of the cellular and molecular plasticity of neuronal connections, the assumed basis for learning and memory (see Chapters 8 and 23–26). *Memory* is evident as recovered experiences that can be brought into consciousness (e.g., remembering what you ate for breakfast), or more often as changes in behavior (e.g., testing how much better you played a piece of music today as a result of yesterday's practice). For non-human animals, changes in behavior must suffice as evidence of memory, although behavioral changes are often taken to reflect conscious remembering, as in many non-human primate studies. Pathological loss of previously stored information (retrograde amnesia) and/or the inability to store new information (anterograde amnesia) have been especially instructive in understanding the neurological underpinnings of memory, a major challenge in modern neuroscience that has yet to be fully met. This chapter reviews the organization of human memory systems, surveys memory disorders and their implications, and considers some key questions about memory that remain unanswered.

Qualitative Categories of Human Memory

Humans have at least two qualitatively different ways of storing information, generally referred to as *declarative memory* and *nondeclarative memory* (Figure 30.1). **Declarative memory** is the storage and retrieval of material that is available to consciousness and can be expressed by language (i.e., "declared"). Examples of declarative memory are the ability to remember a phone number, the words to a song, or a past event. **Nondeclarative memory** (also referred to as **procedural memory**) is not available to consciousness, at least not in any detail. Such memories involve skills and associations that are generally acquired and retrieved at an unconscious level. Remembering how to shoot a basket or how to play the piano are examples of nondeclarative memories. It is difficult or impossible to describe exactly how we do these things, and thinking about how to carry out automatic activities may actually disrupt the ability to perform them efficiently. As discussed later in the chapter, the distinction between declarative and nondeclarative memory is well supported by anatomical, clinical, and other evidence.

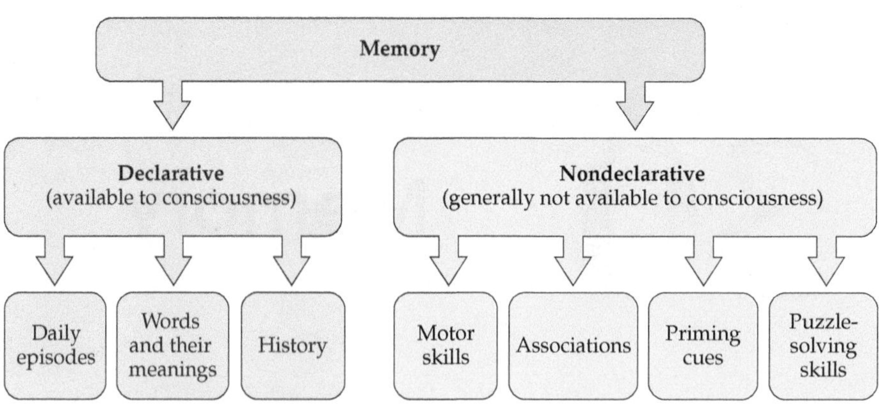

FIGURE 30.1 The major qualitative categories of human memory. Declarative memory includes those memories that can be brought to consciousness and expressed as remembered events, images, sounds, and so on. Nondeclarative, or procedural, memory includes motor skills, cognitive skills, simple classical conditioning, priming effects, and other information that is acquired and retrieved unconsciously.

BOX 30A ■ Phylogenetic Memory

A category of information storage not usually considered in standard accounts is that of memories arising from the experiences of a species over the eons, established by natural selection acting on the cellular and molecular mechanisms of neural development. Such stored information does not depend on postnatal experience but on what a given species typically encountered in its environment over evolutionary time. These "memories" are no less consequential than those acquired by individual experience and are likely to have much underlying biology in common with the memories established during an individual's lifetime; after all, both phylogenetic and ontogenetic memories are based on neuronal connectivity. In the former case, changing connectivity depends on natural selection, and in the latter, on the mechanisms of plasticity discussed in Chapters 8 and 25.

Information about the experience of the species, as expressed by endogenous or "instinctive" behavior, can be quite sophisticated, as is apparent in examples collected by ethologists for a wide range of animals, including primates. The most thoroughly studied instances of such behaviors are those occurring in young birds. Hatchlings arrive in the world with an elaborate set of innate behaviors. First is the complex behavior that allows the young bird to emerge from the egg. Once a bird has hatched, a variety of additional behaviors indicate how much of its early life is dependent on inherited information. Hatchlings of precocial species "know" how to preen, peck, gape their beaks, and carry out a variety of other complex acts immediately. In some species, hatchlings automatically crouch down in the nest when a hawk passes overhead but are oblivious to the overflight of an innocuous bird. Konrad Lorenz and Niko Tinbergen used handheld silhouettes to explore this phenomenon in naïve herring gulls, as illustrated in the figure shown here. "It soon became obvious," wrote Tinbergen, "that ... the reaction was mainly one to shape. When the model had a short neck so that the head protruded only a little in front of the line of the wings, it released alarm, independent of the exact shape of the dummy." Evidently, the memory of what the shadow of a predator looks like is built into the nervous system of this species. Examples in primates include the innate fear that newborn monkeys have of snakes and looming objects.

Despite the relatively scant attention paid to this aspect of memory, evolved and inherited neural associations are the most important component of the stored information in the brain that determines whether or not an individual survives long enough to reproduce.

(A)

(B)

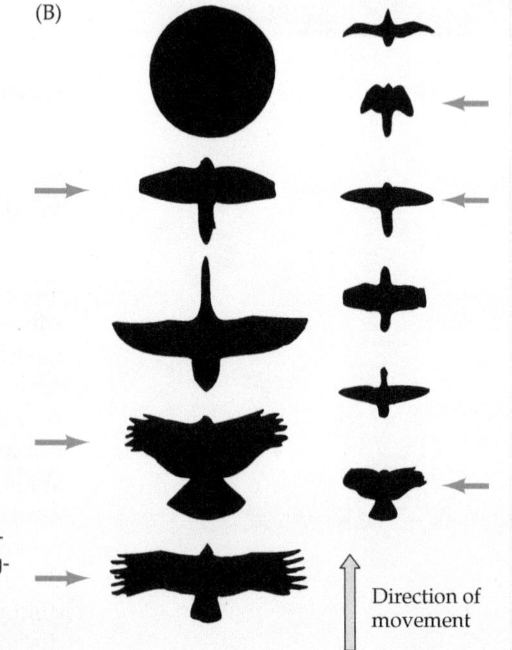

(A) Niko Tinbergen at work. (B) Silhouettes used to study alarm reactions in hatchlings. The shapes that were similar to the shadow of the bird's natural predators (red arrows) when moving in the appropriate direction elicited escape responses (i.e., crouching, crying, seeking cover); silhouettes of songbirds and other innocuous species (or geometrical forms) elicited no obvious response. (After Tinbergen, 1969.)

Although it makes good sense to divide human learning and memory into categories based on the accessibility of stored information to consciousness, this distinction becomes problematic when considering learning and memory processes in non-human animals. From an evolutionary point of view, it is of course unlikely that declarative memory arose de novo in humans with the development of language. Although some researchers favor different classification systems for humans as opposed to other animals, studies suggest that similar memory processes operate in all mammals and that these functions are carried out by homologous neural circuitry. In non-human mammals, declarative memory typically refers to information that they are aware of, and could be "declared" if the species in question had this ability. Another criterion of declarative memory in non-human animals is its dependence on the integrity of the medial temporal lobes (see below). Nondeclarative memory, in humans and other animals alike, can be thought of as the acquisition and storage of neural associations that are not available to consciousness and not dependent on the medial temporal lobes.

Finally, and perhaps most important of all, are the memories we have as a result of the evolution of our species (Box 30A). Although such memories are not often considered, they represent changes in brain connectivity that have been wrought by millions of years of experience with behaviors that work and those that don't. Thus, we each come into the world with a vast array of inherited behaviors that far outweigh in value what we learn in the course of an individual lifetime.

Temporal Categories of Memory

Memory can also be categorized according to the time over which it is effective. Although the details are debated by both psychologists and neurobiologists, three temporal classes of memory are generally accepted (Figure 30.2). The first of these is **immediate memory**. By definition, immediate memory is the ability of the brain to hold onto ongoing experience for a second or so. For example, we normally make saccadic eye movements three or four times a second, thus continuously sending new "snapshots" of the visible environment to the visual system. Although we perceive the visual scene as stable, the snapshots are forgotten absent something arresting. Nonetheless, if you close your eyes at some random moment, immediately you can recall a good deal of information from the image you last saw.

Short-term memory, the second temporal category, is the ability to hold and manipulate information in the mind for seconds to minutes while it is being used to achieve a particular goal (such processes are also referred to as *working memory*). An everyday example is searching for a lost object; short-term memory allows the hunt to proceed efficiently, avoiding places already inspected. A conventional way of testing the integrity of short-term memory at the bedside is to present a string of random numbers, which the patient is then asked to repeat. The normal digit memory span is seven to nine numbers. Because short-term memory is limited in both duration and capacity, the relevant information must be rehearsed if it is to persist for a very long.

Short-term memory is also closely related to attention; indeed, it is sometimes considered to be a special category of attention that operates on internal representations rather than on sensory input as such. Short-term memory is also pertinent to language, reasoning, and problem solving, as the example of searching for a lost item makes plain. Thus, memory influences and is influenced by many other aspects of brain function. Although it is typically studied in the context of declarative memory, short-term memory also operates in the acquisition and ultimate storage of nondeclarative information.

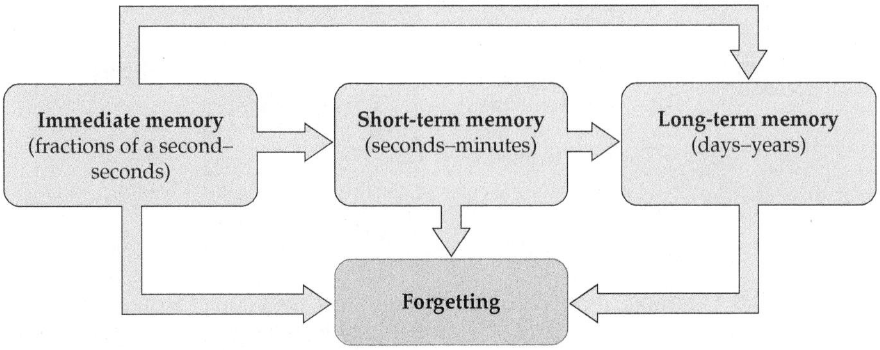

The third temporal category is **long-term memory** and entails retaining information in a more permanent form of storage for days, weeks, or even a lifetime. Information of particular significance in immediate and short-term memory can enter into long-term memory by conscious or unconscious rehearsal or practice. There is general agreement that the so-called **engram**—the physical embodiment of any memory in neuronal machinery—depends on changes in the efficacy of synaptic connections and/or the actual growth and reordering of such connections. As discussed in Chapter 23, there is ample

FIGURE 30.2 The major temporal categories of human memory. Information in both immediate and short-term memory can enter long-term memory, although most information is promptly forgotten.

evidence that mechanisms of synaptic change can and do act over each of the temporal intervals pertinent to the different endurances of memories. The term *consolidation* (Latin, "to make firm") refers to the progressive stabilization of memories that follows the initial encoding of memory "traces." Consolidation involves changes in gene expression, protein synthesis, and other mechanisms of synaptic plasticity that allow the persistence of memories at the cellular level and can be disrupted by interfering with these processes (see Chapters 8 and 23).

TABLE 30.1 ■ The Fallibility of Human Memory[a]

(A) Initial list of words		(B) Subsequent test list
candy	honey	taste
sour	soda	point
sugar	chocolate	sweet
bitter	heart	chocolate
good	cake	sugar
taste	eat	nice
tooth	pie	
nice		

[a]After hearing list A read aloud, subjects were asked to identify which items in list B had also been on list A. See text for the results.

Priming

Another way of exploring the transfer of information from immediate and short-term memory to long-term memory is in terms of priming. **Priming** is defined as a change in the processing of a stimulus due to a previous encounter with the same or a related stimulus with or without conscious awareness of the original encounter. The phenomenon is typically demonstrated by presenting subjects with a set of items to which they are exposed under false pretenses. For example, a list of words can be given with the instruction that the subjects are to identify some feature that is actually extraneous to the experiment (e.g., identifying the words as verbs, adjectives, or nouns). Sometime thereafter (often the next day), the same individuals are given a different test in which they are asked to fill in the missing letters of words with the letters of whatever words come to mind (Figure 30.3) The test list actually includes fragments of words that were presented in the first test, mixed among fragments of words that were not. Subjects tend to fill in the letters to make the words that were presented earlier at a higher rate than expected by chance, and fill them in more quickly than they do new words, even though they may have little or no conscious memory of seeing the words from the earlier list.

The information stored by priming, however, is not particularly reliable. Consider the list of words in Table 30.1A. If the list is read to a group of students who are immediately asked to identify which of several items were on the original list and which were not (Table 30.1B), the result is surprising. Typically, about half the students report that the word *sweet* was included in the list in Table 30.1A; moreover, they are quite certain about it. The mechanism of such erroneous "recognition" is presumably the strong associations that have previously been made between the words on the list in Table 30.1A and the word *sweet*, which biases the students to think that *sweet* was a member of the original set. Clearly, memories, even those we feel quite confident about, are often false.

Priming is resistant to brain injury, aging, and dementia. As a result, its contributions are less obvious (and less

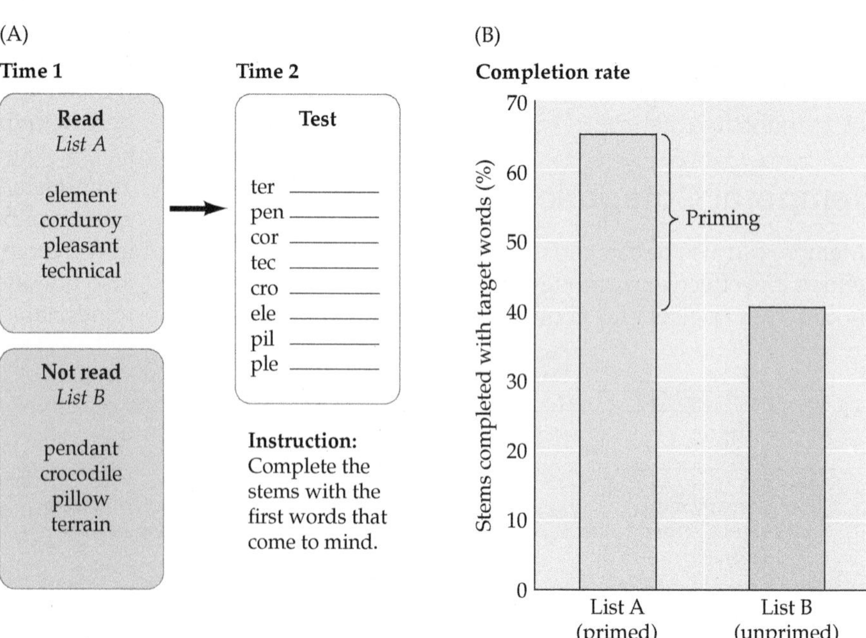

FIGURE 30.3 **Priming.** (A) In a commonly used test, the subject is presented at time 1 with a list of words to study (list A) and is later tested using a word-stem completion task (time 2). The stems could also be completed from list B, which comprises words the subject did not see during the initial session. (B) Subjects typically complete the stems with about 25% more studied than unstudied words; this percentage represents the effect of priming.

easily studied) than other forms of memory that are compromised by specific brain insults, such as impaired declarative memory following damage to the medial temporal lobes (see below). Among other things, priming shows that information previously presented is always influential, even though we are entirely unaware of its effect on subsequent behavior. The significance of priming is well known—at least intuitively—to advertisers, teachers, spouses, and others who want to influence the way we think and act.

The Importance of Association in Information Storage

The normal human capacity for remembering relatively meaningless information is surprisingly limited (as noted, a string of seven to nine numbers or other arbitrary items). This stated capacity, however, is misleading. People can remember 14 or 15 items in a briefly presented 5 × 5 matrix of 25 numbers or other objects if the experimenter points to specific boxes in the blank matrix during recall testing. Moreover, a person's digit memory span can be increased dramatically with practice. For example, a college student who for some months spent an hour each day being paid to successfully remember randomly presented numbers was able to recall a string of up to about 80 digits (Figure 30.4). He did this by making subsets of the string of numbers

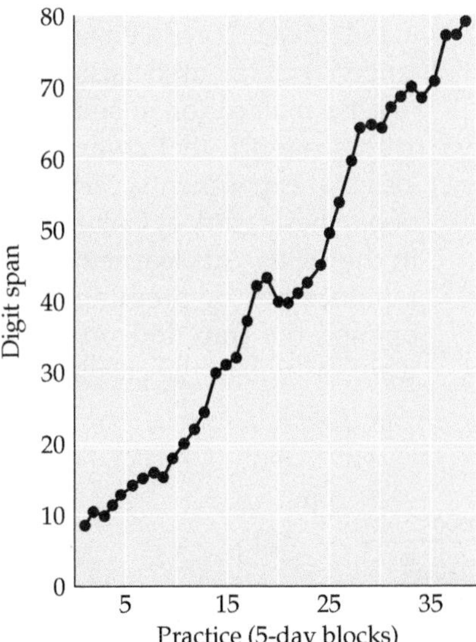

FIGURE 30.4 Increasing the digit span by practice and the development of associational strategies. During many months involving 1 hour of practice each day for 3 to 5 days a week, the subject increased his digit memory span from 7 to 79 numbers. Random digits were read to him at the rate of 1 per second. If a sequence was recalled correctly, 1 digit was added to the next sequence. (After Ericsson et al., 1980.)

he was given signify dates or times at track meets (he was a competitive runner)—in essence, giving meaningless items a meaningful context. This same strategy of enhancing associations is used by professional "mnemonists" who amaze audiences by prodigious feats of memory. A challenge for some mnemonists is memorizing as many as possible of the infinite number of digits in π (3.1416...*n*). The current world record is over 67,000 decimal places. The mnemonists who hold such records use a variety of strategies, a common one being associating the ten digits involved with musical notes and singing the number string.

The capacity of memory very much depends on what the information in question means to the individual and how readily it can be associated with information that has already been stored. A good chess player can remember the position of many more pieces on a briefly examined board than an inexperienced player, presumably because the positions have much more significance for individuals who understand the intricacies of the game (Figure 30.5). Arturo Toscanini, the late conductor of the NBC Philharmonic Orchestra, allegedly kept in his head the complete scores of more than 250 orchestral works, as well as the music and librettos for some 100 operas. Once, just before a concert in St. Louis, the first bassoonist approached Toscanini in some consternation, having just discovered that one of the keys on his bassoon was broken. After a minute or two of deep concentration, the story goes, Toscanini turned to the alarmed bassoonist and informed him that there was no need for concern, since that note did not appear in any of the bassoon parts for the evening's program. Such feats of memory are not achieved by rote learning but are a result of the fascination that aficionados bring to their special interests, sometimes in a pathological way (Box 30B).

Such examples indicate that motivation also plays an important role in memory. In one study of this issue, experimenters asked subjects to study a set of photographs that depicted either pieces of furniture or pieces of food (Figure 30.6). The subjects were later tested with a much larger set of photographs that included images from the previously studied set along with new ones; the subjects were asked to indicate whether a picture was "old" or "new." In one condition, the experimenters increased subjects' hunger by depriving them of food for several hours. Predictably, subjects were much more likely to remember more pictures of food when they were hungry than when they were not. There was no effect of motivation on memory for pictures of furniture.

Although few of us can boast the mnemonic prowess of a π enthusiast or a Toscanini, the human ability to remember the things that deeply interest us—whether baseball statistics, television show, or the details of brain structure—is nothing short of amazing.

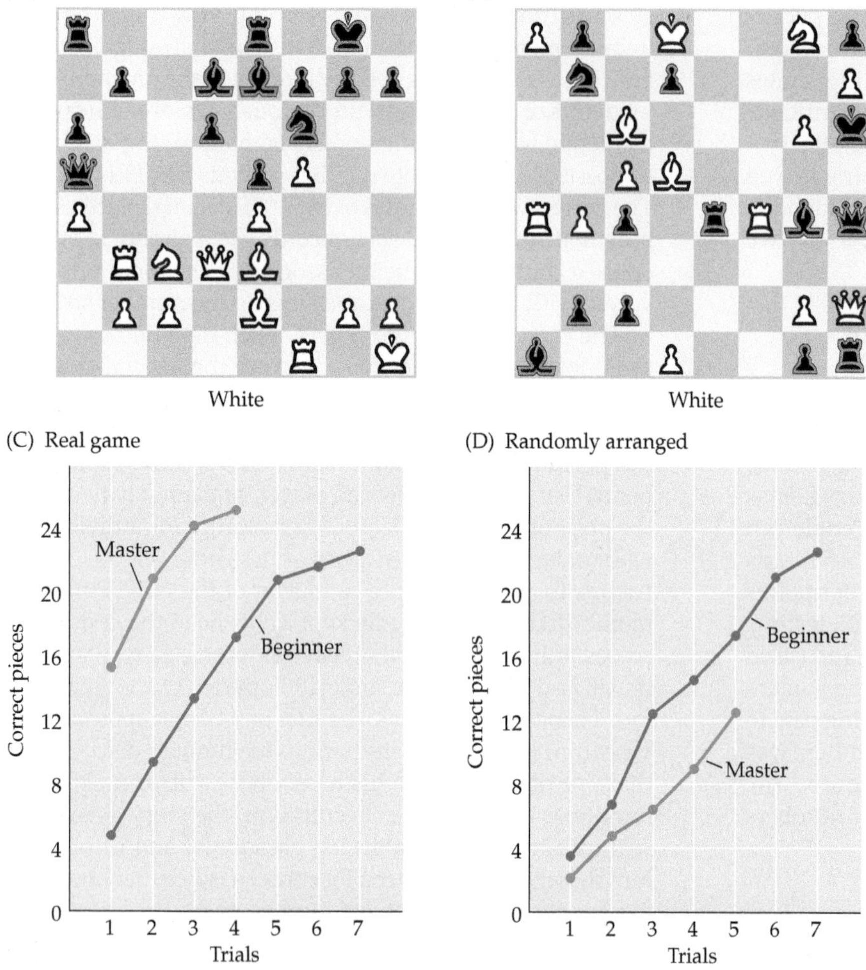

FIGURE 30.5 Retention of briefly presented information depends on past experience, context, and perceived importance. (A) Board position after white's 21st move in game 10 of the 1985 World Chess Championship between A. Karpov (white) and G. Kasparov (black). (B) A random arrangement of the same 28 pieces. (C,D) After briefly viewing the board from the real game, master players reconstruct the positions of the pieces with much greater efficiency than beginning players. With a randomly arranged board, however, beginners perform as well as or better than accomplished players. (After Chase and Simon, 1973.)

Conditioned Learning

Conditioned learning is defined as the generation of a novel response that is gradually elicited by repeatedly pairing a novel stimulus with a stimulus that normally elicits the response being studied. **Classical conditioning** occurs when an innate reflex is modified by associating its normal trigger with an unrelated stimulus; by virtue of the repeated association, the unrelated stimulus eventually triggers the original response. This type of conditioning was famously studied by the Russian psychologist Ivan Pavlov in experiments with dogs and other animals early in the twentieth century. The dogs' innate reflex was salivation (the *unconditioned response*) in reaction to the sight and/or smell of food (the *unconditioned stimulus*). The association was elicited in the animals by repeatedly pairing the sight and smell of food with the sound of a bell (the *conditioned stimulus*). The conditioned reflex was considered established when the conditioned stimulus (the sound of the bell) elicited salivation by itself (the *conditioned response*).

Operant conditioning refers to the altered probability of a behavioral response engendered by associating the response with a reward (or in some instances, a punishment). In Edward Thorndike's original experiments, carried out as part of his thesis work at Columbia University in the 1890s, cats learned to escape from a puzzle box by pressing the lever that opened the trap door to get a food reward. Although the cats initially pressed

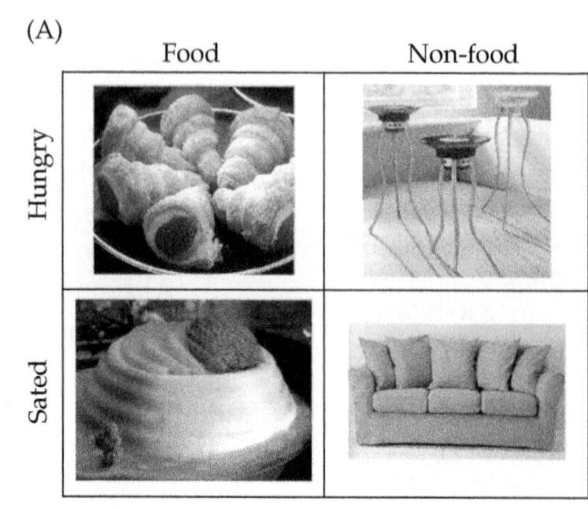

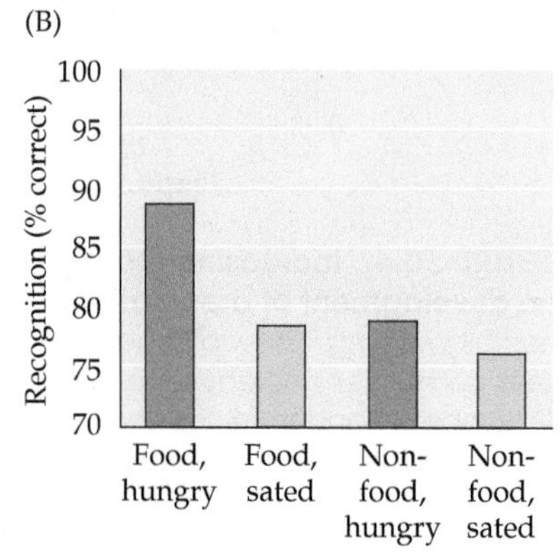

FIGURE 30.6 Motivated memory. (A) Subjects studied a set of pictures of food and non-food (i.e., furniture) items and were later tested for their ability to discriminate the pictures they had seen from a new set of pictures. In one condition, subjects were made hungry by withholding food for several hours. (B) Memory for food items was significantly enhanced when subjects were hungry, but there was no significant effect of hunger on memory for non-food pictures. Results like these emphasize the importance of motivation and interest for memory performance. (From Morris and Dolan, 2001.)

BOX 30B ■ Savant Syndrome

A fascinating developmental anomaly of human memory is seen in rare individuals who until recently were referred to as *idiot savants*; the current literature tends to use the less pejorative phrase *savant syndrome*. Savants are people who, for a variety of poorly understood reasons (typically brain damage in the perinatal period), are severely restricted in most mental activities but extraordinarily competent and mnemonically capacious in one particular domain. The grossly disproportionate skill compared with the rest of their limited mental life can be striking. Indeed, these individuals—whose special talent may be in calculation, history, art, language, or music—are usually diagnosed as being severely impaired.

Many examples could be cited, but a summary of one such case suffices to make the point. The individual whose history is summarized here was given the fictitious name "Christopher" in a detailed study carried out by psychologists Neil Smith and Ianthi-Maria Tsimpli. Christopher was discovered to be severely brain damaged at just a few weeks of age (perhaps as the result of rubella during his mother's pregnancy or anoxia during birth; the record is uncertain in this respect). He had been institutionalized since childhood because he was unable to care for himself, could not find his way around, had poor hand-eye coordination, and had a variety of other deficiencies. Tests on standard IQ scales were low, consistent with his general inability to cope with daily life. Scores on the Wechsler Scale were, on different occasions, 42, 67, and 52.

Despite his severe mental incapacitation, Christopher took an intense interest in books from the age of about 3, particularly those providing factual information and lists (e.g., telephone directories and dictionaries). At about age 6 or 7 he began to read technical papers that his sister sometimes brought home from work, and he showed a surprising proficiency in foreign languages. His special talent in the acquisition and use of language (an area in which savants are often especially limited) grew rapidly. As an early teenager, Christopher could translate from—and communicate in—a variety of languages in which his skills were described as ranging from rudimentary to fluent; these included Danish, Dutch, Finnish, French, German, modern Greek, Hindi, Italian, Norwegian, Polish, Portuguese, Russian, Spanish, Swedish, Turkish, and Welsh. This extraordinary level of linguistic accomplishment is all the more remarkable since he had no formal training in language even at the elementary school level, and could not play tic-tac-toe or checkers because he was unable to grasp the rules needed to make moves in these games.

The neurobiological basis for such extraordinary individuals is not understood. It is fair to say, however, that savants are unlikely to have ability in their areas of expertise that exceeds the competency of normally intelligent individuals who focus passionately on a particular subject. Presumably, the savant's intense interest in a particular cognitive domain is due to one or more brain regions that continue to work reasonably well. Whether because of social feedback or self-satisfaction, savants clearly spend a great deal of their mental time and energy practicing the skill they can exercise more or less normally. The result is that the relevant associations they make become especially rich, as Christopher's case demonstrates.

the lever only occasionally—and more or less by chance—the probability of their doing so increased as the animals learned to associate this action with escape and reward. In Frederick Skinner's far more complete and better-known experiments performed a few decades later at Harvard, pigeons or rats learned to associate pressing a lever with receiving a food pellet in a widely used device that came to be known as a Skinner box (Figure 30.7). In both classical and operant conditioning, it takes a number of trials for the conditioning to become established. If the conditioned animal performs the desired response but the reward is no longer provided, the conditioning gradually disappears, a phenomenon called *extinction*.

It should be apparent that many of our habits and rituals are learned through unconscious conditioning. Habits can be efficient responses to frequently occurring situations. However, habits can be maladaptive as well, as occurs in addiction or obsessive–compulsive disorder.

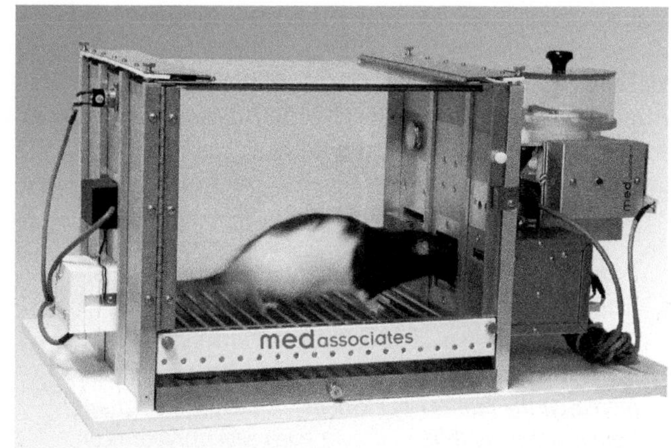

FIGURE 30.7 Modern example of a Skinner box. This apparatus is the most widely used method for studying operant conditioning.

Forgetting

Some years ago, a poll showed that 84% of psychologists agreed with the statement "Everything we learn is permanently stored in the mind, although sometimes particular details are not accessible." The 16% who thought otherwise, however, were correct. Common sense indicates that,

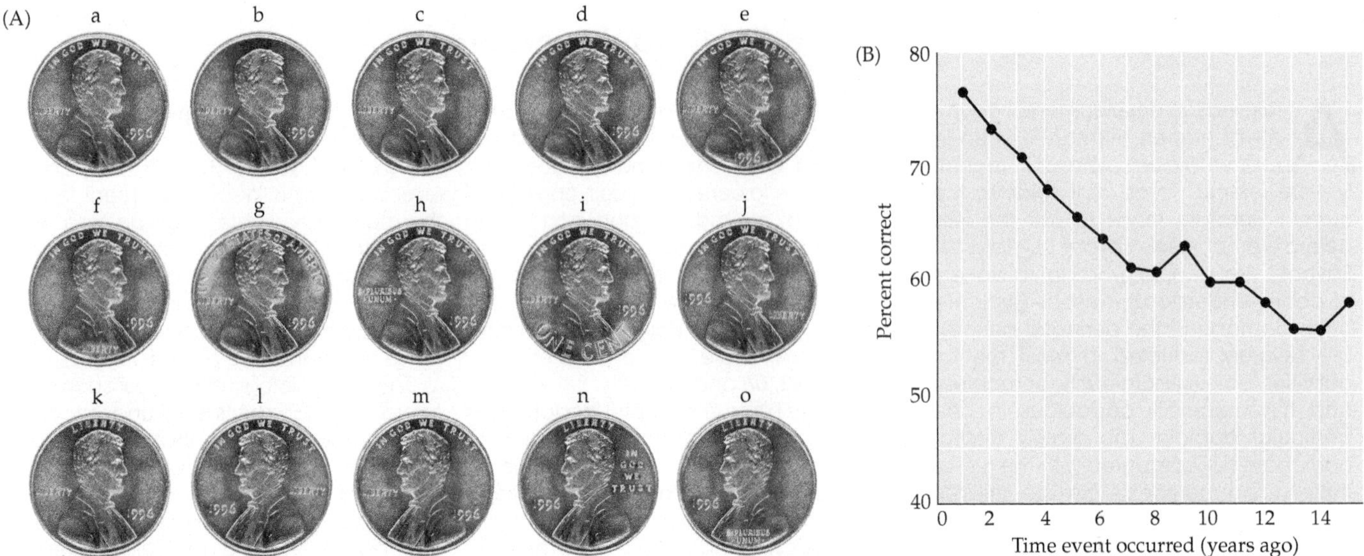

FIGURE 30.8 Forgetting. (A) Different versions of the "heads" side of a penny. Despite innumerable exposures to this familiar design, few people are able to select (a) as the authentic version. Clearly, repeated information is not necessarily retained. (B) The deterioration of long-term memories was evaluated in this example by a multiple-choice test in which the subjects were asked to recognize the names of television programs that had been broadcast for only one season during the past 15 years. Forgetting of stored information that is no longer used evidently occurs gradually and progressively over the years (chance performance = 25%). (A after Rubin and Kontis, 1983; B after Squire, 1989.)

were it not for forgetting, our brains would be impossibly burdened with the welter of useless information that is briefly encoded in our immediate, short-term and even long term memories. In fact, the human brain is very good at forgetting. Like the unreliable performance on tests such as the one shown in Table 30.1, Figure 30.8 shows that our memory of the appearance of a penny (an icon seen thousands of times since childhood) is uncertain at best, and that people tend to gradually forget what they have encoded in long-term memory. Clearly, we forget things that have no particular importance, and unused or unrehearsed memories deteriorate over time.

The ability to forget unimportant information may be as critical for normal life as the ability to retain information that is significant. Evidence for this presumption is demonstrated in rare individuals who have difficulty with the normal erasure of information. The best-known case is a subject studied over several decades by the Russian psychologist Alexander Luria, who referred to the subject simply as "S." Luria's description of an early encounter gives some idea why S, then a newspaper reporter, was so interesting:

> *I gave S a series of words, then numbers, then letters, reading them to him slowly or presenting them in written form. He read or listened attentively and then repeated the material exactly as it had been presented. I increased the number of elements in each series, giving him as many as thirty, fifty, or even seventy words or numbers, but this too, presented no problem for him. He did not need to commit any of the material to memory; if I gave him a series of words or numbers, which I read slowly and distinctly, he would listen attentively, sometimes ask me to stop and enunciate a word more clearly, or, if in doubt whether he had heard a word correctly, would ask me to repeat it. Usually during an experiment he would close his eyes or stare into space, fixing his gaze on one point; when the experiment was over, he would ask that we pause while he went over the material in his mind to see if he had retained it. Thereupon, without another moment's pause, he would reproduce the series that had been read to him.*
>
> A. R. Luria (1987), *The Mind of a Mnemonist*, pp. 9–10

S's phenomenal memory did not always serve him well, however. He had difficulty ridding his mind of the trivial information that he tended to focus on, sometimes to the point of incapacitation. As Luria put it:

> *Thus, trying to understand a passage, to grasp the information it contains (which other people accomplish by singling out what is most important) became a tortuous procedure for S, a struggle against images that kept rising to the surface in his mind. Images, then, proved an obstacle as well as an aid to learning in that they prevented S from concentrating on what was essential. Moreover, since these images tended to jam together, producing still more images, he was carried so far adrift that he was forced to go back and rethink*

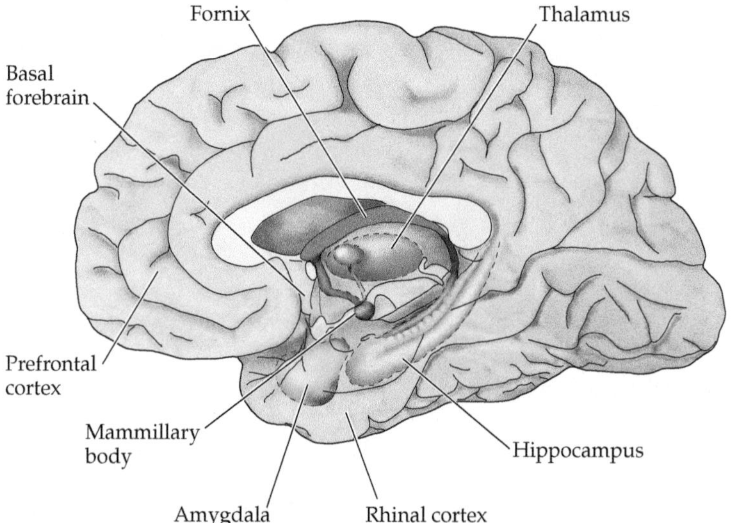

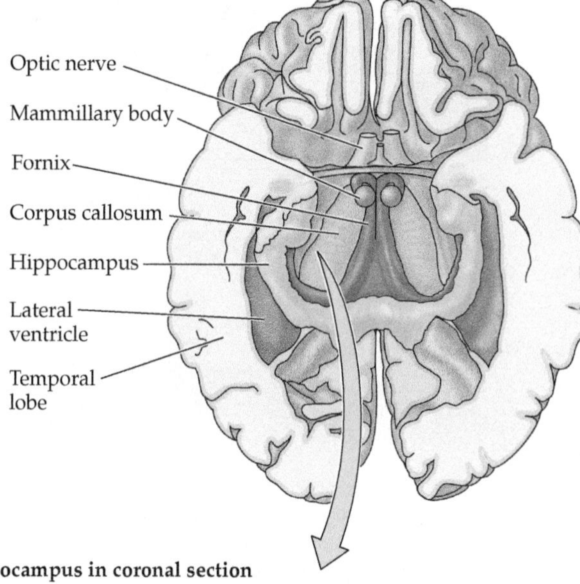

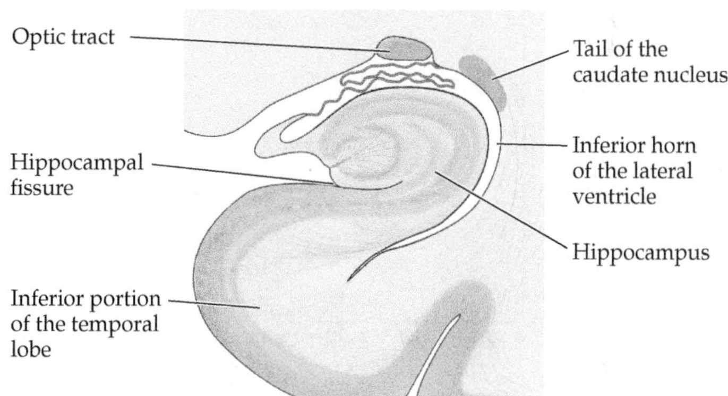

FIGURE 30.9 Brain areas that support declarative memory. By inference from the results of damage to these structures, declarative memory is based on their physiological activity. (A) Studies of amnesic patients have shown that the formation of declarative memories depends on the integrity of the hippocampus and its subcortical connections to the mammillary bodies and dorsal thalamus. (B) Location of the hippocampus as seen in a cutaway view in the horizontal plane. (C) The hippocampus as it would appear in a histological section in the coronal plane, at approximately the level indicated by the line in (B).

> *the entire passage. Consequently, a simple passage—a phrase, for that matter—would turn out to be a Sisyphean task.*
>
> *Ibid.*, p. 113

Luria's patient presumably represents one extreme of a continuum. A number of otherwise normal individuals have what has come to be referred to as *superior autobiographical memory*, the best known of whom is the television actress Marilu Henner. Although not negatively afflicted like Luria's patient, these individuals remember far more details about their daily lives than most of us. Even if years in the past, Henner and others can apparently remember the day of the week pertinent to a given date and much of what occurred that day.

Amnesia

Although forgetting is a normal (and essential) process, it can also be pathological, a condition called **amnesia**. An inability to establish new memories following neurological insult is called **anterograde amnesia**, whereas difficulty retrieving memories established prior to the precipitating neuropathology is called **retrograde amnesia**. Anterograde and retrograde amnesia are often present together, but can be dissociated under various circumstances. Amnesias following bilateral lesions of the temporal lobe and diencephalon have given particular insight into where and how at least some categories of memory are formed and where they are stored, as discussed in the following section.

Brain Systems Underlying Declarative Memory Acquisition and Storage

Several extraordinary clinical cases of amnesia have been especially revealing about the neural systems responsible for the short-term storage and consolidation of declarative information. Taken together, these cases provide dramatic evidence of the importance of midline diencephalic and medial temporal lobe structures—the hippocampus, in particular—in establishing new declarative memories (Figure 30.9). These patients also demonstrate that there

(A)

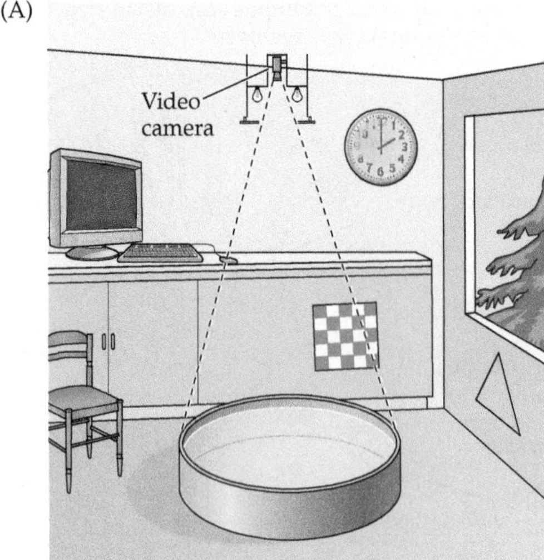

(B)

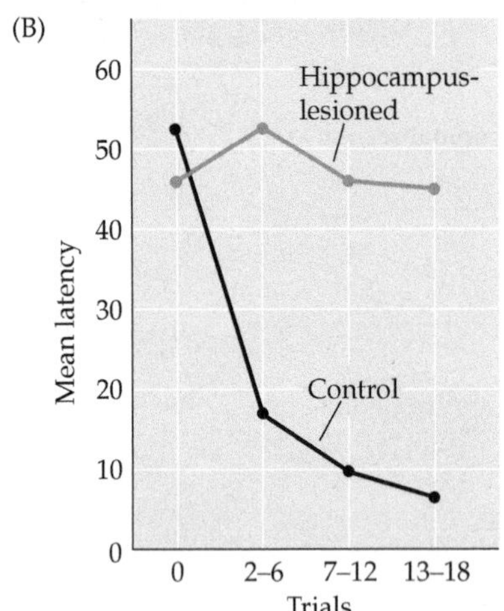

(C) Control rat

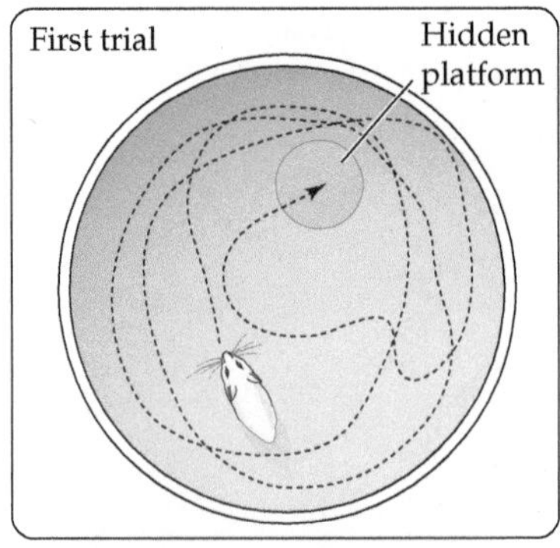

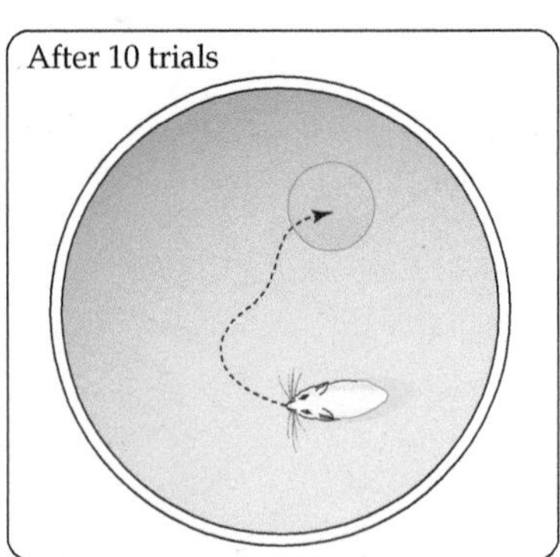

(D) Rat with hippocampal lesions

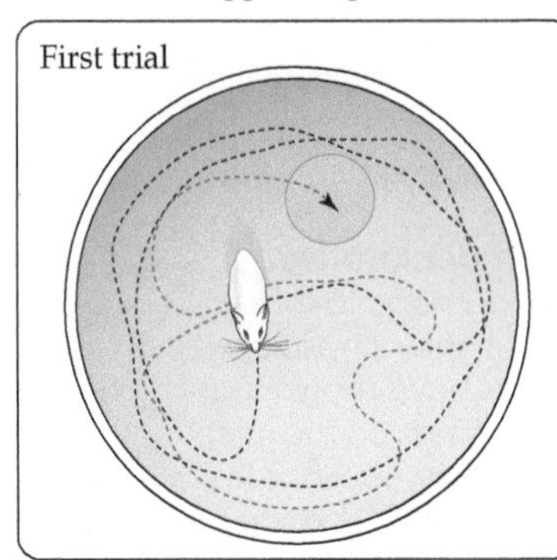

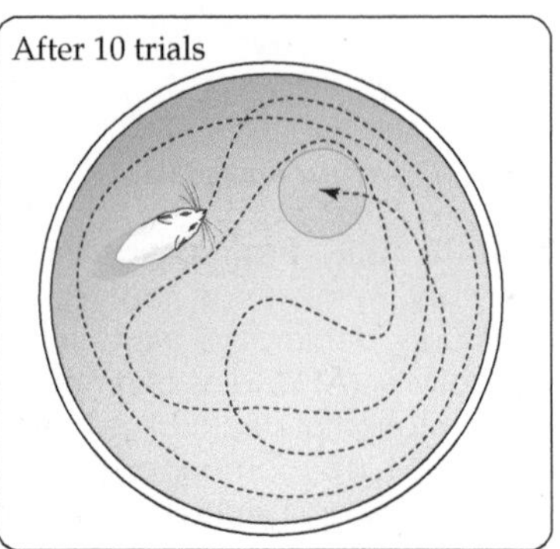

FIGURE 30.10 Spatial learning and memory in rodents depends on the hippocampus. (A) Rats are placed in a circular tank about the size and shape of a child's wading pool, filled with opaque (milky) water. The surrounding environment contains visual cues such as windows, doors, a clock, and so on. A small platform is located just below the surface. As rats search for this resting place, the pattern of their swimming (indicated by the traces in C and D) is monitored by a video camera. (B) After a few trials, normal rats rapidly reduce the time required to find the platform, whereas rats with hippocampal lesions do not. Sample swim paths of normal (C) and hippocampus-lesioned (D) rats on the first and tenth trials. Rats with hippocampal lesions are unable to remember where the platform is located. (B after Eichenbaum, 2000; C,D after Schenk and Morris, 1985.)

is a different anatomical substrate for anterograde and retrograde amnesia, since memories for events and other information acquired before the brain damage they suffered was largely retained. Thus, this sort of injury produces primarily anterograde amnesia.

Studies of animals with lesions of the medial temporal lobe have largely corroborated these findings in human patients. For example, one test of the presumed equivalent of declarative memory formation in animals involves placing rats into a pool filled with opaque water, thus concealing a submerged platform. Surrounding the pool are prominent visual landmarks (Figure 30.10). Normal rats at first search randomly until they find the submerged platform. After repeated testing, however, they learn to swim directly to the platform no matter where they are initially placed in the pool by orienting to the landmarks. Rats with lesions of the hippocampus and nearby structures cannot learn to find the platform, suggesting that remembering its location relative to visual landmarks depends on the same neural structures critical to declarative memory formation in humans. Likewise, destruction of the hippocampus and parahippocampal gyrus in monkeys severely impairs their ability to perform delayed response tasks. These studies suggest that primates and other mammals depend on medial temporal structures to encode and initiate the consolidation of memories of events, just as humans use these same brain regions for the initial encoding and consolidation of declarative memories.

Consistent with the evidence from studies of humans and other animals with lesions to the medial temporal lobe—in particular, to the hippocampus and parahippocampal cortex—recent studies have shown that neurons in these areas are selectively recruited by tasks that involve declarative memory. For example, neuroimaging studies using positron emission tomography show increased metabolism in the hippocampus of human subjects studying information they would later be asked to recall. Studies using fMRI have also shown that the hippocampus and parahippocampal gyrus are activated in human subjects studying a list of items to be remembered. Moreover the amount of activity measured in these areas was higher for items that subjects subsequently remembered compared with the activity measured for items they later forgot (Figure 30.11).

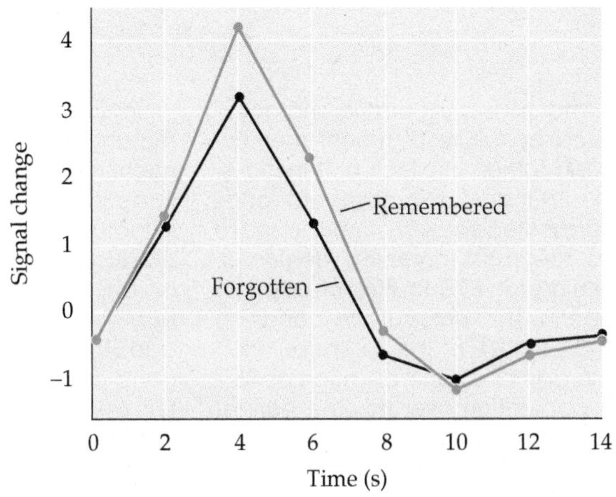

FIGURE 30.11 Activation of the hippocampus and adjacent parahippocampal cortex predicts memory performance. Activation in these areas was much stronger for items that were later remembered. (After Wagner et al., 1998.)

Another example of the importance of medial temporal lobe structures in the formation and consolidation of declarative memories is cab drivers. Anyone who has ridden a taxi in a large city can appreciate the difficulty of negotiating the labyrinth of streets to arrive at a specified destination. A much discussed study showed that the posterior hippocampus, which appears to be particularly useful in remembering spatial information, is larger in London taxi drivers than in age-matched control subjects (Figure 30.12A). Confirming the role of experience in performance, the size of the posterior hippocampus in cab drivers scales positively with the number of months spent driving a cab (Figure 30.12B). Together, such findings support the idea that neuronal activation within the hippocampus and closely allied cortical areas of the medial temporal lobe largely determines the transfer of declarative information into long-term memory, and that the robustness with which such memories are encoded depends on structural and functional changes of neural connections that occur as a result of experience.

In contrast, retrograde amnesia—the loss of memory for events preceding an injury or illness—is more typical of the generalized lesions associated with head trauma and neurodegenerative disorders, such as Alzheimer's disease (Box 30C). Although a degree of retrograde amnesia can occur with the more focal lesions that cause anterograde amnesia, the long-term storage of memories is presumably distributed throughout the brain (see the following sections). Thus, the hippocampus and related diencephalic structures indicated in Figure 30.9 are critical for the initial formation and consolidation of declarative memories that are ultimately stored elsewhere. Together, these observations have lead researchers to think of the hippocampus as providing a *cognitive map*, an idea that is consistent with studies in rodents (Box 30D).

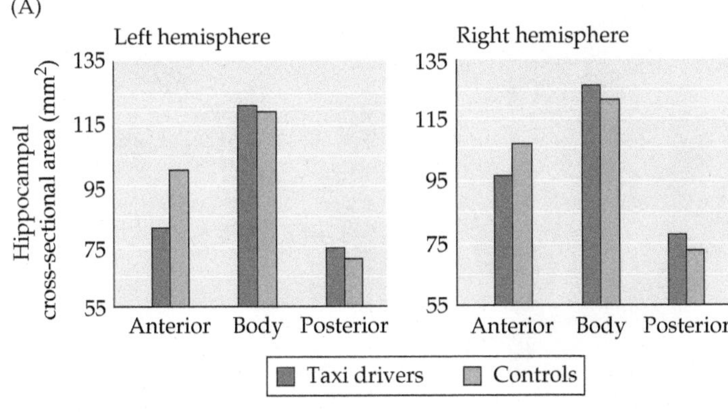

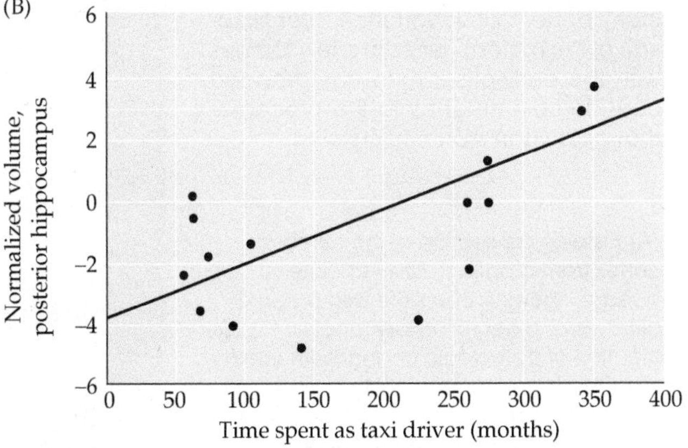

FIGURE 30.12 The hippocampus in London taxi drivers. (A) Structural brain scans show that the posterior hippocampus, a region specialized for remembering spatial information, is larger in taxi drivers than in age-matched controls. (B) Hippocampus size scales positively with experience as a cabbie. (After Maguire et al., 2000.)

BOX 30C ■ Alzheimer's Disease

Dementia is a syndrome characterized by failure of recent memory and other intellectual functions. It is usually insidious in onset but tends to progress steadily. Alzheimer's disease (AD) is the most common dementia, accounting for 60% to 80% of cases in the elderly. This unfortunate condition afflicts 5% to 10% of the U.S. population over the age of 65 and as much as 45% of the population over 85. The earliest signs are an impairment of recent memory function and attention, followed by failure of language skills, visual-spatial orientation, abstract thinking, and judgment. Alterations of personality inevitably accompany these defects.

A tentative diagnosis of AD is based on these characteristic clinical features and can be confirmed only by the distinctive cellular pathology evident on post mortem examination of the brain (Figure A). These histopathological changes consist of three principal features: (1) collections of intraneuronal cytoskeletal filaments called *neurofibrillary tangles*; (2) extracellular deposits of an abnormal protein (called amyloid) in so-called *senile plaques*; and (3) a diffuse loss of neurons. These changes are most apparent in neocortex, limbic structures (hippocampus, amygdala, and their associated cortices), and some brainstem nuclei (typically, the basal forebrain nuclei) (Figure B).

The vast majority of AD cases are "late-onset," arising after age 60 without an obvious cause. In contrast, relatively rare early-onset forms appear in middle life and are caused by monogenic defects consistent with an autosomal dominant pattern of inheritance. Identification of the mutant genes in a few families with the early-onset form has provided considerable insight into the processes that go awry in AD.

Investigators long suspected that a mutant gene responsible for familial AD might reside on chromosome 21, primarily because clinical and neuropathologic features similar to AD often occur in individuals with Down syndrome (caused by an extra copy of chromosome 21), but with a much earlier onset (about age 30, in most cases). A mutation of the gene encoding amyloid precursor protein (APP) emerged as an attractive candidate both because of the prominent amyloid deposits in AD together with isolation of a fragment of APP, Aβ peptide, from amyloid plaques. The gene that encodes APP was subsequently cloned by Dmitry Goldgaber and his colleagues and found to reside on chromosome 21. This discovery eventually led to the identification of mutations of the *APP* gene in almost 20 families with the early-onset, autosomal dominant form of AD. It should be noted, however, that only a few of the early-onset families (and none of the late-onset families) exhibited these particular mutations.

The mutant genes underlying two additional autosomal dominant forms of AD have been subsequently identified as *presenilin 1* and *presenilin 2*. Mutations of these two genes modify processing of APP and result in increased amounts of a particularly toxic form of Aβ peptide, Aβ42. Thus, mutation of any one of several genes appears to be sufficient to cause a heritable form of AD, and these converge on abnormal processing of APP.

In the far more common late-onset form of AD, the disease is clearly not inherited in any simple sense (although the relatives of affected individuals are at a greater risk, for reasons that are not clear). The central role of APP in the families with early-onset forms of the disease nonetheless suggested that APP might be linked to the chain of events culminating in the sporadic forms of AD. Biochemists Warren Strittmatter and Guy Salvesen theorized that pathologic deposition of proteins complexed with Aβ peptide might be responsible.

To test this idea, Strittmatter and Salvesen immobilized Aβ peptide on nitrocellulose paper and searched for proteins in the cerebrospinal fluid of patients with AD that bound with high affinity. One of the proteins they detected was apolipoprotein E (ApoE), a molecule that normally chaperones cholesterol through the bloodstream. This discovery was especially provocative in light of another discovery, this one made by Margaret Pericak-Vance, Allen Roses, and their colleagues, who found that affected members of some families with the late-onset form of AD exhibited an association with genetic

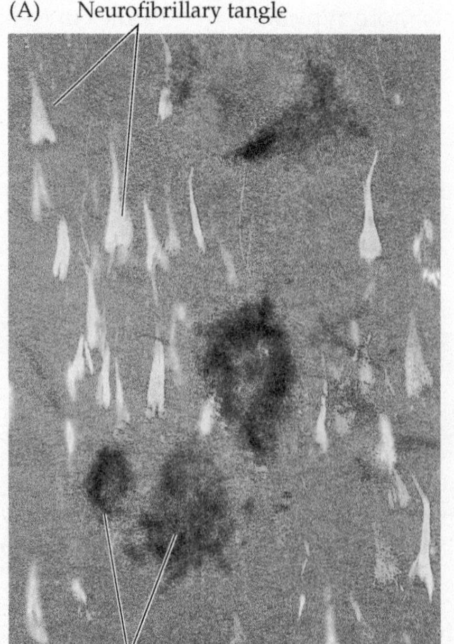

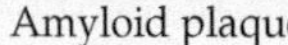

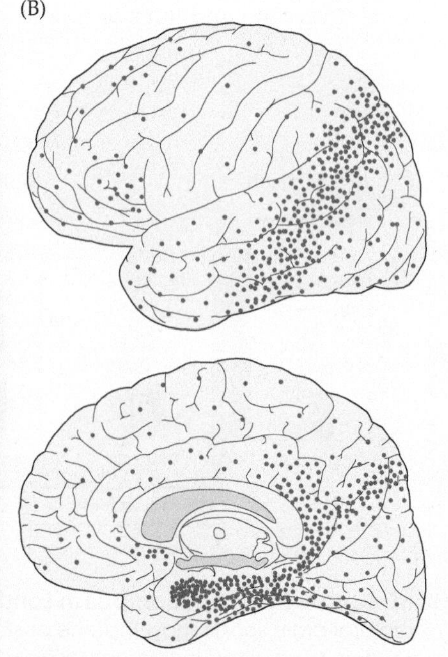

(A) Histological section of the cerebral cortex from a patient with Alzheimer's disease, showing characteristic amyloid plaques and neurofibrillary tangles. (B) Distribution of pathologic changes (including plaques, tangles, neuronal loss, and gray matter shrinkage) in Alzheimer's disease. Dot density indicates severity of pathology. (A courtesy of Gary W. Van Hoesen; B after Blumenfeld, 2002, based on Brun and Englund, 1981.)

BOX 30C ■ (continued)

markers on chromosome 19. This finding was of particular interest because a gene encoding an isoform of ApoE is located in the same region of chromosome 19 implicated by the association studies. As a result, these researchers began to explore the relationship of the different alleles of *ApoE* with individuals with a sporadic, late-onset form of AD.

There are three major alleles of *ApoE*: *e2*, *e3*, and *e4*. The frequency of allele *e3* in the general population is 0.78, and the frequency of allele *e4* is 0.14. The frequency of the *e4* allele in late-onset AD patients, however, is 0.52—almost four times higher than in the general population. Thus, the inheritance of the *e4* allele is a risk factor for late-onset AD. In fact, people homozygous for *e4* are about eight times more likely to develop AD compared with individuals homozygous for *e3*. Among individuals with no copies of *e4*, only 20% develop AD by age 75, compared with 90% of individuals with two copies of *e4*.

In contrast to the mutations of *APP* or *presenilin 1* and *presenilin 2* that cause early-onset familial forms of AD, inheriting the *e4* form of ApoE is not sufficient to cause AD; rather, inheriting this gene simply increases the risk of developing AD. The cellular and molecular mechanisms by which the *e4* allele of ApoE increases susceptibility to late-onset AD are not understood, and elucidating these mechanisms is clearly an important goal.

Clearly, AD has a complex pathology and probably reflects a variety of related molecular and cellular abnormalities. So far, the most apparent common denominator seen in this complex disease is abnormal APP processing. In particular, accumulation of the toxic Aβ42 peptide is thought to be a key factor. This conclusion has led to efforts to develop therapies aimed at inhibiting formation or facilitating clearance of this toxic peptide. It is unlikely that this important problem will be understood without a great deal more research, the hyperbole in the lay press notwithstanding.

BOX 30D ■ Place Cells and Grid Cells

Nearly 70 years ago psychologist Edward Tolman suggested that the brain must possess a cognitive map that represents remembered places in the environment. In 2014 the Nobel Prize in Physiology or Medicine was awarded to three neurophysiologists (John O'Keefe, May-Britt Moser, and Edvard Moser) for their studies of navigation in rodents that confirmed Tolman's idea.

This work began in the late 1960s with O'Keefe's observation that some neurons in the rat hippocampus fire robustly when and only when freely moving animals in an arena occupy a specific place (Figure A). Based on a series of studies by O'Keefe and his colleagues, it became clear that the activity of different combinations of these *place cells* constituted the sort of learned cognitive map that Tolman had imagined. Although recording from neurons in behaving animals is now widely practiced, O'Keefe's group was one of the pioneers in this methodology.

The Mosers (husband and wife) added to this understanding of animal navigation by further exploring the activity

(A)

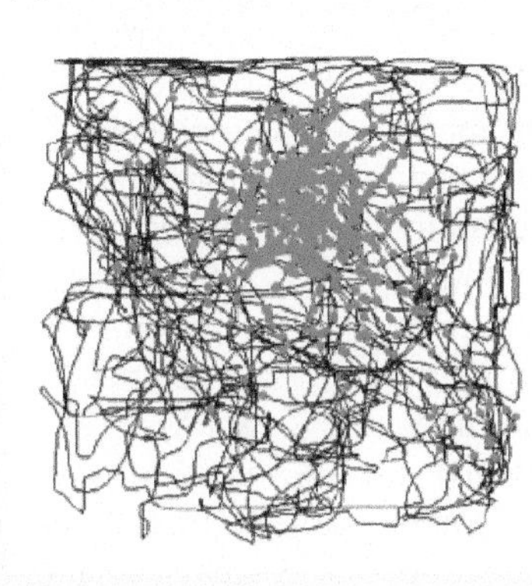

(B)

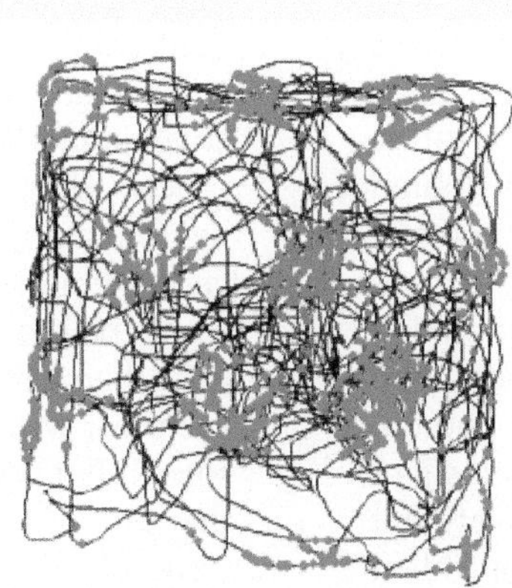

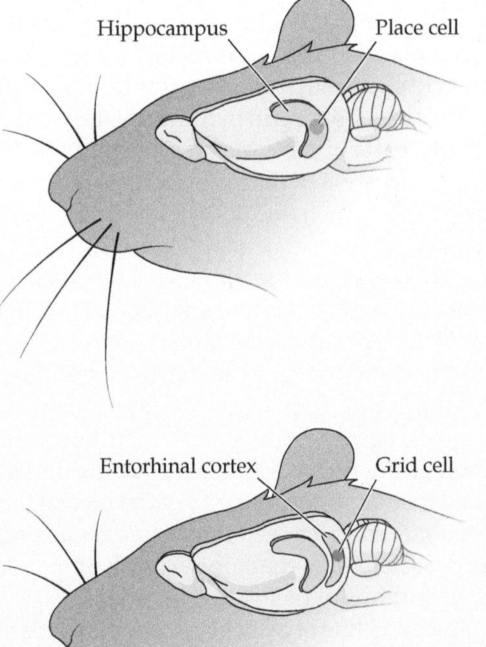

(A) Hippocampal place cells. The picture on the right shows the location of the hippocampus in a rat brain; the panel on the left illustrates place cells (orange dots) that fire only when the rat traverses a specific locus as it moves in an arena. (B) Entorhinal grid cells. The picture on the right shows the location of the entorhinal cortex adjacent to the hippocampus; the panel on the left represents the activity of a single grid cell as the rat traverses points in the arena that form a hexagonal grid (From Moser et al., 2008.).

Continued on the next page

BOX 30D ■ (*continued*)

of neurons in the entorhinal cortex, a region adjacent to the hippocampus and whose neurons project to it. They found that cells in this area also code for place, but other neurons they named grid cells showed quite different patterns of activity. Remarkably, each of these grid cells fired when the rat was in multiple loci that formed a hexagonal grid (Figure B), thus mapping every point in the arena over distances ranging from centimeters to a few meters. The implied significance of *grid cells* is providing measurements of the arena which place cells in the hippocampus then use to couple environmental cues with distance and direction.

Although the link from these findings in rats and mice to human hippocampal and entorhinal functions remains largely speculative, it is not difficult to see how the learning deficits seen in H. M. and other patients with medial temporal lobe damage are related to these basic studies. Indeed, recordings from hippocampal neurons in patients undergoing surgery to relieve intractable epilepsy support this connection.

Sites of Long-Term Memory Storage

Revealing though they have been, clinical studies of amnesic patients have provided relatively little insight into the long-term storage of declarative information in the brain, other than to indicate that such information is *not* stored in the midline diencephalic and medial temporal lobe structures that are affected in anterograde amnesia. A good deal of evidence accumulated over the years implies that the cerebral cortex is the major long-term repository for many aspects of declarative memory.

The idea that memory traces are distributed over the cortex began with the work of the American neuroscientist Karl Lashley in the 1920s. Lashley made cuts that disconnected various regions of the cortex in rats, performing this procedure either before or after the animals had learned to run mazes of varying difficulty. When the cuts he made failed to show much effect on the animals' memory of how to find the food reward in the maze, he went on to actually remove parts of the cortex (Figure 30.13A). Lashley found that the location of the lesions did not matter much—only the extent of the tissue destruction and the difficulty of the task seemed consequential (Figure 30.13B,C). He summarized his findings in terms of what he called the *mass action principle,* which states that any degradation in learning and memory depends on the amount of cortex destroyed; and that the more complex the learning task, the more disruptive the lesion. Only when the damage is widespread does network performance show a significant decline. These findings imply that, whereas acquiring declarative memories depends on the integrity of the medial temporal lobes, storing them over the long term depends on distributed cortical networks that are seriously impaired only when large portions of them are destroyed.

A second line of evidence supporting this interpretation comes from patients with severe depression who undergo electroconvulsive therapy (ECT). The passage of enough electrical current through the brain causes the equivalent of a full-blown seizure. This remarkably useful treatment (which is performed under anesthesia in well-controlled circumstances) was discovered because depression in epileptics often remitted after a spontaneous seizure. However, ECT often causes both anterograde and retrograde amnesia. Patients typically do not remember the treatment itself or the events of the preceding days, and their recall of events over the previous 1 to 3 years can be affected. Animal studies (e.g., rats tested for maze learning) have confirmed the amnesic consequences of ECT. To mitigate this side effect (which may be the result of excitotoxicity and tends to resolve over a few months), ECT is often delivered to only one hemisphere at a time. The nature of amnesia following ECT supports the conclusion that long-term declarative memories are widely stored in the cerebral cortex.

Another sort of evidence comes from patients with damage to regions of the association cortex. Since different cortical regions have significantly different if overlapping functions (see Chapter 27), it is not surprising that these sites store information that reflects their function. For example, the region that links speech sounds and their symbolic significance is located in the association cortex of the superior temporal lobe, and damage to this area typically results in an inability to link words and meanings (Wernicke's aphasia; see Chapter 33). Presumably, the widespread connections of the hippocampus to the language areas serve to transfer declarative information to these and other language-related cortical sites (Figure 30.14). By the same token, the inability of patients with temporal lobe lesions to remember and thus recognize objects and/or faces suggests that such memories are stored in these cortical sites.

One more line of support for the hypothesis that declarative memories are stored in cortical areas specialized for processing particular types of information comes from neuroimaging of human subjects during the recollection of vivid memories. In one such study, subjects first examined words paired with either pictures or sounds. Their brains were then scanned while they were asked to recall whether each test word was associated with either a picture or a sound. Functional images based on these scans showed that the cortical areas activated when subjects viewed pictures or heard sounds were reactivated when these percepts were vividly recalled. In fact, this sort of reactivation can be quite specific. Thus, different classes of visual images—for

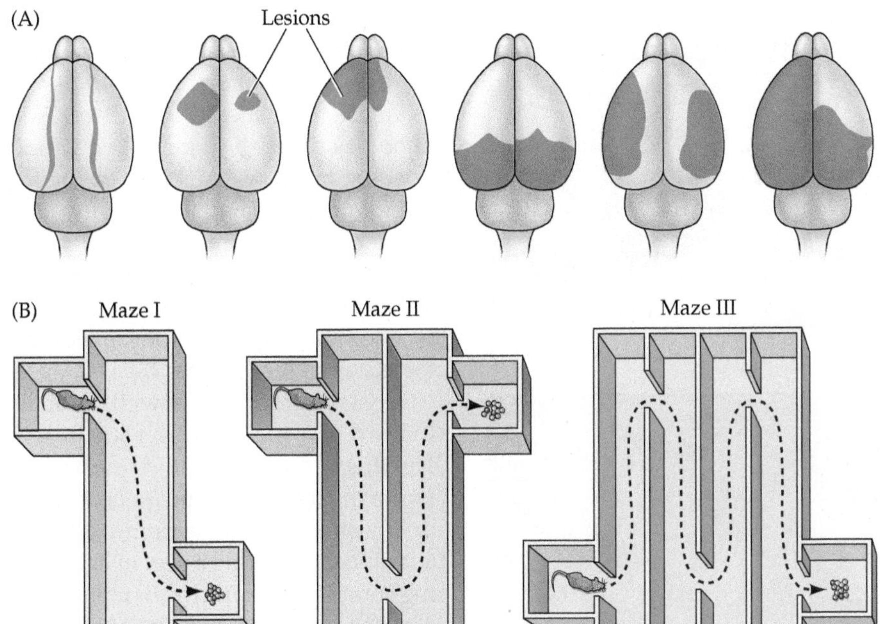

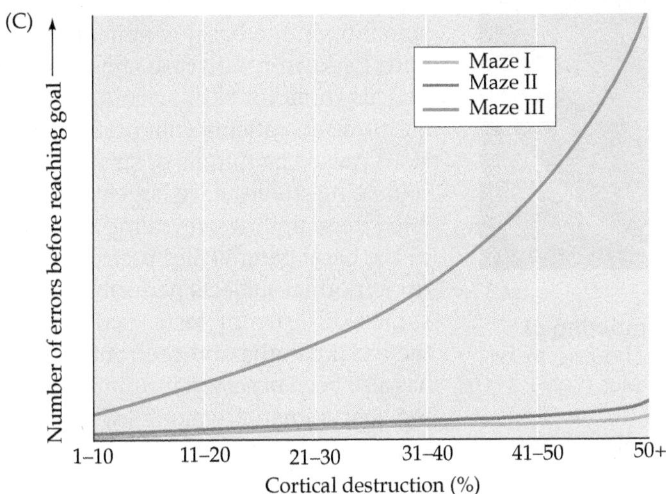

FIGURE 30.13 Lashley's experiments in search of the engram. (A) Lesions of varying size and location (red) were made in rat brains either before or after the animals had learned to run mazes (B) of varying complexity. (C) The reduction in learning Lashley observed was proportional to the amount of tissue destroyed; the lesion locations appear inconsequential. The more complex the learning task, however, the more the lesions affected performance. (After Lashley and Wiley, 1933; Lashley, 1944; and the University of Rome Psychology Lab Website.)

example, faces, houses, or chairs—tend to reactivate the same regions of the visual association cortex that were activated when the objects were actually perceived (Figure 30.15).

Finally, whereas the ability of patients such as H. M. to remember facts and events from the period of their lives preceding their lesions demonstrates that the medial temporal lobe is not necessary for retrieving declarative information held in long-term memory, other studies suggest that these structures may be important for recalling declarative memories during the early stages of consolidation and storage in the cerebral cortex.

Brain Systems Underlying Nondeclarative Memory Acquisition and Storage

The fact that patients such as H. M., N. A., and R. B. had no difficulty establishing or recalling nondeclarative memories implies that such information is laid down using a different anatomical substrate from that used in declarative memory formation. Nondeclarative memory apparently involves the basal ganglia, prefrontal cortex, amygdala, sensory association cortices, and

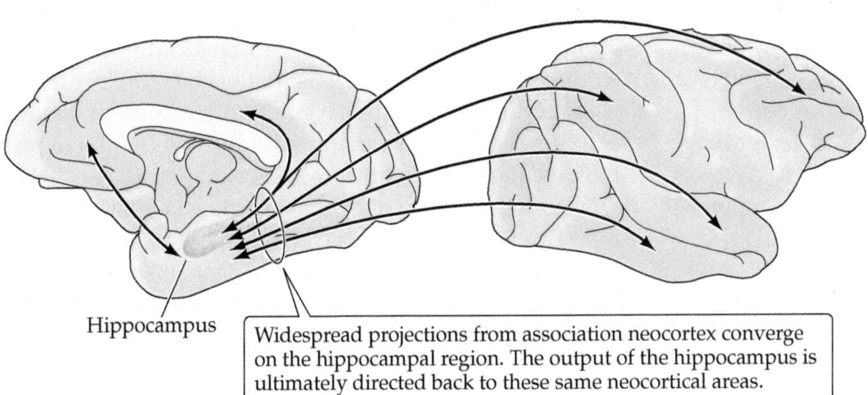

Hippocampus
Widespread projections from association neocortex converge on the hippocampal region. The output of the hippocampus is ultimately directed back to these same neocortical areas.

FIGURE 30.14 The hippocampus and possible declarative memory storage sites. The rhesus monkey brain is shown because these connections are much better documented in non-human primates than in humans. Projections from numerous cortical areas converge on the hippocampus and the related structures known to be involved in human memory; most of these sites also send projections to the same cortical areas. Medial and lateral views are shown, the latter rotated 180° for clarity. (After Van Hoesen, 1982.)

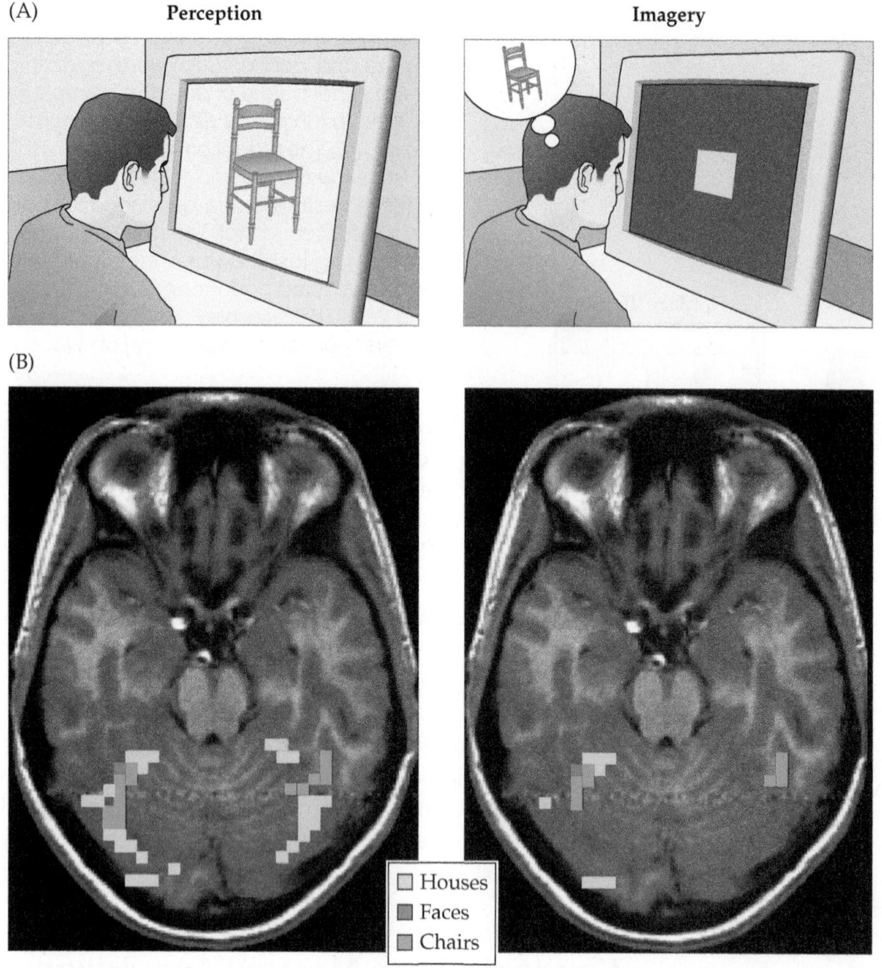

FIGURE 30.15 Reactivation of visual cortex during vivid remembering of visual images. (A) Subjects were instructed either to view images of houses, faces, and chairs (left) or to imagine the objects in the absence of the stimulus (right). (B) At left, bilateral regions of ventral temporal cortex are specifically activated during perception of houses (yellow), faces (red), and chairs (blue). At right, when subjects recall these objects, the same regions preferentially activated during the perception of each object class are reactivated. (From Buckner and Wheeler, 2001.)

cerebellum—but not the medial temporal lobe or midline diencephalon. In support of this view, perceptual priming (the unconscious influence of previously studied information on subsequent performance; see above) depends critically on the integrity of sensory association cortex. For example, lesions of the visual association cortex produce profound impairments in visual priming but leave declarative memory formation intact. Likewise, simple sensorimotor conditioning, such as learning to blink following a tone that predicts a puff of air directed at the eye, relies on the normal activation of neural circuits in the cerebellum. Ischemic damage to the cerebellum following infarcts of the superior cerebellar artery or the posterior inferior cerebellar artery causes profound deficits in classical eye-blink conditioning but does not interfere with the ability to lay down new declarative memories. Evidence from such *double dissociations* endorses the idea that relatively independent brain systems govern the formation and storage of declarative and nondeclarative memories.

The connections between the basal ganglia and prefrontal cortex appear to be especially important for complex motor learning (see Chapter 18). Damage to either structure interferes with the ability to learn new motor skills. Patients with Huntington's disease, which causes atrophy of the caudate and putamen, perform poorly on motor skill learning tests such as manually tracking a spot of light, tracing curves using a mirror, or reproducing sequences of finger movements. Because the loss of dopaminergic neurons in the substantia nigra interferes with normal signaling in the basal ganglia, patients with Parkinson's disease show similar deficits in motor skill learning (Figure 30.16), as do patients with prefrontal lesions caused by tumors or strokes. Neuroimaging studies have largely corroborated these findings, revealing activation of the basal ganglia and prefrontal cortex in normal subjects performing these same skill-learning tests. Activation of the basal ganglia and prefrontal cortex has also been observed in animals carrying out rudimentary motor learning and sequencing tasks.

The dissociation of memory systems supporting declarative and nondeclarative memory suggests the scheme for long-term information storage diagrammed in Figure 30.17. The generality of the diagram emphasizes the rudimentary state of present thinking about exactly how and where long-term memories are stored. A reasonable guess is that each complex memory is embodied in an extensive network of neurons whose activity depends on synaptic weightings that have been molded and modified by previous experience.

In sum, a variety of evidence indicates that long-term memories, whether declarative or nondeclarative, are stored throughout the brain. This conclusion, however, does not imply that individual memory traces are randomly distributed over the cortex. The current view is that memories are stored primarily within the brain regions originally involved in processing each kind of information. That is,

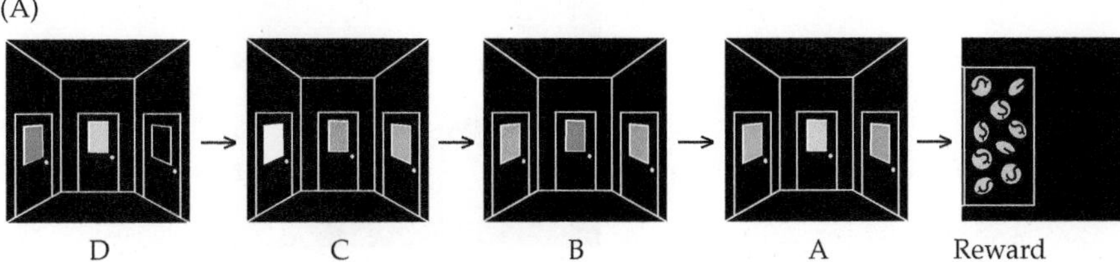

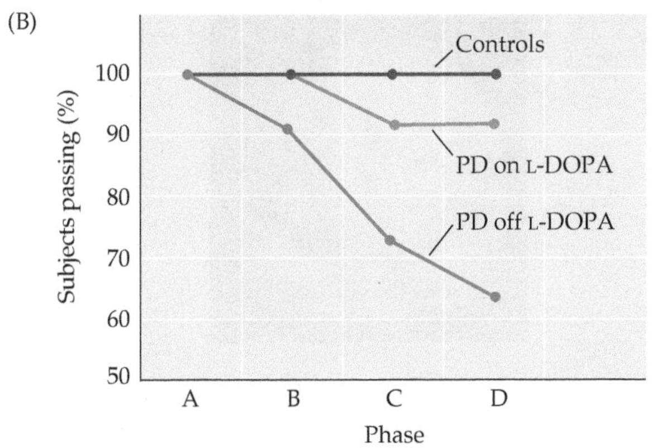

FIGURE 30.16 Parkinson's disease reveals a role for basal ganglia in nondeclarative memory. (A) Subjects performed a probabilistic learning task that had four levels. They first learned that selecting a door of one color (e.g., pink) in condition A led to a reward. Subjects then learned that selecting a differently colored door (e.g., red) in condition B would permit them to proceed to condition A, where they could select the rewarded door. This procedure was continued until subjects made choices from D → C → B → A → reward. (B) Parkinson's patients (PD) who were taking medication to replace depleted dopamine in the midbrain performed nearly as well as age-matched controls. However, Parkinson's patients who were not on dopamine replacement were impaired in their ability to learn the task. (After Shohamy et al., 2005.)

the striate and extrastriate visual cortices store memory traces for visual information, auditory cortices store memory traces for auditory information, and so on. Moreover, some brain-damaged patients are impaired in very specific semantic or object categories, such as information about animals, and some forms of memory storage have been associated with memory mechanisms in restricted brain regions, such as the localization of eye-blink conditioning in the cerebellum, or fear conditioning in the amygdala.

Memory and Aging

Although it is all too obvious that our outward appearance changes with age, most of us would like to believe that the brain is more resistant to the ravages of time. Unfortunately, the evidence suggests that this optimistic view is not justified. The average weight of the normal human brain determined at autopsy steadily decreases from early adulthood onward (Figure 30.18). In elderly individuals, this effect can also be observed with noninvasive imaging as a slight but nonetheless significant shrinkage of the brain. Counts of synapses in the cerebral cortex generally decrease in old age (although the number of neurons probably does not change very much), suggesting that it is mainly the connections between neurons that are lost as humans grow old, consistent with the idea that the networks of connections that represent memories—the engrams—gradually deteriorate.

These observations accord with the difficulty older people have in making associations (e.g., remembering names, or the details of recent experiences) and with declining scores on tests of memory as a function of age. The normal loss of some memory function with age means that clinicians must deal with a large "gray area" in distinguishing individuals subject to normal aging from those suffering from Alzheimer's disease (see Box 30C).

Just as regular exercise slows the deterioration of the neuromuscular system with age, age-related neurodegeneration and the associated cognitive decline may be slowed in elderly individuals who make a special effort to continue using the full range of both declarative and nondeclarative memory. Although cognitive decline with age is ultimately inevitable, neuroimaging studies suggest that

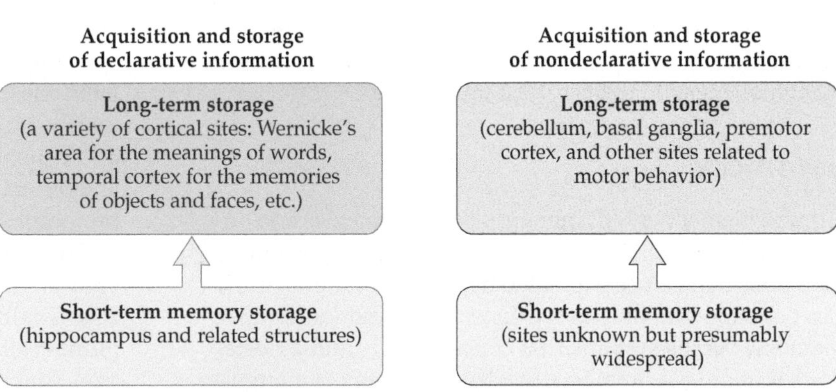

FIGURE 30.17 Summary diagram of the acquisition and storage of declarative versus nondeclarative information.

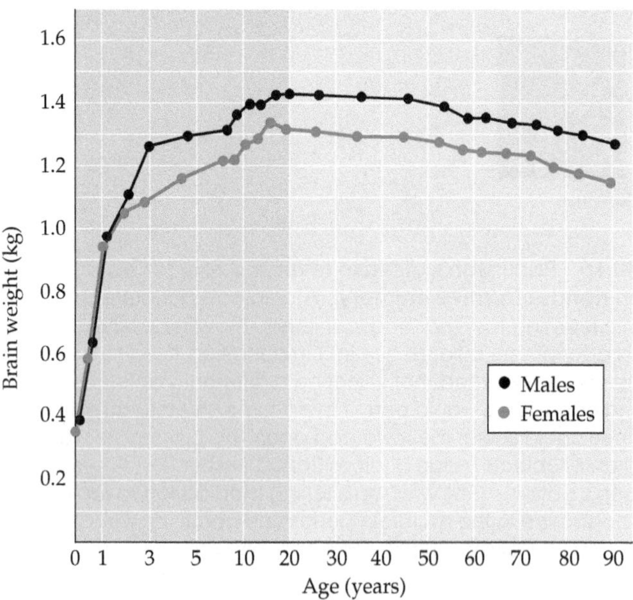

FIGURE 30.18 Brain size as a function of age. The human brain reaches its maximum size (measured by weight in this case) in early adult life and decreases progressively thereafter. This decrease evidently represents the gradual loss of neural circuitry in the aging brain, which presumably underlies the progressively diminished memory function in older individuals. (After Dekaban and Sadowsky, 1978.)

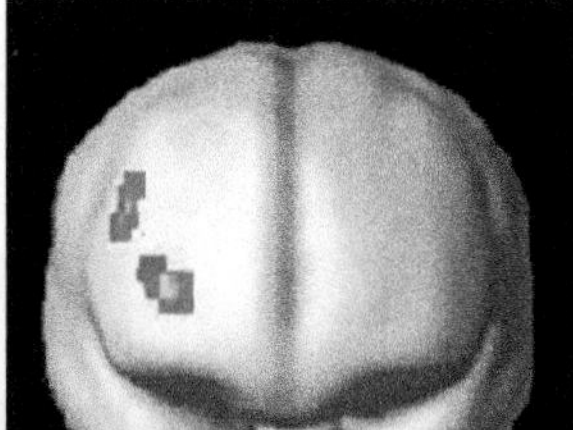

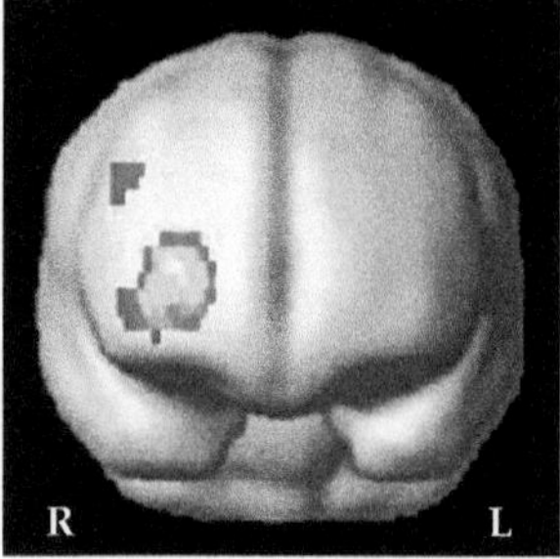

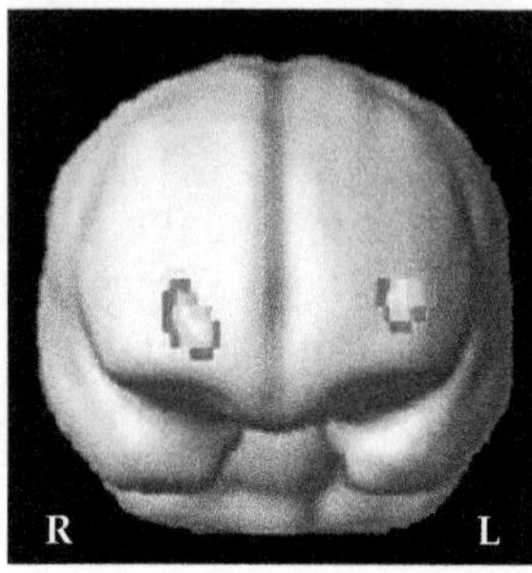

FIGURE 30.19 Compensatory activation of memory areas in high-functioning older adults. During remembering, activity in prefrontal cortex was restricted to the right prefrontal cortex (following radiological conventions, the brain images are left–right reversed) in both young participants and elderly subjects with poor recall. In contrast, elderly subjects with relatively good memory showed activation in both right and left prefrontal cortex. (After Cabeza et al., 2002.)

high-performing older adults may to some degree offset declines in processing efficacy (Figure 30.19).

The inevitable loss of some memory ability in old age raises an ethical issue that most readers will be aware of. Many commercial offerings today advertise that their "brain exercises" will forestall or reverse this effect of aging, often with statements to the effect that the promised results are backed by "neuroscientific evidence." Such claims are true in the sense that any form or physical or mental exercise is beneficial in remaining as healthy as possible as the years pass, as physicians have long known. Whether these products are any better than the mental activity that comes from any interest, hobby, or other form of mental engagement is at best unproven.

Summary

Human memory entails many biological strategies and anatomical substrates. Primary among these are a system for memories that can be expressed by means of language and can be made available to the conscious mind (declarative memory) versus systems that concern skills and associations that are essentially nonverbal, operating at a largely unconscious level (nondeclarative or procedural memory). Based on evidence from amnesic patients and knowledge about normal patterns of neural connections in the human brain, the hippocampus and associated midline diencephalic and medial temporal lobe structures are critically important in acquiring and consolidating declarative memories, although not in storing them, which occurs primarily in the cerebral cortices. In contrast, the acquisition and consolidation of nondeclarative memories for motor and other unconscious skills depend on the integrity of the premotor cortex, basal ganglia, and cerebellum, and are not affected by lesions that impair the declarative memory system. The common denominator of stored information is generally thought to be alterations in the strength and number of the synaptic connections in the cerebral cortices that mediate associations between stimuli and the behavioral responses to them, which include perceptions, thoughts, and emotions as well as motor actions.

ADDITIONAL READING

Reviews

Buckner, R. L. (2000) Neuroimaging of memory. In *The New Cognitive Neurosciences*, 2nd Edition, M. Gazzaniga (ed.). Cambridge, MA: MIT Press, pp. 817–840.

Buckner, R. L. (2002) The cognitive neuroscience of remembering. *Nature Rev. Neurosci.* 2: 624–634.

Cabeza, R. (2001) Functional neuroimaging of cognitive aging. In *Handbook of Functional Neuroimaging of Cognition*, R. Cabeza and A. Kingstone (eds.). Cambridge, MA: MIT Press, pp. 329–377.

Curtis, C. E. and M. D'Esposito (2006) Functional neuroimaging of working memory. In *Handbook of Functional Neuroimaging of Cognition*, 2nd Edition, R. Cabeza and A. Kingstone (eds.). Cambridge, MA: MIT Press, pp. 269–306.

Eichenbaum, H., A. R. Yonelinas and C. Ranganath (2007) The medial temporal lobe and recognition memory. *Annu. Rev. Neurosci.* 30: 123–152.

Erickson, C. A., B. Jagadeesh and R. Desimone (2000) Learning and memory in the inferior temporal cortex of the macaque. In *The New Cognitive Neurosciences*, 2nd Edition, M. Gazzaniga (ed.). Cambridge, MA: MIT Press, pp. 743–752.

Gallistel, C. R. and L. D. Matzel (2013) The neuroscience of learning: Beyond the Hebbian synapse. *Annu. Rev. Psychol.* 64: 169–200. doi: 10.1146/anurev-psych-113011-143807

LeDoux J. E. (2007) Consolidation: Challenging the traditional view. In, H. L. Roediger, Y. Dudai. and S. M. Fitzpatrick (eds.) *Science of Memory: Concepts*. New York: Oxford University Press, pp. 171–175.

Mishkin, M. and T. Appenzeller (1987) The anatomy of memory. *Sci. Am.* 256: 80–89.

Poldrack, R. A. and D. T. Willingham (2006) Functional neuroimaging of skill learning. In *Handbook of Functional Neuroimaging of Cognition*, 2nd Edition, R. Cabeza and A. Kingstone (eds.). Cambridge, MA: MIT Press.

Schacter, D. L. and R. L. Buckner (1998) Priming and the brain. *Neuron* 20: 185–195.

Schiller, D. and 6 others (2015) Memory and Space: Towards an Understanding of the Cognitive Map. *J. Neurosci.* 35: 13904–13911.

Squire, L. R. and B. J. Knowlton (2000) The medial temporal lobe, the hippocampus, and the memory systems of the brain. In *The New Cognitive Neurosciences*, 2nd Edition, M. Gazzaniga (ed.). Cambridge, MA: MIT Press, pp. 765–779.

Squire, L. R., C. E. Stark And R. E. Clark (2004) The medial temporal lobe. *Annu. Rev. Neurosci.* 27: 279–306.

Thompson, R. F. (2005) In search of memory traces. *Annu. Rev. Psychol.* 56: 1–23.

Important Original Papers

Cabeza, R., N. D. Anderson, J. K. Locantore and A. R. McIntosh (2002) Aging gracefully: Compensatory brain activity in high-performing older adults. *NeuroImage* 17: 1394–1402.

Clark, R. E. And L. R. Squire (1998) Classical conditioning and brain systems: The role of awareness. *Science* 280: 77–81.

Gobet, F. and H. A. Simon (1998) Expert chess memory: Revisiting the chunking hypothesis. *Memory* 6: 225–255.

Ishai, A., L. G. Ungerleider and J. V. Haxby (2000) Distributed neural systems for the generation of visual images. *Neuron* 28: 979–990.

LaBar K. S. and R. Cabeza (2006) Cognitive neuroscience of emotional memory. *Nature Rev. Neurosci.* 7: 54–64. doi: 10.1038/nrn1825

Scoville, W. B. and B. Milner (1957) Loss of recent memory after bilateral hippocampal lesions. *J. Neurol. Neurosurg. Psychiat.* 20: 11–21.

Squire, L. R. (1989) On the course of forgetting in very long-term memory. *J. Exp. Psychol.* 15: 241–245.

Wheeler, M. A., D. T. Stuss and E. Tulving (1995) Frontal lobe damage produces episodic memory impairment. *J. Internatl. Neuropsychol. Soc.* 1: 525–536.

Zola-Morgan, S. M. and L. R. Squire (1990) The primate hippocampal formation: Evidence for a time-limited role in memory storage. *Science* 250: 288–290.

Books

Baddeley, A. (1982) *Your Memory: A User's Guide*. New York: Macmillan.

Craik, F. I. M. and T. A. Salthouse (eds.) (1999) *The Handbook of Aging and Cognition*. Mahwah, NJ: Lawrence Erlbaum Associates.

Eichenbaum, H. and N. J. Cohen (2001) *From Conditioning to Conscious Recollection: Memory Systems of the Brain*. New York: Oxford University Press.

Gazzaniga, M. S. (2000) *The New Cognitive Neurosciences*, 2nd Edition. Cambridge MA: MIT Press.

Gazzaniga, M. S., R. B. Ivry and G. R. Mangun (2008) *Cognitive Neuroscience: The Biology of the Mind*, 3rd Edition. New York: W. W. Norton & Company.

Luria, A. R. (1987) *The Mind of a Mnemonist* (translated by L. Solotaroff). Cambridge, MA: Harvard University Press.

Miyake, A. and P. Shah (2001) *Models of Working Memory*. Cambridge: Cambridge University Press.

Neisser, U. (1982) *Memory Observed: Remembering in Natural Contexts*. San Francisco: W. H. Freeman.

Purves, D. and 5 others (2013) *Principles of Cognitive Neuroscience*, 2nd Edition. Sunderland, MA: Sinauer Associates.

Saper, C. B. and F. Plum (1985) *Handbook of Clinical Neurology*, Vol. 1 (45): *Clinical Neuropsychology*, P. J. Vinken, G. S. Bruyn and H. L. Klawans (eds.). New York: Elsevier, pp. 107–128.

Schacter, D. L. (2001) *The Seven Sins of Memory: How the Mind Forgets and Remembers*. Boston: Houghton Mifflin Co.

Squire, L. R. And E. R. Kandel (2000) *Memory: From Mind to Molecules*. New York: Holt.

Go to the NEUROSCIENCE 6e Companion Website at **www.oup.com/uk/Purves6e** for Web Topics, Animations, Flashcards, and more.

CHAPTER

31 Emotion

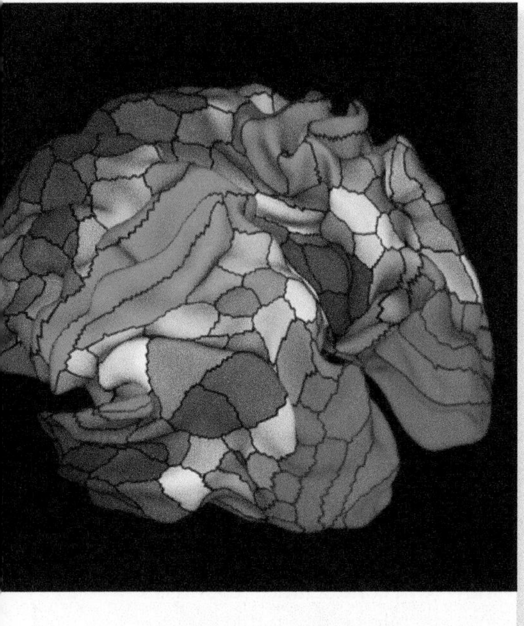

Overview

THE SUBJECTIVE FEELINGS WE CALL EMOTIONS are critically important for human behaviors, as well as those of many other animals. In addition to feelings, emotions are associated with visceral (autonomic) motor effects and stereotyped somatic motor responses, especially movements of the facial muscles and the complex of skeletal muscles underlying posture. Historically, the neural centers that coordinate emotional responses have been grouped under the rubric of the limbic system. More recently, however, other brain regions have been shown to play a role in emotional processing, including the amygdala and several cortical areas in the orbital and medial aspects of the frontal lobe. This broader constellation of cortical and subcortical regions encompasses not only the central components of the visceral motor system but regions in the forebrain and diencephalon that motivate lower motor neuron pools concerned with the somatic expression of emotional behavior. These same forebrain structures participate in a variety of complex brain functions, including the regulation of goal-directed behavior, decision making, social behavior, and even moral judgments. These brain regions are also subject to maladaptive changes due to drug abuse and factors that underlie psychiatric illnesses.

Defining Emotion

In everyday use, the word *emotion* refers to conscious feelings as varied as happiness, anger, fear, surprise, disgust, jealousy, and much more (Figure 31.1A). Because consciousness defies a clear neurobiological explanation, however, defining emotions in terms of subjective states is problematic. Among other difficulties, a definition in subjective terms implies that organisms with less self-awareness do not experience or use emotions in the same way humans do. But as Charles Darwin noted in the late nineteenth century, many non-human animal species—think of a dog—produce affective displays much like those in humans. In any event, neuroscientists today conceptualize emotional states as a composite of subjective feelings, physiological responses, and behaviors that allow humans and other animals to react adaptively to internal and external stimuli (Figure 31.1B).

Physiological Changes Associated with Emotion

The most obvious signs of emotional arousal involve changes in the activity of the visceral motor (autonomic) system (see Chapter 21). Thus, increases or decreases in heart rate, cutaneous blood flow (blushing or turning pale), skin temperature, sweating, piloerection, pupil size, and gut motility can all accompany various emotions. These responses are brought about by changes in activity in the sympathetic,

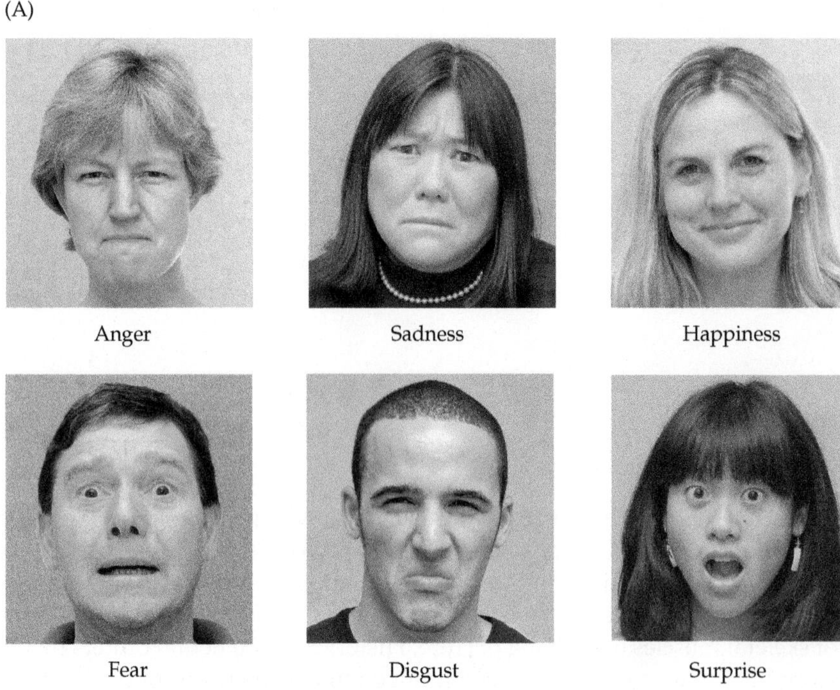

(B)

Behavior

Physiology

Emotion

Feeling

FIGURE 31.1 Emotions. (A) Facial expressions of some major emotional categories. (B) Emotions typically have three components: behavioral manifestations, a subjective feeling, and a physiological state.

parasympathetic, and enteric components of the visceral motor system, which governs smooth muscle, cardiac muscle, and glands throughout the body. As discussed in Chapter 21, Walter Cannon argued that intense activity of the sympathetic division of the visceral motor system prepares the animal to fully utilize metabolic and other resources in challenging or threatening situations. Conversely, activity of the parasympathetic division promotes a building up of metabolic reserves. Cannon further suggested that the natural opposition of the expenditure versus the storage of resources is reflected in a parallel opposition of the emotions associated with these different physiological states. As he pointed out, "The desire for food and drink, the relish of taking them, all the pleasures of the table are naught in the presence of anger or great anxiety."

Measuring the physiological variables associated with emotion is straightforward. In addition to the obvious measures of heart rate, respiratory rate, and blood pressure, two psychophysiological indices that have been particularly useful for evaluating emotional reactions are *skin conductance* and the *startle response*. **Skin conductance** is an index of sweating measured by electrodes on the palmar surface of the hands and feet. Because fear and anxiety are linked to high arousal states, the skin conductance is a good measure of these emotions, and it is widely used in "lie detector" (polygraph) tests in criminal cases. However, skin conductance is modulated by other emotions, such as sexual arousal, as well as attentional orienting responses to novel stimuli. Skin conductance has also been used to identify stimulus-evoked arousal responses that arise unconsciously. For instance, individuals with prosopagnosia often show skin conductance responses to pictures of family members, despite being unable to consciously recognize the individuals by sight.

These and other psychophysiological measures can be combined to measure two fundamental dimensions of emotion: **arousal** and **valence**, the former term referring to level of excitement and the latter referring to whether the subjective emotion is a positive or a negative one.

The Integration of Emotional Behavior

These various manifestations of emotional responses must be integrated in some way. In 1928, Philip Bard reported a series of experiments that pointed to the hypothalamus as a critical center for coordination of both the visceral and somatic motor components of emotional behavior (see Box 21A). Bard removed both cerebral hemispheres (including the cortex, underlying white matter, and basal ganglia) in a series of cats. When the anesthesia had worn off, the animals behaved as if they were enraged. The angry behavior occurred spontaneously and included the usual autonomic correlates of this emotion: increased blood

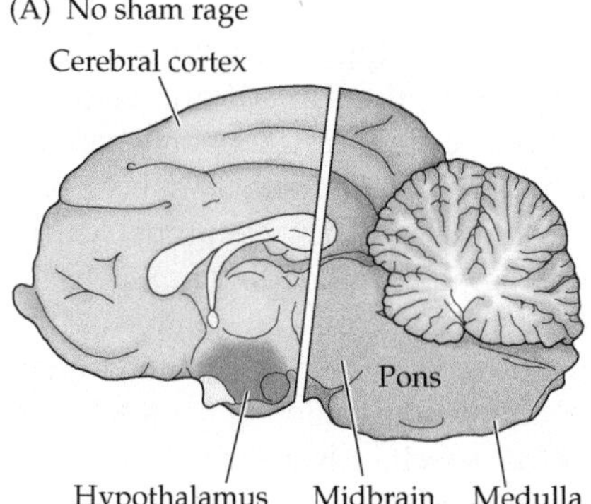

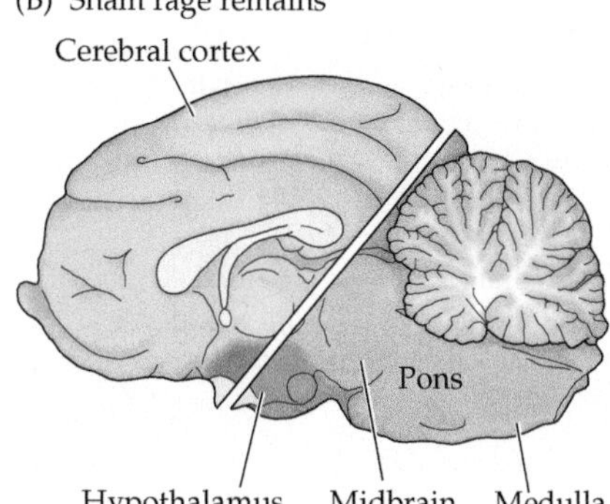

FIGURE 31.2 Midsagittal view of a cat's brain, illustrating the regions sufficient for the expression of emotional behavior. (A) Transection through the midbrain, disconnecting the hypothalamus and brainstem, abolishes sham rage. (B) The integrated emotional responses associated with sham rage survive removal of the cerebral hemispheres as long as the caudal hypothalamus remains intact. (After LeDoux, 1987.)

pressure and heart rate, retraction of the nictitating membranes (the thin connective tissue sheets associated with feline eyelids), dilation of the pupils, and erection of the hairs on the back and tail. The cats also exhibited somatic motor components of anger, such as arching the back, extending the claws, lashing the tail, and snarling. This behavior was called **sham rage** because it had no obvious target. Bard showed that a complete response occurred as long as the caudal hypothalamus was intact (Figure 31.2). Sham rage could not be elicited, however, when the brain was transected at the junction of the hypothalamus and midbrain (although some uncoordinated components of the response were still apparent). Bard suggested that whereas the subjective experience of emotion might depend on an intact cerebral cortex, the expression of coordinated emotional behaviors does not necessarily entail cortical processes. The functional importance of emotions in all mammals is consistent with the involvement of phylogenetically older parts of the nervous system. Bard also emphasized that emotional behaviors are often directed toward self-preservation, as Darwin had pointed out in his work on the evolution of emotion.

Complementary results were reported by Walter Hess, who showed that electrical stimulation of discrete sites in the hypothalamus of awake, freely moving cats could also lead to a rage response, and even to subsequent attack behavior. Moreover, stimulation of other sites in the hypothalamus caused a defensive posture that resembled fear. In 1949, a share of the Nobel Prize in Physiology or Medicine was awarded to Hess "for his discovery of the functional organization of the interbrain [hypothalamus] as a coordinator of the activities of the internal organs." Experiments like those of Bard and Hess led to the important conclusion that the basic circuits for organized behaviors accompanied by emotion are in the diencephalon and the brainstem structures connected to it. Furthermore, their work emphasized that the control of the involuntary motor system is not entirely separable from the control of the voluntary pathways, an important consideration in understanding the motor aspects of emotion, as discussed below.

The routes by which the hypothalamus and other forebrain structures influence the visceral and somatic motor systems are complex. The major targets of the hypothalamus lie in the **reticular formation**, the tangled web of nerve cells and fibers in the core of the brainstem (see Box 17A). This structure contains more than 100 identifiable cell groups, including some of the nuclei that control the brain states associated with sleep and wakefulness described in Chapter 28. Other important circuits in the reticular formation control cardiovascular function, respiration, urination, vomiting, and swallowing. The reticular neurons receive hypothalamic input from and feed into both somatic and autonomic effector systems in the brainstem and spinal cord. Their activity can therefore produce widespread visceral motor and somatic motor responses, often overriding reflex functions and sometimes involving almost every organ in the body (as implied by Cannon's dictum about the sympathetic preparation of the animal for fight or flight). In addition to the hypothalamus, other sources of descending projections from the forebrain to the brainstem reticular formation contribute to the expression of emotional behavior. Collectively, these additional centers in the forebrain are considered part of the limbic system and arise outside the classic motor cortical areas in the posterior frontal lobe.

The descending control of emotional expression thus entails two parallel systems that are anatomically and functionally distinct (Figure 31.3). The voluntary motor component described in detail in Chapters 16–19 comprises the classic motor areas of the posterior frontal lobe and related circuitry in the basal ganglia and cerebellum. The descending pyramidal and extrapyramidal projections from the motor cortex and brainstem ultimately convey the impulses responsible for voluntary somatic movements. In addition to the descending systems that govern volitional movements, several cortical and subcortical structures in the medial frontal lobe and ventral parts of the forebrain, including related circuitry in the ventral part of the basal ganglia and hypothalamus, give rise to separate descending projections that run parallel to the pathways of the volitional motor system. These descending projections of the

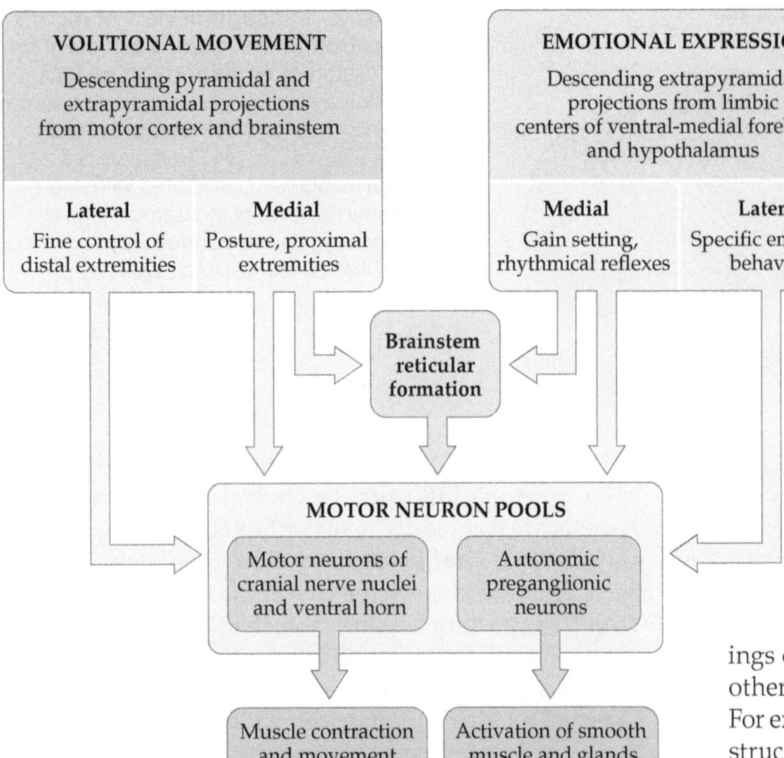

FIGURE 31.3 **Components of the nervous system that organize the expression of emotional experience.** Diagram of the descending systems that control somatic and visceral motor effectors. Motor cortical areas in the posterior frontal lobe give rise to descending projections that, together with secondary projections arising in the brainstem, are organized into medial and lateral components. As described in Chapter 17, these descending projections account for volitional somatic movements. Functionally and anatomically distinct centers in the forebrain govern the expression of non-volitional somatic motor and visceral motor functions, which are coordinated to meditate emotional behavior. Limbic centers in the ventral-medial forebrain and hypothalamus also give rise to medial and lateral descending projections. For both systems of descending projections, the lateral components elicit specific behaviors (e.g., volitional digit movements and emotional facial expressions), while the medial components support and modulate the execution of such behaviors. The descending projections of both systems terminate in several integrative centers in the brainstem reticular formation, as well as the motor neuron pools of the brainstem and spinal cord. In addition, the limbic forebrain centers innervate components of the visceral motor system that govern preganglionic autonomic neurons in the brainstem and spinal cord.

medial and ventral forebrain terminate on visceral motor centers in the brainstem reticular formation, preganglionic autonomic neurons, and certain somatic premotor and motor neuron pools that also receive projections from volitional motor centers. The two types of facial paresis illustrated in Box 31A underscore this dual nature of descending motor control.

In summary, the somatic and visceral activities associated with emotional behavior are mediated by the activity of both the somatic and visceral motor neurons, which integrate parallel, descending inputs from a constellation of forebrain sources. The remainder of the chapter is devoted to the organization and function of the forebrain centers that specifically govern the experience and expression of emotional behavior.

Is the Motor Activity Underlying Emotion Cause or Effect?

The concerted action of the visceral and somatic motor systems in response to the diverse brain regions that control them is, in effect, an "emotional motor system." But do the subjective feelings of an emotion initiate this motor activity, or is it the other way around? Some evidence favors the latter view. For example, if individuals are given muscle-by-muscle instructions that result in facial expressions recognizable as anger, disgust, fear, happiness, sadness, or surprise without being told which emotion they are simulating, each pattern of facial muscle activity is accompanied by specific and reproducible differences in visceral motor activity (as measured by indices such as heart rate, skin conductance, and skin temperature). Moreover, autonomic responses are strongest when the facial expressions are judged to most closely resemble actual emotional expression and are often accompanied by the subjective experience of that emotion. One interpretation of these findings is that when voluntary facial expressions are produced, signals in the brain engage not only the motor cortex but also some of the circuits that produce emotional states. Perhaps this relationship helps explain how good actors can be so convincing, and why we are adept at recognizing the difference between a contrived facial expression and the spontaneous smile that accompanies a positive emotional state (see Box 31A).

This sort of evidence indicates that a major source of emotion (but certainly not the only source) is feedback from muscles and internal organs that are activated reflexively by external circumstances. However, physiological responses can also be elicited by complex and idiosyncratic stimuli mediated by the forebrain. For example, an anticipated tryst with a lover, a suspenseful episode in a novel or film, stirring patriotic or religious music, or dishonest accusations can all lead to autonomic activation and strongly felt emotions. The neural activity evoked by such complex stimuli is relayed from the forebrain to visceral and somatic motor nuclei via the hypothalamus and brainstem reticular

BOX 31A ■ Determination of Facial Expressions

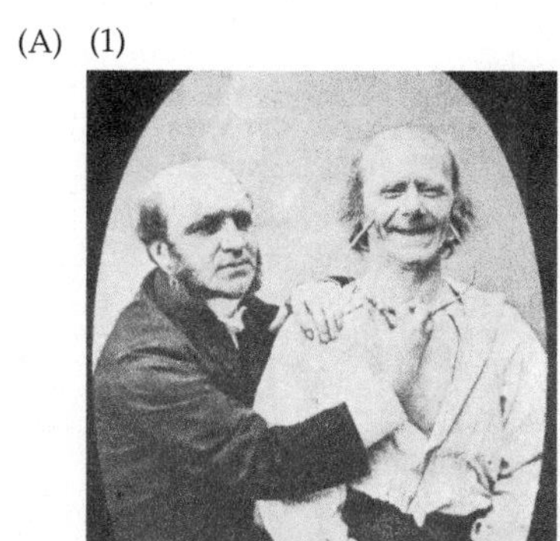

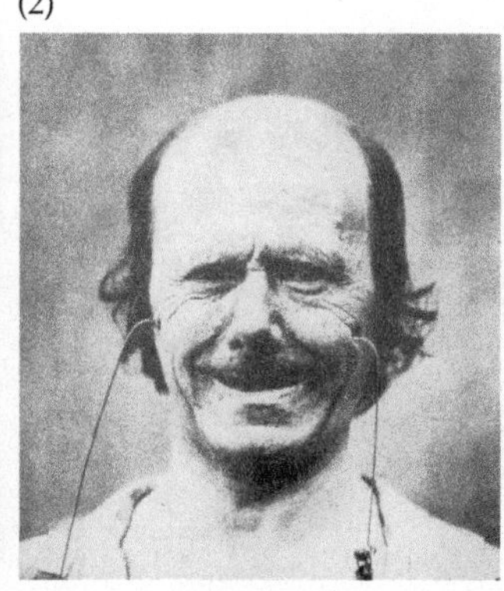

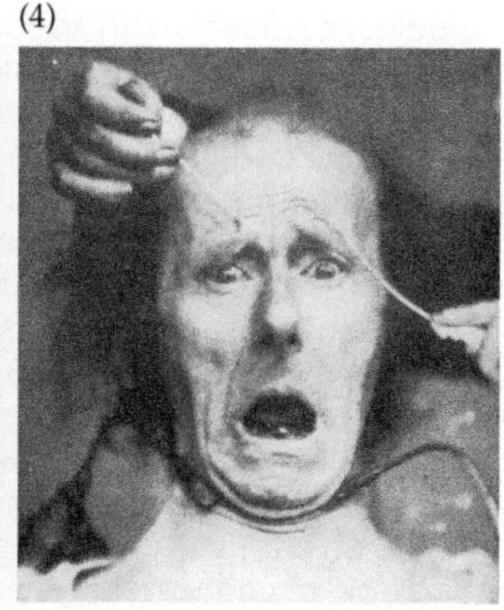

(A) Duchenne made use of early photography to study human facial expressions. (1) Duchenne and one of his subjects (a Parisian shoemaker) undergoing "faradization" of the facial muscles. (2) Bilateral electrical stimulation of the zygomaticus major mimicked a genuine expression of happiness, although closer examination shows insufficient contraction of the obicularis oculi (surrounding the eyes) compared with spontaneous laughter (3). In (4), stimulation of the brow and neck produced an expression of "terror mixed with pain, torture ... that of the damned"; however, the man reported no discomfort or emotional experience consistent with the evoked contractions. (Photos 2 and 3 from the Wellcome Library, London.)

In 1862, the French neurologist and physiologist G.-B. Duchenne de Boulogne published a remarkable treatise on facial expressions. His work was the first to systematically examine the contributions of small groups of cranial muscles to the expressions that communicate the richness of human emotion. Duchenne reasoned that "one would be able, like nature herself, to paint the expressive lines of the emotions of the soul on the face of man." In so doing, he sought to understand how the coordinated contractions of groups of muscles express distinct, pan-cultural emotional states. To achieve this goal, he pioneered the use of transcutaneous electrical stimulation (then called *faradization*, after the British chemist and physicist Michael Faraday) to activate single muscles and small groups of muscles in the face, dorsal surface of the head, and neck.

Duchenne also documented the faces of his subjects with another technological innovation: photography (Figure A). His seminal contribution was the identification of muscles and muscle groups, such as the obicularis oculi, that are not easily controlled by force of the will, but are mainly "put into play by the sweet emotions of the soul." Duchenne concluded that the emotion-driven contraction of these muscle groups surrounding the eyes, together with the zygomaticus major, conveys the genuine experience of happiness, joy, and laughter. In recognition of these insights, psychologists sometimes refer to this facial expression as the "Duchenne smile."

In typical individuals, such as the Parisian shoemaker shown here, the difference between a forced smile (produced by voluntary contraction or electrical stimulation of facial muscles) and a spontaneous (emotional) smile testifies to the convergence of descending motor signals from different forebrain centers onto premotor and motor neurons in the brainstem that control the facial musculature. In contrast to the Duchenne smile, the contrived smile of volition (sometimes called a *pyramidal smile*) is driven by the motor cortex, which communicates with the brainstem and spinal cord via the pyramidal tracts. The Duchenne smile is motivated by motor areas in the anterior cingulate gyrus that access facial motor circuitry via multisynaptic, extrapyramidal pathways through the brainstem reticular formation.

Continued on the next page

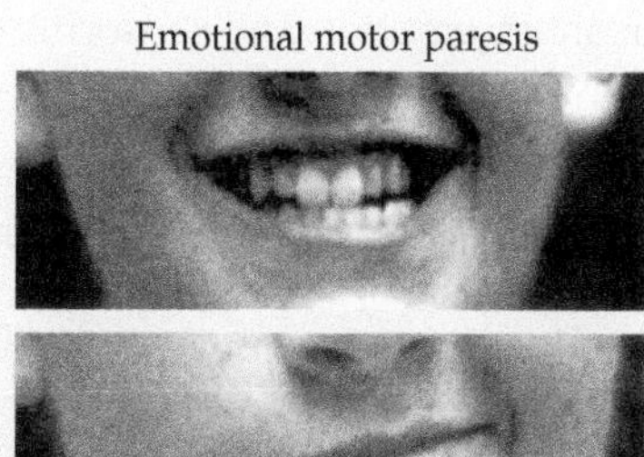
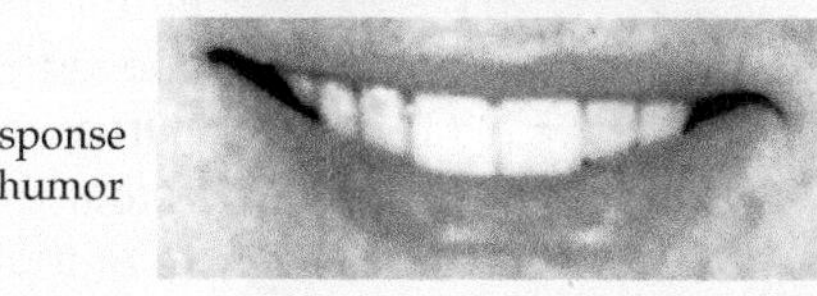

(B) Left panels: Mouth of a patient with a lesion that destroyed descending fibers from the right motor cortex displaying voluntary facial paresis. When asked to show her teeth, the patient was unable to contract the muscles on the left side of her mouth (upper left), yet her spontaneous smile in response to a humorous remark was nearly symmetrical (lower left). Right panels: Face of a child with a lesion of the left forebrain that interrupted descending pathways from non-classic motor cortical areas, producing emotional facial paresis. When the child was asked to smile volitionally, the contractions of the facial muscles were nearly symmetrical (upper right). In spontaneous response to a humorous comment, however, the right side of the child's face failed to express emotion (lower right). (Left photos from G. Holstege et al., 1996; right photos from Trosch et al., 1990.)

BOX 31A ■ *(continued)*

Studies of patients with specific neurological injury to these separate descending systems of control have further differentiated the forebrain centers responsible for control of the muscles of facial expression (Figure B). Patients with unilateral facial paralysis due to damage of descending pathways from the motor cortex (upper motor neuron syndrome; see Chapter 17) have considerable difficulty moving their lower facial muscles on one side, either voluntarily or in response to commands, a condition called voluntary facial paresis (see Figure B, left panels). Nonetheless, many such individuals produce symmetrical *involuntary* facial movements when they laugh, frown, or cry in response to amusing or distressing stimuli. In such patients, pathways from regions of the forebrain other than the classic motor cortex in the posterior frontal lobe remain available to activate facial movements in response to stimuli with emotional significance.

A much less common form of neurological injury, called *emotional facial paresis*, demonstrates the opposite set of impairments—that is, loss of the ability to express emotions by using the muscles of the face without loss of volitional control (see Figure B, right panels). Such individuals are able to produce symmetrical pyramidal smiles, but fail to display spontaneous emotional expressions involving the facial musculature contralateral to the lesion. These two systems are diagrammed in Figure C.

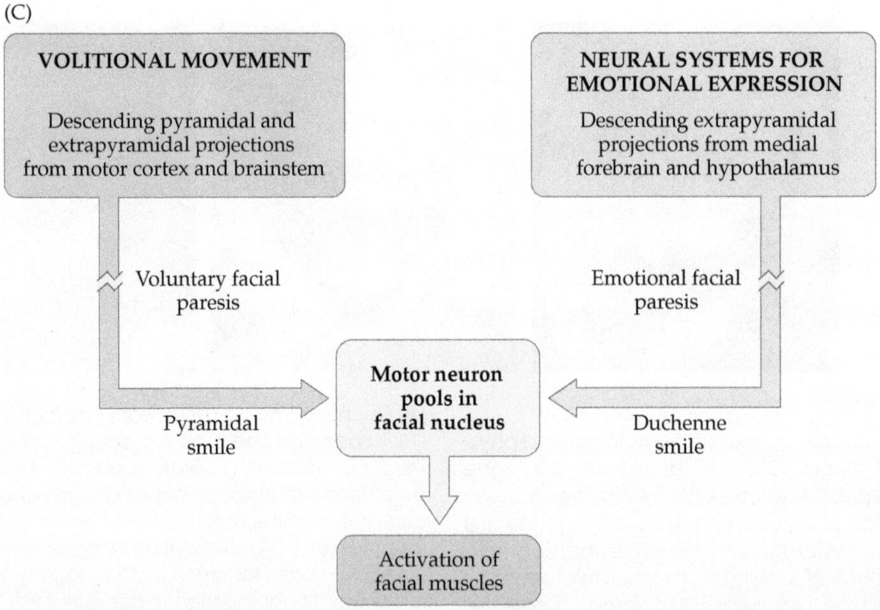

(C) The complementary deficits demonstrated in Figure B are explained by selective lesions of one of two anatomically and functionally distinct sets of descending projections that motivate the muscles of facial expression.

formation, the major structures that coordinate the expression of emotional behavior.

In short, emotion and sensorimotor behavior are inextricably linked. As William James put it more than a century ago:

> *What kind of an emotion of fear would be left if the feeling neither of quickened heart-beats nor of shallow breathing, neither of trembling lips nor of weakened limbs, neither of goose-flesh nor of visceral stirrings, were present, it is quite impossible for me to think … I say that for us emotion dissociated from all bodily feeling is inconceivable.*
>
> William James, 1893 (*Psychology*, p. 379)

The Limbic System

Attempts to rationalize the effector systems that control emotional behavior have a long history. In 1937, James Papez (pronounced *Papes*) first proposed that specific brain circuits are devoted to emotional experience and expression (much as the occipital cortex is devoted to vision, for instance). In seeking to understand what parts of the brain serve this function, he began to explore the medial aspects of the cerebral hemisphere. In the 1850s, Paul Broca used the term **limbic lobe** to refer to the part of the cerebral cortex that forms a rim (*limbus* is Latin for "rim") around the corpus callosum and diencephalon on the medial face of the hemispheres (Figure 31.4). Two prominent components of this region are the **cingulate gyrus**, which lies above the corpus callosum, and the **parahippocampal gyrus**, which lies in the medial temporal lobe. For many years, these structures, along with the olfactory bulbs, were thought to be concerned primarily with the sense of smell. Indeed, Paul Broca considered the olfactory bulbs to be the principal source of input to the limbic lobe. Papez, however, speculated that the function of the limbic lobe might be more related to emotions. He knew from the work of Bard and Hess that the hypothalamus influences the expression of emotion; he also knew, as everyone does, that emotions reach consciousness and that higher cognitive functions affect emotional behavior. Ultimately, Papez showed that the cingulate cortex and hypothalamus are interconnected via projections from the **mammillary**

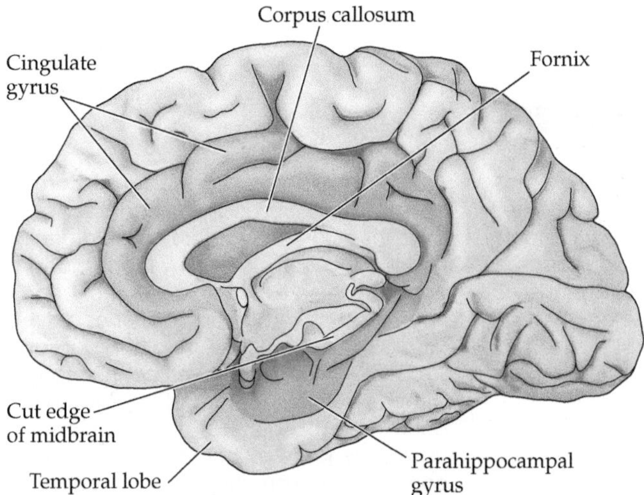

FIGURE 31.4 The limbic lobe. This constellation of regions includes the cortex on the medial aspect of the cerebral hemisphere that forms a rim around the corpus callosum and diencephalon, including the cingulate gyrus (lying above the corpus callosum) and the parahippocampal gyrus. Historically, the olfactory bulb and olfactory cortex (not illustrated here) have also been considered to be important elements of the limbic lobe.

bodies (part of the posterior hypothalamus) to the anterior nucleus of the dorsal thalamus, which projects in turn to the cingulate gyrus. The cingulate gyrus (and many other cortical regions as well) projects to the hippocampus. Finally, Papez showed that the hippocampus projects via the **fornix** (a large fiber bundle) back to the hypothalamus. Papez suggested that these pathways provided the connections necessary for cortical control of emotional expression, and they became known as the *Papez circuit.*

Over time, the concept of a forebrain circuit for the control of emotional expression, first elaborated by Papez, has been revised to include parts of the orbital and medial prefrontal cortex, ventral parts of the basal ganglia, mediodorsal nucleus of the thalamus (a different thalamic nucleus than the one emphasized by Papez), and a large nuclear mass in the temporal lobe anterior to the hippocampus, the **amygdala**. This set of structures, together with the parahippocampal gyrus and cingulate cortex, is generally referred to as the **limbic system** (Figure 31.5). Thus, some of the structures that Papez originally described (e.g., the hippocampus) now appear to have little to do with emotional behavior, whereas the amygdala, which Papez hardly mentioned, clearly plays a major role in the experience and expression of emotion (Box 31B).

About the same time that Papez proposed that these structures were important for the integration of emotional behavior, Heinrich Klüver and Paul Bucy were carrying out a series of experiments on rhesus monkeys in which they removed a large part of both medial temporal lobes, thus

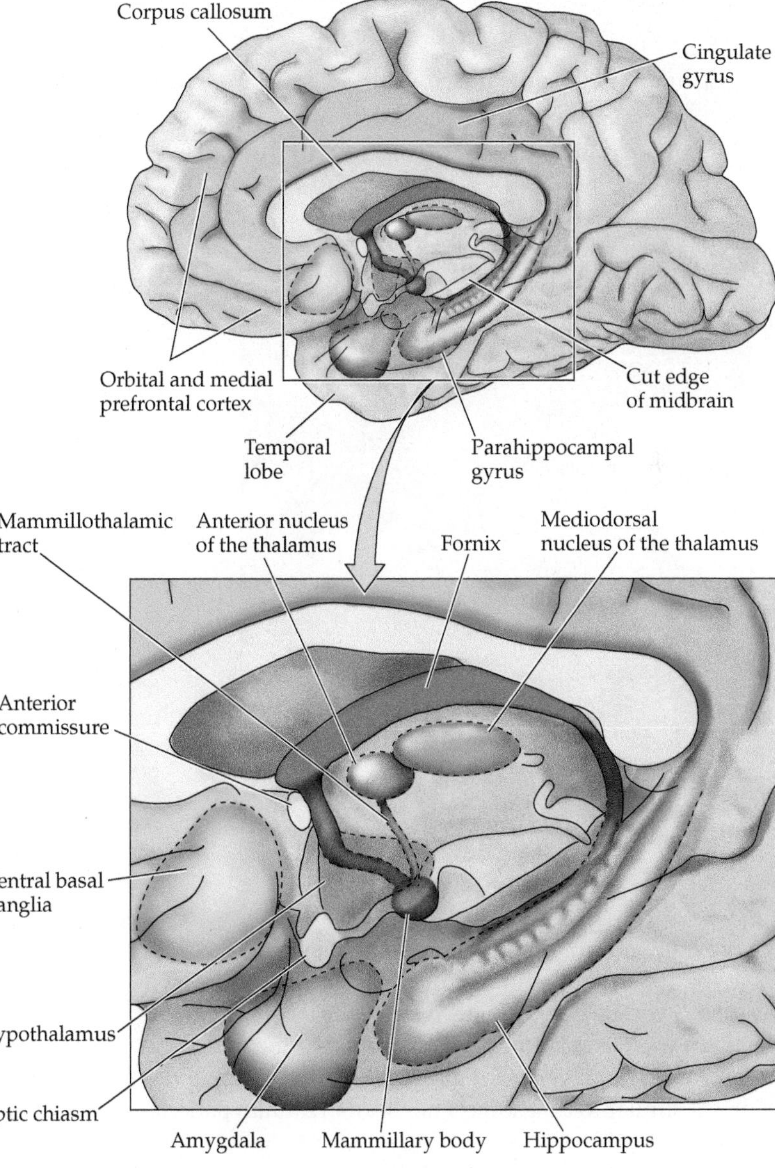

FIGURE 31.5 Modern conception of the limbic system. Two especially important components of the limbic system not emphasized in early anatomical accounts are the orbital and medial prefrontal cortex and the amygdala. These two telencephalic regions, together with related structures in the thalamus, hypothalamus, and ventral striatum, are especially important in the experience and expression of emotion (green). Other parts of the limbic system, including the hippocampus and the mammillary bodies of the hypothalamus, are no longer considered important neural centers for processing emotion (blue).

BOX 31B ■ The Amygdala

The amygdala is a complex mass of gray matter buried in the anterior-medial portion of the temporal lobe, just rostral to the hippocampus (Figure A). It comprises multiple, distinct subnuclei and cortical regions that are richly connected to nearby cortical areas on the ventral and medial aspect of the hemispheric surface.

The amygdala (or amygdaloid complex, as it is often called) can best be thought of in terms of three major functional and anatomical subdivisions, each of which has a unique set of connections with other parts of the brain (Figures B and C). The medial group of subnuclei has extensive connections with the olfactory bulb and the olfactory cortex. The basolateral group, which is especially large in humans, has major connections with the cerebral cortex, especially the orbital and medial prefrontal cortex of the frontal lobe and the associational cortex of the anterior temporal lobe. The central and anterior group of nuclei is characterized by connections with the hypothalamus and brainstem, including such visceral sensory structures as the nucleus of the solitary tract and the parabrachial nucleus (see Chapter 21).

The amygdala thus links cortical regions that process sensory information with hypothalamic and brainstem effector systems. Cortical inputs provide information about highly processed visual, somatosensory, visceral sensory, and auditory stimuli. These pathways from

(A) Amygdala

(B) Central group, Medial group, Basolateral group

(A) Coronal section through the forebrain at the level of the amygdala. (B) Histological section through the human amygdala (boxed area in panel A), stained with silver salts to reveal the presence of myelinated fiber bundles. These bundles subdivide major nuclei and cortical regions within the amygdaloid complex. (B courtesy of Joel Price.)

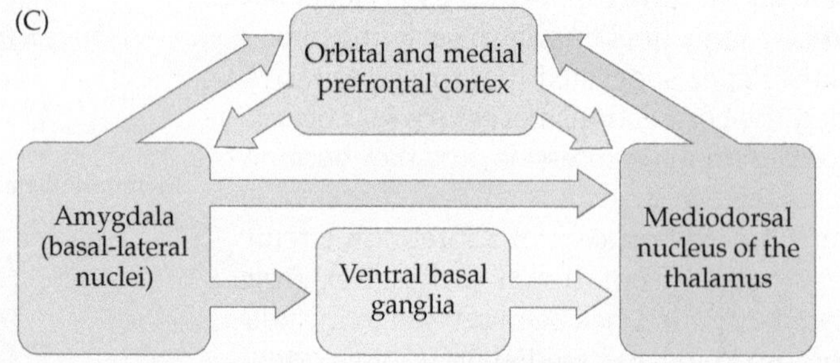

(C) The amygdala (specifically, the basolateral group of nuclei) participates in a "triangular" circuit linking the amygdala, the thalamic mediodorsal nucleus (directly and indirectly via the ventral parts of the basal ganglia), and the orbital and medial prefrontal cortex. These complex interconnections allow direct interactions between the amygdala and prefrontal cortex, as well as indirect modulation via the circuitry of the ventral basal ganglia.

destroying much of the limbic system. They reported a set of atypical behaviors in these animals that is now known as the Klüver–Bucy syndrome. Among the most prominent changes was visual agnosia: The animals appeared to be unable to recognize objects, although they were not blind, a deficit similar to that sometimes seen in human patients following lesions of the temporal cortex. In addition, the monkeys displayed bizarre oral behaviors, putting objects into their mouths that typical monkeys would not. They exhibited hyperactivity and hypersexuality, approaching and making physical contact with virtually anything in their environment; most important, they showed marked

BOX 31B ■ (continued)

sensory cortical areas distinguish the amygdala from the hypothalamus, which receives relatively unprocessed sensory inputs. The amygdala also receives sensory input directly from some thalamic nuclei, the olfactory bulb, and visceral sensory relays in the brainstem. Thus, many neurons in the amygdala respond to visual, auditory, somatosensory, visceral sensory, gustatory, and olfactory stimuli.

Physiological studies have confirmed this convergence of sensory information. Moreover, highly complex stimuli are often required to evoke a neuronal response. For example, there are neurons in the basolateral group of nuclei that respond selectively to the sight of faces, very much like the "face" cells in the inferior temporal cortex (see Chapter 27). In addition to sensory inputs (e.g., visual), the prefrontal and temporal cortical connections of the amygdala give it access to more overtly cognitive neocortical circuits, which integrate the emotional significance of sensory stimuli and guide complex behavior.

Finally, projections from the amygdala to the hypothalamus and brainstem (and possibly as far as the spinal cord) allow the amygdala to play an important role in the expression of emotional behavior by influencing activity in both the somatic and visceral motor efferent systems.

changes in emotional behavior. Because they had been caught in the wild, the monkeys had typically reacted with hostility and fear to humans before their surgery. Postoperatively, however, they were virtually tame. Motor and vocal reactions generally associated with anger or fear were no longer elicited by the approach of humans, and the animals showed little or no excitement when the experimenters handled them. Nor did they show fear when presented with a snake—a strongly aversive stimulus for a typical rhesus monkey. Klüver and Bucy concluded that this remarkable change in behavior was at least partly due to the interruption of the pathways described by Papez. A similar syndrome has been described in humans who have suffered bilateral damage of the temporal lobes.

When it was later demonstrated that the emotional disturbances of the Klüver–Bucy syndrome could be elicited by removal of the amygdala alone, attention turned more specifically to the role of this structure in the control of emotional behavior.

The Importance of the Amygdala

Experiments first performed in the late 1950s by John Downer at University College London vividly demonstrated the importance of the amygdala in aggressive behavior. Downer removed one amygdala in rhesus monkeys, at the same time transecting the optic chiasm and the commissures that link the two hemispheres (principally, the corpus callosum and anterior commissure). In so doing, he produced animals with a single amygdala that had access only to visual inputs from the eye on the same side of the head. Downer found that the animals' behavior depended on which eye was used to view the world. When the monkeys were allowed to see with the eye on the side of amygdala lesion, they behaved in some respects like those described by Klüver and Bucy (for example, they were relatively placid in the presence of humans). If, however, they were allowed to see only with the eye on the side of the intact amygdala, they reverted to their usual fearful and often aggressive behavior. Thus, in the absence of the amygdala, a monkey does not interpret the significance of the visual stimulus presented by an approaching human in the same way as a normal animal. Importantly, only visual stimuli presented to the eye on the side of the ablation produced this abnormal state; thus, if the animal was touched on either side, a full aggressive reaction occurred, implying that somatosensory information about both sides of the body had access to the remaining amygdala. These anecdotal data, taken together with what is now a rich trove of empirical results and clinical observations in both experimental animals and humans, show that the amygdala mediates neural processes that invest sensory experience with emotional significance.

To better understand the role of the amygdala in evaluating stimuli, and to define more precisely the specific circuits and mechanisms involved, several other animal models of emotional behavior have since been developed. One of the most useful is based on conditioned fear responses in rats. Conditioned fear develops when an initially neutral stimulus is repeatedly paired with an inherently aversive one. Over time, the animal begins to respond to the neutral stimulus with behaviors similar to those elicited by the threatening stimulus (i.e., it learns to attach a new meaning to the neutral stimulus). Studies of the parts of the brain involved in the development of conditioned fear in rats have begun to shed some light on this process. Joseph LeDoux and his colleagues at New York University trained rats to associate a tone with a mildly aversive foot shock delivered shortly after onset of the sound. To assess the animals' responses, they measured blood pressure and the length of time the animals crouched without moving (a fearful reaction called *freezing*). Before training, the rats did not react to the tone, nor did their blood pressure change when the tone was presented. After training, however, the onset of the tone caused a marked increase in blood pressure and prolonged periods of behavioral freezing. Using this paradigm, LeDoux and his colleagues worked out the neural circuitry that established the association between the tone and fear (Figure 31.6). First, they demonstrated that the medial geniculate complex (MGC) is necessary for the development of the conditioned fear response. This result is not surprising, since all auditory information that reaches the forebrain travels through the MGC of the dorsal thalamus (see Chapter 13). They went on to show, however, that the responses were still elicited if the connections between the MGC and auditory cortex were severed, leaving only a direct projection between the MGC and the basolateral group of nuclei in the amygdala. Furthermore, if the part of the MGC that projects to the amygdala was also destroyed, the fear responses were abolished. Subsequent work in LeDoux's laboratory established that projections from the central group of nuclei in the amygdala to the midbrain reticular formation are critical in the expression of freezing behavior, while other projections from this group to the hypothalamus control the rise in blood pressure.

Since the amygdala is a site where neural activity produced by both tones and shocks can be processed, it is reasonable to suppose that the amygdala is also the site where learning about fearful stimuli occurs. This supposition led to the broader hypothesis that the amygdala participates in establishing associations between neutral sensory stimuli, such as a mild auditory tone or the sight of inanimate objects in the environment, and other stimuli that have some primary reinforcement value (Figure 31.7). The neutral sensory input can be stimuli in the external environment, stimuli communicated centrally via the special sensory

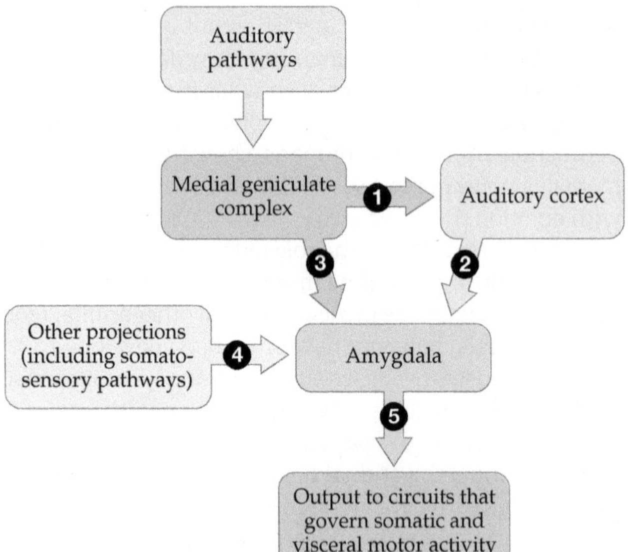

FIGURE 31.6 Pathways in the rat brain that mediate the association of auditory and aversive somatosensory stimuli. Information processed by the auditory centers in the brainstem is relayed to the auditory cortex via the medial geniculate complex (1). The amygdala receives auditory information indirectly via the auditory cortex (2) and directly from one subdivision of the medial geniculate (3). The amygdala also receives sensory information about other sensory modalities, including pain (4). Thus, the amygdala is in a position to associate diverse sensory inputs, leading to new behavioral and autonomic responses to stimuli that were previously devoid of emotional content (5).

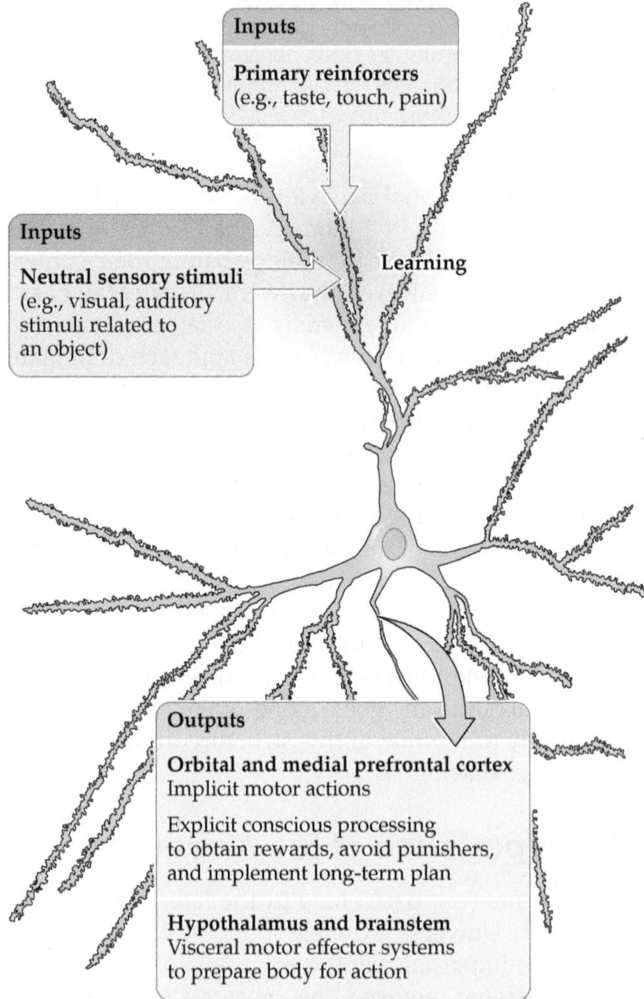

FIGURE 31.7 Model of associative learning in the amygdala relevant to emotional function. Most neutral sensory inputs are relayed to principal neurons in the amygdala by projections from sensory processing areas that represent objects (e.g., faces). If these sensory inputs depolarize neurons at the same time as inputs that represent other sensations with primary reinforcing value, then associative learning would presumably occur by strengthening synaptic linkages between the previously neutral inputs and the neurons of the amygdala.. The output of the amygdala would then inform a variety of integrative centers responsible for the somatic and visceral motor expression of emotion, and for modifying behavior relevant to seeking rewards and avoiding punishment. (After Rolls, 1999.)

afferent systems, or internal stimuli derived from activation of visceral sensory receptors. The stimuli with primary reinforcement value include sensory stimuli that are inherently rewarding, such as the sight, smell, and taste of food, or stimuli with negative valences such as an aversive taste, loud sounds, or painful mechanical stimulation. This associative learning presumably strengthens the connections relaying the information about the neutral stimulus, provided that the connections activate the postsynaptic neurons in the amygdala at the same time as inputs pertaining to the primary reinforcer. The discovery that long-term potentiation (LTP) occurs in the amygdala provides further support for this hypothesis. Indeed, the acquisition of conditioned fear in rats is blocked by infusion into the amygdala of NMDA antagonists, which prevents the induction of LTP. Finally, the behavior of patients with selective damage to the anterior-medial temporal lobe indicates that the amygdala plays a similar role in the human experience of fear (Box 31C).

BOX 31C ■ Fear and the Human Amygdala

Studies of fear conditioning in rodents show that the amygdala plays a critical role in the association of an innocuous auditory tone with an aversive mechanical sensation. Does this finding imply that the human amygdala is similarly involved in the experience of fear and the expression of fearful behavior? Reports of at least one extraordinary patient support the idea that the amygdala is indeed a key brain center for the experience of fear.

The patient (S. M.) suffers from a rare autosomal recessive condition called Urbach–Wiethe disease, a disorder that causes bilateral calcification and atrophy of the anterior-medial temporal lobes. As a result, both of S. M.'s amygdalas are extensively damaged, with little or no detectable injury to the hippocampal formation or nearby temporal neocortex (Figure A). She has no motor or sensory impairment, and no notable deficits in intelligence, memory, or language function. However, when asked to rate the intensity of emotion in a series of photographs of facial expressions, she cannot recognize the emotion of fear (Figure B). Indeed, S. M.'s ratings of emo-

Continued on the next page

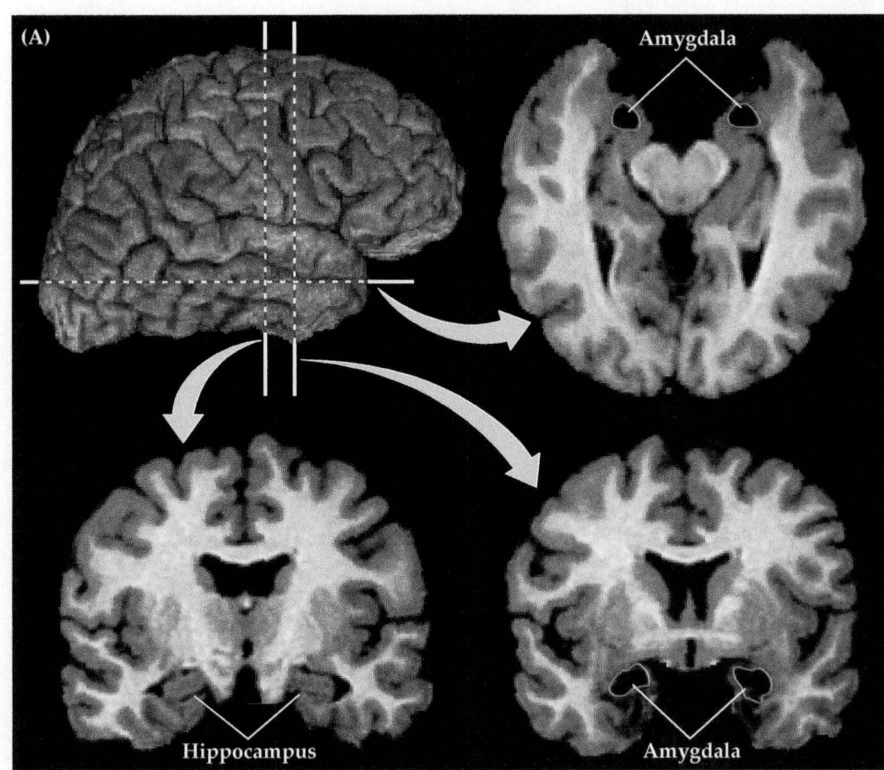

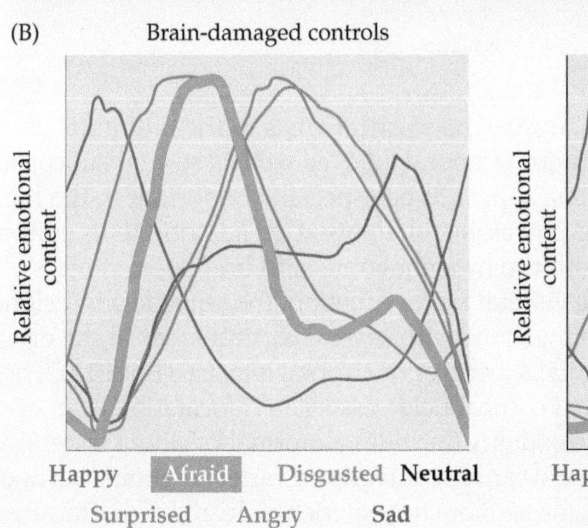

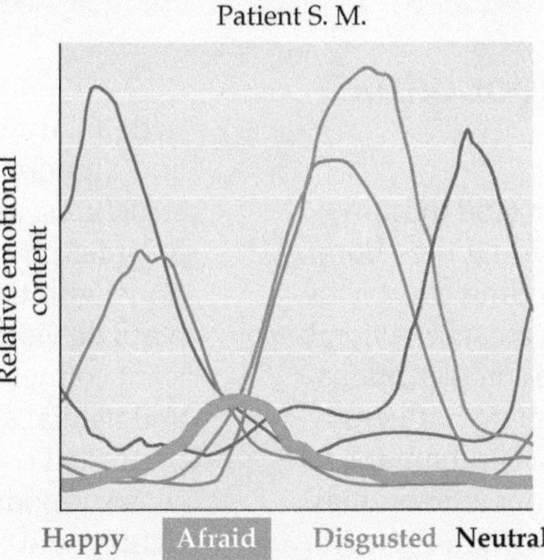

(A) MRI showing the extent of brain damage in patient S. M.; note the bilateral destruction of the amygdala and the preservation of the hippocampus. (B) Patients with brain damage outside the anterior-medial temporal lobe and patient S. M. rated the emotional content of a series of facial expressions. Each colored line represents the intensity of the emotions judged in the face. S. M. recognized happiness, surprise, anger, disgust, sadness, and neutral qualities in facial expressions about as well as controls did. However, she failed to recognize fear (orange lines). (A courtesy of R. Adolphs; B after Adolphs et al., 1995.)

BOX 31C (continued)

tional content in fearful facial expressions were several standard deviations below the ratings of control patients who had suffered brain damage outside the anterior-medial temporal lobe.

The investigators next asked S. M. (and brain-damaged control patients) to draw facial expressions of the same set of emotions from memory. Although the individuals obviously differed in artistic ability and the details of their renderings, S. M. (who has some artistic experience) produced skillful pictures of each emotion, except for fear (Figure C). At first, she could not produce a sketch of a fearful expression and, when prodded to do so, explained that she "did not know what an afraid face would look like." After several failed attempts, she produced the sketch of a cowering figure with hair standing on end, evidently because she knew these clichés about the expression of fear. In short, S. M. has a severely limited concept of fear and, consequently, fails to recognize the emotion of fear in facial expressions, in part because she fails to seek out salient social information from the eye regions of human faces. Studies of other individuals with bilateral destruction of the amygdala are consistent with this account. As might be expected, S. M.'s deficiency also limits her ability to experience fear in situations where this emotion is appropriate.

Despite the admonition "have no fear," to truly live without fear is to be deprived of a crucial neural mechanism that facilitates appropriate social behavior, helps make advantageous decisions in critical circumstances, and ultimately promotes survival.

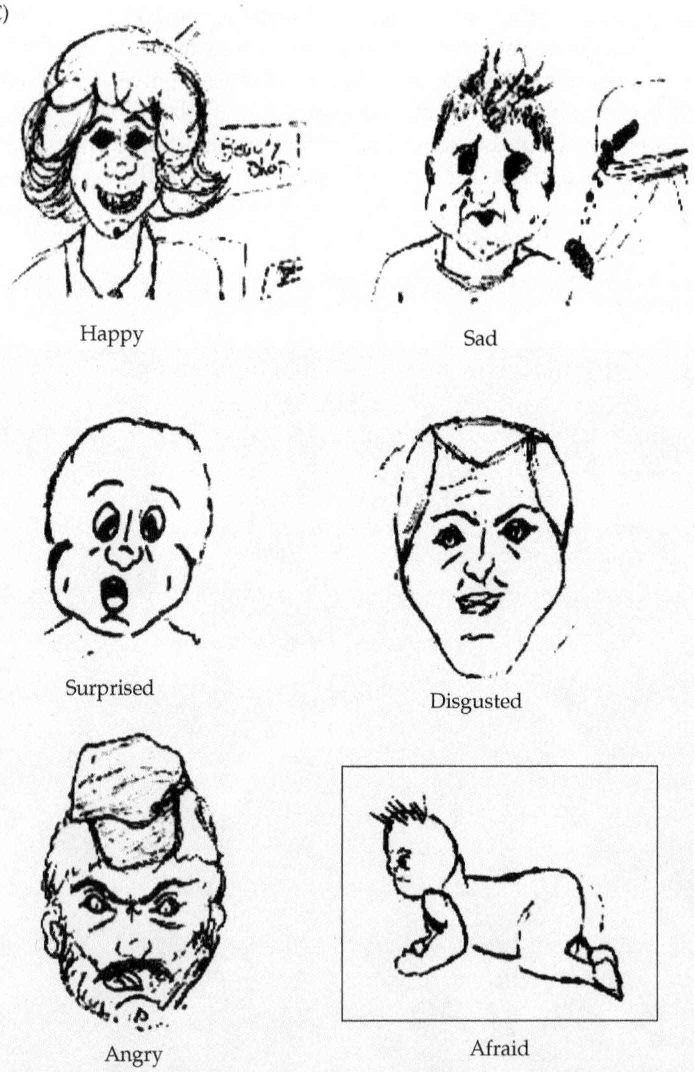

(C) Sketches made by S. M. when asked to draw facial expressions of emotion. (From Adolphs et al., 1995.)

The Relationship between Neocortex and Amygdala

As these observations on the limbic system (and the amygdala in particular) make plain, understanding the neural basis of emotions also requires understanding the role of the cerebral cortex. In animals like the rat, most behavioral responses are highly stereotyped. In more complex brains, however, individual experience is increasingly influential in determining responses to stimuli. Thus, in humans, a stimulus that evokes fear or sadness in one person may have little or no effect on the emotions of another. Although the pathways underlying such responses are not well understood, the amygdala and its interconnections with an array of neocortical areas in the prefrontal cortex and anterior temporal lobe, as well as several subcortical structures, appear to be especially important in the higher-order processing of emotion. In addition to its connections with the hypothalamus and brainstem centers that regulate visceral motor function, the amygdala has significant connections with several cortical areas in the orbital and medial aspects of the frontal lobe (see Box 31B). These prefrontal cortical fields associate information from every sensory modality (including information about visceral activities) and can thus integrate a variety of inputs pertinent to moment-to-moment experience. In addition, the amygdala projects to the thalamus (specifically, the mediodorsal nucleus), which projects in turn to these same cortical areas. Finally, the amygdala innervates neurons in the ventral portions of the basal ganglia that receive the major corticostriatal projections from the regions of the prefrontal cortex thought to process emotions. Considering all these connections, the amygdala emerges as a nodal point in a network that links together the cortical and subcortical brain regions involved in emotional processing.

Clinical evidence concerning the significance of this circuitry linked through the amygdala has come from functional imaging studies of individuals suffering from depression. In such patients this set of interrelated forebrain structures shows atypical patterns of cerebral blood flow, especially in the left hemisphere. More generally, the amygdala and its connections to the prefrontal cortex and basal ganglia are likely to influence the selection and initiation of behaviors aimed at obtaining rewards and avoiding punishments (recall that the process of motor program selection and initiation is an important function of basal ganglia circuitry; see Chapter 18). The parts of the prefrontal cortex interconnected with the amygdala are also involved in organizing and planning future behaviors; thus, the amygdala may provide emotional input to overt (and covert) deliberations of this sort (see the section "Emotion, Reason, and Social Behavior" below).

Finally, it is likely that interactions between the amygdala, the neocortex, and related subcortical circuits account for what is perhaps the most enigmatic aspect of emotional experience: the highly subjective "feelings" that attend most emotional states. Although the neurobiology of such experience is not understood, it is reasonable to assume that emotional feelings arise as a consequence of a more general cognitive capacity for self-awareness. In this conception, feelings entail both the immediate conscious experience of implicit emotional processing (arising from amygdala–neocortical circuitry) and explicit processing of semantically based thought (arising from hippocampal–neocortical circuitry). Thus, feelings can be plausibly conceived as the product of an emotional working memory that sustains neural activity related to the processing of these various elements of emotional experience. Given the evidence for working memory functions in the prefrontal cortex (see Chapter 30), this portion of the frontal lobe—especially the orbital and medial sector—is the likely substrate when such associations are conscious.

Cortical Lateralization of Emotional Functions

Since functional asymmetries of complex cortical processes are commonplace, it should come as no surprise that the two hemispheres make different contributions to the governance of emotion. Emotion is lateralized in the cerebral hemispheres in at least two ways. First, as discussed in Chapter 33, the right hemisphere is especially important for the expression and comprehension of the affective aspects of speech. Thus, patients with damage to the supra-Sylvian portions of the posterior frontal and anterior parietal lobes on the right side may lose the ability to express emotion by modulation of their speech patterns (this loss of emotional expression is referred to as *aprosody* or *aprosodia,* and that similar lesions in the left hemisphere give rise to Broca's aphasia). Patients with aprosodia tend to speak in a monotone, no matter what the circumstances or meaning of what is said. For example, one such patient, a teacher, had trouble maintaining discipline in the classroom. Because her pupils (and even her own children) couldn't tell when she was angry or upset, she had to resort to adding phrases such as "I am angry and I really mean it" to indicate the emotional significance of her remarks. The wife of another patient felt her husband no longer loved her because he could not imbue his speech with cheerfulness or affection. Although such individuals cannot express emotion in speech, they nonetheless experience typical emotional feelings.

A second way in which the hemispheric processing of emotionality is asymmetrical concerns mood. Both clinical and experimental studies indicate that the left hemisphere is more importantly involved with what can be thought of as positive emotions, whereas the right hemisphere is more involved with negative ones. For example, the incidence and severity of depression are significantly higher in individuals with lesions of the left anterior hemisphere compared with any other location. In contrast, patients with lesions of the right anterior hemisphere are often described as unduly cheerful. These observations suggest that lesions of the left hemisphere result in a relative loss of positive feelings, facilitating depression, whereas lesions of the right hemisphere result in a loss of negative feelings, leading to inappropriate optimism. Hemispheric asymmetry related to emotion is also apparent in normal individuals. For instance, auditory experiments that introduce sound into one ear or the other indicate a right-hemisphere superiority in detecting the emotional nuances in speech. Moreover, when facial expressions are specifically presented to either the right or the left visual hemifield, the depicted emotions are more readily and accurately identified from the information in the left hemifield (i.e., the hemifield perceived by the right hemisphere; see Chapter 12). Finally, kinematic studies of facial expressions show that most individuals more quickly and fully express emotions with the left facial musculature than with the right (recall that the left lower face is controlled by the right hemisphere, and vice versa) (Figure 31.8).

Taken together, this evidence is consistent with the idea that the right hemisphere is more intimately concerned with both the perception and expression of emotions than is the left hemisphere. However, as in the case of other

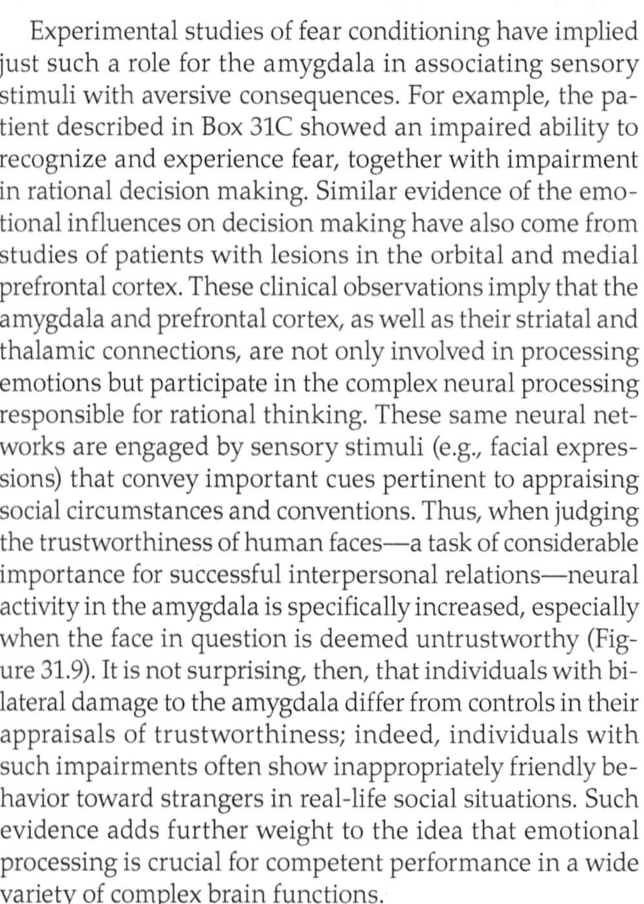

FIGURE 31.8 Asymmetrical smiles on some famous faces. Studies of typical individuals show that facial expressions are often more quickly and fully expressed by the left facial musculature than the right, as suggested by examination of these examples (try covering one side of the faces and then the other). Because the left lower face is governed by the right hemisphere, some psychologists have suggested that the majority of humans are "left-faced," in the same general sense that most of us are right-handed. (Top row, left to right: Courtesy U.S. Department of State; Musée du Louvre, Paris; © AF archive/Alamy. Bottom row, left to right: © D. Levenson/Alamy; © Everett Collection/ Alamy; courtesy of Pete Souza.)

lateralized behaviors (language, for instance), both hemispheres participate in processing emotion.

Emotion, Reason, and Social Behavior

The experience of emotion—even on a subconscious level—has a powerful influence on other brain functions, including the neural faculties responsible for making rational decisions and the interpersonal judgments that guide social behavior. Evidence for this statement has come principally from studies of patients with damage to parts of the orbital and medial prefrontal cortex, as well as patients with injury or disease involving the amygdala. Such individuals often have impaired emotional processing, especially of emotions engendered by complex personal and social situations. As a result they may have difficulty making advantageous decisions.

Antonio Damasio and his colleagues at the University of Southern California have suggested that such decision making entails the rapid evaluation of a set of possible outcomes with respect to the future consequences associated with each course of action. It seems plausible that the generation of conscious or subconscious mental images that represent the consequences of each contingency trigger emotional states that involve either actual alterations of somatic and visceral motor function, or the activation of neural representations of such activity. Whereas William James proposed that we are "afraid because we tremble," Damasio and his colleagues suggest a vicarious representation of motor action and sensory feedback in the neural circuits of the frontal and parietal lobes. It is these vicarious states, according to Damasio, that give mental representations of contingencies the emotional valence that helps an individual identify favorable or unfavorable outcomes.

Experimental studies of fear conditioning have implied just such a role for the amygdala in associating sensory stimuli with aversive consequences. For example, the patient described in Box 31C showed an impaired ability to recognize and experience fear, together with impairment in rational decision making. Similar evidence of the emotional influences on decision making have also come from studies of patients with lesions in the orbital and medial prefrontal cortex. These clinical observations imply that the amygdala and prefrontal cortex, as well as their striatal and thalamic connections, are not only involved in processing emotions but participate in the complex neural processing responsible for rational thinking. These same neural networks are engaged by sensory stimuli (e.g., facial expressions) that convey important cues pertinent to appraising social circumstances and conventions. Thus, when judging the trustworthiness of human faces—a task of considerable importance for successful interpersonal relations—neural activity in the amygdala is specifically increased, especially when the face in question is deemed untrustworthy (Figure 31.9). It is not surprising, then, that individuals with bilateral damage to the amygdala differ from controls in their appraisals of trustworthiness; indeed, individuals with such impairments often show inappropriately friendly behavior toward strangers in real-life social situations. Such evidence adds further weight to the idea that emotional processing is crucial for competent performance in a wide variety of complex brain functions.

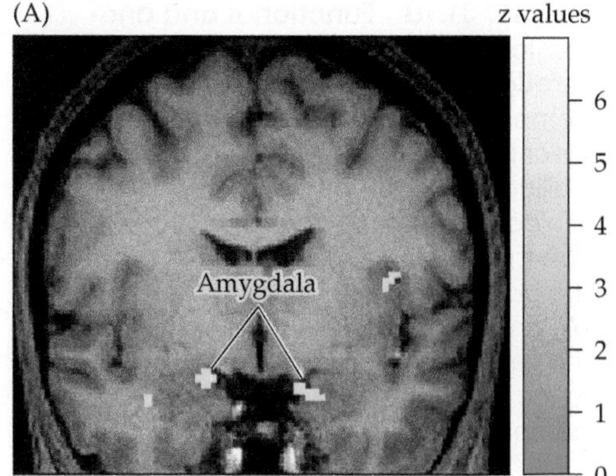

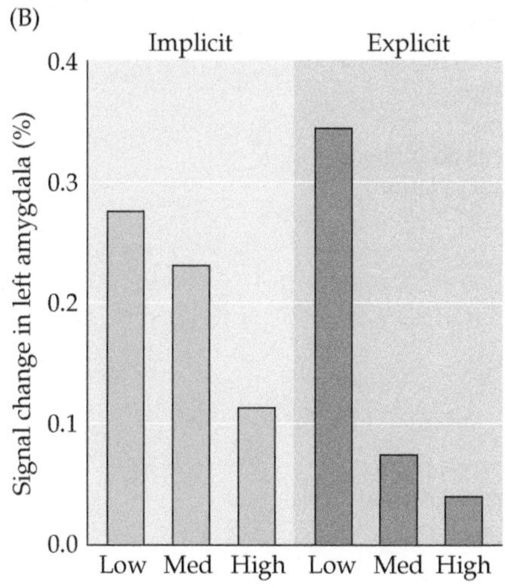

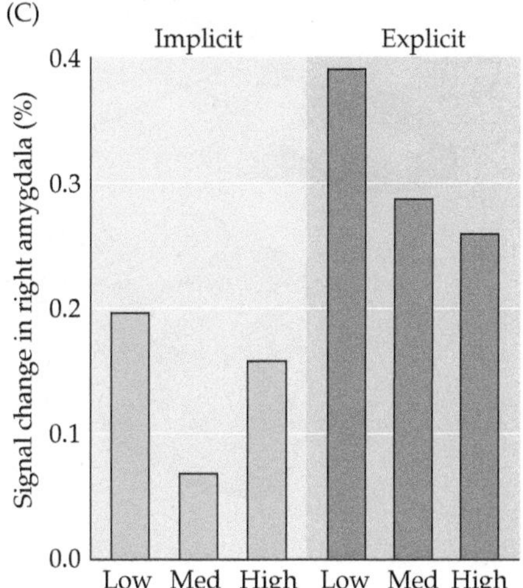

FIGURE 31.9 Activation of the amygdala during judgments of trustworthiness. (A) Functional MRI shows increased neural activation bilaterally in the amygdala when normal individuals appraise the trustworthiness of human faces; activity is also increased in the right insular cortex. (B,C) The degree of activation is greatest when individuals evaluate faces that are considered untrustworthy (low, med, and high indicate ratings of trustworthiness; low = untrustworthy). The same effect was observed when individuals were instructed to evaluate the trustworthiness of the faces (explicit condition) or whether the faces were those of high school or university students (implicit condition). (After Winston et al., 2002.)

Emotion and Addiction

Understanding the neurobiological basis of emotion and the contributions of affective neural processing to higher brain functions remains an important goal of neuroscience. The urgency of achieving this goal is highlighted by the pervasiveness and impact of drug abuse and addiction, behaviors that reflect the vulnerability of emotional neural circuits when exposed to drugs that alter their normally adaptive operations in goal-directed behavior. Under physiological conditions, emotional processing in the limbic system can signal the presence of or prospect for reward and punishment, as well as promote the activation of motor programs aimed at procuring beneficial rewards and avoiding punishment. With this in mind, it is not surprising that most known drugs of abuse—including heroin and other opiates, cocaine, alcohol, marijuana, nicotine, amphetamines, and their synthetic analogs—act on elements of limbic circuitry (see Figure 31.11). What may be surprising is that such diverse natural and synthetic compounds do so by altering the neuromodulatory influence of **dopamine** on the processing of reinforcement signals in the ventral divisions of the basal ganglia, which in turn leads to the consolidation of addictive behavior in limbic circuitry.

Recall from Chapter 18 that the dorsal divisions of the basal ganglia (dorsal caudate, putamen, and globus pallidus) are instrumental in gating the activation of thalamocortical circuits that initiate volitional movement. Also mentioned briefly in Chapter 18 is the existence of other, parallel processing streams that similarly gate the activation of non-motor programs, including those that pertain to cognition and affective processing in limbic circuits. The organization of these non-motor processing streams is fundamentally comparable to the "direct pathway" for volitional movement: There are major excitatory inputs from cortex to striatum, neuromodulatory projections from midbrain dopaminergic neurons to striatum, internuclear connections from striatum to pallidum, and output projections from pallidum to thalamus. What distinguishes the "limbic loop" from the "motor loop" discussed in Chapter 18 is the source and nature of the cortical input, the relevant divisions of striatum and pallidum that process this input, the source of dopaminergic projections from the midbrain, and the thalamic target of the pallidal output (Figure 31.10).

Central to the organization and function of the limbic loop are inputs from the amygdala, the subiculum (a ventral division of the hippocampal formation), and orbital-medial prefrontal cortex that convey signals relevant to emotional reinforcement to ventral divisions of the anterior striatum, the largest component of which is called the *nucleus accumbens*. Like the caudate and putamen, the nucleus accumbens contains medium spiny neurons that integrate excitatory telencephalic inputs under the modulatory influence of dopamine. However, unlike the larger dorsal division of the striatum, the nucleus accumbens receives its dopaminergic projections from a collection of neurons that lies just dorsal and medial to the substantia nigra, in a region of the midbrain called the **ventral tegmental area**

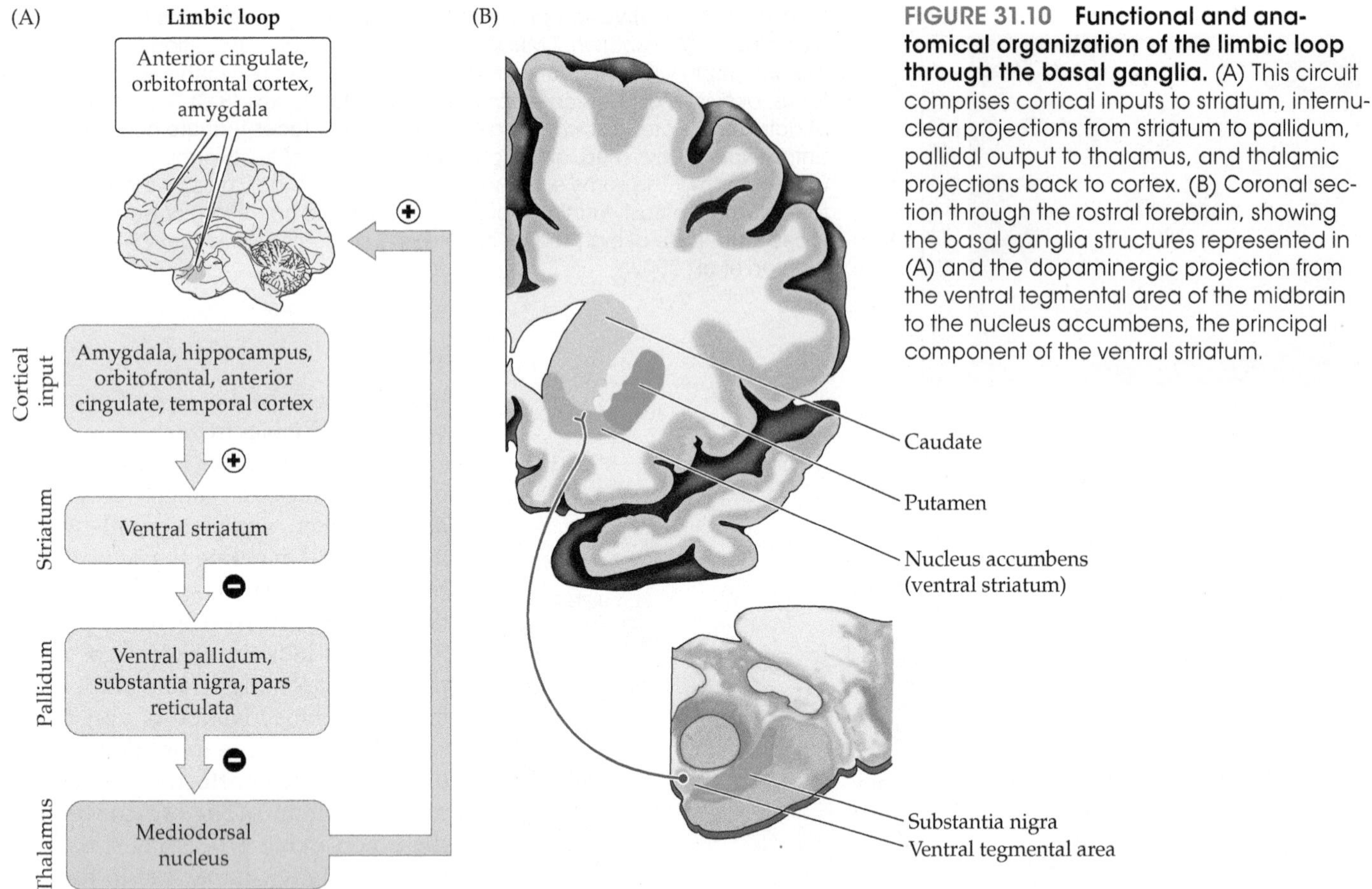

FIGURE 31.10 Functional and anatomical organization of the limbic loop through the basal ganglia. (A) This circuit comprises cortical inputs to striatum, internuclear projections from striatum to pallidum, pallidal output to thalamus, and thalamic projections back to cortex. (B) Coronal section through the rostral forebrain, showing the basal ganglia structures represented in (A) and the dopaminergic projection from the ventral tegmental area of the midbrain to the nucleus accumbens, the principal component of the ventral striatum.

(**VTA**). The nucleus accumbens and the ventral tegmental area are primary sites where drugs of abuse interact with the processing of neural signals related to emotional reinforcement; they do so by prolonging the action of dopamine in the nucleus accumbens or by potentiating the activation of neurons in the ventral tegmental area and nucleus accumbens (Figure 31.11).

Under normal conditions, these dopaminergic neurons are only phasically active; when they do fire a barrage of action potentials, however, dopamine is released in the nucleus accumbens, and medium spiny neurons are much more responsive to coincident excitatory input from telencephalic structures such as the amygdala and orbital-medial prefrontal cortex. These activated striatal neurons in turn project to and inhibit pallidal neurons in a region just below the globus pallidus called the **ventral pallidum**, as well as in the pallidal division of the substantia nigra (pars reticulata). The suppression of tonic activity in the pallidum then disinhibits the thalamic target of the limbic loop, which is the mediodorsal nucleus, the thalamic nucleus that innervates cortical divisions of the limbic forebrain (see Figure 31.5). Activation of these cortical regions via the mediodorsal nucleus is reinforced by direct cortical projections of dopaminergic neurons in the ventral tegmental area and glutamatergic projections from the basolateral group of nuclei in the amygdala that target these same regions (see Box 31B).

Activation of these complex limbic circuits is believed to cause the rewarding effects of natural agents and experiences such as food, water, sex, and more complex social rewards. However, the phasic release of dopamine is also subject to experience-dependent plasticity. During associative learning, for example, the activity of ventral tegmental neurons comes to signal the presence of the reward-predicting stimulus with diminished responsiveness to the presence of the primary reward itself (Figure 31.12A,B). Interestingly, if a stimulus is not followed by delivery of the predicted reward (e.g., because of an inappropriate behavioral response in an instrumental learning paradigm), ventral tegmental neurons are suppressed at precisely the same time that the neuronal response would have signaled the presence of a reward (Figure 31.12C). These observations suggest that the phasic release of dopamine signals the presence of reward relative to its prediction, rather than the unconditional presence of reward. The integration of such signals in the nucleus accumbens, orbital-medial

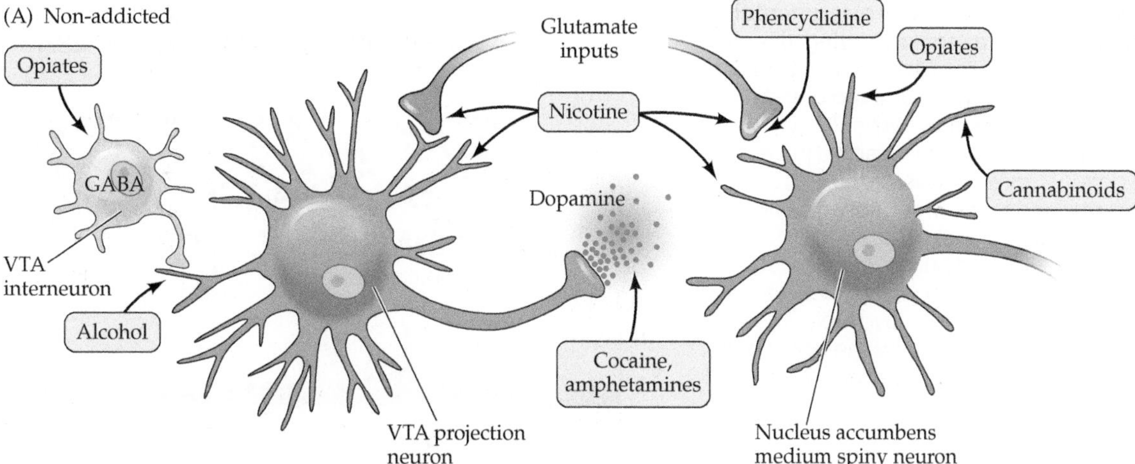

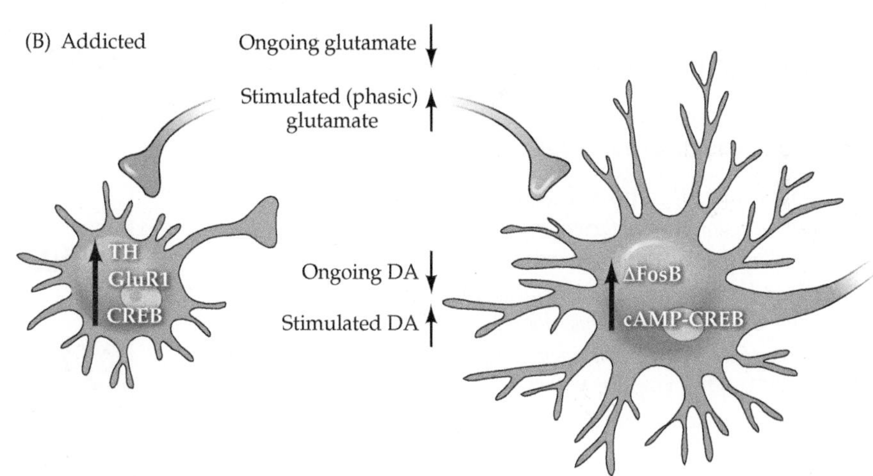

FIGURE 31.11 Schematic representation of how drugs of abuse affect dopamine projections from the ventral tegmental area to the nucleus accumbens. (A) Most drugs of abuse potentiate the activity of dopamine by interacting directly with dopamine synapses in the nucleus accumbens, or indirectly by modulating the activity of neurons in the ventral tegmental area (VTA). Other drugs may act directly on accumbens neurons to increase their responsiveness to telencephalic input. (B) Drug addiction is associated with cellular and molecular adaptations of this circuit (see text for details). The net effect of addiction is a chronic decrease in basal activity and an increase in the intensity of phasic activity in the presence of abusive drugs. (After Nestler, 2005.)

prefrontal cortex, and amygdala leads to the activation of behaviors directed at obtaining and consolidating the benefits of the rewarding event.

Unfortunately, the plastic potential of these limbic circuits can be co-opted by chronic exposure to drugs of abuse, leading to cellular and molecular changes that promote abnormal regulation (see Figure 31.11B). In the ventral tegmental area of addicted individuals, the activity of the dopamine-synthesizing enzyme tyrosine hydroxylase increases, as does the ability of VTA neurons to respond to excitatory inputs. The latter effect is secondary to increases in the activity of the transcription factor CREB (see Chapter 7) and the upregulation of GluR1, an important subunit of AMPA receptors for glutamate (see Chapter 6). In the nucleus accumbens, addiction is characterized by increases in another transcription factor, ΔFosB, in addition to induction of CREB with chronic exposure to at least some classes of addictive drugs. Activation of these molecular signaling pathways leads to a generalized reduction in the responsiveness of accumbens neurons to glutamate released by telencephalic inputs by regulation of different AMPA receptor subunits and/or changes in postsynaptic density proteins that alter the dynamics of receptor trafficking. However, during the phasic release of dopamine, the responsiveness of these striatal neurons is intensified, mediated in part by a shift in the expression of D1 and D2 classes of dopamine receptors and a coordinated upregulation of cAMP–PKA signaling pathways.

The cellular and molecular maladaptations of ventral tegmental, striatal, and cortical neurons to chronic exposure of abusive drugs remain poorly understood. Nevertheless, the net effect of these and other changes is that addiction dampens the response of this emotional reinforcement circuitry to less potent natural rewards, while intensifying the response to addictive drugs. At a systems level, these changes are likely reflected in the "hypofrontality"—reduced baseline activity in orbital-medial prefrontal cortex—commonly seen in addicts. Taken together, these alterations in neural processing could account for the waning influence of adaptive emotional signals in the operation of decision-making faculties as drug-seeking and drug-taking behaviors become habitual and eventually compulsive.

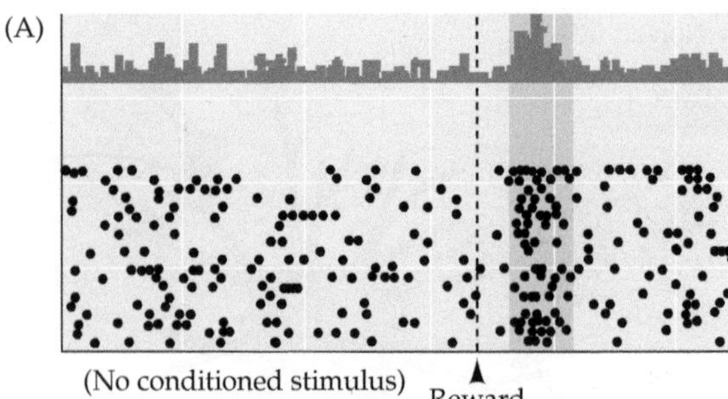

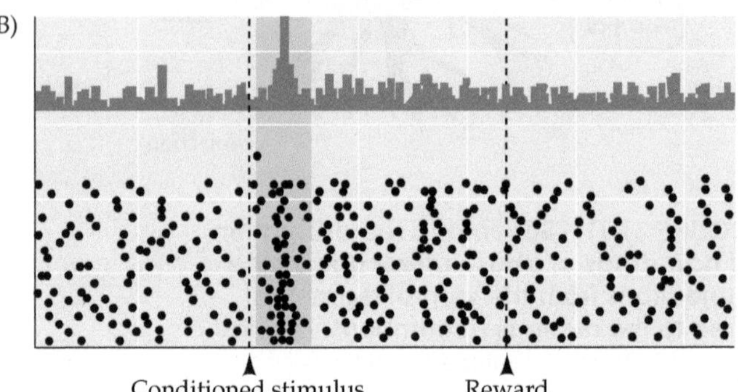

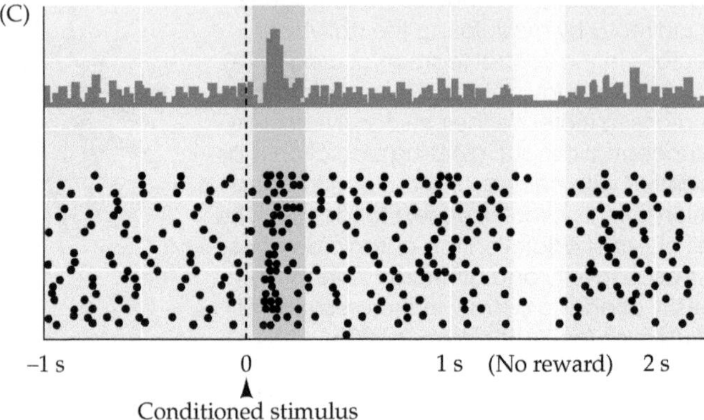

FIGURE 31.12 **Changes in the activity of ventral tegmental area dopamine neurons in an awake monkey during stimulus-reward learning.** In each panel, poststimulus time histograms (blue, above) and data plots (below) report summed activity across trials and individual spikes within trials, respectively. (A) Before learning, the presentation of an unexpected juice reward evokes a burst of activity (darker-shaded zone). (B) After learning trials, the neurons respond to the presentation of visual or auditory cues (conditioned stimuli), but not to the reward itself. (C) In trials when a reward was predicted but never delivered, dopaminergic neurons become suppressed (light-shaded zone) at the time when the reward would have been available. These results show that the ventral tegmental area signals the occurrence of reward relative to its prediction. (After Schultz et al., 1997.)

Posttraumatic Stress Disorder

The following description gives some sense of what **posttraumatic stress disorder** (**PTSD**) entails:

> *For months after the attack, I couldn't close my eyes without envisioning the face of my attacker. I suffered horrific flashbacks and nightmares. For four years after the attack I was unable to sleep alone in my house. I obsessively checked windows, doors, and locks ... I lost all ability to concentrate or even complete simple tasks. Normally social, I stopped trying to make friends or get involved in my community.*
>
> P. K. Philips, www.adaa.org

PTSD typically emerges following exposure to a traumatic stressor, such as rape, robbery, or combat, that elicits feelings of fear, horror, or helplessness to forestall bodily injury or threat of death. Community-based studies in the United States estimate that 50% of people will have a traumatic experience during their lifetime, and an estimated 5% of men and 9% of women will develop PTSD as a result. Symptoms include persistently re-experiencing the traumatic event, avoiding reminders of the event, numbed responsiveness, and heightened arousal. PTSD is often accompanied by depression and substance abuse, each of which complicates treatment and recovery. While cognitive–behavioral therapies and antianxiety and antidepressant medications often help, there is no cure for this debilitating condition that can persist for decades.

Some of the structural abnormalities associated with PTSD are reductions in hippocampal and amygdala volume and altered dendritic remodeling in these structures (Figure 31.13A). Hippocampal atrophy has been linked to declarative memory deficits in some patients with PTSD, and functional impairments in the amygdala are associated with hyperarousal symptoms and exaggerated responses to threats in others. Problems with fear reduction are further exacerbated by hyporesponsiveness in the rostral anterior cingulate and ventromedial prefrontal cortex, which provide inhibitory control over neurons in the amygdala (Figure 31.13B).

Treatment with serotonin uptake inhibitors (e.g., Prozac) may partially reverse hippocampal volume differences and alleviate anxiety symptoms, but no single treatment "cures" this complex disorder. A major focus of ongoing research is to determine whether the brain alterations in PTSD are causal, or whether they are a consequence of the chronic stress associated with the syndrome.

Researchers and clinicians working with PTSD patients face other challenges as well. Since it is generally considered unethical to induce physical or psychological trauma in the laboratory, the topic is difficult to approach experimentally. For instance, is it ethical to have PTSD patients

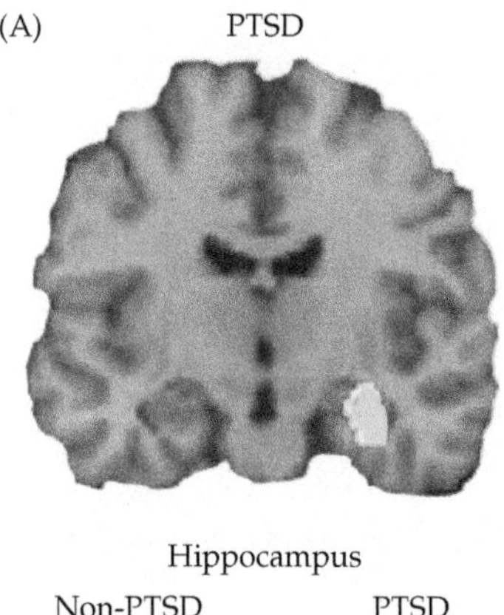

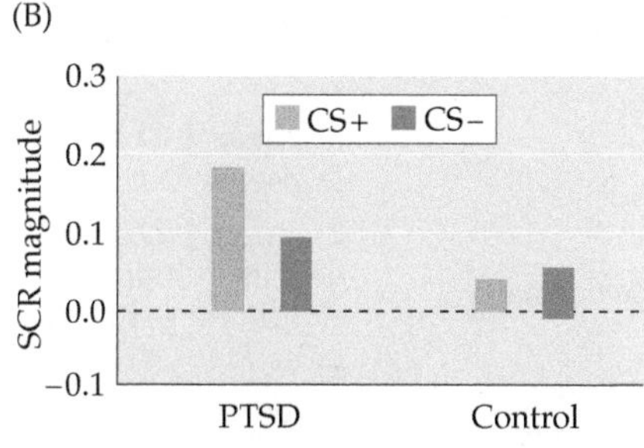

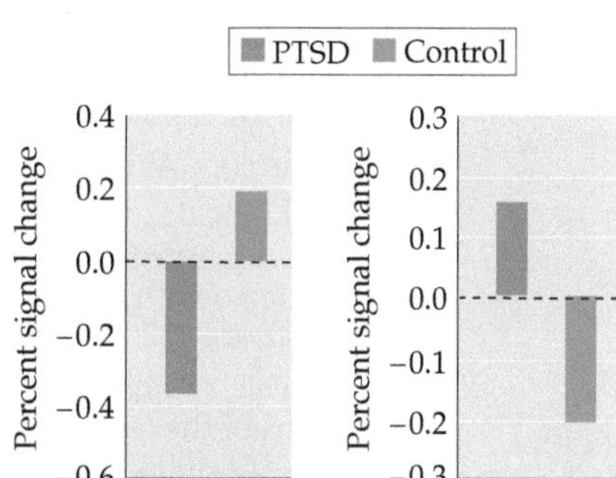

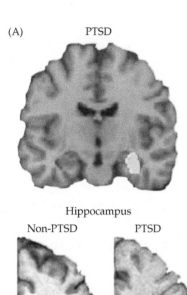

FIGURE 31.13 Brain abnormalities associated with posttraumatic stress disorder (PTSD). (A) The volume of the hippocampus is often reduced in adult patients with PTSD and childhood abuse, compared with both abused individuals who never developed PTSD and normal individuals. (B) Compared with trauma-exposed control individuals, PTSD patients have a difficult time extinguishing fear responses to cues that previously predicted a threat. During fear conditioning, one cue (CS+) predicted delivery of a mild shock whereas another cue (CS–) did not. Individuals then underwent an extinction procedure in which the shock was removed and the CS+ was now "safe." Despite the fact that the cue no longer predicted a shock, the PTSD patients continued to show greater skin conductance responses (SCR), a measure of sympathetic arousal, and greater amygdala activity to the CS+ during the extinction test. In addition, they failed to engage the ventromedial prefrontal cortex (vmPFC) during the extinction test. This functional activity pattern is indicative of persistent hyperreactivity to threats and difficulty engaging executive control processes to suppress acquired fears when they are not appropriate to express. (A from Parsons and Ressler, 2013 after Carrión et al., 2010; B after Milad et al., 2009.)

relive their painful past experiences for the purpose of studying these extreme emotions in the laboratory? As new treatments develop, additional dilemmas have emerged. For example, if a pharmacological agent selectively blocks emotional memories, should it be routinely administered to rape victims? If a genetic variant of a molecular marker is discovered to be a risk factor for developing PTSD, should military recruits be screened for it? Is it ethical to expose research animals to chronic stress to investigate the neurobiological mechanisms?

Questions like these raise concerns not only for PTSD, but for emotion research generally. A primary goal of such research is to alleviate the suffering of patients afflicted with affective disorders. On the one hand, researchers must study emotional phenomena that mimic or elicit the emotions associated with the disorder of interest. On the other hand, if emotions are evoked only weakly in the laboratory, the mechanisms uncovered may bear little resemblance to those that operate in the real world. Evoking stronger emotions, however, may do harm to a patient or study volunteer. As with much scientific research, the risks posed to patients and volunteers must be weighed against the potential benefits.

Summary

The word *emotion* covers a wide range of states that have in common the association of visceral motor responses, somatic behavior, and powerful subjective feelings. The visceral motor responses are mediated by the visceral motor nervous system, which is itself regulated by inputs from many other parts of the brain. The organization of the somatic motor behavior associated with emotion is governed by circuits in the limbic system, which includes the hypothalamus, the amygdala, and several regions of the cerebral cortex. Although a good deal is known about the neuroanatomy and transmitter chemistry of the different parts of the limbic system and how they are affected by drugs, there is still a dearth of information about how this complex circuitry mediates specific emotional states. Similarly, psychologists, neurologists, and psychiatrists have only recently begun to study the role of emotional processing in decision making and social behavior. The prevalence and social significance of human emotions and their disorders ensure that the neurobiology of emotion will be an increasingly important theme in modern neuroscience.

ADDITIONAL READING

Reviews

Adolphs, R. (2003) Cognitive neuroscience of human social behavior. *Nat. Rev. Neurosci.* 4: 165–178.

Appleton, J. P. (1993) The contribution of the amygdala to normal and abnormal emotional states. *Trends Neurosci.* 16: 328–333.

Borod, J. C., R. L. Bloom, A. M. Brickman, L. Nakhutina and E. A. Curko (2002) Emotional processing deficits in individuals with unilateral brain damage. *Appl. Neuropsychol.* 9: 23–36.

Campbell, R. (1986) Asymmetries of facial action: Some facts and fancies of normal face movement. In *The Neuropsychology of Face Perception and Facial Expression*, R. Bruyer (ed.). Hillsdale, NJ: Erlbaum, pp. 247–267.

Craig, A. D. (2007) Interoception and emotion: A neuroanatomical perspective. In *Handbook of Emotions*, 3rd Edition, M. Lewis, J. M. Haviland-Jones and L. F. Barrett (eds.). New York: Guilford Press, pp. 395–408.

Davis, M. (1992) The role of the amygdala in fear and anxiety. *Annu. Rev. Neurosci.* 15: 353–375.

Dolan, R. J. (2002) Emotion, cognition, and behavior. *Science* 308: 1191–1194.

Harmon-Jones, E., P. A. Gable and C. K. Peterson (2010) The role of asymmetric frontal cortical activity in emotion-related phenomena: A review and update. *Biol. Psychol.* 84: 451–462.

Jones, C. L., J. Ward and H. D. Critchley (2010) The neuropsychological impact of insular cortex lesions. *J. Neurol. Neurosurg. Psychiatry* 81: 611–618.

Kauer, J. A. (2004) Learning mechanisms in addiction: Synaptic plasticity in the ventral tegmental area as a result of exposure to drugs of abuse. *Annu. Rev. Physiol.* 66: 447–475.

LeDoux, J. E. (1987) Emotion. In V. B. Mountcastle, F. Plum, S. R. Geiger (eds.), *Handbook of Physiology, Section 1. The Nervous System (Vol. 5). Higher Function of the Brain, Part 1.* Bethesda, MD: American Physiological Society, pp. 419–459.

LeDoux, J. E. (1991) The limbic system concept. *Concepts Neurosci.* 2: 169–199.

LeDoux, J. E. (2012) Rethinking the emotional brain. *Neuron* 73(4): 653–676.

Mayberg, H. S. (1997) Limbic-cortical dysregulation: A proposed model of depression. *J. Neuropsychiatry Clin. Neurosci.* 9: 471–481.

Nestler, E. J. (2005) Is there a common molecular pathway for addiction? *Nature Neurosci.* 8: 1445–1449.

Ochsner K. N. and J. J. Gross (2005) The cognitive control of emotion. *Trends Cogn. Sci.* 9: 242–249.

Viulleumier, P. and J. Driver (2007) Modulation of visual processing by attention and emotion: Windows on causal interactions between human brain regions. *Phil. Trans. R. Soc. B: Biol. Sci.* 362: 837–855.

Important Original Papers

Anderson, A. K. and E. A. Phelps (2001) Lesions of the human amygdala impair enhanced perception of emotionally salient events. *Nature* 411: 305–309.

Bard, P. (1928) A diencephalic mechanism for the expression of rage with special reference to the sympathetic nervous system. *Am. J. Physiol.* 84: 490–515.

Bremner, J. D. (2006) Traumatic stress: effects on the brain. *Dialogues Clin. Neurosci.* 8: 445–461.

Bremner, J. D. and 13 others (2003) MRI and PET study of deficits in hippocampal structure and function in women with childhood sexual abuse and posttraumatic stress disorder. *Am. J. Psychiatry* 160: 924–932.

Calder, A. J., J. Keane, F. Manes, N. Antoun and A. W. Young (2000) Impaired recognition and experience of disgust following brain injury. *Nat. Neurosci.* 3: 1077–1078.

Canli, T. and Z. Amin (2002) Neuroimaging of emotion and personality: Scientific evidence and ethical considerations. *Brain Cogn.* 50: 414–431.

Critchey, H. D., S. Wiens, P. Rotshtein, A. Öhman and R. J. Dolan (2004) Neural systems supporting interoceptive awareness. *Nat. Neurosci.* 7: 189–195.

Downer, J. L. de C. (1961) Changes in visual agnostic functions and emotional behaviour following unilateral temporal pole damage in the "split-brain" monkey. *Nature* 191: 50–51.

Ekman, P., R. W. Levenson and W. V. Friesen (1983) Autonomic nervous system activity distinguishes among emotions. *Science* 221: 1208–1210.

Glannon, W. (2006) Neuroethics. *Bioethics* 20: 37–52.

Kalin, N. H., C. Larson, S. E. Shelton and R. J. Davidson (1998) Asymmetric frontal brain activity, cortisol, and behavior associated with fearful temperament in rhesus monkeys. *Behav. Neurosci.* 112: 286–302.

Klüver, H. and P. C. Bucy (1939) Preliminary analysis of functions of the temporal lobes in monkeys. *Arch. Neurol. Psychiat.* 42: 979–1000.

MacLean, P. D. (1949) Psychosomatic disease and the "visceral brain": Recent developments bearing on the Papez theory of emotion. *Psychosom. Med.* 11: 338–353.

MacClean, P. D. (1964) Psychosomatic disease and the "visceral brain": Recent developments bearing on the Papez theory of emotion. In *Basic Readings in Neuropsychology*, R. L. Isaacson (ed.). New York: Harper and Row, pp. 181–211.

Milad, M. R. and 9 others (2009) Neurobiological basis of failure to recall extinction memory in posttraumatic stress disorder. *Biol. Psychiatry* 66: 1075–1082.

Papez, J. W. (1937) A proposed mechanism of emotion. *Arch. Neurol. Psychiat.* 38: 725–743.

Phillips, R. G. and LeDoux, J. E. (1992) Differential contribution of amygdala and hippocampus to cued and contextual fear conditioning. *Behav. Neurosci.* 106: 274–285.

Robinson, T. E. and K. C. Berridge (1993) The neural basis of drug craving: an incentive-sensitization theory of addiction. *Brain Res. Rev.* 18: 247–301.

Ross, E. D. and M.-M. Mesulam (1979) Dominant language functions of the right hemisphere? Prosody and emotional gesturing. *Arch. Neurol.* 36: 144–148.

Shackman A. J., B. W. McMenamin, J. S. Maxwell, L. L. Greischar and R. J. Davidson (2010) Right dorsolateral prefrontal cortical activity and behavioral inhibition. *Psychol. Sci.* 20: 1500–1506.

Shin, L. M. and I. Liberzon (2010) The neurocircuitry of fear, stress, and anxiety. *Neuropsychopharmacol. Rev.* 35: 169–191.

Vuilleumier, P., M. Richardson, J. Armony, J. Driver and R. J. Dolan (2004) Distant influences of amygdala lesion on visual cortical activation during emotional face processing. *Nat. Neurosci.* 7: 1271–1278.

Whalen, P. J. and 9 others (2004) Human amygdala responsivity to masked fearful eye whites. *Science* 306: 2061.

Williams, M. A., A. P. Morris, F. McGlone, D. F. Abbott and J. B. Mattingley (2004) Amygdala responses to fearful and happy facial expressions under conditions of binocular suppression. *J. Neurosci.* 24: 2898–3004.

Books

Appleton, J. P. (ed.) (1992) *The Amygdala: Neurobiological Aspects of Emotion, Memory and Mental Dysfunction*. New York: Wiley-Liss.

Armony, J. and P. Vuilleumier (2013) *The Cambridge Handbook of Human Affective Neuroscience*. Cambridge, UK: Cambridge University Press.

Barrett, L. F., M. Lewis and J. M. Haviland-Jones (2016) *Handbook of Emotions,* 4th Edition. New York: Guilford Press.

Borod, J. C. (2000) *The Neuropsychology of Emotion*. New York: Oxford University Press.

Damasio, A. R. (1994) *Descartes Error: Emotion, Reason, and the Human Brain*. New York: Avon Books.

Darwin, C. (1890) *The Expression of Emotion in Man and Animals,* 2nd Edition. In *The Works of Charles Darwin*, vol. 23, London: William Pickering, 1989.

Ekman, P. and R. J. Davidson (1994) *The Nature of Emotions.* New York: Oxford University Press.

Gross, J. J. (2007) *Handbook of Emotion Regulation*. New York: Guilford Press.

Hellige, J. P. (1993) *Hemispheric Asymmetry: What's Right and What's Left*. Cambridge, MA: Harvard University Press.

James, W. (1890) *The Principles of Psychology*, vols. 1 and 2. New York: Dover Publications (1950).

LeDoux, J. (1998) *The Emotional Brain: The Mysterious Underpinnings of Emotional Life*. New York: Simon and Schuster.

Rolls, E. T. (1999) *The Brain and Emotion*. Oxford: Oxford University Press.

Go to the NEUROSCIENCE 6e Companion Website at **www.oup.com/uk/Purves6e** for Web Topics, Animations, Flashcards, and more.

CHAPTER

32 Thinking, Planning, and Deciding

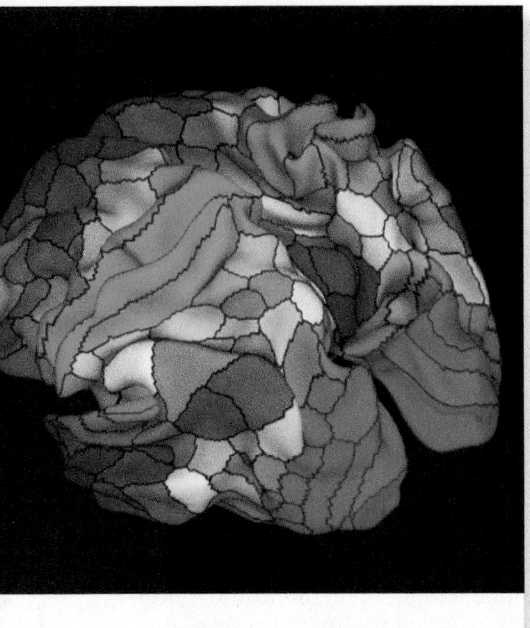

Overview

THINKING, PLANNING, AND DECIDING are some of the most advanced functions of human (and presumably many other) brains. Although people sometimes react in stereotyped ways to the environment, the human brain has an astonishing capacity to respond flexibly and to anticipate the outcome of behavior. Thus, behavior depends not just on sensory input, but on remembered information, goals, and predictions about what might happen. This flexibility gives humans the ability to play chess, write novels, and do science, among other complex behaviors.

Because of this complexity, such brain functions are difficult to study with the same precision used to understand other neural processes. Consequently, thinking, planning, and deciding are as closely tied to psychology and philosophy as to neurobiology. Nevertheless, recent advances in EEG, PET, and fMRI methods have enabled neuroscientists to provide a rough sketch of the brain networks associated with these key abilities. Complementing this work, studies of non-human animals have revealed some of the underlying circuitry and cellular mechanisms. Finally, the advent of high-speed computing and the development of computational models have allowed neuroscientists to test biologically plausible simulations of these brain functions.

Adding their importance, many neurological and psychiatric diseases are characterized by impairments of thinking, planning, and deciding. Addiction, depression, schizophrenia, and obsessive-compulsive disorder are all linked to dysregulation of cognitive flexibility, including impairments in long-term planning, the ability to learn from mistakes, motivation, and decision making. The impact of these impairments on patients and their families, not to mention on society and the economy, underscores the importance of understanding them.

A Sketch of the Relevant Circuitry

Thinking, planning, and deciding draw on a suite of related mental activities that are loosely described under the rubric of *executive functions*. Although this term alludes to managing corporate activities, it would be wrong to imagine that the brain actually works this way. It seems more apt to consider executive functions as part of a *control system* (a term borrowed from engineering) that adapts cognitive functions to the current environment and state of the organism—just as a thermostat adjusts heating and cooling systems to maintain the temperature. A complex neural control system would entail several elements, including short-term memory, reward evaluation, conflict resolution, and response inhibition. Because several anatomical areas in the prefrontal cortex have been linked to these functions, it is tempting to conclude that each of these elements maps directly onto a specific brain region. In fact,

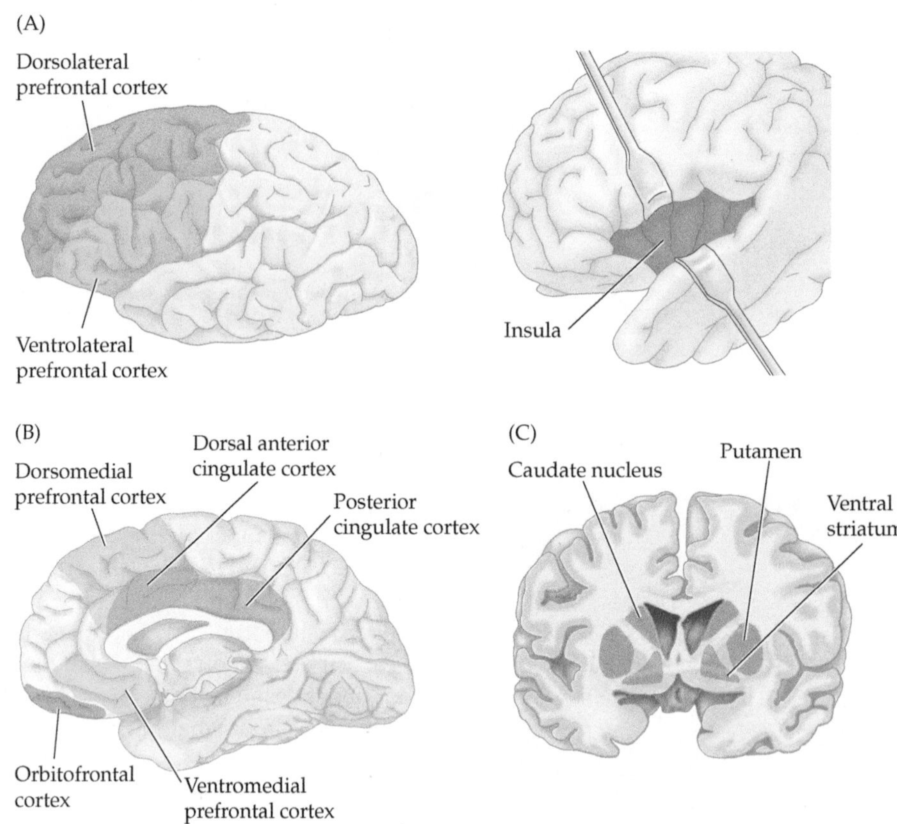

FIGURE 32.1 Major brain regions involved in thinking, planning, and deciding. Three views of the human brain: lateral (A), medial (B), and coronal (C). Highlighted are the dorsolateral prefrontal cortex, ventrolateral prefrontal cortex, insula, dorsomedial prefrontal cortex, ventromedial prefrontal cortex, dorsal anterior cingulate cortex, orbitofrontal cortex, and posterior cingulate cortex and their major targets in the striatum: the caudate nucleus, putamen, and ventral striatum.

complex mental operations such as thinking, planning, and deciding are distributed across multiple, interacting brain regions (Figure 32.1), including the prefrontal cortex.

The **prefrontal cortex** (**PFC**) is the portion of the frontal lobe anterior to the motor cortex in both humans and non-human primates (Figure 32.2), and it is especially prominent in humans. Although the existence of Brodmann cytoarchetectonic areas is well established (see Figure 27.1), histological boundaries defining subregions of the PFC are not agreed on (Figure 32.3). Nor are these regions clearly defined electrophysiologically, as they are, for example, in the visual cortex in the occipital lobe. Thus, it is difficult if not impossible to reliably identify the flow of information through the PFC; connections between different cortical regions are often reciprocal, and it is hard to differentiate which connections are feedback and which are feedforward. This robust connectivity is a conundrum that applies to the association cortices generally, as indicated in Chapter 27.

Despite this complexity, a rough path from input to output can be identified. Information about sensory stimuli is conveyed to the orbitofrontal cortex (that is, the orbital portion of the PFC), where representations of the *values* of various options may be represented. Value signals, and probably much other information, then flow rostrally and laterally to the lateral and medial PFC, where information that influences decision making comes into play. One interpretation is that the resulting signals then flow dorsally to other PFC regions that use this information to plan possible responses. From there, signals propagate to the premotor and parietal cortices, and finally to the motor and other cortical regions that give rise to behavior (remember that behavior is not limited to motor actions but also includes perception, attention, emotion, memory, and more) (Figure 32.4). These pathways and their targets are influenced by neuromodulatory transmitters such as dopamine, serotonin, and acetylcholine; by specialized cortico-basal ganglia loops; and by emotional and memory processes in the amygdala and hippocampus, respectively. The remainder of the chapter outlines some of these interactions, with the caveat that their complexity, and the difficulty of modeling these functions in non-human animals, limit what can be concluded with confidence.

Orbitofrontal Cortex and the Evaluation of Options

Reward is a basic influence on the decisions individuals make and the responses they make to stimuli. Generally speaking, the reward value of an option refers to the benefit it provides the decision maker, either in the short term (such as the relief provided by a cold drink on a hot day) or in the long term (such as a retirement annuity or the increased opportunities provided by enrolling one's child in a better school). Estimating the value of an option

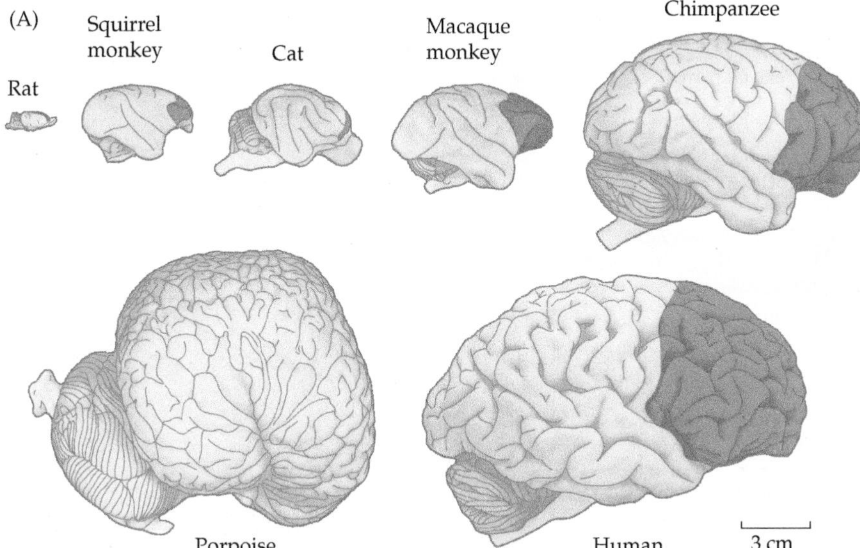

FIGURE 32.2 Size of the cerebral cortex and prefrontal cortex in mammals. (A) Among the seven species shown here, humans have the largest cerebral cortex but have a larger PFC (blue) relative to the other (non-primate) species, even controlling for brain size. The porpoise brain is provided for size comparison; its PFC is not indicated because the borders of it are not known. (B) Within the order of primates, the size of the PFC is roughly proportional to that of the rest of the neocortex. Brodmann's work in the early twentieth century had suggested that humans and other great apes have a disproportionately large PFC. (C) Later work has indicated that relative size of the PFC is roughly constant within the order of primates. (B after Brodmann, 1912; C after Semendeferi et al., 1997, 2002.)

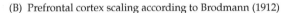

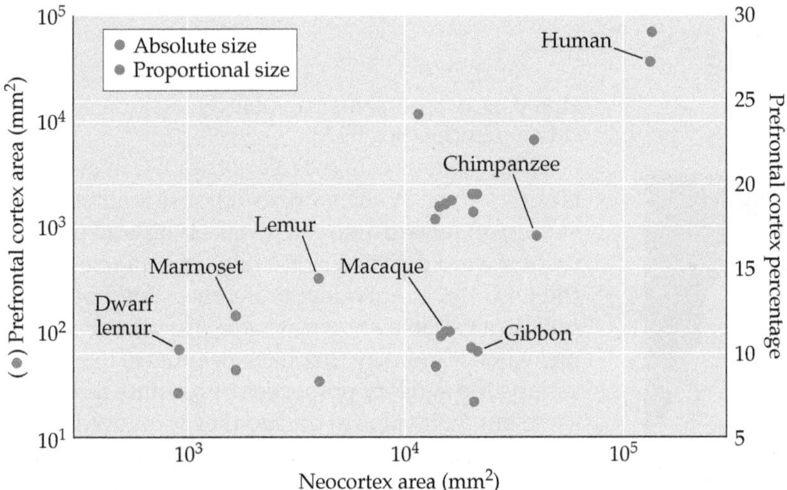

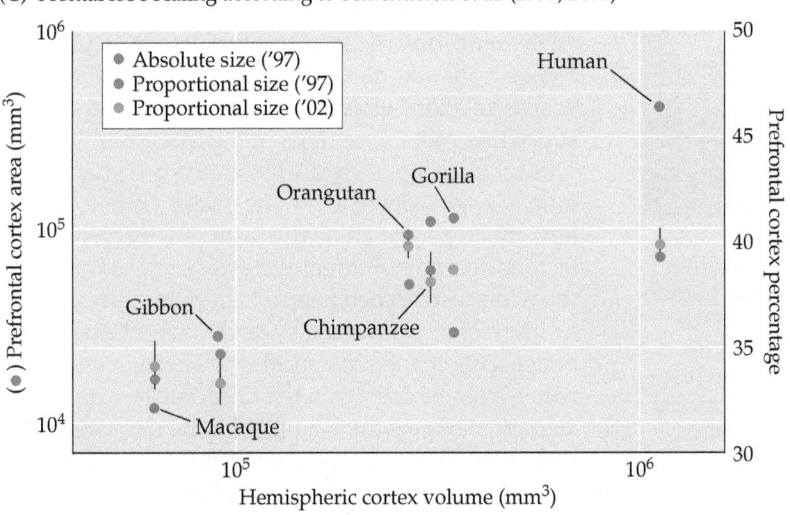

involves perceiving its sensory properties, identifying the situation, and retrieving information about past experiences with similar contexts from memory. For example, a diner might be attracted by a particular entrée but may recall a prior experience with food poisoning after eating that dish and decide to pass. Such details are combined to estimate a value that can be compared with the values of other options to guide decisions. **Evaluation**—the estimation of the value of an option based on both past and present information—has been linked to the **orbitofrontal cortex** (**OFC**) (see Figures 32.3 and 32.4) by virtue of the OFC's anatomical connections, the behavioral disruptions that follow damage to this area, and its activation during learning and decision making.

The OFC receives input from all of the major sensory modalities (vision, audition, somatic sensation, olfaction, and gustation), giving it access to the information necessary to identify options. Unlike other prefrontal regions, however, the OFC has few motor connections, consistent with the idea that the OFC provides inputs to systems that themselves inform the selection and execution of behavior. Furthermore, the OFC receives inputs from the hippocampus and adjacent regions in the medial temporal lobe that are involved in memory storage and retrieval (see Chapter 30). These inputs presumably provide information about prior experiences to improve estimates of value. Finally, the OFC receives inputs from reward-related dopamine neurons in the midbrain that help

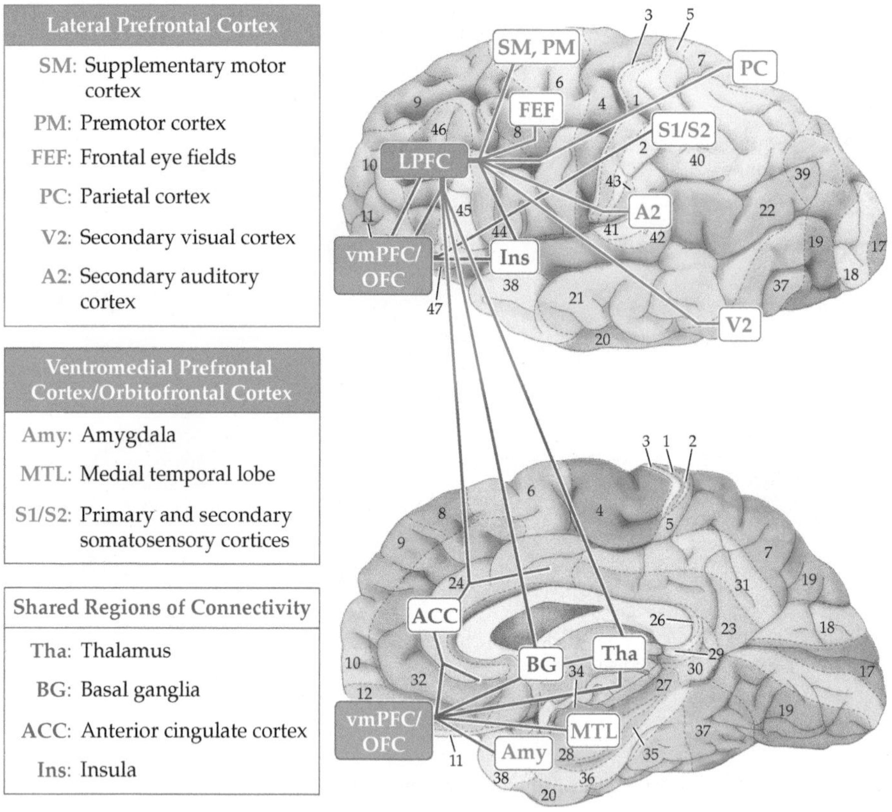

FIGURE 32.3 Connectivity of the prefrontal cortex. Neurons in the PFC project to and receive input from secondary sensory cortices, motor preparatory structures, and parietal cortex. This schematic diagram shows some of the major connections for the lateral prefrontal cortex (LPFC) and for the ventromedial prefrontal cortex and orbitofrontal cortex (vmPFC/OFC, combined here for simplicity). All indicated connections are bidirectional, with the important exception of a unidirectional projection from the LPFC to the basal ganglia (which projects back to the LPFC via the thalamus).

Lateral Prefrontal Cortex

- SM: Supplementary motor cortex
- PM: Premotor cortex
- FEF: Frontal eye fields
- PC: Parietal cortex
- V2: Secondary visual cortex
- A2: Secondary auditory cortex

Ventromedial Prefrontal Cortex/Orbitofrontal Cortex

- Amy: Amygdala
- MTL: Medial temporal lobe
- S1/S2: Primary and secondary somatosensory cortices

Shared Regions of Connectivity

- Tha: Thalamus
- BG: Basal ganglia
- ACC: Anterior cingulate cortex
- Ins: Insula

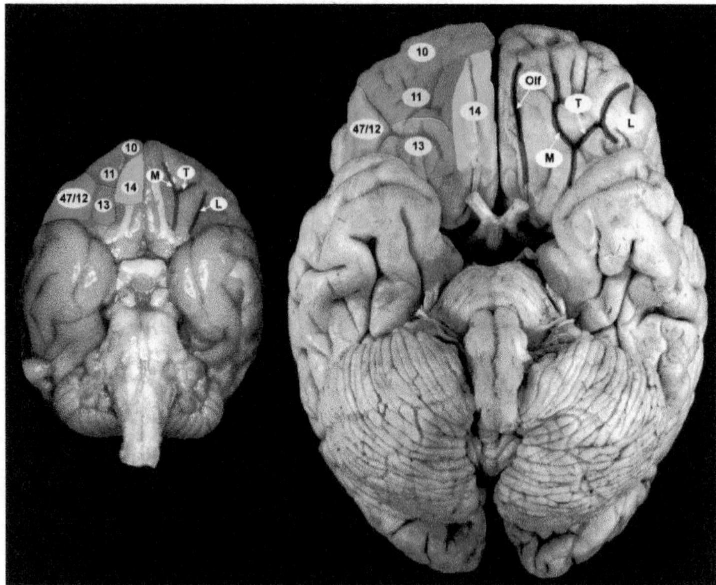

FIGURE 32.4 Gross anatomy of the orbitofrontal cortex. The ventral surfaces of the brains of a rhesus monkey (left) and a human (right) are shown for comparison. In spite of the different sizes of the brains, the structure of the OFC (the colored regions) is largely preserved across the two species. Numbers here refer to Brodmann areas, which are the major anatomical subdivisions of the OFC. The majority of research on the OFC in both species has been focused on Brodmann's area 13. (Olf = olfactory sulcus, M = medial orbital sulcus, T = transverse orbital sulcus, L = lateral orbital sulcus.) (From Wallis, 2007.)

shape associations among objects, actions, and their consequences (Box 32A).

Direct evidence for the role of the OFC in evaluation is fairly strong. When monkeys choose among options varying in reward amount, reward type, and probability of reward delivery, firing rates of some neurons in the OFC track individual preferences for a particular option, a variable known as *subjective value*. For example, when a monkey that likes peanuts is fed them to satiety, the sensory properties of peanuts remain the same but their value to the monkey is reduced. Satiety is accompanied by a reduction in the firing rates of OFC neurons in response to more peanuts, implying that the OFC encodes the subjective value of a food and not just its sensory properties.

Evidence for the evaluative role of the OFC also comes from neuroimaging studies, which show a robust correlation between hemodynamic activity and the subjective value of options in decision-making tasks. This relationship is vividly illustrated in a study on the human enjoyment of wine. The investigators found that how much people enjoy the taste of wine depends on factors that do not affect its taste, such as how much the wine costs. When people believe they are drinking a more expensive wine, they tend to report that it tastes better. This change in appeal is reflected in changes in hemodynamic activity in the OFC, suggesting that the activity mediates the change in enjoyment associated with price. The relationship between activity in the OFC and personal preferences is so robust that measures of

BOX 32A ■ Dopamine and Reward Prediction Errors

Perhaps no molecule has permeated popular culture as much as dopamine. References to dopamine can be found on late-night TV, in newspaper editorials, and in casual conversation. Dopamine has come to prominence in part through its role in drug addiction. Nearly all addictive drugs exert their effects through their ability to alter dopamine release or reuptake from the relevant synapses. This basic understanding of the molecular basis of drug addiction has led to the view that dopamine signaling is the basis of pleasure. The popular view has some truth to it but is misleading.

One factor that has encouraged public awareness of dopamine is the series of self-stimulation studies carried out in rats in the 1950s. James Olds and Peter Milner implanted rats with electrodes in the medial forebrain bundle, a structure that, when activated, causes the release of dopamine in the nucleus accumbens of the striatum. Olds and Milner's innovation was allowing rats to control the stimulation. Faced with the choice between self-stimulation and other activities, the rats chose to self-stimulate, to the exclusion of eating and drinking, even to the point of death. These studies emphasize the direct influence of dopamine circuitry over an individual's basic drives. Along with its demonstrated importance in addiction, this work led to the idea that dopamine release is the cause of pleasure.

That inference is wrong, however. In further studies on rats, Kent Berridge and colleagues showed that the motivation to seek rewards and the pleasure obtained from them are distinct, with different chemical and neuroanatomical substrates. The terms *liking* and *wanting* are often used to describe these two processes. An individual can enjoy (*like*) things without having a drive to seek more of them, just as one can be driven (*want*) to pursue things, often with great vigor, without pleasure. Consider a smoker trying to quit who is miserable and hates stepping outside to smoke, but is compelled by his addiction to do it anyway. It is unlikely the smoker would describe the smoking experience as pleasurable. Or consider a sufferer of OCD, hand washing repeatedly; even though motivated to do it, the sufferer would not describe this process as enjoyable. It is this drive—the wanting, not the liking—that dopamine regulates. Berridge argues that a different brain system, the μ-opioid system, is responsible for pleasure.

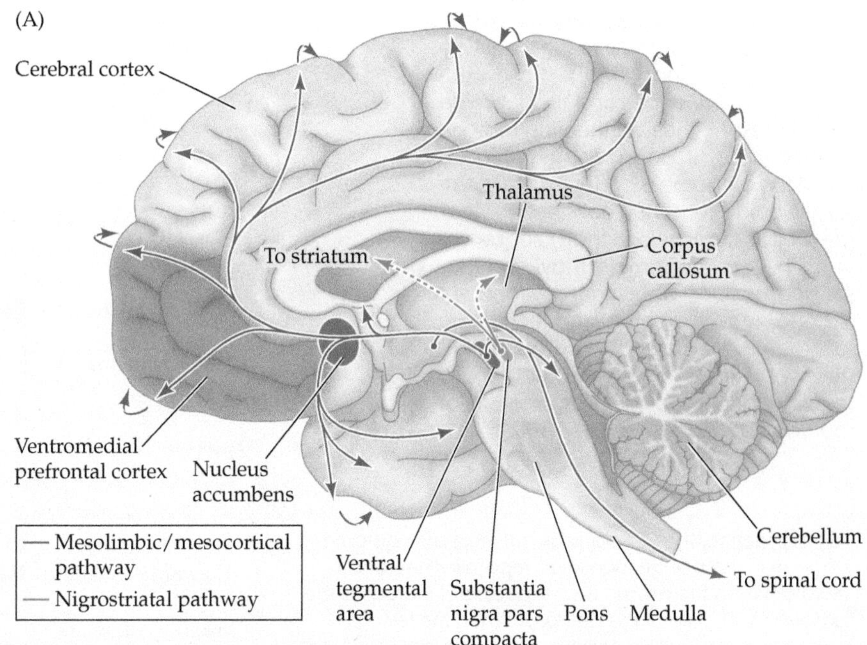

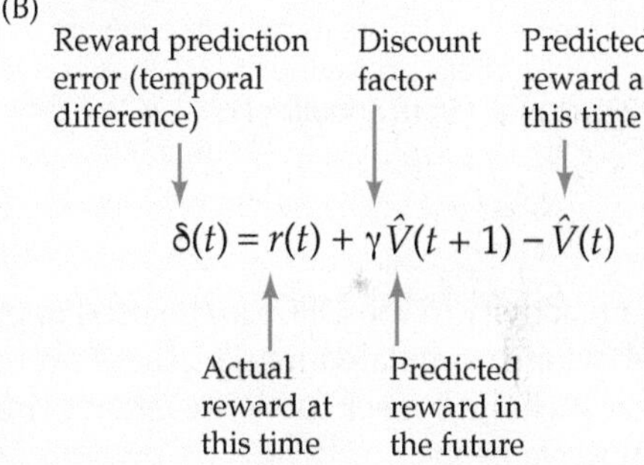

(A) Schematic of the dopamine system of the brain. Two of the major sources of dopamine in the brain are the ventral tegmental area (VTA) and the substantia nigra pars compacts (SNPc). These two regions house the cell bodies of neurons that project to much of the brain and serve as the source of the neuromodulator dopamine. (B) The equation of the reward prediction error (RPE) signal—the hallmark of the response of the dopamine neuron. (After Schultz et al., 1997.)

The most completely studied dopaminergic neurons are those whose cell bodies are in the ventral tegmental area (VTA) of the midbrain and the substantia nigra pars compacta (SNPc) (Figure A). These neurons project widely throughout the brain, but especially to the PFC and ventral striatum, where they are thought to regulate neural activity. Recordings from dopamine neurons show that their responses can be described as a **reward prediction error** (**RPE**). Whenever individuals execute some behavior, they predict the likely outcome (*reward*). RPE is simply the difference between what was predicted and what actually materialized (Figure B). If what materialized is better than predicted, the RPE is positive; if what materialized is worse than predicted, the RPE is negative. The RPE is important for learning: If one option is better than expected, individuals update their estimate of its value and choose it more often in the future—as predicted by formal learning models developed in psychology and computer science.

Continued on the next page

BOX 32A ■ *(continued)*

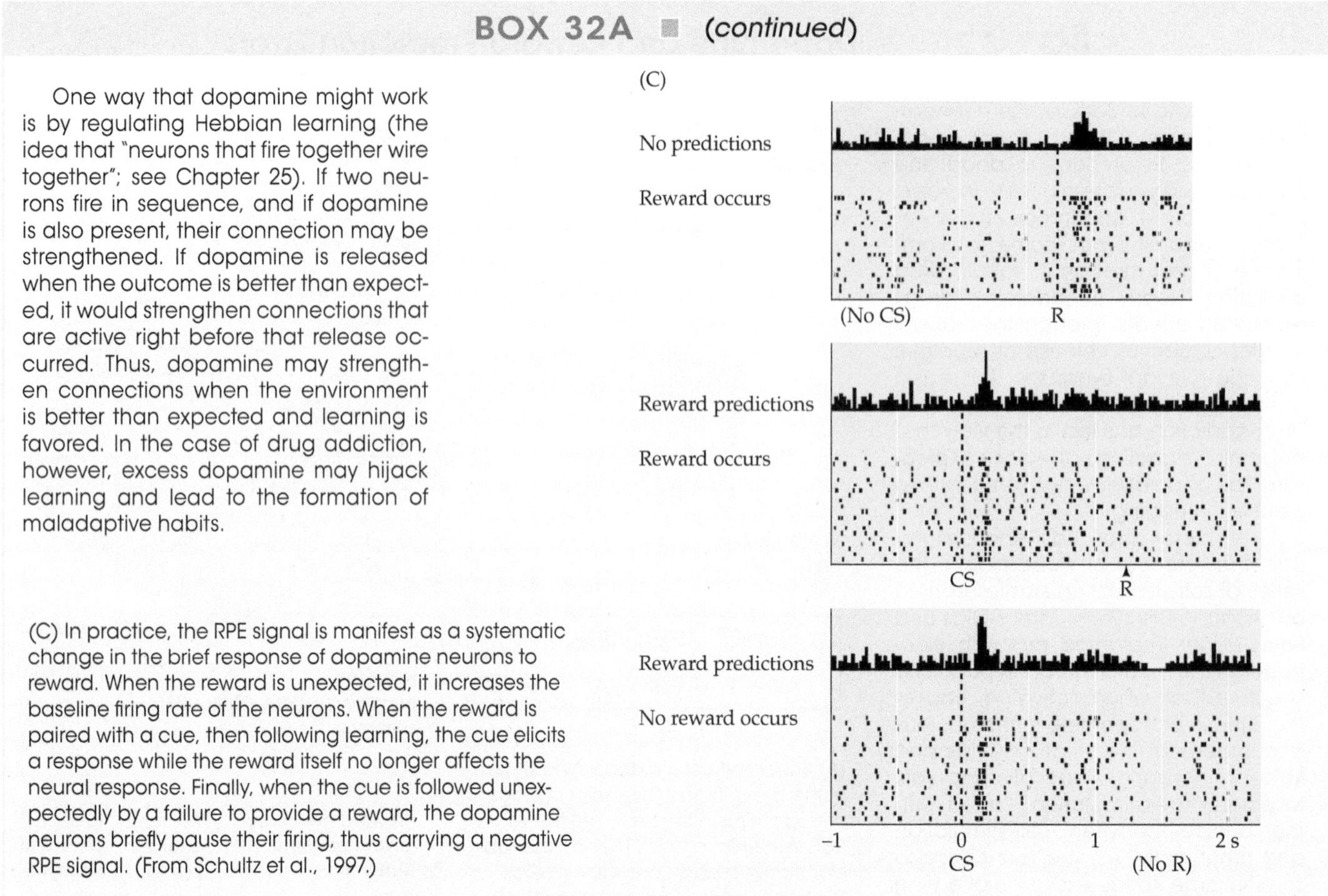

One way that dopamine might work is by regulating Hebbian learning (the idea that "neurons that fire together wire together"; see Chapter 25). If two neurons fire in sequence, and if dopamine is also present, their connection may be strengthened. If dopamine is released when the outcome is better than expected, it would strengthen connections that are active right before that release occurred. Thus, dopamine may strengthen connections when the environment is better than expected and learning is favored. In the case of drug addiction, however, excess dopamine may hijack learning and lead to the formation of maladaptive habits.

(C) In practice, the RPE signal is manifest as a systematic change in the brief response of dopamine neurons to reward. When the reward is unexpected, it increases the baseline firing rate of the neurons. When the reward is paired with a cue, then following learning, the cue elicits a response while the reward itself no longer affects the neural response. Finally, when the cue is followed unexpectedly by a failure to provide a reward, the dopamine neurons briefly pause their firing, thus carrying a negative RPE signal. (From Schultz et al., 1997.)

brain activity in the OFC can be used to predict purchasing behavior in simulated marketplaces over an array of products, including snack foods, beverages, and ads that feature attractive people.

Indeed, a remarkable feature of the OFC is its apparent ability to contribute directly to decisions about so many different things, ranging from which soda to buy from a vending machine to which college to attend (see Figure 27.6). This flexibility has led to the proposal that the OFC and/or the **ventromedial prefrontal cortex (vmPFC)** make use of a universal value "format" that allows comparison of any set. This **common currency theory** is supported by hemodynamic response patterns measured with fMRI. However, direct recordings from OFC neurons reveal a much greater diversity of response patterns, suggesting that the common currency theory may be simplistic.

Actually making a choice occurs after evaluation and requires active maintenance of the values of two or more options. This process seems likely to occur in the OFC as well, or in the neighboring vmPFC, a structure with similar neuroanatomy and function as the OFC. Neurons in both regions show systematic changes in firing when the values of multiple options are maintained in short-term memory. Lesions of the vmPFC are associated with deficits in comparing the values of disparate options. For example, when a person chooses between options that differ along multiple dimensions, such as price, styling, and gas mileage of a new car, the brain must make separate comparisons and use the results to make a decision. Lesions to the vmPFC (which often include the OFC as well) impair this process. Individuals with vmPFC lesions are less efficient when comparing options.

This evidence extends the role of the OFC to include **credit assignment**, which is the process of identifying the one stimulus among many in the current context that is responsible for a reward or punishment. The idea of credit assignment was anticipated by the psychologist Edward Thorndike in his classic work on the *law of effect*. His idea was that "responses that produce a satisfying effect in a particular situation become more likely to occur again in that situation, and responses that produce a discomforting effect become less likely to occur again in that situation." More generally, stimuli associated with pleasant events gain pleasant associations themselves, and vice versa. This concept seems simple, but in many cases a large number of stimuli occur at the same time and their consequences may be delayed or provide both plusses and minuses. Brain mechanisms need to assign credit to the actual stimuli that

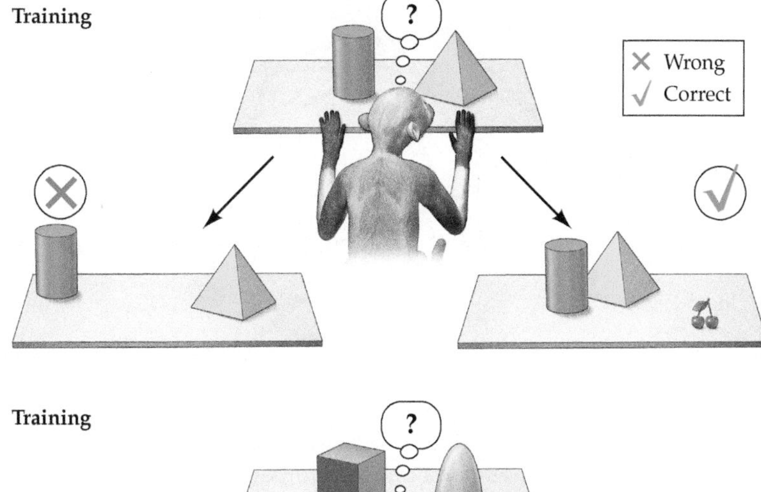

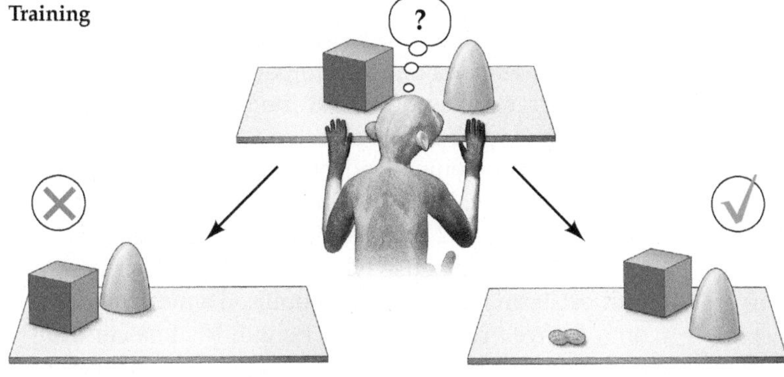

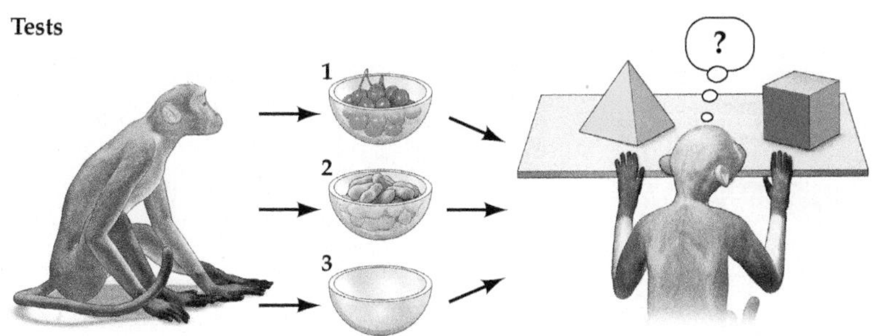

FIGURE 32.5 **Effects of orbitofrontal cortex lesions on the reinforcer devaluation task.** Some of the most important insights into the role of the OFC come from the results of lesion tasks in primates. In the reinforcer devaluation task, a monkey is trained with pairs of stimuli that are associated through training with either a reward or no reward. The rewards themselves differ and are predicted by the stimuli. During the testing phase, preference for the stimulus is experimentally controlled by selective satiation. For example, to make the monkey more motivated to eat cherries, the monkey would be fed to satiation on peanuts (this process devalues the reinforcer). The monkey then chooses between stimuli associated with cherries or peanuts. Monkeys with OFC lesions do not adjust their preference toward the preferred food as well as non-lesioned control monkeys do. (After Baxter and Murray, 2002.)

predict rewards and punishments, and a good deal of evidence supports the idea that the OFC contributes to this process. For example, lesions to the OFC selectively impair the ability to link rewards to events (Figure 32.5). Monkeys with OFC lesions assign positive value to stimuli that predict aversive events as long as the stimuli are surrounded by other positive events. It thus seems that the OFC is critical for accurately linking events with their values.

It would be overly simple, however, to conclude that the OFC and vmPFC are the only brain areas in which values are computed, maintained, and compared. Other evidence supports roles for additional brain regions, including the ventral striatum and the **dorsal anterior cingulate cortex (dACC)**, in these processes. Moreover, a good deal of evidence points to a more sophisticated role for the OFC in behavior. For example, one current theory holds that the OFC maintains a cognitive map of the set of currently relevant behavioral stimuli, their values, and potential outcomes. This idea suggests that the OFC operates much like a switchboard, linking the external world and internal states with the possible outcomes of choices.

Dorsolateral Prefrontal Cortex and the Planning and Organization of Behavior

Imagine taking a regular route to work. If one day the road is blocked, you quickly recalculate and adjust your path to follow the best alternative route. This flexibility is a remarkable but poorly understood feature of human cognition. By its very nature, flexibility must override habits, which are efficient solutions to predictable problems. The simplest habit is a reflex, which is a built-in response to events and outcomes that have occurred repeatedly throughout the lifetime of an organism or across the generations of a species. Reflexive behavior is most closely associated with invertebrates, and more ancient portions of the human nervous system such as the spinal cord and the autonomic nervous system. Flexibility, in contrast, is more often associated with animals that have relatively larger brains, such as primates, carnivores, and some cetaceans. Nonetheless, there is no clear dividing line between reflexes and flexible behavior, other than the number, complexity, and time horizon of processes that intervene between sensory input and motor (or other) output. The most flexible, complex, and future-oriented behaviors produced by humans and other mammals appear to be organized and planned, in part, by processes occurring in the **dorsolateral prefrontal cortex (DLPFC)**.

The DLPFC consists primarily of Brodmann's areas 9 and 46 and is connected to several other cortical regions. It communicates with reward-related regions such as the OFC and anterior cingulate cortex, premotor cortex, and parietal areas concerned with attention. Thus, it has connections well suited to serve as regulators of input-output pathways, using value and other factors to shape this process. Accordingly, the function of the DLPFC is sometimes likened to switching in a railroad yard, rerouting

FIGURE 32.6 Schematic illustration of the function of the dorsolateral prefrontal cortex. The brain can be thought of as a network that transforms inputs to outputs through information propagated along weighted connections. The path that information takes is, in turn, influenced by regulatory units thought to be housed in the DLPFC. These regulatory units receive information from other units in the system. (After Miller, 2000.)

connections between different tracks to align trains from their origins with their intended destinations (Figure 32.6). (This idea is distinguished from the cognitive map model of the OFC because it involves an active regulation of other circuits, not just a linkage of stimuli and their values.) In a similar manner, the DLPFC may control the responses of other groups of neurons, making them more or less responsive to inputs and feedback, thereby producing different responses in different contexts.

A particularly important component of this type of control system is short-term memory, the ability to keep information in mind to guide behavior (see Chapter 30). Short-term memory is distinguished from long-term memory by its duration and purpose. If someone asks you to remember a phone number, you can maintain it in mind for tens of seconds, but the memory quickly fades, especially if you become distracted. Storage of information in short-term memory reflects changes in the firing rates of neurons in various regions of the brain, including the DLPFC (Figure 32.7). Specifically, firing rates of neurons in the DLPFC increase while information is maintained in short-term memory. For example, if a monkey is trained on a delayed response task, he will fixate a central spot and then covertly remember one of eight positions from a set of positions around the spot; he must store that spatial

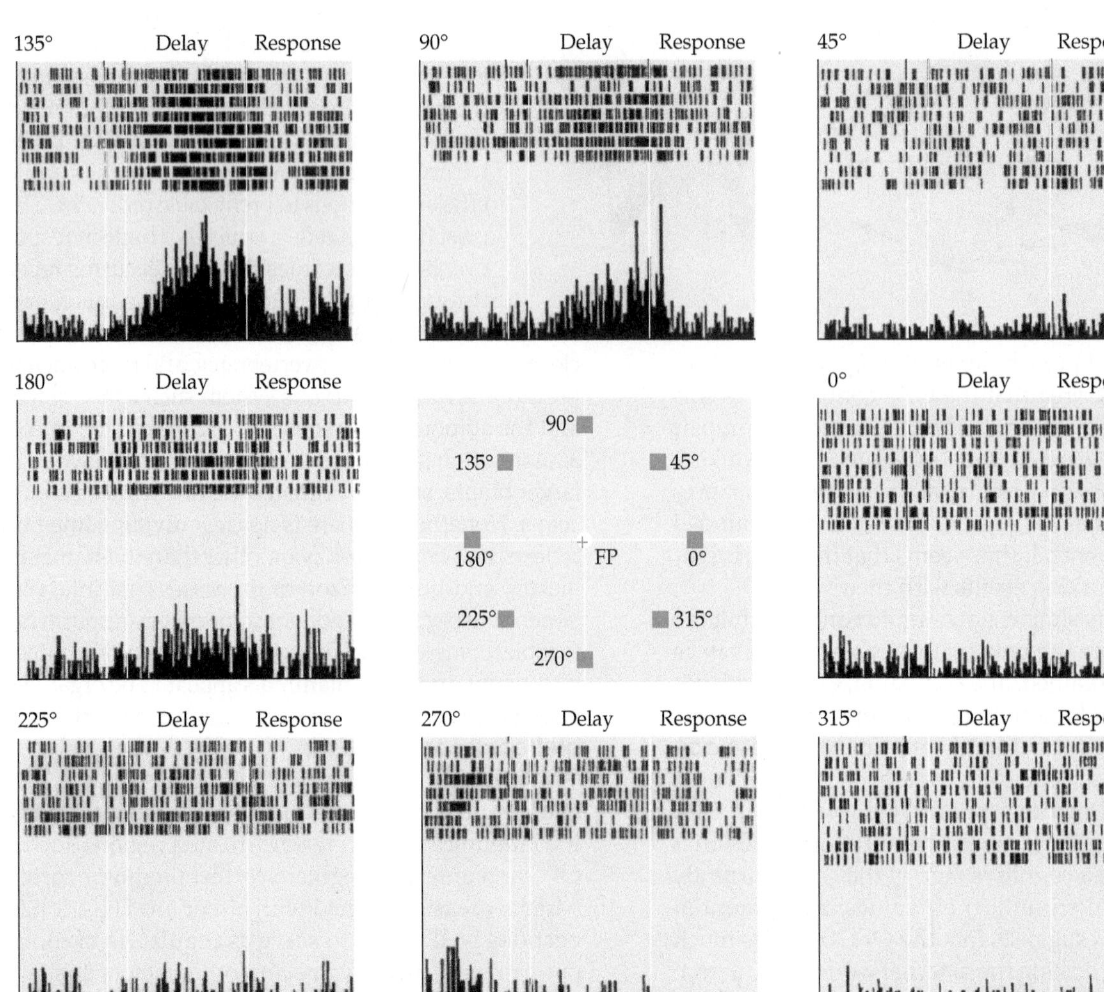

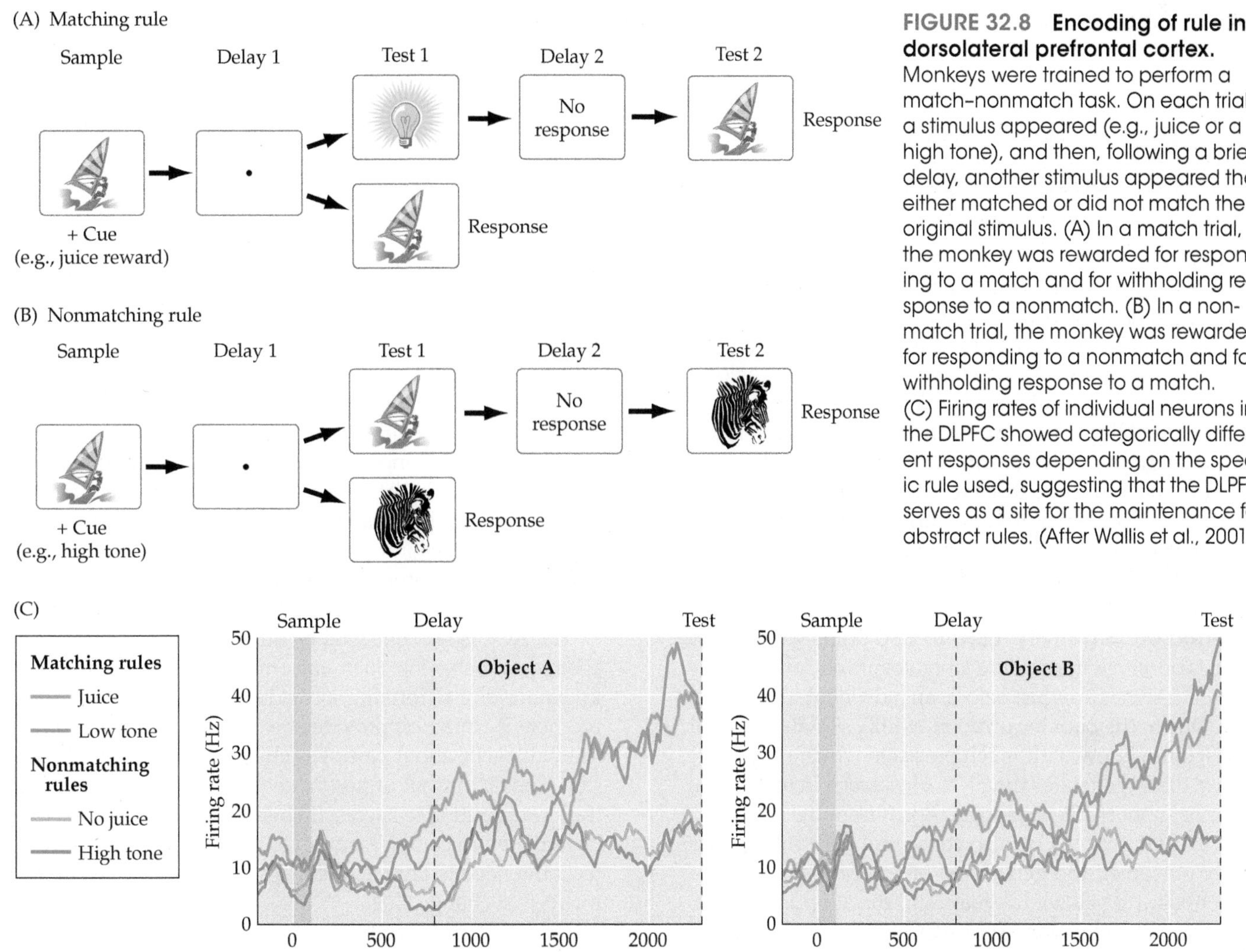

FIGURE 32.8 Encoding of rule in dorsolateral prefrontal cortex. Monkeys were trained to perform a match–nonmatch task. On each trial, a stimulus appeared (e.g., juice or a high tone), and then, following a brief delay, another stimulus appeared that either matched or did not match the original stimulus. (A) In a match trial, the monkey was rewarded for responding to a match and for withholding response to a nonmatch. (B) In a nonmatch trial, the monkey was rewarded for responding to a nonmatch and for withholding response to a match. (C) Firing rates of individual neurons in the DLPFC showed categorically different responses depending on the specific rule used, suggesting that the DLPFC serves as a site for the maintenance for abstract rules. (After Wallis et al., 2001.)

information in this form of short-term memory. Neurons in the DLPFC show specific changes in their firing rate that depend on the position of the remembered dot; that is, their response depends on the contents of short-term memory. Endorsing the link between neuronal activity and short-term memory, damage to the DLPFC is associated with impairments in short-term memory capacity and duration. Such impairments are evident in experimental animals with ablations to the DLPFC as well as in human patients who have suffered damage to this area.

Another important element in cognitive control is maintenance of rules and corresponding changes in behavior when the rules change. Neurons in the DLPFC show systematic patterns of activity that accord with specific rules, suggesting this area also maintains a representation of abstract information that guides complex behavior (Figure 32.8). Moreover, systematic changes in the firing rates of neurons in the DLPFC accompany changes in the rules that govern effective behavior in a particular context. In the **Wisconsin Card Sorting Task** (see Box 27C), an individual is shown a set of cards, each of which has a different number of distinct shapes of varying color. The individual must then place a new card according to an unstated rule, such as shape, color, or number. After a series of trials, the rule is surreptitiously switched. Patients with DLPFC damage can learn to perform this task using the initial rule, but when the rule is changed they tend to continue with the old one and perform poorly. This impairment corresponds to the tendency for patients with DLPFC lesions to become stuck in behavioral routines and not adapt to changing circumstances.

◀ **FIGURE 32.7 Role of the dorsolateral prefrontal cortex in short-term memory.** Evidence for the role of the DLPFC in short-term memory comes from responses of DLPFC neurons in the delayed response task. Monkeys are trained to fixate their gaze on a central spot (FP, fixation cross), and then, while they are fixating, a saccade target appears in the periphery at one of eight eccentric positions (the red squares, 0° through 315°). Monkeys then withhold a saccade, the target disappears, and the monkey is required to remember the position of the eccentric target in short-term memory before a "go" cue indicates the monkey should shift its gaze. Firing rates of individual neurons in the DLPFC depend on the position of the remembered cue, meaning they carry sufficient information to store the short-term memory. This change is observable in averages of the firing rate for each position, and, critically, is observed during the delay period of the task, when there is no information on the screen and short-term memory is the only site of information maintenance. (From Funahashi et al., 1989.)

Cingulate Cortex and Learning from the Consequences of Behavior

Humans and other "intelligent" animals are capable of learning from the consequences of their actions. Doing so requires circuitry that can evaluate the outcomes of decisions and update the control systems that regulate the connections between stimulus inputs and behavioral outputs. Presumably, such feedback signals influence brain regions, like the DLPFC, that are involved in control processes that mediate flexibility. Such **monitoring** is most strongly associated with the **anterior cingulate cortex** (**ACC**). This region is well positioned to serve this role since it has multiple inputs conveying information from a variety of systems, including perception, emotion, attention, and memory.

The ACC consists of Brodmann's area 24 as well as parts of areas 9, 6, and 32 and is thought to be the source of the **error-related negativity** (**ERN**) in EEG studies. The ERN signal is observed in standard laboratory tasks immediately after (or in some cases just before) an individual commits an error. Due to the poor localization of EEG signal sources, it is difficult to know with much precision which part of the brain actually generates the ERN. However, circumstantial evidence supports the idea that it arises from the ACC.

Activity in the ACC is also consistent with the evaluation of outcomes and generation of feedback signals useful in updating behavioral goals and the adoption of new cognitive rules. Responses of ACC neurons are affected not so much by the nature of options or their values, but by the consequences of choosing them. In laboratory studies, the ACC is activated by errors that reduce rewards, which in real life would correspond to disappointment in not achieving a goal. In tasks where the reward depends on the choices individuals make, ACC neurons signal the size of rewards that follow decisions. The ACC thus seems to track the values of outcomes when this information is used to guide future behavior. The ACC also tracks counterfactual or fictive outcomes—in other words, the rewards or punishments associated with options that were not chosen. People and other animals not only monitor the consequences of their actions but also attend to what they might have experienced had they acted differently. There is evidence that the ACC monitors both types of outcomes simultaneously. For example, when stock traders play the market, they respond differently to financial returns depending on how other stocks they could have invested in perform.

The ACC is also engaged by outcomes that lead to rapid changes in behavior. Such outcomes include those that are unexpected or surprising, provide useful information and promote learning, or provide new information about other individuals or about pain, which signals what is or is not safe. The factor that unites these drivers of ACC activity is their contribution to successful behavior. The ACC thus can be regarded as complementary to the DLPFC; the ACC detects the need to change behavior, and the DLPFC implements that change.

Another influence on ACC activity is the **conflict** associated with different action plans activated at the same time. A classic example is the **Stroop effect** (Figure 32.9).

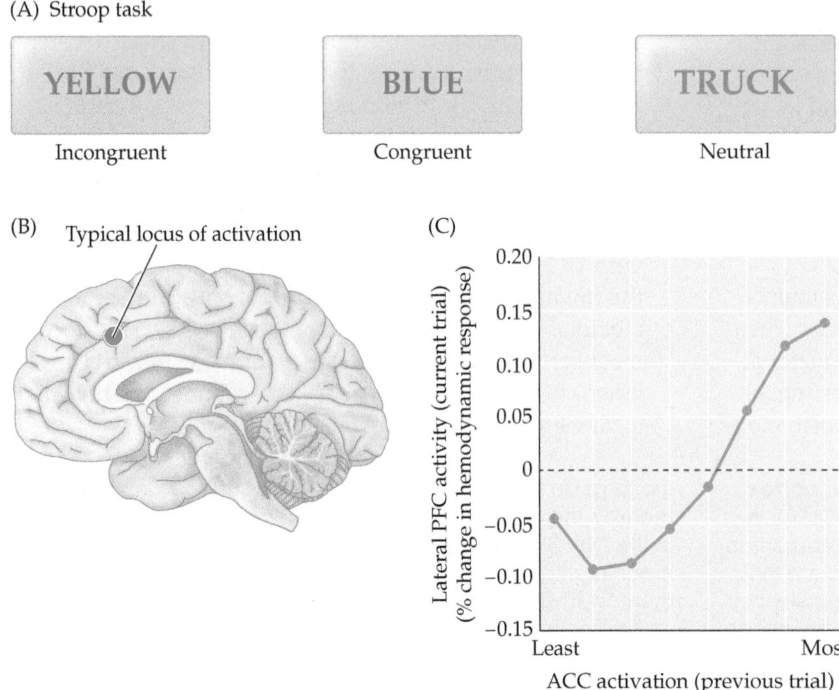

FIGURE 32.9 **Anterior cingulate cortex and the Stroop effect.** (A) In the Stroop task, individuals are asked to say a word printed in a color on a card or computer screen. When the word is incongruent—a different color than the color of the ink—individuals are slower to respond. When the word is neutral (not color related) or congruent (same color), individuals have no difficulty responding. It is thought that the competition between saying the word and saying the ink color come into conflict. (B) During cognitive conflict, responses in the anterior cingulate cortex (ACC) are greatly enhanced. (C) Following conflicting trials, individuals adjust their behavior to improve performance. This adjustment is thought to reflect changes driven by inputs to the dorsolateral prefrontal cortex from the ACC. Activation of the ACC on the previous trial, as measured by fMRI-predicted the response in the DLPFC on the current trial, predicts adjustment. (B after Bush et al., 1998; C after Kerns et al. 2004.)

In a typical Stroop scenario, people are faster saying the word *red* when it is printed in red ink than they are when it is printed in green ink. Presumably they are simultaneously intending to say *red*—the word itself—and *green*—the color in which it is printed. There is some evidence that the ACC is involved in dealing with such conflicts. Direct electrophysiological recordings from the ACC in human neurosurgical patients show evidence of this conflict. If individuals are asked to press a left button when an arrow on the computer screen points left and a right button when an arrow on the screen points right, they can do this task with ease. If the arrows are flanked by other arrows that the individuals are asked to ignore, the individuals are faster and more accurate when the flanking arrows point in the same direction as the central one, and slower and less accurate when they point in the opposite direction. This conflict presumably requires extra control, and this additional demand activates neurons in the ACC, as indicated by a reliable enhancement in activity associated with the hemodynamic response in this area as well.

Consistent with these observations, damage to the ACC leads to impairments in learning from the consequences of actions. An example is obsessive-compulsive disorder (OCD), which is associated with atypical levels of activity in the ACC. Individuals with OCD are overly sensitive to stimuli that would generally be ignored, such as dirt and clutter. OCD is also associated with self-doubt (being unsure the oven is actually off, for example) and failures of self-control. Thus, these individuals feel compelled to perform acts such as counting and washing that they know are illogical and don't want to perform, but can't suppress.

Cingulotomy, or surgical ablation of the ACC, can be an effective treatment for the most severe cases of OCD, for reasons that are unclear. One possibility is that regulatory monitoring activity in the ACC is hyperactive in such individuals, leading to enhanced anxiety and maladaptive behavior. Suppressing this hyperactivity by severing the connections of the ACC may reduce individuals' sensitivity to minor errors, allowing them to move ahead with more adaptive responses. In any case, surgical intervention in OCD is an example where basic science and surgery have worked well together.

Ventrolateral Prefrontal Cortex and Self-Control

After deciding to act and preparing to do so, individuals sometimes change their minds and countermand their planned actions. Patients with damage to the **ventrolateral prefrontal cortex (VLPFC)** are impaired in this override function and respond impulsively. When tested, they respond more quickly but less accurately in timed tasks, and in life make poor decisions in various domains, including purchases, dietary choices, and social interactions. They may reach out and touch or even grab things that come into view or say the first thought that pops into their mind, even if it is inappropriate; and they continue to make bad choices even when they recognize that these actions are harmful.

The VLPFC is well positioned to govern the flow of information as it undergoes transformation from stimulus to behavior, and may directly regulate control processes in the DLPFC. The VLPFC connects with sensory areas in the inferotemporal cortex and the auditory superior temporal gyrus, and has outputs that support a role in regulating the DLPFC. The VLPFC consists of Brodmann's areas 44, 45, and 12/47. The role of the VLPFC in inhibition is well documented (Figure 32.10). For example, disorders associated with reduced ability to inhibit unwanted actions and thoughts, including Tourette syndrome, OCD, and clinical depression (habitual negative thoughts), are associated with damage to this region. When neuronal activity in the VLPFC is enhanced using transcranial magnetic stimulation (TMS), an individual's ability to suppress unwanted actions in laboratory tasks is improved. Consequently, TMS is currently being evaluated as a therapy to help individuals with OCD inhibit unwanted impulses.

Identifying the mechanisms that mediate behavioral inhibition involves using tests that target the ability to countermand learned or habitual actions. The go/no-go task is an example. Individuals are told to perform a specific action when they see a stimulus—for example, press a button when a green light is illuminated. But occasionally a subsequent stimulus—say a red light—supplants the first one, and the individual must ignore the green light and withhold the initially planned action. Such tasks reliably and selectively activate the VLPFC, and damage to the VLPFC impairs performance on this task. Another test entails task switching, in which individuals alternate between performing two different tasks. The role of inhibition in task switching may not be obvious at first, but performing one task for several trials tends to instill the response pattern. Switching from one task to another requires suppressing the practiced patterns, and failure to do so introduces confusion between tasks and conflict. fMRI studies on these tasks reliably show activation of a small number of brain regions, chief among them the VLPFC.

Inhibition is directly related to self-control. Although widely studied, self-control has no agreed-on definition, which makes it difficult to form definitive theories of its function. Nonetheless, most investigators agree that self-control consists of multiple discrete processes, of which inhibition is probably the most critical. Many theories of self-control invoke direct competition between different systems—often simplified as *hot* and *cold*—for control of behavior. One way, then, that self-control is conceptualized, is as an inhibition of the hot system by the cold system, which allows prudent actions to govern

(A)

Training phase

Experimental phase

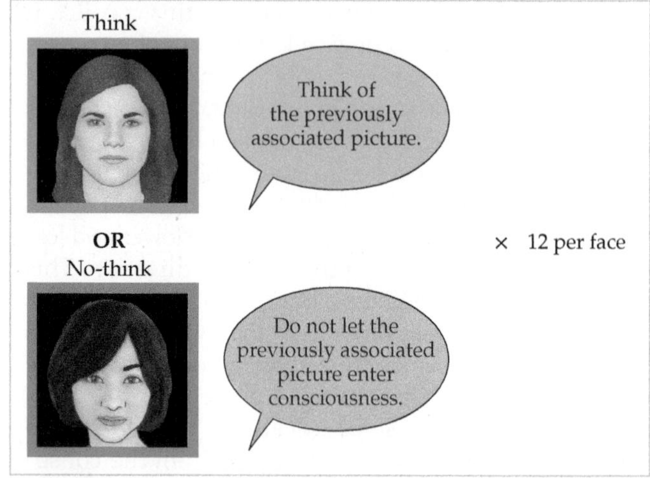

Testing phase

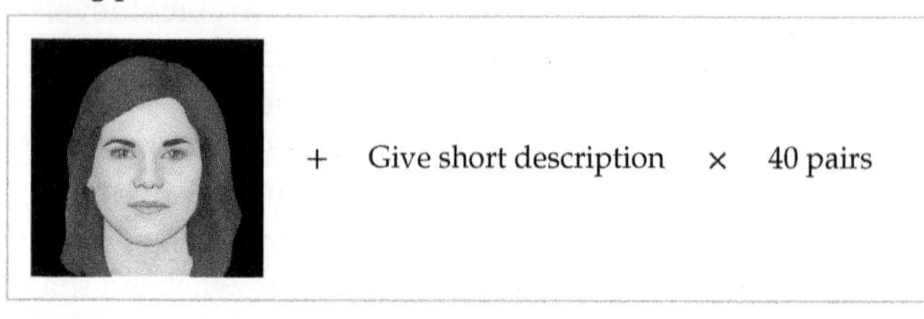

(B)

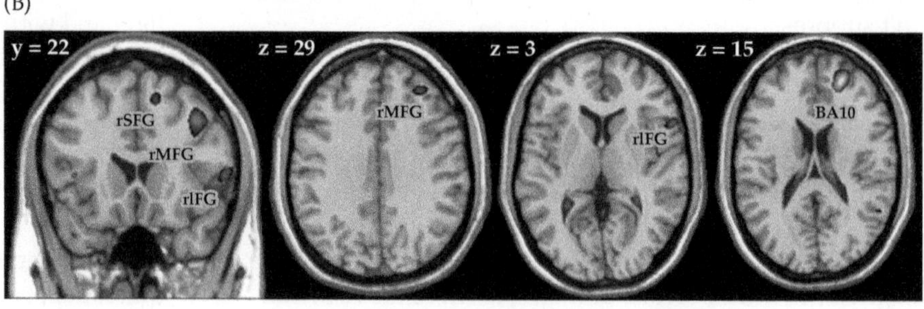

FIGURE 32.10 Role of the ventrolateral prefrontal cortex in behavioral inhibition. (A) In one experiment, individuals learned to associate faces with particular scenes. Following training, presentation of the face naturally led to recall of the associated scene. On some trials, individuals were asked not to think of the associated scene—that is, to inhibit the thought process. (B) Deliberate inhibition of thought led to activation of the right inferior frontal gyrus (rIFG), a major component of the VLPFC. Colored areas indicate regions of hemodynamic activation. Y/Z measures indicate sagittal and coronal positions of brain images. Other regions activated, and thus potentially involved in inhibition, include Brodmann's area 10 (BA 10), the right medial frontal gyrus (rMFG), the right superior frontal gyrus (rSFG), and the right lateral frontal gyrus (rLFG). (From Depue et al., 2007.)

decision-making. In the laboratory, tasks that involve inhibition often involve a competition between the VLPFC and its target structures. Presumably, self-control arises in part from interaction between these cortical regions.

These ideas are echoed by popular science books that characterize decision making as a battle between fast, intuitive behaviors commanded by the lower-level "reptilian brain" and thoughtful, deliberate behaviors controlled by the higher-level "mammalian brain." The implication is that humans are wiser than our ostensibly less advanced ancestors because our mammalian brain confers a greater ability to deliberate and suppress quick but ill-considered actions. Neither neuroscience nor evolutionary biology supports these ideas.

Anterior Insula and the Internal Milieu

Just as individuals use cognitive information to deal with information about reward, error, memory, and surprise, they also use information about basic body states to regulate behavior. These states include hunger, thirst, temperature, pain, itch, fatigue, heart rate, and many more. Although these sensations are often unconscious, they nonetheless affect decision making. These processes are most closely associated with the **anterior insula**, which receives visceral inputs—that is, information about bodily states, largely via subcortical brain areas. The anterior insula is also associated with emotional awareness and expression, relaying this information to the rest of the cerebral cortex as well as subcortical areas involved in the expression of emotions (see Chapter 31).

The anterior insula is not part of the prefrontal cortex but is discussed here because of its role in thinking, planning, and deciding. It lies buried within the lateral sulcus, which separates the temporal lobe from the inferior parietal and frontal lobes (see Figure 32.1A, right). The body states that activate it are also associated with the cingulate cortex; indeed, the anterior insula and the ACC are often co-activated. The anterior insula receives inputs from the ACC, the inferior temporal lobe, the PFC, the OFC, the central nucleus of the amygdala, and the hippocampus. It is also

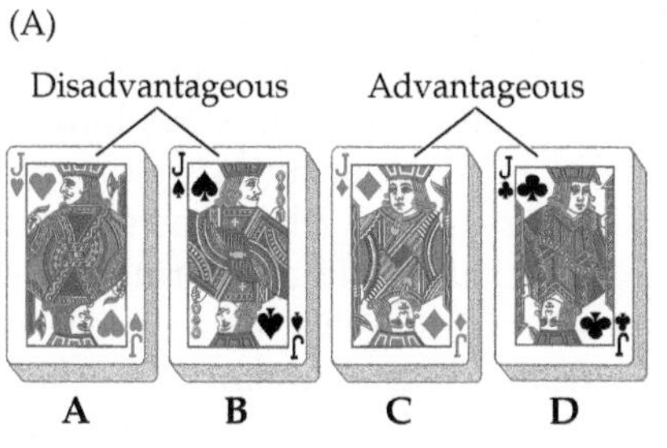

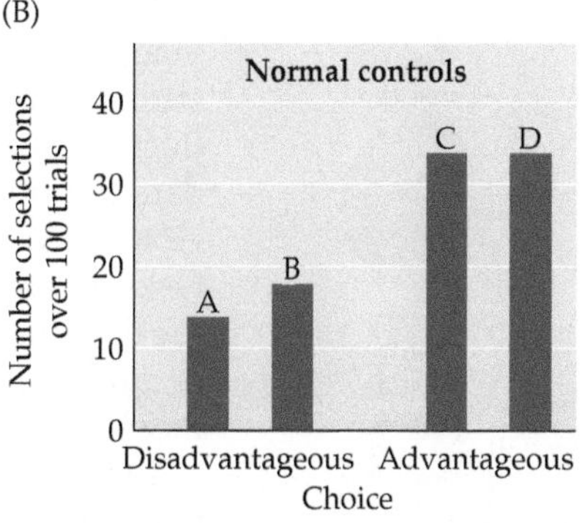

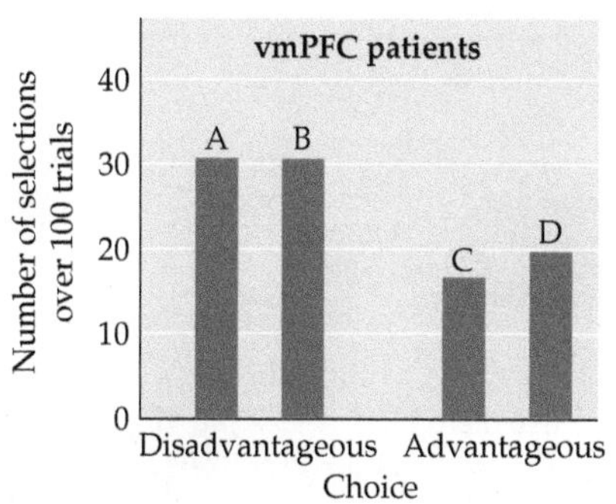

FIGURE 32.11 The Iowa Gambling Task. (A) Across 100 trials, individuals turn over cards from any of four decks. Each card is associated with either winning or losing a certain amount of money. Two of the decks (A and B) provide frequent small wins and occasional large losses; the other two decks (C and D) provide frequent small losses and occasional large wins. (B) Most individuals start by preferring decks A and B but gradually figure out that the optimal strategy is to focus on C and D. Patients with vmPFC lesions generally do not make the transition to choosing the advantageous over the disadvantageous decks. (C) The Galvanic skin response (GSR) reflects low-level bodily processes that drive the change in behavior. These processes may include fear responses that precede choices of the disadvantageous decks. In control individuals, the GSR rises as they learn the payoffs of the decks; the GSR does not rise in vmPFC patients. (B after Bechara et al., 1994, C after Bechara et al., 1996.)

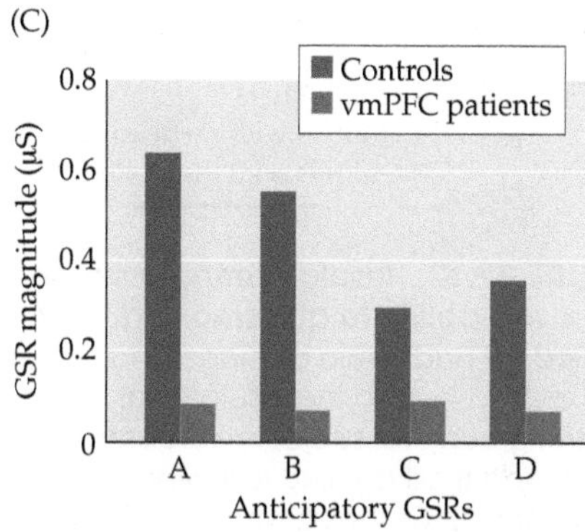

linked to attention, time perception, romantic and maternal love, mood, speech, and music—presumably because at least some of these states are accompanied by bodily states often linked to emotion. Thus, the anterior insula can be seen as part of a larger system that uses information about internal states to help make decisions and guide behavior.

These internal states are so fundamental that they are continuous with and sometimes difficult to distinguish from body states as such. For example, we may know we are anxious by sensing our own reactions, such as shaking and sweating. Conversely, music lovers often describe body states—such as "spine tingling"—to emphasize their reactions to music they like. These responses are sometimes referred to as *somatic markers* and are controlled by and influence a network of brain regions that includes the anterior insula and the vmPFC. The **somatic marker hypothesis** was proposed by Antonio Damasio, who suggested that many decisions, even seemingly deliberative and non-emotional ones, are strongly influenced by somatic markers. Through learning, exposure to stimuli associated with particular outcomes reactivates the same body states that the stimuli originally produced. These reactivations can then forecast the potential consequences of one's actions.

Testing this model has been closely associated with the Iowa Gambling Task (Figure 32.11). In this task, individuals draw cards from one of four decks. Each card is associated, in turn, with winning or losing money, and individuals naturally try to maximize their winnings. Two decks are associated, in turn, with frequent small wins and occasional large losses, while two others are associated with frequent small losses and occasional large wins. The optimal strategy, given these distributions, is to select from the second pair of decks. Typical individuals initially focus on the two disadvantageous decks but soon switch to the advantageous ones. Individuals with damage to the anterior insula or vmPFC, however, persist in choosing the disadvantageous decks.

More specifically, when selecting from the disadvantageous deck, typical individuals will rapidly develop an arousal response immediately before they choose from the disadvantageous deck. This response can be measured as a change in skin conductance (i.e., a change in electrical conductivity associated with sweating), which serves as the basis for the polygraph, or lie detector, test (see Chapter 31). This measure indicates that peripheral systems actively monitor the state of the world around us and provide a measurable indicator of its status, often before we are consciously aware of them,

The insula figures in several disorders, the most prominent being compulsive gambling and drug addiction, both of which affect decision making. Indeed, there is some evidence that drug addiction and compulsive behaviors such as gambling may be the outcome of a vicious cycle in which the addictive substance or behavior begins to take over decision-making, and push the addict toward more of that substance or behavior. One study examined patients who experienced acute damage to the insula (typically through stroke) who then tried to quit smoking (Figure 32.12).

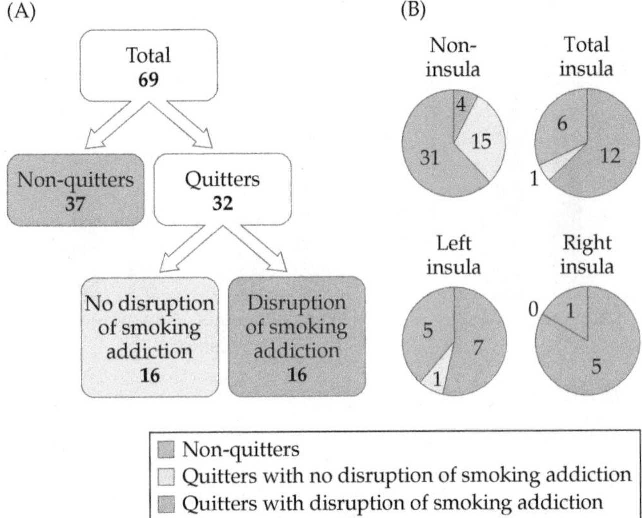

FIGURE 32.12 Insula damage is associated with an improved ability to quit smoking. (A) Among all subjects studied, 19 had insula damage and 50 had other damage that did not involve the insula (that is, they served as the control group). All subjects were smokers; 32 of the total group had tried quitting following their lesion; the remaining 37 did not try. Of the 32 who tried to quit, half experienced a disruption in their smoking addiction; that is, quitting was relatively effortless and the urge to smoke again was low. (B) Critically, the subjects who experienced distrusted smoking addiction were more likely to be the insula-lesioned patients. (After Naqvi et al., 2007.)

Those individuals experienced greater success in their attempts to quit, a lower likelihood of relapse, and reduced cravings for cigarettes. An inability to experience the body states typically associated with nicotine withdrawal may have allowed these individuals to lose the connection between smoking and its immediate consequences, thereby allowing them to break the habit.

Posterior Cingulate Cortex and Self-Awareness

In addition to an awareness of bodily states, humans (and presumably many other animals) are aware of themselves as part of a larger world. We are creative and unpredictable, and we ruminate and muse. These ineffable, but important, aspects of mental life are at best difficult to study (Box 32B). But as neuroscientists have begun to explore these issues, a region that appears to be especially important is the **posterior cingulate cortex** (**PCC**), a midline cortical structure lying at the caudal end of the cingulate sulcus that includes Brodmann's areas 31 and 23.

The PCC, like the anterior insula, it is not part of the prefrontal cortex, and until recently it received relatively little attention from cognitive neuroscientists. When examined with recording electrodes in experimental animals, PCC neurons do not generally respond to sensory stimuli. Instead, they tend to exhibit slow, long-lasting fluctuations in firing rates. Neuroimaging during standard cognitive laboratory tasks further shows that this region shows reduced activity during performance of most tasks and returns to a high baseline activity level during the delay between trials. When individuals are distracted or daydreaming, activity in the PCC is elevated. By some measures, the PCC is the most metabolically active part of the brain, the implication being that it must be doing something vital. But precisely what that is remains a mystery.

These facts are puzzling. Why would the brain include a region whose activity is—or seems to be—antagonistic to the sorts of functions probed in laboratory tasks? One hypothesis is that some brain functions relevant to cognition happen at rest or between the trials of a test. This idea has given rise to the concept of *default-mode function* (Figure 32.13; also see Chapter 28). In this view, brain regions that show greater activity when individuals are not engaged in a task are operating in default mode. The PCC is not the only region in the default-mode network (DMN). Others include the vmPFC, temporal–parietal junction, and cerebellar tonsils. Although the types of contexts that activate the DMN may vary based on task, the PCC is the most reliably and strongly modulated node in the DMN network.

What might these default functions be? One possibility is that the PCC regulates information about the self. One of the most obvious examples of this function is the retrieval of information from autobiographical memory. Humans and presumably other animals are especially interested in information about their past, perhaps subconsciously and even during sleep. This focus would help individuals evaluate past actions so they could make better plans for the future. The anatomical connections between the PCC and the medial temporal lobe—which plays an important role in memory (see Chapter 30)—are consistent with the idea that PCC activation reflects these processes.

Activation of the PCC is also associated with thinking about oneself in the future and considering others in relation to oneself. These processes may involve complex representations of how the self relates to the broader environment, which includes the future and the roles of other people in it. To construct such representations, one would have to draw on memories to formulate a best guess about the present and derive a reasonable prediction about the future. The idea that the PCC helps individuals situate themselves in the larger world can potentially explain the relationship between PCC activity and reward value in economic choice tasks. Economic choices involve evaluation of past encounters with the options available. Such memory retrieval could also explain the antagonistic

BOX 32B ■ What Does Neuroscience Have to Say about Free Will?

Any discussion of thinking, planning, and decision making is bound at some point to come up against the philosophical question of "free will." The debate over free will, which like many philosophical arguments goes back millennia, asks whether individuals govern their own actions, or whether all behaviors are in fact determined.

Neuroscience has been a player in the free will debate since a study by neurophysiologist Benjamin Libet in 1984. Libet took advantage of a classic finding from EEG: that an individual's actions are preceded by an elevation in neural activity (Figure A), generated in the premotor cortex, known as the **readiness potential** (Figure B). This potential likely reflects the responding of neurons in the premotor cortex, which are thought to directly influence motor neurons through feedforward projections. Libet wondered whether this process reflects brain events that occurred before conscious awareness, which would indicate that individuals only become aware of their intentions after the action had already been determined. He asked people to look at a specially designed clock and note, privately, the time on the clock when they first felt the conscious urge to act. (Their task was simply to move their arm when they felt the urge). Individuals were then asked to indicate the time that the event occurred.

Libet found that the readiness potential was statistically detectable up to 300 ms before individuals reported the conscious intention to act. More modern methods using more sophisticated brain measures show that brain activity can predict choices as much as 20 seconds in advance. These findings suggest that consciousness is an effect, not a cause, of the decision to act, which occurs unconsciously. This in turn would imply that one's actions are not generated consciously but instead are elicited by other processes, and that consciousness is simply a passive observer.

Libet accepted this interpretation only in part. He believed that individuals have the ability to override decisions made unconsciously, and that the override function follows conscious awareness and is thus indicative of free will. According to this model, individuals don't freely select their actions, but rather reject some and permit others, and this veto process results in the indeterminacy or unpredictability of behavior. Not surprisingly, this interpretation has attracted a great deal of debate among philosophers.

Other pertinent evidence comes from some very simple psychological experiments looking at the factors that influence putatively free choice. In one important experiment, individuals chose one of two pictures of faces as more attractive (the faces were monochrome and were rated nearly equal in attractiveness by independent raters). The experimenter then handed the participants the picture and asked them to explain why they chose the one they did. Unbeknownst to the individuals, however, the experimenter used sleight of hand to change which picture was handed to the individuals. In most cases, participants did not notice the switch. Then when faced with the need to explain their choice, most individuals made up plausible explanations. This finding suggests that the human brain will often fabricate beliefs and memories to support what it "thinks" occurred. Perhaps our sense of free will is a similar confabulation that we employ to make sense of a determined reality we can't accept.

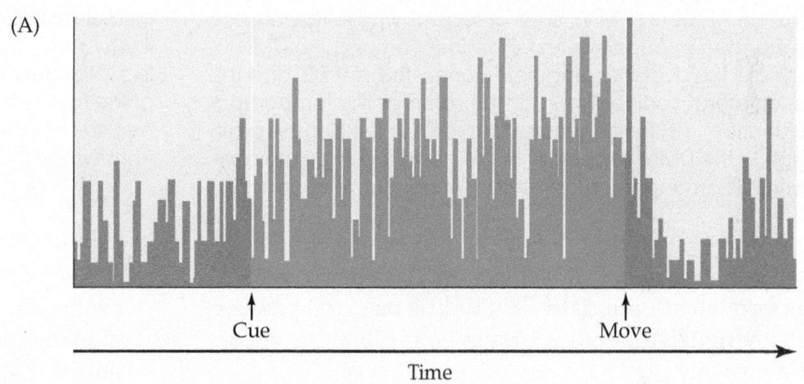

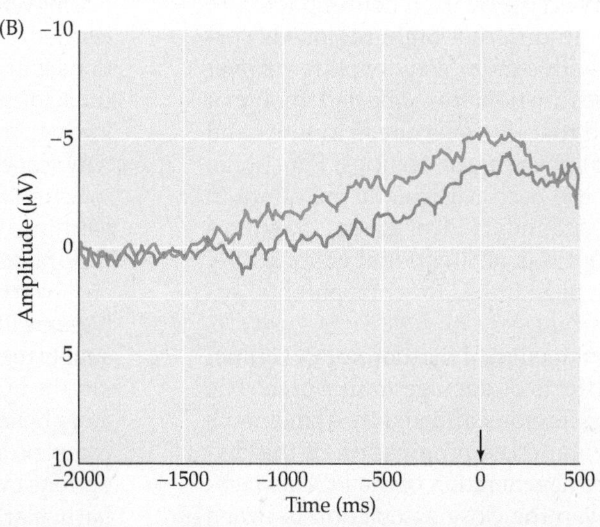

(A) When we move our limbs, our action is preceded by specific patterns of brain activity that seem to have a direct causative role in preparing for the action. One of the most prominent such signals occurs in the premotor cortex, a region of the brain located rostrally to the motor cortex. The intention period that precedes the movement is associated with a gradual increase in firing rate of neurons in the premotor cortex, a region that provides direct inputs to the motor cortex. Once the movement begins, neurons in this region return to baseline level, suggesting that their major contribution is to promote motor intention. (B) The readiness potential. When EEG electrodes are placed over the premotor cortex, they show a consistent ramping up of activity that precedes the performance of the action. This ramping up is greater when individuals attend to the urge to move (red line) rather than attending to the movement itself (blue line), but it is observed in both cases. (After Sirigu et al., 2003.)

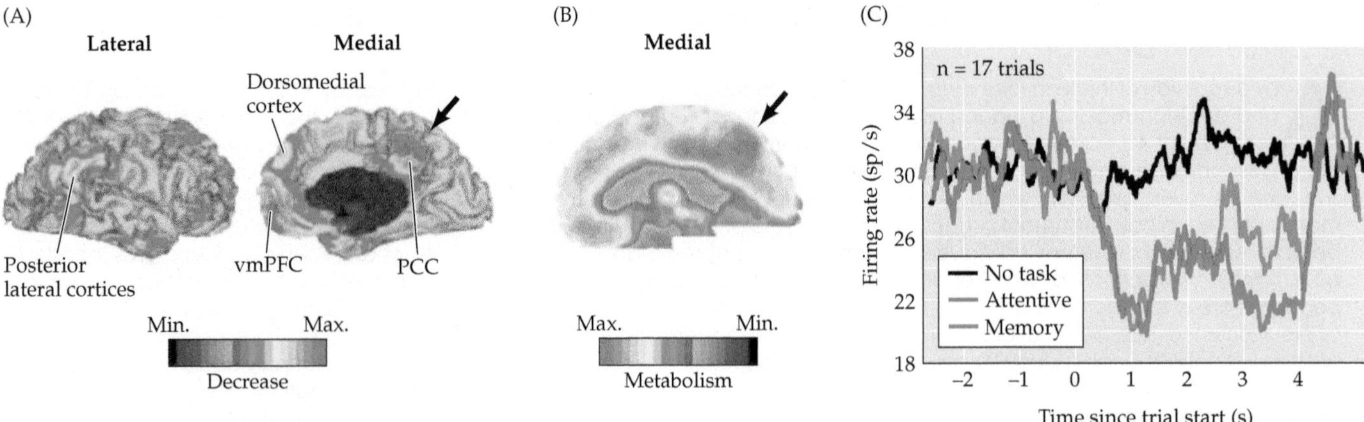

FIGURE 32.13 Default-mode function and the posterior cingulate cortex. (A) Images of the human brain showing metabolic responses as measured by PET. Individuals either performed standard laboratory tasks or viewed the same stimuli passively—that is, without any task requirements. Active task performance led to consistent decreases in activity in the default-mode network (DMN), a set of regions whose responses are enhanced at cognitive rest. Prominent among the DMN regions are the posterior cingulate cortex, the vmPFC, and the posterior lateral cortices (roughly the same as the temporoparietal junction). (B) Resting metabolic rates in the human brain are high in the DMN, especially in the posterior cingulate cortex, which is among the most metabolically active regions of the brain. (C) Responses of single neurons in the monkey PCC show similar patterns as the hemodynamic response. That is, they show sustained reductions in activity during the cognitively demanding parts of rapid laboratory tasks and a return to high baseline levels during the delay between trials. This figure comes from a study in which rhesus monkeys performed either a difficult attention task (blue line) or short-term memory task (red line), or else performed no task (black line). In the task, tonic firing rates during the delay were reduced during the difficult part of the task relative to the rest condition. When information was a help in short-term memory, responses were further reduced. (A,B from Gusnard and Raichle, 2001; C after Hayden et al., 2009.)

relationship between PCC activity and task performance: To perform an attentionally demanding task, the process of retrieving past memories must be inhibited to focus on the present.

Neuroscientists have recently used neuroimaging to measure how activity in different brain regions is correlated. Two regions with similar activity patterns over time are assumed to be functionally connected, though it should be emphasized that this measure is indirect and cannot measure actual physical connections. Functional connectivity studies have been particularly important in understanding the DMN and its involvement in psychiatric disorders. Disruptions of functional connectivity within the DMN have been linked to schizophrenia, autism, and OCD.

Some of the most compelling indications of PCC function come from the effects of damage to this area. The PCC is one of the first regions affected in Alzheimer's disease (see Box 30C), and the progression of the disease is associated with degeneration of the PCC as measured after death. Given the close association between the progression of Alzheimer's disease and the loss of autobiographical memories, this finding provides additional evidence that the PCC in particular, and default mode function in general, is linked with the regulation of autobiographical memory.

One recent study highlights this linkage directly by examining the activity of single neurons in the brains of monkeys performing a difficult learning task. The task required that monkeys learn to pair scenic photographs with eye movements to the left or right. Once an association was learned, the researchers added new picture–action associations for the monkeys to learn. Activation of PCC neurons tracked these associations and was particularly high following errors and when the monkeys were confronted with new scenes. When the investigators inactivated the PCC by injecting a small amount of the GABA agonist muscimol, learning was impaired, providing direct evidence that the PCC plays a causal role in regulating learning.

Given the contexts that activate and deactivate the PCC, it seems likely that its function is much more complex than simply regulating memory retrieval. To give just one example, the PCC appears to play a key role in regulating exploratory behavior (Figure 32.14). Individuals often must make complex decisions in which their choice affects the range of options available to them in the future. Such decisions are particularly common in complex foraging situations and in social interactions. Activation of the PCC is higher during such strategic decisions. Such decisions involve memory, but also emotion, prospection, error monitoring, the delicate trade-off between exploration and exploitation, and the integration of all these factors into a decision.

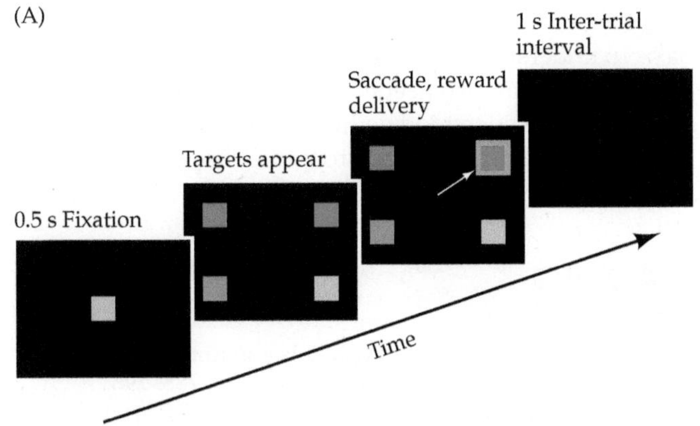

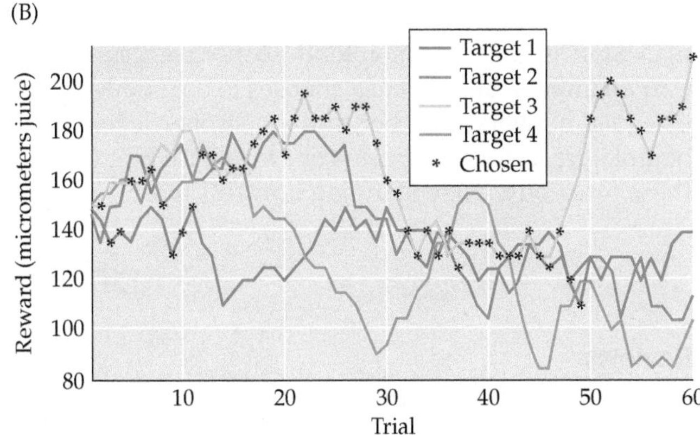

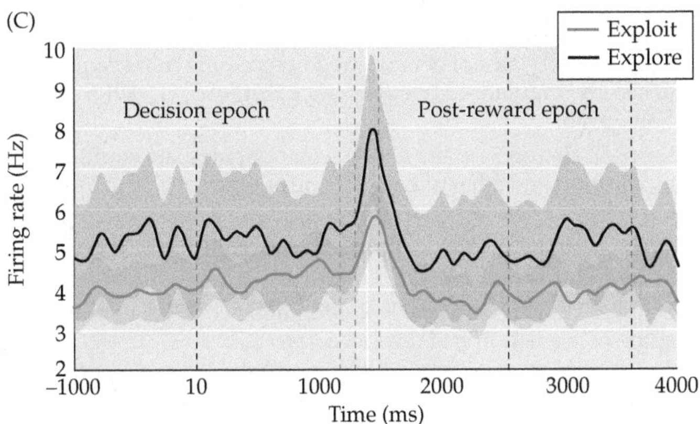

FIGURE 32.14 Exploratory behavior and the posterior cingulate cortex. (A) Exploratory behavior in animals, including monkeys, can be studied with the four-arm bandit task, in which individuals choose among four different options that provide random rewards. In this task, monkeys looked at a computer screen with four squares on it. On each trial, they chose one of the squares and received a reward. (B) The value of the reward for each option was stochastically related to its value on the past trial. That is, the value rose or fell by a small amount at random on each trial. Lines in the panel indicate the value that would be offered for each option (although the monkey did not know this except by learning directly by choosing the option). Stars indicate the option the monkey did choose. As a consequence of the payoff structure, the optimal strategy is to choose the best option most of the time, but, since the value of the other options changes, the monkey must occasionally sample the alternatives to see if it now has a better value than the one he has been choosing. Even though solving the optimal strategy for the task is difficult, monkeys are able to approximate it quite well. In this example session, the stars, which indicate the monkeys' choices, align closely with the optimal strategy although they also deviate a bit. (C) Monkeys alternate between exploitative modes (in which they choose the most rewarding option given their knowledge of the rewards) and exploratory modes (in which they choose inferior options to gain information that could help improve their understanding of the environment, and ultimately increase reward intake). In exploratory modes, firing rates of neurons in the PCC are enhanced throughout the trial. Panel shows the average firing rate of a single neuron from PCC aligned to the start of the trial. This neuron showed elevated firing on explore trials both during the decision epoch (when the stimuli were presented but before the animal had chosen) and following it (the post-reward epoch). (After Pearson et al., 2011.)

Summary

Humans and other intelligent animals stand out because of the number and complexity of the associations that can be made between sensory input and behavioral responses. This flexibility depends, at least in part, on the prefrontal cortex, a region that consists of many areas that work together to produce rich, sophisticated, and even creative behaviors. These functions are not mediated by the prefrontal cortex alone, but recruit an extended network of structures with sometimes overlapping and sometimes conflicting roles. These areas acquire information about both the state of the world and the state of the body, and elicit further associations that can then be used to evaluate options, deal with conflicting possibilities, and regulate the allocation of cognitive resources accordingly. This network incorporates rapidly changing contextual information to generate an effective plan of action, inhibit unwanted or maladaptive plans, and monitor the consequences of whatever an individual ultimately chooses to do. Insults to this extended network, whether through stroke, degenerative neurological disorders, trauma or drug abuse, compromise these functions. Understanding how this mosaic of brain regions contributes to thinking, planning, and deciding is basic to understanding "higher-order" human brain functions, which is in turn needed to properly deal with a range of still poorly understood neurological and psychiatric disorders. It should be apparent from this chapter that there is still a long way to go in reaching this goal.

ADDITIONAL READING

Reviews

Baxter, M. G. and E. A. Murray (2002) The amygdala and reward. *Nat. Rev. Neurosci.* 3: 563–573. doi: 10.1038/nrn875

Eagleman, D. M. (2004) The where and when of intention. *Science* 303: 1144–1146.

Gusnard, D. A. and M. E. Raichle (2001) Searching for a baseline: Functional imaging and the resting human brain. *Nat. Rev. Neurosci.* 2: 685–694.

Miller, E. K. (2000) The prefrontal cortex and cognitive control. *Nat. Rev. Neurosci.* 1: 59–65.

Pearson, J. M., S. R. Heilbronner, D. L. Barack, B. Y. Hayden and M. L. Platt (2011) Posterior cingulate cortex: Adapting behavior to a changing world. *Trends Cogn. Sci.* 15: 143–151.

Wallis, J. D. (2007) Orbitofrontal cortex and its contribution to decision-making. *Annu. Rev Neurosci.* 30: 31–56.

Important Original Papers

Anderson, S. W., A. Bechara, H. Damasio, D. Tranel and A. R. Damasio (1999) Impairment of social and moral behavior related to early damage in the human prefrontal cortex. *Nat. Neurosci.* 2: 1032–1037.

Bechara, A., A. R. Damasio, H. Damasio and S. W. Anderson (1994) Insensitivity to future consequences following damage to the human prefrontal cortex. *Cognition* 50: 7–15.

Bush, G. and 5 others (1998) The counting Stroop: An interference task specialized for functional neuroimaging—Validation study with functional MRI. *Hum. Brain. Mapp.* 6: 270–282.

Depue, B. E., T. Curran and M. T. Banich (2007) Prefrontal regions orchestrate suppression of emotional memories via a two-phase process. *Science* 317: 215–219.

Fellows, L. K. and M. J. Farah (2005) Is anterior cingulate cortex necessary for cognitive control? *Brain* 128: 788–796.

Funahashi, S., C. J. Bruce and P. S. Goldman-Rakic (1989) Mnemonic coding of visual space in the monkey's dorsolateral prefrontal cortex. *J. Neurophysiol.* 61: 331–349.

Hayden, B. Y., D. Smith and M. L. Platt (2009) Electrophysiological correlates of default-mode processing in macaque posterior cingulate cortex. *Proc. Natl. Acad. Sci. USA* 106: 5948–5953.

Kerns, J. G. and 5 others (2004) Anterior cingulate conflict monitoring and adjustments in control. *Science* 303: 1023–1026.

Naqvi, N. H., D. Rudrauf, H. Damasio and A. Bechara (2007) Damage to the insula disrupts addition to cigarette smoking. *Science* 315: 531–534.

Schultz, W., P. Dayan and P. R. Montague (1997) A neural substrate of prediction and reward. *Science* 275: 1593–1599.

Wallis, J. D., K. C. Anderson and E. K. Miller (2001) Single neurons in prefrontal cortex encode abstract rules. *Nature* 411: 953–956.

Books

Damasio, A. (2005) *Descartes' Error: Emotion, Lesion, and the Human Brain.* New York: Penguin Books.

Fuster, J. (2015) *The Prefrontal Cortex,* 5th Edition. New York: Academic Press.

Glimcher, P. (2004) *Decisions, Uncertainty, and the Brain: The Science of Neuroeconomics.* New York: Bradford.

Passingham R. E. and S. P. Wise (2012) *The Neurobiology of the Prefrontal Cortex: Anatomy, Evolution, and the Origin of Insight.* London: Oxford University Press.

Thorndike, E. L. (1911) *Animal Intelligence: Experimental Studies.* New York: Macmillan.

Go to the NEUROSCIENCE 6e Companion Website at **www.oup.com/uk/Purves6e** for Web Topics, Animations, Flashcards, and more.

CHAPTER

33 Speech and Language

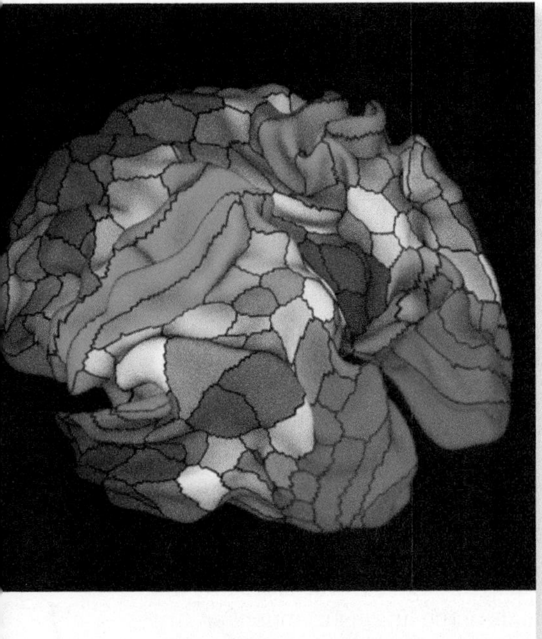

Overview

PERHAPS THE MOST REMARKABLE cognitive function in humans is the ability to associate arbitrary symbols with specific meanings to express ideas, feelings, desires, emotions, and more to ourselves and others by means of language, whether in thought, speech, or writing. The achievements and transmission of human culture rest largely on this kind of communication, and a person who for one reason or another fails to develop a facility for language as a child is severely incapacitated. Studies using electrophysiological methods and/or functional brain imaging of subjects and of patients with damage to specific cortical regions indicate that the linguistic abilities of humans depend on the integrity of several specialized areas located primarily in the association cortices of the left temporal, parietal, and frontal lobes. Understanding functional localization and hemispheric lateralization of language is especially important in clinical practice. The loss of language has such devastating consequences that neurologists and neurosurgeons make every effort to identify and preserve those cortical areas involved in its comprehension and production (the so-called *eloquent* areas). The need to map language functions in patients for the purpose of sparing these brain regions has provided another source of information about the neural organization of this critical human attribute. The linkage between speech sounds and their meanings is mainly represented in the left temporal and parietal cortices, and the circuitry for the motor commands that organize the production of meaningful speech is mainly found in the left frontal cortex. While the left hemisphere typically processes the lexical, grammatical, and syntactic aspects of language, the emotional (affective) content of speech is governed largely by the right hemisphere. Studies of congenitally deaf individuals have shown further that the cortical areas devoted to sign language are generally the same as those that organize spoken and heard communication. The regions of the brain devoted to language are therefore specialized for symbolic representation and communication rather than for heard and spoken language per se. Finally, whether and in what ways language is limited to humans, if it is, remains hotly debated.

Representation of Language in the Brain

It has been known for more than a century that two regions in the frontal and temporal association cortices of the left cerebral hemisphere are especially important for the explicitly verbal aspects of human language. That language abilities are both localized and lateralized is not surprising; ample evidence of the localization and lateralization of other functions has been discussed in earlier chapters. Although lateralization has already been introduced in describing the unequal functions of the parietal lobes in attention and of the temporal lobes in recognizing different categories of objects, it

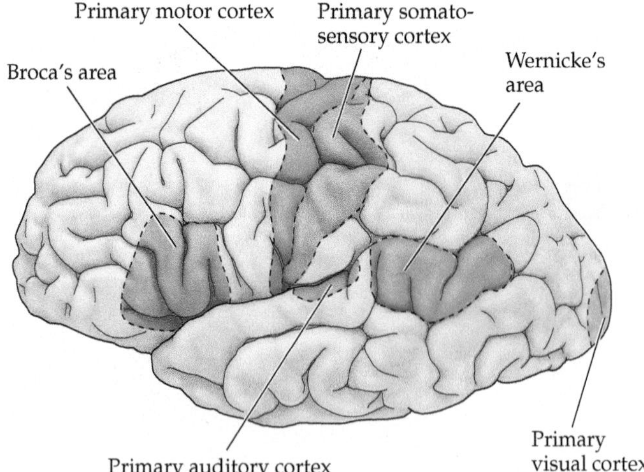

FIGURE 33.1 **Diagram of the major brain areas involved in the comprehension and production of language.** The primary sensory, auditory, visual, and motor cortices are indicated to show the relation of Broca's and Wernicke's language areas to these other areas that are involved in the comprehension and production of speech, albeit in a less specialized way.

is in language that the idea of lateralization has been most thoroughly documented. Because language is so important to humans, its lateralization has given rise to the misleading idea that one hemisphere is "dominant" over the other—namely the hemisphere in which the major capacity for verbal expression resides. But the true significance of lateralization for language (or for any other cognitive ability) lies in the efficient subdivision of complex functions between the two hemispheres rather than in any superiority of one hemisphere over the other. Indeed, pop psychological dogmas about cortical redundancy notwithstanding, it is safe to assume that every region of the brain is being efficiently used for *something* important all the time.

A first step in sorting out the organization of language in the brain is to recognize that the representation of language is distinct from, although related to, the circuitry concerned with the motor planning and control of the larynx, pharynx, mouth, and tongue—the structures that produce speech sounds (Box 33A). The cortical representation of language is also distinct from, although clearly related to, the circuits underlying the auditory perception of spoken words and the visual perception of written words in the primary auditory and visual cortices, respectively (Figure 33.1). Whereas the neural substrates for language depend on sensory and motor functions, the regions of the brain that are specifically devoted to language transcend these more basic demands. The main concern of the areas of cortex that represent language is using symbols for communication—spoken and heard, written and read, or in the case of sign language, gestured and seen (see the section "Sign Language" later in the chapter). Thus, the essential function of the cortical language areas, and indeed of language, is symbolic representation. Obedience to a set of rules for using these symbols (called grammar), ordering them to generate useful meanings (called syntax), and giving utterances the appropriate emotional valence by varying intensity, pitch, stress and rhythm (called **prosody**) are all important and readily recognized elements of communication, regardless of the particular mode of expression.

Aphasias

The distinction between language and the related sensory and motor functions on which it depends was first apparent in patients with damage to specific brain regions. Clinical evidence from such cases showed that the ability to move the muscles of the larynx, pharynx, mouth, and tongue can be compromised without abolishing the ability to use language to communicate (even though a motor deficit may make communication difficult). Similarly, damage to the auditory pathways can impede the ability to hear without interfering with language functions per se (as is obvious in individuals who have become partially or wholly deaf later in life). Damage to specific brain regions, however, can compromise essential language functions while leaving the sensory and motor infrastructure of verbal communication intact. These syndromes, collectively referred to as aphasias, diminish or abolish the ability to comprehend and/or to produce language as a vehicle for communicating meaningful statements, while sparing the ability to perceive the relevant stimuli and to produce intelligible words. Missing in these patients is the capacity to recognize or employ the meaning of words correctly, thus depriving such individuals of the linguistic understanding, grammatical and syntactic organization, and/or appropriate intonation that distinguish language from nonsense.

The first evidence for the localization of language function to a specific region (and to a hemisphere) of the cerebrum is usually attributed to the French neurologist Paul Broca and the German neurologist Carl Wernicke, who made seminal observations in the late 1800s. Both Broca and Wernicke examined the brains of individuals who had become aphasic and later died. Based on correlations of the clinical picture and the location of the brain damage seen in autopsy, Broca suggested that language abilities were localized in the ventroposterior region of the frontal lobe (Figure 33.2). More important, he observed that the loss of the ability to produce meaningful language—as opposed to the ability to move the mouth and produce words—was usually associated with damage to the left hemisphere. *"On parle avec l'hemisphere gauche,"* Broca concluded. The preponderance of aphasic syndromes associated with damage to the left hemisphere has supported Broca's claim that "we speak with the left hemisphere," a conclusion amply

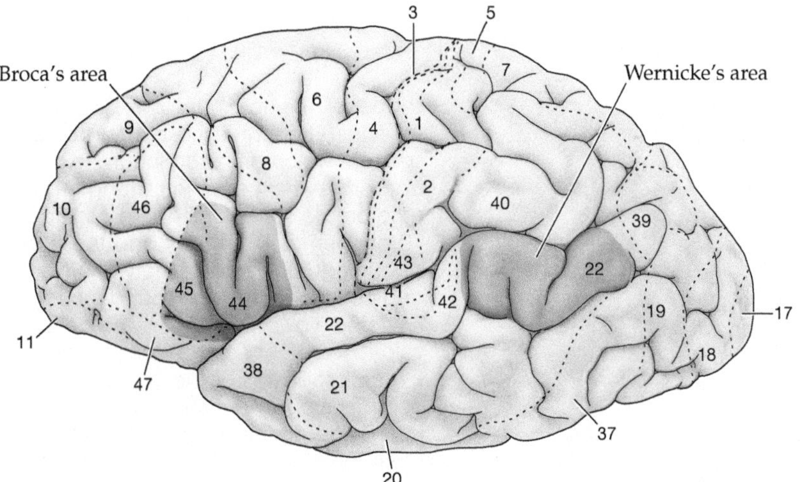

FIGURE 33.2 Relationship of the major language areas to the classic cytoarchitectonic map of the cerebral cortex. As discussed in Chapter 27, about 50 histologically distinct regions (cytoarchitectonic areas) have been described in the human cerebral cortex. The language functions described by Broca and Wernicke are associated with at least three of the cytoarchitectonic areas defined by Brodmann (area 22, at the junction of the parietal and temporal lobes [Wernicke's area]; and areas 44 and 45, in the ventral and posterior region of the frontal lobe [Broca's area]) but are not coextensive with any of them.

BOX 33A ■ The Generation of Speech

The organs that produce speech include the lungs, which serve as a reservoir of air; the larynx, which is the source of the periodic quality of "voiced" sounds; and the pharynx, oral, and nasal cavities and their included structures (e.g., tongue, teeth, and lips), which modify (or filter) the speech sounds that eventually emanate from the speaker (Figure A). This generally accepted *source-filter model* of speech is an old one, having been proposed by Johannes Mueller in the nineteenth century (Figure B).

Although the physiological details are complex, the general operation of the vocal apparatus is simple. Air expelled from the lungs accelerates as it passes through a constricted opening between the **vocal folds** ("vocal cords") called the *glottis*, thus decreasing the pressure in the air stream according to Bernoulli's principle. As a result, the vocal folds come together until the pressure buildup from below forces them open again. The ongoing iteration of this process results in an oscillation of sound wave pressure, the frequency of which is determined primarily by the muscles that control the tension on the vocal cords. The fundamental frequencies of these oscillations range from about 100 to about 400 Hz, depending on the gender, size, and age of the speaker.

The larynx has many other consequential effects on the speech signal, and these create additional speech sounds. For instance, the vocal folds can open suddenly to produce what is called a *glottal stop* (as in the beginning of the exclamation "Idiot!"). Alternatively, the vocal folds can hold an intermediate position for the production of consonants such as *h*, or they can be completely open for "unvoiced" consonants such as *s* or *f* (i.e., speech sounds that don't have the periodic quality derived from vocal fold oscillations). In short, the larynx is important in the production of virtually all vocalizations.

The vocal system as a whole can be thought of as a sort of musical instrument capable of exquisite modulation. As in the sound produced by a musical instrument, however, the primary source of oscillation (e.g., the reed of a clarinet or the vocal folds in speech) is hardly the whole story. The entire pathway between the vocal folds and the lips (and nostrils) is equally critical in determining speech sounds, comparable to the structure of a musical instrument. The most important determinants of the sound that emanates from an instrument are its natural resonances, which shape or filter the sound pressure oscillation. For the vocal tract, the natural resonances that modulate the air stream generated by the larynx are called *formants*. Formants are thus peaks of power in the spectrum of a vocal sound stimulus (see Figure B). The power in the laryngeal source

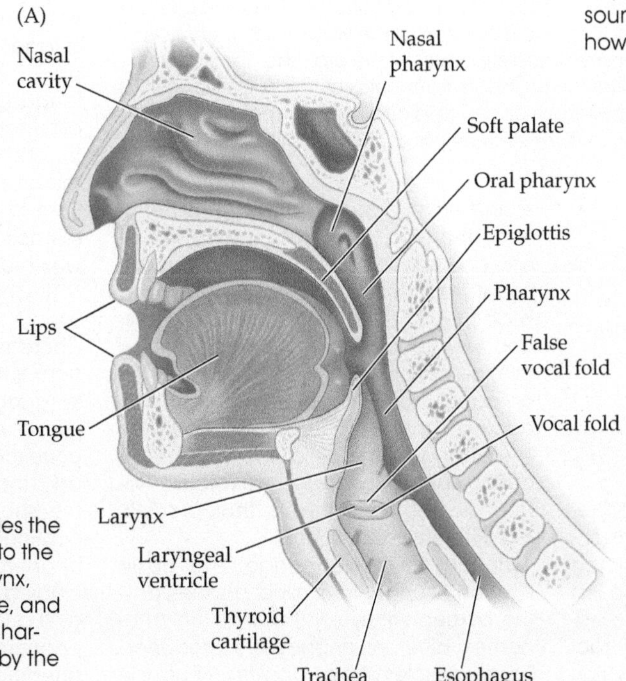

(A) The human vocal tract includes the vocal apparatus from the larynx to the lips. The structures above the larynx, including the pharynx, soft palate, and nasal cavity, shape and filter the harmonic series that are generated by the vocal cords when they vibrate.

Continued on the next page

BOX 33A ■ (continued)

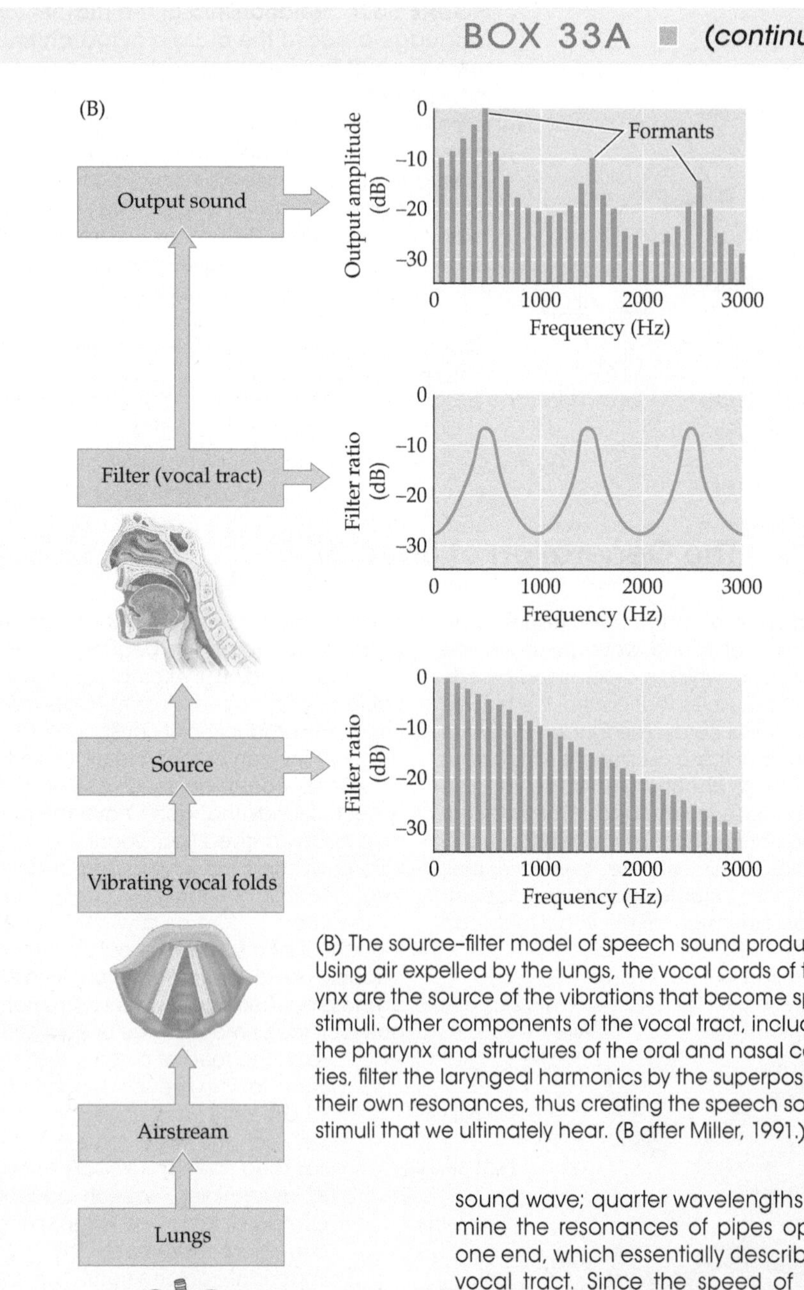

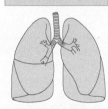

(B) The source–filter model of speech sound production. Using air expelled by the lungs, the vocal cords of the larynx are the source of the vibrations that become speech stimuli. Other components of the vocal tract, including the pharynx and structures of the oral and nasal cavities, filter the laryngeal harmonics by the superposition of their own resonances, thus creating the speech sound stimuli that we ultimately hear. (B after Miller, 1991.)

near the formant frequencies will be reinforced, and power at other frequencies will, in varying degrees, be filtered out. The resonance frequency of the first formant arises from the fact that the approximate length of the adult vocal tract in its relaxed state is about 17 cm, which is the quarter wavelength of a 68-cm sound wave; quarter wavelengths determine the resonances of pipes open at one end, which essentially describes the vocal tract. Since the speed of sound is about 33,500 cm/sec, the lowest resonance frequency of an open tube or pipe of this length will be 33,500/68 or about 500 Hz; additional formants occur at about 1500 Hz and 2500 Hz. The changing shape of the vocal tract during speech changes the relative frequencies of the formants, thus producing different vowel sounds.

In any given language, the basic speech sounds are called *phones* and the percepts they elicit are called *phonemes*; different phones are produced as the muscles of the vocal tract change the tension on the vocal fold and the shape of the resonant cavities above the folds. Phonemes make up syllables in speech, which are used in turn to make up words, which are then strung together to create sentences. There are about 40 phonemes in English, and these are about equally divided between vowel and consonant speech sound percepts. As noted, vowel sounds are the periodic voiced elements of speech generated by the oscillation of the vocal cords. In contrast, consonant sounds involve rapid changes in the sound signal and are more complex. In English, consonants begin and/or end syllables, each of which entails a vowel sound. Consonant sounds are categorized according to the site in the vocal tract that determines them (the *place of articulation*), or the physical way they are generated (the *manner of articulation*). With respect to place, there are labial consonants (such as *p* and *b*), dental consonants (*f* and *v*), palatal consonants (*sh*), and glottal consonants (*h*) (among many others). With respect to manner, there are plosive, fricative, nasal, liquid, and semi-vowel consonants. Plosives are produced by blocking the flow of air somewhere in the vocal tract, fricatives by producing turbulence, nasals by directing the flow of air through the nose, and so on. Another variation on the use of consonants is found in the "click languages" of southern Africa, of which about 30 survive today. Many of these languages have four or five different click sounds. They are produced by the tongue enclosing a pocket of air against the palate and then being sucked down from the roof of the mouth. The tongue release can occur at different positions, which creates different click sounds.

It should be obvious that speech sound stimuli are enormously complex (there are more than 200 different phonemes in the approximately 6000 human languages in the world today). To complicate matters further, there is no simple correspondence between phonemes and the elements in speech. Because the shape of the vocal tract changes continuously in normal speech, one phoneme overlaps with the next, precluding any split into discrete segments, as the abstract concept of phones and phonemes implies. Given these facts, our ability to produce and comprehend speech is truly amazing.

confirmed by a variety of modern studies using functional imaging (albeit with some important caveats discussed later in the chapter).

Although Broca was basically correct, he failed to grasp the limitations of thinking about language as a unitary function localized in a single cortical region. This issue was better appreciated by Wernicke, who distinguished between patients with impaired ability to comprehend language and those with impaired ability to produce language. Wernicke recognized that some aphasic patients who hear normally have great difficulty understanding speech. These patients retain the ability to produce utterances with reasonable grammatical and syntactic fluency, but often without meaningful content. He concluded that lesions of the posterior and superior temporal lobe on the left side tend to result in deficits of this sort. In contrast, other patients continue to comprehend language but lack the ability to produce syntactically complete utterances, even though it is clear that they know what they are trying to say. Thus, they repeat syllables and words and utter agrammatical phrases, even though the meaning eventually gets through (see examples below). This type of deficit is associated with damage to the posterior and inferior region of the left frontal lobe, the area that Broca emphasized as an important substrate for language production (see Figures 33.1 and 33.2).

As a consequence of these early observations, two rules about the localization of language have been taught ever since. The first is that lesions of the left frontal lobe in a region referred to as **Broca's area** affect the ability to *produce* language efficiently. This deficiency is called **motor** or **expressive aphasia**, and is also known as **Broca's aphasia**. (Expressive aphasias must be distinguished from *dysarthria*, which is the impaired ability to move the muscles of the mouth, tongue, and pharynx that mediate speech.) The deficient motor-planning aspects of expressive aphasias accord with the complex motor functions of the posterior frontal lobe and its proximity to the primary motor cortex.

The second rule is that damage to the left temporal lobe causes difficulty *understanding* spoken language, a deficiency referred to as **sensory** or **receptive aphasia**, also known as **Wernicke's aphasia**. (Deficits of reading and writing—*alexias* and *agraphias*—are separate disorders that can arise from damage to related but different brain areas; most aphasics, however, also have difficulty with these closely linked abilities.) Receptive aphasia generally reflects damage to the auditory association cortices in the posterior temporal lobe, a region referred to as **Wernicke's area**.

A final broad category of language deficiency syndromes is **conduction aphasia**. These disorders arise from lesions to the pathways connecting the relevant temporal and frontal regions, such as the arcuate fasciculus that links Broca's and Wernicke's areas. Interruption of these pathways may result in an inability to produce appropriate responses to heard communication, even though the communication is understood.

An Ingenious Confirmation of Language Lateralization

Until the 1960s, observations about language localization and lateralization were based primarily on patients with brain lesions of varying severity, location, and etiology. Up until that time, the inevitable uncertainties of clinical findings had allowed some skeptics to argue that language and other complex cognitive functions might not be localized or even lateralized in the brain. Definitive evidence supporting the inferences from neurological observations came from studies of patients whose corpus callosum and anterior commissure were severed as a treatment for medically intractable epileptic seizures. (Recall that a fraction of severe epileptics are refractory to medical treatment, and that interrupting the connection between the two hemispheres remains an effective way of treating epilepsy in highly selected patients. In such patients, investigators could assess the function of the two cerebral hemispheres *independently* because the major axon tracts that connect them had been interrupted. The first studies of these so-called **split-brain patients** were carried out by Roger Sperry and his colleagues at the California Institute of Technology in the 1960s and 1970s, and established the hemispheric lateralization of language beyond any doubt. Their work also demonstrated many other functional differences between the left and right hemispheres (Figure 33.3) and stands as an extraordinary contribution to the understanding of brain organization.

To evaluate the functional capacity of each hemisphere in split-brain patients, it is essential to provide information to one side of the brain only. Sperry, Michael Gazzaniga (a key collaborator in this work), and others devised several simple ways to do this, the most straightforward of which was to ask the subject to use each hand independently to identify objects without any visual assistance (see Figure 33.3A). Recall from Chapter 9 that somatosensory information from the right hand is processed by the left hemisphere, and vice versa. Therefore, by asking the subject to describe an item being manipulated by one hand or the other, the researchers could test the language capacity of the relevant hemisphere. Such studies showed clearly that the two hemispheres differ in their language ability, in keeping with the postmortem correlations described earlier.

Using the left hemisphere, split-brain patients were able to name objects held in the right hand without difficulty. In contrast, and quite remarkably, an object held in the left hand could not be named. Using the right hemisphere, most split-brain patients could produce only an indirect description of the object that relied on rudimentary words and phrases rather than the precise lexical symbol for the object (for instance, "a round thing" instead of "a ball");

(A)

Left-hemisphere functions	Right-hemisphere functions
Analysis of right visual field	Analysis of left visual field
Stereognosis (right hand)	Stereognosis (left hand)
Lexical and syntactic language	Emotional coloring of language
Writing	Spatial abilities
Speech	Rudimentary speech

(B) Normal individual

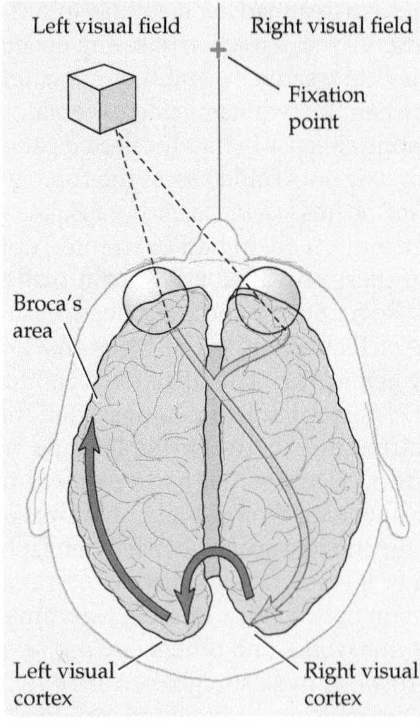

Split-brain individual

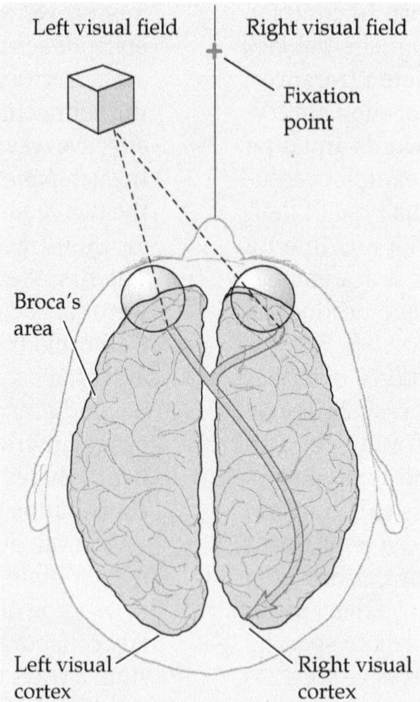

Split-brain individual

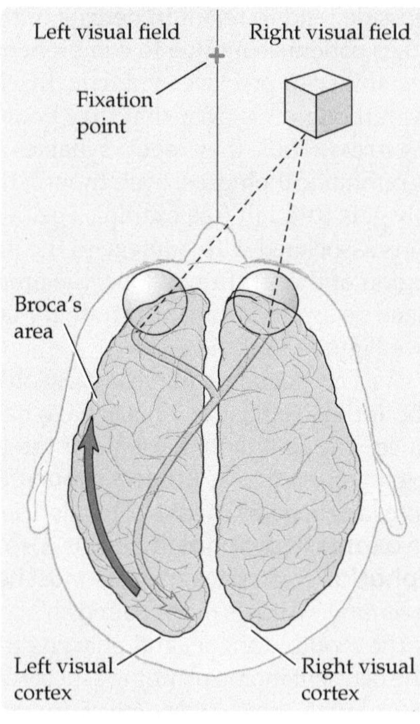

FIGURE 33.3 **Confirmation of hemispheric specialization for language obtained by studying individuals in whom the connections between the right and left hemispheres have been surgically divided.** (A) Single-handed, vision-independent *stereognosis* can be used to evaluate the language capabilities of each hemisphere in split-brain patients. Objects held in the right hand, which provides somatosensory information to the left hemisphere, are easily named; objects held in the left hand, however, are not readily named by most of these patients. (B) Visual stimuli or simple instructions can also be given independently to the right or left hemisphere in intact and split-brain individuals. Since the left visual field is perceived by the right hemisphere (and vice versa; see Chapter 12), a briefly presented (*tachistoscopic*) instruction in the left visual field is appreciated only by the right brain (assuming that the individual maintains fixation on a mark in the center of the viewing screen). In normal subjects, activation of the right visual cortex leads to hemispheric transfer of visual information via the corpus callosum to the left hemisphere. In split-brain patients, information presented to the left visual field cannot reach the left hemisphere, and patients are unable to produce a verbal report regarding the stimuli. However, such patients are able to provide a verbal report of stimuli presented to the right visual field. A wide range of hemispheric functions can be evaluated using this method, even in normal subjects. The list above enumerates some of the different functional abilities of the left and right hemispheres, as deduced from a variety of behavioral tests in split-brain patients.

some could not provide any verbal account of what they held in their left hand. Observations using techniques to present visual information to the hemispheres independently (a method called *tachistoscopic presentation*; see Figure 33.3B) showed further that the left hemisphere responds to written commands, whereas the right hemisphere typically responds only to nonverbal stimuli (e.g., pictorial instructions or, in some cases, rudimentary written commands). These distinctions reflect broader hemispheric differences summarized by the statement that the left hemisphere in most humans is specialized for (among other things) the verbal and symbolic processing important in communication, whereas the right hemisphere is specialized for (among other things) visuospatial and emotional processing.

The ingenious work of Sperry and his colleagues on split-brain patients put an end to the century-long controversy about language lateralization. In most individuals, the left hemisphere is unequivocally the seat of the explicitly verbal language functions. There is significant variation in the degree of lateralization among individuals, however, and it would be wrong to suppose that the right hemisphere has no language capacity. As noted, in some individuals the right hemisphere can produce rudimentary words and phrases, a few individuals have fully right-sided verbal functions, and even for the majority with strongly left-lateralized language semantic abilities, the right hemisphere is normally the source of our emotional coloring of language (see Chapter 31). Moreover, the right hemisphere in many split-brain patients understands language to a modest degree, since these patients can respond to simple visual commands presented tachistoscopically in the left visual field. Consequently, Broca's conclusion that we speak with our left brain is not strictly correct; it would

BOX 33B ■ Language and Handedness

Approximately nine out of ten people are right-handed, a proportion that appears to have been stable over thousands of years and across all cultures in which handedness has been examined. In addition, handedness, or its equivalent, is not peculiar to humans; many studies have demonstrated paw preference in animals ranging from mice to monkeys that is, at least in some ways, similar to human handedness. Unlike in humans, however, which hand is preferred varies about equally among individuals.

Handedness is usually assessed by having individuals answer a series of questions about preferred manual behaviors, such as "Which hand do you use to write?"; "Which hand do you use to throw a ball?"; or "Which hand do you use to brush your teeth?" Each answer is given a value, depending on the preference indicated, providing a quantitative measure of the inclination toward right- or left-handedness. Anthropologists have determined the incidence of handedness in ancient cultures by examining artifacts—the shape of a flint ax, for example, can indicate whether it was made by a right- or left-handed individual. Handedness in antiquity has also been assessed by noting the incidence of people in artistic representations who are using one hand or the other. Based on this evidence, the human species appears always to have been a primarily right-handed one.

Whether an individual is right- or left-handed has several interesting consequences. As will be obvious to left-handers, the world of human artifacts is in many respects a right-handed one. Implements such as can openers, scissors, and power tools are constructed for the right-handed majority (Figure A). Books and magazines are also designed for right-handers (compare turning this page with your left and right hands), as are golf clubs and guitars. By the same token, the challenge of penmanship is different for left- and right-handers by virtue of writing from left to right (Figure B). Perhaps as a consequence of such biases, the accident rate for left-handers in all categories (work, home, sports, traffic fatalities) is higher than for right-handers. However, there are also some advantages to being left-handed. For example, an inordinate number of international fencing champions have been left-handed. The reason for this fact is simply that the majority of any individual's opponents will be right-handed; therefore, the average fencer, whether right- or left-handed, is less practiced at parrying thrusts from left-handers.

Continued on the next page

(A)

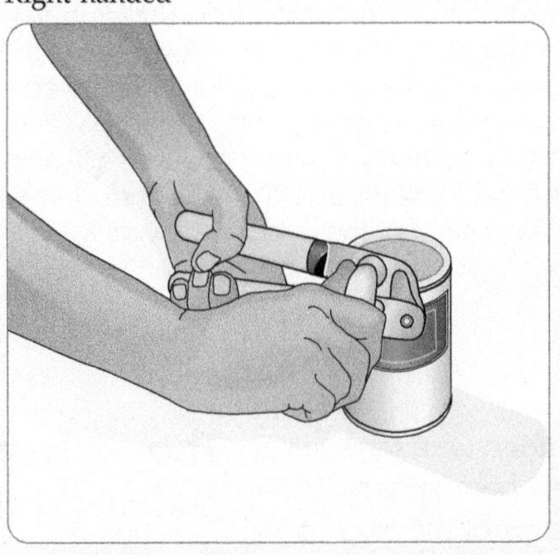

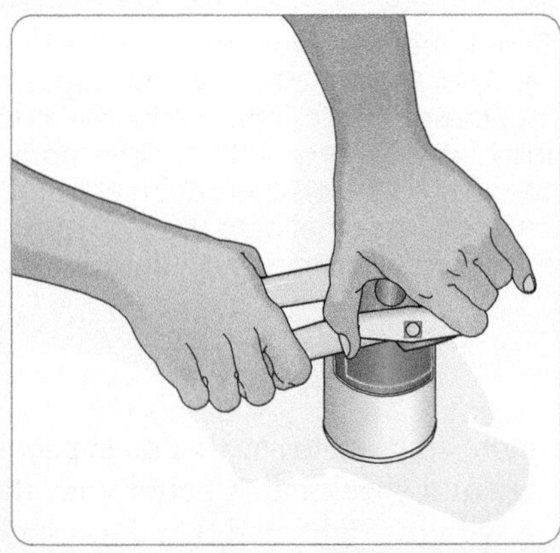

(A) A simple manual can opener is one example of the many common objects designed for use by the right-handed majority.

BOX 33B ■ *(continued)*

(B)

Right-handed writing

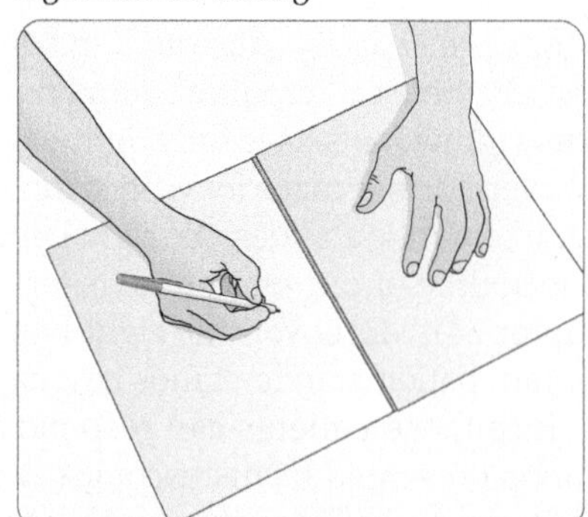

Left-handed writing

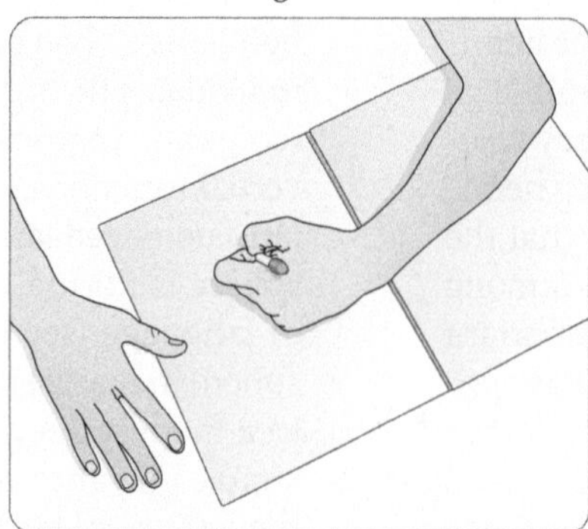

(C)

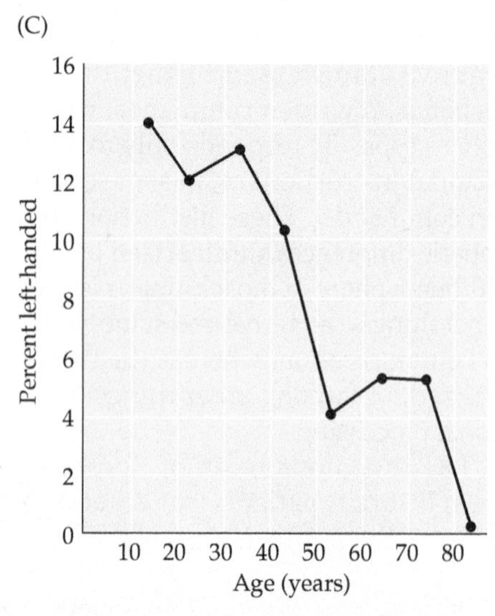

(B) Writing techniques for right- and left-handed individuals. (C) Percentage of left-handers in the U.S. population as a function of age (based on more than 5000 individuals assessed in 1990). Taken at face value, these data indicate that right-handers live longer than left-handers. A more likely possibility, however, is that the paucity of elderly left-handers simply reflects changes in the social pressures on children to become right-handed. (After Coren, 1992.)

Hotly debated over the years have been the related questions of whether being left-handed is in any sense "pathological," and whether being left-handed entails a diminished life expectancy. No one disputes the fact that there is currently a surprisingly small number of left-handers among the elderly (Figure C). These data have come from studies of the general population and have been supported by information gleaned from *The Baseball Encyclopedia* (in which longevity and other characteristics of a large number of healthy left- and right-handers have been recorded because of interest in the U.S. national pastime).

Two explanations for this peculiar finding have been put forward. Stanley Coren and his collaborators at the University of British Columbia have argued that these statistics reflect a higher mortality rate among left-handers, partly as a result of increased accidents, but also because other data show left-handedness to be associated with a variety of pathologies (there is, for instance, a higher incidence of left-handedness among individuals classified as mentally impaired). Coren and others have suggested that left-handedness may arise because of developmental problems in the pre- and/or perinatal period. If this were shown to be true, then a rationale for decreased longevity would have been identified that might combine with greater proclivity for accidents in a right-hander's world.

An alternative explanation, however, is that the diminished number of left-handers among the elderly at present is primarily a reflection of sociological factors—namely a greater acceptance of left-handed children today compared with earlier in the twentieth century. In this view, there are fewer older left-handers because in earlier generations parents, teachers, and other authority figures encouraged (and sometimes insisted on) right-handedness. The weight of the evidence favors the sociological explanation.

The relationship between handedness and other lateralized functions—language in particular—has long been a source of confusion. It is unlikely that there is any direct relationship between language and handedness, despite much speculation to the contrary. The most straightforward evidence on this point comes from the results of the Wada test described in the text. The large number of such tests carried out for clinical purposes indicates that about 97% of humans, including the majority of left-handers, have verbal language functions represented in the left hemisphere (although that right-hemispheric dominance for language is more common among left-handers). Since most left-handers don't have language function on the same side of the brain as the control of their preferred hand, it is hard to argue for any strict relationship between these two lateralized functions. In all likelihood, handedness, like language, is first and foremost an example of the advantage of having any specialized function in one hemisphere or the other to maximize "wiring efficiency."

be more accurate to say that most people understand language and speak much better with the left hemisphere than with the right, and that the contributions of the two hemispheres to the overall goals of communication are different (see Box 33B for further discussion of cerebral "dominance" and handedness).

The Search for Anatomical Differences between the Hemispheres

The differences in language function between the left and right hemispheres have naturally inspired neurologists and neuropsychologists to search for some structural

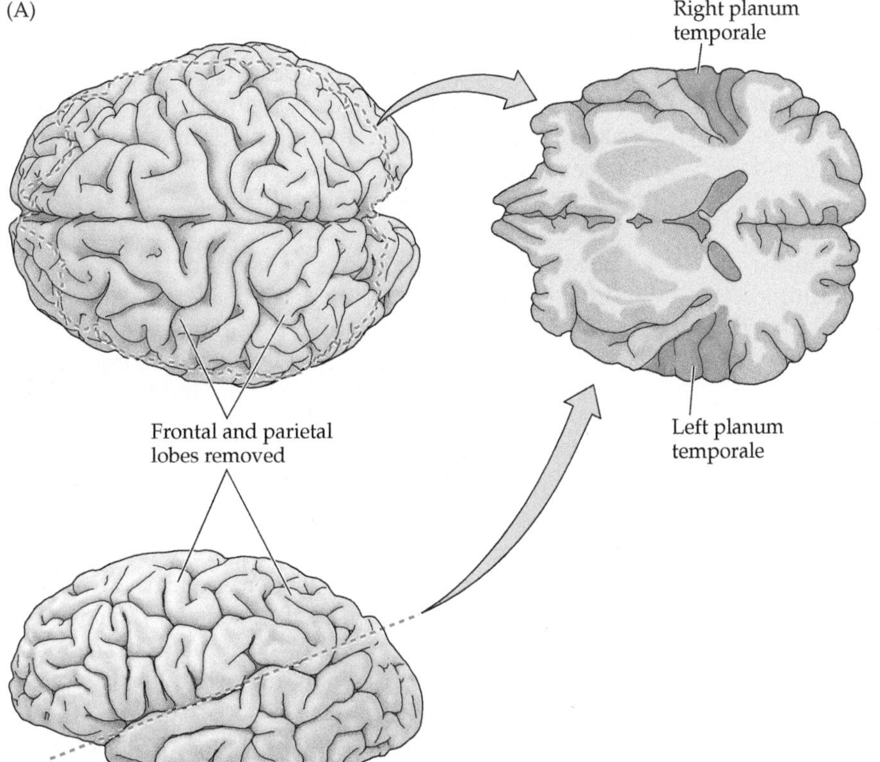

(B)

Mean size of the planum temporale measured in 100 adult and 100 infant brains		
	Left hemisphere	Right hemisphere
Infant	20.7	11.7
Adult	37.0	18.4

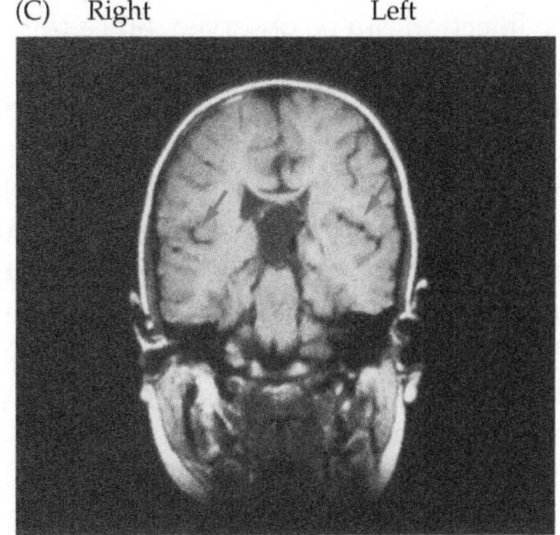

FIGURE 33.4 Asymmetry of the right and left human temporal lobes. (A) The superior portion of the brain has been removed as indicated to reveal the dorsal surface of the temporal lobes in the right-hand diagram (which presents a dorsal view of the horizontal plane). A region of the surface of the temporal lobe called the planum temporale is significantly larger in the left hemisphere of most (but far from all) individuals. (B) Measurements of the planum temporale in adult and infant brains. The mean size of the planum temporale is expressed in arbitrary planimetric units to get around the difficulty of measuring the curvature of the gyri within the planum, which tends to be larger on the left side of the brain. The asymmetry is evident at birth and persists in adults at roughly the same magnitude (on average, the left planum is about 50% larger than the right). (C) An MRI image in the frontal plane, showing this asymmetry (arrows) in a normal brain.

correlate of behavioral lateralization. One hemispheric difference that has received much attention over the years was identified in the late 1960s by Norman Geschwind and his colleagues at Harvard Medical School, who found an asymmetry in the superior aspect of the temporal lobe known as the **planum temporale** (Figure 33.4). This area is significantly larger on the left side in about two-thirds of humans studied postmortem; a similar difference has been reported in great apes but not in other primates.

Because the planum temporale is near (although not congruent with) the regions of the temporal lobe that contain cortical areas essential to language (i.e., Wernicke's area and other auditory association areas), it was initially suggested that this leftward asymmetry reflected the greater involvement of the left hemisphere in language. However, these anatomical differences in the two hemispheres of the brain, which are recognizable at birth, are unlikely to be an anatomical correlate of the language lateralization. The fact that a detectable planum asymmetry is present in only 67% of human brains, whereas the preeminence of language in the left hemisphere is evident in 97% of the population, argues that this association has some other cause. Structural correlates of the functional left–right differences in hemispheric language abilities, if indeed they exist at a gross anatomical level, are simply not clear. This lack of information applies equally to other lateralized brain functions, which evidently depend on more subtle differences in the connectivity of the relevant brain regions.

Mapping Language Functions

The pioneering work of Broca and Wernicke, and later Geschwind and Sperry, clearly established differences in hemispheric function. Several techniques have since been developed that allow hemispheric attributes to be assessed in neurological patients with an intact corpus callosum and in healthy individuals.

One method that has long been used to assess language lateralization in patients was devised in the 1960s by Juhn Wada at the Montreal Neurological Institute. In the so-called **Wada test**, a short-acting anesthetic (e.g., sodium amytal) is injected into the left carotid artery; this procedure transiently "anesthetizes" the left hemisphere and thus tests the functional capabilities of the affected half of the brain. If the left hemisphere is indeed "dominant" for language, then the patient becomes transiently aphasic while carrying out an ongoing verbal task like counting. The anesthetic is rapidly diluted by the circulation, but not before its local effects on the hemisphere on the side of the injection can be observed. Since this test is potentially dangerous, its use is typically limited to neurological and neurosurgical patients in which a surgical procedure is contemplated. Such studies have provided valuable diagnostic tools to determine the location of the speech and language cortex in preparation for neurosurgery. The Wada test confirms which hemisphere is "eloquent." Such confirmation is important: Although most individuals have the major language functions in the left hemisphere, a few—about 3% of the population—do not (the latter are much more often left-handed).

Less invasive (but also less definitive) ways to test the cognitive abilities of the two hemispheres in typical individuals include PET, fMRI, TMS, and the sort of tachistoscopic presentation used so effectively by Sperry and his colleagues (even when the hemispheres are normally connected, participants show delayed verbal responses and other differences when the right hemisphere receives the instruction). Application of these noninvasive approaches has amply confirmed the hemispheric lateralization of language functions, and improvements in the approaches has already lead to the replacement of the Wada test.

Once the "dominant" hemisphere is known using these methods, neurosurgeons typically map language areas more precisely by electrical stimulation of the cortex during surgery. By the 1930s, the neurosurgeon Wilder Penfield and his colleagues at the Montreal Neurological Institute had already carried out a detailed localization of cortical capacities in a large number of patients. Penfield used electrical mapping techniques adapted from neurophysiological work in animals to delineate the language areas of the cortex prior to removing brain tissue in the treatment of tumors or epilepsy (Figure 33.5A). Such intraoperative mapping guaranteed that the cure would not be worse than the disease and has been widely used ever since, with increasingly sophisticated stimulation and recording methods. As a result, a wealth of detailed information about language localization has emerged.

Penfield's observations, together with more recent studies performed by George Ojemann and his group at the University of Washington, and others, have further advanced the conclusions inferred from postmortem correlations and other approaches (Figure 33.5B). As expected, these studies using electrophysiological recording methods during surgery have shown that a large region of the perisylvian cortex of the left hemisphere is clearly involved in language production and comprehension. A surprise, however, has been the variability in language localization from patient to patient. Ojemann found that the brain regions involved in language are only approximately those indicated by textbook treatments, and that their locations

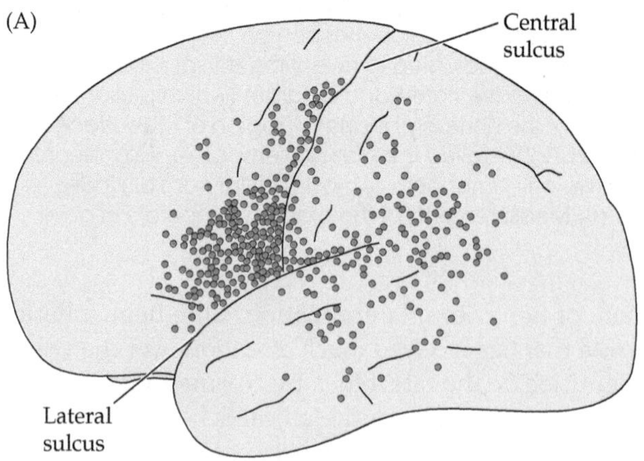

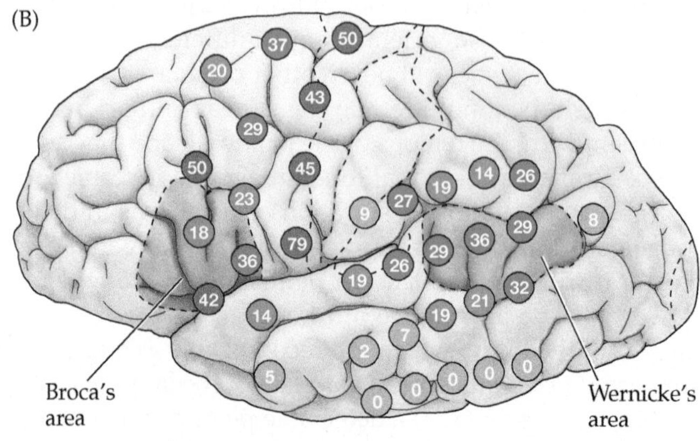

FIGURE 33.5 Evidence for the variability of language representation among individuals, determined by electrical stimulation during neurosurgery. (A) Diagram from Penfield's original study illustrating sites in the left hemisphere at which electrical stimulation interfered with speech. (B) Diagram summarizing data from 117 patients whose language areas were mapped by electrical recording at the time of surgery. The number in each red circle indicates the percentage of patients who showed interference with language in response to stimulation at that site. Note that many of the sites that elicited interference fall outside the classic language areas (Broca's area, shown in purple; Wernicke's area, shown in blue). (A after Penfield and Roberts, 1959; B after Ojemann et al., 1989.)

differ unpredictably among individuals. Equally unexpected, bilingual patients do not necessarily use the same bit of cortex for storing the names of the same objects in two different languages. Moreover, although neurons in the temporal cortex in and around Wernicke's area respond preferentially to spoken words, they do not show preferences for a particular word. Rather, a wide range of words can elicit a response at any given recording site.

Despite these advances, neurosurgical studies are complicated by their intrinsic difficulty, the risk involved, and the fact that the brains of the patients on whom they are carried out are not typical. The advent of PET in the 1980s, and more recently of fMRI, has allowed the investigation of language regions in research participants by noninvasive brain-imaging techniques (Figure 33.6). Recall that these techniques reveal the areas of the brain that are active during a particular task because the related electrical activity increases local metabolic activity and blood flow. Much like Ojemann's studies in neurosurgical patients, the results of these approaches, particularly in the hands of Marc Raichle, Steve Petersen, and their colleagues at Washington University in St. Louis, have challenged excessively rigid views of the localization and lateralization of linguistic function. Although high levels of activity occur in the expected regions, large areas of both hemispheres are activated in word recognition or production tasks.

Finally, several investigators, including Hanna Damasio and her colleagues, then at the University of Iowa, and Alex Martin and his collaborators at the National Institute of Mental Health, have shown that distinct regions of the temporal cortex are activated by tasks requiring study participants to name, view, or match human faces, animals, or tools (Figure 33.7). This observation helps explain the clinical finding that when a relatively limited region of the temporal lobe is damaged (usually but by no means always on the left side), language deficits are sometimes restricted to a particular *category* of objects. These studies are also consistent with Ojemann's electrophysiological studies, which indicate that some aspects of language are organized according to categories of meaning rather than individual words. Taken together, such studies are rapidly augmenting the information available about how language is represented in the brain.

Passively viewing words

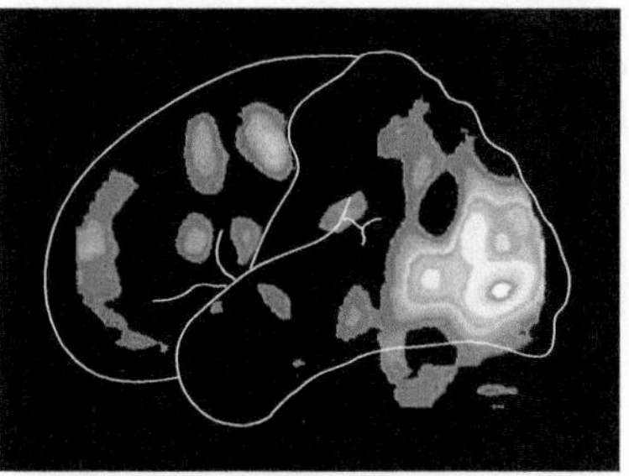

Listening to words

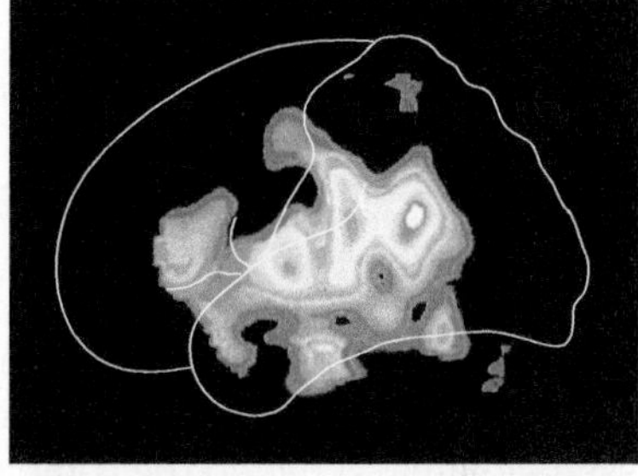

Speaking words

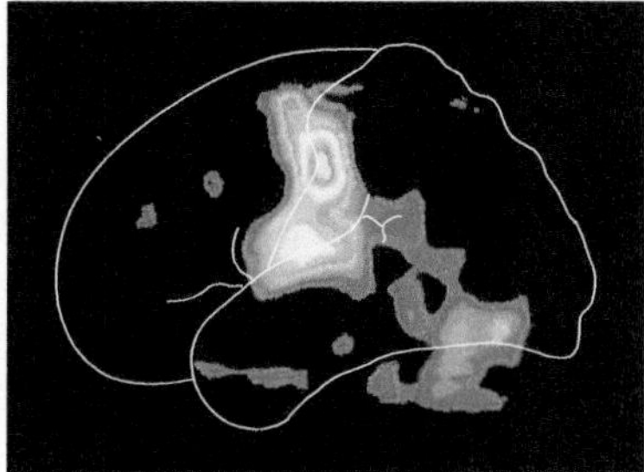

Generating word associations

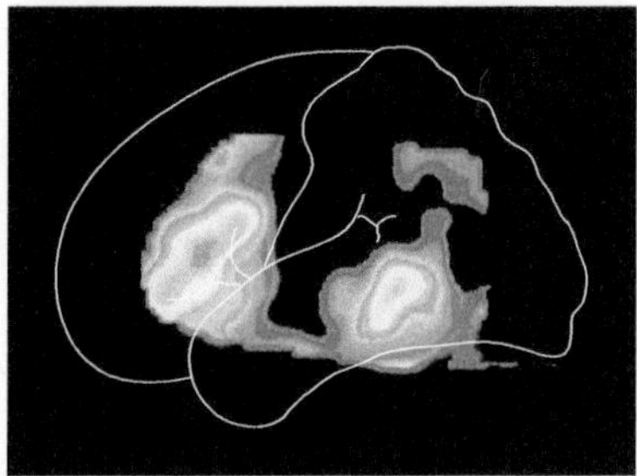

FIGURE 33.6 **Language-related regions of the left hemisphere mapped by positron emission tomography (PET) in intact individuals.** Individuals in the PET scanner followed instructions on a special display (details not illustrated). The left panels indicate the task being practiced prior to scanning. The PET scan images are shown on the right. Language tasks such as listening to words and generating word associations elicit activity in Broca's and Wernicke's areas, as expected. However, there is also activity in primary and association sensory and motor areas for both active and passive language tasks. These observations indicate that language processing involves many cortical regions in addition to the classic language areas. Warm colors indicate greater activation. (From Posner and Raichle, 1994.)

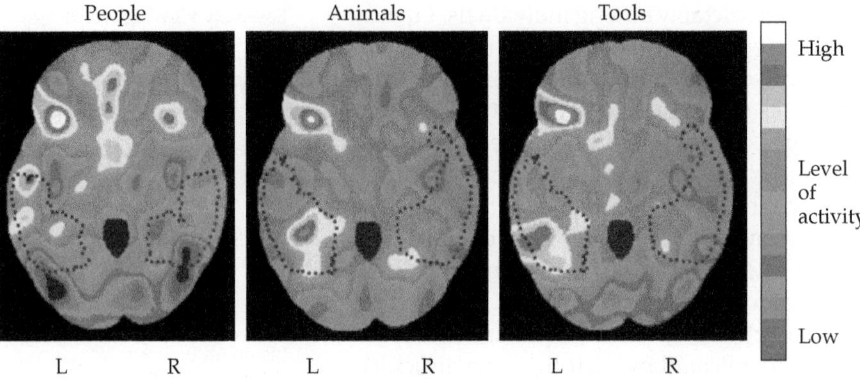

FIGURE 33.7 **Different regions in the temporal lobe are activated by different word categories using PET imaging.** Dotted lines show the locations of the relevant temporal regions in these horizontal views. Note the different patterns of activity in the temporal lobe in response to each stimulus category. (From Damasio et al., 1996.)

The Role of the Right Hemisphere in Language

Because the same gross anatomical and cytoarchitectonic areas exist in the cortex of both hemispheres, a puzzling issue remains. What do the comparable areas in the right hemisphere actually do? In fact, language deficits often *do* occur following damage to the right hemisphere. The most obvious effect of such lesions is an absence of the normal emotional and tonal components of language conveyed by prosody. As mentioned ealier, the prosodic aspects of speech (or prelingual vocal sounds) include stress, intonation, and rhythm, informing listeners about aspects of communication not necessarily encoded in grammar, such as whether an utterance is a question, statement, or command. This "coloring" of speech is critical to the message conveyed; indeed, in some languages, such as Mandarin Chinese, variations in tone are used to change the meaning of the word uttered. Deficiencies in speech prosody, referred to as **aprosodias**, are associated with right-hemisphere damage to the cortical regions that correspond to Broca's and Wernicke's areas and related regions in the left hemisphere. Aprosodias emphasize that although the left hemisphere (or better put, specialized cortical regions in that hemisphere) figures prominently in the comprehension and production of language for most humans, other regions, including corresponding areas in the right hemisphere, are needed to generate the full richness of everyday speech.

In summary, whereas the classically defined regions of the left hemisphere operate more or less as advertised, a variety of more recent studies have shown that other left- and right-hemisphere areas clearly make significant contributions to the generation and comprehension of language.

Genes and Language

The search for genetic contributions to cognitive functions and their disorders is being pursued in many contexts. Because genes play some role in all phenotypic features, and because the propensity for language acquisition by infants is obvious, exploring the genes involved is plausible. Furthermore, the occurrence of language and/or reading problems that run in families makes plain that genetic anomalies can play a role in the normal development of these cognitive functions.

An inherited but quite rare disorder has more specifically raised the question of the genetic determination of language and, in popular accounts, whether there might be a "language gene." The gene of interest, called *FOXP2*, is located on human chromosome 7. It was discovered in 1990 in a family in which about half the members are afflicted. The affected individuals in the pedigree, known in the literature as the K.E. family, are unable to fluently select the movements of the vocal apparatus needed to make appropriate speech sounds. Thus, what they try to say is largely incomprehensible. The impairment, which is caused by a single autosomal recessive mutation, is thus one of motor organization as it pertains to speech rather than one of comprehension. The afflicted family members, however, also have lower IQs than their unafflicted relatives, indicating that the defect is not specific for language.

The mechanism by which the gene defect exerts these effects is not known, but the protein that it encodes is a transcription factor, meaning that the gene product is an agent that binds to the promoter regions of other genes to control their expression. The *FOXP2* gene is strongly expressed in other animals, including mice, where it affects many aspects of development, including the ultrasonic vocalization of these animals.

Interesting though this gene may be, reports about the discovery of a "language gene" were clearly unwarranted. Because it encodes a transcription factor, *FOXP2* affects many other genes with a range of developmental consequences, some evidently influencing the mechanisms that generate neural circuits in those parts of the brain that support the organization and expression of language.

Sign Language

An implication of several aspects of the foregoing account is that the cortical organization of language does not simply reflect specializations for hearing and speaking; the language regions of the brain appear to be more broadly organized for processing symbols pertinent to social communication. Strong support for this conclusion has come from studies of sign language in individuals deaf from birth.

American Sign Language has all the components (grammar, syntax, and emotional tone) of spoken and heard language. Based on this knowledge, Ursula Bellugi and her colleagues at the Salk Institute for Biological Studies examined the cortical localization of sign language abilities in patients who had suffered lesions of either the left or right hemisphere. All of these deaf individuals had never learned verbal language, had been signing throughout their lives, had deaf spouses, were members of the deaf community, and were right-handed. The patients with left-hemisphere lesions, which in each case involved the language areas of the frontal and/or temporal lobes, had measurable deficits in sign production and comprehension when compared with normal signers of similar age (Figure 33.8). In contrast, the patients with lesions in approximately the same areas in the right hemisphere did not have such signing "aphasias." Instead, as predicted from hearing patients with similar lesions, right hemisphere abilities such as visuospatial processing, emotional processing, and the emotional tone evident in signing were impaired. Although the number of people studied was necessarily small (deaf signers with lesions of the language areas are understandably difficult to find), the capacity for signed and seen communication was predominantly represented in the same areas as spoken language in the left hemisphere. This evidence accords with the idea that the language regions of the brain are specialized for the representation of social communication by means of symbols, rather than for heard and spoken language per se.

The capacity for seen and signed communication, like its heard and spoken counterpart, emerges in early infancy. Careful observation of babbling in infants with normal hearing shows the production of a predictable pattern of sound production related to the ultimate acquisition of spoken language. Thus, babbling prefigures mature language, and indicates that an innate capacity for language imitation is a key part of the process by which a full-blown language is ultimately acquired. The hearing offspring of deaf, signing parents "babble" with their hands in gestures that are apparently the forerunners of mature signs. Like verbal babbling, the amount of manual babbling increases with age until the child begins to form accurate, meaningful signs. These observations indicate that the strategy for acquiring the rudiments of symbolic communication from parental or other cues—regardless of the means of expression—is similar.

Patient with signing deficit:

Arrive

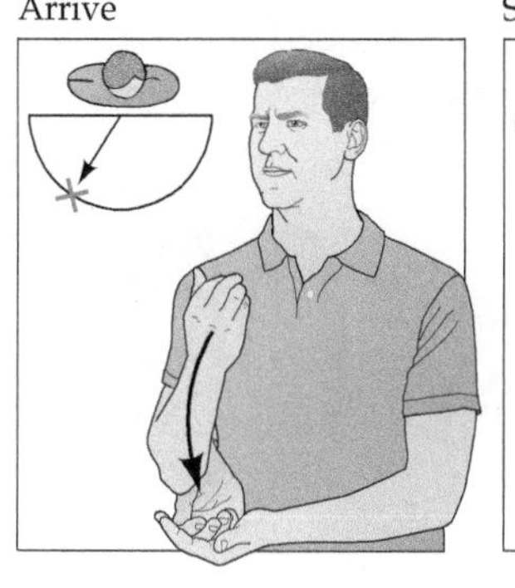

Stay

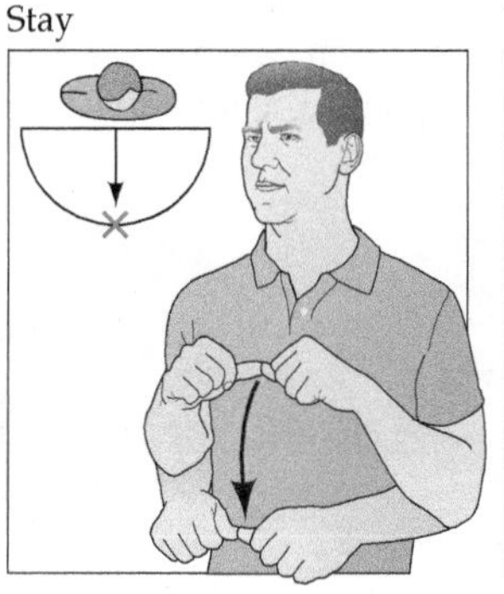

There

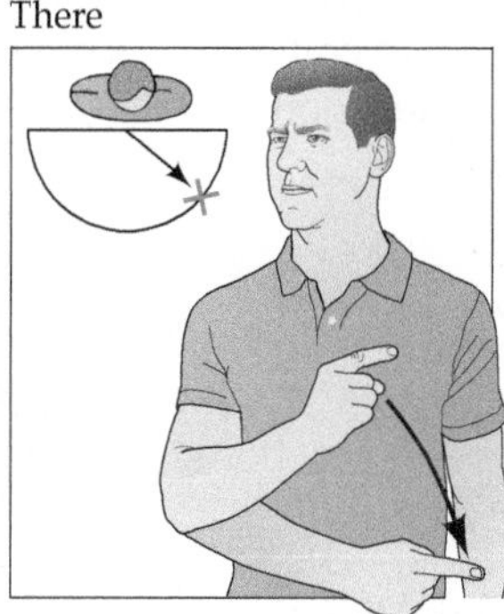

Correct form:

Arrive

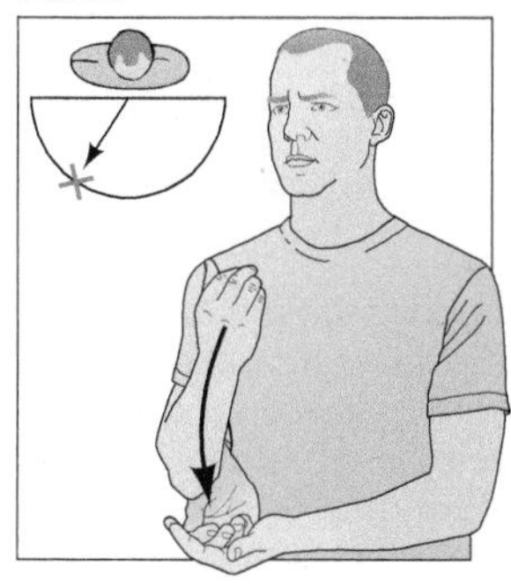

Stay

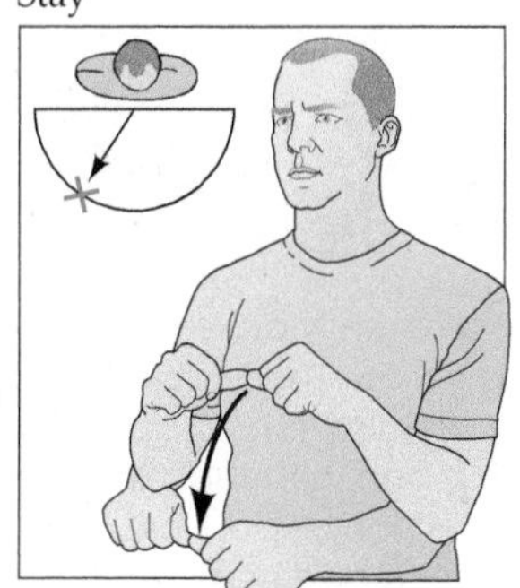

There

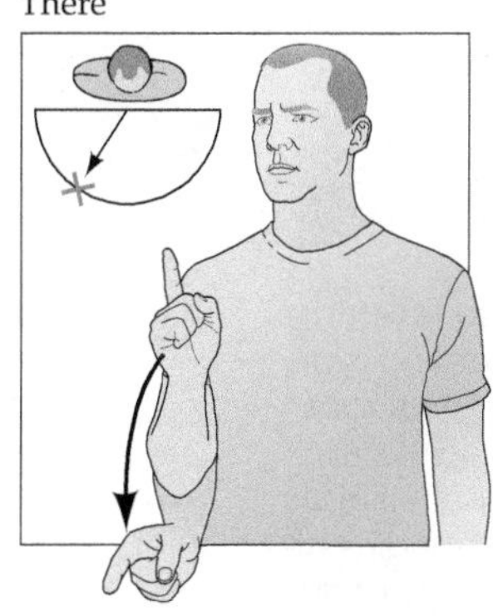

FIGURE 33.8 Signing deficits in congenitally deaf individuals who learned sign language from birth and later suffered lesions of the language areas in the left hemisphere. Left-hemisphere damage produced signing problems in these patients analogous to the aphasias seen after comparable lesions in hearing, speaking patients. In this example, the patient (upper panels) is expressing the sentence "We arrived in Jerusalem and stayed there." Compared with a control (lower panels), the patient cannot properly control the spatial orientation of the signs. The direction of the correct signs and the aberrant direction of the "aphasic" signs are indicated in the upper left-hand corner of each panel. (After Bellugi et al., 1989.)

A Critical Period for Language Acquisition

Although a maturing child begins to lose the ability to hear non-native phonetic distinctions at a remarkably early age, the ability to learn another language fluently persists for some years. As is apparent from the experience of learning a new language in school or from watching friends and family who began learning a second language at various ages, becoming fluent nevertheless requires linguistic experience relatively early in life. This fact reflects a broader generalization about neural development—namely that neural circuitry is especially susceptible to modification during early development, and that this malleability gradually diminishes with maturation. The window for extensive neural modification supporting a behavior is referred to as the **critical period** (also called the sensitive period), and it is evident in the acquisition of language, and in many other cognitive behaviors.

Psycholinguists Jacqueline Johnson and Elissa Newport at the University of Rochester undertook a detailed study of this aspect of language learning, examining the critical period for the acquisition of a second language in Chinese Americans who had come to the United States at various ages (Figure 33.9A). Using a battery of grammatical and other tests of fluency, Johnson and Newport found that learning a second language before about age 7 results in adult performance that is indistinguishable from that of native speakers, although the details for second language learning are debated.

The requirement for experience during a critical period is also apparent in studies of language acquisition in children who become deaf at different ages. The effects on language skills tend to be more marked when the onset of deafness occurs early in life than when the onset occurs later in childhood or in adult life. Younger children who have acquired some speech but then lose their hearing suffer a substantial decline in spoken language because they are unable to hear themselves speak and thus cannot refine their initial efforts at speech by testing the relative adequacy of what they are trying to say through auditory feedback. Differences in brain activation observed in children and adults doing language-based tasks provide some indication of the neural regions pertinent to diminished language-learning skills in adults (Figure 33.9B).

An issue complementary to normal language learning concerns the effects of language deprivation, a topic already touched on in the consideration of the deterioration of language in hearing-deficient individuals as a function of age. A broader question, asked since antiquity, is what would happen if an otherwise typical child were never exposed to language. Would the child remain mute, or could he or she develop some ability to speak and, if so, what sort of language would she or he have?

The closest approximation to a realization of this sort of "thought experiment" is a handful of unfortunate cases in which children have been deprived of significant language exposure as a result of having been raised by deranged parents. In the most fully documented case, a girl in a Los Angeles suburb was raised from infancy until age 13 under conditions of almost total language deprivation (she is known in the literature as Genie). Genie was brought to the attention of social workers in 1970, who found her locked in a small room where she had been isolated and allegedly beaten if she made any noise. She was removed from these

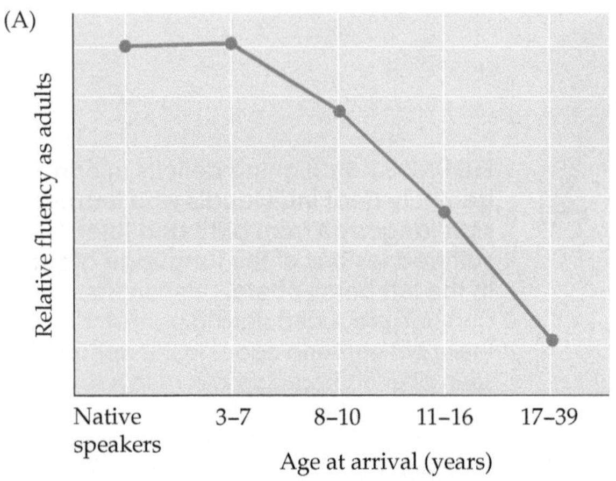

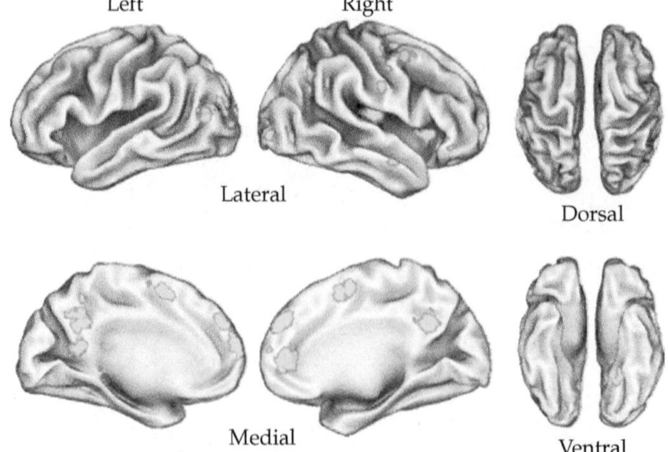

FIGURE 33.9 The critical period for language learning. (A) The critical period for fully fluent language learning is apparent in studies of the fluency of adult Chinese Americans as a function of the age of their arrival in the United States, marking the onset of significant exposure to English. Ultimate fluency starts to drop off when language learning begins after about age 7. (B) Areas in the brains of children and adults that are differently active during language-based tasks are shown in yellow. These differences provide a possible neural basis for the diminishing ability to learn a new language with increasing age, although more specific interpretation of this evidence is not yet possible. (A after Johnson and Newport, 1989; B from Brown et al., 2005.)

desperate conditions and taken to the children's hospital at the University of California, Los Angeles, where she was found to be in adequate general health. Given these highly unusual circumstances, a team of psychologists and linguists at UCLA studied Genie's language and other cognitive skills during the subsequent 5 years. Although Genie had little or no language ability initially, the investigators found no evidence of brain damage or mental retardation in the usual sense, and they described her overall personality as rather docile and generally pleasant. As might be expected, Genie also received extensive remedial training to teach her the language skills that she had never learned as a child. Despite these efforts, as well as daily life in more or less normal conditions in foster homes, Genie never acquired more than rudimentary language skills. Although she eventually learned a reasonable vocabulary, she could not put words together grammatically, saying things like "Applesauce buy store?" when she wanted to ask whether she might buy some applesauce at the store. Genie's case and a few similar examples starkly define the importance of adequate early experience for successfully learning any language, in accord with the more abundant evidence of a critical period for learning a second language.

In sum, researchers agree that the normal acquisition of human language is subject to a critical period of approximately a decade; exposure and practice must occur within this time for a person to achieve full fluency. Of course, some ability to learn language persists into adulthood, but at a reduced level of efficiency and ultimate performance. This generalization is consistent with much other evidence from neural development that underscores the special importance of early experience in the full development of cognitive abilities (see Chapter 25).

Reading and Dyslexia

Reading is obviously an interpretation of visual symbols closely related to language and is subject to some important disorders. Dyslexia, for example, is a common problem that affects a child's ability to read. Despite having normal or above-normal intelligence, dyslexics are poor readers and have difficulty more generally in processing speech sounds and translating visual to verbal information, and vice versa. Thus, in addition to suffering reading problems, dyslexics often have difficulty writing, are poor spellers, and are prone to errors arising from letter transposition. Because there is no specific diagnostic criterion, estimates of the prevalence of dyslexia vary widely; 5% to 15% of children are affected, with greater occurrence in boys. Although dyslexia is generally accepted as a learning disability, its cause is unclear, and some investigators have argued that this disorder is simply the lower tail of the distribution of performance in learning to read. There is, however, a strong tendency for the problem to run in families, implying a genetic basis.

Given these facts, dyslexia is a broad category with several causes. Nonetheless, researchers have understandably focused on areas of the brain concerned with reading. fMRI studies indicate that a specific set of left-hemispheric brain areas is activated during reading. Some of these areas are also activated by spoken language, but one of them, the *visual word form area* (VWFA) located in the region of the left occipito-temporal sulcus, is selectively activated by written characters but not by spoken words or low-level visual stimuli (Figure 33.10). The organization of the VWFA appears to depend on experience, and activation levels in this area in children and adolescents predict word–phoneme decoding abilities. Dyslexics tend to have a weaker BOLD signal in this general area compared with nondyslexics, as well as underdevelopment of the associated cortex and white matter tracts.

Evidence for a functionally specific brain region is surprising, given that until recently few humans have been literate. Thus, the VWFA could not have evolved to support reading per se, and may be better thought of as a brain region with specific processing characteristics rather than a brain region devoted to a specific stimulus class such as words. To make sense of these observations, Stanislas Dehaene and Laurent Cohen have argued that brain circuits "recycle" information for new, culturally specified functions. An extension of this idea is that cultural inventions such as reading and writing are constrained to specific brain circuits across cultures. For example, letters in all alphabets are made up of roughly three strokes, which might in turn be related to the efficient use of receptive fields by neurons at successively higher levels of visual cortex. The argument is that by matching the appearance of letters to the inherent functions of neurons involved in recognizing elementary objects, writing and reading systems are

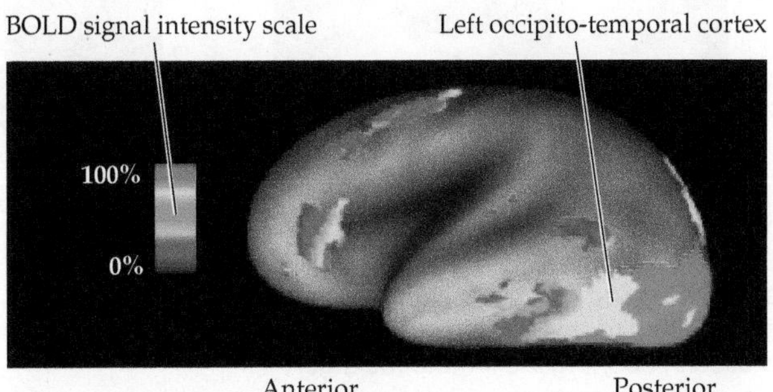

FIGURE 33.10 Hierarchical organization across the extent of the VWFA in the left occipito-temporal cortex. The figure shows an anterior–posterior gradient of BOLD signal intensity as stimuli being read are progressively changed from single letters to complete words. In typical individuals, but less so in dyslexics, increasingly closer approximations to the usual appearance of words (i.e., shifts from false fonts to regular fonts) lead to increasing anterior activation in the left but not the right occipito-temporal cortex. (From Vinckier et al., 2007.)

similarly determined across cultures. In any event, there is no accepted treatment for dyslexia, although identifying the problem early and implementing remediation through extra training and effort are helpful in most dyslexics.

Do Other Animals Have Language?

Over the centuries, theologians, natural philosophers, and a good many modern neuroscientists have argued that language is uniquely human, this extraordinary behavior being seen as setting us apart from our fellow animals. However, the gradual accumulation of evidence during the last 80 years demonstrating highly sophisticated systems of communication in animals as diverse as bees, birds, monkeys, and whales has made this point of view increasingly untenable, at least in a broad sense (Box 33C). Until recently, however, human language *has* appeared to be unique in its ability to associate specific meanings with arbitrary symbols, ad infinitum. In the dance of the honeybee described so beautifully by Karl von Frisch, for

BOX 33C ■ Learned Vocal Communication in Birds

Humans are not the only animals that learn to communicate during a critical period of development, and studies of non-human species have added greatly to a better understanding of social communication by vocalization. Many animal vocalizations are innate in the sense that they require no experience to be correctly produced and interpreted. For example, quails raised in isolation or deafened at hatching so that they never hear conspecific vocal stimuli nonetheless produce the full repertoire of species-specific vocalizations. Some species of birds, however, learn to communicate by vocal sounds, a process that is in some respects similar to the way humans learn language. Particularly well characterized is vocal learning in song sparrows, canaries, and finches. These and other bird species use songs to define their territory and attract mates (Figures A and B).

As with human language, early sensory exposure and practice are key determinants of subsequent perceptual and behavioral capabilities. Furthermore, the developmental period for learning these vocal behaviors, as for learning language, is restricted to early life. (Canaries are exceptional in that they continue to build their song repertoire from season to season, which is one reason these birds have been such popular pets over the centuries.) Song learning in these species entails an initial stage of *sensory acquisition*, when the juvenile bird listens to and memorizes the song of an adult male "tutor" of its own species. This period is followed by a stage of vocal learning through practice, when the young bird matches its song to the memorized tutor model by auditory feedback. This *sensory-motor learning* stage ends with the onset of sexual maturity, when songs become acoustically stable (called *crystallized song*; Figure C).

(A)

(B)

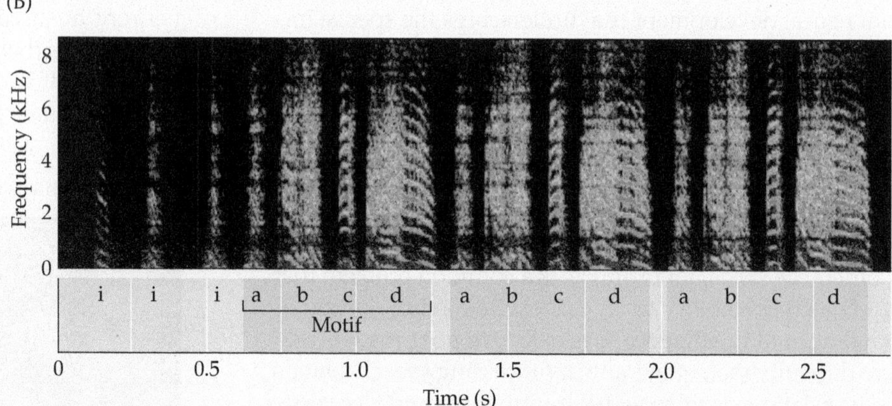

(C)

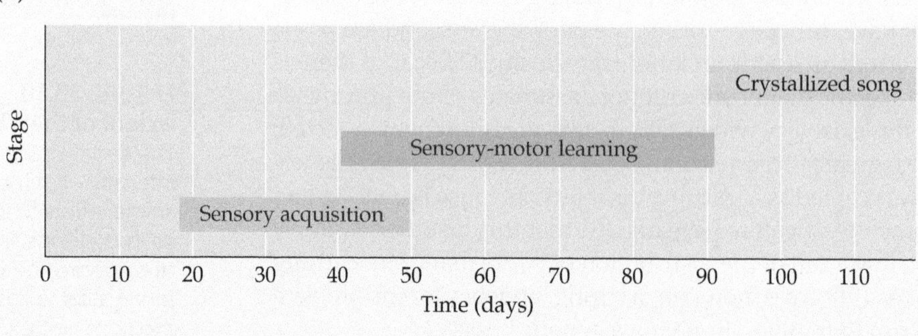

Birdsong learning. The spectrogram in (B) shows the song of an adult male zebra finch—the bird on the right in (A)—that is used in courting the female (as a general rule, only male songbirds sing). The recording plots the frequency of the song against time, showing the syllables and motifs that characterize the song of this species. Color indicates the intensity of the vocal signal, with red representing higher intensity and blue lower. (C) The stages of song learning in the zebra finch (0 indicates the time of hatching). (Courtesy of Rich Mooney.)

BOX 33C (*continued*)

In the species typically studied, young birds are especially impressionable during the first 2 months after hatching and become refractory to further exposure to tutor songs as they age, thus defining a critical (or sensitive) period for song learning. The early exposure to the tutor (typically the father) generates a memory that can remain intact for months (or longer) in some species before the onset of the vocal practice phase. Moreover, juveniles need to hear the tutor song only 10 to 20 times to vocally mimic it months later, and exposure to other songs after the sensory acquisition period does not affect this memory. The songs heard during this time, but not later, are the only ones that the young bird mimics. Songbirds also exhibit learned regional dialects, much as human infants learn the language characteristic of the region in which they are raised.

Other studies indicate that birds have a strong intrinsic predisposition to learning the song of their own species. Thus, when presented during maturation with a variety of recorded songs that include their own species' song, together with that of other species, juvenile birds preferentially learn the song of their own species. This observation shows that juveniles are not really naïve, but are innately biased to learn the songs of their own species in preference to those of other species. Indeed, some evidence suggests that songbirds have a very rough template of their species song that is expressed in the absence of any exposure to that song or any other. Thus, birds raised in isolation produce highly abnormal "isolate" songs that have some characteristics of the song they would normally have learned (unlike Genie in the comparable human example). Such songs, however, are biologically ineffective in that they fail to attract mates.

In sum, the vocally relevant parts of the bird brain are already prepared during early life to learn the specific vocal sounds of the species, much as the brains of human infants are prepared at birth to learn language. Although the similarities with human language acquisition can be exaggerated, at least some aspects of human language have analogues in the vocal communicative abilities of other animals.

example, symbolic movements made by a foraging bee that returns to the hive encode only a single meaning; both expression and its appreciation have been hardwired into the nervous systems of the actor and the respondents.

A series of studies in great apes, however, has claimed that the rudiments of the human symbolic communication are evident in the behavior of our closest relatives. While techniques have varied, most psychologists who study chimps have used some form of manipulable symbols that can be arranged to express ideas in an interpretable manner. For example, chimps can be trained to manipulate tiles or other symbols (such as the gestures of sign language) to represent words and syntactic constructs, allowing them to communicate simple demands, questions, and even spontaneous expressions. The most remarkable results have come from increasingly sophisticated work with chimps using keyboards with a variety of symbols (Figure 33.11). With appropriate training, chimps can choose from as many as 400 different symbols to construct expressions, allowing the researchers to have something resembling a rudimentary conversation with their charges. The more accomplished of these animals are alleged to have "vocabularies" of several thousand words or phrases (how they use these words compared with young children, however, is dramatically less impressive).

Given the challenge this work presents to some long-held beliefs about the uniqueness of human language, it is not surprising that these claims continue to stir up debate. Nonetheless, the issues raised deserve careful consideration by anyone interested in human language abilities and how our remarkable symbolic skills may have evolved from the communicative capabilities of our ancestors. The pressure for the evolution of some form of symbolic communication in great apes seems clear enough. Ethologists studying chimpanzees in the wild have described extensive social communication based on gestures, the manipulation of objects, and facial expressions. In addition, studies

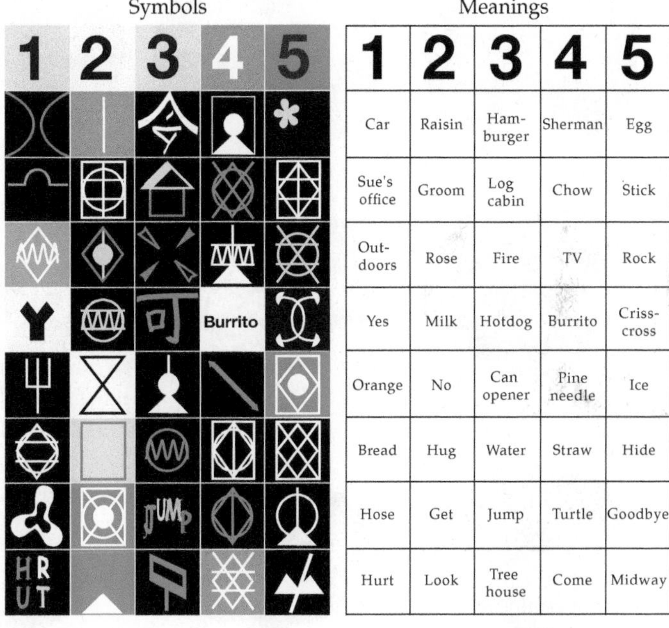

1	2	3	4	5
Car	Raisin	Hamburger	Sherman	Egg
Sue's office	Groom	Log cabin	Chow	Stick
Outdoors	Rose	Fire	TV	Rock
Yes	Milk	Hotdog	Burrito	Crisscross
Orange	No	Can opener	Pine needle	Ice
Bread	Hug	Water	Straw	Hide
Hose	Get	Jump	Turtle	Goodbye
Hurt	Look	Tree house	Come	Midway

FIGURE 33.11 Rudiments of language in non-human primates. Keyboard showing lexical symbols used to study symbolic communication in great apes. (From Savage-Rumbaugh et al., 1998.)

of monkeys have shown that some species normally use a variety of vocalizations in socially meaningful ways, and that these vocalizations may activate regions in the frontal and temporal lobes that are homologous to Broca's and Wernicke's areas in humans (Figure 33.12). This intricate social intercourse by gestures, facial expression, and limited vocalizations in non-human primates is likely to represent the antecedents of human language; one need only think of the importance of gestures, facial expressions, and non-verbal human vocal sounds as ancillary aspects of our own speech to appreciate this point.

In the end, it may turn out to be that human language, for all its differences (e.g., not context-bound; creative; open-ended), is based on the same general scheme of inherent and acquired neural associations between tokens and meanings that appears to be the basis of any animal communication.

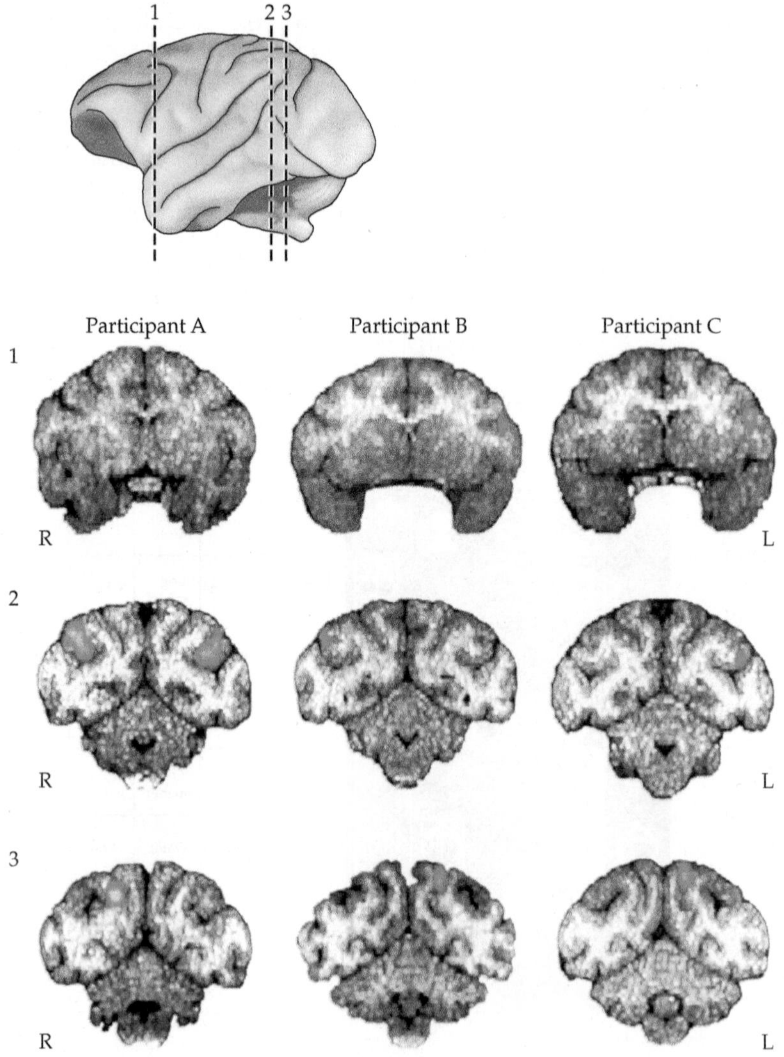

FIGURE 33.12 Activation of areas in the frontal and temporal lobes of three rhesus monkeys responding to conspecific vocal calls. The areas activated are arguably similar to the major language areas in the human brain. (From Gil-da-Costa et al., 2006.)

Summary

A variety of methods have been used to understand the organization of language in the human brain. This effort began in the nineteenth century by correlating clinical signs and symptoms with the location of brain lesions determined postmortem. In the twentieth and now twenty-first centuries, additional clinical observations together with studies of split-brain patients, mapping prior to neurosurgery, transient anesthesia of a single hemisphere, and noninvasive imaging techniques such as PET and fMRI have greatly extended knowledge of the neural substrates of language. Together, these various approaches show that the perisylvian cortices of the left hemisphere are especially important for normal language in the vast majority of humans. The right hemisphere also contributes importantly to language, most obviously by giving it emotional tone. The similarity of the deficits after comparable brain lesions in congenitally deaf individuals and their speaking counterparts has shown further that the cortical representation of language is independent of the means of its expression or perception (spoken and heard versus gestured and seen). The specialized language areas that have been identified are evidently the major components of a widely distributed set of brain regions that allow humans to communicate effectively by means of symbols that can be attached to objects, concepts, and feelings. Unlike social communication in other species, linguistic symbols can be manipulated and organized to create an endless range of meanings that are not tied to the context at hand. These observations raise a host of as-yet unresolved issues, including questions about the origins of human language and the basis of its extraordinary development.

ADDITIONAL READING

Reviews

Bachorowski, J.-A. and M. J. Owren (2003) Sounds of emotion: The production and perception of affect-related vocal acoustics. *Ann. NY Acad. Sci.* 1000: 244–265.

Belin, P., S. Fecteau and C. Bedard (2004) Thinking the voice: Neural correlates of voice perception. *Trends Cogn. Sci.* 8: 129–135.

Bellugi, U., H. Poizner and E. S. Klima (1989) Language, modality, and the brain. *Trends Neurosci.* 12: 380–388.

Binder, J. R., H. Rutvik, W. Desai, W. Graves and L. Conant (2009) Where is the semantic system? A critical review and meta-analysis of 120 functional neuroimaging studies. *Cereb. Cortex* 19: 3267–3296.

Bloomfield, T. C., T. Q. Gentner and D. Margoliash (2012) What birds have to say about language. *Nat. Neurosci.* 14: 947–948.

Buhusi, C. V. and W. H. Meck (2005) What makes us tick? Functional and neural mechanisms of interval timing. *Nat. Rev. Neurosci.* 6: 755–765.

Damasio, A. R. (1992) Aphasia. *New Engl. J. Med.* 326: 531–539.

Damasio, A. R. and N. Geschwind (1984) The neural basis of language. *Annu. Rev. Neurosci.* 7: 132–147.

Evans, N. and S. C. Levinson (2009) The myth of language universals: Language diversity and its importance for cognitive science. *Behav. Brain Sci.* 32: 429–492.

Friederici, A. D. (2009) Pathways to language: fiber tracts in the human brain. *Trends Cogn. Sci.* 13: 175–181.

Gazzaniga, M. S. (1998) The split brain revisited. *Sci. Am.* 329 (1): 50–55.

Gazzaniga, M. S. and R. W. Sperry (1967) Language after section of the cerebral commissures. *Brain* 90: 131–147.

Ghazanfar, A. A. and M. D. Hauser (2001) The auditory behavior of primates: A neuroethological perspective. *Curr. Opin. Neurobiol.* 12: 712–720.

Gibbon, J., C. Malapani, C. L. Dale and C. R. Gallistel (1997) Toward a neurobiology of temporal cognition: Advances and challenges. *Curr. Opin. Neurobiol.* 7: 170–184.

Hauser, M. D., N. Chomsky and W. T. Fitch (2002) The faculty of language: What is it, who has it, and how did it evolve? *Science* 298: 1569–1579.

Kuhl, P. K. (2000) A new view of language acquisition. *Proc. Natl. Acad. Sci. USA* 97: 12850–12857.

Kutas, M., C. K. Van Petten and R. Kluender (2006) Psycholinguistics electrified II (1924–2005). In *Handbook of Psycholinguistics*, 2nd Edition, M. A. Gernsbacher and M. Traxler (eds.). New York: Elsevier, pp. 659–724.

Mayberry, R. I. (2010) Early language acquisition and adult language ability: What sign language reveals about the critical period for language. In *The Oxford Handbook of Deaf Studies, Language, and Education,* M. Marschark and P. E. Spencer (eds.). Oxford: Oxford University Press, pp. 281–291.

Miles, H. L. W. and S. E. Harper (1994) "Ape language" studies and the study of human language origins. In *Hominid Culture in Primate Perspective*, D. Quiatt and J. Itani (eds.). Niwot: University Press of Colorado, pp. 253–328.

Newport, E. L., D. Bavelier and H. J. Neville (2001) Critical Thinking about Critical Periods: Perspectives on a Critical Period for Language Acquisition. In *Language, Brain, and Cognitive Development: Essays in Honor of Jacques Mehler*, E. Dupoux (ed.). Cambridge, MA: MIT Press, pp. 481–502.

Ojemann, G. A. (1983) The intrahemispheric organization of human language, derived with electrical stimulation techniques. *Trends Neurosci.* 4: 184–189.

Pollick, A. S. and F. B. M. de Waal (2007) Ape gestures and language evolution. *Proc. Natl. Acad. Sci. USA* 104: 8184–8189.

Seyfarth, D. M. and D. I. Cheney (1984) The natural vocalizations of non-human primates. *Trends Neurosci.* 7: 66–73.

Spelke, E. S. (2003) What makes us smart? Core knowledge and natural language. In *Language in Mind: Advances in the Study of Language and Thought*, vol. 8, M. S. E. Gazzaniga (ed.). Cambridge, MA: MIT Press, pp. 277–311.

Zuberbuhler, K. (2005) Linguistic prerequisites in the primate lineage. In *Language Origins: Perspectives on Evolution*, M. Tallerman (ed.). New York: Oxford University Press, pp. 262–282.

Important Original Papers

Abe, K. and D. Watanabe (2012) Songbirds possess the spontaneous ability to discriminate syntactic rules. *Nat. Neurosci.* 14: 1067–1074.

Bagley, W. C. (1900–1901) The apperception of the spoken sentence: A study in the psychology of language. *Am. J. Psychol.* 12: 80–130.

Belin, P., R. J. Zatorre, P. Lafaille, P. Ahad and B. Pike (2000). Voice- selective areas in human auditory cortex. *Nature* 403: 309–312.

Berwick, R. C., A. D. Friederici, N. Chomsky and J. J. Bolhuis (2013) Evolution, brain, and the nature of language. *Trends Cogn. Sci.* 17: 89–98.

Brown, T. T. and 5 others (2005) Developmental changes in human cerebral functional organization for word generation. *Cereb. Cortex* 15: 275–290.

Chang, E. F., C. A. Niziolek, R. T. Knight, S. S. Nagarajan and J. F. Houde (2013) Human cortical sensorimotor network underlying feedback control of vocal pitch. *Proc. Nat. Acad. Sci. USA* 110: 2653–2658.

Chao, L. L., J. V. Haxby and A. Martin (1999) Attribute-based neural substrates in temporal cortex for perceiving and knowing about objects. *Nat. Neurosci.* 2: 913–919.

DeLong, K., T. Urbach and M. Kutas (2005) Probabilistic word pre-activation during language comprehension inferred from electrical brain activity. *Nat. Neurosci.* 8: 1217–1221.

Fromkin, V., S. Krashen, S. Curtis, D. Rigler and M. Rigler (1974) The development of language in Genie: A case of language acquisition beyond the "critical period." *Brain Lang.* 1: 81–107.

Gentner, T. Q., K. M. Fenn, D. Margoliash and H. C. Nusbaum (2006) Recursive syntactic pattern learning by songbirds. *Nature* 440: 1204–1207.

Ghazanfar, A. A. and N. Logothetis (2003) Facial expressions linked to monkey calls. *Nature* 423: 937.

Gil-da-Costa, R. and 5 others (2006) Species-specific calls activate homologs of Broca's and Wernicke's areas in the macaque. *Nat. Neurosci.* 9: 1064–1070.

Johnson, J. S. and E. L. Newport (1989) Critical period effects in second language learning: The influence of maturational state on the acquisition of English as a second language. *Cogn. Psychol.* 21: 60–99.

Kuhl, P. K., B. T. Conboy, D. Padden, T. Nelson and J. Pruitt (2005) Early speech perception and later language development: Implications for the "critical period." *Lang. Learn. Dev.* 1: 237–264.

Kuhl, P. K., K. A. Williams, F. Lacerda, K. N. Stevens and B. Lindblom (1992) Linguistic experience alters phonetic perception in infants 6 months of age. *Science* 255: 606–608.

Kutas, M. and S. A. Hillyard (1980) Reading senseless sentences: Brain potentials reflect semantic incongruity. *Science* 207: 203–205.

Leonard, M. K. and E. F. Chang (2014) Dynamic speech representations in the human temporal lobe. *Trends Cogn. Sci.* 18: 472–479.

Miller, G. A. and J. C. R. Licklider (1950) The intelligibility of interrupted speech. *J. Acoust. Soc. Am.* 22: 167–173.

Nieder, A. and E. K. Miller (2004) A parieto-frontal network for visual numerical information in the monkey. *Proc. Natl. Acad. Sci. USA* 101: 7457–7462.

Ojemann, G. A. and H. A. Whitaker (1978) The bilingual brain. *Arch. Neurol.* 35: 409–412.

Pica, P., C. Lemer, W. Izard and S. Dehaene (2004) Exact and approximate arithmetic in an Amazonian indigene group. *Science* 306: 499–503.

Pollick, A. S. and F. B. M. de Waal (2007) Ape gestures and language evolution. *Proc. Natl. Acad. Sci. USA* 104: 8184–8189.

Rilling, J. K. (2014) Comparative primate neurobiology and the evolution of brain language systems. *Curr. Opin. Neurobiol.* 28: 10–14.

Schlaggar, B. L. and 5 others (2002) Functional neuroanatomical differences between adults and school-age children in the processing of single words. *Science* 296: 1476–1479.

Shulman, G. L., J. M. Ollinger, M. Linenweber, S. E. Petersen and M. Corbetta (2001) Multiple neural correlates of detection in the human brain. *Proc. Natl. Acad. Sci. USA* 98: 313–318.

Tomasello, M. (2004) What kind of evidence could refute the UG hypothesis? *Stud. Lang.* 28: 642–644.

Vinckier, F. and 5 others (2007) Hierarchical coding of letter strings in the ventral stream: Dissecting the inner organization of the visual work-form system. *Neuron* 55: 143–156

Whiten, A. and 8 others (1999) Cultures in chimpanzees. *Nature* 399: 682–685.

Xu, F. and E. Spelke (2000) Large number discrimination in 6-month-old infants. *Cognition* 74: B1–B12.

Books

Anderson, S. (2012) Languages: *A Very Short Introduction.* Oxford: Oxford University Press.

Bloom, P. (2002) *How Children Learn the Meanings of Words.* Cambridge, MA: MIT Press.

Chomsky, N. (1957) *Syntactic Structures.* The Hague: Elsevier.

Darwin, C. (1872) *The Expression of Emotion in Man and Animals.* Reprint, Chicago: University of Chicago Press, 1965.

Gelman, R. and C. Gallistel (1978) *The Child's Understanding of Number.* Cambridge, MA: Harvard University Press.

Goodall, J. (1990) *Through a Window: My Thirty Years with the Chimpanzees of Gombe.* Boston: Houghton Mifflin.

Griffin, D. R. (1992) *Animal Minds.* Chicago: University of Chicago Press.

Hauser, M. (1996) *The Evolution of Communication.* Cambridge MA: MIT Press.

Lenneberg, E. (1967) *The Biological Foundations of Language.* New York: Wiley.

Liberman, A. M. (1996) *Speech: A Special Code.* Cambridge, MA: MIT Press.

McNeil, D. (2000) *Language and Gesture.* Cambridge: Cambridge University Press.

Miller, G. A. (1991) *The Science of Words.* New York: Scientific American Library.

Plomp, R. (2002) *The Intelligent Ear: On the Nature of Sound Perception.* Mahwah, NJ: Erlbaum.

Posner, M. I. and M. E. Raichle (1994) *Images of Mind.* New York: Scientific American Library.

Provine, R. R. (2000) *Laughter: A Scientific Investigation.* New York: Penguin.

Rogers, T. T. and J. L. McClelland (2004) *Semantic Cognition: A Parallel Distributed Processing Approach.* Cambridge, MA: MIT Press.

Savage-Rumbaugh, S., S. G. Shanker and T. J. Taylor (1998) *Apes, Language, and the Human Mind.* New York: Oxford University Press.

Schnupp, J., I. Nelken and A. King (2012) *Auditory Neuroscience: Making Sense of Sound.* Cambridge, MA: MIT Press.

Tomasello, M. (2008) *Origin of Human Communication.* Cambridge MA: MIT Press.

von Frisch, K. (1993) *The Dance Language and Orientation of Bees.* (Translated by Leigh E. Chadwick.) Cambridge, MA: Harvard University Press.

Winchester, S. (2003) *The Meaning of Everything: The Story of the Oxford English Dictionary.* Oxford: Oxford University Press.

Go to the NEUROSCIENCE 6e Companion Website at **www.oup.com/uk/Purves6e** for Web Topics, Animations, Flashcards, and more.

APPENDIX

Survey of Human Neuroanatomy

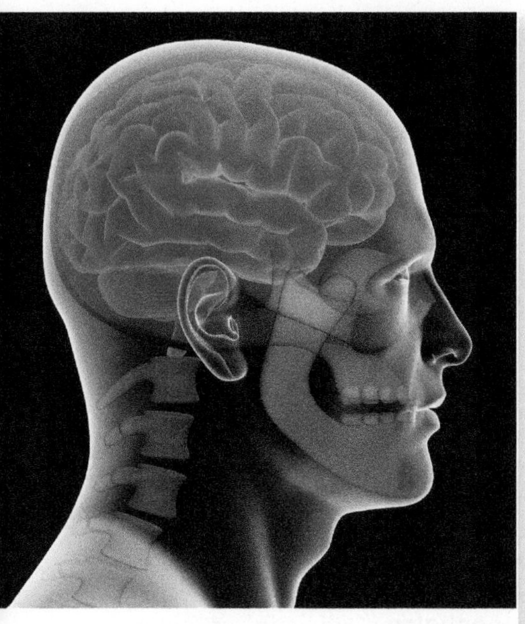

Overview

PERHAPS THE MAJOR REASON that neuroscience remains such an exciting field is the wealth of unanswered questions about the fundamental organization and function of the human brain. To understand this remarkable organ (and its interactions with the body that it governs), the myriad cell types that constitute the nervous system must be identified, their mechanisms of excitability and plasticity characterized, their interconnections traced, and the physiological role of the resulting neural circuits defined in behaviorally meaningful contexts. These challenges have been at the forefront of the five units of this textbook, where a broad range of questions about how nervous systems are organized and how they generate behavior has been addressed (albeit leaving many questions unanswered, especially those that pertain to distinctly human behaviors). This appendix provides an anatomical framework for the integration of this knowledge and its application to the human nervous system. It reviews the basic terms and anatomical conventions used in discussing human neuroanatomy, and provides a general picture of the organization of the human forebrain, brainstem, and spinal cord. The appendix is followed by an atlas of surface and sectional images of the human central nervous system on which relevant neuroanatomical structures are identified.

Neuroanatomical Terminology

The terms used to specify *location* in the CNS are the same as those used for the gross anatomical description of the rest of the body (Figure A1A). Thus, *anterior* and *posterior* indicate, respectively, front and behind; *rostral* and *caudal*, nose and "tail" (i.e., the lower spinal region); *dorsal* and *ventral*, top and bottom (back and belly); and *medial* and *lateral*, at the midline or to the side. But the use of these coordinates in the body versus their use to describe position in the brain can be confusing, especially as these terms are applied to humans. For the entire body, these anatomical terms refer to the long axis, which is straight. The long axis of the human CNS, however, has a bend in it. In humans (and other bipeds), the rostral–caudal axis of the forebrain is tilted forward (because of the cephalic flexure that forms in embryogenesis; see Chapter 22) with respect to the long axis of the brainstem and spinal cord (see Figure A1A). Once this forward tilt is appreciated, the other terms that describe position in the brain and the terms used to identify planes of section can be easily assigned.

The proper assignment of the anatomical axes dictates the standard planes for histological sections or live images that are used to study the internal anatomy of the brain and to localize function (Figure A1B). **Horizontal sections** (also referred to as **axial sections**) are taken parallel to the rostral–caudal axis of the brain; thus, in an

(A)

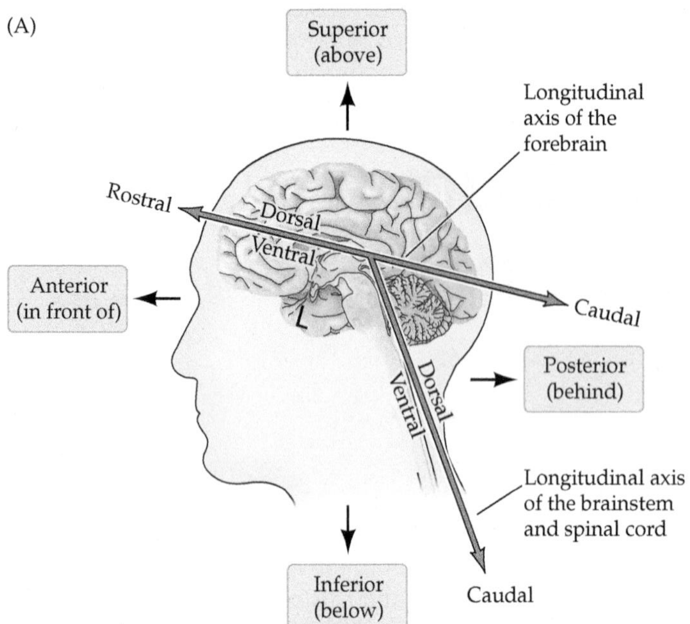

(B)

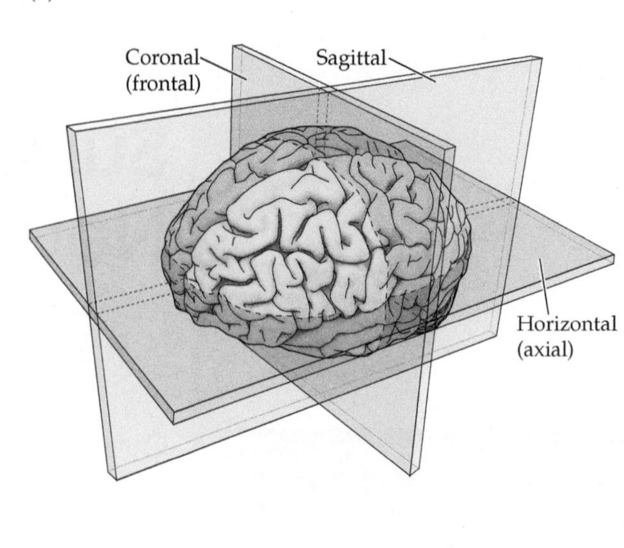

FIGURE A1 Axes of the human nervous system. A flexure in the long axis of the nervous system arose as humans evolved upright posture, leading to an approximately 120° angle between the long axis of the brainstem and that of the forebrain. The consequences of this flexure for anatomical terminology are indicated in (A). The terms *anterior, posterior, superior,* and *inferior* refer to the long axis of the body, which is straight. Therefore, these terms indicate the same direction for both the forebrain and the brainstem. In contrast, the terms *dorsal, ventral, rostral,* and *caudal* refer to the long axis of the CNS. The dorsal direction is toward the back for the brainstem and spinal cord, but toward the top of the head for the forebrain. The opposite direction is ventral. The rostral direction is toward the top of the head for the brainstem and spinal cord, but toward the face for the forebrain. The opposite direction is caudal. (B) The major planes of section used in cutting or imaging the brain.

individual standing upright, such sections are parallel to the ground. Sections taken in the plane dividing the two hemispheres are **sagittal** and can be further categorized as **midsagittal** or **parasagittal**, according to whether the section is at the midline (midsagittal) or is more lateral (parasagittal). Sections in the plane of the face are called **coronal** or **frontal**.

Different terms are usually used to refer to sections of the brainstem and spinal cord. The plane of section orthogonal to the long axis of the brainstem and spinal cord is the **transverse**, whereas sections parallel to this axis are **longitudinal**. In a transverse section through the human brainstem and spinal cord, the dorsal–ventral axis and the posterior–anterior axis indicate the same directions (see Figure A1A). This terminology is essential for understanding the basic subdivisions of the nervous system and for discussing the locations of brain structures in a common frame of reference.

Basic Subdivisions of the Central Nervous System

As detailed in Chapter 22, the four embryological divisions of the CNS arise in early brain development after neurulation, as three swellings appear at the cephalic end of the neural tube (see Figure 22.3); together, these swellings develop into the brain, while the rest of the neural tube gives rise to the spinal cord. The most rostral of the three swellings, the **prosencephalon** ("forward brain" or "front brain"), divides into two parts: the **telencephalon** ("end brain" or "outer brain"), which gives rise to the cerebral hemispheres, and the **diencephalon** ("between brain" or "through brain"), from which are derived the thalamus, the hypothalamus, and also the retina (via the optic vesicle). These structures together make up the adult **forebrain**. The **mesencephalon** is the middle swelling in the embryonic brain, and it does not divide further; this division becomes the **midbrain** of the adult. The **rhombencephalon** is also known as the **hindbrain**; it is the third of the three cephalic swellings, and it develops just caudal to the mesencephalon. The rhombencephalon further divides into the **metencephalon**, which becomes the pons and the overlying cerebellum, and the **myelencephalon**, which becomes the medulla oblongata (or simply, the medulla). The term **brainstem** is used commonly to refer to the midbrain, pons, and medulla as a collective structure, despite their distinct embryological origins. The neural tube caudal to the three cephalic swellings becomes the spinal cord.

Because the nervous system starts out as a simple tube, the lumen of the tube remains in the adult brain as a series of connected, fluid-filled spaces. These spaces, known as

Embryonic brain		Adult brain derivatives	Associated ventricular space
Prosencephalon (forebrain)	Telencephalon	Cerebral cortex	Lateral ventricles
		Cerebral nuclei (basal ganglia, amygdala, basal forebrain)	
	Diencephalon	Thalamus	Third ventricle
		Hypothalamus	
		Retina	
Mesencephalon (midbrain)		Superior and inferior colliculi Red nucleus Substantia nigra	Cerebral aqueduct
Rhombencephalon (hindbrain)	Metencephalon	Cerebellum	Fourth ventricle
		Pons	
	Myelencephalon	Medulla oblongata	Fourth ventricle
Spinal cord		Spinal cord	Central canal

FIGURE A2 Representative relationships between the embryonic and adult forms of the central nervous system. See Chapter 22 for an account of brain development that more fully explains the formation of regional identity in the developing CNS, including the origin of the ventricular spaces.

the **ventricles**, are filled with **cerebrospinal fluid** (**CSF**) and provide important landmarks on sectional images of the nervous system. As the brain grows, the shape of the ventricles changes from that of a simple tube to that of the complex adult form (see the section "The Ventricular System" below). The ventricles, although continuous, acquire different names in each of the embryological subdivisions of the CNS. Thus, the spaces inside the hemispheres are known as the lateral ventricles, and the space inside the diencephalon is the third ventricle. The space inside the midbrain is called the cerebral aqueduct. The space inside the developing rhombencephalon (between the cerebellum and the pons and rostral medulla) is called the fourth ventricle. In embryos and young children, the opening in the spinal cord is patent and is known as the central canal.

Figure A2 accounts for the conserved relationships among the parts of the developing brain and their adult brain derivatives, including the components of the ventricular system. Figure A3 shows the subdivisions of the CNS as they are situated in the human body, including illustration of the relationship among the spinal cord, the spinal nerves, and the vertebrae. The same embryonic

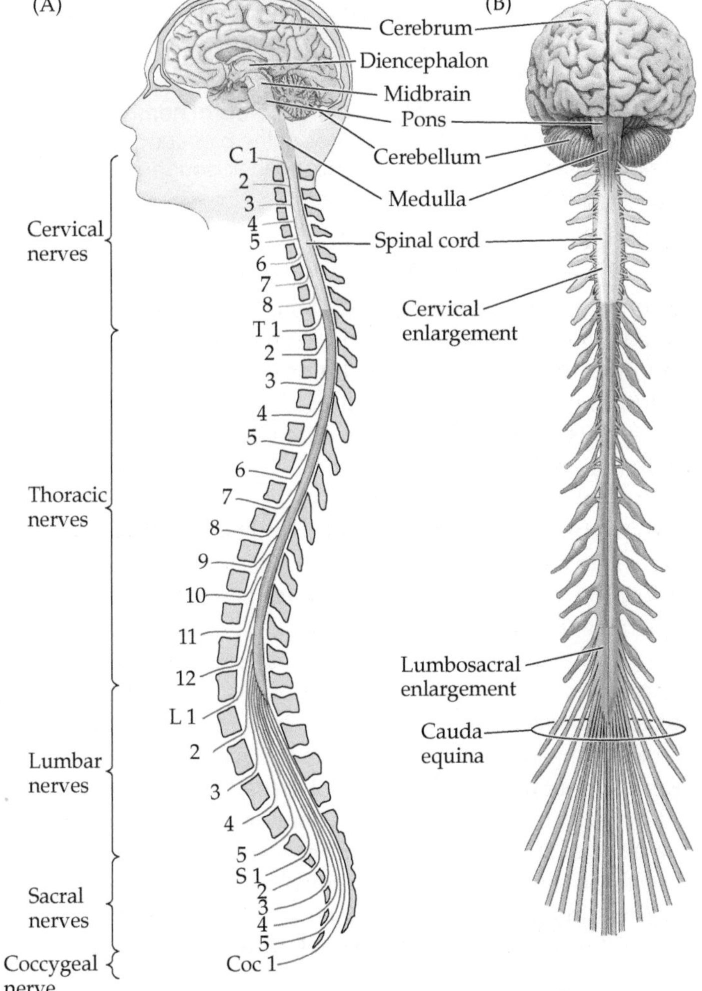

FIGURE A3 Subdivisions and components of the central nervous system. (A) A lateral view indicating the major subdivisions and components of the CNS. (Note that the position of the brackets on the left side of the figure refers to the location of the spinal nerves as they exit the intervertebral foramina, not the position of the corresponding spinal cord segments.) (B) The CNS in ventral view, indicating the emergence of the spinal nerves, the cervical and lumbar enlargements, and the cauda equina.

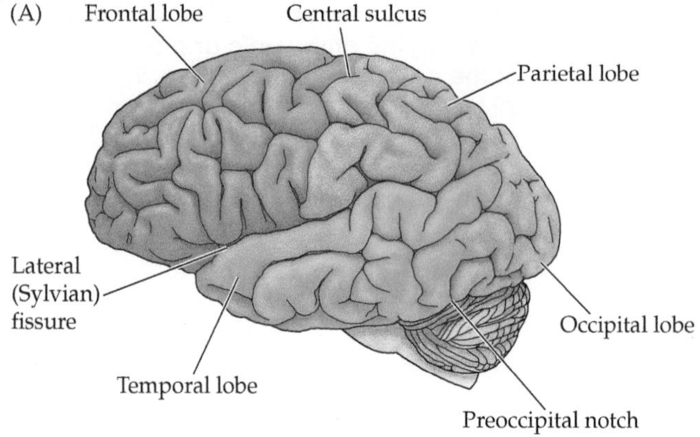

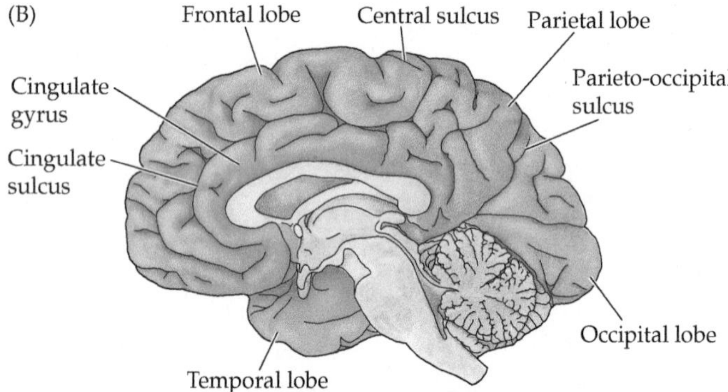

FIGURE A4 Surface anatomy of the cerebral hemisphere. These depictions show the four lobes of the brain and the major fissures and sulci that help define their boundaries. (A) Lateral view. (B) Midsagittal view.

relationships shown in Figure A2 should be discoverable in the adult form in Figure A3, although the relatively greater growth of the **cerebral hemispheres** makes some of these relationships difficult to appreciate since the hemispheres are the largest and most prominent feature of the human brain (Figure A4).

In humans, the cerebral hemispheres (the outermost portions of which are continuous, highly folded sheets of cortex) are characterized by **gyri** (singular: *gyrus*), or crests of folded cortical tissue, and by **sulci** (singular: *sulcus*), which are the grooves or spaces that divide gyri from one another. Although gyral and sulcal patterns vary among individuals, several consistent landmarks divide the **cerebral cortex** in each hemisphere into four **lobes**. The names of the lobes are derived from the cranial bones that overlie them: **occipital**, **temporal**, **parietal**, and **frontal**. A key feature of the surface anatomy of the cerebrum is the **central sulcus** located roughly halfway between the rostral and caudal poles of the hemispheres (see Figure A4). This prominent sulcus divides the frontal lobe in the rostral half of the hemisphere from the more caudal parietal lobe. Other prominent landmarks that divide the cerebral lobes are the **lateral fissure** (also called the **Sylvian fissure**), which divides the temporal lobe inferiorly from the overlying frontal and parietal lobes, and the **parieto-occipital sulcus**, which separates the parietal lobe from the occipital lobe on the medial surface of the hemisphere. The remaining major subdivisions of the forebrain are not visible from the surface; they comprise gray matter and white matter structures that lie deeper in the cerebral hemispheres and can be seen only in sectional views.

Next, we describe the characteristic superficial features of these major subdivisions of the human CNS and their internal organization in more detail from caudal to rostral, beginning with the spinal cord.

External Anatomy of the Spinal Cord

The spinal cord extends caudally from the brainstem, running from the medullary-spinal junction at about the level of the first cervical vertebra to about the level of the first lumbar vertebra in adults (see Figure A3). The vertebral column (and the spinal cord within it) is divided into **cervical**, **thoracic**, **lumbar**, **sacral**, and **coccygeal** regions. The peripheral nerves (called the **spinal** or **segmental nerves**) that innervate much of the body arise from the spinal cord's 31 pairs of spinal nerves. On each side of the midline, the cervical region of the cord gives rise to 8 cervical nerves (C1–C8), the thoracic region to 12 thoracic nerves (T1–T12), the lumbar region to 5 lumbar nerves (L1–L5), the sacral region to 5 sacral nerves (S1–S5), and the coccygeal region to a single coccygeal nerve. The spinal nerves leave the vertebral column through the intervertebral foramina that lie adjacent to the respectively numbered vertebral body. Sensory information carried by the afferent axons of the spinal nerves enters the cord via the **dorsal roots**, and motor commands carried by the efferent axons leave the cord via the **ventral roots** (Figure A5). Once the dorsal and ventral roots join, sensory and motor axons (with some exceptions) travel together in the spinal nerves.

Two regions of the spinal cord are enlarged to accommodate the greater number of nerve cells and connections needed to process information related to the upper and lower limbs. The spinal cord expansion that corresponds to the arms is called the **cervical enlargement** and includes spinal segments C3–T1; the expansion that corresponds to the legs is called the **lumbosacral enlargement** and includes spinal segments L1–S2 (see Figure A3B). Because the spinal cord is considerably shorter than the vertebral column in adults (see Figure A3A), lumbar and sacral nerves run for some distance in the vertebral canal before emerging, thus forming a collection of nerve roots known as the **cauda equina**. The space surrounding the cauda equina is the target for an important clinical

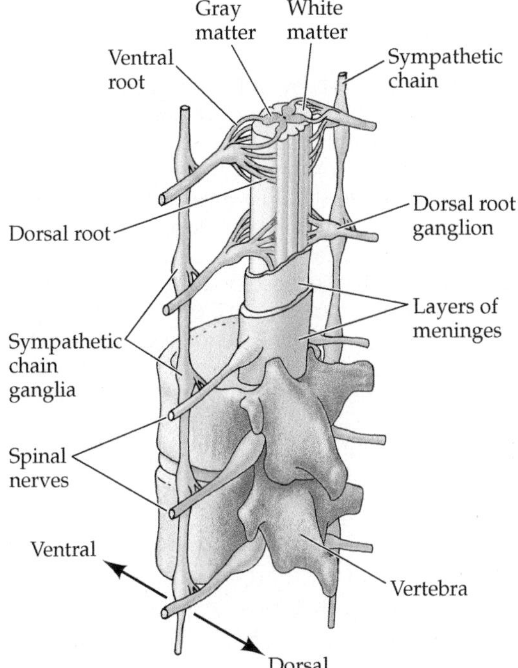

FIGURE A5 Relationship of the spinal cord and spinal nerves in the vertebral column. Sensory information carried by the spinal nerves enters the cord via the dorsal roots of ganglia; motor commands leave the cord via the ventral roots. Once the dorsal and ventral roots join, sensory and motor axons usually travel together in the spinal nerves.

procedure—the *lumbar puncture*—that allows for the collection of CSF by placing a needle into this lumbar cistern to withdraw fluid for analysis. In addition, local anesthetics can be safely introduced into the cauda equina, producing spinal anesthesia; at this level, the risk of damage to the spinal cord with insertion of a needle is minimal.

Internal Anatomy of the Spinal Cord

The arrangement of gray and white matter in the spinal cord is relatively simple: The interior of the cord is formed by gray matter, which is surrounded by white matter (Figure A6A). In transverse sections, the gray matter is conventionally divided into dorsal (posterior) and ventral (anterior) "horns." The neurons of the **dorsal horns** receive sensory information that enters the spinal cord via the dorsal roots of the spinal nerves (Figure A6B). The **lateral horns** are present primarily in the thoracic region and contain the preganglionic visceral motor neurons that project to the sympathetic ganglia (illustrated in Figure A5). The **ventral horns** contain the cell bodies of motor neurons that send axons via the ventral roots of the spinal nerves to terminate on striated muscles. These major divisions of gray matter have been further subdivided according to the distribution of neurons in the dorsal–ventral axis.

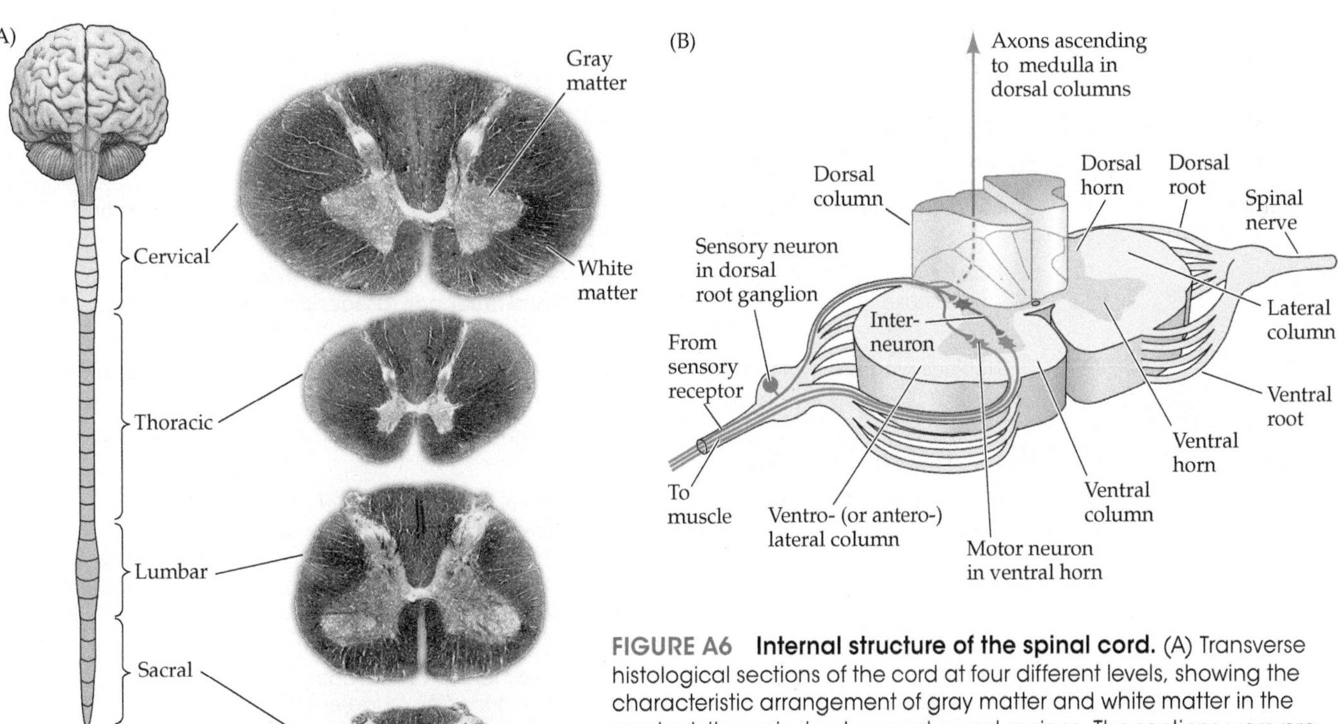

FIGURE A6 Internal structure of the spinal cord. (A) Transverse histological sections of the cord at four different levels, showing the characteristic arrangement of gray matter and white matter in the cervical, thoracic, lumbar, and sacral regions. The sections were processed to simulate myelin staining; thus, white matter appears darker, and gray matter lighter. (B) Diagram of the internal structure of the spinal cord.

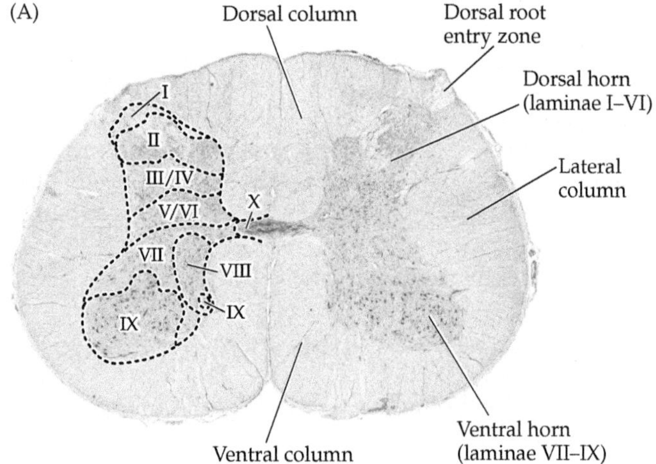

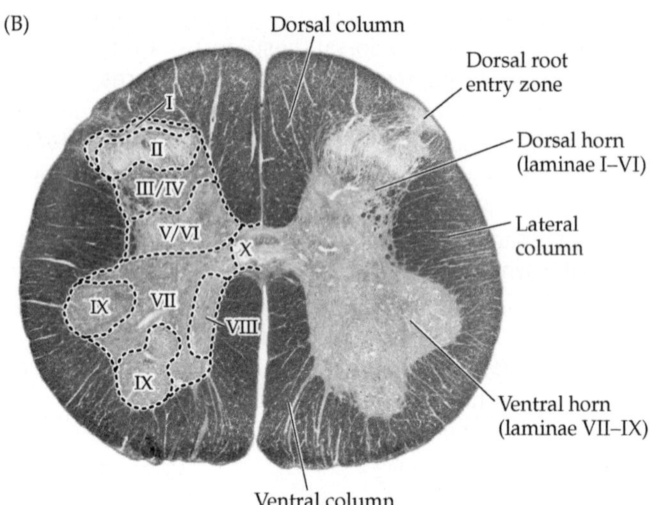

FIGURE A7 Internal histology of the human spinal cord in a lumbar segment. (A) Photomicrograph of a section stained for the demonstration of Nissl substance (showing cell bodies in a blue stain). (B) Photomicrograph of a section that was acquired and processed to simulate myelin staining. On the left side of both images, dotted lines indicate the boundaries between cytoarchitectonic divisions of spinal cord gray matter, known as Rexed's laminae (see Table A1 and Figure 10.3B). Among the more conspicuous divisions are lamina II, which corresponds to the substantia gelatinosa and is important in pain transmission (see Chapter 10), and lamina IX, which contains the columns of lower motor neurons that innervate skeletal muscle (see Chapter 16).

The Swedish neuroanatomist Bror Rexed recognized that neurons in the dorsal horn are organized into layers, and that neurons in the ventral horns (especially in the enlargements) are arranged into longitudinal columns (Figure A7). Rexed proposed a scheme for naming these subdivisions (since termed *Rexed's laminae*) that is still used by neuroanatomists and clinicians, although more descriptive terms are also applied (Table A1).

The white matter of the spinal cord is subdivided into dorsal (or posterior), lateral, and ventral (or anterior) columns, each of which contains axon tracts related to specific functions. The **dorsal columns** carry ascending sensory information mainly from somatic mechanoreceptors (see Figure A6B). The **lateral columns** include axons that

TABLE A1 ■ Subdivisions of Spinal Cord Gray Matter

Division	Rexed lamina	Descriptive term	Significance
Dorsal horn	I	Marginal zone	Projection neurons that receive input from small-diameter afferents; one source of anterolateral system projections
	II	Substantia gelatinosa	Interneurons that receive input mainly from small-diameter afferents; integrates feedforward and feedback (descending) inputs that modulate pain transmission (see Chapter 10)
	III/IV	Nucleus proprius	Interneurons that integrate inputs from small- and large-diameter afferents
	V/VI	Base of dorsal horn	Projection neurons that receive input from both large- and small-diameter afferents and spinal interneurons; another source of anterolateral system projections
Intermediate zone (lateral horn)	VII	Intermediate gray	Mainly interneurons that communicate between dorsal and ventral horns; in the thoracic cord, also contains projection neurons of the dorsal nucleus of Clarke, a spinocerebellar relay (see Chapter 19), and the sympathetic preganglionic visceral motor neurons of the intermediolateral cell column (underlying the lateral horn); in the sacral cord, also contains preganglionic visceral motor neurons (see Chapter 21)
Ventral horn	VIII	Motor interneurons	Interneurons in the medial aspect of ventral horn that coordinate the activities of lower motor neurons (see Chapter 16)
	IX	Motor neuron columns	Columns of lower motor neurons that govern limb musculature (see Chapter 16)
Central zone	X	Central gray	Interneurons surrounding the rudiment of the central canal

extend from the cerebral cortex to interneurons and motor neurons in the ventral horns; this important pathway is called the **lateral corticospinal tract** (see Chapter 17). The lateral columns also convey proprioceptive signals from spinal cord neurons to the cerebellum (see Chapter 19). The **ventral** (and **ventrolateral** or **anterolateral**) **columns** carry both ascending information about pain and temperature, and descending motor information from the brainstem and motor cortex concerned with postural control and gain adjustment.

Brainstem and Cranial Nerves

The brainstem is one of the most complex regions of the CNS. It comprises the midbrain, pons, and medulla and is continuous rostrally with the diencephalon (thalamus and hypothalamus); caudally it is continuous with the spinal cord. Although the medulla, pons, and midbrain participate in myriad specific functions, the integrated actions of these brainstem components give rise to three fundamental functions. First, the brainstem is the target or source for the **cranial nerves** that deal with sensory and motor function in the head and neck, and it provides for local circuits that integrate afferent signals and coordinate or organize efferent signals (Table A2). Second, the brainstem provides a "throughway" for all of the ascending sensory tracts from the spinal cord; the sensory tracts for the head and neck; the descending motor tracts from the forebrain; and local pathways that link eye movement centers. Finally, the brainstem is involved in regulating the level of consciousness, primarily though the extensive forebrain projections of a key modulatory center in the brainstem core, the **reticular formation** (see Box 17C).

Understanding the internal anatomy of the brainstem is generally regarded as essential for neurological diagnosis and the practice of clinical healthcare. Brainstem structures are compressed into a relatively small volume that has a regionally restricted vascular supply. Thus, vascular accidents in the brainstem—which are common—result in distinctive, and often devastating, combinations of functional deficits (see below). These deficits can be used both for diagnosis and for better understanding the intricate anatomy of the medulla, pons, and midbrain.

Unlike the surface appearance of the spinal cord, which is relatively homogeneous along its length, the surface appearance of each brainstem subdivision is characterized by unique bumps and bulges formed by the underlying gray matter (nuclei) or white matter (tracts) (Figure A8). A series of swellings on the dorsal and ventral surfaces of the medulla reflects many of the major structures in this caudal part of the brainstem. One prominent landmark that can be seen laterally is the **inferior olive**. Just medial to the inferior olives are the **medullary pyramids**, prominent swellings on the ventral surface of the medulla that

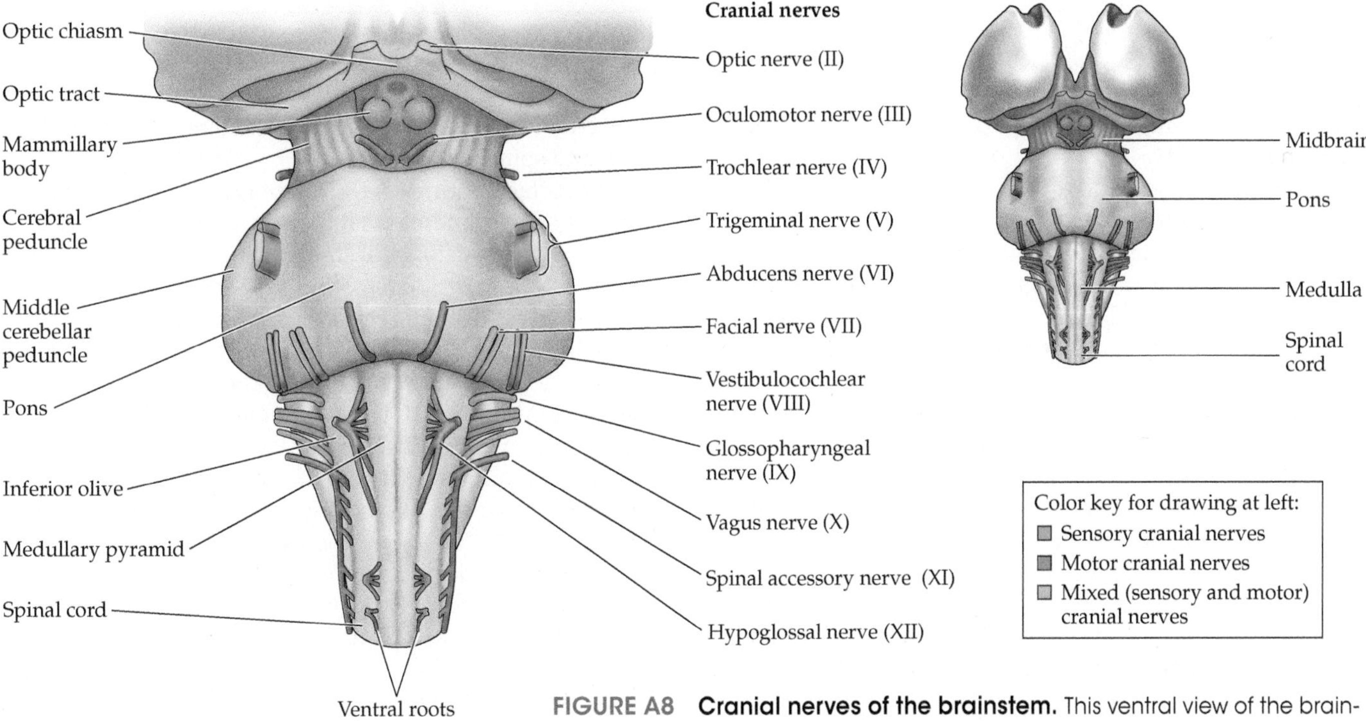

FIGURE A8 Cranial nerves of the brainstem. This ventral view of the brainstem shows the locations of the cranial nerves as they enter or exit each of the brainstem subdivisions (midbrain, pons, and medulla, shown at the right).

TABLE A2 ■ **The Cranial Nerves and Their Primary Functions**

Cranial nerve	Name	Sensory and/or motor	Major function
I	Olfactory nerve	Sensory	Sense of smell
II	Optic nerve	Sensory	Vision
III	Oculomotor nerve	Motor	Eye movements; pupillary constriction and accommodation; muscle of upper eyelid
IV	Trochlear nerve	Motor	Eye movements (intorsion, downward gaze)
V	Trigeminal nerve	Sensory and motor	Somatic sensation from face, mouth, cornea; muscles of mastication
VI	Abducens nerve	Motor	Eye movements (abduction or lateral movements)
VII	Facial nerve	Sensory and motor	Controls the muscles of facial expression; taste from anterior tongue; lacrimal and salivary glands
VIII	Vestibulocochlear (auditory) nerve	Sensory	Hearing; sense of balance
IX	Glossopharyngeal nerve	Sensory and motor	Sensation from posterior tongue and pharynx; taste from posterior tongue; carotid baroreceptors and chemoreceptors; salivary gland
X	Vagus nerve	Sensory and motor	Autonomic functions of gut; cardiac inhibition; sensation from larynx and pharynx; muscles of vocal cords; swallowing
XI	Spinal accessory nerve	Motor	Shoulder and neck muscles
XII	Hypoglossal nerve	Motor	Movements of tongue

reflect the underlying descending corticospinal tracts (see Chapter 17).

The **pons** (Latin, "bridge") is rostral to the medulla and is easily recognized by the mass of decussating fibers that cross (bridge) the midline on its ventral surface, giving rise to the name of this subdivision. The **cerebellum** is attached to the dorsal aspect of the pons by three large white matter tracts: the **superior**, **middle**, and **inferior cerebellar peduncles**. Each of these tracts contains either efferent (superior and inferior) or afferent (inferior and middle) axons from or to the cerebellum (see Chapter 19).

The midbrain contains the **superior** and **inferior colliculi** defining its dorsal surface, or *tectum* (Latin, "roof"). Several midbrain nuclei lie in the ventral portion of the midbrain, including the bipartite structure the **substantia nigra** (pars reticulata and pars compacta; see Chapter 18), and the **red nucleus** (see Chapter 19). Another noteworthy anatomical feature of the midbrain is the presence of the prominent **cerebral peduncles** that are visible from the ventral surface; these structures are formed by massive projections from the cerebral cortex to targets in the brainstem and spinal cord.

The surface features of the midbrain, pons, and medulla can be used as landmarks for locating the source and termination of the majority of cranial nerves in the brainstem. Unlike for the spinal nerves, the entry and exit points of the cranial nerves are not regularly arrayed along the length of the brainstem. Two cranial nerves, the **olfactory nerve** (**I**) and the **optic nerve** (**II**), enter the forebrain directly. The remaining cranial nerves enter and exit at distinct regions of the ventral (and in one case, the dorsal) surface of the midbrain, pons, and medulla (see Figure A8). The **oculomotor nerve** (**III**) exits into the space between the two cerebral peduncles on the ventral surface of the midbrain. The **trochlear nerve** (**IV**) associated with the caudal midbrain is the only cranial nerve to exit on the dorsal surface of the brainstem; it is also the only nerve (cranial or spinal) that supplies contralateral tissue. The **trigeminal nerve** (**V**)—the largest cranial nerve—exits the ventrolateral pons by traversing the fibers of the middle cerebellar peduncle. The **abducens nerve** (**VI**), **facial nerve** (**VII**),

Location of cells whose axons form the nerve	Clinical test of function
Nasal epithelium	Test sense of smell with standard odor
Retina	Assess acuity, pupillary light reflex, and integrity of visual field
Oculomotor nucleus in midbrain; Edinger-Westphal nucleus in midbrain	Test eye movements (individual can't look up, down, or medially if nerve involved); look for ptosis and pupillary dilation; assess pupillary light reflex
Trochlear nucleus in midbrain	Test for downward eye movement when eye adducted
Trigeminal motor nucleus in pons; trigeminal sensory ganglion (the gasserian ganglion)	Test sensation on face; test ability to clamp jaw tightly; palpate masseter muscles and temporal muscle
Abducens nucleus in pons	Test for lateral eye movement
Facial motor nucleus in pons; superior salivatory nuclei in pons; geniculate ganglion	Test facial expression plus taste on anterior tongue
Spiral ganglion; vestibular (Scarpa's) ganglion	Test audition with tuning fork; test vestibular function by assessing gaze fixation during head rotation and balance during perturbation; perform caloric test
Nucleus ambiguus in medulla; inferior salivatory nucleus in pons; glossopharyngeal ganglia	Test swallowing and pharyngeal gag reflex
Dorsal motor nucleus of vagus; nucleus ambiguus; vagal nerve ganglion	Test swallowing and pharyngeal gag reflex; assess hoarseness; observe uvula and posterior pharynx at rest and during phonation
Spinal accessory nucleus in superior cervical cord	Test sternocleidomastoid and trapezius muscles
Hypoglossal nucleus in medulla	Test deviation of tongue during protrusion (points to side of lesion) and symmetry of force when pushing tongue against cheek

and **vestibulocochlear nerve** (**VIII**) emerge in a medial to lateral manner, respectively, at the junction of the pons and medulla with the abducens nerve emerging most medial and the vestibulocochlear nerve emerging most laterally. The **glossopharyngeal nerve** (**IX**) and the **vagus nerve** (**X**) are associated with the lateral medulla, whereas the **hypoglossal nerve** (**XII**) exits the ventromedial medulla between the medullary pyramid and the inferior olive. The **spinal accessory nerve** (**XI**) does not originate in the brainstem but, as its name implies, exits the lateral portion of the upper cervical spinal cord.

Despite the somewhat irregular distribution of these cranial nerves relative to their points of attachment to the brainstem, there is an orderly arrangement of certain groups of nerves. For example, the nerves that convey motor signals to muscles derived from embryological somitomeres—cranial nerves III, IV, VI, and XII—all attach to the brainstem along the same parasagittal plane (see Figure A8). Likewise, the nerves that supply muscles derived from the embryological pharyngeal arches (also known as branchiomeres)—the motor root of cranial nerve V and cranial nerves VII, IX, X, and XI—also emerge from the same parasagittal plane through the brainstem (with some distortion to account for the lateral expansion of the pons). The sensory root of cranial nerve V and cranial nerve VIII attach to the brainstem in a similar but slightly more lateral parasagittal plane; these are the cranial nerves that convey predominantly sensory signals to the brainstem. Table A2 presents a more complete description of the major functions of the cranial nerves.

Cranial nerve nuclei within the brainstem are the targets of cranial sensory nerves or the source of cranial motor nerves (Table A3, Figure A9). These nuclei are located in the **tegmentum** or central core of the brainstem, between the ventricular system dorsally and the division-specific structures and long motor pathways located ventrally. Cranial nerve nuclei that receive sensory input (analogous to the dorsal horns of the spinal cord) are located separately from those that give rise to motor output (which are analogous to the ventral horns). The primary sensory neurons that innervate these nuclei are found in ganglia associated with the cranial nerves—similar to the

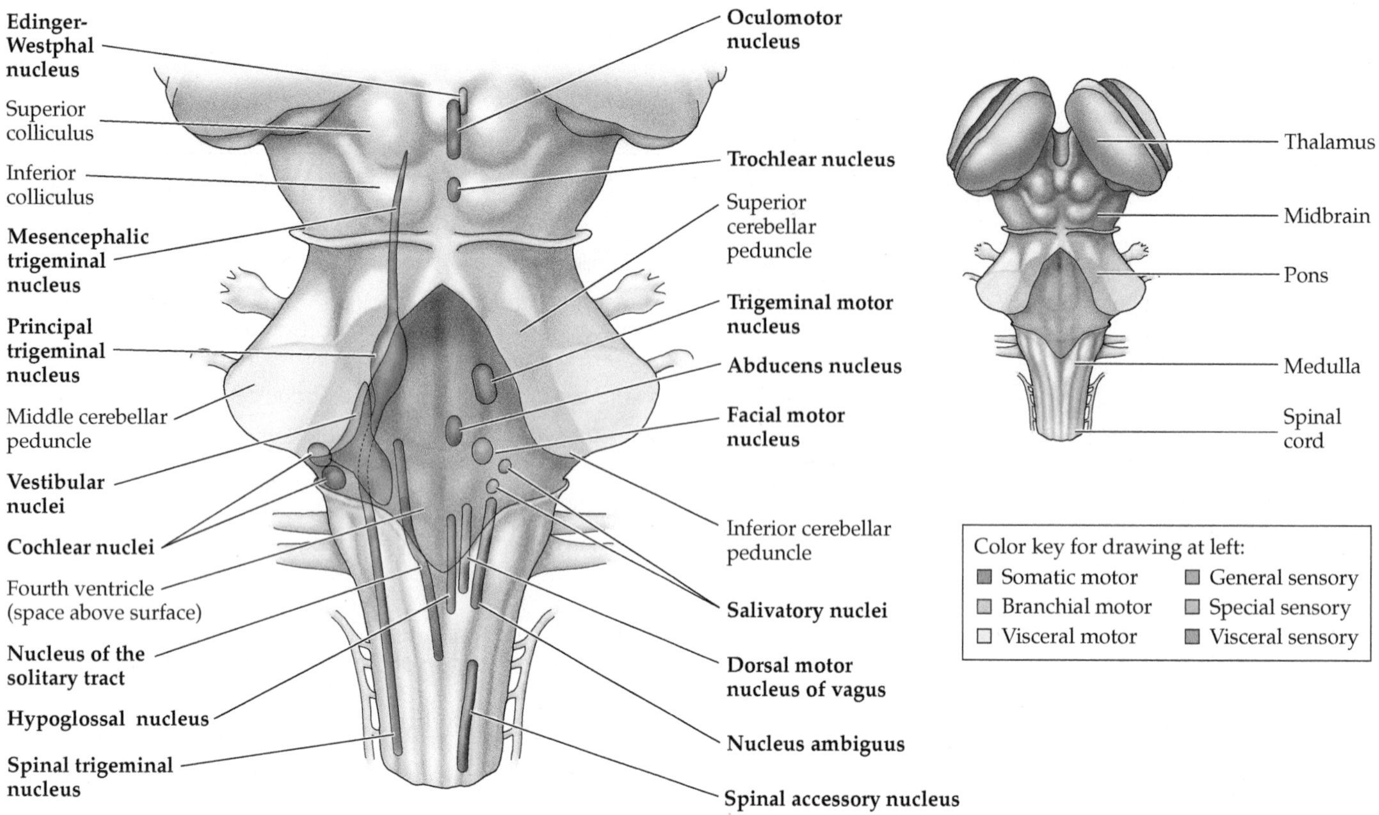

FIGURE A9 **Cranial nerve nuclei of the brainstem.** This "phantom" view of the brainstem's dorsal surface shows the locations of the brainstem nuclei that are either the target or the source of the cranial nerves. (See Table A2 for the relationship between each cranial nerve and cranial nerve nuclei, and Table A3 for a functional scheme that localizes cranial nerve nuclei with respect to brainstem subdivision and sensory or motor function.) With the exception of the nuclei associated with the trigeminal nerve, there is close correspondence between the location of the cranial nerve nuclei in the midbrain, pons, and medulla and the location of the associated cranial nerves. At right, the territories of the major brainstem subdivisions are indicated on the dorsal surface.

TABLE A3 ■ Classification and Location of the Cranial Nerve Nuclei[a]

<table>
<tr><th>LOCATION</th><th>SOMATIC MOTOR</th><th>BRANCHIAL MOTOR</th><th>VISCERAL MOTOR</th><th>GENERAL SENSORY</th><th>SPECIAL SENSORY</th><th>VISCERAL SENSORY</th></tr>
<tr><td>Midbrain</td><td>Oculomotor nucleus (III)
Trochlear nucleus (IV)</td><td></td><td>Edinger–Westphal nucleus (III)</td><td>Trigeminal sensory: mesencephalic nucleus (V, VII)</td><td></td><td></td></tr>
<tr><td>Pons</td><td>Abducens nucleus (VI)</td><td>Trigeminal motor nucleus (V)
Facial nucleus (VII)</td><td>Superior salivatory nucleus (VII)
Inferior salivatory nucleus (IX)</td><td rowspan="2">Trigeminal sensory: principal nucleus (V, VII, IX, X)
Trigeminal sensory: spinal nucleus (V, VII, IX, X)</td><td rowspan="2">Vestibular nuclei (VIII)
Cochlear nuclei (VIII)</td><td rowspan="2">Nucleus of the solitary tract (VII, IX, X)</td></tr>
<tr><td>Medulla</td><td>Hypoglossal nucleus (XII)</td><td>Nucleus ambiguus (IX, X)
Spinal accessory nucleus (XI)</td><td>Nucleus ambiguus (X)
Dorsal motor nucleus of vagus (X)</td></tr>
</table>

[a]Associated cranial nerves are shown in parentheses.

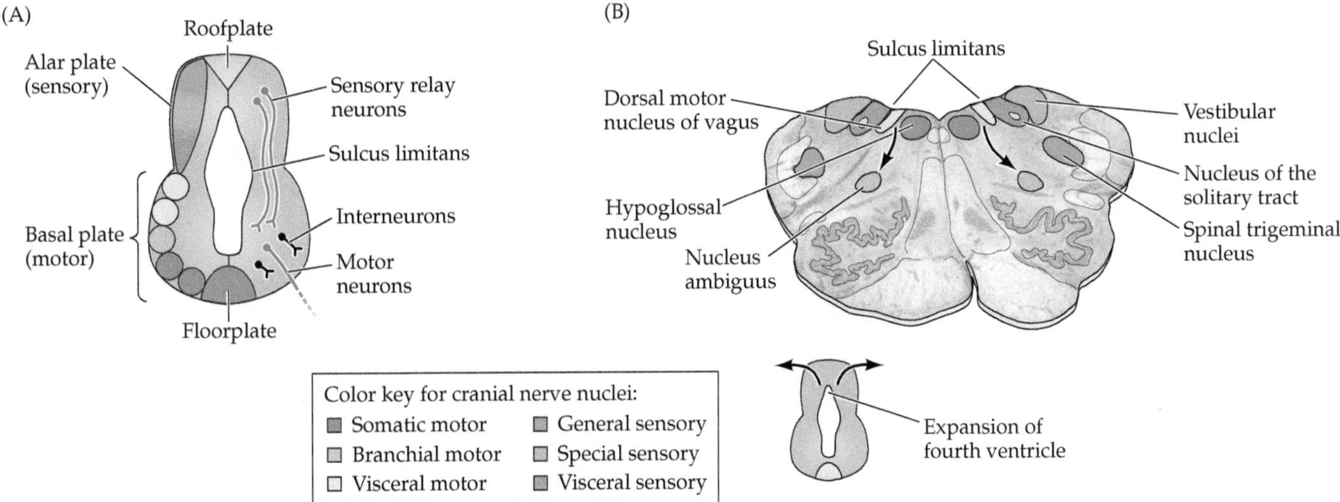

FIGURE A10 Embryological derivation of internal structure in the brainstem. (A) Illustration of a transverse section through the developing neural tube demonstrating the division of the alar plate from the basal plate by the sulcus limitans. The alar plate differentiates into the dorsal horn of the spinal cord and the sensory nuclei of the brainstem. The basal plate differentiates into the ventral horn of the spinal cord and the motor nuclei of the brainstem. (B) Representative transverse section from the brainstem (middle medulla) illustrating the location and identity of alar and basal plate derivatives. With expansion of the fourth ventricle, the alar plate derivatives develop lateral to the basal plate derivatives, like the opening of a book, with the floorplate in the position of the binding of the book (see inset). Note the secondary migration (curved arrows) of branchial motor nuclei, such as the nucleus ambiguus, to an intermediate position in the brainstem tegmentum.

relationship between dorsal root ganglia and the spinal cord. In general, sensory nuclei are found laterally in the brainstem, whereas motor nuclei are located more medially (see Figure A9).

A more precise accounting of the location of cranial nerve nuclei should be appreciated in relation to the embryological origins of the nuclei and the target tissues they innervate (Figure A10). Early in the development of the CNS, the neural tube establishes regional identity in the rostral–caudal axis (the divisions of the brainstem and spinal cord discussed above). The neural tube also gives rise to an important differentiation of dorsal and ventral identity, with the dorsal gray matter establishing an **alar plate** and the ventral gray matter establishing a **basal plate** (see Chapter 22). The alar and basal plates are separated by a shallow longitudinal groove called the **sulcus limitans**, which extends the length of the spinal cord through the mesencephalon. The alar plates gives rise to the dorsal horn of the spinal cord and the sensory nuclei of the cranial nerves, and the basal plate gives rise to the ventral horn of the spinal cord and the motor nuclei of the cranial nerves.

In the spinal cord, the division of sensory (dorsal) and motor (ventral) gray matter is straightforward and easy to appreciate in transverse section. In the brainstem, the enlargement of the ventricular system that generates the fourth ventricle (see below) contributes to the lateral displacement of the alar plate (see Figure A10B). Thus, in the tegmentum of the metencephalon and myelencephalon, the derivatives of the alar plate (the sensory nuclei) are located lateral to the derivatives of the basal plate (the motor nuclei). The derivatives of the basal plate differentiate further into three types of motor nuclei that are arranged in a medial to lateral progression, with respect to the embryological derivatives they innervate. Along the dorsal midline of the tegmentum are the **somatic motor nuclei** that project to striated muscles derived from somitomeres (extraocular muscles and extrinsic muscles of the tongue). Next, in a slightly more lateral column, are the **visceral motor nuclei** that project to peripheral ganglia innervating smooth muscle or glandular targets, similar to preganglionic motor neurons in the spinal cord that innervate autonomic ganglia. The **branchial motor nuclei** project to muscles derived from the pharyngeal or branchial arches, which give rise to the muscles (and bones) of the jaws, larynx, pharynx, and other craniofacial structures. The branchial motor nuclei migrate away from the dorsal aspect of the tegmentum and occupy a more central position just lateral to the visceral motor column, but still medial to the sensory nuclei (see curved arrows in Figure A10B).

The rostrocaudal organization of the cranial nerve nuclei (all of which are bilaterally symmetric) reflects the rostrocaudal distribution of head and neck structures (Figure A11). The more caudal the nucleus, the more caudally located are the target structures in the periphery. For example, the spinal accessory nucleus in the cervical spinal cord

FIGURE A11 Internal organization of the brainstem. Transverse sections through the brainstem along the rostral–caudal axis show the locations of the cranial nerve nuclei in six representative sections. The vascular territories for these brainstem sections are illustrated in Figure A21.

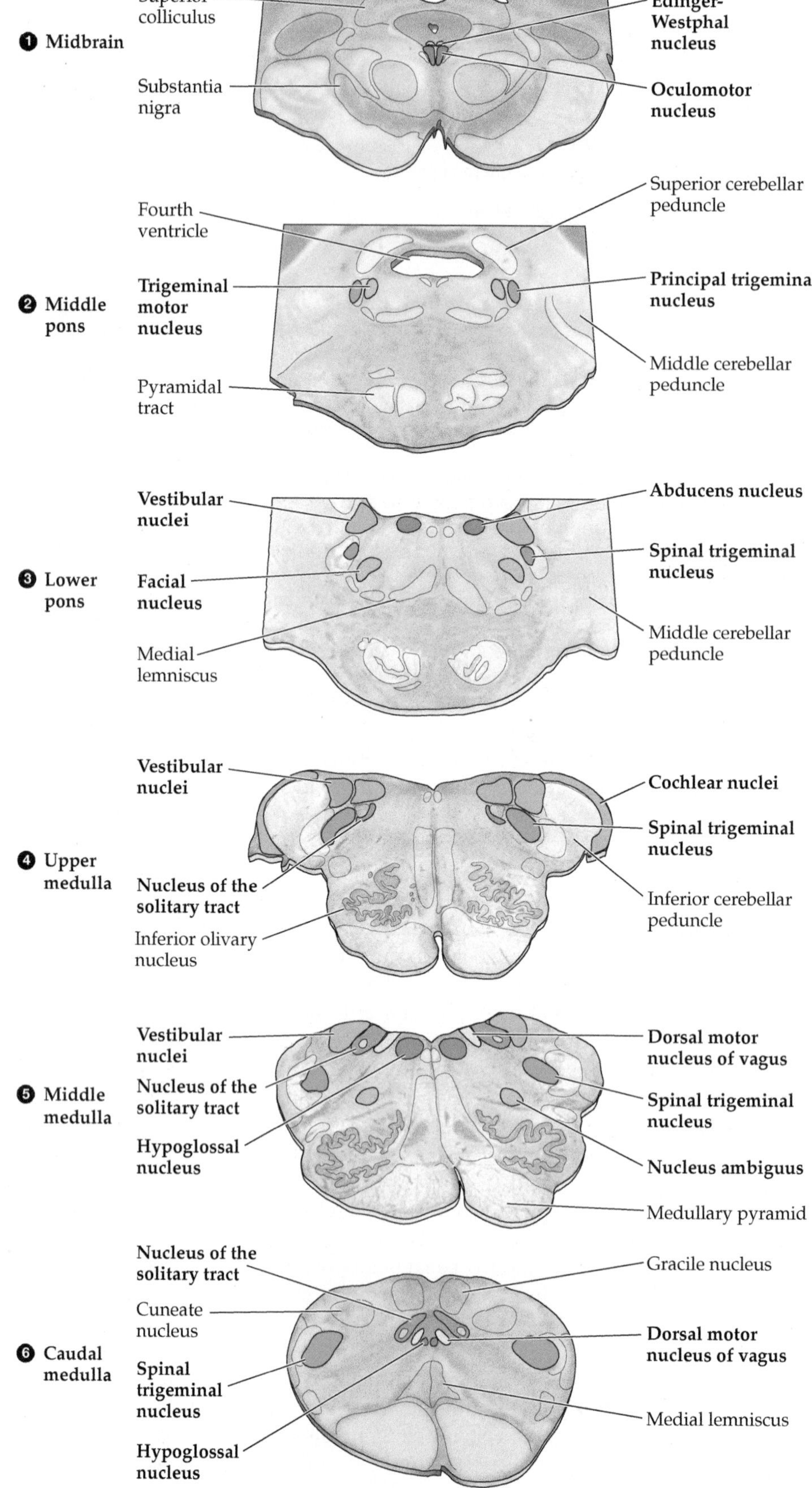

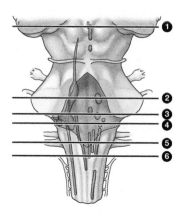

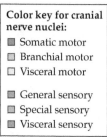

Color key for cranial nerve nuclei:

- Somatic motor
- Branchial motor
- Visceral motor
- General sensory
- Special sensory
- Visceral sensory

and caudal medulla provides branchial motor innervation for neck and shoulder muscles, and the motor nucleus of the vagus nerve provides preganglionic (parasympathetic) innervation for many enteric and visceral targets. In the pons, the sensory and motor nuclei are concerned primarily with somatic sensation from the face (the principal trigeminal nuclei) as well as movement of the jaws and the muscles of facial expression (the trigeminal motor and facial nuclei). Further rostrally, in the mesencephalic portion of the brainstem, are nuclei concerned primarily with eye movements (the oculomotor and trochlear nuclei) and preganglionic parasympathetic innervation of the iris (the Edinger–Westphal nucleus). While this list is not complete, it indicates the basic order of the rostrocaudal organization of the brainstem.

Healthcare professionals assess combinations of cranial nerve deficits to infer the location of brainstem lesions, or to place the source of brain dysfunction either in the spinal cord or forebrain. The most common brainstem lesions reflect the vascular territories that supply subsets of cranial nerve nuclei as well as ascending and descending tracts, which are located generally in the tegmentum (sensory) or basal (motor) regions of the brainstem (see below). For example, an occlusion of the posterior inferior cerebellar artery (PICA), a branch of the vertebral artery that supplies the dorsolateral region of the middle and rostral medulla, results in damage to several cranial nerve nuclei and tracts (see the "Upper medulla" section in Figure A11). Accordingly, there are functional deficits that reflect the loss of the spinal trigeminal nucleus, the vestibular and cochlear nuclei, and the nucleus ambiguus (which contains branchial motor neurons that project to the larynx and pharynx) on the same side as the lesion. In addition, ascending pathways from the spinal cord that relay pain and temperature from the contralateral body surface are disrupted, leading to a contralateral loss of these functions (see Chapters 9 and 10). Finally, the inferior cerebellar peduncle, which contains projections that relay information about body position to the cerebellum for postural control, is damaged. This loss results in ataxia (clumsiness) on the side of the lesion (see Chapter 19).

Anatomical relationships and shared vascularization, rather than any functional principle, unite these deficits and allow clinical localization of brainstem damage. For both clinicians and neurobiologists, understanding the brainstem requires integrating regional anatomical information with knowledge about functional organization and pathology.

Lateral Surface of the Brain

A lateral view of the human brain is the best perspective from which to appreciate all four lobes of the cerebral hemisphere (see Figure A4A). In this view, the two most salient landmarks are the deep lateral fissure that separates the temporal lobe from the overlying frontal and parietal lobes, and the central sulcus, which serves as the boundary between the frontal and parietal lobes (Figure A12). A particularly important feature of the frontal lobe is the **precentral gyrus**. (The prefix *pre-*, when used anatomically, refers to a

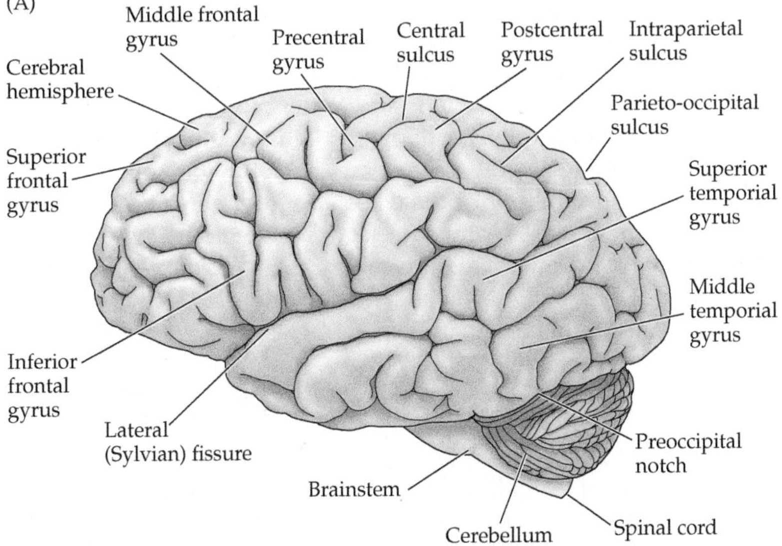

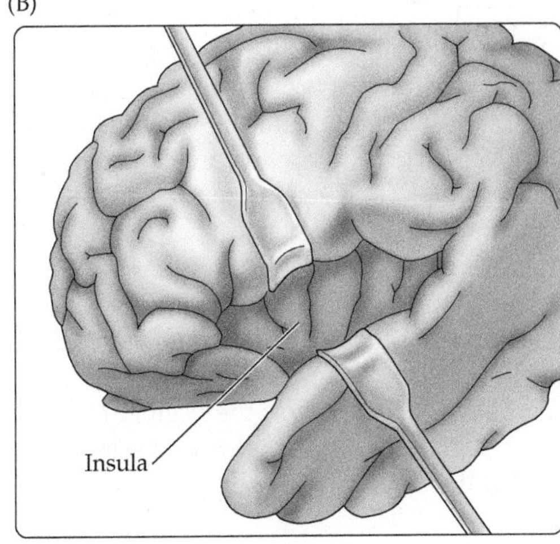

FIGURE A12 Lateral view of the human brain. (A) Illustration of some of the major gyri and sulci from this perspective. (B) The banks of the lateral (Sylvian) fissure have been pulled apart to expose the insula.

structure that is in front of or anterior to another.) The cortex of the precentral gyrus is referred to as the **motor cortex** and contains neurons whose axons project to the lower somatic and branchial motor neurons in the brainstem and spinal cord (see Chapter 17). Anterior to the precentral gyrus are three long parallel gyri called the **superior, middle**, and **inferior frontal gyri**. The posterior portion of the left inferior frontal gyrus is the typical localization of Broca's area, which is involved in the expression of language (see Chapter 33).

The lateral surface of the temporal, like the frontal, lobe features three long parallel gyri termed the **superior, middle**, and **inferior temporal gyri**. The superior aspect of the temporal lobe contains cortex concerned with audition and language reception, and inferior portions of the lobe deal with highly processed visual information. Hidden beneath the frontal and temporal lobes, the **insular cortex**, or **insula**, can be seen only if these two lobes are pulled apart or removed (see Figure A12B). The posterior portion of the insula is concerned largely with visceral and autonomic function, including taste. More rostral portions of the insula are involved in implicit feelings and their impact on social cognition (see Chapter 32). In the anterior parietal lobe just posterior to the central sulcus is the **postcentral gyrus**; this gyrus harbors cortex that is concerned with somatic (bodily) sensation and is therefore referred to as the **somatosensory cortex** (see Chapter 9). Posterior to the postcentral gyrus are two gyral formations separated by the **intraparietal sulcus** called the **superior** and **inferior parietal lobules**. These cortical regions associate somatosensory, visual, auditory, and vestibular signals and generate a neural construct of the body, the position of its parts, and its movements (body image or schema).

The boundary between the parietal lobe and the occipital lobe, the most posterior of the hemispheric lobes, is a somewhat arbitrary line from the parieto-occipital sulcus to the preoccipital notch. The occipital lobe, only a small part of which is apparent from the lateral surface of the brain, is concerned primarily with vision and visualization (even when the eyelids are closed). In addition to its role in primary and sensory processing, each cortical lobe participates in complex brain functions related to one or more dimensions of cognition (see Chapter 27). Thus, the parietal lobe is critical for attending to stimuli, the temporal lobe is used to recognize stimuli, and the frontal lobe is critical in planning responses to stimuli and in the future organization of behavior; the occipital lobe is involved in all aspects of visual perception and may also participate in multimodal sensory processing.

Dorsal and Ventral Surfaces of the Brain

Although the primary subdivisions of the cerebral hemispheres can be appreciated from a lateral view, other key landmarks are better seen from the dorsal and ventral surfaces (Figure A13). When viewed from the dorsal surface, the approximate bilateral symmetry of the cerebral hemispheres is apparent. Although there is some variation, major landmarks such as the central sulci and intraparietal sulci are usually very similar in arrangement on the two sides. If the cortical hemispheres are spread slightly apart from the dorsal midline, another major structure, the **corpus callosum**, can be seen bridging the two hemispheres (see Figure A13C). This structure contains axons that originate from pyramidal neurons in the cerebral cortex of both hemispheres. Callosal axons interconnect neurons in opposite (homotypical) cortical regions.

The external features of the brain that are best seen on its ventral surface are shown in Figure A13B. Extending along the inferior surface of the frontal lobe near the midline are the **olfactory tracts**, which arise from enlargements at their anterior ends called the **olfactory bulbs**. The olfactory bulbs receive input from neurons in the epithelial lining of the nasal cavity whose axons make up the first cranial nerve (cranial nerve I is therefore called the olfactory nerve; see Table A2 and Chapter 15). The olfactory bulbs and tracts lie on the medial margins of the **orbital gyri**, so named because these complex and highly variable gyri of the ventral frontal lobe are situated just superior to the orbits of the skull. On the ventromedial surface of the temporal lobe, the **parahippocampal gyrus** conceals the amygdala and also the **hippocampus**, a highly convoluted cortical structure that consolidates memory (see Chapter 30). A prominent medial protrusion of the parahippocampal gyrus is the **uncus**, which includes the cortical divisions of the amygdala. Between the parahippocampal and the inferior temporal gyri is the **occipitotemporal gyrus**, the posterior portion of which is sometimes called the **fusiform gyrus** (hidden by the cerebellum in Figure A13B). At the most central aspect of the ventral surface of the forebrain is the **optic chiasm**, and immediately posterior, the ventral surface of the **hypothalamus**, including the **infundibulum** (also called the **pituitary stalk**, at the base of the pituitary gland) and the **mammillary bodies**. Posterior to the hypothalamus, the paired cerebral peduncles are located on either side of the ventral midline of the midbrain. Finally, the ventral surfaces of the pons, medulla, and cerebellar hemispheres (see Figure A8) can be seen in this ventral view.

Midsagittal Surface of the Brain

When the brain is hemisected in the midsagittal plane, all of its major subdivisions plus several additional structures are visible on the cut surface. In this view, the cerebral hemispheres, because of their greater size, are still the most obvious structures. The frontal lobe of each hemisphere extends forward from the central sulcus, the medial end of which can just be seen terminating in the

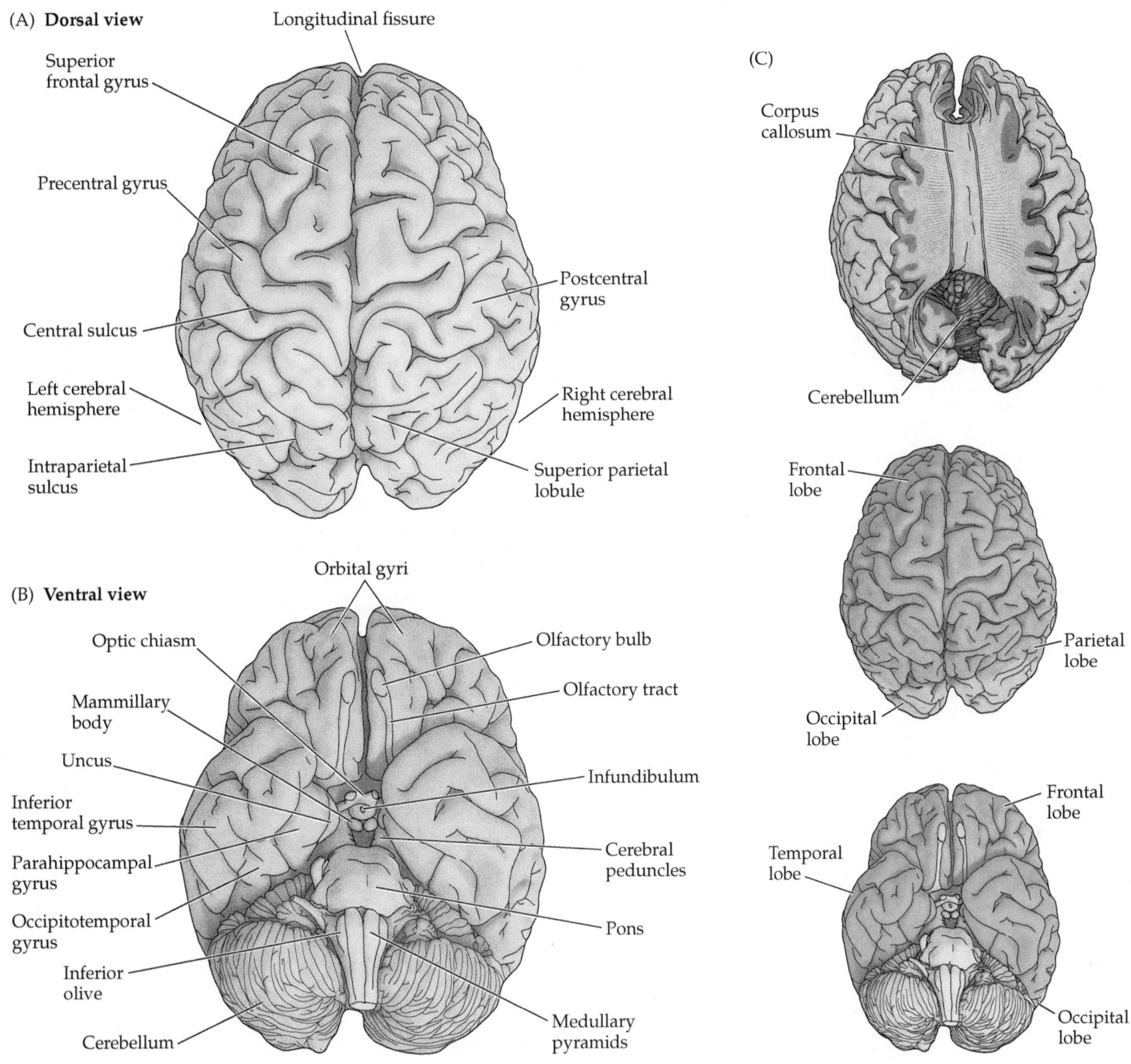

FIGURE A13 Dorsal and ventral views of the human brain. (A) Dorsal view. (B) Ventral view. Both views indicate some of the major features visible from these perspectives. (C) In the upper image (dorsal view), the cerebral cortex has been removed to reveal the underlying corpus callosum. The two lower images highlight the four lobes of the cerebral cortex. (C from Rohen et al., 1993.)

paracentral lobule (Figure A14A,B). The parieto-occipital sulcus, running from the superior to the inferior aspect of the hemisphere, is most obvious in this view of the hemisphere as it separates the **precuneus gyrus** in the parietal lobe from two major gyri in the occipital lobe. The **calcarine sulcus** divides the medial surface of the occipital lobe, running at nearly a right angle from the parieto-occipital sulcus and marking the location of the **primary visual cortex** (see Chapter 12). The upper bank of the calcarine sulcus is formed by the **cuneus gyrus** and the lower bank by the **lingual gyrus**. A long sulcus that follows the curvature of the corpus callosum, the **cingulate sulcus**, extends across the medial surface of the frontal and parietal lobes, ending in a dorsal ramus

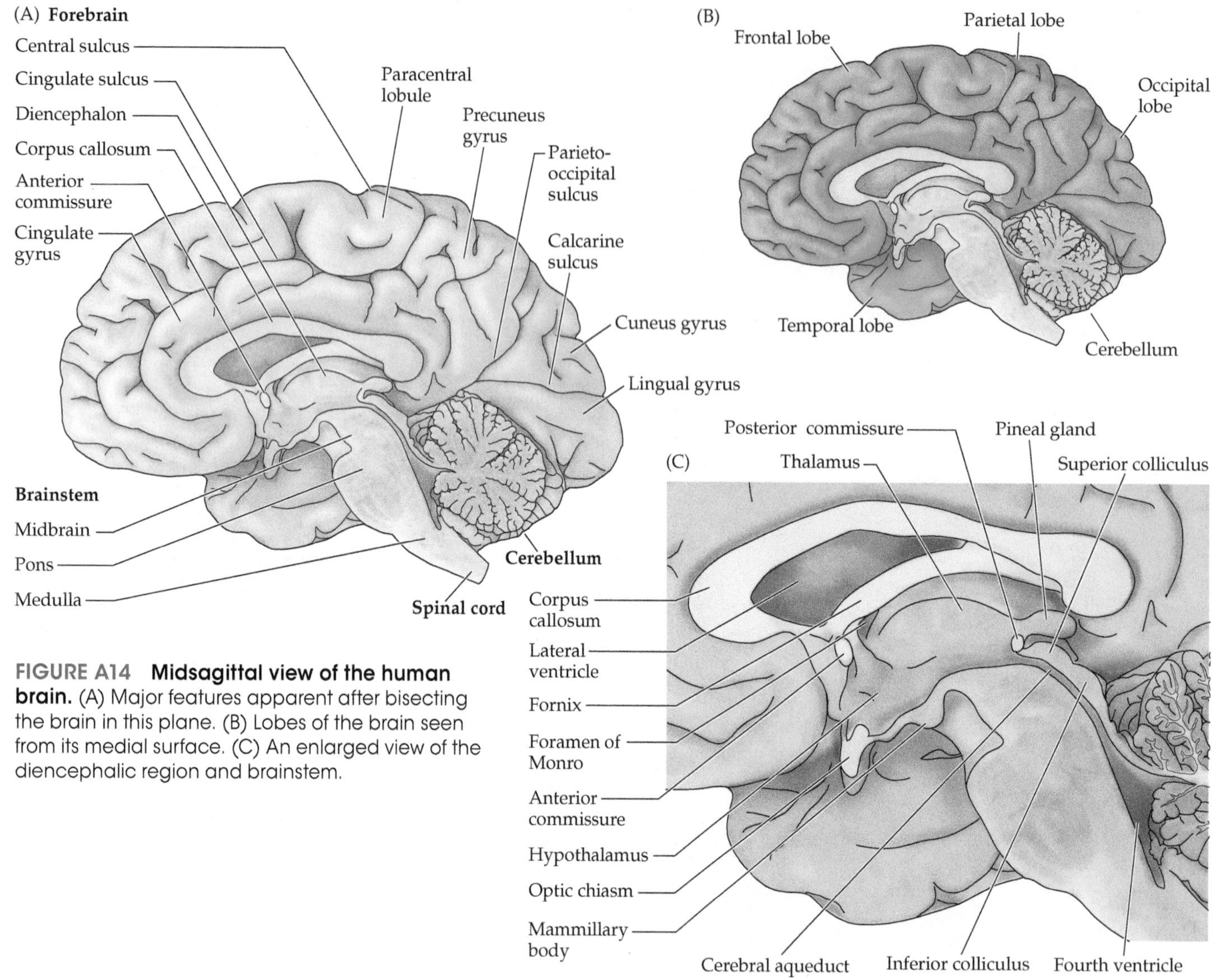

FIGURE A14 Midsagittal view of the human brain. (A) Major features apparent after bisecting the brain in this plane. (B) Lobes of the brain seen from its medial surface. (C) An enlarged view of the diencephalic region and brainstem.

that marks the posterior boundary of the paracentral lobule. Below the cingulate sulcus is the **cingulate gyrus**, a prominent component of the **limbic forebrain**, which comprises cortical and subcortical structures in the frontal and temporal lobes that form a medial rim of cerebrum roughly encircling the corpus callosum and diencephalon (*limbic* means "border" or "rim"). The limbic forebrain is important in the experience and expression of emotion, as well as the regulation of attending visceral motor activity (see Chapter 31). Finally, ventral to the cingulate gyrus is the cut, midsagittal surface of the corpus callosum.

Although parts of the diencephalon, brainstem, and cerebellum are visible at the ventral surface of the brain, their overall structure is especially clear from the midsagittal surface (Figure A14C). From this perspective, the diencephalon can be seen to consist of two parts. The **thalamus**, the largest component of the diencephalon, comprises several subdivisions, all of which relay information to the cerebral cortex from other parts of the brain (Box A). The hypothalamus—a small but crucial part of the diencephalon—is devoted to the control of homeostatic and reproductive functions, among other diverse activities (see Box 21A). The hypothalamus is intimately related, both structurally and functionally, to the pituitary gland, a critical endocrine organ whose posterior part is connected to the hypothalamus by the infundibulum.

The midbrain lies caudal to the thalamus, and the pons is caudal to the midbrain. The cerebellum lies over the pons and rostral medulla just beneath the occipital lobe of the cerebral hemispheres. From the midsagittal surface, the most visible feature of the cerebellum is the **cerebellar cortex**, a continuous layered sheet of cells folded into small convolutions called **folia**. The most caudal structure seen

BOX A ■ Thalamus and Thalamocortical Relations

With one notable exception, studies of the sensory and motor systems in Units II and III have all included descriptions of important connections between the thalamus and some circumscribed division of the cerebral cortex. Here, we draw together these descriptions into a brief consideration of the thalamus and its anatomical relations to the cerebral cortex.

The thalamus is a large mass of gray matter in the dorsal aspect of the diencephalon, superior to the hypothalamus (the hypothalamus is described in Box 21A) and medial to the massive collections of fibers that form the genu and posterior limb of the internal capsule. Conventionally, the thalamus comprises three main parts: the *epithalamus*, a small strip of tissue on the dorsomedial aspect of the thalamus to which the pineal gland is attached; the *subthalamus*, a region just above and slightly posterior to the hypothalamus containing nuclei that modulate basal ganglia output (including the subthalamic nucleus discussed in Chapter 18); and the *dorsal thalamus*, which is the largest and most complex of the three parts. The dorsal thalamus is the part that is now simply called the *thalamus* (Greek, "inner chamber" or "marriage bed," a Galenic reference to the thalamus's central position in the brain and possibly also a sexual reference implying the regeneration or procreation of sensory signals). It is this part of the dorsal diencephalon that is most closely associated anatomically and functionally with the circuitry of the cerebral cortex.

The mammalian thalamus is a complex (in the sense that the amygdala is a complex; see Box 31B) comprising some 50 or so nuclear subdivisions that maintain distinct connections of inputs and outputs (Figure A). Despite the complexity of the thalamus, it is possible to understand its structure in broad terms based on the locations of its subdivisions and the patterns of their projections to the cerebral cortex. Broadly speaking, the thalamus is divided into medial, lateral, and anterior sectors by a Y-shaped band of white matter called the internal medullary lamina. Thus, there are groups of medial nuclei, lateral nuclei, and anterior nuclei; there are nuclei embedded in the internal medullary lamina itself; and there are nuclei along the midline of the thalamus. There is also a thin, shell-shaped nucleus that envelops the thalamus laterally, called the reticular nucleus (not to be confused with the brainstem reticular formation), which has a profound influence over the firing patterns of thalamocortical projection neurons (see Figure 28.11).

The thalamus is usually considered a major relay station for sending sensory and motor signals to specific areas of the cerebral cortex (with the exception of the olfactory cortex, which receives sensory signals directly from the olfactory bulb). These sensory and motor signals arise mainly from nuclei in the lateral group. The thalamus also conveys less well understood but nonetheless specific signals to associational areas of the cerebral cortex; these arise from medial and anterior nuclei and from certain nuclei in the posterior pole of the thalamus. Thus, every area of the cerebral cortex receives incoming signals that are topographically organized from some particular subdivision of the thalamus. Indeed, the degree to which the functional—and even the anatomical—identify of cortical areas is determined by specific connections with thalamic nuclei remains an active area of research and debate in developmental neuroscience.

Continued on the next page

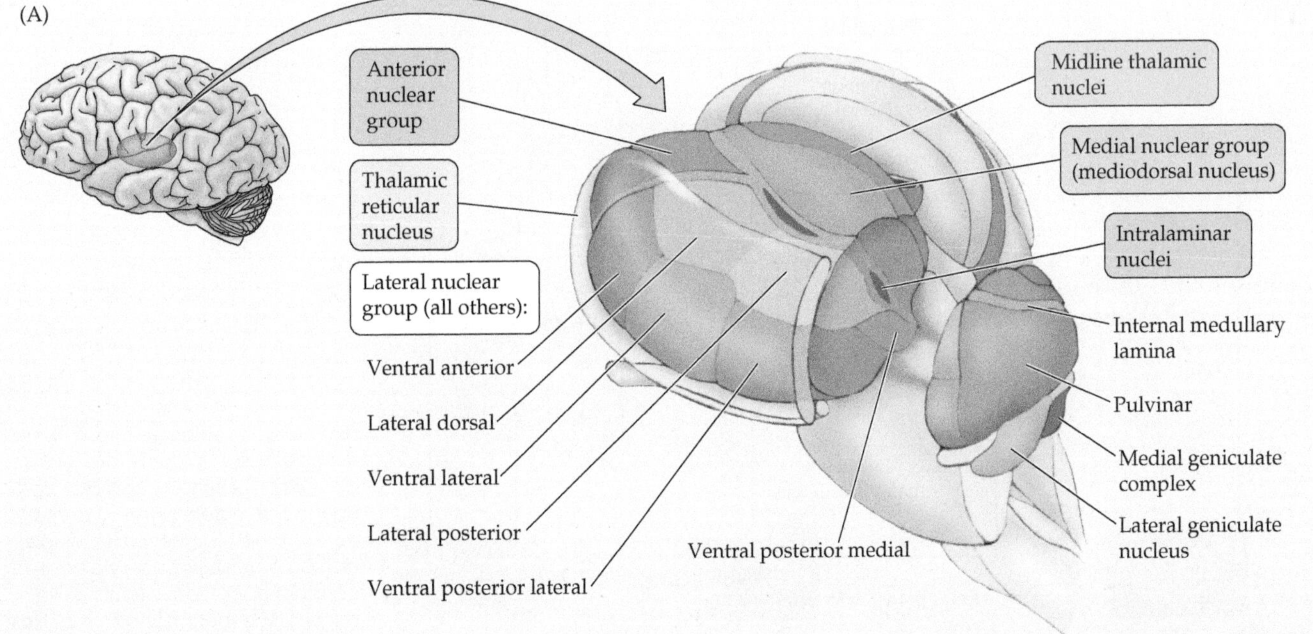

(A) Subdivisions of the thalamus in the human brain. (From Blumenfeld, 2010.)

BOX A ■ (continued)

These so-called specific thalamic projections are targeted to the middle layers of the cerebral cortex, where they serve to drive or sharply modulate activity in local columnar circuits of cortical neurons (Figure B). Examples include thalamocortical projections from the *somatosensory thalamus* (ventral posterior lateral and ventral posterior medial nuclei to the postcentral gyrus; see Chapters 9 and 10), the *visual thalamus* (lateral geniculate nucleus projections to the lingual and cuneus gyri; see Chapter 12), the *auditory thalamus* (medial geniculate complex projections to superior, transverse temporal gyri; see Chapter 13), or the *motor thalamus* (ventral anterior and ventrolateral nuclei projections to the precentral gyrus; see Chapters 18 and 19). However, there are prominent thalamic nuclei—such as the pulvinar in the posterior pole of the thalamus—that are themselves driven primarily by inputs from layer 5 of the cerebral cortex rather than by lower-order sensory or motor centers. Such thalamic nuclei, in turn, provide higher-order input that drives activity in other (non-primary) cortical areas (see Figure B). Thus, the thalamus may serve as both a relay of first-order sensory and motor input signals to relevant primary cortical areas, and a distributer of higher-order output signals from one cortical area to another. Evidently, it is the timely activation of such thalamocortical and cortico-thalamocortical projections that triggers sensory processing and the execution of behavioral programs that move body, mind, and emotion.

In contrast to these "specific" thalamocortical relations, there are much more diffuse projections arising from intralaminar and midline nuclei that terminate diffusely in the upper layers of the cerebral cortex. Rather than conveying specific sensory or motor signals, these so-called nonspecific thalamic projections have widespread modulatory influences over vast networks of cortical neurons—the sort of influences that could mediate attention, mood change, behavioral arousal, and transitions in sleep and wakefulness. Unfortunately, it is also these nonspecific projections of the thalamus that synchronize paroxysmal activity in generalized seizures, accounting for the nearly simultaneous and rhythmical discharge of cortical neurons across the cerebral hemispheres.

Finally, thalamocortical projections are reciprocated by massive systems of inputs from layer 6 of the cerebral cortex that appear morphologically and physiologically to serve as feedback modulators of the same thalamic neurons that provide driving signals to cortical networks (descending dashed arrows in Figure B). In fact, for certain thalamic nuclei such as the lateral geniculate nucleus, the number of presumptive modulatory inputs derived from the cerebral cortex is several-fold larger than the number of synaptic connections received from lower-order processing centers (the retina, in the case of the lateral geniculate nucleus). Despite the preeminence of these corticothalamic inputs, the precise role of feedback modulation in thalamic function remains poorly understood. Clearly, there remains much to be discovered concerning the neural computations hosted by this "inner chamber."

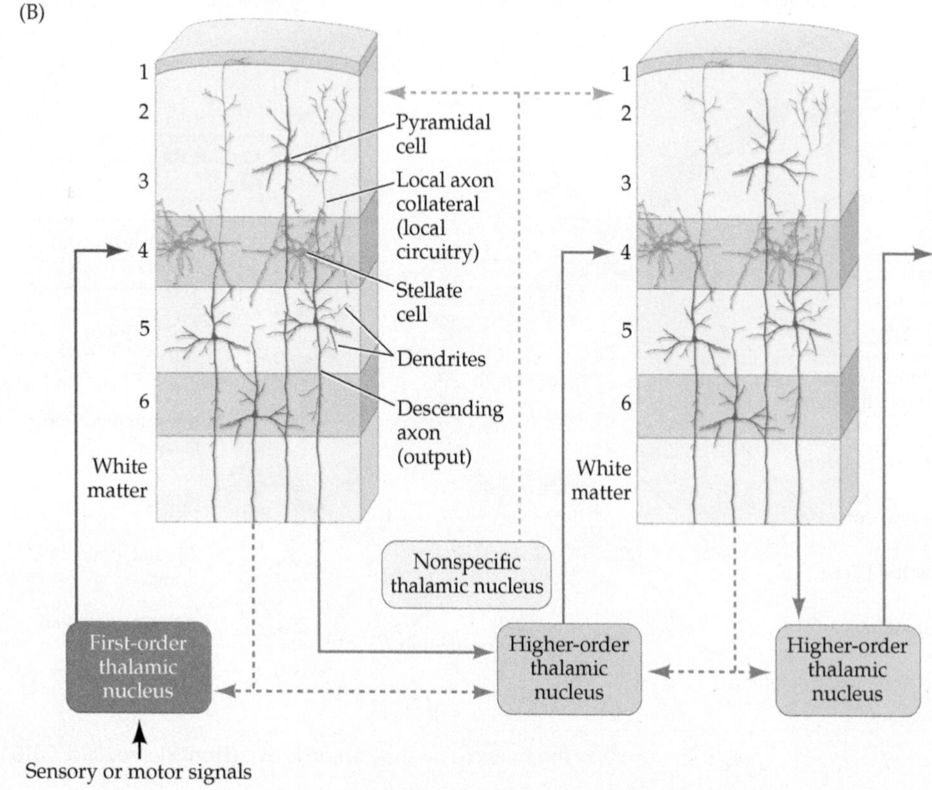

(B) Thalamocortical relations. Some specific thalamic nuclei are "first-order" relays of sensory or motor signals to the middle layers of primary cortex; other, "higher-order" nuclei distribute output signals via cortico-thalamocortical circuits (solid arrows). The thalamus also receives and distributes modulatory signals (dashed lines). (After Sherman and Guillery, 2011.)

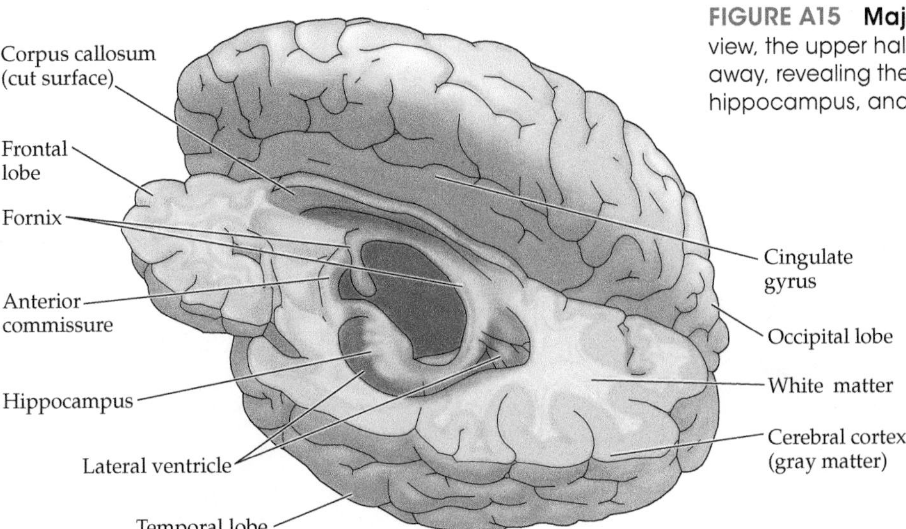

FIGURE A15 Major internal structures of the brain. In this view, the upper half of the left hemisphere has been dissected away, revealing the temporal horn of the lateral ventricle, the hippocampus, and the fornix.

from the midsagittal surface of the brain is the medulla, which merges into the spinal cord.

Internal Anatomy of the Forebrain

A much more detailed neuroanatomical picture of the forebrain is apparent in gross or histological slices. In these slices (or sections), deep structures that are not visible from any brain surface can be identified. In addition, relationships between brain structures seen from the surface can be appreciated more fully. The major challenge to understanding the internal anatomy of the brain is to integrate the rostral–caudal, dorsal–ventral, and medial–lateral landmarks seen on the brain surface with the position of structures seen in brain sections taken in the horizontal (axial), frontal, and sagittal planes. This challenge is not only important for understanding brain function; it is essential for interpreting noninvasive images of the brain, most of which are displayed as sections (see Atlas).

In any plane of section through the forebrain, the cerebral cortex is evident as a thin layer of neural tissue that covers the entire cerebrum. Most cerebral cortex is made up of six layers and is referred to as **neocortex** (see Box 27A). Phylogenetically older cortex (**paleocortex**) with fewer cell layers occurs on the inferior and medial aspect of the temporal lobe within the parahippocampal gyrus and in the pyriform cortex (a major division of the olfactory cortex near the junction of the temporal and frontal lobes). The simplest and most primitive division of the cortex, the **archicortex**, occurs in the hippocampus. The hippocampal cortex is folded into the medial aspect of the temporal lobe and therefore is visible only in dissected brains or in sections (Figure A15).

Embedded within the cerebral hemispheres are the **cerebral nuclei**, the largest of which are the components of the **basal ganglia**: the **caudate** and **putamen nuclei** (together referred to as **striatum**) and the **globus pallidus** (Figure A16). (The term *ganglia* does not usually refer to nuclei in the brain; the usage here is an exception.) The basal ganglia are visible in sections through the forebrain that also contain the lateral ventricles. The anterior "head" and central "body" of the caudate nucleus forms the lateral wall of the lateral ventricle, and the tail of the caudate may be found in the temporal lobe in the roof of the temporal horn of the lateral ventricle. The neurons of these large nuclei receive input from the cerebral cortex and participate in the organization and guidance of complex motor functions (see Chapter 18). In the base of the forebrain, ventral and medial to the basal ganglia are several smaller clusters of nerve cells known as the **septal** or **basal forebrain nuclei**. These nuclei are of particular interest because they modulate neural activity in the cerebral cortex and hippocampus, and they are among the forebrain systems that degenerate in Alzheimer's disease. The other clearly discernible structure visible in sections through the cerebral hemispheres at the level of the uncus is the **amygdala**, a complex of nuclei and cortical divisions that lies in front of the hippocampus just behind the anterior pole of the temporal lobe.

In addition to these cortical and nuclear structures, several important axon tracts are localized to the internal anatomy of the forebrain. As already mentioned, cerebral cortex in the two cerebral hemispheres is interconnected by the corpus callosum; in some anterior sections, the smaller **anterior commissure** that interconnects cortex in the anterior temporal lobes and ventral frontal lobes can also be seen (see Figure A16A). Axons descending from (and ascending to) the cerebral cortex assemble into another large fiber bundle tract called the **internal capsule** (see Figure A16A,B). The internal capsule lies just

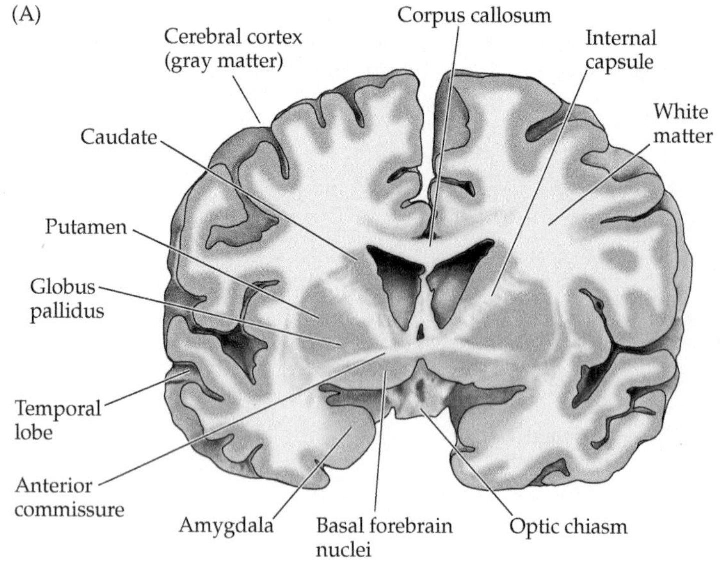

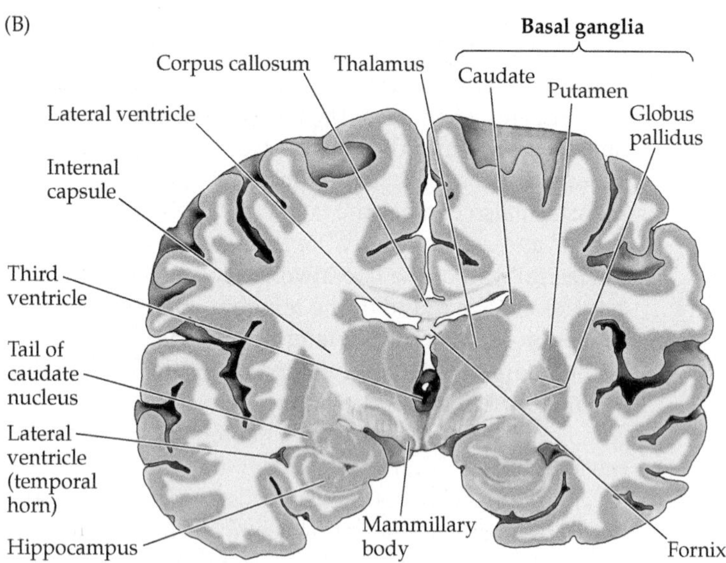

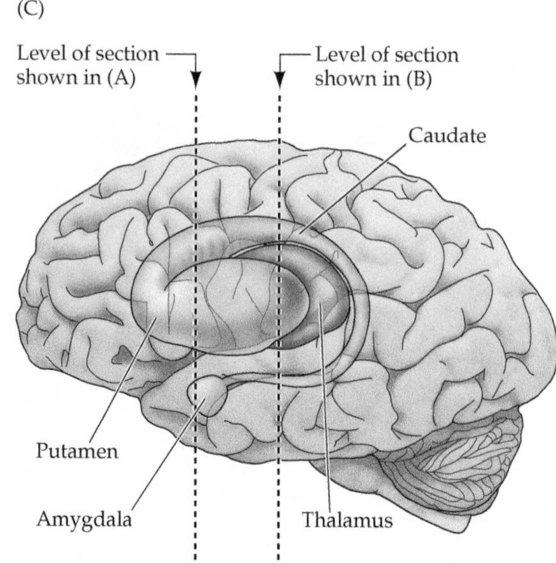

FIGURE A16 Internal structures of the brain seen in coronal section. (A) This section passes through the basal ganglia. (B) A more posterior section also includes the thalamus. (C) A transparent view of the cerebral hemisphere showing the approximate locations of the sections in (A) and (B) relative to deep gray matter (the basal ganglia, thalamus, and amygdala are represented). Notice that because the caudate nucleus has a "tail" that arcs into the temporal lobe, it appears twice in section (B); the same is true of other brain structures, including the lateral ventricle. (Also see Figure A20 and Atlas Plate 2.) (After Blumenfeld, 2010.)

lateral to the diencephalon (forming a "capsule" around it), and many of its axons arise from or terminate in the thalamus. The internal capsule is seen most clearly in frontal sections through the middle one-third of the rostrocaudal extent of forebrain, or in horizontal sections through the level of the thalamus. Other axons descending from the cortex in the internal capsule continue past the diencephalon to enter the cerebral peduncles of the midbrain. Axons in these corticobulbar and corticospinal tracts project to several targets in the brainstem and spinal cord, respectively (see Chapter 17). Thus, the internal capsule is the major pathway linking the cerebral cortex to the rest of the brain and spinal cord. Strokes or other injury to this structure interrupt the flow of ascending and descending nerve impulses, often with devastating consequences. The internal capsule is also a useful landmark for understanding the distribution of major cerebral nuclei, including the basal ganglia and the thalamus. The caudate nucleus and the thalamus lie on the medial aspect of the internal capsule, while the globus pallidus and putamen are found on its lateral aspect. Finally, a smaller fiber bundle within each of the hemispheres, the **fornix**, interconnects the hippocampus and the hypothalamus and the septal region of the basal forebrain.

Blood Supply of the Brain and Spinal Cord

Understanding the blood supply of the brain and spinal cord is crucial for neurological diagnoses and the practice of medicine, particularly for neurology and neurosurgery.

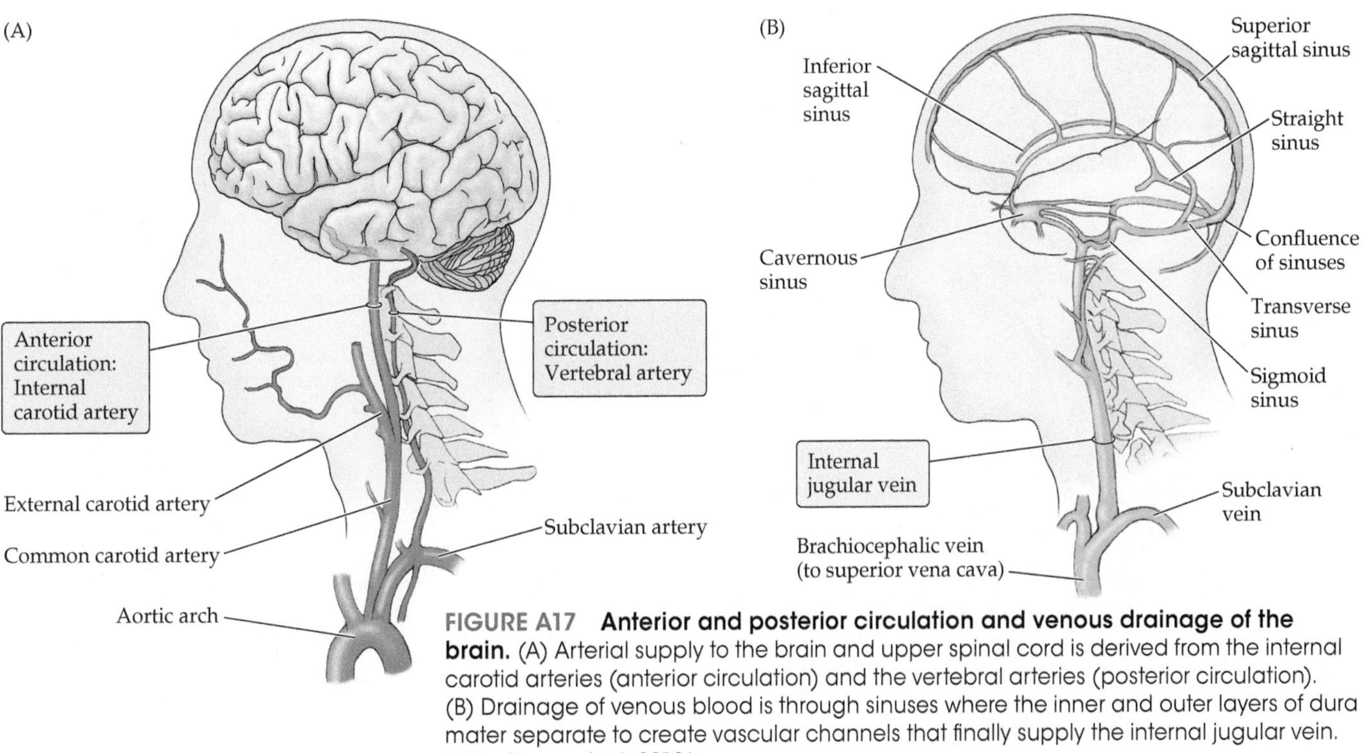

FIGURE A17 Anterior and posterior circulation and venous drainage of the brain. (A) Arterial supply to the brain and upper spinal cord is derived from the internal carotid arteries (anterior circulation) and the vertebral arteries (posterior circulation). (B) Drainage of venous blood is through sinuses where the inner and outer layers of dura mater separate to create vascular channels that finally supply the internal jugular vein. (After Blumenfeld, 2010.)

Damage to major blood vessels by trauma or stroke results in combinations of functional deficits that reflect both local cell death and the disruption of axons passing through the region compromised by the vascular damage. Thus, a firm knowledge of the major cerebral blood vessels and the neuroanatomical territories they perfuse facilitates the initial diagnoses of a broad range of brain damage and disease.

The entire blood supply of the brain and spinal cord depends on two sets of branches from the dorsal aorta (Figures A17 and A18). The **internal carotid arteries** are branches of the common carotid arteries, while the **vertebral arteries** arise from the subclavian arteries. These two major sets of arterial branches and the vascular distributions supplied by them are often conceptualized as the **anterior** and **posterior circulation**, with the anterior circulation derived from the internal carotid arteries and the posterior circulation from the vertebral arteries (see Figure A17A). Generally speaking, the anterior circulation supplies the forebrain (the cerebral hemispheres and diencephalon), while the posterior circulation supplies the brainstem, cerebellum, and upper portion of the spinal cord. However, the posterior cerebral artery supplies the posterior forebrain, including some deep structures. Thus, as indicated by listing this artery twice in Table A4, the posterior cerebral artery contributes to both the anterior and posterior circulations.

Figure A19 shows the major arteries of the brain. Anterior to the spinal cord and brainstem, the internal carotid arteries branch to form two major cerebral arteries, the **anterior** and **middle cerebral arteries**. The right and left vertebral arteries come together at the level of the pons on the ventral surface of the brainstem to form the midline **basilar artery**. The basilar artery joins the blood supply from the internal carotids in an arterial ring at the base of the brain (in the vicinity of the hypothalamus and cerebral peduncles) called the **circle of Willis** (see Figure A19A). The **posterior cerebral arteries** arise at this confluence, as do three small bridging arteries, the single **anterior** and paired **posterior communicating arteries**. In most humans, the posterior cerebral artery receives its blood supply from the vertebral/basilar system. In some people, the posterior communicating artery is quite large, and the posterior cerebral artery may be perfused by the carotid artery. Conjoining the major sources of cerebral vascular supply via the circle of Willis presumably improves the chances of any region of the brain continuing to receive blood if one of the major arteries becomes occluded.

Each of the major arterial branches that comprise the anterior circulation (i.e., those branches derived from the internal carotid artery plus the posterior cerebral artery) gives rise to superficial vessels that supply cortical structures and deep vessels that penetrate the ventral surface of the brain and supply internal structures. An extensive region of the central and lateral cerebral hemispheres is supplied by the middle cerebral artery (shaded green in Figure A19B). Included in this region are the sensorimotor

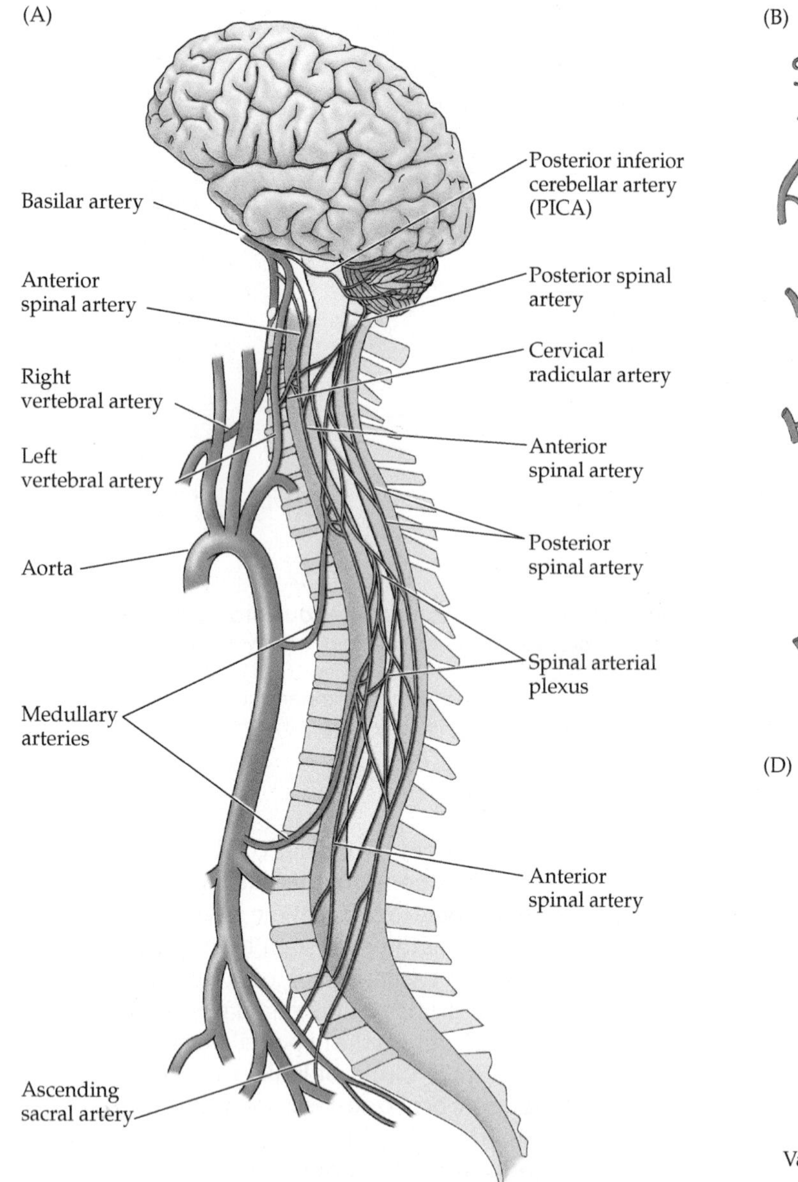

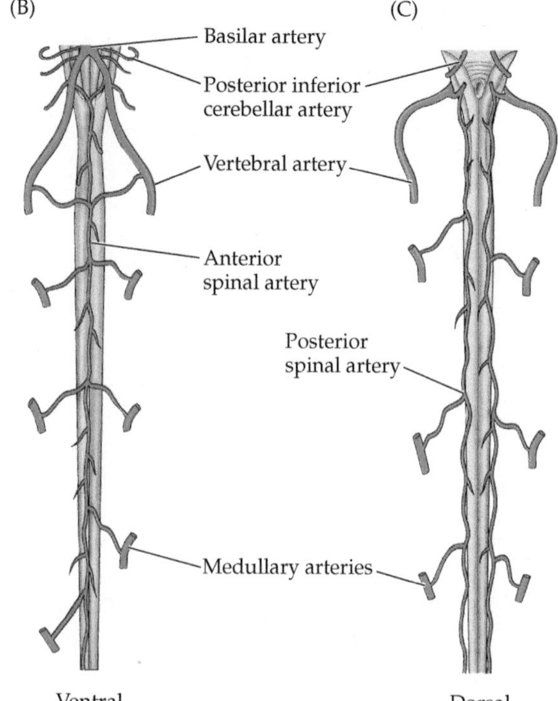

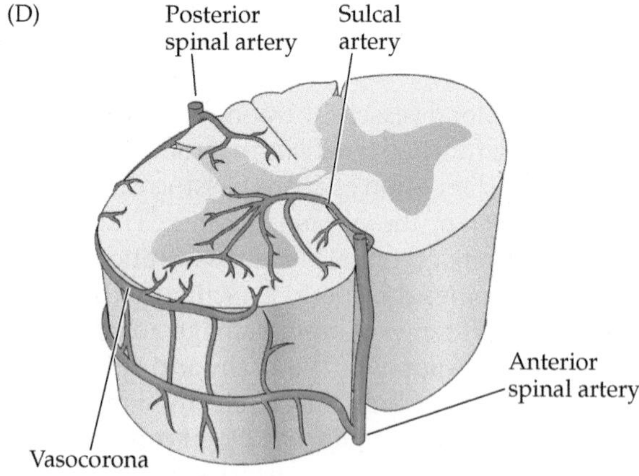

FIGURE A18 Blood supply of the spinal cord. (A) View of the left side of blood supply to the brainstem and spinal cord in relation to the aorta from which the supply is derived. (B) View of the ventral (anterior) surface of the spinal cord. At the level of the medulla, the vertebral arteries give off branches that merge to form the anterior spinal artery. Approximately six to ten segmental arteries (which arise from various branches of the aorta) join the anterior spinal artery along its course. These segmental arteries are known as medullary arteries. (C) The vertebral arteries (or the posterior inferior cerebellar artery) give rise to paired posterior spinal arteries that run along the dorsal (posterior) surface of the spinal cord. (D) Cross section through the spinal cord, illustrating the distribution of the anterior and posterior spinal arteries. The anterior spinal artery gives rise to numerous sulcal branches that supply the anterior two-thirds of the spinal cord in alternating fashion (one side and then the other) along the length of the cord. The posterior spinal arteries supply much of the dorsal horn and the dorsal columns. A network of vessels known as the vasocorona connects these two sources of supply and sends branches into the white matter around the margin of the spinal cord.

areas that govern the upper extremities and face, and the language areas of the left hemisphere (Broca's area and Wernicke's area; see Chapter 33). The anterior cerebral artery supplies regions in the medial aspect and dorsal and orbital margins of the frontal lobe, and the medial aspect and dorsal margin of the anterior parietal lobe (yellow area in Figure A19B). Included in this extended territory are sensorimotor areas in the paracentral lobule that govern the lower extremity, accessory motor areas in the cingulate gyrus that govern the upper face, and limbic areas in the medial frontal lobe. The posterior cerebral artery supplies regions in the posterior parietal lobe, inferior temporal

TABLE A4 ■ Organization of Blood Supply to Brain and Spinal Cord

Circulation	Supply	Cerebrospinal artery
Anterior	Internal carotid arteries	Anterior cerebral arteries
		Middle cerebral arteries
		Anterior choroidal arteries
		Posterior communicating arteries
		Posterior cerebral arteries
Posterior	Vertebral/basilar arteries	Posterior cerebral arteries
		Superior cerebellar arteries
		Anterior inferior cerebellar arteries
		Posterior inferior cerebellar arteries
		Anterior spinal artery (upper portion)
		Posterior spinal arteries (upper portion)

lobe, and occipital lobe (lavender in Figure A19B). Included in this region are primary and associational (higher-order) visual areas in each lobe and limbic regions in the posterior cingulate and parahippocampal gyri.

As Figure A20 illustrates, an anterior-to-posterior pattern of deeply penetrating branches arises from the anterior circulation. The anterior cerebral artery supplies the anterior caudate and putamen and the anterior limb of the internal capsule. The middle cerebral artery supplies the body of the caudate and most of the putamen, most of the globus pallidus, the middle part (or genu) of the internal capsule, and the anterior hypothalamus. These deep-penetrating branches of the middle cerebral artery are usually called the **lenticulostriate arteries** (see Figure A19D). The **anterior choroidal artery**, which arises from the middle cerebral artery just distal to the circle of Willis, supplies the amygdala, hippocampus, anterior part of the thalamus, part of the globus pallidus, posterior limb of the internal capsule, and choroid plexus of the lateral ventricle. The posterior communicating and posterior cerebral arteries supply the posterior hypothalamus, most of the thalamus, and the choroid plexus of the third ventricle. (Branches of the posterior cerebral artery also supply the midbrain, as described below.) Thus, the deep structures of the forebrain are divided approximately into four sectors progressing from anterior to posterior, and each sector is perfused by a different artery.

Blood from the anterior circulation makes its passage through the brain from the arterial vasculature back to the heart via the internal **jugular veins** through a series of venous sinuses (see Figure A17B). The major venous sinuses inside the cranium are formed by a separation of the two layers of dura mater, the tough outer component of the meninges that surrounds the brain (see below). Thus, the more superficial veins of the cerebrum drain into the **superior sagittal sinus** along the dorsal midline of the hemisphere, or the **cavernous sinus** in the base of the cranium. The superior sagittal sinus and the deeper sinuses drain into the **confluence of sinuses** at the posterior end of the longitudinal fissure, before giving rise to the **transverse sinuses**, which are oriented roughly in the horizontal plane. Venous blood then passes a short distance in the anterior direction before the transverse sinuses turn in the inferior direction, curving into a sigmoid shape. Finally, venous blood exits the cranial vault as the **sigmoid sinuses** pass through the skull bass and join the internal jugular veins.

The posterior circulation of the brain supplies the posterior cerebral cortex, thalamus, and brainstem. The pattern of arterial distribution is similar for all the subdivisions of the brainstem: midline arteries supply medial structures, lateral arteries supply the lateral brainstem, and dorsolateral arteries supply dorsolateral brainstem structures and the cerebellum (Figure A21). Among the most important dorsolateral arteries (also called **long circumferential arteries**) are the **posterior inferior cerebellar artery** (**PICA**) and the **anterior inferior cerebellar artery** (**AICA**), which supply distinct regions of the medulla and pons along their way to the cerebellum. These arteries, as well as branches of the posterior cerebral and basilar arteries that penetrate the brainstem from its ventral and lateral surfaces—the **paramedian** and **short circumferential arteries**—are especially common sites of occlusion and result in specific functional deficits of cranial nerve, somatosensory, and motor function. Most vascular lesions of the brainstem are

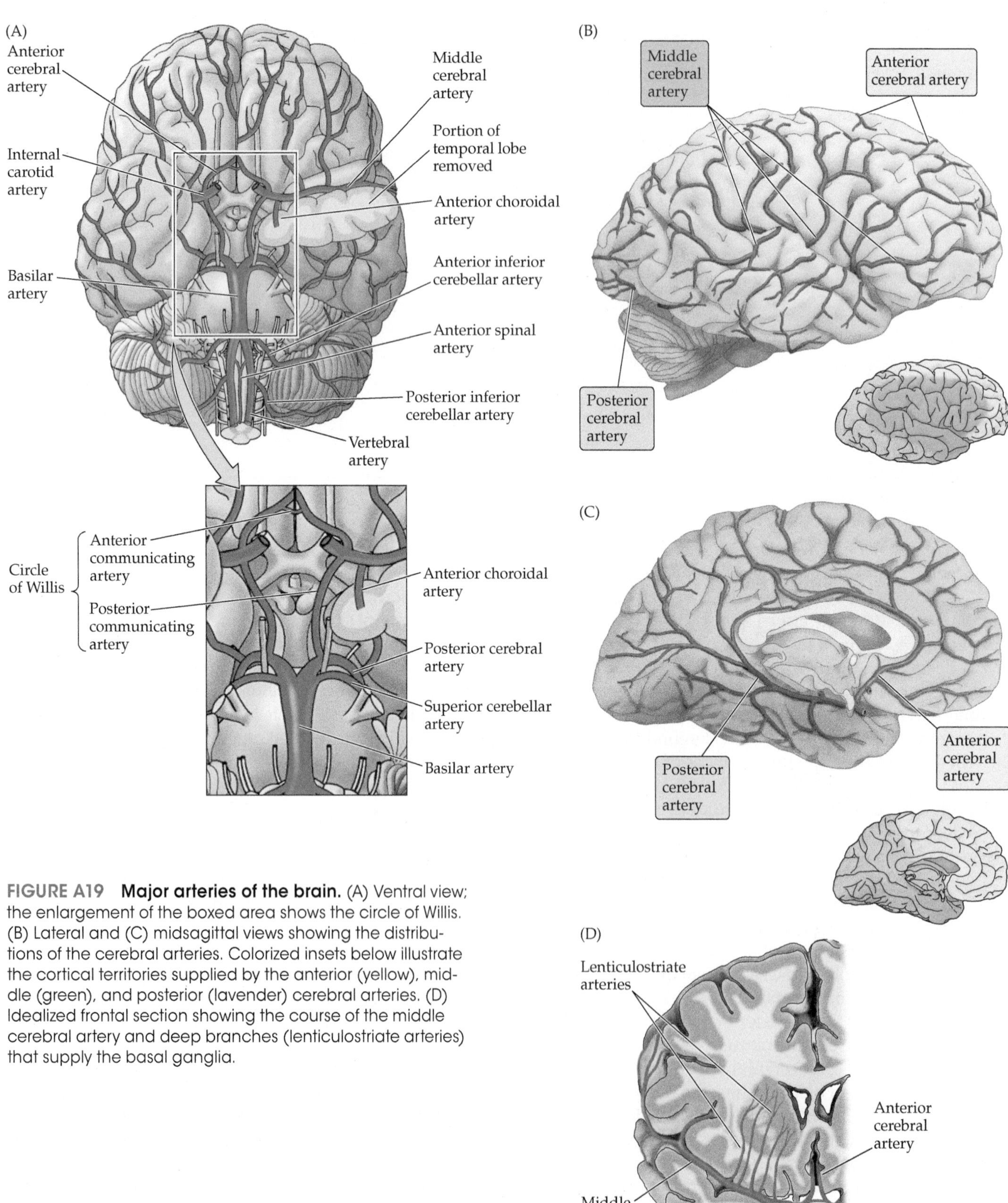

FIGURE A19 Major arteries of the brain. (A) Ventral view; the enlargement of the boxed area shows the circle of Willis. (B) Lateral and (C) midsagittal views showing the distributions of the cerebral arteries. Colorized insets below illustrate the cortical territories supplied by the anterior (yellow), middle (green), and posterior (lavender) cerebral arteries. (D) Idealized frontal section showing the course of the middle cerebral artery and deep branches (lenticulostriate arteries) that supply the basal ganglia.

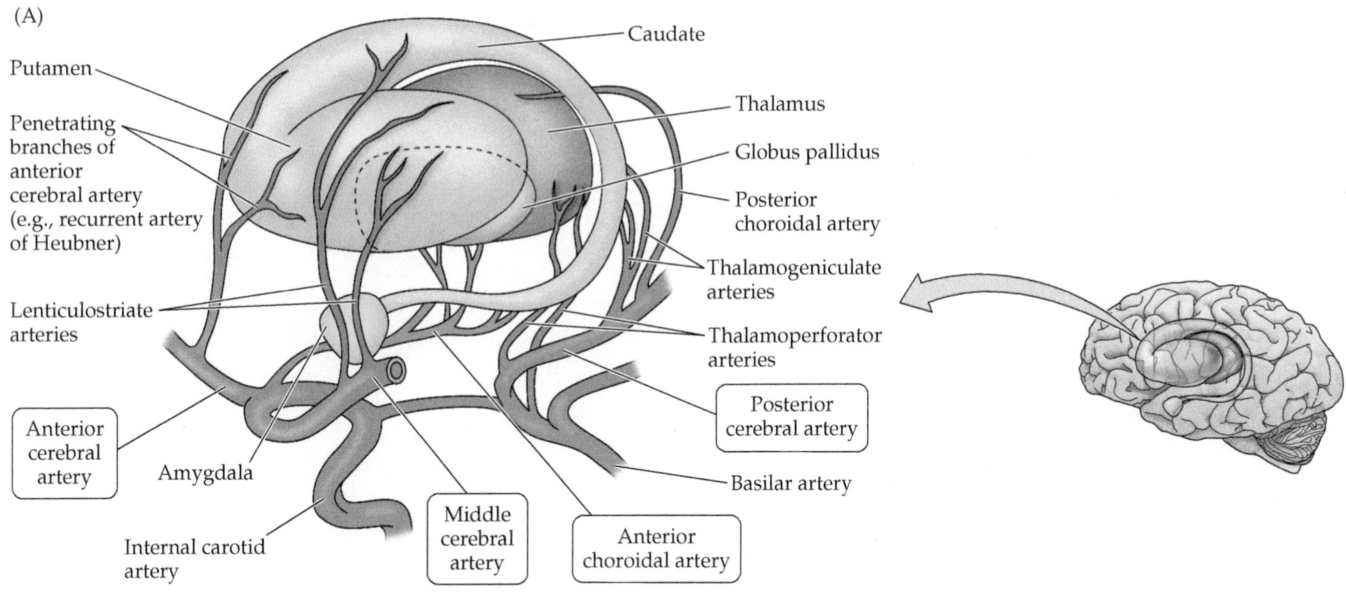

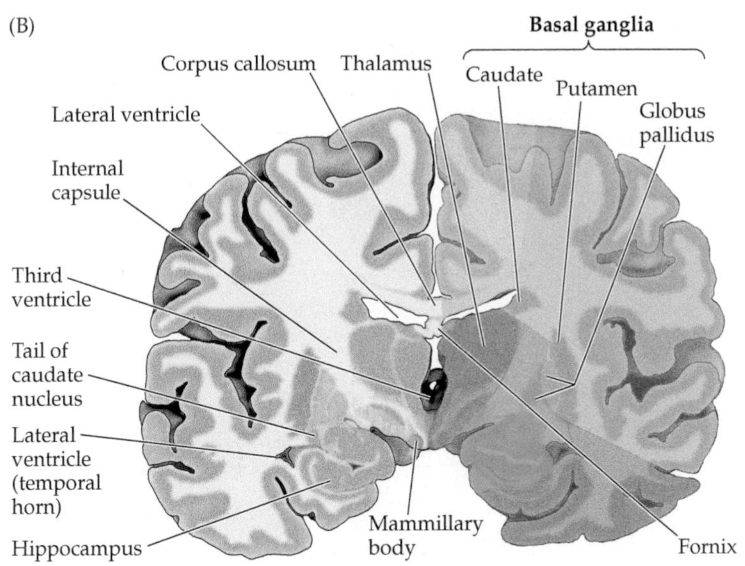

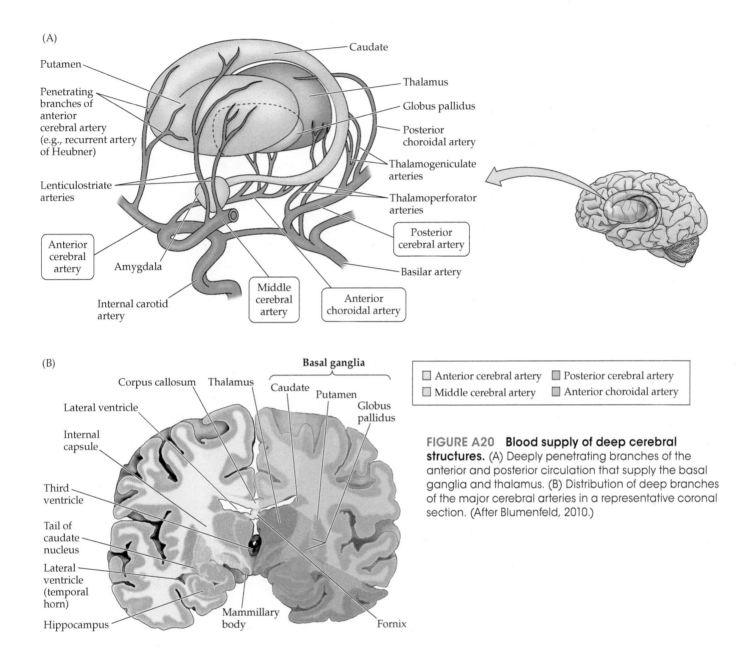

FIGURE A20 Blood supply of deep cerebral structures. (A) Deeply penetrating branches of the anterior and posterior circulation that supply the basal ganglia and thalamus. (B) Distribution of deep branches of the major cerebral arteries in a representative coronal section. (After Blumenfeld, 2010.)

unilateral, since each side of the brainstem is supplied by different sets of circumferential vessels. However, this may not be the case if the basilar artery itself is blocked, since it gives rise to vessels that supply both sides.

Blood is supplied to the spinal cord by the vertebral arteries and the six to ten **medullary arteries** that arise from segmental branches of the aorta (see Figure A18). These medullary arteries join to form a single **anterior spinal artery** and a pair of **posterior spinal arteries**. An anastomotic network of vessels known as the **vasocorona** connects these two sources of supply and sends branches into a narrow zone of white matter around the margin of the spinal cord. The vasocorona may be sufficient to supply the most lateral white matter in cases in which the anterior spinal artery is occluded. Nevertheless, if any of the medullary arteries are obstructed or damaged (during abdominal surgery, for example), the blood supply to specific parts of the spinal cord may be compromised. The pattern of resulting neurological damage differs according to whether supply to a posterior spinal artery or the anterior spinal artery is interrupted. As might be expected from the arrangement of ascending and descending neural pathways in the spinal cord described above, loss of the posterior supply generally leads to loss of sensory functions, whereas loss of the anterior supply more often causes motor deficits.

The physiological demands on the brain's blood supply are particularly significant because neurons are more

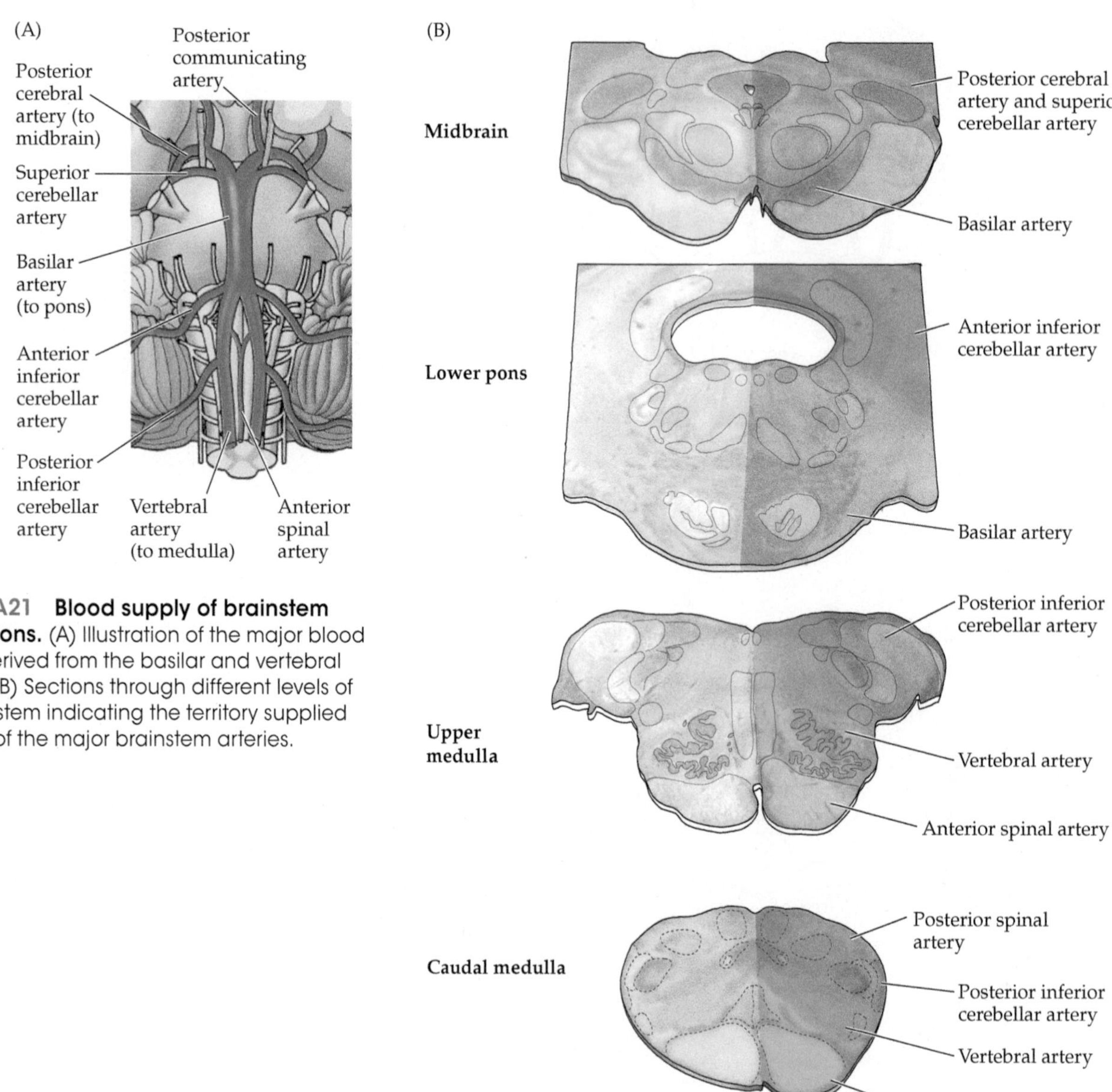

FIGURE A21 **Blood supply of brainstem subdivisions.** (A) Illustration of the major blood supply derived from the basilar and vertebral arteries. (B) Sections through different levels of the brainstem indicating the territory supplied by each of the major brainstem arteries.

sensitive to oxygen and glucose deprivation than are cells with lower rates of metabolism. The high metabolic rate of neurons means that brain tissue deprived of oxygen and glucose as a result of compromised blood supply is likely to sustain transient or permanent damage. Even brief loss of blood supply (referred to as ischemia) can cause cellular changes that, if not quickly reversed, can lead to cell death through the mechanisms of excitotoxicity). Sustained loss of blood supply leads much more directly to death and degeneration of the deprived cells. Stroke—an anachronistic term that refers to the death or dysfunction of brain tissue due to vascular disease—often follows the occlusion of (or hemorrhage from) the brain's arteries. Historically, studies of the functional consequences of strokes, and their relation to vascular territories in the brain and spinal cord, provided information about the location of various brain functions. The location of the major language functions in the left hemisphere, for instance, was discovered in this way in the latter part of the nineteenth century (see Chapter 33). Now, noninvasive functional imaging techniques based on blood flow have largely supplanted the correlation of clinical signs and symptoms with the location of tissue damage observed at autopsy (see Chapter 1).

The Blood–Brain Barrier

In addition to their susceptibility to oxygen and glucose deprivation, brain cells are at risk from toxins circulating in the bloodstream. The brain is specifically protected in this respect, however, by the **blood–brain barrier**. The interface between the walls of capillaries and the surrounding tissue is important throughout the body, as it keeps vascular and extravascular concentrations of ions and molecules at appropriate levels in these two compartments. In the brain, this interface is especially significant—hence its unique and alliterative name. The special properties of the blood–brain barrier were first observed by the nineteenth-century bacteriologist Paul Ehrlich, who noted that intravenously injected dyes leaked out of capillaries in most regions of the body to stain the surrounding tissues; brain tissue, however, remained unstained. Ehrlich wrongly concluded that the brain had a low affinity for the dyes. It was his student, Edwin Goldmann, who showed that in fact such dyes do not traverse the specialized walls of brain capillaries.

The restriction of large molecules such as Ehrlich's dyes (and many smaller molecules) to the vascular space is the result of tight junctions between neighboring capillary endothelial cells in the brain (Figure A22). Such junctions are not found in capillaries elsewhere in the body, where the spaces between adjacent endothelial cells allow much more ionic and molecular traffic. The structure of tight junctions was first demonstrated in the 1960s by Tom Reese, Morris Karnovsky, and Milton Brightman. Using electron microscopy after the injection of electron-dense intravascular agents such as lanthanum salts, they showed that the close apposition of the endothelial cell membranes prevented such ions from passing (see Figure A22B). Substances that traverse the walls of brain capillaries must move *through* the endothelial cell membranes. Accordingly, molecular entry into the brain should be determined by an agent's solubility in lipids, the major constituent of cell membranes. Nevertheless, many ions and molecules not readily soluble in lipids *do* move quite readily from the vascular space into brain tissue. A molecule such as glucose, the primary source of metabolic energy for neurons and glial cells, is an obvious example. This paradox is explained by the presence of specific transporters in the endothelial plasma membrane for glucose and other critical molecules and ions.

In addition to tight junctions, astrocytic *end feet* (the terminal regions of astrocytic processes) surround the outside of capillary endothelial cells (see Figure A22A). The reason for this endothelial–glial allegiance is unclear, but may reflect an influence of astrocytes on the formation and maintenance of the blood–brain barrier and/or the passage of cerebrospinal fluid from perivascular space through aqueous channels in the astrocytic end feet (see below).

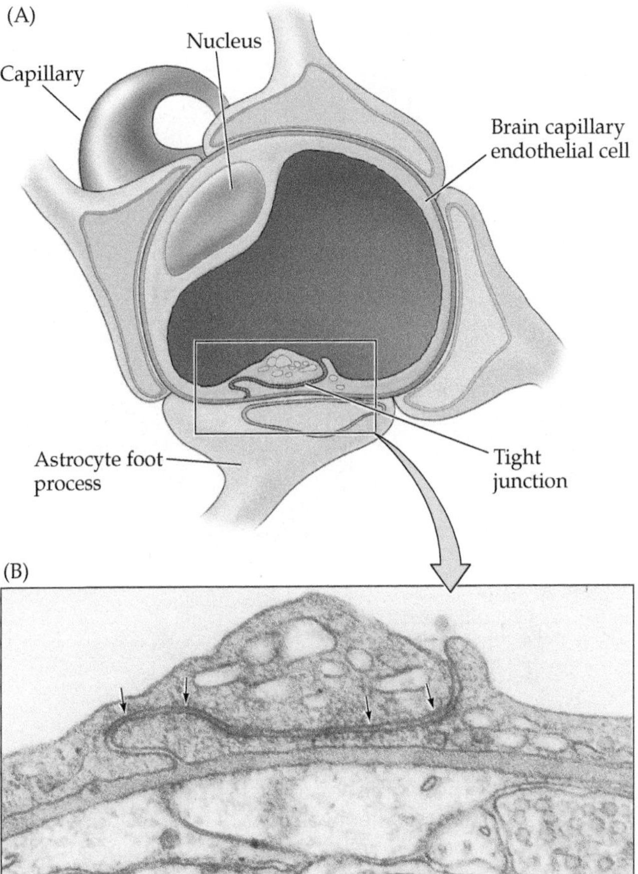

FIGURE A22 **Cellular basis of the blood–brain barrier.** (A) Diagram of a brain capillary in cross section and reconstructed views, showing endothelial tight junctions and the investment of the capillary by astrocytic end feet. (B) Electron micrograph of boxed area in (A), showing the appearance of tight junctions between neighboring endothelial cells (arrows). (A after Goldstein and Betz, 1986; B from Peters et al., 1991.)

The brain, more than any other organ, must be carefully shielded from abnormal variations in its ionic milieu, as well as from the potentially toxic molecules that find their way into the vascular space by ingestion, infection, or other means. The blood–brain barrier is thus crucial for protection and homeostasis. It also presents a significant problem for the delivery of drugs to the brain. Large (or lipid-insoluble) molecules can be introduced to the brain only by transiently disrupting the blood–brain barrier with hyperosmotic agents such as the sugar mannitol.

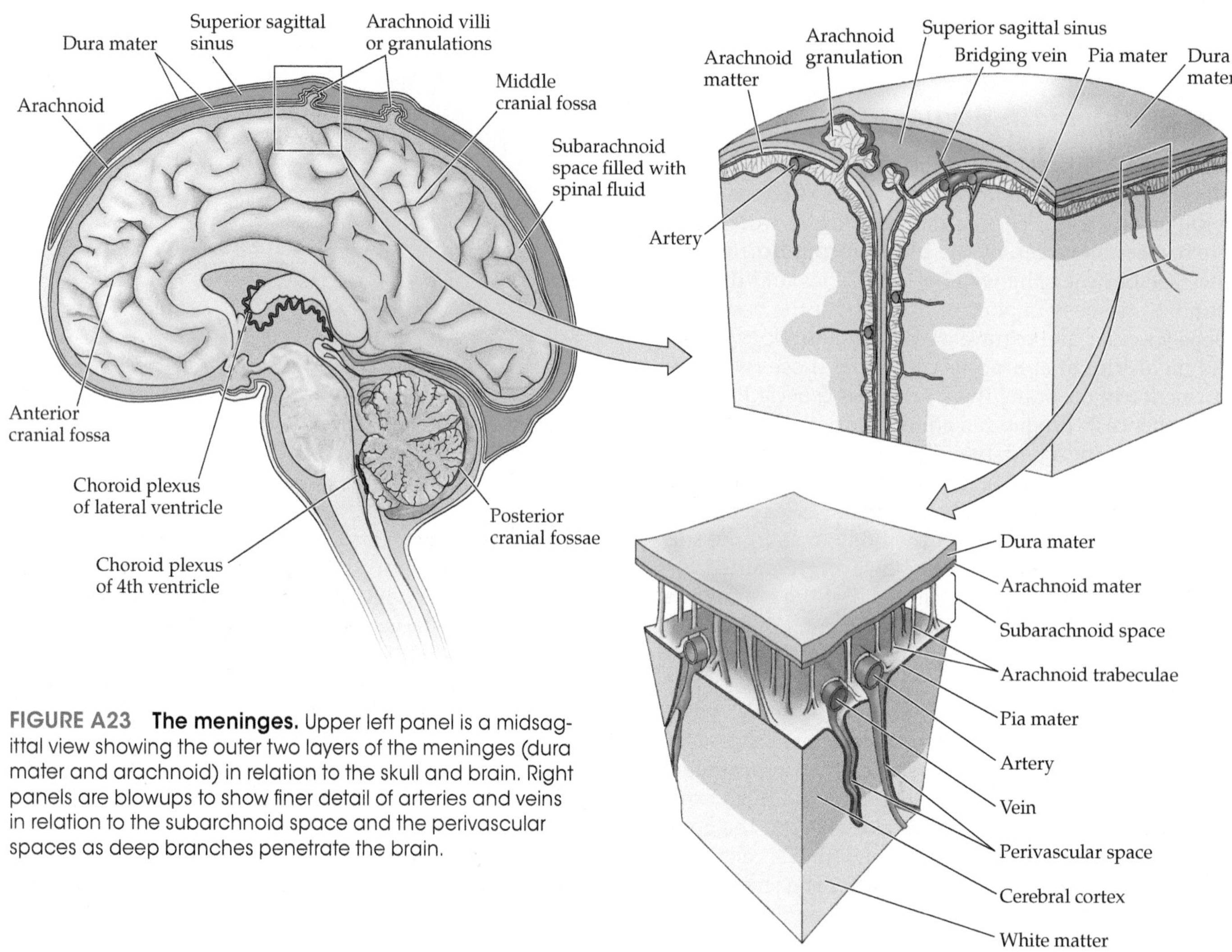

FIGURE A23 The meninges. Upper left panel is a midsagittal view showing the outer two layers of the meninges (dura mater and arachnoid) in relation to the skull and brain. Right panels are blowups to show finer detail of arteries and veins in relation to the subarchnoid space and the perivascular spaces as deep branches penetrate the brain.

The Meninges

The cranial cavity is conventionally divided into three regions called the anterior, middle, and posterior cranial fossae. Surrounding and supporting the brain within this cavity are three protective tissue layers, which also extend down the brainstem and the spinal cord. Together these layers are called the **meninges** (Figure A23). The outermost layer of the meninges is called the **dura mater** ("hard mother," referring to its thick and tough qualities). The middle layer is called the **arachnoid mater** because of spiderweb-like processes called arachnoid trabeculae, which extend from it toward the third layer, the **pia mater** ("tender mother"), a delicate layer of cells that envelopes subarachnoid vessels and apposes the basement membrane on the outer glial surface of the brain. Because the pia closely adheres to the brain as its surface curves and folds whereas the arachnoid does not, there are places, called **cisterns**, where the subarachnoid space enlarges to form significant collections of cerebrospinal fluid (the fluid that fills the ventricles; see the next section). Since the major arteries supplying the brain course through the subarachnoid space on the surface of the cerebrum, this space is a frequent site of bleeding following trauma. A collection of blood between the meningeal layers is referred to as a subdural or subarachnoid hemorrhage (or hematoma), as distinct from bleeding within the brain itself.

The Ventricular System

The cerebral ventricles are a series of interconnected, fluid-filled spaces that lie in the core of the forebrain and brainstem (Figures A24 and A25). These spaces are filled with cerebrospinal fluid (CSF) produced by a modified vascular structure called the **choroid plexus**, which is present in each ventricle. CSF percolates through the ventricular system and flows into the subarachnoid space through

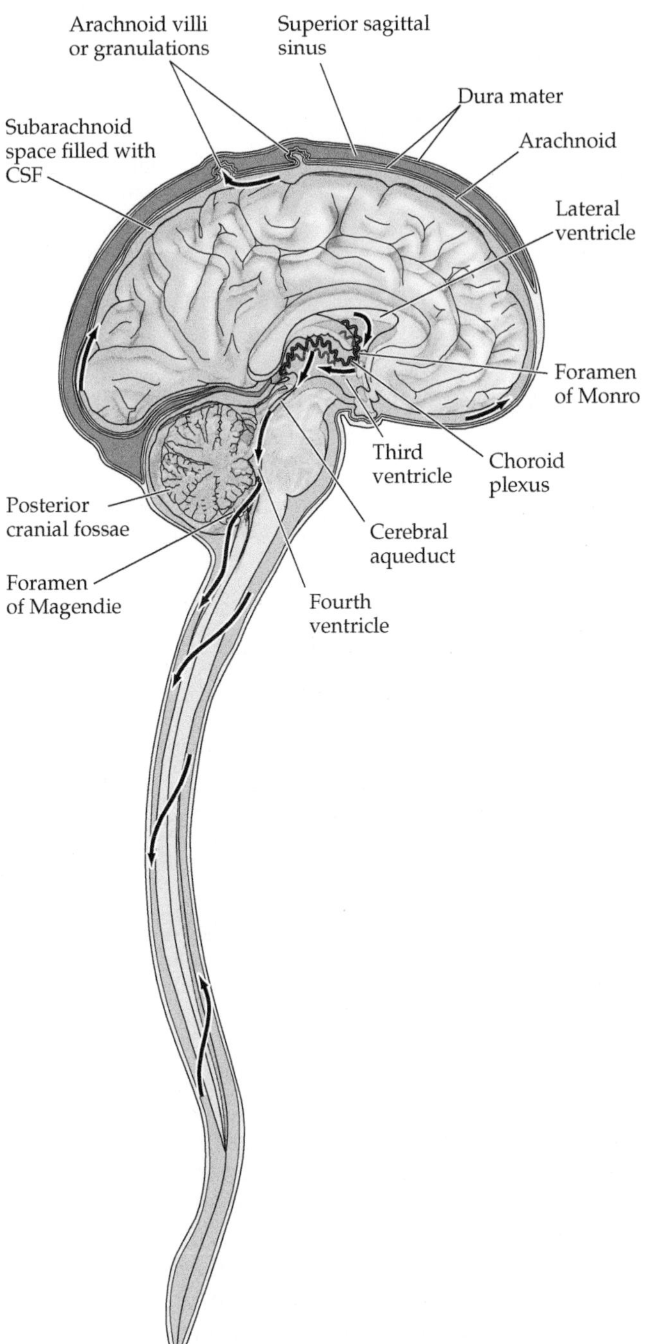

FIGURE A24 Circulation of cerebrospinal fluid. Cerebrospinal fluid (CSF) is produced by the choroid plexus and flows from the lateral ventricles through the paired interventricular foramina (singular: foramen; foramina of Monro) into the third ventricle, through the cerebral aqueduct, and into the fourth ventricle. CSF exits the ventricular system through several foramina associated with the fourth ventricle (e.g., foramen of Magendie along the midline) into the subarachnoid space surrounding the CNS. CSF is eventually passed through the arachnoid granulations and returned to the venous circulation in the superior sagittal sinus.

perforations in the thin covering of the fourth ventricle (midline foramen of Magendie and two lateral foramina of Luschka; see Figure A24); it is eventually passed through specialized structures called **arachnoid villi** or **granulations** along the dorsal midline of the forebrain (see Figure A23) and returned to the venous circulation via the superior sagittal sinus.

The presence of ventricular spaces in the various subdivisions of the brain reflects the fact that the ventricles are the adult derivatives of the open space, or lumen, of the embryonic neural tube (see Chapter 22). Although they have no unique function, the ventricular spaces present in sections through the brain provide another useful guide to location (see Figure A2). The largest of these spaces are the **lateral ventricles** (formerly called the first and second ventricles), one within each of the cerebral hemispheres. These particular ventricles are best seen in frontal sections, where their ventral and lateral surfaces are usually defined by the basal ganglia, their dorsal surface by the corpus callosum, and their medial surface by the **septum pellucidum**, a membranous tissue sheet that forms part of the midline sagittal surface of the cerebral hemispheres. The lateral ventricles, like several telencephalic structures, possess a C shape. This pattern results from the non-uniform growth of the cerebral hemispheres and the formation of the temporal lobes during embryonic development. CSF flows from the lateral ventricles through small openings (called the **interventricular foramina**, or the **foramina of Monro**) into a narrow midline space between the right and left diencephalon, the **third ventricle**. The third ventricle is continuous caudally with the **cerebral aqueduct** (also referred to as the **aqueduct of Sylvius**), which runs though the midbrain. At its caudal end, the aqueduct opens into the **fourth ventricle**, a larger space dorsal to the pons and medulla. The fourth ventricle, covered on its dorsal aspect by the cerebellum, narrows caudally to form the central canal of the spinal cord, which normally does not remain patent beyond the early postnatal period.

Recent studies have demonstrated that, in addition to the bulk flow of CSF through the ventricular system, the subarachnoid space, and into the superior sagittal sinus, CSF also passes through the interstitial spaces of brain tissue itself (i.e., brain parenchyma). Maiken Nedergaard and Steven Goldman and their colleagues at the University of Rochester and the University of Copenhagen used chemical dyes and advanced in vivo microscopy to observe the passage of CSF through the parenchyma. Some quantity of CSF enters the perivascular space that surrounds the arterial branches penetrating deep into the brain from the subarachnoid compartment. Propelled by the pumping of arterial blood, this CSF moves into brain tissue by passing through water channels comprising aquaporin-4 proteins in the astrocytic end feet. As the CSF passes through the parenchyma and mixes with extracellular fluid in the

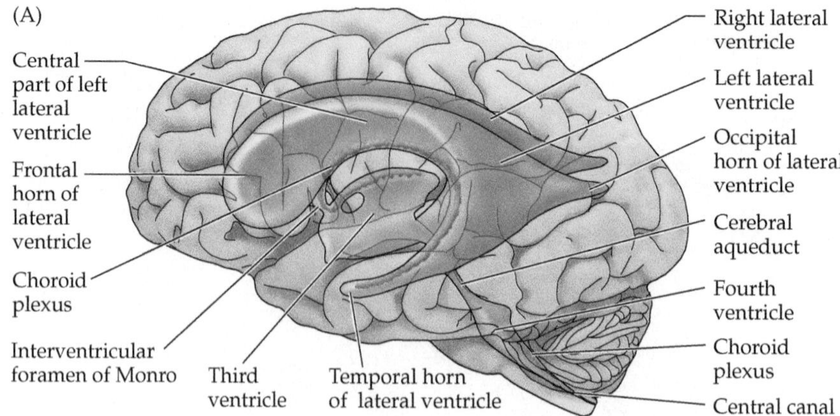

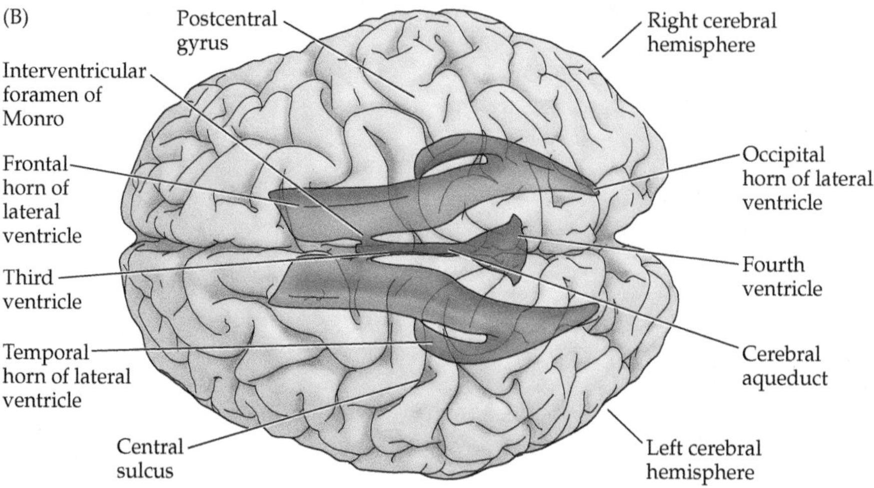

FIGURE A25 Ventricular system of the human brain. (A) Location of the ventricles as seen in a transparent left lateral view. (B) Dorsal view of the ventricles.

interstitial spaces, metabolic waste and discarded proteins are carried away (Figure A26). This fluid eventually passes into the perivascular spaces surrounding small veins and flows back into the subarachnoid space or into newly discovered lymphatic vessels that course along the superior sagittal sinus. It is estimated that this system, termed the brain's **glymphatic system** due to the participation of glial cells in a lymphatic-like system, is responsible for removing nearly the brain's own weight in waste material over the course of a year.

Not surprisingly, the discovery of this glymphatic system has led to keen interest in its role in brain health and neurological disease. One intriguing observation is that the rate of glymphatic flow increases during sleep (see Chapter 28), when the brain's interstitial spaces are thought to expand by some 50% or so. This expansion helps create convective flow of interstitial fluids through the parenchyma and a significant increase in the efficiency of waste removal. The finding of beta-amyloid and synuclein proteins (proteins implicated in Alzheimer's disease and Parkinson's disease, respectively) in fluids flowing through the glymphatic system suggests that this system may serve to remove potentially toxic substances from the brain. Furthermore, it raises the possibility that disruption of this cleansing function might contribute to the onset or progression of neurological disease. Perhaps this mechanism of circadian waste removal is responsible for the association of poor sleep in middle age and an increased risk of cognitive decline in later years. It may also help rationalize what would seem to be an excessively high rate of daily CSF production: The normal total volume of CSF in the ventricular system and subarachnoid space is approximately 150 mL, while the choroid plexus produces approximately 500 mL of CSF per day. Thus, the entire volume of CSF present in the ventricular system is turned over several times a day. However, this high rate of CSF production and clearance poses a risk if there is a blockage of CSF flow through the ventricular spaces or the arachnoid granulations. Obstruction results in an excess of CSF in the intracranial cavity, a dangerous condition called **hydrocephalus** (literally, "water head") that can lead to enlargement of the ventricles and compression of the brain.

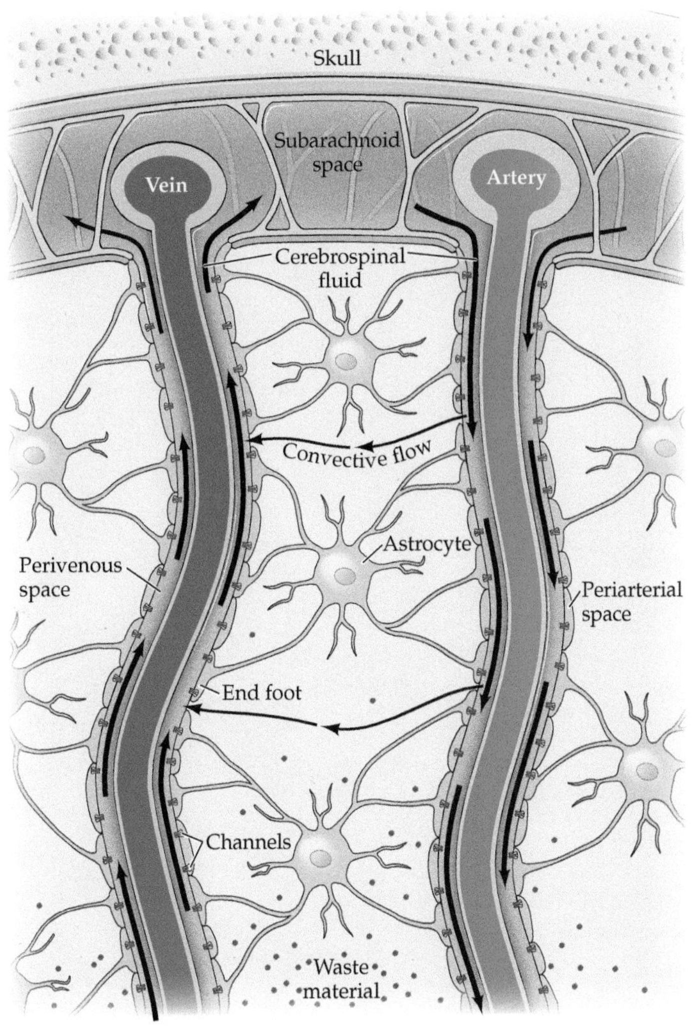

FIGURE A26 Glymphatic system of the brain. Cerebrospinal fluid (CSF) passes from arterial perivascular space through the substance of the brain. As it does so, metabolic wastes and discarded proteins are rinsed from the parenchyma and pass out of the brain via the perivascular spaces surrounding veins. This convective flow of CSF and interstitial fluid increases during sleep, when extracellular spaces expand. (After Nedergaard and Goldman, 2016.)

REFERENCES

Blumenfeld, H. (2010) *Neuroanatomy through Clinical Cases*, 2nd Edition. Sunderland, MA: Sinauer Associates.

Brightman, M. W. and T. S. Reese (1969) Junctions between intimately opposed cell membranes in the vertebrate brain. *J. Cell Biol.* 40: 648–677.

Brodal, P. (2010) *The Central Nervous System: Structure and Function*, 4th Edition. New York: Oxford University Press.

England, M. A. and J. Wakely (1991) *Color Atlas of the Brain and Spinal Cord: An Introduction to Normal Neuroanatomy.* St. Louis, MO: Mosby Yearbook.

Goldstein, G. W. and A. L. Betz (1986) The blood–brain barrier. *Sci. Am.* 255: 74–83.

Haines, D. E. (2007) *Neuroanatomy: An Atlas of Structures, Sections, and Systems*, 7th Edition. Baltimore: Lippincott Williams & Wilkins.

Mai, J. K. and G. Paxinos (2012) *The Human Nervous System*, 3rd Edition. New York: Elsevier.

Martin, J. H. (2012) *Neuroanatomy: Text and Atlas*, 4th Edition. New York: McGraw-Hill Medical.

Nedergaard, M. and S. A. Goldman (2016) Brain drain. *Sci. Am.* 314: 44–49.

Netter, F. H. (1983) *The CIBA Collection of Medical Illustrations*, Vols. I and II. West Caldwell, NJ: CIBA Pharmaceutical Co.

Parent, A. and M. B. Carpenter (1996) *Carpenter's Human Neuroanatomy*, 9th Edition. Baltimore: Williams & Wilkins.

Peters, A., S. L. Palay and H. deF. Webster (1991) *The Fine Structure of the Nervous System: Neurons and Their Supporting Cells*, 3rd Edition. Oxford University Press, New York.

Reese, T. S. and M. J. Karnovsky (1967) Fine structural localization of a blood–brain barrier to exogenous peroxidase. *J. Cell Biol.* 34: 207–217.

Rexed, B. (1952) The cytoarchitectonic organization of the spinal cord of the cat. *J. Comp. Neurol.* 96: 414–495.

Schmidley, J. W. and E. F. Maas (1990) Cerebrospinal fluid, blood–brain barrier and brain edema. In *Neurobiology of Disease*, A. L. Pearlman and R. C. Collins (eds.). New York: Oxford University Press, pp. 380–398.

Tarasoff-Conway, J. M. and 15 others (2015) Clearance systems in the brain—implications for Alzheimer disease. *Nat. Rev. Neurol.* 11: 457–470.

ATLAS

The Human Central Nervous System

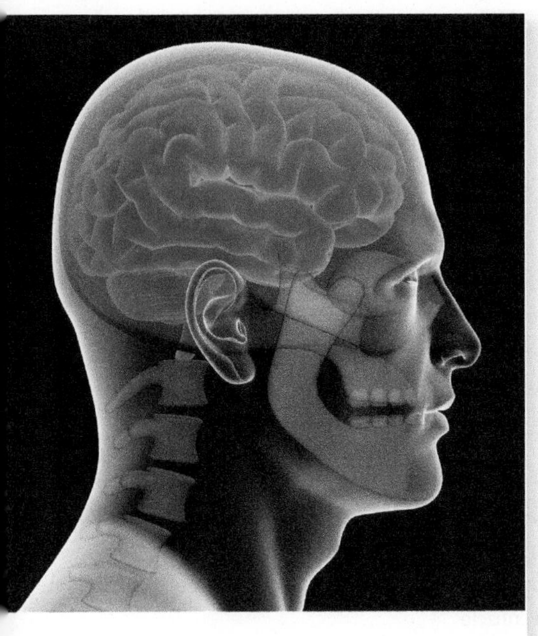

This series of seven plates presents labeled images of the human brain and spinal cord. The surface features of the brain are shown in photographs of a postmortem specimen after removal of the meninges and superficial blood vessels (Plate 1). Sectional views of the forebrain in each of three standard anatomical planes (see Figure A1) are derived from T1-weighted magnetic resonance imaging of a living subject (Plates 2–4). In these images, compartments filled with aqueous fluids, such as the ventricles, appear dark; tissues that are enriched with lipids, such as white matter, appear bright; and tissues that are relatively poor in lipid (myelin) and high in water content, such as gray matter, appear in intermediate shades of gray. Thus, the appearance of gray matter and white matter in the T1-weighted series is similar to what would be observed when dissecting a brain specimen obtained postmortem. Plate 5 presents reconstructions of fiber tracts in the white matter of the human brain in a living subject obtained by means of diffusion tensor imaging (see Chapter 1 and Box 23B). The final images are transverse sections obtained from the major subdivisions of the brainstem (Plate 6) and spinal cord (Plate 7). Each of these histological images was acquired and processed to simulate myelin staining; thus, white matter appears dark, while gray matter and poorly myelinated fibers appear light. Note the small insets that show the actual, typical size of cross sections through the human brainstem and spinal cord.

PLATE 1 PHOTOGRAPHIC ATLAS: BRAIN SURFACE

(A)

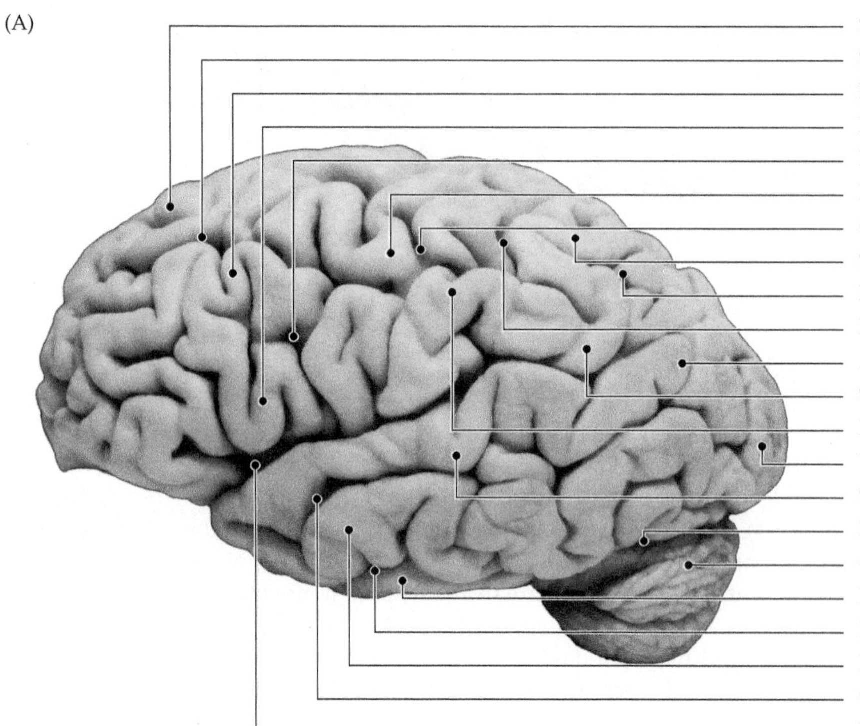
Superior frontal gyrus
Superior frontal sulcus
Middle frontal gyrus
Inferior frontal gyrus
Inferior frontal sulcus
Precentral gyrus
Central sulcus
Superior parietal lobule
Intraparietal sulcus
Postcentral sulcus
Angular gyrus
Supramarginal gyrus
Postcentral gyrus
Lateral occipital gyri
Superior temporal gyrus
Preoccipital notch
Cerebellar hemisphere
Inferior temporal gyrus
Inferior temporal sulcus
Middle temporal gyrus
Superior temporal sulcus
Lateral (Sylvian) fissure

(B)

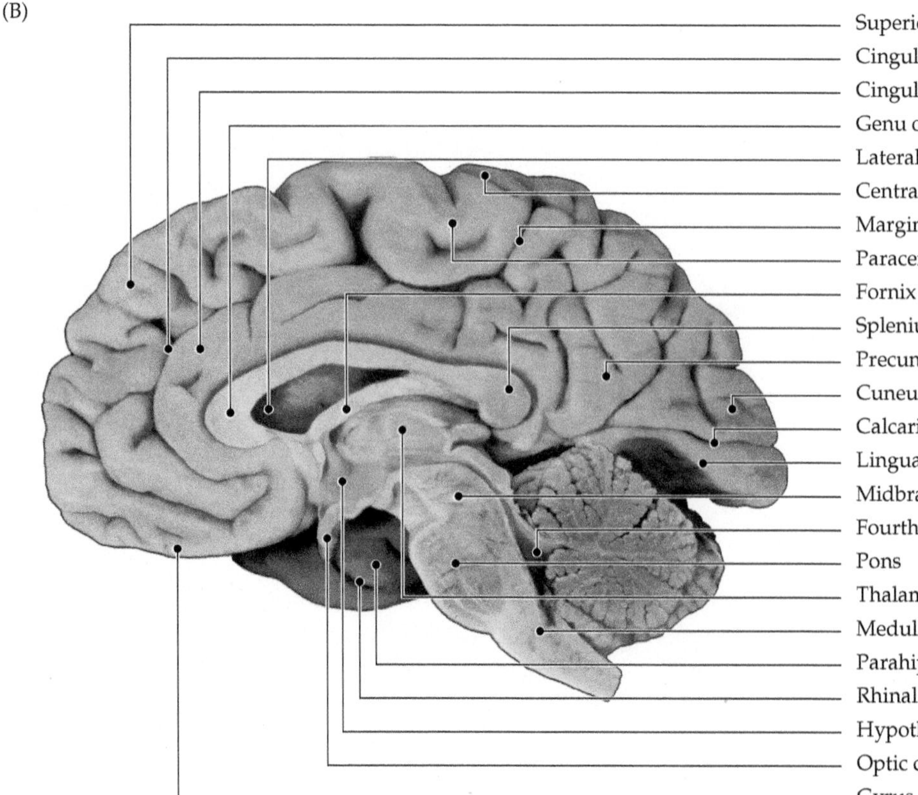
Superior frontal gyrus
Cingulate sulcus
Cingulate gyrus
Genu of corpus callosum
Lateral ventricle
Central sulcus
Marginal branch of cingulate sulcus
Paracentral lobule
Fornix
Splenium of corpus callosum
Precuneus gyrus
Cuneus gyrus
Calcarine sulcus
Lingual gyrus
Midbrain
Fourth ventricle
Pons
Thalamus
Medulla oblongata
Parahippocampal gyrus
Rhinal sulcus
Hypothalamus
Optic chiasm
Gyrus rectus

Surface features of a human brain specimen. (A) Lateral view of the left hemisphere. (B) Midsagittal view of right hemisphere. (C) Dorsal view. (D) Ventral view.

(C)

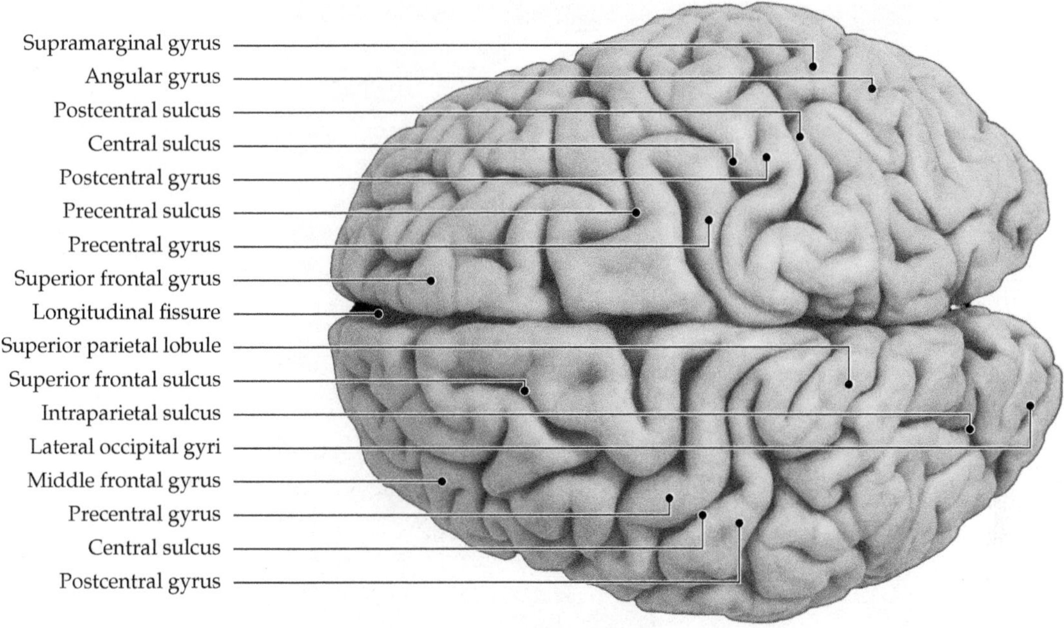

(D)

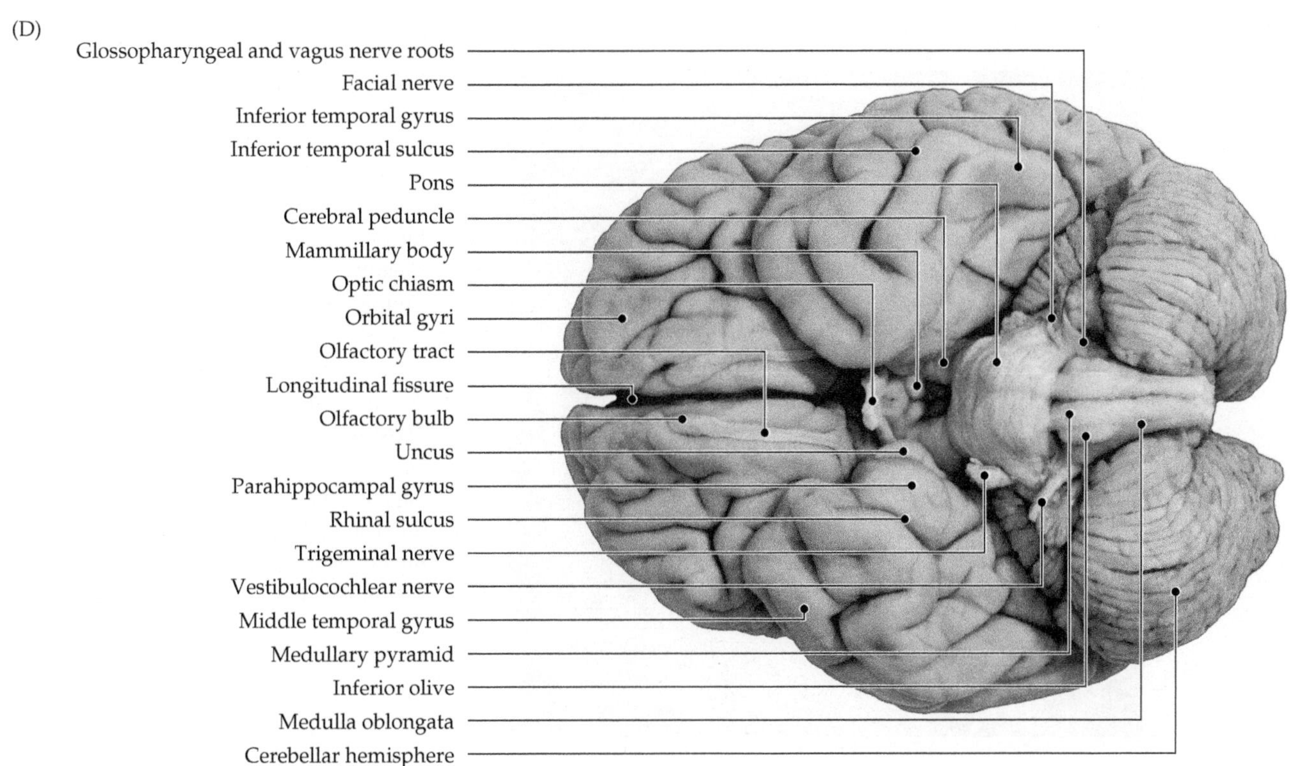

PLATE 2 CORONAL MR ATLAS

(A)

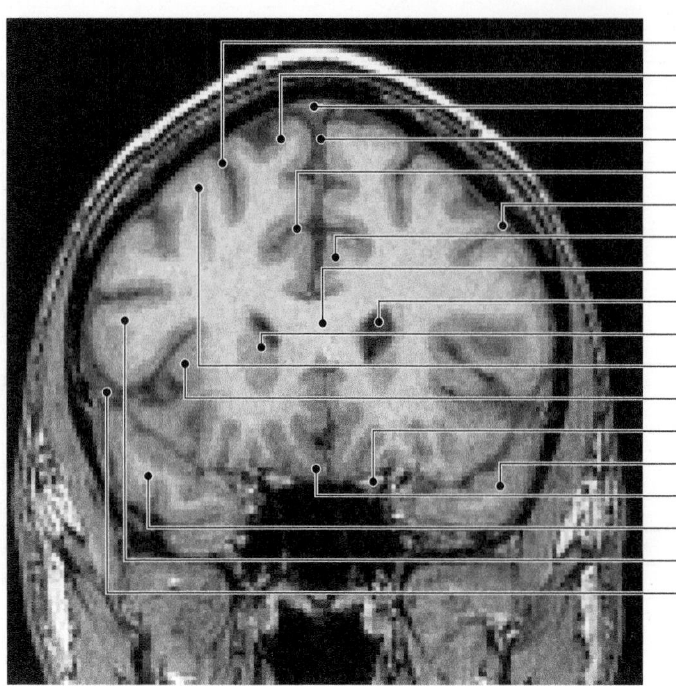
Superior frontal sulcus
Superior frontal gyrus
Superior sagittal sinus
Longitudinal fissure
Cingulate sulcus
Inferior frontal sulcus
Cingulate gyrus
Corpus callosum, genu
Lateral ventricle, anterior horn
Caudate
Middle frontal gyrus
Insular gyri
Optic nerve
Temporal pole
Gyrus rectus
Middle temporal gyrus
Inferior frontal gyrus
Lateral (sylvian) fissure

(B)

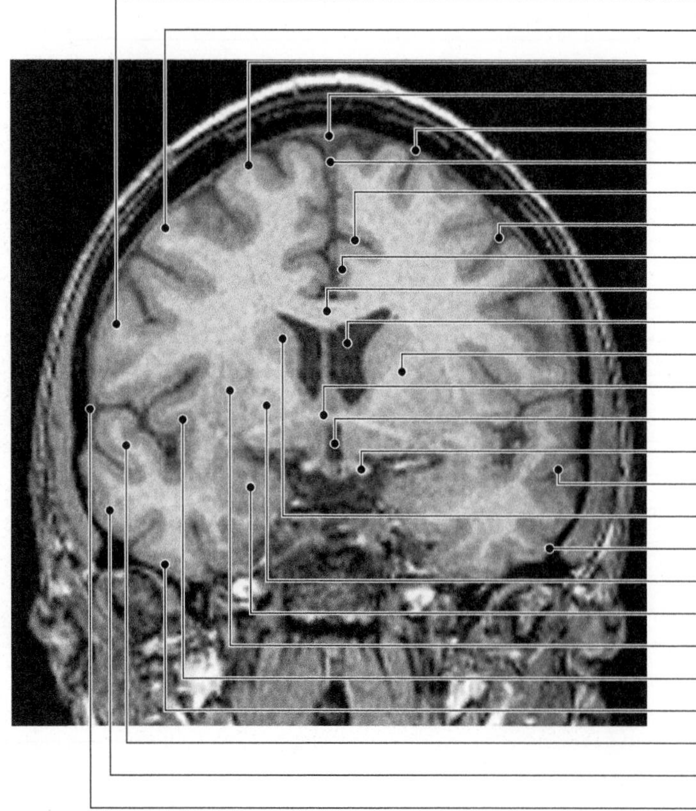
Inferior frontal gyrus
Middle frontal gyrus
Superior frontal gyrus
Superior sagittal sinus
Superior frontal sulcus
Longitudinal fissure
Cingulate sulcus
Inferior frontal sulcus
Cingulate gyrus
Corpus callosum, body
Lateral ventricle
Internal capsule
Anterior commissure
Third ventricle
Optic tract
Superior temporal sulcus
Caudate
Inferior temporal sulcus
Globus pallidus
Amygdala
Putamen
Insular gyri
Inferior temporal gyrus
Superior temporal gyrus
Middle temporal gyrus
Lateral (sylvian) fissure

Coronal sections of the human brain demonstrating internal forebrain structures in magnetic resonance images; images in (A)–(D) are arranged from rostral to caudal.

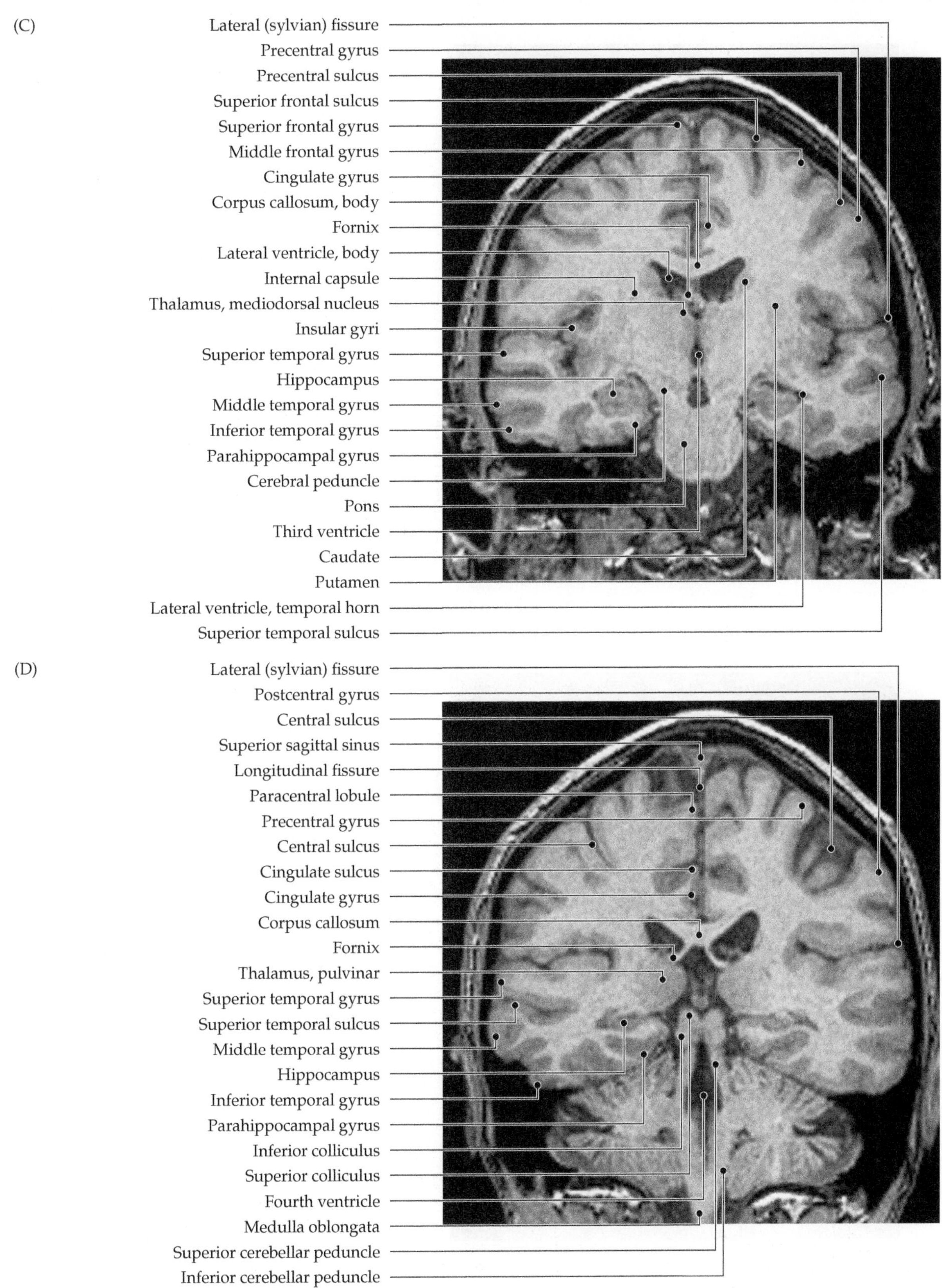

PLATE 3 AXIAL MR ATLAS

(A)

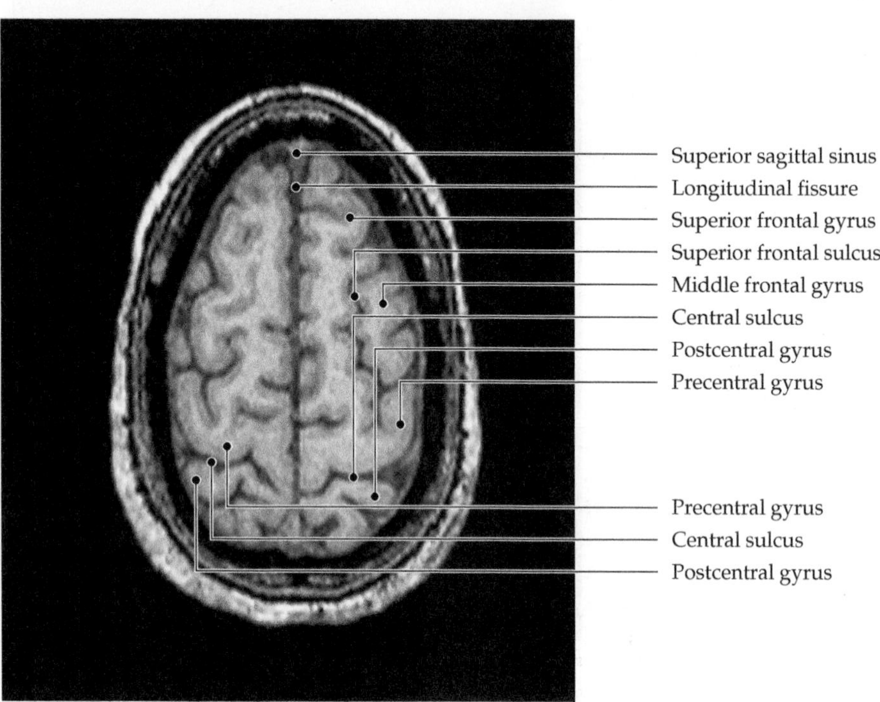
Superior sagittal sinus
Longitudinal fissure
Superior frontal gyrus
Superior frontal sulcus
Middle frontal gyrus
Central sulcus
Postcentral gyrus
Precentral gyrus
Precentral gyrus
Central sulcus
Postcentral gyrus

(B)

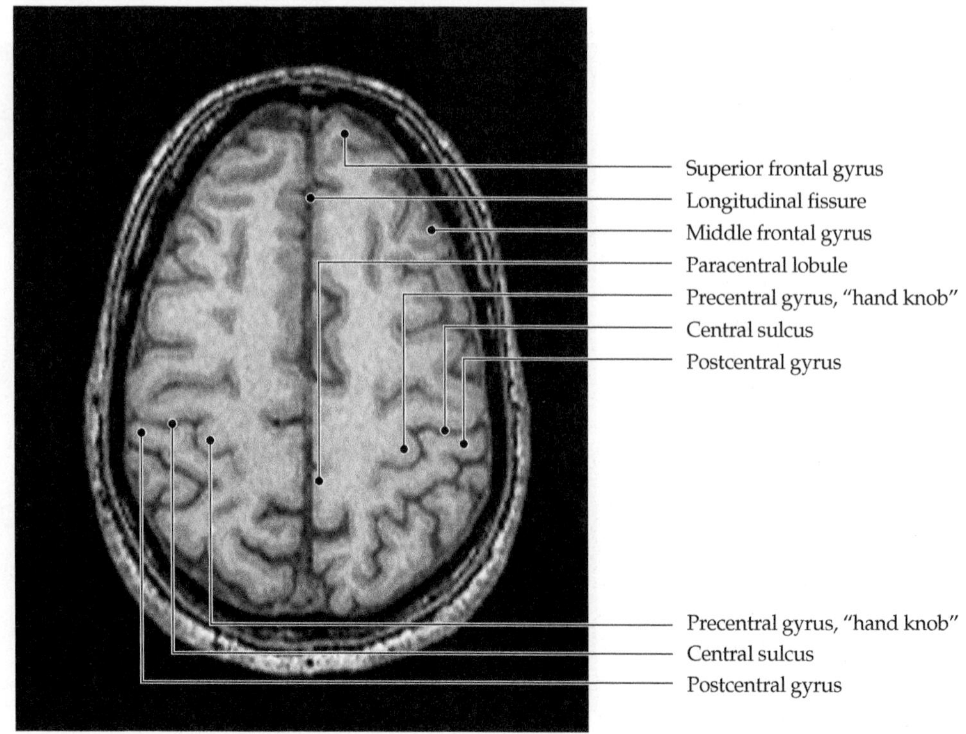
Superior frontal gyrus
Longitudinal fissure
Middle frontal gyrus
Paracentral lobule
Precentral gyrus, "hand knob"
Central sulcus
Postcentral gyrus
Precentral gyrus, "hand knob"
Central sulcus
Postcentral gyrus

Axial sections of the human brain demonstrating internal forebrain structures in T1-weighted magnetic resonance images; images in (A)–(H) are arranged from superior to inferior.

(C)

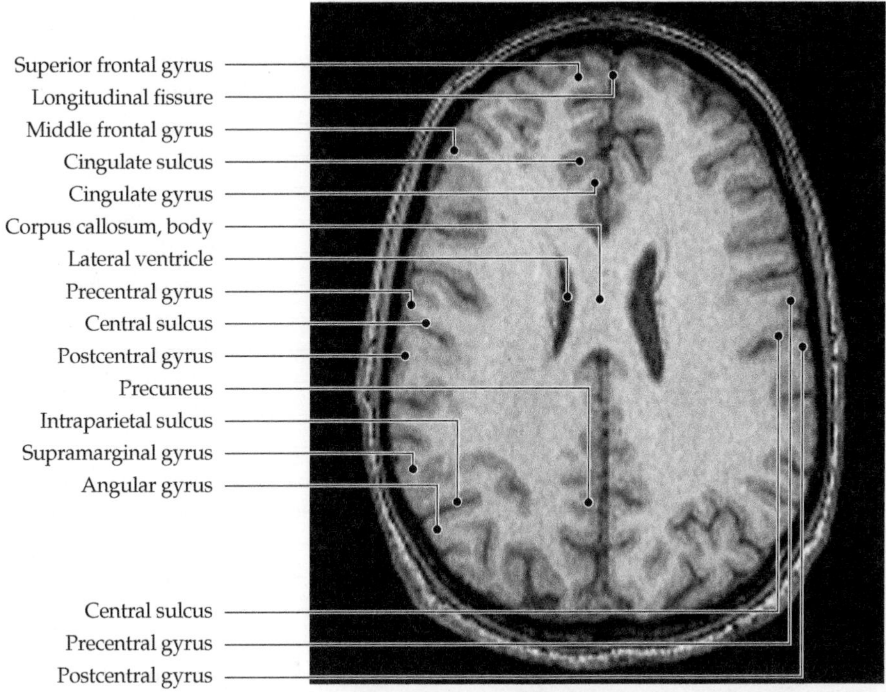

(D)

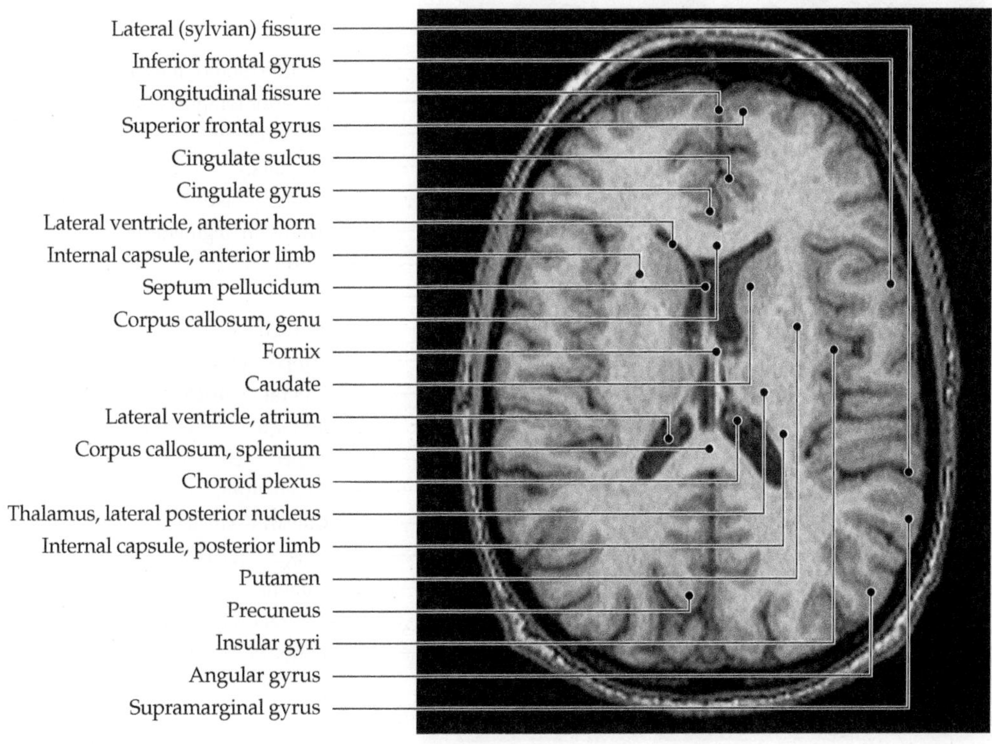

PLATE 3 AXIAL MR ATLAS (CONTINUED)

(E)

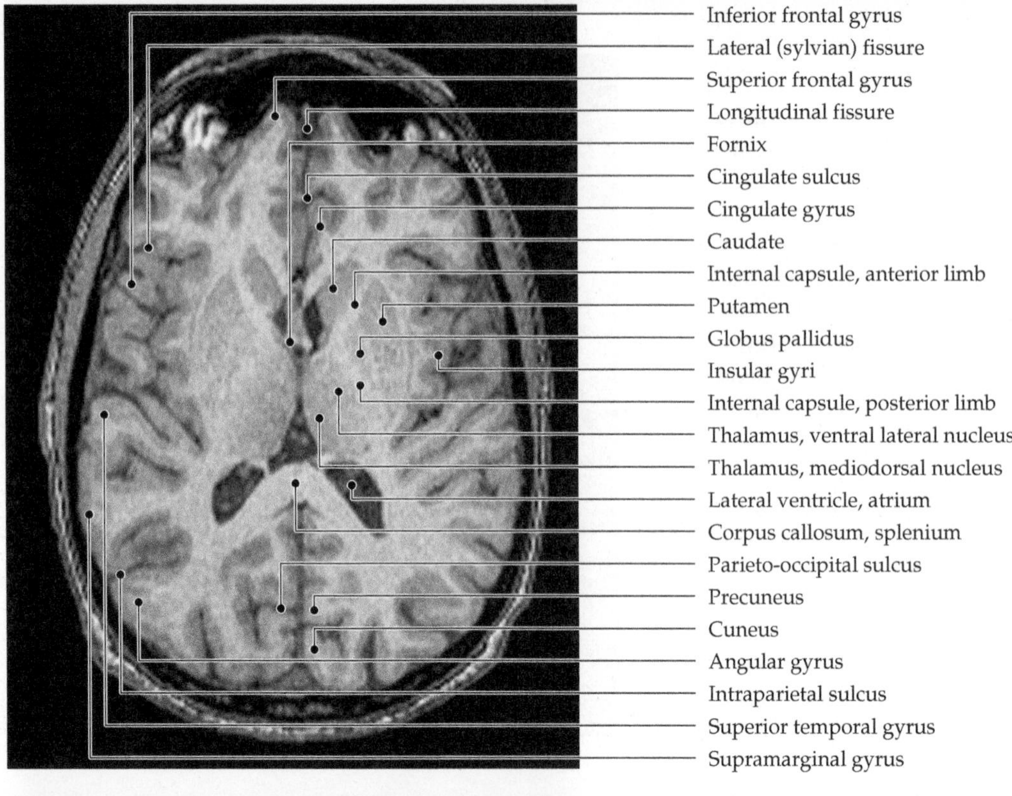
Inferior frontal gyrus
Lateral (sylvian) fissure
Superior frontal gyrus
Longitudinal fissure
Fornix
Cingulate sulcus
Cingulate gyrus
Caudate
Internal capsule, anterior limb
Putamen
Globus pallidus
Insular gyri
Internal capsule, posterior limb
Thalamus, ventral lateral nucleus
Thalamus, mediodorsal nucleus
Lateral ventricle, atrium
Corpus callosum, splenium
Parieto-occipital sulcus
Precuneus
Cuneus
Angular gyrus
Intraparietal sulcus
Superior temporal gyrus
Supramarginal gyrus

(F)

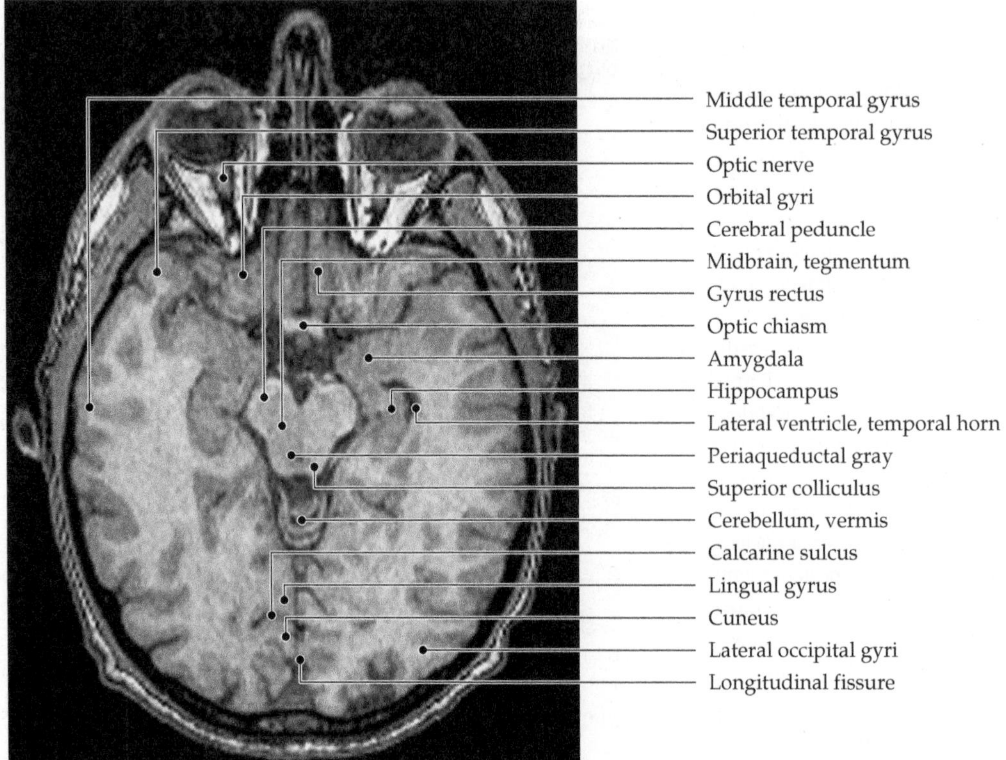
Middle temporal gyrus
Superior temporal gyrus
Optic nerve
Orbital gyri
Cerebral peduncle
Midbrain, tegmentum
Gyrus rectus
Optic chiasm
Amygdala
Hippocampus
Lateral ventricle, temporal horn
Periaqueductal gray
Superior colliculus
Cerebellum, vermis
Calcarine sulcus
Lingual gyrus
Cuneus
Lateral occipital gyri
Longitudinal fissure

Axial sections of the human brain demonstrating internal forebrain structures in T1-weighted magnetic resonance images; images in (A)–(H) are arranged from superior to inferior.

(G)

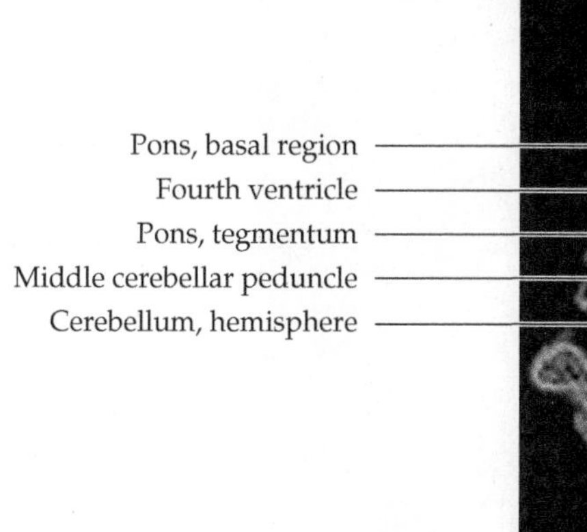

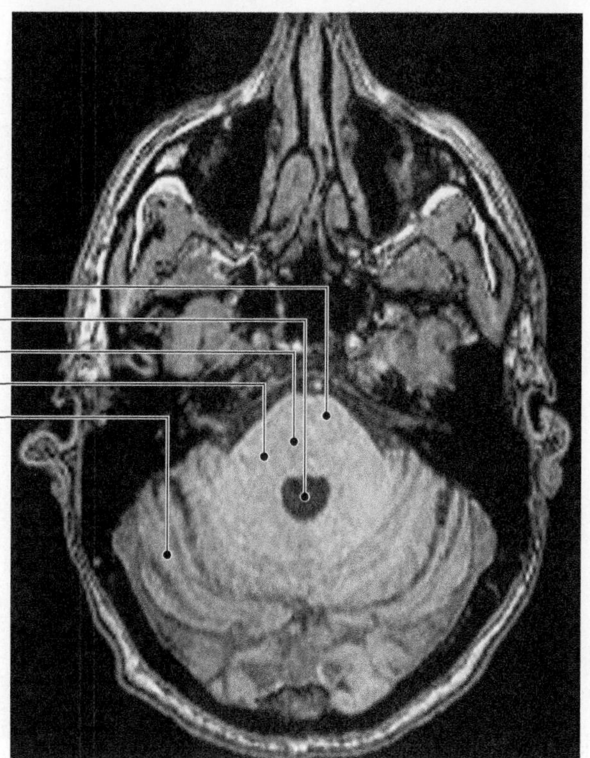

(H)

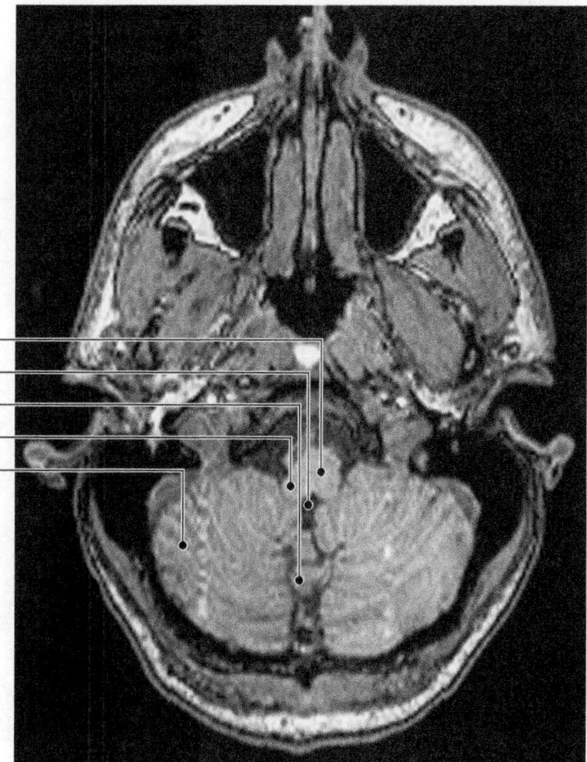

PLATE 4 SAGITTAL MR ATLAS

(A)

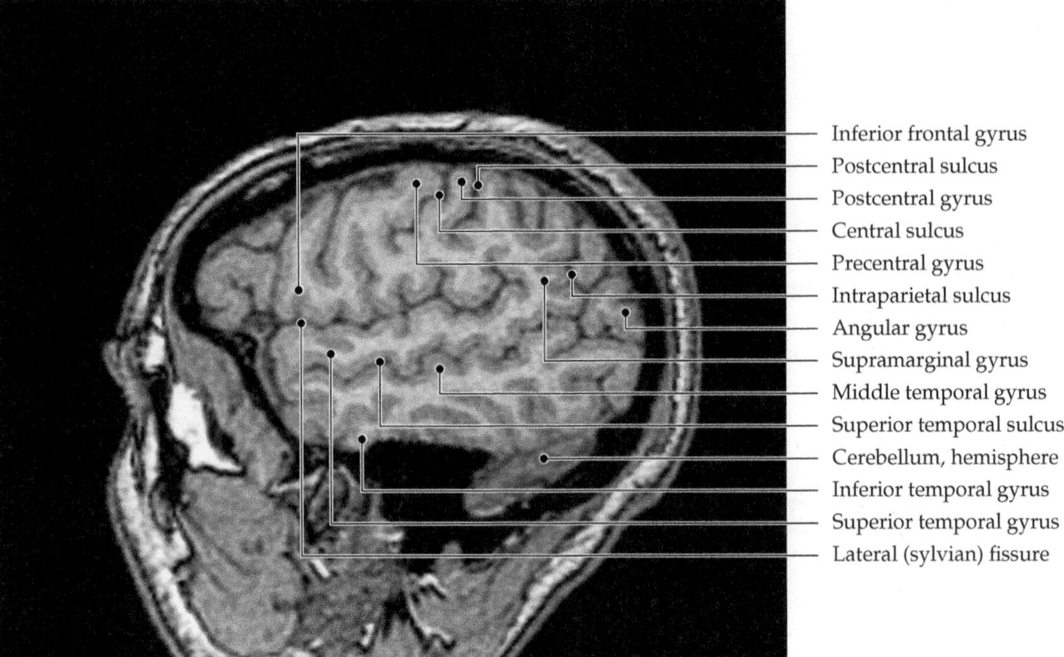
Inferior frontal gyrus
Postcentral sulcus
Postcentral gyrus
Central sulcus
Precentral gyrus
Intraparietal sulcus
Angular gyrus
Supramarginal gyrus
Middle temporal gyrus
Superior temporal sulcus
Cerebellum, hemisphere
Inferior temporal gyrus
Superior temporal gyrus
Lateral (sylvian) fissure

(B)

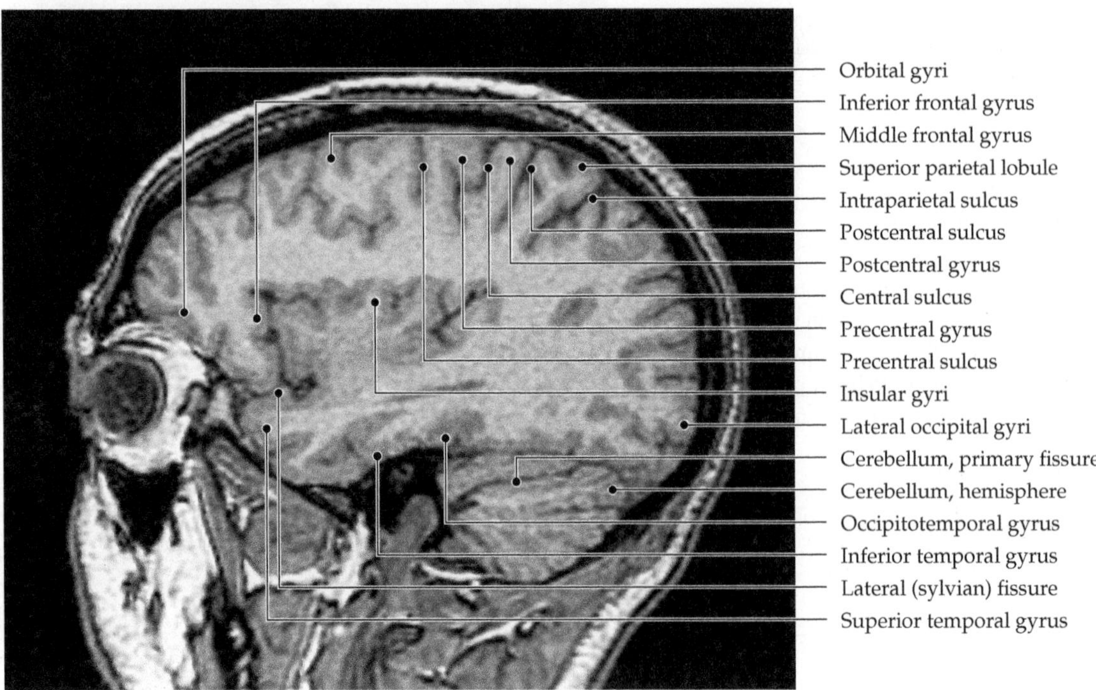
Orbital gyri
Inferior frontal gyrus
Middle frontal gyrus
Superior parietal lobule
Intraparietal sulcus
Postcentral sulcus
Postcentral gyrus
Central sulcus
Precentral gyrus
Precentral sulcus
Insular gyri
Lateral occipital gyri
Cerebellum, primary fissure
Cerebellum, hemisphere
Occipitotemporal gyrus
Inferior temporal gyrus
Lateral (sylvian) fissure
Superior temporal gyrus

Sagittal sections of the human brain demonstrating internal forebrain structures in T1-weighted magnetic resonance images; images in (A)–(D) are arranged from lateral to medial.

(C)

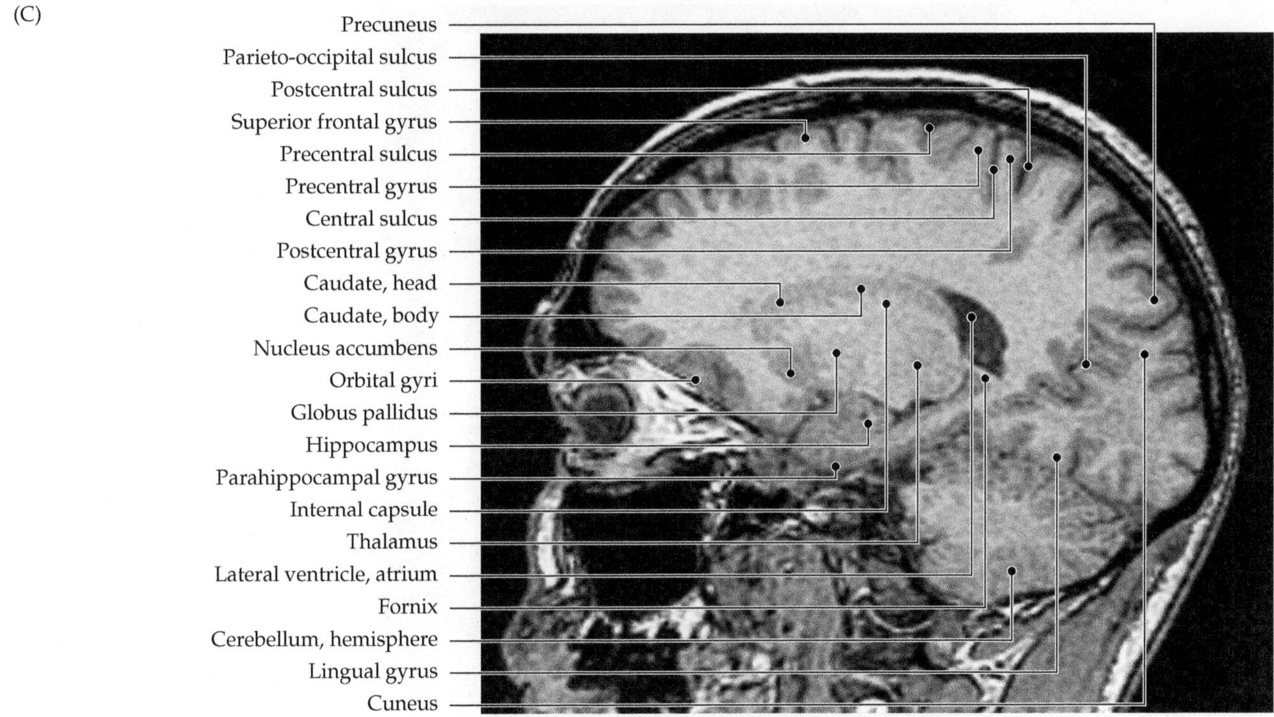

(D)

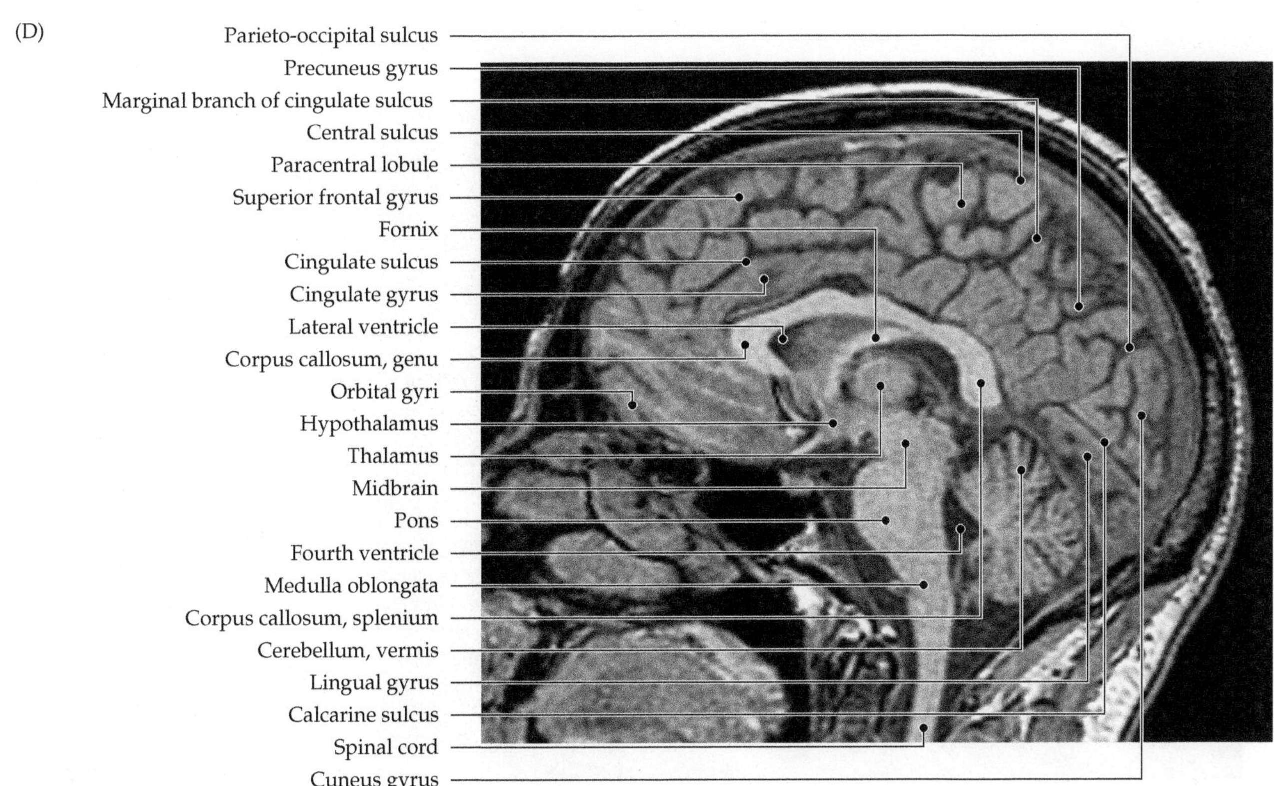

PLATE 5 DIFFUSION TENSOR IMAGING

(A)

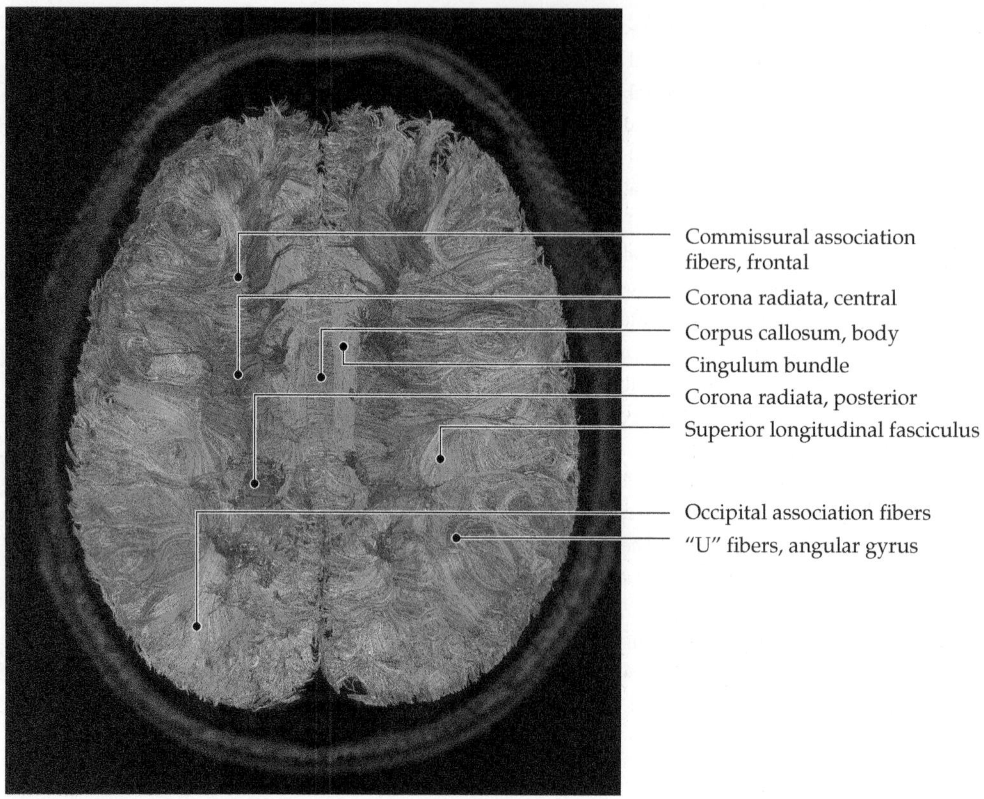
Commissural association fibers, frontal
Corona radiata, central
Corpus callosum, body
Cingulum bundle
Corona radiata, posterior
Superior longitudinal fasciculus
Occipital association fibers
"U" fibers, angular gyrus

(B)

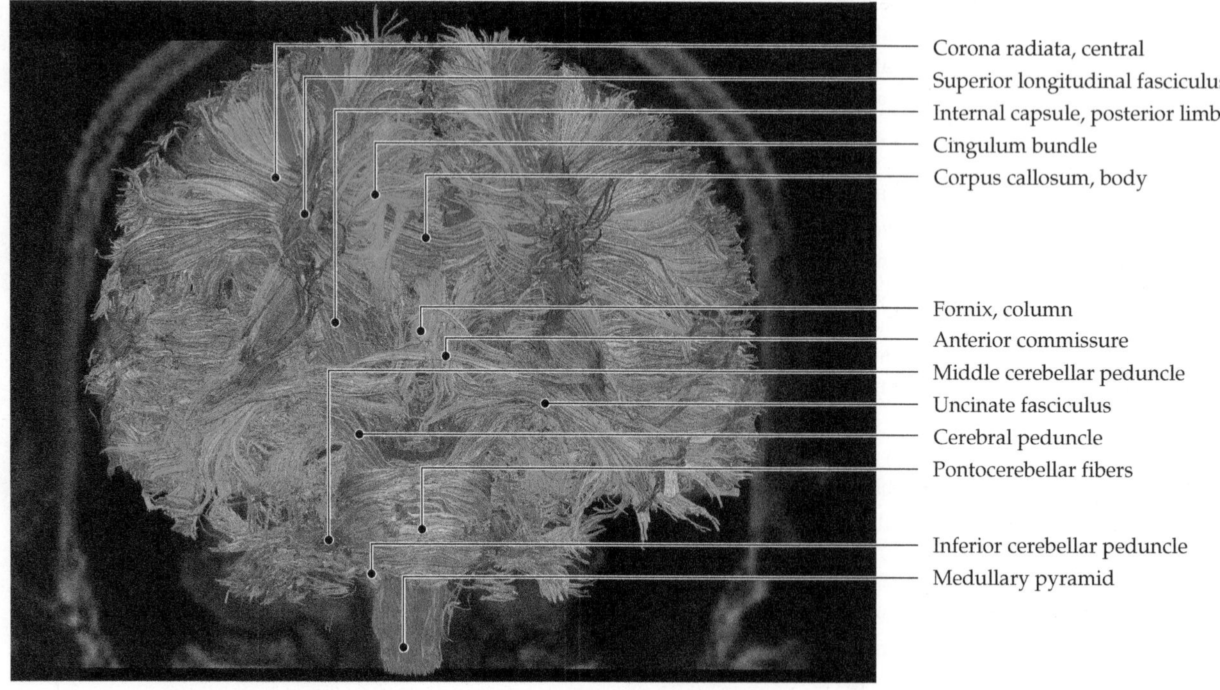
Corona radiata, central
Superior longitudinal fasciculus
Internal capsule, posterior limb
Cingulum bundle
Corpus callosum, body
Fornix, column
Anterior commissure
Middle cerebellar peduncle
Uncinate fasciculus
Cerebral peduncle
Pontocerebellar fibers
Inferior cerebellar peduncle
Medullary pyramid

(A) Axial, (B) coronal, and (C) sagittal sections through a diffusion tensor imaging dataset used to compute fiber tracts, which represent the structure of white matter fibers in a human brain. Lower panel shows color code for spatial orientation of fiber tracts. (Images courtesy of Allen W. Song and Iain Bruce, Duke-UNC Brain Imaging and Analysis Center.)

(C)

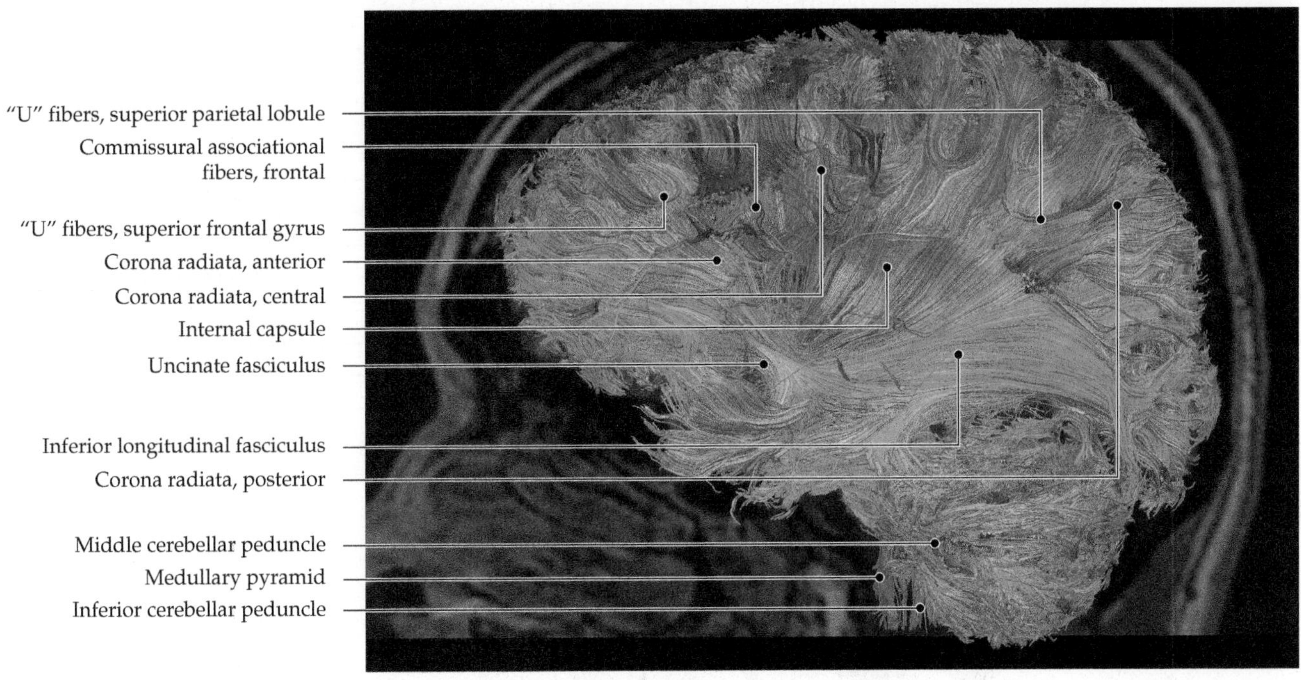

(D)

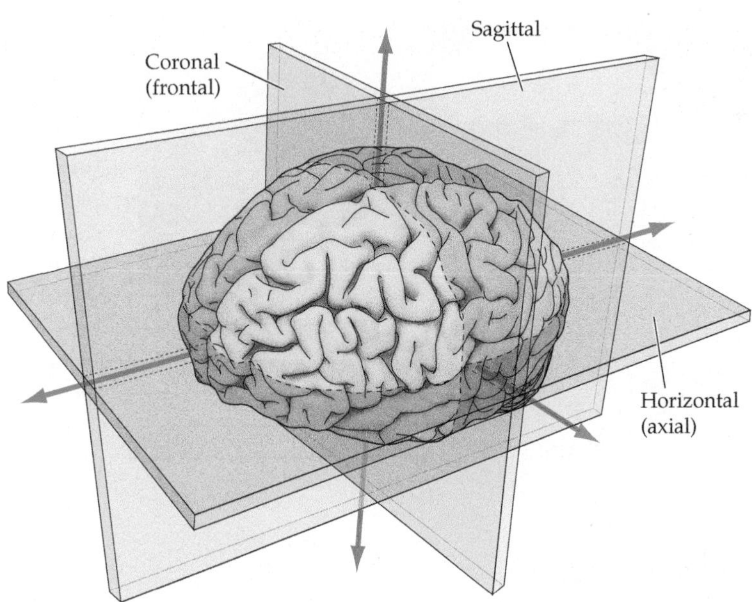

PLATE 6 BRAINSTEM ATLAS

(A)

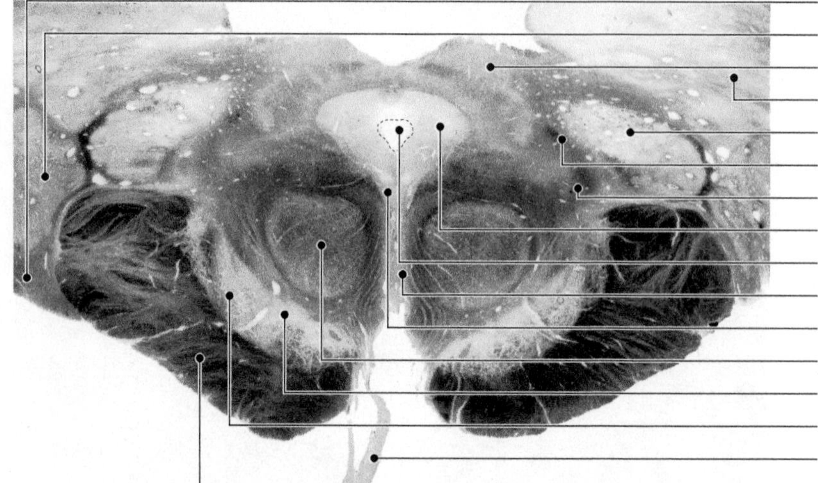
Optic tract
Lateral geniculate nucleus
Superior colliculus
Pulvinar
Medial geniculate nucleus
Anterolateral system
Medial lemniscus
Periaqueductal gray
Cerebral aqueduct
Raphe nuclei
Oculomotor complex
Red nucleus
Substantia nigra, pars compacta
Substantia nigra, pars reticulata
Oculomotor nerve
Cerebral peduncle

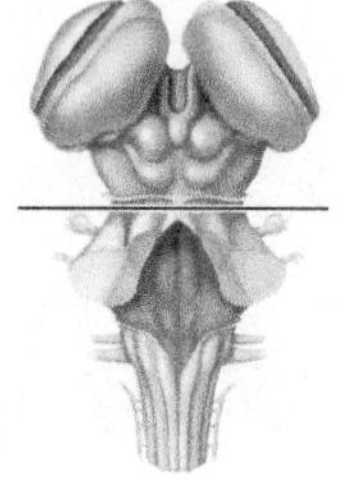

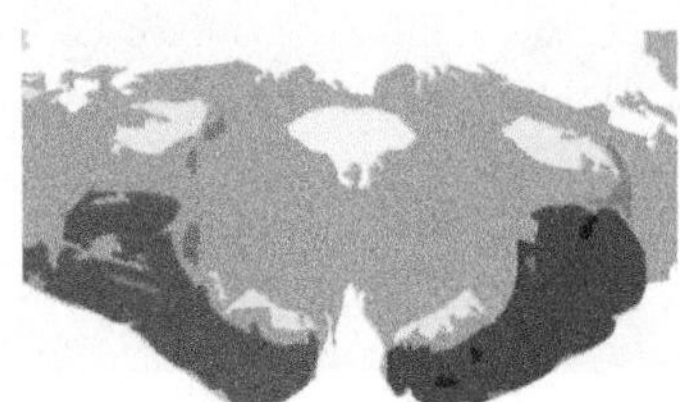

(B)

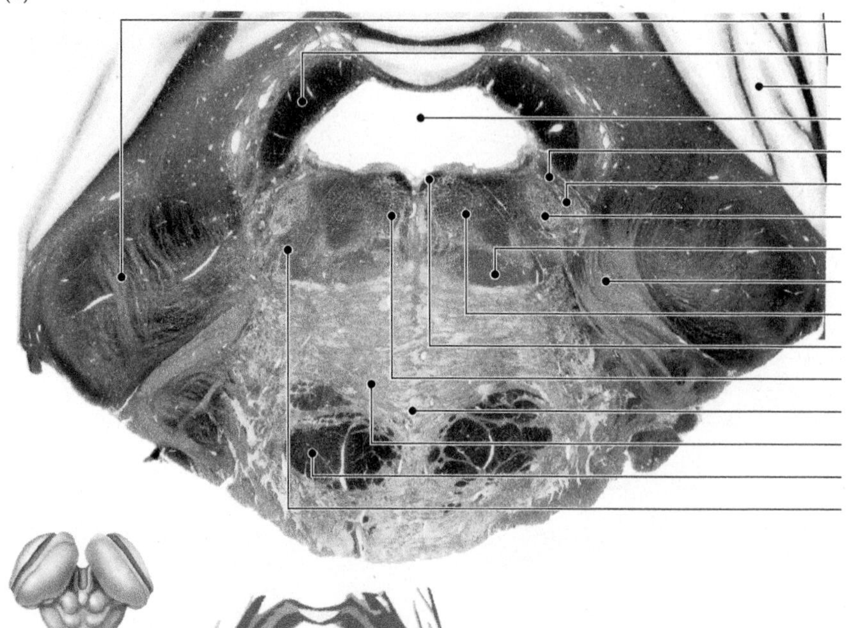
Middle cerebellar peduncle
Superior cerebellar peduncle
Cerebellum, cortex
Fourth ventricle
Mesencephalic trigeminal tract and nucleus
Chief sensory nucleus of the trigeminal complex
Trigeminal motor nucleus
Medial lemniscus
Trigeminal nerve roots
Central tegmental tract
Medial longitudinal fasciculus
Tectospinal fibers
Pontine nuclei
Pontocerebellar fibers
Corticobulbar and corticospinal fibers
Anterolateral system

Transverse sections of the human brainstem acquired and prepared to simulate myelin staining. (A) Midbrain. (B) Pons. (C) Medulla oblongata. (D) Caudal medulla oblongata. Sections in insets printed at actual size.

(C)

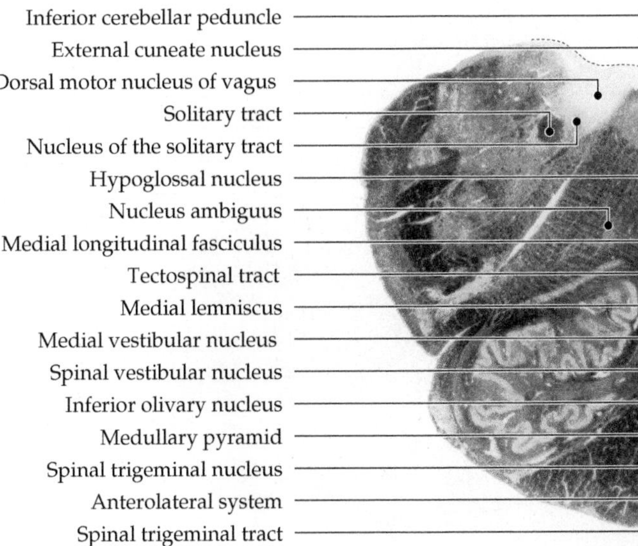

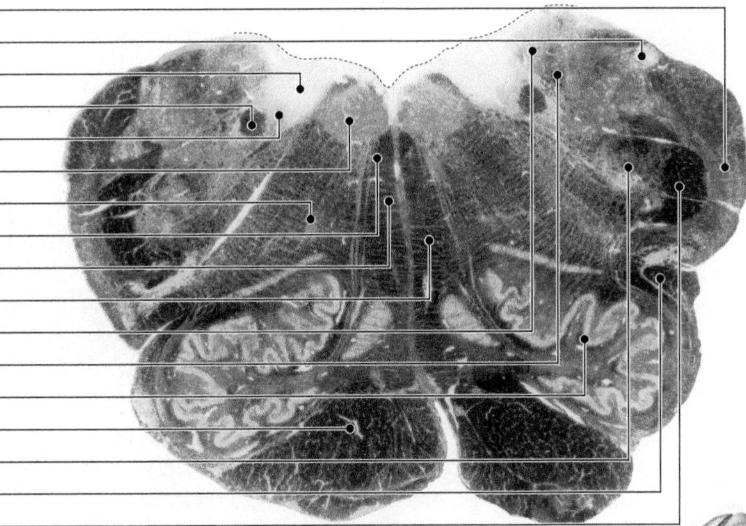

(D)

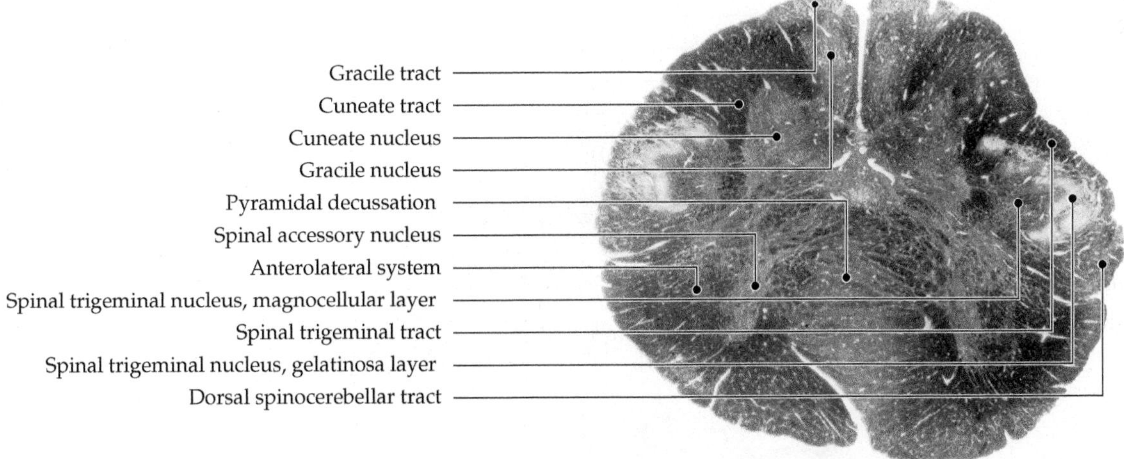

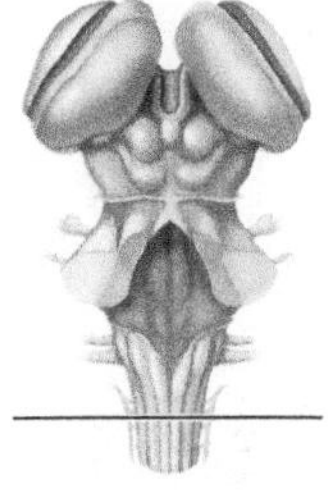

PLATE 7 SPINAL CORD ATLAS

(A)

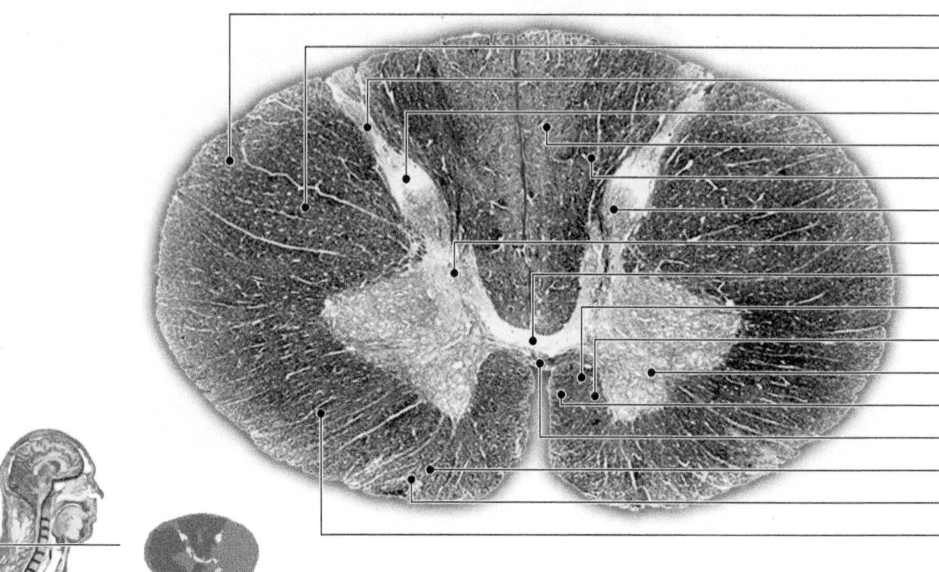

(B)

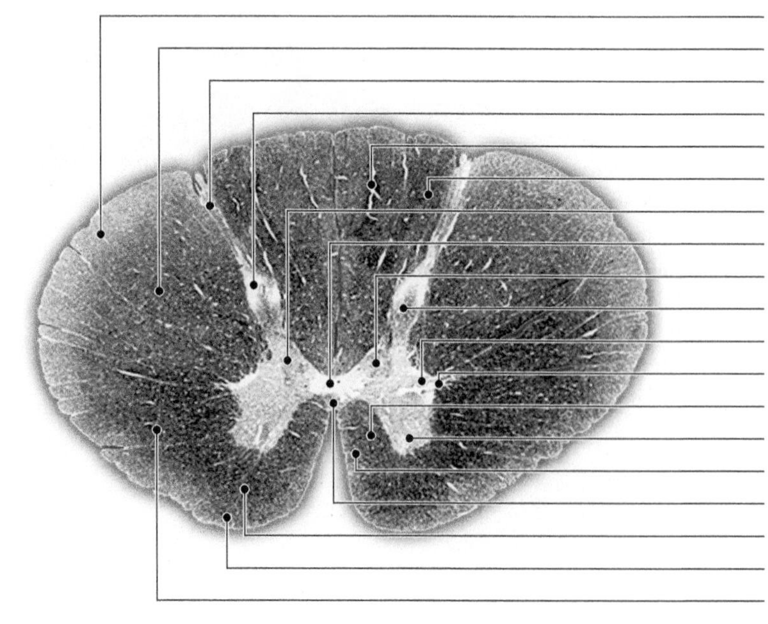

Transverse sections of the human spinal cord acquired and prepared to simulate myelin staining. (A) Cervical segment. (B) Thoracic segment. (C) Lumbar sement. (D) Sacral segment. Sections in insets printed at actual size.

(C)

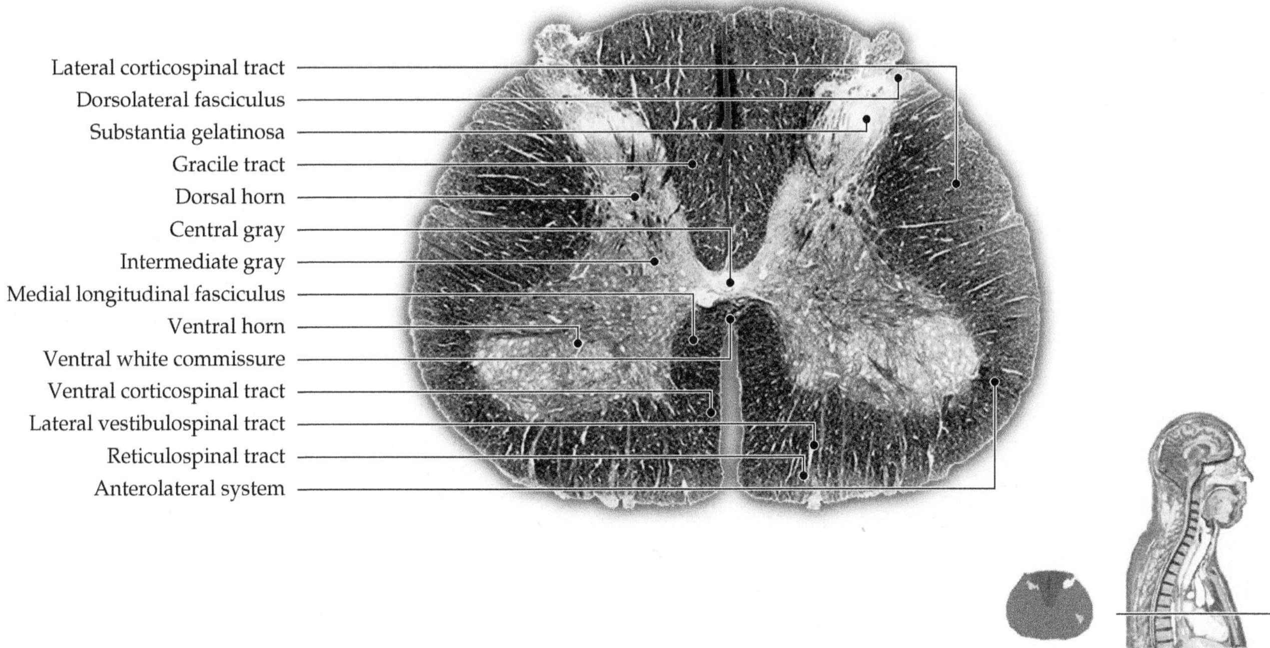

(D)

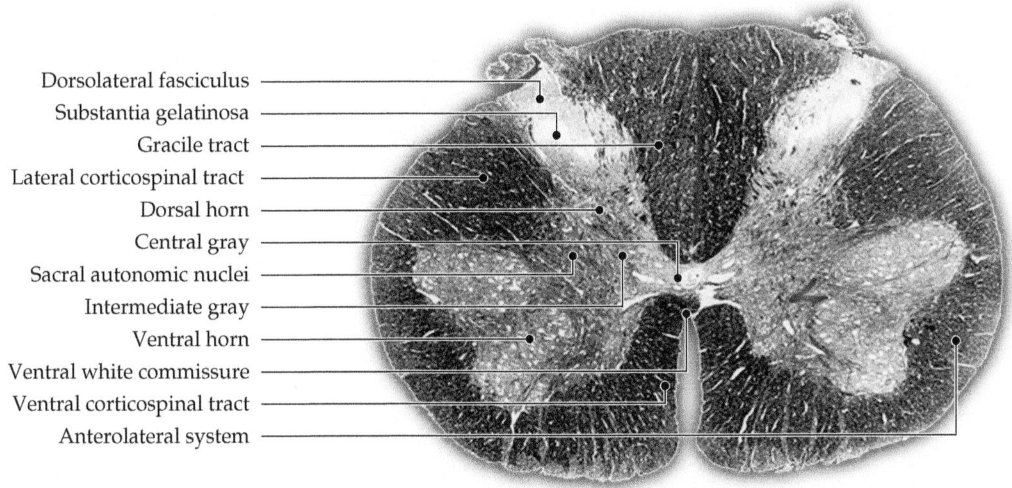

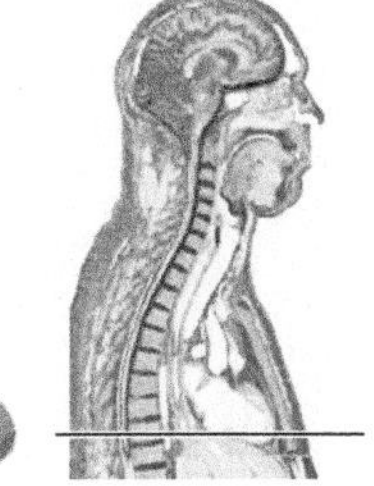

Glossary

A

abducens nerve (VI) Cranial nerve VI, an efferent nerve that controls the lateral rectus muscle of the eye.

accessory olfactory bulb (AOB) The target of axons from the VNO, adjacent to the main olfactory bulb that relays vomeronasal information to the hypothalamus and other basal forebrain regions.

accommodation 1. Dynamic changes in the lens of the eye that enable the viewer to focus. When viewing distant objects, the lens is made relatively thin and flat; for near vision, the lens becomes thicker and rounder and has more refractive power. 2. Term used by Piaget (along with assimilation) to describe how children might react to a new person, event, or object by modifying their scheme of thought.

acetylcholine (Ach) Neurotransmitter at motor neuron synapses, in autonomic ganglia, and in a variety of central synapses. Binds to two types of **acetylcholine receptors (AChRs)**, either ligand-gated ion channels (nicotinic receptors) and G-protein-coupled receptors (muscarinic receptors).

acetylcholinesterase (AChE) Enzyme in the synaptic cleft that clears the cleft of acetylcholine released by the presynaptic cell. AChE hydrolyzes ACh into acetate and choline; the choline is then transported back into nerve terminals, where it is used to resynthesize ACh.

actin A cytoskeletal protein involved in maintaining cell shape and organelle movement.

actin cytoskeleton The meshwork of polymers of the fibrillary protein actin that forms a flexible, but strong scaffold to allow for localization of key proteins and organelles in the cytoplasm, and maintains the integrity of the cell membrane so that the cell retains its volume. The actin cytoskeleton is also essential for force generation in motile cells or motile cell extensions like axonal or dendritic growth cones.

action potential The electrical signal generated and conducted along axons (or muscle fibers) by which information is conveyed from one place to another in the nervous system (or within muscle fibers).

action tremors See intention tremors.

activation The time-dependent opening of ion channels in response to a stimulus, typically membrane depolarization.

active transporters Transmembrane proteins that actively move ions into or out of cells against their concentration gradients. Their source of energy may be ATP or the electrochemical gradients of various ions. See also co-transporters; ion exchangers.

active zone The location within the presynaptic terminal where synaptic vesicles fuse with the presynaptic plasma membrane to discharge their neurotransmitters into the synaptic cleft.

adaptation In the context of evolution, moving animal phenotypes closer to the demands of their environments. Roughly synonymous with evolving "fitness."

adenylyl cyclase III (ACIII) Membrane-bound enzyme that can be activated by G-proteins to catalyze the synthesis of cyclic AMP from ATP.

adrenaline See epinephrine.

affective–motivational The fear, anxiety, and autonomic nervous activation that accompany exposure to a noxious stimulus.

afferent neurons Neurons or axons that conduct action potentials from the periphery toward the central nervous system.

agnosias The inability to name objects; literally means "not knowing."

alar plate Embryonic gray matter structure in the brainstem and spinal cord that gives rise to sensory nuclei of the brainstem and the dorsal horn of the spinal cord.

allelic silencing A genomic mechanism that renders one of the two copies of each gene (paternal and maternal copies on each chromosome) unable to be expressed. Allelic silencing is thought to be due to a combination of direct binding of transcription regulators to regulatory sequences of one but not the other allele, as well as local histone modifications. Allelic silencing is a more general example of specific forms of transcriptional selection of one copy versus another of a particular gene. Other examples include parental imprinting when either the paternal or maternal allele of a gene is selectively expressed and the other suppressed, and X-inactivation in females when one of two copies of identical genes on each X chromosome is silenced to maintain appropriate gene dosage.

allodynia The induction of pain by a normally innocuous stimulus.

allometry Measurement of one or more body parts in relation to the body as a whole.

α-fetoprotein A protein that actively sequesters circulating estrogens, preventing maternal estrogen from affecting the sexual differentiation of the fetus.

alpha (α) motor neurons Neurons in the ventral horn of the spinal cord that innervate force-generating, extrafusal fibers of skeletal muscle.

amacrine cells Retinal neurons that mediate lateral interactions between bipolar cell terminals and the dendrites of ganglion cells.

amblyopia Diminished visual acuity as a result of the failure to establish appropriate visual cortical connections in early life.

amnesia The pathological inability to remember or establish memories; retrograde amnesia is the inability to recall existing memories, whereas anterograde amnesia is the inability to lay down new memories.

AMPA receptors See ionotropic glutamate receptors.

ampullae The juglike swellings at the base of the semicircular canals that contain the hair cells and cupulae. See also cupulae.

amygdala A nuclear and corticoid complex in the anterior-medial temporal lobe that forms part of the limbic forebrain; its major functions concern implicit processing with respect to autonomic, emotional, and sexual behavior.

androgen insensitivity syndrome (AIS) A condition in which, due to a defect in the gene that codes for the androgen receptor, testosterone cannot act on its target tissues. Also called *testicular feminization.*

anesthesia Procedures that reduce sensation during surgical procedures, most often to alleviate pain or to create a state of unconsciousness.

anesthetics Drugs that produce anesthesia.

anomalous trichromats Individuals with one or more atypical cone opsins.

anopsias A large deficit in the visual field resulting from pathological changes in some component of the primary visual pathway.

anosmias Loss of the sense of smell; can be total or restricted to a single odorant.

anosmin The protein encoded by the gene mutated in Kallman's syndrome, also known as KAL1. Anosmin is a transmembrane Ca^{2+} independent cell adhesion molecule. In the absence of anosmin protein, olfactory receptor neurons do not extend their axons to the olfactory bulb nor do GNRH neurons migrate to the hypothalamus. Thus, individual lacking anosmin because of a specific gene mutation are both unable to smell things (anosmia) and also sterile due to a lack of gonadotrophin releasing hormone secretion.

anterior cerebral arteries Major vessels derived from the internal carotid arteries that supply the anterior and medial aspects of the frontal and parietal lobes, including associated deep structures.

anterior chamber The part of the eye that is just behind the cornea and in front of the lens.

anterior choroidal artery Branch of the proximal middle cerebral artery that supplies blood to the medial temporal lobe and deep white matter and gray matter, including parts of the basal ganglia and the internal capsule.

anterior cingulate cortex (ACC) The portion of the midline frontal lobe comprising the anterior extent of the cingulate gyrus and adjacent cortex; its dorsal regions are associated with executive functions.

anterior circulation Vasculature derived from the internal carotid arteries that supplies blood to the forebrain.

anterior commissure A small commissural fiber tract that lies anterior to the third ventricle and inferior to the genu of the corpus callosum; like the callosum, it serves to connect the two hemispheres, but its origins and terminations are mainly in the ventral frontal lobe, olfactory bulb, and anterior temporal lobe.

anterior communicating arteries Small vessels that cross the midsagittal plane joining the two anterior cerebral arteries, forming the anterior aspect of the circle of Willis.

anterior inferior cerebellar artery (AICA) Long circumferential branch of the basilar artery that supplies dorsolateral aspects of the caudal pons and anterior and inferior aspects of the cerebellum

anterior insula A functional division of the cerebral cortex located within the lateral sulcus between the temporal lobe and the frontal and parietal lobes. It is associated with emotion and homeostatic regulation.

anterior spinal artery Principal artery on the anterior aspect of the spinal cord supplied by the vertebral arteries and the medullary arteries; gives rise to some 200 unilateral sulcal branches that alternate left and right along the length of the spinal cord supplying blood to the anterior two-thirds of the cord.

anterograde Signals or impulses that travel "forward," e.g., from the cell body to the axon terminal, from the presynaptic terminal to the postsynaptic cell, or from the CNS to the periphery.

anterograde amnesia The inability to lay down new memories.

anterolateral columns See ventral columns.

anterolateral system Ascending sensory pathway in the spinal cord and brainstem that carries information about pain and temperature to the thalamus.

anteroventral paraventricular nucleus (AVPV) A sexually dimorphic hypothalamic nucleus that regulates cyclical ovulation in female mammals.

antiporters Active transporters that use the energy from ionic gradients to carry multiple ions across the membrane in opposite directions.

apoptosis Cell death resulting from a programmed pattern of gene expression; also known as *programmed cell death.*

aprosodias The inability to infuse language with its normal emotional content. See also prosody.

aqueduct of Sylvius See cerebral aqueduct.

aqueous humor A clear, watery liquid that supplies nutrients to the cornea and lens of the eye.

arachnoid mater One of the three coverings of the brain that make up the meninges; lies between the dura mater and the pia mater and forms a spiderweb-like network of trabeculae (arachnoid mater means "spiderweb-like mother) that allows for the flow of cerebrospinal fluid and the distribution of superficial blood vessels in the subarachoid space.

arachnoid villi Protrusions of arachnoid mater into the superior sagittal sinus that allow for the passage of cerebrospinal fluid from the subarachnoid space into the venous drainage.

archicortex Phylogenetically, the simplest and most primitive division of the cerebral cortex, which occurs in the hippocampus.

aromatase The enzyme that converts testosterone, and some additional steroids, to the active form of estrogen, 17β estradiol.

arousal 1. A global state of the brain (or the body) reflecting an overall level of responsiveness. 2. The degree of intensity of an emotion.

arrestin A protein that binds to rhodopsin.

assimilation Term used by Piaget (along with accommodation) to describe how children adapt to their environment during development.

associational systems Neural cell circuits that are not part of the relatively defined sensory (input) and motor (output) systems; they mediate the most complex and least well-defined brain functions that require the integration or association of signals from multiple sensory and/or motor systems.

associativity A mechanism that serves to link together two or more independent processes. For example, associative learning results from the pairing of unconditioned and conditioned stimuli presented to an experimental subject. In the hippocampus, associativity allows a weakly activated group of synapses to undergo long-term potentiation when a nearby group of synapses is strongly activated.

astrocytes One of the three major classes of glial cells found in the central nervous system; important in maintaining and regulating, in a variety of ways, an appropriate chemical environment for neuronal signaling; also involved in the formation of the blood-brain barrier, the secretion of substances that influence the construction of new synaptic connections, and the proliferation of new cells in the adult brain that retain characteristics of stem cells.

ATPase pumps Membrane pumps that use the hydrolysis of ATP to translocate ions against their electrochemical gradients.

auditory area 1 (A1) The cortical target of the neurons in the medial geniculate nucleus; the terminus of the primary auditory pathway.

auditory meatus Opening of the external ear canal.

Auerbach's plexus See myenteric plexus.

augmentation An activity-dependent form of short-term synaptic plasticity that enhances synaptic transmission over a time course of a few seconds. Augmentation is caused by an increase in the amount of neurotransmitter released in response to presynaptic action potentials and results from persistent calcium signaling within presynaptic terminals, perhaps due to actions on the SNARE-regulatory protein, munc13.

Australopithecines The first ancestors in the human lineage about 3–4 million years ago. The subsequent lineage includes *Homo habilis, H. erectus,* Neanderthals, and *H. sapiens.*

autonomic ganglia Collections of autonomic motor neurons outside the central nervous system that innervate visceral smooth muscles, cardiac muscle, and glands.

autonomic motor division See visceral motor division.

autonomic nervous system The components of the nervous system (peripheral and central) concerned with the regulation of smooth muscle, cardiac muscle, and glands. Also known as the visceral motor system; sometimes called the "involuntary" nervous system. Consists of sympathetic and parasympathetic divisions, and a semi-autonomous division in the gut, the enteric nervous system.

autophagy A cellular state in which proteins and other macromolecules within a cell are isolated and eventually trafficked to lysosomes for degradation. Autophagy is regulated by a number of genes, many of which respond to cellular damage, infection, or oxidative stress.

autosomes Any chromosome other than the X and Y sex chromosomes.

auxilin An accessory protein that promotes the actions of HSC70 during vesicle uncoating after endocytosis.

axial sections See horizontal sections.

axon The neuronal process (typically, much longer than any dendrite) that conveys the action potential from the nerve cell body to its terminals.

axonal transport The process that allows cellular components, such as proteins and organelles, to move within the axons of neurons.

Aδ group Myelinated, faster conducting pain fibers.

B

bacteriorhodopsin A protein that, in response to light of the proper wavelength, acts as a proton pump transporting protons from inside the cell to outside; in its native host, the resulting proton gradient is subsequently converted into chemical energy; when engineered into a neuron for optogenetics, it hyperpolarizes the neuron when exposed to light.

Balint's syndrome A neurological syndrome, caused by bilateral damage to the posterior parietal and lateral occipital cortex, that has three hallmark symptoms: (1) *simultanagnosia,* the inability to attend to and/or perceive more than one visual object at a time; (2) *optic ataxia,* the impaired ability to reach for or point to an object in space under visual guidance; and (3) *oculomotor apraxia,* difficulty voluntarily directing the eye gaze toward objects in the visual field with a saccade. Simultanagnosia is the sign most closely associated with the syndrome, and the one most studied from a cognitive neuroscience standpoint.

basal cells Basal cells are found in the region of the olfactory epithelium adjacent to the lamina propria, where blood vessels and connective tissue that support the olfactory epithelium are found. They retain neural stem cell identity and can generated new olfactory receptor neurons throughout life.

basal forebrain nuclei Cerebral nuclei anterior to the hypothalamus and ventral to the basal ganglia; give rise to widespread modulatory projections to diverse targets in the cerebral hemispheres.

basal ganglia Cerebral nuclei lying deep in the subcortical white matter of the cerebral hemispheres lateral and central to the lateral ventricle. The caudate, putamen, and globus pallidus are the major components of the basal ganglia; together with the subthalamic nucleus and substantia nigra, these structures modulate the initiation and suppression of behavior.

basal lamina A thin layer of extracellular matrix material (primarily collagen, laminin, and fibronectin) that surrounds muscle cells and Schwann cells. Also underlies all epithelial sheets. Also called the basement membrane.

basal plate Embryonic gray matter structure in the brainstem and spinal cord that gives rise to motor nuclei of the brainstem and the ventral horn of the spinal cord.

basilar artery Major vessel formed by the fusion of the two vertebral arteries that lies along the ventral midline of the pons; gives rise to the anterior inferior and superior cerebellar arteries before bifurcating and giving rise to the paired posterior cerebral arteries at the midbrain.

basilar membrane The membrane that forms the floor of the cochlear duct, on which the cochlear hair cells are located.

basket cells Inhibitory interneurons in the cerebellar cortex whose cell bodies are located within the Purkinje cell layer and whose axons make basket-like terminal arbors around Purkinje cell bodies, providing lateral inhibition that focuses the spatial distribution of Purkinje cell activity.

belt and parabelt regions Regions of the auditory cortex that surround the core region.

bHLH (basic helix-loop-helix) Neurogenic transcription factors (named for a shared *b*asic *h*elix-*l*oop-*h*elix amino acid motif that defines their DNA-binding domain) that have emerged as central to the differentiation of distinct neural and glial fates.

binocular field The two symmetrical, overlapping visual hemifields. The left hemifield includes the nasal visual field of the right eye and the temporal visual field of the left eye; the right hemifield includes the temporal field of the right eye and the nasal field of the left eye.

biogenic amines Category of small-molecule neurotransmitters; includes the catecholamines (epinephrine, norepinephrine, dopamine), serotonin, and histamine.

bipolar cells Retinal neurons that provide a direct link between photoreceptor terminals and ganglion cell dendrites.

bitter One of the five basic tastes; the taste quality, generally considered unpleasant, produced by substances like quinine or caffeine. Compare *salty*, *sour*, and *sweet*. Bitter is transduced by taste cells via the T2R G-protein-coupled taste receptors.

blindsight A pathological phenomenon in which patients with damage to the primary visual cortex are blind in the affected area of the contralateral visual field.

blood oxygenation level-dependent (BOLD) Endogenous signals reflecting the oxygenation of hemoglobin in blood that are modulated by changes in the local level of neural activity; for example, when neural activity in a local brain region increases, more oxygen is consumed and within seconds the local microvasculature responds by increasing the flow of oxygen-rich blood to the active region, thus constituting a BOLD signal that may be detected by fMRI.

blood–brain barrier A diffusion barrier between the cerebral vasculature and the substance of the brain formed by tight junctions between capillary endothelial cells and the surrounding astrocytic endfeet.

bone morphogenetic proteins (BMPs) Peptide hormones that play important roles in neural induction and differentiation.

Bowman's glands Mucous producing specializations composed of secretory cells surrounding a lumen that is continuous with the surface of the olfactory epithelium.

brachium conjunctivum See cerebellar peduncles.

brachium pontis See cerebellar peduncles.

brain The rostral (supraspinal) portion of the central nervous system comprised of the cerebral hemispheres, diencephalon, cerebellum, and brainstem.

brain-derived neurotrophic factor (BDNF) One member of a family of neutrophic factors, the best-known constituent of which is nerve growth factor (NGF).

brainstem The portion of the brain that lies between the diencephalon and the spinal cord; comprises the midbrain, pons, and medulla.

branchial motor nuclei Brainstem nuclei (derived from the basal plate) that give rise to efferent fibers innervating striated muscle fibers derived from embryonic branchial (pharyngeal) arches; located in an intermediate position in the tegmentum after embryonic migration from a more dorsal location.

Broca's aphasia Difficulty producing speech as a result of damage to Broca's area in the left frontal lobe. Also called *motor, expressive*, or *production aphasia*.

Broca's area An area in the left inferior frontal lobe specialized for the production of speech and the expression of language in non-vocal forms.

Brodmann's area 17 Another name for the primary visual cortex (V1) in the occipital lobe; major cortical target of the retinal sensory cells. Also called striate cortex because the prominence of layer 4 in myelin-stained sections gives this region a striped (striated) appearance.

C

C fiber group Unmyelinated, slower conducting pain fibers.

c-fos A transcription factor, originally isolated from *c*ellular *f*eline *os*teosarcoma cells, that binds as a heterodimer, thus activating gene transcription.

Ca^{2+}/calmodulin-dependent protein kinase, type II A protein kinase that is activated by the second messenger, calcium ions, binding to the calcium-binding protein calmodulin. Once activated by calcium and calmodulin, this protein kinase can phosphorylate numerous substrate proteins to alter their signaling properties.

CACNA genes Ca^{2+} channel genes.

cadherins A family of calcium-dependent cell adhesion molecules found on the surfaces of growth cones and the cells over which they grow.

calbindin A protein that slows transient changes in intracellular calcium concentration by reversibly binding calcium ions.

calcarine sulcus Major sulcus on the medial aspect of the occipital lobe that divides the cuneus and lingual gyri; the primary visual (striate) cortex lies largely within this sulcus.

calcium imaging Method of monitoring by optical means the levels of calcium within cells using calcium-sensitive fluorescent dyes; calcium dynamics within the cytoplasm of neurons reflect the integration of synaptic inputs and the generation of postsynaptic electrical activity.

calcium pump ATPases that remove calcium ions from the cytoplasm of cells. Calcium pumps are found both on the plasma membrane and intracellular membranes, such as the endoplasmic reticulum.

calmodulin A calcium-binding protein that serves as a sensor for many calcium-regulated intracellular signaling processes.

cAMP response element binding protein (CREB) A protein activated by cyclic AMP that binds to specific regions of DNA, thereby increasing the transcription rates of nearby genes.

CAMs The general abbreviation for all *c*ell *a*dhesion *m*olecules.

cataracts Opacities in the lens of the eye that cause a loss of transparency and, ultimately, degrading vision.

catecholamines A term referring to molecules containing a catechol ring and an amino group; examples are the neurotransmitters epinephrine, norepinephrine, and dopamine.

cauda equina The collection of segmental ventral and dorsal roots that attach to the lumbosacral enlargement through the caudal segments of the spinal cord and pass out of the spinal canal through caudal intervertebral foramina and the sacrum.

caudate One of the three major components of the striatum (the other two are the putamen and nucleus accumbens)

cavernous sinus Dural venous sinus in the anterior middle cranial fossa; drains venous blood from ventral cerebral hemisphere and the face. Cranial nerves III, IV, two divisions of V, and VI, and the internal carotid artery all pass through the cavernous sinus.

cell-associated signaling molecules Chemical signals that are attached to, or embedded in, the plasma membrane.

cell-impermeant molecules Chemical signals that are incapable of permeating the plasma membrane, either because they are too hydrophobic or are attached to the membranes of other nearby cells. Such molecules activate intracellular signaling by activating receptors on the plasma membrane of their target cells.

cell-permeant molecules Hydrophobic chemical signals that are capable of permeating the plasma membranes of their target cells. Such molecules activate intracellular signaling by activating receptors in the cytoplasm or nucleus of their target cells.

central autonomic network Collection of nuclei and cortical regions in the brain that integrate visceral sensory signals, distribute those signals to more widespread brain regions, and give rise to signals that govern visceral motor activity.

central nervous system (CNS) The brain and spinal cord of vertebrates (by analogy, the central nerve cord and ganglia of invertebrates).

central pattern generators Oscillatory spinal cord or brainstem circuits responsible for programmed, rhythmic movements such as locomotion.

central sensitization Increased excitability of neurons in the dorsal horn following high levels of activity in nociceptive afferents and that can result in hyperalgesia and allodynia.

central sulcus A major sulcus on the dorsolateral aspect of the cerebral hemispheres that forms the boundary between the frontal and parietal lobes. The anterior bank of the sulcus contains the primary motor cortex; the posterior bank contains the primary sensory cortex.

cephalic flexure Sharp bend in the neural tube which during early neurulation balloons out to form the prosencephalon, which in turn will give rise to the forebrain and later to the cerebral hemispheres.

cerebellar ataxia A pathological inability to make coordinated movements, associated with lesions or congenital malformation of the cerebellum.

cerebellar cortex Laminated, superficial gray matter of the cerebellum.

cerebellar peduncles Three bilateral pairs of tracts that convey axons to and from the cerebellum. The **superior cerebellar peduncle**, or **brachium conjunctivum**, is primarily an efferent motor pathway; the **middle cerebellar peduncle**, or **brachium pontis**, is an afferent pathway arising from the pontine nuclei. The smallest but most complex is the **inferior cerebellar peduncle**, or **restiform body**, which encompasses multiple afferent and efferent pathways.

cerebellum Prominent hindbrain structure concerned with motor coordination, posture, and balance; derived from the embryonic metencephalon. Composed of a three-layered cortex and deep nuclei; attached to the brainstem by the cerebellar peduncles.

cerebral achromatopsia Loss of color vision as a result of damage to extrastriate visual cortex.

cerebral akinetopsia A rare disorder in which one is unable to appreciate the motion of objects.

cerebral angiography An x-ray based means for imaging blood vessels in the brain involving injection of a contrast agent into the systemic circulation.

cerebral aqueduct Narrow channel derived from the lumen of the neural tube in the dorsal mesencephalon that connects the third and fourth ventricles. Also called the aqueduct of Sylvius.

cerebral cortex The superficial gray matter of the cerebral hemispheres derived from the outer aspect of the telencephalic vesicles.

cerebral hemispheres The two symmetrical halves of the cerebrum derived from the telencephalic vesicles.

cerebral nuclei Masses of gray matter located in the deep or basal regions of the cerebral hemispheres, including the basal ganglia, basal forebrain nuclei, septal nuclei, and nuclear components of the amygdala.

cerebral peduncles Paired "stalks" (peduncle means stalk) of white matter that define the ventral aspect of the midbrain; contain major axon tracts that originate in the cerebral cortex and terminate in the brainstem and spinal cord, including the corticopontine, corticobulbar and corticospinal tracts.

cerebrocerebellum Lateral part of the cerebellar hemisphere, greatly expanded in humans, that receives input from the cerebral cortex via axons from pontine relay nuclei and sends output to the premotor and prefrontal cortex via the thalamus; concerned with the planning and execution of complex spatial and temporal sequences of skilled movement.

cerebrospinal fluid (CSF) Clear and cell-free fluid that fills the ventricular system of the central nervous system produced by choroid plexus in the ventricles.

cervical Rostral region of the spinal cord related to the upper trunk and upper extremities.

cervical enlargement The spinal cord expansion that relates to the upper extremities; includes spinal segments C3–T1.

channel-linked receptors Receptors that are ligand-gated ion channels; binding of ligands leads to channel opening.

channelrhodopsin Typically, a protein that, in response to light of the proper wavelength, opens a channel that is permeable to cations; when engineered into a neuron for optogenetics, it depolarizes the neuron when exposed to light; anion-conducting channelrhodopsins have also been discovered, which would have inhibitory effects when activated in mature neurons.

characteristic frequency The lowest threshold of a tuning curve.

chemical synapses Synapses that transmit information via the secretion of chemical signals (neurotransmitters).

chemoaffinity hypothesis The idea that nerve cells bear chemical labels that determine their connectivity.

cholinergic nuclei Nuclei in which synaptic transmission is mediated by acetylcholine.

Chordin An endogenous antagonist of Bmps that acts in combination with Noggin.

choroid plexus Specialized, highly vascularized epithelium in the ventricular system that produces cerebrospinal fluid.

chromosomal sex The sex of an individual organism based upon the sex chromosomes in its genome.

chronic traumatic encephalopathy (CTE) A neurological syndrome that is caused by repeated concussive force delivered to the head. This syndrome is common in athletes who sustain constant forceful blows to the head, including boxers, hockey players and football players. In addition, soldiers exposed to repeated explosive blasts at close range are at risk.

While alive, individuals with CTE suffer a gradual dementia like decline in cognition and social behaviors. At autopsy, their brains have several signs of neurodegenerative and neuroinflammatory damage, and resemble those of individuals, often far older in age, with Alzheimer's disease or other age-related neurodegenerative disorders.

ciliary body Two-part ring of tissue encircling the lens of the eye. The muscular component is important for adjusting the refractive power of the lens. The vascular component produces the fluid that fills the front of the eye.

ciliary muscle Muscle that controls the shape of the lens.

cingulate gyrus Prominent gyrus on the medial aspect of each cerebral hemisphere, lying just superior to the corpus callosum; a major component of the limbic forebrain.

cingulate sulcus Prominent sulcus on the medial aspect of each cerebral hemisphere.

circadian Refers to variations in physiological functions that occur on a daily basis.

circle of Willis Arterial anastomosis on the ventral aspect of the midbrain; connects the posterior and anterior cerebral circulation.

circumvallate papillae Circular structures that form an inverted *V* on the rear of the tongue (three to five on each side, with the largest in the center). Circumvallate papillae are moundlike structures surrounded by a trench (like a moat). These papillae are much larger than fungiform papillae. Compare *foliate papillae* and *fungiform papillae.*

cisterns Large, cerebrospinal fluid-filled spaces that lie within the subarachnoid space.

Clarke's nucleus A group of relay neurons (also called the dorsal nucleus of Clarke) located in the medial aspect of the intermediate gray matter of the spinal cord (lamina VII) in spinal levels T1 through L2–3; conveys proprioceptive signals originating in the lower body to the ipsilateral cerebellum and the dorsal column nuclei via the dorsal spinocerebellar tract.

classical conditioning Also called *conditioned reflex.* The modification of an innate reflex by associating its normal triggering stimulus with an unrelated stimulus. The unrelated stimulus comes to trigger the original response by virtue of this repeated association. Compare *operant conditioning.*

clathrin The most important protein for endocytotic budding of vesicles from the plasma membrane; its three-pronged "triskelia" attach to the vesicular membrane to be retrieved.

CLCN genes Cl^- channel genes.

climbing fibers Axons that originate in the inferior olivary nuclei, ascend through the inferior cerebellar peduncle, and make terminal arborizations that invest the proximal dendritic trees of Purkinje cells; induce complex spikes and long-term depression in cerebellar Purkinje neurons.

co-transmitters Two or more types of neurotransmitters within a single synapse; may be packaged into separate populations of synaptic vesicles or co-localized within the same synaptic vesicles.

co-transporters Active transporters that use the energy from ionic gradients to carry multiple ions across the membrane in the same direction.

coccygeal Most caudal region of the spinal cord.

cochlea The coiled structure in the inner ear where vibrations caused by sound are transduced into neural impulses.

cochlear microphonics The generation of sound by the cochlea, thought to be driven by the cochlear amplifier.

cognitive neuroscience The field of neuroscience devoted to studying and understanding cognitive functions.

coincidence detector A device that detects the simultaneous presence of two or more signals. In the context of long-term synaptic plasticity, a mechanism for detecting the coioncidence of two or more synaptic signals; for example, NMDA receptors detect the simultaneous occurrence of presynaptic glutamate release and postsynaptic depolarization during long-term synaptic potentiation.

coincidence detectors A neuron that detects simultaneous events, as in sound localization.

collagens Fibrillary extracellular matrix proteins with multiple binding domains for a variety of cell surface receptors.

columns Term used to describe an elongated gray matter structure (e.g., the motor neuronal pool in the ventral horn of the spinal cord that innervates a muscle) or a subdivision of white matter (e.g., a region of white matter in the spinal cord containing long axon tracts).

coma A pathological state of profound and persistent unconsciousness.

commissures Axon tracts that cross the midline of the brain or spinal cord.

common currency theory A hypothesis about the function of the reward system. In this theory, the brain uses a single scale to compare the values of all goods; the scale is consistent for different items regardless of their type.

competitive interaction The struggle among nerve cells, or nerve cell processes, for limited resources essential to survival or growth.

computational map An assembly of neural circuits in a specific brain region that represent inputs that do not have a direct correspondence to a topographic map, such as those in the somatosensory or visual systems. Some cognitive capacities, including language and declarative memory, are thought to depend on computational maps.

computerized tomography (CT) Radiographic procedure in which a three-dimensional image of a body structure is constructed by computer from a series of cross-sectional X-ray images.

concha A component of the external ear.

conditional mutations A genetic engineering approach, typically reliant upon the Cre/lox system, whereby an exogenous recombinase enzyme recognizes unique DNA excision sequences (loxP sequences) introduced at the 5′ and 3′ ends of an endogenous gene and eliminates the intervening sequence.

conditioned learning The generation of a novel response that is gradually elicited by repeatedly pairing a novel stimulus with a stimulus that normally elicits the response being studied.

conduction aphasia Difficulty producing speech as a result of damage to the connection between Wernicke's and Broca's language areas.

conduction velocity The speed at which an action potential is propagated along an axon.

conductive hearing loss Diminished sense of hearing due to the reduced ability of sounds to be mechanically transmitted to the inner ear. Common causes include occlusion of the ear canal, perforation of the tympanic membrane, and arthritic degeneration of the middle ear ossicles. Contrast with sensorineural hearing loss.

cones Photoreceptor cells specialized for high visual acuity and the perception of color.

conflict A psychological process that arises when multiple competing demands compete for control of behavior or attention. It is usually associated with increased error rates and/or reaction times.

confluence of sinuses Dural venous sinus formed by the junction of the superior sagittal sinus with the transverse sinuses.

congenital adrenal hyperplasia (CAH) Genetic deficiency that leads to overproduction of androgens and a resultant masculinization of external genitalia in genotypic females.

conjugate eye movements The paired movements of the two eyes in the same direction, as occurs in saccades, smooth pursuit eye movements, optokinetic movements, and the vestibulo-ocular reflex.

connexins Transmembrane proteins that serve as the subunits of connexons, the transcellular channels that permit electrical and metabolic coupling between cells at electrical synapses. See gap junctions.

connexons Precisely aligned, paired transmembrane channels that form gap junctions between cells. They are formed from **connexins**, members of a specialized family of channel proteins.

contralateral neglect syndrome Neurological condition in which the patient does not acknowledge or attend to the left visual hemifield or the left half of the body. The syndrome typically results from lesions of the right parietal cortex.

convergence Innervation of a target cell by axons from more than one neuron. In vision refers specifically to the convergence of both rod and cone photoreceptor cells onto retinal ganglion cells.

cornea The transparent surface of the eyeball in front of the lens; the major refractive element in the optical pathway.

coronal Standard anatomical planes of section; any vertical plane passing parallel to the medial-to-lateral axis through the head (in humans, parallel to the face), dividing the head into anterior (front) and posterior (rear) segments. Also known as *frontal*.

corpus callosum The large medial fiber bundle that connects the cortices of the two cerebral hemispheres.

cortex (pl. cortices) The superficial mantle of gray matter (a sheet-like array of nerve cells) covering the cerebral hemispheres and cerebellum, where most of the neurons in the brain are located.

corticobulbar tract Pathway carrying motor information from the motor cortex to brainstem nuclei.

corticocortical connections Connections made between cortical areas in the same hemisphere, or between corresponding areas in the two hemispheres via the cerebral commissures.

corticospinal tract Pathway carrying motor information from the motor cortex to the spinal cord. Essential for the performance of discrete voluntary movements, especially of the hands and feet.

corticostriatal pathway Excitatory (glutamateric) projections from deep layers of the cerebral cortex to the striatum. Projections are organized topographically with distinct cortical areas projecting to distinct divisions of the striatum.

covert attention The focusing of visual attention toward a location or item in the visual field without shifting the direction of gaze. Can apply to other sensory modalities or to attentional paradigms. Compare *overt attention*.

cranial nerve ganglia The sensory ganglia associated with the cranial nerves; these correspond to the dorsal root ganglia of the segmental nerves of the spinal cord.

cranial nerves The 12 pairs of sensory, motor, and mixed sensorimotor nerves that innervate targets in the head and neck.

cranial placodes The local thickening of the non-neural surface ectoderm of the mid-gestation embryo that undergo a form of neural induction so that they can generate cranial peripheral sensory neurons (mechanoreceptor cells in the cranial sensory ganglia, olfactory and vomeronasal receptor neurons in the nose, and hair cells in the inner ear) as well as the lens in the eye.

Cre recombinase A viral DNA cutting enzyme used to excise a floxed exons. See Cre/lox.

Cre/lox A genetic engineering system for achieving conditional mutations of endogenous mammalian genes using introduced loxP sequences, which are not found in mammalian genomes but occur in bacterial genomes and are targeted by certain viruses, and a viral DNA cutting enzyme, Cre recombinase. With expression of the Cre DNA introduced into host genome, the resulting Cre recombinase engages the loxP binding sites, and the intervening endogenous exon targeted for elimination (the so-called floxed sequence) is excised.

credit assignment When multiple events occur and are associated with different values, the decision-maker must determine which event produces which value. This process, known as credit assignment, is trivially easy in many cases, but in other cases, can be quite difficult.

cribiform plate A bony structure of the facial portion of the skull comprising many small fensetra (tiny holes), at the level of the eyebrows, that separates the olfactory epithelium from the brain. The axons from the olfactory sensory neurons pass through the tiny holes of the cribriform plate to enter the brain.

crista The hair cell-containing sensory epithelium of the semicircular canals.

critical periods Restricted developmental periods during which the nervous system is particularly sensitive to the effects of experience.

cuneate nucleus Somatosensory relay nucleus in the lower medulla containing second-order sensory neurons that relay mechanosensory information originating in peripheral receptors in the upper body (excluding the face) to the contralateral thalamus.

cuneate tract Lateral division of the dorsal column in the upper half of the spinal cord containing the central processes of first-order afferents and the postsynaptic dorsal column projection; conveys mechanosensory signals derived from the upper body excluding the face.

cuneus gyrus Gyral structure on the superior aspect of the medial occipital lobe forming the upper bank of the calcarine sulcus; portion of the primary visual cortex that represents the inferior quadrant of the contralateral visual hemifields.

cupula Gelatinous structure in the semicircular canals in which the hair cell bundles are embedded.

cyclic nucleotide-gated channels A class of ion channels that are activated and inactivated by second messenger cascades. These second messenger cascades usually involve the activation of a G-protein that is coupled to a G-protein-coupled receptor leading to increased phosphorylation capacity of adenylyl or guanyl cyclases: enzymes that can phosphorylate the channels and modify their permeability to ions.

cytoarchitectonic areas Distinct regions of the neocortical mantle identified by differences in cell size, packing density, and laminar arrangement (layering). Most prominent in humans is the 6-layered neocortex. The evolutionary older archicortex (or hippocampal cortex) has 3–4 layers, and the ancient paleocortex has 3 layers.

D

decerebrate rigidity Excessive tone in extensor muscles as a result of damage to descending motor pathways at the level of the brainstem.

declarative memory Memories available to consciousness that can be expressed by language.

decussation of the superior cerebellar peduncle Midline crossing of fibers in the superior cerebellar peduncles as they converge in the tegmentum of the rostral pons.

deep cerebellar nuclei Subcortical nuclei at the base of the cerebellum that give rise to output from the cerebellum to the thalamus and brainstem; integrate afferent signals to the cerebellum and cortical processing conveyed by Purkinje neurons.

delayed response genes Genes whose protein products are not produced rapidly after a triggering stimulus. The delay in gene expression is caused by the requirement for transcriptional regulators that must first be synthesized in response to the stimulus.

delayed response task A behavioral paradigm used to test cognition and memory.

delta ligands Transmembrane proteins whose ectodomain (the region of the protein that extends beyond the cell's outer membrane) binds to receptors (notch proteins) on the surfaces of adjacent cells to initiate a signaling cascade that promotes local cellular differentiation.

delta waves Slow (<4 Hz) electroencephalographic waves that characterize stage IV (slow-wave) sleep.

dendrites Neuronal processes (typically, much shorter than the axon) arising from the nerve cell body that receive synaptic input.

dendritic polarization The process by which the dendrite of a nerve cell, which is the primary site for synaptic input at post-synaptic specializations becomes distinguished from the axon, which is the primary site for synaptic output at presynaptic terminals.

dentate nucleus Largest and most lateral of the deep cerebellar nuclei; source of output from the cerebrocerebellum to the premotor and prefrontal cortex via the thalamus and to the parvocellular red nucleus.

depolarization Displacement of a cell's membrane potential toward a less negative value.

dermatomes The area of skin supplied by the sensory axons of a single dorsal root ganglion.

dichromatic The color vision that arises when animals have only two cone types.

diencephalon Portion of the brain derived from the posterior part of the embryonic forebrain vesicle that lies just rostral to the midbrain; comprises the thalamus and hypothalamus.

diffusion tensor imaging (DTI) A type of magnetic resonance imaging used in live humans that allows for the selective visualization of large axon tracts in the brain based upon the alignment of the water molecules in myelinated axons bundled together and extending in the same direction.

dimorphic Having two different forms depending on genotypic or phenotypic sex.

diplopia Double vision.

disconjugate (disjunctive) eye movements Movements of the two eyes in opposite directions, such as in vergence eye movements.

dissociated sensory loss Loss of mechanosensation on one side of the body accompanied by pain and temperature deficits on the other side of the body, often caused by lateral hemisection of the spinal cord.

divergence The branching of a single axon to innervate multiple target cells.

dopamine A catecholamine neurotransmitter that is involved many brain functions, including motivation, reward and motor control.

dorsal anterior cingulate cortex (dACC) A functional division of the prefrontal cortex that surrounds the corpus callosum and includes the cingulate gyrus; important in emotional and visceral motor behavior.

dorsal columns Major ascending tracts in the spinal cord that carry mechanosensory information from first-order sensory neurons in dorsal root ganglia and second-order, postsynaptic dorsal column projection neurons to the dorsal column nuclei; also called the posterior funiculi.

dorsal horns The dorsal portions of the spinal cord gray matter derived from the alar plate; populated by neurons that process somatosensory information.

dorsal motor nucleus of the vagus nerve Visceral motor nucleus of the rostral medulla containing parasympathetic preganglionic neurons that innervate thoracic and upper abdominal visceral, mediating a range of autonomic functions.

dorsal nucleus of Clarke Column of relay neurons in the medial aspect of the intermediate gray matter of the spinal cord from T1–L3; receives first-order sensory signals from proprioceptors that supply the lower body and gives rise to ipsilateral dorsal spinocerebellar tract. Also called Clarke's nucleus.

dorsal root ganglia The segmental sensory ganglia of the spinal cord; they contain the cell bodies of the first-order neurons of all somatic sensory and visceral sensory pathways arising in the spinal cord.

dorsal roots The bundle of axons that runs from the dorsal root ganglia to the dorsal horn of the spinal cord, carrying somatosensory information from the periphery.

dorsal spinocerebellar tract Axonal projection arising from Clarke's nucleus to the ipsilateral cerebellum and the dorsal column nuclei; conveys proprioceptive signals originating in the lower body.

dorsolateral geniculate nucleus The portion of the thalamus that sends (and receives) axons to the cerebral cortex via the internal capsule.

dorsolateral prefrontal cortex (DLPFC) A functional division of the prefrontal cortex roughly corresponding to the middle and superior frontal gyri, as located anterior to motor cortex and the frontal eye fields. Compare ventrolateral prefrontal cortex.

dorsolateral tract of Lissauer A small bundle of mostly unmyelinated axons that is situated on the posterior margin of the dorsal horn and that conveys pain information to second order neurons in Rexed's laminae 1, 2, and 5.

dreaming A unique state of awareness that entails some features of memory and hallucinations in the sense that the experience of dreams is not related to corresponding sensory stimuli arising from the present environment.

DSCAM A class of transmembrane cell adhesion molecules encoded by the DSCAM genes. In *Drosophila*, the DSCAM gene structure allows for a total of 37,000 possible splice variants based upon its complex structure of exons for alternative splicing. The DSCAMs can either bind homophilically (binding between the same isoforms) or heterophilically (binding between different isoforms) to initiate avoidance of dendrites or axon branches from the same neuron, or recognition and apposition of dendrites or axons from different cells.

dura mater The thick external covering (dura mater means "tough mother") of the brain and spinal cord; one of the three components of the meninges, the other two being the pia mater and arachnoid mater

dynamin A GTP-hydrolyzing enzyme involved in the fission of membranes during endocytosis.

dynorphins A class of endogenous opioid.

dysdiadochokinesia Difficulty performing rapid alternating movements.

dysmetria Inaccurate movements due to faulty judgment of distance, especially over- or underreaching; characteristic of cerebellar pathology.

E

ectoderm The most superficial of the three embryonic germ layers; gives rise to the nervous system and epidermis.

Edinger–Westphal nucleus Midbrain visceral motor nucleus containing the parasympathetic preganglionic neurons that constitute the efferent limb of the pupillary light reflex.

efferent neurons Neurons or axons that conduct information away from the central nervous system toward the periphery.

electrical synapse Synapses that transmit information via the direct flow of electrical current at gap junctions.

electrochemical equilibrium The condition in which no net ionic flux occurs across a membrane because ion concentration gradients and opposing transmembrane potentials are in exact balance.

electroencephalography (EEG) The study of electrical potentials generated in the brain recorded from electrodes placed on the scalp.

electrophysiological recording Measure of the electrical activity across the membrane of a nerve cell by use of electrodes.

embryonic stem cells (ES cells) Cells derived from pre-gastrula embryos that have the potential for infinite self-renewal and can give rise to *all* tissue and cell types of the organism. See also glial stem cells; neural stem cells.

emmetropic Having normal vision.

end plate current (EPC) A macroscopic postsynaptic current resulting from the summed opening of many ion channels; produced by neurotransmitter release and binding at the motor end plate.

end plate potential (EPP) Depolarization of the membrane potential of skeletal muscle fiber, caused by the action of the transmitter acetylcholine at the neuromuscular synapse.

end plates The complex postsynaptic specializations at the site of nerve contact on skeletal muscle fibers.

endocannabinoids A family of endogenous signals that participate in several forms of synaptic transmission, interacting cannabinoid receptors. These receptors are the molecular targets of the psychoactive component of the marijuana plant, *Cannabis*.

endocrine Referring to the release of signaling molecules whose effects are made widespread by distribution in the general circulation.

endoderm The innermost of the three embryonic germ layers. Gives rise to the digestive and respiratory tracts and the structures associated with them.

endogenous antagonists The endogenous antagonists are secreted proteins from the notochord and other sources that bind to Bmp ligands and inactivate them. The specification of the neuroectoderm relys upon the activity of the endogenous antagonists to prevent the undifferentiated ectoderm from becoming epidermal ectoderm.

endogenous attention A form of attention in which processing resources are directed voluntarily to specific aspects of the environment; typically prompted by experimental instructions or, more normally, by an individual's goals, expectations, and/or knowledge. Compare *exogenous attention*.

endogenous opioids Peptide neuritransmitters in the central nervous system that have the same pharmacological effects as morphine and other derivatives of opium, being agonists at opioid receptors, virtually all of which contain the sequence Tyr-Gly-Gly-Phe. There are three classes: dynorphins, endorphins, and enkephalins.

endolymph The potassium-rich fluid filling both the cochlear duct and the membranous labyrinth; bathes the apical end of the hair cells.

endorphins A family of neuropeptides first identified as endogenous mimics of morphine. These neuropeptides, as well as morphine, act by activating opioid receptors.

engram A term used to describe the physical basis of a stored memory.

enkephalins A class of endogenous opioid.

enteric nervous system (ENS) A subsystem of the visceral motor system, made up of small ganglia and individual neurons scattered throughout the wall of the gut; influences gastric motility and secretion. Also called the *enteric system*.

enzyme-linked receptors Receptors that activate intracellular signaling processes via their enzymatic activity. Most of these receptors are protein kinases that phosphorylate intracellular target proteins.

ephrin A large family of transmembrane cell surface adhesion molecules, also referred to as *Eph* ligands that can bind to and activate the protein kinase activity of ephrin receptors. Ephrins also can initiate retrograde signaling in the cell whose membrane they are embedded in.

ephrin receptors Ephrin or Eph receptors are a large family of transmembrane receptor tyrosine kinases (RTKs). Upon binding of an ephrin (also known as an eph ligand) embedded in the membrane of a neighboring cell or process, the tyrosine kinase domain of the receptor is activated leading to phosphorylation of multiple protein targets and initiation of signaling that changes the cytoskeleton, cell motility and gene expression.

epibrachial placodes The cranial placodes found in the branchial arches that contribute neural precursors to cranial sensory ganglia to generate mechanoreceptor sensory neurons.

epinephrine Catecholamine hormone and neuro-transmitter that binds to adrenergic G-protein-coupled receptors.

epithelial-to-mesenchymal transition The dissolution of tight junctions and other molecular specializations that maintain cells in a sheet like (epithelial) arrangement so that the liberated cells can acquire motile capacity and migrate to distal locations. The delamination and migration of the neural crest is the best-known example of an epithelial-to-mesenchymal transition in the embryo. These changes, resulting in migratory cells, can also be pathological in mature tissues, leading to metastatic tumor formation when transformed cells escape epithelial constraints.

equilibrium potential The membrane potential at which a given ion is in electrochemical equilibrium.

error-related negativity (ERN) An electrophysiological marker that occurs when participants make errors in cognitive tasks.

esotropia A type of ocular misalignment in which one or both eyes of an individual turn inward (toward the nose).

estradiol The principal estrogen, secreted by ovarian follicles.

estrogen A steroid hormone (including estradiol) that affects sexual differentiation during development and reproductive function and behavior in mature adults.

evaluation Assigning a specific value to an option or possible action. Usually based on learned associations with past experiences with similar options or actions.

event related potential (ERP) Averaged EEG recordings measuring time-locked brain responses to repeated presentations of a stimulus or repeated execution of a motor task.

excitatory Postsynaptic potentials that increase the probability of firing a postsynaptic action potential (**EPSPs).**

executive control The cognitive functions that allow flexible and goal-directed control of thought and behavior.

exocytosis A form of cell secretion resulting from the fusion of the membrane of a storage organelle, such as a synaptic vesicle, with the plasma membrane.

exogenous attention Also called *reflexive attention.* A form of attention in which processing resources are directed to specific aspects of the environment in response to a sudden stimulus change, such as a loud noise or sudden movement, that attracts attention automatically. Compare *endogenous attention.*

exotropia A type of ocular misalignment in which one or both eyes of an individual turn outward (toward the temporal bones, or the ears).

express saccades A reflexive type of saccade in response to the sudden appearance of a sensory stimulus; mediated by a direct pathway from the retina (and auditory and somatosensory centers in the brainstem) to the superior colliculus.

expressive aphasia See Broca's aphasia.

external cuneate nucleus A group of relay neurons just lateral to the cuneate nucleus in the caudal medulla; conveys proprioceptive signals originating in the upper body, excluding the face, to the ipsilateral cerebellum via the cuneocerebellar tract.

external segment A lateral subdivision of the globus pallidus.

exteroception The modality of touch and pressure.

extracellular matrix cell adhesion molecules Fibrillar proteins usually secreted by epithelial cells that form a macromolecular meshwork to provide a substrate for binding and stability of epithelial cells (via the basal lamina, which is composed of extracellular matrix adhesion molecules) or signals for cell motility and force generation. Fibronectin, laminins, and collagens are the main subtypes of extracellular matrix (ECM) cell adhesion molecules.

extracellular recording Recording the electrical potentials in the extracellular space near active neurons. Compare *intracellular recording.*

extrafusal fibers Fibers of skeletal muscles that generate primary biomechanical force during muscle contraction; a term that distinguishes ordinary muscle fibers from the specialized intrafusal fibers associated with muscle spindles.

F

facial nerve (VII) Cranial nerve VII, a mixed sensorimotor nerve that conveys visceral afferents (taste from anterior two-thirds of tongue) and somatosensory afferents (mechanosensation from skin on or near pinna) to the brainstem, and branchial motor efferents from the brainstem to the muscles of mastication and visceral motor efferents to lacrimal and salivary glands. Middle of three cranial nerves that attaches to the brainstem at the junction of the pons and medulla.

far cells Visual cortical neurons that change their rate of firing in response to retinal disparities that arise from points beyond the plane of fixation.

fast fatigable (FF) motor units Large motor units comprising large, pale muscle fibers that generate large amounts of force; however, these fibers have sparse mitochondria and are therefore easily fatigued. FF motor units are especially important for brief exertions that require large forces, such as running or jumping.

fast fatigue-resistant (FR) motor units Motor units of intermediate size comprising muscle fibers that are not as fast as FF motor units, but generate about twice the force of S motor units and are resistant to fatigue.

fastigial nucleus Most medial of the deep cerebellar nuclei; source of output from the median spinocerebellum to brainstem upper motor neurons.

fibroblast growth factor (FGF) A peptide growth factor, originally defined by its mitogenic effects on fibroblasts; also acts as an inducer during early brain development.

fibronectin A large cell adhesion molecule that binds integrins.

filopodia Slender protoplasmic projections, arising from the growth cone of an axon or a dendrite, that explore the local environment.

first pain A category of pain perception described as sharp.

flocculus Lateral portion of the vestibulocerebellum that receives input from the vestibular nuclei in the brainstem and the vestibular nerve; coordinates the vestibulo-ocular reflex and movements that maintain posture and equilibrium.

floorplate A specialized region of columnar neuroepithelial cells at the midline of the ventral neural tube, just above the notochord, that becomes a source for secreted signals, particularly Sonic hedgehog (Shh) that establish the ventral-dorsal pattern of the neural tube.

folia The gyral formations of the cerebellar cortex.

foliate papillae Folds of tissue containing taste buds. Foliate papillae are located on the rear of the tongue lateral to the circumvallate papillae, where the tongue attaches to the mouth. Compare *circumvallate papillae* and *fungiform papillae*.

foramina of Monro See interventricular foramina.

forebrain The anterior portion of the brain derived from the prosencephalon that includes the diencephalon and telencephalon.

fornix Axon tract, best seen from the medial surface of the divided brain, that interconnects the septal nuclei and hypothalamus with the hippocampus.

fourth ventricle The ventricular space derived from the lumen of the neural tube that lies between the pons and rostral medulla and the cerebellum.

fovea Area of the retina specialized for high acuity in the center of the macula; contains a high density of cones and few rods.

foveation Aligning the foveae with a visual target and maintaining fixation.

foveola Capillary-free and rod-free zone in the center of the fovea.

free nerve endings Afferent fibers that lack specialized receptor cells; they are especially important in the sensation of pain and tempertaure.

frontal See coronal.

frontal eye fields A region of the frontal lobe that lies in a rostral portion of the premotor cortex and that contains cells that respond to visual and motor stimuli.

frontal lobe The hemispheric lobe that lies anterior to the central sulcus and superior to the lateral fissure.

frontal-parietal attention network A postulated cortical network for controlling attention.

functional magnetic resonance imaging (fMRI) Magnetic resonance imaging that detects changes in blood flow and therefore identifies regions of the brain that are particularly active during a given task.

fundus The inner surface of the retina.

fungiform papillae Mushroom-shaped structures (maximum diameter 1 millimeter) that are distributed most densely on the edges of the tongue, especially the tip. Taste buds (an average of six per papilla) are buried in the surface.

fusiform gyrus See occipitotemporal gyrus.

G

G-protein-coupled receptors A large family of neurotransmitter or hormone receptors, characterized by seven transmembrane domains; the binding of these receptors by agonists leads to the activation of intracellular G-proteins. See also metabotropic receptors.

G-proteins Proteins that are activated by exchanging bound GDP for bound GTP (and thus also known as GTP-binding proteins).

gamma (γ) motor neurons Class of spinal motor neurons specifically concerned with the regulation of muscle spindle length; these neurons innervate the contractile elements of intrafusal muscle fibers in muscle spindles.

ganglia (sing. ganglion) Collection of hundreds to thousands of neurons found outside the brain and spinal cord along the course of peripheral nerves.

ganglion cells Neurons located in a ganglion.

ganglionic eminences The bilaterally symmetric accumulations of forebrain neural precursors at the ventromedial aspect of the telencephalon that give rise to the nuclei of the basal ganglia (caudate, putamen, globus pallidus, subthalamic nucleus) as well as the majority of GABAergic interneurons that migrate into the cerebral cortex.

gap junctions Specialized intercellular contacts formed by channels that directly connect the cytoplasm of two cells.

gastrulation The cell movements (invagination and spreading) that transform the embryonic blastula into the gastrula.

gate theory of pain The idea that the flow of nociceptive information through the spinal cord is modulated by the concomitant activation of low threshold mechanoreceptors.

gaze centers Collections of local circuit neurons in the reticular formation that organize the output of cranial nerves III, IV, and VI to control eye movements along the horizontal or vertical axis.

gene Hereditary unit located on the chromosomes; genetic information is carried by linear sequences of nucleotides in DNA that code for corresponding sequences of amino acids.

generator potential See receptor potential.

genetic analysis The analysis of the relationship between single genes and the phenotypes to which each gene contributes.

genetic engineering A methodological means for inducing mutations in genes or otherwise editing or altering the structure and/or the function of targeted genes for experimental or therapeutic benefit. Also called *reverse genetics*.

genome-wide association studies (GWAS) A statistical correlation of likely associated genes drawn from analyses of large cohorts of individuals with the same phenotype or clinical diagnoses.

genomics The comprehensive analysis of nuclear DNA sequences within or between species or individuals. Genomic analyses include "whole exome" evaluation in which all nuclear regions that code RNA transcripts are assessed or compared, and "whole genome" evaluation in which the entire DNA sequence of an individual or organism is assessed or compared.

germ layers The three layers—ectoderm, mesoderm, and endoderm—of the developing embryo from which all adult tissues arise. Neural cells and structures arise from the ectoderm and from the mesodermally-generated notochord.

ghrelin Peptide hormone secreted by the stomach prior to feeding thought to signal hunger.

glaucoma Condition in which the eye's aqueous humor is not adequately drained, resulting in increased intraocular pressure, reduced blood supply to the eye, and eventual damage to the retina.

glial cells (glia) The support cells associated with neurons (astrocytes, oligodendrocytes, and microglial cells in the central nervous system; Schwann cells in peripheral nerves; and satellite cells in ganglia).

glial scarring Local proliferation of glial precursors and extensive growth of processes from existing glia within or around the site of a brain injury.

glial stem cells Neural precursor cells in the adult brain that retain the capacity to proliferate and generate both additional precursor cells and differentiated glial cells (and, in some cases, differentiated neurons).

globus pallidus Principal component of the pallidum. External division is involved with the indirect pathway from striatum to pallidum. Internal division provides major output from the basal ganglia to motor circuits in the thalamus and brainstem.

glomeruli Characteristic collections of neuropil in the olfactory bulb; formed by dendrites of mitral cells and terminals of olfactory receptor cells, as well as processes from local interneurons.

glossopharyngeal nerve (IX) Cranial nerve IX, a mixed sensorimotor nerve that conveys visceral afferents (taste from posterior one-third of tongue and sensation from oropharynx and middle ear) and somatosensory afferents (mechanosensation from skin on or near pinna) to the brainstem, and branchial motor efferents from the brainstem to muscles of the soft palate and pharynx and visceral motor efferents to a salivary (parotid) gland. Attaches to the rostral medulla in a cleft between the inferior olive and the inferior cerebellar peduncle, just superior to the roots of the vagus nerve.

glutamate–glutamine cycle A metabolic cycle of glutamate release and resynthesis involving both neuronal and glial cells.

glymphatic system Vascular, glial, and lymphatic system that allows for the passage of cerebrospinal fluid (CSF) from arterial perivascular space through the substance of the brain and back into venous perivascular space; with this flow of CSF, metabolic wastes and discarded proteins are rinsed from the parenchyma and pass out of the brain, especially during sleep when extracellular spaces in the brain expand.

Goldman equation Mathematical formula that permits membrane potential to be calculated for case where a membrane is permeable to multiple ions.

G_{olf} A G-protein found uniquely in olfactory receptor neurons.

Golgi cells Inhibitory interneurons in the granular cell layer of the cerebellar cortex that provide inhibitory feedback from parallel fibers to granule cells, regulating the temporal properties of the granule cell input to the Purkinje cells.

Golgi tendon organs Receptors at the interface of muscle and tendon that provide mechanosensory information to the central nervous system about muscle tension.

gracile nucleus Somatosensory nucleus in the caudal medulla containing second-order sensory neurons that relay mechanosensory information originating in peripheral receptors in the lower body to the contralateral thalamus via the medial lemniscus.

gracile tract Medial division of the dorsal column containing the central processes of first-order afferents and the postsynaptic dorsal column projection; conveys mechanosensory signals derived from the lower body (also conveys visceral pain signals from the lower abdominal viscera).

granulations See arachnoid villi.

gray matter General term that describes regions of the central nervous system rich in neuronal cell bodies and neuropil; includes the cerebral and cerebellar cortices, the nuclei of the brain, and the central portion of the spinal cord.

group Ib afferents Axons of primary sensory neurons that innervate Golgi tendon organs.

growth cone The specialized end of a growing axon (or dendrite) that generates the motive force for elongation.

GTP-binding protein (G-protein) Proteins that are activated by exchanging bound GDP for bound GTP.

gustatory nucleus The portion of the solitary nucleus in the brainstem that receives input from sensory neurons that are activated by taste receptors cells in taste buds in the tongue. The neurons of the gustatory nucleus then send their axons to the ventral posterior nucleus of the thalamus to relay taste information to the insular cortex.

gustducin A specialized G-protein that is activated by binding of specific tastants to a subclass of G-protein coupled taste receptors.

gyri (sing. gyrus) Folds in the cerebral cortex. Also called *convolutions*.

gyrification index Measure of cortical folding (in relation to brain weight or volume).

H

habituation Reduced behavioral responsiveness to the repeated occurrence of a sensory stimulus.

hair cells The sensory cells in the inner ear that transduce mechanical displacement into neural impulses.

halorhodopsin A protein that, in response to light of the proper wavelength, opens a channel that is selectively permeable to chloride ions; when engineered into a mature neuron for optogenetics, it inhibits the neuron when exposed to light.

haptics The exploration and perception of somatosensory stimuli using active touching and proprioception.

Hebb's postulate The idea that when pre- and postsynaptic neurons fire action potentials at the same time, the synaptic association between those cells strengthens. Sometimes phrased as "cells that fire together wire together," the postulate provides one explanation for the formation of certain neural networks.

helicotrema The opening at the apex of the cochlea that joins the perilymph-filled cavities of the scala vestibuli and scala tympani.

hemianopsia A loss of vision confined to either the nasal or temporal visual fields of each eye. Also called *heteronomous hemianopsia*.

hemiballismus A basal ganglia syndrome resulting from damage to the subthalamic nucleus and characterized by involuntary, ballistic movements of the limbs.

hemispatial (or **contralateral) neglect syndrome)** Neurological condition in which the patient does not acknowledge or attend to the left visual hemifield or the left half of the body. The syndrome typically results from lesions of the right parietal cortex.

heteronomous hemianopsia A loss of vision confined to the temporal visual field of each eye. Also called *bitemporal hemianopsia*. Due to lesions at the optic chiasm.

heterotrimeric G-proteins A large group of proteins consisting of three subunits (α, β, and γ) that can be activated by exchanging bound GDP for GTP, resulting in the liberation of two signaling molecules—αGTP and the $\beta\gamma$ dimer.

hindbrain See rhombencephalon.

hippocampus A cortical structure in the dorsomedial margin of the parahippocampal gyrus; in humans, concerned with short-term declarative memory, among many other functions.

histamine A biogenic amine neurotransmitter, derived from the amino acid histidine, that is involved in arousal, attention, and other central and peripheral functions.

histamine-containing neurons Neurons in the tuberomammillary nucleus (TMN) of the hypothalamus that contribute to the neuronal basis of wakefulness.

homeobox genes The family of genes that encode transcription factor proteins with a homeodomain DNA binding domain (the homeobox). The homeobox genes are essential for early anterior-posterior patterning in most animal embryos.

homeotic genes Genes that determine the developmental fate of an entire segment of an animal. Mutations in these genes drastically alter the characteristics of the body segment (as when wings grow from a fly body segment that should have produced legs).

hominins Term for ancestral humans.

homologous recombination An endogenous cellular mechanism for DNA replication and repair involving DNA polymerases and ligases; may be used in genetic engineering to replace ("recombine") a native sequence of nucleotides in a gene with an exogenous sequence.

homonymous hemianopsia A loss of vision in both left and right hemifields due to lesions of the optic tract.

homonymous quadrantanopsia A loss of vision in both left of right quadrants of the visual field due to lesions of the optic radiation.

horizontal cells Retinal neurons that mediate lateral interactions between photoreceptor terminals and the dendrites of bipolar cells.

horizontal gaze center See paramedian pontine reticular formation.

horizontal sections Standard anatomical planes of section; when standing, horizontal sections are parallel to the ground. Also known as *axial sections.*

Hox genes A group of conserved genes characterized by a specific DNA sequence—the homeobox—and that specify body axis (particularly the anterior–posterior axis) and regional identity in the developing vertebrate embryo.

Hsc70 An ATP-hydrolyzing enzyme involved in dissociation of clathrin-coats on vesicles following endocytosis.

Huntington's disease An autosomal dominant genetic disorder in which a single gene mutation results in personality changes, progressive loss of control of voluntary movement, and eventually death. Primary target early in the disease is medium spiny neurons of the striatum that participate in the indirect pathway.

hydrocephalus Enlarged ventricles that can compress the brain and expand the entire cranium as a result of increased cerebrospinal fluid pressure (typically due to a mechanical outflow blockage).

hyperacusis A painful sensitivity to moderate or even low-intensity sounds.

hyperalgesia Increased perception of pain.

hyperopic Far sighted.

hyperpolarization The displacement of a cell's membrane potential toward a more negative value.

hypnotic Drug that induces sleep.

hypocretin Another name for orexin.

hypoglossal nerve (XII) Cranial nerve XII, a somatic motor nerve that conveys efferents from the brainstem (hypoglossal nucleus) to the extrinsic muscles of the tongue. Attaches to the rostral medulla in a cleft between the medullary pyramid and the inferior olive.

hypothalamus Heterogeneous collection of small nuclei in the base of the diencephalon that plays an important role in the coordination and expression of visceral motor activity, as well as neuroendocrine and somatomotor activities that promote homeostasis and allostasis. The diverse functions in which hypothalamic involvement is at least partially understood include: the control of blood flow, the regulation of energy metabolism, the regulation of reproductive activity, and the coordination of responses to threatening conditions.

I

immediate early gene A gene whose protein product is produced rapidly, within 30-60 minutes, after a triggering stimulus. These often serve as transcriptional activators for delayed response genes. *C-fos* is one of the best-known examples of an immediate early gene.

immediate memory Sensory impressions that last only a few seconds.

inactivation The time-dependent closing of ion channels in response to a stimulus, typically membrane depolarization.

inferior cerebellar peduncles See cerebellar peduncles.

inferior colliculi (sing. colliculus) Paired hillocks on the dorsal surface of the midbrain; concerned with auditory processing.

inferior division Referring to the region of the visual field of each eye that corresponds to the bottom half of the retina.

inferior frontal gyri Inferior of three, parallel longitudinal gyri anterior to the precentral sulcus; part of the prefrontal cortex and anterior premotor cortex, including Broca's area in the left hemisphere.

inferior oblique muscles Extraocular muscles that extort the eyeballs when in the primary position (eyes straight ahead) and rotate upward when in adduction.

inferior olivary nucleus Prominent nucleus in the ventral medulla; source of climbing fiber input to the contralateral cerebellum; induces complex spikes and long-term depression in cerebellar Purkinje neurons. Also called *inferior olive.*

inferior olive See inferior olivary nucleus.

inferior parietal lobules Gyral formations of the lateral (inferior) parietal lobes that are involved in associating somatosensory, visual, auditory, and vestibular signals and generating a neural construct of the body, the position of its parts, and its movements (body image or schema).

inferior rectus muscles Extraocular muscles that rotate the eyeballs downward.

inferior salivatory nuclei Visceral motor nucleus of the caudal pons containing parasympathetic preganglionic neurons that mediate salivation.

inferior temporal gyri Inferior of three, parallel longitudinal gyri that define the inferior-lateral margin of the temporal lobe; involved in associational functions pertaining to stimulus recognition.

infundibulum The connection between the hypothalamus and the pituitary gland; also known as the *pituitary stalk.*

inhibitory Postsynaptic potentials that decrease the probability that a postsynaptic cell will generate an action potential (**IPSPs**).

inositol trisphosphate (IP_3) receptor A ligand-gated ion channel in the endoplasmic reticulum membrane. This receptor binds to the second messenger, IP_3, and elevates cytoplasmic calcium concentration by mediating flux of calcium out of the lumen of the endoplasmic reticulum.

insular cortex (insula) The portion of the cerebral cortex that is buried within the depths of the lateral fissure by the growth inferior frontal and parietal lobes and the superior temporal lobe. The posterior portion of the insula is concerned largely with visceral and autonomic function, including taste, while more rostral portions are involved in implicit feelings and their impact on social cognition. Also called the insula.

insular taste cortex A neocortical region located with the gyri of the insula, a region covered by dorsal region of the temporal lobe and the ventral region of the parietal lobe. Taste information from the tongue.

integrins A family of receptor molecules found on growth cones that bind to cell adhesion molecules such as laminin and fibronection.

intelligence quotient (IQ) A widely used but highly controversial measure of human intelligence based on standard tests.

intention tremors Tremor that occurs while performing a voluntary motor act; characteristic of cerebellar pathology. Also called *action tremor.*

interhemispheric connections The corpus callosum and anterior commissure, together. Mediate corticocortical connections between cortical regions in the opposite hemispheres.

intermediolateral cell column Rod-shaped distribution of sympathetic preganglionic neurons in the lateral, intermediate gray matter of the spinal cord; in thoracic segments, accounts for a lateral protrusion of gray matter into the white matter known as the lateral horn.

internal arcuate fibers Axons of dorsal column nuclei in the caudal brainstem that sweep across the midline and turn in the rostral direction forming the medial lemniscus.

internal capsule Large, fan-shaped white matter tract that lies between the diencephalon and the basal ganglia, formed by the growth of axons supplying the cerebral cortex (mainly from the thalamus) and axons originating in the cerebral cortex and terminating in subcortical targets; features an anterior limb, a genu, and a posterior limb.

internal carotid arteries Large arteries, one on each side of the head, that carry blood to the head. They divide into an external branch (supplying the neck and face), and an internal branch (supplying the brain and eye).

internal segment A medial subdivision of the globus pallidus.

interneurons Technically, a neuron in the pathway between primary sensory and primary effector neurons; more generally, a neuron whose relatively short axons branch locally to innervate other neurons. Also known as *local circuit neuron.*

interoception The sense of the internal state of the organism.

interphotoreceptor retinoid binding protein (IRBP) A critical protein in the retinoid cycle.

interposed nuclei Intermediate deep cerebellar nuclei; source of output from the paramedian spinocerebellum to the motor cortex via the thalamus and to brainstem upper motor neurons.

intersexuality Having a biologically ambiguous or intermediate sex based either upon indeterminate chromosomal sex (XXY males), primary sex characteristics (differentiation of gonads, genitalia), or secondary sex characteristics.

interstitial nuclei of the hypothalamus (INAH) Four cell groups located slightly lateral to the third ventricle in the anterior hypothalamus of primates; thought to play a role in sexual behavior.

interventricular foramina Narrow channels that allow for the passage of cerebrospinal fluid from the paired lateral ventricles into the single third ventricle; form in each hemisphere between the fornix and medial aspect of the anterior thalamus. Also known as the *foramina of Monro.*

intracellular receptors Receptors that participate in signal transduction by binding to cell-permeant chemical signals. The activate form of these receptors typically interact with nuclear DNA and produce new mRNA and protein within target cells.

intracellular recording Recording the potential between the inside and outside of a neuron with a microelectrode. Compare *extracellular recording.*

intracellular signal transduction A process that converts binding of ligands to plasma membrane receptors to intracellular signaling processes.

intrafusal muscle fibers Specialized muscle fibers found in muscle spindles.

intraparietal sulcus Prominent longitudinal sulcus of the posterior parietal lobe that divides the superior and inferior parietal lobules.

ion channels Integral membrane proteins possessing pores that allow only certain ions to diffuse across cell membranes, thereby conferring selective ionic permeability.

ion exchangers Membrane transporters that exchange intracellular and extracellular ions against their concentration gradient by using the electrochemical gradient of other ions as an energy source. See also antiporters and co-transporters.

ion selectivity The ability of channels to discriminate between different ions.

ionotropic receptors Receptors in which the ligand binding site is an integral part of the receptor molecule.

J

joint receptors Mechanoreceptors found in and around joints; especially important for monitoring finger movements during fine manual manipulations.

jugular veins Principal means for draining venous blood from the cranium; arise from the sigmoid sinuses as they pass through the jugular foramina in the base of the skull.

K

kainate receptors See ionotropic glutamate receptors.

kairomones Volatile chemicals from other species indicating predator, prey, or symbiotic status that bind specifically to vomeronasal receptor proteins localized on subsets of vomeronasal sensory receptor cells.

KAL1 See anosmin.

KCN genes K^+ channel genes.

KIF21A A member of the Kinesin family of molecular motors that interact with the microtubule cytoskeleton as well as with macromolecules and organelles referred to as cargo so that the cargo is transported from the cell body to the periphery of the cell. KIF21A and related kinesins are essential for moving proteins and organelles from the cell body to the axon as well as transporting cargos back to the cell body, often as endosomes for signaling to the cell body, or as membrane bound compartments for protein turnover and degradation.

kinocilium A true ciliary structure which, along with the stereocilia, comprises the hair bundle of vestibular and fetal cochlear hair cells in mammals (it is not present in the adult mammalian cochlear hair cell).

koniocellulary (K-cell) pathway A third poorly understood pathway from retina to cortex characterized by the anatomical location of it cells in the lateral geniculate nucleus that process of short wavelength light.

L

L1 A member of the Ca^{2+} independent family of transmembrane cell surface adhesion molecules. L1 binds homophillically (thus to other L1 molecules on adjacent cells) It signals primarily through non-receptor tyrosine kinases Fyn and Src. It is particularly essential for the fasciculation (binding together) of axons within growing nerves.

labyrinth A set of interconnected chambers in the internal ear, comprising the cochlea, vestibular apparatus, and the bony canals in which these structures are housed.

lamellipodium A sheetlike extension, rich in actin filaments, on the leading edge of a motile cell or growth cone.

laminae (sing. lamina) Cell layers that characterize the neocortex, hippocampus, and cerebellar cortex. The gray matter of the spinal cord is also arranged in laminae.

laminins Large cell adhesion molecules that bind integrins. Laminin is a major component of the extracellular matrix.

large dense-core vesicles A type of synaptic vesicle characterized by a large diameter (typically 90–250 nanometers) and the presence of an electron-dense core. These vesicles typically contain and release neuropeptides.

lateral columns The lateral regions of spinal cord white matter that convey motor information from the brain to the ventral horn via the lateral corticospinal tract and convey proprioceptive signals from spinal cord neurons to the cerebellum via the spinocerebellar tracts.

lateral corticospinal tract Spinal portion of the corticospinal tract in the lateral column of the spinal cord derived from the contralateral motor cortex; governs skilled movements of the extremities.

lateral fissure The cleft on the lateral surface of the brain that separates the temporal lobe below from the frontal and parietal lobes above. Also called the *Sylvian fissure.*

lateral horns Lateral protrusion of intermediate gray matter into the adjacent white matter known; characteristic feature of the thoracic segments of the spinal cord. See intermediolateral cell column.

lateral olfactory tract The bundle of mitral and tufted cell axons that relay olfactory information to the accessory olfactory nuclei, the olfactory tubercle, the pyriform and entorhinal cortices, and portions of the amygdala.

lateral rectus muscles Extraocular muscles that rotate the eyeballs laterally.

lateral superior olive (LSO) A structure in the auditory brainstem that aids in sound localization that aids in sound localization by computing interaural intensity differences.

lateral ventricles Largest of the ventricles derived from the lumen of the neural tube that expanded in the formation of the paired telencephalic vesicles; components of the lateral ventricles are present in each lobe of the cerebral hemisphere: anterior (frontal) horn, body, atrium, posterior (occipital) horn, and temporal horn.

lateral vestibulospinal tract Ipsilateral projection of the anterior white matter of the spinal cord from the lateral vestibular nuclei to the medial ventral horn; mediates reflexes that activate extensor (anti-gravity) muscles with rapid lateral roll of the head, as when jostled to one side when standing on a moving train.

lenticulostriate arteries Numerous small branches of the middle and anterior cerebral arteries that penetrate deep into the anterior cerebral hemispheres supplying blood to deep white matter and cerebral nuclei, including the basal ganglia.

leptin Peptide hormone secreted by adipocytes following feeding thought to signal satiety.

lesion studies The method of observing and documenting change in function following damage (lesion) of a distinct brain region, nerve, or tract; damage may be acquired in humans or induced experimentally in non-human models; predominant method of studying the human nervous system prior to the advent of modern neurophysiological and brain imaging tools.

ligand-gated ion channels Ion channels that respond to chemical signals rather than to the changes in membrane potential generated by ionic gradients. The term covers a large group of neurotransmitter receptors that combine receptor and ion channel functions into a single molecule.

light adaptation Gain control of vision according to the prevailing level of illumination.

limbic forebrain Constellation of cortical and subcortical structures in the frontal and temporal lobes that form a medial rim of cerebrum roughly encircling the corpus callosum and diencephalon (limbic means "border" or "rim"). Comprises an olfactory division that processes olfactory cues; a parahippocampal division that generates cognitive maps in spatial frameworks that facilitate the acquisition of episodic and declarative memory; and an amygdaloid/orbital cortical division that is important in the experience and expression of emotion.

limbic lobe Cortex that lies superior to the corpus callosum on the medial aspect of the cerebral hemispheres; forms the cortical component of the limbic system.

limbic system Term that refers to those cortical and subcortical structures concerned with the emotions; the most prominent components are the cingulate gyrus, the hippocampus, and the amygdala.

lingual gyrus Gyral structure on the inferior aspect of the occipital lobe forming the lower bank of the calcarine sulcus; portion of the primary visual cortex that represents the superior quadrant of the contralateral visual hemifields.

lobes The four major divisions of the cerebral hemispheres named for the overlying cranial bones (frontal, parietal, occipital, and temporal).

local circuit neurons General term referring to a neuron whose activity mediates interactions among other neurons in the CNS; exemplified by short-axon neurons in the spinal cord that mediate transmission of signals from sensory neurons to motor neurons. Interneuron is often used as a synonym.

long circumferential arteries Long branches of the vertebral and basilar arteries that supply dorsolateral aspects of the brainstem and cerebellum.

long-term depression (LTD) A form of long-term synaptic plasticity that produces a persistent, activity-dependent weakening of synaptic transmission.

long-term memory Memories that last days, weeks, months, years, or a lifetime.

long-term potentiation (LTP) A form of long-term synaptic plasticity that produces a persistent, activity-dependent strengthening of synaptic transmission.

longitudinal Standard anatomical planes of section that pass through the CNS parallel to its long axis.

loudness The sensory quality elicited by the intensity of sound stimuli.

lumbar Caudal region of the spinal cord between the thoracic and sacral regions related to the lower extremities.

lumbosacral enlargement The spinal cord expansion that relates to the lower extremities; includes spinal segments L1–S2.

luminance The physical measure of light intensities.

luminance The physical measure of light intensities.

luminance contrast The difference between the level of illumination that falls on the receptive field center and the level of illumination that falls on the surround.

M

macroscopic currents Ionic currents flowing through large numbers of ion channels distributed over a substantial area of membrane.

macula The sensory epithelium of the otolith organs, comprising hair cells and associated supporting cells.

macula lutea The central region of the retina that contains the fovea (the term derives from the yellowish appearance of this region in ophthalmoscopic examination); also, the sensory epithelia of the otolith organs.

macular sparing The loss of vision throughout wide areas of the visual field, with the exception of foveal vision.

magnetic resonance imaging (MRI) A noninvasive technique that uses magnetic energy and radiofrequency pulses to generate images that reveal structural and/or functional information in the living brain.

magnetic source imaging (MSI) A non-invasive means for localizing brain activity that combines magnetoencephalography with structural magnetic resonance imaging.

magnetoencephalography (MEG) A passive and noninvasive functional brain-imaging technique that measures the tiny magnetic fields produced by active neurons, in order to identify regions of the brain that are particularly active during a given task.

magnocellular layers A component of the primary visual pathway specialized for the perception of motion; so named because of the relatively large ("magno") cells involved.

major histocompatibility complex (MHC) A macromolecular assembly of immune proteins that mediate recognition of antigens derived from an individual organism itself ("self") and those derived from external sources ("other"). The MHC provides a unique molecular signature for an individual, and is thought to act as a stimulus in for the vomeronasal system.

mammillary bodies Small prominences on the ventral surface of the posterior diencephalon; anatomically and functionally part of the caudal hypothalamus that is related to the hippocampus and its role in memory formation.

mechanoreceptors Receptors specialized to sense mechanical forces.

mechanosensitive Ion channels that respond to mechanical distortion of the plasma membrane.

medial geniculate complex (MGC) The thalamic nucleus in the primary auditory pathway. Compare *dorsolateral geniculate nucleus.*

medial lemniscus Axon tract in the brainstem that carries mechanosensory information from the dorsal column nuclei to the ipsilateral thalamus.

medial longitudinal fasciculus Axon tract that carries excitatory projections from the abducens nucleus to the contralateral oculomotor nucleus; important in coordinating conjugate eye movements.

medial nucleus of the trapezoid body (MNTB) A structure that provides inhibitory input to the lateral superior olive.

medial rectus muscles Extraocular muscles that rotate the eyeballs medially.

medial superior olive (MSO) A structure in the auditory brainstem that aids in sound localization by computing interaural time differences.

medial vestibulospinal tract Bilateral projection of the anterior-medial white matter of the spinal cord from the medial vestibular nuclei to the cervical cord; mediates reflexes that extend the arms and dorsiflex the neck with rapid downward pitch of the head, as when falling forward.

medium spiny neurons The principal projection neurons of the striatum.

medullary arteries Segmental branches of the descending aorta that supply blood to the vertebral column and the spinal cord.

medullary pyramids Longitudinal bulges on the ventral aspect of the medial medulla formed by the corticospinal tract and a small remnant of the corticobulbar tract.

Meissner afferents Encapsulated cutaneous mechanosensory receptors in the tips of dermal papillae specialized for the detection of fine touch and pressure.

Meissner's plexus See submucous plexus.

melanopsin A photopigment located in the retinal ganglion cells that help to set the biological clock.

melatonin Sleep-promoting neurohormone produced in the pineal gland.

membrane conductance The reciprocal of membrane resistance. Changes in membrane conductance result from, and are used to describe, the opening or closing of ion channels.

meninges The external covering of the brain and spinal cord; includes the pia, arachnoid, and dura mater.

Merkel cell afferents Peripheral axons of primary sensory neurons that make synaptic connections with Merkel cells forming Merkel cell-neurite complexes; Merkel afferents transduce the dynamic aspects of stimuli giving rise to fine touch sensations.

Merkel cell–neurite complexes Specializations at the boundary of the epidermis and dermis comprising Merkel cells in the tips of the primary epidermal ridges (coinciding with fingerprint ridges on the skin surface) and terminals of primary sensory afferents in the dermis; transduce light touch and pressure stimuli with high spatial resolution.

Merkel cells Specialized cells in the basal epidermis that contact Merkel afferents forming Merkel cell-neurite complexes;

Merkel cells signal the static aspect of a touch stimulus, such as light pressure, and release peptide neurotransmitters onto the terminals of Merkel afferents.

mesencephalic locomotor region Collection of neurons in the reticular formation of the midbrain tegmentum that can trigger locomotion and change the speed and pattern of the movement by changing the level of activity delivered to the spinal cord through reticulospinal projections originating in the pons and medulla.

mesencephalic trigeminal nucleus Array of pseudounipolar proprioceptive neurons in the rostral pons and midbrain (mesencephalon) on the ventral-lateral margin of the periaqueductal gray; mediates sensory limb of myotatic reflexes for jaw muscles and other striated muscles of the anterior cranium.

mesencephalon See midbrain.

mesoderm The middle of the three embryonic germ layers; gives rise to muscle, connective tissue, skeleton, and other structures.

mesopic vision Vision in light levels at which both the rods and cones are active.

metabotropic receptors Receptors that are indirectly activated by the action of neurotransmitters or other extracellular signals, typically through the aegis of G-protein activation. Also called *G-protein-coupled receptors.*

metencephalon The part of the embryonic hindbrain (the entire rhombencephalon at earliest stages) that generates the pons and the cerebellum, and the trigeminal (V), abducens (VI), facial (VII) and vestibulocochlear (VIII) cranial nerves and surrounds the fourth ventricle in the mature brainstem.

Meyer's loop That part of the optic radiation that runs in the caudal portion of the temporal lobe.

microglial cells One of the three major classes of glial cells found in the central nervous system derived primarily from hematopoietic precursor cells; function as scavenger cells that remove cellular debris from sites of injury or normal cell turnover, and secrete signaling molecules that modulate local inflammatory responses.

microscopic currents Ionic currents flowing through single ion channels.

microtubule cytoskeleton The polymers of a variety of tubulin proteins that form a framework of parallel tubes that provide stability and resilience to long cellular processes, especially dendrites and axons. Polymerization of tubulins occurs as an axon or dendrite extends from the cell body of a differentiating neuron. Once the process is stable, this arrangement of parallel microtubules maintains the volume of the process and also acts as a set of tracks upon which molecules or organelles are transported from the cell body to distal domains of the processes using the molecular "motor" proteins kinesin or dynein.

midbrain The most rostral portion of the brainstem; identified by the superior and inferior colliculi on its dorsal surface, and the cerebral peduncles on its ventral aspect. Also known as the *mesencephalon.*

middle cerebellar peduncles See cerebellar peduncles.

middle cerebral arteries Major vessels derived from the internal carotid arteries that supply the lateral aspects of the frontal, parietal and temporal lobes, including associated deep structures.

middle frontal gyri Middle of three, parallel longitudinal gyri anterior to the precentral sulcus; part of the prefrontal cortex and anterior premotor cortex.

middle temporal area (MT) Region of the extrastriate cortex in which neurons respond mainly to movement without regard to color.

middle temporal gyri Middle of three, parallel longitudinal gyri that define the lateral aspect of the temporal lobe; part of the lateral temporal network that encodes language content and, at its posterior margin, contains visual areas involved in motion discrimination.

midsagittal Standard anatomical plane of section; the vertical plane passing from anterior to posterior through the midline dividing the body (and brain) into right and left sections.

miniature end plate potentials (MEPPs) Small, spontaneous depolarization of the membrane potential of skeletal muscle cells, caused by the release of a single quantum of acetylcholine.

mirror motor neurons Neurons in the posterior frontal and inferior parietal lobes that respond during the execution of goal-oriented action and the observation of the same actions, even when such actions are not executed.

mitral cells The major output neurons of the olfactory bulb.

monitoring The process that evaluates the appropriateness of a given behavior for the current context; examples include evaluating the accuracy of answers generated during a memory test or the adequacy of a response rule in an executive function paradigm.

monomeric G-proteins GTP-binding proteins, also called small G-proteins, that relay signals from activate cell surface receptors to intracellular targets. In contrast to heterotrimeric G-proteins, monomeric G-proteins consist of a single protein. Also called *small G-proteins.*

mosaic brain evolution Brain evolution considered in terms of proposed functional modules.

mossy fibers Afferent axons to the cerebellum from all sources except for the inferior olivary nuclei; the vast majority enter the cerebellum via the inferior and middle cerebellar peduncles.

motor aphasia See Broca's aphasia.

motor cortex Region of the cerebral cortex in the posterior frontal lobe that gives rise to corticobulbar and corticospinal projections and is concerned with motor behavior. Includes the primary motor cortex in the anterior bank of the central sulcus that is essential for the voluntary control of movement, and the premotor cortex (anterior to the primary motor cortex) that is involved in planning and programming voluntary movements.

motor neurons By common usage, nerve cells that innervate and send efferent signals to skeletal muscle.

motor systems A broad term used to describe all the central and peripheral structures that support motor behavior.

motor unit A motor neuron and the skeletal muscle fibers it innervates.

muscarinic ACh receptors (mAChRs) Metabootropic ACh receptors that can be pharmacologically identified by their selective activation by muscarine.

muscle spindles Highly specialized mechanosensory organs found in most skeletal muscles; provide proprioceptive information about muscle length.

muscle tone The normal, ongoing tension in a muscle; measured by resistance of a muscle to passive stretching.

myelencephalon The part of the hindbrain that gives rise to the medulla and the glossopharyngeal (IX), vagal (X), Spinal accessory (XI), and hypoglossal (XII) cranial nerves (motor neurons and neural crest that will form related cranial ganglia).

myelin The multilaminated wrapping around many axons formed by oligodendrocytes or Schwann cells.

myelination Process by which glial cells (oligodendrocytes or Schwann cells) wrap around axons to form multiple layers of glial cell membrane, thus insulating the axonal membrane and increasing conduction velocity.

myenteric plexus Network of neurons in the enteric division of the visceral motor system concerned with regulating the musculature of the gut. Also called *Auerbach's plexus.*

myopic Near sighted.

myotatic reflex A fundamental spinal reflex that is generated by the motor response to afferent sensory information arising from muscle spindles; also called a "stretch" or "deep tendon" reflex. The knee jerk reaction is a common example.

N

N-acetyl aspartate (NAA) An abundant metabolite in the neurons synthesized in mitochondria from the amino acid aspartic acid and acetyl-coenzyme A.

Na^+/Ca^{2+} exchanger An active transport protein that removes calcium from the cytoplasm of cells by exchanging intracellular calcium ions for extracellular sodium ions.

nasal division Referring to the region of the visual field of each eye in the direction of the nose. See also binocular field.

nasal mucosa The nasal mucosa is the general term for the entire epithelial lining of the nasal cavities. It includes both the non-neural respiratory epithelium and the neural olfactory sensory epithelium and their constituent cells. It is named based upon the layer of mucous that coats the entire outer surface of the respiratory and olfactory epithelia in the nose.

Neanderthals A group in the human lineage that evolved from *Homo erectus* about 50,000 years ago in what is now Europe and the Middle East.

near cells Visual cortical neurons that respond to retinal disparities that arise from points in front of the plane of fixation.

near reflex triad Reflexive response induced by changing binocular fixation to a closer target; comprises convergence, accommodation, and pupillary constriction.

neocortex The six-layered cortex that forms the surface of most of the cerebral hemispheres (all cerebral cortex lateral and dorsal to the rhinal sulcus).

Nernst equation A mathematical formula that predicts the electrical potential generated ionically across a membrane at electrochemical equilibrium.

nerve cells See neurons.

nerve growth factor (NGF) A neurotrophic protein factor required for survival and differentiation of sympathetic ganglion cells and certain sensory neurons. Preeminent member of the neurotrophin family of growth factors.

nerves A collection of peripheral axons that are bundled together and travel a common route.

netrins A family of diffusible molecules that act as attractive or repulsive cues to guide growing axons.

neural circuits A collection of interconnected neurons mediating a specific function.

neural crest cells Cells that migrate to become a variety of cells types and structures, including peripheral sensory neurons, enteric neurons, and glial cells.

neural crest A transient region where the edges of the folded neural plate come together, at the dorsalmost limit of the neural tube. Gives rise to **neural crest cells** that migrate to become a variety of cells types and structures, including peripheral sensory neurons, enteric neurons, and glial cells as well as facial bones, teeth, and melanocytes in the skin.

neural induction The mechanism by which ectodermal cells, in response to local signals available in the embryo, acquire neural stem cell identity.

neural plate The thickened region of the dorsal ectoderm of a neurula that gives rise to the neural tube.

neural precursor cells Undifferentiated stem cells in the embryonic neural tube.

neural stem cells The neuroectodermal cells, established immediately after gastrulation via signals from the notochord, that have the capacity to give rise to all neuronal and glial cell types of the CNS and PNS, plus the non-neural derivatives of the neural crest.

neural tube The primordium of the brain and spinal cord; derived from the neural ectoderm.

neuregulin1 (Nrg1) Neuregulin 1 is a member of the broader class of secreted neuregulin ligands. Via binding to the Erb family of neuregulin receptors, neuregulin1 and related neuregulins can mediate cell motility, local receptor clustering or a variety of growth responses.

neurexins Adhesion molecules of the presynaptic membrane in developing synapses. Neurexins bind to neuroligin in the postsynaptic membrane, promoting adhesion, and help localize synaptic vesicles, docking proteins, and fusion molecules.

neuroblasts Dividing cells, the progeny of which develop into neurons; immature nerve cells.

neuroectoderm The portion of the outermost embryonic germ layer (the ectoderm) that based upon its proximity to the mesodermally derived notochord differentiates into a field of multipotent neural stem cells that give rise to the entire nervous system (CNS and PNS).

neuroectodermal precursor cells The cells in the neuroectoderm that can give rise via proliferation to a variety of neural and glial cell classes.

neuroethology The field of study devoted to using evolutionary and comparative approaches for observing complex behaviors of animals in their native environments (e.g., social communication in birds and non-human primates) and inferring underlying mechanisms for nervous system regulation.

neuroglia See glial cells.

neuroligins Postsynaptic binding partners of the presynaptic adhesion molecule neurexin. Promote the clustering of receptors and channels of the postsynaptic density as the synapse matures.

neuromeres The repeating units of the neural tube.

neurons Also called *nerve cells.* Cells specialized for the generation, conduction, and transmission of electrical signals in the nervous system.

neuropathic pain A chronic, intensely painful experience that is difficult to treat with conventional analgesic medications.

neuropeptides A general term describing a large number of peptides that are synthesized by neurons and function as neurotransmitters or neurohormones.

neuropil The dense tangle of axonal and dendritic branches, the synapses between them, and associated glia cell processes that lies between neuronal cell bodies in the gray matter of the brain and spinal cord.

neurotransmitters Substances released by synaptic terminals for the purpose of transmitting information from one cell (the **presynaptic cell**) to another (the **postsynaptic cell**).

neurotrophic factors Chemical substances, secreted by cells in a target tissue, that promote the growth and survival of neurons.

neurotrophin 4 or **5 (NT-4/5)** NT-4 and NT-5 are names used for the same member of the neurotrophin family of secreted growth and survival signaling molecules. NT-4/5 signals through the TrkB receptor as the LNGFR/P75.

neurotrophin-3 (NT-3) A member of the neurotrophin family of secreted growth and survival signaling molecules. NT-3 binds with high affinity to both the TrkB and TrkC neurotrophin receptor-kinases, as well as with low affinity to the low affinity neurotrophin receptor (LNGFR, also known as P75). NT-3 has trophic effects on a variety of peripheral and central neurons.

neurotrophins A family of trophic factor molecules that promote the growth and survival of several different classes of neurons.

neurulation The process by which the neuroectoderm rounds into the neural tube. This process establishes the midline of the neural plate as the ventral midline of the CNS and the lateral edges of the neural plate as the dorsal, or alar regions that give rise to the neural crest at the margins, or the dorsal CNS.

niche The tissue environment necessary to support the retention, quiescence, and when elicited, proliferation of tissue specific stem cells in an adult organism.

nicotinic ACh receptor (nAChR) Ionotropic ACh receptors that can be pharmacologically identified by their selective activation by nicotine.

NMDA receptors See ionotropic glutamate receptors.

nociceptors Cutaneous and subcutaneous receptors (especially free nerve endings) specialized for the detection of harmful (noxious) stimuli.

nodes of Ranvier Periodic gaps in the myelination of axons where action potentials are generated.

nodulus Medial portion of the vestibulocerebellum that receives input from the vestibular nuclei in the brainstem and the vestibular nerve; coordinates the vestibulo-ocular reflex and movements that maintain posture and equilibrium.

Noggin An endogenous antagonist of the Bmp family of Tgfβ ligands that binds secreted Bmps and inactivates them, preventing the acquisition of epidermal fate for ectodermal cells that go on to become neural plate neural stem cells. Noggin is secreted by the notochord, in combination with positive signals like Shh that drive neuronal differentiation.

non-rapid eye movement (non-REM) sleep Collectively, those phases of sleep (stages I–IV) characterized by the absence of rapid eye movements.

nondeclarative memory Unconscious memories such as motor skills and associations. Also called *procedural memory*.

noradrenaline See norepinephrine.

noradrenergic neurons Neurons that contribute to the neuronal basis of wakefulness.

norepinephrine Catecholamine hormone and neurotransmitter that binds to α- and β-adrenergic receptors, both of which are G-protein-coupled receptors. Also known as *noradrenaline*.

Notch cell surface receptors These transmembrane proteins transduce signals upon binding a delta ligand on the surface of a neighboring cell. Delta binding to Notch activates a local cascade at the inner surface of the cell membrane that recruits a protease to cleave the intracellular domain of the Notch protein. This Notch intracellular domain (NICD) then complexes with other cytoplasmic proteins and is translocated to the nucleus to regulate gene expression.

notochord A transient, cylindrical structure of mesodermal cells underlying the neural plate (and later the neural tube) in vertebrate embryos. Source of important inductive signals for neural development.

NSF *N*EM-*s*ensitive *f*usion protein. An enzyme responsible for dissociating complexes of SNARE proteins.

nuclei of the lateral lemniscus A brainstem nucleus in the primary auditory pathway.

nucleus (pl. nuclei) Collection of nerve cells in the brain and spinal cord that are anatomically discrete, and which typically serve a particular function.

nucleus ambiguus Branchial motor nucleus of the rostral medulla containing somatic motor neurons that innervate striated muscles of the larynx and pharynx; also contains a visceral motor division with parasympathetic preganglionic neurons that mediate slowing of the heart rate.

nucleus of the solitary tract Nucleus of the caudal pons and rostral medulla that contains a rostral gustatory division and a caudal visceral sensory division; integrates inputs relayed from the rostral division and several primary and secondary visceral sensory afferents that are relevant to the autonomic control of the gut, the cardiovascular system, and other target organs; receives visceral and taste information via several cranial nerves and relays this information to the thalamus.

nutrient A chemical that is needed for growth, maintenance, and repair of the body but is not used as a source of energy.

nystagmus Repetitive rotational movements of the eyes normally elicited by large-scale motion of the visual field (optokinetic nystagmus), with each cycle involving a slower phase driven by central circuits in the brainstem and higher brain centers and a faster, reflexive phase resetting the position of the eye in the orbit; in the absence of physiological visual or vestibular stimuli, nystagmus may indicate brainstem or cerebellar pathology.

O

occipital lobe The posterior lobe of the cerebral hemisphere; primarily devoted to vision.

occipitotemporal gyrus Longitudinal gyrus of the inferior temporal lobe between the parahippocampal gyrus and the inferior temporal gyrus; part of the ventral "what" visual processing stream concerned with object recognition. Posterior portion also known as the *fusiform gyrus*.

ocular dominance columns The segregated termination patterns of thalamic inputs representing the two eyes in the primary visual cortex of some mammalian species.

oculomotor nerve (III) Cranial nerve III, a mixed efferent nerve with somatic motor components that controls the superior rectus, the inferior rectus, the medial rectus and the inferior oblique eye muscles, as well as the levator palpebrae superioris—a muscle that retracts the upper eyelid; also contains a parasympathetic component that constricts the pupil.

odorant receptor gene family The largest set of genes in most animals' genomes devoted to a singular function: the binding and transduction of volatile chemicals in the air that are sensed as odors. All odorant receptors are members of the 7 transmembrane G-protein coupled receptor class of cell surface receptors. There numbers in vertebrates are between 600 and 2000 individual genes.

odorants Molecules capable of eliciting responses from receptors in the olfactory mucosa.

OFF-center neuron A visual neuron whose receptive field center is inhibited by light.

olfactory bulbs Telencephalic structures that lie on the orbital surface of the frontal lobe and receive axons from cranial nerve I; contain local circuit neurons and project neurons that transmit olfactory signals to the olfactory cortex via the olfactory tracts.

olfactory cilia Actin based protrusions from the apical domain of an olfactory sensory receptor neuron. The olfactory cilia are the site of concentration of odorant receptor molecules and the cytoplasmic signaling intermediates necessary for odor transduction and the initiation of odor processing in the olfactory pathway.

olfactory ensheathing cells The glial cells that surround the unmyelinated axons of the peripheral portion of the olfactory nerve. These cells have several properties of Schwann cells that fulfill a similar function for peripheral somatosensory, motor and autonomic axons in peripheral nerves throughout the body.

olfactory epithelium Pseudostratified epithelium that contains olfactory receptor cells, supporting cells, and mucus-secreting glands.

olfactory nerve (I) The set of bundles of unmyelinated axons originating from the olfactory receptor neurons in the olfactory epithelium of the nose that project through the cribiform plate and terminate in the olfactory bulb.

olfactory receptor neurons (ORNs) Bipolar neurons in olfactory epithelium that contain receptors for odorants.

olfactory tracts The projections from the olfactory bulb to the various divisions of the olfactory cortex in the ventromedial forebrain. See lateral olfactory tracts.

oligodendrocytes One of the three major classes of glial cells found in the central nervous system; their major function is to lay down myelin, which facilitates the efficient generation and rapid conduction of action potentials; also produce signaling molecules that modulate growth cone activity in regenerating axons.

ON-center neuron A visual neuron whose receptive field center is excited by light.

Onuf's nucleus Sexually dimorphic nucleus in the human spinal cord that innervates striated perineal muscles mediating contraction of the bladder in males, and vaginal constriction in females.

operant conditioning A form of conditioning shaped by reward rather that pairing a reflex response with an arbitrary signal.

opsins Proteins in photoreceptors that absorb light (in humans, rhodopsin and the three specialized cone opsins).

optic chiasm The junction of the two optic nerves on the ventral aspect of the diencephalon, where axons from the nasal divisions of each retina cross the midline.

optic disk The region of the retina where the axons of retinal ganglion cells exit to form the optic nerve and where the ophthalmic artery and vein enter the eye. Also called the *optic papilla*.

optic nerve (II) The nerve (cranial nerve II) containing the axons of retinal ganglion cells; extends from the eye to the optic chiasm.

optic papilla The region of the retina where the axons of retinal ganglion cells exit to form the optic nerve and where the ophthalmic artery and vein enter the eye. Also called the *optic disk*.

optic radiation Portion of the internal capsule that comprises the axons of lateral geniculate neurons that carry visual information to the striate cortex.

optic tectum The first central station in the visual pathway of many non-mammalian vertebrates (analogous to the superior colliculus in mammals).

optic tract The axons of retinal ganglion cells after they have passed through the region of the optic chiasm en route to the lateral geniculate nucleus of the thalamus.

optic vesicles The evagination of the forebrain vesicles that generates the retina and induces lens formation in the overlying ectoderm.

optogenetics The use of genetic tools to induce neurons to become sensitive to light, such that experimenters can excite or inhibit a cell by exposing it to light.

optokinetic eye movements Movements of the eyes that compensate for head movements; the stimulus for optokinetic movements is large-scale motion of the visual field.

optokinetic nystagmus Repeated reflexive responses of the eyes to ongoing large-scale movements of the visual scene.

orbital gyri Gyral formations on the ventral aspect of the frontal lobes that lie superior to the orbits in the anterior cranial fossae; part of the prefrontal cortex that is involved in implicit processing, including emotion, bodily feeling, and related aspects of cognition.

orbitofrontal cortex (OFC) The division of the prefrontal cortex that lies above the orbits in the most rostral and ventral extension of the sagittal fissure; important in emotional processing and decision making.

orexin A peptide secreted by the hypothalamus, which promotes waking. Also called *hypocretin*.

orthologous genes Genes expressed in model organisms that are identical or similar to target genes (typically expressed in humans and associated with disease) based on sequence and chromosomal location.

orthostatic hypotension Fall in blood pressure upon standing up as a result of blood pooling in the lower extremities.

oscillations Rhythmic patterns of either sub-threshold or spike related electrical activity in the brain that continues over extended periods of time, and can influence the capacity for neurons to engage in plastic changes in synaptic connections.

oscillopsia Inability, as a result of vestibular damage, to fixate visual targets while the head is moving.

ossicles The bones of the middle ear.

otoconia The calcium carbonate crystals that rest on the otolithic membrane overlying the hair cells of the sacculus and utricle.

otolith organs The two organs in the labyrynth of the inner ear—the utricle and saccule—that respond to linear accelerations of the head and static head position relative to the gravitational axis.

otolithic membrane The gelatinous and fibrous membrane on which the otoconia lie and in which the tips of the hair bundles are embedded.

oval window Site where the middle ear ossicles transfer vibrational energy to the cochlea.

overshoot phase The peak, positive-going phase of an action potential, caused by high membrane permeability to a cation such as Na^+ or Ca^{2+}.

overt attention The focusing of attention (typically visual) by voluntarily shifting gaze. Compare *covert attention*.

P

P2X receptors A family of ionotropic purinergic neurotransmitter receptors.

p75 The alternate name for the low affinity neurotrophin receptor (LNGFR). P75/LNGFR interacts with the AKT kinase pathway, and is particularly important for neurotrophin signaling that influences neuronal survival or death.

Pacinian afferents Encapsulated cutaneous mechanosensory receptors in the deep dermis (also found in other tissues) specialized for the detection of high-frequency vibrations.

pain matrix A broad array of brain areas, including the somatosensory cortex, insular cortex, amygdala, and anterior cingulate cortex, whose activity is associated with the experience of pain.

paleocortex Phylogenetically primitive cortex with few cell layers on the inferior and medial aspect of the temporal lobe within the parahippocampal gyrus and the junction of the temporal and frontal lobes.

pallidum Division of the basal ganglia that receives input from the striatum and provides inhibitory (GABAergic) output to the thalamus and brainstem.

parabrachial nucleus Nucleus of the rostral pons that relays visceral sensory information to the hypothalamus, amygdala, thalamus, and medial prefrontal and insular cortex.

paracentral lobule Gryal formation on the medial aspect of the cerebral hemisphere formed by the fusion of the pre- and post-central gyri surrounding the medial termination of the central sulcus; comprises the somatic sensorimotor representation of the contralateral foot.

paracrine Referring to the secretion of hormone-like agents whose effects are mediated locally rather than by the general circulation.

parahippocampal gyrus Medial-most gyral structure in the inferior temporal lobe; part of the medial temporal lobe memory system that generates cognitive maps in spatial frameworks that facilitate the acquisition of episodic and declarative memory.

parallel fibers The bifurcated axons of cerebellar granule cells that extend along the length of the folia in the molecular layer of the cerebellar cortex where they synapse on dendritic spines of Purkinje cells.

parallel pathways Afferent pathways that carry distinct submodalities of sensory information centrally at the same time along anatomically distinct projections.

paramedian circumferential arteries Shorter branches of the vertebral and basilar arteries that supply lateral aspects of the brainstem.

paramedian pontine reticular formation (PPRF) Neurons in the reticular formation of the pons that coordinate the actions of motor neurons in the abducens and oculomotor nuclei to generate horizontal movements of the eyes; also called the horizontal gaze center.

parasagittal Standard anatomical planes of section; any vertical plane passing from anterior to posterior that is parallel to the sagittal plane.

parasympathetic A division of the visceral motor system (division) in which the effectors are cholinergic ganglion cells located near target organs and the central preganglion neurons that innervate them.

parasympathetic ganglia Locus of primary parasympathetic motor neurons; unlike sympathetic ganglia, which are relatively close to the spinal column, parasympathetic ganglia are further removed and typically embedded in or very near the end organs they innervate.

paravertebral sympathetic chain Chain of cervical and thoracic sympathetic ganglia located lateral to the spinal column.

parietal lobe The hemispheric lobe that lies between the frontal lobe anteriorly, and the occipital lobe posteriorly.

parieto-occipital sulcus Prominent sulcus on the medial surface of the cerebral hemisphere that divides the parietal and occipital lobes.

Parkinson's disease Progressive neurodegenerative disease of the substantia nigra pars compacta that results in a characteristic tremor at rest and a general paucity of movement.

parvocellular layers A component of the primary visual pathway specialized for the detection of detail and color; so named because of the relatively small cells involved.

passive electrical responses Responses to applied electrical currents that do not require activation of voltage-gated ion channels.

patch clamp An extraordinarily sensitive voltage clamp method that permits the measurement of ionic currents flowing through individual ion channels.

peptide neurotransmitters See neuropeptides.

perilymph The potassium-poor fluid that bathes the basal end of the cochlear hair cells.

peripheral nervous system (PNS) All nerves and neurons that lie outside the brain and spinal cord.

peripheral sensitization Increased responsiveness of peripheral pain-sensing neurons following tissue damage that is one source of hyperalgesia.

persistent vegetative state A state that results from profound damage to the brain, perhaps by injury or disease, that is characterized by a lack of awareness. A patient with persistant vegetative state typically can still react to stimuli and exhibit degrees of wakefulness and quiescence.

phenotypic sex The visible somatic features that define one sex from the other. Phenotypic sex can variably include body hair patterns, body size and musculature, voice, as well as the primary sex characteristics associated with males versus females.

pheromones Species-specific odorants that play important roles in behavior in some animals, including many mammals.

photopic vision Vision at high light levels, which is mediated almost entirely by cone cells. Contrast with *scotopic vision*.

photoreceptors The specialized neurons in the eye—rods and cones—that are sensitive to light.

phototransduction The process by which light is converted in electrical signals in the retina.

pia mater The innermost of the three layers of the meninges; a delicate layer (pia mater means "tender or delicate mother") closely applied to the surface of the brain.

pineal gland Midline neural structure lying on the dorsal surface of the midbrain; important in the control of circadian rhythms (and, incidentally, considered by Descartes to be the seat of the soul).

pinna A component of the external ear.

pitch The sensory quality roughly corresponding to periodic vibrations of sound stimuli.

pituitary stalk See infundibulum.

placebo effect The physiological effects resulting from administration of an inert substance (a placebo).

planum temporale Region on the superior surface of the temporal lobe posterior to Heschl's gyrus; notable because it is larger in the left hemisphere in about two-thirds of humans.

PLCβ2 Phospholipase Cβ2, a membrane associated enzyme activated by gustducin, a G-protein specifically associated with G-protein-coupled taste receptor proteins in taste receptors cells. PLCβ2 enzymatically converts the membrane lipid phospho-inositol-biphosphate into second messenger lipids inositol triphosphate (IP_3) and DiacylGlycerol (DAG). IP_3/DAG signaling leads to Ca^{2+} release and activation of TRP channels.

plexus A complex network of nerves, blood vessels, or lymphatic vessels.

pneumoencephalography An x-ray based means for brain imaging involving the displacement of cerebrospinal fluid by injection of air into the subarachnoid space to increase signal contrast.

point of fixation The point in visual space that falls on the fovea of each eye.

polarized epithelial cells Cells arranged in sheets or layers, that have an apical (top) and basal (bottom) domain that are distinguished by molecular differences. The apical domain of an epithelial cell is specialized for interactions and transduction of signals from the environment. The basal domain is specialized for secretion.

polyneuronal innervation A state in which neurons or muscle fibers receive synaptic inputs from multiple, rather than single, axons.

pons One of the three major divisions of the brainstem, lying between the midbrain rostrally and the medulla oblongata caudally; derived from the embryonic metencephalon.

pontine nuclei Collections of neurons in the base of the pons that receive input from the ipsilateral cerebral cortex and send their axons across the midline to the contralateral cerebellum via the middle cerebellar peduncle.

pontine reticular formation Collections of neurons in the pons that receive input from the cerebral cortex and send their axons across the midline to the cerebellar cortex via the middle cerebellar peduncle.

pontine-geniculo-occipital (PGO) waves Characteristic encephalographic waves that signal the onset of rapid eye movement sleep.

pore Structural feature of an ion channel that allows ions to diffuse through the channel.

pore loop An extracellular domain of amino acids, found in certain ion channels, that lines the channel pore and allows only certain ions to pass.

positron emission tomography (PET) A technique for examining brain function following injection of unstable, positron-emitting isotopes that are then incorporated into bioactive molecules or metabolites; the emission of positrons are detected by gamma ray detectors and tomographic images are computed that indicate the localization and concentration of the isotopes.

post-tetanic potentiation (PTP) An enhancement of synaptic transmission resulting from high-frequency trains of action potentials. See synaptic potential.

postcentral gyrus The gyrus that forms the posterior bank of the central sulcus; contains the primary somatic sensory cortex.

posterior cerebral arteries Major vessels derived from the basilar artery that supply the ventral midbrain and the posterior, inferior, and medial aspects of the occipital, parietal, and temporal lobes, including associated deep structures.

posterior chamber The region of the eye between the lens and the iris.

posterior cingulate cortex (PCC) A functional division of the cerebral cortex located on its midline surface caudal to the central sulcus that surrounds the corpus callosum. It is associated with task-negative cognition, including mind-wandering, and reward.

posterior circulation Vasculature derived from the vertebral and basilar arteries that supplies blood to the hindbrain and posterior forebrain.

posterior communicating arteries Small vessels that join the internal carotid arteries to the posterior cerebral arteries, forming the lateral aspects of the circle of Willis.

posterior funiculi See dorsal columns.

posterior inferior cerebellar artery (PICA) Long circumferential branch of the vertebral artery that supplies dorsolateral aspects of the medulla and posterior and inferior aspects of the cerebellum.

posterior spinal arteries Principal arteries on the posterior aspect of the spinal cord supplied by the vertebral or posterior inferior cerebellar arteries; supplies blood to the posterior one-third of the spinal cord.

postganglionic axons Axons that link visceral motor neurons in autonomic ganglia to their targets.

postsynaptic Referring to the compartment of a neuronal process (typically, a dendritic spine or shaft) or a location on a cell body that is specialized for transmitter reception; downstream at a synapse.

postsynaptic current (PSC) The current produced in a postsynaptic neuron by the binding of neurotransmitter released from a presynaptic neuron.

postsynaptic density A cytoskeletal junction in developing synapses that may serve to organize postsynaptic receptors and speed their response to neurotransmitter.

postsynaptic dorsal column projection Axonal projection arising from dorsal horn neurons through the dorsal columns terminating in dorsal column nuclei; conveys mechanosensory information in parallel with central processes of first-order afferents.

postsynaptic potential (PSP) The potential change produced in a postsynaptic neuron by the binding of neurotransmitter released from a presynaptic neuron.

posttraumatic stress disorder (PTSD) A clinical condition that emerges following the experience of one or more traumatic, stressful events. Symptoms include heightened arousal, emotional numbness, avoidance of event reminders, and persistent reexperiencing of the traumatic event(s).

potentiation An activity-dependent form of short-term synaptic plasticity that enhances synaptic transmission. Potentiation is caused by an increase in the amount of neurotransmitter released in response to presynaptic action potentials and results from persistent calcium actions within presynaptic terminals. Because potentiation acts over a time course of seconds to minutes, it often outlasts the high-frequency trains of action potentials that evoke it, leading to the phenomenon of post-tetanic potentiation.

pre-propeptides The first protein translation products synthesized in a cell. These polypeptides are usually much larger than the final, mature peptide and often contain signal sequences that target the peptide to the lumen of the endoplasmic reticulum.

precentral gyrus The gyrus that forms the anterior bank of the central sulcus; contains the primary motor cortex.

precuneus gyrus Gyral structure on the medial aspect of the parietal lobe between the dorsal ramus of the cingulate sulcus and the parieto-occipital sulcus; a component of the default-mode network.

prefrontal cortex (PFC) Cortical regions in the frontal lobe that are anterior to the primary and association motor cortices; thought to be involved in planning complex cognitive behaviors and in the expression of personality and appropriate social behavior.

preganglionic neurons Visceral motor neurons in the spinal cord and brainstem that innervate autonomic ganglia.

premotor cortex Motor association areas in the frontal lobe anterior to the primary motor cortex; involved in planning or programming of voluntary movements and a source of descending projections to motor neurons in the spinal cord and cranial nerve nuclei.

presbyopia The condition in which aging affects the accommodative ability of the eye.

presynaptic Referring to the compartment of a neuronal process (typically, a terminal of an axon) at a synapse specialized for transmitter release; upstream at a synapse.

pretectum A group of nuclei located at the junction of the thalamus and the midbrain; these nuclei are important in the pupillary light reflex, relaying information from the retina to the Edinger–Westphal nucleus.

prevertebral ganglia Sympathetic ganglia that lie anterior to the spinal column (distinct from the sympathetic chain ganglia).

primary motor cortex A major source of descending projections to motor neurons in the spinal cord and cranial nerve nuclei; located in the precentral gyrus (Brodmann's area 4) and essential for the voluntary control of movement.

primary sex characteristics The distinguishing body features related to chromosomal sex: male versus female gonads and genitalia.

primary somatosensory cortex (SI) Functional division of the cerebral cortex in the postcentral gyrus, corresponding to Brodmann's areas 3, 1, and 2, that receives somatosensory projections from the ventral posterior complex of thalamic nuclei; processes somatosensory information from the body surface, subcutaneous tissues, muscles, and joints.

primary visual cortex (V1) Brodmann's area 17 in the medial occipital lobe; major cortical target of the thalamic lateral geniculate nucleus. Also called *striate cortex* because of the prominence of a heavily myelinated stripe in layer 4 (called the stria of Gennari) that gives this region a striped (striated) appearance.

primary visual pathway Pathway from the retina via the lateral geniculate nucleus of the thalamus to the primary visual cortex; carries the information that allows conscious visual perception. Also known as the *retinogenticulocortical pathway.*

priming A change in the processing of a stimulus due to a previous encounter with the same or a related stimulus, with or without conscious awareness of the original encounter.

primitive pit An important source of neural inductive signals during gastrulation.

primitive streak Axial indentation in the gastrulas of birds and mammals that generates the notochord and defines the embryonic midline. At the thickened anterior end of the streak is an indentation, the **primitive pit**, which is an important source of neural inductive signals during gastrulation.

principal nucleus Main nuclear division of the trigeminal nuclear complex of the brainstem in the pons (also known as the chief *sensory nucleus*) that receives mechanosensory afferents from the trigeminal nerve; gives rise to the trigeminal lemniscus the supplies the contralateral ventral posterior medial nucleus.

procedural memory See nondeclarative memory.

projection neurons Neurons with long axons that project to distant targets.

promoter DNA sequence (usually within 35 nucleotides upstream of the start site of transcription) to which the RNA polymerase and its associated factors bind to initiate transcription.

propeptide Partially processed forms of proteins containing peptide sequences that play a role in the correct folding of the final protein.

prosencephalon The part of the brain that includes the diencephalon and telencephalon derived from the embryonic forebrain vesicle.

prosody The normal rhythm, stress, and tonal variation of speech that give it emotional meaning.

prosopagnosia The inability to recognize faces; usually associated with lesions to the right inferior temporal cortex.

protein kinases Enzymes that participate in intracellular signal transduction by phosphorylating their target proteins, thereby altering the function of these targets.

protein phosphatases A family of enzymes that participate in intracellular signal transduction by removing phosphate groups from their target proteins, thereby altering the function of these targets.

protocadherins A large family of the cadherin class of Ca^{2+} dependent cell adhesion molecules. There are at least 50 or more genes in the mammalian genome that encode protocadherins, and most protocadherin genes are organized so that their exons can be alternatively spliced to generate multiple protocadherin isoforms from the same gene.

pseudounipolar Morphology of a somatosensory ganglion neuron whose peripheral and central components of an afferent fiber are continuous, attached to the cell body by a single process.

ptosis A drooping of the upper eyelid.

pupil The perforation in the center of the iris that allows light to enter the eye. The pupillary light reflex mediates pupillary constriction in full light and expansion (dilation) in dim light; these responses can also be induced by chemicals and by certain emotional states, and thus can be clinically important.

pupillary light reflex The reduction in the diameter of the pupil that occurs when sufficient light falls on the retina.

putamen One of the three major components of the striatum (the other two are the caudate and the nucleus accumbens). Also called *putamen nuclei.*

pyriform cortex Component of cerebral cortex in the temporal lobe pertinent to olfaction; so named because of its pearlike shape.

R

radial glia Glial cells that contact both the luminal and pial surfaces of the neural tube, providing a substrate for neuronal migration.

rapid eye movement (REM) sleep The phase of sleep characterized by low-voltage, high-frequency electroencephalographic activity accompanied by rapid eye movements.

rapidly adapting afferents Afferents that fire transiently in response to stimulus onset or offset.

ras The first monomeric G-protein discovered. It is involved in many types of neuronal signaling and also controls differentiation and proliferation of non-neuronal cells.

readiness potential An electrical potential, recorded from the motor and premotor cortices with EEG electrodes, that signals the intention to initiate a voluntary movement well in advance of actual production of the movement.

receptive aphasia See Wernicke's aphasia.

receptive field The region of a receptive surface (e.g., the body surface, or a specialized structure such as the retina) within which a specific stimulus elicits the greatest action potential response from a sensory cell in a sensory ganglion or within the CNS.

receptive field properties Neurophysiological characteristics the define the particular responses of sensory neurons to stimuli in their receptive fields; examples for a neuron in the primary visual cortex include the preferred location, orientation, motion direction, and spatial frequency of stimuli that yields the greatest neural response.

receptor molecule A molecule that binds to chemical signals and transduces these signals into a cellular response.

receptor potentials The membrane potential change elicited in receptor neurons during sensory transduction. Also called *generator potentials.* Compare *synaptic potential.*

reciprocal innervation Pattern of connectivity in local circuits of the spinal cord involving excitatory and inhibitory interneurons arranged to ensure that contraction of agonistic muscles produce forces that are opposite to those generated by contraction of antagonistic muscles; thus, reciprocal innervation mediates the simultaneous relaxation of antagonists during contraction of agonists.

red nucleus Prominent parvocellular nucleus of the midbrain tegmentum involved in regulating activity in the inferior olivary nucleus; integrates input from the cerebral cortex and the contralateral dentate nucleus of the cerebrocerebellum. In non-human mammals, also features a magnocellular division that gives rise to the rubrospinal tract, which participates in upper motor neuronal control of the distal musculature of the upper limbs or forelimbs.

reentrant (or **recurrent) neural activation)** Following a stimulus or event, a process in which neural activity is fed back to the same brain region activated earlier in the processing sequence.

refractory period The brief period after the generation of an action potential during which a second action potential is difficult or impossible to elicit.

regenerative A process that is self-sustaining. For example, action potential propagation is regenerative because an action potential produced at one location depolarizes downstream regions, thereby activating voltage-gated ion channels to generate an action potential in these regions.

restiform body See cerebellar peduncles.

resting membrane potential The inside-negative electrical potential that is normally recorded across all cell membranes.

reticular activating system Region in the brainstem tegmentum that, when stimulated, causes arousal; involved in modulating sleep and wakefulness.

reticular formation A network of neurons and axons that occupies the core of the brainstem, giving it a reticulated ("netlike") appearance in myelin-stained material; major functions are modulatory (e.g., regulating states of consciousness) and premotor (e.g., coordinating eye movements, posture, and the regulation of respiratory and cardiac rhythms).

retina Laminated neural component of the eye that contains the photoreceptors (rods and cones) and the initial processing machinery for the primary (and other) visual pathways.

retinal A light-absorbing chromophore; the aldehyde form of vitamin A.

retinal waves A type of oscillatory activity established in the developing mammalian retina, usually prenatally or before eye opening, that is independent of visual input. These oscillations are established by subthreshold activity of subsets of amacrine cells leading to rhythmic firing of subsets of retinal ganglion cells in a spatially distinct manner.

retinogeniculostriate pathway Another term for the primary visual pathway.

retinohypothalamic pathway The route by which variation in light levels influences the broad spectrum of functions that are entrained to the day–night cycle.

retinoic acid (RA) A derivative of vitamin A that acts as an inductive signal during early brain development.

retinoid cycle Process in which retinal is restored to a form capable of signaling photon capture.

retinoid receptors The family of nuclear transcription factor proteins that form heterodimers and bind isomers of retinoic acid to initiate recognition of response element DNA binding sequences and subsequently influence gene expression. Retinoid receptors are a subclass of the broader steroid/thyroid receptor-transcription factor family.

retrograde Signals or impulses that travel "backward," e.g., from the axon terminal toward the cell body, or from the postsynaptic cell to the presynaptic terminal, or from the periphery to the CNS.

retrograde amnesia The inability to recall existing memories.

reversal potential Membrane potential of a postsynaptic neuron (or other target cell) at which the action of a given neurotransmitter causes no net current flow.

reward prediction error (RPE) A quantity given by the difference between the reward that was expected and what actually occurs; the activity of some dopaminergic neurons seems to convey this quantity.

reward A poorly defined term that generally refers to a sense of pleasure following a successful response to some challenge. Often taken to entail dopaminergic neural circuitry.

rhodopsin The photopigment found in rods.

rhombencephalon The caudal part of the brain between the mesencephalon and the spinal cord derived from the embryonic hindbrain vesicle; includes the pons, cerebellum, and medulla. Also known as the hindbrain.

rising phase The initial, depolarizing, phase of an action potential, caused by the regenerative, voltage-dependent influx of a cation such as Na^+ or Ca^{2+}.

robo The transmembrane receptor for Slit that activates signaling via interactions with non-receptor tyrosine kinases, RhoGTPases and other cytoplasmic signaling molecules to initiate the repulsion/avoidance of an axon or dendrite for a specific direction or location.

ROBO3 An additional ROBO receptor, encoded by a separate gene. ROBO3 is particularly essential for the guidance of axons through the ventral commissure at the midline spinal cord, and ensuring that these axons do not turn around and cross back at the midline.

rods Photoreceptor cells specialized for operating at low light levels.

roofplate The thinned dorsal-most medial region of the neural tube, where the two edges of the lateral/alar neuroectoderm fuse during neurulation. This neuroectodermal cells of this region, like the floorplate at the ventral midline, provide secreted signals to specify the neural crest as well as dorsal cell types in the neural tube.

rostral (R) Anterior, or "headward."

rostral interstitial nucleus Cluster of neurons in the reticular formation that coordinates the actions of neurons in the oculomotor nuclei to generate vertical movements of the eye; also called the *vertical gaze center*.

rostral migratory stream A specific migratory route, defined by a distinct subset of glial cells, that facilitates migration of newly generated neurons from the stem cell niche of the anterior subventricular zone to the olfactory bulb.

rostrotemporal (RT) One of the divisions of the core region of the auditory cortex in non-human primates.

round window Along with the oval window, a region at the base of the cochlea where the overlying bone is absent.

rubrospinal tract In non-human mammals, the pathway from the magnocellular divisions of the red nucleus of the midbrain to the spinal cord; participates with the lateral corticospinal tract in governing the distal extremities. In humans, however, the corticospinal tract serves this function and the rubrospinal tract is vestigial (perhaps even nonexistent).

Ruffini afferents Encapsulated cutaneous mechanosensory receptors in the deep dermis (also found in other tissues) specialized for the detection of cutaneous stretching produced by digit or limb movements.

ryanodine receptor A ligand-gated ion channel in the endoplasmic reticulum membrane. This receptor binds to the drug ryanodine and elevates cytoplasmic calcium concentration by mediating flux of calcium out of the lumen of the endoplasmic reticulum.

S

saccades Ballistic, conjugate eye movements that change the point of foveal fixation.

saccule The otolith organ that detects linear accelerations and head tilts in the vertical plane.

sacral Caudal region of the spinal cord between the lumbar and coccygeal regions related to the lower extremities and pelvic visceral motor outflow.

sagittal Standard anatomical plane of section; the vertical plane passing from anterior to posterior through the midline dividing the body (and brain) into right and left sections.

saliency maps A theoretical construct of visual attention in which the importance of different stimuli in the visual field is set by a combination of top-down processes based on behavioral goals and bottom-up processes resulting from how distinctive the different elements of a stimulus are compared to the background.

salt One of the five basic tastes; the taste quality produced by the cations of salts (e.g., the sodium in sodium chloride produces the salty taste). Some cations also produce other taste qualities (e.g., potassium tastes bitter as well as salty). The purest salty taste is produced by sodium chloride (NaCl), common table salt. Salt taste is transduced by taste cells via an ameloride sensitive Na^+ channel.

saltatory Mechanism of action potential propagation in myelinated axons; so named because action potentials "jump" from one node of Ranvier to the next due to generation of action potentials only at these sites.

scala media The fluid-filled chamber within the cochlea that sits on the basilar membrane and that lies between the scala vestibuli and the scala tympani.

scala tympani The fluid-filled chamber within the cochlea at the base of which is located the round window.

scala vestibuli The fluid-filled chamber within the cochlea at the base of which is located the oval window.

Scarpa's ganglion See vestibular nerve ganglion.

Schwann cells Glial cells in the peripheral nervous system that lay down myelin, which facilitates the efficient generation and rapid conduction of action potentials; also facilitate axon regeneration in damaged nerves (named after the nineteenth-century anatomist and physiologist Theodor Schwann).

sclera The external connective tissue coat of the eyeball.

SCN genes Na^+ channel genes. These genes produce proteins that differ in their structure, function, and distribution in specific tissues.

scotoma A small deficit in the visual field resulting from pathological changes in some component of the primary visual pathway.

scotopic vision Vision in dim light, where the rods are the operative receptors.

second pain A category of pain perception described as more delayed, diffuse, and longer-lasting than first pain.

secondary sex characteristics Anatomical characteristics, such as breasts and facial hair, that generally differ between the sexes but are not necessarily concordant with the chromosomal sex of the individual

secondary somatosensory cortex (SII) Functional division of the cerebral cortex in the parietal operculum just posterior to the postcentral gyrus; processes somatosensory information received from the primary somatosensory cortex.

sedative Drug that calms patient, in some cases inducing sleep.

segmental nerves See spinal nerves.

segmentation The anterior–posterior division of animals into roughly similar repeating units.

selective serotonin reuptake inhibitors (SSRIs) A class of drugs that work by inhibiting the ability of the SERT serotonin transporter to take serotonin up into presynaptic terminals.

selectivity filter Structure within an ion channel that allows selected ions to permeate, while rejecting other types of ions.

semaphorins A family of diffusible, growth-inhibiting molecules.

semicircular canals The vestibular end organs in the inner ear that sense rotational accelerations of the head.

sensitization Increased sensitivity to stimuli in an area surrounding an injury. Also, a generalized aversive response to an otherwise benign stimulus when it is paired with a noxious stimulus.

sensorineural hearing loss Diminished sense of hearing due to damage of the inner ear or its related central auditory structures. Contrast with *conductive hearing loss.*

sensory aphasia See Wernicke's aphasia.

sensory systems Term sometimes used to describe all the components of the central and peripheral nervous system concerned with sensation.

sensory transduction Process by which the energy of a stimulus is converted into electrical signals by peripheral sensory receptors and then processed by the central nervous system.

sensory–discriminative The aspect of pain that allows one to distinguish the location, intensity, and quality of a noxious stimulation.

septal forebrain nuclei Cerebral nuclei at the anterior base of the septum pellucidum; give rise to widespread modulatory projections to diverse targets in the cerebral hemispheres.

septum pellucidum Non-neural tissue that forms the medial wall along the anterior horns, bodies, and atria of the paired lateral ventricles.

serotonergic neurons Neurons that contribute to the neuronal basis of wakefulness.

serotonin A biogenic amine neurotransmitter, derived from the amino acid tryptophan, that is involved in a wide range of behaviors, including emotional states and mental arousal.

sex chromosomes Either of a pair of chromosomes (XX in female or XY in male mammals) that differ between the sexes.

sexual identity The subjective perception one feels about one's phenotypic sex.

sexual orientation The direction of an individual's sexual feelings: sexual attraction toward persons of the opposite sex (heterosexual), the same sex (homosexual), or both sexes (bisexual).

sexually dimorphic nucleus of the preoptic area (SDN-POA) A hypothalamic nucleus that in humans and several other mammals differs in size in males versus females: usually the male SDN-POA is larger than that of the female. The SDN-POA is thought to regulate sexual behaviors directly involved with reproduction as well as partner selection.

sham rage An emotional reaction elicited in cats by electrical stimulation of the hypothalamus, characterized by hissing, growling, and attack behaviors directed randomly toward innocuous targets.

short circumferential arteries The shortest branches of the vertebral and basilar arteries that supply medial aspects of the brainstem.

short-term memory Memories held briefly in mind that enable a particular task to be accomplished (e.g., efficiently searching a room for a lost object). Also called *working memory.*

sigmoid sinuses Dural venous sinuses that convey venous blood from the transverse sinuses through the jugular foramina and into the jugular veins.

signal amplification A consequence of intercellular or intracellular signal transduction, resulting from the involvement of reactions that generate a much larger number of products than the number of molecules required to initiate the process. Signal amplification is one of the most important advantages of chemical signaling.

signaling endosome A vesicular structure, internalized into the cell, usually at an axonal or dendritic process, via the invagination and then constriction of the plasma membrane. These vesicles include transmembrane receptor kinases that have bound their activating ligand while still on the cell surface. The ligand remains bound to the receptor, but as an internalized vesicle the ligand activated kinase domain faces the cytoplasm of the cell. The vesicle can then be transported retrogradely to the cell body and continue to transduce the signal detected at its point of origin throughout the entire cell.

single-photon emission computerized tomography (SPECT) A technique for examining brain function following injection or inhalation of radiolabeled compounds, which produce photons that are detected by a gamma camera moving rapidly around the head and used to generate tomographic images indicating the localization and concentration of the isotopes.

size principle The orderly recruitment of motor neurons by size to generate increasing amounts of muscle tension.

skin conductance A stimulus-induced increase in the electrical conductance of the skin due to increased hydration.

sleep spindles Periodic bursts of activity at about 10 to 12 Hz that generally last 1 to 2 seconds and arise as a result of interactions between thalamic and cortical neurons; intermittent high-frequency EEG spike clusters characteristic of stage II sleep.

slit A secreted signaling molecule that acts a repulsive cue for growing axons or dendrites.

slow (S) motor units Small motor units comprising small muscle fibers that contract slowly and generate relatively small forces; but because of their rich myoglobin content, plentiful mitochondria, and rich capillary beds, these small red fibers are resistant to fatigue. S motor units are especially important for activities that require sustained muscular contraction, such as maintaining an upright posture.

slow-wave sleep The component of sleep characterized by delta waves.

slowly adapting afferents Afferents that continue to fire, with only modest decrement in firing, in response to the sustained presence of a stimulus.

small clear-core vesicles A type of synaptic vesicle characterized by a small diameter (typically on the order of 50 nanometers) and an absence of an electron-dense core. These vesicles typically contain and release small-molecular neurotransmitters, such as acetylcholine, glutamate and GABA.

small G-proteins See monomeric G-proteins.

small-molecule neurotransmitters The non-peptide neurotransmitters such as acetylcholine, the amino acids glutamate, aspartate, GABA, and glycine, as well as the biogenic amines.

smooth pursuit movements Slow, tracking movements of the eyes designed to keep a moving object aligned with the foveae.

SNAP-25 A SNARE associated with the plasma membrane. This protein forms a SNARE complex with synaptobrevin and syntaxin that mediates fusion of synaptic vesicles with the presynaptic plasma membrane.

SNAPs *S*oluble *N*SF-*a*ttachment *p*roteins. A protein that attaches the enzyme NSF to SNARE complexes, to allow NSF to dissociate the SNARE complexes.

SNAREs *SNAP re*ceptors. Proteins that are found on two membranes and are responsible for fusing the two membranes together.

somatic marker hypothesis A theory that motivated behavior is influenced by neural representations of body states (the "somatic markers"), whose re-experiencing can shape behavior positively or negatively; the hypothesis that evaluation of one's own body states makes important contributions to decision making.

somatic motor division The components of the motor system that support skeletal movements mediated by the contraction of skeletal muscles that are derived from embryonic somites or somitomeres.

somatic motor nuclei Brainstem nuclei (derived from the basal plate) that give rise to efferent fibers innervating striated muscle fibers derived from embryonic somitomeres; located in the dorsal tegmentum alongside the midline.

somatic stem cells Cells that can divide to give rise to more cells like itself, but also can divide to give rise to a new stem cell plus one or more differentiated cells of the relevant tissue type (e.g., a hematopoietic stem cell can give rise to all types of blood cells, neural stem cells give rise to all neuronal types, and glial stem cells to glia). Contrast with *embryonic stem cell.*

somatosensory cortex Functional division of the cerebral cortex in the postcentral gyrus and anterior parietal lobe that receives somatosensory projections from the ventral posterior complex of thalamic nuclei; processes somatosensory information from the body surface, subcutaneous tissues, muscles, and joints.

somatotopic maps Cortical or subcortical arrangements of sensory inputs and local circuits that reflect the topological organization of the body.

Sonic hedgehog (Shh) An inductive signaling hormone essential for development of the mammalian nervous system; believed to be particularly important for establishing the identity of neurons in the ventral portion of the developing spinal cord and hindbrain.

sour One of the five basic tastes; the taste quality produced by the hydrogen ion in acids. Sour tastes are transduced by taste cells via a H^+ selective TRP channel.

spike timing-dependent plasticity (STDP) Changes in synaptic transmission that depend upon the precise temporal relationship between presynaptic action potentials and postsynaptic responses.

spinal accessory nerve (XI) Cranial nerve XI, a branchial motor nerve that conveys efferents from the rostral cervical spinal cord (spinal accessory nucleus) to the upper trapezius and sternocleidomastoid muscles. Attaches to the rostral cervical cord in a cleft medial to the inferior cerebellar peduncle; enters the cranial vault through the foramen magnum and exits via the jugular foramen.

spinal cord The caudal (post cranial) portion of the central nervous system (CNS) that extends from the lower end of the brainstem (the medulla) to the cauda equina; mediates the transmission of afferent and efferent neural signals between the CNS and the body.

spinal nerves Mixed sensory and motor nerves that arise in bilaterally symmetrical pairs from each of 31 segments of the spinal cord.

spinal nucleus Component of the trigeminal nuclear complex of the brainstem in the caudal pons and medulla oblongata that receives afferents from the trigeminal nerve concerning pain and temperature (also receives collateral of mechanosensory afferents); comprises several subdivisions each of which gives rise to the trigeminothalamic tract that supplies the contralateral ventral posterior medial nucleus.

spinal nucleus of the bulbocavernosus (SNB) Sexually dimorphic collection of neurons in the lumbar region of the rodent spinal cord that innervate striated perineal muscles.

spinal shock The initial, short-lived period of flaccid paralysis that accompanies damage to upper motor neurons or their descending pathways to lower motor neurons.

spindle cells Neurons in the insula and anterior cingulate cortex that may support cognition in great apes and humans.

spinocerebellum Medial part of the cerebellum that receives proprioceptive input from the spinal cord and sends output to the motor cortex via the thalamus and to brainstem upper motor neurons; includes paramedian zones that coordinate movements of distal muscles, and a median zone, called the vermis, that coordinates movements of proximal muscles, including eye movements.

splice variants Variable messenger RNA transcripts derived from the same gene that are typically produced by including or excluding certain exons from a gene; the result such alternative splicing is the production of a diverse set of related protein products.

split-brain patients Individuals who have had the cerebral commissures divided in the midline to control epileptic seizures.

Sprague effect Hemispatial neglect induced by a parietal lesion in humans can be mostly compensated by a lesion of the superior colliculus on the other side.

***SRY* (sex-reversal gene on the *Y* chromosome)** Gene on the Y chromosome whose expression triggers a signaling and transcriptional regulatory cascade that masculinizes the developing fetus.

stellate cells Inhibitory interneurons in the cerebellar cortex that receive parallel fiber input and provide inhibitory output to Purkinje cell dendrites.

stereocilia The actin-rich processes that, along with the kinocilium, form the hair bundle extending from the apical surface of the hair cell; site of mechanotransduction.

stereognosis Responses to object shapes.

stereopsis The perception of depth that results from the fact that the two eyes view the world from slightly different angles.

steroid–thyroid nuclear receptors A large class of receptor transcription factors that selectively bind different members of the steroid-thyroid family of hormones. This class of receptor-transcription factors includes estrogen and androgen receptors.

strabismus Developmental misalignment of the two eyes; may lead to binocular vision being compromised.

stria vascularis Specialized epithelium lining the cochlear duct that maintains the high potassium concentration of the endolymph.

striate cortex See primary visual cortex.

striatum General term applied to the caudate, putamen, nucleus accumbens and other minor divisions of the ventral basal forebrain. The name derives from the bridges ("striations") of gray matter that unite the caudate and putamen around which course fibers of the anterior limb of the internal capsule. Principal component of the corpus striatum, an historical term that has been used collectively to refer to the striatum and the globus pallidus.

striola A line found in both the sacculus and utricle that divides the hair cells into two populations with opposing hair bundle polarities.

Stroop effect A slowing of response time when a stimulus harbors inherently conflicting information (e.g., responding to the word *red* printed in green ink versus the word printed in red ink).

submucous plexus Network of neurons in the enteric division of the visceral motor system just beneath the mucus membranes of the gut and is concerned with chemical monitoring and glandular secretion. Also called *Meissner's plexus.*

substance P An 11-amino acid neuropeptide; the first neuropeptide to be discovered.

substantia nigra Bipartite gray matter structure in the ventral midbrain; contains a pallidal division, called the **pars reticulata**, defined by a network (reticulum) of cells that provide inhibitory (GABAergic) output from the basal ganglia to the thalamus and brainstem, and a compact division, called the **pars compacta**, comprising densely packed neurons that synthesize and release dopamine in the caudate and putamen.

substantia nigra pars compacta Compact cell division of the substantia nigra in the ventral midbrain featuring densely packed neurons that synthesize and release dopamine in the caudate and putamen.

substantia nigra pars reticulata Pallidal division of the substantia nigra in the ventral midbrain featuring a network (reticulum) of cells that provide inhibitory (GABAergic) output from the basal ganglia to the thalamus and brainstem.

subthalamic nucleus A nucleus in the ventral thalamus that receives input from the cerebral cortex and external segment of the globus pallidus and sends excitatory (glutamatergic) projections to the internal segment of the globus pallidus. A component of the indirect pathway from striatum to pallidum.

sulci (sing. sulcus) Spaces between gyri; the largest of these spaces are called fissures.

sulcus limitans Shallow longitudinal groove in the lateral wall of the lumen of the neural tube that defines a boundary between the alar and basal plates.

summation The addition in space and time of sequential synaptic potentials to generate a postsynaptic response larger than that produced by a single synaptic potential.

superior cerebellar peduncle See cerebellar peduncles.

superior colliculus (pl. colliculi) Laminated gray matter structure that forms part of the roof of the midbrain; plays an important role in orienting movements of the head and eyes.

superior division Referring to the region of the visual field of each eye that corresponds to the top half of the retina.

superior frontal gyri Superior of three, parallel longitudinal gyri that define the dorsomedial margin of frontal lobe anterior to the precentral sulcus; part of the prefrontal cortex and anterior premotor cortex.

superior oblique muscles Extraocular muscles that intort the eyeballs when in the primary position (eyes straight ahead) and rotate downward when in adduction.

superior parietal lobules Gyral formations of the dorsal (superior) parietal lobes that are involved in associating somatosensory, visual, auditory, and vestibular signals and generating a neural construct of the body, the position of its parts, and its movements (body image or schema).

superior rectus muscles Extraocular muscles that rotate the eyeballs upward.

superior sagittal sinus Large dural venous sinus along the dorsal aspect of the longitudinal fissure; provides for the drainage of venous blood from the dorsal cerebral hemisphere and the return of cerebrospinal fluid via the arachnoid granulations.

superior salivatory nuclei Visceral motor nucleus of the rostral pons containing parasympathetic preganglionic neurons that mediate tearing and salivation.

superior temporal gyri Superior of three, parallel longitudinal gyri that define the dorsolateral margin of temporal lobe; part of the auditory cortex and lateral temporal network that encodes language content.

suprachiasmatic nucleus (SCN) Hypothalamic nucleus lying just above the optic chiasm that receives direct input from the retina; involved in light entrainment of circadian rhythms.

supramodal attention The focusing of attention on stimulus information across multiple modalities at the same time.

sustentacular cells The primary support cells of the olfactory epithelium. Sustentacular cells help to maintain appropriate ionic milieu and epithelial integrity for the olfactory sensory neurons and their basal cell precursors throughout life.

sweet One of the five basic tastes; the taste quality produced by some sugars, such as glucose, fructose, and sucrose. These three sugars are particularly biologically useful to us, and our sweet receptors are tuned to them. Some other compounds (e.g., saccharin, aspartame) are also sweet. Sweet is transduced by taste cells via the T1R class of G-protein-coupled taste receptors.

Sylvian fissure See lateral fissure.

sympathetic A division of the visceral motor system (division) in vertebrates comprising, for the most part, adrenergic ganglion cells located relatively far from the related end organs and the central preganglion neurons that innervate them.

sympathetic ganglia Locus of primary sympathetic motor neurons. See paravertebral sympathetic chain and prevertebral ganglia.

synapse elimination The developmental process by which the number of axons innervating some classes of target cells is diminished. Also called input elimination.

synapses The junctions between neurons where information is passed from one to the other; typically refers to chemical synapses where a physical cleft exists between communicating neurons, but could also refer to electrical synapses mediated by gap junctions.

synapsin A protein which reversibly binds to synaptic vesicles and is responsible for tethering these vesicles within a reserve pool.

synaptic cleft The space that separates pre- and postsynaptic neurons at chemical synapses.

synaptic depression A short-term decrease in synaptic strength resulting from the depletion of synaptic vesicles at active synapses.

synaptic facilitation An increase in synaptic strength that occurs when two or more action potentials invade the presynaptic terminal within a few milliseconds of each other. Facilitation is typically caused by an increase in the amount of neurotransmitter released by a presynaptic action potential and lasts for a fraction of a second.

synaptic potentials A membrane potential change (or a conductance change) generated by the action of a chemical transmitter agent. Synaptic potentials allow the transmission of information from one neuron to another. Compare *receptor potential.*

synaptic transmissions The chemical and electrical process by which the information encoded by action potentials is passed from a presynaptic (initiating) cell to a postsynaptic (target) cell.

synaptic vesicle cycle Sequence of budding and fusion reactions that occurs in presynaptic terminals to maintain the supply of synaptic vesicles.

synaptic vesicles Spherical, membrane-bound organelles in presynaptic terminals that store neurotransmitter molecules and associated molecular machinery that facilitates exocytosis.

synaptobrevin A SNARE protein located in the membrane of synaptic vesicles. This protein forms a SNARE complex with syntaxin and SNAP-25 that mediates fusion of synaptic vesicles with the presynaptic plasma membrane.

synaptojanin A protein involved in uncoating of synaptic vesicles. It works by modifying a vesicular lipid, which serves as a cue for vesicle uncoating by Hsc70.

synaptotagmins A family of calcium-binding proteins found in the membrane of synaptic vesicles and elsewhere. Synaptotagmins 1 and 2 serve as the calcium sensors that trigger the rapid release of neurotransmitters.

synesthesia A sensory anomaly in which individuals conflate experiences in one sensory domain with those in another.

syntaxin A SNARE protein found primarily in the plasma membrane. This protein forms a SNARE complex with synaptobrevin and SNAP-25 that mediates fusion of synaptic vesicles with the presynaptic plasma membrane.

T

taste buds Onion-shaped structures in the mouth and pharynx that contain taste cells.

taste cells The sensory receptors cells that transduce water or lipid soluble molecules with chemical identities that are perceived as one of the five taste categories: sweet, sour, bitter, salty, and umami (savory, associated with amino acids in foods). Taste cells transduce these five categories of taste stimuli either via homomonomeric or dimeric G-protein coupled taste receptors or through ion-selective channels for Na^+ (salty) and H^+ (sour, which reflects acidity in food).

taste papillae Multicellular protuberances on the tongue, along which taste buds are distributed.

tectorial membrane The fibrous sheet overlying the apical surface of the cochlear hair cells; produces a shearing motion of the stereocilia when the basilar membrane is displaced.

tegmentum A general term that refers to the central core of the brainstem.

telencephalon The part of the brain derived from the anterior part of the embryonic forebrain vesicle; includes the cerebral hemispheres (cerebral cortex and cerebral nuclei).

temporal division Referring to the region of the visual field of each eye in the direction of the temple.

temporal lobe The hemispheric lobe that lies inferior to the lateral fissure.

testis-determining factor (TDF) The original name for the gene product of the *SRY* gene (see *SRY*).

testosterone The principal androgen, synthesized in the testes and, in lesser amounts, in the ovaries and adrenal glands.

thalamus A collection of nuclei that forms the major component of the dorsal diencephalon. Although its subdivisions and functions are many, a primary role of the thalamus is to interact with neural circuits in the cerebral cortex through reciprocal, topographically organized interconnections.

thermosensitive Ion channels that respond to heat.

third ventricle Narrow, slit-like ventricle derived from the lumen of the neural tube between the paired diencephalon, which form the lateral wall of the third ventricle.

thoracic Intermediate region of the spinal cord related to the trunk and sympathetic outflow.

threshold potential The level of membrane potential at which an action potential is generated.

tinnitus A pathological condition characterized by spontaneous ringing or rushing noises, which can be either peripheral or central in origin.

tonotopy The topographic mapping of sound frequency across the surface of a structure, which originates in the cochlea and is preserved in ascending auditory structures, including the auditory cortex.

topographic maps Point-to-point correspondence between neighboring regions of the sensory periphery (e.g., the visual field or the body surface) and neighboring neurons within the central components of the system (e.g., in the brain and spinal cord).

totipotent The diploid cells of the early vertebrate embryo that can give rise to all cell classes in the organism including the germ cells (gametes: egg/sperm) are considered to be totipotent.

tracrRNA Small, trans-encoding RNA that combines with a specific guide RNA species to form an RNA duplex, which then acts to guide a bacterial excision/repair enzyme (endonuclease Cas9) to a genomic location targeted for excision. Following Following Cas9 excision, the DNA may be repaired by non-homologous end joining, yielding a microdeletion mutation; alternatively, a donor DNA sequence can be inserted following Cas9 cleavage via a mechanism similar to homologous recombination.

tracts Bundles of fasciculated axons in the central nervous system that are gathered into compact structures and typically share a common origin and termination; more or less analogous to nerves in the periphery.

transcranial magnetic stimulation (TMS) Localized, noninvasive stimulation of cortical neurons through the induction of electrical current by the application of strong, focal magnetic fields.

transcription factors See transcriptional activator proteins.

transcriptional activator proteins DNA-binding proteins that attach near the start site of a gene, thereby activating transcription of the gene. Also called *transcription factors.*

transducin G-protein involved in the phototransduction cascade.

transient receptor potential channel (TRP) The transient receptor potential (TRP) family of ion channels constitute approximately 28 individual genes and the proteins that they encode. All are transmembrane cation selective channels that mediate depolarization in response primarily to various sensory stimuli. These include taste stimuli that interact with the T1R/T2R GPCR taste receptors to activate the TRPM5 channel. There also TRP channels involved in transducing mechanical displacement/stretch across cellular membranes.

transit amplifying cell A precursor cell, capable of rapid asymmetric divisions, descended from a stem cell. Transit amplifying cells can generate a large number of post-mitotic cells via a series of asymmetric divisions that yield one post-mitotic cell that goes on to differentiate and one precursor. These cells, however, are not self-renewing like stem cells—eventually, there is a terminal symmetric division in which both progeny cease to divide.

translational movements Linear motion along the X, Y, and Z axes.

transverse Standard anatomical planes of section that pass through the CNS orthogonal to its long axis.

transverse pontine fibers Axons of pontine nuclei that cross the midline and form the middle cerebellar peduncles.

transverse sinuses Dural venous sinuses that convey venous blood in the anterior direction from the confluence of sinuses at the back of the cranium to the sigmoid sinuses.

traveling wave The sound-evoked propagation of motion from the basal toward the apical end of the basilar membrane.

trichromatic Referring to the presence of three different cone types in the human retina, which generate the initial steps in color vision by differentially absorbing long, medium, and short wavelength light.

trigeminal brainstem complex Nuclei of the brainstem that receive or give rise to sensory or motor axons in the trigeminal nerve; comprises the mesencephalic trigeminal nucleus in the midbrain and rostral pons, the principal (chief sensory) nucleus in the pons, the trigeminal motor nucleus in the pons, and the spinal trigeminal nucleus, which itself contains several subdivisions, in the caudal pons and medulla oblongata.

trigeminal (cranial nerve V) ganglion Cranial nerve conveying somatosensory information from the face to the trigeminal nuclear complex of the brainstem; also conveys motor signals from the brainstem to the muscles of mastication.

trigeminal lemniscus Axonal projection arising from the principal nucleus of the trigeminal nuclear complex of the brainstem and terminating in the contralateral ventral posterior medial nucleus of the thalamus; conveys mechanosensory signals derived from the face.

trigeminal nerve (V) See trigeminal (cranial nerve V) ganglion.

tripartite synapse A three-way junction involving a presynaptic terminal, a postsynaptic process, and neighboring glia.

trochlear nerve (IV) Cranial nerve IV, an efferent motor nerve that controls the superior oblique muscle of the eye.

trophic interaction Referring to the long-term interdependence between nerve cells and their targets.

tuned zero Optic neurons that respond selectively to points that lie on the plane of fixation.

tuning curves The function obtained when a neuron's receptive field is tested with stimuli at different orientations; its peak defines the maximum sensitivity of the neuron in question.

two-point discrimination Distance between caliper tips needed to distinguish one versus two points of stimulation.

tympanic membrane The eardrum.

tyrosine kinase receptors (Trk) The receptor proteins for neurotrophins have extracellular domains with specific affinities for one or at most two of the neurotrophin ligands, and intracellular domains with enzymatic activity that can catalyze phosphorylation of a number of target proteins specifically on tyrosine amino acids in those target proteins. There are three neurotrophin tyrosine kinase receptors: TrkA (ligand:NGF), TrkB (ligands: BDNF, NT3, NT4/5), and TrkC (NT-3).

U

umami The last of five basic tastes: umami is taste detected in response to amino acids in proteins like meat. Umami is also referred to as "savory" taste. It is transduced via the T1R class of G-protein- coupled taste receptors.

uncus Medial protrusion of the anterior parahippocampal gyrus formed by cortical division of the amygdala.

undershoot The final, hyperpolarizing phase of an action potential, typically caused by the voltage-dependent efflux of a cation such as K^+.

upper motor neuron syndrome Signs and symptoms that result from damage to descending motor pathways; these include weakness, spasticity, clonus, hyperactive reflexes, and a positive Babinski sign.

upper motor neurons Neurons in the motor cortex or brainstem that give rise to descending projections that govern the activity of lower motor neurons in the brainstem and spinal cord.

utricle The otolith organ that senses linear accelerations and head tilts in the horizontal plane.

uveal tract A layer of eye tissue adjacent to the retina that includes three distinct but continuous structures: the choroid, the ciliary body, and the iris.

V

V1Rs A sub-class of vomeronasal receptors that interact with the G-protein Gαi2 to transduce vomeronasal sensory stimuli.

V2Rs A sub-class of vomeronasal receptors that interact with the G-protein Gαo to transduce vomeronasal signals.

V4 Region in the extrastriate cortex that contains a high percentage of neurons that respond selectively to color without regard to motion.

vagal crest The region of the neural crest that arises from the posterior rhombencephalon, migrates into the visceral endoderm and eventually gives rise to the enteric nervous system as well as most of the parasympathetic ganglia.

vagus nerve (X) Cranial nerve X, a mixed sensorimotor nerve that conveys visceral afferents (taste from posterior oropharynx) and somatosensory afferents (mechanosensation from skin on or near pinna) to the brainstem, and branchial motor efferents from the brainstem to muscles of the larynx and pharynx and visceral motor efferents to widely distributed targets in the thorax and upper abdomen. Attaches to the rostral medulla in a cleft between the inferior olive and the inferior cerebellar peduncle, just inferior to the glossopharyngeal nerve.

valence The degree of pleasantness of a stimulus.

vasocorona A network of blood vessels on the lateral and ventrolateral margins of the spinal cord connecting circumferential branches of the posterior and anterior spinal arteries.

vection The sense of self-motion created by visual flow.

ventral (anterior) corticospinal tract Spinal portion of the corticospinal tract in the anterior–medial column of the spinal cord derived from the ipsilateral motor cortex; contributes to postural control.

ventral anterior nuclei Nuclei in the ventral tier of the thalamus that receive input from the basal ganglia and cerebellum and project to the motor cortex.

ventral columns The ventral (anterior) and ventrolateral (anterolateral) regions of spinal cord white matter that convey both ascending information about pain and temperature, and descending motor information from the brainstem and motor cortex concerned with postural control and gain adjustment. Also known as *ventrolateral columns* or *anterolateral columns.*

ventral horns The ventral portion of the spinal cord gray matter derived from the basal plate; populated by interneurons and primary motor neurons.

ventral lateral nuclei Nuclei in the ventral tier of the thalamus that receive input from the basal ganglia and cerebellum and project to the motor cortex.

ventral pallidum A structure within the basal ganglia whose fibers project to thalamic nuclei, such as the mediodorsal nucleus.

ventral posterior complex Group of thalamic nuclei that receive the somatosensory projections from the dorsal column nuclei and the trigeminal nuclear complex of the brainstem.

ventral posterior lateral nucleus (VPL) Component of the ventral posterior complex of thalamic nuclei that receives brainstem projections via the medial lemniscus carrying somatosensory information from the body (excluding the face), and gives rise to projections that pass through the posterior limb of internal capsule and terminate somatotopically in the dorsal two-thirds of the postcentral gyrus.

ventral posterior medial (VPM) nucleus A component of the ventral posterior complex of thalamic nuclei that receives brainstem projections carrying somatic sensory information from the face including the inputs from the facial, glossopharyngeal and vagal nerve that innervate the taste buds in the tongue peripherally and the gustatory nucleus portion of the solitary nucleus in the brainstem.

ventral roots The collection of nerve fibers containing motor axons that exit ventrally from the spinal cord and contribute the motor component of each segmental spinal nerve.

ventral tegmental area (VTA) A part of the midbrain that contains many dopaminergic neurons and is important for reward and learning.

ventricles The spaces in the vertebrate brain that are filled with cerebrospinal fluid and represent the lumen of the embryonic neural tube.

ventrolateral columns See ventral columns.

ventrolateral prefrontal cortex (VLPFC) A functional division of the prefrontal cortex roughly corresponding to the inferior frontal gyrus and surrounding sulci, as located anterior to motor cortex. Compare dorsolateral prefrontal cortex.

ventromedial prefrontal cortex (vmPFC) The ventral portion of the prefrontal cortex surrounding the hemispheric midline; plays a key role in the control of emotions and social behavior.

vergence movements Disjunctive movements of the eyes (convergence or divergence) that align the fovea of each eye with targets located at different distances from the observer.

vermis Median zone of the spinocerebellum that receives proprioceptive input from the spinal cord and sends output to brainstem upper motor neurons; coordinates movements of proximal muscles, including eye movements.

vertebral arteries Major source of posterior circulation to hindbrain and posterior forebrain.

vertical gaze center See rostral interstitial nucleus.

vestibular nerve ganglion Contains the bipolar afferent neurons that innervate the semicircular canals and otolith organs of the auditory vestibule. Also called *Scarpa's ganglion.*

vestibular nuclei Clusters of neurons in the medulla that receive direct innervation from the vestibular nerve.

vestibulo-ocular movements Involuntary movement of the eyes in response to displacement of the head; this reflex allows retinal images to remain stable during head movement.

vestibulo-ocular reflex (VOR) Involuntary movement of the eyes in response to displacement of the head. This reflex allows retinal images to remain stable while the head is moved.

vestibulocerebellum Caudal-inferior lobes of the cerebellum, including the flocculus and nodulus, that receives input from the vestibular nuclei in the brainstem and the vestibular nerves; concerned with the vestibulo-ocular reflex and the coordination of movements that maintain posture and equilibrium.

vestibulocochlear nerve (VIII) Cranial nerve VIII, a sensory nerve that conveys vestibular afferents from the various components of the vestibular labyrinth and auditory afferents from the cochlea to the vestibular and cochlear nuclei, respectively, in the dorsolateral caudal pons and rostral medulla. Lateral-most of three cranial nerves that attaches to the brainstem at the junction of the pons and medulla.

visceral motor division The components of the nervous system (peripheral and central) concerned with the regulation of smooth muscle, cardiac muscle, and glands; organized anatomically and physiologically into sympathetic, parasympathetic, and enteric divisions. Also known as the *autonomic nervous system* or *autonomic motor division*.

visceral motor nuclei Nuclei (derived from the basal plate) in the brainstem and spinal cord that give rise to efferent fibers innervating smooth muscle, cardiac muscle or glands. In the brainstem, located in the dorsal tegmentum just lateral to somatic motor nuclei; in the spinal cord, located in the intermediolateral cell column of thoracic and sacral segments.

visceral motor system See autonomic nervous system.

visual field The area in the external world normally seen by one or both eyes (referred to, respectively, as the monocular and binocular fields).

vitreous humor A gelatinous substance that fills the space between the back of the lens and the surface of the retina.

vocal folds Source of vocal vibration in the larynx. Synonymous with *vocal cords*.

voltage clamp method A technique that uses electronic feedback to simultaneously control the membrane potential of a cell and measure the transmembrane currents that result from the opening and closing of ion channels.

voltage gated Term used to describe ion channels whose opening and closing is sensitive to membrane potential.

voltage sensor Charged structure within a membrane-spanning domain of an ion channel that confers the ability to sense changes in transmembrane potential.

voltage-gated ion channels Ion channels that are are opened or closed in response to changes in the transmembrane potential.

vomeronasal organs (VNO) A pair of chemical sensing organs in the septum (medial process) of the olfactory epithelium that are specialized for the detection and transduction of specific classes of volatile chemicals, pheromones and kairomones. The sensory neurons of the VNO are bipolar vomeronasal sensory receptor neurons that resemble olfactory receptor neurons in the olfactory epithelium. The VNO is the site of expression of a distinct family of GPCR chemosensory receptors that specifically bind pheromones and kairomones.

vomeronasal receptor neurons (VRNs) A class of bipolar chemosensory neurons found in the vomeronasal organ that uniquely express vomeronasal receptors, and whose axons project to the accessory olfactory bulb.

vomeronasal receptors (VRs) A large class of 7-transmembrane G-protein coupled receptor proteins (GPCRs) that bind and transduce phermonal and kairomonal signals. Vomeronasal receptor proteins are localized only to vomeronasal receptor neurons in the vomeronasal organ.

vomeronasal system A specialized chemical detection system that detects pheromones—volatile chemicals released into the air from conspecifics to regulate social interactions—or kairomones—volatile chemicals from other species indicating predator, prey, or symbiotic status. The vomeronasal system includes the peripheral sensory organ, the vomeronasal organ adkacent to the olfactory epithelium in the nose, and its primary target in the forebrain, the accessory olfactory bulb.

W

Wada test A procedure sometimes used as a diagnostic tool to determine the location of the speech and language cortex in preparation for neurosurgery. Involves carotid injection of an anesthetic agent.

Wernicke's aphasia Difficulty comprehending speech as a result of damage to Wernicke's language area. Also called *sensory* or *receptive aphasia*.

Wernicke's area Region of cortex in the superior and posterior region of the left temporal lobe that helps mediate language comprehension. Named after the nineteenth-century neurologist Carl Wernicke.

white matter A general term that refers to regions of the brain and spinal cord containing large axonal tracts; the phrase derives from the fact that axonal tracts have a whitish cast when viewed in the freshly cut material due to the abundance of myelin.

wide-dynamic-range neurons Multimodal lamina V neurons that receive converging inputs from nociceptive and non-nociceptive afferents, a quality which makes them a likely substrate for referred pain.

Wisconsin Card Sorting Task A cognitive test that involves classifying a set of cards, each showing one or more images of a simple shape, into categories based on rules that periodically change throughout the session.

Wnt A large family of secreted ligands that regulate stem and precursor cell proliferation, transcriptional activation/repression, and differentiation within and beyond the nervous system.

Z

zonule fibers Radially arranged connective tissue bands that hold the lens of the eye in place.

Box References

CHAPTER 1 Studying the Nervous System

BOX 1A Model Organisms in Neuroscience

Bockamp, E. and 7 others (2002) Of mice and models: Improved animal models for biomedical research. *Physiol. Genomics* 11: 115–132.

Muquit, M. M. and M. B. Feany (2002) Modelling neurodegenerative diseases in *Drosophila*: A fruitful approach? *Nat. Rev. Neurosci.* 3: 237–243.

Rinkwitz, S., P. Mourrain and T. S. Becker (2011) Zebrafish: An integrative system for neurogenomics and neurosciences. *Prog. Neurobiol.* 93: 231–243.

Sengupta, P. and A. D. Samuel (2009) *Caenorhabditis elegans*: A model system for systems neuroscience. *Curr. Opin. Neurobiol.* 19: 637–643.

CHAPTER 2 Electrical Signals of Nerve Cells

BOX 2A The Remarkable Giant Nerve Cells of Squid

Llinás, R. (1999) *The Squid Synapse: A Model for Chemical Transmission*. Oxford, UK: Oxford University Press.

Young, J. Z. (1939) Fused neurons and synaptic contacts in the giant nerve fibres of cephalopods. *Phil. Trans. R. Soc. Lond.* 229: 465–503.

BOX 2B Action Potential Form and Nomenclature

Barrett, E. F. and J. N. Barrett (1976) Separation of two voltage-sensitive potassium currents, and demonstration of a tetrodotoxin-resistant calcium current in frog motoneurones. *J. Physiol. (Lond.)* 255: 737–774.

Chen, S., G. J. Augustine and P. Chadderton (2016) The cerebellum linearly encodes whisker position during voluntary movement. *eLife* 5: e10509.

Dodge, F. A. and B. Frankenhaeuser (1958) Membrane currents in isolated frog nerve fibre under voltage clamp conditions. *J. Physiol. (Lond.)* 143: 76–90.

Hodgkin, A. L. and A. F. Huxley (1939) Action potentials recorded from inside a nerve fibre. *Nature* 144: 710–711.

Llinás, R. and Y. Yarom (1981) Electrophysiology of mammalian inferior olivary neurones *in vitro*. Different types of voltage-dependent ionic conductances. *J. Physiol. (Lond.)* 315: 549–567.

CHAPTER 3 Voltage-Dependent Membrane Permeability

BOX 3A The Voltage Clamp Method

Cole, K. S. (1968) *Membranes, Ions and Impulses: A Chapter of Classical Biophysics*. Berkeley, CA: University of California Press.

CHAPTER 4 Ion Channels and Transporters

BOX 4A The Patch Clamp Method

Dunlop, J., M. Bowlby, R. Peri, D. Vasilyev and R. Arias (2008) High-throughput electrophysiology: An emerging paradigm for ion-channel screening and physiology. *Nat. Rev. Drug Discov.* 7: 358–368.

Hamill, O. P., A. Marty, E. Neher, B. Sakmann and F. J. Sigworth (1981) Improved patch-clamp techniques for high-resolution current recording from cells and cell-free membrane patches. *Pflügers Arch.* 391: 85–100.

Levis, R. A. and J. L. Rae (1998) Low-noise patch-clamp techniques. *Meth. Enzym.* 293: 218–266.

Sakmann, B. and E. Neher (1995) *Single-Channel Recording*, 2nd Edition. New York: Plenum Press.

BOX 4B Toxins That Poison Ion Channels

Cahalan, M. (1975) Modification of sodium channel gating in frog myelinated nerve fibers by *Centruroides sculpturatus* scorpion venom. *J. Physiol. (Lond.)* 244: 511–534.

Catterall, W. A. and 5 others (2007) Voltage-gated ion channels and gating modifier toxins. *Toxicon* 49: 124–141.

Dutertre, S. and R. J. Lewis (2010) Use of venom peptides to probe ion channel structure and function. *J. Biol. Chem.* 285: 13315–13320.

Green, B. R. and 9 others (2016) Structural basis for the inhibition of voltage-gated sodium channels by conotoxin μO§-GVIIJ. *J. Biol. Chem.* 291: 7205–7220.

Narahashi, T. (2008) Tetrodotoxin: A brief history. *Proc. Jpn. Acad. Ser. B Phys. Biol. Sci.* 84: 147–154.

Schmidt, O. and H. Schmidt (1972) Influence of calcium ions on the ionic currents of nodes of Ranvier treated with scorpion venom. *Pflügers Arch.* 333: 51–61.

CHAPTER 5 Synaptic Transmission

BOX 5A The Tripartite Synapse

Cornell-Bell, A. H., S. M. Finkbeiner, M. S. Cooper and S. J. Smith (1990) Glutamate induces calcium waves in cultured astrocytes: Long-range glial signaling. *Science* 247: 470–473.

Fiacco, T. A., C. Agulhon and K. D. McCarthy (2009) Sorting out astrocyte physiology from pharmacology. *Annu. Rev. Pharmacol. Toxicol.* 49: 151–174.

Han, X. and 13 others (2013) Forebrain engraftment by human glial progenitor cells enhances synaptic plasticity and learning in adult mice. *Cell Stem Cell* 12: 342–353.

Haydon, P. G. and M. Nedergaard (2014) How do astrocytes participate in neural plasticity? *Cold Spring Harb. Perspect. Biol.* 7: a020438.

Jahromi, B. S., R. Robitaille and M. P. Charlton (1992) Transmitter release increases intracellular calcium in perisynaptic Schwann cells in situ. *Neuron* 8: 1069–1077.

Lee, S. and 7 others (2010) Channel-mediated tonic GABA release from glia. *Science* 330: 790–796.

Olsen, M. L. and 5 others (2015) New insights on astrocyte ion channels: Critical for homeostasis and neuron-glia signaling. *J. Neurosci.* 35: 13827–13835.

Perea, G. and A. Araque (2007) Astrocytes potentiate transmitter release at single hippocampal synapses. *Science* 317: 1083–1086.

Perea, G., M. Navarrete and A. Araque (2009) Tripartite synapses: astrocytes process and control synaptic information. *Trends Neurosci.* 32: 421–431.

Witcher, M. R., S. A. Kirov and K. M. Harris (2007) Plasticity of perisynaptic astroglia during synaptogenesis in the mature rat hippocampus. *Glia* 55: 13–23.

CHAPTER 6 Neurotransmitters and Their Receptors

BOX 6A Neurotoxins That Act on Neurotransmitter Receptors

Han, T. S., R. W. Teichert , B. M. Olivera and G. Bulaj (2008) Conus venoms: A rich source of peptide-based therapeutics. *Curr. Pharm. Des.* 14: 2462–2479.

Lebbe, E. K. M., S. Peigneur, I. Wijesekara and J. Tytgat (2014) Conotoxins targeting nicotinic acetylcholine receptors: An overview. *Marine Drugs* 12: 2970–3004.

Lewis, R. L. and L. Gutmann (2004) Snake venoms and the neuromuscular junction. *Seminars Neurol.* 24: 175–179.

Tsetlin, V. I. (2015) Three-finger snake neurotoxins and Ly6 proteins targeting nicotinic acetylcholine receptors: Pharmacological tools and endogenous modulators. *Trends Pharmacol. Sci.* 36: 109–123.

BOX 6B Excitatory Actions of GABA in the Developing Brain

Berglund, K. and 8 others (2006) Imaging synaptic inhibition in transgenic mice expressing the chloride indicator, Clomeleon. *Brain Cell Biol.* 35: 207–228.

Cherubini, E., J. L. Gaiarsa andY. Ben-Ari (1991) GABA: An excitatory transmitter in early postnatal life. *Trends Neurosci.* 14: 515–519.

Glykys, J. and 7 others (2009) Differences in cortical versus subcortical GABAergic signaling: a candidate mechanism of electroclinical uncoupling of neonatal seizures. *Neuron* 63: 657–672.

Obata, K., M. Oide and H. Tanaka (1978) Excitatory and inhibitory actions of GABA and glycine on embryonic chick spinal neurons in culture. *Brain Res.* 144: 179–184.

Owens, D. F. and A. R. Kriegstein (2002) Is there more to GABA than synaptic inhibition? *Nat. Rev. Neurosci.* 3: 715–727.

Payne, J. A., C. Rivera, J. Voipio and K. Kaila (2003) Cation-chloride co-transporters in neuronal communication, development and trauma. *Trends Neurosci.* 26: 199–206.

Rivera, C. and 8 others (1999) The K^+/Cl^- co-transporter KCC2 renders GABA hyperpolarizing during neuronal maturation. *Nature* 397: 251–255.

BOX 6C Marijuana and the Brain

Adams, A. R. (1941) Marihuana. *Harvey Lecture* 37: 168.

Freund, T. F., I. Katona and D. Piomelli (2003) Role of endogenous cannabinoids in synaptic signaling. *Physiol. Rev.* 83: 1017–1066.

Gerdeman, G. L., J. G. Partridge, C. R. Lupica and D. M. Lovinger (2003) It could be habit forming: Drugs of abuse and striatal synaptic plasticity. *Trends Neurosci.* 26: 184–192.

Howlett, A. C. (2005) Cannabinoid receptor signaling. *Handbook Exp. Pharmacol.* 168: 53–79.

Iversen, L. (2003) *Cannabis* and the brain. *Brain* 126: 1252–1270.

Mechoulam, R. (1970) Marihuana chemistry. *Science* 168: 1159–1166.

Onaivi, E. S. (2009) Cannabinoid receptors in brain: Pharmacogenetics, neuropharmacology, neurotoxicology, and potential therapeutic applications. *Int. Rev. Neurobiol.* 88: 335–369.

Shao, Z. and 6 others (2016) High-resolution crystal structure of the human CB1 cannabinoid receptor. *Nature* 540: 602–606.

CHAPTER 7 Molecular Signaling within Neurons

BOX 7A Dynamic Imaging of Intracellular Signaling

Chalfie, M., Y. Tu, G. Euskirchen, W. W. Ward and D. C. Prasher (1994) Green fluorescent protein as a marker for gene expression. *Science* 263: 802–805.

Connor, J. A. (1986) Digital imaging of free calcium changes and of spatial gradients in growing processes in single mammalian central nervous system cells. *Proc. Natl. Acad. Sci. USA* 83: 6179–6183.

Finch, E. A. and G. J. Augustine (1998) Local calcium signaling by IP_3 in Purkinje cell dendrites. *Nature* 396: 753–756.

Grynkiewicz, G., M. Poenie and R.Y. Tsien (1985) A new generation of Ca^{2+} indicators with greatly improved fluorescence properties. *J. Biol. Chem.* 260: 3440–3450.

Livet, J. and 7 others (2007) Transgenic strategies for combinatorial expression of fluorescent proteins in the nervous system. *Nature* 450: 56–62.

Rodriguez, E. A. and 8 others (2017) The growing and glowing toolbox of fluorescent and photoactive proteins. *Trends Biochem. Sci.* 42: 111–129.

Shimomura, O. (2009) Discovery of green fluorescent protein (GFP) (Nobel Lecture). *Angew Chem. Int. Ed. Engl.* 48: 5590–5602.

Tsien, R.Y. (2010) Nobel lecture: constructing and exploiting the fluorescent protein paintbox. *Integr. Biol.* (*Camb.*). 2: 77–93.

Vidal, G. S., M. Djurisic, K. Brown, R. W. Sapp and C. J. Shatz (2016) Cell-autonomous regulation of dendritic spine density by PirB. *eNeuro* 3: 1–15. doi: 10.1523/ENEURO.0089-16.2016

BOX 7B Dendritic Spines

Bhatt, D. H., S. Zhang and W. B. Gan (2009) Dendritic spine dynamics. *Ann. Rev. Physiol.* 71: 261–282.

Goldberg, J. H., G. Tamas, D. Aronov and R. Yuste (2003) Calcium microdomains in aspiny dendrites. *Neuron* 40: 807–821.

Harnett, M. T., J. K. Makara, N. Spruston, W. L. Kath and J. C. Magee (2012) Synaptic amplification by dendritic spines enhances input cooperativity. *Nature* 491: 599–602.

Harris, K. M. and R. J. Weinberg (2012) Ultrastructure of synapses in the mammalian brain. *Cold Spring Harb. Perspect. Biol.* 4: a005587.

Nishiyama, J. and R. Yasuda (2015) Biochemical computation for spine structural plasticity. *Neuron* 87: 63–75.

Noguchi, J., M. Matsuzaki, G. C. Ellis-Davies and H. Kasai (2005) Spine-neck geometry determines NMDA receptor-dependent Ca^{2+} signaling in dendrites. *Neuron* 46: 609–622.

Penzes, P., M. E. Cahill, K. A. Jones, J. E. VanLeeuwen and K. M. Woolfrey (2011) Dendritic spine pathology in neuropsychiatric disorders. *Nat. Neurosci.* 14: 285–293.

Popovic, M. A., N. Carnevale, B. Rozsa and D. Zecevic (2015) Electrical behaviour of dendritic spines as revealed by voltage imaging. *Nat. Commun.* 6: 8436.

Sabatini, B. L., T. G. Oertner and K. Svoboda (2002) The life cycle of Ca^{2+} ions in dendritic spines. *Neuron* 33: 439–452.

Santamaria, F., S. Wils, E. De Schutter and G. J. Augustine (2006) Anomalous diffusion in Purkinje cell dendrites caused by spines. *Neuron* 52: 635–648.

Sheng, M. and E. Kim (2011) The postsynaptic organization of synapses. *Cold Spring Harb. Perspect. Biol.* 3: a00567.

CHAPTER 8 Synaptic Plasticity

BOX 8A Genetics of Learning and Memory in the Fruit Fly

Androschuk, A., B. Al-Jabri and F.V. Bolduc (2015) From learning to memory: What flies can tell us about intellectual disability treatment. *Front. Psychiatry* 6: 85.

Davis, R. L. (2004) Olfactory learning. *Neuron* 44: 31–48.

Quinn, W. G., W. A. Harris and S. Benzer (1974) Conditioned behavior in *Drosophila melanogaster*. *Proc. Natl. Acad. Sci. USA* 71: 708–712.

Tully, T. (1996) Discovery of genes involved with learning and memory: An experimental synthesis of Hirshian and Benzerian perspectives. *Proc. Natl. Acad. Sci. USA* 93: 13460–13467.

Waddell, S. and W. G. Quinn (2001) Flies, genes, and learning. *Annu. Rev. Neurosci.* 24: 1283–1309.

Weiner, J. (1999) *Time, Love, Memory: A Great Biologist and His Quest for the Origins of Behavior.* New York: Knopf.

BOX 8B Silent Synapses

Derkach, V. A., M. C. Oh, E. S. Guire and T. R. Soderling (2007) Regulatory mechanisms of AMPA receptors in synaptic plasticity. *Nat. Rev. Neurosci.* 8: 101–113.

Gomperts, S. N., A. Rao, A. M. Craig, R. C. Malenka and R. A. Nicoll (1998) Postsynaptically silent synapses in single neuron cultures. *Neuron* 21: 1443–1451.

Huang, Y. H. and 12 others (2009) In vivo cocaine experience generates silent synapses. *Neuron* 63: 40–47.

Liao, D., N. A. Hessler and R. Malinow (1995) Activation of postsynaptically silent synapses during pairing-induced LTP in CA1 region of hippocampal slice. *Nature* 375: 400–404.

Luscher, C., R. A. Nicoll, R. C. Malenka and D. Muller (2000) Synaptic plasticity and dynamic modulation of the postsynaptic membrane. *Nat. Neurosci.* 3: 545–550.

Petralia, R. S. and 6 others (1999) Selective acquisition of AMPA receptors over post–natal development suggests a molecular basis for silent synapses. *Nat. Neurosci.* 2: 31–36.

CHAPTER 9 The Somatosensory System: Touch and Proprioception

BOX 9A Patterns of Organization within the Sensory Cortices: Brain Modules

da Costa, N. M. and K. A. Martin (2010) Whose cortical column would that be? *Front. Neuroanat.* 4 (May 31): 16.

Horton, J. C. and D. L. Adams (2005) The cortical column: A structure without a function. *Philos. Trans. R. Soc. Lond. B* 360: 837–862.

Hubel, D. H. (1988) *Eye, Brain, and Vision.* Scientific American Library. New York: W. H. Freeman.

Lorente de Nó, R. (1949) The structure of the cerebral cortex. *Physiology of the Nervous System*, 3rd Edition. New York: Oxford University Press.

Mountcastle, V. B. (1957) Modality and topographic properties of single neurons of cat's somatosensory cortex. *J. Neurophysiol.* 20: 408–434.

Mountcastle, V. B. (1998) *Perceptual Neuroscience: The Cerebral Cortex.* Cambridge, MA: Harvard University Press.

Purves, D., D. Riddle and A. LaMantia (1992) Iterated patterns of brain circuitry (or how the cortex gets its spots). *Trends Neurosci.* 15: 362–369.

Woolsey, T. A. and H. Van der Loos (1970) The structural organization of layer IV in the somatosensory region (SI) of mouse cerebral cortex. The description of a cortical field composed of discrete cytoarchitectonic units. *Brain Res.* 17: 205–242.

CHAPTER 10 Pain

BOX 10A Capsaicin

Caterina, M. J. and 5 others (1997) The capsaicin receptor: A heat-activated ion channel in the pain pathway. *Nature* 389: 816–824.

Caterina, M. J. and 8 others (2000) Impaired nociception and pain sensation in mice lacking the capsaicin receptor. *Science* 288: 306–313.

Szallasi, A. and P. M. Blumberg (1999) Vanilloid (capsaicin) receptors and mechanisms. *Pharm. Reviews* 51: 159–212.

Tominaga, M. and 8 others (1998) The cloned capsaicin receptor integrates multiple pain-producing stimuli. *Neuron* 21: 531–543.

Zygmunt, P. M. and 7 others (1999) Vanilloid receptors on sensory nerves mediate the vasodilator action of anandamide. *Nature* 400: 452–457.

BOX 10B Referred Pain

Capps, J. A. and G. H. Coleman (1932) *An Experimental and Clinical Study of Pain in the Pleura, Pericardium, and Peritoneum.* New York: Macmillan.

Head, H. (1893) On disturbances of sensation with special reference to the pain of visceral disease. *Brain* 16: 1–32.

Kellgren, J. H. (1939–1942) On the distribution of pain arising from deep somatic structures with charts of segmental pain areas. *Clin. Sci.* 4: 35–46.

BOX 10C A Dorsal Column Pathway for Visceral Pain

Al-Chaer, E. D., N. B. Lawand, K. N. Westlund and W. D. Willis (1996) Visceral nociceptive input into the ventral posterolateral nucleus of the thalamus: a new function for the dorsal column pathway. *J. Neurophysiol.* 76: 2661–2674.

Al-Chaer, E. D., N. B. Lawand, K. N. Westlund and W. D. Willis (1996) Pelvic visceral input into the nucleus gracilis is largely mediated by the postsynaptic dorsal column pathway. *J. Neurophysiol.* 76: 2675–2690.

Becker, R., S. Gatscher, U. Sure and H. Bertalanffy (2001) The punctate midline myelotomy concept for visceral cancer pain control—case report and review of the literature. *Acta Neurochir.* (Suppl.) 79: 77–78.

Hitchcock, E. R. (1970) Stereotactic cervical myelotomy. *J. Neurol. Neurosurg. Psychiatry* 33: 224–230.

Kim, Y. S. and S. J. Kwon (2000) High thoracic midline dorsal column myelotomy for severe visceral pain due to advanced stomach cancer. *Neurosurgery* 46: 85–90.

Nauta, H. and 8 others (2000) Punctate midline myelotomy for the relief of visceral cancer pain. *J. Neurosurg.* (*Spine 2*) 92: 125–130.

Willis, W. D., E. D. Al-Chaer, M. J. Quast and K. N. Westlund (1999) A visceral pain pathway in the dorsal column of the spinal cord. *Proc. Natl. Acad. Sci. USA* 96: 7675–7679.

CHAPTER 11 Vision: The Eye

BOX 11A Myopia and Other Refractive Errors

Bock, G. and K. Widdows (1990) *Myopia and the Control of Eye Growth*. Ciba Foundation Symposium 155. Chichester: Wiley.

Coster, D. J. (1994) *Physics for Ophthalmologists*. Edinburgh: Churchill Livingston.

Kaufman, P. L. and A . Alm (eds.) (2002) *Adler's Physiology of the Eye: Clinical Application*, 10th Edition St. Louis, MO: Mosby Year Book.

Wallman, J., J. Turkel and J. Tractman (1978) Extreme myopia produced by modest changes in early visual experience. *Science* 201: 1249–1251.

Wallman, J. and J. Winawer (2004) Homeostasis of eye growth and the question of myopia. *Neuron* 43: 447–468.

Wiesel, T. N. and E. Raviola (1977) Myopia and eye enlargement after neonatal lid fusion in monkeys. *Nature* 266: 66–68.

BOX 11B Retinitis Pigmentosa

Rivolta, C., D. Sharon, M. M. DeAngelis and T. P. Dryja (2002) Retinitis pigmentosa and allied diseases: Numerous diseases, genes and inheritance patterns. *Hum. Molec. Genet.* 11: 1219–1227.

Weleber, R. G. and K. Gregory-Evans (2006) Retinitis pigmentosa and allied disorders. In *Retina*, 4th Edition, Vol. 1: *Basic Science and Inherited Retinal Diseases,* S. J. Ryan (ed.-in-chief). Philadelphia: Elsevier, pp. 395–498.

Wright, A. F., C. F. Chakarova, M. M. Abd El-Aziz and S. S. Bhattacharya (2010) Photoreceptor degeneration: Genetic and mechanistic dissection of a complex trait. *Nat. Rev. Genet.* 11: 273–284.

BOX 11C The Importance of Context in Color Perception

Land, E. (1986) Recent advances in Retinex theory. *Vision Research* 26: 7–21.

Purves, D. and R. B. Lotto (2011) *Why We See What We Do Redux: An Empirical Theory of Vision*, Chapters 2 and 3. Sunderland, MA: Sinauer Associates, pp. 15–91.

BOX 11D The Perception of Light Intensity

Adelson, E. H. (1999) Light perception and lightness illusions. In *The Cognitive Neurosciences*, 2nd Edition, M. Gazzaniga (ed.). Cambridge, MA: MIT Press, pp. 339–351.

Purves, D. and R. B. Lotto (2011) *Why We See What We Do Redux: An Empirical Theory of Vision*, Chapters 2 and 3. Sunderland, MA: Sinauer Associates, pp. 15–91.

Purves, D., Y. Morgenstern and W. T. Wojtach (2015) Perception and reality: Why a wholly empirical paradigm is needed to understand vision. *Front. Syst. Neurosci*. 9: 156. doi: 10.3389/fnsys.2015.00156

CHAPTER 12 Central Visual Pathways

BOX 12A Random Dot Stereograms and Related Amusements

Julesz, B. (1971) *Foundations of Cyclopean Perception*. Chicago: University of Chicago Press.

Julesz, B. (1995) *Dialogues on Perception*. Cambridge, MA: MIT Press.

N. E. Thing Enterprises (1993) *Magic Eye: A New Way of Looking at the World*. Kansas City: Andrews and McMeel.

CHAPTER 13 The Auditory System

BOX 13A The Sweet Sound of Distortion

Jaramillo, F., V. S. Markin and A. J. Hudspeth (1993) Auditory illusions and the single hair cell. *Nature* 364: 527–529.

Planchart, A. E. (1960) A study of the theories of Giuseppe Tartini. *J. Music Theory* 4: 32–61.

Robles, L., M. A. Ruggero and N. C. Rich (1991) Two-tone distortion in the basilar membrane of the cochlea. *Nature* 439: 413–414.

BOX 13B Representing Complex Sounds in the Brains of Bats and Humans

Ehret, G. (1987) Left hemisphere advantage in the mouse brain for recognizing ultrasonic communication calls. *Nature* 325: 249–251.

Esser, K.-H., C. J. Condon, N. Suga and J. S. Kanwal (1997) Syntax processing by auditory cortical neurons in the FM-FM area of the mustached bat, *Pteronotus parnellii*. *Proc. Natl. Acad. Sci. USA* 94: 14019–14024.

Hauser, M. D. and K. Andersson (1994) Left hemisphere dominance for processing vocalizations in adult, but not infant, rhesus monkeys: Field experiments. *Proc. Natl. Acad. Sci. USA* 91: 3946–3948.

Kanwal, J. S., J. Kim and K. Kamada (2000) Separate, distributed processing of environmental, speech and musical sounds in the cerebral hemispheres. *J. Cog. Neurosci.* (Supp.): 32.

Kanwal, J. S., J. S. Matsumura, K. Ohlemiller and N. Suga (1994) Acoustic elements and syntax in communication sounds emitted by mustached bats. *J. Acoust. Soc. Amer.* 96: 1229–1254.

Wang, X., D. Bendor and E. Bartlett (2008) Neural coding of temporal information in auditory thalamus and cortex. *Neuroscience* 157: 484–494.

CHAPTER 14 The Vestibular System

BOX 14A Mauthner Cells in Fish

Eaton, R. C., R. A. Bombardieri and D. L. Meyer (1977) The Mauthner-initiated startle response in teleost fish. *J. Exp. Biol.* 66: 65–81.

Furshpan, E. J. and T. Furukawa (1962) Intracellular and extracellular responses of the several regions of the Mauthner cell of the goldfish. *J. Neurophysiol.* 25: 732–771.

Jontes, J. D., J. Buchanan and S. J. Smith (2000) Growth cone and dendrite dynamics in zebrafish embryos: Early events in synaptogenesis imaged in vivo. *Nat. Neurosci.* 3: 231–237.

Korn, H. and D. S. Faber (2005) The Mauthner cell half a century later: A neurobiological model for decision-making? *Neuron* 47: 13–28.

O'Malley, D. M., Y. H. Kao and J. R. Fetcho (1996) Imaging the functional organization of zebrafish hindbrain segments during escape behaviors. *Neuron* 17: 1145–1155.

CHAPTER 15 The Chemical Senses

BOX 15A The "Dogtor" Is In

Church, J. and H. Williams (2001) Another sniffer dog for the clinic? *Lancet* 358: 930.

McCulloch, M. and 5 others (2006) Diagnostic accuracy of canine scent detection in early- and late-stage lung and breast cancers. *Integ. Cancer Therap.* 5: 30–39.

Phillips, M. and 7 others (2003) Detection of lung cancer with volatile markers in the breath. *Chest* 123: 2115–2123.

Willis, C. M. and 7 others (2004) Olfactory detection of human bladder cancer by dogs: Proof of principle study. *BMJ* 329: 712.

CHAPTER 16 Lower Motor Neuron Circuits and Motor Control

BOX 16A Motor Unit Plasticity

Brownstone, R. M., T. V. Bui and N. Stifani (2015) Spinal circuits for motor learning. *Curr. Opin. Neurobiol.* 33: 166–173.

Buller, A. J., J. C. Eccles and R. M. Eccles (1960a) Differentiation of fast and slow muscles in the cat hind limb. *J. Physiol.* 150: 399–416.

Buller, A. J., J. C. Eccles and R. M. Eccles (1960b) Interactions between motoneurones and muscles in respect of the characteristic speeds of their responses. *J. Physiol.* 150: 417–439.

Close, R. (1965) Effects of cross-union of motor nerves to fast and slow skeletal muscles. *Nature* 206: 831–832.

Duchateau, J., J. G. Semmler and R. M. Enoka (2006) Training adaptations in the behavior of human motor units. *J. Appl. Physiol.* 101: 1766–1775.

Gordon, T., N. Tyreman, V. F. Rafuse and J. B. Munson (1997) Fast-to-slow conversion following chronic low-frequency activation of medial gastrocnemius muscle in cats. I. Muscle and motor unit properties. *J. Neurophysiol.* 77: 2585–2604.

Lieber, R. L. (2002) *Skeletal Muscle Structure, Function, and Plasticity,* 2nd Edition. Baltimore, MD: Lippincott Williams & Wilkins.

Munson, J. B., R. C. Foehring, L. M. Mendell and T. Gordon (1997) Fast-to-slow conversion following chronic low-frequency activation of medial gastrocnemius muscle in cats. II. Motoneuron properties. *J. Neurophysiol.* 77: 2605–2615.

Van Cutsem, M., J. Duchateau and K. Hainaut (1998) Changes in single motor unit behaviour contribute to the increase in contraction speed after dynamic training in humans. *J. Physiol.* 513: 295–305.

BOX 16B Locomotion in the Leech and the Lamprey

Alford, S. T. and M. H. Alpert (2014) A synaptic mechanism for network synchrony. *Front. Cell. Neurosci.* 8: 290. doi:10.3389/fncel.2014.00290

Grillner, S., P. Wallén, K. Saitoh, A. Kozlov and B. Robertson (2008) Neural bases of goal-directed locomotion in vertebrates—an overview. *Brain Res. Rev.* 57: 2–12.

Kristan, Jr., W. B., R. L. Calabrese and W. O. Friesen (2005) Neuronal control of leech behavior. *Prog. Neurobiol.* 76: 279–327.

Marder, E. and R. L. Calabrese (1996) Principles of rhythmic motor pattern generation. *Physiol. Rev.* 76: 687–717.

Mullins, O. J., J. T. Hackett, J. T. Buchanan and W. D. Friesen (2011) Neuronal control of swimming behavior: Comparison of vertebrate and invertebrate model systems. *Prog. Neurobiol.* 93: 244–269.

Sharples, S. A., K. Koblinger, J. M. Humphreys and P. J. Whelan (2014) Dopamine: a parallel pathway for the modulation of spinal locomotor networks. *Front. Neural Circuits* 8: 55. doi:10.3389/fncir.2014.00055

CHAPTER 17 Upper Motor Neuron Control of the Brainstem and Spinal Cord

BOX 17A What Do Motor Maps Represent?

Barinaga, M. (1995) Remapping the motor cortex. *Science* 268: 1696–1698.

Graziano, M. S. A. (2016) Ethological action maps: a paradigm shift for the motor cortex. *Trends. Cog. Sci.* 20: 121–132.

Graziano, M. S. A., T. N. S. Aflalo and D. F. Cooke (2005) Arm movements evoked by electrical stimulation in the motor cortex of monkeys. *J. Neurophysiol.* 94: 4209–4223.

Lemon, R. (1988) The output map of the primate motor cortex. *Trends Neurosci.* 11: 501–506.

Penfield, W. and E. Boldrey (1937) Somatic motor and sensory representation in the cerebral cortex of man studied by electrical stimulation. *Brain* 60: 389–443.

Schieber, M. H. and L. S. Hibbard (1993) How somatotopic is the motor cortex hand area? *Science* 261: 489–491.

Woolsey, C. N. (1958) Organization of somatic sensory and motor areas of the cerebral cortex. In *Biological and Biochemical Bases of Behavior*, H. F. Harlow and C. N. Woolsey (eds.). Madison: University of Wisconsin Press, pp. 63–81.

BOX 17B Minds and Machines

Chaudhary, U., N. Birbaumer and A. Ramos-Murguialday (2016) Brain–computer interfaces for communication and rehabilitation. *Nat. Rev. Neurol.* 12: 513–525.

Donati, A. R. C. and 19 others (2016) Long-term training with a brain-machine interface-based gait protocol induces partial neurological recovery in paraplegic patients. *Sci. Rep.* 6: 30383. doi: 10.1038/srep30383

Krucoff, M. O., S. Rahimpour, M. W. Slutzky, V. R. Edgerton and D. A. Turner (2016) Enhancing nervous system recovery through neurobiologics, neural interface training, and neurorehabilitation. *Front. Neurosci.* 10: 584. doi: 10.3389/fnins.2016.00584

Nicolelis, M. A. L. (2012) Mind in motion. *Sci. Am.* 307: 58–63.

Figure A After Dr. Eric C. Leuthardt, Professor of Neurological Surgery, Washington University School of Medicine, Director of The Center for Innovation in Neuroscience and Technology.

BOX 17C The Reticular Formation

Blessing, W. W. (1997) Inadequate frameworks for understanding bodily homeostasis. *Trends Neurosci.* 20: 235–239.

Holstege, G., R. Bandler and C. B. Saper (eds.) (1996) *Progress in Brain Research*, vol. 107. Amsterdam: Elsevier.

Loewy, A. D. and K. M. Spyer (eds.) (1990) *Central Regulation of Autonomic Functions.* New York: Oxford University Press.

Mason, P. (2001) Contributions of the medullary raphe and ventromedial reticular region to pain modulation and other homeostatic functions. *Annu. Rev. Neurosci.* 24: 737–777.

Moruzzi, G. and H. W. Magoun (1949) Brain stem reticular formation and activation of the EEG. *EEG Clin. Neurophys.* 1: 455–476.

CHAPTER 18 Modulation of Movement by the Basal Ganglia

BOX 18A Making and Breaking Habits

Desrochers, T. M., K. Amemori and A. M. Graybiel (2015) Habit learning by naive macaques is marked by response sharpening of striatal neurons representing the cost and outcome of acquired action sequences. *Neuron* 87: 853–868.

O'Hare, J. K. and 6 others (2016) Pathway-specific striatal substrates for habitual behavior. *Neuron* 89: 472–479.

Smith, K. S. and A. M. Graybiel (2016) Habit formation. *Dialog. Clin. Neurosci.* 18: 33–43.

BOX 18B Basal Ganglia Loops and Non-Motor Brain Functions

Alexander, G. E., M. R. DeLong and P. L. Strick (1986) Parallel organization of functionally segregated circuits linking basal ganglia and cortex. *Annu. Rev. Neurosci.* 9: 357–381.

Drevets, W. C. and 6 others (1997) Subgenual prefrontal cortex abnormalities in mood disorders. *Nature* 386: 824–827.

Jahanshahi, M., I. Obeso, J. C. Rothwell and J. A. Obeso (2015) A fronto–striato–subthalamic–pallidal network for goal-directed and habitual inhibition. *Nat. Rev. Neurosci.* 16: 719–732.

Middleton, F. A. and P. L. Strick (2000) Basal ganglia output and cognition: Evidence from anatomical, behavioral, and clinical studies. *Brain Cogn.* 42: 183–200.

Shepherd, G. M. G. (2013) Corticostriatal connectivity and its role in disease. *Nat. Rev. Neurosci.* 14: 278–291.

Smith, K. S. and A. M. Graybiel (2016) Habit formation. *Dialog. Clin. Neurosci.* 18: 33–43.

CHAPTER 19 Modulation of Movement by the Cerebellum

BOX 19A Genetic Analysis of Cerebellar Function

Caviness, V. S., Jr. and P. Rakic (1978) Mechanisms of cortical development: A view from mutations in mice. *Annu. Rev. Neurosci.* 1: 297–326.

D'Arcangelo, G. and 5 others (1995) A protein related to extracellular matrix proteins deleted in the mouse mutation *reeler*. *Nature* 374: 719–723.

Kloth, A. D. and 16 others (2015) Cerebellar associative sensory learning defects in five mouse autism models. *eLife* 4: e06085.

Patil, N. and 5 others (1995) A potassium channel mutation in *weaver* mice implicates membrane excitability in granule cell differentiation. *Nat. Genet.* 11: 126–129.

Rakic, P. (1977) Genesis of the dorsal lateral geniculate nucleus in the rhesus monkey: Site and time of origin, kinetics of proliferation, routes of migration and pattern of distribution of neurons. *J. Comp. Neurol.* 176: 23–52.

Rakic, P. and V. S. Caviness, Jr. (1995) Cortical development: A view from neurological mutants two decades later. *Neuron* 14: 1101–1104.

Taroni, F. and S. DiDonato (2004) Pathways to motor incoordination: the inherited ataxias. *Nat. Rev. Neurosci.* 5: 641–655.

Wang, S. S., A. D. Kloth and A. Badura (2014) The cerebellum, sensitive periods, and autism. *Neuron* 83: 518–532.

CHAPTER 20 Eye Movements and Sensorimotor Integration

BOX 20A The Perception of Stabilized Retinal Images

Barlow, H. B. (1963) Slippage of contact lenses and other artifacts in relation to fading and regeneration of supposedly stable retinal images. *Q. J. Exp. Psychol.* 15: 36–51.

Coppola, D. and D. Purves (1996) The extraordinarily rapid disappearance of entopic images. *Proc. Natl. Acad. Sci. USA* 96: 8001–8003.

Heckenmueller, E. G. (1965) Stabilization of the retinal image: A review of method, effects and theory. *Psychol. Bull.* 63: 157–169.

Krauskopf, J. and L. A. Riggs (1959) Interocular transfer in the disappearance of stabilized images. *Amer. J. Psychol.* 72: 248–252.

Martinez-Conde, S., J. Otero-Millan and S. L. Macknik (2013) The impact of microsaccades on vision: towards a unified theory of saccadic function. *Nat. Rev. Neurosci.*14: 83–96.

Riggs, L. A., F. Ratliff, J. C. Cornsweet and T. N. Cornsweet (1953) The disappearance of steadily fixated visual test objects. *J. Opt. Soc. Am.* 43: 495–501.

Rucci, M. and J. D. Victor (2014) The unsteady eye: an information-processing stage, not a bug. *Trends Neurosci.* 38: 194–206.

BOX 20B Sensorimotor Integration in the Superior Colliculus

Isa, T. and W. C. Hall (2009) Exploring the superior colliculus in vitro. *J. Neurophysiol.* 102: 2581–2593.

Lee, P. H., M. C. Helms, G. J. Augustine and W. C. Hall (1997) Role of intrinsic synaptic circuitry in collicular sensorimotor integration. *Proc. Natl. Acad. Sci. USA* 94: 13299–13304.

Ozen, G., G. J. Augustine and W. C. Hall (2000) Contribution of superficial layer neurons to premotor bursts in the superior colliculus. *J. Neurophysiol.* 84: 460–471.

Sparks, D. L. and J. S. Nelson (1987) Sensory and motor maps in the mammalian superior colliculus. *Trends Neurosci.* 10: 312–317.

Wurtz, R. H. and J. E. Albano (1980) Visual-motor function of the primate superior colliculus. *Annu. Rev. Neurosci.* 3: 189–226.

BOX 20C From Place Codes to Rate Codes

Fuchs, A. F. and E. S. Luschei (1970) Firing patterns of abducens neurons of alert monkeys in relationship to horizontal eye movement. *J. Neurophysiol.* 33: 382–392.

Groh, J. M. (2001) Converting neural signals from place codes to rate codes. *Biol. Cybern.* 85: 159–165.

Sparks, D. L. (1975) Response properties of eye movement-related neurons in the monkey superior colliculus. *Brain Res.* 90: 147–152.

CHAPTER 21 The Visceral Motor System

BOX 21A The Hypothalamus

Saper, C. B. (2012) Hypothalamus. In *The Human Nervous System*, 3rd Edition. J. K. Mai and G. Paxinos (eds.). Amsterdam: Elsevier, pp. 548–583.

Swanson, L. W. and P. E. Sawchenko (1983) Hypothalamic integration: Organization of the paraventricular and supraoptic nuclei. *Annu. Rev. Neurosci.* 6: 269–324.

BOX 21B Obesity and the Brain

Horvath, T. L. and S. Diano (2004) The floating blueprint of hypothalamic feeding circuits. *Nat. Rev. Neurosci.* 5: 662–667.

Huxing, C., M. López and K. Rahmouni (2017) The cellular and molecular bases of leptin and ghrelin resistance in obesity. *Nat. Rev. Endocrin.*, doi: 10.1038/nrendo.2016.222.

Kaye, W. H., J. L. Fudge and M. Paulus (2009) New insights in symptoms and neurocircuit function of anorexia nervosa. *Nat. Rev. Neurosci.* 10: 573–584.

Marx, J. (2003) Cellular warriors at the battle of the bulge. *Science* 299: 846–849.

Morton, G. J., T. H. Meek and M. W. Schwartz (2014) Neurobiology of food intake in health and disease. *Nat. Rev. Neurosci.* 15: 367–378.

O'Rahilly, S., I. S. Farooqi, G. S. H. Yeo and B. G. Challis (2003) Human obesity—lessons from monogenic disorders. *Endocrinology* 144: 3757–3764.

Saper, C. B., T. C. Chou and J. K. Elmquist (2002) The need to feed: homeostatic and hedonic control of eating. *Neuron* 36: 199–201.

Schwartz, M. W., S. C. Woode, D. Porte, R. J. Seely and D. G. Baskin (2000) Central nervous system control of food intake. *Nature* 404: 661–671.

Ziauddeen, H., I. S. Farooqi and P. C. Fletcher (2012) Obesity and the brain: how convincing is the addiction model? *Nat. Rev. Neurosci.* 13: 279–286.

CHAPTER 22 Early Brain Development

BOX 22A Stem Cells: Promise and Peril

Brustle, O. and 7 others (1999) Embryonic stem cell derived glial precursors: A source of myelinating transplants. *Science* 285: 754–756.

Castro, R. F., K. A. Jackson, M. A. Goodell, C. S. Robertson, H. Liu and H. D. Shine (2002) Failure of bone marrow cells to transdifferentiate into neural cells in vivo. *Science* 297: 1299.

Dolmetsch, R. and D. H. Geschwind (2011) The human brain in a dish: The promise of iPSC-derived neurons. *Cell* 145: 831–834.

Seaberg, R. M. and D. Van Der Kuoy (2003) Stem and progenitor cells: The premature desertion of rigorous definition. *Trends Neurosci.* 26: 125–131.

Wichterle, H., I. Lieberam, J. A. Porter and T. M. Jessell (2002) Directed differentiation of embryonic stem cells into motor neurons. *Cell* 110: 385–397.

Wu, S. M. and K. Hochedlinger (2011) Harnessing the potential of induced pluri-potent stem cells for regenerative medicine. *Nat. Cell Biol.*13: 497–505.

CHAPTER 23 Construction of Neural Circuits

BOX 23A Choosing Sides: Axon Guidance at the Optic Chiasm

Guillery, R. W. (1974) Visual pathways in albinos. *Sci. Amer.* 230: 44–54.

Guillery, R. W., C. A. Mason and J. S. Taylor (1995) Developmental determinants at the mammalian optic chiasm. *J. Neurosci.* 15: 4727–4737.

Herrera, E. and 8 others (2003) Zic2 patterns binocular vision by specifying the uncrossed retinal projection. *Cell* 114: 545–557.

Rasband, K., M. Hardyv and C. B. Chien (2003) Generating X: Formation of the optic chiasm. *Neuron* 39: 885–888.

Williams, S. E. and 9 others (2003) Ephrin-B2 and EphB1 mediate retinal axon divergence at the optic chiasm. *Neuron* 39: 919–935.

BOX 23B Why Do Neurons Have Dendrites?

Hume, R. I. and D. Purves (1981) Geometry of neonatal neurons and the regulation of synapse elimination. *Nature* 293: 469–471.

Purves, D. and R. I. Hume (1981) The relation of postsynaptic geometry to the number of presynaptic axons that innervate autonomic ganglion cells. *J. Neurosci.* 1: 441–452.

Purves, D. and J. W. Lichtman (1985) Geometrical differences among homologous neurons in mammals. *Science* 228: 298–302.

Purves, D., E. Rubin, W. D. Snider and J. W. Lichtman (1986) Relation of animal size to convergence, divergence and neuronal number in peripheral sympathetic pathways. *J. Neurosci.* 6: 158–163.

Snider, W. D. (1988) Nerve growth factor promotes dendritic arborization of sympathetic ganglion cells in developing mammals. *J. Neurosci.* 8: 2628–2634.

CHAPTER 24 Circuit Development: Intrinsic Factors and Sex Differences

BOX 24A The Science of Love (or, Love As a Drug)

Acevedo, B. P., A. Aron, H. E. Fisher and L. L. Brown (2011) Neural correlates of long-term intense romantic love. *Soc. Cogn. Affect. Neurosci.* 7: 145–159.

Aron, A. and 5 others (2005) Reward, motivation, and emotion systems associated with early-stage intense romantic love. *J. Neurophysiol.* 94: 327–337.

Bartels, A. and S. Zeki (2000) The neural basis of romantic love. *NeuroReport* 11: 3829–3834.

Bartels, A. and S. Zeki (2004) The neural correlates of maternal and romantic love. *Neuroimage* 21: 1155–1166.

Fisher, H. E., A. Aron, and L. L. Brown (2005) Romantic love: An fMRI study of neural mechanisms for mate choice. *J. Comp. Neurol.* 493: 58–62.

Fisher, H. E., L. L. Brown, A. Aron, G. Strong and D. Mashek (2010) Reward, addiction, and emotion regulation systems associated with rejection in love. *J. Neurophysiol.* 104: 51–60.

Insel, T. R. and L. J. Young (2001) The neurobiology of attachment. *Nat. Rev. Neurosci.* 2: 129–136.

Young, L. J. and Z. Wang (2004) The neurobiology of pair bonding. *Nat. Neurosci.* 7: 1048–1054.

Zeki, S. (2007) The neurobiology of love. *FEBS Lett.* 581: 2575–2579.

CHAPTER 25 Experience-Dependent Plasticity in the Developing Brain

BOX 25A Built-In Behaviors

Harlow, H. F. (1959) Love in infant monkeys. *Sci. Amer.* 2: 68–74.

Harlow, H. F. and R. R. Zimmerman (1959) Affectional responses in the infant monkey. *Science* 130: 421–432.

Lorenz, K. (1970) *Studies in Animal and Human Behaviour.* (Translated by R. Martin.) Cambridge, MA: Harvard University Press.

Macfarlane, A. J. (1975) Olfaction in the development of social preferences in the human neonate. *Ciba Found. Symp.* 33: 103–117.

Schaal, B. E. and 5 others (1980) Les stimulations olfactives dans les relations entre l'enfant et la mère. *Reprod. Nutr. Dev.* 20: 843–858.

Tinbergen, N. (1953) *Curious Naturalists*. Garden City, NY: Doubleday.

CHAPTER 26 Repair and Regeneration in the Nervous System

BOX 26A Specific Regeneration of Synaptic Connections in Autonomic Ganglia

Landmesser, L. and G. Pilar (1970) Selective reinnervation of two cell populations in the adult pigeon ciliary ganglion. *J. Physiol. (Lond.)* 211: 203–216.

Langley, J. N. (1897) On the regeneration of pre-ganglionic and post-ganglionic visceral nerve fibres. *J. Physiol. (Lond.)* 22: 215–230.

Purves, D. and J. W. Lichtman (1983) Specific connections between nerve cells. *Annu. Rev. Physiol.* 45: 553–565.

Purves, D., W. Thompson and J. W. Yip (1981) Re-innervation of ganglia transplanted to the neck from different levels of the guinea-pig sympathetic chain. *J. Physiol. (Lond.)* 313: 49–63.

BOX 26B Nuclear Weapons and Neurogenesis

Au, E. and G. Fishell (2006) Adult cortical neurogenesis: Nuanced, negligible, or nonexistent? *Nat. Neurosci.* 9: 1086–1088.

Bhardwaj, R. D. and 10 others (2006) Neocortical neurogenesis in humans is restricted to development. *Proc. Natl. Acad. Sci. USA* 103: 12564–12568.

Gould, E., A. J. Reeves, M. S. Graziano and C. G. Gross (1999) Neurogenesis in the neocortex of adult primates. *Science* 286: 548–552.

Koketsu, D., A. Mikami, Y. Miyamoto and T. Hisatsune (2003) Nonrenewal of neurons in the cerebral neocortex of adult macaque monkeys. *J. Neurosci.* 23: 937–942.

Kornack, D. R. and P. Rakic (2001) Cell proliferation without neurogenesis in adult primate neocortex. *Science* 294: 2127–2130.

Rakic, P. (2006) No more cortical neurons for you. *Science* 313: 928–929.

CHAPTER 27 Cognitive Functions and the Organization of the Cerebral Cortex

BOX 27B Synesthesia

Baron-Cohen, S. and J. E. Harrison (eds.) (1997) *Synesthesia: Classic and Contemporary Readings*. Malden, MA: Blackwell Scientific.

Bridgeman, B., D. Winter and P. Tseng (2010) Dynamic phenomenology of grapheme-color synesthesia. *Perception* 39: 671–676.

Palmeri, T. J., Blake, R. B., Marois, R., Flanery, M.A. and Whetsell, W. O. (2002) The perceptual reality of synesthetic color. *Proc. Natl. Acad. Sci. USA* 99, 4127–4131

Ramachandran, V. S. and E. M. Hubbard (2001) Psychophysical investigations into the neural basis of synaesthesia. *Proc. R. Soc. Lond. B* 368: 979–983.

Ramachandran, V. S. and E. M. Hubbard (2005) Neurocognitive mechanisms in synesthesia. *Neuron* 48: 509–520.

BOX 27C Neuropsychological Testing

Berg, E. A. (1948) A simple objective technique for measuring flexibility in thinking. *J. Gen. Psychol.* 39: 15–22.

Lezak, M. D. (1995) *Neuropsychological Assessment*, 3rd Edition. New York: Oxford University Press.

Milner, B. (1963) Effects of different brain lesions on card sorting. *Arch. Neurol.* 9: 90–100.

Milner, B. and M. Petrides (1984) Behavioural effects of frontal-lobe lesions in man. *Trends Neurosci.* 4: 403–407.

Stoet, G. and L. H. Snyder (2009) Neural correlates of executive control functions in the monkey. *Trends Cog. Sci.* 13: 228–234.

CHAPTER 28 Cortical States

BOX 28A Electroencephalography

Adrian, E. D. and K. Yamagiwa (1935) The origin of the Berger rhythm. *Brain* 58: 323–351.

Andersen, P. and S. A. Andersson (1968) *Physiological Basis of the Alpha Rhythm*. New York: Appleton-Century-Crofts.

Caton, R. (1875) The electrical currents of the brain. *Brit. Med. J.* 2: 278.

Da Silva, F. H. and W. S. Van Leeuwen (1977) The cortical source of the alpha rhythm. *Neurosci. Letters* 6: 237–241.

Dempsey, E. W. and R. S. Morrison (1943) The electrical activity of a thalamocortical relay system. *Amer. J. Physiol.* 138: 273–296.

Niedermeyer, E. and F. L. Da Silva (1993) *Electroencephalography: Basic Principles, Clinical Applications, and Related Fields.* Baltimore: Williams & Wilkins.

Nuñez, P. L. (1981) *Electric Fields of the Brain: The Neurophysics of EEG*. New York: Oxford University Press.

BOX 28B Sleep and Memory

Ji, D.Y. and M. A. Wilson (2007) Coordinated memory replay in the visual cortex and hippocampus during sleep. *Nat. Neurosci.* 10: 100–107.

Rudoy, J. D., J. L. Voss, C. E. Westerberg and K. A. Paller (2009) Strengthening individual memories by reactivating them during sleep. *Science* 326: 1079.

Tambini, A., N. Ketz and L. Davachi (2010) Enhanced brain correlations during rest are related to memory for recent experiences. *Neuron* 65: 280–290.

BOX 28C Dreaming

Foulkes, D. (1999) *Children's Dreaming and the Development of Consciousness*. Cambridge, MA: Harvard University Press.

Hobson, J. A. (1990) Sleep and dreaming. *J. Neurosci.* 10: 371–382.

Hobson, J. A. (2002) *Dreaming*. New York: Oxford University Press.

Hobson, J. A., R. Strickgold and E. F. Pace-Schott (1998) The neuropsychology of REM sleep and dreaming. *NeuroReport* 9: R1–R14.

CHAPTER 29 Attention

BOX 29A: Attention and the Frontal Eye Fields

Moore, T., K. M. Armstrong and M. Fallah (2003) Visuomotor origins of covert spatial attention. *Neuron* 40: 671–683.

Thompson, K. G., K. L. Biscoe and T. R. Sato (2005) Neuronal basis of covert spatial attention in the frontal eye field. *J. Neurosci.* 25: 9479–9487.

CHAPTER 30 Memory

BOX 30A Phylogenetic Memory

Dukas, R. (1998) *Cognitive Ecology*. Chicago: University of Chicago Press.

Lorenz, K. (1970) *Studies in Animal and Human Behaviour* (translated by R. Martin). Cambridge, MA: Harvard University Press.

Tinbergen, N. (1953) *The Herring Gull's World*. New York: Harper & Row.

Tinbergen, N. (1969) *Curious Naturalists*. Garden City, NY: Doubleday.

BOX 30B Savant Syndrome

Howe, M. J. A. (1989) *Fragments of Genius: The Strange Feats of Idiots Savants*. Routledge, NY: Chapman and Hall.

Miller, L. K. (1989) *Musical Savants: Exceptional Skill in the Mentally Retarded*. Hillsdale, NJ: Lawrence Erlbaum Associates.

Smith, N. and I.-M. Tsimpli (1995) *The Mind of a Savant: Language Learning and Modularity*. Oxford, U.K.: Basil Blackwell Ltd.

Smith, S. B. (1983) *The Great Mental Calculators: The Psychology, Methods, and Lives of Calculating Prodigies, Past and Present*. New York: Columbia University Press.

BOX 30C Alzheimer's Disease

Citron, M. and 8 others (1992) Mutation of the β-amyloid precursor protein in familial Alzheimer's disease increases β-protein production. *Nature* 360: 672–674.

Corder, E. H. and 8 others (1993) Gene dose of apolipoprotein E type 4 allele and the risk of Alzheimer's disease in late-onset families. *Science* 261: 921–923.

Goldgaber, D., M. I. Lerman, O. W. McBride, U. Saffiotti and D. C. Gajdusek (1987) Characterization and chromosomal localization of a cDNA encoding brain amyloid of Alzheimer's disease. *Science* 235: 877–880.

Gotz, J. and L. M. Ittner (2008) Animal models of Alzheimer's disease and frontotemporal dementia. *Nat. Rev. Neurosci.* 9: 532–534.

Murrell, J., M. Farlow, B. Ghetti and M. D. Benson (1991) A mutation in the amyloid precursor protein associated with hereditary Alzheimer's disease. *Science* 254: 97–99.

Rogaev, E. I. and 20 others (1995) Familial Alzheimer's disease in kindreds with missense mutations in a gene on chromosome 1 related to the Alzheimer's disease type 3 gene. *Nature* 376: 775–778.

Sherrington, R. and 33 others (1995) Cloning of a gene bearing missense mutations in early-onset familial Alzheimer's disease. *Nature* 375: 754–760.

Whitehouse P. J., George D. (2008) *The Myth of Alzheimer's*. New York NY: St. Martin's Press.

BOX 30D Place Cells and Grid Cells

Brun, V. H. and 6 others (2002) Place cells and place recognition maintained by direct entorhinal-hippocampal circuitry. *Science* 296: 2243–2246.

Fyhn, M., S. Molden, M. P. Witter, E. I. Moser and M. B. Moser (2004) Spatial representation in the entorhinal cortex. *Science* 305: 1258–1264.

Jacobs, J. and 10 others (2013) Direct recordings of grid-like neuronal activity in human spatial navigation. *Nat. Neurosci.* 6: 1188–1190.

Moser, E. I., E. Kropff and M.-B. Moser (2008) Place cells, grid cells, and the brain's spatial representation system. *Ann. Rev. Neurosci.* 31: 69–89.

O'Keefe, J. (1976) Place units in the hippocampus of the freely moving rat. *Exp. Neurol.* 51: 78–109.

O'Keefe, J. and L. Nadel (1978) *The Hippocampus as a Cognitive Map*. Oxford, UK: Oxford University Press.

Tolman, E. C. (1948) Cognitive maps in rats and men. *Psychol. Rev.* 55: 189–208.

CHAPTER 31 Emotion

BOX 31A Determination of Facial Expressions

Duchenne de Boulogne, G.-B. (1862) *Mecanisme de la Physionomie Humaine*. Paris: Editions de la Maison des Sciences de l'Homme. Edited and translated by R. A. Cuthbertson (1990) Cambridge, UK: Cambridge University Press.

Hopf, H. C., W. Müller-Forell and N. J. Hopf (1992) Localization of emotional and volitional facial paresis. *Neurology* 42: 1918–1923.

Trosch, R. M., G. Sze, L. M. Brass and S. G. Waxman (1990) Emotional facial paresis with striatocapsular infarction. *J. Neurol. Sci.* 98: 195–201.

Waxman, S. G. (1996) Clinical observations on the emotional motor system. In *Progress in Brain Research*, vol. 107, G. Holstege, R. Bandler and C. B. Saper (eds.). Amsterdam: Elsevier, pp. 595–604.

BOX 31B The Amygdala

Price, J. L., F. T. Russchen and D. G. Amaral (1987) The limbic region II: The amygdaloid complex. In *Handbook of Chemical Neuroanatomy*, vol. 5, *Integrated Systems of the CNS*, part I, *Hypothalamus, Hippocampus, Amygdala, Retina*, A. Björklund and T. Hökfelt (eds.). Amsterdam: Elsevier, pp. 279–388.

Phelps, E. A. and P. J. Whalen (2009) *The Human Amygdala*. New York: Guilford Press.

BOX 31C Fear and the Human Amygdala

Adolphs, R., D. Tranel, H. Damasio and A. R. Damasio (1995) Fear and the human amygdala. *J. Neurosci.* 15: 5879–5891.

Adolphs, R. and 5 others (2005) A mechanism for impaired fear recognition after amygdala damage. *Nature* 433: 68–72.

Bechara, A., H. Damasio, A. R. Damasio and G. P. Lee (1999) Differential contributions of the human amygdala and ventromedial prefrontal cortex to decision-making. *J. Neurosci.* 19: 5473–5481.

CHAPTER 32 Thinking, Planning, and Deciding

BOX 32A Dopamine and Reward Prediction Errors

Murayama, K., M. Matsumoto, K. Izuma and K. Matsumoto (2010) Neural basis of the undermining effect of monetary reward on intrinsic motivation. *Proc. Natl. Acad. Sci. USA* 107: 20911–20916.

Schultz, W., P. Dayan and P. R. Montague (1997) A neural substrate of prediction and reward. *Science* 275: 1593–1599.

BOX 32B What Does Neuroscience Have to Say about Free Will?

Eagleman, D. M. (2004) The where and when of intention. *Science* 303: 1144–1146.

Pearson, J. M., S. R. Heilbronner, D. L. Barack, B.Y. Haden and M. L. Platt (2011) Posterior cingulate cortex: Adapting behavior to a changing world. *Trends Cogn. Sci.* 15: 143–151.

CHAPTER 33 Speech and Language

BOX 33A The Generation of Speech

Gardner, H. (1974) *The Shattered Mind: The Person after Brain Damage*. New York: Vintage.

Liberman, A. M. (1996) *Speech: A Special Code.* Cambridge, MA: MIT Press.

Miller, G. A. (1991) *The Science of Words*, chapter 4, The spoken word. New York: Scientific American Library.

Plomp, R. (2002) *The Intelligent Ear: On the Nature of Sound Perception*. Mahwah, NJ: Erlbaum.

Sandrone, S. (2013) Norman Geschwind (1926–1984) *J. Neurol.* 260: 3197–3198. doi: 10.1007/s00415-013-6871-9

Warren, R. M. (1999) *Auditory Perception: A New Analysis and Synthesis*, Chapter 7, Speech. Cambridge: Cambridge University Press.

BOX 33B Language and Handedness

Bakan, P. (1975) Are left-handers brain damaged? *New Scientist* 67: 200–202.

Coren, S. (1992) *The Left-Hander Syndrome: The Causes and Consequence of Left-Handedness.* New York: The Free Press.

Davidson, R. J. and K. Hugdahl (eds.) (1995) *Brain Asymmetry*. Cambridge, MA: MIT Press.

Salive, M. E., J. M. Guralnik and R. J. Glynn (1993) Left-handedness and mortality. *Am. J. Pub. Health* 83: 265–267.

BOX 33C Learned Vocal Communication in Birds

Brenowitz, E. A. and M. D. Beecher (2005) Song learning in birds: diversity and plasticity, opportunities and challenges. *Trends Neurosci.* 28: 127–132.

Doupe, A. and P. Kuhl (1999) Birdsong and human speech: Common themes and mechanisms. *Annu. Rev. Neurosci.* 22: 567–631.

Marler, P. and H. W. Slabbekoorn (2004) *Nature's Music: The Science of Birdsong*. New York: Academic Press.

APPENDIX Survey of Human Neuroanatomy

BOX A Thalamus and Thalamocortical Relations

Jones, E. G. (2007) *The Thalamus*, 2nd Edition. Cambridge, UK: Cambridge University Press.

Sherman, S. M. and R. W. Guillery (2006) *Exploring the Thalamus and Its Role in Cortical Function*. Cambridge, MA: MIT Press.

Sherman, S. M. and R. W. Guillery (2011) Distinct functions for direct and transthalamic corticocortical connections. *J. Neurophysiol*. doi:10.1152/jn.00429.2011

Illustration Credits

CHAPTER 1 Studying the Nervous System

Opening Image Courtesy of Allen W. Song, Duke–UNC Brain Imaging and Analysis Center.

Figure 1.1A Ramsköld, D., E. T. Wang, C. B. Burge and R. Sandberg (2009) An abundance of ubiquitously expressed genes revealed by tissue transcriptome sequence data. *PLoS* 5:12 e1000598. **Figure 1.1C** Bond, J. and 11 others (2002) ASPM is a major determinant of cerebral cortical size. *Nat. Genet.* 32: 316–320. **Figure 1.3** Peters, A., S. L. Palay and H. deF. Webster (1991) *The Fine Structure of the Nervous System: Neurons and Their Supporting Cells*, 3rd Edition Oxford University Press, New York. **Figure 1.4B** Kalil, K., G. Szebenyi and E.W. Dent (2000) Common mechanisms underlying growth cone guidance and axon branching. *Dev. Neurobio.* 44: 145–158. **Figure 1.4E,F** Matus, A. (2000) Actin dynamics and synaptic plasticity. *Science* 290: 754–758. **Figure 1.5A–C** Jones, E. G. and M. W. Cowan (1983) The nervous tissue. In *The Structural Basis of Neurobiology,* E. G. Jones (ed.). New York: Elsevier, Chapter 8. **Figure 1.5D,E** Nishiyama, A., M. Komitova, R. Suzuki and X. Zhu (2009) Polydendrocytes (NG2 cells): Multifunctional cells with lineage plasticity. *Nat. Rev. Neurosci.* 10: 9–22. **Figure 1.5H** Bhat, M.A. and 11 others (2001) Axon-Glia Interactions and the Domain Organization of Myelinated Axons Requires Neurexin IV/Caspr/Paranodin. *Neuron* 30: 369–383. **Figure 1.10A** Mank, M. and 12 others (2008) A genetically encoded calcium indicator for chronic in vivo two-photon imaging. *Nat. Meth.* 5: 805–811. **Figure 10B,C** Ohki, K., S. Chung, Y. H. Ch'ng, P. Kara and R. C. Reid (2005) Functional imaging with cellular resolution reveals precise micro-architecture in visual cortex. *Nature* 433: 597–603. **Figure 1.11A** Zhang, F. and 12 others (2011) The microbial opsin family of optogenetic tools. *Cell* 147: 1446–1457. **Figure 1.11C,D** Kravitz, A. V. and 6 others (2010) Regulation of parkinsonian motor behaviours by optogenetic control of basal ganglia circuitry. *Nature* 466: 622–626. **Figure 1.13A** Stewart, T. A. and B. Mintz (1981) Successive generations of mice produced from an established culture line of euploid teratocarcinoma cells. *Proc. Nat. Acad. Sci. USA* 78: 6314–6318. **Figure 1.15** Zylka, M. J., F. L. Rice and D. J. Anderson (2005) Topographically distinct epidermal nociceptive circuits revealed by axonal tracers targeted to Mrgprd. *Neuron* 46: 17–25. **Figure 1.17A** Hoeffner, E. G., S. K. Mukherji, A. Srinivasan and D. J. Quint. (2012) Neuroradiology back to the future: Brain imaging. *Am. J. Neuroradiol.* 33: 5–11. **Figure 1.19C** Khairy, S. and 5 others (2015) Duodenal obstruction as first presentation of metastatic breast cancer. *Case Reports in Surgery* 2015: 605719. **Figure 1.20B,C** Seiger, R. and 10 others (2015) Voxel-based morphometry at ultra-high fields. a comparison of 7T and 3T MRI data. *Neuroimage* 113: 207–216. **Figure 1.21** Goodyear, B., E. Liebenthal and V. Mosher (2014) Active and passive fMRI for presurgical mapping of motor and language cortex. In *Advanced Brain Neuroimaging Topics in Health and Disease—Methods and Applications*, D. Duric (Ed.), *InTech*. doi: 10.5772/58269

CHAPTER 2 Electrical Signals of Nerve Cells

Opening Image, Unit 1 Structure of a chemical synapse within the cerebral cortex. A presynaptic terminal (top) forms a synapse with a dendritic spine of the postsynaptic neuron (bottom). Colors indicate different organelles found within these structures. Courtesy of Alain Burette and Richard Weinberg. **Figure 2.3** Hodgkin, A. L. and W. A. Rushton (1946) The electrical constants of a crustacean nerve fibre. *Proc. R. Soc. Lond. B* 133: 444–479. **Figures 2.8 & 2.9** Hodgkin, A. L. and B. Katz (1949) The effect of sodium ions on the electrical activity of the giant axon of the squid. *J. Physiol. (Lond.)* 108: 37–77. **Box 2B Figure A** Hodgkin, A. L. and A. F. Huxley (1939) Action potentials recorded from inside a nerve fibre. *Nature* 144: 710–711. **Box 2B Figure B** Dodge, F. A. and B. Frankenhaeuser (1958) Membrane currents in isolated frog nerve fibre under voltage clamp conditions. *J. Physiol. (Lond.)* 143: 76–90. **Box 2B Figure C** Barrett, E. F. and J. N. Barrett (1976) Separation of two voltage-sensitive potassium currents, and demonstration of a tetrodotoxin-resistant calcium current in frog motoneurones. *J. Physiol. (Lond.)* 255: 737–774. **Box 2B Figure D** Llinás, R. and Y. Yarom (1981) Electrophysiology of mammalian inferior olivary neurones in vitro. Different types of voltage-dependent ionic conductances. *J. Physiol. (Lond.)* 315: 549–567. **Box 2B Figure E** Chen, S., G. J. Augustine and P. Chadderton (2016) The cerebellum linearly encodes whisker position during voluntary movement. *eLife* 5: e10509.

CHAPTER 3 Voltage-Dependent Membrane Permeability

Figures 3.1, 3.2, & 3.3 Hodgkin, A. L., A. F. Huxley and B. Katz (1952) Measurements of current–voltage relations in the membrane of the giant axon of *Loligo. J. Physiol.* 116: 424–448. **Figures 3.4** Hodgkin, A. L. and A. F. Huxley (1952a) Currents carried by sodium and potassium ions through the membrane of the giant axon of *Loligo. J. Physiol.* 116: 449–472. **Figure 3.5** Armstrong, C. M. and L. Binstock (1965) Anomalous rectification in the squid giant axon injected with tetraethylammonium chloride. *J. Gen. Physiol.* 48: 859–872; Moore, J. W., M. P. Blaustein, N. C. Anderson and T. Narahashi (1967) Basis of tetrodotoxin's selectivity in blockage of squid axons. *J. Gen. Physiol.* 50: 1401–1410. **Figures 3.6 & 3.7** Hodgkin, A. L. and A. F. Huxley (1952b) The components of membrane conductance in the giant axon of Loligo. *J. Physiol.* 116: 473–496. **Figure 3.8** Hodgkin, A. L. and A. F. Huxley (1952d) A quantitative description of membrane current and its application to conduction and excitation in nerve. *J. Physiol.* 116: 507–544. **Figure 3.11B** Chen, C. and 17 others (2004) Mice lacking sodium channel beta1 subunits display defects in neuronal excitability, sodium channel expression, and nodal architecture. *J. Neurosci.* 24: 4030–4042.

CHAPTER 4 Ion Channels and Transporters

Figure 4.1B,C Bezanilla, F. and A. M. Correa (1995) Single-channel properties and gating of Na^+ and K^+ channels in the squid giant axon. In *Cephalopod Neurobiology*, N. J. Abbott, R. Williamson and L. Maddock (eds.). New York: Oxford University Press, pp. 131–151. **Figure 4.1D** Vanderberg, C. A. and F. Bezanilla (1991) A sodium channel model based on single channel, macroscopic ionic, and gating currents in the squid giant axon. *Biophys. J.* 60: 1511–1533. **Figure 4.1E** Correa, A. M. and F. Bezanilla (1994) Gating of the squid sodium channel at positive potentials. II. Single channels reveal two open states. *Biophys. J.* 66: 1864–1878. **Figure 4.2B–D** Augustine, C. K. and F. Bezanilla (1990) Phosphorylation modulates potassium conductance and gating current of perfused giant axons of squid. *J. Gen. Physiol.* 95: 245–271. **Figure 4.2E** Perozo, E., D. S. Jong and F. Bezanilla (1991) Single-channel studies of the phosphorylation of K^+

channels in the squid giant axon. II. Nonstationary conditions. *J. Gen. Physiol.* 98: 19–34. **Figure 4.4A,B** Doyle, D. A. and 7 others (1998) The structure of the potassium channel: Molecular basis of K^+ conduction and selectivity. *Science* 280: 69–77. **Box 4B Figure A** Schmidt, O. and H. Schmidt (1972) Influence of calcium ions on the ionic currents of nodes of Ranvier treated with scorpion venom. *Pflügers Arch.* 333: 51–61. **Box 4B Figure B** Cahalan, M. (1975) Modification of sodium channel gating in frog myelinated nerve fibers by Centruroides sculpturatus scorpion venom. *J. Physiol.* (*Lond.*) 244: 511–534. **Figure 4.5A,B** Long, S. B., E. B. Campbell and R. Mackinnon (2005b) Voltage sensor of Kv1.2: Structural basis of electromechanical coupling. *Science* 309: 903–908. **Figure 4.5C** Tao, X., A. Lee, W. Limapichat, D.A. Dougherty and R. MacKinnon (2010) A gating charge transfer center in voltage sensors. *Science* 328: 67-73. **Figure 4.5D** Lee, A. G. (2006) Ion channels: A paddle in oil. *Nature* 444: 697. **Figure 4.6A** Ahuja, S. and 34 others (2015) Structural basis of Nav1.7 inhibition by an isoform-selective small-molecule antagonist. *Science* 350: aac5464. **Figure 4.6B** Wu, J. and 6 others (2015) Structure of the voltage-gated calcium channel CaV1.1 complex. *Science* 350: aad2395. **Figure 4.6C** Long, S. B., E. B. Campbell and R. Mackinnon (2005) Crystal structure of a mammalian voltage-dependent Shaker family K^+ channel. *Science* 309: 897–903. **Figure 4.6D** Dutzler, R., E. B. Campbell, M. Cadene, B. T. Chait and R. Mackinnon (2002) X-ray structure of a ClC chloride channel at 3.0 A reveals the molecular basis of anion selectivity. *Nature* 415: 287–294. **Figure 4.8A** Sobolevsky, A. I., M. P. Rosconi and E. Gouaux (2009) X-ray structure, symmetry and mechanism of an AMPA-subtype glutamate receptor. *Nature* 462: 745–756. **Figure 4.8B** Gonzalez, E. B., T. Kawate and Eric Gouaux (2009) Pore architecture and ion sites in acid sensing ion channels and P2X receptors. *Nature* 460: 599–604. **Figure 4.8C** Hite, R.K., X. Tao and R. MacKinnon (2017) Structural basis for gating the high-conductance Ca^{2+}-activated K^+ channel. *Nature* 541: 52–57. **Figure 4.8D** Li, M. and 8 others (2017) Structure of a eukaryotic cyclicnucleotide-gated channel. *Nature* doi: 10.1038/nature20819 **Figure 4.9A** Gao,Y., E. Cao, D. Julius andY. Cheng (2016) TRPV1 structures in nanodiscs reveal mechanisms of ligand and lipid action. *Nature* 534: 347–351. **Figure 4.9B** Ge, J. and 9 others (2015) Architecture of the mammalian mechanosensitive Piezo1 channel. *Nature* 527: 64–69. **Figure 4.10A** Shinoda, T., H. Ogawa, F. Cornelius and C. Toyoshima (2009) Crystal structure of the sodium-potassium pump at 2.4 Å resolution. *Nature* 459: 446–450. **Figure 4.10B** Toyoshima, C., H. Nomura and T. Tsuda (2004) Luminal gating mechanism revealed in calcium pump crystal structures with phosphate analogues. *Nature* 432: 361–368. **Figure 4.11** Hodgkin, A. L. and R. D. Keynes (1955) Active transport of cations in giant axons from *Sepia* and *Loligo*. *J. Physiol.* 128: 28–60. **Figure 4.12A** Lingrel, J. B.and 5 others (1994) Structure-function studies of the Na, K-ATPase. *Kidney Internat. Suppl.* 44: S32–S39. **Figure 4.12B** Nyblom, M. and 7 others (2013) Crystal structure of Na^+, K^+-ATPase in the Na^+-bound state. *Science* 342: 123–127.

CHAPTER 5 Synaptic Transmission

Figure 5.2A,B Sotelo, C., R. Llinas and R. Baker (1974) Structural study of inferior olivary nucleus of the cat: morphological correlates of electrotonic coupling. *J. Neurophysiol.* 37: 541–559. **Figure 5.2D** Maeda, S. and 6 others (2009) Structure of the connexin 26 gap junction channel at 3.5 Å resolution. *Nature* 458: 597–602. **Figure 5.3A** Furshpan, E. J. and D. D. Potter (1959) Transmission at the giant motor synapses of the crayfish. *J. Physiol.* (*Lond.*) 145: 289–324. **Figure 5.3B** Waxman, S. G and G. W. Zamponi (2014) Regulating excitability of peripheral afferents: emerging ion channel targets. *Nat. Neurosci.* 17: 153–163. **Figure 5.4A,B** Burette, A. C. and 6 others (2012) Electron tomographic analysis of synaptic ultrastructure. *J. Comp. Neurol.* 520: 2697–2711. **Figure 5.5** Fatt, P. and B. Katz (1952) Spontaneous subthreshold activity at motor nerve endings. *J. Physiol.* (*Lond.*) 117: 109–127. **Figure 5.6** Boyd, I. A. and A. R. Martin (1955) Spontaneous subthreshold activity at mammalian neuromuscular junctions. *J. Physiol.* 132: 61–73. **Figure 5.7C** Heuser, J. E. and 5 others (1979) Synaptic vesicle exocytosis captured by quick freezing and correlated with quantal transmitter release. *J. Cell Biol.* 81: 275–300. **Figure 5.8** Heuser, J. E. and T. S. Reese (1973) Evidence for recycling of synaptic vesicle membrane during transmitter release at the frog neuromuscular junction. *J. Cell Biol.* 57: 315–344. **Figure 5.9** Augustine, G. J. and R. Eckert (1984) Divalent cations differentially support transmitter release at the squid giant synapse. *J. Physiol.* 346: 257–271. **Figure 5.10A** Smith, S. J., J. Buchanan, L. R. Osses, M. P. Charlton and G. J. Augustine (1993) The spatial distribution of calcium signals in squid presynaptic terminals. *J. Physiol.* (*Lond.*) 472: 573–593. **Figure 5.10B** Miledi, R. (1973) Transmitter release induced by injection of calcium ions into nerve terminals. *Proc. R. Soc. Lond. B* 183: 421–424. **Figure 5.10C** Adler, E. M. Adler, G. J. Augustine, M. P. Charlton and S. N. Duffy (1991) Alien intracellular calcium chelators attenuate neurotransmitter release at the squid giant synapse. *J. Neurosci.* 11: 1496–1507. **Figure 5.11A** Takamori, S. and 21 others (2006) Molecular anatomy of a trafficking organelle. *Cell* 127: 831–846. **Figure 5.12A** Sutton, R. B., D. Fasshauer, R. Jahn and A. T. Brünger (1998) Crystal structure of a SNARE complex involved in synaptic exocytosis at 2.4 Å resolution. *Nature* 395: 347–353; Madej, T. and 6 others (2014) MMDB and VAST$^+$: Tracking structural similarities between macromolecular complexes. *Nucleic Acids Res.* 42: D297–D303. **Figure 5.12B** Zhou, Q. and 19 others (2015) Architecture of the synaptotagmin-SNARE machinery for neuronal exocytosis. *Nature* 525: 62–67. **Figure 5.13A** Fotin, A. and 6 others (2004) Molecular model for a complete clathrin lattice from electron cryomicroscopy. *Nature* 432: 573–579. **Figure 5.13B** Reubold, T. F. and 12 others (2015) Crystal structure of the dynamin tetramer. *Nature* 525: 404–408. **Figure 5.13C** Shupliakov, O. and L. Brodin (2010) Recent insights into the building and cycling of synaptic vesicles. *Exp. Cell Res.* 316: 1344–1350. **Figures 5.16A–C & 5.17** Takeuchi, A. and N. Takeuchi (1960) On the permeability of end-plate membrane during the action of transmitter. *J. Physiol.* 154: 52–67. **Box 5A Figures A,B** Witcher, M. R., S. A. Kirov and K. M. Harris (2007). Plasticity of perisynaptic astroglia during synaptogenesis in the mature rat hippocampus. *Glia* 55: 13-23. **Box 5A Figure C** Cornell-Bell, A. H., S. M. Finkbeiner, M. S. Cooper, and S. J. Smith (1990). Glutamate induces calcium waves in cultured astrocytes: Long-range glial signaling. *Science* 247: 470-473. **Box 5A Figure D** Perea, G. and A. Araque (2007) Astrocytes potentiate transmitter release at single hippocampal synapses. *Science* 317: 1083–1086.

CHAPTER 6 Neurotransmitters and Their Receptors

Box 6A Figure B Tsetlin, V. I. (2015) Three-finger snake neurotoxins and Ly6 proteins targeting nicotinic acetylcholine receptors: Pharmacological tools and endogenous modulators. *Trends Pharmacol. Sci.* 36: 109–123. **Box 6A Figure D** Lebbe, E. K. M., S. Peigneur, I. Wijesekara, and J. Tytgat (2014) Conotoxins targeting nicotinic acetylcholine receptors: An overview. *Marine Drugs* 12: 2970–3004. **Figure 6.3A–C** Unwin, N. (2005) Refined structure of the nicotinic acetylcholine receptor at 4 Å resolution. *J. Mol. Biol.* 346: 967–989. **Figure 6.3D,E** Miyazawa, A., Y. Fujiyoshi and N. Unwin (2003) Structure and gating mechanism of the acetylcholine receptor pore. *Nature* 423: 949–955. **Figure 6.4A,B** Haga, K. and 10 others (2012) Structure of the human M2 muscarinic acetylcholine receptor bound to an antagonist. *Nature* 482: 547–551. **Figure 6.6A** Watanabe, J., A. Rozov and L. P. Wollmuth (2005) Target-specific regulation of synaptic amplitudes in the neocortex. *J. Neurosci.* 25: 1024–1033. **Figure 6.6B**

Mott, D. D., M. Benveniste and R. J. Dingledine (2008) pH-dependent inhibition of kainate receptors by zinc. *J. Neurosci.* 28: 1659–1671. **Figure 6.7A,E** Traynelis, S. F. and 9 others (2010) Glutamate receptor ion channels: Structure, regulation, and function. *Pharmacol. Rev.* 62: 405–496. **Figure 6.7B–D,F** Sobolevsky, A. I., M. P. Rosconi and E. Gouaux (2009) X-ray structure, symmetry and mechanism of an AMPA-subtype glutamate receptor. *Nature* 462: 745–756. **Figure 6.8 C–E** Karakas, E. and H. Furukawa (2014) Crystal structure of a heterotetrameric NMDA receptor ion channel. *Science* 344: 992–997. **Figure 6.8F** Zhu, S. and 6 others (2016) Mechanism of NMDA receptor inhibition and activation. *Cell* 165: 704-714. **Figure 6.9** Pin, J.-P. and B. Bettler (2016) Organization and functions of mGlu and GABAB receptor complexes. *Nature* 540: 60–68. **Figure 6.11A** Chavas, J. and A. Marty (2003) Coexistence of excitatory and inhibitory GABA synapses in the cerebellar interneuron network. *J. Neurosci.* 23: 2019–2030. **Figure 6.11B–C** Miller, P. S. and A. R. Aricescu (2014) Crystal structure of a human GABAA receptor. *Nature* 512: 270–275. **Figure 6.11D** Puthenkalam, R. and 6 others (2016) Structural studies of GABAA receptor binding sites: Which experimental structure tells us what? *Front. Mol. Neurosci.* 9: 44. **Box 6B Figure B** Berglund, K. and 8 others (2006) Imaging synaptic inhibition in transgenic mice expressing the chloride indicator, Clomeleon. *Brain Cell Biol.* 35: 207–228. **Box 6B Figure C** Obata, K., M. Oide and H. Tanaka (1978) Excitatory and inhibitory actions of GABA and glycine on embryonic chick spinal neurons in culture. *Brain Res.* 144: 179–184. **Figure 6.12** Pin, J.-P. and B. Bettler (2016) Organization and functions of mGlu and GABAB receptor complexes. *Nature* 540: 60-68. **Figure 6.13** Du, J., W. Lü, S. Wu, Y. Cheng and E. Gouaux (2015) Glycine receptor mechanism elucidated by electron cryo-microscopy. *Nature* 526: 224–229. **Figure 6.16A** Chien, E.Y. and 10 others (2010) Structure of the human dopamine D3 receptor in complex with a D2/D3 selective antagonist. *Science* 330: 1091–1095. **Figure 6.16B** Betke, K. M., C. A. Wells and H. E. Hamm (2012) GPCR mediated regulation of synaptic transmission. *Prog. Neurobiol.* 96: 304–321. **Figure 6.19A** Wacker D. and 12 others (2017) Crystal structure of an LSD-bound human serotonin receptor. *Cell* 168: 377–389. **Figure 6.19B** Hassaine, G, and 14 others (2014) X-ray structure of the mouse serotonin 5-HT3 receptor. *Nature* 512: 276–281. **Figure 6.20A–C** Kawate, T., J. C. Michel, W. T. Birdsong and E. Gouaux (2009) Crystal structure of the ATP-gated P2X(4) ion channel in the closed state. *Nature* 460: 592–598. **Figure 6.20D** Jaakola, V. P. and A. P. Ijzerman (2010) The crystallographic structure of the human adenosine A2A receptor in a high-affinity antagonist-bound state: Implications for GPCR drug screening and design. *Curr. Opin. Struct. Biol.* 20: 401–414. **Box 6C Figure C and Figure 6.23C** Iversen, L. (2003) Cannabis and the brain. *Brain* 126: 1252–1270. **Box 6C Figure D** Shao, Z. and 6 others (2016) High-resolution crystal structure of the human CB1 cannabinoid receptor. *Nature* 540: 602–606. **Figure 6.23A,B** Freund, T. F., I. Katona and D. Piomelli (2003) Role of endogenous cannabinoids in synaptic signaling. *Physiol Rev.* 83: 1017–1066. **Figure 6.24** Ohno-Shosaku, T., T. Maejima and M. Kano (2001) Endogenous cannabinoids mediate retrograde signals from depolarized postsynaptic neurons to presynaptic terminals. *Neuron* 29: 729–738.

CHAPTER 7 Molecular Signaling within Neurons

Figure 7.9A The Protein Data Bank. H. M. Berman and 7 others (2000) *Nucleic Acids Research* 28: 235–242. doi:10.1093/nar/28.1.235; rcsb.org; https://pdb101.rcsb.org/motm/152. **Figure 7.9B** Craddock, T. J. A., J. A. Tuszynski, and S. Hameroff (2012) Cytoskeletal signaling: Is memory encoded in microtubule lattices by CaMKII phosphorylation? *PLoS Comput. Biol.* 8: e1002421. **Figure 7.9C** Leonard, T. A., B. Różycki, L. F. Saidi, G. Hummer and J. H. Hurley (2011) Crystal structure and allosteric activation of protein kinase C βII. *Cell* 144: 55–66. **Figure 7.9D** Turk, B. E. (2007). Manipulation of host signalling pathways by anthrax toxins. *Biochem. J.* 402: 405–417. **Figure 7.10A** Bollen, M., W. Peti, M. J. Ragusa and M. Beullens (2010) The extended PP1 toolkit: Designed to create specificity. *Trends Bio-chem. Sci.* 35: 450–458. **Figure 7.10B** Cho, U. S. and W. Xu (2007) Crystal structure of a protein phosphatase 2A heterotrimeric holoenzyme. *Nature* 445: 53–57. **Figure 7.10C** Li, H., A. Rao and P.G. Hogan (2011) Interaction of calcineurin with substrates and targeting proteins. *Trends Cell Biol.* 21: 91–103. **Box 7A Figure A** Grynkiewicz, G., M. Poenie and R.Y. Tsien (1985) A new generation of Ca^{2+} indicators with greatly improved fluorescence properties. *J. Biol. Chem.* 260: 3440–3450. **Box 7A Figure B** Finch, E. A. and G. J. Augustine (1998) Local calcium signalling by inositol-1,4,5-trisphosphate in Purkinje cell dendrites. *Nature* 396: 753–756. **Box 7A Figure D** Vidal, G. S. M. Djurisic, K. Brown, R. W. Sapp and C. J. Shatz (2016) Cell-autonomous regulation of dendritic spine density by PirB. *eNeuro* 3: 1–15 ENEURO.0089-16.2016. **Box 7A Figure E** Livet, J. and 7 others (2007) Transgenic strategies for combinatorial expression of fluorescent proteins in the nervous system. *Nature* 450: 56–62. **Box 7B Figure B** Harris, K. M. and R. J. Weinberg (2012) Ultrastructure of synapses in the mammalian brain. *Cold Spring Harb. Perspect. Biol.* 4: a005587. **Box 7B Figure C** http://synapseweb.clm.utexas.edu/atlas; reprinted with permission from J. Spacek.

CHAPTER 8 Synaptic Plasticity

Figure 8.1A,B Charlton, M. P. and G. D. Bittner (1978) Presynaptic potentials and facilitation of transmitter release in the squid giant synapse. *J. Gen. Physiol.* 72: 487–511. **Figure 8.1C** Swandulla, D., M. Hans, K. Zipser and G. J. Augustine (1991) Role of residual calcium in synaptic depression and posttetanic potentiation: Fast and slow calcium signaling in nerve terminals. *Neuron* 7: 915–926. **Figure 8.1D** Betz, W.J. (1970) Depression of transmitter release at the neuromuscular junction of the frog. *J. Physiol.* (*Lond.*) 206: 629–644. **Figure 8.1E** Lev-Tov, A., M. J. Pinter and R. E. Burke (1983) Posttetanic potentiation of group Ia EPSPs: Possible mechanisms for differential distribution among medial gastrocnemius motoneurons. *J. Neurophysiol.* 50: 379–398. **Figure 8.2A** Katz, B. (1966) *Nerve, Muscle, and Synapse.* New York: McGraw-Hill. **Figure 8.2B** Malenka, R. C. and S. A. Siegelbaum (2001) Synaptic plasticity: Diverse targets and mechanisms for regulating synaptic efficacy. In *Synapses*. W. M. Cowan, T. C. Sudhof and C. F. Stevens (eds.). Baltimore: John Hopkins University Press, pp. 393–413. **Figures 8.3, 8.4 & 8.5** Squire, L. R. and E. R. Kandel (1999) *Memory: From Mind to Molecules*. New York: Scientific American Library. **Box 8A** Tully, T. (1996) Discovery of genes involved with learning and memory: An experimental synthesis of Hirshian and Benzerian perspectives. *Proc. Natl. Acad. Sci. USA* 93: 13460–13467. Copyright (1996) National Academy of Sciences, U.S.A. **Figure 8.7A–C** Malinow, R., H. Schulman, and R. W. Tsien (1989) Inhibition of postsynaptic PKC or CaMKII blocks induction but not expression of LTP. *Science* 245: 862–866. **Figure 8.7D** Abraham, W. C., B. Logan, J. M. Greenwood and M. Dragunow (2002) Induction and experience-dependent consolidation of stable long-term potentiation lasting months in the hippocampus. *J. Neurosci.* 22: 9626–9634. **Figure 8.8** Gustafsson, B., H. Wigstrom, W. C. Abraham, and Y.Y. Huang (1987) Long-term potentiation in the hippocampus using depolarizing current pulses as the conditioning stimulus to single volley synaptic potentials. *J. Neurosci.* 7: 774–780. **Figure 8.10** Nicoll, R. A., J. A. Kauer and R. C. Malenka (1988) The current excitement in long-term potentiation. *Neuron* 1: 97–103. **Figure 8.11A,B** Matsuzaki, M., N. Honkura, G.C. Ellis-Davies and H. Kasai (2004) Structural basis of long-term potentiation in single dendritic spines. *Nature* 429: 761–766. **Figure 8.11C and Box 8B Figure A** Liao, D., N. A. Hessler and R. Malinow (1995) Activation of postsynaptically silent synapses during pairing-induced LTP in CA1 region of hippocampal slice. *Nature* 375: 400–404. **Box 8B Figure C** Petralia, R. S. and 6 others (1999) Selective acquisition of AMPA receptors over post-natal development suggests a molecular basis for silent synapses. *Nat. Neurosci.* 2: 31–36. **Figure 8.12** Lee, S. J., Y. Escobedo-Lozoya, E. M. Szatmari and R. Yasuda (2009) Activation of CaMKII in single dendritic spines during long-term potentiation. *Nature* 458: 299–304. **Figure 8.14** Frey, U. and R. G. Morris (1997) Synaptic tagging and long-term potentiation. *Nature* 385: 533–536. **Figure 8.15A** Squire, L. R. and E. R. Kandel (1999) *Memory: From Mind to Molecules.* New York: Scientific American Library. **Figure 8.15B** Engert, F. and T. Bonhoeffer (1999) Dendritic spine changes associated with hippocampal long-term synaptic plasticity. *Nature* 399: 66–70. **Figure 8.16B** Mulkey, R. M., C. E. Herron and R. C. Malenka (1993) An essential role for protein phosphatases in hippocampal long-term depression. *Science* 261: 1051–1055. **Figure 8.17B** Sakurai, M. (1987) Synaptic modification of parallel fibre-Purkinje cell transmission in in vitro guinea-pig cerebellar slices. *J. Physiol. (Lond.)* 394: 463–480. **Figure 8.18** Bi, G. Q. and M. M. Poo (1998) Synaptic modifications in cultured hippocampal neurons: dependence on spike timing, synaptic strength, and postsynaptic cell type. *J. Neurosci.* 18: 10464–10472.

CHAPTER 9 The Somatosensory System: Touch and Proprioception

Opening Image, Unit II Scanning electron micrograph of outer hair cells (stereocilia) in the inner ear. © Steve Gschmeissner/ Science Photo Library. **Figure 9.3C** Weinstein, S. (1968) Intensive and extensive aspects of tactile sensitivity as a function of body part, sex, and laterality. In D.R. Kenshalo (Ed.), *The Skin Senses.* Springfield, IL: Charles C. Thomas, pp. 195–222. **Figure 9.5A** Johansson, R. S. and A. B. Vallbo (1983) Tactile sensory coding in the glabrous skin of the human. *Trends Neurosci.* 6: 27–32. **Figure 9.5B** Abraira, V. E. and D. D. Ginty (2013) The sensory neurons of touch. *Neuron* 79: 618–639. **Table 9.2** Johnson, K. O. (2002) Neural basis of haptic perception. In *Seven's Handbook of Experimental Psychology*, 3rd Edition. Vol 1: Sensation and Perception. H. Pashler and S. Yantis (eds.). New York: Wiley, pp. 537–583. **Figure 9.6** Phillips, J. R., R. S. Johansson and K. O. Johnson (1990) Representation of Braille characters in human nerve fibres. *Exp. Brain Res.* 81: 589–592. **Figure 9.7A** Matthews, P. B. C. (1964) Muscle spindles and their motor control. *Physiol. Rev.* 44: 219–289. **Figure 9.10** Brodal, P. (1992) *The Central Nervous System: Structure and Function.* New York: Oxford University Press, p. 151; Jones, E. G. and D. P. Friedman (1982) Projection pattern of functional components of thalamic ventrobasal complex on monkey somatosensory cortex. *J. Neurophys.* 48: 521–544. **Figure 9.11** Penfield, W. and T. Rasmussen (1950) *The Cerebral Cortex of Man: A Clinical Study of Localization of Function.* New York: Macmillan; Corsi, P. (1991) *The Enchanted Loom: Chapters in the History of Neuroscience*, P. Corsi (ed.). New York: Oxford University Press. **Box 9A** Purves, D., D. Riddle and A. LaMantia (1992) Iterated patterns of brain circuitry (or how the cortex gets its spots). *Trends Neurosci.* 15: 362–369. **Figure 9.13A** Kaas, J. H. (1993) The functional organization of somatosensory cortex in primates. *Ann. Anat.* 175: 509–518. **Figure 9.13C** Sur, M. (1980) Receptive fields of neurons in areas 3b and 1 of somatosensory cortex in monkeys. *Brain Res.* 198: 465–471. **Figure 9.14** Merzenich, M. M., R. J. Nelson, M. P. Stryker, M. S. Cynader, A. Schoppmann and J. M. Zook (1984) Somatosensory cortical map changes following digit amputation in adult monkeys. *J. Comp. Neurol.* 224: 591–605. **Figure 9.15** Jenkins, W. M., M. M. Merzenich, M. T. Ochs, T. Allard and E. Guic-Robles (1990) Functional reorganization of primary somatosensory cortex in adult owl monkeys after behaviorally controlled tactile stimulation. *J. Neurophysiol.* 63: 82–104.

CHAPTER 10 Pain

Figure 10.1 Fields, H. L. (1987) *Pain.* New York: McGraw-Hill. **Figure 10.2** Fields, H. L. (ed.) (1990) *Pain Syndromes in Neurology.* London: Butterworths. **Box 10C Figure B** Willis, W. D., E. D. Al-Chaer, M. J. Quast and K. N. Westlund (1999) A visceral pain pathway in the dorsal column of the spinal cord. *Proc. Natl. Acad. Sci. USA* 96: 7675–7679. **Box 10C Figure C** Hirshberg, R. M., E. D. Al-Chaer, N. B. Lawand, K. N. Westlund and W. D. Willis (1996) Is there a pathway in the posterior funiculus that signals visceral pain? *Pain* 67: 291–305; drawing after Nauta, H. J. W., E. Hewitt, K. N. Westlund and W. D. Willis, Jr. (1997) Surgical interruption of a midline dorsal column visceral pain pathway: Case report and review of the literature. *J. Neurosurg.* 86(3): 538–542.

CHAPTER 11 Vision: The Eye

Box 11A Figure D Westheimer, G. (1974) In *Medical Physiology*, 13th Edition V. B. Mountcastle (ed.) St. Louis: Mosby. **Figure 11.4A–C** Hilfer, S. R. and J. J. W. Yang (1980) Accumulation of CPC-precipitable material at apical cell surfaces during formation of the optic cup. *Anat. Rec.* 197: 423–433. **Figure 11.6A** Oyster, C. W. (1999) *The Human Eye.* Sunderland, MA: Sinauer Associates. **Figure 11.6B,C** Young, R. W. (1971) Shedding of discs from rod outer segments in the rhesus monkey. *J. Ultrastruc. Res.* 34: 190–203. **Box 11B** Hamel, C. (2006) Retinitis pigmentosa. *Orphanet Journal of Rare Diseases* 2006 1:40. CC-BY 2.0 https://creativecommons.org/licenses/by/2.0/. **Figure 11.7** Baylor, D. A. (1987) Photoreceptor signals and vision. *Invest. Ophthalmol. Vis. Sci.* 28: 34–49. **Figure 11.9A** Oyster, C. W. (1999) *The Human Eye.* Sunderland, MA: Sinauer Associates; Stryer, L. (1986) Cyclic GMP cascade of vision. *Annu. Rev. Neurosci.* 9: 87–119. **Figure 11.9B** Oyster, C. W. (1999) *The Human Eye.* Sunderland, MA: Sinauer Associates; Stryer, L. (1987) The molecules of visual excitation. *Sci. Am.* 257: 42–50. **Figure 11.12** Baylor, D. A. (1987) Photoreceptor signals and vision. Invest. *Ophthalmol. Vis. Sci.* 28: 34–49. **Figure 11.13A** Curcio, C.A., K. R. Sloan, R. E. Kalina, A. E. Hendrickson (1990) Human photoreceptor topography. *J. Comp. Neurol.* 292: 497–523; Purves, D. and R. B. Lotto (2011) *Why We See What We Do Redux: An Empirical Theory of Vision*, Appendix. Sunderland, MA: Sinauer Associates. **Figure 11.14A** Schnapf, J. L., T. W. Kraft, and D. A. Baylor (1987) Spectral sensitivity of human cone photoreceptors. *Nature* 325: 439–441. **Figure 11.14B** Hofer, H., J. Carroll, J. Neitz, M. Neitz and D. R. Williams (2005) Organization of the human trichromatic cone mosaic. *J. Neurosci.* 25: 9669–9679. **Box 11C** Purves, D. and R. B. Lotto (2011) *Why We See What We Do Redux: An Empirical Theory of Vision*, Chapter 3. Sunderland, MA: Sinauer Associates. **Figure 11.16A** Nathans, J. (1987) Molecular biology of visual pigments. *Annu. Rev. Neurosci.* 10: 163–194. **Figure 11.16B,C** Deeb, S. S. (2005) The molecular basis of variation in human color vision. *Clin. Genet.* 67: 369–377. **Figure 11.19** Sakmann, B. and O. D. Creutzfeldt (1969) Scotopic and mesopic light adaptation in the cat's retina. *Pflügers Arch.* 313: 168–185. **Figure 11.20B** Hubel, D. and T.N. Wiesel (1961) Integrative action in the cat's lateral geniculate body, *J. Physiol.* 155: 385–398.

CHAPTER 12 Central Visual Pathways

Figure 12.8A Hubel, 1988. *Eye, Brain, and Vision.* New York: Scientific American Library. **Figure 12.10B** Ohki, K., S. Ghung, P. Kara, M. Hubener, T. Bonhoeffer and R. C. Reid (2006) Highly ordered arrangement of single neurons in orientation pinwheels. *Nature* 442: 925–928. **Figure 12.11D** Horton and E. T. Hedley-Whyte (1984) Mapping of cytochrome oxidase patches and ocular dominance columns in human visual cortex. *Philos. Trans.* 304: 255–272. **Box 12A Figure A** Wandell, B. A. (1995) *Foundations of Vision*. Sunderland, MA: Sinauer Associates. **Box 12A Figure B** Julesz, B. (1964) Binocular depth perception without familiarity cues. *Science* 45: 356–362. **Figure 12.13A** Watanabe, M. and R. W. Rodieck (1989) Parasol and midget ganglion cells of the primate retina. *J. Comp. Neurol.* 289: 434–454. **Figure 12.13B** Andrews, T. J., S. D. Halpern and D. Purves (1997) Correlated size variations in human visual cortex, lateral geniculate nucleus, and optic tract. *J. Neurosci.* 17: 2859–2868. **Figure 12.14A** Maunsell, J. H. R. and W.T. Newsome (1987) Visual processing in monkey extrastriate cortex. *Annu. Rev. Neurosci.* 10: 363–401. **Figure 12.14B** Felleman, D. J. and D. C. Van Essen (1991) Distributed hierarchical processing in primate cerebral cortex. *Cereb. Cortex* 1: 1–47. **Figure 12.15** Sereno, M. I. and 7 others (1995) Borders of multiple visual areas in humans revealed by functional magnetic resonance imaging. *Science* 268: 889–893.

CHAPTER 13 The Auditory System

Figure 13.5 (inset) Kessel, R. G. and R. H. Kardon (1979) *Tissue and Organs: A Text-Atlas of Scanning Electron Microscopy*. San Francisco: W.H. Freeman. **Figure 13.6** Dallos, P. (1992) The active cochlea. *J. Neurosci.* 12: 4575–4585; von Bèkèsy, G. (1960) *Experiments in Hearing*. New York: McGraw-Hill. **Figure 13.6** von Bekesy, G. (1960) *Experiments in Hearing*. New York: McGraw-Hill. (A collection of von Bekesy's original papers.) **Figure 13.8A** Hudspeth, A. J. (2014) Integrating the active process of hair cells with cochlear function. *Nat. Rev. Neurosci.* 15: 600–614. **Figure 13.8B** Hudspeth, A. J. (1983) The hair cells of the inner ear. *Sci. Amer.* 248: 54–64. **Figure 13.8D** Kachar, B., M. Parakkal, M. Kurc, Y. Zhao and P. G. Gillespie (2000) High-resolution structure of hair-cell tip links. *Proc. Natl. Acad. Sci. USA* 97: 13336–13341. Copyright (2000) National Academy of Sciences, U.S.A. **Figure 13.9** Lewis, R. S. and A. J. Hudspeth (1983) Voltage- and ion-dependent conductances in solitary vertebrate hair cells. *Nature* 304: 538–541. **Figure 13.10A** Shotwell, S. L., R. Jacobs, and A. J. Hudspeth (1981) Directional sensitivity of individual vertebrate hair cells to controlled deflection of their hair bundles. *Ann. NY Acad. Sci.* 374: 1–10. **Figure 13.10B** Hudspeth, A. J. and D. P. Corey (1977) Sensitivity, polarity and conductance change in the response of vertebrate hair cells to controlled mechanical stimuli. *Proc. Natl. Acad. Sci. USA* 74: 2407–2411. **Figure 13.10C** Palmer, A. R. and I. J. Russell (1986) Phase-locking in the cochlear nerve of the guinea-pig and its relation to the receptor potential of inner hair cells. *Hear. Res.* 24: 1–14. **Figure 13.12A** Kiang, N.Y. and E. C. Moxon (1972) Physiological considerations in artificial stimulation of the inner ear. *Ann. Otol. Rhinol. Laryngol.* 81: 714–729. **Figure 13.12C** Kiang, N.Y. S. (1984) Peripheral neural processing of auditory information. In *Handbook of Physiology: A Critical, Comprehensive Presentation of Physiological Knowledge and Concepts,* Section 1: The Nervous System, Vol. III. Sensory Processes, Part 2, J. M. Brookhart, V. B. Mountcastle, I. Darian-Smith and S. R. Geiger (eds.). Bethesda, MD: American Physiological Society, pp. 639–674. **Figure 13.14** Jeffress, L. A. (1948) A place theory of sound localization. *J. Comp. Physiol. Psychol.* 41: 35–38.

CHAPTER 14 The Vestibular System

Figure 14.3 Dickman, J. D., Huss, D., & Lowe, M. (2004). Morphometry of otoconia in the utricle and saccule of developing Japanese quail. *Hear. Res.* 188: 89–103. **Figure 14.6 & 14.9** Fernández, C. and J. M. Goldberg, J. M. and (1976) Physiology of peripheral neurons innervating otolith organs of the squirrel monkey, Parts 1, 2, 3. *J. Neurophys.* 39: 970–1008. **Box 14A Figure A** Eaton, R. C., R. A. Bombardieri and D. L. Meyer (1977) The Mauthner-initiated startle response in teleost fish. *J. Exp. Biol.* 66: 65–81. **Box 14A Figure C** Furshpan, E. J. and T. Furukawa (1962) Intracellular and extracellular responses of the several regions of the Mauthner cell of the goldfish. *J. Neurophysiol.* 25: 732–771.

CHAPTER 15 The Chemical Senses

Figures 15.1E & 15.4D Rolls, E. T., M. L. Kringelbach, and I. E. T. de Araujo, (2003) Different representations of pleasant and unpleasant odours in the human brain. *Eur. J. Neurosci.* 18: 695–703. **Figure 15.2A** Shier, D., J. Butler, and R. Lewis (2004) *Hole's Human Anatomy and Physiology*. Boston: McGraw-Hill. **Figure 15.3** Porter, J. and 8 others (2007) Mechanisms of scent-tracking in humans. *Nat. Neurosci.* 10: 27–29. **Figure 15.4A** Cain, W. S., R. Schmidt and P. Wolkoff (2007) Olfactory detection of ozone and d-limonene: Reactants in indoor spaces. *Indoor Air* 17: 337–347. **Figure 15.5A** Cain, W. S. and J. F. Gent (1986) Use of odor identification in clinical testing of olfaction. In *Clinical Measurement of Taste and Smell,* H. L. Meiselman and R. S. Rivlin (eds.). New York: Macmillan, pp. 170–186. **Figure 15.5B** Murphy, C. (1986) Taste and smell in the elderly. In *Clinical Measurement of Taste and Smell,* H. L. Meiselman and R. S. Rivlin (eds.). New York: Macmillan, pp. 343–371. **Figure 15.5C** Wang, J., P. Eslinger, M. B. Smith, and Q. X. Yang (2005) Functional magnetic resonance imaging study of human olfaction and normal aging. *J. Gerontol. Med. Sci.* 60A: 510–514. **Figure 15.6A** Anholt, R. R. H. (1987) Primary events in olfactory reception. *Trends Biochem. Sci.* 12: 58–62. **Figure 15.6B** Rawson, N. E. and A.-S. LaMantia (2006) Once and again: Retinoic acid signaling in the developing and regenerating olfactory pathway. *J. Neurobiol.* 66: 653–676. **Figure 15.6C** Leung C. T., P. A. Coulombe and R. R. Reed (2007) Contribution of olfactory neural stem cells to tissue maintenance and regeneration. *Nat. Neurosci.* 10: 720–726. **Figure 15.7** Firestein, S., F. Zufall and G. M. Shepherd (1991) Single odor-sensitive channels in olfactory receptor neurons are also gated by cyclic nucleotides. *J. Neurosci.* 11: 3565–3572. **Figures 15.8A & 15.10A** Menini, A. (1999) Calcium signalling and regulation in olfactory neurons. *Curr. Opin. Neurobiol.* 9: 419–425. **Figure 15.8B** Dryer, L. (2000) Evolution of odorant receptors. *BioEssays* 22: 803–809. **Figure 15.9B–D** Bozza, T., P. Feinstein, C. Zheng and P. Mombaerts (2002) Odorant receptor expression defines functional units in the mouse olfactory system. *J. Neurosci.* 22: 3033–3043. **Figure 15.10B** Wong, S. T. and 8 others (2000) Disruption of the type III adenylyl cyclase gene leads to peripheral and behavioral anosmia in transgenic mice. *Neuron* 27: 487–497; Belluscio, L., G. H. Gold, A. Nemes, and R. Axel (1998) Mice deficient in Golf are anosmic. *Neuron* 20: 69–81; Brunet, L., G. H. Gold and J. Ngai (1996) General anosmia caused by a targeted disruption of the mouse olfactory cyclic nucleotide–gated cation channel. *Neuron* 17: 681–693. **Figure 15.11** Firestein S., G. M. Shepherd. Neurotransmitter antagonists block some odor responses in olfactory receptor neurons. *Neuroreport*. 3: 661-664. **Figure 15.12 graphs** Bozza, T., P. Firestein, C. Zheng and P. Mombaerts (2002) Odorant receptor expression defines functional units in the mouse olfactory system. *J. Neurosci.* 22: 3033–3043. **Figure 15.12 A micrograph** Bozza, T. and J. Kauer. Odorant response properties of convergent olfactory receptor neurons. *J. Neurosci.* 18.12 (1998): 4560-4569. **Figure 15.13A inset** Wang, J. W., A. M. Wong, J. Flores, L. B. Vosshall and R. Axel (2003) Two-photon calcium imaging reveals an odor-evoked map of activity in the fly brain. *Cell* 112: 271–282. **Figure 15.13C** Pomeroy, S. L., A.-S. LaMantia and D. Purves (1990) Postnatal construction of neural activity in the mouse olfactory bulb. *J. Neurosci.* 10: 1952–1966. **Figure 15.13E inset** Tadenev, A. L., Kulaga, H. M., May-Simera, H. L., Kelley, M. W., Katsanis, N., and Reed. R, R. 2011. Loss of Bardet-Biedl syndrome protein-8 (BBS8) perturbs olfactory function, protein localization, and axon targeting. *Proc.*

Natl. Acad. Sci. USA 108: 10320–10325. **Figure 15.13E** Mombaerts, P. and 7 others (1996) Visualizing an olfactory sensory map. *Cell* 87: 675–686. **Figure 15.14A** Wang, J. W., A. M. Wong, J. Flores, L. B. Vosshall and R. Axel (2003) Two-photon calcium imaging reveals an odor-evoked map of activity in the fly brain. *Cell* 112: 271–282. **Figure 15.14B** Fleischmann, A. and 10 others (2008) Mice with a "monoclonal nose": Perturbations in an olfactory map impair odor discrimination. *Neuron* 60: 1068–1081. **Figure 15.15A** Sosulski, D. L., M. L. Bloom, T. Cutforth, R. Axel and S. R. Datta (2011) Distinct representations of olfactory information in different cortical centers. *Nature* 472: 213–216. **Figure 15.15B** Zhan, C. and M. Luo (2010) Diverse patterns of odor representation by neurons in the anterior piriform cortex of awake mice. *J. Neurosci.* 30: 16662–16672. **Figure 15.16** Stettler, D. D. and R. Axel (2009) Representations of odor in the piriform cortex. *Neuron* 63: 854–864. **Figure 15.17B** Pantages, E. and C. Dulac (2000) A novel family of candidate pheromone receptors in mammals. *Neuron* 28: 835–845. **Figure 15.18** Savic, I., H. Berglund, B. Gulyas and P. Roland (2001) Smelling of odorous sex hormone-like compounds causes sex-differentiated hypothalamic activations in humans. *Neuron* 31: 661–668. **Figure 15.19A** Dulac, C. and A. T. Torello (2003) Molecular detection of pheromone signals in mammals: from genes to behaviour. *Nat. Rev. Neurosci.* 4: 551–562. **Figure 15.19B** Isogai, Y. and 5 others (2011) Molecular organization of vomeronasal chemoreception *Nature* 478: 241–245. **Figures 15.20C & 15.22B** Schoenfeld, M. A. and 6 others (2004) Functional magnetic resonance tomography correlates of taste perception in human primary taste cortex. *Neuroscience* 127: 347–353. **Figure 15.25A,B** Nelson, G. and 5 others (2001) Mammalian sweet taste receptors. *Cell* 106: 381–390. **Figure 15.25C** Adler, E. and 5 others (2000) A novel family of mammalian taste receptors. *Cell* 100: 693–702.

CHAPTER 16 Lower Motor Neuron Circuits and Motor Control

Opening Image, Unit III Corticospinal tract, DTI MRI scan. © Sherbrooke Connectivity Imaging Lab/Science Photo Library. **Figure 16.2** Burke, R. E., P. L. Strick, K. Kanda, C. C. Kim and B. Walmsley (1977) Anatomy of medial gastrocnemius and soleus motor nuclei in cat spinal cord. *J. Neurophys.* 40: 667–680. **Figure 16.6** Burke, R. E., D. N. Levine, P. Tsairis, and F. E., III, Zajac (1973) Physiological types and histochemical profiles in motor units of the cat gastrocnemius. *J. Physiol.* 234: 723–748. **Box 16A Figure B** Gordon, T., N. Tyreman, V. F. Rafuse and J. B. Munson (1997) Fast-to-slow conversion following chronic low-frequency activation of medial gastrocnemius muscle in cats. I. Muscle and motor unit properties. *J. Neurophysiol.* 77: 2585–2604. **Box 16A Figure C** Munson, J. B., R. C. Foehring, L. M. Mendell and T. Gordon (1997) Fast-to-slow conversion following chronic low-frequency activation of medial gastrocnemius muscle in cats. II. Motoneuron properties. *J. Neurophysiol.* 77: 2605–2615. **Box 16A Figure E** Van Cutsem, M., J. Duchateau and K. Hainaut (1998) Changes in single motor unit behaviour contribute to the increase in contraction speed after dynamic training in humans. *J. Physiol.* 513: 295–305. **Figure 16.7** Walmsley, B., J. A. Hodgson and R. E. Burke (1978) Forces produced by medical gastrocnemius and soleus muscles during locomotion in freely moving cats. *J. Neurophys.* 41: 1203–1216. **Figure 16.9** Monster, A. W. and H. Chan (1977) Isometric force production by motor units of extensor digitorum communis muscle in man. *J. Neurophys.* 40: 1432–1443. **Figure 16.11** Hunt, C. C. and S. W. Kuffler (1951) Stretch receptor discharges during muscle contraction. *J. Physiol.* (Lond.) 113: 298–314. **Figure 16.13** Patton, H. D. (1965) Reflex regulation of movement and posture. In *Physiology and Biophysics*, 19th Edition, T. C. Ruch and H. D. Patton (eds.). Philadelphia: Saunders, pp. 181–206. **Figure 16.15A–C** Pearson, K. (1976) The control of walking. *Sci. Amer.* 235: 72–86. **Figure 16.15D** Kiehn, O. (2016) Decoding the organization of spinal circuits that control locomotion. *Nat. Rev. Neurosci.* 17: 224-238. **Figure 16.16** Drew, T. and D. S. Marigold (2015) Taking the next step: cortical contributions to the control of locomotion. *Curr. Opin. Neurobiol.* 33: 25–33; Kiehn, O. (2016) Decoding the organization of spinal circuits that control locomotion. *Nat. Rev. Neurosci.* 17: 224-238.

CHAPTER 17 Upper Motor Neuron Control of the Brainstem and Spinal Cord

Box 17A & Figure 17.7 Graziano, M. S. A., T. N. S. Aflalo and D. F. Cooke (2005) Arm movements evoked by electrical stimulation in the motor cortex of monkeys. *J. Neurophysiol.* 94: 4209–4223. **Figure 17.6** Porter, R. and R. Lemon (1993) *Corticospinal Function and Voluntary Movement.* Oxford: Oxford University Press. **Figure 17.8 A–C** Georgopoulos, A. P., J. F. Kalaska, R. Caminiti and J. T. Massey (1982) On the relations between the direction of two-dimensional arm movements and cell discharge in primate motor cortex. *J. Neurosci.* 2: 1527–1537. **Figure 17.8D** Georgopoulos, A. P., R. Caminit, J. F. Kalaska and J. T. Massey (1983) Spatial coding of movement: A hypothesis concerning the coding of movement direction by motor cortical populations. In *Neural Coding of Motor Performance,* J. Massion, J. Paillard, W. Schultz and M. Wiesendanger M (eds.). *Exp. Brain. Res. (Suppl.)* 7: 327–336. **Figure 17.9** Geyer, S., M. Matelli and G. Luppino (2000) Functional neuroanatomy of the primate isocortical motor system. *Anat. Embryol.* 202: 443–474. **Box 17B Figure A** Dr. Eric C. Leuthardt, Professor of Neurological Surgery, Washington University School of Medicine, Director of The Center for Innovation in Neuroscience and Technology. **Box 17B Figure B** Donati, A. and 19 others (2016) Long-term training with a brain-machine interface-based gait protocol induces partial neurological recovery in paraplegic patients. *Sci. Rep.* 6: 30383. doi: 10.1038/srep30383 **Figure 17.10** Rizzolatti, G., L. Fadiga, V. Gallese and L. Fogassi (1996) Premotor cortex and the recognition of motor actions. *Cogn. Brain Res.* 3: 131–141. **Figure 17.11A** Rizzolatti, G. and C. Sinigaglia (2016) The mirror mechanism: a basic principle of brain function. *Nature Rev. Neurosci.* 17: 757–765. **Figure 17,14** Nashner, L. M. (1979). Organization and programming of motor activity during posture control. In *Progress in Brain Research, Volume 50, Reflex Control of Posture and Movement*, R. Granit and O. Pompeiano (eds.). Amsterdam: Elsevier/North Holland Biomedical Press, pp. 177–184.

CHAPTER 18 Modulation of Movement by the Basal Ganglia

Figure 18.6A1 Hikosaka, O. and R. H. Wurtz (1986) Cell activity in monkey caudate nucleus preceding saccadic eye movements. *Exp. Brain Res.* 63: 659–662. **Figure 18.6A2–3,B** Hikosaka, O. and R. H. Wurtz (1983) Visual and oculomotor functions of monkey substantia nigra pars reticulata. IV. Relation of substantia nigra to superior colliculus. *J. Neurophysiol.* 49: 1285–1301. **Figures 18.9B & 18.10B** DeLong, M. R. (1990) Primate models of movement disorders of basal ganglia origin. *Trends Neurosci.* 13: 281–285. **Box 18A Figure A–C** Desrochers, T. M., K. Amemori and A. M. Graybiel (2015) Habit learning by naive macaques is marked by response sharpening of striatal neurons representing the cost and outcome of acquired action sequences. *Neuron* 87: 853–868.

CHAPTER 19 Modulation of Movement by the Cerebellum

Figure 19.10A Stein, J. F. (1986) Role of the cerebellum in the visual guidance of movement. *Nature* 323: 217–220. **Figure 19.11** Cerminara, N. L., E. J. Lang, R. V. Sillitoe and R. Apps (2015) Redefining the cerebellar cortex as an assembly of non-uniform Purkinje cell microcircuits. *Nat. Rev. Neurosci.* 16: 79–93. **Figure 19.12** Thach, W. T. (1968) Discharge of Purkinje and cerebellar nuclear neurons during rapidly alternating arm movements in the monkey. *J. Neurophys.* 31: 785–797. **Figure 19.13** Optican, L. M. and D. A. Robinson (1980) Cerebellar-dependent adaptive control of primate saccadic system. *J. Neurophys.* 44: 1058–1076. **Box 19A** Caviness, V. S., Jr. and P. Rakic (1978) Mechanisms of cortical development: A view from mutations in mice. *Annu. Rev. Neurosci.* 1: 297–326.

CHAPTER 20 Eye Movements and Sensory Motor Integration

Figure 20.1 Yarbus, A. L. (1967) *Eye Movements and Vision*. Basil Haigh, trans. New York: Plenum Press. **Box 20A Figure A** Pritchard, R. M. (1961) Stabilized images on the retina. *Sci. Amer.* 204 (June): 72–78. **Box 20A Figure B** Riggs, L. A., F. Ratliff, J. C. Cornsweet and T. N. Cornsweet (1953) The disappearance of steadily fixated visual test objects. *J. Opt. Soc. Am.* 43: 495–501. **Figures 20.4 & 20.5** Fuchs, A. F. (1967) Saccadic and smooth pursuit eye movements in the monkey. *J. Physiol. (Lond.)* 191: 609–630. **Figure 20.6** Baarsma, E. and H. Collewijn (1974) Vestibulo-ocular and optokinetic reactions to rotation and their interaction in the rabbit. *J. Physiol.* 238: 603-625. **Figure 20.7** Fuchs, A. F. and E. S. Luschei (1970) Firing patterns of abducens neurons of alert monkeys in relationship to horizontal eye movements. *J. Neurophys.* 33: 382–392. **Figure 20.9** Schiller, P. H. and M. Stryker (1972) Single unit recording and stimulation in superior colliculus of the alert rhesus monkey. *J. Neurophys.* 35: 915–923. **Box 20B Figure B** Wurtz, R. H. and J. E. Albano (1980) Visual-motor function of the primate superior colliculus. *Annu. Rev. Neurosci.* 3: 189–226. **Box 20B Figure C** Ozen, G., G. J. Augustine and W. C. Hall (2000) Contribution of superficial layer neurons to premotor bursts in the superior col-liculus. *J. Neurophysiol.* 84: 460–471. **Figure 20.10** Sparks, D. L. and L. E. Mays (1983) Spatial localization of saccade targets. I. Compensation for stimulation-induced perturbations in eye position. *J. Neurophysiol.* 49: 45–63. **Box 20C Figure A** Sparks, D. L. (1975) Response properties of eye movement-related neurons in the monkey superior colliculus. *Brain Res.* 90: 147–152. **Box 20C Figure B** Fuchs, A. F. and E. S. Luschei (1970) Firing patterns of abducens neurons of alert monkeys in relationship to horizontal eye movement. *J. Neurophysiol.* 33: 382–392. **Figure 20.12** Schall, J. D. (1995) Neural basis of saccade target selection. *Rev. Neurosci.* 6: 63–85. **Figure 20.13** Krauzlis, R. J. (2005) The control of voluntary eye movements: New perspectives. *Neuroscientist* 11: 124–137.

CHAPTER 21 The Visceral Motor System

Box 21B Figure A Marx, J. (2003) Cellular warriors at the battle of the bulge. *Science* 299: 846–849; Morton, G. J., T. H. Meek and M. W. Schwartz (2014) Neurobiology of food intake in health and disease. *Nature Rev. Neurosci.* 15: 367–378. **Box 21B Figure B** Yaswen, L., N. Diehl, M. B. Brennan and U. Hochgeschwender (1999) Obesity in the mouse model of pro-opiomelanocortin deficiency responds to peripheral melanocortin. *Nat. Med.* 5: 1066–1070. **Box 21B Figure C** O'Rahilly, S., S. Farooqi, G. S. H. Yeo and B. G. Challis (2003) Human obesity: Lessons from monogenic disorders. *Endocrinology* 144: 3757–3764.

CHAPTER 22 Early Brain Development

Opening Image, Unit IV A trigeminal ganglion from an E10.5 mouse embryo grown in cell culture for 24 hours. The green is neuron-selective label (b-tubulin III), the red is a transcription factor marker (Six1) and the blue is a nuclear DNA label (DAPI). Courtesy of Anthony-Samuel LaMantia. **Box 22A Figure B** Wichterle, H., I. Lieberam, J. A. Porter and T. M. Jessell (2002) Directed differentiation of embryonic stem cells into motor neurons. *Cell* 110: 385–397. **Figure 22.2** Sanes, J. R. (1989) Extracellular matrix molecules that influence neural development. *Annu. Rev. Neurosci.* 12: 491–516. **Figure 22.4A** Gilbert, S. F. (1994) *Developmental Biology*, 4th Edition. Sunderland, MA: Sinauer Associates; Lawrence, P. A. (1992) *The Making of a Fly: The Genetics of Animal Design.* Oxford: Blackwell Scientific Publications. **Figure 22.4B** Ingham, P. (1988) The molecular genetics of embryonic pattern formation in *Drosophila*. *Nature* 335: 25–34. **Figure 22.4C** Veraksa, A. and W. McGinnis (2000) Developmental patterning genes and their conserved functions: From model organisms to humans. *Molec. Genet. Metab.* 69: 85-100. **Figure 22.7A** Dodd, J., T. M. Jessell and M. Placzek (1998) The when and where of floor plate induction. *Science* 282: 1654–1657. **Figure 22.7B** Fausett, S. R., L. J. Brunet and J. Klingensmith (2014) BMP antagonism by *Noggin* is required in presumptive notochord cells for mammalian foregut morphogenesis. *Dev. Biol.* 391: 111–124. **Figure 22.8B** Noctor, S. C., V. A. Martinez-Cedeno and A. R. Kriegstein (2008) Distinct behaviors of neural stem and progenitor cells underlie cortical neurogenesis. *J. Comp. Neurol.* 508: 28–44. **Figures 22.9** Rakic, P. (1974) Neurons in rhesus monkey visual cortex: Systematic relation between time of origin and eventual disposition. *Science* 183: 425–427. **Figure 22.11** Kintner C. (2002) Neurogenesis in embryos and in adult neural stem cells. *J. Neurosci.* 22: 639–643. **Figures 22.13A & 22.14A** Cowan, W. M. (1979) The development of the brain. *Sci. Am.* 241(3): 112–133; Rakic, P. (1971) Guidance of neurons migrating to the fetal monkey neocortex. *Brain Res.* 33: 471–476. **Figure 22.13B** Noctor, S. C., A. C. Flint, T. A. Weissman, R. S. Dammerman and A. R. Kriegstein (2001) Neurons derived from radial glial cells establish radial units in neocortex. *Nature* 409: 714–720. **Figure 22.14B,C** Hong, S. E. and 7 others (2000) Autosomal recessive lissencephaly with cerebellar hypoplasia (LCH) is associated with human reelin gene mutations. *Nat. Genet.* 26: 93–96. **Figure 22.15** Lumsden, A. and R. Keynes (1989) Segmental patterns of neuronal development in the chick hindbrain. *Nature* 337: 424–428. **Figure 22.16C, left** Carmona, F. D., R. Jiménez and J. M. Collinson (2008) The molecular basis of defective lens development in the Iberian mole. *BMC Biol.* 6: 44. doi: 10.1186/1741-7007-6-44. © Carmona et al; licensee BioMed Central Ltd. 2008. **Figure 22.16D, left** Birol, O., T. Ohyama, R. K. Edlund, K. Drakou, P. Georgiades, A. K. Groves (2016) The mouse Foxi3 transcription factor is necessary for the development of posterior placodes *Dev. Biol.* 409(1): 139–151 **Figure 22.16D, right** Hudspeth, A. J. (1985) The cellular basis of hearing: the biophysics of hair cells. *Science* 230: 745–752.

CHAPTER 23 Construction of Neural Circuits

Figure 23.1D Shi, S. H., L. Y. Jan and Y. N. Jan (2003) Hippocampal neuronal polarity specified by spatially localized mPar3/mPar6 and PI 3-kinase activity. *Cell* 112: 63–75. **Figure 23.2A (1st part)** Takahashi, M., M. Narushima and Y. Oda (2002) In vivo imaging of functional inhibitory networks on the Mauthner cell of larval zebrafish. *J. Neurosci.* 22: 3929–3938. **Figure 23.2A (other 2 parts)** Jontes, J. D., J. Buchanan and S. Smith (2000) Growth cone and dendrite dynamics in zebrafish embryos: early events in synaptogenesis imaged in vivo. *Nat. Neurosci.* 3: 231–237. **Box 23A** Herrera, E. and 8 others (2003) Zic2 patterns binocular vision by specifying the uncrossed retinal projection. *Cell* 114: 545–557. **Figure 23.3A** Dent, E.W. and F.B. Gertler (2003) Cytoskeletal dynamics and transport in growth cone motility and axon guidance. *Neuron* 40 :209–227. **Figure 23.3B** Dent, E. W. and K. Kalil (2001) Axon branching requires interactions between dynamic microtubules and actin filaments. *J. Neurosci.* 21: 9757–9769. **Figure 23.23C** Huber, A. B., A. L. Kolodkin, D. D. Ginty and J.-F. Cloutier (2003) Signaling at the growth cone: Ligand-receptor complexes and the control of axon growth and guidance. *Ann. Rev. of Neurosci.* 26: 509–563. **Figure 23.3D** Gomez, T. M. and

J. O. Zheng (2006) The molecular basis for calcium-dependent axon pathfinding. *Nat. Rev. Neurosci.* 7: 115–117. **Figure 23.5A** Huber, A. B., A. L. Kolodkin, D. D. Ginty and J. F. Cloutier (2003) Signaling at the growth cone: Ligand-receptor complexes and the control of axon growth and guidance. *Annu. Rev. Neurosci.* 26: 509–563. **Figure 23.5D** Gomez, T. M. and J. O. Zheng (2006) The molecular basis for calcium-dependent axon pathfinding. *Nat. Rev. Neurosci.* 7: 115–117; Dontchev, V. D. and P. C. Letourneau (2002) Nerve growth factor and semaphorin 3A signaling pathways interact in regulating sensory neuronal growth cone motility. *J. Neurosci.* 22: 6659–6669. **Figure 23.6A** Serafini, T. and 5 others (1994) The netrins define a family of axon outgrowth-promoting proteins homologous to *C. elegans* UNC-6. *Cell* 78: 409–423. **Figure 23.6B** Dickinson, B. J. (2001) Moving on. *Science* 291: 1910–1911. **Figure 23.6C** Serafini, T. and 6 others (1996) Netrin-1 is required for commissural axon guidance in the developing vertebrate nervous system. *Cell* 87: 1001–1014. **Figure 23.7A,C** Whitford, K. L., P. Dijkhuizen, F. Polleux and A. Ghosh (2002) Molecular control of cortical dendrite development. *Ann. Rev. Neurosci.* 25: 127–149. **Figure 23.7B** Choi, J. H., M.Y. Law, C. B. Chien, B. A. Link, and R. O. Wong (2010) In vivo development of dendritic orientation in wild-type and mislocalized retinal ganglion cells. *Neural Dev.* 5: 29. © Choi et al.; licensee BioMed Central Ltd. 2010. **Figure 23.8A,B,D** Hattori, D., S. S. Millard, W. M. Wojtowicz and S. L. Zipursky (2008) Dscam-mediated cell recognition regulates neural circuit formation. *Annu. Rev. Cell Dev. Biol.* 24: 597–620. **Figure 23.8C** Lefebvre, J. L., J. R. Sanes and J. N. Kay (2015) Development of dendritic form and function. *Ann. Rev. Cell Dev. Biol.* 31: 741–777. **Figure 23.9A,B** Sperry, R. W. (1963) Chemoaffinity in the orderly growth of nerve fiber patterns and connections. *Proc. Natl. Acad. Sci. USA* 50: 703–710. **Figure 23.9C** Walter, J., S. Henke-Fahle and F. Bonhoeffer (1987) Avoidance of posterior tectal membranes by temporal retinal axons. *Development* 101: 909–913. **Figure 23.9D** Wilkinson, D. G. (2001) Multiple roles of EPH receptors and ephrins in neural development. *Nat. Rev. Neurosci.* 2: 155–164. **Figure 23.10A,B** Waites, C. L., A. M. Craig and C. C. Garner (2005) Mechanisms of vertebrates synaptogenesis. *Annu. Rev. Neurosci.* 28: 251–274. **Figure 23.10C** Dean, C. and T. Dresbach (2006) Neuroligins and neurexins: Linking cell adhesion, synapse formation, and cognitive function. *Trends Neurosci.* 29: 21–29 **Figure 23.11A** Wang, X., H. Su and A. Bradley (2002) Molecular mechanisms governing *Pcdh-γ* gene expression: Evidence for a multiple promoter and *cis*-alternative splicing model. *Genes Dev.* 16: 1890–1905; Hamada, S. and T. Yagi (2001) The cadherin-related receptor family: A novel diversified cadherin family at the synapse. *Neurosci. Res.* 41: 207–215. **Figure 23.11B** Phillips, G. R. and 6 others (2003) Gamma-protocadherins are targeted to subsets of synapses and intracellular organelles in neurons. *J. Neurosci.* 23: 5096–5104. **Figure 23.12** Hamburger, V. (1958) Regression versus peripheral controls of differentiation in motor hypoplasia. *Amer. J. Anat.* 102: 365–409; Hamburger, V. (1977) The developmental history of the motor neuron. The F. O. Schmitt Lecture in Neuroscience, 1976, *Neurosci. Res. Prog. Bull.* 15, Suppl. III: 1–37. Hollyday, M. and V. Hamburger (1976) Reduction of the naturally occurring motor neuron loss by enlargement of the periphery. *J. Comp. Neurol.* 170: 311–320. **Bxo 23B** Purves, D. and R. I. Hume (1981) The relation of postsynaptic geometry to the number of presynaptic axons that innervate autonomic ganglion cells. *J. Neurosci.* 1: 441–452. **Figure 23.13** Purves, D. and J. W. Lichtman (1980) Elimination of synapses in the developing nervous system. *Science* 210: 153–157. **Box 23C** Horch, H. W., A. Kruittgen, S. D. Portbury and L. C. Katz (1999) Destabilization of cortical dendrites and spines by BDNF. *Neuron* 23: 353–364. **Figure 23.14A** Walsh, M. K. and J. W. Lichtman (2003) In vivo time-lapse imaging of synaptic takeover associated with naturally occurring synapse elimination. *Neuron* 37: 67–73. **Figure 23.14B** Hashimoto, K., R. Ichikawa, K. Kitamura, M. Watanabe and M. Kano (2009) Translocation of a"winner"climbing fiber to the Purkinje cell dendrite and subsequent elimination of"losers"from the soma in developing cerebellum. *Neuron* 63: 106–118. **Figure 23.15A,B** Purves, D. and J. W. Lichtman (1985) *Principles of Neural Development.* Sunderland, MA: Sinauer Associates. **Figure 23.15C** Chun, L. L. and P. H. Patterson (1977) Role of nerve growth factor in the development of rat sympathetic neurons in vitro. III; Effect on acetylcholine production. *J. Cell Biol.* 75: 712–718. **Figure 23.15D** Levi-Montalcini, R. (1972) The morphological effects of immunosympathectomy. In *Immonosympathectomy*, G. Steiner and E. Schönbaum (eds.). Amsterdam: Elsevier. **Figure 23.16A** Maisonpierre, P. C. and 6 others (1990) Neurotrophin-3: A neurotrophic factor related to NGF and BDNF. *Science* 247: 1446–1451. **Figure 23.16B** Bibel, M. and Y.-A. Barde (2000) Neurotrophins: Key regulators of cell fate and cell shape in the vertebrate nervous system. *Genes Dev.* 14: 2919–2937. **Figure 23.17A** Campenot, R. B. (1981) Regeneration of neurites on long-term cultures of sympatic neurons deprived of nerve growth factor. *Science* 214: 579–581. **Figure 23.17B** Li, Y. and 6 others (2005) Essential role of TRPC channels in the guidance of nerve grwoth cones by brain-derived neurotrophic factor. *Nature* 434: 894–898. **Figure 23.17C** Zweifel, L. S., R. Kuruvilla and D. D. Ginty (2005) Functions and mechanisms of retrograde neurotrophin signalling. *Nat. Rev. Neurosci.* 6: 615–625.

CHAPTER 24 Circuit Development: Intrinsic Factors and Sex Differences

Figure 24.2B Arnold, A. P. 1980. Sexual differences in the brain. *Am. Sci.* 68: 165–173. **Figure 24.2C** Grisham, W. and 6 others (2011) Using digital images of the zebra finch song system as a tool to teach organizational effects of steroid hormones: A free downloadable module. *CBE—Life Sciences Education* 10: 222–230. **Figure 24.3B** Moore, K. L. (1977) *The Developing Human*, 2nd Edition. Philadelphia: W. B. Saunders, p. 241. **Box 24A Figures A–C** Young, L. J. and Z. Wang (2004) The neurobiology of pair bonding. *Nat. Neurosci.* 7: 1048–1054. **Box 24A Figure D** Fisher, H. E., A. Aron, and L. L. Brown (2005) Romantic love: An fMRI study of neural mechanisms for mate choice. *J. Comp. Neurol.* 493: 58–62. **Box 24A Figure E** Bartels, A. and S. Zeki (2000) The neural basis of romantic love. *Neuroreport.* 11: 3829–3834. **Figure 24.4B** Gustafson, M. L. and P. K. Donahoe (1994) Male sex determination: Current concepts of male sexual differentiation. *Annu. Rev. Med.* 45: 505–524. **Figure 24.5A** McEwen, B. S., P. G. Davis, B. Parsons and D. W. Pfaff (1979) The brain as a target for steroid hormone action. *Annu. Rev. Neurosci.* 2: 65–112. **Figure 24.5B** McEwen, B. S. (1976) Interactions between hormones and nerve tissue. *Sci. Am.* 235: 48–58. **Figure 24.6A** Breedlove, S. M. and A. P. Arnold (1981) Sexually dimorphic motor nucleus in the rat lumbar spinal cord: Response to adult hormone manipulation, absence in androgen-insensitive rats. *Brain Res.* 225: 297–307. **Figure 24.6B** Diagram after Morris, J. A., C. L. Jordan and S. M. Breedlove (2004) Sexual differentiation of the vertebrate nervous system. Nat, Neurosci. 7: 1034–1039.; micrographs from Breedlove, S. M. and A. P. Arnold (1983) Hormonal control of a developing neuromuscular system. II. Sensitive periods for the androgen-induced masculinization of the rat spinal nucleus of the bulbocavernosus. *J. Neurosci.* 3: 424–432. **Figure 24.6C** Forger, N. G. and S. M. Breedlove (1986) Sexual dimorphism in human and canine spinal cord: Role of early androgen. *Proc. Natl. Acad. Sci. USA* 83: 7527–7530. **Figure 24.7B** Arnold, A. P. and R. A. Gorski (1984) Gonadal steroid induction of structural sex differences in the central nervous system. *Ann. Rev. Neurosci.* 7: 413–442; Gorski, R. A. (1983) Steroid-induced sexual characteristics in the brain. In *Neuroendocrine Perspectives, Vol. 2.* E. E. Muller and R. M. MacLeod (eds.), Amsterdam: Elsevier/North Holland. **Figure 24.8A** Toran-Allerand, C. D. (1976) Sex steroids and the development of the newborn mouse hypothalamus and preoptic area in vitro: Implications for sexual differentiation. *Brain Res.* 106:

407–412. **Figure 24.8B** Woolley, C. S. and B. S. McEwen (1993) Roles of estradiol and progesterone in regulation of hippocampal dendritic spine density during the estrous cycle in the rat. *J. Comp. Neurol.* 336: 293–306. **Figure 24.8C** Meusburger, S. M. and J. R. Keast (2001) Testosterone and nerve growth factor have distinct but interacting effects on neurotransmitter expression of adult pelvic ganglion cells in vitro. *Neuroscience* 108: 331–340. **Figure 24.9** Wooley, C. (2007) Acute effects of estrogen on neuronal physiology. *Annu. Rev. Pharmacol. Toxicol.* 47: 5.1–5.24. **Figure 24.10** Patchev, V. K., J. Schroeder, F. Goetz, W. Rhode and A.V. Patchev (2004) Neurotropic actions of androgens: principles, mechanisms and novel targets. *Exp. Gerontol.* 39: 1651–1660. **Figure 24.11** Savic, I., H. Berglund and P. Lindström (2005) Brain response to putative pheromones in homosexual men. Proc. Nat. Acad. Sci. USA 102: 7356–7361. **Figure 24.12** Cahill, L. (2006) Why sex matters for neuroscience. *Nature Rev. Neurosci.* 7: 477–484. **Figure 24.13** Cahill, L., M. Uncapher, L. Kilpatrick, M. T. Aikire and J. Turner (2004) Sex-related hemispheric lateralization of amygdala function in emotionaly influenced memory: An fMRI investitation. *Learn. Mem.* 11: 261–266.

CHAPTER 25 Experience-Dependent Plasticity in the Developing Brain

Figure 25.2B Conel, J. L. (1939–1967). *The Postnatal Development of the Human Cerebral Cortex: Vols. 1–8*. Cambridge, MA: Harvard University Press. **Figure 25.2C** Huttenlocher, P. R., C. De Courten, L. J. Garey and H. Van der Loos (1982) Synaptogeneis in human visual cortex: Evidence for synapse elimination during normal development. *Neurosci. Lett.* 33: 247–252. **Figure 25.3A,B** Feller, M. B., D. P. Wellis, D. Stellwagen, F. S. Werblin and C. J. Shatz (1996) Requirement for cholinergic synaptic transmission in the propagation of spontaneous retinal waves. *Science* 272: 1182–1187. **Figure 25.3C** Hanganu, I. L., Y. Ben-Ari and R. Khazipov (2006) Retinal waves trigger spindle bursts in the neonatal rat visual cortex. *J. Neurosci.* 26: 6728–6736. **Figure 25.4** LeVay, S., T. N. Wiesel and D. H. Hubel (1980) The development of ocular dominance columns in Sillnormal and visually deprived monkeys. *J. Comp. Neurol.* 191: 1–51. **Figure 25.5A** Hubel, D. H. and T. N. Wiesel (1962) Receptive fields, binocular interaction and functional architecture in the cat's visual cortex. *J. Physiol.* 160: 106–154. **Figure 25.5B** Wiesel, T. N and D. H. Hubel (1963) Single-cell responses in striate cortex of kittens deprived of vision in one eye. *J. Neurophysiol.* 26: 1003–1017. **Figures 25.5C & 25.6** Hubel, D. H. and T. N. Wiesel (1970) The period of susceptibility to the physiological effects of unilateral eye closure in kittens. *J. Physiol.* 206: 419–436. **Figure 25.7A** Horton, J. C. and D. R. Hocking (1999) An adult-like pattern of ocular dominance columns in striate cortex of newborn monkeys prior to visual experience. *J. Neurosci.* 16: 1791–1807. **Figure 25.7B** Hubel, D. H., T. N. Wiesel and S. LeVay (1977) Plasticity of ocular dominance columns in monkey striate cortex. *Phil. Trans. R. Soc. Lond. B.* 278: 377–409. **Figure 25.8** Antonini, A. and M. P. Stryker (1993) Rapid remodeling of axonal arbors in the visual cortex. *Science* 260: 1819–1821. **Figure 25.9** Hubel, D. H. and T. N. Wiesel (1965) Binocular interaction in striate cortex of kittens reared with artificial squint. *J. Neurophysiol.* 28: 1041–1059. **Table 25.1** Hensch, T. K. (2004) Critical period regulation. *Annu. Rev. Neurosci.* 27: 549–579. **Figure 25.10** Wang, B.-S., R. Sarnaik and J. Cang (2010) Critical period plasticity matches binocular orientation preference in the visual cortex. *Neuron* 65: 246–256. **Figure 25.11A** Ebert, D. H. and M. E. Greenberg (2013) Activity-dependent neuronal signaling and autism spectrum disorder. *Nature* 493: 327–337. **Figure 25.11B,C** Wong, R. O. and A. Ghosh (2002) Activity-dependent regulation of dendritic growth and patterning. *Nat. Rev. Neurosci.* 3: 803–812. **Figure 25.12** Petitto, L. A. and P. F. Marentette (1991) Babbling in the manual mode: Evidence for the ontogeny of language. *Science* 251: 1493–1496. **Figure 25.13A** Johnson, J. S. and E. I. Newport (1989) Critical period effects in second language learning: the influences of maturational state on the acquisition of English as a second language. *Cog. Psychol.* 21. **Figure 25.13B** Schlaggar, B. L. and 5 others (2002) Functional neuroanatomical differences between adults and school-age children in the processing of single words. *Science* 296: 1476–1479. **Figure 25.14** Rakic, P. J. P. Bourgeois, M. F. Eckenhoff, N. Zecevic and P. S. Goldman-Rakic (1986) Concurrent overproduction of synapses in diverse regions of the primate cerebral cortex. *Science* 232: 232–235. **Figure 25.15A** Gogtay, N. and 11 others (2004) Dynamic mapping of human cortical development during childhood through early adulthood. *Proc. Natl. Acad. Sci. USA* 101: 8174–8179. **Figure 25.15B** Lenroot, R. K. and 11 others (2007) Sexual dimorphism of brain development trajectories during childhood and adolescence. *NeuroImage* 36: 1065–1073. **Figure 25.16A** Shaw, P. and 9 others (2007) Attention-deficit/hyperactivity disorder is characterized by a delay in cortical maturation. *Proc. Natl. Acad. Sci. USA* 104: 19649–19654. Copyright (2007) National Academy of Sciences, U.S.A. **Figure 25.16B** Shaw, P. and 8 others (2006) Longitudinal mapping of cortical thickness and clinical outcome in children and adolescents with attention-deficit/hyperactivity disorder. *Arch. Gen. Psychiatry* 63: 540–549.

CHAPTER 26 Repair and Regeneration in the Nervous System

Figures 26.1 Case, L. C. and M. Tessier-Lavigne (2005) Regeneration of the adult central nervous system. *Curr. Biol.* 15: R749–R753. **Figure 26.2** Ward, N. S., M. M. Brown, A. J. Thompson, and R. S. J. Frackowiak (2003a) Neural correlates of outcome after stroke: A cross-sectional fMRI study. *Brain* 126: 1430–1448; Ward, N. S., M. M. Brown, A. J. Thompson, and R. S. J. Frackowiak (2003b) Neural correlates of motor recovery after stroke: A longitudinal fMRI study. *Brain* 126: 2476–2496. **Figure 26.4** Head, H., W. H. R. Rivers and J. Sherren. (1905) The afferent nervous system from a new aspect. *Brain* 28: 99–111. **Figure 26.5B,C** Pan, Y. A., T. Misgeld, J. W. Lichtman and J. R. Sanes (2003) Effects of neurotoxic and neuroprotective agents on peripheral nerve regeneration assayed by time-lapse imaging in vivo *J. Neurosci.* 23: 11479–11488. **Figure 26.5D** Sabatier, M. J., N. Redmon, G. Schwartz and A. W. English (2008) Treadmill training promotes axon regrowth in injured peripheral neurons. *Exp. Neurol.* 211: 489–493. **Figure 26.7A** So, K. F. and A. J. Aguayo (1985) Lengthy regrowth of cut axons from ganglion cells after peripheral nerve transplantation into the retina of adult rats. *Brain Res.* 359: 402–406. **Figure 26.7B** Aguayo, A. J., D. A. Carter, T. Zwimpfer, M. Vidal-Sanz and G. M. Bray. 1990. Axonal regeneration and synapse formation in the injured CNS of adult mammals. In *Brain Repair.* A. Bjorklund, A. J. Aguayo and D. Ottoson (eds.). Wenner-Gren International Symposium Series. Vol. 56: 252–272. New York: Stockton Press. **Box 26A** Purves, D., W. Thompson and J. W. Yip (1981) Re-innervation of ganglia transplanted to the neck from different levels of the guinea-pig sympathetic chain. *J. Physiol. (Lond.)* 313: 49–63. **Figure 26.8B** Pitts, E. V., S. Potluri, D. M. Hess and R. J. Balice-Gordon (2006) Neurotrophin and Trk-mediated signaling in the neuromuscular system. *Int. Anesthes. Clin.* 44: 21–76. **Figure 26.8C** Nguyen, Q. T., J. R. Sanes and J. W. Lichtman (2002) Pre-existing pathways promote precise projection patterns. *Nat. Neurosci.* 5: 861–867. **Figure 26.8D** Campanari, M.-L., M.-S. Garcia-Ayllon, S. Clura, J. Saez-Valero and E. Kabashi. (2016) Neuromuscular junction impairment in amyotrophic lateral sclerosis: Reassessing the role of acetylcholinesterase. *Front. Mol. Neurosci.* 9: article 160. **Figure 26.9A** Manabat, C. and 8 others (2003) Reperfusion differentially induces caspase-3 activation in ischemic core and penumbra after stroke in immature brain. *Stroke* 34: 207–213. **Figure 26.10 (top)** McGraw, J., G. W. Hiebert and J. D. Stevens (2001) Modulating astrogiosis after neurotrauma. *J. Neurosci. Res.* 63: 109–115. **Figure 26.10 (center)** Tan, A. M., W. Zhang and J. M. Levine (2005) NG2: A component of the glial scar that inhibits axon growth. *J. Anat.* 207: 717–725. **Figure 26.10 (bottom)** Ladeby, R. and 6 others (2005) Microglial cell population dynamics in the injured adult CNS. *Brain Res. Rev.* 48: 196–206. **Figure 26.11B** Kolodkin, A. L. and M. Tessier-Lavigne (2011) Mechanisms and molecules of neuronal wiring: A primer. *Cold Spring Harb. Perspect. Biol.* (epub) **Figure 26.12A,B** Waisman, A., R. S. Liblau and B. Becher (2015) Innate and adaptive immune responses in the CNS. *Lancet Neurol.* 14: 945–955. **Figure 26.12C & Table 26.1** Mckee, C. A. and J. R. Lukens (2016) Emerging roles for the immune system in traumatic brain injury. *Front. Immunol.* 7: 556. **Figure 26.13A** Otteson, D. C. and P. F. Hitchcock (2003) Stem cells in the teleost retina: Persistent neurogenesis and injury-induced regeneration. *Vis. Res.* 43: 927–936. **Figure 26.13B** Goldman, S. A. (1998) Adult neurogenesis: From canaries to the clinic. *Trends Neurosci.* 21: 107–114. **Figure 26.14A,B** Gage, F. H. (2000) Mammalian neural stem cells. *Science* 287: 1433–1438. **Figure 26.14C** Kelsch, W., S. Sim and C. Lois (2010) Watching synaptogenesis in the adult brain. *Annu. Rev. Neurosci.* 33: 131–149. **Figure 26.15A** Alvarez-Buylla, A. and D. A. Lim (2004) For the long run: maintaining germinal niches in the adult brain. *Neuron* 41: 683–686. **Figure 26.15B** Councill, E. S. and 7 others (2006) Limited influence of olanzapine on adult forebrain neural precursors in vitro. *Neuroscience* 140: 111–122. **Figure 26.16A** Ghashghaei, H. T. and 9 others (2006) The role of neuregulin–ErbB4 interactions on the proliferation and organization of cells in the subventricular zone. *Proc. Natl. Acad. Sci. USA* 103: 1930–1935. Copyright © 2006 by The National Academy of Sciences of the USA. **Figure 26.16C** Peretto, P. A. Merighi, A. Fasolo and L. Bonfanti (1999) The subependymal layer in rodents: A site of structural plasticity and cell migration in the adult mammalian brain. *Brain Res. Bull.* 49: 221–243. **Figure 26.16B** Ghashghaei, H. T., C. Lai and E. S. Anton (2007) Neuronal migration in the adult brain: are we there yet? *Nat. Rev. Neurosci.* 8: 141–151. **Box 26B Figure A** Au, E. and G. Fishell (2006) Adult cortical neurogenesis: Nuanced, negligible, or nonexistent? *Nat. Neurosci.* 9: 1086–1088. **Box 26B Figure B,C** Bhardwaj, R. D. and 10 others (2006) Neocortical neurogenesis in humans is restricted to development. *Proc. Natl. Acad. Sci. USA* 103: 12564–12568. Copyright (2006) National Academy of Sciences, U.S.A.

CHAPTER 27 Cognitive Functions and the Organization of the Cerebral Cortex

Opening Image, Unit V Glasser, M. F. and 11 others (2016) A multi-modal parcellation of human cerebral cortex. *Nature* 536: 171–178 **Box 27B Figure C** Palmeri, T.J., R. B. Blake, R. Marois, M. A. Flanery and W. O. Whetsell (2002) The perceptual reality of synesthetic color. *Proc. Natl. Acad. Sci. USA* 99: 4127–4131. Copyright (2002) National Academy of Sciences, U.S.A. **Figure 27.5A** Lynch, J. C.,V. B. Mountcastle, W. H. Talbot and T. C.Yin (1977) Parietal lobe mechanisms for directed visual attention. *J. Neurophys.* 40: 362–369. **Figure 27.5B** Platt, M. L. and P. W. Glimcher (1999) Neural correlates of decision variables in parietal cortex. *Nature* 400: 233–238. **Figure 27.6** McClure, S. M. and 5 others (2004) Neural correlates of behavioral preference for culturally familiar drinks. *Neuron* 44: 379–387. **Figure 27.7** Goldman-Rakic, P. S. (1987) Circuitry of the prefrontal cortex and the regulation of behavior by representational memory. In *Handbook of Physiology*. Section 1, The Nervous System. Vol. 5, Higher Functions of the Brain, Part I. F. Plum (ed.). Bethesda: American Physiological Society, pp. 373–417.

CHAPTER 28 Cortical States

Figure 28.1A Aschoff, J. (1965) Circadian rhythms in man. *Science* 148: 1427–1432. **Figure 28.3** Okamura, H. and 8 others (1999) Photic induction of mPer1 and mPer2 in Cry-deficient mice lacking a biological clock. *Science* 276: 2531–2534. **Box 28A Figures B,C** Bear, M., M. A. Paradiso and B. Connors (2001) *Neuroscience: Exploring the Brain*, 2nd Ed. Philadelphia: Williams & Wilkins/Lippincott. **Figures 28.1B, 28.4, 28.5, 28.8, and 28.12** Hobson, J. A. (1989) *Sleep*. New York: Scientific American Library. **Figure 28.6A** Woods, R. and H. B. Greenhouse (1974) *The New World of Dreams*. NewYork: Macmillan. **Figure 28.6B,C** Funkhouser, A. (2014) funkhouser.dreamunit.net/Dream-notes; Jovanovic, U. J. (1971) *Normal Sleep in Man*. Stuttgart: Hippocrates. **Figure 28.7** Magoun, H. W. (1952) An ascending reticular activating system in the brain stem. AMA Arch. *Neurol. Psychiat.* 67: 145–154. **Figure 28.10** McCormick, D. A. and H. C. Pape (1990) Properties of a hyperpolarization-activated cation current and its role in rhythmic oscillation in thalamic relay neurones. *J. Physiol.* 431: 291–318. **Figure 28.11** Steriade, M., D. A. McCormick and T. J. Sejnowski (1993) Thalamocortical oscillations in the sleeping and aroused brain. *Science* 262: 679–685. **Figure 28.14** Fox, M. D. and 5 others (2005) The human brain is intrinsically organized into dynamic, anticorrelated function networks. *Proc. Natl. Acad. Sci. USA* 102: 9673–9678.

CHAPTER 29 Attention

Figure 29.1 Cherry, E. C. (1953) Some experiments on the recognition of speech, with one and with two ears. *J. Acoust. Soc. Am.* 25: 975–979. **Figure 29.2** Posner, M. I., C. R. R. Snyder and B. J. Davidson (1980) Attention and the detection of signals. *J. Exp. Psychol. Gen.* 59: 160–174. **Figure 29.3** Klein, R. M. (2000) Inhibition of return. *Trends Cogn. Sci.* 4:138–147; Posner, M. I. and Y. Cohen (194) Componenets of visual orienting. In *Attention and Performance*, vol. 10: Control of Language Porcesses, H. Bouma and D. Bouwhuis (eds.). London: Erlbaum, pp. 531–556. **Figure 29.4** Heilman, H. and E. Valenstein (1985) *Clinical Neurospychology*, 2nd Edition. NewYork: Oxford University Press. **Figure 29.5A,C** Posner, M. I. and M. E. Raichle (1994) *Images of Mind*. NewYork: Scientific American Library. **Figure 29.5B** Blumenfeld, H. (2010) *Neuroanatomy through Clinical Cases*, 2nd Edition. Sunderland, MA: Sinauer. **Figure 29.5D** Grabowecky, M., L. C. Robertson and A. Treisman (1993) Preattentive processes guide visual search: Evidence from patients with unilateral visual neglect. *J. Cogn. Neurosci.* 5: 288–302. **Box 29A** Moore, T., K. M. Armstrong and M. Fallah (2003) Visuomotor origins of covert spatial attention. *Neuron* 40: 671–683. **Figure 29.6** Corbetta, M. and G. L. Shulman (2002) Control of goal-directed and stimulus-driven attention in the brain. *Nat. Rev. Neurosci.* 3: 201–215. **Figure 29.7** Moran, J. and R. Desimone (1985) Selective attention gates visual processing in the extrastriate cortex. *Science* 229: 782–784.

CHAPTER 30 Memory

Box 30A Tinbergen, N. (1969) *Curious Naturalists*. Garden City, NY: Doubleday. Photo © Nina Leen/Getty Images. **Figure 30.4** Ericsson, K. A., W. G. Chase, and S. Faloon (1980) Acquisition of a memory skill. *Science*. 208: 1181–1182. **Figure 30.5** Chase W. G. and H. A. Simon (1973) *The Mind's Eye in Chess in Visual Information Processing*, W. G. Chase, ed. NewYork: Academic Press, pp. 215–281. **Figure 30.6** Morris, J. S. and R. J. Dolan (2001) Involvement of human amygdala and orbitofrontal cortex in hunger-enhanced memory for food stimuli. *J. Neurosci.* 21: 5304–5310. **Figure 30.8A** Rubin, D. C. and T. C. Kontis (1983) A schema for common cents. *Mem. Cog.* 11: 335–341. **Figure 30.8B** Squire, L. R. (1989) On the course of forgetting in very long-term memory. *J. Exp. Psychol.* 15: 241–245. **Figure 30.10B** Eichenbaum, H. (2000) A cortical-hippocampal system for declarative memory. *Nat. Rev. Neurosci.* 1: 41–50. **Figure 30.10C,D** Schenk, F. and R. G. Morris (1985) Dissociation between components of spatial memory in rats after recovery from the effects of retrohippocampal lesions. *Exp. Brain Res.* 58: 11–28. **Figure 30.11**

Wagner, A. D. and 7 others (1998) Building memories: Remembering and forgetting of verbal experiences. *Science* 281: 1188. **Figure 30.12** Maguire, E. A. and 6 others (2000) Navigation-related structural change in the hippocampi of taxi drivers. *Proc. Natl. Acad. Sci. USA* 97: 4398–4403. Copyright (2000) National Academy of Sciences, U.S.A. **Box 30C Figure B** Blumenfeld, H. (2002) *Neuroanatomy through Clinical Cases.* Sunderland, MA: Sinauer Associates; Brun, A. and E. Englund (1981) Regional pattern of degeneration in Alzheimer's disease: Neuronal loss and histopathalogical grading. *Histopathology* 5: 459–564. **Box 30D** Moser, E. I., E. Kropff and M.-B. Moser (2008) Place cells, grid cells, and the brain's spatial representation system. *Annu. Rev. Neurosci.* 31: 69–89. **Figure 30.13** Lashley, K. S. and Wiley, L. E. (1933) Studies of cerebral function in learning IX. Mass action in relation to the number of elements in the problem to be learned. *J. Comp. Neurol.*, 57: 3–55. doi:10.1002/cne.900570102; Lashley, K. S. (1944) Studies of cerebral function in learning. XIII. Apparent absence of transcortical association in maze learning. *J. Comp. Neurol.*, 80: 257–281. doi:10.1002/cne.900800207; University of Rome Psychology Lab Website. **Figure 30.14** Van Hoesen, G. W. (1982) The parahippocampal gyrus. *Trends Neurosci.* 5: 345–350. **Figure 30.15** Buckner, R. L. And M. E. Wheeler (2001) The cognitive neuroscience of remembering. *Nat. Rev. Neurosci.* 2: 624–634. **Figure 30.16** Shohamy, D., C. E. Myers, S. Grossman, J. Sage and M. A. Gluck (2005) The role of dopamine in cognitive sequence learning: Evidence from Parkinson's disease. *Behav. Brain Res.* 156: 191–199. **Figure 30.18** Dekaban, A. S. and D. Sadowsky (1978) Changes in brain weights during the span of human life: Relation of brain weights to body heights and body weights. *Ann. Neurol.* 4: 345–356. **Figure 30.19** Cabeza, R., N. D. Anderson, J. K. Locantore and A. R. McIntosh (2002) Aging gracefully: compensatory brain activity in high-performing older adults. *Neuroimage* 17: 1394–1402.

CHAPTER 31 Emotion

Figure 31.2 LeDoux, J. E. (1987) Emotion. In *Handbook of Physiology*, Section 1, The Nervous System, vol. 5. F. Blum, S. R. Geiger, and V. B. Mountcastle (eds.). Bethesda, MD: American Physiological Society, pp. 419–459. **Box 31A Figure A** Part A (2,3) from Wellcome Library, London. http://wellcomeimages.org **Box 31A Figure B** Left photos: G. Holstege, R. Bandler and C.B. Saper (1996) The Emotional Motor System. In *Progress in Brain Research,* vol. 107. G. Holstege, R. Bandler and C. B. Saper (eds.). Amsterdam: Elsevier, pp. 3-6.
Right photos: Trosch, R. M., G. Sze, L. M. Brass and S. G. Waxman (1990) Emotional facial paresis with striatocapsular infarction. *J. Neurol. Sci.* 98: 195–201. **Figure 31.7** Rolls, E. T. (1999) *The Brain and Emotion*. Oxford: Oxford University Press. **Box 31C Figures B,C** Adolphs, R., D. Tranel, H. Damasio and A. R. Damasio (1995) Fear and the human amygdala. *J. Neurosci.* 15: 5879–5891. **Figure 31.8** Moscovitch, M. and J. Olds (1982) Asymmetries in spontaneous facial expressions and their possible relation to hemispheric specialization. *Neuropsychologia* 20: 71–81. **Figure 31.9** Winston, J. S., B. A. Strange, J. O'Doherty and R. J. Dolan (2002) Automatic and intentional brain responses during evaluation of trustworthiness of faces. *Nat. Neurosci.* 5: 277–283. **Figure 31.11** Nestler, E. J. (2005) Is there a common molecular pathway for addiction? *Nat. Neurosci.* 8: 1445–1449. **Figure 31.12** Schultz, W., P. Dayan and R. P. Montague (1997) A neural substrate of prediction and reward. *Science* 275: 1593–1599. **Figure 31.13A** Parsons, R. G. and K. J. Ressler (2013) Implications of memory modulation for post-traumatic stress and fear disorders. *Nat. Neurosci.* 16: 146–153; Carrión, V. G., B. W. Haas, A. Garrett, S. Song, and A. L. Reiss (2010) Reduced hippocampal activity in youth with posttraumatic stress symptoms: An fMRI study. *J. Ped. Psych.* 35: 559–569.

CHAPTER 32 Thinking, Planning, and Deciding

Figure 32.2B Brodmann, K. (1912) Neue Ergebnisse über die vergleichende histologische Lokalisation der Grosshirnrinde mit besonderer Berücksichtigung des Stirnhirns. *Anat. Anzeiger* 41: 157–216. **Figure 32.2C** Semendeferi, K., H. Damasio, R. Frankand G. W. Van Hoesen (1997) The evolution of the frontal lobes: A volumetric analysis based on three-dimensional reconstructions of magnetic resonance scans of human and ape brains. *J. Hum. Evol.* 32: 375–388; Semendeferi, K., A. Lu, N. Schenker and H. Damasio (2002) Humans and great apes share a large frontal cortex. *Nat. Neurosci.* 5: 272–276. **Figure 32.4** Wallis, J. D. (2007) Orbitofrontal cortex and its contribution to decision-making. *Annu. Rev Neurosci.* 30: 31–56. **Box 32A Figures B,C** Schultz, W., P. Dayan and P. R. Montague (1997) A neural substrate of prediction and reward. *Science* 275: 1593–1599. **Figure 32.5** Baxter, M. G. and E. A. Murray (2002) The amygdala and reward. *Nat. Rev. Neurosci.* 3:563–573. doi: 10.1038/nrn875 **Figure 32.6** Miller, E. K. (2000) The prefrontal cortex and cognitive control. *Nat. Rev. Neurosci.* 1: 59–65. **Figure 32.7** Funahashi, S., C. J. Bruce and P. S. Goldman-Rakic (1989) Mnemonic coding of visual space in the monkey's dorsolateral prefrontal cortex. *J. Neurophysiol.* 61: 331–349. **Figure 32.8** Wallis, J. D., K. C. Anderson and E. K. Miller (2001) Single neurons in prefrontal cortex encode abstract rules. *Nature* 411: 953–956. **Figure 32.9B** Bush, G., P. J. Whalen, B. R. Rosen, M. A. Jenike, S. C. McInerney and S. L. Rauch (1998) The counting Stroop: An interference task specialized for functional neuroimaging—Validation study with functional MRI. *Hum. Brain Mapp.* 6: 270–282. **Figure 32.9C** Kerns, J. G., J. D. Cohen, A. W. MacDonald III, R.Y. Cho, V. A. Stenger and C. S. Carter (2004) Anterior cingulate conflict monitoring and adjustments in control. *Science* 303: 1023–1026. **Figure 32.10** Depue, B. E., T. Curran and M. T. Banich (2007) Prefrontal regions orchestrate suppression of emotional memories via a two-phase process. *Science* 317: 215–219. **Figure 32.11B** Bechara, A., A. R. Damasio, H. Damasio and S. W. Anderson (1994) Insensitivity to future consequences following damage to the human prefrontal cortex. *Cognition* 50: 7–15. **Figure 32.11C** Bechara, A., D. Tranel, H. Damasio and A. R. Damasio (1996) Failure to respond autonomically to anticipated future outcomes following damage to prefrontal cortex. *Cereb. Cortex* 6: 215–225. **Box 32B Figure B** Sirigu, A., E. Daprati, S. Ciancia, P. Giraux, N. Nighoghossian, A. Posada and P. Haggard (2003) Altered awareness of voluntary action after damage to the parietal cortex. *Nat. Neurosci.* 7: 80–84. **Figure 32.12** Naqvi, N. H., D. Rudrauf, H. Damasio and A. Bechara (2007) Damage to the insula disrupts addition to cigarette smoking. *Science* 315: 531–534. **Figure 32.13A,B** Gusnard, D. A. and M. E. Raichle (2001) Searching for a baseline: Functional imaging and the resting human brain. *Nat. Rev. Neurosci.* 2: 685–694. **Figure 32.13C** Hayden, B.Y., D. Smith and M. L. Platt (2009) Electrophysiological correlates of default-mode processing in macaque posterior cingulate cortex. *Proc. Natl. Acad. Sci. USA* 106: 5948–5953. **Figure 32.14** Pearson, J. M., S. R. Heilbronner, D. L. Barack, B.Y. Hayden, and M. L. Platt (2011) Posterior cingulate cortex: Adampting behavior to a changing world. *Trends Cogn. Sci.* 15: 143–151.

CHAPTER 33 Speech and Language

Box 33A Figure B Miller, G. A. (1991) *The Science of Words*, chapter 4, The spoken word. New York: Scientific American Library. **Box 33B Figure B** Coren, S. (1992) *The Left-Hander Syndrome: The Causes and Consequence of Left-Handedness.* New York: The Free Press. **Figure 33.5A** Penfield, W. and L. Roberts (1959) *Speech and Brain Mechanisms.* Princeton, NJ: Princeton University Press. **Figure 33.5B** Ojemann, G. A., I. Fried and E. Lettich (1989) Electrocorticographic (EcoG) correlates of language. *Electroencephalo. Clin. Neurophys.* 73: 453–463. **Figure 33.6** Posner, M. I. and M. E. Raichle (1994) *Images of Mind.*

New York: Scientific American Library. **Figure 33.7** Damasio, H., T. J. Grabowski, D. Tranel, R. D. Hichwa and A. Damasio (1996) A neural basis for lexical retrieval. *Nature* 380: 499–505. **Figure 33.8** Bellugi, U., H. Poizner and E. S. Klima (1989) Language, modality, and the brain. *Trends Neurosci.* 12: 380–388. **Figure 33.9 A** Johnson, J. S. and E. L. Newport (1989) Critical period effects in second language learning: The influence of maturational state on the acquisition of English as a second language. *Cogn. Psychol.* 21: 60–99. **Figure 33.9B** Brown, T. T., H. M. Lugar, R. S. Coalson, F. M. Miezin, S. E. Peteresen and B. L. Schlagger (2005) Developmental changes in human cerebral functional organization for word generation. *Cereb. Cortex* 15: 275–290. **Figure 33.10** Vinckier, F. and 5 others (2007) Hierarchical coding of letter strings in the ventral stream: Dissecting the inner organization of the visual word-form system. *Neuron* 55: 143–156. **Figure 33.11** Savage-Rumbaugh, S., S. G. Shanker and T. J. Taylor (1998) *Apes, Language, and the Human Mind.* New York: Oxford University Press. **Figure 33.12** Gil-da-Costa, R. and 5 others (2006) Species-specific calls activate homologs of Broca's and Wernicke's areas in the macaque. *Nat. Neurosci.* 9: 1064–1070.

APPENDIX Survey of Human Neuroanatomy

Opening Image Copyright © Max Delson/istockphoto.com. **Figure A13C** Rohen, J. W. and C. Yokochi (1993) *Color Atlas of Anatomy.* New York: Igaku-Shoin. **Figure A16, A17, A20A** Blumenfeld, H. (2010) *Neuroanatomy through Clinical Cases*, 2nd Edition. Sunderland, MA: Sinauer Associates. **Figure A26** Nedergaard, M. and S. A. Goldman (2016) Brain drain. *Sci. Am.* 314: 44–49.

Index

Entries with an italic *f* next to the page number indicate that the information will be found in a figure; entries with an italic *t* next to the page number indicate that the information will be found in a table; and entries with an italic *b* next to the page number indicate that the information will be found in a box.

D

E

U

V

W

ABOUT THE BOOK

Editor: Sydney Carroll

Production Editor: Martha Lorantos

Copyeditor: Lou Doucette

Indexer: Hughes Analytics

Production Manager: Christopher Small

Book Design: Jefferson Johnson

Cover Design: Jefferson Johnson

Book Production: Jefferson Johnson and Joanne Delphia

Illustration Program: Dragonfly Media Group